Antenna
Handbook

Antenna Handbook

THEORY, APPLICATIONS, AND DESIGN

Edited by
Y. T. Lo
Electromagnetics Laboratory
Department of Electrical and Computer Engineering
University of Illinois–Urbana

S. W. Lee
Electromagnetics Laboratory
Department of Electrical and Computer Engineering
University of Illinois–Urbana

 VAN NOSTRAND REINHOLD COMPANY
New York

Copyright © 1988 by Van Nostrand Reinhold Company Inc.

Library of Congress Catalog Card Number 87-16833

ISBN 0-442-25843-7

Printed in the United States of America

Van Nostrand Reinhold Company Inc.
115 Fifth Avenue
New York, New York 10003

Van Nostrand Reinhold Company Limited
Molly Millars Lane
Wokingham, Berkshire RG11 2PY, England

Van Nostrand Reinhold
480 La Trobe Street
Melbourne, Victoria 3000, Australia

Macmillan of Canada
Division of Canada Publishing Corporation
164 Commander Boulevard
Agincourt, Ontario M1S 3C7, Canada

16 15 14 13 12 11 10 9 8 7 6 5 4 3 2 1

Library of Congress Cataloging-in-Publication Data

Antenna handbook: theory, applications, and design/
  edited by Y. T. Lo and S. W. Lee.
      p.   cm.
    Includes bibliographies and index.
    ISBN 0-442-25843-7:
    1. Antennas (Electronics)   I. Lo, Y. T.   II. Lee, S. W.
TK7871.6.A495 1988
621.38′028′3—dc19   87-16833
                                                CIP

Contents

Appendixes

PART A

Fundamentals and Mathematical Techniques

Chapter 1

Basics

S. W. Lee
University of Illinois

CONTENTS

 **Shung-Wu Lee** was born in Kiangsi, China. He received his BS degree in electrical engineering from Cheng Kung University in Tainan, Taiwan, in 1961. After one year of military service in Taiwan, he came to the United States in 1962. At the University of Illinois in Urbana, he received his MS and PhD degrees in electrical engineering and has been on the faculty since 1966. Currently he is a professor of electrical and computer engineering, and an associate director of the Electromagnetics Laboratory.

While on leave from the University of Illinois, Dr. Lee was with Hughes Aircraft Company, Fullerton, California, in 1969–1970, and with the Technical University at Eindhoven, The Netherlands, and the University of London, England, in 1973–74. Dr. Lee received several professional awards, including the 1968 Everitt Teaching Excellence Award from the University of Illinois, 1973 NSF NATO Senior Scientist Fellowship, 1977 Best Paper Award from IEEE Antennas and Propagation Society, and the 1985 Lockheed Million Dollar Award.

Dr. Lee has published more than 100 papers in technical journals on antennas and electromagnetic theory. He is the coauthor of a book on guided waves published by Macmillan in 1971, and a coeditor of this book. He is a Fellow of IEEE.

1. The Maxwell Equations and Time-Harmonic Fields

An electric field $\mathscr{E}$ is generally a function of spatial variable $\mathbf{r}$ and time variable t, and is denoted $\mathscr{E}(\mathbf{r}, t)$. Unless specifically mentioned otherwise, the time variation is assumed to be sinusoidal:

$$\mathscr{E}(\mathbf{r}, t) = \sqrt{2}\, \text{Re}\{\mathbf{E}(\mathbf{r})\, \epsilon^{+j\omega t}\} \tag{1}$$

where $\omega = 2\pi f$ is the angular frequency in radians per second, and f is the frequency in hertz. We will work with the complex phasor $\mathbf{E}(\mathbf{r})$ throughout this book.*

In a continuous medium the behavior of the time-harmonic electromagnetic field is governed by the Maxwell equations. Using the International System of Units (SI units), the Maxwell equations for phasors are

$$\nabla \times \mathbf{E} = -j\omega\mathbf{B} - \mathbf{K} \tag{2a}$$

$$\nabla \times \mathbf{H} = j\omega\mathbf{D} + \mathbf{J} \tag{2b}$$

$$\nabla \cdot \mathbf{B} = \varrho_m \tag{2c}$$

$$\nabla \cdot \mathbf{D} = \varrho \tag{2d}$$

where

$\mathbf{E}$ = electric field in volts per meter (V/m)

$\mathbf{H}$ = magnetic field in amperes per meter (A/m)

$\mathbf{D}$ = electric flux density in coulombs per square meter (C/m^2)

$\mathbf{B}$ = magnetic flux density in teslas (T) or in webers per square meter (Wb/m^2)

$\mathbf{J}$ = electric current density in amperes per square meter, (A/m^2)

$\mathbf{K}$ = magnetic current density in volts per square meter (V/m^2)

ϱ = electric charge density in coulombs per cubic meter (C/m^3)

---

*Most antenna terms and definitions used in this chapter are consistent with those in *IEEE Standard Definitions of Terms for Antennas*, IEEE Std. 145–1983, appeared in *IEEE Trans. Antennas Propag.*, vol. AP-31, no. 8, November 1983.

ϱ_m = magnetic charge density in webers per cubic meter (Wb/m^3)

From (2) we can deduce the equations of continuity:

$$\nabla\cdot\mathbf{J} + j\omega\varrho = 0 \tag{3a}$$

$$\nabla\cdot\mathbf{K} + j\omega\varrho_m = 0 \tag{3b}$$

which state the conservation of charges.

In an isotropic medium the constitutive relation is

$$\mathbf{D} = \epsilon\mathbf{E}, \quad \mathbf{B} = \mu\mathbf{H}, \quad \mathbf{J} = \sigma\mathbf{E} \tag{4}$$

The conductivity σ is given in siemens per meter (S/m) or in mhos per meter (℧/m) and is generally real, while the other two parameters, the dielectric constant (permittivity) ϵ and the permeability μ, are complex:

$$\epsilon = \epsilon' - j\epsilon'' \quad \text{in farads per meter (F/m),} \tag{5a}$$

$$\mu = \mu' - j\mu'' \quad \text{in henrys per meter (H/m)} \tag{5b}$$

where ϵ', ϵ'', μ', and μ'' are positive real. The factors $(\epsilon''/\epsilon', \mu''/\mu')$ are called (*dielectric, magnetic*) *loss tangents*, respectively. They become vanishingly small if the medium is lossless. The *wave number k* in (meter)$^{-1}$ is defined by

$$k = k' - jk'' = \omega\sqrt{\mu\epsilon(1 - j\sigma/\omega\epsilon)} \tag{6}$$

where k' and k'' are positive real. The wavelength in meters is

$$\lambda = \frac{2\pi}{k'} \tag{7}$$

The intrinsic wave impedance Z in ohms and admittance Y in siemens of the medium are

$$Z = \frac{1}{Y} = \frac{\sqrt{\mu}}{\sqrt{\epsilon(1 - j\sigma/\omega\epsilon)}} \tag{8}$$

Here the square roots should be taken such that, if $a = \sqrt{b}$, Re $\{a\} \gtreqless 0$ and Im $\{a\} \lesseqgtr 0$. It follows that the phase angle of the complex number Z is between $-\pi/4$ and $+\pi/4$.

Free space or a vacuum is an isotropic medium in which*

---

*The approximate value of ϵ_0 in (9a) is obtained from $\epsilon_0 = \mu_0^{-1}c^{-2}$ with the speed of light $c \cong 3 \times 10^8$ m/s. If the more exact value $c \cong 2.997\,925 \times 10^8$ m/s is used, then $\epsilon_0 \cong 8.854\,185 \times 10^{-12}$ F/m.

$$\epsilon = \epsilon_0 \cong \frac{1}{36\pi \times 10^9} \quad (\text{F/m}) \tag{9a}$$

$$\mu = \mu_0 = 4\pi \times 10^{-7} \quad (\text{H/m}) \tag{9b}$$

$$\sigma = 0 \quad (\text{S/m}) \tag{9c}$$

For other isotropic media it is convenient to introduce the dimensionless ratios

$$\epsilon_r = \epsilon/\epsilon_0, \qquad \mu_r = \mu/\mu_r \tag{10}$$

which are known as the *relative dielectric constant* and *relative permeability*, respectively. In a lossless medium the wave number k and wavelength λ are given explicitly by

$$k = 0.209\,44 \times f \times \sqrt{\mu_r \epsilon_r} \quad (\text{cm}^{-1})$$
$$= 0.531\,98 \times f \times \sqrt{\mu_r \epsilon_r} \quad (\text{in}^{-1})$$
$$\lambda = \frac{30}{f \times \sqrt{\mu_r \epsilon_r}} \quad (\text{cm})$$
$$= \frac{11.811\,02}{f \times \sqrt{\mu_r \epsilon_r}} \quad (\text{in})$$

in which f is given in gigahertz.

2. The Poynting Theorem

Consider a volume V enclosed by the boundary surface $S = \partial V$ in an isotropic medium characterized by constitutive parameters (ϵ, μ, σ). In general, (ϵ, μ) are functions of ω (dispersive medium). The complex power in watts supplied by sources inside V is

$$P_s = -\iiint_V (\mathbf{E} \cdot \mathbf{J}^* + \mathbf{H}^* \cdot \mathbf{K}) \, dv \tag{11}$$

[As described in (1) we use rms values for phasors. Therefore the factor 1/2 does not appear in (11).] The time-averaged power supplied by sources inside V is equal to the real part of P_s. The complex power leaving V across S is

$$P = \iint_S \mathbf{E} \times \mathbf{H}^* \cdot d\mathbf{s} \tag{12}$$

The direction of $d\mathbf{s}$ is in the outward normal direction of surface S (pointing away from V). The time-averaged power dissipated (conduction loss) inside V is

$$P_c = \iiint_V \sigma |\mathbf{E}|^2 \, dv \tag{13}$$

The time-averaged energy in joules stored in electric and magnetic fields inside V is

$$\overline{W}_e = \frac{1}{2} \iiint_V \overline{\epsilon} |\mathbf{E}|^2 \, dv \tag{14a}$$

$$\overline{W}_m = \frac{1}{2} \iiint_V \overline{\mu} |\mathbf{H}|^2 \, dv \tag{14b}$$

where

$$\overline{\epsilon} = \epsilon + \omega \frac{d\epsilon}{d\omega} \tag{14c}$$

$$\overline{\mu} = \mu + \omega \frac{d\mu}{d\omega} \tag{14d}$$

For the special case in which (ϵ, μ) are not functions of ω (nondispersive medium) the quantities in (14) reduce to

$$(\overline{\epsilon}, \overline{\mu}) \rightarrow (\epsilon, \mu) \tag{15a}$$

$$(\overline{W}_e, \overline{W}_m) \rightarrow (W_e, W_m) \tag{15b}$$

where

$$W_e = \frac{1}{2} \iiint_V \epsilon |\mathbf{E}|^2 \, dv \tag{15c}$$

$$W_m = \frac{1}{2} \iiint_V \mu |\mathbf{H}|^2 \, dv \tag{15d}$$

Note that in a dispersive medium (W_e, W_m) do not represent time-averaged energy stored in the electromagnetic field, as (ϵ, μ) may be negative. The stored energies are given by $(\overline{W}_e, \overline{W}_m)$, which are always positive. The Poynting theorem for the time-harmonic fields is

$$P_s = P + P_c + j\omega(W_m - W_e) \tag{16}$$

which holds for both dispersive and nondispersive media.

3. Boundary, Radiation, and Edge Conditions

At a point at the interface of two media we define a surface normal $\hat{\mathbf{n}}_{21}$ pointing into medium 1 (Fig. 1). Then the fields in the two media at this point are related by boundary conditions:

$$\hat{\mathbf{n}}_{21} \times (\mathbf{E}_1 - \mathbf{E}_2) = -\mathbf{K}_s \tag{17a}$$

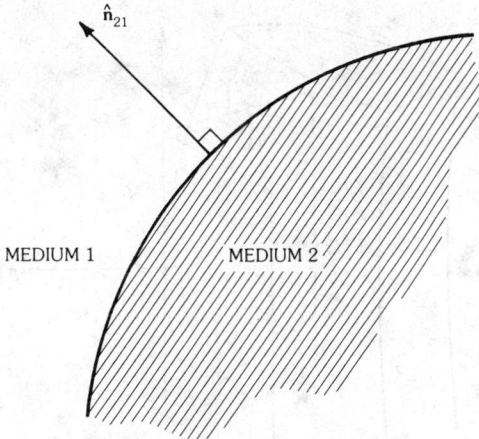

Fig. 1. Interface between two media.

$$\hat{\mathbf{n}}_{21} \times (\mathbf{H}_1 - \mathbf{H}_2) = \mathbf{J}_s \tag{17b}$$

$$\hat{\mathbf{n}}_{21} \cdot (\mathbf{D}_1 - \mathbf{D}_2) = \varrho_s \tag{17c}$$

$$\hat{\mathbf{n}}_{21} \cdot (\mathbf{B}_1 - \mathbf{B}_2) = \varrho_{ms} \tag{17d}$$

where $(\mathbf{J}_s, \varrho_s; \mathbf{K}_s, \varrho_{ms})$ are possible surface current and charge densities at the interface.

In an unbounded space the solution of the Maxwell equations may not be unique. To make it unique we impose an additional constraint to the Maxwell equations, namely, the radiation condition. For definiteness let us assume that all sources are contained in a finite region near $r = 0$. If the medium in space is lossy, the radiation condition requires that the fields vanish as $r \to \infty$. If the medium is lossless and isotropic, it requires that all field components have a phase progressively outward, and have an amplitude that decreases at least as rapidly as r^{-1}.

In problems where geometrical singularities (edges or tips) of perfect conductors are present, the solutions of the Maxwell equations must satisfy yet another constraint, namely, the edge condition [1]. It requires that the electric and magnetic energies stored in any finite neighborhood of a geometrical singularity be finite; that is,

$$\int_V |\mathbf{E}|^2 \, dv \to 0, \qquad \int_V |\mathbf{H}|^2 \, dv \to 0 \tag{18}$$

as the volume V contracts to the geometrical singularity. For the conducting wedge in Fig. 2 the upper bounds of the fields near an edge point O are, as $\varrho \to 0$,

$$E_\varrho, E_\phi, H_\varrho, H_\phi = O(\varrho^{(\beta - \pi)/(2\pi - \beta)}) \tag{19a}$$

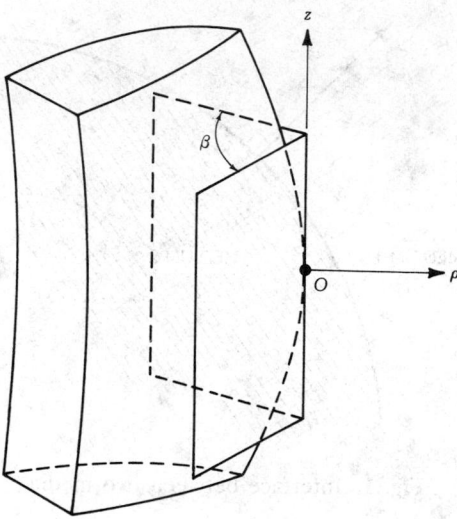

Fig. 2. A conducting wedge with interior wedge angle β and tangent planes at point O.

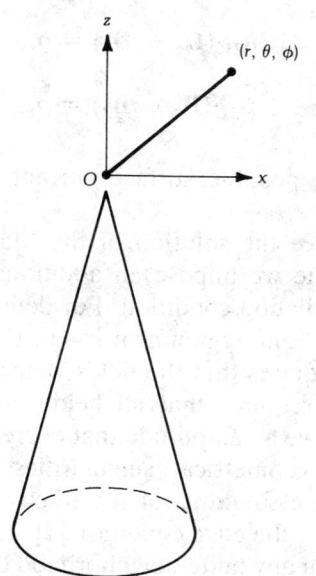

Fig. 3. A conducting cone with tip O.

$$E_z = O(\varrho^{\pi/(2\pi-\beta)}) \tag{19b}$$

$$H_z = O(1) \tag{19c}$$

where β is the interior wedge angle at point O. For the conducting cone in Fig. 3 all field components must grow slower than $r^{-3/2}$ as $r \to 0$.

4. Radiation from Sources

A source described by current densities $(\mathbf{J}, \mathbf{K})$ radiates in an unbounded homogeneous space characterized by constitutive parameters (ϵ, μ). See Fig. 4. The radiation field at an arbitrary observation point $\mathbf{r}$ can be calculated from two vector potentials $(\mathbf{A}, \mathbf{F})$ by the relations [2]

$$\mathbf{E}(\mathbf{r}) = -\nabla \times \mathbf{F} + \frac{1}{j\omega\epsilon} (\nabla \times \nabla \times \mathbf{A} - \mathbf{J}) \tag{20a}$$

$$\mathbf{H}(\mathbf{r}) = \nabla \times \mathbf{A} + \frac{1}{j\omega\mu} (\nabla \times \nabla \times \mathbf{F} - \mathbf{K}) \tag{20b}$$

where

$$\mathbf{A}(\mathbf{r}) = \iiint_V \mathbf{J}(\mathbf{r}') g(\mathbf{r} - \mathbf{r}') \, dv'$$

$$\mathbf{F}(\mathbf{r}) = \iiint_V \mathbf{K}(\mathbf{r}') g(\mathbf{r} - \mathbf{r}') \, dv'$$

$$g(\mathbf{r} - \mathbf{r}') = \frac{e^{-jk|\mathbf{r}-\mathbf{r}'|}}{4\pi|\mathbf{r} - \mathbf{r}'|}$$

$$k = \omega\sqrt{\mu\epsilon} = 2\pi/\lambda$$

The integrals above are over the source region V, and $\mathbf{r}'$ is a typical point in V.

Now consider an important special case of the radiation problem sketched in Fig. 4: the source region V is situated near the origin O, whereas the observation point $\mathbf{r} = (r, \theta, \phi)$ is far away $(kr \gg 1)$. Then Fig. 4 becomes Fig. 5, and the

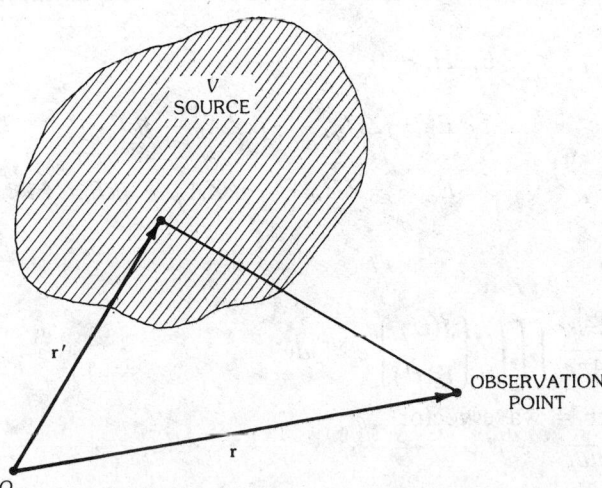

Fig. 4. Calculation of radiation field at an observation point $\mathbf{r}$.

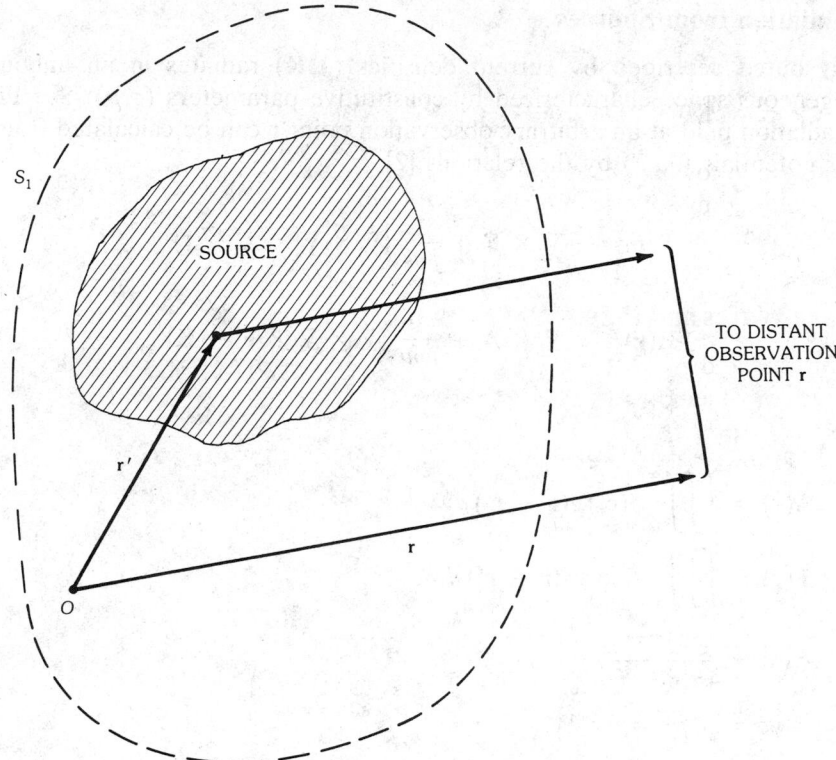

Fig. 5. Calculation of radiation field with **r** in the far zone and S_1 an arbitrary surface enclosing the source.

formulas in (20) can be simplified. Neglecting terms of order r^{-2} and higher in amplitude and phase, (20) is reduced to, as $r \to \infty$, the asymptotic equalities

$$E_r, H_r \sim 0 \tag{21a}$$

$$E_\theta \sim ZH_\phi \sim -jkZA_\theta - jkF_\phi \tag{21b}$$

$$E_\phi \sim -ZH_\theta \sim -jkZA_\phi + jkF_\theta \tag{21c}$$

where

$$\begin{Bmatrix} \mathbf{A(r)} \\ \mathbf{F(r)} \end{Bmatrix} = \frac{e^{-jkr}}{4\pi r} \iiint_V \begin{Bmatrix} \mathbf{J(r')} \\ \mathbf{K(r')} \end{Bmatrix} e^{j\mathbf{k}\cdot\mathbf{r'}} \, dv'$$

$\mathbf{k} = k\hat{\mathbf{r}}$ = wave vector

$Z = \sqrt{\mu/\epsilon}$

$\mathbf{r'}$ = source point with coordinates (x', y', z'), (ϱ', ϕ', z'), or (r', θ', ϕ')

$$\mathbf{k}\cdot\mathbf{r}' = k[(x'\cos\phi + y'\sin\phi)\sin\theta + z'\cos\theta]$$
$$= k[\varrho'\sin\theta\cos(\phi' - \phi) + z'\cos\theta]$$
$$= kr'[\sin\theta'\sin\theta\cos(\phi' - \phi) + \cos\theta'\cos\theta]$$

We note the following characteristics of the far field (21) of a finite source: (*a*) the field is an outgoing spherical wave, (*b*) both **E** and **H** are transverse to the direction of propagation, and (*c*) **E** and **H** satisfy the impedance relation $\mathbf{E} \sim Z\mathbf{H} \times \hat{\mathbf{r}}$.

5. Plane Waves and Polarization

The far field in (21) is locally a plane wave. A plane wave in an isotropic homogeneous medium has the following representation:

$$\begin{Bmatrix} \mathbf{E}(\mathbf{r}) \\ \mathbf{H}(\mathbf{r}) \end{Bmatrix} = C \begin{Bmatrix} \sqrt{Z}\,\mathbf{u} \\ \sqrt{Y}\,\hat{\mathbf{k}} \times \mathbf{u} \end{Bmatrix} e^{-j\mathbf{k}\cdot\mathbf{r}} \qquad (22)$$

Here $\mathbf{k} = k\hat{\mathbf{k}}$, with spherical coordinates (k,θ,ϕ) describing the direction of propagation (Fig. 6). The term $Z = Y^{-1} = \sqrt{\mu/\epsilon}$ is the wave impedance of the medium. The complex vector **u** satisfies two conditions:

(*a*) $\mathbf{u}\cdot\mathbf{u}^* = 1$ (unitary)   (23a)

(*b*) $\mathbf{u}\cdot\mathbf{k} = 0$ (transverse to direction of propagation)   (23b)

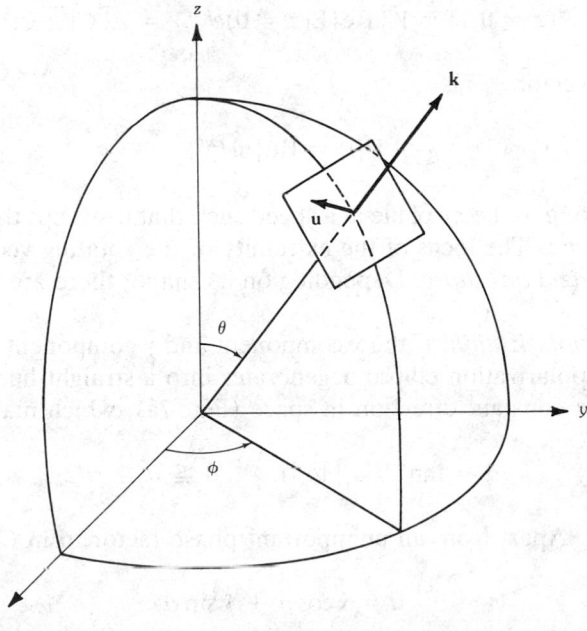

Fig. 6. A plane wave propagating in the direction $\mathbf{k} = (k,\theta,\phi)$ and with the polarization **u**.

The amplitude C in (18) is, in general, a complex number and has the unit $(\text{watt})^{1/2}$ $(\text{meter})^{-1}$. The power density of the plane wave is $|C|^2$ W/m^2.

The unitary vector $\mathbf{u}$ describes the polarization of the plane wave. It is defined only within a phase angle. Thus $\mathbf{u}$ and $\mathbf{u}e^{j\delta}$ describe the same polarization. To study $\mathbf{u}$ in further detail it is convenient to fix the direction of propagation in the z direction ($\mathbf{k} = k\hat{\mathbf{z}}$). Then (22) becomes

$$\left\{\begin{matrix} \mathbf{E}(\mathbf{r}) \\ \mathbf{H}(\mathbf{r}) \end{matrix}\right\} = C \left\{\begin{matrix} \sqrt{Z}\,\mathbf{u} \\ \sqrt{Y}\,\hat{\mathbf{z}} \times \mathbf{u} \end{matrix}\right\} e^{-jkz} \tag{24}$$

Because of (23), a general expression for $\mathbf{u}$ has the form

$$\mathbf{u} = (\hat{\mathbf{x}}a_x + \hat{\mathbf{y}}a_y) \tag{25}$$

Here a_x and a_y are two complex constants

$$a_x = |a_x|\,e^{j\delta_x}, \qquad a_y = |a_y|\,e^{j\delta_y} \tag{26a}$$

subject to the constraint

$$\sqrt{|a_x|^2 + |a_y|^2} = 1 \tag{26b}$$

In the time domain the electric-field vector at a reference plane (say $z = 0$) derived from (24) is

$$\mathscr{E}(z = 0, t) = \sqrt{2}\,\text{Re}\{\mathbf{E}(z = 0)e^{j\omega t}\} = \sqrt{2}\,C\sqrt{Z}\,\mathbf{U}(t) \tag{27a}$$

Here the real vector

$$\mathbf{U}(t) = \text{Re}\{\mathbf{u}e^{j\omega t}\} \tag{27b}$$

is a vector rotating in the xy plane at a speed such that it sweeps through a constant area per unit time. The locus of the extremity of the rotating vector is an ellipse, called the *polarization ellipse*. Depending on its shape, there are three cases to be considered:

(*a*) *Linear polarization.* If the x component and y component of $\mathbf{u}$ are in phase ($\delta_x = \delta_y$), the polarization ellipse degenerates into a straight-line segment. Thus, $\mathbf{U}(t)$ points to a constant direction in space (Fig. 7a), which makes an angle

$$\alpha = \tan^{-1}(|a_y|/|a_x|), \qquad 0 \le \alpha \le \pi/2 \tag{28}$$

with the x axis. Apart from an unimportant phase factor, $\mathbf{u}$ in (25) is reduced to

$$\mathbf{u} = \hat{\mathbf{x}}\cos\alpha + \hat{\mathbf{y}}\sin\alpha \tag{29}$$

which describes a linear polarization.

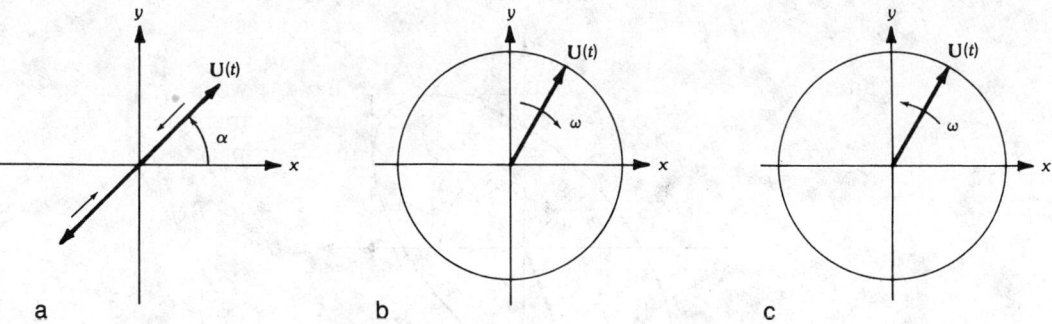

Fig. 7. Rotating vector $\mathbf{U}(t)$ for linear and circular polarizations. (*a*) Linear. (*b*) Left-hand circular. (*c*) Right-hand circular.

(*b*) *Circular polarization.* If the x component and y component of $\mathbf{u}$ are equal in magnitude ($|a_x| = |a_y|$) and 90° out of phase ($\delta_y - \delta_x = \pm\pi/2$), then the polarization ellipse becomes a circle (Figs. 7b and 7c). Apart from an unimportant phase factor, $\mathbf{u}$ in (25) is reduced to one of the following two unitary vectors:

$$\mathbf{L} = (\hat{\mathbf{x}} + j\hat{\mathbf{y}})/\sqrt{2} \qquad \text{(LHCP)} \tag{30a}$$

$$\mathbf{R} = (\hat{\mathbf{x}} - j\hat{\mathbf{y}})/\sqrt{2} \qquad \text{(RHCP)} \tag{30b}$$

Vector $\mathbf{L}$ (vector $\mathbf{R}$) represents a vector in the time domain which rotates in the left-hand (right-hand) sense with respect to the direction of propagation, and is called a left-hand (right-hand) circular polarization vector.

(*c*) *Elliptical polarization.* For general values of a_x and a_y, $\mathbf{u}$ in (25) represents an elliptical polarization. The extremity of the time-domain vector $\mathbf{U}(t)$ traces out an ellipse (Fig. 8). An alternative presentation of (25) is obtained by decomposing $\mathbf{u}$ into two circular polarizations, viz.,

$$\mathbf{u} = \hat{\mathbf{x}}a_x + \hat{\mathbf{y}}a_y \tag{31a}$$
$$= \mathbf{L}a_L + \mathbf{R}a_R$$

where

$$a_L = |a_L|\, e^{j\delta_L} = \mathbf{u}\cdot\mathbf{L}^* = \frac{1}{\sqrt{2}}(a_x - ja_y) \tag{31b}$$

$$a_R = |a_R|\, e^{j\delta_R} = \mathbf{u}\cdot\mathbf{R}^* = \frac{1}{\sqrt{2}}(a_x + ja_y) \tag{31c}$$

The polarization ellipse in Fig. 8 is characterized by three parameters described as follows. (*a*) The axis ratio AR is defined by the ratio of maximum length and minimum length of $\mathbf{U}(t)$, or the ratio of the semimajor and semiminor axes of the polarization ellipse. It may be calculated from

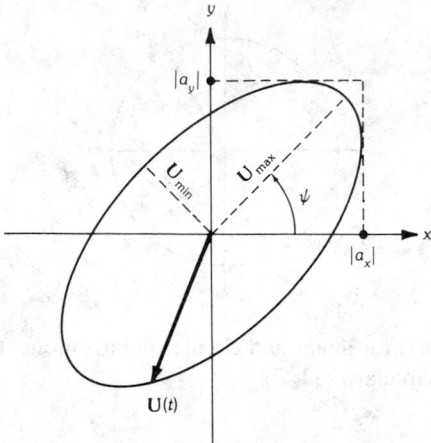

Fig. 8. Elliptical polarization: the tip of the rotating vector $\mathbf{U}(t)$ traces out an ellipse.

$$AR = \frac{U_{\max}}{U_{\min}} = \left| \frac{|a_L| + |a_R|}{|a_L| - |a_R|} \right| \tag{32}$$

For a linear polarization, $AR = \infty$. For a circular polarization, $AR = 1$. (b) The tilt angle ψ is measured from the x axis to a semimajor axis of the ellipse. It may be calculated from

$$\psi = \frac{1}{2}(\delta_R - \delta_L) + m\pi, \qquad m = 0, \pm 1, \pm 2, \ldots \tag{33}$$

Customarily, we choose the integer m so that ψ falls in the range $0 \leqq \psi < \pi$. (c) The sense of rotation of $\mathbf{U}(t)$ is determined by the comparison of $|a_L|$ and $|a_R|$. If $|a_L| > |a_R|$, $\mathbf{u}$ in (31a) represents a left-hand elliptical polarization. If $|a_L| = |a_R|$, $\mathbf{u}$ represents a linear polarization. If $|a_L| < |a_R|$, $\mathbf{u}$ represents a right-hand elliptical polarization.

The plane wave in (22) is characterized by its propagation direction $\mathbf{k}$ and its polarization $\mathbf{u}$. We say that the plane wave is in a state $(\mathbf{k}, \mathbf{u})$. In Chapter 2 we will use the time-reversed field of (22), namely, the plane wave propagates in the opposite direction as the original one, and has the *same* polarization. This time-reversed plane wave is in the state $(-\mathbf{k}, \mathbf{u}^*)$.

6. Antenna Near and Far Fields

The radiation field from a transmitting antenna can be roughly decomposed into two components: (a) radiating field in which the complex Poynting vector $\mathbf{E} \times \mathbf{H}^*$ is real, and $(\mathbf{E}, \mathbf{H})$ decay as r^{-1}, where r is the distance from the antenna; and (b) reactive field in which $\mathbf{E} \times \mathbf{H}^*$ is imaginary, and $(\mathbf{E}, \mathbf{H})$ decay more rapidly than r^{-1}. These two field components dominate in different regions in the space around the antenna. Based on this we can divide the space into three regions:

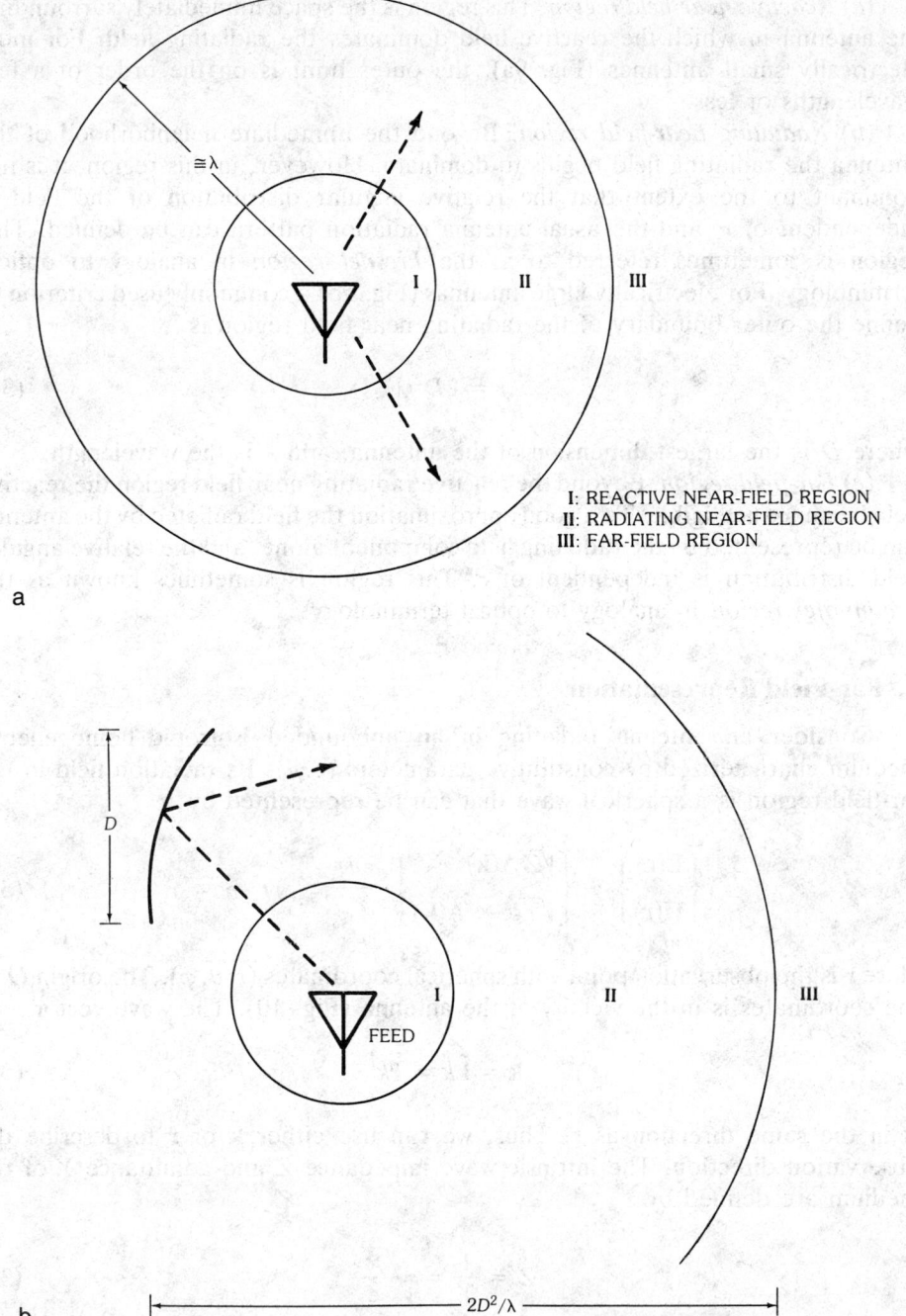

I: REACTIVE NEAR-FIELD REGION
II: RADIATING NEAR-FIELD REGION
III: FAR-FIELD REGION

Fig. 9. Near and far fields. (*a*) Electrically small antennas. (*b*) Electrically large reflector.

(a) *Reactive near-field region*. This region is the space immediately surrounding the antenna in which the reactive field dominates the radiating field. For most electrically small antennas (Fig. 9a), the outer limit is on the order of a few wavelengths or less.

(b) *Radiating near-field region*. Beyond the immediate neighborhood of the antenna the radiating field begins to dominate. However, in this region, it is not dominant to the extent that the relative angular distribution of the field is independent of r, and the usual antenna radiation pattern can be defined. This region is sometimes referred to as the *Fresnel region* in analogy to optical terminology. For electrically large antennas (Fig. 9b) a commonly used criterion to define the outer boundary of the radiating near-field region is

$$r = 2D^2/\lambda \tag{34}$$

where D is the largest dimension of the antenna, and λ is the wavelength.

(c) *Far-field region*. Beyond the reactive/radiating near-field region the reactive field becomes negligible. To a good approximation the field radiated by the antenna can be represented by the radiating field component alone, and the relative angular field distribution is independent of r. This region is sometimes known as the *Fraunhofer region* in analogy to optical terminology.

7. Far-Field Representation

Consider an antenna radiating in an unbounded isotropic homogeneous medium characterized by constitutive parameters (ϵ, μ). Its radiation field in the far-field region is a spherical wave that can be represented by*

$$\begin{Bmatrix} \mathbf{E}(\mathbf{r}) \\ \mathbf{H}(\mathbf{r}) \end{Bmatrix} \sim \begin{Bmatrix} \sqrt{Z}\,\mathbf{A}(\mathbf{k}) \\ \sqrt{Y}\,\hat{\mathbf{k}} \times \mathbf{A}(\mathbf{k}) \end{Bmatrix} \frac{e^{-jkr}}{r}, \qquad r \to \infty \tag{35}$$

Here $\mathbf{r}$ is the observation point with spherical coordinates (r, θ, ϕ). The origin O of the coordinates is in the vicinity of the antenna (Fig. 10). The wave vector

$$\mathbf{k} = \hat{\mathbf{k}}k = \hat{\mathbf{r}}k \tag{36}$$

is in the same direction as $\mathbf{r}$. Thus, we can use either $\hat{\mathbf{k}}$ or $\hat{\mathbf{r}}$ to describe the observation direction. The intrinsic wave impedance Z and admittance Y of the medium are defined by

$$Z = \frac{1}{Y} = \sqrt{\frac{\mu}{\epsilon}} \tag{37}$$

If the medium is free space, then Z has the numerical value

---

*Do not confuse amplitude $\mathbf{A}(\mathbf{k})$ with vector potential $\mathbf{A}(\mathbf{r})$ used in (20). The present representation of the far field was described in G. A. Deschamps [3].

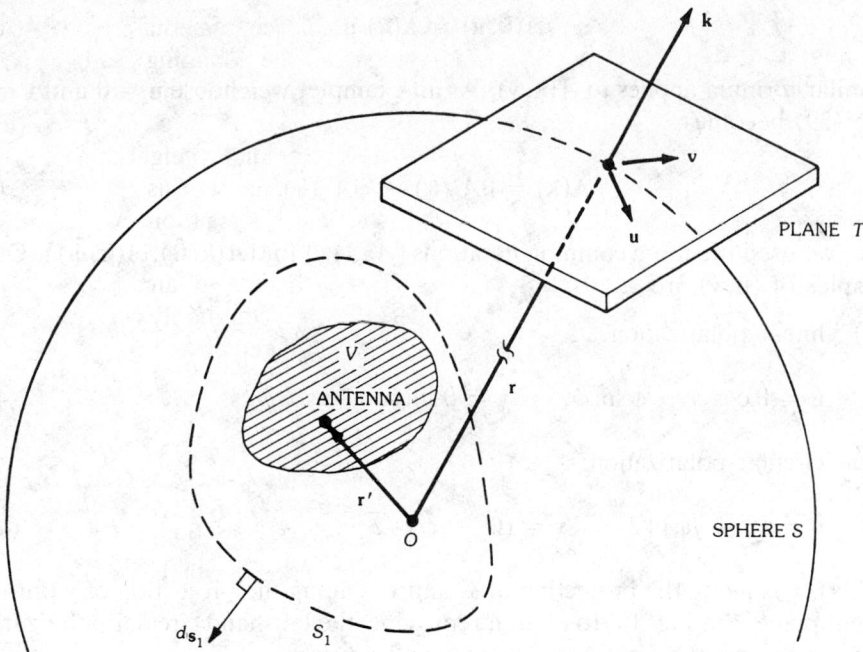

Fig. 10. An antenna located inside V radiates in unbounded space, with observation point $\mathbf{r}$ on a large sphere S, whose tangent plane at $\mathbf{r}$ is T.

$$Z_0 = Y_0^{-1} \cong 120\pi \quad \text{ohms} \tag{38}$$

The amplitude vector $\mathbf{A}(\mathbf{k})$ in (35) is generally complex, and has the unit $(\text{watt})^{1/2}$. We will discuss it in detail below.

Referring to Fig. 10, we draw a sphere S, called the *radiation sphere*, passing through the observation point $\mathbf{r}$ and centered at O. A property of $\mathbf{A}(\mathbf{k})$ is that it is transverse to $\mathbf{k}$. Thus $\mathbf{A}$ lies in the tangent plane T of sphere S. Therefore the set of $\mathbf{A}$s forms a four-dimensional manifold (the tangent bundle of S). We decompose $\mathbf{A}$ into two orthogonal components:

$$\mathbf{A}(\mathbf{k}) = \mathbf{u}A(\mathbf{k},\mathbf{u}) + \mathbf{v}A(\mathbf{k},\mathbf{v}) \tag{39}$$

Here $(\mathbf{u},\mathbf{v})$ satisfy the following restrictions:

$$|\mathbf{u}|^2 = \mathbf{u}\cdot\mathbf{u}^* = |\mathbf{v}|^2 = 1 \tag{40a}$$

$$\mathbf{u}^*\cdot\mathbf{v} = \mathbf{u}\cdot\mathbf{v}^* = 0 \tag{40b}$$

$$\mathbf{u}\cdot\mathbf{k} = \mathbf{v}\cdot\mathbf{k} = 0 \tag{40c}$$

Unitary vectors $(\mathbf{u},\mathbf{v})$ describe the polarization of the radiated field (*cf.* Section 5). The scalar $A(\mathbf{k},\mathbf{u})$ is the component of $\mathbf{A}(\mathbf{k})$ in the direction $\mathbf{k}$ and with polarization $\mathbf{u}$, or the component in the state $(\mathbf{k},\mathbf{u})$. It is calculated by

$$A(\mathbf{k}, \mathbf{u}) = \mathbf{A}(\mathbf{k}) \cdot \mathbf{u}^* \tag{41}$$

A similar formula applies to $A(\mathbf{k}, \mathbf{v})$. As an example, we choose $\mathbf{u} = \hat{\boldsymbol{\theta}}$ and $\mathbf{v} = \hat{\boldsymbol{\phi}}$. Then (39) becomes

$$\mathbf{A}(\mathbf{k}) = \hat{\boldsymbol{\theta}} A_\theta(\mathbf{k}) + \hat{\boldsymbol{\phi}} A_\phi(\mathbf{k}) \tag{42}$$

where we used the more common notations (A_θ, A_ϕ) for $(A(\mathbf{k}, \hat{\boldsymbol{\theta}}), A(\mathbf{k}, \hat{\boldsymbol{\phi}}))$. Other examples of $(\mathbf{u}, \mathbf{v})$ are

(*a*) linear polarization:

$$\mathbf{u} = \hat{\boldsymbol{\theta}} \cos \phi - \hat{\boldsymbol{\phi}} \sin \phi, \qquad \mathbf{v} = \hat{\boldsymbol{\theta}} \sin \phi + \hat{\boldsymbol{\phi}} \cos \phi \tag{43a}$$

(*b*) circular polarization:

$$\mathbf{u} = (\hat{\boldsymbol{\theta}} + j\hat{\boldsymbol{\phi}})/\sqrt{2}, \qquad \mathbf{v} = (\hat{\boldsymbol{\theta}} - j\hat{\boldsymbol{\phi}})/\sqrt{2} \tag{43b}$$

In (43a), $\mathbf{u}$ is along the projection of $\hat{\mathbf{x}}$, and $\mathbf{v}$ is along the projection of $\hat{\mathbf{y}}$ onto the tangent plane T in Fig. 10. In (43b) $\mathbf{u}$ represents the left-hand circular polarization, and $\mathbf{v}$ represents the right-hand circular polarization.

For the far field in (35) the radiation intensity in watts per steradian (power per solid angle) in the state $(\mathbf{k}, \mathbf{u})$ is defined by

$$I(\mathbf{k}, \mathbf{u}) = (\mathbf{A} \cdot \mathbf{u}^*)(\mathbf{A} \cdot \mathbf{u}^*)^* = |A(\mathbf{k}, \mathbf{u})|^2 \tag{44}$$

A plot of $I(\mathbf{k}, \mathbf{u})$ as a function of observation direction $\hat{\mathbf{k}} = (\theta, \phi)$ is known as the antenna pattern. A typical pattern is shown in Fig. 11, where $I(\mathbf{k}, \hat{\boldsymbol{\phi}})$ is plotted as a function of θ for a fixed $\phi = \phi_0$. The maximum intensity for all observation directions occurs at a direction $\mathbf{k} = \mathbf{k}_0 = (k, \theta_0, \phi_0)$. The lobe in this direction is called the *main beam*, while all other lobes are called *side lobes*. The vertical axis of Fig. 11 shows the decibel (dB) value of the normalized intensity defined by

$$10 \log_{10} \left[\frac{I(\mathbf{k}, \mathbf{u})}{I(\mathbf{k}_0, \mathbf{u})} \right]$$

Thus the main beam is at the 0-dB level. The highest side lobe in Fig. 11 is -10 dB. It is customary to say that the side lobe level in $\phi = \phi_0$ cut is 10 dB down from the main beam. The width of the main beam at -3 dB level is called the *half-power beamwidth*, and in the cut $\phi = \phi_0$ is given by $\theta_2 - \theta_1$ in radians or in degrees.

For the far field in (35) the radiation intensity in all (both) polarizations is

$$|\mathbf{A}(\mathbf{k})|^2 = |A(\mathbf{k}, \mathbf{u})|^2 + |A(\mathbf{k}, \mathbf{v})|^2 \tag{45}$$

Alternatively, we can rewrite (39) as

$$\mathbf{A}(\mathbf{k}) = |A(k)|\mathbf{U} \tag{46}$$

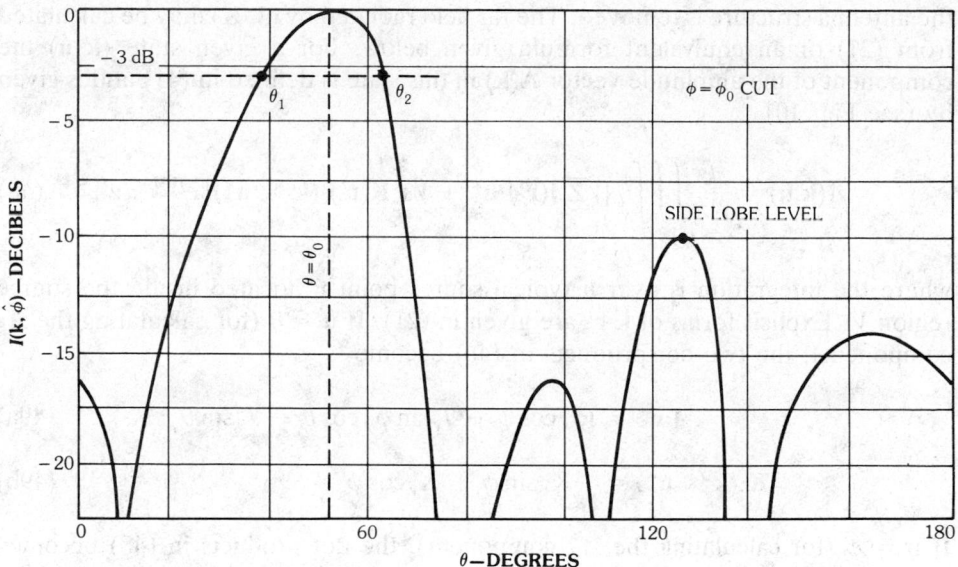

Fig. 11. An antenna pattern: the normalized intensity in the polarization $\hat{\phi}$ versus observation polar angle θ for a fixed $\phi = \phi_0$.

where the unitary vector **U** is given by

$$\mathbf{U} = \frac{1}{|\mathbf{A}(\mathbf{k})|} [\mathbf{u}A(\mathbf{k}, \mathbf{u}) + \mathbf{v}A(\mathbf{k}, \mathbf{v})] \tag{47}$$

Then the polarization of the far field in the direction **k** is described by a single unitary vector **U**.

In summary, an antenna far field can be represented by the spherical wave in (35). To describe amplitude vector **A(k)** we need to specify an observation direction **k**, and a polarization **u**, or a state **(k, u)**. The component of **A(k)** in this state is $A(\mathbf{k}, \mathbf{u})$ in (41), and the radiation intensity is $I(\mathbf{k}, \mathbf{u})$ in (44). There are two degrees of freedom in the polarization of **A(k)**. We customarily describe them by two orthogonal unitary vectors **(u, v)** in the manner stated in (39). A common choice is $(\mathbf{u} = \hat{\theta}, \mathbf{v} = \hat{\phi})$; another one is (**u** = reference polarization, **v** = cross polarization), which will be described in Section 9.

8. Calculation of the Far Field

We will look at the calculation of the far field (*a*) from given currents and (*b*) from given fields over a closed surface.

From Given Currents

The antenna far field is represented by the expression in (35). We will now consider its calculation. For a given transmitting antenna we may replace it by a mathematical current source **(J, K)** which radiates in the unbounded medium after the antenna structure is removed. The far field radiated by $(\mathbf{J}, \mathbf{K})$ may be calculated from (21) or an equivalent formula given below. For a given state $(\mathbf{k}, \mathbf{u})$ the component of the amplitude vector $\mathbf{A}(\mathbf{k})$ in this state is defined in (41) and is given by (see Fig. 10)

$$A(\mathbf{k}, \mathbf{u}) = \frac{k}{j4\pi} \iiint_V [\sqrt{Z}\mathbf{J}(\mathbf{r}') \cdot \mathbf{u}^* + \sqrt{Y}\mathbf{K}(\mathbf{r}') \cdot (\hat{\mathbf{k}} \times \mathbf{u}^*)] e^{j\mathbf{k} \cdot \mathbf{r}'} \, dv' \qquad (48)$$

where the integration is over a typical source point $\mathbf{r}'$ located inside the source region V. Explicit forms of $\mathbf{k} \cdot \mathbf{r}'$ are given in (21). If $\mathbf{u} = \hat{\boldsymbol{\theta}}$ (for calculating the A_θ component), the two dot products in (48) become

$$\mathbf{J} \cdot \mathbf{u}^* = (J_x \cos\phi + J_y \sin\phi) \cos\theta - J_z \sin\theta \qquad (49a)$$

$$\mathbf{K} \cdot (\hat{\mathbf{k}} \times \mathbf{u}^*) = -K_x \sin\phi + K_y \cos\phi \qquad (49b)$$

If $\mathbf{u} = \hat{\boldsymbol{\phi}}$ (for calculating the A_ϕ component), the dot products in (48) become

$$\mathbf{J} \cdot \mathbf{u}^* = -J_x \sin\phi + J_y \cos\phi \qquad (50a)$$

$$\mathbf{K} \cdot (\hat{\mathbf{k}} \times \mathbf{u}^*) = (K_x \cos\phi + K_y \sin\phi)(-\cos\theta) + K_z \sin\theta \qquad (50b)$$

From Given Fields over a Closed Surface

Referring to the radiation problem in Fig. 10, let us denote the (exact) radiation field from the antenna over a closed surface S_1 enclosing the antenna by $(\mathbf{E}_1, \mathbf{H}_1)$. If $(\mathbf{E}_1, \mathbf{H}_1)$ are given, we can calculate the far field (35) from them. The component of the amplitude vector $\mathbf{A}(\mathbf{k})$ in the state $(\mathbf{k}, \mathbf{u})$ is calculated from

$$A(\mathbf{k}, \mathbf{u}) = \frac{k}{j4\pi} \iint_{S_1} e^{j\mathbf{k} \cdot \mathbf{r}'_1} [\sqrt{Y}(\hat{\mathbf{k}} \times \mathbf{u}^*) \times \mathbf{E}_1(\mathbf{r}'_1) - \sqrt{Z}\mathbf{u}^* \times \mathbf{H}_1(\mathbf{r}'_1)] \cdot d\mathbf{s}'_1 \qquad (51)$$

where the integration is over the closed surface S_1, and $\mathbf{r}'_1$ is a typical point on S_1, and the differential surface $d\mathbf{s}'_1$ is in the direction of outward normal of S_1. For the special case that S_1 is the infinite plane $z = 0$ (Fig. 12), the two spherical components of $\mathbf{A}(\mathbf{k})$ deduced from (51) are

$$\begin{Bmatrix} A_\theta(\mathbf{k}) \\ A_\phi(\mathbf{k}) \end{Bmatrix} = \frac{k}{j4\pi} \iint_{-\infty}^{\infty} \begin{Bmatrix} B_\theta \\ B_\phi \end{Bmatrix} e^{jk\sin\theta\,(x'\cos\phi + y'\sin\phi)} \, dx' \, dy' \qquad (52)$$

where

$$B_\theta = \sqrt{Y}(-1)(E_x \cos\phi + E_y \sin\phi) + \sqrt{Z}(H_x \sin\phi - H_y \cos\phi) \cos\theta \qquad (53a)$$

$$B_\phi = \sqrt{Y}(E_x \sin\phi - E_y \cos\phi) \cos\theta + \sqrt{Z}(H_x \cos\phi + H_y \sin\phi) \qquad (53b)$$

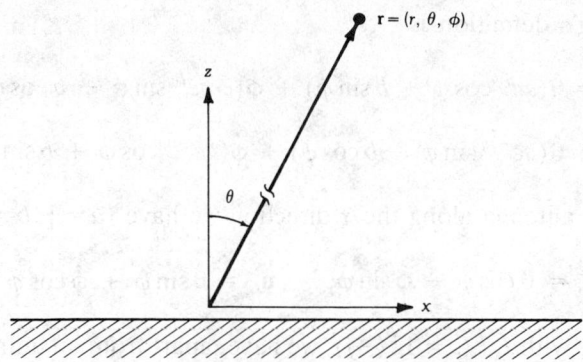

Fig. 12. Calculation of far field at **r** from a given field distribution over $z = 0$ plane.

In (53), (E_x, E_y, H_x, H_y) are the components of $(\mathbf{E}_1, \mathbf{H}_1)$ evaluated at a typical source point $(x', y', z' = 0)$ on surface S_1. The formula in (51) is related to that in (48) by Huygens' principle, which is discussed in Chapter 2.

9. Reference and Cross Polarizations

Consider a transmitting antenna with its far field given by (35). The amplitude vector $\mathbf{A}(\mathbf{k})$ can be expressed in terms of two polarization vectors $(\mathbf{u}, \mathbf{v})$ in the manner described in (39). Among many possible choices for $(\mathbf{u}, \mathbf{v})$, we often select a particular set $(\mathbf{u}_R, \mathbf{u}_C)$ and call them

$\mathbf{u}_R$ = polarization vector for the reference polarization

$\mathbf{u}_C$ = polarization vector for the cross polarization

(Alternative names for reference polarization are copolarization and principal polarization.) The unitary vectors $(\mathbf{u}_R, \mathbf{u}_C)$ are, in general, functions of observation direction **k**, and satisfy the constraints in (40).

There is no standard way to define $(\mathbf{u}_R, \mathbf{u}_C)$. This difficulty may be traced to the geometrical property that there is no unique way to transport a tangent vector at one point on the sphere S in Fig. 10 to another point on S. Consequently, if $(\mathbf{u}_R, \mathbf{u}_C)$ are defined in one direction, say the antenna main beam direction, there is no unique way to extend into other directions.

Nevertheless, a popular definition for $(\mathbf{u}_R, \mathbf{u}_C)$ is given by Ludwig [4]. His definition is useful for a class of antennas in which:

(*a*) the antenna main beam is nearly in the z direction, and
(*b*) the antenna current (or equivalent current) has the form

$$\mathbf{J}(\mathbf{r}) = J(\mathbf{r})(\hat{x}ae^{j\psi} + \hat{y}b) \tag{54}$$

where (a, b, ψ) are real (independent of **r**), and $a^2 + b^2 = 1$.

Then Ludwig's definition is

$$\mathbf{u}_R = \hat{\boldsymbol{\theta}}(ae^{j\psi}\cos\phi + b\sin\phi) + \hat{\boldsymbol{\phi}}(-ae^{j\psi}\sin\phi + b\cos\phi) \tag{55a}$$

$$\mathbf{u}_C = \hat{\boldsymbol{\theta}}(ae^{-j\psi}\sin\phi - b\cos\phi) + \hat{\boldsymbol{\phi}}(ae^{-j\psi}\cos\phi + b\sin\phi) \tag{55b}$$

For a linear wire antenna along the x direction we have $(a = 1, b = 0, \psi = 0)$ and

$$\mathbf{u}_R = \hat{\boldsymbol{\theta}}\cos\phi - \hat{\boldsymbol{\phi}}\sin\phi, \qquad \mathbf{u}_C = \hat{\boldsymbol{\theta}}\sin\phi + \hat{\boldsymbol{\phi}}\cos\phi$$

For a cross dipole with $J_x : J_y = 1 : (-j)$, we have, apart from an unimportant phase factor,

$$\mathbf{u}_R = \frac{1}{\sqrt{2}}(\hat{\boldsymbol{\theta}} - j\hat{\boldsymbol{\phi}}), \qquad \mathbf{u}_C = \frac{1}{\sqrt{2}}(\hat{\boldsymbol{\theta}} + j\hat{\boldsymbol{\phi}})$$

which describe circular polarizations as expected.

10. Antenna Gains

There are several different terms relating to the gain of a transmitting antenna, depending on the incident power normalization, the direction of observation, and polarization of its far field.

Incident Powers on an Antenna

A transmitting antenna is connected to a generator through a transmission line (or a waveguide) as sketched in Fig. 13. The (time-averaged) power in watts incident from the generator to the antenna is denoted by P_1. Due to the mismatch between the transmission line and the antenna, only P_2 is accepted by the antenna, where

$$P_2 = (1 - |\Gamma|^2)P_1 \tag{56}$$

and $|\Gamma|^2$ is the power reflection coefficient. If the antenna is lossless, P_2 is radiated

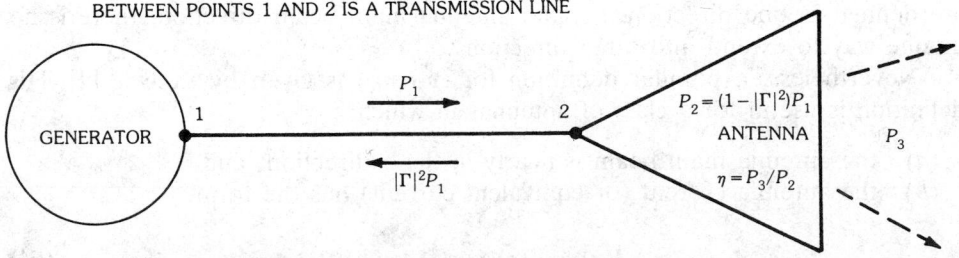

Fig. 13. The power incident to the antenna is P_1, the power accepted by the antenna is P_2, and that radiated is P_3.

into the space. If the antenna is lossy, the radiated power P_3 is only a portion of P_2. The ratio

$$\eta = P_3/P_2 \tag{57}$$

is called the *radiation efficiency*.

Three Gains in State $(\mathbf{k}, \mathbf{U})$

As indicated in (39) the antenna radiates in two orthogonal polarizations $(\mathbf{u}, \mathbf{v})$. Alternatively, its polarization can be represented by (within a phase factor) a single unitary vector $\mathbf{U}$ in the manner described in (46) and (47). Then the radiation intensity in watts per steradian in state $(\mathbf{k}, \mathbf{U})$ is

$$I(\mathbf{k}) = |\mathbf{A}(\mathbf{k})|^2 = |A(\mathbf{k}, \mathbf{u})|^2 + |A(\mathbf{k}, \mathbf{v})|^2 \tag{58}$$

which includes radiation in all (both) polarizations. The radiated power P_3 by the antenna in space is

$$P_3 = \int_0^\pi d\theta \int_0^{2\pi} d\phi \, [I(\mathbf{k}) \sin \theta] \tag{59}$$

The general definition of a gain (dimensionless) in the direction $\mathbf{k}$ and for all polarizations is

$$G(\mathbf{k}) = \frac{4\pi}{P_n} I(\mathbf{k}) = \frac{4\pi}{P_n} |\mathbf{A}(\mathbf{k})|^2 = 4\pi \frac{\text{intensity of the antenna in direction } \mathbf{k}}{\text{a reference power}} \tag{60}$$

where P_n is a reference power. Depending on P_n, there are three commonly used gains:

(a)   realized gain $G_1(\mathbf{k})$ if $P_n = P_1 = $ power incident at the antenna,
(b)   gain $G_2(\mathbf{k})$ if $P_n = P_2 = $ power accepted by the antenna,
(c)   directivity $D(\mathbf{k})$ if $P_n = P_3 = $ power radiated by the antenna.

The relations among the three gains are

$$G_1(\mathbf{k}) = (1 - |\Gamma|^2)G_2(\mathbf{k}) = \eta(1 - |\Gamma|^2)D(\mathbf{k}) \tag{61}$$

The three gains are graphically illustrated in the lower half of Fig. 14.

Three Gains in State $(\mathbf{k}, \mathbf{u})$

The antenna has a polarization $\mathbf{U}$ as defined in (47). In defining gains we may specify a preferred polarization $\mathbf{u}$. Then the radiation intensity in state $(\mathbf{k}, \mathbf{u})$ is

$$I(\mathbf{k}, \mathbf{u}) = |A(\mathbf{k}, \mathbf{u})|^2 = |\mathbf{A}(\mathbf{k}) \cdot \mathbf{u}^*|^2 \tag{62}$$

in watts per steradian. The ratio

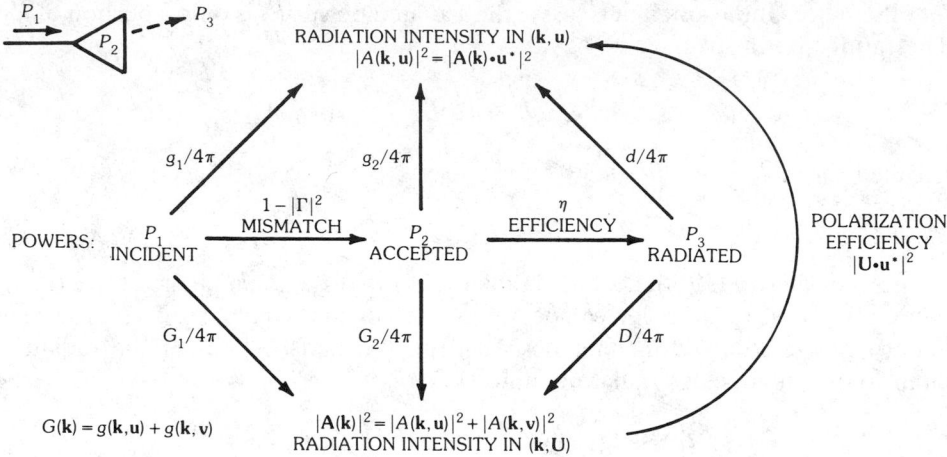

Fig. 14. Six definitions of gain ($a \xrightarrow{t} b$ means $at = b$). (*Courtesy G. A. Deschamps*)

$$p = \frac{I(\mathbf{k}, \mathbf{u})}{I(\mathbf{k})} = |\mathbf{U} \cdot \mathbf{u}^*|^2 \tag{63}$$

is called the *polarization efficiency in state* $(\mathbf{k}, \mathbf{u})$ *for an antenna that has polarization* **U**. With respect to state $(\mathbf{k}, \mathbf{u})$ we define a gain

$$g(\mathbf{k}, \mathbf{u}) = \frac{4\pi}{P_n} I(\mathbf{k}, \mathbf{u}) = \frac{4\pi}{P_n} |\mathbf{A}(\mathbf{k}) \cdot \mathbf{u}^*|^2 = pG(\mathbf{k}) \tag{64}$$

$$= 4\pi \frac{\text{intensity of the antenna in state } (\mathbf{k}, \mathbf{u})}{\text{a reference power}}$$

Again, depending on the reference power P_n, there are three gains g_1, g_2, and d, which are analogous to the cases associated with (60). These three gains are illustrated in the upper half of Fig. 14.

The gain defined in (64) is called the *partial gain* for a specific polarization **u**. The (total) gain $G(\mathbf{k})$ in (60) is the sum of partial gains for any two orthogonal polarizations:

$$G(\mathbf{k}) = g(\mathbf{k}, \mathbf{u}) + g(\mathbf{k}, \mathbf{v})$$

which follows from (58).

Peak Gains

All of the six gains depend on the observation direction $\hat{\mathbf{k}} = (\theta, \phi)$. In applications we often use the maximum of a gain as a function of $\hat{\mathbf{k}}$. Thus, corresponding to (60), we define a *peak gain* for all polarizations by

$$G = \max G(\mathbf{k}) = \frac{4\pi}{P_n} \max |\mathbf{A}(\mathbf{k})|^2 \tag{65}$$

In a similar manner we define a peak gain for a specific polarization **u**. When a gain is given without a specified observation direction, it is customarily assumed to be the peak gain.

Gain-Related Terms

When defining antenna terms related to gain, we can use any one of the six definitions of gain. If the radiation in all polarizations is of interest, we use $G(\mathbf{k})$ in (60). If the radiation in a specific polarization is of interest, we use $g(\mathbf{k}, \mathbf{u})$ in (62). With this understanding we use $G(\mathbf{k})$ to represent a typical gain in the discussion below.

An isotropic radiator is a hypothetical lossless antenna having equal radiation intensity in all directions, i.e., $I(\mathbf{k}) =$ a constant. If an input power P_n were fed to an isotropic radiator, its radiation intensity in watts per steradian would be

$$I_{\text{iso}}(\mathbf{k}) = \frac{P_n}{4\pi} \tag{66}$$

Here P_n can be any one of three powers explained at the beginning of this section. In terms of (66) the gain definition in (60) has the following interpretation:

$$G(\mathbf{k}) = \frac{I(\mathbf{k})}{I_{\text{iso}}(\mathbf{k})}$$

$$= \frac{\text{intensity of the antenna in direction } \mathbf{k}}{\text{intensity of an isotropic radiator fed by the same power}} \tag{67}$$

We often express the dimensionless G by its decibel (dB) value: $10 \log_{10} G$. Sometimes we write dB as dBi, where the letter "i" emphasizes that the gain is over an isotropic radiator.

Consider an antenna with gain $G(\mathbf{k})$ fed by an input power P_n. Its radiation intensity $I(\mathbf{k})$ would be the same as that for an isotropic radiator if the latter were fed with an input power given in watts by

$$\text{EIRP} = P_n G(\mathbf{k}) \tag{68}$$

where EIRP stands for equivalent (effective) isotropically radiated power. We often express EIRP by $10 \log_{10} P_n G$ in decibels referred to 1 W (dBW).

When an antenna is used for receiving, a figure of merit of the antenna is defined by

$$\frac{G}{T_a} = \frac{\text{peak gain of the antenna}}{\text{noise temperature of the antenna}} \tag{69}$$

T_a, which is usually given in kelvins (K), is discussed in Chapter 2.

11. Pattern Approximation by $(\cos\theta)^q$

The far-field patterns of many aperture-type antennas have two characteristics: (*a*) A single major lobe exists in the forward half-space $z > 0$ and the lobe maximum is in the $+z$ direction. (*b*) The radiation in the backward half-space $z < 0$ is negligible. For these types of antennas their far fields can be approximated by simple analytical functions described in this section. For an *x*-polarized antenna, such as the flanged waveguide shown in Fig. 15, we have (for $\theta \leqq \pi/2, 0 \leqq \phi \leqq 2\pi$, and $r \to \infty$)

$$\mathbf{E}(\mathbf{r}) = A_0[\hat{\boldsymbol{\theta}}\, C_E(\theta)\cos\phi - \hat{\boldsymbol{\phi}}\, C_H(\theta)\sin\phi]\,\sqrt{Z}\,\frac{e^{-jkr}}{r} \qquad (70\text{a})$$

For a *y*-polarized antenna we have

$$\mathbf{E}(\mathbf{r}) = A_0[\hat{\boldsymbol{\theta}}\, C_E(\theta)\sin\phi + \hat{\boldsymbol{\phi}}\, C_H(\theta)\cos\phi]\,\sqrt{Z}\,\frac{e^{-jkr}}{r} \qquad (70\text{b})$$

Here A_0 is a complex constant in (watts)$^{1/2}$, and

$$C_E(\theta) = (\cos\theta)^{q_E} = E\text{-plane pattern} \qquad (71\text{a})$$

$$C_H(\theta) = (\cos\theta)^{q_H} = H\text{-plane pattern} \qquad (71\text{b})$$

The shape of the pattern is controlled by indices (q_E, q_H). For a given linearly polarized antenna pattern we can approximate it by (70), with (q_E, q_H) determined by matching the given pattern and (70) at two selected directions. A proper superposition of (70a) and (70b) gives a circularly polarized far field, namely,

$$\mathbf{E}(\mathbf{r}) = A_0 e^{j\tau\phi}[\hat{\boldsymbol{\theta}}\, C_E(\theta) + \hat{\boldsymbol{\phi}}\, j\tau\, C_H(\theta)]\,\sqrt{Z}\,\frac{e^{-jkr}}{r} \qquad (72)$$

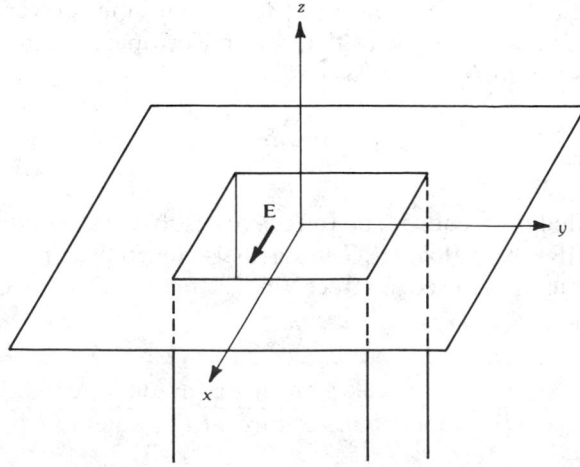

Fig. 15. Radiation from a flanged rectangular waveguide.

where $\tau = +1$ for left-handed circular polarization, and $\tau = -1$ for right-handed circular polarization. Note that (72) represents a perfectly circularly polarized wave only in the main beam direction ($\theta = 0$). Away from this direction it is generally elliptically polarized (unless $q_E = q_H$).

The directivity of the pattern in (70) or (72) can be calculated from (60) and (59). The result for the peak directivity in the main beam direction is

$$D(\theta = 0) = \frac{2(2q_E + 1)(2q_H + 1)}{q_E + q_H + 1} \tag{73}$$

which is valid for both linear and circular polarizations. It should be noted that (73) applies only if the antenna has no radiation in the backward half-space ($z < 0$ in Fig. 15). For an antenna which has a symmetrical pattern in the forward and backward half-spaces, the directivity is one half of the value calculated from (73). For example, an electric dipole oriented in the x direction has a pattern described by (70a), with $q_E = 1$ and $q_H = 0$ for all $\theta \leq \pi$. Its directivity is 3/2, one half of the value given by (73).

The formula in (73) can be expressed in a different form. The half-power beamwidths (θ_E, θ_H) in the (E, H) planes are related to (q_E, q_H) by

$$\left(\cos \frac{\theta_n}{2}\right)^{q_n} = \frac{1}{\sqrt{2}}, \qquad \text{for } n = E, H \tag{74}$$

The use of (74) in (73) leads to

$$D(\theta = 0) = \frac{4}{\dfrac{1}{1 + 0.3/|\log \cos(\theta_E/2)|} + \dfrac{1}{1 + 0.3/|\log \cos(\theta_H/2)|}} \tag{75}$$

where the log is of base 10. If θ_E and θ_H are much less than one radian, (75) is simplified to read

$$D(\theta = 0) \cong \frac{36\,300}{(\theta_E^2 + \theta_H^2)/2} \tag{76}$$

where θ_E and θ_H are expressed in degrees.

Another directivity formula, which is similar to (76), is based on the approximation that $D \cong 4\pi/\Omega$, where Ω is the solid angle extended by the main beam at the half-power level. Using a further approximation that $\Omega \cong \theta_E \theta_H$, we obtain

$$D \cong \frac{41\,300}{\theta_E \theta_H} \tag{77}$$

where θ_E and θ_H are expressed in degrees. Formulas similar to (76) and (77) exist in the literature, and the coefficients in the numerator usually fall in the range from 27 900 to 41 300 (see [5, 6]).

12. TE and TM Field Representations

Consider an isotropic, homogeneous medium characterized by the constitutive parameters (ϵ, μ). In a source-free region of the medium, the field can be decomposed into two components: TE (transverse electric) field and TM (transverse magnetic) field with respect to the z direction. For the TE component the fields are derived from an electric vector potential $\mathbf{F(r)}$ by

$$\mathbf{E} = -\nabla \times \mathbf{F} \tag{78a}$$

$$\mathbf{H} = \frac{k}{jZ}\left[\mathbf{F} + \frac{1}{k^2}\nabla(\nabla\cdot\mathbf{F})\right] \tag{78b}$$

where

$$\mathbf{F} = \hat{\mathbf{z}}\psi(\mathbf{r}) \tag{79a}$$

$$(\nabla^2 + k^2)\psi(\mathbf{r}) = 0 \tag{79b}$$

$$k = \omega\sqrt{\mu\epsilon}, \qquad Z = \sqrt{\mu/\epsilon} \tag{79c}$$

For the TM field the corresponding equations are

$$\mathbf{E} = -jkZ\left[\mathbf{A} + \frac{1}{k^2}\nabla(\nabla\cdot\mathbf{A})\right] \tag{80a}$$

$$\mathbf{H} = \nabla \times \mathbf{A} \tag{80b}$$

where

$$\mathbf{A} = \hat{\mathbf{z}}\frac{1}{Z}\overline{\psi}(\mathbf{r}) \tag{81a}$$

$$(\nabla^2 + k^2)\overline{\psi}(\mathbf{r}) = 0 \tag{81b}$$

We introduce Z^{-1} in (81a) so that ψ and ψ have the same dimensions in volts. In rectangular coordinates, (78) and (80) may be written explicitly:

Transverse Electric

$$E_x = -\frac{\partial\psi}{\partial y} \qquad\qquad H_x = \frac{1}{jkZ}\frac{\partial^2\psi}{\partial x\partial z}$$

$$E_y = \frac{\partial\psi}{\partial x} \qquad\qquad H_y = \frac{1}{jkZ}\frac{\partial^2\psi}{\partial y\partial z} \tag{82}$$

$$E_z = 0 \qquad\qquad H_z = \frac{1}{jkZ}\left(\frac{\partial^2}{\partial z^2} + k^2\right)\psi$$

Transverse Magnetic

$$E_x = \frac{1}{jk}\frac{\partial^2\overline{\psi}}{\partial x\partial z} \qquad\qquad H_x = \frac{1}{Z}\frac{\partial\overline{\psi}}{\partial y}$$

$$E_y = \frac{1}{jk}\frac{\partial^2\overline{\psi}}{\partial y\partial z} \qquad\qquad H_y = -\frac{1}{Z}\frac{\partial\overline{\psi}}{\partial x} \qquad (83)$$

$$E_z = \frac{1}{jk}\left(\frac{\partial^2}{\partial z^2} + k^2\right)\overline{\psi} \qquad\qquad H_z = 0$$

In cylindrical coordinates, (78) and (80) may be written explicitly:

Transverse Electric

$$E_\varrho = -\frac{1}{\varrho}\frac{\partial\psi}{\partial\phi} \qquad\qquad H_\varrho = \frac{1}{jkZ}\frac{\partial^2\psi}{\partial\varrho\partial z}$$

$$E_\phi = \frac{\partial\psi}{\partial\varrho} \qquad\qquad H_\phi = \frac{1}{jkZ\varrho}\frac{\partial^2\psi}{\partial\phi\partial z} \qquad (84)$$

$$E_z = 0 \qquad\qquad H_z = \frac{1}{jkZ}\left(\frac{\partial^2}{\partial z^2} + k^2\right)\psi$$

Transverse Magnetic

$$E_\varrho = \frac{1}{jk}\frac{\partial^2\overline{\psi}}{\partial\varrho\partial z} \qquad\qquad H_\varrho = \frac{1}{Z\varrho}\frac{\partial\overline{\psi}}{\partial\phi}$$

$$E_\phi = \frac{1}{jk\varrho}\frac{\partial^2\overline{\psi}}{\partial\phi\partial z} \qquad\qquad H_\phi = -\frac{1}{Z}\frac{\partial\overline{\psi}}{\partial\varrho} \qquad (85)$$

$$E_z = \frac{1}{jk}\left(\frac{\partial^2}{\partial z^2} + k^2\right)\overline{\psi} \qquad\qquad H_z = 0$$

Two remarks are in order. First, in a conducting medium with conductivity σ all the formulas in this section remain valid if ϵ is replaced by $\epsilon(1 - j\sigma/\omega\epsilon)$. Second, to decompose the field into TE and TM with respect to the x direction, we make the following changes in (82) and (83): $x \to y$, $y \to z$, and $z \to x$.

13. Plane-Wave Spectrum Representation

A general solution of the wave equations in (79b) or (81b) is

$$\left\{\begin{matrix}\psi(r)\\ \overline{\psi}(r)\end{matrix}\right\} = \int_{-\infty}^{\infty} dk_x \int_{-\infty}^{\infty} dk_y \left\{\begin{matrix}\Psi(k_x, k_y)\\ \overline{\Psi}(k_x, k_y)\end{matrix}\right\} e^{-j(k_x x + k_y y + k_z z)} \qquad (86)$$

Here (k_x, k_y) are two integration variables, and

$$k_z = \sqrt{k^2 - (k_x^2 + k_y^2)} \tag{87a}$$

where $k = \omega\sqrt{\mu\epsilon}$. The square root in (87a) is taken such that

$$\text{Re}\,\{k_z\} \geqq 0 \quad \text{and} \quad \text{Im}\,\{k_z\} \leqq 0 \tag{87b}$$

The integral in (86) represents a superposition of plane waves and constitutes the plane-wave spectrum representation (or Fourier transform representation) of fields in a source-free region.

The amplitudes of the plane waves are related to potentials via the inverse Fourier theorem, namely,

$$\Psi(k_x, k_y) = \left(\frac{1}{2\pi}\right)^2 e^{jk_z z_0} \int_{-\infty}^{\infty} dx_0 \int_{-\infty}^{\infty} dy_0\, \psi(x_0, y_0, z_0)\, e^{+j(k_x x_0 + k_y y_0)} \tag{88}$$

Thus $\Psi(k_x, k_y)$ can be determined from the values of $\psi(\mathbf{r})$ over a reference plane P_0, on which a typical point is $(x_0, y_0, z_0) = $ a constant (Fig. 16).

The explicit field components of $(\mathbf{E}, \mathbf{H})$ in the plane-wave representation can be obtained by substituting (86) into (82) and (83). Including both TE and TM components, the $\mathbf{E}$ field is

$$\begin{Bmatrix} E_x(\mathbf{r}) \\ E_y(\mathbf{r}) \\ E_z(\mathbf{r}) \end{Bmatrix} = \int_{-\infty}^{\infty} dk_x \int_{-\infty}^{\infty} dk_y\, e^{-j(k_x x + k_y y + k_z z)}$$

$$\times \left[\begin{Bmatrix} jk_y \\ -jk_x \\ 0 \end{Bmatrix} \Psi(k_x, k_y) + \begin{Bmatrix} -k_x k_z \\ -k_y k_z \\ k_x^2 + k_y^2 \end{Bmatrix} \frac{1}{jk} \overline{\Psi}(k_x, k_y) \right] \tag{89}$$

A similar formula exists for the $\mathbf{H}$ field. We note that any rectangular component of $(\mathbf{E}, \mathbf{H})$ has the same form as $\psi(\mathbf{r})$ given in (86). In the following we will give several formulas related to the integral of $\psi(\mathbf{r})$. These formulas apply equally well to any rectangular components of $(\mathbf{E}, \mathbf{H})$.

(a) *Far field due to finite sources.* If the sources that produce the field in (86) are confined in a finite region near $\mathbf{r} = 0$, then amplitude $\Psi(k_x, k_y)$ is an entire function of three independent complex variables k_x, k_y, and k_z. In the far-field region the integral in (88) can be asymptotically evaluated, with the result

$$\psi(\mathbf{r}) \sim \frac{e^{-jkr}}{r} (2\pi j\, k \cos\theta)\, \Psi(k_x = k \sin\theta \cos\phi,\ k_y = k \sin\theta \sin\phi) \tag{90}$$

valid for $r \to \infty$. Here (r, θ, ϕ) are the spherical coordinates of the observation point $\mathbf{r}$.

(b) *Huygens-Fresnel principle.* Substituting (88) into (86) and making use of (90), we obtain another far-field formula valid for $R \to \infty$ (see Fig. 16):

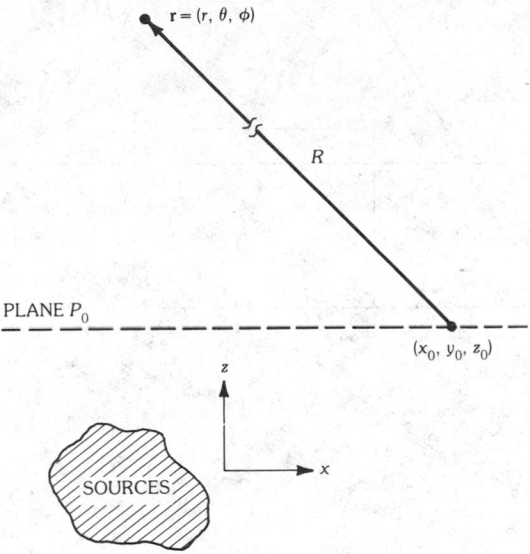

Fig. 16. Far field at **r** is calculated from the field over a plane P_0.

$$\psi(\mathbf{r}) \sim \left(\frac{jk}{2\pi}\right) \int_{-\infty}^{\infty} dx_0 \int_{-\infty}^{\infty} dy_0\, \psi(x_0, y_0, z_0)\, \frac{z - z_0}{R}\, \frac{e^{-jkR}}{R} \qquad (91a)$$

where

$$R = \sqrt{(x - x_0)^2 + (y - y_0)^2 + (z - z_0)^2} \qquad (91b)$$

Due to finite sources near plane P_0, the factor $(z - z_0)/R$ in (91a) is close to unity. Apart from this factor, (91a) agrees with a corresponding formula derived by using the Huygens-Fresnel principle in optics [7].

(c) *Field with azimuthal symmetry.* If the field is independent of ϕ, the representation in (86) can be rewritten as

$$\psi(\varrho, z) = 2\pi \int_0^{\infty} dk_\varrho\, [k_\varrho J_0(k_\varrho \varrho)\, \Psi(k_\varrho)]\, e^{-j\sqrt{k^2 - k_\varrho^2}\, z} \qquad (92)$$

The representation in (92) is valid for any finite ϱ (including $\varrho = 0$). In a region which is unbounded in the ϱ direction and excludes $\varrho = 0$, however, we must replace $J_0(\cdot)$ in (92) by $H_0^{(2)}(\cdot)$ on account of the radiation condition.

14. Periodic Structure

Consider an infinitely large periodic structure of thickness τ situated in an unbounded, isotropic, homogeneous space as sketched in Fig. 17. In terms of geometry and material the structure is repetitive from cell to cell, with its periodic lattice specified by three parameters (a, b, Ω). The structure is illuminated by an

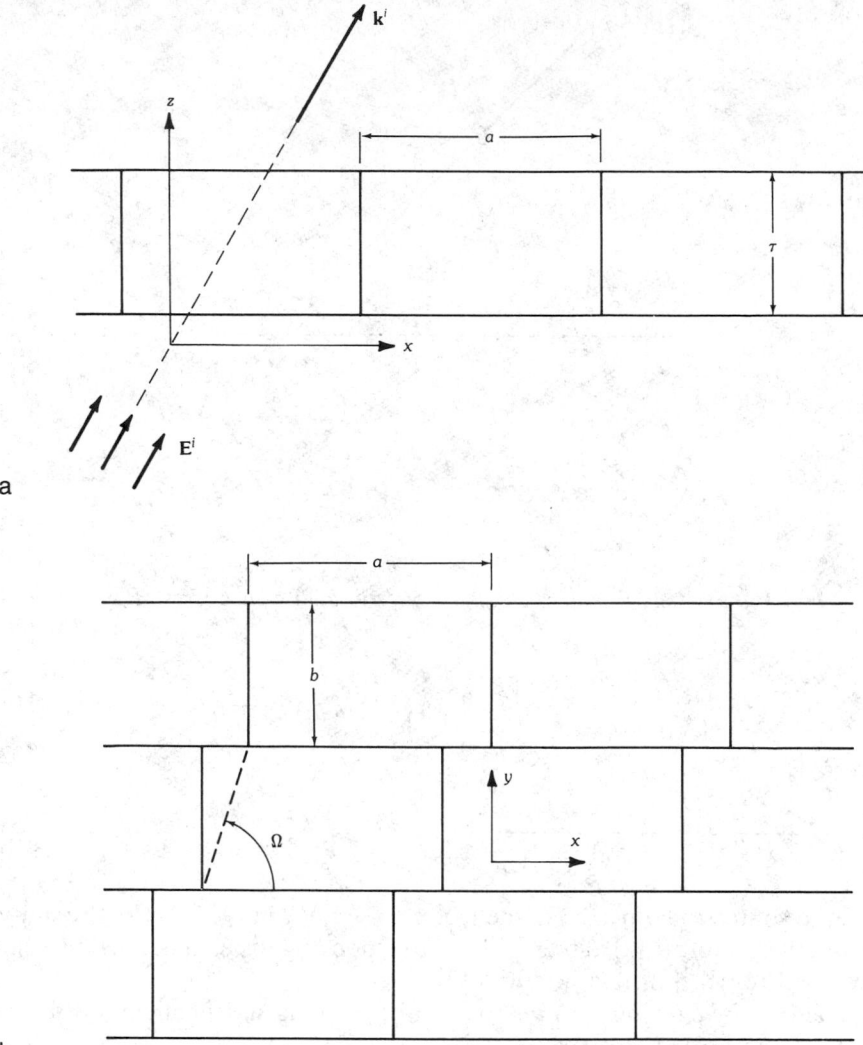

Fig. 17. A periodic structure illuminated by an incident plane wave. (*a*) Side view. (*b*) Top view.

incident plane wave described in (22). The plane wave propagates in the direction $\mathbf{k} = \mathbf{k}^i$, where

$$\mathbf{k}^i = k(\hat{\mathbf{x}} \sin \theta_0 \cos \phi_0 + \hat{\mathbf{y}} \sin \theta_0 \sin \phi_0 + \hat{\mathbf{z}} \cos \theta_0) \qquad (93)$$

The problem at hand is to find a field representation for the scattered field in the unbounded space (outside of the periodic slab).

Due to the periodic nature of the structure and the uniformity of the incident field, the scattered field can be expressed in terms of Floquet space harmonics. For the (p, q)th space harmonics, its transverse field variation is given by

$$Q_{pq}(x, y) = \exp[-j(u_{pq}x + v_{pq}y)], \qquad p, q = 0, \pm 1, \pm 2, \ldots \qquad (94a)$$

where

$$u_{pq} = (2p\pi/a) + k \sin\theta_0 \cos\phi_0 \qquad (94b)$$

$$v_{pq} = (-2p\pi/a)\cot\Omega + (2q\pi/b) + k \sin\theta_0 \sin\phi_0 \qquad (94c)$$

Its z variation is $\exp(\pm j\gamma_{pq}z)$, where the propagation constant is

$$\gamma_{pq} = \sqrt{k^2 - (u_{pq}^2 + v_{pq}^2)} \qquad (95)$$

The square root in (95) is taken such that

$$\mathrm{Re}\{\gamma_{pq}\} \geqq 0 \quad \text{and} \quad \mathrm{Im}\{\gamma_{pq}\} \leqq 0 \qquad (96)$$

In a lossless medium (with real k), only a finite number of $\{\gamma_{pq}\}$s are real, corresponding to propagating space harmonics. The remaining $\{\gamma_{pq}\}$s are negative imaginary, corresponding to attenuating space harmonics. In phased array terminology, the dominant space harmonic with $(p = 0, q = 0)$ is called the *main beam*, whereas all others are *grating lobes*. In most applications we wish to choose an array lattice so that only the main beam is propagating and all grating lobes are attenuating. To achieve this for the incident polar angle in the range $0 \leqq \theta_0 \leqq \theta_1$ and for $0 \leqq \phi_0 \leqq 2\pi$, the lattice parameters must be such that

$$1 + \sin\theta_1 < \Gamma \qquad (97a)$$

where

$$\Gamma = \min\left\{\frac{\lambda}{a\sin\Omega}, \frac{\lambda}{b}, \left[\left(\frac{\lambda}{a}\right)^2 + \left(\frac{\lambda}{b} - \frac{\lambda}{a}\cot\Omega\right)^2\right]^{1/2}\right\} \qquad (97b)$$

In other words, when (97a) is satisfied, only γ_{00} in (95) is real in the range $0 \leqq \theta_0 \leqq \theta_1$ and $0 \leqq \phi_0 \leqq 2\pi$. For several commonly used lattices, (97b) can be simplified to the following:

(a) For a rectangular lattice ($\Omega = 90°$),

$$\Gamma = \min\left\{\frac{\lambda}{a}, \frac{\lambda}{b}\right\} \qquad (98a)$$

(b) For an exact triangular lattice ($b = \sqrt{3}a/2$, $\Omega = 60°$),

$$\Gamma = \frac{2\lambda}{\sqrt{3}a} \qquad (98b)$$

(c) For an isosceles triangular lattice ($\cot\Omega = 0.5a/b$),

$$\Gamma = \min\left\{\frac{\lambda \csc \Omega}{a}, \frac{\lambda}{b}\right\} \tag{98c}$$

which reduces to $\Gamma = \sqrt{5}\lambda/(2a)$ if $a = b$ ($\Omega = 63.435°$).

The scattered field in the unbounded space can be decomposed into TE_z and TM_z components in the manner described in Section 12. The two potentials are expressed in terms of Floquet space harmonics:

$$\left\{\begin{matrix} \psi(\mathbf{r}) \\ \overline{\psi}(\mathbf{r}) \end{matrix}\right\} = \sum_{p=-\infty}^{\infty} \sum_{q=-\infty}^{\infty} \left[\left\{\begin{matrix} C_{pq} \\ \overline{C}_{pq} \end{matrix}\right\} e^{-j\gamma_{pq}z} + \left\{\begin{matrix} D_{pq} \\ \overline{D}_{pq} \end{matrix}\right\} e^{+j\gamma_{pq}z}\right] Q_{pq}(x, y) \tag{99}$$

The fields $(\mathbf{E}, \mathbf{H})$ can be calculated from (99), (82), and (83). Retaining only the component traveling in the $+z$ direction, we have

$$\left\{\begin{matrix} \mathbf{E}_t \\ E_z \\ \mathbf{H}_t \\ H_z \end{matrix}\right\} = \sum_p \sum_q \left[\left\{\begin{matrix} \boldsymbol{\alpha}_{pq} \\ 0 \\ (\gamma_{pq}/kZ)\boldsymbol{\beta}_{pq} \\ -w_{pq}^2/kZ \end{matrix}\right\} C_{pq} + \left\{\begin{matrix} (\gamma_{pq}/k)\boldsymbol{\beta}_{pq} \\ -w_{pq}^2/k \\ -Z^{-1}\boldsymbol{\alpha}_{pq} \\ 0 \end{matrix}\right\} \overline{C}_{pq}\right]$$

$$\times [jQ_{pq}(x, y)\, e^{-j\gamma_{pq}z}] \tag{100}$$

where

$$\boldsymbol{\alpha}_{pq} = (\hat{\mathbf{x}}v_{pq} - \hat{\mathbf{y}}u_{pq})$$

$$\boldsymbol{\beta}_{pq} = (\hat{\mathbf{x}}u_{pq} + \hat{\mathbf{y}}v_{pq}) = \hat{\mathbf{z}} \times \hat{\boldsymbol{\alpha}}_{pq}$$

$$w_{pq}^2 = u_{pq}^2 + v_{pq}^2$$

$$Z = \sqrt{\mu/\epsilon}$$

The field component traveling in the $-z$ direction associated with coefficients $(D_{pq}, \overline{D}_{pq})$ is also given by (100) after changing the sign of γ_{pq}. The time-averaged power carried by the total field in the $+z$ direction in each unit cell $a \times b$ is defined by

$$\overline{P} = \iint_{\text{a cell}} \text{Re}\{(\mathbf{E} \times \mathbf{H}^*)\cdot\hat{\mathbf{z}}\}\, dx\, dy \tag{101a}$$

If the medium is lossless, we have

$$\overline{P} = \frac{ab}{kZ} \sum_p \sum_q^{\text{prop.}} w_{pq}^2 \gamma_{pq}\{|C_{pq}|^2 + |\overline{C}_{pq}|^2 - |D_{pq}|^2 - |\overline{D}_{pq}|^2\} \tag{101b}$$

in which the summation includes only propagating space harmonics with real γ_{pq}s.

The constant coefficients C_{pq} in (99) are to be determined by matching

boundary conditions at $z = 0$ and $z = \tau$. This can be done only after a more detailed description of the periodic slab is given. Of particular interest is the dominant
space harmonic with $(p = 0, q = 0)$. The fields with coefficients $(C_{00}, \overline{C}_{00})$ are the
(perpendicular, parallel) components of a plane wave propagating in the direction
of $\mathbf{k}^i$:

$$\mathbf{E}(\mathbf{r}) = (jk \sin \theta_0)[(-\hat{\boldsymbol{\phi}}^i)C_{00} + \hat{\boldsymbol{\theta}}^i \overline{C}_{00}] e^{-j\mathbf{k}^i \cdot \mathbf{r}} + \cdots \tag{102a}$$

$$\mathbf{H}(\mathbf{r}) = \frac{1}{Z} (jk \sin \theta_0)[\hat{\boldsymbol{\theta}}^i C_{00} + \hat{\boldsymbol{\phi}}^i \overline{C}_{00}] e^{-j\mathbf{k}^i \cdot \mathbf{r}} + \cdots \tag{102b}$$

where $\mathbf{k}^i$ is given in (93), and

$$\hat{\boldsymbol{\theta}}^i = (\hat{\mathbf{x}} \cos \phi_0 + \hat{\mathbf{y}} \sin \phi_0) \cos \theta_0 - \hat{\mathbf{z}} \sin \theta_0 = \hat{\boldsymbol{\phi}}^i \times \hat{\mathbf{k}}^i \tag{102c}$$

$$\hat{\boldsymbol{\phi}}^i = -\hat{\mathbf{x}} \sin \phi_0 + \hat{\mathbf{y}} \cos \phi_0 = \hat{\mathbf{k}}^i \times \hat{\boldsymbol{\theta}}^i \tag{102d}$$

The fields with coefficients $(D_{00}, \overline{D}_{00})$ are the (perpendicular, parallel) components
of a plane wave propagating in the direction of $\mathbf{k}^r$, where $\mathbf{k}^r = \mathbf{k}^i - 2\hat{\mathbf{z}}(\hat{\mathbf{z}} \cdot \mathbf{k}^i)$ is the
reflected wave vector. By "perpendicular" component we mean that the $\mathbf{E}$ vector of
the field is linearly polarized and is perpendicular to the plane of incidence defined
by $\hat{\mathbf{z}}$ and $\mathbf{k}^i$.

15. Rectangular Waveguide

Consider a metallic rectangular waveguide filled with an isotropic homogeneous medium (Fig. 18). The field in the waveguide can be decomposed into TE_z
and TM_z. They are derivable from two potential functions (for the component
traveling in the $+z$ direction):

$$\psi = \sum_{m=0}^{\infty} \sum_{n=0}^{\infty}{}' C_{mn} \cos\left(\frac{m\pi}{a} x\right) \cos\left(\frac{n\pi}{b} y\right) e^{-j\gamma_{mn}z} \tag{103a}$$

$$\overline{\psi} = \sum_{m=1}^{\infty} \sum_{n=1}^{\infty} \overline{C}_{mn} \sin\left(\frac{m\pi}{a} x\right) \sin\left(\frac{n\pi}{b} y\right) e^{-j\gamma_{mn}z} \tag{103b}$$

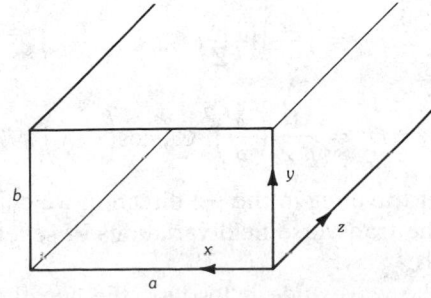

Fig. 18. A rectangular waveguide.

In (103a) the prime signifies that the term with $m = n = 0$ is omitted in the summation. The propagation constants γ_{mn} for either TE_{mn} or TM_{mn} are given by

$$\gamma_{mn} = \sqrt{k^2 - (m\pi/a)^2 - (n\pi/b)^2} \qquad (104)$$

The square-root convention stated in (96) should be observed. The field components for the (m, n)th mode calculated from (82) and (83) are

$$\mathbf{E}_t = C_{mn}\boldsymbol{\alpha}_{mn} + \overline{C}_{mn}\boldsymbol{\beta}_{mn} \qquad (105a)$$

$$E_z = \frac{1}{jk}\left[\left(\frac{m\pi}{a}\right)^2 + \left(\frac{n\pi}{b}\right)^2\right]\overline{\psi} \qquad (105b)$$

$$\mathbf{H}_t = C_{mn}\frac{\gamma_{mn}}{kZ}\hat{\mathbf{z}} \times \boldsymbol{\alpha}_{mn} + \overline{C}_{mn}\frac{k}{\gamma_{mn}Z}\hat{\mathbf{z}} \times \boldsymbol{\beta}_{mn} \qquad (105c)$$

$$H_z = \frac{1}{jkZ}\left[\left(\frac{m\pi}{a}\right)^2 + \left(\frac{n\pi}{b}\right)^2\right]\psi \qquad (105d)$$

where

$$\boldsymbol{\alpha}_{mn} = \left[\hat{\mathbf{x}}\left(\frac{n\pi}{b}\right)\cos\left(\frac{m\pi}{a}x\right)\sin\left(\frac{n\pi}{b}y\right)\right.$$
$$\left. - \hat{\mathbf{y}}\left(\frac{m\pi}{a}\right)\sin\left(\frac{m\pi}{a}x\right)\cos\left(\frac{n\pi}{b}y\right)\right]e^{-j\gamma_{mn}z} \qquad (106a)$$

$$\boldsymbol{\beta}_{mn} = \left(-\frac{\gamma_{mn}}{k}\right)\left[\hat{\mathbf{x}}\left(\frac{m\pi}{a}\right)\cos\left(\frac{m\pi}{a}x\right)\sin\left(\frac{n\pi}{b}y\right)\right.$$
$$\left. + \hat{\mathbf{y}}\left(\frac{n\pi}{b}\right)\sin\left(\frac{m\pi}{a}x\right)\cos\left(\frac{n\pi}{b}y\right)\right]e^{-j\gamma_{mn}z} \qquad (106b)$$

In particular, the nonzero field components of the dominant TE_{10} mode are

$$\text{TE}_{10}: \quad E_y = -\left(\frac{\pi}{a}\right)C_{10}\sin\left(\frac{\pi}{a}x\right)e^{-j\gamma_{10}z} \qquad (107a)$$

$$H_x = \left(\frac{\pi}{a}\right)\left(\frac{\gamma_{10}}{kZ}\right)C_{10}\sin\left(\frac{\pi}{a}x\right)e^{-j\gamma_{10}z} \qquad (107b)$$

$$H_z = \frac{1}{jkZ}\left(\frac{\pi}{a}\right)^2 C_{10}\cos\left(\frac{\pi}{a}x\right)e^{-j\gamma_{10}z} \qquad (107c)$$

For the field component traveling in the $-z$ direction we change the sign of γ_{mn} in (103) through (107). The transverse field variations of several lower-order modes are sketched in Fig. 19.

If the medium in the waveguide is lossless, the cutoff frequency of TE_{mn} or TM_{mn} is

$$f_c = \frac{1}{2\sqrt{\epsilon\mu}} \sqrt{\left(\frac{m}{a}\right)^2 + \left(\frac{n}{b}\right)^2} \tag{108}$$

which is plotted in Fig. 20 for several lower-order modes. The propagation constant of TE_{mn} or TM_{mn} is

$$\gamma_{mn} = \begin{cases} k\sqrt{1 - (f_c/f)^2} = 2\pi/\lambda_g & \text{if } f > f_c \\ -jk\sqrt{(f_c/f)^2 - 1} & \text{if } f < f_c \end{cases} \tag{109}$$

where λ_g is the waveguide wavelength. The time-averaged power carried by the field in (105) is

$$\bar{P} = \frac{ab}{4Z} \sum_m^{\text{prop.}} \sum_n \Delta_m \Delta_n \frac{\gamma_{mn}}{k} \left[\left(\frac{m\pi}{a}\right)^2 + \left(\frac{n\pi}{b}\right)^2\right] (|\bar{C}_{mn}|^2 + |C_{mn}|^2) \tag{110}$$

when the summation includes only propagating modes with real γ_{mn}s, and $\Delta_m = 2$ if $m = 0$, and $\Delta_m = 1$ if $m \neq 0$.

There are two common sources of waveguide loss. One is the dielectric loss in the medium inside the waveguide: $k = k' - jk''$. The other is the conduction loss due to the finite conductivity σ of the metallic waveguide walls. When both losses are small, the propagation constant of a propagating mode sufficiently above cutoff is approximately given by

$$\gamma_{mn} = \beta - j\alpha, \qquad \beta \gg \alpha \tag{111}$$

where

$$\beta = \sqrt{(k')^2 - (m\pi/a)^2 - (n\pi/b)^2} \tag{112}$$

$$\alpha = \alpha_d + \alpha_c \tag{113}$$

$$\alpha_d = k'k''/\beta \tag{114}$$

with β in radians per meter and α in nepers per meter. The attenuation constant due to conduction loss is approximately given by

$$TE_{mo}: \quad \alpha_c = v\left(\frac{1}{b} u^2 + \frac{1}{2a}\right) \tag{115a}$$

$$TE_{on}: \quad \alpha_c = v\left(\frac{1}{a} u^2 + \frac{1}{2b}\right) \tag{115b}$$

$$TE_{mm}: \quad \alpha_c = v\left[\left(\frac{1}{a} + \frac{1}{b}\right) u^2 + (1 - u^2)\frac{m^2 b + n^2 a}{m^2 b^2 + n^2 a^2}\right] \tag{115c}$$

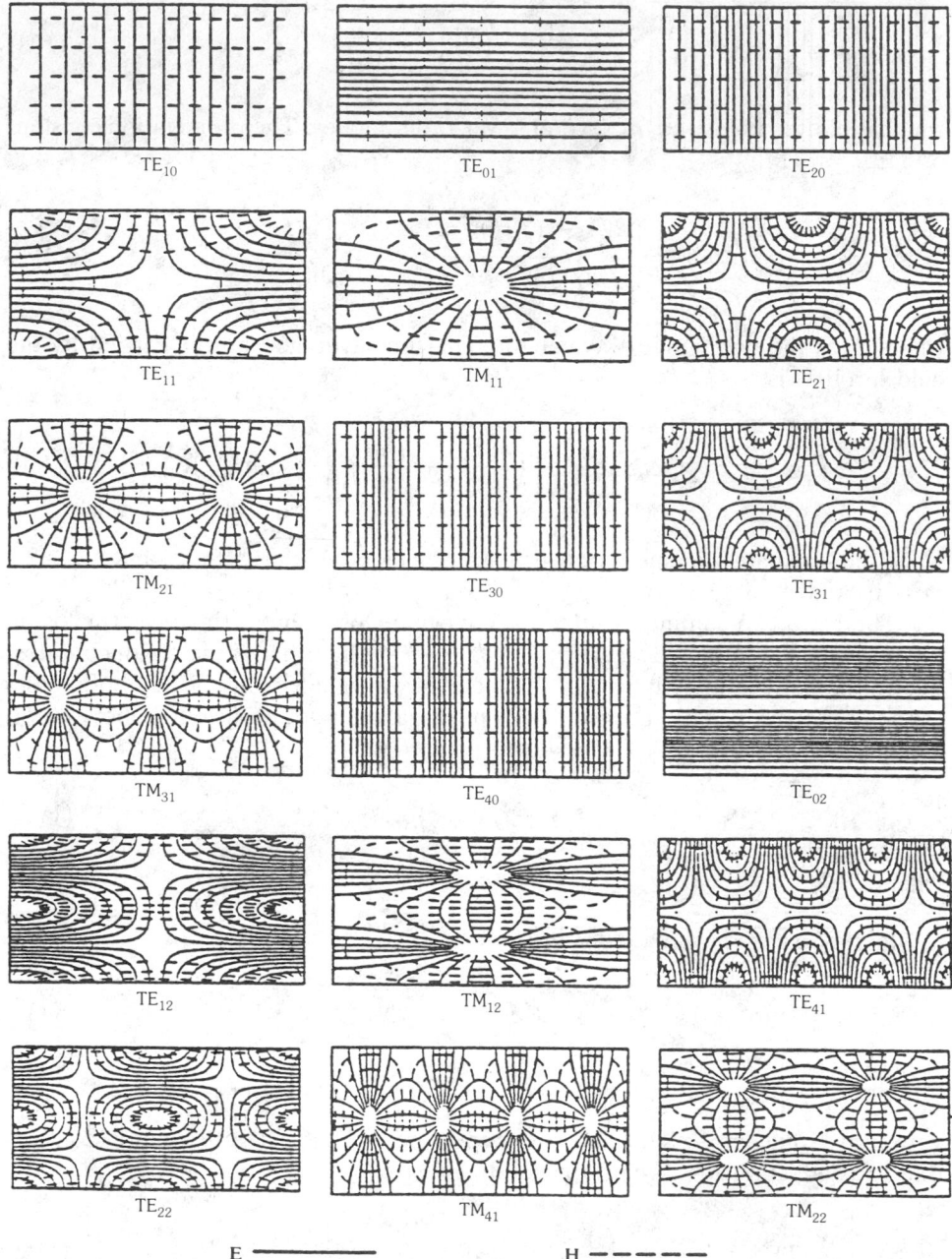

Fig. 19. Modal transverse field distributions in a rectangular waveguide. (*After C. S. Lee, S. W. Lee, and S. L. Chuang [8], © 1985 IEEE*)

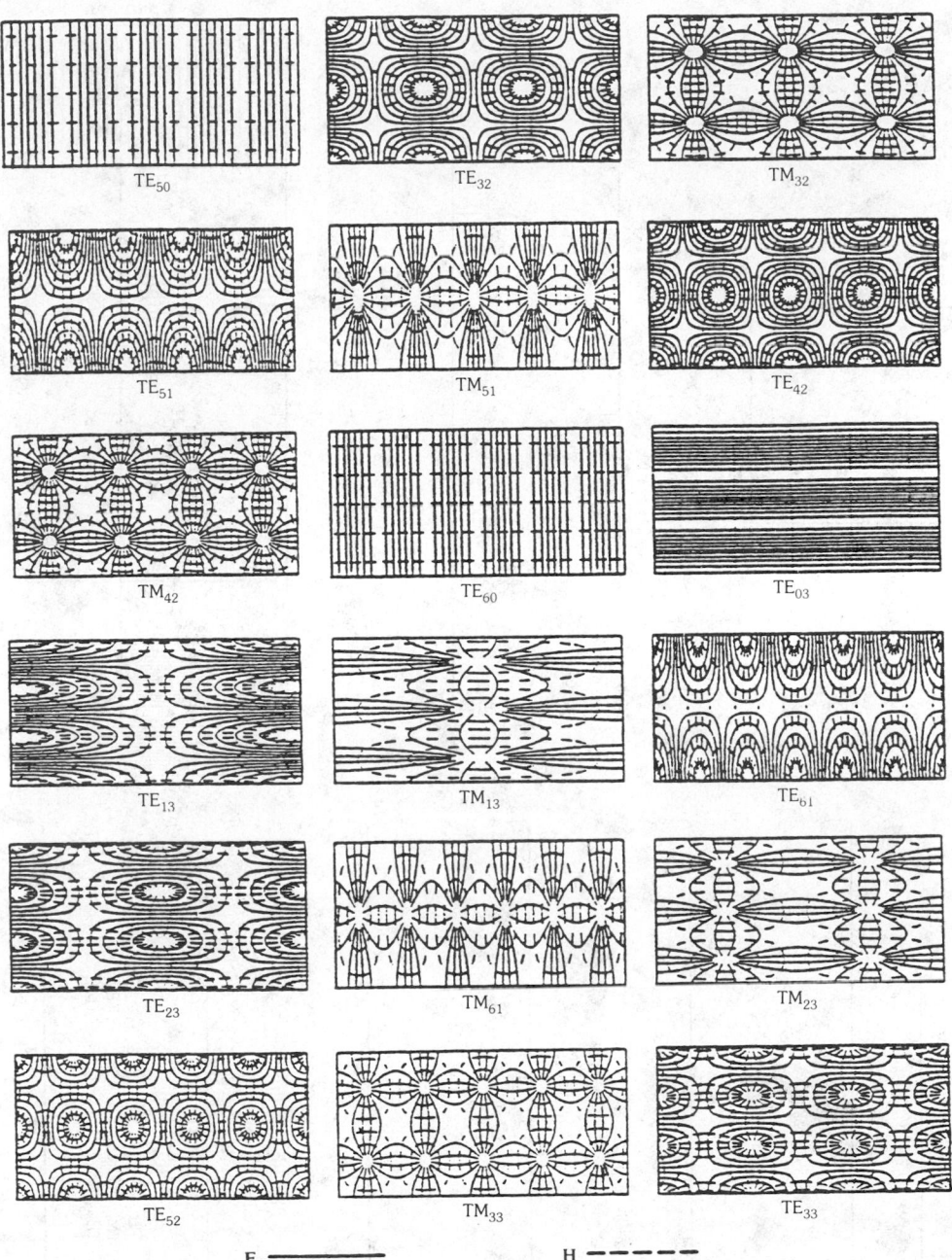

$$E \text{ ———} \qquad H \text{ ------}$$

Fig. 19, *continued.*

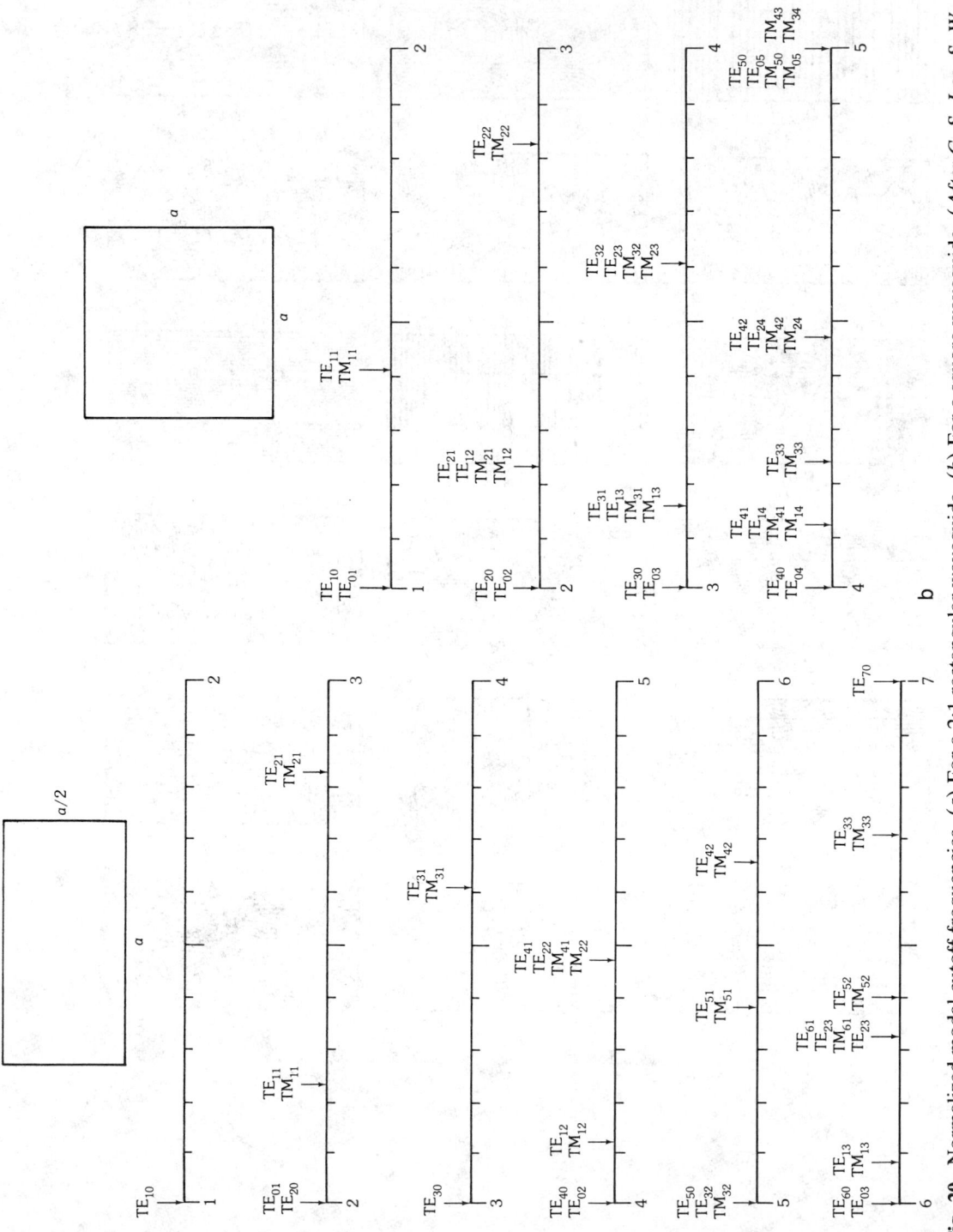

Fig. 20. Normalized modal cutoff frequencies. (*a*) For a 2:1 rectangular waveguide. (*b*) For a square waveguide. (*After C. S. Lee, S. W. Lee, and S. L. Chuang [8], © 1985 IEEE*)

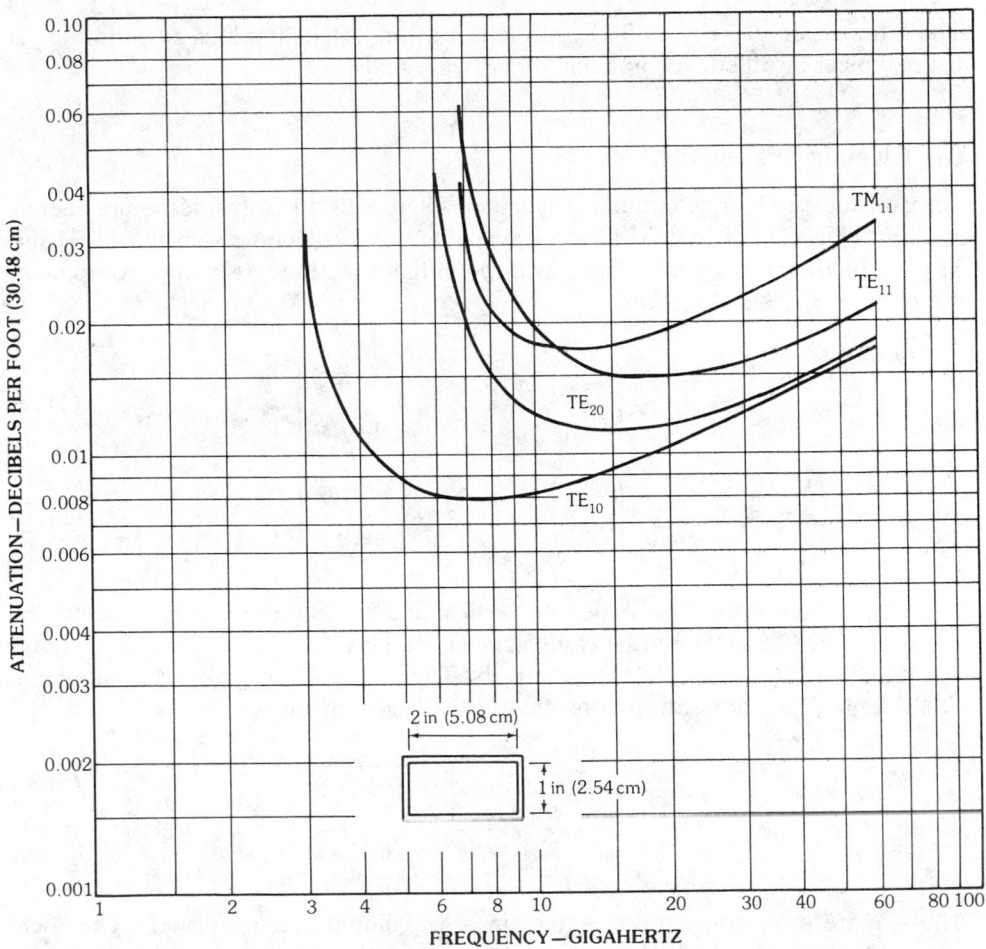

Fig. 21. Attenuation versus frequency curves for lower-order modes in a typical rectangular brass waveguide. (*After Jordan and Balmain [9], © 1968 Prentice-Hall, Inc.; reprinted by permission of Prentice-Hall, Inc., Englewood Cliffs, NJ.*)

$$\text{TM}_{mn}: \quad \alpha_c = v\left[\frac{m^2 b^3 + n^2 a^3}{ab(m^2 b^2 + n^2 a^2)}\right] \tag{115d}$$

where

$$u = f_c/f \tag{116a}$$

$$v = \frac{2R_s}{Z\sqrt{1 - u^2}} \tag{116b}$$

$$R_s = (\omega\mu/2\sigma)^{1/2} = \text{skin-effect surface resistance of waveguide wall} \tag{116c}$$

The attenuation of TE_{mn} or TM_{mn} is $8.686\,\alpha$ dB/m. An example is given in Fig. 21 for a typical air-filled, rectangular, brass waveguide.

16. Circular Waveguide

Consider a metallic circular waveguide filled with an isotropic homogeneous medium (Fig. 22). The field in the waveguide can be decomposed into TE_z and TM_z, which are derivable from two potential functions (for the component traveling in the $+z$ direction):

$$\psi = \sum_m \sum_n \left\{ \begin{matrix} C_{mn}^v \\ C_{mn}^h \end{matrix} \right\} J_m(\xi'_{mn}\varrho/a) \left\{ \begin{matrix} \cos m\phi \\ \sin m\phi \end{matrix} \right\} e^{-j\gamma_{mn}z} \tag{117a}$$

$$\overline{\psi} = \sum_m \sum_n \left\{ \begin{matrix} \overline{C}_{mn}^v \\ \overline{C}_{mn}^h \end{matrix} \right\} J_m(\xi_{mn}\varrho/a) \left\{ \begin{matrix} \sin m\phi \\ \cos m\phi \end{matrix} \right\} e^{-j\overline{\gamma}_{mn}z} \tag{117b}$$

Here the superscripts v and h denote vertical and horizontal modes, respectively. For example, C_{mn}^v is the modal coefficient of the TEV_{mn} mode, and (ξ_{mn}, ξ'_{mn}) are the nth roots of $(J_m(x), J'_m(x))$, respectively. Their lower-order values are tabulated in Table 1. The propagation constants in (117) are given by

$$\text{TE}_{mn}: \quad \gamma_{mn} = \sqrt{k^2 - (\xi'_{mn}/a)^2} \tag{118a}$$

$$\text{TM}_{mn}: \quad \overline{\gamma}_{mn} = \sqrt{k^2 - (\xi_{mn}/a)^2} \tag{118b}$$

The square-root convention stated in (96) should be observed. The field components of TE_{mn} calculated from (117a) and (84) are

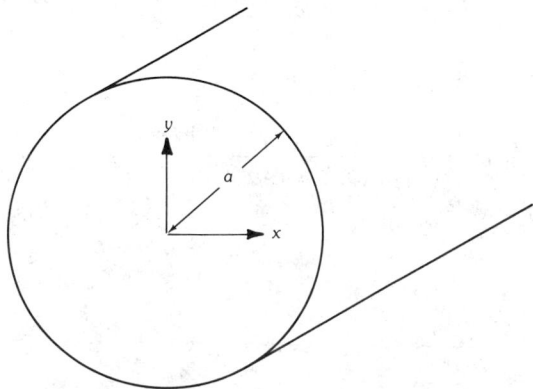

Fig. 22. A circular waveguide.

Table 1. Zeros of Bessel Functions*

Number	Transverse Electric		Transverse Magnetic	
	m, n	ξ'_{mn}	m, n	ξ_{mn}
1	1, 1	1.841 18	0, 1	2.404 83
2	2, 1	3.054 24	1, 1	3.831 71
3	0, 1	3.831 71	2, 1	5.135 62
4	3, 1	4.201 19	0, 2	5.520 08
5	4, 1	5.317 55	3, 1	6.380 16
6	1, 2	5.331 44	1, 2	7.015 59
7	5, 1	6.415 62	4, 1	7.588 34
8	2, 2	6.706 13	2, 2	8.417 24
9	0, 2	7.015 59	0, 3	8.653 73
10	6, 1	7.501 27	5, 1	8.771 48
11	3, 2	8.015 24	3, 2	9.761 02
12	7, 1	8.577 84	6, 1	9.936 11

*The nth zeros of $[J_m(x), J'_m(x)]$ are (ξ_{mn}, ξ'_{mn}), respectively.

$$\text{TE}_{mn}: \quad \mathbf{E} = C^v_{mn}\boldsymbol{\alpha}^v_{mn} + C^h_{mn}\boldsymbol{\alpha}^h_{mn} \tag{119a}$$

$$\mathbf{H}_t = \frac{\gamma_{mn}}{kZ}\hat{\mathbf{z}} \times \mathbf{E} \tag{119b}$$

$$H_z = \frac{1}{jkZ}\left(\frac{\xi'_{mn}}{a}\right)^2 \psi \tag{119c}$$

where

$$\begin{Bmatrix}\alpha^v_{mn}\\\alpha^h_{mn}\end{Bmatrix} = \left[\hat{\boldsymbol{\varrho}}\,\frac{m}{\varrho}\,J_m(\xi'_{mn}\varrho/a)\begin{Bmatrix}\sin m\phi\\-\cos m\phi\end{Bmatrix}\right.$$

$$\left. + \hat{\boldsymbol{\phi}}\left(\frac{\xi'_{mn}}{a}\right)J'_m(\xi'_{mn}\varrho/a)\begin{Bmatrix}\cos m\phi\\\sin m\phi\end{Bmatrix}\right]e^{-j\gamma_{mn}z} \tag{119d}$$

The field components of TM_{mn} calculated from (117b) and (85) are

$$\text{TM}_{mn}: \quad \mathbf{E}_t = (\overline{C}^v_{mn}\boldsymbol{\beta}^v_{mn} + \overline{C}^h_{mn}\boldsymbol{\beta}^h_{mn})(\overline{\gamma}_{mn}/k) \tag{120a}$$

$$\mathbf{E}_z = \frac{1}{jk}\left(\frac{\xi_{mn}}{a}\right)^2 \overline{\psi} \tag{120b}$$

$$\mathbf{H} = \frac{k}{\overline{\gamma}_{mn}Z}\hat{\mathbf{z}} \times \mathbf{E}_t \tag{120c}$$

where

$$\left\{\begin{matrix}\beta_{mn}^v \\ \beta_{mn}^h\end{matrix}\right\} = \left[-\hat{\varrho}\left(\frac{\xi_{mn}}{a}\right)J_m'(\xi_{mn}\varrho/a)\left\{\begin{matrix}\sin m\phi \\ \cos m\phi\end{matrix}\right\}\right.$$

$$\left. + \hat{\phi}\frac{m}{\varrho}J_m(\xi_{mn}\varrho/a)\left\{\begin{matrix}-\cos m\phi \\ \sin m\phi\end{matrix}\right\}\right]e^{-j\gamma_{mn}z} \tag{120d}$$

In particular, the nonzero field components of the dominant TE_{11} are

$$TE_{11}: \quad E_\varrho = \left\{\begin{matrix}C_{11}^v \\ C_{11}^h\end{matrix}\right\}\frac{1}{\varrho}J_1(\xi_{11}'\varrho/a)\left\{\begin{matrix}\sin\phi \\ -\cos\phi\end{matrix}\right\}e^{-j\gamma_{11}z}$$

$$E_\phi = \left\{\begin{matrix}C_{11}^v \\ C_{11}^h\end{matrix}\right\}\frac{\xi_{11}'}{a}J_1'(\xi_{11}'\varrho/a)\left\{\begin{matrix}\cos\phi \\ \sin\phi\end{matrix}\right\}e^{-j\gamma_{11}z}$$

$$H_\varrho = -\frac{\gamma_{11}}{kZ}E_\phi$$

$$H_\phi = \frac{\gamma_{11}}{kZ}E_\varrho$$

$$H_z = \frac{1}{jkZ}\left\{\frac{\xi_{11}'}{a}\right\}^2 J_1(\xi_{11}'\varrho/a)\left\{\begin{matrix}\cos\phi \\ \sin\phi\end{matrix}\right\}e^{-j\gamma_{11}z} \tag{121}$$

The transverse field variations of several lower-order modes are sketched in Fig. 23.

If the medium in the guide is lossless, the modal cutoff frequencies are

$$TE_{mn}: \quad f_c = \frac{\xi_{mn}'}{2\pi a\sqrt{\epsilon\mu}} \tag{122a}$$

$$TM_{mn}: \quad f_c = \frac{\xi_{mn}}{2\pi a\sqrt{\epsilon\mu}} \tag{122b}$$

which are plotted in Fig. 24 for several lower-order modes. In terms of f_c, the propagation constants $(\gamma_{mn}, \overline{\gamma}_{mn})$ can be again written in the form of (109). The time-averaged power carried by the modal fields is

$$\overline{P} = \frac{\pi}{2Z}\sum_m^{prop.}\sum_n \Delta_m\frac{1}{k}\{\gamma_{mn}[(\xi_{mn}')^2 - m^2][J_m(\xi_{mn}')]^2(|C_{mn}^v|^2 + |C_{mn}^h|^2)$$

$$+ \gamma_{mn}[\xi_{mn}J_{m+1}(\xi_{mn})]^2[|\overline{C}_{mn}^v|^2 + |\overline{C}_{mn}^h|^2]\} \tag{123}$$

Fig. 23. Transverse modal field distributions for a circular waveguide. (*After C. S. Lee, S. W. Lee, and S. L. Chuang [8]*, © *1985 IEEE*)

Fig. 23, *continued.*

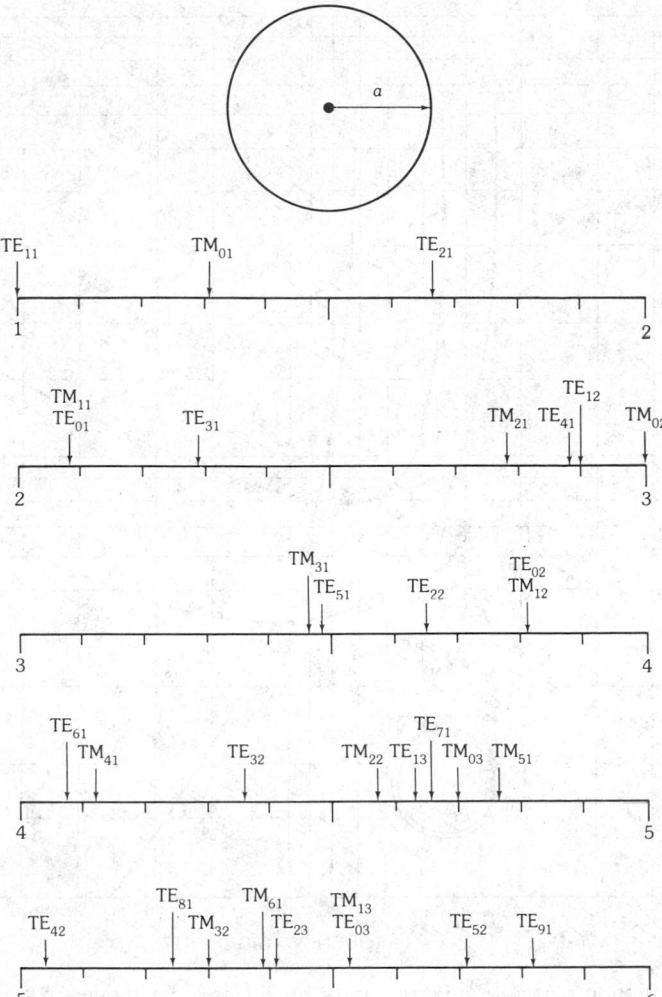

Fig. 24. Normalized modal cutoff frequencies for a circular waveguide. (*After C. S. Lee, S. W. Lee, and S. L. Chuang [8], © 1985 IEEE*)

where the summation includes only propagating modes with real γ_{mn} or $\bar{\gamma}_{mn}$s, and $\Delta_m = 2$ if $m = 0$, and $\Delta_m = 1$ if $m \neq 0$.

The loss in a circular waveguide is similar to that in a rectangular waveguide. Thus the formulas in (111), (113), and (114) still hold. The attenuation constant due to the conduction loss in a circular waveguide is

$$\text{TE}_{mn}: \quad \alpha_c = \frac{v}{2a}\left[\frac{m^2}{(\xi'_{mn})^2 - m^2} + u^2\right] \tag{124a}$$

$$\text{TM}_{mn}: \quad \alpha_c = \frac{v}{2a} \tag{124b}$$

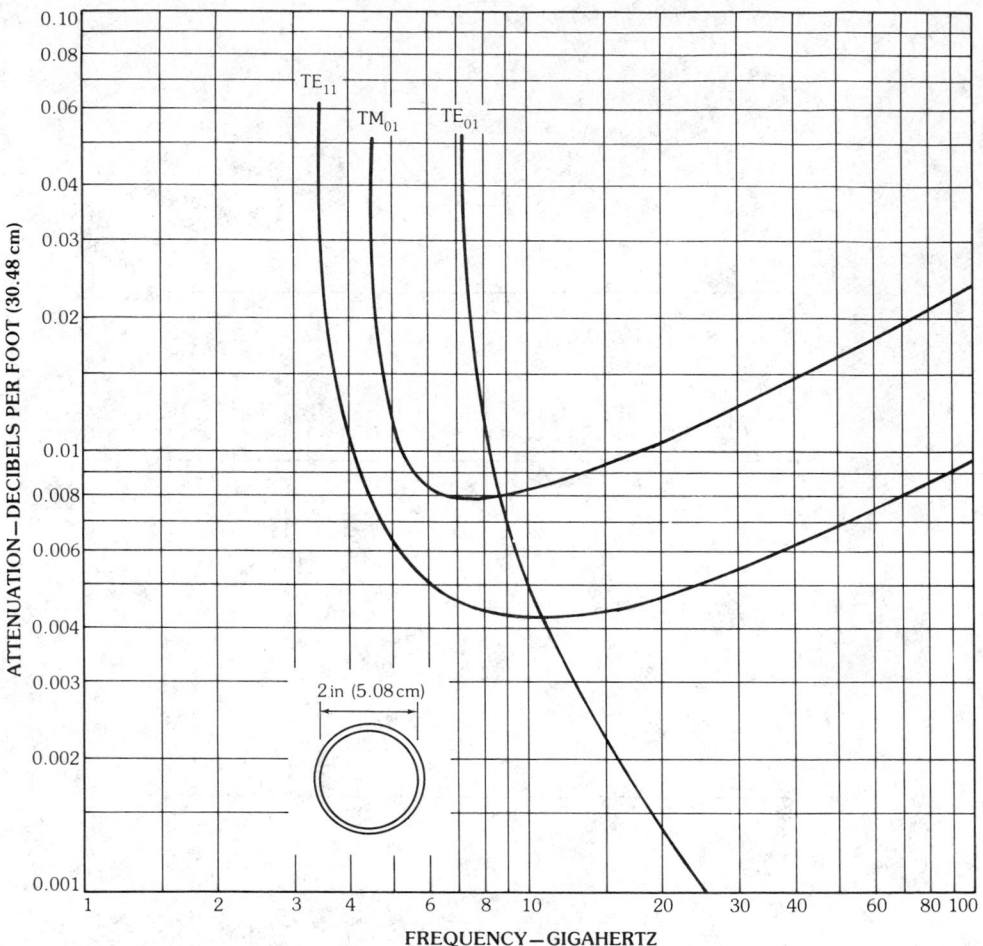

Fig. 25. Attenuation versus frequency curves for lower-order modes in a typical circular brass waveguide. (*After Jordan and Balmain [9], © 1968 Prentice-Hall Inc.; reprinted by permission of Prentice-Hall, Inc., Englewood Cliffs, NJ.*)

where u and v are defined in (116), and f_c is given in (122). Note that the conduction loss of TE_{0n} decreases without limit as $f \to \infty$. An example of the conduction loss in a typical air-filled circular brass waveguide is shown in Fig. 25.

17. References

[1] D. S. Jones, *The Theory of Electromagnetism*, New York: Macmillan Co., 1964, pp. 566–569.

[2] R. F. Harrington, *Time-Harmonic Electromagnetic Fields*, New York: McGraw-Hill Book Co., 1961, pp. 99–100.

[3] G. A. Deschamps, "I. Le principe de réciprocité en électromagnétisme. II. Application du principe de réciprocité aux antennes et aux guides d'ondes," *Revue du CETHEDEC* (Paris) N° 8–4°, pp. 71–101, 1966.

[4]   A. C. Ludwig, "The definition of cross polarization," *IEEE Trans. Antennas Propag.*, vol. AP-21, pp. 116–119, 1973.

[5]   R. J. Stegen, "The gain-beamwidth product of an antenna," *IEEE Trans. Antennas Propag.*, vol. AP-21, pp. 505–506, 1964.

[6]   C. T. Tai and C. S. Pereira, "An approximate formula for calculating the directivity of an antenna," *IEEE Trans. Antennas Propag.*, vol. AP-24, pp. 235–236, 1976.

[7]   M. Born and E. Wolf, *Principles of Optics*, 5th ed., New York: Pergamon Press, 1975, p. 436.

[8]   C. S. Lee, S. W. Lee, and S. L. Chuang, "Plot of modal field distribution in rectangular and circular waveguides," *IEEE Trans. Microwave Theory Tech.*, vol. MTT-33, pp. 271–274, 1985.

[9]   E. C. Jordan and K. G. Balmain, *Electromagnetic Waves and Radiating Systems*, 2nd ed., Englewood Cliffs: Prentice-Hall, 1968, p. 271.

Chapter 2

Theorems and Formulas

S. W. Lee
University of Illinois

CONTENTS

**Shung-Wu Lee** was born in Kiangsi, China. He received his BS degree in electrical engineering from Cheng Kung University in Tainan, Taiwan, in 1961. After one year of military service in Taiwan, he came to the United States in 1962. At the University of Illinois in Urbana, he received his MS and PhD degrees in electrical engineering and has been on the faculty since 1966. Currently he is a professor of electrical and computer engineering, and an associate director of the Electromagnetics Laboratory.

While on leave from the University of Illinois, Dr. Lee was with Hughes Aircraft Company, Fullerton, California, in 1969–1970, and with the Technical University at Eindhoven, The Netherlands, and the University of London, England, in 1973–74. Dr. Lee received several professional awards, including the 1968 Everitt Teaching Excellence Award from the University of Illinois, 1973 NSF NATO Senior Scientist Fellowship, 1977 Best Paper Award from IEEE Antennas and Propagation Society, and the 1985 Lockheed Million Dollar Award.

Dr. Lee has published more than 100 papers in technical journals on antennas and electromagnetic theory. He is the coauthor of a book on guided waves published by Macmillan in 1971, and a coeditor of this book. He is a Fellow of IEEE.

1. Duality

Consider the two radiation problems sketched in Fig. 1. In problem 1, the source $(\mathbf{J}_1, \mathbf{K}_1)$ radiates in medium described by $\epsilon_1(\mathbf{r})$ and $\mu_1(\mathbf{r})$, and in the presence of two typical scatterers: a perfect electric conductor $(\text{PEC})_1$, and a perfect magnetic conductor $(\text{PMC})_1$. The total (radiation) field in space is $(\mathbf{E}_1, \mathbf{H}_1)$. A similar description holds for problem 2. Now, these two problems are "dual" if*

$$\mathbf{J}_1(\mathbf{r}) \rightarrow \mathbf{K}_2(\mathbf{r}) \tag{1a}$$

$$\mathbf{K}_1(\mathbf{r}) \rightarrow -\mathbf{J}_2(\mathbf{r}) \tag{1b}$$

$$\epsilon_1(\mathbf{r}) \rightarrow \mu_2(\mathbf{r}) \tag{1c}$$

$$\mu_1(\mathbf{r}) \rightarrow \epsilon_2(\mathbf{r}) \tag{1d}$$

$$(\text{PEC})_1 \rightarrow (\text{PMC})_2 \tag{1e}$$

$$(\text{PMC})_1 \rightarrow (\text{PEC})_2 \tag{1f}$$

For example, (1e) indicates that the PEC in problem 1 is replaced by a PMC in problem 2. A consequence of (1) is

$$\mathbf{E}_1(\mathbf{r}) \rightarrow \mathbf{H}_2(\mathbf{r}) \tag{2a}$$

$$\mathbf{H}_1(\mathbf{r}) \rightarrow -\mathbf{E}_2(\mathbf{r}) \tag{2b}$$

Thus, knowing $(\mathbf{E}_1, \mathbf{H}_1)$, we can immediately write down, with the help of (2), the field solution $(\mathbf{E}_2, \mathbf{H}_2)$ of the dual problem. In other words, dual sources and media imply dual fields.

In some dual scattering problems the sources are not explicitly specified. We are given instead the incident field $(\mathbf{E}_1^i, \mathbf{H}_1^i)$ in problem 1 and field $(\mathbf{E}_2^i, \mathbf{H}_2^i)$ in problem 2. Then, these two problems remain dual provided that (1a) and (1b) are replaced by

$$\mathbf{E}_1^i(\mathbf{r}) \rightarrow \mathbf{H}_2^i(\mathbf{r}) \tag{3a}$$

$$\mathbf{H}_1^i(\mathbf{r}) \rightarrow -\mathbf{E}_2^i(\mathbf{r}) \tag{3b}$$

---

*The symbol $A \rightarrow B$ means replacing A in problem 1 by B in problem 2.

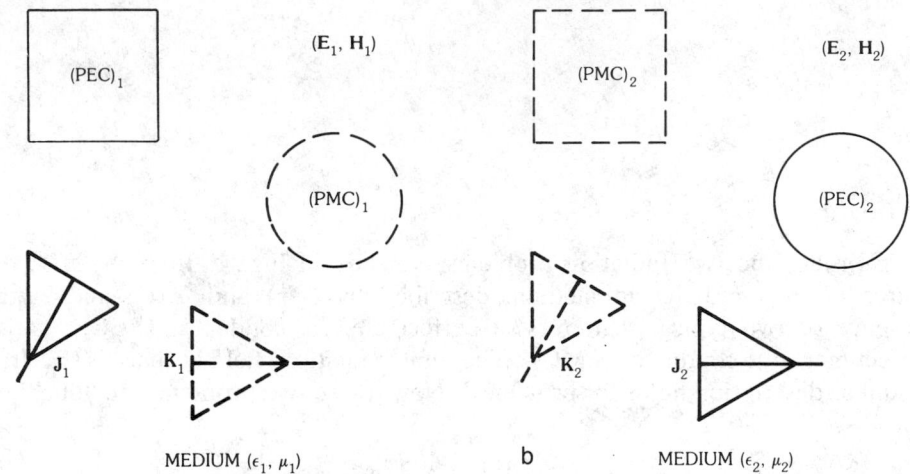

Fig. 1. Dual radiation problems. (*a*) Problem 1. (*b*) Problem 2.

After this replacement the satisfaction of (1) again implies the validity of (2). The fields in (2) can be either total fields or the scattered fields.

2. Green's Function in an Unbounded Space

A Green's function is the field due to a point source described by a delta function. Once it is known, the field due to an arbitrary source can be calculated by a convolution integral involving the source distribution and the Green's function.

Scalar Wave Equation

A Green's function $G(\mathbf{r}, \mathbf{r}')$ involves two points:

$$\mathbf{r} = (x_1, x_2, x_3) = \text{observation point} \tag{4a}$$

$$\mathbf{r}' = (x_1', x_2', x_3') = \text{source point} \tag{4b}$$

In an unbounded isotropic medium, G depends only the distance

$$R = |\mathbf{r} - \mathbf{r}'| = \sqrt{(x_1 - x_1')^2 + (x_2 - x_2')^2 + (x_3 - x_3')^2} \tag{4c}$$

Hence we often write $G(\mathbf{r}, \mathbf{r}')$ as $G(R)$.

In a three-dimensional space filled with a homogeneous isotropic medium, the scalar wave equation with a point source is

$$(\nabla^2 + k^2)g(R) = -\delta(\mathbf{r} - \mathbf{r}') \tag{5}$$

where $\delta(\cdot)$ is the delta function. The solution of (5) subject to the radiation condition is [1]

$$g(R) = \frac{e^{-jkR}}{4\pi R} \qquad (6)$$

For a two-dimensional case the corresponding Green's function is

$$2D: \quad g(R_2) = \frac{1}{4j} H_0^{(2)}(kR_2) \qquad (7)$$

$$\sim \frac{e^{-j(kR_2 + \pi/4)}}{2\sqrt{2\pi kR_2}}, \qquad kR_2 \gg 1 \qquad (8)$$

where $R_2 = \sqrt{(x_1 - x_1')^2 + (x_2 - x_2')^2}$. For a one-dimensional case it is

$$1D: \quad g(R_1) = \frac{1}{2jk} e^{-jkR_1} \qquad (9)$$

where $R_1 = |x_1 - x_1'|$.

It is important to realize that Green's function $g(R)$ is singular or has discontinuous derivatives when the observation point is at the source point ($\mathbf{r} = \mathbf{r}'$). We must properly interpret their meanings when we perform integrations/differentiations involving $g(R)$. In the following manipulations we consider $g(R)$ a generalized function. We manipulate $g(R)$ formally and explain its singular terms later.

Vector Wave Equation

A vector wave equation with a point source oriented in the x_n direction reads

$$(\nabla \times \nabla \times -k^2)\mathbf{G}^{(n)}(R) = \hat{\mathbf{x}}_n \delta(\mathbf{r} - \mathbf{r}'), \qquad \text{for } n = 1, 2, 3 \qquad (10)$$

The solution of (10) is

$$\mathbf{G}^{(n)} = \left[1 + \frac{1}{k^2} \nabla \nabla \cdot \right] [\hat{\mathbf{x}}_n g(R)], \qquad \text{for } n = 1, 2, 3 \qquad (11)$$

Using the dyadic notation we may rewrite (10) and (11) as

$$(\nabla \times \nabla \times -k^2)\bar{\bar{\mathbf{G}}}(R) = \bar{\bar{\mathbf{I}}}\delta(\mathbf{r} - \mathbf{r}') \qquad (12)$$

$$\bar{\bar{\mathbf{G}}}(R) = \left(1 + \frac{1}{k^2} \nabla \nabla \cdot \right) \bar{\bar{\mathbf{I}}} g(R) \qquad (13a)$$

$$= \left(\bar{\bar{\mathbf{I}}} + \frac{1}{k^2} \nabla \nabla \right) g(R) \qquad (13b)$$

where $\bar{\bar{\mathbf{I}}}$ is the unit dyad defined by

$$\bar{\mathbf{I}} = \hat{\mathbf{x}}_1\hat{\mathbf{x}}_1 + \hat{\mathbf{x}}_2\hat{\mathbf{x}}_2 + \hat{\mathbf{x}}_3\hat{\mathbf{x}}_3 \tag{14}$$

Since $g(R)$ is symmetrical in $\mathbf{r}$ and $\mathbf{r}'$, an alternative expression of (13b) reads

$$\bar{\bar{\mathbf{G}}}(R) = \left(\bar{\bar{\mathbf{I}}} + \frac{1}{k^2}\nabla'\nabla'\right)g(R) \tag{15}$$

Note the difference in (13b) and (15):

$$\nabla = \hat{\mathbf{x}}_1\frac{\partial}{\partial x_1} + \hat{\mathbf{x}}_2\frac{\partial}{\partial x_2} + \hat{\mathbf{x}}_3\frac{\partial}{\partial x_3} \tag{16}$$

$$\nabla' = \hat{\mathbf{x}}_1\frac{\partial}{\partial x_1'} + \hat{\mathbf{x}}_2\frac{\partial}{\partial x_2'} + \hat{\mathbf{x}}_3\frac{\partial}{\partial x_3'} \tag{17}$$

In matrix notation, (15) may be explicitly written as

$$\bar{\bar{\mathbf{G}}}(R) = \begin{bmatrix} 1 + \frac{1}{k^2}\left(\frac{\partial}{\partial x_1'}\right)^2 & \frac{1}{k^2}\frac{\partial^2}{\partial x_1'\partial x_2'} & \frac{1}{k^2}\frac{\partial^2}{\partial x_1'\partial x_3'} \\[2mm] \frac{1}{k^2}\frac{\partial^2}{\partial x_2'\partial x_1'} & 1 + \frac{1}{k^2}\left(\frac{\partial}{\partial x_2'}\right)^2 & \frac{1}{k^2}\frac{\partial^2}{\partial x_2'\partial x_3'} \\[2mm] \frac{1}{k^2}\frac{\partial^2}{\partial x_3'\partial x_1'} & \frac{1}{k^2}\frac{\partial^2}{\partial x_3'\partial x_2'} & 1 + \frac{1}{k^2}\left(\frac{\partial}{\partial x_3'}\right)^2 \end{bmatrix} \frac{e^{-jkR}}{4\pi R} \tag{18}$$

The differentiations in (18) formally yield the following results:

$$\frac{\partial^2 g}{\partial x_m'^2} = k^2\left[-\cos^2\theta_m + \frac{j}{kR}\left(1 - \frac{j}{kR}\right)(3\cos^2\theta_m - 1)\right]g(R),$$

$$m = 1, 2, 3 \tag{19a}$$

$$\frac{\partial^2 g}{\partial x_m'\partial x_n'} = k^2\cos\theta_m\cos\theta_n\left[-1 + \frac{3j}{kR}\left(1 - \frac{j}{kR}\right)\right]g(R),$$

$$m \neq n \tag{19b}$$

where

$$\cos\theta_n = (x_n' - x_n)/R, \quad \text{for } n = 1, 2, 3 \tag{19c}$$

Note that $\bar{\bar{\mathbf{G}}}(R)$ contains the R^{-3} singularity.

The Electric Field

In an unbounded, isotropic, homogeneous medium the electric field $\mathbf{E}$ due to a current $\mathbf{J}$ satisfies the following wave equation:

$$(\nabla^2 + k^2)\left(\mathbf{E} + \frac{1}{j\omega\epsilon}\mathbf{J}\right) = \frac{-1}{j\omega\epsilon}\nabla \times \nabla \times \mathbf{J} \tag{20}$$

For a given $\mathbf{J}$, we wish to calculate $\mathbf{E}(\mathbf{r})$ by using the Green's functions defined above. We shall give three formulas. The first formula is derived from (5) and (20), namely,

$$\mathbf{E}(\mathbf{r}) = \frac{1}{j\omega\epsilon}\int_V [\nabla' \times \nabla' \times \mathbf{J}(\mathbf{r}')]\,g(R)\,dv' - \frac{1}{j\omega\epsilon}\mathbf{J}(\mathbf{r}) \tag{21}$$

where V is the support of $\mathbf{J}$ (outside volume V, the current $\mathbf{J}$ is identically zero). The second formula makes use of $\bar{\bar{\mathbf{G}}}$ defined in (13b), namely,

$$\mathbf{E}(\mathbf{r}) = (-j\omega\mu)\left(\bar{\bar{\mathbf{I}}} + \frac{1}{k^2}\nabla\nabla\right)\int_V \mathbf{J}(\mathbf{r}')\,g(R)\,dv' \tag{22}$$

The third formula makes use of $\bar{\bar{\mathbf{G}}}$ defined in (15), namely,

$$\mathbf{E}(\mathbf{r}) = (-j\omega\mu)\int_V \left[\left(\bar{\bar{\mathbf{I}}} + \frac{1}{k^2}\nabla'\nabla'\right)g(R)\right]\mathbf{J}(\mathbf{r}')\,dv' \tag{23}$$

Note the difference between (22) and (23). In (22) the differentiation with respect to the observation point coordinates (x_1, x_2, x_3) takes place after the integration over the source is performed, whereas in (23) the differentiation with respect to the source point coordinates (x_1', x_2', x_3') takes place before. If the integral is to be evaluated numerically, (23) is preferred to (22). The problem with (23) is how to integrate over the R^{-3} singularity. This is discussed next.

Integration Involving the $\mathbf{R}^{-3}$ Singularity

In evaluating the integral in (23), we encounter the following typical integral:

$$I_{mn}(\mathbf{r}) = \int_V J(\mathbf{r}')\frac{\partial^2 g}{\partial x_m' \partial x_n'}\,dv', \qquad \text{for } m, n = 1, 2, 3 \tag{24}$$

As shown in (19) the second derivatives of g have an R^{-3} singularity at $R = 0$ (when observation point $\mathbf{r}$ is at a source point $\mathbf{r}'$). This singularity is not generally integrable. Thus (24) may be a divergent integral. We shall regularize (24) so that $I_{mn}(\mathbf{r})$ is a well-defined function of $\mathbf{r}$ for all $\mathbf{r}$. The result of a particular regularization is [2]

$$I_{mn}(\mathbf{r}) = A_{mn} + B_{mn} + C_{mn}, \qquad \text{for } m, n = 1, 2, 3 \tag{25a}$$

where

$$A_{mn} = \int_{V-V_\epsilon} J(\mathbf{r}')\frac{\partial^2 g}{\partial x_m' \partial x_n'}\,dv' \tag{25b}$$

$$B_{mn} = \int_{V_\epsilon} \left[J(\mathbf{r}') \frac{\partial^2 g}{\partial x_m' \partial x_n'} - J(\mathbf{r}) \frac{\partial^2 g_0}{\partial x_m' \partial x_n'} \right] dv' \tag{25c}$$

$$C_{mn} = J(\mathbf{r}) \left(\frac{-1}{4\pi} \right) \int_{\partial V_\epsilon} \frac{(\hat{\mathbf{x}}_m \cdot \hat{\mathbf{N}})(\hat{\mathbf{x}}_n \cdot \hat{\mathbf{R}})}{R^2} \, ds' \tag{25d}$$

The various notations in (25) are explained below. Volume V is the support of current $J(\mathbf{r}')$ (see Fig. 2). Volume V_ϵ is an arbitrary volume inside V and contains the observation point $\mathbf{r}$. We emphasize that:

(a) V_ϵ need not be small, and
(b)   the value of I_{mn} is independent of the choice of V_ϵ.

Term A_{mn} in (25b) is convergent because the region V_ϵ is excluded from the domain of integration. The static Green's function g_0 in (25c) is given by

$$g_0(R) = \frac{1}{4\pi R} \tag{26}$$

which is obtained by setting $k = 0$ in the (dynamic) Green's function in (6). Note that g and g_0 have the same R^{-1} singularity at $R = 0$. This fact ensures the convergence of term B_{mn} in (25c). In term C_{mn} in (25d), there is a surface integral over the boundary surface of V_ϵ, which is denoted by ∂V_ϵ (Fig. 3). In this figure $\hat{\mathbf{N}}$ is the unit-outward normal of ∂V_ϵ at a point $\mathbf{r}'$, and $\hat{\mathbf{R}}$ is the unit vector along $\mathbf{R} = \mathbf{r}' - \mathbf{r}$. In summary, we regularize the divergent integral (24) so that it is defined by (25a). With this particular regularization it can be shown that I_{mn} is identical with

$$I_{mn}(\mathbf{r}) = \frac{\partial^2}{\partial x_m \partial x_n} \int_V J(\mathbf{r}') g(R) \, dv', \qquad \text{for } m, n = 1, 2, 3 \tag{27}$$

In other words, the convergent integral in (27) can be alternatively evaluated via (25). This establishes the equivalence of two Green's functions in (13b) and (15) when the latter is regularized in the manner described in (25).

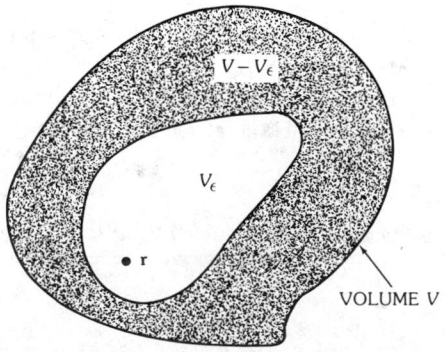

Fig. 2. Arbitrary volume V_ϵ is inside V and contains observation point $\mathbf{r}$.

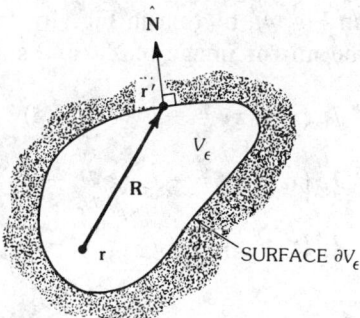

Fig. 3. Symbols for the surface integral in (25d): $\mathbf{r}$ is the observation point and $\mathbf{r}'$ is the integration point on surface ∂V_ϵ.

Explicit Expressions for $\mathbf{C}_{mn}$

The term C_{mn} defined in (25d) can be explicitly evaluated for several special V_ϵs [3].

(a) V_ϵ is a sphere of radius a with the observation point $\mathbf{r}$ at an arbitrary point inside V_ϵ:

$$C_{mn} = -\frac{1}{3} J(\mathbf{r})\delta_{mn} \tag{28}$$

(b) V_ϵ is a cube with center at $\mathbf{r}$ but is arbitrarily oriented, i.e., the faces of the cube need not be orthogonal to the x_1, x_2, x_3 axes:

$$C_{mn} = -\frac{1}{3} J(\mathbf{r})\delta_{mn} \tag{29}$$

(c) V_ϵ is a cylinder with center at $\mathbf{r}$, radius a, and height $2b$, and its axis coincides with the x_3 axis:

$$C_{11} = C_{22} = -\frac{b}{2\sqrt{a^2 + b^2}} J(\mathbf{r}) \tag{30}$$

$$C_{33} = \left(\frac{b}{\sqrt{a^2 + b^2}} - 1\right) J(\mathbf{r}) \tag{31}$$

$$C_{mn} = 0, \quad \text{if } m \neq n \tag{32}$$

3. Image Theory

Consider the radiation problem sketched in Fig. 4a with source $(\mathbf{J}_1, \mathbf{K}_1)$ and scatterer Σ_1, both situated above an infinitely large, planar, perfect electric conductor at $z = 0$. To calculate the field in the upper half-space $z > 0$, we may replace the configuration in Fig. 4a by that in Fig. 4b. In Fig. 4b the conductor is removed, scatterer Σ_2 is the mirror image of Σ_1, and source $(\mathbf{J}_2, \mathbf{K}_2)$ is given by:

$$J_{1x}(x, y, z) = -J_{2x}(x, y, -z) \tag{33a}$$

$$J_{1y}(x, y, z) = -J_{2y}(x, y, -z) \tag{33b}$$

$$J_{1z}(x, y, z) = +J_{2z}(x, y, -z) \tag{33c}$$

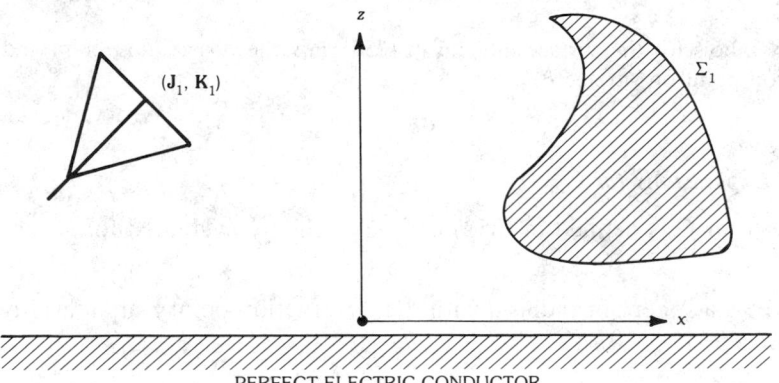

a

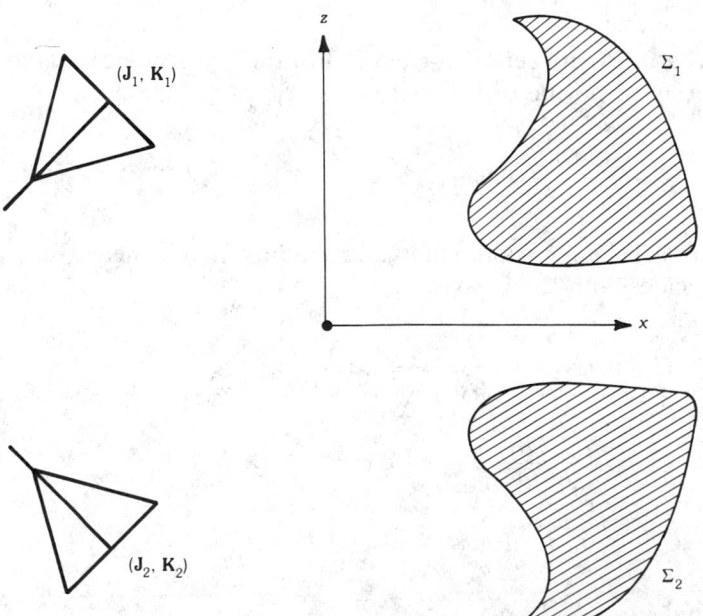

b

Fig. 4. By the image theory the problem in (*a*) may be replaced by that in (*b*) for calculating fields in the upper half-space $z > 0$. (*a*) Given radiation problem. (*b*) Calculating field in $z > 0$.

$$K_{1x}(x,y,z) = +K_{2x}(x,y,-z) \tag{34a}$$

$$K_{1y}(x,y,z) = +K_{2y}(x,y,-z) \tag{34b}$$

$$K_{1z}(x,y,z) = -K_{2z}(x,y,-z) \tag{34c}$$

The above image theory can be also applied if sources $(\mathbf{J}_1, \mathbf{J}_2)$ in (33) are replaced by incident fields $(\mathbf{E}_1^i, \mathbf{E}_2^i)$, and $(\mathbf{K}_1, \mathbf{K}_2)$ in (34) by $(\mathbf{H}_1^i, \mathbf{H}_2^i)$.

Now if the perfect electric conductor in Fig. 4a is replaced by a perfect magnetic conductor, the relations in (33) and (34) must be replaced by:

$$J_{1x}(x,y,z) = +J_{2x}(x,y,-z) \tag{35a}$$

$$J_{1y}(x,y,z) = +J_{2y}(x,y,-z) \tag{35b}$$

$$J_{1z}(x,y,z) = -J_{2z}(x,y,-z) \tag{35c}$$

$$K_{1x}(x,y,z) = -K_{2x}(x,y,-z) \tag{36a}$$

$$K_{1y}(x,y,z) = -K_{2y}(x,y,-z) \tag{36b}$$

$$K_{1z}(x,y,z) = +K_{2z}(x,y,-z) \tag{36c}$$

4. The Babinet Principle

The Babinet principle relates the field solutions of two problems with complementary configurations and dual sources.

Scattering Problems

Let us first consider the scattering version of the Babinet principle involving the following two scattering problems:

(a)   As sketched in Fig. 5a, an infinitely large, thin, perfectly conducting plane Σ_1 with apertures lies in the plane $z = 0$. It is illuminated by an incident field, from $z < 0$, described by

$$\mathbf{E}_1^i = \mathbf{E}_0, \qquad \mathbf{H}_1^i = \mathbf{H}_0 \tag{37a}$$

We express the total electric field by

$$\mathbf{E}_1^{\text{total}} = \begin{cases} \mathbf{E}_1^i + \mathbf{E}_1^s & \text{for } z > 0 \\ \mathbf{F}_1^i + \mathbf{E}_1^r + \mathbf{E}_1^d & \text{for } z < 0 \end{cases} \tag{37b}$$

Here $\mathbf{E}_1^r$ is the reflected field from an infinite conducting plane (without aperture) at $z = 0$ due to the incidence of (37a), and $\mathbf{E}_1^d$ is the diffracted

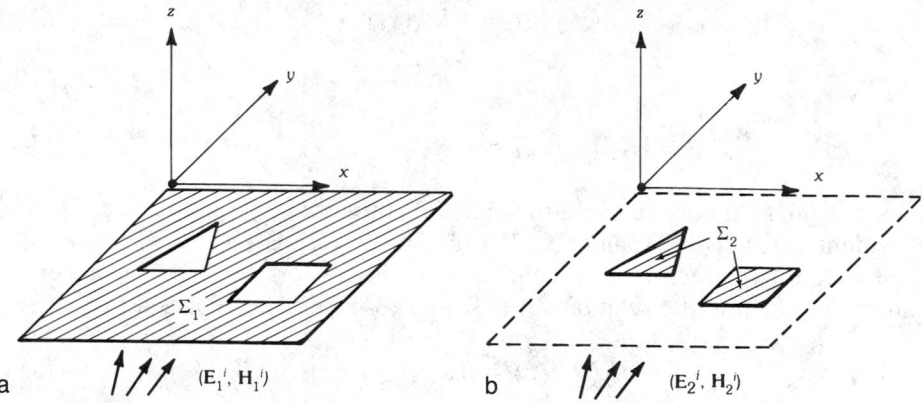

Fig. 5. Two scattering problems related by the Babinet principle. (*a*) Perfectly conducting plane Σ_1 is infinitely large. (b) Planes Σ_2 are complementary to Σ_1.

field. (If apertures were absent, $\mathbf{E}_1^d$ would be zero.) A similar expression holds for the total magnetic field $\mathbf{H}_1^{\text{total}}$.

(*b*)   Complementary to the infinite plane Σ_1 in Fig. 5a, we have two thin, perfectly conducting planes sketched in Fig. 5b. They are denoted by Σ_2. Let Σ_2 be illuminated by an incident field from $z < 0$ described by

$$\mathbf{E}_2^i = -Z\mathbf{H}_0, \qquad \mathbf{H}_2^i = Y\mathbf{E}_0 \tag{38a}$$

where $Z = Y^{-1} = \sqrt{\mu/\epsilon}$. We express the total electric field everywhere by

$$\mathbf{E}_2^{\text{total}} = \mathbf{E}_2^i + \mathbf{E}_2^s, \qquad \text{for all } z \tag{38b}$$

A similar expression holds for $\mathbf{H}_2^{\text{total}}$.

In the transmitted half-space the Babinet principle states that

$$\left.\begin{aligned}
\mathbf{E}_1^{\text{total}} + Z\mathbf{H}_2^{\text{total}} - \mathbf{E}_0 &= 0 \\
\mathbf{H}_1^{\text{total}} - Y\mathbf{E}_2^{\text{total}} - \mathbf{H}_0 &= 0
\end{aligned}\right\} \quad \text{for } z > 0 \tag{39a}$$

when stating in terms of total fields; or

$$\left.\begin{aligned}
\mathbf{E}_1^s + Z\mathbf{H}_2^s + \mathbf{E}_0 &= 0 \\
\mathbf{H}_1^s - Y\mathbf{E}_2^s + \mathbf{H}_0 &= 0
\end{aligned}\right\} \quad \text{for } z > 0 \tag{39b}$$

when stating in terms of scattered fields. In the reflected half-space the Babinet principle states that

$$\left.\begin{aligned}
\mathbf{E}_1^{\text{total}} - Z\mathbf{H}_2^{\text{total}} - \mathbf{E}_1^r &= 0 \\
\mathbf{H}_1^{\text{total}} + Y\mathbf{E}_2^{\text{total}} - \mathbf{H}_1^r &= 0
\end{aligned}\right\} \quad \text{for } z < 0 \tag{40a}$$

when stating in terms of total fields; or

$$\left.\begin{array}{l} \mathbf{E}_1^d - Z\mathbf{H}_2^s = 0 \\ \mathbf{H}_1^d + Y\mathbf{E}_2^s = 0 \end{array}\right\} \quad \text{for } z < 0 \tag{40b}$$

when stating in terms of scattered/diffracted fields. In either problem (a) or (b), the following relations hold for the field in the aperture (nonmetal part) at the $z = 0$ plane:

$$E_z^{\text{total}} = E_z^i \tag{41}$$

$$H_x^{\text{total}} = H_x^i, \qquad H_y^{\text{total}} = H_y^i \tag{42}$$

This is due to the fact that the induced surface current on the conducting screen Σ_1 or Σ_2 produces symmetric (E_x^s, E_y^s, H_z^s), and antisymmetric (E_z^s, H_x^s, H_y^s) on the two sides of the $z = 0$ plane.

An Impedance Problem

Consider two complementary planar antennas: the slot antenna excited by a voltage source across the small gap ab in Fig. 6a, and the strip antenna excited by a voltage source across the small gap cd in Fig. 6b. Their impedances are related by

$$Z_{\text{slot}} Z_{\text{strip}} = \frac{1}{4} Z^2 \tag{43}$$

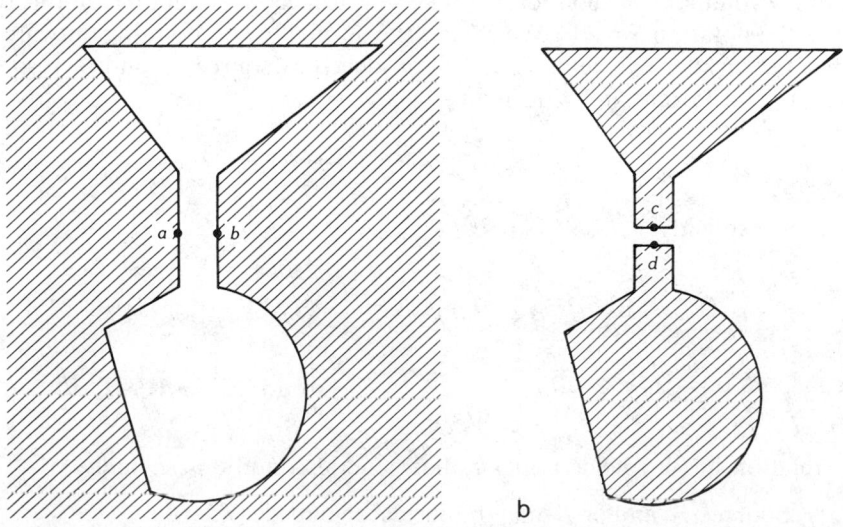

Fig. 6. Two complementary planar antennas. (a) Slot excited at (a, b). (b) Strip excited at (c, d).

where $Z = \sqrt{\mu/\epsilon}$ is the intrinsic wave impedance of the unbounded medium where both antennas are situated.

5. Reciprocity

The reciprocity theorem is useful in formulating problems as well as checking final answers. It "removes" the explicit dependence of media from a radiation or scattering problem.

Bra-Ket Notation

A source A generally consists of an electric current described by its density $\mathbf{J}_a(\mathbf{r})$ and a magnetic current described by its density $\mathbf{K}_a(\mathbf{r})$. In a given environment the field produced by A is $(\mathbf{E}_a, \mathbf{H}_a)$, denoted by F_a. Over a volume V the "reaction" between a source A and a field $F_b = (\mathbf{E}_b, \mathbf{H}_b)$ is defined by [4]

$$\langle AVF_b \rangle = \int_V (\mathbf{J}_a \cdot \mathbf{E}_b - \mathbf{K}_a \cdot \mathbf{H}_b)\, dv \tag{44}$$

which has the unit of watt. Through a surface S (not necessarily closed) the *cross-flux* of two fields F_a and F_b is defined by [4]

$$\langle F_a S F_b \rangle = \int_S (\mathbf{E}_a \times \mathbf{H}_b - \mathbf{E}_b \times \mathbf{H}_a) \cdot \hat{\mathbf{N}}\, ds \tag{45}$$

which again has the unit of watt. The surface S in (45) is oriented, meaning a unit normal $\hat{\mathbf{N}}$ is specified.

The Lorentz Relation

For a volume V, its boundary is a closed surface S. To emphasize this relation we use the notation $S = \partial V$. We choose the normal $\hat{\mathbf{N}}$ of S pointing outward (away from V). Consider two fields F_a and F_b produced by sources A and B, respectively (Fig. 7). The Lorentz relation reads

$$\langle F_a S F_b \rangle = \langle AVF_b \rangle - \langle BVF_a \rangle \tag{46a}$$

or, more explicitly,

$$\int_S (\mathbf{E}_a \times \mathbf{H}_b - \mathbf{E}_b \times \mathbf{H}_a) \cdot \hat{\mathbf{N}}\, ds = \int_V (\mathbf{J}_a \cdot \mathbf{E}_b - \mathbf{K}_a \cdot \mathbf{H}_b)\, dv$$
$$- \int_V (\mathbf{J}_b \cdot \mathbf{E}_a - \mathbf{K}_b \cdot \mathbf{H}_a)\, dv \tag{46b}$$

The relation in (46) holds under rather general conditions, namely:

(a) Sources A and/or B may be inside or outside V.
(b) Fields F_a and F_b may be produced in different media provided that these two media coincide inside V.

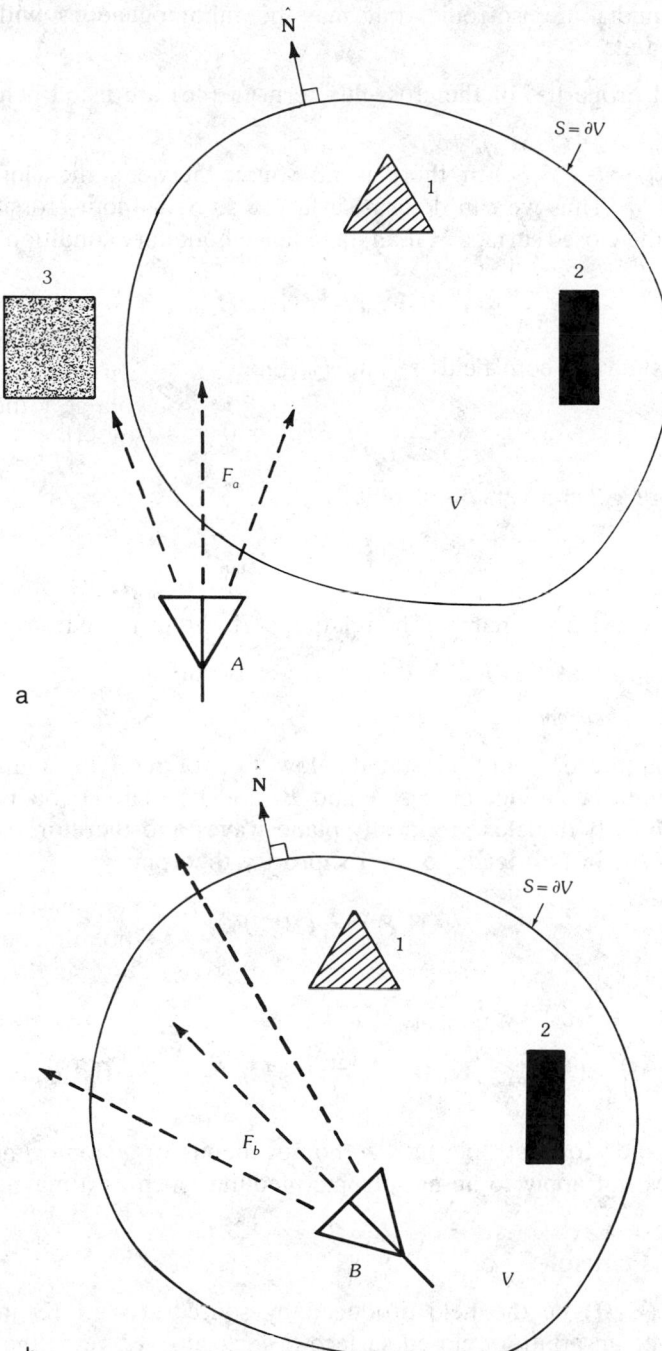

Fig. 7. Fields (F_a, F_b) are produced by sources (A, B), respectively, where blocks $(1, 2, 3)$ are scatterers and the medium in (a) is identical with the medium in (b) only inside V, not necessarily outside. (a) Field produced by A. (b) Field produced by B.

(c)   The media are isotropic, but may be inhomogeneous with scatterers presented.

Several useful properties of the cross-flux term in (46) are listed below:

(a) $\langle F_a S F_b \rangle = -\langle F_b S F_a \rangle$

(b) $\langle F_a S F_b \rangle = \langle F_a S_1 F_b \rangle$ if there is no source between the closed surfaces S and S_1. Thus we can deform surface S to S_1 without crossing sources.

(c)   Over the closed surface S, if an impedance boundary condition of the form

$$\mathbf{E}_{\tan} = Z \hat{\mathbf{N}} \times \mathbf{H}_{\tan} \tag{47a}$$

is satisfied for both fields F_a and F_b, then

$$\langle F_a S F_b \rangle = 0 \tag{47b}$$

A more general version of (47a) is

$$\mathbf{E}_{\tan} = \bar{\bar{\mathbf{Z}}} \mathbf{H}_{\tan} \tag{47c}$$

where $\bar{\bar{\mathbf{Z}}}$ is a 2×2 matrix. The relation (47b) holds if the trace of $\bar{\bar{\mathbf{Z}}}$ is zero.

We also mention that $\langle A V F_b \rangle = 0$ if source A is outside V.

The Reciprocity Theorem

A useful application of (47) is stated below. Let volume V be an infinitely large spherical volume enclosing sources A and B, and be denoted by V_∞. Over the surface $S = \partial V_\infty$ both fields are locally plane waves and therefore satisfy (47a). The use of (47b) in (46) leads to the reciprocity theorem:

$$\langle A V_\infty F_b \rangle = \langle B V_\infty F_a \rangle \tag{48a}$$

or, more explicitly,

$$\int_{V_\infty} (\mathbf{J}_a \cdot \mathbf{E}_b - \mathbf{K}_a \cdot \mathbf{H}_b)\, dv = \int_{V_\infty} (\mathbf{J}_b \cdot \mathbf{E}_a - \mathbf{K}_b \cdot \mathbf{H}_a)\, dv \tag{48b}$$

Reciprocity holds for isotropic media, homogeneous or inhomogeneous, but it generally does not apply to an anisotropic medium, such as a magnetoplasma.

6. Huygens' Principle

Let $F = (\mathbf{E}, \mathbf{H})$ be the field produced by source $A = (\mathbf{J}, \mathbf{K})$ in medium a (Fig. 8a). Over an arbitrary closed surface S enclosing A, we define a Huygens' source described by surface current densities as

$$\left. \begin{array}{l} \mathbf{J}_s = \hat{\mathbf{N}} \times \mathbf{H} \\[4pt] \mathbf{K}_s = \mathbf{E} \times \hat{\mathbf{N}} \end{array} \right\} \quad \text{over } S \tag{49}$$

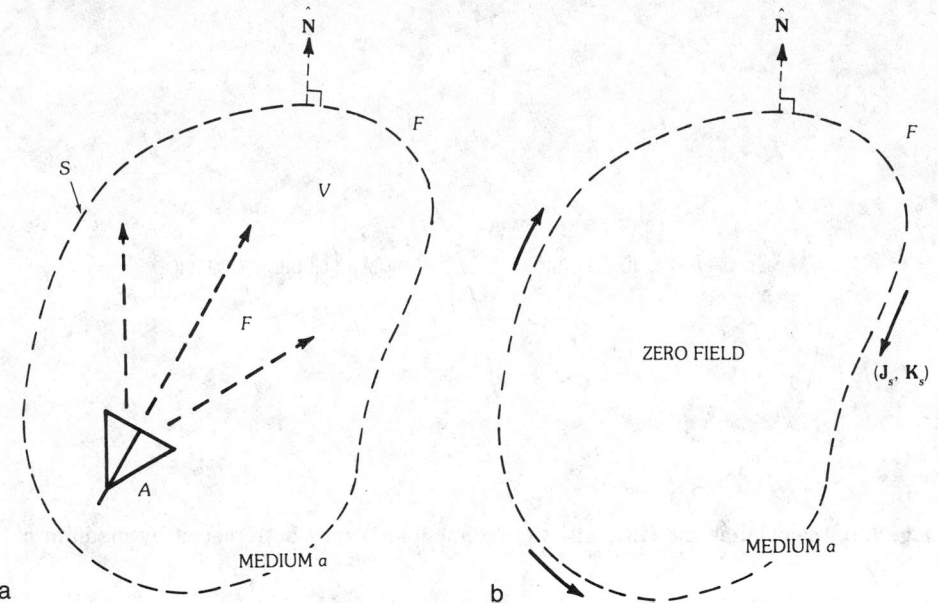

Fig. 8. Huygens' source $(\mathbf{J}_s, \mathbf{K}_s)$ produces field F outside V and zero field inside V. (*a*) Field F produced by source. (*b*) Huygens' source.

It may be shown that, when radiating in medium a, the Huygens' source produces

(*a*) zero field inside V, and

(*b*) the original field F outside V.

Thus, for observation points outside V, source A and its Huygens' source are equivalent. From $(\mathbf{J}_s, \mathbf{K}_s)$ in (49) we may calculate its exact radiation field by using (20) in Chapter 1, or its far field by using (21) or (48) in Chapter 1.

The Huygens' source is not unique. In addition to the one given in (49), an alternative Huygens' source is

$$\left.\begin{array}{l} \mathbf{J}_s' = \hat{\mathbf{N}} \times \mathbf{H}' \\ \mathbf{K}_s' = \mathbf{E}' \times \hat{\mathbf{N}} \end{array}\right\} \quad \text{over } S \qquad (50)$$

Here $(\mathbf{E}', \mathbf{H}')$ is the field produced by source A in medium b, which coincides with medium a inside V and may be different from medium a outside V (Fig. 9). The source $(\mathbf{J}_s', \mathbf{K}_s')$ radiating in medium a (not b) produces the original field F outside V.

7. The Kirchhoff Approximation

Consider an infinitely large perfectly conducting screen Σ_1, with a finite aperture Σ_2 (Fig. 10a). For a given source $A = (\mathbf{J}, \mathbf{K})$, the problem is to find the transmitted field $F = (\mathbf{E}, \mathbf{H})$ in the upper half-space $z > 0$. If the dimension of Σ_2 is

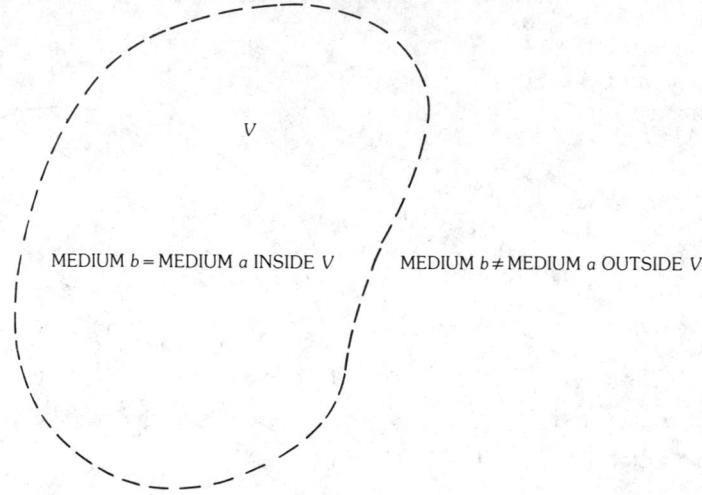

Fig. 9. In calculating the Huygens' source, medium a may be replaced by medium b.

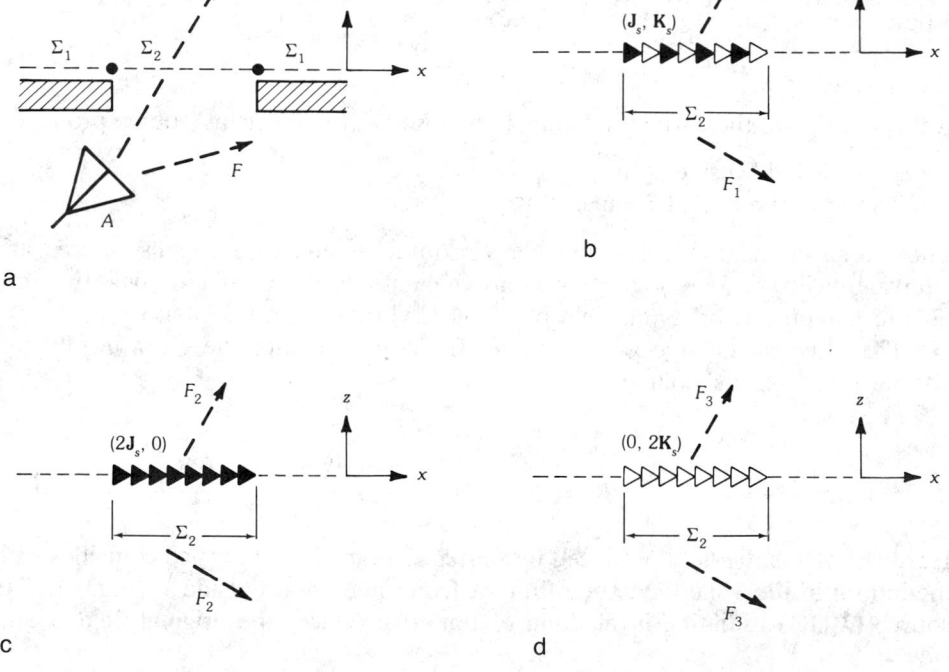

Fig. 10. Three versions of the Kirchhoff approximation for the field in the upper half-space $z > 0$. (*a*) Radiation problem. (*b*) Approximating electric and magnetic currents. (*c*) Approximating electric current alone. (*d*) Approximating magnetic current alone.

large in terms of wavelength, the field F may be approximately calculated by using one of the following three versions of Kirchhoff's approximations:

(a) *Electric and magnetic currents* (Fig. 10b). We replace source A by an equivalent surface source distributed over the aperture Σ_2:

$$\left.\begin{array}{l} \mathbf{J}_s = \hat{\mathbf{N}} \times \mathbf{H}^i \\ \mathbf{K}_s = \mathbf{E}^i \times \hat{\mathbf{N}} \end{array}\right\} \quad \text{over } \Sigma_2 \qquad (51)$$

Here $\hat{\mathbf{N}} = \hat{\mathbf{z}}$ is the outward normal of Σ_2. Field $(\mathbf{E}^i, \mathbf{H}^i)$ is that produced by source A in the absence of screen Σ_1. The field produced by $(\mathbf{J}_s, \mathbf{K}_s)$ in the absence of Σ_1 is F_1, which is an approximation of F in the upper half-space $z > 0$. Note that F_1 is not a good approximation of F for $z < 0$.

(b) *Electric current alone* (Fig. 10c). The equivalent source in (51) consists of both electric and magnetic surface currents. An alternative equivalent source is $(2\mathbf{J}_s, 0)$, which doubles the electric current in (51) and contains no magnetic current. The field produced by $(2\mathbf{J}s, 0)$ in the absence of Σ_1 is F_2, which is the second approximation of F for $z > 0$.

(c) *Magnetic current alone* (Fig. 10d). Another equivalent source is $(0, 2\mathbf{K}_s)$, which doubles the magnetic current in (51) and contains no electric current. The field produced by it in the absence of Σ_1 is F_3, which is the third approximation of F for $z > 0$.

Version (a) above is sometimes known as the Stratton-Chu formula. Field F_1 is the average of F_2 and F_3. Kirchhoff's approximation has been applied to transmission through aperture problems in which the aperture is not necessarily in a planar screen, such as radiation from an open-ended waveguide.

8. Scattering by an Obstacle

Consider the scattering problem sketched in Fig. 11. The obstacle (scatterer) is illuminated by an incident plane wave F^i given by

$$\left\{\begin{array}{l} \mathbf{E}^i(\mathbf{r}) \\ \mathbf{H}^i(\mathbf{r}) \end{array}\right\} = C \left\{\begin{array}{l} \sqrt{Z}\,\mathbf{u}_1 \\ \sqrt{Y}\,\hat{\mathbf{k}}_1 \times \mathbf{u}_1 \end{array}\right\} e^{-j\mathbf{k}_1 \cdot \mathbf{r}} \qquad (52)$$

where

$\mathbf{k}_1$ = wave vector with magnitude $k = \omega\sqrt{\mu\epsilon}$ and pointing in the direction of propagation of F^i

$Z = Y^{-1} = \sqrt{\mu/\epsilon}$

$\mathbf{u}_1$ = a unitary vector which describes the polarization of F^i and is orthogonal to $\mathbf{k}_1$

C = amplitude of F^i in $(\text{watt})^{1/2}$ $(\text{meter})^{-1}$

We say that F^i is in state $(\mathbf{k}_1, \mathbf{u}_1)$. In the presence of the obstacle the total field

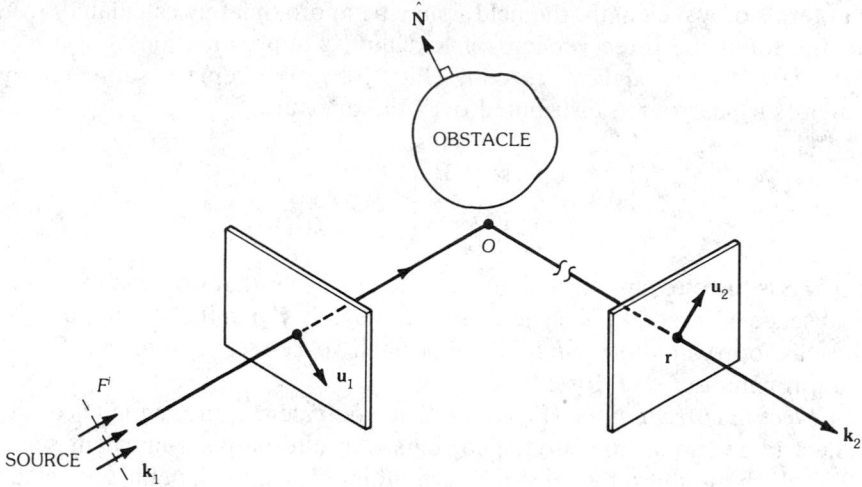

Fig. 11. We are interested in the scattered far field at **r** with state $(\mathbf{k}_2, \mathbf{u}_2)$ when an obstacle is illuminated by an incident plane wave F^i with state $(\mathbf{k}_1, \mathbf{u}_1)$.

everywhere is the sum of F^i and a scattered field F. At a far-field observation point **r**, field F is represented by a spherical wave

$$\begin{Bmatrix} \mathbf{E}(\mathbf{r}) \\ \mathbf{H}(\mathbf{r}) \end{Bmatrix} \sim \begin{Bmatrix} \sqrt{Z}\,\mathbf{A}(\mathbf{k}_2) \\ \sqrt{Y}\,\hat{\mathbf{k}}_2 \times \mathbf{A}(\mathbf{k}_2) \end{Bmatrix} \frac{e^{-jkr}}{r}, \qquad r \to \infty \tag{53}$$

The distance r is measured from a reference point O in the vicinity of the obstacle. Vector $\mathbf{k}_2$ has a magnitude equal to k and is in the direction of $O\mathbf{r}$. The amplitude vector $\mathbf{A}(\mathbf{k}_2)$ given in $(\text{watt})^{1/2}$ may be decomposed into two orthogonal components, as discussed in Section 7, Chapter 1. Let us concentrate on a particular component of $\mathbf{A}(\mathbf{k}_2)$ with polarization $\mathbf{u}_2$, namely, $\mathbf{A} \cdot \mathbf{u}_2^*$. We introduce the notation (after G. A. Deschamps)

$$A(2,1) = \mathbf{A}(\mathbf{k}_2) \cdot \mathbf{u}_2^* \tag{54}$$

which represents the scattering amplitude in state 2 (direction $\mathbf{k}_2$ and polarization $\mathbf{u}_2$) due to an incident plane wave in state 1. Using this notation, we will introduce some definitions and theorems for the scattering problem in Fig. 11.

Bistatic Cross Section

The bistatic cross section (BCS) in square meters from state 1 to state 2 is defined by

$$\text{BCS} = \frac{4\pi |A(2,1)|^2}{|C|^2} = \frac{4\pi |\mathbf{A}(\mathbf{k}_2) \cdot \mathbf{u}_2^*|^2}{|C|^2} \tag{55a}$$

$$= \frac{4\pi \ (\text{intensity of scattered field in state 2})}{\text{power density of incident plane wave in state 1}} \qquad (55b)$$

Radar Cross Section

The plane wave in (52) is in state 1 described by $(\mathbf{k}_1, \mathbf{u}_1)$. Its time-reversed counterpart is in state 1' described by $(-\mathbf{k}_1, \mathbf{u}_1^*)$. Note that these two fields have the *same* polarization.* The radar cross section (RCS) in square meters is a special case of the bistatic cross section BCS(2, 1) with state 2 equal to state 1', namely,

$$\text{RCS} = \frac{4\pi |A(1', 1)|^2}{|C|^2} = \frac{4\pi |\mathbf{A}(-\mathbf{k}_1) \cdot \mathbf{u}_1|^2}{|C|^2} \qquad (56a)$$

$$= \frac{4\pi \ (\text{intensity of scattered field in state 1'})}{\text{power density of incident plane wave in state 1}} \qquad (56b)$$

We will consider two examples of RCS. (*a*) For a smooth conductor whose dimension is large in terms of wavelength and is illuminated by a linearly polarized incident field, its RCS is approximately independent of polarization and is given by

$$\text{RCS} \cong \pi |R_1 R_2| \qquad (57a)$$

where R_1 and R_2 are two principal radii of curvature of the conducting body at the specular point P (Fig. 12). The point P is determined by the relation in which the surface normal $\hat{\mathbf{N}}$ is in the opposite direction of $\mathbf{k}_1$. We assume that there is only one such specular point. If the conductor is a sphere of radius a, the use of (57a) leads to RCS $\cong \pi a^2$, a well-known result. (*b*) For a rectangular conducting plate of dimension $a \times b$ in the $z = 0$ plane, the RCS is again approximately independent of polarization and is given by

$$\text{RCS} \cong \frac{1}{\pi} \left(kab \cos \theta_0 \frac{\sin \alpha}{\alpha} \frac{\sin \beta}{\beta} \right)^2 \qquad (57b)$$

where

$(\theta_0, \phi_0) = $ spherical angles of back-scattered wave vector $\mathbf{k}_2$ or $-\mathbf{k}_1$

$\alpha = ka \sin \theta_0 \cos \phi_0$

$\beta = kb \sin \theta_0 \sin \phi_0$

The expression in (57b) is approximately valid when dimensions (a, b) are large in terms of wavelength, and incident angle θ_0 is not too large (near normal incidence).

The RCS defined in (56) is proportional to the scattered intensity in the same polarization as the incident one. It is also known as the RCS for the copolarization. We can define an RCS in square meters for the cross polarization by

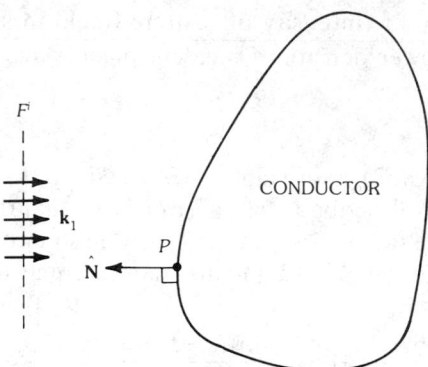

Fig. 12. The radar cross section (RCS) of a large, smooth conductor.

$$\text{RCS for cross polarization} = \frac{4\pi |\mathbf{A}(-\mathbf{k}_1) \cdot \mathbf{v}_1|^2}{|C|^2} \tag{58a}$$

Here $\mathbf{v}_1$ is the unitary vector describing the cross polarization satisfying

$$\mathbf{v}_1 \cdot \mathbf{k}_1 = 0, \qquad \mathbf{v}_1 \cdot \mathbf{v}_1^* = 1, \qquad \mathbf{v}_1 \cdot \mathbf{u}_1^* = 0 \tag{58b}$$

A common choice for the copolarization and cross polarization is $\mathbf{u}_1$ describing vertical polarization and $\mathbf{v}_1$ horizontal polarization, or vice versa. This results in four values for RCS, with two of them being equal.

Reciprocity

In terms of the notation in (54), the reciprocity for the scattering amplitude may be stated as

$$A(2,1) = A(1',2') \tag{59}$$

The amplitude $A(2,1)$ is explained in Fig. 11, and amplitude $A(1',2')$ in Fig. 13.

Scattering Cross Section

For the scattered field in (53), the scattering cross section (SCS) in square meters is defined by

$$\text{SCS} = |C|^{-2} \int_{4\pi} |\mathbf{A}|^2 \, d\Omega \tag{60a}$$

$$= \frac{\text{total scattered power in all directions and all polarizations}}{\text{power density of incident plane wave in state 1}} \tag{60b}$$

Clearly, SCS is a function of the incident state 1. It can be shown that SCS of a lossless obstacle is related to the forward scattering amplitude by

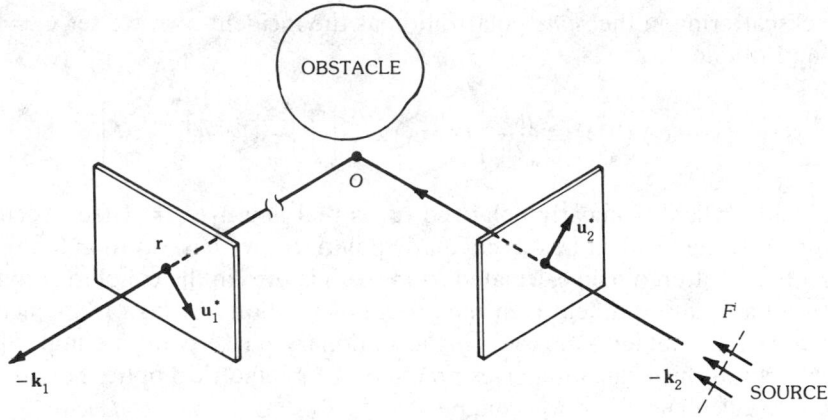

Fig. 13. Reciprocity applies to the above situation and that in Fig. 11.

$$\text{SCS} = \frac{4\pi}{k} \, \text{Im}\{A(1,1)\} = \frac{4\pi}{k} \, \text{Im}\{\mathbf{A}(\mathbf{k}_1)\cdot\mathbf{u}_1^*\} \tag{61}$$

which is known as the *scattering cross section theorem*.

Physical Optics Approximation for Scattering by a Conductor

Consider the scattering problem in Fig. 11, with the incident field described by (52) and the scattered field by (53). If the obstacle in Fig. 11 is a perfect conductor, the scattered field is due to the radiation of a surface electric current $\mathbf{J}_s$ on the very surface of the conductor. At high frequencies we may use the so-called physical optics approximation for $\mathbf{J}_s$, namely,

$$\mathbf{J}_s(\mathbf{r}) \cong \begin{cases} 2\hat{\mathbf{N}} \times \mathbf{H}^i & \text{for } \mathbf{r} \text{ in the lit region} \\ 0 & \text{for } \mathbf{r} \text{ in the shadow} \end{cases} \tag{62}$$

For a convex obstacle the lit region is defined by relation $\mathbf{k}_1\cdot\hat{\mathbf{N}} < 0$, and shadow region by $\mathbf{k}_1\cdot\hat{\mathbf{N}} > 0$. The scattered field defined in (53) is the radiation field of $\mathbf{J}_s$ in the unbounded space. The scattering amplitude in state $(\mathbf{k}_2, \mathbf{u}_2)$ may be calculated from (62) above and (48) in Chapter 1. The final result is

$$\mathbf{A}(\mathbf{k}_2)\cdot\mathbf{u}_2^* \cong \frac{jkC}{2\pi} \left[\mathbf{u}_2^* \times (\hat{\mathbf{k}}_1 \times \mathbf{u}_1) \right] \cdot \left[\iint_{\text{lit}} \hat{\mathbf{N}} \, e^{j(\mathbf{k}_2-\mathbf{k}_1)\cdot\mathbf{r}'} \, ds' \right] \tag{63a}$$

where the surface integration is over the source point $\mathbf{r}'$ in the lit portion of the conductor surface. For backscattering, $\mathbf{k}_2 = -\mathbf{k}_1$, and (63a) is reduced to

$$\mathbf{A}(-\mathbf{k}_1)\cdot\mathbf{u}_2^* \cong \frac{jkC}{2\pi} \left[(\mathbf{u}_2^*\cdot\mathbf{u}_1) \, \hat{\mathbf{k}}_1 \right] \cdot \left[\iint_{\text{lit}} \hat{\mathbf{N}} \, e^{-j2\mathbf{k}_1\cdot\mathbf{r}'} \, ds' \right] \tag{63b}$$

For backscattering in the *same* polarization as the incident one, we set $\mathbf{u}_2 = \mathbf{u}_1^*$ in (63b) and obtain

$$\mathbf{A}(-\mathbf{k}_1)\cdot(\mathbf{u}_1^*)^* \cong \frac{jkC}{2\pi}\left[(\mathbf{u}_1\cdot\mathbf{u}_1)\,\hat{\mathbf{k}}_1\right]\cdot\left[\iint_{\text{lit}}\hat{\mathbf{N}}\,e^{-j2\mathbf{k}_1\cdot\mathbf{r}'}\,ds'\right] \qquad (63c)$$

If the incident field is linearly polarized, $\mathbf{u}_1$ is real and $\mathbf{u}_1\cdot\mathbf{u}_1 = 1$ (see Section 5, Chapter 1). If the incident field is circularly polarized, we have $\mathbf{u}_1\cdot\mathbf{u}_1 = 0$, implying that the backscattered field calculated from (63c) is zero in the copolarization, and all of the backscattered field is in the cross polarization. At high frequencies we may evaluate the integral in (63a) by the stationary phase point method [5]. The leading term of this evaluation gives precisely the geometrical optics field that can be directly calculated by ray techniques.

Two Other Versions of the Physical Optics Approximation

Consider the scattering problem sketched in Fig. 14. An electrically large perfect conductor is illuminated by an incident field $(\mathbf{E}^i, \mathbf{H}^i)$ from a source at point A. The scattered field $F(B)$ at B is to be found. According to the physical optics approximation described above, we may approximate $F(B)$ by $F_1(B)$. Here $F_1(B)$ is the field produced by a surface current $\mathbf{J}_s$ on the conductor's surface radiating in the free space, and $\mathbf{J}_s$ is given by

$$\mathbf{J}_s(\mathbf{r}) = \begin{cases} 2\hat{\mathbf{N}} \times \mathbf{H}^i & \text{for } \mathbf{r} \text{ in the surface portion visible from } A \\ 0 & \text{elsewhere on the conductor's surface} \end{cases} \qquad (64)$$

The surface portion with nonzero $\mathbf{J}_s$ is indicated by a zigzag line in Fig. 14a. The second version of the physical optics approximation is to approximate $F(B)$ by $F_2(B)$. The latter is produced by a magnetic surface current $\mathbf{K}_s$ radiating in the free space, and $\mathbf{K}_s$ is given by

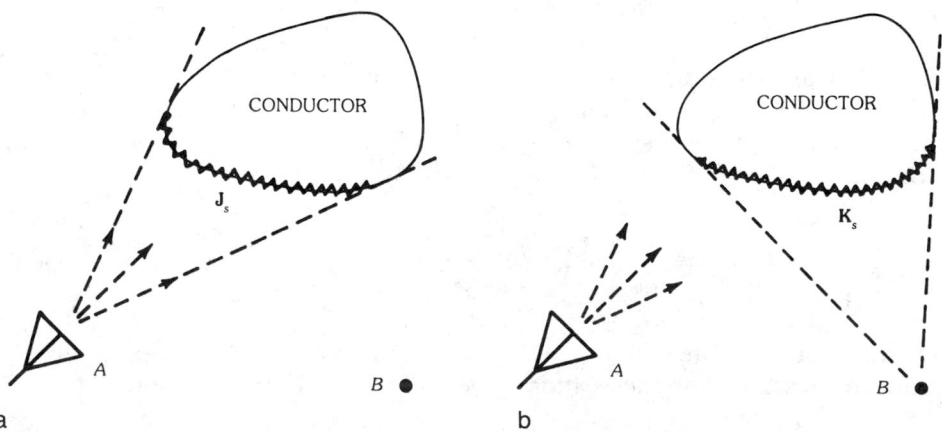

Fig. 14. Scattering by a large conductor: two versions of the physical optics approximation. (*a*) Using electric surface current $\mathbf{J}_s$. (*b*) Using magnetic surface current $\mathbf{K}_s$.

$$\mathbf{K}_s(\mathbf{r}) = \begin{cases} 2\mathbf{E}^i \times \hat{\mathbf{N}} & \text{for } \mathbf{r} \text{ in the surface portion visible from } B \\ 0 & \text{elsewhere on the conductor's surface} \end{cases} \tag{65}$$

The surface portion with nonzero $\mathbf{K}_s$ is indicated by a zigzag line in Fig. 14b. Note that $\mathbf{J}_s$ and $\mathbf{K}_s$ are distributed over different portions of the conductor's surface. A third version of the physical optics approximation is to approximate $F(B)$ by the average of $F_1(B)$ and $F_2(B)$. When the positions of the source and the observation points are interchanged, the reciprocity holds in the third version, but not in the first two. It is also interesting to mention that the three versions of physical optics are the counterparts of the three Kirchhoff approximations discussed in Section 7.

9. The Antenna as a One-Port Device

An antenna is usually fed through a transmission line or a waveguide, as the two examples sketched in Fig. 15. We assume that the transmission line (waveguide) supports only one propagating mode. The total field along the line (inside the waveguide) is a superposition of two traveling waves, one in the $+z$ direction and one in the $-z$ direction:

$$\mathbf{E}(\mathbf{r}) = \mathbf{e}(x, y)[ae^{-j\beta z} + be^{+j\beta z}] \tag{66a}$$

$$\mathbf{H}(\mathbf{r}) = \mathbf{h}(x, y)[ae^{-j\beta z} - be^{+j\beta z}] \tag{66b}$$

The reference plane $z = 0$ can be arbitrarily chosen, and need not be exactly at the antenna–transmission-line junction. Here β is the propagation constant of the dominant mode, and (a, b) are wave amplitudes in (watts)$^{1/2}$. Transverse field variations are described by $(\mathbf{e}, \mathbf{h})$, whose units are $(\Omega^{1/2}\mathrm{m}^{-1}, \Omega^{-1/2}\mathrm{m}^{-1})$, respectively, and satisfy the normalization condition

$$\iint (\mathbf{e} \times \mathbf{h}^*)\cdot\hat{\mathbf{z}}\, dx\, dy = 1 \tag{67}$$

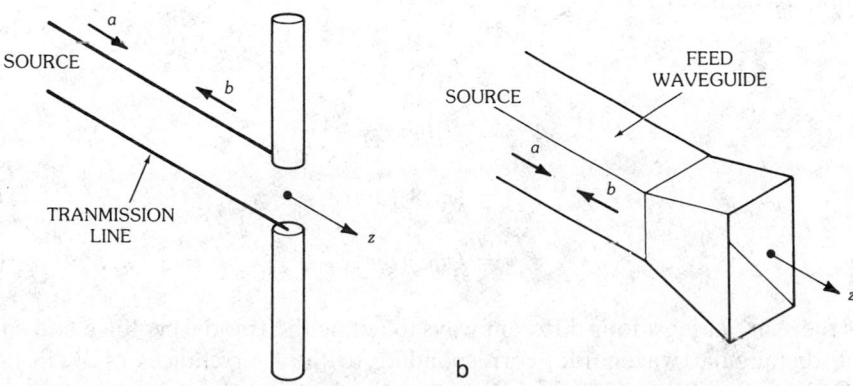

a   b

Fig. 15. Antennas are fed via transmission lines or waveguides. (*a*) Dipole. (*b*) Horn.

where the integration is over the infinite plane transverse to the transmission line, or the transverse cross section of the waveguide. The ratio

$$\Gamma = b/a \tag{68}$$

is called the E-field (voltage) reflection coefficient at a reference plane $z = 0$. The H-field (current) reflection coefficient is $-\Gamma$.

The above description of the field in the waveguide is from the wave viewpoint. Alternatively, it may be described from the circuit viewpoint. To do so, we are forced to introduce a so-called characteristic impedance $Z_c = Y_c^{-1}$, and then define the (modal) voltage and current by

$$V^+ = \sqrt{Z_c}\, a, \qquad I^+ = \sqrt{Y_c}\, a \tag{69}$$

where the superscript $+$ signifies a wave traveling in the $(+z)$ direction. We emphasize that in a general waveguide there is no obvious (unique) way to define Z_c because the ratio of transverse electric and magnetic fields varies from point to point. As an example, for a rectangular-feed waveguide having cross section $(c \times d)$ with $c > d$, we have

$$\mathbf{e}(x, y) = \hat{\mathbf{y}} \left(Z \frac{k}{\beta} \frac{2}{cd} \right)^{1/2} \cos\left(\frac{\pi}{c} x \right) \tag{70a}$$

$$\mathbf{h}(x, y) = (Z k/\beta)^{-1} \hat{\mathbf{z}} \times \mathbf{e} \tag{70b}$$

where

$$Z = Y^{-1} = (\mu/\epsilon)^{1/2}$$
$$\beta = k[1 - (\pi/kc)^2]^{1/2} \tag{71}$$

As for the characteristic impedance, there are at least four commonly used definitions, namely,

$$Z_{c1} = Zk/\beta \tag{72a}$$

$$Z_{c2} = 2(d/c)Z_{c1} \tag{72b}$$

$$Z_{c3} = (\pi^2/8)(d/c)Z_{c1} \tag{72c}$$

$$Z_{c4} = (\pi/2)(d/c)Z_{c1} \tag{72d}$$

Thus there are at least four different ways to define the (modal) voltage and current in the rectangular waveguide, corresponding to the four choices of Z_c in (72).

Once a choice of Z_c is made, we find the (total) voltage and current at the reference plane $z = 0$ are given by

$$V = V^+ + V^- = \sqrt{Z_c}\,(a + b) \tag{73}$$

$$I = I^+ + I^- = \sqrt{Y_c}\,(a - b) \tag{74}$$

The input impedance of the antenna is defined by

$$Z_{in} = \frac{1}{Y_{in}} = \frac{V}{I} \tag{75}$$

The relations among Γ, Z_{in}, and Y_{in} are

$$\frac{Z_{in}}{Z_c} = \frac{Y_c}{Y_{in}} = \frac{1 + \Gamma}{1 - \Gamma} \tag{76}$$

$$\Gamma = \frac{(Z_{in}/Z_c) - 1}{(Z_{in}/Z_c) + 1} = \frac{1 - (Y_{in}/Y_c)}{1 + (Y_{in}/Y_c)} \tag{77}$$

It is clear from (76) that antenna impedance Z_{in} also depends on the choice of Z_c.

The power transmitted (radiated) from the source into the free space via the antenna is given by

$$P_t = \mathrm{Re}\left\{ \int\!\!\int (\mathbf{E} \times \mathbf{H}^*)\cdot\hat{z}\, dx\, dy \right\} \tag{78a}$$

$$= |a|^2 - |b|^2 = |a|^2(1 - |\Gamma|^2) \tag{78b}$$

$$= \mathrm{Re}\{VI^*\} \tag{79}$$

The reflected power back toward the source (due to the mismatch of Z_c and Z_{in}) is $|b|^2$.

10. Three Ideal Sources for Transmitting Antennas

As discussed in the previous section, an antenna as a one-port device can be described by either wave amplitudes (a, b), or by circuit parameters (V, I). When the antenna is used for transmitting, a source is connected to the transmission line (or feed waveguide). Following Deschamps [6], we will introduce three principal ideal sources:

    (a)   unit amplitude source (defined by $a = 1$ (watt)$^{1/2}$)
    (b)   unit voltage source ($V = 1$ volt)
    (c)   unit current source ($I = 1$ ampere)

Their graphical representation is shown in Fig. 16. Note that the dot in each source indicates the propagation direction of the incident wave leaving the dot, the positive voltage terminal at the dot, or the current out of the dot. The term "ideal" is used because those sources have special internal impedance Z_s, viz.,

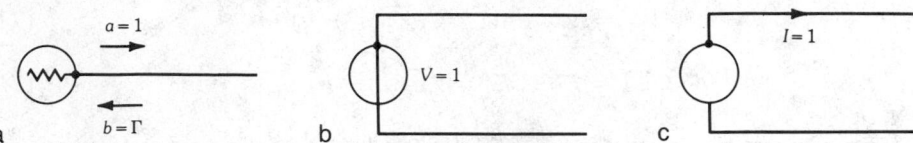

Fig. 16. Three ideal sources for transmitting antennas.

(a) $Z_s = Z_c$ for the amplitude source, in which Z_c is the characteristic impedance of the transmission line. Thus the source is matched to the transmission line, and there is no reflection at the source–transmission-line junction. (However, the mismatch at the antenna–transmission-line junction may still exist; Z_c may or may not be equal to Z_{in}.)

(b) $Z_s = 0$ for the voltage source so that there is no internal voltage drop within the source.

(c) $Z_s \to \infty$ for the current source so that the total source current enters the transmission line.

In Fig. 16 we use the zigzag line, straight line, and gap inside the circles to indicate the source impedances.

For a transmitting antenna any one of the three ideal sources may be used as the excitation. For each case, we list in Table 1 the internal feed waveguide quantities (V, I, a, b) and the external radiation quantities (radiated field F and its power P_t). Here F stands for vector fields ($\mathbf{E}, \mathbf{H}$). The subscripts $(1, 2, 3)$ are used to identify excitations due to a unit (amplitude, voltage, current) source. For example, a_1 is the value of a when the excitation is a unit amplitude source, and F_3 is the value of F when the excitation is a unit current source. The three fields (F_1, F_2, F_3) in Table 1 are related by

$$F_2 = \frac{\sqrt{Y_c}}{1 + \Gamma} \frac{V_2}{a_1} F_1 = \frac{1}{2}\sqrt{Y_c}\left(1 + \frac{Y_{in}}{Y_c}\right)\frac{V_2}{a_1} F_1 \tag{80}$$

$$F_3 = \frac{\sqrt{Z_c}}{1 - \Gamma} \frac{I_3}{a_1} F_1 = \frac{1}{2}\sqrt{Z_c}\left(1 + \frac{Z_{in}}{Z_c}\right)\frac{I_3}{a_1} F_1 \tag{81}$$

where $a_1 = 1$ (watt)$^{1/2}$, $V_2 = 1$ volt, and $I_3 = 1$ ampere.

11. Three Ideal Meters for Receiving Antennas

When an antenna is used for receiving, the source is replaced by a receiver or, for our present purpose, a meter. Again, following Deschamps we introduce three principal ideal meters:

(a) Amplitude meter, which measures the incoming traveling-wave amplitude b at a reference plane $z = 0$, and has internal source impedance $Z_s = Z_c$ such that the meter is matched to the feed waveguide. (The antenna impedance Z_{in} may or may not be matched to Z_c.) Note that b is a complex number, including both magnitude and phase information.

Table 1. Relationships for Transmitting Antennas

Quantity \ Source	Amplitude	Voltage	Current
V (volt)	$\sqrt{Z_c}(1+\Gamma)a_1$	$V_2 = 1$	$Z_{in}I_3$
I (ampere)	$\sqrt{Y_c}(1-\Gamma)a_1$	$Y_{in}V_2$	$I_3 = 1$
a (watt)$^{1/2}$	$a_1 = 1$	$\frac{1}{2}\sqrt{Y_c}\left(1+\frac{Y_{in}}{Y_c}\right)V_2$	$\frac{1}{2}\sqrt{Z_c}\left(\frac{Z_{in}}{Z_c}+1\right)I_3$
b (watt)$^{1/2}$	Γa_1	$\frac{1}{2}\sqrt{Y_c}\left(1-\frac{Y_{in}}{Y_c}\right)V_2$	$\frac{1}{2}\sqrt{Z_c}\left(\frac{Z_{in}}{Z_c}-1\right)I_3$
P_t (watt)	$(1-|\Gamma|^2)a_1^2$	$V_2^2\,\mathrm{Re}\,Y_{in}$	$I_3^2\,\mathrm{Re}\,Z_{in}$
Field	F_1	F_2	F_3
Source impedance	Z_c (matched)	0 (short)	∞ (open)

(b)   Voltmeter, which measures the voltage V at a reference plane $z = 0$, and has an infinite internal source impedance ($Z_s \to \infty$) such that V is the open-circuit voltage.

(c)   Ammeter, which measures the current I at a reference plane $z = 0$, and has a zero internal source impedance ($Z_s = 0$) such that I is the short-circuit current.

The graphical representation of meters is given in Fig. 17. Note our convention that the circles represent sources, while squares represent meters (compare Figs. 16 and 17).

When the receiving antenna is illuminated by an incident field we may connect any of the three ideal meters in Fig. 17 to its feed waveguide, corresponding to matched-load, open-circuit, and short-circuit situations. Table 2 lists the internal feed waveguide quantities (V, I, a, b). The subscripts (4, 5, 6) are used to identify the use of an (amplitude meter, voltmeter, ammeter). For the same incident field the three situations are related by

$$V_5 = \sqrt{Z_c}\,\frac{2b_4}{1-\Gamma} = b_4\sqrt{Z_c}\left(1 + \frac{Z_{in}}{Z_c}\right) \tag{82}$$

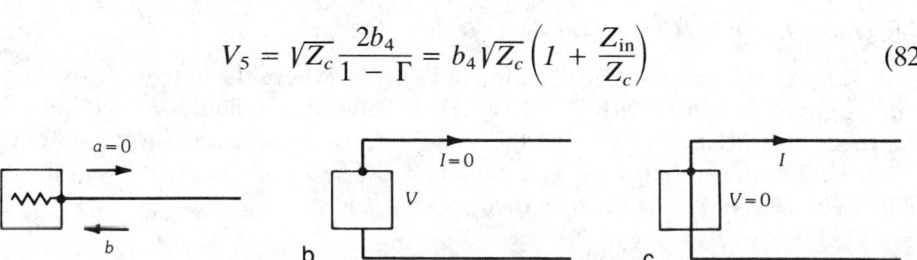

Fig. 17.   Three ideal meters for receiving antennas. (a) Amplitude. (b) Voltage. (c) Current.

Table 2. Relationships for Receiving Antennas

Quantity \ Meter	Amplitude	Voltmeter	Ammeter
V (volt)	$\sqrt{Z_c}\,b_4$	V_5	0
I (ampere)	$-\sqrt{Y_c}\,b_4$	0	I_6
a (watt)$^{1/2}$	0	$\dfrac{1}{2}\sqrt{Y_c}\,V_5$	$\dfrac{1}{2}\sqrt{Z_c}\,I_6$
b (watt)$^{1/2}$	b_4	$\dfrac{1}{2}\sqrt{Y_c}\,V_5$	$-\dfrac{1}{2}\sqrt{Z_c}\,I_6$
Source impedance	Z_c (matched)	∞ (open)	0 (short)

$$I_6 = \sqrt{Y_c}\frac{(-2b_4)}{1+\Gamma} = -b_4\sqrt{Y_c}\left(1 + \frac{Y_{\text{in}}}{Y_c}\right) \tag{83}$$

Commonly, V_5 is known as the *open-circuit voltage*, and I_6 as the *short-circuit current* of a receiving antenna.

12. Reciprocity between Antenna Transmitting and Receiving

The reciprocity theorem in a circuit is well known. Its application to antennas, however, is not simple for the reason explained below. If viewed from the transmission line, the antenna looks like a circuit element whose transmitting and receiving properties are describable by two (complex) numbers: (a, b) or (V, I), as discussed in Section 9. Outside the antenna in the free-space region, either the radiated field of the transmitting antenna or the incident field on the receiving antenna is more complex. They are vector fields characterized by polarization and spatial variation, which are not describable by circuit quantities. Hence the reciprocity for an antenna cannot be simply stated by the usual exchange of sources and meters. In this section we will give two reciprocity relations for antennas.

Reciprocity Involving General Incident Fields

Consider the transmitting situation in Fig. 18a, where the antenna is excited by a unit amplitude source with $a_1 = 1$ (watt)$^{1/2}$. The radiated field is $F_1 = (\mathbf{E}_1, \mathbf{H}_1)$. In the receiving situation (Fig. 18b) the same antenna is connected to an amplitude meter with matched impedance (so that $a_4 = 0$), and is illuminated by an incident field $F_4 = (\mathbf{E}_4, \mathbf{H}_4)$. Then a reciprocity states [6]

$$b_4 = \frac{1}{2a_1}\int_S (\mathbf{E}_4 \times \mathbf{H}_1 - \mathbf{E}_1 \times \mathbf{H}_4)\cdot\hat{\mathbf{N}}\,dS \tag{84}$$

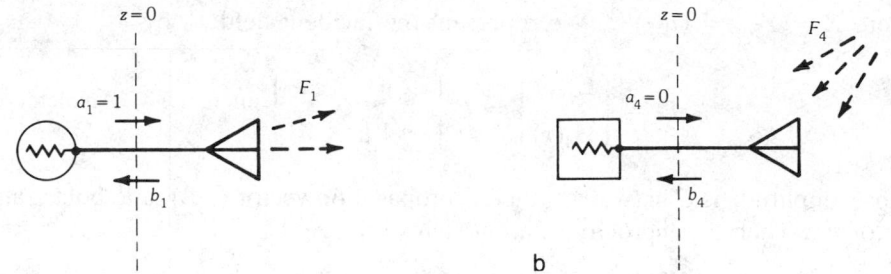

Fig. 18. An antenna in transmitting excited with a unit-amplitude source, and in receiving connected with an amplitude meter. (*a*) Transmitting. (*b*) Receiving.

$$= \frac{1}{2a_1} \langle F_4 S F_1 \rangle$$

Here S is an arbitrary closed surface which encloses the antenna but excludes the source of F_4. Its outward unit normal is $\hat{N}$. It is emphasized that F_4 is the incident field that would exist in the absence of the antenna in Fig. 18b, and does not include the scattered field F_4' from the antenna. It can be shown, however, that (84) remains valid if F_4 is replaced by $F_4 + F_4'$, because the cross flux $\langle F_4' S F_1 \rangle = 0$.

Reciprocity Involving Plane Waves

In many applications we are interested in a special case of Fig. 18, namely, the radiated field F_1 in the transmitting situation is known in the far-field zone (Fig. 19a), and the incident field F_4 in the receiving situation is a plane wave (Fig. 19b). We express the radiated field F_1 by

$$F_1 = \begin{Bmatrix} \mathbf{E}_1(\mathbf{r}) \\ \mathbf{H}_1(\mathbf{r}) \end{Bmatrix} \sim \begin{Bmatrix} \sqrt{Z}\,\mathbf{A}(\mathbf{k}) \\ \sqrt{Y}\,\hat{\mathbf{k}} \times \mathbf{A}(\mathbf{k}) \end{Bmatrix} \frac{e^{-jkr}}{r}, \qquad r \to \infty \qquad (85)$$

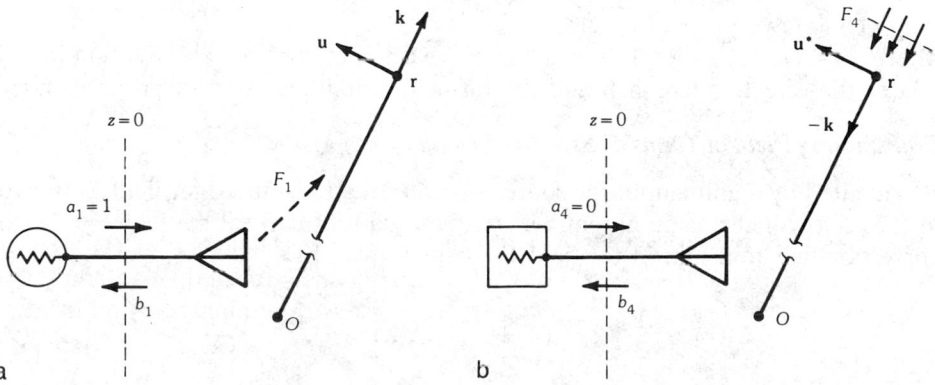

Fig. 19. Same antenna as Fig. 18 except incident field F_4 is a plane wave. (*a*) Transmitting. (*b*) Receiving.

where $Z = Y^{-1} = (\mu/\epsilon)^{1/2}$. We represent the incident field F_4 by

$$F_4 = \left\{\begin{array}{c} \mathbf{E}_4(\mathbf{r}) \\ \mathbf{H}_4(\mathbf{r}) \end{array}\right\} = C \left\{\begin{array}{c} \sqrt{Z}\,\mathbf{u}^* \\ -\sqrt{Y}\hat{\mathbf{k}} \times \mathbf{u}^* \end{array}\right\} e^{j\mathbf{k}\cdot\mathbf{r}} \tag{86}$$

whose amplitude is C in $\text{(watt)}^{1/2}/\text{meter}$, propagation vector $(-\mathbf{k})$, and polarization vector $\mathbf{u}^*$. Then a reciprocity relation states [6]

$$b_4 = -j\lambda \left(\frac{C}{a_1}\right)[\mathbf{A}(\mathbf{k})\cdot\mathbf{u}^*] \tag{87}$$

where $\lambda = 2\pi/k$ is the free-space wavelength. If the open-circuit voltage and short-circuit current in the receiving situation are of interest, we may use (87) or (84) in conjunction with (82) and (83). The interpretation of (87) is as follows: The received amplitude b_4 of an antenna under the matched condition due to the incidence of a plane wave in state $(-\mathbf{k}, \mathbf{u}^*)$ is proportional to the antenna radiation far-field amplitude in state $(\mathbf{k}, \mathbf{u})$. [Remember that $(-\mathbf{k}, \mathbf{u}^*)$ and $(\mathbf{k}, \mathbf{u})$ have the same polarization.]

Antenna Effective Length

Consider the receiving situation in Fig. 19b, where the antenna is connected to a matched meter and is illuminated by an incident plane wave F_4 defined in (86). The effective length $\mathbf{h}$ is defined by the relation

$$V_5 = \mathbf{h}\cdot\mathbf{E}_4(\mathbf{r} = \mathbf{0}) \tag{88}$$

where V_5 is the open-circuit voltage of the transmission line at a reference plane $z = 0$, and $\mathbf{E}_4(\mathbf{r} = \mathbf{0})$ is the electric field at a reference point O of the incident plane wave given in (86). From (88), (87), and (82) we conclude that

$$\mathbf{h} = \sqrt{\frac{Z_c}{Z}}\left(1 + \frac{Z_{\text{in}}}{Z_c}\right)\left(-j\frac{\lambda}{a_1}\right)\mathbf{A}(\mathbf{k}) \tag{89}$$

where $Z = (\mu/\epsilon)^{1/2}$. Clearly $\mathbf{h}$ is a complex vector in meters. The relation in (89) relates the effective length $\mathbf{h}$ and the far-field amplitude $\mathbf{A}(\mathbf{k})$ of an antenna.

Transmitting Field in Terms of Effective Length

Excited by a unit amplitude source $(a = 1)$, the transmitted far field $\mathbf{E}_1$ is given in (85). Now if the same antenna is excited by a unit current source $(I = 1)$, the corresponding transmitted far field $\mathbf{E}_3$ is then given by

$$\mathbf{E}_3(\mathbf{r}) \sim \frac{1}{2}\sqrt{Z_c Z}\left(1 + \frac{Z_{\text{in}}}{Z_c}\right)\mathbf{A}(\mathbf{k})\frac{e^{-jkr}}{r}, \qquad r \to \infty \tag{90}$$

where we have used the formula (81) in relating $\mathbf{E}_1$ and $\mathbf{E}_3$. Replacing $\mathbf{A}$ by $\mathbf{h}$ in accordance with (89), we rewrite (90) as

$$E_3(r) \sim \frac{jZ}{2\lambda} \mathbf{h} \frac{e^{-jkr}}{r} \tag{91}$$

which may be used as an alternative definition of the effective length $\mathbf{h}$.

Receiving Cross Section

Consider the receiving situation in Fig. 19b, where the antenna is connected to a matched amplitude meter. The incident plane wave is given in (86), with amplitude C and state $(-\mathbf{k}, \mathbf{u}^*)$. The received wave amplitude is b_4, given in (87). Then we define the receiving cross section (effective area) σ in state $(\mathbf{k}, \mathbf{u})$ of the antenna by

$$\begin{aligned}
\sigma(\mathbf{k}, \mathbf{u}) &= \left| \frac{b_4}{C} \right|^2 \\
&= \frac{\text{received power of the antenna under matched condition}}{\text{power density of incident plane wave in state } (-\mathbf{k}, \mathbf{u}^*)}
\end{aligned} \tag{92}$$

in square meters. Making use of (87), we have

$$\sigma(\mathbf{k}, \mathbf{u}) = \lambda^2 \left| \frac{1}{a_1} \mathbf{A}(\mathbf{k}) \cdot \mathbf{u}^* \right|^2 \tag{93}$$

A partial gain of the antenna in state $(\mathbf{k}, \mathbf{u})$ is defined by (Section 10, Chapter 1)

$$g_1(\mathbf{k}, \mathbf{u}) = 4\pi \left| \frac{1}{a_1} \mathbf{A}(\mathbf{k}) \cdot \mathbf{u}^* \right|^2 \tag{94}$$

From (93) and (94) we have

$$g_1(\mathbf{k}, \mathbf{u}) = \frac{4\pi}{\lambda^2} \sigma(\mathbf{k}, \mathbf{u}) \tag{95}$$

where $\sigma_\lambda = \lambda^2/4\pi$ is sometimes known as the *receiving cross section of the fictitious isotropic radiator*. As discussed in Section 10, Chapter 1, the partial gain g_1 in a preferred polarization $\mathbf{u}$ is related to the (total) gain G_1 in all (both) polarizations by

$$g_1(\mathbf{k}, \mathbf{u}) = G_1(\mathbf{k}) |\mathbf{u}^* \cdot \mathbf{U}|^2 \tag{96}$$

where $\mathbf{U}$ describes the polarization of the antenna in direction $\mathbf{k}$. Substituting (96) in (95) gives

$$\sigma(\mathbf{k}, \mathbf{u}) = \frac{\lambda^2}{4\pi} |\mathbf{u}^* \cdot \mathbf{U}|^2 G_1(\mathbf{k}) \tag{97}$$

which relates the receiving cross section and the gain of an antenna. The factor

$$p = |\mathbf{u}^* \cdot \mathbf{U}|^2 \tag{98}$$

is called the *polarization efficiency* (or *polarization mismatch factor*). We emphasize that unitary vector $\mathbf{u}^*$ (not $\mathbf{u}$) describes the polarization of the incoming plane wave in direction $(-\mathbf{k})$, while $\mathbf{U}$ describes the polarization of the receiving antenna in the outgoing direction $(+\mathbf{k})$. For example, let $\mathbf{k} = \hat{\mathbf{z}}$ and $\mathbf{U} = (\hat{\mathbf{x}} - j\hat{\mathbf{y}})/\sqrt{2}$ for a right-hand circularly polarized antenna. If the incoming plane wave is also right-hand circularly polarized, we have $\mathbf{u}^* = (\hat{\mathbf{x}} + j\hat{\mathbf{y}})/\sqrt{2}$ and $p = 1$.

13. The Radar Equation and Friis Transmission Formula

The configuration of a bistatic radar is sketched in Fig. 20. The obstacle (target) in the vicinity of point 0 is illuminated by an incident wave from the transmitting antenna at point 1. A part of the scattered energy is received by a receiving antenna at point 2. Both distances R_1 and R_2 are large in terms of wavelength so that the scatterer is in the far-field zones of the antennas. The problem at hand is to determine the power ratio P_2/P_1, where

P_1 = power incident from the generator to the transmitting antenna,

P_2 = power received by the receiver via the receiving antenna.

To this end we need to define the antennas and the obstacle more precisely, as below.

(*a*) In the vicinity of point 0, the radiated field of the transmitting antenna is in

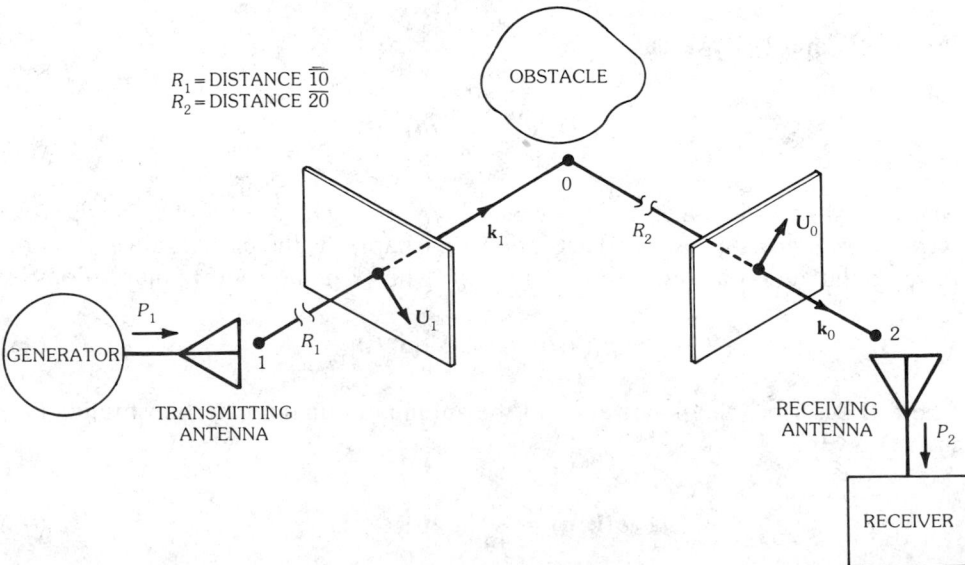

Fig. 20. A bistatic radar.

the direction $\mathbf{k}_1$ and has a polarization described by a unitary vector $\mathbf{U}_1$. In other words, the state of the transmitting antenna at point 0 is $(\mathbf{k}_1, \mathbf{U}_1)$. The gain of the transmitting antenna in the direction $\mathbf{k}_1$ is $G_1(\mathbf{k}_1)$, which is related to the directivity $D_1(\mathbf{k}_1)$ by

$$G_1(\mathbf{k}_1) = \eta_1(1 - |\Gamma|^2)D_1(\mathbf{k}_1) \tag{99}$$

Here η_1 is the transmitting antenna efficiency accounting for the conductor and dielectric losses, and Γ_1 is the reflection coefficient accounting for the impedance mismatch between the transmitting antenna and its feed (Section 10, Chapter 1).

(b) If the receiving antenna were used for transmitting (Fig. 21), the state of its radiated field at point 0 would be $(\mathbf{k}_2, \mathbf{U}_2)$, and its gain would be $G_2(\mathbf{k}_2)$. A relation similar to (99) holds for G_2 and D_2.

(c) Due to the illumination from the transmitting antenna the obstacle produces a scattered field at point 2, with state $(\mathbf{k}_0 = -\mathbf{k}_2, \mathbf{U}_0)$. Its bistatic cross section from state $(\mathbf{k}_1, \mathbf{U}_1)$ to state $(\mathbf{k}_0, \mathbf{U}_0)$ is denoted by the BCS.

With the above description we may now calculate various power quantities. The power density W_0 of the radiated field at point 0 from the transmitting antenna is

$$W_0 = \frac{P_1}{4\pi R_1^2} G_1(\mathbf{k}_1) \tag{100}$$

in watts per square meter. The power density W_2 of the scattered field at point 2 from the obstacle is

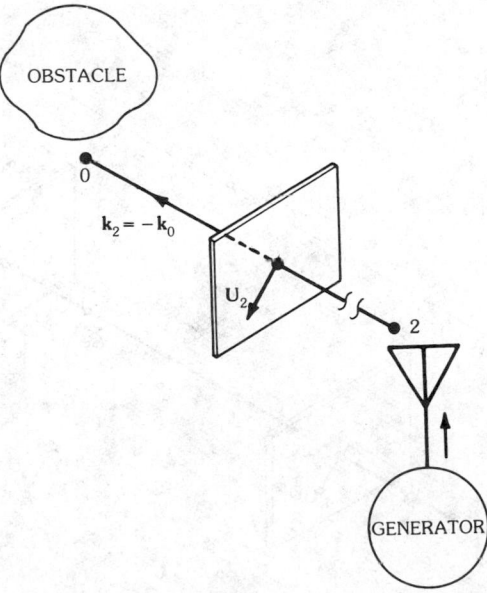

Fig. 21. The receiving antenna of Fig. 20 is now used for transmitting.

$$W_2 = W_0 \frac{\text{BCS}}{4\pi R_2{}^2} \tag{101}$$

The received power P_2 by the receiver at point 2 via the receiving antenna may be calculated from (92) and (97) with the result

$$P_2 = W_2 \frac{\lambda^2}{4\pi} |\mathbf{U}_0 \cdot \mathbf{U}_2|^2 G_2(\mathbf{k}_2) \tag{102}$$

Combining (100) through (102), we have the power ratio for the bistatic radar in Fig. 20, namely,

$$\frac{P_2}{P_1} = \frac{\text{BCS}}{4\pi} \left(\frac{\lambda}{4\pi R_1 R_2}\right)^2 |\mathbf{U}_0 \cdot \mathbf{U}_2|^2 G_1(\mathbf{k}_1) G_2(\mathbf{k}_2) \tag{103}$$

which is known as the *radar (range) equation.*

    Next, we will consider a special case of (103). Let us remove the obstacle in Fig. 21, and study the direct power transmission from point 1 to point 2 (Fig. 22). Making use of the following relations:

$$\mathbf{k}_2 = -\mathbf{k}_1, \qquad \mathbf{U}_0 = \mathbf{U}_1, \qquad R_1 = R$$

$$W_0 = W_2, \qquad \text{BCS} = 4\pi R_2{}^2$$

then (103) becomes the Friis transmission formula for the far-field transmission between two antennas, namely,

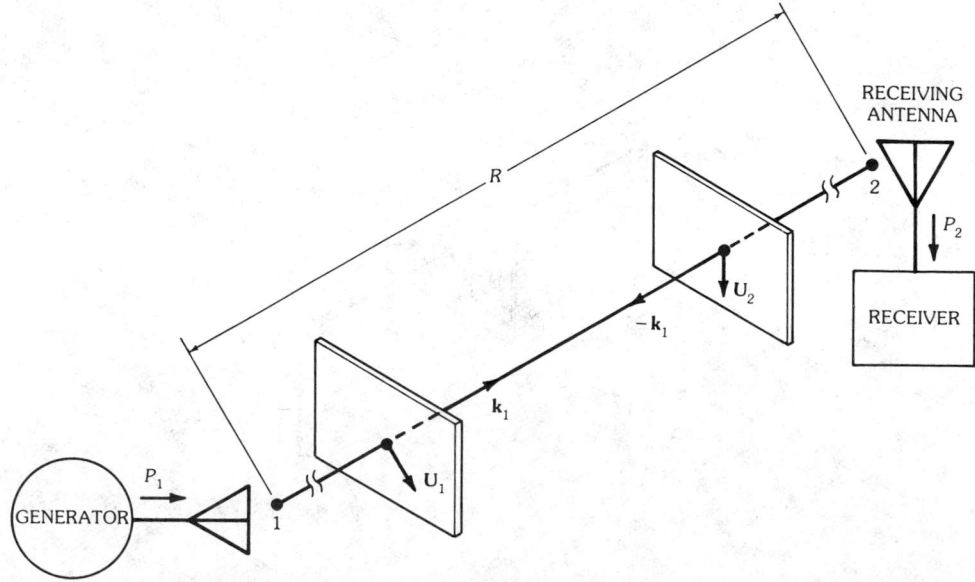

Fig. 22. Power transmission between two antennas.

$$\frac{P_2}{P_1} = \left(\frac{\lambda}{4\pi R}\right)^2 |\mathbf{U}_1\cdot\mathbf{U}_2|^2 G_1(\mathbf{k}_1)\,G_2(-\mathbf{k}_1) \tag{104}$$

The factor $(\lambda/4\pi R)^2$ is called the *free-space loss factor*, and it accounts for the loss due to the spherical spread of the transmitted field. The polarization efficiency

$$p = |\mathbf{U}_1\cdot\mathbf{U}_2|^2 \tag{105}$$

accounts for the loss due to polarization mismatch between the transmitting and the receiving antennas. If both antennas have the same polarization, $\mathbf{U}_1 = \mathbf{U}_2^*$ and $p = 1$. (The conjugate on $\mathbf{U}_2^*$ is due to the fact that $\mathbf{U}_2$ refers to a propagation direction opposite that of $\mathbf{U}_1$.) Note the symmetry between 1 and 2 in the right-hand side of (104), implying that the same formula applies if the roles of the receiving and transmitting antennas are interchanged in Fig. 22.

14. Noise Temperature of an Antenna

For a high-resolution antenna a crucial factor that limits its ability to detect a weak signal is the antenna noise, which is the subject of the present section.

Antenna Noise Temperature

The receiving antenna sketched in Fig. 23 receives the desired signal as well as noise. We denote the available noise power at terminal Σ by P_a in watts. It is a common practice to express P_a in terms of an (effective) antenna noise temperature T_a in kelvins via the relation

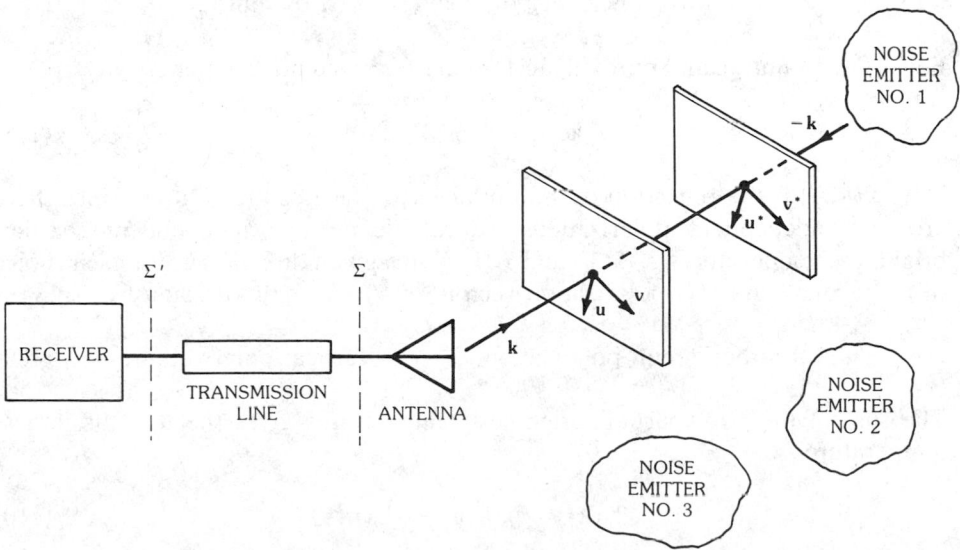

Fig. 23. A receiving antenna receives thermal noises from external noise emitters.

$$T_a = P_a/(k'\Delta f) \tag{106}$$

where

k' = Boltzmann's constant = 1.38×10^{-23} J/K

Δf = bandwidth of the antenna receiving system in hertz

The interpretation of (106) is that antenna noise power P_a is equal to that of a matched resistor whose physical temperature is T_a. Hereafter we will use P_a and T_a interchangeably. Clearly, T_a depends on the antenna receiving characteristics and external noise emitter.

Brightness Temperature of an Emitter

Every object with its physical temperature above the absolute zero (0 K) is an emitter of thermal energy in the form of electromagnetic waves. Within a narrow frequency band the amount of energy radiated in direction **k** and polarization **u** or simply in state (**k**, **u**) is proportional to a parameter called the *brightness temperature* $T_b(\mathbf{k}, \mathbf{u})$. The latter is related to the physical temperature T_p of the emitter via the relation

$$T_b(\mathbf{k}, \mathbf{u}) = \varepsilon(\mathbf{k}, \mathbf{u})T_p \tag{107}$$

where the dimensionless parameter $\varepsilon(\mathbf{k}, \mathbf{u})$ is called the *emissivity in state* (**k**, **u**) of the emitter. To determine $\varepsilon(\mathbf{k}, \mathbf{u})$ we illuminate the emitter by an incident plane wave in state ($-\mathbf{k}, \mathbf{u}^*$). See Fig. 24. Then

$$\varepsilon(\mathbf{k}, \mathbf{u}) = \frac{\text{incident power absorbed by emitter}}{\text{incident power intercepted by emitter}}$$

For a large emitter an approximate formula for calculating its emissivity is

$$\varepsilon(\mathbf{k}, \mathbf{u}) \cong 1 - |\Gamma(\mathbf{k}, \mathbf{u})|^2 \tag{108}$$

Here $\Gamma(\mathbf{k}, \mathbf{u})$ is the reflection coefficient at the surface of the emitter. Since there are two independent polarizations we can define two independent (partial) brightness temperatures $T_b(\mathbf{k}, \mathbf{u})$ and $T_b(\mathbf{k}, \mathbf{v})$ in a given direction **k**, for each choice of orthogonal, unitary, polarization vectors (**u**, **v**). A common choice of (**u**, **v**) is

$$\mathbf{u} = \text{horizontal polarization}, \qquad \mathbf{v} = \text{vertical polarization} \tag{109}$$

The sum of the two (partial) brightness temperatures gives the total brightness temperature

$$T_b(\mathbf{k}) = T_b(\mathbf{k}, \mathbf{u}) + T_b(\mathbf{k}, \mathbf{v}) \tag{110}$$

which is proportional to the total energy radiated by the emitter in direction **k** in

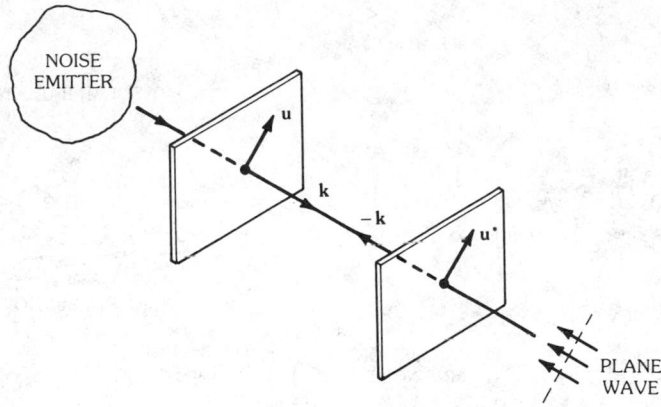

Fig. 24. To determine the emissivity $\epsilon(\mathbf{k}, \mathbf{u})$ of a noise emitter we illuminate the emitter by an incident plane wave of state $(-\mathbf{k}, \mathbf{u}^*)$.

both polarizations. A blackbody absorbs all the incoming energy impinging on it (a perfect absorber). Its emissivity is unity (a perfect emitter), and $\frac{1}{2}T_b(\mathbf{k}) = T_b(\mathbf{k}, \mathbf{u}) = T_b(\mathbf{k}, \mathbf{v}) = T_p$ for any direction $\mathbf{k}$.

Calculation of Antenna Noise Temperature

The power received by a receiving antenna can be traced to three sources: the desired signal, interference from other coherent radiators, and incoherent noise from noise emitters. Fig. 25 shows some important noise emitters in the free-space environment. Let us concentrate on a typical noise emitter (No. 1 in Fig. 23). Its contribution to the antenna noise temperature can be calculated from the following formula:

$$T_a = \frac{1}{4\pi} \int_0^\pi \sin\theta \, d\theta \int_0^{2\pi} d\phi \left[T_b(-\mathbf{k}, \mathbf{u}^*) \, d(\mathbf{k}, \mathbf{u}) + T_b(-\mathbf{k}, \mathbf{v}^*) \, d(\mathbf{k}, \mathbf{v}) \right] \quad (111)$$

where

$(\mathbf{u}, \mathbf{v})$ = unitary vectors describing two orthogonal polarizations

$d(\mathbf{k}, \mathbf{u})$ = (partial) directivity of the antenna in state $(\mathbf{k}, \mathbf{u})$

$T_b(-\mathbf{k}, \mathbf{u}^*)$ = (partial) brightness temperature of the emitter in state $(-\mathbf{k}, \mathbf{u}^*)$. Note that states $(\mathbf{k}, \mathbf{u})$ and $(-\mathbf{k}, \mathbf{u}^*)$ have the opposite directions but the same polarization

If there is more than one noise emitter in space, the superposition principle applies to the calculation of T_a. This is so because noise emitters are incoherent and the superposition of powers (temperatures) is permissible. For an idealized omni-directional antenna, which radiates equally in both polarizations, we have $d(\mathbf{k}, \mathbf{u}) = d(\mathbf{k}, \mathbf{v}) = 1/2$, and

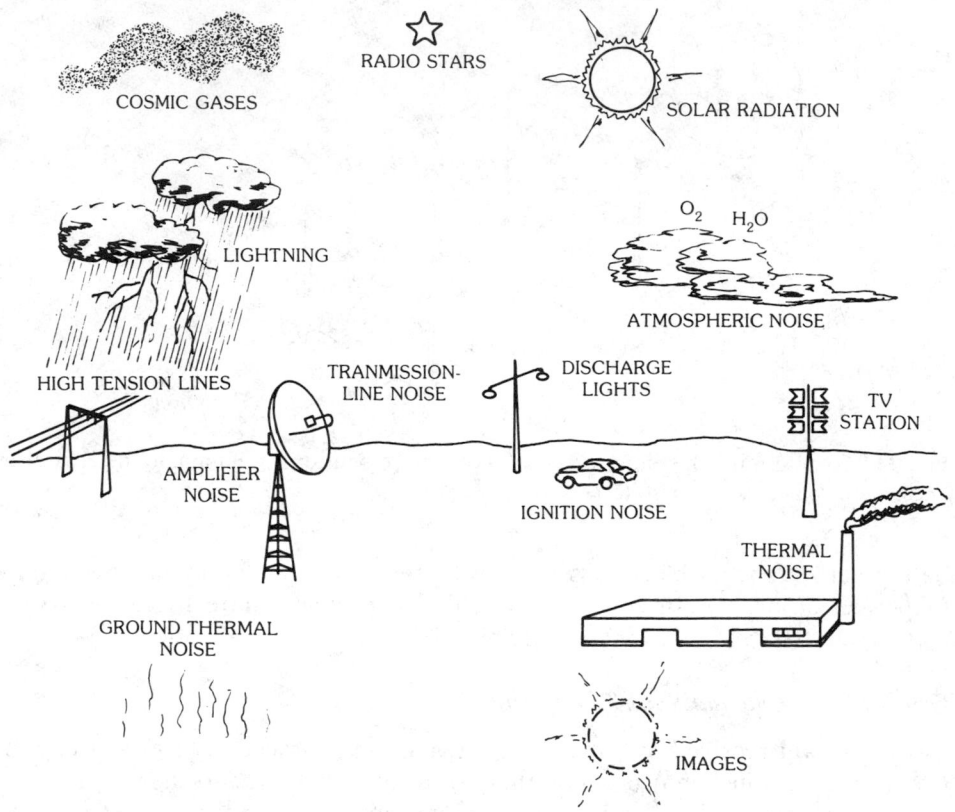

COSMIC GASES

RADIO STARS

SOLAR RADIATION

LIGHTNING

O_2 H_2O

ATMOSPHERIC NOISE

HIGH TENSION LINES

TRANSMISSION-
LINE NOISE

DISCHARGE
LIGHTS

TV
STATION

AMPLIFIER
NOISE

IGNITION NOISE

THERMAL
NOISE

GROUND THERMAL
NOISE

IMAGES

Fig. 25. Important noise emitters for a receiving antenna. (*Courtesy Y. T. Lo*)

$$T_a = T_{a0} = \frac{1}{8\pi} \int_0^\pi \sin\theta \, d\theta \int_0^{2\pi} d\phi \, [T_b(-\mathbf{k}, \mathbf{u}^*) + T_b(-\mathbf{k}, \mathbf{v}^*)] \qquad (112)$$

Fig. 26 presents some typical values of T_{a0} of an omnidirectional antenna, together with the noise temperature of a typical receiver.

Noise Power at the Receiver's Terminal

Corresponding to the antenna noise temperature T_a in (111), the available noise power at terminal Σ in Fig. 23 is given by P_a according to (106). This incoming power propagates through the transmission line, and arrives at the receiver's terminal Σ' with its value in watts equal to

$$P'_a = k' \, \Delta f [\eta T_a + (1 - \eta) T_0] \qquad (113)$$

where

η = power transmission efficiency between terminals Σ and Σ'

T_0 = physical temperature of the transmission line

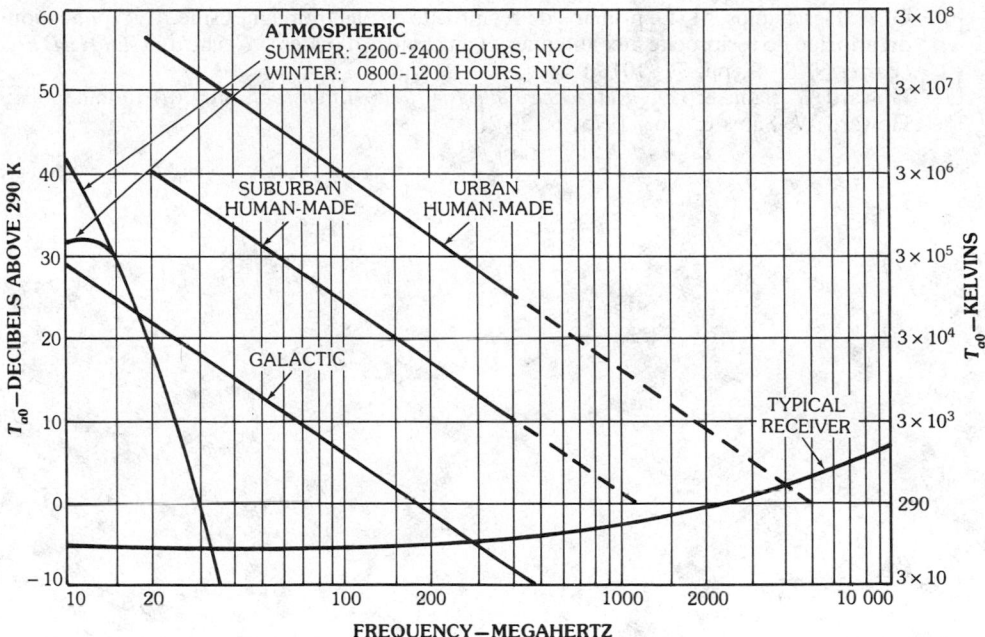

Fig. 26. Median values of average antenna noise temperature for an omnidirectional antenna near the earth's surface. (*After Sams [7], © 1975 Howard W. Sams & Company, Indianapolis; reprinted with permission*)

The power transmission efficiency is defined by

$$\eta = \frac{\text{output power at terminal } \Sigma'}{\text{input power at terminal } \Sigma} \tag{114}$$

It includes the power loss due to the mismatches at Σ and Σ', and conductor/dielectric losses of the transmission line.

15. References

[1] C. T. Tai, *Dyadic Green's Functions in Electromagnetic Theory*, Scranton: Intext Educational, 1971.

[2] S. W. Lee, J. Boersma, C. L. Law, and G. A. Deschamps, "Singularity in Green's function and its numerical evaluation," *IEEE Trans. Antennas Propag.*, vol. AP-28, pp. 311–317, 1980.

[3] A. D. Yaghjian, "Electric dyadic Green's functions in the source region," *Proc. IEEE*, vol. 68, pp. 248–263, 1980.

[4] G. A. Deschamps, "Scattering diagrams in electromagnetic theory," in *Electromagnetic Theory and Antennas*, Part I, ed. by E. C. Jordan, New York: Pergamon, 1963, pp. 235–251.

[5] R. D. Kodis, "A note on the theory of scattering from an irregular surface," *IEEE Trans. Antennas Propag.*, vol. AP-14, pp. 77–82, 1966.

[6] G. A. Deschamps, "I. Le principe de réciprocité en électromagnétisme. II. Application du principe de réciprocité aux antennes et aux guides d'ondes," *Revue du CETHEDEC* (Paris) N° 8–4°, pp. 71–101, 1966.

[7] Howard W. Sams & Co., *Reference Data for Radio Engineers*, 6th ed., Indianapolis: Howard W. Sams & Co., 1975, p. 29-2.

Chapter 3

Techniques for Low-Frequency Problems

A. J. Poggio
Lawrence Livermore National Laboratory

E. K. Miller
Rockwell International Science Center

CONTENTS

Andrew J. Poggio was born in New York City in 1941. He received a PhD in electrical engineering from the University of Illinois in 1969. His doctoral research was directed toward the numerical solution of integral equations for dipole and slot antennas. As a research associate at MBAssociates he performed research in numerical techniques and mathematical modeling in electromagnetic wave theory and was instrumental in developing computer methods for solving boundary value problems associated with electromagnetic scatterers and antennas. At the Cornell Aeronautical Laboratory he was a research engineer studying the propagation of electromagnetic waves in random media. Now a staff member of the Lawrence Livermore National Laboratory, he is at present involved in microwave and millimeter-wave analytical and experimental programs and leads the Electromagnetic Diagnostic Systems Group and the Microwave Engineering Facility.

Edmund K. Miller was born in Milwaukee in 1935. He received a PhD in electrical engineering from the University of Michigan in 1965. At the Radiation Laboratory, and later at the High Altitude Research Laboratory, of the University of Michigan, he conducted experimental and analytical research in plasma–electromagnetic-wave interaction. In 1968 he joined MBAssociates, where he worked on integral-equation methods for modeling antennas in both the time and frequency domains. His involvement in numerical techniques continued at the Lawrence Livermore National Laboratory, which he joined in 1971 and where he served as leader of the Engineering Research Division and the Nuclear Energy Systems Division until 1985. He was Regent's Distinguished Professor of Electrical and Computer Engineering at the University of Kansas at Lawrence from 1985 to 1987 and is now Manager of Electromagnetics at Rockwell International Science Center. He has lectured widely on computer methods in electromagnetics and has taught numerous short courses on the subject. Dr. Miller has published more than 60 articles on numerical methods, their electromagnetic applications, signal processing, graphics, and related topics.

Introduction to Low-Frequency Techniques

In this chapter we survey various techniques that can be used to evaluate antennas in the regime where their physical size is a maximum of a few wavelengths in extent. There are two basic approaches besides experimentation that can be considered for such problems: analytical and numerical.

Analytical procedures are the older of the two, representing the only practical way to design and characterize new antennas prior to the advent of the digital computer. These techniques are discussed in Part 1 of this chapter, beginning with the problem of finding the fields of a prescribed current for various geometries. The presentation follows a sequence of problems of increasing complexity, including expressions for radiation resistance and near and far fields. Also considered are a variety of tools that can be used to simplify or extend analytical techniques, based for example on image theory, duality, and the like.

A common denominator of these analytical techniques, with the exception of the few problems that can be solved using boundary-value formulations, is that of assuming a current distribution. That approach, adequate for a surprising range of problems, can be extended in applicability by using variational techniques. But for really general problems more elaborate numerical computations are necessary. Part 1 concludes with presentation of several integral equations suitable for such numerical modeling, specialized to the wire geometries which make up the majority of low-frequency problems.

The emphasis of Part 2 of the chapter is on numerical techniques. It extends the treatment of Part 1 to situations where the current distribution cannot be assumed, but must be solved as part of a generalized boundary-value problem. A discussion of the issues needing consideration in developing and applying numerical techniques is given. These include the formulation, numerical implementation, computation, and validation. Both frequency-domain and time-domain techniques are considered. Numerous tables are used to summarize salient points, and a number of existing computer codes are outlined to give more concrete examples of what can be accomplished.

Part 1
Selected Analytical Issues for Antenna Engineering

In this part on analysis we will cover the electromagnetic field equations used in antenna analysis, some antenna characteristics, and integral equations in antenna analysis.

1. Theory

This section treats the basic theory of electromagnetic fields as applied to the characterization of antennas.

The Electromagnetic Field Equations for Antenna Analysis

The fundamental equations for electromagnetics can be written in their time-dependent form or specialized to time-harmonic behavior as in Chapter 1, (1) through (3b). For the purposes of this chapter and the definition of antenna behavior in the frequency domain, we will confine our attention to the time-harmonic Maxwell equations given by

$$\nabla \times \mathbf{E} = -j\omega\mu\mathbf{H} - \mathbf{K} \tag{1}$$

$$\nabla \times \mathbf{H} = j\omega\varepsilon\mathbf{E} + \mathbf{J} \tag{2}$$

$$\nabla \cdot \mathbf{E} = \varrho/\epsilon \tag{3}$$

$$\nabla \cdot \mathbf{H} = \varrho_m/\mu \tag{4}$$

with the continuity equations

$$\nabla \cdot \mathbf{J} = -j\omega\varrho \tag{5}$$

$$\nabla \cdot \mathbf{K} = j\omega\varrho_m \tag{6}$$

The solution of the coupled differential equations for **E** and **H** when driven by the forcing functions **J** and **K** has been the subject of numerous books and is beyond the scope of this work. Rather, we will employ the expressions in Chapter 1— (20a) and (20b)—which are written in terms of vector potentials [1], viz., the magnetic vector potential, $\mathbf{A}(\mathbf{r})$, and the electric vector potential, $\mathbf{F}(\mathbf{r})$. These expressions are presented in Chart 1. Also shown in that chart are alternate representations obtained from direct integration of the field equations [2, 3, 4, 5, 6, 7]. In the equations, ∇' implies operations in the source or primed coordinates. The equations in Chart 1 are the general representations for the electromagnetic fields due to volumetric distributions of electric and magnetic sources in an unbounded homogeneous medium.

The Chart 1 representations are most useful when evaluating the fields due to assumed (or approximate) current distributions. While this may often be adequate, the current distributions on complex structures can be difficult to accurately predict or approximate. In such cases it may be possible to solve Maxwell's equations directly subject to appropriate boundary conditions and driving conditions for the fields and the induced surface currents. This approach is particularly useful when the antenna geometry conforms to a separable coordinate system, but since the class of antennas fitting separable coordinate systems is limited, the direct solution of Maxwell's equations (or the corresponding wave equation) has also been limited. An alternative to solving Maxwell's equations is to cast the field equations into

Chart 1. Field Equation Pairs

E Field

$$\mathbf{E(r)} = -\nabla \times \mathbf{F} + \frac{1}{j\omega\epsilon}(\nabla \times \nabla \times \mathbf{A} - \mathbf{J}) \qquad (7a)$$

$$\mathbf{E(r)} = -\nabla \times \mathbf{F} + \frac{1}{j\omega\epsilon}(\nabla\nabla\cdot + k^2)\mathbf{A} \qquad (7b)$$

H Field

$$\mathbf{H(r)} = \nabla \times \mathbf{A} + \frac{1}{j\omega\mu}(\nabla \times \nabla \times \mathbf{F} - \mathbf{K}) \qquad (8a)$$

$$\mathbf{H(r)} = \nabla \times \mathbf{A} + \frac{1}{j\omega\mu}(\nabla\nabla\cdot + k^2)\mathbf{F} \qquad (8b)$$

$$\mathbf{A(r)} = \frac{1}{4\pi}\int_v \mathbf{J(r')}\,\varphi(\mathbf{r,r'})\,dv' \qquad (9)$$

$$\mathbf{F(r)} = \frac{1}{4\pi}\int_v \mathbf{K(r')}\,\varphi(\mathbf{r,r'})\,dv' \qquad (10)$$

$$\varphi(\mathbf{r,r'}) = \frac{e^{-jk|\mathbf{r}-\mathbf{r'}|}}{|\mathbf{r}-\mathbf{r'}|}$$

$$\mathbf{E(r)} = -\frac{1}{4\pi}\int_v \left(j\omega\mu\mathbf{J}\varphi + \mathbf{K}\times\nabla'\varphi + \frac{\nabla'\cdot\mathbf{J}}{j\omega\epsilon}\nabla'\varphi\right)dv' \qquad (11)$$

$$\mathbf{E(r)} = -\frac{1}{4\pi}\int_v (j\omega\mu\mathbf{J}\varphi + \mathbf{K}\times\nabla'\varphi)\,dv' + \frac{1}{4\pi j\omega\epsilon}\nabla\int_v \nabla'\cdot\mathbf{J}\varphi\,dv'$$

$$\mathbf{E(r)} = \frac{1}{4\pi j\omega\epsilon}\int_v (\mathbf{J}\cdot(k^2\overline{\overline{\mathbf{I}}} + \nabla\nabla)\varphi - j\omega\epsilon\mathbf{K}\times\nabla'\varphi)\,dv'$$

$$\mathbf{H(r)} = \frac{1}{4\pi}\int_v \left(-j\omega\epsilon\mathbf{K}\varphi + \mathbf{J}\times\nabla'\varphi + \frac{\nabla'\cdot\mathbf{K}}{j\omega\mu}\nabla'\varphi\right)dv' \qquad (12)$$

$$\mathbf{H(r)} = \frac{1}{4\pi}\int_v (-j\omega\epsilon\mathbf{K}\varphi + \mathbf{J}\times\nabla'\varphi)\,dv' + \frac{1}{4\pi j\omega\mu}\nabla\int_v \nabla'\cdot\mathbf{K}\rho\,dv'$$

$$\mathbf{H(r')} = \frac{1}{4\pi j\omega\mu}\int_v (\mathbf{K}\cdot(k^2\overline{\overline{\mathbf{I}}} + \nabla\nabla)\varphi + j\omega\mu\mathbf{J}\times\nabla'\varphi)\,dv'$$

$$\overline{\overline{\mathbf{I}}} = \text{unit dyad}$$

integral equations in which the induced source distributions in the form of **J** and **K** on the conducting surface are the unknowns driven by the exciting source. The radiated fields are then computed using equations of the form given in Chart 1. This particular approach, which has found significant usage in recent times, will be discussed in a later section.

The complexity of the solution processes discussed above is justified for general geometries since the current distributions or the fields which result from the computations are consistent, i.e., they satisfy Maxwell's equations with the appropriate boundary and driving conditions. As a result a great deal of confidence can be placed in the results of the computation. On the other hand, for geometries where the induced distributions are reasonably well known so that confidence in the results can be maintained, the fields can be directly evaluated. Fig. 1 attempts to capture the essence of the above discussion.

Integral Representations for Far Fields

The evaluation of fields can progress directly once the current distribution over the radiating structure is known. Using (7) through (10) in Chart 1, it is rather straightforward though sometimes demanding to perform this computation. The integral representations are a complete and precise description of the field based on

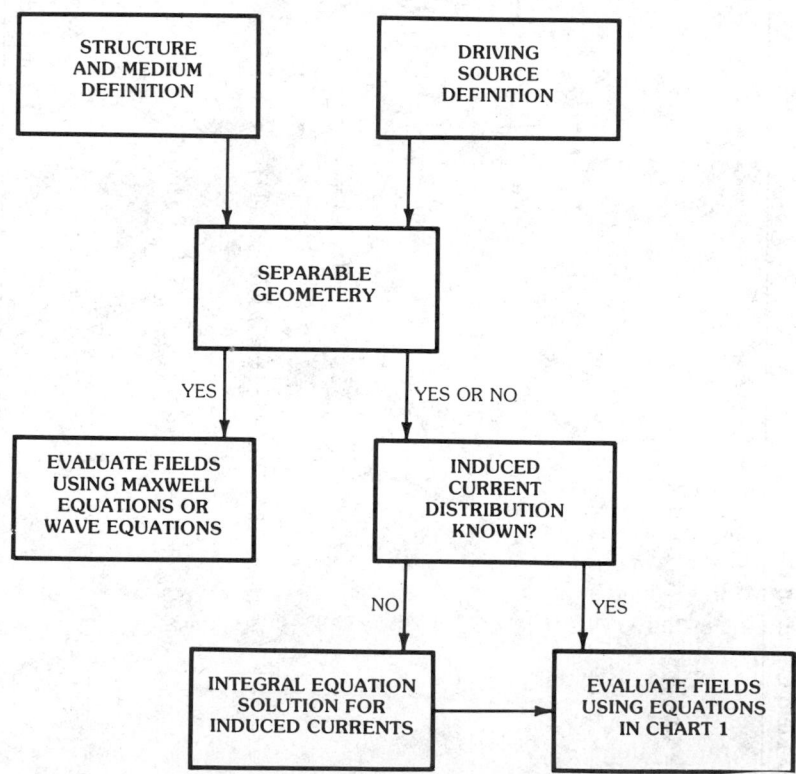

Fig. 1. A flow graph for the evaluation of radiated fields.

the source distributions. In keeping with the discussion of Chapter 1, Section 6, it is often convenient to compute only the portion of the complete field which dominates over a particular region of space. In the following we focus our attention on the far field since the development of expressions for the near-zone fields is most conveniently pursued for specific structures.

The fields at large distances from source distributions can be written in simplified forms by investigation of the vector potential in (9) and (10). In the far zone $|\mathbf{r}| \gg |\mathbf{r}'|$ and $|kr| \gg 1$, so that

$$|\mathbf{r} - \mathbf{r}'| = \rightarrow r - \mathbf{r} \cdot \mathbf{r}'/r = r - \hat{\mathbf{r}} \cdot \mathbf{r}' \text{ and}$$

$$\varphi(\mathbf{r}, \mathbf{r}') = \frac{e^{-jk|\mathbf{r}-\mathbf{r}'|}}{|\mathbf{r} - \mathbf{r}'|} \rightarrow \frac{e^{-jkr}}{r} e^{jk\hat{\mathbf{r}} \cdot \mathbf{r}'} \tag{13}$$

$$\nabla'\varphi = \nabla' \frac{e^{-jk|\mathbf{r}-\mathbf{r}'|}}{|\mathbf{r} - \mathbf{r}'|} \rightarrow \hat{\mathbf{r}} jk \frac{e^{-jkr}}{r} e^{jk\hat{\mathbf{r}} \cdot \mathbf{r}'}$$

Hence the far-zone representations of the equations in Chart 1 can be written by using (13) and letting the ∇ operator be expressed as $\nabla \rightarrow -jk\hat{\mathbf{r}}$. The equations for the far-zone fields (14 through 21) are presented in Chart 2. Note that the equations are related through the far-zone plane-wave relation

$$\mathbf{H}(\mathbf{r}) = \frac{1}{Z_0} \hat{\mathbf{r}} \times \mathbf{E}(\mathbf{r}), \qquad Z_0 = \sqrt{\mu/\epsilon} \tag{22}$$

and furthermore that the far-zone fields satisfy the conditions imposed in Chapter 1, Section 6, namely, that $\mathbf{E} \times \mathbf{H}^*$ be real and that the fields decay as r^{-1}.

Duality

The similarity in the form of Maxwell's equations for $\mathbf{E}(\mathbf{r})$ and $\mathbf{H}(\mathbf{r})$ allows them to be referred to as dual equations. Two examples of this duality are captured in Chart 3, where the indicated changes will lead to a set of Maxwell's equations for the dual case which are identical with those for the original [1]. Thus, given the fields that satisfy Maxwell's equations with the sources $[\mathbf{J}(\mathbf{r}), \mathbf{K}(\mathbf{r})]$ and constitutive parameters (ϵ, μ), the solution for the fields satisfying Maxwell's equations with the sources $[\mathbf{K}'(\mathbf{r}) = \mathbf{J}(\mathbf{r}), \mathbf{J}'(\mathbf{r}) = -\mathbf{K}'(\mathbf{r})]$ and constitutive parameters $(\epsilon' = \mu, \mu' = \epsilon)$ will be given by $[\mathbf{E}'(\mathbf{r}) = -\mathbf{H}(\mathbf{r}), \mathbf{H}'(\mathbf{r}) = \mathbf{E}(\mathbf{r})]$. Note that in this case we have dealt with dual problems having "dual" conditions achieved by interchanging μ and ϵ. Also shown in Chart 3 is a case where the medium remains unchanged. Thus the solution to a problem in the dual space can be related to a solution of a problem in the original space with the same constitutive parameters by the definitions in example 2 of Chart 3. The usefulness of the "duality" approach will be illustrated later for the small loop antenna.

It should be noted that the described duality relationships, while being the most widely used, are not exhaustive.

Chart 2. Some Convenient Expressions for Far-Zone Fields

E Field

$$\mathbf{F} = \frac{1}{4\pi}\frac{e^{-jkr}}{r}\int_V \mathbf{K}(\mathbf{r}')\,e^{jk\hat{\mathbf{r}}\cdot\mathbf{r}'}\,dv' \qquad (14)$$

$$\mathbf{A} = \frac{1}{4\pi}\frac{e^{-jkr}}{r}\int_V \mathbf{J}(\mathbf{r}')\,e^{jk\hat{\mathbf{r}}\cdot\mathbf{r}'}\,dv' \qquad (15)$$

$$\mathbf{E}(\mathbf{r}) = jk\hat{\mathbf{r}}\times\mathbf{F} + j\omega\mu\hat{\mathbf{r}}\times\hat{\mathbf{r}}\times\mathbf{A} \qquad (16)$$

$$\mathbf{E}(\mathbf{r}) = -\frac{1}{4\pi}\frac{e^{-jkr}}{r}\int_V \left(j\omega\mu\mathbf{J} + jk\mathbf{K}\times\hat{\mathbf{r}} + \sqrt{\mu/\epsilon}\,\nabla'\cdot\mathbf{J}\hat{\mathbf{r}}\right)e^{jk\hat{\mathbf{r}}\cdot\mathbf{r}'}\,dv' \qquad (18)$$

$$\mathbf{E}(\mathbf{r}) = \frac{j\omega\mu}{4\pi}\frac{e^{-jkr}}{r}\int_V \left[(\hat{\mathbf{r}}\cdot\mathbf{J})\hat{\mathbf{r}} - \mathbf{J} - \sqrt{\epsilon/\mu}\,\mathbf{K}\times\hat{\mathbf{r}}\right]e^{jk\hat{\mathbf{r}}\cdot\mathbf{r}'}\,dv' \qquad (20)$$

H Field

$$\mathbf{H}(\mathbf{r}) = -jk\hat{\mathbf{r}}\times\mathbf{A} + j\omega\epsilon\hat{\mathbf{r}}\times\hat{\mathbf{r}}\times\mathbf{F} \qquad (17)$$

$$\mathbf{H}(\mathbf{r}) = \frac{1}{4\pi}\frac{e^{-jkr}}{r}\int_V \left(-j\omega\epsilon\mathbf{K} + jk\mathbf{J}\times\hat{\mathbf{r}} - \sqrt{\epsilon/\mu}\,\nabla'\cdot\mathbf{K}\hat{\mathbf{r}}\right)e^{jk\hat{\mathbf{r}}\cdot\mathbf{r}'}\,dv' \qquad (19)$$

$$\mathbf{H}(\mathbf{r}) = \frac{j\omega\epsilon}{4\pi}\frac{e^{-jkr}}{r}\int_V\int_V \left[(\hat{\mathbf{r}}\cdot\mathbf{K})\hat{\mathbf{r}} + \mathbf{K} - \sqrt{\mu/\epsilon}\,\mathbf{J}\times\hat{\mathbf{r}}\right]e^{jk\hat{\mathbf{r}}\cdot\mathbf{r}'}\,dv' \qquad (21)$$

Chart 3. Examples of Duality Relationships

Initial Quantities		Dual Quantities
Example 1	$\mathbf{E}(\mathbf{r})$	$\mathbf{H}(\mathbf{r})$
	$\mathbf{H}(\mathbf{r})$	$-\mathbf{E}(\mathbf{r})$
	$\mathbf{J}(\mathbf{r})$	$\mathbf{K}(\mathbf{r})$
	$\mathbf{K}(\mathbf{r})$	$-\mathbf{J}(\mathbf{r})$
	$\mu(\mathbf{r})$	$\epsilon(\mathbf{r})$
	$\epsilon(\mathbf{r})$	$\mu(\mathbf{r})$
Example 2	$\mathbf{E}(\mathbf{r})$	$\sqrt{\mu/\epsilon}\,\mathbf{H}(\mathbf{r})$
	$\mathbf{H}(\mathbf{r})$	$-\sqrt{\epsilon/\mu}\,\mathbf{E}(\mathbf{r})$
	$\mathbf{J}(\mathbf{r})$	$\sqrt{\epsilon/\mu}\,\mathbf{K}(\mathbf{r})$
	$\mathbf{K}(\mathbf{r})$	$-\sqrt{\mu/\epsilon}\,\mathbf{J}(\mathbf{r})$
	$\mu(\mathbf{r})$	$\mu(\mathbf{r})$
	$\epsilon(\mathbf{r})$	$\epsilon(\mathbf{r})$

Radiated Power

The time-average power exiting a region of space or equivalently crossing the bounding surface can be found by integrating the complex Poynting vector (power flux density) over the surface. Thus

$$P = \mathrm{Re}\left\{ \oiint \mathbf{E} \times \mathbf{H}^* \cdot d\mathbf{s} \right\} \tag{23}$$

where $d\mathbf{s}$ has a unit normal in the outward direction and Re denotes the real part.

In a similar manner the time-average power supplied by sources is given by

$$P_{\text{sources}} = Re\left\{ \int_V (\mathbf{E}\cdot\mathbf{J}^* + \mathbf{H}^*\cdot\mathbf{K})\, dv \right\} \tag{24}$$

The power P_r radiated by an antenna is defined with (23) evaluated using the far-zone fields. Similarly, it can be evaluated using (23) and then taking its limiting value as the observation point $\mathbf{r}$ approaches infinity.

The time-average power balance for an antenna is given by

$$P_r + P_{\text{losses}} = P_{\text{sources}} \tag{25}$$

Thus the power supplied by sources is either radiated or dissipated through loss mechanisms in the antenna or the medium.

For a lossless antenna composed of perfect conductors, the losses vanish so that the input power is equal to the radiated power. Over a small antenna source region, which we can treat in an analogous network manner,

$$\mathrm{Re}\left\{ \int_{\substack{\text{source} \\ \text{region}}} \mathbf{E}\cdot\mathbf{J}^*\, dv \right\} = \mathrm{Re}\{VI^*\} \tag{26}$$

Thus the antenna time-average input power can lead to the evaluation of antenna radiation resistance R_r:

$$\text{Re}\{VI^*\} = \frac{1}{R_{\text{in}}}|I_{\text{in}}|^2 \tag{27a}$$

and

$$R_r = \frac{1}{|I_{\text{in}}|^2} \text{Re}\left\{ \oiint (\mathbf{E} \times \mathbf{H}^*) \cdot d\mathbf{s} \right\} \tag{27b}$$

where I_{in} is the current at input terminals.

Directive Gain, Directivity, Efficiency, and Gain

A number of useful quantities are defined for the characterization of antennas. These are dealt with in Chapter 1 and in several texts [3, 8, 9, 10, 11, 12, 13].

The *directive gain* of an antenna $G_D(\theta, \varphi)$ is defined as the ratio of real power flux density in the far zone in a given direction (θ, φ) to its average value over the entire radiation sphere. Thus

$$G_D(\theta, \varphi) = \frac{\text{Re}\{(\mathbf{E} \times \mathbf{H}^*) \cdot \hat{\mathbf{r}}\}}{\lim_{r \to \infty} (1/4\pi r^2) \oint (\mathbf{E} \times \mathbf{H}^*) \cdot d\mathbf{s}} = \frac{\text{Re}\{(\mathbf{E} \times \mathbf{H}^*) \cdot \hat{\mathbf{r}}\}}{P_r/4\pi r^2} \tag{28}$$

The quantity $P_r/4\pi r^2$ is equivalent to the power density of an isotropic antenna radiating the power P_r. The *directivity* D of an antenna is defined as the maximum value of the directive gain or simply

$$D = \max\{G_D(\theta, \varphi)\} \tag{29}$$

For lossy antennas, i.e., antennas for which the total input power at the antenna terminals is not radiated, additional quantities of interest are defined, namely, radiation efficiency η and gain $G(\theta, \varphi)$. The ratio of power P_r radiated to total input power P_{in} is the *radiation efficiency* η. Thus η can be related to the radiation resistance through the radiated power and, through the total input power P_{in}, to the total input resistance composed of the radiation resistance and resistances accounting for the other losses R_L [9]. Thus

$$\eta = \frac{P_r}{P_{\text{in}}} = \frac{R_r}{R_r + R_L} \tag{30}$$

The *gain* of an antenna $G(\theta, \varphi)$ is defined in a manner similar to directive gain but with the inclusion of losses. It is the ratio of the real power flux density in the far zone in a given direction to the power flux density which would be obtained if the input power were radiated isotropically. Thus

$$G(\theta, \varphi) = \frac{\text{Re}\{(\mathbf{E} \times \mathbf{H}^*)\cdot\hat{\mathbf{r}}\}}{P_{\text{in}}/4\pi r^2} = \eta G_D(\theta, \varphi) \tag{31}$$

The descriptions above can be related to the concepts in Fig. 14 in Chapter 1 by the substitution of P_{in} for P_1 and P_r for P_3. The consequences of input impedance mismatch to a feed line have not been included but can be readily taken into account as in Chapter 1 and Fig. 14 of Chapter 1.

2. Characteristics of Some Classical Antennas

In the following we will tabulate the characteristics of several simple antennas which are computed using the equations established in the preceding section. The examples illustrate how, with a certain amount of approximation (e.g., an assumed current distribution), one can obtain realistic estimates of antenna characteristics. In subsequent sections we will deal with more complicated structures and more precise techniques.

The Electric Dipole Antenna

The linear dipole antenna has received treatment in a very large number of works on antennas. Here we will consider the dipole antenna in a progression of models, from the very short to the capacitively loaded and then to the somewhat longer dipole, for each of which we can estimate a realistic current distribution.

The Point Source—The electric point-current source (sometimes called a Hertzian source), while possessing little realism, permits an individual to conceptualize the radiation due to electric currents. In a coordinate system as shown in Chart 4 the point source is defined as

$$\mathbf{J}(\mathbf{r}) = \hat{\mathbf{z}}\,\delta(\mathbf{r}) \tag{32}$$

where $\delta(\mathbf{r})$ is the Dirac delta function. The utility of this current source is that all other sources can be constructed as a superposition of these infinitesimal elements. Using (7), (8), (9), (16), (17), and (23) we can construct the relevant characteristics for the electric point-current source given in Chart 4. From the expressions for $\mathbf{E}(\mathbf{r})$ and $\mathbf{H}(\mathbf{r})$ one can identify [3] the far-zone terms behaving as $1/r$ and the near-zone terms composed of the "static" portion behaving as $1/r^3$, similar to that of a static dipole distribution, and the "induction" field behaving as $1/r^2$.

The Short Dipole Antenna—The short electric dipole antenna [3, 8, 11, 13, 14, 15] is of interest since it represents the behavior of dipole antennas at low frequencies where $L/\lambda \ll 1$. Two electric current distributions are of interest for the short electric dipole driven at its center. These distributions and the relevant antenna characteristics are presented in Chart 5. The first case, possessing a current distribution similar to the unit current element, i.e., a constant current over its length, can be constructed in reality by capacitively loading a short dipole with end caps. The second is the triangular distribution with maximum current at its center

Chart 4. Characteristics for the Electric Point-Current Source

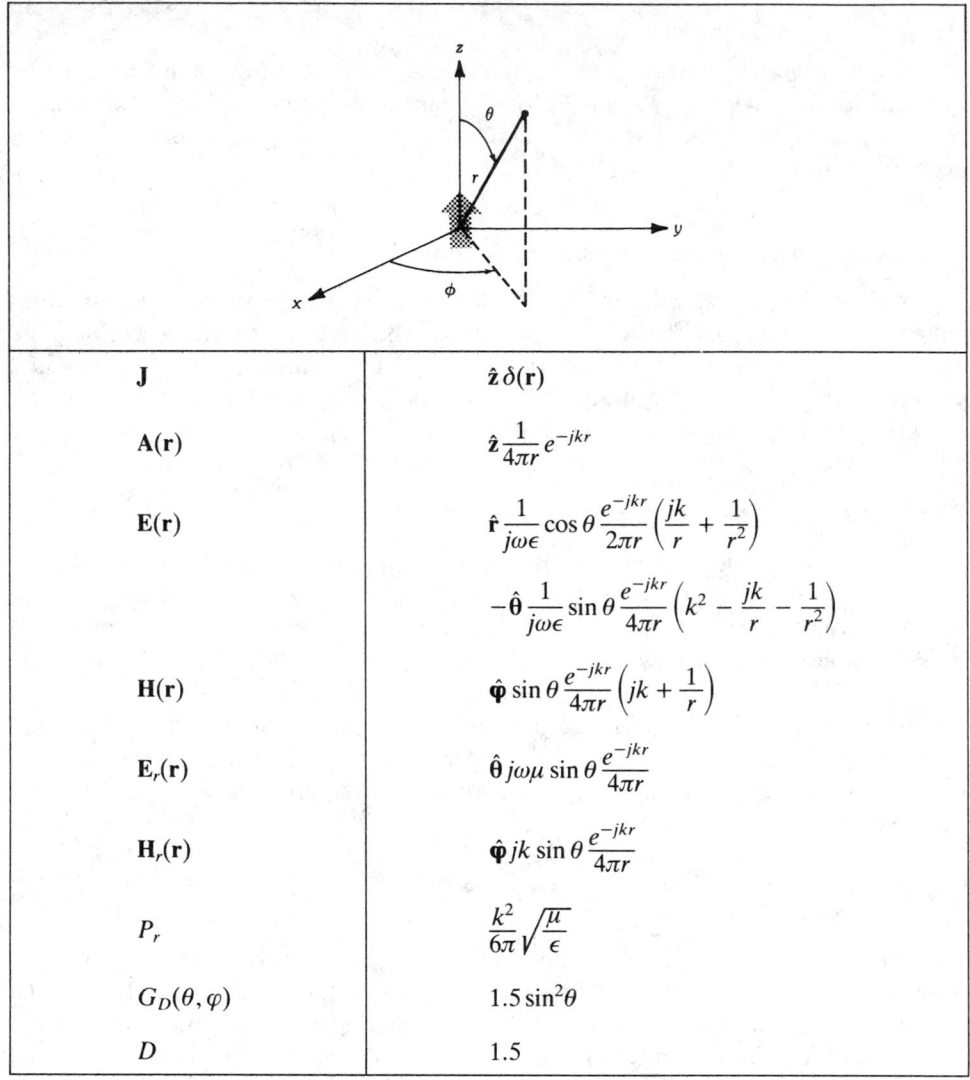

J	$\hat{z}\,\delta(\mathbf{r})$
$\mathbf{A(r)}$	$\hat{z}\,\dfrac{1}{4\pi r}\,e^{-jkr}$
$\mathbf{E(r)}$	$\hat{r}\,\dfrac{1}{j\omega\epsilon}\cos\theta\,\dfrac{e^{-jkr}}{2\pi r}\left(\dfrac{jk}{r}+\dfrac{1}{r^2}\right)$
	$-\hat{\theta}\,\dfrac{1}{j\omega\epsilon}\sin\theta\,\dfrac{e^{-jkr}}{4\pi r}\left(k^2-\dfrac{jk}{r}-\dfrac{1}{r^2}\right)$
$\mathbf{H(r)}$	$\hat{\varphi}\,\sin\theta\,\dfrac{e^{-jkr}}{4\pi r}\left(jk+\dfrac{1}{r}\right)$
$\mathbf{E}_r(\mathbf{r})$	$\hat{\theta}\,j\omega\mu\,\sin\theta\,\dfrac{e^{-jkr}}{4\pi r}$
$\mathbf{H}_r(\mathbf{r})$	$\hat{\varphi}\,jk\,\sin\theta\,\dfrac{e^{-jkr}}{4\pi r}$
P_r	$\dfrac{k^2}{6\pi}\sqrt{\dfrac{\mu}{\epsilon}}$
$G_D(\theta,\varphi)$	$1.5\sin^2\theta$
D	1.5

and zero current at its ends. This current realistically approximates the current on a very short center-driven dipole.

The Sinusoidal Current Distribution

Current distribution measurements have indicated that for thin cylindrical antennas of diameter less than approximately $\lambda/100$, the current distribution can be well approximated by a sinusoid [3, 9, 10, 11, 12, 14, 15]. Theoretical analyses have supported these observations. Thus the sinusoidal current distribution whose characteristics are presented in Chart 6 occupies an important place in antenna

Chart 5. Characteristics for the Short Electric Dipole Antenna

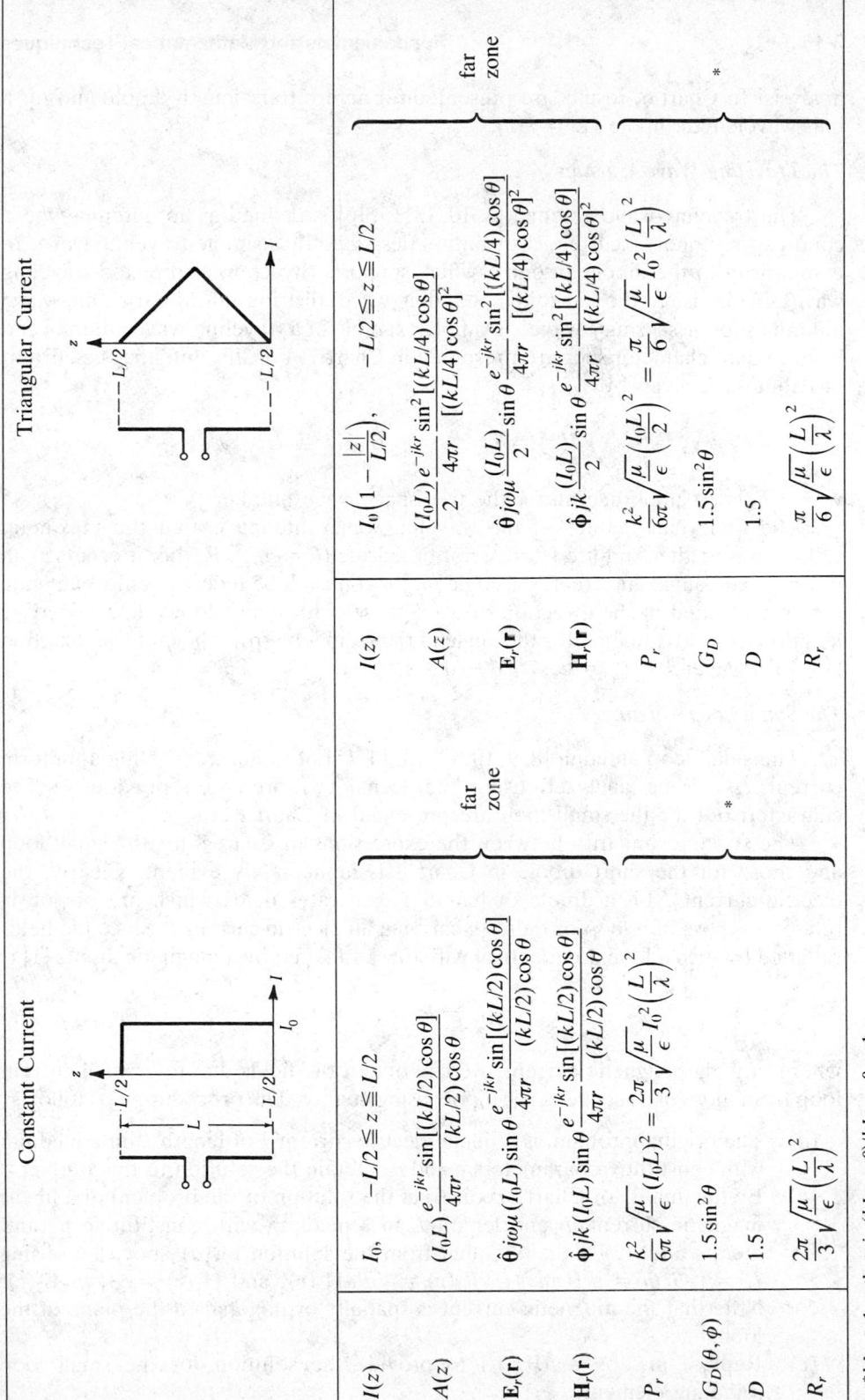

	Constant Current	Triangular Current
$I(z)$	$I_0,\quad -L/2 \leqq z \leqq L/2$	$I_0\left(1-\dfrac{|z|}{L/2}\right),\quad -L/2 \leqq z \leqq L/2$
$A(z)$	$(I_0L)\dfrac{e^{-jkr}}{4\pi r}\sin\theta\dfrac{\sin[(kL/2)\cos\theta]}{(kL/2)\cos\theta}$	$\dfrac{(I_0L)}{2}\dfrac{e^{-jkr}}{4\pi r}\dfrac{\sin^2[(kL/4)\cos\theta]}{[(kL/4)\cos\theta]^2}$
$\mathbf{E}_r(\mathbf{r})$	$\hat{\boldsymbol{\theta}}\,j\omega\mu\,(I_0L)\sin\theta\,\dfrac{e^{-jkr}}{4\pi r}\dfrac{\sin[(kL/2)\cos\theta]}{(kL/2)\cos\theta}$	$\hat{\boldsymbol{\theta}}\,j\omega\mu\,\dfrac{(I_0L)}{2}\sin\theta\,\dfrac{e^{-jkr}}{4\pi r}\dfrac{\sin^2[(kL/4)\cos\theta]}{[(kL/4)\cos\theta]^2}$
$\mathbf{H}_r(\mathbf{r})$	$\hat{\boldsymbol{\phi}}\,jk\,(I_0L)\sin\theta\,\dfrac{e^{-jkr}}{4\pi r}\dfrac{\sin[(kL/2)\cos\theta]}{(kL/2)\cos\theta}$	$\hat{\boldsymbol{\phi}}\,jk\,\dfrac{(I_0L)}{2}\sin\theta\,\dfrac{e^{-jkr}}{4\pi r}\dfrac{\sin^2[(kL/4)\cos\theta]}{[(kL/4)\cos\theta]^2}$
P_r	$\dfrac{k^2}{6\pi}\sqrt{\dfrac{\mu}{\epsilon}}(I_0L)^2 = \dfrac{2\pi}{3}\sqrt{\dfrac{\mu}{\epsilon}}I_0^2\left(\dfrac{L}{\lambda}\right)^2$	$\dfrac{k^2}{6\pi}\sqrt{\dfrac{\mu}{\epsilon}}\left(\dfrac{I_0L}{2}\right)^2 = \dfrac{\pi}{6}\sqrt{\dfrac{\mu}{\epsilon}}I_0^2\left(\dfrac{L}{\lambda}\right)^2$
$G_D(\theta,\phi)$	$1.5\sin^2\theta$	$1.5\sin^2\theta$
D	1.5	1.5
R_r	$\dfrac{2\pi}{3}\sqrt{\dfrac{\mu}{\epsilon}}\left(\dfrac{L}{\lambda}\right)^2$	$\dfrac{\pi}{6}\sqrt{\dfrac{\mu}{\epsilon}}\left(\dfrac{L}{\lambda}\right)^2$

(Constant Current: far zone applies to $\mathbf{E}_r(\mathbf{r})$, $\mathbf{H}_r(\mathbf{r})$, P_r; R_r marked *)

(Triangular Current: far zone applies to $\mathbf{E}_r(\mathbf{r})$, $\mathbf{H}_r(\mathbf{r})$, P_r; R_r marked *)

*$kL \ll 1$ so that $\sin(kL\cos\theta)/kL\cos\theta \to 1$.

analysis. In Chart 6, results are presented for an arbitrary-length dipole and for a half-wavelength dipole ($L = \lambda/2$).

The Traveling-Wave Antenna

The traveling-wave antenna [8, 10, 13, 15, 16] is defined as an antenna whose conductors support a current distribution described by a simple traveling wave. In comparison to the dipole antenna, which supports two counter-directed traveling waves giving rise to the sinusoidal (standing-wave) distribution, the traveling-wave antenna supports a single wave. A simple example of a traveling-wave antenna and its attendant characteristics are presented in Chart 7. For this antenna the current distribution is given by

$$I(z) = I_0 e^{-jkz}, \qquad 0 \leqq z \leqq L$$

where I_0 is its amplitude and k the free-space wave number.

The important feature of the traveling-wave antenna is that the maximum radiation is neither end-fire ($\theta = 0$) nor broadside ($\theta = \pi/2$). Rather it occurs at an intermediate angle and creates a single major conical lobe independent of antenna length and tilted in the direction of wave travel with minor lobes determined by length. An approximation for the angle of the main lobe from the end-fire direction [13, 16] is given by $\theta_m = \cos^{-1}[1 - 0.371/(L/\lambda)]$.

The Small Loop Antenna

The small loop antenna [8, 9, 10, 11, 13, 14, 15] of radius a, carrying a uniform current I, can be analyzed using the techniques presented previously. The characteristics for the small loop are presented in Chart 8.

The striking similarity between the expressions in Chart 8 for the small loop and those for the short dipole in Chart 4 is immediately evident. Clearly, the electric current I on a dipole of length L generates fields which are obviously related to those of a loop of radius a carrying an electric current I. Since the fields radiated by such a loop are identical with those radiated by a magnetic dipole [1] if

$$mL = j\omega\mu I(\pi a^2)$$

where m is the magnetic current, we can obtain the fields due to a small current loop from those of a short electric dipole using duality. The procedure is as follows:

(a) The original problem is a linear electric current I of length L in a medium with constitutive parameters ϵ and μ. Obtain the solution to this problem.

(b) By the duality of Chart 3, construct the solution of the problem of a linear magnetic current m and length L in a medium with constitutive parameters ϵ and μ. This is obtained from the solution in (a) above by setting $I_1 = \sqrt{\epsilon/\mu}\, m = jkI(\pi a^2)/L$, $\mathbf{E}_1(\mathbf{r}) = \sqrt{\mu/\epsilon}\, \mathbf{H}_2(\mathbf{r})$, and $\mathbf{H}_1(\mathbf{r}) = -\sqrt{\epsilon/\mu}\, \mathbf{E}_2(\mathbf{r})$. Note that the magnetic current is spatially orthogonal to the plane of the loop.

(c) Replace mL by $j\omega\mu I(\pi a^2)$ to provide the solution for the small loop radiating element.

Chart 6. Characteristics for a Dipole with Sinusoidal Current Distribution

$I(z)$	$I_0 \sin k_0(L/2 - |z|), \quad -L/2 \leqq z \leqq L/2$

Arbitrary-Length Dipole

$A_z(\mathbf{r})$	$I_0 \dfrac{e^{-jkr}}{2\pi r} \dfrac{\cos\left[(kL/2)\cos\theta\right] - \cos(kL/2)}{k\sin^2\theta}$
$\mathbf{E}_r(\mathbf{r})$	$\hat{\theta}\, j\sqrt{\dfrac{\mu}{\epsilon}}\, I_0 \dfrac{e^{-jkr}}{2\pi r} \dfrac{\cos\left[(kL/2)\cos\theta\right] - \cos(kL/2)}{\sin\theta}$
$\mathbf{H}_r(\mathbf{r})$	$\hat{\phi}\, j I_0 \dfrac{e^{-jkr}}{2\pi r} \dfrac{\cos\left[(kL/2)\cos\theta\right] - \cos(kL/2)}{\sin\theta}$

far zone (brace spanning A_z, $\mathbf{E}_r$, $\mathbf{H}_r$)

P_r [11, 12]	$\sqrt{\dfrac{\mu}{\epsilon}}\, I_0^2 \dfrac{1}{2\pi}\Big\{ \sin kL\,[Si(kL) - \tfrac{1}{2}Si(2kL)]$
	$\qquad + (1 + \cos kL)[\ln(kL\gamma) - Ci(kL)]$
	$\qquad - \dfrac{\cos kL}{2}[\ln(2kL\gamma) - Ci(2kL)] \Big\}$
	where
	$Si(x) = \displaystyle\int_0^x \frac{\sin x}{x}\,dx, \quad Ci(x) = -\displaystyle\int_x^\infty \frac{\cos x}{x}\,dx, \quad \ln\gamma = 0.5772$

Half-Wavelength Dipole ($L = \lambda/2$)

$A_z(\mathbf{r})$	$I_0 \dfrac{e^{-jkr}}{2\pi r} \dfrac{\cos\left[(\pi/2)\cos\theta\right]}{k\sin^2\theta}$
$\mathbf{E}_r(\mathbf{r})$	$\hat{\theta}\, j\sqrt{\dfrac{\mu}{\epsilon}}\, I_0 \dfrac{e^{-jkr}}{2\pi r} \dfrac{\cos\left[(\pi/2)\cos\theta\right]}{\sin\theta}$
P_r	$\sqrt{\dfrac{\mu}{\epsilon}}\, I_0^2 \dfrac{1}{4\pi}[\ln(2\pi\gamma) - Ci(2\pi)] = 0.194\sqrt{\mu/\epsilon}\, I_0^2$
R_r	$\sqrt{\dfrac{\mu}{\epsilon}}\, \dfrac{1}{4\pi}[\ln(2\pi\gamma) - Ci(2\pi)] = 0.194\sqrt{\mu/\epsilon}\; (R_r = 73\ \Omega \text{ in free space})$
$G_D(\theta,\phi)$	$\dfrac{4\cos^2\left[(\pi/2)\cos\theta\right]}{\sin^2\theta\,[\ln(2\pi\gamma) - Ci(2\pi)]} = 1.644\,\dfrac{\cos^2\left[(\pi/2)\cos\theta\right]}{\sin^2\theta}$
D	$\dfrac{4}{\ln(2\pi\gamma) - Ci(2\pi)} = 1.644$

Chart 7. Characteristics for the Traveling-Wave Antenna

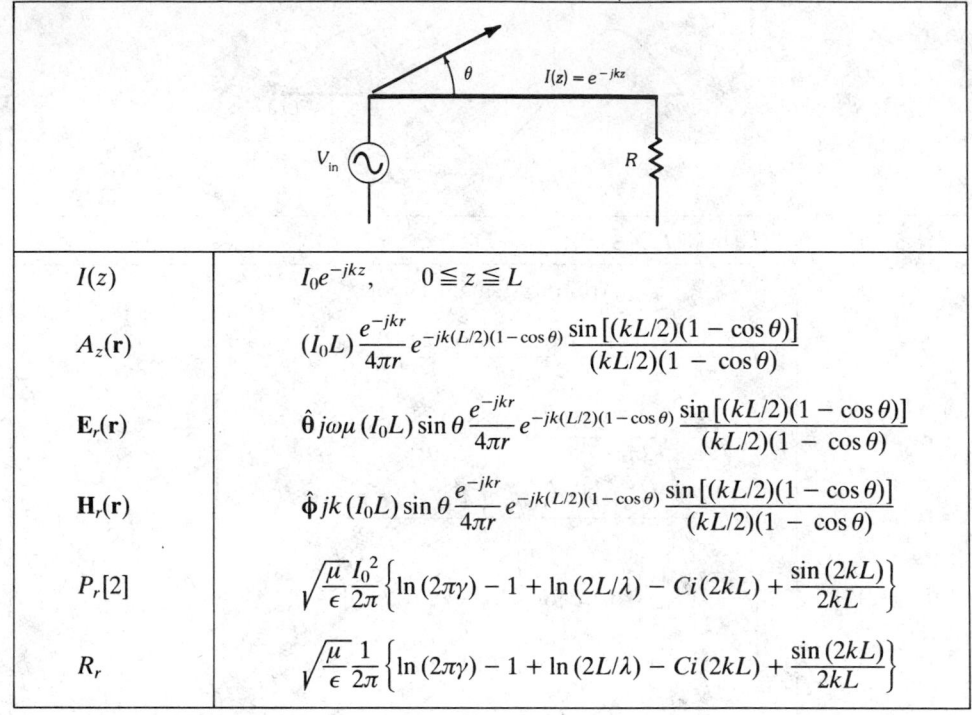

$I(z)$	$I_0 e^{-jkz}, \qquad 0 \leqq z \leqq L$
$A_z(\mathbf{r})$	$(I_0 L)\dfrac{e^{-jkr}}{4\pi r} e^{-jk(L/2)(1-\cos\theta)} \dfrac{\sin\left[(kL/2)(1-\cos\theta)\right]}{(kL/2)(1-\cos\theta)}$
$\mathbf{E}_r(\mathbf{r})$	$\hat{\boldsymbol{\theta}}\, j\omega\mu\,(I_0 L)\sin\theta\,\dfrac{e^{-jkr}}{4\pi r} e^{-jk(L/2)(1-\cos\theta)} \dfrac{\sin\left[(kL/2)(1-\cos\theta)\right]}{(kL/2)(1-\cos\theta)}$
$\mathbf{H}_r(\mathbf{r})$	$\hat{\boldsymbol{\phi}}\, jk\,(I_0 L)\sin\theta\,\dfrac{e^{-jkr}}{4\pi r} e^{-jk(L/2)(1-\cos\theta)} \dfrac{\sin\left[(kL/2)(1-\cos\theta)\right]}{(kL/2)(1-\cos\theta)}$
$P_r[2]$	$\sqrt{\dfrac{\mu}{\epsilon}}\dfrac{I_0^2}{2\pi}\left\{\ln(2\pi\gamma) - 1 + \ln(2L/\lambda) - Ci(2kL) + \dfrac{\sin(2kL)}{2kL}\right\}$
R_r	$\sqrt{\dfrac{\mu}{\epsilon}}\dfrac{1}{2\pi}\left\{\ln(2\pi\gamma) - 1 + \ln(2L/\lambda) - Ci(2kL) + \dfrac{\sin(2kL)}{2kL}\right\}$

The Perfect Ground Plane

The perfect ground plane, i.e., the planar, perfectly conducting surface with conductivity $\sigma \to \infty$, is often found in the environment of antenna structures. Since the previous characterizations have been for antennas in homogeneous media with real ϵ and μ, the introduction of conducting bodies such as infinite planes will require a modification.

The boundary conditions at any perfectly conducting surface require that the tangential electric field satisfy, with $\hat{\mathbf{n}}(\mathbf{r})$ the outward normal to the surface and $\mathbf{r}$ on the surface,

$$\hat{\mathbf{n}}(\mathbf{r}) \times \mathbf{E}(\mathbf{r}) = 0 \qquad (33)$$

and that the magnetic field satisfy

$$\hat{\mathbf{n}}(\mathbf{r}) \cdot \mathbf{H}(\mathbf{r}) = 0 \qquad (34)$$

The exclusion of fields from the interior of the surface leads to the generation of surface charge and current densities given by

$$\hat{\mathbf{n}}(\mathbf{r}) \cdot \mathbf{E}(\mathbf{r}) = \varrho_s(\mathbf{r})/\epsilon$$

$$\hat{\mathbf{n}}(\mathbf{r}) \times \mathbf{H}(\mathbf{r}) = \mathbf{J}_s(\mathbf{r}) \qquad (35)$$

Chart 8. Characteristics of the Small Loop Antenna

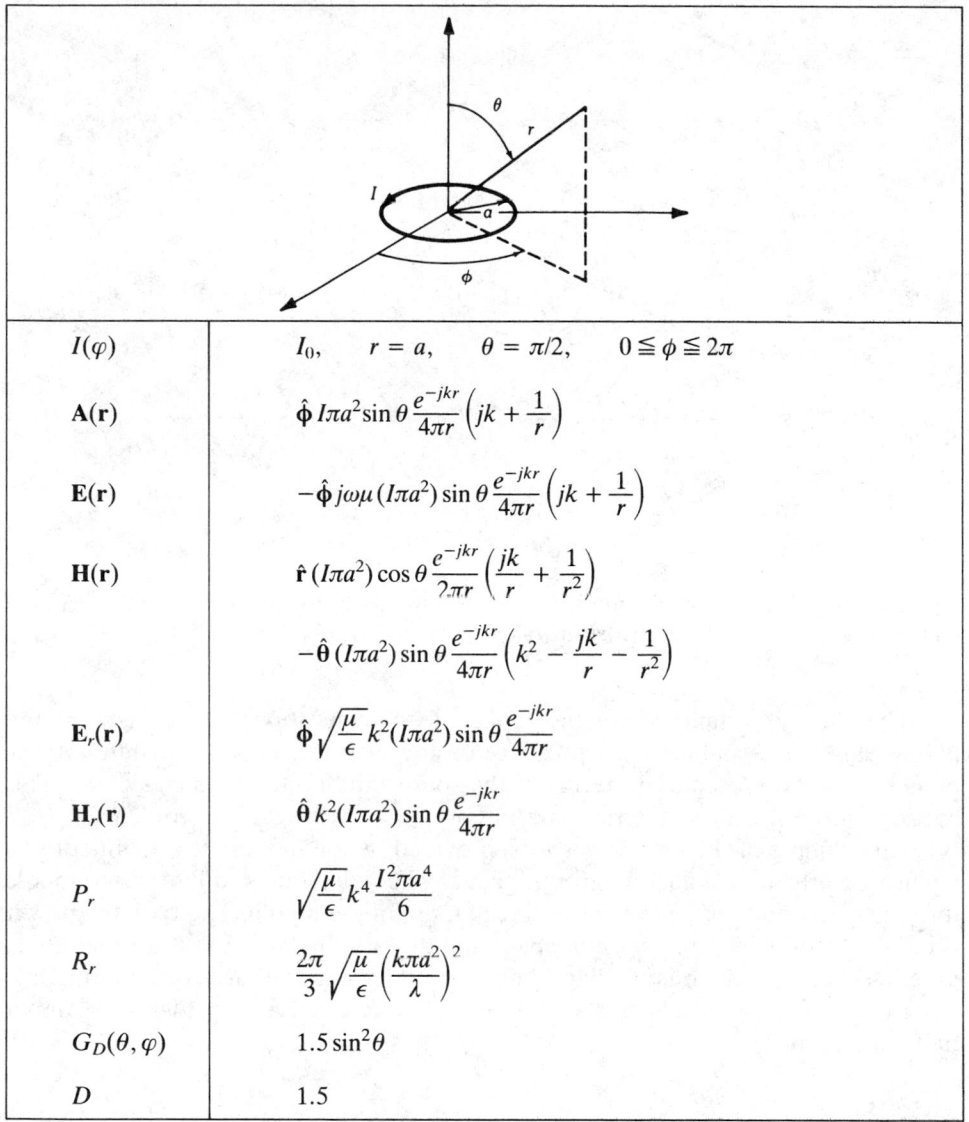

$I(\varphi)$	$I_0, \quad r = a, \quad \theta = \pi/2, \quad 0 \le \phi \le 2\pi$
$\mathbf{A(r)}$	$\hat{\phi}\, I\pi a^2 \sin\theta\, \dfrac{e^{-jkr}}{4\pi r}\left(jk + \dfrac{1}{r}\right)$
$\mathbf{E(r)}$	$-\hat{\phi}\, j\omega\mu\,(I\pi a^2)\sin\theta\, \dfrac{e^{-jkr}}{4\pi r}\left(jk + \dfrac{1}{r}\right)$
$\mathbf{H(r)}$	$\hat{\mathbf{r}}\,(I\pi a^2)\cos\theta\, \dfrac{e^{-jkr}}{2\pi r}\left(\dfrac{jk}{r} + \dfrac{1}{r^2}\right)$
	$-\hat{\boldsymbol{\theta}}\,(I\pi a^2)\sin\theta\, \dfrac{e^{-jkr}}{4\pi r}\left(k^2 - \dfrac{jk}{r} - \dfrac{1}{r^2}\right)$
$\mathbf{E_r(r)}$	$\hat{\phi}\,\sqrt{\dfrac{\mu}{\epsilon}}\,k^2(I\pi a^2)\sin\theta\, \dfrac{e^{-jkr}}{4\pi r}$
$\mathbf{H_r(r)}$	$\hat{\boldsymbol{\theta}}\,k^2(I\pi a^2)\sin\theta\, \dfrac{e^{-jkr}}{4\pi r}$
P_r	$\sqrt{\dfrac{\mu}{\epsilon}}\,k^4\,\dfrac{I^2\pi a^4}{6}$
R_r	$\dfrac{2\pi}{3}\sqrt{\dfrac{\mu}{\epsilon}}\left(\dfrac{k\pi a^2}{\lambda}\right)^2$
$G_D(\theta,\varphi)$	$1.5\sin^2\theta$
D	1.5

An arbitrarily oriented antenna located above a perfectly conducting ground plane must generate a vanishing electric field at the surface in order to satisfy the boundary condition on the electric field. Such a condition can be satisfied by creating an equivalent problem where the ground plane is removed and an image source is introduced to produce, when combined with the original source, a vanishing tangential electric field at the location of the plane. The appropriate images for electric and magnetic currents are shown in Fig. 2. A close inspection will show that the tangential electric and normal magnetic fields vanish at the plane surface and that the fields are identical in the upper half-space.

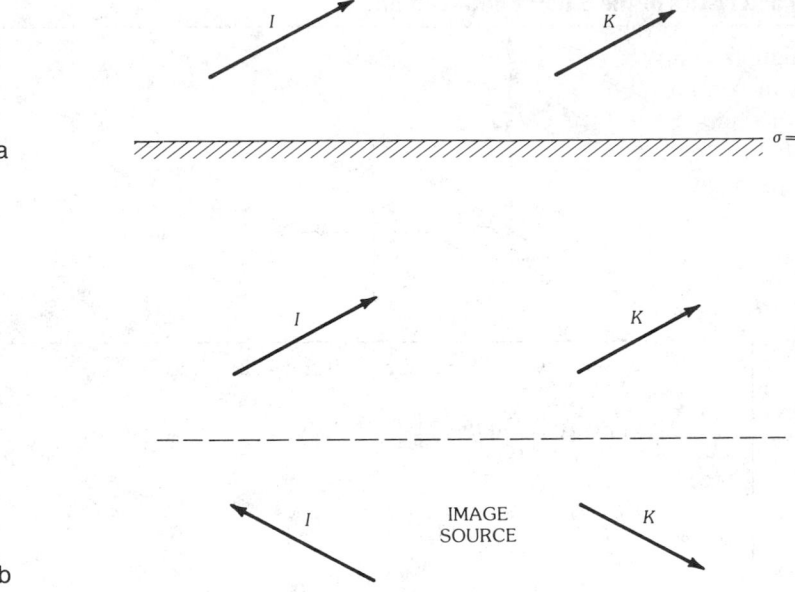

Fig. 2. Equivalent source distributions for sources above perfectly conducting ground planes. (*a*) Original problem. (*b*) Equivalent problem.

The theory of images can be applied to configurations other than infinite planes. In fact, a source in the presence of any surface composed of intersecting planes can be represented in terms of the source and multiple images [1, 8]. This concept applies to both exterior and interior (e.g., waveguide) problems.

An example of the use of images is provided by the monopole antenna driven against a perfectly conducting ground plane. The fields radiated by the monopole into the upper half-space are identical with those of the dipole. The radiated power is contained only in the upper half-space and thus is only half of that for the dipole antenna. Hence we can immediately infer that the input voltage to the monopole need only be one-half that of the dipole to produce the same fields in the upper half-space. Thus

$$P_r = \sqrt{\frac{\mu}{\epsilon}}\, I_0^2 \frac{1}{4\pi} \{[\sin{(kL)}][Si(kL) - \tfrac{1}{2}Si(2kL)]$$

$$+ [1 - \cos{(kL)}][\ln{(kL\gamma)} - Ci(kL)] - [\cos{(kL)}][\ln{(2kL\gamma)} - Ci(2kL)]\}$$

The input resistance is given by $R_r = P_r/I_0^2$, and for a quarter-wavelength monopole

$$R_r = \sqrt{\frac{\mu}{\epsilon}} \frac{1}{8\pi} \{\ln{(2\pi\gamma)} - Ci(2\pi)\} = 36.5 \ \Omega$$

The Rectangular-Aperture Antenna

The rectangular-aperture antenna [1, 3, 8, 9, 10, 11, 13, 15], shown in Fig. 3, has many features in common with the electric dipole. In the following we will illustrate the use of equivalence and images to provide a convenient mechanism for evaluating its fields. In each case the electric-field distribution in the aperture will be assumed to be known. Then the fields produced by the aperture can be obtained using equivalence, images, and the field representations presented earlier.

For an assumed field distribution in the aperture in a perfectly conducting plane, the pictorial representation of Fig. 4a is appropriate. Using equivalence, the situation in Fig. 4b applies and the fields in the half-space $z > 0$ are identical with those in the original problem but zero for $z < 0$. In this figure $\mathbf{J}_a = \hat{\mathbf{z}} \times \mathbf{H}$ and $\mathbf{K}_a = -\hat{\mathbf{z}} \times \mathbf{E}$. As a result of the vanishing fields for $z < 0$, the perfectly conducting plane is completed through the aperture as shown in Fig. 4c. Using images, the plane can be removed and the situation in Fig. 4d holds. Note that the fields are the correct fields for the half-space $z > 0$ and not for $z < 0$.

The fields radiated by the aperture distribution can be evaluated using equations in Charts 1 and 2 with only a magnetic current source or an equivalent electric vector potential. In the following we tabulate the expressions necessary to construct the radiation fields and certain characteristics for some commonly encountered aperture distributions defined in general as

$$\mathbf{E}^a(x, y, z) = \hat{\mathbf{y}} E^a(x, y)\delta(z) \qquad -L_x/2 < x < L_x/2, \quad -L_y/2 < y < L_y/2 \quad (36)$$

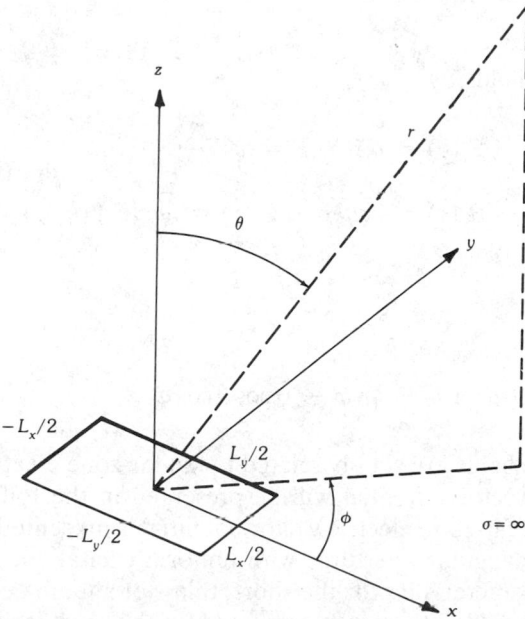

Fig. 3. The rectangular aperture or slot antenna in a perfectly conducting plane.

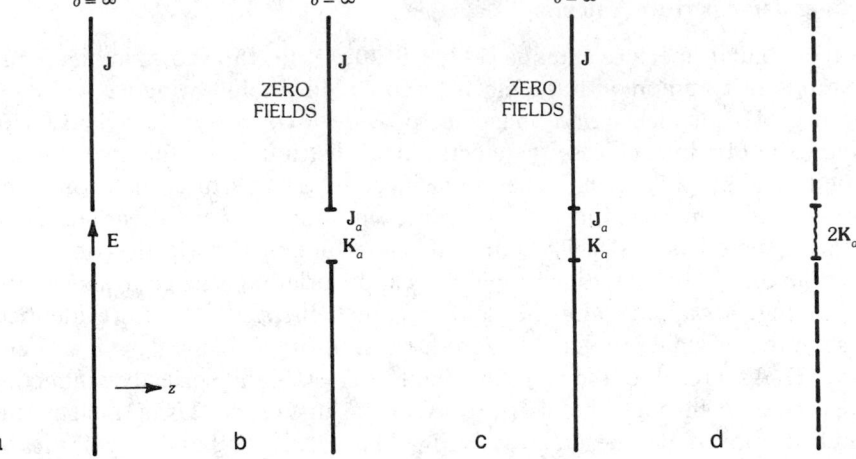

Fig. 4. Equivalent source distributions and their environments for fields in the right half-space. (*a*) Aperture field distribution in a perfectly conducting plane. (*b*) Using equivalence with (*a*). (*c*) Perfectly conducting plane is completed through aperture. (*d*) Using images to (*c*).

In the far zone the electric vector potential is given by

$$\mathbf{F} = \hat{\mathbf{x}} \frac{1}{2\pi} \frac{e^{-jkr}}{r} \int_{-\infty}^{\infty} dz'\, \delta(z') \int_{-L_y/2}^{L_y/2} dy'\, e^{jky' \sin\theta \sin\varphi} \int_{-L_x/2}^{L_x/2} dx'\, E^a(x',y')\, e^{jkx' \sin\theta \cos\varphi}$$

$$(37)$$

and the radiated fields by

$$\mathbf{E}_r(\mathbf{r}) = jk\hat{\mathbf{r}} \times \mathbf{F} = jk|\mathbf{F}|\hat{\mathbf{e}}(\theta,\varphi) \tag{38}$$

$$\mathbf{H}_r(\mathbf{r}) = \sqrt{\epsilon/\mu}\,\hat{\mathbf{r}} \times \mathbf{E}_r(\mathbf{r}) = j\omega\epsilon|\mathbf{F}|\hat{\mathbf{h}}(\theta,\varphi) \tag{39}$$

where

$$\hat{\mathbf{e}}(\theta,\varphi) = \hat{\boldsymbol{\varphi}} \cos\theta \cos\varphi + \hat{\boldsymbol{\theta}} \sin\varphi$$
$$\hat{\mathbf{h}}(\theta,\varphi) = \hat{\mathbf{r}} \times \hat{\mathbf{e}}(\theta,\varphi) = \hat{\boldsymbol{\varphi}} \sin\varphi - \hat{\boldsymbol{\theta}} \cos\theta \cos\varphi$$

Since the far-zone fields are simply related to the far-zone electric vector potential, only the electric vector potential will be presented in the following charts.

In Chart 9 the far-zone electric vector potential is presented for various lengths and widths of rectangular apertures with uniform excitation. Also presented are some radiation characteristics for the short, thin-slot antenna excited by a uniform transverse electric field. The results presented for the short, thin slot are consistent with the relationship between the impedances of complementary structures [13], i.e.,

Chart 9. Far-Zone Characteristics for Apertures with Uniform Aperture Distribution

Uniform Aperture Distribution

$$\mathbf{E}^a(x,y,z) = \hat{\mathbf{y}} E_0 \delta(z), \qquad -L_x/2 < x < L_x/2, \quad -L_y/2 < y < L_y/2$$

Aribtrary Length: L_x, L_y

$$\mathbf{F}(\mathbf{r}) = \hat{\mathbf{x}} \frac{e^{-jkr}}{2\pi r} E_0 L_x L_y \frac{\sin\left[(kL_x/2)\sin\theta\cos\varphi\right]}{(kL_x/2)\sin\theta\cos\varphi} \frac{\sin\left[(kL_y/2)\sin\theta\sin\varphi\right]}{(kL_y/2)\sin\theta\sin\varphi}$$

Thin Slot: $kLy \ll 1$, L_x Arbitrary

$$E_0 L_y \to V_0$$

$$\mathbf{F}(\mathbf{r}) = \hat{\mathbf{x}} \frac{e^{-jkr}}{2\pi r} V_0 L_x \frac{\sin\left[(kL_x/2)\sin\theta\cos\varphi\right]}{(kL_x/2)\sin\theta\cos\varphi}$$

Short, Thin Slot: $kL_x \ll 1$, $kL_y \ll 1$

$$E_0 L_y \to V_0$$

$$\mathbf{F}(\mathbf{r}) = \hat{\mathbf{x}} \frac{e^{-jkr}}{2\pi r} V_0 L_x$$

$$\mathbf{E}(\mathbf{r}) = jk V_0 L_x \frac{e^{-jkr}}{2\pi r} \hat{\mathbf{e}}(\theta,\varphi)$$

$$\mathbf{H}(\mathbf{r}) = j\omega\epsilon V_0 L_x \frac{e^{-jkr}}{2\pi r} \hat{\mathbf{h}}(\theta,\varphi)$$

$$P_r = \frac{8\pi}{3}\sqrt{\frac{\epsilon}{\mu}} V_0 \left(\frac{L_x}{\lambda}\right)^2$$

$$R_r = V_0^2/P_r = \left[\frac{8\pi}{3}\sqrt{\frac{\epsilon}{\mu}}\left(\frac{L_x}{\lambda}\right)^2\right]^{-1}$$

$$G_D(\theta,\varphi) = \frac{3}{2}(1 - \cos^2\varphi\sin^2\theta), \qquad D = 1.5$$

$$Z_S Z_D = \frac{1}{4}\left(\frac{\mu}{\epsilon}\right) \tag{40}$$

where

Z_S = the slot impedance (Chart 9)

Z_D = the dipole impedance (Chart 5)

The far-zone representations for the electric vector potentials for triangular and sinusoidal aperture distributions are presented in Chart 10.

Chart 10. Far-Zone Vector Potentials for Triangular and Sinusoidal Aperture Distributions

---

Triangular Aperture Distribution

$$E^a(x, y, z) = \hat{\mathbf{y}} E_0(1 - 2|x|/L_x)\delta(z), \qquad -L_x/2 < x < L_x/2, \quad -L_y/2 < y < L_y/2$$

$$\mathbf{F}(\mathbf{r}) = \hat{\mathbf{x}}\, \frac{e^{-jkr}}{4\pi r} E_0 L_x L_y\, \frac{\sin\left[(kL_y/2)\sin\theta\sin\varphi\right]}{(kL_y/2)\sin\theta\sin\varphi}\, \frac{\sin^2\left[(kL_x/2)\sin\theta\cos\varphi\right]}{\left[(kL_x/4)\sin\theta\cos\varphi\right]^2}$$

$$E_r(\mathbf{r}) = jk\,|\mathbf{F}|\,\hat{\mathbf{e}}(\theta, \varphi), \qquad H_r(\mathbf{r}) = j\omega\epsilon\,|\mathbf{F}|\,\hat{\mathbf{h}}(\theta, \varphi)$$

Sinusoidal Aperture Distribution

$$E^a(x, y, z) = \hat{\mathbf{y}} E_0 \sin\left[k(L_x/2 - |x|)\right], \qquad -L_x/2 < x < L_x/2, \quad -L_y/2 < x < L_y/2$$

$$\mathbf{F}(\mathbf{r}) = \hat{\mathbf{x}}\, \frac{e^{-jkr}}{2\pi r} E_0 L_x L_y\, \frac{\sin\left[(kL_y/2)\sin\theta\sin\varphi\right]}{(kL_y/2)\sin\theta\sin\varphi}\, \frac{\cos\left[(kL_x/2)\sin\theta\sin\varphi\right] - \cos(kL_x/2)}{(kL_x/2)(1 - \sin^2\theta\sin^2\varphi)}$$

$$E_r(\mathbf{r}) = jk\,|\mathbf{F}|\,\hat{\mathbf{e}}(\theta, \varphi), \qquad H_r(\mathbf{r}) = j\omega\epsilon\,|\mathbf{F}|\,\hat{\mathbf{h}}(\theta, \varphi)$$

The Biconical Antenna

The solution for antenna characteristics is simplified when the antenna surface coincides with coordinate surfaces. While there are only a few such coordinate systems and furthermore only one of finite dimensionality, the spheroidal, the method of separation of variables has been used to study some specific antennas. By the "perturbation" of the separable geometries, such as the spheroidal and conical shapes, other shapes such as the cylindrical antenna have been investigated.

The infinite biconical antenna shown in Fig. 5 has been studied extensively [4, 5, 8, 9, 11, 13, 14, 15, 17]. Such a structure coincides with spherical coordinate surfaces and serves as a guide for spherical waves. By solving the wave equation in the spherical coordinate system, the components of the assumed TEM wave are

$$H_\varphi = \frac{1}{r\sin\theta}\, H_0 e^{-jkr} \tag{41}$$

$$E_\theta = \sqrt{\mu/\epsilon}\, H_\varphi \tag{42}$$

The voltage and current for the biconical transmission line are

$$V(r) = \int_{\theta_c}^{\pi - \theta_c} E_\theta r\, d\theta = 2\sqrt{\mu/\epsilon}\, H_0 e^{-jkr} \ln\left[\cot(\theta_c/2)\right] \tag{43}$$

$$I(r) = \int_0^{2\pi} H_\varphi r\sin\theta\, d\varphi = 2\pi H_0 e^{-jkr} \tag{44}$$

The characteristic impedance of the biconical transmission line is then

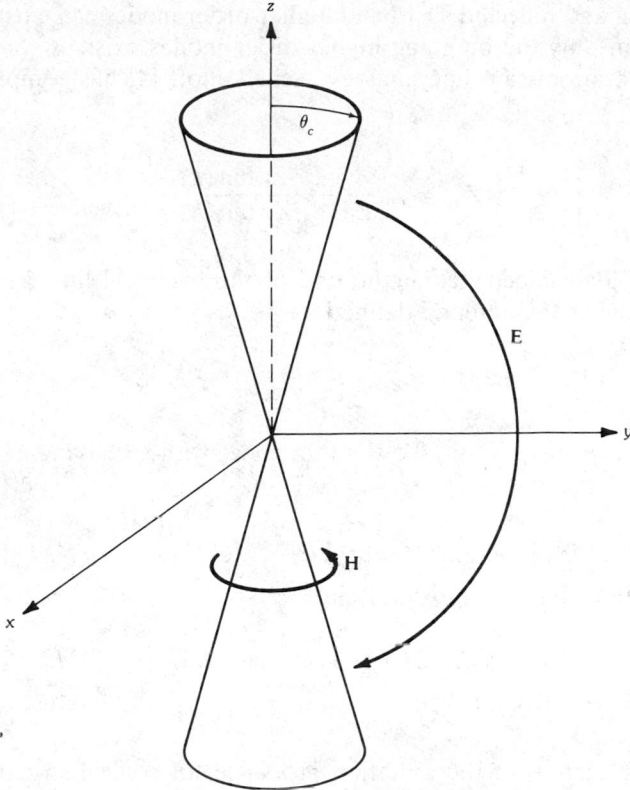

Fig. 5. The biconical antenna.

$$Z_0 = \frac{V(r)}{I(r)} = \frac{\sqrt{\mu/\epsilon}}{\pi} \ln\left[\cot\left(\theta_c/2\right)\right] \tag{45}$$

The radiated power is

$$P_r = 4\pi\sqrt{\mu/\epsilon}\, H_0^2 \ln\left[\cot\left(\theta_c/2\right)\right] \tag{46}$$

and the directive gain is

$$G_D = \frac{1}{\sin^2\theta \ln\left[\cot\left(\theta_c/2\right)\right]} \tag{47}$$

Note that because of the TEM fields and the structure's infinite length there is no frequency dependence exhibited in the preceding quantities.

A realistic biconical antenna is not infinite in length as required by the model created above. For the practical antenna of finite length ($r = L$), the space is divided into two regions. One is the modal or TEM region containing the bounded wave descriptions ($r < L$), and the other is the radiation region ($r > L$). In the modal region outgoing and reflected TEM and higher-order modes can exist, while in the radiation region only the outgoing higher-order modes exist.

Using the transmission line analogy, Schelkunoff [4] has computed an input impedance

$$Z_{in} = Z_0 \frac{Z_L + jZ_0 \tan(kL)}{Z_0 + jZ_L \tan(kL)} \tag{48}$$

where Z_L is a load impedance at the end of the biconical line representing the transition region of the antenna defined as

$$Z_L = Z_0^2/Z_m \tag{49}$$

where $Z_m = R_m + jX_m$. Then, for the thin cone, with $Ci(x)$ and $Si(x)$ defined in Chart 6,

$$
\begin{aligned}
R_m &= 60Ci(2kL) + 30[0.577 + \ln(kL) - 2Ci(kL) + Ci(4kL)]\cos(2kL) \\
&\quad + 30[Si(4kL) - 2Si(2kL)]\sin(2kL) \\
X_m &= 60Si(2kL) + 30[Ci(4kL) - \ln(kL) - 0.577]\sin(2kL) \\
&\quad - 30Si(4kL)\cos(2kL)
\end{aligned}
\tag{50}
$$

A detailed analysis of the radiation properties of conical structures has been pursued and presented in many works [4, 5, 11, 17, 18, 19, 20]. A formal solution is presented in "Electromagnetic radiation from conical structures" by J. R. Wait in [11] and by Schelkunoff and Friis in [5], where expansions for the fields in the two regions are obtained and a formalism for determining the unknown coefficients is established. The limiting cases of a spherical antenna ($\theta_c \to \pi/2$) and the thin-wire antenna ($\theta_c \to 0$) are considered as well.

Other antennas which have received theoretical attention because of their coincidence with coordinate surfaces are the spherical [8, 16] and spheroidal [2, 4, 8, 15] antennas.

An Antenna Reference Chart

A number of works have characterized many antenna systems. The work represented in Jasik [10], *Antenna Engineering Handbook*, is monumental and extremely useful. Here we will augment the approximate analysis methods useful for simple antennas presented earlier with an antenna reference chart useful for qualitative and approximate quantitative analysis [21].

The antenna reference chart is duplicated with minor corrections in Fig. 6. The chart includes the antenna name, some physical characteristics such as size, a diagram with coordinate system, its resistance at the lowest resonance frequency, the half-power bandwidth in percent, the antenna gain in decibels relative to a half-wavelength dipole and isotropic source, the polarization characteristic of the antenna, and the principal-plane radiation patterns characterizing the antenna. In the text, mention is made of another valuable reference on antenna characteristics which has, as well, a handy section of tabulated antenna characteristics [22].

Imperfect Grounds

The introduction of a ground plane, such as the earth, with realistic electrical parameters in the vicinity of an antenna can modify the antenna characteristics. The current distribution is affected through near-field interactions between the antenna and ground, and the radiated field is modified by the altered antenna currents and the ground reflection of the radiation field. The former effect, where near-field interactions perturb the current distribution, is considered in a following section. The latter effect can be included in the solution process by using plane-wave reflection techniques where the radiated fields are constructed from a direct and a ground-reflected wave [9, 13, 23].

The use of image and reflection coefficients is rather straightforward, being only somewhat more involved than that for the perfect ground plane as exhibited previously. The approach uses the image as induced in a perfect ground plane but with a modified strength which is proportional to the appropriate Fresnel plane-wave reflection coefficient. The relevant geometry for an electric current source and ground with parameters ϵ, μ_0, and σ is shown in Fig. 7.

The reflection of the incident **E** and **H** fields at the interface will depend on the polarization of the field with respect to the plane of incidence (the plane containing the surface normal and the propagation vector for the wave). Thus the Fresnel reflection coefficients, which are strictly true for an infinite plane-wave field, will exhibit this dependency on polarization.

The two cases of interest are illustrated in Fig. 8, where the wave with **E(r)** in the plane of incidence is termed vertically polarized and **E(r)** normal to the plane of incidence as horizontally polarized.

The Fresnel reflection coefficients for the vertically and horizontally polarized waves are

$$R_v = \frac{\epsilon' \cos\theta - \sqrt{\epsilon' - \sin^2\theta}}{\epsilon' \cos\theta + \sqrt{\epsilon' - \sin^2\theta}} \tag{51}$$

and

$$R_h = \frac{\cos\theta - \sqrt{\epsilon' - \sin^2\theta}}{\cos\theta + \sqrt{\epsilon' - \sin^2\theta}} \tag{52}$$

where

$$\cos\theta = -\hat{\mathbf{k}}\cdot\hat{\mathbf{z}} = \hat{\mathbf{k}}_R\cdot\hat{\mathbf{z}}$$

$$\epsilon = \frac{\epsilon}{\epsilon_0} - j\frac{\sigma}{\omega\epsilon_0} \tag{53}$$

The determination of the radiated field $\mathbf{E}(\theta, \varphi)$ for an arbitrarily oriented antenna over a finitely conducting ground can be readily computed. First, the

Type	Configuration	Impedance: Resistive at f_r, $R\,(\Omega)$	−3-dB Bandwidth Percent	Gain: dB above		Polarization	Pattern Number
				Isotropic	Dipole		
Isotropic radiator (theoretical)		—	—	0	−2.14	none	A
Small dipole $L > \lambda/2$		—	—	1.74	−0.4	h	B
Thin dipole $L = \lambda/2$ $L/D = 276$		60	34	2.14	0	h	B
Thick dipole $L = \lambda/2$ $L/D = 51$		49	55	2.14	0	h	B
Cylindrical dipole $L = \lambda/2$ $L/D = 10$		37	100	2.14	0	h	B
Cylindrical dipole $L = \lambda$ $L/D = 9.6$		150	130	3.64	1.5	h	B

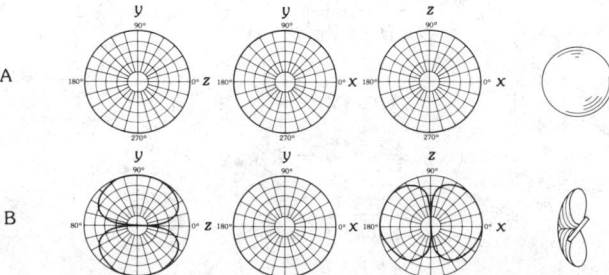

Fig. 6. An antenna chart. (*After Salati [21]*)

Type	Configuration	Impedance: Resistive at f_r, $R\,(\Omega)$	−3-dB Bandwidth Percent	Gain: dB above		Polarization	Pattern Number
				Isotropic	Dipole		
Folded dipole $L = \lambda/4$ $L/d = 13$		6000	5	1.64	−0.5	h	B
Folded dipole $L = \lambda/2$ $L/d = 25.5$		300	45	2.14	0	h	B
Biconical $L = \lambda/2$		72	100	2.14	0	h	B
Biconical $L = \lambda$		350	200	2.14	0	h	B
Turnstile $L = \lambda/2$ $L/d = 25.5$		150	50	−0.86	−3	h	C
Folded dipole over reflecting sheet $L = \lambda/2$ $L/d = 25.5$ $\lambda/8$ above sheet		150	20	7.14	5	h	D

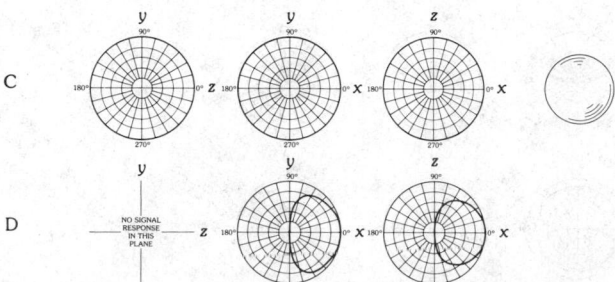

Fig. 6, *continued.*

Type	Configuration	Impedance: Resistive at f_r, $R(\Omega)$	−3-dB Bandwidth Percent	Gain: dB above		Polarization	Pattern Number
				Isotropic	Dipole		
Dipole over small ground plane $L=\lambda/4$ $L/D=53$ $\ell=2\lambda$		28	40	2.14	0	v	B
Folded unipole over small ground plane $L=\lambda/4$ $L/D=53$ $\ell=2\lambda$ $L/d=13$		150	45	2.14	0	v	B
Coaxial dipole $L=\lambda/4$ $L/D=40$		50	16	2.14	0	v	B
Biconical coaxial dipole $L=\lambda/2$ $d=\lambda/8$ $D=3\lambda/8$		72	200	2.14	0	v	B
Disc-cone or rod disc-cone $L=\lambda/4$ $\ell=\lambda$		50	300	2.14	0	v	B
Biconical horn $L=9\lambda/2$ $D=14\lambda$		20	25	14.14	12	v	B

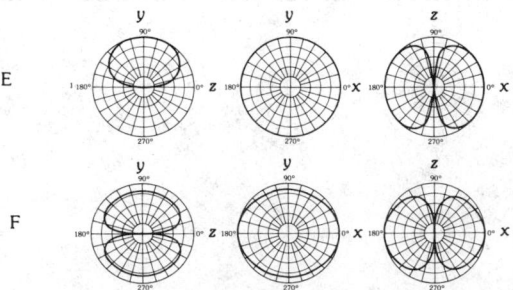

Fig. 6, *continued.*

Type	Configuration	Impedance: Resistive at f_r, R (Ω)	−3-dB Bandwidth Percent	Gain: dB above		Polarization	Pattern Number
				Isotropic	Dipole		
Slot in large ground plane $L = \lambda/2$ $l/d = 29$		350	70	2.14	0	h	E
Vertical full-wave loop $D = \lambda/\pi$ $D/d = 36$		45	13	3.14	1	h	F
Helical over reflector screen, tube 6λ long coiled into 6 turns $\lambda/4$ apart		130	200	10.14	8	circ.	G
Rhombic $L = 9\lambda$ $l = 9\lambda/2$		600	100	16.74	14.5	h	H
Parabolic with folded dipole feed ($\lambda/2$) $D = 5\lambda/2$		300	30	14.74	12.5	h	H
Horn, coaxial feed $L = 3\lambda$ $l = 3\lambda$		50	35	15.14	13	h	H

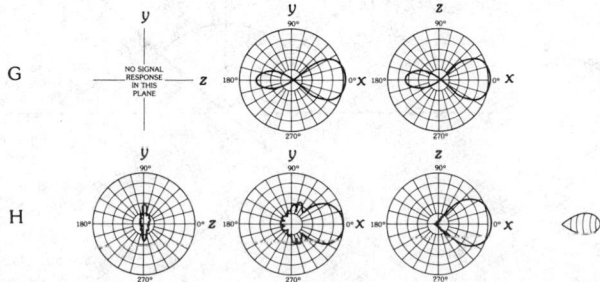

Fig. 6, *continued.*

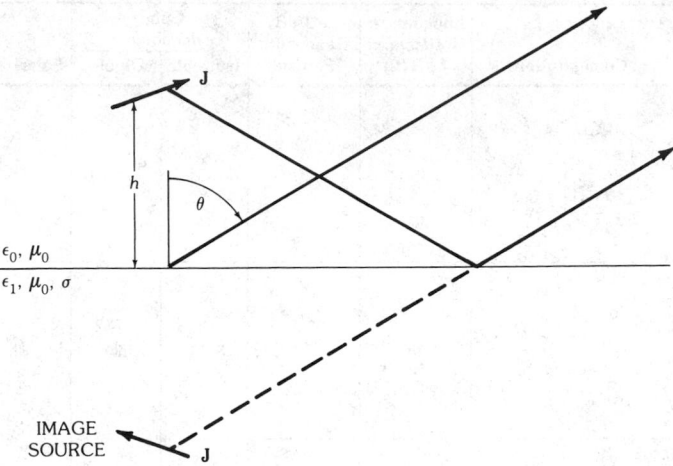

Fig. 7. Images for imperfect ground analysis.

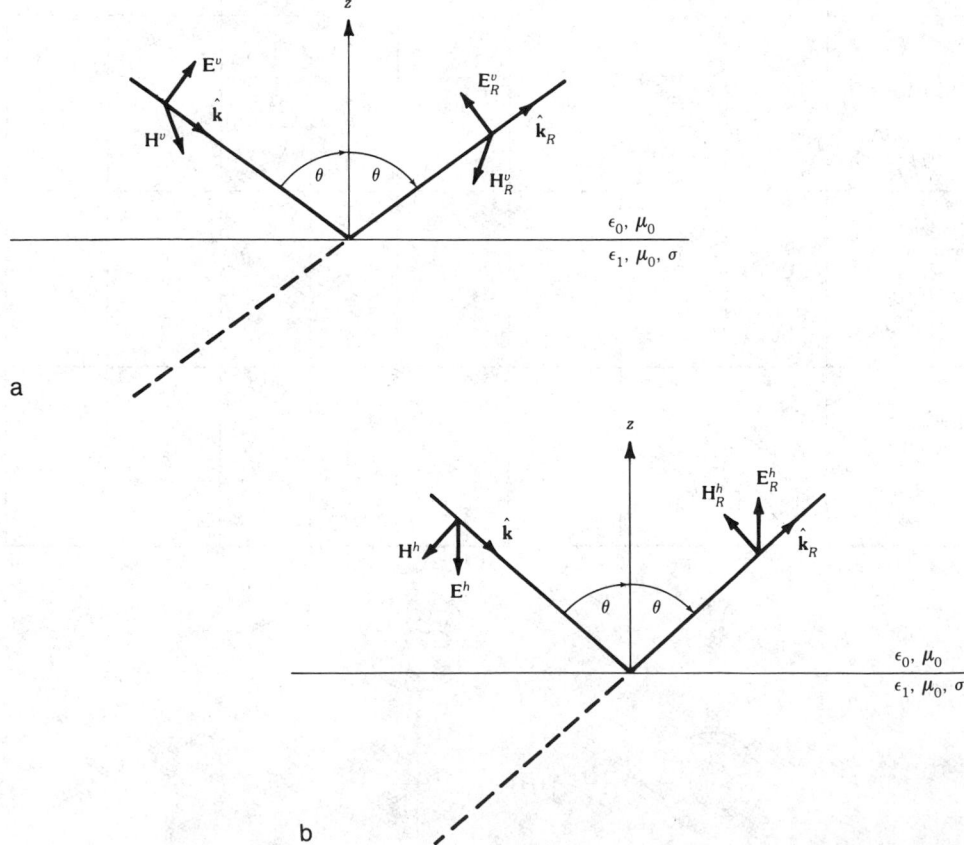

Fig. 8. Plane-wave reflection at an interface.

radiated field $\mathbf{E}_r(\theta, \varphi)$ due to the given antenna is computed for the case when it is located in an infinite homogeneous medium at its original location. Then the reflected field $\mathbf{F}_R(\theta, \varphi)$ is calculated from

$$\mathbf{E}_R(\theta, \varphi) = R_v\mathbf{E}_I(\theta, \varphi) + (R_h - R_v)[\mathbf{E}_I(\theta, \varphi)\cdot\hat{\mathbf{p}}]\hat{\mathbf{p}} \qquad (54)$$

where $\mathbf{E}_I(\theta, \varphi)$ is the field due to the image of the original source in a perfect ground plane and $\hat{\mathbf{p}}$ is the unit vector normal to the plane of incidence ($\mathbf{p} = \hat{\mathbf{z}} \times \hat{\mathbf{k}}$). Finally, the total field $\mathbf{E}(\theta, \varphi)$ is the sum of the two contributions:

$$\mathbf{E}(\theta, \varphi) = \mathbf{E}_r(\theta, \varphi) + \mathbf{E}_R(\theta, \varphi) \qquad (55)$$

Chart 11 illustrates the steps and the results of a procedure for evaluating the fields of a vertical and horizontal short dipole carrying a constant current I_0. The procedure for analyzing an arbitrary antenna, though more involved, can proceed in a similar manner.

The Fresnel reflection coefficients are strictly valid only for plane waves and are therefore not rigorously valid for antennas near ground planes [13, 23, 24, 25].

Arrays

The subject of antenna arrays has been extensively documented and will be discussed in Chapters 13, 14, and 17. Previous work is described in the writings of Bach and Hansen [26], Kraus [14], Jasik [10], Stutzman and Thiele [13], Weeks [9], and Ma [27]. Here we will merely show the relationship of some aspects of array analysis to previously discussed subjects in this chapter.

In the following we will present expressions which are useful for expressing the fields due to an array of identical antenna elements, each having an identical current distribution differing by, at most, a complex scaling constant. The similarity of current distribution implies identical orientation of the elements. The field due to an electric current source $\mathbf{J}(\mathbf{r})$ can be evaluated using the equations in Charts 1 and 2.

Some elements in an antenna array are shown in Fig. 9. Using element number 1 as the reference element in an N-element array we establish our coordinate system definitions so that a linear shift of the nth element will cause it to be coincident with element 1. In this case such a shift between coincident points is $\mathbf{r}_n - \mathbf{r}_1$ or $\mathbf{r}'_n - \mathbf{r}'_1$. The relationship between the element current distributions is given by

$$J_r(\mathbf{r}'_n) = \alpha_n\mathbf{J}_1(\mathbf{r}'_1) \qquad (56)$$

The magnetic vector potential in the far zone for the nth element is

$$\mathbf{A}_n(\mathbf{r}) = \alpha_n e^{jk\hat{\mathbf{r}}\cdot(\mathbf{r}_n - \mathbf{r}_1)} \frac{1}{4\pi} \frac{e^{-jkr}}{\mathbf{r}} \int_V \mathbf{J}_1(\mathbf{r}')e^{jk\hat{\mathbf{r}}\cdot\mathbf{r}'}dv' \qquad (57)$$

or equivalently,

Chart 11. Evaluation of Field Components for a Short Dipole of Constant Current in Presence of a Finitely Conducting Ground

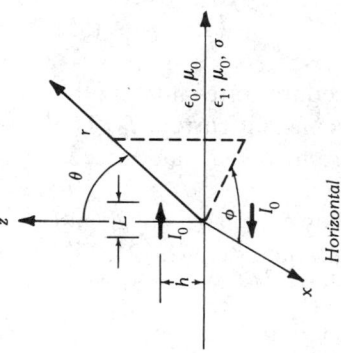

Vertical

$$\mathbf{E}_r(\theta, \varphi) = \hat{\boldsymbol{\theta}}\, j\omega\mu_0 I_0 L \sin\theta\, \frac{e^{-jk(r-h\cos\theta)}}{4\pi r}$$

$$\mathbf{E}_I(\theta, \varphi) = \hat{\boldsymbol{\theta}}\, j\omega\mu_0 I_0 L \sin\theta\, \frac{e^{-jk(r+h\cos\theta)}}{4\pi r}$$

$$\mathbf{E}_R(\theta, \varphi) = R_v \mathbf{E}_I(\theta, \varphi)$$

$$\mathbf{E}(\theta, \varphi) = \hat{\boldsymbol{\theta}}\, j\omega\mu_0 I_0 L \sin\theta\, \frac{e^{-jkr}}{4\pi r}\, \left(e^{jkh\cos\theta} + R_v e^{-jkh\cos\theta} \right)$$

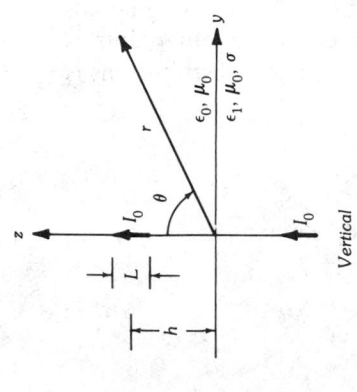

Horizontal

Step 1. The field, due to a primary source in free space, is evaluated:

$$A_y = I_0 L\, \frac{e^{-jk(r-h\cos\theta)}}{4\pi r}$$

$$\mathbf{E}_r(\theta, \varphi) = j\omega\mu_0\hat{\mathbf{r}} \times \mathbf{A}$$

$$= j\omega\mu_0 I_0 L\, \frac{e^{-jk(r-h\cos\theta)}}{4\pi r}\, (-\cos\varphi\, \hat{\boldsymbol{\varphi}} - \cos\theta\sin\varphi\, \hat{\boldsymbol{\theta}})$$

Step 2. The field, due to the image source, is evaluated:

$$\mathbf{E}_I(\theta, \varphi) = j\omega\mu_0 I_0 L\, \frac{e^{-jk(r+h\cos\theta)}}{4\pi r}\, (\cos\varphi\, \hat{\boldsymbol{\varphi}} + \cos\theta\sin\varphi\, \hat{\boldsymbol{\theta}})$$

Step 3. Evaluate the ground-reflected field using (54):

$$\mathbf{E}_R(\theta, \varphi) = R_v \mathbf{E}_I(\theta, \varphi) + (R_h - R_v)(\mathbf{E}_I(\theta, \varphi)\cdot\hat{\boldsymbol{\varphi}})\hat{\boldsymbol{\varphi}}$$

$$= j\omega\mu_0 I_0 L\, \frac{e^{-jk(r+h\cos\theta)}}{4\pi r}\, (-R_h\cos\varphi\, \hat{\boldsymbol{\varphi}} + R_v\cos\theta\sin\varphi\, \hat{\boldsymbol{\theta}})$$

Step 4. Evaluate the total field using (55):

$$\mathbf{E}(\theta, \varphi) = j\omega\mu_0 I_0 L\, \frac{e^{-jkr}}{4\pi r}\, \left[-\hat{\boldsymbol{\varphi}}\cos\theta\, (e^{jkh\cos\theta} + R_h e^{-jkh\cos\theta}) \right.$$

$$\left. -\hat{\boldsymbol{\theta}}\cos\theta\sin\varphi\, (e^{jkh\cos\theta} - R_v e^{-jkh\cos\theta}) \right]$$

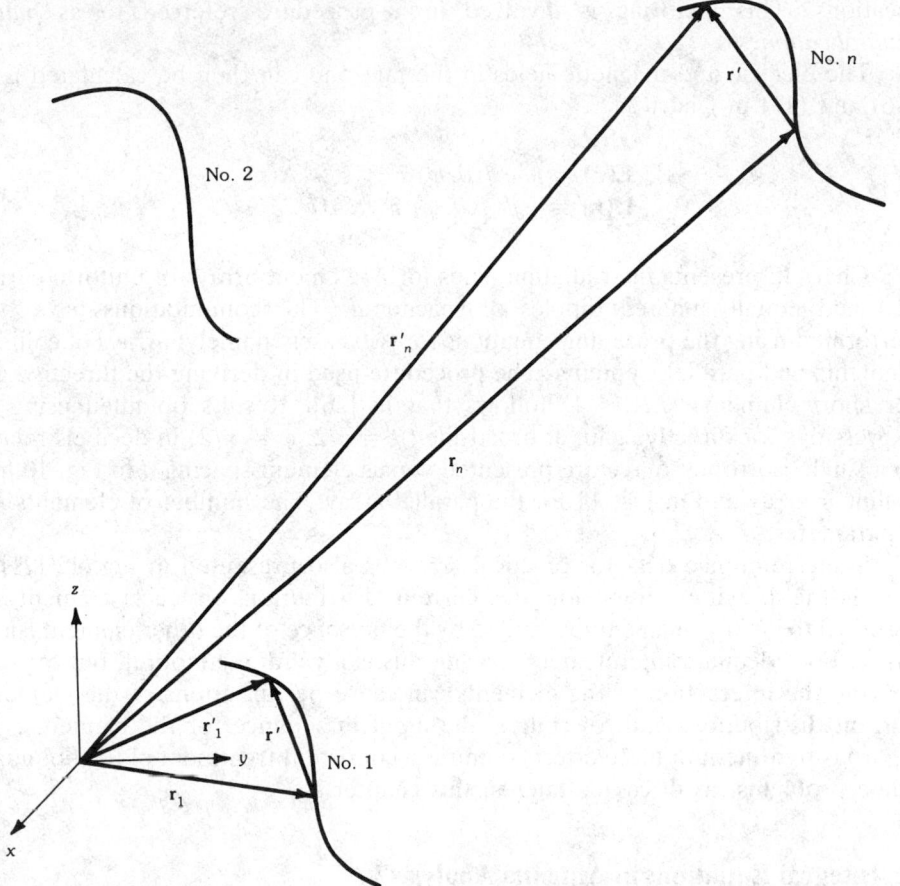

Fig. 9. Antenna array geometry.

$$\mathbf{A}_n(\mathbf{r}) = \alpha_n e^{jk\hat{\mathbf{r}}\cdot(\mathbf{r}_n - \mathbf{r}_1)}\mathbf{A}_1(\mathbf{r}) \tag{58}$$

The far-zone magnetic vector potential for the array is then

$$\mathbf{A}(\mathbf{r}) = \mathbf{A}_1(\mathbf{r})\sum_{n=1}^{N} \alpha_n e^{jk\hat{\mathbf{r}}\cdot(\mathbf{r}_n - \mathbf{r}_1)} \tag{59}$$

where the summation, denoted by $f(\theta, \varphi)$, is referred to as the *array factor*:

$$f(\theta, \varphi) = \sum_{n=1}^{N} \alpha_n e^{jk\hat{\mathbf{r}}\cdot(\mathbf{r}_n - \mathbf{r}_1)} \tag{60}$$

This factorization separates the element contribution or element pattern $\mathbf{A}_1(\mathbf{r})$ from the array pattern $f(\theta, \varphi)$, which depends only on the relative source strengths and locations. This factoring is involved in a procedure referred to as *pattern multiplication*.

The electric and magnetic fields in the far zone can then be calculated using (16) and (17) in Chart 2:

$$\mathbf{E}(\mathbf{r}) = j\omega\mu\, f(\theta,\varphi)\, \hat{\mathbf{r}} \times \hat{\mathbf{r}} \times \mathbf{A}(\mathbf{r})$$
$$\mathbf{H}(\mathbf{r}) = -jk\, f(\theta,\varphi)\, \hat{\mathbf{r}} \times \mathbf{A}(\mathbf{r}) \tag{61}$$

Chart 12 presents the radiation fields for N-element arrays of uniform current (I_0) and equally phased dipoles of spacing d. The computations have been performed using the preceding equations for two cases, namely, arrays of collinear elements and parallel elements. The procedure used in deriving the directive gain for short elements ($kL \ll 1$) follows that in [26]. Results obtained using the expressions for directive gain at broadside ($\theta = \pi/2$, $\varphi = \pi/2$) in decibels relative to a single isotropic source are presented versus element spacing d in Fig. 10 for a collinear array and in Fig. 11 for the parallel array. The number of elements N is a parameter.

Some reference data for practical arrays is also presented in Fig. 12 [21].

In the previous discussion the current distribution on each element was assumed to be known and unperturbed by the presence of the other elements in the array. For adequate interelement spacing this is a valid assumption, but for close spacing the interaction of the elements can cause perturbations of the elemental current distributions and, of course, the input impedance for the elements. The rigorous treatment of these effects requires a consistent treatment of the boundary-value problems, as discussed later in this chapter.

3. Integral Equations in Antenna Analysis

In previous discussions the current distributions on radiating elements were assumed to be known. This, however, is generally not the case since the precise description of the current distribution on a metallic structure such as an antenna in the presence of an exciting source, such as a voltage generator at its terminals, involves the solution of a complicated boundary-value problem. In the following we will describe the integral equations which can be solved for the unknown source distributions induced by specified excitations, and in Part 2 we will consider the numerical solution of these equations and some associated issues.

Perfectly Conducting Wires and Bodies

Here we will focus our attention on radiating structures composed of perfect electric conductors over which the boundary conditions given by (33), (34), and (35) must hold. Furthermore, we will devote our attention mainly to wire structures, with conducting bodies touched on briefly.

In order to facilitate ensuing discussions concerning integral equations, questions concerning validity of each specific equation, the existence or uniqueness of solutions, and various features of the limiting process which reduce the integral representations for radiated fields to integral equations for unknown source dis-

Chart 12. Radiation Fields for Vertical Collinear and Parallel Arrays of Short Dipoles Supporting Equal Amplitude and Equal Phase Currents

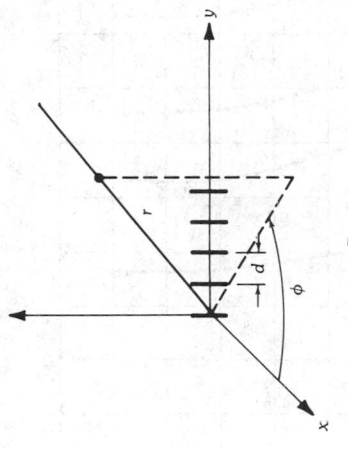

Collinear

$$\mathbf{A}_1(\mathbf{r}) = \hat{\mathbf{z}}\frac{e^{-jkr}}{4\pi r}(I_0 L)\frac{\sin[(kL/2)\cos\theta]}{(kL/2)\cos\theta}$$

$$f(\theta,\varphi) = \sum_{n=1}^{N} e^{jk(n-1)d\cos\theta} = e^{j[k(N-1)d/2]\cos\theta}\frac{\sin[(kNd/2)\cos\theta]}{\sin[(kd/2)\cos\theta]}$$

$$\mathbf{E}_r(\mathbf{r}) = \hat{\boldsymbol{\theta}}j\omega\mu(I_0 L)\sin\theta\frac{e^{-jkr}}{4\pi r}\frac{\sin[(kL/2)\cos\theta]}{(kL/2)\cos\theta}\frac{\sin[(kNd/2)\cos\theta]}{\sin[(kd/2)\cos\theta]}$$
$$\times e^{j[k(N-1)d/2]\cos\theta}$$

$$G_D(\theta,\varphi) = $$
$$\frac{\sin^2\theta\sin^2[(kNd/2)\cos\theta]\sin^2[(kd/2)\cos\theta]}{2N/3 + 4\sum_{m=1}^{N-1}(N-m)[(\sin mkd)/(mkd)^3 - (\cos mkd)/(mkd)^2]}$$

Elemental Vector Potential

$$\mathbf{A}_1(\mathbf{r}) = \hat{\mathbf{z}}\frac{e^{-jkr}}{4\pi r}(I_0 L)\frac{\sin[(kL/2)\cos\theta]}{(kL/2)\cos\theta}$$

Array Factor ($\alpha_n = 1$)

$$f(\theta,\varphi) = e^{j[k(N-1)d/2]\sin\theta\sin\varphi}\frac{\sin[(kNd/2)\sin\theta\sin\varphi]}{\sin[(kd/2)\sin\theta\sin\varphi]}$$

Far-Zone Field

$$\mathbf{E}_r(\mathbf{r}) = \hat{\boldsymbol{\theta}}j\omega\mu(I_0 L)\sin\theta\frac{e^{-jkr}}{4\pi r}\frac{\sin[(kL/2)\cos\theta]}{(kL/2)\cos\theta}\frac{\sin[(kNd/2)\sin\theta\sin\varphi]}{\sin[(kd/2)\sin\theta\sin\varphi]}$$
$$\times e^{j[k(N-1)d/2]\sin\theta\sin\varphi}$$

Directive Gain

$$G_D(\theta,\varphi) = $$
$$\frac{\sin^2\theta\sin^2[(kNd/2)\sin\theta\sin\varphi]\sin^2[(kd/2)\sin\theta\sin\varphi]}{2N/3 + 2\sum_{m=1}^{N-1}(N-m)(\{[(mkd)^2-1]\sin mkd\}/(mkd)^3 + (\cos mkd)/(mkd)^2)}$$

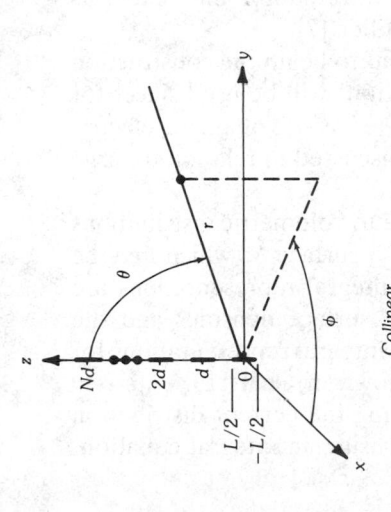

Parallel

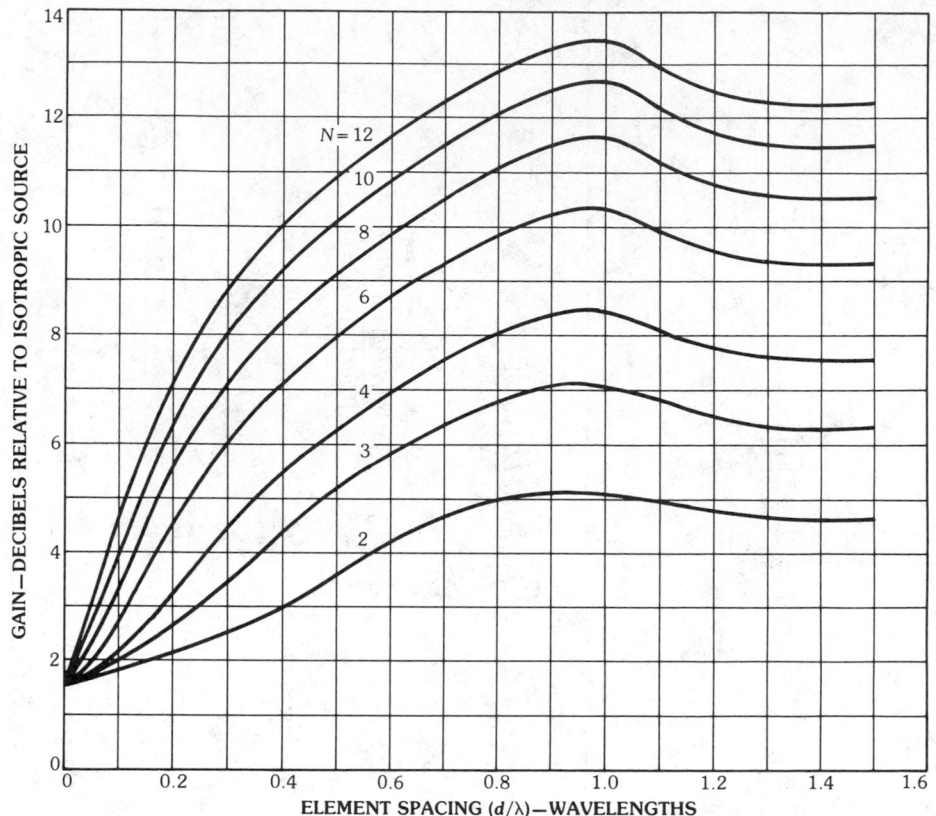

Fig. 10. Gain of a collinear array of short dipoles relative to isotropic source.

tributions are necessarily glossed over. For further information the reader is referred to Stratton [2], Silver [3], and Poggio and Miller [7].

The field equations in Chart 1 are a convenient point to begin the construction of the integral equations. The electric-field representations will be used since, for antennas, the driving source is most easily specified in terms of voltage or electric field. Later, the magnetic-field representation will be discussed in relation to large conducting surfaces.

The representations in Chart 1 are for the fields due to volumetric distributions of sources. For electric current sources constrained to a surface S (which may be considered to be the boundary of V in Chart 1) the integral representations are simply modified in that the volume densities become surface densities and the volume integral becomes a surface integral over S. The integral representations for the electric field due to electric sources over S are shown in Chart 13.

The general boundary-value problem of determining the current distribution on a perfect electric conducting surface is approached using an integral equation. The boundary condition on S is stated as

$$\hat{\mathbf{n}}(\mathbf{r}) \times \mathbf{E}_t(\mathbf{r}) = 0, \qquad \mathbf{r} \in S \tag{62}$$

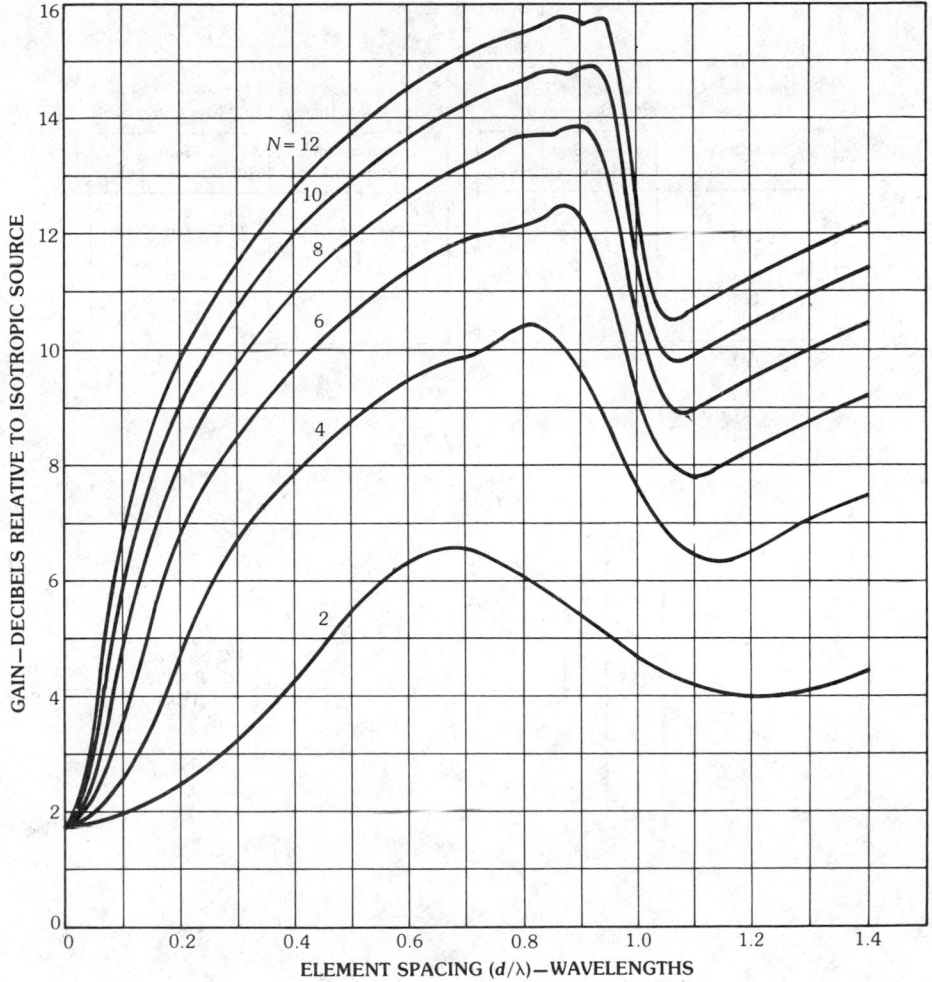

Fig. 11. Gain of equiphased short-dipole array relative to isotropic source.

with $\hat{\mathbf{n}}(\mathbf{r})$ the outwardly pointing normal to S and $\mathbf{E}_t(\mathbf{r})$, the total electric field at the surface. The total field is composed of an incident or driven portion $\mathbf{E}^i(\mathbf{r})$ and a portion generated by the induced surface sources $\mathbf{J}_s(\mathbf{r}')$ and $\mathbf{p}_s(\mathbf{r}')$ referred to as $\mathbf{E}(\mathbf{r})$. The boundary condition then requires

$$\hat{\mathbf{n}}(\mathbf{r}) \times \mathbf{E}(\mathbf{r}) = -\hat{\mathbf{n}}(\mathbf{r}) \times \mathbf{E}^i(\mathbf{r}), \qquad \mathbf{r} \in S \qquad (63)$$

The field component $\mathbf{E}(\mathbf{r})$ is given by the integral representations in Chart 5 but with the observation point $\mathbf{r}$ on the surface. The procedure of taking the observation point to the surface must be performed delicately due to the singularity in $\varphi(\mathbf{r}, \mathbf{r}')$ when $|\mathbf{r} - \mathbf{r}'| \to 0$. These issues are dealt with in Stratton [2], Silver [3], Poggio and Miller [7], and Maue [28]. The integral equations most widely used and derived from the electric-field integral representations are presented in Chart

Broadside Array
$L = \lambda/2$
Polarization: Vertical

Theoretical Gain of Broadside $\frac{1}{2}\lambda$ Elements at Different Spacings a		Theoretical Gain of Broadside $\frac{1}{2}\lambda$ Elements for Different Numbers of Elements	
Spacing a (wavelengths)	Gain (dB above dipole)	Number of Elements	Gain (dB above dipole)
5/8	4.8	2	4.0
3/4	4.6	3	5.5
1/2	4.0	4	7.0
3/8	2.4	5	8.0
1/4	1.0	6	9.0
1/8	0.3		

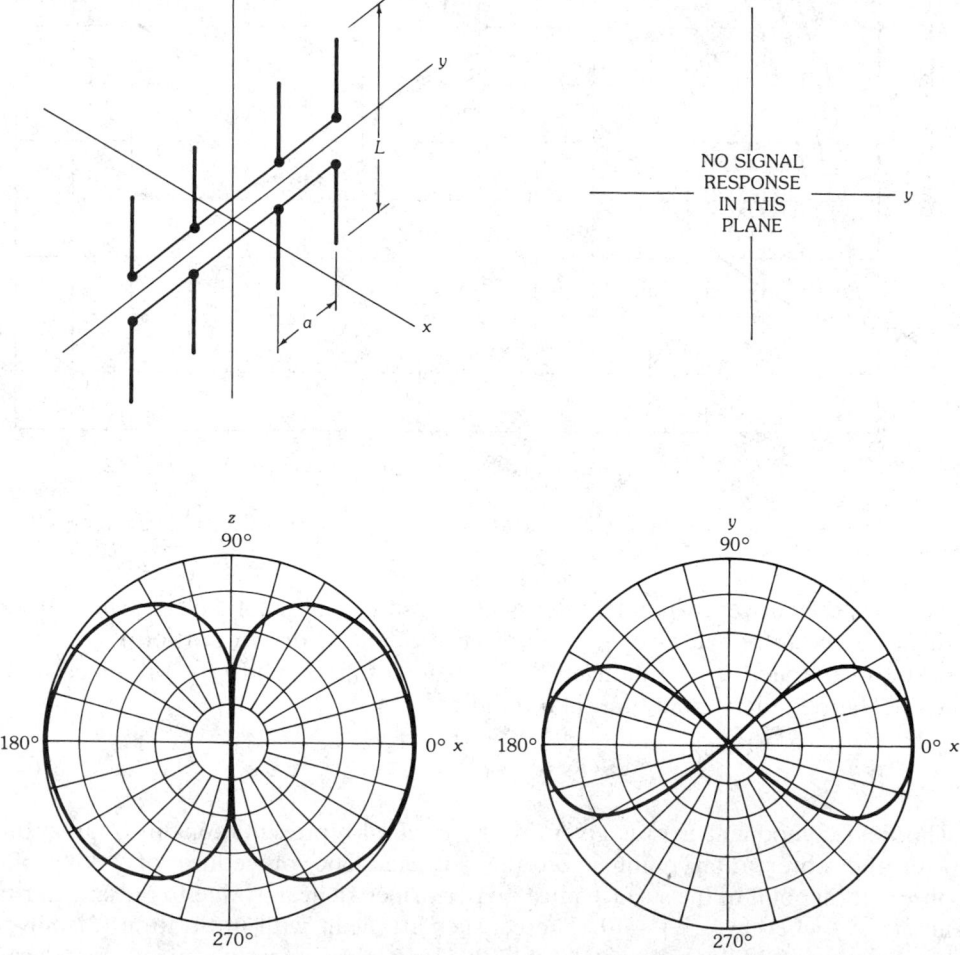

Fig. 12. Reference data for practical arrays. (*After Salati [21]*)

End-Fire Array
$L = \lambda/2$
Polarization: Vertical

Theoretical Gain of Two End-Fire $\frac{1}{2}\lambda$
Elements for Various Spacings a

a	Gain (dB above dipole)
5/8	1.7
1/2	2.2
3/8	3.0
1/4	3.8
1/20	4.1
1/8	4.3

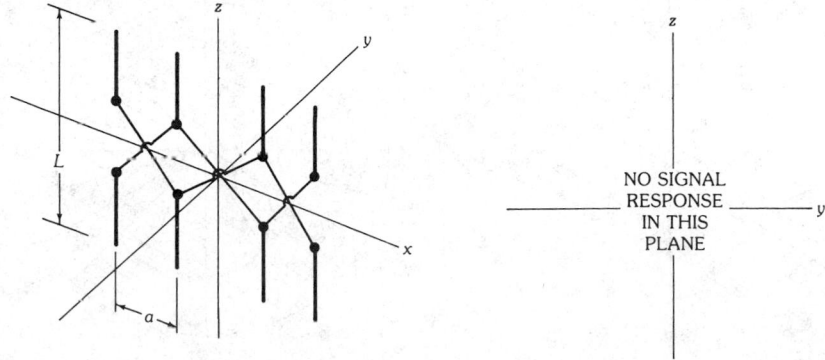

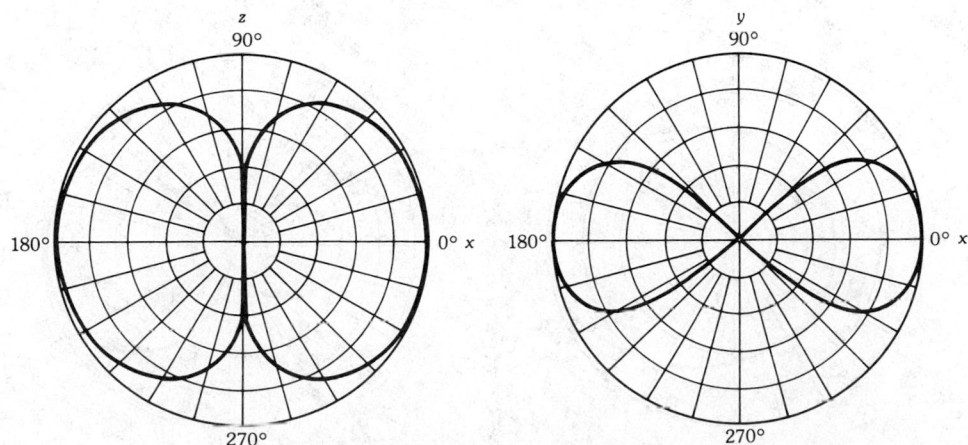

Fig. 12, *continued.*

Parasitic Array
$L = \lambda/2$
Polarization: Horizontal

Number of Elements	Gain (dB above dipole)	Front-to-Back Ratio (dB)
2	4 to 5	10 to 15
3	6 to 7	15 to 25
4	7 to 9	20 to 30
5	9	

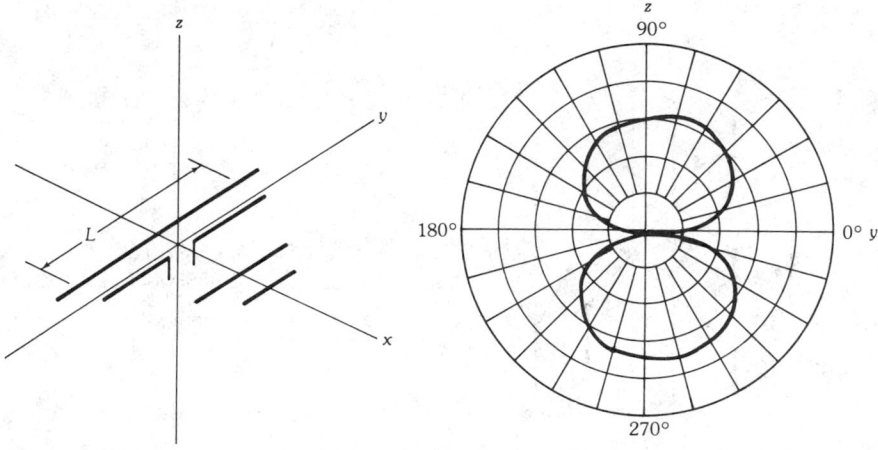

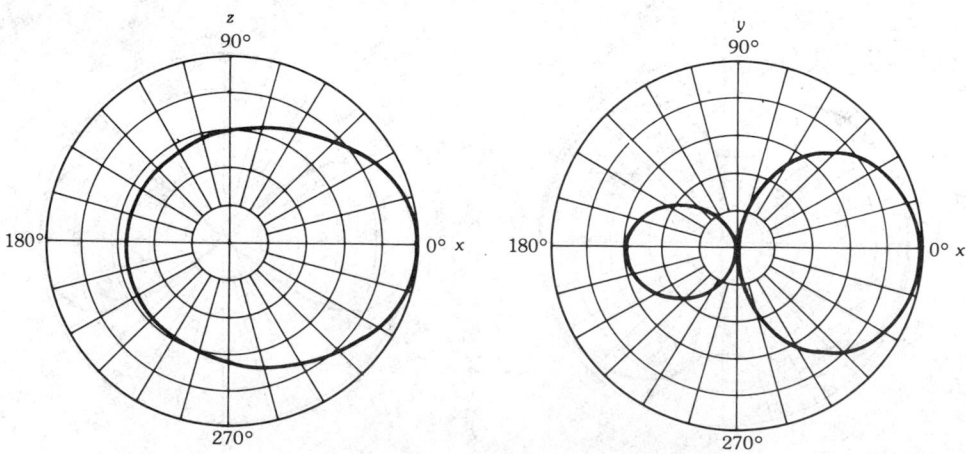

Fig. 12, *continued.*

Collinear Array
$L = \lambda/2$
$b = \lambda/4$

Spacing a Between Centers of Adjacent $\frac{1}{2}\lambda$ Elements	Number of $\frac{1}{2}\lambda$ Elements in Array Versus Gain in dB Above a Reference Dipole				
	2	3	4	5	6
$a = \frac{1}{2}\lambda$	1.8	3.3	4.5	5.3	6.2
$a = \frac{3}{4}\lambda$	3.2	4.8	6.0	7.0	7.8

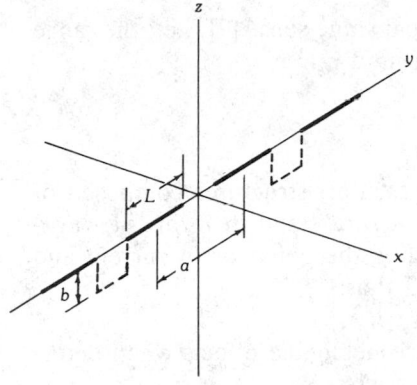

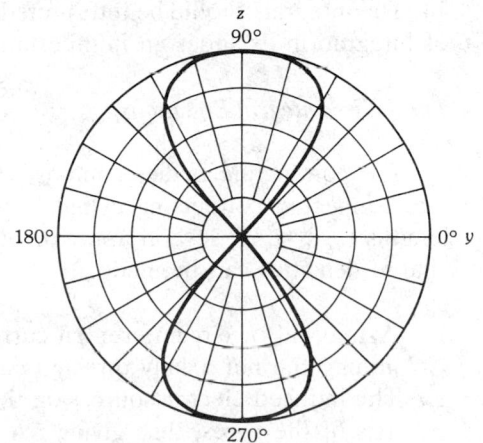

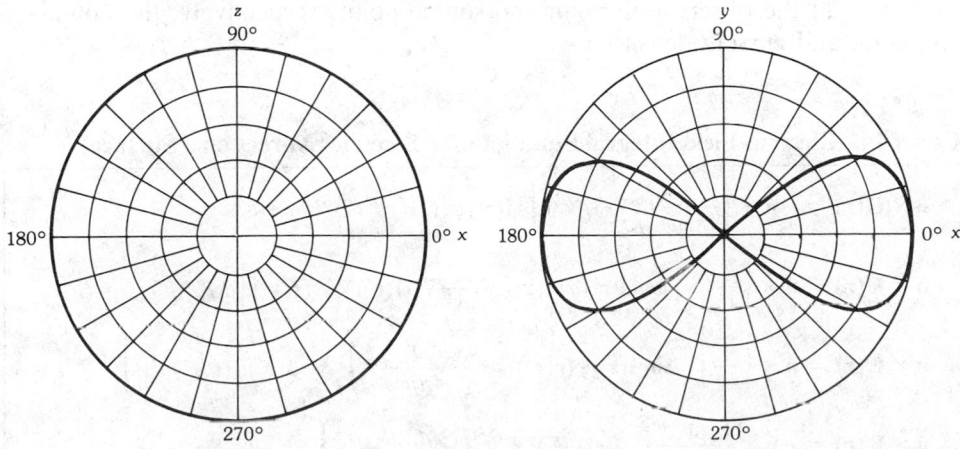

Fig. 12, *continued.*

Chart 13. Electric-Field Integral Representations for Electric Sources

$$E(r) = \frac{1}{4\pi j\omega\epsilon}(\nabla\nabla\cdot + k^2)\int_S J(r')\varphi(r,r')\,d^2r'$$

$$E(r) = -\frac{1}{4\pi}\int_S \left[j\omega\mu J(r')\varphi(r,r') + \frac{1}{j\omega\epsilon}\nabla'\cdot J(r')\nabla'\varphi(r,r')\right]d^2r'$$

$$E(r) = -\frac{1}{4\pi}\int_S j\omega\mu J(r')\varphi(r,r')\,d^2r' + \frac{1}{4\pi j\omega\epsilon}\nabla\int_S \nabla'\cdot J(r')\,\varphi(r,r')\,d^2r'$$

$$E(r) = \frac{1}{4\pi j\omega\epsilon}\int_S J(r')\cdot(\nabla\nabla + k^2\overline{\overline{I}})\,\varphi(r,r')\,d^2r'$$

14. The integrals should be interpreted in a principal value sense [7], i.e., the range of integration excludes an infinitesimal region around $r = r'$.

Thin-Wire Integral Equations

For conducting bodies composed of thin wires, i.e., structures composed of interconnected conducting cylinders whose radii a_i are small in terms of wavelengths ($a_i/\lambda \ll 1$), several approximations regarding the behavior of current and charge densities can be made. It can be assumed that:

1. Azimuthal or circumferential currents produce negligible effects when determining the net axially directed current on the wires.
2. The induced electric sources on the surface of the wires can be located on the axis of the wires, thus giving rise to a filamentary source representation.
3. The boundary condition on the electric field can be enforced on the surface of the conductor.

Using $\hat{s}(r)$ and $\hat{s}'(r)$ to denote unit vectors tangent to the conductor and parallel to its axis at the observation point and source point, respectively, the boundary condition and current density are

Chart 14. Electric-Field Integral Equations for Electric Sources on a Surface

$$\hat{n}\times E^i(r) = -\hat{n}\times\left\{\frac{1}{4\pi j\omega\epsilon}(\nabla\nabla\cdot + k^2)\int_S J(r')\varphi(r,r')\,d^2r'\right\}, \qquad r\in S$$

$$\hat{n}\times E^i(r) = \hat{n}\times\left\{\frac{1}{4\pi}\int_S \left[j\omega\mu J(r')\varphi(r,r') + \frac{1}{j\omega\epsilon}\nabla'\cdot J(r')\nabla'\varphi(r,r')\right]d^2r'\right\}, \qquad r\in S$$

$$\hat{n}\times E^i(r) = \hat{n}\times\left\{\frac{1}{4\pi}\int_S j\omega\mu J(r')\varphi(r,r')\,d^2r' - \frac{1}{4\pi j\omega\epsilon}\nabla\int_S \nabla'\cdot J(r')\varphi(r,r')\,d^2r'\right\}, \qquad r\in S$$

$$\hat{n}\times E^i(r) = -\hat{n}\times\left\{\frac{1}{4\pi j\omega\epsilon}\int_S J(r')\cdot(\nabla\nabla + k^2\overline{\overline{I}})\,\varphi(r,r')\,d^2r'\right\}, \qquad r\in S$$

$$\hat{s}(\mathbf{r}) \cdot \mathbf{E}_t(\mathbf{r}) = 0, \qquad \mathbf{r} \in s \tag{64}$$

$$\mathbf{J}(\mathbf{r}') = \hat{s}'(\mathbf{r}') I(\mathbf{r}') / 2\pi a \tag{65}$$

Because of the assumptions of the locations of the source and observation points, the distance variable $|\mathbf{r} - \mathbf{r}'|$ in $\varphi(\mathbf{r}, \mathbf{r}')$ is approximated by a distance R, which can never be zero since it is the distance from a point on the axis of the wire to a point on the surface and is thus never less than a. For example, for a z-aligned straight wire, $R = [(z - z')^2 + a^2]^{1/2}$. Such an approximation leads to the widely used thin wire or reduced kernel.

With s and s' denoting the axial coordinates at the observation and source points and $C(\mathbf{r})$ denoting the range of integration over the wires, the integral equations in Chart 14 can be written in their thin-wire forms as shown in Chart 15. Some of the integral equations have been widely used for thin wires. The "mixed-potential integral equation" has been used in the analysis of many structures [29]. Another equation which has found wide usage was derived from the first of the equations in Charts 14 and 15 by solving the differential equation for a

Chart 15. Thin-Wire Integral Equations

$$\hat{s} \cdot \mathbf{E}^i(s) = -\frac{1}{4\pi j\omega\epsilon} \left(\frac{\partial}{\partial s} \nabla \cdot + k^2 \hat{s} \cdot \right) \int_{C(\mathbf{r})} \hat{s}' \, I(s') \frac{e^{-jkR}}{R} ds', \qquad s \in C(\mathbf{r})$$

$$\hat{s} \cdot \mathbf{E}^i(s) = \frac{1}{4\pi} \int_{C(\mathbf{r})} \left[\hat{s} \cdot \hat{s}' j\omega\mu \, I(s') \frac{e^{-jkR}}{R} - \frac{1}{j\omega\epsilon} \frac{\partial I(s')}{\partial s'} \hat{s} \cdot \nabla \frac{e^{-jkR}}{R} \right] ds', \qquad s \in C(\mathbf{r})$$

$$\hat{s} \cdot \mathbf{E}^i(s) = \frac{1}{4\pi} \int_{C(\mathbf{r})} \hat{s} \cdot \hat{s}' j\omega\mu \, I(s') \frac{e^{-jkR}}{R} ds' - \frac{1}{4\pi j\omega\epsilon} \frac{\partial}{\partial s} \int_{C(\mathbf{r})} \frac{\partial I(s')}{\partial s'} \frac{e^{-jkR}}{R} ds', \qquad s \in C(\mathbf{r})$$

Pocklington's Integral Equation

$$\hat{s} \cdot \mathbf{E}^i(s) = -\frac{1}{4\pi j\omega\epsilon} \int_{C(\mathbf{r})} I(s') \left[k^2 \hat{s} \cdot \hat{s}' - \frac{\partial}{\partial s} \frac{\partial}{\partial s'} \right] \frac{e^{-jkR}}{R} ds', \qquad s \in C(\mathbf{r})$$

Magnetic Vector Potential Integral Equation for Arbitrarily Curved Wires

$$\int_{C(\mathbf{r})} I(s') \left\{ \frac{e^{-jkR}}{R} \hat{s} \cdot \hat{s}' + \frac{1}{2} \int_{C(\mathbf{r})} d\xi \, \theta(\xi - s) \left[\frac{\partial}{\partial s'} \frac{e^{-jkR}}{R} (\hat{\xi} \cdot \hat{s}') + \frac{\partial}{\partial \xi} \left[(\hat{\xi} \cdot \hat{s}') \frac{e^{-jkR}}{R} \right] \right] e^{-jk|s-\xi|} \right\} ds$$

$$= Ae^{-jks} + Be^{jks} + \frac{1}{2\sqrt{\mu/\epsilon}} \int_{C(\mathbf{r})} \hat{s}' \cdot \mathbf{E}^i(s') e^{-jk|s-s'|} ds', \qquad s \in C(\mathbf{r}), \quad \begin{array}{l} \theta(u) = 1, \quad u \geq 0 \\ \theta(u) = 0, \quad u < 0 \end{array}$$

Magnetic Vector Potential Integral Equation for Straight Wires
(Hallen's Integral Equation)

$$\int_{C(\mathbf{r})} I(s') \frac{e^{-jkR}}{R} ds' = Ae^{-jks} + Be^{jks} + \frac{1}{2\sqrt{\mu/\epsilon}} \int_{C(\mathbf{r})} \hat{s}' \cdot \mathbf{E}^i(s') e^{-jk|s-s'|} ds', \qquad s \in C(\mathbf{r})$$

straight wire by Hallen [30] and generalized to curved wires by Mei [31]. Also, "Pocklington's integral equation" has found widespread usage [7, 13, 23].

Integral Equations for Solid Bodies

The integral equations in Chart 14 in terms of electric field have been reduced to forms appropriate for thin-wire analysis as shown in Chart 15. Of course, the equations in Chart 14 can be used for the analysis of the radiating characteristics of nonwire bodies. To this end the magnetic-field integral equation has been developed and discussed in detail [7, 28] and, with the various forms of the electric-field integral equation, leads to useful representations for describing the interaction of electromagnetic waves with conducting bodies.

The two integral equations for use with perfect electric conductors are given by

$$\hat{\mathbf{n}}(\mathbf{r}) \times \mathbf{E}^i(\mathbf{r}) = \frac{1}{4\pi}\hat{\mathbf{n}} \times \int_S \left[j\omega\mu \mathbf{J}_s(\mathbf{r}')\varphi(\mathbf{r},\mathbf{r}') + \frac{\nabla'\cdot\mathbf{J}(\mathbf{r}')}{j\omega\epsilon}\nabla'\varphi \right] d^2r' \qquad (66)$$

and

$$\mathbf{J}_s(\mathbf{r}) = 2\hat{\mathbf{n}}(\mathbf{r}) \times \mathbf{H}^i(\mathbf{r}) + \frac{1}{2\pi}\hat{\mathbf{n}} \times \int_S \mathbf{J}_s(\mathbf{r}) \times \nabla\varphi d^2r' \qquad (67)$$

where $\int$ represents the principal value integral. These equations are referred to as the *electric-field integral equation* and *magnetic-field integral equation*, respectively.

In a shorthand mathematical notation the electric-field integral equation is written as

$$\hat{\mathbf{n}} \times \mathbf{E}^i(\mathbf{r}) = -\hat{\mathbf{n}} \times \left\{ \frac{1}{4\pi j\omega\epsilon}\int_S \mathbf{J}_s(\mathbf{r})\cdot\bar{\bar{\mathbf{G}}}(\mathbf{r},\mathbf{r}')d^2r' \right\} \qquad (68)$$

and the magnetic-field integral equation as

$$\mathbf{J}_s(\mathbf{r}) = 2\hat{\mathbf{n}} \times \mathbf{H}^i(\mathbf{r}) + \frac{1}{2\pi}\hat{\mathbf{n}} \times \int \mathbf{J}_s(\mathbf{r})\cdot\bar{\bar{\Gamma}}(\mathbf{r},\mathbf{r}')d^2r' \qquad (69)$$

where

$$\bar{\bar{\mathbf{G}}}(\mathbf{r},\mathbf{r}') = (\nabla\nabla + k^2\bar{\bar{\mathbf{I}}})\varphi$$

$$\bar{\bar{\Gamma}}(\mathbf{r},\mathbf{r}') = \bar{\bar{\mathbf{I}}} \times \nabla'\varphi(\mathbf{r},\mathbf{r}')$$

$$\varphi(\mathbf{r},\mathbf{r}') = \frac{e^{-jk|\mathbf{r}-\mathbf{r}'|}}{|\mathbf{r}-\mathbf{r}'|}$$

and where $\bar{\bar{\mathbf{G}}}(\mathbf{r},\mathbf{r}')$ and $\bar{\bar{\Gamma}}(\mathbf{r},\mathbf{r}')$ are referred to as Green's dyads for electric and magnetic fields due to electric current sources.

The electric-field integral equation, as seen previously, is widely used for wire antenna analysis in part because of the driving-source definition in terms of electric field. On the other hand, the magnetic-field integral equation has been used extensively for the analysis of nonwirelike structures such as closed surfaces. In part this is promoted by the fact that it is a Fredholm integral equation of the second kind with the unknown outside and within the integral. For flat surfaces of infinite extent it has a trivial solution:

$$\mathbf{J}_s(\mathbf{r}) = 2\hat{\mathbf{n}} \times \mathbf{H}^i(\mathbf{r})$$

so that one can infer that the integral is a correction term accounting for body size limitations or curvature. Examples of the application of the magnetic-field integral equation are provided in [7] while the inclusion of this equation in a widely used computer program is well documented by Burke and Poggio [23]. The electric-field integral equation, a Fredholm integral equation of the first kind, has also been applied to arbitrary bodies [45].

The Imperfectly Conducting Ground

The integral equations presented above have dealt with surface integrals over the surface of the antenna radiating structure. Of course, the presence of a conducting ground plane can, in principle, be treated in a manner similar to any conducting body in the system of interest (i.e., by solving for the sources induced on that plane). However, a simplification can be introduced into the solution of the problem by the use of appropriately modified kernels in the integral equation which take into account the effect of the ground plane. For a perfectly conducting ground plane the principle of images as illustrated previously for simple antenna models can be used to construct a rigorous modified kernel. For a ground plane of arbitrary electrical parameters, a rigorous treatment requires a more involved approach, although an approximate treatment can be achieved by a modified image theory via reflection coefficients (the reflection coefficient approximation). In the following we outline the various approaches, progressing from the implementation of image theory for a perfect ground to the reflection coefficient approximation and finally to the Sommerfeld integral approach.

Perfect Ground—Implementation of images for perfectly conducting grounds in the integral equation approach is straightforward [23]. The Green's function for a perfectly conducting ground is the sum of the free-space Green's function of the source current element and the negative of the free-space Green's function of the image of the source reflected in the ground plane. For the electric field the Green's dyad for a perfect ground in the $z = 0$ plane is

$$\bar{\bar{\mathbf{G}}}_{pg}(\mathbf{r}, \mathbf{r}') = \bar{\bar{\mathbf{G}}}(\mathbf{r}, \mathbf{r}') + \bar{\bar{\mathbf{G}}}_I(\mathbf{r}, \mathbf{r}') \tag{70}$$

where

$$\bar{\bar{\mathbf{G}}}_I(\mathbf{r}, \mathbf{r}') = -\bar{\bar{\mathbf{I}}}_R \cdot \bar{\bar{\mathbf{G}}}(\mathbf{r}, \bar{\bar{\mathbf{I}}}_R \cdot \mathbf{r}')$$

$$\bar{\bar{\mathbf{I}}}_R = \hat{\mathbf{x}}\hat{\mathbf{x}} + \hat{\mathbf{y}}\hat{\mathbf{y}} - \hat{\mathbf{z}}\hat{\mathbf{z}}$$

and $\overline{\overline{\mathbf{I}}}_R$ is a dyad that produces a reflection in the $z = 0$ plane when used in a dot product. For the magnetic field with free-space Green's dyad $\overline{\overline{\Gamma}}(\mathbf{r}, \mathbf{r}')$ given in (69), the Green's dyad over a perfect ground is

$$\overline{\overline{\Gamma}}_{pg} = \overline{\overline{\Gamma}}(\mathbf{r}, \mathbf{r}') + \overline{\overline{\Gamma}}_I(\mathbf{r}, \mathbf{r}') \tag{71}$$

where

$$\overline{\overline{\Gamma}}_I(\mathbf{r}, \mathbf{r}') = -\overline{\overline{\mathbf{I}}}_R \cdot \overline{\overline{\Gamma}}(\mathbf{r}, \overline{\overline{\mathbf{I}}}_R \cdot \mathbf{r}') \tag{72}$$

The introduction of these dyads into (68) and (69) yields the integral equations for perfectly conducting bodies over perfectly conducting ground planes.

Imperfect Ground: Modified Image Theory or Reflection Coefficient Approximation
—The Green's dyads for electric and magnetic fields over an imperfectly conducting ground resulting from the reflection coefficient approximation are, in keeping with the development in a previous section [23],

$$\overline{\overline{\mathbf{G}}}_g(\mathbf{r}, \mathbf{r}') = \overline{\overline{\mathbf{G}}}(\mathbf{r}, \mathbf{r}') + R_v \overline{\overline{\mathbf{G}}}_I(\mathbf{r}, \mathbf{r}') + (R_h - R_r)[\overline{\overline{\mathbf{G}}}_I(\mathbf{r}, \mathbf{r}') \cdot \hat{\mathbf{p}}]\hat{\mathbf{p}} \tag{73}$$

and

$$\overline{\overline{\Gamma}}_g(\mathbf{r}, \mathbf{r}') = \overline{\overline{\Gamma}}(\mathbf{r}, \mathbf{r}') + R_h \overline{\overline{\Gamma}}_I(\mathbf{r}, \mathbf{r}') + (R_v - R_h)[\overline{\overline{\Gamma}}_I(\mathbf{r}, \mathbf{r}') \cdot \hat{\mathbf{p}}]\hat{\mathbf{p}} \tag{74}$$

where

$$\hat{\mathbf{p}} = \frac{(\mathbf{r} - \mathbf{r}') \times \hat{\mathbf{z}}}{|(\mathbf{r} - \mathbf{r}') \times \hat{\mathbf{z}}|}$$

The reflection coefficient approximation for finitely conducting grounds uses image fields modified by Fresnel plane-wave reflection coefficients as described earlier for simple antenna models. These reflection coefficients are strictly correct only for an infinite plane-wave field but have been used in the integral equation approach and have been shown to provide useful results for structures that are not too near the ground [23, 24, 25].

In the integral equation formulation a ground plane changes the solution in three ways: (1) by modifying the current distribution through the fields reflected from the ground; (2) by changing the field illuminating the structure; and (3) by changing the reradiated field. Effects 2 and 3 are easily analyzed by plane-wave reflection as a direct ray and a ray reflected from the ground. The reradiated field is not a plane wave when it reflects from the ground, but, as can be seen from reciprocity, plane-wave reflection gives the correct far-zone field. Analysis of the near-field interaction effect is, however, much more difficult in general. In the following we consider the rigorous treatment based on the work of Sommerfeld [32] and Banos [33].

Imperfect Ground: Rigorous Sommerfeld Treatment—The integral equation for an arbitrarily oriented straight wire antenna over an imperfectly conducting ground of complex relative dielectric constant ϵ_E has been derived in a rigorous manner in Miller et al. [24, 25]. For a straight wire of radius a and length L, the integral equation for the current excited by a field $\mathbf{E}^A$ can be written as

$$\frac{j\omega\mu}{4\pi}\int_L I(s')ds'\left[\left(1 + \frac{1}{k^2}\frac{\partial}{\partial s^2}\right)g_0 + \left(\cos 2\beta' + \frac{1}{k^2}\frac{\partial}{\partial s\partial s^*}\right)g_i\right.$$

$$+ \left(\cos\beta' + \frac{1}{k^2}\frac{\partial^2}{\partial s\partial z}\right)\left(g_{sHz}\sin\beta' - g_{sVz}\cos\beta'\right) \tag{75}$$

$$\left. + \sin\beta'\left(\sin\beta + \frac{1}{k^2}\frac{\partial^2}{\partial s\partial s'}\right)g_{sHs}\right] = -E^A(s), \qquad s \in L$$

where

$$\varrho' = [(x - x')^2 + (y - y')^2 + a^2]^{1/2}$$

$$g_{sHs} = 2\int_0^\infty \frac{\lambda}{\mu + \mu_E}J_0(\lambda\varrho')e^{-\mu(z + z')}\,d\lambda$$

$$g_{sHz} = \frac{\cos(\phi' - \alpha')}{k^2}\int_0^\infty \frac{\mu - \mu_E}{\epsilon_E\mu + \mu_E}J'_0(\lambda\varrho')\,e^{-\mu(z + z')}\,\lambda^2 d\lambda$$

$$g_{sVz} = 2\int_0^\infty \frac{\mu_E}{\epsilon_E\mu + \mu_0}J_0(\lambda\varrho')e^{-\mu(z + z')}\frac{\lambda}{\mu}\,d\lambda$$

$$g_0 = \frac{e^{-jkR}}{R}, \qquad R = [(x - x')^2 + (y - y')^2 + (z - z')^2 + a^2]^{1/2}$$

$$g_i = \frac{e^{-jkR^*}}{R^*}, \qquad R^* = [(x - x')^2 + (y - y')^2 + (z + z')^2]^{1/2}$$

$$\phi' = \tan^{-1}[(y - y')/(x - x')]$$

$$\mu = (\lambda^2 - k^2)^{1/2}, \qquad k = \omega(\mu_o\epsilon_0)^{1/2}$$

$$\mu_E = (\lambda^2 - k_E^2)^{1/2}, \qquad k_E = (\epsilon_E)^{1/2}k$$

In the preceding, α and β are the direction angles as shown in Fig. 13 with a prime used when referring to source coordinates, s and s' are observation and source coordinates of the wire, $\hat{s}^*$ is a unit vector in the direction of the image current with s^* the corresponding coordinate for the image current. The zeroth order Bessel function and its derivative with respect to argument are denoted by J_0 and J'_0. The description of the origin of the terms g_{sHs}, g_{sHz}, and g_{sVz} is beyond the scope of this chapter but is provided in the literature [24, 25].

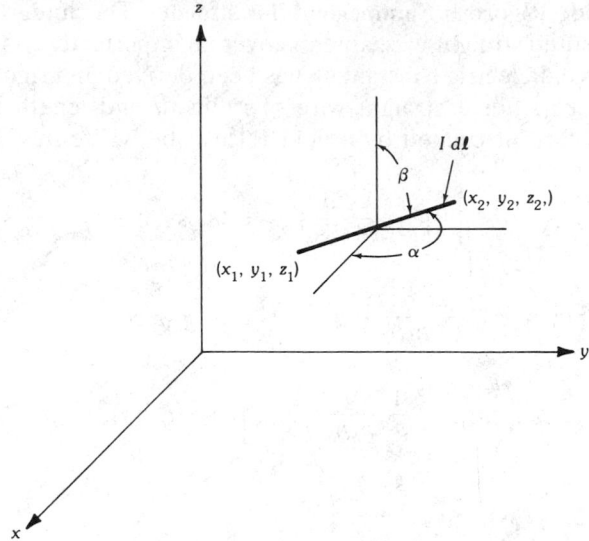

Fig. 13. Geometry for a current element above an imperfectly conducting ground.

For a wire antenna of arbitrary geometry and orientation, denote the wire contour $C(\mathbf{r})$, and $\alpha(\mathbf{r})$, $\beta(\mathbf{r})$, and $\hat{s}(\mathbf{r})$ the direction angles with respect to the x and z axes and the tangent vector to the wire, respectively. Further, letting $\alpha = \alpha(\mathbf{r})$, $\alpha' = \alpha(r')$, etc., we obtain the rigorous integral equation for the antenna current $I(s')$ as a function of position s' along the wire as

$$
\frac{j\omega\mu}{4\pi} \int_{C(\mathbf{r})} I(s')ds' \left[\left(\hat{s}\cdot\hat{s} + \frac{1}{k^2}\frac{\partial^2}{\partial s \partial s'} \right) g_0 + \left(\hat{s}\cdot\hat{s}^* + \frac{1}{k^2}\frac{\partial^2}{\partial s \partial s^*} \right) g_i \right.
$$

$$
+ \left(\cos\beta + \frac{1}{k^2}\frac{\partial^2}{\partial s \partial z} \right) \left(g_{sHz}\sin\beta' - g_{sVz}\cos\beta' \right) \tag{76}
$$

$$
\left. + \sin\beta' \left(\sin\beta \cos(\alpha - \alpha') + \frac{1}{k^2}\frac{\partial^2}{\partial s \partial s'} \right) g_{sHs} \right] = -E^i(s), \qquad s \in C(\mathbf{r})
$$

where $\int_{C(\mathbf{r})} ds'$ implies integration along the wire length over the contour $C(\mathbf{r})$.

The preceding discussion concerning integral equations for antennas over lossy grounds is condensed and necessarily sketchy since the related work has been widespread. The indicated references will, however, serve as convenient starting points which will ultimately lead the reader to more involved and detailed developments.

Integral Equations in the Time Domain

In the previous discussions a frequency-domain approach has been emphasized. All analyses were performed for time-harmonic fields so that the analyses were strictly valid only for the frequency for which they were performed.

It is sometimes desirable to consider the performance of systems such as antennas under transient conditions, i.e., when excited by either a transient local source or when receiving a nonsinusoidal electromagnetic wave. Analyses of the transient problem can proceed along two different avenues. The first is by performing the time-harmonic analyses over a broad spectrum of frequencies so that the spectral characterization can be used to construct the temporal characterization as one would when using the Fourier or Laplace transform to convert frequency-domain information into the time domain. The other is to perform the analyses directly in the time domain. While a detailed discussion of the issues associated with a direct solution in the time domain is beyond the scope of this text, a brief discussion is appropriate.

The theoretical development of time-domain integral equations is covered in Poggio and Miller [7], Mittra [34], and Sengupta and Tai [34]. Further, the application of the integral equation approach to radiation and scattering is presented in the first two references above, whereas Sengupta and Tai delve more deeply and with broader scope into the transient radiation and reception properties of linear antennas.

The time-domain equivalents to the integral equations presented earlier for electromagnetic wave interactions with perfect electric conductors are [7]

Electric-Field Integral Equation

$$
\hat{\mathbf{n}} \times \mathbf{E}^i(\mathbf{r}, t) = \frac{1}{4\pi} \hat{\mathbf{n}} \times \oint_S \left\{ \mu \frac{\partial}{\partial \tau} \mathbf{J}_s(\mathbf{r}', \tau) \frac{1}{|\mathbf{r} - \mathbf{r}'|} \right. \\
\left. - \left[\frac{1}{|\mathbf{r} - \mathbf{r}'|} + \frac{1}{c} \frac{\partial}{\partial \tau} \right] \frac{\varrho_s}{\epsilon} \frac{\mathbf{r} - \mathbf{r}'}{|\mathbf{r} - \mathbf{r}'|^2} \right\} d^2 r'
$$

(77)

Magnetic-Field Integral Equation

$$
J_s(\mathbf{r}, t) = 2\hat{\mathbf{n}} \times \mathbf{H}^i(\mathbf{r}, t) = \frac{1}{2\pi} \hat{\mathbf{n}} \times \oint_S \left[\frac{1}{|\mathbf{r} - \mathbf{r}'|} \right. \\
\left. + \frac{1}{c} \frac{\partial}{\partial \tau} \right] \mathbf{J}_s(\mathbf{r}', \tau) \times \frac{(\mathbf{r} - \mathbf{r}')}{|\mathbf{r} - \mathbf{r}'|^2} d^2 r'
$$

(78)

where $\mathbf{r} \in S$ and $\tau = t - |\mathbf{r} - \mathbf{r}'|/c$.

Integral equations based on the magnetic vector potential for curved and straight wires are described in [7]. These later equations have not found extensive application as have (77) and (78).

The specialization of the electric-field integral equation to thin wires, which has found extensive applications, is detailed in [36] and [37]. In these works the thin-wire time-domain integral equation is derived for a geometry as described for the frequency domain earlier and is given as

$$\hat{\mathbf{s}}\cdot\mathbf{E}^i(\mathbf{r},t) = \frac{\mu_0}{4\pi} \int_{c(\mathbf{r})} \left[\frac{\hat{\mathbf{s}}\cdot\hat{\mathbf{s}}'}{R} \frac{\partial}{\partial\tau} I(s',\tau) \right.$$

$$\left. + c\frac{\hat{\mathbf{s}}\cdot\mathbf{R}}{R^2} \frac{\partial}{\partial s'} I(s',\tau) - c^2\frac{\hat{\mathbf{s}}\cdot\mathbf{R}}{R^3} q(s',\tau) \right] ds' \tag{79}$$

with $\mathbf{r} \in C(\mathbf{r})$, $\mathbf{R} = \mathbf{r} - \mathbf{r}'$, $R = |\mathbf{R}|$, and $\tau = t - R/c$.

As in the frequency domain, the thin-wire approximation is used so that the distance R is always greater than zero with the integration path $C(\mathbf{r})$ along the wire contour displaced from the observation point path by the wire radius.

The numerical solution of (77), (78), and (79) will be discussed in a later section. To date, time-domain analysis for antenna systems has not received the attention allocated to frequency-domain analysis. But, increasing attention is being given because of the expanding interest in pulsed applications. Fast pulse radar and inverse scattering applications have led to interest in and numerical solution of (78), whereas pulse signal transmission and reception lead to interest in (79).

Part 2

Numerical Issues Involved in Integral Equations for Antenna Analysis

Now we turn to the numerical treatment of the integral equations that we have developed in Part 1.

1. Introduction

The advent of the digital computer has opened up new vistas in antenna analysis. In this section we present a summary of computational procedures for solving antenna problems of a more general nature than can be handled by some of the analytically based approaches described in the previous part.

We must caution the reader that in cataloging the various specific computer codes that might be considered, there can exist great differences among them with respect to documentation, validation, availability, support, and the like. A qualitative assessment of these aspects of such computer tools is given in the summary tables included at the end of this section. Before considering such specifics, however, we make a brief tour through the analytical and numerical issues which precede the actual development and use of a computer model.

2. Preliminary Discussion

The process of developing a numerical procedure for solving electromagnetic-field problems may be conveniently divided into the steps outlined in Fig. 14 and discussed below. Since approximations are an intrinsic part of all the steps, we summarize some of those most widely used in Chart 16.

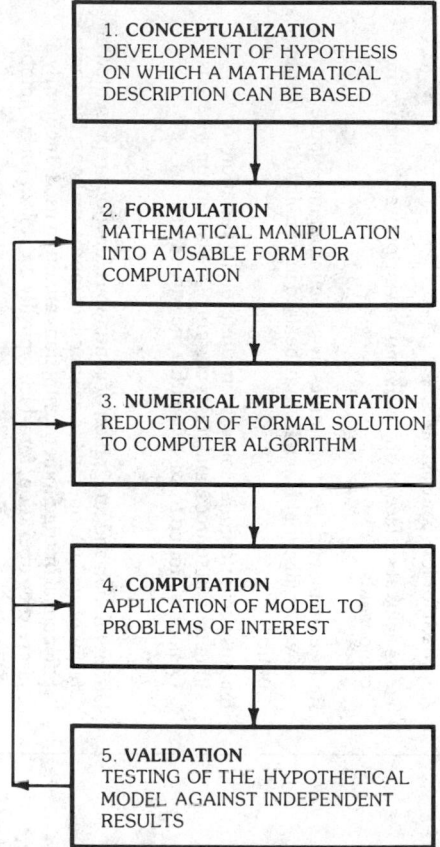

Fig. 14. The basic steps involved in developing a computer model of an electromagnetic field.

Step 1. Conceptualization

It is at this step where physical principles, experimental observations, and so on, are used to form hypotheses from which mathematical descriptions of the relevant phenomena can be devised.

Step 2. Formulation

This involves the evolution of the physical idea or mathematical description from its elementary form into one suitable for numerical or analytical evaluation. Various approximations may be utilized to make the subsequent analysis and/or computation easier.

Step 3. Numerical Implementation

At this stage the formulation is reduced to a form suitable for computation, leading to a computer code or algorithm. For the problems considered here, this

Chart 16. Examples of Approximations Used in the Various Steps of Developing a Computer Solution

Step	Approximation	Limitations/Impact
Conceptualization	Physical optics	Best for backscatter and main lobe region of reflector antennas, from resonance region ($ka > 1$, where a is the object radius) and up in frequency. Fields propagated via Green's function with amplitude established from tangential **H** on surface
	Geometrical theory of diffraction	Generally applicable for $ka > 2$ to 5, where a is the object radius. Fields propagated via a divergence factor with amplitude obtained from diffraction coefficient. Can involve complicated ray tracing
	Geometrical optics	Ray tracing without diffraction. Improves with increasing frequency
	Compensation theorem	Solution obtained in terms of perturbation from a reference, known solution
	Born-Rytov	Approach for low-contrast objects
	Rayleigh	Treats fields at surface of object as only having outward propagating components
Formulation	Surface impedance	Reduces number of field quantities by assuming an impedance relation between tangential **E** and **H**. May be used in combination with physical optics
	Thin-wire	Ignores circumferential sources, circumferential variations of longitudinal sources, and treats current as a filament on wire axis. Generally limited to $ka < 1$, with a the wire radius
Numerical implementation	$\dfrac{\partial f}{\partial x} \to \dfrac{\Delta f}{\Delta x}$ $\int f dx \to \sum f_i \Delta x_i$	Differentiation and integration of continuous functions represented in terms of analytic operations on sampled approximations, for which polynomial or trigonometric functions are often used. Inherently a discretizing operation, for which typically $\Delta x < \lambda/2\pi$ for acceptable accuracy
	$I(s') \to \sum\limits_{n=1}^{N} a_n f_n(s')$	Representation of unknown is also approximated in terms of basis functions, the number and form of which affect solution accuracy and efficiency
	$L(s,s')f(s') = g(s) \to$ $\sum\limits_{n=1}^{N} a_n \langle w_m(s), L(s,s')f_n(s') \rangle$ $= \langle w_m(s), g(s) \rangle$	Inner product operation with weight functions determines the manner in which the original equation is satisfied by the numerical model
Computation	Deviation of numerical model from physical reality	The greatest source of uncertainty for most problems. Model details affect solution in ways that are difficult to quantify
	Nonconverged solution	Discretized solutions usually converge using a global measure in proportion to $\exp(-AN)$ with A determined by the problem. At least two solutions using different values of N are needed to estimate A

step almost always involves developing a linear system of equations which are solved using matrix techniques.

Step 4. Computation/Application

In the computation/application the limitations and "bugs" are uncorked and accuracy measures are established. Computation also involves approximation, but in a more ambiguous way than the previous steps, because a model is employed by the user to represent the reality of interest, and there is rarely a simple or obvious way to model most real problems. Furthermore, the numerical model itself may not be solved exactly (see Section 4, under "Modeling Errors," below).

Step 5. Validation

This step is probably the most crucial as it establishes the degree to which the code can eventually be relied upon. It is an open-ended process since a code is usually applied to an expanding variety of problems. Several kinds of validation can be used, which include internal checks for self-consistency, and external checks, which can include independent analytical and numerical results and experimental measurement.

It can be seen that code developers are most intensively concerned with steps 1–3, while code users are more involved with steps 4 and 5.

3. Numerical Implementation

Maxwell's equations can be written in either integral or differential form and either as a function of frequency or time. Each of these can be solved numerically using the method of moments (MOM). This involves approximating the unknowns in a set of basis or expansion functions and satisfying the governing equations using a set of testing or weight functions. Solution of the problem is thus reduced to finding the coefficients of the expansion from a set of linear equations. For various reasons as discussed below, integral equations are more widely useful for antenna modeling, and after some preliminary discussion subsequent detailed attention will be limited to them.

The General Idea

A differential equation (or a set, e.g., Maxwell's equations) relates field and source quantities in a local, pointwise sense (see Chart 17). Field propagation through a medium is represented as a continuum of local interactions which are approximated in a discretized sequence (grid) by the numerical model. Boundary conditions are enforced by specifying the values of the relevant quantities at the appropriate places in the solution continuum in the analytical case or in a solution grid in the numerical case. Analytical solutions are obtainable for only a few separable geometries. Since numerical values must be found throughout the solution volume, the number of unknowns $N \propto A(D/\Delta D)^d$, where D is a characteristic dimension of the solution volume, ΔD is the spatial resolution that is sought, d is the spatial dimensionality (1, 2, or 3), and A is a problem-specific constant determined by the number of field quantities being found at each grid point.

Chart 17. Demonstration of the Difference between an Integral and Differential Formulation of an Electromagnetic Field in the Frequency Domain

Infinite Cylinder in z Direction (Normal to Paper)

(Geometry is two-dimensional for simplicity with TM polarization (E_z, H_x, and H_y components)

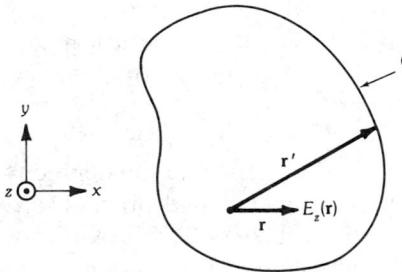

$C(\mathbf{r})$, boundary between electrically dissimilar media. For simplicity the outer medium is assumed to be perfectly conducting

Integral Description

$$E_z(\mathbf{r}) = \int_{C(\mathbf{r})} I_z(\mathbf{r}')K(\mathbf{r},\mathbf{r}')dr'$$

$$\cong \sum_{i=1}^{I} I_z(\mathbf{r}_i)K(\mathbf{r},\mathbf{r}_i)\Delta r_i$$

Field at $\mathbf{r}$ determined from knowledge of equivalent sources on boundary [at most ($\hat{\mathbf{n}} \times \mathbf{E}$, $\hat{\mathbf{n}} \times \mathbf{H}$)], in this case $I_z\hat{\mathbf{z}} = \hat{\mathbf{n}} \times \mathbf{H}$

Differential Description

$$E_z(\mathbf{r}) = -\frac{j}{\omega}\left[\frac{\partial}{\partial x}H_y(\mathbf{r}) - \frac{\partial}{\partial y}H_x(\mathbf{r})\right]$$

$$\cong -\frac{j}{\omega}\left\{\left[\frac{H_y(\mathbf{r} + \hat{\mathbf{x}}\Delta x/2) - H_y(\mathbf{r} - \hat{\mathbf{x}}\Delta x/2)}{\Delta x}\right]\right.$$

$$\left. -\left[\frac{H_x(\mathbf{r} + \hat{\mathbf{y}}\Delta y/2) - H_x(\mathbf{r} - \hat{\mathbf{y}}\Delta y/2)}{\Delta y}\right]\right\}$$

E_z field at $\mathbf{r}$ requires displaced $\mathbf{H}$ fields at $\mathbf{r} \pm \hat{\mathbf{x}}\Delta x/2$ and $\mathbf{r} \pm \hat{\mathbf{y}}\Delta y/2$, which requires in turn the displaced $\mathbf{E}$ fields

An integral equation, on the other hand, relates field and source quantities in a global sense (see Chart 17). Field propagation through a medium is described by a source-field relationship, which is commonly called a Green's function. Whereas analytical solutions can be found for a few separable boundaries (e.g., planes, cylinders, spheres, etc.) using special Green's functions, the integral representations and equations presented previously use an infinite, homogeneous-medium Green's function for the most part (an exception is the interface problem discussed previously), whatever the geometry. The evaluation of fields is then achieved by integrating the contribution of all sources within the volume of interest (known) and the contributions of all equivalent sources over the boundary (see Charts 1, 2, and 17). While the latter contributions are difficult to specify due to the unknown equivalent sources, the coincidence of this boundary with a surface whose boundary conditions are specified permits an integral equation to be written.

Obviously, the fields in space are then evaluated simply by summing the contributions from bounding surfaces. Since numerical values are needed only on these surfaces, the number of unknowns $N \propto A(D/\Delta D)^{d-1}$. (This relationship applies to objects having extended surfaces. Wire objects represent a special case with $N \propto (D/\Delta D)$, with D the total wire length and ΔD the desired resolution.)

An integral equation model for a given problem thus has a factor of approximately $\Delta D/D$ fewer unknowns than its differential equation counterpart. This can represent a significant difference in computer storage and time requirements, especially for wire objects. On the other hand, a differential-equation approach can be applied to anisotropic, inhomogeneous, nonlinear, and time-varying media, problems for which a Green's function is not easily derivable, if at all. In spite of these advantages, differential equation–based models have been less widely used, primarily because most problems of practical interest do not involve such media. Except for occasional comments the following presentation is therefore limited to an integral-equation formulation. A brief summary of differential and integral equation characteristics is given in Table 1, where N is the number of unknown field values (samples), $d = 1$, 2, or 3 is the problem dimensionality, D and ΔD the problem spatial size and resolution, and N_t the number of steps in a time-domain solution.

It was brought out earlier that various kinds of integral equations can be formulated even for the same problem. When the problems vary significantly, ranging from perfectly conducting to penetrable dielectric bodies for example, the variety can become even greater [7]. The integral equations for all such problems share certain characteristics, however.

First, they involve integrals over prescribed regions of known kernel (or Green's functions) functions operating on unknown field quantities. Second, they include the known source field outside the integral, and sometimes the unknown as well. If the latter is the case, they are called Fredholm integral equations of the second kind; otherwise, they are first-kind integral equations. Third, the required boundary conditions are satisfied implicitly by virtue of their enforcement during the construction of the integral equation. Finally, the unknown often, but not always, is subject to differential operators as a result of the kernel function.

Frequency-Domain Method of Moments

The method of moments (MOM) is an intuitively logical approach for solving operator equations numerically (see Table 2) [7, 38, 39]. Following commonly used notation, let a generic form of the integral equation of interest be written as

$$L(s, s')f(s') = g(s) \tag{80}$$

where

$L =$ the integral operator (in the frequency domain)

$f =$ the unknown (response, e.g., a current) at source coordinate s'

$g =$ the known (source or forcing function, e.g., a tangential electric field) at observation coordinate s

Table 1. Comparison of Some of the Characteristics of Differential- and Integral-Based Computer Models

Characteristic	Differential Form	Integral Form
Field propagator	Defining differential equation	Green's function, e.g., $e^{j\mathbf{k}\cdot\mathbf{R}}/4\pi R$
Boundary treatment: Radiation condition	Local "lookback" (approximate) via outward propagating Green's function Global "lookback" (rigorous) via match to model expansion or integral equation solution	Green's function
Boundary condition	Appropriate field values specified on grid boundaries to obtain staircase or piecewise linear approximation to boundary	Appropriate field values specified on object contour which can in principle be a general curvilinear surface, although this possibility seems to be seldom used
Medium properties that can be handled: Linear	Yes	Yes
Anisotropic	Yes	Yes
Inhomogeneous	Yes	Feasible at expense of increasing dimensionality of unknown (e.g., from surface to volume)
Nonlinear	Yes	Generally inapplicable
Time-varying	Yes	Generally inapplicable
Combination of above	Yes	Generally inapplicable (Except linear anisotropic)
Solution characteristics: Number of unknowns	$N \propto (D/\Delta D)^d$	$N \propto (D/\Delta D)^{d-1}$
Linear system	Sparse, but larger	Dense, but smaller
Approximate solution time: Time domain	$\propto NN_t \propto (D/\Delta D)^{d+1}$	$\propto N^2 N_t \propto (D/\Delta D)^{p(d-1)}$
Frequency domain	$\propto N^2 \propto (D/\Delta D)^{2d}$	$\propto N^p, \quad 2 \leqq p \leqq 3$

Note that the equations in Charts 14 and 15 can, with suitable interpretations, be written in this form.

Now expand f using a set of basis functions $\{f_n(s')\}$, $n = 1, \ldots, N$:

$$f(s') = \sum_{n=1}^{N} a_n f_n(s') \tag{81}$$

Table 2. Basic Steps in Developing a Moment-Method Solution (*After Harrington [38] and Miller [39]*)

Operation	Differential Equation	Integral Equation
Sampling of unknowns via basis-function expansion	Subdomain basis functions, usually of low order, are used for differential equations. When the basis functions are pulses, the differential-equation approach is referred to as a finite-difference procedure. When the basis functions are linear, the approach is referred to as a finite-element procedure	Integral equations can use either entire-domain or subdomain basis functions. Use of the former is generally confined to bodies of rotation for sampling orthogonal to the axis of rotation. Subdomain sampling usually is of low order, with piecewise linear or sinusoidal being the maximum variation employed
Matching of equations via integration with weight functions	Pointwise matching is commonly employed, using a delta-function weight. Pulse and linear matching can also be used	Pointwise matching is commonly employed, using a delta-function weight. For wires, pulse, linear, and sinusoidal weight functions are frequently used as well
Numerical approximation of derivatives and integrals. Determined in part at least by the basis and/or weight functions	$\dfrac{d}{ds}f(s) \rightarrow [f(s_i + \Delta s_i/2)$ $- f(s_i - \Delta s_i/2)]/\Delta s_i$	$\displaystyle\int f(s)ds \rightarrow \sum f(s_i)\Delta s_i$
Solving for the unknown coefficients using matrix manipulation	$\mathbf{ZA = B}$	$\mathbf{ZA = B}$

which may be either a complete domain (each f_n is defined over the entire object or domain of interest as, for example, in a Fourier series) or subdomain expansion (each f_n defined over only a part of the object). From these equations there follows

$$L(s,s') \sum_{n=1}^{N} a_n f_n(s') = \sum_{n=1}^{N} a_n L(s,s') f_n(s') = g(s) \tag{82}$$

On then forming an inner product* of (82) with a set of weight functions

$$\{w_m(s)\}, \qquad m = 1, \ldots, M$$

there results

---

*An inner product of two functions $p(\mathbf{r}), q(\mathbf{r})$ over a surface S is defined as $\langle p(\mathbf{r}), q(\mathbf{r}) \rangle = \int_S p(\mathbf{r})q(\mathbf{r}) d^2r$.

$$\sum_{n=1}^{N} a_n \langle w_m(s), L(s,s') f_n(s') \rangle = \langle w_m(s), g(s) \rangle, \qquad m = 1, \ldots, M \qquad (83)$$

As for the basis function expansion, each weight function w_m can have either a complete domain or subdomain support. Some of the more commonly used basis-weight combinations are given in Table 3. The approaches most often used for wire problems involve a subdomain procedure of the form of subsectional collocation or Galerkin's method.

The form of (83) allows the representation

$$\sum_{n=1}^{N} Z_{mn} a_n = b_m, \qquad m = 1, \ldots, M \qquad (84)$$

with

$$Z_{mn} = \langle w_m(s), L(s,s') f_n(s') \rangle \quad \text{and} \quad b_m = \langle w_m(s), g(s) \rangle$$

Note that this equation represents a set of M simultaneous equations in N unknowns and can be written in matrix equation form as

$$\mathbf{ZA} = \mathbf{B} \qquad (85)$$

where

$\mathbf{Z} = [Z_{mn}]$, an $M \times N$ matrix

$\mathbf{B} = [b_m]$, an $M \times 1$ vector

$\mathbf{A} = [a_n]$, an $N \times 1$ vector

Table 3. Representative Pairs of General Basis and Weight Functions Commonly Used in the Moment Method (*After Poggio and Miller [7], © 1973 Pergamon Press*)

Method	nth Term of Basis Function	mth Term of Weight Function
Galerkin	$a_n f_n(x)$	$f_m(x)$
Least square	$a_n f_n(x)$	$Q(x) \dfrac{\delta \epsilon(x)}{\delta a_m}$
General collocation	$a_n f_n(x)$	$\delta(x - x_m)$
Point matching	$a_n \delta(x - x_n)$	$\delta(x - x_m)$
Subsectional collocation*	$U(x_n) \sum\limits_{p=1}^{P} a_{np} f_p(x)$	$\delta(x - x_m)$
Subsectional Galerkin*	$U(x_n) \sum\limits_{p=1}^{P} a_{np} f_p(x)$	$U(x_m) \sum\limits_{p=1}^{P} f_{mp}(x)$

*Here $U(x_n) = 1$ for x in the subsection (segment) Δx_n, and $U(x_n) = 0$ otherwise.

A formal solution of (85) can be written as

$$\mathbf{A} = \mathbf{YB} \tag{86}$$

or

$$a_n = \sum_{m=1}^{M} Y_{nm} b_m, \qquad n = 1, \ldots, N \tag{87}$$

where $\mathbf{Y} = [Y_{nm}] = [Z_{mn}]^{-1}$ for $M = N$. For $M > N$, a solution can be achieved by using the generalized inverse

$$\mathbf{A} = (\mathbf{Z}^T\mathbf{Z})^{-1}\mathbf{Z}^T\mathbf{B} \tag{88}$$

In the above the superscript T implies conjugate transpose.

Since the matrix $\mathbf{Z}$ relates a current to its electric field when this representation is used with the integral equations in Charts 14 and 15, it is called an *impedance matrix*. Similarly, the matrix $\mathbf{Y}$ is known as an *admittance matrix* and possesses some especially useful features. First, it provides a solution for arbitrary excitation. Second, it may be stored for subsequent reuse. Third, and most important, $\mathbf{Y}$ represents a complete electromagnetic characterization of a structure, within the limits of the approximations involved in its derivation. In this sense it has some characteristics that are similar to a hologram. The general expression for the field E_r radiated by the object obtained using reciprocity highlights this property:

$$E_r = \sum_{n=1}^{N} E_n^o \sum_{m=1}^{N} Y_{nm} E_m^i$$

with $E_n^o = \hat{s}_n \cdot \mathbf{E}^o(s_n)$ and $E_m^i = \hat{s}_m \cdot \mathbf{E}^i(s_m)$. Here, E^i is the field of the exciting source, and $\mathbf{E}^o$ is the field produced at the object by a point source located at the observation point. Clearly, it is possible to construct the spatial field produced anywhere in space by the object.

The computer time T_f involved in obtaining a frequency-domain solution of an integral equation at N_f frequencies for a general object (nonsymmetric) in free space can be approximated as*

$$T_f \cong (T_{f_1} + T_{f_2}N)N^2N_f \tag{89a}$$

while the variable storage for a single frequency is approximately

$$S_f \cong (S_{f_1} + 2N)N \tag{89b}$$

where

---

*These approximate formulas include only the dominant terms and ignore symmetry effects.

$$N \gtrsim \begin{cases} 2\pi L/\lambda & \text{for wires of length } L/\lambda \\ 2A(2\pi/\lambda)^2 & \text{for surfaces of area } A/\lambda^2 \end{cases}$$

The T_{f_1} term accounts for computation of the impedance matrix and the T_{f_2} term its solution via factorization or inversion where T_{f_1} and T_{f_2} are algorithm- and computer-dependent coefficients. It can be deduced that the computer time can vary with frequency over the range f^2 (wire object with impedance matrix computation time dominating) to f^6 (surface object with matrix solution time dominating). The constant S_{f_1} in (89b) accounts for geometry and other structure-dependent storage.

Time-Domain Method of Moments

A time-domain counterpart to the generic frequency-domain integral equation (80) can be written as

$$\tilde{L}(s,t; s',t')\tilde{f}(s',t') = \tilde{g}(s,t) \tag{90}$$

with $\tilde{g}(s,t)$ the time- and space-dependent forcing function, $\tilde{f}(s',t')$ the resulting response, and $\tilde{L}(s,t; s't')$ the integral operator which can be defined using (77), (78), and (79). It is important to observe that while $\tilde{L}$ is apparently a function of the space-time coordinates of the observation point (s,t) and source point (s',t'), the spatial integration over the source coordinate eliminates both source-point variables. The reason is that the spatial integration over s' for a given observation point (s,t) in space-time has an additional constraint imposed by $t' = t - R(s,s')/c$, where $R(s,s')$ represents the distance between the spatial source and observation points. Thus, as the integration over s' is performed the dependency on t' also vanishes. In essence, $\tilde{L}$ performs a spatial integration over the space-time cone with due regard paid to the retarded time restriction [7].

Because there are two independent variables, we approximate the unknown in a separable form:

$$\tilde{f}(s',t') = \sum_{n=1}^{N} \sum_{m=1}^{N_t} a_{nm}\tilde{f}_n(s')\tilde{q}_m(t') \tag{91}$$

which, with (90), leads to

$$\sum_{n=1}^{N} \sum_{m=1}^{N_t} a_{nm}\tilde{L}(s,t; s',t')\tilde{f}_n(s')\tilde{q}_m(t') = g(s,t) \tag{92}$$

On employing the weight function expansion

$$\{w_{pq}(s,t)\} = \{w_q(t)v_p(s)\}, \qquad p = 1, \ldots, N, \quad q = 1, \ldots, N_t \tag{93}$$

there follows

$$\sum_{n=1}^{N} \sum_{m=1}^{N_t} a_{nm} \langle w_q(t), \langle v_p(s), \bar{L}(s,t;s',t')\tilde{f}_n(s')\tilde{q}_m(t')\rangle\rangle$$
$$= \langle w_q(t), \langle v_p(s), g(s,t)\rangle\rangle \tag{94}$$

Most numerical solutions to wire problems in the time domain employ subsectional collocation [see a description of TWTD (thin-wire time domain) in a following section]. Then

$$\tilde{f}(s',t') = \sum_{n=1}^{N} \sum_{m=1}^{N_t} a_{nm}p_{nm}(s',t')\tilde{f}_n(s')\tilde{q}_m(t') \tag{95}$$

and

$$w_{pq}(s,t) = \delta(s - s_p)\delta(t - t_q) \tag{96}$$

where

$$p_{nm}(s',t') = \begin{bmatrix} 1 & \text{if } s' \in \Delta s'_n \text{ and } t' \in \Delta t'_m \\ 0 & \text{otherwise} \end{bmatrix}$$

The general form of the current [the most commonly used unknown as in (77), (78), and (79)] can be shown to be

$$I_{nm} = I(s_n, t_m) = Y_{nn}E^t_{nm} = Y_{nn}(E^i_{nm} + E^s_{nm})$$
$$= Y_{nn}(E^i_{nm} + \sum_{n'=1}^{N} A_{nn'}I_{n', m-f(n,n')}) \tag{97}$$

where E^t_{nm} = total E_{nm}, E^s_{nm} = scattered E_{nm}, and $f(n,n') = |n - n'|$ for a straight wire, but in general is a more complicated function of object geometry, and $[A_{nn'}]$ is an object-dependent interaction matrix. This form is similar to a multi-input, multi-output linear predictor. In the general case, the general solution is developed by time-stepping beginning with a set of initial (known) conditions. Many of the details of applying the technique to solid bodies and wires using the magnetic-field and electric-field integral equations in space-time domain are found in References 7, 34, 36, and 37.

The computer time and storage involved in obtaining a time-domain solution having N_t time steps are given approximately by

$$T_t \cong (T_{t_3}N + T_{t_4}N_a)NN_tN_i \tag{98a}$$

and

$$S_t \cong (N + N_t)N \tag{98b}$$

with N_i the number of incident fields or excitations, and N_a the number of angles at which the far field is evaluated. In (98a) the T_{t_3} term accounts for current-charge computation, and T_{t_4} for far-field evaluation.

The N-Port Analogy

Solutions based on the MOM in either the frequency domain or time domain can be viewed as leading to an N-port equivalent network because in either case an $N \times N$ matrix can be developed as an approximation to the actual boundary-value problem of interest. Thus, finding the response of the structure to an arbitrary incident field variation in either the time domain or frequency domain requires determining the currents induced at each of the N ports for as many specific exciting source distributions as are necessary to adequately describe the incident field variation. This viewpoint can provide useful physical insight [7]; an idea of the amount of computational effort required is shown in Table 4. In this table, N is the number of unknowns, N_f the number of frequency samples, and N_i the number of independent spatial source distributions (incident fields). Note that N_f and N_i can be related to N for specific problems based on a sampling theorem argument. A complex interaction term is considered to be composed of two real interaction terms.

Table 4. Comparison of Frequency-Domain and Time-Domain Models in Terms of the Number of Real Interaction Terms Involved (*After Poggio and Miller [7]*, © *1973 Pergamon Press*)

Computation*	Equivalent Real Interaction Terms for N-Port Structure		
	Frequency Domain (FD) N_f Frequencies	Time Domain (TD) N_t Time Steps	FD/TD > 1 favors TD < 1 favors FD
Monostatic (N_i angles of incidence)			
Single frequency ($N_f = 1$)	$2N^2$	NN_tN_i	$2N/N_tN_i$
Time response** ($N_f = N_t/2$)	$2N^2N_f$	NN_tN_i	N/N_i
Bistatic (one angle of incidence)			
Single frequency ($N_f = 1$)	$2N^2$	NN_t	$2N/N_t$
Time response** ($N_f = N_t/2$)	$2N^2N_f$	NN_t	N

*A monostatic computation involves finding the far field for enough incident sources (N_i) to define the angle-dependent back-scattering pattern, while a bistatic computation involves finding the scattered field for one incident source. The latter is the situation most relevent to antenna analysis where only a few antenna feed points would be of interest.

**For equivalent information, $N_f = N_t$ in transforming between the frequency domain and time domain. But because the admittance samples for plus and minus frequencies are complex conjugate, only the plus-frequency values are needed so that $N_f = N_t/2$.

Comparison of Frequency- and Time-Domain Approaches

It is useful to compare the steps involved in performing frequency-domain and time-domain computations, as is done in Table 5. In this table the various constants for storage and time are computer and algorithm dependent, and n is a symmetry measure given by 2^d for objects having $d = 1, 2,$ or 3 planes of mirror symmetry, and by the number of sectors for objects having discrete rotational symmetry. (Symmetry effects are discussed more thoroughly in Table 7.)

The most important points to note in Table 5 are that a frequency-domain approach yields a solution valid at a single frequency but for an arbitrary spatial source distribution. The time-domain approach, on the other hand, yields a solution valid over a wide band of frequencies but for only a single spatial source distribution. In addition, it is relevant to mention that the time-domain approach can handle nonlinear effects, while dispersive and/or dissipative media can be handled more readily in the frequency domain using integral equations.

Comparison of the computer-time and -storage expressions in (89) and (98) reveals further interesting differences. A summary of computer-time dependencies for several kinds of common problems is given in Table 6, which was developed assuming the dependence on L and C of the various sampling densities as indicated in the footnote. This table includes only the dominant terms of the computer-time dependencies; i.e., as frequency increases without limit, the solution time should eventually exhibit the indicated behavior. The multiplying factor needed to estimate the actual time, which is computer, algorithm, and problem dependent, is not included. It is enlightening, however, to see the trends that characterize these different problems, permitting the potentially most efficient approaches to be identified.

Benefits of Symmetry

Symmetry can substantially reduce both storage and computation time, essentially by eliminating redundant operations. Consider, for example, the case of the center-fed, straight dipole. Its current is symmetric about the feed point, so that finding the current on just one half of the antenna defines the current on the whole antenna. In addition, the impedance matrix for equal-sized segments exhibits a symmetry about the antenna's midpoint, a fact that can reduce the matrix fill time by one half. Actually, a further symmetry holds for this structure in that only the first row is needed to define the entire impedance matrix. This is because the structure is translation invariant, i.e., the interaction between any two points a given distance apart on the wire is independent of their actual positions.

There are three different kinds of symmetries that can be exploited: plane, rotational, and translation symmetry. Each is discussed briefly below and summarized in Table 7, which was constructed considering only multiply and divide operations.

Plane Symmetry—An object can exhibit up to three orthogonal symmetry planes. The dipole example discussed above possessed one. Each plane halves both the storage needed for the impedance matrix and the computer time needed to fill it. Solution time is reduced because the admittance matrix possesses the same kind of symmetry as the original impedance matrix. Finally, if the exciting field is either

Table 5. Comparison of the Steps Involved in Developing and Using Frequency-Domain and Time-Domain Computer Models (After Miller [40])

Frequency Domain			Development of Solution	Time Domain		
Time	Storage	Operation		Operation	Storage	Time
		$f(s,t) = e^{i\omega t}f(s)$	Maxwell's equations	Time dependent		
		$L(s,s')f(s') = g(s)$	Plus BC, etc.	$\tilde{L}(s,s')\tilde{f}(s',t') = g(s,t)$ $\quad t' = t - R/c$		
$T_{f_1}N^2/n$	$2N^2/n + S_{f_1}N/n$	$\displaystyle\sum_{n=1}^{N} Z_{mn}f_n = g_m$	Apply MOM to get Nth-order system ($m = 1, \ldots, N$)	$\displaystyle\sum_{n=1}^{N} \tilde{Z}_{mn}\tilde{f}_{np} = \tilde{g}_{mp}$ $\quad p = 1, \ldots, N_t$	$S_{t_1}N^2$	$T_{t_1}N^2$
$T_{f_2}N^3/n^2$		$(Z_{mn})^{-1} \to Y_{mn}$	Solve matrix	$[\tilde{Z}_{mn}]^{-1} \to \tilde{Y}_{mn}$		$T_{t_2}N^3/n^2$
$T_{f_3}N^2/n$	$2S_{f_3}N$	$\displaystyle f_m = \sum_{n=1}^{N} Y_{mn}g_n$	Obtain induced current	$\displaystyle \tilde{f}_{mp} = \sum_{n=1}^{N} \tilde{Y}_{mn}\tilde{g}_{np}$	NN_t	$T_{t_3}N^2N_t$
$T_{f_4}NN_a$	$2N_a$	$\displaystyle E_0 = \sum_{m=1}^{N} E_{0m}f_m$ $\quad \times N_I$	Compute far field for $0 = 1, \ldots, N_a$ observation angles	$\displaystyle \tilde{E}_{0p} = \sum_{m=1}^{N} E_{0m}\tilde{f}_{mp}$ $\quad \times N_I$	N_aN_t	$T_{t_4}NN_tN_a$
		$\times N_f$	Repeat for additional spatial sources (total N_I) Repeat for additional frequencies $q = 1, \ldots, N_f$ Observe that the two solutions are related by a Fourier transform			
			$f_{mq} \overset{FT}{\Longleftrightarrow} \tilde{f}_{mp}$			

Table 6. Dependence of Computation Time on Object Size for Wires of Length L and Surfaces with Enclosing Spheres of Circumference C Enclosing Them (*After Poggio and Miller [7]*)

Calculation*	Frequency Domain	Time Domain
	Transient	
Monostatic:		
Surface	$(C/\lambda)^7$	$(C/\lambda)^6$
Wire	$(L/\lambda)^4$	$(L/\lambda)^4$
Bistatic:		
Surface	$(C/\lambda)^7$	$(C/\lambda)^5$
Wire	$(L/\lambda)^4$	$(L/\lambda)^3$
	Monochromatic	
Monostatic:		
Surface	$(C/\lambda)^6$	$(C/\lambda)^6$
Wire	$(L/\lambda)^3$	$(L/\lambda)^4$
Bistatic:		
Surface	$(C/\lambda)^6$	$(C/\lambda)^5$
Wire	$(L/\lambda)^3$	$(L/\lambda)^3$

*These dependencies on object size are arrived at by determining the highest-order terms in (89a) and (98a), assuming that the various sampling densities depend on object size in a systematic way. For wires we use $N \propto L/\lambda$, the number of incidence angles $N_i \propto N$, the number of time steps $N_t \propto N$, and the number of frequencies $N_f \propto N$. Similar values for surfaces are $N \propto (C/\lambda)^2$, $N_i \propto C/\lambda$, $N_t \propto C/\lambda$, and $N_f \propto C/\lambda$, respectively.

Table 7. Effect of Various Kinds of Symmetry on Impedance Matrix Storage and Fill Time and Admittance Matrix Solution Time Relative to Same Number of Unknowns and Without Symmetry

Type of Symmetry	Impedance Matrix Storage and Fill Time	Admittance Matrix Solution Time	
		General Source	One Symmetric Source
None	1	1	1
Orthogonal plane (d planes)	$1/2^d$	$1/2^{2d}$	$1/2^{3d}$
Rotational (n sectors)	$1/n$	$1/n^2$	$1/n^3$
Translational	$1/N$	$1/N$	$1/N$

even or odd about the symmetry plane(s), only that part of the admittance matrix which gives an even or odd current is needed.

Rotational Symmetry—An object can be either continuously rotational about an axis (a circular loop, for example) or discretely rotational (an *n*-sided polygon). Since a wire code normally models curved objects using piecewise linear (or straight) segments, discrete rotational symmetry is the one usually encountered in wire modeling. Storage and impedance matrix fill time are reduced in proportion to the number of identical sectors (n) comprising the model. Because the admittance matrix is circulant in the same way as the impedance matrix, its solution time is decreased due to needing only a part of the factored or inverted matrix. Furthermore, if the exciting source has symmetries in common with the object, additional savings are possible. There is a close correspondence between continuously and discretely rotational objects, in that the continuous Fourier transform applicable to the former is analogous to the discrete Fourier transform applicable to the latter.

Translational Symmetry—This is the symmetry exhibited by an object that is space invariant along some line. It can be continuous or discrete, as demonstrated by a continuous shell of arbitrary contour that is modeled using a wire grid. Impedance matrix storage and fill time are reduced in this case by the fact that interactions along the structure are invariant with respect to absolute coordinates. The admittance matrix is a product of two matrices having the same structure as the impedance matrix, yielding a further, very substantial time savings. This kind of symmetry produces what is termed a *Toeplitz matrix*.

As a concluding comment it should be observed that the above symmetries can occur in combination. When this is the case, even more reductions of computer resources are possible.

4. Computation

Computation is, of course, the focal point of modeling, and is the purpose to which all the other efforts are ultimately directed. It is where a user first encounters a given code and may experience difficulties. Because these are often due to deficiencies in documentation, maintenance, and assistance for the EM codes, we discuss each of these areas below.

1. *Documentation*: At a minimum the documentation should include a written description of the code and a reasonable amount of comment statements within it and directions on how to use it. Also highly desirable are applications guidelines, a discussion of the code's limitations, and examples of its use to provide check cases for the new user. As an option (highly desirable) the documentation could also provide the theoretical basis for the code and a discussion of the numerical treatment.

2. *Maintenance and Updating*: Most codes that receive any significant amount of use benefit greatly from periodic updating. One of these benefits is to correct "bugs" and identify unanticipated or unknown limitations in the code. The other is to inform the users of new capabilities that have been added to it.

3. *User Assistance*: Perhaps the most crucial factor in making the user-code encounter a successful one is that of user assistance since the best engineered and documented codes can leave unanswered questions and produce inexplicable results. Access to an experienced user, or the code developer, can save a user much time and anxiety. User assistance is also beneficial as a feedback mechanism for the developer.

Modeling Errors

Although a variety of errors can arise in computer modeling, they can be conveniently put into just two categories defined according to their cause as physical modeling errors and numerical modeling errors [41]. The former, which we denote by ϵ_p, provides a measure of the mismatch which may occur between the numerical model and the physical reality it is intended to represent. The latter, which we denote by ϵ_N, provides a measure of the mismatch between the solution obtained from the computer model and an exact one for the numerical model being used. In Table 8 are listed examples of some commonly encountered physical and numerical modeling errors.

As a general rule, ϵ_N can be made acceptably small by simply increasing the number of unknowns (number of segments N in subsectional models) used for the numerical model. This is because various measures of ϵ_N (see Table 9, for example) demonstrate that it tends to be an exponential function of the number of samples per wavelength. For geometrically complex structures, unfortunately, the value of N needed before ϵ_N exhibits this behavior may be unaffordably large so that ϵ_N is not reducible to an acceptable level. Some results of increasing N are shown in Figs. 15 and 16.

In Fig. 15 the results are based on the backscatter radar cross section averaged over 4π steradians of incidence angles using the value obtained from a model having the maximum feasible number of segments used as a reference for convergence purposes. All objects tend to yield the same slope past some minimum sampling density per wavelength, N_λ, with $\epsilon_N \propto \exp(-AN_\lambda)$, where A is a value in the range from 0.15 to 0.30.

In Fig. 16a the rms current $I_{rms}(N)$ for N segments is shown, while Fig. 16b shows the rms current error relative to a reference: $I_{rms}(N)/I_{rms}(\text{ref})$. Models using sinusoidal bases are seen to be superior where a traveling-wave current can exist.

There are other ways of estimating ϵ_N besides increasing N, as discussed in the next section.

Physical modeling errors are intrinsically the more difficult to establish because they almost invariably require experimental measurements for their direct assessment. But because experimentation is subject to its own errors and uncertainties, differences between a measured physical model and a computed numerical one cannot necessarily be attributed to the computation alone. One way to handle this problem is to use two or more experimental models, with one intended to model the real physical problem as closely as possible and another which matches the numerical model. Assuming systematic errors affect both measured results in the same way, a comparison of the two measurements should provide an accurate indication of ϵ_p. Two (or more) numerical models might be used in a similar fashion to estimate ϵ_p. The important point to keep in mind is that comparing either two (or more) measurements or two (or more) calculations can provide a way to estimate ϵ_p. The value of such comparisons is not limited to this application, however. Other uses are discussed below.

Limitations

Every code without exception is subject to limitations of various kinds. These limitations may be known or unknown, intrinsic or imposed, and can arise from the formulation, numerical treatment, or other choices made by the developer. Whatever the cause, it is important that the user be informed about known and anticipated limitations which affect the code's applications.

Information on limitations should be included in the documentation and can be compiled by listing the modeling capabilities incorporated in the code or, less preferably, by listing the things it cannot or was not designed to do.

Table 8. Generic Examples of Physical and Numerical Modeling Errors and Specific Examples for Representative Problems

Physical Modeling Errors	
Cause	Effect
Smoothly curved object represented numerically by a piecewise linear contour (e.g., a circular loop modeled as a polygon)	Introduces a nonphysical discretization that can obscure the needed relationship between the physical and numerical models. One criterion is to use equal lengths for wire models and areas for surface models that equal those quantities for the object being modeled.
Surface (closed or open) represented numerically by a wire grid or mesh	Provides a nonphysical mechanism for fields to "leak through" the surface. A modeling criterion sometimes used here is to choose the grid parameters such that its surface area equals that of the actual object.
Neglect of end cap in thin-wire numerical model	Produces a negligible effect on the solution for the most part, except near the end of a wire
Omission of fine object detail in developing the numerical model (sometimes referred to as *model order reduction*)	Reduces geometry-dependent correlation of computed results. Object complexity can impose more severe sampling requirements than can size, which is why this error source can be so important.
Numerical Modeling Errors	
Thin-wire approximation	Limits minimum segment length to approximately the wire diameter, or sharply oscillatory solution will result. Also ignores possible circumferential current variation near bends and junctions.
Under sampling of the unknown and boundary conditions	Unpredictable, but can generally be expected to most affect near-field quantities
Basis and weight function incompatibility with formulation	Can lead to anomalous-appearing non-physical near-field behavior, due to discontinuities in the basis functions that are not smoothed by the weight functions

Table 8, *continued.*

Specific Examples of Modeling Errors		
Object	Physical Modeling Error	Numerical Modeling Error
Straight wire	—	Thin-wire approximation and neglect of end caps
Circular loop	Piecewise linear model is discrete representation of curved object	Thin-wire approximation and neglect of possible circumferential variation at the bends
Wire-grid model of continuous surface	"Lumped" filamentary model of continuous surface	Thin-wire approximation and neglect of possible circumferential variation at the junctions
Wire attached to wire-grid model of continuous surface	Lumped filamentary model of continuous surface plus constrained flow onto surface at connection point	Thin-wire approximation and neglect of possible circumferential variation at the junctions
Surface-patch model of continuous surface	None necessarily. Usually however, the surface is modeled with piecewise linear patches which neglect surface curvature.	Variation of source within surface patch is not always included
Wire attached to surface-patch model of continuous surface	None necessarily. Usually however, the surface is modeled with piecewise linear patches which neglect surface curvature.	Variation of source within surface patch is not always included; thin-wire approximation and neglect of possible circumferential variation at the junctions
Box or other object	As above for either wire-grid or surface-patch model	Variation of source within surface patch is not always included; for example, edge singularities in the induced sources are usually not explicitly modeled.
Wire attached to box near edge or corner	As above for either wire-grid or surface-patch model	Variation of source within surface patch is not always included; for example, edge singularities in the induced sources are usually not explicitly modeled.

Table 9. Examples of Convergence Measures for Evaluating Numerical Modeling Errors

ϵ_N Measure	Form	Demonstrates	Properties
Local convergence: Current	$\lim\limits_{N \to N\max} I(s)$	Convergence of input impedance, current, fields, etc.	Can yield non-monotonic measure of convergence
Field	$\lim\limits_{N \to N\max} E(\mathbf{r})$		
Global convergence: Current	$\int_{C(\mathbf{r})} I(s')I^*(s')ds'$ $\cong \sum I_n I_n^* \Delta s'_n$	Convergence over entire object $[C(\mathbf{r})]$	A more complete measure of convergence
Field	$\int_{R^n} \mathbf{E}(\mathbf{r}) \cdot \mathbf{E}^*(\mathbf{r}) dx^n$	Convergence of field over $n = 1$, 2, or 3 dimensions	Can be used for near or far field. If a near-field quantity, can be expensive computation
Random convergence (local or global)	$\dfrac{1}{R} \sum\limits_{n=1}^{R} F(\mathbf{r}_n)$ With $F(\mathbf{r}_n)$ a field quantity which is a function of a random variable $\mathbf{r}_n$	Convergence of any field quantity measured by a random observation variable	Permits estimation of convergence and uncertainty of convergence estimates

Information concerning limitations should also be incorporated in the code itself. This can be as simple as providing internal checks that prohibit unacceptable parameter values from being used or inappropriate problems from being run, at least without warning the user. Further checks might be performed at subsequent steps in the computation to determine whether other limitations have been violated.

5. Validation

Unless a code can be used with some acceptable degree of confidence that it produces reliable results, it will have little value to prospective users. The process of validation is a vital and ongoing one. Here we discuss several ways to validate a code.

Experimental Validation

In spite of the difficulties which attend experimental validation already alluded to in the section on modeling errors, measurements still remain the most satisfying

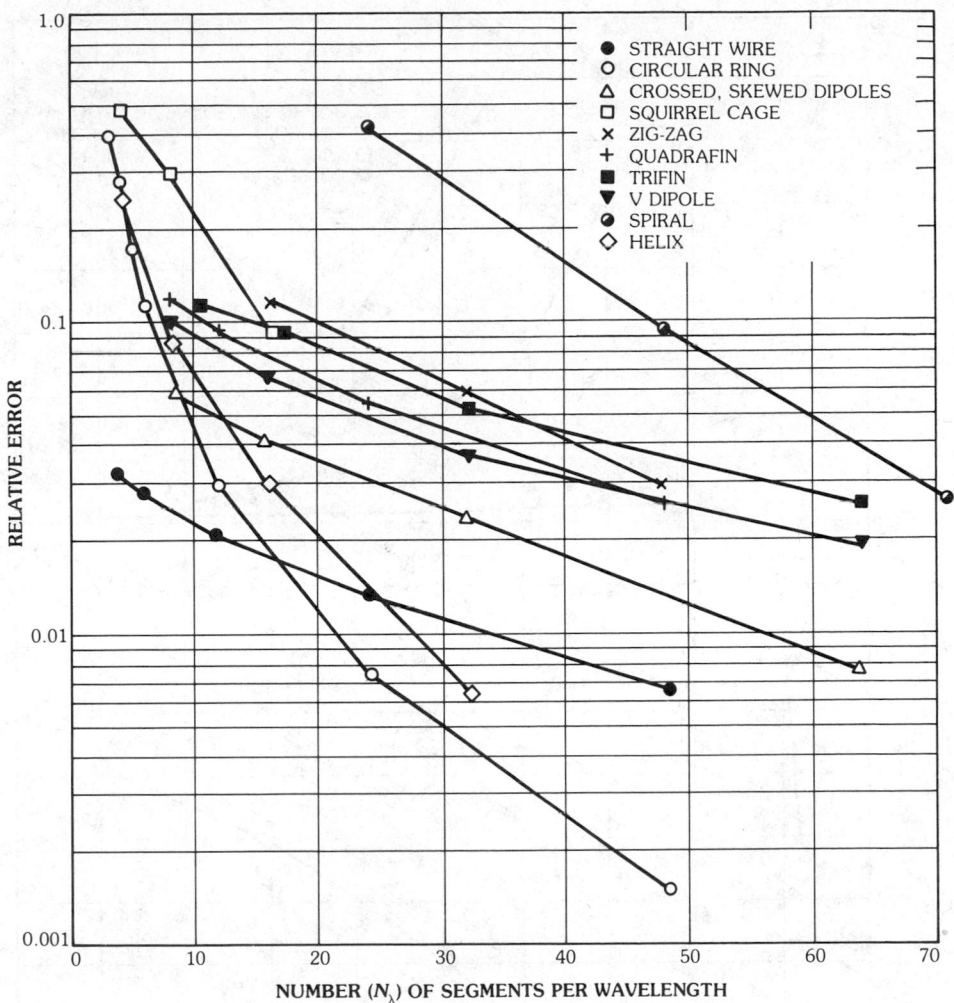

Fig. 15. Convergence behavior of various wire objects. (*After Miller, Burke, and Selden [42], © 1975 IEEE*)

way to demonstrate a code's validity. Experimental validation is most suitable where the measurement and computation can be made wholly congruent to reduce or eliminate any associated ambiguity that affects their comparison. If this is the case, it is then possible to make meaningful, absolute comparisons as opposed to the relative ones previously discussed.

The quality of the experimental validation can depend on the kinds of quantities being compared, specifically the relative degree of difficulty their measurement represents. Progressing in order of increasing difficulty, common quantities for validating a code are far fields (patterns and frequency dependence), radiation resistance (computed from integrating the far field), current and charge distribution, impedance resonances, and input reactance. An example of a resonance check is given in Fig. 17. Shown here is the resonance in the input impedance of a center-fed antenna consisting of a variable number of long zigzags

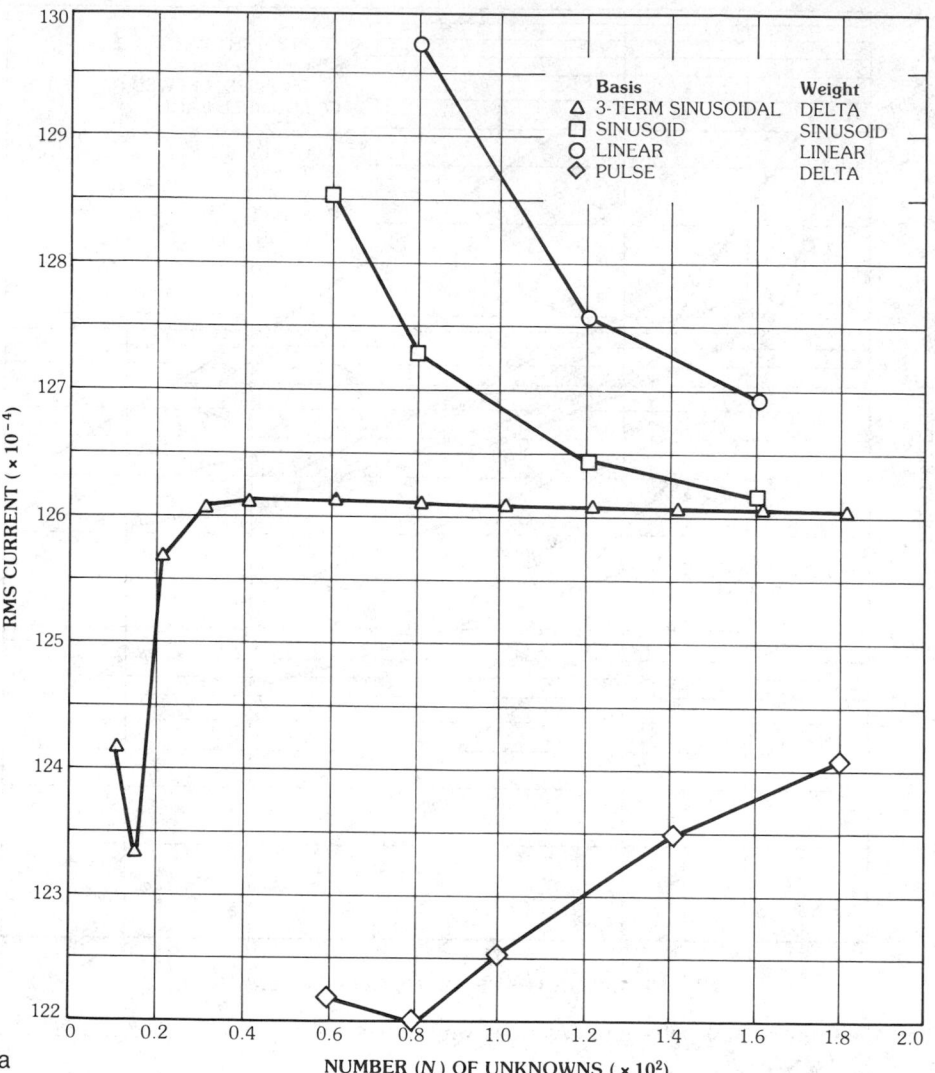

Fig. 16. Convergence results for scattering from a wire 5.8λ long, based on an rms-current measure. (*a*) Showing the rms current. (*b*) Showing the rms-current relative error. (*After Poggio, Bevensee, and Miller [43], © 1974 IEEE*)

as a function of the bend angle α, which is 180° when the antenna is a straight wire. The tip-to-tip length of the antenna is kept fixed at 29.6 in (175.18 cm). Good agreement is observed between the measured and computed results.

Because input reactance is so dependent on feed-region details, both in the experimental and numerical models, it is probably the most sensitive quantity to use for validation. But it is extremely common to observe shifts in frequency between measured and computed impedances. Shifts in far-field pattern details are also frequently observed both as a function of angle and frequency, as depicted graphically in Fig. 18. It is for this reason that relative comparisons, i.e., comparing a measurement with a measurement, or a computation with a computation, are

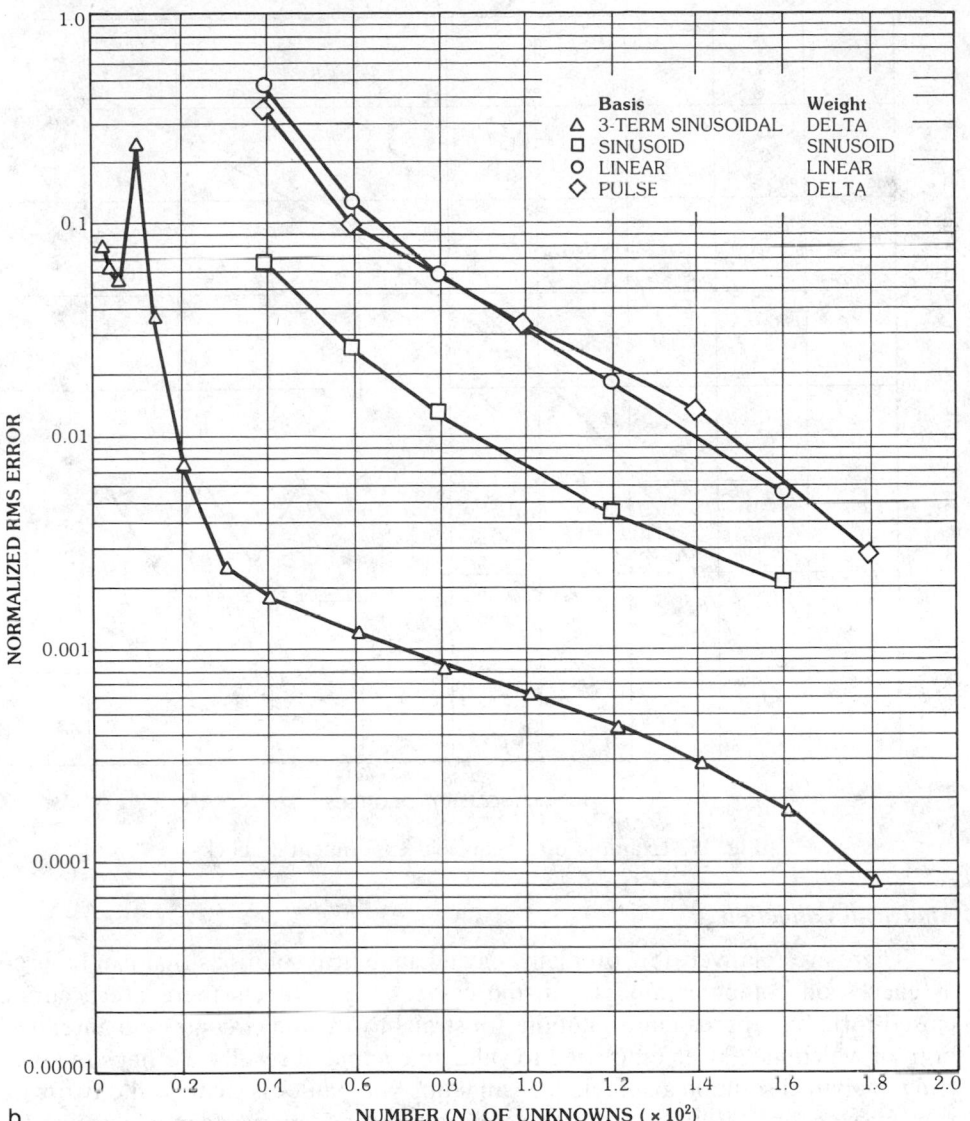

b

NUMBER (N_t) OF UNKNOWNS ($\times 10^2$)

Fig. 16, *continued.*

often more appropriate. Besides their use in validating a code, experimental results can be invaluable in developing an appropriate model for a given problem. It may seem to be reasonable to conclude that a computer model can be developed merely from a description of the problem, but it often happens that the physical modeling error depends unpredictably on the details of the computer model. Without the insight provided by the experimental data, there is little to guide the modeler in selecting the model parameters best suited to the particular problem. The outcome may be that valid numerical results are obtained, but for a significantly different physical problem than the one of interest. Only by using the experimental data can the most appropriate model be chosen.

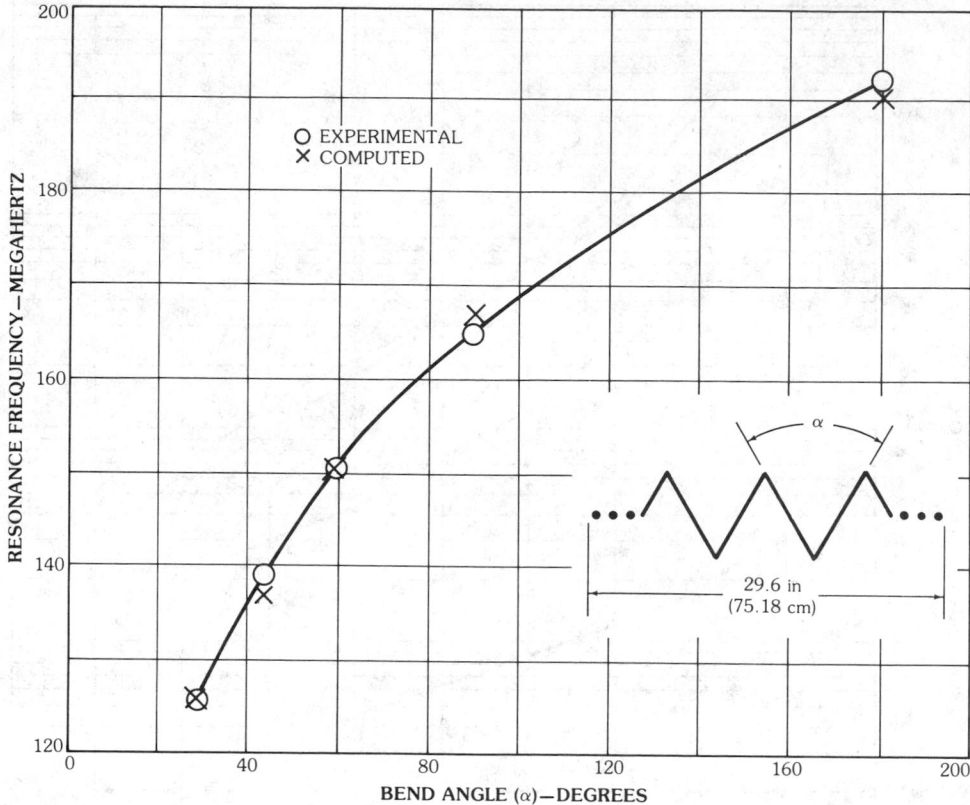

Fig. 17. Example of a near-field experimental check.

Analytical Validation

There are relatively few problems having analytical solutions that can be used as checks on computer models. In the case of wire objects there are accurate closed-form but approximate solutions for straight wire and circular loop antennas, both of which have been employed to validate computed results. Perhaps the only truly analytical solution available for validating wire codes is that for the two-wire transmission line. But since the transmission-line equations do not account for radiation, this check is also an approximate one. For surface objects, the sphere, spheroid, and ellipsoid provide useful check cases.

Unfortunately, all these problems tend to provide relatively undemanding test cases, as straight wires and spheres seem to be the simplest of problems to solve numerically. As checks on computer models they may be viewed more as necessary rather than sufficient conditions for establishing code validity. Analytical concepts, however, do provide the basis for other checks as described below under internal validation.

Numerical Validation

There are essentially two kinds of computations that can be used for numerical validation of a code. These are external checks, in which one code is used to vali-

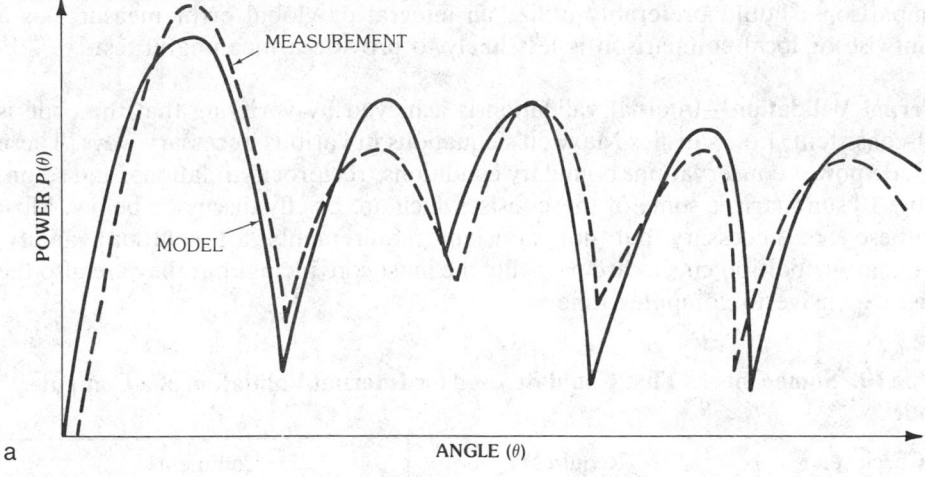

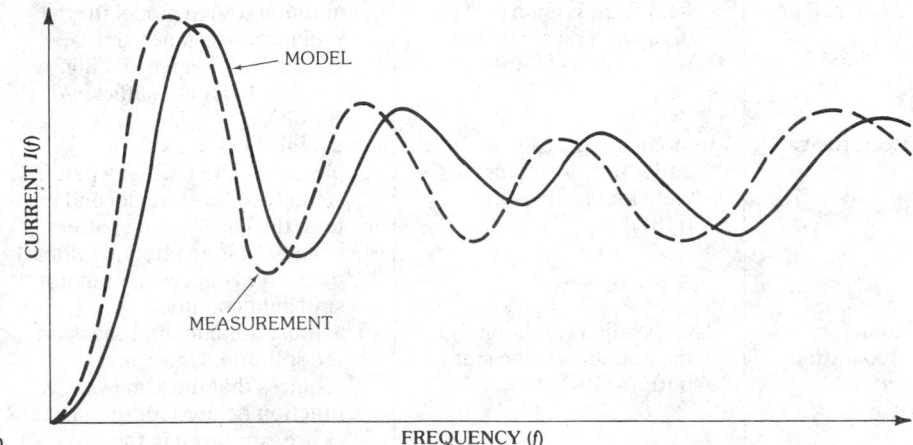

Fig. 18. Generic example of shifts frequently found between computed and measured results. (*a*) As a function of angle. (*b*) As a function of frequency.

date another, and internal checks, wherein various conditions required of a valid solution, but perhaps not routinely examined, are computed. Each kind of check is discussed further below.

External Validation—One of the most reassuring ways of validating a code's numerical performance is to find that, using some standard error measure, it agrees with a different code within some acceptable error bound, ϵ_E, when they are both applied to the same problem. To the extent that these codes are independent in their formulation and numerical treatment, and ϵ_E has a meaningfully small value, agreement between them will imply their mutual validity. However, the value of ϵ_E may signify nothing quantitative about ϵ_N, because neither code can be necessarily assumed to have converged. In addition, we emphasize that such comparisons should preferably utilize an integral or global error measure, as a pointwise or local comparison is less likely to provide a meaningful test.

Internal Validation—Internal validation is achieved by verifying that the code is self-consistent, i.e., satisfies Maxwell's equations in various necessary ways. These include power conservation, boundary conditions, reciprocity relations, and so on. Table 10 summarizes some of these tests, which are briefly discussed below. Most of these are necessary but not sufficient requirements for solution validity. Boundary-condition checks are probably the most convincing, but they are also the most expensive in computer time.

Table 10. Some Checks That Could Be Used for Internal Validation of a Computer Model

Check	Requires	Comments
Power balance	Power supplied by incident field to equal sum of dissipated plus radiated power [(99)–(100)]	Provides a good check on the antenna source model for the radiation resistance and validity of resistive loading. A necessary but not sufficient condition
Reciprocity	Interchanging observation and source locations to yield identical results [(101)]	A useful check for antenna patterns. The receiving pattern is likely to be more dependable than the transmitting pattern because it is an integral (global) quantity. A necessary but not sufficient condition
Boundary condition	The specified conditions on the boundary to be met [(103)–(104)]	The most fundamental check on the solution. Consistency requires that the same weight function be used for this check as are employed in the numerical model. Can be computationally expensive. A necessary and sufficient condition
Convergence	Computed result to approach a limiting value as N is increased	Exhibits behavior of a given quantity as number of unknowns is increased. Can be applied on a local or global basis to any observable. Near-field results provide a more sensitive measure especially on a local or pointwise basis
"Nonphysical behavior"	Computed result to exhibit a physically plausible behavior	Can be a subjective check. One example is provided by spatial oscillation of current over distance much less than the wavelength. Other examples are negative antenna resistance and anomalous near fields

Power Conservation—For lossless media, all the power supplied to an object must equal the sum of its resistive losses and integrated far-field power flow. This condition can be stated mathematically as [see (26)]

$$
\text{Re}\left\{\int_{C(\mathbf{r})} \mathbf{I}(s')\cdot\mathbf{E}^{*i}(s')ds'\right\} = \text{Re}\left\{\int_{C(\mathbf{r})} \mathbf{I}(s')\cdot\mathbf{I}^*(s')R(s')ds'\right\}
$$
$$
+ \text{Re}\left\{\lim_{R\to\infty}\oint_S \mathbf{E}(\mathbf{r})\times\mathbf{H}^*(\mathbf{r})\cdot d^2\mathbf{r}'\right\} \tag{99}
$$

where the term on the left is the supplied power, and the right-side terms account for the dissipated and far-field power, respectively. Note that the supplied-power term arises from integrating the inward Poynting's vector normal to the object's surface due to the exciting field. Equation 99 applies to both antenna and scattering problems, but for the special case of an antenna excited at a single point (segment), the supplied power simplifies to

$$
\text{Re}\left\{\int_{C(\mathbf{r})} \mathbf{I}^*(s')\cdot\mathbf{E}^i(s')d^2\mathbf{r}'\right\} = \text{Re}\{I_f^* V_{\text{in}}\} = |I_f|^2 R_{\text{in}} \tag{100}
$$

Thus (100) provides an explicit test on the computed input resistance when the object is excited as an antenna.

Reciprocity Relations—Another general requirement of a valid solution to Maxwell's equations is that of reciprocity, for which a general mathematical statement is

$$
\int_{C_2(\mathbf{r})} \mathbf{E}_1(s')\cdot\mathbf{I}_2(s')ds' = \int_{C_1(\mathbf{r})} \mathbf{E}_2(s')\cdot\mathbf{I}_1(s')ds' \tag{101}
$$

A numerical model can be tested for reciprocity in various ways. Both far-field and near-field tests are useful. An example of the former is

$$
\mathbf{E}(\theta_0, \theta_i) = \mathbf{E}(\theta_i, \theta_0) \tag{102}
$$

i.e., the field scattered in direction θ_0 due to a plane wave incident from direction θ_i is the same as when the two angles are interchanged.

A near-field test is provided by subjecting the admittance matrix to a test defined by

$$
\Delta_j Y_{ij} \cong \Delta_i Y_{ji}
$$

where the source segment length is $\Delta_{(\cdot)}$. Because of reciprocity the impedance matrix is symmetric when a Galerkin's technique is used, but with point matching (delta-function weights) this is not the case unless the two segments in question are of equal length.

Boundary Conditions—Although the integral equation is derived with boundary conditions explicitly satisfied everywhere on the object, the numerical solution can only approximate that condition. It is therefore useful to compute the surface fields over the object using the currents (and charges) that have been computed. However, this check cannot be applied using the same testing-function locations as were used to generate the impedance matrix since the boundary conditions at these locations should be satisfied to computer accuracy. In between these locations, though, it can provide a demanding test of solution accuracy.

Such a check is best done in an integral sense, as a few isolated field values are not likely to provide a meaningful test. Furthermore, the errors have a quantitative significance only relative to the exciting field value. One logical measure of boundary-condition mismatch is given by

$$\epsilon_B = \frac{\int_{C(\mathbf{r})} [\mathbf{E}(s') - \mathbf{E}^i(s')]\cdot\hat{\mathbf{s}}' \ [\mathbf{E}^*(s') - \mathbf{E}^{*i}(s')]\cdot\hat{\mathbf{s}}' \ ds'}{\int_{C(\mathbf{r})} [\mathbf{E}^i(s')\cdot\hat{\mathbf{s}}'] \ [\mathbf{E}^{*i}(s')\cdot\hat{\mathbf{s}}'] \ ds'} \tag{103}$$

This error functional does not include explicitly the possible role of the current in establishing the importance of boundary-condition mismatch. It can happen, for example, that large errors in the field are less important in places where the current is small than where it is large. This is because the field error when multiplied by the current is equivalent to a nonzero power flow normal to the object. Thus an alternative to (103) is

$$\epsilon'_B = \frac{\int_{C(\mathbf{r})} \{\mathbf{I}(s')\cdot[\mathbf{E}(s') - \mathbf{E}^i(s')] \ \mathbf{I}^*(s')\cdot[\mathbf{E}(s') - \mathbf{E}^i(s')]^*\} \ ds'}{\int_{C(\mathbf{r})} [\mathbf{I}(s')\cdot\mathbf{E}^i(s')] \ [\mathbf{I}(s')\cdot\mathbf{E}^i(s')]^* ds'} \tag{104}$$

Any validation procedure which requires evaluation of near fields can be computationally expensive to implement. Thus, even though boundary-condition checking may be a quantitatively meaningful way to estimate solution validity, it is not viable for routine use. Such checks are generally most valuable when employed selectively, for example when a new kind of problem is being modeled.

Check Cases—It is frequently useful to have results available for standard check cases to use in establishing that a code is continuing to produce valid results. This is especially true when the code has been adapted to a new machine, an updated operating system is introduced, the user has made some changes, and the like. At the very least, some problems which exercise various of the code's features should be described and their output included in the documentation.

6. A Guided Tour of Some Codes and Their Features

In this section we will discuss some of the issues and trade-offs to be considered in developing and selecting a modeling code, and we will present a detailed catalog of representative examples.

The General Versus the Specific

Compromises must be made in developing any modeling code. Perhaps most immediate is that of deciding on the degree of generality (or complexity) versus specificity (or simplicity) the code is designed for in terms of problems to be treated and features it is to contain. By and large a code intended to handle a wide variety of problems will not be as efficient or accurate for a given problem as a code designed for that particular problem alone. Furthermore, other measures of code "usability," such as storage requirements, user ease, modification ease, and other user resource requirements, may favor the specific code over the general one, even at the expense of the limited scope [44]. Nevertheless, the general code can be an essential and important tool for a wide spectrum of users. The point is that there is a need for both kinds of codes, and the user should be aware of their relative advantages and disadvantages.

Generic Characteristics

Formulation—Although Maxwell's equations provide the starting point for any formulation, various significant differences in capability can arise in the process of developing a suitable mathematical description. Rather than reviewing this issue in detail, instead a few of the major options and their consequences that might be considered at this step are summarized in Chart 18.

Numerical Treatment—Once the formulation has been developed, the needs of the numerical treatment are fairly clear. Furthermore, whether a differential equation, integral equation, hybrid, or some other approach has been selected, these needs are quite similar, and involve the basic steps given in Table 2.

An issue of importance comparable to that of the formulation is the choice of basis and testing functions. We will discuss each briefly here, and summarize in Table 11 examples of some of the more commonly used combinations. Sinusoidal bases are preferable for objects having traveling-wave currents.

Basis Functions—The goal of the basis function is that it provide a match to the unknown to the degree of accuracy being sought using an acceptable number of unknown coefficients. There is quite a degree of latitude provided by this general goal, as several possibly conflicting factors may affect the choice, in a specific situation.

Perhaps the most obvious factor is that of the physical variation the basis function is intended to represent. Other things being equal, that basis function which can provide the closest match to the physical quantity being modeled while using the smallest number of unknowns would be preferable. The situation is hardly ever this simple, however, since the trade-off between impedance-matrix fill time and the subsequent solution time must also be considered. While the latter is smaller the fewer unknowns there are, the former may more than offset this savings by increasing the fill time because of the increasing basis-function complexity. Thus the spatial variability provided by the basis function and evaluation of its fields are its important characteristics.

Another factor is associated with the nature of the integral or differential

Chart 18. A Ranking of Time-Domain and Frequency-Domain Approaches Based on Integral Equation and Differential Equation Formulations with Respect to Various Kinds of Solution Capabilities and Issues

Time Domain		Issue	Frequency Domain	
Differential Equation	Integral Equation		Differential Equation	Integral Equation
		Medium		
√	√	Linear	√	√
–	–	Dispersive	√	√
√	–	Lossy	√	√
√	~	Anisotropic	√	√
√	–	Inhomogeneous	√	–
√	–	Nonlinear	–	–
√	–	Time-varying	–	–
		Object		
~	√	Line (Wire)	~	√
√	√	Closed surface	√	√
√	√	Volume	√	√
~	√	Open surface	~	√
		Boundary Conditions		
√	√	Interior problem	√	√
~	√	Exterior problem	~	√
√	√	Linear	√	√
√	√	Nonlinear	–	–
√	√	Time-varying	–	–
–	–	Half-space	~	√
		Other Aspects		
~	~	Symmetry exploitation	√	√
~	√	Far-field evaluation	~	√
–	~	Number of unknowns	~	√
√	~	Length of code	~	–
		Suitability for Combining with Other		
–	√	Numerical procedures	√	√
–	–	Analytical techniques	–	√
–	–	Geometrical theory of diffraction	–	√
		Legend		

√ = Highly suited or most advantageous
~ = Moderately suited or neutral
– = Unsuited or least advantageous

Table 11. Specific Examples of Basis and Weight Function Combinations Used for Wire Models (*After Miller and Deadrick [41]*)

nth Basis	mth Weight	Comments
Subdomain		
Pulse $A_n P_n(s')$	Delta function $\delta(s - s_m)$	Produces point-matched fields. Current and charge are discontinuous at domain junctions
Linear $A_n \ell_n^{(+)} + B_n \ell^{(-)}$	Linear $\ell_m^{(+)} + \ell_m^{(-)}$	A Galerkin's procedure. Provides continuous current
Sinusoidal $A_n s_n^{(+)} + B_n s^{(-)}$	Sinusoidal $s_m^{(+)} + s_m^{(-)}$	A Galerkin's procedure. Provides continuous current. Especially good for longer $(L > 2\lambda)$ wires
Parabolic $A_n + B_n \ell_n^{(+)} + C_n \ell_n^{(+)2}$	Delta function $\delta(s - s_m)$	Produces point-matched fields. Can provide continuous current and charge
Three-term sinusoidal $A_n + B_n s_n + C_n c_n$	Delta function $\delta(s - s_m)$	Produces point-matched fields. Can provide continuous current and charge. Especially good for longer $(L > 2\lambda)$ wires
Entire Domain		
Fourier series $I(s') = \sum A_n e^{jns'}$	Delta function $\delta(s - s_m)$	Used for axisymmetric objects
Legend		
$\Delta s_n = $ length of nth segment $s_n = $ center coordinate of Δs_n along $C(\mathbf{r})$	$\ell_n^{(+)} = s' - s_n + \Delta s_n,$ $s_n = \sin[k(s' - s_n)]$ $\ell_n^{(-)} = s' - s_n - \Delta s_n,$ $c_n = \cos[k(s' - s_n)]$	$s_n^{(+)} = \sin k\ell_n^{(+)},$ $s_n^{(-)} = \sin k\ell_n^{(-)},$

operator which generates the field from the sources. Care must be taken, for example, if the order of a differential operator exceeds the differentiability of the basis function. When this happens a nonphysical field behavior can occur, unless an extended interpretation of the differentiation is employed, as is done when using a pulse current basis and a pulse charge basis. If the differentiation is the more subtle kind that occurs from the multiplication of a source by its differentiated Green's function, it is no less a problem. Consequently, a basis function must possess the degree of continuity required by the formulation and numerical treatment.

Weight Functions—A third factor which affects the suitability of a given basis function is the weight function being considered. The weight functions may vary from a simple delta function (this is called *point matching*), to being the same as the basis function (this is called *Galerkin's Method*), to providing even more variability than the basis function. A weight function (other than a delta function) can be viewed, in general, as smoothing out the fields produced by the combination of the basis function and operator. For a given operator, it is the total variability (this might be quantified by the number of degrees of freedom) of the basis-weight function combination that is important. Thus, using a lower-order (fewer degrees of freedom) testing function might be compensated for by using a higher-order basis [45].

As a specific example of this possibility we observe that the combination of piecewise sinusoidal basis and testing functions seems closely comparable to the use of a three-term basis with a delta-function weight in solving the thin-wire electric-field integral equation. Curiously enough, it can also be shown that interchanging the basis and weight functions does not change the impedance matrix elements [45]. This emphasizes the need to evaluate the fields using a testing function that is compatible with the basis function being employed.

Features—While it is the formulation and numerical treatment that establish the features a code is fundamentally capable of providing, it is often the add-ons that make it truly useful. Summarized here, and in Table 12, are some of the more commonly included features to be found, divided into three categories as input, run time, and output features, respectively. Included are what could be expected as a minimum, as well as extended features for more fully developed codes.

Input Features—Aside from the task of developing the parameters of the numerical model to be used, the most time-consuming part of the modeling process for the user is that of actually putting the model data (which consists of both geometrical and electrical information) into the computer.

The problem to be modeled may be described to the computer via input data, stored modules, and computation. The simplest, but most laborious, method for inputting object geometry is to give the required parameters (for example, center coordinates, length, radius, and direction angles) of each individual wire segment (or surface patch) in the object. This segment-oriented and inefficient procedure can be significantly improved by adopting a wire-oriented approach, wherein only the geometry (straight or curved wire, etc.), end points, radius, and number of segments of each wire are given. The computer then computes the data required to specify each segment. An additional enhancement can be realized by storing geometry modules that may occur as parts of more complex objects.

For objects having planar or rotational symmetry the description can be further simplified. Reflections about one to three symmetry plans can expedite model development, as well as shorten the computation time. Rotation about an axis can be similarly used, either for axisymmetric objects or sectors of such objects.

Displaying the object geometry on a display terminal or plotting it out in hard copy is a check that is always worthwhile doing, even for simple objects. At the least a hard-copy plot provides a record that is valuable for interpreting the output data and keeping with it as a permanent record. More than that, these plots can reveal errors in the input data, or verify its validity. A tabular printout of the model data is also useful for quantitative values.

Besides object geometry, model data also includes (for most codes) information about electrical connectivity of the geometrical segments which make up the object. This information is needed to establish current reference directions, to relate the basis function on each segment to those segments connected to it, and to identify open-ended wires where the current goes to zero. Because segments that

Table 12. Examples of Generically Useful Code Features

Function	Minimum	Extended
Input		
Object definition	Specify object coordinates on a segment-by-segment basis	Provide a selection of "building blocks" from which a relatively complex object can be developed
Exciting source description	Specify on a segment basis	Develop exciting field over object from menu of user-selected options
Data check	None	Automatic checking of input data and parameters to identify user violation of recommended guidelines or inconsistent data
Run Time		
Admittance matrices	Verify that their product with the impedance matrix yields the identity matrix	Perform extended reciprocity check using admittance matrix
Current and charge evolution in the time domain	Produce error flag if the computed quantity becomes divergent	Estimate the exponential (SEM) parameters of the computed quantity to predict behavior and terminate computation as soon as possible
Output		
Induced current and charge distributions	Numerical printout	Monitor display or hard copy plotting. Could be interactive as user varies exciting field
Far fields	Numerical printout	Monitor display or hard copy plotting. Could be interactive as user varies exciting field
Near fields	Numerical printout	Monitor display or hard copy plotting. Could be interactive as user varies exciting field

are specified to be electrically connected must have their physical end points at common locations, electrical-connectivity data can be derived from the geometry. But, by providing both kinds of object data independently, a self-consistency check can then be made to validate the input data.

This object data is all that is needed to develop a model, as the N-port admittance matrix can be obtained from such data alone. Additional data is eventually required to specify the excitation, to select whatever run-time options the code provides, and to specify the output desired.

Run-Time Features—Until now, most codes have been designed for batch processing. Little or no control or monitoring of the actual computation is available to the user. The increasing interactivity and speed provided by present and future computers offers some interesting possibilities. Some of the more obvious are given below.

One run-time feature that would give the modeler more control of the computation is an interrupt and restart capability to enable the computation to be stored at an intermediate point. This could be done in order to examine its accuracy, for example, so that an otherwise expensive and valueless run could be terminated and appropriate changes made in the input parameters to redo the computation. Such a capability could be especially valuable when new problems are being modeled.

Another run-time feature of potential value is one which would enable the modeler to make changes in the model parameters during the computation. If an accuracy check revealed that some error measure exceeded a desired threshold value, then having the option of increasing the sampling in that area of the model could perhaps rectify the problem. In essence, the modeler becomes part of a feedback loop, making decisions about modeling that would be difficult to anticipate a priori. After gaining experience with that kind of problem, the feasibility of automating such decisions for similar problems could be established.

Output Features—Even the most elementary code can produce an overwhelming volume of output. Unless this output can be presented to the user in an understandable way, it is likely that not only will important results be missed, but also that wrong conclusions will be reached. At the least this makes the computation less useful than it should be, and at the worse, makes it counterproductive.

Output options provided by most models include a choice by the user of a variety of EM phenomena. For antennas these range from (most often) far-field patterns, input impedance, transfer or mutual impedances, current and charge distributions, and (least often) near fields. Auxiliary quantities derivable from the above include radiated power, input power, antenna loss, antenna efficiency, absolute and directive gain, polarization, field impedance, and insulator and impedance-load voltage drops. This is a broad range of quantities, many of which are not conveniently available from measurement.

The Importance of the Details

It is fair for the prospective user to ask why all the details of the formulation, numerical treatment, etc., should be of interest to him or her. After all, the average computer user does not need comparably detailed information about most of the software provided by a computer's operating system. The difference between these situations is that the typical EM modeling code is neither as simple nor as easy to use as the system routines. Therefore it is important for the user at least to be aware that there are trade-offs involved in designing the various codes that could be considered, and the implications thereof for his or her applications.

For example, consider the specific issue of the basis function used in a particular code. As has been discussed above, one criterion for the basis-function selection is that it provide a reasonable match to the expected current behavior so as to minimize the number of unknowns. In modeling long, traveling-wave structures, such as the Beverage and sloping vee antennas, a basis which includes a sinusoid is preferable to one which uses polynomials [43].

A Code Catalog

Since even two codes designed for application to the same kinds of problems may differ greatly, it is worthwhile to identify a set of generic features by which they and other codes can be usefully compared. In this section we outline the properties of a representative collection of codes in these categories: (1) code description; (2) analytical foundation; (3) numerical treatment; and (4) user features. The codes summarized in Table 13 were chosen to illustrate the variability that may typically be encountered between such modeling codes. Equivalent information is not available in all cases so that one-to-one comparisons are not always possible. Nevertheless, information of the kind presented has been found to be useful and may provide a model for the interested reader in comparing codes being considered for his or her own use. Another source of code information, in this case for EMP applications, is given in [46].

A Closer Examination of Two Specific Codes

Two widely available computer codes are NEC (Numerical Electromagnetic Code) [23] and TWTD (Thin-Wire Time Domain) [47, 48, 49]. Their characteristics are summarized in Table 14. These are not the only computer tools that are applicable to wire modeling in the frequency and time domains, but they can be considered representative of this general kind of electromagnetic code. Also, NEC is one of the best documented and most widely used codes, and TWTD is almost the only one available for time-domain modeling of general wire objects.

For simplicity, and because of problem-caused variability, the computer and storage estimates listed in Table 14 do not include the effects of symmetry. Also, these aspects have been discussed previously.

In Table 15 are listed several representative problems to illustrate application of these two codes. Specific problems can always be found which reveal that one approach may have a computational advantage over another. It can be expected, however, that when each produces the equivalent information, without redundant or unneeded results, they generally will exhibit similar efficiency when developed using similar numerical and programming techniques.

Application Guidelines

It should be clear that the successful application of any EM modeling code depends on the user's adhering to the limitations that particular code may be subject to. These limitations may either be intrinsic (or generic) properties of the formulation, numerical implementation, and so on, or extrinsic constraints (specific) introduced by the developer. In either case it is vital that the user be made aware of guidelines appropriate for a given code to increase the probability of its producing valid results. While selecting a numerical model of a physical problem involves a high degree of art, choosing the parameters of the computer model that gets the best performance from a given code is hardly less artful. Table 16 summarizes some generic guidelines for using EM modeling codes.

Table 13. A Catalog of Some Representative Modeling Codes with Respect to Overall Description, Analytical Foundation, Numerical Treatment, and User Features

Code Description*

Issue	FDTD, A. Taflove, Northwestern U., Chicago, IL	Surface Patch Code, E. H. Newman, Ohio State U., Columbus, OH	Main Scattering Program, P. Barber, Clarkson Inst. of Tech.	EM Scattering, Wilton, Rao & Glisson, U. of Miss., University, MS	LASMN3, Hagman, Grandi & Durney, U. of Utah, Salt Lake City, UT	GEMACS, Balestri et al., BDM Corp., Albq., NM	NEC, Burke et al., LLNL, Livermore, CA	MFIE/Aircraft Real Earth Code, Sancer & Siegel, RDA, Santa Monica, CA	COBRA, Medgyesi-Mitchang, McDonnell Douglas, St. Louis, MO	ROTSY, C. L. Bennett, Raytheon, Sudbury, MA	TWTD, J. A. Landt, Amtech, Los Alamos, NM
Machines	CDC STAR 100 CYBER 203 CYBER 205	IBM 370-165	UNIVAC 1108 CDC 7600 CRAY 1 Minicomp. LSI-11/2	DEC-10	PDP-10	HONEYWELL 6180	CDC 6600 7600 CRAY IBM 360 3033 VAX	CDC 6600 7600 PRIM 750	CDC 6600 7600	UNIVAC 1106 HONEYWELL 635	CDC 6600 CDC 7600 IBM 3601
Availability†	Author Research Code	Author	Author	Author	Author	Author	Author or LLNL	Author	Restricted	Author or LLNL	Author
Length (lines) Storage B = bytes W = words	350 $3(10^6)$W	4000 26 KW $+ N(N+1)$	1500 50 KB	3000 85 KW $N < 140$	950 ≈ 9(no. cells)2 complex W	19000 93 KW	8000 83 KW (7600)	4000 52 KW (7600)	4600 Variable	1000 29 KW	1000–2000 30 KW
Time	$0.3\ \mu s$/cell per time step (CYBER 203)	$\lambda/4$ wire on $1-\lambda^2$ plate $\cong 20$s (IBM 370-165)	23 min for prolate spheroid of $ka = 5.6$ (TERAK-8510/H)	15–20 min $N + 90$ (DEC-10)	$\cong 20$ hr for $N = 540$ (PDP-10)	3 hr for $N = 1000$ (HON. 6180)	$AN^2 + BN^3$ $A = 3(10^{-4})s$ $B = 2(10^{-6})s$ free space (CDC-7600)	$AN^2 + BN^3$ $A = 2.4(10^{-3})s$ $B = 2(10^{-5})s$ (CDC-7600)	$AN^2 + BN^3$ $A = 1.2(10^{-3})s$ $B = 2(10^{-5})s$ (CDC-7600)	Problem dependent (UNIVAC, HON)	$AN^2N + BNN$, $A = 2.5(10^{-5})s$ $B = 5(10^{-9})s$ (CDC 7600)
Documentation	Report	J. Article	J. Article	Report	Report	Report	Very thorough 3-part report	Report	Report	Report	Report
Scope, goals	Sinusoidal steady state penetration studies for 2D & 3D objects, using transient approach	Wires and plates (connected)	Scat for axisym dielectric	Conducting body with apertures OK	Absorption of EM by man	General wire modeling	General antennas and scatterers. Wires and surfaces, including ground.	EMP aircraft on real earth	Wire and slot radiators on a body of revolution	Time-domain scattering from closed, rotationally symmetric bodies	Wire objects in time domain

Analytical Formulation

Category	FDTD, A. Taflove, Northwestern U. Chicago, IL	Surface Patch Code, E. H. Newman, Ohio State U., Columbus, OH	Main Scattering Program, P. Barber, Clarkson Inst. of Tech.	EM Scattering, Wilton, Rao & Glisson, U. of Miss., University, MS	LASMN3, Hagman, Grandi & Durney, U. of Utah, Salt Lake City, UT	GEMACS, Balestri et al., BDM Corp., Albq., NM	NEC, Burke et al., LLNL, Livermore, CA	MFIE/Aircraft Real Earth Code, Sancer & Siegel, RDA., Santa Monica, CA	COBRA, Medgyesi-Mitchgang, McDonnell Douglas, St. Louis, MO	ROTSY, C. L. Bennett, Raytheon, Sudbury, MA	TWTD, J. A. Landt, Amtech, Los Alamos, NM
Domain (T = time. F = frequency)	T	F	F	F	F	F	F	F	F	T	T
Equations: EFIE = electric integral eq. / MFIE = magnetic field integral eq.	Maxwell 3D	Reaction Integral	EF-MF hybrid integral equation.	EFIE mixed potential	EFIE	EFIE wires, MFIE for surfaces	EFIE wires, MFIE for surfaces	MFIE	EFIE	MFIE	EFIE
Media type: Lossless	✓	✓	✓	✓	✓						
Lossy	✓		✓			✓	✓	✓	✓	✓	✓
Dispersive	✓							✓	✓		
Inhomogeneous	✓										
Nonlinear	✓										✓
Object impedance	Volume and surface impedance	Lumped or distorted RLC on wires	Lossy	None	Lossy dielectric	Lumped or distributed RLC on wires	Lumped or distributed RLC on wires	None	Lumped or distributed RLC on wires	Surface impedance	Lumped or distributed RLC. lumped nonlinear
Geometry: Wire	Cables	✓	✓			✓	✓		✓		✓
Surface‡	Closed	Open	Closed	Open		Closed	Closed	Closed	Closed	Closed	
Penetrable volume	✓	✓	✓	✓	✓		✓	✓	✓	✓	✓
Discontinuities: Wire bends	✓	✓		✓		✓	✓		✓	✓	✓
Wire junctions	✓	✓				✓	✓	✓	✓		✓
Edges	✓	✓		✓	✓		✓	✓	✓	✓	✓
Apertures	✓	✓							✓		✓
Approximations	Linear approx. of curved surfaces	$\geqq 24$ basis functions/λ^2		Triangle patch edges $\leqq \lambda/8$	Cell size $\leqq \lambda/5$	Essentially same as NEC	Thin wire approx; pulse basis functions on surface plates	$k \cdot$(body length) < 10–20	Norton fields for earth	Object size < 2 pulse-widths	Linear approx. of curved wire

Table 13, *continued.*

Numerical Treatment

Issue	FDTD, A. Taflove, Northwestern U., Chicago, IL	Surface Patch Code, E. H. Newman, Ohio State U., Columbus, OH	Main Scattering Program, P. Barber, Clarkson Inst. of Tech.	EM Scattering, Wilton, Rao & Glisson, U. of Miss., University, MS	LASMN3, Hagman, Grandi & Durney, U. of Utah, Salt Lake City, UT	GEMACS, Balestri et al., BDM Corp., Albq., NM	NEC, Burke et al., LLNL, Livermore, CA	MFIE/Aircraft Real Earth Code, Sancer & Siegel, Santa Monica, CA	COBRA, Medgyesi-Mitchang, McDonnell Douglas, St. Louis, MO	ROTSY, C. L. Bennett, Raytheon, Sudbury, MA	TWTD, J. A. Landt, Amtech, Los Alamos, NM
Method	Finite difference	MOM	MOM, Volume	MOM	MOM, Volume	MOM	MOM	MOM	MOM	MOM Time Step	MOM Time Step
Basis functions	Pulse	Piecewise sine on wires and surfaces	Spherical expansion	Linear on pair of triangles	Pulse, interpolated	3-term trig, on wires, impulse on surfaces	3-term trig, on wires, impulse on surfaces	Pulse	Triangle and $e^{jn\phi}$	Pulse in space, polyn. in time	9-term quadratic in space and time
Weight functions	Delta functions	Same	Same	Same	Impulse	Impulse	Impulse	Pulse	Same	Impulse	Impulse
Boundary conditions	Dirichlet, Neumann, impedance, radiation	Object surface	Extended bc	Object surface	Extended bc	E on wires, H on surfaces	Object surface	Object surface and earth interface	Object surface	Object surface	Wire surface
Symmetry exploitation	No	No	Plane, rotational	No	Plane	Plane rotational translation	Plane rotational	Plane	Rotational	Rotational discrete	Rotational discrete
Integration	N/A	Numerical	Numerical	Numerical	Numerical	Numerical	Analytical, adaptive numerical	Numerical self term	Numerical	Numerical self term	Analytical closed form
Antenna radiation	No	Yes	No	No	No	Yes	Yes	No	Yes	No	Yes
Scattering	Yes	Yes	Yes	Yes	Yes	Yes	Yes	Yes	No	Yes	Yes
Ground	No	No	No	No	No	Perfectly conducting	RCA, Sommerfeld	Sommerfeld	RCA	No	Perfect ground, dielectric plane
Other		Basis functions provide current cont. at edges	—	—	—	Banded matrix technique	—	—	—	—	

User Features

Feature	FDTD, A. Taflove, Northwestern U., Chicago, IL	Surface Patch Code, E. H. Newman, Ohio State U., Columbus, OH	Main Scattering Program, P. Barber, Clarkson Inst. of Tech.	EM Scattering, Wilton, Rao & Glisson, U. of Miss., University, MS	LASMN3, Hagman, Grandi & Durney, U. of Utah, Salt Lake City, UT	GEMACS, Balestri et al., BDM Corp., Albq., NM	NEC, Burke et al., LLNL, Livermore, CA	MFIE/Aircraft Real Earth Code, Sancer & Siegel, Santa Monica, CA	COBRA, Medgyesi-Mitchgang, McDonnell Douglas, St. Louis, MO	ROTSY, C. L. Bennett, Raytheon, Sudbury, MA	TWTD, J. A. Landt, Amtech, Los Alamos, NM
Input data	Manual	Gross plate geometry	Cards (large computers); interactive (minicomp.)	Cards	Manual	Field-free language	Batch	Batch	Computer assisted	Manual	Manual batch
Output	Tabular, primitive graphics	Tab., orthographic plot of geometry	Tabular post-plotter	Tabular	Tabular	Line printer	Tabular	Menu sel. tabular	Menu selected	Tabular	Tabular plots
Input error checking	No	No	Yes	No	No	Yes	Input data checks	Minimal	On input data and during runs	Input checks, run-time checks	Input checks
Run-time checking	No	No	Yes	No	No	Yes	Integrated $E \times H$	No	No	Some	Some
Test problems	3-dimensional cylinders	Wires, wire on plates intersec. plates	Prolate spheroids	Sphere	Block man-model on ground plane	Wire grid sphere, reflectors	Yes	Sphere, cyl. aircraft model, cyl. slab	Monopoles on sphere; cone-sphere	Sphere	Dipole loop spiral
Debug help	No	No	No	Yes	?	Extensive, EMC/ IAP ctr	LLNL	Yes	No	Yes	Yes
Documentation within code	Yes	?	?	No	?	?	Some	?	Yes	Yes	Some

*Information supplied here is based primarily on responses given to a questionnaire. It is intended more to be illustrative of what code characteristics might be of interest than to be definitive about the codes actually listed.

†Codes selected for inclusion here are thought to be available from the authors, or can be obtained from LLNL as indicated.

‡An open surface is a conducting plate or shell which can become closed.

Table 14. Closer Comparison of NEC and TWTD Wire Codes (*After Miller [40]*)

Feature	NEC	TWTD
Language	FORTRAN IV	FORTRAN IV
Machines	CDC-6800, 7600; UNIVAC-1110; IBM-360/65 VAX11 CRAY 1, IBM 3033, XEROX Sigma 9	CDC-6600, 7600; IBM-360
Number of statements (cards)	8000	1000–2000 (depending on features)
Storage required (decimal, words)		
Program	$\cong 83\,000$	$\cong 30\,000$
Variable	$\cong 30N + 2N^2$	$\cong (12)N^2 + 2NN_t$
Time required (CDC-7600, seconds)	$\cong (10^{-4}N^2 + 2 \times 10^{-6}N^3)N_f$	$\cong 2.5 \times 10^{-5}N^2N_tN_i + 5 \times 10^{-6}NN_tN_iN_a$
Objects treated	Wires; Surfaces (continuous or wire grid); Wire-surface combination	Wires; Wire-grid surface
MOM technique used	Subsectional collocation	Subsectional collocation
Expansion (basis) functions	3-term trigonometric in space	9-term quadratic in space-time
Junction treatment, 2-wire	Amplitude and slope matching at segment juncture	Matches adjacent current samples
Junction treatment, multiwire	Charge density $\alpha \log (\alpha a)$ center	Matches equivalent segment at center
Testing (weight) function	Delta (point match)	Delta (point match)
Exciting (incident) field:		
Antenna	Tangential electric field, charge discontinuity, current source	Tangential electric field
Scatterer	Plane wave, local point source	Plane wave, local point source
Impedance loading:		
Lumped	Linear R, L, C	Nonlinear R, L, C
Distributed	Linear R, L, C; imperfectly conducting wire	Nonlinear R, L, C
Circuits	Nonradiating networks (including transmission lines)	Nonlinear networks

Symmetry included		
Planar	1 to 3 orthogonal planes	1 plane (storage and current computation)
Rotational	Up to 16 sectors	Up to n-fold (storage only)
Environments	Infinite medium (free-space or lossy). Lossy ground, wires on either side of or penetrating the interface	Free space. Perfect ground. Lossless, nondispersive, dielectric half-space
Input data preparation	Specify end points and connecting path (straight or circular arc), number of segments and radii. Provides for reflection or discrete rotation for object symmetry — Excitation — Options for output data	Specify end points and connecting (straight only), number of segments and radii. Also, special curved shapes (loop, spiral). Provides for reflection about ground plane. Time-variation of exciting field (functional or tabular). — Options for output data
Output data options	Impedance, admittance, input power, current and charge distributions, efficiency, radiation patterns; directive gain, power gain, average gain, "worst case" coupling, scattering cross section, near **E** & **H** fields, receiving pattern — NGF — file which can be used to add to the model and implement a partitioned matrix solution	Current and charge distributions near electric field, far fields in time domain and, via fast Fourier transform, impedance, admittance, far field, radiation pattern in frequency domain
Diagnostics	Data card error flags (geometry, dimension) — Run-time error messages (integration, pivot element)	Checks for time-step consistency ($\Delta tc \leqq \Delta_{max}$)
Documentation	Extensive, 3-volume set [23]: 1. Program description — theory 2. Program description — code 3. User's guide	Moderate: two somewhat redundant user's manuals [47, 48]

Note: N is number of space samples $\quad(>2\pi L/\lambda)$
N_t is number of time steps $\quad(\cong 2N)$
N_f is number of frequency steps $\quad(\cong N)$
N_i is number of incident fields $\quad(\cong 2\pi N)$
N_a is number of observation angles $\quad(\cong 2\pi N)$

These are sampling estimates based on the need to resolve spatial, temporal, or spectral variations to resolutions approximately 2π samples per wavelength or hertz. For N_i and N_a, there are assumed to be two principle-plane pattern cuts of 2π radian coverage with a total number of lobes of approximately (4 to 8)L/λ. Thus, with approximately 2π samples per lobe, $N_i \cong N_a \cong 2\pi N$. Factors of 2 variation in any of these sampling rates would not be uncommon.

Table 15. A Set of Representative Problems and Computer Times Using NEC and TWTD (After Miller [40])

Structure	Number N of Space Samples (NEC and TWTD)	Number N_t of Time Samples (TWTD)	Computer Time* CDC-7600 (s) TWTD ($N_i = 1$)	Computer Time CDC-7600 (s) NEC ($N_f = 1$)
Straight wire	48	480 (Well-converged solution)	$\cong 60$	$\cong 1$
Conical spiral	60	600 (Persistent ringing without resistive loading)	$\cong 100$	$\cong 1.5$
LP antenna	120	120	$\cong 45$	$\cong 8$
Truck (over perfect ground)	70	250	$\cong 60$	$\cong 4$
General			$\cong 2.5 \times 10^{-5} N^2 N_t N + 5 \times 10^{-6} N N_t N_a$	$\cong (5 \times 10^{-4} N^2 + 2 \times 10^{-6} N^3) N_f$

*Time is quite sensitive to i/o; a factor of 2 variation is not uncommon.

Table 16. Generic Modeling Guides

Parameter or Issue	Nominal value or Range	Reason
Wire length (L)	$L > 10d$	Neglect of end caps in thin-wire treatment
Wire diameter* (d)	$\lambda > \pi d$	Neglect of circumferential variation in current
Wire segment length* (Δ):		
As related to diameter	$\Delta > d$	Use fo thin-wire kernel in integral equation
As related to wavelength	$\lambda > 2\pi\Delta \qquad (\Delta < \lambda/2\pi)$	Necessity of sampling current densely enough in wavelengths
Step change in wire radius* (σa).	$\Delta > 10\sigma a$	Neglect of treatment for sources on stepped surface (similar to end-cap problem)
Source location	Do not place on last segment of open-ended wire.	Avoids nonphysical situation of driving wire at open end
Angle of wire bend (α).	$a > 6a/k\Delta$	Keep adjacent wires from occupying too large a common volume
Axial separation of parallel wires.		Neglect of circumferential variation of current
With match points aligned	$r \gtrsim 5a$	Avoids placing one match point in error field of other wire
> otherwise	$r \gtrsim 10a$	
Surface patch area (Δ_s):		Need to sample surface currents densely enough in wavelengths
Frequency domain	$\lambda > 6\sqrt{}\Delta_s$	
Time domain	$\lambda > 4\sqrt{}\Delta_s$	
Piecewise linear model of curved wire or curved surface of radius of curvature R	$R > \Delta \quad$ or $\quad \sqrt{}\Delta_s$	Necessity of sampling a circular arc at least six times per 2π radians
Starting time in time-domain solution (t_{st})	$E_v^i(t)\big|_{max} \gtrsim 100\, E(t_{st})$	Achieves smooth buildup of exciting field
Stopping time† in time-domain solution (t_{sp})	$E(t)\big|_{max} \gtrsim 100\lvert E(t) - E(t_{sp})\rvert$ or $I(t)$ reaches steady state	Permits final current or field value to stabilize
Time step (ϱ) in time-domain solution (c = velocity of light)	$c\varrho \lesssim \Delta$	Satisfies Courant stability condition. Required for MFIE, but not for EFIE
Maximum frequency of transient source in time-domain solution (assume Gaussian source $e^{-g^2+2} \Rightarrow e^{-\omega^2/4g^2}$	$g \propto 2f_{max} = \omega_{max}/\pi$	Ensures source spectrum does not exceed upper frequency for model validity

*Wavelength of frequency-domain solution or approximate minimum wavelength for valid time-domain solution.
†If steady-state response is not needed, then to t_{sp} could be substantially shorter.

Acknowledgments

The authors are indebted to the following individuals for their gracious support in the preparation of this chapter:

Marian Holten

Rose O'Brien

Margaret Poggio

Diane Ray

7. References

[1] R. F. Harrington, *Time-Harmonic Electromagnetic Fields*, New York: McGraw-Hill Book Co., 1961.

[2] J. A. Stratton, *Electromagnetic Theory*, New York: McGraw-Hill Book Co., 1941.

[3] S. Silver, *Microwave Antenna Theory and Design*, New York: McGraw-Hill Book Co., 1949.

[4] S. A. Schelkunoff, *Advanced Antenna Theory*, New York: John Wiley & Sons, 1952.

[5] S. A. Schelkunoff and H. T. Friis, *Antennas: Theory and Practice*, New York: John Wiley & Sons, 1952.

[6] B. B. Baker and E. T. Copson, *The Mathematical Theory of Huygens' Principle*, 2nd ed., London: Oxford University Press, 1953.

[7] A. J. Poggio and E. K. Miller, "Integral equation solutions for three-dimensional scattering problems," in *Computer Techniques for Electromagnetics*, ed. by R. Mittra, New York: Pergamon Press, pp. 159–264, 1973.

[8] E. A. Wolff, *Antenna Analysis*, New York: John Wiley & Sons, 1966.

[9] W. L. Weeks, *Antenna Engineering*, New York: McGraw-Hill Book Co., 1968.

[10] H. Jasik, ed., *Antenna Engineering Handbook*, New York: McGraw-Hill Book Co., 1961.

[11] R. E. Collin and F. J. Zucker, eds., *Antenna Theory*, pt. 1, New York: McGraw-Hill Book Co., 1969.

[12] R. E. Collin and F. J. Zucker, eds., *Antenna Theory*, pt. 2, New York: McGraw-Hill Book Co., 1969.

[13] W. L. Stutzman and G. A. Thiele, *Antenna Theory and Design*, New York: John Wiley & Sons, 1981.

[14] J. D. Kraus, *Antennas*, New York: McGraw-Hill Book Co., 1950.

[15] S. Ramo and J. R. Whinnery, *Fields and Waves in Modern Radio*, New York: John Wiley & Sons, 1960.

[16] C. H. Walker, *Traveling-Wave Antennas*, New York: Dover Publications, 1965.

[17] R. M. Bevensee, *A Handbook of Conical Antennas and Scatterers*, New York: Gordon and Breach Science Publishers, 1973.

[18] C. H. Papas, "Input impedance of wide-angle conical antennas," *Cruft Lab Tech. Rep. 52*, Harvard Univ., 1948.

[19] P. D. P. Smith, "The conical dipole of wide angle," *J. Appl. Phys.*, vol. 19, 1948.

[20] C. T. Tai, "On the theory of biconical antennas," *J. Appl. Phys.*, vol. 20, 1949.

[21] O. M. Salati, "Antenna chart for system designers," *Electron. Engineer*, January 1968.

[22] A. B. Bailey, *TV and Other Receiving Antennas*, New York: J. F. Ryder Publisher, 1950.

[23] G. J. Burke and A. J. Poggio, "Numerical electromagnetic code (NEC)—method of moments, vol. 1, pt. I: program description—theory; vol. 1, pt. II: program description—code; vol. 2, pt. III: user guide," *Tech. Doc. 116*, San Diego, CA 92152: Naval Ocean Systems Center, 1977.

[24] E. K. Miller, A. J. Poggio, G. J. Burke, and E. S. Selden, "Analysis of wire antennas

in the presence of a conducting half space, pt. I: the vertical antenna in free space," *Can J. Phys.*, vol. 50, 1972.

[25] E. K. Miller, A. J. Poggio, G. J. Burke, and E. S. Selden, "Analysis of wire antennas in the presence of a conducting half space, pt. II: the horizontal antenna in free space," *Can. J. Phys.*, vol. 50, 1972.

[26] H. Bach and J. E. Hansen, "Uniformly spaced arrays," in *Antenna Theory*, pt. I, ed. by R. E. Collin and F. J. Zucker, New York: McGraw-Hill Book Co., 1969.

[27] M. T. Ma, *Theory and Application of Antenna Arrays*, New York: John Wiley & Sons, 1974.

[28] A. W. Maue, "The formulation of a general diffraction problem by an integral equation," *Z. für Phys.*, Bd. 126, pp. 601–618.

[29] H. H. Chao and B. J. Strait, "Radiation and scattering by configurations of bent wires with junctions," *IEEE Trans. Antennas Propag.*, vol. AP-19, no. 5, 1971.

[30] E. Hallen, "Theoretical investigation into transmitting and receiving antennae," *Nova Acta Regiae Societatis Scientiarum Upsaliensis* (Sweden), 1938.

[31] K. K. Mei, "On the integral equation of thin wire antennas," *IEEE Trans. Antennas Propag.*, vol. AP-13, 1965.

[32] A. Sommerfeld, *Partial Differential Equations*, New York: Academic Press, 1969.

[33] A. Banos, *Dipole Radiation in the Presence of a Conducting Half Space*, New York: Pergamon Press, 1966.

[34] R. Mittra, "Integral equation methods for transient scattering," in *Transient Electromagnetic Fields*, ed. by L. B. Felsen, New York: Springer-Verlag, 1973.

[35] D. L. Sengupta and C.-T. Tai, "Radiation and reception of transients by linear antennas," in *Transient Electromagnetic Fields*, New York: Springer-Verlag, 1973.

[36] E. K. Miller, A. J. Poggio, and G. J. Burke, "An integro-differential equation technique for the time-domain analysis of thin-wire structures, I: the numerical method," *J. Comput. Phys.*, vol. 12, no. 1, 1973.

[37] A. J. Poggio, E. K. Miller, and G. J. Burke, "An integro-differential equation technique for the time-domain analysis of thin-wire structures, II: numerical results," *J. Comput. Phys.*, vol. 12, no. 2, 1973.

[38] R. F. Harrington, *Field Computation by Moment Methods*, New York: Macmillan Co., 1968.

[39] E. K. Miller, "Some computational aspects of transient electromagnetics," *Rept. UCRL 51276*, Lawrence Livermore Lab., 1972.

[40] E. K. Miller, "Time-domain modeling of wires," in *Application of the Method Moments to Electromagnetic Fields*, ed. by B. J. Strait, Kissimmee, Florida: SCEEE Press, 1980.

[41] E. K. Miller and F. J. Deadrick, "Some computational aspects of thin-wire modeling," in *Numerical and Asymptotic Techniques in Electromagnetics*, ed. by R. Mittra, Berlin: Springer-Verlag, 1975.

[42] E. K. Miller, G. J. Burke, and E. S. Selden, "Accuracy modeling guidelines for integral-equation evaluation of thin wire structures," *IEEE Trans. Antennas Propag.*, vol. AP-19, pp. 534–536, 1975.

[43] A. J. Poggio, R. M. Bevensee, and E. K. Miller, "Evaluation of some thin-wire computer programs," *1974 Intl. IEEE/AP-S Symp. Dig.*, Georgia Tech, Atlanta, GA.

[44] E. K. Miller and A. J. Poggio, "Moment method techniques in electromagnetics from an applications viewpoint," in *Electromagnetic Scattering*, ed. by P. O. E. Uslenghi, New York: Academic Press, 1978.

[45] D. R. Wilton and C. M. Butler, "Effective methods for solving integral and integro-differential equations," *Electromagnetics*, vol. 1, pp. 289–308, 1981.

[46] R. M. Bevensee, J. N. Brittingham, F. J. Deadrick, T. H. Lehman, E. K. Miller, and A. J. Poggio, "Computer codes for EMP interaction and coupling," *IEEE Trans. Antennas Propag.*, vol. AP-26, pp. 156–165, 1978.

[47] M. Van Blaricum and E. K. Miller, "TWTD: a computer program for the time-domain analysis of thin-wire structures," *Rept. UCRL 5127*, Lawrence Livermore Laboratory, 1972.

[48] J. A. Landt, E. K. Miller, and M. L. Van Blaricum, "WT-MBA/LLL1B: a computer program for the time-domain response of thin-wire structures," *Rept. UCRL 51585*, Lawrence Livermore Laboratory, 1974.

[49] E. K. Miller and J. A. Landt, "Direct time-domain techniques for transient radiation and scattering from wires," *Proc. IEEE*, vol. 68, pp. 1396–1423, 1980.

Chapter 4

Techniques for High-Frequency Problems

P. H. Pathak
Ohio State University ElectroScience Laboratory

CONTENTS

Prabhakar H. Pathak received the BSc degree in physics from the University of Bombay, India, in 1962, the BS degree in electrical engineering from Louisiana State University, Baton Rouge, in 1965, and the MS and PhD degrees in electrical engineering from Ohio State University, Columbus, in 1970 and 1973, respectively.

From 1965 to 1966 he was an instructor in the Department of Electrical Engineering at the University of Mississippi, Oxford. During the summer of 1966, he worked as an electronics engineer with the Boeing Company in Renton, Washington. Since 1968 he has been with the Ohio State University ElectroScience Laboratory, where his research interests have centered around mathematical methods, electromagnetic antenna and scattering problems, and uniform ray techniques. He is also an associate professor in the Department of Electrical Engineering at Ohio State University, where he teaches courses in electromagnetics, antennas, and linear systems.

Dr. Pathak has participated in invited lectures, and several short courses on the uniform geometrical theory of diffraction, both in the United States and abroad. He has also coauthored chapters on ray methods for four books. Dr. Pathak is a member of Commission B of the International Scientific Radio Union (URSI), and of Sigma Xi. He is a Fellow of the IEEE.

1. Introduction

Techniques based on the method of modal expansions, the Rayleigh-Stevenson expansion in inverse powers of the wavelength, and also the method of moments solution of integral equations are essentially restricted to the analysis of electromagnetic radiating structures which are small in terms of the wavelength. It therefore becomes necessary to employ approximations based on "high-frequency techniques" for performing an efficient analysis of electromagnetic radiating systems that are large in terms of the wavelength.

One of the most versatile and useful high-frequency techniques is the *geometrical theory of diffraction* (GTD), which was developed around 1951 by J. B. Keller [1, 2, 3]. A class of diffracted rays are introduced systematically in the GTD via a generalization of the concepts of classical geometrical optics (GO). According to the GTD these diffracted rays exist in addition to the usual incident, reflected, and transmitted rays of GO. The diffracted rays in the GTD originate from certain "localized" regions on the surface of a radiating structure, such as at discontinuities in the geometrical and electrical properties of a surface, and at points of grazing incidence on a smooth convex surface as illustrated in Fig. 1. In particular, the diffracted rays can enter into the GO shadow as well as the lit regions. Consequently, the diffracted rays entirely account for the fields in the shadow region where the GO rays cannot exist. Thus the GTD overcomes the failure of GO in the shadow region; it also improves the GO solution in the lit region. In this sense the GTD constitutes a significant improvement over GO. The initial amplitude of a diffracted ray is given in terms of a diffraction coefficient just as the initial values of GO reflected and transmitted rays are given in terms of reflection and transmission coefficients. Away from the points of diffraction the diffracted rays in the GTD propagate according to the laws of ordinary GO. The diffraction coefficients can be found from the asymptotic solutions to appropriate canonical problems. As a result, the GTD provides an efficient high-frequency solution to problems that cannot be solved rigorously. For example, an analysis of complex radiating systems such as antennas on aircraft, missiles, or ships can be performed efficiently by simulating those structures with approximate mathematical models which are "built up" from simpler shapes. These simpler shapes carefully simulate parts of the actual structure that dominate the reflection and diffraction effects, as is described later in Chapter 20 dealing with the engineering applications of GTD.

The ability of the GTD to accurately solve the problems of electromagnetic radiation and scattering from complex structures in a relatively simple, physically appealing, and efficient manner makes it a very powerful tool for antenna engineers. Moreover, the GTD can also provide information on ways to control the radiation and scattering properties of a structure. Although the GTD is a high-

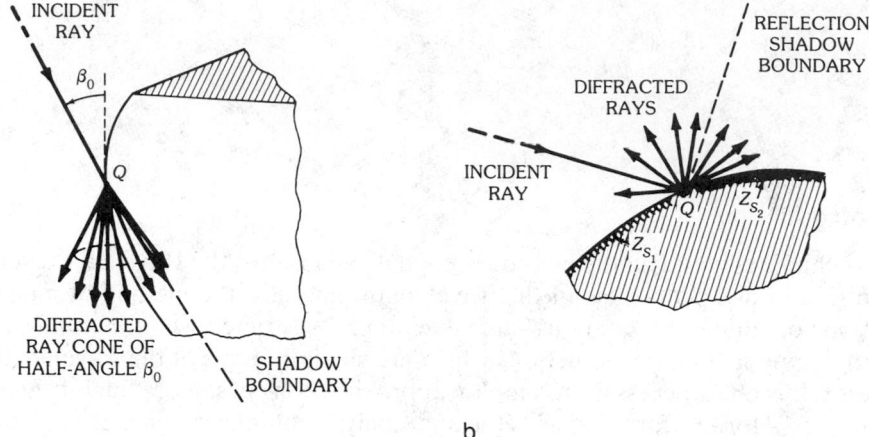

a   b

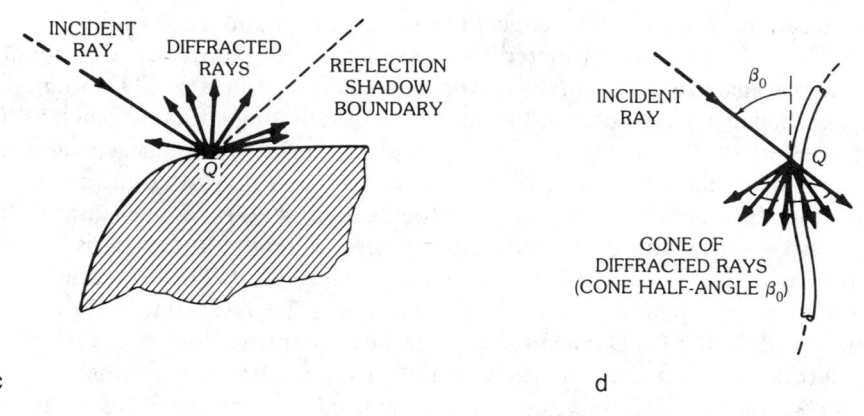

c   d

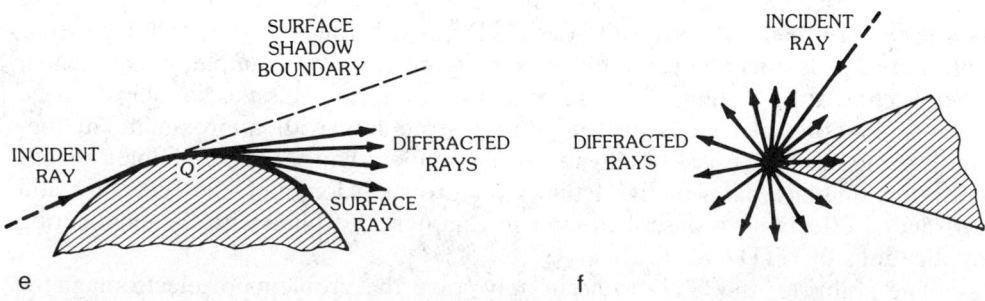

e   f

Fig. 1. Examples of diffracted rays. (*a*) Diffraction by a curved wedge. (*b*) Diffraction at a discontinuity in surface impedance (Z_{s_1} and Z_{s_2}). (*c*) Diffraction at a discontinuity in surface curvature. (*d*) Diffraction by a thin, curved wire. (*e*) Diffraction by a smooth, convex surface. (*f*) Diffraction by a vertex in a plane screen.

frequency technique it works surprisingly well in many situations, even for radiating objects almost as small as a wavelength in extent. Even though the GTD is not rigorous it yields the leading terms in the asymptotic expansions of many diffraction problems. However, because the GTD is a purely ray optical theory it fails within the transition regions adjacent to the GO incident and reflection shadow boundaries that divide the space surrounding an illuminated structure into the lit and shadow zones; it also fails at and near caustics* of diffracted rays just as GO fails at caustics of GO rays.

The failure of GTD within the incident and reflection shadow boundary transition regions can be overcome through the use of uniform ray techniques based on the *uniform geometrical theory of diffraction* (UTD) [4,5], and the *uniform asymptotic theory* (UAT) [6]. The UTD and UAT automatically reduce to the GTD outside the shadow boundary transition regions where the latter theory becomes valid. In engineering applications it is essential to use the uniform version of the GTD, namely the UTD or UAT, because the diffracted field generally assumes its strongest value at the shadow boundary where the GTD fails. The GTD, UTD, and UAT fail within the diffracted-ray caustic regions; in these regions the GTD, UTD, and UAT may be augmented by an *equivalent current method* (ECM) [7,8,9], which indirectly employs the GTD far from caustics to obtain equivalent currents that radiate fields valid at caustics. The ECM, which generally reduces to the GTD outside the caustic regions, is good only when the GO shadow boundary transition regions do not overlap with the diffracted-ray caustic regions. The GTD, UTD, UAT, and ECM all generally fail in regions where there is an overlap of the GO shadow boundary transition regions with the diffracted and/or GO ray caustic regions, especially if the illumination is nonuniform; in such regions the fields may be calculated via the *physical theory of diffraction* (PTD) [10,11] and its modifications. The PTD was developed by P. Ya. Ufimtsev in the Soviet Union at about the same time as Keller's GTD, and it constitutes a systematic extension of the method of physical optics (PO) just as GTD is an extension of GO. The PO field is calculated from a radiation integral [12] which employs a GO approximation for the surface currents on the structure. At the present time the PTD is developed only for edged bodies and is therefore not as general as the GTD and its uniform versions.

The format of this chapter is as follows. First the GO and PO high-frequency methods are reviewed in Sections 2 and 3. Then a discussion of the GTD, UTD, and UAT is presented in Section 4. Subsequently the ECM and PTD are discussed in Sections 5 and 6. The treatment in this chapter is restricted to perfectly conducting structures in an isotropic, homogeneous medium. Not included in this chapter are some recent generalizations of the ray methods, such as those dealing with Gaussian beams [13], and the hybrid combinations of ray methods with the moment method or model techniques [14,15,16]. In this chapter the fields are assumed to be time harmonic with an $e^{j\omega t}$ time dependence, which is suppressed. Only here does ω refer to the angular frequency and t refer to time.

---

*Ray caustics or focii occur whenever a family of rays (i.e., ray congruences) merge or intersect to form a focal surface, a focal line, or a focal point.

2. Wavefronts, Rays, and the Geometrical Optics Field

In this section some background material is introduced briefly to indicate the connection between the wavefronts and rays at high frequencies, and to also provide examples of ray optical fields. In addition, the basic ideas of the geometrical optics (GO) ray technique are reviewed.

The Ray Concept

A wavefront is an equiphase surface. At high frequencies the electromagnetic energy flow in an isotropic medium is associated primarily with the propagation of its wavefront along curved paths which are everywhere normal to the wavefront. Such highly localized paths of wave propagation which are directed along the normals to the wavefronts are called *rays*; these ray paths are straight lines in a homogeneous medium. The family of rays associated with the propagation of a wavefront is usually referred to as a *normal congruence of rays*.

The above connection between wavefronts and rays may be established via an application of the method of stationary phase to the radiation integral over the equivalent (Huygens') sources which are associated with the field distribution on the wavefront; this procedure constitutes a straightforward generalization of the scalar treatment of rays and wavefronts given by Silver [12]. Let $\mathbf{E}(\mathbf{r}')$ and $\mathbf{H}(\mathbf{r}')$ refer to the electric and magnetic fields at any point $\mathbf{r}'$ on the wavefront surface S', and let dS' be an elemental area of S' at the point $\mathbf{r}'$ as shown in Fig. 2. Let $d\mathbf{E}(\mathbf{r})$ denote the field associated with a spherical wave at P which originates from dS'. Thus the total electric field $\mathbf{E}(\mathbf{r})$ at P is then obtained by superposing the spherical wave contributions from each elemental dS' over the entire surface S'; namely,

$$\mathbf{E}(\mathbf{r}) = \iint_{S'} d\mathbf{E}(\mathbf{r}) \tag{1a}$$

where

$$d\mathbf{E}(\mathbf{r}) = \frac{jkZ_0}{4\pi}[\hat{\mathbf{R}} \times \hat{\mathbf{R}} \times \mathbf{J}_s(\mathbf{r}') + Y_0\hat{\mathbf{R}} \times \mathbf{M}_s(\mathbf{r}')]\frac{e^{-jkR}}{R} dS' \tag{1b}$$

in which the equivalent (or Huygens') electric and magnetic current sources on S' are given by $\mathbf{J}_s(\mathbf{r}') = \hat{\mathbf{n}}' \times \mathbf{H}(\mathbf{r}')$ and $\mathbf{M}_s(\mathbf{r}') = \mathbf{E}(\mathbf{r}') \times \hat{\mathbf{n}}'$. It is noted that $Z_0 = (Y_0)^{-1}$ denotes the wave impedance in the medium, k constitutes the wave number of the medium, and $\hat{\mathbf{n}}'$ denotes the unit outward normal vector to S' at $\mathbf{r}'$.

For convenience the point O is chosen in the present development to correspond to a point on S' which is nearest to a given observation point P such that $OP = \hat{\mathbf{z}}|OP|$, and also such that the unit outward normal vector $\hat{\mathbf{n}}'$ to the surface S' is given by $\hat{\mathbf{n}}' = \hat{\mathbf{z}}$ at O, as in Fig. 2. Generally, there exists at least one such point O. To start with, it is assumed that there exists only one such point O on S'; otherwise, the point P is said to be a *caustic* or *focal point*.

It can be shown via the principle of stationary phase [12] that e^{-jkR} in the integrand of (1b) oscillates rapidly for large k to produce a destructive interference or cancellation between the different spherical wave contributions to P in (1a) that

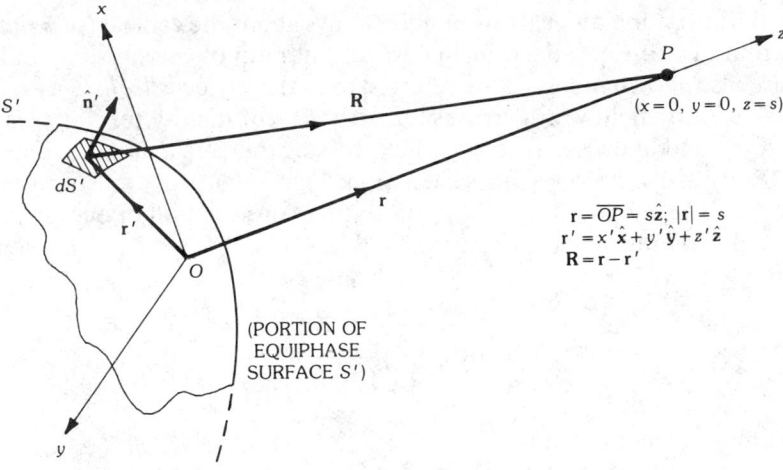

Fig. 2. Wavefront geometry.

arise from each of the equivalent sources on dS' at points on S' which do not lie in the neighborhood of O, whereas e^{-jkR} changes slowly to produce a constructive interference between the various spherical wave contributions to the radiation integral of (1a) from points on S' which lie in the neighborhood of O. The point O is thus referred to as the *stationary phase point*. Therefore, at high frequencies (or large k) the dominant contribution to the integral in (1a) comes from the *stationary* point at O.

In particular, an evaluation of (1a) via the method of stationary phase [12] yields

$$\mathbf{E}(P) \sim \mathbf{E}(O) \sqrt{\frac{\varrho_1}{(\varrho_1 + s)} \frac{\varrho_2}{(\varrho_2 + s)}} \, e^{-jks}, \qquad |OP| = s \qquad (2)$$

where ϱ_1 and ϱ_2 are the principal radii of curvature of the wavefront surface S' at the stationary point O. It is important to note that

$$\sqrt{\frac{\varrho}{\varrho + s}} = \left| \sqrt{\frac{\varrho}{\varrho + s}} \right| e^{j\{^0_{\pi/2}\}} \qquad \text{if} \quad \frac{\varrho}{\varrho + s} \begin{cases} > 0 \\ < 0 \end{cases} \qquad (3)$$

in which ϱ can be either ϱ_1 or ϱ_2. The above result in (2) indicates that the electric field $\mathbf{E}(P)$ arrives at P along a "ray" from O to P, i.e., this field is a continuation of the electric field $\mathbf{E}(O)$ at the point O on the initial wavefront S' to the wavefront containing the observation point P. It is noted that the amplitude of $\mathbf{E}(P)$ varies as $\sqrt{\varrho_1\varrho_2/(\varrho_1 + s)(\varrho_2 + s)}$ along the ray, and e^{-jks} represents the phase path delay along that ray. Furthermore, $\mathbf{E}(O)$ is polarized transverse to the ray path OP, and $\mathbf{E}(P)$ also has the same polarization as $\mathbf{E}(O)$ because the ray path is straight in a homogeneous medium. Since energy in the high-frequency electromagnetic (EM) field is transported along rays, it is evident from geometrical considerations that energy must be conserved in a tube of rays (or a ray bundle). This may be verified by considering a tube or a narrow bundle of rays about the central (or axial) ray OP as shown in Fig. 3. First, the principal wavefront radii of curvature ϱ_1 and ϱ_2 at O, which are also shown in Fig. 3, are referred to as the *ray caustic distances*. Now, the energy flux in the field which crosses the area dA_O of the ray tube at O is given by $|\mathbf{E}(O)|^2 dA_O$, and likewise, the energy flux crossing the area dA of the same ray tube at P is $|\mathbf{E}(P)|^2 dA_P$; however, it is clear from Fig. 3 that $dA_O \cong |(\varrho_1 d\psi_1)(\varrho_2 d\psi_2)|$, and $dA_P \cong |[(\varrho_1 + s)d\psi_1][(\varrho_2 + s)d\psi_2]|$, so that conservation of energy flux within the ray tube requires one to satisfy $|\mathbf{E}(P)|^2 dA_P = |\mathbf{E}(O)|^2 dA_O$. It then follows immediately that

$$|\mathbf{E}(P)| = |\mathbf{E}(O)| \left| \sqrt{\frac{\varrho_1 \varrho_2}{(\varrho_1 + s)(\varrho_2 + s)}} \right| \tag{4}$$

which is in agreement with (2).

The field in (2) associated with the ray OP of Fig. 3 is referred to as an *arbitrary ray optical field* since ϱ_1 and ϱ_2 can be arbitrary. Also, the ray tube in Fig. 3 is commonly referred to as an *astigmatic ray tube* or a *quadratic ray pencil* because of the quadratic wavefront surface approximation at O which is employed in the stationary phase method to obtain the result in (2). It can be easily verified that plane, cylindrical, conical, and spherical wave fields are special cases of an arbitrary ray optical field, i.e., each of those wave fields is also a ray optical field.

The Geometrical Optics Field

Geometrical optics (GO) is a high-frequency approximation which employs rays to describe the fields that are directly incident from the source, and to describe the fields which are reflected and refracted (or transmitted) at an interface between two different media. According to classical GO, the high-frequency electromagnetic field is assumed to propagate along ray paths which satisfy Fermat's principle, and the family of rays is everywhere orthogonal to the wavefronts in an isotropic medium. The ray paths are straight lines in a homogeneous medium, but they can

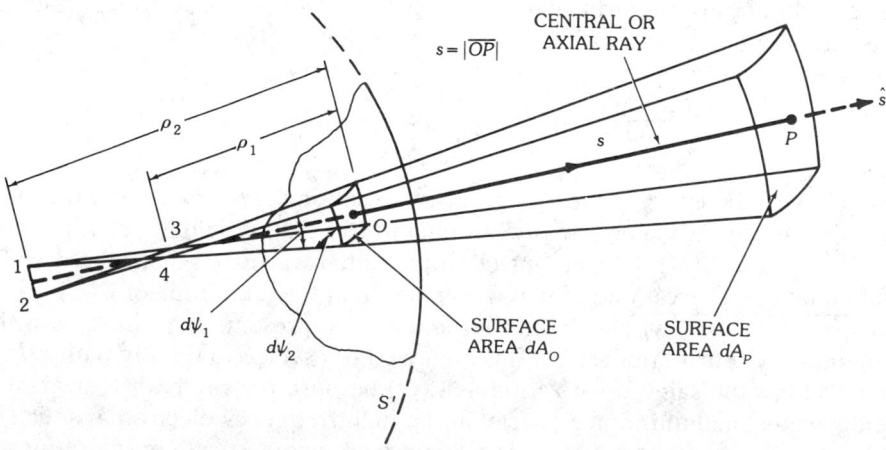

Fig. 3. Geometrical (ray tube) interpretation of the result in (2).

change directions at an interface between two different media according to the well-known (Snell's) laws of reflection and refraction, which can be deduced from Fermat's principle. The amplitude variation of the classical GO field is obtained by requiring conservation of energy flux in an astigmatic ray tube associated with a central (or axial) ray as indicated previously with the aid of Fig. 3. These considerations lead to an expression identical with that in (4), as might be expected. The information on the field polarization (as being transverse to the ray) and on the wave nature of the field is then introduced heuristically in the classical GO approximation of (4) to arrive at a more complete expression for the GO field. The latter, more complete expression is found to be identical with that obtained in (2) via stationary phase considerations. Indeed, the surface S' in Fig. 3 could be associated with any wavefront; thus, if S' represents the wavefront of an incident or reflected type ray congruence, then (2) would represent the GO incident or reflected electric field, respectively.

An alternative development for the GO field is based on an asymptotic high-frequency solution to Maxwell's equations in which the electromagnetic fields are expanded in inverse powers of the wave number k, as done by Luneberg and Kline [17, 18]. The leading term in their expansion corresponds to the GO field. According to Luneberg and Kline the electric field $\mathbf{E}$ in a source-free, homogeneous, isotropic medium can be expressed at high frequencies by

$$\mathbf{E}(\mathbf{r}, k) \sim e^{-jk\psi(\mathbf{r})} \sum_{n=0}^{\infty} \frac{\mathbf{E}_n(\mathbf{r})}{(-jk)^n} \tag{5}$$

in which $\mathbf{r}$ is the position vector of the field point. Substituting (5) into the vector Helmholtz equation satisfied by $\mathbf{E}$, namely $(\nabla^2 + k^2)\mathbf{E} = 0$ (in which ∇^2 is the Laplacian operator), leads to the usual eikonal and transport equations $|\nabla\psi|^2 = 1$ and $[\nabla^2\psi + 2\nabla\psi\cdot\nabla]\mathbf{E}_n = -\nabla^2\mathbf{E}_{n-1}$ (with $\mathbf{E}_{-1} \equiv 0$), respectively. The surfaces of constant ψ are defined as the wavefronts. Integrating the transport equation from some reference point $\mathbf{r}_O$ to $\mathbf{r}$ for the $n = 0$ case yields the leading term in the Luneberg-Kline expansion as

$$\mathbf{E}(\mathbf{r}) \sim \mathbf{E}(\mathbf{r}_O) \sqrt{\frac{\varrho_1\varrho_2}{(\varrho_1 + s)(\varrho_2 + s)}} \, e^{-jks} \tag{6}$$

in which

$$\mathbf{E}(\mathbf{r}_O) \equiv \mathbf{E}_0(\mathbf{r}_O) e^{-jk\psi(\mathbf{r}_o)}$$

where $\mathbf{E}_0$ implies $\mathbf{E}_n$ for $n = 0$ in (5). The quantities ϱ_1, ϱ_2, and s in (6) have the same meaning as in Fig. 3, with $\mathbf{r}$ being the position vector of P and $\mathbf{r}_O$ being the position vector of O. Furthermore, from the requirement $\nabla\cdot\mathbf{E}(\mathbf{r}) = 0$ for source-free regions, one obtains $\hat{\mathbf{s}}\cdot\mathbf{E}(\mathbf{r}) = 0$, which implies that $\mathbf{E}(\mathbf{r})$ in (6) is polarized transverse to the ray direction $\hat{\mathbf{s}}$ (see Fig. 3). It is thus evident that (6) is the same as (2). A Luneberg-Kline expansion for $\mathbf{H}$ yields

$$\mathbf{H}(\mathbf{r}) \sim \hat{\mathbf{s}} \times Y_0 \mathbf{E}(\mathbf{r}) \tag{7}$$

In (6) the distance s is measured as positive in the direction of ray propagation. The caustic distances ϱ_2 and ϱ_1 are positive if the associated caustics at $1-2$ and $3-4$ in Fig. 3 occur before the reference point O as one propagates along the ray; otherwise, they are negative. The positive branch of the square roots is chosen in (6). Therefore, if $\varrho_{1,2} < 0$ and $s > -|\varrho_2|$, or $s > -|\varrho_1|$, then a caustic is crossed at $1-2$ or $3-4$ (see Fig. 3), respectively, and $(\varrho_2 + s)$ or $(\varrho_1 + s)$ changes sign within the square root of (6) so that a phase jump of $\pi/2$ results naturally in each case; namely,*

$$\sqrt{\frac{\varrho_i}{\varrho_i + s}} = \left|\frac{\varrho_i}{\varrho_i + s}\right| e^{j\pi/2}, \qquad \text{if } s > -|\varrho_i| \quad \text{for } i = 1, 2$$

The above result is consistent with (3), which is implied in (2). The GO field in (6) is therefore completely consistent with the expression in (2) which was obtained via the method of stationary phase. At a ray caustic, $s = -|\varrho_i|$ for $i = 1, 2$, and the GO field in (6) and (2) becomes infinite. Consequently, the GO field representation fails at and near the GO ray caustics.

For a perfectly conducting surface illuminated by a source (antenna), only the fields directly incident from the source and the fields reflected by that surface in just the specular direction can be described by GO. These GO incident and reflected fields are discussed next for the two-dimensional (2-D) and three-dimensional (3-D) cases.

The GO Incident Field (3-D Case)—The GO incident field is the field directly radiated by the source (antenna) to the observation point; this incident ray optical (or GO) field exists "in the presence of" the surface which it illuminates. Part of the GO incident ray system (or congruence) which strikes an impenetrable surface is blocked by that surface; as a result the surface creates a shadow region behind it where the incident rays cannot exist, and consequently GO predicts a zero field in the shadow zone! A shadow boundary is created naturally by the presence of the shadow region; this boundary divides the space surrounding the surface into the lit and shadow zones corresponding to regions where the source is directly visible and where it is not, respectively. In particular, there is an incident shadow boundary (ISB) for a perfectly conducting structure with an edge, such as a wedge as in Fig. 4a, and likewise there is a surface shadow boundary (SSB) associated with a smooth, perfectly conducting, convex surface as shown in Fig. 4b. Thus the GO incident field is truncated to zero at the incident shadow boundary (ISB or SSB) so that it vanishes within the shadow region of an obstacle illuminated by that field.

If the incident wavefront is characterized by two different radii of curvature ϱ_1^i and ϱ_2^i, then the GO incident field is given by (6) as

$$\mathbf{E}^i(\mathbf{r}) = \mathbf{E}^i(\mathbf{r}_O) \sqrt{\left(\frac{\varrho_1^i}{\varrho_1^i + s^i}\right)\left(\frac{\varrho_2^i}{\varrho_2^i + s^i}\right)} \, e^{-jks^i} \qquad (8)$$

---

*Actually, if both caustics (at $1-2$ and $3-4$) are crossed, then a total phase jump of $\pi/2 + \pi/2 = \pi$ results.

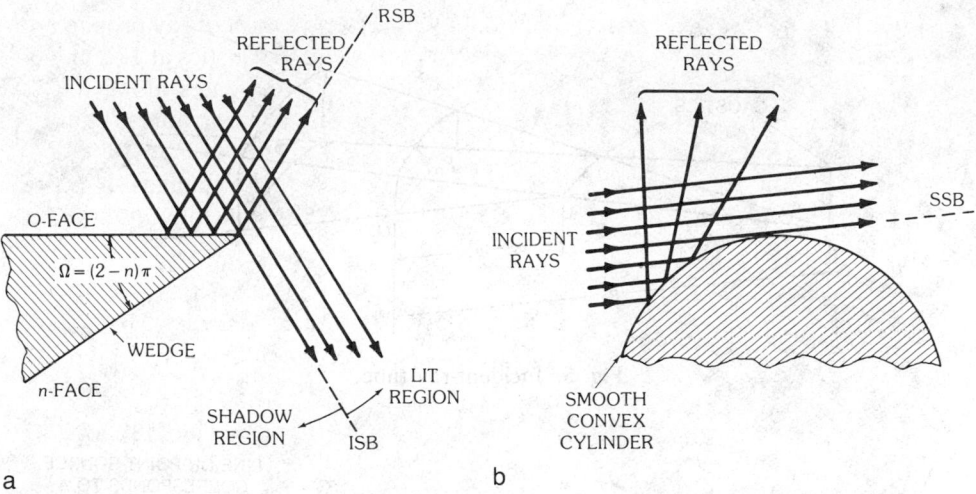

Fig. 4. Shadow boundaries ISB and SSB. (*a*) Incident shadow boundary. (*b*) Surface shadow boundary.

where the superscript i is employed to denote quantities associated with the incident ray as in Fig. 5. In the case of plane-wave illumination, ϱ_1^i and ϱ_2^i are infinite, so that (8) reduces to

$$\mathbf{E}^i(\mathbf{r}) = \mathbf{E}^i(\mathbf{r}_O)e^{-jks^i} \tag{9}$$

for a *plane-wave* incidence. Likewise, in the case of spherical-wave illumination for which $\varrho_1^i = \varrho_2^i \equiv \varrho^i$, and for which the reference point P_O in Fig. 5 is moved to the point caustic (so that $\varrho^i \to 0$), it is seen that for a *spherical-wave* or *point-source* type illumination, (8) reduces to

$$\mathbf{E}^i(\mathbf{r}) = \mathbf{C}^i \frac{e^{-jks^i}}{s^i} \tag{10}$$

with

$$\mathbf{C}^i \equiv \lim_{\varrho^i \to 0} \varrho^i \mathbf{E}^i(\mathbf{r}_O)$$

in which $\mathbf{C}^i$ is related to the strength of the point source. The plane-wave and spherical-wave type of GO incident fields are illustrated in Fig. 6.

It is noted that the incident magnetic field $\mathbf{H}^i(\mathbf{r})$ associated with $\mathbf{E}^i(\mathbf{r})$ of (8) through (10) is simply given by

$$\mathbf{H}^i(\mathbf{r}) \sim Y_0 \hat{\mathbf{s}}^i \times \mathbf{E}^i(\mathbf{r}) \tag{11}$$

The GO Incident Field (2-D Case)—In the case of cylindrical wave or 2-D line

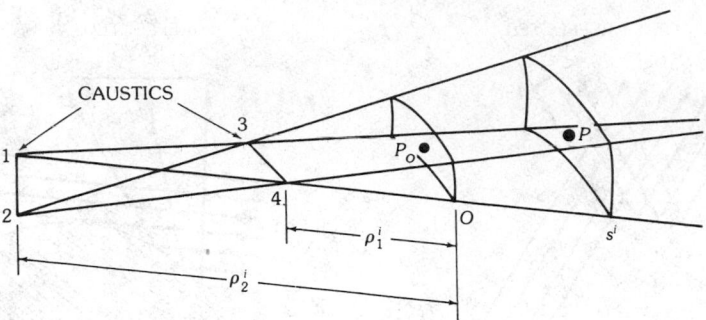

Fig. 5. Incident-ray tube.

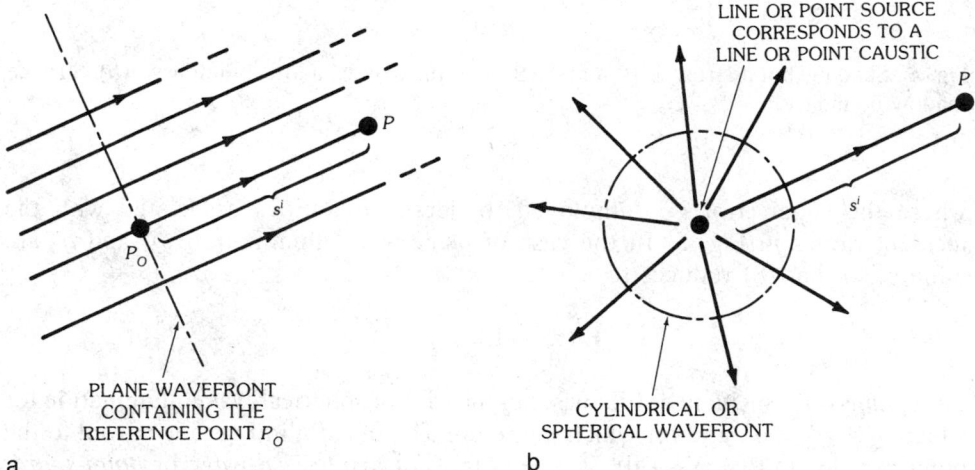

a   b

Fig. 6. Plane-wave and cylindrical- or spherical-wave type GO incident fields. (*a*) Plane-wave GO incident field. (*b*) Cylindrical- or spherical-wave GO incident field.

source illumination, one may let $\varrho_1^i \to \infty$ in (8), and allow the reference point P_O (at $\mathbf{r}_O$) in Fig. 5 be moved to the caustic 1–2 for convenience so that $\varrho_2^i \to 0$, together with the condition that

$$\lim_{\varrho_2^i \to 0} \sqrt{\varrho_2^i\, \mathbf{E}^i(\mathbf{r}_O)} \equiv \mathbf{C}^i$$

where $\mathbf{C}^i$ is finite and related to the strength of the line source. Incorporating this limit in (8) yields

$$\mathbf{E}^i(\mathbf{r}) = \mathbf{C}^i \frac{e^{-jks^i}}{\sqrt{s^i}} \tag{12}$$

for a *cylindrical wave* or *line source* type illumination.

The magnetic field $\mathbf{H}^i(\mathbf{r})$ can be obtained from $\mathbf{E}^i(\mathbf{r})$ via (11).

The GO Reflected Field (3-D Case)—When a family of rays incident from a source strike a perfectly conducting surface, they are transformed at the surface into a family of reflected rays as shown in Figs. 4a and 4b. Such a reflected ray congruence is discontinuous across the reflection shadow boundary (RSB) for the edge type structure shown in Fig. 4a; on the other hand, the incident and reflection shadow boundaries merge into the surface shadow boundary (SSB) for the convex surface as shown in Fig. 4b. Furthermore, the GO reflected field vanishes at the SSB and also within the shadow region for the problem in Fig. 4b. The GO field associated with a reflected ray has the form indicated in (6). In particular, the electric field $\mathbf{E}^r(\mathbf{r})$ associated with a GO reflected ray is given by

$$\mathbf{E}^r(P) = \mathbf{E}^r(Q_R) \sqrt{\frac{\varrho_1^r}{(\varrho_1^r + s^r)} \frac{\varrho_2^r}{(\varrho_2^r + s^r)}} \, e^{-jks^r} \tag{13}$$

where the reference position Q_R on the ray is chosen here to be at the point of reflection on the surface as shown in Fig. 7. The distances ϱ_1^r, ϱ_2^r, and s^r are associated with the reflected ray from Q_R to P. The reflected field $\mathbf{E}^r(Q_R)$ is related to the incident field $\mathbf{E}^i(Q_R)$ at the point of reflection Q_R by the boundary condition

$$\hat{\mathbf{n}} \times [\mathbf{E}^i(Q_R) + \mathbf{E}^r(Q_R)] = 0 \tag{14}$$

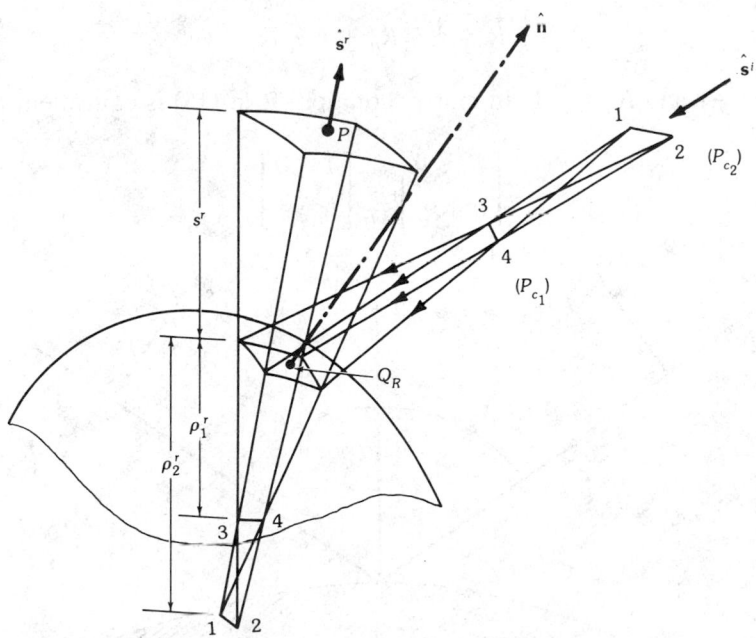

Fig. 7. Reflected-ray tube.

which requires the total tangential electric field to vanish on the perfectly conducting surface. In (14), $\hat{n}$ is the unit normal vector to the surface at Q_R. As a consequence of (14), the following relationship holds:

$$\mathbf{E}^r(Q_R) = \overline{\overline{\mathbf{R}}} \cdot \mathbf{E}^i(Q_R) \tag{15}$$

where $\overline{\overline{\mathbf{R}}}$ denotes the dyadic surface reflection coefficient at Q_R.
Incorporating (15) into (13) yields

$$\mathbf{E}^r(P) = \mathbf{E}^i(Q_R) \cdot \overline{\overline{\mathbf{R}}} \sqrt{\frac{\varrho_1^r}{\varrho_1^r + s^r} \frac{\varrho_2^r}{\varrho_2^r + s^r}} \, e^{-jks^r} \tag{16}$$

Likewise, the associated reflected magnetic field $\mathbf{H}^r(P)$ is simply given by

$$\mathbf{H}^r(P) \sim Y_0 \hat{s}^r \times \mathbf{E}^r(P) \tag{17}$$

The expression for $\overline{\overline{\mathbf{R}}}$ simplifies if one expresses the fields in terms of an appropriate set of unit vectors which are fixed in the incident and reflected rays. Thus, let $\hat{\mathbf{e}}_{\parallel}^i$ be a unit vector transverse to $\hat{s}^i$ such that it lies in the plane of incidence defined by $\hat{n}$ and $\hat{s}^i$ as in Fig. 8. It is noted that the reflected ray along $\hat{s}^r$ also lies in the plane of incidence. Likewise, let $\hat{\mathbf{e}}_{\parallel}^r$ be a unit vector in the plane of incidence but transverse to the reflected ray direction $\hat{s}^r$ as in Fig. 8, and let $\hat{\mathbf{e}}_{\perp}$ be a unit vector perpendicular to the plane of incidence. The $\hat{\mathbf{e}}_{\perp}$ is then automatically transverse to the directions of incidence and reflection, respectively. In this ray coordinate system,

$$\overline{\overline{R}} = \hat{\mathbf{e}}_{\parallel}^i \hat{\mathbf{e}}_{\parallel}^r R_h + \hat{\mathbf{e}}_{\perp} \hat{\mathbf{e}}_{\perp} R_s \tag{18}$$

with $R_{\hat{s}} = -1$ and $R_{\hat{h}} = +1$. In matrix notation, $\overline{\overline{\mathbf{R}}}$ in (18) is equivalent to

$$\begin{bmatrix} R_h & 0 \\ 0 & R_s \end{bmatrix} = \begin{bmatrix} 1 & 0 \\ 0 & -1 \end{bmatrix}$$

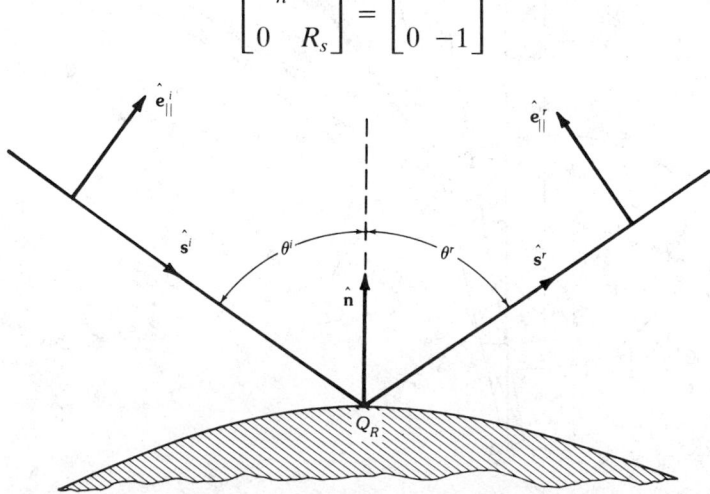

Fig. 8. Ray unit vectors associated with the reflection problem.

or (16) is equivalent via (18) to

$$
\begin{bmatrix} E_\parallel^r(P) \\ E_\perp^r(P) \end{bmatrix} = \begin{bmatrix} 1 & 0 \\ 0 & -1 \end{bmatrix} \begin{bmatrix} E_\parallel^i(Q_R) \\ E_\perp^i(Q_R) \end{bmatrix} \sqrt{\frac{\varrho_1^r \varrho_2^r}{(\varrho_1^r + s^r)(\varrho_2^r + s^r)}}\, e^{-jks^r}
$$

where $\mathbf{E}^i(Q_R) = E_\parallel^i(Q_R)\hat{\mathbf{e}}_\parallel^i + E_\perp^i(Q_R)\hat{\mathbf{e}}_\perp$ and $\mathbf{E}^r(P) = E_\parallel^r(P)\hat{\mathbf{e}}_\parallel^r + E_\perp^r(P)\hat{\mathbf{e}}_\perp$ in the ray coordinates of Fig. 8.

The principal radii of curvature $(\varrho_1^i, \varrho_2^i)$ and $(\varrho_1^r, \varrho_2^r)$ of the incident and reflected wavefronts and their associated principal directions $(\hat{\mathbf{X}}_1^i, \hat{\mathbf{X}}_2^i)$ and $(\hat{\mathbf{X}}_1^r, \hat{\mathbf{X}}_2^r)$ are given in [19], and $(\varrho_1^r, \varrho_2^r)$ are expressed more compactly here as*

$$
\frac{1}{\varrho_{1,2}^r} = \frac{1}{\varrho_m^i} + \frac{f}{\varrho_g(Q_R)\cos\theta^i}\left(1 \pm \left\{\frac{\varrho_g^2(Q_R)\cos^2\theta^i}{4f^2} \cdot \left(\frac{1}{\varrho_1^i} - \frac{1}{\varrho_2^i}\right)^2\right.\right.
$$

$$
+ \frac{\varrho_g^2(Q_R)\cos\theta^i}{f^2}\left(\frac{1}{\varrho_1^i} - \frac{1}{\varrho_2^i}\right) \cdot \left[\frac{g\cos 2\alpha_0}{\varrho_g(Q_R)} - \sin 2\alpha_0 \sin 2\omega_0 \cos\theta^i\right.
$$

$$
\times \left.\left.\left(\frac{1}{R_1} - \frac{1}{R_2}\right)\right] + 1 - \frac{4\varrho_g^2(Q_R)\cos^2\theta^i}{f^2 R_1 R_2}\right\}^{1/2}\Bigg) \tag{19a}
$$

where $R_1 > 0$ for a convex surface and $R_2 < 0$ for a concave surface, and

$$
\frac{1}{\varrho_m^i} = \frac{1}{2}\left(\frac{1}{\varrho_1^i} + \frac{1}{\varrho_2^i}\right) \tag{19b}
$$

$$
\begin{Bmatrix} f \\ g \end{Bmatrix} = \left\{1 \pm \frac{\varrho_g(Q_R)}{\varrho_t(Q_R)}\cos^2\theta^i\right\} \tag{19c}
$$

The quantities R_1 and R_2 constitute the principal radii of curvatures of the surface at Q_R, and $\hat{\mathbf{U}}_1$ and $\hat{\mathbf{U}}_2$ denote the corresponding principal surface directions at Q_R. The quantity ϱ_g denotes the radius of curvature of the surface at Q_R in the plane of incidence which contains $\hat{\mathbf{s}}^i$, $\hat{\mathbf{n}}$, and $\hat{\mathbf{t}}$, where $\hat{\mathbf{t}}$ is tangent to the surface. Also, ϱ_t is the radius of curvature of the surface at Q_R in the plane containing $\hat{\mathbf{n}}$ and the binormal vector $\hat{\mathbf{b}}$. The unit vectors $\hat{\mathbf{t}}$, $\hat{\mathbf{n}}$, $\hat{\mathbf{b}}$, $\hat{\mathbf{U}}_1$, and $\hat{\mathbf{U}}_2$ are shown in Fig. 9a together with the angle ω_0 between $\hat{\mathbf{t}}$ and $\hat{\mathbf{U}}_2$.

Note that ω_0 corresponds to the angle between the plane of incidence and one of the principal surface directions $(\hat{\mathbf{U}}_2)$ at the point of reflection.

The principal directions $(\hat{\mathbf{X}}_1^i, \hat{\mathbf{X}}_2^i)$ of the incident wavefront and α_0 are shown in Fig. 9b. The angle of incidence θ^i is defined by $\hat{\mathbf{n}}\cdot\hat{\mathbf{s}}^i = -\cos\theta^i = -\hat{\mathbf{n}}\cdot\hat{\mathbf{s}}^r$.

A matrix formulation to calculate $(\varrho_1^r, \varrho_2^r)$ is given in [20, 21, 22]; in particular, [20, 22] contain results for both reflected and transmitted ray caustics in the case of penetrable media.

---

*The principal radii of curvature $\varrho_\ell^{i,r} = \pm|\varrho_\ell^{i,r}|$ for a diverging (converging) pair of adjacent rays. Here, $\ell = 1, 2$.

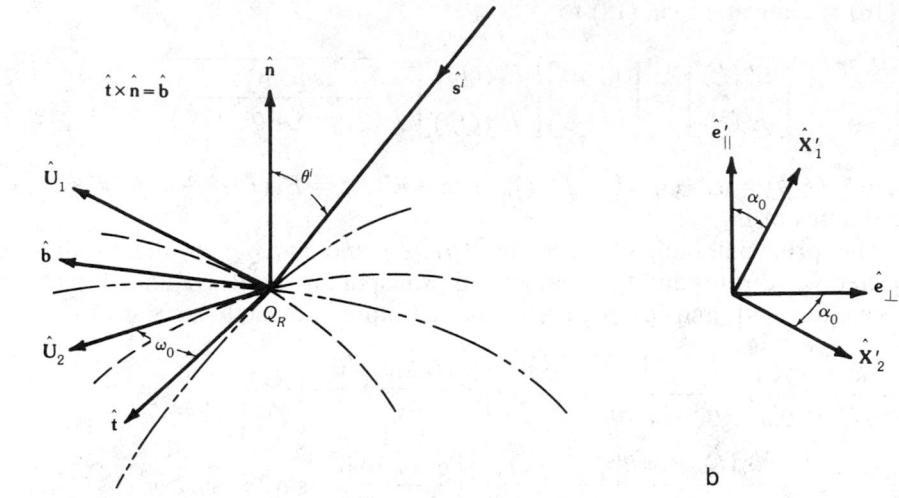

Fig. 9. Geometry for description of wavefront reflected from a curved surface. (*a*) Showing unit vectors. (*b*) Showing principal directions. (*After Sitka, Burnside, Chu, and Peters [27],* © *1983 IEEE*)

The incident spherical wavefront is commonly encountered; for this case it can be shown that

$$\frac{1}{\varrho^r_{1,2}} = \frac{1}{s^i} + \frac{1}{\cos\theta^i}\left(\frac{\sin^2\theta_2}{R_1} + \frac{\sin^2\theta_1}{R_2}\right) \pm \sqrt{\frac{1}{\cos^2\theta^i}\left(\frac{\sin^2\theta_2}{R_1} + \frac{\sin^2\theta_1}{R_2}\right)^2 - \frac{4}{R_1 R_2}}$$

in which s^i is the radius of curvature of the incident spherical wavefront at Q_R, θ_1 is the angle between $\hat{s}^i$ and $\hat{U}_1$, and θ_2 is the angle between $\hat{s}^i$ and $\hat{U}_2$.

It is easily verified that

$$\sin^2\theta_1 = \cos^2\omega_0 + \sin^2\omega_0\cos^2\theta^i$$

$$\sin^2\theta_2 = \sin^2\omega_0 + \cos^2\omega_0\cos^2\theta^i$$

Thus a further simplification is possible in the case of spherical wave incidence if $\omega_0 = 0$ so that the preceding expressions for $(\varrho^r_1)^{-1}$ and $(\varrho^r_2)^{-1}$ become

$$\frac{1}{\varrho^r_1} = \frac{1}{s^i} + \frac{2\cos\theta^i}{R_1}, \qquad \frac{1}{\varrho^r_2} = \frac{1}{s^i} + \frac{2}{R_2\cos\theta^i}, \qquad \text{if } \omega_0 = 0$$

It is noted that $\omega_0 = 0$ implies that the plane of incidence is aligned with one of the principal directions ($\hat{U}_2$) on the surface at Q_R.

Explicit results are given in [23] for the GO field reflected from an arbitrary surface due to an incident spherical wave.

It is clear that the GO representation of (16) fails at caustics which are the intersection of the paraxial rays (associated with the ray tube or pencil) at the lines 1–2 and 3–4 as shown in Figs. 5 and 7. On crossing a caustic in the direction of propagation, $(\varrho^{i,r} + s^{i,r})$ changes sign under the radical in (16), and a phase jump of $+\pi/2$ results. Furthermore, the reflected field $\mathbf{E}^r$ of (16) fails in the transition region adjacent to SSB of Fig. 4b. It is important to note that near the SSB (i.e., as $\theta^i \to \pi/2$), ϱ_1^r and ϱ_2^r of (19a) approach the following limiting values:

$$\varrho_2^r \to \left(\frac{\varrho_g(Q_R) \cos \theta^i}{2} \right) \to 0 \quad \left. \right\} \quad \text{(20a)}$$
$$\left. \text{for } \theta^i \to \pi/2 \right.$$
$$\varrho_1^r \to \varrho_b^i \quad \text{(20b)}$$

where

$$1/\varrho_b^i = (\sin^2\alpha_0/\varrho_1^i) + (\cos^2\alpha_0/\varrho_2^i) \quad \text{(21)}$$

and ϱ_b^i is the radius of curvature of the incident wavefront in the $(\hat{\mathbf{t}}\hat{\mathbf{b}})$ plane (i.e., in the plane tangent to the surface) at Q_R for $\theta^i \to \pi/2$. Furthermore, the principal directions $\hat{\mathbf{X}}_1^r$ and $\hat{\mathbf{X}}_2^r$ of the reflected wavefront approach the following values for grazing incidence:

$$\hat{\mathbf{X}}_1^r \to \hat{\mathbf{b}} \quad \text{(at } Q_R) \quad \left. \right\} \quad \text{(22a)}$$
$$\left. \text{for } \theta^i \to \pi/2 \right.$$
$$\hat{\mathbf{X}}_2^r = (-\hat{\mathbf{s}}^r \times \hat{\mathbf{X}}_1^r) \to \hat{\mathbf{n}} \text{ (at } Q_R) \quad \text{(22b)}$$

The limiting values in (22a) and (22b) are independent of α_0. The total GO electric field $\mathbf{E}^{GO}$ at P_L in the lit region is the sum of the incident and reflected ray optical fields; hence

$$\mathbf{E}^{GO}(P_L) \sim \mathbf{E}^i(P_L) + \mathbf{E}^i(Q_R) \cdot \bar{\bar{\mathbf{R}}} \sqrt{\frac{\varrho_1^r \varrho_2^r}{(\varrho_1^r + s^r)(\varrho_2^r + s^r)}} \, e^{-jks^r} \quad \text{(23)}$$

In summary, it is noted that the GO incident and reflected fields are *discontinuous* across their associated shadow boundaries, such as the ISB, RSB, and SSB in Figs. 4a and 4b. These GO field discontinuities cannot be removed by including additional terms in the Luneberg-Kline expansion, which loses validity when calculating the fields at and near such boundaries, even though these additional terms may be useful for improving the GO approximation elsewhere in the deep lit zone. The failure of GO to account for a proper nonzero field within the shadow region behind an impenetrable obstacle can be overcome through the GTD and its uniform versions, the UTD and UAT, which are discussed in Section 4. Nevertheless, GO generally yields the dominant contribution to the total high-frequency fields, and, as will be seen in Section 4, it constitutes the leading term in the GTD solution.

The GO Reflected Field (2-D Case)—The GO reflected field $\mathbf{E}^r(P_L)$ for the 2-D case can be deduced directly from the 3-D case by allowing ϱ_1^r to approach infinity. Thus one may let $\varrho_2^r \equiv \varrho^r$ and $\varrho_1^r \to \infty$ in (16) to arrive at the 2-D reflected GO field $\mathbf{E}^r(P_L)$ as

$$\mathbf{E}^r(P_L) = \mathbf{E}^i(Q_R) \cdot \overline{\overline{\mathbf{R}}} \sqrt{\frac{\varrho^r}{\varrho^r + s^r}} \, e^{-jks^r} \tag{24}$$

in which the incident ray optical field $\mathbf{E}^i(Q_R)$ is a cylindrical wave at Q_R in the 2-D case, and the caustic distance ϱ^r in (24) for the 2-D case is given by

$$\frac{1}{\varrho^r} = \frac{1}{s^i} + \frac{2}{\varrho_g(Q_R)\cos\theta^i} \tag{25}$$

where θ^i has the same meaning as before, and s^i is the radius of curvature of the incident cylindrical wavefront at Q_R. If the cylindrical wave is produced by a 2-D line source, then s^i in (25) can be chosen to be the distance from that line source to the point Q_R of reflection on the 2-D boundary. The quantity $\varrho_g(Q_R)$ in (25) denotes the radius of curvature of the 2-D boundary at the point Q_R of reflection.

3.   The Physical Optics Field

The total electromagnetic field of a source (antenna) which radiates in the presence of a perfectly conducting surface as in Fig. 10 may be expressed as a superposition of the incident fields $(\mathbf{E}^i, \mathbf{H}^i)$ and the fields $(\mathbf{E}^s, \mathbf{H}^s)$ which are scattered by the surface. The incident fields referred to above are chosen here to denote the electric and magnetic fields of the source which exist everywhere, i.e., they exist as if the scatterer was "absent"; this is unlike the GO incident field, which exists in the presence of the surface of the scatterer (and which hence becomes discontinuous at the ISB or SSB, as in Figs. 4a and 4b). The scattered

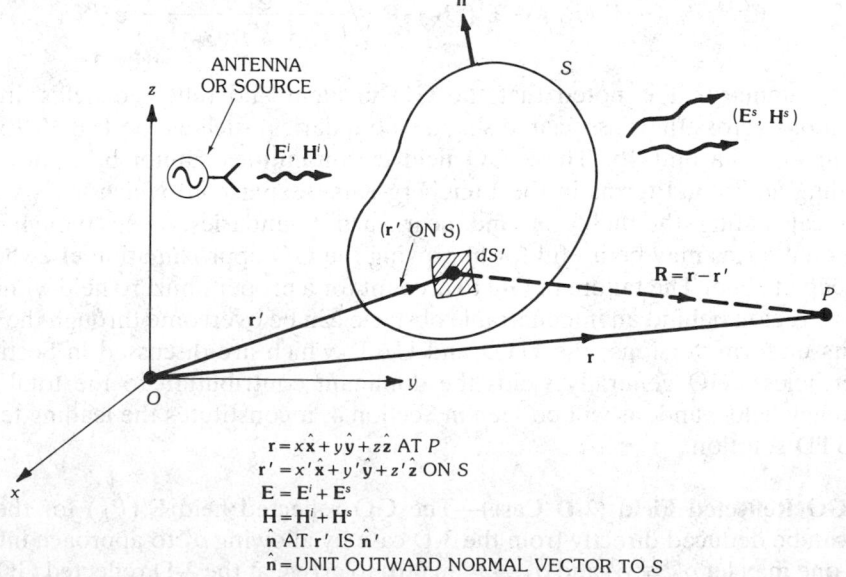

Fig. 10. Geometry pertaining to a 3-D obstacle (surface S) illuminated by an antenna.

fields in this case can be expressed in terms of the radiation integrals [12] over the actual currents induced on the surface S of the scatterer. These currents also radiate the scattered fields in the absence of the scatterer, i.e., these currents are now viewed as equivalent sources which are impressed over the mathematical boundary of S but with the perfectly conducting scatterer removed. The medium internal and external to the mathematical surface S in this equivalent problem is the same as that which exists external to S in the original problem. Such an equivalent configuration produces the same scattered fields as in the original problem outside the mathematical surface S, whereas it produces scattered fields which exactly cancel the incident fields inside S (to yield a zero total field inside S).

In the physical optics (PO) method, that radiation integral for the scattered field is calculated by employing a GO approximation for the currents induced on S, which is assumed to be electrically large; hence the PO method is also a high-frequency method. The PO approximation for the 3-D and 2-D cases is discussed below.*

Three-Dimensional Case

The total electric field $\mathbf{E}(\mathbf{r})$ at a point P exterior to S which may be evaluated in the near zone of S but not extremely close to S (in the problem of Fig. 10) is given by

$$\mathbf{E}(\mathbf{r}) = \mathbf{E}^i(\mathbf{r}) + \mathbf{E}^s(\mathbf{r}) \tag{26a}$$

where

$$\mathbf{E}^s(\mathbf{r}) \cong \frac{jkZ_0}{4\pi} \iint_S \hat{\mathbf{R}} \times \hat{\mathbf{R}} \times d\mathbf{p}_e(\mathbf{r}') \frac{e^{-jkR}}{R} \tag{26b}$$

as in [12]. Likewise, the total magnetic field $\mathbf{H}(\mathbf{r})$ at P is

$$\mathbf{H}(\mathbf{r}) = \mathbf{H}^i(\mathbf{r}) + \mathbf{H}^s(\mathbf{r}) \tag{27a}$$

where

$$\mathbf{H}^s(\mathbf{r}) \cong \frac{-jk}{4\pi} \iint_S \hat{\mathbf{R}} \times d\mathbf{p}_e(\mathbf{r}') \frac{e^{-jkR}}{R} \tag{27b}$$

In (26) and (27) the source distribution $d\mathbf{p}_e(\mathbf{r}')$ at each point $(\mathbf{r}')$ on S is given by

$$d\mathbf{p}_e(\mathbf{r}') = \mathbf{J}_s(\mathbf{r}')dS' = \hat{\mathbf{n}} \times \mathbf{H}(\mathbf{r}')dS' \tag{28}$$

The $\mathbf{J}_s(\mathbf{r}')$ in (28) is the induced electric current density at $\mathbf{r}'$ on S, and dS' is an elemental area of S at $\mathbf{r}'$. If P (at $\mathbf{r}$) is extremely close to S, then it is necessary to include additional terms which depend on R^{-2} and R^{-3} in the integrands of (26b)

---

*Additional important comments on the PO method are given after the 3-D and 2-D cases are discussed.

and (27b) as indicated in [12]; these additional terms are neglected in the present development. If $\mathbf{J}_s(\mathbf{r}')$ in (28) is approximated by its value $\mathbf{J}_s^{\text{GO}}(\mathbf{r}')$ as predicted by geometrical optics (GO), then

$$\mathbf{J}_s(\mathbf{r}') \cong \mathbf{J}_s^{\text{GO}}(\mathbf{r}') = \begin{cases} 2\hat{\mathbf{n}}' \times \mathbf{H}^i(\mathbf{r}') & \text{on the lit portion of } S \\ 0 & \text{on the shadowed portion of } S \end{cases} \tag{29}$$

Here, $\mathbf{E}^i$ and $\mathbf{H}^i$ refer to the "incident" electric and magnetic fields which are produced by the source (antenna) in the "absence" of the surface S. If the above GO approximation of (29) is employed in (28), then the resulting $\mathbf{E}$ and $\mathbf{H}$ so obtained from (26b) and (27b) are defined as the physical optics (PO) fields $\mathbf{E}^{\text{PO}}$ and $\mathbf{H}^{\text{PO}}$, respectively. Thus

$$\mathbf{E}^{\text{PO}}(\mathbf{r}) = \mathbf{E}^i(\mathbf{r}) + \frac{jkZ_0}{4\pi} \iint_{S_{\text{lit}}} \hat{\mathbf{R}} \times \hat{\mathbf{R}} \times [2\hat{\mathbf{n}}' \times \mathbf{H}^i] \frac{e^{-jkR}}{R} \, dS' \tag{30}$$

and

$$\mathbf{H}^{\text{PO}}(\mathbf{r}) = \mathbf{H}^i(\mathbf{r}) + \frac{-jk}{4\pi} \iint_{S_{\text{lit}}} \hat{\mathbf{R}} \times [2\hat{\mathbf{n}}' \times \mathbf{H}^i] \frac{e^{-jkR}}{R} \, dS' \tag{31}$$

where S_{lit} in the integrals of (30) and (31) refers to the part of S which is directly illuminated by the source, i.e., to the area of only the lit portion of S in accordance with (29). Let $\mathbf{E}_{\text{PO}}^s$ and $\mathbf{H}_{\text{PO}}^s$ denote the integrals in (30) and (31); thus they denote the scattered fields based on the PO approximation. In the event that the observation point P at $(\mathbf{r})$ is in the far zone of S, then $\hat{\mathbf{R}}$ can be replaced by $\hat{\mathbf{r}}$, which can be removed outside the integrals of (30) and (31); furthermore,

$$\frac{e^{-jkR}}{R} \cong \frac{e^{-jkr}}{r} e^{+jk\hat{\mathbf{r}}\cdot\mathbf{r}'}$$

in the far zone as usual, and the e^{-jkr}/r can also be factored outside those integrals. In the far zone the scattered fields in (26), (27), (30), and (31) are related by

$$\mathbf{E}^s(\mathbf{r}) \sim -\hat{\mathbf{r}} \times Z_0 \mathbf{H}^s(\mathbf{r})$$
$$\mathbf{H}^s(\mathbf{r}) \sim Y_0 \hat{\mathbf{r}} \times \mathbf{E}^s(\mathbf{r}) \tag{32}$$

and

$$\mathbf{E}_{\text{PO}}^s(\mathbf{r}) \sim -\hat{\mathbf{r}} \times Z_0 \mathbf{H}_{\text{PO}}^s(\mathbf{r})$$
$$\mathbf{H}_{\text{PO}}^s(\mathbf{r}) \sim Y_0 \hat{\mathbf{r}} \times \mathbf{E}_{\text{PO}}^s(\mathbf{r}) \tag{33}$$

Two-Dimensional Case

Next, consider the two-dimensional (2-D) geometry illustrated in Fig. 11, where a perfectly conducting scatterer bounded by a contour C is illuminated by a

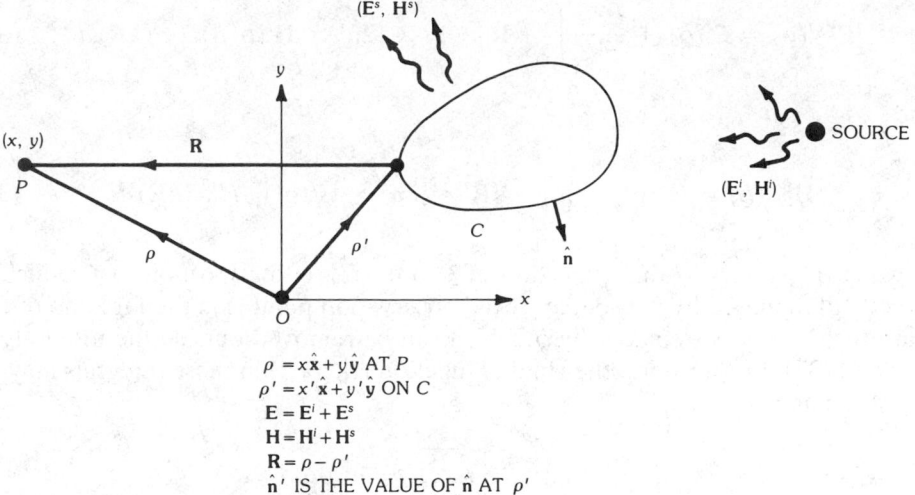

$$\rho = x\hat{\mathbf{x}} + y\hat{\mathbf{y}} \text{ AT } P$$
$$\rho' = x'\hat{\mathbf{x}} + y'\hat{\mathbf{y}} \text{ ON } C$$
$$\mathbf{E} = \mathbf{E}^i + \mathbf{E}^s$$
$$\mathbf{H} = \mathbf{H}^i + \mathbf{H}^s$$
$$\mathbf{R} = \rho - \rho'$$
$$\hat{\mathbf{n}}' \text{ IS THE VALUE OF } \hat{\mathbf{n}} \text{ AT } \rho'$$

Fig. 11. Geometry pertaining to a 2-D obstacle (boundary C) illuminated by a line source.

2-D source. The total fields $(\mathbf{E}, \mathbf{H})$ exterior to the boundary C are given in a manner analogous to (26b) and (27b) as follows:

$$\mathbf{E}(\varrho) \cong \mathbf{E}^i(\varrho) + \frac{kZ_0}{4} \int_C [\hat{\mathbf{R}} \times \hat{\mathbf{R}} \times \mathbf{J}(\ell')] H_0^{(2)}(kR)\, d\ell' \tag{34}$$

and

$$\mathbf{H}(\varrho) \cong \mathbf{H}^i(\varrho) - \frac{k}{4} \int_C [\hat{\mathbf{R}} \times \mathbf{J}(\ell')] H_0^{(2)}(kR)\, d\ell' \tag{35}$$

where $\mathbf{R} = \varrho - \varrho'$ in the 2-D case, and $H_0^{(2)}(kR)$ corresponds to the cylindrical Hankel function of the second kind and of order zero. The argument of this Hankel function is $kR = k|\varrho - \varrho'|$. Following (26) and (27) for the 3-D case, the expressions in (34) and (35) for the 2-D case are likewise valid if the observation point P (at ϱ) is not too close to the boundary C. In the PO approximation the electric current density $\mathbf{J}(\ell') = \mathbf{J}(\varrho')$ in (34) and (35) must be replaced by its GO approximation,

$$\mathbf{J}(\ell') \cong \mathbf{J}^{GO}(\ell') = \begin{cases} 2\hat{\mathbf{n}}' \times \mathbf{H}^i(\varrho') & \text{in the lit zone of } C \\ 0 & \text{in shadowed zone of } C \end{cases}$$

as indicated previously in (29). With this GO approximation for $\mathbf{J}(\ell')$ in (34) and (35), one obtains the total fields $(\mathbf{E}^{PO}, \mathbf{H}^{PO})$ based on PO for the 2-D problem of Fig. 11 as

$$\mathbf{E}^{PO}(\varrho) = \mathbf{E}^i(\varrho) + \frac{kZ_0}{4} \int_{C_{\text{lit}}} (\hat{\mathbf{R}} \times \hat{\mathbf{R}} \times [2\hat{\mathbf{n}}' \times \mathbf{H}^i(\varrho')]) H_0^{(2)}(kR)\, d\ell' \quad (36)$$

and

$$\mathbf{H}^{PO}(\varrho) = \mathbf{H}^i(\varrho) - \frac{k}{4} \int_{C_{\text{lit}}} \{\hat{\mathbf{R}} \times [2\hat{\mathbf{n}}' \times \mathbf{H}^i(\varrho')]\} H_0^{(2)}(kR)\, d\ell' \quad (37)$$

in which C_{lit} refers to the lit portion of C, i.e., C_{lit} is that portion of C which is directly illuminated by the source. If the observation point is in the far zone of the scatterer, $\hat{\mathbf{R}}$ may be replaced by $\hat{\varrho}$, which can be removed outside the integrals in (36) and (37); furthermore, the Hankel function $H_0^{(2)}(kR)$ in those integrals may be approximated by

$$H_0^{(2)}(kR) \cong \sqrt{\frac{2j}{\pi k \varrho}}\, e^{-jk\varrho}\, e^{jk\hat{\varrho}\cdot\varrho'}$$

in the far-zone case. The quantity $\sqrt{2j/\pi k\varrho}\, e^{-jk\varrho}$ in the preceding far-zone approximation can also be removed outside the integrals of (36) and (37).

In general, the PO integrals in (30) and (31) [and also in (36) and (37)] must be evaluated numerically. If one can approximate the PO integral asymptotically via the method of stationary phase, then one obtains the GO reflected field which is associated with the stationary phase point on S_{lit} (or C_{lit}). In addition, there is a contribution to the integral arising from the boundary of S_{lit} (or C_{lit}), and also from regions where $\mathbf{J}_s^{GO}$ (or $\mathbf{J}^{GO}$) is discontinuous; the latter contribution can provide an approximate description for the fields diffracted by edged bodies which are otherwise smooth. The edge-diffracted fields given by PO are accurate only near the GO shadow boundary (or ISB and RSB directions in Fig. 4). Furthermore, the fields in the PO approximation do not satisfy reciprocity except in the direction of specular reflection [24]. The above limitations are to be anticipated from the GO current approximation inherent in the PO procedure. Thus the PO approximation is expected to be inaccurate in regions where the GO current does not produce the dominant scattering, e.g., for grazing angles of incidence, and for the fields in the deep shadow region behind S or C. It is noted that the truncation of the GO currents at the shadow boundary on S (i.e., at the boundary of S_{lit}) is one of the main sources of error in the PO approximation. On the other hand, PO does yield useful estimates in directions where the GO currents produce the dominant contribution to the scattered fields, e.g., as in the calculation of the main beam and the first few side lobes of a parabolic reflector antenna. Also, unlike GO, the PO integral provides field values even in directions away from specular reflection. Furthermore, the GO field amplitude is independent of the wave number k, whereas the PO integral is dependent on k and it may therefore provide an improvement over GO. The PO approximation can be improved by including a diffraction correction to the GO currents as in the PTD procedure, which is described later in Section 6. It is important to recall once again that in the PO calculation described here, the incident fields ($\mathbf{E}^i, \mathbf{H}^i$) exist everywhere as if the scatterer were absent; this is in contrast to the GO calculation, in which the incident fields are truncated at the incident shadow boundaries (ISB/SSB of Fig. 4) so that they vanish in the geometrical shadow region of the scatterer.

4. The Geometrical Theory of Diffraction and Its Uniform Versions

As pointed out in Section 1, the geometrical theory of diffraction (GTD) is a systematic extension of the ideas of GO in which diffracted rays are introduced via a generalization of Fermat's principle. While the GO rays exist only in the lit zone the diffracted rays, in general, enter into both the shadow as well as the lit zones.

The total GTD field consists of a superposition of the GO (incident and reflected) field and the field of all the diffracted rays which can reach the observation point. In particular, the total electric and magnetic fields as predicted by GTD are denoted here by $\mathbf{E}^{\mathrm{GTD}}$ and $\mathbf{H}^{\mathrm{GTD}}$, respectively, and

$$\mathbf{E}^{\mathrm{GTD}} = \mathbf{E}^{\mathrm{GO}} + \mathbf{E}_k^d \qquad (38)$$

$$\mathbf{H}^{\mathrm{GTD}} = \mathbf{H}^{\mathrm{GO}} + \mathbf{H}_k^d \qquad (39)$$

In (38) and (39) the $\mathbf{E}^{\mathrm{GO}}$ and $\mathbf{H}^{\mathrm{GO}}$ denote the GO component of the electric and magnetic fields; likewise, $\mathbf{E}_k^d$ and $\mathbf{H}_k^d$ refer to the corresponding diffracted-ray field components as given by Keller's GTD. The general form of $\mathbf{E}_k^d$ and $\mathbf{H}_k^d$ is discussed below.

General Form of the GTD Diffracted-Ray Fields E_k^d and H_k^d

It is appropriate to begin a discussion of diffracted-ray fields after summarizing the postulates on which the GTD is based, namely:

(*a*) Diffraction, like reflection, is a local phenomenon at high frequencies.
(*b*) The diffracted rays satisfy a generalized Fermat's principle.
(*c*) Away from the point of diffraction the diffracted rays behave according to the laws of conventional geometrical optics (GO).

It was mentioned in Section 1 that the initial value of the diffracted-ray field is given in terms of a diffraction coefficient in the GTD, and as a result of postulate *a*, the relevant diffraction coefficients can be found from the asymptotic solutions to simpler canonical problems which model the geometrical and electrical properties of the original configuration only in the local vicinity of the points of diffraction.

As a result of postulate *b*, the rays diffracted from a line of discontinuity (e.g., an edge), as in Figs. 1 and 12, lie on a cone about the tangent to the line of discontinuity with a cone half-angle β_0, which is identical with the angle that the incident ray makes with that same tangent at the point of diffraction. Likewise, postulate *b* requires that an incident ray which grazes a smooth convex surface must launch rays which follow geodesic paths on that surface. As an additional consequence of postulate *b*, rays are shed along the forward tangents to each geodesic surface ray path, thereby giving rise to surface-diffracted rays as in Fig. 13. Thus the surface-ray field attenuates as it propagates due to the continual shedding of energy from the surface-ray field. The example in Fig. 13 illustrates the GO

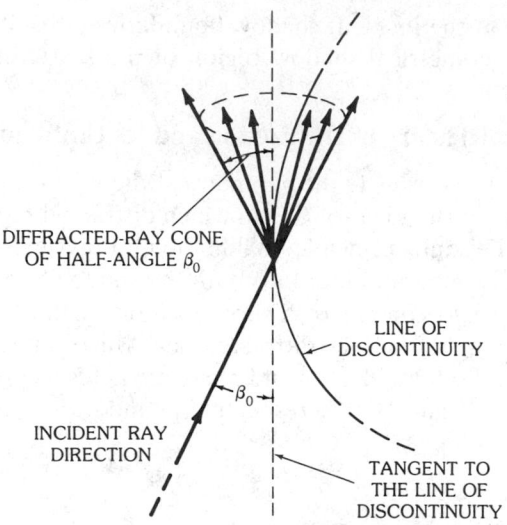

Fig. 12. Diffracted-ray cone from a line of discontinuity.

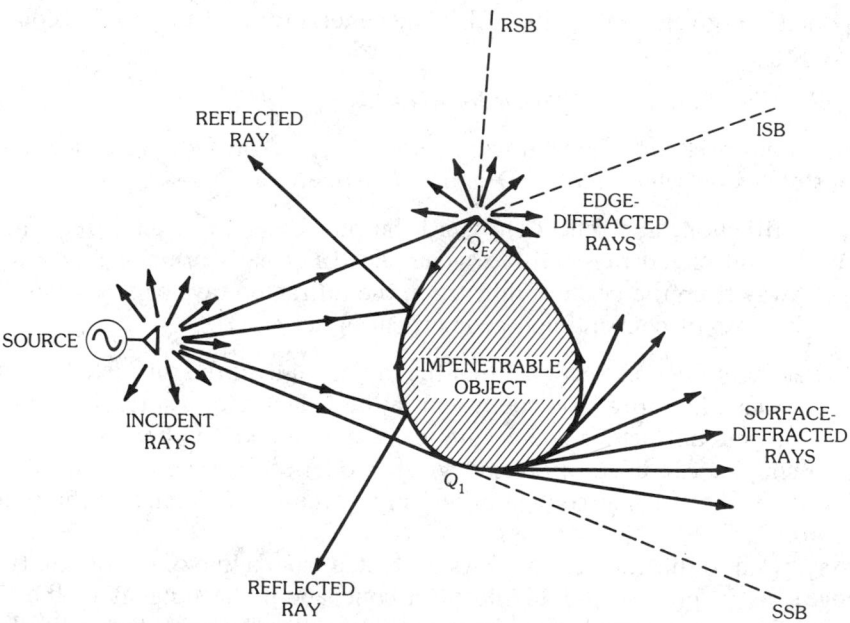

Fig. 13. Rays associated with the reflection and diffraction from an impenetrable surface.

incident and reflected rays as well as the edge- and surface-diffracted rays that can contribute to the total field at any exterior point when a source (antenna) radiates in the presence of an impenetrable structure consisting of an edge in an otherwise smooth convex body. It is shown in Fig. 13 that the incident ray at the edge can also excite surface rays and, conversely, surface rays can diffract from the edge. Through postulate *b*, the diffracted-ray optical path length is an extremum—usually a minimum. As mentioned previously the total field is the sum of the fields of all the individual rays that reach an observation point. The contribution from multiply interacting rays is usually much weaker and may thus be ignorable in such cases. In some cases it may be sufficient to include second-order interactions if they are important in comparison with the remaining higher-order multiply interacting rays, whose effects may be negligible. More is said about these interactions in Chapter 20. Next, the general form of the GTD diffracted fields $(\mathbf{E}_k^d, \mathbf{H}_k^d)$ associated with the diffraction by edges, convex surfaces, and vertices will be obtained. However, the explicit forms of the various diffraction functions present in the expressions for these GTD diffracted fields will be indicated later in the second part of Section 4, which deals with the uniform versions of the GTD.

GTD for Edges—Consider a perfectly conducting wedge which is illuminated by a source at O as shown in Fig. 14. The total electric field $\mathbf{E}^{\mathrm{GTD}}(P)$ at any point P exterior to the wedge is given by

$$\mathbf{E}^{\mathrm{GTD}}(P) = \mathbf{E}^{\mathrm{GO}}(P) + \mathbf{E}_k^d(P) \tag{40}$$

where the GO field component $\mathbf{E}^{\mathrm{GO}}(P)$ is given as

$$\mathbf{E}^{\mathrm{GO}}(P) = \mathbf{E}^i(P)U_i + \mathbf{E}^r(P)U_r \tag{41}$$

in which the GO incident and reflected fields $\mathbf{E}^i$ and $\mathbf{E}^r$ exist only in the lit regions. The domains of existence of these incident and reflected GO fields are indicated by the step functions U_i and U_r, respectively. In the case of the wedge, U_i and U_r are defined by:

$$U_i = \begin{cases} 1 & \text{if } 0 < \phi < \pi + \phi' \\ 0 & \text{if } \pi + \phi' < \phi < n\pi \end{cases}$$

and

$$U_r = \begin{cases} 1 & \text{if } 0 < \phi < \pi - \phi' \\ 0 & \text{if } \pi - \phi' < \phi < n\pi \end{cases}$$

The GO incident and reflected fields have been discussed previously in Section 2.

The diffracted field $\mathbf{E}_k^d(P)$ (and likewise $\mathbf{H}_k^d$) exists everywhere exterior to the wedge, i.e., for $0 < \phi < n\pi$. The diffracted rays lie on cones about the edge as in Figs. 12 and 14. Away from the edge, $\mathbf{E}_k^d(P)$ behaves as a ray optical (or GO) type field via Keller's postulate (*c*) above; the validity of this postulate can be observed directly via the discussion in Section 2, under "The Ray Concept," by considering S' (of Fig. 3) to be the wavefront associated with the edge-diffracted field (since S' could represent any wavefront). Thus, from (2) one may write

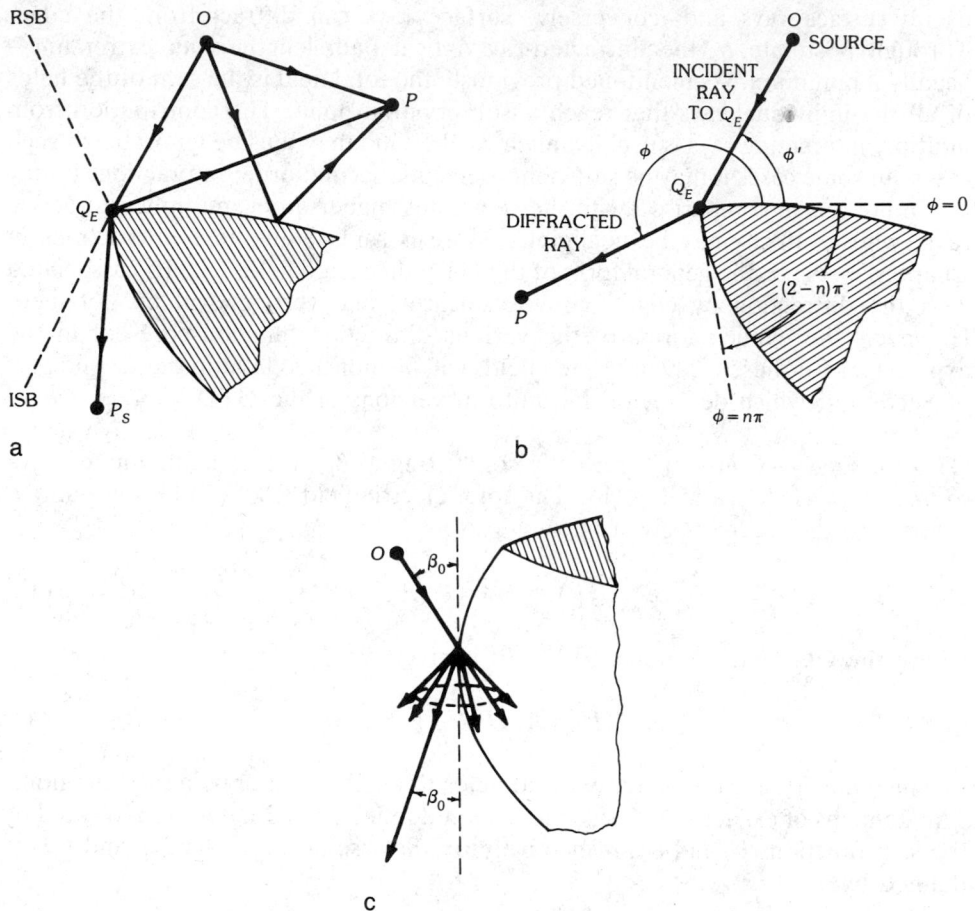

Fig. 14. Rays associated with the problem of edge diffraction. (*a*) Incident, reflected, and diffracted rays. (*b*) Side view, for edge-diffracted ray. (*c*) Perspective view, for edge-diffracted ray.

$$\mathbf{E}_k^d(P) \sim \mathbf{E}_k^d(P_0) \sqrt{\frac{\varrho_1^d \, \varrho_2^d}{(\varrho_1^d + s_0^d)(\varrho_2^d + s_0^d)}} \; e^{-jks_0^d} \qquad (42a)$$

The diffracted-ray tube corresponding to (42) is shown in Fig. 15, where the superscript d on ϱ_1^d, ϱ_2^d, s_0^d, and s^d denotes that these quantities are associated with the diffracted-ray field component. In order to relate $\mathbf{E}_k^d(P)$ to the incident field at the point Q_E of edge diffraction, one moves the reference P_0 in Fig. 15 to the point Q_E of diffraction on the edge (i.e., by letting $\varrho_1^d \to 0$) so that

$$\mathbf{E}_k^d(P) = \lim_{\varrho_1^d \to 0} [\sqrt{\varrho_1^d} \, \mathbf{E}_k^d(P_0)] \sqrt{\frac{\varrho_2^d}{(\varrho_1^d + s_0^d)(\varrho_2^d + s_0^d)}} \; e^{-jks_0^d} \qquad (42b)$$

Since $\mathbf{E}_k^d(P)$ is independent of the reference point P_0, the above limit exists and one defines it as

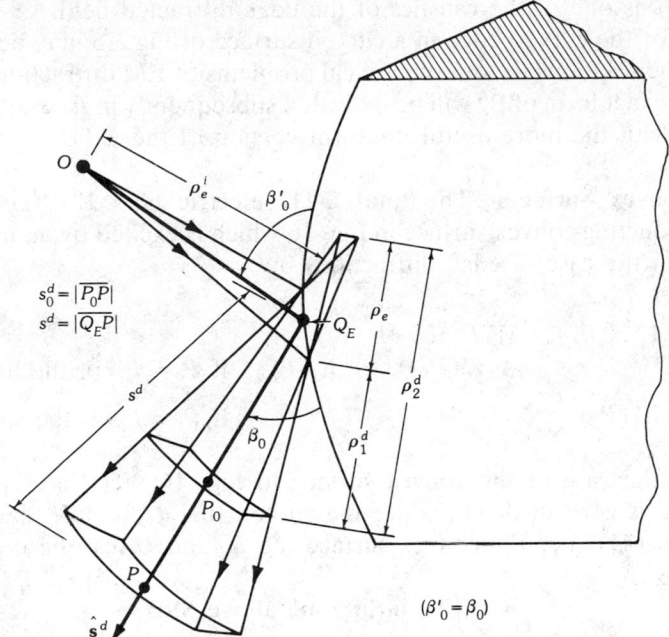

Fig. 15. Edge-diffracted ray tube.

$$\lim_{\varrho_1^d \to 0} \sqrt{\varrho_1^d}\, \mathbf{E}_k^d(P_0) \equiv \mathbf{E}^i(Q_E) \cdot \bar{\bar{\mathbf{D}}}_e^k \tag{43}$$

where $\bar{\bar{\mathbf{D}}}_e^k = \bar{\bar{\mathbf{D}}}_e^k(\phi, \phi', \beta_0; k)$ is the *dyadic edge-diffraction coefficient*, which indicates how the energy is distributed in the diffracted field as a function of the angles ϕ, ϕ', and β_0; the dyadic $\bar{\bar{\mathbf{D}}}_e^k$ also depends on n and the wave number k. From (42) and (43) one obtains

$$\mathbf{E}_k^d(P) \sim \mathbf{E}^i(Q_E) \cdot \bar{\bar{\mathbf{D}}}_e^k(\phi, \phi', \beta_0; k) \sqrt{\frac{\varrho_e}{s^d(\varrho_e + s^d)}}\, e^{-jks^d} \tag{44}$$

where

$$\lim_{\varrho_1^d \to 0} \varrho_2^d \equiv \varrho_e$$

is the edge-diffracted ray caustic distance, and likewise

$$\lim_{\varrho_1^d \to 0} s_0^d \equiv s^d$$

as shown in Fig. 15. The field $\mathbf{E}_k^d(P)$ is polarized transverse to the diffracted-ray direction $\hat{\mathbf{s}}^d$ *since* $\mathbf{E}_k^d(P)$ is ray optical; thus the associated magnetic field can be expressed as

$$\mathbf{H}_k^d(P) \sim Y_0 \hat{\mathbf{s}}^d \times \mathbf{E}_k^d(P) \tag{45}$$

In Fig. 15 Q_E is one of the caustics of the edge-diffracted field.

The $\bar{\bar{\mathbf{D}}}_e^k$ for the curved edge in a curved surface of Fig. 15 may be found from the asymptotic solutions to some canonical problems of EM diffraction by a wedge [25, 26]. The exact form of $\bar{\bar{\mathbf{D}}}_e^k$ will be indicated subsequently in the part of Section 4 which deals with the more useful, uniform version of the GTD.

GTD for Convex Surfaces—The total GTD electric field $\mathbf{E}^{\text{GTD}}$ exterior to a perfectly conducting convex surface in Fig. 16 which is excited by an incident wave is given (as in the case of edge diffraction) by

$$
\mathbf{E}^{\text{GTD}}(P) =
\begin{cases}
\mathbf{E}^{\text{GO}}(P_L) = \mathbf{E}^i(P_L)U_i \\
\qquad\qquad + \mathbf{E}^r(P_L)U_r + \mathbf{E}_k^d(P_L) & \text{if } P = P_L \text{ in the lit zone} \\
\mathbf{E}_k^d(P_S) & \text{if } P = P_S \text{ in the shadow zone}
\end{cases}
$$

However, in the case of the convex surface in Fig. 16, $\mathbf{E}_k^d(P_L)$ as predicted by Keller's GTD is zero in the lit zone. It is noted that U_i and U_r have the same meaning as in (41). For the convex surface the U_i and U_r can be defined by:

$$
U_i = U_r =
\begin{cases}
1 & \text{in lit zone above SSB} \\
0 & \text{in shadow zone below SSB}
\end{cases}
$$

since the ISB and RSB merge into the SSB. Therefore, for the convex surface in Fig. 16 the above expression for $\mathbf{E}^{\text{GTD}}$ becomes

$$
\mathbf{E}^{\text{GTD}}(P) =
\begin{cases}
\mathbf{E}^i(P_L) + \mathbf{E}^r(P_L) & \text{if } P = P_L \text{ in the lit zone} \\
\mathbf{E}_k^d(P_S) & \text{if } P = P_S \text{ in the shadow zone}
\end{cases}
\tag{46}
$$

The ray paths associated with $\mathbf{E}^i$, $\mathbf{E}^r$, and $\mathbf{E}_k^d$ have been discussed previously in Sections 1 (see Fig. 1) and 2. The surface shadow boundary (SSB) is an extension of the incident ray which grazes the surface at Q_1; it divides the exterior space into the lit and shadow zones. The explicit forms of $\mathbf{E}^i(P_L)$ and $\mathbf{E}^r(P_L)$ have been

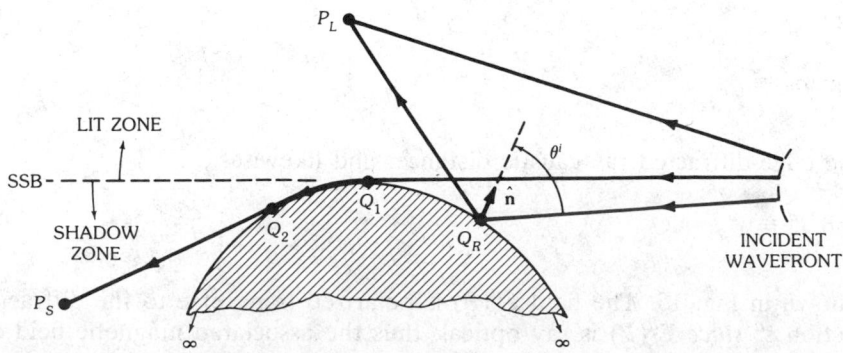

Fig. 16. Rays associated with the scattering and diffraction by a convex surface.

introduced earlier in Section 2, "The Geometrical Optics Field"; thus it remains only to consider the form of the diffracted-ray field. The field $\mathbf{E}_k^d(P_S)$ can be expressed ray optically via Keller's postulate c (or by making use of the arguments indicated in Section 2, under "The Ray Concept," which allow one to consider S' of Fig. 3 to be the wavefront of the surface-diffracted ray from Q_2 to P_S); therefore $\mathbf{E}_k^d(P_S)$ becomes

$$\mathbf{E}_k^d(P_S) \sim \mathbf{E}_k^d(P_0) \sqrt{\frac{\varrho_1^d \, \varrho_2^d}{(\varrho_1^d + s_0^d)(\varrho_2^d + s_0^d)}} \, e^{-jks_0^d} \tag{47}$$

The usual distances ϱ_1^d, ϱ_2^d, s_0^d, and s^d associated with a surface-diffracted ray tube corresponding to (47) are illustrated in Fig. 17. In order to relate $\mathbf{E}_k^d(P_0)$ to the incident field at the point of grazing Q_1, one moves the reference point P_0 to the point Q_2 from where the surface-diffracted ray is shed tangentially to arrive at the point P_S. Since $\mathbf{E}_k^d(P_S)$ is independent of the reference point P_0, one may write

$$\mathbf{E}_k^d(P_S) = \lim_{\varrho_1^d \to 0} \left[\sqrt{\varrho_1^d} \, \mathbf{E}_k^d(P_0) \right] \sqrt{\frac{\varrho_2^d}{(\varrho_1^d + s_0^d)(\varrho_2^d + s_0^d)}} \, e^{-jks_0^d} \tag{48}$$

together with

$$\lim_{\varrho_1^d \to 0} \sqrt{\varrho_1^d} \, \mathbf{E}_k^d(P_0) \equiv \mathbf{E}^i(Q_1) \cdot \overline{\overline{\mathbf{T}}}^k(Q_1, Q_2) \, e^{-jkt} \sqrt{\frac{d\eta(Q_1)}{d\eta(Q_2)}} \tag{49}$$

so that (47) becomes

$$\mathbf{E}_k^d(P_S) \sim \mathbf{E}^i(Q_1) \cdot \overline{\overline{\mathbf{T}}}^k(Q_1, Q_2) \, e^{-jkt} \sqrt{\frac{d\eta(Q_1)}{d\eta(Q_2)}} \sqrt{\frac{\varrho_s}{s^d(\varrho_s + s^d)}} \, e^{-jks^d} \tag{50}$$

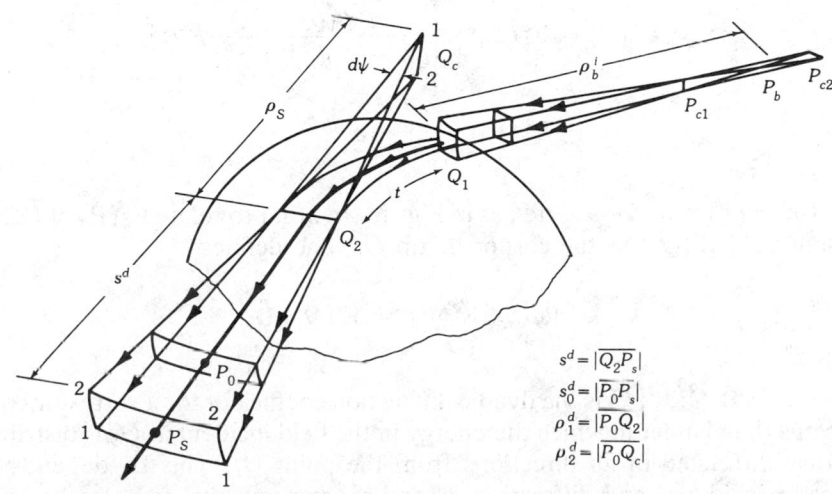

$$s^d = |\overline{Q_2 P_s}|$$
$$s_0^d = |\overline{P_0 P_s}|$$
$$\rho_1^d = |\overline{P_0 Q_2}|$$
$$\rho_2^d = |\overline{P_0 Q_c}|$$

Fig. 17. Surface-diffracted ray tube.

In arriving at (50), use is also made of the definitions

$$\lim_{\varrho_1^d \to 0} \varrho_2^d \equiv \varrho_s \quad \text{and} \quad \lim_{\varrho_1^d \to 0} s_0^d = s^d$$

In (49) and (50) $d\eta(Q_1)$ and $d\eta(Q_2)$ refer to the widths of the surface-ray strip at Q_1 and Q_2, respectively, as illustrated in Fig. 17. Furthermore, the factor $\sqrt{d\eta(Q_1)/d\eta(Q_2)}\,e^{-jkt}$ indicates a conservation of the energy flux in the surface-ray strip from Q_1 to Q_2, in which e^{-jkt} denotes the dominant phase delay along that surface ray path. However, since rays are shed along the forward tangents to the geodesic surface-ray paths to give rise to the surface-diffracted rays, energy is lost from the surface rays and the surface-ray field attenuates. Thus the dyadic transfer function $\bar{\bar{\mathbf{T}}}^k(Q_1, Q_2)$ in (48) and (49) is introduced to indicate the launching of the surface-ray field at Q_1, the attenuation of the surface-ray field between Q_1 and Q_2, and also the amount of diffraction of the surface-ray field from Q_2. The explicit form of $\bar{\bar{\mathbf{T}}}^k(Q_1, Q_2)$ will be indicated subsequently in Section 4, under "Uniform Version of the GTD." The shadowed part of the surface in Fig. 16 is a caustic of the surface-diffracted rays. The diffracted-ray optical field $\mathbf{E}_k^d(P_S)$ has an associated magnetic field $\mathbf{H}_k^d(P_S)$ given as usual by

$$\mathbf{H}_k^d(P_S) \sim Y_0 \hat{s}^d \times \mathbf{E}_k^d(P_S) \tag{51}$$

GTD for Vertices—A vertex is formed, for example, at the tip of a cone, the tip of a pyramid, or by the corner of a plane angular sector. From Keller's postulate *a*, an incident ray which strikes a perfectly conducting vertex produces a continuum of diffracted rays which emanate in all directions from the vertex as shown in Fig. 1f. One such vertex-diffracted ray to P is shown in Fig. 18. Using postulate *c*, as before, allows one to express the electric field $\mathbf{E}_k^d(P)$ diffracted by the corner (or vertex) as in (2) or (42a); thus,

$$\mathbf{E}_k^d(P) \sim \mathbf{E}_k^d(P_0) \sqrt{\frac{\varrho_1^d \varrho_2^d}{(\varrho_1^d + s_0^d)(\varrho_2^d + s_0^d)}}\, e^{-jks_0^d}$$

$$\sim \mathbf{E}_k^d(P_0) \frac{\varrho_v^d}{\varrho_v^d + s_0^d}\, e^{-jks_0^d} \tag{52}$$

since $(\varrho_1^d = \varrho_2^d) \equiv \varrho_v^d$ for a vertex as in Fig. 18. In order to relate $\mathbf{E}_k^d(P)$ of (52) to the incident field $\mathbf{E}^i(Q_v)$ at the corner or tip Q_v, one defines

$$\lim_{\varrho_v^d \to 0} \varrho_v^d \mathbf{E}_k^d(P_0) = \mathbf{E}^i(Q_v) \cdot \bar{\bar{\mathbf{D}}}_v \tag{53}$$

where $\bar{\bar{\mathbf{D}}}_v = \bar{\bar{\mathbf{D}}}_v(\hat{s}^d, \hat{s}^i; k)$ is the dyadic diffraction coefficient for a vertex or corner; it indicates the manner in which the energy in the field incident at Q_v is distributed in the rays diffracted in all directions from the point Q_v. This $\bar{\bar{\mathbf{D}}}_v$ depends on the angles of incidence and diffraction, $\hat{s}^i$ and $\hat{s}^d$, respectively, and also on the wave number k besides depending on the local geometry of the corner at Q_v. The limit in

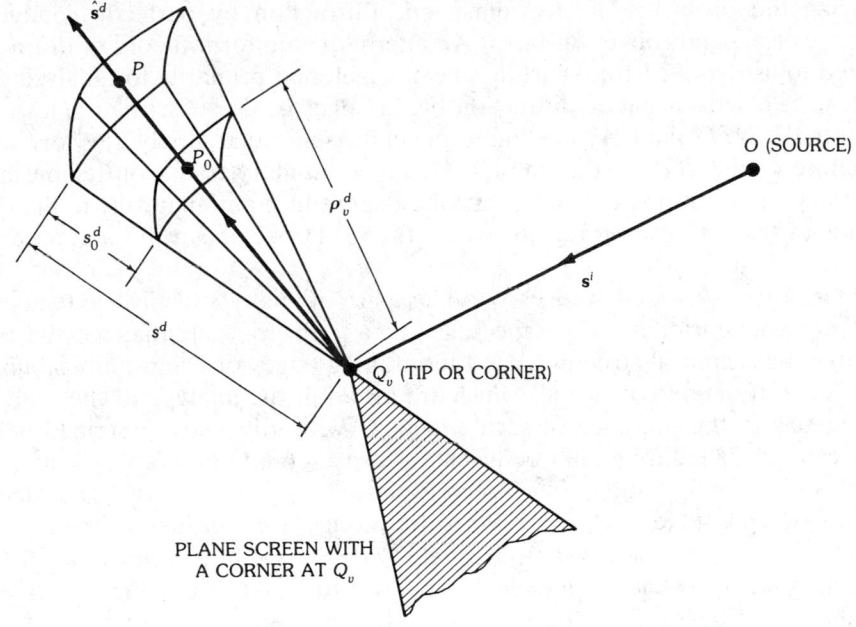

Fig. 18. Corner- or vertex-diffracted ray tube.

(53) exists because $\mathbf{E}_k^d(P)$ in (52) is independent of the position of the reference point at P_0. From Fig. 18 it is evident that

$$\lim_{\varrho_v^d \to 0} s_0^d = s^d$$

Using this information and (53) allows one to write (52) as

$$\mathbf{E}_k^d(P) \sim \mathbf{E}^i(Q_v) \cdot \bar{\bar{\mathbf{D}}}_v(\hat{\mathbf{s}}^i, \hat{\mathbf{s}}^d, k) \frac{e^{-jks^d}}{s^d} \tag{54}$$

when $\varrho_v^d \to 0$. The vertex- or corner-diffracted ray optical field in (54) is polarized transverse to the ray direction $\hat{\mathbf{s}}^d$ and, as before, the associated magnetic field $\mathbf{H}_k^d(P)$ is given by

$$\mathbf{H}_k^d(P) \sim Y_0 \hat{\mathbf{s}}^d \times \mathbf{E}_k^d(P) \tag{55}$$

The point Q_v is a caustic of the vertex-diffracted rays. Also, the total GTD field surrounding the vertex or corner is the sum of the GO field and the corner-diffracted field, as in (38). The explicit form of $\bar{\bar{\mathbf{D}}}_v$ will be indicated later in the following section dealing with the uniform version of the GTD.

Uniform Version of the GTD

A uniform version of the GTD which is referred to as the *uniform geometrical theory of diffraction*, and which is abbreviated as *UTD* [4, 5], is described below for analyzing the problems of electromagnetic diffraction by perfectly conducting edges, vertices, and convex surfaces. An alternative uniform theory of diffraction referred to as the *UAT* [6], which has been developed primarily for analyzing the problem of electromagnetic diffraction by an edge, is also described below.

Both the *UTD* and *UAT* are "uniform" in the sense that, firstly, they overcome the failure of the GTD within the GO shadow boundary transition regions away from the point of diffraction and, secondly, they reduce automatically to the GTD exterior to these transition regions where the GTD becomes valid and accurate.

UTD for Edges—Curved wedges occur as part of many practical antenna and scattering configurations, e.g., the edge of a reflector antenna, the bases of cylindrical and conical structures, and the trailing edges of wings and stabilizers. The UTD diffraction coefficients which are useful in the analysis of the radiation by antennas in the presence of such structures with edges are described below. Specifically, UTD diffraction coefficients are presented for both 3-D and 2-D curved wedge configurations which serve to locally model 3-D and 2-D structures with edges. Only field points exterior to the wedge are considered here.

3-D Case—The total UTD electric field $\mathbf{E}(P)$ at a point exterior to a perfectly conducting wedge, which is illuminated by a source at O as in Fig. 14, is given according to UTD by

$$\mathbf{E}(P) = \mathbf{E}^{\mathrm{GO}}(P) + \mathbf{E}^d(P) \tag{56}$$

The field $\mathbf{E}^{\mathrm{GO}}(P)$ is as defined earlier in (41). The UTD field $\mathbf{E}(P)$ in (56) remains valid even within the transition regions adjacent to the incident and reflection shadow boundaries (ISB and RSB of Fig. 4a), respectively, where the GTD fails. Furthermore, exterior to these transition regions, $\mathbf{E}(P)$ reduces automatically to $\mathbf{E}^{\mathrm{GTD}}(P)$ of (40), where the GTD becomes valid; i.e., the uniform edge-diffracted field $\mathbf{E}^d(P)$ in (56) reduces properly to Keller's edge-diffracted field $\mathbf{E}_k^d(P)$ [in (40)] exterior to the transition regions. Since $\mathbf{E}^{\mathrm{GO}}(P)$ has been described earlier in Section 2, under "The Geometrical Optics Field," it only remains to give explicit expressions for the diffracted field $\mathbf{E}^d(P)$ in (56). Firstly, the edge-diffracted field $\mathbf{E}^d(P)$ is associated with the edge-diffracted rays of the GTD. If an incident ray strikes the edge obliquely, making an angle β'_0 with the unit vector $\hat{\mathbf{e}}$ which is tangent to the edge at the point of diffraction, then the diffracted rays lie on a Keller cone whose half-angle is $\beta_0 = \beta'_0$ (law of edge diffraction) as shown in Figs. 12, 14, and 15. Secondly, the UTD expression for $\mathbf{E}^d(P)$ associated with an edge-diffracted ray can also be expressed in the format of the GTD as in (44) for $\mathbf{E}_k^d(P)$; in particular,

$$\mathbf{E}^d(P) = \mathbf{E}^i(Q_E) \cdot \bar{\bar{\mathbf{D}}}_e(\phi, \phi', \beta_0; k) \sqrt{\frac{\varrho_e}{s^d(\varrho_e + s^d)}} \, e^{-jks^d} \tag{57}$$

Thus the GTD dyadic edge-diffraction coefficient $\bar{\bar{\mathbf{D}}}_e^k$ in (44) is replaced with the uniform dyadic edge-diffraction coefficient $\bar{\bar{\mathbf{D}}}_e$ in the UTD expression of (57) for the edge-diffracted field. In addition, the UTD dyadic edge-diffraction coefficient $\bar{\bar{\mathbf{D}}}_e$ can be expressed conveniently in terms of unit vectors fixed in the incident and diffracted rays as follows.* Note that the Keller cone is defined by the law of edge diffraction, $\hat{s}^i \cdot \hat{e} = \hat{s}^d \cdot \hat{e}$, which results from Keller's generalization of Fermat's principle (Section 4, under "General Form of the GTD Diffracted-Ray Fields $\mathbf{E}_k^d$ and $\mathbf{H}_k^d$"). Let the incident-ray direction $\hat{s}^i$ and $\hat{e}$ define an edge-fixed plane of incidence; likewise, let the direction of the diffracted ray $\hat{s}^d$ and $\hat{e}$ define an edge-fixed plane of diffraction. The unit vectors $\hat{\beta}_0'$ and $\hat{\beta}_0$ are parallel to the edge-fixed plane of incidence and diffraction, respectively, and

$$\hat{\beta}_0' = \hat{s}^i \times \hat{\phi}' \tag{58a}$$

$$\hat{\beta} = \hat{s}^d \times \hat{\phi} \tag{58b}$$

as shown in Fig. 19a. Here, $\hat{\phi}'$ and $\hat{\phi}$ point in the direction of increasing angles ϕ' and ϕ, respectively, which indicate the azimuthal (or circumferential) angular coordinates of the incident and diffracted rays about the edge tangent $\hat{e}$ as illustrated in Fig. 19b. The field $\mathbf{E}^d(P)$ can now be expressed invariantly in terms of these triads of unit vectors $(\hat{s}^i, \hat{\beta}_0', \hat{\phi}')$ and $(\hat{s}^d, \hat{\beta}_0, \hat{\phi})$, respectively, which are fixed in the incident and diffracted rays. In this ray coordinate system [25, 26] one may write

$$\mathbf{E}^d(P) = \hat{\beta}_0 E_{\beta_0}^d + \hat{\phi} E_\phi^d \tag{59a}$$

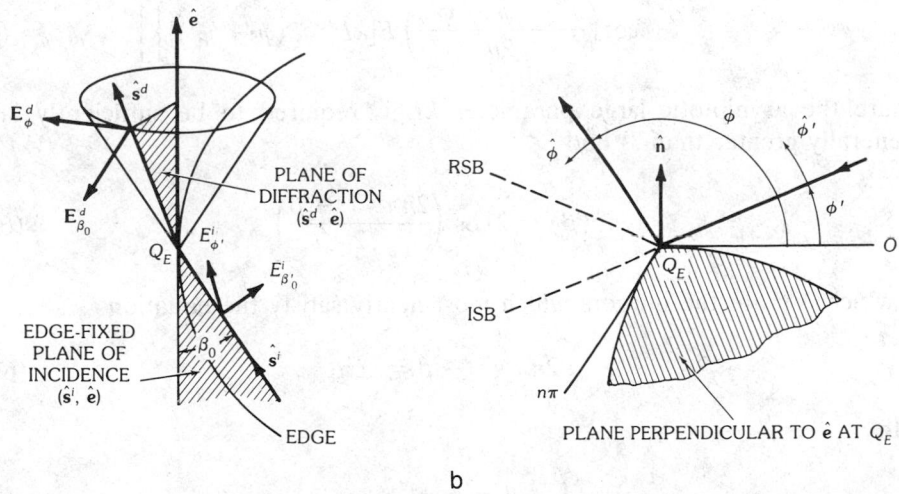

Fig. 19. Edge-fixed planes of incidence and diffraction. (*a*) Planes of incidence and diffraction. (*b*) Angular coordinates about $\hat{e}$. (*After Kouyoumjian and Pathak [26], © 1974 IEEE*)

---

*The main difference between $\bar{\mathbf{D}}_e$ and $\bar{\mathbf{D}}_e^k$ is that the former is range dependent, whereas the latter is not. As a result, the expression in (57) is not strictly ray optical within the shadow boundary transition regions; outside these regions, (57) reduces to (44), as shown later.

and

$$\mathbf{E}^i(Q_E) = \hat{\beta}_0' E_{\beta_0}^i + \hat{\phi}' E_{\phi'}^i \tag{59b}$$

with

$$\bar{\bar{\mathbf{D}}}_e = -\hat{\beta}_0'\hat{\beta}_0 D_{es} - \hat{\phi}'\hat{\phi} D_{eh} \tag{59c}$$

A matrix representation for (57) using (59a)–(59c) is given by

$$\begin{bmatrix} E_{\beta_0}^d \\ E_{\phi}^d \end{bmatrix} = \begin{bmatrix} -D_{es} & 0 \\ 0 & -D_{eh} \end{bmatrix} \begin{bmatrix} E_{\beta_0}^i \\ E_{\phi'}^i \end{bmatrix} \sqrt{\frac{\varrho_e}{s^d(\varrho_e + s^d)}} e^{-jks^d} \tag{60}$$

in which [5, 26]

$$D_{es,eh}(\phi, \phi'; \beta_0) = \frac{-e^{-j\pi/4}}{2n\sqrt{2\pi k}\sin\beta_0} \left[\cot\left(\frac{\pi + (\phi - \phi')}{2n}\right) F[kL^i a^+(\phi - \phi')] \right.$$

$$+ \cot\left(\frac{\pi - (\phi - \phi')}{2n}\right) F[kL^i a^-(\phi - \phi')]$$

$$\mp \left\{ \cot\left(\frac{\pi + (\phi + \phi')}{2n}\right) F[kL^{rn} a^+(\phi + \phi')] \right.$$

$$\left. \left. + \cot\left(\frac{\pi - (\phi + \phi')}{2n}\right) F[kL^{ro} a^-(\phi + \phi')] \right\} \right] \tag{61}$$

where the asymptotic large parameter kL is required to be sufficiently large (generally greater than 3) and

$$a^{\pm}(\beta) = 2\cos^2\left(\frac{2n\pi N^{\pm} - \beta}{2}\right) \tag{62a}$$

in which $N^{\pm}$ are the integers which most nearly satisfy the equation

$$2n\pi N^{\pm} - \beta = \pm\pi \tag{62b}$$

with

$$\beta = \phi \pm \phi' \tag{62c}$$

In (61) [and (62a, 62b)] n defines the exterior wedge angle as in Fig. 19b; hence $n = 2$ for a half-plane, and $n = 3/2$ for an exterior right-angle wedge, etc. The two faces of the wedge are thus referred to as the "o" face at $\phi = 0$ and the "n" face at $\phi = n\pi$, respectively, as in Fig. 19b.

For exterior edge diffraction, $N^+ = 0$ or 1, and $N^- = -1, 0$, or 1. The values of $N^{\pm}$ at the shadow and reflection boundaries as well as their associated transition regions are given in Table 1 for exterior wedge angles ($1 < n < 2$).

Table 1. Values of N for Exterior Wedge Angles

	The Cotangent Is Singular When	Value of N at the Boundary
$\cot\left[\dfrac{\pi + (\phi - \phi')}{2n}\right]$	$\phi = \phi' - \pi$, an SB Surface $\phi = 0$ is shadowed	$N^+ = 0$
$\cot\left[\dfrac{\pi - (\phi - \phi')}{2n}\right]$	$\phi = \phi' + \pi$, an SB Surface $\phi = n\pi$ is shadowed	$N^- = 0$
$\cot\left[\dfrac{\pi + (\phi + \phi')}{2n}\right]$	$\phi = (2n - 1)\pi - \phi'$, an RB Reflection from surface $\phi = n\pi$	$N^+ = 1$
$\cot\left[\dfrac{\pi - (\phi + \phi')}{2n}\right]$	$\phi = \pi - \phi'$, an RB Reflection from surface $\phi = 0$	$N^- = 1$

For a point-source (or spherical-wave) type illumination the distance parameter L^i is

$$L^i = \frac{s^i s^d}{s^i + s^d}\,\sin^2\beta_0 \tag{63}$$

in which s^i and s^d are the distances from the point of edge diffraction at Q_E to the source and observation points, respectively. Only for a straight wedge with planar faces that is illuminated by a point source does

$$L^{ro} = L^{rn} = L^i = \frac{s^i s^d}{s^i + s^d}\,\sin^2\beta_0$$

as in (63). In the case of an arbitrary ray-optical type of illumination which is characterized by two distinct principal wavefront radii of curvature, ϱ_1^i and ϱ_2^i, the above L^i must be modified as shown below in the general expressions for L^{ro} and L^{rn} pertaining to a curved wedge; thus

$$L^i = \left[\frac{s^d(\varrho_e^i + s^d)\varrho_1^i\varrho_2^i\sin^2\beta_0}{\varrho_e^i(\varrho_1^i + s^d)(\varrho_2^i + s^d)}\right]_{\text{ISB}} \tag{64a}$$

$$L^r = \left[\frac{s^d(\varrho_e^r + s^d)\varrho_1^r\varrho_2^r\sin^2\beta_0}{\varrho_e^r(\varrho_1^r + s^d)(\varrho_2^r + s^d)}\right]_{\text{RSB}} \tag{64b}$$

Here, L^{ro} and L^{rn} are the values of L^r associated with the o and n faces of the wedge, respectively. Furthermore, ϱ_e^r is given by

$$\frac{1}{\varrho_e^r} = \frac{1}{\varrho_e^i} - \frac{2(\hat{\mathbf{n}}\cdot\hat{\mathbf{n}}_e)(\hat{\mathbf{s}}^i\cdot\hat{\mathbf{n}})}{a\,\sin^2\beta_0} \tag{65a}$$

Also ϱ_e in (57) is given by

$$\frac{1}{\varrho_e} = \frac{1}{\varrho_e^i} - \frac{\hat{\mathbf{n}}_e \cdot (\hat{\mathbf{s}}^i - \hat{\mathbf{s}}^d)}{a \sin^2\beta_0} \tag{65b}$$

The unit vector $\hat{\mathbf{n}}$ is defined in Fig. 19b, whereas $\hat{\mathbf{n}}_e$ is a unit vector normal to the edge which is directed away from the center of edge curvature at Q_E. The radius of edge curvature is denoted by a in (65). The term ϱ_e^i is the radius of curvature of the incident wavefront at Q_E which lies in the edge-fixed plane of incidence. In the far zone when $s^d \gg \varrho_{1,2}^i$, $s^d \gg \varrho_{1,2}^r$, and $s^d \gg \varrho_e$, then the L^i and L^r in (64a) and (64b) simplify to $L \cong (\varrho_1\varrho_2 \sin^2\beta_0)/\varrho_e$ in which the appropriate superscripts on L, ϱ_1, and ϱ_2 are omitted for convenience. It is noted that L^i and L^r in (64a) and (64b) are calculated on the appropriate shadow boundaries. The transition function F which appears in (61) contains a Fresnel integral; it is defined by

$$F(x) = 2j\sqrt{x}\, e^{jx} \int_{\sqrt{x}}^{\infty} e^{-j\tau^2}\, d\tau \tag{66}$$

A plot of the above $F(x)$ is illustrated in Fig. 30. In (66), $\sqrt{x} = |\sqrt{x}|$ if $x > 0$ and $\sqrt{x} = -j|\sqrt{x}|$ if $x < 0$. If $x < 0$, then $F(x)|_{x<0} = F^*(|x|)$ where $*$ denotes the complex conjugate. Exterior to the ISB and RSB transition regions, x becomes large and $F(x) \to 1$ so that the uniform $D_{es,eh}$ in (61) then reduces to Keller's form [1] as it should; namely,

$\mathbf{E}^d(P) \to \mathbf{E}_k^d(P)$, outside the transition region, or

$\bar{\bar{\mathbf{D}}}_e \to \bar{\bar{\mathbf{D}}}_e^k$, outside the transition region, where

$$\bar{\bar{\mathbf{D}}}_e^k = -\hat{\boldsymbol{\beta}}_0'\hat{\boldsymbol{\beta}}_0 D_{es}^k - \hat{\boldsymbol{\phi}}'\hat{\boldsymbol{\phi}}\, D_{eh}^k \tag{67}$$

$$D_{es,eh}^k(\phi, \phi'; \beta_0) = \frac{e^{-j\pi/4}\sin(\pi/n)}{n\sqrt{2\pi k}\,\sin\beta_0} \left\{ \frac{1}{\cos(\pi/n) - \cos[(\phi - \phi')/n]} \right.$$

$$\left. \mp \frac{1}{\cos(\pi/n) - \cos[(\phi + \phi')/n]} \right\} \tag{68}$$

Note that $D_{es,eh}^k$ in (68) becomes singular at the shadow boundaries ISB ($\phi - \phi' = \pi$) and RSB ($\phi + \phi' = \pi$) for the o face. On the other hand, $D_{es,eh}$ in (61) is well behaved at these shadow boundaries. Indeed, at these (ISB and RSB) boundaries the small-argument approximation for $F(x)$ may be employed (since $x = 0$ on the ISB and RSB), namely, one may insert

$$F(x) \underset{x \to 0}{\to} \sqrt{\pi x}\, e^{j(\pi/4 + x)} \tag{69}$$

into (61) to arrive at the following result:

$$\mathbf{E}^d|_{\text{ISB;RSB}} = \mp \frac{1}{2}\, \mathbf{E}^{i;r} + \text{continuous higher-order terms in } k,$$

$$\text{if } \left\{ \begin{array}{l} \text{on just the lit side of ISB;RSB} \\ \text{on just the shadow side of ISB;RSB} \end{array} \right\} \tag{70}$$

The above result in (70) ensures the continuity of the total high-frequency field in (56) at the ISB and RSB. The field contributions arising from the edge-excited "surface-diffracted rays" are not included in (56); these may be important for observation points close to the surface shadow boundaries (SSB) associated with the tangents to the o and n faces of a curved wedge at Q_E if the o and n faces are convex boundaries. The result in (56) and (57) together with (61) is valid away from any diffracted-ray caustics and away from the edge caustic at Q_E.

In the case of grazing angles of incidence on a wedge with planar faces, $D_{es} = 0$, and D_{eh} must be replaced by $(1/2) D_{eh}$. The reason for the factor 1/2 in the latter case may be argued as follows. The incident and reflected GO fields tend to combine into a single "total incident field" as one approaches grazing angles of incidence; consequently, only half of this "total field" illuminating the edge at grazing constitutes the incident GO field, while the other half constitutes the reflected GO field. The case of grazing angles of incidence at an edge on a curved surface cannot be handled as easily as the case of a wedge with planar faces. In fact, work needs to be done to extend the UTD to grazing angles of incidence on a curved surface forming the wedge; at present one can only treat angles of incidence that are greater than $[2/k\varrho_g(Q_E)]^{1/3}$, where $\varrho_g(Q_E)$ is the radius of curvature of the surface in the direction of the incident ray at the point Q_E of edge diffraction.

Under the above restrictions the general result in (61) for $D_{es,eh}$ simplifies in the case of a plane or curved screen ($n = 2$ case) to

$$D_{es,eh}(\phi, \phi'; \beta_0) = \frac{-e^{-j\pi/4}}{2\sqrt{2\pi k}\,\sin\beta_0}\left\{\sec\left(\frac{\phi - \phi'}{2}\right)F[kL^i a(\phi - \phi')]\right.$$
$$\left. \mp \sec\left(\frac{\phi + \phi'}{2}\right)F[kL^r a(\phi + \phi')]\right\}$$

where $a(\beta) \equiv 2\cos^2(\beta/2)$ and $L^{i,r}$ are as in (64a, 64b) with the understanding that L^r is evaluated at the RSB corresponding to the face which is illuminated; hence the superscripts o and n in L^r are dropped for this $n = 2$ case.

2-D Case—As in the 3-D case the total electric field $\mathbf{E}(P)$ at a point exterior to a 2-D perfectly conducting wedge illuminated by a 2-D line source is described by (56) in which the GO field $\mathbf{E}^{GO}(P_L)$ is given by (41), and $\mathbf{E}^i$ and $\mathbf{E}^r$ are found via (12) and (24). The 2-D edge-diffracted field $\mathbf{E}^d(P)$ can be obtained from (57) by allowing ϱ_e to approach infinity and by requiring $\beta_0 = \pi/2$; thus, for the 2-D case,

$$\mathbf{E}^d(P) = \mathbf{E}^i(Q_E) \cdot \bar{\bar{\mathbf{D}}}_e(\phi, \phi', \pi/2; k) \frac{e^{-jks^d}}{\sqrt{s^d}} \tag{71}$$

The $\bar{\bar{\mathbf{D}}}_e$ in (71) for the 2-D case is available from (59c) and (61), with $\beta_0 = \pi/2$ (or $\sin\beta_0 = 1$). Also, L^i for the 2-D case is as in (63), with $\beta_0 = \pi/2$; in particular,

$$L^i = \frac{s^i s^d}{s^i + s^d} \tag{72}$$

Likewise, L^r is obtained from (64b), with $\beta_0 = \pi/2, \varrho_1^r \to \infty, \varrho_2^r \equiv \varrho^r$ [as in (24)], and $\varrho_e^r \to \infty$; therefore, in the 2-D case,

$$L^r = \frac{\varrho^r s^d}{\varrho^r + s^d} \tag{73}$$

Note that ϱ^r in (73) is the same as the one in (24); however, ϱ^r is in general different for the o and n faces of the wedge, with L^{ro} and L^{rn} denoting the values of L^r for these two different faces. While the expression for L^r in (64b) is fixed to its value on the RSB for convenience [26], the one in (73) can be evaluated as a function of the observation point with almost the same ease as if one had approximated the value of L^r by its value at the RSB. The values of L^i and L^r in (64a) and (64b) involve several quantities for the general 3-D case, and since these parameters L^i and L^r are slowly varying within the ISB and RSB transition regions, it is therefore convenient in the 3-D case to approximately fix their values to those at the ISB and RSB, respectively, as done in (64a) and (64b). Outside their respective transition regions, $kL^i a^i$ and $KL^{rn} a^{rn}$ and $kL^{ro} a^{ro}$ become large anyway so that $F(kL^i a^i) \to 1$ and $F(kL^{rn} a^{rn}) \to 1$ and $F(kL^{ro} a^{ro}) \to 1$, unaffected by the aforementioned approximations.

The comments under (70) pertaining to grazing angles of incidence on a wedge also apply to the 2-D case.

Slope Diffraction by Edges Based on UTD—The UTD results for the edge-diffracted field $\mathbf{E}^d(P)$ in (57) and (71) for the 3-D and 2-D cases are accurate if the incident field $\mathbf{E}^i(Q_E)$ in those expressions exhibits a slow spatial variation (in its amplitude but not necessarily in its phase along the incident ray) at Q_E. If the field $\mathbf{E}^i(Q_E)$ is not slowly varying, then a higher-order term, referred to as the *slope diffraction term*, must be added to (57) and (71), respectively, as indicated in [4, 5]. This slope diffraction contribution pertaining to 3-D and 2-D cases is given below.

3-D Case—The UTD edge-diffracted field $\mathbf{E}^{dt}$ produced by a rapidly varying field $\mathbf{E}^i(Q_E)$ which is incident on a wedge may be expressed as

$$\mathbf{E}^{dt}(P) = \mathbf{E}^d(P) + \mathbf{E}^{sd}(P) \tag{74}$$

in which $\mathbf{E}^d(P)$ is as in (57) for the 3-D case, and $\mathbf{E}^{sd}(P)$ is the additional higher-order slope diffraction contribution. The total UTD field exterior to the wedge is as usual given by

$$\mathbf{E}(P) = \mathbf{E}^i(P)U_i + \mathbf{E}^r(P)U_r + \mathbf{E}^{dt}(P) \tag{75}$$

The field $\mathbf{E}^{sd}(P)$ in (74) may be expressed as in (59a) by

$$\mathbf{E}^{sd}(P) = \hat{\boldsymbol{\beta}}_0 E^{sd}_{\beta_0} + \hat{\boldsymbol{\phi}} E^{sd}_\phi \tag{76}$$

where

$$\begin{bmatrix} E^{sd}_{\delta_0} \\ E^{sd}_\phi \end{bmatrix} = \left\{ \begin{bmatrix} -d^i & 0 \\ 0 & -d^i \end{bmatrix} \begin{bmatrix} \dfrac{\partial E^i_{\beta'_0}}{\partial n^i} \\ \dfrac{\partial E^i_{\phi'}}{\partial n^i} \end{bmatrix} + \begin{bmatrix} -d^r_s & 0 \\ 0 & -d^r_h \end{bmatrix} \begin{bmatrix} \dfrac{\partial E^r_{\beta'_r}}{\partial n^r} \\ \dfrac{\partial E^r_{\phi'_r}}{\partial n^r} \end{bmatrix} \right\} \sqrt{\dfrac{\varrho_e}{s^d(\varrho_e + s^d)}}\, e^{-jks^d}$$

following the matrix notation of (60). In (77)

$$d^i = \frac{1}{jk \sin \beta_0} \left\{ A_0 \left[\csc^2\left(\frac{\pi + \beta^-}{2n}\right) F_s[kL^i a^+(\beta^-)] \right. \right.$$

$$\left. \left. - \csc^2\left(\frac{\pi - \beta^-}{2n}\right) F_s[kL^i a^-(\beta^-)] \right] \right\} \tag{78}$$

and

$$d^r_{s,h} \cong \frac{(\pm 1)}{jk \sin \beta_0} \left\{ A_0 \left[\csc^2\left(\frac{\pi + \beta^+}{2n}\right) F_s[kL^{rn} a^+(\beta^+)] \right. \right.$$

$$\left. \left. - \csc^2\left(\frac{\pi - \beta^+}{2n}\right) F_s[kL^{ro} a^-(\beta^+)] \right] \right\} \tag{79}$$

with $\beta^{\mp} = \phi \mp \phi'$, and $a^{\pm}(\beta)$ as in (62a). Also,

$$A_0 \equiv \frac{-e^{-j\pi/4}}{4n^2 \sqrt{2\pi k} \, \sin \beta_0} \tag{80}$$

and

$$F_s(x) \equiv 2jx[1 - F(x)] \tag{81}$$

with $F(x)$ as in (66).

The terms within the curly braces in (78) and (79) are the ϕ' derivatives of those present in $D_{es,eh}$ of (61). The partial derivative $\partial/\partial n^i$ in (77) is evaluated at Q_E and $n^i = \varrho^i_{\varrho'} \cdot \hat{\phi}'$, where $\varrho^i_{\phi'}$ is the radius of curvature of the incident wavefront in the $\hat{\phi}'$ direction. For a point source, $\varrho^i_{\phi'} = s^i(Q_E)$, where $s_i(Q_E)$ is simply the distance from the point Q_E of edge diffraction to the source point. The terms involving d^i in the first matrix on the right-hand side of (77) ensure that the slope of the total field is continuous across the ISB, in addition to the total field being continuous there. The continuity of the total field is already ensured by the term $\mathbf{E}^d$ in (74), as discussed previously. If the field $\mathbf{E}^i$ exhibited a null in the direction of incidence at Q_E, then $\mathbf{E}^d$ of (57) would vanish [since $\mathbf{E}^d$ is proportional to $\mathbf{E}^i(Q_E)$]; in this special case the higher-order slope diffraction term $\mathbf{E}^{sd}$ becomes the main contributor to the edge-diffracted field. The terms involving d^i are also important near the ISB if the incident field has a pattern which varies rapidly near Q_E. Likewise, the terms involving $d^r_{s,h}$ in the second matrix on the right-hand side of (77) are important at and near the RSB, where they ensure that the slope of the total field is continuous there. In order to evaluate these terms involving $d^r_{s,h}$ in (77), the reflected field $\mathbf{E}^r$ may be expressed for convenience as in (59b) by

$$\mathbf{E}^r = \hat{\mathbf{B}}'_r E_{\hat{\beta}'_r} + \hat{\phi}'_r E^r_{\phi'_r} \tag{82}$$

so as to view the reflected field $\mathbf{E}^r$ as being "incident" from virtual or image space. The direction of incidence from image space is of course $\hat{s}'$. Then $\hat{\beta}'_r$ and $\hat{\phi}'_r$ may be given the same meaning with respect to $\hat{s}^r$ (the direction of incidence from image space) as $\hat{\beta}'_0$ and $\hat{\phi}'$ have with respect to $\hat{s}^i$. Thus $\hat{\beta}'_r = \hat{s}^r \times \hat{\phi}'_r$, and $\mathbf{n}^r = \varrho^r_{\phi'_r} \hat{\phi}'_r$ (for evaluating $\partial/\partial n'$) in which $\varrho^r_{\phi'_r}$ is the radius of curvature of the reflected wavefront in the direction $\hat{\phi}'_r$, etc. In the case of a very blunt wedge for which the RSB corresponding to the two faces may be close, the results in (77)–(79) must be appropriately modified; that modification is not given here.

In the special case of a *half-plane* illuminated at the edge by a rapidly varying spherical wave, (77) simplifies to

$$
\begin{bmatrix} E^{sd}_{\beta_0} \\ E^{sd}_{\phi} \end{bmatrix} = \begin{bmatrix} -d_s & 0 \\ 0 & -d_h \end{bmatrix} \begin{bmatrix} \dfrac{\partial}{\partial n'} E^i_{\beta'_0} \\ \dfrac{\partial}{\partial n'} E^i_{\phi'} \end{bmatrix} \sqrt{\frac{\varrho_e}{s^d(\varrho_e + s^d)}}\, e^{-jks^d} \tag{83}
$$

where

$$
\begin{aligned}
d_{s,h} = \frac{e^{-j\pi/4}}{4\sqrt{2\pi k}\,\sin\beta_0} &\left\{ \frac{\sin(\beta^-/2)}{\cos^2(\beta^-/2)} F_s[2kL^i\cos^2(\beta^-/2)] \right. \\
&\left. \pm \frac{\sin(\beta^+/2)}{\cos^2(\beta^+/2)} F_s[2kL^r\cos^2(\beta^+/2)] \right\}
\end{aligned} \tag{84}
$$

in which $L^i = L^r$ as given previously in (63).

The UTD result in (75) is valid only if $\mathbf{E}^i$ is a ray optical field. If $\mathbf{E}^i$ is not a ray optical field, then one may have to decompose it into its ray optical components so that (74) can be employed to each of these incident-ray optical components, and these results may then be superposed. For example, if one is dealing with a large discrete antenna array near Q_E, then the near field of this array may not be ray optical at Q_E; however, each element of that array could still yield a ray optical field at Q_E. Likewise, a large-aperture antenna near Q_E can be quantized so that each quantized source in the aperture could yield a ray optical field at Q_E.

2-D Case—The results in (77) for the 3-D case can be reduced directly to treat the corresponding 2-D situation by requiring $\beta_0 = \pi/2$ and $\varrho_e \to \infty$, etc., exactly as done earlier to obtain (71) from (57). Thus, for the 2-D case the factor $\sqrt{\varrho_e/s^d(\varrho_e + s^d)}$ in (77) is replaced by $1/\sqrt{s^d}$. Also, the result in (84) for the special case of a *half-plane* illuminated at the edge by a rapidly varying spherical wave can be employed to deal with rapidly varying cylindrical- and plane-wave illumination if $\beta_0 = \pi/2$ and $\varrho_e \to \infty$ in (84), and if $s^i \to \infty$ for the case of an incident plane wave.

UAT for Edges—From the preceding discussions one may recall that the GTD solution for edge diffraction is expressed as $\mathbf{E}^{\mathrm{GTD}} = \mathbf{E}^{\mathrm{GO}} + \mathbf{E}^d_k$ [see (40)], while the corresponding UTD solution [3, 4] is expressed as $\mathbf{E} = \mathbf{E}^{\mathrm{GO}} + \mathbf{E}^d$ [see (56)]. It is therefore evident that $\mathbf{E}^d_k$ in the GTD solution is replaced by $\mathbf{E}^d$ in the UTD, whereas the $\mathbf{E}^{\mathrm{GO}}$ present in the GTD solution is left unchanged in the UTD. Also, as noted earlier, $\mathbf{E}^d_k$ in the GTD solution for edge diffraction becomes singular at the GO shadow boundaries (ISB and RSB) and it is therefore not valid within the transition regions adjacent to these boundaries; on the other hand, the $\mathbf{E}^d$ in the corresponding UTD solution is bounded and it properly compensates for the discontinuities of $\mathbf{E}^{GO}$ (at ISB and RSB) to yield a total field which is continuous everywhere away from the edge and other diffracted-ray caustics. Consequently the UTD overcomes the failure of the GTD within the GO shadow boundary transition regions; furthermore, outside these transition regions, $\mathbf{E}^d$ of the UTD reduces automatically to $\mathbf{E}_k^d$ of the GTD so that the total UTD solution in (56) also reduces uniformly to the GTD solution in (40), where the latter is valid and accurate.

In an alternative solution based on the uniform asymptotic theory (UAT) for edge diffraction [6], the total field $\mathbf{E}$ exterior to a perfectly conducting wedge is expressed as

$$\mathbf{E} = \mathbf{E}^G + \mathbf{E}_k^d \qquad \text{(in the UAT)} \tag{85}$$

in which $\mathbf{E}^{GO}$ of the GTD is modified to $\mathbf{E}^G$ in the UAT, while $\mathbf{E}_k^d$ is left unchanged. The field $\mathbf{E}^G$ in the UAT now contains the necessary Fresnel integrals to ensure the uniform nature of $\mathbf{E}$ in (85), just as $\mathbf{E}^d$ contains those Fresnel integrals in the F type transition functions of the UTD solution of (60) and (61). Thus, exterior to the GO shadow boundary transition regions, $\mathbf{E}^G$ reduces automatically to $\mathbf{E}^{GO}$, thereby allowing $\mathbf{E}$ of the UAT to also reduce properly to $\mathbf{E}^{GTD}$ outside those transition regions.

Since $\mathbf{E}_k^d$ is as defined previously in (44) [together with (67) and (68)], it is only necessary to define the remaining quantity $\mathbf{E}^G$ in the UAT expression of (85). In particular,

$$\mathbf{E}^G(P) = [f(\xi^i) - \tilde{f}(\xi^i)]\mathbf{E}^i(P) + [f(\xi^r) - \tilde{f}(\xi^r)]\mathbf{E}^r(P) \tag{86}$$

where $\mathbf{E}^i(P)$ represents the unperturbed "incident" field which exists in the absence of the wedge; in contrast, the GO incident field $\mathbf{E}^i U_i$ [see (41)] exists only in the lit zone. Likewise, $\mathbf{E}^r(P)$ in (86) is interpreted here as the unperturbed "reflected" field which is incident from "image space." The $\mathbf{E}^r$ plays the same role as $\mathbf{E}^i$ (as opposed to the truncated GO reflected field $\mathbf{E}^r U_r$). It is important to note that $\mathbf{E}^i$ and $\mathbf{E}^r$ in the UAT expression for $\mathbf{E}^G$ in (86) are not discontinuous at the ISB and RSB, respectively, whereas the $\mathbf{E}^i U_i$ and $\mathbf{E}^r U_r$ in $\mathbf{E}^{GO}$ [see (41)] are of course discontinuous there as explicitly indicated by the step functions U_i and U_r. It is quite easily seen that $\mathbf{E}^i$ in (86) is a continuous vector field since it exists as if the wedge were absent. On the other hand, $\mathbf{E}^r$ is the field reflected from the o or the n face of the wedge which is illuminated by $\mathbf{E}^i$; consequently it is necessary to construct a smooth extension of $\mathbf{E}^r$ beyond the edge of the illuminated surface of the wedge to ensure that $\mathbf{E}^r$ does not become discontinuous across the RSB. For the wedge in Fig. 19b, if the o face alone is illuminated, this face is extended smoothly past the edge into a surface o from which $\mathbf{E}^i$ continues to be reflected to yield an $\mathbf{E}^r$ past the edge. The same arguments apply if the n face alone is illuminated, in which case the surface n is constructed as a smooth extension of the n face past the edge. If both faces of the wedge are illuminated, then (86) should be replaced by

$$\mathbf{E}^G(P) = [f(\xi^i) - \tilde{f}(\xi^i)]\mathbf{E}^i(P) + [f(\xi^{ro}) - \tilde{f}(\xi^{ro})]\mathbf{E}^{ro}(P)$$
$$+ [f(\xi^{rn}) - \tilde{f}(\xi^{rn})]\mathbf{E}^{rn}(P) \tag{87}$$

where ξ^{ro} and ξ^{rn} are the values of ξ^r pertaining to the o and n faces of the wedge, respectively, and $\mathbf{E}^{ro}$ and $\mathbf{E}^{rn}$ likewise refer to the values of $\mathbf{E}^r$ for these two faces (and their smooth extensions).

The function $f(\xi^{i,r})$ in (86) is defined by

$$f(\xi^{i,r}) = \frac{e^{j\pi/4}}{\sqrt{\pi}} \int_{\xi^{i,r}}^{\infty} e^{-j\tau^2} d\tau \tag{88}$$

and $\tilde{f}(\xi^{i,r})$ represents the first term in the large-argument approximation of $f(\xi^{i,r})$, namely,

$$\tilde{f}(\xi^{i,r}) = \frac{e^{-j\pi/4}}{2\xi^{i,r}\sqrt{\pi}} e^{-j(\xi^{i,r})^2} \tag{89}$$

The $\xi^{i,r}$ are referred to as the *detour parameters* defined by

$$\xi^i(P) = \varepsilon^i(P)|\sqrt{k[s^d(P) + s^i(Q_E) - s^i(P)]}| \tag{90}$$

and

$$\xi^r(P) = \varepsilon^r(P)|\sqrt{k[s^d(P) + s^i(Q_E) - s^i(Q_R) - s^r(P)]}| \tag{91}$$

where the $s^d(P)$, $s^i(P)$, and $s^r(P)$ are the distances s^d, s^i, and s^r to the observation point P as defined previously. Also, $s^i(Q_R)$ and $s^i(Q_E)$ are the values of s^i at the point of reflection Q_R, and the point of edge diffraction Q_E as in Fig. 20.

$$\varepsilon^i(P) = \begin{cases} +1 & \text{if } P \text{ is on the shadow side of the ISB} \\ -1 & \text{if } P \text{ is on the lit side of the ISB} \end{cases} \tag{92}$$

and

$$\varepsilon^r(P) = \begin{cases} +1 & \text{if } P \text{ is on the shadow side of the RSB} \\ -1 & \text{if } P \text{ is on the lit side of the RSB} \end{cases} \tag{93}$$

In the UTD solution for the field in the presence of a wedge expressed as $\mathbf{E} = \mathbf{E}^{GO} + \mathbf{E}^d$ [see (56)], both $\mathbf{E}^{GO}$ and $\mathbf{E}^d$ are finite everywhere (except at caustics and the edge) including the ISB and RSB, and hence $\mathbf{E}$ is also finite there as indicated previously. In contrast, $\mathbf{E}_k^d$ in the UAT solution of (85) becomes singular at the ISB and RSB; however, $\mathbf{E}^G$ in (85) also becomes singular at these shadow boundaries such that it exactly cancels the singularity in $\mathbf{E}_k^d$, thereby keeping the total UAT solution in (85) finite and continuous at these boundaries. Like the UTD the UAT edge-diffraction solution is also valid away from the edge and ray caustics. Care must be exercised in evaluating the UAT field of (85) in the vicinity of the image sources [i.e., the virtual sources or caustics of $\mathbf{E}^r$ in (86)]; otherwise a spurious singularity may result at those points. Note that the general form in (85) is valid for both 3-D and 2-D cases. The $\mathbf{E}_k^d$ in (85) is given by (44) and (67) in the 3-D

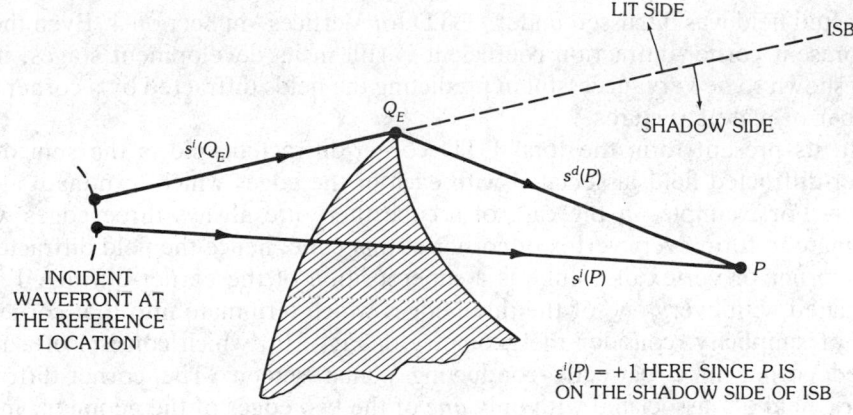

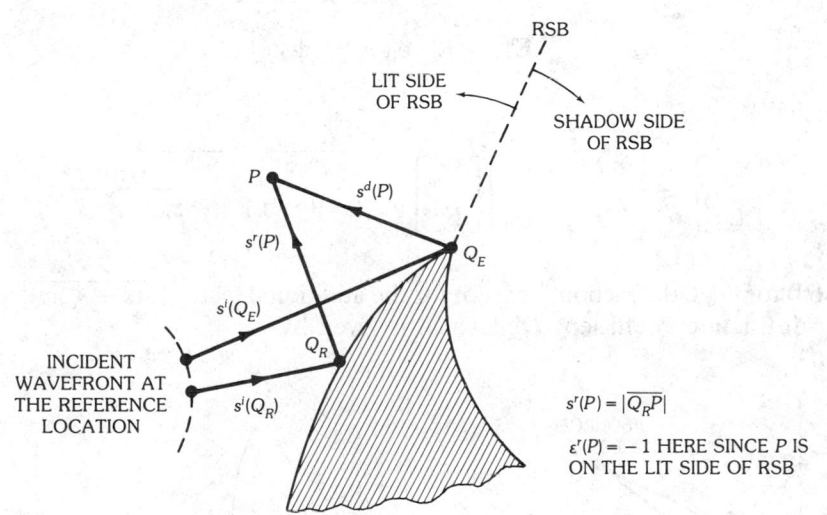

Fig. 20. Path lengths in the calculation of the detour parameters. (*a*) Calculation of $\varepsilon^i(P)$. (*b*) Calculation of $\varepsilon^r(P)$.

case; whereas $\sqrt{\varrho_e/s^d(\varrho_e + s^d)}$ in (44) is replaced by $1/\sqrt{s^d}$ for $\mathbf{E}_k^d$ in the 2-D case (since $\varrho_e \to \infty$ in the 2-D case), just as (71) is obtained from (57) for the 2-D situation in evaluating $\mathbf{E}^d$ of the UTD.

Explicit results are given in [23] for the UAT solution to the problem of the diffraction by an arbitrary smooth surface bounded by an arbitrary smooth edge.

UTD for Vertices—A vertex or a corner can be formed by the truncation of an edge. Figs. 21a and 21b illustrate a corner formed by the truncation of edges in planar as well as nonplanar geometries. An empirical UTD corner-diffraction solution proposed in [27] is described below. The general form of the corner-diffracted field was discussed under "GTD for Vertices" in Section 4. Even though the present corner-diffraction coefficient is still in its development stages, it has been shown to be very successful in predicting the fields diffracted by a corner for a number of plate structures.

In its present form the total UTD corner-diffracted field is the sum of the corner-diffracted fields associated with each of the edges which terminate at that corner. For example, in the case of a cube there are always three edges which terminate to form every vertex or corner of that cube; hence the field diffracted by each corner or vertex of a cube is a superposition of the corner-diffracted fields associated with every one of the three edges which terminate into that corner.

For simplicity, consider the geometry of Fig. 21a, which consists of a right-angled corner in a perfectly conducting planar screen. The corner-diffracted electric field $\mathbf{E}^{dc}$ associated with only *one* of the two edges of the geometry in Fig. 21a, when the illumination is due to a spherical wave (or point source), may be expressed as

$$\mathbf{E}^{dc} = \mathbf{E}^c_{\beta_0} \,\hat{\boldsymbol{\beta}}_0 + E^c_\phi \,\hat{\boldsymbol{\phi}} \tag{94}$$

in which

$$\begin{bmatrix} E^c_{\beta_0} \\ E^c_\phi \end{bmatrix} = \begin{bmatrix} -D^c_s & 0 \\ 0 & -D^c_h \end{bmatrix} \begin{bmatrix} E^i_{\beta'_0} \\ E^i_{\phi'} \end{bmatrix} \sqrt{\frac{s'}{s''(s' + s'')}} \sqrt{\frac{s(s + s_c)}{s_c}} \frac{e^{-jks}}{s} \tag{95}$$

as in (60) for edge diffraction,* except for the additional factor $\sqrt{s(s + s_c)/s_c}$, and the corner-diffraction coefficient $D^c_{s,h}$ which is given by

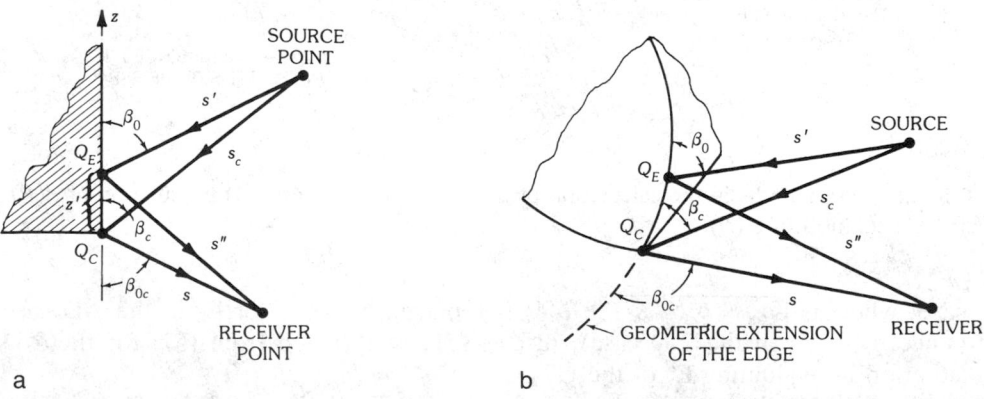

Fig. 21. Geometry for corner-diffraction problem. (*a*) Corner in a planar surface. (*b*) Corner in a nonplanar surface. (*After Sitka, Burnside, Chu, and Peters [27], © 1983 IEEE*)

---

*The incident field in (95) is evaluated at Q_c. Moreover, the corner-diffracted field in (94) and (95) is in terms of $\hat{\beta}_0$ and $\hat{\phi}$ associated with the ray reaching the receiver from Q_c (see Fig. 21).

$$D_{s,h}^c = \frac{e^{-j\pi/4}}{\sqrt{2\pi k}} C_{s,h}(Q_E) \frac{\sqrt{\sin\beta_c \sin\beta_{0c}}}{\cos\beta_{0c} - \cos\beta_c} F[kL_c a(\pi + \beta_{0c} - \beta_c)] \tag{96}$$

For the case of a corner in a planar geometry as in Fig. 21a, $C_{s,h}(Q_E)$ simplifies to

$$C_{s,h}(Q_E) = \frac{-e^{-j\pi/4}}{2\sqrt{2\pi k}\sin\beta_0} \left\{ \frac{F[kLa(\beta^-)]}{\cos(\beta^-/2)} \middle| F\left[\frac{kLa(\beta^-)/2\pi}{kL_c a(\pi + \beta_{0c} - \beta_c)}\right] \middle| \right.$$
$$\left. \mp \frac{F[kLa(\beta^+)]}{\cos(\beta^+/2)} \middle| F\left[\frac{kLa(\beta^+)/2\pi}{kL_c a(\pi + \beta_{0c} - \beta_c)}\right] \middle| \right\} \tag{97}$$

where

$$a(\psi) \equiv 2\cos^2(\psi/2) \tag{98}$$

Note that F in (97) is the same as in (66) and $\beta^{\mp} = \phi \mp \phi'$ is as in (62c). In addition,

$$L = \frac{s's''}{(s' + s'')}\sin^2\beta_0 \tag{99a}$$

and

$$L_c = \frac{s_c s}{s_c + s} \tag{99b}$$

for spherical-wave incidence. Since the corner in Fig. 21a is formed by the intersection of *two* edges, a corner-diffracted field contribution of the type in (95) and (96), which is associated with the *other* edge, must also be included; it is found in a similar fashion. The total UTD corner-diffracted field ensures the continuity of the total high-frequency field across the edge-diffraction shadow boundary (Fig. 22), where the edge-diffracted field becomes discontinuous past the corner.

Note that the corner-diffracted field should be added to any edge-diffracted and geometrical optics field components which can exist at a given observation point. In Fig. 22 only one edge-diffracted field component due to diffraction from $Q_E^{(b)}$ on edge (b) contributes, because the diffraction from $Q_E^{(a)}$ on the geometric extension of edge (a) past the corner does not fall within the physical limits of that edge. Another situation is shown in Fig. 23, where both $Q_E^{(a)}$ and $Q_E^{(b)}$ lie on their edge extensions; therefore no edge-diffracted field component is present. Fig. 24 shows the case where edge- and corner-diffracted fields from both edges are incident on the observation point or the receiver.

In order to treat corners formed by the truncation of nonplanar edged structures as in Fig. 21b, it is necessary only to generalize the $C_{s,h}(Q_E)$ and L_c for a planar corner in (97) and (99b). Such a generalization of $C_{s,h}(Q_E)$ and L_c for any *one* of the edges terminating into a vertex in a *nonplanar* geometry with spherical-wave illumination is given by

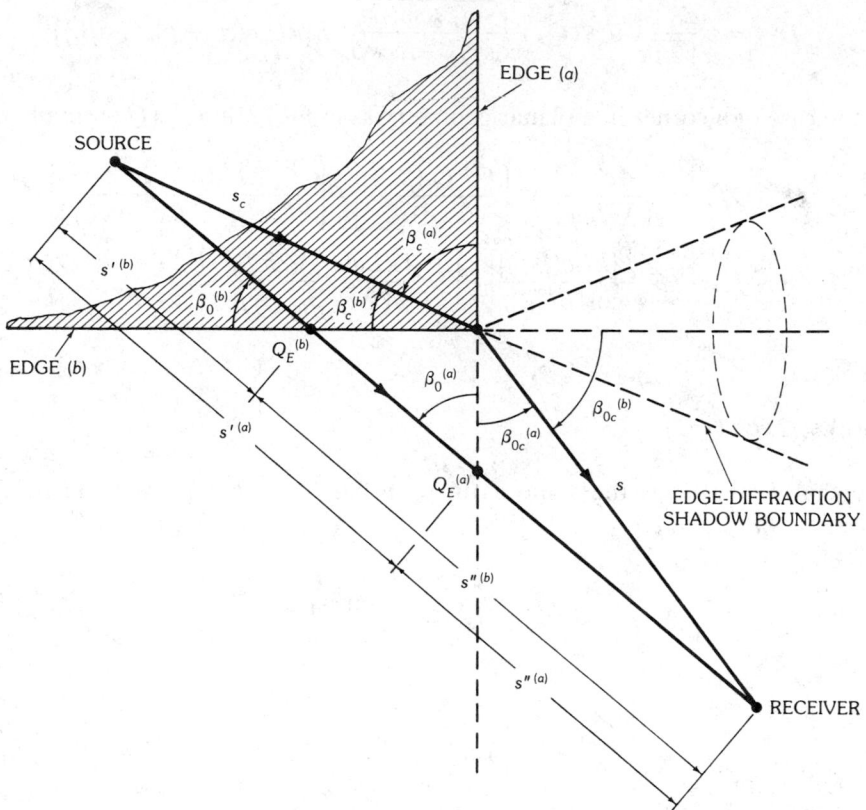

Fig. 22. Case where only one edge-diffracted field and both corner fields are received. (*After Sitka, Burnside, Chu, and Peters [27],* © *1983 IEEE*)

$$
C_{s,h} = \frac{-e^{-j\pi/4}}{2n\sqrt{2\pi k}\,\sin\beta_0} \left(\left\{ \cot\left[\frac{\pi + (\phi - \phi')}{2n}\right] F[kL^i a^+(\phi - \phi')] \right.\right.
$$

$$
\times \left| F\left[\frac{L^i a^+(\beta^-)/\lambda}{kL_c a(\pi + \beta_{0c} - \beta_c)}\right]\right| + \cot\left[\frac{\pi - (\phi - \phi')}{2n}\right] F[kL^i a^-(\phi - \phi')]
$$

$$
\times \left| F\left[\frac{L^i a^-(\beta^-)/\lambda}{kL_c a(\pi + \beta_{0c} - \beta_c)}\right]\right| \right\} \mp \left\{ \cot\left[\frac{\pi + (\phi + \phi')}{2n}\right] F[kL^{rn} a^+(\phi + \phi')] \right.
$$

$$
\times \left| F\left[\frac{L^{rn} a^+(\beta^+)/\lambda}{kL_c a(\pi + \beta_{0c} - \beta_c)}\right]\right| + \cot\left[\frac{\pi - (\phi + \phi')}{2n}\right] F[kL^{ro} a^-(\phi + \phi')]
$$

$$
\times \left.\left.\left| F\left[\frac{L^{ro} a^-(\beta^+)/\lambda}{kL_c a(\pi + \beta_{0c} - \beta_c)}\right]\right| \right\} \right) \tag{100}
$$

where

$$
L_c = \frac{\varrho_c s}{\varrho_c + s} \tag{101a}
$$

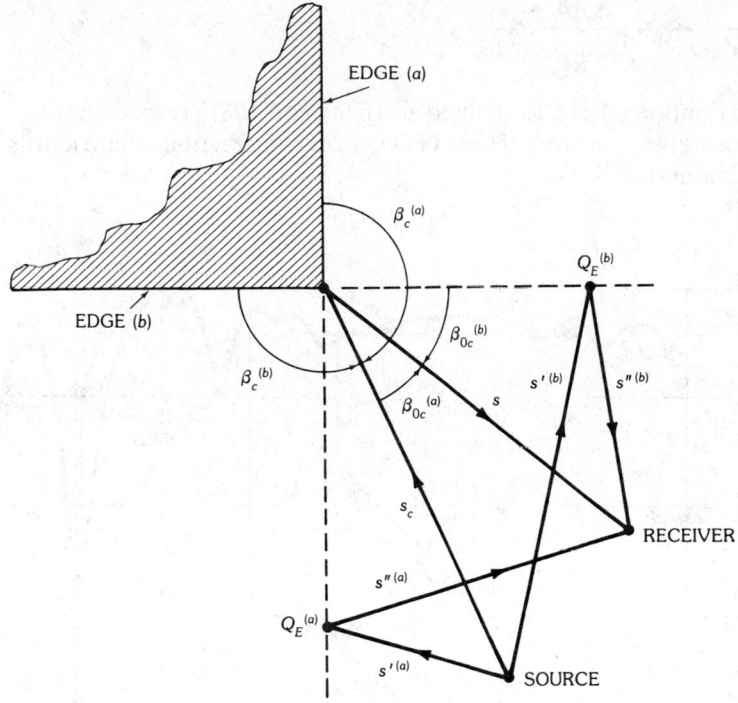

Fig. 23. Case where only corner fields are received. (*After Sitka, Burnside, Chu, and Peters [27], © 1983 IEEE*)

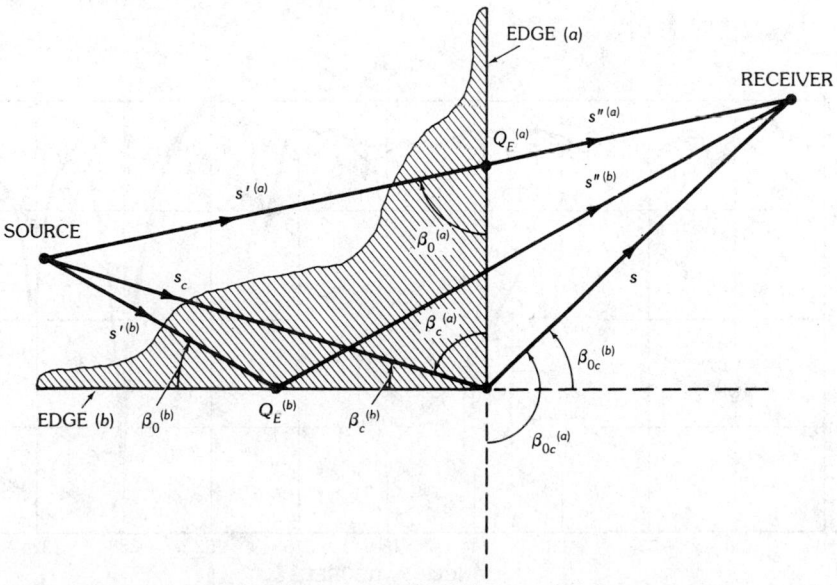

Fig. 24. Case where all edge- and corner-diffracted fields are received. (*After Sitka, Burnside, Chu, and Peters [27], © 1983 IEEE*)

$$\frac{1}{\varrho_c} = \frac{1}{\varrho_e^i(Q_c)} - \frac{\hat{\mathbf{n}}_e \cdot (\hat{\mathbf{s}}_c - \hat{\mathbf{s}})}{a_c \sin \beta_c \sin \beta_{0c}} \tag{101b}$$

and $a^{\pm}(\beta)$ and $a(\psi)$ are as defined in (62a) and (98), respectively.

The usefulness of the present UTD corner-diffraction coefficient is illustrated in Figs. 25a and 25b.

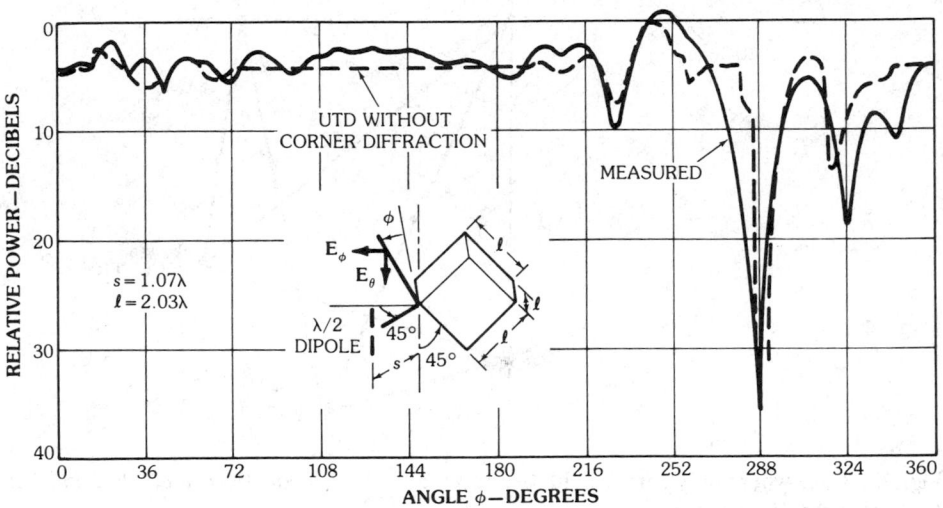

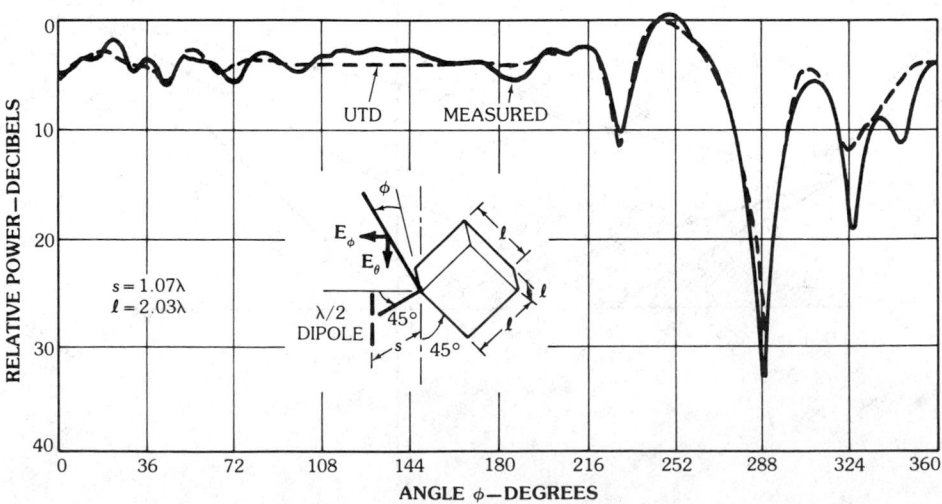

Fig. 25. Comparison of measured and calculated E_ϕ radiation pattern for a dipole near a box in the indicated plane.

UTD for Analyzing Problems of the Radiation by Antennas Either On or Off a Smooth, Perfectly Conducting, Convex Surface—The UTD solutions to the problems associated with the prediction of high-frequency radiation from sources which are located either near or directly on smooth convex surfaces are useful in analyzing a variety of antenna problems. For example, these solutions are useful for the following problems: (*a*) in predicting the scattering from the smooth convex portions of an aircraft fuselage, a ship mast, or a satellite shape, which is illuminated by a nearby antenna (that is located on some other parts of these structures); and (*b*) in predicting the radiation patterns of conformal antennas which are mounted directly on the smooth convex portions of an aircraft, a missile, or a spacecraft, etc.

The situation that arises in dealing with problems of type *a* above is depicted in Fig. 26. The problem indicated in Fig. 26 will henceforth be referred to as the "scattering problem"; in this problem, the source and observer are both located off the convex surface. The observer in Fig. 26 may be located either at a point P_L in the lit region or at a point P_S in the shadow region.

The situation that typically arises in dealing with problems of type *b* indicated above is illustrated in Fig. 27, and it defines what will henceforth be referred to as the "radiation problem." The source is positioned directly on the convex surface in this case, whereas the observation point is always located off the surface, either at P_L in the lit zone or at P_S in the shadow zone as indicated in Fig. 27. Note that the radiation problem of Fig. 27 is directly related to finding the electric current density, or the charge density, that is induced at the point Q' on a perfectly conducting convex surface by an appropriate source at P_L or P_S via the reciprocity theorem for electromagnetic fields. This reciprocal problem of determining the currents and charges induced on a perfectly conducting convex surface by an external source is of importance, for example, in electromagnetic pulse applications.

Another situation of interest which arises in dealing with problems of type *b* above is shown in Fig. 28; this problem in Fig. 28 will henceforth be referred to as the "mutual coupling problem." The source and the observation points are both positioned directly on the surface in this situation. It is clear that the problem in Fig. 28 is related directly to the calculation of the mutual coupling between a pair of

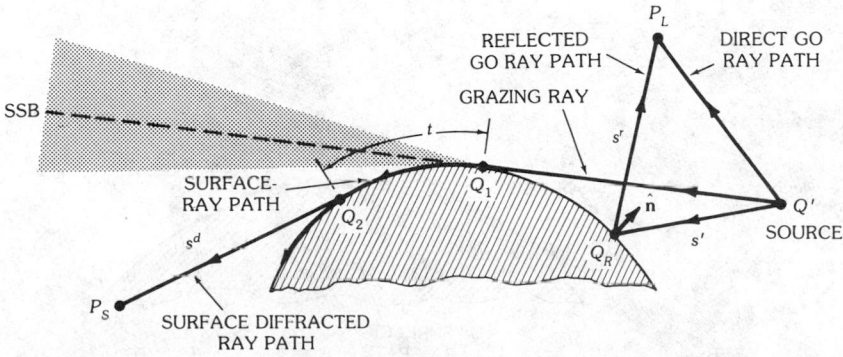

Fig. 26. Geometry associated with the scattering problem.

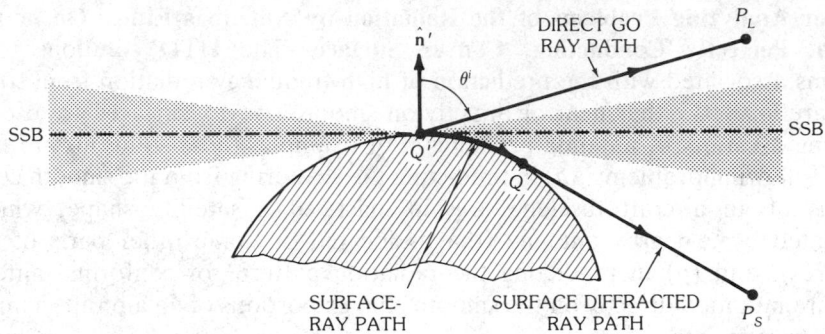

Fig. 27. Geometry associated with the radiation problem.

conformal antennas on a convex surface; such a calculation essentially reduces to finding the "surface fields" at Q due to a source at Q' on the same surface.

The scattering problem will be treated first, to be followed by similar treatments for the radiation and mutual coupling problems. It is of course implied that the principal surface radii of curvature at points of reflection and diffraction are sufficiently large in terms of the wavelength so that these high-frequency solutions remain valid. Some examples are provided which shed information as to the approximate bounds on how large in terms of the wavelength the surface has to be for obtaining accurate results based on these solutions. Note that the UTD solutions to be described below for the scattering, radiation, and mutual coupling associated with a convex surface all employ the rays of the GTD.

UTD Analysis of the Scattering by a Smooth, Convex Surface (3-D Case)— A uniform GTD (or UTD) solution which is convenient and accurate for engineering applications is presented below for the scattering problem in Fig. 26 such that it overcomes the limitations of the GTD within the SSB transition region (shaded in Fig. 26); furthermore, this UTD solution automatically reduces to the GTD solution exterior to the SSB transition region, where the latter solution is indeed valid. The angular extent of the SSB transition region is of the order $[2/k\varrho_g(Q_1)]^{1/3}$ radians, where $\varrho_g(Q_1)$ is the surface radius of curvature along the incident-ray

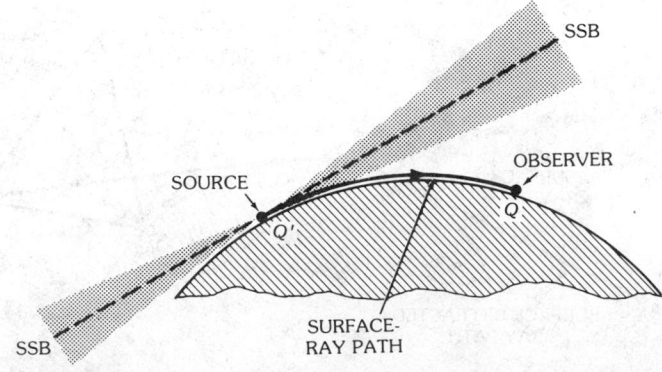

Fig. 28. Geometry associated with the mutual coupling problem.

direction at Q_1. The incident field can be an arbitrary ray optical type field as in (8); it is noted that plane- and spherical-wave fields are special cases of this arbitrary ray optical field (see "The GO Incident Field" in Section 2). The UTD solution given below is applicable, provided that the field point and the caustics of the incident-ray system are not in the close vicinity of the surface. It is also assumed that the amplitude (or the pattern) of the incident field does not exhibit a rapid spatial variation in the vicinity of Q_1, otherwise additional slope diffraction contributions to the total field need to be included; expressions for the latter have not yet been developed for the arbitrary convex surface. Two additional minor restrictions on the solution are indicated later on, when the explicit field expressions are given.

The total electric field $\mathbf{E}$ for the problem in Fig. 26 is given as

$$
\mathbf{E}(P) = \begin{cases} \mathbf{E}^i(P_L)U + \mathbf{E}^r(P_L)U + \mathbf{E}^d(P_L) & \text{if } P = P_L \text{ in the lit zone} \\ \mathbf{E}^d(P_S)[1 - U] & \text{if } P = P_S \text{ in the shadow zone} \end{cases}
$$

$$(102)$$

The incident and reflected fields $\mathbf{E}^i$ and $\mathbf{E}^r$ are associated with the incident and reflected GO rays shown in Fig. 26. The step function U in (102) is defined as

$$
U = \begin{cases} 1 & \text{in the lit region which lies above the SSB} \\ 0 & \text{in the shadow region which lies below the SSB} \end{cases}
$$

$$(103)$$

where the surface shadow boundary (SSB) is as illustrated in Fig. 26. The field $\mathbf{E}^d(P_S)$ in (102) within the shadow region follows the surface-diffracted ray path also shown in Fig. 26, whereas the field $\mathbf{E}^d(P_L)$ in (102) which is diffracted into the lit region follows the reflected-ray path (of $\mathbf{E}^r$) in this solution. Therefore it is convenient in this problem to combine the GO reflected field $\mathbf{E}^r(P_L)U$ and the diffracted field $\mathbf{E}^d(P_L)$ into a "generalized reflected field" $\mathbf{E}^{gr}(P_L)U$ in the lit region so that (102) becomes

$$
\mathbf{E}(P) = \begin{cases} \mathbf{E}^i(P_L)U + \mathbf{E}^{gr}(P_L)U & \text{if } P = P_L \text{ in the lit zone} \\ \mathbf{E}^d(P_S)[1 - U] & \text{if } P = P_S \text{ in the shadow zone} \end{cases}
$$

$$(104)$$

The surface-diffracted ray path (see Fig. 26) associated here with $\mathbf{E}^d(P_S)$ was described earlier under "GTD for Convex Surfaces" in Section 4. Also, the incident GO field $\mathbf{E}^i(P_L)$ was described previously under "The GO Incident Field" in Section 2. The fields $\mathbf{E}^{gr}(P_L)$ and $\mathbf{E}^d(P_S)$ are obtained in [28, 29] and are given by

$$
\mathbf{E}^{gr}(P_L) \sim \mathbf{E}^i(Q_R) \cdot [\mathcal{R}_s \hat{\mathbf{e}}_\perp \hat{\mathbf{e}}_\perp + \mathcal{R}_h \hat{\mathbf{e}}_\parallel^i \hat{\mathbf{e}}_\parallel^r] \sqrt{\frac{\varrho_1^r \varrho_2^r}{(\varrho_1^r + s^r)(\varrho_2^r + s^r)}} e^{-jks^r} \quad (105)
$$

$$
\mathbf{E}^d(P_S) \sim \mathbf{E}^i(Q_1) \cdot [\mathcal{D}_s \hat{\mathbf{b}}_1 \hat{\mathbf{b}}_2 + \mathcal{D}_h \hat{\mathbf{n}}_1 \hat{\mathbf{n}}_2] \sqrt{\frac{\varrho_s}{s^d(\varrho_s + s^d)}} e^{-jks^d} \quad (106)
$$

where the points Q_R and Q_1 and the distances s^r and s^d are indicated in Fig. 26. The quantities ϱ_1^r and ϱ_2^r have the same meaning as in (16) for the ordinary reflected GO field $\mathbf{E}^r(P_L)$, which is discussed under "The GO Reflected Field" in Section 2. Also, ϱ_s in (106) is the same as in (50) of Section 4. The quantities within brackets involving $\mathscr{R}_{s,h}$ in (105) and $\mathscr{D}_{s,h}$ in (106) may be viewed as generalized dyadic coefficients for surface reflection* and diffraction, respectively. It is noted that (105) and (106) are expressed invariantly in terms of the unit vectors fixed in the reflected and surface-diffracted ray coordinates. The unit vectors $\hat{\mathbf{e}}_\parallel^i$, $\hat{\mathbf{e}}_\parallel^r$, and $\hat{\mathbf{e}}_\perp$ in (105) have been defined in "The GO Reflected Field" in Section 2 (see Fig. 8). At Q_1 let $\hat{\mathbf{t}}_1$ be the unit vector in the direction of incidence, $\hat{\mathbf{n}}_1$ be the unit outward normal vector to the surface, and $\hat{\mathbf{b}}_1 = \hat{\mathbf{t}}_1 \times \hat{\mathbf{n}}_1$; likewise at Q_2 let a similar set of unit vectors $(\hat{\mathbf{t}}_2, \hat{\mathbf{n}}_2, \hat{\mathbf{b}}_2)$ be defined with $\hat{\mathbf{t}}_2$ in the direction of the diffracted ray as in Fig. 29. In the case of surface rays with zero torsion, $\hat{\mathbf{b}}_1 = \hat{\mathbf{b}}_2$. It is noted that ϱ_s in (106) is the wavefront radius of curvature of the surface-diffracted ray evaluated in the $\hat{\mathbf{b}}_2$ direction at Q_2. The $\mathscr{R}_{s,h}$ and $\mathscr{D}_{s,h}$ in (105) and (106) are given by [28, 29]:

$$\mathscr{R}_{s,h} = -\left(\sqrt{\frac{-4}{\xi^L}}\, e^{-j(\xi^L)^3/12} \left\{ \frac{e^{-j\pi/4}}{2\sqrt{\pi}\xi^L} [1 - F(X^L)] + \tilde{P}_{s,h}(\xi^L) \right\} \right) \qquad (107)$$

and for the lit region

$$\mathscr{D}_{s,h} = -\left\{ \sqrt{m(Q_1)m(Q_2)}\, \sqrt{\frac{2}{k}}\, \frac{e^{-j\pi/4}}{2\sqrt{\pi}\xi} [1 - F(X^d)] + \tilde{P}_{s,h}(\xi) \right\} \cdot \sqrt{\frac{d\eta(Q_1)}{d\eta(Q_2)}}\, e^{-jkt} \qquad (108)$$

for the shadow region. The function F appearing above has been introduced in the section dealing with edge diffraction [see (66)]. The Fock type surface-reflection functions $\tilde{P}_s$ and $\tilde{P}_h$ are related to the soft and hard Pekeris functions p^* and q^* by [28, 29] (note that $\delta = 0$ at SSB):

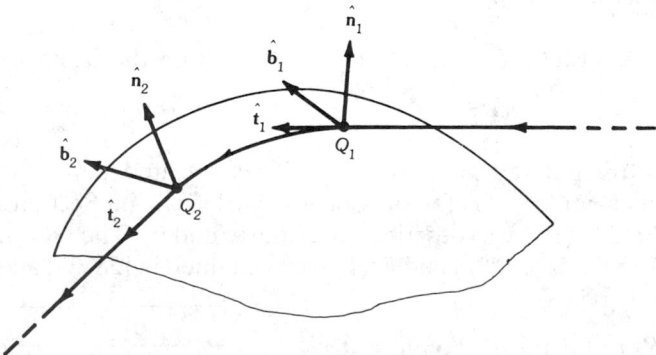

Fig. 29. Surface-ray unit vectors $\hat{\mathbf{t}}$, $\hat{\mathbf{n}}$, and $\hat{\mathbf{b}}$.

---

*Actually, cross terms exist in the above generalized dyadic reflection coefficient; but in general their effect is seen to be weak within the SSB transition region. Also these terms vanish in the deep lit region and on the SSB, hence they have been ignored in (105).

$$\tilde{P}_{s,h}(\delta) = \begin{Bmatrix} p^*(\delta) \\ q^*(\delta) \end{Bmatrix} e^{-j\pi/4} - \frac{e^{-j\pi/4}}{2\sqrt{\pi}\delta}$$

$$(109a)$$
$$(109b)$$

where p^* and q^* are finite and well behaved even when $\delta = 0$; these universal functions are plotted in Figs. 30, 31, and 32. Also,

$$\tilde{P}_{s,h}(\delta) = \frac{e^{-j\pi/4}}{\sqrt{\pi}} \int_{-\infty}^{\infty} \frac{\tilde{Q}V(\tau)}{\tilde{Q}W_2(\tau)} e^{-j\delta\tau} d\tau$$

where

$$\tilde{Q} = \begin{cases} 1 & \text{soft case} \\ \partial/\partial\tau & \text{hard case} \end{cases}$$

$$(110a)$$
$$(110b)$$

in which the Fock type Airy functions $V(\tau)$ and $W_2(\tau)$ are [28, 29]

$$2jV(\tau) = W_1(\tau) - W_2(\tau) \tag{111a}$$

$$W_1(\tau) = \frac{1}{\sqrt{\pi}} \int_{\infty \exp(-j2\pi/3)}^{\infty} e^{\tau t - t^3/3} \, dt \tag{111b}$$

$$W_2(\tau) = \frac{1}{\sqrt{\pi}} \int_{\infty \exp(+j2\pi/3)}^{\infty} e^{\tau t - t^3/3} \, dt \tag{111c}$$

The remaining quantities occurring in (107) and (108) are*

$$\xi^L = -2m(Q_R)\cos\theta^i \tag{112}$$

$$\xi = \int_{Q_1}^{Q_2} \frac{m(t')}{\varrho_g(t')} dt' \tag{113}$$

$$m(\cdot) = \left[\frac{k\varrho_g(\cdot)}{2} \right]^{1/3} \tag{114}$$

$$t = \int_{Q_1}^{Q_2} dt' \tag{115}$$

$$X^L = 2kL \cos^2\theta^i \tag{116}$$

and

---

*In [29] ξ^l is more precisely given by $\xi^L = 2m(Q_R)f^{-1/3}\cos\theta^i$, where $f^{-1/3}$ depends on the principal surface radii of curvatures at Q_R and θ^i in a complicated manner. However, it appears that replacing $f^{-1/3}$ by unity as is done in (112) for all θ^i does not impair the accuracy of the solution. It is noted that $f = 1$ when $\theta^i = \pi/2$ (i.e., at SSB), and it differs from unity as $\theta^i \to 0$; however, as $\theta^i \to 0$, $\xi^L \ll 0$ and $\tilde{P}_{s,h} \to \pm\sqrt{-\xi^L/4} \, e^{j(\xi^L)^{3/2}}$ so that $\mathcal{R}_{s,h} \to \mp 1$ as it should with either definition for ξ^L.

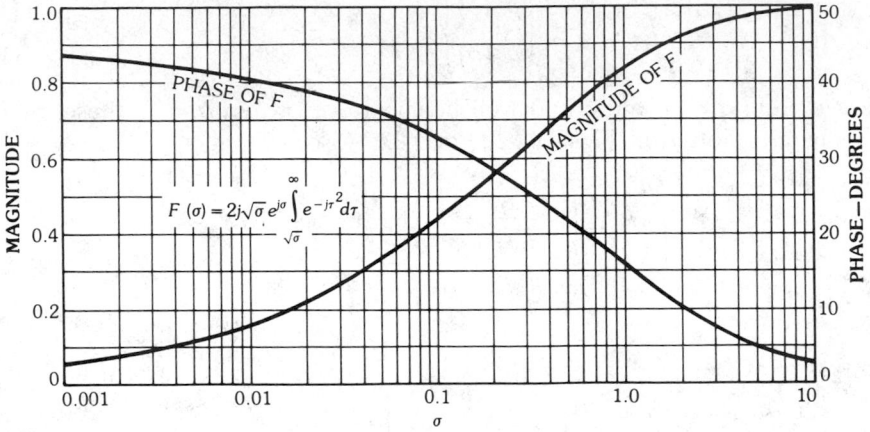

Fig. 30. A plot of $F(\sigma)$ versus σ. (*After Pathak [28]*, © *1979 American Geophysical Union*)

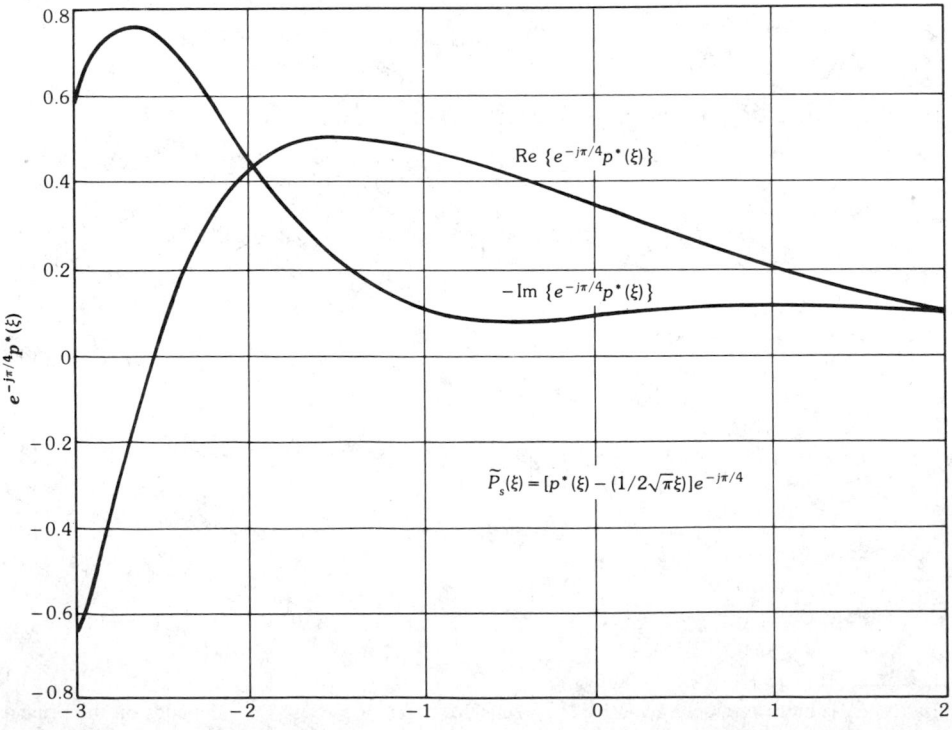

Fig. 31. A plot of $e^{-j\pi/4}p^*(\xi)$ versus ξ. (*After Pathak [28]*, © *1979 American Geophysical Union*)

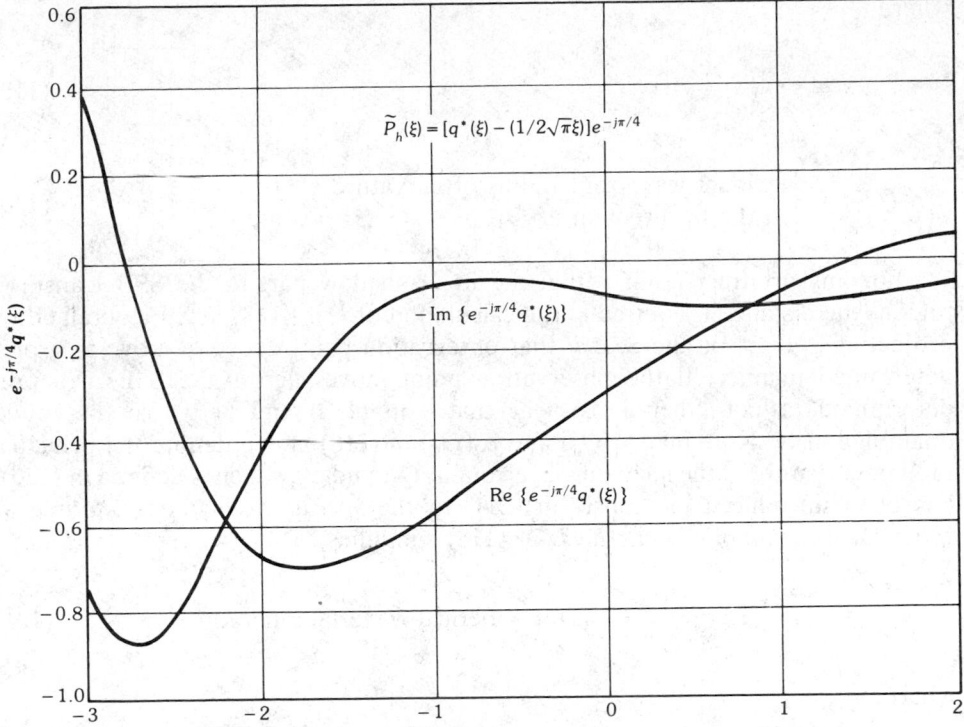

Fig. 32. A plot of $e^{-j\pi/4}q^*(\xi)$ versus ξ. (*After Pathak [28], © 1979 American Geophysical Union*)

$$X^d = \frac{kL(\xi)^2}{2m(Q_1)m(Q_2)} \tag{117}$$

The $\varrho_g(Q_R)$ in $m(Q_R)$ is the surface radius of curvature at Q_R in the plane of incidence, whereas $\varrho_g(Q_i)$ for $i = 1$ or 2 is the surface radius of curvature at Q_i in the $\hat{t}_i$ direction. Here, dt' is an incremental arc length along the surface-ray path. The angle of incidence θ^i is shown in Fig. 8. The $d\eta(Q_1)$ and $d\eta(Q_2)$ denote the widths of the surface-ray tube at Q_1 and Q_2, respectively; the surface-ray tube is formed by considering a pair of rays adjacent to the central ray as in Fig. 17. It is noted that the geodesic surface-ray paths are easy to find on cylinders, spheres, and cones. For example, the geodesic paths on a convex cylinder are helical, whereas they are great circle paths on a sphere. The geodesic surface-ray paths on cylinders and cones are illustrated later in Figs. 45 and 46. For more general types of convex surfaces the geodesic surface-ray paths must be found numerically with the aid of differential geometry. The distance parameter L in (116) and (117) is given by

$$L = \frac{\varrho_1^i(Q_1)\,\varrho_2^i(Q_1)}{[\varrho_1^i(Q_1) + s][\varrho_2^i(Q_1) + s]} \cdot \frac{s[\varrho_b^i(Q_1) + s]}{\varrho_b^i(Q_1)} \tag{118}$$

where

$$s \equiv s^r \bigg|_{\text{SSB}} = s^d \bigg|_{\text{SSB}} \tag{119}$$

$$\varrho_b^i(Q_1) = \begin{cases} \text{incident wavefront radius of curvature} \\ \text{in the } \hat{\mathbf{b}}_1 \text{ direction at } Q_1 \end{cases} \tag{120}$$

For any arbitrary point within the lit or shadow part of the SSB transition region, the distance s required in the calculation of L in (118) may be obtained by projecting s^r or s^d on the SSB *if* that observation point does *not* move in a pre-determined manner. If the observation point moves across the SSB in a pre-determined fashion, then it is clear that s in (118) and (119) can be found unambiguously. Note that $\varrho_1^i(Q_1)$ and $\varrho_2^i(Q_1)$ in (118) above denote the principal radii of curvature of the incident wavefront at Q_1, and ϱ_b^i, which is defined in (120), has been introduced previously in (21). In the *special case of point-source or spherical-wave illumination*, the L in (118) simplifies to*

$$L = \frac{s's}{s' + s} \qquad \text{for spherical-wave illumination} \tag{121}$$

where

$$s' \equiv [\varrho_1^i(Q_1) = \varrho_2^i(Q_1) = \varrho_b^i(Q_1)] = \begin{cases} \text{distance from the point} \\ \text{source to the point of} \\ \text{grazing incidence at } Q_1 \end{cases} \tag{122}$$

In the case of *plane-wave illumination*, $s' \to \infty$ and hence (121) above simplifies to

$$L = s \qquad \text{for plane-wave illumination} \tag{123}$$

If the incident wavefront is of the converging ($\varrho_{1,2}^i < 0$) or converging-diverging ($\varrho_1^i > 0$ and $\varrho_2^i < 0$; or $\varrho_1^i < 0$ and $\varrho_2^i > 0$) type, then the parameter L in (118) can become negative. It has not been investigated in detail how the general solution can be completed when L becomes negative. However, if one of the principal directions of the incident wavefront coincides with one of the principal planes of the surface at grazing, then one can treat a converging or converging-diverging type wavefront for which $L < 0$ by replacing $F(X^{L,d})$ with $F^*(|X^{L,d}|)$. Note that the asterisk on F^* denotes the complex conjugate operator. The use of $F^*(|X^{L,d}|)$ when $L < 0$ leads to a continuous total field at the SB in this case.

The above UTD result remains accurate outside the paraxial (i.e., near axial) regions of quasi-cylindrical or elongated convex surfaces; a different solution is required in these regions and it has not yet been completed.

---

*In general the present solution appears to be accurate even for kL as small as 3; in some special cases kL can be made as small as 1.

The surface-diffracted field of the type $\mathbf{E}^d(P_S)$ can also be present in the lit zone if the surface is closed; this may be visualized by noting that the field of the type $\mathbf{E}^d$ can propagate around the closed surface. Also, additional contributions to $\mathbf{E}^d(P_S)$ can be present in the shadow zone for a closed surface because surface-diffracted rays can be initiated at all points of grazing incidence on that closed surface; furthermore, these surface rays can undergo multiple encirclements around the closed body. These additional surface-diffracted ray contributions, however, are generally quite weak in comparison with the $\mathbf{E}^{gr}$ contribution within the lit zone for surfaces which are quite large in terms of the wavelength; hence their contributions may be neglected in such cases. Fig. 33 indicates the rays associated with these additional contributions for a closed surface.

It is clear that the parameters ξ^L, ξ, X^L, and X^d become small as one approaches the surface shadow boundary, SSB, from both the lit and shadow regions. As one approaches the SSB the small-argument limiting form of the transition function $F(X)$ which has been introduced earlier in (66) becomes helpful for verifying the continuity of the total high-frequency field at the SSB. On the other hand, the above parameters become large as one moves outside the SSB transition region; in this case $F(X) \to 1$ for large X and

$$\tilde{P}_{s,h}(\delta)\big|_{\delta \ll 0} \sim \pm \sqrt{\frac{-\delta}{4}}\, e^{j\delta^3/12} \tag{124}$$

$$\tilde{P}_{s,h}(\delta)\big|_{d \gg 0} = \begin{cases} -\dfrac{e^{-j\pi/4}}{\sqrt{\pi}} \sum\limits_{n=1}^{N} \dfrac{e^{j\pi/6} e^{\delta q_n \exp(-j5\pi/6)}}{2[Ai'(-q_n)]^2} \\[2em] -\dfrac{e^{-j\pi/4}}{\sqrt{\pi}} \sum\limits_{n=1}^{N} \dfrac{e^{j\pi/6} e^{\delta \bar{q}_n \exp(-j5\pi/6)}}{2\bar{q}_n [Ai(-\bar{q}_n)]^2} \end{cases} \tag{125}$$

where $N = 2$ is generally sufficient to compute $\tilde{P}_{s,h}(\delta)$ accurately for $\delta \gg 0$ via (125). In (125) the Miller type Airy function $Ai(\tau) = V(\tau)/\sqrt{\pi}$, and $Ai'(\tau) =$

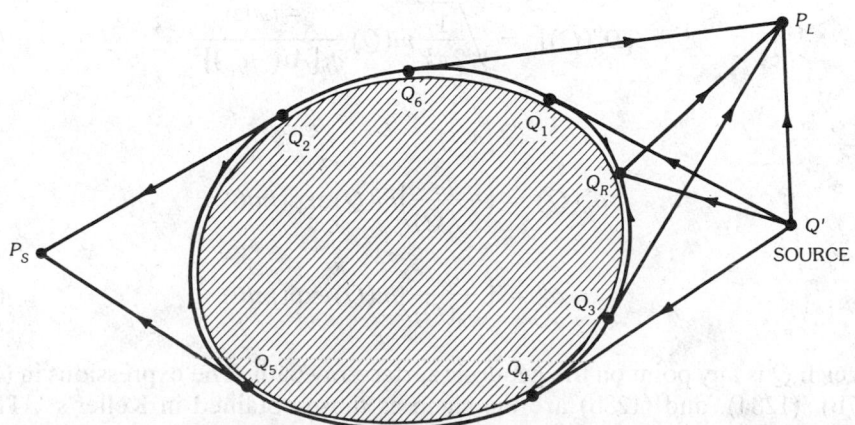

Fig. 33. Multiply encircling surface rays and contribution to P_L from surface-diffracted rays.

$dAi(\tau)/d\tau$. The parameters q_n and $\bar{q}_n$ in (125) are defined by $Ai(-q_n) = 0$ and $Ai'(-\bar{q}_n) = 0$, respectively. Explicit values of q_n and $\bar{q}_n$ are available in [30]; they are also given for $n = 1$ and $n = 2$ in Chart 1. Thus on incorporating the limiting values of (124) and (125), which are valid outside the SSB transition region, into (107) and (108) and replacing $F(X)$ by its asymptotic value of unity, it is clear that (105) and (106) properly reduce to $\mathbf{E}^{\mathrm{GTD}}$ of (46). Indeed, $\mathcal{R}_{s,h}$ in (107) reduces to $\mathcal{R}_{s,h} = \mp 1$ outside the SSB transition region so that $\mathbf{E}^{gr}(P_L) \to \mathbf{E}^r(P_L)$, and likewise $\mathbf{E}^d(P_S) \to \mathbf{E}_k^d(P_S)$ therein, respectively, in which $\mathbf{E}^r(P_L)$ and $\mathbf{E}_k^d(P_S)$ of (46) are as given by (16) and (50). The $\bar{\bar{\mathbf{T}}}^k(Q_1, Q_2)$ appearing in (50) for $\mathbf{E}_k^d$ is seen to be

$$\bar{\bar{\mathbf{T}}}^k(Q_1, Q_2) = \hat{\mathbf{b}}_1 \hat{\mathbf{b}}_2 T_s + \hat{\mathbf{n}}_1 \hat{\mathbf{n}}_2 T_h \tag{126a}$$

where

$$T_{s,h} = \sum_{n=1}^{N} D_n^{s,h}(Q_1) \exp\left[-\int_{Q_1}^{Q_2} \alpha_n^{s,h}(t')\, dt'\right] D_n^{s,h}(Q_2) \tag{126b}$$

The $D_n^{s,h}$ and $\alpha_n^{s,h}$ are the Keller's GTD diffraction coefficients and attenuation constants for the nth soft (s) or hard (h) surface-ray mode. Thus, in the GTD the surface-ray field consists of surface-ray modes which propagate independently of one another. Also, this surface-ray field is not the true field on the surface; it is a boundary layer field which gives rise to the interpretation of $\bar{\bar{\mathbf{T}}}^k$ in (50) and (126a) as a dyadic transfer function to indicate the launching of the surface-ray field at Q_1 [via $D_n^{s,h}(Q_1)$], the attenuation of the surface-ray field between Q_1 and Q_2 $\{$via $\exp[-\int_{Q_1}^{Q_2} \alpha_n^{s,h}(t)dt]\}$, and the diffraction of the surface-ray field at Q_2 [via $D_n^{s,h}(Q_2)$] to arrive at P_S via the surface-diffracted ray path $\overline{Q_2 P_s}$ as in Fig. 26. From (125) it can be seen that $D_n^{s,h}$ and $\alpha_n^{s,h}$ are given by

$$[D_n^s(Q)]^2 = \sqrt{\frac{1}{2\pi k}}\, m(Q)\, \frac{e^{-j(\pi/12)}}{[Ai'(-q_n)]^2} \tag{127a}$$

$$[D_n^h(Q)]^2 = \sqrt{\frac{1}{2\pi k}}\, m(Q)\, \frac{e^{-j(\pi/12)}}{\bar{q}_n[Ai(-\bar{q}_n)]^2} \tag{127b}$$

and

$$\alpha_n^s(Q) = \left(\frac{q_n}{\varrho_g(Q)}\right) m(Q) \exp(j\pi/6) \tag{128a}$$

$$\alpha_n^h = \left(\frac{\bar{q}_n}{\varrho_g(Q)}\right) m(Q) \exp(j\pi/6) \tag{128b}$$

in which Q is any point on the geodesic surface-ray path. The expressions in (127a), (127b), (128a), and (128b) are the same as those obtained in Keller's GTD [3].

The GTD result of (46) in terms of (127) and (126) is not valid within the SSB transition region; it is valid only outside that transition region, as pointed out under

Chart 1. Zeroes of the Airy Function and Its Derivative

Zeroes of the Airy Function		Zeroes of the Derivative of the Airy Function	
$Ai(-q_p) = 0$		$Ai'(-\bar{q}_p) = 0$	
q_1	$= 2.338\,11$	$\bar{q}_1$	$= 1.018\,79$
q_2	$= 4.087\,95$	$\bar{q}_2$	$= 3.248\,20$
$Ai'(-q_1)$	$= 0.701\,21$	$Ai(-\bar{q}_1)$	$= 0.535\,66$
$Ai'(-q_2)$	$= -0.803\,11$	$Ai(-\bar{q}_2)$	$= -0.419\,02$

"GTD for Convex Surfaces" in Section 4. Therefore one must use the more general UTD result of (104) together with (105) and (106), which not only remains valid within the SSB transition region but which also reduces automatically and uniformly to the usual GTD result of (46) exterior to that transition region. Some examples indicating the usefulness of the UTD result for the scattering problem are shown in Figs. 34, 35, and 36. In Fig. 34 the UTD solution is compared with an independent moment method (MM)–based solution for the same configuration.

Note that the magnetic field $\mathbf{H}(P)$ can be obtained as usual from the corresponding $\mathbf{E}(P)$ using the local plane-wave approximation to each component of $\mathbf{E}$ as in (11), (17), and (51).

UTD Analysis of the Scattering by a Smooth, Convex Surface (2-D Case)—The above UTD result for the 3-D scattering configuration can be simply modified to recover the corresponding UTD result for the 2-D case by allowing the caustic distances ϱ_1^r and ϱ_s in (105) and (106) to recede to infinity. Then, let

$$\varrho_2^r \equiv \varrho^r, \quad \text{if } \varrho_1^r \to \infty \text{ and } \varrho_s \to \infty \tag{129}$$

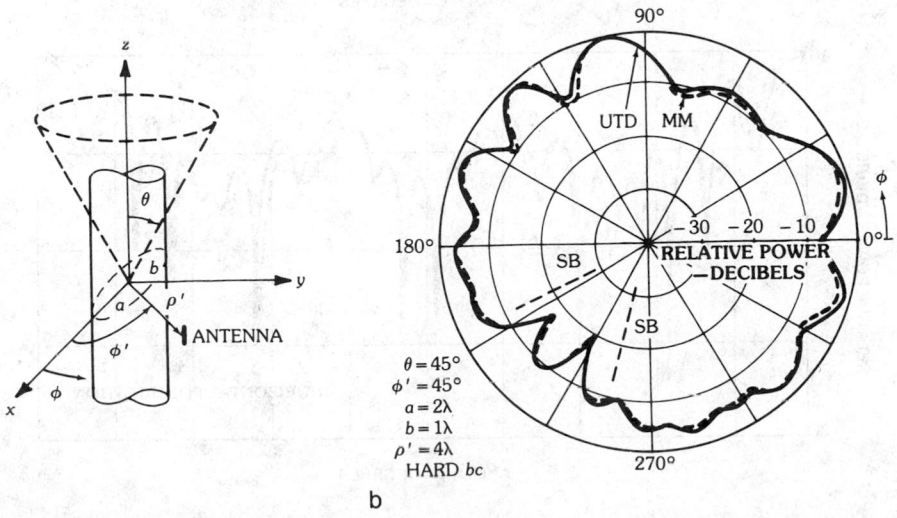

Fig. 34. Radiation pattern of a magnetic dipole located parallel to the axis of an elliptic cylinder. (*a*) Cylinder and antenna. (*b*) Radiation pattern. (*After Pathak, Burnside, and Marhefka [29],* © *1980 IEEE*)

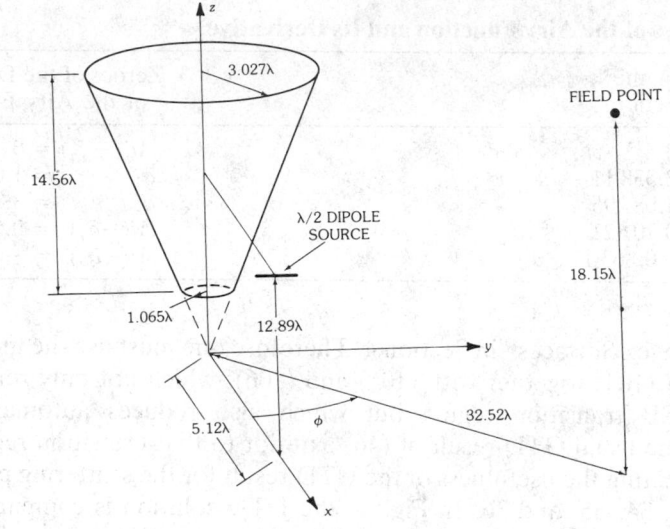

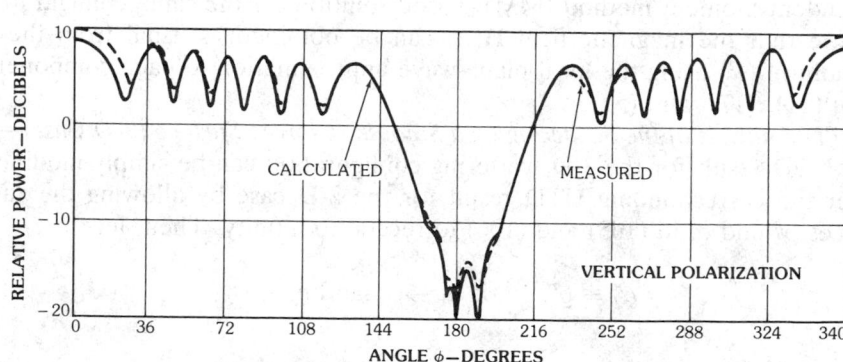

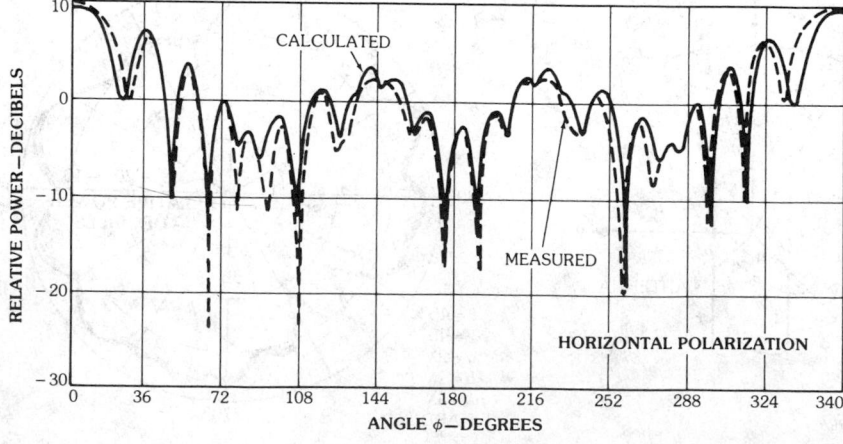

Fig. 35. Radiation patterns of an electric dipole mounted near a circular cone (see Fig. 37b). (*a*) Geometry for circular cone. (*b*) Comparison of measured and calculated radiation patterns. (*After Pathak, Burnside, and Marhefka [29], © 1980 IEEE*)

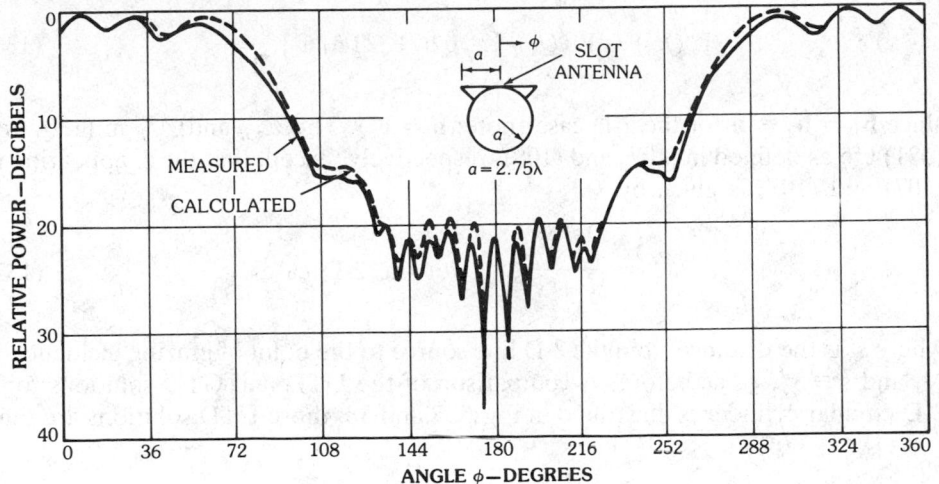

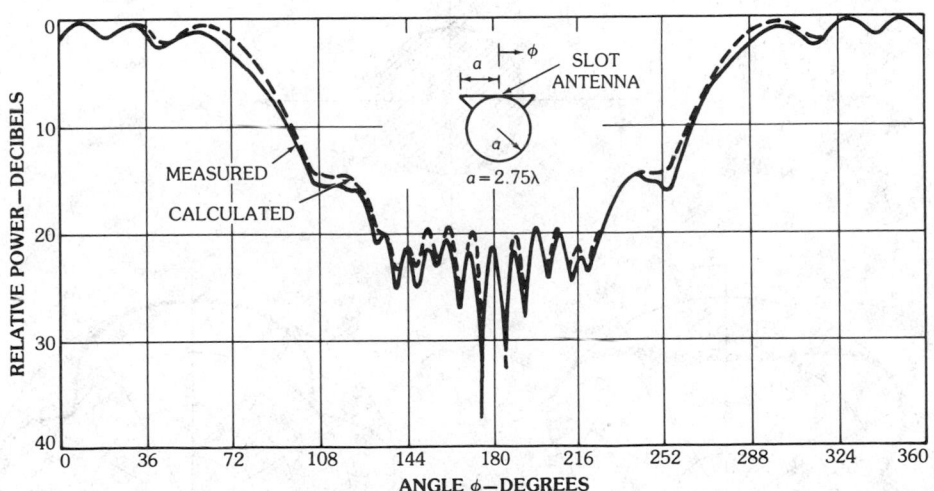

Fig. 36. Comparison of measured and calculated radiation patterns for a slot mounted in a plate-cylinder configuration with the slot parallel to the cylinder axis. (*After Pathak, Burnside, and Marhefka [29], © 1980 IEEE*)

so that $\mathbf{E}(P)$ is still given by (104), with (105) and (106) now changed to the following for the 2-D case:

$$\mathbf{E}^{gr}(P_L) \sim \mathbf{E}^i(Q_R) \cdot [\mathscr{R}_s \hat{\mathbf{e}}_\perp \hat{\mathbf{e}}_\perp + \mathscr{R}_h \hat{\mathbf{e}}_\parallel^i \hat{\mathbf{e}}_\parallel^r] \sqrt{\frac{\varrho^r}{\varrho^r + s^r}} e^{-jks'} \qquad (130)$$

in which ϱ^r is discussed under "The GO Reflected Field" in Section 2, and

$$\mathbf{E}^d(P_S) \sim \mathbf{E}^i(Q_1) \cdot [\mathscr{D}_s \,\hat{\mathbf{b}}\,\hat{\mathbf{b}} + \mathscr{D}_h \,\hat{\mathbf{n}}_1 \,\hat{\mathbf{n}}_2] \frac{e^{-jks^d}}{\sqrt{s^d}} \tag{131}$$

since $\hat{\mathbf{b}}_1 = \hat{\mathbf{b}}_2 \equiv \hat{\mathbf{b}}$ for the 2-D case (note: $\hat{\mathbf{b}} = \hat{\mathbf{e}}_\perp$). The $\mathscr{R}_{s,h}$ and $\mathscr{D}_{s,h}$ in (130) and (131) are as defined in (107) and (108), respectively, except that the L appearing in (107) and (108) is given by

$$L = \frac{s's^d}{s' + s^d}, \qquad \text{for the 2-D case} \tag{132}$$

where s' is the distance from the 2-D line source to the point of grazing incidence at Q_1 and $s \equiv s^d|_{\text{SSB}}$ as before. A comparison of the UTD and GTD solutions for a 2-D circular cylinder is illustrated in Figs. 37 and 38; those UTD solutions are then

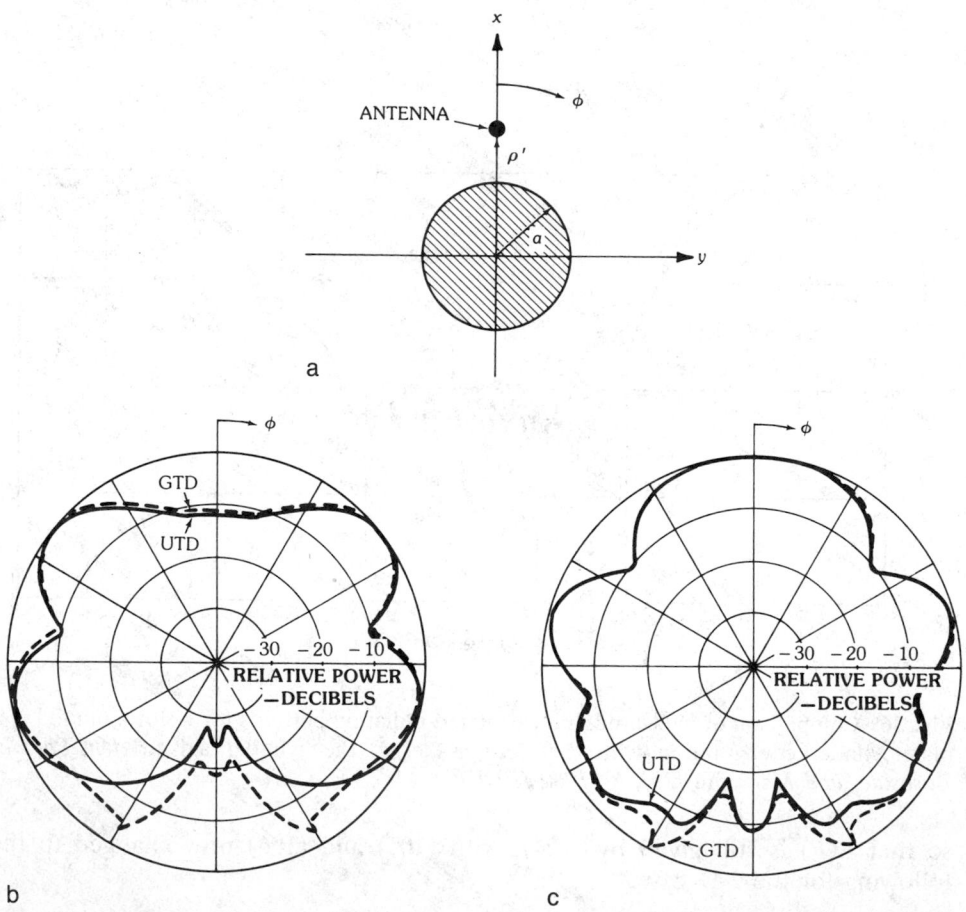

Fig. 37. Comparison of patterns calculated by uniform GTD (UTD) and ordinary GTD solutions for radiation by dipoles parallel to the axis of a perfectly conducting cylinder with a = 1λ and $\varrho' = 2\lambda$. (*a*) Geometry. (*b*) Electric line source case. (*c*) Magnetic line source case. (*After Pathak, Burnside, and Marhefka [29], © 1980 IEEE*)

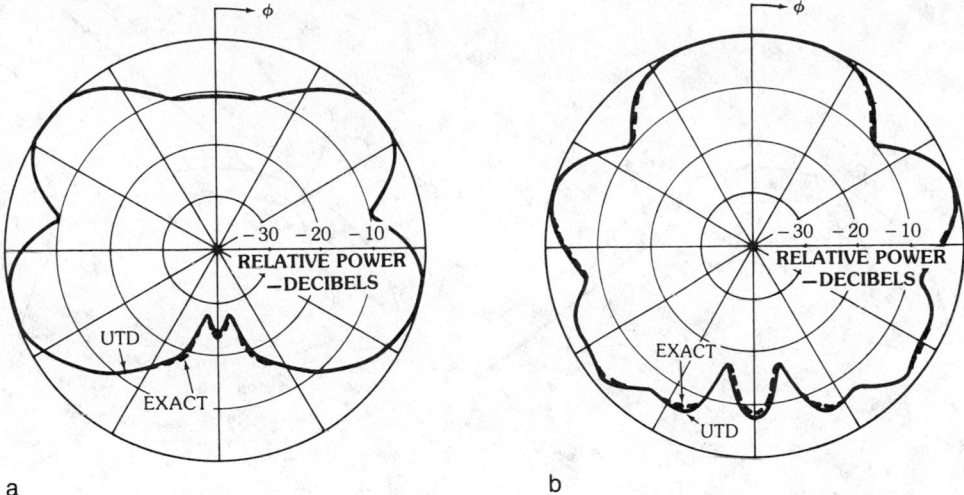

Fig. 38. Comparison of UTD solution of Fig. 37a with exact model series solution. (*a*) Radiation by electric line source. (*b*) Radiation by magnetic line source. (*After Pathak, Burnside, and Marhefka [29],* © *1980 IEEE*)

compared with the corresponding exact eigenfunction solutions, indicating the better accuracy of the UTD result over the ordinary GTD result, especially in the SSB transition region.

The magnetic field $\mathbf{H}(P)$ can be found from $\mathbf{E}(P)$ as in (11), (17), and (51) via the local plane-wave approximation.

UTD Analysis of the Radiation by Antennas on a Smooth, Convex Surface (3-D Case)—A compact uniform GTD (or UTD) solution is described below for the problem in Fig. 27 such that it remains uniformly valid across the SSB transition region. This UTD solution for the radiation problem employs the ray coordinates of the GTD. In the shadow region the field radiated by a source at Q' propagates along Keller's surface-diffracted ray path to the point P_S (Fig. 39a), whereas in the lit region the field propagates along the GO ray path directly from the source at Q' to the field point P_L (Fig. 39b). These ray fields reduce to the GO field in the deep lit region and remain uniformly valid across the SSB transition region into the deep shadow region. Surface-ray torsion, which affects the radiated field in both the shadow and the SSB transition regions, appears "explicitly" in the solution as a torsion factor. The solution to the scattering problem described earlier did not explicitly contain such a torsion factor to the first order; thus, explicit effects of surface-ray torsion appear to be localized only to regions in the neighborhood of sources on a convex surface. Since the field in the deep lit region is essentially that obtained from GO, and the field in the deep shadow region is relatively weak, the practical importance of the UTD solution described here is its ability to predict complex, surface-dependent field and polarization effects in the transition region adjacent to the shadow boundary (SSB) of a convex surface (Fig. 27).

The UTD solution for the radiated field is given below for the lit and shadow regions. These solutions for the two regions join smoothly at the shadow boundary

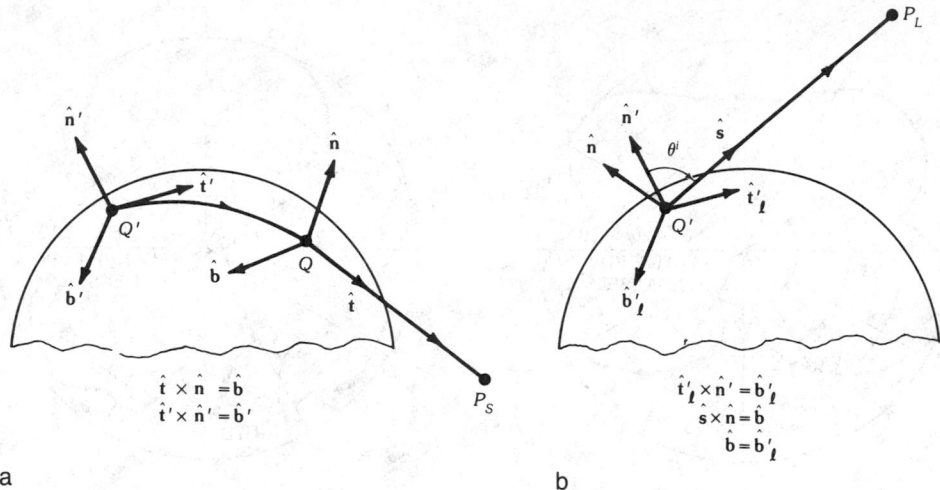

a   b

Fig. 39. Ray paths in the shadow and lit regions. (*a*) Field point in shadow region. (*b*) Field point in lit region. (*After Pathak, Wang, Burnside, and Kouyoumjian [31], © 1981 IEEE*)

(SSB). The UTD solution is presented here for the electromagnetic field radiated by an aperture in or by a short, thin monopole on a smooth, perfectly conducting, convex surface surrounded by a homogeneous isotropic medium. For an aperture in a convex surface it is convenient to define an infinitesimal magnetic current moment $d\mathbf{p}_m(Q')$ at any point Q' in the aperture as

$$d\mathbf{p}_m(Q') = \mathbf{E}^a(Q') \times \hat{\mathbf{n}}' \, dA' \tag{133}$$

where $\mathbf{E}^a(Q')$ is the electric field, $\hat{\mathbf{n}}'$ is the outward unit normal vector to the surface, and dA' is an area element at Q'. The tangential electric field in the aperture is assumed to be known. The $d\mathbf{p}_m(Q')$ radiating in the presence of the perfectly conducting convex surface, which now covers the aperture as well, constitutes the equivalent source of the electric field $d\mathbf{E}_m(P|Q')$ produced at any point P exterior to the surface. The total radiated electric field $\mathbf{E}_m(P)$ is then found by integrating the incremental field $d\mathbf{E}_m(P|Q')$ over the total area A of the aperture. Thus

$$\mathbf{E}_m(P) = \iint_A d\mathbf{E}_m(P|Q') \tag{134}$$

In the present context one notes that an aperture antenna frequently occurs in the form of a slot which is cut in a conducting surface. Slot radiators of rectangular, circular, or annular shapes are commonly employed in practice. Following the above development for the equivalent sources in the aperture radiation problem, one may similarly define an infinitesimal electric current moment $d\mathbf{p}_e(\ell')$ in dealing with the radiation by a monopole on a convex surface as

$$d\mathbf{p}_e(Q') = I(\ell') d\ell' \, \hat{\mathbf{n}}' \tag{135}$$

where $I(\ell')$ denotes the electric current distribution on the monopole, and ℓ' is the distance along its length measured from the base at Q'. It is assumed here that the monopole is a short, thin wire whose total length h is such that $h \ll R_{1,2}(Q')$, where $R_1(Q')$ and $R_2(Q')$ denote the principal surface radii of curvatures at Q'. It is also assumed that the current distribution $I(\ell')$ is known. The current moment $d\mathbf{p}_e(\ell')$ radiating in the presence of the perfectly conducting surface then constitutes the equivalent source of the electric field $d\mathbf{E}_e(P|\ell')$ produced at P exterior to the surface. The total electric field $\mathbf{E}_e(P)$ radiated by the short monopole can be approximately calculated from a knowledge of only the field $d\mathbf{E}_e(P|Q')$ upon simply replacing the source strength $d\mathbf{p}_e(Q')$, which occurs in the latter solution, by $\int_0^h d\mathbf{p}_e(\ell')\cos(k\ell'\cos\theta^i)$ if P is in the lit region, or by $\int_0^h d\mathbf{p}_e(\ell')$ if P is in the shadow region.* The above integrals serve to properly incorporate the effects of the wire length and the current distribution of the short monopole into the radiation calculation based only on $d\mathbf{E}_e(P|Q')$. It is noted that the term $\cos\theta^i$ inside the integral is defined by $\cos\theta^i = \hat{\mathbf{n}}' \cdot \hat{\mathbf{s}}$, where $\hat{\mathbf{s}}$ is the radiation direction in the lit zone from any point ℓ' on the monopole, with $0 \leqslant \ell' \leqslant h$. For calculating the field radiated by monopoles which are longer than the one being considered here, one must employ the integral $\int_0^h d\mathbf{E}_e(P|\ell')$ over the length of the monopole. The latter integral may also be employed to calculate the field radiated by arbitrary curved wire monopoles; however, in this case the $d\mathbf{E}_e(P|\ell')$ is produced by an arbitrarily oriented $d\mathbf{p}_e(\ell')$ above the convex surface. The cases of long, straight, or curved wire monopoles will not be considered here. From the preceding discussion note that it suffices to present a uniform GTD (or UTD) formulation for only the incremental fields $d\mathbf{E}_m(P|Q')$ and $d\mathbf{E}_e(P|Q')$ radiated by the sources $d\mathbf{p}_m(Q')$ and $d\mathbf{p}_e(Q')$, respectively. The present formulation or ansatz of the uniform GTD solution leads to separate representations for the radiated field $d\mathbf{E}_{m,e}(P|Q')$ in the shadow and lit regions, respectively. As mentioned earlier, however, these different representations match exactly in polarization, amplitude, and phase, at the shadow boundary. The shadow region field representation for $d\mathbf{E}_{m,e}(P|Q')$ is presented first; a corresponding lit region field representation for $d\mathbf{E}_{m,e}(P|Q')$ is given next.

The field $d\mathbf{E}_{m,e}(P_S|Q')$ for the field at P_S in the shadow region arrives from Q' along the surface-diffracted ray path shown in Fig. 39a; it is given by [31]

$$dE_{m,e}(P_S|Q') = dp_{m,e}(Q') \cdot [\bar{\bar{\mathbf{T}}}_{m,e}(Q'|Q) + \bar{\bar{\Delta}}^{m,e}] \sqrt{\frac{\varrho^d}{s^d(\varrho^d + s^d)}}\, e^{-jks^d} \tag{136}$$

with

$$\bar{\bar{\mathbf{T}}}_m(Q'|Q) = \frac{-jk}{4\pi}\, [\hat{\mathbf{b}}'\hat{\mathbf{n}}T_1(Q')H + \hat{\mathbf{t}}'\hat{\mathbf{b}}T_2(Q')S + \hat{\mathbf{b}}'\hat{\mathbf{b}}T_3(Q')S$$
$$+ \hat{\mathbf{t}}'\hat{\mathbf{n}}T_4(Q')H]\, e^{-jkt} \sqrt{\frac{d\psi_0}{d\eta(Q)}} \left[\frac{\varrho_g(Q)}{\varrho_g(Q')}\right]^{1/6} \tag{137}$$

---

*Here, $d\mathbf{E}_e(P|Q')$ is the value of $d\mathbf{E}_e(P|\ell')$ at $\ell' = 0$.

for the slot or $d\mathbf{p}_m$ case, and

$$\bar{\bar{\mathbf{T}}}_e(Q'|Q) = \frac{-jkZ_0}{4\pi}[\hat{\mathbf{n}}'\hat{\mathbf{n}}T_5(Q')H + \hat{\mathbf{n}}'\hat{\mathbf{b}}T_6(Q')S]\,e^{-jkt}\,\sqrt{\frac{d\psi_0}{d\eta(Q)}}\left[\frac{\varrho_g(Q)}{\varrho_g(Q')}\right]^{1/6}$$

(138)

for the monopole or $d\mathbf{p}_e$ case. The $\bar{\bar{\mathbf{\Delta}}}_{m,e}$ constitute the higher-order terms which are needed in special cases where the contribution from the leading terms in $\bar{\bar{\mathbf{T}}}_{m,e}$ is weak. The $\bar{\bar{\mathbf{\Delta}}}_{m,e}$ will be given below after the terms in $\bar{\bar{\mathbf{T}}}_{m,e}$ are defined in Table 2.

Let $\hat{\mathbf{\tau}}_1'$ and $\hat{\mathbf{\tau}}_2'$ denote the principal directions on the surface at Q' as shown in Fig. 40 along with the angle α' between $\hat{\mathbf{t}}'$ and $\mathbf{\tau}_1'$. Then R_1 and R_2 in Table 2 denote the principal surface radii of curvatures in the $\hat{\mathbf{\tau}}_1'$ and $\hat{\mathbf{\tau}}_2'$ directions at Q'. For the sake of definiteness one chooses $R_1 \geqslant R_2$ in this development. One notes that the expression for torsion $T(Q')$ given in Table 2 becomes negative if $\pi/2 < \alpha' < \pi$ or if $3\pi/2 < \alpha' < 2\pi$. The ϱ^d is expressed in terms of the quantities E and G in Table 2 which denote two of the three coefficients E, F, and G that occur in the "first fundamental form" representing the differential arc length along a curve on a surface (see, for example, D. J. Struik, *Differential Geometry*, 2nd ed., Addison-Wesley Publishing Co., Reading, Massachusetts, 1961). The functions H and S in $\bar{\bar{\mathbf{T}}}_{m,e}$ of (137) and (138) are

$$H = g(\xi) \tag{139}$$

$$S = \frac{-j}{m(Q')}\,\tilde{g}(\xi) \tag{140}$$

Here $g(\xi)$ and $\tilde{g}(\xi)$ denote the acoustic hard and soft Fock functions (or Fock integrals); they are defined as

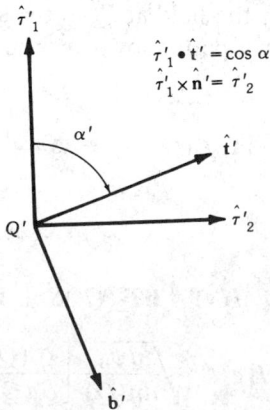

Fig. 40. Principal surface directions at the source. (*After Pathak, Wang, Burnside, and Kouyoumjian [31], © 1981 IEEE*)

Table 2. Terms for the Shadow Region

Type of Convex Surface	Slot or $d\mathbf{p}_m$ Case			Monopole or $d\mathbf{p}_m$ Case			Surface-Ray Torsion $T(Q')$	Surface Radius of Curvature in $\hat{\mathbf{t}}'$ Direction $\varrho_g(Q')$	Surface-Diffracted Ray Caustic Distance ϱ_c
	$T_1(Q')$	$T_2(Q')$	$T_3(Q')$	$T_4(Q')$	$T_5(Q')$	$T_6(Q')$			
Sphere	1	1	0	0	1	0	0	a	$a\tan(t/a)$
Circular cylinder	1	1	$\dfrac{\sin 2\alpha'}{2a}\dfrac{a}{\sin^2\alpha'}$	0	1	$\dfrac{\sin 2\alpha'}{2a}\dfrac{a}{\sin^2\alpha'}$	$\dfrac{\sin 2\alpha'}{2a}$	$\dfrac{a}{\sin^2\alpha'}$	t
Arbitrary convex surface	1	1	$T(Q')\varrho_g(Q')$	0	1	$T(Q')\varrho_g(Q')$	$\dfrac{\sin 2\alpha'}{2}\left[\dfrac{1}{R_2(Q')}-\dfrac{1}{R_1(Q')}\right]$ with $R_1(Q')\geq R_2(Q')$	$\left[\dfrac{\cos^2\alpha'}{R_1(Q')}+\dfrac{\sin^2\alpha'}{R_2(Q')}\right]^{-1}$	$\dfrac{2\sqrt{E}\,G}{\partial G/\partial t}$

$$g(\xi) = \frac{1}{\sqrt{\pi}} \int_{\infty \, \exp(-j2\pi/3)}^{\infty} e^{-j\tau\xi} \, [W_2'(\tau)]^{-1} \, d\tau \qquad (141)$$

and

$$\tilde{g}(\xi) = \frac{1}{\sqrt{\pi}} \int_{\infty \, \exp(-j2\pi/3)}^{\infty} e^{-j\tau\xi} \, [W_2(\tau)]^{-1} \, d\tau \qquad (142)$$

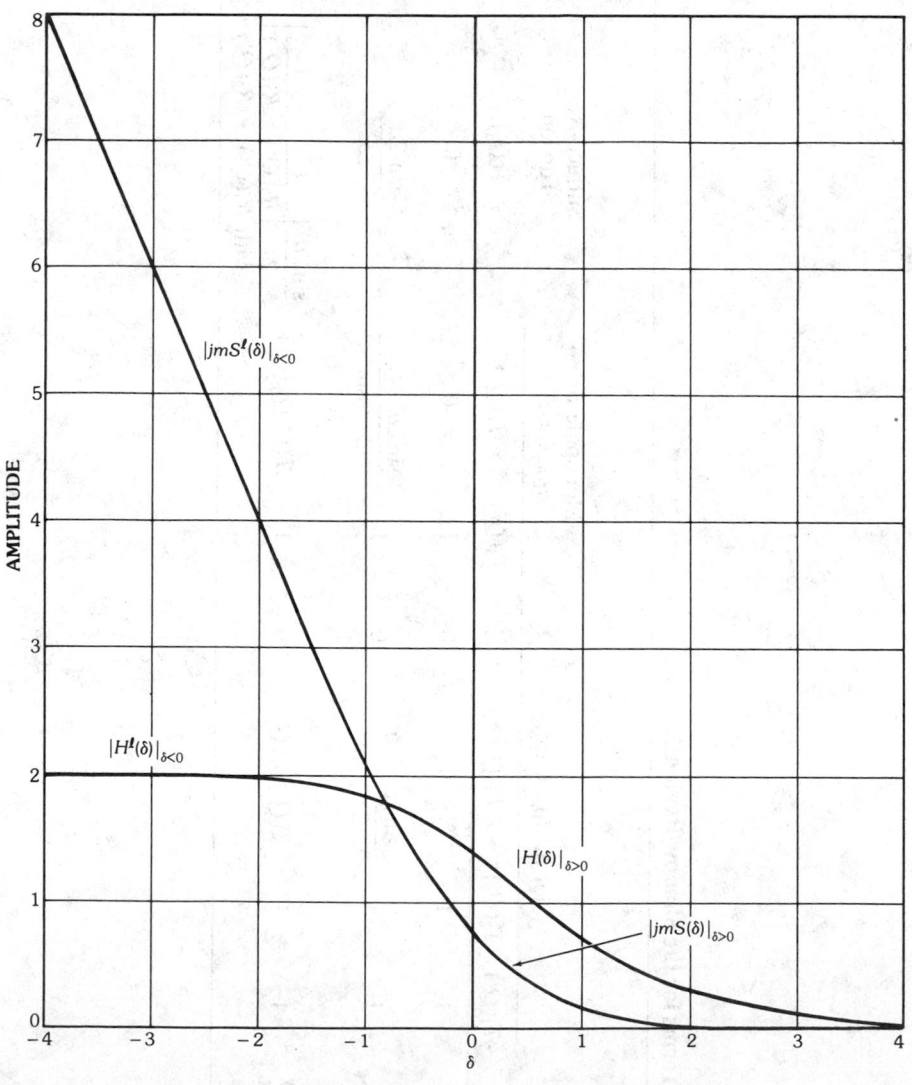

a

Fig. 41. Plots of the Fock radiation functions. (*a*) Amplitude. (*b*) Phase.

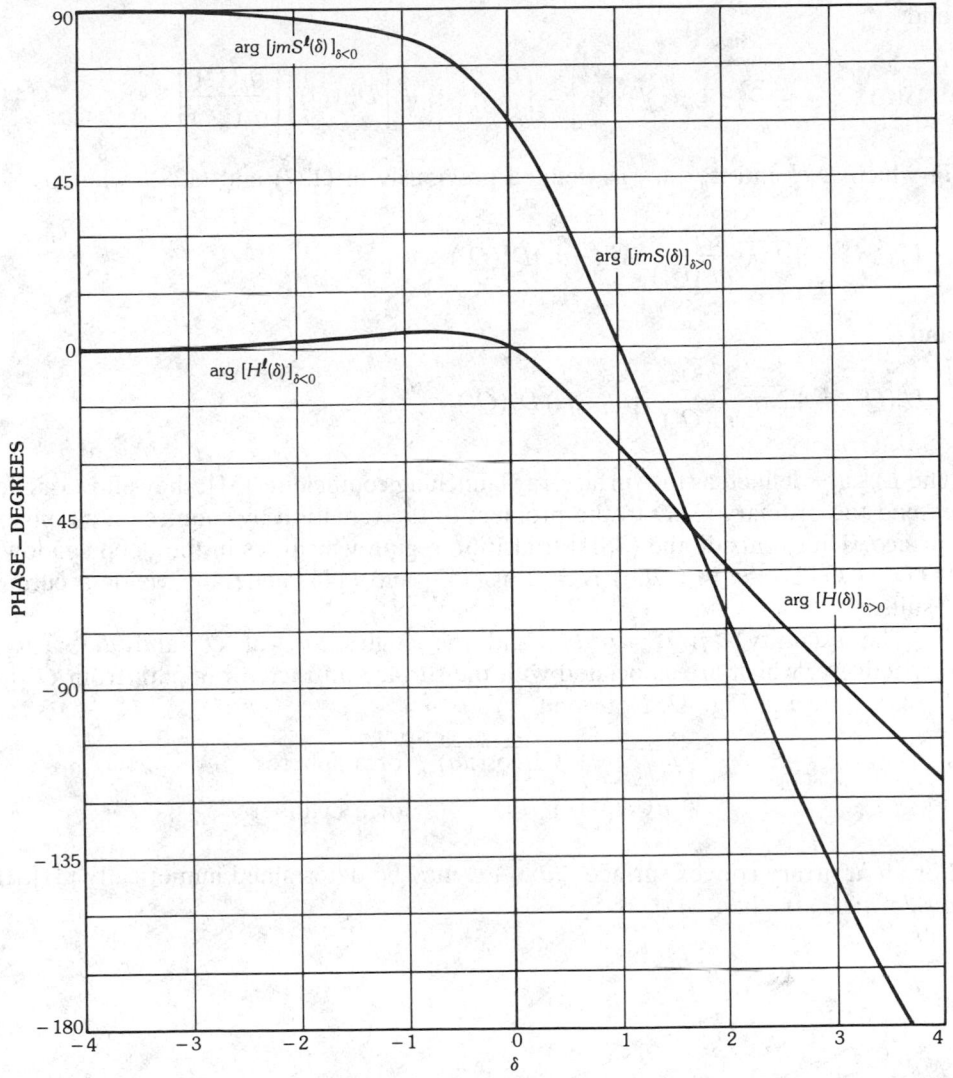

b

Fig. 41, *continued*

in which the Fock type Airy functions W_2 and W_2' are given by (111c) and $dW_2/d\tau$, respectively. The Fock parameter ξ for the "shadow zone" is defined as in (113) together with (114), but with Q_1 and Q_2 in (113) replaced by Q' and Q, respectively. Noted that $\xi > 0$ in the shadow zone. Also the distance t in e^{-jkt} of (137) and (138) is the length of the surface-ray geodesic path from Q' to Q. The functions $H(\xi)$ and $S(\xi)$ are plotted in Fig. 41 for $\xi > 0$. It is interesting to note that in the deep shadow zone, where $\xi \gg 0$, H and S can alternatively be expressed as

$$H(\xi)|_{\xi \gg 0} \sim \sum_{n=1}^{N} \left\{ L_n^h(Q') \exp\left[-\int_{Q'}^{Q} \alpha_n^h(t')dt'\right] D_n^h(Q) \right\} \left[\frac{\varrho_g(Q)}{\varrho_g(Q')}\right]^{-1/6} \quad (143)$$

and

$$S(\xi)|_{\xi \gg 0} \sim \sum_{n=1}^{N} \left\{ L_n^s(Q') \exp\left[-\int_{Q'}^{Q} \alpha_n^s(t')dt'\right] D_n^s(Q) \right\} \left[\frac{\varrho_g(Q)}{\varrho_g(Q')}\right]^{-1/6} \tag{144}$$

in which $D_n^{s,h}$ and $\alpha_n^{s,h}$ are as defined previously in (127) and (128), whereas

$$L_n^s(Q') = \sqrt{2\pi k} \, \frac{e^{-j\pi/12}}{m^2(Q')} \, Ai'(-q_n) D_n^s(Q')$$

and

$$L_n^h(Q') = \sqrt{2\pi k} \, \frac{e^{j\pi/12}}{m(Q')} \, Ai'(-\bar{q}_n) \, D_n^h(Q')$$

The $L_n^{s,h}$ are defined as the surface-ray launching coefficients [31]; they allow one to extend the ordinary GTD to the problem of the radiation by sources on a convex surface. Thus, outside the (SSB) transition region which lies in the deep shadow, UTD $\rightarrow$ GTD. As in (126b) $N = 2$ in (143) and (144) generally yields accurate results.

The quantity $d\eta(Q) = \varrho^d d\psi$, and the angles $d\psi_0$ at Q' and $d\psi$ at Q, respectively, which are associated with the surface-diffracted ray path from Q' to P_S, are shown in Fig. 42. Note that

$$\sqrt{\frac{d\psi_0}{d\psi}} = \begin{cases} \sqrt{1/\cos(t/a)} & \text{for a sphere} \\ 1 & \text{for a cylinder} \end{cases}$$

For an arbitrary convex surface, $\sqrt{d\psi_0/d\psi}$ may be determined numerically [31]. If $d\psi_0/\varrho^d d\psi < 0$, then

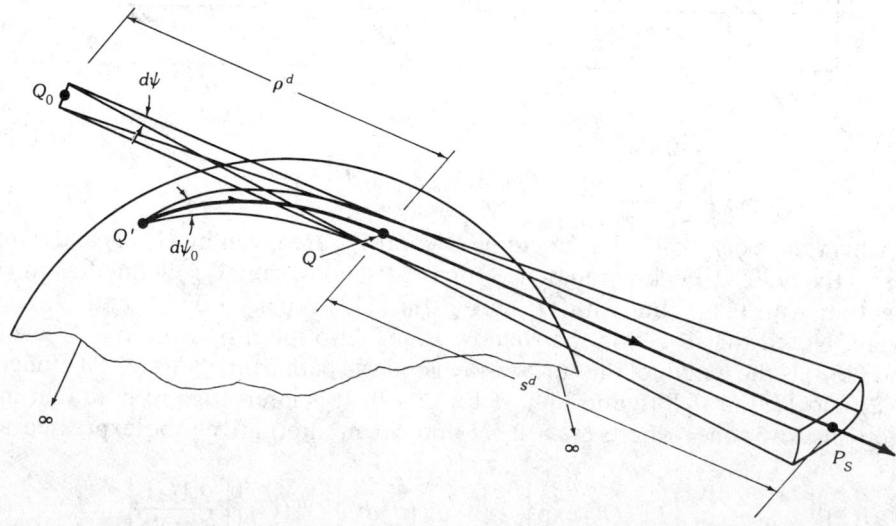

Fig. 42. Surface-diffracted ray tube associated with the surface-diffracted ray from Q' to P_s.

$$\sqrt{d\psi_0/\varrho^d d\psi} = \sqrt{d\psi_0/d\eta(Q)} = +j|\sqrt{d\psi_0/\varrho^d d\psi}|$$

Finally, the higher-order terms $\bar{\bar{\Delta}}_{m,e}$ in (136) are given by [31]

$$\bar{\bar{\Delta}}_m = \left(\frac{-jk}{4\pi}\right)\left\{(\hat{\mathbf{b}}'\hat{\mathbf{n}}T_0 + \hat{\mathbf{t}}'\hat{\mathbf{n}})\left[T_0\frac{jH'}{2m^2(Q')}\sqrt{\frac{d\psi_0}{d\psi}}\right] + \hat{\mathbf{t}}'\hat{\mathbf{b}}\frac{jH}{2m^3(Q')}\Lambda_s\left(\frac{d\psi_0}{d\psi}\right)\right\}$$
$$\times \left[\frac{\varrho_g(Q)}{\varrho_g(Q')}\right]^{1/6}\sqrt{\frac{d\psi_0}{d\eta(Q)}}\,e^{-jkt} \tag{145}$$

and

$$\bar{\bar{\Delta}}_e = \left(\frac{-jkZ_0}{4\pi}\right)\hat{\mathbf{n}}'\hat{\mathbf{n}}\Lambda_c(1 + T_0^2)\frac{jH'}{2m^2(Q')}\sqrt{\frac{d\psi_0}{d\psi}}\left[\frac{\varrho_g(Q)}{\varrho_g(Q')}\right]^{1/6}\sqrt{\frac{d\psi_0}{d\eta(Q)}}\,e^{-jkt} \tag{146}$$

where $T_0 = T(Q')\varrho_g(Q')$, in which $T(Q')$ and $\varrho_g(Q')$ are defined in Table 2. Also, the function H' is defined as $dH/d\xi$, in which H has been defined previously in (139). A reasonable choice for Λ_s and Λ_c is given by [31]

$$\Lambda_s = R_2(Q')/R_1(Q') \tag{147a}$$

and

$$\Lambda_c = 1 - \Lambda_s \tag{147b}$$

with $R_1(Q') \gneqq R_2(Q')$, in order to interpolate smoothly between the higher-order terms in the canonical circular cylinder and sphere solutions.

Next, the field $\mathbf{E}_{m,e}(P_L|Q')$ at a point P_L in the lit region arrives from the point Q' along the direct-ray path as shown in Fig. 39b; it is given by [31]

$$d\mathbf{E}_{m,e}(P_L|Q') = d\mathbf{p}_{m,e}(Q') \cdot (\bar{\bar{\mathbf{T}}}^\ell_{m,e} + \bar{\bar{\Delta}}^\ell_{m,e})\,e^{-jks} \tag{148}$$

with

$$\bar{\bar{\mathbf{T}}}^\ell_m = \frac{-jk}{4\pi}\,(\hat{\mathbf{b}}'_\ell\hat{\mathbf{n}}A + \hat{\mathbf{t}}'_\ell\hat{\mathbf{b}}B + \hat{\mathbf{b}}'_\ell\hat{\mathbf{b}}C + \hat{\mathbf{t}}'_\ell\hat{\mathbf{n}}D) \tag{149}$$

for the slot or $d\mathbf{p}_m$ case, and

$$\bar{\bar{\mathbf{T}}}^\ell_e = \frac{-jkZ_0}{4\pi}\,(\hat{\mathbf{n}}'\hat{\mathbf{n}}M + \hat{\mathbf{n}}'\hat{\mathbf{b}}N) \tag{150}$$

for the monopole or $d\mathbf{p}_e$ case. Again the $\bar{\bar{\Delta}}^\ell_{m,e}$ in (148) constitute the higher-order terms in the lit zone; whereas the $\bar{\bar{\mathbf{T}}}^\ell_{m,e}$ are the leading terms, which are generally more important than $\bar{\bar{\Delta}}^\ell_{m,e}$. The terms in $\bar{\bar{\mathbf{T}}}^\ell_{m,e}$ are defined below in Table 3. It is noted that $T(Q')$ and $\varrho_g(Q')$ in Table 3 have the same meaning as in Table 2 for the shadow zone. However, the angle α' in the definition of $T(Q')$ for the *lit zone* is now the angle between $\hat{\mathbf{t}}_\ell$ and $\hat{\mathbf{t}}'_1$. The functions H^ℓ and S^ℓ in Table 3 are

Table 3. Terms for the Lit Region

	Slot or $d\mathbf{p}_m$ Case		
A	B	C	D
$H^\ell + T_0^2 \Upsilon \cos\theta^i$	$S^\ell - T_0^2 \Upsilon \cos^2\theta^i$	$T_0\Upsilon$	$T_0\Upsilon \cos\theta^i$

	Monopole or $d\mathbf{p}_{\hat{e}}$ Case		
M	N	T_0	T
$\sin\theta^i(H^\ell + T_0^2 \Upsilon \cos\theta^i)$	$T_0\Upsilon \sin\theta^i$	$T(Q')\varrho_g(Q')$	$\dfrac{S^\ell - H^\ell \cos\theta^i}{1 + T_0^2\cos\theta^i}$

$$H^{\ell} = g(\xi_{\ell})e^{-j(\xi_{\ell}^3/3)} \qquad (151)$$

$$S^{\ell} = \frac{-j}{m_{\ell}(Q')}\tilde{g}(\xi_{\ell})e^{-j(\xi_{\ell}^3/3)} \qquad (152)$$

in which the hard and soft Fock functions g and $\tilde{g}$ have been defined previously in (141) and (142). The Fock parameter ξ_{ℓ} for the lit region is given by

$$\xi_{\ell} = -m_{\ell}(Q')\cos\theta^i \qquad (153)$$

where

$$m_{\ell}(Q') = \frac{m(Q')}{(1 + T_0^2\cos^2\theta^i)^{1/3}} \qquad (154)$$

The angle θ^i is defined by $\hat{\mathbf{n}}' \cdot \hat{\mathbf{s}} = \cos\theta^i$; this angle is also shown in Fig. 43. It is noted that the Fock parameter $\xi_{\ell} < 0$ in the lit region. In the deep lit region it can be shown from the asymptotic properties of g and $\tilde{g}$ for $\xi_{\ell} \ll 0$ that $\overline{\Upsilon} \to 0$. The factor $\overline{\Upsilon}$ has been defined in Table 3. One may view $T_0\overline{\Upsilon}$ as the "surface depolarization factor" since it plays an important role within the lit portion of the shadow boundary transition region, where it serves as a measure of the extent to which the polarization of the radiated field is affected by the surface geometry near Q'. Clearly, in the deep lit region where the geometrical optics approximation becomes valid, the polarization of the radiated field is dictated by the source rather than the surface, and $T_0\overline{\Upsilon} \to 0$ as one would expect.

The functions $H^{\ell}(\xi_{\ell})$ and $S^{\ell}(\xi_{\ell})$ are plotted in Fig. 41 for $\xi_{\ell} < 0$. It is easily verified that the $d\mathbf{E}_{m,e}$ of (148) reduces to the geometrical optics field solution in the deep lit region, where $T_0\overline{\Upsilon} \to 0$, $H^{\ell} \to 2$, and $S^{\ell} \to 2\cos\theta^i$ for $\xi_{\ell} \ll 0$, i.e., the leading terms yield

$$d\mathbf{E}_m(P_L|Q') \sim \frac{-jk}{4\pi}d\mathbf{p}_m(Q') \cdot (\hat{\mathbf{b}}'_{\ell}\hat{\mathbf{n}}2 + \hat{\mathbf{t}}'_{\ell}\hat{\mathbf{b}}2\cos\theta^i)\frac{e^{-jks}}{s}, \qquad \text{for } \xi_{\ell} \ll 0 \qquad (155)$$

and

$$d\mathbf{E}_e(P_L|Q') \sim \frac{-jkZ_0}{4\pi}d\mathbf{p}_e(Q') \cdot (\hat{\mathbf{n}}'\hat{\mathbf{n}}2\sin\theta^i)\frac{e^{-jks}}{s}, \qquad \text{for } \xi_{\ell} \ll 0 \qquad (156)$$

Finally, the higher-order terms $\overline{\overline{\Delta}}^{\ell}_{m,e}$ in (148) are given by [31]

$$\overline{\overline{\Delta}}^{\ell}_m = \frac{-jk}{4\pi}\left\{(\hat{\mathbf{b}}'_{\ell}\hat{\mathbf{n}}\,T_0 + \hat{\mathbf{t}}'_{\ell}\hat{\mathbf{n}})\left[T_0\frac{\sin^2\theta^i}{(1 + T_0^2\cos^2\theta^i)^2}\frac{jH^{\ell'}}{2m_{\ell}^2(Q')}\right]\right.$$
$$\left. + \hat{\mathbf{t}}'_{\ell}\hat{\mathbf{b}}\left[\frac{jH^{\ell}}{2m_{\ell}^3(Q')}\Lambda_s - \frac{T_0^2\cos\theta^i\sin^2\theta^i}{(1 + T_0^2\cos^2\theta^i)^2}\frac{jH^{\ell'}}{2m_{\ell}^2(Q')}\right]\right\} \qquad (157)$$

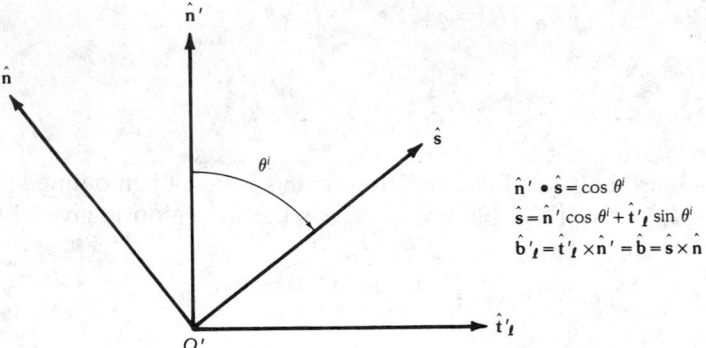

Fig. 43. Unit vectors for the lit region. (*After Pathak, Wang, Burnside, and Kouyoumjian [31], © 1981 IEEE*)

and

$$\bar{\bar{\Delta}}_e^\ell = \left(\frac{-jkZ_0}{4\pi}\right)\left[\hat{\mathbf{n}}'\hat{\mathbf{n}}\, \frac{\sin\theta^i\Lambda_c}{1+T_0^2\cos^2\theta^i}\frac{1+T_0^2}{1+T_0^2\cos^2\theta^i}\frac{jH^{\ell'}}{2m_\ell^2}\right.$$

$$\left. + \hat{\mathbf{n}}'\hat{\mathbf{b}}\,\frac{(-\sin\theta^i\cos\theta^i T_0)\Lambda_c}{1+T_0^2\cos^2\theta^i}\frac{1+T_0^2}{1+T_0^2\cos^2\theta^i}\frac{jH^{\ell'}}{2m_\ell^2}\right]\tag{158}$$

where T_0 is specified in Table 3. Also $H^{\ell'} = dH^\ell/d\xi_\ell$ in which H^ℓ is as defined in (151). The remaining quantities in (157) and (158) are identical with those defined earlier.

At the shadow boundary (SSB) it can be readily shown that $\bar{\bar{\Delta}}_{m,e}^\ell = \bar{\bar{\Delta}}_{m,e}$, so that the higher-order terms, like the leading terms, are continuous at $\theta^i = \pi/2$.

The solution presented above is employed here to compute the radiation from slots in cylinders, spheres, cones, and spheroids, and also from a monopole on a spheroid. In these computations the pertinent rays which pass through the field point in the lit and shadow regions are depicted in Fig. 44. The ray which is launched at Q' and then propagates in a clockwise direction around the convex surface before it sheds to the field point P_L in the lit region is not shown in Fig. 44a. The field of this ray, as well as that of rays shed from multiple encirclements, can be neglected for electrically large surfaces. The fields of rays which traverse long distances on surfaces with radii of curvature large in terms of the wavelength become highly attenuated and hence their effect on the resulting pattern is negligible. One notes that in the case of a prolate spheroidal surface there are two additional surface-diffracted ray contributions besides the ones shown in Fig. 44a; however, these contribute as significantly as the two rays in Fig. 44b only in the deep shadow region. Elsewhere, their contribution is negligible.

The various functions contained in the present solution are rather easy to compute; on the other hand, the associated geodesic surface-ray paths can be easily determined only in the case of some simple shapes, such as planes, cylinders, spheres, and cones; otherwise, the surface geodesics must be determined

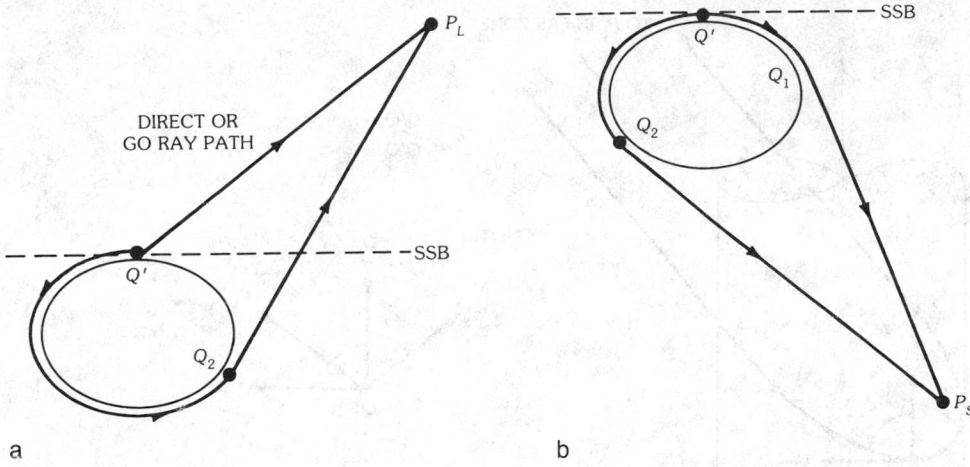

Fig. 44. Ray paths in the shadow and lit regions for a closed convex body. (*a*) Lit region. (*b*) Shadow region. (*After Pathak, Wang, Burnside, and Kouyoumjian [31],* © *1981 IEEE*)

numerically from their governing differential equations (or from an alternative integral expression). Fig. 45 indicates the dominant helical geodesic surface-ray paths for a convex cylinder. In Fig. 45a angles α'_1 and α'_2 stay the same with respect to the generator (or the $\hat{z}$ direction) of the cylinder. For a spherical surface the geodesic surface-ray paths are great circle paths. A discussion of the numerical computation of geodesic paths on circular and elliptic cylinders may be found in [32], whereas the geodesics on a cone may be computed as described in [33, 34]. A typical geodesic surface-ray path on a cone is illustrated in Fig. 46. Also, a procedure for finding the geodesic paths corresponding to a "given" radiation direction ($\hat{t}$ or $\hat{s}$) is discussed in [33] for the more general prolate spheroidal surface.

Radiation pattern calculations using only the leading terms of this asymptotic (UTD) solution are presented in Figs. 47b, 47c, and 47d for circumferential and axial slots in circular and elliptic cylinders. The rectangular slots in these cases are short and thin so that the dominant mode cosine distribution can be assumed in the slot. The cylinder geometry is shown in Fig. 47a. It is observed that the agreement between the asymptotic (UTD) calculations and the exact calculations in Figs. 47b, 47c, and 47d is very good even though the cylinders are not too large in terms of the wavelength. The exact results in Figs. 47c and 47d for the elliptic cylinder case have been obtained from the work of [35].

The next two radiation pattern calculations presented in Figs. 48 and 49 illustrate the importance of the higher-order terms in the UTD solution.* While the inclusion of the higher-order terms is in general not essential for accuracy, there are

---

*A caustic of the surface-diffracted rays is present at $\theta = 0°$ and $180°$ for the spherical geometry. The field of this caustic is weak in comparison with the direct or geometrical optics field of the source at $\theta = 0°$; however, it is the only field which exists at $\theta = 180°$. The present UTD solution is not directly valid at caustics, hence the pattern is not shown close to $\theta = 180°$ in Fig. 48; more will be said about modifications of the UTD solution at caustics later.

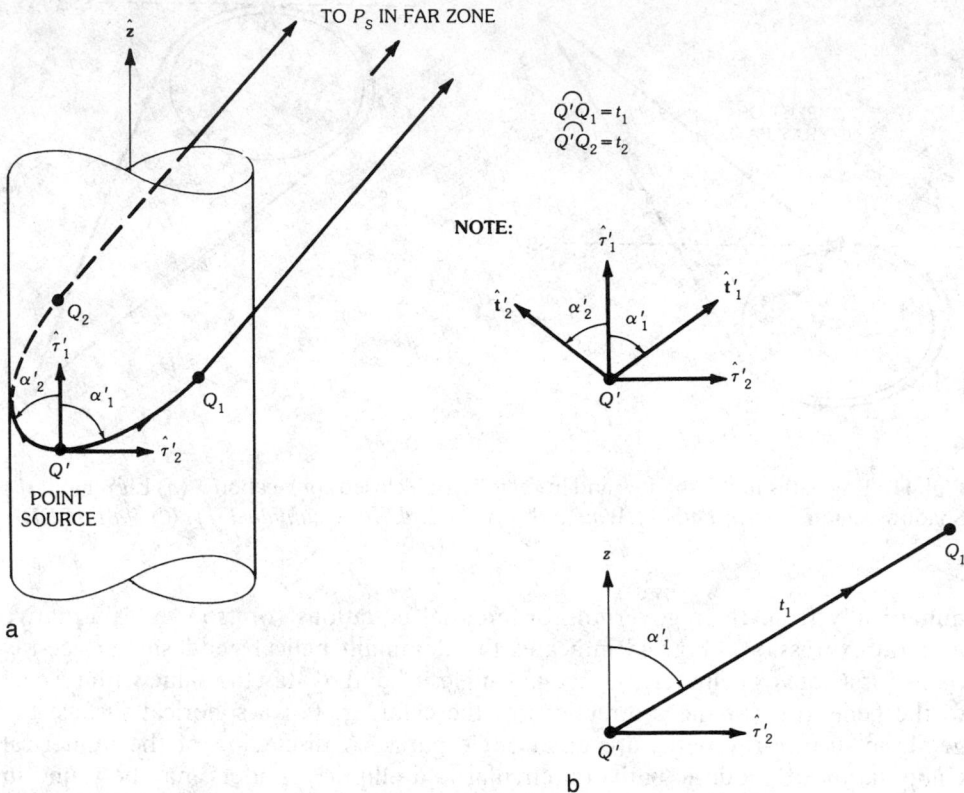

Fig. 45. Helical geodesic surface-ray paths from Q'. (*a*) Dominant path to points Q_1 and Q_2 of diffraction on a convex cylinder. (*b*) Helical geodesic surface ray path from Q' to point Q_1 of Fig. 45a on a developed cylinder.

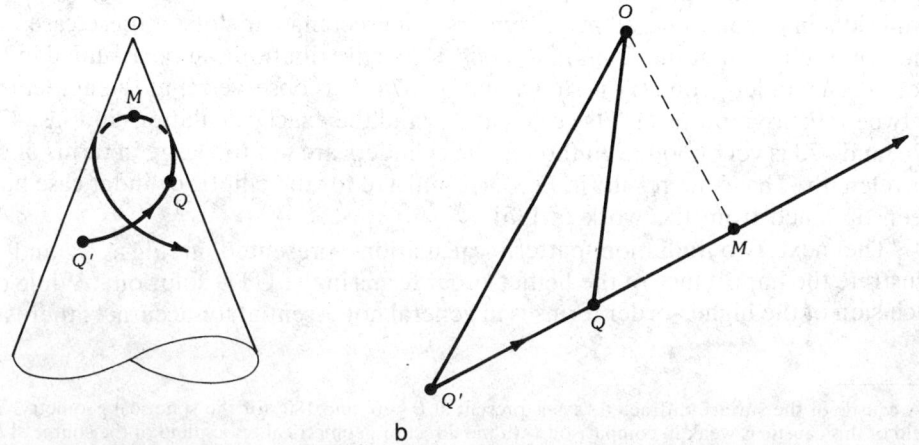

Fig. 46. Typical surface-ray path on a semi-infinite cone. (*a*) Geodesic surface-ray path on a cone. (*b*) Geodesic path on a developed cone.

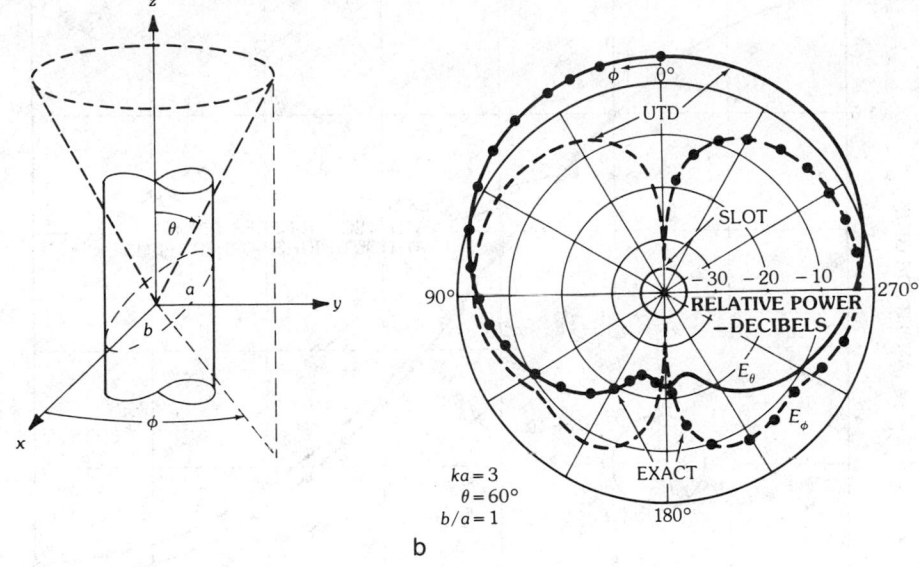

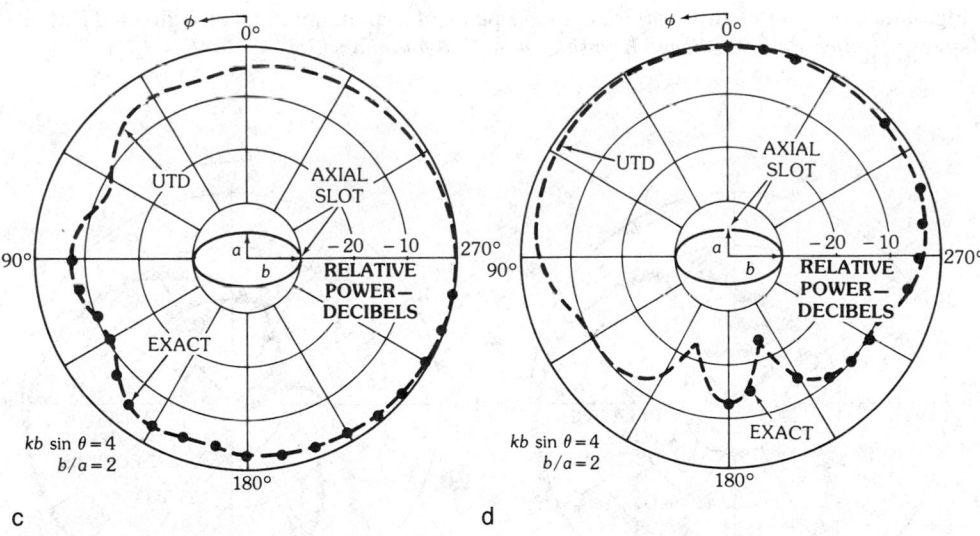

Fig. 47. Radiation pattern calculations using only the leading terms of UTD solution for circumferential and axial slots in circular and elliptic cylinders. (*a*) Elliptic cylinder geometry. (*b*) E_θ and E_ϕ radiation patterns of circumferential slot on circular cylinder. (*c*) $|E_\phi|$ radiation patterns of axial slot at major axis of elliptic cylinder. (*d*) $|E_\phi|$ radiation patterns of axial slot at minor axis of cylinder. (*After Pathak, Wang, Burnside, and Kouyoumjian [31], © 1981 IEEE*)

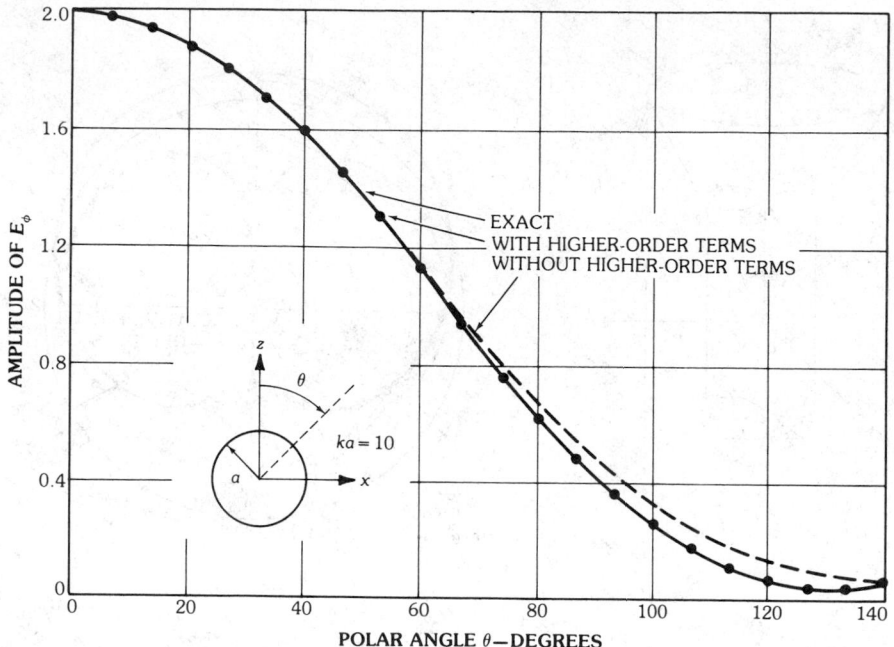

Fig. 48. The $|E_\phi|$ radiation pattern in the xz plane of a circumferential or $\hat{\mathbf{x}}$-directed slot in a sphere. (*After Pathak, Wang, Burnside, and Kouyoumjian [31], © 1981 IEEE*)

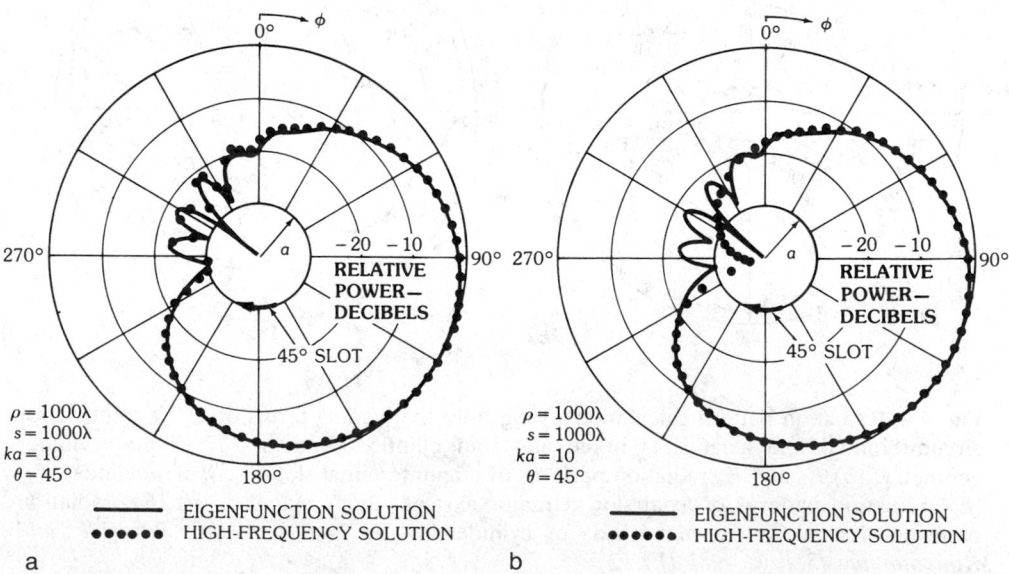

Fig. 49. The $|E_\phi|$ radiation pattern of a 45° (tilted) slot in a circular cylinder. (*a*) High-frequency solution with higher-order terms. (*b*) High-frequency solution without higher-order terms. (*After Pathak, Wang, Burnside, and Kouyoumjian [31], © 1981 IEEE*)

special situations where these terms become as important as the leading terms and they must then be included to maintain accuracy.

The UTD radiation pattern calculations for short, thin, radial slots in a cone are illustrated in Figs. 50 and 51. These calculations are compared with the corresponding exact (eigenfunction) results and measurements which are given in [36]. In Fig. 50 the patterns corresponding to both the $\hat{\theta}$ and the $\hat{\phi}$ polarized components of the electric field are presented for $\theta = 80°$ and $0° \leq \phi \leq 360°$. In Fig. 51 the $\hat{\phi}$ component of the electric field is plotted as a function of θ in the $\phi = 40°$ and $\phi = 220°$ planes. It is noted for the cone configuration of Fig. 51 that the field point in the $\phi = 40°$ plane lies in the lit region, whereas for the $\phi = 220°$ case it lies primarily in the shadow region.

Figs. 52a–52d illustrate the UTD radiation pattern calculation for antennas on prolate spheroids together with measured patterns for comparison. The geometry of this problem is shown in Fig. 53. This close agreement between calculations

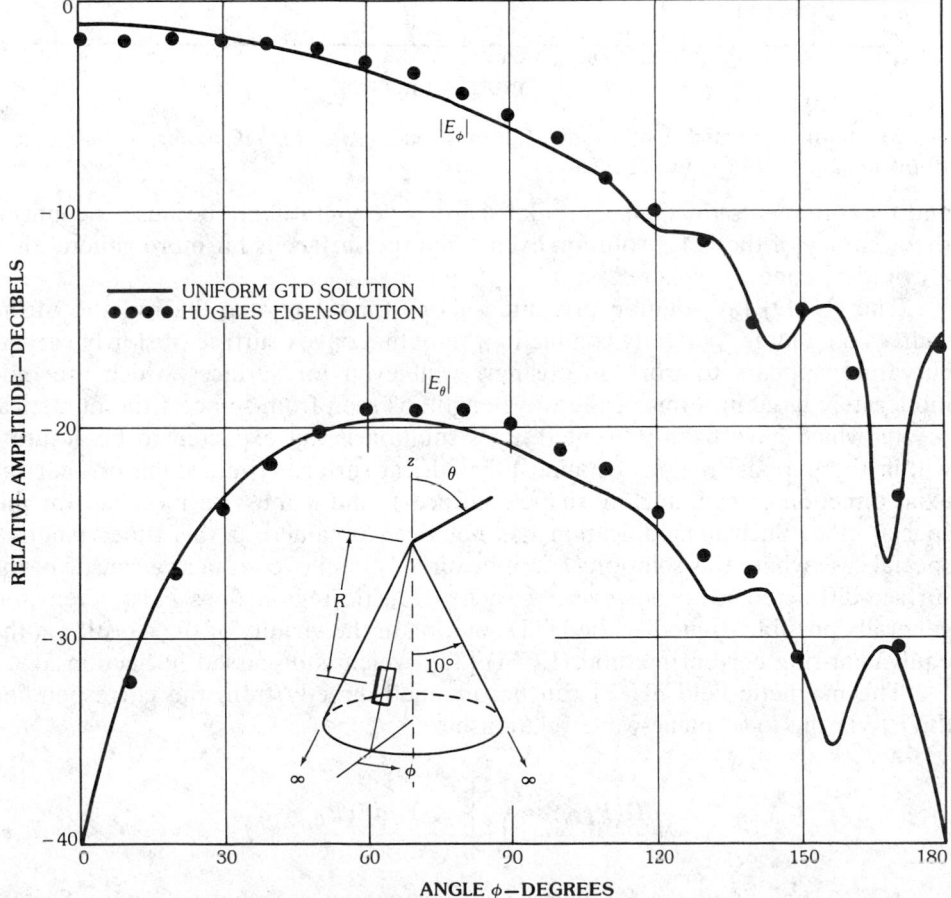

Fig. 50. The $|E_\phi|$ radiation pattern of a radial slot in a cone. (*After Pathak, Wang, Burnside, and Kouyoumjian [31], © 1981 IEEE*)

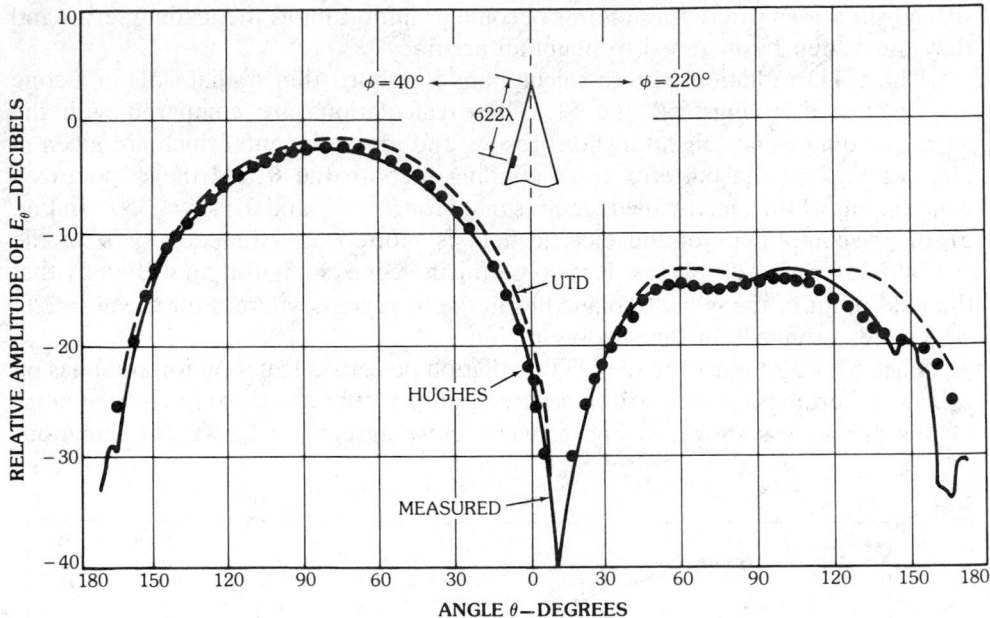

Fig. 51. Radiation patterns of a radial slot in a cone. (*After Pathak, Wang, Burnside, and Kouyoumjian [31],* © *1981 IEEE*)

and measurements for the spheroidal shape is very gratifying because it confirms the accuracy of the UTD solution even when the surface is far more general than a cylinder, cone, or sphere.

The (UTD) ray solution presented above for calculating the radiation from sources on a large, perfectly conducting, smooth, convex surface of slowly varying curvature appears to work surprisingly well even for surfaces which are only moderately large in terms of the wavelength, as seen from some of the numerical results which have been presented. This solution is not expected to be accurate within the paraxial regions of almost cylindrical surfaces (i.e., along or near the axial direction of cylindrically shaped surfaces), and it must be modified for this special case. Such a modification has not been obtained at this time. Another special case where this solution cannot be directly employed is near a caustic of the surface-diffracted rays; however, if such a caustic region does exist, then it is generally possible to modify the UTD solution in the vicinity of the caustic via the equivalent-ring current method (ECM) [37], which is discussed in Section 5.

The magnetic field $d\mathbf{H}(P)$ can be obtained directly from the corresponding $d\mathbf{E}(P)$ via the local plane-wave relationship

$$d\mathbf{H}(P_{L,S}) = \begin{Bmatrix} \hat{\mathbf{s}} \\ \hat{\mathbf{t}} \end{Bmatrix} \times Y_0 \, d\mathbf{E}(P_{L,S})$$

UTD Analysis of the Radiation by Antennas on a Smooth, Convex Surface (2-D Case)—The UTD solution for analyzing the radiation from 2-D sources on a 2-D convex surface can be obtained from the 3-D solution by noting that $\hat{\mathbf{b}}' = \hat{\mathbf{b}}$

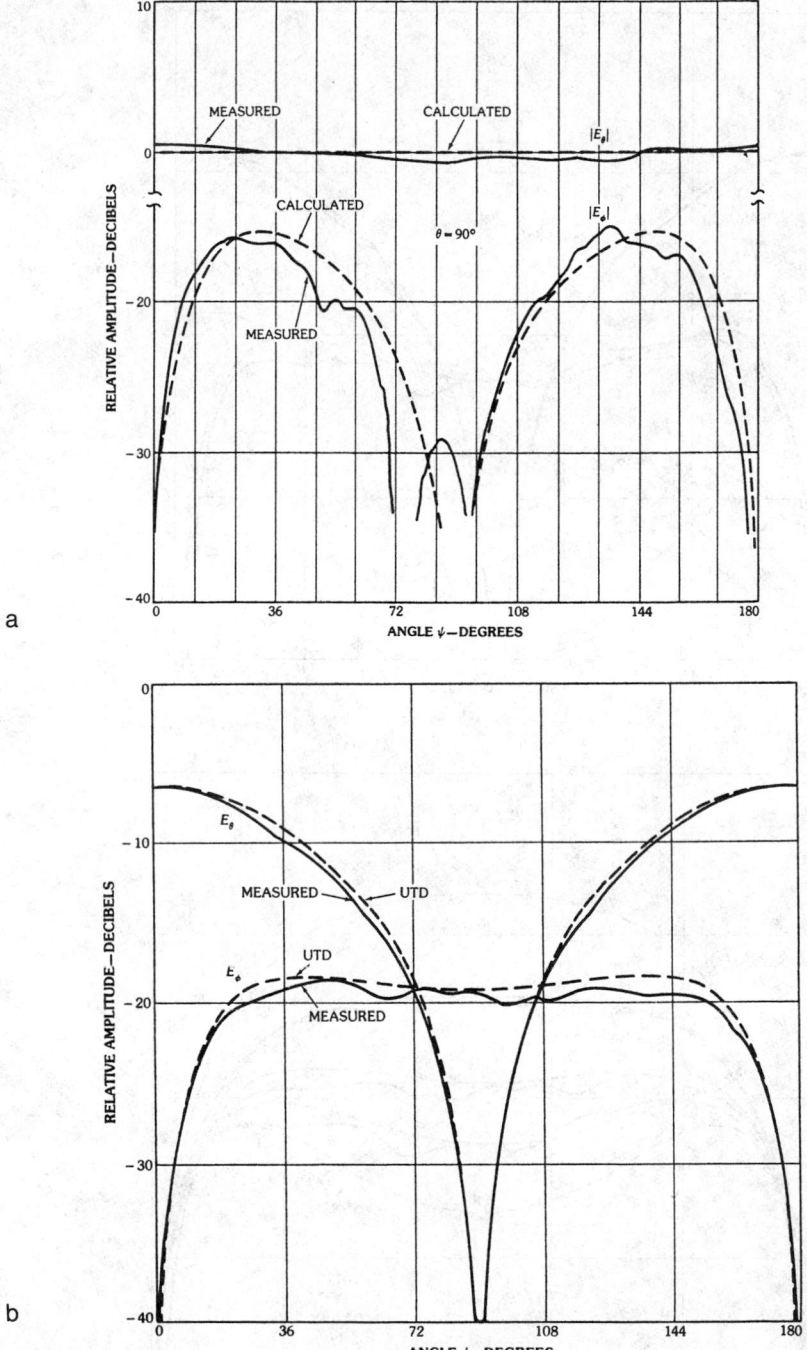

Fig. 52. Radiation patterns of antennas on a prolate spheroid (at the source location in Fig. 54). (*a*) Of **n̂**-directed monopole. (*b*) Of **x̂**-directed rectangular slot with $\theta = 90°$. (*c*) Of **x̂**-directed rectangular slot with $\theta = 100°$. (*d*) Of **x̂**-directed rectangular slot with various θ. (*After Pathak, Wang, Burnside, and Kouyoumjian [31], © 1981 IEEE*)

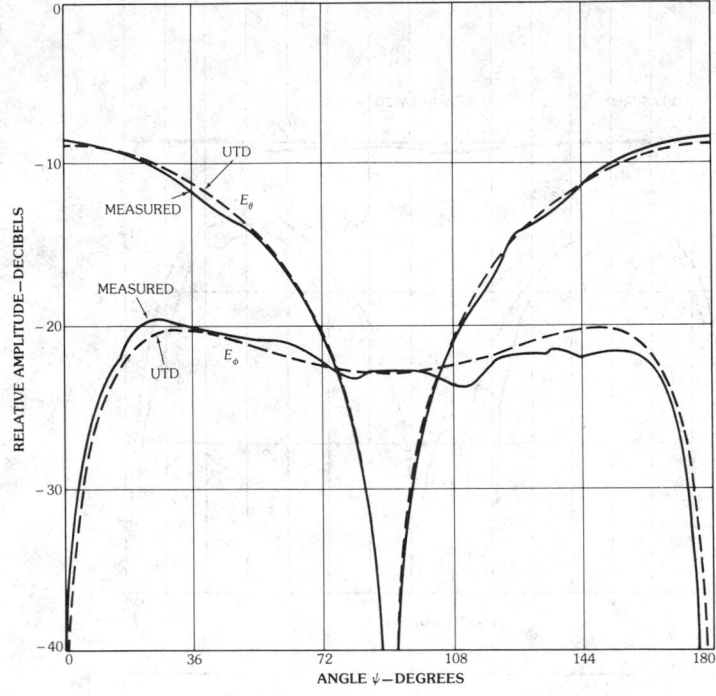

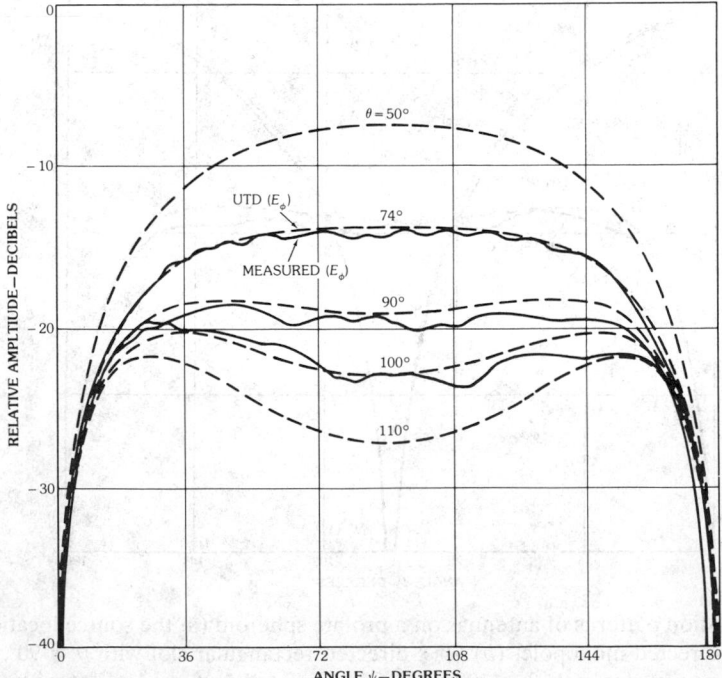

Fig. 52, *continued*

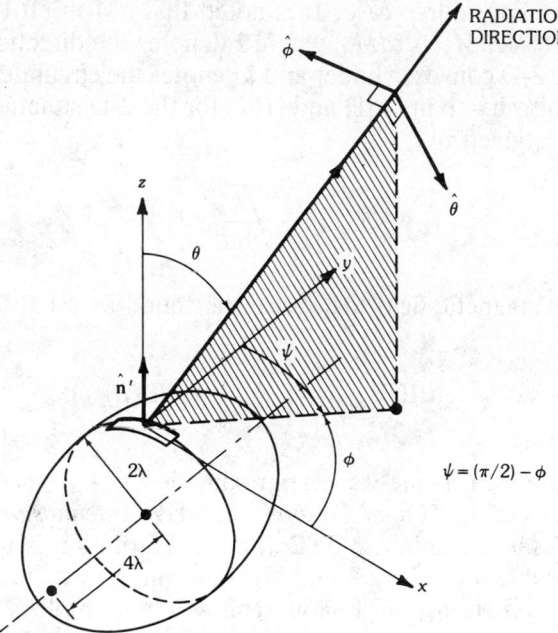

Fig. 53. Prolate spheroidal geometry. (*After Pathak, Wang, Burnside, and Kouyoumjian [31], © 1981 IEEE*)

and $\varrho^d \to \infty$ in the 2-D case, and by allowing the cross terms in the dyads $\bar{\bar{\mathbf{T}}}_{e,m}$ and $\bar{\bar{\mathbf{T}}}^\ell_{e,m}$ to vanish [31, 38]. Thus the electric field radiated by 2-D electric and magnetic line currents of strength $dI_m\hat{\mathbf{n}}'$ and $d\mathbf{M}$ at Q', respectively, on a 2-D convex surface are given by

$$dE(P_L) = C_0 Z_0 \, dI_m(Q')\hat{n}' \cdot [\hat{\mathbf{n}}'\hat{\mathbf{n}}_\ell \, H^\ell(\xi_\ell)] \frac{e^{-jks}}{\sqrt{s}} \sin\theta^i \qquad (159)$$

and

$$dE(P_S) = C_0 Z_0 \, dI_m(Q')\hat{\mathbf{n}}' \cdot [\hat{\mathbf{n}}'\hat{\mathbf{n}} \, H(\xi)] \left[\frac{\varrho_g(Q)}{\varrho_g(Q')}\right]^{1/6} \frac{e^{-jkt-jks^d}}{\sqrt{s^d}} \qquad (160)$$

for the electric line monopole source case, and by

$$dE(P_L) = C_0 \, d\mathbf{M}(Q') \cdot [\hat{\mathbf{b}}'\hat{\mathbf{n}}_\ell H^\ell(\xi_\ell) + \hat{\mathbf{t}}'\hat{\mathbf{b}}S^\ell(\xi_\ell)] \frac{e^{-jks}}{\sqrt{s}} \qquad (161)$$

and

$$dE(P_S) = C_0 \, d\mathbf{M}(Q') \cdot [\hat{\mathbf{b}}'\hat{\mathbf{n}}H(\xi) + \hat{\mathbf{t}}'\hat{\mathbf{b}}S(\xi)] \left[\frac{\varrho_g(Q)}{\varrho_g(Q')}\right]^{1/6} \frac{e^{-jkt-jks^d}}{\sqrt{s^d}} \qquad (162)$$

for the magnetic line source case. It is noted that $d\mathbf{M}$ in (161) and (162) can be expressed as $d\mathbf{M} = \hat{\mathbf{z}}dM_z + \hat{\mathbf{t}}dM_d$, where $\hat{\mathbf{z}}$ denotes the direction along the axis or generator of the 2-D convex cylinder and $\hat{\mathbf{t}}$ denotes the circumferential direction on that cylinder;* also $\hat{\mathbf{b}} = \hat{\mathbf{b}}$ in (161) and (162) for the 2-D situation. The quantity C_0 in (159)–(162) is given by

$$C_0 = -\sqrt{\frac{k}{8\pi}}\, e^{+j\pi/4} \tag{163}$$

Finally, the magnetic field $d\mathbf{H}(P)$ can be found as usual from $d\mathbf{E}(P)$ via

$$d\mathbf{H}(P_{L,S}) = \left\{ \begin{matrix} \hat{\mathbf{s}} \\ \hat{\mathbf{t}} \end{matrix} \right\} \times Y_0\, d\mathbf{E}(P_{L,S})$$

on employing the local plane-wave approximation.

UTD Analysis of the Mutual Coupling Between Antennas on a Smooth, Convex Surface (3-D Case)—A uniform GTD (or UTD) solution is presented below for analyzing the problem in Fig. 28, and it is uniform in the sense that it can also be used within the SSB transition region. This solution employs surface rays of the GTD, and it is assumed, once again, that the convex surface is electrically large. In particular, this UTD solution specifically describes in a simple fashion the surface field at Q which is excited by an infinitesimal magnetic current moment $d\mathbf{p}_m(Q')$, or an infinitesimal electric current moment $d\mathbf{p}_e(Q')$, respectively, at Q' on the same convex surface.

The electromagnetic surface fields excited by an aperture in or a short, thin monopole on a convex conducting surface may be obtained in terms of the surface fields due to infinitesimal magnetic or electric current moments on the same surface as follows. In the aperture case, let $\mathbf{E}^a(Q')$ denote the electric field at any point Q' in the aperture. One may then define an infinitesimal magnetic current moment $d\mathbf{p}_m(Q')$ as in (133) for the radiation problem. The $d\mathbf{p}_m(Q')$ radiating in the presence of the perfectly conducting surface which covers the aperture as well now constitutes an equivalent source in the incremental electric and magnetic fields $d\mathbf{E}_m(Q|Q')$ and $d\mathbf{E}_m(Q|Q')$, respectively, at Q on the surface. The surface fields $\mathbf{E}_m(Q)$ and $\mathbf{H}_m(Q)$ due to the aperture are then found via an integration of $d\mathbf{E}_m$ and $d\mathbf{H}_m$ over the aperture area S_a. Thus,

$$\mathbf{E}_m(Q) = \iint_{S_a} d\mathbf{E}_m(Q|Q') \tag{164a}$$

$$\mathbf{H}_m(Q) = \iint_{S_a} d\mathbf{H}_m(Q|Q') \tag{164b}$$

---

*Note that $d\mathbf{M}(Q') = \lim_{dt'\to 0} \mathbf{E}^a(Q') \times \hat{\mathbf{n}}'\, dt'$, where dt' is the infinitesimal circumferential width of an infinitesimal 2-D aperture and $\mathbf{E}^a(Q')$ is the electric field at Q' in the aperture.

Let t be the arc length of the geodesic surface-ray path from Q' to Q as in Fig. 54. The expressions in (164a) and (164b) and those for $d\mathbf{E}_m$ and $d\mathbf{H}_m$ obtained here are valid even for $t \to 0$. One may likewise define an infinitesimal electric current moment $d\mathbf{p}_e(\ell')$ at Q' again, as done previously in (135) for the radiation problem. The $d\mathbf{p}_e(\ell')$ now constitutes an equivalent source of the incremental electric and magnetic fields $d\mathbf{E}_e(Q|\ell')$ and $d\mathbf{H}_e(Q|\ell')$, respectively, on the surface at Q. The total fields $\mathbf{E}_e(Q)$ and $\mathbf{H}_e(Q)$ are obtained by integrating the incremental fields over the total length h of the monopole as follows:

$$\mathbf{E}_e(Q) = \int_0^h d\mathbf{E}_e(Q|\ell') \tag{165a}$$

and

$$\mathbf{H}_e(Q) = \int_0^h d\mathbf{H}_e(Q|\ell') \tag{165b}$$

If the distance t of the observation point Q is far enough away from Q' such that the monopole is not directly visible at Q, then

$$\mathbf{E}_e(Q)\big|_{t>t_0} \cong \frac{d\mathbf{E}_e(Q|Q')}{dp_e(Q')}\left[\int_0^h dp_e(\ell')\right]; \tag{166a}$$

$$\mathbf{H}_e(Q)\big|_{t>t_0} \cong \frac{d\mathbf{H}_e(Q|Q')}{dp_e(Q')}\left[\int_0^h dp_e(\ell')\right] \tag{166b}$$

where $dp_e(Q')$ is the value of $dp_e(\ell')$ at the base of the monopole, i.e., at $\ell' = 0$ or at Q'. Since $dp_e(\ell') = I(\ell')d\ell'$, it follows that $dp_e(Q') = I(Q')d\ell'$, and $I(Q')$ is the current at the base of the monopole. The fields $d\mathbf{E}_e(Q|Q')$ and $d\mathbf{H}_e(Q|Q')$ are directly proportional to $dp_e(Q')$, and the distance t_0 at which (166a) and (166b) may be used is given by $t_0 \cong \varrho_g(Q')\cos^{-1}\{\varrho_g(Q')/[\varrho_g(Q') + h]\}$. Clearly t_0 can be made small only if h is sufficiently small. Even though the $\mathbf{E}_e$ and $\mathbf{H}_e$ in

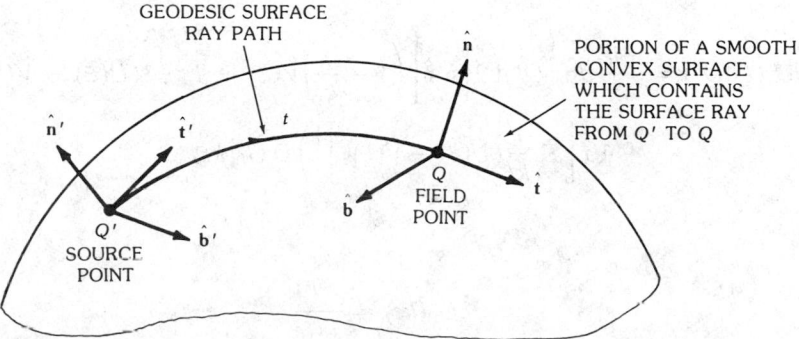

Fig. 54. Surface-ray coordinates. (*After Pathak and Wang [39], © 1981 IEEE*)

(166a) and (166b) are valid for $t > t_0$, the $d\mathbf{E}_e(Q|Q')$ and $d\mathbf{H}_e(Q|Q')$ due to $d\mathbf{p}_e(Q')$ on the surface are valid even for $t \to 0$ (i.e., $Q \to Q'$).

Explicit expressions for the surface fields $d\mathbf{H}_m$ and $d\mathbf{E}_m$ due to a source on an arbitrary convex surface are as follows [39]:

$$
d\mathbf{H}_m(Q|Q') = \frac{-jk}{4\pi} d\mathbf{p}_m(Q') \left(2Y_0 \left\{ \hat{\mathbf{b}}'\hat{\mathbf{b}} \left[\left(1 - \frac{j}{kt} \right) \bar{V}(\xi) \right. \right. \right.
$$
$$
\left. + \mathsf{D}^2 \left(\frac{j}{kt} \right)^2 (\Lambda_s \bar{U}(\xi) + \Lambda_c \bar{V}(\xi)) + \bar{T}_0^2 \frac{j}{kt} (\bar{U}(\xi) - \bar{V}(\xi)) \right]
$$
$$
+ \hat{\mathbf{t}}'\hat{\mathbf{t}} \left[\mathsf{D}^2 \frac{j}{kt} \bar{V}(\xi) + \frac{j}{kt} \bar{U}(\xi) - 2 \left(\frac{j}{kt} \right)^2 (\Lambda_s \bar{U}(\xi) + \Lambda_c \bar{V}(\xi)) \right]
$$
$$
\left. \left. + (\hat{\mathbf{t}}'\hat{\mathbf{b}} + \hat{\mathbf{b}}'\hat{\mathbf{t}}) \left[\frac{j}{kt} \bar{T}_0 (\bar{U}(\xi) - \bar{V}(\xi)) \right] \right\} \right) \mathsf{D}\, G_0(kt) \qquad (167)
$$

and

$$
d\mathbf{E}_m(Q|Q') = \frac{-jk}{4\pi} d\mathbf{p}_m(Q') \left(2 \left\{ \hat{\mathbf{b}}'\hat{\mathbf{n}} \left[\left(1 - \frac{j}{kt} \right) \bar{V}(\xi) + \bar{T}_0^2 \frac{j}{kt} (\bar{U}(\xi) - \bar{V}(\xi)) \right] \right. \right.
$$
$$
\left. \left. + \hat{\mathbf{t}}'\hat{\mathbf{n}} \left[\bar{T}_0 \frac{j}{kt} (\bar{U}(\xi) - \bar{V}(\xi)) \right] \right\} \right) \mathsf{D}\, G_0(kt) \qquad (168)
$$

Similarly, the surface fields $d\mathbf{H}_e$ and $d\mathbf{E}_e$ due to $d\mathbf{p}_e$ on an arbitrary convex surface are given by [39]:

$$
d\mathbf{E}_e(Q|Q') = \frac{-jk}{4\pi} d\mathbf{p}_e(Q') \left\{ 2Z_0\, \hat{\mathbf{n}}'\hat{\mathbf{n}} \left[\bar{V}(\xi) - \frac{j}{kt} \bar{V}(\xi) \right. \right.
$$
$$
\left. \left. + \left(\frac{j}{kt} \right)^2 (\Lambda_s \bar{V}(\xi) + \Lambda_c \bar{U}(\xi)) + \bar{T}_0^2 \frac{j}{kt} (\bar{U}(\xi) - \bar{V}(\xi)) \right] \right\} \mathsf{D}\, G_0(kt) \quad (169)
$$

and

$$
d\mathbf{H}_e(Q|Q') = \frac{-jk}{4\pi} d\mathbf{p}_e(Q') \left(2 \left\{ \hat{\mathbf{n}}'\hat{\mathbf{b}} \left[\left(1 - \frac{j}{kt} \right) \bar{V}(\xi) + \bar{T}_0^2 \frac{j}{kt} (\bar{U}(\xi) - \bar{V}(\xi)) \right] \right. \right.
$$
$$
\left. \left. + \hat{\mathbf{n}}'\hat{\mathbf{t}} \left[\bar{T}_0 \frac{j}{kt} (\bar{U}(\xi) - \bar{V}(\xi)) \right] \right\} \right) \mathsf{D}\, G_0(kt) \qquad (170)
$$

Here,

$$
G_0(kt) = \frac{e^{-jkt}}{t} \qquad (171)
$$

and ξ is as defined previously in (113) together with (114). Furthermore,

$$D = \sqrt{t \, d\psi_0 / d\eta(Q)} \tag{172a}$$

$$d\eta(Q) = \varrho^d \, d\psi \tag{172b}$$

In Fig. 55 a pair of infinitesimally separated surface rays adjacent to the central ray from Q' at Q is shown; these adjacent rays constitute a surface-ray strip (or a surface-ray tube), and $d\psi_0$ in (172a) then refers to the angle between these adjacent surface rays at the source point Q'. On the other hand, $d\psi$ is the angle between the backward tangents to the same pair of adjacent surface rays at the field point Q. The distance ϱ^d between Q and Q_c is the geodesic circle at Q. Thus, $d\eta(Q)$ in (172b) denotes the width of the surface-ray strip at Q as discussed previously in the radiation configuration (see Table 3 and Fig. 42). In the case of a sphere of radius a, one obtains the following simplifications: $\xi = mt/a$, with $m = (ka/2)^{1/3}$, and $D = \sqrt{(t/a)/\sin(t/a)}$. Also, $\varrho^d = a \tan(t/a)$ for a sphere. For a circular cylinder of radius a, one obtains $\xi = mt/\varrho_g$ along a helical geodesic surface-ray path with $m = (k\varrho_g/2)^{1/3}$ and $\varrho_g = a/\sin^2 \alpha'$, with α' constant along a given helical geodesic path on a convex cylinder.* Also, $D = 1$, and $T_0 = T\varrho_g$ in which $T = (\sin 2\alpha')/2a$ for a circular cylinder. The generalized torsion factor $\tilde{T}_0$ for the arbitrary convex surface is given by

$$\tilde{T}_0 = \mp |\sqrt{T_0(Q')T_0(Q)}| \tag{173}$$

where the negative sign in (173) is chosen if $T_0(Q') < 0$ and/or $T_0(Q) < 0$; otherwise, the positive sign is chosen. Note that for a general convex surface, one employs

$$T_0(Q') = T(Q')\varrho_g(Q') \tag{174a}$$

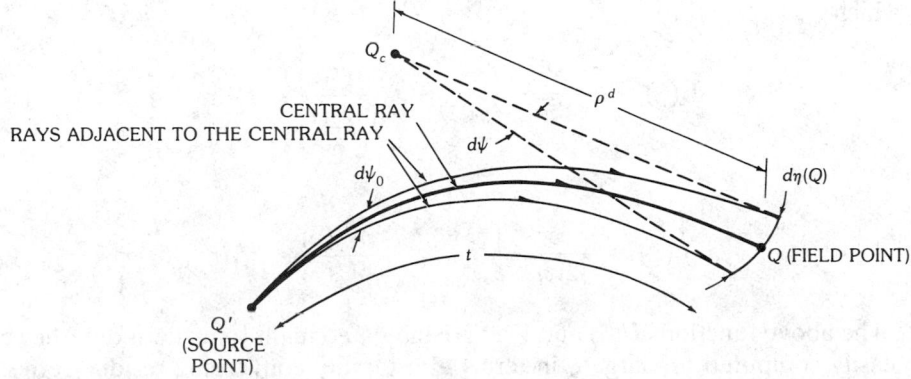

Fig. 55. Surface-ray strip (or tube). (*After Pathak and Wang [39], © 1981 IEEE*)

---

*Here α' could be either α'_1 or α'_2, as in Fig. 45a.

$$T(Q') = \frac{\sin 2\alpha'}{2}\left[\frac{1}{R_2(Q')} - \frac{1}{R_1(Q')}\right] \tag{174b}$$

and

$$\frac{1}{\varrho_g(Q')} = \frac{\cos^2\alpha'}{R_1(Q')} + \frac{\sin^2\alpha'}{R_2(Q')} \tag{174c}$$

as in Table 2, where α' is shown in Fig. 40, and $R_1(Q') \geqq R_2(Q')$ as in the radiation problem. Note that $T_0(Q)$ is given by (174a)–(174c) on replacing Q' by Q therein. A reasonable choice for Λ_s and Λ_c is [39]

$$\Lambda_s = \left|\sqrt{\frac{R_2(Q')}{R_1(Q')}\frac{R_2(Q)}{R_1(Q)}}\right| \tag{175a}$$

$$\Lambda_c = 1 - \Lambda_s \tag{175b}$$

in order to interpolate smoothly between the higher-order terms in the canonical circular cylinder and sphere solutions. The above choice of Λ_s and Λ_c is also essential for recovering the field solution for the planar surface case when $R_{1,2} \to \infty$. The generalized Fock integrals $\tilde{U}(\xi)$ and $\tilde{V}(\xi)$ for the surface fields on an arbitrary convex surface which are present in (167)–(170) are given by

$$\tilde{U}(\xi) = \left[\frac{kt}{2m(Q')m(Q)\xi}\right]^{3/2} U(\xi) \tag{176a}$$

$$\tilde{V}(\xi) = \left[\frac{kt}{2m(Q')m(Q)\xi}\right]^{1/2} V(\xi) \tag{176b}$$

in which

$$U(\xi) \equiv \frac{\xi^{3/2}e^{j3\pi/4}}{\sqrt{\pi}} \int_{\infty\exp(-j2\pi/3)}^{\infty} \frac{W_2'(\tau)}{W_2(\tau)} e^{-j\xi\tau}d\tau \tag{177}$$

and

$$V(\xi) \equiv \frac{\xi^{1/2}e^{j\pi/4}}{2\sqrt{\pi}} \int_{\infty\exp(-j2\pi/3)}^{\infty} \frac{W_2(\tau)}{W_2'(\tau)} e^{-j\xi\tau}d\tau \tag{178}$$

The above functions $U(\xi)$ and $V(\xi)$ are tabulated in [40]; furthermore, they can be easily computed for large ξ in terms of a rapidly converging residue series as follows [40]:

$$V(\xi) = e^{-j\pi/4}\sqrt{\pi}\,\xi^{1/2}\sum_{n=1}^{\infty}(\tau_n')^{-1}\,e^{-j\xi\tau_n'} \tag{179}$$

$$U(\xi) = e^{j\pi/4} 2\sqrt{\pi}\, \xi^{3/2} \sum_{n=1}^{\infty} e^{-j\xi\tau_n} \tag{180}$$

$$V_1(\xi) = e^{j\pi/4} 2\sqrt{\pi}\, \xi^{3/2} \sum_{n=1}^{\infty} e^{-j\xi\tau_n'} \tag{181}$$

$$V'(\xi) = \frac{1}{2} e^{-j\pi/4} \sqrt{\pi}\, \xi^{-1/2} \sum_{n=1}^{\infty} (1 - j2\xi\tau_n')(\tau_n')^{-1}\, e^{-j\xi\tau_n'} \tag{182}$$

$$U'(\xi) = e^{j\pi/4} 3\sqrt{\pi}\, \xi^{1/2} \sum_{n=1}^{\infty} (1 - j2\xi\tau_n/3)e^{-j\xi\tau_n} \tag{183}$$

where τ_n and τ_n' are zeros of $W_2(\tau)$ and $W_2'(\tau)$, respectively; they are tabulated in Chart 2. Note that the sums in (179)–(183) can be truncated at $n = 10$ without losing accuracy. The additional functions $U'(\xi)$, $V'(\xi)$, and $V_1(\xi)$ in (183), (182), and (181) are defined by $U'(\xi) = dU/d\xi$, $V'(\xi) = dV(\xi)/d\xi$, and

$$V_1(\xi) = \frac{e^{j3\pi/4}\, \xi^{3/2}}{\sqrt{\pi}} \int_{\infty \exp(-j2\pi/3)}^{\infty} \frac{W_2(\tau)}{W_2'(\tau)} \tau e^{-j\xi\tau}\, d\tau$$

If ξ is small and positive, one may employ a small-argument asymptotic expansion for the Fock functions as follows [40]:

$$V(\xi) \sim 1 - \frac{\sqrt{\pi}}{4} e^{j\pi/4} \xi^{3/2} + \frac{7j}{60} \xi^3 + \frac{7\sqrt{\pi}}{512} e^{-j\pi/4} \xi^{9/2} - \cdots \tag{184}$$

$$U(\xi) \sim 1 - \frac{\sqrt{\pi}}{2} e^{j\pi/4} \xi^{3/2} + \frac{5j}{12} \xi^3 + \frac{5\sqrt{\pi}}{64} e^{-j\pi/4} \xi^{9/2} - \cdots \tag{185}$$

$$V_1(\xi) \sim 1 + \frac{\sqrt{\pi}}{2} e^{j\pi/4} \xi^{3/2} - \frac{7j}{12} \xi^3 - \frac{7\sqrt{\pi}}{64} e^{-j\pi/4} \xi^{9/2} + \cdots \tag{186}$$

Chart 2. Zeros of $W_2(\tau)$ and $W_2'(\tau)$

n	$\tau_n = \lvert\tau_n\rvert\, e^{-j\pi/3}$ $\lvert\tau_n\rvert$	$\tau_n' = \lvert\tau_n'\rvert\, e^{-j\pi/3}$ $\lvert\tau_n'\rvert$
1	2.338 11	1.018 79
2	4.087 95	3.248 19
3	5.520 56	4.820 10
4	6.786 61	6.163 31
5	7.944 13	7.372 18
6	9.022 65	8.488 49
7	10.040 2	9.535 45
8	11.008 5	10.527 7
9	11.930 0	11.475 1
10	12.828 8	12.384 8

$$V'(\xi) \sim \frac{3\sqrt{\pi}}{8} e^{-j3\pi/4} \xi^{1/2} + \frac{7j}{20} \xi^2 + \frac{63\sqrt{\pi}}{1024} e^{-j\pi/4} \xi^{7/2} - \dots \tag{187}$$

$$U'(\xi) \sim \frac{3}{4} \sqrt{\pi} e^{-j3\pi/4} \xi^{1/2} + \frac{5j}{4} \xi^2 + \frac{45\sqrt{\pi}}{128} e^{-j\pi/4} \xi^{7/2} - \dots \tag{188}$$

For $\xi \geq \xi_0$ the residue series representation with the first ten terms in the summation may be used. For $\xi \leq \xi_0$ the small-argument asymptotic expression with the first three terms may be used. Here, ξ_0 is taken to be 0.6, as shown in [40].

The above results in (167)–(170), which are available from [39], can be modified for even greater accuracy in the paraxial (near axial) regions of quasicylindrical (or elongated) convex surfaces by adding higher-order terms in m^{-1}. In particular, these higher-order terms to be added to (167), (168), (169), and (170) are

$$\frac{-jk}{4\pi} d\mathbf{p}_m \cdot [2Y_0 \hat{\mathbf{b}}' \hat{\mathbf{b}} C] DG_0(kt)$$

$$\frac{-jk}{4\pi} d\mathbf{p}_m \cdot [2\hat{\mathbf{b}}' \hat{\mathbf{n}} C] DG_0(kt)$$

$$\frac{-jk}{4\pi} d\mathbf{p}_e \cdot [2Z_0 \hat{\mathbf{n}}' \hat{\mathbf{n}} C] DG_0(kt)$$

and

$$\frac{-jk}{4\pi} d\mathbf{p}_e \cdot [2Y_0 \hat{\mathbf{n}}' \hat{\mathbf{b}} C] DG_0(kt)$$

respectively, in which

$$C = j(4/3)(\sqrt{2} k\tilde{\varrho}_g)^{-2/3} (\tilde{\varrho}_g / \tilde{\varrho}_b) \tilde{\tau} V'(\xi)$$

with $V'(\xi) = dV/d\xi$, and with $\tilde{\tau}$ and $\tilde{\varrho}_g$ as defined later in (189b) and (189c).

The expressions given above for the surface fields of infinitesimal magnetic and electric current moments on an arbitrary convex surface are employed to calculate the mutual coupling between a pair of antennas on such a surface. Before proceeding to an illustration of the numerical results based on these calculations, it is useful to also give an alternative expression for $d\mathbf{H}_m(Q|Q')$ developed in [34] as compared with the one in (167) developed in [39], namely

$$d\mathbf{H}_m(Q|Q') = \frac{k^2 Y_0}{2\pi j} d\mathbf{p}_m(Q') \cdot \hat{\mathbf{b}}' \hat{\mathbf{b}} \left\{ (1 - j/kt) \ \bar{V}(\xi) - (1/kt)^2 \ \bar{U}(\xi) \right.$$

$$\left. + j(\sqrt{2} k\tilde{\varrho}_g)^{-2/3} \left[\tilde{\tau} V'(\xi) + \frac{\tilde{\varrho}_g}{\tilde{\varrho}_b} \tilde{\tau}^3 \ U'(\xi) \right] \right\} + \hat{t}' \hat{t} (j/kt)[\bar{V}(\xi)$$

$$+ (1 - 2j/kt) \ \bar{U}(\xi) + j(\sqrt{2} k\tilde{\varrho}_g)^{-2/3} \ \tilde{\tau}^3 U'(\xi)] \ DG_0(t) \tag{189}$$

with

$$\tilde{\tau} = \left[\frac{kt}{2m(Q')m(Q)\xi}\right]^{1/2} \tag{189b}$$

$$\tilde{\varrho}_g = [\varrho_g(Q')\varrho_g(Q)]^{1/2} \tag{189c}$$

$$\tilde{\varrho}_b = [\varrho_b(Q')\varrho_b(Q)]^{1/2} \tag{189d}$$

Here ϱ_g and ϱ_b are the surface radii of curvature in the $\hat{\mathbf{t}}$ and $\hat{\mathbf{b}}$ directions as usual, which are associated with the surface-ray path. Alternative expressions for $d\mathbf{E}_m$, $d\mathbf{H}_e$, and $d\mathbf{E}_e$ are not available in [34]. It is noted that these alternative expressions for $d\mathbf{H}_m$ in (167) and (189) yield almost the same accuracy.

Consider a pair of rectangular slots in a smooth, perfectly conducting, convex surface. Let the electric field in the aperture of the first slot antenna be $\mathbf{E}_1^a$; likewise, let $\mathbf{E}_2^a$ denote the aperture distribution in the second slot antenna. If the slots are sufficiently short and thin, the dominant mode approximation may be employed for $\mathbf{E}_1^a$ and $\mathbf{E}_2^a$. Thus,

$$\mathbf{E}_1^a = V_{11}\mathbf{e}_1, \qquad \mathbf{E}_2^a = V_{22}\mathbf{e}_2 \tag{190}$$

where V_{11} and V_{22} are the modal voltages associated with the dominant electric vector mode functions $\mathbf{e}_1$ and $\mathbf{e}_2$ [41]. If the two slots have identical dimensions, then $\mathbf{e}_1 = \mathbf{e}_2$. The expression for the mutual admittance Y_{21} between the two slots with the first slot transmitting and the second slot short-circuited is [42]

$$Y_{21} = -\iint_{S_2} \iint_{S_1} d\mathbf{H}_m(Q|Q') \cdot d\mathbf{p}_m(Q)/V_{11}V_{22} \tag{191}$$

where Q' is any point in the aperture of the first slot, and Q is any point in the aperture of the second slot which is short-circuited. As before, $d\mathbf{H}_m(Q|Q')$ is the surface-magnetic field at Q due to the equivalent source $d\mathbf{p}_m(Q')$ at Q'; furthermore,

$$d\mathbf{p}_m(Q') = \mathbf{E}_1^a(Q') \times \hat{\mathbf{n}}' \, dS_1 \quad \text{and} \quad d\mathbf{p}_m(Q) = \mathbf{E}_2^a(Q) \times \hat{\mathbf{n}} \, dS_2$$

The double integrals in (191) are evaluated over the surface areas S_1 and S_2 of the two apertures. One notes that $Y_{21} = Y_{12}$. The result in (191) is also applicable to a pair of slots in an array environment provided that the field distribution in each of the slot array elements can be approximated by the dominant mode as in (190) and the array elements are not too closely spaced. Equation (191) is employed here to calculate the isolation S_{12} (which can be expressed in terms of Y_{12}) between a pair of rectangular slots in a perfectly conducting cone; the developed cone is shown in Fig. 56 and the results of the calculations are shown in Figs. 57a and 57b. The results for $d\mathbf{H}_m(Q|Q')$ based on (167) are employed in (191) to obtain the curves designated by OSU in Figs. 57a and 57b, whereas those based on (189) are

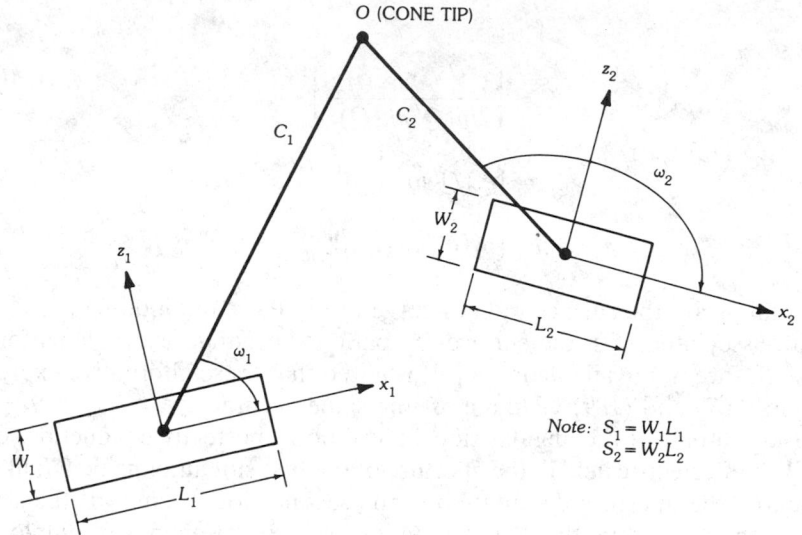

Fig. 56. Two rectangular slots on a developed cone.

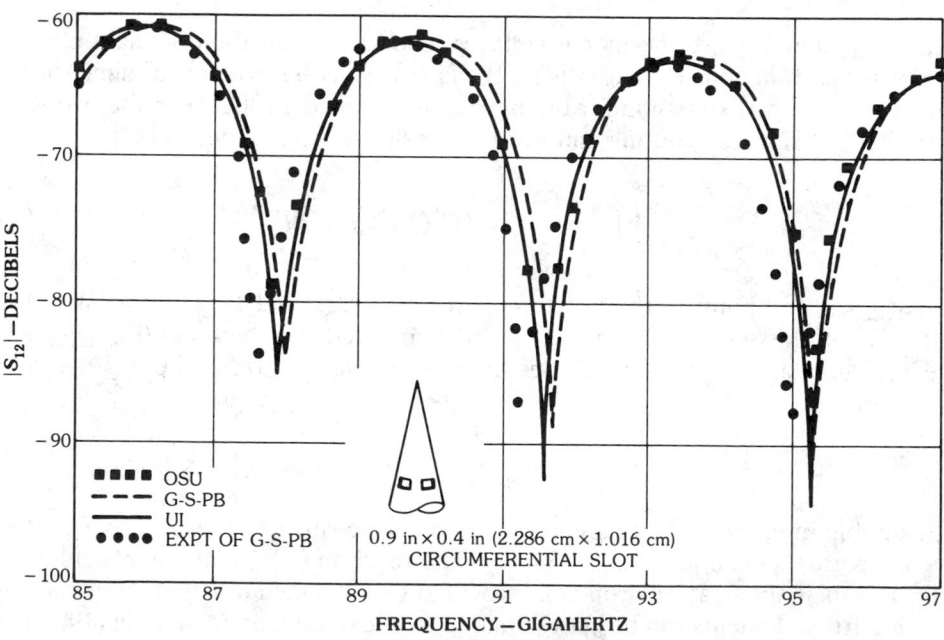

a

Fig. 57. Coupling coefficient S_{12} between two circumferential slots on a cone versus frequency. (*After Pathak and Wang [39]*) (*a*) Radial separation between slots is $|C_1 - C_2| = |45.53 - 45.53| = 0$ cm, angular separation is $\phi_0 = 60.8°$, and cone half-angle is $\theta_0 = 12.2°$. (*b*) Radial separation between slots is $|C_1 - C_2| = |27.03 - 25.88| = 1.15$ cm, angular separation is $\phi_0 = 80°$, and cone half-angle is $\theta_0 = 11°$. (*After Pathak and Wang [39]*, © 1981 IEEE)

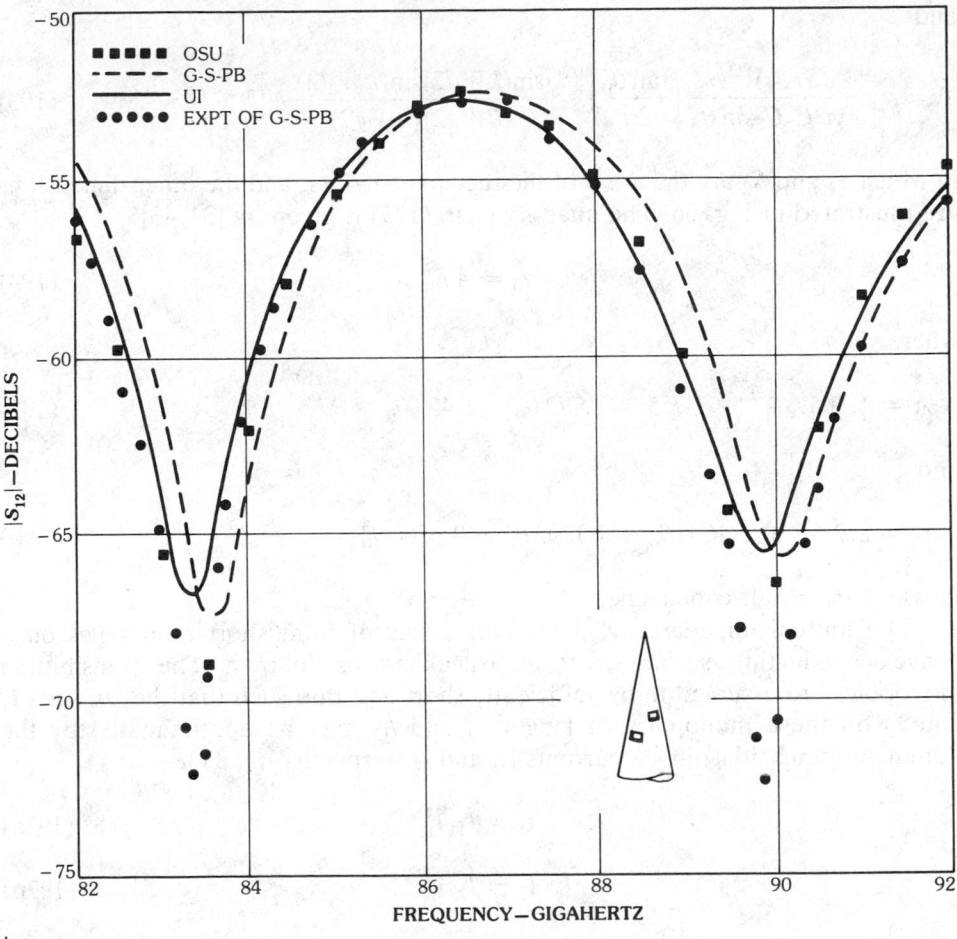

b

Fig. 57, *continued*

designated by UI in these figures. In addition, the OSU and UI solutions are compared in those figures with the measured results and also with calculations based on an equivalent cylinder model, both of which were obtained by Golden, Stewart, and Pridmore-Brown [43]. The latter measured results are designated by EXPT of G-S-PB, and the approximate cylinder model–based calculations are designated by G-S-PB in these figures. The diffraction by the cone tip has been included in the calculations designated as OSU and UI, respectively. This tip-diffraction contribution is essential for obtaining the interference pattern which is present in these figures, and it has been calculated via the interpolation formula given in [34] pertaining to the cone tip diffraction contribution available from [43]. In particular, $Y_{12}^{\text{total}} = Y_{12}$ [of (190)] $+ Y_{12}^{\text{tip}}$, for the tipped cone geometry, where [34, 43]

$$Y_{12}^{\text{tip}} = \tilde{\Upsilon} \sin \omega_1 \sin \omega_2 \qquad (192)$$

and

$$\tilde{Y} = \sigma_0 \frac{(S_1 S_2)^{1/2}}{30\pi^4 C_1 C_2 \sin\theta_0} \left(\frac{\tan\theta_0}{2\pi}\right)^{1/2} \cdot \frac{\sin(kW_1/2)\sin(kW_2/2)}{(kW_1/2)(kW_2/2)} e^{j(\pi/4 - kC_1 - kC_2)} \tag{193}$$

in which S_1 and S_2 are the areas of the rectangular slots, and the other dimensions are illustrated in Fig. 56. The quantity σ_0 in (193) is given by [34, 43]

$$\sigma_0 = Ae^{jB} \tag{194}$$

where

$$A = 1.3057 \theta_0^{-1} - 1.755 + 2.772 \theta_0 - 1.459 \theta_0^2 \tag{195}$$

and

$$B = 2.7195 + 1.4608 \theta_0 - 1.1295 \theta_0^2 + 0.6566 \theta_0^3 \tag{196}$$

in which $\theta_0 = $ half-cone angle.

The mutual impedance Z_{21} between a pair of thin, short monopoles on a convex conducting surface may be calculated as follows. The transmitting monopoles are assumed to be sufficiently short and thin such that the currents $\mathbf{I}_1$ and $\mathbf{I}_2$ on these monopoles of length h_1 and h_2 may be approximated by the dominant (sinusoidal) mode currents ($\mathbf{i}_1$ and $\mathbf{i}_2$, respectively). Thus,

$$\mathbf{I}_1 = I_{11}\mathbf{i}_1 \tag{197a}$$

$$\mathbf{I}_2 = I_{22}\mathbf{i}_2 \tag{197b}$$

where I_{11} and I_{22} are the monopole base currents. Let the first monopole be $\hat{\mathbf{n}}'$-directed on the surface Q'; likewise, let the second monopole be $\hat{\mathbf{n}}$-directed on the surface at Q. Then Z_{21} for the case when the first monopole is transmitting and the second one is open circuited is given by

$$Z_{21} = -\int_0^{h_2}\int_0^{h_1} d\mathbf{E}_e(Q|Q')\cdot d\mathbf{p}_e(Q)/I_{11} I_{22} \tag{198}$$

where $d\mathbf{E}_e(Q|Q')$ is the field at Q due to the equivalent source $d\mathbf{p}_e(Q')$ at Q' and

$$d\mathbf{p}_e(Q') = I_{11}i_1(Q') d\ell_1(Q')\hat{\mathbf{n}}' \tag{199a}$$

$$d\mathbf{p}_e(Q) = I_{22}i_2(Q) d\ell_2(Q)\hat{\mathbf{n}} \tag{199b}$$

in which $d\ell_1(Q')$ and $d\ell_2(Q)$ are the incremental lengths of the two monopoles at points Q' and Q, respectively. It is assumed that the open-circuited monopole at Q scatters negligibly if it is short.

UTD Analysis of the Mutual Coupling Between Antennas on a Smooth, Convex Surface (2-D Case)—The tangential surface magnetic field **H** at a point Q [or (x,y)] on a perfectly conducting convex cylinder due to a magnetic line source of strength $d\mathbf{M} = \hat{\mathbf{z}}dM_z + \hat{\mathbf{t}}dM_d$ [see (161) and (162)] which is located at Q' [or (x',y')] on the same surface is given in terms of the surface-ray coordinates as [38]

$$d\mathbf{H}(Q) \sim C_0 Y_0 d\mathbf{M} \cdot [\hat{\mathbf{b}}'\hat{\mathbf{b}}F_s + \hat{\mathbf{t}}'\hat{\mathbf{t}}G_s]e^{-jkt} \tag{200}$$

in which $\hat{\mathbf{b}}' = \hat{\mathbf{b}}$ for the 2-D case, and t denotes the surface-ray distance from Q' to Q. The functions G_s and F_s for the TM_z (or dM_d excitation) and TE_z (or dM_z excitation) cases, respectively, describe the behavior of the surface field. These functions are related to the mutual coupling Fock functions, U and V, respectively, as follows:

$$F_s = \left(\frac{jk}{2}\right)^{1/2}\left\{m(Q')\left[\frac{\varrho_g(Q)}{\varrho_g(Q')}\right]^{1/6}\right\}^{-1} 2\xi^{-1/2} e^{-j\pi/4} V(\xi) \tag{201}$$

and

$$G_s = -\left(\frac{jk}{2}\right)^{1/2}\left\{m(Q')\left[\frac{\varrho_g(Q)}{\varrho_g(Q')}\right]^{1/6}\right\}^{-3} \xi^{-3/2} e^{-j3\pi/4} U(\xi) \tag{202}$$

in which $U(\xi)$ and $V(\xi)$ are defined earlier in (177) and (178). Near the source, $Q \to Q'$ (i.e., $\xi \to 0$ and $t \to 0$); in this case one may use the small-argument approximation for $U(\xi)$ and $V(\xi)$ as in (185) and (184) to obtain

$$\hat{\mathbf{b}}\cdot d\mathbf{H}(Q)\Big|_{\xi\,\text{small}} \cong -\frac{kY_0 dM_z}{2}\sqrt{\frac{2j}{\pi kt}}\, e^{-jkt}\left(1 - e^{j\pi/4}\frac{\sqrt{\pi}}{2}\xi^{3/2}\right.$$
$$\left. + j\frac{7}{60}\xi^3 + \frac{7\sqrt{\pi}}{512}e^{-j\pi/4}\xi^{9/2}\dots\right) \tag{203}$$

and

$$\hat{\mathbf{t}}\cdot d\mathbf{H}(Q)\Big|_{\xi\,\text{small}} \cong \frac{kY_0 dM_d}{2}\frac{j}{kt}\frac{2j}{\sqrt{\pi kt}}\, e^{-jkt}\left(1 - e^{j\pi/4}\frac{\sqrt{\pi}}{2}\xi^{3/2}\right.$$
$$\left. + j\frac{5}{12}\xi^3 + \frac{5\sqrt{\pi}}{64}e^{-j\pi/4}\xi^{9/2} - \dots\right) \tag{204}$$

A more accurate representation for $\xi \to 0$ (and $t \to 0$) is possible if one replace the large-argument asymptotic form of the Hankel function identified as $\sqrt{2j/\pi kt}\, e^{-jkt}$ in (203) by $H_0^{(2)}(kt)$, and likewise if one replaces the large-argument asymptotic form $(j/kt)\sqrt{2j/\pi kt}\, e^{-jkt}$ in (204) by its corresponding original function $H_1^{(2)}(kt)/kt$.

Such a heuristic modification allows one to obtain the proper singular behavior for $\hat{\mathbf{b}} \cdot d\mathbf{H}(Q)$ and $\hat{\mathbf{t}} \cdot d\mathbf{H}(Q)$ in the immediate neighborhood of the source at Q'.

Far from the source, where $\xi > 0.6$, the rapidly convergent residue series or the surface-ray modal expansion representation for V and U in (184) and (185) may be used to obtain

$$F_s \bigg|_{\xi \gg 0} \cong \sum_{n=1}^{10} L_n^h(Q') \exp\left[-\int_{Q'}^{Q} \alpha_n^h(t')dt' \right] A_n^h(Q) \tag{205}$$

and

$$G_s \bigg|_{\xi \gg 0} \cong \sum_{n=1}^{10} L_n^s(Q') \exp\left[-\int_{Q'}^{Q} \alpha_n^s(t')dt' \right] A_n^s(Q) \tag{206}$$

in which $L_n^{s,h}$ and $\alpha_n^{s,h}$ have been employed earlier in this section [see (143) and (144)], whereas $A_n^{s,h}$ are referred to as the *attachment coefficients* [37]. It can be shown that [37]

$$L_n^h = A_n^h \tag{207a}$$

$$L_n^s = -A_n^s \tag{207b}$$

The attachment coefficient transforms the boundary layer GTD surface-ray field to the true field on the surface. The above relationships in (207a) and (207b) are consistent with reciprocity. Incorporating (205) and (206) into (200) yields a GTD representation for $d\mathbf{H}(Q)$ which is valid outside the SSB transition region (see Fig. 28).

If an electric line monopole current source of strength $\hat{\mathbf{n}}'dI_m$ as in (159) and (160) is placed at Q', then the $\hat{\mathbf{n}}$-directed electric field of this source which is observed at Q may be expressed as [38]

$$d\mathbf{E}(Q) \sim C_0 Z_0 (dI_m \hat{\mathbf{n}}') \cdot [\hat{\mathbf{n}}' \hat{\mathbf{n}} F_s] e^{-jkt} \tag{208}$$

5. The Equivalent Current Method

The equivalent current method (ECM) is primarily useful for evaluating the fields at and near the caustics of diffracted rays since the GTD and its uniform versions (UTD, UAT, etc., which remain valid within the GO shadow boundary transition regions) predict infinite fields at these caustics and are therefore not valid there. It is important, however, to note at the outset that the ECM which employs GTD to calculate its equivalent sources (or currents) is a valid procedure for evaluating the fields only if the caustic region is not close to the GO shadow boundary transition regions; otherwise, one may have to resort to the use of PO or its refined version, the PTD. Concepts based on the ECM are also useful for correcting the PO solution in a simple manner via a slight modification of the original PTD ansatz, as will be described in the next section, which deals with the

PTD. In addition, an approach based on the ECM is useful in analyzing many other special problems, such as in estimating the modal reflection coefficients of an open-ended waveguide, in the formulation of the UTD solution for edge-excited surface rays, and in formulating a UTD solution to the problem of the diffraction by a vertex which is given under "UTD for Vertices" in Section 4. A detailed description of the ECM may be found in [7, 8, 9].

Even though the GTD fails at and near the caustics of diffracted rays, it can be employed far from the caustics to calculate the necessary equivalent sources (or currents) of the ECM; these equivalent sources are then incorporated into the radiation integral [12] to estimate the fields at and near the caustics of diffracted rays. Outside the shadow boundary transition regions where this ECM is valid, the UTD and UAT reduce to the GTD. Thus the *ECM employs GTD* to find the equivalent sources which radiate the fields at the caustics of diffracted rays in a manner analogous to PO, which employs GO to find the source distribution of the scattered fields; in fact, the PO method is useful for calculating the fields at the caustics of GO rays. Also, the ECM requires an integration over the equivalent line sources (or currents) similar to the PO source integration. However, the ECM integration is over a line current, whereas the PO integral is over a surface current distribution for three-dimensional (3-D) problems. On the other hand, the ECM requires no integration for 2-D applications, whereas the PO surface integral reduces to a line integral in the 2-D case. It is important to note that away from the caustics an asymptotic evaluation of the integration in the ECM yields results which generally blend into the GTD solution. Since the ECM corrects the GTD only through the caustic regions of the diffracted rays, it is thus preferable to switch from the ECM solution to the GTD ray solution away from the diffracted-ray caustic regions, where the integration in the ECM becomes inefficient and unnecessary.

ECM for Edge-Diffracted Ray Caustic Field Analysis

Consider a curved edge of a general curved wedge structure which is illuminated by an incident electric field $\mathbf{E}^i$, as shown in Fig. 58a. Let $\mathbf{H}^i$ denote the incident magnetic field which is associated with $\mathbf{E}^i$. As mentioned previously, it is assumed in the GTD approximation that the curved wedge (containing the curved edge of Fig. 58a) can be modeled locally by a straight wedge whose faces are tangent to the surfaces forming the curved wedge at the point of diffraction, and whose edge is tangent to the curved edge at that point. Let $\hat{\mathbf{e}}$ denote the unit vector tangent to the edge at the point Q of diffraction as in Fig. 58a. The $\hat{\mathbf{e}}$-directed components of the GTD edge-diffracted electric and magnetic fields ($\mathbf{E}_k^d$ and $\mathbf{H}_k^d$) at P are given by (44), (45), (67), and (68) as

$$\hat{\mathbf{e}} \cdot \mathbf{E}_k^d(P) = [\hat{\mathbf{e}} \cdot \mathbf{E}^i(Q)] \, D_{es}^k(\phi, \phi'; \beta_0) \sqrt{\frac{\varrho_e}{s^d(\varrho_e + s^d)}} \, e^{-jks^d} \qquad (209)$$

and

$$\hat{\mathbf{e}} \cdot \mathbf{H}_k^d(P) = [\hat{\mathbf{e}} \cdot \mathbf{H}^i(Q)] \, D_{eh}^k(\phi, \phi'; \beta_0) \sqrt{\frac{\varrho_e}{s^d(\varrho_e + s^d)}} \, e^{-jks^d} \qquad (210)$$

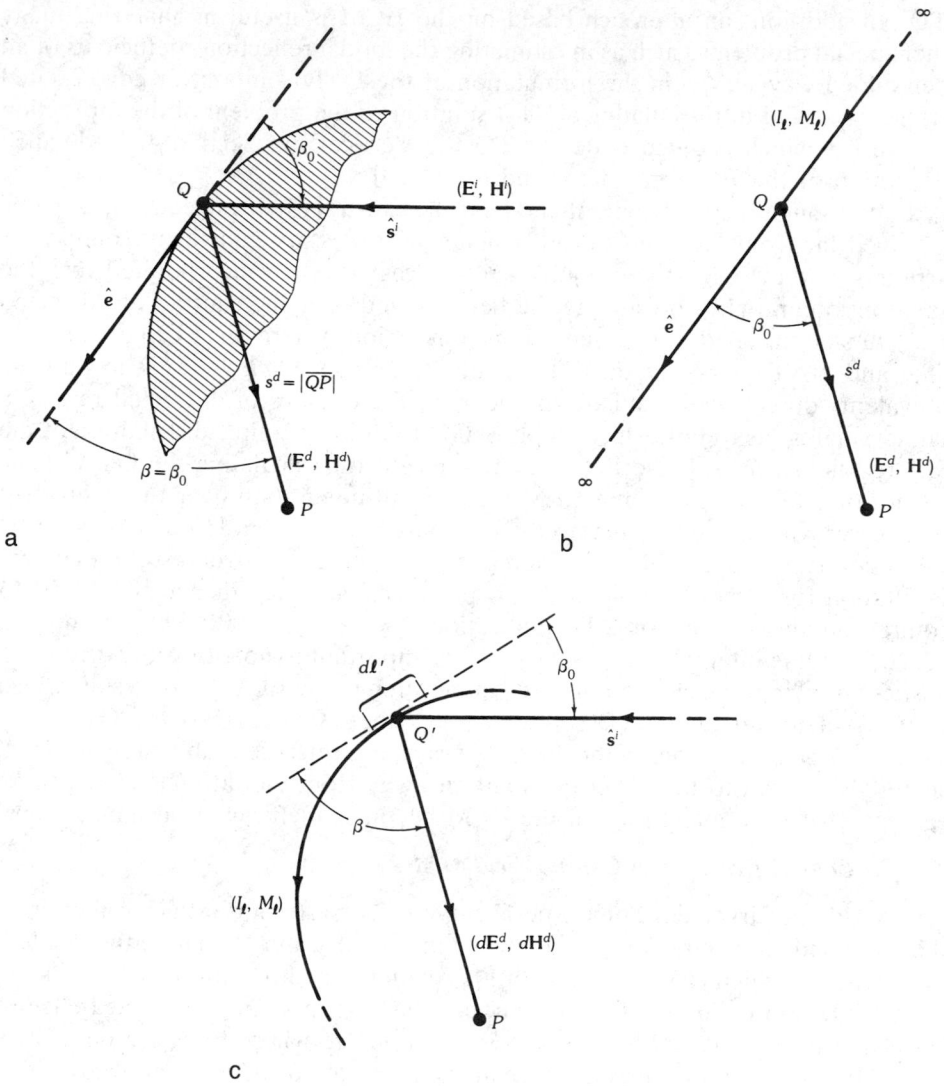

Fig. 58. Illustration of the equivalent edge current concept. (*a*) Diffraction by a curved wedge. (*b*) Equivalent line currents I_ℓ and M_ℓ at Q which generate edge-diffracted ray fields $(\mathbf{E}^d, \mathbf{H}^d)$ as in (*a*). (*c*) Contribution to P from an element $d\ell'$ at any point is Q' on the equivalent edge current.

In the case of a uniform wedge it can be proved readily that a knowledge of these two independent solutions $(\hat{\mathbf{e}} \cdot \mathbf{E}^d)$ and $(\mathbf{e} \cdot \mathbf{H}^d)$ is sufficient for obtaining all the remaining field components (which are transverse to $\hat{\mathbf{e}}$) [25]; this concept can be generalized to hold true asymptotically for the curved edge (in a curved wedge) via the principle of locality of high-frequency diffraction. If $s^d \ll \varrho_e$, even though s^d may be large in terms of the wavelength, then the two independent edge ($\hat{\mathbf{e}}$) directed field components in (209) and (210) can be approximated as

$$\hat{\mathbf{e}} \cdot \mathbf{E}_k^d(P) \cong [\hat{\mathbf{e}} \cdot \mathbf{E}^i(Q)] \, D_{es}^k \, \frac{e^{-jks^d}}{\sqrt{s^d}} \tag{211}$$

and

$$\hat{\mathbf{e}} \cdot \mathbf{H}_k^d(P) \cong [\hat{\mathbf{e}} \cdot \mathbf{H}^i(Q)] \, D_{eh}^k \, \frac{e^{-jks^d}}{\sqrt{s^d}} \tag{212}$$

Next, consider a traveling-wave electric or magnetic line current of strength I_ℓ or M_ℓ, respectively, which is located along the edge tangent ($\hat{\mathbf{e}}$) at Q as illustrated in Fig. 58b. Also, let I_ℓ and M_ℓ be of constant magnitude and let them possess a constant traveling-wave propagation factor equal to $k \cos \beta_0$ (in the $\hat{\mathbf{e}}$ direction), i.e., let $I_\ell = |I_\ell| e^{-jk\ell\cos\beta_0}$ and $M_\ell = |M_\ell| e^{-jk\ell\cos\beta_0}$, where the path ℓ is measured in the $\hat{\mathbf{e}}$ direction. Each of these traveling-wave currents radiates a conical wave of cone half-angle β_0. Furthermore, the field at P can be shown asymptotically to be associated with a ray from the point Q on the line source to the point P as shown in Fig. 58b. In particular, the line current strengths $I_\ell(Q)$ and $M_\ell(Q)$, at Q, generate the following $\hat{\mathbf{e}}$-directed components of the conical-wave electric and magnetic fields $\mathbf{E}^I(P)$ and $\mathbf{H}^M(P)$ at P, respectively:

$$\hat{\mathbf{e}} \cdot \mathbf{E}^I(P) \sim \left[\frac{-kZ_0 I_\ell(Q) e^{j\pi/4} \sin\beta_0}{\sqrt{8\pi k}} \right] \frac{e^{-jks^d}}{\sqrt{s^d}} \tag{213}$$

$$\hat{\mathbf{e}} \cdot \mathbf{H}^M(P) \sim \left[\frac{-kY_0 M_\ell(Q) e^{j\pi/4} \sin\beta_0}{\sqrt{8\pi k}} \right] \frac{e^{-jks^d}}{\sqrt{s^d}} \tag{214}$$

On comparing (211) and (212) with (213) and (214) outside the shadow boundary transition regions,* respectively, one obtains readily the strengths of the electric and magnetic line currents $I_\ell(Q)$ and $M_\ell(Q)$ which will generate the edge-diffracted fields $\mathbf{E}_k^d(P)$ and $\mathbf{H}_k^d(P)$ of (211) and (212); in particular,

$$I_\ell(Q) = \left(-\frac{1}{Z_0} \sqrt{\frac{8\pi}{k}} \, e^{-j\pi/4} \right) \frac{[\hat{\mathbf{e}} \cdot \mathbf{E}^i(Q)]}{\sin\beta_0} \, D_{es}^k(\phi, \phi'; \beta_0) \tag{215}$$

and

$$M_\ell(Q) = \left(-\frac{1}{Y_0} \sqrt{\frac{8\pi}{k}} \, e^{-j\pi/4} \right) \frac{[\hat{\mathbf{e}} \cdot \mathbf{H}^i(Q)]}{\sin\beta_0} \, D_{eh}^k(\phi, \phi'; \beta_0) \tag{216}$$

---

*The D_{es}^k and D_{eh}^k are not valid within the ISB and RSB (see Fig. 4a) transition regions; they must be replaced by the D_{es} and D_{eh} of the UTD. These D_{es} and D_{eh} are range (s^d) dependent because they contain Fresnel integrals which involve s^d; consequently, the edge-diffracted field within the shadow boundary transition regions is not a conical wave as in (211) and (212), and a comparison with (213) and (214) is therefore not possible within such regions.

It is evident from the above expressions that the currents I_ℓ and M_ℓ not only depend on the position (Q) on the edge, but they also depend on the angles of incidence (ϕ') and diffraction (ϕ) at the point of diffraction (Q). On the other hand, true currents do not depend on the angle of observation; consequently, the above aspect-dependent I_ℓ and M_ℓ are referred to as the strengths of "equivalent" currents which generate the edge-diffracted fields. It was mentioned earlier that a knowledge of $\hat{\mathbf{e}} \cdot \mathbf{E}_k^d$ and $\hat{\mathbf{e}} \cdot \mathbf{H}_k^d(P)$ is sufficient for generating all the remaining field components (which are then transverse to $\hat{\mathbf{e}}$) via Maxwell's equations. Similarly, when both the equivalent currents $I_\ell(Q)$ and $M_\ell(Q)$, of (215) and (216) above, are incorporated into the radiation integrals [12] for obtaining the fields radiated by these currents, then they also yield a complete description of the high-frequency edge-diffracted fields. It follows that one can employ both the equivalent edge currents I_ℓ and M_ℓ to radiate from every point Q on a curved edge as illustrated in Fig. 58c. Unlike the edge-diffracted fields as given by GTD or its uniform versions (UTD and UAT), the fields radiated by the "equivalent edge currents" $[I_\ell(Q)$ and $M_\ell(Q)]$ remain valid even within the caustic regions of the edge-diffracted rays; such a procedure constitutes the ECM for edge-diffraction.

While the equivalent edge currents I_ℓ and M_ℓ radiate the edge-diffracted fields everywhere exterior to the curved wedge (via the radiation integral [12]), it is evident from (215) and (216) that these currents are defined only for observation points lying on the Keller edge-diffracted ray cone of half-angle β_0 as in Fig. 58b. In order to extend the above I_ℓ and M_ℓ to include observation points which are not restricted to lie on the Keller cone of edge-diffracted rays, one must introduce an additional angle (of diffraction), β, into the expressions in (215) and (216) for I_ℓ and M_ℓ. The angle β is shown in Fig. 58c.* It is noted that β and β_0 are in general different (except on the Keller edge-diffracted ray cone which is defined by $\beta = \beta_0$). A generalization of I_ℓ and M_ℓ to include information on β as well as β_0 can be performed heuristically, and in a manner consistent with reciprocity, by replacing the factor $\sin\beta_0$ in the previous definition of I_ℓ and M_ℓ with the new factor $\sqrt{\sin\beta_0 \sin\beta}$. With this change the previous I_ℓ and M_ℓ of (215) and (216) are replaced by the new edge currents $\tilde{I}_\ell$ and $\tilde{M}_\ell$, as follows:

$$\tilde{I}_\ell(Q) = \left(-\frac{1}{Z_0}\sqrt{\frac{8\pi}{k}}\,e^{-j\pi/4}\right)\frac{[\hat{\mathbf{e}} \cdot \mathbf{E}^i(Q)]}{\sqrt{\sin\beta_0 \sin\beta}}\,\tilde{D}_{es}^k(\phi,\phi';\beta_0,\beta) \qquad (217)$$

and

$$\tilde{M}_\ell(Q) = \left(-\frac{1}{Y_0}\sqrt{\frac{8\pi}{k}}\,e^{-j\pi/4}\right)\frac{[\hat{\mathbf{e}} \cdot \mathbf{H}^i(Q)]}{\sqrt{\sin\beta_0 \sin\beta}}\,\tilde{D}_{eh}^k(\phi,\phi';\beta_0,\beta) \qquad (218)$$

Furthermore,

$$\mathbf{H}^i(Q) = \hat{\mathbf{s}}^i \times Y_0\mathbf{E}^i(Q)$$

---

*Note: $\beta = \beta_0$ if $Q' = Q$. Also, β_0 at Q may not be the same as β_0 at Q'. In general β does not have to equal β_0 for any point Q on the line current.

The above relation is true since $\mathbf{E}^i$ is assumed to be a ray optical field. It is observed that $D_{es,eh}^k$ in the old definitions of I_ℓ and M_ℓ contains the factor $(\sin\beta_0)^{-1}$ [see (67) and (68)] which must also be replaced by $(\sqrt{\sin\beta_0\sin\beta})^{-1}$. Thus, one includes $\tilde{D}_{es,eh}^k$ in the new definitions for $\tilde{I}_\ell$ and $\tilde{M}_\ell$ of (217) and (218); the $\tilde{D}_{es,eh}^k$ contains $(\sqrt{\sin\beta_0\sin\beta})^{-1}$ instead of just $(\sin\beta_0)^{-1}$ as in $D_{es,eh}^k$. In particular, $\tilde{D}_{es,eh}^k$ is simply related to $D_{es,eh}^k$ as follows:

$$\tilde{D}_{es,eh}^k(\phi,\phi';\beta_0,\beta) \equiv \frac{\sin\beta_0 \, D_{es,eh}^k(\phi,\phi';\beta_0,\beta)}{\sqrt{\sin\beta_0\sin\beta}} \qquad (219)$$

The electric and magnetic fields $\mathbf{E}_c^d$ and $\mathbf{H}_c^d$ radiated by the equivalent edge currents of (217) and (218) are found via the radiation integrals [12]; thus,

$$\mathbf{E}_c^d(P) \approx \frac{jKZ_0}{4\pi} \oint_{\mathscr{L}} [\hat{\mathbf{R}} \times \hat{\mathbf{R}} \times \tilde{I}_\ell(\ell')\hat{\boldsymbol{\ell}}' + Y_0\hat{\mathbf{R}} \times \tilde{M}_\ell(\ell')\hat{\boldsymbol{\ell}}'] \frac{e^{-jkR}}{R} \, d\ell' \qquad (220)$$

and

$$\mathbf{H}_c^d(P) \approx \frac{-jk}{4\pi} \oint_{\mathscr{L}} [\hat{\mathbf{R}} \times \tilde{I}_\ell(\ell')\hat{\boldsymbol{\ell}}' - Y_0\hat{\mathbf{R}} \times \hat{\mathbf{R}} \times \tilde{M}_\ell(\ell')\hat{\boldsymbol{\ell}}'] \frac{e^{-jkR}}{R} \, d\ell' \qquad (221)$$

The integrations in (220) and (221) are over a closed path $\mathscr{L}$ formed by a line of diffraction as in Fig. 59, i.e., by a ring type of edge discontinuity on the surface of the scatterer from which the edge-diffracted rays are produced. If the contour $\mathscr{L}$ is not closed, then the end points of the path $\mathscr{L}$ will contribute to the integrals in (220) and (221); these end point contributions may, in some cases, result from incorrectly truncating the currents $\tilde{I}_\ell$ and $\tilde{M}_\ell$ on those portions of the edge which are shadowed by the surface of the scatterer for a certain range of aspects. In the latter case, however, the effect of the rays launched on the surface (i.e., surface rays) which

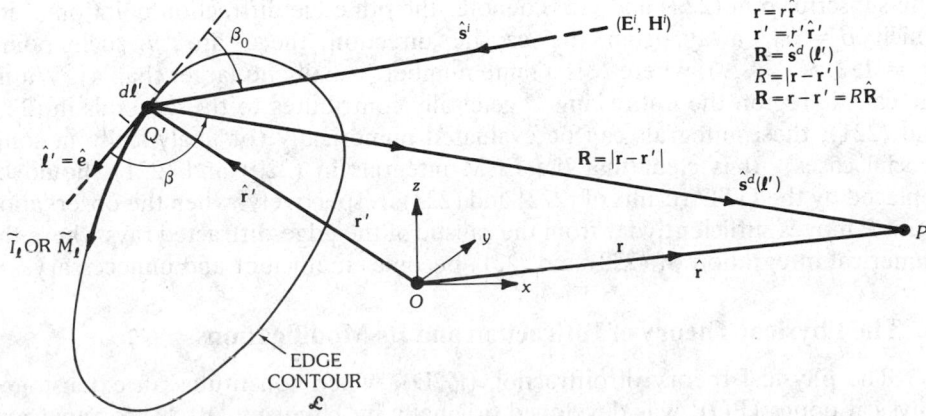

Fig. 59. Geometry associated with the radiation by the equivalent edge currents I_ℓ and M_ℓ on $\mathscr{L}$.

then undergo edge diffraction must also be taken into consideration so that $\tilde{I}_\ell$ and $\tilde{M}_\ell$ are not truncated, and spurious diffraction effects arising from such improper truncation can thereby be avoided. For a point P in the far zone of the scatterer one may employ the usual approximations in (220) and (221), namely,

$$\hat{\mathbf{R}} \cong \hat{\mathbf{r}}$$

$$\frac{e^{-jkR}}{R} \cong \frac{e^{-jkr}}{r} e^{+jk\hat{\mathbf{r}}\cdot\mathbf{r}'}$$

and

$$\mathbf{H}_c^d(P) \sim Y_0 \hat{\mathbf{r}} \times \mathbf{E}_c^d(P)$$

Here, $\hat{\mathbf{r}}$ denotes the radiation direction. The quantities $\hat{\mathbf{r}}$, $\hat{\mathbf{R}}$, $\hat{\ell}'$, β_0, and β are shown in Fig. 59. Far from the caustic directions, it can be shown that the dominant contribution to each of the integrals in (220) and (221) occurs from a few isolated stationary points corresponding to points of edge diffraction on $\mathscr{L}$ for which the law of edge diffraction ($\beta = \beta_0$) holds true, and in that case (220) and (221), in general, reduce to the GTD result given by

$$\mathbf{E}_c^d(P)\Big|_{\substack{\text{far} \\ \text{from} \\ \text{caustics}}} \sim \sum_{p=1}^{N} \mathbf{E}^i(Q_p) \cdot \bar{\bar{\mathbf{D}}}_e^k(Q_p) \sqrt{\frac{\varrho_{ep}}{s_p^d(\varrho_{ep} + s_p^d)}} e^{-jks_p^d} \qquad (222)$$

and

$$\mathbf{H}_c^d(P)\Big|_{\substack{\text{far} \\ \text{from} \\ \text{caustics}}} \sim Y_0 \sum_{p=1}^{N} \hat{\mathbf{s}}_p^d \times \left[\mathbf{E}^i(Q_p) \cdot \bar{\bar{\mathbf{D}}}_e^k(Q_p) \sqrt{\frac{\varrho_{ep}}{s_p^d(\varrho_{ep} + s_p^d)}} e^{-jks_p^d} \right] \qquad (223)$$

The subscript p in (222) and (223) denotes the pth edge diffraction point on $\mathscr{L}$ for which $\beta = \beta_0$; away from the caustic direction there are N such points ($p = 1, 2, 3, \ldots, N$), where N is a finite number (usually no larger than 4). Within the caustic region the entire ring $\mathscr{L}$ generally contributes to the integrals in (220) and (221); these integrals can be evaluated numerically (or analytically in some special cases). It is clear that the ECM integrals in (220) and (221) should be replaced by the GTD results of (222) and (223), respectively, when the observation point P moves sufficiently far from the caustic of the edge-diffracted rays where the numerical integration of (220) and (221) becomes inefficient and unnecessary.

6. The Physical Theory of Diffraction and Its Modifications

The physical theory of diffraction (PTD), which constitutes an extension of physical optics (PO), was developed originally by Ufimtsev [10, 11] for analyzing the high-frequency scattering from conducting surfaces. In particular, the PTD refines the PO field approximation just as the GTD refines the GO field approximation. In the PO technique the currents induced on the surface of a scatterer are approximated according to GO as described earlier in Section 3; however, it is clear that the GO approximation for the currents would be accurate only on the portion of the scatterer which is strongly illuminated by the source, whereas it would be totally inaccurate in the shadowed portion of a smooth convex surface, where, the GO yields a zero value for the surface current. The GO approximation for the surface current is also erroneous within the transition regions adjacent to the shadow boundary which divides a smooth convex surface into its illuminated (or lit) and shadowed portions; furthermore, it is expected to be inaccurate in the regions on the surface where a discontinuity (such as an edge) could exist. In the PTD approach, the GO current approximation is improved by including a correction which Ufimtsev refers to as a "nonuniform" component of the current. This nonuniform component of the current in the PTD formulation is supposed to include the effects not accounted for in the GO approximation to the surface current. In his original work Ufimtsev [10] considers the effects of the nonuniform component of the current only for conducting surfaces with edge type discontinuities; in addition, he neglects the effects of surface rays on smooth, convex bodies containing edges. Ufimtsev also does not give any expressions for the nonuniform component of the current nor does he explicitly integrate over these currents to obtain the fields radiated by them; instead, he obtains those fields via indirect considerations that are much simpler. The latter considerations involve an asymptotic high-frequency analysis of the canonical problem of the diffraction of an electromagnetic plane wave by a perfectly conducting wedge. As a result the fields of the nonuniform component of the current, as originally obtained by Ufimtsev, exhibit a ray optical character and hence become infinite at and near caustics, even though, unlike the GTD, they maintain their validity at and near the optical shadow boundaries except for grazing angles of incidence. Thus Ufimtsev introduced caustic matching functions in an *ad hoc* manner into the fields radiated by the nonuniform component of the current to correct for the infinite field behavior along the edge diffracted–ray caustic directions. That *ad hoc* procedure, which employs caustic matching functions, can be circumvented via an application of the ECM, which employs equivalent line currents deduced from Ufimtsev's expressions for the fields that are radiated far from the caustic by the nonuniform component of the current, rather than from GTD as done in Section 5; of course, these fields of nonuniform component of the current emanate from a line of discontinuity (such as an edge) on the surface of the scatterer. The latter modification of Ufimtsev's work based on the concept of the ECM was also suggested in [9], and it will be employed in the present development to modify the original version of the PTD.

The PTD field will be presented here as the superposition of the PO field and the field of the ECM which employs Ufimtsev's diffraction correction to PO instead of the GTD for deducing the equivalent line currents. In those situations where the PO and ECM type integrals occurring in this PTD approach for edged bodies can be evaluated asymptotically, one generally recovers the GTD solution; of course, if that asymptotic evaluation is performed in a uniform manner, then the corresponding uniform GTD (i.e., UTD or UAT) solution which remains valid even within the GO shadow boundary transition regions may be recovered from the

PTD. The GTD (and UTD or UAT) constitutes a far more efficient and physically appealing solution than the PTD, which in general requires an integration over the GO currents and also over the Ufimtsev-based equivalent line currents. Nevertheless, the PTD is very useful for estimating the fields in regions where there is a confluence of the transition regions associated with ray caustics and shadow boundaries, respectively; it is noted that the GTD, UTD, and UAT all fail in this special situation. The PTD also automatically remains valid in the neighborhood of the caustics of GO rays, since it contains the PO solution which provides a good estimate for the dominant fields near the caustics of GO rays associated with edge bodies. The PTD also remains valid in the neighborhood of the caustics of edge-diffracted rays. However, if the diffracted-ray caustic lies far from the GO shadow boundaries, then it is more efficient to employ the ECM of Section 5, which employs GTD based equivalent line currents, rather than to use the PTD (which involves the PO surface integral).

PTD for Edged Bodies

A Canonical Problem in the PTD Formulation—First, consider the 3-D canonical problem of the diffraction of an obliquely incident electromagnetic plane wave by a perfectly conducting wedge. A PO solution to this problem can be expressed formally in terms of the PO integral (over the GO currents) that is evaluated only on the illuminated faces of the wedge. Thus, the total PO electric field $\mathbf{E}^{PO}$ external to the wedge becomes

$$\mathbf{E}^{PO} = \mathbf{E}^i + \mathbf{E}^s_{PO} \tag{224}$$

as in (30), where $\mathbf{E}^i$ is the incident field in the absence of the wedge, and the scattered field $\mathbf{E}^s_{PO}$ is given by the integral in (30). For the present it is assumed that the observation point lies outside the GO shadow boundary transition regions. An asymptotic high-frequency evaluation of the integral for $\mathbf{E}^s_{PO}$ far from the GO shadow boundaries yields

$$\mathbf{E}^s_{PO} \sim - \mathbf{E}^i(1 - U_i) + \mathbf{E}^r U_r + \mathbf{E}^d_{PO} \tag{225}$$

Note that the step functions U_i and U_r have been defined previously under "GTD for Edges" in Section 4.

Combining (224) and (225) one obtains

$$\mathbf{E}^{PO} \sim \mathbf{E}^i U_i + \mathbf{E}^r U_r + \mathbf{E}^d_{PO} \tag{226}$$

where $\mathbf{E}^d_{PO}$ is viewed as the edge-diffracted field, as predicted by PO, at an observation point P; it can be expressed as in (57) by

$$\mathbf{E}^d_{PO}(P) = \mathbf{E}^i(Q) \cdot \bar{\bar{\mathbf{D}}}^{PO}_e(Q) \sqrt{\frac{\varrho_e}{s^d(\varrho_e + s^d)}} \, e^{-jks^d} \tag{227}$$

The PO dyadic edge-diffraction coefficient $\bar{\bar{\mathbf{D}}}^{PO}_e$ at the point of diffraction Q on the edge is of the form

$$\bar{\bar{D}}_e^{PO} = -\hat{\beta}_0'\hat{\beta}_0 D_{es}^{PO} - \hat{\phi}'\hat{\phi} D_{eh}^{PO} \tag{228}$$

which is similar to $\bar{\bar{D}}_e$ of (59c). The precise form of $D_{es,eh}^{PO}$ will be indicated shortly. The PO edge-diffracted field $\mathbf{E}_{PO}^d$ arises from the truncation of the GO current by the edge. On the other hand, a GTD solution to the same canonical problem yields [see (40) and (41) of Section 4]

$$\mathbf{E}_{GTD} \sim \mathbf{E}^i U_i + \mathbf{E}^r U_r + \mathbf{E}_k^d \tag{229}$$

where

$$\mathbf{E}_k^d(P) = \mathbf{E}^i(Q) \cdot \bar{\bar{D}}_e^k(Q) \sqrt{\frac{\varrho_e}{s^d(\varrho_e + s^d)}} e^{-jks^d} \tag{230}$$

$$\bar{\bar{D}}_e^k = -\hat{\beta}_0'\hat{\beta}_0 D_{es}^k - \hat{\phi}'\hat{\phi} D_{eh}^k \tag{231}$$

The GTD result in (229) together with (230) and (231) constitutes the leading terms in the direct asymptotic approximation of the exact solution to this canonical problem [25, 26], whereas the corresponding asymptotic PO field expression in (226) together with (227) and (228) constitutes an asymptotic approximation of a PO based solution which in itself is approximate. Thus the diffraction correction to $\mathbf{E}^{PO}$ should now be readily evident from (226) and (229); it is denoted here by $\mathbf{E}_u^d$ and is simply obtained by subtracting (226) from (229). In particular,

$$\mathbf{E}_u^d = \mathbf{E}_{GTD} - \mathbf{E}^{PO} = \mathbf{E}_k^d - \mathbf{E}_{PO}^d \tag{232}$$

or

$$\mathbf{E}_u^d(P) = \mathbf{E}^i(Q) \cdot \bar{\bar{D}}_e^u \sqrt{\frac{\varrho_e}{s^d(\varrho_e + s^d)}} e^{-jks^d} \tag{233}$$

with

$$\bar{\bar{D}}_e^u = -\hat{\beta}_0'\hat{\beta}_0(D_{es}^k - D_{es}^{PO}) - \hat{\phi}'\hat{\phi}(D_{eh}^k - D_{eh}^{PO}) \tag{234a}$$

or, more compactly,

$$\bar{\bar{D}}_e^u = \bar{\bar{D}}_e^k - \bar{\bar{D}}_e^{PO} \tag{234b}$$

Ufimtsev's PTD ansatz is then essentially based on adding the required diffraction correction $\mathbf{E}_u^d$ to $\mathbf{E}^{PO}$ to arrive at an expression $\mathbf{E}_{PTD}$ for the electric field which constitutes a refinement over PO; thus

$$\mathbf{E}_{PTD} = \mathbf{E}^{PO} + \mathbf{E}_u^d \tag{235a}$$

or

$$\mathbf{E}_{PTD} = (\mathbf{E}^i + \mathbf{E}_{PO}^s) + \mathbf{E}_u^d \tag{235b}$$

It is important to note the $\mathbf{E}^s_{\mathrm{PO}}$ of (235b) in the PTD formulation is left formally as an integral over the GO surface current approximation; this is in contrast to $\mathbf{E}_{\mathrm{GTD}}$ of (229), which employs rays and requires no integration over the currents. Although the $\mathbf{E}^d_u$ above was developed initially for observation points sufficiently far from the GO shadow boundaries, it is easily verified that the singularities in the nonuniform edge-diffracted fields $\mathbf{E}^d_k$ and $\mathbf{E}^d_{\mathrm{PO}}$, which occur at the GO shadow boundaries, exactly cancel each other in (232), thereby making $\mathbf{E}^d_u$ bounded, continuous, and valid at these boundaries (and their associated transition regions).

PTD for 3-D Edged Bodies—A PTD analysis of the problem of EM radiation from an antenna in the presence of perfectly conducting edge bodies, which are otherwise smooth and convex, can be developed directly from the canonical PTD wedge solution of (234) due to the local nature of $\mathbf{E}^d_u$. Thus the PTD electric field $\mathbf{E}_{\mathrm{PTD}}$ for edged bodies can be expressed as in (235a) and (235b), namely,

$$\mathbf{E}_{\mathrm{PTD}}(P) \sim \mathbf{E}^{\mathrm{PO}}(P) + \mathbf{E}^d_u(P) \tag{236a}$$

where

$$\mathbf{E}^{\mathrm{PO}}(P) = \mathbf{E}^i(P) + \frac{jkZ_0}{4\pi} \iint_{S_{\mathrm{lit}}} \hat{\mathbf{R}} \times \hat{\mathbf{R}} \times [2\hat{\mathbf{n}}' \times \mathbf{H}^i] \frac{e^{-jkR}}{R} \, dS' \tag{236b}$$

as in (30), and

$$\mathbf{E}^d_u(P) = \sum_{p=1}^{N} \mathbf{E}^i(Q_p) \cdot \bar{\bar{\mathbf{D}}}^u_e(Q_p) \sqrt{\frac{\varrho_{ep}}{s^d_p(\varrho_{ep} + s^d_p)}} \, e^{-jks^d_p} \tag{236c}$$

as in (233), except that the above expression in (236c) accounts for all the N edge-diffraction points which give rise to N diffracted rays that reach the point P. These ($p = 1, 2, \ldots, N$) points of edge diffraction obey Keller's law of edge diffraction so that the field $\mathbf{E}^d_u$ propagates along the same paths as the GTD edge-diffracted rays, i.e., the ray field $\mathbf{E}^d_u$ is associated with the Keller cone of edge-diffracted rays as in Figs. 1 and 12. It is important to note once again that $\mathbf{E}^d_u$ represents only a correction to the edge-diffracted field as predicted by PO. Since $\mathbf{E}^d_u$ is a ray optical type field, it becomes singular at the caustics of edge-diffracted rays; therefore a more useful form of $\mathbf{E}^d_u$ which circumvents this problem is available from (220) via the concept of the ECM. Thus, more generally, $\mathbf{E}^d_u$ is given by

$$\mathbf{E}^d_u \cong \frac{jkZ_0}{4\pi} \oint_{\mathscr{L}} [\hat{\mathbf{R}} \times \hat{\mathbf{R}} \times \bar{I}^u_\ell(\ell')\hat{\ell}' + Y_0 \hat{\mathbf{R}} \times \bar{M}^u_\ell(\ell')\hat{\ell}'] \frac{e^{-jkR}}{R} \, d\ell' \tag{237}$$

in which the equivalent Ufimtsev type electric and magnetic edge currents $\bar{I}^u_\ell$ and $\bar{M}^u_\ell$ on the edge contour are now defined as in (217) and (218), but with $\bar{D}^u_{es}$ and $\bar{D}^u_{eh}$ replacing $\bar{D}^k_{es}$ and $\bar{D}^k_{eh}$, respectively, in those equations. Hence,

$$\tilde{I}_{\ell}^{u}(\ell') = \left(-\frac{1}{Z_0}\sqrt{\frac{8\pi}{k}}\ e^{-j\pi/4}\right)\frac{[\hat{\mathbf{e}}\cdot\mathbf{E}^{i}(\ell')]}{\sqrt{\sin\beta_0\sin\beta}}\ \tilde{D}_{es}^{u}(\phi,\phi';\beta_0,\beta) \tag{238}$$

and

$$\tilde{M}_{\ell}^{u}(\ell') = \left(-\frac{1}{Y_0}\sqrt{\frac{8\pi}{k}}\ e^{-j\pi/4}\right)\frac{[\hat{\mathbf{e}}\cdot\mathbf{H}^{i}(\ell')]}{\sqrt{\sin\beta_0\sin\beta}}\ \tilde{D}_{eh}^{u}(\phi,\phi';\beta_0,\beta) \tag{239}$$

It is clear from (219) that $\tilde{D}_{es,eh}^{u}$ are related to $D_{es,eh}^{u}$ via

$$\tilde{D}_{es,eh}^{u}(\phi,\phi';\beta_0,\beta) \equiv \frac{\sin\beta_0\ D_{es,eh}^{u}(\phi,\phi';\beta_0)}{\sqrt{\sin\beta_0\sin\beta}} \tag{240}$$

The contour $\mathscr{L}$ in (237) is shown to be closed; however, if it is nonclosed, then the discussion under (221) applies also to this case. The result in (237) remains valid along the diffracted-ray caustic directions even if these occur within GO shadow boundary transition regions, since $\tilde{I}_{\ell}^{u}$ and $\tilde{M}_{\ell}^{u}$ are valid there. On the other hand, recall that $\tilde{I}_{\ell}$ and $\tilde{M}_{\ell}$ of (217) and (218) are not valid near the GO shadow boundaries. Away from the caustic directions it is more efficient to replace (237) by its asymptotic approximation given previously in (236c).

It only remains to give explicit expressions for $D_{es}^{u}(\phi,\phi';\beta_0)$ and $D_{eh}^{u}(\phi,\phi';\beta_0)$, which occur in $\bar{\bar{\mathbf{D}}}_{e}^{u}$ of (234a) and also in the definition of the equivalent Ufimtsev type edge currents of (238) and (239), to complete the PTD solution. From (234a) and (234b) it is clear that

$$D_{es,eh}^{u}(\phi,\phi';\beta_0) \equiv D_{es,eh}^{k}(\phi,\phi';\beta_0) - D_{es,eh}^{PO}(\phi,\phi';\beta_0) \tag{241}$$

The $D_{es,eh}^{PO}$ is defined as follows:

$$D_{es,eh}^{PO}(\phi,\phi';\beta_0) = \begin{cases} -\dfrac{e^{-j\pi/4}}{2\sqrt{2\pi k}}\dfrac{1}{\sin\beta_0}\left[\tan\left(\dfrac{\phi-\phi'}{2}\right)\mp\tan\left(\dfrac{\phi+\phi'}{2}\right)\right] \\ \quad\text{if } 0\leqslant\phi'\leqslant(\Omega-\pi)\ \text{(or only } \phi=0 \text{ face is illuminated)} \\[2mm] -\dfrac{e^{-j\pi/4}}{2\sqrt{2\pi k}}\dfrac{1}{\sin\beta_0}\left[\mp\tan\left(\dfrac{\phi+\phi'}{2}\right)\mp\tan\left[\dfrac{2\Omega-(\phi+\phi')}{2}\right]\right] \\ \quad\text{if } (\Omega-\pi)\leqslant\phi'\leqslant\pi\ \text{(or both faces are illuminated)} \\[2mm] -\dfrac{e^{-j\pi/4}}{2\sqrt{2\pi k}}\dfrac{1}{\sin\beta_0}\left[-\tan\left(\dfrac{\phi-\phi'}{2}\right)\mp\tan\left[\dfrac{2\Omega-(\phi+\phi')}{2}\right]\right] \\ \quad\text{if } \pi\leqslant\phi'\leqslant\Omega\ \text{(or only } \phi=\Omega \text{ face is illuminated)} \end{cases} \tag{242a}$$

with

$$\Omega = n\pi \tag{242b}$$

(see Fig. 19). Note that $n = 2$ for a half-plane.

The $D_{es,eh}^k$ was defined earlier in (68). It is clear from (219), (240), and (234) that

$$\bar{D}_{es,eh}^{PO}(\phi, \phi'; \beta_0, \beta) \equiv \frac{\sin\beta_0 \, D_{es,eh}^{PO}(\phi, \phi'; \beta_0, \beta)}{\sqrt{\sin\beta_0 \sin\beta}} \tag{243}$$

and

$$\bar{D}_{es,eh}^u(\phi, \phi'; \beta_0, \beta) = \bar{D}_{es,eh}^k(\phi, \phi'; \beta_0, \beta) - \bar{D}_{es,eh}^{PO}(\phi, \phi'; \beta_0, \beta) \tag{244}$$

PTD for 2-D Edged Bodies—The PTD analysis of the problem of the radiation by a line source in the presence of a perfectly conducting 2-D structure with edges is similar to that in (236a), (236b), and (236c) except that (236b) becomes the following in the 2-D case [see (36)]:

$$\mathbf{E}^{PO}(P) = \mathbf{E}^i(P) + \frac{kZ_0}{4} \int_{C_{\text{lit}}} [\hat{\mathbf{R}} \times \hat{\mathbf{R}} \times (2\hat{\mathbf{n}}' \times \mathbf{H}^i)] H_0^{(2)}(kR) \, d\ell' \tag{245}$$

and (236c) reduces in the 2-D case to

$$\mathbf{E}_u^d(P) \cong \sum_{p=1}^N \mathbf{E}^i(Q_p) \cdot \bar{\bar{\mathbf{D}}}_e^u(Q_p) \frac{e^{-jks_p^d}}{\sqrt{s_p^d}} \tag{246}$$

It is noted that there is no diffracted-ray caustic in the 2-D case; hence there is no need for any integration as in (237) for calculating $\mathbf{E}_u^d$.

Results based on an application of the PTD for analyzing some simple antenna problems are illustrated in Figs. 60 through 63. Fig. 60 shows a comparison of the PTD and UTD based far-zone radiation pattern calculations for a 2-D parabolic reflector excited by a magnetic line source at the focus; furthermore, an independent formally exact moment method solution to this problem is also included in that figure for comparison. Outside the diffracted-ray caustic regions and exterior to the regions of confluence of diffracted-ray caustics with the GO shadow boundaries and/or GO ray caustics, a uniform asymptotic evaluation of the PO integral in the PTD solution generally allows the PTD solution to yield the leading terms of the UTD solution. However, an exact or numerical evaluation of the PO integral in the PTD solution inherently includes an additional contribution not included within the UTD approximation. For example, this additional contribution present in the exact PO integral tends to indicate that the field in the shadow zone is accounted for not only by diffraction but also by what may equivalently be viewed as leakage through the surface. It is evident, however, from the GTD/UTD/UAT that at high frequencies the field in the shadow zone is accounted for only by diffraction. Thus the additional contribution from the exact PO integration (which is based on approximate GO currents) appears to be erroneous; furthermore, the $\mathbf{E}_u^d$ in the PTD is not able to compensate for this error in the $\mathbf{E}^{PO}$; such an error in $\mathbf{E}^{PO}$ also exists to a lesser degree in the lit region. Consequently, the PTD solution does not agree as closely with the moment method solution in Fig. 60 as does the UTD solution except in the main beam direction ($\theta \cong 0°$), where the reflection shadow boundary and the reflected-ray caustic directions merge, rendering the UTD invalid in this region. Both the UTD and the PTD deviate by roughly the same amount from the moment method solution for $60° < \theta < 100°$ because the diffraction of creeping or surface rays by the edges and also the doubly diffracted rays are ignored in these UTD and PTD calculations. Note that double and higher-order multiple edge diffractions can be handled more easily by the UTD; hence, the first-order PTD result can be modified to include the multiple diffraction effects via the UTD. Figs. 61 through 63 illustrate a comparison of UTD and PTD results with those based on the corresponding moment method solutions for the far-zone radiation pattern of a magnetic line source near a planar reflector. Again, the PTD result exhibits a small leakage effect in the shadow zone in contrast to the pure UTD interference pattern resulting from only the two edge-diffracted rays. In general, this error in the PTD may not be too serious in most applications where extreme pattern detail is not required in the shadow (and also the lit) zone;

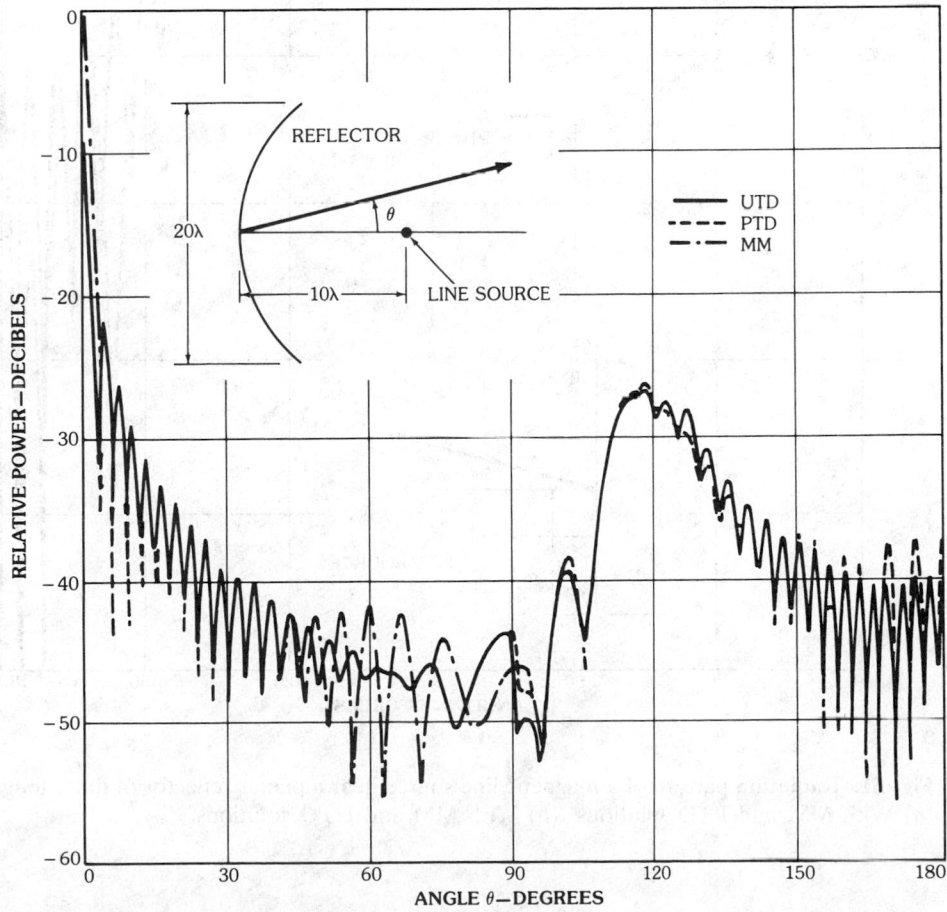

Fig. 60. Radiation pattern of a magnetic line source near a parabolic cylinder reflector.

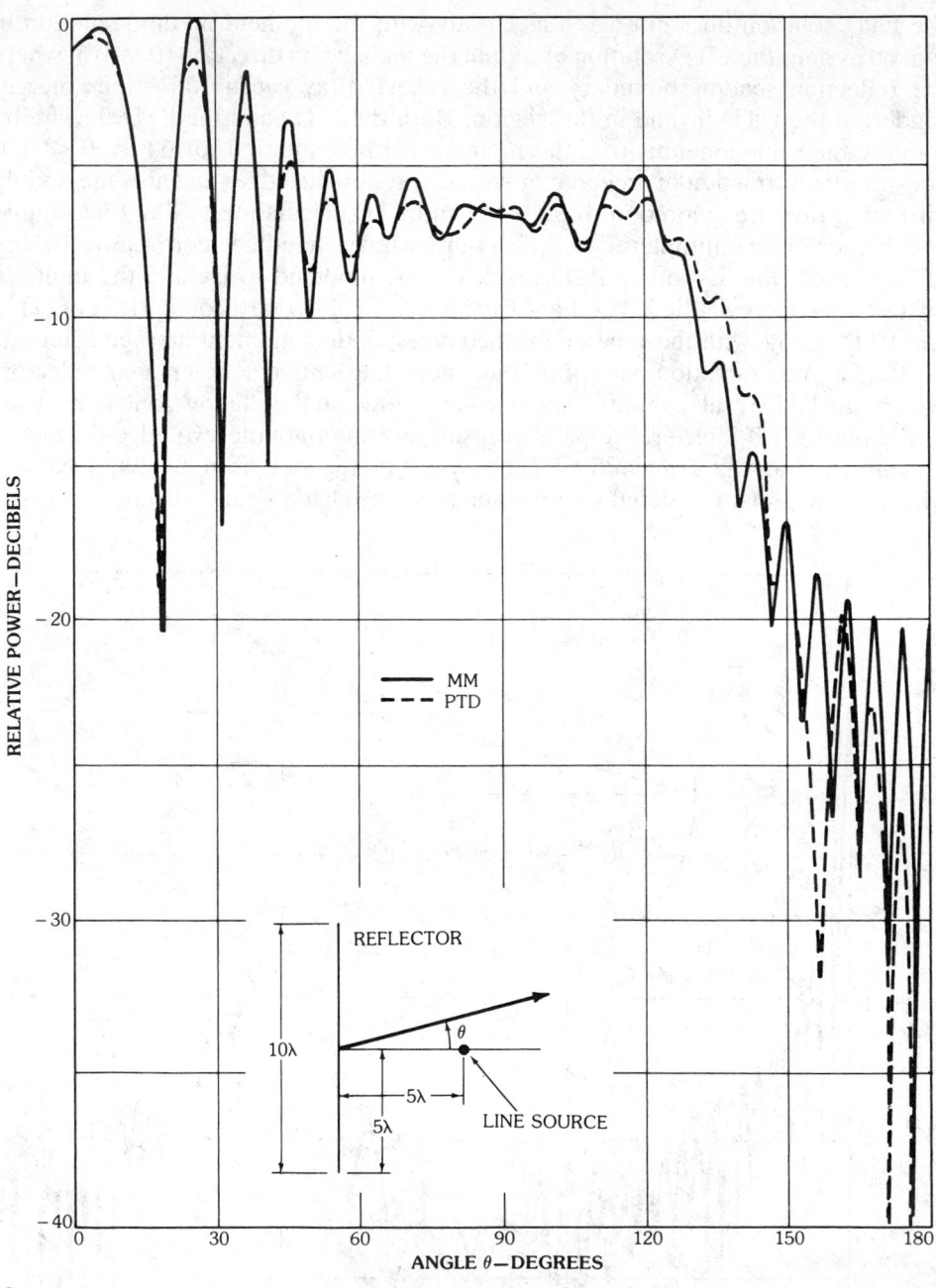

a

Fig. 61. Radiation pattern of a magnetic line source near a planar reflector of finite length. (*a*) With MM and PTD solutions. (*b*) With MM and UTD solutions.

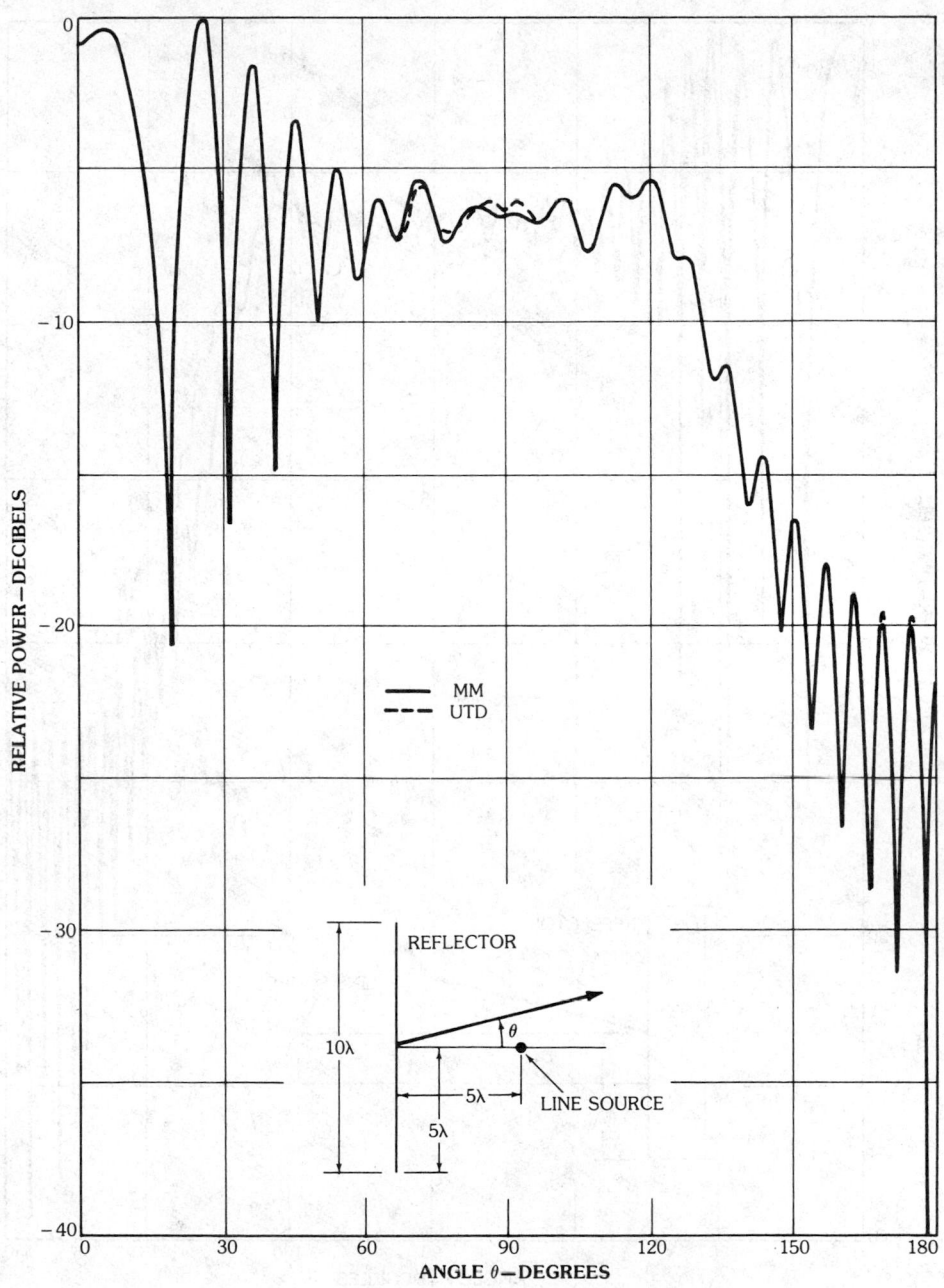

b

Fig. 61, *continued*

in fact, this error is expected to decrease as the size of the reflector increases (so PO becomes more accurate) and as the source is moved farther away from the reflector as in Figs. 62 and 63. Of course, PTD is most useful for obtaining the pattern for 0° < θ < 10° in Fig. 60 pertaining to the parabolic reflector, since the UTD is not

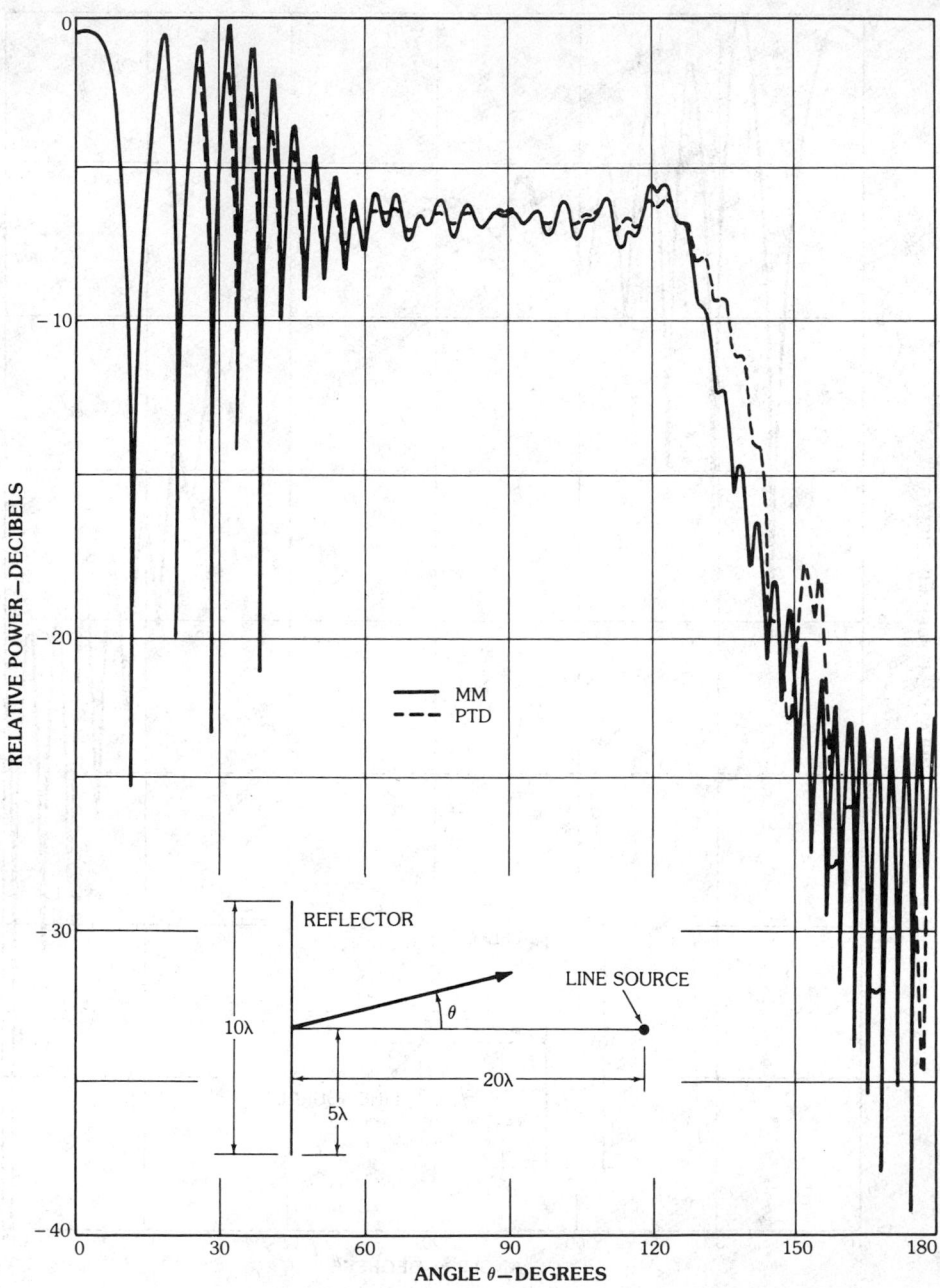

a

Fig. 62. Radiation pattern of a magnetic line source near a planar reflector of finite length. (*a*) With MM and PTD solutions. (*b*) With MM and UTD solutions.

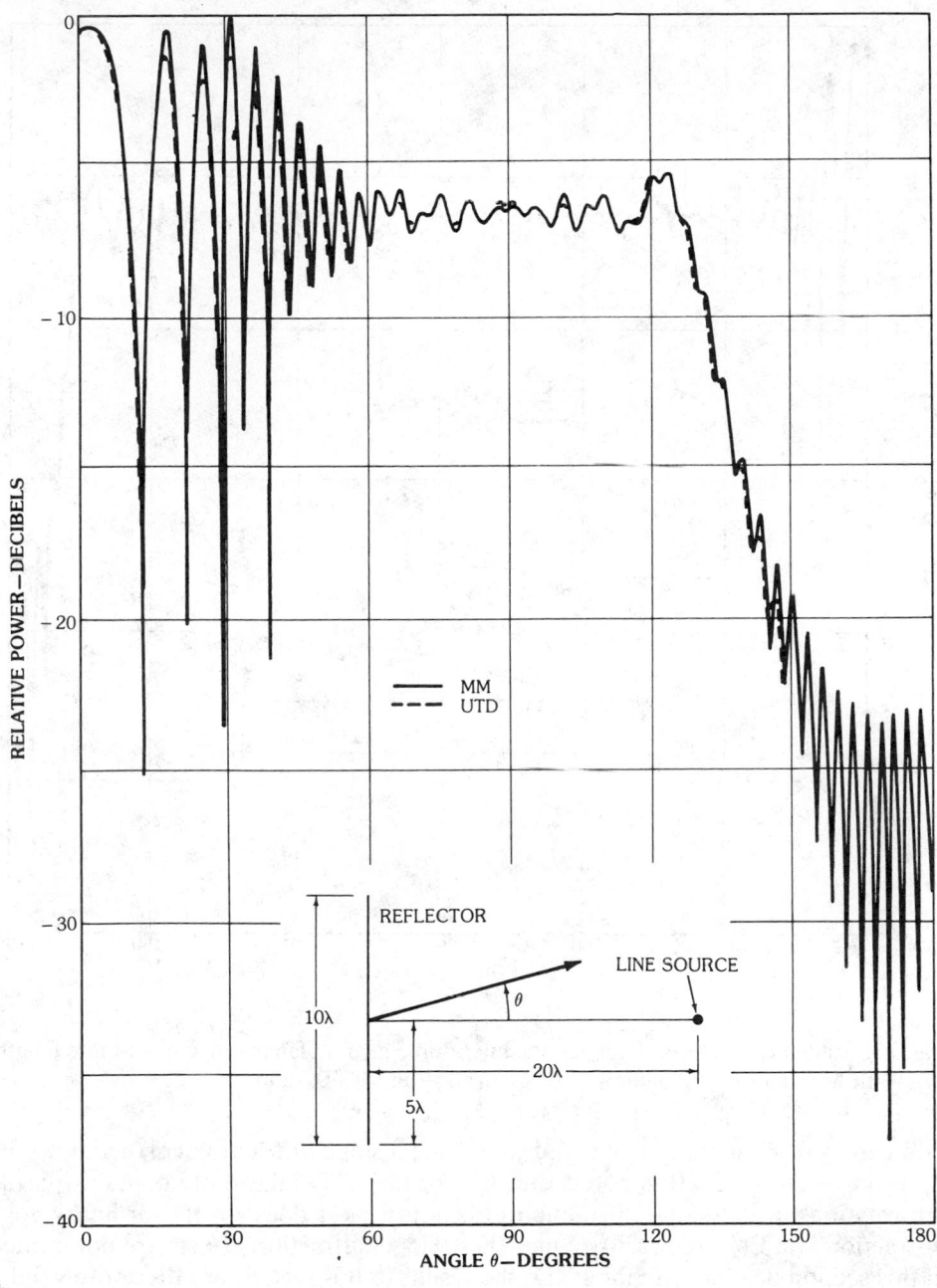

b

Fig. 62, *continued*

applicable at those aspects, as mentioned earlier. For $10° < \theta < 180°$ the UTD may be employed instead of PTD for greater efficiency.

Finally, it may be mentioned that effects of slope edge diffraction can be incorporated in the equivalent currents of the ECM of Section 5; such a modifica-

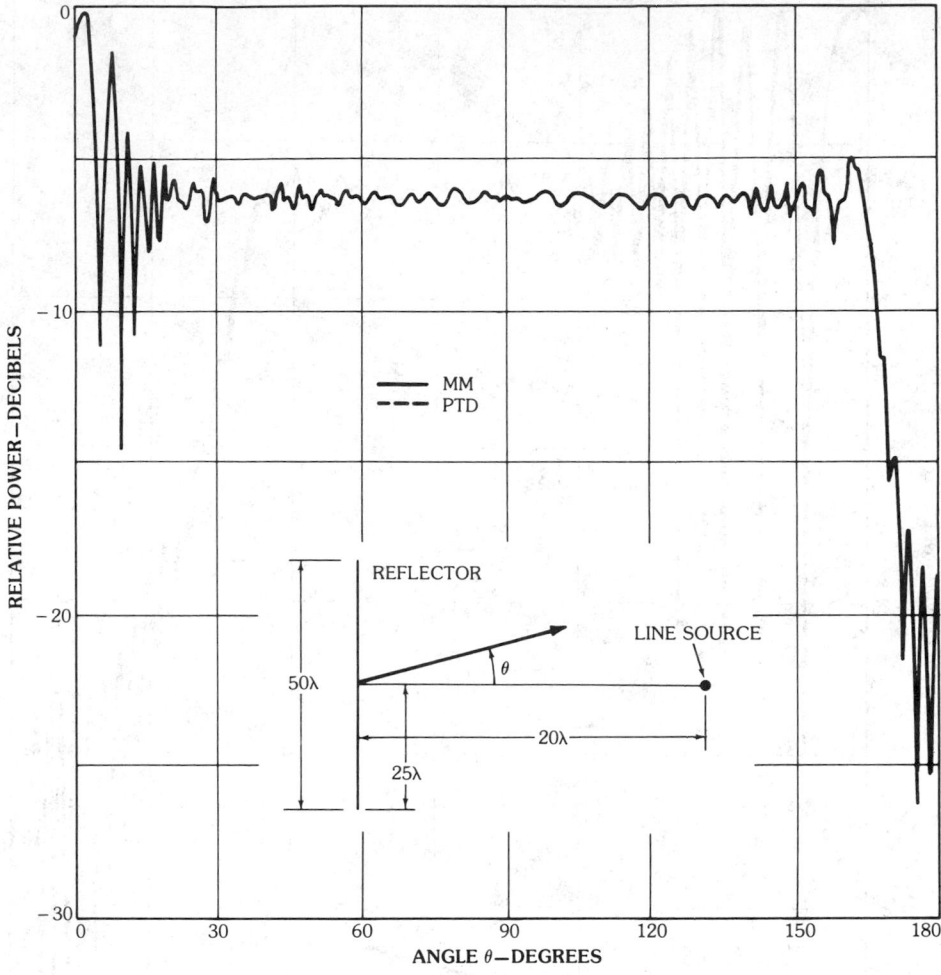

a

Fig. 63. Radiation pattern of a magnetic line source near a planar reflector of finite length. (*a*) With MM and PTD solutions. (*b*) With MM and UTD solutions.

tion can also be similarly employed to include a slope diffraction correction to the $\mathbf{E}_u^d$ term of the PTD. It is noted that $\mathbf{E}^{PO}$ of the PTD inherently contains partial information on the slope diffraction effects just as it does on the ordinary edge diffraction [via $\mathbf{E}_{PO}^d$; see (226)]. Since these slope diffraction effects are not treated in this section dealing with the PTD, the results of this section are therefore valid if those effects are negligible; otherwise, their effects must be included.

Acknowledgments

The author wishes to thank Dr. Ayhan Altintas of the Ohio State University ElectroScience Laboratory for his review of this chapter and for his helpful comments.

The work in this chapter was supported in part by Contract No. ONR N00014-

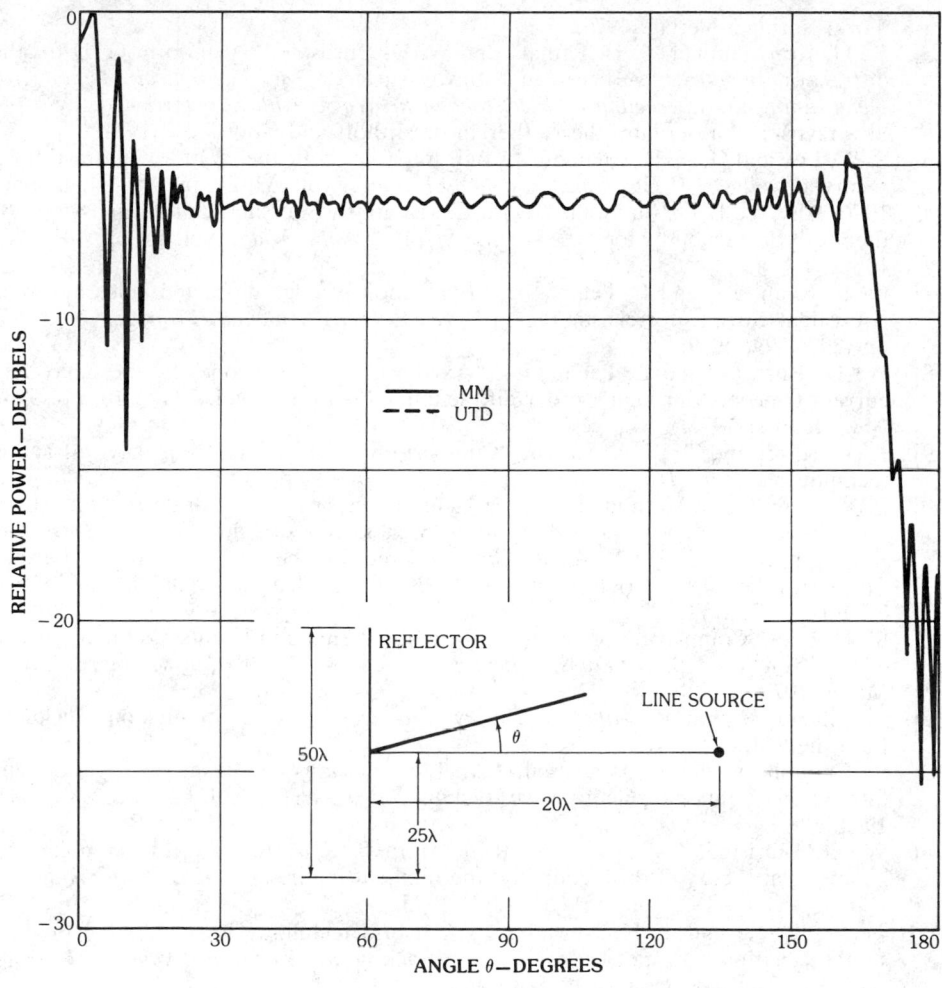

Fig. 63, *continued*

78-C-0049, between the Office of Naval Research and the Ohio State University Research Foundation.

7. References

[1]   J. B. Keller, "Geometrical theory of diffraction," *J. Opt. Soc. Am.*, vol. 52, pp. 116–130, 1962.

[2]   J. B. Keller, "A geometrical theory of diffraction," in *Calculus of Variations and Its Applications*, ed. by L. M. Graves, New York: McGraw-Hill Book Co., 1958, pp. 27–52.

[3]   B. R. Levy and J. B. Keller, "Diffraction by a smooth object," *Commun. Pure Appl. Math.*, vol. 12, pp. 159–209, 1959.

[4]   R. G. Kouyoumjian, "The geometrical theory of diffraction and its applications," in *Numerical and Asymptotic Techniques in Electromagnetics*, ed. by R. Mittra, New

York: Springer-Verlag, 1975.

[5] R. G. Kouyoumjian, P. H. Pathak, and W. D. Burnside, "A uniform GTD for the diffraction by edges, vertices, and convex surfaces," in *Theoretical Methods for Determining the Interaction of Electromagnetic Waves with Structures*, ed. by J. K. Skwirzynski, Amsterdam, the Netherlands: Sijthoff and Noordhoff, 1981.

[6] S. W. Lee and G. A. Deschamps, "A uniform asymptotic theory of EM diffraction by a curved wedge," *IEEE Trans. Antennas Propag.*, vol. AP-24, pp. 25–34, January 1976. Also see D. S. Ahluwalia, R. M. Lewis, and J. Boersma, "Uniform asymptotic theory of diffraction by a plane screen," *SIAM J. Appl. Math.*, vol. 16, pp. 783–807, 1968.

[7] C. E. Ryan, Jr., and L. Peters, Jr., "Evaluation of edge diffracted fields including equivalent currents for caustic regions," *IEEE Trans. Antennas Propag.*, vol. AP-7, pp. 292–299, 1969.

[8] W. D. Burnside and L. Peters, Jr., "Axial RCS of finite cones by the equivalent current concept with higher-order diffraction," *Radio Sci.*, vol. 7, no. 10, pp. 943–948, October 1972.

[9] E. F. Knott and T. B. A. Senior, "Comparison of three high-frequency diffraction techniques," *Proc. IEEE*, vol. 62, pp. 1468–1474, 1974.

[10] P. Ya. Ufimtsev, "Method of edge waves in the physical theory of diffraction" (from the Russian "Method Krayevykh voin v fizicheskoy teorii difraktsii," *Izd-Vo Sov. Radio*, pp. 1–243, 1962), translation prepared by the U.S. Air Force Foreign Technoloy Division, Wright-Patterson AFB, Ohio; released for public distribution September 7, 1971.

[11] S. W. Lee, "Comparison of uniform asymptotic theory and Ufimtsev's theory of EM edge diffraction," *IEEE Trans. Antennas Propag.*, vol. AP-25, no. 2, pp. 162–170, March 1977.

[12] S. Silver, *Microwave Antenna Theory and Design*, Boston: Boston Technical Publishers, Inc., 1964.

[13] S. Choudhary and L. B. Felsen, "Analysis of Gaussian beam propagation and diffraction by inhomogeneous wave tracking," *Proc. IEEE*, vol. 62, pp. 1530–1541, 1974.

[14] W. D. Burnside, C. L. Yu, and R. J. Marhefka, "A technique to combine the geometrical theory of diffraction and the moment method," *IEEE Trans. Antennas Propag.*, vol. AP-23, pp. 551–558, July 1975.

[15] G. A. Thiele and T. H. Newhouse, "A hybrid technique for combining moment methods with the geometrical theory of diffraction," *IEEE Trans. Antennas Propag.*, vol. AP-23, no. 1, January 1975.

[16] L. B. Felsen and A. H. Kamel, "Hybrid ray-mode formulation of parallel plane waveguide green's functions," *IEEE Trans. Antennas Propag.*, vol. AP-29, no. 4, pp. 637–649, July 1981.

[17] R. K. Luneberg, *Mathematical Theory of Optics*, Providence: Brown Univ., 1964.

[18] M. Kline, "An asymptotic solution of Maxwell's equation," *Commun. Pure Appl. Math.*, vol. 4, pp. 225–263, 1951.

[19] R. G. Kouyoumjian and P. H. Pathak, "The dyadic diffraction coefficient for a curved edge," *Rep. 3001-3*, ElectroScience Laboratory, Dept. Electr. Eng., Ohio State Univ., Columbus. Prepared under Grant NGR 36-008-144 for NASA, Langley Research Center, Hampton, Va. (see appendix III), August 1973.

[20] G. A. Deschamps, "Ray techniques in electromagnetics," *Proc. IEEE*, vol. 60, pp. 1022–1035, September 1972.

[21] S. W. Lee, "Electromagnetic reflection from a conducting surface: geometrical optics solution," *IEEE Trans. Antennas Propag.*, vol. AP-23, pp. 184–191, 1975.

[22] S. W. Lee, M. S. Sheshadari, V. Jamnejad, and R. Mittra, "Refraction at a curved dielectric interface: geometrical optics solution," *IEEE Trans. Microwave Theory Tech.*, MTT-30, no. 1, pp. 12–19, January 1982.

[23] S. W. Lee, P. Cramer, Jr., K. Woo, and Y. Rahmat-Samii, "Diffraction by an arbitrary subreflector: GTD solution," *IEEE Trans. Antennas Propag.*, vol. AP-27, pp. 305–316, 1979.

[24] R. G. Kouyoumjian, "Asymptotic high-frequency methods," *Proc. IEEE*, vol. 53, pp. 864–876, August 1965.

[25] P. H. Pathak and R. G. Kouyoumjian, "The dyadic diffraction coefficient for a perfectly conducting wedge," *Rep. 2183-4*, ElectroScience Lab., Dept. Elec. Eng., Ohio State Univ., Columbus. Prepared under Contract AF 19(628)-5929 for AF Cambridge Res. Labs. (AFCRL-69-0546), also *ASTIA Doc. AD 707 827*, June 5, 1970.

[26] R. G. Kouyoumjian and P. H. Pathak, "A uniform geometrical theory of diffraction for an edge in a perfectly conducting surface," *Proc. IEEE*, vol. 62, pp. 1448–1461, November 1974.

[27] F. A. Sikta, W. D. Burnside, T. T. Chu, and L. Peters, Jr., "First-order equivalent current and corner-diffraction scattering from flat-plate structures," *IEEE Trans. Antennas Propag.*, vol. 31, no. 4, pp. 584–589, July 1983.

[28] P. H. Pathak, "An asymptotic result for the scattering of a plane wave by a smooth convex cylinder," *Radio Sci.*, vol. 14, no. 3, pp. 419–435, May–June 1979.

[29] P. H. Pathak, W. D. Burnside, and R. J. Marhefka, "A uniform GTD analysis of the diffraction of electromagnetic waves by a smooth convex surface," *IEEE Trans. Antennas Propag.*, vol. AP-28, no. 5, pp. 631–642, September 1980.

[30] M. Abramowitz and I. A. Stegun, eds., *Handbook of Mathematical Functions* (Applied Mathematics Series 55). Washington, D.C.: National Bureau of Standards, p. 478, 1964.

[31] P. H. Pathak, N. Wang, W. D. Burnside, and R. G. Kouyoumjian, "Uniform GTD solution for the radiation from sources on a smooth convex surface," *IEEE Trans. Antennas Propag.*, vol. AP-29, no. 4, pp. 609–621, July 1981.

[32] W. D. Burnside, "Analysis of on-aircraft antenna patterns," PhD dissertation, Ohio State Univ., Columbus, 1972.

[33] N. Wang and W. D. Burnside, "An efficient-geodesic path solution for prolate spheroids," *Quarterly Report 711305-2*, The Ohio State Univ. ElectroScience Lab., Dept. Elec. Eng., July 1979.

[34] S. W. Lee, "Mutual admittance of slots on a cone: solution by ray technique," *IEEE Trans. Antennas Propag.*, vol. AP-26, no. 6, pp. 768–773, November 1978.

[35] J. R. Wait and W. E. Mientka, "Calculated patterns of slotted elliptic-cylinder antennas," *Appl. Sci. Res.*, Sec. B, vol. 7, pp. 449–462, 1959.

[36] P. C. Bargeliotes, A. T. Villeneuve, and W. H. Kummer, "Pattern synthesis of conformal arrays," Radar Microwave Lab., Aerospace Groups, Hughes Aircraft Co., Culver City, Calif., 1975.

[37] P. H. Pathak and R. G. Kouyoumjian, "The radiation from apertures in curved surfaces," *Rep. 3001-2*, ElectroScience Lab., Dept. Elec. Eng., Ohio State Univ. Prepared under Grant NGR 36-008-144 for NASA Langley Research Center, Hampton, Va., December 1972.

[38] P. H. Pathak, "Uniform GTD solutions for a class of problems associated with the diffraction by smooth convex surfaces," in volume 1 of the Ohio State University short course notes on *The Modern Geometrical Theory of Diffraction*, June 1983.

[39] P. H. Pathak and N. N. Wang, "Ray analysis of mutual coupling between antennas on a convex surface," *IEEE Trans. Antennas Propag.*, vol. AP-29, no. 6, pp. 911–922, November 1981.

[40] S. W. Lee and S. Safavi-Naini, "Approximate asymptotic solution of surface field due to a magnetic dipole on a cylinder," *IEEE Trans. Antennas Propag.*, vol. AP-26, no. 4, pp. 593–598, July 1978.

[41] R. F. Harrington, *Time Harmonic Electromagnetic Fields*, New York: McGraw-Hill Book Co., 1961.

[42] J. H. Richmond, "A reaction theorem and its application to antenna impedance calculations," *IRE Trans. Antennas Propag.*, vol. AP-9, no. 6, pp. 515–520, November 1961.

[43] K. E. Golden, G. E. Stewart, and D. C. Pridmore-Brown, "Approximation techniques for the mutual admittances of slot antennas on metallic cones," *IEEE Trans. Antennas Propag.*, vol. AP-22, pp. 43–48, 1974.

PART B

Antenna Theory

Chapter 5

Radiation from Apertures

E. V. Jull
University of British Columbia

CONTENTS

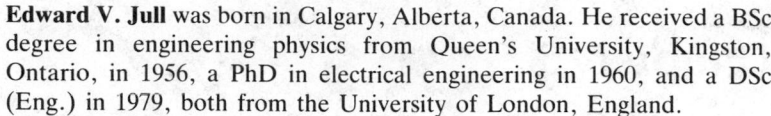

Edward V. Jull was born in Calgary, Alberta, Canada. He received a BSc degree in engineering physics from Queen's University, Kingston, Ontario, in 1956, a PhD in electrical engineering in 1960, and a DSc (Eng.) in 1979, both from the University of London, England.

In 1956–57 and 1961–72 he was a research officer in the microwave section and later the antenna engineering section of the Division of Electrical Engineering of the National Research Council of Canada Laboratories in Ottawa. During 1963–65 he was a guest worker in the Electromagnetics Institute of the Technical University of Denmark and the Microwave Institute of the Royal Institute of Technology, Stockholm, Sweden. In 1972 he joined the University of British Columbia, Vancouver, Canada, where he is now a professor in the Department of Electrical Engineering.

In 1964 Dr. Jull was a joint winner of the IEEE Antennas and Propagation Society Best Paper Award. He has been chairman of Canadian Commission VI for the International Union of Radio Science (URSI), chairman of the Canadian National Committee for URSI, an associate editor of *Radio Science*, and an international director of the Electromagnetics Society. He is currently a vice president of URSI and is the author of *Aperture Antennas and Diffraction Theory* (Peter Peregrinus, 1981).

1. Alternative Formulations for Radiation Fields

An aperture antenna is an opening in a surface designed to radiate. Examples are radiating slots, horns, and reflectors. It is usually more convenient to calculate aperture radiation patterns from the electromagnetic fields of the aperture rather than from the currents on the antenna. There are now basically two methods for doing this. Traditionally the pattern has been derived from the tangential electric and magnetic fields in the aperture. This aperture field method is an electromagnetic formulation of the Huygens-Kirchhoff method of optical diffraction. In application it is convenient and accurate for the forward pattern of large apertures. More recently the pattern has also been derived from fields associated with rays which pass through the aperture and rays diffracted by the aperture edges. Its origins can be traced to the early ideas on optical diffraction of Young as more recently formulated by Keller [1] in his geometrical theory of diffraction. It is particularly useful in deriving the radiation pattern in the lateral and rear directions and is described in Chapter 4.

This chapter deals only with the derivation of radiation patterns from the tangential fields in the aperture. Two methods of formulating the radiation integrals are given in this section. They lead to the same result but differ in their concepts of radiation from apertures.

Plane-Wave Spectra

This approach has the advantages of conceptual simplicity for the radiative fields and completeness in its inclusion of the reactive fields of the aperture [2–5]. It uses the fact that any radiating field can be represented by a superposition of plane waves in different directions. The amplitude of the plane waves in the various directions of propagation, or the spectrum function, is determined from the tangential fields in the aperture. This spectrum function is the far-field radiation pattern of the aperture for radiation in real direction angles. The reactive aperture fields are represented by the complex directions of propagation in the total spectrum function.

The method is most appropriate for planar apertures. If the aperture lies in the $z = 0$ plane of Fig. 1 and radiates into $z > 0$, components E_x and E_y of the electric field in $z \geqq 0$ can be written in terms of their corresponding spectrum functions P_x and P_y as

$$
\begin{Bmatrix} E_x(x,y,z) \\ E_y(x,y,z) \end{Bmatrix} = \frac{1}{(2\pi)^2} \int_{-\infty}^{\infty} \int_{\infty}^{\infty} \begin{Bmatrix} P_x(k_x,k_y) \\ P_y(k_x,k_y) \end{Bmatrix} e^{-j(k_x x + k_y y + k_z z)} \, dk_x \, dk_y \qquad (1)
$$

If α, β are the directions of propagation of each plane wave in Fig. 1, then $k_x = k \sin\alpha \cos\beta$, $k_y = k \sin\alpha \sin\beta$, and $k_z = k \cos\alpha$ are the components of the propagation vector $\mathbf{k}$ for each plane wave. It is necessary to specify

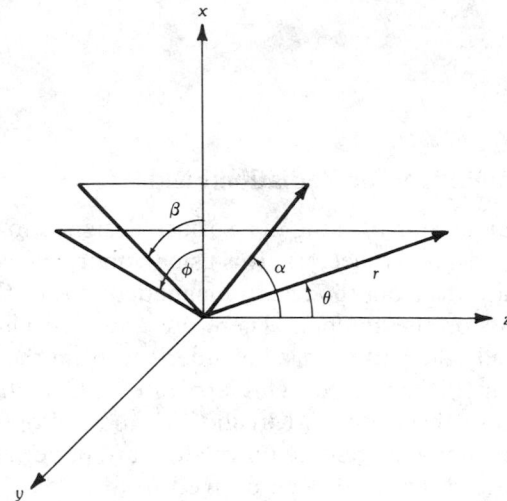

Fig. 1. Radiation from an aperture in the $z = 0$ plane: a plane wave radiates in a direction defined by the angles α and β, and the coordinates of a far-field point are (r, θ, ϕ).

$$
k_z = \begin{cases}
\sqrt{k^2 - (k_x{}^2 + k_y{}^2)} & \text{when } k_x{}^2 + k_y{}^2 \leqq k^2 \\
-j\sqrt{(k_x{}^2 + k_y{}^2) - k^2} & \text{when } k_x{}^2 + k_y{}^2 > k^2
\end{cases}
\tag{2}
$$

so for real angles α, β (with $k_x{}^2 + k_y{}^2 \leqq k^2$) there is radiation, and for complex angles α, β (with $k_x{}^2 + k_y{}^2 > k^2$), the fields decay exponentially outwards from the aperture.

Putting $z = 0$ in (1) and inverting the transforms shows the relation between the spectrum functions and the aperture fields:

$$
\begin{Bmatrix} P_x(k_x, k_y) \\ P_y(k_x, k_y) \end{Bmatrix} = \int_{-\infty}^{\infty} \int_{-\infty}^{\infty} \begin{Bmatrix} E_x(x, y, 0) \\ E_y(x, y, 0) \end{Bmatrix} e^{j(k_x x + k_y y)} \, dx \, dy
\tag{3}
$$

Each plane wave has a z component associated with its x and y components in (1) and their relative magnitudes follow from $\mathbf{k} \cdot \mathbf{E} = 0$ for each plane wave. Thus the total field is

$$
\mathbf{E}(x, y, z) = \frac{1}{(2\pi)^2} \int_{-\infty}^{\infty} \int_{-\infty}^{\infty} [(\hat{\mathbf{x}} k_z - \hat{\mathbf{z}} k_x) P_x(k_x, k_y)
$$
$$
+ (\hat{\mathbf{y}} k_z - \hat{\mathbf{z}} k_y) P_y(k_x, k_y)] e^{-j(k_x x + k_y y + k_z z)} k_z{}^{-1} \, dk_x \, dk_y
\tag{4}
$$

To evaluate (4) at large distances from the aperture it is convenient to first convert to spherical coordinates, then apply stationary phase integration to the two double integrals. The stationary point is $\alpha = \theta$, $\beta = \phi$, and the final result for $kr \gg 1$ is

$$\mathbf{E}(r,\theta,\phi) \simeq \frac{je^{-jkr}}{\lambda r} [\hat{\boldsymbol{\theta}}(P_x\cos\phi + P_y\sin\phi) - \hat{\boldsymbol{\phi}}\cos\theta(P_x\sin\phi - P_y\cos\phi)] \quad (5)$$

where

$$\begin{Bmatrix} P_x \\ P_y \end{Bmatrix} = \int_{-\infty}^{\infty}\int_{-\infty}^{\infty} \begin{Bmatrix} F_x(x,y,0) \\ E_y(x,y,0) \end{Bmatrix} e^{jk(x\sin\theta\cos\phi + y\sin\theta\sin\phi)}\, dx\, dy \quad (6)$$

The far magnetic field follows from

$$\mathbf{H}(r,\theta,\phi) = Y_0\hat{\mathbf{r}} \times \mathbf{E}(r,\theta,\phi) \quad (7)$$

where $Y_0 = \sqrt{\epsilon_0/\mu_0}$ is the free-space wave admittance.

Equation 5 provides the complete far-field radiation pattern from the Fourier transforms (6) of the tangential electric field in the aperture. It is also possible to use instead the tangential magnetic field in the aperture. Then, in terms of the spectrum functions

$$\begin{Bmatrix} Q_x \\ Q_y \end{Bmatrix} = \int_{-\infty}^{\infty}\int_{-\infty}^{\infty} \begin{Bmatrix} H_x(x,y,0) \\ H_y(x,y,0) \end{Bmatrix} e^{jk(x\sin\theta\cos\phi + y\sin\theta\sin\phi)}\, dx\, dy \quad (8)$$

the total electric field for $kr \gg 1$ is

$$\mathbf{E}(r,\theta,\phi) \simeq \frac{-jZ_0e^{-jkr}}{\lambda r} [\hat{\boldsymbol{\theta}}(Q_x\sin\phi - Q_y\cos\phi)\cos\theta + \hat{\boldsymbol{\phi}}(Q_x\cos\phi + Q_y\sin\phi)] \quad (9)$$

where $\simeq$ means asymptotically equal and $Z_0 = \sqrt{\mu_0/\epsilon_0}$ is the free-space wave impedance.

An equivalent result in terms of both electric and magnetic tangential components of the aperture field is half the sum of (5) and (9), i.e.,

$$\mathbf{E}(r,\theta,\phi) \simeq \frac{je^{-jkr}}{2\lambda r} \{\hat{\boldsymbol{\theta}}[P_x\cos\phi + P_y\sin\phi) - Z_0\cos\theta(Q_x\sin\phi - Q_y\cos\phi)]$$

$$- \hat{\boldsymbol{\phi}}[(P_x\sin\phi - P_y\cos\phi)\cos\theta + Z_0(Q_x\cos\phi + Q_y\sin\phi)]\} \quad (10)$$

in which P_x, P_y and Q_x, Q_y are defined by (6) and (8), respectively. This superposition of electric and magnetic current sources provides a field which satisfies Huygens' principle in that radiation into $z < 0$ is suppressed.

The three expressions (5), (9), and (10) all yield the exact far-field pattern from the exact aperture field integrated over the entire aperture plane. Usually electric fields are more convenient to measure and calculate than magnetic fields, so (9) is rarely used. The choice between (5) and (10) should depend on how well the true boundary conditions are satisfied by whichever approximations are used. For example, it is convenient to assume that the field vanishes in the aperture plane outside the aperture. Then (5) should be used for apertures mounted in a large conducting plane as the boundary conditions on the conductor are rigorously satisfied. For apertures not in a conducting plane, however, (5) with this assumption generally yields less accuracy near the aperture plane than (10).

For apertures which are large in wavelengths the pattern is well predicted away from the aperture plane by all of these expressions. Then the aperture electric and magnetic fields are related by essentially free-space conditions, i.e., $E_x(x, y, 0) = Z_0 H_y(x, y, 0)$, $E_y(x, y, 0) = -Z_0 H_x(x, y, 0)$ and $Q_x = -Z_0^{-1} P_y$, $Q_y = Z_0^{-1} P_x$. Equations 9 and 10 become, respectively,

$$\mathbf{E}(r, \theta, \phi) \simeq \frac{je^{-jkr}}{\lambda r} [\hat{\theta}(P_x \cos \phi + P_y \sin \phi) \cos \theta - \hat{\phi}(P_x \sin \phi - P_y \cos \phi)] \quad (11)$$

and

$$\mathbf{E}(r, \theta, \phi) \simeq \frac{je^{-jkr}}{2\lambda r} (1 + \cos \theta) [\hat{\theta}(P_x \cos \phi + P_y \sin \phi) - \hat{\phi}(P_x \sin \phi - P_y \cos \phi)]$$

$$(12)$$

For small angles of θ, $\cos \theta \cong 1$ and the three expressions (5), (11), and (12) yield essentially identical results in that region of the pattern where their accuracy is highest. All have less precision at angles far off the beam axis, where (12) yields values which are the average of (5) and (11). As (12) is in terms of an aperture electric field which vanishes in the rear ($\theta = \pi$) direction, it is most commonly used for larger apertures, such as horns or reflectors, which are not in a conducting screen.

Equivalent Currents

The fields of sources within a surface S of Fig. 2 can be calculated from the tangential electric and magnetic fields $\mathbf{E}_s$ and $\mathbf{H}_s$ of the sources on S. Alternatively, these surface fields may be replaced by equivalent currents [3, 6, 7]. An electric surface current density $\mathbf{J}_s = \hat{\mathbf{n}} \times \mathbf{H}_s$, where $\hat{\mathbf{n}}$ is a unit vector normally outward from S, represents the tangential magnetic fields. Their contribution to the magnetic vector potential

$$\mathbf{A} = \frac{1}{4\pi} \int_S \mathbf{J}_s \frac{e^{-jkR}}{R} \, dS \quad (13)$$

gives a magnetic field $\mathbf{H} = \nabla \times \mathbf{A}$ and an electric field

$$\mathbf{E} = \frac{1}{j\omega\epsilon_0} \nabla \times \nabla \times \mathbf{A} \quad (14)$$

Similarly, a magnetic surface current density $\mathbf{K}_s = \mathbf{E}_s \times \hat{\mathbf{n}}$ represents the tangential electric fields on S and the resulting electric vector potential

$$\mathbf{F} = \frac{1}{4\pi} \int_S \mathbf{K}_s \frac{e^{-jkR}}{R} \, dS \quad (15)$$

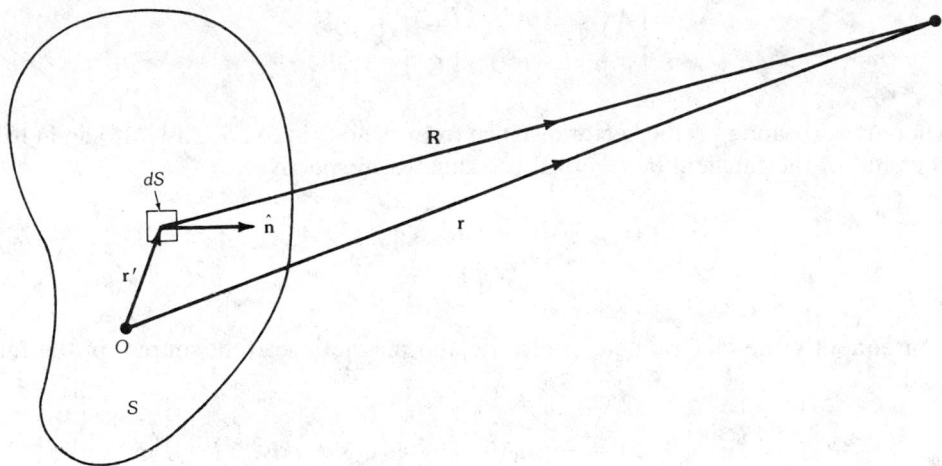

Fig. 2. Coordinates for calculation of radiation from a surface S.

provides an electric field

$$\mathbf{E} = -\nabla \times \mathbf{F} \tag{16}$$

and a magnetic field $\mathbf{H} = \nabla \times \nabla \times \mathbf{F}/(j\omega\mu_0)$.

Contributions from both electric and magnetic current sources give the total electric and magnetic fields outside S:

$$\mathbf{E} = -\nabla \times \mathbf{F} + \frac{1}{j\omega\epsilon_0}\nabla \times \nabla \times \mathbf{A} \tag{17}$$

$$\mathbf{H} = \nabla \times \mathbf{A} + \frac{1}{j\omega\mu_0}\nabla \times \nabla \times \mathbf{F} \tag{18}$$

With this superposition of the fields of electric and magnetic surface current densities radiation into the region enclosed by S is suppressed and Huygens' principle satisfied. There will also be no fields inside S if that region is a perfect conductor. Then $\mathbf{K}_s = 0$ on S and the total electric field outside S can be calculated from (14) with $2\mathbf{K}_s$ in (15).

As discussed previously, under "Plane-Wave Spectra," the fields can be calculated from the tangential electric or magnetic fields on S or from a superposition of the two sets of fields and all three methods yield the same result if the boundary conditions are rigorously satisfied. Usually $\mathbf{E}_s$ and $\mathbf{H}_s$ are approximated by incident fields, and scattered fields on S are neglected. The choice between the three methods should depend on the convenience of accurately approximating the true boundary conditions.

At distances r much larger than the maximum dimension of S in Fig. 2, $R \cong r - \hat{\mathbf{r}} \cdot \mathbf{r}'$ and (13) and (15) become

$$\begin{Bmatrix} \mathbf{A} \\ \mathbf{F} \end{Bmatrix} = \frac{e^{-jkr}}{4\pi r} \int_S \begin{Bmatrix} \mathbf{J}_s \\ \mathbf{K}_s \end{Bmatrix} e^{jk\hat{\mathbf{r}}\cdot\mathbf{r}'} \, dS \tag{19}$$

where $\hat{\mathbf{r}} = \mathbf{r}/r$ and $\mathbf{r}'$ is the vector distance from the origin to dS. Also (14) and (16) become, in the far field in spherical coordinates, respectively,

$$\mathbf{E}(r,\theta,\phi) = -j\omega\mu_0(\hat{\boldsymbol{\theta}}A_\theta + \hat{\boldsymbol{\phi}}A_\phi) \tag{20}$$

$$= -jk\mathbf{F} \times \hat{\mathbf{r}} \tag{21}$$

Consequently the electric field of electric and magnetic current sources in the far field is

$$\mathbf{E}(r,\theta,\phi) = -j\omega\mu_0(\hat{\boldsymbol{\theta}}A_\theta + \hat{\boldsymbol{\phi}}A_\phi) - jk\mathbf{F} \times \hat{\mathbf{r}} \tag{22}$$

and the corresponding magnetic fields are given by (7). Again, if electric or magnetic currents alone are used to calculate the total fields from (20) or (21), a factor of 2 must be included in the right side of (19).

With the surface S in the plane $z = 0$ of Fig. 2 and radiation into $z > 0$, $\hat{\mathbf{n}} = \hat{\mathbf{z}}$ in (19), which becomes

$$\mathbf{A} = \frac{e^{-jkr}}{4\pi r}(-\hat{\mathbf{x}}Q_y + \hat{\mathbf{y}}Q_x) \tag{23a}$$

$$\mathbf{F} = \frac{-e^{-jkr}}{4\pi r}(-\hat{\mathbf{x}}P_y + \hat{\mathbf{y}}P_x) \tag{23b}$$

in which P_x, P_y and Q_x, Q_y are defined by (6) and (8). Using (23a) in (20) with a factor of 2 gives (9). Equation 23b in (21) with a factor of 2 gives (5), and (23a) and (23b) in (22) yields (10). Thus the equivalent-current and plane-wave spectrum formulations provide identical results for the radiation fields of an aperture. The equivalent current method is simpler mathematically, but does not account for the reactive fields of the aperture.

2. Radiation Patterns of Planar Aperture Distributions

The expressions (5), (8), and (10) for the radiating far fields of an aperture in terms of the Fourier transforms of the tangential electric and magnetic aperture fields (6) and (8) are exact, but approximations are required in their application.

Approximations

In obtaining the approximations the usual assumptions are the following:

(a) The integration limits in (6) and (8) are the antenna aperture dimensions, i.e., fields in the aperture plane outside the aperture are assumed negligible.

(b) The aperture field is assumed to be the incident field from the antenna feed, i.e., scattered fields in the aperture are assumed negligible.

(c) Aperture electric and magnetic fields are assumed related by free-space conditions, i.e., the aperture is assumed large in wavelengths. This is so in (11) and (12) but not in (5), (9), and (10).

(d) Aperture fields are assumed separable in the coordinates of the aperture. Fortunately this assumption applies in many antenna designs; otherwise numerical integration of double integrals is usually required.

Clearly, the accuracy of the final result will depend on the degree to which all of the above assumptions are satisfied.

Rectangular Apertures

The Uniform Aperture Distribution—If the rectangular aperture of Fig. 3 is large and not in a conducting plane, its far field is conveniently calculated from (12) (assumption c above). Each component of aperture electric field can be dealt with separately. From (12) with $P_y = 0$, the x component of aperture field produces the far field

$$\mathbf{E}(r, \theta, \phi) = \mathbf{A} \int_{-a/2}^{a/2} \int_{-b/2}^{b/2} E_x(x, y, 0) e^{j(k_1 x + k_2 y)} \, dx \, dy \tag{24}$$

where

$$\mathbf{A} = \mathbf{A}(r, \theta, \phi) = j \frac{e^{-jkr}}{2\lambda r} (1 + \cos\theta)(\hat{\boldsymbol{\theta}} \cos\phi - \hat{\boldsymbol{\phi}} \sin\phi) \tag{25}$$

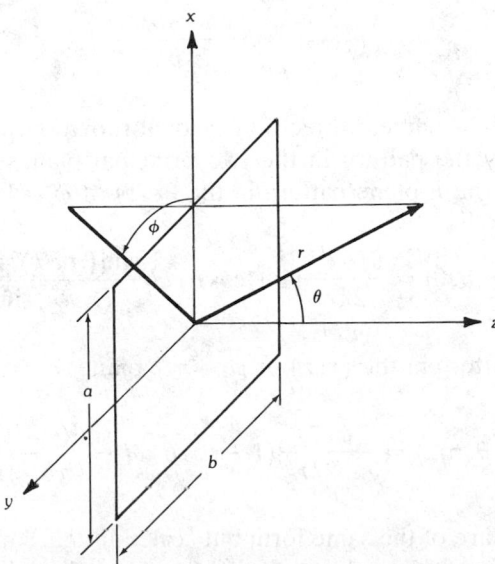

Fig. 3. Coordinates of a rectangular aperture in the $z = 0$ plane.

and

$$k_1 = k \sin\theta\cos\phi$$
$$k_2 = k \sin\theta\sin\phi \tag{26}$$

Assumption a has been used in (24). For separable aperture fields (assumption d) $E_x(x, y, 0) = E_0 E_1(x) E_2(y)$, where $E_1(x)$ and $E_2(y)$ are the field distributions normalized to the field E_0. Then (24) becomes

$$\mathbf{E}(r, \theta, \phi) = \mathbf{A}E_0 F_1(k_1) F_2(k_2) \tag{27}$$

where

$$F_1(k_1) = \int_{-a/2}^{a/2} E_1(x) e^{jk_1 x}\, dx \tag{28}$$

$$F_2(k_2) = \int_{-b/2}^{b/2} E_2(y) e^{jk_2 y}\, dy \tag{29}$$

It is now necessary to choose an aperture field (assumption b). For the ideal case of a field uniform in amplitude and phase across the aperture, $E_1(x) = E_2(y) = 1$ and

$$F_1(k_1) = a\frac{\sin(k_1 a/2)}{k_1 a/2} \tag{30}$$

$$F_2(k_2) = b\frac{\sin(k_2 b/2)}{k_2 b/2} \tag{31}$$

This specifies the complete three-dimensional radiation pattern. For practical reasons it is usually the pattern in the two principal planes which are of major interest. These are the E-plane pattern in the (x, z) or $\phi = 0$ plane,

$$\mathbf{E}(r, \theta, 0) = \frac{\hat{\boldsymbol{\theta}} j e^{-jkr}}{2\lambda r}(1 + \cos\theta)\, ab\, \frac{\sin[(\pi a/\lambda)\sin\theta]}{(\pi a/\lambda)\sin\theta} \tag{32}$$

and the H-plane pattern in the (y, z) or $\phi = \pi/2$ plane

$$\mathbf{E}(r, \theta, \pi/2) = -\frac{\hat{\boldsymbol{\phi}} e^{-jkr}}{2\lambda r}(1 + \cos\theta)\, ab\, \frac{\sin[(\pi b/\lambda)\sin\theta]}{(\pi b/\lambda)\sin\theta} \tag{33}$$

These patterns are of the same form but scaled in θ according to the aperture dimensions in their respective planes. If the aperture is large ($a, b \gg \lambda$), the main beam and first side lobes are contained in a small angle θ, so $1 + \cos\theta \cong 2$ and the function $(\sin u)/u$, where $u = (\pi a/\lambda) \sin \theta$, determines the pattern. This function is plotted in Fig. 4.

The first null in the pattern occurs at $u = \pi$ or $\theta = \sin^{-1}(\lambda/a)$. Hence the full width of the main beam is $2 \sin^{-1}(\lambda/a) \cong 2\lambda/a$ radians for $a \gg \lambda$. At $u = 1.39$ the field is 0.707 its peak value. Thus the half-power beamwidth in radians is

$$\Delta\theta_{HP} = 2\sin^{-1}\left(\frac{1.39\lambda}{\pi a}\right) \cong 0.88\lambda/a \tag{34}$$

for $a \gg \lambda$. The first side lobes are at $u = \pm 1.43\pi$ and are $20 \log_{10}(0.217) = -13.3$ dB below the peak value of the main beam.

Simple Distributions—Radiation patterns are usually characterized by their principal plane half-power beamwidths and first side lobe levels. These parameters are given for several simple symmetrical aperture distributions in Table 1. The pattern functions there are derived from the Fourier cosine transforms of the aperture distributions, i.e., if $E_1(-x) = E_1(x)$, then (28) becomes

$$F_1(k_1) = 2\int_0^{a/2} E_1(x)\cos k_1 x\, dx \tag{35}$$

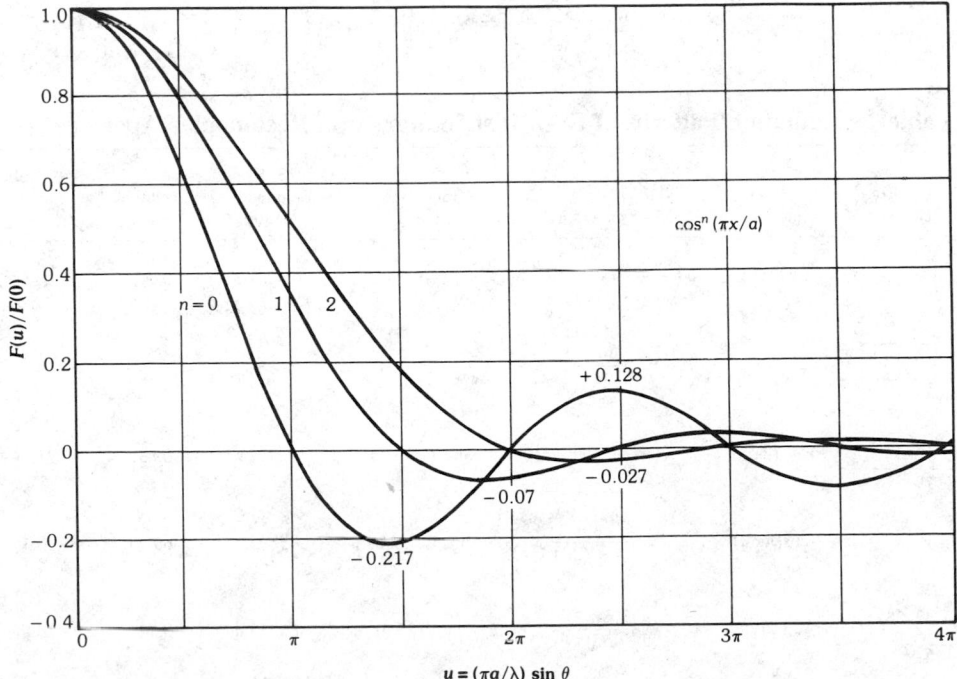

Fig. 4. Pattern functions of in-phase symmetrical field distributions in a rectangular aperture: uniform ($n = 0$), cosinusoidal ($n = 1$), and cosine-squared ($n = 2$).

The results in Table 1 are arranged in order of increasing beamwidths and decreasing first side lobe levels. For in-phase distributions the uniform aperture field has the highest gain but it also has a high side lobe level. The more the distribution decreases toward the aperture edges, the broader is the main beam and the lower the side lobe levels.

If the aperture distribution is an odd function, i.e., if $E_1(-x) = -E_1(x)$ in (28), its pattern is a Fourier sine transform,

$$F(k_1) = 2j \int_0^{a/2} E_1(x) \sin k_1 x \, dx \tag{36}$$

and is itself an odd function $[F(-k_1) = -F(k_1)]$. Several examples are shown in Table 2. These patterns have nulls at $\theta = 0$ and consequently their beamwidths and side lobe levels are unspecified. Instead, the angle at which the first and main lobe of the pattern appears is given, and the examples are arranged in order of increasing values of this angle. Such patterns may arise from the cross-polarized fields of a paraboloidal reflector, or they may be used for tracking on the pattern null, but otherwise they are rarely encountered.

The two-dimensional patterns of Tables 1 and 2 combine in (27) to give three-dimensional patterns of rectangular apertures. For example, an open-ended rectangular waveguide with the TE_{01} mode has an electric-field distribution in the aperture of Fig. 3

$$E_x(x, y, 0) = E_0 \cos(\pi y/b) \tag{37}$$

Table 1. Radiation Patterns of Even Distributions in a Rectangular Aperture

Aperture Distribution	$E(x)$ ($|x| < a/2$)	$F(u)$ $u = (\pi a/\lambda) \sin \theta$	First Null (rad)	3-dB Beamwidth (rad)	Relative Gain	Side Lobe Level (dB)
	$\dfrac{2}{a}|x|$	$a\left(\dfrac{\sin u}{u} - \dfrac{1 - \cos u}{u^2}\right)$	$\sin^{-1}(0.75\lambda/a)$	$0.796\lambda/a$	0.742	-4.6
	1	$a\dfrac{\sin u}{u}$	$\sin^{-1}(\lambda/a)$	$0.886\lambda/a$	1.0	-13.3
	$1 - (2x/a)^2$	$\dfrac{2a}{u^2}\left(\dfrac{\sin u}{u} - \cos u\right)$	$\sin^{-1}(1.43\lambda/a)$	$1.179\lambda/a$	0.833	-21.3
	$\cos(\pi x/a)$	$2\pi a\dfrac{\cos u}{\pi^2 - (2u)^2}$	$\sin^{-1}(1.5\lambda/a)$	$1.189\lambda/a$	0.810	-23.1
	$1 - 2|x|/a$	$\dfrac{a}{2}\dfrac{\sin^2(u/2)}{(u/2)^2}$	$\sin^{-1}(2\lambda/a)$	$1.273\lambda/a$	0.742	-26.5
	$\cos^2(\pi x/a)$	$\dfrac{a}{2}\dfrac{\sin u}{u}\dfrac{\pi^2}{\pi^2 - u^2}$	$\sin^{-1}(2\lambda/a)$	$1.441\lambda/a$	0.667	-31.5

Table 2. Radiation Patterns of Odd Distributions in a Rectangular Aperture

Aperture Distribution	$E(x)$ $(\|x\| < a/2)$	$F(u)$ $u = (\pi a/\lambda) \sin \theta$	First Peak (rad)
RECTILINEAR	$\dfrac{2x}{a}$	$-j\dfrac{a}{u^2}(u \cos u - \sin u)$	$\sin^{-1}(0.6\lambda/a)$
ANTIPHASE CONSTANT	± 1 $(x \neq 0)$	$-ja\dfrac{1 - \cos u}{u}$	$\sin^{-1}(0.7\lambda/a)$
SINE	$\sin(2\pi x/a)$	$-ja\dfrac{\pi \sin u}{\pi^2 - u^2}$	$\sin^{-1}(0.81\lambda/a)$
ANTIPHASE SINE-SQUARED	$\pm \sin^2(2\pi x/a)$ $(x \neq 0)$	$-j\dfrac{a}{2}\dfrac{1 - \cos u}{u}\dfrac{4\pi^2}{4\pi^2 - u^2}$	$\sin^{-1}(0.9\lambda/a)$
ANTIPHASE TRIANGULAR	$\pm 1 - 2x/a$ $(x \neq 0)$	$-ja\dfrac{u - \sin u}{u^2}$	$\sin^{-1}(\lambda/a)$

and a radiation pattern, from (27) and Table 1,

$$\mathbf{E}(r, \theta, \phi) = j\frac{e^{-jkr}}{2\lambda r} (1 + \cos \theta)(\hat{\boldsymbol{\theta}} \cos \phi - \hat{\boldsymbol{\phi}} \sin \phi) 2\pi ab\, E_0 \frac{\sin[(\pi a/\lambda) \sin \theta \cos \phi]}{(\pi a/\lambda) \sin \theta \cos \phi}$$

$$\times \frac{\cos[(\pi b/\lambda) \sin \theta \sin \phi]}{\pi^2 - [(2\pi b/\lambda) \sin \theta \sin \phi]^2}, \quad \text{for } -\frac{\pi}{2} < \theta < \frac{\pi}{2} \tag{38}$$

This is not a very accurate representation of the pattern of an open-ended rectangular waveguide which supports only the dominant mode since the dimensions a, b are less than $\lambda/2$ and the assumptions above, under "Approximations," are invalid. The inaccuracy is largest near the aperture plane.

For the same rectangular waveguide aperture set in a conducting plane in $z = 0$ of Fig. 3, the far field follows from (5) with $P_y = 0$, i.e.,

$$\mathbf{E}(r, \theta, \phi) = j\frac{e^{-jkr}}{\lambda r} (\hat{\boldsymbol{\theta}} \cos \phi - \hat{\boldsymbol{\phi}} \cos \theta \sin \phi) 2\pi ab\, E_0 \frac{\sin[(\pi a/\lambda) \sin \theta \cos \phi]}{(\pi a/\lambda) \sin \theta \cos \phi}$$

$$\times \frac{\cos[(\pi b/\lambda) \sin \theta \sin \phi]}{\pi^2 - [(2\pi b/\lambda) \sin \theta \sin \phi]^2} \tag{39}$$

This result satisfies the aperture plane boundary conditions and so is accurate even for narrow rectangular slots in a conducting plane.

Compound Distributions—The patterns of aperture distributions which are linear combinations of the simple distributions of Tables 1 and 2 are the same linear combinations of their patterns. That is, if

$$\int_{-\infty}^{\infty} E_n(x) e^{jk_1 x} dx = F_n(k_1), \qquad n = 1, 2, \ldots m \tag{40}$$

are the patterns of m simple aperture distributions, the pattern of a *compound distribution* is

$$\int_{-\infty}^{\infty} \sum_{n=1}^{m} a_n E_n(x) e^{jk_1 x} dx = \sum_{n=1}^{m} a_n F_n(k_1) \tag{41}$$

For example, the compound distribution

$$E(x) = \begin{cases} C + (1 - C) \cos^2(\pi x/a) & \text{for } |x| < a/2 \\ 0 & \text{for } |x| > a/2 \end{cases} \tag{42}$$

has the radiation pattern

$$F(u) = a \frac{\sin u}{u} \left[C + \left(\frac{1 - C}{2} \right) \frac{\pi^2}{\pi^2 - u^2} \right] \tag{43}$$

in which $u = k_1 a/2$. The half-power beamwidth of (43) lies between those of uniform and cosine-squared distributions, according to the value of C, as indicated by the solid curve in Fig. 5. Also shown are the half-power beamwidths of compound uniform and cosinusoidal and uniform and parabolic distributions. The side lobe levels of these compound distributions are also between those of the component parts.

Another practical example is that of aperture blockage by a reflector feed or subreflector of width $\delta \ll a$. The blocked aperture distribution is then

$$E'(x) = \begin{cases} E(x), & \delta/2 < |x| < a/2 \\ 0 & , \quad |x| < \delta/2, \quad |x| > a/2 \end{cases} \tag{44}$$

and the resulting pattern is

$$F'(k_1) = F(k_1) - \delta \frac{\sin(k_1 \delta/2)}{k_1 \delta/2} \tag{45}$$

Here $F(k_1)$ is the radiation pattern of the unblocked distribution $E(x)$. If $k_1 \delta \ll 1$, then $F'(k_1) \cong F(k_1) - \delta$. The blocked pattern is uniformly reduced by δ. The effect is a narrower main beam and higher side lobe levels than the unblocked pattern.

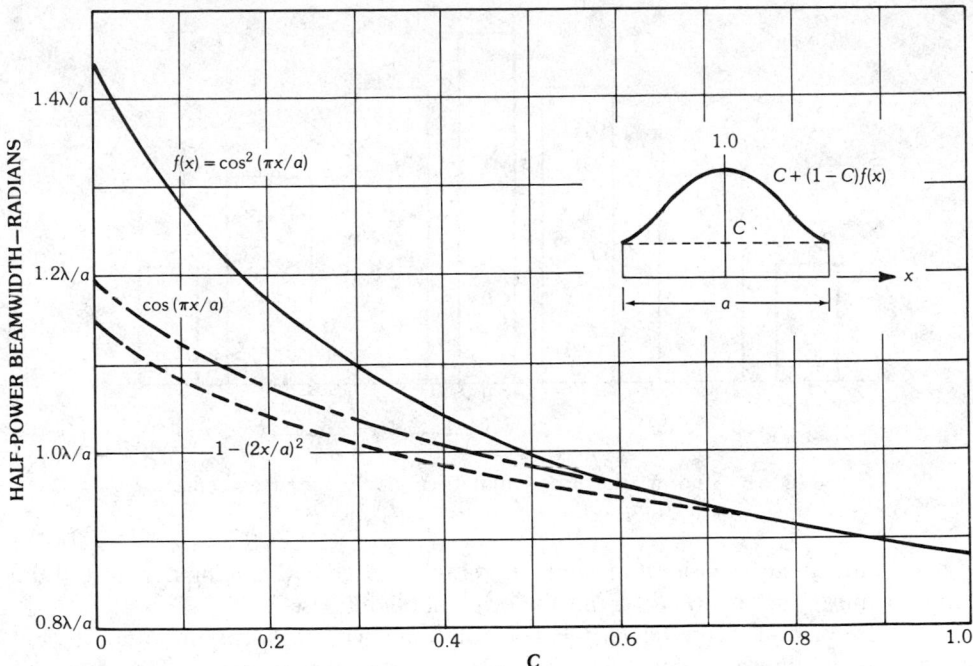

Fig. 5. Half-power beamwidths of compound symmetrical distributions in a rectangular aperture.

Displaced and Phase-Shifted Distributions—If the aperture distributions $E_n(x)$ of (41) are laterally displaced by x_n in the aperture plane, the radiation pattern is

$$\int_{-\infty}^{\infty} \sum_{n=1}^{m} a_n E_n(x - x_n) e^{jk_1 x} \, dx = \sum_{n=1}^{m} a_n F_n(k_1) e^{jk_1 x_n} \qquad (46)$$

For example, the radiation pattern of an array of $2m + 1$ uniform, in-phase distributions of symmetrical amplitudes $a_n = a_{-n}$ and widths w_n, symmetrically placed about the array center as in Fig. 6, is

$$F(k_1) = a_0 w_0 \frac{\sin(k_1 w_0/2)}{k_1 w_0/2} + 2 \sum_{n=1}^{m} a_n w_n \frac{\sin(k_1 w_n/2)}{k_1 w_n/2} \cos(k_1 x_n) \qquad (47)$$

Thus the patterns of arbitrary distributions can be obtained by summing contributions from segments of the aperture of essentially uniform amplitude and phase.

If the aperture distributions of (41) have a progressive linear phase shift exponent $(-j2\pi n x/a)$ across the aperture of width a, the pattern is

$$\int_{-\infty}^{\infty} \sum_{n=1}^{m} a_n E_n(x) e^{j[k_1 - (2\pi n/a)]x} \, dx = \sum_{n=1}^{m} a_n F_n[k_1 - (2\pi n/a)] \qquad (48)$$

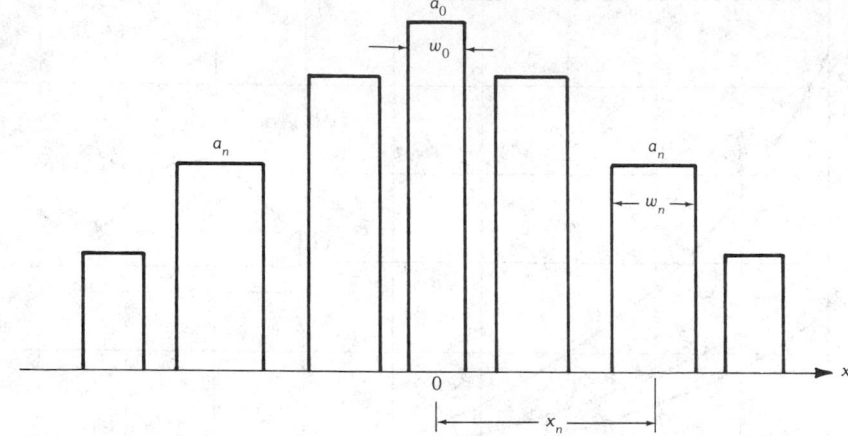

Fig. 6. Uniform apertures symmetrical about the array center.

This is a linear superposition of radiation patterns, each with an angular shift of the main beam $\theta_n = \sin^{-1}(n\lambda/a)$ from the original beam axis.

As an example, consider $2m + 1$ distributions with uniform amplitude b_n but with a linear phase variation of $2\pi n$ radians across the aperture. The radiation pattern is

$$F(k_1) = \int_{-a/2}^{a/2} \sum_{n=-m}^{m} b_n e^{j[k_1 - (2\pi n/a)]x}\, dx = a \sum_{n=-m}^{m} b_n \frac{\sin[(k_1 a/2) - n\pi]}{(k_1 a/2) - n\pi} \quad (49)$$

Equation 49 can be used to synthesize an antenna pattern $F_1(k_1)$ by equating b_n with $F_1(2\pi n/a)$. The accuracy of this approximation will be limited by the aperture width a since the largest value of m for real angles $\theta = \sin^{-1}(n\lambda/a)$ is a/λ.

Optimum Pattern Distributions—Generally antenna radiation patterns with narrow beams and low side lobe levels are sought. This involves a compromise, for in reducing side lobe levels by symmetrically tapering the aperture distribution the main beam broadens. An *optimum distribution* is one which has the lowest side lobe levels for a given width of main beam, or the narrowest main beam for a given side lobe level. Optimizing in this way results in a pattern with a maximum number of side lobes, all of the same level and width. The Chebyshev polynomials conveniently provide this as pattern functions for arrays of a finite number of elements (see Chapter 11) and the corresponding current distributions are rather similar to those of a cosine-squared function on a pedestal.

For a continuous field distribution in an aperture of width a, an ideal pattern function, which may be derived from an approximation to the Chebyshev polynomials of large order, is

$$F(u) = \begin{cases} \cosh\sqrt{(\pi A)^2 - u^2} & \text{for } u < \pi A \\ \cos\sqrt{u^2 - (\pi A)^2} & \text{for } u > \pi A \end{cases} \quad (50)$$

with $u = k_1 a/2 = (\pi a/\lambda) \sin \theta$.

As the side lobe levels are all unity the main beam to side lobe ratio is

$$R = F(0) = \cosh \pi A \tag{51}$$

Pattern nulls occur for

$$
\begin{aligned}
u_n &= \pm \pi \sqrt{A^2 + (n - 1/2)^2}, \quad n = 1, 2, \dots \\
&\cong \pm (n - 1/2)\pi, \quad u \gg \pi A
\end{aligned}
\tag{52}
$$

Hence the main beamwidth is determined by the side lobe levels and the far side lobes are equispaced. This ideal pattern is impractical, however, because the aperture distribution required to produce it is infinite at the aperture edges.

For a realizable pattern the far side lobe levels must decrease. This occurs in the normalized pattern [9]

$$F(u, A, \bar{n}) = \frac{\sin u}{u} \prod_{n=1}^{\bar{n}-1} \frac{1 - (u/u_n)^2}{1 - (u/n\pi)^2} \tag{53}$$

in which the nulls are

$$
u_n = \begin{cases} \pm \pi \sigma \sqrt{A^2 + (n - 1/2)^2}, & 1 \leqq n < \bar{n} \\ \pm n\pi & , \quad \bar{n} \leqq n < \infty \end{cases}
\tag{54}
$$

with

$$\sigma = \frac{\bar{n}}{\sqrt{A^2 + (\bar{n} - 1/2)^2}} \tag{55}$$

This pattern has a first side lobe level of $1/R$ and gradually decreasing near side lobes with null positions determined by the product in (53). Beyond the nth null the pattern is approximately that of the uniform distribution $(\sin u)/u$ depressed by the factor $(\pi/u_1)^2$, with side lobes of equal width and decreasing height. The scaling parameter σ ensures that the nulls of the two patterns coincide for $n = \bar{n}$.

This optimum pattern (53) may be synthesized by the method indicated by (49). The normalized aperture distribution is expressed as a Fourier series,

$$E(x) = 1 + 2 \sum_{p=1}^{\bar{n}} f(p, A, \bar{n}) \cos (2\pi px/a) \tag{56}$$

in which the coefficients are samples of the pattern (53) for $u = m\pi$ and $n < \bar{n}$ or

$$f(p, A, \bar{n}) = \frac{[(\bar{n} - 1)!]^2}{(\bar{n} - 1 + p)!(\bar{n} - 1 - p)!} \prod_{m=1}^{\bar{n}-1} [1 - (n\pi/u_m)^2] \tag{57}$$

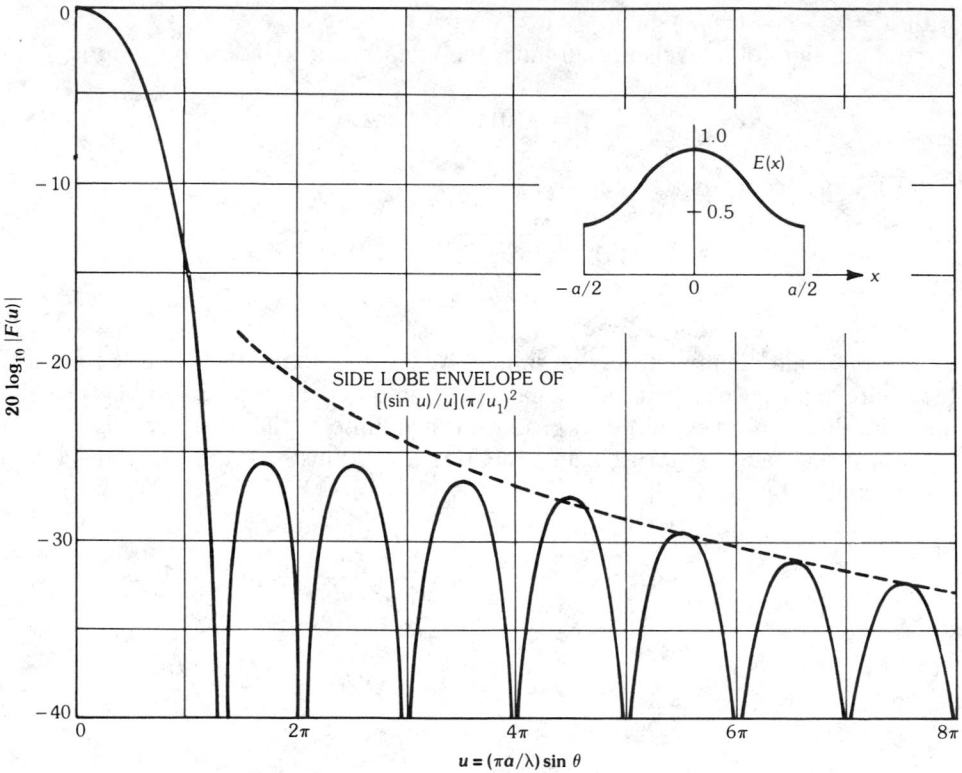

Fig. 7. Optimum aperture distribution and radiation pattern for first side lobe levels down 25 dB and $\overline{n} = 5$.

Tables of these coefficients are available [10]. The half-power beamwidth of the pattern is approximately ([10], p. 56)

$$\Delta\theta_{\mathrm{HP}} \cong 2\sin^{-1}\left\{\frac{\lambda\sigma}{\pi a}[(\cosh^{-1}R)^2 - (\cosh^{-1}R/\sqrt{2})^2]^{1/2}\right\} \tag{58}$$

with σ given by (55). The half-power beamwidth of the ideal pattern (50) is slightly smaller and given precisely by (58) with $\sigma = 1$.

In the numerical example of Fig. 7 the first side lobe of the pattern is 25 dB below the main beam ($R = 17.78$) and the pattern nulls are those of $(\sin u)/u$ for $u \gtrsim 5\pi$ ($\overline{n} = 5$). Hence from (51), (54), and (55), $A = 1.1365$, $\sigma = 1.0773$, and the first four nulls are at $u/\pi = 1.34, 2.03, 2.96$, and 3.96.

Circular Apertures

For a circular aperture of diameter a as in Fig. 8 the far field of the x component of aperture field $E_x(\varrho', \phi')$ is, from (24) with $x = \varrho' \cos\phi'$ and $y = \varrho' \sin\phi'$,

$$\mathbf{E}(r, \theta, \phi) = \mathbf{A} \int_0^{2\pi}\int_0^{a/2} E_x(\varrho', \phi')\, e^{jk\varrho'\sin\theta\cos(\phi - \phi')}\varrho'\, d\varrho'\, d\phi' \tag{59}$$

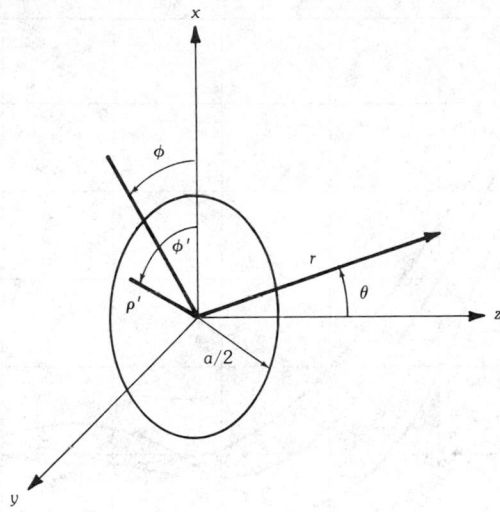

Fig. 8. Coordinates for radiation from a circular aperture.

where

$$\mathbf{A} = \frac{je^{-jkr}}{2\lambda r}(1 + \cos\theta)(\hat{\boldsymbol{\theta}}\cos\phi - \hat{\boldsymbol{\phi}}\sin\phi) \tag{60}$$

for an aperture not in a conducting plane and

$$\mathbf{A} = \frac{je^{-jkr}}{2\lambda r}(\hat{\boldsymbol{\theta}}\cos\phi - \hat{\boldsymbol{\phi}}\cos\theta\sin\phi) \tag{61}$$

for a conducting plane in $z = 0$, $\varrho' > a/2$.

If the aperture distribution is independent of ϕ' then $E_x(\varrho',\phi') = E_x(\varrho')$. Integration of (59) in ϕ' gives the Bessel function J_0 and

$$\mathbf{E}(r,\theta,\phi) = \mathbf{A}2\pi\int_0^{a/2} E_x(\varrho')J_0(k\varrho'\sin\theta)\varrho'\,d\varrho' \tag{62}$$

The family of aperture distributions

$$E_x(\varrho') = E_0[1 - (2\varrho'/a)^2]^n, \qquad n = 0, 1, 2, \ldots \tag{63}$$

in (62) yield the radiation patterns

$$\mathbf{E}(r,\theta,\phi) = \mathbf{A}E_0\frac{\pi a^2}{2}2^n n! \frac{J_{n+1}[(\pi a/\lambda)\sin\theta]}{[(\pi a/\lambda)\sin\theta]^{n+1}} \tag{64}$$

The normalized patterns are illustrated in Fig. 9 for the uniform distribution ($n = 0$), the parabolic distribution ($n = 1$), and the parabolic-squared distribution

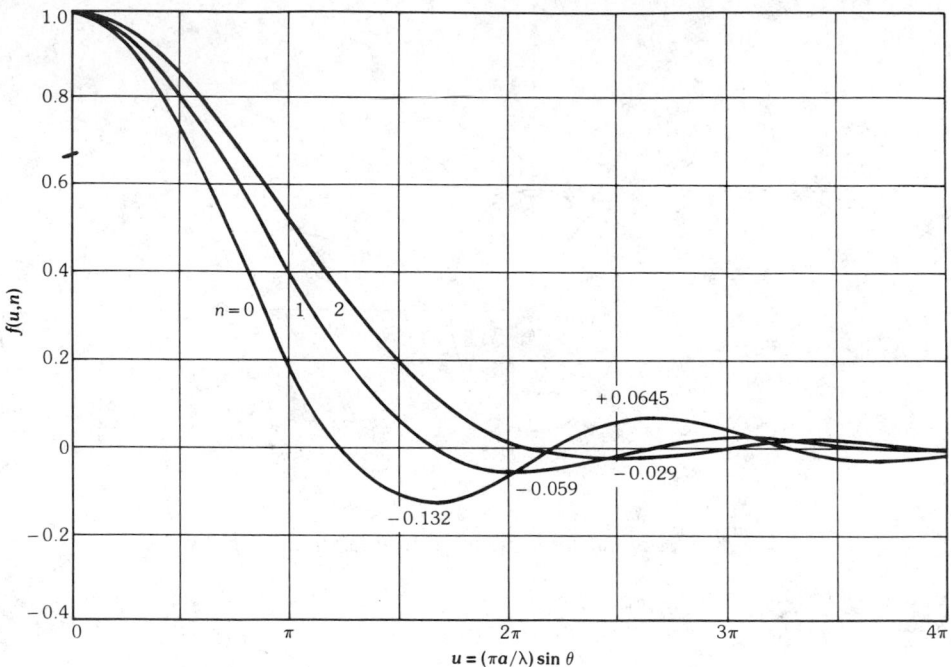

Fig. 9. Pattern functions of in-phase symmetrical field distributions in a circular aperture: uniform ($n = 0$), parabolic ($n = 1$), and parabolic-squared ($n = 2$).

($n = 2$). Values of the angular position off the beam axis of the first pattern null, the half-power beamwidth, and first side lobe levels are given in Table 3. Also given are corresponding values for the H-plane pattern for the dominant (TE_{11}) mode in the aperture of an open-ended circular waveguide. The E-plane pattern characteristics in this situation are those of a uniform aperture distribution in a circular aperture.

Compound distributions in a circular aperture are analyzed as above, under "Compound Distributions." For the distributions

$$E(\varrho') = C + (1 - C)[1 - (2\varrho'/a)^2]^n, \qquad n = 0, 1, 2 \tag{65}$$

in a circular aperture, the normalized patterns are

$$\frac{Cf(u, 0) - (1 - C)/(n + 1) f(u, n)}{C + (1 - C)/(n + 1)} \tag{66}$$

where

$$f(u, n) = 2^{n+1}(n + 1)! \frac{J_{n+1}(u)}{u^{n+1}} \tag{67}$$

are the normalized patterns of (64) in which $u = (\pi a/\lambda) \sin \theta$. The half-power beamwidths of patterns of a parabolic taper on a pedestal distribution for $n = 1$ and

Table 3. Radiation Patterns of Simple Distributions in a Circular Aperture

Aperture Distribution	$E(r)$ $\varrho' < a/2$	$F(u)$ $u = (\pi a/\lambda)\sin\theta$	First Null (rad)	3-dB Beamwidth (rad)	Relative Gain	First Side Lobe Level (dB)
UNIFORM	1	$\dfrac{\pi a^2}{2}\dfrac{J_1(u)}{u}$	$\sin^{-1}(1.22\lambda/a)$	$1.016\lambda/a$	1.0	-17.6
PARABOLA	$1 - (2\varrho'/a)^2$	$\pi a^2 \dfrac{J_2(u)}{u^2}$	$\sin^{-1}(1.63\lambda/a)$	$1.267\lambda/a$	0.75	-24.6
PARABOLA-SQUARED	$[1 - (2\varrho'/a)^2]^2$	$4\pi a^2 \dfrac{J_3(u)}{u^3}$	$\sin^{-1}(2.03\lambda/a)$	$1.47\lambda/a$	0.55	-30.6
H-PLANE PATTERN OF TE$_{11}$ MODE	$\dfrac{\partial}{\partial\varrho'}[J_1(3.682\varrho'/a)]$	$\dfrac{\pi a^2}{2}\dfrac{J_0(u) - J_1(u)/u}{1 - (u/1.841)^2}$	$\sin^{-1}(1.71\lambda/a)$	$1.29\lambda/a$		-26.2

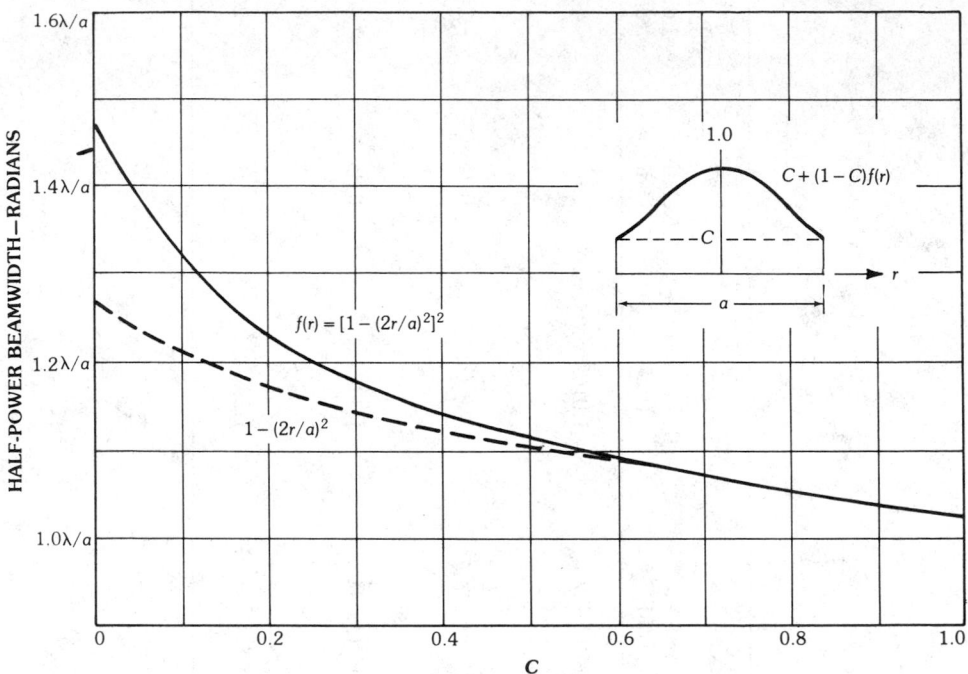

Fig. 10. Half-power beamwidths of compound symmetrical distributions in a circular aperture.

$n = 2$ in the above expressions are plotted in Fig. 10 for various values of C.

Data for the design of optimum pattern distributions in circular apertures is also available [11, 12].

Near-Field Patterns

The pattern expressions preceding are far field, i.e., the distance is very large compared with the aperture dimensions. There are linear phase differences in the radiation reaching the field point from different parts of the aperture. The first-order *near-field* effect is an additional quadratic phase difference in the radiation from the aperture. Thus for an x-polarized separable field in the aperture of Fig. 3,

$$\mathbf{E}(r, \theta, \phi) = \mathbf{A}E_0 D_1(k_1)D_2(k_2) \tag{68}$$

where

$$D_1(k_1) = \int_{-a/2}^{a/2} E_1(x)\, e^{-j(\beta x^2 - k_1 x)}\, dx \tag{69}$$

$$D_2(k_2) = \int_{-b/2}^{b/2} E_2(y)\, e^{-j(\beta y^2 - k_2 u)}\, dy \tag{70}$$

with $\beta = \pi/r\lambda$ and k_1 and k_2 defined by (26). As $r \to \infty$, $\beta \to 0$ and the Fresnel transforms (69) and (70) become the Fourier transforms (28) and (29).

For a uniform distribution $E_1(x) = 1$, so

$$D_1(k_1) = \sqrt{\frac{r\lambda}{2}}\, e^{jr\lambda k_1^2/4\pi} \{C(u_2) - C(u_1) - j[S(u_2) - S(u_1)]\} \tag{71}$$

where the *Fresnel integrals** $C(u)$ and $S(u)$ are defined by

$$C(u) - jS(u) = \int_0^u e^{-j\pi v^2/2}\, dv \tag{72}$$

and have arguments

$$\begin{Bmatrix} u_2 \\ u_1 \end{Bmatrix} = \pm\frac{a}{\sqrt{2r\lambda}} - \frac{k_1}{2\pi}\sqrt{2r\lambda} \tag{73}$$

For the cosinusoidal distribution $E_2(y) = \cos(\pi y/b)$, so

$$D_2(k_2) = \frac{1}{2}\sqrt{\frac{r\lambda}{2}}\, (e^{j(r\lambda/4\pi)(k_2+\pi/b)^2}\{C(v_2) - C(v_1) - j[S(v_2) - S(v_1)]\}$$

$$+ e^{j(r\lambda/4\pi)(k_2-\pi/b)^2}\{C(w_2) - C(w_1) - j[S(w_2) - S(w_1)]\}) \tag{74}$$

where

$$\begin{Bmatrix} v_2 \\ v_1 \end{Bmatrix} = \pm\frac{b}{\sqrt{2r\lambda}} - \frac{k_2}{2\pi}\sqrt{2r\lambda} - \frac{1}{b}\sqrt{\frac{r\lambda}{2}}$$

$$\begin{Bmatrix} w_2 \\ w_1 \end{Bmatrix} = \pm\frac{b}{\sqrt{2r\lambda}} - \frac{k_2}{2\pi}\sqrt{2r\lambda} + \frac{1}{b}\sqrt{\frac{r\lambda}{2}} \tag{75}$$

The above expressions together give the near-field pattern of an open-ended rectangular waveguide and reduce to the far-field pattern expressions (38) and (39) as $r \to \infty$.

Some near-field patterns in the $\phi = \pi/2$ plane ($k_1 = k\sin\theta$, $k_2 = 0$) for a cosinusoidal distribution $E_2(y) = \cos(\pi y/b)$ in a rectangular aperture are shown in Fig. 11. The near-field effects on the pattern are a broadening of the main beam, a filling in of the nulls, and a raising of the side lobe levels. Similar effects are observed in the near-field patterns of circular apertures [14].

---

*Computer subroutines and tabulated values are available. For example, see [13].

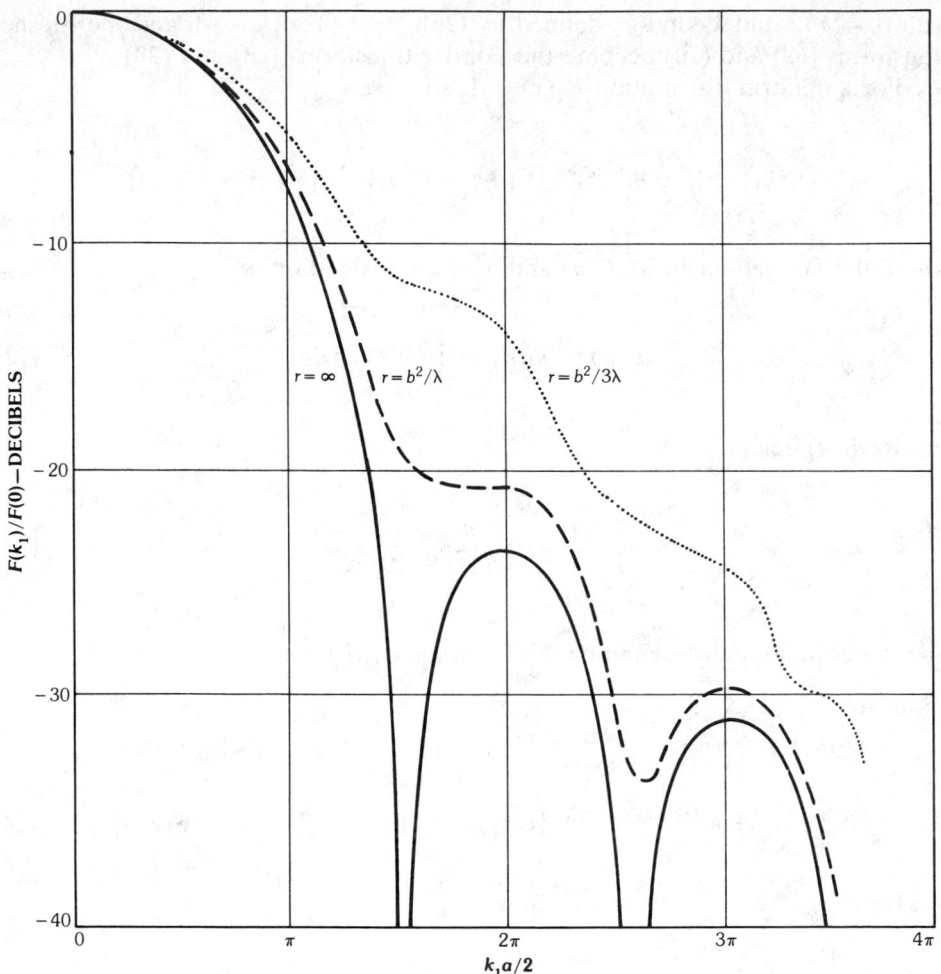

Fig. 11. Normalized radiation patterns of a cosinusoidal distribution. (*After Jull [5], © 1981 Peter Peregrinus Ltd.*)

3. Aperture Gain

The transmitting properties of an antenna are usually characterized by its gain. The *directive gain* $G(\theta, \phi)$ of an antenna in the (θ, ϕ) direction is the ratio of the radiated power density from it at (r, θ, ϕ) to its average radiated power density about the antenna at the range r. Thus

$$G(\theta, \phi) = \frac{\frac{1}{2}Y_0\{|E_\theta(r, \theta, \phi)|^2 + |E_\phi(r, \theta, \phi)|^2\}}{P_r/4\pi r^2} \tag{76}$$

Here P_r is the total power radiated and can be calculated by integrating the numerator of (76) over a sphere of radius r surrounding the antenna. For planar antennas it is usually simpler to integrate the time-averaged complex Poynting vector over the aperture. For a rectangular aperture in the $z = 0$ plane

$$P_r = \frac{1}{2} \text{Re} \left\{ \int_{\text{ap.}} (E_x H_y^* - E_y H_x^*) \, dx \, dy \right\}$$

$$= \frac{1}{2} Y_0 \int_{\text{ap.}} (|E_x|^2 + |E_y|^2) \, dx \, dy \tag{77}$$

where the integration is over the aperture with the aperture dimensions sufficiently large in wavelengths that electric and magnetic fields in it are related by essentially free-space conditions.

For in-phase symmetrical aperture distributions the maximum directive gain, or directivity, is in the $\theta = 0$ direction. Then, from (12), (76), and (77), this is

$$G = \frac{4\pi}{\lambda^2} \frac{\left| \int E_x \, dx \, dy \right|^2 + \left| \int E_y \, dx \, dy \right|^2}{\int (|E_x|^2 + |E_y|^2) \, dx \, dy} \tag{78}$$

For a uniform in-phase distribution the aperture fields $E_x = E_y$ are a constant and

$$G = 4\pi A/\lambda^2 \tag{79}$$

where A is the aperture area. It can readily be shown that for in-phase distributions this uniform distribution yields the highest gain, i.e., $G \leqq 4\pi A/\lambda^2$. "Supergain" antennas with nonuniform phase distributions can in principle have gains higher than (79) but are almost invariably impractical for apertures large in wavelengths.

4. Effective Area and Aperture Efficiency

The receiving properties of an antenna are characterized by its *effective area*. This is defined as the ratio of the power available from an antenna to the power density incident on it which is polarization matched to it. Simple reciprocity arguments show that the ratio of gain to effective area is a constant for all antennas. The value of this constant is, from (79), $4\pi/\lambda^2$. Hence the effective area A_{eff} of an antenna is related to the antenna gain G by

$$A_{\text{eff}} = \lambda^2 G/4\pi \quad \text{or} \quad G = 4\pi A_{\text{eff}}/\lambda^2 \tag{80}$$

From (78) and (80) for an in-phase distribution in a large rectangular aperture, the effective area can be calculated from

$$A_{eff} = \frac{\left|\int\int E_x \, dx \, dy\right|^2 + \left|\int\int E_y \, dx \, dy\right|^2}{\int (|E_x|^2 + |E_y|^2) \, dx \, dy} \tag{81}$$

For a uniform aperture distribution with $E_x = E_y$, a constant, (81) gives an effective area equal to the physical area of the aperture. This is the maximum effective area for in-phase distributions. If the aperture field is polarized in the x direction ($E_y = 0$) and separable in the aperture coordinates $E_x = E_1(x)E_2(y)$, (81) may be written as

$$A_{eff} = (A_{eff})_1 (A_{eff})_2 \tag{82}$$

where

$$(A_{eff})_1 = \left|\int E_x \, dx\right|^2 \Big/ \int |E_x|^2 \, dx$$

$$(A_{eff})_2 = \left|\int E_x \, dy\right|^2 \Big/ \int |E_x|^2 \, dy \tag{83}$$

The *aperture efficiency* is the ratio of the effective area to the physical area of the aperture. For separable distributions in a rectangular aperture, as above,

$$\varepsilon = \frac{A_{eff}}{ab} = \varepsilon_1 \varepsilon_2 = \frac{(A_{eff})_1}{a} \frac{(A_{eff})_2}{b} \tag{84}$$

The two-dimensional aperture efficiency A_{eff}/a for simple aperture distributions is given under "Relative Gain" in Table 1. For compound distributions in a rectangular aperture the two-dimensional aperture efficiencies, or gain relative to a uniform distribution, are as follows:

For a parabola on a pedestal of height C, as in Fig. 12, the one-dimensional aperture efficiency is

$$\varepsilon_1 = \frac{(2 + C)^2}{9[1 - (2/3)(1 - C) + (1/5)(1 - C)^2]} \tag{85}$$

for a cosine on a pedestal,

$$\varepsilon_1 = \frac{[C + (2/\pi)(1 - C)]^2}{C^2 + (1/2)(1 - C) + (4C/\pi)(1 - C)} \tag{86}$$

for a cosine-squared on a pedestal,

$$\varepsilon_1 = \frac{[(C + 1)/2]^2}{C^2\{1 + (1 - C)/C + (3/8)[(1 - C)/C]^2\}} \tag{87}$$

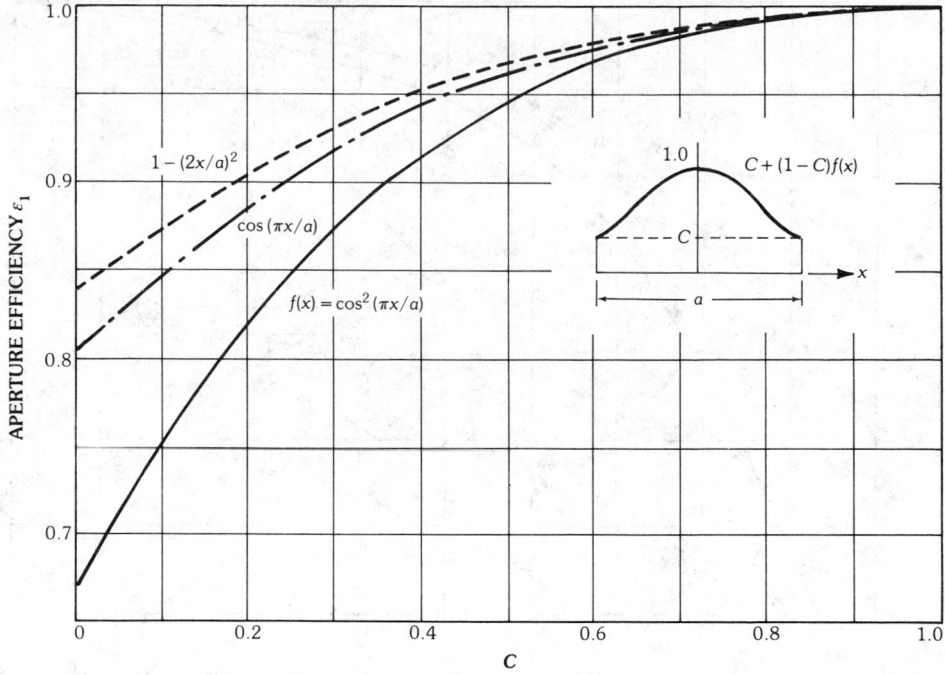

Fig. 12. One-dimensional aperture efficiencies of compound symmetrical distributions in a rectangular aperture.

These are plotted as dashed, broken, and solid curves, respectively, in Fig. 12. The gain of a rectangular aperture with separable distributions is then

$$G = \frac{4\pi}{\lambda^2} \varepsilon_1 \varepsilon_2 \, ab \tag{88}$$

Circular apertures with simple distributions have the aperture efficiency or relative gain given in Table 3. For the compound distributions of a parabola and a parabola-squared on a pedestal, the aperture efficiencies are plotted as the dashed and solid curves, respectively, in Fig. 13.

5. Near-Field Axial Gain and Power Density

On the beam axis ($\theta = 0$) the gain of an x-polarized aperture distribution is, from (76),

$$G = \frac{\frac{1}{2} Y_0 |E_\theta(r,0)|^2}{P_r/4\pi r^2} \tag{89}$$

For a uniform x-polarized distribution in the aperture of Fig. 3 the axial near-field follows from (68) and (71):

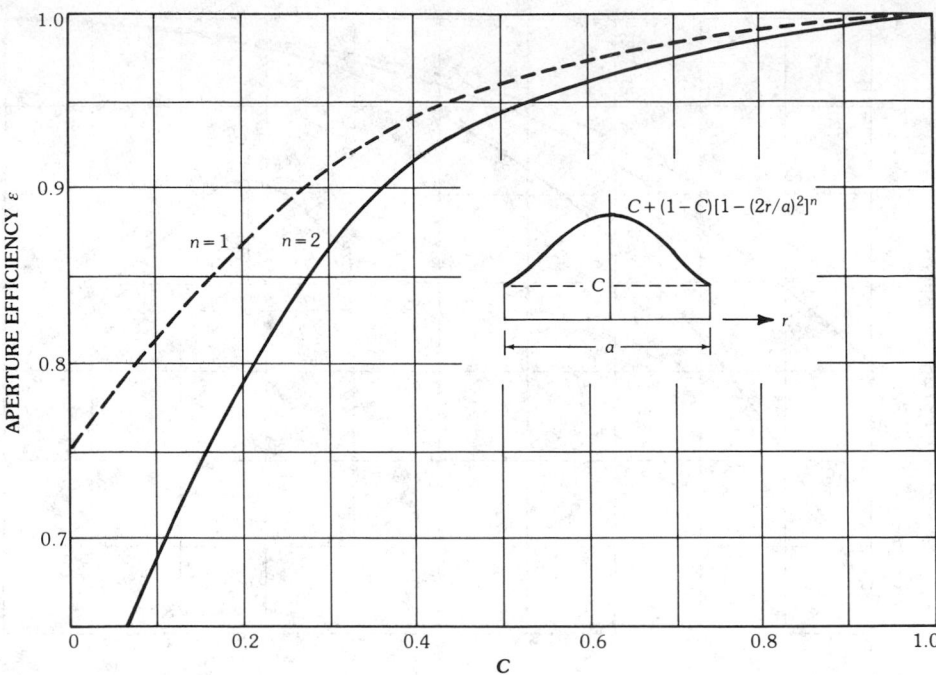

Fig. 13. Aperture efficiencies of compound symmetrical distributions in a circular aperture.

$$\mathbf{E}(r,0) = \hat{\boldsymbol{\theta}}\, 2j\, E_0 e^{-jkr}[C(u) - jS(u)][C(u') - jS(u')] \tag{90}$$

where $C(u)$ and $S(u)$ are the Fresnel integrals defined by (72), $u = a/\sqrt{2r\lambda}$, and $u' = b/\sqrt{2r\lambda}$. The power radiated is $P_r = \frac{1}{2} Y_0 E_0^2 ab$ and the axial near-field gain can be written as

$$G = \frac{4\pi ab}{\lambda^2} R_E(u)\, R_E(u') \tag{91}$$

where

$$R_E(u) = \frac{C^2(u) + S^2(u)}{u^2} \tag{92}$$

is the near-field gain reduction factor for a one-dimensional uniform distribution. This factor is plotted in Fig. 14.

For a uniform and cosinusoidal distribution linearly polarized in the aperture of Fig. 4, the axial near-field is, from (68) with (71) and (74),

$$\mathbf{E}(r,0) = \hat{\boldsymbol{\theta}}\, j E_0\, e^{-jkr+j(\pi r\lambda/4b^2)}[C(u) - jS(u)]\{C(v) - C(w) - j[S(v) - S(w)]\} \tag{93}$$

where, from (75),

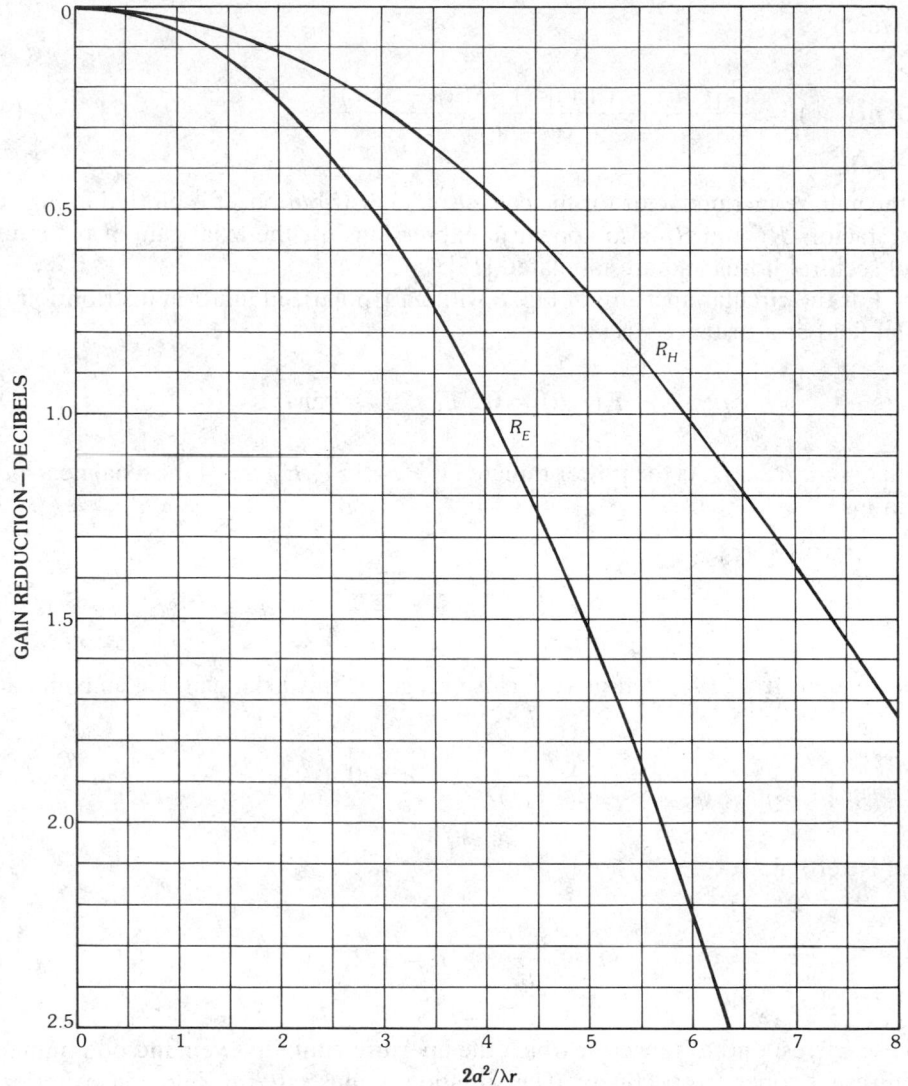

Fig. 14. Near-field axial gain reduction factors in decibels: R_E, (92), for a uniform distribution, and R_H, (96), for a cosinusoidal distribution in a rectangular aperture. (*After Jull [5], © 1981 Peter Peregrinus Ltd.*)

$$\left\{ \begin{matrix} v \\ w \end{matrix} \right\} = \pm \frac{b}{\sqrt{2r\lambda}} + \frac{1}{b} \sqrt{\frac{r\lambda}{2}} \tag{94}$$

The power radiated is now $P_r = \frac{1}{4} Y_0 E_0^2 ab$ and the axial gain can be written as

$$G = \frac{32ab}{\pi \lambda^2} R_E(u) R_H(v, w) \tag{95}$$

in which

$$R_H(v, w) = \frac{\pi^2}{4} \frac{[C(u) - C(w)]^2 + [S(v) - S(w)]^2}{(v - w)^2} \tag{96}$$

is the gain reduction factor for the cosinusoidal distribution. It is plotted in Fig. 14. The factors R_E and R_H also appear in expressions for the axial gain of pyramidal and sectoral horns and are tabulated [15].

For the circular aperture of Fig. 8 with an x-polarized uniform distribution the axial field at a distance r is

$$\mathbf{E}(r, \theta) = \hat{\boldsymbol{\theta}} 2jE_0 e^{-j(kr+t)} \sin t \tag{97}$$

with $t = \pi a^2/(8\lambda r)$. As the power radiated is $P_r = \frac{1}{2} Y_0 E_0^2 \pi a^2/4$ the axial near-field gain is

$$G = \frac{4\pi}{\lambda^2} \frac{\pi a^2}{4} \frac{\sin^2 t}{t^2} \tag{98}$$

It is evident from (98) that at very close ranges r, this axial gain has maxima at

$$r = \frac{a^2}{4(2n + 1)\lambda}, \qquad n = 0, 1, 2, \ldots \tag{99}$$

and is zero at

$$r = \frac{a^2}{8n\lambda}, \qquad n = 1, 2, \ldots \tag{100}$$

These correspond to ranges at which the aperture contains even and odd numbers of Fresnel zones, respectively. The situation is illustrated in Fig. 15a, where the axial near-field power density of a uniform circular aperture of diameter a,

$$2Y_0 E_0^2 \sin^2 t \tag{101}$$

is plotted.

Fig. 15b shows corresponding results for a uniform square aperture of side a. The axial power density

$$2Y_0 E_0^2 [C^2(u) + S^2(u)] \tag{102}$$

is not completely canceled at any range because the Fresnel zones are annular and incomplete in a square aperture.

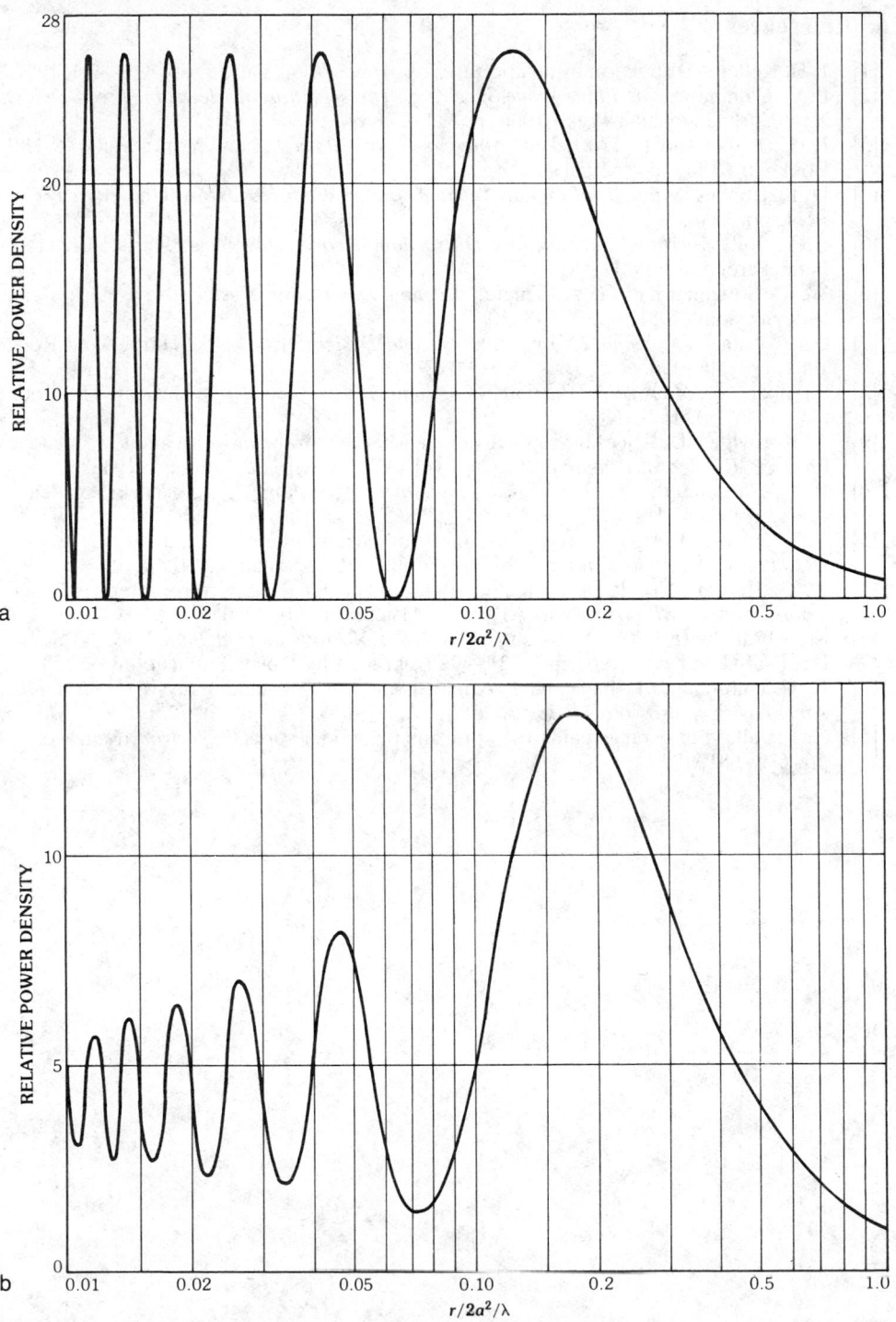

Fig. 15. Axial power density in the near field of uniform distributions. (*a*) Circular aperture of diameter *a*; (*b*) Square aperture of side *a*.(*After Jull [5], © 1981 Peter Peregrinus Ltd.*)

6. References

[1] J. B. Keller, "Diffraction by an aperture," *J. Appl. Phys.*, vol. 28, pp. 426–444, 1957.

[2] P. C. Clemmow, *The Plane Wave Spectrum Representation of Electromagnetic Fields*, New York: Pergamon Press, 1966, pp. 11–37.

[3] R. E. Collin and F. J. Zucker, *Antenna Theory, Part 1*, New York: McGraw-Hill Book Co., 1969, pp. 62–74.

[4] D. R. Rhodes, *Synthesis of Planar Antenna Sources*, Oxford: Oxford University Press, 1974, pp. 9–42.

[5] E. V. Jull, *Aperture Antennas and Diffraction Theory*, Stevenage, Herts., England: Peter Peregrinus, 1981, pp. 7–17.

[6] W. L. Stutzman and G. A. Thiele, *Antenna Theory and Design*, New York: John Wiley & Sons, 1981, pp. 375–384.

[7] C. A. Balanis, *Antenna Theory: Analysis and Design*, New York: Harper and Row, 1982, pp. 446–456.

[8] S. Silver, ed., *Microwave Antenna Theory and Design*, New York: McGraw-Hill Book Co., 1946, pp. 159–168.

[9] T. T. Taylor, "Design of line-source antennas for narrow beamwidth and low side lobe levels," *IRE Trans. Antennas Propag.*, vol. AP-3, pp. 16–28, January 1955.

[10] R. C. Hansen, ed., *Microwave Scanning Antennas, Volume 1*, New York: Academic Press, 1964, pp. 419, 427.

[11] T. T. Taylor, "Design of circular apertures for narrow beamwidth and low side lobes," *IRE Trans. Antennas Propag.*, vol. AP-8, pp. 17–22, January 1960.

[12] R. C. Hansen, "Table of Taylor distributions for circular aperture antennas," *IRE Trans. Antennas Propag.*, vol. AP-8, pp. 23–26, January 1960.

[13] M. Abramowitz and I. A. Stegun, *Handbook of Mathematical Functions*, NBS, US Dept. of Commerce, 1964, pp. 321–322 (reprinted by Dover Publications, 1965).

[14] R. C. Hansen and L. L. Baillin, "A new method of near-field analysis," *IRE Trans. Antennas Propag.*, vol. AP-7, pp. 458–467, 1959.

[15] E. V. Jull, "Finite range gain of sectoral and pyramidal horns," *Electron. Lett.*, vol. 6, pp. 680–681, 1970.

Chapter 6

Receiving Antennas

P. K. Park
Hughes Aircraft Company

C. T. Tai
University of Michigan

CONTENTS

 **Pyong Kiel Park** was born in Pyong-An-Do, Korea. He received his BSEE degree from In-Ha University, Korea, in 1962, the MSEE degree from Yon Sei University, Korea, in 1965, and the PhD in electrical engineering from UCLA in 1979. He is a senior staff engineer at Hughes Aircraft Company, Missile Systems Division, which he joined in 1979. Currently he is responsible for designing a monopulse concurrent low side-lobe antenna, a coherent side lobe canceller antenna of the AIM-54, and their monopulse feed networks. His prior experience includes lecturer at Yon Sei University, assistant professor at Kwang Woon University, member of the technical staff at TRW, and senior engineer at Ford Aeronatronics. His main interests are in electromagnetic theory, antennas, and microwave circuits. He is a Senior Member of IEEE.

 **Chen-To Tai** received his BS degree in physics from Tsing-Hua University in 1937 and his DSc degree in communication engineering from Harvard University in 1947. He has been a professor of electrical engineering at the University of Michigan since 1964. He has been a visiting professor at several universities in the United States, Europe, and the Far East, and held an honorary professorship from the Shanghai Normal University. A Life Fellow of IEEE and a past chairman of the Antennas and Propagation Society (1971), he received a Distinguished Faculty Award from the University of Michigan in 1975, and several awards from the Department of Electrical and Computer Engineering and the College of Engineering at that university.

1. Equivalent Circuit of a Receiving Antenna

By applying either Thevenin's theorem or Norton's theorem to a receiving antenna placed in a harmonically oscillating incident electromagnetic field, the load current can be calculated from the corresponding equivalent circuit (Fig. 1), i.e.,

$$I_L = \frac{V_{oc}}{Z_{in} + Z_L} \tag{1}$$

and

$$I_L = \frac{Z_{in}}{Z_{in} + Z_L} I_{sc} \tag{2}$$

where

V_{oc} = the open-circuit voltage measured at the receiving terminals

Z_{in} = the input impedance of the antenna when it is operating in its transmitting mode

Z_L = the load impedance

I_{sc} = the short-circuit current through the receiving terminals

The open-circuit voltage and the short-circuit current are related by

$$V_{oc} = Z_{in} I_{sc} \tag{3}$$

2. Vector Effective Height of an Antenna

When an antenna is operating in its transmitting mode the far-zone electric and magnetic fields can be written in the form

$$\mathbf{E} = \frac{-jkZ_0 \mathbf{N} e^{-jkr}}{4\pi r} \tag{4}$$

and

$$\mathbf{H} = \frac{\hat{\mathbf{r}} \times \mathbf{E}}{Z_0} \tag{5}$$

where

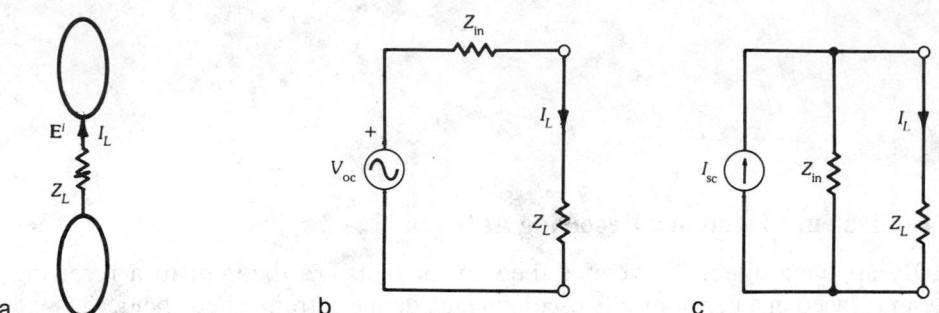

Fig. 1. Illustration for calculating load current with Thevenin's and Norton's theorems. (*a*) Receiving antenna in incident electromagnetic field. (*b*) Thevenin's equivalent circuit for load current. (*c*) Norton's equivalent circuit for load current.

$$k = 2\pi/\lambda = \omega(\mu_0 \epsilon_0)^{1/2}$$
$$Z_0 = (\mu_0/\epsilon_0)^{1/2}$$

$\hat{\mathbf{r}}$ = the unit vector along $\mathbf{r}$, the distance vector from antenna terminal to point of observation

r = the spherical radial distance from the antenna terminal to the point of observation of the field

The function $\mathbf{N}$ in (4) is related to the amplitude vector $\mathbf{A}$ defined by (39) in Chapter 1, namely,

$$\mathbf{N} = j\frac{4\pi}{k\sqrt{Z_0}}\mathbf{A} \tag{6}$$

It is understood that the medium under consideration is free space with constitutive constants μ_0 and ϵ_0. The physical dimension of the vector $\mathbf{N}$ is measured in ampere-meters while that of $\mathbf{A}$ is measured in (watts)$^{1/2}$.

One of the most useful parameters in antenna theory was introduced by G. Sinclair [1]. It is designated as the *vector effective height function* or simply the *vector effective height*. It is defined by

$$\mathbf{h} = \frac{\mathbf{N}}{I_{\text{in}}} \tag{7}$$

where I_{in} denotes the input current to the antenna when it is operating in its transmitting mode.* The dimension of $\mathbf{h}$ is measured in meters, hence it is identified as the "height" function. The function $\mathbf{h}$, like the function $\mathbf{N}$ or $\mathbf{A}$, is a function of the spherical angular variables (θ, ϕ) only, being independent of the radial variable r. A plot of $|\mathbf{h}|$ yields the field pattern and that of $|\mathbf{h}|^2$ yields the power pattern.

---

*Sinclair's original definition includes a negative sign, i.e., $\mathbf{h} = -\mathbf{N}/I_{\text{in}}$.

The vector effective height is particularly useful in relating the open-circuit voltage or the short-circuit current in the theory of receiving antennas. By means of the reciprocity theorem one finds

$$V_{oc} = Z_{in} I_{sc} = \mathbf{h} \cdot \mathbf{E}^i \tag{8}$$

where $\mathbf{E}^i$ denotes the incident electric field measured at the site of the antenna terminals. The expressions for the effective height of several simple antennas are listed in Table 1. It is assumed that the dipoles are pointed in the z direction and θ denotes the polar angle measured from the z axis. For the small loop, its axis is assumed to coincide with the z axis and $\hat{\mathbf{t}}$ denotes the unit tangent vector to the loop. Its shape, however, could be arbitrary as long as its largest linear dimension is electrically small.

For loop type antennas it is convenient to introduce a parameter designated as the *vector effective area*, which is defined by

$$\mathbf{S} = \frac{j}{k} (\hat{\mathbf{r}} \times \mathbf{h}) \tag{9}$$

where $\mathbf{r}$ is the unit vector pointed in the radial direction. In terms of $\mathbf{S}$ the open-circuit voltage V_{oc} given by (8) can be written in the form

$$
\begin{aligned}
V_{oc} = \mathbf{h} \cdot \mathbf{E}^i &= \mathbf{h} \cdot (-\hat{\mathbf{r}} \times \mathbf{H}^i) Z_0 \\
&= \frac{-\mathbf{h} \cdot (\hat{\mathbf{r}} \times \mathbf{B}^i) Z_0}{\mu_0} = -j\omega \left[\frac{j}{k} \mathbf{B}^i \cdot (\hat{\mathbf{r}} \times \mathbf{h}) \right] \\
&= -j\omega (\mathbf{S} \cdot \mathbf{B}^i)
\end{aligned}
\tag{10}
$$

where $\mathbf{H}^i$ and $\mathbf{B}^i$ denote, respectively, the incident magnetic field and the incident magnetic induction. Thus the open-circuit voltage can be calculated using the vector effective area and the incident magnetic induction field, instead of the vector effective height and the incident electric field. This alternative formula conforms to the expression of the induced emf in a loop deduced from Faraday's law. For a small loop the vector effective area is given by

Table 1. Vector Effective Height

Antenna Type	Current Distribution		$\mathbf{h}$
Hertzian dipole	$\mathbf{I} = I\hat{z}$,	$|z| \leqq \ell$	$-2\ell \sin\theta \, \hat{\theta}$
Abraham dipole	$\mathbf{I} = I\left(1 - \dfrac{|z|}{\ell}\right)\hat{z}$,	$|z| \leqq \ell$	$-\ell \sin\theta \, \hat{\theta}$
Half-wave dipole	$\mathbf{I} = I\cos\left(\dfrac{2\pi}{\lambda} z\right)\hat{z}$,	$|z| \leqq \lambda/4$	$\dfrac{\lambda}{\pi} \dfrac{\cos[(\pi/2)\cos\theta]}{\sin\theta} \, \hat{\theta}$
Small loop of area A	$\mathbf{I} = I\hat{t}$		$jkA \sin\theta \, \hat{\phi}$

$$S = \frac{j}{k}(\hat{\mathbf{r}} \times \mathbf{h}) = A \sin \theta \,\hat{\boldsymbol{\theta}} \tag{11}$$

where θ is the angle between $\hat{\mathbf{r}}$ and $\hat{\mathbf{h}}$, and A is the area of the loop.

3. Receiving Cross Section, Impedance-Matching Factor, and Polarization-Matching Factor

The *receiving cross section* of an antenna is defined by

$$\sigma_R = \frac{\text{power received by an antenna}}{\text{incident power density}}$$

$$= \frac{1}{2}|I_L|^2 R_L \,\Big/\, \frac{1}{2}\frac{|\mathbf{E}^i|^2}{Z_0} \tag{12}$$

Using (1), (3), and (8) we have

$$\sigma_R = \frac{|\mathbf{h} \cdot \mathbf{E}^i|^2 R_L Z_0}{|Z_L + Z_{\text{in}}|^2 |\mathbf{E}^i|^2} \tag{13}$$

The receiving cross section is a measure of the effective "capture" area of an antenna with respect to the incident power density. As far as the dependence on the load impedance is concerned, the maximum value of σ_R occurs when Z_L is equal to Z_{in}^*, the conjugate of Z_{in}. Thus it is convenient to introduce an impedance-matching factor defined by

$$q = 1 - \left|\frac{Z_L - Z_{\text{in}}^*}{Z_L + Z_{\text{in}}}\right|^2 = \frac{4R_L R_{\text{in}}}{|Z_L + Z_{\text{in}}|^2} \tag{14}$$

whose value lies between zero and unity. Equation 13 can then be written as

$$\sigma_R = \frac{q Z_0 |\mathbf{h} \cdot \mathbf{E}^i|^2}{4 R_{\text{in}} |\mathbf{E}^i|^2} \tag{15}$$

The quantity Z_0/R_{in} can be expressed in terms of the effective height and the directivity of the antenna.

We start with the expression for the power input to an antenna:

$$P_{\text{in}} = 1/2 \, |I_{\text{in}}|^2 R_{\text{in}} \tag{16}$$

The radiated power P_r is related to the input power by

$$P_r = \eta P_{\text{in}} \tag{17}$$

where η denotes the efficiency of radiation to account for the ohmic loss of the antenna. Furthermore, the *directivity* or the *directive gain* of the antenna in an arbitrary direction (θ, ϕ) is defined by

$$D(\theta, \phi) = \frac{4\pi r^2 S(\theta, \phi)}{P_r} \tag{18}$$

where $S(\theta, \phi)$ denotes the power density in the far-zone region in the (θ, ϕ) direction, i.e.,

$$S(\theta, \phi) = \frac{|\mathbf{E}(\theta, \phi)|^2}{2Z_0} = \frac{k^2 Z_0 |I_{in} \mathbf{h}|^2}{32\pi^2 r^2} \tag{19}$$

in view of (4) and (7). Hence

$$\begin{aligned} D(\theta, \phi) &= \frac{k^2 Z_0 |I_{in}|^2 |\mathbf{h}|^2}{8\pi P_r} \\ &= \frac{k^2 Z_0 |\mathbf{h}|^2}{4\pi \eta R_{in}} \end{aligned} \tag{20}$$

on account of (16) and (17).

Eliminating R_{in}/Z_0 between (15) and (20), we obtain

$$\sigma_R = \frac{\lambda^2}{4\pi} \eta q p D(\theta, \phi) \tag{21}$$

where

$$p = \frac{|\mathbf{h} \cdot \mathbf{E}^i|^2}{|\mathbf{h}|^2 |\mathbf{E}^i|^2} \tag{22}$$

is designated as the *polarization-matching factor*, and its value lies between zero and unity. Equation 21 is the most general formula to characterize the receiving capability of an antenna [2].

To compute the value of p, it is convenient to introduce the complex polarization ratios for $\mathbf{h}$ and $\mathbf{E}^i$ defined by

$$\frac{h_\theta}{h_\phi} = t e^{j\beta} \tag{23}$$

$$\frac{F_\theta^i}{E_\phi^i} = s e^{j\alpha} \tag{24}$$

Then

$$p = \frac{1 + 2st\cos(\alpha + \beta) + s^2t^2}{(1 + s^2)(1 + t^2)} \tag{25}$$

The maximum value of p, equal to unity, occurs when $s = t$ and $\alpha = -\beta$, and corresponds to two identical elliptically polarized states with opposite sense of rotation. When $\alpha = \beta = 0$ both $\mathbf{h}$ and $\mathbf{E}^i$ are linearly polarized. A circularly polarized incident field corresponds to $s = 1$ and $\alpha = \pm\pi/2$. An antenna which radiates a circularly polarized far-zone field corresponds to $t = 1$ and $\beta = \pm\pi/2$. The polarization-matching factor between a circularly polarized incident field and a linearly polarized antenna is 1/2. The matching factor can also be displayed graphically on the Poincaré sphere or alternatively on a Smith chart [3].

4. Generalized Friis Transmission Formula

When we are dealing with two distant antennas, one transmitting, designated as no. 1, and another receiving, designated as no. 2, the pertinent parameters involved would be the power transfer ratio and the mutual impedance between the two antennas. Both parameters can most conveniently be expressed in terms of the vector effective heights of these two antennas. The power transfer ratio between the two antennas can be written in the form

$$\frac{P_2}{P_1} = \frac{\sigma_2 \eta_1 D_1(\theta, \phi)}{4\pi r^2} \tag{26}$$

where

P_2 = the power received by antenna no. 2

P_1 = the input power to antenna no. 1

σ_2 = the receiving cross section of antenna no. 2

η_1 = the radiation efficiency of antenna no. 1

D_1 = the directivity of antenna no. 1

r = the distance between the terminals of the two antennas

Using the expression for σ_2 given by (21) we obtain

$$\frac{P_2}{P_1} = \left(\frac{\lambda}{4\pi r}\right)^2 \eta_1 \eta_2 q_2 p D_1(\theta, \phi) D_2(\theta, \phi) \tag{27}$$

where

η_2 = the radiation efficiency of antenna no. 2 when it is operating in its transmitting mode

q_2 = the impedance-matching factor of antenna no. 2

$$= 1 - \left| \frac{Z_L - Z_{in,2}^*}{Z_L + Z_{in,2}} \right|^2$$

$Z_{in,2}$ = the input impedance of antenna no. 2

Z_L = the load impedance connected to the terminals of antenna no. 2

$D_2(\theta, \phi)$ = the directivity of antenna no. 2 in the direction (θ, ϕ)

$p = |\mathbf{h}_1 \cdot \mathbf{h}_2|^2 / |\mathbf{h}_1|^2 |\mathbf{h}_2|^2$ = polarization-matching factor of the two antennas

$\mathbf{h}_1, \mathbf{h}_2$ = the vector effective heights of the two antennas

Equation 27 is the generalized version of Friis' transmission formula [4] originally formulated for two linear antennas without considering the radiation efficiency, the impedance-matching factor, or the polarization-matching factor.

5. Mutual Impedance between Distant Antennas

The mutual impedance between two antennas, one of which lies in the far-zone field of the other, can be calculated by using the vector effective height of these two antennas. By considering the relation

$$Z_{12} = \frac{(V_2)_{oc}}{I_1} \tag{28}$$

where

$(V_2)_{oc}$ = the open-circuit voltage excited at the terminals of antenna no. 2

I_1 = the input current to the terminals of antenna no. 1

one finds

$$Z_{12} = \frac{jkZ_0 e^{-jkr}}{4\pi r} (\mathbf{h}_1 \cdot \mathbf{h}_2) \tag{29}$$

This expression has been tested for two half-wave dipoles. The result shows that the expression is quite accurate where r, the distance between the two antennas, is greater than half of a wavelength. The formula could be used to estimate the coupling between antennas since the vector effective heights of most antennas can be calculated using some reasonable assumption of their current distributions or aperture field distribution.

6. Small Antennas

Most receiving antennas in practical use at low frequencies are small in wavelength. An antenna is said to be *small* if its physical dimensions do not exceed approximately one eighth of the wavelength. The small transmitting antenna is never satisfactory from the standpoint of efficiency because much of the generated power is wasted in heating the ohmic resistance. On the other hand, small receiving antennas often are very satisfactory. Sensitivity rather than power efficiency is the important factor in reception, and sensitivity is limited by noise. Schelkunoff and Friis ([5], Chap. 10) are suggested for a complete discussion of small antennas. There are two types of small antennas: the short dipole and the small loop. In the limiting case of vanishingly low frequency the short dipole is just a capacitor and the small loop is an inductor.

The Short Dipole

Radiation Pattern and Gain—The far fields of a short dipole oriented in the $\hat{z}$ direction (see Fig. 2a) are given by

$$E_\theta = j\frac{60\pi \sin\theta}{r\lambda}M \tag{30}$$

$$H_\phi = \frac{E_\theta}{120\pi} \tag{31}$$

where

M = the moment of current distribution defined as $M = \int_{-\ell}^{\ell} I(z)\,dz$

r = distance in meters to the observation point

λ = wavelength in meters

k = wave number $(2\pi/\lambda)$

The radiation pattern of a short dipole $(\sin\theta)$ is shown in solid lines in Fig. 3. For comparison purposes the radiation pattern of a half-wave dipole is shown in dotted lines. The directivity of a short dipole with respect to the isotropic radiation is

$$D(\theta,\phi) = \frac{4\pi}{\int_0^{2\pi}\int_0^{\pi} \Psi \sin\theta\, d\theta\, d\phi} = 1.5 \tag{32}$$

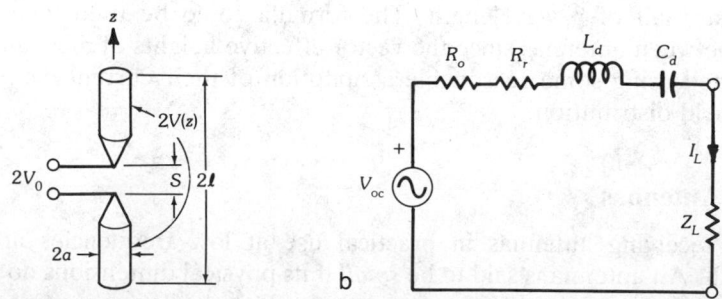

Fig. 2. Short dipole antenna. (*a*) Incremental dipole. (*b*) Equivalent circuit of (*a*).

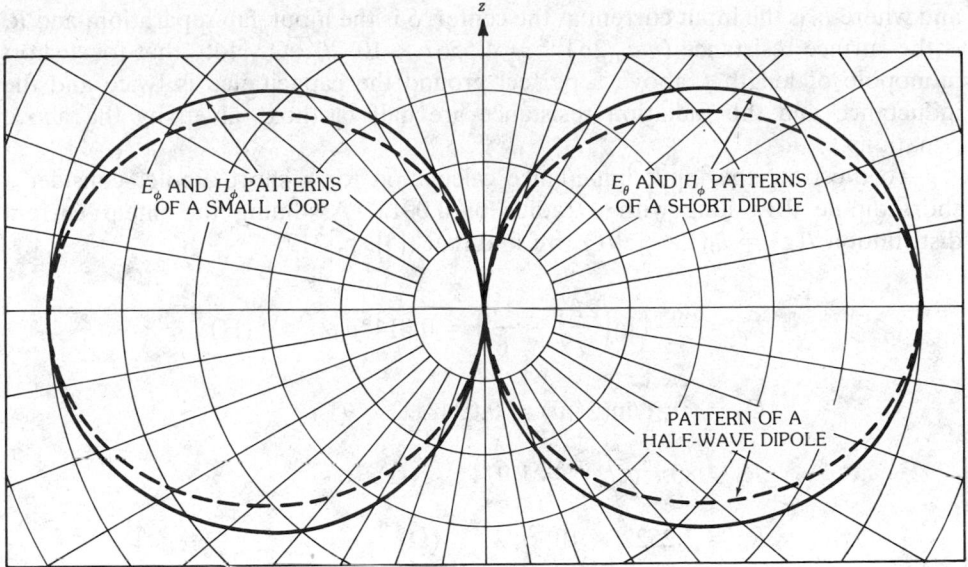

Fig. 3. Radiation patterns of small antennas.

where $\theta = 90°$ and the radiated power pattern Ψ is proportional to $\sin^2\theta$. Note that its directivity is only slightly less than that (1.64) of a half-wave dipole ([6], p. 54).

Impedance—The input impedance of a short dipole as shown in Fig. 2b is composed of the series inductance, the capacitance, the radiation resistance, and the ohmic resistance, which are expressed as

$$L_d = \frac{1}{|I_0|^2} \int_0^\ell L(z) |I(z)|^2 \, dz \tag{33}$$

where

$L(z) = (\mu_0/\pi) \ln(2z/a)$ = the inductance per unit length

$$C_d = \frac{1}{|V_0|^2} \int_{S/2}^\ell C(z) |V(z)|^2 \, dz$$

$$= \int_{S/2}^\ell C(z) \, dz, \qquad V_0 \cong V(z) \tag{34}$$

$C(z) = \pi\epsilon_0/\ln(2z/a)$ = the capacitance per unit length

$$R_r = \frac{80\pi^2}{I_0^2} \left(\frac{M}{\lambda}\right)^2 \tag{35}$$

$$R_0 = \frac{2\ell}{2\pi a} R_s \tag{36}$$

and where I_0 is the input current at the center, S is the input gap separation, and R_s is the surface resistance $(\omega\mu_0/2\sigma)^{1/2} = 4.5567 \times 10^{-3}/\sqrt{\lambda}\,\Omega$. Note that for a short monopole of length ℓ above a perfect ground the capacitance is twice and the inductance and the radiation resistance are half of those given by the above equations.

To illustrate the input impedance calculation for a short dipole, consider a short dipole 0.1λ long with a radius of 0.001λ. Assuming the linear current distribution $I(z) = I_0(1 - |z|/\ell)$, we find that ([5], p. 312)

$$L_d = \frac{\mu_0 \ell}{3\pi}\left(\ln\left(\frac{2\ell}{a}\right) - \frac{11}{6}\right) = 0.0147\mu_0\lambda \qquad \text{(H)}$$

$$C_d = \pi\epsilon_0\ell/\ln(2\ell/a) = 0.034\epsilon_0\lambda \qquad \text{(F)}$$

$$R_r = 80\pi^2(\ell/\lambda)^2 = 1.974 \qquad (\Omega)$$

$$R_o = 7.2522 \times 10^{-2}/\sqrt{\lambda} \qquad (\Omega)$$

The capacitive reactance $(X_{C_d} = 1/j\omega C_d = -j1763.5\,\Omega)$ for this short dipole is much greater than the inductive reactance $(X_{L_d} = j\omega L_d = j34.769\,\Omega)$. This example demonstrates that a short dipole is just a capacitor in the low-frequency limit.

Maximum Effective Height—The magnitude of the maximum effective height of a straight linear receiving antenna in (8) is equal to the effective length of a straight linear transmitting antenna, which is defined as the moment of its current distribution divided by the input current ([5], p. 301)

$$h_{max} = \frac{M}{I_0} = \int_{-\ell}^{\ell} I(z)\,dz/I_0 \qquad (37)$$

Maximum Receiving Cross Section $(\sigma_R)_{max}$—From (15) the maximum receiving cross section $(\sigma_R)_{max}$ occurs when the matching factor q becomes 1, and the vector effective height $\mathbf{h}$ is at its maximum (at $\theta = 90°$):

$$(\sigma_R)_{max} = \frac{Z_0(h_{max})^2(E^i)^2}{4R_r(E^i)^2} = \frac{3}{8\pi}\lambda^2 \qquad (38)$$

where $Z_0 = 120\pi$ is the free-space impedance and R_r is the radiation resistance of the antenna. In Table 2 the radiation resistance, the maximum effective height, and the maximum receiving cross section are listed for various short dipoles. Note that the half-wave dipole is also included in Table 2 for comparison. The receiving cross section of a half-wave dipole is $0.13\lambda^2$ ([6], p. 51), which is only slightly greater than the cross section of the short dipole $0.119\lambda^2$. (See Table 2.) This interesting condition exists because the radiation resistance of the antenna decreases with antenna size as rapidly as the square of the induced voltage, so the power available from the antenna remains constant. In practice the available power cannot be obtained because the antenna and its matching network are generally lossy, and also the high antenna reactance generally limits the bandwidth over which an effective match can be made with a simple network.

To induce resonance an inductance may be added in series with the antenna and the generator. The loading inductor keeps the current distribution nearly constant from the feed to the load point, with a linear decrease from the load to the end as shown in row 2 of Table 2. The inductive loading is advantageous since the loading increases the current moment, and the receiving parameter—effective length—varies as the current moment. The location of the inductor in the radiating structure affects the value of the radiation resistance in the sense that the current distribution along the antenna is modified.

The optimum location for this inductance with regard to efficiency is about four-tenths of the length of the antenna above the ground plane [7]. The radiation efficiency of this inductively loaded monopole is still small (in practice, on the order of 10 percent at the lower end of the band).

Top loading as shown in row 3 of Table 2 is another traditional approach toward achieving a fairly high efficiency in a relatively small size antenna. Note that the current distribution in these structures is almost constant. Fig. 4 shows plots of theoretical radiation resistance R_r of a top-loaded monopole antenna for an assumed linear current distribution. Further improvement of radiation efficiency can be achieved by combining the inductive loading and the top loading as shown in row 4 of Table 2.

The Small Loop

Radiation Pattern and Gain—The radiation pattern of a small loop is identical with that of a short dipole oriented normal to the plane of the loop with **E** and **H** fields interchanged. If the normal direction to the plane of the loop is the z axis (see Fig. 5a), the radiated electromagnetic fields are given by

$$E_\phi = 120\pi \frac{\pi N}{r} \frac{A}{\lambda^2} I \sin \theta \tag{39}$$

and

$$H_\theta = -\frac{\pi N}{r} \frac{A}{\lambda^2} I \sin \theta = -\frac{E_\phi}{120\pi} \tag{40}$$

where

r = distance from the antenna

I = antenna current

N = number of turns of the loop

A = area of loop

λ = wavelength

Table 2. Short Dipole Parameters

Parameter / Dipole	Current Distribution	Current Moment $M = \int_{-L}^{L} I(z)\, dz$	Radiation Resistance $R_r = 80\pi^2 \dfrac{(M/\lambda)^2}{I_0^2}$	Maximum Effective Height $h_{\max} = \dfrac{M}{I_0}$	Maximum Receiving Cross Section $\sigma R_{\max} = \dfrac{Z_0 h^2}{4R_r}$
Short (or Abraham)		$I_0 L$	$80\pi^2 \dfrac{L^2}{\lambda^2}$	L	$\dfrac{3}{8\pi}\lambda^2 = 0.119\lambda^2$
Inductively loaded short		$I_0(2L - L_1)$	$80\pi^2 \dfrac{(2L - L_2)^2}{\lambda^2}$	$2L - L_1$	$\dfrac{3}{8\pi}\lambda^2$

Top-loaded short		$(I_0 + I_1)L$	$80\pi^2\dfrac{(I_0+I_1)^2}{\lambda^2 I_0^2}$	$\left(1+\dfrac{I_1}{I_C}\right)L$	$\dfrac{3}{8\pi}\lambda^2$
Top-loaded and inductively loaded short (or Hertzian)		$2I_0 L$	$320\pi^2\dfrac{L^2}{\lambda^2}$	$2L$	$\dfrac{3}{8\pi}\lambda^2$
Half-wave		$I_0\dfrac{\lambda}{\pi}$	$73\,\Omega$	$\dfrac{\lambda}{\pi}$	$\dfrac{30}{73\pi}\lambda^2 = 0.13\lambda^2$

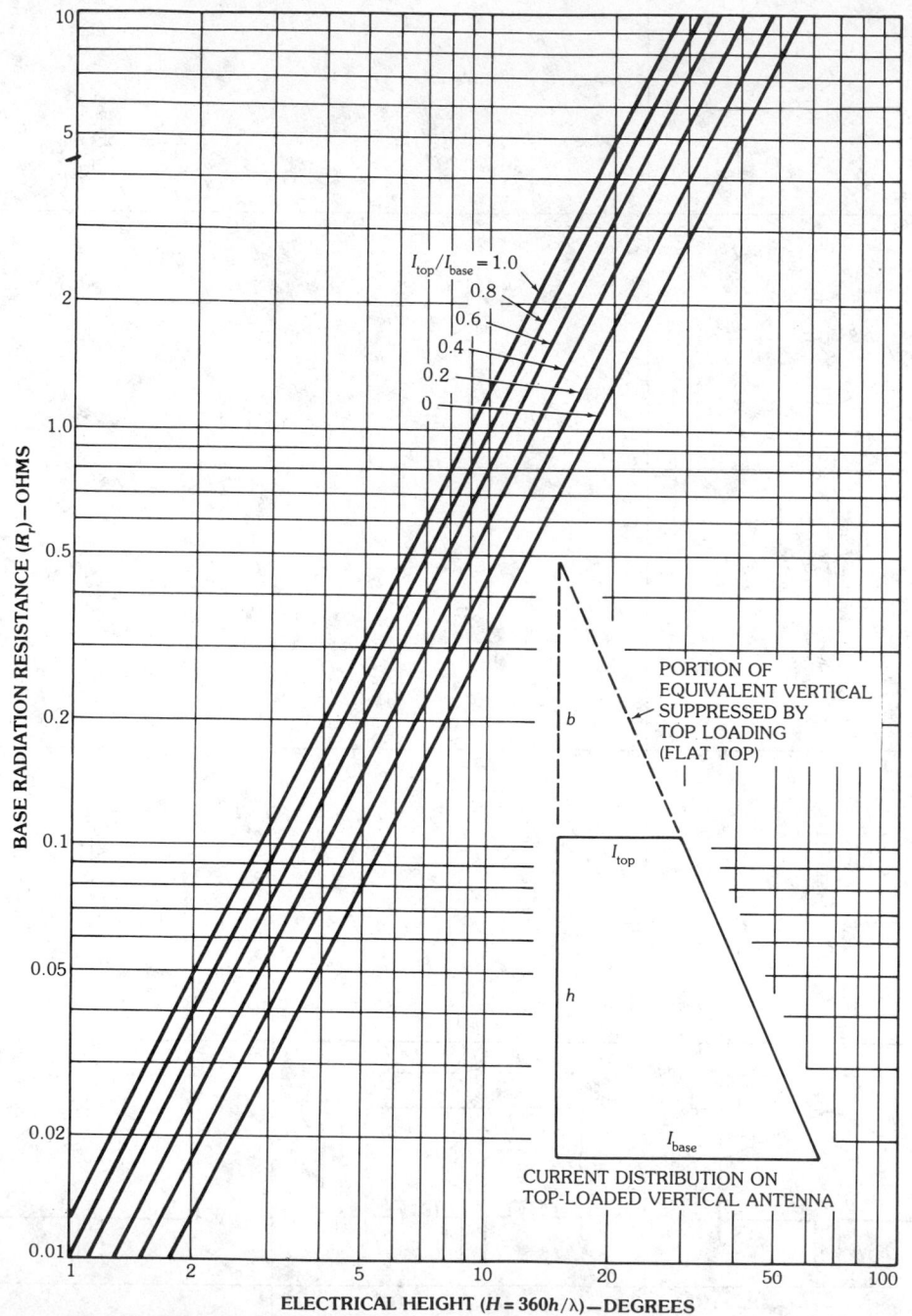

Fig. 4. Theoretical radiation resistance of vertical antenna for assumed linear current distribution. (*After Laport [15]; reprinted with permission of McGraw-Hill Book Company*)

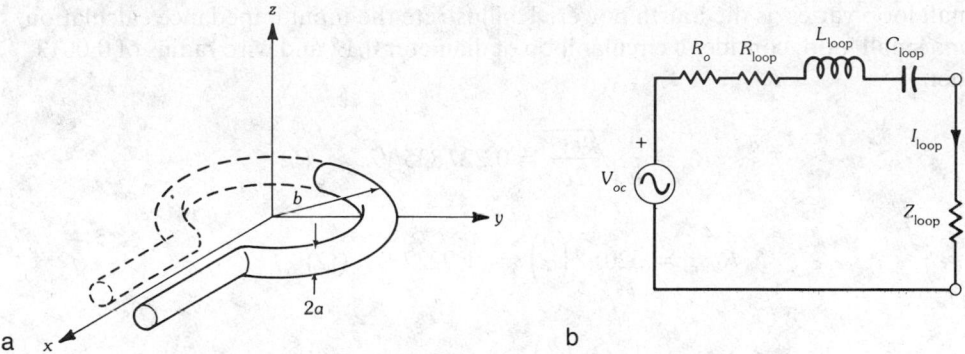

Fig. 5. Magnetic dipole. (*a*) Incremental magnetic dipole. (*b*) Equivalent circuit of (*a*).

The above equations are valid for arbitrary loop cross sections, provided that the loop diameter is small (i.e., loop radius $a < \lambda/2$) and uniform so that the loop current is uniform. The radiation pattern is shown in Fig. 3. The gain of a small loop with respect to the isotropic radiator is again 1.5 (= 1.76 dB).

Impedance—The input impedance of a small loop of one turn shown in Fig. 5a is composed of the ohmic resistance R_o, the series inductance L_{loop}, and the radiation resistance R_{loop}, which may be expressed as

$$R_o = \frac{\text{loop length} \times R_s}{\text{perimeter of wire cross section}} = \frac{b}{a} R_s \tag{41}$$

$$L_{\text{loop}} = \mu_0 b \ln(b/a) \tag{42}$$

$$R_{\text{loop}} = 20(k^2 S)^2 = 20[(h_{\text{max}})^2/S]^2 = 320\pi^4 \frac{S^2}{\lambda^4} \tag{43}$$

where

    b = the loop radius

    a = the wire radius

    S = the loop area

    R_s = the surface resistance = $\sqrt{\omega\mu/2\sigma}$, with σ the conductivity of the metal that the loop is made of

A satisfactory formula for the series capacitance of a coil is not available. However, since the capacitive reactance is much smaller than the inductive reactance of a multiple-turn small loop, one may ignore the series capacitance. If there are N turns, the total circulating current is the current per turn multiplied by N; hence R_{loop} and L_{loop} are to be multiplied by N^2. Note that while the radiation resistance of a dipole varies as the second power of the frequency, the radiation resistance of a small loop varies as the fourth power. To illustrate the input impedance calculation for a small loop, consider a circular loop of diameter 0.1λ and wire radius of 0.001λ. Then

$$R_o = \frac{b}{a} \sqrt{\frac{\mu_0 \omega}{2\sigma}} = 0.227\,835/\sqrt{\lambda} \quad (\Omega)$$

$$R_{\text{loop}} = 320\pi^6 \left(\frac{b}{\lambda}\right)^4 = 1.9227 \quad (\Omega)$$

$$L_{\text{loop}} = \mu_0 b \ln\left(\frac{b}{a}\right) = 0.1956\mu_0\lambda \quad (\text{H})$$

The inductive reactance ($X_L = j\omega L = j463\,\Omega$) for this small loop is much greater than R_o and R_{loop}, indicating that a small loop is just an inductor in the low-frequency limit.

Maximum Effective Height—For a loop placed in a medium of its relative permeability μ_r where a uniform field exists, the maximum induced voltage is given by

$$V_{\text{max}} = -\frac{\partial\phi}{\partial t} = -j\omega\mu_0\mu_r HS \tag{44}$$

where ϕ is the magnetic flux ($\phi = \int_{\text{loop}} \mathbf{B}\cdot d\mathbf{S} = \mu_r\mu_0 HS$) and S is the area of the loop. Since the magnetic field in free space is related to the electric field as $H = \sqrt{\epsilon_0/\mu_0}\, E$, the maximum induced voltage can be expressed in terms of the incident electric-field vector $\mathbf{E}^i$ as

$$V_{\text{max}} = -j\mu_r kSE^i, \qquad k = \omega\sqrt{\mu_0\epsilon_0} \tag{45}$$

or the magnitude of the maximum vector effective height h_{max} becomes $h_{\text{max}} = -j\mu_r kS$ (see Table 1).

Maximum Receiving Cross Section $(\sigma_R)_{\text{max}}$—The maximum receiving cross-section for a small loop placed in a free space ($\mu_r = 1$) is identical with that for a short dipole, that is,

$$(\sigma_R)_{\text{max}} = \frac{Z_0 h_{\text{max}}}{4R_r} \frac{120\pi k^2 S^2}{4 \times 320\pi^4 S^2/\lambda^2} = \frac{3}{8\pi}\lambda^2 \tag{46}$$

Table 3 lists the receiving parameters for various small loops.

7. Ferrite Loop Antennas

For a given current a loop wound on a magnetic core produces a stronger field than the loop alone. If the loop and its core are small, the directive pattern is still of

Table 3. Radiation Resistance of Various Small Loops

Loop	Shape	Area	Radiation Resistance $31\,000\ S^2/\lambda^4$
Circular		πA^2	$31\,000\left(\dfrac{\pi^2 A^4}{\lambda^4}\right)$
Ellipse		πAB	$31\,000\left(\dfrac{\pi^2 A^2 B^2}{\lambda^4}\right)$
Rectangular		AB	$31\,000\left(\dfrac{A^2 B^2}{\lambda^4}\right)$
N-turn loop		$N\pi A^2$	$N^2\left(31\,000\ \dfrac{\pi^2 A^4}{\lambda^4}\right)$
Ferrite-core N-turn coil		$N\pi A^2$	$N^2(\bar{\mu}_{\text{eff}})^2\left(31\,000\ \dfrac{\pi^2 A^4}{\lambda^4}\right)$

the figure-8 shape; hence the directivity of the loop is not affected by the core. Since for the given current the field and hence the radiated power are increased by the core, the radiation resistance must become larger.

Mean Effective Permeability of a Ferrite Rod—The permeability of a core $\mu_c(x)$ of any configuration (see Fig. 6), with the exception of an ellipsoid, is a function of the core cross-section shape and the location of the coil in the core's longitudinal axis. Fig. 7 shows the curve μ_c as a function of the axial coordinate of the cross section.

The distribution of μ_c along the rod may be approximated by a trinomial [8]

$$\mu_c(x) = \mu_{cs}(1 + 0.106\bar{x} - 0.988\bar{x}^2) \tag{47}$$

where

$$\bar{x} = \frac{|x|}{\ell/2}$$

$$\mu_{cs} = \begin{cases} \dfrac{\mu}{1 + (\mu - 1)(D/\ell)^2(\ln(\ell/D)\{0.5 + 0.7[1 - \exp(-\mu \times 10^{-3})]\} - 1)} \\ \qquad\qquad\qquad\qquad\qquad\qquad\qquad\qquad\text{for cylindrical cross section} \\ \dfrac{\mu}{1 + (\mu - 1)(4ab/\pi\ell^2)\{\ln[k\ell/(a + b)] - 1\}} \\ \qquad\qquad\qquad\qquad\qquad\qquad\qquad\text{for rectangular cross section} \end{cases}$$

$k = 4 - 0.732[1 - \exp(-5.5b/a)] - 1.23\exp(-\mu \times 10^{-3})$

μ = initial or reversible permeability of the core material

ℓ = the core length

D = the cylinder diameter

a = the height of the rectangular cross section

b = the width of the rectangular cross section

The empirical equation (47) has very little error for permeabilities of from 50 to 1000 and for relative core lengths (core length divided by core diameter) of from 10 to 40. These ranges embrace practically all ferrite antenna construction. The mean

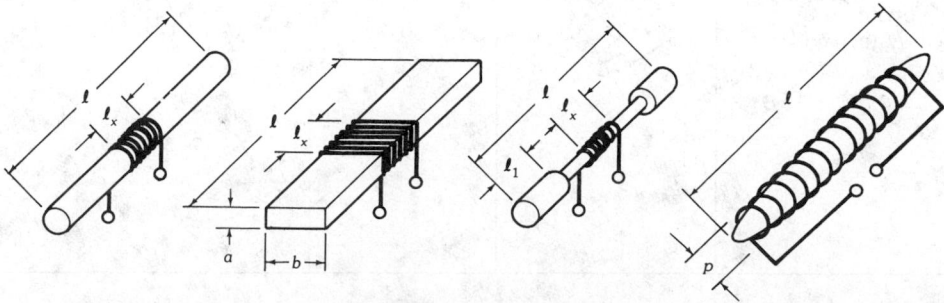

Fig. 6. Ferrite antennas.

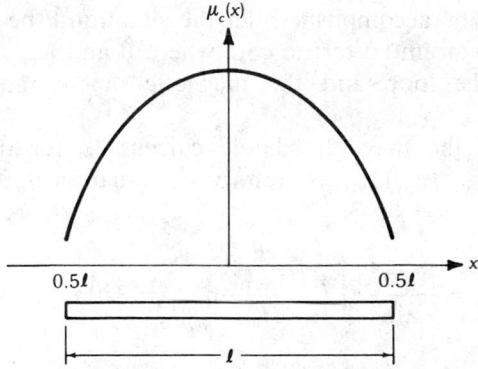

Fig. 7. Variation of permeability along ferrite rod.

effective core permeability $\bar{\mu}_{\text{eff}}$ for a coil length ℓ_{coil} enclosed between coordinates x_1 and x_2 is defined as

$$\bar{\mu}_{\text{eff}} = \frac{1}{\ell_{\text{coil}}} \int_{x_1}^{x_2} \mu_c(x)\, dx \tag{48}$$

It was reported that the mean effective permeability $\bar{\mu}_{\text{eff}}$ for a hyperboloid core was about 1.8 times larger than that for a cylindrical core having the same weight. With the dumbbell shape, however, the reduction of weight was accomplished with a factor less than 1.2.

Impedance—An equivalent circuit of a ferrite antenna (which is a coil of N turns around a ferrite core) consists of an antenna coil inductance L_{coil}, radiation resistance R_r, an ohmic resistance R_o, an equivalent resistance of losses in the ferrite R_{loss}, a parasitic antenna capacitance C_p, and an equivalent emf V_{oc}, the effective value of which equals

$$V_{\text{oc}} = \mathbf{h} \cdot \mathbf{E}^i \tag{49}$$

where

$$\mathbf{h} = \bar{\mu}_{\text{eff}} k A \sin\theta\, \boldsymbol{\theta}$$
$$R_o = N(b/a)\, R_s$$
$$L_{\text{coil}} = N^2 \bar{\mu}_{\text{eff}}\, \mu_0 b \ln(b/a) = \bar{\mu}_{\text{eff}} L_{\text{loop}} = \bar{\mu}_{\text{eff}} N^2 \mu_0 b \ln(b/a)$$

The radiation resistance of a small loop around the ferrite core can be found by applying the reciprocity theorem, given in the form

$$\int \mathbf{B}^b \cdot \mathbf{K}_m^a\, dV = \int \mathbf{B}^a \cdot \mathbf{K}_m^b\, dV \tag{50}$$

To see how this can be accomplished, let the situation *a* be an air loop and the situation *b* be a loop around a ferrite core where **B** and **K**$_m$ are the magnetic flux density inside of the loop and the magnetic dipole moment of the loop, respectively.

Then from (50) the magnetic dipole current J_m^b for the ferrite-core loop becomes $J_m^b = (H_{\text{ferrite}}/H_{\text{air}}) J_m^a$. Therefore the radiation resistance of the ferrite loop becomes [9]

$$R_{\text{loop}} = \frac{320\pi^4}{\lambda^4} \left| \iint_S \frac{H_{\text{ferrite}}}{H_{\text{air}}} dS \right|^2 = 320\pi^4 (\mu_{\text{eff}})^2 \frac{S^2}{\lambda^4} \tag{51}$$

It is noted that since no satisfactory formulas for the inductance and the capacitance of a ferrite antenna are available, the quality factor of the ferrite antenna cannot be determined theoretically. It was observed that the quality factor increases with a shifting of the coil toward the end of the core, and in addition by making the coil diameter larger than the diameter of the core (the optimum diameter ratio is 4/3).

It is easy to verify that the maximum receiving cross section for a ferrite loop antenna remains the same as for a small loop antenna.

8. Bandwidth and Efficiency

A certain relationship exists between antenna size and its theoretical maximum bandwidth and efficiency. Chu [10] obtained the quality factor Q for the electrically small antenna as

$$Q = \frac{1 + 3k^2 a^2}{k^3 a^3 (1 + k^2 a^2)} \tag{52}$$

where a is the radius of the smallest sphere which encloses the longest antenna dimension and k is the wave number in free space.

This Q is based on the lowest TM mode. When both a TM mode and a TE mode are excited, the value of Q is halved. Note that for $ka \ll 1$ the Q varies inversely as the cube of sphere radius in wavelengths. The importance of the Chu result is that it relates the lowest achievable Q to the largest dimension of an electrically small antenna, and the result is independent of the art that is used to construct the antenna within the hypothetical sphere, except in determining whether a pure TE or TM, or both modes, is excited. The above equation is for a lossless antenna. If the antenna is lossy, its ohmic resistance can be added in series with the radiation resistance, so the effect on Q is apparent. Fig. 8 plots single-mode Q for various efficiencies. The radiation efficiency is defined as

$$\eta = \frac{R_r}{R_r + R_L} \tag{53}$$

where R_r is the radiation resistance of the small antenna and R_L is all the losses in the circuit. As Fig. 8 shows, improvement of the bandwidth for an electrically small

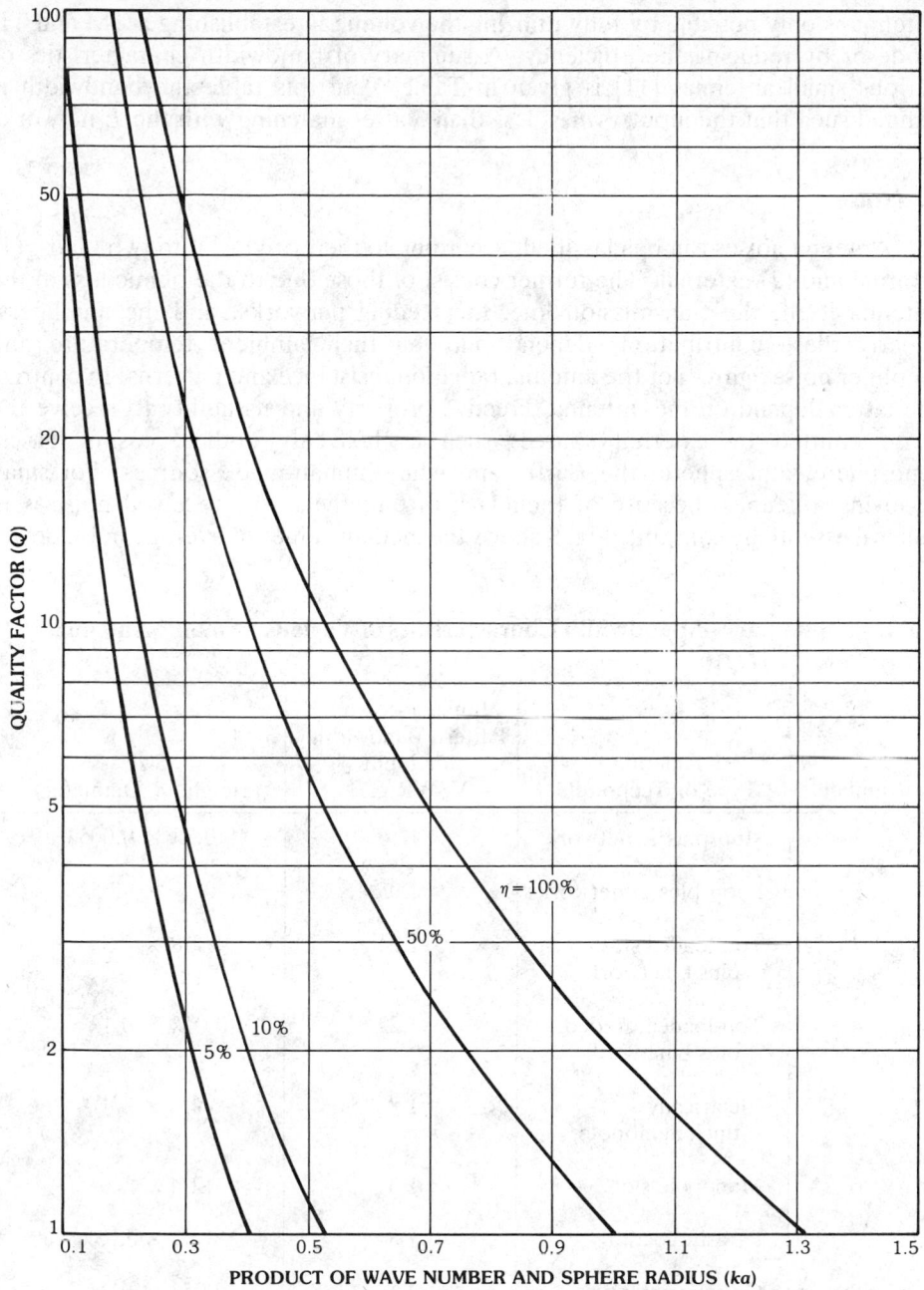

Fig. 8. Chu-Harrington fundamental limitations for single-mode antenna versus efficiency. (*After Hansen [16], © 1981 IEEE*)

antenna is only possible by fully utilizing the volume in establishing a TM and TE mode or by reducing the efficiency. A summary of bandwidth characteristics of various small antennas [11] is given in Table 4. In this table the bandwidth is defined such that the input vswr is less than 3 after matching with the L network.

9. Noise

Antenna noises can be classified, according to their origins, into two kinds: (1) internal and (2) external. The former consist of those due to the ohmic loss of the antenna itself, the transmission line, the feeding networks, and the amplifiers. Clearly these contributions depend only on their ambient temperature and amplifier noise figure, not the antenna radiation resistance and patterns. In contrast the latter depend on the antenna directive property and its ability to receive the noise emitted by external sources such as heavenly bodies, cosmic gases, ionosphere, atmosphere, the Earth, and many human-made sources. For small receiving antennas, because of their lack of directivity, the received noise is in general essentially constant. Fig. 9 shows the median values of average noise power

Table 4. Summary of Bandwidth Characteristics of Various "Small" Antennas (*After Desantis [11]*)

Number	Antenna Type or Technique	Impedance and Pattern Bandwidth for Input VSWR $\leqq$ 3	Size (Height × Diameter)
1	Stub plus L network	1.16	$0.1\lambda \times 0.005\lambda$
2	Loop plus L network	1.05	$0.1\lambda \times 0.05\lambda$
3	Top-loaded stub plus L network	$\cong 1.24$	$\lambda/8 \times \lambda/8$
4	Top-loaded, folded, plus L network	$\cong 1.22$	$0.07\lambda \times 0.1\lambda$
5	Electrically thick monopole	$\cong 1.8$	$\lambda/2 \times \lambda/4$
6	Monopole-slot	1.3	$\lambda/4 \times \tfrac{3}{8}\lambda$
7	Parasite-loading	1.8	$\lambda/2 \times 0.05\lambda$
8	Goubau antenna	2	$0.05\lambda \times 0.2\lambda$
9	Electrically small, complementary pair	>2.5	$\lambda/9 \times \lambda/4$
10	Slotted-cone antenna	>3	$\lambda/8 \times 0.44\lambda$
11	Hallen	>3	$\lambda/2 \times 0.03\lambda$

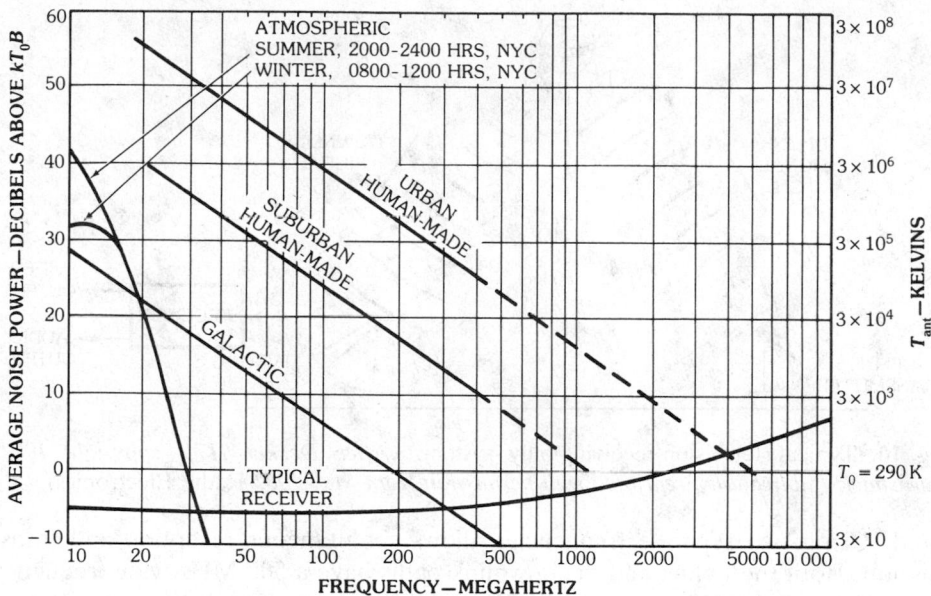

Fig. 9. Median values of average antenna noise temperature for an omnidirectional antenna near the Earth surface. (*Reprinted with permission of Howard W. Sams*)

expected from various sources. In this figure $k = 1.38 \times 10^{-23}$ J/B, $T_0 = 290$ K, and B is the receiver bandwidth in hertz. External human-made noise levels decrease with frequency but can increase rapidly from a suburban to an urban area. Furthermore, for frequencies below the gigahertz range, the sky noise increases sharply due to contributions from what is called atmospheric noise and galactic noise, and it is desirable to maximize the desired signal-to-noise ratio. This problem is considered in Chapter 11. For frequencies above the gigahertz range, the internal noise may become significant. It has been shown that for the internal noise a proper amount of mismatch between the antenna and the amplifier may actually improve the signal-to-noise ratio [12, 13, 14].

10. Satellite TV Earth Station Receiving Antenna*

The design of an Earth station antenna (which is not electrically small) requires a broad variety of system considerations which include microwave antenna, satellite link, and receiver characteristics. There are a number of factors that contribute to the overall performance of any satellite system. A typical tvro (television receiving only) system is shown in Fig. 10. The function of the uplink transmitter is to send a frequency modulated video signal and an fm audio signal to the satellite using a carrier in the 5.9- to 6.5-GHz band. Then the satellite retransmits those signals back to Earth on a lower carrier in the 3.7- to 4.2-GHz

---

*The material contained in this section is based mainly on the private note "Earth station considerations," dated June 1983, by James F. Corum of West Virginia University and Basil F. Pinzone, Jr., president of Pinzone Communication Products, Inc. Their assistance is duly acknowledged.

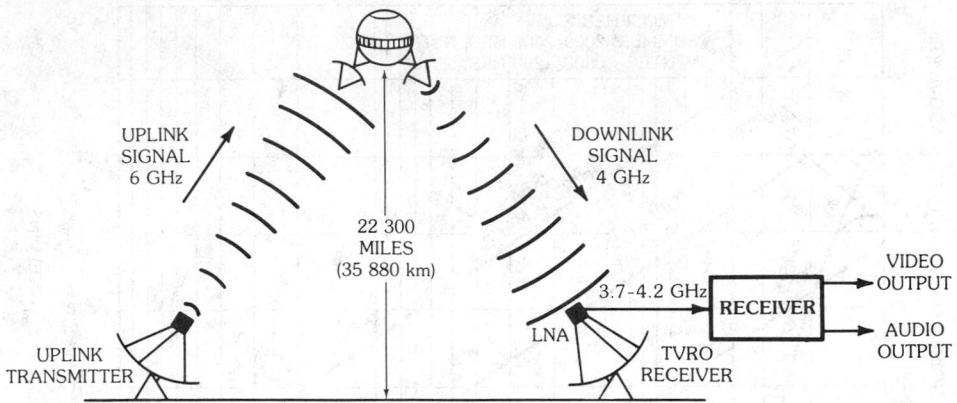

Fig. 10. Typical television-receiving-only system. (*After Decker [17], copyright 1982, Gernsback Publications; reprinted with permission from May 1982* Radio Electronics)

band. Using two separate frequencies allows simultaneous reception and transmission. Both the uplink and the downlink paths have a 500-MHz-wide frequency range (5.9 to 6.4 GHz and 3.7 to 4.2 GHz), which means that there is room for 12 channels, each 40 MHz wide. However, 24 channels are allocated in the same space by overlapping the channels with cross-polarized signals as shown in Fig. 11. The downlink signal transmitted by the satellite has an output power of 36 dBW, including the onboard antenna gain (approximately 30 dB). Fig. 12 shows an antenna pattern "footprint" on the Earth. This signal travels 22 300 miles to Earth, losing 196 dB of its initial strength. Therefore the ground receiving station is not left with much to work with. Once the signal is received by the receiving antenna, the lna (low-noise amplifier) amplifies the 4-GHz signal by about 40 dB while adding only 1 or 2 dB of noise. It is important to introduce as little noise at this stage as possible, since any that is introduced will be carried and amplified through the rest of the system. Then it is fed to a receiver where the 4-GHz rf signal transforms into standard composite video and audio. A summary of the frequency allocations for each television delivery technology is shown in Fig. 13.

Satellite Location

Communications satellites have geosynchronous orbits, which are directly above the equator and have an angular velocity the same as the angular velocity of the Earth's rotation. An observer on Earth sees the satellite as a stationary point in the sky and wants to locate the satellite. Figs. 14 and 15 are self-explanatory. Given the satellite point F (the point along with Earth's equator directly below the satellite) and the Earth station's coordinates (latitude and longitude), the standard spherical trigonometric quantities provide the observer with the azimuth and elevation angles of the desired satellite.

Satellite Link Calculation

Since the satellite is line of sight for the Earth station the received power P_R at the Earth receiving antenna can be expressed in a form of the Friis transmission formula

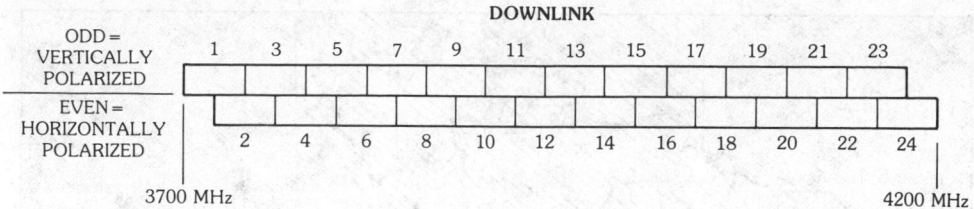

Fig. 11. How to fit 24 channels in a 12-channel bandwidth. (*After Decker [17], copyright 1982, Gernsback Publications; reprinted with permission from May 1982* Radio Electronics)

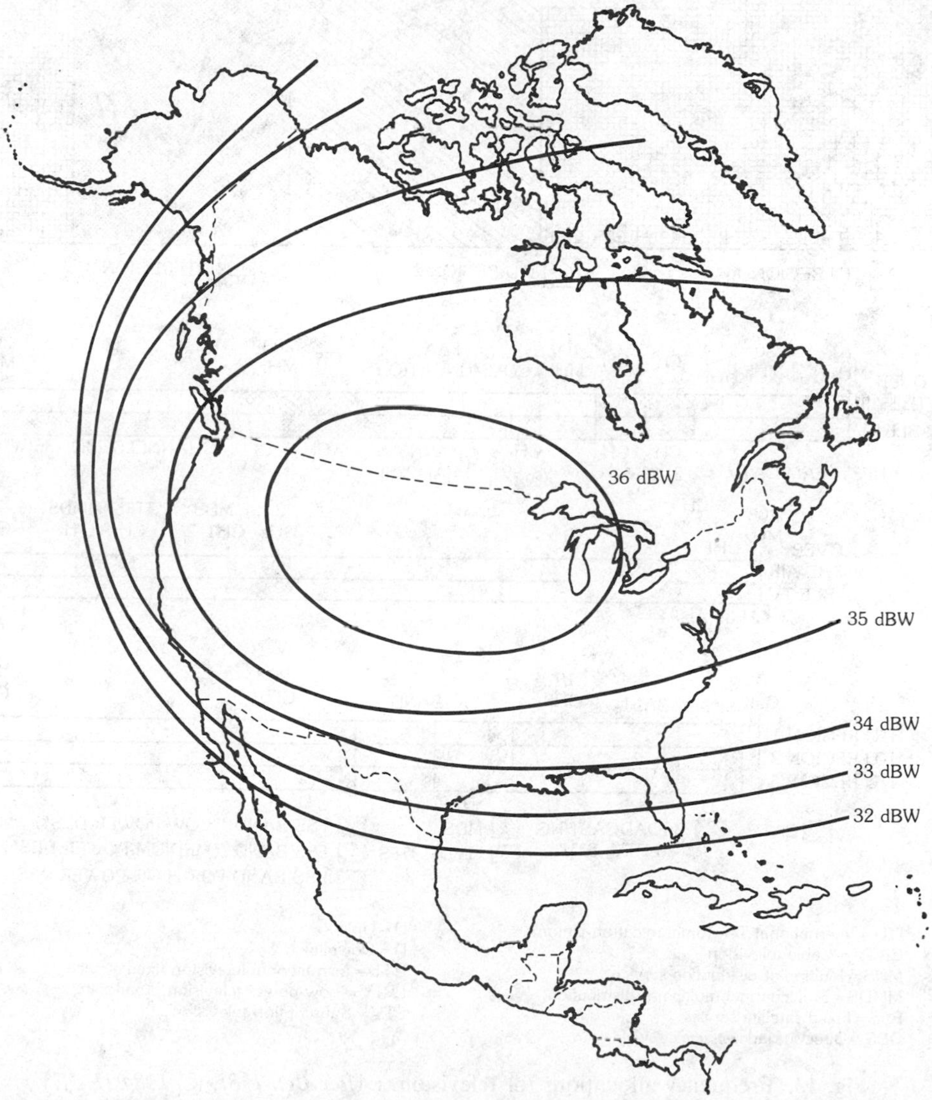

36 dBW

35 dBW

34 dBW

33 dBW

32 dBW

Fig. 12. Pattern footprint (contour of EIRP) on North America. (*Courtesy James Corum and Basil Pinzone*)

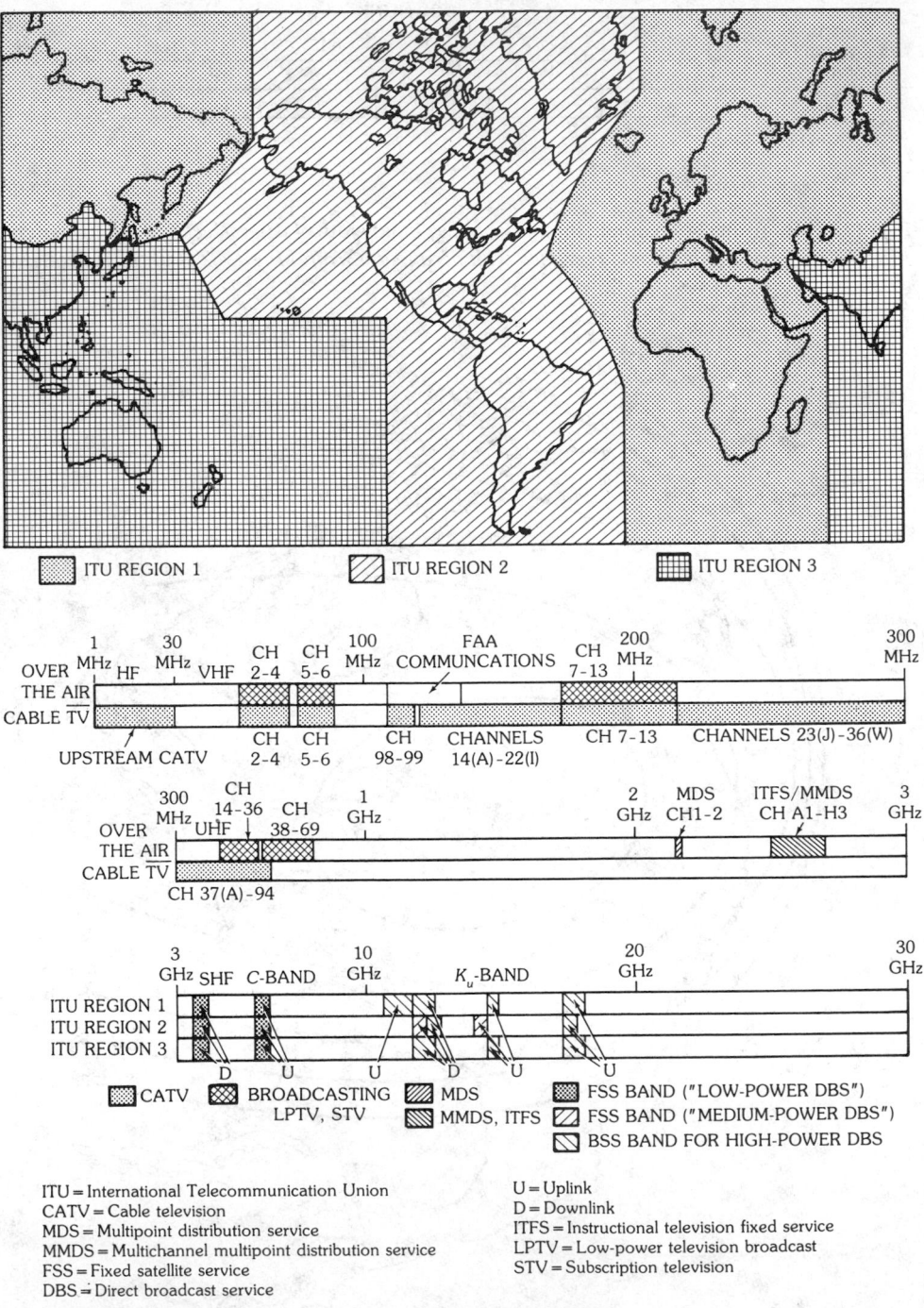

Fig. 13. Frequency allocations for television. (*After Bell [18], © 1984 IEEE*)

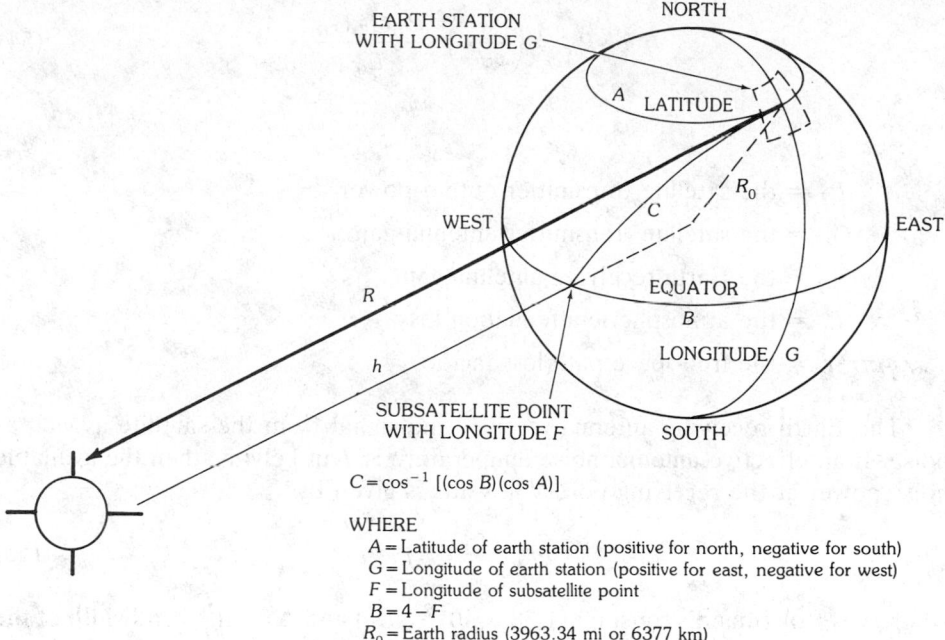

$$C = \cos^{-1}[(\cos B)(\cos A)]$$

WHERE

A = Latitude of earth station (positive for north, negative for south)
G = Longitude of earth station (positive for east, negative for west)
F = Longitude of subsatellite point
$B = 4 - F$
R_0 = Earth radius (3963.34 mi or 6377 km)

Fig. 14. Standard spherical trigonometric quantities required for locating a desired satellite from a specified Earth station. (*Courtesy James Corum and Basil Pinzone*)

$$a_N = 180° + \tan^{-1}(\tan B / \sin A)$$
$$R = \sqrt{R_0{}^2 + (R_0 + h)^2 - 2(R_0 + h)R_0 \cos C}$$
$$e = \tan^{-1}\{[\cos C - (R_0/R)]/\sqrt{1 - \cos^2 C}\}$$

EXAMPLE:

Site: Morgantown ($A = 39°$, $G = -79°$)
Satellite: Statcom I ($0°$, $-135°$, $h = 22\ 282$ mi or 35 852 km)
Great Circle Arc: $C = 64.24°$
Azimuth: $a_N = 247° = 113°$ west of north
Elevation: $e = 16.81°$
Slant Range: $R = 24\ 396$ mi (39 253 km)

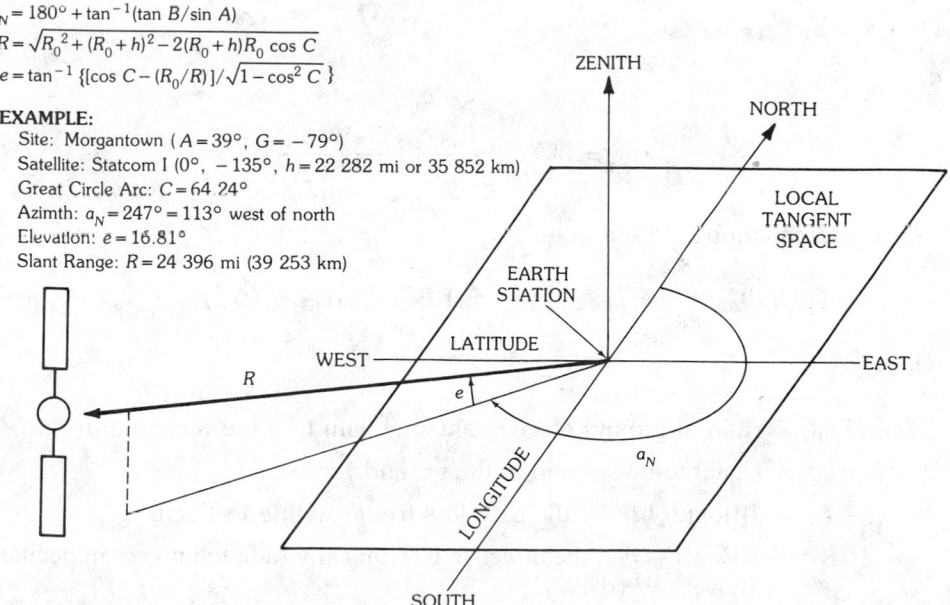

Fig. 15. Earth station look angles for satellite antennas. (*Courtesy James Corum and Basil Pinzone*)

$$P_R = \left(\frac{\lambda}{4\pi r^2}\right)^2 \frac{P_T G_T G_R}{L_0} \tag{54}$$

where

P_T = the satellite transmitter output power

G_T = the satellite transmitter antenna gain

G_R = the Earth receiving antenna gain

L_0 = the atmospheric attenuation loss

$(\lambda/4\pi r^2)^2$ = the free-space path loss factor

The Earth receiving antenna received the signal from the satellite as well as noise. If an effective antenna noise temperature is T in kelvins, then the available noise power at the receiving port N in watts is given by

$$P_n = kT\Delta f \tag{55}$$

where k is Boltzmann's constant (1.38×10^{-23} J/K) and Δf is the bandwidth of the receiving antenna system in hertz. The ratio of signal power to the noise power appearing at the antenna's terminal is given by

$$\text{snr} = \frac{P_R}{P_n} = \frac{(\lambda/4\pi r^2)^2(P_T G_T G_R/L_0)}{kT\Delta f}$$

or

$$\text{snr} = \left(\frac{\lambda}{4\pi r^2}\right)^2 \frac{P_T G_T}{L_0}\left(\frac{G_R}{T}\right)\frac{1}{k\Delta f} \tag{56}$$

This equation can be written as

$$(G_R/T)_{\text{dB}} = (\text{snr})_{\text{dB}} - P_L - \text{EIRP} + (L_0)_{\text{dB}} + (\Delta f)_{\text{dB}} + k_{\text{dB}} \tag{57}$$

where

$(G_R/T)_{\text{dB}}$ = ratio of ground receiver antenna gain to noise temperature

$(\text{snr})_{\text{dB}}$ = signal-to-noise ratio in the ground receiver

$P_L = 10\log(\lambda/4\pi r^2)^2$, the path loss from satellite to Earth

$\text{EIRP} = 10\log(P_T G_T)$, the effective isotropically radiated power in decibels above 1 W (dBW)

$(L_0)_{\text{dB}}$ = atmospheric loss in decibels

$(\Delta f)_{\text{dB}} = 10\log(\Delta f)$, channel bandwidth in decibels

$k_{\text{dB}} = 10\log(1.38 \times 10^{-23}) = -228.6$ dB

The G_R/T quantity above is a figure of merit for the Earth station. Typically the sky temperature is on the order of 3 to 20 K. The contribution to the antenna temperature due to the side lobes facing the Earth will decrease as the antenna elevation is increased. Typically the antenna temperature will be on the order of 25 K for elevation angle above 45° and will rise to perhaps 90 K as the antenna elevation is lowered to about 5° above the horizon as shown in Fig. 16. Consequently one would provide a high antenna gain and a vanishingly small side lobe in order to increase the G_R/T figure of merit for the Earth station.

Earth Station Receiving Antenna—From (57) one can determine the antenna type and size. As an example, let us consider low-power, direct-broadcast, satellite services which are those available in the *C* band between 3.7 to 4.2 GHz; the satellite transmits an output power of 36 dBW (an output power P_T on the order of 5 to 10 W and the onboard antenna gain G_T of 30 dB), namely, the EIRP of 36 dBW. The path loss of −196 dB (22 300 mi or 35 880 km from satellite to Earth at 4 GHz), an atmospheric attenuation of 2 dB, and the channel bandwidth of 74.7 dB (for assuming a bandwidth of 30 MHz) are estimated.

To ensure reasonably good picture quality the signal-to-noise ratio (snr)$_{dB}$, which depends on the receiver quality, needs to be about 12 dB. The ground receiver antenna gain-to-noise ratio $(G_R/T)_{dB}$ becomes 20.1 dB by substituting all the estimated values into the above equation. Assuming the effective antenna temperature of 40 K, the minimum required receiving antenna gain G_R becomes 36 dB. Once the antenna gain is defined, the aperture size, aperture taper, and antenna type can be selected. The design procedure of the antenna for a specified

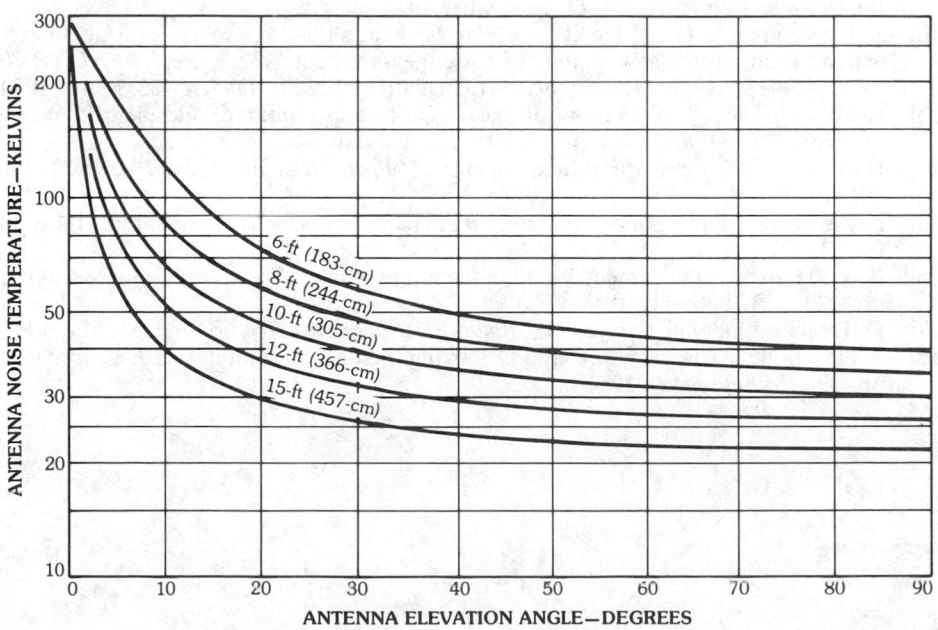

Fig. 16. Typical variation of antenna noise temperature with elevation angle, for five popular antenna sizes. (*Courtesy James Corum and Basil Pinzone*)

gain, side lobe level, and bandwidth are described in the chapters on reflector antennas and arrays.

11. References

[1] G. Sinclair, "The transmission and the reception of elliptically polarized waves," *Proc. IRE*, vol. 38, pp. 148–151, February 1950.

[2] C. T. Tai, "On the definition of the effective aperture of antennas," *IEEE Trans. Antennas Propag.*, vol. AP-9, pp. 224–225, March 1961.

[3] G. A. Deschamps, "Geometrical representation of the plane electromagnetic wave," *Proc. IRE*, vol. 39, pp. 540–544, 1961.

[4] H. T. Friis, "A note on a simple transmission formula," *Proc. IRE*, vol. 34, pp. 254–256, 1946; see also [5], pp. 388–394.

[5] S. A. Schelkunoff and H. T. Friis, *Antenna Theory and Practice*, New York: John Wiley & Sons, 1952.

[6] J. D. Kraus, *Antennas*, New York: McGraw-Hill Book Co., 1950.

[7] R. C. Hansen, "Efficiency and matching trade-offs for inductively loaded short antennas," *IEEE Trans. Commun.*, vol. COM-23, no. 4, April 1975.

[8] E. A. Mel'nikov and L. N. Mel'nikova, "Receiving induction ferrite antennas," *Izvestiya Vysshikh Uchebnykh Zavedeniy: Radioelectronica*, tr. Leo Kanner Associates, vol. 17, no. 10, 1974. Obtainable from the Department of the Army, US Army Foreign Science and Technology Center, 230 Seventh St. NE, Charlottesville, Va 22901.

[9] V. H. Rumsey and W. L. Weeks, "Electrically small loaded long antennas," *IRE Natl. Con. Rec.*, vol. 4, pt. 1, pp. 165–170, 1956.

[10] L. J. Chu, "Physical limitations of omnidirectional antennas," *J. Appl. Phys.*, vol. 19, p. 1163, December 1948.

[11] C. M. Desantis, "Low profile antenna performance study, pt. III: bibliography," *R&D Tech. Rep ECOM-4547*, November 1977.

[12] J. J. Sweeny and G. A. Deschamps, "A wave amplitude approach to noise analysis with application to a new method of noise measurement," *Tech. Rep. AFAL-TR-70-20*, Air Force Avionics Laboratory, Wright-Patterson AFB, Ohio 45433, March 1970.

[13] E. W. Harold, "An analysis of the signal-to-noise ratio of ultrahigh-frequency receivers," *RCA Rev.*, vol. 6, pp. 302–322, January 1942.

[14] H. T. Friis, "Noise figure of radio receivers," *Proc. IRE*, vol. 32, pp. 419–422, July 1944.

[15] E. A. Laport, *Radio Antenna Engineering*, chapter 1, New York: McGraw-Hill Book Co., 1952.

[16] R. C. Hansen, "Fundamental limitations in antennas," *Proc. IEEE*, vol. 69, no. 2, pp. 170–173, February 1981.

[17] D. Decker, "Satellite tv receiver," *Radio Electronics*, vol. 53, no. 5, p. 51, May 1982.

[18] T. E. Bell, "The new television: looking behind the tube," *IEEE Spectrum*, pp. 52–53, September 1984.

Chapter 7

Wire and Loop Antennas

L. W. Rispin
MIT Lincoln Laboratory

D. C. Chang
University of Colorado

CONTENTS

Lawrence W. Rispin received the BSEE degree from Akron University, in Akron, Ohio, in 1971, and an MSE degree from Arizona State University, in Tempe, Arizona, in 1973. In 1982 Dr. Rispin was awarded a PhD in electrical engineering from the University of Colorado, Boulder, Colorado. His doctoral dissertation involved research in cylindrical antenna theory.

His industrial experience includes cooperative student employment with Diebold, Inc., Canton, Ohio, and staff positions with Hughes Aircraft Company, Fullerton, California, and Motorola, Inc., Phoenix, Arizona, where he helped develop microwave power transistors. Currently he is in the Communication Antennas Group at MIT Lincoln Laboratory, in Lexington, Massachusetts, where he has been engaged in building an earth terminal for experimental EHF satellite communications.

He is a member of the IEEE, Eta Kappa Nu, and Sigma Tau. His major interests lie in analytical methods in antenna theory.

David C. Chang received his PhD in applied physics from Harvard University, Cambridge, Massachusetts, in 1967. He joined the faculty in the Department of Electrical and Computer Engineering, University of Colorado, Boulder, Colorado, in September 1967 and has been a professor of electrical and computer engineering since 1975, and Chairman of the Department since 1982.

Dr. Chang has been active in electromagnetic theory, antennas, and microwave circuits research. He has served, on various occasions, as the associate editor (1980–82), coordinator for the Distinguished Lecturers Program (1982–85), Chair of the Ad Hoc Committee for Basic Research (1985–present), and a member of the Administrative Committee (1985–present) of the IEEE Professional Society on Antennas and Propagation; as a member of the Technical Subcommittee on Microwave Field Theory of the IEEE Professional Society on Microwave Theory and Techniques (1975–85); as a member-at-large, US National Committee (1982–85), Chair of Technical Program Committee, Commission B on Fields and Waves (1983–86) of the URSI (International Union of Radio Science).

Dr. Chang is a Fellow of the IEEE.

1. Introduction

Wire and loop antennas are widely used in communication systems from low to ultra-high frequencies, either in the form of individual elements or arranged with other similar elements to form phased arrays. They are also frequently used as probes to sense unknown environments or as bases for modeling more complex systems and structures. In this chapter we shall be concerned with the important properties of wire and loop antennas as isolated elements. Our emphasis will be to develop simple expressions with sufficient accuracy for some basic antenna forms so that readers can generalize them to more complicated, composite structures pertaining to their particular needs without excessive reformulation and computing. For this reason our approach will be substantially different from those computationally more demanding methods reported in the literature. It is important for readers to recognize the physical description of our solution process in the later sessions in order to fully benefit from this approach.

The Thin-Wire Antenna

A general thin-wire antenna structure having the radius a and of length $2h$ is shown in Fig. 1a. The wire is assumed to be perfectly conducting and satisfies the following conditions:

$$a \ll \lambda_0 \quad \text{and} \quad a \ll h \tag{1}$$

where λ_0 is the free-space wavelength of a plane wave at an angular operating frequency ω in radians per second. The electrical properties of this antenna can be described by the axial current on its surface. Choosing the thin-wire conductor axis to coincide with the curvilinear coordinate s and assuming the axially directed current $I(s)$ to be uniformly distributed about the conductor surface, the average induced tangential electric field $\langle E_s(s) \rangle$ at the conductor surface generated by this current can be written in the general manner [1]

$$\langle E_s(s) \rangle = \int_C I(s') \, K(s;s') \, ds' \tag{2}$$

where C is the line contour along the antenna from $s = -h$ to h. The kernel $K(s;s')$ represents the tangential electric field produced by an elemental dipole moment, $I(s') \, ds'$. The specific form of the kernel $K(s;s')$ depends on the particular antenna structure at hand. For example, in the case of a linear dipole antenna, i.e., a straight wire as shown in Fig. 1b where the natural coordinate system is a cylindrical coordinate system, it is given as [2], [3],

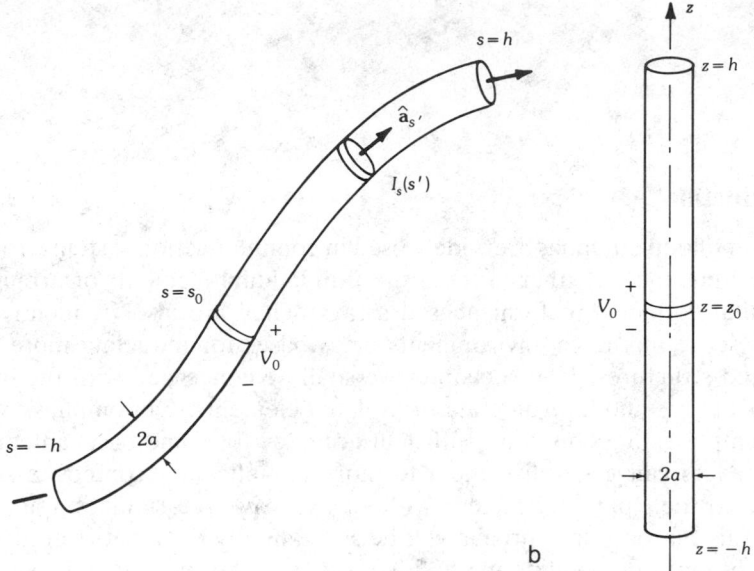

Fig. 1. The thin-wire antenna. (*a*) General form. (*b*) Straight wire.

$$K(z;z') = -j\frac{\xi_0}{4\pi}\left[\frac{d^2}{dz^2} + k_0^2\right] G(z;z') \tag{3}$$

where

$$G(z;z') = \frac{1}{2\pi}\int_{-\pi}^{\pi}\frac{e^{-jk_0R}}{k_0R}d\phi, \quad R = [(z - z')^2 + 4a^2\sin^2(\phi/2)]^{1/2} \tag{4}$$

and $\xi_0 = 120\pi$ ohms is the characteristic impedance in free space, k_0 is the wave number, and λ_0 is the free-space wavelength.

For a transmitting antenna we usually assume the antenna is excited by a source potential V_0 maintained across an infinitesimally small gap (i.e., a so-called delta-function voltage generator) in the thin-wire conductor at $s = s_0$ so that

$$\langle E_s(s)\rangle = -V_0\delta(s - s_0), \quad -h < s < h \tag{5}$$

Equating (5) with (2) yields an integral equation for finding the transmitting current.

In the receiving situation the induced electric field on the conductor must cancel the axially directed electric-field component associated with an incident plane wave, i.e.,

$$\langle E_s(s)\rangle = -\hat{\mathbf{a}}_s\cdot\mathbf{E}^i e^{-jk_0\hat{\mathbf{a}}_i\cdot\mathbf{x}}, \quad \mathbf{x}\in C \tag{6}$$

where $\mathbf{E}^i$ in volts per meter is the amplitude of the plane-wave field and $\hat{\mathbf{a}}_i$ is the unit vector in the direction of propagation of the plane wave; $\mathbf{x}$ is the position vector at an observation point s along the antenna and $\hat{\mathbf{a}}_s$ is the axially directed unit vector at the same point. Equating (6) with (2) gives us the desired integral equation for finding the receiving current distribution.

Various approximate methods have been employed to solve the appropriate form of the integral equation in (2). Many of these methods [2–4] rely on the fact that the value of the kernel function $K(s;s')$ becomes extremely large at $s = s'$. Numerical techniques may also be employed to obtain numerically "exact" solutions for specific thin-wire structures. These methods will be consulted for the purposes of comparison in the sections to follow. The particular method we have adopted for our discussion, except in the case of loop antennas, is derived from the observation that the current distribution on a finite-length thin-wire antenna is essentially that of a standing wave resulting from multiply reflected currents which "bounce" back and forth between discontinuities (such as the ends, conductor radius changes, lumped impedance elements, etc.) in the thin-wire structures. Thus, by characterizing the reflected and transmitted current waves for each type of discontinuity, we can construct the standing-wave current distributions by superposing these effects.

Before we proceed with this method of constructing a solution for any particular thin-wire antenna, we need to first address some of the general issues regarding thin-wire antenna structures.

Input Admittance or Impedance

As mentioned earlier, the excitation of a thin-wire transmitting antenna is usually modeled mathematically by a finite voltage source V_0 maintained across an infinitesimal gap along the thin-wire structure. In practice, however, the antenna must be fed by some sort of transmission line, such as coaxial cable or open two-wire line. The mathematical model is therefore only a convenient way to separate the antenna problem from the circuit analysis of the transmission line. We now designate $J^T(s;s_0)$ as the current distribution of a transmitting antenna due to a *unit* voltage source at a feed point s_0, i.e., $I^T(s) = V_0 J^T(s;s_0)$. The superscript T refers to a transmitting antenna. The input admittance in siemens is then given by

$$Y_{in} = I^T(s_0)/V_0 \quad \text{or} \quad J^T(s_0;s_0) \tag{7}$$

and the input admittance in ohms is

$$Z_{in} = Y_{in}^{-1} \tag{8}$$

which acts as a load terminating the connecting transmission line. We should caution the readers, however, that because the electromagnetic coupling between the antenna and the transmission line usually cannot be completely avoided, an "end-correction" network usually has to be added if a high degree of accuracy is desired. The specific form of this network obviously depends on the particular way the antenna is fed. This topic is beyond the scope of this chapter and readers are referred to [2] for more detail.

An additional complication which often arises in thin-wire antenna analyses concerns the gap capacitance associated with the highly idealized delta-function voltage source. Because this capacitance is an integral part of the antenna admittance, we usually cannot separate one from the other. This means that the input susceptance calculation has an inherent error which can be avoided by detailed modeling of the actual source region, which is usually impractical.

Far-Field Radiation from a Thin-Wire Transmitting Antenna

The radiation field at an observation point (r, θ, ϕ) in the far zone or the Fresnel zone, i.e., $r \gg h$ and λ_0, can be expressed approximately in terms of a vector potential function $\mathbf{A}(r, \theta, \phi)$ as follows:

$$\mathbf{H} = -jk_0 \hat{\mathbf{a}}_r \times \mathbf{A} \quad \text{and} \quad \mathbf{E} = \xi_0 \mathbf{H} \times \hat{\mathbf{a}}_r \tag{9}$$

where $\hat{\mathbf{a}}_r$ is a radially directed unit vector at the observation point. Provided the normalized current distribution $J^T(s; s_0)$ is known, we can write the expression for $\mathbf{A}$ approximately as

$$\mathbf{A}(r, \theta, \phi) = V_0 \mathbf{F}(\theta, \phi; s_0) \frac{e^{-jk_0 r}}{4\pi r} \tag{10}$$

where

$$\mathbf{F}(\theta, \phi; s_0) = \int_C \hat{\mathbf{a}}_s J^T(s'; s_0) e^{jk_0 \hat{\mathbf{a}}_r \cdot \mathbf{x}'} ds' \tag{11}$$

in siemens-meters and $\mathbf{x}'$ again is the position vector at the source point s' and $\hat{\mathbf{a}}_s$ is the axial unit vector at that point. The function $\mathbf{F}$, which varies only angularly, is the so-called vector far-field pattern. In the case of a linear dipole antenna, $\mathbf{F}$ readily reduces to the more familiar form,

$$\mathbf{F}(\theta, \phi; z_0) = \hat{\mathbf{a}}_z F(\theta; z_0) \tag{12}$$

where

$$F(\theta; z_0) = \int_{-h}^{h} J^T(z'; z_0) e^{jk_0 z' \cos\theta} dz' \tag{13}$$

The Receiving Antenna

It is well known that the far-field pattern of a transmitting antenna has the same angular dependence as the current of an unloaded receiving antenna at the location corresponding to the feed point of the transmitting case. This can be readily observed when we compare (5) with (6), and conclude that the receiving case is actually equivalent to a transmitting antenna, excited by a distributed voltage source of amplitude $\hat{\mathbf{a}}_{s'} \cdot \mathbf{E}^i ds'$ at point s' on the antenna. The receiving current can thus be obtained by integrating over all the elementary sources on the antenna:

$$I^R(s) = \int_C (\mathbf{E}^i \cdot \hat{\mathbf{a}}_{s'}) J^T(s;s') e^{-jk_0 \hat{\mathbf{a}}_i \cdot \mathbf{x}'} ds' \tag{14}$$

where $\mathbf{E}^i = \hat{\mathbf{a}}_e E^i$ is the field intensity vector of an incident plane wave with an amplitude E^i volts per meter and polarized in the direction of the unit vector $\hat{\mathbf{a}}_e$; on the other hand, $\hat{\mathbf{a}}_i$ is the direction of propagation of the plane wave. To relate this current with the far-field pattern of a transmitting antenna fed at s, we first have to recognize that the reciprocity theorem requires that $J^T(s;s') = J^T(s';s)$. Now since the amplitude E^i of a plane-wave field is constant everywhere and the direction of the incident plane wave is opposite to the observation direction in the transmitting case, i.e., $\hat{\mathbf{a}}_r = -\hat{\mathbf{a}}_i$, we have from (11) and (14) the following result:

$$I^R(s) = \mathbf{E}^i \cdot \mathbf{F}(\pi - \theta, \phi; s) \tag{15}$$

The superscript R refers to the receiving current in this case. A normalized receiving current can be similarly defined as

$$J^R(\theta, \phi; s) = I^R(s)/E^i = \hat{\mathbf{a}}_e \cdot \mathbf{F}(\pi - \theta, \phi; s) \tag{16}$$

Loaded Thin-Wire Antennas

The electrical properties of a thin-wire antenna can be drastically altered by introducing lumped impedance elements along the antenna. The effects of such loading can be modeled mathematically by equivalent voltage sources corresponding to the actual voltage drops across individual loads. The overall current distribution on a thin-wire transmitting antenna having N voltage sources V_1, $V_2, \ldots, V_N$ in series with impedances $Z_1, Z_2, \ldots, Z_N$ located at $s = s_1, s_2, \ldots, s_N$ along the antenna can be written as

$$I^T(s) = \sum_{n=1}^{N} [V_n - Z_n I^T(s_n)] J^T(s;s_n), \quad -h < s < h \tag{17}$$

The values of $I^T(s_n)$ can be found from the N linear equations obtained by setting $s = s_1, s_2, \ldots, s_N$ on both sides of (17). If any of the Z_n's are simple passive impedance elements not associated with a voltage source, the corresponding V_n's in (17) are simply set to zero.

The first term in the square bracket of (17) represents the overall unloaded current distribution resulting from the N voltage sources. In the receiving situation this term would be replaced by one representing the unloaded receiving current distribution excited by any number (say M) of incident plane-wave fields. The loaded receiving current distribution is then given by

$$I^R(s) = \sum_{m=1}^{M} E_m^i J^R(\theta_m, \phi_m; s) - \sum_{n=1}^{N} Z_n I^R(s_n) J^T(s;s_n) \tag{18}$$

Again, we can determine $I^R(s_n)$ for $n = 1, 2, \ldots, N$ by setting $s = s_n$ on both sides of (18) and solving the set of N linear equations.

Transient Response

Study of the transient response on thin-wire structures has many contemporary applications, among them the use of these structures as electromagnetic pulse (EMP) simulators. Basically, the homogeneous integral equation [5, 6] associated with the expression in (2) possesses nontrivial solutions, $I(s; \omega_\alpha)$ for some complex frequencies $\omega = \omega_\alpha$, where $\alpha = 1, 2, \ldots$, so that

$$\int_C I(s'; \omega_\alpha) K(s, s'; \omega_\alpha) \, ds' = 0 \tag{19}$$

Each of these solutions can be identified as a resonance of the structure, very much in the same way as a waveguide cavity, where the electromagnetic waves excited by the current distribution on the cavity wall constructively interact with each other in phase. A thin-wire antenna, in fact, can be considered as an open resonator with complex natural frequency ω_α and associated natural mode currents $G_\alpha(s; \omega_\alpha)$, with $\alpha = 1, 2, \ldots$. Transient response of a thin-wire structure is then given by the so-called SEM method [5, 6]:

$$I(t, t_0; s, s_0) = \text{Re}\left\{ \sum_{\alpha=1}^{\infty} A_\alpha G_\alpha(s; \omega_\alpha) G_\alpha(s_0; \omega_\alpha) e^{j\omega_\alpha(t-t_0)} \right\} \tag{20}$$

where s_0 and t_0 are, respectively, the source location and turn-on time, and A_α is the excitation factor, which, of course, depends on the particular pulse shape of the voltage source. According to the time causality principle, such an expression can be used only after the arrival time, which is determined by the observation distance divided by the speed of light.

Equivalent Radius for Noncircular Cylindrical Thin-Wire Conductors

Should a thin-wire antenna be constructed from a noncircular cylindrical conductor, an equivalent radius [7] can be assigned to the antenna, provided that the current can be assumed to be fairly uniform around the noncircular conductor. For the cross section shown in Fig. 2, an approximate equivalent radius can be written as [7]

$$a_e = \exp\left\{ \frac{1}{S^2} \oint_\ell \oint_\ell \ln|\mathbf{w} - \mathbf{w}'| \, dw \, dw' \right\} \tag{21}$$

where S is the peripheral length around the cross section and $|\mathbf{w} - \mathbf{w}'|$ is the distance between two points on the contour ℓ bounding the peripheral surface that are located by the position vectors $\mathbf{w}$ and $\mathbf{w}'$. The equivalent radii for a few common noncircular thin-wire antenna-conductor cross sections are given in [7].

Solid Thin-Wire Antenna Conductors

For the most part the solutions given in this chapter are based on analyses of tubular thin-wire antennas. In practice, however, thin-wire antennas are usually

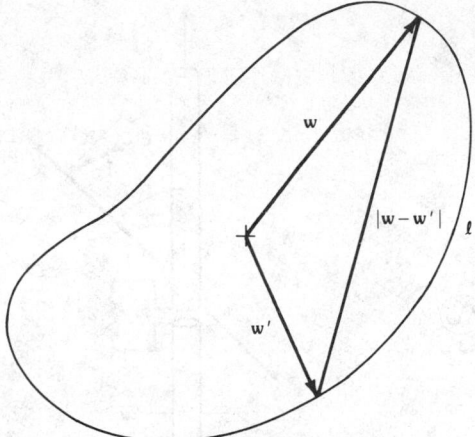

Fig. 2. General noncircular cylindrical conductor.

constructed from solid conductors. The essential effect is that an antenna made from a solid thin-wire conductor appears to be electrically slightly longer. The apparent increase in overall length has been shown to be approximately $0.26a$, which is indeed negligible for practical thin-wire structures [8].

Antenna Parameters

Before we conclude this section we should consider two important parameters frequently used to characterize a thin-wire antenna:

$$\Omega = 2\ln(2h/a), \qquad C_a = -\ln(k_0 a) - \gamma \tag{22}$$

where $\gamma = 0.577$ is the Euler constant. For thin-wire antennas that satisfy the conditions in (1), either Ω or C_a can be used as the large expansion parameter for developing asymptotic solutions in antenna problems.

2. The Linear Dipole Antenna

The linear dipole antenna depicted in Fig. 3 is one of the most basic antenna forms. By assuming the current is concentrated along the axis of the antenna, the integral equation given in (2) becomes more readily tractable. In fact, based on the observation that the kernel function $K(z; z')$ is at its peak when $z = z'$, iterative solutions and a few trigonometric-term solutions for the current distribution have been successfully developed and a large amount of data compiled [2, 3, 9, 10, 11]. Accurate numerical computation of this current can also be achieved using moment methods [4, 12, 13], sometimes without the constraint of the thin-wire approximation [14–15]. The approach we will adopt in this chapter, however, involves the construction of a current standing wave on the linear thin-wire antenna based upon a hollow cylindrical model [16–22]. Although this approach was generally regarded as a long-antenna theory in the past, recent improvements in the analysis [23–24] have removed many of the restrictions. The formulas to be presented in the

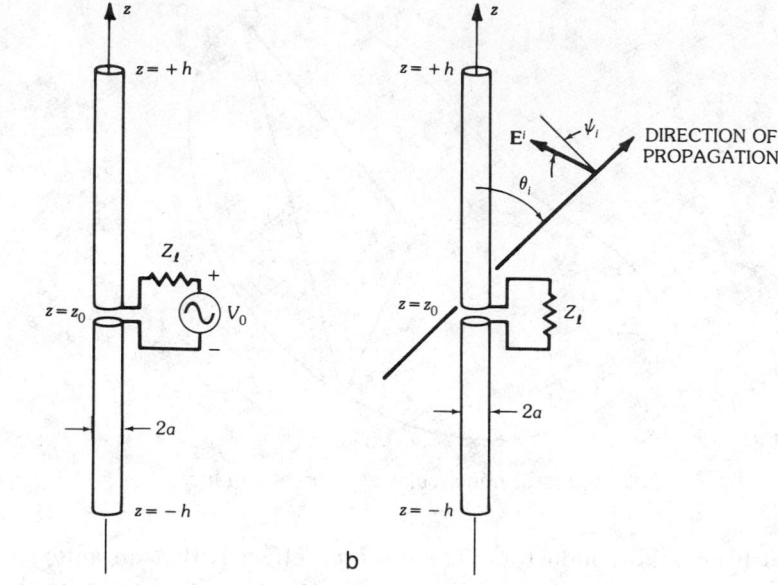

Fig. 3. Linear thin-wire dipole antenna. (*a*) Transmitting. (*b*) Receiving.

later sections of this chapter are well within the acceptable accuracy in engineering practice provided that

$$k_0 a \leqq 0.1 \quad \text{and} \quad \Omega \geqq 10 \tag{23}$$

which, of course, encompasses almost all the thin-wire antenna structures commonly used. As we mentioned earlier, our emphasis is to develop a simple procedure which will allow readers to extend the results in this chapter to more complicated, composite structures.

Unloaded Transmitting Antennas

Consider the linear thin-wire transmitting antenna illustrated in Fig. 3a. The voltage source V_0 applied across the electrically small gap at z_0 excites currents which travel (i.e., progress in phase) outward away from the source along the thin wire in a manner analogous to the propagation of current along a transmission line. Unlike the transmission-line current, however, the antenna current is subject to attenuation due to radiation. This initial, or "primary," current can be obtained from the current distribution existing between $-h < z < h$ on a similarly excited, infinitely long, linear, thin-wire antenna. An approximate expression for this current, here normalized to V_0, is given by [20]

$$U_a^s(|z-z_0|) = \frac{-j}{\xi_0} e^{-jk_0|z-z_0|}$$

$$\times \ln\left\{1 + \frac{j2\pi}{2C_a + \gamma - j3\pi/2 + \ln[k_0|z-z_0| + (k_0^2|z-z_0|^2 + e^{-2\gamma})^{1/2}]}\right\} \tag{24}$$

where $\xi_0 = 120\pi$ ohms and $\gamma = 0.577$. Originally derived for $k_0|z - z_0| > 1$, the above expression gives an accurate value for the real part of the input current even when it is evaluated at the feed point [11].

As depicted in Fig. 4a, the primary current emanating from the source at z_0 impinges on the ends of the antenna and is reflected back in the opposite direction. The form of these initial "secondary" currents can be obtained from a semi-infinitely long antenna of the same radius, excited by waves incident at the angle π with respect to a particular end. Determined from a Wiener-Hopf analysis, the initial secondary currents shown in Fig. 4b from both ends are denoted by the terms $-V_0 U_a^s(h - z_0) R_a U_a(h - z)$ and $-V_0 U_a^s(h + z_0) R_a U_a(h + z)$, respectively, where $V_0 U_a^s(h \mp z_0)$ is the value of the primary current at $z = \pm h$. The *reflection coefficient* R_a is given approximately by

$$R_a = \frac{\xi_0}{2\pi}(2C_a - j\pi)/(1 + \delta_a)^2 \tag{25}$$

where

$$\delta_a = \mathrm{Re}\left\{\frac{-j}{2\pi}\ln\left(1 + \frac{j2\pi}{2C_a - j3\pi/2}\right) - \frac{1}{2C_a - j\pi + \ln 2}\right\} \tag{26}$$

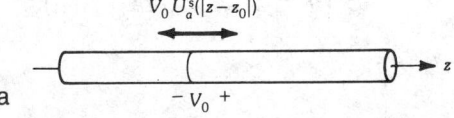

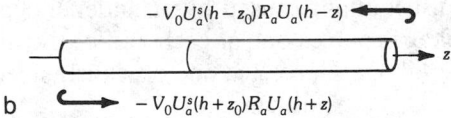

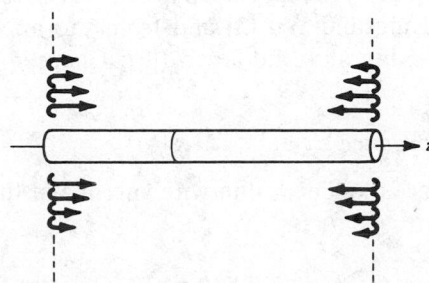

Fig. 4. The multiple reflection concept as applied to a finite-length thin-wire transmitting antenna. (*a*) Primary current emanating from the source. (*b*) Initial reflected currents from both antenna ends. (*c*) Subsequent multiple reflected currents from antenna ends.

The expression for the current distribution of the reflected wave is given by

$$U_a(z) = \frac{2\pi}{\xi_0} \frac{e^{-jk_0z}}{2C_a + \gamma - j\pi/2 + \ln(2k_0z) + e^{j2k_0z}E_1(j2k_0z)} \qquad (27)$$

where E_1 is the exponential integral of the first kind ([25], Chapter 5). We note that the primary current U_a^s and the reflected distribution U_a are essentially the same for $k_0z \gtrsim 1$. As mentioned earlier, the small kz behavior of the former expression was manipulated in order to achieve an accurate real component of input current.

Continuing, the two secondary currents travel (progress in phase and attenuate) on the antenna in opposite directions until they impinge on the other ends, whereupon another secondary current is generated. This process continues on and on as depicted in Fig. 4c, leading to two sets of infinite, though summable, series of multiply reflected waves. Assembling the constituent components gives the total current distribution per unit volt on an unloaded thin-wire transmitting antenna as

$$J_a^T(z; z_0) = U_a^s(|z - z_0|) - Q_a^T(h)R_aU_a(h - z) - Q_a^T(-h)R_aU_a(h + z),$$

$$\text{for } -h < z < h \qquad (28)$$

where

$$Q_a^T(\pm h) = [U_a^s(h \mp z_0) - U_a^s(h \pm z_0)R_aU_a(2h)]/\Delta(h) \qquad (29)$$

$$\Delta(h) = 1 - R_a^2 U_a^2(2h) \qquad (30)$$

The tacit equivalence between the primary transmitting current $U_a^s(z)$ and the reflected current $U_a(z)$ means that the total current distribution may be thought of as the superimposed distributions caused by an independent unit voltage source and two dependent voltage sources one at each end of the antenna. To obtain the actual current due to a voltage source V_0, one obviously needs only to multiply by V_0 to obtain $I_a^T(z) = V_0 J_a^T(z; z_0)$ in amperes.

The current distribution on a center-driven, half-wave ($k_0h = \pi/2$) linear antenna where $\Omega = 2\ln(2h/a) = 10$ is shown in Fig. 5. Corresponding data from the three-term theory of King and Wu [3] and from the approximate second-order iteration procedure of King and Middleton ([2], Chapter 1, Section 22) are also shown in the figure.

Input Admittance or Impedance

The input admittance of a linear thin-wire antenna at the position z_0 is given in siemens by

$$Y_{\text{in}} = U_a^s(0) - Q_a^T(h)R_aU_a(h - z_0) - Q_a^T(-h)R_aU_a(h + z_0) \qquad (31)$$

Figs. 6a and 6b show the input conductance and susceptance, respectively, of a center-fed antenna as a function of half-length k_0h. Agreement with the results in

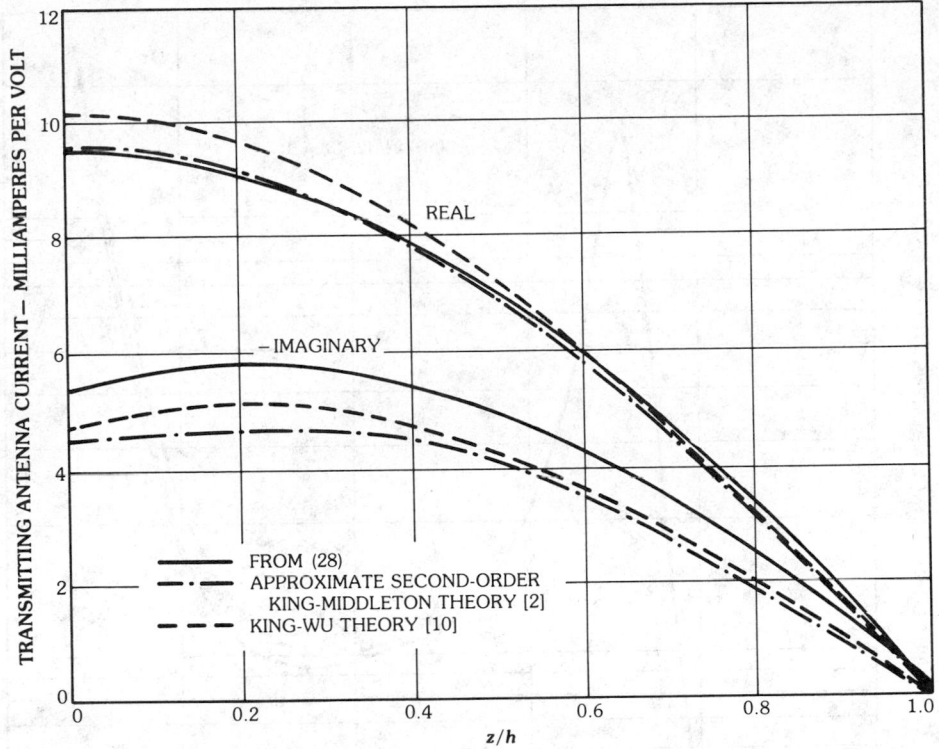

Fig. 5. Current distribution on a linear thin-wire transmitting antenna with $kh = \pi/2$, $\Omega = 10$.

[2] and [3] is indeed very good in the case of input conductance for all lengths. The shift in the input susceptance is closely related to the issue of how to determine the gap capacitance inherent in the idealized voltage source as we addressed earlier [26].

A close look at the input conductance near the first resonance $k_0 h = \pi/2$ for several values of the antenna parameter Ω is given in Fig. 7. Note that as the antenna becomes thinner (Ω increases), the bandwidth as well as the maximum value of G decreases and the resonance peak approaches $k_0 h = \pi/2$.

Far-Field Radiation from a Linear Thin-Wire Transmitting Antenna

Using the current distribution in (28), the far-field pattern function of a thin-wire antenna can be obtained approximately from (11) as

$$
F(\theta; z_0) = \left(\frac{j2\pi}{k_0 \xi_0}\right) \Big\{ e^{jk_0 z_0 \cos\theta} [W_a(h + z_0; \theta) + W_a(h - z_0; \pi - \theta)]
$$
$$
- R_a [e^{jk_0 h \cos\theta} Q_a^T(h) W_a(2h; \theta)
$$
$$
+ e^{-jk_0 h \cos\theta} Q_a^T(-h) W_a(2h; \pi - \theta)] \Big\} \tag{32}
$$

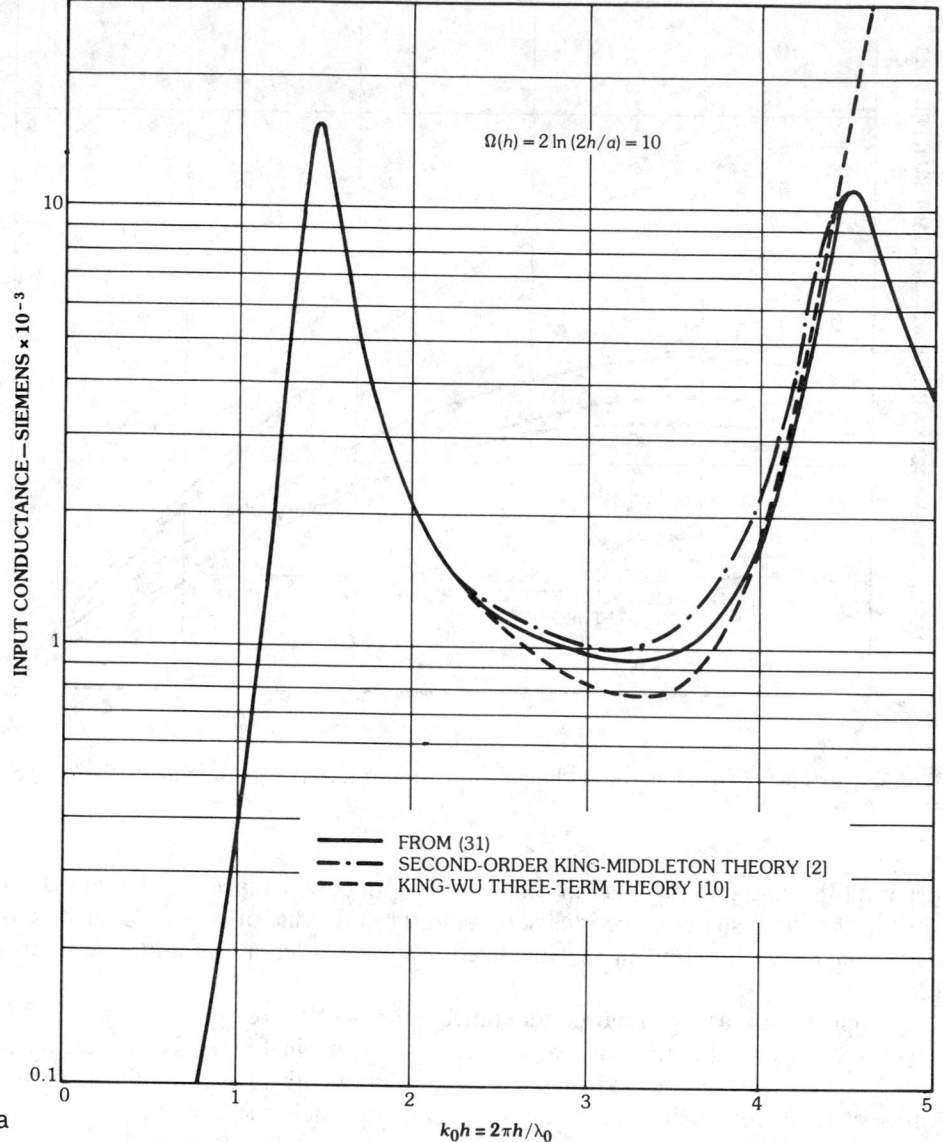

Fig. 6. Input impedance of a linear thin-wire antenna with $0 < kh < 5$, $\Omega = 10$. (a) Input conductance. (b) Input susceptance.

where, based on a similar evaluation in [27],

$$
W_a(z;\theta) = \frac{1}{1 + \cos\theta}\left[\frac{e^{-jk_0z(1+\cos\theta)}}{2C_a - j\pi/2 + \gamma + P(\theta;z)}\right.
$$

$$
\left. - \frac{1}{2C_a - j\pi - 2\csc^2(\theta/2)\ln\cos(\theta/2)}\right]
$$

(33a)

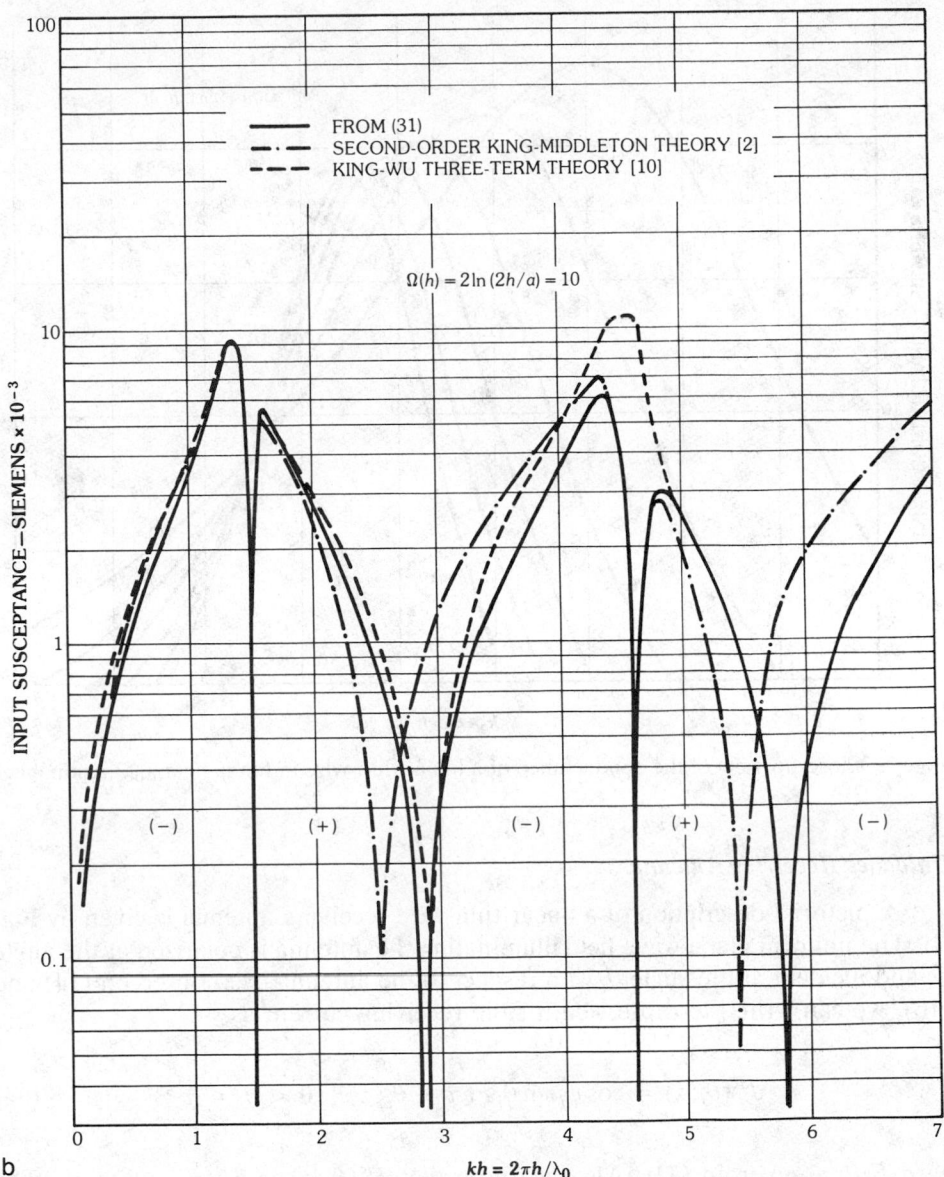

Fig. 6, *continued.*

$$P(\theta; z) = \ln(2k_0 z) + \csc^2(\theta/2) e^{jk_0 z(1+\cos\theta)} E_1(jk_0 z[1 + \cos\theta])$$
$$- \cot^2(\theta/2) e^{j2k_0 z} E_1(j2k_0 z) \tag{33b}$$

Here $\gamma = 0.577$ is the Euler constant and E_1 again is the exponential integral of the first kind. We note that for an infinitely thin antenna, i.e., $C_a \to \infty$, the pattern function readily reduces to the well-known expression for a sinusoidal current given earlier in Chapter 1.

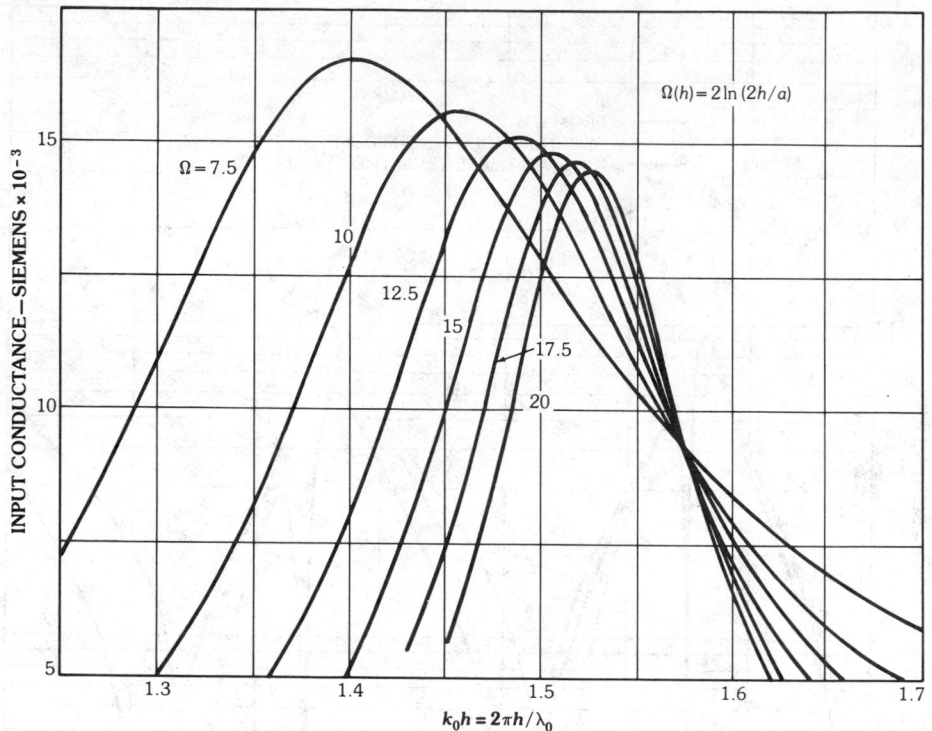

Fig. 7. Close-up view of the conductance of a linear thin-wire antenna near $\lambda_0/2$ resonance.

Unloaded Receiving Antennas

A pictorial description of a linear thin-wire receiving antenna is given by Fig. 3b. The uniform plane-wave field illuminating the antenna is polarized at the angle ψ_i and incident at the angle θ_i with respect to the antenna axis. Based on (14) and (16), we can write the expression for the receiving current as

$$J_a^R(\theta_i; z) = \cos \psi_i \sin \theta_i \, F(\pi - \theta_i; z), \quad 0 < \theta_i < \pi \qquad (34)$$

with $F(\theta; z)$ given in (32). Also, we note that $J_a^R(\theta_i; z) = J_a^R(\pi - \theta_i; -z)$. The actual receiving current is $I^R(z) = E^i J_a^R(\theta_j; z)$, where E^i is the complex amplitude of the incident plane wave in volts per meter.

The current distributions on a half-wave receiving antenna for several incident angles are shown in Fig. 8. Corresponding data for the normal incidence ($\theta_i = \pi/2$) case as determined from the first-order King-Middleton theory ([2], Section IV.7) and the three-term theory of King [9] are also shown.

The receiving current antenna pattern at three different positions is shown in Fig. 9. Note the almost sinusoidal (as it is often assumed) behavior with respect to θ_i. This figure also corresponds to the far-field pattern (when θ_i is replaced by $\pi - \theta_i$) of a linear thin-wire transmitting antenna fed at the same points indicated.

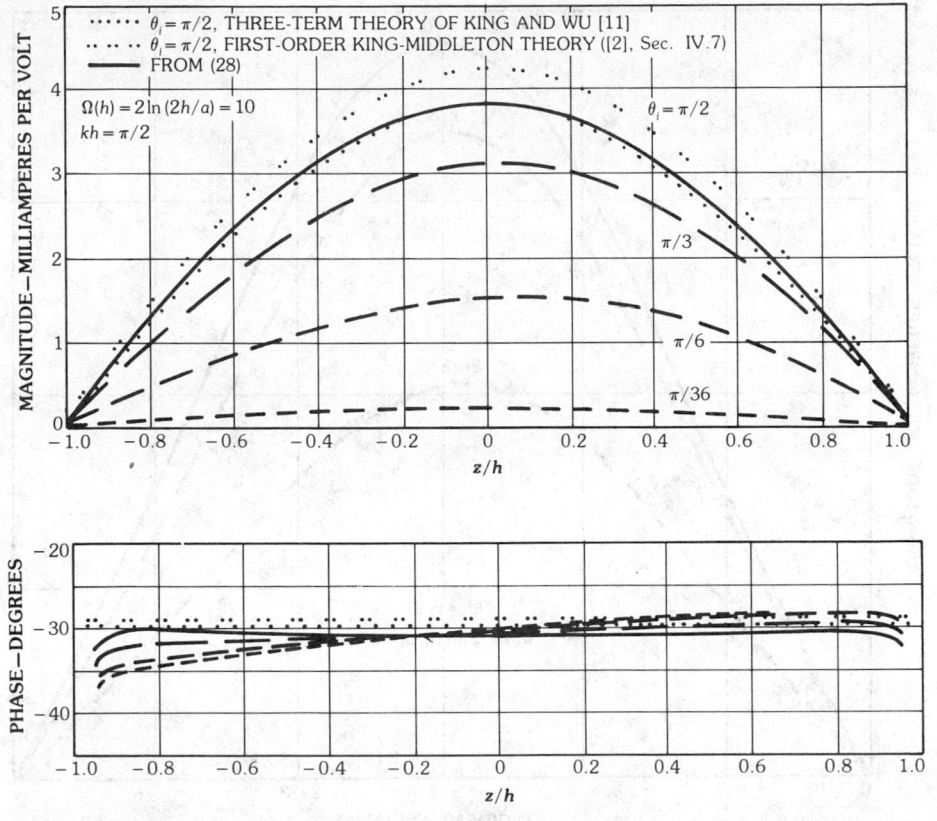

Fig. 8. Receiving current distribution (normalized to the incident electric field and the wavelength, i.e., $I_a^R(\theta_i; z)/\lambda_0 E^i$) with $kh = \pi/2$, $\Omega = 10$.

Impedance-Loaded Antennas

Expressions for the total current on a dipole antenna with an impedance loading of Z_ℓ at the location z_0 is immediately available from (17) and (18):

$$I_a^T(z) = V_0 \left(\frac{Y_\ell}{Y_{\text{in}} + Y_\ell} \right) J_a^T(z; z_0) \tag{35}$$

for a transmitting antenna, and

$$I_a^R(\theta_i; z) = E^i \left\{ J_a^R(\theta_i; z) - \frac{J_a^R(\theta_i; z_0)}{Y_{\text{in}} + Y_\ell} J_a^T(z; z_0) \right\} \tag{36}$$

for a receiving antenna, where $Y_\ell = Z_\ell^{-1}$. A multiple impedance loading can be handled by following a similar procedure as in (17) and (18).

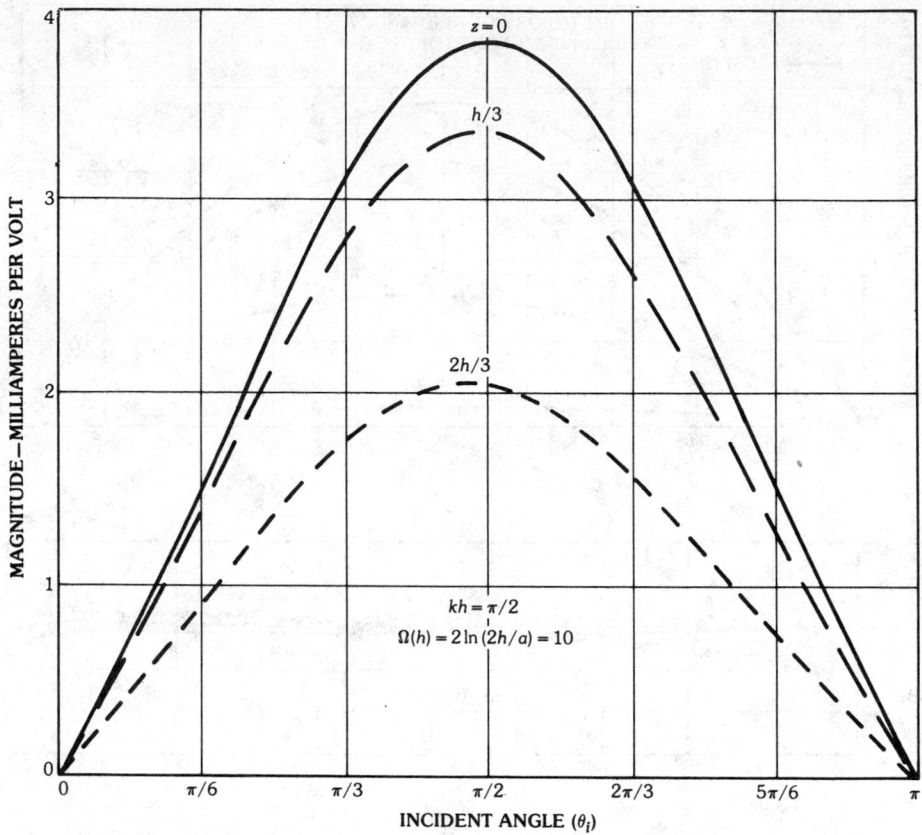

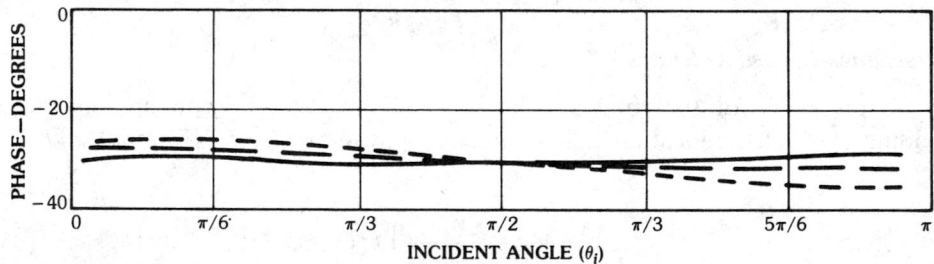

Fig. 9. Receiving current (normalized to the incident electric field and the wavelength, i.e., $I_a^R(\theta_i;z)/\lambda_0 E^i$) with $0 < \theta_i < \pi$, $kh = \pi/2$, $\Omega = 10$.

An Impedance-Loaded Monopole Antenna

A monopole antenna is a thin-wire antenna mounted vertically above a large conducting ground plane and driven at or near its base. Use of the image theorem allows us to replace it by a dipole antenna with two voltage sources at $\pm z_0$ and the associated load impedances. We can therefore write, with the help of (17) and (18),

$$I^T(z) = \frac{Y_\ell V_0}{Y_\ell + Y_{\text{in}} + J_a^T(z_0; -z_0)}[J_a^T(z; z_0) + J_a^T(z; -z_0)] \qquad (37)$$

for the transmitting antenna and

$$I^R(z) = E^i\left\{ J_a^R(\theta_i; z) + J_a^R(\pi - \theta_i; z) \right.$$

$$\left. - \frac{[J_a^R(\theta_i; z_0) + J_a^R(\pi - \theta_i; z_0)]}{Y_\ell + Y_{\text{in}} + J_a^T(z_0; -z_0)}[J_a^T(z; z_0) + J_a^T(z; -z_0)] \right\} \qquad (38)$$

for a receiving antenna. Here the input admittance $Y_{\text{in}} = J_a^T(z_0; z_0)$, and the currents J_a^R and J_a^T refer to those on a linear antenna having an overall length of $2h$. Allowing the feed point to approach the conducting plane, i.e., $z_0 \to 0$, yields the well-known relationships between monopole and dipole antennas.

Transient Behavior of a Dipole Antenna

As we mentioned in the general discussion, transient behavior of a thin-wire dipole antenna acting as an open resonator can be described by a set of natural frequencies and natural modes. According to (19), these modes can be determined by seeking the nontrivial solution of the current distribution in the absence of a voltage source. From the expression that $I^T(z) = V_0 J_a^T(z; z_0)$ with the normalized current J_a^T given by (28) through (30), we can readily establish the resonant condition is

$$\Delta(h; \omega_\alpha) \equiv 1 - R_a^2(\omega_\alpha) U_a^2(2h; \omega_\alpha) = 0, \qquad \alpha = 1, 2, \dots \qquad (39)$$

where $R_a(\omega_\alpha)$, $U_a(2h; \omega_\alpha)$ and $\Delta(h; \omega_\alpha)$ are given by (25), (27), and (30), with k_0 replaced everywhere by ω_α/c.

Now since R_a and U_a in (25) through (27) are given by simple functions, searching for the roots of (39) in the complex plane is a relatively simple task. The transient current response of a transmitting antenna with a step-function voltage source, $V(t) = V_0$ for $t > t_0$ and 0 for $t < t_0$, is derived in [27] as

$$I^T(t, t_0; z, z_0) = 2V_0 \operatorname{Re}\left\{ \sum_{\alpha=1}^{\infty} \frac{R_a(\omega_\alpha)}{\omega_\alpha \dfrac{\partial \Delta(h; \omega_\alpha)}{\partial \omega_\alpha}} G_\alpha(z; \omega_\alpha) G_\alpha(z_0; \omega_\alpha) \right.$$

$$\left. \times \exp[j\omega_\alpha(t - t_0 - |z - z_0|/c)] \right\} \qquad (40a)$$

after the arrival time $t > t_0 + |z - z_0|/c$, where c is the speed of light, and z_0 and t_0 are respectively the location and the turn-on time of the source. In a similar manner the transient current response of a receiving antenna due to a step-function uniform plane wave $\mathbf{E}^i(t:\mathbf{r}) = \mathbf{E}^i \exp[j\omega(t - z\cos\theta_i/c)]$ for $t > [(h + z)/c]\cos\theta_i$, and 0 for $t < [(h + z)/c]\cos\theta_i$ can be written as [27]

$$-I^R(t:\theta_i;z) = 2E^i\cos\psi_i\sin\theta_i \operatorname{Re}\left\{ \sum_{\alpha=1}^{\infty} \frac{R_\alpha(\omega_\alpha)}{\omega_\alpha \dfrac{\partial\Delta(h;\omega_\alpha)}{\partial\omega_\alpha}} G_\alpha(z;\omega_\alpha) \right.$$

$$\times \left(\frac{j2\pi}{\xi_0\omega_\alpha/c}\right)\{\exp(j\omega_\alpha h\cos\theta_i/c)[W_\alpha(z_2;\theta_i;\omega_\alpha)$$

$$- W_\alpha(z_1;\theta_i;\omega_\alpha)] - (-1)^\alpha \exp(-j\omega_\alpha h\cos\theta_i/c)$$

$$\times [W_\alpha(2h - z_2;\pi - \theta_i;\omega_\alpha)$$

$$\left. - W_\alpha(2h - z_1;\pi - \theta_i;\omega_\alpha)]\} \exp(j\omega_\alpha t) \right\} \tag{40b}$$

after the arrival time $t > (h + z)\cos\theta_i/c$ where $W_\alpha(z;\theta;\omega_\alpha)$ is based on (33) with k_0 replaced everywhere by ω_α/c and

$$z_1 = \begin{cases} \left(\dfrac{h + z - ct}{1 - \cos\theta_i}\right) & \text{for } (h + z)\cos\theta_i/c < t < (h + z)/c \\ 0, & \text{otherwise} \end{cases}$$

$$z_2 = \begin{cases} 0 & \text{for } t < (h + z)\cos\theta_i/c \\ \left(\dfrac{h + z + ct}{1 + \cos\theta_i}\right) & \text{for } (h + z)\cos\theta_i/c < t < (2h\cos\theta_i + h - z)/c \\ 2h & \text{for } t > (2h\cos\theta_i + h - z)/c \end{cases}$$

In both transmitting and receiving cases the natural-mode current $G_\alpha(z;\omega_\alpha)$ is simply

$$G_\alpha(z;\omega_\alpha) = U_\alpha(h + z;\omega_\alpha) - (-1)^\alpha U_\alpha(h - z;\omega_\alpha) \tag{41}$$

and has either a basically sinusoidal (α even) or a cosinusoidal (α odd) distribution. Here, $U_\alpha(z;\omega_\alpha)$ is obtained from (27) for $U_\alpha(z)$ with k_0 everywhere replaced by ω_α/c.

Transient response of the feed-point current of a center-fed dipole antenna with $\Omega = 2\ln(2h/a) = 10$ due to a step-function voltage source is shown in Fig. 10 and compared to the numerical results in [29] and as previously derived as a closed-form result in [28]. The transient response for the midpoint current of a receiving antenna ($\Omega = 10$) illuminated by a plane wave at an angle of 90° with respect to the antenna, and with a step-function electric field in the plane of incidence, is shown in Fig. 11 and again compared with other numerical [29] and analytical [28] results.

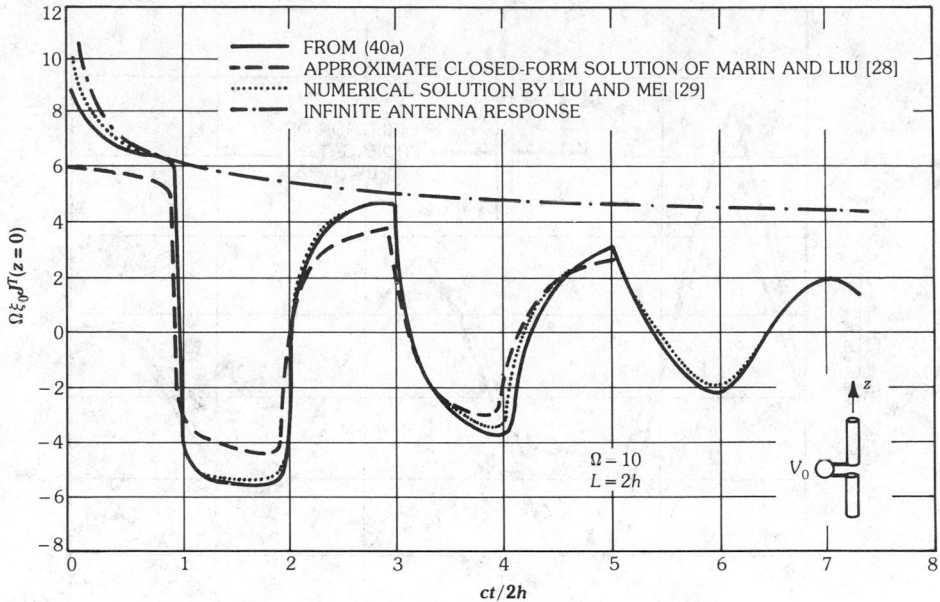

Fig. 10. Time-domain response of the driving-point current on a linear antenna excited by a step-function voltage applied at $z_0 = 0$ at $t_0 = 0$. (*After Hoorfar and Chang [27], © 1982 IEEE*)

3. The Sleeve Antenna

A thin-wire sleeve, or, as it is sometimes called, a "coaxial" antenna, basically consists of two coaxial tubular cylinders, the thinner, longer one fitting partially within the larger (in radius) shorter one, such as depicted in Fig. 12. In normal applications this type of antenna is vertically oriented and is quite similar in appearance as well as electrical characteristics to the linear thin-wire antenna described in the previous section. An important practical advantage of the sleeve antenna is the coaxial line contained within the structure, which can be used as a means of feeding the antenna. Utilization of this type of feed, however, requires that the feed line exiting the antenna be decoupled from the antenna itself.

Approximate simulations [2, 30] of the sleeve antenna have been formulated using the linear thin-wire antenna as a basis. A direct approach [31] based on the same concepts of constructing a standing-wave solution as in the last section will be discussed here. Our method is similar to the one pursued by Hurd [32].

The Junction Effect

We need to first examine the effect of the junction created by the truncation of the outer conductor of the thin-wire coaxial line shown in Figs. 13a through 13c. There are three possible current waves that can impinge on this junction: an antenna current $U_a(z_a - z)$ emanating from a source at z_a on the extended portion of the inner conductor of radius a which acts as an antenna, an antenna current $U_b(z - z_b)$ emanating from a source at z_b on the outer surface $\varrho = b^+$ of the outer

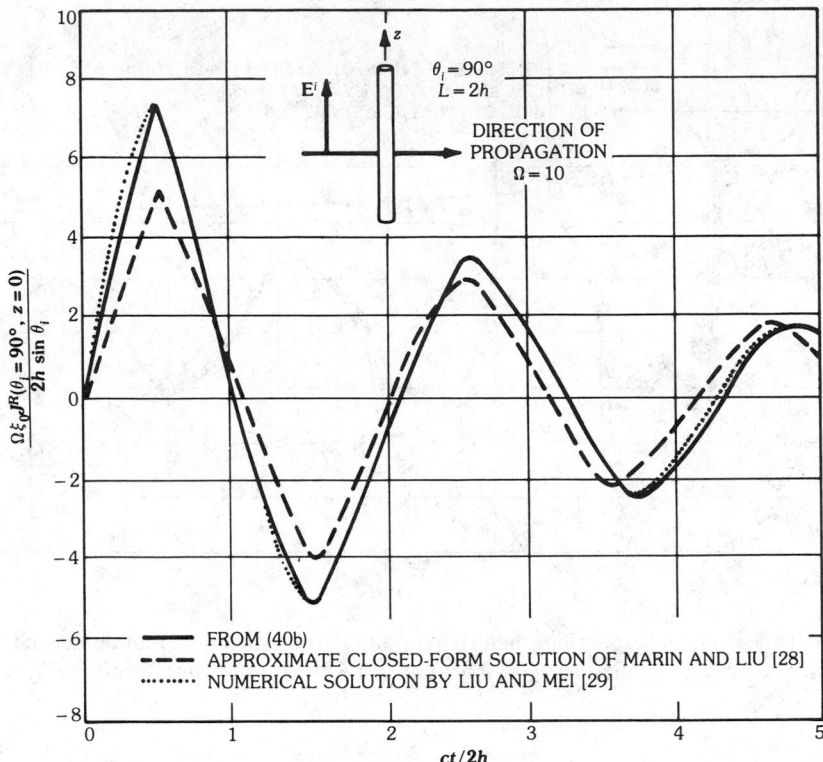

Fig. 11. Time-domain response of the current at the center of a linear antenna excited by a normally incident uniform plane wave arriving at $t_0 = 0$. (*After Hoorfar and Chang [27]*, © *1982 IEEE*)

conductor of radius b which also acts as an antenna, and finally a transmission-line current $Z_0^{-1} \exp[-jk_0(z - z_c)]$ amperes from a source at z_c inside the coaxial region with a characteristic impedance $Z_0 = (\xi_0/2\pi) \ln(b/a)$ ohms where $\xi_0 = 120\pi$ ohms. For convenience we should only consider a "one-sided" voltage source [33] for the excitation of the antenna currents U_b so that we can ignore any other current excitation on the inner surface of the same conductor. Reflections and transmission of these current waves have been previously determined by [31] using a Wiener-Hopf method. The results given in [31] are summarized as follows:

1. The junction has essentially no effect on an antenna current incident from the left. The current wave $U_a(z_a - z)$ in this case continues unperturbed from conductor a to the outer surface of conductor b. In doing so, transmission-line currents $U_a(z_a) \exp(jk_0 z)$ on conductor a and $-U_a(z_a) \exp(jk_0 z)$ on the inner surface of conductor b are transmitted into the coaxial region, thereby providing a smooth transition between the antenna current $U_a(z_a - z)$ and the transmission-line current $U_a(z_a) \exp(jk_0 z)$ at the junction.

2. When an antenna current $U_b(z - z_b)$ is incident from the right, it also continues to propagate across the junction from conductor b to conductor a. How-

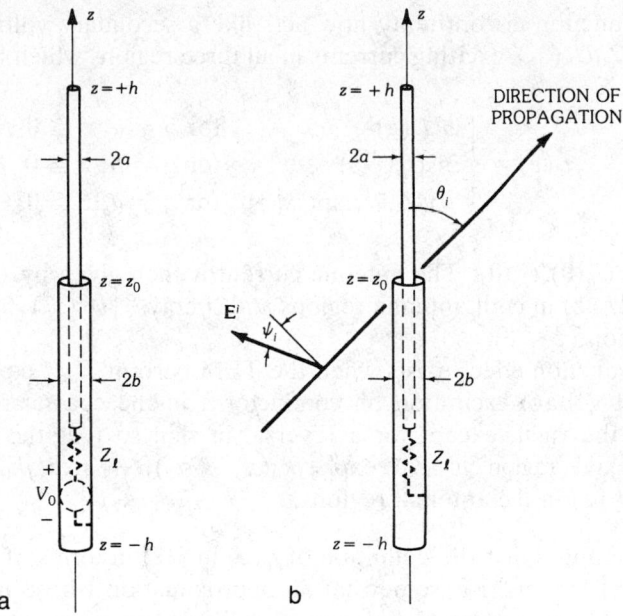

Fig. 12. Sleeve antenna illustration. (*a*) Transmitting. (*b*) Receiving.

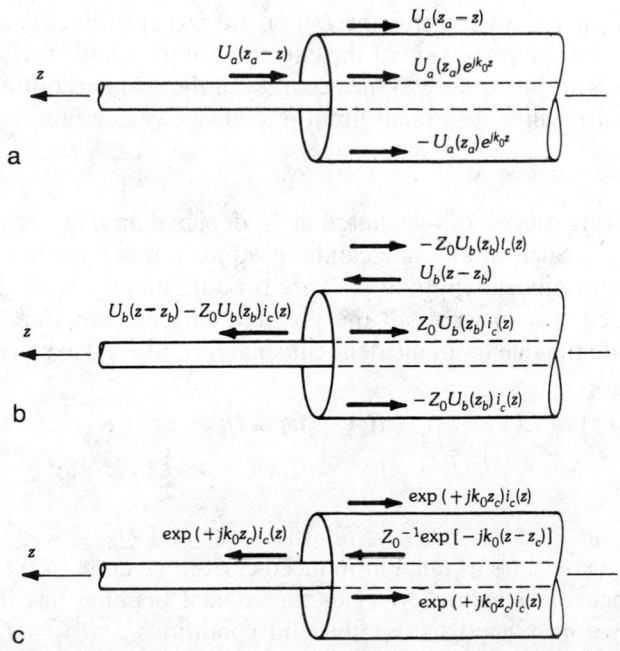

Fig. 13. Coaxial sleeve junction effect. (*a*) Antenna current $U_a(z_a - z)$ impinging on junction. (*b*) Antenna current $U_b(z - z_b)$ impinging on junction. (*c*) Transmission-line current $Z_0^{-1} \exp[-jK_0(z - z_c)]$ impinging on junction. Here, junction location is taken to be $z = 0$.

ever, the junction discontinuity now acts like a secondary voltage source of amplitude $Z_0 U_b(z_b)$, exciting currents in all three regions which have the form

$$i_c(z) = \begin{cases} 2\,U_a^s(z) & \text{for } \varrho = a,\ z \gtreqqless 0 \\ 2v\,U_b^s(-z) & \text{for } \varrho = b,\ z \lesseqqgtr 0 \\ -v\,Z_0^{-1}\exp(jk_0z) & \text{for } \varrho = a,\ z \lesseqqgtr 0 \end{cases} \qquad (42)$$

where $v = U_a^s(0)/U_b^s(0)$. The antenna current is then given by $U_b(z + z_b) - [U_b(z_b)\,Z_0]\,i_c(z)$ in both antenna regions and simply $-[U_b(z_b)\,Z_0)]\,i_c(z)$ in the coaxial region ($z < 0$).

3. The same junction effect exists when the TEM current $Z_0^{-1}\exp[jk_0(z + z_c)]$ due to unit voltage excitation on conductor a in the coaxial region is incident from the right, except for a reversal in sign so that the total current in the coaxial region is $Z_0^{-1}\exp[-jk_0(z - z_c)] + \exp(jk_0z_c)\,i_c(z)$ and $\exp(jk_0z_c)\,i_c(z)$ in the antenna regions.

It should be noted that the definition of $i_c(z)$ in (42) in terms of the functions $U_a^s(z)$ and $U_b^s(z)$ from (24) is somewhat an approximation whose intention is to yield (in a manner analogous to the case of a linear thin-wire antenna) a more accurate value for the real part of the input current at the junction. As a result of this approximation one can determine that the sum of the currents entering the junction at $z = 0$ is only approximately equal to the sum of the currents leaving the junction. If the expressions for the $U_a(z)$ and $U_b(z)$ from (27) are used in the definition of $i_c(z)$ in (42), the sum of the currents entering and leaving the junction is zero. But this is at the expense of the accuracy of the real part of the input current and the accuracy of any subsequent input impedance calculations.

The Sleeve Dipole Antenna

A transmitting sleeve dipole antenna is depicted in Fig. 12a with built-in voltage source V_0. Such an antenna can be used for remote sensing and telemetry purposes. To determine the input impedance (or admittance) defined at the coaxial aperture we need first to write all the possible current components on the two halves of the antenna due to an incident current wave of $Z_0^{-1}\exp[-jk_0(z - z_0)]$ as

$$J^T(z) = i_c(z - z_0) - R_a Q_a^T(h)\,U_a(h - z) - R_b Q_b^T(-h)$$
$$\times\ [U_b(h + z) - Z_0 U_b(h + z_0)\,i_c(z - z_0)] \qquad (43)$$

for $z > z_0$, $\varrho = a$, and $z < z_0$, $\varrho = b$, where Q_a^T and Q_b^T are yet undetermined constants and $i_c(z - z_0)$ is a junction-induced current given in (42) for each region of interest. Since the junction effect at the coaxial opening has now been fully incorporated, we only need to use the end conditions $J^T(h) = J^T(-h) = 0$ to determine the two unknown constants. At $z = +h$, the reflected current is $U_a(h_1 - z)$ and the incident currents are $i_c(z)$ and $U_b(h + z)$. Hence

$$Q_a^T(h) = i_c(h - z_0) - R_b Q_b^T(-h)[U_b(2h) - Z_0 U_b(h + z_0)\,i_c(h - z_0)] \qquad (44a)$$

Likewise, at the other end, $z = -h$, the reflected current is $U_b(h + z)$ and the incident current is $i_c(z - z_0)$, and $U_a(h - z)$ so that

$$Q_b^T(-h) = i_c(-h - z_0) - R_a Q_a^T(h) U_a(2h)$$
$$+ R_b Q_b^T(-h) Z_0 U_b(h + z_0) i_c(-h - z_0) \quad \text{(44b)}$$

The two equations now allow us to obtain explicit expressions for Q_a^T and Q_b^T as follows:

$$Q_a^T(h) = [i_c(h - z_0) - R_b U_b(2h) i_c(-h - z_0)]/\Delta \quad \text{(45a)}$$

$$Q_b^T(-h) = [i_c(-h - z_0) - R_a U_a(2h) i_c(h - z_0)]/\Delta \quad \text{(45b)}$$

where

$$\Delta = 1 - R_a R_b U_a(2h) U_b(2h) - Z_0 R_b U_b(h + z_0)$$
$$\times [i_c(-h - z_0) - R_a U_a(2h) i_c(h - z_0)] \quad \text{(46)}$$

and R_a and R_b are given in (25) for radii a and b. We note that, interestingly enough, the formulas for a sleeve dipole antenna as given in (43) through (46) reduce immediately to those of a simple dipole antenna in (28) through (30) when we set $a = b$ and replace the incident current $i_c(z)$ with the transmitting antenna current $U_a^s(|z|)$.

To find the input impedance we only need to know that each component current in (43) has to continue into the coaxial region so that the total coaxial current for an incident voltage amplitude V_0 is given by

$$I_c = \left(\frac{V_0}{Z_0}\right)[e^{-jk_0(z - z_0)} - \Gamma_c e^{jk_0(z - z_0)}] \quad \text{(47)}$$

where

$$\Gamma_c = v + Z_0 R_a Q_a^T(h) U_a(h - z_0) + v Z_0 R_b Q_b^T(-h_2) U_b(h + z_0) \quad \text{(48)}$$

is actually the (voltage) reflection coefficient for the reflected TEM wave. The input impedance is then defined as

$$Z_{\text{in}} = Z_0(1 + \Gamma_c)/(1 - \Gamma_c) \quad \text{(49)}$$

The far-field pattern of such an antenna can be obtained in exactly the same manner as in the simple dipole case. Using the current expression in (43) and the far-field formula in (11) we obtain

$$\left(\frac{j2\pi}{k_0\xi_0}\right)^{-1} F(\theta; z_0) = [1 + Z_0 R_b U_b(h + z_0) Q_b^T(-h)] e^{jk_0 z_0 \cos\theta}$$

$$\times [W_a(h - z_0; \pi - \theta) + \nu W_b(h + z_0; \theta)]$$

$$- R_a Q_a^T(h) e^{jk_0 h \cos\theta} W_a(2h; \theta) - R_b Q_b^T(-h)$$

$$\times e^{-jk_0 h \cos\theta} W_b(2h; \pi - \theta) \qquad (50)$$

and W_a (W_b) is given in (33) for radius a (b).

Fig. 14 shows the input impedance as determined by using (48) and (49) for an

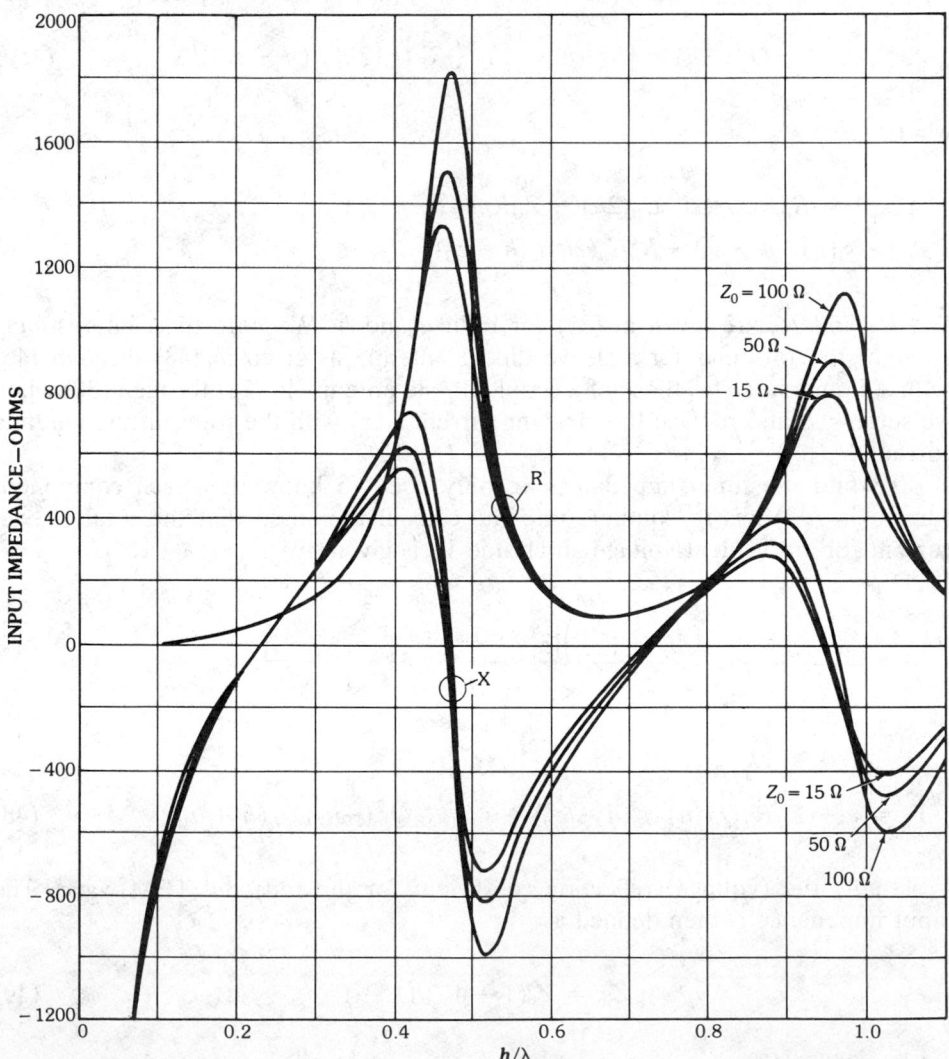

Fig. 14. Input impedance of isolated center-fed sleeve dipoles, where $\Omega = 2\ln(2h/b) = 10$ and $Z_0 = (\xi_0/2\pi)\ln(b/a)$.

isolated thin-wire $\Omega = 2\ln(2h/b) = 10$ sleeve dipole antenna as a function of its half-length kh for several ratios of b/a. These data show the input impedance to be fairly insensitive to the characteristic impedance of the internal coaxial line. In fact, the nominal input impedance in all the cases is quite similar to that of a linear thin-wire antenna having a constant radius (where $\Omega = 2\ln(2h/\sqrt{ab})$) such as discussed in Section 2. Furthermore, the current distributions on both transmitting and receiving sleeve dipoles would be essentially quite similar to those on equivalently sized linear thin-wire dipoles.

The Sleeve Monopole Antenna

Fig. 15 depicts a transmitting sleeve antenna mounted on a ground plane, which is often used to experimentally study the effect of the location of feed point on the excitation of a monopole antenna [2, 30]. To construct a solution for such an antenna we first have to use the image theorem to determine the kind of current components that can exist on the structure, as shown in Fig. 15. We then incorporate the required conditions at the junctions into our solution to obtain

$$J^T(z) = i_c(z - z_0) - R_a Q_a^T U_a(h - z) + A i_c(-z - z_0) + B U_a(h + z)$$
$$- Z_0 [A i_c(-2z_0) + B U_a(h + z_0)] i_c(z - z_0), \quad \text{for } 0 < z < h \quad (51)$$

Here $i_c(z - z_0)$ as given in (42) represents the primary current, due to an incident current wave of $Z_0^{-1} \exp[-jk_0(z - z_0)]$ inside the coaxial region; $U_a(h - z)$ is the reflected current from the end at $z = h$, and, as we know, the junction has no effect on this current; $i_c(-z - z_0)$ and $U_a(h + z)$ are the corresponding image currents originating from the "junction" at $z = -z_0$ and the other "end" at $-h$,

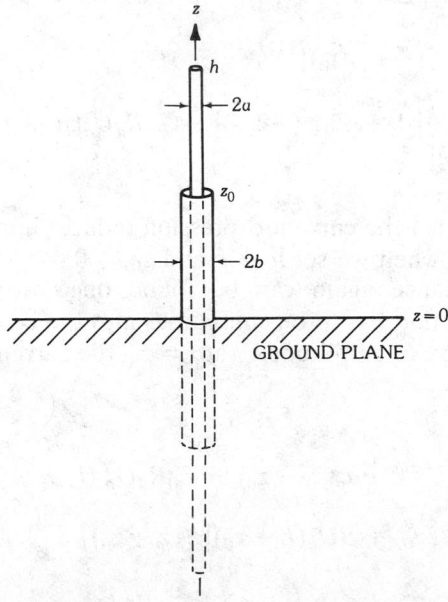

Fig. 15. Sleeve monopole illustration.

respectively. Together these two image currents impinge onto the junction at $z = z_0$ and produce yet another scattered current represented by the last term in the square bracket. In order to determine the unknown constants Q_a^T, A, and B, we invoke the symmetry requirement with respect to the ground plane $J^T(z) = J^T(-z)$ to obtain

$$B = -R_a Q_a^T$$
$$A = 1 - Z_0[Ai_c(-2z_0) + BU_a(h + z_0)]$$

and because of the end condition at $J^T(h) = 0$, we know the reflected current amplitude is related to the (total) incident current amplitude by $-R_a$ so that

$$Q_a^T = i_c(h - z_0) + Ai_c(-h - z_0) + BU_a(2h)$$
$$- Z_0[Ai_c(-2z_0) + BU_a(h + z_0)i_c(h - z_0)]$$

These equations can then be used to obtain an explicit expression for the current distribution:

$$J^T(z) = A[i_c(z - z_0) + i_c(-z - z_0)]$$
$$- R_a Q_a^T[U_a(h - z) + U_a(h + z)], \qquad 0 \leqq z < h \qquad (52)$$

where

$$A = [1 + R_a U_a(2h)]/\Delta \qquad (53a)$$

$$Q_a^T = [i_c(h - z_0) + i_c(-h - z_0)]/\Delta \qquad (53b)$$

$$\Delta = [1 + R_a U_a(2h)][1 + Z_0 i_c(-2z_0)] - Z_0 R_a U_a(h + z_0)[i_c(h - z_0)$$
$$+ i_c(-h - z_0)] \qquad (53c)$$

The reader can show that the current expression reduces immediately to the result of a simple monopole when we set $b \rightarrow a$ and $z_0 \rightarrow 0$.

The input impedance again can be found once we recognize that each component at the junction has to be continuous into the coaxial region. Thus, for an incident voltage wave of amplitude V_0 at $z = z_0$, the current in the coaxial region is given from (51) as

$$I_c = \left(\frac{V_0}{Z_0}\right)\{e^{-jk_0(z-z_0)} + i_c(z - z_0) - Z_0 R_a Q_a^T U_a(h - z_0)e^{-jk_0(z_0-z)}$$
$$- Z_0[Ai_c(-2z_0) + BU_a(h + z_0)]i_c(z - z_0)\} \qquad (54)$$

which, on the substitution of all the relevant constants, provides us an explicit expression for the reflection coefficient Γ_c:

$$I_c = \frac{V_0}{Z_0}[e^{-jk_0(z-z_0)} - \Gamma_c e^{jk_0(z-z_0)}] \tag{55}$$

where $\Gamma_c = \nu A + Z_0 R_a Q_a^T U_a(h - z_0)$, with A and Q_a^T given in (53). The input impedance is then given by the expression $Z_0(1 + \Gamma_c)/(1 - \Gamma_c)$. Likewise, using the far-field formula in (11) and the current expression in (52), we can obtain the far-field pattern function

$$\left(\frac{j2\pi}{k_0\xi_0}\right)^{-1} F(\theta; z_0) = Ae^{jk_0 z_0 \cos\theta}[W_a(h - z_0; \pi - \theta)$$
$$+ \nu W_b(h + z_0; \theta)] + 2Ae^{-jk_0 z_0 \cos\theta}$$
$$\times [\nu W_b(h + z_0; \pi - \theta) + W_a(h - z_0; \theta)]$$
$$- R_a Q_a^T[e^{jk_0 h \cos\theta} W_a(2h; \theta)$$
$$+ e^{-jk_0 h \cos\theta} W_a(2h; \pi - \theta)] \tag{56}$$

Figs. 16 and 17 show the current distributions on two quite different monopole sleeve antennas. In Fig. 16 the currents on the larger, b conductor are near resonance while those on the smaller, a conductor are not. In Fig. 17 the opposite situation is the case. The current distributions on the antennas depend not only on the overall length (including image) but on the position of the coaxial junction(s) as well, the currents emanating from the junction(s) being excited by the internal source or by the effect of the junction on the external currents. The data for these figures were calculated by using (52) and assuming an incident current of $1/Z_0 = 1/90 = 11.1$ mA. Such a current could be generated by a matched (source impedance $Z_\ell = Z_0$) source having an open circuit voltage of 2 V. Furthermore, these figures are consistent with the experimentally determined distributions obtained by Taylor [30] and readily available in the book *The Theory of Linear Antennas* ([2], Section III.30) by King.

The input impedance to a monopole sleeve antenna for several different overall lengths is shown in Fig. 18 as a function of the position of the coaxial junction. The same conductor radii ($ka = 0.02$ and $kb = 0.09$) were chosen here as they were used in the preceding figures. Note that the input resistance becomes very large in all cases when $kz_0 = 2\pi z_0/\lambda_0 \cong \pi/4$ and $3\pi/4$, corresponding to situations in which the junctions are $\lambda_0/2$ and $3\lambda_0/2$ apart. At these positions the source and its image are opposing one another. For the present theory to yield results comparable to the experimental data measured by Taylor (see [2], Section III.30) we found that it was necessary to include a shunt susceptance of $j2 \times 10^{-3}$ siemens in Fig. 18 in order to account for the finite thickness of the outer conductor in the experiment.

The Coaxial Sleeve Antenna with a Decoupling Choke

In most practical applications, sleeve antennas are fed via the inherent coaxial line within them with the feed line exciting the larger conductor in the manner illustrated in Fig. 19. Normally of coaxial construction itself, the outer sheath of the

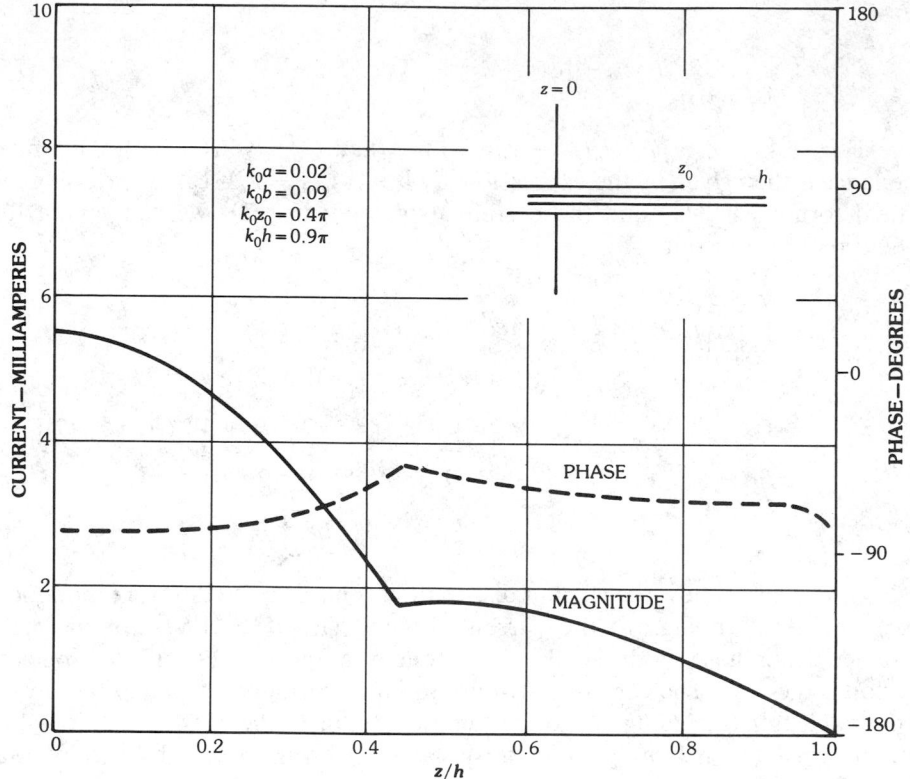

Fig. 16. Current distribution on a monopole sleeve antenna due to an incident coaxial current of $Z_0^{-1}\exp(-jk_0 z)$.

feed line and the hollow larger conductor form another sleeve-type junction. With the proper choice of terminating impedance $Z_t = 1/Y_t$ for the b-c coaxial line indicated in the figure, the sleeve antenna itself may be virtually isolated from its feed line at a single frequency. In the most practical situation, Z_t is chosen so that the equivalent of an open circuit appears at the lower truncation of conductor b. If physically permissible, this can be accomplished by shorting the b-c coaxial line at a distance of $\lambda_0/4$ from the opening. Obviously this type of decoupling has a strong dependence on the operating frequency. In the forthcoming analysis, current waves traveling up the c conductor toward the b and a conductors, which form the intended antenna structure, are not considered. Such waves could result if the antenna is not mounted sufficiently high enough above the surrounding environment (especially the ground) and operated too far from the choke resonance point.

To determine the input impedance of such an antenna we again first write all the component currents existing on the two halves of the antenna, i.e., conductors a and b, due to an incident current wave in the coaxial region:

$$J^T(z; z_0) = i_c(z - z_0) - R_a Q_a^T U_a(h - z) + A[i_{ch}(-h - z)$$
$$- Z_0 i_{ch}(-h - z_0) i_c(z - z_0)], \quad \text{for } -h < z < +h \qquad (57)$$

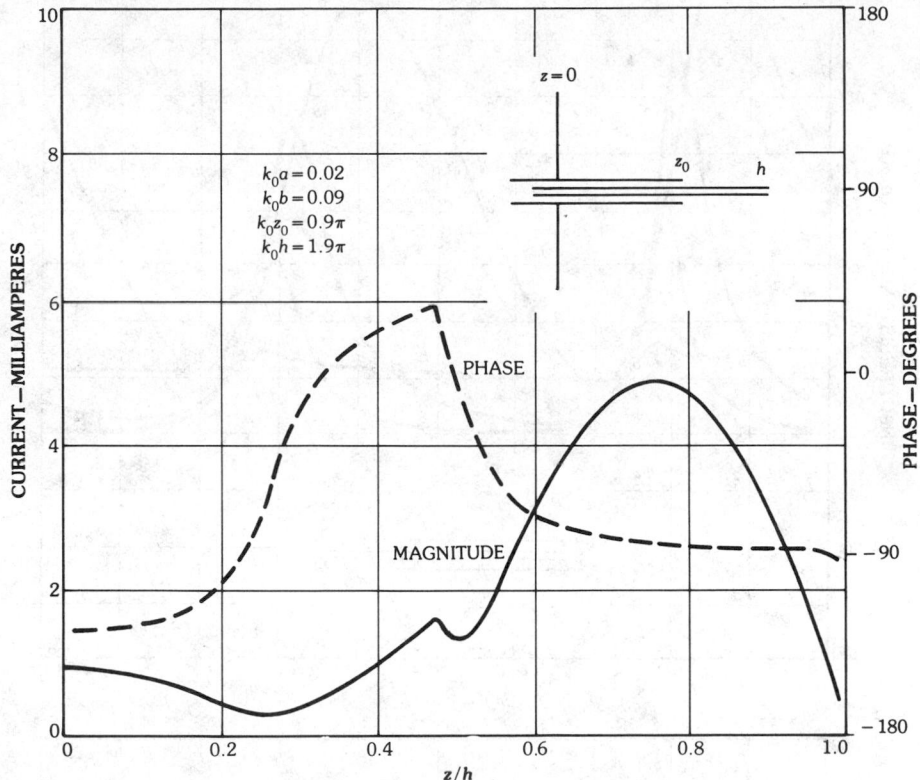

Fig. 17. Current distribution on a monopole sleeve antenna due to an incident coaxial current of $Z_0^{-1}\exp(-jk_0z)$.

Here i_{ch} is the junction current due to the choke at $z = -h$. The expression for i_{ch} is the same as i_c in (42) except that radii (a, b) are replaced by (c, b). Here we note that the choke current $i_{ch}(-h - z)$ is a result of multiple bounces of current waves in the choke region $-h < z < z_{ch}$ back to conductor b. Since for each subsequent bounce a factor of $-v_{ch}\exp(-j2k_0\ell)$ is introduced, where ℓ is the length of the choke, i.e., $\ell = z_{ch} + h$ and $v_{ch} = U_c^s(0)/U_b^s(0)$, the amplitude of the total current wave scattered back from the choke at $z = -h$ can be summed together as

$$-Z_{ch} + Z_{ch}v_{ch}e^{-j2k_0\ell}[1 - v_{ch}e^{-j2k_0\ell} + v_{ch}^2e^{-j4k_0\ell} - \cdots]$$
$$= -Z_{ch}(1 + v_{ch}e^{-j2k_0\ell})^{-1} \tag{58}$$

for an incident current wave of unity on conductor b. Here $Z_{ch} = (\xi_0/2\pi)\ln(b/c)$ is the characteristic impedance of the coaxial choke section. Using the above equation to relate the incident wave and reflected wave at $z = -h$, we have

$$A = -Z_{ch}(1 + v_{ch}e^{-j2k_0\ell})^{-1}$$
$$\times [i_c(-h - z_0) - R_aQ_a^TU_a(2h) - AZ_0i_{ch}(-h - z_0)i_c(-z_0 - h)]$$

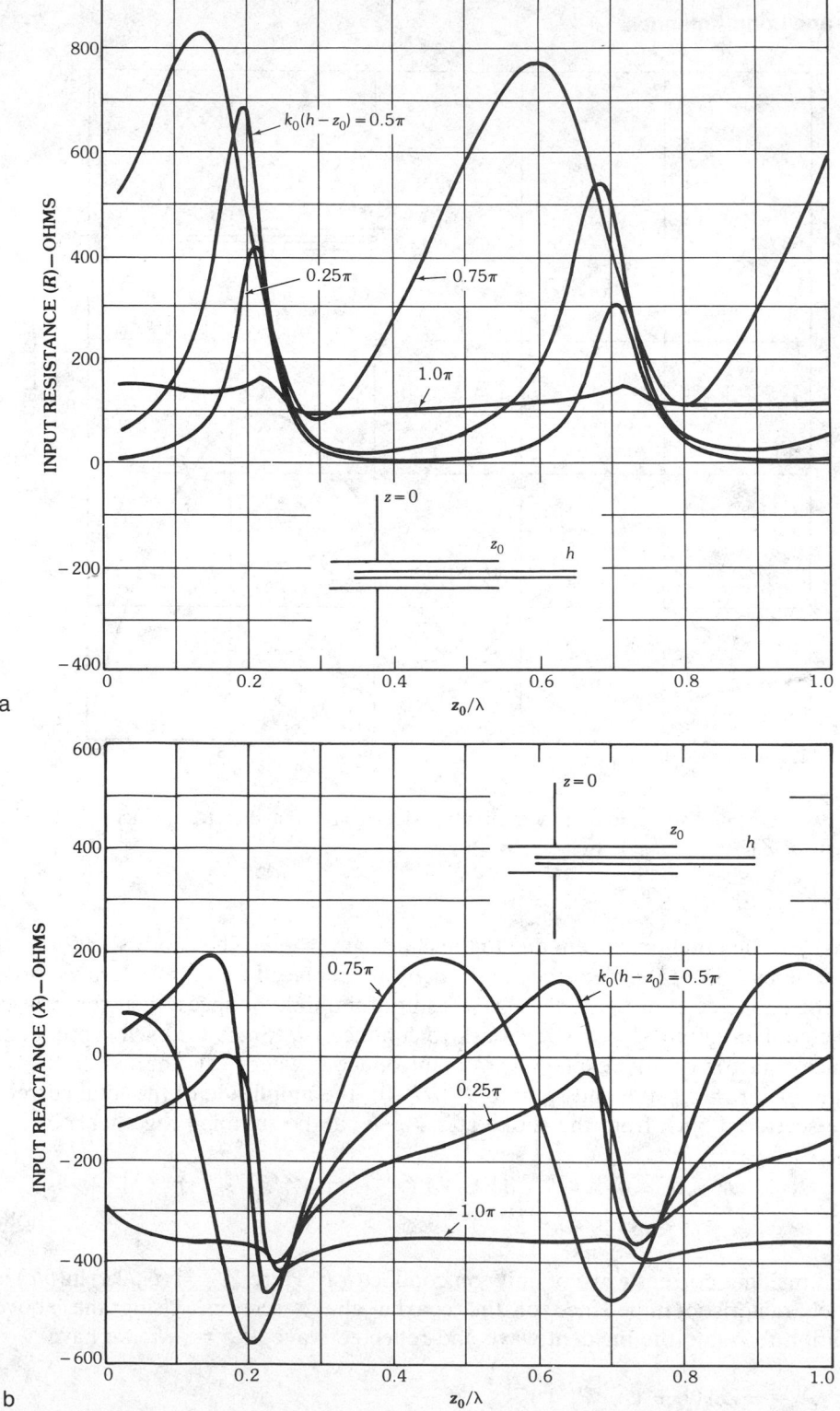

Fig. 18. Input impedance of a monopole sleeve antenna. (*a*) Input resistance. (*b*) Input reactance.

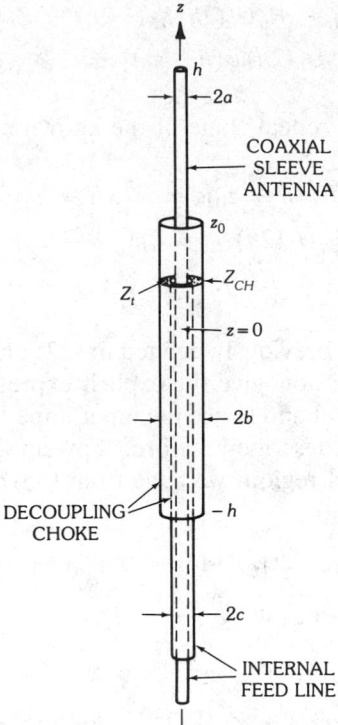

Fig. 19. Decoupled sleeve antenna.

We note that the sum of the incident and the reflected current at the choke is proportional to

$$1 - Z_{ch}(1 + v_{ch}e^{-j2k_0\ell})^{-1}i_{ch}(0)$$

which can be shown to vanish when the length of the choke is a quarter-wave long, i.e., $[1 - Z_{ch}i_{ch}(0)/(1 - v_{ch})] \to 0$. Thus the current is forced to zero at $z = h$ by the choke and remains vanishingly small beyond that point, along the conductor c. In any event, if we now employ the condition at the end of the upper half of the antenna at $J^T(h; z_0) = 0$

$$Q_a^T = i_c(h - z_0) + A[i_{ch}(-2h) - Z_0 i_{ch}(-h - z_0)i_c(h - z_0)]$$

we can solve for the two unknown constants A and Q_a^T:

$$A = [-i_c(-h - z_0) + R_a U_a(2h)i_c(h - z_0)]/\Delta \tag{59}$$

$$Q_a^T = [-i_c(-h - z_0)i_{ch}(-2h) + (1 + v_{ch}e^{-j2k_0\ell})i_c(h - z_0)/Z_{ch}]/\Delta \tag{60}$$

where

$$\Delta = (1 + v_{ch}e^{-j2k_0\ell})/Z_{ch} - R_a U_a(2h)\, i_{ch}(-2h) - Z_0 i_{ch}(-h - z_0)$$
$$\times [i_c(-h - z_0) - R_a U_a(2h)\, i_c(h - z_0)] \tag{61}$$

For easy reference we will repeat some of the known expressions here:

$$i_c(-h - z_0) = 2v U_b(h + z_0), \qquad i_c(h - z_0) = 2U_a(h - z_0)$$
$$i_{ch}(-2h) = 2v_{ch}U_b(2h), \qquad i_{ch}(-h - z_0) = 2v_{ch}U_b(h + z_0)$$
$$v = U_a^s(0)/U_b^s(0), \qquad\qquad v_{ch} = U_c^s(0)/U_b^s(0)$$

and $U_a(z)$ and $U_a^s(z)$ were previously defined in (27) and (24). These expressions, when substituted into (57), now give the explicit expression for the current.

The reflection coefficient and hence the input impedance of the antenna can be obtained in the manner described before. For an incident voltage wave of amplitude V_0 in the coaxial region, we have from (55)

$$I_c = \frac{V_0}{Z_0}[e^{-jk_0(z - z_0)} - \Gamma_c e^{jk_0(z - z_0)}]$$

where

$$\Gamma_c = v - v A Z_0 i_{ch}(-h - z_0) + Z_0 R_a Q_a^T U_a(h - z_0) \tag{62}$$

Fig. 20 shows the input impedance to a decoupled sleeve dipole antenna as a function of its electrical half-length $k_0 h$, which is proportional to frequency, i.e., $k_0 = 2\pi f/c$. The electrical length of the choke is taken to be the same as the antenna half-length, $k\ell = kh$. A somewhat hypothetical case, to be sure (unless the b-c coaxial line is filled with dielectric so that the physical choke length and half-length are appreciably different, with $\ell < h$), but one which illustrates the frequency dependence of this type of antenna quite well. At the resonances of $kh = \pi/2$ and $3\pi/2$, the choke makes the antenna behave essentially like an isolated sleeve dipole. The bandwidths about these resonant points, however, are seen to be quite narrow. For single-frequency operation, though, this does not present a problem.

4. The Folded Dipole Antenna

Folded dipole antennas, such as the ones depicted in Fig. 21, offer performance similar to that of the linear thin-wire antenna (discussed in Section 2) with the added advantage of a certain measure of control over their input admittance near resonance. A simplified analysis [34] of this type of antenna is possible through the consideration of a length of uniform two-conductor transmission line shorted at both ends (which the folded dipole resembles) with the aid of the equivalent radius concept mentioned in Section 1.

The currents excited on each arm of an unloaded, folded dipole antenna can be approximately determined through a decomposition of the source voltage into a

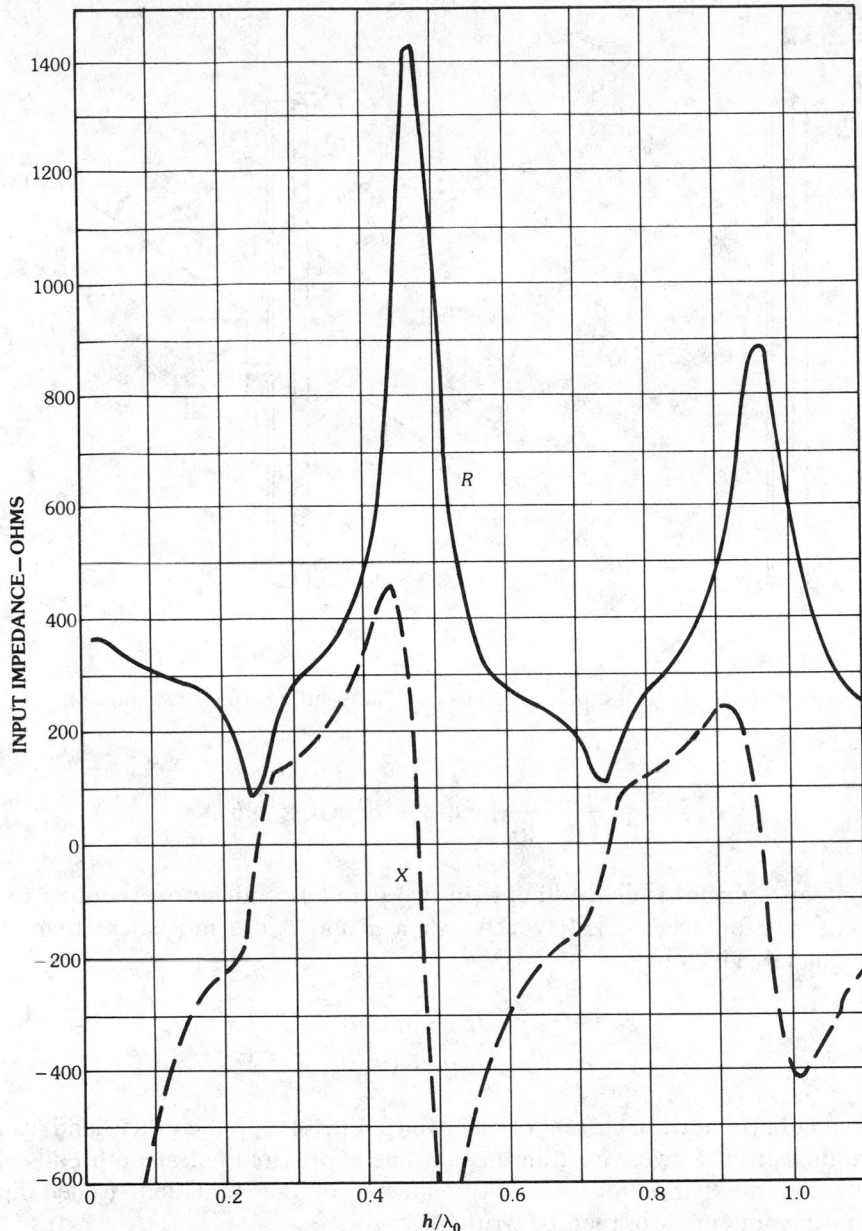

Fig. 20. Input impedance to a decoupled sleeve dipole antenna, where $\Omega = 2 \ln (2h/b) = 10$, $Z_c = (\xi_0/2\pi) \ln (b/a) = 50$ ohms, $Z_{ch} = (\xi_0/2\pi) \ln (c/b) = 72$ ohms, and $k_0\ell = \pi/2$.

symmetrical and an antisymmetrical arrangement [2] as shown in Figs. 22a and 22b. Requiring that $d \ll \lambda_0$, the strong mutual coupling between the two closely spaced parallel conductors allows the symmetrical current (sometimes called *antenna current*) to be approximated by one corresponding to a single antenna of the same overall length $2h$ and an equivalent radius (see the discussion in Section 1) equal to

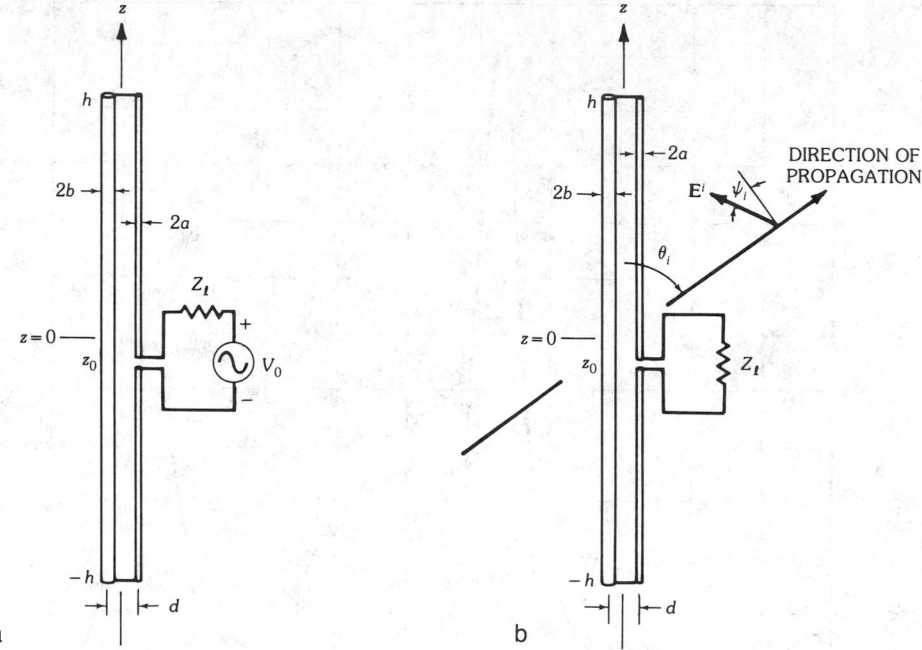

Fig. 21. Folded dipole antenna. (*a*) Transmitting. (*b*) Receiving.

$$a_e \approx \exp\left\{\frac{1}{(a+b)^2}[a^2\ln a + b^2\ln b + 2ab\ln d]\right\} \tag{63}$$

The antenna current is divided between two parallel conductors according to the "current division factor" [34] which, when a and b are much less than d, is approximately given by

$$\eta = \frac{J_b^T}{J_a^T} \approx \frac{\ln(d/a)}{\ln(d/b)} \tag{64}$$

The antisymmetrical current on the two conductors, on the other hand, can be determined from a two-wire transmission line short-circuited at both ends. The overall current distribution on each conductor of the unloaded, folded-dipole transmitting antenna can then be written as

$$I_{a,b}^T(z;z_0) = V_0 \begin{cases} \dfrac{1}{2}\left(\dfrac{\eta}{1+\eta}\right)J_{a_c}^T(z;z_0) - J_{tl}(z;z_0), & b \text{ conductor} \\[3mm] \dfrac{1}{2}\left(\dfrac{1}{1+\eta}\right)J_{a_c}(z;z_0) + J_{tl}(z;z_0), & a \text{ conductor} \end{cases} \tag{65}$$

where $J_{a_c}^T$ is given by (28). Now since the transmission-line current also can be formulated in terms of bouncing waves, we can still use (28) for J_{tl} with the

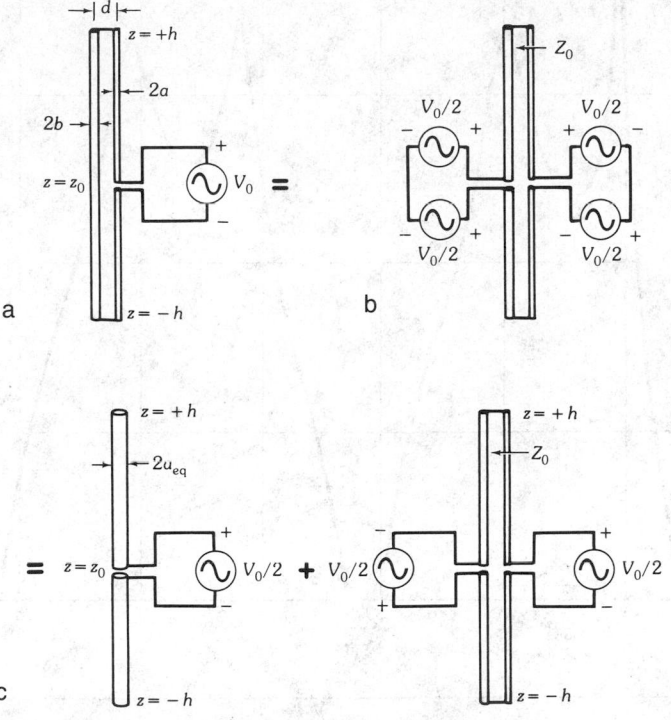

Fig. 22. Approximate model for a folded dipole transmitting antenna. (*a*) Folded dipole transmitting antenna. (*b*) Decomposition of voltage source into symmetric and unsymmetric parts. (*c*) Equivalent symmetrical and antisymmetrical problems.

following replacements:

$$U_a^s(z - z_0) \rightarrow (2Z_0)^{-1} e^{-jk_0|z - z_0|}$$

$$U_a(z) \rightarrow (2Z_0)^{-1} e^{-jk_0 z}$$

$$R_a \rightarrow -2Z_0$$

and Z_0 is the characteristic impedance of the two-wire line, i.e.,

$$Z_0 = (\xi_0/\pi) \cosh^{-1}\left(\frac{d}{2\sqrt{ab}}\right), \qquad \xi_0 = 120\pi \text{ ohms}$$

The input admittance at the feed point $z = z_0$ in conductor a is easily deduced from the input current in (65), i.e.,

$$Y_{in}(z_0) = \frac{1}{2} \frac{1}{(1 + \eta)} J_{a_c}^T(z_0; z_0) + J_{tl}(z_0; z_0), \qquad -h < z_0 < +h$$

where

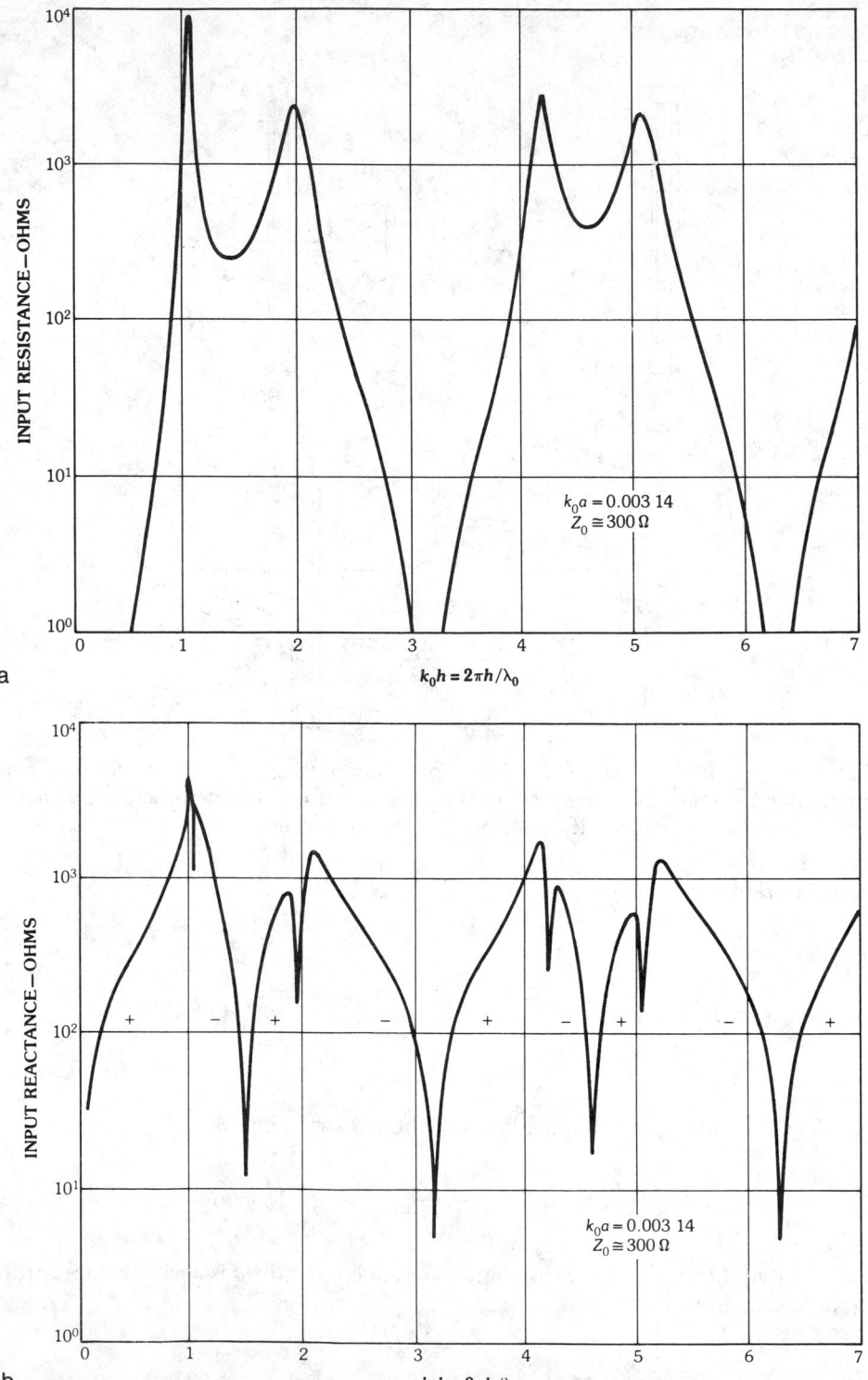

Fig. 23. Input resistance and reactance of a folded dipole antenna, with $a = b$ and $d = 12.5a$. (a) Input resistance. (b) Input reactance.

$$J_{tl}(z_0; z_0) = Y_{tl}(z_0) = \frac{1}{jZ_0\{\tan[k_0(h + z_0)] + \tan[k_0(h - z_0)]\}}$$

is the input admittance of the transmission-line circuit.

The far-field radiation of a folded dipole, however, comes only from the in-phase antenna current on the two conductors. Again, if we use the equivalent radius concept, the far-field pattern function is then given by (32) and (33) in Section 2, with the exception that a be replaced by a_e and that the actual field strength should be multiplied by $V_0/2$ instead of V_0.

Figs. 23a and 23b show the input resistance and reactance, respectively, to a center-fed folded dipole antenna as a function of its electrical half-length, $k_0h = 2\pi h/\lambda_0$. The electrical radius of both conductors is constant at $k_0a = 0.001\pi$ and their separation remains fixed at $d = 12.5a$, thereby yielding a nominal transmission-line impedance of $Z_0 = 300$ ohms.

Fig. 24 affords a close-up view of the input resistance to a center-fed folded dipole antenna near half-wave resonance. The conductors have the electrical radii $k_0a = 0.001\pi$ and $k_0b = k_0a$, $2k_0a$, $3k_0a$, and $4k_0a$. The separation of the conductors has been appropriately chosen so that the nominal transmission-line characteristic impedance is $Z_0 \cong 300$ ohms in each case, i.e., $d = 12.5\sqrt{ab}$.

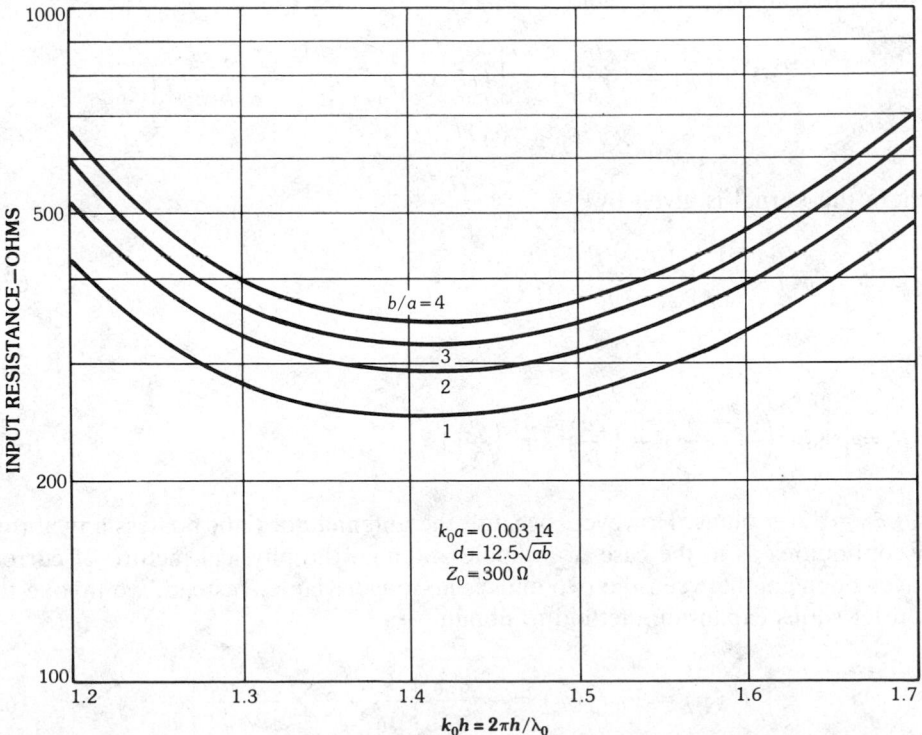

Fig. 24. Close-up view of the input resistance of thin-wire folded dipoles near $\lambda_0/2$ resonance.

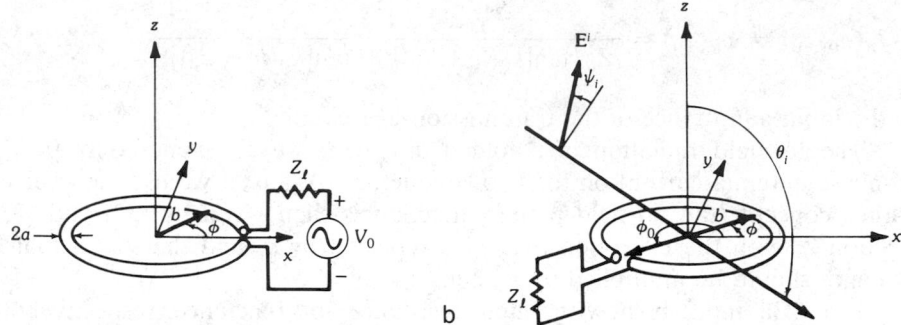

Fig. 25. Circular thin-wire loop antennas. (*a*) Transmitting. (*b*) Receiving.

5. The Thin-Wire Loop Antenna

The thin-wire loop antenna discussed in this section is constructed from a single turn of electrically thin wire (as shown in Fig. 25). Such antennas find application in direction-finding systems and uhf communications, and serve as probes for magnetic intensity measurement.

Following the analysis given by King [35], the basic characteristics and properties of a transmitting loop antenna can be inferred from solving the integral equation given as

$$-V_0\delta(\phi) = -j\frac{\xi_0}{4\pi}\int_{-\pi}^{+\pi}\left[k_0b\cos(\phi - \phi') + \frac{1}{k_0b}\frac{\partial^2}{\partial\phi^2}\right]$$
$$\times G(\phi - \phi')I^T(\phi')\,d\phi' \tag{66}$$

where the kernel is given by

$$G(\phi - \phi') = \frac{1}{2\pi}\int_{-\pi}^{+\pi}\frac{e^{-jk_0bR}}{R}\,d\psi \tag{67a}$$

with

$$R = \left[4\sin^2\left(\frac{\phi - \phi'}{2}\right) + \left(\frac{2a}{b}\right)^2\sin^2\left(\frac{\psi}{2}\right)\right]^{1/2} \tag{67b}$$

and $\xi_0 = 120\pi$ ohms. However, because the antenna does not possess any abrupt discontinuities as in the case of a dipole antenna, the physical picture of current waves bouncing between the two ends is no longer viable. Instead, we invoke the Fourier series expansion method to obtain

$$I^T(\phi) = V_0J^T(\phi) = -j\frac{V_0}{\pi\xi_0}\left\{\frac{1}{A_0} + 2\sum_{n=1}^{\infty}\frac{\cos n\phi}{A_n}\right\} \tag{68}$$

and likewise, for the kernel function G,

$$G(\phi - \phi') = K_0 + 2 \sum_{n=1}^{\infty} K_n \cos n(\phi - \phi') \tag{69}$$

Substitution of these expressions into the integral equation in (66) permits the solution of the A_n coefficients in terms of

$$A_n = \frac{k_0 b}{2}(K_{n+1} + K_{n-1}) - \frac{n^2}{k_0 b} K_n, \qquad K_{-n} = K_{+n} \tag{70}$$

The K_n coefficients above are expressible in terms of integrals involving Bessel and Lommel-Weber (Anger) functions [36].

In most practical situations, the electrical radius $k_0 b = 2\pi b/\lambda_0$ of the loop is seldom very much greater than unity. Keeping only the most important terms, the coefficients A_0, A_1, and A_2 can be approximated as

$$A_0 = \frac{k_0 b}{\pi}\left[\ln\left(\frac{8b}{a}\right) - 2\right] + \frac{1}{\pi}[0.667(k_0 b)^3 - 0.267(k_0 b)^5]$$
$$- j[0.167(k_0 b)^4 - 0.033(k_0 b)^6] \tag{71a}$$

$$A_1 = \left(k_0 b - \frac{1}{k_0 b}\right)\frac{1}{\pi}\left[\ln\left(\frac{8b}{a}\right) - 2\right] + \frac{1}{\pi}[-0.667(k_0 b)^3 + 0.207(k_0 b)^5]$$
$$- j[0.333(k_0 b)^2 - 0.133(k_0 b)^4 + 0.026(k_0 b)^6] \tag{71b}$$

$$A_2 = \left(k_0 b - \frac{4}{k_0 b}\right)\frac{1}{\pi}\left[\ln\left(\frac{8b}{a}\right) - 2.667\right]$$
$$+ \frac{1}{\pi}[-0.40(k_0 b) + 0.21(k_0 b)^3 - 0.086(k_0 b)^5]$$
$$- j[0.050(k_0 b)^4 - 0.012(k_0 b)^6] \tag{71c}$$

Above $k_0 b = 1.3$, the accuracy of the above expressions rapidly deteriorates. Figs. 26a and 26b show the real and imaginary components of the inverses of the above approximate Fourier coefficients for loops where $\Omega = 8$, 9, 10, 11, and 12 as a function of the electrical radius $k_0 b$. Over the limited range of $k_0 b$ shown, these approximate coefficients are consistent with the more exact numerically evaluated data given by King [35]. The higher-order ($n \geq 3$) coefficients are negligible compared with the A_0, A_1, and A_2 coefficients (on an individual basis) for $k_0 b \leq 1.3$. Near the feed point, however, the contributions of these higher-order terms are cumulative and their neglect leads primarily to an error in the determination of the imaginary component of the current in this region. An error in the determination of the input susceptance will also result.

The current distributions on transmitting loop antennas where $\Omega = 2\ln(2\pi b/b) = 10$ for $k_0 b = 0.2, 0.4, 0.6, 0.8, 1.0$, and 1.2 as determined from (68) and (71) are shown in Figs. 27a and 27b. These distributions are consistent with the more exact results (arrived at by the numerical calculation of the A_n coefficients up to $n = 20$) given in [35], except for the imaginary current component near the feed point.

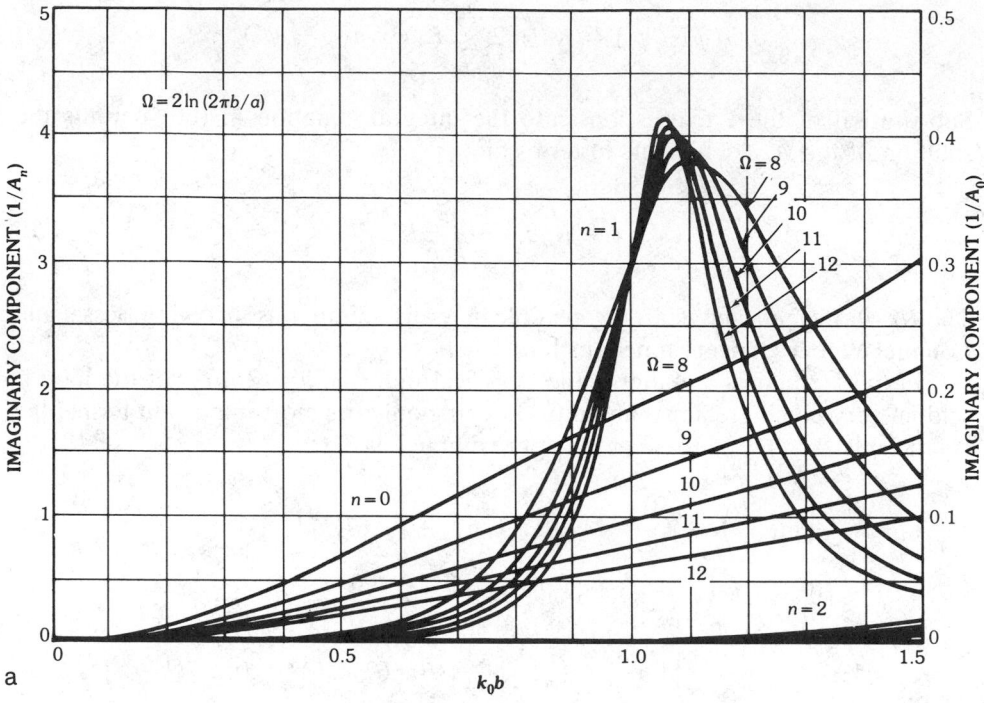

Fig. 26. Approximate real and imaginary components of the inverses of the Fourier coefficients for the current on circular thin-wire loop antennas. (*a*) Approximate imaginary component, with the a_0 components shown ten times actual value. (*b*) Approximate real component.

The input admittance to an electrically small, thin-wire loop can be determined from (68) in the manner:

$$Y_{\text{in}} = I^T(0)/V_0 = J^T(0), \qquad Z_{\text{in}} = 1/Y_{\text{in}} \tag{72}$$

The input resistance and reactance of thin-wire loop antennas where $\Omega = 2\ln(2\pi b/a) = 8$, 10, and 12 are shown in Figs. 28a and 28b, respectively, as functions of the electrical radius $k_0 b = 2\pi b/\lambda_0$.

Fair-Field Radiation from a Circular Loop Antenna

The far-field radiation from a thin-wire loop antenna can be formulated using the general approach outlined in Section 1. In the case of an electrically small loop ($k_0 b \leqq 1.3$) with a voltage source located at $\phi = \phi_0$, the vector far-field pattern is sufficiently well approximated by

$$\mathbf{F}(\theta, \phi; \phi_0) = -j\frac{2b}{\xi_0}\left\{\left[\frac{f_0(\theta)}{A_0} + 2\frac{f_1(\theta)}{A_1}\cos(\phi - \phi_0) + 2\frac{f_2(\theta)}{A_2}\cos 2(\phi - \phi_0)\right]\hat{\mathbf{a}}_\phi\right.$$
$$\left. + \cos\theta\left[2\frac{g_1(\theta)}{A_1}\sin(\phi - \phi_0) + 2\frac{g_2(\theta)}{A_2}\sin 2(\phi - \phi_0)\right]\hat{\mathbf{a}}_\theta\right\} \tag{73}$$

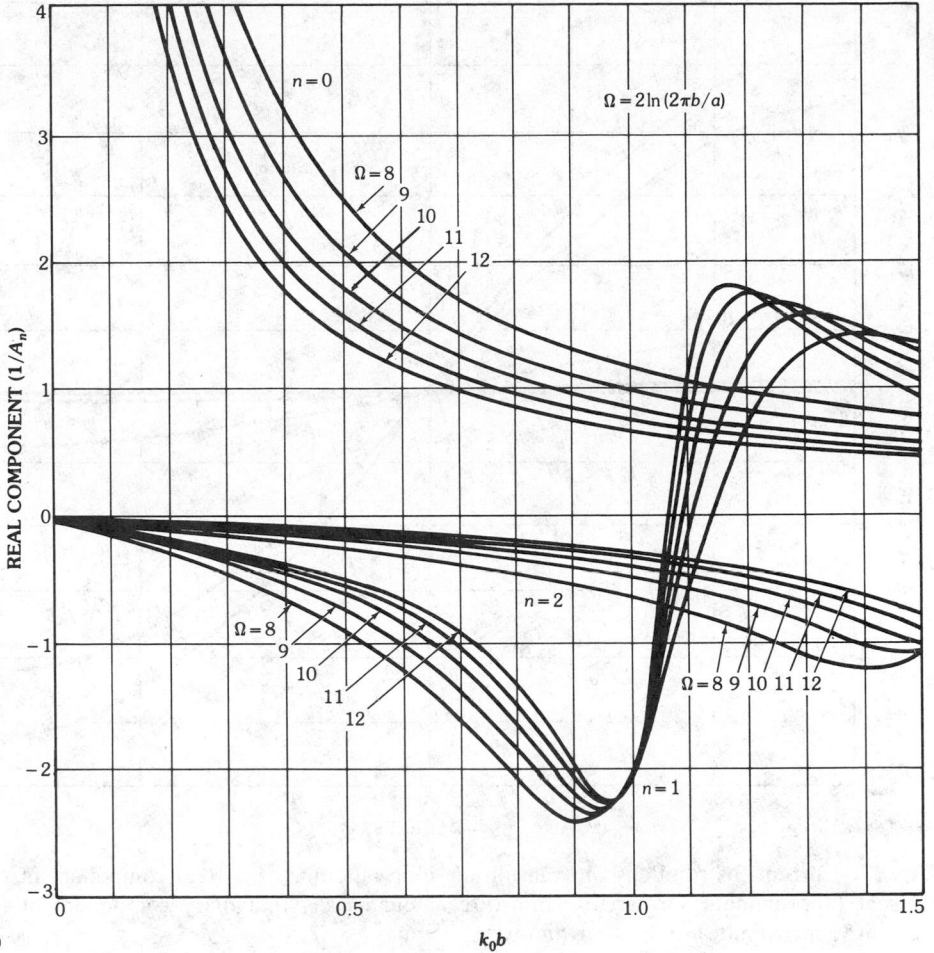

Fig. 26, *continued.*

where

$$f_n(\theta) = (j)^{n-1} J'_n(k_0 b \sin \theta) \tag{74a}$$

$$g_n(\theta) = (j)^{n-1} \frac{n J_n(k_0 b \sin \theta)}{k_0 b \sin \theta} \tag{74b}$$

Here the Bessel function J_n and its derivative J'_n (with respect to its argument) can be easily evaluated using the truncated series

$$J_n(k_0 b \sin \theta) \cong \sum_{\ell=0}^{3} \frac{(-1)^{\ell}}{\ell!(\ell + n)!} \left(\frac{k_0 b}{2} \sin \theta \right)^{2\ell+n} \tag{75}$$

and

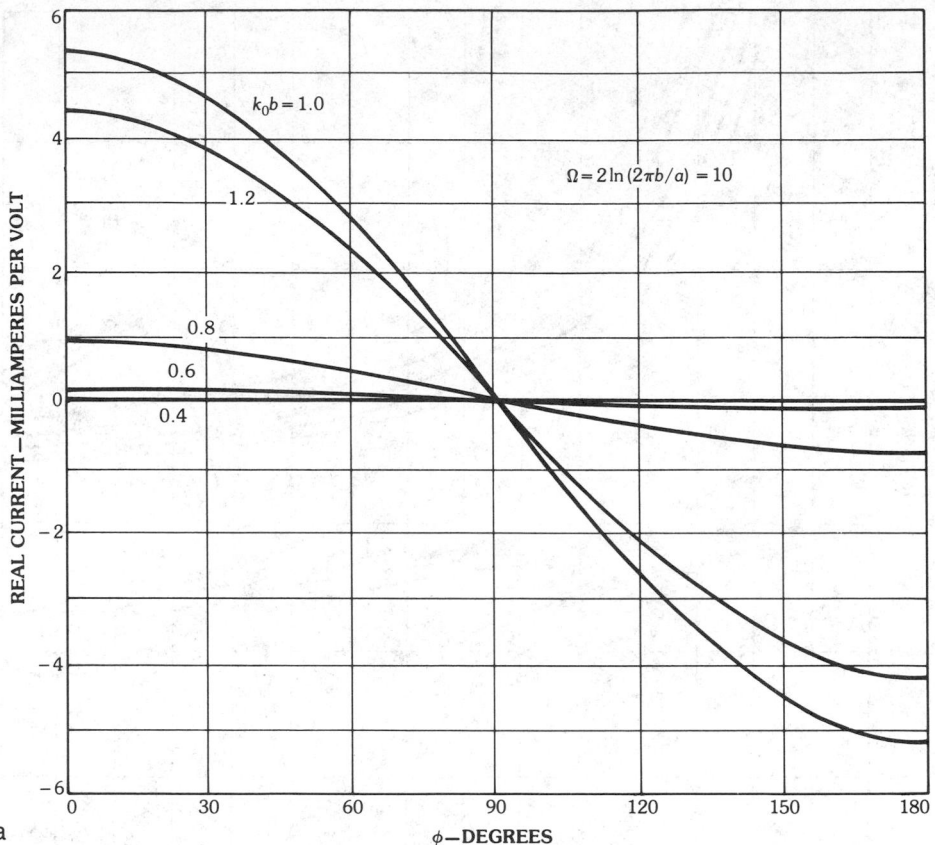

Fig. 27. Current distributions on transmitting loop antennas. (*a*) Real component of the current (approximate) on circular thin-wire antennas. (*b*) Imaginary component of the current (approximate) on circular thin-wire antennas.

$$J'_n(k_0 b \sin \theta) \cong \frac{1}{2} \sum_{\ell=0}^{3} (-1)^\ell \frac{(2\ell + n)}{\ell!(\ell + n)!} \left(k_0 \frac{b}{2} \sin \theta \right)^{2\ell+n-1} \tag{76}$$

which are useful for $k_0 b \leqq 1.3$.

The Electrically Small Receiving Loop Antenna

Consider the electrically small ($k_0 b = 2\pi b/\lambda_0 \leqq 1.3$) circular thin-wire loop antenna illuminated by a uniform plane wave as illustrated in Fig. 25b. This plane wave is incident at an angle θ_i with respect to the axis of the loop (z axis) and polarized at an angle ψ_i with respect to the y axis. The electric-field vector of the incident plane-wave field is expressible as

$$\mathbf{E}^i = E^i(-\sin \psi_i \cos \theta_i \hat{\mathbf{a}}_x + \cos \psi_i \hat{\mathbf{a}}_y + \sin \psi_i \sin \theta_i \hat{\mathbf{a}}_z) \tag{77}$$

The incident azimuthal angle is taken here as $\phi_i = 0$ without any loss of generality.

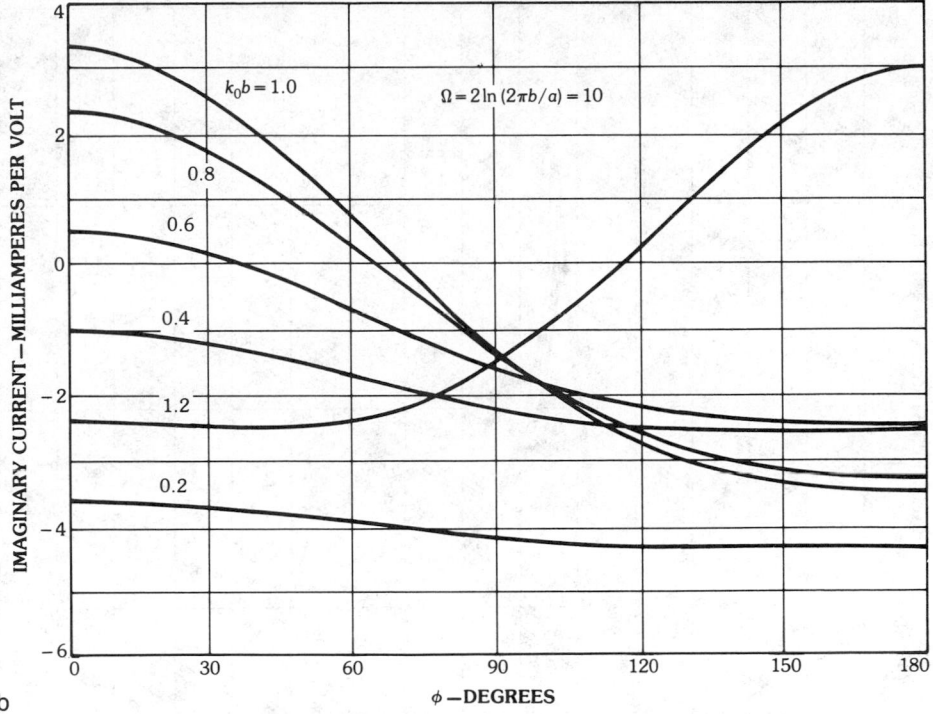

b

Fig. 27, *continued.*

Based on the discussion in Section 1, using (15), (73), and (77), the receiving current distribution can be simply expressed as

$$I^R(\theta_i, \phi_0) = E^i j \frac{2b}{\xi_0} \left\{ \cos \psi_i \left[\frac{f_0(\theta_i)}{A_0} + 2 \frac{f_1(\theta_i)}{A_1} \cos \phi_0 + 2 \frac{f_2(\theta_i)}{A_2} \cos 2\phi_0 \right] \right.$$
$$\left. - \sin \psi_i \cos \theta_i \left[2 \frac{g_1(\theta_i)}{A_1} \sin \phi_0 + 2 \frac{g_2(\theta_i)}{A_2} \sin 2\phi_0 \right] \right\} \tag{78}$$

where ϕ_0 specifies the observation point on the loop.

Loaded Loop Antennas

The introduction of a source or load impedance in a circular transmitting or receiving loop antenna is easily handled in the manner described in Section 1. The current distribution about a thin-wire transmitting loop antenna driven by a voltage source V_0 at $\phi = 0$ having a source admittance $Y_\ell = 1/Z_\ell$ can be written as

$$I^T(\phi) = V_0 \left\{ \frac{Y_\ell}{Y_{\text{in}} + Y_\ell} \right\} J^T(\phi) \tag{79}$$

And the current distribution on a receiving loop antenna having a load admittance $Y_\ell = 1/Z_\ell$ located at an azimuth of ϕ_0 on the loop can be written as

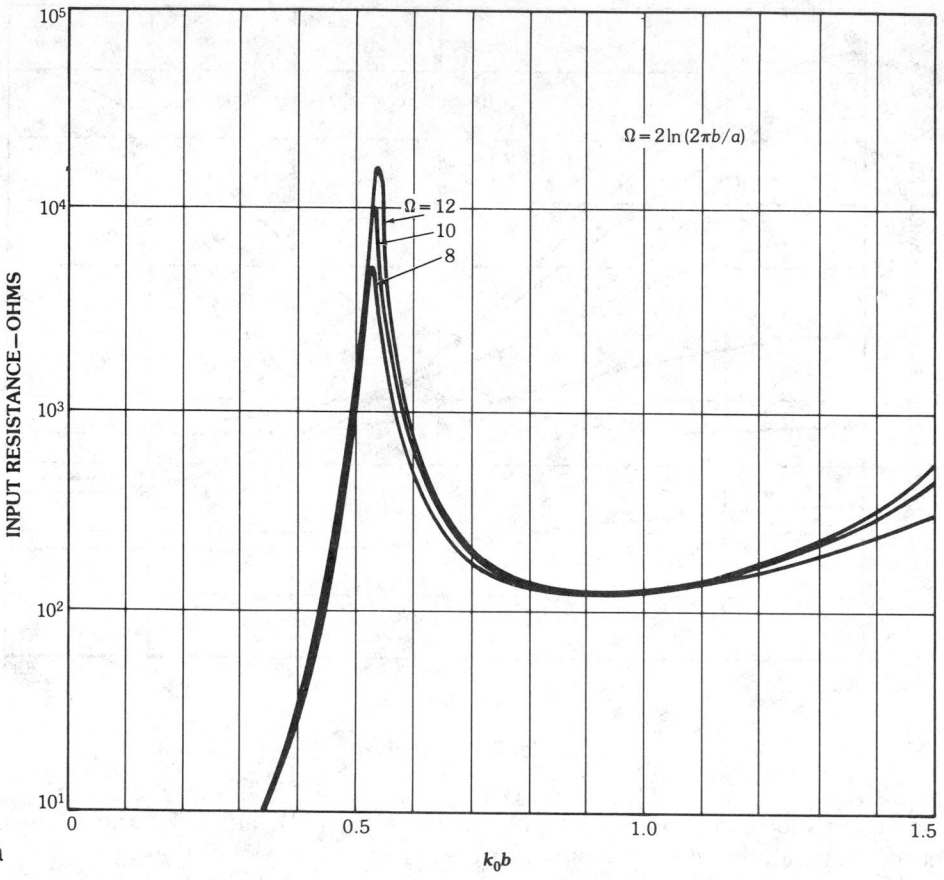

Fig. 28. Approximate input impedance of circular thin-wire loop antennas. (*a*) Input resistance. (*b*) Input reactance.

$$I^R(\phi) = E^i\left\{J^R(\theta_i;\phi) - \frac{J^R(\theta_i;\phi_0)}{Y_{\text{in}} + Y_\ell}J^T(\phi - \phi_0)\right\} \qquad (80)$$

where Y_{in} is the input admittance to the loop $Y_{\text{in}} = 1/Z_{\text{in}}$.

6. Concluding Remarks

Various basic thin-wire antenna structures have been discussed and simple mathematical expressions given that permit the calculation of the electrical properties of these antennas with relative ease. A host of composite antenna structures which embody some of these more basic structures can be analyzed through an appropriate combination of the individual analyses. For example, a circular folded loop antenna could be analyzed by combining the elements of the folded dipole and circular loop antenna analyses in Sections 4 and 5, respectively. Partially shielded, coaxially fed loop antennas could be handled by incorporating the "junction effect" in Section 3 into the loop antenna discussion of Section 5.

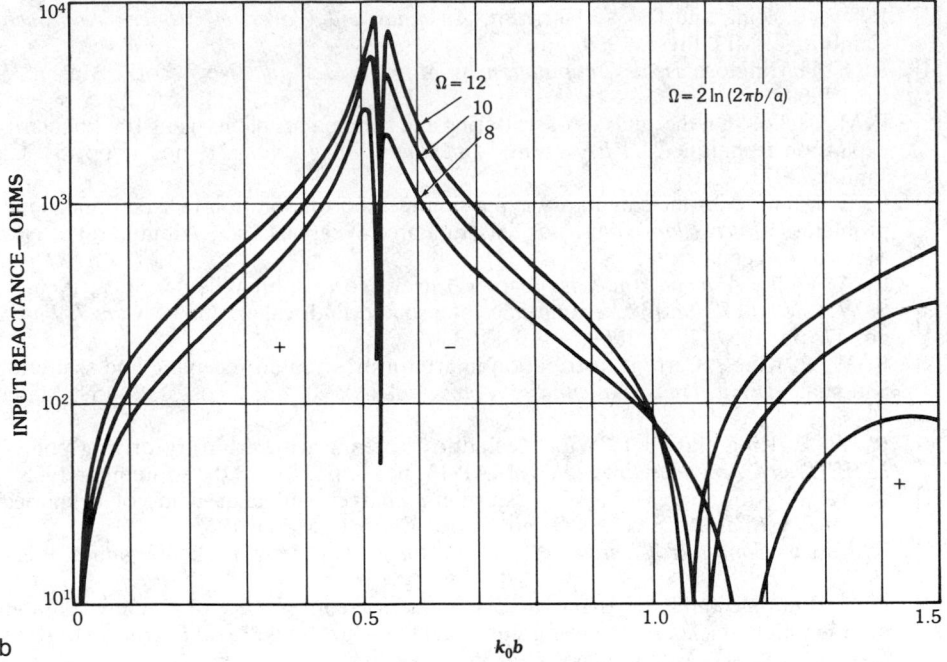

b

Fig. 28, *continued.*

Other composite structures could be adequately modeled using the basic elements described here.

Based on the general equations for the current on thin-wire antennas given in this chapter, many other related antenna parameters can be obtained. For instance, the charge distribution is easily obtained from the derivative of the current distribution with respect to the axial coordinate. The open-circuit terminal voltage of a receiving antenna is obtained from the product of the short-circuit current and the antenna impedance. Furthermore, the equations given here for the current distributions on thin-wire antennas are, for the most part, analytic with respect to frequency, thereby lending themselves to approximate transient analyses as described in Section 2 for the thin-wire linear dipole.

Acknowledgments

The authors wish to thank Dr. Ahmad Hoorfar for his participation in several helpful discussions related to the transient analysis of linear antennas.

7. References

[1] K. K. Mei, "On the integral equations of thin wire antennas," *IEEE Trans. Antennas Propag.*, vol. 13, no. 3, pp. 374–378, May 1965.

[2] R. W. P. King, *The Theory of Linear Antennas*, Cambridge: Harvard University Press, 1956.

[3] R. W. P. King and C. W. Harrison, *Antennas and Waves: A Modern Approach*, Cambridge: MIT Press, 1969.

[4] R. F. Harrington, *Field Computation by Moment Methods*, New York: Macmillan Co., 1968.

[5] F. M. Teche, "On the analysis of scattering and antenna problems using the singularity expansion technique," *IEEE Trans. Antennas Propag.*, vol. 21, no. 1, pp. 53–62, January 1973.

[6] C. E. Baum, "On the singularity expansion method for solution of electromagnetics problems," *Interaction Notes*, no. 88, Air Force Weapons Lab, Albuquerque, New Mexico, December 1971.

[7] E. A. Wolff, *Antenna Analysis*, chapter 3, New York: John Wiley & Sons, 1966.

[8] S. W. Lee and R. Mittra, "Admittance of a solid cylindrical antenna," *Can. J. Phys.*, no. 47, pp. 1959–1970, 1969.

[9] R. W. P. King, "Current distribution in arbitrarily oriented receiving and scattering antenna," *IEEE Trans. Antennas Propag.*, vol. AP-20, no. 2, pp. 152–159, March 1979.

[10] R. W. P. King and T. T. Wu, "Cylindrical antenna with arbitrary driving point," *IEEE Trans. Antennas Propag.*, vol. AP-13, no. 5, pp. 711–718, September 1965.

[11] R. W. P. King and T. T. Wu, "Currents, charges and near fields of cylindrical antennas," *Radio Sci.*, vol. 69D, no. 3, pp. 429–446, March 1965.

[12] R. Mittra, *Computer Techniques for Electromagnetics*, New York: Pergamon Press, 1973.

[13] W. A. Imbriale and P. G. Ingerson, "On numerical convergence of moment solutions of moderately thick antenna using sinusoidal basis functions," *IEEE Trans. Antennas Propag.*, vol. AP-21, no. 3, pp. 363–366, May 1973.

[14] D. C. Chang, "On the electrically thick monopole, part I," *IEEE Trans. Antennas Propag.*, vol. AP-16, no. 1, pp. 58–64, January 1968.

[15] C. C. Kao, "Electromagnetic scattering from a finite cylinder, numerical solution," *Radio Sci.*, vol. 5, no. 3, pp. 617–624, March 1970.

[16] L. A. Weinstein, *The Theory of Diffraction and the Factorization Method*, Boulder: Golem Press, 1969.

[17] R. Mittra and S. W. Lee, *Analytical Techniques in the Theory of Guided Waves*, New York: Macmillan Co., 1971.

[18] J. B. Anderson, "Admittance of infinite and finite cylindrical metallic antenna," *Radio Sci.*, vol. 3 (New Series), no. 6, June 1968.

[19] C.-L. Chen, "On the scattering of electromagnetic waves from a long wire," *Radio Sci.*, vol. 3 (New Series), no. 7, June 1968.

[20] L. C. Shen, T. T. Wu, and R. W. P. King, "A simple formula for current in dipole antennas," *IEEE Trans. Antennas Propag.*, vol. 16, no. 5, pp. 543–547, September 1968.

[21] O. Einarsson, "Electromagnetic scattering by a thin wire," *Acta Polytech. Scandinavia*, Electrical Engineering Series 23, Stockholm, 1969.

[22] L. C. Shen, "A simple theory of receiving and scattering antennas," *IEEE Trans. Antennas Propag.*, vol. 18, no. 1, pp. 112–114, January 1970.

[23] D. C. Chang, S. W. Lee, and L. W. Rispin, "Simple formula for current on a receiving antenna," *IEEE Trans. Antennas Propag.*, vol. 26, no. 5, pp. 683–690, September 1978.

[24] L. W. Rispin and D. C. Chang, "A unified theory for thin-wire antennas of arbitrary length," *Sci. Rep. No. 38* (N00014-76-C-0318), Department of Electrical Engineering, University of Colorado, Boulder, February 1980.

[25] M. Abramowitz and A. Segun, *Handbook of Mathematical Functions*, New York: Dover Publications, 1972.

[26] E. K. Miller, "Admittance dependence of the infinite cylindrical antenna upon exciting gap thickness," *Radio Sci.*, vol. 2 (New Series), no. 12, pp. 1431–1435, December 1967.

[27] A. Hoorfar and D. C. Chang, "Analytic determination of the transient response of a

thin-wire antenna based upon an SEM representation," *IEEE Trans. Antennas Propag.*, vol. AP-30, no. 6, pp. 1145–1152, November 1982.

[28] L. Marin and T. K. Liu, "A simple way of solving transient thin-wire problems," *Radio Sci.*, vol. 11, no. 2, pp. 149–155, February 1976.

[29] T. K. Liu and K. K. Mei, "A time-domain integral equation for linear antennas and scatterers," *Radio Sci.*, vol. 8, no. 9, pp. 797–804, September 1973.

[30] J. Taylor, *The Sleeve Antenna*, doctoral dissertation, Harvard University, Cambridge, Massachusetts, 1950.

[31] D. C. Chang, "Junction effect of two thin, coaxial cylinders of dissimilar radius," p. 103, *Nat. Radio Sci. Mtg. Dig.*, Seattle, June 18–22, 1979.

[32] A. Hurd, private communication.

[33] R. W. P. King and T. T. Wu, "The thick tubular transmitting antenna," *Radio Sci.*, vol. 2, no. 9, pp. 1061–1066, September 1967.

[34] S. Uda and Y. Mushiake, *Yagi-Uda Antennas*, Tokyo: Maruzen Co., p. 19, 1954.

[35] R. W. P. King, "The loop antenna for transmission and reception," chapter 11 of *Antenna Theory*, *Part I*, ed. by R. E. Collin and F. J. Zucker, New York: McGraw-Hill Book Co., 1969.

[36] T. T. Wu, "Theory of the thin-circular loop antenna," *J. Math. Phys.*, vol. 3, no. 6, pp. 1301–1304, November–December, 1962.

Chapter 8

Horn Antennas

Constantine A. Balanis
Arizona State University

CONTENTS

Constantine A. Balanis was born October 1938 in Trikala, Greece. He received his PhD in electrical engineering from Ohio State University in 1969.

From 1964 to 1970 he was with NASA at Langley Research Center. In 1970 he joined the Department of Electrical Engineering of West Virginia University as a visiting associate professor and held the positions of associate and full professor. Since 1983 he has been a full professor in the Department of Electrical and Computer Engineering at Arizona State University, where he teaches graduate and undergraduate courses in electromagnetic theory, microwave circuits, and antennas.

Formerly he was an associate editor of the *IEEE Transactions on Antennas and Propagation*, and of the *IEEE Transactions on Geoscience and Remote Sensing*, and editor of the *Newsletter* of the IEEE Geoscience and Remote Sensing Society. Dr. Balanis is a Fellow of the IEEE and the author of *Antenna Theory: Analysis and Design* (Harper & Row, 1982). His research interests are in high-frequency asymptotic methods (such as GTD and PTD), radar cross section (RCS), electromagnetic geotomography, wave propagation in microstrip lines, and electromagnetic-wave multipath.

1. Introduction

One of the simplest and probably the most widely used microwave antennas is the horn. Its existence and early use date back to the late 1800s. Although neglected somewhat in the early 1900s its revival began in the late 1930s from the interest in microwaves and waveguide transmission lines during World War II. Since that time a number of articles have been written describing its radiation mechanism, optimization design methods, and applications. Many of the articles published since 1939 which deal with the fundamental theory, operating principles, and designs of a horn as a radiator can be found in a book of reprinted papers [1].

The horn is widely used as a feed element for large radioastronomy, satellite-tracking, and communication dishes found installed throughout the world. In addition to its utility as a feed for reflectors and lenses it is a common element of phased arrays and serves as a universal standard for calibration and gain measurements of other high-gain antennas. Its widespread applicability stems from its simplicity in construction, ease of excitation, versatility, large gain, and preferred overall performance.

An electromagnetic horn can take many different forms, four of which are shown in Fig. 1. The horn is nothing more than a hollow pipe of different cross sections which has been tapered to a larger opening. The type, direction, and amount of taper can have a profound effect on the overall performance of the element as a radiator.

The total field radiated by a conventional horn is a combination of the direct field and the diffractions of it from the edges of the aperture, which can be accounted for using diffraction techniques [2–5]. Techniques found in Chapter 11 of [6] can be used to determine both the direct field and the diffractions from the edges. The edge diffractions, especially those that occur at edges where the electric field is normal to them, influence the antenna pattern structure especially in the back lobe region. The diffractions provide undesirable radiation in the minor lobe structure of the pattern, as well as in that of the main lobe. However, they dominate in low-intensity regions.

A conventional optimum gain horn is usually designed so that its on-axis gain is maximum. This is accomplished by controlling the dimensions of the horn (length and/or flare angle) in such a way that diffractions from the aperture edges add in phase with the direct radiation. In addition, a number of other schemes have been introduced to minimize the effect of diffractions, provide a better pattern symmetry in all planes, and reduce the side lobe intensity. These include the

(a)   introduction of corrugations on the inside walls of the horns (pyramidal or conical),

(b)   curving of the walls of the horn at its aperture,

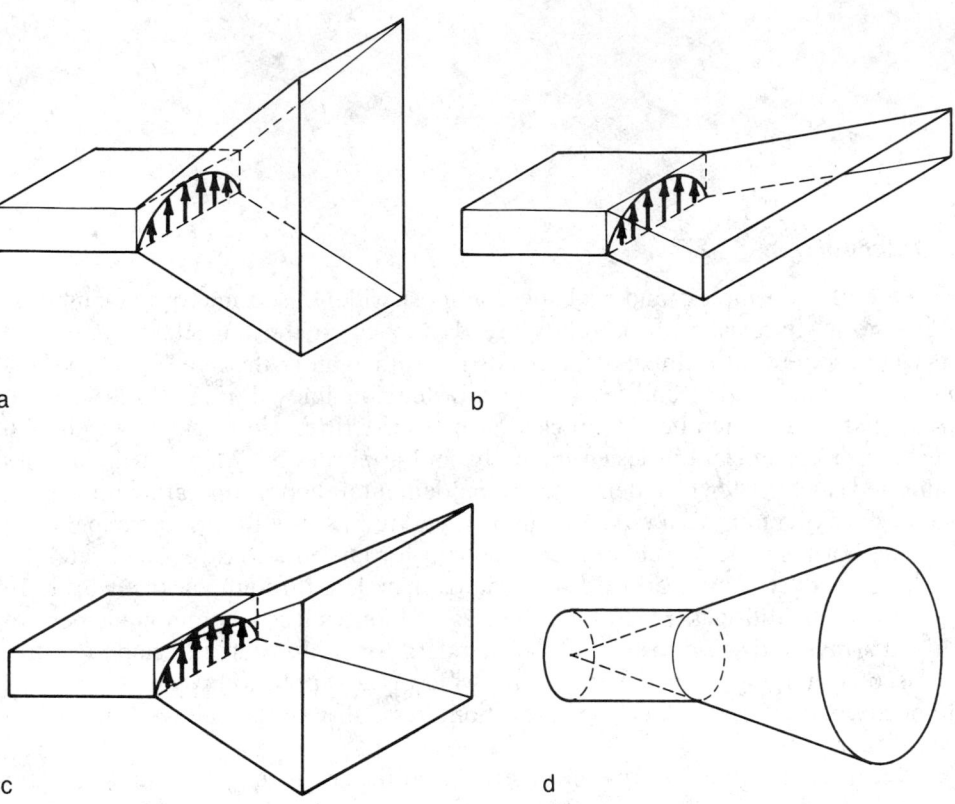

Fig. 1. Typical electromagnetic horn antenna configurations. (*a*) *E*-plane. (*b*) *H*-plane. (*c*) Pyramidal. (*d*) Conical. (*After Balanis [6], © 1982; reprinted by permission of Harper & Row Publishers, Inc.*)

(*c*)   incorporation of corrugations with the curving of the walls at its aperture, and

(*d*)   introduction of higher-order modes within the horn so that the field at the aperture edges is very weak.

   In addition to their wide utilization as single-element radiators, horns have also been used frequently as the basic elements of arrays [7–9]. Techniques have been introduced which can be used to suppress grating lobes for limited-scan applications [7, 8]. In addition, a method has been presented, and verified experimentally, which can be used to eliminate blind spots which can possibly exist in rectangular horn arrays with oversized (overmoded) apertures [9]. The blind spots are a result of forced aperture resonances, and they have been verified both analytically and experimentally at 14 GHz in rectangular-grid arrays of tapered rectangular horns. The techniques can possibly be extended to arrays of conical horns and other tapered structures.

   Much of the material in this chapter is drawn from an antenna textbook this author has written [6].

2. The *E*-Plane Horn

The *E*-plane horn is formed by flaring the walls of a rectangular waveguide in the direction of the **E** field, as shown in Fig. 1a. A more detailed geometry is shown in Figs. 2a and 2b. In this section we want to present, in a summary form, the most pertinent equations related to the aperture fields, radiated fields, and directivity, and to outline a procedure for optimum horn design.

Aperture Fields

The fields at the aperture of an *E*-plane horn will be assumed to be the same as those of the dominant TE_{10} mode of a rectangular waveguide whose dimensions are the same as those of the horn aperture. The only difference is that for the fields of the horn antenna a phase term must also be included to account for the difference in phase that the fields exhibit across the aperture. The phase term is necessary to account for the differences in path that the waves travel from the throat to the different points at the aperture of the horn, as shown in Fig. 2b.

It can be shown [6] that if the (1) fields of the feed waveguide are those of its dominant TE_{10} mode and (2) horn length is large compared with the aperture dimensions, the lowest-order-mode fields at the aperture of the horn are given by

$$E'_z = E'_x = H'_y = 0 \tag{1a}$$

$$E'_y(x', y') \cong E_1 \cos\left(\frac{\pi}{a} x'\right) e^{-j(k/2)(y')^2/\varrho_1} \tag{1b}$$

$$H'_z(x', y') \cong j E_1 \left(\frac{\pi}{ka\eta}\right) \sin\left(\frac{\pi}{a} x'\right) e^{-j(k/2)(y')^2/\varrho_1} \tag{1c}$$

$$H'_x(x', y') \cong -\frac{E_1}{\eta} \cos\left(\frac{\pi}{a} x'\right) e^{-j(k/2)(y')^2/\varrho_1} \tag{1d}$$

$$\varrho_1 = \varrho_e \cos\psi_e \tag{1e}$$

where E_1 is a constant. The primes are used to indicate the fields at the aperture of the horn. The expressions are similar to the fields of a TE_{10} mode for a rectangular waveguide with aperture dimensions of a and b_1 (with $b_1 > a$). The only difference is the complex exponential term which is used here to represent the quadratic phase variations of the fields over the aperture of the horn.

The necessity of the quadratic phase term in (1b)–(1d) can be illustrated geometrically. Referring to Fig. 2b, let us assume that at the imaginary apex of the horn (shown dashed) there exists a line source radiating cylindrical waves. As the waves travel in the outward radial direction the constant phase fronts are cylindrical. At any point y' at the mouth of the horn the phase of the field will not be the same as that at the origin ($y' = 0$). The phase is different because the wave has traveled different distances from the apex to the aperture. The difference in travel path, designated as $\delta(y')$, can be obtained by referring to Fig. 2b. For any point y'

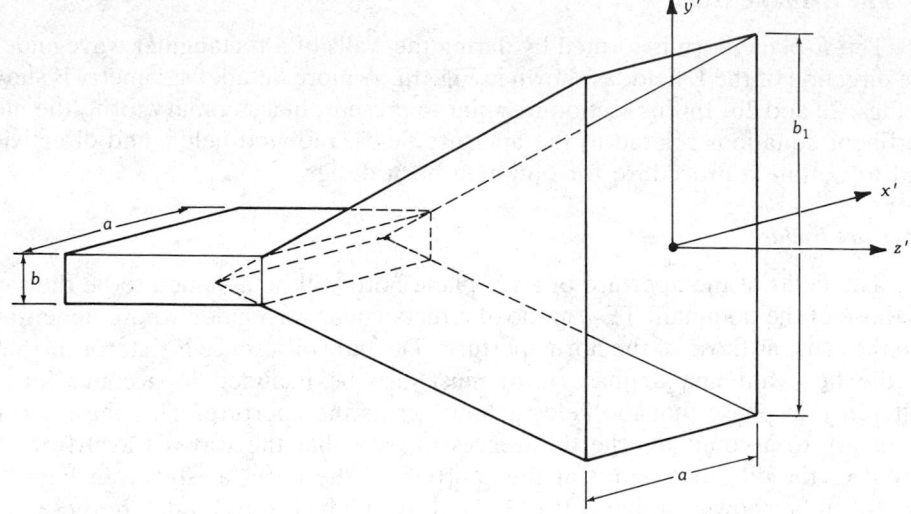

a

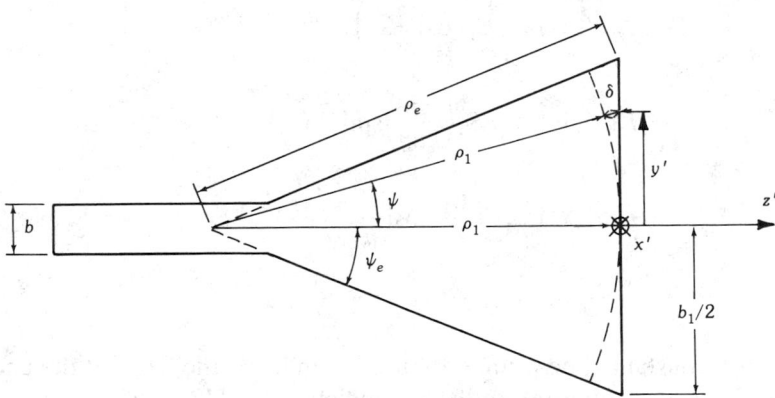

b

Fig. 2. *E*-plane horn and coordinate system. (*a*) *E*-plane sectoral horn. (*b*) *E*-plane view. (*After Balanis [6], © 1982; reprinted by permission of Harper & Row Publishers, Inc.*)

$$[\varrho_1 + \delta(y')]^2 = \varrho_1^2 + (y')^2 \tag{2a}$$

or

$$\delta(y') = -\varrho_1 + [\varrho_1^2 + (y')^2]^{1/2} = -\varrho_1 + \varrho_1[1 + (y'/\varrho_1)^2]^{1/2} \tag{2b}$$

Using the binomial expansion and retaining only the first two terms of (2b) reduces it to

$$\delta(y') \cong -\varrho_1 + \varrho_1\left[1 + \frac{1}{2}\left(\frac{y'}{\varrho_1}\right)^2\right] = \frac{1}{2}\frac{(y')^2}{\varrho_1} \tag{2c}$$

When (2c) is multiplied by the phase factor k, the result is identical with the quadratic phase term in (1b)–(1d).

Radiated Fields

To find the fields radiated by the horn, only the tangential components of the **E** and/or **H** fields over a closed surface need to be known [6]. The closed surface is chosen to coincide with an infinite plane passing through the aperture of the horn. To solve for the fields the approximate equivalent of Section 11.5.2 of [6] is used. That is, we assume that the equivalent current densities $\mathbf{J}_s$ and $\mathbf{M}_s$ exist over the horn aperture, and they are zero elsewhere. Doing this yields

$$\mathbf{J}_s = \hat{\mathbf{n}} \times \mathbf{H}_a = \hat{\mathbf{a}}_y J_y = -\hat{\mathbf{a}}_y\frac{E_1}{\eta}\cos\left(\frac{\pi}{a}x'\right)e^{-jk\delta(y')}, \qquad -a/2 \leqq x' \leqq a/2$$

$$\mathbf{M}_s = -\hat{\mathbf{n}} \times \mathbf{E}_a = \hat{\mathbf{a}}_x M_x = \hat{\mathbf{a}}_x E_1\cos\left(\frac{\pi}{a}x'\right)e^{-jk\delta(y')}, \qquad -b_1/2 \leqq y' \leqq b_1/2 \tag{3}$$

where $\mathbf{E}_a$ and $\mathbf{H}_a$ represent, respectively, the electric and magnetic fields at the horn aperture, as given by (1a)–(1e), and $\hat{\mathbf{n}} = \hat{\mathbf{a}}_z$ is a unit vector normal to the horn aperture.

Using (3) and the equations of Section 11.3 of Reference 6 it can be shown that the far-zone radiated fields are given by [6]

$$E_r = 0 \tag{4a}$$

$$E_\theta = -j\frac{a\sqrt{\pi k\varrho_1}E_1 e^{-jkr}}{8r}\left\{e^{jk_y^2(\varrho_1/2k)}\sin\phi(1 + \cos\theta)\right.$$

$$\left.\left[\frac{\cos(k_x a/2)}{(k_x a/2)^2 - (\pi/2)^2}\right]F(t_1, t_2)\right\} \tag{4b}$$

$$E_\phi = -j\frac{a\sqrt{\pi k\varrho_1}E_1 e^{-jkr}}{8r}\left\{e^{jk_y^2(\varrho_1/2k)}\cos\phi(\cos\theta + 1)\right.$$

$$\left.\left[\frac{\cos(k_x a/2)}{(k_x a/2)^2 - (\pi/2)^2}\right]F(t_1, t_2)\right\} \tag{4c}$$

where

$$F(t_1, t_2) = [C(t_2) - C(t_1)] - j[S(t_2) - S(t_1)] \tag{4d}$$

$$C(x) = \int_0^x \cos\left(\frac{\pi}{2}t^2\right)dt \tag{4e}$$

$$S(x) = \int_0^x \sin\left(\frac{\pi}{2} t^2\right) dt \tag{4f}$$

$$t_1 = \sqrt{\frac{1}{\pi k\varrho_1}} \left(-\frac{kb_1}{2} - k_y\varrho_1\right) \tag{4g}$$

$$t_2 = \sqrt{\frac{1}{\pi k\varrho_1}} \left(-\frac{kb_1}{2} - k_y\varrho_1\right) \tag{4h}$$

$$k_x = k \sin\theta \cos\phi \tag{4i}$$

$$k_y = k \sin\theta \sin\phi \tag{4j}$$

$C(x)$ and $S(x)$ are known as the cosine and sine Fresnel integrals and are well tabulated [6]. Computer subroutines are also available for efficient numerical evaluation of each [10, 11].

In the principal E- and H-planes the electric field reduces to

$$E\text{-Plane } (\phi = \pi/2)$$

$$E_r = E_\phi = 0 \tag{5a}$$

$$E_\theta = -j\frac{a\sqrt{\pi k\varrho_1}E_1 e^{-jkr}}{8r}\left[-e^{j(k\varrho_1\sin^2\theta)/2} (2/\pi)^2(1 + \cos\theta) F(t_1', t_2')\right] \tag{5b}$$

$$t_1' = \sqrt{\frac{k}{\pi\varrho_1}} \left(-\frac{b_1}{2} - \varrho_1 \sin\theta\right) \tag{5c}$$

$$t_2' = \sqrt{\frac{k}{\pi\varrho_1}} \left(+\frac{b_1}{2} - \varrho_1 \sin\theta\right) \tag{5d}$$

$$H\text{-Plane } (\theta = 0)$$

$$E_r = E_\theta = 0 \tag{6a}$$

$$E_\phi = -j\frac{a\sqrt{\pi k\varrho_1}E_1 e^{-jkr}}{8r}\left\{(1 + \cos\theta)\left[\frac{\cos[(1/2) ka \sin\theta]}{[(1/2) ka \sin\theta]^2 - (\pi/2)^2}\right]F(t_1'', t_2'')\right\} \tag{6b}$$

$$t_1'' = -\frac{b_1}{2}\sqrt{\frac{k}{\pi\varrho_1}} \tag{6c}$$

$$t_2'' = +\frac{b_1}{2}\sqrt{\frac{k}{\pi\varrho_1}} \tag{6d}$$

To better understand the performance of an E-plane sectoral horn and gain some insight into its performance as an efficient radiator, a three-dimensional normalized field pattern has been plotted in Fig. 3 utilizing (4a)–(4j). As expected, the E-plane pattern is much narrower than the H-plane because of the flaring and larger dimensions of the horn in that direction. Fig. 3 provides an excellent visual view of the overall radiation performance of the horn. To display additional details the corresponding normalized E- and H-plane patterns (in decibels) are illustrated in Fig. 4. These patterns also illustrate the narrowness of the E-plane and provide information on the relative levels of the pattern in those two planes.

To examine the behavior of the pattern as a function of flaring, the E-plane patterns for a horn antenna with $\varrho_1 = 15\lambda$ and with flare angles of $20° \leqq 2\psi_e \leqq 35°$ are plotted in Fig. 5a and for $40° \leqq 2\psi_e \leqq 55°$ in Fig. 5b. In each figure a total of four patterns are illustrated. Since each pattern is symmetrical, only half of each pattern is displayed. For small included angles the pattern becomes narrower as the flare increases. Eventually the pattern begins to widen, to become flatter around the main lobe, and the phase tapering at the aperture is such that even the main maximum does not occur on axis. This is illustrated in Fig. 5a by the pattern with $2\psi_e = 35°$. As the flaring is extended beyond that point the flatness (with certain allowable ripple) increases and eventually the main maximum returns again on axis as shown in Fig. 5b by the pattern with $2\psi_e = 45°$. It is also observed that as the flaring increases, the pattern exhibits much sharper cutoff characteristics. In practice, to compensate for the phase taper at the opening a lens is usually placed at the aperture, making the pattern of the horn always narrower as its flare increases.

Similar pattern variations occur as the length of the horn is varied while the flare angle is held constant. This is illustrated in Figs. 6a and 6b for an included

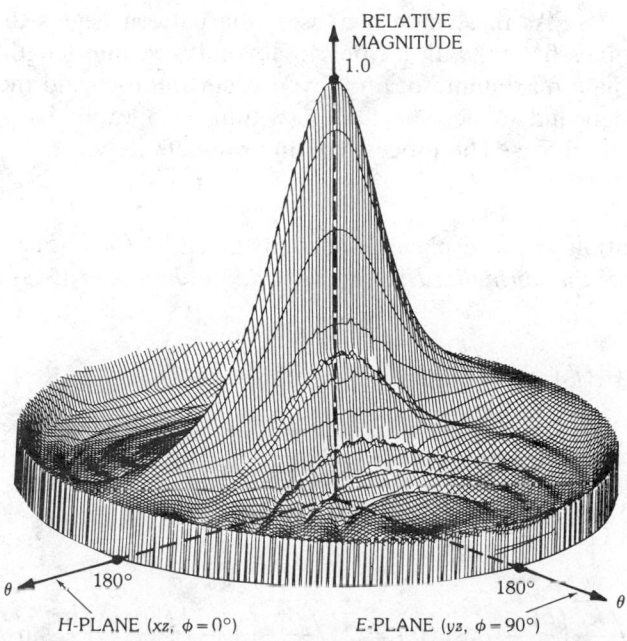

Fig. 3. Three-dimensional field pattern of E-plane sectoral horn ($\varrho_1 = 11.773\lambda$, $b_1 = 4.9269\lambda$, $a = 0.7849\lambda$).

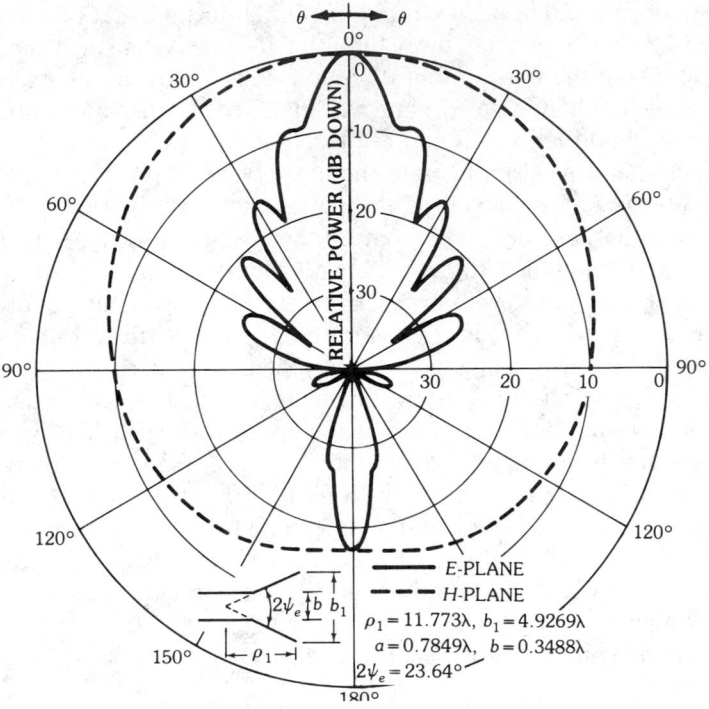

Fig. 4. *E*- and *H*-plane patterns of an *E*-plane sectoral horn.

angle of $2\psi_e = 35°$. As the length increases, the pattern begins to broaden and eventually becomes flatter (with a ripple). Beyond a certain length (see Fig. 6b, $\varrho_1 = 15\lambda$) the main maximum does not even occur on axis, and the pattern continues to broaden and to become flatter (within an allowable ripple) until the maximum return on axis. The process continues indefinitely.

Universal Curves

An observation of the *E*-plane pattern, as given by (5a)–(5d), indicates that the *magnitude of the normalized pattern, excluding the factor* $(1 + \cos\theta)$, can be written as

$$E_{\theta n} = F(t_1', t_2') = [C(t_2') - C(t_1')] - j[S(t_2') - S(t_1')] \tag{7a}$$

$$t_1' = \sqrt{\frac{k}{\pi\varrho_1}}\left(-\frac{b_1}{2} - \varrho_1\sin\theta\right) = 2\sqrt{\frac{b_1^2}{8\lambda\varrho_1}}\left[-1 - \frac{1}{4}\left(\frac{8\varrho_1\lambda}{b_1^2}\right)\left(\frac{b_1}{\lambda}\sin\theta\right)\right]$$

$$= 2\sqrt{s}\left[-1 - \frac{1}{4}\left(\frac{1}{s}\right)\left(\frac{b_1}{\lambda}\sin\theta\right)\right] \tag{7b}$$

$$t_2' = \sqrt{\frac{k}{\pi\varrho_1}}\left(\frac{b_1}{2} - \varrho_1\sin\theta\right) = 2\sqrt{\frac{b_1^2}{8\lambda\varrho_1}}\left[1 - \frac{1}{4}\left(\frac{8\varrho_1\lambda}{b_1^2}\right)\left(\frac{b_1}{\lambda}\sin\theta\right)\right]$$

$$= 2\sqrt{s}\left[1 - \frac{1}{4}\left(\frac{1}{s}\right)\left(\frac{b_1}{\lambda}\sin\theta\right)\right] \tag{7c}$$

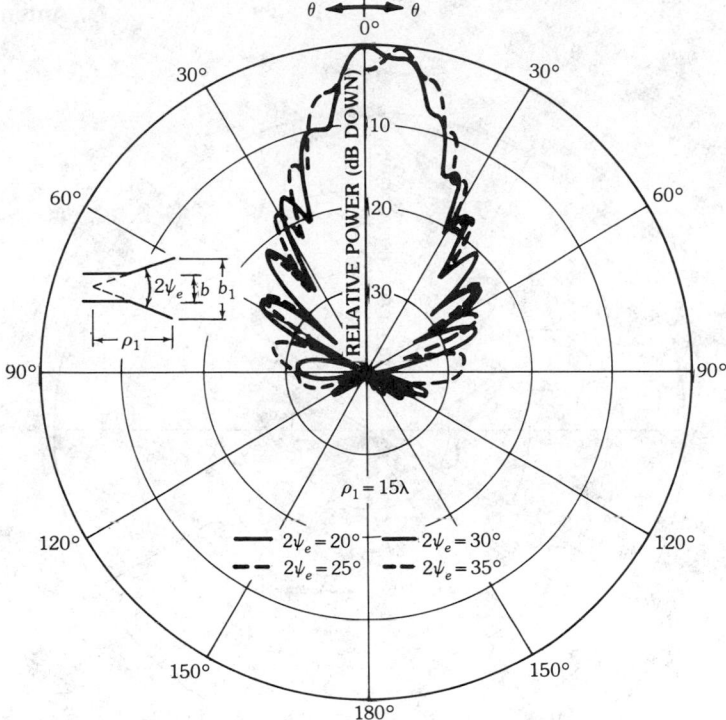

a

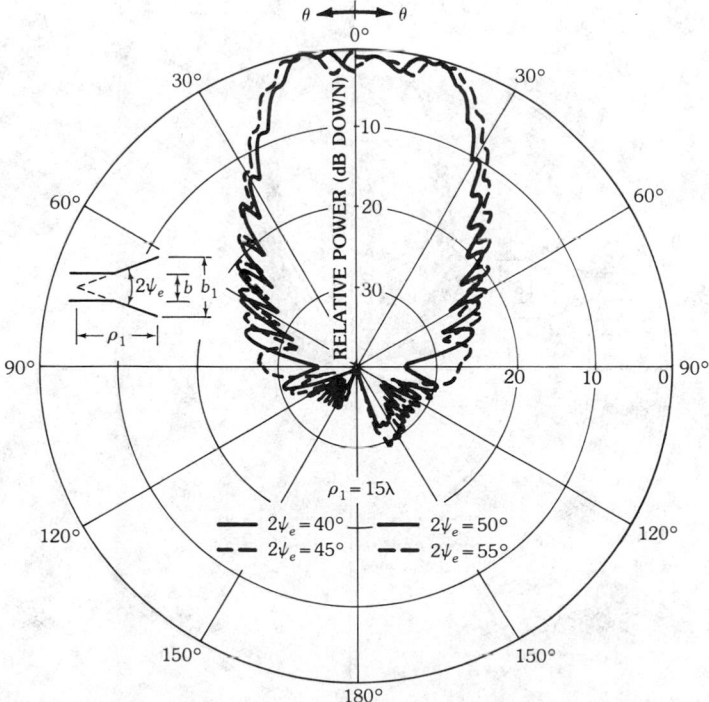

b

Fig. 5. *E*-plane patterns of *E*-plane sectoral horns with constant length and different included angles. (*a*) $2\psi_e = 20°$ to $35°$. (*b*) $2\psi_e = 40°$ to $55°$. (*After Balanis [6], © 1982; reprinted by permission of Harper & Row Publishers, Inc.*)

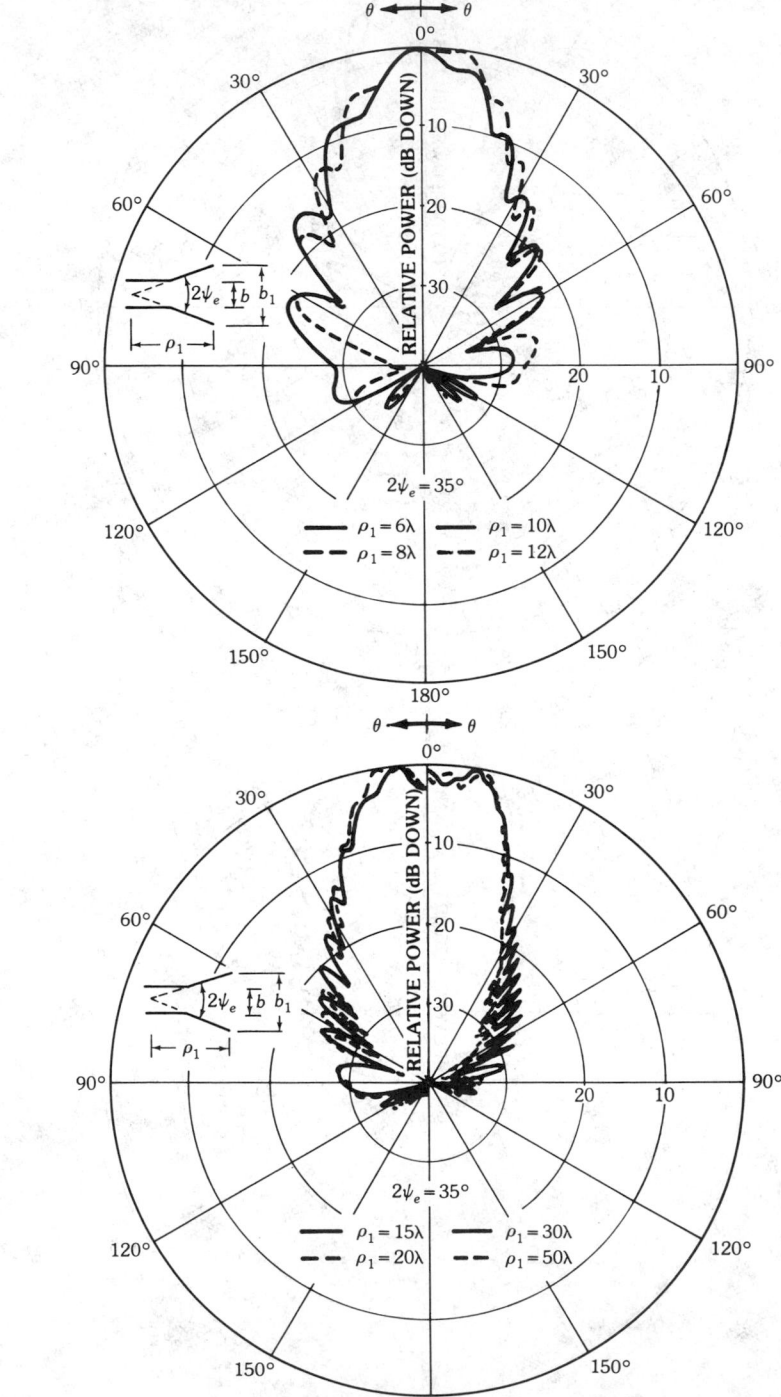

Fig. 6. *E*-plane patterns of *E*-plane sectoral horns for constant included angle and different lengths. (*a*) $\varrho_1 = 6\lambda$ to 12λ. (*b*) $\varrho_1 = 15\lambda$ to 50λ.

$$s = \frac{b_1^2}{8\lambda\varrho_1} \tag{7d}$$

For a given value of s the field of (7a) can be plotted as a function of $(b_1/\lambda)\sin\theta$, as shown in Fig. 7 for $s = 1/64, 1/8, 1/4, 1/2, 3/4$, and 1. These plots are usually referred to as *universal curves*, because from them the normalized E-plane pattern of any E-plane sectoral horn can be obtained. This is accomplished by first determining the value of s from a given b_1 and ϱ_1 by using (7d). For that value of s the field strength (in decibels) as a function of $(b_1/\lambda)\sin\theta$ (or as a function of θ for a given b_1) is obtained from Fig. 7. Finally the value of $(1 + \cos\theta)$, normalized to 0 dB and written as $20\log_{10}[(1 + \cos\theta)/2]$, is added to that number to arrive at the required field strength.

Example 1—An E-plane horn has dimensions of $a = 0.7849\lambda$, $b = 0.3488\lambda$, $b_1 = 4.9269\lambda$, and $\varrho_1 = 11.773\lambda$. Find its E-plane normalized field intensity (in decibels *and* as a voltage ratio) at an angle of $\theta = 30°$ using the universal curves of Fig. 7.

Solution—Using (7d)

$$s = \frac{b_1^2}{8\lambda\varrho_1} = \frac{(4.9269)^2}{8(11.773)} = 0.2577 \cong \frac{1}{4}$$

At $\theta = 30°$

$$\frac{b_1}{\lambda}\sin\theta = 4.9269 \sin 30° = 4.9269(0.5) = 2.463 \cong 2.5$$

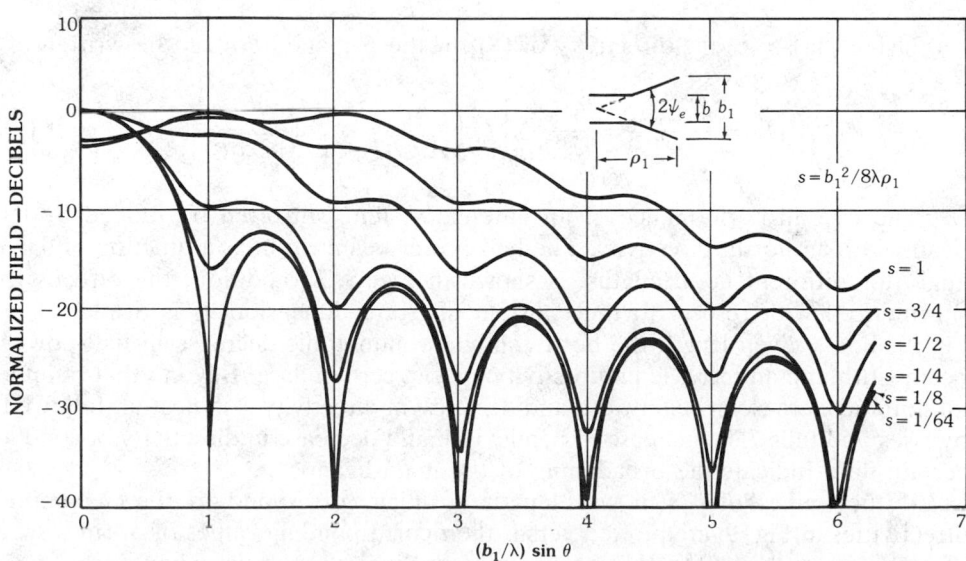

Fig. 7. *E*-plane universal patterns for *E*-plane sectoral and pyramidal horns. (*After Balanis [6], © 1982; reprinted by permission of Harper & Row Publishers, Inc.*)

At that point, using the $s = 1/4$ curve of Fig. 7, the field intensity is about -15.6 dB. Therefore the total field intensity at $\theta = 30°$ is equal to

$$E_\theta = -15.6 + 20\log_{10}\left(\frac{1 + \cos 30°}{2}\right) = -15.6 - 0.6 = -16.2 \text{ dB}$$

or as a normalized voltage ratio of

$$E_\theta = 0.155$$

which closely agrees with the results of Fig. 4.

Directivity

The directivity is a parameter which most often is used as the figure of merit to describe the performance of a horn antenna. By definition the directivity can be written as [6]

$$D_0 = \frac{U_{\max}}{U_0} = \frac{4\pi U_{\max}}{P_r} \tag{8}$$

where

$U_{\max}$ = maximum radiation intensity

U_0 = radiation intensity of isotropic source

P_r = total radiated power

Using (4a)–(4c) the directivity of (8) for the E-plane horn can be written as

$$D_E = \frac{4\pi U_{\max}}{P_r} = \frac{64a\varrho_1}{\pi\lambda b_1}\left[C^2\left(\frac{b_1}{\sqrt{2\lambda\varrho_1}}\right) + S^2\left(\frac{b_1}{\sqrt{2\lambda\varrho_1}}\right)\right] \tag{9}$$

The overall performance of an antenna system can often be judged by its beamwidth and/or its directivity. The half-power beamwidth, as a function of flare angle for different horn lengths, is shown in Fig. 8. In addition, the directivity (normalized with respect to the constant aperture dimension a) is displayed in Fig. 9. For a given length the horn exhibits a monotonic decrease in half-power beamwidth and an increase in directivity up to a certain flare. Beyond that point a monotonic increase in beamwidth and decrease in directivity is indicated, followed by rises and falls. The increase in beamwidth and decrease in directivity beyond a certain flare indicate the broadening of the main beam.

If the values of b_1 (in wavelengths), which correspond to the maximum directivities in Fig. 9, are plotted versus their corresponding values of ϱ_1 (in wavelengths), it can be shown that each optimum directivity occurs when

$$b_1 \cong \sqrt{2\lambda\varrho_1} \tag{10a}$$

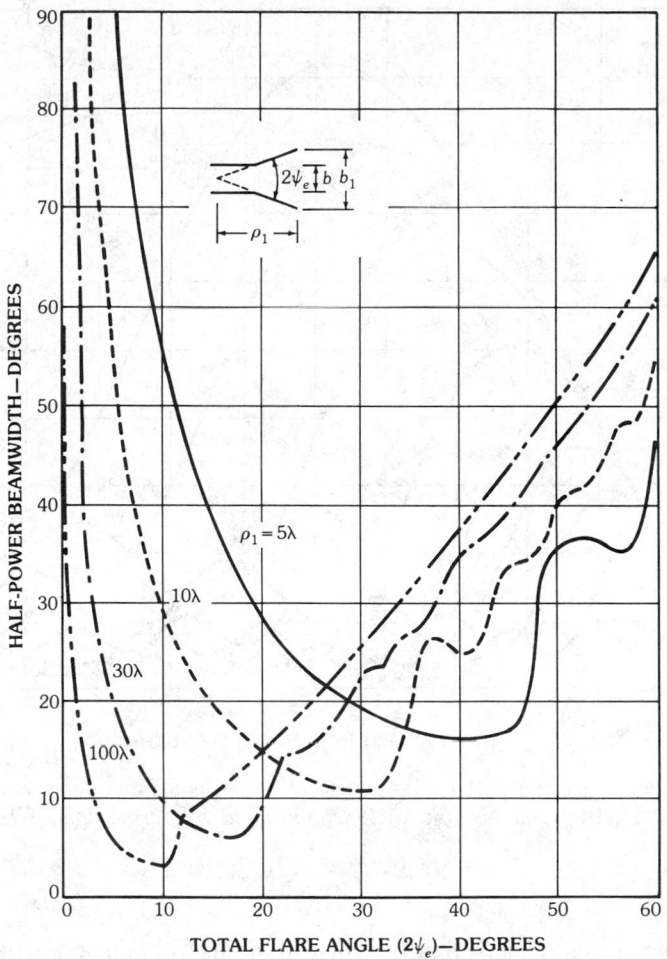

Fig. 8. Half-power beamwidth of *E*-plane sectoral horns as a function of included angle and for different lengths. (*After Balanis [6], © 1982; reprinted by permission of Harper & Row Publishers, Inc.*)

with a corresponding value of *s* equal to

$$s\bigg|_{b_1 = \sqrt{2\lambda\varrho_1}} = s_{\mathrm{op}} = \frac{b_1{}^2}{8\lambda\varrho_1}\bigg|_{b_1 = \sqrt{2\lambda\varrho_1}} = \frac{1}{4} \tag{10b}$$

The directivity of an *E*-plane sectoral horn can also be computed by using the following procedure [12].

(*a*)   Calculate *B* by

$$B = \frac{b_1}{\lambda}\sqrt{\frac{50}{\varrho_e/\lambda}} \tag{11a}$$

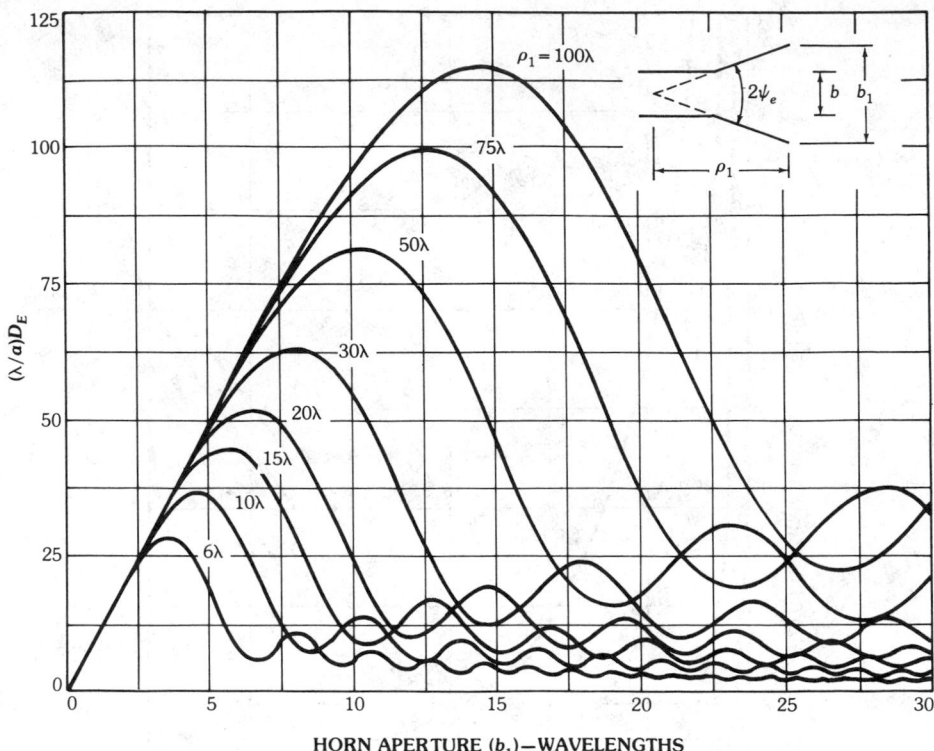

Fig. 9. Normalized directivity of *E*-plane sectoral horns as a function of aperture size for different lengths. (*After Balanis [6], © 1982; reprinted by permission of Harper & Row Publishers, Inc.*)

(*b*) Using this value of *B*, find the corresponding value of G_E from Fig. 10. If, however, the value of *B* is smaller than 2, compute G_E using

$$G_E = \frac{32}{\pi} B \tag{11b}$$

(*c*) Calculate D_E by using the value of G_E from Fig. 10 or from (11b). Thus

$$D_E = \frac{a}{\lambda} \frac{G_E}{\sqrt{50\lambda/\varrho_e}} \tag{11c}$$

It has been found [13] through comparisons with experimental data that the expression of (9) gives values which are about 25 percent below those measured on fairly large horns. The discrepancy will be larger for smaller horns. A convenient formula, whose values pass through the median of experimental values, has been proposed [13] and is given by

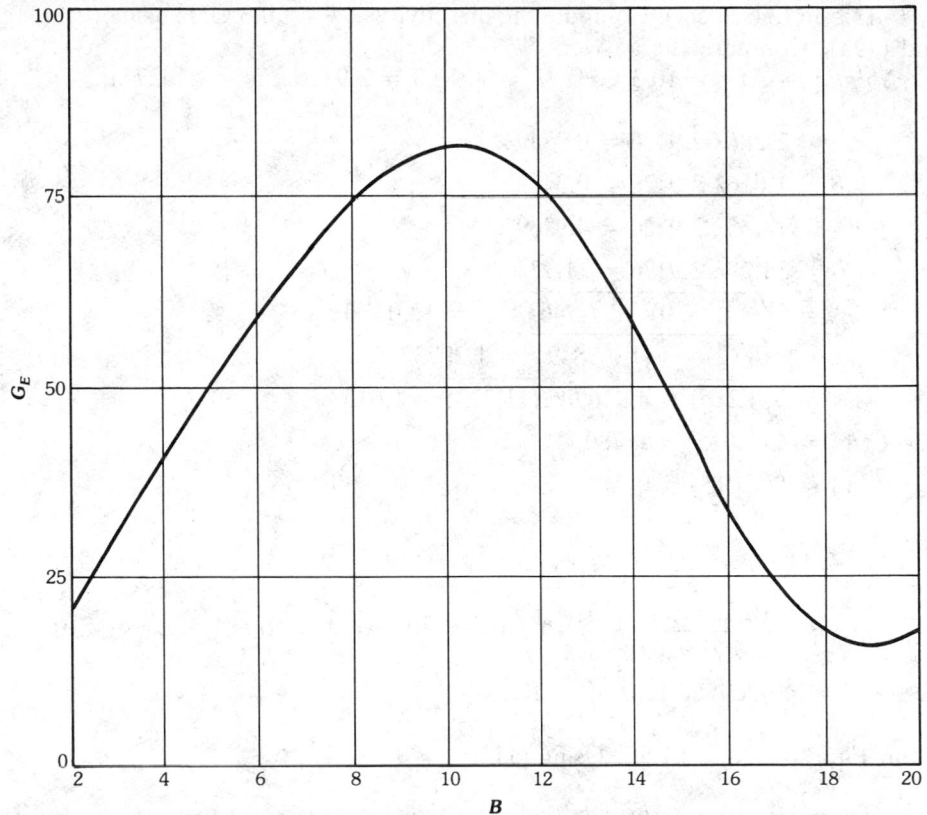

Fig. 10. Term G_E as a function of B. (*Adapted from data by E. H. Braun [12], © 1956 IEEE*)

$$D_E = \frac{16ab_1}{\lambda^2(1 + \lambda_g/\lambda)}\left\{\frac{2\lambda\varrho_1}{b_1^2}\left[C^2\left(\frac{b_1}{\sqrt{2\lambda\varrho_1}}\right) + S^2\left(\frac{b_1}{\sqrt{2\lambda\varrho_1}}\right)\right]\right\}e^{(\pi a/\lambda)(1-\lambda/\lambda_g)} \quad (12a)$$

where

$$\lambda_g = \frac{2\pi}{\beta} = \frac{2\pi}{\sqrt{k^2 - (\pi/a)^2}} = \frac{2}{\sqrt{(2/\lambda)^2 - (1/a)^2}} \quad (12b)$$

Two reasons for the discrepancy in the values as given by (9) are provided by the assumptions that the aperture fields are the incident fields with a propagation constant equal to that of free space and the gain formula is based on the Kirchhoff approximation. These are good approximations for horns with large apertures, such as pyramidal horns; however, typical *E*-plane sectoral horns have closely spaced edges which do not satisfy these approximations well.

Example 2—An *X*-band (8.2–12.4 GHz) *E*-plane sectoral horn has dimensions of $a = 0.9$ in (2.286 cm), $b = 0.4$ in (1.016 cm), $b_1 = 5.65$ in (14.35 cm), and

$\varrho_1 = 13.5$ in (34.29 cm). Compute the directivity at $f = 10.3$ GHz using (9), (11c), and (12a). Compare the answers.

Solution—At $f = 10.3$ GHz, $\lambda = 30/10.3 = 2.9126$ cm $= 1.1427$ in.

$$a = 2.286\lambda/2.9126 = 0.7849\lambda$$
$$b = 1.016\lambda/2.9126 = 0.3488\lambda$$
$$b_1 = 14.35\lambda/2.9126 = 4.9269\lambda$$
$$\varrho_1 = 34.29\lambda/2.9126 = 11.7730\lambda$$
$$\varrho_e = \lambda\sqrt{(11.7730)^2 + (2.463\,45)^2} = 12.027\,9\lambda$$
$$\lambda_g = 2\lambda/\sqrt{(2)^2 - (1/0.7849)^2} = 1.2973\lambda$$
$$w = b_1/\sqrt{2\lambda\varrho_1} = 4.9269/\sqrt{2(11.773)} = 1.015$$
$$C(w) = C(1.015) = 0.780$$
$$S(w) = S(1.015) = 0.440$$

Using (9)

$$D_E = \frac{64(0.7849)(11.773)}{\pi(4.9269)}[(0.780)^2 + (0.440)^2] = 30.643 = 14.863 \text{ dB}$$
$$B = 4.9269\sqrt{50/12.0279} = 10.045$$

From Fig. 10, $G_E = 81.55$. Using (11c)

$$D_E = 0.7849(81.55)/\sqrt{50/12.0279} = 31.394 = 14.968 \text{ dB}$$

Using (12a)

$$D_E = \frac{16(0.7849)(4.9269)}{(1 + 1.2973)}\left\{\frac{2(11.773)}{(4.9269)^2}[(0.780)^2 + (0.440)^2]\right\}e^{\pi(0.7849)(1-1/1.2973)}$$
$$= 41.168 = 16.146 \text{ dB}$$

From a comparison of the three values of D_E, it is evident that those computed using (12a) is about 32 percent higher than those of (9) and (11c).

Gain

A gain G_0 of a horn, as for any other antenna, is related to the directivity D_0 by the total antenna efficiency e_t [6]. Thus we can write that

$$G_0 = e_t D_0 = e_{cd}e_r D_0 = e_{cd}(1 - |\Gamma|^2)D_0 \tag{13}$$

where

e_t = total antenna efficiency

e_{cd} = conduction-dielectric (radiation) efficiency

e_r = reflection efficiency

Γ = reflection coefficient at transmission line–antenna connection

For most well-matched horns, $e_t \cong 1$.

Design Procedure

In designing an optimum directivity E-plane horn, the usual specifications are the following:

Given: (a)   Desired optimum directivity (or gain)
         (b)   Center frequency of operation f
         (c)   Dimensions a, b of the feed waveguide

Desired: Dimensions ϱ_1, b_1, ϱ_e, and angle ψ_e

Procedure: The design procedure is as follows:
         (a)   Substitute (10a) into (9).
         (b)   Knowing D_E (as a dimensionless quantity), a, and λ, find ϱ_1 using (9). The term ϱ_1 will have the same dimensions as a and λ.
         (c)   Determine b_1 using (10a). The term b_1 will have the same dimensions as ϱ_1 and λ.
         (d)   Find ϱ_e and angle ψ_e using the geometry of Fig. 2.

Example 3—Design an optimum gain X-band E-plane sectoral horn such that its directivity at $f = 10$ GHz is 14.437 dB. The dimensions of the feed rectangular waveguide are $a = 0.9$ in (2.286 cm) and $b = 0.4$ in (1.016 cm).

Solution—At $f = 10$ GHz

$$\lambda = \frac{30 \times 10^9}{10 \times 10^9} = 3 \text{ cm}$$

$$a = \frac{2.286}{3}\lambda = 0.762\lambda$$

$$b = \frac{1.016}{3}\lambda = 0.3387\lambda$$

Substituting (10a) into (9) yields

$$D_E = \frac{64a\varrho_1}{\pi\lambda\sqrt{2\lambda\varrho_1}}[C^2(1) + S^2(1)]$$

Since $D_E = 14.437$ dB $= 27.779$,

$$C(1) = 0.77989$$
$$S(1) = 0.43826$$

Then

$$27.779 = \frac{64(0.762)\,\varrho_1}{\pi\sqrt{2\lambda\varrho_1}}[(0.779\,89)^2 + (0.438\,26)^2]$$

$$27.779 = 8.7846\sqrt{\frac{\varrho_1}{\lambda}}$$

so that

$$\varrho_1 = 10\lambda = 30 \text{ cm}$$
$$b_1 = \sqrt{2\lambda(10\lambda)} = 4.472\lambda = 13.416 \text{ cm}$$
$$\psi_e = \tan^{-1}\left(\frac{2.236}{10}\right) = 12.6°$$

The results of this design agree with the data of the $\varrho_1 = 10\lambda$ curve of Fig. 9.

3. The *H*-Plane Horn

The *H*-plane horn is formed by flaring the rectangular waveguide walls in the direction of the **H** field, as shown in Fig. 1b. A more detailed geometry is shown in Figs. 11a and 11b. The aperture fields, radiated fields, and directivity formulas will be summarized in this section.

Aperture Fields

Using a procedure similar to that of an *E*-plane horn and assuming a TE_{10}-mode field distribution, the electric and magnetic field components at the aperture of the *H*-plane sectoral horn of Figs. 11a and 11b can be written as

$$E'_x = H'_y = 0 \tag{14a}$$

$$E'_y(x') = E_2 \cos\left(\frac{\pi}{a_1}x'\right)e^{-jk\delta(x')} \tag{14b}$$

$$H'_x(x') = -\frac{E_2}{\eta}\cos\left(\frac{\pi}{a_1}x'\right)e^{-jk\delta(x')} \tag{14c}$$

$$\delta(x') = \frac{1}{2}\frac{(x')^2}{\varrho_2} \tag{14d}$$

$$\varrho_2 = \varrho_h \cos\psi_h \tag{14e}$$

As with the *E*-plane sectoral horn the $\delta(x')$ function in the exponential term is introduced to account for the phase variation (tapering) across the aperture of the horn.

Radiated Fields

The fields radiated by the *H*-plane sectoral horn are found, as with the *E*-plane sectoral horn, by formulating the equivalent current densities **J**$_s$ and **M**$_s$ over the

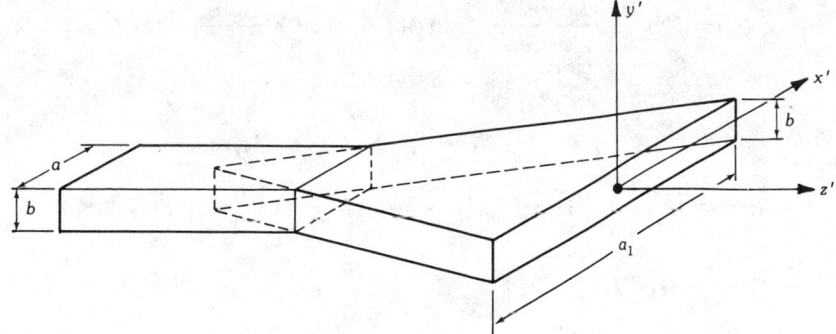

a

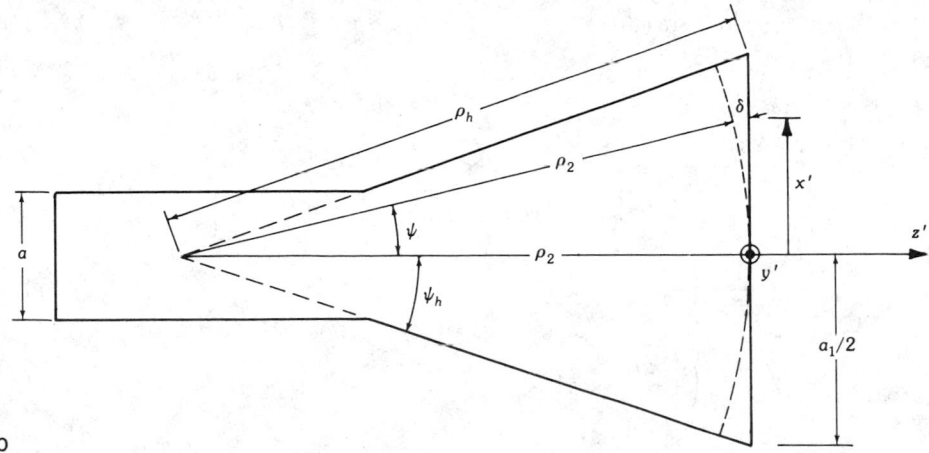

b

Fig. 11. *H*-plane sectoral horn and coordinate system. (*a*) *H*-plane sectoral horn. (*b*) *H*-plane view. (*After Balanis [6],* © *1982; reprinted by permission of Harper & Row Publishers, Inc.*)

horn aperture and assuming they are zero outside it. That is,

$$\mathbf{J}_s = \hat{\mathbf{n}} \times \mathbf{H}_a = \hat{\mathbf{a}}_y J_y = -\hat{\mathbf{a}}_y \frac{E_2}{\eta} \cos\left(\frac{\pi}{a_1} x'\right) e^{-jk\delta(x')}, \qquad -a_1/2 \leqq x' \leqq a_1/2$$

$$\mathbf{M}_s = -\hat{\mathbf{n}} \times \mathbf{E}_a = \hat{\mathbf{a}}_x M_x = \hat{\mathbf{a}}_x E_2 \cos\left(\frac{\pi}{a_1} x'\right) e^{-jk\delta(x')}, \qquad -b/2 \leqq y' \leqq b/2$$

(15)

where $\mathbf{E}_a$ and $\mathbf{H}_a$ are the horn aperture fields given by (14a)–(14e) and $\hat{\mathbf{n}} = \hat{\mathbf{a}}_z$ is a unit vector normal to the horn aperture. Using (15) and the equations of Section 11.3 of Reference 6, it can be shown that the far-zone radiated fields can be written as [6]

$$E_r = 0 \tag{16a}$$

$$E_\theta = jE_2 \frac{b}{8}\sqrt{\frac{k\varrho_2}{\pi}} \frac{e^{-jkr}}{r} \left\{ \sin\phi(1 + \cos\theta) \frac{\sin Y}{Y} [e^{jf_1}F(t_1', t_2') \right.$$

$$\left. + e^{jf_2}F(t_1'', t_2'')] \right\} \tag{16b}$$

$$E_\phi = jE_2 \frac{b}{8}\sqrt{\frac{k\varrho_2}{\pi}} \frac{e^{-jkr}}{r} \left\{ \cos\phi(\cos\theta + 1) \frac{\sin Y}{Y} [e^{jf_1}F(t_1', t_2') \right.$$

$$\left. + e^{jf_2}F(t_1'', t_2'')] \right\} \tag{16c}$$

where

$$Y = \frac{kb}{2} \sin\theta \sin\phi \tag{16d}$$

$$f_1 = \frac{(k_x')^2 \varrho_2}{2k} \tag{16e}$$

$$F(t_1, t_2) = [C(t_2) - C(t_1)] - j[S(t_2) - S(t_1)] \tag{16f}$$

$$t_1' = \sqrt{\frac{1}{\pi k \varrho_2}} \left(-\frac{ka_1}{2} - k_x'\varrho_2 \right) \tag{16g}$$

$$t_2' = \sqrt{\frac{1}{\pi k \varrho_2}} \left(+\frac{ka_1}{2} - k_x'\varrho_2 \right) \tag{16h}$$

$$k_x' = k \sin\theta \cos\phi + \frac{\pi}{a_1} \tag{16i}$$

$$f_2 = \frac{(k_x'')^2 \varrho_2}{2k} \tag{16j}$$

$$t_1'' = \sqrt{\frac{1}{\pi k \varrho_2}} \left(-\frac{ka_1}{2} - k_x''\varrho_2 \right) \tag{16k}$$

$$t_2'' = \sqrt{\frac{1}{\pi k \varrho_2}} \left(+\frac{ka_1}{2} - k_x''\varrho_2 \right) \tag{16l}$$

$$k_x'' = k \sin\theta \cos\phi - \frac{\pi}{a_1} \tag{16m}$$

and $C(x)$ and $S(x)$ are the cosine and sine Fresnel integrals of (4e) and (4f). The electric field in the principal E- and H-planes reduces to

E-Plane ($\phi = \pi/2$)

$$E_r = E_\phi = 0 \tag{17a}$$

$$E_\theta = jE_2 \frac{b}{8} \sqrt{\frac{k\varrho_2}{\pi}} \frac{e^{-jkr}}{r} \left\{ (1 + \cos\theta) \frac{\sin Y}{Y} [e^{jf_1} F(t'_1, t'_2) + e^{jf_2} F(t''_1, t''_2)] \right\} \quad \text{(17b)}$$

$$Y = \frac{kb}{2} \sin\theta \quad \text{(17c)}$$

$$k'_x = \frac{\pi}{a_1} \quad \text{(17d)}$$

$$k''_x = -\frac{\pi}{a_1} \quad \text{(17e)}$$

$$H\text{-Plane } (\phi = 0)$$

$$E_r = E_\theta = 0 \quad \text{(18a)}$$

$$E_\phi = jE_2 \frac{b}{8} \sqrt{\frac{k\varrho_2}{\pi}} \frac{e^{-jkr}}{r} \{(\cos\theta + 1)[e^{jf_1} F(t'_1, t'_2) + e^{jf_2} F(t''_1, t''_2)]\} \quad \text{(18b)}$$

$$k'_x = k\sin\theta + \frac{\pi}{a_1} \quad \text{(18c)}$$

$$k''_x = k\sin\theta - \frac{\pi}{a_1} \quad \text{(18d)}$$

with f_1, f_2, $F(t'_1, t'_2)$, $F(t''_1, t''_2)$, t'_1, t'_2, t''_1, and t''_2 as defined above.

Computations similar to those for the E-plane sectoral horn were also performed for the H-plane sectoral horn. A three-dimensional field pattern of an H-plane sectoral horn is shown in Fig. 12. Its corresponding E- and H-plane patterns are displayed in Fig. 13. This horn exhibits narrow pattern characteristics in the flared H-plane.

Normalized H-plane patterns for a given length horn ($\varrho_2 = 12\lambda$) and different flare angles are shown in Figs. 14a and 14b. A total of four patterns is illustrated in each figure. Since each pattern is symmetrical, only half of each pattern is displayed. As the included angle is increased, the pattern begins to become narrower up to a given flare. Beyond that point the pattern begins to broaden, attributed primarily to the phase taper (phase error) across the aperture of the horn. To correct this a lens is usually placed at the horn aperture which would yield narrower patterns as the flare angle is increased. Similar pattern variations are evident when the flare angle of the horn is maintained fixed while its length is varied, as shown in Figs. 15a and 15b.

Universal Curves

The *universal curves* for the H-plane sectoral horn are based on (18b), in the absence of the factor $(1 + \cos\theta)$. Neglecting the $(1 + \cos\theta)$ factor the normalized H-plane electric field of the H-plane sectoral horn can be written as

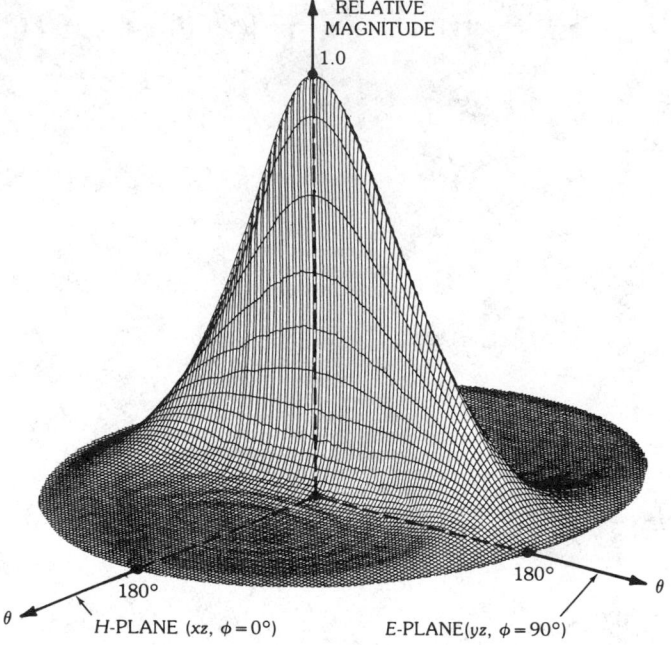

Fig. 12. Three-dimensional field pattern of an *H*-plane sectoral horn ($\varrho_2 = 12.3841\lambda$, $a_1 = 6.6710\lambda$, $b = 0.3488\lambda$).

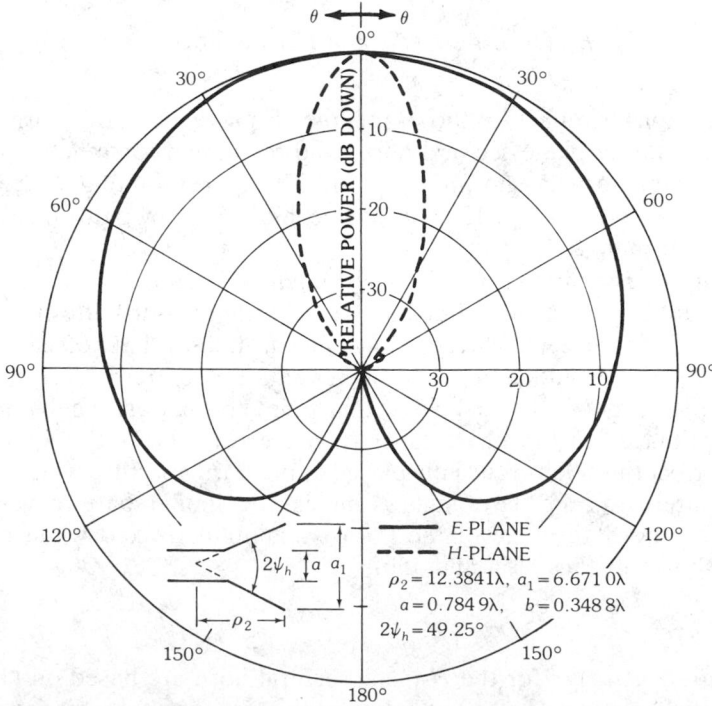

Fig. 13. *E*- and *H*-plane patterns of an *H*-plane sectoral horn.

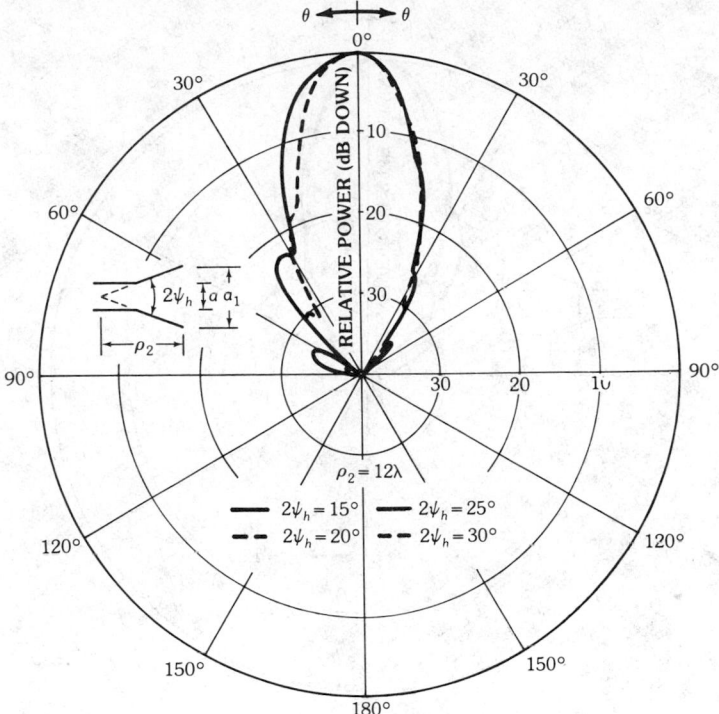

a

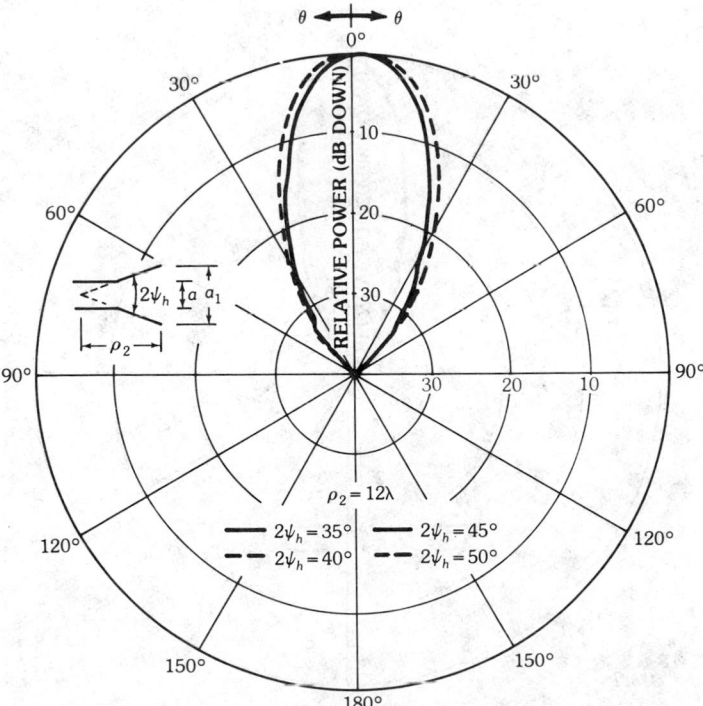

b

Fig. 14. *H*-plane patterns of *H*-plane sectoral horns for constant length and different included angles. (a) $2\psi_h = 15°$ to $30°$. (*After Balanis [6], © 1982; reprinted by permission of Harper & Row Publishers, Inc.*) (b) $2\psi_h = 35°$ to $50°$.

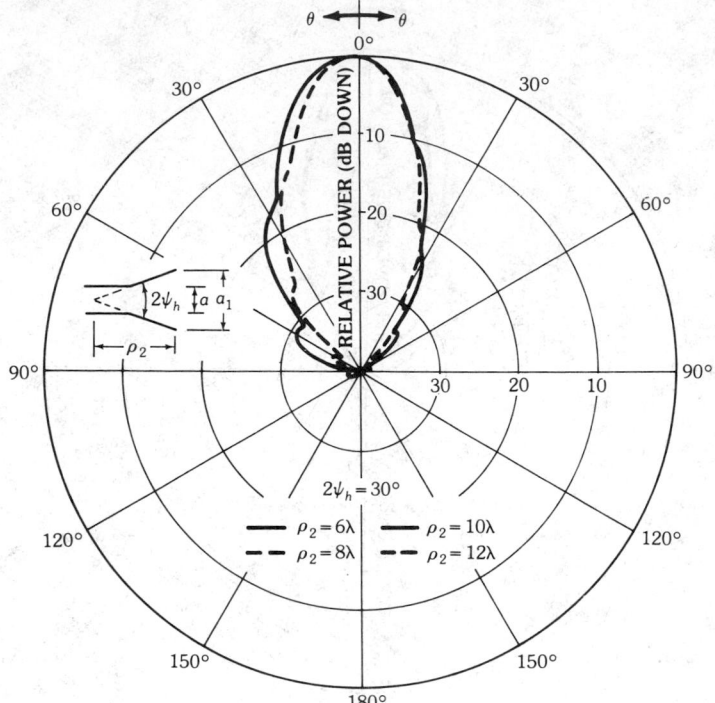

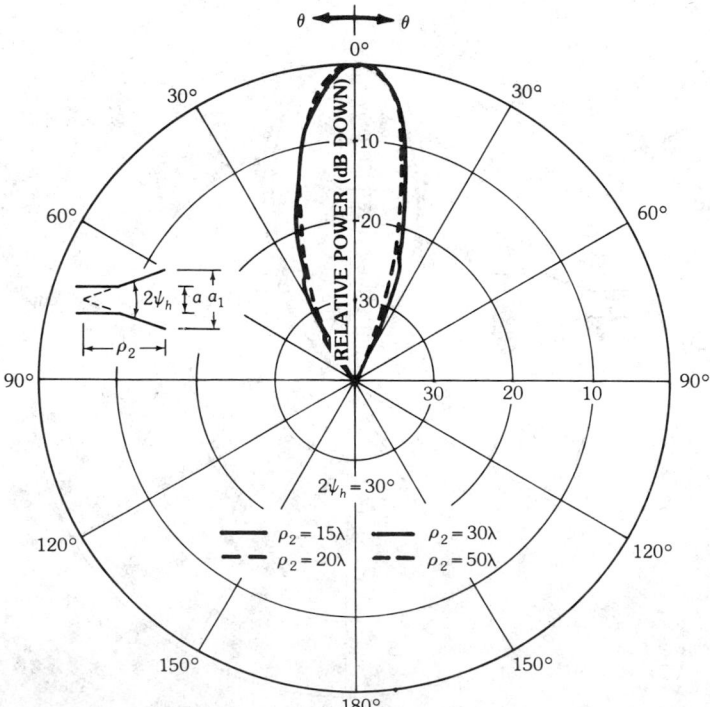

Fig. 15. *H*-plane patterns of *H*-plane sectoral horns of constant included angle and different lengths. (*a*) $\varrho_2 = 6\lambda$ to 12λ. (*b*) $\varrho_2 = 15\lambda$ to 50λ.

$$E_{\phi n} = [e^{jf_1} F(t_1', t_2') + e^{jf_2} F(t_1'', t_2'')] \tag{19a}$$

$$F(t_1, t_2) = [C(t_2) - C(t_1)] - j[S(t_2) - S(t_1)] \tag{19b}$$

$$f_1 = \frac{(k_x')^2 \varrho_2}{2k} = \frac{\varrho_2}{2k}\left(k \sin\theta + \frac{\pi}{a_1}\right)^2$$

$$= \frac{\pi}{8}\left(\frac{1}{t}\right)\left(\frac{a_1}{\lambda}\sin\theta\right)^2\left[1 + \frac{1}{2}\left(\frac{\lambda}{a_1\sin\theta}\right)\right]^2 \tag{19c}$$

$$f_2 = \frac{(k_x'')^2 \varrho_2}{2k} = \frac{\varrho_2}{2k}\left(k \sin\theta - \frac{\pi}{a_1}\right)^2$$

$$= \frac{\pi}{8}\left(\frac{1}{t}\right)\left(\frac{a_1}{\lambda}\sin\theta\right)^2\left[1 - \frac{1}{2}\left(\frac{\lambda}{a_1\sin\theta}\right)\right]^2 \tag{19d}$$

$$t_1' = \sqrt{\frac{1}{\pi k \varrho_2}}\left(-\frac{ka_1}{2} - k_x'\varrho_2\right)$$

$$= 2\sqrt{t}\left[-1 - \frac{1}{4}\left(\frac{1}{t}\right)\left(\frac{a_1}{\lambda}\sin\theta\right) - \frac{1}{8}\left(\frac{1}{t}\right)\right] \tag{19e}$$

$$t_2' = \sqrt{\frac{1}{\pi k \varrho_2}}\left(+\frac{ka_1}{2} - k_x'\varrho_2\right)$$

$$= 2\sqrt{t}\left[+1 - \frac{1}{4}\left(\frac{1}{t}\right)\left(\frac{a_1}{\lambda}\sin\theta\right) - \frac{1}{8}\left(\frac{1}{t}\right)\right] \tag{19f}$$

$$t_1'' = \sqrt{\frac{1}{\pi k \varrho_2}}\left(-\frac{ka_1}{2} - k_x''\varrho_2\right)$$

$$= 2\sqrt{t}\left[-1 - \frac{1}{4}\left(\frac{1}{t}\right)\left(\frac{a_1}{\lambda}\sin\theta\right) + \frac{1}{8}\left(\frac{1}{t}\right)\right] \tag{19g}$$

$$t_2'' = \sqrt{\frac{1}{\pi k \varrho_2}}\left(+\frac{ka_1}{2} - k_x''\varrho_2\right)$$

$$= 2\sqrt{t}\left[+1 - \frac{1}{4}\left(\frac{1}{t}\right)\left(\frac{a_1}{\lambda}\sin\theta\right) + \frac{1}{8}\left(\frac{1}{t}\right)\right] \tag{19h}$$

$$t = \frac{a_1^2}{8\lambda\varrho_2} \tag{19i}$$

For a given value of t, as given by (19i), the normalized field of (19a) is plotted in Fig. 16 as a function of $(a_1/\lambda)\sin\theta$ for $t = 1/64$, $1/8$, $1/4$, $1/2$, $3/4$, and 1. Following a procedure identical with that for the E-plane sectoral horn, the H-plane pattern of any H-plane sectoral horn can be obtained from these curves. The normalized

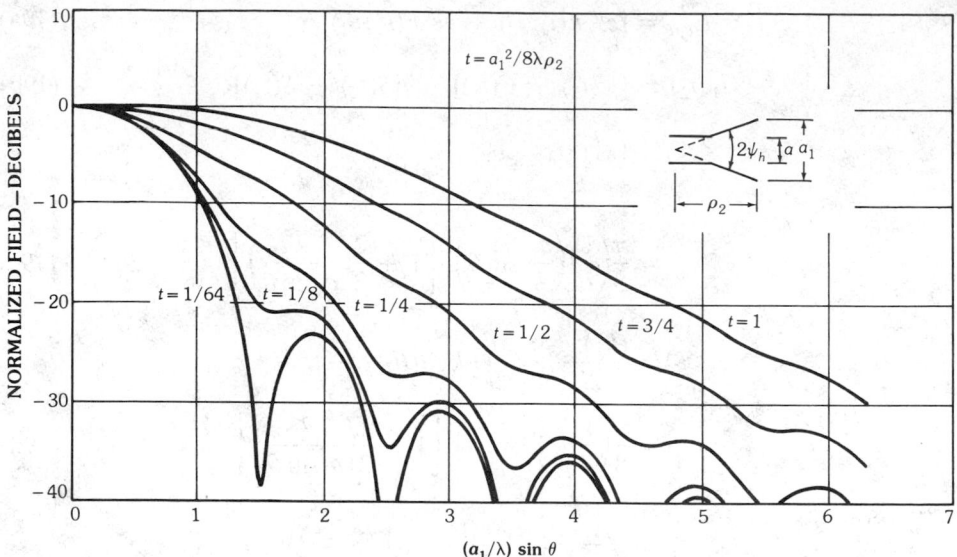

Fig. 16. *H*-plane universal patterns for *H*-plane sectoral and pyramidal horns. (*After Balanis [6], © 1982; reprinted by permission of Harper & Row Publishers, Inc.*)

value of the $(1 + \cos\theta)$ factor in decibels, written as $20 \log_{10}[(1 + \cos\theta)/2]$, must also be included.

Example 4—An *H*-plane horn has dimensions of $a = 0.7849\lambda$, $b = 0.3488\lambda$, $a_1 = 6.6710\lambda$, and $\varrho_2 = 12.3841\lambda$. Find the *H*-plane normalized field intensity (in decibels *and* as a voltage ratio) at an angle of $\theta = 30°$ using the universal curves of Fig. 16.

Solution—Using (19i)

$$t = \frac{a_1^2}{8\lambda\varrho_2} = \frac{(6.6710)^2}{8(12.3841)} = 0.449 \cong 0.45$$

None of the curves in Fig. 16 represents $t = 0.45$. However, the curve of $t = 1/2$ is one which should yield very approximate results. More accurate results could be obtained by interpolating between curves.

At $\theta = 30°$

$$\frac{a_1}{\lambda}\sin\theta = 6.6710 \sin 30° = 6.6710(0.5) = 3.3355 \cong 3.34$$

At that point, using the $t = 1/2$ curve of Fig. 16, the field intensity is about -25 dB. Therefore the total field intensity at $\theta = 30°$ is equal to

$$E_\theta = -25 + 20 \log_{10}\left(\frac{1 + \cos 30°}{2}\right) = 25 + 0.6 = 25.6 \text{ dB}$$

or as a normalized voltage ratio of

$$E_\theta = 0.0525$$

which closely agrees with the results of Fig. 13.

Directivity

Using the field expressions of (16a)–(16m) we can write the directivity as

$$D_H = \frac{4\pi U_{max}}{P_r} = \frac{4\pi b \varrho_2}{a_1 \lambda} \{[C(u) - C(v)]^2 + [S(u) - S(v)]^2\} \qquad (20a)$$

where

$$u = \frac{1}{\sqrt{2}} \left(\frac{\sqrt{\lambda \varrho_2}}{a_1} + \frac{a_1}{\sqrt{\lambda \varrho_2}} \right) \qquad (20b)$$

$$v = \frac{1}{\sqrt{2}} \left(\frac{\sqrt{\lambda \varrho_2}}{a_1} - \frac{a_1}{\sqrt{\lambda \varrho_2}} \right) \qquad (20c)$$

The half-power beamwidth as a function of flare angle is plotted in Fig. 17. The normalized directivity (relative to the constant aperture dimension b) for different horn lengths, as a function of the flare angle ψ_h, is displayed in Fig. 18. As for the E-plane sectoral horn the half-power beamwidth exhibits a monotonic decrease and the directivity a monotonic increase up to a given flare; beyond that the trends are reversed.

If the values of a_1 (in wavelengths), which correspond to the maximum directivities in Fig. 18, are plotted versus their corresponding values of ϱ_2 (in wavelengths), it can be shown that each optimum directivity occurs when

$$a_1 \cong \sqrt{3\lambda \varrho_2} \qquad (21a)$$

with a corresponding value of t equal to

$$t \Big|_{a_1 = \sqrt{3\lambda \varrho_2}} = t_{op} = \frac{a_1^2}{8\lambda \varrho_2} \Big|_{a_1 = \sqrt{3\lambda \varrho_2}} = \frac{3}{8} \qquad (21b)$$

The directivity of an H-plane sectoral horn can also be computed by using the following procedure [12].

(a) Calculate A by

$$A = \frac{a_1}{\lambda} \sqrt{\frac{50}{\varrho_h/\lambda}} \qquad (22a)$$

(b) Using this value of A, find the corresponding value of G_H from Fig. 19. If the value of A is smaller than 2, then compute G_H using

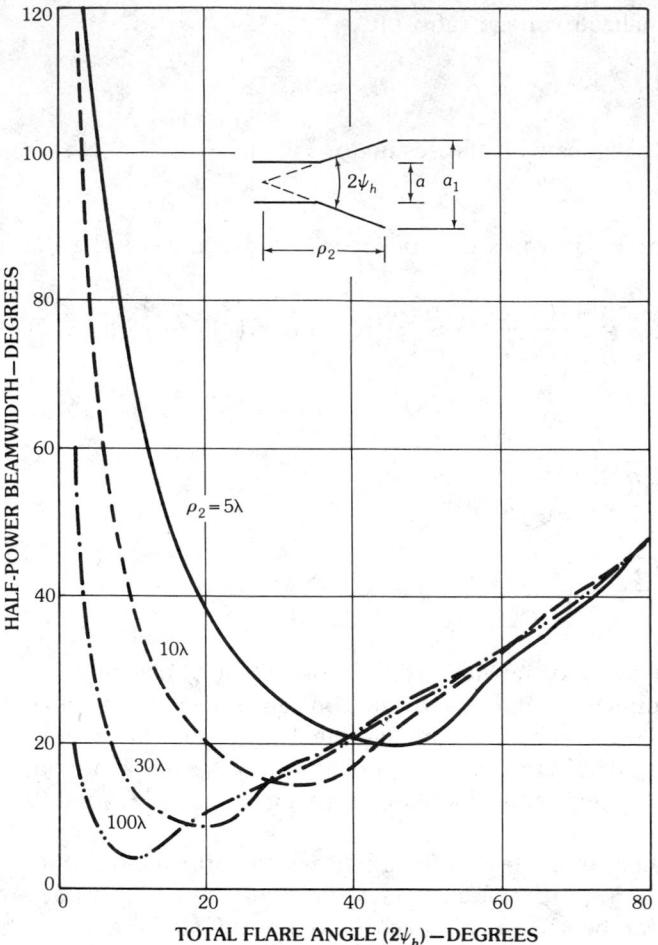

Fig. 17. Half-power beamwidth of *H*-plane sectoral horns as a function of included angle and for different lengths. (*After Balanis [6], © 1982; reprinted by permission of Harper & Row Publishers, Inc.*)

$$G_H = \frac{32}{\pi} A \qquad (22b)$$

(*c*) Calculate D_H by using the value of G_H from Fig. 19 or from (22b). Thus

$$D_H = \frac{b}{\lambda} \frac{G_H}{\sqrt{50\lambda/\varrho_h}} \qquad (22c)$$

This is the actual directivity of the horn.

Example 5—An *H*-plane sectoral horn has dimensions of $a = 0.9$ in (2.286 cm), $b = 0.4$ in (1.016 cm), $a_1 = 7.65$ in (19.431 cm), and $\varrho_2 = 14.2$ in (36.07 cm).

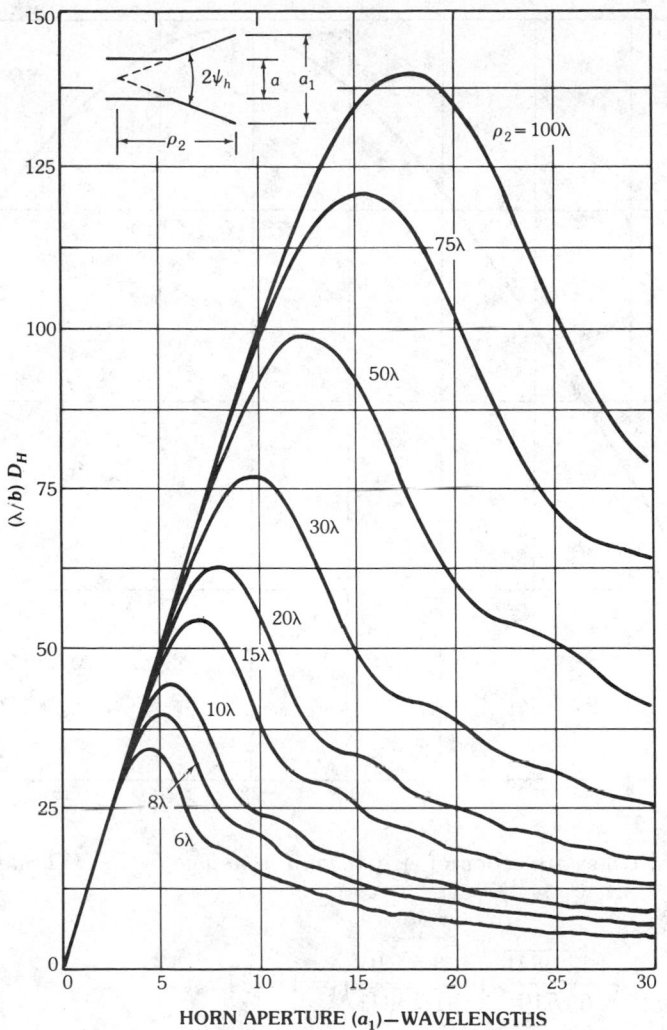

Fig. 18. Normalized directivity of *H*-plane sectoral horns as a function of aperture size and for different lengths. (*After Balanis [6], © 1982; reprinted by permission of Harper & Row Publishers, Inc.*)

Compute the directivity at $f = 10.3$ GHz using (20a) and (22c). Compare the answers.

> *Solution*—At $f = 10.3$ GHz, $\lambda = 30/10.3 = 2.9126$ cm $= 1.1467$ in.

$$a = 2.286\lambda/2.9126 = 0.7849\lambda$$

$$b = 1.016\lambda/2.9126 = 0.3488\lambda$$

$$a_1 = 19.431\lambda/2.9126 = 6.6710\lambda$$

$$\varrho_2 = 36.07\lambda/2.9126 = 12.3841\lambda$$

$$\varrho_h = \lambda\sqrt{(12.384)^2 + (3.3355)^2} = 12.8254\lambda$$

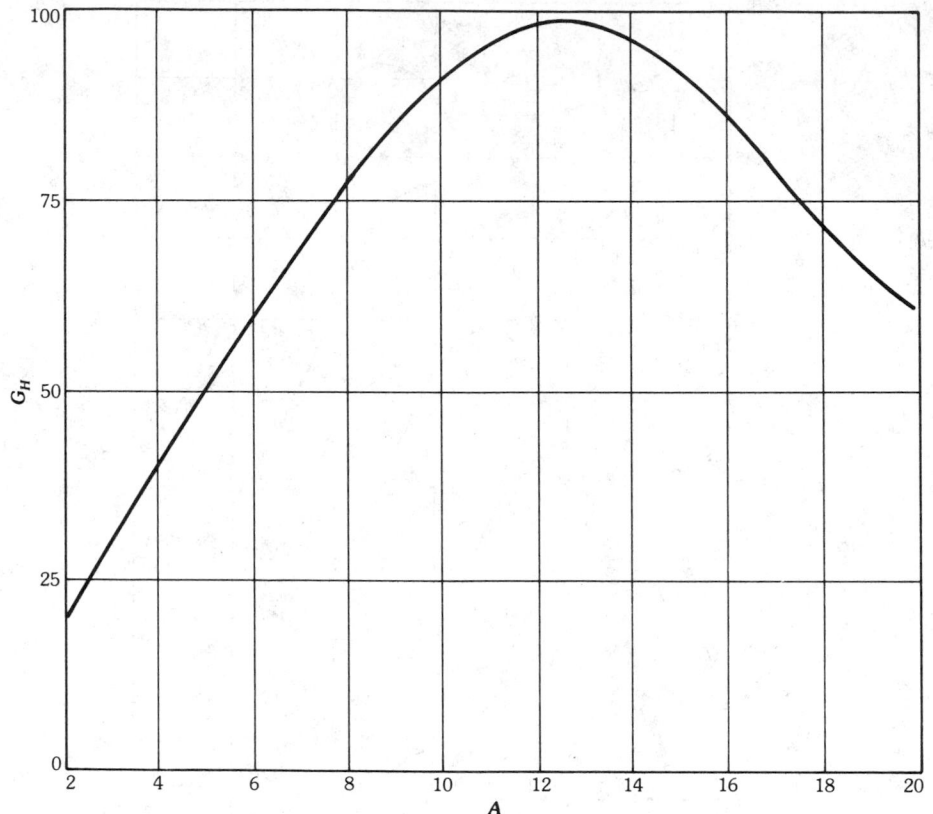

Fig. 19. Term G_H as a function of A. (*Adapted from data by E. H. Braun [12],* © *1956 IEEE*)

$$u = \frac{1}{\sqrt{2}}\left(\frac{\sqrt{12.3841}}{6.6710} + \frac{6.6710}{\sqrt{12.3841}}\right) = 1.7134$$

$$v = \frac{1}{\sqrt{2}}\left(\frac{\sqrt{12.3841}}{6.6710} - \frac{6.6710}{\sqrt{12.3841}}\right) = -0.9674$$

$$C(u) = C(1.7134) = 0.3252$$
$$C(v) = C(-0.9674) = -C(0.9674) = -0.7750$$
$$S(u) = S(1.7134) = 0.5359$$
$$S(v) = S(-0.9674) = -S(0.9674) = -0.4063$$

Using (20a)

$$D_H = \frac{4\pi(0.3488)(12.3841)}{6.6710}\left[(0.3252 + 0.7750)^2 + (0.5359 + 0.4063)^2\right]$$

$$= 17.073 = 12.323 \text{ dB}$$

$$A = 6.6710\sqrt{50/12.8254} = 13.172$$

so that

$$G_H = 98.673$$

From Fig. 19, $G_H = 98.673$. Using (22c)

$$D_H = 0.3488(98.673)/\sqrt{50/12.8254} = 17.431 = 12.413 \text{ dB}$$

A comparison of the values using the two different formulas indicates a very good agreement. For the H-plane sectoral horn, there is no need for a formula comparable to (12) because the propagation constant of the dominant mode in the horn is the same as that of free space and because the Kirchhoff methods yield the exact results for the on-axis gain of the TEM mode in an open-ended parallel-plate waveguide. Also the edges of the H-plane walls are farther apart than those of the E-plane; therefore the interaction between these walls will be weaker.

Design Procedure

In designing an optimum directivity H-plane horn the usual specifications are similar to those of an E-plane and are the following:

Given: (a)   Desired optimum directivity (or gain)
            (b)   Center frequency f of operation
            (c)   Dimensions a, b of the feed waveguide
Desired: Dimensions ϱ_2, a_1, ϱ_h, and angle ψ_h
Procedure: The design procedure is very similar to that for the E-plane, and it is as follows:
            (a)   Substitute (21a) into (20a)–(20c).
            (b)   Knowing D_H (as a dimensionless quantity), b, and λ, determine ϱ_2 using (20a). The term ϱ_2 will have the same dimensions as b and λ.
            (c)   Find a_1 using (21a). The term a_1 will have the same dimensions as ϱ_2 and λ.
            (d)   Determine ϱ_h and ψ_h using the geometry of Fig. 11.

Example 6—Design an optimum gain H-plane sectoral horn such that its directivity at $f = 10$ GHz is 11.884 dB. The dimensions of the feed rectangular waveguide are $a = 0.9$ in (2.286 cm) and $b = 0.4$ in (1.016 cm).

Solution—At $f = 10$ GHz

$$\lambda = \frac{30 \times 10^9}{10 \times 10^9} = 3 \text{ cm}$$

$$a = \frac{2.286}{3}\lambda = 0.762\lambda$$

$$b = \frac{1.016}{3}\lambda = 0.3387\lambda$$

Substituting (21a) into (20b) and (20c), we have

$$u = \frac{1}{\sqrt{2}}\left(\frac{1}{\sqrt{3}} + \sqrt{3}\right) = 1.633$$

$$v = \frac{1}{\sqrt{2}}\left(\frac{1}{\sqrt{3}} - \sqrt{3}\right) = -0.8165$$

Since $D_E = 11.884$ dB $= 15.4312$

$$C(1.633) = 0.35172$$
$$S(1.633) = 0.638\,89$$
$$C(-0.8165) = -C(0.8165) = -0.729\,77$$
$$S(-0.8165) = -S(0.8165) = -0.264\,26$$

(20a) can be written as

$$15.4312 = \frac{4\pi(0.3387)\varrho_2}{\sqrt{3\lambda\varrho_2}}\,[(0.351\,72 + 0.729\,77)^2 + (0.638\,99 + 0.264\,26)^2]$$

$$15.4312 = 4.879\sqrt{\frac{\varrho_2}{\lambda}}$$

so that

$$\varrho_2 = 10\lambda = 30 \text{ cm}$$
$$a_1 = \sqrt{3\lambda\varrho_2} = \sqrt{3\lambda(10\lambda)} = 5.477\lambda = 16.432 \text{ cm}$$
$$\psi_h = \tan^{-1}\left(\frac{8.216}{30}\right) = 15.3°$$

The results of this design agree with the data of the $\varrho_2 = 10\lambda$ curve of Fig. 18.

4. The Pyramidal Horn

The most widely used horn is the one which is flared in both directions, as shown in Fig. 1c. It is widely referred to as a *pyramidal horn*, and its radiation characteristics are essentially a combination of the E- and H-plane sectoral horns. A more detailed geometry of it is shown in Fig. 20.

Aperture and Radiated Fields

To simplify the analysis and to maintain a modeling which leads to computations which have been shown to correlate well with experimental data, the tangential components of the E- and H-fields over the aperture of the horn are approximated by

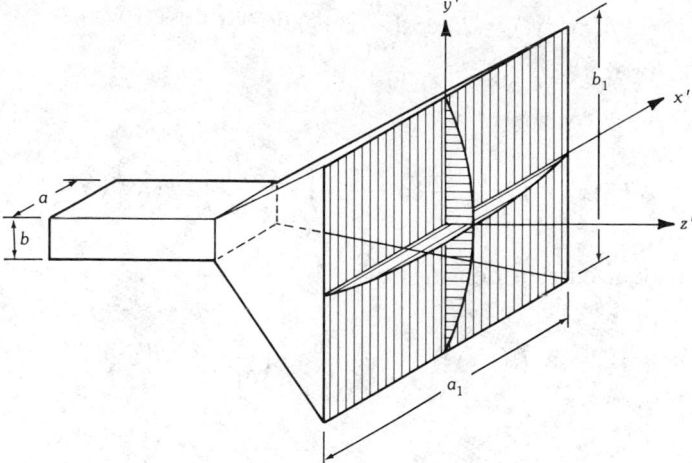

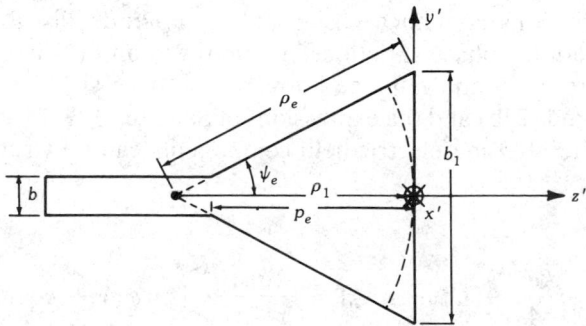

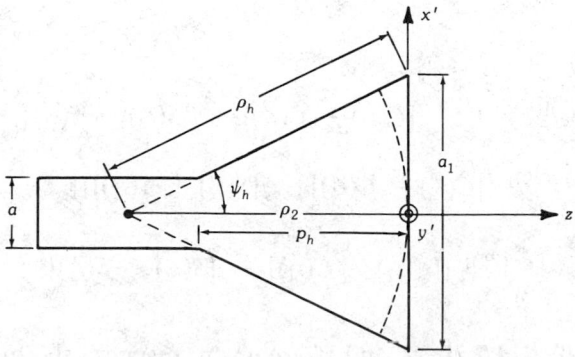

Fig. 20. Pyramidal horn and coordinate system. (*a*) Pyramidal horn. (*b*) *E*-plane view. (*c*) *H*-plane view. (*After Balanis [6], © 1982; reprinted with permission of Harper & Row Publishers, Inc.*)

$$E'_y(x', y') = E_0 \cos\left(\frac{\pi}{a_1} x'\right) e^{-j(k/2)[(x')^2/\varrho_2 + (y')^2/\varrho_1]} \tag{23a}$$

and

$$H'_x(x', y') = -\frac{E_0}{\eta} \cos\left(\frac{\pi}{a_1} x'\right) e^{-j(k/2)[(x')^2/\varrho_2 + (y')^2/\varrho_1]} \tag{23b}$$

and the equivalent current densities by

$$\mathbf{J}_s = \hat{\mathbf{n}} \times \mathbf{H}_a = \hat{\mathbf{a}}_y J_y(x', y') = -\hat{a}_y \frac{E_0}{\eta} \cos\left(\frac{\pi}{a_1} x'\right) e^{-j(k/2)[(x')^2/\varrho_2 + (y')^2/\varrho_1]} \tag{24a}$$

and

$$\mathbf{M}_s = -\hat{\mathbf{n}} \times \mathbf{E}_a = \hat{\mathbf{a}}_x M_x(x', y') = \hat{a}_x E_0 \cos\left(\frac{\pi}{a_1} x'\right) e^{-j(k/2)[(x')^2/\varrho_2 + (y')^2/\varrho_1]} \tag{24b}$$

The above expressions contain a cosinusoidal amplitude distribution in the x' direction and quadratic phase variations in both the x' and y' directions, similar to those of the sectoral E- and H-plane horns.

Using (24a) and (24b) and the expressions of Section 11.3 of Reference 6, it can be shown that the far-zone electric-field components can be written as follows:

$$E_r = 0 \tag{25a}$$

$$E_\theta = -j\frac{ke^{-jkr}}{4\pi r}[L_\phi + \eta N_\theta] = j\frac{kE_0 e^{-jkr}}{4\pi r}[\sin\phi(1 + \cos\theta) I_1 I_2] \tag{25b}$$

$$E_\phi = +j\frac{ke^{-jkr}}{4\pi r}[L_\theta + \eta N_\phi] = j\frac{kE_0 e^{-jkr}}{4\pi r}[\cos\phi(\cos\theta + 1) I_1 I_2] \tag{25c}$$

where

$$I_1 = \frac{1}{2}\sqrt{\frac{\pi\varrho_2}{k}} \left[e^{j[(k'_x)^2 \varrho_2/(2k)]} \{[C(t'_2) - C(t'_1)] - j[S(t'_2) - S(t'_1)]\} \right.$$
$$\left. + e^{j[(k''_x)^2 \varrho_2/(2k)]} \{[C(t''_2) - C(t''_1)] - j[S(t''_2) - S(t''_1)]\} \right] \tag{25d}$$

$$I_2 = \sqrt{\frac{\pi\varrho_1}{k}} e^{j[(k_y)^2 \varrho_1/(2k)]} \{[C(t_2) - C(t_1)] - j[S(t_2) - S(t_1)]\} \tag{25e}$$

Expressions for t'_1, t'_2, k'_x, t''_1, t''_2, and k''_x are given, respectively, by (16g)–(16i) and (16k)–(16m). Similarly t_1, t_2, and k_y can be evaluated, respectively, using (4g), (4h), and (4j).

The fields radiated by a pyramidal horn, as given by (25a)–(25c), are valid for all angles of observation. An examination of these equations reveals that the principal E-plane pattern ($\phi = \pi/2$) of a pyramidal horn, aside from a normalization factor, is identical with the E-plane pattern of an E-plane sectoral horn. Similarly the H-plane ($\phi = 0$) is identical with the H-plane of an H-plane sectoral horn. Therefore the pattern of a pyramidal horn is very narrow in both principal planes and, in fact, in all planes. This is illustrated in Fig. 21a. The corresponding E-plane pattern is shown in Fig. 4 and the H-plane pattern in Fig. 13.

To demonstrate that the maximum radiation for a pyramidal horn is not necessarily directed along its axis, the three-dimensional field pattern for a horn with $\varrho_1 = \varrho_2 = 6\lambda$, $a_1 = 12\lambda$, $b_1 = 6\lambda$, $a = 0.50\lambda$, and $b = 0.25\lambda$ is displayed in Fig. 21b. The corresponding two-dimensional E- and H-plane patterns are shown in Fig. 22. The maximum does not occur on axis because the phase error taper at the aperture is such that the rays emanating from the different parts of the aperture toward the axis are not in phase.

To physically construct a pyramidal horn the dimension p_e of Fig. 20b given by

$$p_e = (b_1 - b)\left[\left(\frac{\varrho_e}{b_1}\right)^2 - \frac{1}{4}\right]^{1/2} \tag{26a}$$

should be equal to the dimension p_h of Fig. 20c given by

$$p_h = (a_1 - a)\left[\left(\frac{\varrho_h}{a_1}\right)^2 - \frac{1}{4}\right]^{1/2} \tag{26b}$$

The dimensions chosen for Figs. 21a and 21b do satisfy these requirements. For the horn of Fig. 21a, $\varrho_e = 12.0279\lambda$, $\varrho_h = 12.8247\lambda$, and $p_e \cong p_h \cong 10.9\lambda$, while for that of Fig. 21b, $\varrho_e = 6.7082\lambda$, $\varrho_h = 8.4853\lambda$, and $p_e = p_h = 5.75\lambda$.

The fields of (25a)–(25c) provide accurate patterns for angular regions near the main lobe and its closest minor lobes. To accurately predict the field intensity of the pyramidal and other horns, especially in the minor lobes, diffraction techniques can be utilized [2–5]. These methods take into account diffractions that occur near the aperture edges of the horn. The diffraction contributions become more dominant in regions where the radiation of (25a)–(25c) is of very low intensity.

Directivity

As for the E- and H-plane sectoral horns the directivity of the pyramidal configuration is vital to the antenna designer and practicing engineer. The maximum radiation of the pyramidal horn is directed nearly along the z axis ($\theta = 0$). See Fig. 22. Using (25a)–(25c) it can be shown that the directivity for the pyramidal horn defined by (8) can be written as

$$D_p = \frac{4\pi U_{max}}{P_r} = \frac{8\pi\varrho_1\varrho_2}{a_1 b_1}\{[C(u) - C(v)]^2 + [S(u) - S(v)]^2\}$$

$$\times \left\{C^2\left(\frac{b_1}{\sqrt{2\lambda\varrho_1}}\right) + S^2\left(\frac{b_1}{\sqrt{2\lambda\varrho_1}}\right)\right\} \tag{27}$$

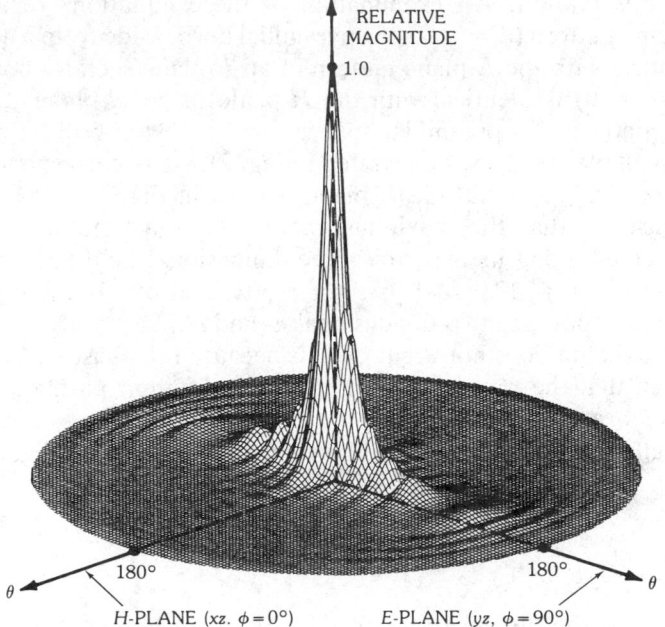

a

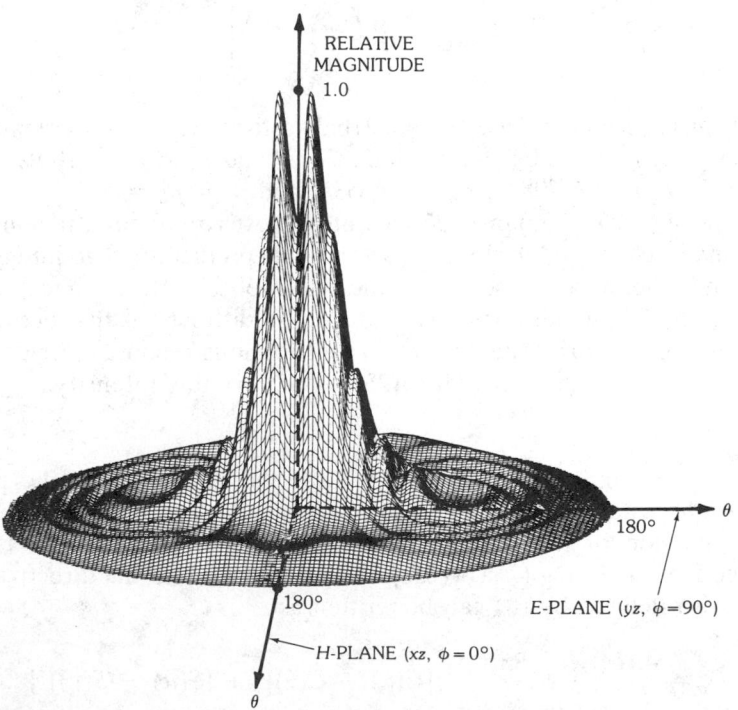

b

Fig. 21. Three-dimensional patterns of pyramidal horns. (*a*) With maximum on axis: $\varrho_1 = 11.773\lambda$, $\varrho_2 = 12.3841\lambda$, $a_1 = 6.6710\lambda$, $b_1 = 4.9269\lambda$, $a = 0.7849\lambda$, $b = 0.3488\lambda$. (*b*) With maximum not on axis: $\varrho_1 = \varrho_2 = 6\lambda$, $a_1 = 12\lambda$, $b_1 = 6\lambda$, $a = 0.5\lambda$, $b = 0.25\lambda$.

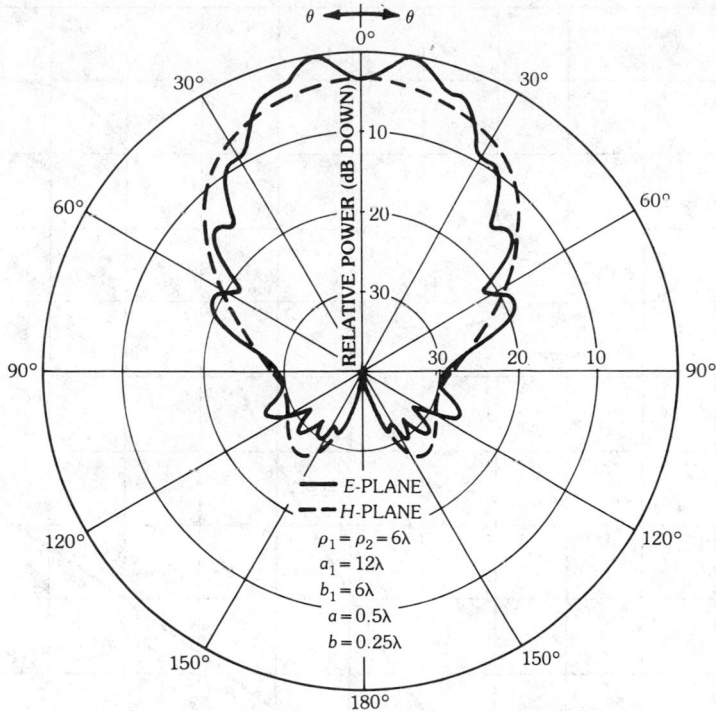

Fig. 22. *E*- and *H*-plane amplitude patterns of a pyramidal horn with maximum not on axis. (*After Balanis [6], © 1982; reprinted with permission of Harper & Row Publishers, Inc.*)

where u and v are defined, respectively, by (20b) and (20c). Another form of (27) is to write it as

$$D_p = \frac{\pi\lambda^2}{32ab} D_E D_H \tag{28}$$

where D_E and D_H are the directivities of the *E*- and *H*-plane sectoral horns as given, respectively, by (9) or (11c) or (12a) and (20a) or (22c). This is a well-known relationship and has been used extensively in the design of pyramidal horns.

The directivity (in decibels) of a pyramidal horn, over isotropic, can also be approximated by [14]

$$D_0(\text{dB}) = 10\left[1.008 + \log_{10}\left(\frac{a_1 b_1}{\lambda^2}\right)\right] - (L_E + L_H) \tag{29}$$

where L_E and L_H represent, respectively, the losses (in decibels) due to phase errors in the *E*- and *H*-planes of the horn which are found plotted in Fig. 23.

The directivity of a pyramidal horn can also be calculated by doing the following [12].

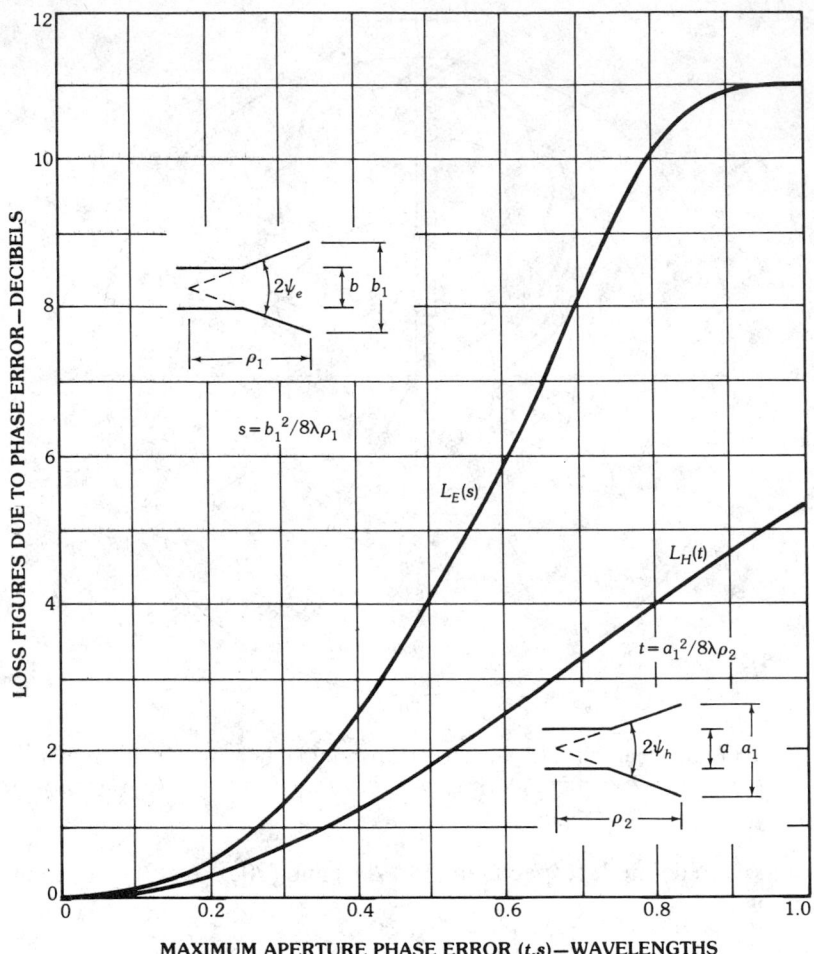

Fig. 23. Loss figures for *E*- and *H*-planes of pyramidal horn due to phase errors. (*After W. C. Jakes [14], © 1961; reprinted with permission of McGraw-Hill Book Company*)

(*a*)   Calculate

$$A = \frac{a_1}{\lambda} \sqrt{\frac{50}{\varrho_h/\lambda}} \tag{30a}$$

$$B = \frac{b_1}{\lambda} \sqrt{\frac{50}{\varrho_e/\lambda}} \tag{30b}$$

(*b*)   Using *A* and *B*, find G_H and G_E, respectively, from Figs. 19 and 10. If the values of either *A* or *B* or both are smaller than 2, then calculate G_E and/or G_H by

$$G_E = \frac{32}{\pi} B \tag{30c}$$

$$G_H = \frac{32}{\pi} A \qquad (30d)$$

(c) Calculate D_p by using the values of G_E and G_H from Figs. 10 and 19 or from (30c) and (30d). Thus

$$D_p = \frac{G_E G_H}{\frac{32}{\pi} \sqrt{50/(\varrho_e/\lambda)} \sqrt{50/\varrho_h/\lambda}} = \frac{G_E G_H}{10.1859 \sqrt{50/(\varrho_e/\lambda)} \sqrt{50/\varrho_h/\lambda}} = \frac{\lambda^2 \pi}{32ab} D_E D_H \qquad (30e)$$

where D_E and D_H are, respectively, the directivities of (9) or (11c) or (12a) and (20a) or (22c). This is the actual directivity of the horn. The above procedure has led to results accurate to within 0.01 dB for a horn with $\varrho_e = \varrho_h = 50\lambda$.

A typical X-band (8.2–12.4 GHz) horn is that shown in Fig. 24. It is a precision standard gain horn, which can be used as a

(a) standard for calibrating other antennas,
(b) feed for reflectors and lenses,
(c) pickup horn for sampling power, and
(d) receiving and/or transmitting antenna.

Example 7—A pyramidal horn has dimensions of $\varrho_1 = 11.7730\lambda$, $\varrho_2 = 12.3841\lambda$, $a_1 = 6.6710\lambda$, $b_1 = 4.9269\lambda$, $a = 0.7849\lambda$, and $b = 0.3488\lambda$.

(a) Check to see if such a horn can be constructed physically.
(b) Compute the directivity using (28), (29), and (30e).

Solution—From Examples 2 and 4

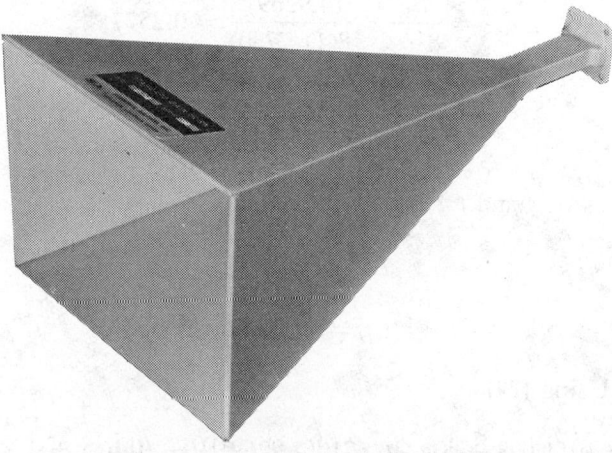

Fig. 24. Commercial X-band (8.2- to 12.4-GHz) pyramidal horn. (*Courtesy Scientific-Atlanta, Atlanta, Georgia*)

$$\varrho_e = 12.0279\lambda$$

$$\varrho_h = 12.8247\lambda$$

Thus

$$p_e = (4.9269 - 0.3488)\lambda \sqrt{\left(\frac{12.0279}{4.9269}\right)^2 - \frac{1}{4}} = 10.939\lambda$$

$$p_h = (6.6710 - 0.7849)\lambda \sqrt{\left(\frac{12.8247}{6.6710}\right)^2 - \frac{1}{4}} = 10.926\lambda$$

Since $p_e \cong p_h \cong 10.9\lambda$, the horn can be constructed physically. The directivity can be computed by utilizing the results of Examples 2 and 4. Using (28) with the values of D_E and D_H computed using (9) and (20a), respectively, gives

$$D_p = \frac{\pi\lambda^2}{32ab}D_ED_H = \frac{\pi}{32(0.7849)(0.3488)}(34.217)(17.073)$$

$$= 209.49 = 23.21 \text{ dB}$$

Utilizing the values of D_E and D_H computed using (11c) and (22c), respectively, the directivity of (30e) is equal to

$$D_p = \frac{\pi\lambda^2}{32ab}D_ED_H = \frac{\pi}{32(0.7849)(0.3488)}(31.394)(17.431)$$

$$= 196.24 = 22.93 \text{ dB}$$

For this horn

$$s = \frac{b_1^2}{8\lambda\varrho_1} = \frac{(4.9269)^2}{8(11.7730)} = 0.2577$$

$$t = \frac{a_1^2}{8\lambda\varrho_2} = \frac{(6.6710)^2}{8(12.3841)} = 0.4492$$

For these values of s and t

$$L_E = 0.625 \text{ dB}$$

$$L_H = 1.50 \text{ dB}$$

from Fig. 23. Using (29)

$$D_0(\text{dB}) = 10\{1.008 + \log_{10}[6.6710(4.9269)]\} - (0.625 + 1.50) = 23.12$$

The agreement is best between the directivities of (28) and (29).

5. The Design Procedure for the Pyramidal Horn

The pyramidal horn is widely used as a standard to make gain measurements of other antennas [6], and as such it is often referred to as a *standard-gain horn*. To design a pyramidal horn one usually knows the desired gain G_0 and the dimensions a, b of the rectangular-feed waveguide. The objective of the design is to determine the remaining dimensions (a_1, b_1, ϱ_e, ϱ_h, p_e, and p_h) that will lead to an optimum gain. The procedure that follows can be used to accomplish this [6].

The design equations are derived by first selecting values of b_1 and a_1 that lead, respectively, to optimum directivities for the *E*- and *H*-plane sectoral horns using (10a) and (21a). Since the overall efficiency (including both the antenna and aperture efficiencies) of a horn antenna is about 50 percent, the gain of the antenna can be related to its physical area. Thus, it can be written using (10a) and (21a) as [6]

$$G_0 = \frac{4\pi}{\lambda^2} A_{em} = \varepsilon_{ap} \frac{4\pi}{\lambda^2} A_p \cong \frac{1}{2} \frac{4\pi}{2} (a_1 b_1)$$

$$= \frac{2\pi}{\lambda^2} \sqrt{3\lambda\varrho_1} \sqrt{2\lambda\varrho_2} = \frac{2\pi}{\lambda^2} \sqrt{3\lambda p_h} \sqrt{2\lambda p_e} \tag{31}$$

since for long horns $\varrho_2 \cong p_h$ and $\varrho_1 \cong p_e$. In (31) ε_{ap} is the aperture efficiency and A_p is the physical area of the horn aperture. In order for a pyramidal horn to be physically realizable, p_e and p_h of (26a) and (26b) must be equal. Using this equality it can be shown that (31) reduces to

$$(\sqrt{2\chi} - b/\lambda)^2 (2\chi - 1) = \left(\frac{G_0}{2\pi} \sqrt{\frac{3}{2\pi}} \frac{1}{\sqrt{\chi}} - \frac{a}{\lambda}\right)^2 \left(\frac{G_0^2}{6\pi^3} \frac{1}{\chi} - 1\right) \tag{32a}$$

where

$$\frac{\varrho_e}{\lambda} = \chi \tag{32b}$$

$$\frac{\varrho_h}{\lambda} = \frac{G_0^2}{8\pi^3} \left(\frac{1}{\chi}\right) \tag{32c}$$

Equation 32a is the horn design equation.

1. As a first step of the design find the value of χ which satisfies (32a) for a desired gain G_0 (dimensionless). Use an iterative technique and begin with a trial value of

$$\chi(\text{trial}) = \chi_1 = \frac{G_0}{2\pi\sqrt{2\pi}} \tag{33}$$

2. Once the correct χ has been found, determine ϱ_e and ϱ_h using (32b) and (32c), respectively.

3. Find the corresponding values of a_1 and b_1 using (10a) and (21a) or

$$a_1 = \sqrt{3\lambda \varrho_2} \cong \sqrt{3\lambda \varrho_h} = \frac{G_0}{2\pi} \sqrt{\frac{3}{2\pi \chi}} \lambda \tag{34a}$$

$$b_1 = \sqrt{2\lambda \varrho_1} \cong \sqrt{2\lambda \varrho_e} = \sqrt{2\chi} \lambda \tag{34b}$$

4. The values of p_e and p_h can be found using (26a) and (26b).

Example 8—Design an optimum gain X-band (8.2–12.4 GHz) pyramidal horn so that its gain (above isotropic) at $f = 11$ GHz is 22.6 dB. The horn is fed by a WR 90 rectangular waveguide with inner dimensions of $a = 0.9$ in (2.286 cm) and $b = 0.4$ in (1.016 cm).

Solution—Convert the gain G_0 from decibels to a dimensionless quantity. Thus

$$G_0(\text{dB}) = 22.6 = 10 \log_{10} G_0$$

implies that

$$G_0(\text{dimensionless}) = 10^{2.26} = 181.97$$

Since $f = 11$ GHz,

$\lambda = 2.7273$ cm

$a = 0.8382\lambda$

$b = 0.3725\lambda$

1. The initial value of χ is taken, using (33), as

$$\chi_1 = \frac{181.97}{2\pi \sqrt{2\pi}} = 11.5539$$

which does not satisfy (32a) for the desired design specifications. After a few iterations a more accurate value is $\chi = 11.1157$.

2. Using (32b) and (32c)

$$\varrho_e = 11.1157\lambda = 30.316 \text{ cm} = 11.935 \text{ in}$$
$$\varrho_h = 12.0094\lambda = 32.753 \text{ cm} = 12.895 \text{ in}$$

3. The corresponding values of a_1 and b_1 are

$$a_1 = 6.002\lambda = 16.370 \text{ cm} = 6.445 \text{ in}$$
$$b_1 = 4.715\lambda = 12.859 \text{ cm} = 5.063 \text{ in}$$

4. The values of p_e and p_h are equal to

$$p_e = p_h = 10.005\lambda = 27.286 \text{ cm} = 10.743 \text{ in}$$

The derived design parameters agree closely with those of a commercial gain horn available in the market.

As a check, the gain of the designed horn was computed using (28) and (29), assuming an antenna efficiency e_t of 100 percent, and (31). The values were

$$G_0 \cong D_0 = 22.4 \text{ dB} \quad \text{for (28)}$$
$$G_0 \cong D_0 = 22.1 \text{ dB} \quad \text{for (29)}$$
$$G_0 = 22.5 \text{ dB} \quad \text{for (31)}$$

All three computed values agree closely with the designed value of 22.6 dB.

Design curves of pyramidal horns with a TE_{10}-mode field distribution, synthesized to maximize the aperture efficiency or to produce maximum power transmission when used as feed elements for reflectors, are shown in Fig. 25 [15]. These curves were derived by expanding the focal plane and feed-horn aperture field distributions into finite-term power series whose coefficients were determined using collocation techniques.

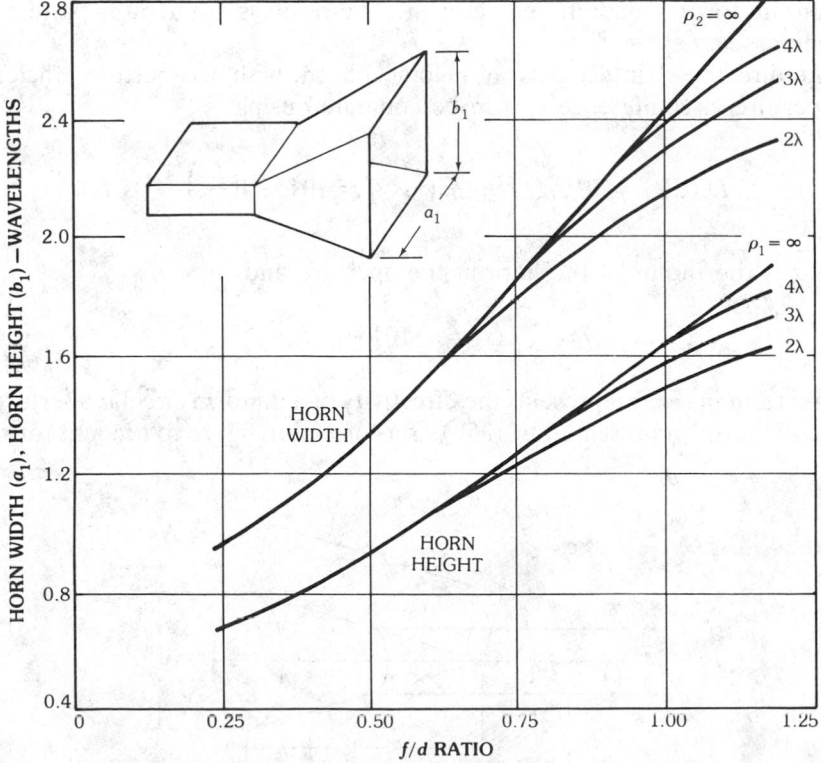

Fig. 25. Optimum pyramidal horn dimensions versus f/d ratio for various horn lengths. (*After Truman and Balanis [15], © 1974 IEEE*)

6. The Conical Horn

Another very practical microwave antenna is the conical horn shown in Fig. 26. While the pyramidal, E-, and H-plane sectoral horns are usually fed by a rectangular waveguide, the feed of a conical horn is often a circular waveguide.

The first rigorous treatment of the fields radiated by a conical horn is that of Schorr and Beck [16]. The modes within the horn are found by introducing a spherical coordinate system and are in terms of spherical Bessel functions and Legendre polynomials. The analysis is too involved and will not be attempted here. Data, however, in the form of curves [17] will be presented that give a qualitative description of the performance of a conical horn.

Referring to Fig. 27 it is apparent that the behavior of a conical horn is similar to that of a pyramidal or a sectoral horn. As the flare angle increases, the directivity for a given length horn increases until it reaches a maximum, beyond which it begins to decrease. The decrease is a result of the dominance of the quadratic phase error at the aperture. In the same figure an optimum directivity line is indicated.

The results of Fig. 27 behave as those of Figs. 9 and 18. When the horn aperture (d_m) is held constant and its length (L) is allowed to vary, the maximum directivity is obtained when the flare angle is zero ($\psi_c = 0$ or $L = \infty$). This is equivalent to a circular waveguide of diameter d_m. As for the pyramidal and sectoral horns a lens is usually placed at the aperture of the conical horn to compensate for its quadratic phase error. The result is a narrower pattern as the flare increases.

The directivity (in decibels) of a conical horn, with an aperture efficiency ε_{ap} and aperture circumference C, can be computed using

$$D_c(\text{dB}) = 10\log_{10}[\varepsilon_{ap}\frac{4\pi}{\lambda^2}(\pi a^2)] = 10\log_{10}\left(\frac{C}{\lambda}\right)^2 - L(s) \qquad (35a)$$

where a is the radius of the horn at the aperture and

$$L(s) = -10\log_{10}\varepsilon_{ap} \qquad (35b)$$

The first term in (35a) represents the directivity of a uniform circular aperture while the second term, represented by (35b), is a correction figure to account for the loss

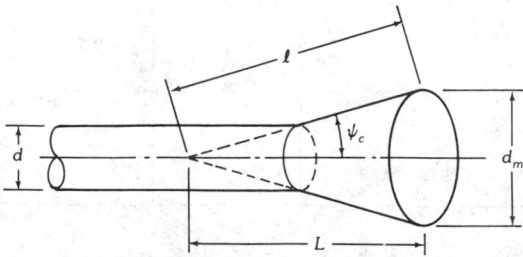

Fig. 26. Geometry of conical horn. (*After Balanis [6], © 1982; reprinted with permission of Harper & Row Publishers, Inc.*)

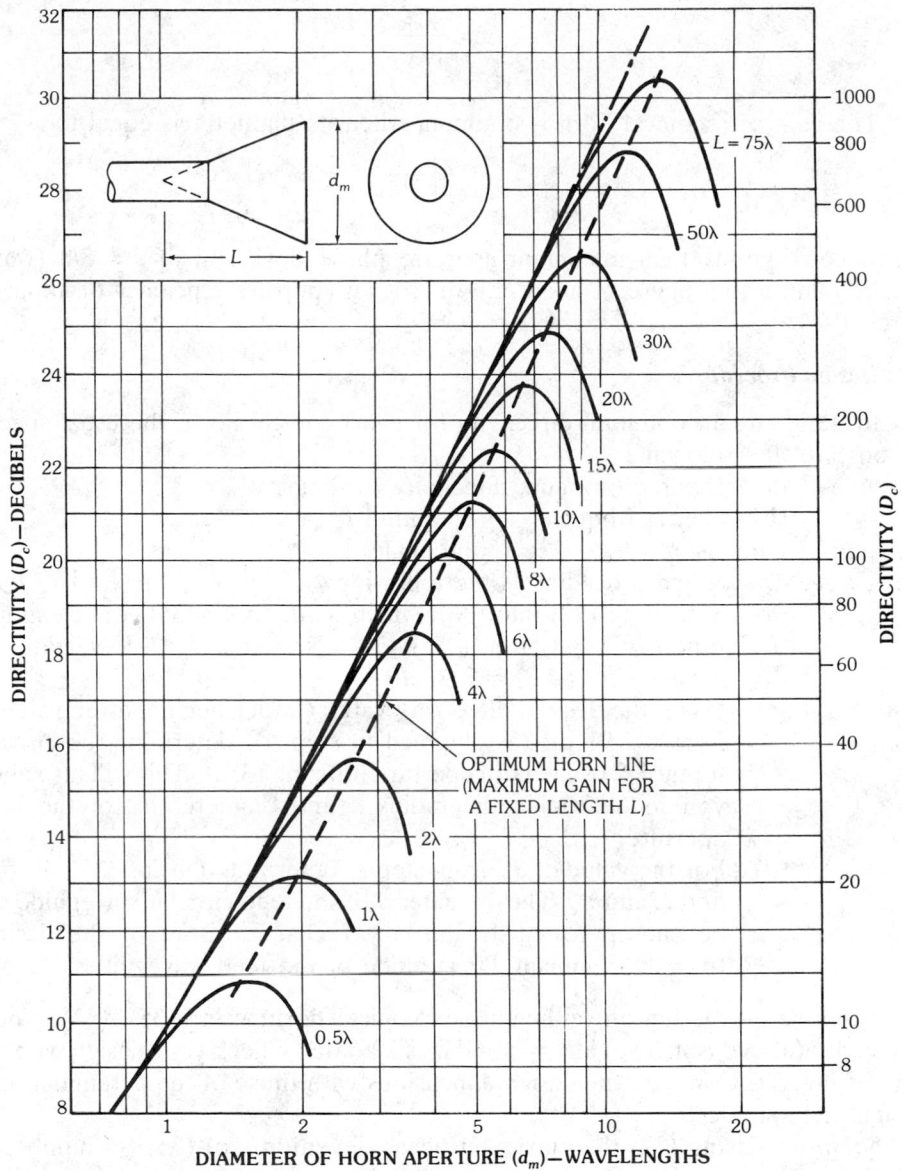

Fig. 27. Directivity of conical horns as a function of aperture diameter and for different horn lengths. (*After King [17], © 1950 IEEE*)

in directivity due to the aperture efficiency. Usually the term in (35b) is referred to as the *loss figure*, which can be computed in decibels using [14]

$$L(s) \cong (0.8 - 1.71s + 26.25s^2 - 17.79s^3)$$ (35c)

where s is the maximum phase deviation (in number of wavelengths), and it is given by

$$s = \frac{d_m^2}{8\lambda\ell} \tag{36}$$

The gain of a conical horn is optimum when its diameter is equal to

$$d_m \cong \sqrt{3\ell\lambda} \tag{37}$$

which corresponds to a maximum aperture phase deviation of $s = 3/8$ (wavelengths) and a loss figure of about 2.9 dB (or an aperture efficiency of about 51 percent).

The Design Procedure

In designing an optimum directivity (or gain) conical horn, the usual specifications are the following:

Given: (*a*)   Desired optimum directivity (or gain)
             (*b*)   Center frequency of operation f

Desired: Dimensions d_m, L, ℓ, d, and angle ψ_c

Procedure: The design procedure is the following:

       (*a*)   Substitute (37) into (36), which leads to $s = 3/8$ (wavelengths).

       (*b*)   Find $L(s = 3/8)$ using (35c). For this value of s, the loss figure L is always $L(s = 3/8) = 2.91$ dB.

       (*c*)   Using the desired directivity value (in decibels) and the value of $L(s) = 2.91$ dB (as obtained in step *b*), determine the circumference C of the horn aperture utilizing (35a). This will also allow you to determine the radius a and diameter d_m of the horn aperture.

       (*d*)   For the value of d_m from step *c*, determine ℓ using (37). In turn find L and ψ_c. The diameter d of the feed circular waveguide can be chosen to satisfy the cutoff characteristics of the desired (usually dominant TE_{11}) mode of the feed waveguides.

Example 9—Design an optimum gain conical horn, using (35)–(37), whose directivity (above isotropic) at $f = 11$ GHz is 22.6 dB. Check your design with the data of Fig. 27. Compare the design dimensions with those of the pyramidal horn design of Example 8.

Solution—Using (37) the maximum phase deviation s of (36) (in number of wavelengths) is equal to

$$s = \frac{d_m^2}{8\lambda\ell} = \frac{3\ell\lambda}{8\ell\lambda} = \frac{3}{8}$$

For $s = 3/8$ the loss figure $L(s = 3/8)$ of (35c) is equal to

$$L(s = 3/8) = 2.91 \text{ dB}$$

Now using (35a) we have

$$22.6 = 10 \log_{10}(C/\lambda)^2 - 2.91$$

so that

$$(C/\lambda)^2 = 10^{2.551} = 355.795$$

and

$$\frac{C}{\lambda} = \frac{\pi d_m}{\lambda} = \sqrt{355.795} = 18.8625$$

means

$$d_m = 6\lambda$$

Since $\lambda = 30 \times 10^9/11 \times 10^9 = 2.7273$ cm, then $d_m = 6\lambda = 16.3636$ cm $= 6.44$ in. Using (37) we have that

$$d_m = \sqrt{3\ell\lambda}$$

and hence

$$\ell = d_m^2/3\lambda = 12\lambda = 32.728 \text{ cm} = 12.88 \text{ in}$$

Thus

$$L = \sqrt{\ell^2 - (d_m/2)^2} = \lambda\sqrt{(12)^2 - (3)^2} = 11.619\lambda = 31.688\lambda = 12.476 \text{ in}$$

and

$$\psi_c = \tan^{-1}\left(\frac{d_m/2}{L}\right) = \tan^{-1}\left(\frac{3}{11.619}\right) = 14.48°$$

which implies that

$$2\psi_c = 28.96°$$

Using Fig. 27 and interpolating between curves, the diameter and length for an optimum directivity of 22.6 dB are equal to

$$d_m \cong 5.8\lambda = 15.818 \text{ cm} = 6.228 \text{ in}$$
$$L \cong 11.5\lambda = 31.364 \text{ cm} = 12.348 \text{ in}$$

The length $\ell = 12\lambda$ and diameter $d_m = 6\lambda$ of the conical horn are contrasted with the lengths $\varrho_e = 11.1157\lambda$, $\varrho_h = 12.0094\lambda$ and the aperture dimensions $a_1 = 6.002\lambda$, $b_1 = 4.715\lambda$ of the pyramidal horn of Example 8.

7. Special Horns

The large emphasis placed on horn antenna research in the 1960s was inspired by the need to reduce spillover efficiency and cross-polarization losses and increase aperture efficiencies of large reflectors used in radioastronomy and satellite communications. In the 1970s high-efficiency and rotationally symmetric antennas were needed in microwave radiometry. Using conventional feeds, aperture efficiencies of 50 to 60 percent were obtained. However, efficiencies on the order of 75 to 80 percent can be obtained with improved feed systems utilizing corrugated horns. In addition, during the same period there was a need for circularly polarized radiators for radar systems, radioastronomy, and ionospheric studies. Most known designs, however, provided circularly polarized waves over a limited angular region, and many efforts were attempted to improve the axial ratio [18–26]. Many of these designs were limited in bandwidth and led to relatively long horns which were delicate in construction.

To overcome some of the above deficiencies and improve the overall radiation characteristics (pattern symmetry, low cross polarization, and low side lobes) of a horn some special horn designs were attempted. Two such designs are the *corrugated* horn [18–25] and the *aperture-matched* horn [26]. The basic concept in each of these designs was to reduce the diffractions at the edges of the aperture. For the corrugated horn aperture diffractions were minimized by reducing the magnitude of the incident field at the edge, while for the aperture-matched horn aperture diffractions were diminished by modifying the horn structure at the aperture so that the magnitude of the diffraction coefficient was reduced.

Corrugated Horns

The unequal *E*- and *H*-plane patterns in a conventional horn result from unfavorable boundary conditions at the top and side walls. The conductive side walls force the tangential components of the electric field to vanish, thus creating for the dominant mode a cosine distribution. The top and bottom walls, however, do not affect the tangential magnetic field, and a uniform distribution is formed between them. If the tangential magnetic field were forced to vanish at the top and bottom walls, the asymmetry of the *E*- and *H*-plane patterns could be corrected.

In 1964 Kay [19] realized that short-circuited quarter-wavelength grooves on the walls of a horn antenna would present the same boundary conditions to all polarizations and would taper the field distribution at the aperture in all the planes. The creation of the same boundary conditions on the walls perpendicular to the **E** field, as those of the *H*-plane, minimizes the spurious diffractions at the edges of the aperture by reducing the incident field at the edge of the aperture. For a square aperture this would lead to an almost rotationally symmetric pattern with equal *E*- and *H*-plane beamwidths over about 60 percent bandwidth and with the 10-dB beamwidth being virtually independent of frequency. In addition, the *E*- and *H*-plane phase centers were found to coincide and be located inside the horn near the throat within a distance of one wavelength from the aperture plane. The bandwidths can be improved by dielectric loading.

For sufficiently long corrugated horns superior pattern performance can be expected over a 2:1 bandwidth. Because the corrugated surface forces the energy away from the horn walls in the E-plane, as the boundary conditions accomplish the same thing in the H-plane, it should be expected that such a horn would have almost identical principal plane patterns. Therefore a corrugated horn can be used as an excellent circularly polarized radiator with axial ratios of 1.05 or smaller over almost the entire beamwidth and with patterns whose maximum radiation is along the axis and with practically no minor lobes.

The Corrugated Pyramidal Horn—A corrugated (grooved) pyramidal horn, with corrugations in the E-plane walls, is shown in Fig. 28a with a side view in Fig. 28b. Since diffractions at the edges of the aperture in the H-plane are minimal, corrugations are usually not placed on the walls of that plane. To form a very effective corrugated surface it usually requires ten or more slots (corrugations) per wave-

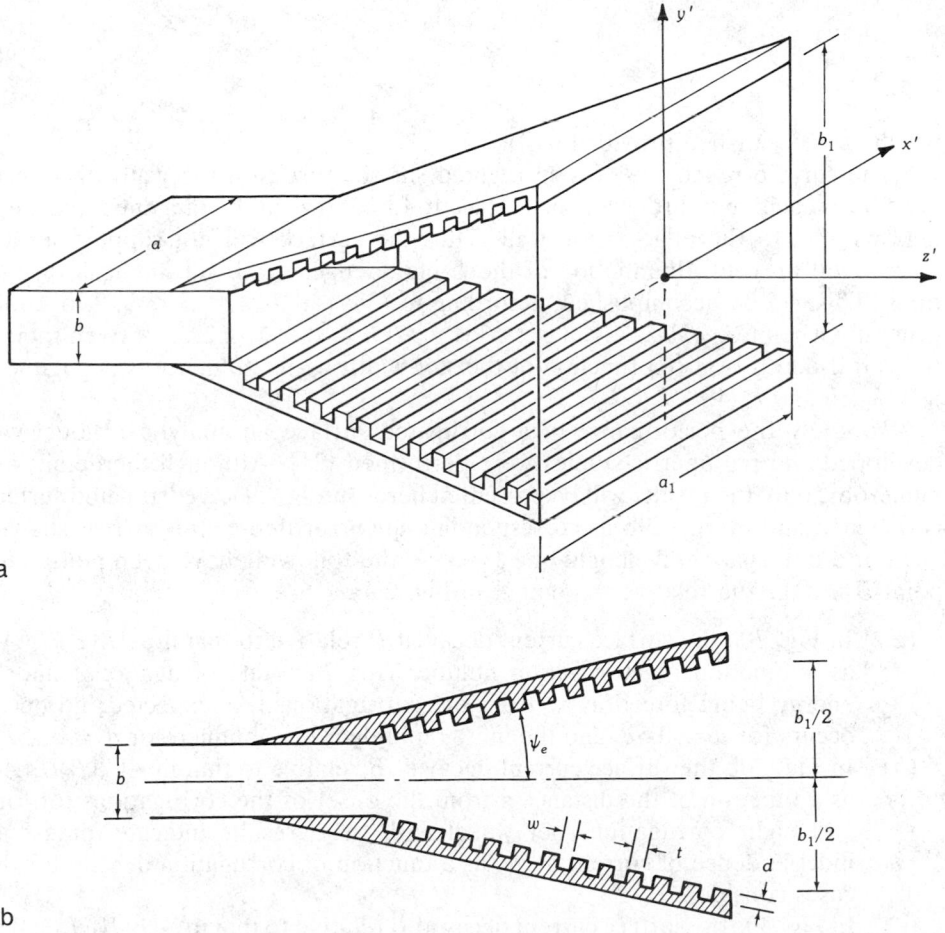

Fig. 28. Pyramidal horn with corrugations in the E-plane. (*a*) Corrugated horn. (*b*) E-plane view. (*After Balanis [6], © 1982; reprinted with permission of Harper & Row Publishers, Inc.*)

length [19]. To simplify the analysis of an infinite corrugated surface, the following assumptions are usually required:

(a) The teeth of the corrugations are vanishingly thin
(b) Reflections from the base of the slot are only those of a TEM mode

The second assumption is satisfied provided the width w of the corrugation is small compared with the free-space wavelength λ_0 and the slot depth d (usually $w < \lambda_0/10$). For a corrugated surface satisfying the above assumptions, its approximate surface reactance is given by [27]

$$X = \frac{w}{w + t} \sqrt{\frac{\mu_0}{\varepsilon_0}} \tan k_0 d \qquad (38a)$$

when

$$\frac{w}{w + t} \cong 1 \qquad (38b)$$

which can be satisfied provided $t \lesssim w/10$.

The surface reactance of a corrugated surface, used on the walls of a horn, must be capacitive in order for the surface to force to zero the tangential magnetic field parallel to the edge at the wall. Thus the surface will not support surface waves, will prevent illumination of the E-plane edges, and will diminish diffractions. This can be accomplished, according to (38a), if $\lambda_0/4 < d < \lambda_0/2$, or more generally when $(2n + 1)\lambda_0/4 < d < (n + 1)\lambda_0/2, n = 0, \pm 1, \pm 2, \ldots$. Even though the cutoff depth is also a function of the slot width w, its influence is negiigible if $w < \lambda_0/10$ and $\lambda_0/4 < d < \lambda_0/2$.

To study the performance of a corrugated surface an analytical model was developed and parametric studies were performed [21]. Although the details are numerous, only the results will be presented here. In Fig. 29a, a corrugated surface is sketched and in Fig. 29b its corresponding uncorrugated counterpart is shown.

For a free-space wavelength of $\lambda_0 = 8$ cm the following have been plotted for point B in Fig. 29a relative to point A in Fig. 29b:

(a) In Fig. 30a the surface current decay at B relative to that at $A[J_s(B)/J_s(A)]$ as a function of corrugation number (for 20 total corrugations) due to energy being forced away from the corrugations. As expected, no decay occurs for $d = 0.5\lambda_0$ and the most rapid decay is obtained for $d = 0.25\lambda_0$.
(b) In Fig. 30b the surface current decay at B relative to that in $A[J_s(B)/J_s(A)]$ as a function of the distance z from the onset of the corrugations for four and eight corrugations per wavelength. The results indicate almost an independence of current decay as a function of corrugation density for the cases considered.
(c) In Fig. 30c the surface current decay at B relative to that in $A[J_s(B)/J_s(A)]$ as a function of the distance z from the onset of the corrugations for $w/(w + t)$ ratios ranging from 0.5 to 0.9. For $z < 4$ cm $= \lambda_0/2$, thinner corrugations [larger $w/(w + t)$ ratios] exhibit larger rates of decay. Approximately

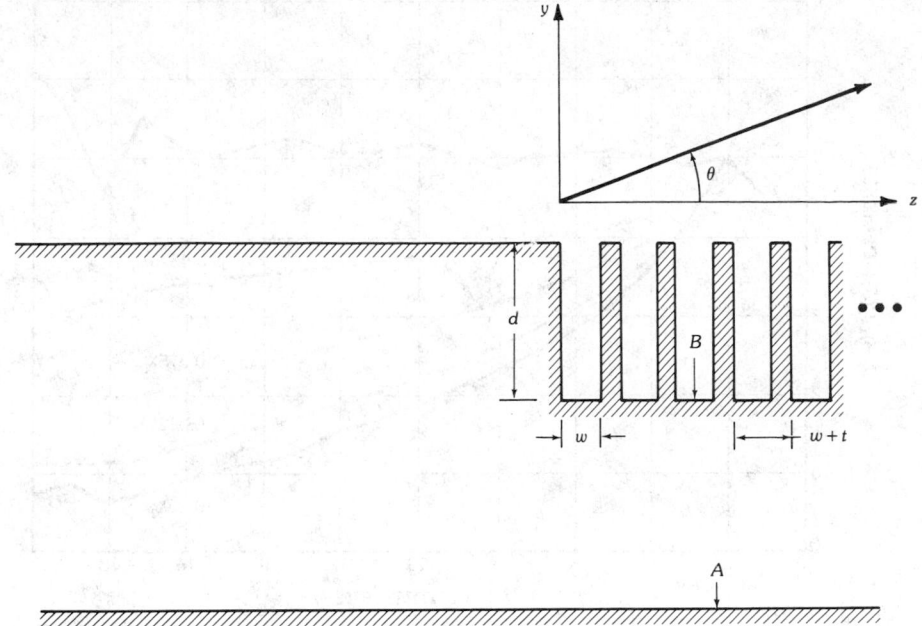

Fig. 29. Geometry of corrugated and plane surfaces. (*a*) Corrugated surface. (*b*) Uncorrugated surface. (*After Mentzer and Peters [21]*, © *1974 IEEE*)

beyond that point the rate of decay is constant. This would indicate that in a practical design thinner corrugations can be used at the onset followed by thicker ones, which are easier to construct.

The effect of the corrugations on the walls of a horn is to modify the electric field distribution in the *E*-plane from uniform (at the waveguide-horn junction) to cosine (at thc aperture). Through measurements it has been shown that the transition from uniform to cosine distribution takes place almost at the onset of the corrugations. For a horn of about 45 corrugations the cosine distribution has been established by the fifth corrugation from the onset and the spherical phase front by the fifteenth [22]. The *E*- and *H*-plane amplitude and phase distributions at the aperture of the horn with 45 corrugations are shown in Figs. 31a and 31b. It is clear that the cosine distribution is well established.

Referring to Fig. 28a, the field distribution at the aperture can be written as

$$E'_y(x', y') = E_0 \cos\left(\frac{\pi}{a_1} x'\right) \cos\left(\frac{\pi}{b_1} y'\right) e^{-j(k/2)[(x')^2/\varrho_2 + (y')^2/\varrho_1]} \tag{39a}$$

$$H'_x(x', y') = -\frac{E_0}{\eta} \cos\left(\frac{\pi}{a_1} x'\right) \cos\left(\frac{\pi}{b_1} y'\right) e^{-j(k/2)[(x')^2/\varrho_2 + (y')^2/\varrho_1]} \tag{39b}$$

corresponding to (23a) and (23b) of the uncorrugated pyramidal horn. Using the above distributions, the fields radiated by the horn can be computed in a manner

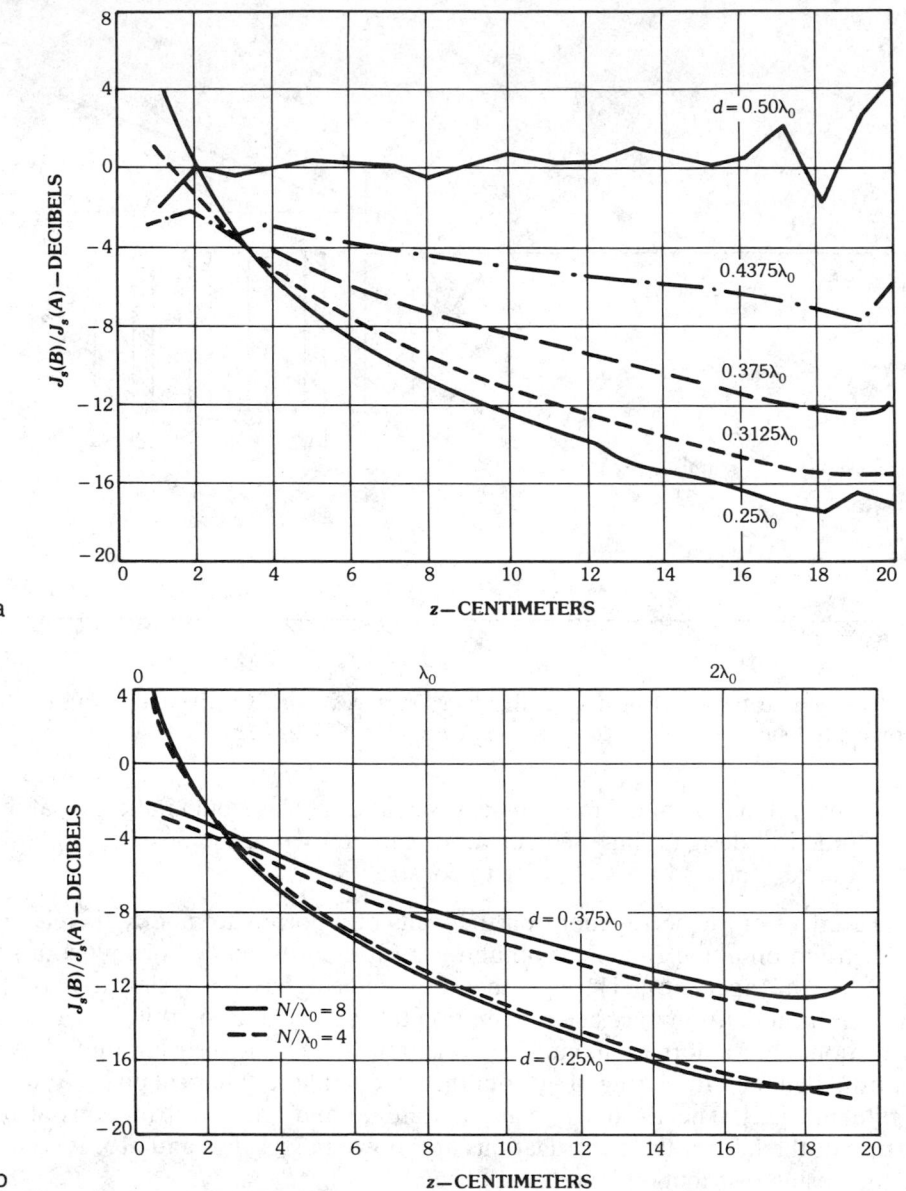

Fig. 30. Surface current decays on corrugated surface. (*a*) Surface current decay due to energy forced away from corrugations. (*b*) Surface current decay as a function of corrugation density. (*c*) Surface current decay on corrugation as a function of corrugation shape. (*After Mentzer and Peters [21], © 1974 IEEE*)

analogous to that of the conventional pyramidal horn of Section 4. Patterns have been computed and compare very well with measurements [22].

In Fig. 32a the measured *E*-plane patterns of an uncorrugated square pyramidal horn (referred to as the *control horn*) and a corrugated square pyramidal

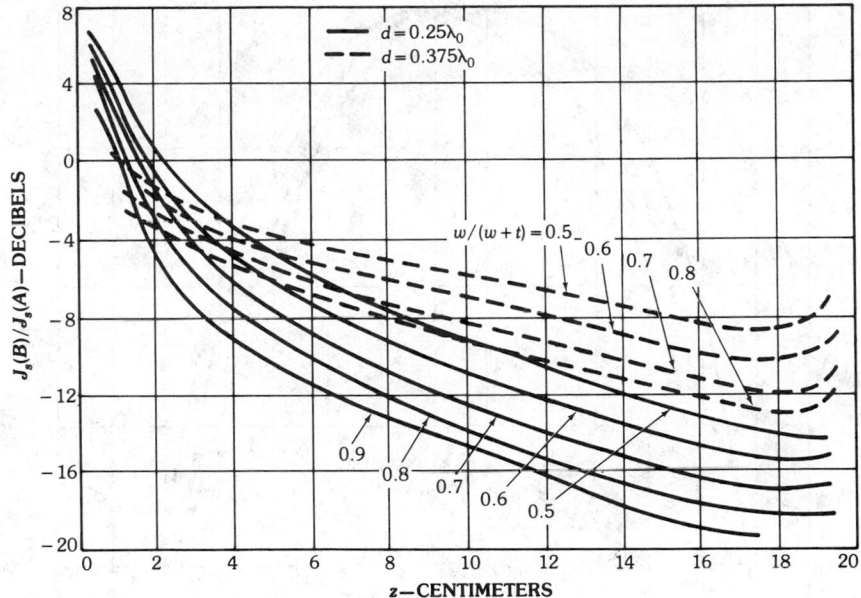

C

Fig. 30, *continued*

horn are shown. The aperture size on each side was 3.5 in or 8.89 cm ($2.96\lambda_0$ at 10 GHz) and the total flare angle in each plane was 50°. It is evident that the levels of the side lobes and back lobes are much lower for the corrugated horn than those of the control horn. However, the corrugated horn also exhibits wider main beam for small angles, and thus a larger 3-dB beamwidth (half-power beamwidth) but a smaller 10-dB beamwidth. This is attributed to the absence of the diffracted fields from the edges of the corrugated horn which, for nearly on-axis observations, add to the direct wave contribution because of their in-phase relationship. The fact that the on-axis far fields of the direct and diffracted fields are nearly in phase is also evident from the pronounced on-axis maximum of the control horn. The E- and H-plane patterns of the corrugated horn are almost identical with those of Fig. 32a over the frequency range from 8 to 14 GHz. These suggest that the main beam in the E-plane can be obtained from known H-plane patterns of horn antennas.

In Fig. 32b the measured E-plane patterns of larger control and corrugated square pyramidal horns, having an aperture of 9.7 in (24.64 cm) on each side ($8.2\lambda_0$ at 10 GHz) and included angles of 34° and 31° in the E- and H-planes, are shown. For this geometry the pattern of the corrugated horn is narrower and its side and back lobes are much lower than those of the corresponding control horn. The saddle formed on the main lobe of the control horn is attributed to the out-of-phase relations between the direct and diffracted rays. The diffracted rays are nearly absent from the corrugated horn and the minimum on-axis field is eliminated. The control horn is a thick-edged horn which has the same interior dimensions as the corrugated horn. The H-plane pattern of the corrugated horn is almost identical with the H-plane pattern of the corresponding control horn.

In Figs. 32c and 32d the back lobe level and the 3-dB beamwidth for the smaller

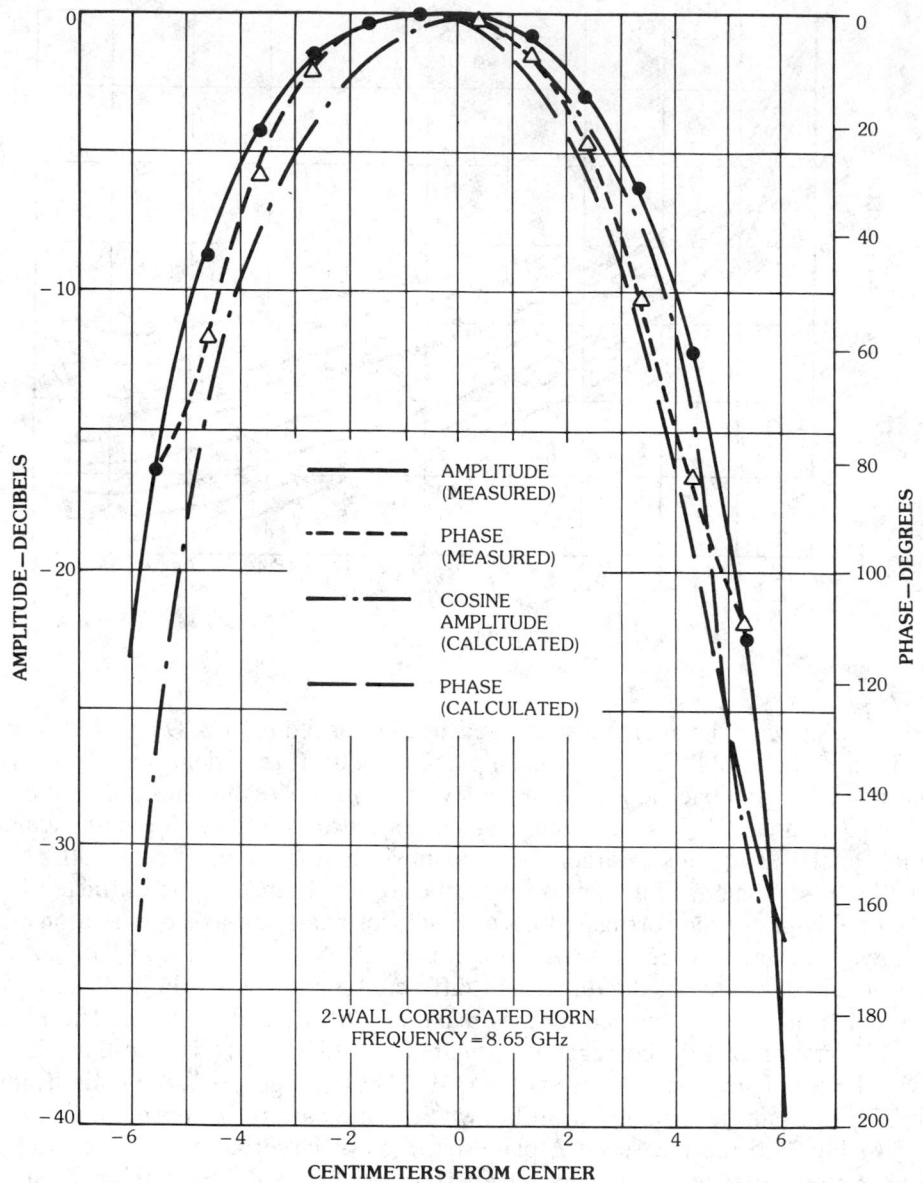

Fig. 31. Amplitude and phase distributions in *H*- and *E*-planes. (*a*) Position in *H*-plane. (*b*) Position in *E*-plane. (*After Mentzer and Peters [22], © 1976 IEEE*)

size control and corrugated horns, whose *E*-plane patterns are shown in Fig. 32a, are plotted as a function of frequency. All the observations made previously for that horn are well evident in these figures.

The presence of the corrugations, especially near the waveguide-horn junction, can affect the impedance and vswr of the antenna. The usual practice is to begin the corrugations at a small distance away from the junction. This leads to low vswr's

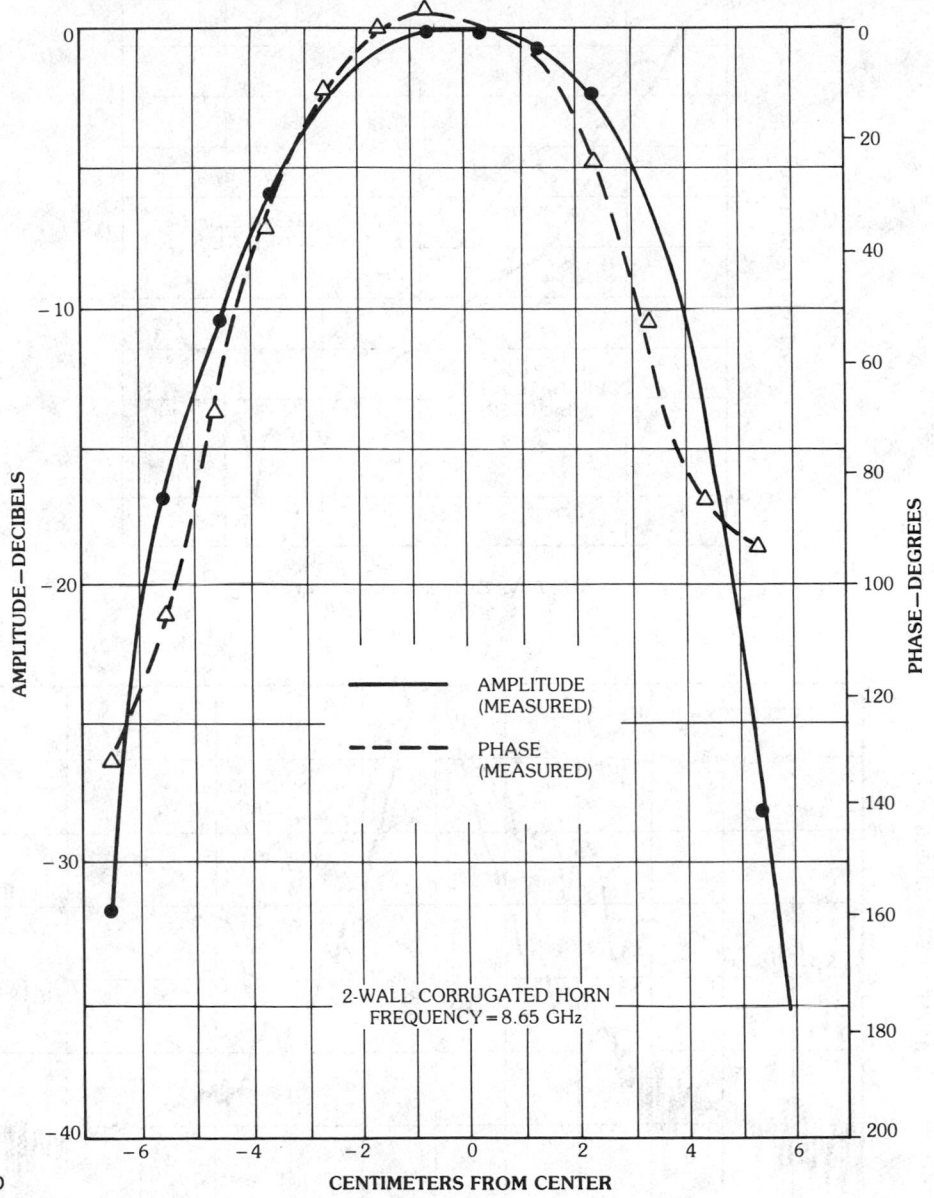

b

CENTIMETERS FROM CENTER

Fig. 31, *continued*

over a broadband. Previously it was indicated that the width *w* of the corrugations must be small (usually $w < \lambda_0/10$) to approximate a corrugated surface. This would cause corona and other breakdown phenomena. However, the large corrugated horn, whose *E*-plane pattern is shown in Fig. 32b, has been used in a system whose peak power was 20 kW at 10 GHz with no evidence of any breakdown phenomena.

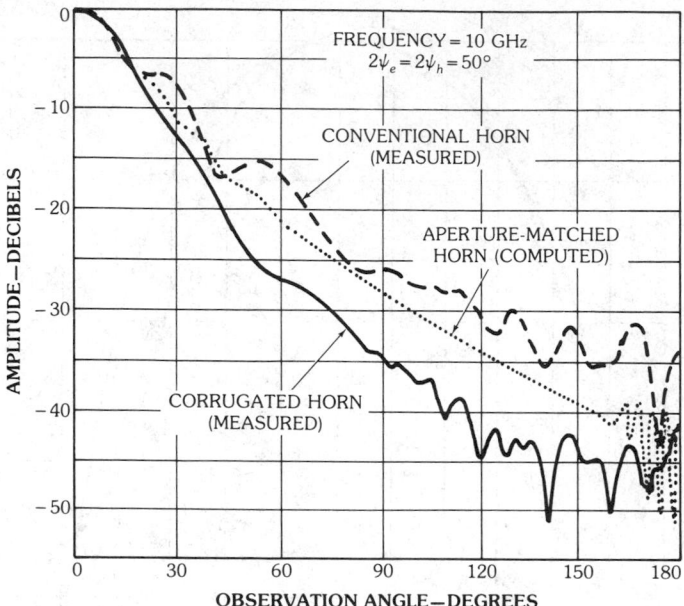

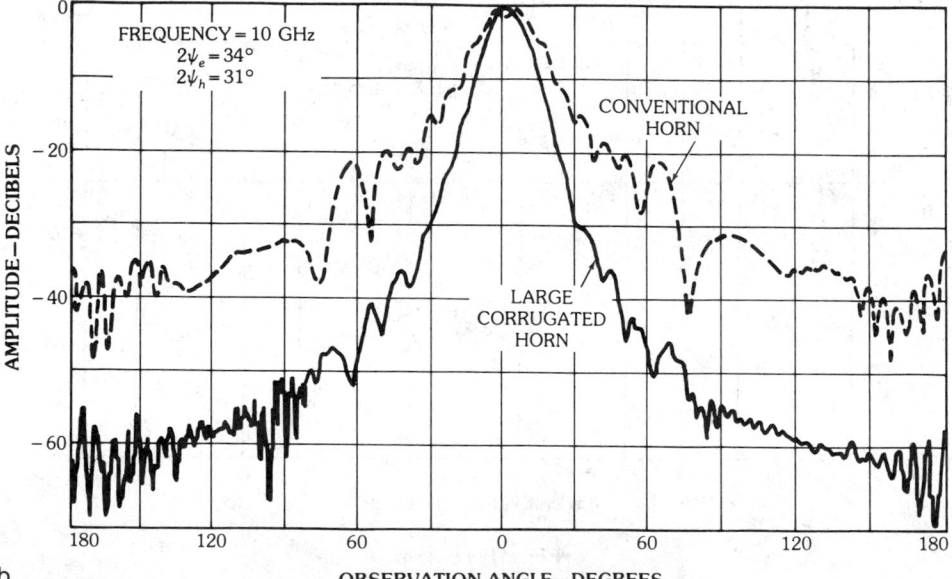

Fig. 32. Radiation characteristics of conventional (control), corrugated, and aperture-matched horns. (*a*) *E*-plane patterns of $2.96\lambda_0 \times 2.96\lambda_0$ pyramidal horns. (*After Burnside and Chuang [26], © 1982 IEEE*) (*b*) Measured *E*-plane patterns of $8.2\lambda_0 \times 8.2\lambda_0$ pyramidal horns. (*After Lawrie and Peters [20], © 1974 IEEE*) (*c*) Back lobe to main lobe *E*-plane level of $2.96\lambda_0 \times 2.96\lambda_0$ pyramidal horns. (*After Burnside and Chuang [26], © 1982 IEEE*) (*d*) Half-power beamwidth of $2.96\lambda_0 \times 2.96\lambda_0$ pyramidal horns. (*After Burnside and Chuang [26], © 1982 IEEE*)

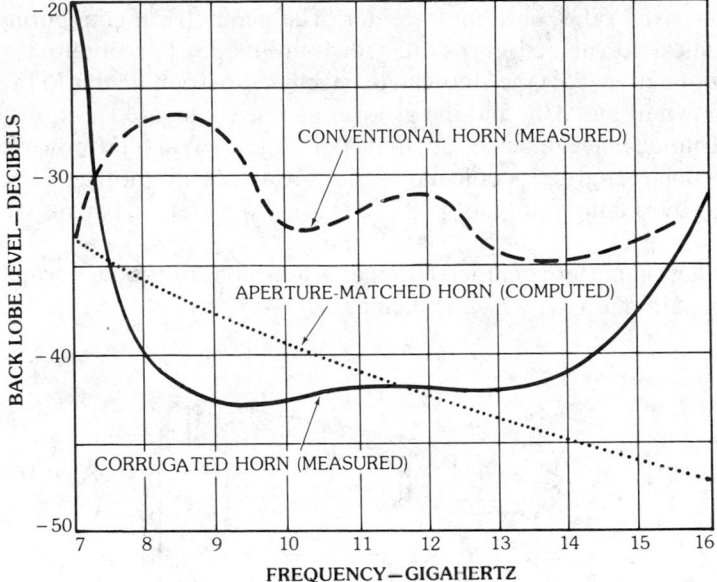

c

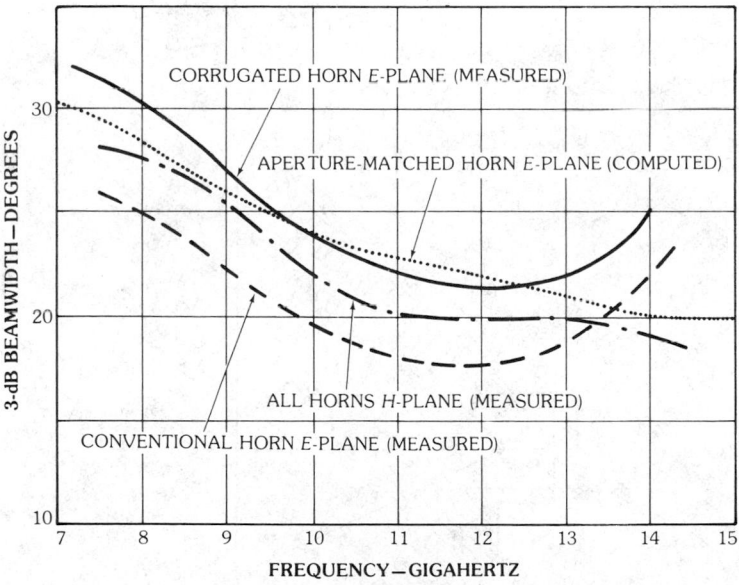

d

Fig. 32, *continued*

The Corrugated Conical Horn—The design concepts of the pyramidal corrugated horn can be extended to include circumferentially corrugated conical horns. Several designs of conical corrugated horns were investigated [18, 19, 23–25] in terms of pattern symmetry, low cross polarization, low side lobe levels, circular polarization, axial ratio, and phase center. The geometrical configurations of two possible conical corrugated horns are shown in Fig. 33. For small flare angles (ψ_c less than about 20° to 25°) the slots can be machined perpendicular to the axis of the horn, as shown in Fig. 33a, and the grooves can be considered sections of parallel-plate TEM-mode waveguides of depth d. For large flare angles, however, the slots should be constructed perpendicular to the surface of the horn, as shown in Fig. 33b. The groove configuration of Fig. 33a is usually preferred because it is easier to fabricate.

The radiation pattern characteristics of a conical horn can be normalized by the dimensions parameter Δ, which is defined as

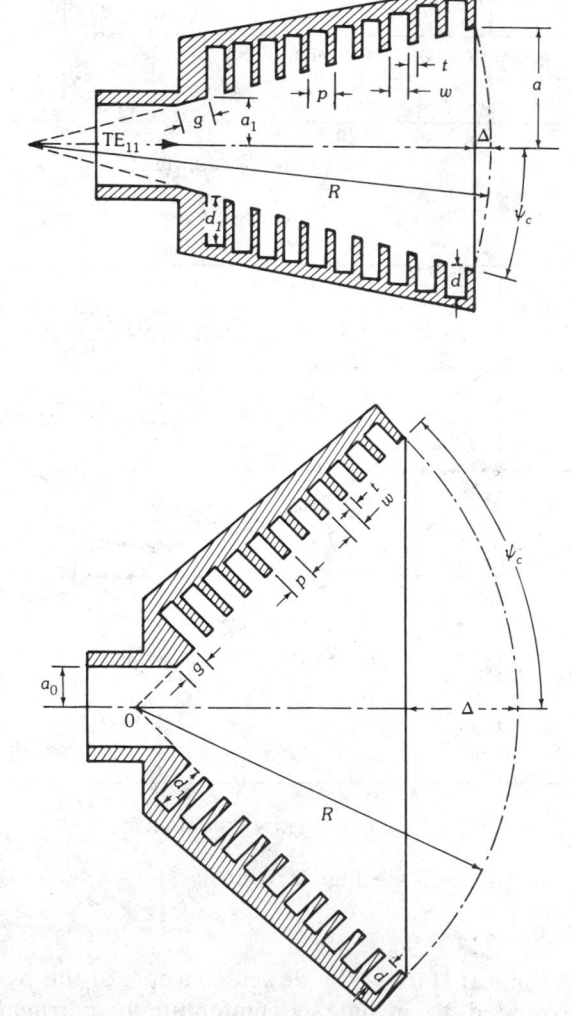

Fig. 33. Geometries of conical corrugated horns of small and large flare angles. (*a*) Small flare angle. (*b*) Large flare angle. (*After Thomas [23], © 1978 IEEE*)

$$\Delta = \frac{a}{\lambda_0} \tan\left(\frac{\psi_c}{2}\right) = \frac{R}{\lambda_0} \sin\psi_c \tan\left(\frac{\psi_c}{2}\right) \qquad (40)$$

As illustrated graphically in Fig. 33, Δ is the difference (in wavelengths) between the spherical wavefront and the plane aperture, and λ_0 is the free-space wavelength.

Depending on the value of Δ, the conical horns can be classified as either "narrowband" or "wideband." Horns with $\Delta < 0.4$ are usually referred to as narrowband, because their characteristics are frequency dependent. For example, it has been found that the beamwidth is determined by the aperture size ka and their phase center moves toward the throat as Δ increases, as shown in Fig. 34. Wideband horns are usually those with $\Delta > 0.75$, because their beamwidth is dependent primarily on the flare angle ψ_c and their phase center is mostly near the throat of the horn. Hence their characteristics are nearly frequency independent, provided $\Delta > 0.75$. In the literature these horns are often referred to as "scalar" horns [19]. Beam efficiencies [23] for narrowband and wideband conical corrugated horns are displayed in Figs. 35a and 35b, respectively.

If the circular waveguide feeding a smooth-surface conical horn operates on its dominant TE_{11} mode and forms a large-diameter (much greater than a wavelength) discontinuity, the first-order forward scattered fields required to match the curved phase front of the TE_{11} mode are those of the TM_{11} and TE_{12} modes, while the second order are those of the TM_{12} and TE_{13} modes [27]. It is assumed that backward scattered modes are negligible. The use of a large diameter (compared to

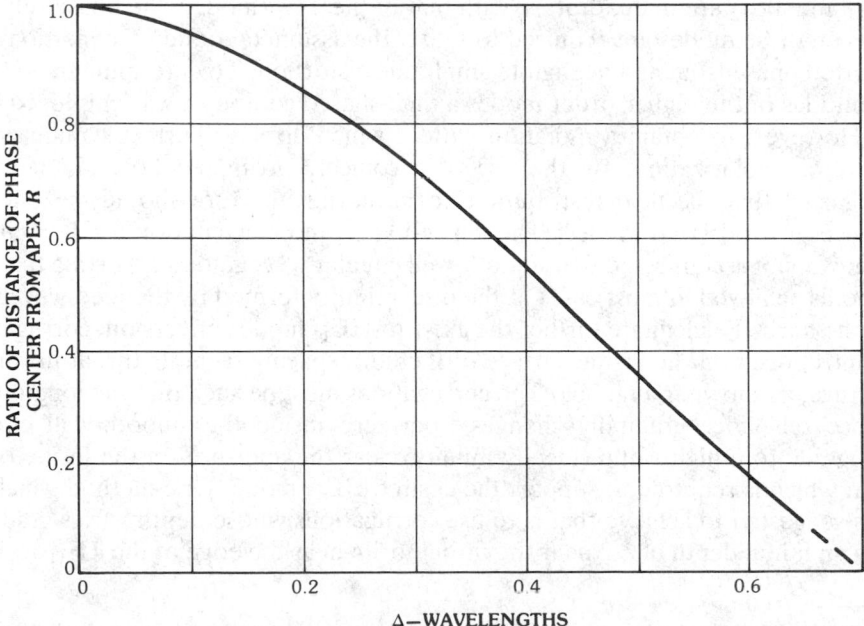

Fig. 34. Normalized phase-center distance (measured from horn apex) of narrowband conical corrugated horn. (*After Thomas [23], © 1978 IEEE*)

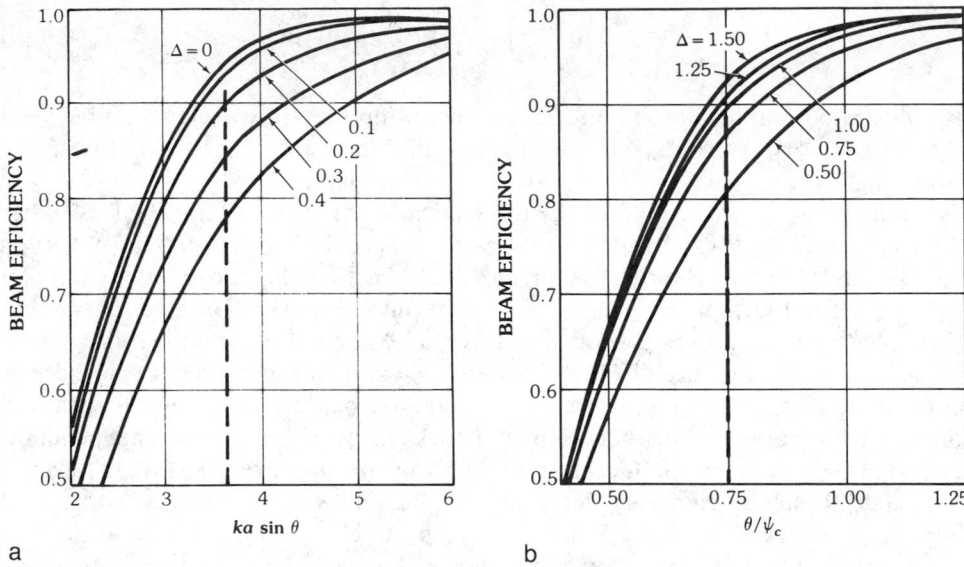

Fig. 35. Beam efficiencies for narrowband and wideband conical corrugated horns. (*a*) Narrowband horn. (*b*) Wideband horn. (*After Thomas [23], © 1978 IEEE*)

the wavelength) ensures the propagation of all of these four higher-order modes. At the center of the junction the electric fields of the TE_{11}, TM_{11}, and TE_{12} modes are oriented in the same direction, while the phases of the electric fields of the TM_{11} and TE_{12} are in quadrature with that of the TE_{11} mode. The phase relations between these modes are required to satisfy the assumed boundary conditions of a distorted phase front and negligible amplitude distortion. To determine the relative amplitudes of the higher-order modes a time-shared computer was employed [27].

However, to obtain a radiation pattern which displays perfect symmetry and zero cross polarization, by the use of a conical corrugated horn, it has been recognized that the field distribution at the horn's aperture should be that of a balanced hybrid HE_{11} mode.* The desired HE_{11}-mode excitation inside a conical horn, when it is connected to a smooth-wall circular waveguide supporting the TE_{11} mode, is achieved in most cases at the discontinuity formed by the feed waveguide and the horn. To facilitate further the TE_{11}-to-HE_{11} mode conversion, corrugations are introduced on the inside surface of the horn spacing from its throat toward its aperture, as shown in Fig. 36a. The corrugations must be such that the longitudinal surface reactance gradually changes from zero inside the smooth-wall circular waveguide to a high value (ideally infinity) near the aperture on the inside of the horn, which is required to support the desired HE_{11} mode. One method which has been suggested to achieve that is to use corrugations whose depth varies gradually from an initial depth of $\lambda/2$ near the throat to $\lambda/4$ near the edge of the TE_{11}-to-HE_{11}

---

*To designate the hybrid mode a mode-content factor γ is defined as the ratio of the longitudinal fields of the TE_{mn} and TM_{mn} components [23]. The term HE_{mn} is used to indicate that the two components are in phase (γ positive) and EH_{mn} when they are out of phase (γ negative). The TM_{mn} modes are represented by $\gamma = 0$ and the TE_{mn}s by $\gamma = \infty$.

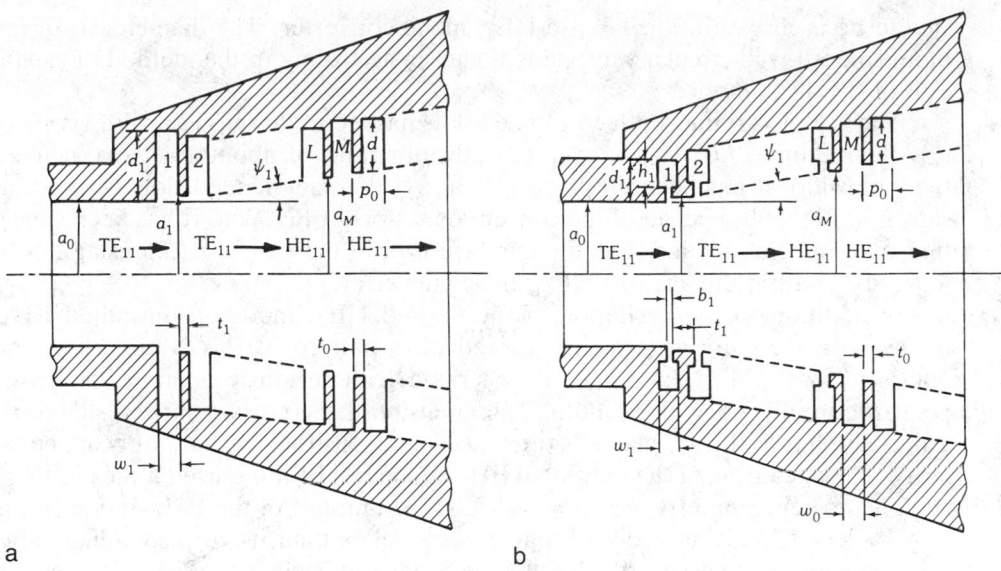

Fig. 36. Geometries of corrugated conical horns with conventional and ring-loaded slots. (*a*) Conventional slots. (*b*) Ring-loaded slots. (*After James [25], © 1982 IEEE*)

mode converter closest to the aperture. The remaining corrugations maintain a constant depth of $\lambda/4$.

The use of $\lambda/2$ deep slots near the throat provides a good match (low vswr) between the TE_{11} and HE_{11} modes over a wide band. To minimize the cross-polarized side lobes, however, the slots should be $\lambda/4$ deep at the center (design) frequency of the bandwidth. In addition, at the discontinuity between the smooth horn and the corrugated surface the EH_{11} surface wave can be introduced [29]. To avoid the generation of it the surface reactance should be negative over the entire desired bandwidth. This can be accomplished by maintaining the depth d_1 of the first slot nearest the throat as $\lambda/4 < d_1 < \lambda/2$ over the desired bandwidth. The design procedure [23] then, for optimum performance, is to choose the depth of the first slot to be slightly less than $\lambda_{max}/2$ at the highest frequency f_{max} of the band. The depth of each successive slot within the TE_{11}-to-HE_{11} converter of Fig. 36a decreases linearly until a depth of $\lambda'/4$ at the center frequency f' is achieved. Even though the surface reactance is negative the undesired EH_{12} mode can still be excited if the diameter of the horn at the first corrugation is too large. For narrow-band horns the radius a_1 of the horn at the first slot should be such that $ka_1 < 4$ at $f = f_{max}$.

The pitch p of the slots for wide-angle "broadband" (scalar) horns should be less than $\lambda_{max}/2$ at $f = f_{max}$, while the wall thickness t between the slots should be very thin so that the width-to-pitch ratio $\delta = w/p$ is near the optimum value of unity. This is a requirement for low cross polarization for wideband operation. For narrow-angle horns the pitch p should be chosen smaller than about $\lambda_{max}/4$ at $f = f_{max}$, which for small ka will avoid increased cross polarization which occurs when wider slots are utilized. For an effective mode conversion ten or more slots should be used within the TE_{11}-to-HE_{11} mode converter. The diameter a_0 of the input smooth-wall circular waveguide is chosen according to the method of excitation of the TE_{11} mode.

The use of varying depth corrugated slots has led to designs [25] with vswr's of less than about 1.065:1 only over a bandwidth ratio of about 1.45. To increase the bandwidth it has been proposed [25, 30, 31] that ring-loaded slots be used, as shown in Fig. 36b, instead of the conventional slots of Fig. 36a. It has been shown that by using ring-loaded slots the bandwidth ratio increases to 1.55 using ten or more slots within the TE_{11}-to-HE_{11} mode converter.

In addition to the excitation of the desired HE_{11} mode an unwanted EH_{12} mode is also usually generated. The radiation pattern of this is entirely cross-polarized in the 45° plane and, if generated, can seriously degrade the cross-polarization efficiency of the horn. The intensity of the cross-polarized side lobe, especially that at the 45° plane, is often used as a sensitive measure of beam circularity. The excitation of the unwanted EH_{12} mode can be minimized if the half-flare angle ψ_c is kept small (typically $\psi_c < 8°$). If the amount of the EH_{12} mode (using ring-loaded slots) is an order of magnitude smaller than its corresponding value using conventional varying depth slots, the bandwidth ratio increases to at least 2.0. The performance of horns having large flare angles toward the aperture can be improved, provided an additional flared section is used between the aperture region and the mode converter [25].

Several other designs of circularly polarized conical corrugated horns were investigated [18] at several frequencies in terms of E- and H-plane patterns, axial ratio, and phase centers. To achieve circular polarization it was recommended that identical dipoles be placed inside the feed guide a $\lambda/4$ guide-wavelength apart and oriented 90° relative to each other. For simplicity in the experiment, however, only one dipole was utilized and the E- and H-plane patterns were compared.

Typical E- and H-plane patterns of one design, at $f = 5.5$ and 6.5 GHz, are shown in Fig. 37. It is evident that the agreement between the E- and H-plane patterns is excellent. To compare better the closeness between the E- and H-plane patterns the axial ratios at several power levels and different frequencies were plotted. The results, measured at -10 and -20 dB for the design whose patterns are displayed in Fig. 37, are shown in Fig. 38. This design exhibits better axial ratios at levels down to -10 dB but deteriorate at lower levels. Additional data for this and other designs can be found in [18].

Aperture-Matched Horns

A horn which provides significantly better performance than an ordinary horn (in terms of pattern, impedance, and frequency characteristics) is that shown in Fig. 39a, which is referred to as an *aperture-matched* horn [26]. The main modification to the ordinary (conventional) horn, which we refer to here as the *control* horn, consists of the attachment of curved surface sections to the outside of the aperture edges, which reduces the diffractions which occur at the sharp edges of the aperture and provides smooth matching sections between the horn modes and the free-space radiation.

In contrast to the corrugated horn, which is complex and costly and reduces the diffractions at the edges of the aperture by minimizing the incident field, the

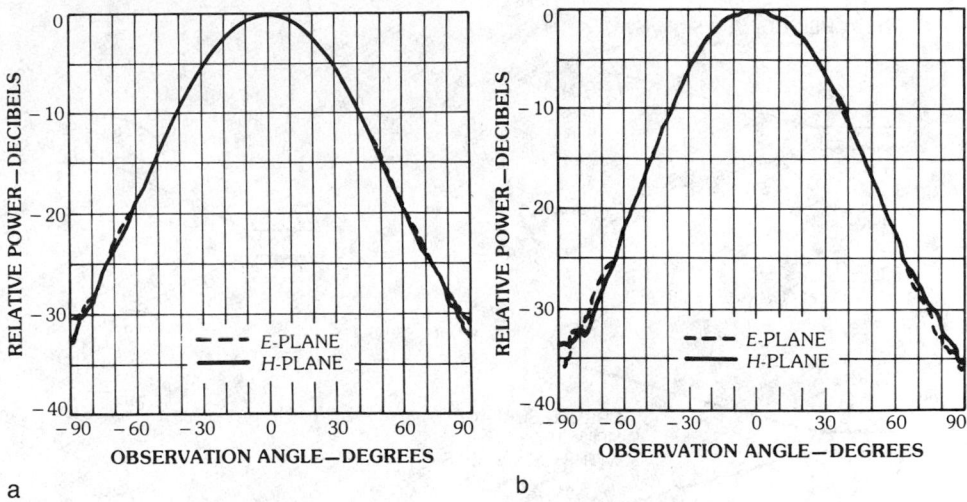

Fig. 37. Measured radiation patterns of a corrugated conical horn. (*a*) Frequency $f = 5.5$ GHz. (*b*) Frequency $f = 6.5$ GHz. (*After Al-Hakkak and Lo [18]*)

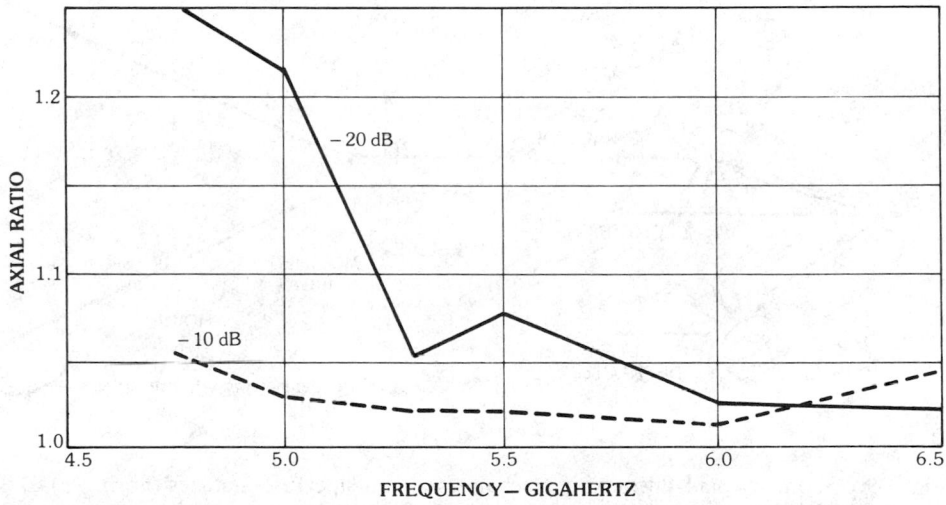

Fig. 38. Axial ratio versus frequency, for a horn whose patterns are shown in Fig. 37, at power levels of -10 dB and -20 dB. (*After Al-Hakkak and Lo [18]*)

aperture-matched horn reduces the diffractions by modifying the structure (without sacrificing size, weight, bandwidth, and cost) so that the diffraction coefficient is minimized. The basic concepts were originally investigated using elliptic cylinder sections, as shown in Fig. 39b; however, other convex curved surfaces, which smoothly blend to the ordinary horn geometry at the attachment point, will lead to similar improvements. This modification in geometry can be used in a wide variety of horns, and includes *E*-plane, *H*-plane, pyramidal, and conical horns. Bandwidths of 2:1 can easily be attained with aperture-matched horns having elliptical,

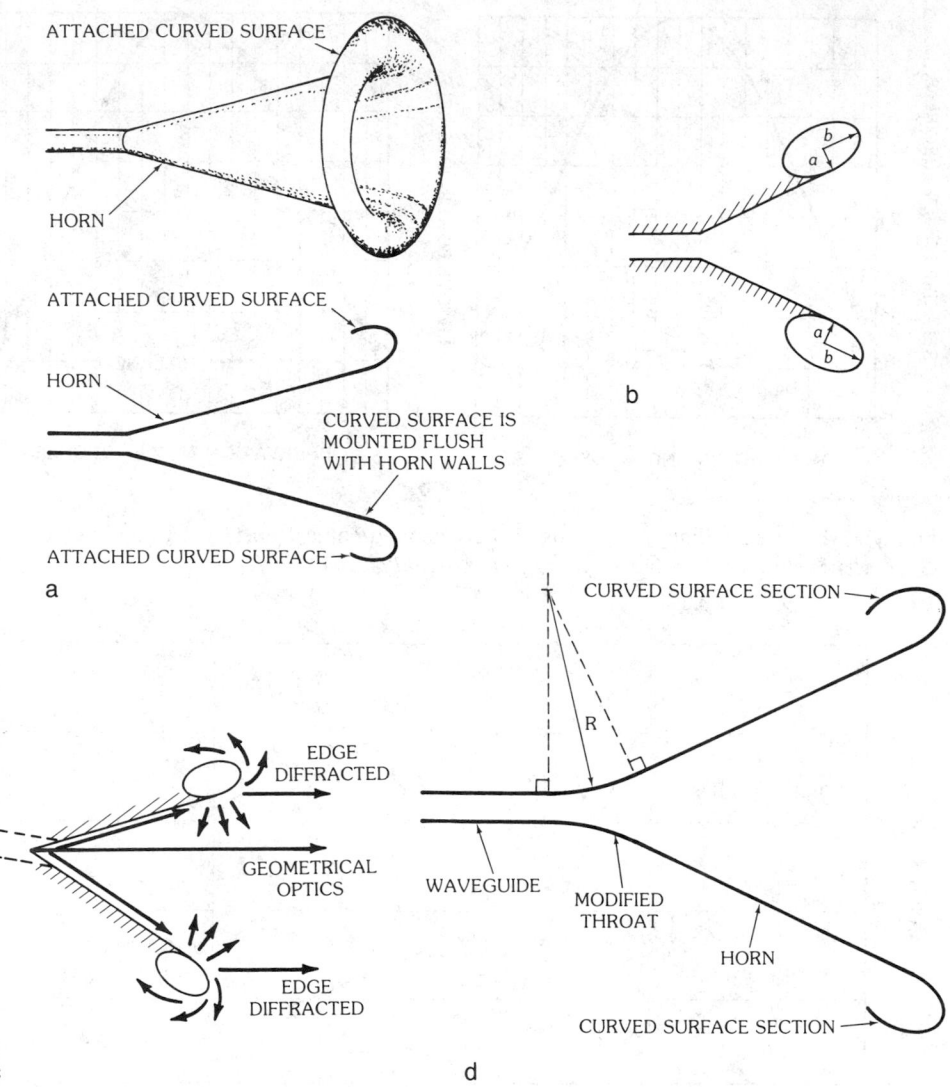

Fig. 39. Geometry and diffraction mechanism of an aperture-matched horn. (*a*) Basic geometry. (*b*) Elliptical cylinder curved surface. (*c*) Diffraction mechanism. (*d*) Modified throat. (*After Burnside and Chuang [26], © 1982 IEEE*)

circular, or other curved surfaces. The radii of curvature of the curved surfaces used in experimental models [26] ranged over $1.69\lambda \leqq a \leqq 8.47\lambda$ with $a = b$ and $b = 2a$. Good results can be obtained by using circular cylindrical surfaces with $2.5\lambda \leqq a \leqq 5\lambda$.

The basic radiation mechanism of such a horn is shown in Fig. 39c. The introduction of the curved sections at the edges does not eliminate diffractions; instead it substitutes edge diffractions by curved-surface diffractions which have a tendency to provide an essentially undisturbed energy flow across the junction, around the curved surface, and into free space. Compared with conventional horns, this radiation mechanism leads to smoother patterns with greatly reduced back lobes and negligible reflections back into the horn. The size, weight, and construction costs of the aperture-matched horn are usually somewhat larger and can be held to a minimum if half (one-half sections of an ellipse) or quadrant (one-fourth sections of an ellipse) sections are used instead of the complete closed surfaces.

To illustrate the improvements provided by the aperture-matched horns, the E-plane pattern, back lobe level, and half-power beamwidth of a pyramidal $2.96\lambda_0 \times 2.96\lambda_0$ horn were computed and were compared with the measured data of corresponding control and corrugated horns. The data are shown in Figs. 32a, 32c, and 32d. It is evident by examining the patterns of Fig. 32a that the aperture-matched horn provides a smoother pattern and lower back lobe level than conventional horns (referred to here as control horns); however, it does not provide, for the wide minor lobes, the same reduction as the corrugated horn. To achieve nearly the same E-plane pattern for all three horns the overall horn size would have to be increased. If the modifications for the aperture-matched and corrugated horns were only made in the E-plane edges, the H-plane patterns for all three horns would be virtually the same except that the back lobe level of the aperture-matched and corrugated horns would be greatly reduced.

The back lobe level of the same three horns (control, corrugated, and aperture-matched) are shown in Fig. 32c. The corrugated horn has lower back lobe intensity at the lower end of the frequency band, while the aperture-matched horn exhibits superior performance at the high end. However, both the corrugated and aperture-matched horns exhibit superior back lobe level characteristics to the control (conventional) horn throughout almost the entire frequency band. The half-power beamwidth characteristics of the same three horns are displayed in Fig. 32d. Because the control (conventional) horn has uniform distribution across the complete aperture plane, compared with the tapered distributions for the corrugated and aperture-matched horns, it possesses the smallest beamwidth almost over the entire frequency band.

In a conventional horn the vswr and antenna impedance are primarily influenced by the throat and aperture reflections. Using the aperture-matched horn geometry of Fig. 39a the aperture reflection toward the inside of the horn is greatly reduced. Therefore the only remaining dominant factor is the throat reflection. To reduce the throat reflections it has been suggested that a smooth curved surface be used to connect the waveguide and horn walls, as shown in Fig. 39d. Such a transition has been applied in the design and construction of a commercial X-band (8.2–12.4 GHz) pyramidal horn (see Fig. 12.23 of [6]), whose tapering is of exponential nature. The vswr's measured in the 8- to 12-GHz frequency band using the conventional exponential X-band horn (shown in Fig. 12.23 of [6]), with and without curved sections at its aperture, are shown in Fig. 40.

The matched sections used to create the aperture-matched horn were small cylinder sections. The vswr's for the conventional horn are very small (less than 1.1) throughout the frequency band because the throat reflection is negligible compared with the aperture reflection. It is evident, however, that the vswr's of the corresponding aperture-matched horn are much superior to those of the conventional horn because both the throat and aperture reflections are very minimal.

The basic design of the aperture-matched horn can be extended to include

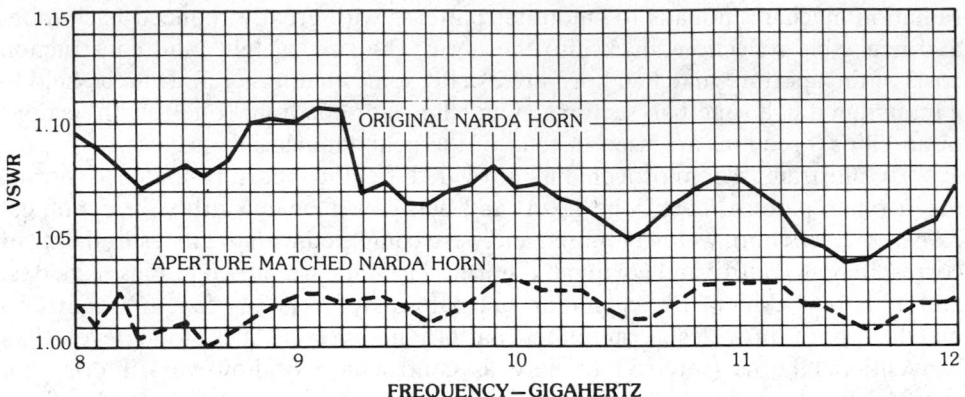

Fig. 40. Measured vswr for exponentially tapered pyramidal horns (conventional and aperture-matched). (*After Burnside and Chuang [26], © 1982 IEEE*)

corrugations on its inside surface. A typical configuration [24] of a conical aperture-matched corrugated horn is shown in Fig. 41. This type of design enjoys the advantages presented by both the aperture-matched and corrugated horns with cross-polarized components of less than −45 dB over a significant part of the bandwidth. Because of its excellent cross-polarization characteristics this horn is

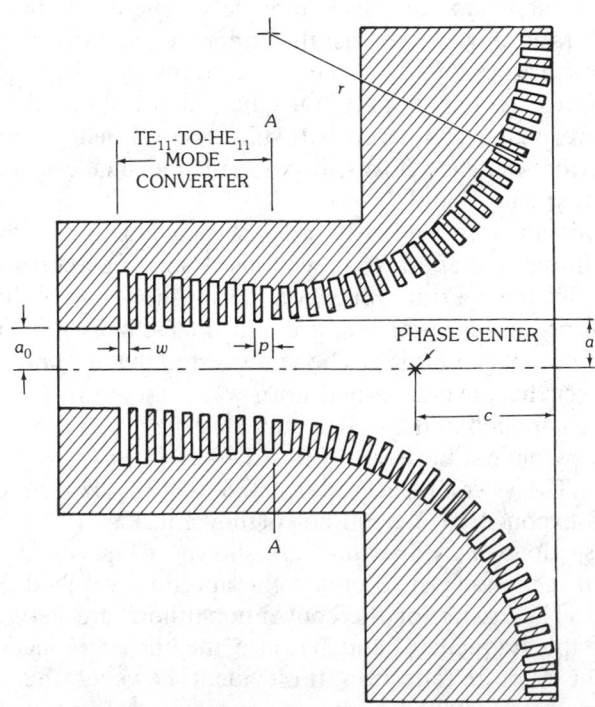

Fig. 41. Cross-section geometry of an aperture-matched corrugated conical horn. (*After Thomas and Greene [24], © 1982 IEEE*)

recommended for use as a reference and for frequency reuse applications in both satellite and terrestrial applications.

The corrugations are designed in the same manner as for the conventional corrugated conical horns, with the depth of the first being $\lambda/2$ at the highest frequency f_{max} of the bandwidth and that of the remaining decreasing linearly through the TE_{11}-to-HE_{11} mode converter. The last groove in the mode converter should have a depth of $\lambda/4$ at the center frequency f', and the remaining corrugations outside the converter toward the aperture should have a constant depth of $\lambda/4$ at $f = f'$. To achieve very low cross-polarization components the normalized radius of curvature of the curved section $\bar{r}$ (normalized with respect to the radius a at the throat, $\bar{r} = r/a$) should be large. Although horns with $\bar{r} = 1.3$ have been designed and tested, cross-polarized components of -45 dB or less have been obtained over a significant part of the bandwidth. The measured $45°$ plane patterns (principal and cross-polarized) for a horn with $\bar{r} = 4.3$ operating at $f = 15$ GHz is shown in Fig. 42. The cross-polarization component of this horn is about -50 dB or smaller. The bandwidth is extended as $\bar{r}$ increases until a value of $\bar{r}$ is reached when the undesired EH_{12} mode is generated.

Multimode Horns

Over the years there has been a need in many applications for horn antennas which provide symmetric patterns in all planes, phase center coincidence for the electric and magnetic planes, and side lobe suppression. All of these are attractive features for designs of optimum reflector systems and monopulse radar systems.

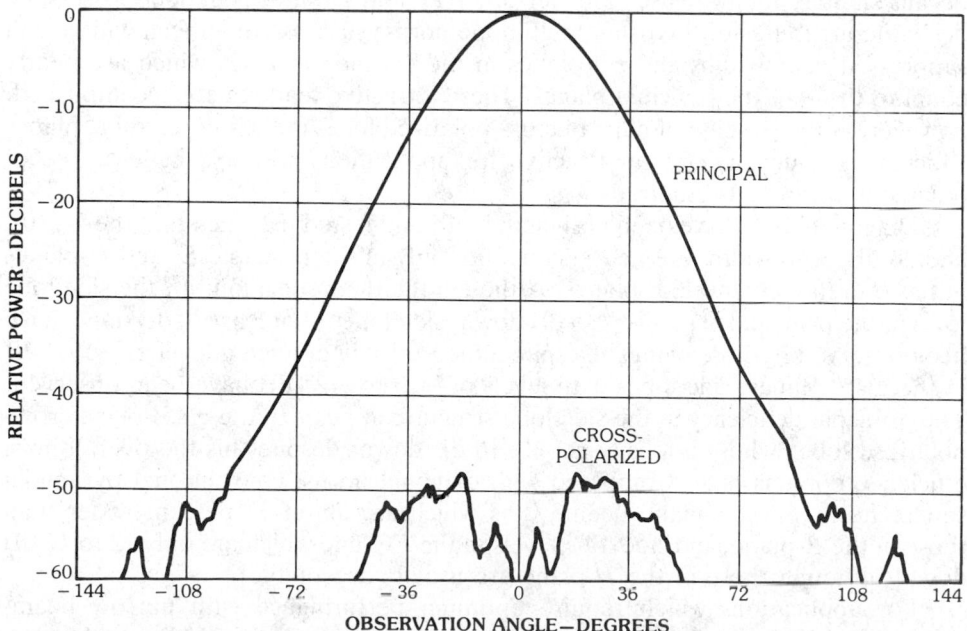

Fig. 42. Principal and cross-polarized $45°$ plane patterns of a corrugated conical horn with $\bar{r} = 4.3$ at $f = 15$ GHz. (*After Thomas and Greene [24], © 1982 IEEE*)

Side lobe reduction is a desired attribute for horn radiators utilized in antenna range, anechoic chamber, and standard gain applications, while pattern plane symmetry is a valuable feature for polarization diversity.

Pyramidal horns have traditionally been used over the years, with good success, in many of these applications. Such radiators, however, possess non-symmetric beamwidths and undesirable side lobe levels, especially in the E-plane. Conical horns, operating in the dominant TE_{11} mode, have a tapered aperture distribution in the E-plane. Thus they exhibit more symmetric electric- and magnetic-plane beamwidths and reduced side lobes than do the pyramidal horns. One of the main drawbacks of a conical horn is its relative incompatibility with rectangular waveguides.

To remove some of the deficiencies of pyramidal and conical horns and further improve some of their attractive characteristics, corrugations were introduced on the interior walls of the waveguides, which lead to the corrugated horns that were discussed in a previous section of this chapter. In some other cases designs were suggested to improve the beamwidth equalization in all planes and reduce side lobe levels by utilizing horn structures with multiple-mode excitations. These have been designated as *multimode* horns, and some of the designs will be discussed briefly here. For more information the reader should refer to the cited references.

One design of a multimode horn is the "diagonal" horn [32], shown in Fig. 43, all of whose cross sections are square and whose internal fields consist of a super-position of TE_{10} and TE_{01} modes in a square waveguide. For small flare angles the field structure within the horn is such that the **E**-field vector is parallel to one of the diagonals. Although it is not a multimode horn in the true sense of the word because it does not make use of higher-order TE and TM modes, it does possess the desirable attributes of the usual multimode horns, such as equal beamwidths and suppressed beamwidths and side lobes in the E- and H-planes which are nearly equal to those in the principal planes. These attractive features are accomplished, however, at the expense of pairs of cross-polarized lobes in the intercardinal planes which make such a horn unattractive for applications where a high degree of polarization purity is required.

Diagonal horns have been designed, built, and tested [32] such that the 3-, 10-, and 30-dB beamwidths are nearly equal not only in the principal E- and H-planes but also in the 45° and 135° planes. Although the theoretical limit of the side lobe level in the principal planes is 31.5 dB down, side lobes of at least 30 dB down have been observed in those planes. Despite a theoretically predicted level of -19.2 dB in the $\pm 45°$ planes, side lobes with levels of -23 to -27 dB have been observed. The principal deficiency in the side lobe structure appears in the $\pm 45°$-plane cross-polarized lobes whose intensity is only 16 dB down; despite this the overall horn efficiency remains high. Compared with diagonal horns, conventional pyramidal square horns have H-plane beamwidths which are about 35 percent wider than those in the E-plane, and side lobe levels in the E-plane which are only 12 to 13 dB down (although those in the H-plane are usually acceptable).

For applications which require optimum performance with narrow beam-widths, lenses are usually recommended for use in conjunction with diagonal horns. Diagonal horns can also be converted to radiate circular polarization by inserting a differential phase shifter inside the feed guide whose cross section is circular and adjusted so that it produces phase quadrature between the two orthogonal modes.

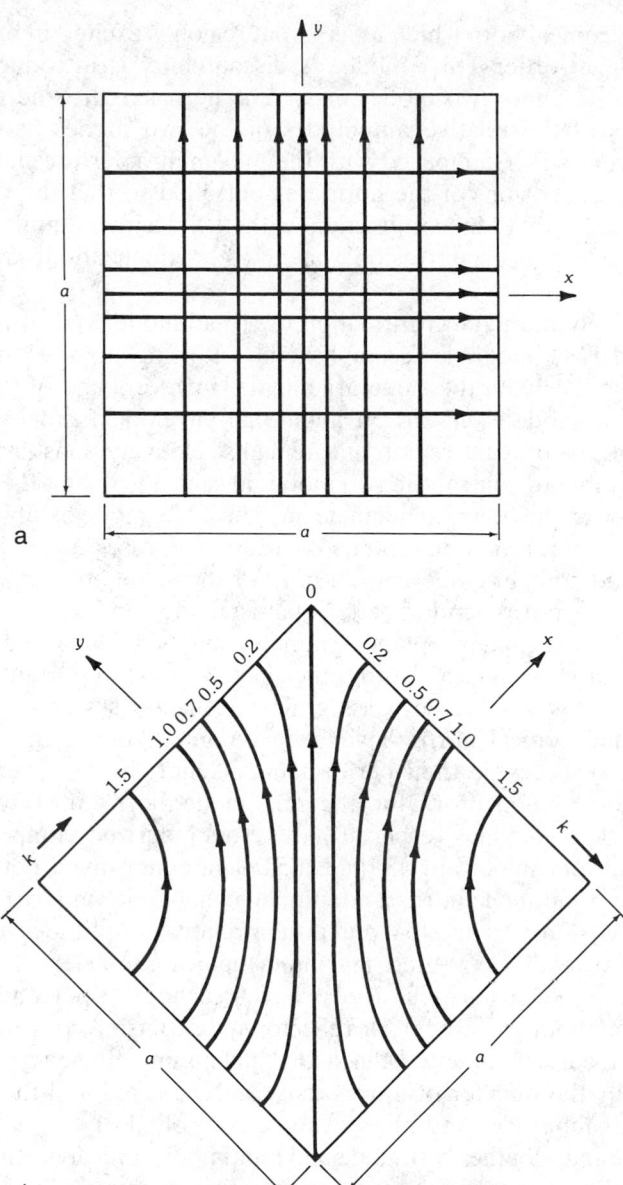

Fig. 43. Electric-field configuration inside square diagonal horn. (*a*) Two coexisting equal orthogonal modes. (*b*) Result of combining the two modes shown in (*a*). (*After Love [32], reprinted with permission of* Microwave Journal, *from March 1962 issue,* © *1962 Horizon House–Microwave, Inc.*)

Another multimode horn which exhibits suppressed side lobes, equal beam-widths, and reduces cross polarization is the *dual-mode* conical horn [33]. Basically this horn is designed so that diffractions at the aperture edges of the horn, especially those in the *E*-plane, are minimized by reducing the fields incident on the aperture edges and consequently the associated diffractions. This is accomplished by utilizing a conical horn which at its throat region is excited in both the dominant TE_{11} and higher-order TM_{11} mode. A discontinuity is introduced at a position within the horn where two modes exist. The horn length is adjusted so that the superposition of the relative amplitudes of the two modes at the edges of the aperture is very small compared with the maximum aperture field magnitude. In addition, the dimensions of the horn are controlled so that the total phase at the aperture is such that, in conjunction with the desired amplitude distribution, it leads to side lobe suppression, beamwidth equalization, and phase center coincidence.

Qualitatively the pattern formation of a dual-mode conical horn operating in the TE_{11} and TM_{11} modes is accomplished by utilizing a pair of modes which have radiation functions with the same argument. However, one of the modes, in this case the TM_{11} mode, contains an additional envelope factor which varies very rapidly in the main beam region and remains relatively constant at large angles. Thus it is possible to control the two modes in such a way that their fields cancel in all directions except within the main beam. The TM_{11} mode exhibits a null in its far-field pattern. Therefore a dual-mode conical horn possesses less axial gain than a conventional dominant-mode conical horn of the same aperture size. Because of that, dual-mode horns render better characteristics and are more attractive for applications where pattern plane symmetry and side lobe reduction are more important than maximum aperture efficiency. A most important application of a dual-mode horn is as a feed of Cassegrainian reflector systems.

Dual-mode conical horns have been designed, built, and tested [33] with relatively good success in their performance. Generally, however, diagonal horns would be good competitors for the dual-mode horns if it were not for the undesirable characteristics (especially the cross-polarized components) that they exhibit in the very important 45° and 135° planes. Improved performance can be obtained from dual-mode horns if additional higher-order modes (such as the TE_{12}, TE_{13}, and TM_{12}) are excited [34] and if their relative amplitudes and phases can be properly controlled. Computed maximum aperture efficiencies of paraboloidal reflectors, using such horns as feeds, have reached 90 percent contrasted with efficiencies of about 76 percent for reflector systems using conventional dominant-mode horn feeds. In practice the actual maximum efficiency achieved will be determined by the number of modes that can be excited and the degree to which their relative amplitudes and phases can be controlled.

The techniques of the dual-mode and multimode conical horns can be extended to the design of horns with rectangular cross sections. In fact a multimode pyramidal horn design has been designed, built, and tested to be used as a feed in a low-noise Cassegrain monopulse system [35]. This rectangular pyramidal horn utilizes additional higher-order modes to provide monopulse capability, side lobe suppression in both the *E*- and *H*-planes, and beamwidth equalization. Specifically the various pattern modes for the monopulse system are formed in a single horn as follows:

(*a*) *Sum*: Utilizes $TE_{10} + TE_{30}$ instead of only TE_{10}. When the relative amplitude and phase excitations of the higher-order TE_{30} mode are properly adjusted, they provide side lobe cancellation at the second minor lobe of the TE_{10}-mode pattern

(b)   E-*Plane Difference*: Utilizes $TE_{11} + TM_{11}$ modes

(c)   H-*Plane Difference*: Utilizes TE_{20} mode

In its input the horn of [35] contained a four-guide monopulse bridge circuitry, a multimode matching section, a difference mode phasing section, and a sum mode excitation and control section. To illustrate the general concept Figs. 44a, 44b, and 44c are plots of three-dimensional patterns of the sum, *E*-plane difference, and *H*-plane difference modes which utilize, respectively, the $TE_{10} + TE_{30}$, $TE_{11} + TM_{11}$, and TE_{20} modes. The relative excitation between the modes has been controlled so that each pattern utilizing multiple modes in its formation displays its most attractive features for its function.

Dielectric-Loaded Horns

Over the years much effort has been devoted in enhancing the antenna and aperture efficiencies of aperture antennas, especially for those that serve as feeds for reflectors (such as the horn). One technique that was proposed and was investigated was to use dielectric guiding structures, referred to as *Dielguides* [36], between the primary feed and the reflector (or subreflector). The technique is simple and inexpensive to implement and provides broadband, highly efficient, and low-noise antenna feeds. The method negates the compromise between taper and spillover efficiencies, and it is based on the principle of internal reflections, which has been utilized frequently in optics. Its role bears a very close resemblance to that of a lens, and it is an extension of the classical parabolic-shaped lens to other geometrical shapes.

Another method which has been used to control the radiation pattern of electromagnetic horns is to insert totally within them various shapes of dielectric material (wedges, slabs, etc.) [37–39] to control in a predictable manner not only the phase distribution over the aperture, as is usually done by using the classical parabolic lenses, but also to change the power (amplitude) distribution over the aperture. The control of the amplitude and phase distributions over the aperture are very essential in the design of very low side lobe antenna patterns.

Symmetrical loading of the *H*-plane walls has also been utilized, by proper parameter selection, to create a dominant longitudinal section electric (LSE) mode and to enhance the aperture efficiency and pattern-shaping capabilities of symmetrically loaded horns [38]. The method is simple and inexpensive, and it can also be utilized to realize high efficiency from small horns which can be used in limited scan arrays. Aperture efficiencies on the order of 92 to 96 percent have been attained, in contrast to values of 81 percent for unloaded horns.

A similar technique has been suggested to symmetrically load the *E*-plane walls of rectangular horns [39], and eventually to line all four of its walls with dielectric slabs. Other similar techniques have been suggested, and a summary of these and other classical papers dealing with dielectric-loaded horns can be found in [1].

8.   Phase Center

Each of the far-zone field components radiated by an antenna can be written, in general, as

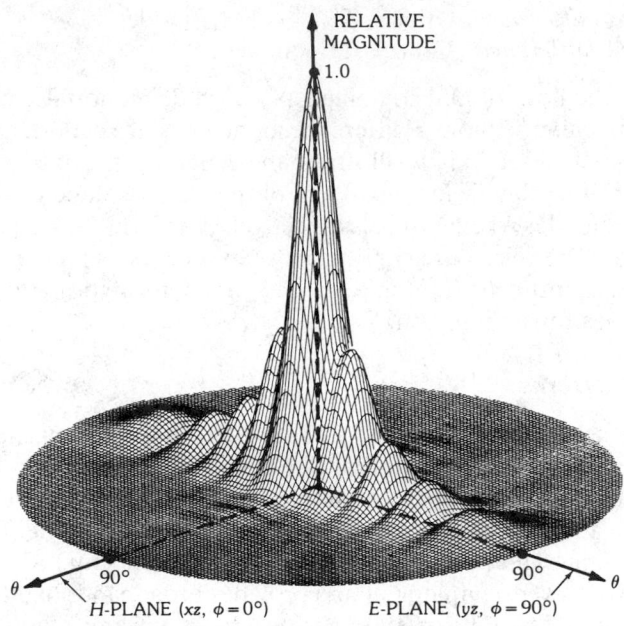

a

H-PLANE (*xz*, $\phi = 0°$)   *E*-PLANE (*yz*, $\phi = 90°$)

b

Fig. 44. Three-dimensional patterns of the sum, *E*-plane difference, and *H*-plane difference patterns. (*a*) Sum field pattern of a monopulse pyramidal horn operating in the TE_{10} and TE_{30} modes. (*b*) *E*-plane difference field pattern of a monopulse pyramidal horn operating in the TE_{11} and TM_{11} modes. (*c*) *H*-plane field pattern of a monopulse pyramidal horn operating in the TE_{20} mode.

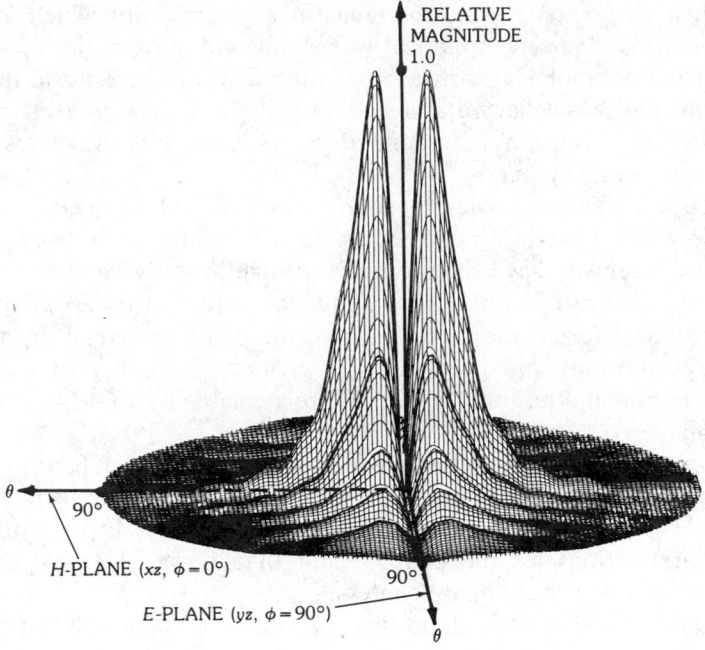

RELATIVE
MAGNITUDE
1.0

θ

90°

H-PLANE (*xz*, $\phi = 0°$)

90°

E-PLANE (*yz*, $\phi = 90°$)

θ

c

Fig. 44, *continued*

$$\mathbf{E}_u = \hat{\mathbf{u}}\, E(\theta, \phi) e^{j\psi(\theta, \phi)} \frac{e^{-jkr}}{r} \tag{41}$$

where $\hat{\mathbf{u}}$ is a unit vector. The terms $E(\theta, \phi)$ and $\psi(\theta, \phi)$ represent, respectively, the (θ, ϕ) variations of the amplitude and phase.

In navigation, tracking, homing, landing, and other aircraft and aerospace systems it is usually desirable to assign to the antenna a reference point such that for a given frequency, $\psi(\theta, \phi)$ of (41) is independent of θ and ϕ [i.e., $\psi(\theta, \phi) =$ constant]. The reference point which makes $\psi(\theta, \phi)$ independent of θ and ϕ is known as the *phase center* of the antenna [40–44]. When referenced to the phase center the fields radiated by the antenna are spherical waves with ideal spherical wavefronts or equiphase surfaces. Therefore a phase center is a reference point from which radiation is said to emanate, and radiated fields measured on the surface of a sphere whose center coincides with the phase center have the same phase.

For practical antennas such as arrays, reflectors, and others, a single unique phase center valid for all values of θ and ϕ does not exist; for most, however, their phase center moves along a surface, and its position depends on the observation point. However, in many antenna systems a reference point can be found such that $\psi(\theta, \phi) = $ constant, or nearly so, over most of the angular space, especially over the main lobe. When the phase center position variation is sufficiently small, that point is usually referred to as the *apparent phase center*.

The need for the phase center can best be explained by examining the radiation characteristics of a paraboloidal reflector (parabola of revolution). Plane waves incident on a paraboloidal reflector focus at a single point which is known as the *focal point*. Conversely, spherical waves emanating from the focal point are reflected by the paraboloidal surface and form plane waves. Thus in the receiving mode all the energy is collected at a single point. In the transmitting mode, ideal plane waves are formed if the radiated waves have spherical wavefronts and emanate from a single point.

In practice, no antenna is a point source with ideal spherical equiphases. Many of them, however, contain a point from which their radiation, over most of the angular space, seems to have spherical wavefronts. When such an antenna is used as a feed for a reflector its phase center must be placed at the focal point.

The analytical formulations for locating the phase center of an antenna are usually very laborious and exist only for a limited number of configurations [40–42]. Experimental techniques [43, 44] are available to locate the phase center of an antenna.

The horn is a microwave antenna which is widely used as a feed for reflectors. To perform as an efficient feed for reflectors it is imperative that its phase center is known and it is located at the focal point of the reflector. Instead of presenting analytical formulations for the phase center of a horn, graphical data will be included to illustrate typical phase centers.

Usually the phase center of a horn is not located at its mouth (throat) or at its aperture but mostly between its imaginary apex point and its aperture. The exact location depends on the dimensions of the horn, especially on its flare angle. For large flare angles the phase center is closer to the apex. As the flare angle of the horn becomes smaller, the phase center moves toward the aperture of the horn.

Computed phase centers for an *E*-plane and an *H*-plane sectoral horn are displayed in Figs. 45a and 45b. It is apparent that for small flare angles the *E*- and *H*-plane phase centers are identical. Although each specific design has its own phase center the data of Figs. 45a and 45b are typical. If the *E*- and *H*-plane phase centers of a pyramidal horn are not identical, its phase center can be taken to be the average of the two. Phase centers for narrowband corrugated horns are also displayed in Fig. 34.

Phase center nomographs for conical corrugated and uncorrugated (TE_{11}-mode) horns are available [42], and they are displayed, respectively, in Figs. 46a and 46b. The procedure to use these is documented in [42], and it is repeated here.

Procedure to Locate Phase Center

Given: (*a*) a/λ = radius of the horn aperture (in wavelengths)

           (*b*) R_0/λ = distance from the aperture, along the horn axis, to the observation point (in wavelengths)

           (*c*) ℓ/λ = distance from the aperture to the horn apex (in wavelengths)

Find: The phase center location Z_0 (in wavelengths). This is the phase center location measured from the aperture along the horn axis toward the apex (positive Z_0s indicate the phase center is within the horn).

Solution: The procedure based on the results of Figs. 46a and 46b, follows:

           (*a*)   Determine $\alpha = a^2/(\lambda\ell)$ and locate the α curve on the appropriate figure.

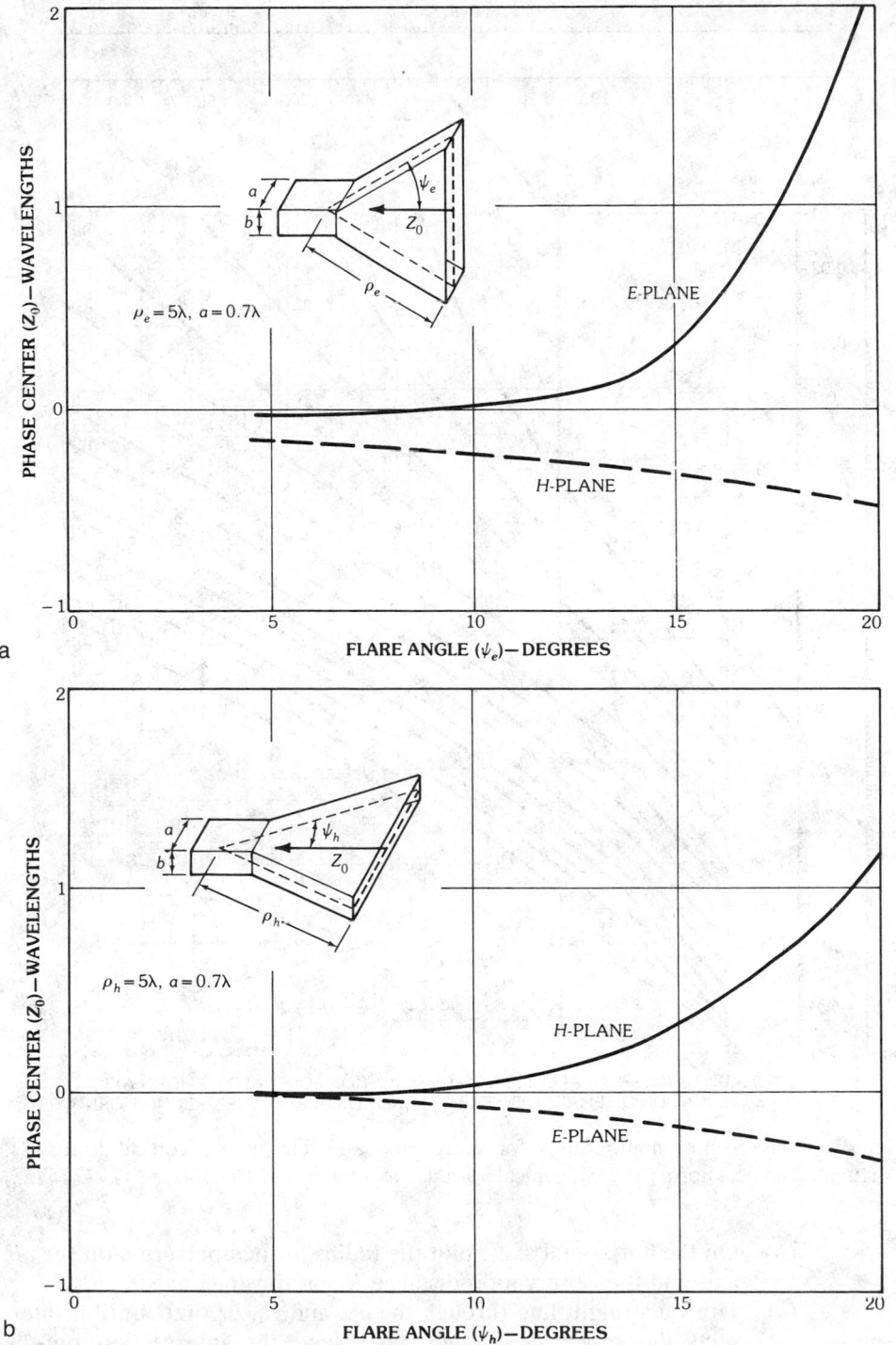

Fig. 45. Phase center location, as a function of flare angle, for *E*- and *H*-plane sectoral horns. (*a*) *E*-plane sectoral horn. (*b*) *H*-plane sectoral horn. (*Adapted from Hu [40]*)

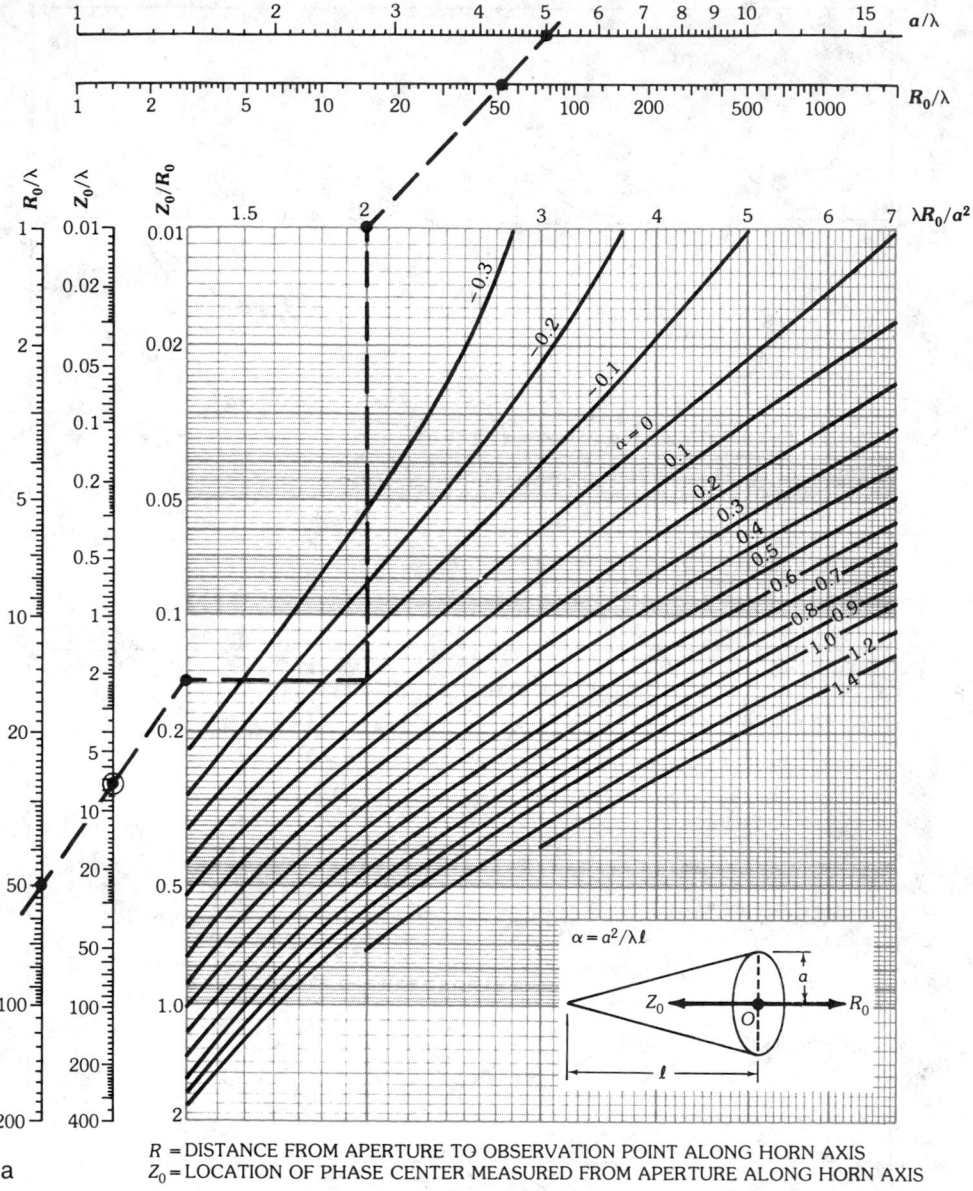

R = DISTANCE FROM APERTURE TO OBSERVATION POINT ALONG HORN AXIS

a Z_0 = LOCATION OF PHASE CENTER MEASURED FROM APERTURE ALONG HORN AXIS

Fig. 46. Phase center nomographs for corrugated and TE_{11}-mode conical horns. (*a*) Corrugated conical horn. (*b*) TE_{11} conical horn. (*After Ohtera and Ujiie [42],* © *1975 IEEE*)

(*b*) On the horizontal axes, plot the radius of the aperture *a* on the a/λ axis and the observation distance R_0 on the R_0/λ axis.

(*c*) Draw a straight line through the a/λ and R_0/λ points until it intersects the reference $\lambda R_0/a^2$ axis. Read the intersection on the reference axis.

(*d*) Through the $\lambda R_0/a^2$ axis intersection, draw a vertical line until it

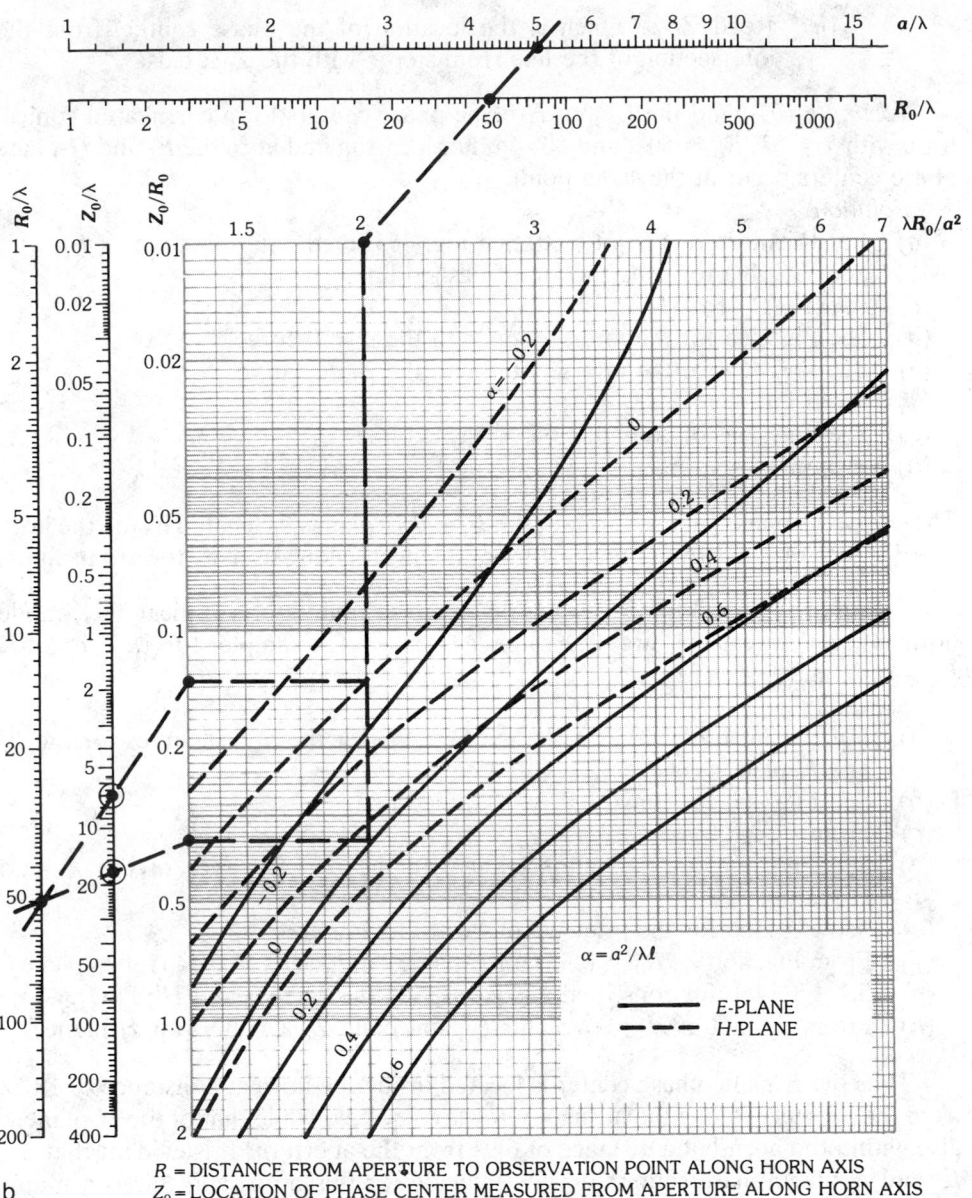

R = DISTANCE FROM APERTURE TO OBSERVATION POINT ALONG HORN AXIS
Z_0 = LOCATION OF PHASE CENTER MEASURED FROM APERTURE ALONG HORN AXIS

b

Fig. 46, *continued*

intersects the α curve found in step *a*.

(*e*) Through the α curve intersection point, draw a horizontal line until it intersects the Z_0/R_0 vertical axis.

(*f*) On the vertical axes, read the intersection point on the reference Z_0/R_0 axis.

(*g*) From the intersection point on the Z_0/R_0 axis, draw a straight line to the given point on the vertical R_0/λ axis.

(h)   Read Z_0/λ, which is the location of the phase center, from the intersection of the line from step g with the Z_0/λ axis.

Example 10—Find the E- and H-plane phase centers for a corrugated conical horn with $a = 5\lambda$, $R_0 = 50\lambda$, and $\ell = \infty$. For a corrugated horn the E- and H-plane phase centers occur at the same point.

Solution:

(a)   For this horn $\alpha = a^2/\lambda\ell = 0$. See Fig. 46a for the $\alpha = 0$ curve.

(b)   See Fig. 46a for construction of dashed line.

(c)   From Fig. 46a, $\lambda R_0/a^2 = 2$.

(d)   See Fig. 46a for the intersection with the $\alpha = 0$ curve.

(e)   See Fig. 46a for the intersection with the Z_0/R_0 curve.

(f)   From Fig. 46a, $Z_0/R_0 = 0.15\lambda$.

(g)   See Fig. 46a for construction of dashed line.

(h)   From Fig. 46a $Z_0/\lambda = 7.47$.

Therefore the phase center, for both the E- and H-planes, is located inside the horn at a distance of 7.47λ from the aperture of the horn, along its axis, toward its apex.

Example 11—Find the E- and H-plane phase centers of a conical TE_{11}-mode horn whose dimensions are the same as those of Example 10 (i.e., $a = 5\lambda$, $R_0 = 50\lambda$, and $\ell = \infty$).

Solution:

(a)   For this horn $\alpha = a^2/\lambda\ell = 0$. See Fig. 46b for the $\alpha = 0$ curves for the E- and H-planes.

(b)   See Fig. 46b for construction of dashed line.

(c)   From Fig. 46b, $\lambda R_0/a^2 = 2$.

(d)   See Fig. 46b for the intersection with the $\alpha = 0$ curves of the E- and H-planes.

(e)   See Fig. 46b for the intersections with the Z_0/R_0 axis.

(f)   From Fig. 46b $Z_0/R_0 = 0.35$ (for E-plane) and $Z_0/R_0 = 0.14$ (for H-plane).

(g)   See Fig. 46b for constructions of dashed lines (for E- and H-plane).

(h)   From Fig. 46b $Z_0/\lambda = 18.5$ (for E-plane) and $Z_0/\lambda = 6.9$ (for H-plane).

Thus the E-plane phase center is located inside the horn at a distance of 18.5λ from the aperture, along the horn axis, toward its apex, while that of the H-plane is also within the horn but a distance of 6.9λ from the aperture. It is evident that the E- and H-plane phase centers do not coincide for this horn, and this is a major deficiency in many applications.

An experimental technique for measuring the phase center of an antenna, based on the work by Dyson [43] and reported also in [18], will be repeated here. The antenna under test is placed in the far field of a transmitting horn, on a positioner that is capable of precise placement of an antenna relative to a rotation axis, as shown in Fig. 47. The relative phase ψ of the unmodulated rf signal e_2 from the test antenna is compared with that of the coherent unmodulated reference signal e_1 in a network analyzer. The output signal e_3, which is proportional to ψ, is then recorded by an X–Y recorder whose X movement is synchronized with the rotation of the antenna under test. A delay line and an attenuator are included in

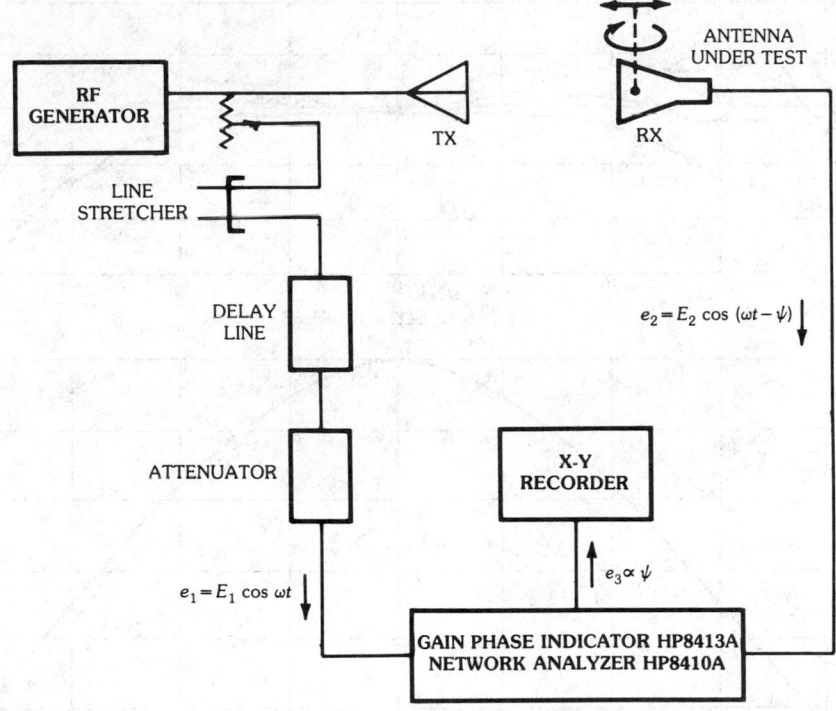

Fig. 47. Circuit diagram for phase center measurement. (*After Al-Hakkak and Lo [18]*)

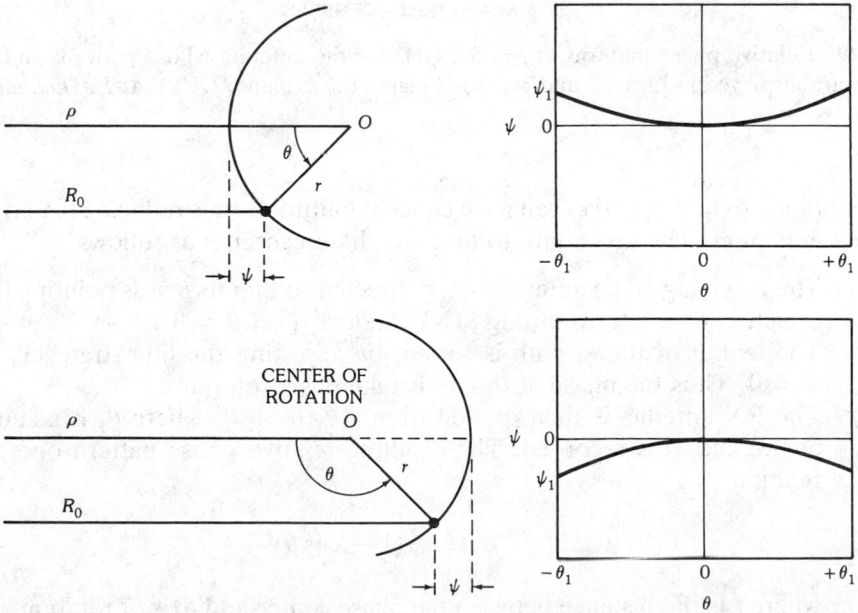

Fig. 48. Geometry and phase change as a displaced antenna is rotated about a given axis. (*After Al-Hakkak and Lo [18]*)

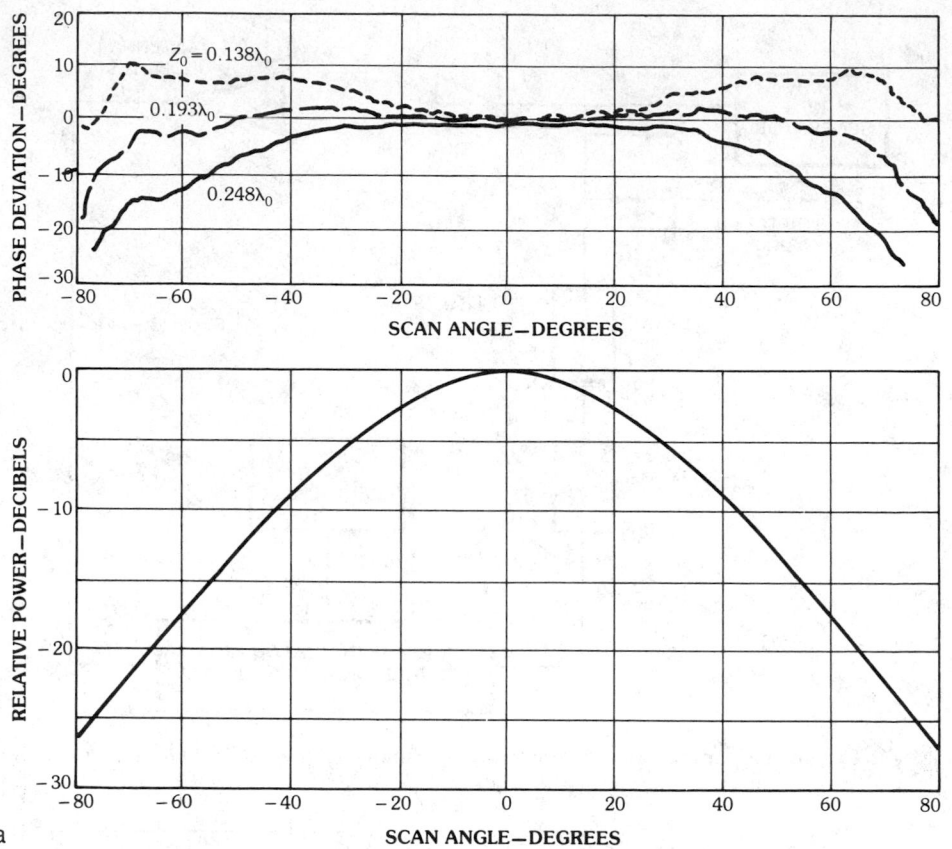

Fig. 49. Relative phase patterns at $f = 5.2$ GHz for the antenna whose patterns and axial ratios are displayed in Figs. 37 and 38. (*a*) *H*-plane. (*b*) *E*-plane. (*After Al-Hakkak and Lo [18]*)

the e_1 branch so that e_1 and e_2 can have equal amplitudes, thus reducing the error in e_3 to a minimum. The procedure to find the phase center is as follows:

(*a*) The receiving (RX) antenna is first directed so that its axis is pointing in the direction of the transmitting (TX) antenna (i.e., $\theta = 0°$).

(*b*) The length of the e_1 path is varied, by adjusting the line stretcher, until $\psi = 0$. Thus the phase at $\theta = 0°$ is taken as a reference.

(*c*) The RX antenna is then rotated from $-\theta_1$ to $+\theta_1$, where θ_1 is a suitable angle, and ψ is recorded. The resulting relative phase pattern obeys the relation

$$\psi \cong k_0 r(1 - \cos\theta) \tag{42}$$

where r is the distance between the phase center and axis of rotation and θ as shown in Fig. 48. If the phase pattern is a straight line, the center of rotation coincides with the phase center of the antenna. If the pattern is not

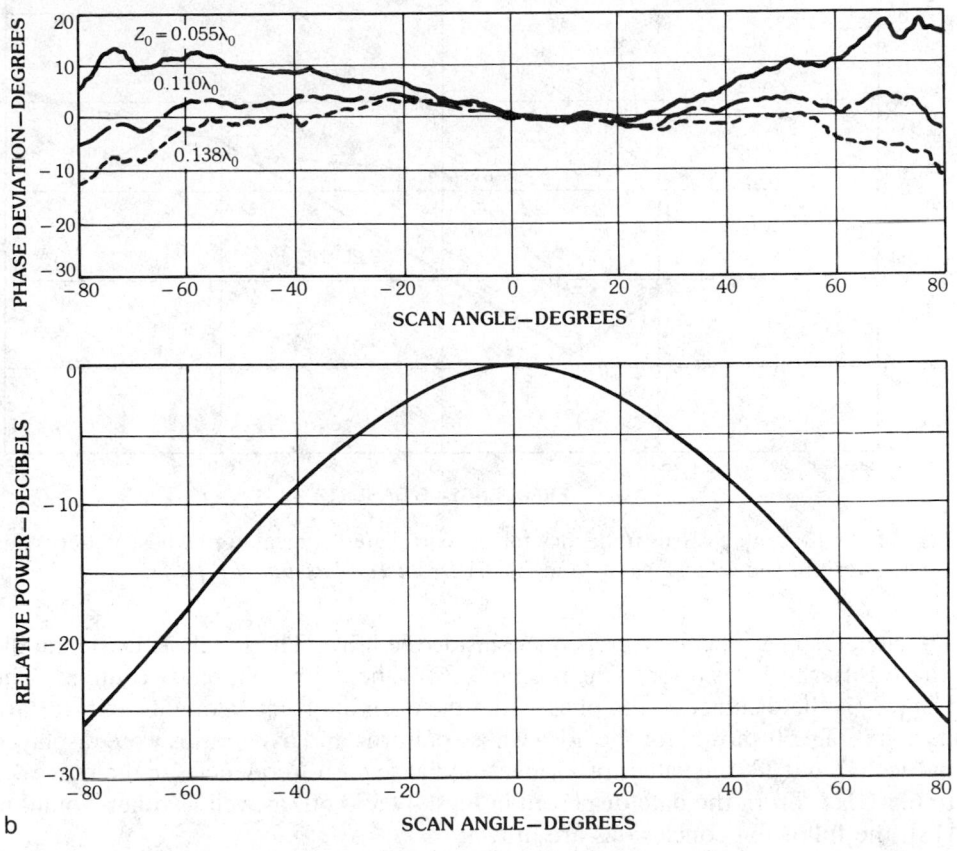

b

Fig. 49, *continued*

straight, then in theory the position of the phase center can be determined from this pattern by finding r from (42) as

$$r = \frac{\lambda_0}{2\pi} \frac{\psi_1}{1 - \cos\theta_1} \qquad (43)$$

where ψ_1 is the relative phase at θ_1. In practice, however, it may be necessary, because of both pattern and experimental anomalies, to record several patterns as the antenna is repositioned along its axis, repeating steps *a–c*. The position that corresponds to the flattest pattern is taken as the phase center.

Using the procedure described above, the phase centers of the conical corrugated horn antenna whose patterns and axial ratios are displayed, respectively, in Figs. 37 and 38 have been determined in both the *E*- and *H*-planes. Typical phase and power patterns measured at 5.2 GHz are shown in Fig. 49. The numbers on the phase patterns refer to the locations of axes of rotation measured in

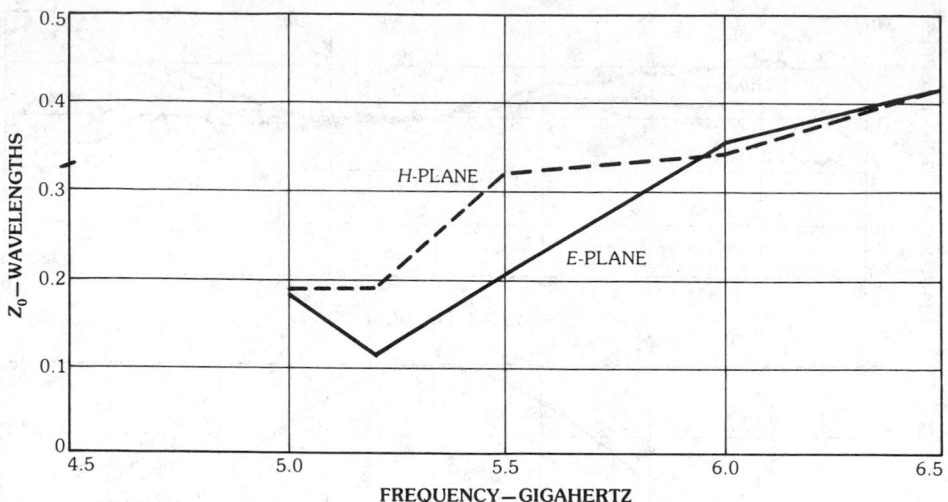

Fig. 50. Phase center versus frequency for the corrugated conical horn whose patterns and axial ratios are shown in Figs. 37 and 38. (*After Al-Hakkak and Lo [18]*)

wavelengths from the aperture's plane inside the horn. The numbers in the middle phase pattern curves refer to the positions Z_0 of the "apparent phase centers." The term Z_0 is the distance of the phase from the horn aperture, along its axis, toward its apex. Fig. 50 shows, for the horn whose patterns and axial ratios were displayed in Figs. 37 and 38, the values of Z_0 measured at several frequencies in the range 4.5 to 6.5 GHz. From the data displayed in Figs. 49 and 50, as well as others found in [18], the following conclusions are drawn:

(*a*)   The apparent phase centers are located inside the horns at distances Z_0 from the aperture plane, that are within one wavelength, in general. However, the longer the horn, the larger Z_0.

(*b*)   As the frequency increases, Z_0 (in wavelengths) becomes larger.

(*c*)   The values of Z_0 for the *E*- and *H*-planes generally differ by less than 0.1 wavelength.

(*d*)   For shorter horns a constant phase pattern can be found over wider angular regions.

A technique which allows the determination of phase centers of horn antennas at *X*-band, with a very high degree of accuracy, has also been documented [44]. The results indicate that for horn antennas as large as 5 in (12.7 cm), the phase centers can be determined to within ±0.25 electrical degrees. Phase center positions at *X*-band were determined to within 0.002 in (0.00508 cm).

Acknowledgments

Portions of this chapter were first published in *Antenna Theory: Analysis and Design* by C. A. Balanis [6], copyright 1982, reprinted by permission of Harper & Row Publishers. Inc.

9. References

[1]   A. W. Love, *Electromagnetic Horn Antennas*, New York: IEEE Press, 1976.

[2]   P. M. Russo, R. C. Rudduck, and L. Peters, Jr., "A method for computing E-plane patterns of horn antennas," *IEEE Trans. Antennas Propag.*, vol. AP-13, no. 2, pp. 219–224, March 1965.

[3]   J. S. Yu, R. C. Rudduck, and L. Peters, Jr., "Comprehensive analysis for E-plane of horn antennas by edge diffraction theory," *IEEE Trans. Antennas Propag.*, vol. AP-14, no. 2, pp. 138–149, March 1966.

[4]   M. A. K. Hamid, "Diffraction by a conical horn," *IEEE Trans. Antennas Propag.*, vol. AP-16, no. 5, pp. 520–528, September 1966.

[5]   M. S. Narasimhan and M. S. Shehadri, "GTD analysis of the radiation patterns of conical horns," *IEEE Trans. Antennas Propag.*, vol. AP-26, no. 6, pp. 774–778, November 1978.

[6]   C. A. Balanis, *Antenna Theory: Analysis and Design*, New York: Harper & Row, Publishers, 1982.

[7]   R. J. Mailloux and G. R. Forbes, "An array technique with grating-lobe suppression for limited-scan applications," *IEEE Trans. Antennas Propag.*, vol. AP-21, no. 5, pp. 597–602, September 1973.

[8]   R. J. Mailloux, L. Zahn, A. Martinez III, and G. R. Forbes, "Grating lobe control in limited scan arrays," *IEEE Trans. Antennas Propag.*, vol. AP-27, no. 1, pp. 79–85, January 1979.

[9]   N. Amitay and M. J. Gans, "Design of rectangular horn arrays with oversized aperture elements," *IEEE Trans. Antennas Propag.*, vol. AP-29, no. 6, pp. 871–884, November 1981.

[10] J. Boersma, "Computation of Fresnel integrals," *Math. Comp.*, vol. 14, p. 380, 1960.

[11] Y.-B. Cheng, "Analysis of aircraft antenna radiation for microwave landing system using geometrical theory of diffraction," MSEE thesis, Dept. of Electr. Eng., West Virginia Univ., pp. 208–211.

[12] E. H. Braun, "Some data for the design of electromagnetic horns," *IRE Trans. Antennas Propag.*, vol. AP-4, no. 1, pp. 29–31, January 1956.

[13] E. V. Jull, "Gain of an E-plane sectoral horn—a failure of the Kirchhoff theory and a new proposal," *IEEE Trans. Antennas Propag.*, vol. AP-22, no. 2, pp. 221–226, March 1974.

[14] W. C. Jakes, "Horn Antennas," Chapter 10 in *Antenna Engineering Handbook*, ed. by H. Jasik, New York: McGraw-Hill Book Co., 1961.

[15] W. M. Truman and C. A. Balanis, "Optimum design of horn feeds for reflector antennas," *IEEE Trans. Antennas Propag.*, vol. AP-22, no. 4, pp. 585–586, July 1974.

[16] M. G. Schorr and F. J. Beck, Jr., "Electromagnetic field of a conical horn," *J. Appl. Phys.*, vol. 21, pp. 795–801, August 1950.

[17] A. P. King, "The radiation characteristics of conical horn antennas," *Proc. IRE*, vol. 38, pp. 249–251, March 1950.

[18] M. J. Al-Hakkak and Y. T. Lo, "Circular waveguides and horns with anisotropic and corrugated boundaries," *Antenna Laboratory Report No. 73-3*, Depart. Electr. Eng., Univ. of Illinois, Urbana, January 1973.

[19] A. F. Kay, "The scalar feed," *AFCRL Rep. 65-347*, AD601609, March 1964.

[20] R. E. Lawrie and L. Peters, Jr., "Modifications of horn antennas for low side lobe levels," *IEEE Trans. Antennas Propag.*, vol. AP-14, no. 5, pp. 605–610, September 1966.

[21] C. A. Mentzer and L. Peters, Jr., "Properties of cutoff corrugated surfaces for corrugated horn design," *IEEE Trans. Antennas Propag.*, vol. AP-22, no. 2, pp. 191–196, March 1974.

[22] C. A. Mentzer and L. Peters, Jr., "Pattern analysis of corrugated horn antennas," *IEEE Trans. Antennas Propag.*, vol. AP-24, no. 3, pp. 304–309, May 1976.

[23] B. MacA. Thomas, "Design of corrugated conical horns," *IEEE Trans. Antennas Propag.*, vol. AP-26, no. 2, pp. 367–372, March 1978.

[24] B. MacA. Thomas and K. J. Greene, "A curved-aperture corrugated horn having very low cross-polar performance," *IEEE Trans. Antennas Propag.*, vol. AP-30, no. 6, pp. 1068–1072, November 1982.

[25] G. L. James, "TE_{11}-to-HE_{11} mode converters for small-angle corrugated horns," *IEEE Trans. Antennas Propag.*, vol. AP-30, no. 6, pp. 1057–1062, November 1982.

[26] W. D. Burnside and C. W. Chuang, "An aperture-matched horn design," *IEEE Trans. Antennas Propag.*, vol. AP-30, no. 4, pp. 790–796, July 1982.

[27] K. Tomiyasu, "Conversion of TE_{11} mode by a large-diameter conical junction," *IEEE Trans. Microwave Theory Tech.*, vol. MTT-17, pp. 277–279, May 1969.

[28] B. MacA. Thomas, "Mode conversion using circumferentially corrugated cylindrical waveguide," *Electron. Lett.*, vol. 8, pp. 394–396, 1972.

[29] J. K. M. Jansen and M. E. J. Jeuken, "Surface waves in corrugated conical horn," *Electron. Lett.*, vol. 8, pp. 342–344, 1972.

[30] Y. Tacheichi, T. Hashimoto, and F. Takeda, "The ring-loaded corrugated waveguide," *IEEE Trans. Microwave Theory Tech.*, vol. MTT-19, pp. 947–950, December 1971.

[31] F. Takeda and T. Hashimoto, "Broadbanding of corrugated conical horns by means of the ring-loaded corrugated waveguide structure," *IEEE Trans. Antennas Propag.*, vol. AP-24, pp. 786–792, 1976.

[32] A. W. Love, "The diagonal horn antenna," *Microwave J.*, vol. V, pp. 117–122, March 1962.

[33] P. D. Potter, "A new horn antenna with suppressed side lobes and equal beamwidths," *Microwave J.*, pp. 71–78, June 1963.

[34] P. D. Potter and A. C. Ludwig, "Beamshaping by use of higher-order modes in conical horns," *Northeast Electron. Res. and Eng. Mtg*, pp. 92–93, November 1963.

[35] P. A. Jensen, "A low-noise multimode Cassegrain monopulse with polarization diversity," *Northeast Electron. Res. and Eng. Mtg*, pp. 94–95, November 1963.

[36] H. E. Bartlett and R. E. Moseley, "Dielguides—highly efficient low-noise antenna feeds," *Microwave J.*, vol. 9, pp. 53–58, December 1966.

[37] L. L. Oh, S. Y. Peng, and C. D. Lunden, "Effects of dielectrics on the radiation patterns of an electromagnetic horn," *IEEE Trans. Antennas Propag.*, vol. AP-18, no. 4, pp. 533–556, July 1970.

[38] G. N. Tsandoulas and W. D. Fitzgerald, "Aperture efficiency enhancement in dielectrically loaded horns," *IEEE Trans. Antennas Propag.*, vol. AP-20, no. 1, pp. 69–74, January 1972.

[39] R. Baldwin and P. A. McInnes, "Radiation patterns of dielectric loaded rectangular horns," *IEEE Trans. Antennas Propag.*, vol. AP-21, no. 3, pp. 375–376, May 1973.

[40] Y. Y. Hu, "A method of determining phase centers and its applications to electromagnetic horns," *Franklin Inst.*, vol. 271, pp. 31–39, January 1961.

[41] E. R. Nagelberg, "Fresnel region phase centers of circular aperture antennas," *IEEE Trans. Antennas Propag.*, vol. AP-13, no. 3, pp. 479–480, May 1965.

[42] I. Ohtera and H. Ujiie, "Nomographs for phase centers of conical corrugated and TE_{11}-mode horns," *IEEE Trans. Antennas Propag.*, vol. AP-23, no. 6, pp. 858–859, November 1975.

[43] J. D. Dyson, "Determination of the phase center and phase patterns of antennas," in "Radio Antennas for Aircraft and Aerospace Vehicles" (W. T. Blackband, ed.), *AGARD Conf. Proc.*, no. 15, Technivision Services, Slough, England, 1967.

[44] M. Teichman, "Precision phase center measurements of horn antennas," *IEEE Trans. Antennas Propag.*, vol. AP-18, no. 5, pp. 689–690, September 1970.

Chapter 9

Frequency-Independent Antennas

Paul E. Mayes
University of Illinois

CONTENTS

 **Paul E. Mayes** was born in Frederick, Oklahoma, on December 21, 1928. He received his PhD in electrical engineering from Northwestern University in 1955.

From 1950 to 1954 he was employed as a graduate assistant and research associate in the Microwave Laboratory at Northwestern, where his research was on electromagnetic-wave propagation along dielectric-rod waveguides and reflection of electromagnetic waves from curved conducting surfaces. Since 1954 he has been on the faculty of the Department of Electrical Engineering, University of Illinois at Urbana, where he is now a full professor teaching courses in electromagnetic theory and antennas and supervising research in the Electromagnetics Laboratory. His research at Illinois has been concerned with slot antennas, numerical electromagnetic analysis, microwave transmission lines, and frequency-independent antennas.

Dr. Mayes was awarded Certificates of Achievement in 1968 and 1969 for papers published in the *IEEE Transactions on Antennas and Propagation*. He was elected IEEE Fellow in 1975 for "contributions to the theory and development of the log-periodic antennas." He has served as a technical consultant to industry and has eleven patents on antenna inventions.

1. Basic Types

Antennas which theoretically have no limitation on the bandwidth are called *frequency independent*. In practice, the lower frequency limit is determined by the size of the antenna; the upper frequency limit, by the precision of construction. Actually, the electrical performance is not strictly independent of frequency, rather it is periodic with the logarithm of the frequency. Hence these antennas are called *logarithmically periodic*, *log-periodic*, or simply *LP* antennas. Some, which have the shape of equiangular spirals, are called *logarithmic spirals*, or *log-spirals*.

Geometrically, frequency-independent antennas are composed of a multiplicity of adjoining cells, each cell being scaled in dimensions relative to the adjacent cell by a factor which remains fixed throughout the structure. Two examples of planar LP geometries, having only a few cells, are shown in Fig. 1. In Fig. 1a, the nth cell is the annular region between concentric circles having radii R_n and R_{n+1}. In Fig. 1b the nth cell is the region between two concentric squares with sides of length L_n and L_{n+1}. If D_n represents some dimension of the nth cell and D_{n+1} the corresponding dimension of the $(n + 1)$st cell, then the relation

$$D_n/D_{n+1} = \tau$$

holds for all values of the integer n.

While it is true that the first successful, i.e., frequency-independent, LP structures were constructed from planar sheets of thin metal conductor, several of the principles of frequency-independent design are not restricted to structures with planar cells. Fig. 1c shows a conical log-spiral, for example, wherein a cell can be defined as the part of a conical surface bounded by two spheres of radii R_n and R_{n+1}. Fig. 1d shows an LP geometry in which each cell is the surface of a truncated pyramid.

The LP geometry is used to lay out an antenna by first configuring an electrical conductor within any one of the cells. The same configuration of conductor, properly scaled, is then reproduced in the other cells. If we presume this process to be repeated infinitely many times for the smaller cells, the resulting structure will converge to a point. Infinite repetition of the larger cells causes the size of the structure to increase without bound. An important observation can be made about such geometry. If any scale factor, τ^n, where n is an arbitrary integer, is applied to this geometric figure, the result is the identical geometric figure. That is to say, any LP geometry which extends to the limit point and to infinite size scales into itself whenever a scale factor of the form τ^n is applied. If we further presume that a frequency-independent point source of electromagnetic waves is located at the limit point of a perfectly conducting antenna having such geometry, the fields associated with the antenna must be identical at all frequencies that are related by τ^n. This

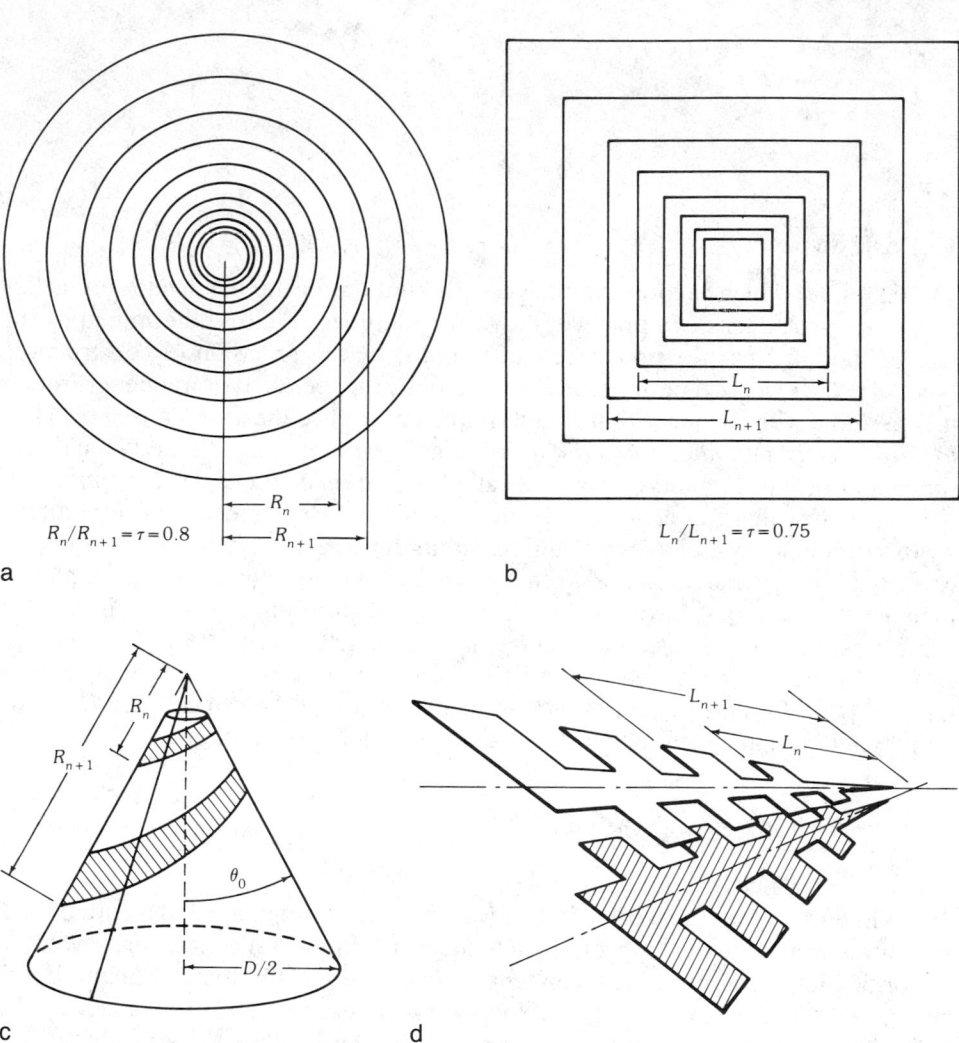

Fig. 1. Some planar and nonplanar log-periodic geometries. (*a*) Circular. (*b*) Square. (*c*) Conical. (*d*) Pyramidal.

demonstrates that the electrical performance of the antenna will be periodic with the logarithm of the frequency, as previously mentioned.

Practically speaking, there are, of course, two impossibilities in the above plan for achieving LP performance: (*a*) cells of near-zero dimension cannot be built, and (*b*) cells of near-infinite size are too large. The trick to achieving log-periodic electrical performance from an antenna with a *finite* LP geometry is associated with the choice of a cell configuration that will minimize the effects of the inevitable truncations of both large and small cells. During the early stages of development of LP antennas, many antennas having finite LP geometry were found not to have LP performance because of large truncation effects. To reduce the effect of the truncation of the large cells, the wave traveling from the excitation point along the structure must be attenuated before it encounters the end. Attenuation due to radiation is preferred over reflective attenuation in order to provide a near-constant impedance versus frequency characteristic. This requires that the cell dimensions must increase to a size sufficiently large compared with the wavelength so that appreciable radiation can occur. Generally, at least one dimension of the largest cell on a frequency-independent antenna must be approximately one-half wavelength in order to substantially eliminate the effect of the large-end truncation. It does not follow, however, that any finite LP which contains such a half-wave cell will display LP performance (zero truncation effect).

In order to approximate the condition of excitation at the limit point it is necessary to accurately scale the small cells until their dimensions are a small fraction of the wavelength at the highest frequency of operation. The conductor configuration in each cell must be capable of providing means of propagating the electromagnetic energy from the feed point toward the larger cells at all frequencies in the intended operating band.

Even though all the above conditions for self-scaling and minimal truncation effects are satisfied by a given structure, there is still no guarantee that the resulting antenna will perform with any desired degree of independence of frequency. Some means must be found to control the variations in performance over a period in frequency. For planar antennas, self-complementary geometry can be used to eliminate, at least theoretically, variations in the input impedance. An example of self-complementary geometry is shown in Fig. 2.

In Fig. 2 the planar cells are partitioned by four sinuous radial lines, *OA*, *OB*, *OC*, *OD*, identical except for rotation by multiples of 90°. The areas between *OA*

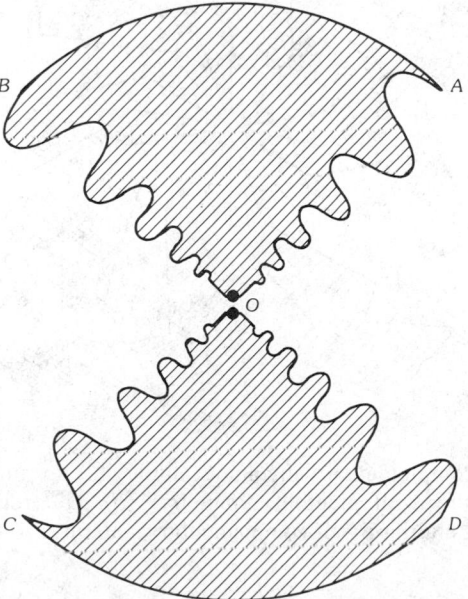

Fig. 2. Example of self-complementary geometry.

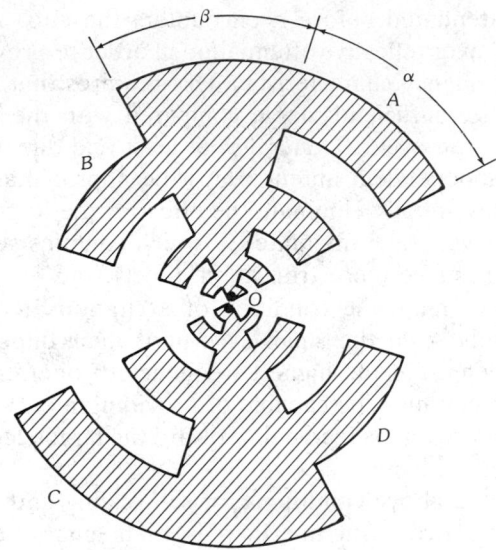

Fig. 3. Self-complementary geometry with flat-top radial lines. (*Adapted from Carrel [8]*)

a

Fig. 4. Multiarm self-complementary antenna example and terminal impedances. (*a*) Self-complementary structure with ninefold symmetry. (*b*) Theoretical values of terminal impedance for several multiarm self-complementary antennas. (*After Deschamps [2],* © *1959 IEEE*)

and OB and between OC and OD are conducting sheet. The areas between OD and OA and between OB and OC have the identical shape but are free space. The shape of the radial lines can take many different forms. Fig. 3 shows a case where the four radial lines, OA, OB, OC, and OD, are properly scaled flat-topped pulses.

The self-complementary property of the structures of Figs. 2 and 3 results from the 90° separation between the radial boundary lines of the cells. Using four such lines produces two conducting regions emanating from the origin. Introducing a small gap at the limit point provides two input terminals for the antenna. The conductor is thus divided into two arms that are symmetrically driven by a generator connected to the terminals. The theoretical value for the input impedance of a two-arm, self-complementary antenna with no truncation effect is $60\pi = 189\ \Omega$ for any frequency [1]. Log-periodic antennas can also be constructed with more than two arms as shown in Fig. 4a. The theoretical values of the terminal impedance for several multiarm self-complementary antennas with various inter-

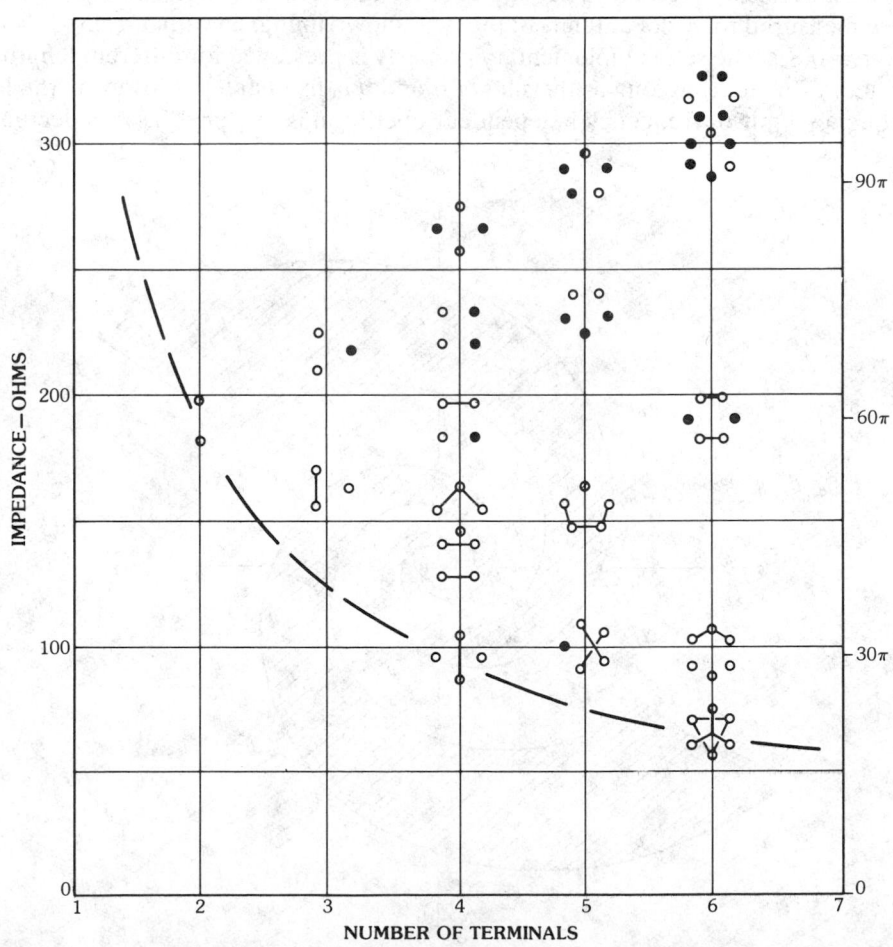

b

Fig. 4, *continued.*

connections are given in Fig. 4b [2]. In this figure, for each configuration the two groups of terminals connected to the source are represented by small circles. The floating terminals are represented by black dots. Measured values differ somewhat from the theoretical ones due to the finite thickness of the conductors, presence of a feed line, etc. [3].

Fig. 5 shows a different sort of boundary for the cells. In this case the edges of the conductor are defined by logarithmic spirals, each rotated 90° with respect to its neighbor. Usually log-spiral antennas are circularly polarized whereas other log-periodic antennas are linearly polarized. Log-spiral antennas are discussed in detail in Section 4.

Planar LP antennas can be truncated either by eliminating all conductor beyond a certain distance from the limit point or by filling the plane with conductor beyond that distance. In the latter case the antenna can be considered to be a slot antenna.

Symmetry dictates that the radiation occur in the same way on both sides of planar antennas. Patterns typical of planar LP antennas are shown in Fig. 6. These were measured for a slot antenna of the type shown in Fig. 6a with $\alpha = 45°$, $\beta = 45°$, and $\tau = 0.81$. The self-complementary property is preserved for different lengths of the teeth (angle α) as long as the sum of α and β is fixed at 90°. However, the low-frequency limit of frequency-independent operation is increased as α is decreased

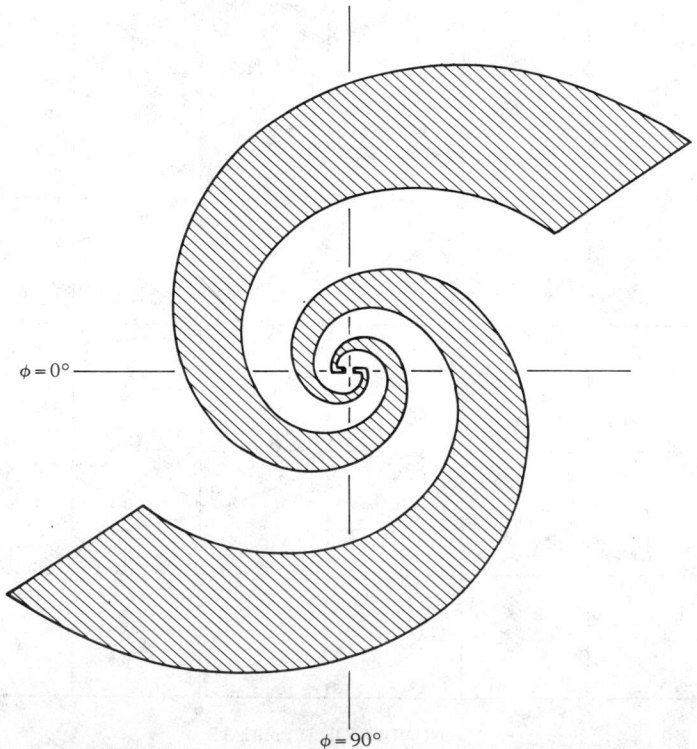

Fig. 5. Logarithmic-spiral geometry. (*After Carrel [8]*)

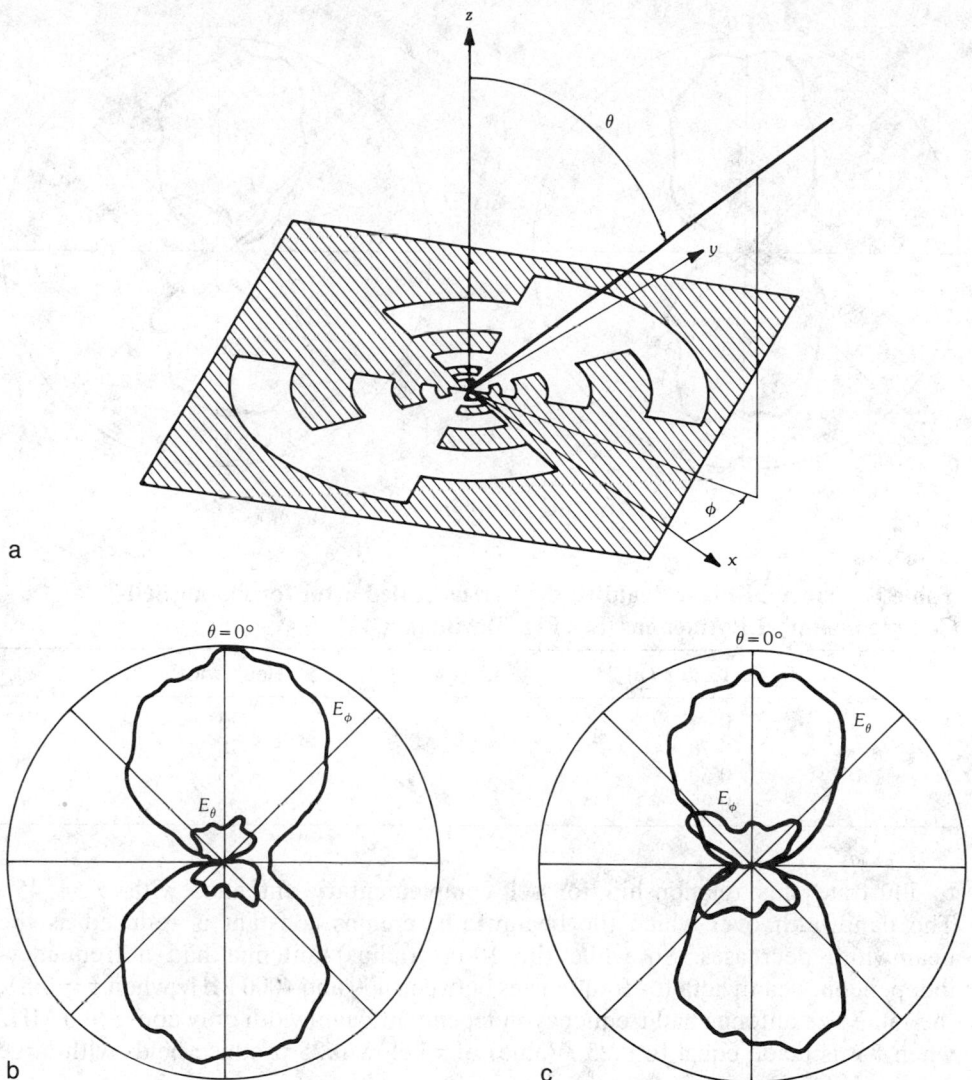

Fig. 6. Typical principal plane radiation patterns. (*a*) Antenna orientation. (*b*) For $\phi = 0°$ and $f = 1530$ MHz. (*c*) For $\phi = 90°$ and $f = 1530$ MHz. (*d*) For $\phi = 0°$ and $f = 1700$ MHz. (*e*) For $\phi = 90°$ and $f = 1700$ MHz. (*After DuHamel and Isbell [3], © 1957 IEEE*)

for a given maximum antenna radius. For example, an antenna with a maximum radius of 10 in had a low-frequency limit of approximately 400 MHz when $\alpha = 45°$, but the low-frequency limit was raised to 800 MHz when α was changed to 20°. It is thus established that the low-frequency limit is determined by the length of the longest tooth.

The principal-plane beamwidths of antennas of the type shown in Fig. 3 can be controlled to some extent by changing the scale factor τ. Table 1 gives data

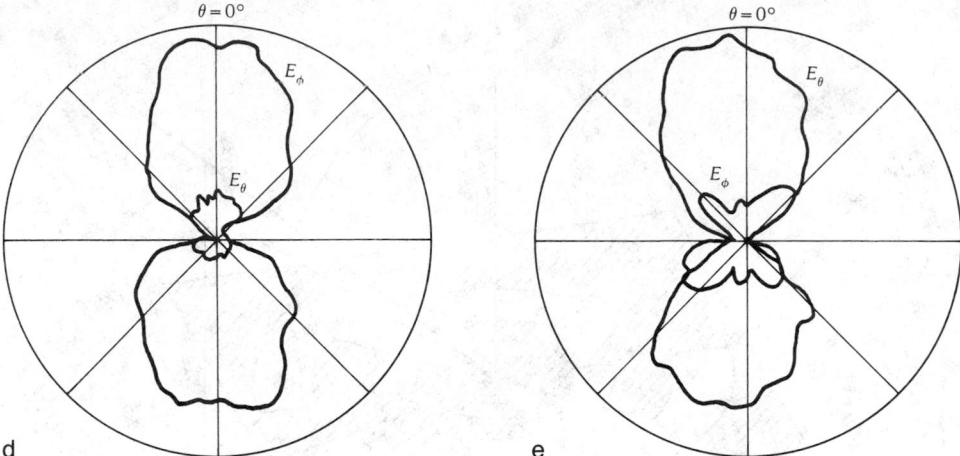

Fig. 6, *continued.*

Table 1. Principal-Plane Beamwidths Versus Scale Factor for Planar Self-Complementary LP Antennas (See Fig. 3) with $\alpha = 45°$

Scale Factor (τ)	Beamwidth
0.81	73°
0.70	70°
0.50	55°
0.25	38°

to illustrate this relationship for self-complementary antennas with $\alpha = 45°$. The bandwidth over which the beamwidth remains constant is reduced as the beamwidth decreases. So while the 10-in (radius) antenna had a frequency-independent beamwidth for frequencies between 400 and 4000 MHz when $\tau = 0.81$, the same size antenna had frequency-independent beamwidth only above 900 MHz when τ was made equal to 0.25. Values of τ below 0.25 produce fields with large cross polarization.

The level of the mean impedance can be reduced by increasing the angle α, e.g., for $\alpha = 75°$ the mean impedance was observed to be approximately 100 Ω. However, the variation of impedance with frequency for such antennas, which are not self-complementary, is greater than it is for antennas which are self-complementary.

For most applications bidirectional patterns, such as those of Fig. 6, are not acceptable. The requirement instead is for a single beam. The symmetry of planar antennas must be eliminated to accomplish this. Planar LP slot antennas can be made unidirectional in a frequency-independent manner by placing an absorbing cavity on one side of the ground plane. For cavities which are sufficiently large the impedance and the patterns in the remaining hemisphere are affected very little by the cavity. Of course, the gain is less by about 3 dB than that of a unidirectional antenna having the same pattern.

When the two arms of a metal-arm LP antenna are inclined toward each other as shown in Fig. 7, unidirectional patterns can result [4]. Fig. 8 shows E-plane patterns measured for the antenna shown in Fig. 7 with $\tau = 0.81$, $\alpha = \beta = 45°$, and $\phi = 90°$. As the angle between the planar elements, ψ, is reduced, one lobe of the pattern decreases in magnitude. For $\psi < 50°$ this lobe is practically nonexistent. The lobe which remains is in the direction from the larger to the smaller end of the structure. Behavior of the nonplanar LP antenna is thus much different than for wire-type vee antennas or horns that radiate in the direction of increasing size. Fig. 9 shows that the mean impedance drops and the variation in impedance over the band increases as the angle between the two arms decreases.

Log-periodic antennas with arms constructed from planar sheet metal may be practical for frequencies high enough so that the physical structure is satisfactory mechanically. However, there are many applications for frequency-independent antennas in long-distance communications which are conducted in the hf (3–30 MHz) band. In this band the antenna must be so large that construction from planar sheet metal is no longer mechanically feasible. Fortunately, much of the conductor can be removed from the arms without appreciably affecting the radiation patterns. The proper procedure is to retain the conductor on the edges of the teeth in the manner illustrated in Fig. 10 [5]. Not only does this result in a structure that is practical for use at high frequency, it also reduces the capacitance between the arms and lessens the amount that the impedance varies with frequency within the operating band. This makes it possible to bring the two arms of the

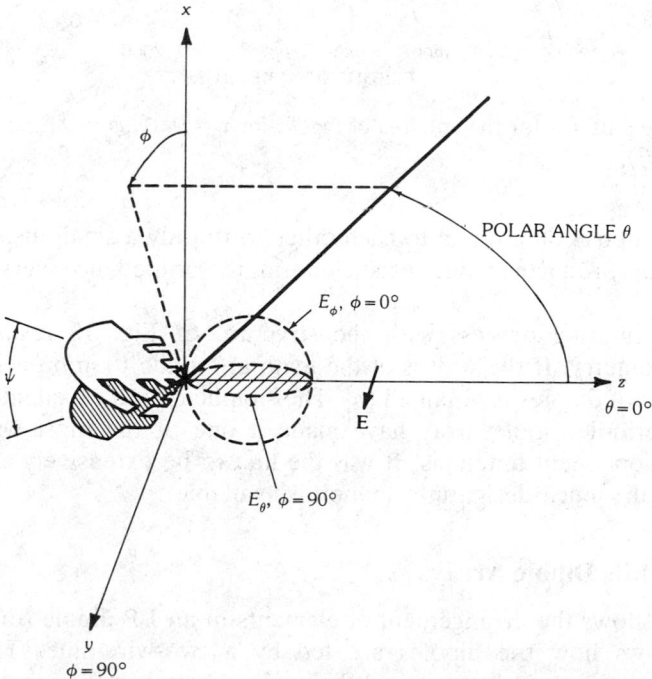

Fig. 7. The first unidirectional LP antenna showing the backfire beam. (*After Isbell [4]*)

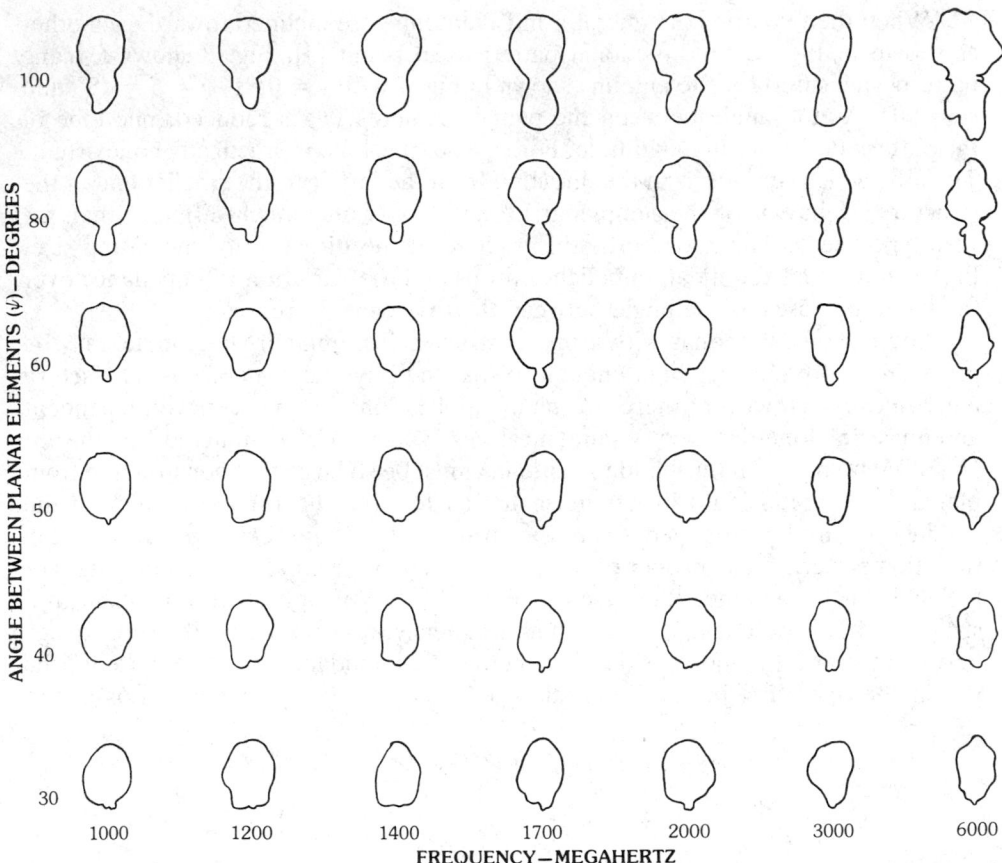

Fig. 8. *E*-plane patterns for the antenna of Fig. 7, for $\tau = 0.81$, $\phi = 90°$, and $\alpha = \beta = 45°$. (*After Isbell [4]*)

antenna into a position parallel to each other, with only a small distance between them, without producing wide excursions in the impedance versus frequency behavior.

Another limiting process with the structure of Fig. 10 results in a very important geometry. If the widths of the teeth are made to approach zero, a log-periodic array of dipoles is obtained [6]. The simplicity and excellent performance of the log-periodic dipole array have made it one of the most widely used of frequency-independent antennas. It was the first to be extensively analyzed [7, 8] and, as a result, much design information is available.

2. Log-Periodic Dipole Arrays

Fig. 11a shows the arrangement of elements in an LP dipole (or LPD) array. Fig. 11b shows how the dipoles are fed by a two-wire line. The two basic parameters are the scale factor τ and the angle α between the centerline and the tips of the dipoles. The alternative parameter, σ, is an "aspect ratio" for each cell

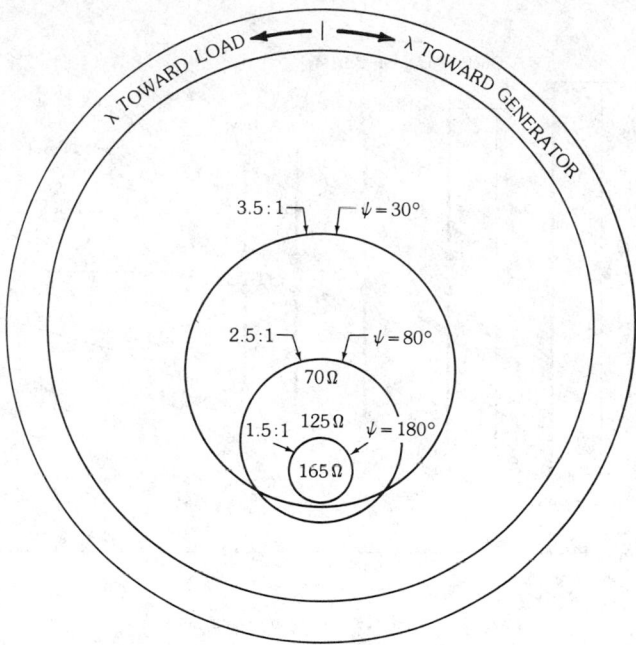

Fig. 9. Boundary circles of the impedance loci as a function of ψ. [*After Isbell [4]*)

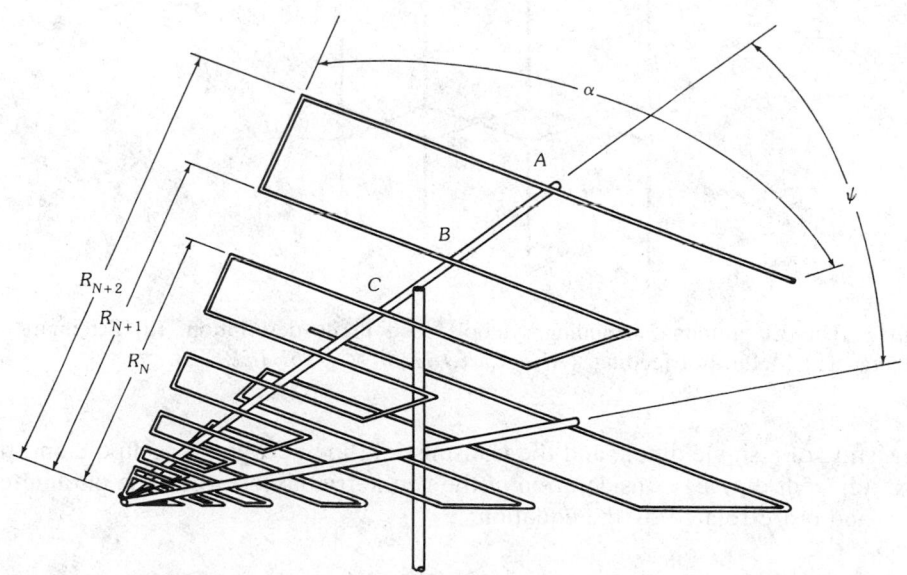

Fig. 10. A wire-outline LP antenna. (*Courtesy Collins Defense Communications, Rockwell International Corporation*)

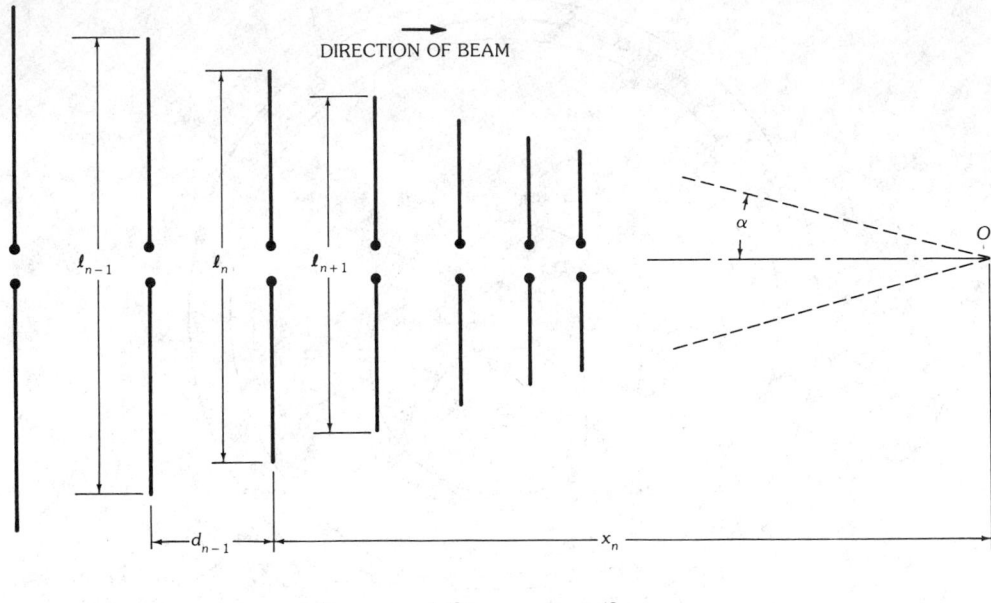

$$x_n/x_{n-1} = \ell_n/\ell_{n-1} = \tau \qquad d_n/2\ell_n = \sigma \qquad h_n = \ell_n/2$$

a

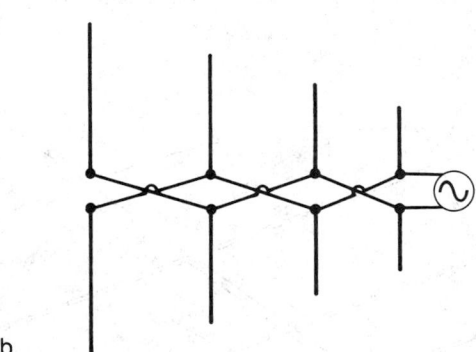

b

Fig. 11. The LP antenna, including symbols used in its description. (*a*) Schematic of antenna. (*b*) Method of feeding. (*After Carrel [7, 8], © 1961 IEEE*)

consisting of a single dipole and the transmission line between that dipole and the next adjacent dipole. Consideration of the geometry discloses that the parameters τ, α, and σ are related by the equation

$$\sigma = \tfrac{1}{4}(1 - \tau)\cot \alpha \tag{1}$$

The nomograph given in Fig. 12 provides a convenient method for transforming from one set of parameters to another, which may occur frequently during the design process.

The transposition of conductors of the feed line is essential to the frequency-

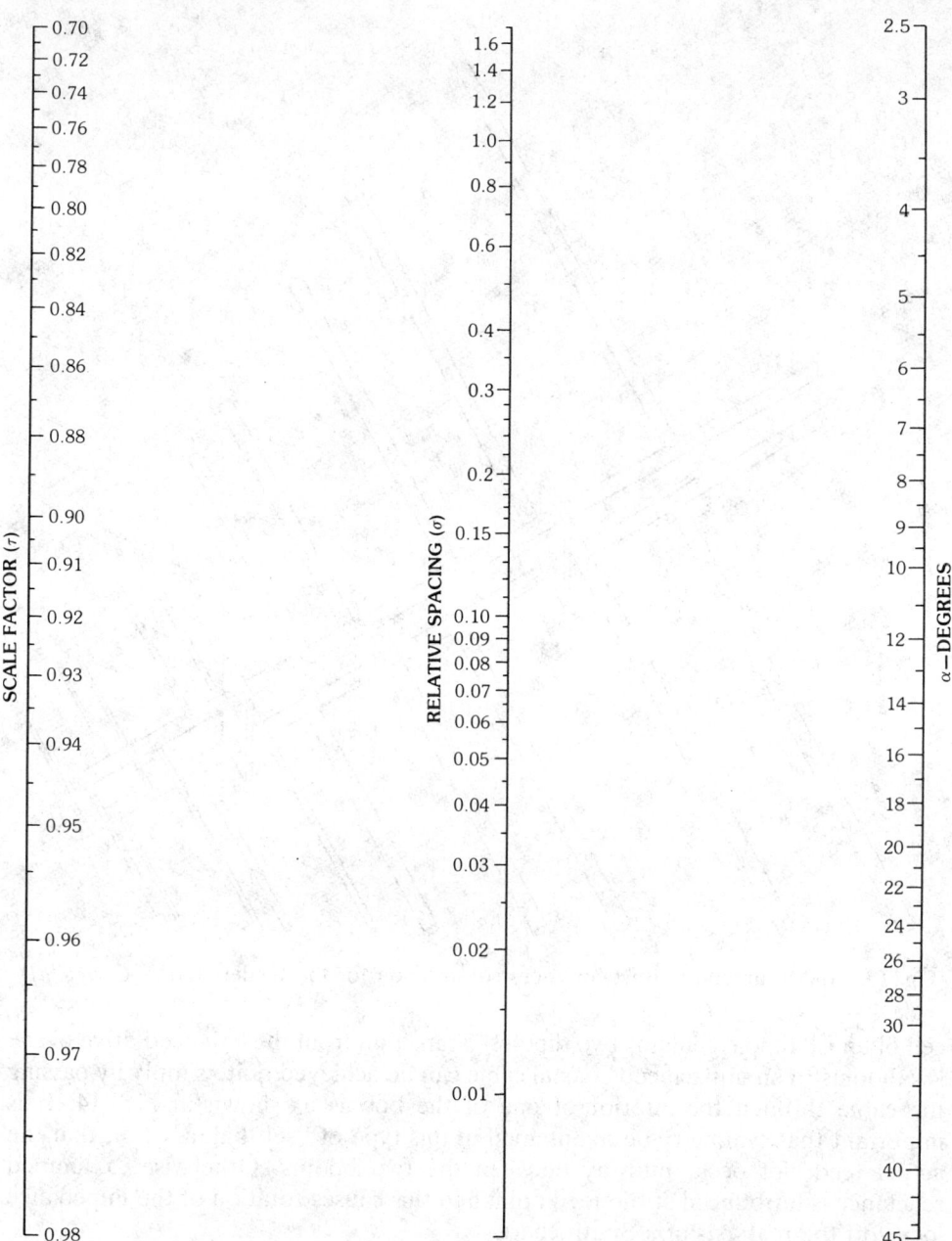

Fig. 12. Nomograph for $\sigma = (1/4)(1 - \tau)\cot\alpha$. (*After Carrel [7, 8], © 1961 IEEE*)

independent performance of LP dipole arrays. One way to achieve the transposition, and provide at the same time mechanical support for the dipoles, is to use twin-boom construction for the feeder, as shown in Fig. 13. It should be noted that, following Carrel, the definition of τ for an LP dipole entails the ratio of the lengths of two adjacent dipoles. However, when the twin-boom construction is used, each

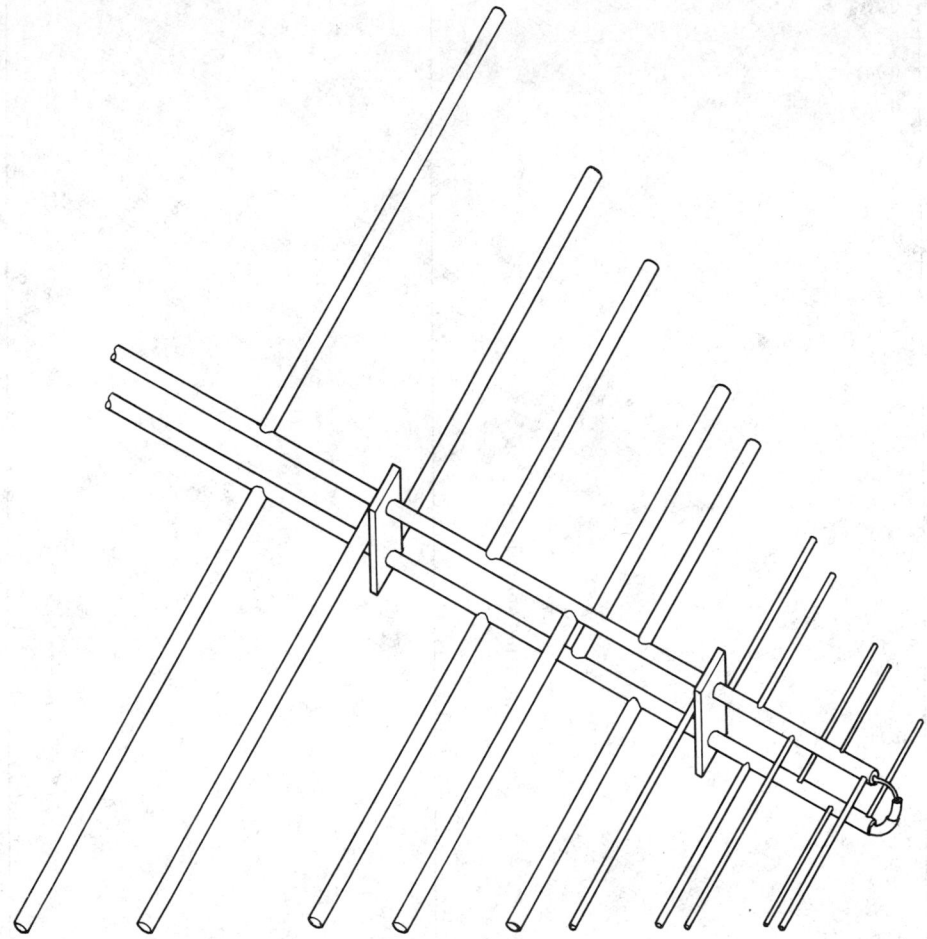

Fig. 13. An LP antenna with twin-boom construction for the feeder. (*After Carrel [8]*)

cell of an LP dipole contains two dipoles. Transition from the balanced drive of the two booms to an unbalanced coaxial cable can be achieved quite simply by passing the cable through the interior of one of the booms as shown in Fig. 14. It is important that symmetry be maintained in this type of "self-balun," i.e., that the actual feed slot occur midway between the two booms. Otherwise, a lumped reactance is introduced at the feed point and this causes rotation of the impedance locus off the real axis of a Smith chart.

    Analysis of LP dipole arrays proceeds by separating the dipoles and transmission line as shown in Fig. 15 [7,8]. The terminal properties of the N-element network of dipoles can be represented by an $N \times N$ impedance matrix. Carrel used the formulas of the induced emf method to calculate the impedance matrix. Later work [9–13] used moment methods. The results of the several techniques are not appreciably different [14].

    The N-terminal-pair network of the feed line can be readily represented by an admittance matrix. By enforcing continuity of the current at all except the

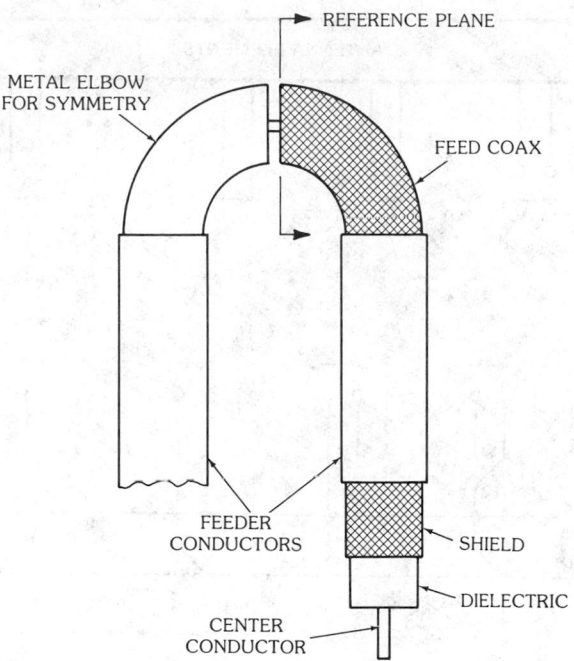

Fig. 14. Details of the symmetrical feed point, showing the reference plane for impedance measurements. (*After Carrel [8]*)

generator terminals when the feeder and antenna networks are paralleled, a set of equations for analysis of the LP dipole results. Fig. 16 shows amplitude and phase of the terminal currents in the several dipoles of an LPD array at the frequency f_3 where the third dipole (counting from the large end) is one-half wavelength. It is to be noted that the input currents of a few dipoles are significantly larger than those of any of the others. These dipoles with highest current are collectively called the *active region*, and they produce most of the radiated field.

When frequency is changed, the active region will move along the axis of the LP dipole in such a way that the dimensions of the active region in wavelengths remain almost constant. For this reason the radiation pattern is insensitive to the frequency changes. Fig. 17 shows the relative amplitude of the input currents at several different frequencies, illustrating the movement of the active region. It is apparent from these results that the pattern will begin to change when the active region encounters either of the two truncation points. The band of frequency-independent patterns is therefore somewhat less than the ratio of longest-to-shortest dipole lengths. This ratio, called the *structure bandwidth*, is given by

$$B_s = h_1/h_N = \tau^{1-N} \tag{2}$$

The operating bandwidth B and the structure bandwidth B_s are related by

$$B = B_s/B_{ar} \tag{3}$$

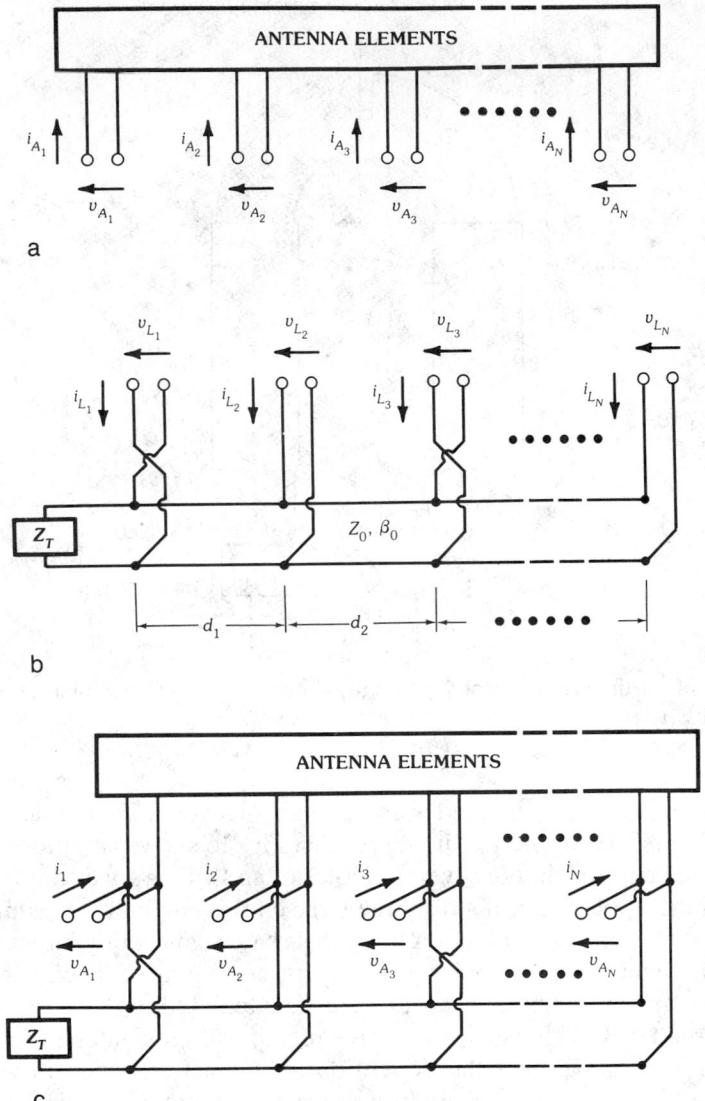

Fig. 15. Schematic circuits for the LP interior problem. (*a*) Element circuit. (*b*) Feeder circuit. (*c*) Complete circuit. (*After Carrel [7, 8], © 1961 IEEE*)

where the term B_{ar} is called the bandwidth of the active region and is dependent on the parameters of the dipole array. Fig. 18 summarizes the results of analyzing many different LP dipole arrays. For $\tau > 0.85$, B_{ar} is almost a linear function of τ. The lines on Fig. 18 show the excellent agreement between the computed data and the empirical formula

$$B_{ar} = 1.1 + 7.7(1 - \tau)^2 \cot \alpha \qquad (4)$$

The nomograph of this relation in Fig. 19 can be used to facilitate design.

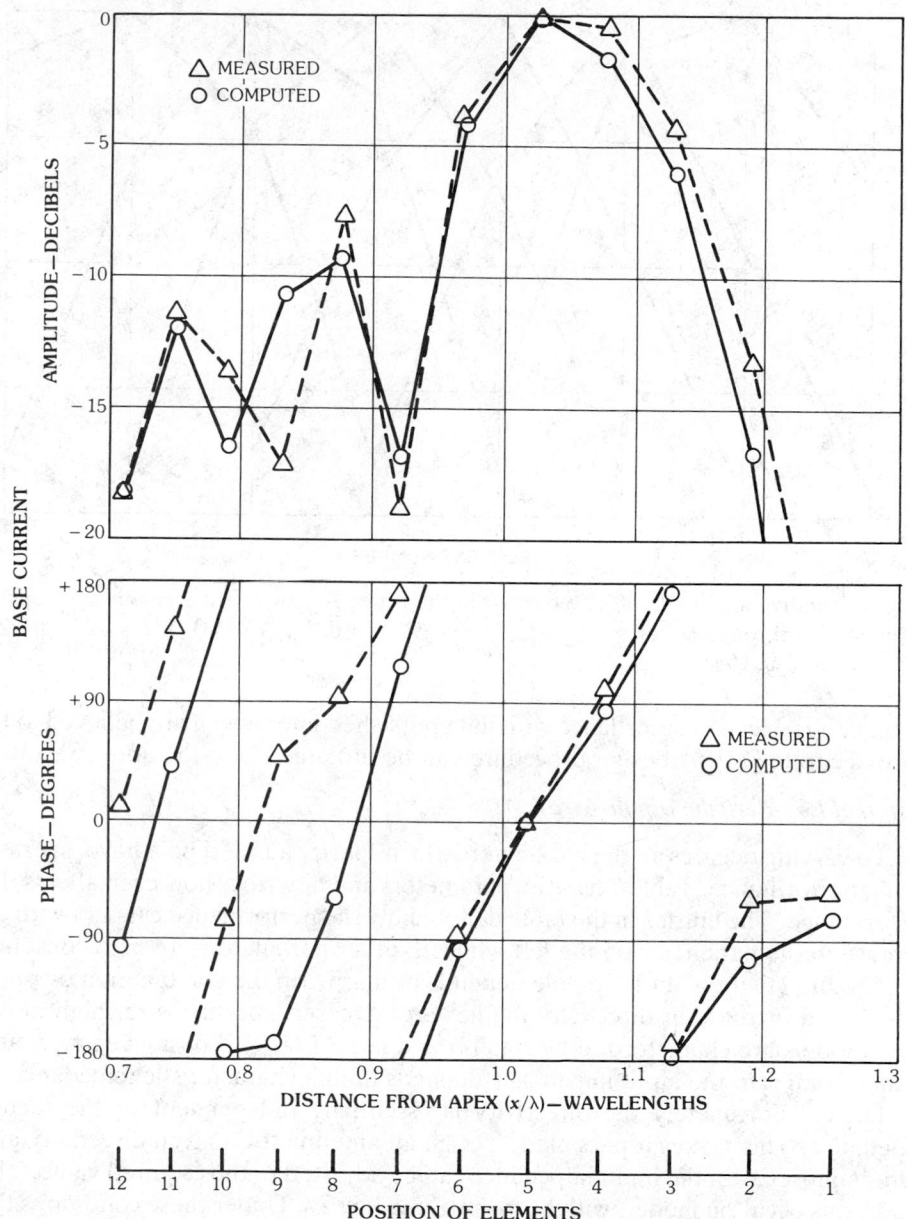

Fig. 16. Amplitude and phase of the element base current versus relative distance from the apex, at frequency f_3, for $\tau = 0.95$, $\sigma = 0.0564$, $Z_0 = 100\ \Omega$, $h/a = 177$, and Z_T a short circuit at $h_1/2$. (*After Carrel [8]*)

Once the dipole base currents have been determined, the far fields can be calculated and, from them, the directivity. These data can be represented as plots of constant directivity contours on τ and σ axes as presented in Fig. 20. The contours shown are those first reported by Carrel. The directivity values, however,

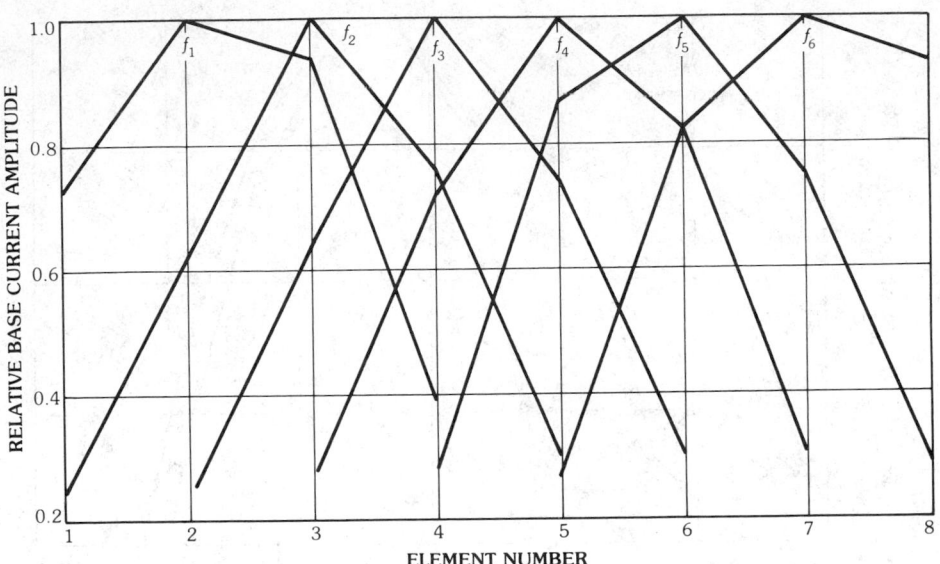

Fig. 17. Relative amplitude of base current in the active region versus element number for frequencies f_1 through f_6, for $\tau = 0.888$, $\sigma = 0.089$, $N = 8$, $Z_0 = 100 \ \Omega$, $h/a = 125$, and Z_T a short at $h_1/2$. (*After Carrel [8]*)

have been reduced in accordance with data published later which are believed to be more accurate [14]. A design procedure can be initiated using the above results.

Design of Log-Periodic Dipole Arrays

To varying degrees all the parameters which specify an LP dipole have an effect on the performance. Table 2 lists the parameters and describes how each affects the performance. The entries in the table denote how the performance changes with an increase in the parameter on the left while all other parameters are held constant.

The directivity of an LP dipole depends primarily on the combination of τ and σ. Since an increase in directivity implies an increased aperture size, high-directivity models are characterized by small α and large L/λ_{max}. For a given τ, σ, and element radius a, the input impedance depends on the characteristic impedance of the feeder. Fortunately the directivity is essentially independent of the feeder impedance. This makes it possible to design an antenna for a given directivity and then, in most cases, the input impedance can be adjusted to the required value. The exceptions occur on models with both low τ and low Z_0. Under these conditions the radiating efficiency of the active region is low, and end effects appear. Input impedances from 50 Ω (for high values of τ) to 200 Ω (for all values of τ) have been obtained.

For most applications the objective is to achieve a given power gain and input impedance over a given frequency band. These specifications do not determine a unique design. The relative importance of minimizing the number of elements or the size of the antenna must also be considered. The number of elements is determined by τ; as τ increases, the number of elements increases. The antenna size is determined by the boom length (the distance between the longest and the

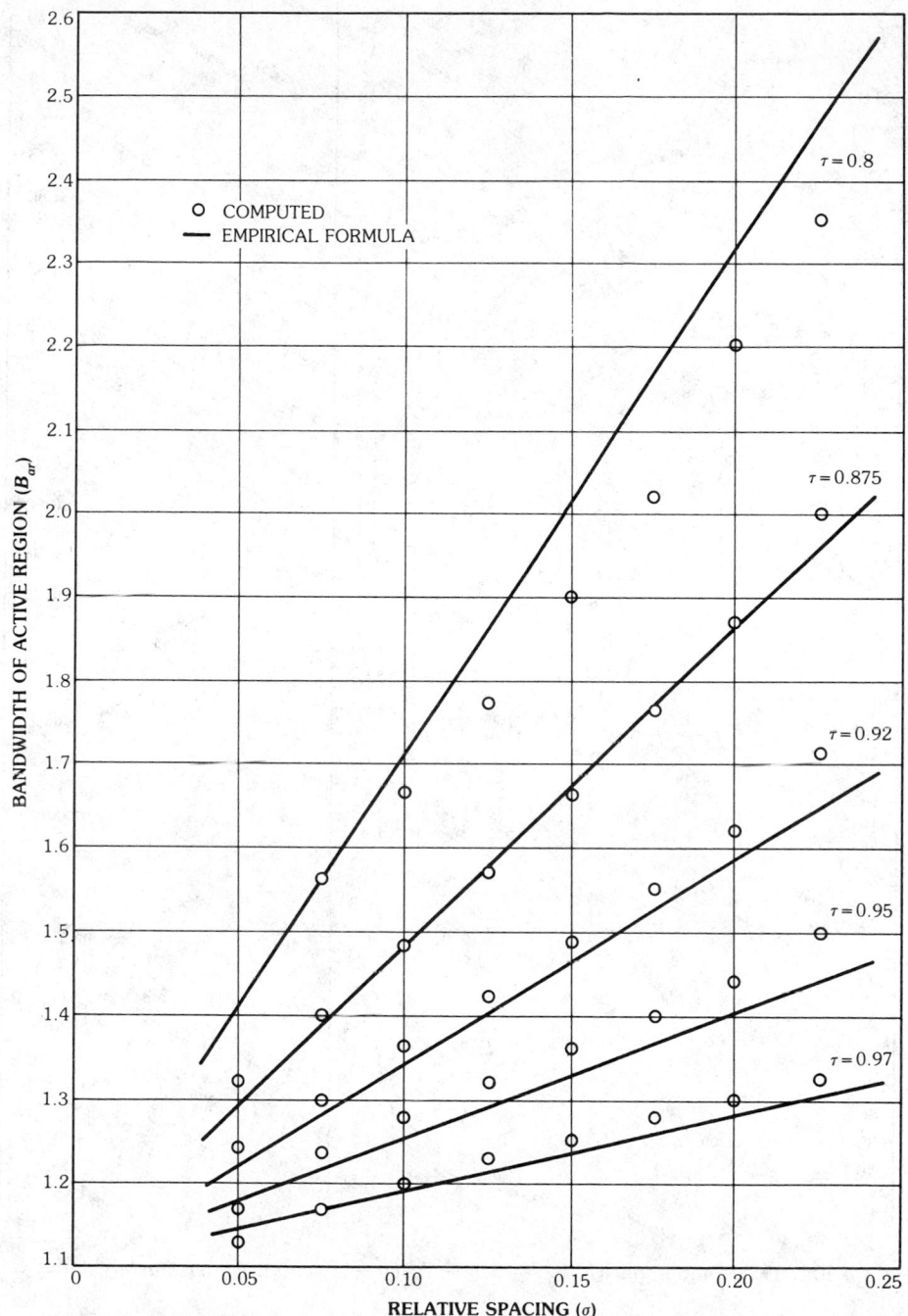

Fig. 18. Bandwidth of the active region versus σ and τ. (*After Carrel [8]*)

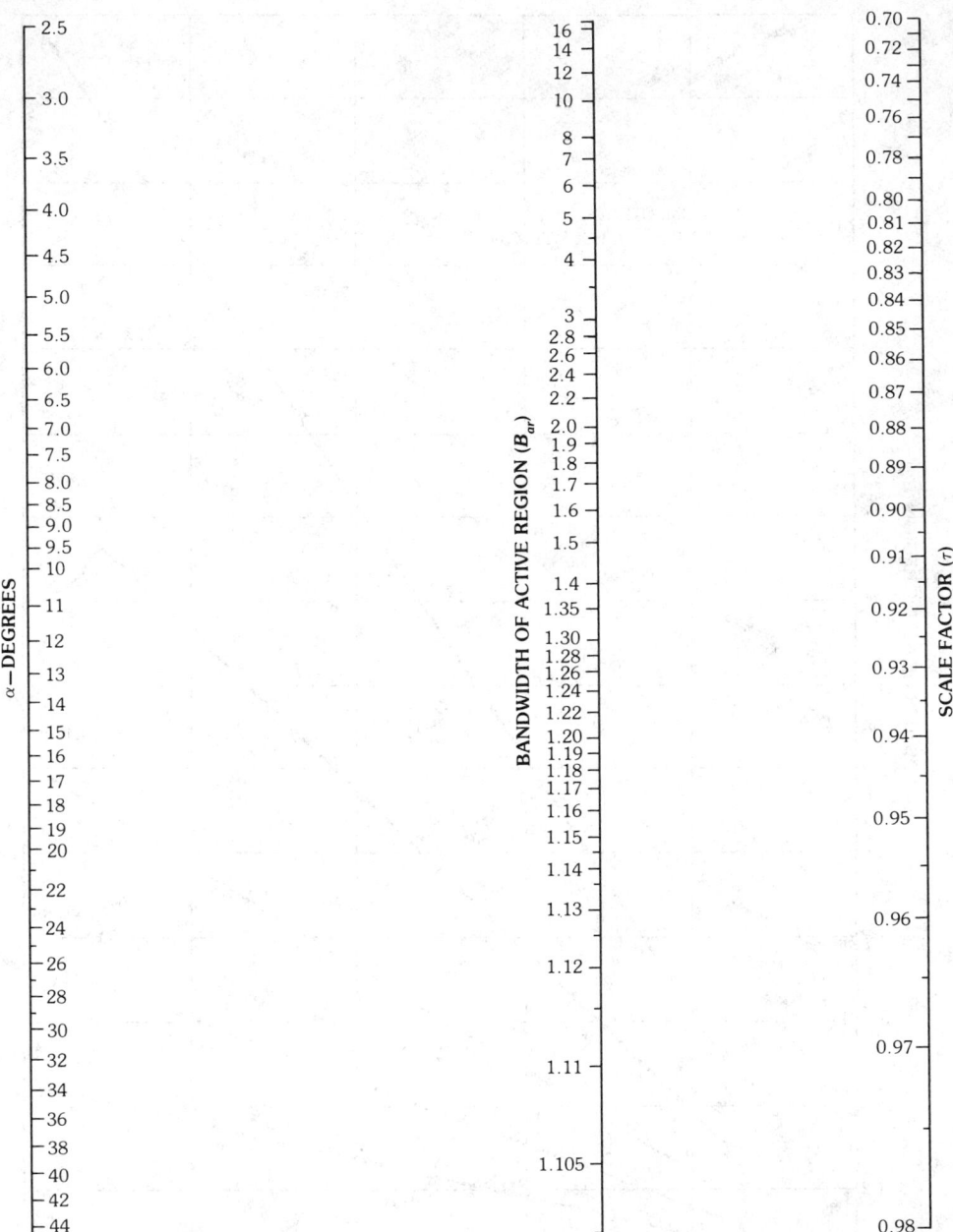

Fig. 19. Nomograph for $B_{ar} = 1.1 + 7.7(1 - \tau)^2 \cot \alpha$. (*After Carrel [8]*)

shortest dipoles), which depends primarily on α. As α decreases, the length increases.

(*a*) Given a value of directivity (desired maximum power gain divided by a reasonable value for efficiency), a set of values for τ and σ can be determined from Fig. 20. Since there are many combinations of τ and σ that will work, it is well to

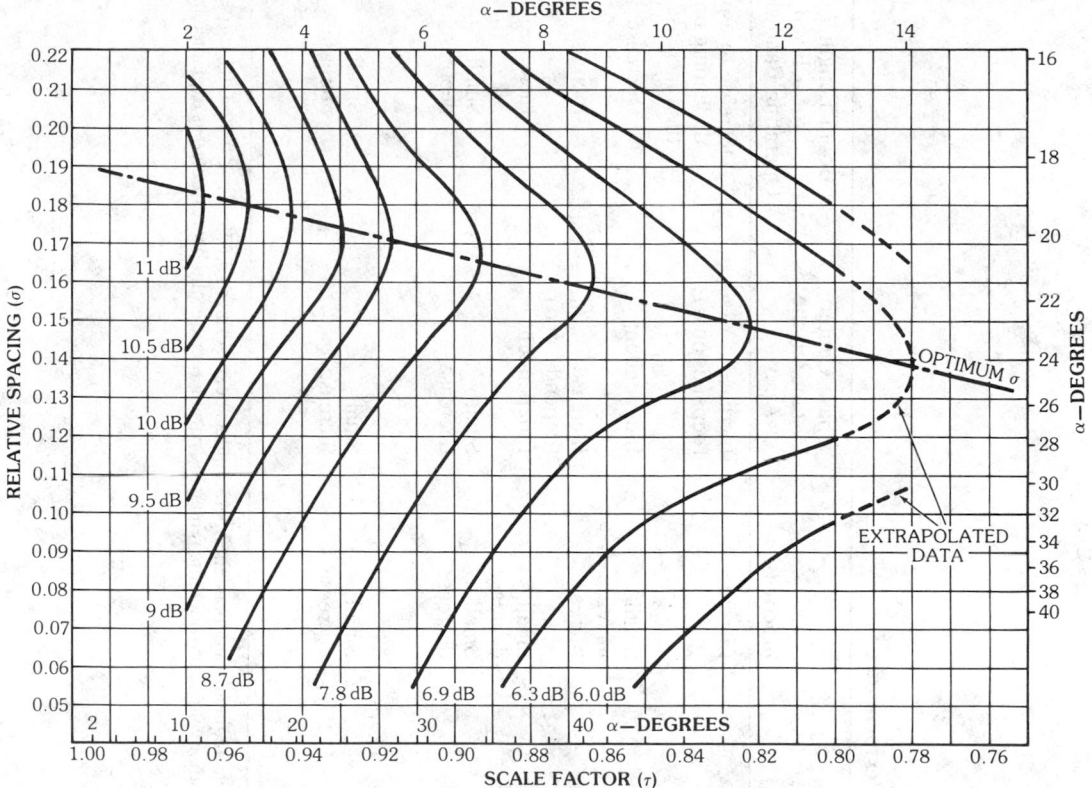

Fig. 20. Computed contours of constant directivity versus τ, σ, and α, for $Z_0 = 100\ \Omega$, $h/a = 177$, and Z_T a short at $h_1/2$. (*Adapted from Carrel [7,8], © 1961 IEEE*)

keep in mind that, for a given bandwidth, a large value σ (small α) will give a very long antenna, and that a large value of τ will give many elements.

(*b*) The value of α for the selected values of τ and σ can be determined from the formula

$$\tan\alpha = (1 - \tau)/4\sigma \tag{5}$$

or read from the nomograph in Fig. 12.

(*c*) Fig. 18 or the nomograph in Fig. 19 can be used to determine the bandwidth of the active region.

(*d*) The required structure bandwidth is next found by using (3).

(*e*) The nomograph of Fig. 21 or the formula

$$L/\lambda_{\max} = \tfrac{1}{4}(1 - 1/B_s)\cot\alpha \tag{6}$$

gives the required length of the array in terms of the wavelength $\lambda_{\max}$ of the low-frequency limit.

(*f*) The formula

Table 2. LP Dipole Parameters and Their Effect on the Observed Performance

LP Dipole Parameter*	Bandwidth of Active Region (B_{ar})	Input Impedance (always less than Z_0)	Directivity	Phase Center Distance to the Apex x_ρ	Boom Length L/λ_{max} for a Fixed Operating Bandwidth B
τ (σ constant)	Decrease	Small decrease	Increase	Increase (depends on α)	Decrease to a point depending on B, then increase
τ (α constant)	Decrease	Small decrease	Small increase	Independent	Decrease
σ (τ constant)	Increase	Increase	Increase	Increase (depends on α)	Increase
σ (α constant)	Increase	Increase	Small decrease	Independent	Increase
z_0	Independent but location of AR moves toward apex	Increase	Small decrease	Small decrease	Small decrease
h/a	Independent, but location of AR moves away from apex	Increase	Small decrease	Small increase	Small increase

*The table entries hold true over the following range of parameters for which frequency-independent operation has been verified: $0.875 < \tau < 0.98$, $0.05 < \sigma < \sigma_{optimum}$, $100 < Z_0 < 500$, and $20 < h/a < 10000$. Any one of τ, σ, or Z_0 may take on other values, provided the remaining parameters are suitably restricted as explained in the text.

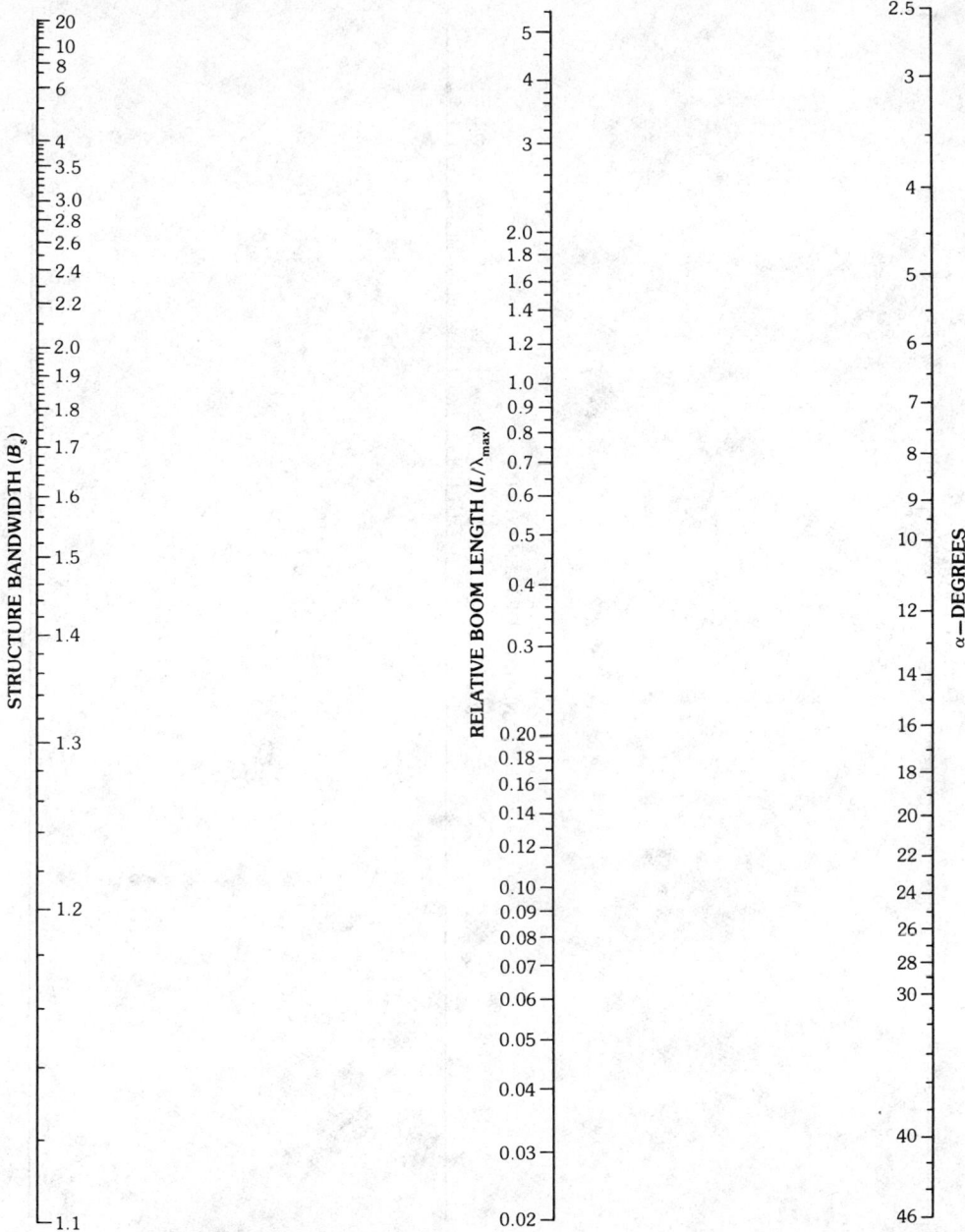

Fig. 21. Nomograph for $L/\lambda_{max} = (1/4)(1 - 1/B_s)\cot\alpha$. (*After Carrel [7, 8], © 1961 IEEE*)

$$N = 1 + (\log B_s)/[\log(1/\tau)] \qquad (7)$$

or the nomograph of Fig. 22 can be used to determine the number of elements required to cover the desired band.

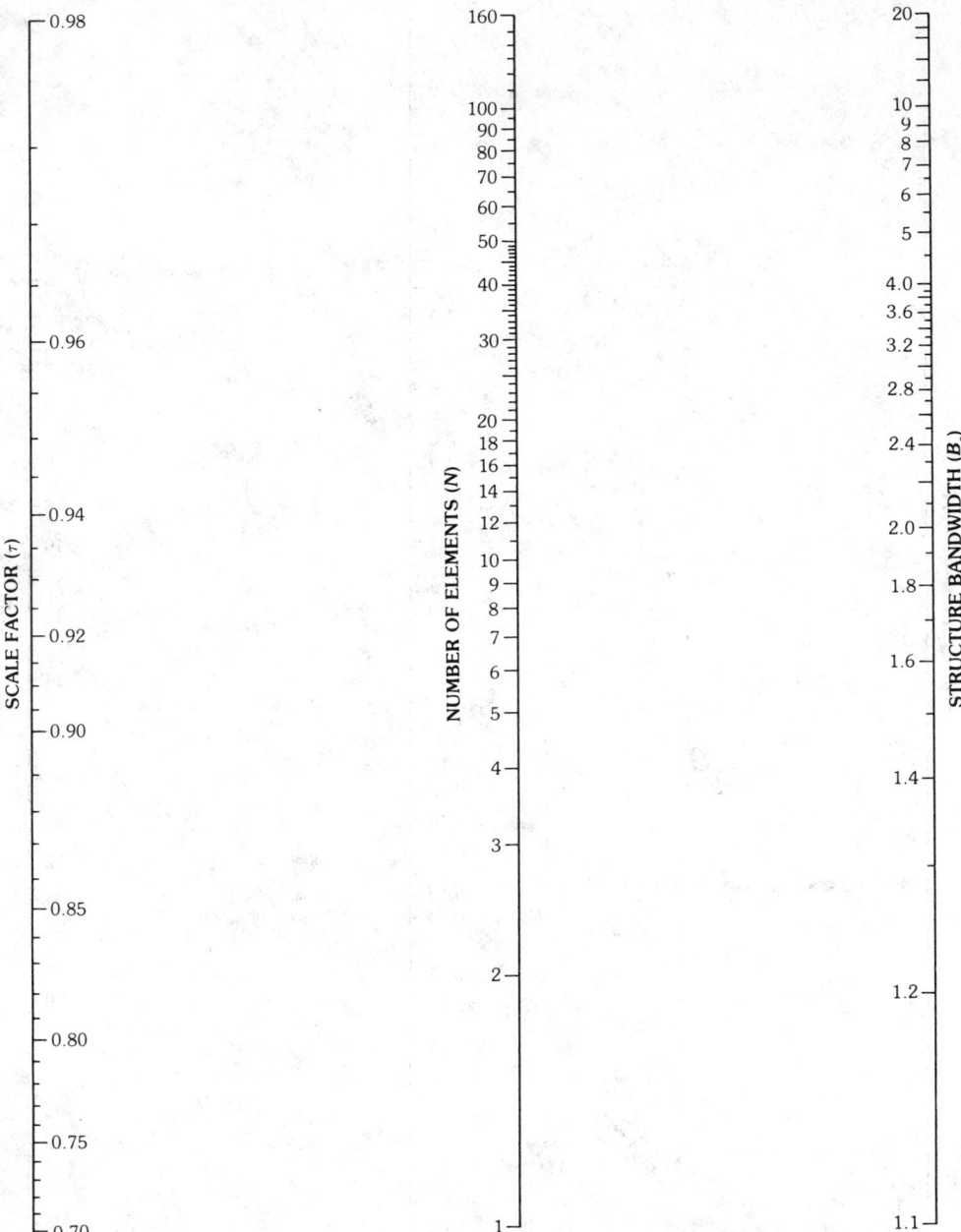

Fig. 22. Nomograph for $N = 1 + (\log B_s)/(\log \tau^{-1})$. (*After Carrel [7, 8]*, © 1961 IEEE)

(g) If either the boom length or the number of elements, but not both, is too large, return to step *a* and select a different τ, σ pair. Repeating the process for different values of τ and σ will establish the trend. For some values of B_s minimum boom length cannot be attained for values of τ and σ within the range of the graph of Fig. 20. In these cases a compromise will have to be made. If a choice of τ and σ exists and there is no apparent basis for a specific selection, it should be noted that the swr increases and the front-to-back decreases as σ departs from the optimum values designated on Fig. 20.

(h) Since the graph of Fig. 20 is based on $Z_0 = 100 \ \Omega$ and $h/a = 177$, an adjustment should be made if it is anticipated that Z_0 and h/a will depart from these values by more than a factor of 2. The exact value of Z_0 is yet to be determined. However, it is known that the feeder impedance is always greater than the input impedance, so if R_0 is greater than $100 \ \Omega$, the directivity contours of Fig. 20 will read a fraction of a decibel high. If the ratio h/a is much different from 177, another adjustment must be made. According to the curve of Fig. 23 the directivity decreases by about 0.1 dB for each doubling of h/a; for $h/a > 177$ the constant directivity contours of Fig. 20 will read high.

(i) Once a design is achieved that has satisfactory patterns, attention can be given to obtaining a desired value for the input impedance. The achievable minimum swr will be dependent on the scale factor τ as shown in Fig. 24. If the required value of swr is lower than achievable with the value of τ already picked for the desired gain, it will be necessary to start the design procedure again with a new value of τ picked on the basis of impedance performance rather than directivity. It should also be noted from Fig. 24 that low values of τ produce low swr only for a narrow range of σ.

(j) Fortunately, the mean value of the impedance of an LP dipole is adjustable over a rather wide range by merely changing the characteristic impedance of the feeder. Fig. 25 illustrates this point for an LP dipole having $\tau = 0.888$, $\sigma = 0.089$, and $h/a = 125$. As Fig. 25 shows, the mean value of the input impedance, R_0, is less than the characteristic impedance of the feeder, Z_0. This is due to the shunt capacitance that is added to the feeder by the presence of the short (compared to

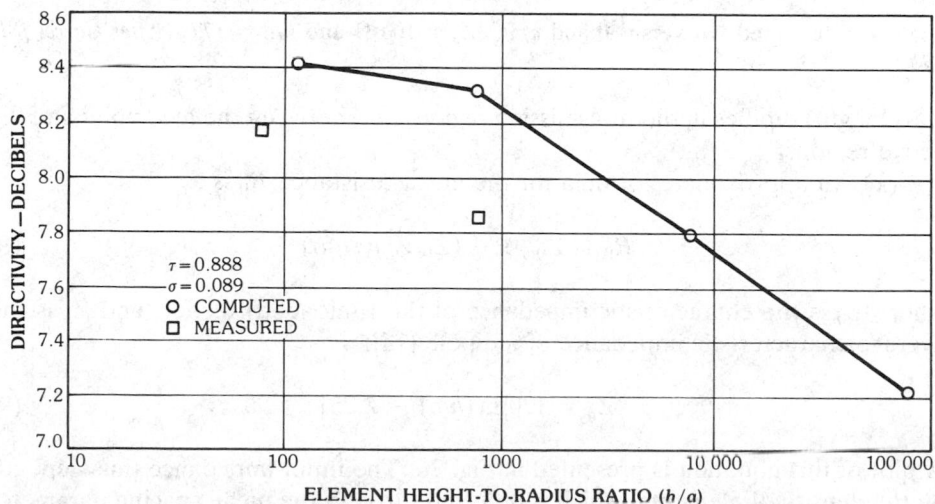

Fig. 23. Example of computed and measured directivity versus height-to-radius ratio. (*After Carrel [8]*)

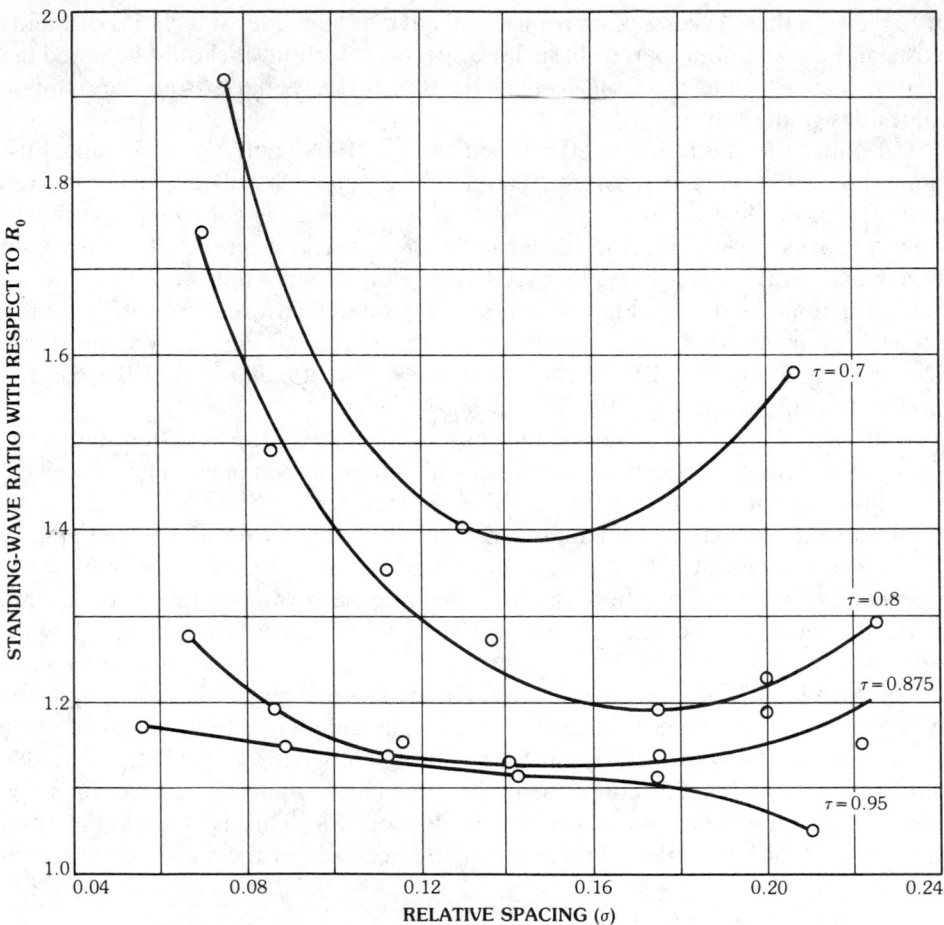

Fig. 24. Computed swr versus σ and τ, for $Z_0 = 100\ \Omega$ and $h/a = 177$. (*After Carrel [8]*)

wavelength) dipoles in the transmission region, i.e., between the feed point and the active region.

(k) An approximate formula for the mean resistance R_0 is

$$R_0 = Z_0/\sqrt{1 + (Z_0/Z_a)(\tau/4\sigma)} \qquad (8)$$

where Z_0 is the characteristic impedance of the (unloaded) feeder, and Z_a is the average characteristic impedance of a dipole [15]:

$$Z_a = 120[\ln(h/a) - 2.25] \qquad (9)$$

A plot of this equation is presented in Fig. 26. The input impedance thus depends on the density of elements, which is reflected through the mean spacing parameter $\sigma' = \sigma/\sqrt{\tau}$. Using Fig. 26 to determine Z_a for a typical value of h/a, normalizing Z_a to the desired mean resistance R_0, evaluating σ' as determined from the above

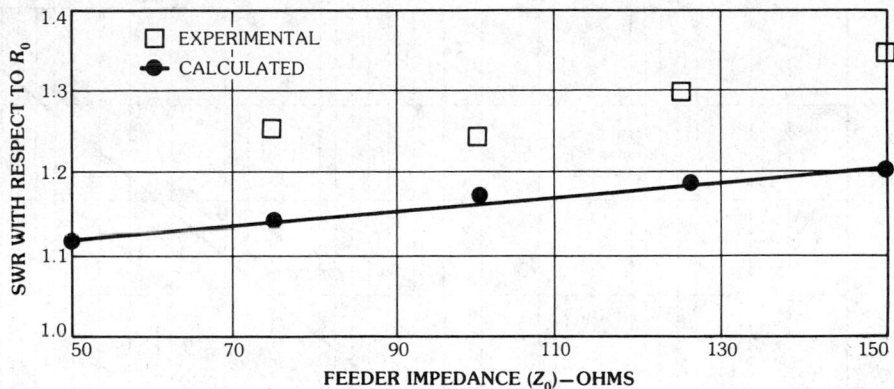

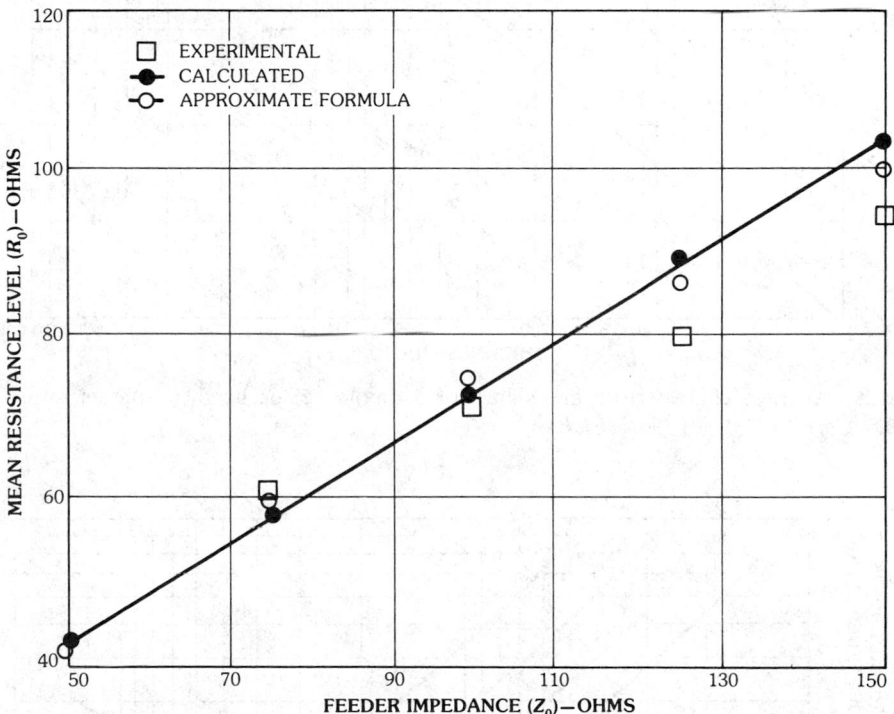

Fig. 25. Input impedance versus feeder impedance, for $\tau = 0.888$, $\sigma = 0.089$, $N = 8$, and $h/a = 125$. (*After Carrel [7, 8], © 1961 IEEE*)

steps related to achieving specified gain, Fig. 27 provides the value of characteristic impedance required for the feeder. For twin-boom or transposed two-wire feeders the spacing-to-diameter ratio of the two conductors determines the characteristic impedance through the relation

$$Z_0 = 120 \cosh^{-1}(b/D) \qquad (10)$$

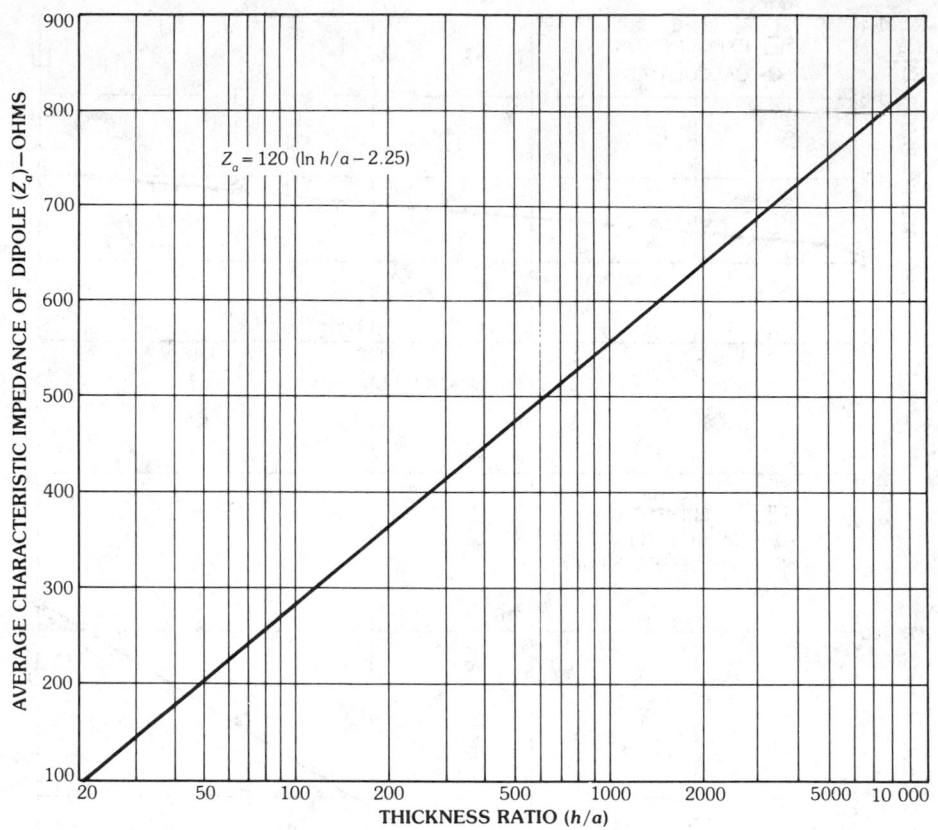

Fig. 26. Average characteristic impedance of a dipole versus height-to-radius ratio *h/a*. (*After Carrel [7, 8],* © *1961 IEEE*)

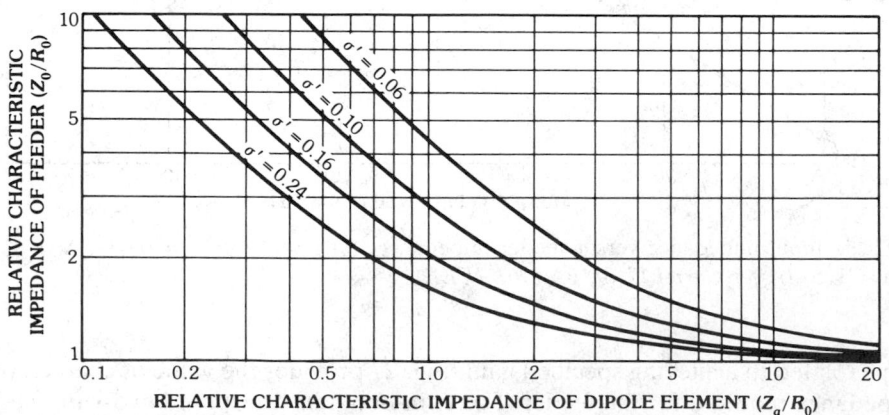

Fig. 27. Relative feeder impedance versus relative dipole impedance from the approximate formula. (*After Carrel [7, 8],* © *1961 IEEE*)

where b is the center-to-center spacing between the parallel conducting cylinders and D is the diameter of each of them.

To determine the physical lengths of the elements the length of one element must be related to a limit on the operating band. The approximation that the longest dipole must be ½ wavelength at the low-frequency limit can be refined to show a slight dependence on the feeder impedance as illustrated in Fig. 28. The length of the longest dipole is then given by

$$\ell_1 = S(\lambda_{\max}/2) = 2h_1 \tag{11}$$

and the lengths of all other dipoles are obtained by multiplying the length of each adjacent longer dipole by the scale factor τ. The chosen value of σ provides means

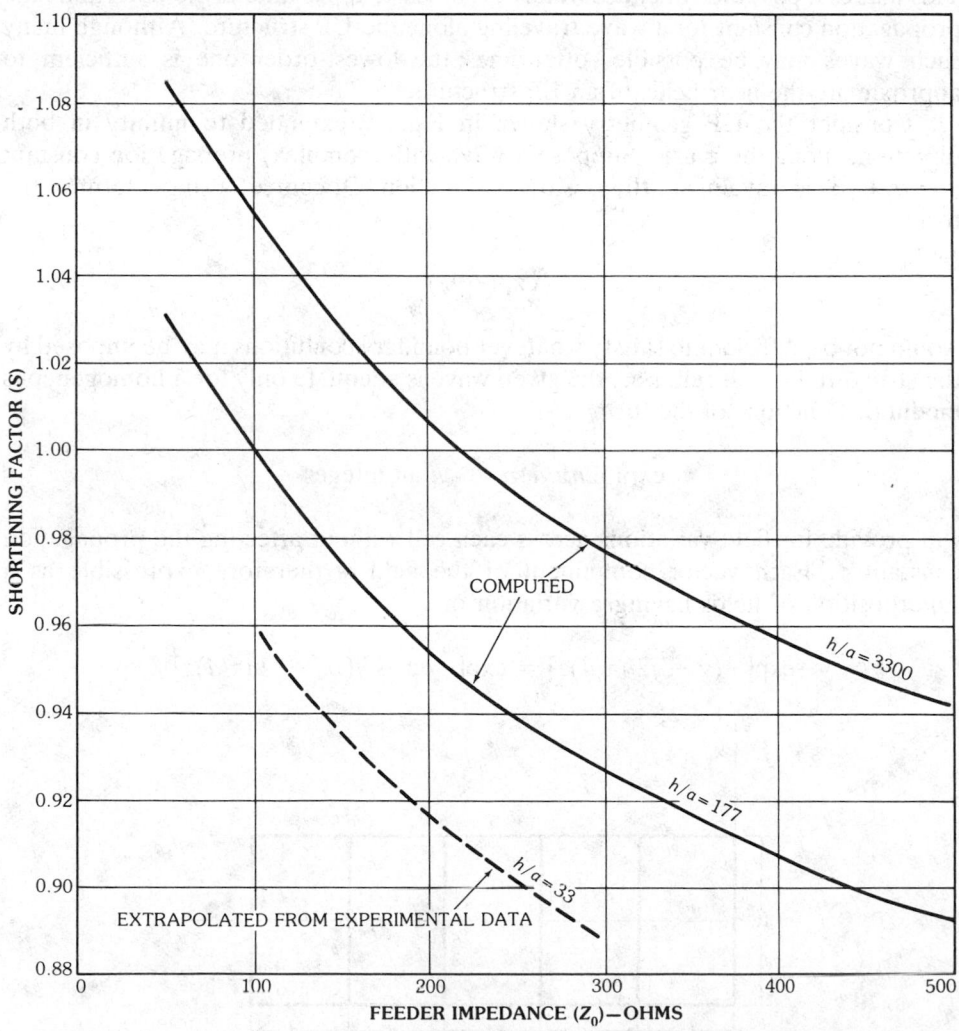

Fig. 28. Shortening factor versus feeder impedance and height-to-radius ratio. (*After Carrel [8]*)

of determining the spacing between each pair of adjacent dipoles. Thus the determination of all dimensions of the antenna has been accomplished and the design procedure is completed.

3. Periodic Structure Theory

The analysis of the log-periodic dipole array, being based on approximations that are valid for thin linear elements, is not readily extendable to other LP structures. Hence it is very useful to have a theory of more general scope. Modifications of the theory of uniform-periodic (UP) structures have provided such a tool [16]. For τ near unity, an LP structure can be considered as a perturbation of a UP structure. The powerful theory of Floquet can be applied to UP structures. This makes it possible either accurately or, at least, approximately to determine the propagation constant for a wave traveling along the UP structure. Although many such waves may be possible, oftentimes the lowest-order one is sufficient to approximate the near fields of an LP structure.

Consider the UP geometry shown in Fig. 29 extended to infinity in both directions along the z axis. Suppose a wave with (complex) propagation constant $\gamma = \alpha + j\beta_0$ is traveling in the positive z direction. Of course, a single term of the form

$$\exp(-\gamma z)$$

would not be sufficient to satisfy whatever boundary conditions may be imposed by the structure in each cell, i.e., the given wave is adequate only for a homogeneous medium. Functions of the form

$$\exp(j2n\pi/d)z, \qquad n \text{ an integer}$$

will provide for field variations across each cell without affecting the propagation constant γ. Each vector component of the field is therefore expressible as a superposition of fields having z variation of

$$\exp[-(\gamma - j2n\pi/d)z] = \exp[-\alpha z - j(\beta_0 - 2n\pi/d)z]$$

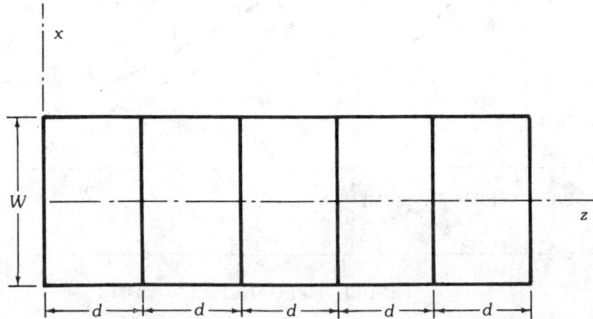

Fig. 29. A periodic structure with five cells.

In keeping with the terminology of Fourier analysis the wave with

$$\gamma_0 = \alpha + j\beta_0 \tag{12}$$

is called the *fundamental wave* and those with

$$\gamma_n = j\beta_n = j(\beta_0 - 2n\pi/d) \tag{13}$$

are called *space harmonics*. The amplitudes and phases of the space harmonics relative to the fundamental are determined by the structure of a specific cell. The concept that the near fields of a periodic structure can be represented as a superposition of waves having the variations given by (12) and (13) is very useful even when the determination of their relative amplitudes and phases is not pursued. In fact, these basic results of periodic structure theory, when combined with some experimental data, provide considerable insight into the operation of LP antennas. This is particularly true for a class of LP antennas that do not involve the excitation of resonant elements. One such antenna is the zigzag.

Consider the zigzag wire shown in Fig. 30. For thin, highly conducting wires, the results of Pocklington [17] show that the current in the wire propagates with the intrinsic phase velocity of the surrounding (homogeneous) medium. It might be expected that a reflected wave would be generated at each of the corners of the zigzag. However, experimental data taken with the practical realizations of the zigzag, which necessarily involve a nonzero radius of curvature for the wire at each corner, fail to display evidence of such reflections. Assuming that the current along the wire travels with free-space phase velocity, the near fields of the zigzag wire will have a fundamental-wave phase constant given by

$$\beta_0 = k \csc \psi \tag{14}$$

where ψ is the pitch angle defined in Fig. 30. The phase constants of fundamental waves for several zigzags with various pitch angles are shown in a plot of frequency versus phase constant (k-β) diagram in Fig. 31. To the degree that (14) holds, the phase constants on the zigzag wire are directly proportional to frequency, i.e., no dispersion occurs. Normalizing all the space-harmonic phase constants with respect to the free-space wave number k,

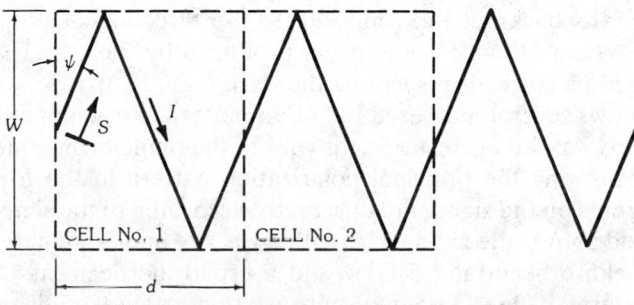

Fig. 30. Basic zigzag wire.

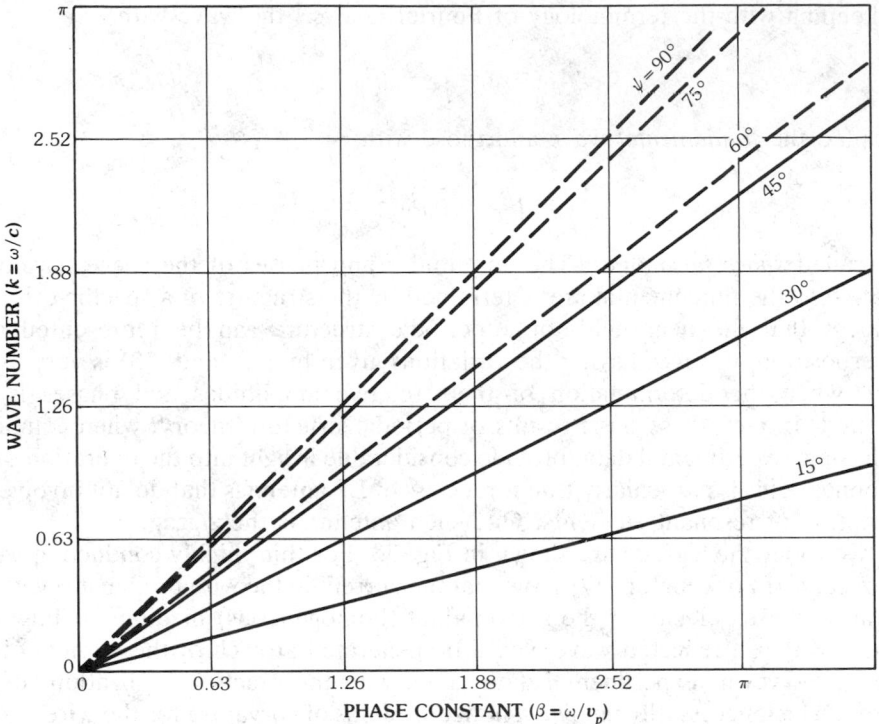

Fig. 31. Approximate phase constants of fundamental waves for several UP zigzag wires.

$$\beta_n/k = \beta_0/k - 2n\pi/kd = \csc\psi + n(\lambda/d) \qquad (15)$$

Fig. 32 shows the normalized space spectrum for a nondispersive periodic structure with $\beta_0/k = 2$. As long as $|\beta_n/k| > 1$ the space harmonics are slow waves which produce no radiation. However, $|\beta_n/k| < 1$ corresponds to the visible range of the space spectrum. A wave with normalized phase constant within the visible range is a fast wave which contributes radiation having maximum value in the direction obtained by projecting from the normalized phase constant upward to the unit circle as shown in Fig. 32c. Note that there are no fast waves when $\lambda/d = 4$. The first space harmonic enters the visible range when $\lambda/d = 3$, and the direction of the beam produced is backfire, i.e., opposite to the direction of propagation of the fundamental. When $\lambda/d = 2.5$, the beam produced by the $n = 1$ space harmonic makes an angle of 60° with respect to the axis.

Fig. 33 shows several measured radiation patterns for a monofilar zigzag wire. The zigzag wire was fed against a linear wire in the plane orthogonal to that of the zigzag. By measuring the principal polarization pattern in the E-plane, only the effects of currents on the zigzag are observed. According to the simple approximate theory outlined above, the zigzag wire with $\csc\psi = 4$ and cell length of 4 cm should produce a backfire beam at 1.5 GHz and a broadside beam at 1.875 GHz. The measured patterns in Fig. 33 agree quite well with these predictions. They also indicate that backfire radiation with narrower beamwidths occurs at frequencies

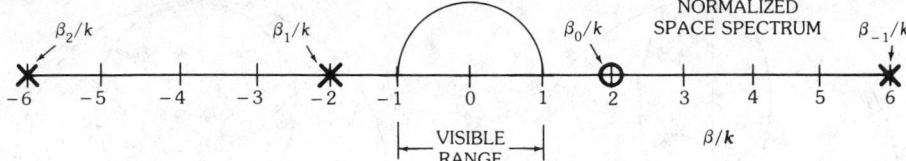

a

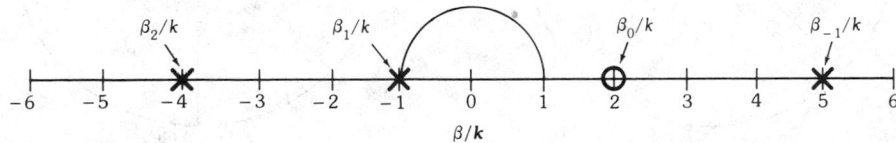

b

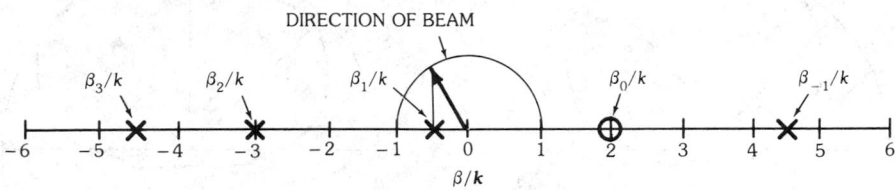

c

Fig. 32. Variation of space harmonics with frequency, where $\beta_n/k = \beta_0/k - n(\lambda/d)$. (a) For $\lambda/d = 4$. (b) For $\lambda/d = 3$. (c) For $\lambda/d = 2.5$.

below 1.5 GHz. The excess cell-to-cell phase shift at the lower frequencies moves the principal lobe away from the visible region, producing the reduced beamwidth. This implies that the active region on an LP zigzag will begin at cells that are smaller than those phased for backfire. When those cells radiate most of the incident energy, the LP zigzag will have a narrow beam. On the other hand, if energy penetrates to the region where the cells are phased for broadside radiation, the pattern can be expected to be much broader.

The k-β diagrams of Fig. 34 were constructed from data measured for several balanced, bifilar UP zigzags with different pitch angles. In the slow-wave regions, $\beta > k$, the wavelength along the axis can be determined by probing the standing-wave field pattern established by reflection of the incident wave from the open end. As one or more space harmonics enter the visible region, however, the incident wave along the zigzag is attenuated and the wave reflected from the open end soon becomes negligibly small. The $n = 1$ space-harmonic phase constant is then determined from the direction of the peak of the beam using the construction of Fig. 32c.

The variations with frequency observed for the UP zigzag correspond to variations with the cell dimensions for an LP zigzag. The region of small cells in Fig. 35 would correspond to the low-frequency conditions for the UP antenna. The fundamental wave and all space harmonics are slow waves and, therefore, little radiation occurs. The region where the first reverse traveling space harmonic

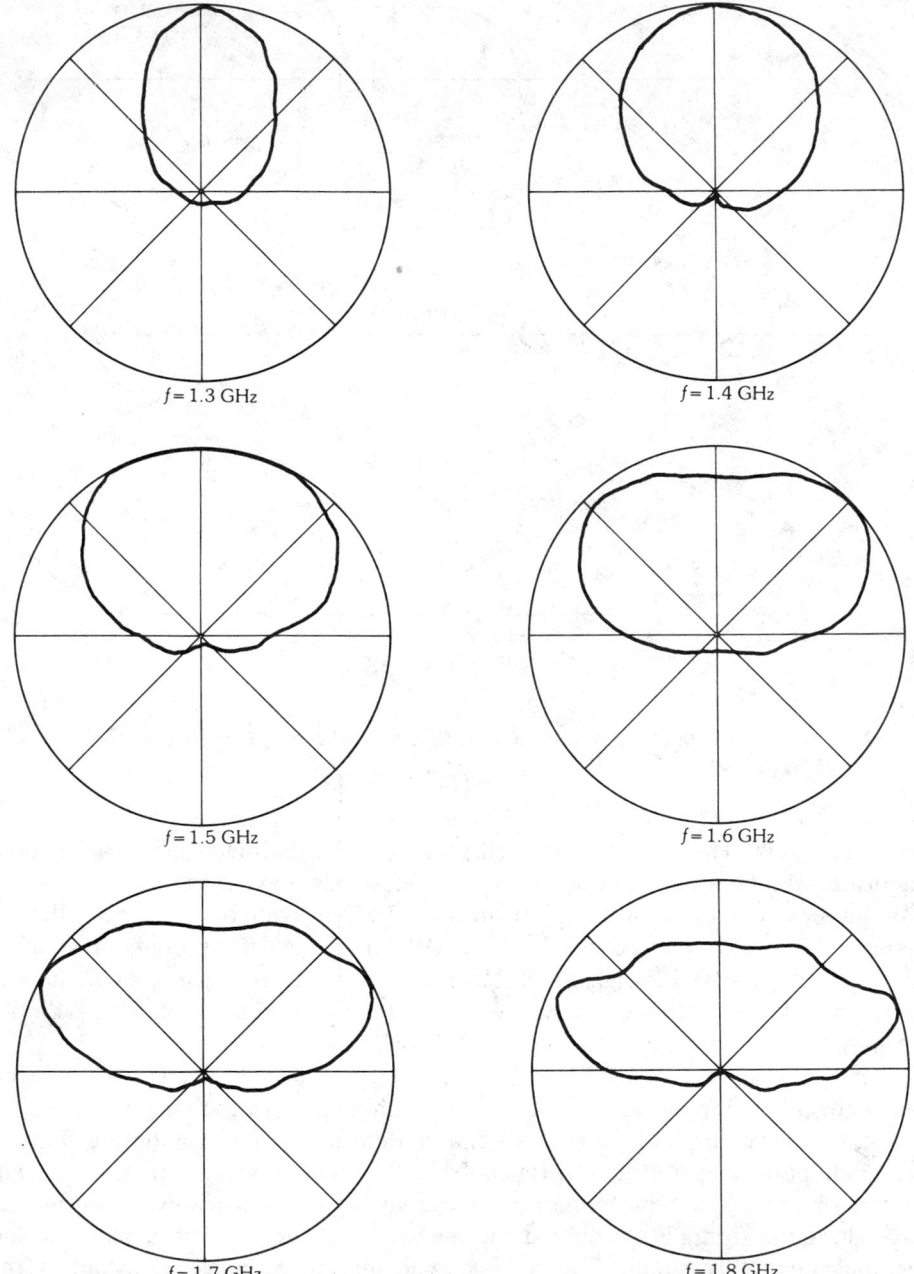

$f=1.3$ GHz $f=1.4$ GHz

$f=1.5$ GHz $f=1.6$ GHz

$f=1.7$ GHz $f=1.8$ GHz

Fig. 33. *H*-plane radiation patterns of a monofilar zigzag antenna.

approaches the backfire condition ($\beta = -k$) produces appreciable radiation and corresponds to the active region. If the radiation is sufficient, the larger cells will be unexcited. This is usually desirable to avoid radiation in directions far from backfire.

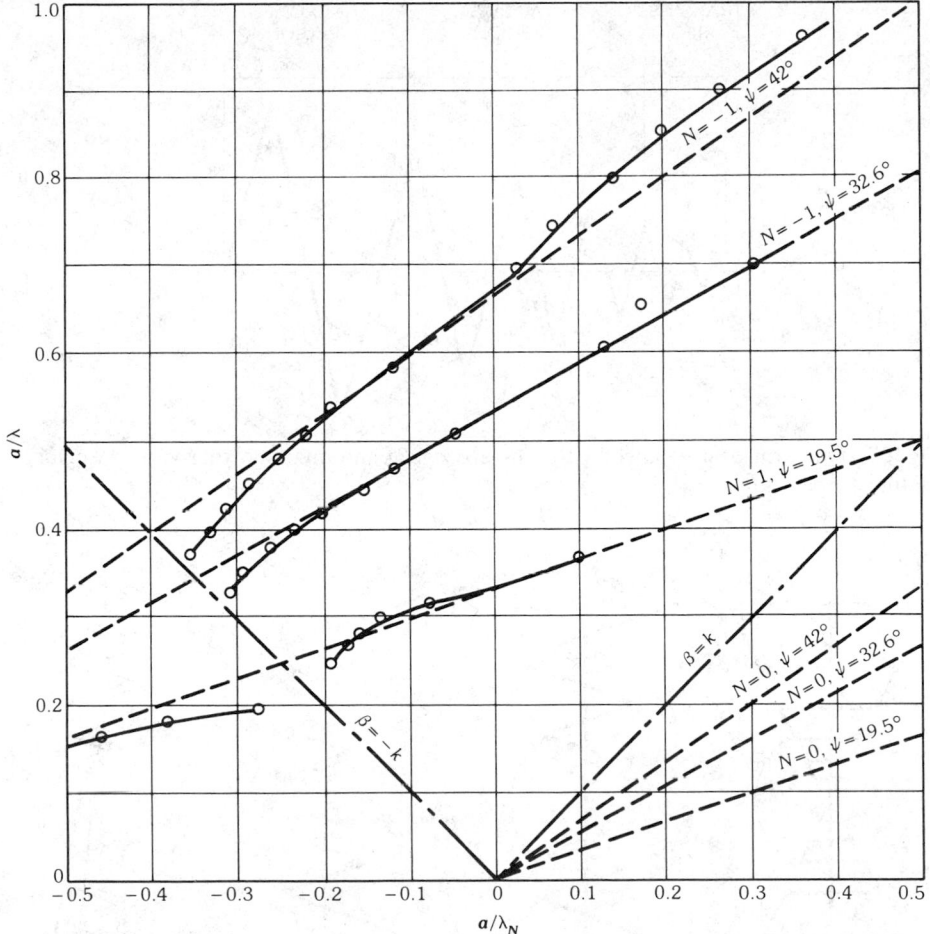

Fig. 34. A k-β diagram for uniform balanced zigzag antennas, showing variation of dispersive curve with pitch angle.

Design of Log-Periodic Zigzag Antennas

The geometric relationships among several types of balanced, two-arm, log-periodic antennas are illustrated in Fig. 36. The wire-outline trapezoidal-tooth LP is shown again in Fig. 36a. A similar antenna with triangular, instead of trapezoidal, teeth is shown in Fig. 36b. The performance of triangular-tooth and trapezoidal-tooth log-periodics of the same parameters (τ, α, and ψ) is very nearly the same [5]. Removing the central boom, the axial conductor of Fig. 36b, from a triangular-tooth, wire-outline LP produces the thin-wire, log-periodic zigzag of Fig. 36c. This antenna does not scale exactly when made using constant-diameter wire. Replacing the wire by a thin metal sheet with tapered width gives the tapered-arm LP zigzag in Fig. 36d. The angle between the center line and the extremities of each arm is called α, as was done for the LPD (see Fig. 11a). However, the angle β is used to define the width of the flat conductor at the tapered-arm zigzag.

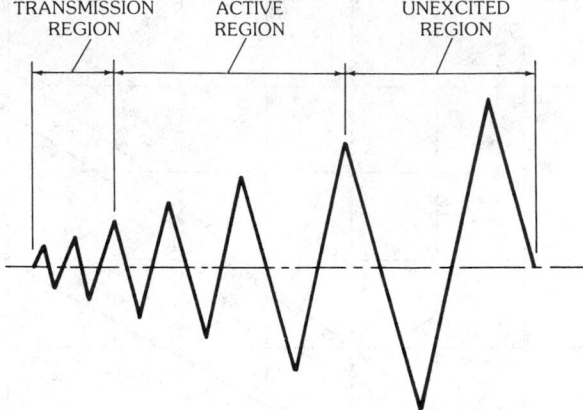

Fig. 35. One arm of a balanced LP thin-wire zigzag antenna, showing various regions for midband operation.

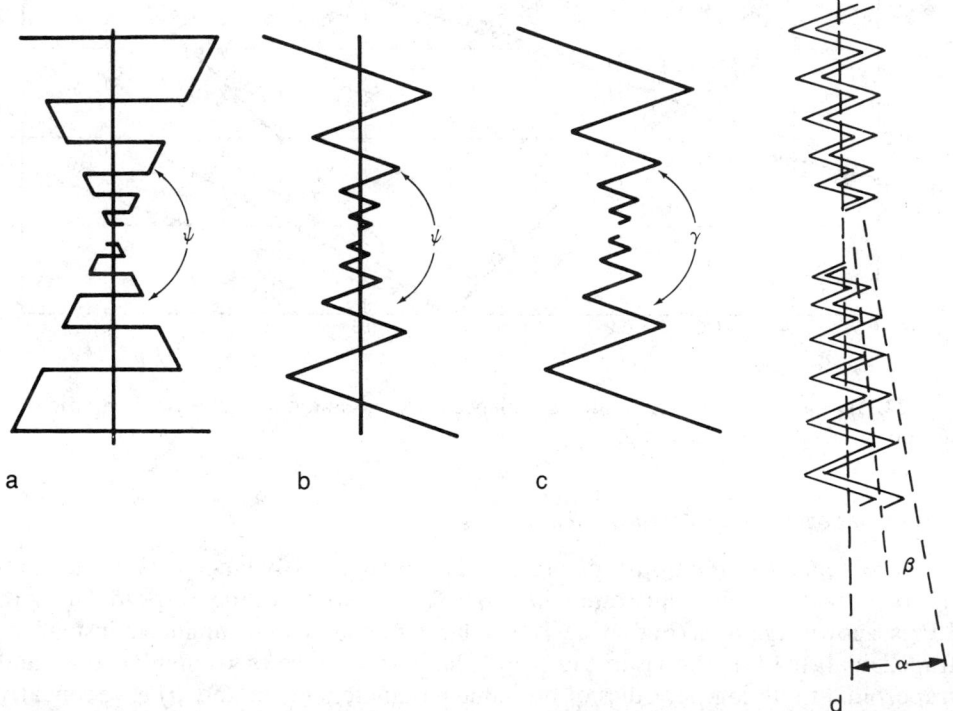

Fig. 36. A comparison of the LP zigzag antenna and the wire, nonplanar trapezoidal-tooth, and triangular-tooth antennas. (*a*) Wire-outline trapezoidal-tooth LP. (*b*) Triangular-tooth LP. (*c*) Thin-wire LP zigzag. (*d*) Tapered-arm LP zigzag. (*After Kuo and Mayes [20]*)

Analysis of zigzag wire antennas like those of Fig. 36c is feasible [18, 19] and provides guides for design that will be presented later. Using the tapered-width arms as in Fig. 36d gives an additional parameter for control of the antenna performance, but the analysis is much more difficult and has not been reported. However, an extensive program of measurements for tapered-arm antennas, including thin-wire arms as a special case, provides some data on which a design procedure can be based [20]. To varying degrees all the parameters which are necessary to specify an LP zigzag antenna have an effect on the observed performance. Table 3 lists the parameters of tapered-arm LP zigzag antennas and qualitatively describes how each affects performance. (The table entries denote the change in performance for an increase in the parameters of the first column.)

The radiation pattern of an LP zigzag antenna depends primarily on the combination of α and β when τ is held constant. When the angle γ between the planes of the zigzag elements is set equal to 2α, the E-plane and H-plane beamwidths will be almost equal. The H-plane beamwidth can be decreased by increasing γ, but the gain increase that can be obtained in this way is limited by the appearance of side lobes for values of γ that are much larger than 2α. The following design procedure is based on $\tau = 0.9$ and $\gamma = 2\alpha$.

(*a*) Desired values of directivity can be reduced to required beamwidths by using the nomograph of Fig. 37.

(*b*) The pattern beamwidths depend primarily on α and β. Values of these two parameters that will give realizable beamwidths are given in the graphs of Fig. 38.

Table 3. Log-Periodic Zigzag Parameters and Their Effect on the Observed Performance*

LP Zigzag Parameters[†]	Input Impedance	Half-Power Beamwidth	Pitch Angle (ψ)	Boom Length for a Fixed Frequency Band
α (β and τ constant)	Increase	Increase	Decrease	Decrease
β (α and τ constant)	Decrease	Decrease	Increase	Independent
τ (α and β constant)	Increase	Decrease and then increase after τ is greater than 0.9	Decrease	Independent
ψ	Decrease	Depends on α and β		Independent

*Table entries denote the change in performance for an increase in the parameters of the first column.
[†]The table entries hold true over the following range of parameters for which frequency-independent operation has been verified: $0.85 < \tau < 0.95$, $5° \leqq \alpha \leqq 15°$, and $0° < \beta \leqq \alpha$.

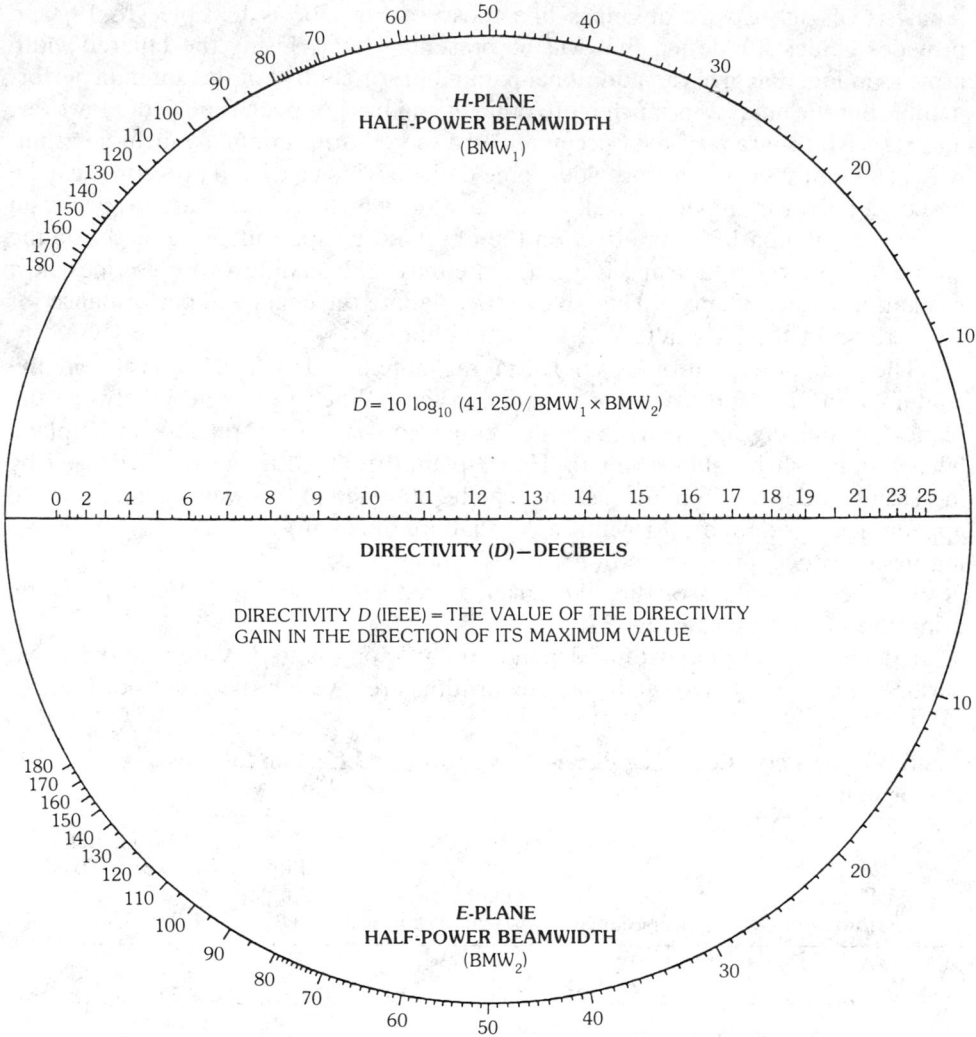

Fig. 37. Nomograph for relating directivity to half-power beamwidths.

The choice of α and β will have an effect on the input impedance which will be given later. Thus if the desired value of impedance cannot be obtained with the chosen values of α and β, a different choice may be required that gives the same beamwidths. The size of antenna that results from certain choices of α and β should also be kept in mind. A small value of α means a long antenna is required to cover a specified frequency band.

(c) The mean input resistance for each combination of α and β can be obtained from Figs. 39 and 40. Combinations of α and β which yield the required beamwidth, even without restrictions on α to meet a limitation on size, will not always produce the desired input resistance. Fig. 41 illustrates the mean input resistance as a function of τ when α and β are held constant. The resistance increases very

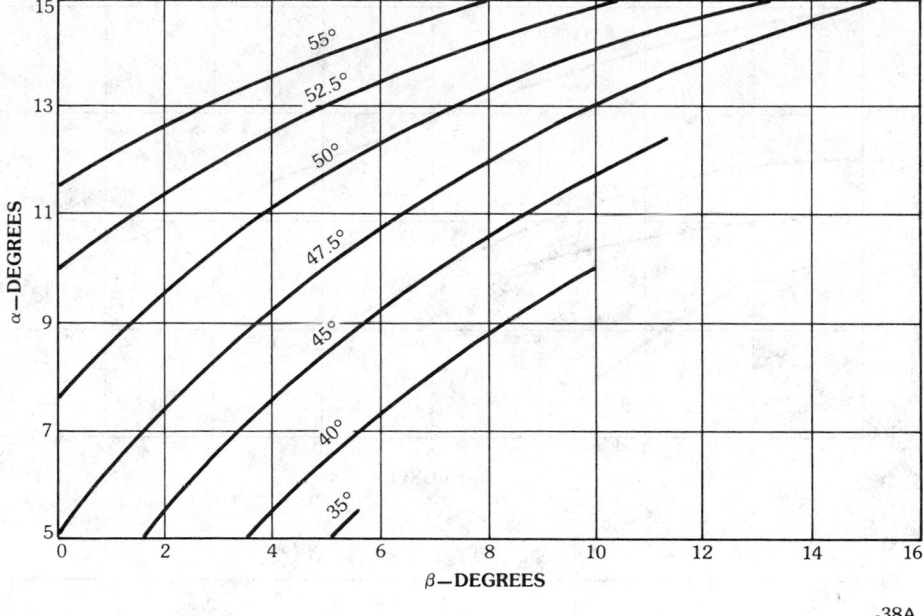

a

-38A

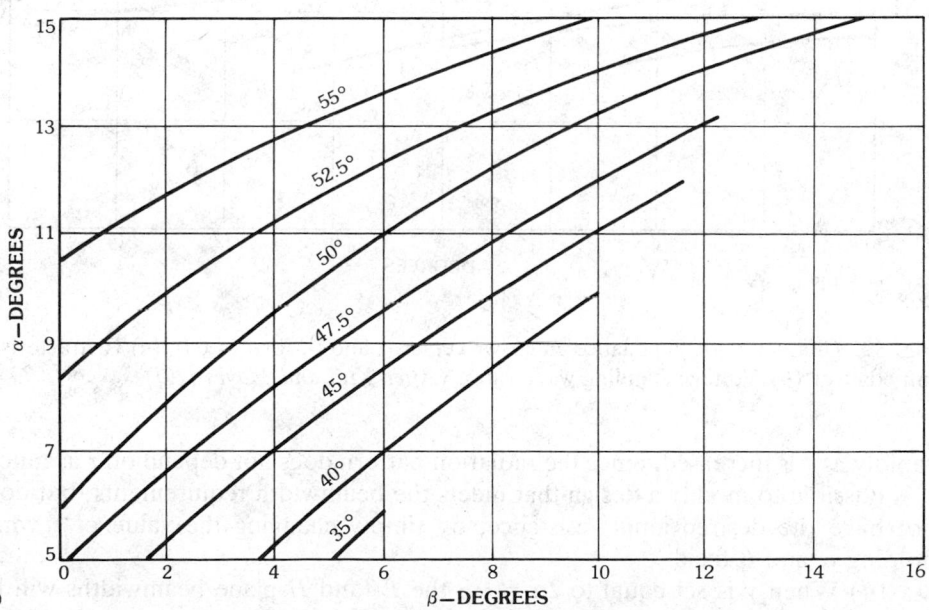

b

Fig. 38. Experimental contours of constant half-power beamwidth versus α and β, for $\tau = 0.9$. (a) H-plane. (b) E-plane. (After Kuo and Mayes [20])

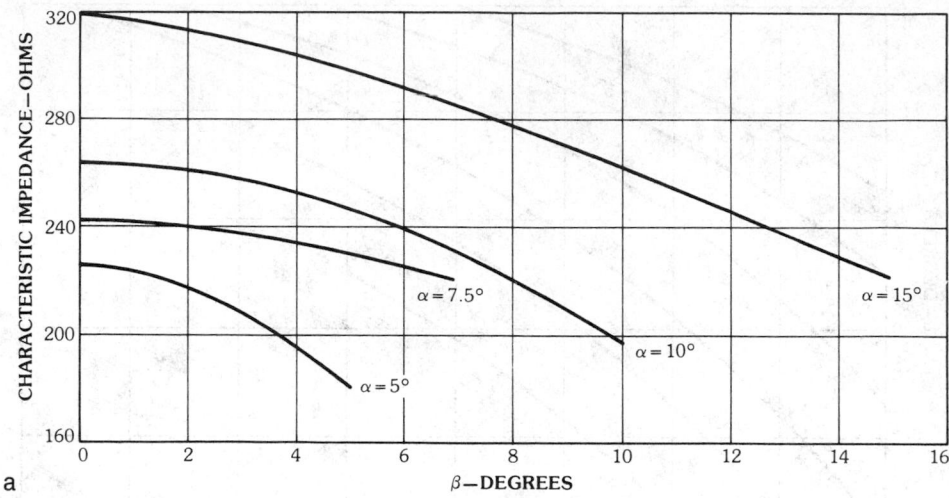

a

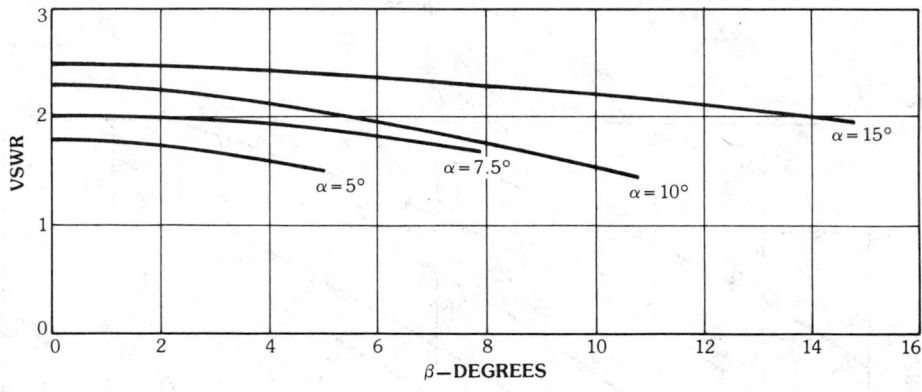

b

Fig. 39. Characteristic impedance and vswr versus α and β, for $\tau = 0.9$. (*a*) Characteristic impedance. (*b*) Voltage standing-wave ratio. (*After Kuo and Mayes [20]*)

rapidly as τ is increased. Since the radiation pattern does not depend on τ as much, it is possible to modify a design that meets the beamwidth requirements, but does not have the desired input resistance, by simply changing the value of τ while holding α and β fixed.

(*d*) When γ is set equal to $2\alpha = \gamma_0$, the *E*- and *H*-plane beamwidths will be almost the same. The *H*-plane beamwidth can be decreased somewhat to achieve higher directivity by increasing γ beyond γ_0. Fig. 42 provides some information about how the beamwidths vary with increasing γ.

(*e*) Fig. 43 shows how the mean input resistance is affected by changing γ while holding α, β, and τ constant.

Log-periodic zigzag antennas with constant-dimension conductors are attractive for simplicity of construction. The experimental data presented above for the

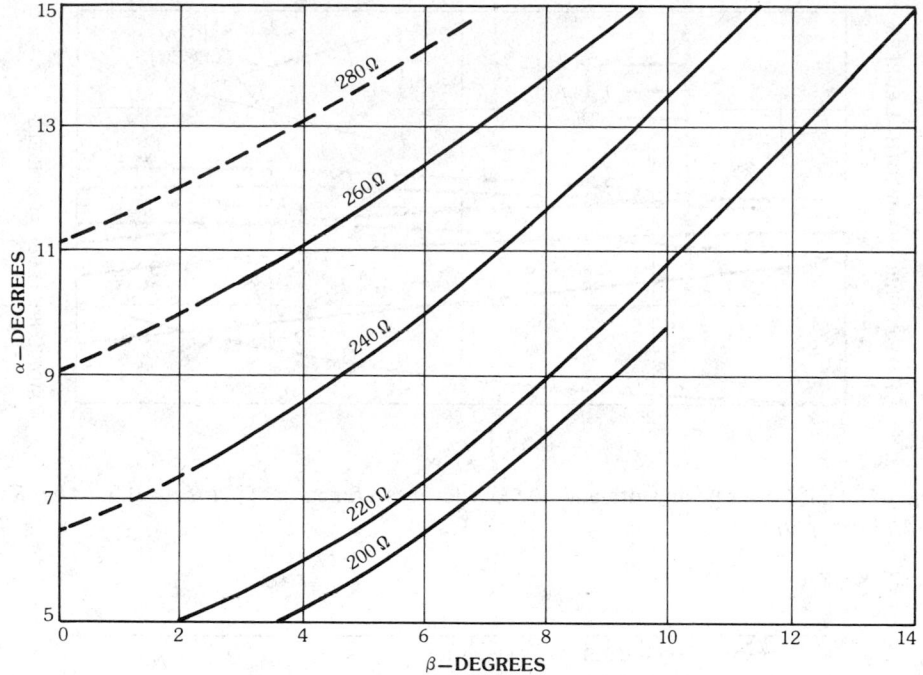

Fig. 40. Experimental contours of constant characteristic impedance versus α and β, for $\tau = 0.9$. (*After Kuo and Mayes [20]*)

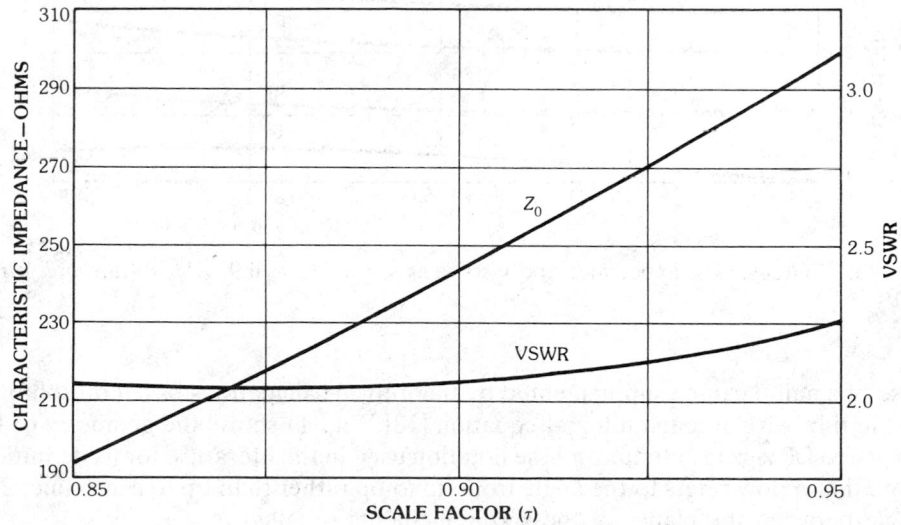

Fig. 41. Characteristic impedance and vswr versus ι, for constants α and β. (*After Kuo and Mayes [20]*)

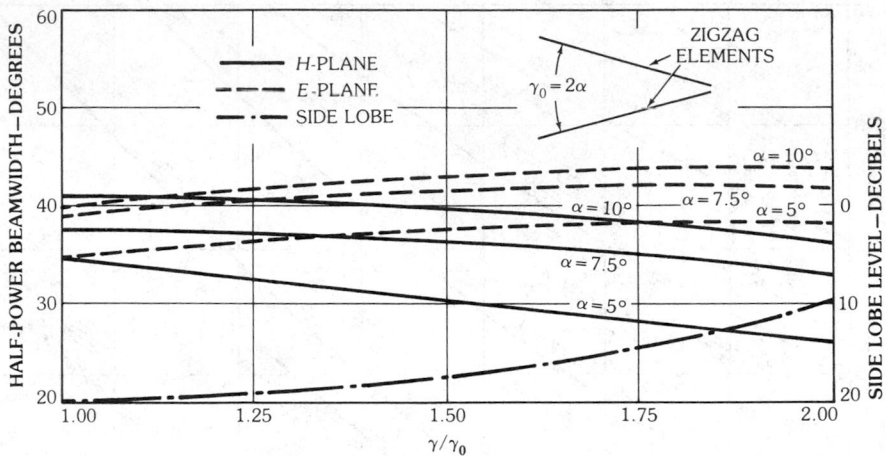

Fig. 42. Half-power beamwidth and side lobe level as a function of γ, for $\tau = 0.9$. (*After Kuo and Mayes [20]*)

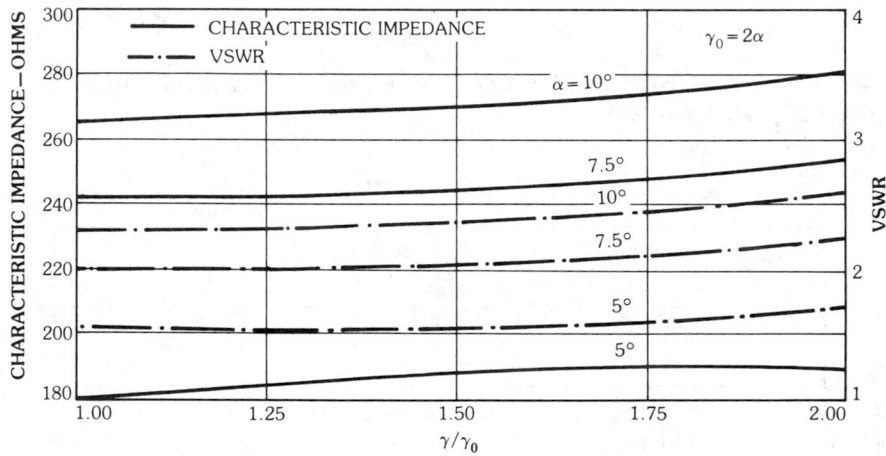

Fig. 43. Characteristic impedance and vswr versus γ, for $\tau = 0.9$. (*After Kuo and Mayes [20]*)

case of small β can be supplemented by theoretical calculations based on solutions of the thin-wire antenna integral equation [18]. Fig. 44 shows the geometry of the thin-wire LP zigzag antenna and the notation used in the literature for its definition. Note that α now refers to the angle from tip to tip rather than tip to centerline. The angle between the planes of the zigzag elements is called ψ.

In Fig. 44 only the zigzag lines designate conductors. The central boom, needed for mechanical support, is nonconducting, and its effects are not included in the analysis. A simple way of fabricating thin-wire LP zigzag antennas is to use small-diameter coaxial cable for the arms. One arm then serves both as a download

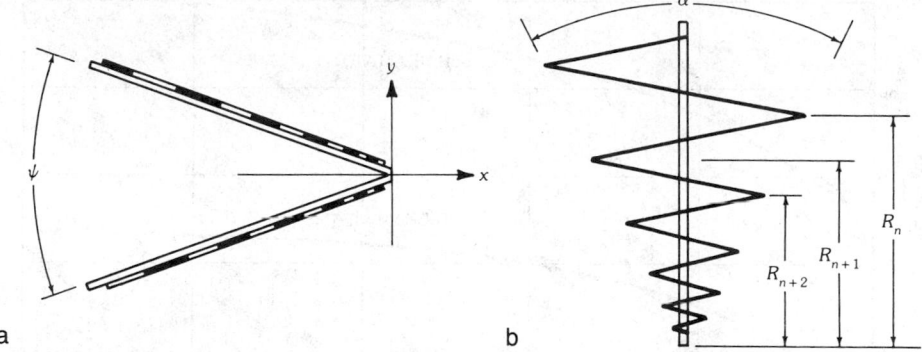

Fig. 44. A balanced LP zigzag antenna. (*a*) Top view. (*b*) Plane view of one half of the antenna. (*After Lee and Mei [19], © 1970 IEEE*)

from the actual feedpoint at the apex and as a radiator. Only the outer conductor is used on the second arm, and it serves only as a radiator.

One advantage of the LP zigzag antenna over the LP dipole array is the control of the *H*-plane beamwidth. When $\psi = \alpha$, the *E*- and *H*-plane beamwidths are nearly equal. This is desirable for some applications, such as a feed for a rotationally symmetrical reflector. However, an even narrower *H*-plane beamwidth would produce greater directivity and gain. Values of ψ much greater than α_0 are not usually advantageous because of the rapid increase in *H*-plane side lobes as shown in Fig. 45.

The design of a thin-wire LP zigzag antenna can proceed from *E*- and *H*-plane pattern beamwidths that will produce a desired gain as approximated using the nomograph of Fig. 37. Figs. 46 and 47 indicate the rather narrow range of beamwidths that can be obtained by changing the values of τ, α, and simultaneously changing ψ so that $\psi = \alpha$. As a consequence of this rather limited range of beamwidths the gain of thin-wire LP zigzag antennas falls between 8 and 10 dB as

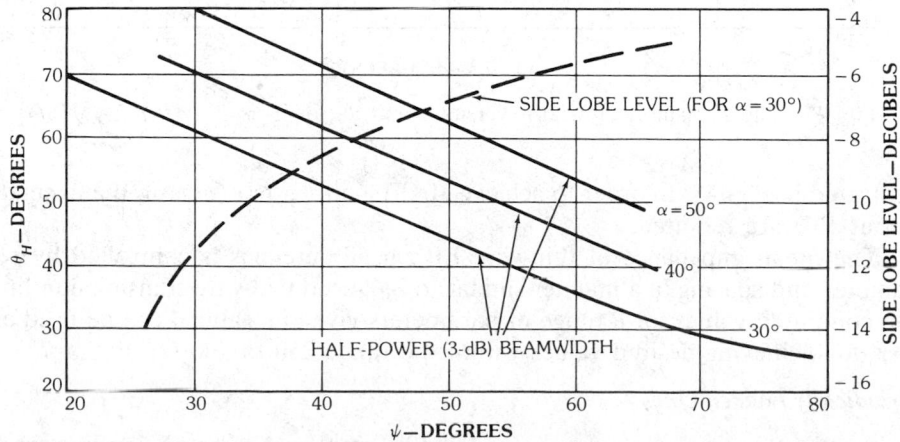

Fig. 45. The *H*-plane beamwidth and side lobe level versus ψ. (*After Lee [18]*)

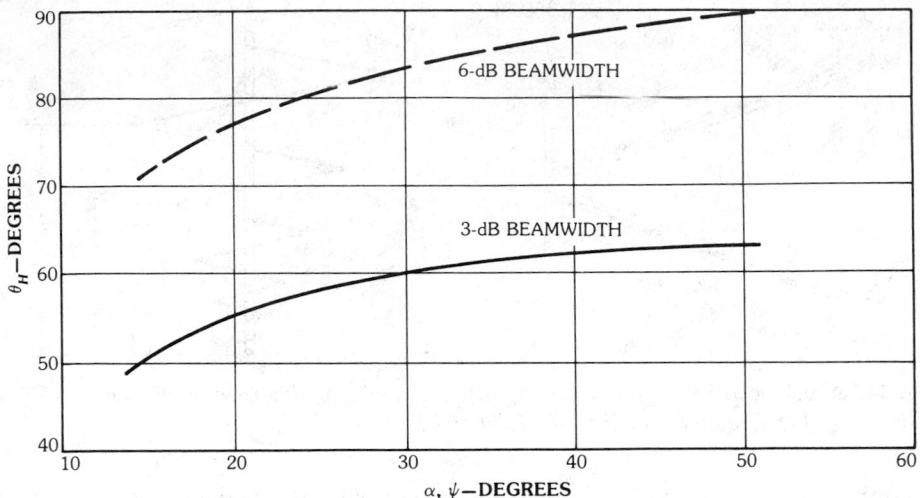

Fig. 46. The *H*-plane beamwidths versus α and ψ, for $\alpha = \psi$. (*After Lee [18]*)

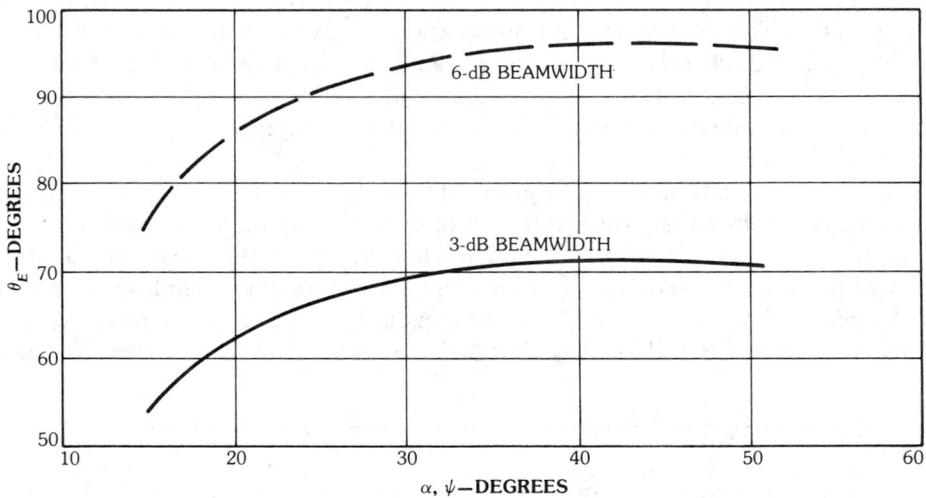

Fig. 47. The *E*-plane beamwidths versus α and ψ, for $\alpha = \psi$. (*After Lee [18]*)

illustrated in Fig. 48. In order to achieve 10 dB gain, large τ (near 0.9) and small α (about 15°) are required.

The mean impedance of thin-wire LP zigzag antennas is controlled by wire diameter and spacing in a manner similar to balanced two-wire transmission lines. The computed values for a range of parameters given in Table 4 can be used as a basis for achieving desired values within the range 250 to 400 Ω.

Periodically Loaded Lines

Some LP antennas are closely modeled by a uniform transmission line loaded with discrete elements. For example, the effect of the dipoles on the waves

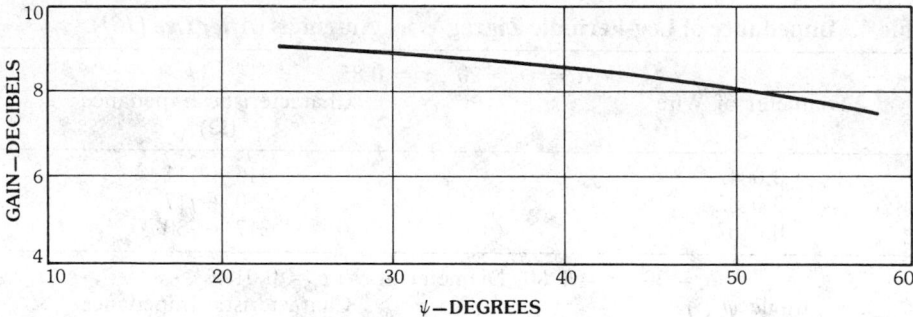

a

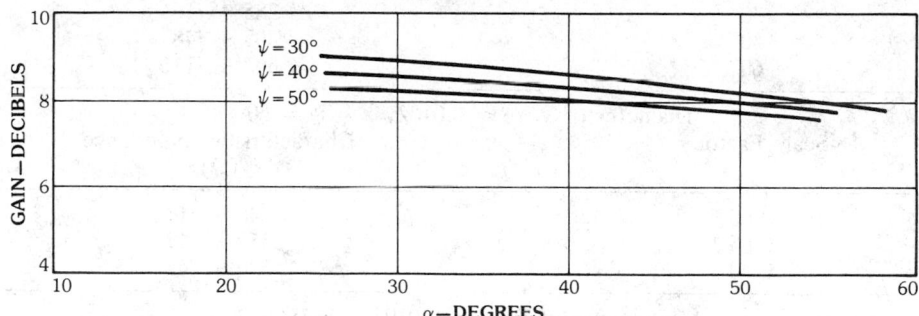

b

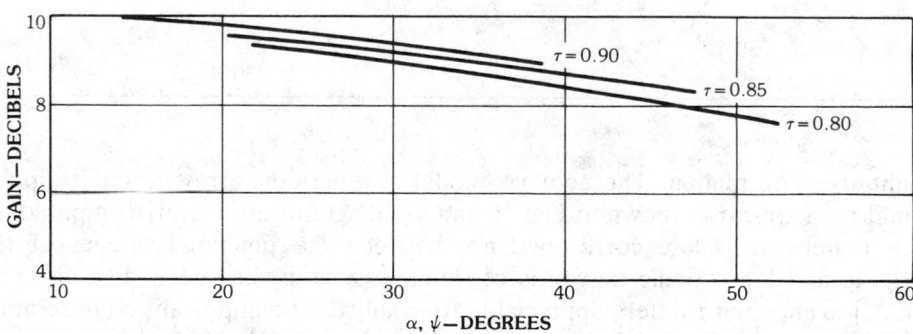

c

Fig. 48. Thin-wire LP zigzag antenna gain. (*a*) Antenna gain versus ψ, for $\tau = 0.80$ and $\alpha = 30°$. (*b*) Antenna gain versus α, for $\tau = 0.80$. (*c*) Antenna gain versus α and ψ, for $\alpha = \psi$. (*After Lee [18]*)

traveling along the feeder of a uniform dipole array can be represented approximately by a discrete shunt admittance as shown in Fig. 49. Of course, this model fails to account for the field interactions (mutual impedance) among the dipoles. The frequency variation of dipole admittance is similar to that of a lossy, open stub. Although the input conductance of the dipole is a function of frequency, a simplified network composed of a lossless stub and a series resistor can give useful

Table 4. Impedance of Log-Periodic Zigzag Wire Antennas (*After Lee [18]*)

$\alpha = \psi = 20°$, $\tau = 0.85$	
Diameter of Wire	Characteristic Impedance (Ω)
0.004λ	$416 + j11$
0.010λ	$350 + j17$
0.016λ	$317 + j30$

$\alpha = 30°$, $\tau = 0.80$, Diameter of Wire $= 0.016\lambda$	
Angle ψ (°)	Characteristic Impedance (Ω)
27	$270 + j12$
30	$285 + j16$
40	$305 + j48$
60	$362 + j116$

Diameter of Wire $= 0.010\lambda$, $\alpha = \psi = 20°$	
Scale Factor τ	Characteristic Impedance (Ω)
0.80	$340 - j11$
0.85	$350 + j17$
0.90	$370 + j45$

Diameter of Wire $= 0.010\lambda$, $\tau = 0.85$	
Angle α (°)	Characteristic Impedance (Ω)
20	$350 + j17$
30	$292 - j3$
40	$236 + j3$

qualitative information. The network model of a periodic array of shunt dipoles would then appear as shown in Fig. 50 and k-β diagrams are readily computed for such a network. Close correspondence between the junction voltages of the periodic and log-periodic networks of shunt lossy stubs has been demonstrated [21, 22] even when τ differs appreciably from unity. The important consideration for this correspondence to hold is that the image impedance of a single symmetric

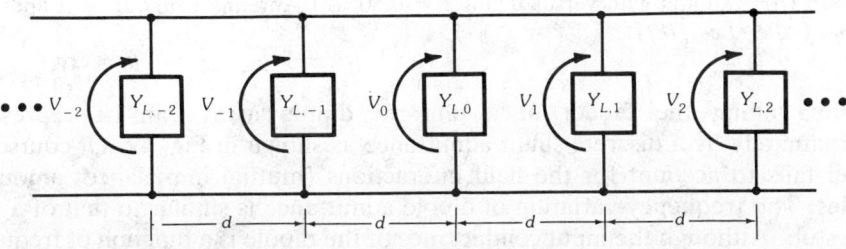

Fig. 49. A transmission line periodically loaded with shunt elements.

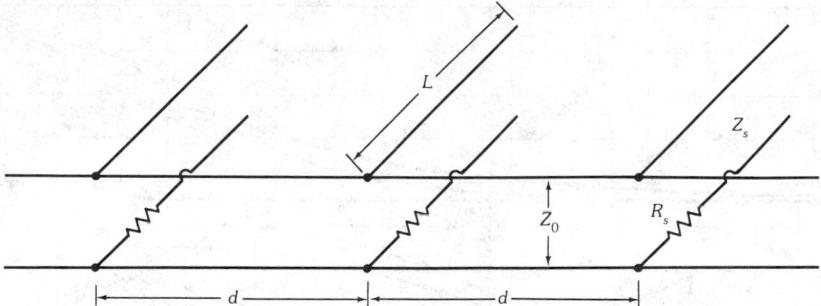

Fig. 50. A transmission line periodically loaded with resistor-stub elements. (*After Ingerson and Mayes [26], © 1968 IEEE*)

cell of the periodic structure be essentially independent of frequency below the first resonance.

A distinctive feature appears in the k-β diagrams of networks like Fig. 50 when the elements have resonant behavior. A stopband occurs in the vicinity of each resonance of the elements regardless of the length of transmission line between the elements. Stopbands associated with resonance of the periodic loads are called *two-terminal stopbands* [23]. Stopbands also occur for nonresonant loads that are spaced approximately one-half wavelength apart. The latter are called *structural* stopbands. Fig. 51 shows two typical plots of complex dispersion data for a system like that shown in Fig. 50 with zero loss. Stopbands are identified in lossless systems by nonzero attenuation. Two stopbands are displayed in Fig. 51. The lowest one is due to the resonance of the stubs, which occurs at 275 MHz. The higher stopband is due to the periodic loading, the stubs being one-half wavelength apart at 397 MHz. Note that the width of the stopbands increases with an increase in the characteristic impedance Z_0 of the main line (feeder).

Fig. 52 shows that adding loss to each cell produces a major effect on the k-β diagram. By increasing the value of R in Fig. 50 from zero to 73 Ω the maximum phase shift per cell in the stopband is reduced to about 40° and the maximum attenuation to about 0.6 Np. Although changing the resistance in each cell affects the phase and attenuation of the cell, it is not a convenient way of controlling either in a log-periodic antenna. The resistance in each cell in the antenna case is desirably due to radiation and is affected by mutual coupling. In Fig. 53 it is seen that the maximum phase shift and attenuation are affected very little by changing the characteristic impedance of the stub, i.e., changing the Q of the loads. The width of the stopband, however, is clearly related to the Q of the loads. An effective way of controlling the phase shift per cell when there is no coupling between the loads is to change the length of line in each cell. This is illustrated in Figs. 54a and 54b, where progressive increases in the line length produce increases in the maximum phase shift occurring just below resonance. It is to be noted, however, that increased line length causes the structural stopband to occur at lower frequency. In particular, the value $d = 67.5$ cm in Fig. 54c produces a structural stopband (180 to 220 MHz) below the first resonance. The effect on attenuation per cell at different line lengths is displayed in Fig. 55. The maximum attenuation first increases with increasing

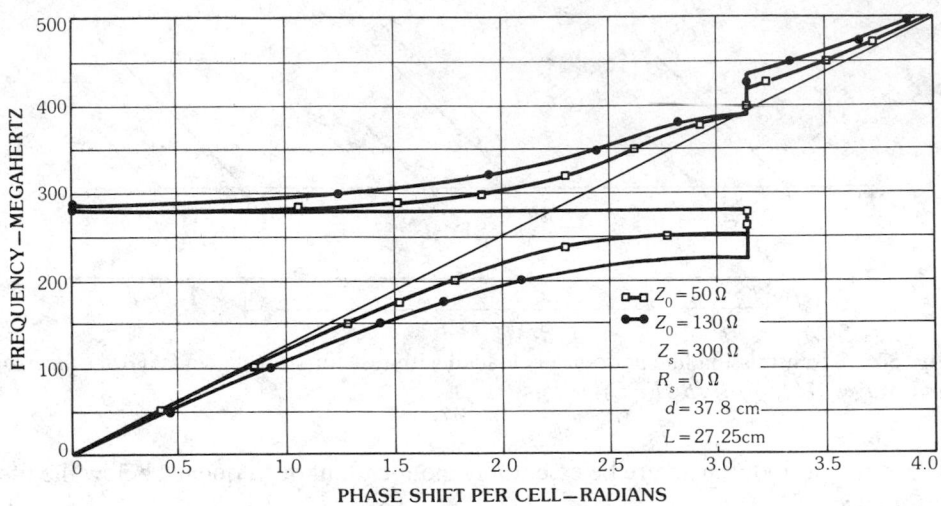

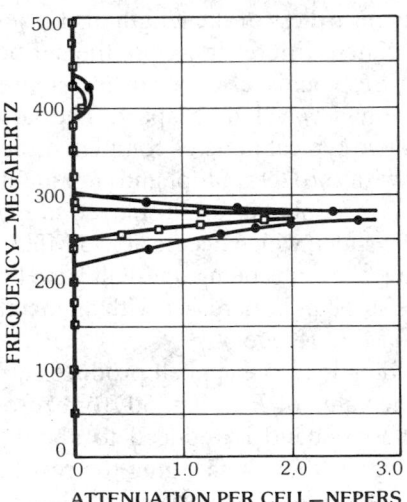

Fig. 51. Dispersion data for a transmission line that is periodically loaded with sections of lossless line, showing dependence on the characteristic impedance.

line length, then decreases as the structural and two-terminal stopbands overlap, increasing again as the structural stopband moves below the two-terminal stopband.

A complete description of symmetrical, reciprocal two-ports requires consideration of image impedances as well as the propagation constant. Fig. 56 shows the image impedance for two examples of cells symmetrically loaded with lossless stubs. The solid line is for the case of smaller spacing ($\sigma = d/4L = 0.25$) and shows how the magnitude of the image impedance changes little until near the first

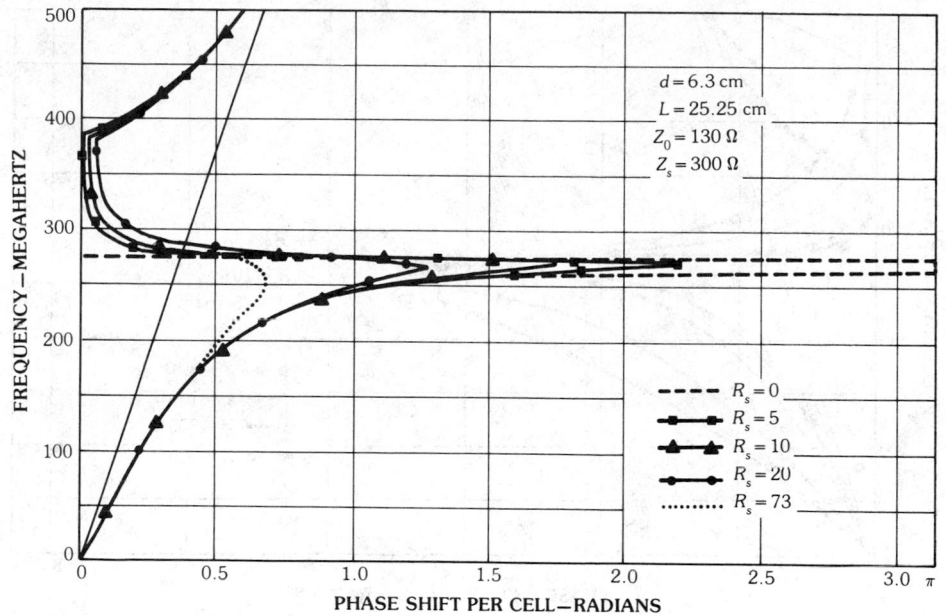

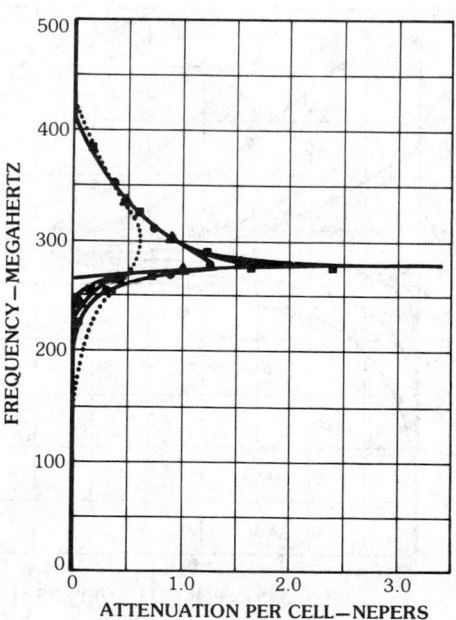

Fig. 52. Dispersion diagram for a transmission line that is periodically loaded with sections of lossy line, showing dependence on loss resistance.

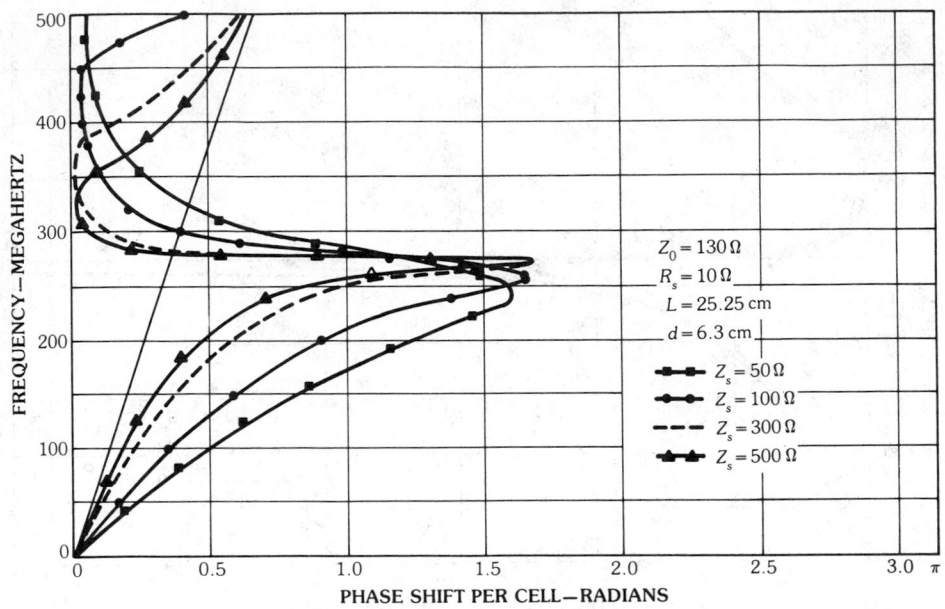

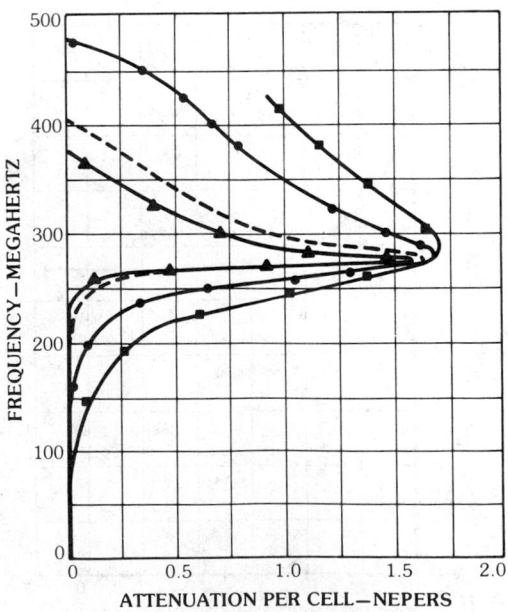

Fig. 53. Computed dispersion curves for stub-loaded transmission line with variable stub characteristic impedance.

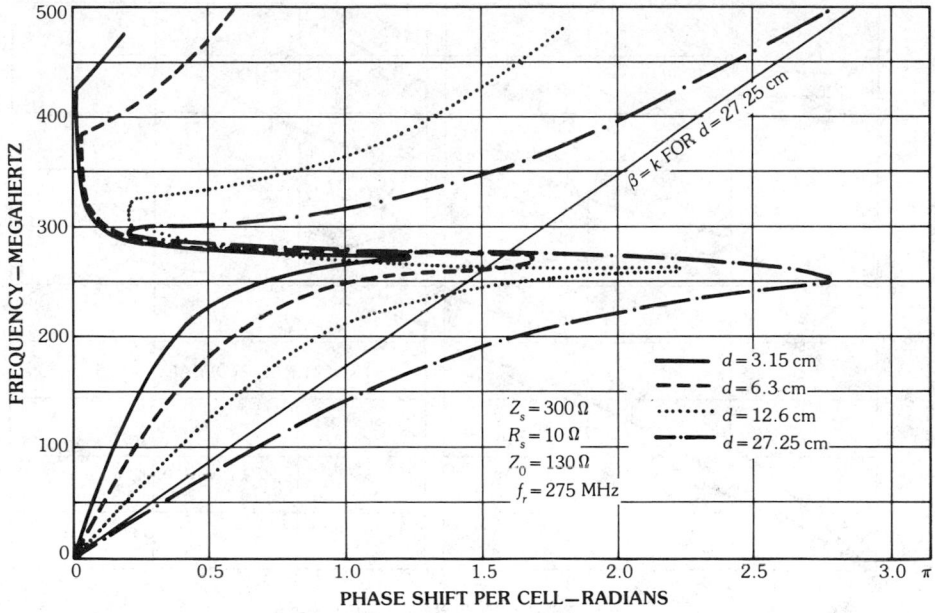

a

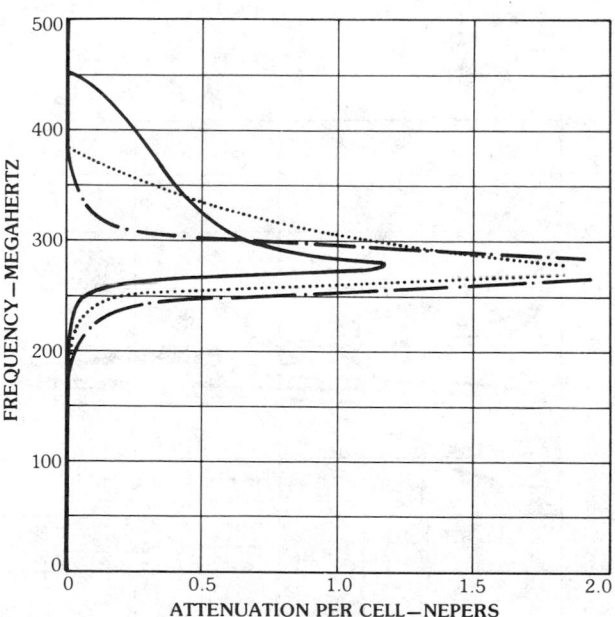

b

Fig. 54. Dispersion data for transmission line periodically loaded with sections of lossless line, showing dependence on spacing. (*a*) Phase shift, 3.15 cm $\leq d \leq$ 27.25 cm. (*b*) Attenuation, 3.15 cm $\leq d \leq$ 27.25 cm. (*c*) Phase shift, 37.5 cm $\leq d \leq$ 67.5 cm. (*After Ingerson [22]*)

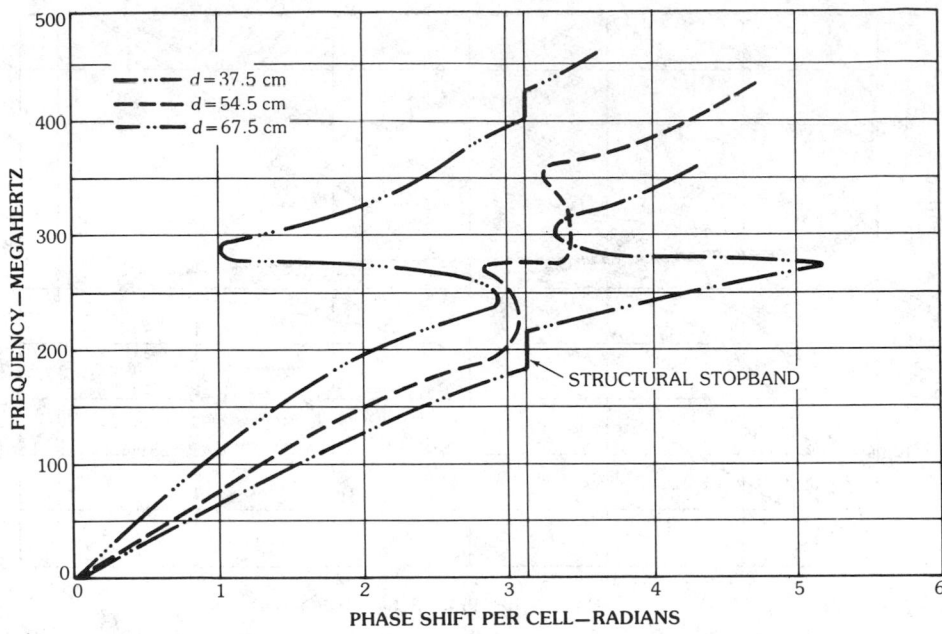

c

Fig. 54, *continued.*

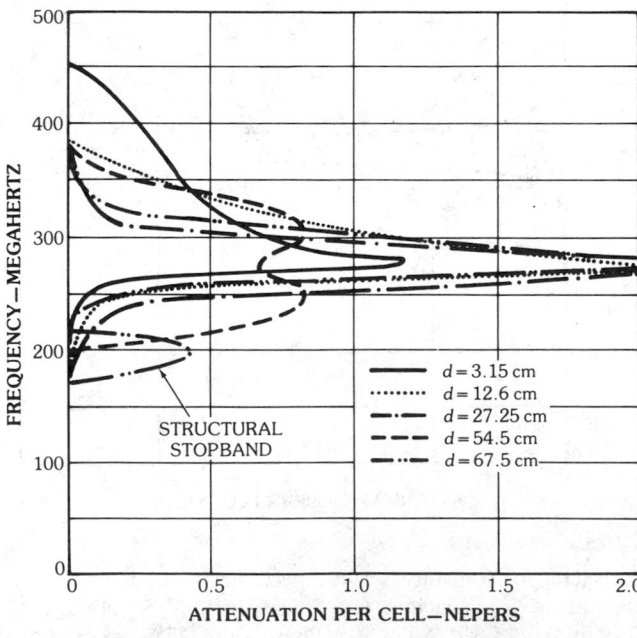

Fig. 55. Computed attenuation curves for stub-loaded transmission line with variable spacing. (*After Ingerson [22]*)

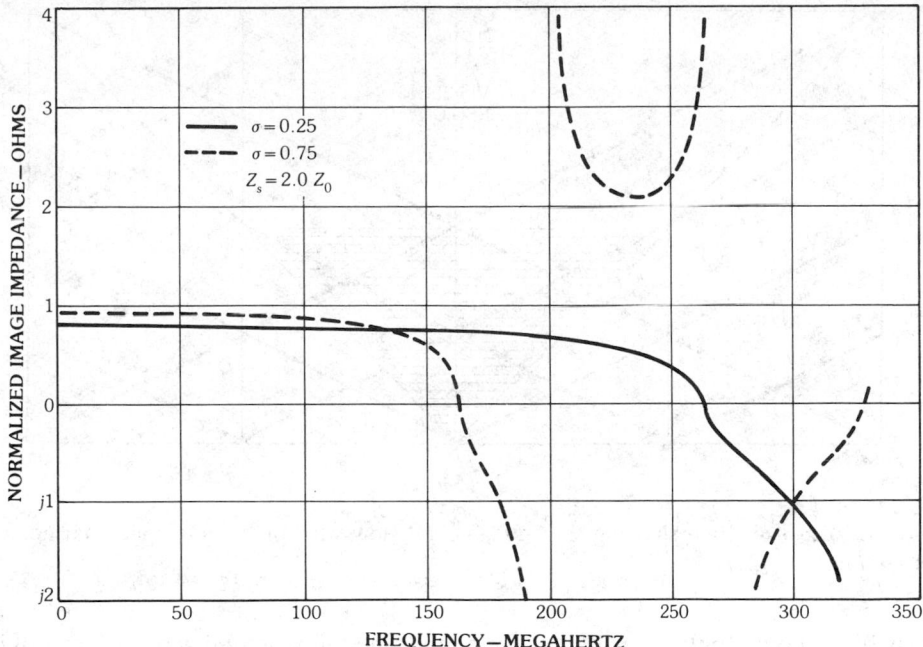

Fig. 56. Image impedance of a symmetrically loaded cell. (*After Ingerson [22]*)

resonance of the loading element. The dashed line ($\sigma = 0.75$) displays the effect of the structural stopband (200 to 275 MHz) wherein the image impedance is purely imaginary. In this case the rapid variation in the magnitude of the image impedance commences just below the stopband (about 160 MHz). When losses are added to the loads the variation of the image impedance with frequency is reduced. A basic principle of achieving low swr in an LP structure is to achieve a minimal variation in the image impedance from one cell to the next, all the way from the feed point to the end of the active region. When this condition is satisfied, the variation in voltage from cell to cell is approximately the same as it would be for individual, image-terminated cells of the corresponding size in terms of wavelengths. Consideration of plots of complex propagation factor and image impedance versus frequency can, therefore, be very helpful in the design of LP antennas. Parameters required to achieve backfire phasing can be determined from the k-β diagrams. Including the attenuation factor in these plots permits one to judge whether appreciable energy will penetrate past the active region and produce truncation effects. Inspection of the variation of image impedance versus frequency for a single cell indicates whether the assumption of minimal reflections at the cell junctions of an LP array of these cells is valid.

Application of the analysis of open-periodic structures to the corresponding log-periodic ones is complicated when appreciable external coupling exists. The base impedance of a radiating element is affected by this coupling. Hence the Q of a resonant radiator in a periodic or log-periodic array may be quite different from its value when the radiator is isolated. Consideration of the k-β diagram is helpful in evaluating this effect. Fig. 57 shows division of the k-β space into visible and

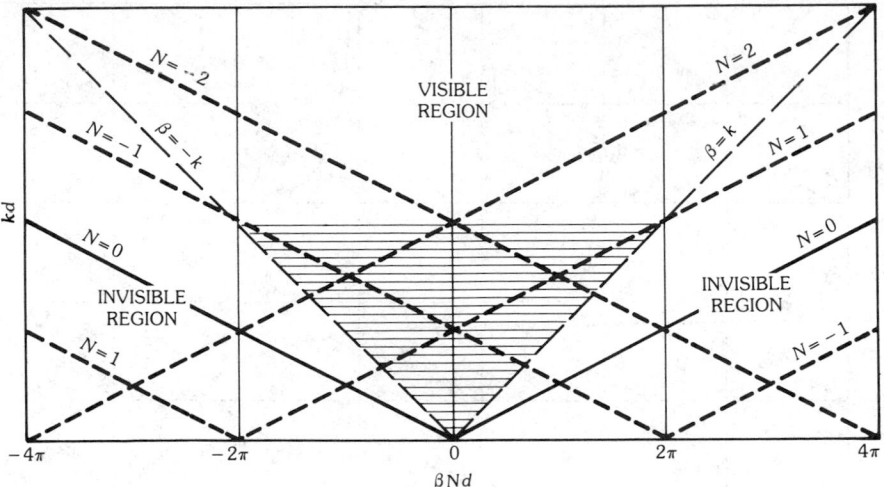

Fig. 57. A k-β diagram showing the fundamental slow wave and some space harmonics. (*After Ingerson [22]*)

invisible regions. In the invisible region neither the fundamental wave nor any of its space harmonics has phase velocity greater than the speed of light. Such waves contribute little to the distant field, and the radiators display high values of Q since the radiative loss is small. On the other hand, a wave in the visible region produces a significant amount of distant field and the radiative loss greatly lowers the Q of elements in an array operating at such points. Hence one important factor when considering the performance of LP antenna arrays with resonant elements is the phasing of the currents on the elements near resonance so that they will be efficient radiators. This requires that the active region phasing correspond to points in or near the visible region of the k-β diagram.

Solving for the modes of propagation on an open structure is also complicated by the mutual coupling. The number of modes on an infinite periodic structure depends on the range of coupling [24]. Interaction limited to adjacent elements produces a single mode; coupling to two adjacent elements produces two modes, and so on. However, experimental measurements on LP structures indicate that only one mode is excited by a source near the tip of such tapered structures.

Fig. 58 shows the k-β diagram of a hypothetical UP dipole array without mutual impedance and the fundamental mode on the same structure when mutual impedances are considered. The main effect of mutual impedance is to change the value of the radiation resistance of the dipoles. To a first approximation the behavior of the dipole in a periodic or log-periodic array can be represented by a lossy stub. When the dipole is short in wavelengths its impedance is approximately that of a stub of characteristic impedance

$$Z_s = 120[\ln(h/a) - 2.25] \tag{16}$$

where h is the dipole length and a is its radius [7,8]. For higher frequencies, through the first resonance, the impedance has appreciable real part which can be included approximately by putting a resistor with value R_s in series with the stub.

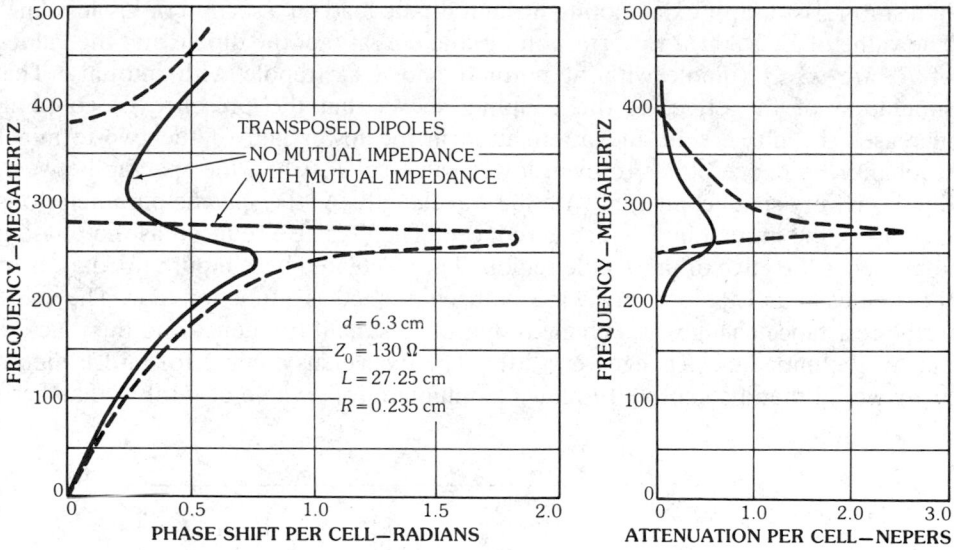

Fig. 58. Dispersion diagram of a periodically loaded transmission line. (*After Ingerson [22]*)

The value of R_s should equal the input resistance of the dipole at resonance. The k-β diagrams for a line symmetrically shunt-loaded by a lossless line in series with the resistor R_s or by a lossy stub with the same characteristic impedance as the lossless one and whose input resistance at resonance is also equal to R_s are found to be essentially identical up to the first antiresonance. Fig. 59 shows the k-β diagrams

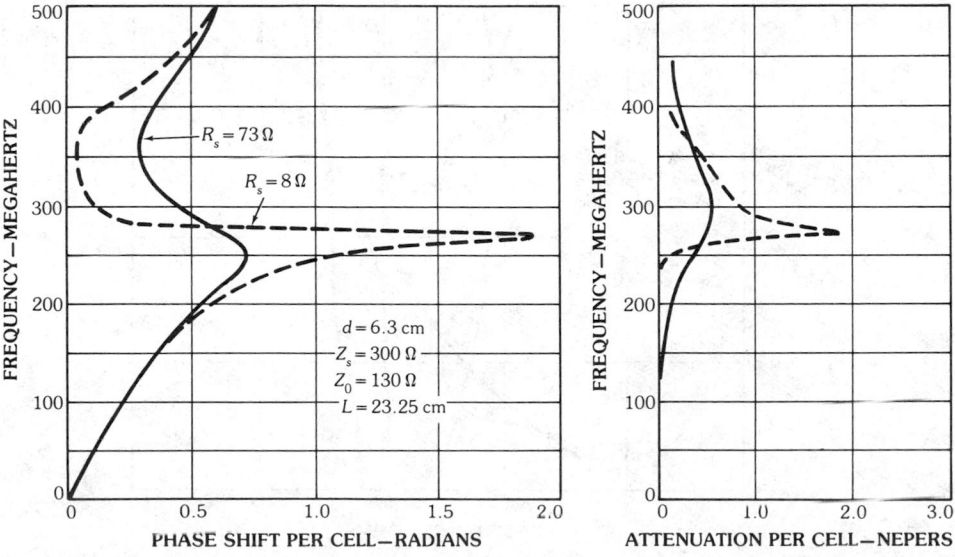

Fig. 59. Dispersion diagram of a periodically stub-loaded transmission line. (*After Ingerson [22]*)

of a stub-resistor approximation to a shunt dipole load on a section of lossless line. The value of Z_s is 300 Ω to correspond to the h/a ratio of the dipole and the values of R_s are 73 Ω (dipole without mutual) and 8 Ω (dipole with mutual). This simulation of the effect of the coupling shows that the presence of coupling increases the phase shift and attenuation in the lower half of the two-terminal stopband. Evidence points to even lower values of R_s when the spacing between dipoles is very small compared with the wavelength. As the spacing parameter σ is increased, the mutual impedance is reduced. The Q is also reduced as the phasing approaches the edge of the visible region. The calculated base input impedances of the dipoles in an LP dipole array are shown in Fig. 60 as a function of σ. The value of the resistance changes as σ changes, but the resonant frequency and the reactive part of the impedance change very little. The swr versus σ curve for an LP dipole array would then be expected to have a minimum for a value of σ other than zero

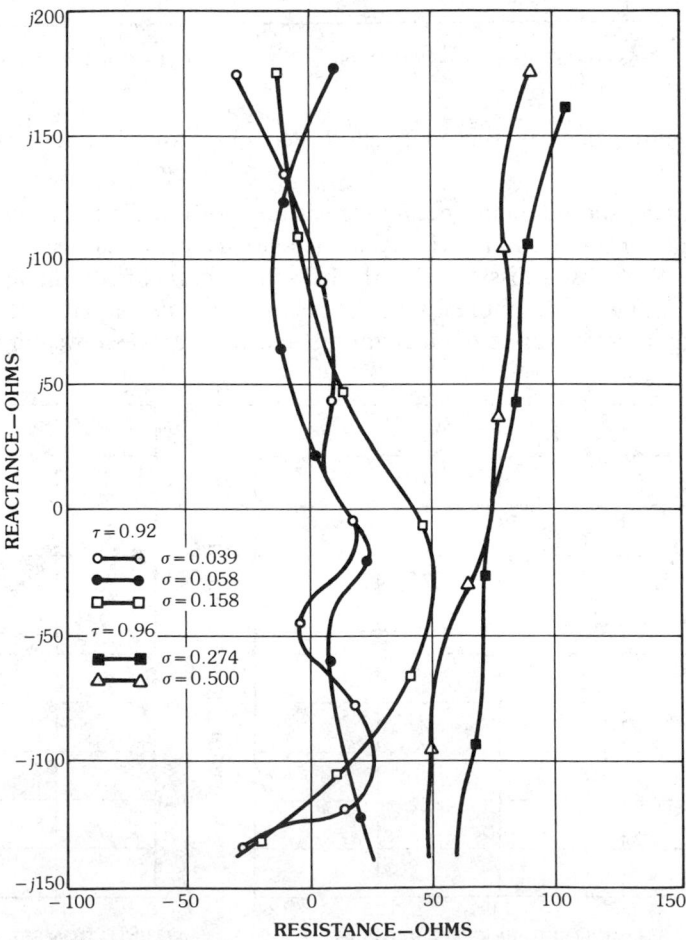

Fig. 60. Calculated base input impedances for dipoles in an LP dipole array. (*After Ingerson [22]*)

(the case without coupling). This is confirmed by the computed swr versus σ curves shown in Fig. 24.

When the dipoles on the feeder of a uniform periodic array are not transposed, a different type of k-β diagram is found for the dominant mode. This is shown in Fig. 61, where two modes are displayed. In the LP structure, however, it again appears as if only the lower wave is excited. The k-β plot for this wave falls inside the invisible region. In the LP structure, then, it would be expected that the dipoles would behave like high-Q stubs and reflect most of the incident energy. A comparison of the transmission-line voltages of the corresponding LP structures with and without transposed dipoles is shown in Fig. 62. It is clearly seen that the untransposed dipoles reflect most of the incident energy producing a high standing wave between the feed point and the resonant dipoles. On the other hand, the transposed dipoles radiate the energy and there is little evidence of a standing wave between the feed point and the active region. The difference between these two cases is attributed to the fact that the currents in the dipoles near resonance are phased close to the visible region when the feeder is transposed.

This view of the dipole excitation being due to waves traveling in opposite directions can be used to relate the swr of LP dipole arrays and the front-to-back ratio of their radiation patterns. It is plausible to expect that the ratio of amplitudes of the energy radiated to the front and rear will be related to the amplitudes of the incident and reflected energy on the feeder. Moreover, except for the scattering due to the larger dipoles, the shape of the patterns would be expected to be similar if τ is not too small. The ratio of incident and reflected power is related to the input vswr by

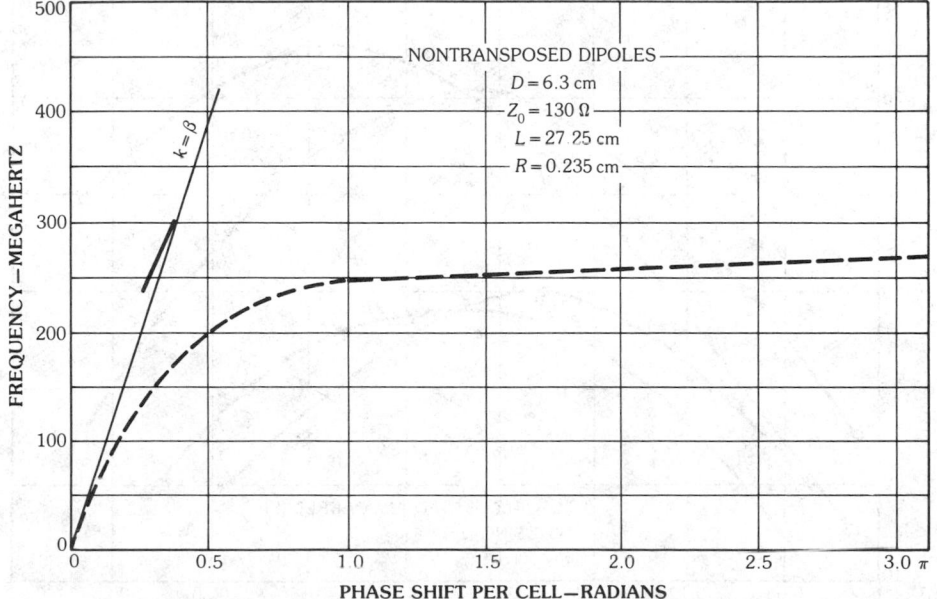

Fig. 61. Dispersion diagram for a periodically loaded transmission line with nontransposed dipoles. (*After Ingerson [22]*)

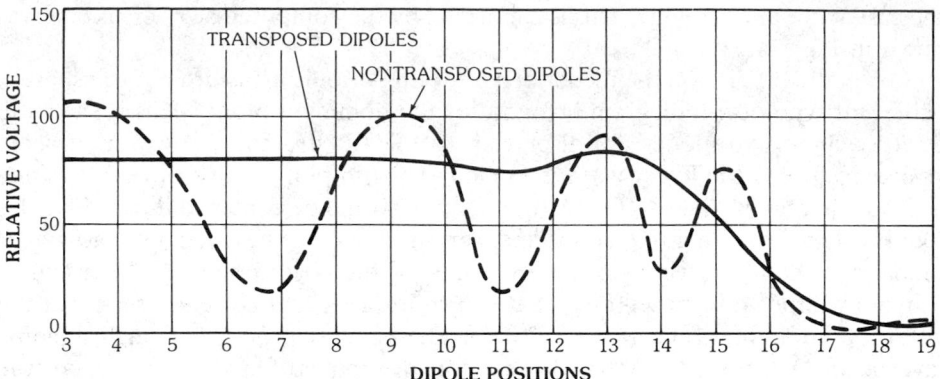

Fig. 62. Relative transmission-line voltage for a transmission line log-periodically loaded with dipoles and with dipoles alternately transposed for additional phase shift. (*After Ingerson [22]*)

$$F/B \cong 10\ln(P_i/P_r) = 20\ln[(\text{vswr} + 1)/(\text{vswr} - 1)] \qquad (17)$$

where P_i and P_r represent the incident and reflected powers. Fig. 63 compares this relationship between front-to-back ratio and the spacing parameter σ with the results previously obtained by computer analysis of many cases [8]. The agreement confirms this as a useful approximation for the front-to-back ratio and the manner in which it will change as a function of antenna parameters. Fig. 64 gives a plot of (17).

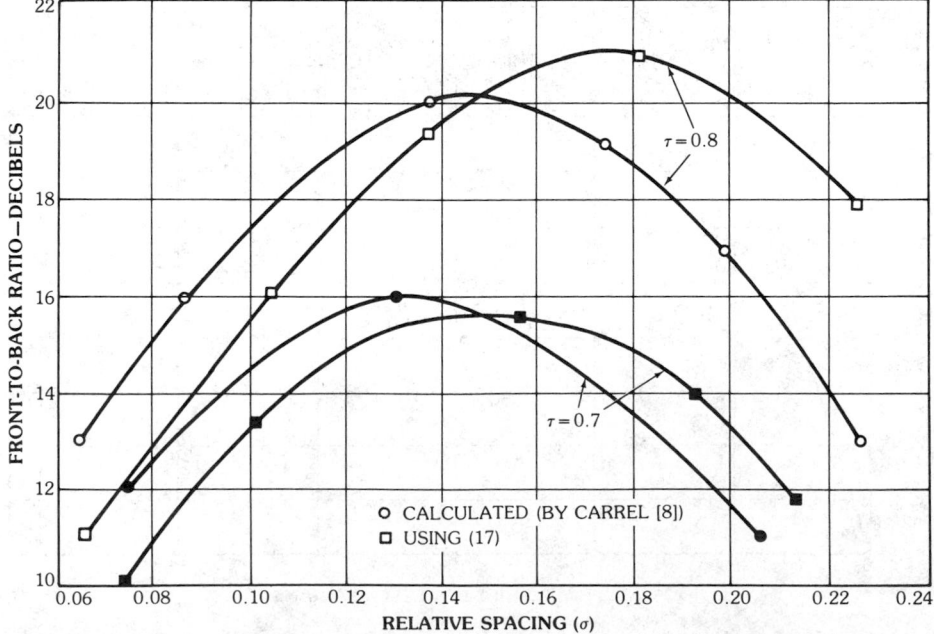

Fig. 63. Calculated front-to-back ratios for LP dipole arrays. (*After Ingerson [22]*)

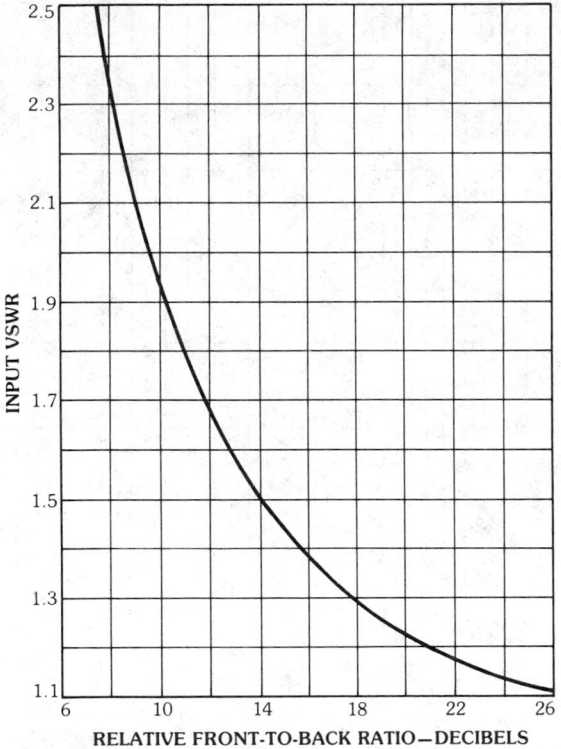

Fig. 64. Simple approximation relating input vswr to front-to-back ratios in LP dipole arrays. (*After Ingerson [22]*)

Several useful design principles can be based on the above arguments about the apparent Q of dipoles in a periodic or log-periodic array environment. For stub-resistor–loaded transmission line sections the swr depends mainly on the Q of the loading elements and not on the characteristic impedance of the feeder. The Q of dipoles in the array depends mainly on σ and only slightly on τ. Hence the vswr of a dipole array is relatively independent of the feeder characteristic impedance. Thus both the vswr and the front-to-back ratio are determined once the minimum Q and τ are known. The directivity, however, is not determined by Q and τ but is a strong function of σ and the characteristic impedance of the feeder.

As mentioned above, the apparent Q of dipoles in a periodic or log-periodic array can be changed by changing σ. For values of σ greater than 0.25 the Q of the dipoles in an array approaches the value for an isolated dipole. The Q of an isolated dipole is primarily dependent on h/a. Fig. 65 shows how the impedance of the equivalent stub depends on h/a. It is clear that to change Q by a factor of 2, h/a must change by a factor of approximately 10. This result substantiates the common practice of using the same radius for several dipole elements having about the same lengths in an LP dipole array. It is also apparent that, unless very thin dipoles have been used, it is not possible to improve the vswr of an LP dipole very much by replacing the dipoles with thicker ones.

In choosing the parameters for an LP antenna it is particularly important that

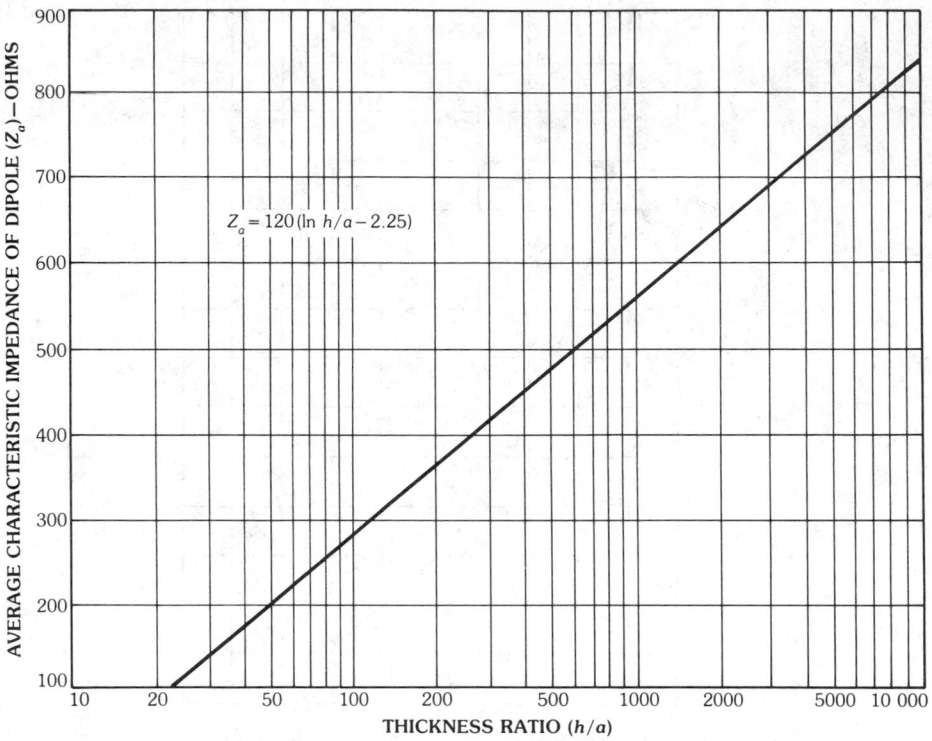

Fig. 65. Average characteristic impedance of a dipole versus height-to-radius ratio h/a. (*After Carrel [8]*)

the stopband which results from a particular choice stop any energy from exciting dipoles which are too long or improperly phased. Since the stopband width is largely determined by the ratio of Z_s/Z_0, the minimum value of τ must be adjusted according to that ratio. To achieve LP dipole designs that have low values of input impedance it is, therefore, necessary to lower Z_s in order to obtain a stopband of sufficient width. Designing LP dipole arrays with low values of input impedance demands that high values of h/a be used. When Z_0 is low compared with Z_s the minimum value of σ must be increased to reduce leakage of energy through the active region.

Log-Periodic Designs Based on Periodic Structure Theory

Parameters for a side-firing LP dipole array can be obtained from study of k-β diagrams. The diagram in Fig. 66 shows that for dipoles with $h/a = 116$, spaced at 50.4 cm, the voltage phasing near resonance approaches 2π radians when a transposed feeder is used between dipoles. Fig. 66 also shows that appreciable attenuation occurs, not only when the dipole is near resonance, but also for shorter dipoles. On an LP structure this would correspond to radiation primarily in the backfire direction. Radiation, however, can be produced in other directions by suppressing radiation in the backfire direction. This is accomplished by using dipoles with very high values of Q in conjunction with a feeder having a low value

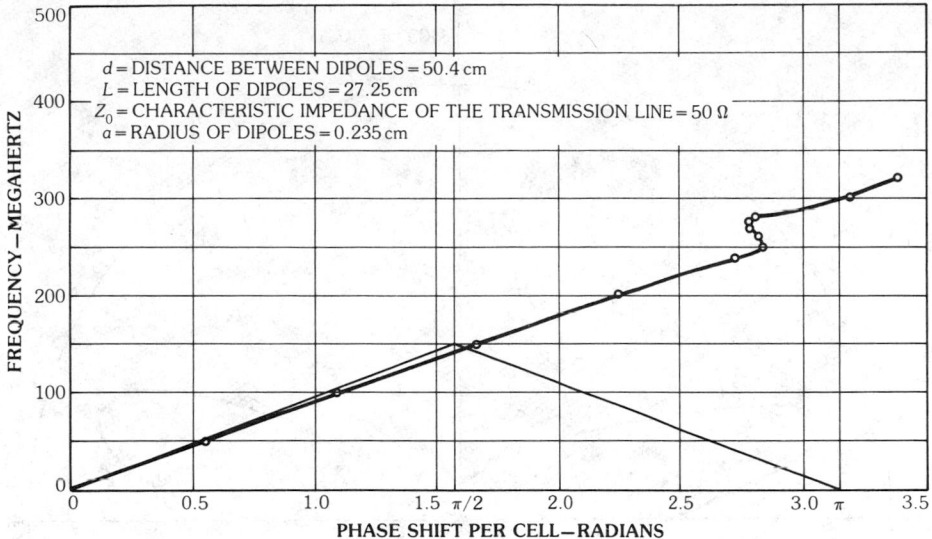

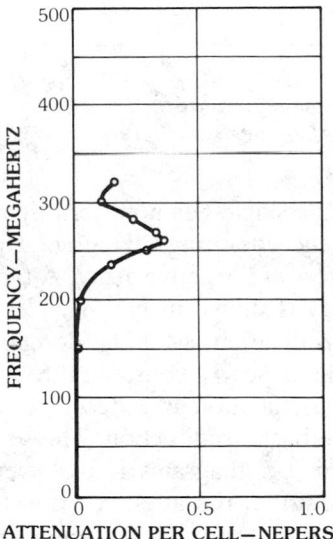

Fig. 66. Dispersion diagram for periodically dipole-loaded transmission line. (*After Ingerson [22]*)

of characteristic impedance. For example, a side-firing LP dipole array has been achieved with $\tau = 0.96$, $\sigma - 0.25$, and $Z_0 = 50\ \Omega$. In addition, dielectric was inserted between the conductors of the feeder in order to increase the phase shift between the dipoles in the active region. Fig. 67 shows a typical *H*-plane radiation pattern measured for the example side-firing antenna. The split-beam pattern of

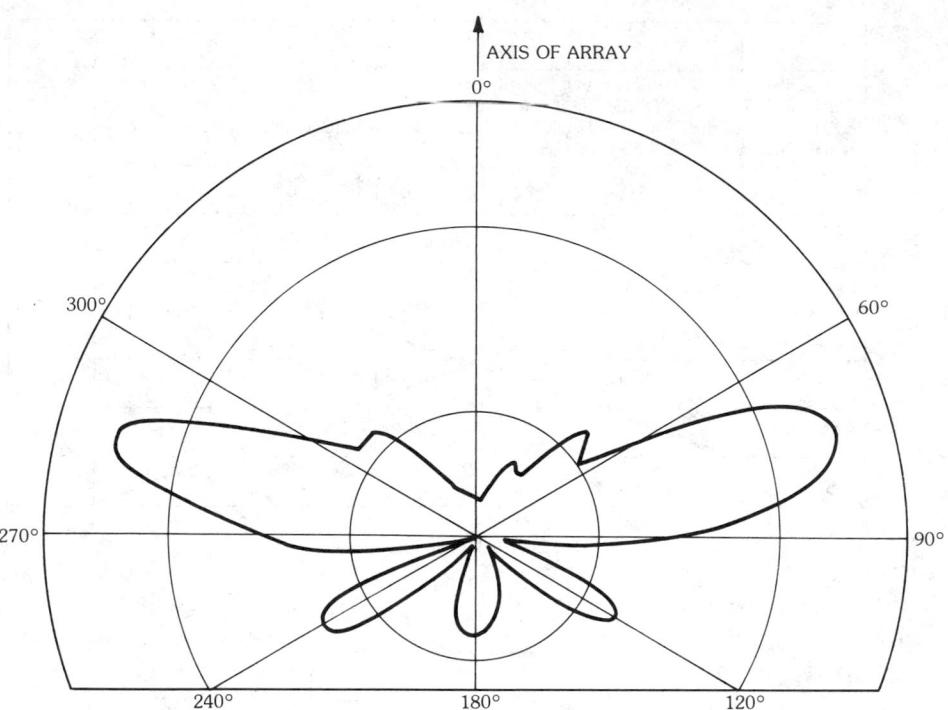

Fig. 67. Typical measured *H*-plane pattern of a side-firing LP dipole array, for $\tau = 0.96$, $\sigma = 0.25$, and $Z_0 \cong 50 \ \Omega$. (*After Ingerson [22]*)

Fig. 67 can be converted to a single-beam pattern by adding an array of parasitic elements along one side of the side-firing LP dipole. The parasitic elements should conform to the LP parameters of the active array. A typical unidirectional *H*-plane pattern obtained in this way is shown in Fig. 68.

Another use of the *k-β* diagram is for the design of high-directivity backfire arrays. If the phasing is held nearly constant through an active region having several elements and the attenuation is reduced, it is possible to obtain more elements that radiate in the backfire direction. The solution indicated by study of the trends indicated by the *k-β* diagrams is to lower the *Q* of the dipoles by increasing *σ* and *a/h*. In addition, the attenuation in the active region should be reduced by lowering the value of the feeder impedance. Fig. 69 shows *E-* and *H*-plane patterns that were measured on an LP dipole array having $\tau = 0.95$, $\sigma = 0.175$, and $Z_0 = 50 \ \Omega$. Dielectric was also used in the feeder to increase the phase shift. In this way it is possible to obtain gain values that are greater than those indicated for the same parameters in Fig. 20.

The dipole arrays discussed above all have transposed feed lines in order to achieve phasing of the active region that is in or near the visible region. Feeding an LP array of monopole elements over ground with a transposed feeder is not possible. Consideration of the *k-β* diagrams of Figs. 54a and 54b shows that backfire phasing could also be achieved by using lengths of feeder that are much longer than the separation between the monopoles. These same *k-β* diagrams, however, show that a structural stopband will occur at frequencies below the first

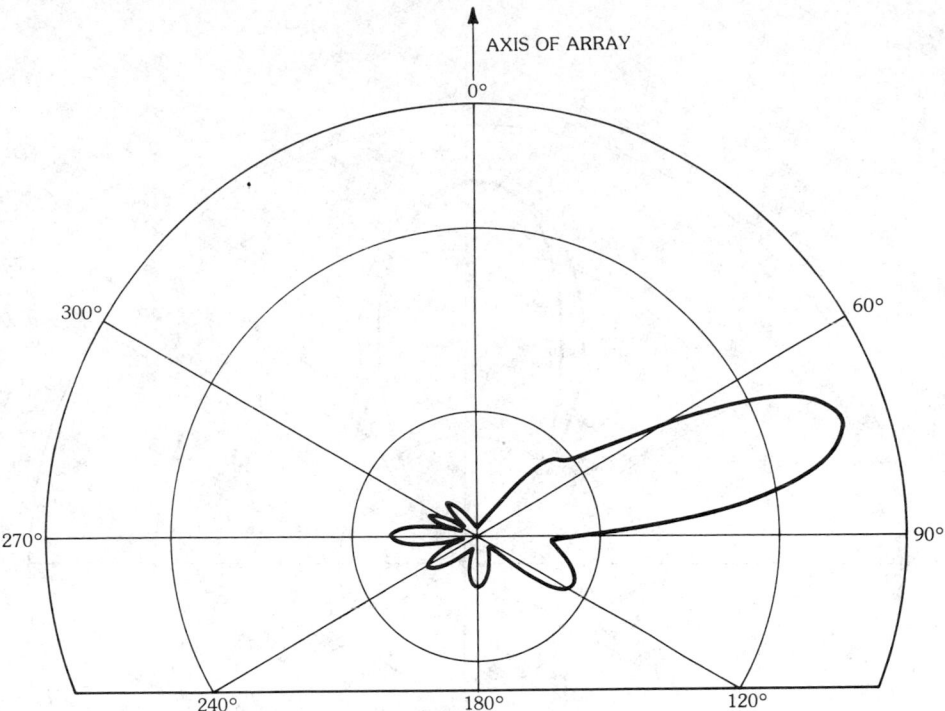

Fig. 68. Measured unidirectional *H*-plane pattern of a side-firing dipole array with an LP array of parasitic reflectors. (*After Ingerson [22]*)

two-terminal stopband when these longer feeders are used. The ability to get energy from the feed point of an LP monopole array with excess feed-line phasing is then dependent on using some technique to eliminate the structural stopband. One such technique is to modulate the impedance of the feeder as shown in Fig. 70 [26].

Except for the feeder, the design of LP monopole arrays with excess feeder follows the procedures previously given for LP dipole arrays. The nontransposed feeder results in a wider active region for LP monopole arrays than is obtained for LP dipole arrays which use a transposed feeder. Design of the modulated impedance feeder can be facilitated by means of a computer-aided technique. The procedure assumes that τ and σ have been chosen by using the LP dipole data. The feed-design steps are then applied to find the parameters of a single cell, say the largest, and the dimensions of the other cells are determined by applying the appropriate power of τ.

The total length L_n of feeder between any two adjacent elements with heights h_n and h_{n+1} is determined from

$$L_n = 4h_n\lambda_r(m - c\sigma), \qquad m = 0, 1, 2, \ldots \tag{18}$$

where λ_r is the ratio of the feed-line wavelength to free-space wavelength and c is a coefficient slightly greater than unity. The parameter m is usually chosen to be unity to avoid overlong feeder sections. No attempt is made in this formula to

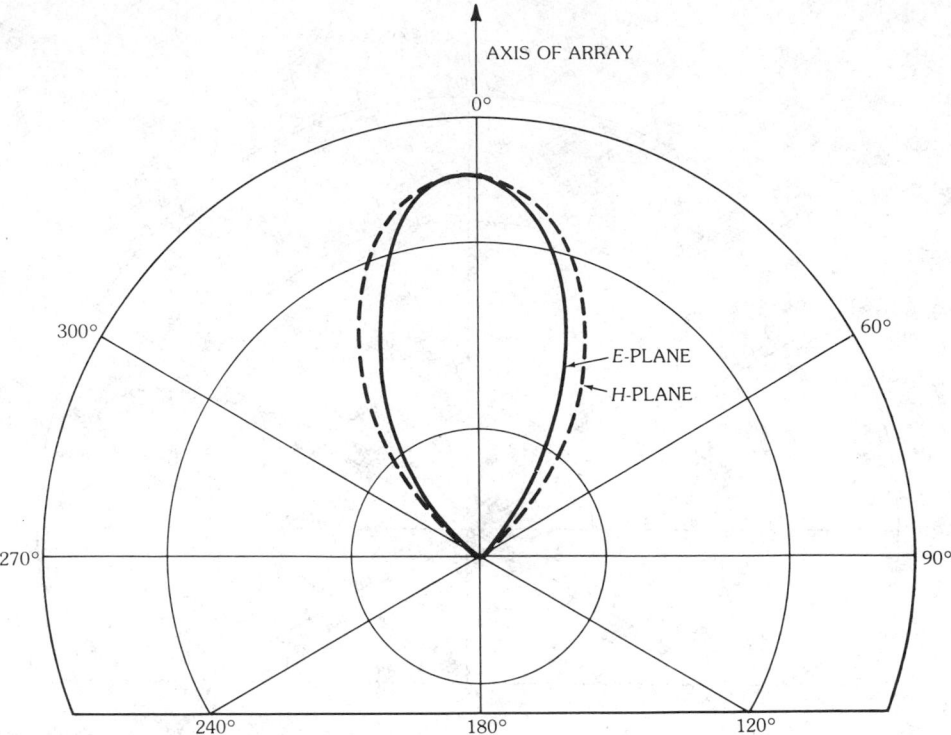

Fig. 69. Measured patterns of a log-periodic dipole array ($\tau = 0.95$, $\sigma = 0.175$), showing increased directivity using low characteristic impedance feeder ($Z_0 \cong 50\ \Omega$). (*After Ingerson [22]*)

consider that the phase velocity along the feeder may be different in sections with different characteristic impedances. Hence, if nonhomogeneous lines, such as microstrip, are used for the feeder, λ_r should be adjusted according to the approximate lengths of line of each different impedance that are expected to be used.

The feeder between any two adjacent monopoles will appear as in Fig. 71. The relative length of the lines with different impedance is

$$r = d_2/d_1 \tag{19}$$

The initial value of r is rather arbitrarily chosen somewhat greater than unity to keep the length of high-impedance line short. For some values of r the image impedance of the subcell may be imaginary and the desired match to R_{01} may not be possible. A different value of r must then be chosen.

Once r has been fixed, d_1 and d_2 can be determined from

$$d_1 = \frac{2L_n}{(1 + r)(1 + \tau)}$$

$$d_2 = rd_1 \tag{20}$$

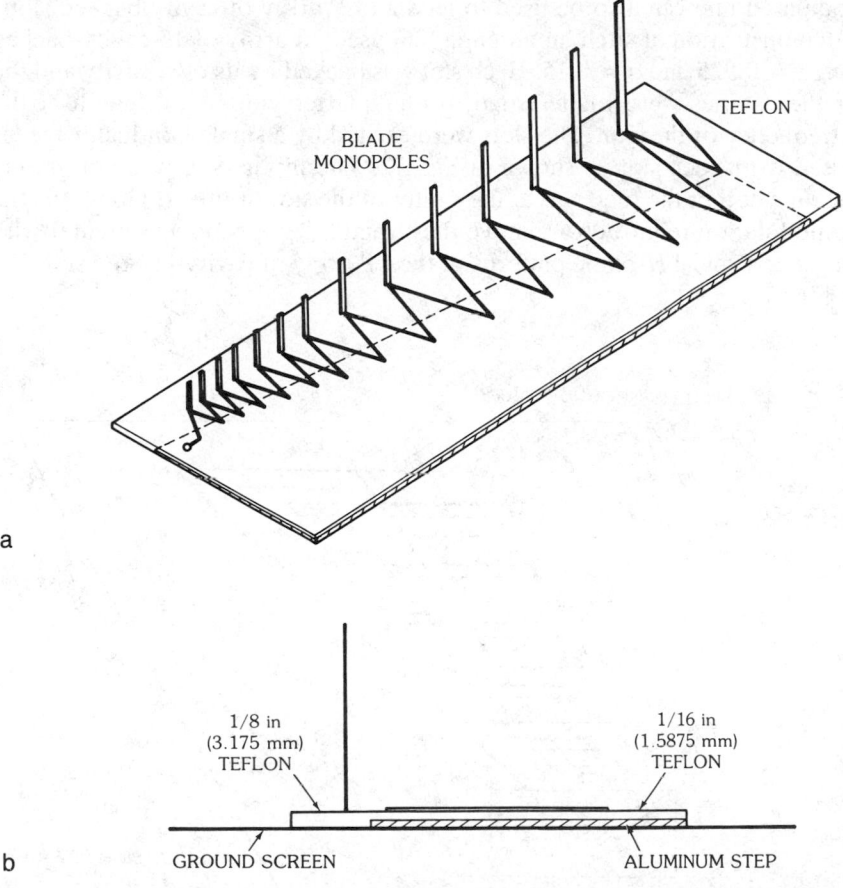

a

b GROUND SCREEN   ALUMINUM STEP

Fig. 70. Log-periodic scaled monopole array fed with a modulated-impedance meandering line. (*a*) Perspective. (*b*) Side view. (*After Ingerson and Mayes [26], © 1968 IEEE*)

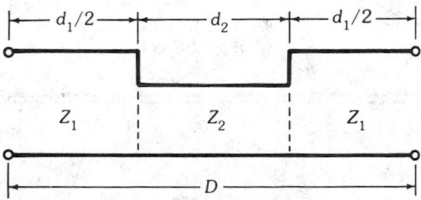

Fig. 71. A section of modulated-impedance feeder for a log-periodic monopole array. (*After Ingerson [22]*)

The dispersion characteristics of the unmodulated cell ($R_{02} = R_{01}$) can be calculated. At the frequency of maximum attenuation in the stopband, set R_{02} equal to the image impedance of the subcell. As a check to ensure that the structural stopband has been minimized, the dispersion characteristics for the modulated cell can be calculated.

A modulated line can also be used to feed an LP array of cavity-backed slots. The first demonstration of such an antenna [26] used an array of 15 cavity-backed slots having $\tau = 0.925$ and $\sigma = 0.15$. Each slot was backed by its own cavity and the depths of the cavities were made equal to one-quarter guide wavelength at the resonant frequency of the slot. The slots were excited by a single-conductor feeder which passed over each slot as shown in Fig. 72. The input vswr was not greatly dependent on whether the feed was at the center of the slot or offset. However, the degree of modulation required to remove the structural stopband was greater with the center feed. Typical *H*-plane patterns of the LP array of cavity-backed slots are shown in Fig. 73.

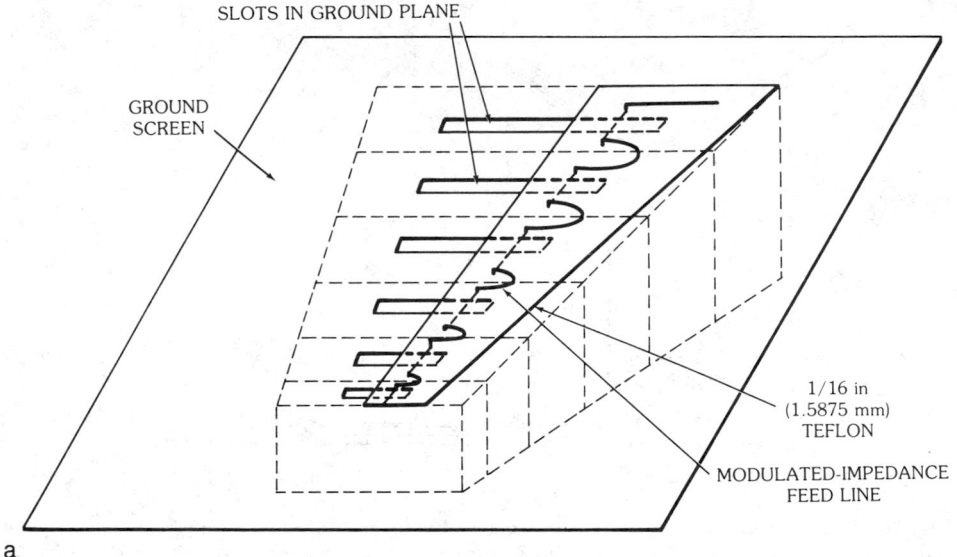

a

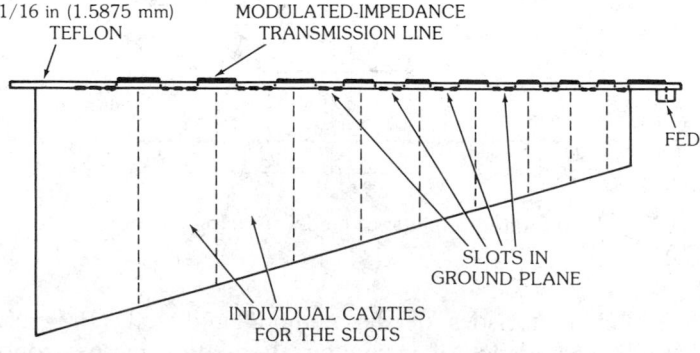

b

Fig. 72. An LP cavity-backed slot array with modulated-impedance feed line. (*a*) Perspective. (*b*) Side view. (*After Ingerson and Mayes [26], © 1968 IEEE*)

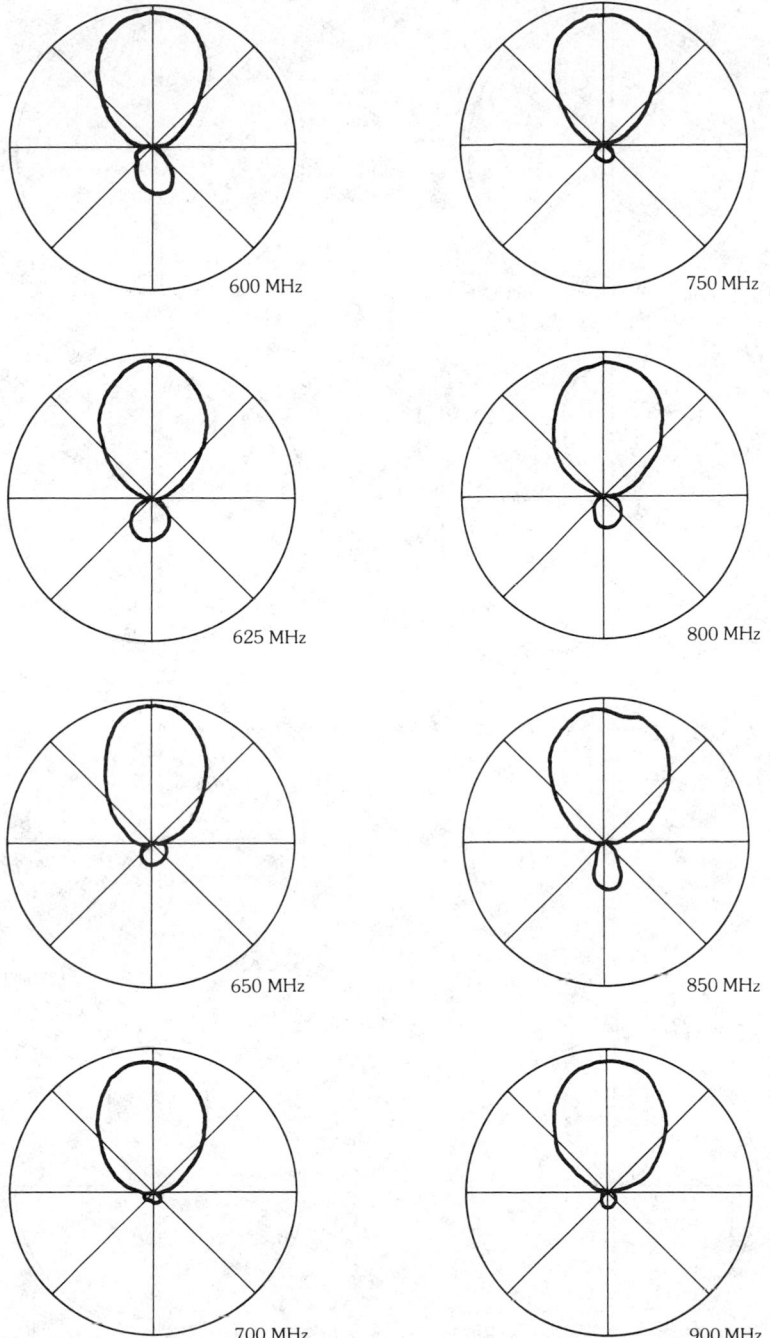

Fig. 73. Typical *H*-plane radiation patterns of a cavity-backed slot array, for $\tau = 0.925$, $\sigma = 0.15$, and $N = 15$. (*After Ingerson [22]*)

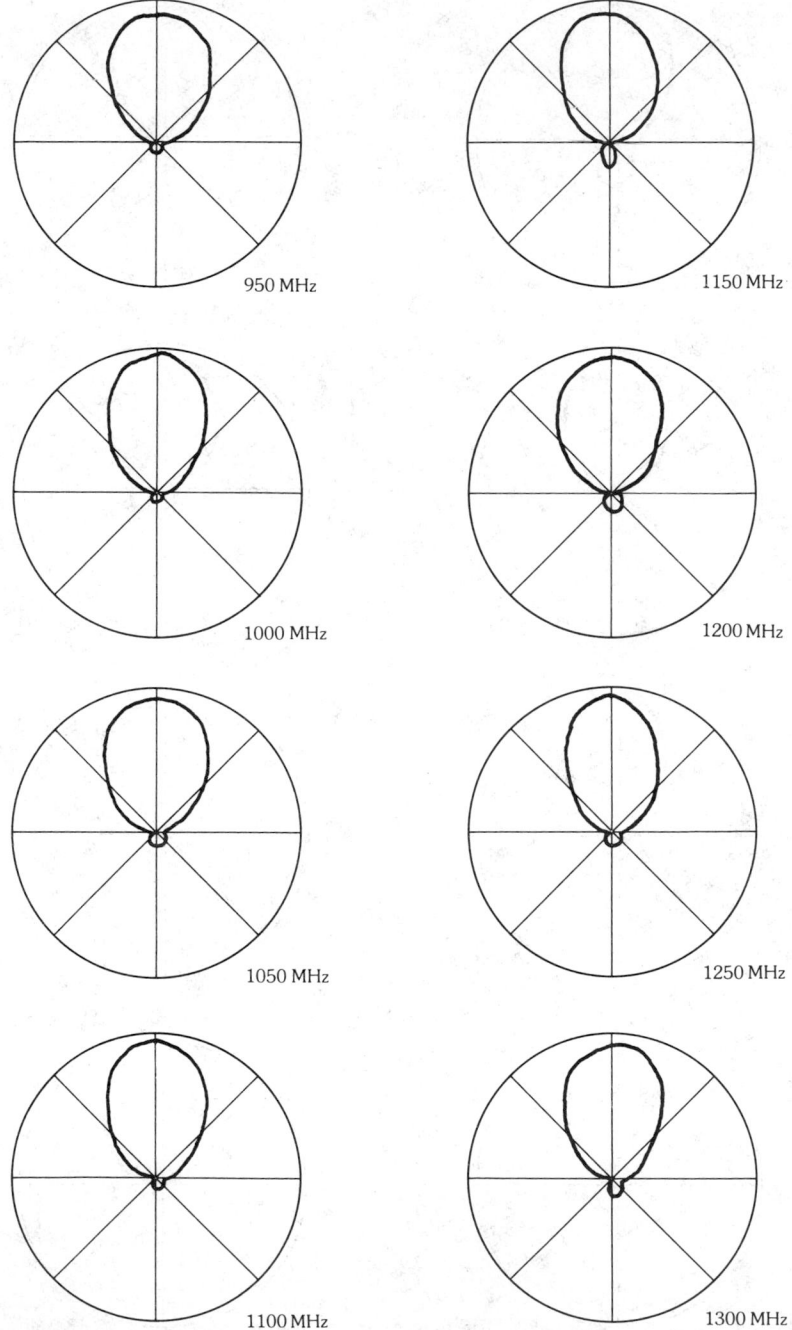

950 MHz

1150 MHz

1000 MHz

1200 MHz

1050 MHz

1250 MHz

1100 MHz

1300 MHz

Fig. 73, *continued.*

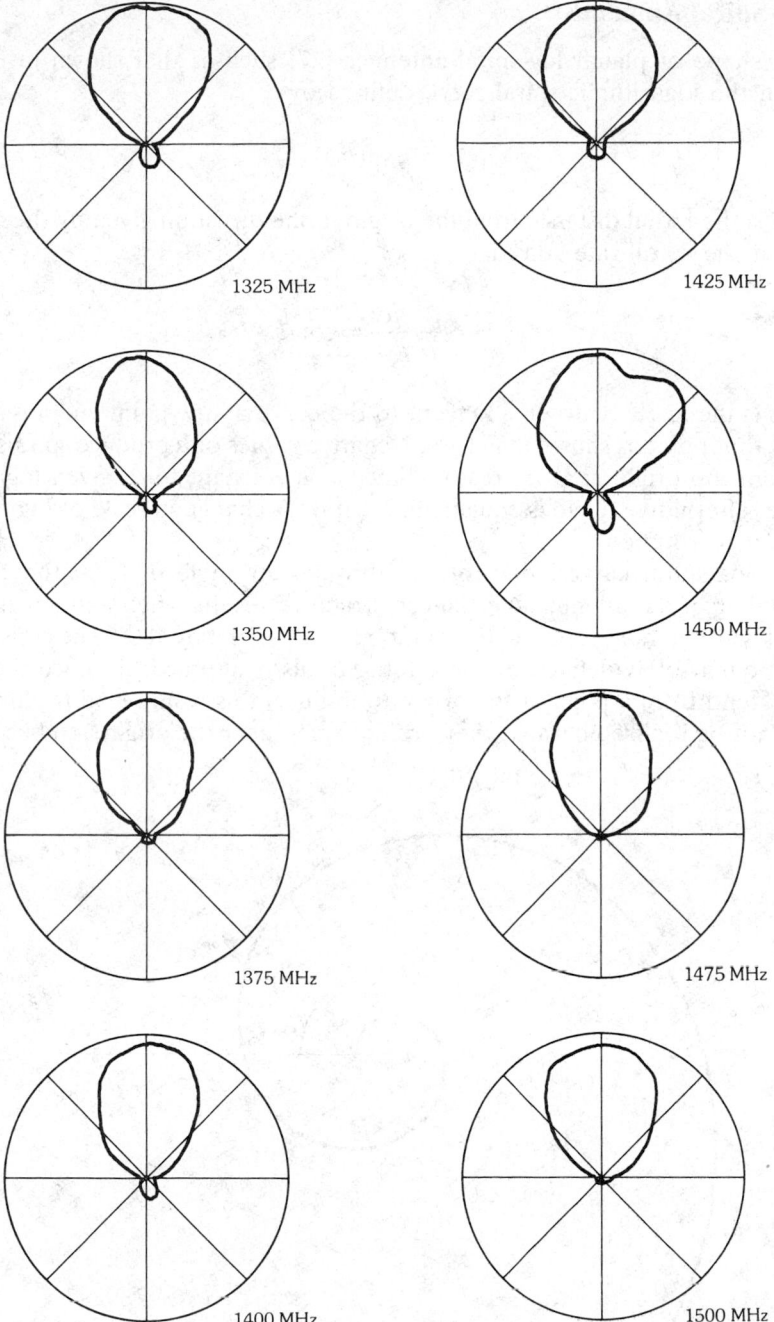

1325 MHz

1425 MHz

1350 MHz

1450 MHz

1375 MHz

1475 MHz

1400 MHz

1500 MHz

Fig. 73, *continued.*

4. Log-Spiral Antennas*

The shape of planar log-spiral antennas [27] such as that shown in Fig. 5 is based on the logarithmic spiral curve defined by

$$\varrho = \exp(b\phi) \tag{21}$$

where ϱ is the radial distance from the origin in the direction given by the angle ϕ. Note that the spiral-rate constant

$$b = \frac{1}{\varrho}\frac{d\varrho}{d\phi} = \cot\alpha \tag{22}$$

where α is the angle between a tangent to the curve at any point and a line to the origin at that point as shown in Fig. 74. Negative values of b produce spirals that go away from the origin as ϕ decreases. Since α is constant for a given logarithmic spiral, an alternative name is equiangular spiral. A change in scale of log-spirals is equivalent to rotation.

Two log-spiral curves, one rotated through an angle δ, form the edges of one member of an antenna. A balanced structure results when another member, identical with the first, is placed by rotation from the first by 180°. The construction of an antenna solely defined by these four spirals is impractical. Since the spirals would extend from a point at the origin to infinity, this would lead to dimensions that cannot be realized because they are too small near the origin and become too

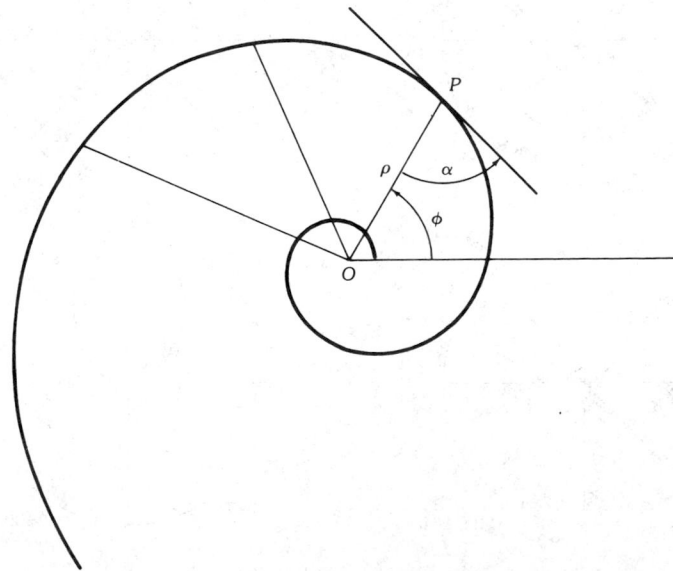

Fig. 74. The equiangular spiral. (*After Dyson [27, 28], © 1959 IEEE*)

*The assistance of John D. Dyson in the preparation of this section is gratefully acknowledged.

large at some distance far short of infinity. Hence the spiral edges must be limited to finite-length curves. These can be conveniently taken as segments of the infinite spirals that lie between two values of ϕ as shown in Fig. 5, or between two values of ϱ as shown in Fig. 75. Using a circle to define the truncation at the large end apparently minimizes the lower limit of frequency-independent performance for a given antenna diameter. A practical, balanced log-spiral antenna is therefore described by the two circles $\varrho_i = d/2$, $\varrho_0 = D/2$, and the four spirals

$$\varrho_3 = \frac{d}{2}\exp(b\phi)$$

$$\varrho_4 = \frac{d}{2}\exp[b(\phi - \delta)]$$

$$\varrho_1 = \frac{d}{2}\exp[b(\phi - \pi)]$$ (23)

$$\varrho_2 = \frac{d}{2}\exp[b(\phi - \pi - \delta)]$$

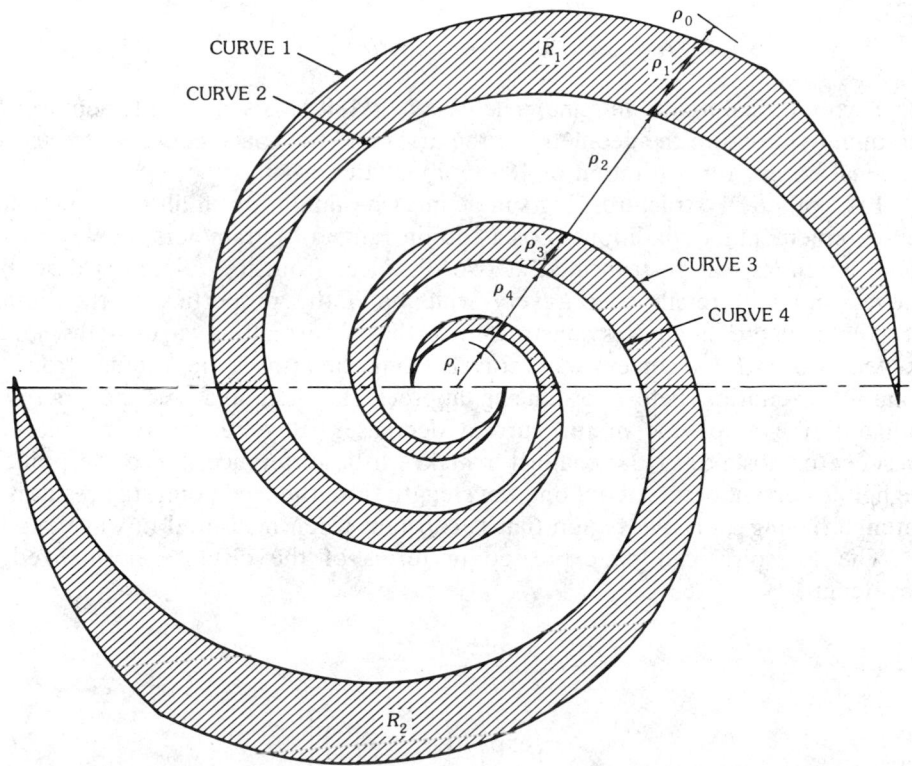

Fig. 75. A balanced, two-arm, log-spiral antenna. (*After Dyson [27]*)

where $d/2 < \varrho_n < D/2$. The regions

$$R_1: \quad \varrho_1 < \varrho < \varrho_2, \qquad \varrho_i < \varrho < \varrho_0$$
$$R_2: \quad \varrho_3 < \varrho < \varrho_4, \qquad \varrho_i < \varrho < \varrho_0$$

shown cross-hatched in Fig. 75, may be covered with conducting sheet material to make a balanced, two-arm, "dipole" antenna, or these same regions may be cut from a large conducting sheet to make a balanced, two-arm, "slot" antenna.

The sense of rotation of the elliptically polarized field radiated by log-spiral antennas is determined by the direction of winding of the spiral arms. Planar spirals radiate one sense of polarization in one half-space and the opposite sense in the other. The sense can be determined from the hand used when pointing the fingers in the direction the current wave travels and the thumb in the direction of the radiated beam. This procedure gives a result which corresponds to the IEEE standard. For right-hand polarization in the positive-z half-space the currents must travel away from the feed point in the direction of increasing ϕ. The spiral in Fig. 75, having a negative value of b, will produce left-hand polarization in the space above the page and right-hand polarization in the space below.

A symmetrical antenna with more than two arms can be constructed by letting the angle equal to π in (23) be $2\pi/n$ instead, where n is the number of arms. Since the angle δ defines the width of the arms,

$$0° < \delta < 360°/n$$

for a symmetrical n-arm antenna. The case $\delta = 180°/n$ is self-complementary; for the infinite structure the geometry of the arms and the space between the arms is identical except for a rotation of $180°/n$ about the origin.

Like other LP structures, log-spirals must be infinite to fulfill the scaling and self-complementary conditions. For a certain range of parameters, however, log-spiral antennas can be truncated at two distances from the origin, as described above, and still retain over a very wide band the properties of the infinite structures. If the antenna is constructed so that d is small in wavelengths, and is excited at $\varrho = d/2$, a current wave travels along the arms some distance from the point of excitation before producing appreciable radiation. As the energy is radiated, the amplitude of the current decreases. Beyond a certain point the presence or absence of the conductor makes little difference. Since the point of negligible current occurs about one wavelength from the feed point, the result is an antenna having an active region that is constant when measured in wavelengths.

The log-spiral can be expressed in terms of the distance normalized to wavelength:

$$\frac{\varrho}{\lambda} = \frac{\exp(b\phi)}{\lambda}$$

$$= \exp\left[b\left(\phi - \frac{\ln\lambda}{b}\right)\right] \qquad (24)$$

$$= \exp(\phi - \phi_0)$$

where $\phi_0 = (\ln \lambda)/b$. The effect of changing the wavelength is equivalent to changing the angle ϕ. Since the active region rotates at a rate dependent on the spiral parameter as the frequency is changed, the radiation pattern also rotates with changing frequency at the same rate. The pattern therefore repeats at frequencies related by integer powers of the scaling factor as for other LP geometries [28, 29]. Except for rotation the radiation characteristics are ideally independent of frequency within the limits imposed by the values of d and D.

A balanced feed is necessary for optimum performance of the balanced antennas. The feed may be brought in perpendicular to the plane of the antenna by a balanced, two-wire line or by an unbalanced (coaxial) line feeding a balanced-to-unbalanced transformer (balun) at the feed point. Of course, the operating bandwidth of the balun must exceed that of the antenna for the latter to be fully realized. The rapid decay of the current along the spiral arms makes possible a convenient frequency-independent method of feeding the balanced antenna with an unbalanced line [27]. The outer conductor (shield) of a coaxial cable may be bonded to the arms of the "dipole" antenna, or to the conductor between the arms of a "slot" antenna. Ideally, this cable would lie along the spiral path midway between the edges of one conductor from $\varrho = D/2$ to $\varrho = d/2$. Only the center conductor crosses the circle $\varrho = d/2$ and is electrically connected to the opposite conductor. A dummy cable, rotated from the feed cable by 180°, can be employed to maintain symmetry. As the frequency is decreased, the diameter of the active region will become greater than D. The current distribution will then be altered by currents on the cable and elsewhere, and the pattern and polarization will differ from that of the frequency-independent mode. The frequency at which the change in performance exceeds some specified value (dependent on the application) becomes the low-frequency limit. The upper frequency limit is determined by the size of the feed region ($\varrho < d/2$) inside which the log-spiral geometry is not continued. Three examples of transitional shapes that have been used in the feed region of log-spiral antennas are shown in Fig. 76. Reducing the dimension of the feed gap in this way will extend somewhat the upper frequency limit. The upper frequency limit is dictated to some degree by the diameter of the feed cable.

The parameters of planar log-spiral antennas are as follows:

b = the rate of expansion (or α, the expansion angle),

$K = \exp(-b\delta)$ = the arm width,

$L = (b^{-2} + 1)^{1/2}(\varrho_2 - \varrho_1)$ = the arm length.

where ϱ_1 and ϱ_2 denote the endpoints. Frequency-independent performance requires that L be approximately equal to one wavelength at the lowest frequency. Hence the maximum diameter, D, is dependent on the rate of spiral, b. The operating characteristics are not strongly dependent on the values of b and K, which typically fall in the ranges 0.2 to 1.2 and 0.375 to 0.97, respectively. The radiation patterns are relatively insensitive to variation in values of b and K, although there are optimum ranges. Good patterns can usually be obtained with only one and one-half turns of the spiral.

Log-spiral antennas radiate two broad lobes in directions perpendicular to the

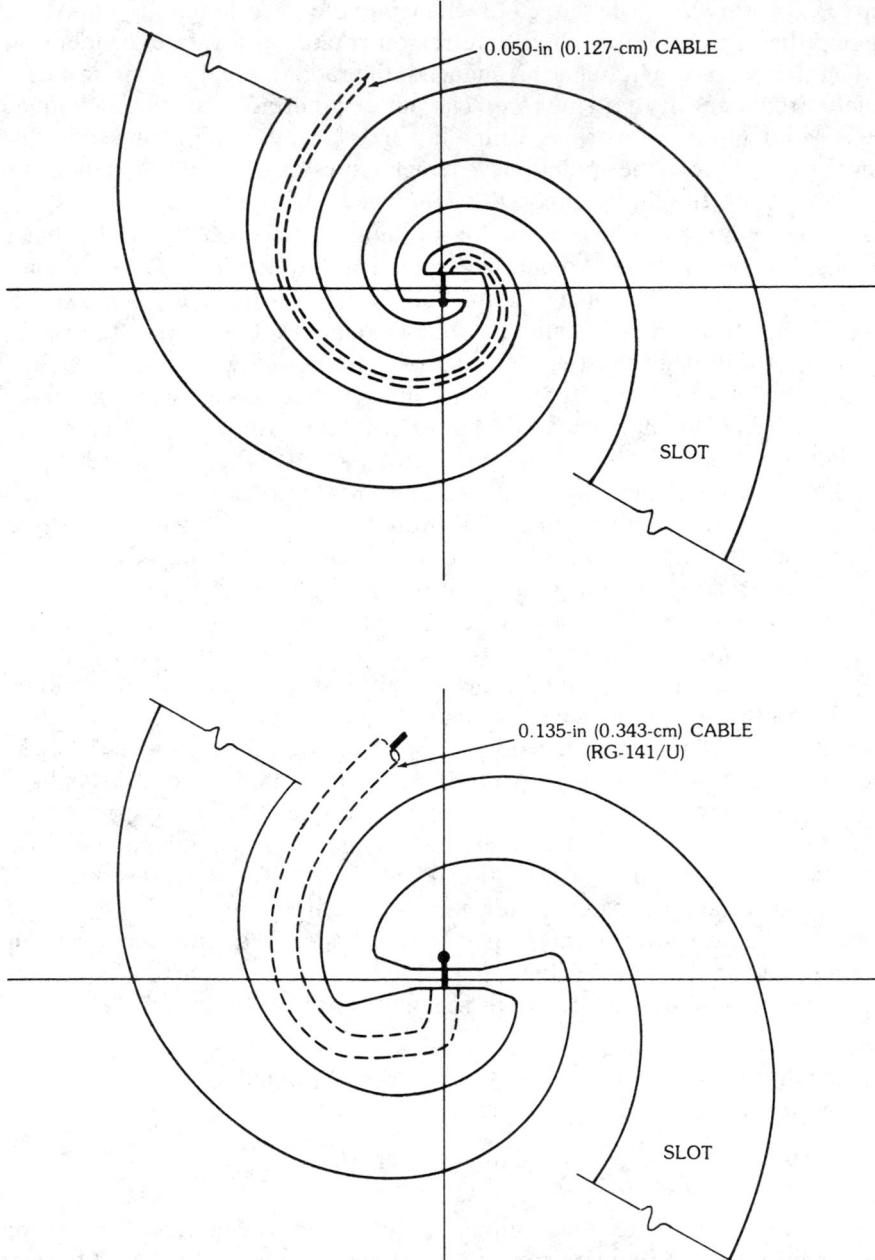

Fig. 76. Examples of feed-gap geometry for balanced, two-arm, planar, log-spiral, slot antennas. (*After Dyson [27]*)

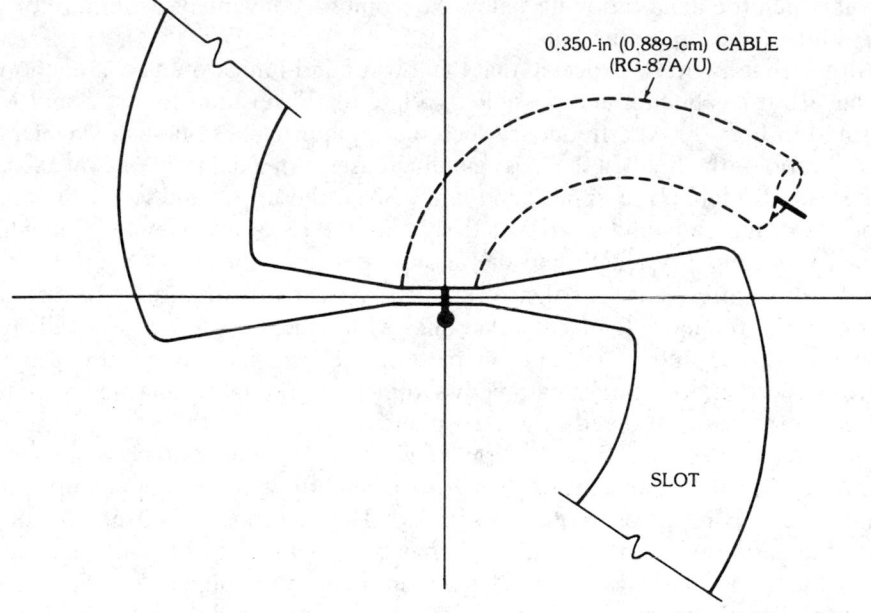

Fig. 76, *continued.*

plane of the antenna. At midband frequencies the beams are circularly polarized on the axis, with axial ratios rising slowly as the off-axis angle increases. Polarization is more sensitive to changes in frequency than is pattern shape. When the antenna arms are very short in wavelengths the polarization is linear, even on the axis. As frequency increases, the axial ratio decreases typically as shown in Fig. 77. The

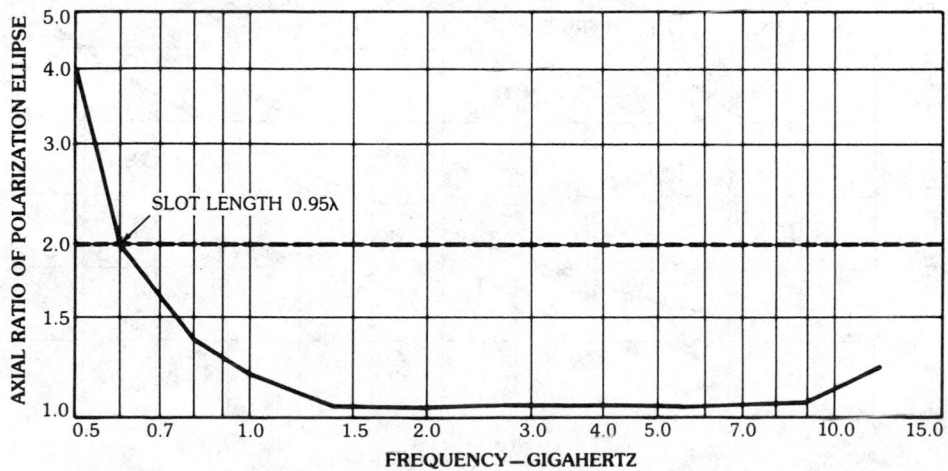

Fig. 77. On-axis polarization of a typical, balanced, two-arm, planar, log-spiral, slot antenna. (*After Dyson [27, 28],* © *1959 IEEE*)

point at which the axial ratio falls below 2 becomes a convenient definition for the lower limit of the operating band.

Although it might be expected that the lower band-limit would be a function of b, δ, and L, it has been found possible to relate the lower limit to just L and K as illustrated in Fig. 78. At a frequency such that ϱ_i approaches one-half wavelength the axial ratio of the field on the axis again increases. An axial ratio of 2 on axis can also be used to define the upper band-limit. Since the upper and lower limits are independent, the only limitations on a design are the required diameter D in which to spiral the necessary length and the allowable size of the feed region.

Although patterns of the balanced spirals are relatively insensitive to variation in parameters, the more tightly spiraled ones with wide arms (b and K small) tend to have smoother and more uniform patterns. These parameters also produce patterns that are more nearly rotationally symmetric, and thus display less variation in beamwidth when observed in a fixed plane.

For frequencies such that the arm lengths of a planar, balanced, log-spiral antenna are greater than one wavelength, or slightly less, the input impedance remains almost constant as frequency is varied. The relationship between the mean impedance and arm width of "slot" antennas is shown in Fig. 79. The data shown pertain directly to antennas made from 1/32-in (0.79-mm) copper, fed with 0.15-in (0.38-mm) diameter coaxial cable bonded to the ground plane midway between the slot arms (no dummy cable). The influence on the standing-wave ratio of using different sizes of feed cables is illustrated in Fig. 80. When $\delta = \pi/2$, the planar log-

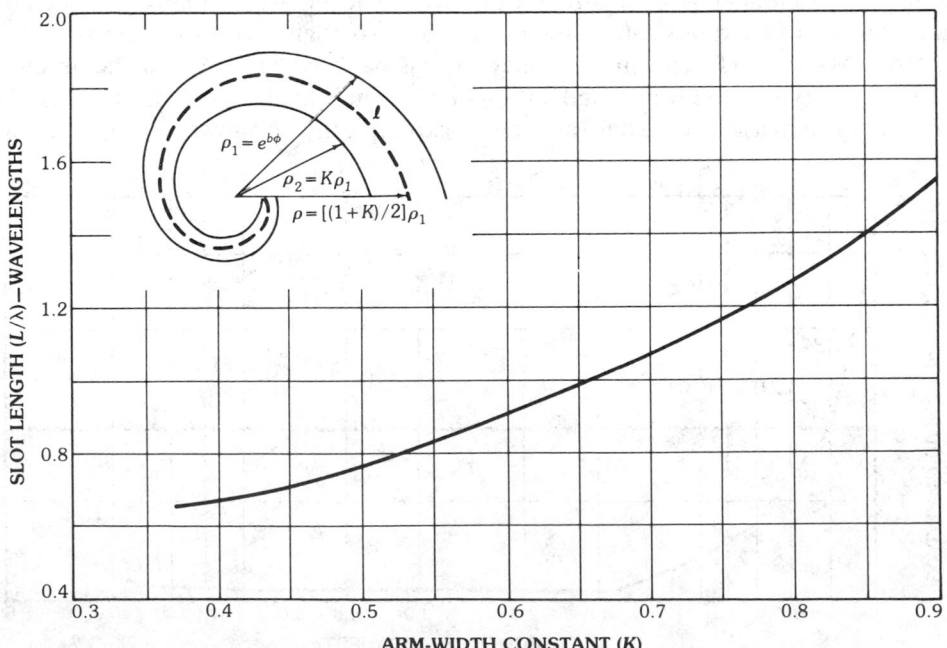

Fig. 78. Minimum slot length necessary to produce a circularly polarized radiated field ($r \leqq 2{:}1$ on axis; $0.2 \leqq a \leqq 0.45$). (*After Dyson [27, 28], © 1959 IEEE*)

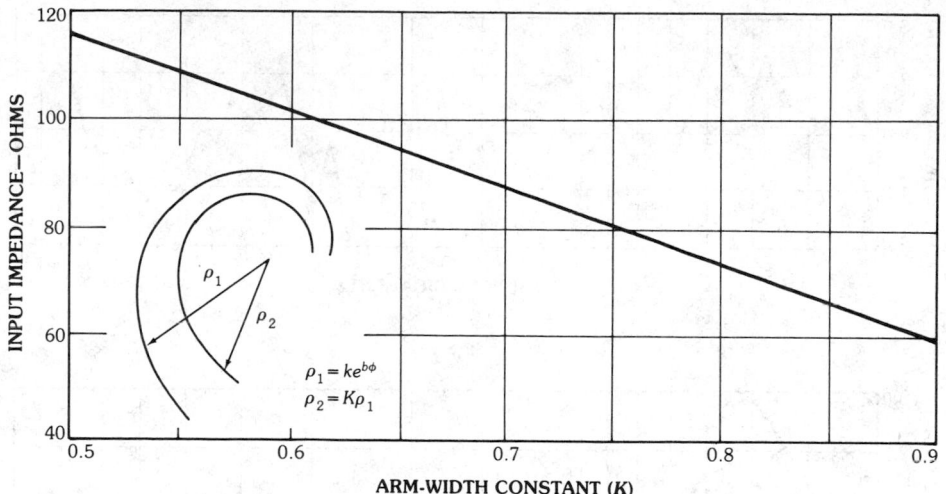

Fig. 79. Input impedance of balanced planar slot antennas. (*After Dyson [27, 28], © 1959 IEEE*)

spirals are self-complementary, and would be expected to have an input impedance of 189 Ω when made of metal with infinitesimal thickness. Self-complementary antennas of finite thickness invariably display lower values of input impedance.

Conical Log-Spirals

Unidirectional patterns can be achieved, in a manner similar to that for other LP structures, by developing log-spirals on a nonplanar surface. A simple, rotationally symmetrical surface is a cone. A log-spiral arm on the surface of a cone is illustrated in Fig. 81. The parameter θ_0 determines the cone angle. As before, the angle δ determines the angular width of the exponentially expanding arms, and the angle α, the rate of wrap of the arms. These angles are constant for any given antenna. The radius vector to any point on the edge of one arm is given by

$$\varrho = \varrho_0 \exp[b(\phi - \delta)] \tag{25}$$

where

$$\varrho_0 = \frac{d/2}{\sin \theta_0} \tag{26}$$

$$b = \frac{\sin \theta_0}{\tan \alpha} \tag{27}$$

In the conical case the parameters d and D define the diameters of the truncated apex and base of the cone. The orientation of the antenna in a spherical coordinate system used to describe the radiation patterns is also indicated in Fig. 81.

Conical log-spirals with small cone angle θ_0 can be treated using the periodic

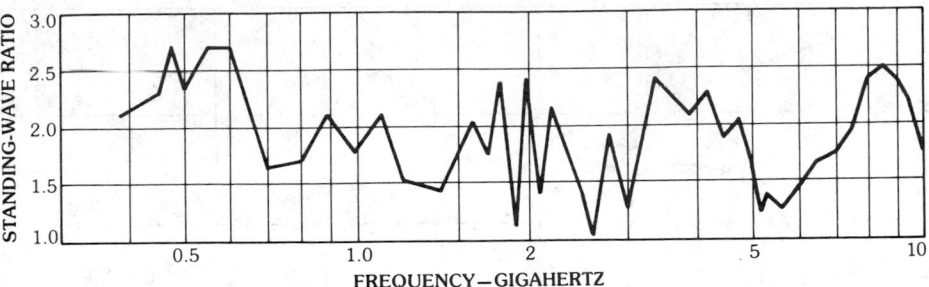

a

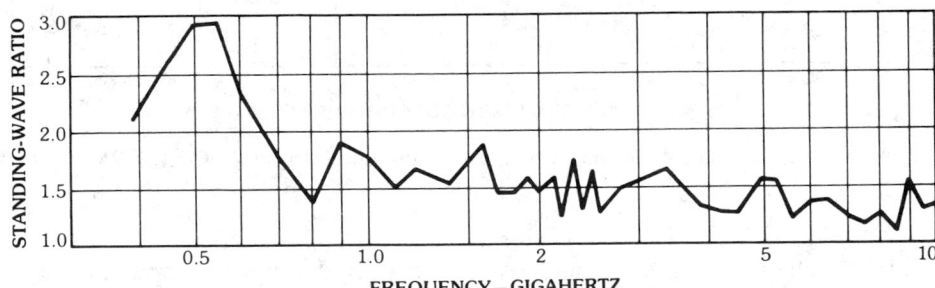

b

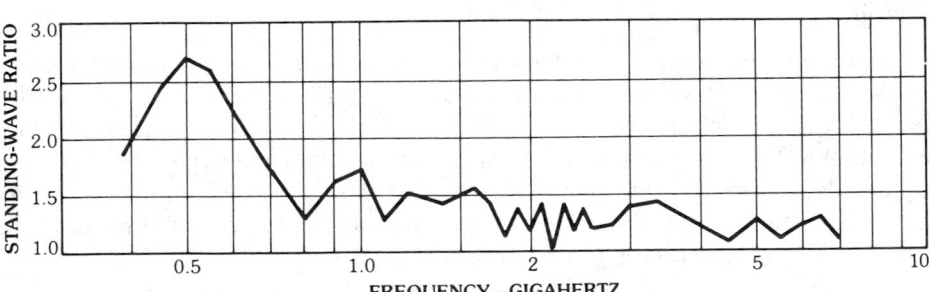

c

Fig. 80. Standing-wave ratios for various feed cable arrangements (50-Ω line). (*a*) Small-diameter cable. (*b*) RG-141/U cable. (*c*) RG-141/U with balancing dummy cable. (*After Dyson [27, 28], © 1959 IEEE*)

structure theory of Section 3. The local propagation constant in a small region along the surface of the conical, two-arm, log-spiral antenna is approximately equal to the propagation constant for the corresponding "average" cylindrical bifilar helix. A study of the propagation constant and other characteristics of the near and far fields leads to identification of the active region [30]. Successful design of the antenna depends on knowledge of the relative position and size of the active region as a function of antenna parameters.

The parameters involved in a comparison of the conical spiral and cylindrical

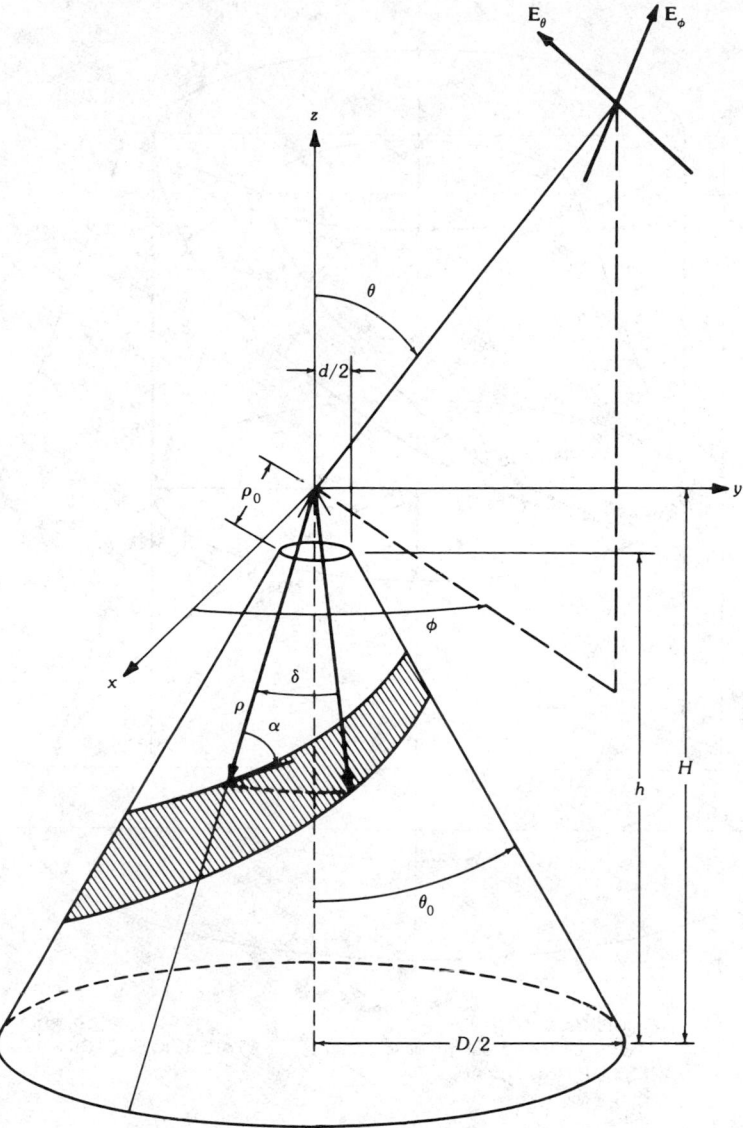

Fig. 81. Conical antenna with associated parameters. (*After Dyson [30], © 1965 IEEE*)

helical geometries are indicated in Fig. 82, which shows one turn (or one cell) of a conical antenna with infinitesimally narrow arms and with the following parameters: ℓ, the length of the turn; ξ, the pitch angle; α, the complementary spiral angle; and θ_0, the half-cone angle. Superimposed on this is one cell of a cylindrical helix with the same pitch angle and with a radius u equal to the geometric mean radius of the conical cell.

For observations parallel to the axis the turn-to-turn phasing of the helix is determined to a first approximation by the ratio of the pitch distance P to the turn

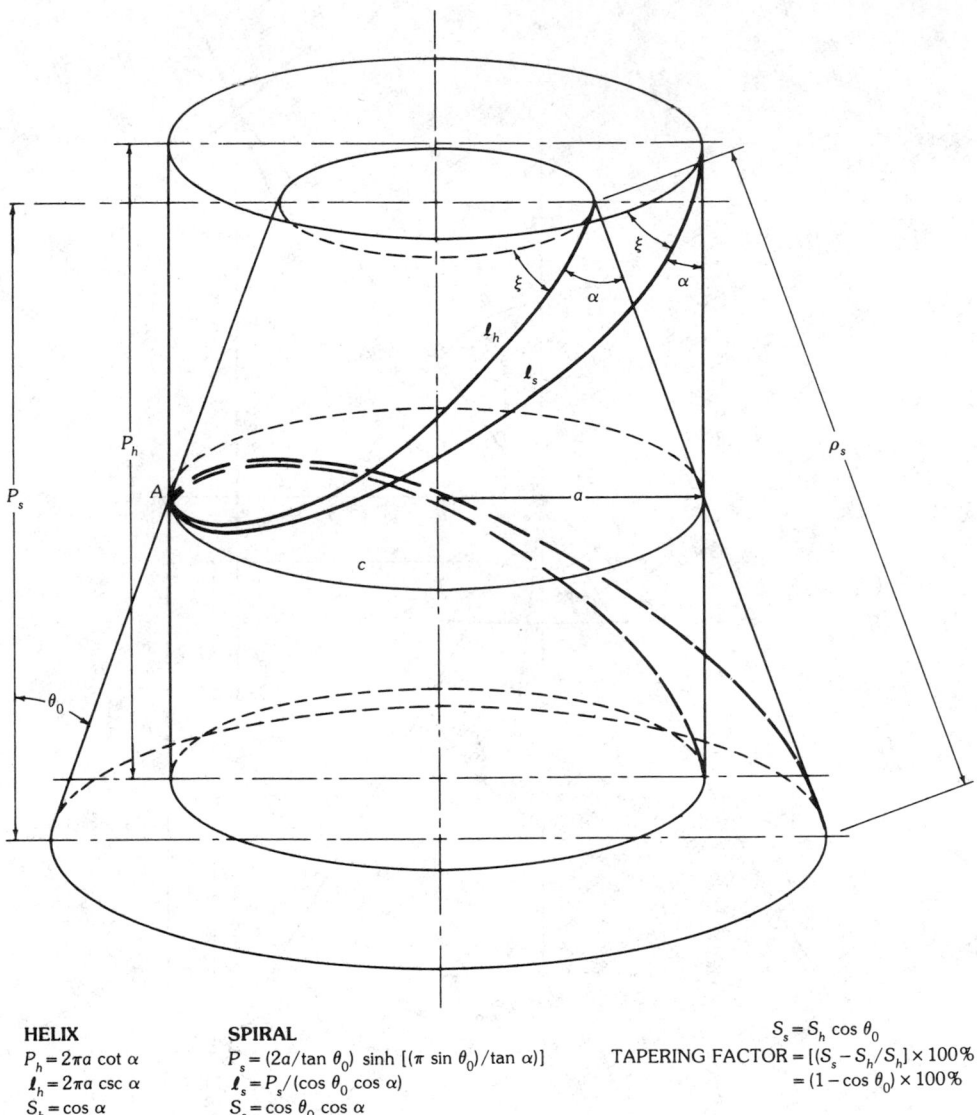

HELIX	SPIRAL	
$P_h = 2\pi a \cot \alpha$	$P_s = (2a/\tan\theta_0)\sinh[(\pi\sin\theta_0)/\tan\alpha)]$	$S_s = S_h \cos\theta_0$
$\mathit{l}_h = 2\pi a \csc \alpha$	$\mathit{l}_s = P_s/(\cos\theta_0 \cos\alpha)$	TAPERING FACTOR $= [(S_s - S_h/S_h] \times 100\%$
$S_h = \cos\alpha$	$S_s = \cos\theta_0 \cos\alpha$	$= (1 - \cos\theta_0) \times 100\%$

Fig. 82. Corresponding conical spiral and helical cells. (*After Dyson [30], © 1965 IEEE*)

length ℓ. This assumes a current wave progressing down the arm at the intrinsic velocity of the surrounding medium. On the helix this ratio is the sine of the pitch angle or $\cos\alpha$. The ratio of the pitch distance (parallel to the axis) to the turn length on the conical structure is equal to $\cos\alpha\cos\theta_0$.

The ratio (S) of pitch to turn length is also the ratio of the propagation constant along the arms, k, to that of a wave propagating along the surface of the cylinder, B. If the difference between the k/B ratios for the helix and the conical spiral are expressed as a function of the helix ratio, the result is simply $1 - \cos\theta_0$ (see Fig. 82). For cones with an included angle of 20° this difference is of the order of

1.5 percent. For all cones which are good unidirectional radiators the difference is only a few percent.

The variation of the propagation constant on periodic antenna structures can be conveniently displayed on the Brillouin (k-β) diagram [16]. One such diagram for the balanced bifilar helix is shown in Fig. 83. The vertical coordinate is given in units $ka/\tan\alpha$ which, since $k = 2\pi/\lambda$, is the pitch distance in free-space wavelengths. The horizontal scale is the pitch distance expressed in the equivalent guide wavelength on the surface of the antenna. For any single helix, since a and α are constant, the only variable involved is the wavelength of operation.

As the frequency of operation is increased, the propagation constant increases. If one considers a single space harmonic to be dominant, and assumes a nondispersive wave along the conductors, there is first a region of slow, closely bound surface waves. As the propagation constant increases still further, there is strong coupling to the first backward space wave, the propagation constant becomes complex, and the structure radiates with a phasing to provide a backfire beam.

If the wavelength is fixed, as a current wave progresses from the point of excitation along a structure with increasing radius, the propagation constant behaves in a manner similar to that observed for the propagation constant for a cylindrical structure as frequency increases. Thus, at a midband frequency the current on the arms near the tip is traveling away from the feed point with constant amplitude. At some distance down the arm, however, where the phasing for backfire radiation is approached, energy is strongly coupled into a radiated wave, and the amplitude of the current diminishes rapidly. The region containing the decaying current of nonnegligible amplitude is the active region. As the frequency is changed, the active region moves on the antenna so that the distance from the apex of the cone to the active region, when expressed in wavelengths, remains constant. The phase center of the radiated field is consistently located in the active region.

Measurements of currents on many conical log-spirals have led to recognition of the region from a point 3 dB below the maximum on the apex side to a point 15 dB below the maximum on the base side as defining the active region. The radii

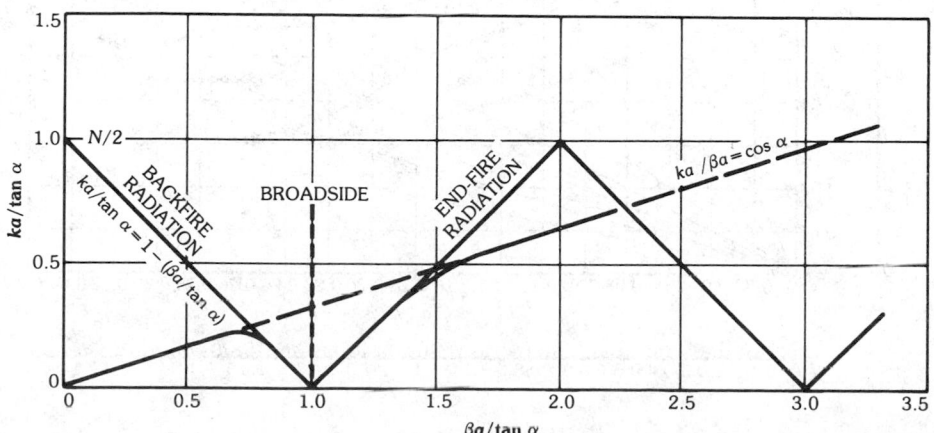

Fig. 83. Brillouin diagram for bifilar helix. (*After Dyson [30], © 1965 IEEE*)

at which these points occur are plotted in Fig. 84 as a function of the included cone angle $2\theta_0$ and the spiral angle α. Since both axes of this graph are normalized to wavelength these curves give the active region bounds. If, however, the vertical axis is considered to be normalized to the shortest wavelength, and the horizontal axis to the longest wavelength, these curves give the required radii of the truncated apex and base of the conical antenna. The circumferences C_3/λ and C_{15}/λ corresponding to these radii are also shown. Fig. 84 also includes a grid indicating the active region bandwidth B_{ar}, previously defined in (3), as a function of the antenna parameters. The data of Fig. 84 are for self-complementary antennas ($\delta = 90°$) and are most accurate for $\theta_0 = 10°$ ($20°$ included cone angle). They tend to become more conservative as the cone angle decreases and slightly more optimistic for larger cone angles.

The rate of decay of the currents decreases as the arm width departs from the self-complementary case, $\delta = 90°$. Fig. 85 gives plots of M, the ratio of maximum radius required for very narrow or very wide arms to that required for arms with $\delta = 90°$. For intermediate values of δ, linear interpolation between the values given in Fig. 85 should be adequate. Since the small-end truncation is affected very little by variations in the arm width, the data for M in Fig. 85 can also be interpreted as the ratio of B_{ar} for other values of δ to that for $\delta = 90°$.

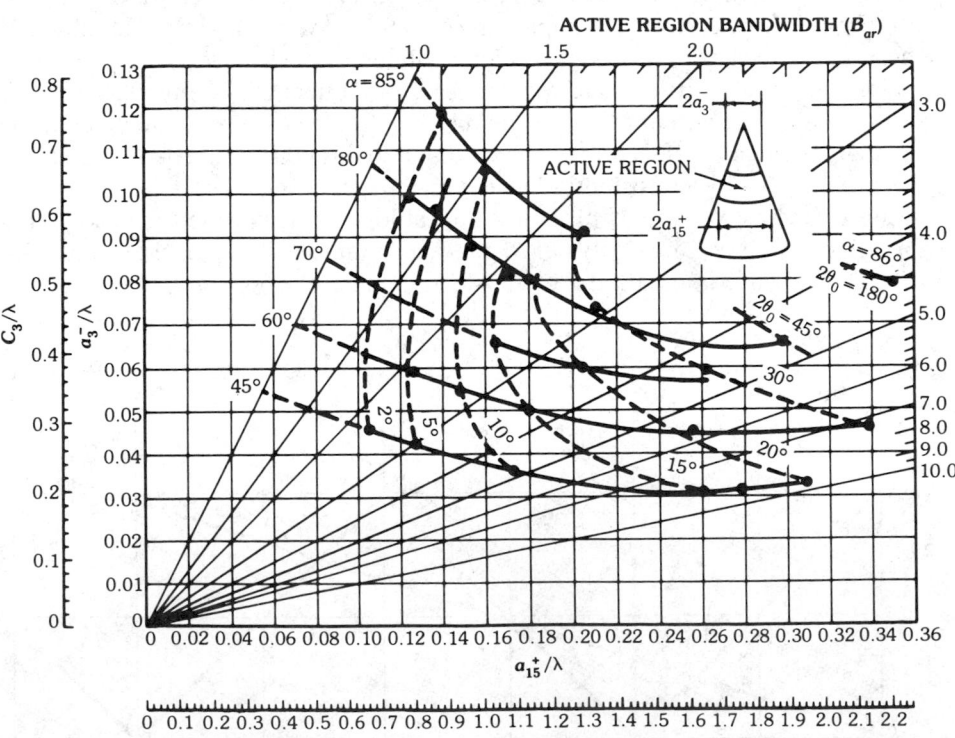

Fig. 84. Bounds of the active region in terms of the radius a or the circumference C of the conical log-spiral antenna. (*After Dyson [30], © 1965 IEEE*)

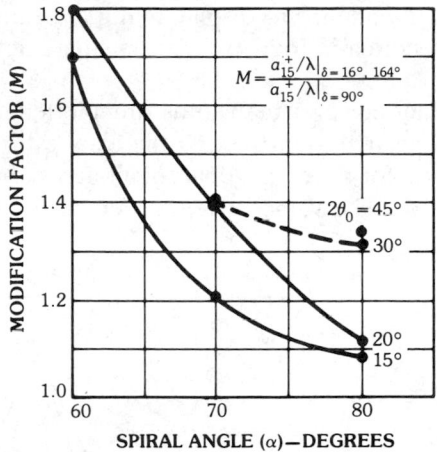

Fig. 85. Modification factor M for very narrow or very wide arm widths. (*After Dyson [30]*, © *1965 IEEE*)

The conical antenna radiates a single lobe which is directed off the apex along the axis of the cone. In Fig. 86 typical radiation patterns are shown as a function of the included angle of the cone $2\theta_0$ and the spiral angle α. The well-formed, relatively narrow beam for small cone angles and $\alpha = 80°$ is indicative of essentially all turns of the active region being phased for backfire radiation. As the spiral angle (α) decreases and the cone angle ($2\theta_0$) increases, the radiation pattern broadens

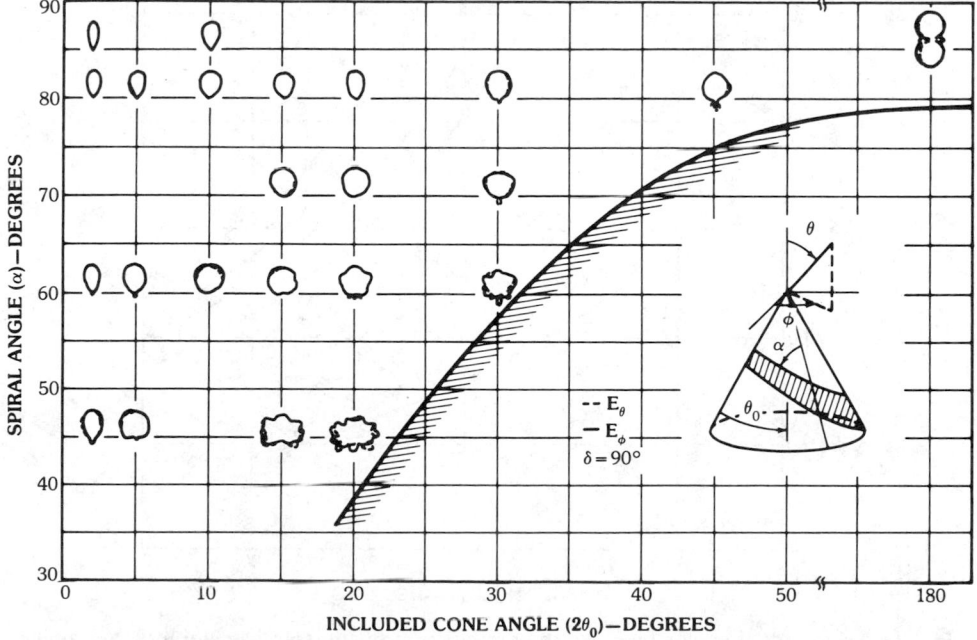

Fig. 86. Electric-field radiation patterns as a function of the spiral angle and cone angle.

and eventually exhibits a tendency to display multiple beams with corresponding irregularities. The shaded area in Fig. 86 represents parameters that are normally unusable.

The approximate half-power beamwidths are plotted in Fig. 87. Although the fields are circularly polarized on the axis the axial ratio increases for off-axis angles. Hence the patterns for the orthogonal components of the electric field differ in beamwidth by 8° to 10°. The values given in Fig. 87 are the averages of these

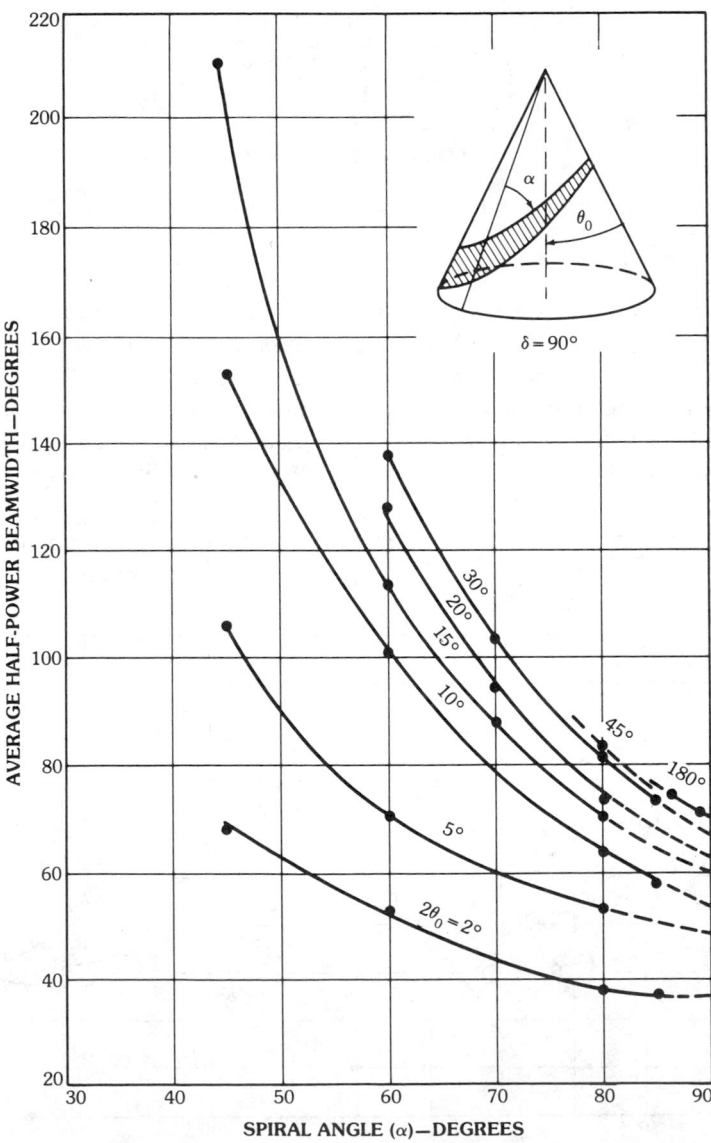

Fig. 87. Average half-power beamwidth as a function of the spiral angle and cone angle. (*After Dyson [30], © 1965 IEEE*)

beamwidths. In addition, the average beamwidth must be interpreted together with the variation in beamwidth with changing frequency, which is shown in Fig. 88. The beamwidth variation is caused by the rotation with frequency of the slightly asymmetrical pattern.

The approximate directivity with respect to a circularly polarized isotropic source is plotted in Fig. 89. These values were calculated using

$$D \cong \frac{32\,600}{\Delta\phi_1 \times \Delta\phi_2} \quad \text{(dB)} \tag{28}$$

where the denominator is the product of the average half-power beamwidths in orthogonal planes. The use of the constant 32 600 is considered conservative, resulting in a directivity that is approximately 1 dB less than that obtained with the value of 41 250 that appears frequently in the literature. For ease in converting to

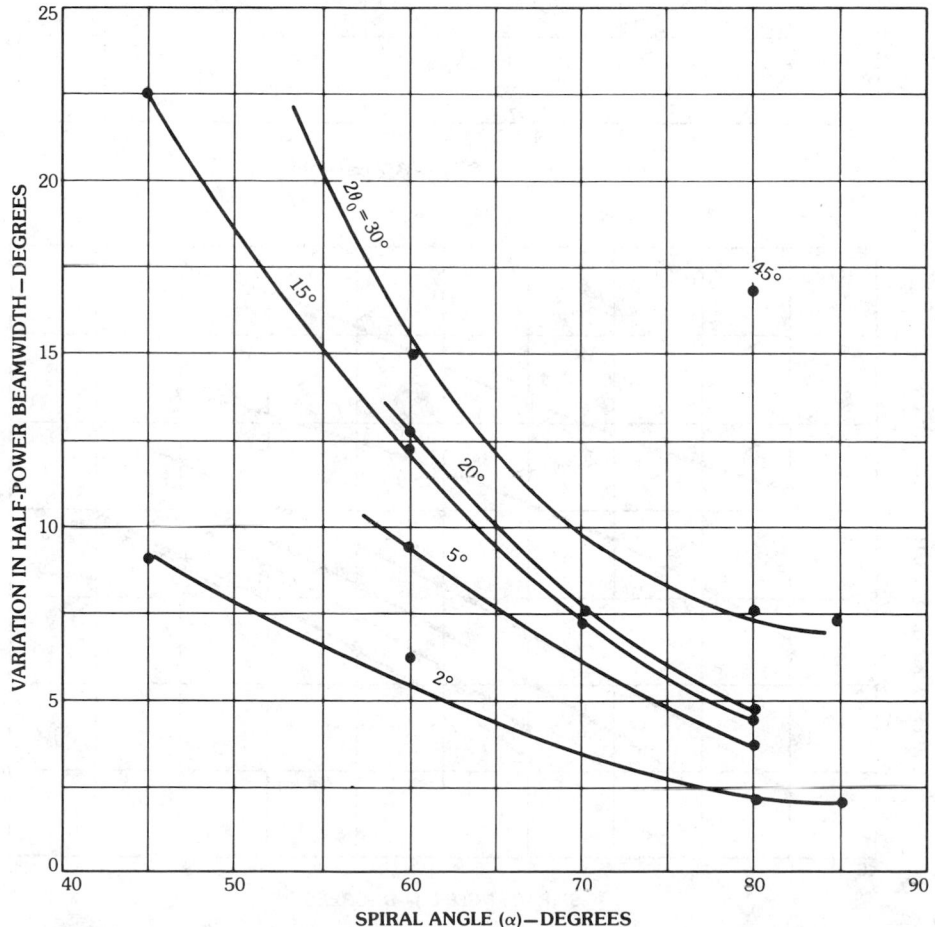

Fig. 88. Variation in average half-power beamwidth ($\delta = 90°$). (*After Dyson [30],* © *1965 IEEE*)

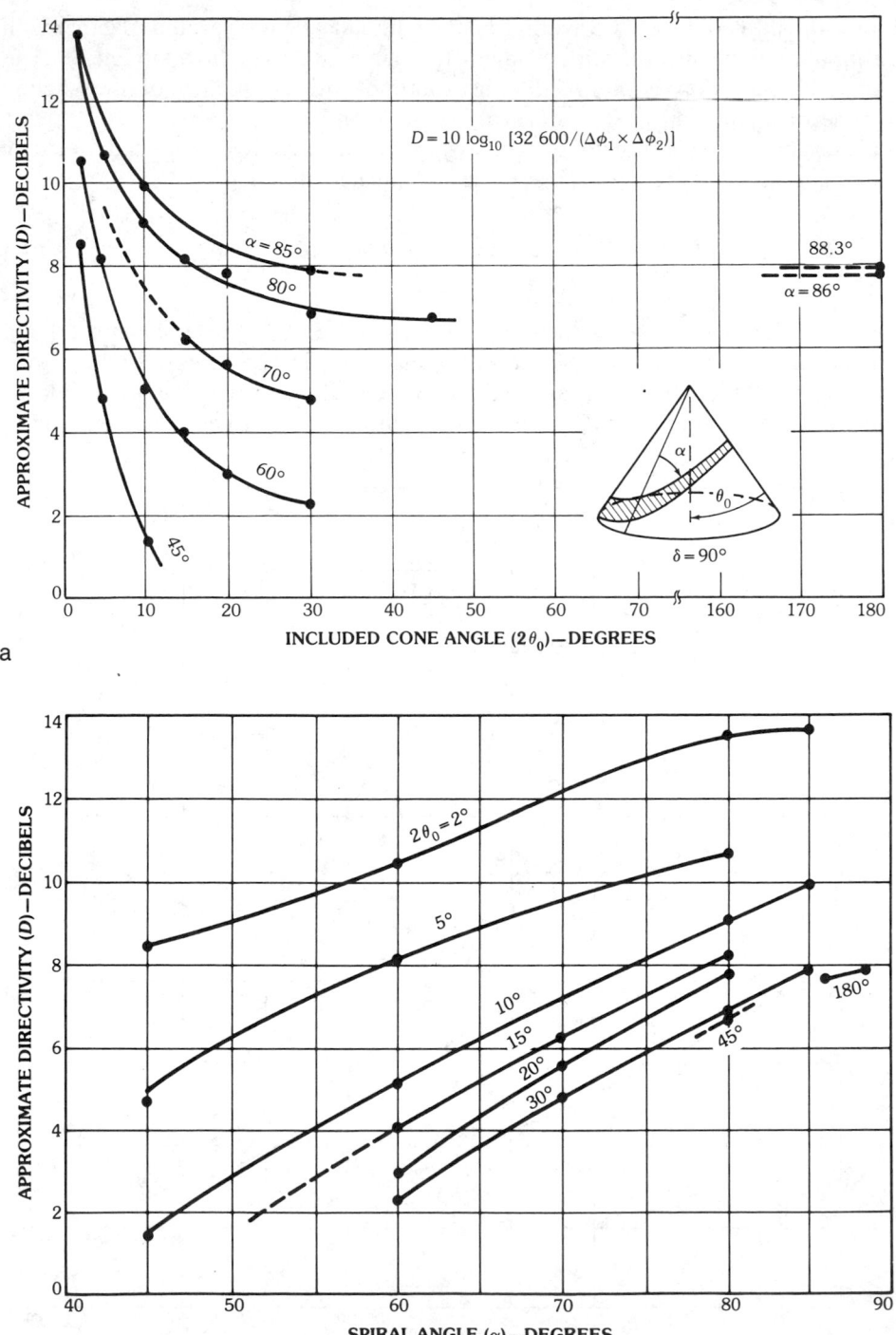

Fig. 89. Directivity as a function of cone angle and spiral angle. (*a*) As a function of cone angle. (*b*) As a function of spiral angle.

the approximate directivity based on the use of other constants in (28), a difference in directivity is plotted in Fig. 90 as a function of the constant. The directivity is directly related to the spiral angle α, rising as α increases. Representative radiation patterns have been plotted in Fig. 91 as a function of the spiral constant b.

In Fig. 92 typical radiation patterns are shown as a function of $2\theta_0$, α, and δ. For each combination of $2\theta_0$ and α, patterns are shown for very narrow arms ($\delta = 16°$), for the self-complementary case ($\delta = 90°$), and for very wide arms ($\delta = 164°$). For each case patterns are shown for two orthogonal polarizations and for two orthogonal planes. The principal change in the radiation patterns with change in arm width from $\delta = 90°$ is an increase in beamwidth.

The patterns shown in Fig. 86 and 92 are typical of those to be expected when the antenna is operated at frequencies such that there is no distortion due to the truncation of the base or tip. As the frequency of operation is decreased and the lower edge of the active region becomes affected by the truncated base, the amplitude of the back lobe will increase rather rapidly. As the leading edge of the active region moves into the truncated tip, the beamwidth may at first decrease and the pattern then becomes rough. Any lack of precision in construction of the apex region will cause pattern tilt and/or distortion. A further increase in frequency may cause the pattern to broaden with a tendency to break into lobes.

Radiation patterns for the very loosely wrapped antennas, $\alpha = 45°$, are shown for self-complementary width arms only. As the arm width deviates from $\delta = 90°$, for antennas with $2\theta_0 \geqq 15°$ and $\alpha \leqq 45°$, the pattern may break into many lobes with a major portion of the energy radiated in the direction of the base. A decrease in the front-to-back ratio may be noted as δ differs from 90°. For $2\theta_0 \geqq 20°$, the variation with δ is less for $\alpha \geqq 75°$. For these small-cone, tightly wrapped antennas it is possible to use thin arms. The constant-width wire or cable arm versions of the antenna with these parameters perform satisfactorily.

The front-to-back ratios are plotted in Fig. 93. These are typical average values

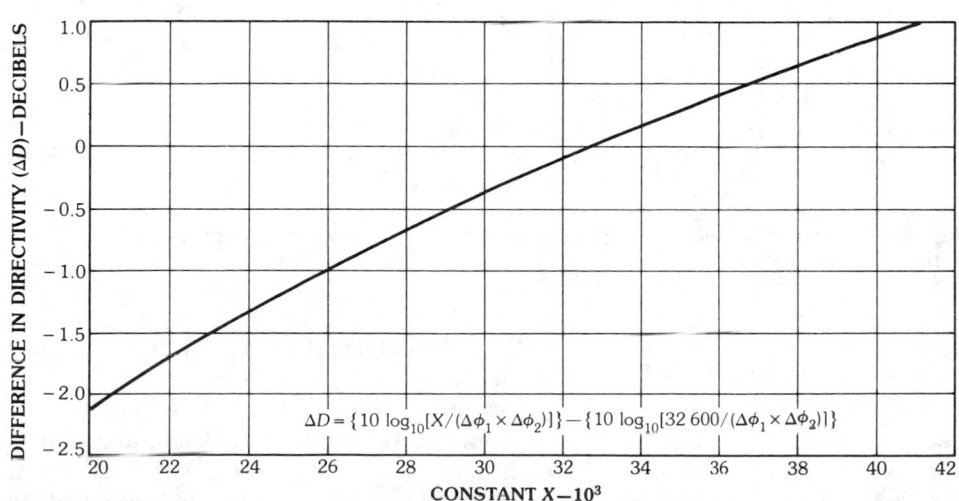

Fig. 90. Change in calculated directivity as the approximate expression is varied.

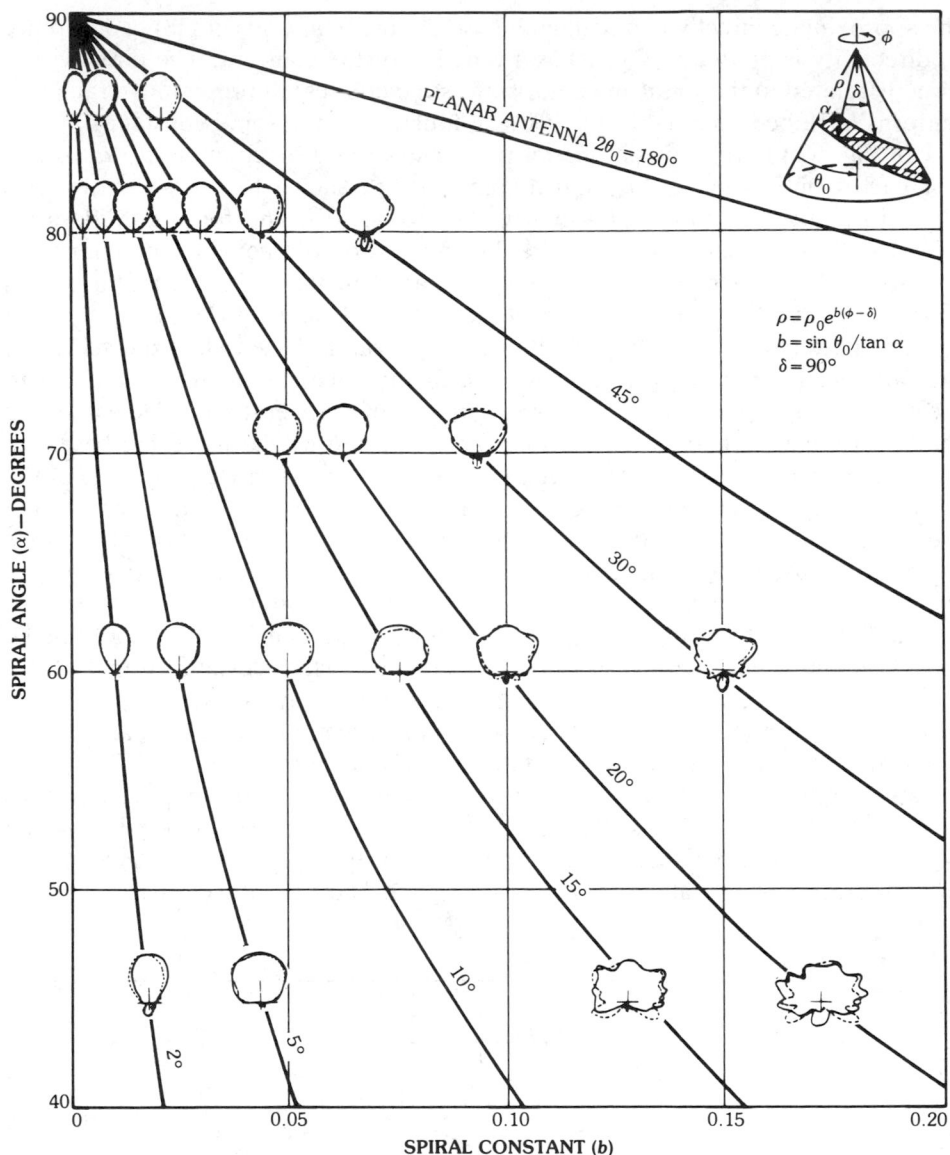

Fig. 91. Electric-field radiation patterns as a function of the spiral constant. (*After Dyson [30], © 1965 IEEE*)

that can be expected from a well-constructed antenna operated at a frequency such that there is no effect due to truncations at the tip or base.

The conical spirals are essentially circularly polarized in any direction where there is substantial radiation. Typical values of axial ratio recorded at angles from the axis of an antenna with $2\theta_0 = 20°$ are shown in Fig. 94. The curves with minimum slope are obtained for the lower values of α. This is indicative of the increased beamwidth and increased energy radiated in directions approaching that

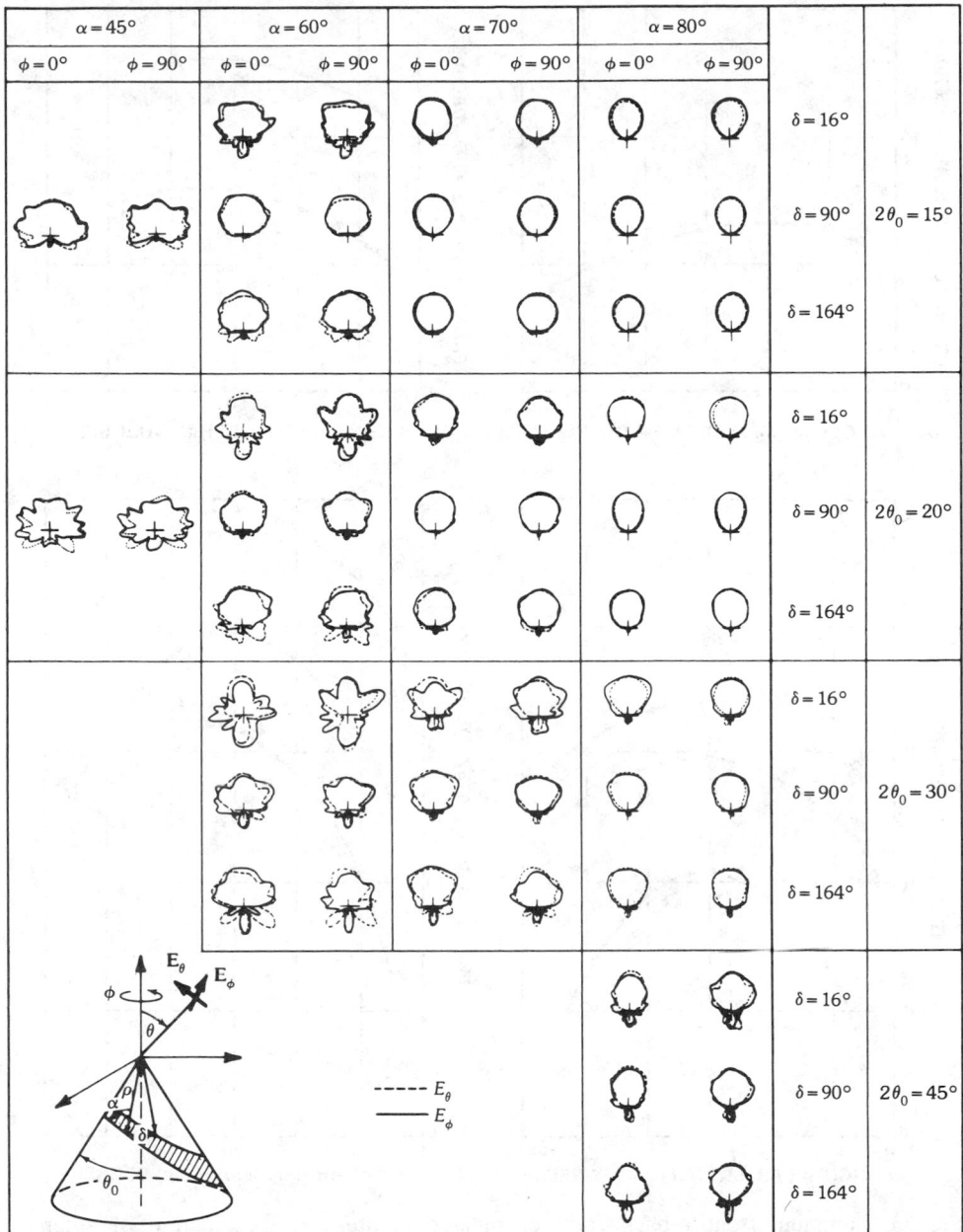

Fig. 92. Electric-field radiation patterns for three angular arm widths δ, for four cone angles $2\theta_0$, and for four spiral angles α. (*After Dyson [30], © 1965 IEEE*)

perpendicular to the axis as the spiral angle α is decreased. As δ departs from 90° the axial ratio rises slightly as a function of θ_0.

Conical log-spirals do not have a unique phase center. However, a phase center can be found for a major portion of the main beam. Fig. 95 shows the distance from

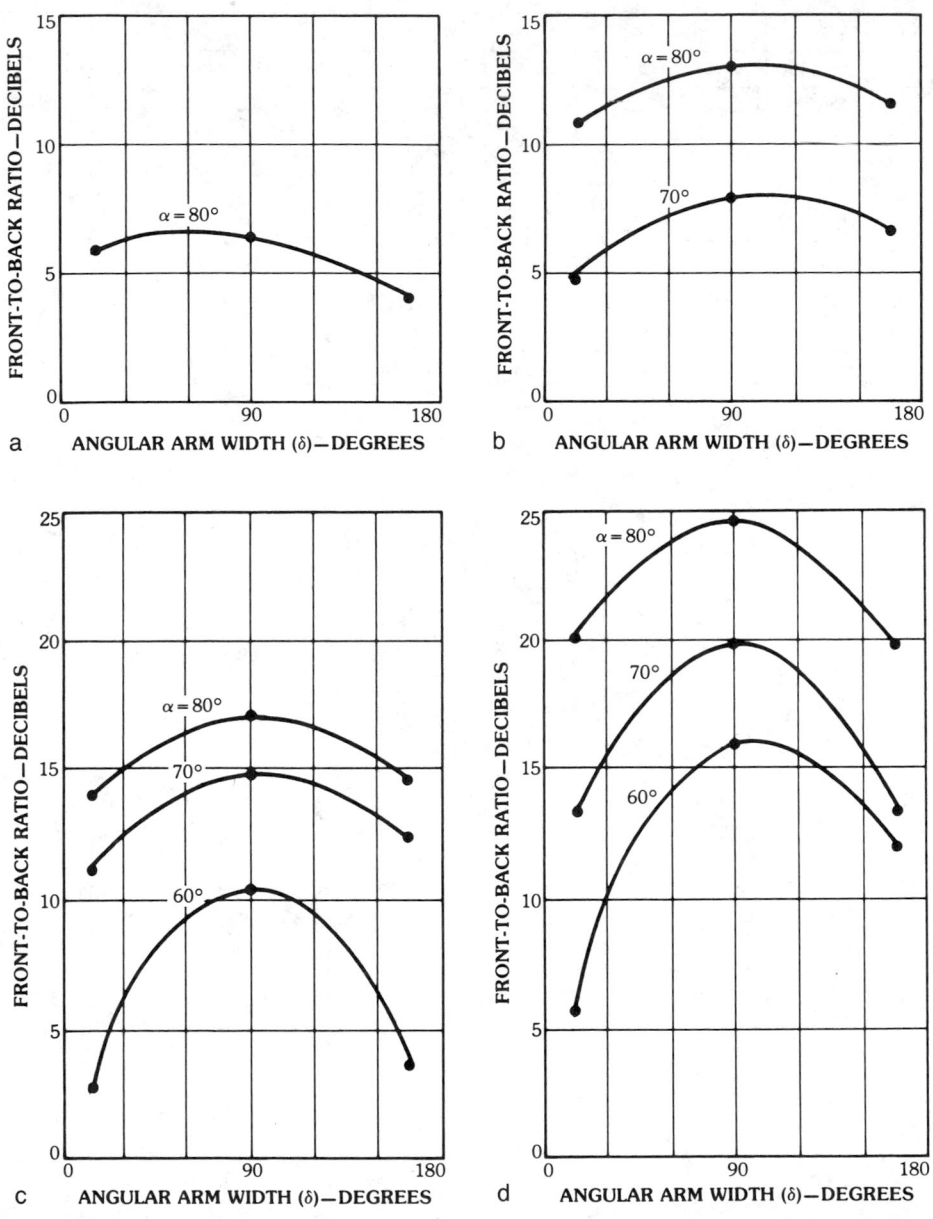

Fig. 93. Minimum front-to-back ratio of radiation patterns as a function of antenna parameters. (*a*) For $2\theta_0 = 45°$. (*b*) for $2\theta_0 = 30°$. (*c*) For $2\theta_0 = 20°$. (*d*) For $2\theta_0 = 15°$. (*After Dyson [30], © 1965 IEEE*)

the virtual apex to the phase centers as measured on several conical log-spirals with $\delta = 90°$. When the conical spirals are used as feeds for parabolic reflectors the antenna should be positioned so that the indicated phase center coincides with the focal point of the reflector. Since the phase center moves as frequency changes, a compromise must be made when operating over a wide frequency band.

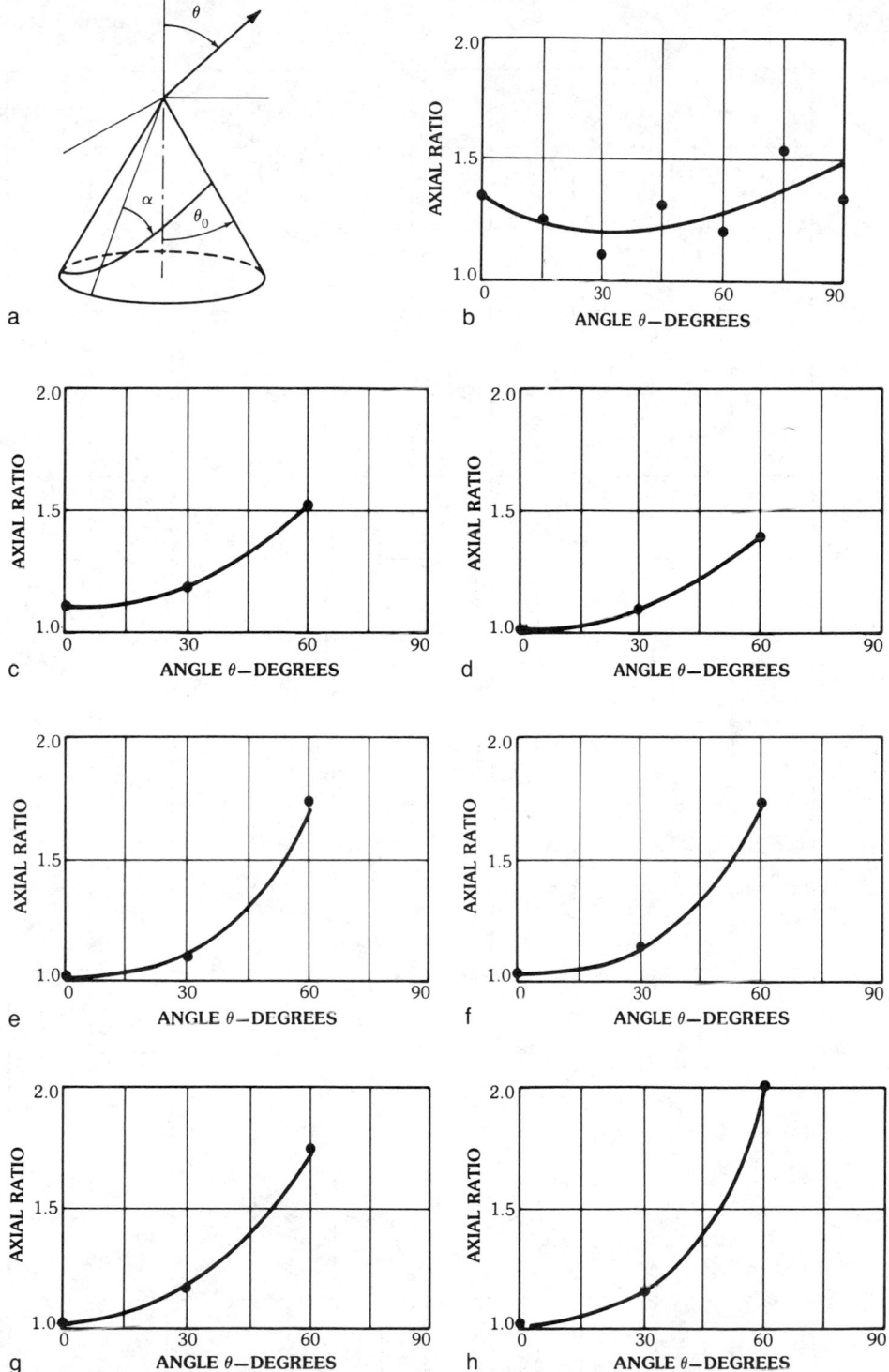

Fig. 94. Typical variation in axial ratio off the axis of the antenna for $2\theta_0 = 20°$. (*a*) Geometry. (*b*) For $\alpha = 45°$, $\delta = 90°$. (*c*) For $\alpha = 70°$, $\delta = 16°$. (*d*) For $\alpha = 70°$, $\delta = 90°$. (*e*) For $\alpha = 70°$, $\delta = 164°$. (*f*) For $\alpha = 80°$, $\delta = 16°$. (*g*) For $\alpha = 80°$, $\delta = 90°$. (*h*) For $\alpha = 80°$, $\delta = 164°$. (*After Dyson [30], © 1965 IEEE*)

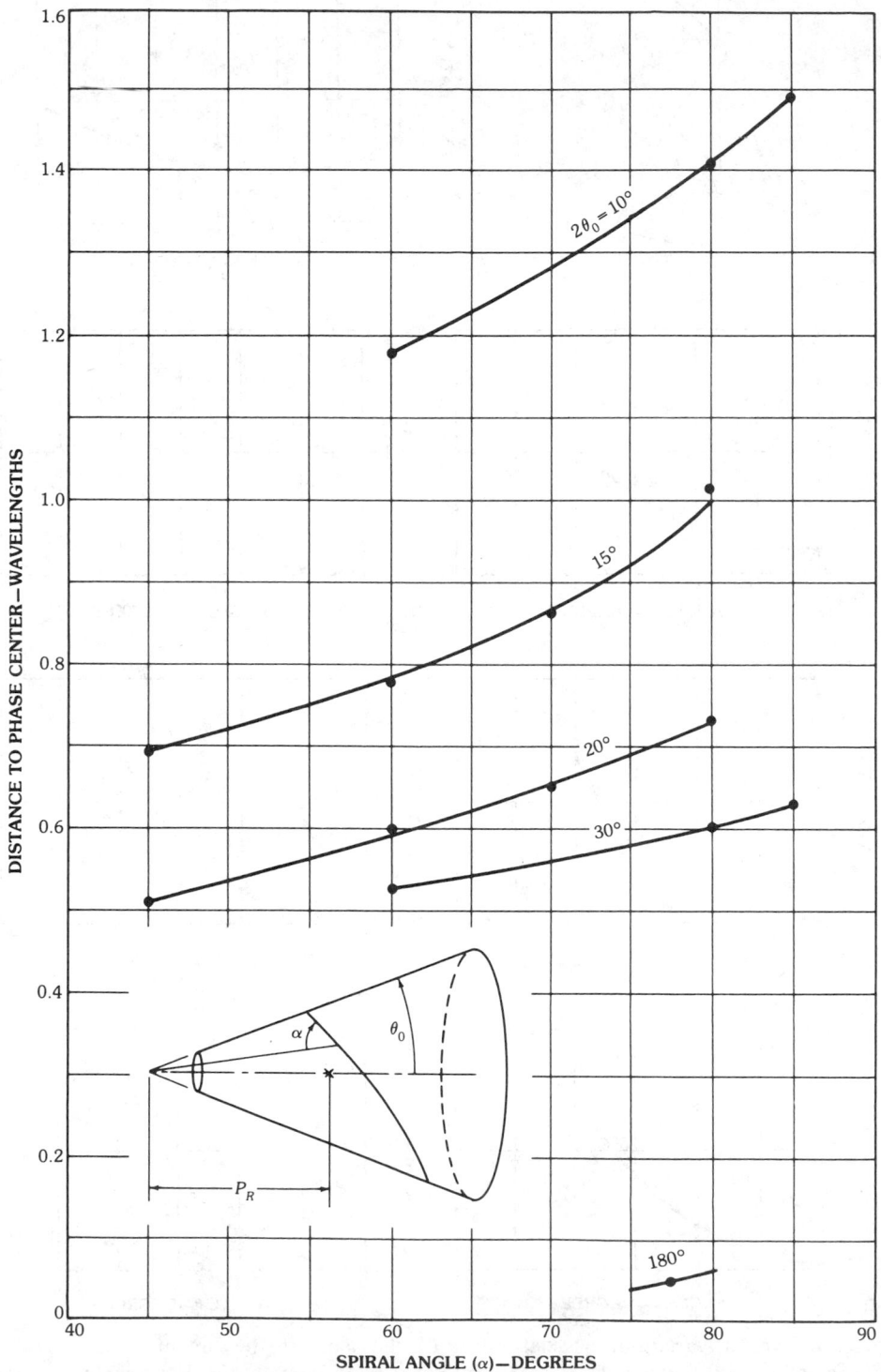

Fig. 95. Location of measured "effective phase center" for balanced, two-arm, conical, log-spiral antennas ($\delta = 90°$).

The two-arm conical log-spiral, like the planar version, is a symmetrical structure and, when excited in a balanced manner, radiates a beam on the axis of the cone without squint or tilt. A tilt in the radiation pattern can be traced to the physical construction or to unbalance in the feed.

Conical log-spirals can be excited by bringing a balanced transmission line along the axis from the large to the small end and connecting one wire to each arm. It is preferable, although not necessary, that this be a shielded line. The presence of metal *on axis* has a minimum effect on the antenna characteristics if the diameter of the metal is no more than one-third the diameter of the antenna at any point on the axis.

The transition from an unbalanced coaxial line to a balanced line may be made by placing a balun inside the antenna if the balun is nonradiating and does not disturb the fields inside the cone. The taperline balun [32] with its extremely wide bandwidth would seem to be ideally suited for this purpose. However, when it is placed inside conical log-spirals a truly balanced feed is seldom realized. For some purposes, however, the degradation of the pattern can be tolerated.

Coaxial 180° hybrids make satisfactory baluns. These units provide equal-amplitude, out-of-phase signals at the side arms. Fifty-ohm coaxial cables connected to these arms may be carried along the axis of the cone, and the two center conductors of the cables thus become a shielded, balanced, 100-Ω transmission line. For maximum symmetry, and to prevent the two outer sheaths from becoming a line for any unbalanced currents, the outer conductors of these cables should be electrically connected along their lengths.

To overcome the possible limiting bandwidths of baluns the conical spirals can also be fed by carrying a coaxial cable along one arm as described before for the planar spirals. To maintain symmetry a similar "dummy" cable must be placed on the other arm. This remains the most satisfactory method if the presence of the cable does not limit the size of the truncated apex region and if the loss in the relatively long length of cable can be tolerated.

The geometry of the truncated tip, including the possible presence of the feed cable on the antenna arms, has a marked effect on the input impedance. Since there are an unlimited number of cable diameter and truncated-tip dimensions, it is dangerous to expect that published data for input impedance will be realized unless care is taken to ensure that the feed-point geometry is duplicated. The use of coaxial hybrids as baluns makes possible a very convenient method of making balanced impedance measurements, and this is the technique that was used to obtain the data reported here [28, 29]. Identical slotted lines were inserted in the coaxial lines between the antenna and hybrid. If the sections are matched for probe insertion and movement, and for detector characteristics, the phase balance of the lines will be preserved.

Measurements were made over a 5:1 bandwidth. The normalized impedances plotted on a Smith chart were enclosed by a circle. The "center" of this circle, chosen to make the hyperbolic distance to all parts of the circle constant, was considered to be the nominal impedance of the antenna. The maximum vswr referred to the nominal impedance was typically less than 1.5.

Fig. 96 indicates the effect that the feed geometry can have on the measured impedance. The data enclosed by the larger circle were obtained when the center conductors of two RG-141/U cables were extended, bent at right angles at the plane

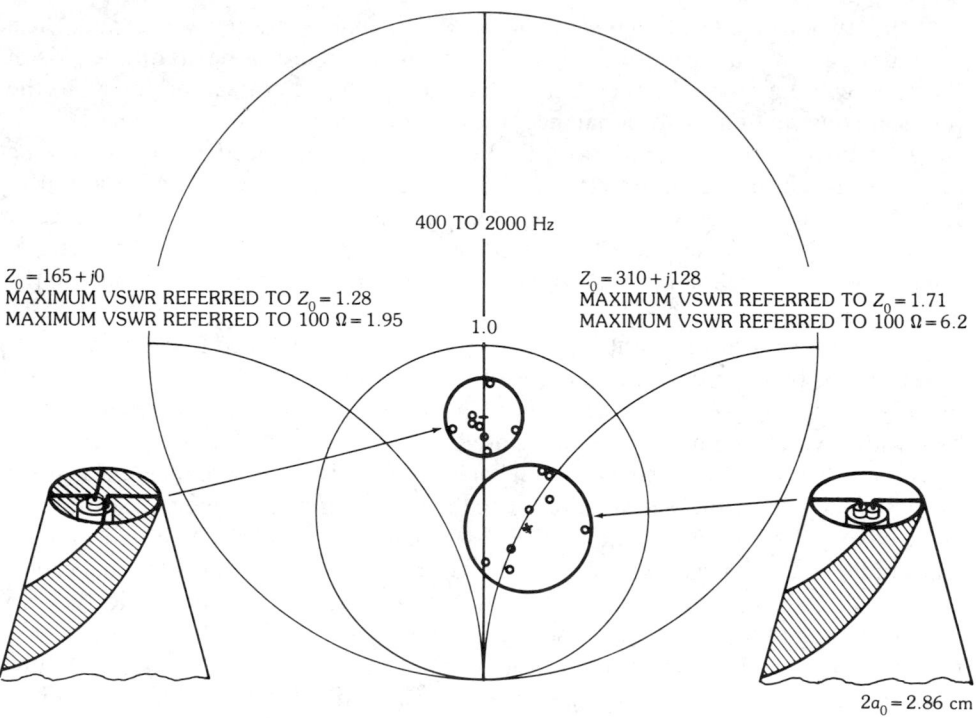

400 TO 2000 Hz

$Z_0 = 165 + j0$
MAXIMUM VSWR REFERRED TO $Z_0 = 1.28$
MAXIMUM VSWR REFERRED TO 100 Ω = 1.95

1.0

$Z_0 = 310 + j128$
MAXIMUM VSWR REFERRED TO $Z_0 = 1.71$
MAXIMUM VSWR REFERRED TO 100 Ω = 6.2

$2a_0 = 2.86$ cm

Fig. 96. Effect of feed region geometry on input impedance. (*After Dyson [30]*, © *1965 IEEE*)

of the tip truncation, and carried perpendicular to the axis of the cone out to the antenna arms. The wires were attached at the sharp end of the spiral arm that results when truncated at the circle $\varrho = d/2$. This provides a tapered transition from the small-diameter wire to the much greater width of the spiral arms. The measured values of input impedance referred to a reference plane established at the termination of the shields of the two coaxial cables fell within a reasonably compact area of the Smith chart, but the data display a positive reactive shift off the real axis.

When the transition between the center conductors of the cables and the antenna arms was made by flat, tapered conductors with angular width equal to δ, the scatter of the impedance data as a function of frequency decreased. In addition, the reactive shift disappeared and the resistive level of the impedance was changed significantly. Symmetry, balance, and tapered leads from the axis to the arms on the surface of the cone are all required for best performance.

The variation of the input impedance with arm width is shown in Fig. 97. The impedance is primarily controlled by the arm width, varying from 320 Ω for very narrow arms to 80 Ω for very wide arms. The impedance increases as the cone angle increases. For self-complementary antennas the impedance approaches 189 Ω, the theoretical value for the planar case. With the infinite-balun feed the presence of cable on the arms tends to give the narrow arms near the apex greater cross section and hence shifts the impedance level. Fig. 97 shows values of the measured

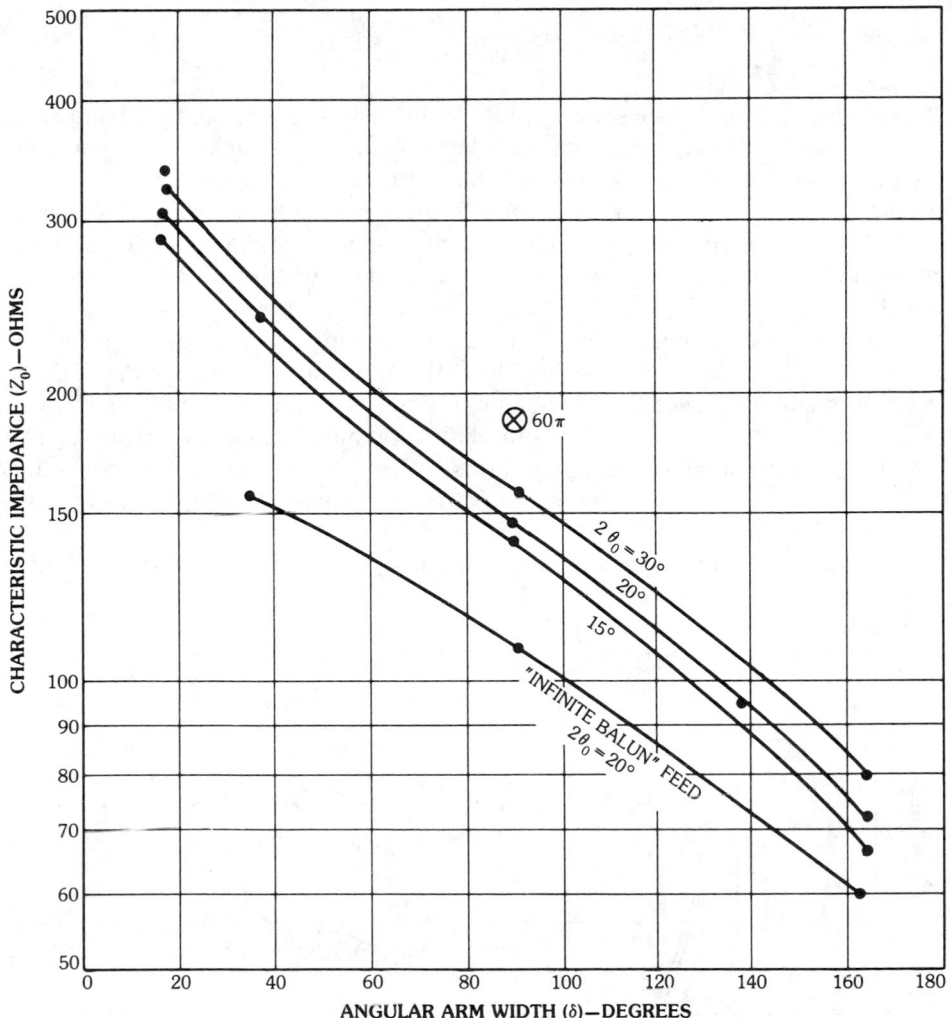

Fig. 97. Average characteristic impedance of conical log-spiral antennas as a function of angular arm width. (*After Dyson [30], © 1965 IEEE*)

impedance for an antenna with $2\theta_0 = 20°$, $\alpha = 60°$, and a tip truncated at 0.75 in, fed with RG-141/U cable of diameter 0.141 in. The maximum vswr with respect to the indicated impedances was about 1.75. The impedance variation with change in the spiral angle α is small, although there is a tendency for the impedance to increase with increasing α. The trend in impedance with cone angle shown in Fig. 97 is not obtained for all cone angles. Impedances for cones with angles less than 15° are not lower than those shown for 15°. For $2\theta_0 = 5°$, $\alpha = 80°$, the impedance is about 300 Ω for $\delta = 16°$, 180 Ω for $\delta = 90°$, and 72 Ω for $\delta = 164°$.

Half-power beamwidths of 40° can be obtained with small values of α. When wide bandwidths are to be covered, however, these antennas will become quite long. The structure length in terms of the longest wavelength λ_L is given by

$$\frac{h}{\lambda_L} = \frac{1}{2 \tan \theta_0} \frac{D}{\lambda_L} - \frac{d}{B\lambda_H} \qquad (29)$$

Plots of this expression are shown in Fig. 98 for 10:1, 5:1, 2:1, and 1:1 bandwidths as a function of the average half-power beamwidth and the cone angle and spiral angle for $\delta = 90°$. No curve is drawn in the 5:1 case except when $\alpha = 45°$. The use of smaller cone angles requires a cone of greater length but, of course, one of smaller diameter. In Fig. 99 the required maximum base diameter at the lowest frequency has been plotted as a function of the beamwidth and the parameters α and θ_0.

Nomographs are a useful aid to the design of conical log-spirals also. Fig. 100 relates the radius vector ϱ in wavelengths to the half-cone angle θ_0 and the diameter D/λ or the circumference (in wavelengths). Fig. 101 relates the total height of the cone from the virtual apex, H/λ, to the half-cone angle θ_0 and the diameter D/λ. Fig. 102 gives the spiral constant b in terms of the spiral angle α and the cone angle θ_0. Fig. 103 relates the spiral constant b, the angular arm width δ, and the arm-width constant K.

The tightly spiraled, self-complementary log-spiral tends to perform best in the

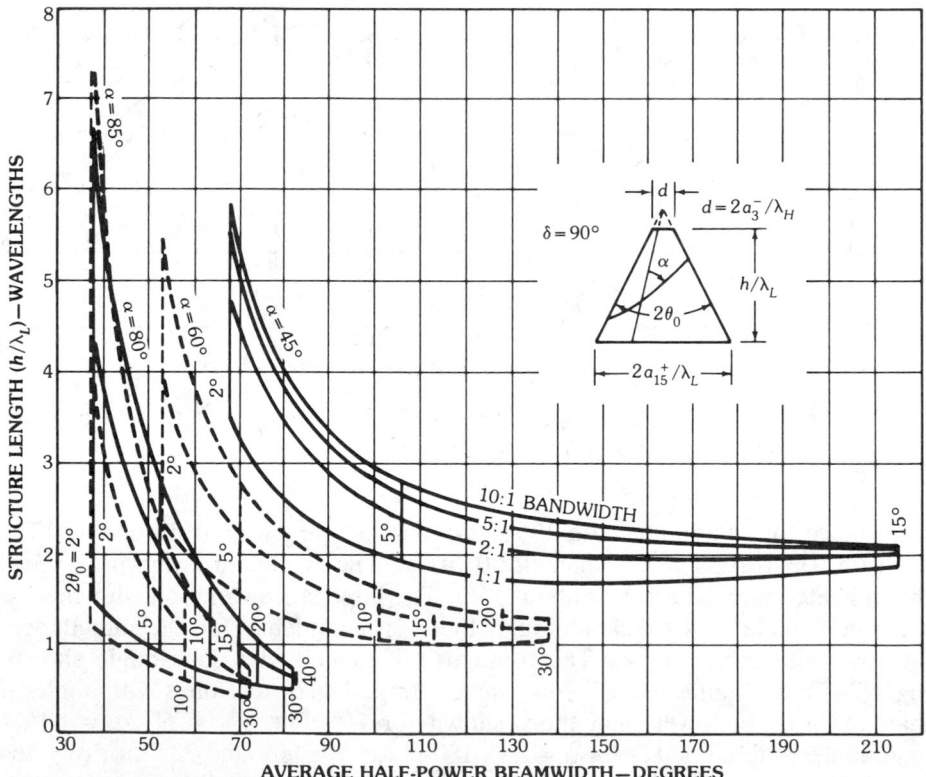

Fig. 98. Approximate cone height normalized to the longest wavelength of operation for bandwidths of 10:1, 5:1, 2:1, and 1:1 when base of cone equals a_{15}^+/λ_L, with $\delta = 90°$.

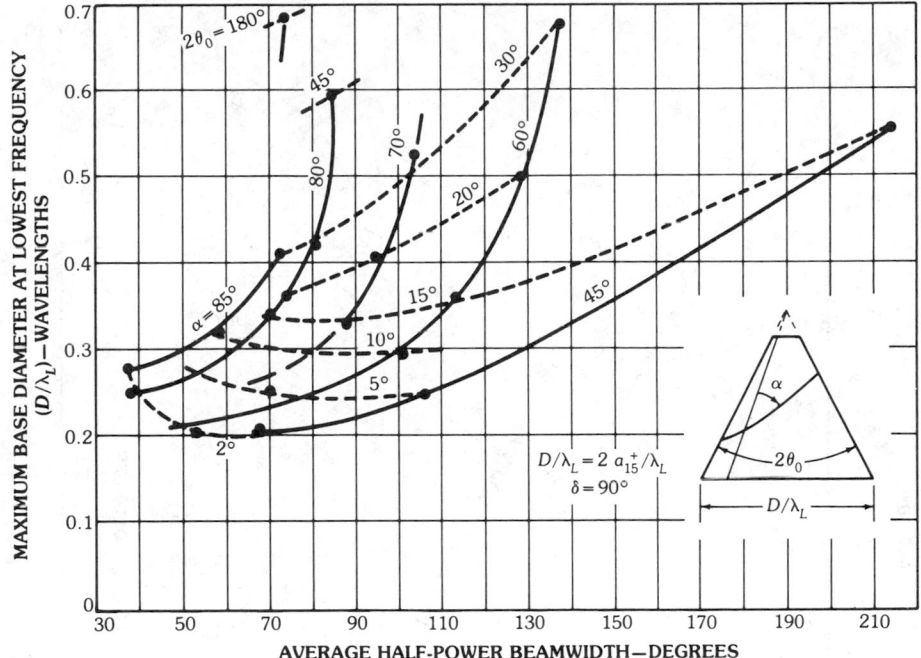

Fig. 99. Maximum diameter of base of cone at longest wavelength of operation when $D/\lambda_L = 2a_{15}^+/\lambda$, with $\delta = 90°$.

sense that it has the narrowest active region. Hence it has the greatest bandwidth for a given physical size. It also tends to have well-formed radiation patterns with the least energy radiated in back lobes. Except for the impedance a change in arm width for cone angles $2\theta_0$ less than 20° and angles of wrap greater than 80° causes only minor changes. Hence the narrow constant-width and cable-arm approximations to the log-spiral antenna can be used effectively. As the cone angle increases and the angle of wrap decreases from these values, a departure from $\delta = 90°$ causes increasingly greater pattern deterioration. The impedance depends primarily on the angular arm width, ranging from 300 Ω for small arm widths to 80 Ω for wide arms. Typical values for the case of $\delta = 90°$ range from 140 to 165 Ω. The impedance bandwidth is consistently greater than the pattern bandwidth.

As shown in Fig. 92 the beamwidth of the two-arm conical spiral can be controlled over a limited range by a suitable choice of α. Typical half-power beamwidths range from 60° to 70° for $\alpha = 82°$, 70° to 80° for $\alpha = 73°$, and 160° to 180° for $\alpha = 60°$. As α is decreased to 45° the beamwidth increases to 180° to 200° and this value of α produces circularly polarized coverage of an entire hemisphere and an omnidirectional pattern in the $\theta = 90°$ plane. To realize this coverage the decrease in α must be accompanied by an increase in δ.

When omnidirectional circular polarization near $\theta = $ constant is desired, it is advantageous to use a four-arm, rather than a two-arm, spiral [31]. When using multiple-arm structures, the number of choices for feeding the antenna increases [2]. There are particular excitations that will produce selected groups from the

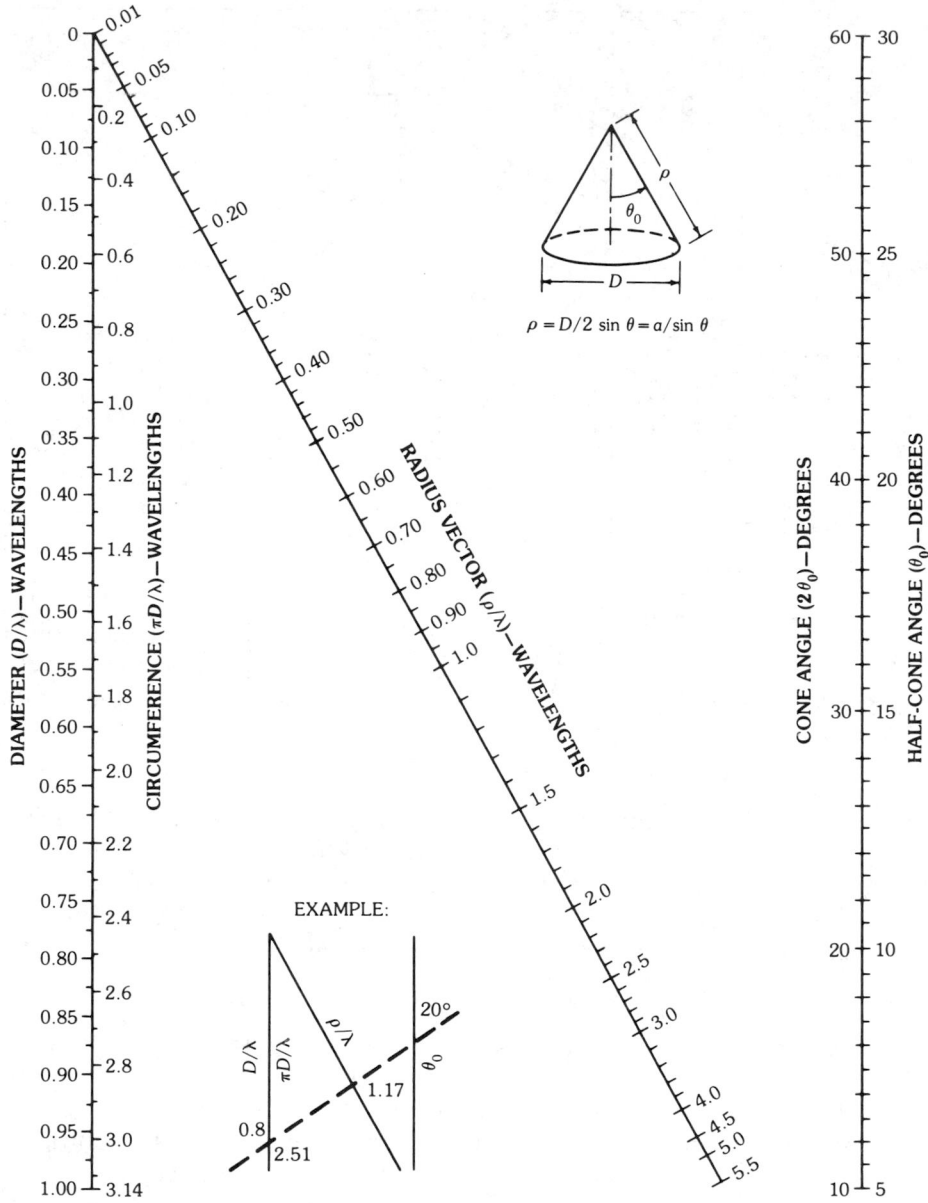

Fig. 100. Nomograph relating radius vector ϱ to diameter D of cone and cone angle $2\theta_0$.

possible azimuthal "modes" having variations $\exp(\pm jm\phi)$, where m is integer. The usual excitation of two-arm spirals shown in Fig. 104a will produce only terms with odd integer values of m since a rotation of 180° is equivalent to 180° phase shift. The remarkable property of the conical log-spirals is that the field is predominantly only the lowest-order term which is consistent with the sense of winding and the excitation. Fig. 105 shows the relative phase of the distant field as a

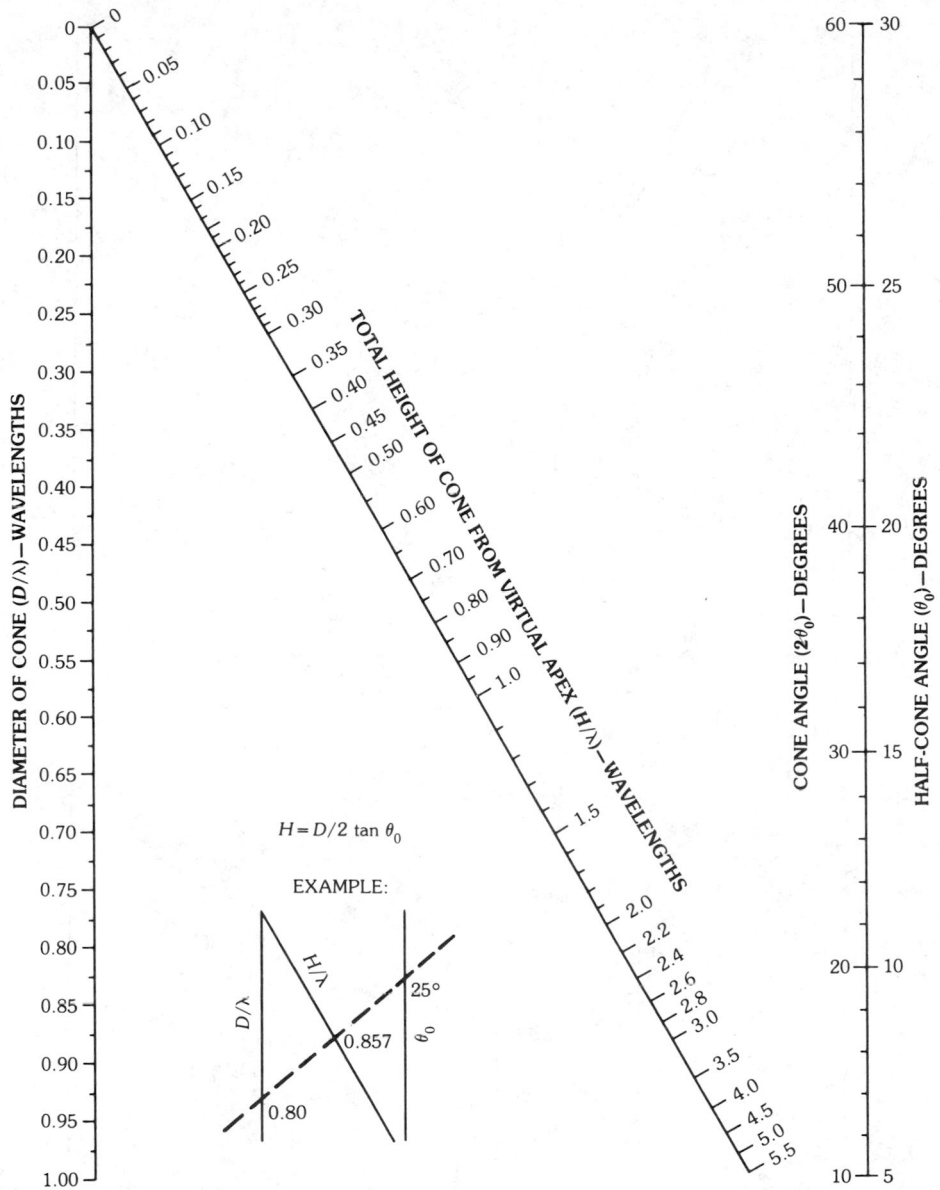

Fig. 101. Nomograph relating full height H of cone to diameter D of cone and cone angle $2\theta_0$.

function of ϕ for two antennas, a two-arm spiral excited with the arms 180° apart corresponding to odd-integer values of m, and a four-arm spiral excited with adjacent arms 180° apart. Note that in each case the phase is almost directly proportional to ϕ, but for the two-arm case the coefficient is 1 while for the four-arm case the coefficient is 2. This indicates that, in each case, there is very little of any of the higher-order terms present.

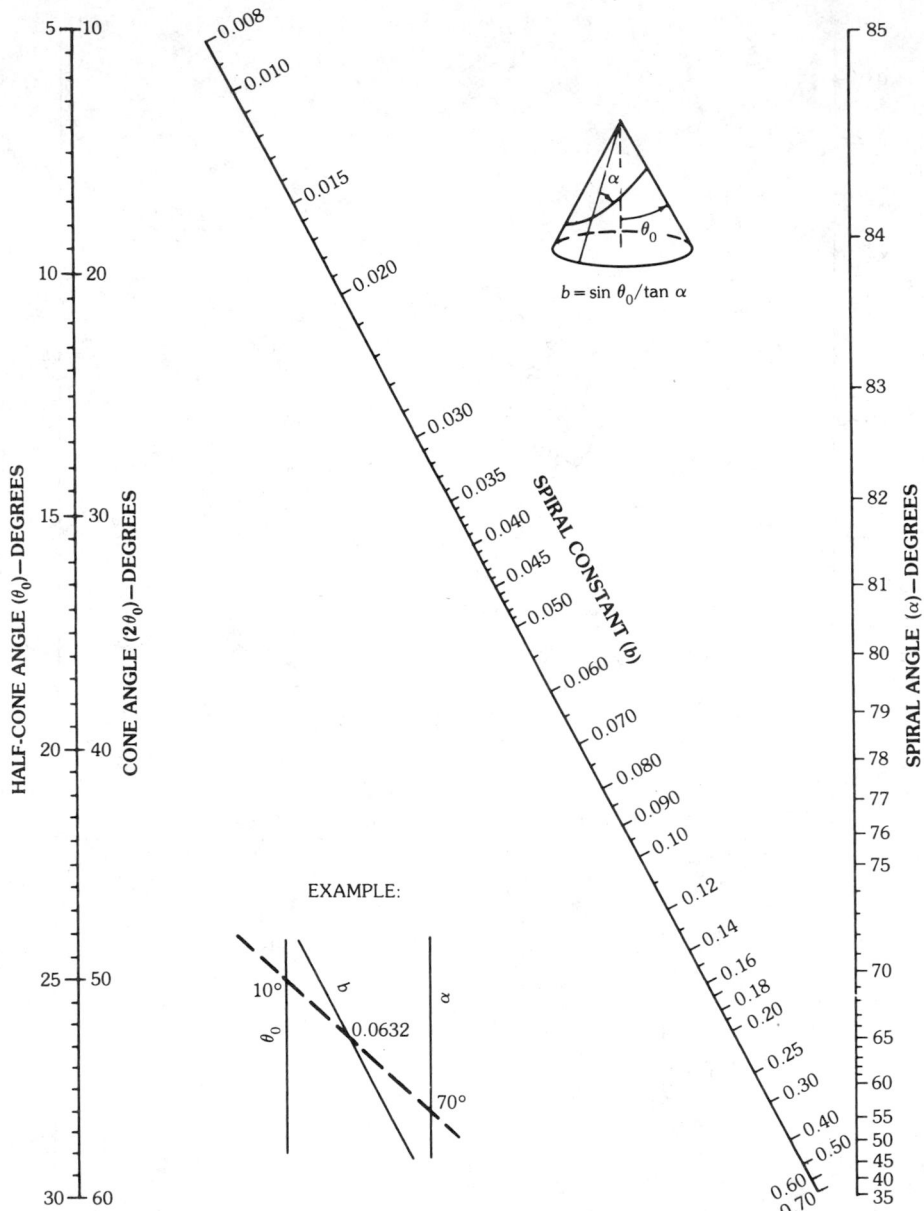

Fig. 102. Nomograph relating spiral constant b to cone angle $2\theta_0$ and spiral angle α.

The Maxwell equations indicate that any field that has m greater than unity will have zero along the polar axis ($\theta = 0$). Any excitation that produces only fields that vary with ϕ according to m greater than one will have a conical pattern with null along the axis and maximum at some nonzero value of θ. The excitation shown in Fig. 104c, producing a minimum value of m of 2, is the simplest case. This excitation is easily achieved by connecting opposite arms together and feeding each

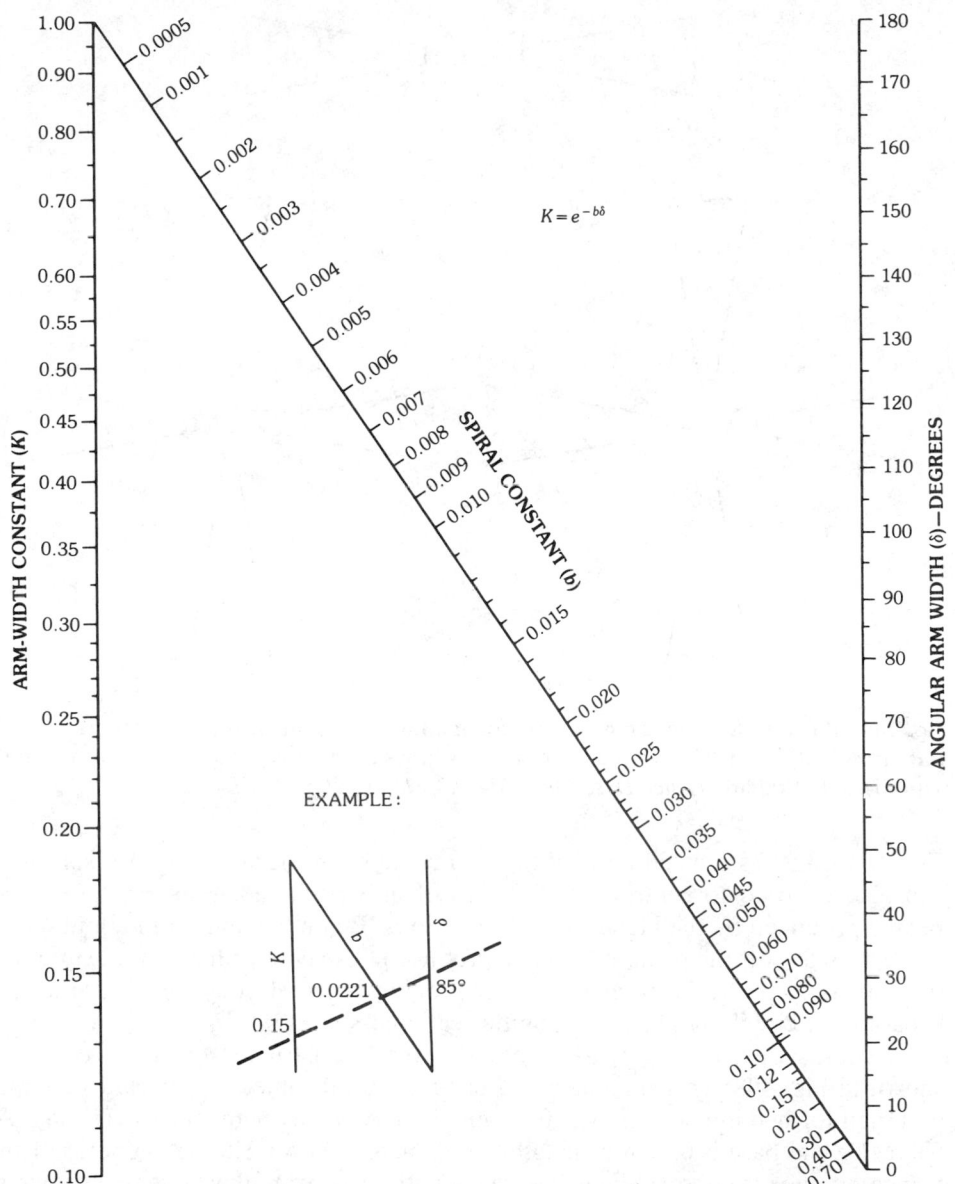

Fig. 103. Nomograph relating arm with constant K to spiral constant b and angular arm width δ.

pair 180° apart. The two terminals created by joining opposite arms can be fed by a balanced line, a coaxial cable connected through a balun, or by carrying a coaxial cable along one of the arms as described previously. Details of the feed-point geometry in the latter case are shown in Fig. 106. Dummy cables on the other arms are required to maintain symmetry.

Fig. 107 shows patterns of symmetrical four-arm spirals fed as shown in

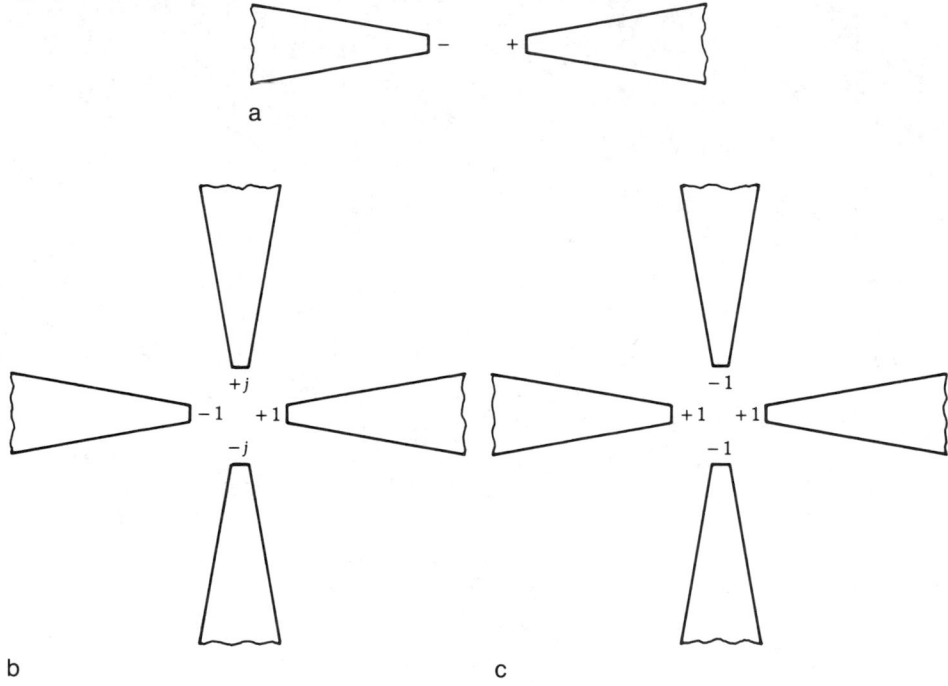

Fig. 104. Possible feeding arrangements for multiarm structures. (*a*) For two-arm spiral, with arms 180° apart. (*b*) For four-arm spiral, with arms 90° apart. (*c*) For four-arm spiral, with arms 180° apart. (*After Dyson and Mayes [31], © 1961 IEEE*)

Fig. 104c. Just as α could be used to control the beamwidth of two-arm spirals, it can also be used to control the direction of maximum radiation for the conical beams produced by the higher-order excitations. Beam maxima from about 40° to more than 90° off the polar axis can be produced. The case with beam maximum at $\theta = 90°$ is of particular interest since it provides a simple way to achieve a very broadband, circularly polarized, omnidirectional source.

Patterns of a four-arm, balanced, log-spiral constructed on a 15° cone are shown in Fig. 108. This antenna was etched from a flexible, copper-clad substrate and then formed into a cone. The feed and dummy cables were RG-141/U coax. At 550 MHz the base is 0.57 wavelengths in diameter. At 4 GHz the diameter of the truncated apex is approximately 0.2 wavelength. The azimuthal coverage shown in the patterns of Fig. 108 is presented in more detail in Fig. 109, where the deviation from omnidirectional is plotted in decibels for the two orthogonal field components. The axial ratio in the $\theta = 90°$ plane is seen to vary somewhat with ϕ. However, over a considerable bandwidth the amplitude deviation from omnidirectional is less than 3 dB and the axial ratio is also less than 3 dB.

The beamwidth of the conical beam patterns in a $\phi = $ constant plane is relatively insensitive to a change in antenna parameters. Antennas with $2\theta_0$ equal to both 15° and 20°, with α between 45° and 73°, and with cable arms or with arms of expanding width all have half-power beamwidths ranging from 35° to 55°.

Four-arm spirals fed in the manner of Fig. 104c, with $2\theta_0$ of 15° or 20°, typically

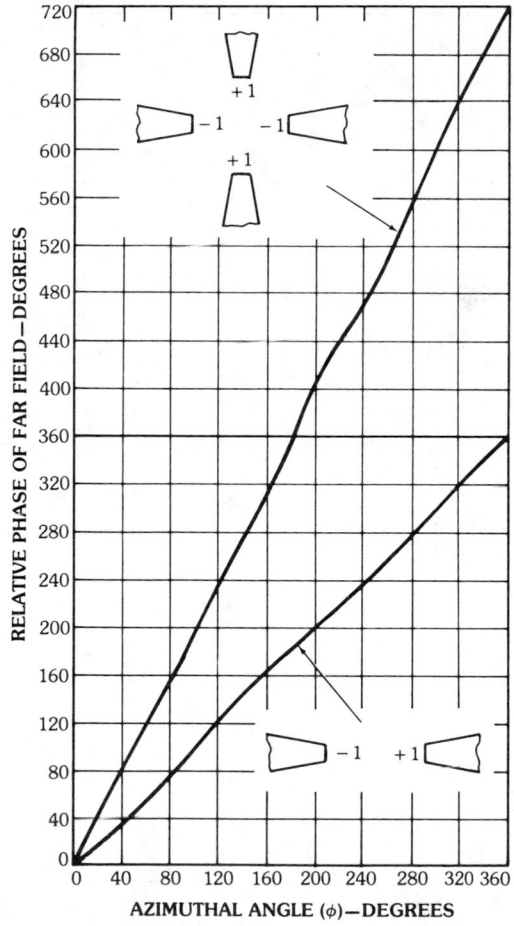

Fig. 105. Measured phase of radiated field as a function of the azimuthal angle ϕ for two antennas, with $\alpha = 45°$ and $\theta = 90°$. (*After Dyson and Mayes [31], © 1961 IEEE*)

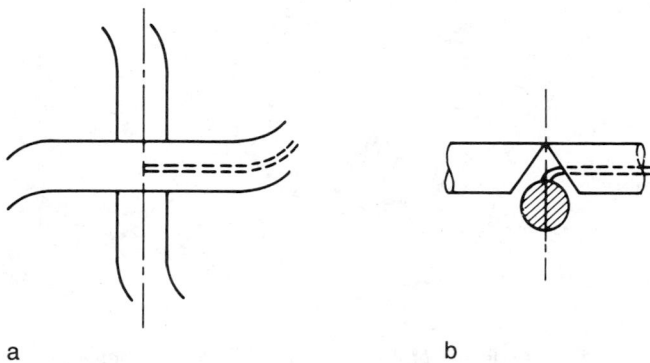

Fig. 106. "Infinite balun" feed used on a four-arm, conical, beam antenna. (*a*) Top view. (*b*) Side view. (*After Dyson and Mayes [31], © 1961 IEEE*)

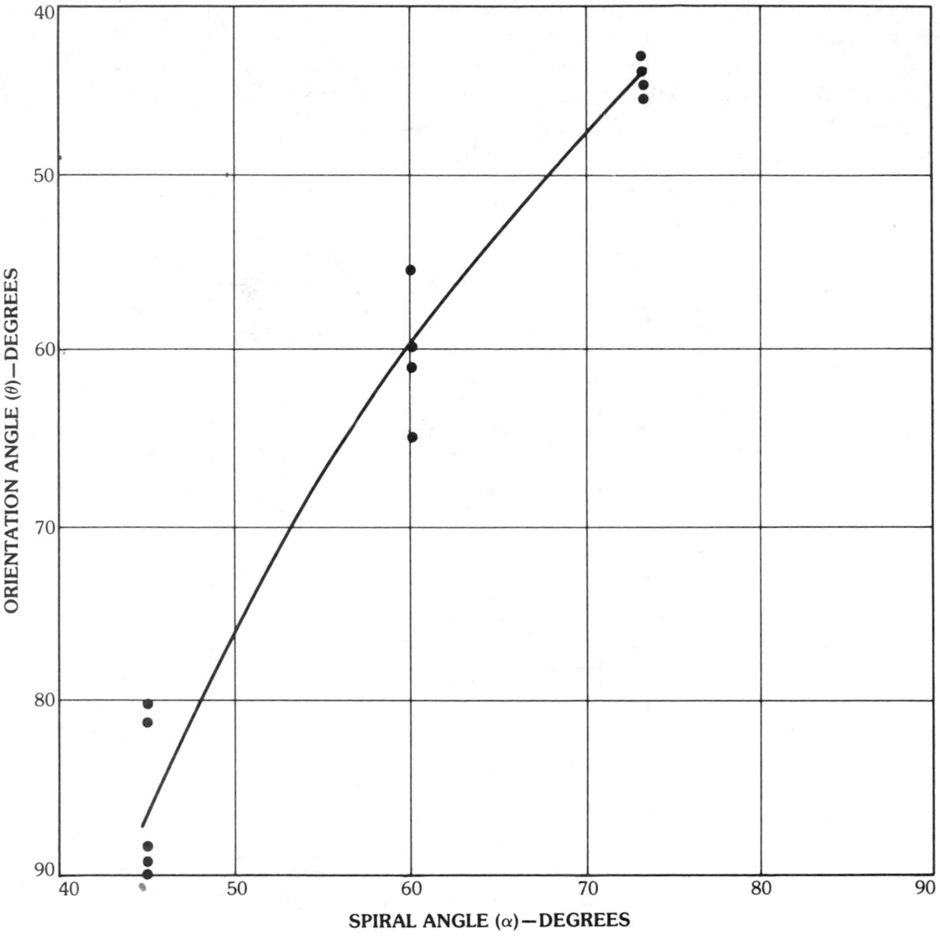

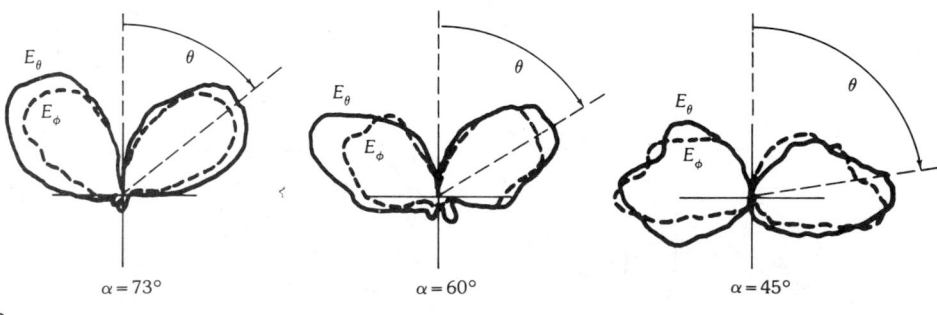

Fig. 107. Typical electric-field patterns and orientation of the conical beam as a function of the spiral angle, with $7.5° \leqq \theta_0 \leqq 10°$. (*a*) Graph. (*b*) Some patterns. (*After Dyson and Mayes [31], © 1961 IEEE*)

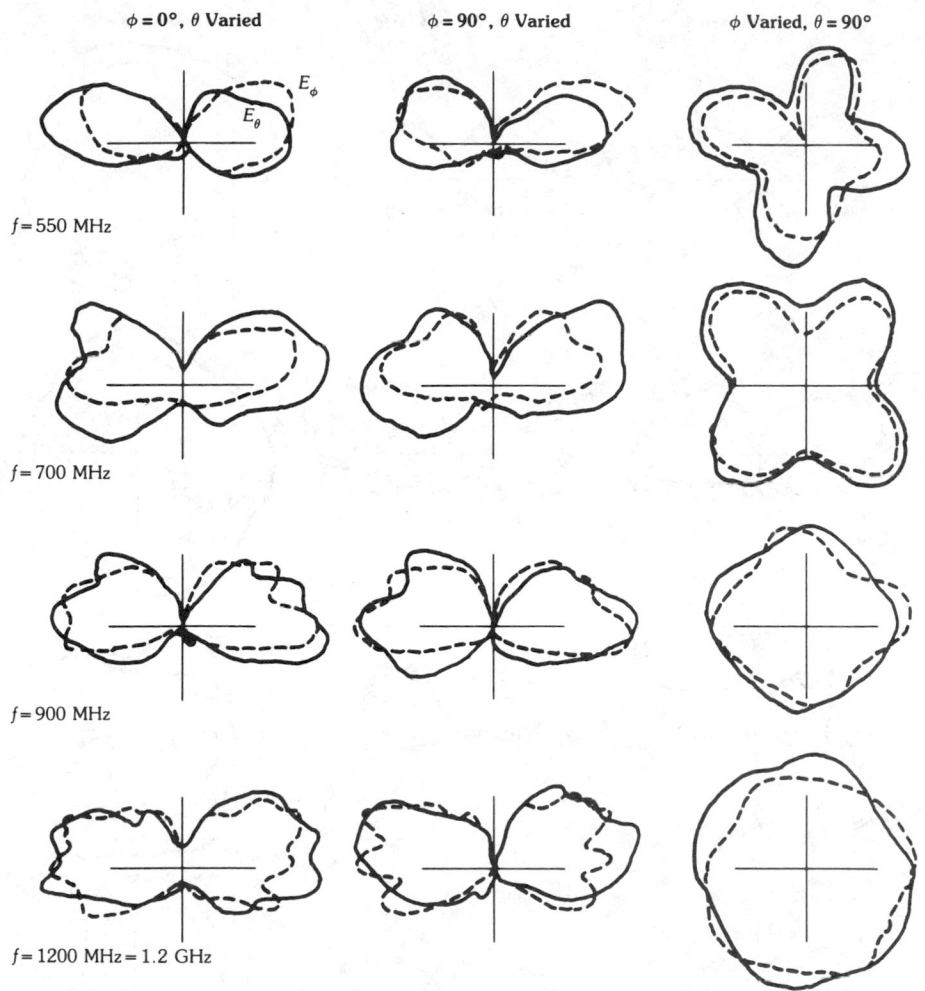

$\phi = 0°$, θ Varied $\phi = 90°$, θ Varied ϕ Varied, $\theta = 90°$

E_ϕ
E_θ

$f = 550$ MHz

$f = 700$ MHz

$f = 900$ MHz

$f = 1200$ MHz $= 1.2$ GHz

Fig. 108. Electric-field patterns of a balanced, symmetrical, four-arm, conical, equiangular-spiral antenna, with $\theta_0 = 7.5°$, $\alpha = 45°$, $K = 0.925$, $D = 31$ cm, and $d = 1.5$ cm. (*After Dyson and Mayes [31], © 1961 IEEE*)

have an impedance from 45 to 55 Ω for α ranging from 45° to 60°. As α is increased to 73° the impedance rises to the neighborhood of 70 Ω. These values are approximately one-half those noted earlier for similar two-arm antennas. The vswr is typically less than 1.5 (referred to the mean impedance) over most of the pattern bandwidth.

The pattern characteristics of log-spiral antennas are relatively constant over an extended frequency band. The beamwidth of two-arm spirals and the beam maximum of four-arm spirals are directly related to the value of α. Archimedean-spiral curves may also be orthogonally projected onto a conical surface as shown in Fig. 110b. However, the value of α is not constant for the Archimedean spirals, but

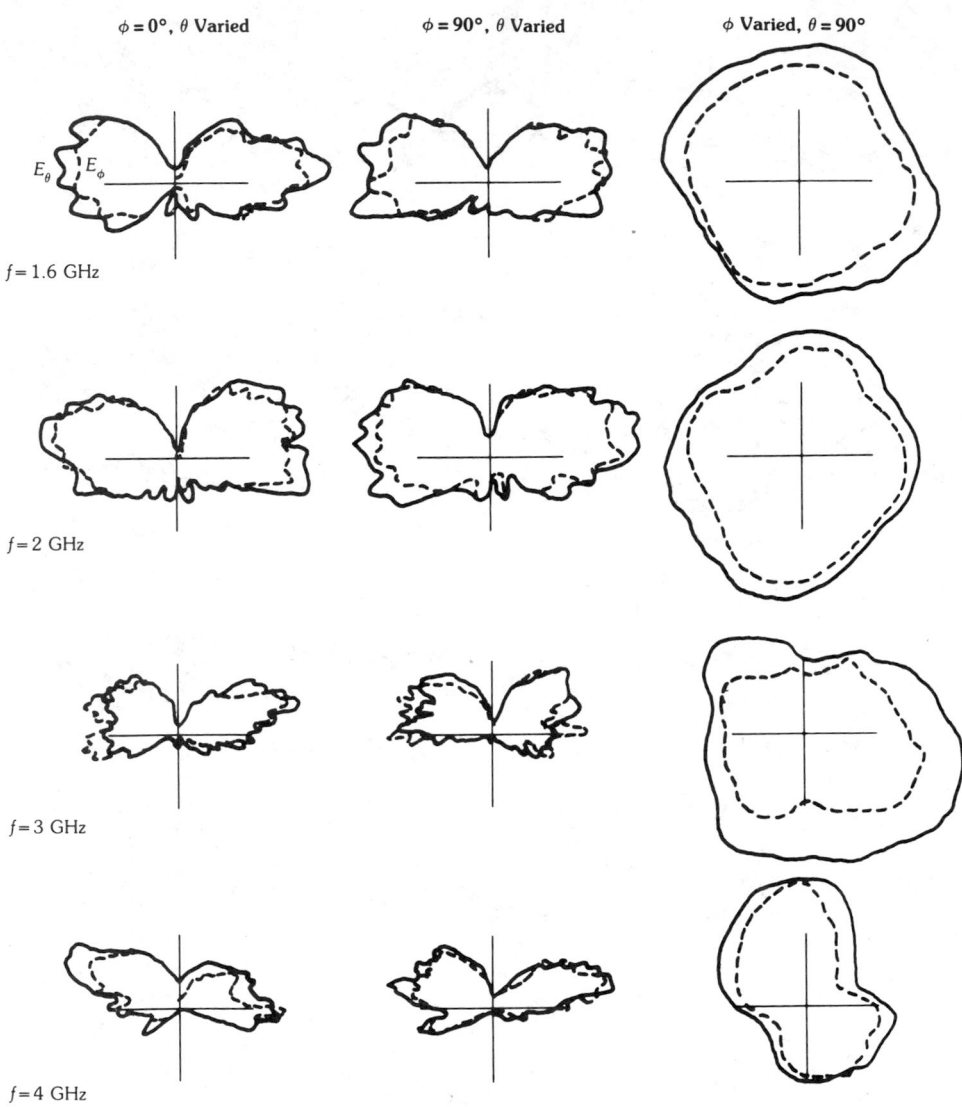

$\phi = 0°, \theta$ Varied $\phi = 90°, \theta$ Varied ϕ Varied, $\theta = 90°$

$E_\theta \quad E_\phi$

$f = 1.6$ GHz

$f = 2$ GHz

$f = 3$ GHz

$f = 4$ GHz

Fig. 108, *continued.*

rather is dependent on the value of ϕ. As the frequency is changed, the active region of a conical Archimedean spiral moves across areas with different values of α. This is manifest in a changing of the beamwidth for a two-arm antenna and a scanning of the beam maximum for the four-arm antennas. Patterns for one particular four-arm Archimedean spiral are shown in Fig. 111a. The value of α for this antenna ranged from about 45° at the apex to 85° at the base. The beam maximum varies from 45° to 90° as the frequency is swept from 1 to 2 GHz. The patterns for a log-spiral with $\alpha = 85°$ are shown in Fig. 111b for comparison.

The various modes of a multiarm spiral could also be selected by carrying an

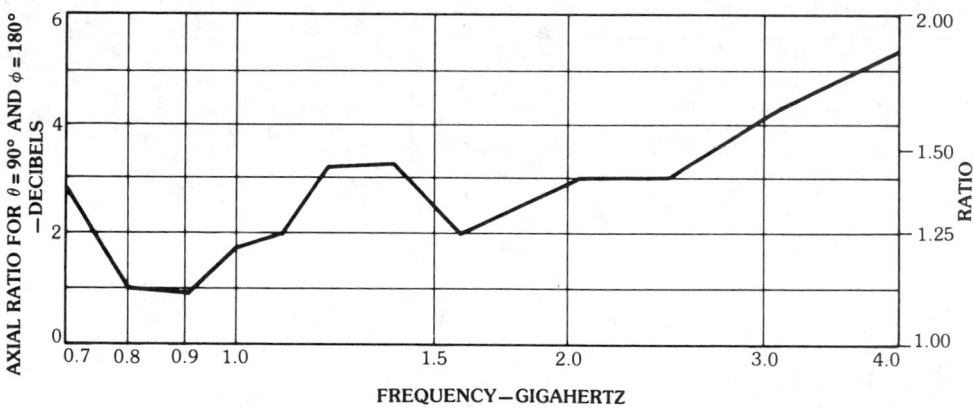

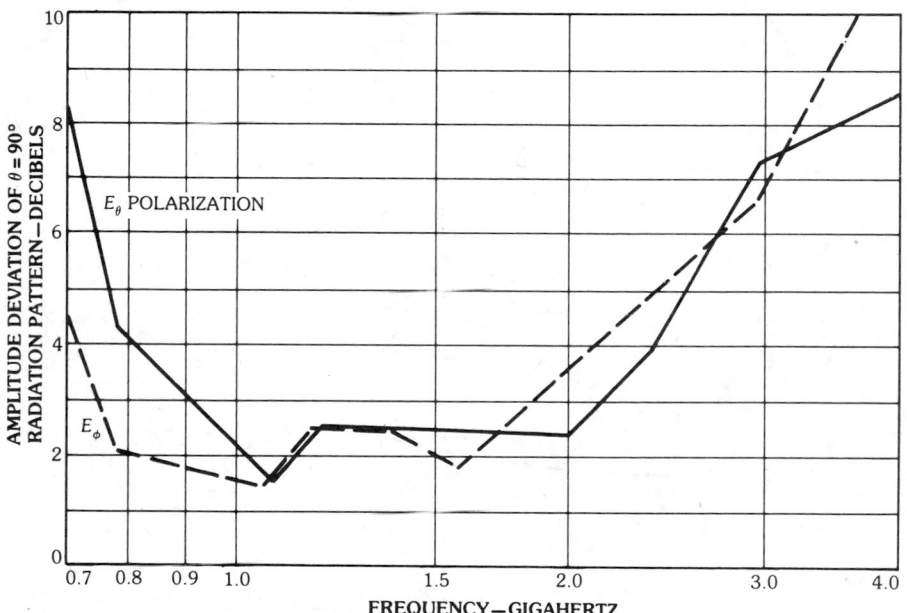

Fig. 109. Azimuthal coverage of the radiation patterns in Fig. 108. (*After Dyson and Mayes [31], © 1961 IEEE*)

active cable for each arm up the axis of the cone and applying the appropriate phase in each line at the base. The phase properties of the far fields for several of the lower-order modes provide a means of determining the direction of arrival of incoming waves. Fields having $\exp(\pm jm\phi)$ far-field behavior have been termed "spiral-phase" fields [33]. In contrast, the fields of a vertical monopole have equiphase contours that are circles, a "circular-phase" field.

Several direction-finding systems can be devised using multiarm log-spirals [34–36]. A system could be based on (*a*) a log-spiral excited in a spiral-phase mode and another antenna operating in the circular-phase mode, (*b*) two log-spiral

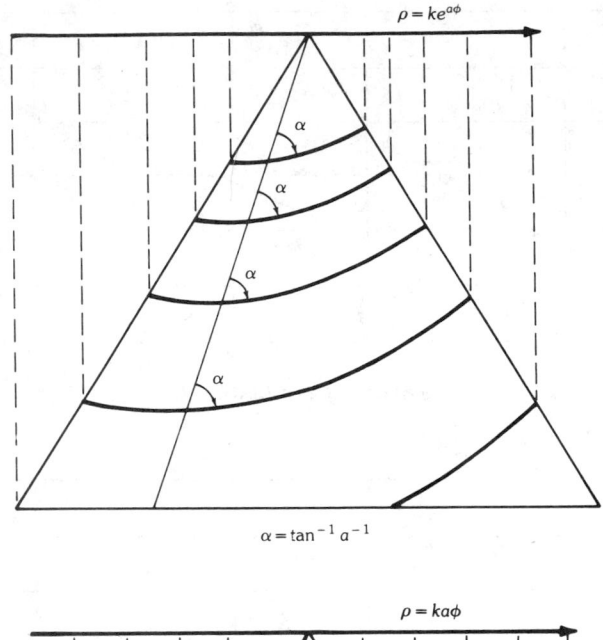

a

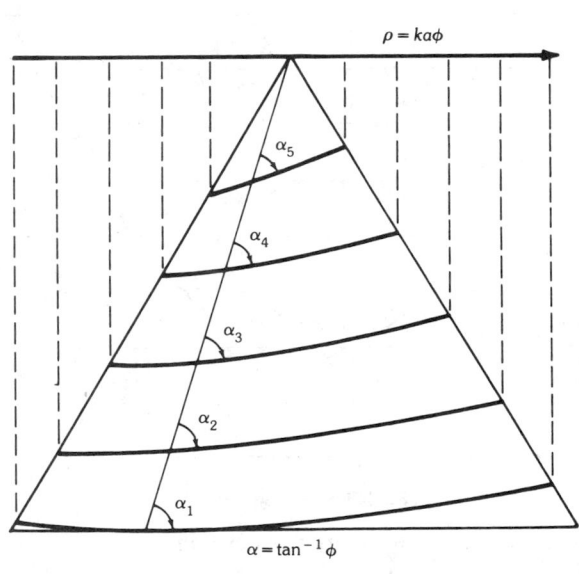

b

Fig. 110. Projection of equiangular-spiral and Archimedean-spiral curves on a conical surface. (*a*) Equiangular spiral. (*b*) Archimedean spiral. (*After Dyson and Mayes [31],* © *1961 IEEE*)

antennas excited in $+m$ and $-m$ modes, or (*c*) a single log-spiral antenna operated successively in two or more modes. The various modes of a multiarm antenna can be excited by carrying a feed cable for each arm up the axis of the cone. Switching among various values of phase shift between arms can then be used to sequentially produce any one of the several possible modes. The technique is illustrated for a four-arm spiral in Fig. 112.

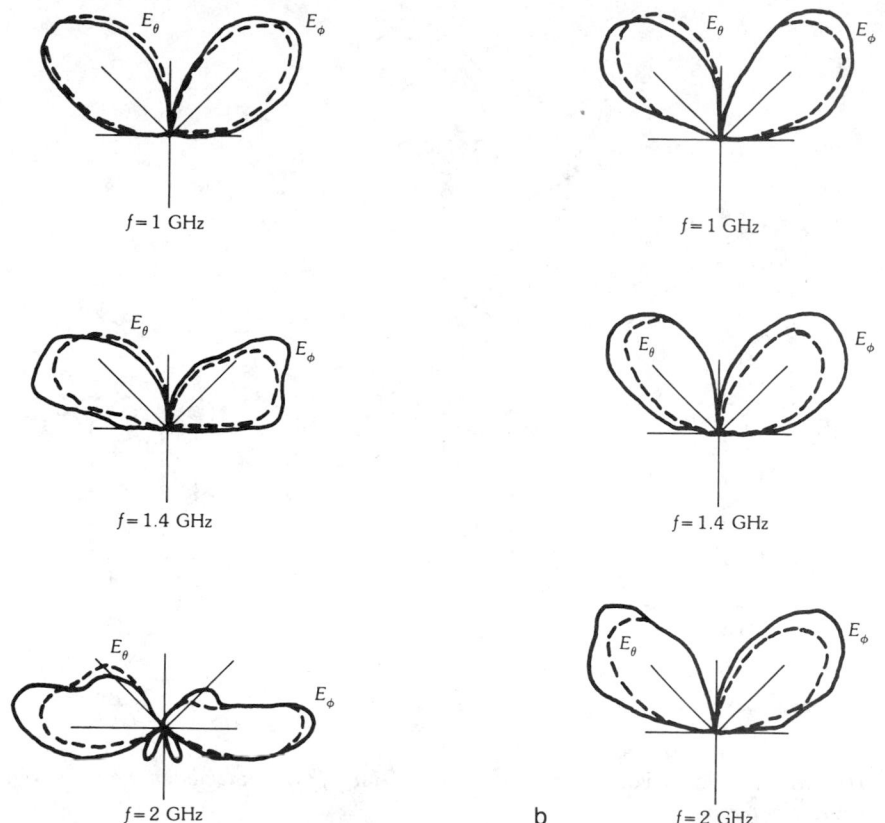

Fig. 111. Electric-field patterns of symmetrical, four-arm, conical antennas, with $\theta_0 = 10°$, $D = 29.5$ cm, and $d = 4.5$ cm ($\phi = 90°$, θ-varied pattern). (a) Archimedean spiral, with $\varrho = 1.026\phi$. (b) Equiangular spiral, with $\varrho = e^{[(\sin 10°)/(\tan 45°)]\phi}$. (*After Dyson and Mayes [31], © 1961 IEEE*)

Construction Techniques

In constructing log-periodic and log-spiral antennas it must be remembered that they are only potentially capable of covering extremely broad bands of frequency. Great care must be taken to use proper techniques if they are to perform with near-constant characteristics over such bandwidths. Generally, all members, both radiating and structural, should scale in size with distance from the apex. The feeders of LP dipole arrays, and other similar transmission means for exciting LP arrays, can be excepted from this rule. Such feeders may require stepped scaling if extremely wide frequency bands are to be covered in a high-performance manner. Element diameters in LP dipole arrays also need not conform exactly to true LP scaling. Symmetry, balance, and precision of construction are all important. An application of log-periodic or log-spiral antennas over extremely wide bandwidths requires all of the precision of construction that can be economically justified. This is particularly true for frequencies above 1 GHz.

Using longitudinal struts for support of the radiating elements has been found

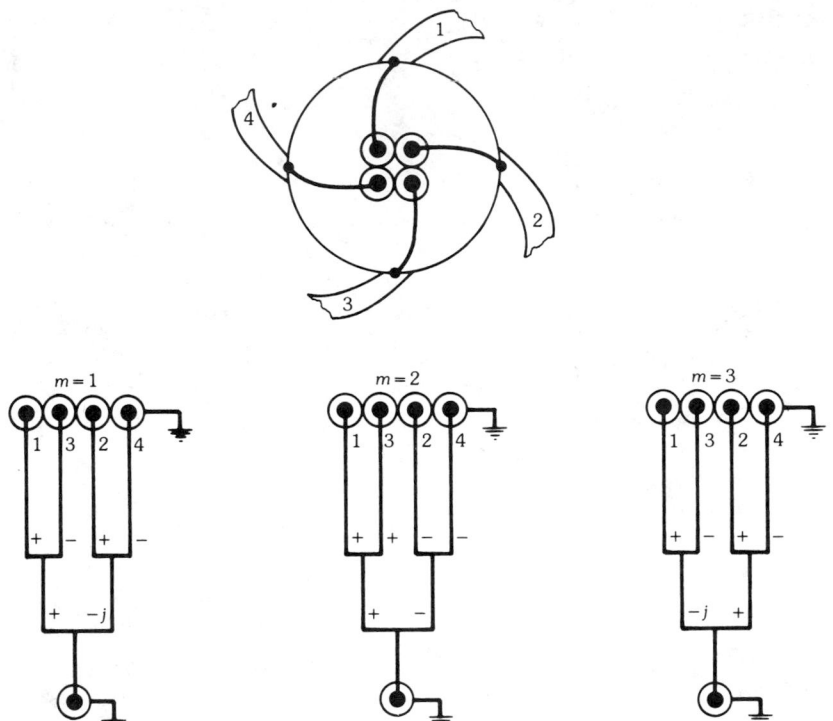

Fig. 112. Multimode feed on axis of cone. (*After Dyson [34], presented at the 1961 National Electronic Conference*)

satisfactory. Ideally, the struts should be tapered. The use of dielectric structural members placed transverse to the axis of the antenna causes some perturbation in performance when the active region passes over these members. Hence such structural members cause greater variation in the performance than members located along a radius vector from the origin (virtual apex).

5. References

[1] Y. Mushiake, *J. Inst. Electron. Eng.*, Japan, vol. 69, pp. 86–88 (in Japanese).
[2] G. A. Deschamps, "Impedance properties of complementary multiterminal planar structures," *IRE Trans. Antennas Propag.*, vol. AP-7, pp. S371–S378, December 1959.
[3] R. H. DuHamel and D. E. Isbell, "Broadband logarithmically periodic antenna structure," *IRE Natl. Conv. Rec.*, pt. I, pp. 119–128, 1957.
[4] D. E. Isbell, "Nonplanar logarithmically periodic antenna structures," *Tech. Rep. 30*, Univ. of Illinois Antenna Lab, Contract AF33(616)-3220, February 1958.
[5] R. H. DuHamel and F. R. Ore, "Logarithmically periodic antenna design," *Tech. Rep. CTR-198*, Collins Radio Co., Cedar Rapids, Iowa, March 1958.
[6] D. E. Isbell, "Log-periodic dipole arrays," *IRE Trans. Antennas Propag.*, vol. AP-8, pp. 260–267, May 1960.
[7] R. L. Carrel, "The design of log-periodic dipole antennas," *IRE Intl. Conv. Rec.*, pt. I, pp. 61–75, 1961.

[8] R. L. Carrel, "Analysis and design of the log-periodic antenna," PhD dissertation, Univ. of Illinois, Urbana, 1961.

[9] W. M. Cheong and R. W. P. King, "Arrays of unequal and unequally spaced dipoles," *Radio Sci.*, vol. 2, no. 11, pp. 1303–1314, November 1967.

[10] W. M. Cheong and R. W. P. King, "Log-periodic dipole antenna," *Radio Sci.*, vol. 2, no. 11, pp. 1315–1325, November 1967.

[11] J. Wolter, "Solution of Maxwell's equations for log-periodic dipole antennas," *IEEE Trans. Antennas Propag.*, vol. AP-18, pp. 734–741, November 1970.

[12] G. DeVito and G. B. Stracca, "Comments on the design of log-periodic dipole antennas," *IEEE Trans. Antennas Propag.*, vol. AP-21, pp. 303–308, May 1973.

[13] G. DeVito and G. B. Stracca, "Further comments on the design of log-periodic dipole antennas," *IEEE Trans. Antennas Propag.*, vol. AP-22, pp. 714–718, September 1974.

[14] P. C. Butson and G. T. Thompson, "A note on the calculation of the gain of log-periodic dipole antennas," *IEEE Trans. Antennas Propag.*, vol. AP-24, pp. 105–106, January 1976.

[15] E. C. Jordan and K. J. Balmain, *Electromagnetic Waves and Radiating Systems*, Englewood Cliffs: Prentice-Hall, 1968, p. 390.

[16] P. E. Mayes, G. A. Deschamps, and W. T. Patton, "Backward-wave radiation from periodic structures and applications to the design of frequency-independent antennas," *Proc. IRE*, vol. 49, pp. 962–963, May 1961.

[17] H. C. Pocklington, "Electrical oscillations in wires," *Camb. Phil. Soc.*, vol. 9, pp. 324–332, October 1897.

[18] S. H. Lee, "Analysis of zigzag antennas," PhD dissertation, Dept. of Electr. Eng., Univ. of California, Berkeley, June 1968.

[19] S. H. Lee and K. K. Mei, "Analysis of zigzag antennas," *IEEE Trans. Antennas Propag.*, vol. AP-18, pp. 760–764, November 1970.

[20] S. C. Kuo and P. E. Mayes, "An experimental study and the design of the log-periodic zigzag antennas," *Tech. Rep. 65-11*, Univ. of Illinois Antenna Lab, Contract AF33(615)-3216, February 1966.

[21] P. G. Ingerson and P. E. Mayes, "Asymmetrical feeders for log-periodic antennas," Seventeenth Annual Symposium, USAF Antenna Res. and Dev. Prog., Allerton Park, Univ. of Illinois, Urbana, October 1967.

[22] P. G. Ingerson, "Analysis of some closed log-periodic antennas with applications to log-periodic antennas," PhD thesis, Univ. of Illinois, Urbana, 1968.

[23] L. Brillouin, *Wave Propagation in Periodic Structures*, New York: Dover Publications, 1953.

[24] E. G. Cristal, "Coupled–transmission-line directional coupler with coupled lines of unequal characteristic impedance," *IEEE Trans. Microwave Theory Tech.*, vol. MTT-14, no. 7, pp. 337–346, July 1966.

[25] R. L. Carrel, "The design of log-periodic dipole antennas," *IRE Intl. Conv. Rec.*, pt. I, pp. 61–75, 1961.

[26] P. G. Ingerson and P. E. Mayes, "Log-periodic antennas with modulated impedance feeders," *IEEE Trans. Antennas Propag.*, vol. AP-16, pp. 633–642, November 1968.

[27] J. D. Dyson, "The equiangular-spiral antenna," Fifth Annual Symposium, USAF Antenna Res. and Dev. Prog., Allerton Park, Univ. of Illinois at Urbana, October 1955.

[28] J. D. Dyson, "The equiangular-spiral antenna," *IRE Trans. Antennas Propag.*, vol. AP-7, pp. 181–187, April 1959.

[29] J. D. Dyson, "Recent developments in spiral antennas," *Proc. IRE Natl. Aeron. Electron. Conf.*, pp. 617–625, 1959.

[30] J. D. Dyson, "The characteristics and design of the conical log-spiral antenna," *IRE Trans. Antennas Propag.*, vol. AP-13, pp. 488–499, July 1965.

[31] J. D. Dyson and P. E. Mayes, "New circularly polarized frequency-independent antennas with conical beam or omnidirectional patterns," *IRE Trans. Antennas Propag.*, vol. AP-19, pp. 334–342, July 1961. Also *Tech. Rep. 60*, Univ. of Illinois

Antenna Lab, Contract AF33(657)-8460, December 1962.

[32] J. W. Duncan and V. P. Minerva, "100:1 bandwidth balun transformer," *Proc. IRE*, vol. 48, pp. 156–164, February 1960.

[33] E. K. Sandeman, "Spiral-phase fields," *Wireless Engineer*, pp. 96–105, March 1949.

[34] J. D. Dyson, "The conical log-spiral antenna in simple arrays," Eleventh Annual Symposium, USAF Res. and Dev. Prog., Allerton Park, Univ. of Illinois, Urbana, October 1961.

[35] J. D. Dyson, "Multimode logarithmic-spiral antennas—possible applications," *Proc. Natl. Electron. Conf.*, pp. 206–213, October 1961.

[36] J. D. Dyson, "The multimode antennas as single-aperture systems," Antenna Forum, Univ. of Illinois, Urbana, January–February 1967.

6. Bibliography

Anders, R., and R. Wohlleben, "Phase velocity on a conical two-armed logarithmic foil-type spiral antenna," *IRE Trans. Antennas Propag.*, vol. AP-17, pp. 233–234, March 1969.

Atia, A. E., and K. K. Mei, "Analysis of multiple-arm conical log-spiral antennas," *IEEE Trans. Antennas Propag.*, vol. AP-19, pp. 320–331, May 1971.

Balmain, K. G., and J. D. Dyson, "The series-fed, log-periodic folded dipole array," *Dig. IEEE PTG-AP Intl. Symp.*, Boulder, Colorado, pp. 143–148, 1963.

Balmain, K. G., and S. W. Mikhail, "Loop coupling to a periodic dipole array," *Electron. Lett.*, vol. 5, no. 11, pp. 228–229, May 29, 1969.

Balmain, K. G., C. C. Bantin, C. R. Oakes, and L. David, "Optimization of log-periodic dipole antennas," *IEEE Trans. Antennas Propag.*, vol. AP-19, pp. 286–288, March 1971.

Balmain, K. G., and J. N. Nkeng, "Asymmetry phenomenon of log-periodic dipole antennas," *IEEE Trans. Antennas Propag.*, vol. AP-24, pp. 402–410, July 1976.

Bantin, C. C., and K. G. Balmain, "Study of compressed log-periodic dipole antennas," *IEEE Trans. Antennas Propag.*, vol. AP-18, pp. 195–203, March 1970.

Barbano, N., "Log-periodic dipole array with parasitic elements," *Microwave J.*, vol. 8, pp. 41–69, October 1965. Also *Tech. Memo EDL-M623*, Electronic Defense Laboratory, Sylvania Electric Products, Inc., Mountain View, Calif., January 1964.

———. "Log-periodic Yagi-Uda array," *IEEE Trans. Antennas Propag.*, vol. AP-14, pp. 235–238, March 1966.

Bell, R. L., C. T. Elfving, and R. E. Franks, "Near-field measurements on a logarithmically periodic antenna," *IRE Trans. Antennas Propag.*, vol. AP-8, pp. 559–567, November 1960.

Berry, D. G., and F. R. Ore, "Log-periodic monopole array," *IRE Intl. Conv. Rec.*, pt. I, pp. 76–85, 1961. Also technical report, Collins Radio Co., Cedar Rapids, Iowa, October 1960.

Besthorn, J. W., "1.0 to 21.0-GHz log-periodic dipole antenna," Eighteenth Annual Symposium, USAF Antenna Res. and Dev. Prog., Allerton Park, Univ. of Illinois, Urbana, October 1968.

Blume, S., and J. Wolter, "Theoretische und experimentelle untersuchungen an logarithmisch periodischen antennen," *Z. Angew. Phys.*, vol. 23, pp. 386–396, 1967.

Brillouin, L., *Wave Propagation in Periodic Structures*, New York: Dover Publications, 1953.

Brooks, G. P., "A conical log-spiral antenna with resistive terminated arms," MS thesis, Dept. of Electr. Eng., Univ. of Illinois at Urbana, 1965.

Butson, P. C., and G. T. Thompson, "A note on the calculation of the gain of log-periodic dipole antennas," *IEEE Trans. Antennas Propag.*, vol. AP-24, pp. 105–106, January 1976.

Campbell, C. K., I. Traboulay, M. S. Suthers, and H. Kneve, "Design of a stripline log-periodic dipole antenna," *IEEE Trans. Antennas Propag.*, vol. AP-25, pp. 718–721, September 1977.

Carr, J. W., "Some variations in log-periodic antenna structures," *IRE Trans. Antennas*

Propag., vol. AP-9, pp. 229–230, March 1961.

Carrel, R. L., "An analysis of log-periodic dipole antennas," Tenth Annual Symposium, USAF Antenna Res. and Dev. Prog., Allerton Park, Univ. of Illinois, Urbana, October 1960.

——. "The design of log-periodic dipole antennas," *IRE Intl. Conv. Rec.*, pt. I, pp. 61–75, 1961.

——. "Analysis and design of the log-periodic dipole antenna," PhD dissertation, Univ. of Illinois, Urbana, 1961. Also *Tech. Rep. 52*, Univ. of Illinois Antenna Lab, Contract AF33(616)-6079, October 1961.

Chan, K. K., and P. Silvester, "Projective analysis of log-periodic dipole antennas," *IEEE Trans. Antennas Propag.*, vol. AP-21, p. 757, September 1973.

——. "Analysis of the log-periodic V-dipole antenna," *IEEE Trans. Antennas Propag.*, vol. AP-23, pp. 397–401, May 1975.

Chatterjee, J. S., and M. N. Roy, "Helical log-periodic array," *IEEE Trans. Antennas Propag.*, vol. AP-16, pp. 592–593, September 1968.

Cheo, B. R-S., V. H. Rumsey, and W. J. Welch, "A solution to the frequency-independent antenna problem," *IRE Trans. Antennas Propag.*, vol. AP-9, pp. 527–534, November 1961.

Cheo, B. R-S., and V. H. Rumsey, "A solution to the equiangular spiral antenna problem —continued," *IEEE Intl. Symp. Antennas Propag.*, pp. 126–130, December 1966.

——. "Surface waves and mode selection on multielement spiral antennas," *Radio Sci.*, vol. 3, no. 3, pp. 267–271, March 1968.

Cheong, W. M., and R. W. P. King, "Arrays of unequal and unequally spaced dipoles," *Radio Sci.*, vol. 2, no. 11, pp. 1303–1314, November 1967.

——. "Log-periodic dipole antenna," *Radio Sci.*, vol. 2, no. 11, pp. 1315–1325, November 1967.

Cornell, D. D., and B. J. Lamberty, "Multimode planar spiral for DF applications," Antenna Applications Symposium, Allerton Park, Univ. of Illinois, Urbana, September 1981.

Craven, J. H., "Dielectric lens for second-mode spiral," *IRE Trans. Antennas Propag.*, vol. AP-9, p. 499, September 1961.

Cristal, E. G., "Coupled-transmission-line directional coupler with coupled lines of unequal characteristic impedance," *IEEE Trans. Microwave Theory Tech.*, vol. MTT-14, no. 7, pp. 337–346, July 1966.

Curtis, J. W., "Spiral antennas," *IRE Trans. Antennas Propag.*, vol. AP-8, pp. 298–306, May 1960.

Deschamps, G. A., "Impedance properties of complementary multiterminal planar structures," *IRE Trans. Antennas Propag.*, vol. AP-7, pp. S371–S378, December 1959. Also *Tech. Rep. 43*, Univ. of Illinois Antenna Lab, Contract AF33(616)-6079, November 1959.

Deschamps, G. A., and J. Dyson, "The logarithmic spiral in a single-aperture multimode antenna system," *IEEE Trans. Antennas Propag.*, vol. AP-19, pp. 90–96, January 1971.

DeVito, G., and G. B. Stracca, "Comments on the design of log-periodic dipole antennas," *IEEE Trans. Antennas Propag.*, vol. AP-21, pp. 303–308, May 1973.

——. "Further comments on the design of log-periodic dipole antennas," *IEEE Trans. Antennas Propag.*, vol. AP-22, pp. 714–718, September 1974.

DeVito, G., "Influence of the earth in the hf transmission of LPD antennas," *IEEE Trans. Antennas Propag.*, vol. AP-25, pp. 891–896, November 1977.

DiFonzo, D. F., "Reduced size log-periodic antennas," *Microwave J.*, vol. 7, pp. 37–42, December 1964.

Donnellan, J. R., "Second-mode operation of the spiral antenna," *IRE Trans. Antennas Propag.*, vol. AP-8, p. 637, November 1960.

DuHamel, R. H., and D. E. Isbell, "Broadband logarithmically periodic antenna structures," *IRE Natl. Conv. Rec.*, pt. I, pp. 119–128, 1957. Also *Tech. Rep. 19*, Univ. of Illinois Antenna Lab, Contract AF33(616)-3220, May 1957.

DuHamel, R. H., and F. R. Ore, "Logarithmically periodic antenna design," *Tech. Rep.*

CTR-198, Collins Radio Co., Cedar Rapids, Iowa, March 1958.

DuHamel, R. H., and D. G. Berry, "Logarithmically periodic antenna arrays," *IRE Wescon Conv. Rec.*, pt. 1, pp. 161–174, August 1958.

DuHamel, R. H., and F. R. Ore, "Logarithmically periodic antenna design," *IRE Natl. Conv. Rec.*, pt. I, pp. 139–151, 1959.

DuHamel, R. H., and D. G. Berry, "A new concept in high-frequency antenna design," *IRE Natl. Conv. Rec.*, pt. I, pp. 42–50, 1959.

DuHamel, R. H., and F. R. Ore, "Log-periodic feeds for lens and reflectors," *IRE Natl. Conv. Rec.*, pt. I, pp. 128–137, 1959.

Duncan, J. W., and V. P. Minerva, "100:1 bandwidth balun transformer," *Proc. IRE*, vol. 48, pp. 156–164, February 1960.

Dyson, J. D., "The equiangular-spiral antenna," Fifth Annual Symposium, USAF Antenna Res. and Dev. Prog., Allerton Park, Univ. of Illinois, Urbana, October 1955. Also *Tech. Rep. 21*, Univ. of Illinois Antenna Lab, Contract AF33(616)-3220, September 1957. Also PhD dissertation, University of Illinois, Urbana, 1957.

———. "The equiangular-spiral antenna," *IRE Trans. Antennas Propag.*, vol. AP-7, pp. 181–187, April 1959.

———. "Recent developments in spiral antennas," *Proc. IRE Natl. Aeron. Electron. Conf.*, pp. 617–625, 1959.

———. "The unidirectional equiangular-spiral antenna," *IRE Trans. Antennas Propag.*, vol. AP-7, pp. 329–334, October 1959. Also *Tech. Rep. 33*, Univ. of Illinois Antenna Lab, Contract AF33(616)3220, July 1958.

Dyson, J. D., and P. E. Mayes, "New circularly polarized frequency independent antennas with conical beam or omnidirectional patterns," *IRE Trans. Antennas Propag.*, vol. AP-9, pp. 334–342, July 1961. Also *Tech. Rep. 46*, Univ. of Illinois, Urbana, Contract AF33(616)-6079, June 1960.

———. "A note of the difference between equiangular and Archimedes spiral antennas," *IRE Trans.*, vol. MTT-9, pp. 203–205, March 1961.

Dyson, J. D., "The conical log-spiral antenna in simple arrays," Eleventh Annual Symposium, USAF Antenna Res. and Dev. Prog., Allerton Park, Univ. of Illinois, Urbana, October 1961.

———. "Multimode logarithmic-spiral antennas—possible applications," *Proc. Natl. Electron. Conf.*, pp. 206–213, October 1961.

———. "A survey of very wideband and frequency independent antennas—1945 to the present," *J. Res. Natl. Bur. Stand.* (US), vol. 66D, pp. 1–6, January–February 1962.

———. "Frequency independent antennas: survey of development," *Electronics*, vol. 35, no. 16, pp. 39–44, April 1962.

———. "An antenna to cover the 220 through 2400 Mc telemetry bands," *Proc. Natl. Telemetering Conf.*, vol. 1, sec. 9-4, May 1962.

———. "The coupling and mutual impedance between balanced wire-arm conical log-spiral antennas," *Tech. Rep. 54*, Univ. of Illinois, Urbana, Contract AF33(657)-8460, June 1962.

———. "The coupling and mutual impedance between conical log-spiral antennas in simple arrays," *IRE Intl. Conv. Rec.*, pt. 1, pp. 165–182, 1962.

———. "On a conical quad-spiral array," *Tech. Rept. 63*, Univ. of Illinois Antenna Lab, Contract AF33(657)-8460, September 1962.

Dyson, J. D., and G. L. Duff, "Near field measurements on the conical logarithmic spiral antenna," *G-AP Intl. Symp. Dig.*, pp. 137–142, 1963.

Dyson, J. D., "Research on conical log-spiral antennas," *Proc. Appl. Forum Antenna Res.*, ed. by P. E. Mayes, Univ. of Illinois, Urbana, pp. 215–265, 1964.

———. "The design of conical log-spiral antennas," *IEEE Intl. Conv. Rec.*, pt. 6, pp. 259–273, 1964.

———. "The characteristics and design of the conical log-spiral antenna," *IEEE Trans. Antennas Propag.*, vol. AP-13, pp. 488–499, July 1965. Also *Tech. Rep. 65-4*, Univ. of Illinois Antenna Lab, Contract AF33(657)-10474, May 1965.

———. "The multimode antennas as single-aperture systems," Antenna Forum, Univ. of

Illinois, Urbana, January–February 1967.

Elfving, C. T., "Foreshortened log-periodic dipole array," *1963 Wescon*, San Francisco, August 1963. Also *Tech. Memo. EDL-M401*, Contract DA36-039-AMC-00088(E), Electronic Defense Laboratories, Sylvania Electric Products, Inc., Mountain View, Calif., September 1963.

Elliott, R. S., "A view of frequency-independent antennas," *Microwave J.*, vol. 5, pp. 61–68, December 1962.

Evans, B. G., "The effects of transverse feed displacements on log-periodic dipole arrays," *IEEE Trans. Antennas Propag.*, vol. AP-18, pp. 124–128, January 1970.

Gans, M. J., D. Kajfez, and V. H. Rumsey, "Frequency independent baluns," *Proc. IEEE*, vol. 53, pp. 647–648, June 1965.

Green, P. B., and P. E. Mayes, "A log-periodic monopole array with a modulated impedance microstrip feeder," *Tech. Rep. 73-2*, Univ. of Illinois Antenna Lab, January 1973.

———. "50-ohm log-periodic monopole array with modulated-impedance microstrip feeder," *IEEE Trans. Antennas Propag.*, vol. AP-22, pp. 332–334, March 1974.

Greiser, J. W., and P. E. Mayes, "Vertically polarized log-periodic zig-zag antennas," *Proc. Natl. Electron. Conf.*, vol. 17, pp. 193–204, 1961.

Greiser, J. W., "The bent log-periodic zigzag antenna," technical report, Univ. of Illinois Antenna Lab, Contract NOBSR85243, May 1962.

———. "The bent log-periodic zigzag antenna," MS thesis, Dept. of Elec. Eng., Univ. of Illinois, Urbana, 1962.

———. "A new class of log-periodic antennas," *Proc. IEEE*, vol. 52, pp. 617–618, May 1964.

Greiser, J. W., and P. E. Mayes, "The bent backfire zigzag—a vertically polarized frequency-independent antenna," *IEEE Trans. Antennas Propag.*, vol. AP-12, pp. 281–290, May 1964.

Hahn, G., and R. Honda, "Conical spiral arrays for passive direction finding," Eighteenth Annual Symposium, USAF Antenna Res. and Dev. Prog., Allerton Park, Univ. of Illinois, Urbana, October 1968.

Hessemer, R. A., Jr., "Backward-wave radiation from an equiangular spiral antenna," *IRE Trans. Antennas Propag.*, vol. AP-9, p. 582, November 1961.

Hong, S., and G. Rassweiler, "Size reduction of a conical log-spiral antenna by loading with magneto-dielectric material," *IEEE Trans. Antennas Propag.*, vol. AP-14, pp. 650–651, September 1966.

Hudock, E., and P. E. Mayes, "Near-field investigation of uniform periodic monopole arrays," *IEEE Trans. Antennas Propag.*, vol. AP-13, pp. 840–855, November 1965.

Ingerson, P. G., and P. E. Mayes, "Design of log-periodic structures using complex dispersion data for periodic lines," *Tech. Rep. 66-10*, Univ. of Illinois Antenna Lab, August 1966.

———. "Design of log-periodic structures using dispersion data for periodic lines," Sixteenth Annual Symposium, USAF Antenna Res. and Dev. Prog., Allerton Park, Univ. of Illinois, Urbana, October 1966.

———. "Asymmetrical feeders for log-periodic antennas," Seventeenth Annual Symposium, USAF Antenna Res. and Dev. Prog., Allerton Park, Univ. of Illinois, Urbana, November 1967.

Ingerson, P. G. "Analysis of some closed log-periodic structures with applications to log-periodic antennas," PhD thesis, Univ. of Illinois, Urbana, 1968.

Ingerson, P. G., and P. E. Mayes, "Log-periodic antennas with modulated impedance feeders," *IEEE Trans. Antennas Propag.*, vol. AP-16, pp. 633–642, November 1968. Also *Tech. Rep. AFAL-TR-69-226*, Univ. of Illinois Antenna Lab, Contract F33615-69-C-1122, July 1969.

Ingerson, P. G., "Modulated arm width (MAW) log-spiral antennas," Twentieth Annual Symposium, USAF Antenna Res. and Dev. Prog., Allerton Park, Univ. of Illinois, Urbana, October 1970.

Isbell, D. E., "Nonplanar logarithmically periodic antenna structures," *Tech. Rep. 30*, Univ.

of Illinois Antenna Lab, Contract AF33(616)-3220, February 1958.

――――. "Multiple terminal log-periodic antennas," Eighth Annual Symposium, USAF Antenna Res. and Dev. Prog., Allerton Park, Univ. of Illinois, Urbana, October 1958.

――――. "A log-periodic reflector feed," *Proc. IRE*, vol. 47, pp. 1152–1153, June 1959.

――――. "Log-periodic dipole arrays," *IRE Trans. Antennas Propag.*, vol. AP-8, pp. 260–267, May 1960. Also *Tech. Rep. 39*, Univ. of Illinois Antenna Lab, Contract AF33(616)-6079, 1959.

Jones, K. E., and R. Mittra, "Some interpretations and applications of the k-beta diagram," *IEEE Intl. Conv. Rec.*, pt. 5, pp. 134–139, March 1965. Also *Tech. Rep. 65-1*, Univ. of Illinois Antenna Lab, Contract AF33(657)-10474, January 1965.

Jones, K. E., and P. E. Mayes, "Continuously scaled transmission lines with applications to log-periodic antennas," *IEEE Trans. Antennas Propag.*, vol. AP-17, pp. 2–9, January 1969.

Jordan, E. C., G. A. Deschamps, J. D. Dyson, and P. E. Mayes, "Developments in broadband antennas," *IEEE Spectrum*, vol. 1, pp. 58–71, April 1964.

Jordan, E. C., and K. J. Balmain, *Electromagnetic Waves and Radiating Systems*, Englewood Cliffs: Prentice-Hall, 1968.

Kaiser, J. A., "The Archimedean two-wire spiral antenna," *IRE Trans. Antennas Propag.*, vol. AP-8, pp. 312–323, May 1960.

――――. "Dual operation with the two-wire spiral antenna," *IRE Trans. Antennas Propag.*, vol. AP-9, pp. 583–584, November 1961.

Kaiser, P., "On the theoretical relations between periodic and log-periodic structures," PhD dissertation, Dept. of Electr. Eng. Univ. of California, Berkeley, September 1966.

――――. "The inclined log-spiral antenna, a new type of unidirectional, frequency independent antenna," *IEEE Trans. Antennas Propag.*, vol. AP-15, pp. 304–305, March 1967.

Keen, K. M., "A planar log-periodic antenna," *IEEE Trans. Antennas Propag.*, vol. AP-22, pp. 489–490, May 1974.

Kieburtz, R. B., "A phase-integral approximation for the current distribution along a log-periodic antenna," *IEEE Trans. Antennas Propag.*, vol. AP-13, pp. 813–814, September 1965.

Kim, O. K., "An experimental investigation of the conical four-arm logarithmic spiral antenna with large cone angle," *Tech. Rep. 66-12*, Univ. of Illinois Antenna Lab, Contract AF33(615)-3216, September 1966. Also MS thesis, Dept. of Electr. Eng., Univ. of Illinois, Urbana, 1966.

Kim, O. K., and J. D. Dyson, "A log-spiral antenna with selectable polarization," *IEEE Trans. Antennas Propag.*, vol. AP-19, pp. 675–677, September 1971.

Klock, P. W., "A study of wave propagation on helices," *Tech. Rep. 68*, Univ. of Illinois Antenna Lab, Contract AF33(657)-10474, March 1963.

Klock, P. W., and R. Mittra, "On the solution of the Brillouin (k-β) diagram of the helix and its application to helical antennas," *G-AP Intl. Symp. Dig.*, pp. 99–103, 1963.

――――. "Complex wave analysis of the backfire bifilar helical antenna," *Fall URSI Meeting*, Univ. of Illinois, p. 67, 1964.

Kosta, S. P., M. D. Singh, N. K. Agarwal, and A. Singh, "A note on the theory of log-periodic dipole antenna," *IEEE Trans. Antennas Propag.*, vol. AP-18, p. 701, September 1970.

Kuo, S. C., and P. E. Mayes, "An experimental study and the design of the log-periodic zigzag antenna," *Tech. Rep. 65-11*, Univ. of Illinois Antenna Lab, Contract AF33(615)-3216, February 1966. Also MS thesis, Dept. of Electr. Eng., Univ. of Illinois, Urbana, 1965.

Kuo, S. C., and C. C. Liu, "Dual polarized center-fed multiarm spiral monopulse antenna," *IEEE/G-AP Intl. Symp.*, University of Michigan, Ann Arbor, p. 233, October 1967.

Kuo, S. C., "Size-reduced log-periodic dipole array," *IEEE/G-AP Intl. Symp.*, pp. 151–158, 1970.

Kuo, S. C., "Size-reduced log-periodic dipole array antenna," *Microwave J.*, vol. 15, no. 12, pp. 27–33, December 1972.

Kyle, R. H., "Mutual coupling between log-periodic antennas," *IEEE Trans. Antennas Propag.*, vol. AP-18, pp. 15–22, January 1970.

Lantz, P. A., "A two-channel monopulse reflector antenna system with a multimode logarithmic spiral feed," Sixteenth Annual Symposium, USAF Antenna Res. and Dev. Prog., Allerton Park, Univ. of Illinois, Urbana, 1966.

Laxpati, S. R., and R. Mittra, "Current distribution on a two-arm thin-wire equiangular spiral antenna," *Electron. Lett.*, vol. 1, pp. 213–215, 1965.

———. "A study of the equiangular spiral antenna," *Tech. Rep. 65-20*, Univ. of Illinois Antenna Lab, Contract AF33(615)-3216, February 1966.

———. "Boundary-value problems associated with source-excited planar-equiangular-spiral antennas," *Proc. IEE* (London), vol. 114, pp. 352–359, March 1967.

Lee, S. H., "Analysis of zigzag antennas," PhD dissertation, Dept. of Electr. Eng., Univ. of California, Berkeley, June 1968.

Lee, S. H., and K. K. Mei, "Analysis of zigzag antennas," *IEEE Trans. Antennas Propag.*, vol. AP-18, pp. 760–764, November 1970.

Lee, W. C., "Analysis of nonplanar spiral antennas," *Tech. Rep. 903-15*, Ohio State Univ., Columbus, November 1960.

Liang, C. S., and Y. T. Lo, "A multipole-field study for the multiarm log-spiral antennas," *IEEE Trans. Antennas Propag.*, vol. AP-16, pp. 656–664, November 1968. Also *Tech. Rep. 69-3*, Univ. of Illinois Antenna Lab, July 1969.

Mast, P. E., "A theoretical study of the equiangular-spiral antenna," PhD dissertation, Department of Electrical Engineering, Univ. of Illinois, Urbana, 1958. Also *Tech. Rep. 35*, Univ. of Illinois Antenna Lab, Contract AF33(616)-3220, September 1958.

Mayes, P. E., and R. L. Carrel, "Logarithmically periodic resonant-V arrays," Tenth Annual Symposium, USAF Antenna Res. and Dev. Prog., Allerton Park, Univ. of Illinois, Urbana, October 1960. Also *IRE Wescon Conv. Rec.*, pt. I, 1961, and *Tech. Rep. 47*, Univ. of Illinois Antenna Lab, Contract AF33(616)-6079, July 1962.

Mayes, P. E., G. A. Deschamps, and W. T. Patton, "Backward-wave radiation from periodic structures and application to the design of frequency independent antennas," *Proc. IRE*, vol. 49, pp. 962–963, May 1961. Also *Tech. Rep. 60*, Univ. of Illinois Antenna Lab, Contract AF33(657)-8460, December 1962.

Mayes, P. E., and P. G. Ingerson, "Near-field measurements on backfire periodic dipole arrays," Twelfth Annual Symposium, USAF Antenna Res. and Dev. Prog., Allerton Park, Univ. of Illinois, Urbana, October 1962.

Mayes, P. E., "Broadband backward-wave antennas," *Microwave J.*, vol. 6, pp. 61–71, January 1963.

———. "Balanced backfire zigzag antennas," *IEEE Intl. Conv. Rec.*, pt. 2, pp. 153–165, 1964. Also *Tech. Rep. 82*, Univ. of Illinois Antenna Lab, Contract AF33(657)-10474, October 1964.

———. "Designing an all-channel tv antenna," *Electron. World*, February 1966.

Mayes, P. E. (ed.), "Wave propagation on smooth and periodic structures and applications to antenna design," *Tech. Rep. 66-11*, Univ. of Illinois Antenna Lab, Contract NASA NGR14-005-043, 1966.

McClelland, O. L., "An investigation of the near fields on the conical equiangular spiral antenna," *Tech. Rep. 55*, Univ. of Illinois Antenna Lab, Contract AF33(657)-8460, May 1962.

Mei, K. K., and D. Johnstone, "A broadside log-periodic antenna," *Proc. IEEE*, vol. 54, no. 6, pp. 889–890, June 1966.

Mittra, R., and K. E. Jones, "How to use k-β diagrams in log-periodic antenna design," *Microwaves*, pp. 18–26, June 1965.

Montague, H., M. J. Horrocks, J. W. Margosian, and J. D. Dyson, "The dual-aperture counterwound log-spiral antenna direction-finder system," *IEEE Trans. Antennas Propag.*, vol. AP-21, pp. 224–226, March 1973.

Mosko, J. A., "Reduced size, dual-mode spiral for two-plane monopulse direction finding," *NAVWEPS Rep. 8758*, U. S. Naval Ordinance Test Station, China Lake, California, 1966.

Murphy, L. R., "A shortened log-spiral antenna by the use of resistive termination," MS thesis, Dept. of Electr. Eng., Univ. of Illinois, Urbana, 1967.

Mushiake, Y., *J. Inst. Electron. Eng.*, Japan, vol. 69, pp. 86–88, 1949 (in Japanese).

Ore, F. R., "A coaxial fed unidirectional log-periodic monopole array," technical report, Collins Radio Company, Cedar Rapids, Iowa, August 1961.

———. "Investigation of the log-periodic coaxial fed monopole array," *Tech. Note No. 1*, Univ. of Illinois Radiolocation Research Lab, Contract NOBSR85243, October 1963.

———. "Log-periodic folded monopole array," *Tech. Rep. 4*, Radiolocation Research Lab, Univ. of Illinois, Urbana, Contract NOBSR89229, June 1964.

———. "A wideband vertical incidence radio location array," *Tech. Rep. 7*, Radiolocation Research Lab, Univ. of Illinois, Urbana, Contract NOBSR89229, September 1964.

Ore, F. R., and P. E. Mayes, "A study of periodic and log-periodic series reactance loading with application to a high-efficiency long-wire antenna," *Tech. Rep. 66-5*, Univ. of Illinois Antenna Lab, Contract N123(953)-51806A, June 1966.

Ore, F. R., "A technical report on the investigation of selected frequency independent arrays," *Tech. Rep. 67-273*, Univ. of Illinois Antenna Lab, Contract AF33(615)-3216, November 1967.

Ostertag, E. L., "Experimental study of a log-periodic cavity backed slot array with computer synthesized 50-ohm modulated impedance feeder," MS thesis, Dept. of Electr. Eng., Univ. of Illinois, Urbana, 1972.

Patton, W. T., "The backfire bifilar helical antenna," *Tech. Rep. 61*, Univ. of Illinois Antenna Lab, Contract AF33(657)-8460, September 1962. Also PhD dissertation, Univ. of Illinois, Urbana, 1963.

———. "The backfire bifilar helical antenna," Twelfth Annual Symposium, USAF Antenna Res. and Dev. Prog., Allerton Park, Univ. of Illinois, Urbana, October 1962.

Ransom, P. L., and J. D. Dyson, "Near-field measurements on the planar four-arm log-spiral antenna," Fourteenth Annual Symposium, USAF Antenna Res. and Dev. Prog., Allerton Park, Univ. of Illinois, Urbana, October 1964.

Ransom, P. L., "An experimental investigation of the four-arm planar logarithmic spiral antenna," *Tech. Rep. 65-5*, Univ. of Illinois Antenna Lab, Contract AF33(657)-10474, May 1965. Also MS thesis, Dept. of Electr. Eng., Univ. of Illinois, Urbana, 1965.

Rumsey, V. H., *Frequency Independent Antennas*, New York: Academic Press, 1966.

Sandeman, E. K., "Spiral-phase fields," *Wireless Engineer*, pp. 96–105, March 1949.

Sinnott, D. H., "Multiple-frequency computer analysis of the log-periodic dipole antenna," *IEEE Trans. Antennas Propag.*, vol. AP-22, pp. 592–594, July 1974.

Sivan-Sussman, R., "Various modes of the equiangular spiral antenna," *IEEE Trans. Antennas Propag.*, vol. AP-11, pp. 533–539, September 1963.

Smith, C. E., *Log-Periodic Antenna Design Handbook*, Cleveland: Smith Electronics, Inc., 1966.

Stephenson, D. T., and P. E. Mayes, "Log-periodic helical dipole arrays," IEEE Wescon, 1963. Also Thirteenth Annual Symposium, USAF Antenna Res. and Dev. Prog., Allerton Park, Univ. of Illinois, Urbana, October 1963.

———. "Broadband arrays of helical dipoles," *Tech. Rep. 2*, Univ. of Illinois Antenna Lab, Contract NEL30508A, January 1964.

———. "Investigations of broadband helical dipole arrays," *Tech. Rep. 6*, Univ. of Illinois Antenna Lab, Contract NEL30508A, October 1964.

Stephenson, D. T., "Broadband helical dipole arrays," *Tech. Rep. 65-19*, Univ. of Illinois Antenna Lab, Contract N123(953)-51806A, October 1965.

Stephenson, D. T., and P. E. Mayes, "Variations of broadband helical dipole arrays," *Tech. Rep. 65-3*, Univ. of Illinois Antenna Lab, Contract N123(953)-51806A, April 1966.

Sussman, R., "The equiangular plane spiral antenna," technical report, Electronics Res. Lab, Univ. of California, Berkeley, September 1961.

Tang, C. H., and O. L. McClelland, "Polygonal spiral antennas," *Tech. Rep. 57*, Univ. of Illinois Antenna Lab, Contract AF33(657)-8460, June 1962.

Tang, C. H., "A class of modified log-spiral antennas," *IEEE Trans. Antennas Propag.*, vol. AP-11, pp. 422–427, July 1963.

Turner, E. M., "Spiral slot antenna," *Note WCLR-55-8*, Wright Air Dev. Center, Dayton,

June 1955.

Wheeler, M. S., "On the radiation from several regions in spiral antennas," *IRE Trans. Antennas Propag.*, vol. AP-9, pp. 100–102, January 1961.

Wickersham, A. F., "Recent developments in very broad-band end-fire arrays," *Proc. IRE*, vol. 48, pp. 794–795, April 1960.

Wickersham, A. F., R. E. Franks, and R. L. Bell, "Further developments in tapered ladder antennas," *Proc. IRE*, vol. 49, p. 378, January 1961.

Wohlleben, R., and B. Schumacher, "Randlinien-abwicklung der zweiarmigen, selbst-komplementaren, konischen, logarithmischen spiralantenne," *NTZ*, pp. 585–590, October 1966.

Wolter, J., "Solution of Maxwell's equations for log-periodic dipole antennas," *IEEE Trans. Antennas Propag.*, vol. AP-18, pp. 734–741, November 1970.

Yeh, Y. S., and K. K. Mei, "Theory of conical equiangular-spiral antennas, part I: numerical technique," *IEEE Trans. Antennas Propag.*, vol. AP-15, pp. 634–639, September 1967.

———. "Theory of conical equiangular-spiral antennas, part II: current distributions and input impedances," *IEEE Trans. Antennas Propag.*, vol. AP-16, pp. 14–21, January 1968.

Chapter 10

Microstrip Antennas

William F. Richards
University of Houston

CONTENTS

 **William F. Richards** was born in Cincinnati, Ohio, in 1950. He received the BS degree in engineering (with a concentration in electrical engineering) in 1970 from Old Dominion University, Norfolk, Virginia, and the MS and PhD degrees in electrical engineering from the University of Illinois in Urbana-Champaign under the direction of Professor Y. T. Lo.

At the University of Illinois he did work on optimization of array designs accounting for the stochastic properties of the excitation network, artificial dielectrics, and did fundamental work with Dr. Lo on the development of the cavity model for microstrip antennas. After obtaining his PhD in 1977, he remained at the University of Illinois until 1980, serving as a visiting assistant professor.

In 1980 he joined the faculty of the Department of Electrical Engineering at the University of Houston, Houston, Texas, where he serves as an associate professor. He has continued his work on microstrip antennas at the University of Houston, developing a theory and a number of applications of loaded microstrip elements. His research interests are microstrip antennas, artificial dielectrics, and numerical techniques in scattering and the on-surface radiation condition. At the time of publication, he was on leave of absence from the University of Houston to attend the University of Miami for its PhD-to-MD program.

Dr. Richards is a member of the IEEE and shared the Antennas and Propagation Society's Best Paper Award in 1979 with his coauthors, Y. T. Lo and D. Solomon, for their paper on the theory and analysis of microstrip antennas.

1. Introduction

A class of antennas that has gained considerable popularity in recent years is the *microstrip antenna*. There are many different varieties of microstrip antennas, but their common feature is that they basically consist of four parts:

(*a*)   a very thin flat metallic region often called the *patch*;
(*b*)   a *dielectric substrate*;
(*c*)   a *ground plane*, which is usually much larger than the patch; and
(*d*)   a *feed*, which supplies the element rf power.

A typical microstrip element is illustrated in Fig. 1. Microstrip elements are often made by etching the patch (and sometimes the feeding circuitry) from a single printed-circuit board clad with conductor on both of its sides.

The longest dimension of the patch is typically about a third to a half of a free-space wavelength (λ_0), while the dielectric thickness is usually in the range of $0.003\lambda_0$ to $0.05\lambda_0$. A commonly used dielectric for such antennas is polytetrafluoral ethylene (PTFE), often set in a reinforcing glass fiber matrix. A relative dielectric constant around 2.5 is typical. Sometimes a low-density cellular "honeycomb" material is used to support the patch. This material has a relative dielectric constant much closer to unity and usually results in an element with better efficiency and larger bandwidth [1] though at the expense of an increased element size. Substrate materials with high dielectric constants can also be used. For the radiation modes most used, however, such substrates result in elements which are electrically small in terms of free-space wavelengths and consequently have relatively smaller bandwidths [1] or low efficiencies [2].

The reasons why this class of antennas has become so popular include the following:

1.   They are low-profile antennas.
2.   They are easily conformable to nonplanar surfaces. Along with their low profile this makes them well suited for use on high-performance airframes.
3.   They are easy and inexpensive to manufacture in large quantities using modern printed-circuit techniques.
4.   When mounted to a rigid surface they are mechanically robust.
5.   They are versatile elements in the sense that they can be designed to produce a wide variety of patterns and polarizations, depending on the mode excited and the particular shape of patch used.
6.   Adaptive elements can be made by simply adding appropriately placed pin or varactor diodes between the patch and ground plane. Using such loaded elements one can vary the antenna's resonant frequency [3,4], polarization [3,5], impedance [6], and even its pattern by simply changing bias voltages on the diodes [7].

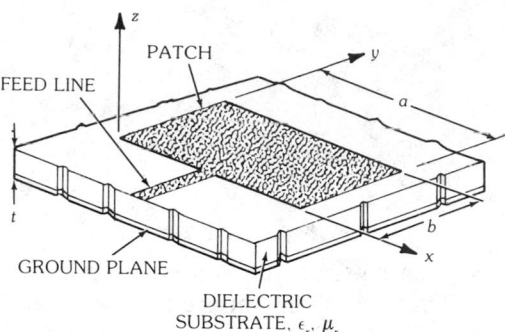

Fig. 1. A typical rectangular microstrip element with dimensional parameters.

These advantages must be weighed against the disadvantages which can be most succinctly stated in terms of the antenna's quality factor, Q. Microstrip antennas are high-Q devices with Qs sometimes exceeding 100 for the thinner elements. High-Q elements have small bandwidths. Also, the higher the Q of an element, the lower is its efficiency. Increasing the thickness of the dielectric substrate will reduce the Q of the microstrip element and thereby increase its bandwidth and its efficiency. There are limits, however. As the thickness is increased, an increasing fraction of the total power delivered by the source goes into a surface wave. This surface-wave contribution can also be counted as an unwanted power loss since it is ultimately scattered at dielectric bends and discontinuities. Such scattered fields are difficult to control and may have a deleterious effect on the pattern of the element [1]. One also needs to be aware that microstrip elements are modal devices. If the band of the element is so large that it encompasses the resonant frequencies of two or more resonant modes, the pattern is likely not to be stable throughout the band even though the vswr at the input could be acceptably low.

Despite the disadvantages the advantages of microstrip antennas have led to their use in many applications in civilian and government systems. In fact, one so-called disadvantage, small bandwidth, is sometimes counted as an advantage instead. For narrow-band applications the antenna itself can act as a filter for unwanted frequency components. Even with the relatively high Q these elements have, a sufficiently thick element with a well-designed external matching circuit can have a bandwidth as large as 35 percent [8]. Finally, if surface-wave loss is not counted as an unwanted power loss (or does not exist because the dielectric is truncated or has a relative dielectric constant near unity), the efficiency of a sufficiently thick element can easily be made larger than 90 percent.

An understanding of the *physical mechanisms* that explain the properties of microstrip antennas is crucial to their creative use in applications. This chapter will introduce the reader to some of the simpler theories and applications of microstrip antennas. No attempt has been made to present a comprehensive review of the literature. Instead, selected additional references are given in the References to assist the reader in obtaining information on theories and applications that exceed the scope of this handbook.

2. Physical Models

In order to design a microstrip element for a given application it is crucial to first have a good understanding of the physical mechanisms that govern these elements. There are several theories for microstrip elements with varying degrees of accuracy and complexity. Among these, two give the best physical insight: (1) the transmission-line model [9, 10] and (2) the cavity model [4, 11]. Of these, the simplest is the transmission-line model. The cavity model, though somewhat more complex, gives a deeper insight into the operation of microstrip antennas. The transmission-line model is considered first.

Transmission-Line Model

The patch element in Fig. 1 can be viewed as a very wide (thus low-impedance) microstrip transmission line of length L. The region between the patch edges and the ground plane at the two ends of the line can be viewed as radiating apertures much like slot antennas. Thus the low-impedance line can be thought of as being loaded at its two ends not by open circuits, but by high-impedance loads. There is also a mutual coupling between the slots. The terminal voltages and currents on the transmission line can be thought of as being coupled through a two-port network representing the propagation of the field in the space exterior to the patch. If the short-circuit parameters of this two-port network are Y_{11} and Y_{12}, then the Π model of the network can be used for the microstrip antenna as illustrated in Fig. 2a.

The self-admittance of the load is $Y_{11} = G_{11} + jB_{11}$. The susceptance B_{11} can be adequately accounted for by extending the length b of the transmission line beyond the physical limits of the patch. One estimate for this extension, ΔL, that has often been used with good success is [12]

$$\Delta L = 0.412t \frac{(\epsilon_{\text{eff}} + 0.3)(a/b + 0.262)}{(\epsilon_{\text{eff}} - 0.258)(a/b + 0.813)} \tag{1}$$

The ϵ_{eff} is the *effective relative dielectric constant* and is given in terms of the substrate relative dielectric constant ϵ_r by [13]

$$\epsilon_{\text{eff}} = \frac{\epsilon_r + 1}{2} + \frac{\epsilon_r - 1}{2(1 + 10t/w)^{1/2}}$$

An estimate for the self-conductance G_{11} that has been commonly used is [14]

$$G_{11} = \frac{a}{120\lambda_0} \left[1 - \frac{\pi^2}{6} (t/\lambda_0)^2 \right]$$

This is based on the conductance of an infinitely long, uniform slot antenna. Its application to the relatively short "slots" associated with microstrip antennas seems dubious. In fact, another formula yielding quite a different conductance is [1]

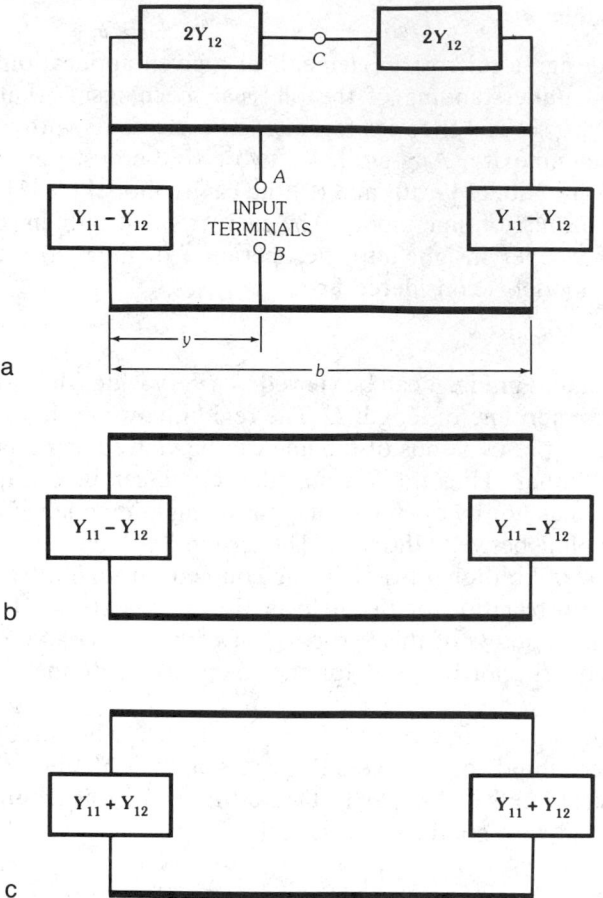

Fig. 2. Transmission-line model. (*a*) Schematic diagram for a rectangular microstrip element including external self and mutual admittances. (*b*) Simplified circuit for symmetric modes. (*c*) Simplified circuit for antisymmetric modes.

$$
G_{11} = \begin{cases}
(a_{\text{eff}})^2/(90\lambda_0)^2 & \text{for } a_{\text{eff}} < 0.35\lambda_0 \\
w_{\text{eff}}/(120\lambda_0) - 1/(60\pi^2) & \text{for } 0.35\lambda_0 < a_{\text{eff}} < 2\lambda_0 \\
w_{\text{eff}}/(120\lambda_0) & \text{for } 2\lambda_0 < a_{\text{eff}}
\end{cases}
$$

The effective width of the microstrip line is given by [15]

$$
a_{\text{eff}} = \frac{120\pi t}{Z_m \sqrt{\epsilon_{\text{eff}}}}
$$

where

$$
Z_m = \frac{60\pi}{\sqrt{\epsilon_{\text{eff}}}} \left[\frac{a}{2t} + 0.441 + \frac{1}{\pi} 1.451 + \ln\left(\frac{a}{2t}\right) + 0.94 \right]^{-1}
$$

The mutual conductance G_{12} can be found by numerically integrating [10]

$$G_{12} = \frac{1}{120\pi^2} \int_0^\pi \frac{\sin^2[(\pi a_{\text{eff}} \cos \theta)/\lambda_0]}{\cos^2 \theta} \sin^3 \theta J_0([2\pi b \sin \theta]/\lambda_0)\, d\theta$$

where J_0 is the Bessel function. The mutual susceptance has been generally ignored; its effect is to slightly alter the element's resonant frequency.

Symmetric and antisymmetric source-free voltage distributions resonate the circuit in Fig. 2a at complex resonant frequencies which can be determined by finding the complex zeros of the admittance and impedance at terminals A and B. Since symmetric voltage distributions produce a virtual open at point C in the circuit, the network can be simplified to that shown in Fig. 2b in this case. For the antisymmetric case, node C becomes a virtual ground node, reducing the equivalent circuit to that illustrated in Fig. 2c.

For both symmetric (even n) and antisymmetric (odd n) resonant voltage distributions, the input impedance at resonance at a distance y from either end of the patch is given approximately by

$$R \cong \frac{1}{G_{11} - (-1)^n G_{12}} \cos^2(n\pi y/b)$$

The resonant frequency (actually the real part of the complex resonant frequency) is

$$f_{\text{res}} \cong \frac{nc}{2\sqrt{\epsilon_{\text{eff}}}(b + 2\Delta L)}$$

where c is the speed of light in a vacuum. The imaginary part of the resonant frequency is its real part divided by twice Q. The Q is the quality factor of the element; it has been approximated by [1]

$$Q \cong \frac{n\pi}{4Z_m[G_{11} - (-1)^n G_{12}]}$$

Once R and Q have been determined, the input impedance at any frequency f near a resonance f_{res} can be approximated by

$$Z_{\text{in}} \cong \frac{R}{1 + j2Q(f - f_{\text{res}})/f_{\text{res}}}$$

The preceding formula is useful for determining the major features of the variation of input impedance with frequency. Its accuracy depends, of course, on the accurary of slot and mutual conductance parameters used. Its accuracy is also limited by its inability to predict an additional inductive term that is associated with the feed although an *ad hoc* correction term can be introduced; see under "Feed Reactance," in Section 4.

The radiation pattern of the element is computed by the same method used for the cavity model (discussed below) except that in the transmission-line model, only the two ends of the patch are considered as radiating apertures. The expression for the far field is

$$\mathbf{E} \cong (\hat{\boldsymbol{\theta}} \sin \phi - \hat{\boldsymbol{\phi}} \cos \phi \cos \theta)$$

$$\times \frac{\sin[(k_0 a/2) \sin \theta \cos \phi] \sin[n\pi/2 + (k_0 b/2) \sin \theta \sin \phi]}{\sin \theta \cos \phi} \frac{e^{-jk_0 r}}{\pi r} V_0$$

where V_0 is the voltage at either of the two ends of the patch, r is the distance from the antenna, and k_0 is the free-space wave number.

The intuitive appeal of the transmission-line model becomes strained when one attempts to adapt it to other than rectangular microstrip elements with a modal field variation only in the y direction. For example, one *could* think of a mode of an annular-sector microstrip antenna as the voltage distribution on a generalized nonuniform transmission line, but there seems to be little advantage in doing so. Instead, the next model to be considered is both conceptually simpler (in this case) and more accurate and comprehensive than the transmission-line model.

Cavity Model

When an oscillating current is injected into a microstrip element a charge distribution is established on the surface of the ground plane and the *two* surfaces (upper and lower) of the patch as illustrated in Fig. 3a. There are two opposing tendencies which shape this charge distribution. (1) There is an *attractive* tendency between the opposite charges at corresponding points on the lower side of the patch and on the ground plane. This attraction tends to keep the patch charge concentrated on the *bottom* of the patch. (2) However, there is also a *repulsive* tendency between like charges on the bottom of the patch. This tends to push some of the charge around the edge of the patch onto its *top* surface. When the element is very thin the first tendency dominates and almost all of the charge on the patch resides on its bottom side. Correspondingly, most of the current flows on the lower side of the patch, with only a small amount flowing around the edge onto its upper surface. Consequently the component of magnetic field *tangential* to the patch edge is small although not exactly zero. Were it *precisely* zero one could introduce a perfect magnetic conductor in the plane between the patch edge and the ground plane without affecting the fields under the patch. The introduction of such a magnetic wall *will* distort the shape of the magnetic field distribution, but not significantly if the element is thin. Thus, to find the *shape* of the magnetic field distribution under the patch, one can replace the antenna by an ideal cavity as illustrated in Fig. 3b. Of course, from the magnetic field distribution the *shape* of the corresponding electric-field distribution can also be found.

In contrast to its shape the *amplitude* of the field under the patch cannot be found by just analyzing the cavity alone. For example, if the dielectric material and the metal parts within the cavity were assumed to be lossless, then the analysis of the *cavity* would yield a *purely reactive* input impedance. The input impedance of the corresponding microstrip *antenna*, of course, is *not* purely reactive; its resistive part accounts for the power radiated by the antenna.

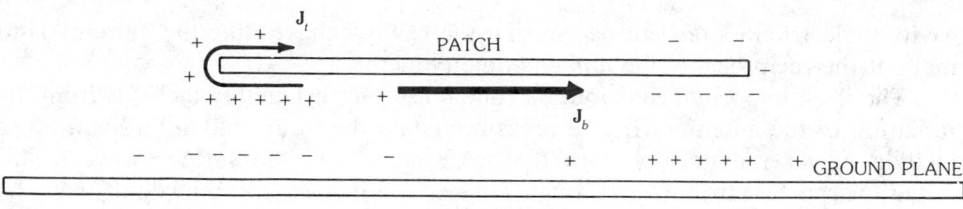

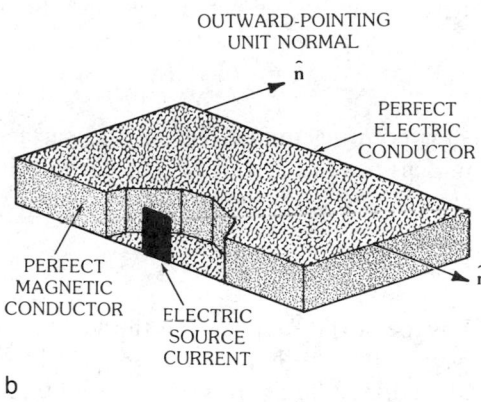

Fig. 3. Development of the cavity model of a microstrip element. (*a*) Charge distributions on the upper and lower sides of the patch and on the ground plane. (*b*) Approximate cavity model of a microstrip antenna.

The impedance function for the ideal *cavity* has only *real* poles. The impedance function for the corresponding *antenna* has *complex* poles. The imaginary parts of these poles account for the power lost by radiation and by dielectric and conduction losses. The real parts of the corresponding cavity and antenna poles are dependent on the *shapes* of their modal field distributions and are consequently almost identical for thin elements. To make the cavity more resemble the antenna it is supposed to model, one can add loss to the cavity dielectric (in one's analysis) by appropriately adjusting the loss tangent of the cavity dielectric. The imaginary parts of the poles of the cavity filled with the lossy dielectric will no longer be zero. The object then becomes to find just how much loss must be added to the cavity so that the imaginary part of its pole near the frequency of interest matches that of the corresponding antenna.

If the dielectric within the cavity (with perfectly conducting electric and magnetic walls) has a dielectric loss tangent δ_{eff}, then at any frequency f near a resonance it has a quality factor of

$$Q = \frac{2\pi f(\text{average total stored energy})}{\text{average power dissipated}} = \frac{1}{2}\frac{\omega_{\text{re}}}{\omega_{\text{im}}} = \frac{1}{\delta_{\text{eff}}}$$

where ω_{re} and ω_{im} are respectively the real and imaginary part of the pole (angular frequency). This result clearly suggests that the appropriate way to choose the cavity dielectric loss tangent δ_{eff} so that the cavity behaves like the antenna is to make it the reciprocal of the *antenna* quality factor.

The most important contribution (one hopes) to this quality factor is from the radiation of the antenna. (In the present context the term "radiation" is used to include power radiated by both the space wave and the surface wave.) This radiation can be attributed, through Huygens' principle, to equivalent magnetic and electric surface currents in the presence of the grounded dielectric slab as illustrated in Fig. 4a. The magnetic surface current density **M** is related to the electric field in the surface between the patch edge and the ground plane by

$$\mathbf{M} = -\hat{\mathbf{n}} \times \mathbf{E}$$

where $\hat{\mathbf{n}}$ is a unit vector pointing outward from the surface of the cavity. The electric surface current densities are $\mathbf{J}_s$ and $\mathbf{J}_t$. The current density $\mathbf{J}_s$ is impressed on the same surface as is **M** and is given by

$$\mathbf{J}_s = \hat{\mathbf{n}} \times \mathbf{H}$$

The current density $\mathbf{J}_t$ is the surface current on the *top* surface of the patch. The preceding discussion argues that both $\mathbf{J}_s$ and $\mathbf{J}_t$ are small, leaving only **M** as the dominant source for the radiated fields. It is emphasized that $\mathbf{J}_t$ is the current *only* on the *top* surface of the patch; it is *not* the *total* patch current. The total patch current, which is the sum of the currents on the top and bottom surfaces of the patch, could be impressed in lieu of all other sources to produce the field radiated by the element [16].

The analysis of the *cavity* yields the *shape* of the electric field **E** under the patch

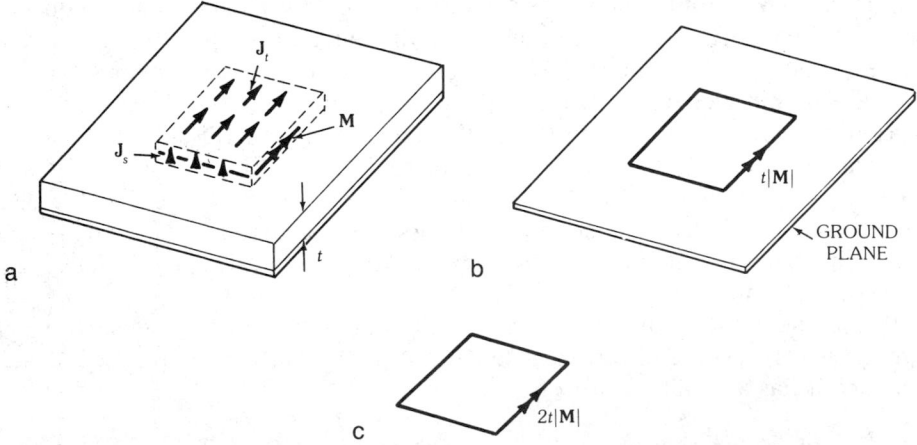

Fig. 4. Source modeling for pattern computations. (*a*) Exact model for the exterior field of a microstrip antenna in terms of equivalent electric and magnetic current. (*b*) Approximate model with the dielectric removed, the electric currents ignored, and the magnetic current condensed to a filamentary magnetic current of strength equal to $t|\mathbf{M}|$. (*c*) Equivalent approximate model with the ground plane removed.

and at its edge. Thus the magnetic current **M** can be determined up to an unknown factor, α; consequently the power P_r it radiates can be determined up to an unknown factor, $|\alpha|^2$. The energy stored in the electric field under the patch is

$$W_E = \frac{1}{2}\epsilon \int_{\text{cavity}} |\mathbf{E}|^2 \, dV$$

which can also be found up to the same unknown factor, $|\alpha|^2$. This unknown amplitude factor cancels in the expression for the quality factor,

$$Q = Q_{\text{ext}} = \frac{4\pi f W_E}{P_r}$$

thus demonstrating that only the shape of the field distribution under the patch is important to find Q.

One can also include the effect of dielectric and conduction losses in the computation of Q. If these are taken into account the expression for antenna quality factor becomes

$$Q = \frac{1}{1/Q_{\text{ext}} + \delta + \Delta/t} \tag{2}$$

where δ is the loss tangent of the *actual* dielectric in the antenna, $\Delta = 0.029(\lambda_0/\sigma)^{1/2}$ is the skin depth in the metal cladding with conductivity σ (with units of siemens per unit length), and t is the thickness of the antenna. Once Q is known, the antenna can be analyzed as if it were a lossy cavity. This is significant since the most commonly used patch shapes correspond to cavities having a separable geometry amenable to simple analytical treatment.

This describes the basic ideas used in the cavity model approximation. A summary of this procedure is given in the flowchart in Fig. 5. A more detailed outline of how to actually carry out a practical analysis is given in the following steps. This analysis has been implemented in FORTRAN and BASIC programs for rectangular, circular-disk, circular-sector, annular, and annular-sector elements. The results show good agreement with experiment for thin elements.

(*a*) *Find the normalized resonant modes of the associated ideal cavity.* Find the z-independent, source-free solutions $\psi_{mn}(u, v)$ to the homogeneous Helmholz equation,

$$[\nabla^2 + (k_{mn})^2]\psi_{mn}(u, v) = 0$$

subject to the homogeneous boundary condition that

$$\nabla\psi_{mn}(u, v)\cdot\hat{\mathbf{n}} = 0$$

for (u, v) on the edge of the patch. The ψ_{mn}s represent the (purely z-directed and z-independent) modal electric fields that resonate in the ideal cavity. It is convenient to normalize these modes so that the integral of their squares over the

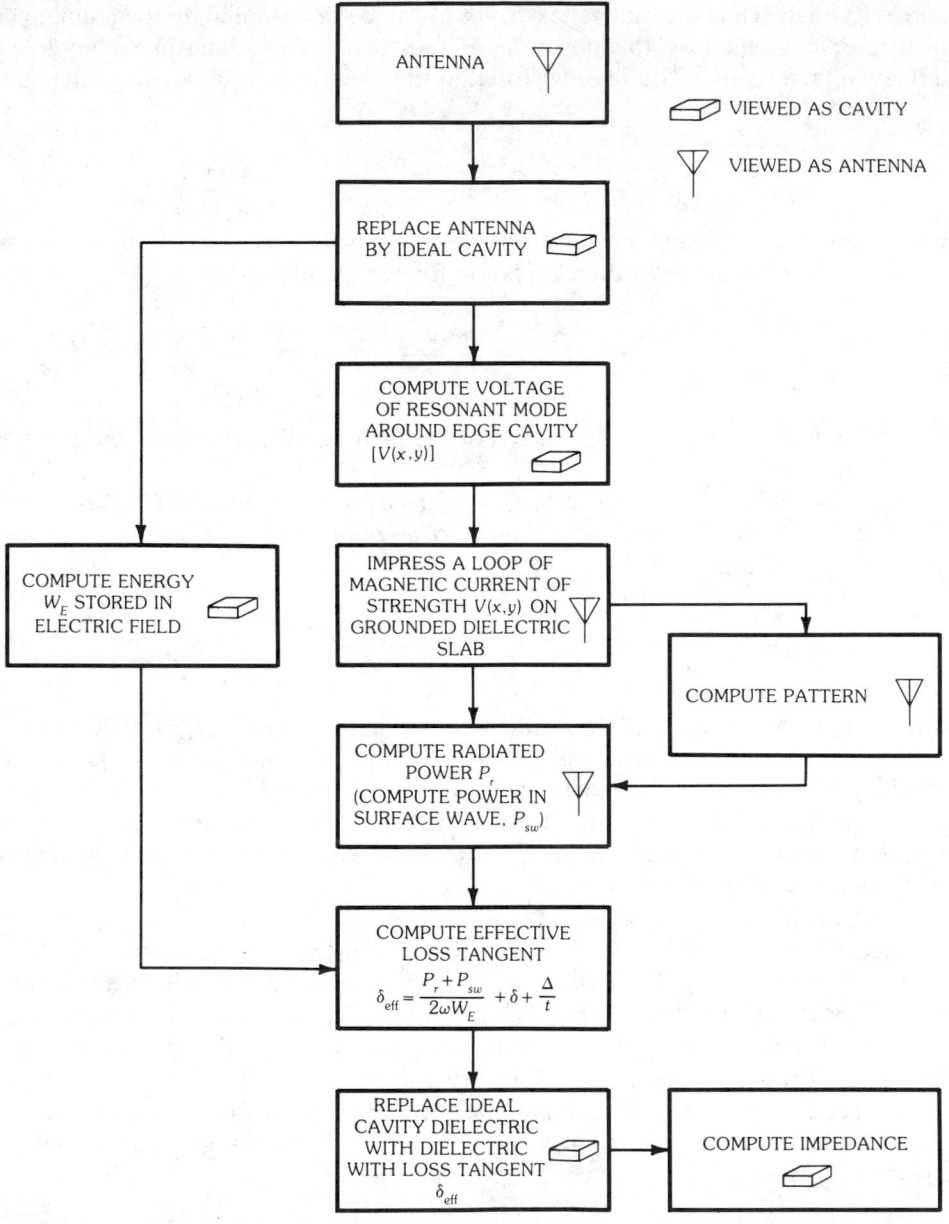

Fig. 5. Flowchart summarizing the steps used in the application of the cavity model.

patch area is 1. The k_{mn}s are the resonant wave numbers of the modes. The point (u, v, z) is a point in the cylindrical coordinate system being used to describe the patch's boundary. The only such systems which are separable are rectangular, circular-cylinder, elliptic-cylinder, and parabolic-cylinder coordinate systems. Table 1 gives the shapes of the patches separable in the rectangular and circular cylinder systems along with their corresponding ψ_{mn}s and characteristic equations for the k_{mn}s. Chart 1 illustrates some of the separable patches in elliptic- and parabolic-cylinder geometries. It also illustrates two triangular patches which can be analyzed in a rectangular system. Little has been done with elliptical disk elements [17] and nothing at all to date is available on other elliptical or parabolic patches. Although limited to only four separable coordinate systems one still has a rich set of geometries available to which the cavity model can be most conveniently applied.

   (b) *Choose a resonant mode.* The next step is to choose a resonant mode ($m = M, n = N$) that produces the desired pattern. A more detailed discussion of how to roughly predict the pattern of the element in order to choose the mode is given in Section 3. How to actually compute the pattern for a given mode is given next.

   (c) *Compute the far-field pattern.* The far-field pattern is obtained by first setting the surface magnetic current density **M** to be

$$\mathbf{M} = \hat{\mathbf{t}}\psi_{mn}(u, v)$$

at the edge of the patch where $\hat{\mathbf{t}}$ is the unit tangent to the edge: $\hat{\mathbf{t}} = \hat{\mathbf{z}} \times \hat{\mathbf{n}}$. The far-field pattern for the (M, N)th mode can then be found as the response to this magnetic current ribbon in the presence of a grounded dielectric slab. A suitable approximation to the far-field pattern can often be obtained by condensing the magnetic current ribbon shown in Fig. 4a into a magnetic current filament acting in the presence of just the ground plane (Fig. 4b). Then, by doubling the magnetic current, the ground plane can also be eliminated. The approximate source obtained is illustrated in Fig. 4c. This approximation will yield an estimate for the space wave. An estimate for the surface wave can be made by extracting the residue of the surface-wave pole in the spectral integral for the fields when the magnetic current filament is acting in the presence of the grounded dielectric substrate. Often, however, the power carried in the surface wave is small compared with that carried by the space wave and can therefore be ignored in the computation of Q. The far-field computation can be done analytically in the case of rectangular, circular-disk, and annular-disk microstrip elements. One can also sample the magnetic current along the edge of the patch at the points $\{(x_i, y_i)\}, i = 0, 1, 2, \ldots, L + 1, x_{L+1} = x_0, y_{L+1} = y_0$. Then, using linear interpolation to reconstruct the magnetic current distribution from the sample points, one obtains the following approximation for the far field:

$$E_\theta = jk_0 t(F_x \sin\phi - F_y \cos\phi)\frac{1}{4\pi r}e^{-jk_0 r}$$

$$E_\phi = jk_0 t(F_x \cos\phi - F_y \sin\phi)\cos\theta\frac{1}{4\pi r}e^{-jk_0 r}$$

Table 1. Separable Patches, Characteristic Equations, Mode Distributions, and Geometric Parameters and Boundary Conditions*

Patch Shape	Characteristic Equation	$\psi_{mn}(u,v)$	Coordinate Transformation	$g(v)$	u_+	v_-	v_+	Boundary Conditions
Rectangular patch	$k_{mn} = \pi\left[\left(\dfrac{m}{a}\right)^2 + \left(\dfrac{n}{b}\right)^2\right]^{1/2}$	$\dfrac{\cos(m\pi x/a)\cos(n\pi y/b)}{\sqrt{ab/(\epsilon_{0m}\epsilon_{0n})}}$ $\epsilon_{0m} = \begin{cases} 1, & m = 0 \\ 2, & m > 0 \end{cases}$	$x = u$ $y = v$	1	a	0	b	$U'_m(u_-) = 0$ $U'_m(u_+) = 0$ $V_m^{+'}(v_+) = 0$ $V_m^{-'}(v_-) = 0$
Circular disk patch	$Z'_{mn}(a) = 0$ $Z_{mn}(\varrho) = J_m(k_{mn}\varrho)$	$\dfrac{J_m(k_{mn}\varrho)\cos m\phi}{D_{mn}(a)}$ $D_{mn}(\varrho) = \sqrt{\dfrac{\phi_0}{2\epsilon_{0m}}}\,|Z_{mn}(\varrho)|\varrho\sqrt{1 - \left(\dfrac{m\alpha}{k_{mn}\varrho}\right)^2}$ $(\alpha = 1;\ \phi_0 = 2\pi)$	$x = ae^v\cos u$ $y = ae^v\sin u$ $\varrho = ae^v$ $\phi = u$	$a^2 e^{2v}$	2π	$-\infty$	0	$U_m(u_-) = U_m(u_+)$ $U'_m(u_-) = U'_m(u_+)$ $|V_m^-(v_-)| < \infty$ $V_m^{+'}(v_+) = 0$

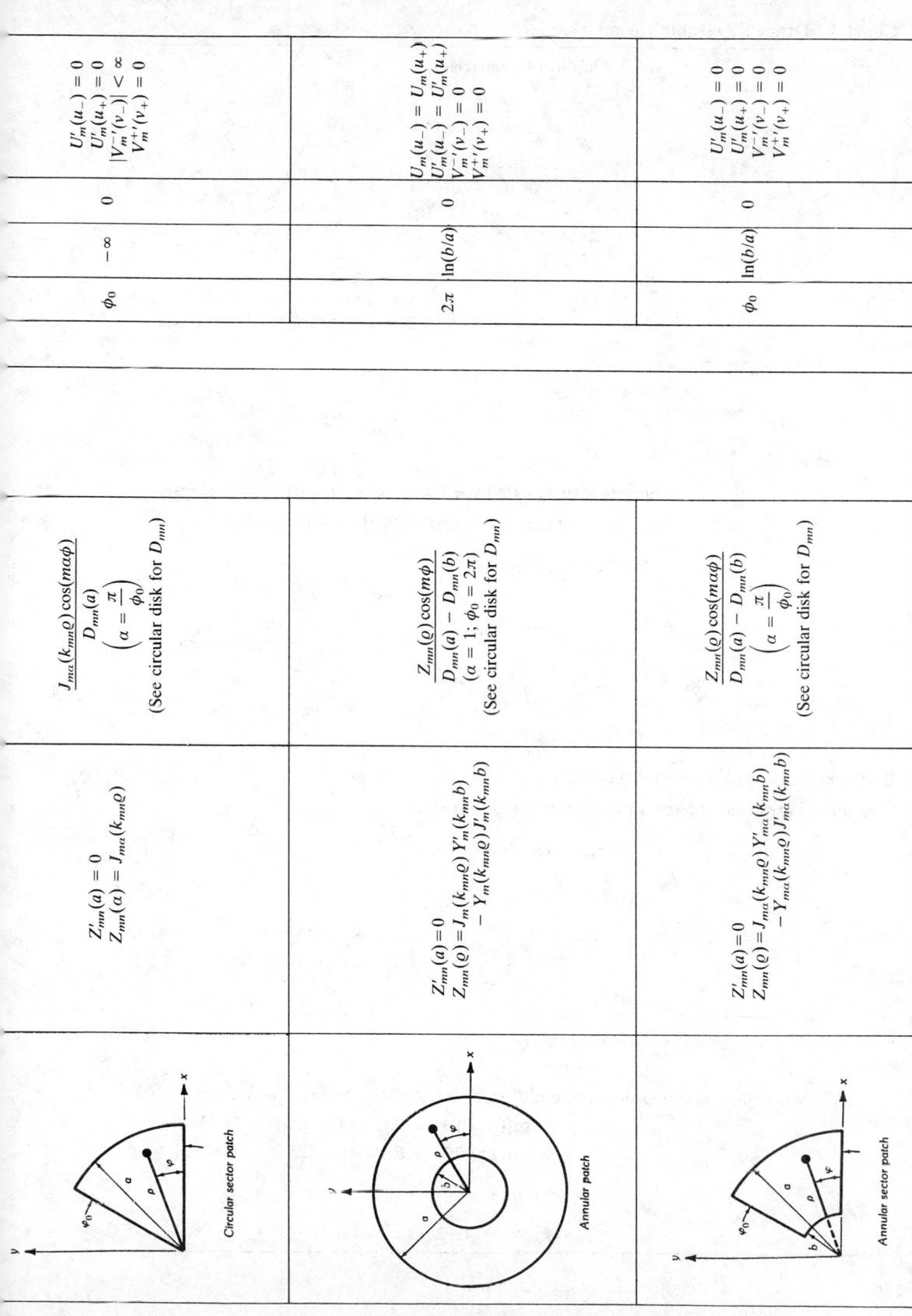

Geometry	Z_{mn} definition	Mode function / D_{mn}				Boundary conditions
Circular sector patch	$Z'_{mn}(a) = 0$ $Z_{mn}(\alpha) = J_{m\alpha}(k_{mn}\varrho)$	$\dfrac{J_{m\alpha}(k_{mn}\varrho)\cos(m\alpha\phi)}{D_{mn}(a)}$ $\left(\alpha = \dfrac{\pi}{\phi_0}\right)$ (See circular disk for D_{mn})	ϕ_0	$-\infty$	0	$U'_m(u_-) = 0$ $U'_m(u_+) = 0$ $|V_m^{-}{}'(v_-)| < \infty$ $V_m^{+}{}'(v_+) = 0$
Annular patch	$Z'_{mn}(a) = 0$ $Z_{mn}(\varrho) = J_m(k_{mn}\varrho)Y'_m(k_{mn}b)$ $\qquad - Y_m(k_{mn}\varrho)J'_m(k_{mn}b)$	$\dfrac{Z_{mn}(\varrho)\cos(m\phi)}{D_{mn}(a) - D_{mn}(b)}$ $(a = 1;\ \phi_0 = 2\pi)$ (See circular disk for D_{mn})	2π	$\ln(b/a)$	0	$U_m(u_-) = U_m(u_+)$ $U'_m(u_-) = U'_m(u_+)$ $V_m^{-}{}'(v_-) = 0$ $V_m^{+}{}'(v_+) = 0$
Annular sector patch	$Z'_{mn}(a) = 0$ $Z_{mn}(\varrho) = J_{m\alpha}(k_{mn}\varrho)Y'_{m\alpha}(k_{mn}b)$ $\qquad - Y_{m\alpha}(k_{mn}\varrho)J'_{m\alpha}(k_{mn}b)$	$\dfrac{Z_{mn}(\varrho)\cos(m\alpha\phi)}{D_{mn}(a) - D_{mn}(b)}$ $\left(\alpha = \dfrac{\pi}{\phi_0}\right)$ (See circular disk for D_{mn})	ϕ_0	$\ln(b/a)$	0	$U'_m(u_-) = 0$ $U'_m(u_+) = 0$ $V_m^{-}{}'(v_-) = 0$ $V_m^{+}{}'(v_+) = 0$

*Note: 1. $u_- = 0$, and $f(u) \equiv 0$ for the rectangular and all circular geometries.
2. The Js and Ys are Bessel functions of the first and second kinds, respectively.

Chart 1. Other Separable Geometries

Elliptical Geometries

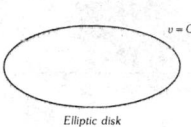

Elliptic disk

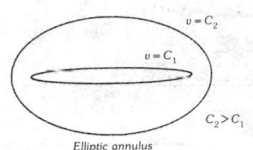

$C_2 > C_1$

Elliptic annulus

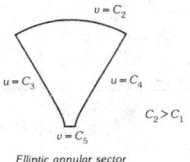

$C_2 > C_1$

Elliptic annular sector

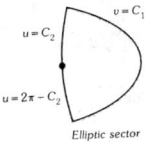

Elliptic sector

$$x = d \cos u \cosh v \quad v \geqq 0$$
$$y = d \sin u \cosh v \quad 0 \leqq u \leqq 2\pi$$

Constant-u curves are hyperbolas; constant-v curves are ellipses

Solutions expressible in terms of Mathieu functions

Parabolic Geometries

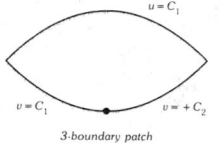

3-boundary patch

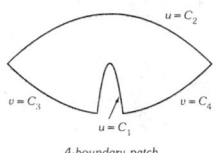

4-boundary patch

$$x = uv$$
$$y = 1/2(u^2 - v^2)$$
$$u \geqq 0$$

$$c_2 > c_1$$
$$c_4 > c_3$$

Both constant-u and constant-v curves are parabolas

Solutions expressible in terms of parabolic cylinder functions

Triangular Geometries

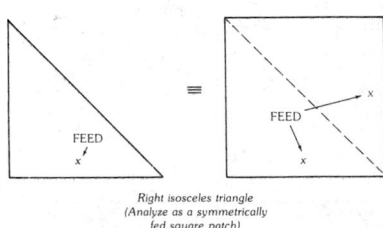

Right isosceles triangle
(Analyze as a symmetrically fed square patch)

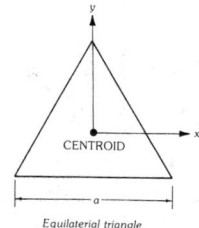

Equilaterial triangle

Mode distribution proportional to $\cos[(2\pi x/\sqrt{3}a + 2\pi/3)\ell] \cos[2\pi(m - n)y/3a]$
$$+ \cos[(2\pi x/\sqrt{3}a + 2\pi/3)m] \cos[2\pi(n - \ell)y/3a]$$
$$+ \cos[(2\pi x/\sqrt{3}a + 2\pi/3)n] \cos[2\pi(\ell - m)y/3a]$$

$$\ell + m + n = 0$$

$$(k_{mn})^2 = (4\pi/3a)^2(m^2 + mn + n^2)$$

where

$$\mathbf{F} = \sum_{\ell=0}^{L} \hat{\tau}_\ell \left[\overline{\psi}_{MN}^{(\ell)} j_0(k_0 \hat{\mathbf{r}} \cdot \Delta \mathbf{r}_\ell/2) + j \frac{\Delta \psi_{MN}^{(\ell)}}{2} j_1(k_0 \hat{\mathbf{r}} \cdot \Delta \mathbf{r}_\ell/2) \right] \Delta s_\ell \, e^{jk_0 \hat{\mathbf{r}} \cdot \mathbf{r}'_\ell}$$

$$\overline{\psi}_{MN}^{(\ell)} = [\psi_{MN}(u_{\ell+1}, v_{\ell+1}) + \psi_{MN}(u_\ell, v_\ell)]/2$$

$$\Delta \psi_{MN}^{(\ell)} = \psi_{MN}(u_{\ell+1}, v_{\ell+1}) - \psi_{MN}(u_\ell, v_\ell)$$

$$\mathbf{r}_l = x_\ell \hat{x} + y_\ell \hat{y}, \qquad \mathbf{r}'_\ell = (\mathbf{r}_{\ell+1} + \mathbf{r}_\ell)/2, \qquad \Delta \mathbf{r}_\ell = \mathbf{r}_{\ell+1} - \mathbf{r}_\ell$$

$$u_\ell = u(x_\ell, y_\ell), \qquad v_\ell = v(x_\ell, y_\ell), \qquad \Delta s_\ell = |\Delta \mathbf{r}_\ell|, \qquad \hat{\tau}_\ell = \frac{\Delta \mathbf{r}_\ell}{\Delta s_\ell}$$

$$\hat{\mathbf{r}} = \hat{x} \cos \phi \sin \theta + \hat{y} \sin \phi \sin \theta + \hat{z} \cos \theta$$

$$j_0(x) = \frac{\sin x}{x}, \qquad j_1(x) = \frac{\sin x}{x^2} - \frac{\cos x}{x}$$

If the presence of the dielectric slab is included, then the E_θ component should be multiplied by e^{TM}, which is given by

$$e^{TM} = \frac{1 + \Gamma^{TM}}{1 + e^{-j2\zeta(k_0 \sin \theta)t} \Gamma^{TM}}$$

where

$$\Gamma^{TM} = \frac{1 - \zeta(k_0 \sin \theta)/(\epsilon_r k_0 \cos \theta)}{1 + \zeta(k_0 \sin \theta)/(\epsilon_r k_0 \cos \theta)}$$

and the E_ϕ component should be multiplied by e^{TE}. The latter coefficient is the same as e^{TM} except that Γ^{TM} is replaced by

$$\Gamma^{TE} = \frac{1 - \mu_r k_0 \cos \theta/\zeta(k_0 \sin \theta)}{1 + \mu_r k_0 \cos \theta/\zeta(k_0 \sin \theta)}$$

where μ_r is the relative permeability of the dielectric substrate. The function $\zeta(\xi)$ is

$$\zeta(\xi) = \sqrt{\epsilon_r \mu_r k_0^2 - \xi^2}$$

This result is based on condensing the magnetic current into a filament and impressing it on the ground plane in the presence of the dielectric slab. The approximation appears to work reasonably well for soft substrates for dielectric thicknesses up to $0.05\lambda_0$.

(d) *Compute the radiated power.* One must next compute the power carried in the space wave and perhaps also the surface wave. This can be done by integrating the far-field power pattern. This must often be done by using a numerical integration. However, since the microstrip elements are not large-aperture elements their patterns are broad and slowly varying and the numerical effort required to perform the integration is small. The integral for power radiated by the space wave is

$$P_r = \frac{1}{\eta_0} \int_0^{2\pi} \int_0^{\pi/2} (|E_\theta|^2 + |E_\phi|^2) r^2 \sin\theta \, d\theta \, d\phi$$

and the power carried by the surface wave is

$$P_{sw} \cong \frac{1}{4\pi\eta_0} (k_0 t)^3 \frac{\epsilon_r - 1}{\epsilon_r} \int_0^{2\pi} |F_x \sin\phi - F_y \cos\phi|^2 \, d\phi$$

where $\eta_0 = 377\,\Omega$. The approximation given above for the surface wave is valid for soft substrates of thickness up to $0.05\lambda_0$.

(e) *Compute the stored energy.* Because of the normalization of the resonant mode the computation of the energy stored in the electric field under the patch is simple. The stored energy is given by

$$W_E = \frac{1}{2}\epsilon t$$

where ϵ is the substrate permittivity.

(f) *Compute the external quality factor.* The external quality factor is obtained from

$$Q_{ext} = \frac{2ck_{mn}W_E}{(\mu_r\epsilon_r)^{1/2}(P_r + P_{sw})}$$

where c is the speed of light in a vacuum. The radiative quality factor Q_r (i.e., the external quality factor obtained ignoring the surface wave) times the substrate thickness t divided by λ_0 is plotted in Fig. 6 for various modes of a rectangular microstrip element. The radiated power is computed ignoring the presence of the dielectric slab.

(g) *Compute the antenna quality factor.* To get the quality factor of the antenna one must include with the external Q the loss contribution due to the actual dielectric loss and the conduction loss. This quality factor is given in (2).

(h) *Compute the effective loss tangent* δ_{eff}. The effective loss tangent is $1/Q$.

(i) *Analyze the lossy cavity model of the antenna to find the impedance and other antenna parameters obtainable from the interior fields.* The voltage between a point (u, v) on the patch and the point on the ground plane directly below it, due to a unit, z-independent, *filamentary* electric current source impressed between the ground plane and the patch at point (u', v'), is [5]

$$G(u, v | u', v') = -jk\eta t \sum_A \frac{\psi_{mn}(u, v)\,\psi_{mn}(u', v')}{k^2(1 - j/Q) - (k_{mn})^2} \qquad (3)$$

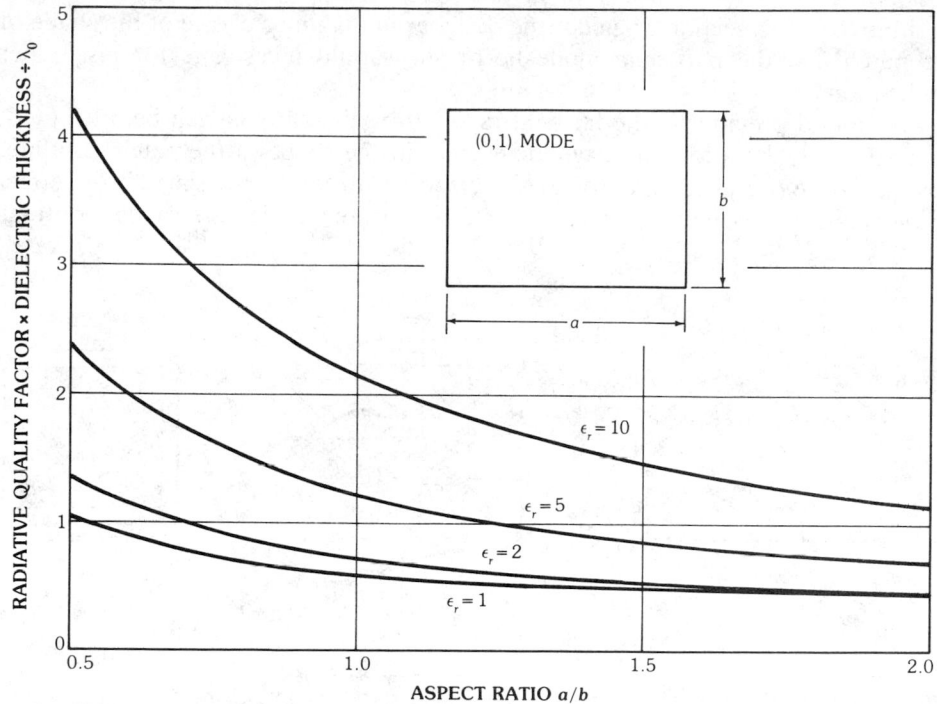

Fig. 6. The dielectric thickness t divided by λ_0 times the radiative quality factor Q for the $(0,1)$ mode versus aspect ratio of a rectangular microstrip element for various relative dielectric constants.

where $\eta = \eta_0(\mu_r/\epsilon_r)^{1/2}$ is the intrinsic wave impedance of the substrate dielectric and $k = 2\pi f(\epsilon_r \mu_r)^{1/2}/c$ is the wave number of the substrate. The summation is over the entire set A of mode indices (m, n). The "$1/Q$" term in the denominator of the summand is the effective loss tangent.

This expression allows one to compute the field everywhere within the cavity due to the filamentary source. If one attempts to set (u', v'), the source point, equal to (u, v), the observation point, to obtain an input impedance, one finds that the series does not converge. This is because there is an infinite inductance associated with a filamentary source. To obtain the correct impedance one must integrate (3) over the perimeter of the actual feed probe or microstrip feed. This integration modifies the summand in (3) by the introduction of an additional factor, $(\xi_{mn})^2$. This factor approaches zero as m and n increase without bound. For the lower-order modes ξ_{mn} is nearly unity. A more detailed discussion of how to actually evaluate the input impedance efficiently is given in Section 4, under "Efficient Computation of Impedance Parameters."

3. Pattern

How to compute the pattern associated with a microstrip element is discussed in Section 2, under (c) "Compute the far-field pattern." The explanation is more qualitative in this section; it guides the designer in making a choice of the shape of the patch and the particular mode he or she should investigate for his or her application.

As noted previously, the far field radiated by the antenna can be attributed, approximately, to a loop of magnetic current in the shape of the patch's outline, flowing in the presence of a grounded dielectric slab, or in free space if the effect of the dielectric is ignored. Fig. 7 illustrates the magnetic current distributions

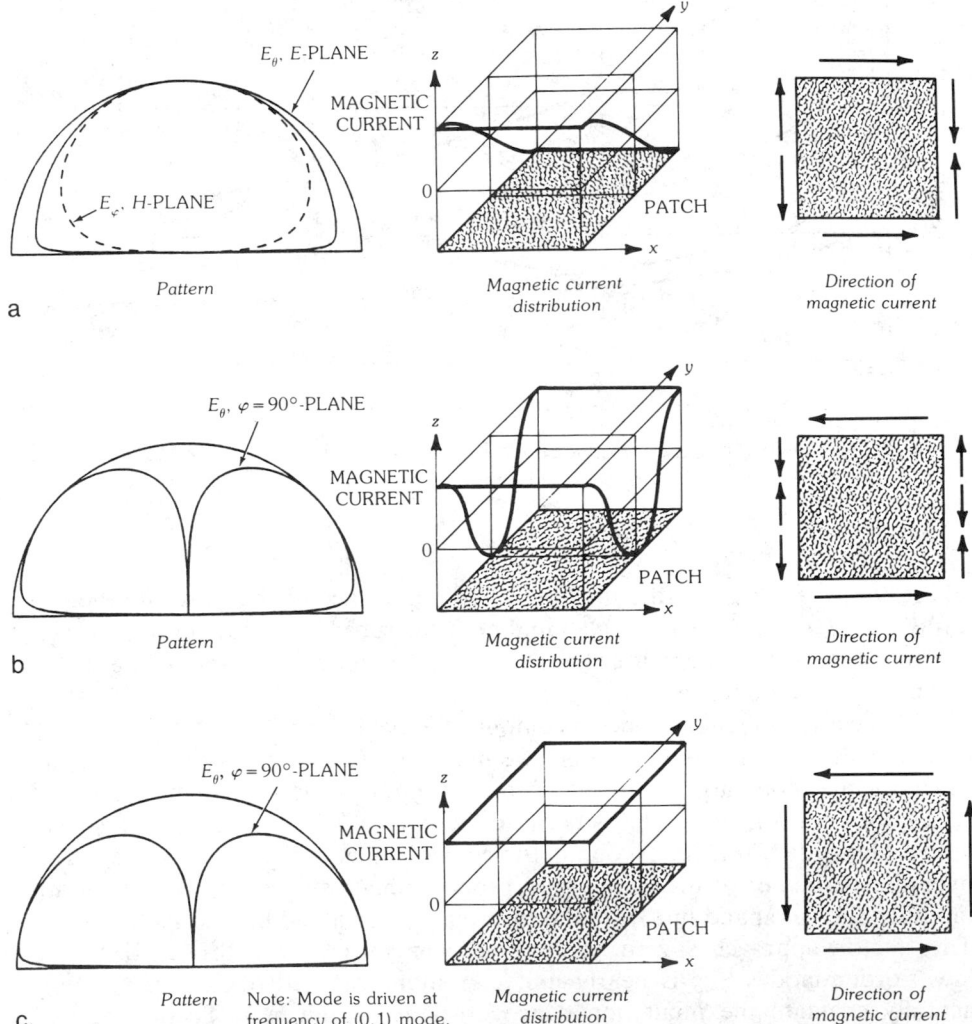

Fig. 7. Principal-plane patterns and magnetic current distributions for the lowest-order modes of rectangular and circular-disk elements. Patterns are plotted on a 40-dB scale and are computed for a $0.01\lambda_0$-thick substrate of relative dielectric constant $\epsilon_r = 2.50$. (*a*) Rectangular element: $(0,1)$ mode. (*b*) Rectangular element: $(0,2)$ mode. (*c*) Rectangular element: dc mode (driven at the resonant frequency of the $(0,1)$ mode). (*d*) Circular-disk element: $(1,1)$ mode. (*e*) Circular-disk element: $(0,1)$ mode.

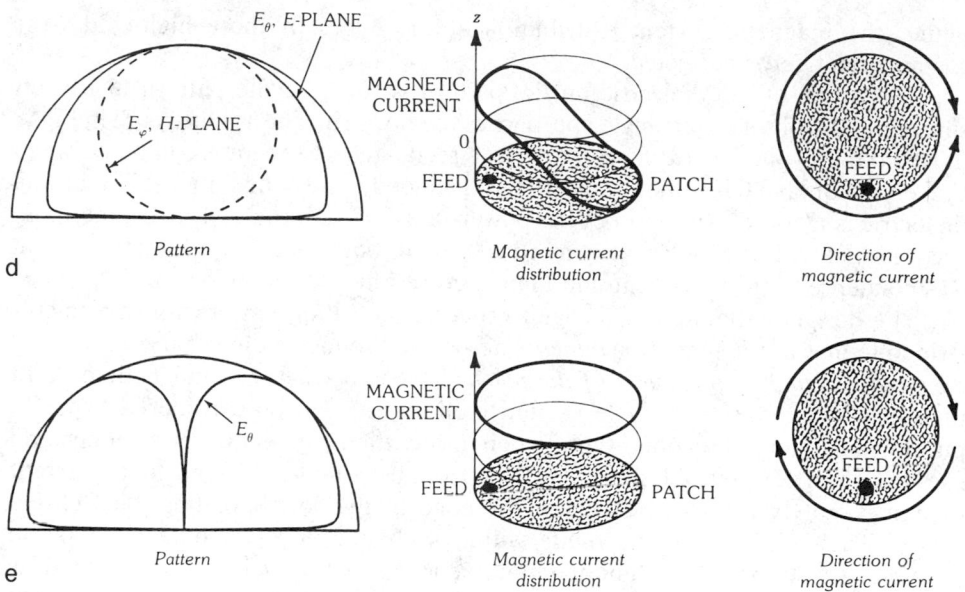

Fig. 7, *continued.*

associated with the lowest few modes of the rectangular and circular disk microstrip antennas. It also includes the corresponding principal-plane patterns. The reader should consult these figures to illuminate the text in the following subsections. These patterns were computed using a magnetic loop current impressed on the ground plane of a $0.01\lambda_0$-thick grounded dielectric slab. The assumed relative dielectric constant of the nonmagnetic substrate was 2.5. Conclusions on the elevation of beam maxima will vary somewhat if different dielectric thicknesses and dielectric constants are assumed. Since the rectangular patch is the most commonly used shape and since its mode functions are the most familiar and the easiest to compute, the discussion in this section centers on this shape of patch. Before considering specific examples, some general discussion is presented on the properties of microstrip-antenna patterns.

General Properties

An important assumption made in the discussion below is that the microstrip elements are mounted on *infinitely large, flat* ground planes. Although some applications could call for microstrip elements mounted on a large, flat, ground plane, it is more likely that the element will be mounted on the curved surface of a missile, or some other curved and more severely abbreviated ground plane. It is clearly impossible to cover all environments in which an element might be used. Thus the designer must be aware of the effects of truncating [18] and bending the ground plane and be ready to compute the pattern radiated by the element in *his or her particular environment*. Fortunately, unless the radius of curvature of the ground plane is small compared with the element's size, one can expect the magnetic current distribution to remain the same when the element is bent [19]. Although some distortion might result from an extreme truncation of the ground plane, the magnetic current distribution at the edges of these high-Q devices remains relatively unaffected.

The effect of the dielectric on the pattern is to cause the pattern to roll off sharply as the observation angle approaches the horizon. This is illustrated in Fig. 8 in which the E-plane pattern produced by a rectangular element is computed when the effect of a $0.05\lambda_0$-thick dielectric is included, and when the effect of the dielectric is ignored. The ratio of the power radiated with the dielectric taken into account and with the dielectric ignored is 0.97. In both cases the magnetic current was condensed into a filament and impressed on the ground plane.

The designer should also recognize that the elevation patterns of thin microstrip antennas are *inherently symmetric* about the broadside direction *regardless of the mode excited or the shape of the patch*. This is because the magnetic current phasor associated with a single excited mode is (or can be considered to be) a purely real-valued function. Only when more than one mode is strongly and simultaneously excited will the magnetic current be complex. The magnetic current is approximately the electric field at the edge of the corresponding ideal closed cavity. The field of a *single mode* within such a cavity can only have phase differences of 0° or 180°. A unidirectional, end-fire pattern cannot be obtained by exciting a *single* mode of any shape of patch. However, by using reactively loaded microstrip elements and exciting an appropriately chosen *pair* of modes, one can, in principle, obtain an end-fire pattern [7].

The (0, 1) Mode

The most commonly used mode of a rectangular patch is the (0, 1) mode. In this mode the magnetic currents on two opposite edges are constant. These edges are called the *radiating edges*. The magnetic currents on the remaining two edges suffer a phase reversal and hence the fields radiated by them largely cancel. The pattern this mode produces is linearly polarized and has a broadside maximum. (In this discussion the term "broadside" is used to indicate a direction normal to the plane of the patch, and "horizon" represents any direction in the plane of the patch.) The "E-plane" is the plane perpendicular to the radiating edges; the electric field for points in this plane is parallel to the plane. The pattern falls in the E-plane by an amount dependent on the dielectric constant of the substrate

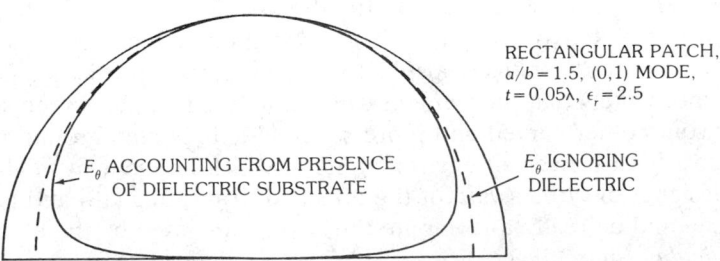

Fig. 8. E-plane far-field pattern of a rectangular element computed when the effect of a $0.05\lambda_0$-thick dielectric slab is taken into account and when it is ignored. Pattern plotted on a 40-dB scale.

material as the observation direction approaches the horizon. For angles very near the horizon, the pattern depends strongly on the truncation of the dielectric.

The (0, 2) Mode

The $(0, 2)$ mode is similar to the $(0, 1)$ mode except that the magnetic currents on the two radiating edges are antiparallel. This mode is commonly called a "higher-order mode." The physical size of the patch in terms of λ_0 is approximately twice that required to resonate a $(0, 1)$ mode at the same frequency. For a substrate with a relative dielectric constant of 2.5, the patch must be approximately $0.63\lambda_0$ long. The pattern produced by this mode has a broadside null and peaks in the E-plane at an angle of 40° above the horizon. The H-plane pattern is null.

The DC Mode

The rectangular patch (as well as all other unloaded patch geometries) supports a dc mode, that is, a mode which "resonates" at a frequency of $f = 0$ Hz. Of course, one wants to excite the dc mode at an rf frequency, not at dc. This requires an external matching network, or it requires that appropriate reactive lumped loads be placed between the patch and the ground plane [7] as discussed in Section 9, under "Frequency-Agile Elements." Under the cavity model approximation the electric field, $\psi_{0,0}(u, v)$, is a constant. The corresponding magnetic current is constant in magnitude. If one recalls the equivalence between a small loop of constant magnetic current and a short, vertical electric dipole, one can see why this mode produces a pattern resembling that of the vertical monopole.

The (1, 1) Mode of a Circular-Disk Element

This mode of the disk element is very similar to the $(0, 1)$ mode of a rectangular element. Both produce broadside nulls. There are two degenerate $(1, 1)$ disk modes. One has an even field distribution about the diameter on which the feed lies. The other has an odd-symmetric field distribution about this diameter. In the absence of anything that perturbs the symmetry of the disk the odd-symmetric mode cannot be excited by the feed. On the other hand, one can strongly excite this odd mode by making the disk slightly elliptical [20] or by loading the disk appropriately. This has applications for producing circularly polarized patterns as discussed in Section 9, under "Single-Feed CP."

The (0, n) Circular-Disk Modes

These modes are independent of the azimuthal variable ϕ. The magnetic current distribution around the perimeter of the patch is a constant and the pattern has a broadside null. For $n = 1$ and for $\epsilon_r = 2.5$, the beam maximum occurs at 53° above the horizon. For $n >, 1$ grating lobes will appear in the pattern for a relative dielectric constant of 2.5.

Other Modes

Some modes such as the $(1, 1)$ mode of the rectangular patch seem to have little practical application although a similar mode, the $(2, 1)$ mode of a circular-disk element, has been used to produce a conical, circularly polarized pattern [21]. When the sum of the mode indices of a rectangular microstrip element is an even number, a broadside null will always occur. When it is odd, a broadside maximum will always occur. As the mode indices increase, the size of the element in terms of λ_0 increases, thus producing grating lobes in the pattern. This typically renders the pattern useless for most applications. Thus one typically only uses one or more of the lowest-order modes of a microstrip element.

4. Impedance and Circuit Models

The general approach for obtaining the input impedance of a microstrip antenna was discussed in Section 2. In this section practical formulas and techniques are given for computing input and mutual impedances of one or more ports on a microstrip element. A physical explanation of the variation of impedance with feed point is given. Simple circuit models for the microstrip element are presented.

General Circuit Model

If one rewrites the factor

$$\frac{-jk}{k^2(1 - j/Q) - (k_{mn})^2} \quad \text{in (3) as} \quad \frac{-j\omega c/\sqrt{\epsilon_r}}{\omega^2 - (k_{mn})^2 c^2/\epsilon_r - j\omega^2/Q}$$

and compares it with the input impedance of the parallel combination of an inductor, capacitor, and resistor, then one concludes that the impedance of a microstrip line as approximated by (3), after integration over the source current distribution, can be modeled as the input impedance to the network illustrated in Fig. 9a. The element values are

$$C_{mn} = \left[\frac{c}{\sqrt{\epsilon_r}} t\eta\psi_{mn}(u',v')^2(\xi_{mn})^2 \right]^{-1}$$

$$L_{mn} = \frac{\epsilon_r}{C_{mn}(k_{mn})^2 c^2}$$

$$R_{mn} = Q/(C_{mn}\omega)$$

The factor $(\xi_{mn})^2$, it is emphasized, arises from the integration of (3) over the feed current distribution and tends toward zero as m and n increase without bound. The inductive susceptance increases much faster than the capacitive susceptance and the conductance as m and n increase. Thus, for a sufficiently high-order mode, the inductance, L_{mn}, dominates its lossy tank circuit and the corresponding resistance and capacitance can be ignored. The inductances corresponding to all sufficiently high-order modes can be combined into the single series inductance indicated in Fig. 9a. The resistances R_{mn} in this model are frequency dependent. However, over the relatively narrow band of a resonant mode, they can be considered approximately constant with the ω replaced by its value at the band center. The network in Fig. 9a is valid for frequencies in the vicinity of the resonance of a *single*

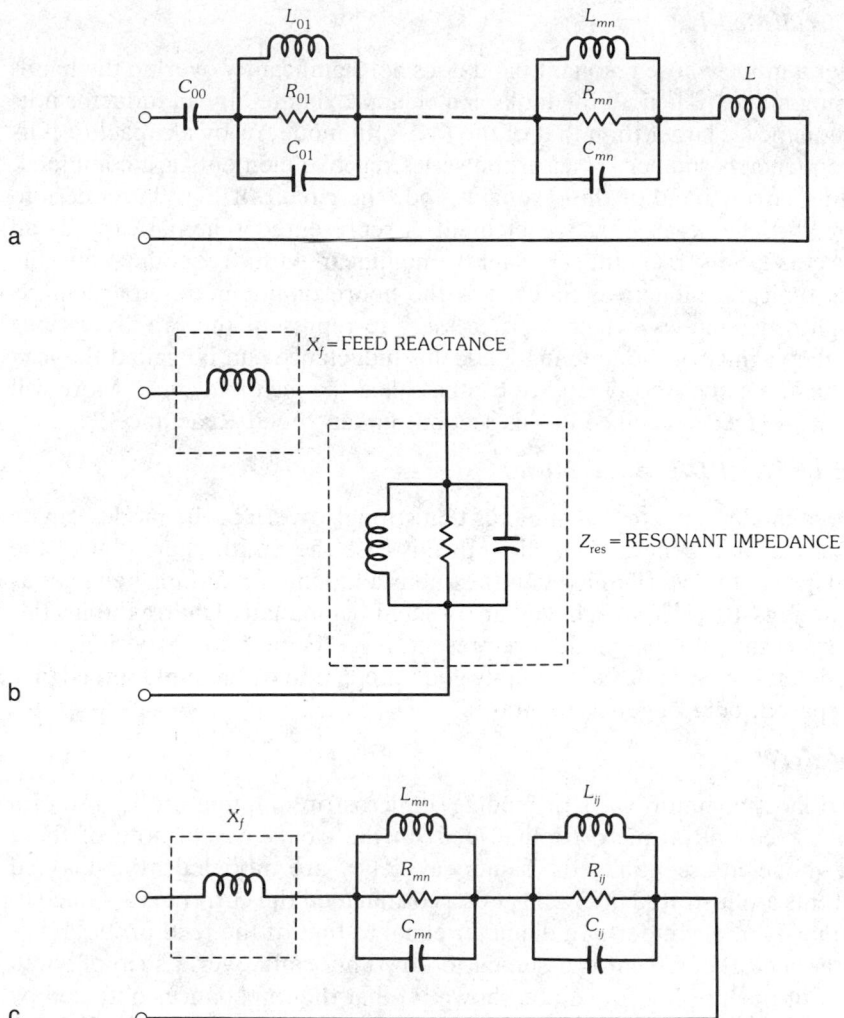

Fig. 9. Circuit models for a microstrip element. (*a*) General circuit model. (*b*) Simplified circuit model valid over the band of a single, isolated mode. (*c*) Circuit model over the overlapping bands of two nearly degenerate modes.

mode of the microstrip element. This is because the effective loss tangent, $1/Q$, was computed for a single [say (M, N)th] mode of the antenna. Since each tank circuit in this network has a relatively high Q, the impedance of any nonresonant tank will be rather insensitive to the exact value of R_{mn}. Thus, one can make the network in Fig. 9a applicable to frequency bands centered about several resonant modes if Q is computed for each of the modes and used to determine the corresponding R_{mn}. This network becomes questionable, however, if one attempts to use it to compute the input impedance of the element at a frequency midway between two well-separated adjacent resonant modes. The problem of analyzing an element far off resonance is considered in Section 8, "Loaded Microstrip Elements."

Simplified Circuit Model

Consider a mode whose resonant band does not significantly overlap the bands of neighboring modes. Then all the tanks can be approximated by an inductor if its resonant frequency is larger than that of the (M, N)th mode, or by a capacitor if its resonant frequency is smaller. If all of the series reactive elements are combined, then over the narrow band of the resonant mode the circuit of Fig. 9a reduces to that of Fig. 9b. The series reactive element is represented schematically as an inductor. Actually, its reactance is slightly nonlinear with frequency, but the reactance is typically inductive. Since it is the nonresonant modes, particularly the very high order modes, which are necessary to represent the rapidly varying currents in the vicinity of the antenna feed, this inductance can be called the *feed inductance* and its associated reactance can be called the *feed reactance*. More will be said about the feed reactance in this section, under "Feed Reactance."

Circuit Model for Near-Degenerate Modes

When two modes have resonant bands that strongly overlap, the modes can be thought of as "near-degenerate" modes. In this case the Smith chart plot of the input impedance in the overlapping bands exhibits a looping or kinking behavior as illustrated in Fig. 10 [11]. Whenever an isolated, unloaded patch exhibits this behavior, expect that muliple modes are present. It has been found experimentally that the model of Fig. 9c yields a resonably good prediction of the input impedance for closely spaced, near-degenerate modes.

Simple Feed Models

The two most common ways of feeding a microstrip antenna are by use of a coaxial-probe feed and a microstripline feed. In the cavity model both of these sources are modeled essentially the same way. They are modeled by z-directed surface currents as illustrated in Fig. 11. For a coaxial feed the current is assumed to flow uniformly over a cylinder of a diameter equal to that of the feed probe. For a microstripline feed the current is assumed to flow uniformly over a strip of width equal to that of the stripline. It can be shown [4] that the impedances obtained by the cavity model for these two source distributions are identical if the cylindrical distribution of diameter d is replaced by a strip distribution of effective feed width $2.24d$.

As the circuit in Fig. 9b suggests, one can view the input impedance of the element as the sum of a resonant component plus a feed-reactive component,

$$Z = Z_{res} + jX_f$$

This decomposition of the impedance into resonant and nonresonant parts has been found quite useful in explaining the behavior of loaded microstrip elements (see Section 8).

Resonant Impedance

If one will operate the element at a frequency in the vicinity of its (M, N)th resonant mode, then the resonant part of the impedance is

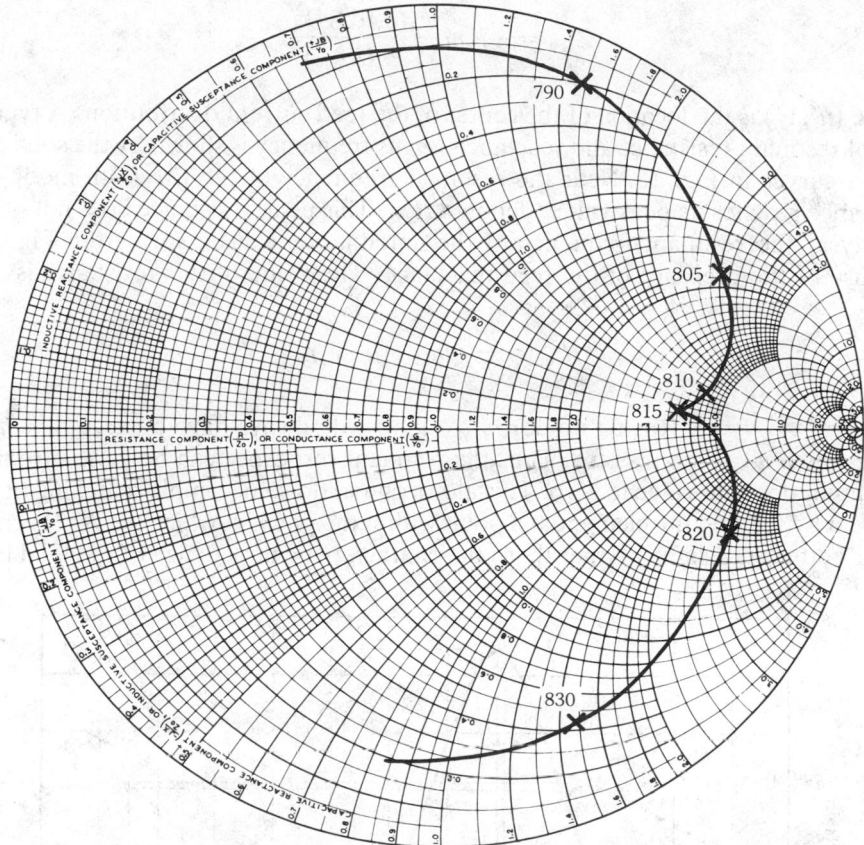

Fig. 10. Typical Smith chart plot of the impedance of an element with two nearly degenerate modes. Measured for an 11.4-cm nearly square patch, with a dielectric 0.16 cm thick and having a relative dielectric constant of 2.62. (*After Lo, Solomon, and Richards [11],* © *1979 IEEE*)

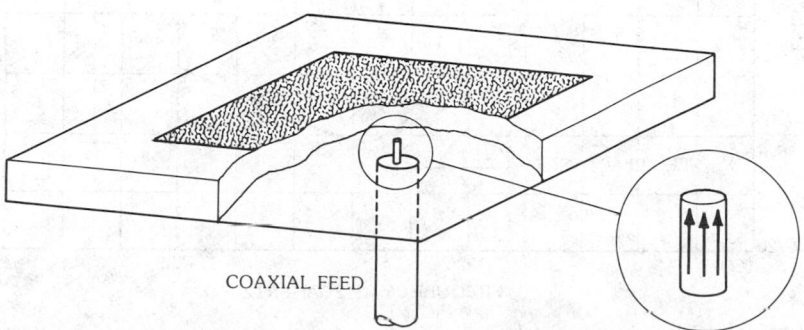

COAXIAL FEED

Fig. 11. Coaxially fed element and the simplified feed model obtained by removing the feed probe, closing the aperture of the coaxial cable, and impressing a uniform electric surface current on the surface corresponding the surface of the feed probe.

$$Z_{\text{res}} = -jk\eta t \frac{\psi_{MN}(u',v')^2}{k^2 - (k_{MN})^2}$$

where (u',v') is the location of the center of the feed current distribution. A typical plot of resonant resistance and reactance *versus* frequency is shown by the solid and dotted curves in Fig. 12. Near the resonant frequency of the (M,N)th mode the resistance reaches its peak value, denoted R_{res}. This peak resistance does not occur exactly at the resonant frequency of the antenna, as can be seen from Fig. 12, because of the presence of the feed reactance. The value of this resistance is

$$R_{\text{res}} = \frac{\eta t}{k} Q\psi_{MN}(u',v')^2 \tag{4}$$

Unless $M = N = 0$ (the dc mode), $\psi_{MN}(u,v)$ will always have a curve $u = h(v)$ along which $\psi_{MN}(u,v)$ is zero. Thus the resonant resistance can be varied from its maximum possible value to zero. This alows one to do some simple matching of the element by feeding the element at a point within the patch at which the R_{res} is equal to (or better yet, slightly larger than) the characteristic impedance of the feed line.

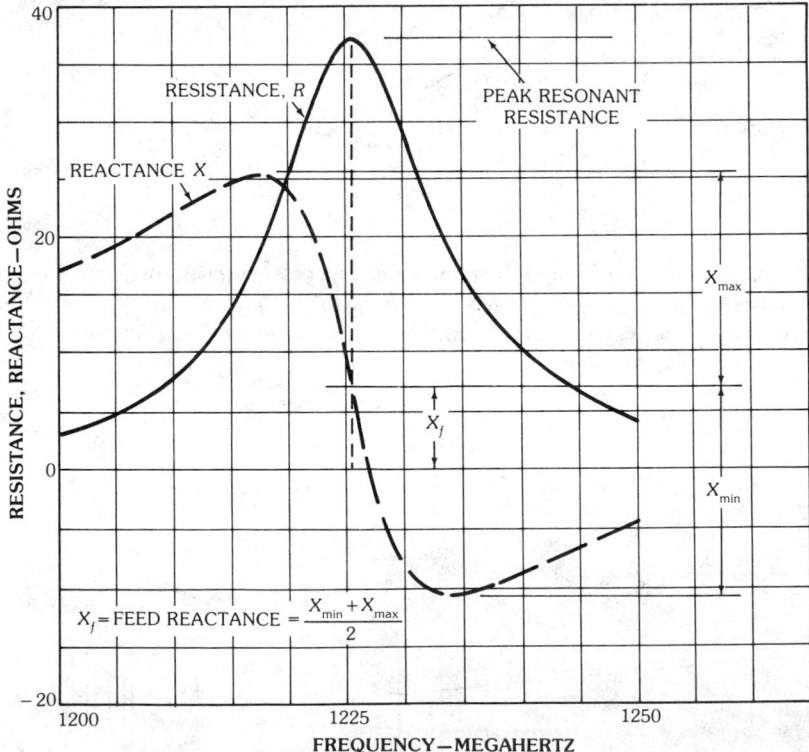

Fig. 12. Typical measured variation of resistance and reactance versus frequency in a 7.62- × 11.43- × 0.16-cm microstrip antenna illustrating one way to empirically determine the feed reactance. (*After Richards et al. [22], reprinted with permission of Hemisphere Publishing Corporation*)

(See Section 7 for more on matching.) Fig. 13 illustrates the theoretical variation of R_{res} with feed location for a rectangular microstrip element in the $(0,1)$ mode.

Feed Reactance

The feed-reactive part [22] X_f of the input impedance represents the contribution to the impedance due to all of the nonresonant modes. The perimeter P of the feed current distribution is an important parameter in determining X_f. If the current were filamentary $(P = 0)$, then the feed reactance would be infinite. The smaller the P, the more inductive the feed reactance will be. The feed reactance is given by

$$\frac{1}{P^2} \int_P \int_P G(u, v | u', v') \, ds \, ds' - Z_{res}$$

where the ds and ds' are elements of arc length associated with the source and observation points (u', v') and (u, v) and the integrals are over the perimeter of the current distribution. How to actually carry out these integrations is considered in this section, under "Multiport Impedance Parameters." Typical plots of feed reactance versus frequency can be found in Section 8, "Loaded Microstrip Elements." Fig. 12 illustrates the feed reactance at the resonant frequency.

The feed reactance may or may not be an important parameter in the design of a microstrip element. For thin elements the feed reactance is typically small compared with the resonant resistance of a well-matched element. For thicker elements this may not be true and the feed reactive component may have to be taken into account in matching the element. For loaded-element applications the feed reactance is a critical parameter since it determines, among other parameters, the resonant frequency of the loaded element. (A discussion of loaded elements is given in Section 8.) For loaded elements an understanding of how the feed reactance varies with feed point can be useful in element design.

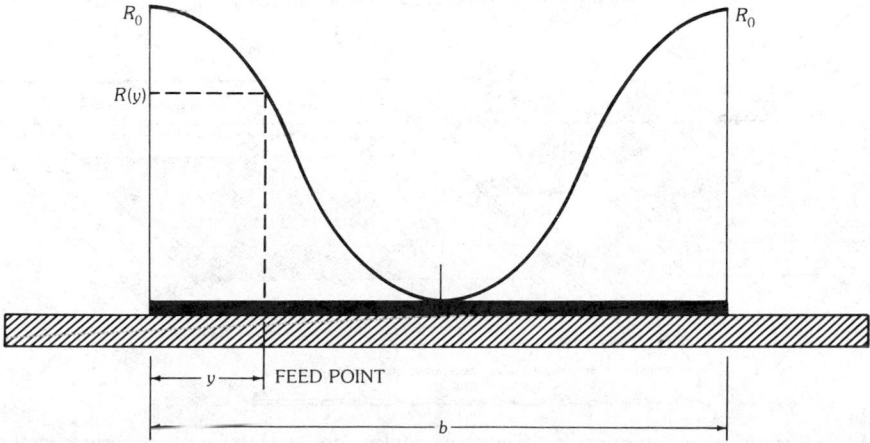

Fig. 13. The theoretical variation of resonant resistance versus feed point on a rectangular microstrip element driven in the $(0,1)$ mode.

To see how feed reactance varies with feed point, one can note that the feed inductance will be roughly proportional to the amount of magnetic energy stored within a small testing volume surrounding the feed (Fig. 14a) [22]. Consider, for example, a rectangular element. The magnetic walls can be removed if the multiple images of the source current are impressed as illustrated in the right-hand side of Fig. 14a. Since these are images through *magnetic* walls, the image currents flow in the same direction as the source current. Suppose the feed current is far from any of the magnetic walls in the cavity model of the element. Then the magnetic fields produced by this current and its multiple images do not strongly overlap (Fig. 14a). However, if the feed current is moved close to one edge, the magnetic fields of the feed current and its nearest image overlap strongly. In the extreme case when the feed current is at the edge of the patch, the feed and image currents coincide and the magnetic field within the testing volume doubles (Fig. 14b). In this case the magnetic energy density in the testing volume quadruples. The volume of the testing volume under the patch, however, is only half as much as it was when the feed was far from a magnetic wall. Thus there is a net doubling of the magnetic energy in the testing volume. One expects from this rough analysis that the feed reactance for a feed near an edge of the patch will be twice that for a feed far from

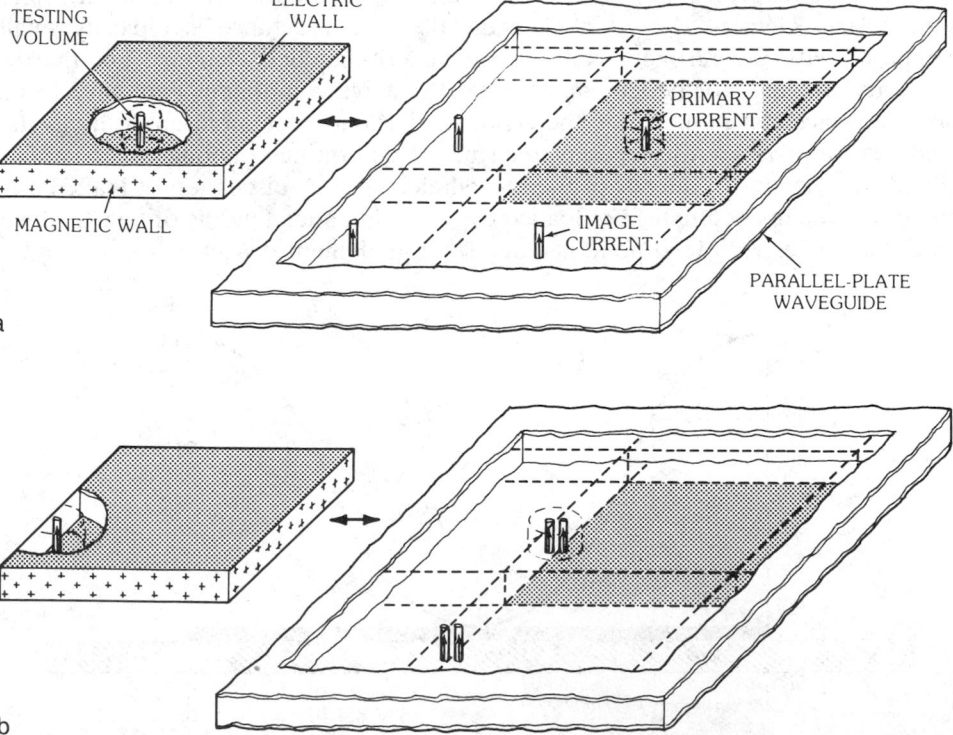

Fig. 14. Multiple-image expansions (right) of cavity model (left) with driving current far from or near to the edge of the patch. (*a*) Feed far from any edge. (*b*) Feed very near one edge. (*After Richards [22], reprinted with permission of Hemisphere Publishing Corporation*)

the edge. If the feed were near a corner of the patch, one would expect a net quadrupling of the inductance because of the presence of three images at the corner.

Although this rough analysis does not prove to be quantitatively accurate, it does precisely predict the trends in feed inductance as a function of feed location. Fig. 15 contains a plot of the feed reactance predicted by the cavity model and the measured feed reactance. The cavity model is able to predict the variation of feed reactance with reasonable accuracy for feeds not too close to the edge. It over-estimates the feed reactance for feeds right on the edge. The reason for this is that the cavity model assumes zero tangential component of magnetic field at the patch edge. Actually, for the cavity model to precisely represent the fields under the patch, one would have to introduce an equivalent electric current which is numerically equal to the small tangential magnetic field at the patch edge. When the feed is very close to the edge the tangential field near the probe is not negligible. A strong equivalent electric current *should be* impressed on the magnetic wall. This equivalent current is in a direction opposite to the feed current, which has the effect of reducing the magnetic energy stored in the testing volume. Hence the *actual* feed reactances for feeds very close to the edge are smaller than those predicted by the cavity model.

A formula which is based on the feed reactance within a parallel-plate waveguide (which ignores all image currents) is given by

$$X_f \cong -\frac{\eta k t}{2\pi}[\ln(kd/4) + 0.577]$$

where d is the diameter of the probe. This formula has been used to estimate the feed reactance. It does not predict the variation of feed reactance versus feed

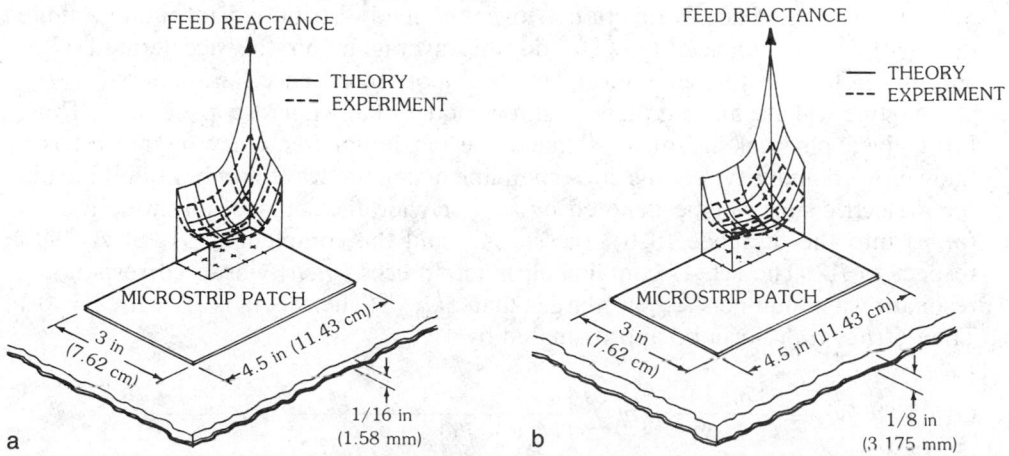

Fig. 15. Feed reactance measured and computed using the cavity model for a 11.43-× 7.62-cm rectangular patch. (*a*) Dielectric thickness 1/16 in (1.6 mm). (*b*) Dielectric thickness 1/8 in (3.2 mm). (*After Richards [22], reprinted with permission of Hemisphere Publishing Corporation*)

position but it is simple and can have some use if only a very rough estimate of the feed reactance is required.

Multiport Impedance Parameters

The discussion above has centered on the computation and the properties of the driving-point or input impedance of a microstrip element. One can also compute the open-circuit parameters of a microstrip element with more than one port. From these open-circuit or z parameters one can compute the corresponding s parameters of the element. The nondiagonal z parameters are essentially independent of the precise feed-current distributions and are given by

$$Z_{ij} = Z_{ji} = G(u_i, v_i | u_j, v_j), \qquad i \ne j \tag{5}$$

The corresponding s-parameter matrix $\mathbf{S}$ is given in terms of the z-parameter matrix $\mathbf{Z}$ by

$$\mathbf{S} = (\mathbf{Z}\mathbf{Y}_0 + \mathbf{I})^{-1}(\mathbf{Z}\mathbf{Y}_0 - \mathbf{I})$$

where $\mathbf{I}$ is the unit matrix and $\mathbf{Y}_0$ is the diagonal matrix containing the characteristic admittances of the feed lines to the element. Typical measured and computed s parameters are illustrated in Fig. 16.

Efficient Computation of Impedance Parameters

Equation 3 for the driving-point impedance and equation 5 for the mutual impedances of a multiple-feed element are not practical to use as is, for two reasons. (1) They both converge very slowly. (2) For all but rectangular elements, computing higher-order modes $\psi_{mn}(u, v)$ becomes a time-consuming process. Fortunately, the equation for $G(u, v | u', v')$ can be accelerated in the way described in this section. As pointed out earlier, under "Feed Reactance," one must average the Green's function G over the feed distribution to obtain a finite driving-point impedance. How to do this averaging on the accelerated G is considered later in this section, under "Multiport Impedance Parameters."

No one will use any but the lowest few modes of any microstrip element. Thus, for a given physical size of an element the maximum frequency of interest is a known, fixed quantity. Let the corresponding maximum wave number of interest in the dielectric substrate be denoted by k_{max}. Divide the set A of all mode indices (m, n) into the dc mode $(0, 0)$, the set A_ξ, and the complement $\overline{A}_\xi$ of A_ξ with respect to A_0. The set A_ξ contains all mode indices (m, n) whose corresponding resonant wave numbers k_{mn} are larger than ξk_{max}. (The set A_0 is A_ξ with $\xi = 0$.) Then $G(u, v | u', v')$ can be approximated by

$$G(u, v | u', v') \cong -\frac{j\eta t}{k} \frac{1}{S} - jk\eta t \sum_{\overline{A}_\xi} \left\{ \frac{1}{k^2(1 - j/Q_{mn}) - (k_{mn})^2} + a_0 \frac{1}{(k_{mn})^2} \right.$$

$$+ k^2 a_1 \frac{1}{(k_{mn})^4} + \dots + k^{2(L-1)} a_{L-1} \frac{1}{(k_{mn})^{2L}} \bigg\} \psi_{mn}(u, v)\, \psi_{mn}(u', v')$$

$$+ \{a_0 G_0(u, v | u', v') + k^2 a_1 G_1(u, v | u', v')$$

$$+ \dots k^{2(L-1)} a_{L-1} G_{L-1}(u, v | u', v')\} \tag{6}$$

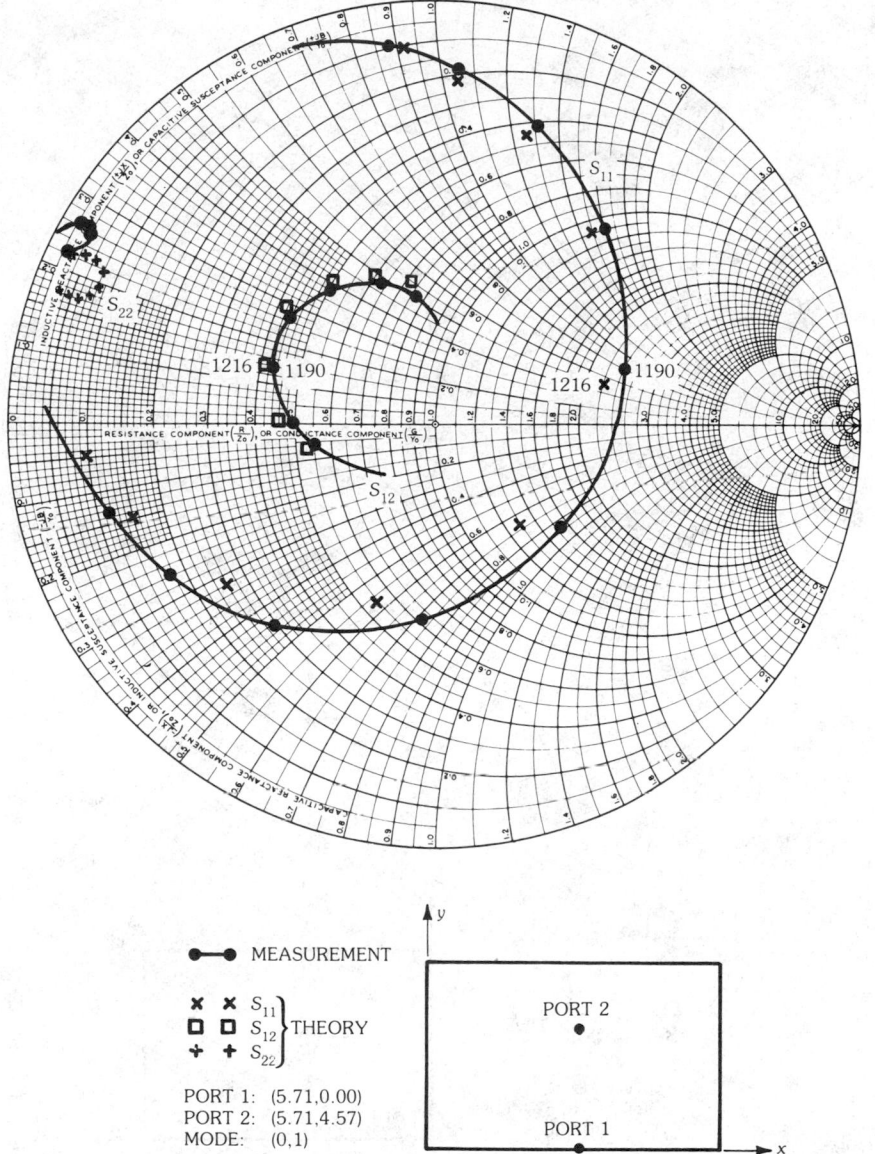

Fig. 16. Typical measured and computed *s*-parameter Smith chart plots for an 11.43- × 7.62- × 0.16-cm rectangular microstrip element with relative dielectric constant of 2.62. (*After Richards, Lo, and Harrison [5], © 1981 IEEE*)

The parameter S is used to represent both the area of the patch and the set of points comprising the patch. The coefficients $a_0, a_1, \ldots, a_{L-1}$ are the coefficients of the polynomial approximation of

$$\frac{1}{1-x} \cong a_0 + a_1 x + a_2 x^2 + \ldots + a_{L-1} x^{(L-1)}$$

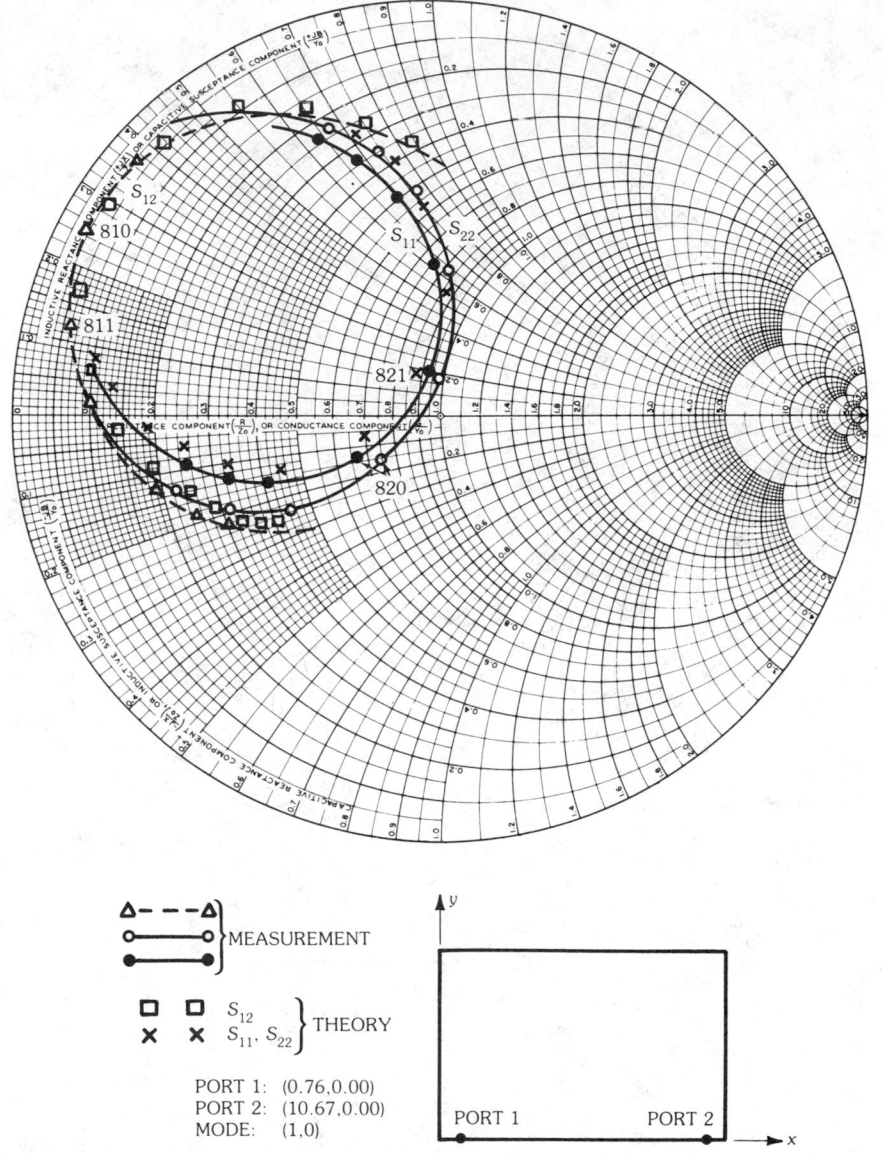

Fig. 16, *continued.*

for $0 \leqq x \leqq 1/\xi^2$. For $L = 2$, $\xi = 2$, $a_0 = 0.989$, and $a_1 = 1.331$, the maximum error in this approximation is about 1 percent. This has been found to be an adequate choice of parameters for the analysis of the microstrip elements. The G_ℓs are quasi-static terms (independent of frequency except for a linear factor) and are given by

$$G_\ell(u, v \,|\, u', v') = jk\eta t \sum_{A_0} \frac{\psi_{mn}(u, v)\,\psi_{mn}(u', v')}{(k_{mn})^{2\ell}} \tag{7}$$

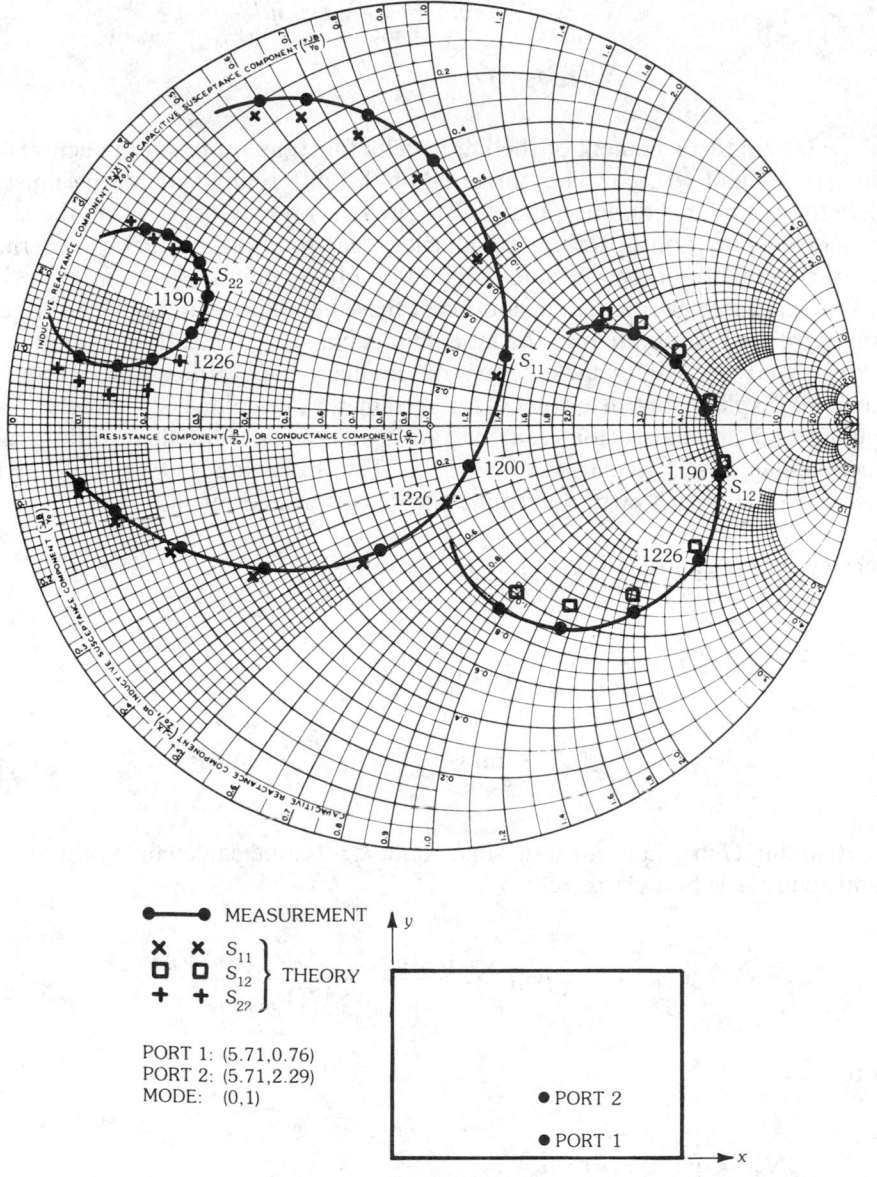

MEASUREMENT

$\left.\begin{array}{l} \times \quad \times \quad S_{11} \\ \square \quad \square \quad S_{12} \\ + \quad + \quad S_{22} \end{array}\right\}$ THEORY

PORT 1: (5.71,0.76)
PORT 2: (5.71,2.29)
MODE: (0,1)

● PORT 2

● PORT 1

Fig. 16, *continued.*

Although the series for G_0 is just as slowly converging as the original series for G, G_0 and G_1 need only be computed *once* for a given pair of observation and feed points, (u, v) and (u', v'). Once they are determined, one need only sum over the *finite* set $\overline{A}_\xi$ in (6) to find the G at any frequency.

The G_ℓs can be computed more efficiently than by summing (7). They satisfy the static boundary value problem,

$$\nabla^2 G_0 = -jk\eta t \delta(r - r') + jk\eta t/S$$

$$\nabla^2 G_1 = -G_0$$

The vectors in the argument of the Dirac delta function are position vectors to the points (u, v) and (u', v'). The solutions of the differential equations must also satisfy the boundary condition that $\nabla G_\ell(u, v \mid u', v') \cdot \hat{n} = 0$ for (u, v) on the edge of the patch. For nonseparable geometries the solutions can be determined numerically by a moment-method solution of a line integral equation.

For separable geometries, solutions of these equations can be found which *do not* require the computation of the modal functions ψ_{mn}. Instead, the solutions can be expressed as single-index series of circular, hyperbolic, and exponential functions. Explicit formulas for G_ℓ for rectangular, circular-disk, circular-sector, annular, and annular-sector elements are given in later in this section, under "Rectangular and Circular Geometries." The following "Overview" gives a general overview of how to determine the quasi-static terms.

Overview—One can obtain G_0 and G_1 by evaluating

$$G_0 = k \lim_{k \to 0} \left(\frac{G}{k} + \frac{j\eta t}{k^2 S} \right)$$

$$(8)$$

$$G_1 = k \lim_{k \to 0} \left\{ \frac{\partial}{\partial k^2} \left(\frac{G}{k} + \frac{j\eta t}{k^2 S} \right) \right\}$$

By expanding G in a Laurent series in k about $k = 0$, one can evaluate these limits. To this end, G is first expressed as

$$G(u, v \mid u', v') = jk\eta t \sum_{m=0}^{\infty} \frac{U_m(u) U_m(u') V_m^-(v_<) V_m^+(v_>)}{N_m W(V_m^-, V_m^+)} \qquad (9)$$

where

$$N_m = \int_{u_-}^{u_+} U_m(u)^2 \, du$$

$$W(V_m^-, V_m^+) = W_m = V_m^-(v_+) V_m^{+\prime}(v_+) - V_m^{-\prime}(v_+) V_m^+(v_+)$$

$$v_< = \begin{cases} v, & \text{for } v < v' \\ v', & \text{for } v > v' \end{cases}$$

$$v_> = \begin{cases} v, & \text{for } v > v' \\ v', & \text{for } v < v' \end{cases}$$

The prime (′) on a dependent variable indicates differentiation with respect to argument. The U_ms and V_ms satisfy the separated Helmholtz equations

$$U_m'' + [(a_m)^2 + k^2 f(u)] U_m = 0$$

$$V_m'' + [-(a_m)^2 + k^2 g(v)] V_m = 0$$

where the + and − superscripts have been suppressed on the V for convenience. The u and v coordinate variables are related to the rectangular x and y coordinate variables through the conformal mapping, $x + jy = w(u + jv)$. This conformal mapping transforms a patch in the xy plane whose edges lie along the curves $u = u_-$, $u = u_+$, $v = v_-$, and $v = v_+$ into a rectangular patch in the uv plane. This is illustrated in Fig. 17.

The square of the common scale factor for the u and v variables is $|w'(u + jv)|^2$ and can be written as $f(u) + g(v)$ for the four cylindrical-coordinate systems in which the Helmholtz equation is separable. These four systems are rectangular, circular-cylinder, elliptic-cylinder, and parabolic-cylinder systems. The $(a_m)^2$ is the separation constant determined by applying the pair of homogeneous boundary conditions that $U_m(u)$ satisfies at $u = u_\pm$. These boundary conditions depend on the geometry of the patch and are listed along with the functions $f(u)$ and $g(v)$ and the $u_\pm$ and $v_\pm$ in Table 1 for the rectangular and circular classes of patch shapes.

These parameters can be expanded in a power series in k^2 as

$$U_m = U_{0m} + k^2 U_{1m} + k^4 U_{2m} + \ldots$$

$$V_m = V_{0m} + k^2 V_{1m} + k^4 V_{2m} + \ldots$$

$$(a_m)^2 = (a_{0m})^2 + k^2 (a_{1m})^2 + k^4 (a_{2m})^2 + \ldots$$

$$N_m = N_{0m} + k^2 N_{1m} + k^4 N_{2m} + \ldots$$

$$W_m = W_{0m} + k^2 W_{1m} + k^4 W_{2m} + \ldots$$

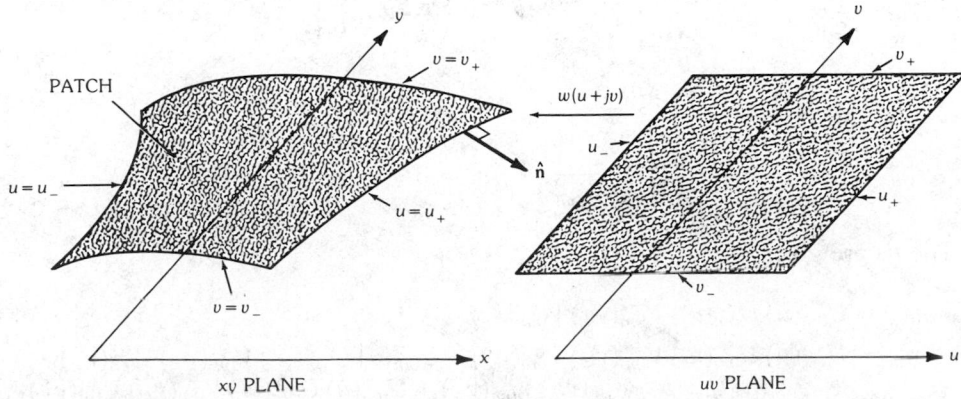

Fig. 17. Conformal mapping of a patch into a rectangular patch for the purpose of obtaining the quasi-static terms.

The U_{nm}s and the $V_{nm}^{\pm}$s satisfy the sequence of differential equations

$$U_{nm}'' + (\alpha_{0m})^2 U_{nm} = -\sum_{\ell=0}^{n-1} [\alpha_{n-\ell,m}]^2 U_{\ell m} - f(u) U_{n-1,m}$$

$$V_{nm}'' - (\alpha_{0m})^2 V_{nm} = \sum_{\ell=0}^{n-1} [\alpha_{n-\ell,m}]^2 V_{\ell m} - g(v) V_{n-1,m}$$

where again the superscript $\pm$ on the V has been dropped for convenience. In addition to the boundary conditions listed in Table 1, the following conditions are applied to obtain a unique solution in the most convenient form:

$$N_{1m} = N_{2m} = \ldots = 0$$

$$V_{nm}^{\pm}(v_{\pm}) = 0, \qquad \text{for } n > 0$$

$$V_{0m}^{\pm}(v_{\pm}) = 1$$

These conditions can be applied since the multiplication of the U and V functions by any arbitrary function of k^2 leaves $G(u, v \mid u', v')$ unchanged, but *does change* the coefficients of the series in k^2 listed above; i.e., the latter expansions are *not* unique.

Using the solutions of these equations in (8) and (9) yields the expression

$$G_{\ell}(u, v \mid u', v') = jk\eta t \left\{ r_{\ell 0} + \sum_{m=0}^{\infty} r_{\ell m} \right\}$$

The r_{ij}s are

$$r_{00} = \frac{1}{S} \left\{ n_{10} - \frac{\Delta u}{S} W_{20} \right\}$$

$$r_{10} = \frac{1}{S} \left\{ n_{20} - n_{10} \frac{\Delta u}{S} W_{20} - \frac{\Delta u}{S} W_{30} + \left(\frac{\Delta u}{S} W_{20} \right)^2 \right\}$$

$$r_{0m} = \frac{2n_{0m}}{\Delta u W_{0m}}, \qquad m > 0$$

$$r_{1m} = \left\{ n_{1m} - \frac{n_{0m} W_{1m}}{W_{0m}} \right\} \frac{2}{\Delta u W_{0m}}, \qquad m > 0$$

The ns are given by

$$n_{0m} = U_{0m}(u) U_{0m}(u') V_{0m}^{-}(v_<) V_{0m}^{+}(v_>)$$

$$n_{1m} = U_{1m}(u) U_{0m}(u') V_{0m}^{-}(v_<) V_{0m}^{+}(v_>) + U_{0m}(u) U_{1m}(u') V_{0m}^{-}(v_<) V_{0m}^{+}(v_>)$$
$$+ U_{0m}(u) U_{0m}(u') V_{1m}^{-}(v_<) V_{0m}^{+}(v_>) + U_{0m}(u) U_{0m}(u') V_{0m}^{-}(v_<) V_{1m}^{+}(v_>)$$

$$n_{20} = U_{20}(u) + U_{20}(u') + V_{20}^{-}(v_<) + V_{20}^{+}(v_>) + U_{10}(u) U_{10}(u') + V_{10}^{-}(v_<) V_{10}^{+}(v_>)$$
$$+ U_{10}(u) V_{10}^{-}(v_<) + U_{10}(u) V_{10}^{+}(v_>) + U_{10}(u') V_{10}^{-}(v_<) + U_{10}(u') V_{10}^{+}(v_>)$$

The range of u is $\Delta u = u_+ - u_-$; the range of v is $\Delta v = v_+ - v_-$, if finite.

The summand of G_0 always contains the term of the form

$$\frac{jk\eta t}{2\pi} \mathcal{R}\left\{\frac{1}{m}e^{-maz}\right\}$$

where $\alpha_{0m} = ma$ and $z = v_> - v_< + j(u - u')$. This term can be summed in closed form to yield

$$G_0(u, v \mid u', v') = \frac{jk\eta t}{2\pi}\left\{\alpha\frac{v_> - v_<}{2} - \ln|\alpha z| - \ln\left|\frac{\sinh(\alpha z/2)}{\alpha z/2}\right|\right\}$$

$$+ \text{ remainder terms}$$

For source points (u', v') not on the edge of the patch the remainder terms of G_0 in the preceding expression are nonsingular as $(u, v) \rightarrow (u', v')$. The logarithmic singularity of the Green's function is explicitly extracted by this method. The remainder terms of G_0 and the terms of G_1 involve sums over m of terms of the form $e^{-maz_i}/(ma)^\ell$, where z_i is a complex number whose real and imaginary parts are linear functions of u, u', v, and v'. The series associated with these terms can be evaluated efficiently using the functions listed in Table 2.

This method has been used to obtain explicit formulas for the rectangular patches and the separable patches in the circular-cylinder coordinate system. This analysis can also be applied to the class of patches separable in parabolic- and elliptic-cylinder coordinate systems.

Rectangular and Circular Geometries—For this class of patches the expressions listed above simplify considerably because $(\alpha_m)^2$ can be chosen as a sequence *independent* of k as can the corresponding sequence of eigenfunctions U_m. In this case $U_{nm} \equiv 0$ and $(\alpha_{nm})^2 = 0$ for $n > 0$. For all patches in this class the large-m form of the summands r_{0m} and r_{1m} can be written as

$$r_{0m} \sim \tilde{r}_{0m} = D_1\frac{1}{m}\left\{\sum_{\ell=1}^{3}e^{-ma\sigma_\ell}\right\}U_m(u, u')$$

$$r_{1m} \sim \tilde{r}_{1m} = \frac{1}{m^2}\left\{\sum_{\ell=1}^{3}\left(D_{2\ell} + \frac{1}{m}D_{3\ell}\right)e^{-ma\sigma_\ell}\right\}U_m(u, u')$$

where

$$D_1 = -\frac{1}{a\Delta u}$$

$$D_{i+1,\ell} = -\frac{4}{a\Delta u}\left[C_{i\ell} + \frac{B_{i-1}}{2a}\right], \qquad i = 1, 2; \quad \ell = 1, 2, 3$$

The C coefficients are given by

Table 2. Definition, Properties, and Approximations of $F_n(z)$

n	A	B	$F_n(z)$		
			$0 \leqq \mathscr{R}z \leqq A$ $0 \leqq \mathscr{I}z \leqq B$	$0 \leqq \mathscr{R}z \leqq A$ $B < \mathscr{I}z \leqq \pi$	$\mathscr{R}z \geqq A$
n			$\displaystyle\sum_{k=1}^{8} e^{-kz}/k^n, \qquad F_n(z+j2\pi) = F_n(z), \qquad F_n^*(z) = F_n(z^*)$		
1			$z/2 - \ln z - \ln\left[\dfrac{\sinh(z/2)}{z/2}\right]$		
2	1.0	2.2	$z\ln z + 1.64493 - z - \dfrac{1}{4}z^2 + \dfrac{1}{72}z^3$ $- \dfrac{1}{14400}z^5$	$-0.82247 + 0.69315w$ $- \dfrac{1}{4}w^2 + \dfrac{1}{24}w^3 - \dfrac{1}{960}w^5;$ $w = z - j\pi$	$e^{-z} + \dfrac{1}{4}e^{-2z} + \dfrac{1}{9}e^{-3z} + \dfrac{1}{16}e^{-4z}$
3	0.6	2.0	$-0.5z^2\ln z + 1.20206 - 1.64493z$ $+ \dfrac{3}{4}z^2 + \dfrac{1}{12}z^3 - \dfrac{1}{288}z^4$	$-0.90154 + 0.82247w - 0.34657w^2$ $+ \dfrac{1}{12}w^3 - \dfrac{1}{96}w^4;$ $w = z - j\pi$	$e^{-z} + \dfrac{1}{8}e^{-2z} + \dfrac{1}{27}e^{-3z} + \dfrac{1}{64}e^{-4z}$

$$C_{11} = \frac{1}{2}[A_1^+ + A_1^-], \qquad C_{21} = \frac{1}{2}[A_2^+ + A_2^-], \qquad C_{12} = \frac{1}{2}[A_1^+ - A_1^-]$$

$$C_{22} = C_{23} = C_{21}, \qquad C_{13} = -C_{12}$$

and the A and B coefficients are listed in Table 3 for rectangular and annular patches. (These coefficients appear in the large-m asymptotic formulas of the Vs and the Ws, respectively.) The function $U_m(u, u')$ is

$$U_m(u, u') = \begin{cases} \cos(mau)\cos(mau') & \text{for rectangular, annular-sector,} \\ & \text{and circular-sector patches} \\ \cos[ma(u - u')] & \text{for circular and annular disk patches} \end{cases}$$

The summation of the r_{ij}s can be performed using

$$\sum_{m=1}^{\infty} r_{0m} = \sum_{m=1}^{\infty} [r_{0m} - \tilde{r}_{0m}] + D_1 \sum_{\ell=1}^{3} \mathscr{S}_{1\ell}$$

$$\sum_{m=1}^{\infty} r_{1m} = \sum_{m=1}^{\infty} [r_{1m} - \tilde{r}_{1m}] + \sum_{\ell=1}^{3} [D_{2\ell}\mathscr{S}_{2\ell} + D_{3\ell}\mathscr{S}_{3\ell}]$$

where

$$\mathscr{S}_{i\ell} = \frac{1}{\tau} \mathscr{R} \sum_{k=1}^{\tau} F_i(az_{k\ell})$$

$$\tau = \begin{cases} 1 & \text{for annular and circular patches} \\ 2 & \text{for rectangular, circular-sector, and annular-sector patches} \end{cases}$$

Table 3. Asymptotic Expansion Coefficients for Rectangular and Annular Patches

	Rectangular	Annular
A_1^+	$-\dfrac{1}{4a}(b - v)$	$-\dfrac{a^2}{8a}(1 - e^{2v})$
A_2^+	0	$\dfrac{1}{a}A_1^+$
A_1^-	$-\dfrac{1}{4a}v$	$\dfrac{1}{8a}(b^2 - a^2 e^{2v})$
A_2^-	0	$-\dfrac{1}{a}A_1^-$
B_0	$\dfrac{b}{4}$	$\dfrac{a^2 - b^2}{8}$
B_1	$\dfrac{1}{4a}$	$\dfrac{a^2 + b^2}{8a}$

$$z_{k\ell} = (\sigma_\ell + j\omega_k)$$

$$\omega_1 = u - u'$$

$$\omega_2 = u + u'$$

$$\sigma_1 = v_> - v_<$$

$$\sigma_2 = v_> + v_< - 2v_-$$

$$\sigma_3 = 2v_+ - (v_> + v_<)$$

$$F_i(z) = \mathcal{R} \sum_{m=1}^{\infty} \frac{e^{-mz}}{m^i}$$

The expressions listed in Table 2 can be used to efficiently evaluate $F_i(z)$. The remaining infinite sums of the difference between the exact and the asymptotic r_{ij}s converge in just a few (three or four) terms. The parameters needed to construct the exact r_{0m} and r_{1m} can be obtained from Table 4 and Table 5. To obtain the appropriate formulas for circular and circular-sector patches, one simply sets b to zero in the formulas for the annular patch, although this must be done taking appropriate limits in the final expression for the quasi-static terms. One could let b/a be a sufficiently small ratio, such as 10^{-6}, to obtain results for circular and circular-sector patches using the same computer program as for the annular and annular-sector patch. The parameter α used in these tables is

$$\alpha = \begin{cases} 1 & \text{for circular and annular disks} \\ \pi/\phi_0 & \text{for circular- and annular-sector patches (ϕ_0 is the sector angle)} \\ \pi/a & \text{for rectangular patches} \end{cases}$$

Multiport Impedance Parameters—Equation 4 can be used for computing Z_{ij} if the accelerated expression for G is used. To find the driving-point impedance one must average the singular part of the Green's function over the source current distribution. Simply setting $u_2 = u_1$ and $v_2 = v_1$ will yield an infinite inductive part of the impedance. On the other hand, averaging over the nonsingular parts of the Green's function yields nearly the same result as evaluating those parts at the center of the source distribution. Suppose the element is being fed with a coaxial probe of diameter d centered at (u_1, v_1). Then the driving-point impedance is obtained by

$$Z_{11} = \overline{G}(u_1, v_1 \,|\, u_1, v_1)$$

where $\overline{G}$ is obtained from G in (6) simply by replacing the z_{11} in $F_1(\alpha z_{11})$ in the expression for G_0 by $d/2$. If the feed distribution is a current strip of width w (such as when one feeds with a microstrip line), replace d by $w/2.24$ in the preceding expression.

5. Resonant Frequency

There are a number of different theories available for computing the resonant frequency of a microstrip element. One of the simplest, though not the most

Table 4. Solutions to the v-Differential Equations for Rectangular and Annular Patches

	Rectangular	Annular
V_{0m}^+	$\cosh ma(b - v)$	$\cosh mav$
V_{1m}^+	$-\dfrac{1}{2ma}(b - v)\sinh[ma(b - v)]$	$\dfrac{a^2}{8}\left\{\left[\dfrac{ma - 2}{ma(ma - 1)} - \dfrac{e^{2v}}{ma + 1}\right]e^{mav} \right.$ $\left. -\left[\dfrac{ma + 2}{ma(ma + 1)} - \dfrac{e^{2v}}{ma - 1}\right]e^{-mav}\right\}, \qquad am \neq 1;$ $-\dfrac{a^2}{16}e^{3v} + \dfrac{a^2}{4}(1 - v)e^v - \dfrac{3a^2}{16}e^{-v}, \qquad am = 1$
V_{10}^+	$-\dfrac{1}{2}(b - v)^2$	$-\dfrac{a^2}{4}e^{2v} + \dfrac{a^2}{4} + v\dfrac{a^2}{2}$
V_{20}^+	$\dfrac{1}{24}(b - v)^4$	$\dfrac{a^4}{8}\left[\dfrac{e^{4v}}{8} + \dfrac{e^{2v}}{2} - ve^{2v} - \dfrac{v}{2} - \dfrac{5}{8}\right]$
V_{30}^+	$-\dfrac{1}{720}(b - v)^6$	$\dfrac{a^6}{64}\left[-\dfrac{e^{6v}}{36} - \dfrac{e^{4v}}{2} + \dfrac{ve^{4v}}{2} + ve^{2v} + \dfrac{e^{2v}}{4} + \dfrac{1}{6}v + \dfrac{5}{18}\right]$
V_{nm}^-	$V_{nm}^+(v + b)$	$\left(\dfrac{b}{a}\right)^{2n}V_{nm}^+[v - \ln(b/a)]$

Table 5. Wronskians for Rectangular and Annular Patches

	Rectangular	Annular
W_{0m}	$-ma\sinh mab$	$ma\sinh[ma\ln(b/a)]$
W_{1m}	$\dfrac{1}{2ma}\sinh(mab) + \dfrac{b}{2}\cosh(mab)$	$-\dfrac{1}{8}\left\{\left[\dfrac{ma - 2}{ma - 1}b^2 - \dfrac{ma + 2}{ma + 1}a^2\right]e^{-ma\ln(b/a)} \right.$ $\left. +\left[\dfrac{ma + 2}{ma + 1}b^2 - \dfrac{ma - 2}{ma - 1}a^2\right]e^{ma\ln(b/a)}\right\}, \qquad ma \neq 1;$ $-\dfrac{1}{4}ab\ln\left(\dfrac{b}{a}\right) + \dfrac{3}{16}\left[\dfrac{a^3}{b} - \dfrac{b^3}{a}\right], \qquad ma = 1$
W_{10}	b	$\dfrac{1}{2}(a^2 - b^2)$
W_{20}	$-\dfrac{1}{6}b^3$	$-\dfrac{a^4 - b^4}{16} - \dfrac{a^2b^2}{4}\ln\left(\dfrac{b}{a}\right)$
W_{30}	$\dfrac{1}{120}b^5$	$\dfrac{a^6 - b^6}{384} + \dfrac{3a^2b^2(a^2 - b^2)}{128} + \dfrac{a^2b^2(a^2 + b^2)\ln(b/a)}{32}$

accurate, is that provided by the cavity model. As discussed in Section 2, under "Cavity Model," the field under the microstrip element is distributed in about the same way as the field in the corresponding magnetic-walled cavity. The resonant wave number k_{mn} of the cavity mode $\psi_{mn}(u, v)$ is approximately the same as that of the corresponding antenna. The characteristic equation for this wave number is obtained from

$$\nabla \psi_{mn}(u, v) \cdot \hat{\mathbf{n}} = 0$$

for (u, v) on the edge of the patch. In terms of the functions used in Section 4, under "Overview," this characteristic equation is

$$\frac{\partial}{\partial v} V_m^-(v_+) = 0$$

where the equation $v = v_+$ describes at least a portion of the edge of the patch in the curvilinear coordinate system (u, v). See Table 1 for a list of characteristic equations for various patch shapes in the rectangular and circular classes.

The resonant frequency predicted by this equation has an accuracy typically in the range of from approximately 0 to 17 percent, depending on the particular shape of patch and mode being used and the thickness of the microstrip element [23]. The thicker the element, the larger the error is. The bandwidth of a thin microstrip antenna is small; designs based solely on the predictions of the cavity model may yield actual operating bands that completely miss the desired resonant frequency. Consequently, the designer must either use a more accurate theory, or actually build a prototype and subsequently adjust the size of the element. Some simple modifications to the cavity-model formulas for rectangular and circular disk elements are listed in the following subsections. More accurate theories for predicting resonant frequency can be found in the References. The option of building a prototype is not difficult. Any complicated feed circuitry can usually be replaced by a very simple feed.

The following five subsections present formulas and results that pertain to rectangular, circular-disk, circular-sector, annular, and annular-sector microstrip elements. The tables, formulas, and graphs are in terms of *normalized resonant frequency* f_n. To obtain a cavity-model estimate of the resonant frequency f_{res} (in megahertz) for your patch, use the formula

$$f_{res} = \frac{4775}{(\epsilon_r \mu_r)^{1/2}} \frac{1}{L} f_n$$

where L is the "characteristic length" (in centimeters) given in the following sections.

Rectangular Patch

See Table 1 for the geometrical parameters for this and the following patches. The characteristic length is $L = a$. The normalized resonant frequency of the (m, n)th mode is

$$f_n = \left[(m\pi)^2 + \left(\frac{n\pi a}{b} \right)^2 \right]^{1/2}$$

One can usually obtain a much better estimate [23] of the resonant frequency if one replaces a and b by $a + 2\Delta L$ and $b + 2\Delta L$, where ΔL is given in (1) in Section 2.

Circular-Disk Patch

The characteristic dimension is the disk radius, a. The normalized resonant frequencies of the (m, n)th modes are listed in Table 6. A better estimate for resonant frequency can often be obtained if one replaces a by [24]

$$a\left[1 + \frac{2t}{\pi a \epsilon_r}\left(\ln\left(\frac{\pi a}{2t}\right) + 1.7726\right)\right]^{1/2}$$

Circular-Sector Patch

The characteristic dimension is the sector radius, a. The normalized resonant frequencies for the (m, n)th modes as a function of sector angle ϕ_0 can be read off of the graph in Fig. 18. The resonant frequencies of the $(0, n)$ modes are the same as those of the circular disk. Note that the resonant frequency of the (sm, n)th mode of a patch with a sector angle of $s\phi_0$, for $s = 1, 2, \ldots, \ell$ such that $\ell\phi_0 < 2\pi$, has the same resonant frequency as the (m, n)th mode of a patch with a sector angle of ϕ_0.

Annular Patch

The characteristic dimension is the outer radius, a. The normalized resonant frequencies for the $(m, 1)$th modes as a function of b/a can be obtained from the graph in Fig. 19.

Annular-Sector Patch

The characteristic dimension is the outer radius, a. The normalized resonant frequency for the $(1, 1)$th mode as a function of b/a and sector angle ϕ_0 can be obtained from the graphs in Fig. 20. The resonant frequency of the $(0, n)$ mode is the same as that for the annulus. Also, the same comment under "Circular-Sector Patch" above about the equivalence involving the (sm, n)th mode holds here as well.

Table 6. Normalized Resonant Frequencies for a Circular-Disk Microstrip Antenna

Mode		Resonant Frequency		
m \ n	0	1	2	3
0	0	3.832	7.016	10.173
1		1.841	5.331	8.536
2		3.054	6.706	9.969
3		4.201	8.015	11.346

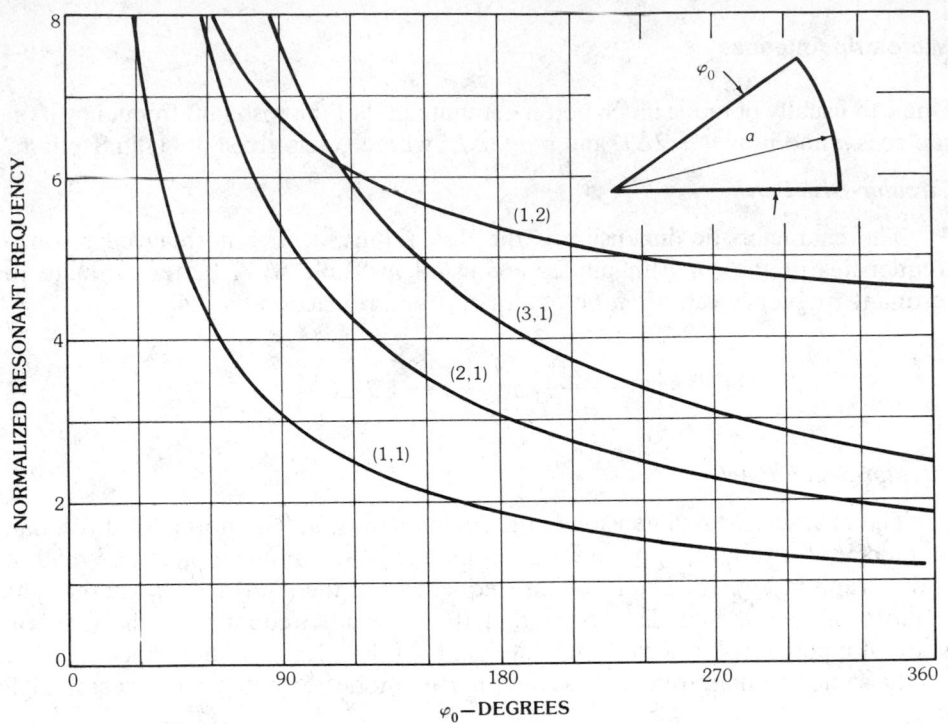

Fig. 18. Normalized resonant frequency of the $(1,1)$, $(2,1)$, $(3,1)$, and $(1,2)$ modes of a circular-sector microstrip element versus sector angle ϕ_0.

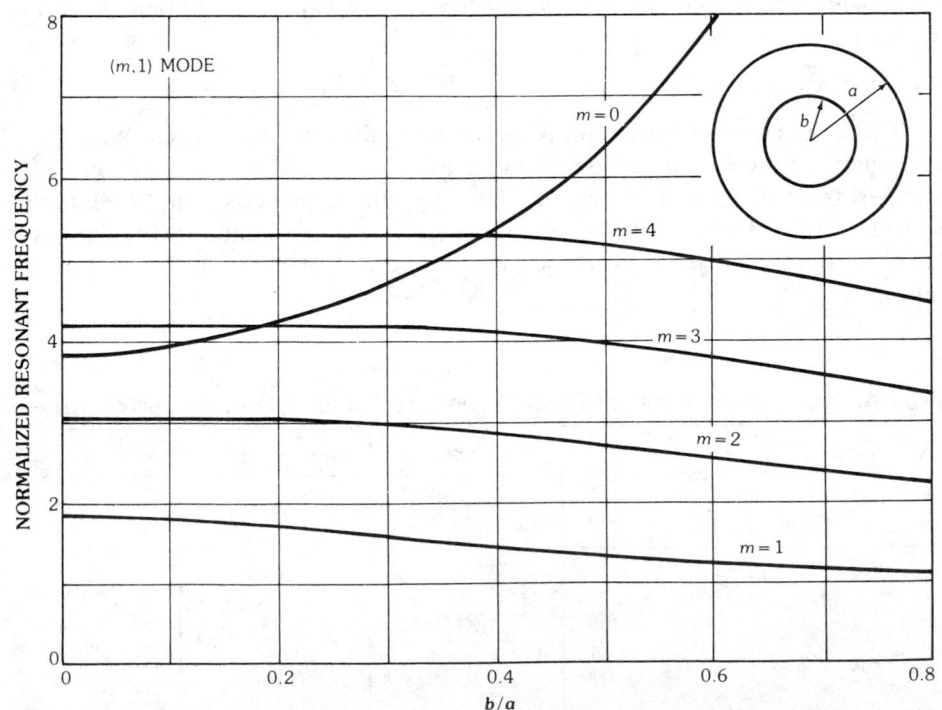

Fig. 19. Normalized resonant frequency of the $(m,1)$ mode of an annular disk microstrip element versus b/a for $m = 0, 1, 2, 3, 4$.

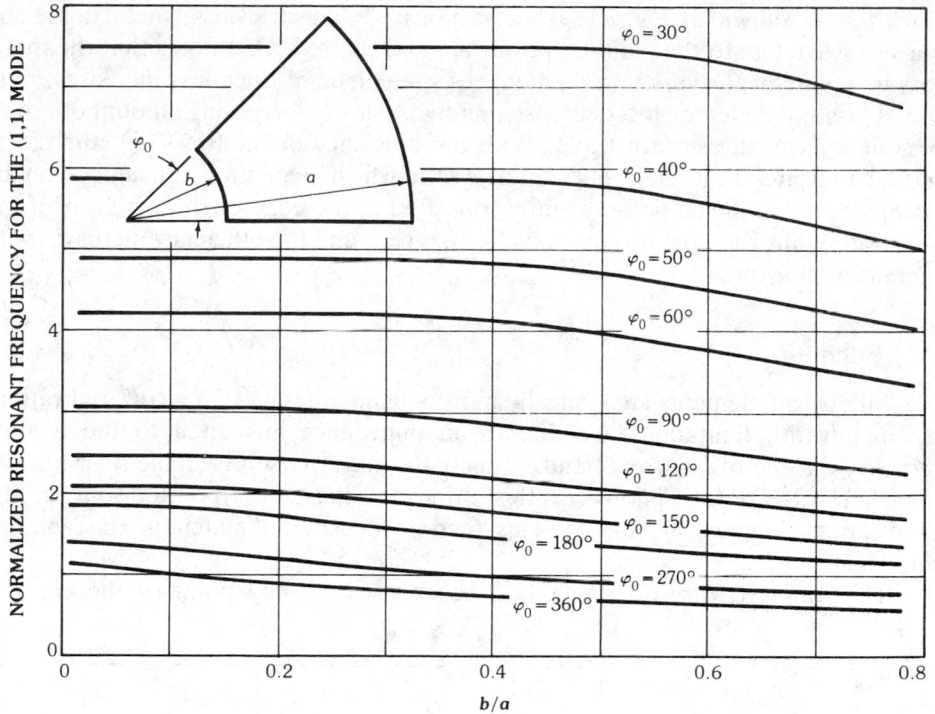

Fig. 20. Normalized resonant frequency of the $(1, 1)$ mode of an annular-sector microstrip element versus b/a for various sector angles ϕ_0.

6. Efficiency

The efficiency of a microstrip element is given by

$$\text{efficiency} = \frac{1/Q_r}{1/Q_r + 1/Q_{sw} + \delta + \Delta/t} \tag{10}$$

where Q_r is the radiative (space-wave) quality factor, Q_{sw} is the surface-wave quality factor, δ is the dielectric loss tangent of the substrate, and Δ is the skin depth in the patch and ground plane. The quality factors can be found in the way described in Section 2, under "Cavity Model." The dielectric loss tangent is usually available from the printed-circuit board manufacturer although the *effective* conductivity of the metal cladding from which Δ can be computed usually is not quoted. An optimistic guess for this conductivity is the conductivity of the pure metal cladding itself. This is usually too high because of the surface roughness of the cladding. A simple experimental method [25] of estimating these parameters from a sample of printed circuit board is available.

The surface-wave quality factor was excluded from the numerator of (10) since the power carried by the surface wave is ultimately scattered at the truncation of the dielectric substrate and combines with the space wave in a way that is not easy to control. A plot based on the cavity model of the efficiency *versus* dielectric thickness is shown in Fig. 21. The fractions of dielectric loss, metal loss, and surface-wave loss to the radiated power are also plotted. One notes that the metal loss is dominant for very thin elements. As the element thickness increases, both metal loss and dielectric loss decrease rapidly. A slowly increasing amount of power is contained in the surface wave. With the efficiency defined by (10), the cavity model indicates that there is a thickness at which maximum efficiency can be obtained. If the dielectric is abruptly truncated at the edge of the patch, or if the substrate is air, then the surface-wave term is zero and the efficiency increases with increasing thickness.

7. Matching

Microstrip elements are typically narrow-band antennas. In most applications the bandwidth limitations are due to an impedance mismatch to the feeding circuitry outside of a narrow band. Usually the pattern itself is stable over a wider band. There are exceptions to this principle; these are pointed out in the applications section, Section 9. This section focuses on matching the antenna impedance.

The simplest form of matching is to choose the feed point of the element

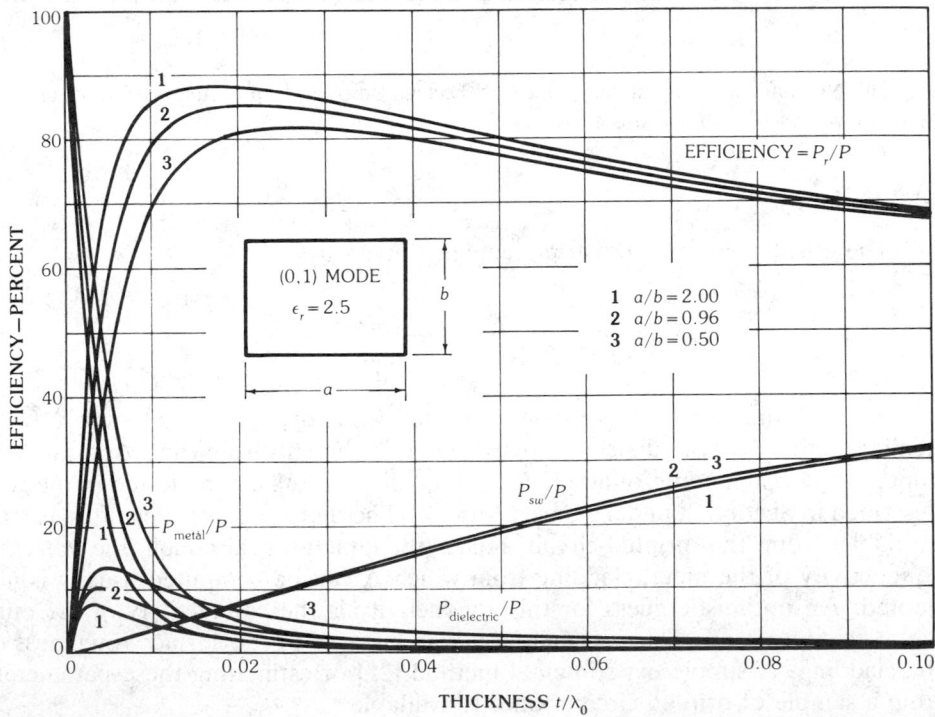

Fig. 21. Plot of efficiency and percent power loss contributions from metal-conduction, dielectric-heating, and surface-wave loss components versus thickness for a rectangular microstrip element with a relative dielectric constant of 2.5 and with three different aspect ratios a/b.

wisely. It was pointed out in Section 4, under "Resonant Impedance," that the resonant resistance of the antenna varies with feed point [see (4)]. By choosing the feed location (u', v') where the resonant resistance is equal to the feed-line impedance (or where it is slightly larger), one can match the element over a limited band. The bandwidth is inversely proportional to the quality factor of the antenna.

This optimum feed location is not unique. For some modes of some elements a suitable feed point can be found on the edge of the patch. For example, one can excite the $(0, 1)$ mode of a rectangular element as illustrated in Fig. 22a and Fig. 22b with the same effect. In other elements, such as a simple, unloaded circular disk, it is impossible to find an optimal feed point on the edge. One must feed at a point interior to the patch. This can be done by using a coaxial feed. If a microstripline feed is desired, then one can feed the element as illustrated in Fig. 22c. Even if it is possible to feed on an edge it still sometimes preferable to feed as in Fig. 22c. Such symmetrical feeding can reduce the excitation of nonresonant modes which can cause a small cross-polarized component of field. The slots cut to inset the microstrip feed should be aligned with the lines of current of the mode being excited to minimize disturbing the modal distribution. The amount that the maximum input resistance is reduced by using this feeding technique will depend on the width of the slots. Other techniques for feeding the element are possible besides coaxial feeds and microstripline feeds. The reader is referred to Section 10, "References," for these.

The matching technique described above does not extract the largest possible bandwidth from the element. It is important for the designer to recognize that there are fundamental theoretical limitations [26] to the bandwidth achievable from

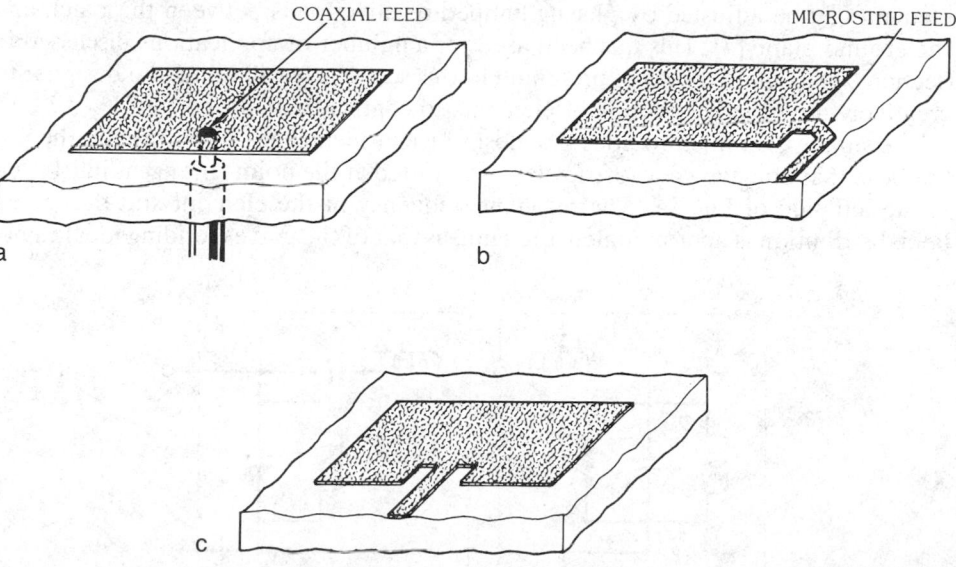

Fig. 22. Three simple matching schemes for a rectangular microstrip element. (*a*) Inset coaxial feed. (*b*) Microstripline feed on nonradiating sides of a patch driven in the $(0, 1)$ mode. (*c*) Inset microstripline feed.

these elements using lossless matching networks. Nevertheless, for a 2:1 vswr a bandwidth improvement by a factor of 3.86 is theoretically possible and an improvement by a factor of 3.18 is practical [8]. This bandwidth improvement can be obtained by feeding through an external matching network. For example, a network as illustrated in Fig. 23, with properly selected element values, will increase the best 2:1 vswr, bandwidth of the patch element alone by a factor of about 2.84. Adding another series *LC* pair can raise this factor to 3.18. There is little point to using more than three *LC* pairs of matching elements.

The section of the network in Fig. 23 that is enclosed by the dashed line represents the model of the input impedance of the microstrip element. One notes that part of the inductance required by the first matching section is supplied by the microstrip element itself. This inductive part can be significant, particularly for thicker elements and elements fed at high-inductance points.

The price for this improvement is a reduction in overall efficiency due to unavoidable losses in the matching filter. A greater price is the space that the matching network must occupy. The network can be realized using microstripline or stripline elements. If the latter is used, the network can be placed in a second layer of the circuit board under the patch.

How to actually choose the element values and how to realize them using microstripline or stripline circuits does not fall within the scope of this chapter. The reader is directed to look under the "Matching" category (17) of the References for publications on matching circuit design.

8. Loaded Microstrip Elements

The behavior (pattern, impedance, resonant frequency) of a microstrip element can be adjusted by placing lumped reactive loads between the patch and the ground plane [4]. This has been used for a number of applications discussed in Section 9. A simple theory is presented in this section. It will allow the designer to creatively use reactive loading of elements to control their properties.

A simple example provides the most efficient means of presenting the theory. Suppose that one has a microstrip element shorted at the point (u_1, v_1) as illustrated at the left side of Fig. 24. The resonant frequency of the element and the modal field distribution is approximately the same as that of the corresponding ideal cavity

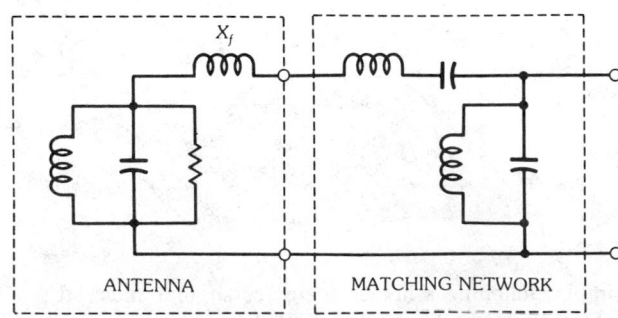

Fig. 23. External matching circuit for a microstrip element using two matching sections.

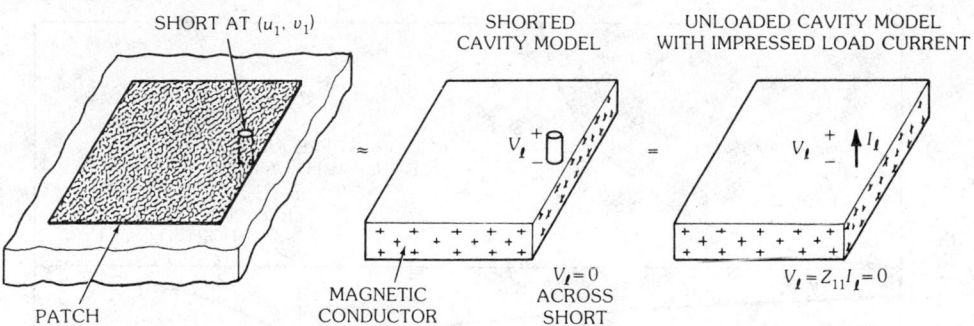

SHORT AT (u_1, v_1) SHORTED CAVITY MODEL UNLOADED CAVITY MODEL WITH IMPRESSED LOAD CURRENT

PATCH   MAGNETIC CONDUCTOR $V_\ell = 0$ ACROSS SHORT $V_\ell = Z_{11} I_\ell = 0$

Fig. 24. Modeling a loaded microstrip element. Shorted element (left) is modeled by shorted cavity (middle) and then by an *empty* cavity (right) driven by the load current I_ℓ.

model of the antenna (the middle of Fig. 24). The modal field is the source-free field of the shorted cavity. Associated with this source-free field is an induced electric current I_ℓ on the surface of the shorting probe. The equivalence principle allows the removal of this probe as long as its surface current is impressed in its place (the right side of Fig. 24). The result is an *empty* (i.e., unloaded cavity) element with an impressed electric source current. The voltage induced by this current must be $Z_{11}(f) I_\ell$, where f is the frequency. Since the cavity was shorted at this point, this voltage must vanish. Thus, at the frequency of the source-free mode, $Z_{11}(f)$ must be zero. That is, the resonant frequencies of the shorted cavity are the zeros of the driving-point impedance.

As discussed in Section 4, under "Simple Feed Models," the driving-point impedance Z_{11} can be written as the sum of a resonant impedance, jX_{res}, and a feed-reactive term, jX_f. (In this analysis the effective loss tangent is not introduced since the modal distribution and resonant frequency are not strongly dependent on this parameter. Thus Z_{11} in this context is purely reactive.) The characteristic equation for the resonant frequency is

$$X_{res} = -X_f$$

Fig. 25 is a plot of X_{res} and $-X_f$ versus frequency. The resonant frequency of the shorted element is at the intersection of these two curves. Since X_f is typically inductive the effect of shorting the cavity is to raise its resonant frequency.

The modal field distribution Φ_{MN} of the loaded element is

$$\Phi_{MN}(u, v) = \left\{ \sum_A \frac{\psi_{mn}(u, v)\, \psi_{mn}(u_1, v_1)}{(k_{\ell MN})^2 - (k_{mn})^2} \right\} \bigg/ \left\{ \sum_A \frac{[\psi_{mn}(u_1, v_1)]^2}{[(k_{\ell MN})^2 - (k_{mn})^2]^2} \right\}^{1/2} \quad (11)$$

where $k_{\ell MN}$ is the resonant wave number of the loaded element. This distribution is normalized so that the integral of the square of $\Phi_{MN}(u, v)$ over the patch is unity. The series in the numerator of (11) is best computed using the accelerated expressions for $G(u, v \,|\, u', v')$ found in Section 4, under "Efficient Computation of Impedance Parameters." The Q of the loaded antenna can be found from Φ_{MN} in the way described in Section 2, under "Cavity Model." From this Q, an effective

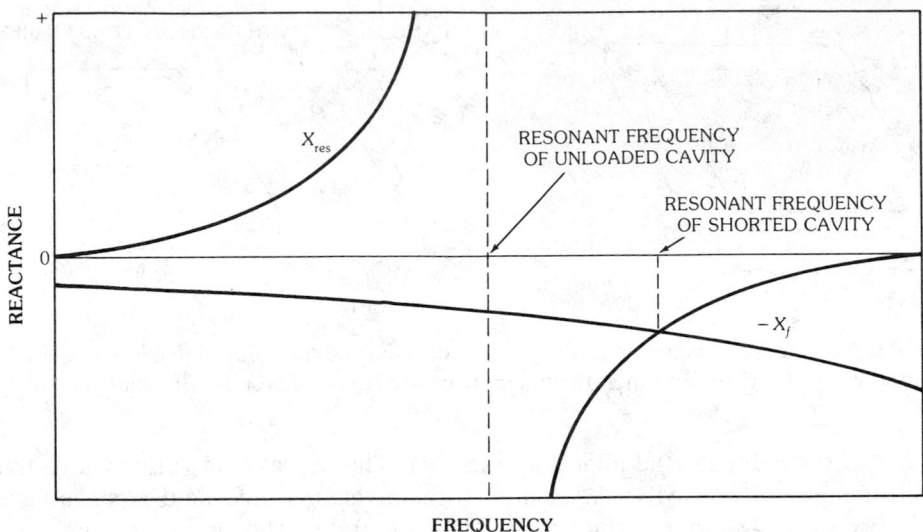

Fig. 25. Plot of X_{res} and $-X_f$ versus frequency. Intersection of the curves shows the resonant frequency of a shorted element.

loss tangent of the loaded element can be found and the resonant part of the impedance of the loaded element can be determined. The expression for this is

$$Z_{\text{res}} = -jk\eta t \frac{[\Phi_{MN}(u_2, v_2)]^2}{k^2(1 - j/Q) - (k_{\ell MN})^2} \tag{12}$$

where (u_2, v_2) is the location of the feed point of the loaded element. The feed-reactive part of the input impedance at the resonant frequency of the loaded element is given by

$$X_f = \text{Im}\left\{ -\frac{(Z_{12})^2}{Z_{11}} + Z_{22} \right\}$$

where Z_{11}, Z_{12}, and Z_{22} are the z parameters of the unloaded cavity. A typical input impedance and plot of the edge magnetic current distribution for a shorted element is shown in Fig. 26.

If the load is not a short but a reactive load, X_ℓ, then the characteristic equation for the resonant frequency is just

$$X_{\text{res}} = -X_f - X_\ell \tag{13}$$

The normalized loaded-element resonant mode is still given by (11) and the corresponding resonant part of the input impedance is still given by (12). The feed-reactive part of the input impedance to the loaded element, $X_{\ell f}$, is given by

$$X_{\ell f} = \text{Im}\left\{ \frac{-(Z_{12})^2}{Z_{11} + jX_\ell} + Z_{22} \right\}$$

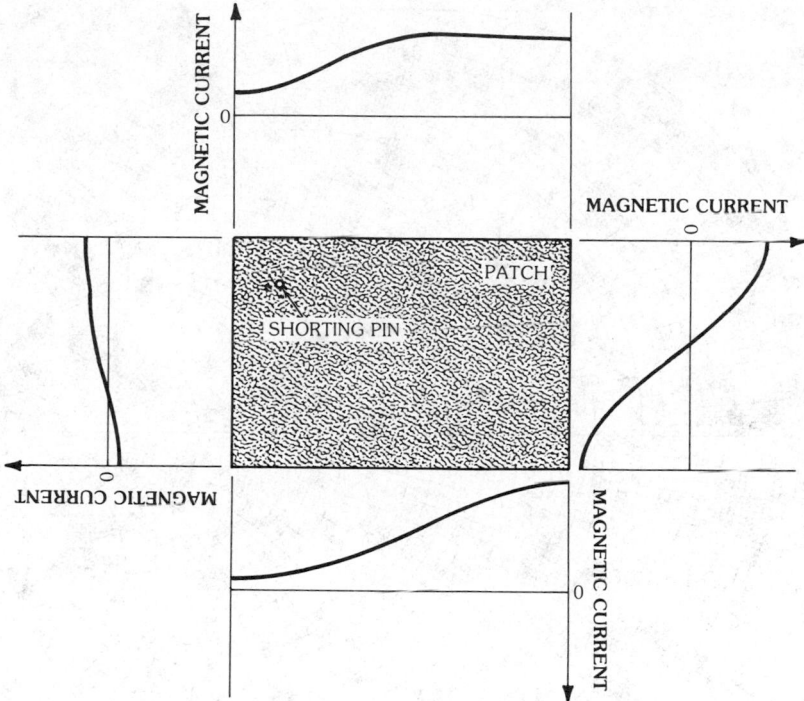

a

Fig. 26. The edge magnetic current distribution and input impedance at two feed locations of a shorted rectangular element of dimensions $11.43 \times 7.62 \times 0.152$ cm with relative dielectric constant of 2.43 (*a*) Computed magnetic current distribution for a short located at coordinates $(1.52, 6.10)$ cm. (*b*) Computed and measured input impedance with a feed located at coordinates $(2.54, 1.27)$ cm. (*c*) Computed and measured input impedance with a feed located at coordinates $(8.89, 6.35)$ cm. (*After Richards and Lo [4], reprinted with permission of Hemisphere Publishing Corporation*)

If there are L loads with reactances $X_1, X_2, \ldots, X_L$, located at (u_1, v_1), $(u_2, v_2), \ldots, (u_L, v_L)$, then the characteristic equation for the resonant frequency is that the determinant of the matrix

$$\mathbf{C} = \begin{bmatrix} jX_1 + Z_{11} & Z_{12} & \ldots & Z_{1L} \\ Z_{12} & jX_2 + Z_{22} & \ldots & Z_{2L} \\ \vdots & & & \\ Z_{1L} & Z_{2L} & \ldots & jX_L + Z_{LL} \end{bmatrix}$$

must vanish. To find the normalized resonant mode of this general loaded element, first compute

$$\mathbf{I} = -\mathbf{A}^{-1}\mathbf{B}$$

where $\mathbf{I}$ is the $(L-1)$-tuple $(I_2, I_3, \ldots, I_L)$, $\mathbf{A}$ is the $(L-1) \times (L-1)$ matrix obtained from $\mathbf{C}$ by removing its first row and column, and $\mathbf{B}$ is the $(L-1)$-tuple $(Z_{12}, Z_{13}, \ldots, Z_{1L})$. Then one must form the series

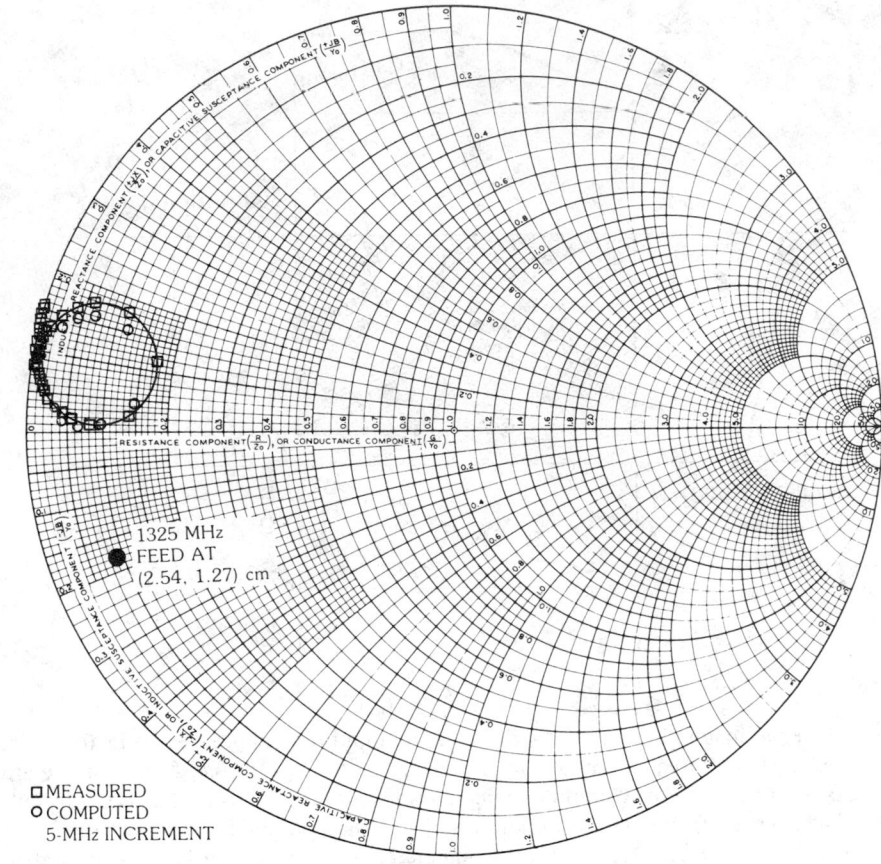

1325 MHz
FEED AT
(2.54, 1.27) cm

□ MEASURED
○ COMPUTED
5-MHz INCREMENT

b

Fig. 26, *continued.*

$$\Psi_{mn} = \psi_{mn}(u_1, v_1) + \sum_{i=2}^{L} I_i \psi_{mn}(u_i, v_i)$$

The normalized resonant mode of the loaded cavity is then given by (11) with the $\psi_{mn}(u_1, v_1)$ replaced by Ψ_{mn}. The resonant part of the input impedance of this multiply loaded element is again given by (12). Its feed-reactive part at resonance is

$$X_f = Z_{L+1,L+1} - \mathbf{Z}^t \mathbf{C}^{-1} \mathbf{Z}$$

where Z is the L-tuple $(Z_{1,L+1}, Z_{2,L+1}, \ldots, Z_{L,L+1})$. The $Z_{i,L+1}$s are the z parameters of the unloaded element with ports at $(u_1, v_1), (u_2, v_2), \ldots, (u_L, v_L)$ and with the feed port at (u_{L+1}, v_{L+1}).

9. Applications

In this section a number of selected applications of microstrip elements are presented. To the extent possible, simple design formulas for the applications are given. More importantly, the physical concepts which support the applications

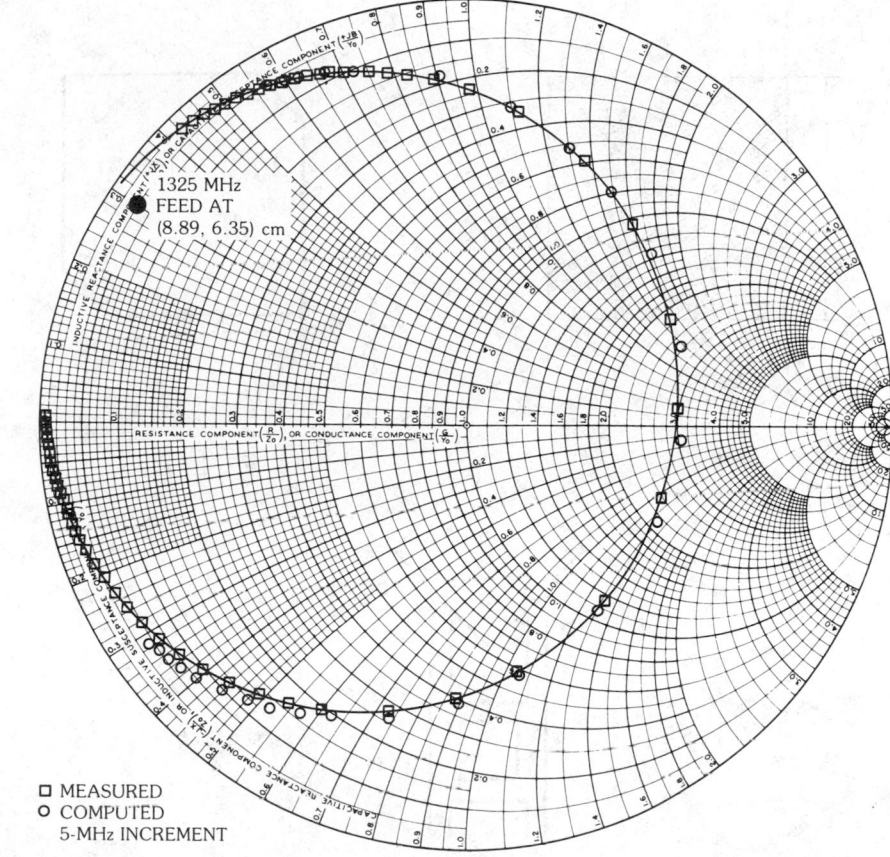

1325 MHz
FEED AT
(8.89, 6.35) cm

□ MEASURED
○ COMPUTED
5-MHz INCREMENT

C

Fig. 26, *continued.*

are presented to aid the reader in developing his or her own designs. Because of the physical insight afforded by it, and the relative simplicity of the theory, the design concepts are derived from the cavity model of the microstrip element.

Circular Polarization

There are two ways of achieving circular polarization (CP) from microstrip elements. The most conventional way, and the way that has the widest CP bandwidth, is to feed the element with two or more feeds in phase quadrature. It is somewhat surprising that under the conditions described in the following subsection one can feed the microstrip element with just a *single* feed and also obtain CP. While no quadrature hybrid circuits are required by this second method, the CP bandwidth is much smaller than even the impedance bandwidth of the element. Common to both methods of producing CP is the requirement that the element be of such a shape that it possesses *two* degenerate modes resonant at the CP operating frequency.

Single-Feed CP—To see how a single-feed, CP patch works, consider the example of a nearly square patch illustrated in Fig. 27. Assume that the dimensions *a* and *b*

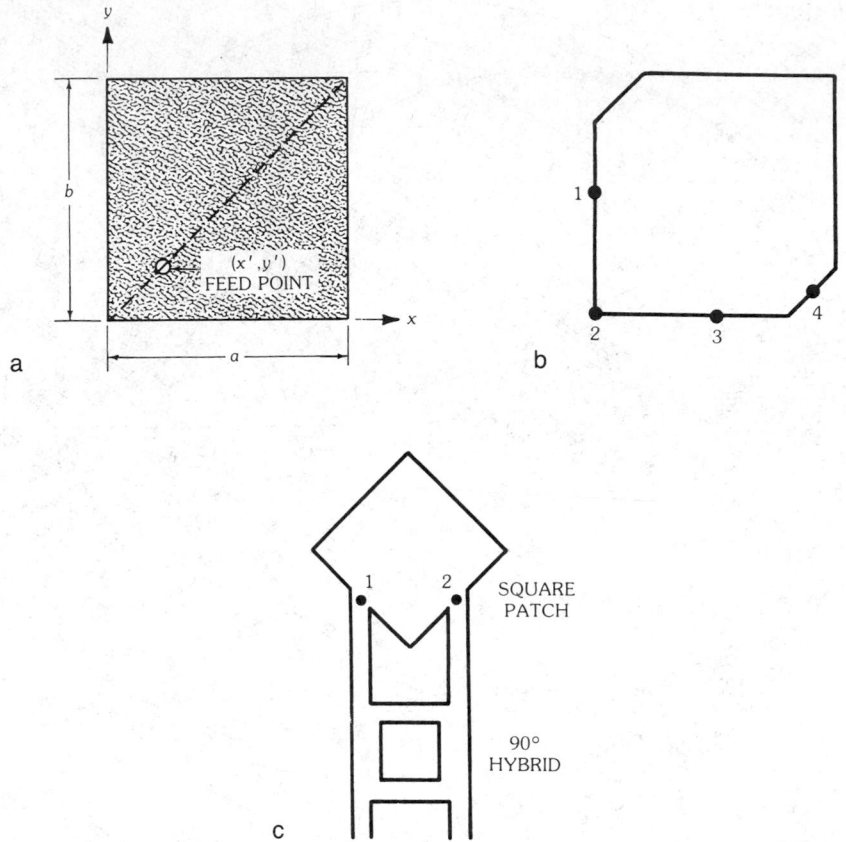

Fig. 27. Circular-polarization elements. (*a*) Nearly square element. (*b*) Square element perturbed at opposite corners. (*After Richards, Lo, and Harrison [5], © 1981 IEEE*) (*c*) Square element with quadrature feed.

of the patch are so nearly the same that the bands of the resonant $(0,1)$ and $(1,0)$ modes overlap significantly. In the broadside direction, the $(0,1)$ mode produces an electric far field E_y polarized in the y direction. The $(1,0)$ mode produces an x-polarized far field E_x. These fields satisfy the approximate proportionalities

$$E_x = \alpha \frac{\sin(\pi y'/b)}{k^2(1 - j/Q) - (k_{01})^2}, \qquad E_y = \alpha \frac{\sin(\pi x'/a)}{k^2(1 - j/Q) - (k_{10})^2}$$

The proportionality constant α is identical in both cases if the fields are measured at the same distance in the broadside direction. The effective loss tangent, $1/Q$, is the same in both cases since one of the two nearly degenerate modes is just the other one rotated by 90°. The terms $k_{01} = \pi/b$ and $k_{10} = \pi/a$ are the resonant wave numbers of the $(0,1)$ and $(1,0)$ modes. Suppose for the time being that the feed point (x',y') is along the diagonal of the element so that $y'/b = x'/a$. Then the ratio of the x and y components of the field is approximately

$$\frac{E_y}{E_x} \cong \frac{k(1 - j/2Q) - k_{10}}{k(1 - j/2Q) - k_{01}} \tag{14}$$

In order to achieve CP, the magnitude of this ratio should be 1 and its phase $\pm 90°$. Since Q is typically large for microstrip elements (at least 10 or larger in most cases), the complex phasor $1 - j/2Q$ is almost parallel to the real axis. A plot of the locus of the point $k(1 - j/2Q)$ in the complex plane as k varies over the very narrow band between k_{10} and k_{01} is illustrated in Fig. 28. One can see from this figure that the condition for CP will be met when the phasors in the numerator and the denominator of (14) have equal length and intersect at right angles. This picture immediately gives in geometrical terms the conditions under which this can occur: (1) The difference $k_{01} - k_{10}$ must be $\bar{k}/Q$. (2) The operating frequency must be midway between the resonant frequencies of the $(0, 1)$ and $(1, 0)$ modes. The first condition yields the simple formula

$$a = b(1 + 1/Q) \tag{15}$$

Feeding the antenna along the lower-left-corner-to-upper-right-corner diagonal will produce left-hand CP. Fed along the opposite diagonal, the element will produce right-hand CP.

This geometrical picture also quickly demonstrates why exciting CP in this way is extremely narrowband. For an axial ratio less than or equal to AR (specified in dB), the CP bandwidth in percent is approximately equal to

$$\text{percent bandwidth} \cong 12 \times AR/Q \tag{16}$$

Equations 15 and 16 have been found to work remarkably well even for elements with as low a value of Q as 10.

The results discussed above pertain to CP along the broadside direction. However, the CP actually remains quite good over a fairly large angular region, as the experimental results shown in Fig. 29 illustrate. These patterns were measured over a very large ground plane. If the ground plane is small, even better CP performance can be obtained.

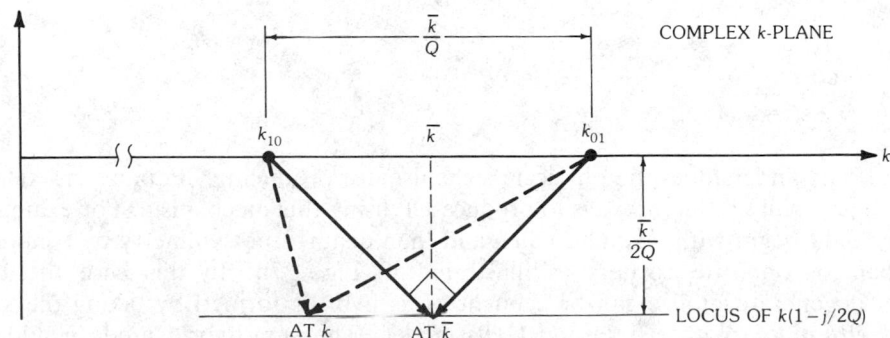

Fig. 28. Phasor diagram illustrating the conditions necessary for single-feed CP for a nearly square patch fed along a diagonal.

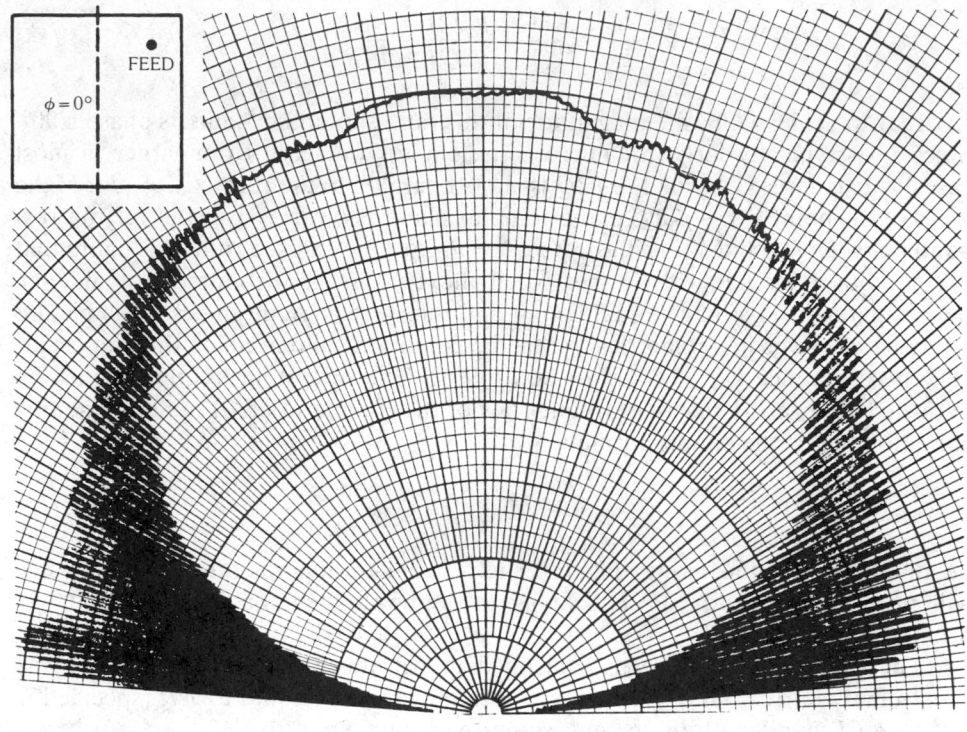

a

Fig. 29. Spinning-dipole CP patterns measured over a very large ground plane for a nearly square microstrip element. (*a*) Elevation pattern in the $\phi = 0°$ plane. (*b*) Elevation pattern in the $\phi = 90°$ plane. (*After Richards, Lo, and Harrison [5], © 1981 IEEE*)

One can also feed the element off the diagonal. In this case the ratio of the dimensions of the element are related by

$$a = b\left(1 + \frac{A + 1/A}{2Q}\right),$$

where

$$A = \frac{\cos(\pi y'/b)}{\cos(\pi x'/a)}$$

If one understands the physical mechanism for producing CP, one can see how there are many different ways to produce CP using this mechanism. For example, one could begin with a square patch and then disturb the symmetry by trimming a pair of opposite corners as illustrated in Fig. 27b. In this case the two near-degenerate modes can be thought of as hybrids formed by taking the sum and difference of the $(0,1)$ and $(1,0)$ modes. The sum-hybrid mode would be unaffected by the trimming while the difference-hybrid mode's resonant frequency would be raised slightly. By feeding at either point 1 or point 3 as indicated in

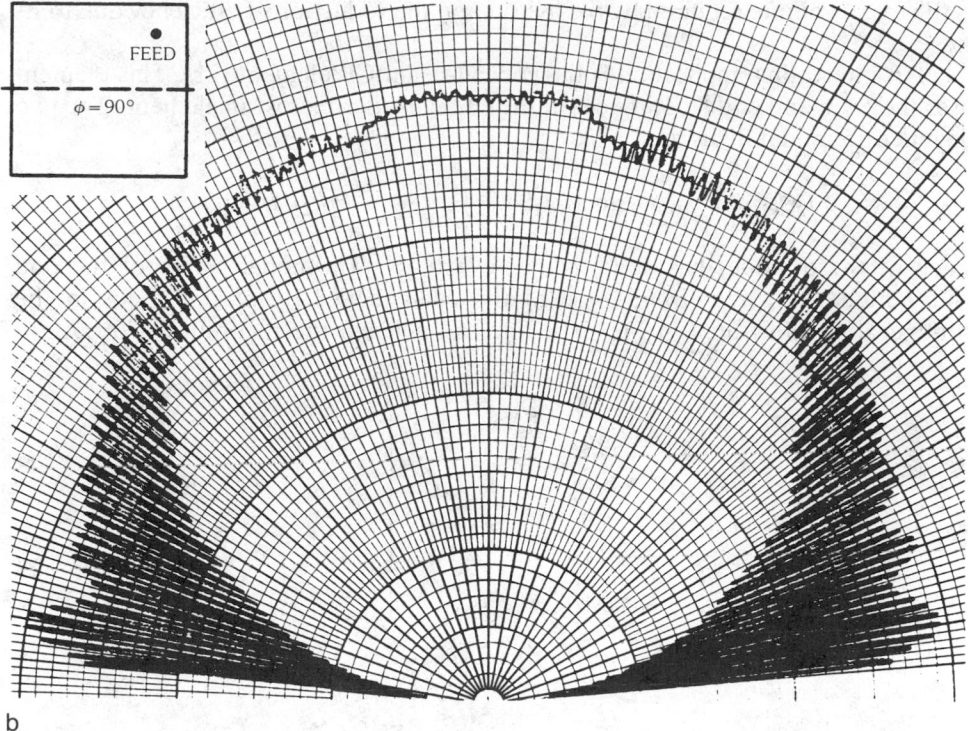

Fig. 29, *continued.*

Fig. 27b, one can achieve CP if the resonant wave numbers of the hybrid modes satisfy the conditions stated above. Similarly, one can achieve CP by disturbing the symmetry of a circular-disk element by adding tabs or by making it slightly elliptical. Adding loads to an otherwise symmetrical element can also be used to disturb the symmetry of otherwise precisely degenerate modes to produce CP. This latter technique is considered in more detail in this section, under "Polarization-Agile Elements."

Multiple-Feed CP Elements—One can generate the required phase quadrature between orthogonal far-field components by using external quadrature hybrids. In this case the element should support exactly degenerate modes instead of the nearly degenerate modes called for by single-feed CP elements. Each mode is excited independently by placing its respective feed at a node of the other mode. For example, Fig. 27c illustrates a dual-feed, square, CP microstrip element. Port 1 is located at a point where mode $(0, 1)$ is zero. Thus port 1 does not load or excite any part of mode $(0, 1)$. Similarly, the second feed port 2 is located where mode $(1, 0)$ is zero.

Because the bandwidth of the external phase shifter is typically larger than even the impedance bandwidth of the microstrip element itself, the CP bandwidth of this type of element will exceed the impedance bandwidth. The price, of course, is the extra volume required by the hybrid and the reduction of efficiency due to its losses.

An interesting application is the conical-beam CP element [21]. This element uses the even and odd $(2, 1)$ modes of the circular-disk microstrip element. The two degenerate modal fields are

$$J_2(k_{21}\varrho)\cos 2\phi \quad \text{and} \quad J_2(k_{21}\varrho)\sin 2\phi, \qquad k_{21} = 3.054/a$$

Each of these modes has nodal planes every $45°$. The nodal planes of the odd (sine) mode are $22\frac{1}{2}°$ from those of the even mode. By placing feeds in the configuration illustrated in Fig. 30a, one can achieve the theoretical CP pattern

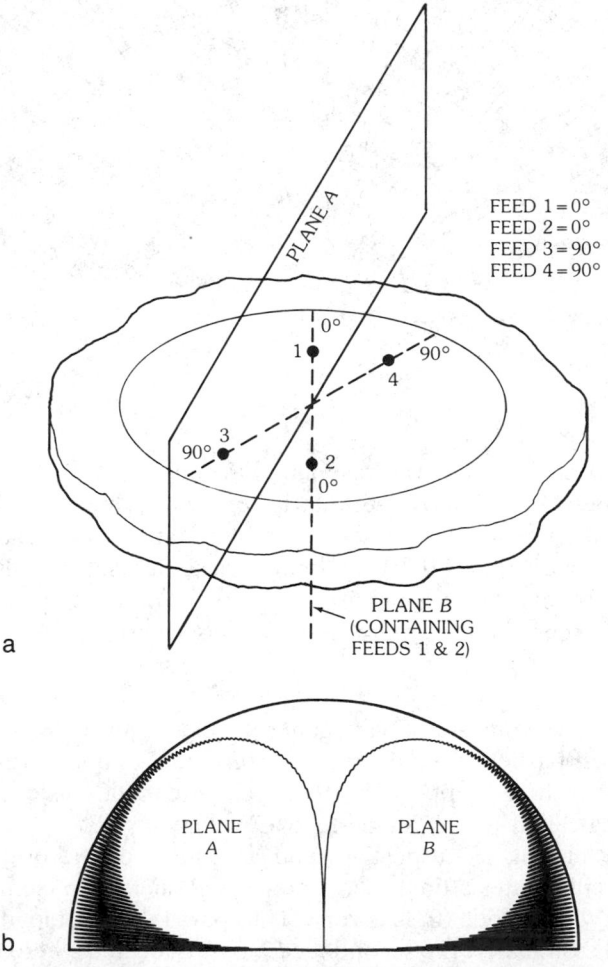

Fig. 30. Circular-disk antenna using degenerate $(2, 1)$ modes to produce a conical CP beam. (*a*) Element and feed phasing to suppress unwanted modes. (*b*) Computed spinning dipole pattern for an element mounted on an infinite grounded $0.01\lambda_0$-thick dielectric slab. The assumed relative dielectric constant is 2.5. Pattern is plotted on a 40-dB scale.

illustrated in Fig. 30b. Four feeds were used in this case to help suppress the excitation of other lower- and higher-order modes.

In principle, one could incorporate these techniques for obtaining CP with the dual-band techniques described next to produce dual-band CP elements.

Dual-Band Elements

This section describes the physical and design principles of dual-band microstrip elements using reactive loads. Other loading techniques using stacked elements have been reported [27] for obtaining dual-band operation. For these the reader is directed to Section 10, "References." The discussion in this section uses the theory of loaded microstrip elements presented in Section 8.

Suppose one loads an element with a lossless resonator of reactance X_ℓ. The characteristic equation for the resonant frequency of the loaded element is (13). This equation is graphically represented in Fig. 31a. The intersection of the two solid curves occurs at the resonant frequency of the loaded element. It is clear from the figure that by adjusting the resonant frequency of the load to be somewhat less than that of the unloaded element, *two* intersections will occur representing *two* resonant frequencies of the loaded element.

By adjusting the resonant frequency of the load, one can shift the locations of the two resonances of the loaded patch. It has been found that there is an optimal pair of frequencies of all that could be achieved by adjusting the resonant frequency of the load. For this optimal pair the impedance characteristics over the dual bands are almost identical as the measured results in Fig. 32 illustrate.

The magnetic current distributions at the two resonant frequencies can be computed using (11) and have been plotted in Fig. 33 for the element corresponding to Fig. 32. In both cases the magnetic current distribution can be viewed as the sum of the magnetic current distribution of the parent $(0, 1)$ unloaded cavity mode and a magnetic current distribution which varies rapidly at points nearest the load. The latter component does not produce a very strong radiated field. Thus the radiation patterns of the two modes of the dual-band elements are expected to be nearly the same, and nearly the same as the pattern of the parent $(0, 1)$ mode of the unloaded cavity. In fact, the measured results verify that this is the case. A typical pattern is given in Fig. 33c.

The resonator load could be an opened or shorted transmission line of an appropriately chosen length. If the transmission line in the resonant load has its characteristic impedance raised, then the reactance plot of Fig. 31a changes to that of Fig. 31b. This illustrates one way to decrease the distance between the dual bands. A simpler way is to simply move the location of the load nearer to a nodal plane of the unloaded cavity mode. When this is done the reactance plot changes to that illustrated in Fig. 31c. By using these two techniques one can produce dual bands spaced as close to each other as one wishes.

The transmission-line resonator is best chosen to be a shorted line since it can be made less than one quarter-wavelength long at the lowest of the two operating bands. This line can be microstripline. It is best located symmetrically so as not to produce a large cross-polarized component of the radiated field. If it is necessary to achieve the desired band separation, the line can be inset as illustrated in Fig. 34.

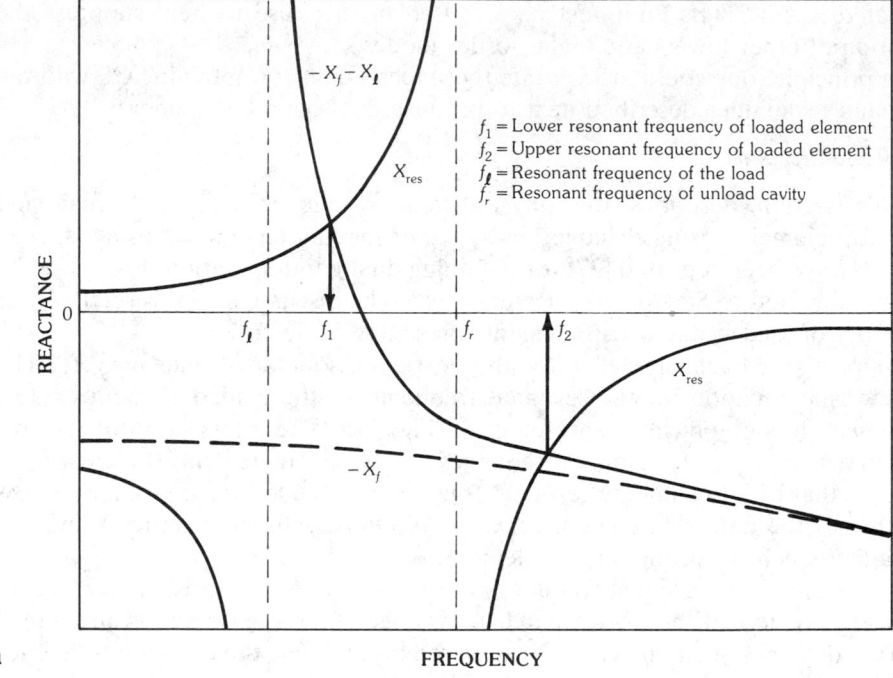

a

b

Fig. 31. Reactance plots for determining the resonant frequency of dual-band elements. (*a*) The X_{res} and $-X_\ell - X_f$ curves for a dual-band element using a shorted stub as its reactive load. (*b*) Same curve but for a higher characteristic impedance shorted stub load. (*c*) Same curve but for the load placed nearer to a nodal curve of the resonant mode of the unloaded cavity.

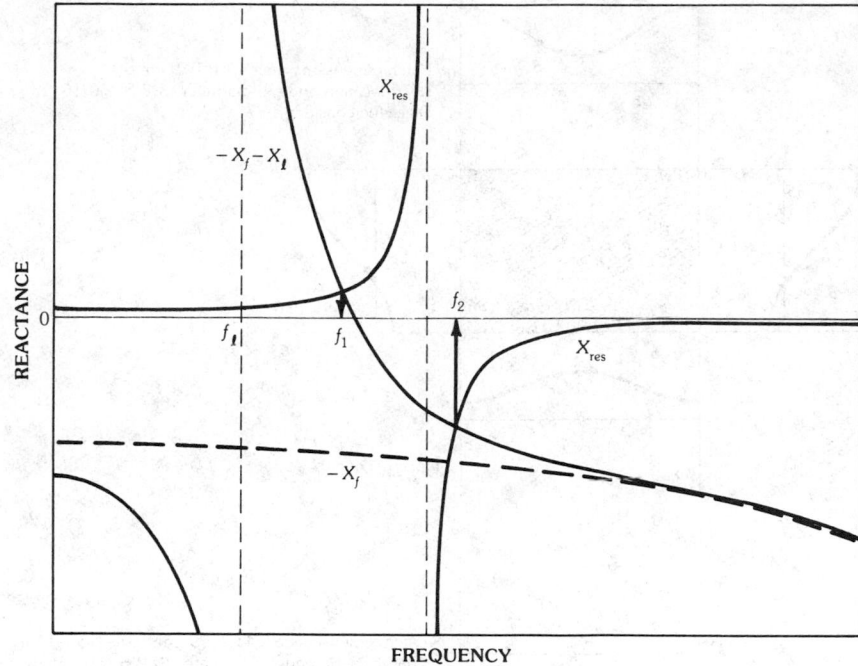

Fig. 31, *continued.*

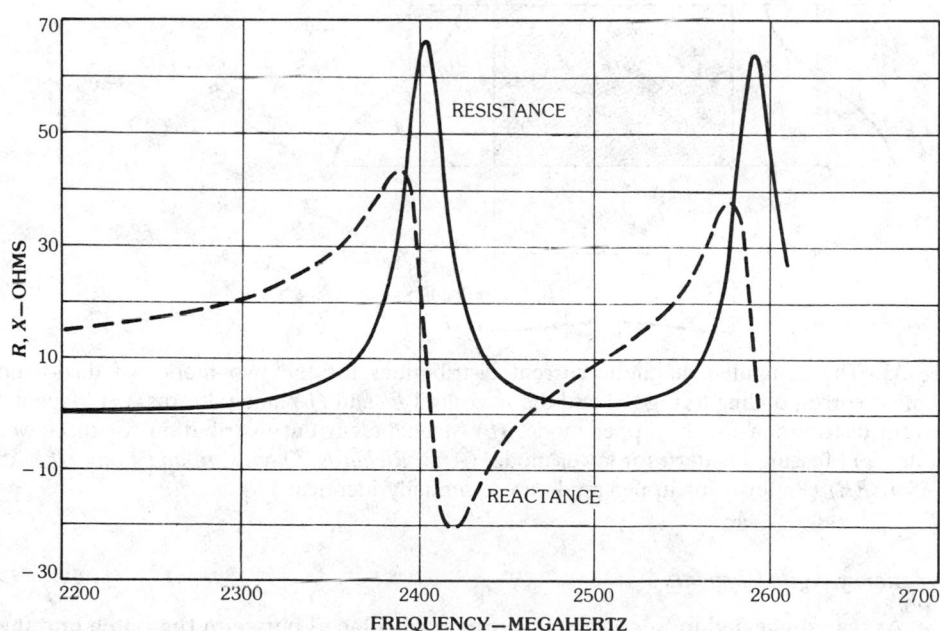

Fig. 32. The measured input resistance and reactance plotted versus frequency for an optimally-loaded rectangular dual-band element. The patch dimensions are 6.0 × 4.0 cm mounted on a 0.079-cm dielectric with relative dielectric constant of 2.17 and loaded by a 5.1-cm, 24.3-Ω shorted stub inset 1.5 cm from the patch edge. (*After Richards, Davidson, and Long [28],* © *1985 IEEE*)

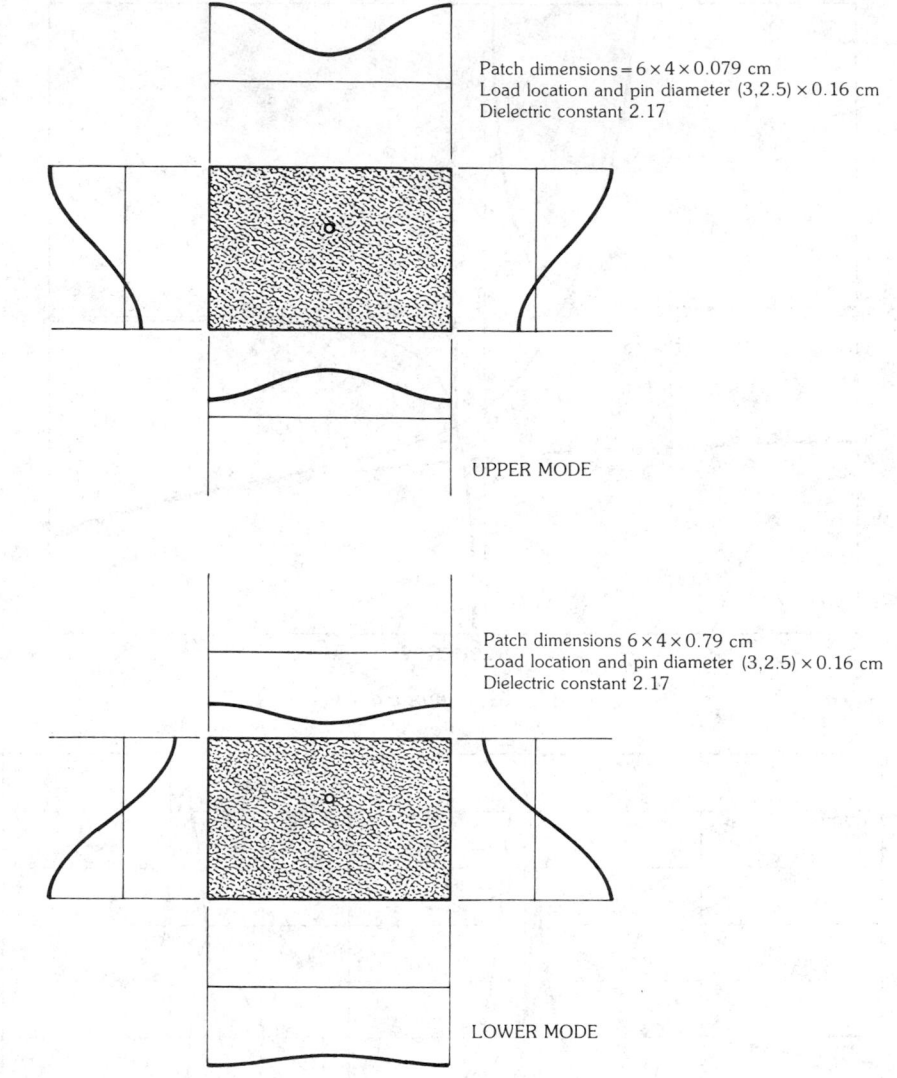

Patch dimensions = 6 × 4 × 0.079 cm
Load location and pin diameter (3,2.5) × 0.16 cm
Dielectric constant 2.17

UPPER MODE

a

Patch dimensions 6 × 4 × 0.79 cm
Load location and pin diameter (3,2.5) × 0.16 cm
Dielectric constant 2.17

LOWER MODE

b

Fig. 33. The computed magnetic current distributions for the two modes of dual-band element corresponding to Fig. 32 and the measured *E*- and *H*-plane patterns. (*a*) Magnetic current distribution for the upper mode. (*b*) Magnetic current distribution for the lower mode. (*c*) Measured pattern for lower mode. (*After Richards, Davidson, and Long [28],* © *1985 IEEE*) (Patterns for upper mode are essentially identical.)

Frequency-Agile Elements

As was suggested in Section 8, shorting posts placed between the patch and the ground plane can be used to alter the resonant frequency of a microstrip element [3]. This can be made dynamic by implementing the shorts as pin diodes which can be biased to represent an rf open or short.

It was seen from Fig. 25 in Section 8 that, because the feed reactance is

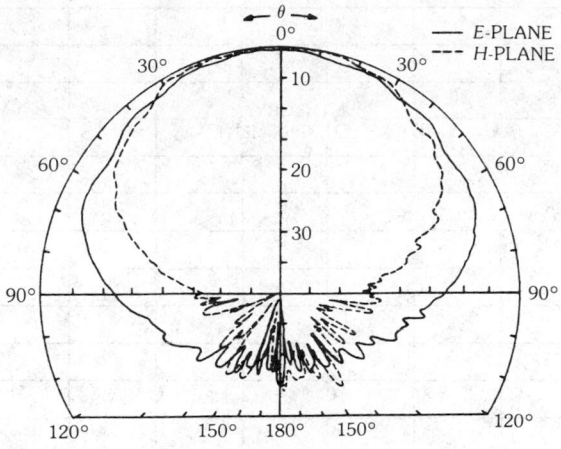

Fig. 33, *continued.*

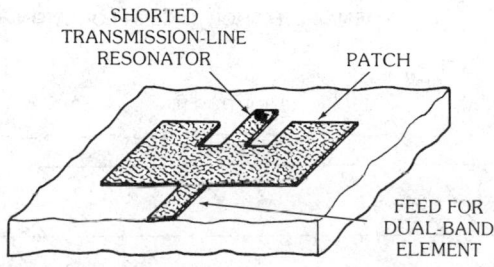

Fig. 34. A monolithic (except for shorting pin) dual-band element with an inset load location to produce narrow-band separation. (Completely monolithic version omits the short and adjusts the line length.)

inductive, shorting the element will raise its resonant frequency. If another short is added, the resonant frequency will be increased even further. In fact, it has been found that one can adjust the resonant frequency of a rectangular element over a 1.7:1 range [3] while still maintaining good vswr and a stable pattern. An implementation of the loaded element theory for multiple loads should give the designer a good quantitative guide for producing a tunable or frequency-agile element.

One problem with this approach is that the diodes should be biased independently to step in small increments through the tuning range. This will require that the diodes be capacitively coupled to the patch, or capacitively coupled to the ground plane. This may require an additional layer of dielectric to support the capacitors and dc biasing lines.

An alternative is to use varactor diodes as the loads and adjust their capacity by adjusting their reverse bias. If all diodes are biased identically, no dc blocking capacitors on the patch are necessary. One need only bias the patch itself through a

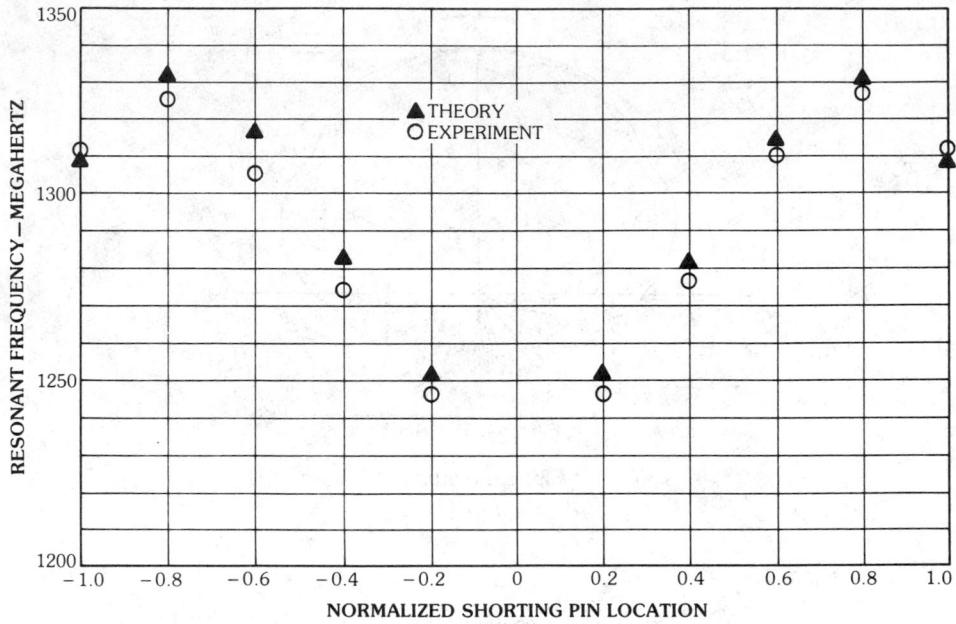

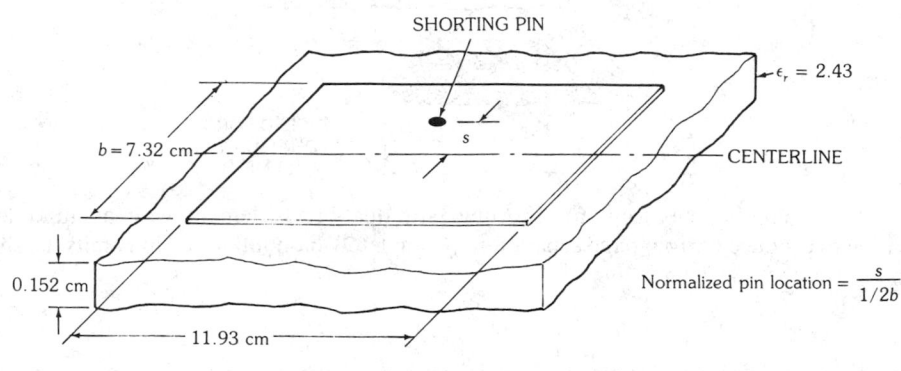

Fig. 35. Measured and computed resonant frequencies of a rectangular element with a single shorting pin versus normalized pin location. (*a*) Frequency variation. (*After Richards and Lo [4], reprinted with permission of Hemisphere Publishing Corporation*) (*b*) Element.

microwave inductive choke. This choke could be on the same level as the patch itself or be placed further back in the microwave feed network.

A good quantitative guide to the design of such frequency-agile elements can be obtained by writing a computer program implementing the theory presented in Section 8. The following qualitative guidelines are also available. The closer to a nodal curve (curve along which the modal field is zero) of the unloaded resonant cavity mode the short is placed, the thinner the plot of the X_{res} versus frequency becomes, as illustrated by the transformation of X_{res} in Fig. 31a to Fig. 31c. Thus the loaded element's resonant frequency will be closer to the unloaded cavity's resonant frequency. This is obvious since a load placed right on a nodal curve has no effect on the unloaded cavity mode. Consequently, the farther away from a nodal curve the load is placed, the higher the resonant frequency of the loaded element will be. However, because the feed reactance increases very rapidly close to the edge of a patch for reasons discussed in Section 4, under "Feed Reactance," the loading effect of a short is diminished somewhat when it is placed very near or right on the patch edge. These trends are illustrated in Fig. 35. These results pertain to a single short; similar qualitative results will also hold for multiple shorts.

In applications where a broadside maximum is not desired, but a monopolelike pattern is sought, one can use the dc mode of a patch element. Of course, the dc mode must somehow be excited (resonated) not at dc but at the desired operating frequency. This can be done by adding shorting pins using exactly the same principles as discussed above.

Polarization-Agile Elements

This application is similar in spirit to the frequency-agile element except the shift in frequency is very small. It was seen in the discussion of single-feed CP that a single-feed CP element could be produced if the ratio of the difference of the two near-degenerate modes to their average value could be made to be $1/Q$. This can easily be done by placing pin diodes as illustrated in Fig. 36a. Based on the cavity model the distance s of the two diodes from the center of the patch should be

$$s = a \left(\frac{a}{t} \frac{X_f}{\eta} \frac{1}{2\pi Q} \right)^{1/2}$$

By turning diode 1 on (rf short) and diode 2 off (rf open), the element will produce right-hand CP. By switching the roles of the two diodes, one produces left-hand CP. By biasing both diodes either on or off, linear polarization is obtained.

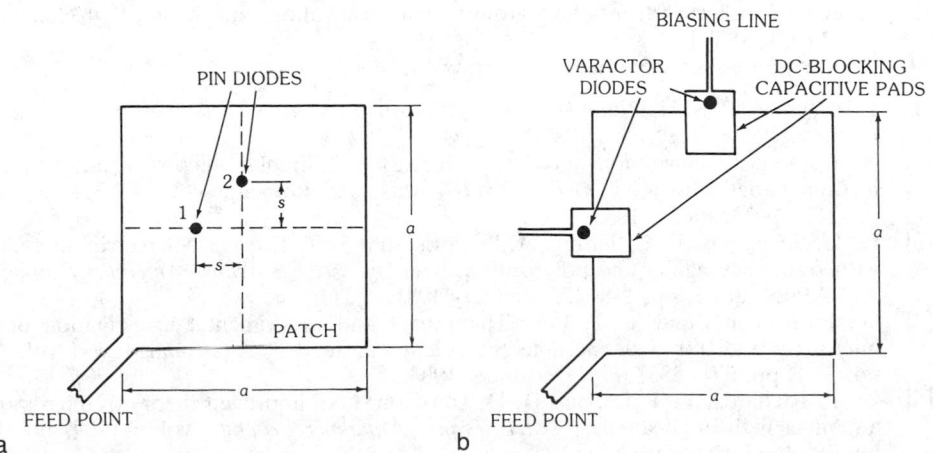

Fig. 36. Polarization-agile elements. (*a*) Square element using pin diodes. (*b*) Square element using varactor diodes.

Another version of the same idea can be accomplished by placing varactor diodes on the edge of the patch as in Fig. 36b. In this case one can sweep the element continuously through whatever axial-ratio elliptical polarization (including circular and linear, of course) one wants. Two diodes are used and biased such that the mean frequency of the two near-degenerate modes is constant. This leaves the operating frequency of the elliptical polarization independent of the selected axial ratio. Clearly, these ideas can be extended to differently shaped patches.

10. References

The following is an index of publications by category. The numbers in square brackets given in each category refer to the numbered references which follow category (19).

(1) Reviews of literature and books on microstrip antennas: [1, 31, 32, 33, 34].
(2) Approximate analytical treatments of microstrip antennas: [1, 3, 4, 5, 10, 11, 17, 22, 24, 28, 29, 31, 34, 47, 48, 49, 50, 56, 62, 63, 64, 69, 73, 74].
(3) Numerical analysis of microstrip elements: [35, 36, 37, 38, 39, 40, 41, 42, 43, 44, 45, 46, 58, 67, 71].
(4) Rectangular microstrip elements: [1, 3, 4, 5, 9, 10, 11, 19, 22, 28, 30, 31, 33, 34, 35, 36, 37, 41, 42, 43, 46, 47, 48, 50, 53, 54, 62, 64, 68, 69, 77, 78, 79, 83].
(5) Circular-disk microstrip patches: [1, 2, 5, 11, 16, 21, 24, 29, 31, 33, 34, 38, 39, 56, 57, 63, 67, 70, 75].
(6) Circular-sector microstrip patches: [11, 74].
(7) Annular-ring microstrip patches: [11, 40, 55, 74].
(8) Annular-sector microstrip patches: [11, 74].
(9) Elliptical microstrip antennas: [11, 17, 20, 54, 59].
(10) Circularly polarized microstrip antennas: [1, 3, 5, 17, 20, 21, 30, 31, 33, 34, 54, 78].
(11) Adaptable microstrip elements: [3, 4, 5, 6, 7].
(12) Dual-band microstrip elements: [27, 28, 31, 33].
(13) Parametric studies (e.g., thickness, dielectric, loss): [1, 23, 31, 34, 51, 52, 57, 60, 61, 63, 65, 68].
(14) Covered microstrip elements: [65, 72].
(15) Feeding microstrip elements: [22, 31, 66, 79, 80].
(16) Mutual coupling between microstrip elements: [30, 43, 67, 68, 73, 76].
(17) External matching of microstrip elements: [8, 81].
(18) Effects of dielectric truncation, ground-plane truncation, and bending of element: [13, 19, 69, 70, 77].
(19) Ferromagnetic substrates and superstrates: [2, 4, 65].

[1] J. R. James, P. S. Hall, and C. Wood, *Microstrip Antenna Theory and Design*, IEE, London: Peter Peregrinus, 1981.
[2] N. D. S. K. Chowdhury, and J. S. Chatterjee, "Circular microstrip antenna on ferrimagnetic substrate," *IEEE Trans. Antennas Propag.*, vol. AP-31, no. 1, pp. 188–190, January 1983.
[3] D. H. Schaubert, F. G. Farrar, A. Sindoris, and S. T. Hayes, "Microstrip antennas with frequency agility and polarization diversity," *IEEE Trans. Antennas Propag.*, vol. AP-29, no. 1, pp. 118–123, January 1981.
[4] W. F. Richards and Y. T. Lo, "Theoretical and experimental investigation of a microstrip radiator with multiple lumped linear loads," *Electromagnetics*, vol. 3, no. 3–4, pp. 371–385, July–December 1983.
[5] W. F. Richards, Y. T. Lo, and D. D. Harrison, "An improved theory of microstrip antennas with applications," *IEEE Trans. Antennas Propag.*, vol. AP-29, no. 1, pp. 38–46, January 1981.
[6] W. F. Richards and S. A. Long, "Impedance control of microstrip antennas utilizing reactive loading," *Proc. Intl. Telemetering Conf.*, pp. 285–290, Las Vegas, 1986.

[7]   W. F. Richards and S. A. Long, "Adaptive pattern control of a reactively loaded, dual-mode microstrip antenna," *Proc. Intl. Telemetering Conf.*, pp. 291–296, Las Vegas, 1986.

[8]   D. A. Pascen, "Broadband microstrip matching techniques," *Proc. 1983 Antenna Appl. Symp.*, EM Laboratory, Univ. of Illinois, Urbana (no page numbers in digest).

[9]   R. E. Munson, "Conformal microstrip antennas and microstrip phased arrays," *IEEE Trans. Antennas Propag.*, vol. AP-22, no. 1, pp. 74–78, January 1974.

[10] A. G. Derneryd, "A theoretical investigation of the rectangular microstrip antenna element," *IEEE Trans. Antennas Propag.*, vol. AP-26, no. 4, pp. 532–535, July 1978.

[11] Y. T. Lo, D. Solomon, and W. F. Richards, "Theory and experiment on microstrip antennas," *IEEE Trans. Antennas Propag.*, vol. AP-27, no. 2, pp. 137–145, March 1979.

[12] E. O. Hammerstad, "Equations for microstrip circuit design," *Proc. Fifth European Microwave Conf.*, pp. 268–272, September 1975.

[13] M. V. Scheider, "Microstrip dispersion," *Proc. IEEE*, vol. 60, no. 1, pp. 144–146, January 1972.

[14] R. F. Harrington, *Time-Harmonic Electromagnetic Fields*, New York: McGraw-Hill Book Co., 1961, p. 183.

[15] H. A. Wheeler, "Transmission-line properties of a strip on a dielectric plane," *IEEE Trans. Microwave Theory Tech.*, vol. MTT-13, pp. 172–185.

[16] S. L. Chuang, L. Tsang, J. A. Kong, and W. C. Chew, "The equivalence of the electric and magnetic surface current approaches in microstrip antenna studies," *IEEE Trans. Antennas Propag.*, vol. AP-28, no. 4, pp. 569–571, July 1980.

[17] L. C. Shen, "The elliptical microstrip antenna with circular polarization," *IEEE Trans. Antennas Propag.*, vol. AP-29, no. 1, pp. 91–94, January 1981.

[18] J. Huang, "The finite ground-plane effect on the microstrip antenna radiation patterns," *IEEE Trans. Antennas Propag.*, vol. AP-31, no. 4, pp. 649–653, July 1983.

[19] C. M. Krowne, "Cylindrical and rectangular microstrip antenna," *IEEE Trans. Antennas Propag.*, vol. AP-31, no. 1, pp. 194–199, July 1983.

[20] Y. T. Lo and W. F. Richards, "Perturbation approach to design of circularly polarized microstrip antennas," *Electron. Lett.*, vol. 17, no. 11, pp. 383–385, May 28, 1981.

[21] J. Huang, "Circularly polarized conical patterns from circular microstrip antennas," *1984 IEEE AP-S Symp. Dig.*, pp. 271–274.

[22] W. F. Richards, J. R. Zinecker, R. D. Clark, and S. A. Long, "Experimental and theoretical investigation of the inductance associated with a microstrip antenna feed," *Electromagnetics*, vol. 3, no. 3–4, pp. 327–346, July–December 1983.

[23] E. Chang, S. A. Long, and W. F. Richards, "An experimental investigation of electrically thick rectangular microstrip antennas," *IEEE Trans. Antennas Propag.*, vol. AP-34, no. 6, pp. 767–772, June 1986.

[24] L. C. Shen, S. A. Long, M. R. Allerding, and M. D. Walton, "Resonant frequency of a circular-disc, printed-circuit antenna," *IEEE Trans. Antennas Propag.*, vol. AP-25, no. 4, pp. 595–596, July 1977.

[25] W. F. Richards, Y. T. Lo, and J. Brewer, "A simple experimental method for separating loss parameters of a microstrip antenna," *IEEE Trans. Antennas Propag.*, vol. AP-29, no. 1, pp. 150–151, January 1981.

[26] R. M. Fano, "Theoretical limitations on the broadband matching of arbitrary impedances," *J. Franklin Inst.*, vol. 249, pp. 57–83, January 1950; pp. 139–154, February 1950.

[27] S. A. Long and M. D. Walton, "A dual-frequency stacked circular-disc antenna," *IEEE Trans. Antennas Propag.*, vol. AP-27, no. 2, pp. 270–273, March 1979.

[28] W. F. Richards, S. Davidson, and S. A. Long, "Dual-band, reactively loaded microstrip antennas," *IEEE Trans. Antennas Propag.*, vol. AP-33 (accepted for publication, 1985).

[29] W. C. Chew, J. A. Kong, and L. C. Shen, "Radiation characteristics of a circular microstrip antenna," *J. Appl. Phys.*, vol. 51, no. 7, pp. 3907–3915, July 1980.

[30] R. E. Munson, "Conformal microstrip antennas and microstrip phased arrays," *IEEE*

Trans. Antennas Propag., vol. AP-22, no. 1, pp. 74–78, January 1974.

[31] K. R. Carver and J. W. Mink, "Microstrip antenna technology," *IEEE Trans. Antennas Propag.*, vol. AP-29, no. 1, pp. 2–24, January 1981.

[32] R. J. Malloux, J. F. McIlvenna, and N. P. Kernweis, "Microstrip array technology," *IEEE Trans. Antennas Propag.*, vol. AP-29, no. 1, pp. 25–37, January 1981.

[33] *Proceedings of the Workshop on Printed-Circuit Antenna Technology* (Keith Carver, Technical Program Chairman), October 17–19, 1979, New Mexico State University, N Mex.

[34] I. J. Bahl and P. Bhartia, *Microstrip Antennas*, Dedham, Mass.: Artech House, 1980.

[35] P. K. Agrawal and M. C. Bailey, "An analysis technique for microstrip antennas," *IEEE Trans. Antennas Propag.*, vol. AP-25, no. 6, pp. 756–759, November 1977.

[36] E. H. Newman and P. Tulyathan, "Analysis of microstrip antennas using moment methods," *IEEE Trans. Antennas Propag.*, vol. AP-29, no. 1, pp. 47–53, January 1981.

[37] T. Itoh and W. Menzel, "A full-wave analysis method for open microstrip structures," *IEEE Trans. Antennas Propag.*, vol. AP-29, no. 1, pp. 63–68, January 1981.

[38] W. C. Chew and J. A. Kong, "Analysis of a circular microstrip disk antenna with a thick dielectric substrate," *IEEE Trans. Antennas Propag.*, vol. AP-29, no. 1, pp. 68–76, January 1981.

[39] K. Araki and T. Itoh, "Hankel transform domain analysis of open circular microstrip radiating structures," *IEEE Trans. Antennas Propag.*, vol. AP-29, no. 1, pp. 84–89, January 1981.

[40] S. M. Ali, W. C. Chew, and J. A. Kong, "Vector Hankel transform analysis of annular-ring microstrip antenna," *IEEE Trans. Antennas Propag.*, vol. AP-30, no. 4, pp. 637–644, July 1982.

[41] M. D. Deshpande and M. C. Baily, "Input impedance of microstrip antennas," *IEEE Trans. Antennas Propag.*, vol. AP-30, no. 4, pp. 645–650, July 1982.

[42] M. C. Baily and M. D. Deshpande, "Integral equation formulation of microstrip antennas," *IEEE Trans. Antennas Propag.*, vol. AP-30, no. 4, pp. 651–656, July 1982.

[43] D. M. Pozar, "Input impedance and mutual coupling of rectangular microstrip antennas," *IEEE Trans. Antennas Propag.*, vol. AP-30, no. 6, pp. 1191–1196, November 1982.

[44] R. Chandra and K. C. Gupta, "Segmentation method using impedance matrices for analysis of planar microwave circuits," *IEEE Trans. Microwave Theory Tech.*, vol. MTT-29, no. 1, pp. 71–74, January 1981.

[45] P. C. Sharma and K. C. Gupta, "Desegmentation method for analysis of two-dimensional microwave circuits," *IEEE Trans. Microwave Theory and Techniques*, vol. MTT-29, no. 10, pp. 1094–1097, October 1981.

[46] D. M. Pozar, "Improved computational efficiency for the moment method solution of printed dipoles and patches," *Electromagnetics*, vol. 3, no. 3–4, pp. 299–307, July–December 1983.

[47] D. C. Chang, "Analytical theory of an unloaded rectangular microstrip patch," *IEEE Trans. Antennas Propag.*, vol. AP-29, no. 1, pp. 54–62, January 1981.

[48] E. F. Kuester, R. T. Johnk, and D. C. Chang, "The thin-substrate approximation for reflection from the end of a slab-loaded parallel-plate waveguide with application to microstrip patch antennas," *IEEE Trans. Antennas Propag.*, vol. AP-30, no. 5, pp. 910–917, September 1980.

[49] E. F. Kuester and D. C. Chang, "A geometrical theory for the resonant frequencies and Q-factors of some triangular microstrip patch antennas," *IEEE Trans. Antennas Propag.*, vol. AP-31, no. 1, pp. 27–34, January 1983.

[50] J. Venkataraman and D. C. Chang, "Imput impedance to a probe-fed rectangular microstrip patch antenna," *Electromagnetics*, vol. 3, no. 3–4, pp. 387–399, July–December 1983.

[51] D. M. Pozar, "Considerations for millimeter-wave printed antennas," *IEEE Trans. Antennas Propag.*, vol. AP-31, no. 5, pp. 740–747, September 1983.

[52] I. J. Bahl, P. Bhartia, and S. S. Stuchly, "Design of microstrip antennas covered with

a dielectric layer," *IEEE Trans. Antennas Propag.*, vol. AP-30, no. 2, pp. 314–318, March 1982.

[53] N. Das and S. K. Chowdhury, "Rectangular microstrip antenna on a ferrite substrate," *IEEE Trans. Antennas Propag.*, vol. AP-30, no. 3, pp. 499–502, May 1982 (plus correction, vol. AP-30, no. 6, p. 1268, November 1982).

[54] S. A. Long, L. C. Shen, D. H. Schaubert, and F. G. Farrar, "An experimental study of the circular-polarized elliptical printed-circuit antenna," *IEEE Trans. Antennas Propag.*, vol. AP-29, no. 1, pp. 95–99, January 1981.

[55] W. C. Chew, "A broadband annular-ring microstrip antenna," *IEEE Trans. Antennas Propag.*, vol. AP-30, no. 5, pp. 918–922, September 1982.

[56] S. Yano and A. Ishimaru, "A theoretical study of the input impedance of a circular microstrip disk antenna," *IEEE Trans. Antennas Propag.*, vol. AP-29, no. 1, pp. 77–83, January 1981.

[57] J. S. Dahele and K. Lee, "Effect of substrate thickness on the performance of a circular-disk microstrip antenna," *IEEE Trans. Antennas Propag.*, vol. AP-31, no. 2, pp. 358–360, March 1983.

[58] J. R. Mosig and F. E. Gardiol, "Analytical and numerical techniques in the Green's function treatment of microstrip antennas and scatterers," *IEEE Proc.*, vol. 130, pt. H, no. 2, pp. 175–182, March 1983.

[59] S. A. Long and M. W. McAllister, "The impedance of an elliptical printed-circuit antenna," *IEEE Trans. Antennas Propag.*, vol. AP-30, no. 6, pp. 1197–1200, November 1982.

[60] A. G. Derneryd and I. Karlsson, "Broadband microstrip antenna element and array," *IEEE Trans. Antennas Propag.*, vol. AP-29, no. 1, pp. 140–141, January 1981.

[61] J. W. Mink, "Sensitivity of microstrip antennas to admittance boundary variations," *IEEE Trans. Antennas Propag.*, vol. AP-29, no. 1, pp. 142–144, January 1981.

[62] P. Hammer, D. Van Bouchaute, D. Verschraeven, and A. Van de Capelle, "A model for calculating the radiation field of microstrip antennas," *IEEE Trans. Antennas Propag.*, vol. AP-27, no. 2, pp. 267–270, March 1979.

[63] A. G. Derneryd, "Analysis of the microstrip disk antenna element," *IEEE Trans. Antennas Propag.*, vol. AP-27, no. 5, pp. 660–664, September 1979.

[64] A. G. Derneryd, "Extended analysis of rectangular microstrip resonator antennas," *IEEE Trans. Antennas Propag.*, vol. AP-27, no. 6, pp. 846–849, November 1979.

[65] N. G. Alexopoulos and D. R. Jackson, "Radiation efficiency optimization for printed-circuit antennas using magnetic superstrates," *Electromagnetics*, vol. 3, no. 3–4, pp. 255–269, July–December 1983.

[66] J. Rivera and T. Itoh, "Analysis of a suspended patch antenna excited by an inverted microstrip line," *Electromagnetics*, vol. 3, no. 3–4, pp. 289–298, July–December 1983.

[67] T. M. Habashy and J. A. Kong, "Coupling between two circular microstrip disk resonators," *Electromagnetics*, vol. 3, no. 3–4, pp. 347–370, July–December 1983.

[68] C. M. Krowne, "Dielectric and width effect on *H*-plane and *E*-plane coupling between rectangular microstrip antennas," *IEEE Trans. Antennas Propag.*, vol. AP-31, no. 1, pp. 39–47, January 1983.

[69] E. Lier and K. R. Jakobsen, "Rectangular microstrip patch antennas with infinite and finite ground-plane dimensions," *IEEE Trans. Antennas Propag.*, vol. AP-31, no. 6, pp. 978–984, November 1983.

[70] S. B. De Assis Fonseca and A. J. Giarola, "Microstrip disk antennas, part I: efficiency of space wave launching," and "Microstrip disk antennas, part II: the problem of surface wave radiation by dielectric truncation," *IEEE Trans. Antennas Propag.*, vol. AP-32, no. 6, pp. 561–573, June 1984.

[71] Y. Suzuki and T. Chiba, "Computer analysis method for arbitrarily shaped microstrip antenna with multiterminals," *IEEE Trans. Antennas Propag.*, vol. AP-32, no. 6, pp. 585–590, June 1984.

[72] N. G. Alexopoulos and D. R. Jackson, "Fundamental superstrate (cover) effects on printed-circuit antennas," *IEEE Trans. Antennas Propag.*, vol. AP-32, no. 8,

pp. 807–815, August 1984.

[73] E. H. Van Lil and A. R. Van de Capelle, "Transmission-line model for mutual coupling between microstrip antennas," *IEEE Trans. Antennas Propag.*, vol. AP-32, no. 8, pp. 816–821, August 1984.

[74] W. F. Richards, J. D. Ou, and S. A. Long, "A theoretical and experimental investigation of annular, annular-sector, and circular-sector microstrip antennas," *IEEE Trans. Antennas Propag.*, vol. AP-32, no. 8, pp. 864–867, August 1984.

[75] K. F. Lee, K. Y. Ho, and J. S. Dahele, "Circular-disk microstrip antenna with an air gap," *IEEE Trans. Antennas Propag.*, vol. AP-32, no. 8, pp. 880–884, August 1984.

[76] P. B. Katehi and N. G. Alexopoulos, "On the modeling of electromagnetically coupled microstrip antennas—the printed-strip dipole," *IEEE Trans. Antennas Propag.*, vol. AP-32, no. 11, pp. 1179–1186, November 1984.

[77] K. R. Jakobsen, "The radiation from rectangular microstrip antennas mounted on two-dimensional objects," *IEEE Trans. Antennas Propag.*, vol. AP-32, no. 11, pp. 1255–1259, November 1984.

[78] Y. T. Lo, B. Engst, and R. Q. H. Lee, "Circularly polarized microstrip antennas," Antenna Applications Symposium, Allerton Park, Univ. of Illinois, September 22–24, 1984.

[79] P. S. Hall, C. Wood, and C. Garrett, "Wide-bandwidth microstrip antennas for circuit integration," *Electron. Lett.*, vol. 15, no. 15, pp. 458–460, July 19, 1979.

[80] J. R. James, P. S. Hall, C. Wood, and A. Henderson, "Some recent developments in microstrip antenna design," *IEEE Trans. Antennas Propag.*, vol. AP-29, no. 1, pp. 124–128, January 1981.

[81] G. L. Matthaei, L. Young, and E. M. T. Jones, *Microwave Filters, Impedance-Matching Networks, and Coupling Structures*, New York: McGraw-Hill Book Co., 1965.

[82] S. S. Zhong and Y. T. Lo, "Single-element rectangular microstrip antenna for dual-frequency operation," *Electron. Lett.*, vol. 19, no. 8, pp. 298–300, April 14, 1983.

[83] S. M. Wright and Y. T. Lo, "Efficient analysis for infinite microstrip dipole array," *Electron. Lett.*, vol. 19, no. 24, pp. 1043–1045, November 24, 1983.

Chapter 11

Array Theory

Y. T. Lo
University of Illinois

CONTENTS

Yuen T. Lo is a professor and the director of the Electromagnetics Laboratory (formerly the Antenna Laboratory) in the Electrical and Computer Engineering Department, University of Illinois at Urbana-Champaign. He is a member of the National Academy of Engineering, a Fellow of IEEE, and a member of the International Union of Radio Science. He received the 1964 IEEE AP-S Bolljahn Memorial Award, and the 1964 IEEE AP-S Best Paper Award, the 1979 IEEE AP-S Best Paper Award, the IEEE Centennial Medal, and the Halliburton Education Leadership award. He served as an AP-S AdCom member, the Chairman of the AP-S Education and Tutorial Papers Committee, and twice (1979–1982 and 1984–1987) as the IEEE AP-S National Distinguished Lecturer. Dr. Lo is an honorary professor of the Northwest Telecommunication Engineering Institute and also the Northwestern Polytechnical University, both at Xian, China. He has published over 100 technical articles in refereed journals covering a wide spectrum, from theoretical to experimental works. His works include large-antenna arrays, radiotelescopes, multiple-beam antennas, multiple scattering, antenna synthesis, antijamming antennas, antenna in plasmas, corrugated guides and horns, artificial dielectrics, and microstrip antennas. He designed the University of Illinois Radiotelescope, considered to be the world's largest antenna, in the early 1960s.

1. Introduction

Single-element antennas are discussed in other chapters in the book but their performance is somewhat limited. To obtain high directivity, narrow beams, low side lobes, steerable beams, particular pattern characteristics, etc., commonly a group of antenna elements, called an *antenna array*, or simply *array*, is used. The design of an array involves mainly first the selection of elements and array geometry, and then the determination of the element excitations required for achieving a particular performance, sometimes under a given constraint. The realization of the desired excitation requires a detailed knowledge of element input impedance characteristics as well as the mutual impedance between any two elements in the given array environment. In general, this is a difficult problem which may be solved approximately for large arrays with an infinite array model and for small arrays in a two-element environment model. While discussions on these models for some simple elements can be found elsewhere in the book, in this chapter we confine our discussion to the problems stated above.

There is no reason why all elements in an array must be of the same type other than simplicity in fabrication and analysis. In fact, radioastronomers have made use of two or more existing radiotelescopes of different types for interferometry measurement. Furthermore, even with the same type of elements, the shape of current or aperture field distribution of the element near the edge of the array can be different from that of the element in the central portion of the array, depending on the array geometry, element spacing and orientation, and, of course, the element type. Generally such a difference may not cause a serious deterioration in the array performance in some applications.

There is also no particular reason why the array structure must be periodic other than the two reasons just stated above. In fact, a periodic structure can result in grating lobes, frequency sensitivity, blind angles, etc. Even a periodic structure of scatterers, such as those used in artificial dielectrics, possesses some interesting but undesirable properties, such as birefringence, anisotropy, and dispersion. For analytic simplicity, however, uniformly spaced arrays, particularly the linear ones, have been studied in a great detail [1–120]. At least for these arrays the performance can be predicted accurately. Therefore we shall consider uniformly spaced linear arrays first. With some understanding of linear arrays, for convenience in later discussions, we then consider a general transformation theory and its application to planar, circular, and elliptical arrays. Finally, a few array synthesis problems are discussed.

2. General Formulation

For radiation characteristics, only the far field is of interest. In all the following discussions the use of the far-field approximation will be understood. Let $\mathbf{A(r)}$ be the magnetic vector potential due to a typical element. Then

$$\mathbf{A}(\mathbf{r}) \cong \frac{\mu}{\pi} \frac{e^{-jkr}}{r} \int_V \mathbf{J}(\mathbf{r}') e^{jkr' \cos \xi} dV' \tag{1}$$

where

$\mathbf{J}(\mathbf{r}')$ = the electrical current distribution in the element

$\mathbf{r}$ = the position vector of an observation point

$\mathbf{r}'$ = the position vector of a typical source point

V = the volume of the source element

dV' = a differential volume element of the source

$\cos \xi = \hat{\mathbf{r}} \cdot \hat{\mathbf{r}}'$

$k = \omega \sqrt{\mu \epsilon} = 2\pi/\lambda$ = free-space wave number

The important consequence of the far-field approximation is that the dependence of $\mathbf{A}(\mathbf{r})$ on $\mathbf{r}$ is separated into two parts: one depending on r only and the other on (θ, ϕ) through $\hat{\mathbf{r}}$ (or $\cos \xi$ in the integral). Thus, for a sphere with fixed r, we can write the factor $(jkr' \cos \xi)$ as $(j\mathbf{k} \cdot \mathbf{r}')$ and $\mathbf{A}(\mathbf{r})$ as $\mathbf{A}(\mathbf{k})$ with $\mathbf{k} = k\hat{\mathbf{r}}$, the direction in which $\mathbf{A}$ is to be evaluated. In practice, only the directional characteristic is of main interest; therefore, we shall focus our attention on the integral and rewrite (1) in the following form:

$$\mathbf{A}(\mathbf{r}) = \hat{\mathbf{r}} A_r + \hat{\boldsymbol{\theta}} A_\theta + \hat{\boldsymbol{\phi}} A_\phi \cong \frac{\mu e^{-jkr}}{4\pi r} [\hat{\mathbf{r}} f_r(\hat{\mathbf{k}}) + \hat{\boldsymbol{\theta}} f_\theta(\hat{\mathbf{k}}) + \hat{\boldsymbol{\phi}} f_\phi(\hat{\mathbf{k}})] \tag{2a}$$

where

$$\begin{Bmatrix} f_\theta(\theta, \phi) \\ f_\phi(\theta, \phi) \end{Bmatrix} = \begin{Bmatrix} f_\theta(\hat{\mathbf{k}}) \\ f_\phi(\hat{\mathbf{k}}) \end{Bmatrix} = \begin{Bmatrix} \hat{\boldsymbol{\theta}} \\ \hat{\boldsymbol{\phi}} \end{Bmatrix} \cdot \int \mathbf{J}(\mathbf{r}') e^{j\mathbf{k} \cdot \mathbf{r}'} dV' \tag{2b}$$

Then the far fields are

$$\mathbf{E}(\mathbf{r}) = -j\omega \mathbf{A} + \frac{1}{j\omega \epsilon} \nabla \nabla \cdot \mathbf{A} \cong -j\omega [\hat{\boldsymbol{\theta}} \hat{\boldsymbol{\theta}} + \hat{\boldsymbol{\phi}} \hat{\boldsymbol{\phi}}] \cdot \mathbf{A}$$

$$\mathbf{H}(\mathbf{r}) \cong \frac{1}{\eta} \hat{\mathbf{r}} \times \mathbf{E} \tag{3}$$

where $\eta = \sqrt{\mu/\epsilon}$.

The angular-dependent parts $f_\theta(\theta, \phi)$ and $f_\phi(\theta, \phi)$ give the directional characteristics of E_θ and E_ϕ. It should be noted that f_θ and f_ϕ are complex functions. Their absolute values, or magnitudes, $|f_\theta(\theta, \phi)|$ and $|f_\phi(\theta, \phi)|$, are generally referred to as the *pattern functions*; in fact, often one of them is termed the major component (or copolarization) pattern function and the other the cross-polarization pattern function. Their phases, namely $\angle f_\theta(\theta, \phi)$ and $\angle f_\phi(\theta, \phi)$, are referred to as the *phase pattern functions*. Obviously the significance of a phase pattern function should be weighted by its associated pattern function $|f(\theta, \phi)|$. The phase pattern is of interest in some applications, such as the determination of the phase center. Sometimes one is also interested in evaluating the patterns in terms of circularly, instead of linearly, polarized components. In that case

$$\mathbf{E}(\hat{r}) \propto \hat{\mathbf{L}} f_L(\theta, \phi) + \hat{\mathbf{R}} f_R(\theta, \phi) \tag{4}$$

where

$$f_L(\theta, \phi) = f_\theta(\theta, \phi) - jf_\phi(\theta, \phi) = \text{LCP pattern}$$
$$f_R(\theta, \phi) = f_\theta(\theta, \phi) + jf_\phi(\theta, \phi) = \text{RCP pattern} \tag{5}$$

Finally, it may also be noted that, in practice, for simplicity, $f(\theta, \phi)$ is often called the pattern function, rather than the precise term *complex* pattern function.

For an array of N arbitrary elements, as shown in Fig. 1, the far field is

$$\mathbf{E}(\mathbf{r}) \cong \frac{-j\omega\mu e^{-jkr}}{4\pi r} \mathbf{f}(\theta, \phi) \tag{6}$$

where

$$\mathbf{f}(\theta, \phi) = \sum_{n=1}^{N} \mathbf{f}_n(\theta, \phi)$$

$$\mathbf{f}_n(\theta, \phi) = (\hat{\boldsymbol{\theta}}\hat{\boldsymbol{\theta}} + \hat{\boldsymbol{\phi}}\hat{\boldsymbol{\phi}}) \int_{\substack{n\text{th} \\ \text{element}}} \mathbf{J}_n(\mathbf{r}'_n) \, e^{jk\hat{\mathbf{r}} \cdot (\mathbf{r}' - \mathbf{r}_n)} \, dV'_n \tag{7}$$

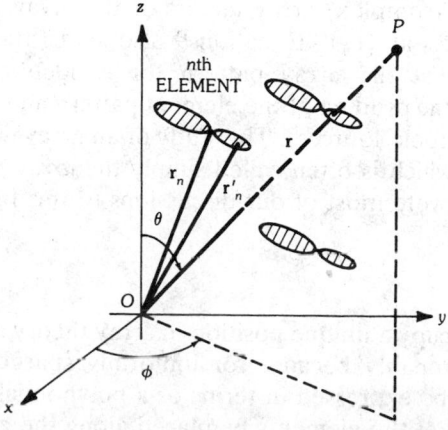

Fig. 1. Geometry of an array with identical elements.

$\mathbf{r}_n$ = a convenient reference point, such as the phase center, of the nth element

$\mathbf{r}'_n$ = a typical point on the nth element

$\mathbf{f}_n(\theta, \phi)$ = the nth-element pattern function

If all elements are identical and also identically oriented, with the assumption that the current distribution of all elements in the array have identical shape except for a constant multiplier, one can write

$$\mathbf{f}_n(\theta, \phi) = I_n \mathbf{f}_0(\theta, \phi) \tag{8}$$

where

$\mathbf{f}_0(\theta, \phi)$ = complex pattern function of a single element

I_n = relative complex excitation to the nth element

It should be noted again that element patterns in different array environments could be significantly different from each other (see Chapters 13 and 14). For large arrays or arrays with simple elements the neglect of this difference may still lead to a useful approximation. Then, using (8), one can rewrite (6) as follows:

$$\mathbf{E}(\mathbf{r}) \cong -\frac{j\omega\mu e^{-jkr}}{4\pi r} \mathbf{f}_0(\theta, \phi) F(\theta, \phi) \tag{9}$$

where

$$F(\theta, \phi) = \sum_n I_n e^{jk\hat{\mathbf{r}} \cdot \mathbf{r}_n} \tag{10}$$

is usually called the (complex) array factor, or the array pattern function for N isotropic point sources at $\{\mathbf{r}_n\}$. Equation 9 also states the *pattern multiplication principle*, namely, that the array pattern for N identical elements, similarly oriented, is equal to the product of the element pattern and the array factor (or the array pattern for isotropic sources). The study of an array usually implies the study of the array factor, which is often called simply the *array pattern function*. In the following we shall devote most of our discussions to this function.

3. Linear Arrays

Linear arrays occupy a unique position in array theory and have received great attention. This is probably because for uniformly spaced elements the pattern function can simply be expressed in terms of a polynomial for which the analytic tool is well-known. Let the elements be placed along the z axis with interelement spacing d; then for the nth element $\mathbf{r}_n = \hat{z}nd$ and for an array of $(N + 1)$ elements

$$F(\theta) = \sum_{n=0}^{N} I_n e^{jknd\cos\theta} = \sum_{n=0}^{N} I_n Z^n \tag{11}$$

where

$$Z = e^{j\psi} \quad \text{and} \quad \psi = kd\cos\theta \tag{12}$$

For the physically observable region $0 \leqq \theta \leqq \pi$, the function $F(\theta)$ is given by the value of the polynomial in (11) with Z only on a unit circle and its phase angle ψ bounded between $-kd$ and $+kd$, which is called the *visible region*. Polynomials have been thoroughly studied and it is therefore not surprising to find a large number of contributions in the literature on linear arrays. A few important results are summarized below.

Arrays with Prescribed Nulls

These are useful for antijamming and interference elimination. Let $\{\theta_n\}$, $n = 1, 2, \ldots, N$, be the set of null angles. Then the desired array pattern function is

$$F(\theta) = c \prod_{n=1}^{N} (Z - Z_n) = c[I_0 + I_1 Z + \cdots + I_N Z^N] \tag{13}$$

where c = a constant, commonly a normalization factor such that

$$\max |F(\theta)| = 1,$$
$$Z_n = e^{j\psi_n},$$
$$\psi_n = kd\cos\theta_n,$$
$$I_n = \text{the required complex excitation for the } n\text{th element}$$

Comments:

(*a*) The magnitude pattern is

$$|c| \prod_{n=1}^{N} |Z - Z_n| = 20 \left(\log|c| + \sum_{n=1}^{N} \log|Z - Z_n| \right)$$

in decibels, where $|Z - Z_n|$ is the length between a typical point $Z(\theta)$ on the unit circle and Z_n, which is also on the circle as shown in Fig. 2.

(*b*) Excitations are given by the coefficients of the polynomial expansion as in (13).

(*c*) Because of the axial symmetry the 3D pattern is the generation of the above pattern function rotated about the array axis.

(*d*) To obtain a deep null at, say, θ_n, multiplicity of the root Z_n can be imposed. In fact, $(Z - Z_n)^p$, with p an integer, implies that all derivatives up to the $(p - 1)$th order vanish at $Z = Z_n$.

(*e*) These excitations are determined only by the desired null directions without

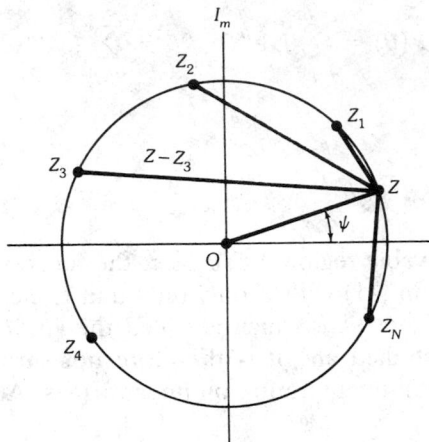

Fig. 2. Zeros of an array polynomial on a unit circle.

considering the desired signal. For a more meaningful solution see Section 6, where signal-to-interference ratio is to be maximized.

(f) Since each element can be a subarray, the element could be designed against the jamming signals while the entire array is designed for receiving the desired signal, or vice versa, provided the desired signal is in a direction different from that of any of the jamming signals.

(g) Null steering can be achieved by varying Z_n on a unit circle, but in general this change will alter excitations of many elements.

(h) Of course, some roots $\{Z_n\}$ need not be on the unit circle nor in the visible region. In that case the array will not have deep nulls at $\{\theta_n\}$.

Binomial Arrays

If all roots $\{Z_n\}$ coincide and are equal to Z_1, then

$$F(\theta) = c(Z - Z_1)^N$$

$$= c[Z^N + N(-Z_1)Z^{N-1} + \frac{N(N-1)}{2}(-Z_1)^2 Z^{N-2}$$

$$+ \cdots + C_k^N(-Z_1)^k Z^{N-k} + \cdots + (-Z_1)^N] \tag{14}$$

where

$Z_1 = e^{jkd\cos\theta_1}$

θ_1 = null angle

C_k^N = binomial coefficient, which can be found more easily by using the Pascal triangle

Thus $I_1 = 1$, $I_2 = NZ_1$, $\ldots$, $I_k = C_k^N(-Z_1)^k$, $\ldots$, $I_N = (-Z_1)^N$.

Comments:

(*a*) The magnitude pattern indecibels is

$$|c| \, |Z - Z_1|^N = 20(\log|c| + N\log|Z - Z_1|)$$

(*b*) This array has no "side lobes" if $d \leq \lambda/2$. For example, if $d = \lambda/2$ and $\theta_1 = 0$, then $Z_1 = -1$ and all element excitations are in phase with weight $\{C_k^N\}$. The pattern has only a broadside beam, irrespective of the array length. If $\theta_1 = \pi/2$, $Z_1 = 1$, all element excitations will be alternatively opposite in phase. The pattern will consist of two separate beams along $\theta = 0$ and $\theta = \pi$, respectively, and a null at $\theta = \pi/2$. Again there are no other lobes, no matter how long the array is. If d is near or larger than λ, grating lobes appear partially or totally.

Uniform Arrays

When all excitations are equal, say 1,

$$F(\theta) = \sum_{n=0}^{N-1} Z^n = \frac{Z^N - 1}{Z - 1} \tag{15}$$

If the array center is chosen as the origin,

$$F(\theta) = \frac{\sin(N\psi/2)}{\sin(\psi/2)} \tag{16}$$

where

$$\psi = kd \cos \theta \tag{17}$$

The normalized pattern function of (16) as a function of ψ for a few values of N is shown in Fig. 3. The graphical method for determining $F(\theta)$ through (17) is demonstrated in Fig. 4 for two cases: broadside and scan angle θ_0.

Comments:

(*a*) As will be seen in Section 6 the uniform array for d nearly equal to or greater than $\lambda/2$ has a directivity close to maximum. For $d < \lambda/2$, a substantial increase in directivity over that of the uniform excitation is only possible in theory but not in practice. Thus the directivity of a uniformly excited array is about the maximum achievable in practice and therefore is often used as a reference for comparing directivities of various designs.

(*b*) The beam maximum of (16) appears at $\theta = \pi/2$ and, therefore, the array is called a *broadside array*. The beam can be steered to any direction θ_0 if the excitation I_n contains a progress phase factor $nkd \cos \theta_0$, for all ns, as will be seen in a general discussion in Section 4, under "Application to Planar Periodic Arrays." This phase shift results in a translation of ψ by $kd \cos \theta_0$. Thus, instead of (17),

$$\psi = kd(\cos \theta - \cos \theta_0) \tag{18}$$

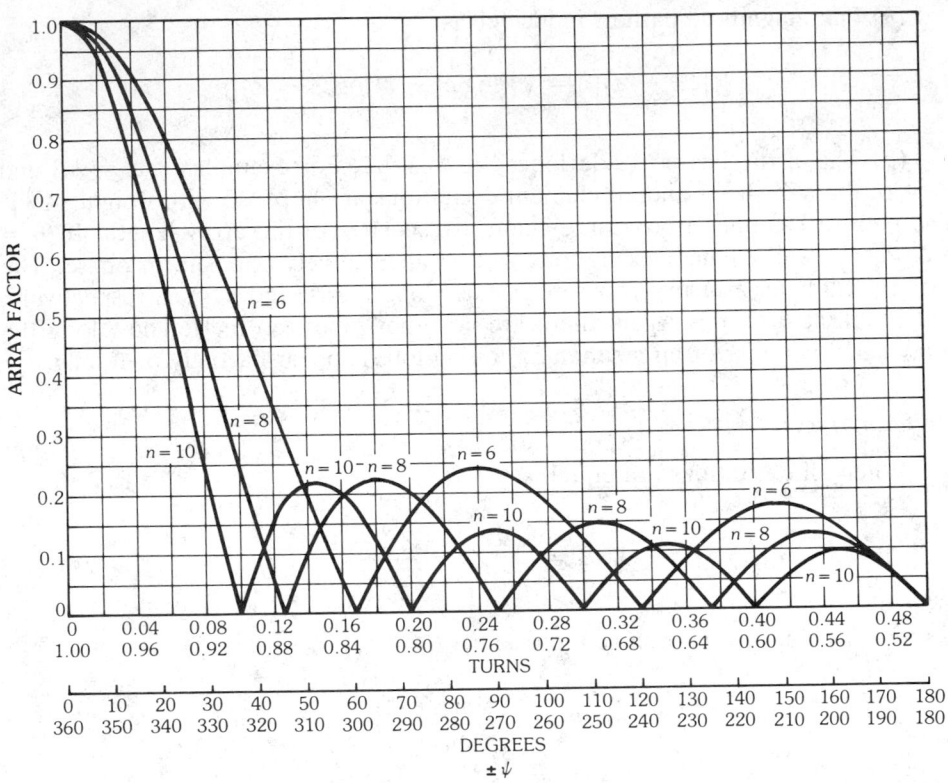

Fig. 3. Plots of $\sin(N\psi/2)/\sin(\psi/2)$ vs ψ for $N = 6$ through 11. (*After Kraus [60], © 1950; reprinted with permission of McGraw-Hill Book Co.*)

In other words, what happened at $\theta = \pi/2$ before introducing the phase shift will happen afterwards at $\theta = \theta_0$. The pattern function does not change with respect to ψ except for a simple translation. But this is not true when plotted against θ due to the nonlinear functional dependence of ψ on θ. Therefore it is often more convenient to study the pattern as a function of ψ, or the so-called u space in Section 4. When $\theta_0 = 0°$, the main beam will be along the array axis and the array will be called *end-fire*.

(c) A few important formulas, such as directivity D, half-power beamwidth, beamwidth between first nulls, null angular position, and side lobe maximum position, for broadside and end-fire uniform arrays are listed in Table 1.

(d) Also listed in Table 1 are the formulas for Hansen-Woodyard end-fire arrays. Hansen and Woodyard [50] found that for a long uniform end-fire array when element spacing is small the directivity in the $\theta = 0°$ direction can be increased from that of the ordinary end-fire if the phase shift per element is [4]

$$\beta = -\left(kd + \frac{2.94}{N}\right) \cong -\left(kd + \frac{\pi}{N}\right) \tag{19}$$

and if

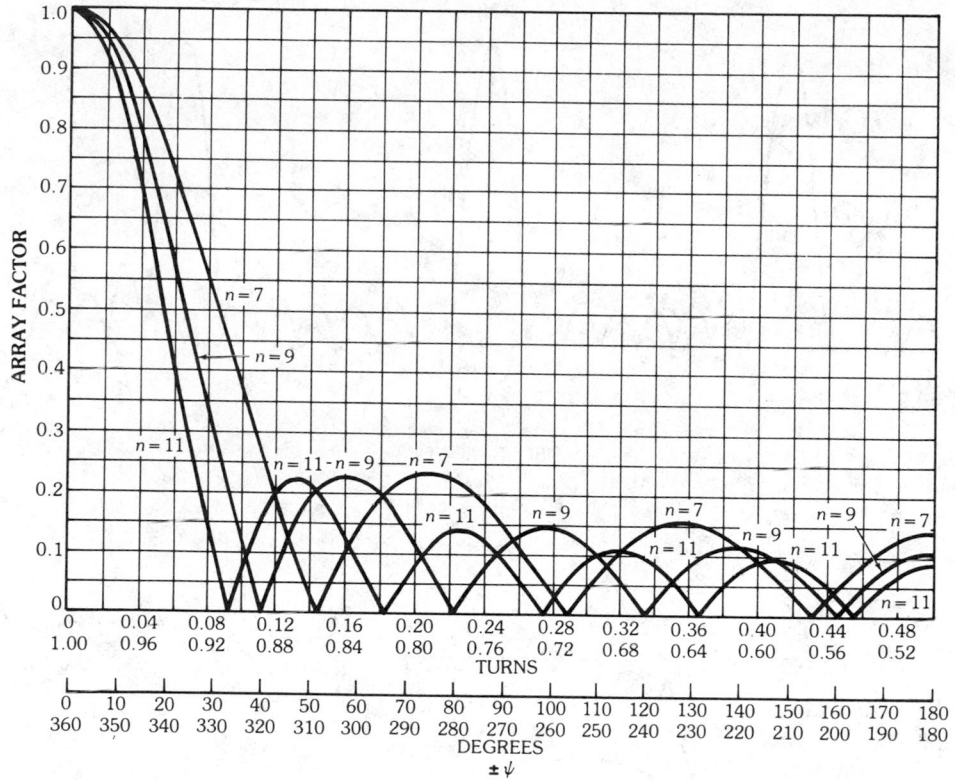

Fig. 3, *continued.*

$$|\psi| = |kd\cos\theta + \beta|_{\theta=0^\circ} \cong \pi/N$$

$$|\psi| = |kd\cos\theta + \beta|_{\theta=180^\circ} \cong \pi \tag{20}$$

It will be seen in Section 6 that this excitation does not give the maximum directivity and, in fact, for $d \cong \lambda/2$ or larger, its directivity is even smaller than that of the ordinary end-fire.

Dolph-Chebyshev Arrays

It is well known that the antenna aperture distribution and pattern function (in the wave vector **k** space, or the *u* space) are a Fourier transform pair. Since for any practical antenna the aperture function must vanish outside a finite region, its Fourier transform, namely the pattern function, is analytic. An analytic function which is zero (or constant) over a finite region must vanish (or be constant) everywhere. This result, if translated into antenna language, implies that any physically realizable antenna pattern must either have a broad beam covering the entire visible region, or have side lobes. From the Parseval's theorem the L_2 norm of the pattern function must be finite. Thus if one side lobe is pushed down, somewhere else the pattern function must go up. Therefore a meaningful optimum

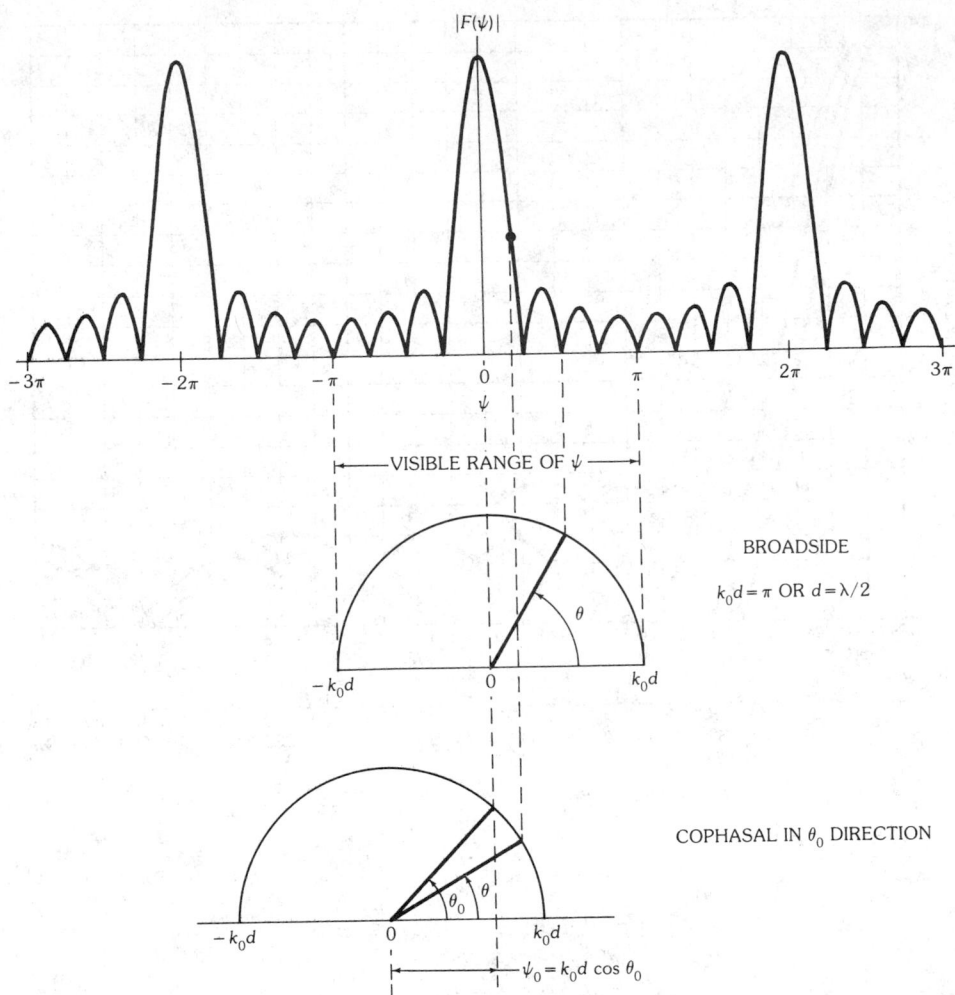

Fig. 4. Graphical method of constructing the radiation pattern from a universal pattern for a uniformly excited broadside array and a uniformly excited cophasal array with the beam in the θ_0 direction.

design would be one in which no one side lobe has a level higher than any others, i.e., equal side lobe levels. For an aperture antenna this is impossible because of the finite L_2 norm of the pattern function for $|u| < \infty$, and the pattern function must approach zero as $|u|$ approaches infinity. Thus the best design one can hope for will be one in which only a *finite* number of side lobes are of equal level. Taylor [100] has provided a solution to this problem which is discussed in Chapter 4.

Since a discrete array can be considered as an aperture antenna sampled at a discrete set of points, all previously stated results are generally applicable. However, when all elements are uniformly spaced, the pattern function becomes periodic (see Chapters 13 and 14). In this case the L_2 norm can be defined over a period, and the Parseval's theorem becomes the well-known power relation. It is thus possible to realize a design of a pattern function with a finite L_2 norm and all

Table 1. Approximate Formulas for a Few Linear Arrays

	Uniform Broadside	Uniform End-fire	Hansen-Woodyard End-fire
Directivity $\cong$ ($Nd \gg \lambda$)	$2N(d/\lambda)$	$4N(d/\lambda)$	$1.79 \times [4N(d/\lambda)]$
Half-power beamwidth $\cong$ ($\pi d/\lambda \ll 1$)	$2\left[\dfrac{\pi}{2} - \cos^{-1}\left(\dfrac{1.391\lambda}{\pi Nd}\right)\right]$	$2\cos^{-1}\left(1 - \dfrac{1.391\lambda}{\pi Nd}\right)$	$2\cos^{-1}\left(1 - 0.1398\dfrac{\lambda}{Nd}\right)$
Beamwidth between nulls	$2\left[\dfrac{\pi}{2} - \cos^{-1}\left(\dfrac{\lambda}{Nd}\right)\right]$	$2\cos^{-1}\left(1 - \dfrac{\lambda}{Nd}\right)$	$2\cos^{-1}\left(1 - \dfrac{\lambda}{2Nd}\right)$
Null angular position $n = 1, 2, \ldots$ $n \neq N, 2N, \ldots$	$\cos^{-1}\left(\pm\dfrac{n}{N}\dfrac{\lambda}{d}\right)$	$\cos^{-1}\left(1 - \dfrac{n\lambda}{Nd}\right)$	$\cos^{-1}\left[1 + (1 - 2n)\dfrac{\lambda}{2Nd}\right]$
Side lobe maximum position $\cong$ ($\pi d/\lambda \ll 1$) ($s = 1, 2, \ldots$)	$\cos^{-1}\left[\pm\dfrac{(2s + 1)\lambda}{2Nd}\right]$	$\cos^{-1}\left[1 - \dfrac{(2s + 1)\lambda}{2Nd}\right]$	$\cos^{-1}\left(1 - \dfrac{s\lambda}{Nd}\right)$

(After Balanis [4], © 1982 Harper & Row, Publishers, Inc.; reprinted by permission of the publisher)

side lobes in one period of equal level. This is achieved by Dolph [22] by making use of the Chebyshev polynomials.

(*a*) For the Dolph-Chebyshev array first consider the Chebyshev polynomials. They can be expressed in either of two equivalent forms, the use of which depends on a particular consideration. For the *m*th degree, they are

$$T_m(x) = \begin{cases} \cos(m\alpha) = \cos(m\cos^{-1}x), & |x| \leqq 1 \\ \cosh(m\alpha) = \cosh(m\cosh^{-1}x), & |x| \geqq 1 \end{cases} \tag{21}$$

or

$$T_m(x) = \text{Re}\{e^{jm\alpha}\} = \text{Re}\{(\cos\alpha + j\sin\alpha)^m\}$$

$$= \cos^m\alpha - \binom{m}{2}(\cos^{m-2}\alpha)(\sin^2\alpha) + \cdots$$

$$+ (-1)^n\binom{m}{2n}(\cos^{m-2n}\alpha)(\sin^{2n}\alpha) + \cdots + \text{Re}\{(j)^m\sin^m\alpha\}$$

$$= x^m - \binom{m}{2}x^{m-2}(1 - x^2) + \cdots + \text{Re}\{(j)^m(1 - x^2)^{m/2}\}$$

$$= A_m\prod_{p=1}^{m}(x - x_p) \tag{22}$$

where

$$x = \cos \alpha \tag{23}$$

$$x_p = \cos \alpha_p = \cos\left(\frac{2p - 1}{2m}\pi\right) = p\text{th root, with } p = 1, 2, \ldots, m \tag{24}$$

$$\binom{m}{2n} = \frac{m!}{(2n)!(m - 2n)!} = \text{binomial coefficient} \tag{25}$$

$$A_0 = A_1 = 1$$
$$A_m = 2A_{m-1}, \qquad m \geqq 2 \tag{26}$$

A few of these polynomials are sketched in Fig. 5. Note that $T_m(x)$ is a

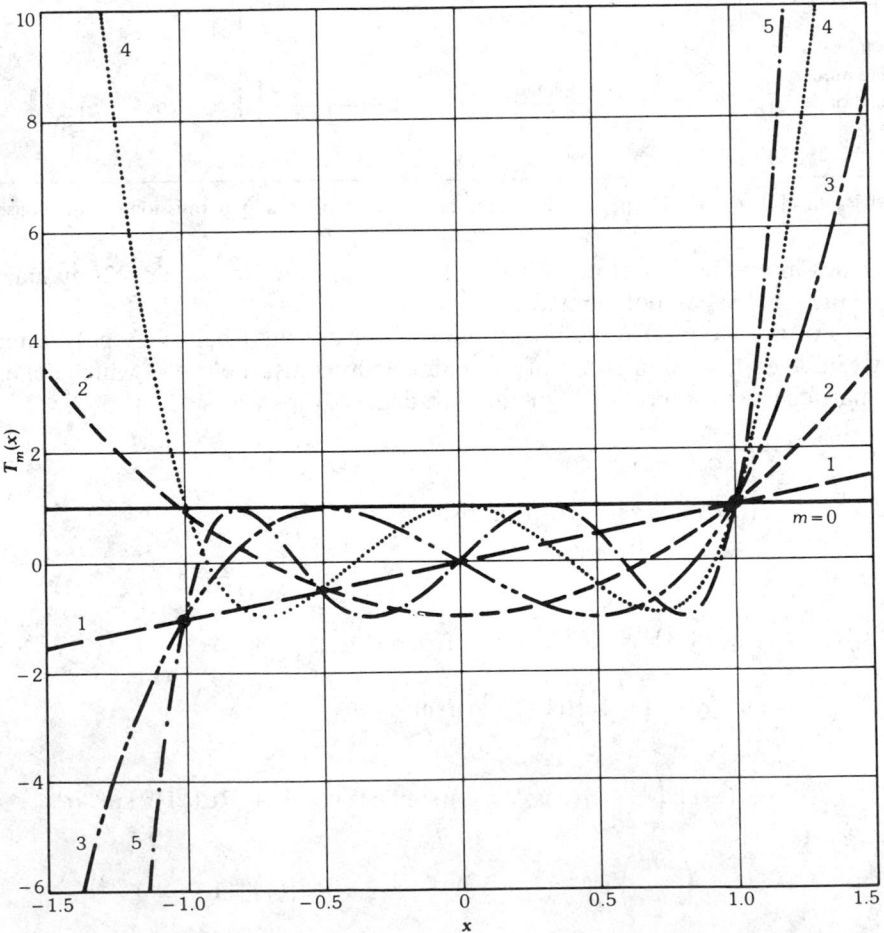

Fig. 5. Chebyshev polynomial $T_m(x)$ for $m = 0, 1, \ldots, 5$. (*After Balanis [4], © 1982, Harper & Row, Publishers, Inc.; reprinted by permission of the publisher*)

polynomial of mth degree, having only even-power terms of x if m is even and only odd-power terms of x if m is odd. The recursion formula, useful for generation, is

$$T_{m+1}(x) + T_{m-1}(x) = 2xT_m(x) \tag{27}$$

with $T_0(x) = 1$ and $T_1(x) = x$. Thus $T_m(x)$ is $1/x$ times the arithmetic mean of its two adjacent neighbors, $T_{m-1}(x)$ and $T_{m+1}(x)$.

(b) Pattern function of an array with N equally spaced and symmetrically excited elements:

If N is odd,

$$\begin{aligned}
F(\theta) &= I_0 + (I_1 e^{j\psi} + I_{-1} e^{-j\psi}) + (I_2 e^{j2\psi} + I_{-2} e^{-j2\psi}) \\
&\quad + \cdots + [I_{N-1/2} e^{j(N-1)\psi/2} + I_{-N-1/2} e^{-j(N-1)\psi/2}] \\
&= I_0 + \sum_{n=1}^{(N-1)/2} I_n(Z^n + Z^{-n})
\end{aligned} \tag{28}$$

If N is even,

$$\begin{aligned}
F(\theta) &= I_1 e^{j\psi/2} + (I_{-1} e^{-j\psi/2}) + (I_2 e^{j3\psi/2} + I_{-2} e^{-j3\psi/2}) \\
&\quad + \cdots + [I_{N/2} e^{j(N-1)\psi/2} + I_{-N/2} e^{-j(N-1)\psi/2}] \\
&= \sum_{n=1}^{N/2} I_n[Z^{(2n-1)/2} + Z^{-(2n-1)/2}]
\end{aligned} \tag{29}$$

where $\psi = kd \cos \theta$, $Z = e^{j\psi}$, and d is the interelement spacing.

(c) To relate (a) and (b), let

$$x = a \cos \psi/2 = a(Z^{1/2} + Z^{-1/2})/2 \tag{30}$$

Then

$$\begin{aligned}
T_{N-1}(x) &= A_{N-1} \prod_{p=1}^{N-1} (x - x_p) \\
&= A_{N-1} \prod_{p=1}^{N-1} a(Z^{1/2} + Z^{-1/2} - Z_p^{1/2} - Z_p^{-1/2})/2
\end{aligned} \tag{31}$$

$$x_p = a \cos(\psi_p/2) = \cos\left[\frac{2p-1}{2(N-1)}\pi\right] \tag{32}$$

$$\begin{aligned}
Z_p &= e^{j\psi}p \\
\psi_p &= kd \cos \theta_p
\end{aligned} \tag{33}$$

$$\theta_p = \text{angular position of the } p\text{th pattern null} \tag{34}$$

Note that the pattern function as given by (28) or (29) is a polynomial in $Z^{1/2}$ of power from $-(N - 1)$ to $(N - 1)$, and so is $T_{N-1}(x)$ as given by (31). They can be equated to determine the excitation $\{I_n\}$ once the spacing d/λ and the parameter a are chosen. To find a, one needs to specify the desired side lobe level. Let the main beam to side lobe level ratio be b. Then

$$b = T_{N-1}(a) = \cosh[(N - 1)\cosh^{-1}a]$$

or

$$a = \cosh\left(\frac{1}{N - 1}\cosh^{-1}b\right) \tag{35}$$

as θ varies from $\qquad\qquad 0 \to \pi/2 \to \pi$

ψ varies from $\qquad 2\pi d/\lambda \to\ 0\ \to -2\pi d/\lambda$

x varies from $a\cos(\pi d/\lambda) \to\ a\ \to a\cos(\pi d/\lambda)$ $\tag{36}$

From Fig. 6 it is seen that to avoid grating lobes

$$a\cos(\pi d/\lambda) \leqq -1 \tag{37}$$

Although tables for $\{I_n\}$ are available [93], today, with the availability of computers, it is simple to compute them according to the formula given by Elliott [28]:

$$I_n = I_{-n} = \sum_{p=n}^{(N-1)/2}(-1)^{(N-1)/2-p}\frac{N - 1}{N - 1 + 2p}\begin{bmatrix}(N - 1)/2 + p\\2p\end{bmatrix}\begin{bmatrix}2p\\p - n\end{bmatrix}a^{2p},$$

$$\text{if } N \text{ is odd} \tag{38}$$

$$I_n = I_{-n} = \sum_{p=n}^{N/2}(-1)^{(N/2)-p}\frac{N - 1}{N + 2(p - 1)}\begin{bmatrix}N/2 + p - 1\\2p - 1\end{bmatrix}\begin{bmatrix}2p - 1\\p - n\end{bmatrix}a^{2p-1},$$

$$\text{if } N \text{ is even} \tag{39}$$

(d) Design procedure: Choose number of elements N and main beam-to-side-lobe ratio b; then successively determine the following quantities, using the equations shown in the boxes, and finally the required excitation $\{I_n\}$

$$\downarrow \quad \boxed{(35)}$$
$$a$$

$$\downarrow \quad \boxed{(37)}$$

the largest permissible value of d

$$\downarrow \boxed{(32)}$$

$$\text{roots } \{x_k\} \text{ of } T_{N-1}(x)$$

$$\downarrow \boxed{(33) \text{ and } (34)}$$

$$\psi_k \text{ and } Z_k$$

$$\downarrow \boxed{(31)}$$

expand $T_{N-1}(x)$ in terms of $Z^{\pm 1/2}$ (ignoring the factor A_{N-1} and a)

$$\downarrow$$

compare with (28) or (29) to obtain $\{I_n\}$, or use (38) and (39).

(e) To find the pattern, use (30) and (31) [or (28) and (29)]. Or, graphically, as shown in Fig. 6, first construct $T_{N-1}(x)$ and a circle with radius a and center at $x = 0$, and then use (36) to find a point on the circle corresponding to a given value θ and project on the x axis to find $T_{N-1}(x)$, which is $F_{N-1}(\theta)$ except for a constant multiplier.

(f) Cophasal Dolph-Chebyshev array. Let θ_0 be the main beam angle and replace ψ in the above discussion by $\psi - \psi_0 = (2\pi d/\lambda)(\cos\theta - \cos\theta_0)$; then what happened at $\theta = \pi/2$ (i.e., the main beam) will occur at $\theta = \theta_0$. However, for the visible region,

$$\theta = \qquad 0 \qquad\qquad \rightarrow \theta_0 \rightarrow \pi$$

$$\psi - \psi_0 = \frac{2\pi d}{\lambda}(1 - \cos\theta_0) \qquad \rightarrow 0 \rightarrow -\frac{2\pi d}{\lambda}(1 + \cos\theta_0)$$

$$x = a\cos[(\psi - \psi_0)/2] = a\cos\left[\frac{\pi d}{\lambda}(1 - \cos\theta_0)\right] \rightarrow a \rightarrow a\cos\left(\frac{\pi d}{\lambda}\right)(1 + \cos\theta_0) \tag{40}$$

To avoid the presence of grating lobes, neither of the end points of x in the above range should be less than -1. This will determine the largest allowable element spacing d.

Comments:

(1) For large arrays and for side lobe levels in the range of -20 to -60 dB, the excitation transformation method of Elliott [28] gives an approximation solution for the half-power beamwidth in terms of the beam-broadening factor f, which is defined as the ratio of the half-power beamwidth of the array with the Dolph-Chebyshev excitation and that with a uniform excitation. Fig. 7a shows the plot of f versus side lobe level in decibels.

(2) A Dolph-Chebyshev array is optimum only in the sense of narrowest beamwidth for a given side lobe level, or lowest side lobe level for a given beamwidth, but not for maximum directivity. Elliott [28] gives an approximate expression for the directivity of a large Dolph-Chebyshev array:

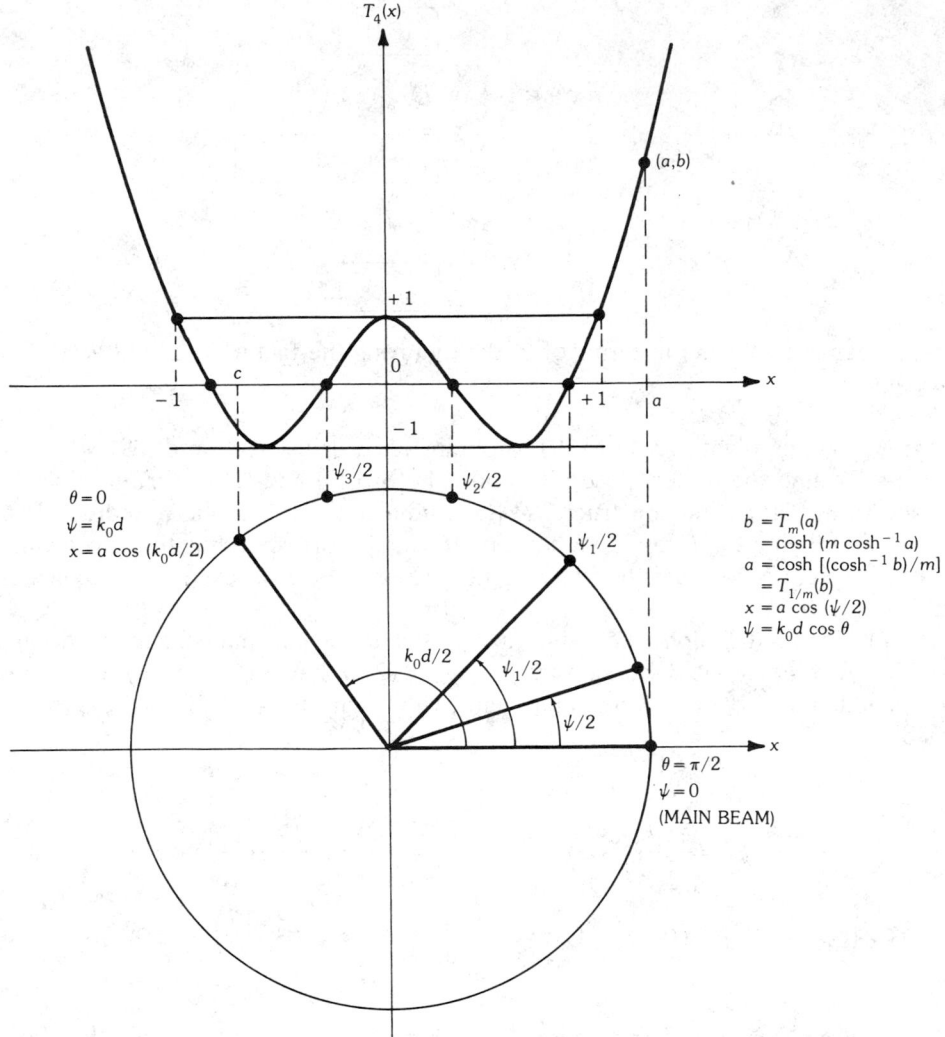

Fig. 6. Graphical method for constructing the radiation pattern of a five-element Dolph-Chebyshev array with side lobe level $20 \log b$ and element spacing d.

$$D \cong \frac{2b^2}{1 + (b^2 - 1)f\lambda/(L + d)}$$

where

   L = the total array length

   f = the beam-broadening factor given in Fig. 7a

   b = determined by the desired side lobe level [see (35)]

Fig. 7b shows the plots of D versus $(L + d)/\lambda$ for several side lobe levels. The directivity is generally lower than that of a uniform array, particularly for large L and high side lcbe levels.

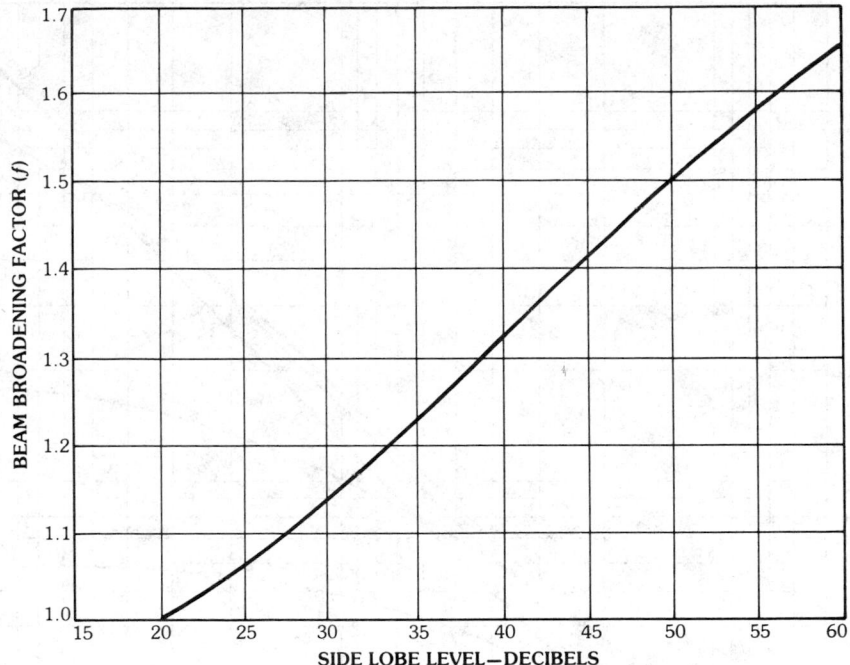

a

Fig. 7. Beam broadening factor and directivity of a Chebyshev array as functions of side lobe level. (*a*) Beam broadening factor *f* versus side lobe level. (*b*) Directivity versus ($L + d$)/λ for various side lobe levels from 15 to 40 dB. (*After Elliott [121], reprinted with permission of* Microwave Journal, *from the December 1963 issue,* © *1963 Horizon House–Microwave, Inc.*)

(*g*) An alternative Dolph-Chebyshev array design. So far the Chebyshev polynomial in (*a*) and the pattern function in (*b*) are related by (30) as shown graphically in Fig. 6, where the center of the circle is at $x = 0$. The visible region as given by (36) may not necessarily reach the point $x = -1$, depending on the value of $a \cos \pi d/\lambda$. To make full use of the interval $(-1, a)$ for x, one may let

$$x = c \cos \psi + h = 2c(Z + Z^{-1}) + h \tag{41}$$

where

$$\psi = kd \cos \theta$$

$$a = \cosh\left(\frac{2 \cosh^{-1} b}{N - 1}\right)$$

$$c = \frac{a + 1}{1 - \cos kd} \tag{42}$$

$$h = -\frac{a \cos kd + 1}{1 - \cos kd}$$

In this case,

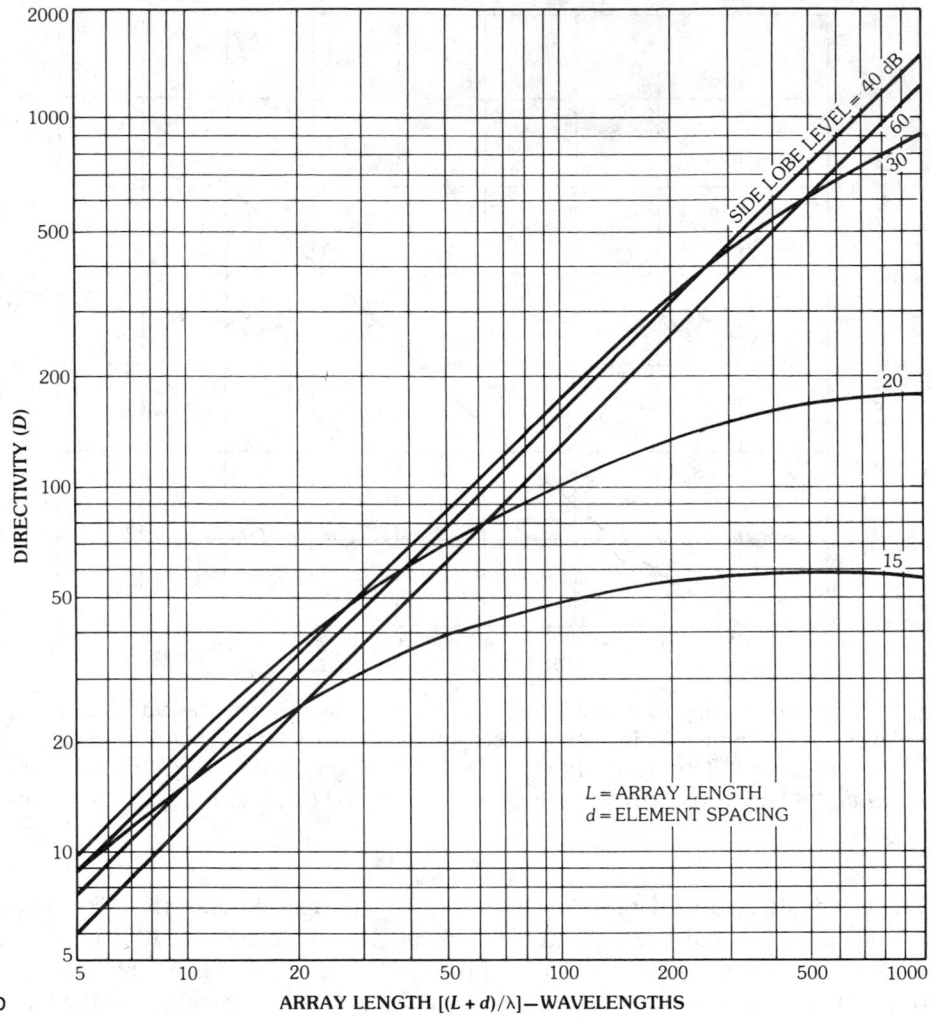

b

ARRAY LENGTH [$(L + d)/\lambda$]—WAVELENGTHS

Fig. 7, *continued.*

$$\theta = 0 \to \pi/2 \to \pi$$

$$\psi = kd \to 0 \to -kd$$

$$x = -1 \to a \to -1$$

To find $\{I_n\}$, the term x defined by (41), instead of (30), must be substituted in (31) and the polynomial is compared with (28) for odd N, since now the polynomials in (28) and (31) are both in powers of Z, not $Z^{1/2}$, and also there is a zero-degree term due to h in (41). Furthermore, the power of Z in (28) ranges from $(N - 1)/2$ to $-(N - 1)/2$ [or similarly from $(N - 1)$ to 0 by multiplying all terms by $Z^{-(N-1)/2}$], whereas that in (31) with x substituted by (41) ranges from $(N - 1)$ to $-(N - 1)$.

Since the number of terms in the polynomial must equal the number of elements, N, only the Chebyshev polynomial of $(N - 1)/2$ degree, i.e., $T_{(N-1)/2}(x)$, is needed for (31). Drane [23] has given the following formula for the excitation of an array with N (odd) number of elements:

$$I_n = \frac{\varepsilon_n}{4M} \sum_{m=0}^{M_1} \varepsilon_m \varepsilon_{M_2-m} T_n(Y_m) S_M^n(c, h, y_m) \tag{43}$$

where

$$\varepsilon_0 = 1 \text{ and } \varepsilon_n = 2 \quad \text{for } n \neq 0,$$
$$y_m = \cos(m\pi/M)$$
$$M = (N - 1)/2$$
$$M_1, M_2 = \text{the integer parts of } M/2 \text{ and } (M + 1)/2, \text{ respectively,}$$
$$S_M^n(c, h, y_m) = T_M(cy_m + h) + (-1)^n T_M(h - cy_m) \tag{44}$$

If M is even,

$$S_M^n(c, h, y_{M/2}) = T_M(cy_{M/2} + h) \tag{45}$$

4. Linear Transformations in Antenna Arrays

A linear transformation can be applied to array pattern functions in two different ways: (1) When applied to element positions it relates the pattern function of an array of one geometry to that of another [64]. In other words, the pattern of one member of a family of arrays determines the patterns of the entire family, if their geometries are linearly transformable. (2) When applied to excitations it can express the pattern function of an array with one type of excitation in terms of the patterns of the same array with different types of excitations, in particular, the patterns for some simple canonical excitations, such as uniform cophasal excitations. Of course, the application of some of these transformations is valid only if the mutual coupling effect among elements can be ignored, since this effect, in general, is not linearly dependent on element spacings. But the transformation can also be used if the feeding network is redesigned to compensate for the coupling effect after transformation. We shall discuss some fundamental results in this section.

Linear Transformations in Array Geometry

Let us consider a Cartesian coordinate system with bases $\{\hat{\mathbf{e}}_1, \hat{\mathbf{e}}_2, \hat{\mathbf{e}}_3\}$ in which all lengths are expressed in terms of free-space wavelength λ. First we define the following vectors in matrix notations:

$$\mathbf{E} = (\hat{\mathbf{e}}_1, \hat{\mathbf{e}}_2, \hat{\mathbf{e}}_3) \tag{46}$$

$$\mathbf{X} = (x_1, x_2, x_3)^t \tag{47}$$

$$\mathbf{U} = (u_1, u_2, u_3)^t \tag{48}$$

where the superscript t, as usual, designates the transpose, and the us are Cartesian components of a unit vector $\hat{\mathbf{u}}$ in spherical coordinates, namely

$$u_1 = \sin\theta\cos\phi, \qquad u_2 = \sin\theta\sin\phi, \qquad u_3 = \cos\theta \tag{49}$$

In the following discussion we shall use either vector notations or matrix notations, depending on which one is more convenient in a particular discussion. Thus

$$\mathbf{x} = x_1\hat{\mathbf{e}}_1 + x_2\hat{\mathbf{e}}_2 + x_3\hat{\mathbf{e}}_3 = \mathbf{EX} \tag{50}$$

$$\hat{\mathbf{u}} = u_1\hat{\mathbf{e}}_1 + u_2\hat{\mathbf{e}}_2 + u_3\hat{\mathbf{e}}_3 = \mathbf{EU} \tag{51}$$

Then from (2a) for each component, the pattern, due to a source called x, with distribution $J(\mathbf{x})$, as a function of (θ, ϕ), or $\hat{\mathbf{u}}$, is

$$F_x(\hat{\mathbf{u}}) = \int_{\text{source } x} J(\mathbf{x})\, e^{j2\pi\hat{\mathbf{u}}\cdot\mathbf{x}}\, d\mathbf{x} \tag{52}$$

where $d\mathbf{x} = dx_1\, dx_2\, dx_3$, the subscript x of $F_x(\cdot)$ denotes the pattern function for source x, and $J(\mathbf{x})$ can be any source distribution function, with continuous, sectionally continuous, or discrete point sources. For the latter case,

$$J(\mathbf{x}) = \sum_n I(\mathbf{x}_n)\, \delta(\mathbf{x} - \mathbf{x}_n) \tag{53}$$

and

$$F_x(\hat{\mathbf{u}}) = \sum_n I(x_n)\, e^{j2\pi\hat{\mathbf{u}}\cdot\mathbf{x}_n} \tag{54}$$

which is simply the array factor for point elements at $\{\mathbf{x}_n\}$ with excitations $\{I_n\}$. Thus the so-called array factor can be regarded as a special case of (52).

Equation 52 indicates that $F_x(\hat{\mathbf{u}})$ and $J(\mathbf{x})$ are a Fourier transform pair. However, in physical space (θ, ϕ) are real, and, because of (49), the observable pattern function is described by only a part of the Fourier transform of $J(x)$ which lies on a unit sphere in the transform space u.

Now let $\mathbf{X}$ and $\mathbf{U}$ undergo linear transformations described by matrices $\mathbf{A}$ and $\mathbf{B}$, respectively, which takes $\mathbf{X}$ to $\mathbf{Y} = \mathbf{AX}$ and $\mathbf{U}$ to $\mathbf{V} = \mathbf{BU}$. If

$$\mathbf{B}^{-1} = \mathbf{A}^t \tag{55}$$

then the scalar product

$$\mathbf{v \cdot y} = \mathbf{V'Y} = (\mathbf{BU})'(\mathbf{AX}) = \mathbf{U'X} = \mathbf{u \cdot x} \tag{56}$$

remains invariant. Hence (52) can be written as

$$F_x(\hat{\mathbf{u}}) = \int_{\text{source } x} J(\mathbf{x}) \, e^{j2\pi \hat{\mathbf{u}} \cdot \mathbf{x}} \, d\mathbf{x}$$

$$= \int_{\text{source } y} J(\mathbf{A}^{-1}\mathbf{Y}) \, e^{j2\pi \mathbf{v} \cdot \mathbf{y}} \, |\mathbf{A}|^{-1} d\mathbf{Y} = F_y(\mathbf{v}) \tag{57}$$

where $|\mathbf{A}|$ is the determinant of $\mathbf{A}$. This relation states in effect that the field $J(\mathbf{x})$ observed in the direction $\hat{\mathbf{u}}$ due to the source x is the same as that in the direction $\mathbf{v}$ due to the source y, that is, $J(\mathbf{y}) = |\mathbf{A}|^{-1} J(\mathbf{A}^{-1}\mathbf{Y})$. Thus the pattern function of a member of a family of arrays which are related to each other by a linear transformation in geometry determines the patterns of all members of the family. However, since in general $\mathbf{u} \neq \mathbf{v}$, part of the invisible region in the u space may become visible in the v space after the transformation, or vice versa. To see this transformation in detail, let us consider first the scalar product invariant transformation, from (55):

$$\mathbf{B'A} = \mathbf{A'B} = \mathbf{E}$$

Let $\mathbf{a}_1, \mathbf{a}_2, \mathbf{a}_3$ be the column vector of $\mathbf{A}$ and $\mathbf{b}_1, \mathbf{b}_2, \mathbf{b}_3$ be those of $\mathbf{B}$. Then $\mathbf{Y} = \mathbf{AX}$ implies that

$$\mathbf{y} = y_1 \hat{\mathbf{e}}_1 + y_2 \hat{\mathbf{e}}_2 + y_3 \hat{\mathbf{e}}_3 = x_1 \mathbf{a}_1 + x_2 \mathbf{a}_2 + x_3 \mathbf{a}_3 \tag{58}$$

This transformation can be regarded as a *mapping* (rotation and linear stretching) of $\mathbf{x}$ into $\mathbf{y}$, or, by comparing the right-hand side of (58) with (50), simply a *relabeling* of the base vectors of $\mathbf{x}$ from $\mathbf{e}_n$ into $\mathbf{a}_n$. Similar interpretations can be given to $\mathbf{V} = \mathbf{BU}$,

$$\mathbf{v} = v_1 \hat{\mathbf{e}}_1 + v_2 \hat{\mathbf{e}}_2 + v_3 \hat{\mathbf{e}}_3 = u_1 \mathbf{b}_1 + u_2 \mathbf{b}_2 + u_3 \mathbf{b}_3 \tag{59}$$

For invariant scalar product, (55) states that

$$\mathbf{b}_i \cdot \mathbf{a}_j = \delta_{ij}, \quad i, j = 1, 2, 3 \tag{60}$$

where the Kronecker delta $\delta_{ij} = 0$ if $i \neq j$ and 1 if $i = j$. Thus $\mathbf{b}_i$ has a projection on $\mathbf{a}_i$ equal to $1/a_i$ and a direction perpendicular to all other $\mathbf{a}_j$s. Specifically,

$$\mathbf{b}_1 = |\mathbf{A}|^{-1}(\mathbf{a}_2 \times \mathbf{a}_3), \qquad \mathbf{b}_2 = |\mathbf{A}|^{-1}(\mathbf{a}_3 \times \mathbf{a}_1), \qquad \mathbf{b}_3 = |\mathbf{A}|^{-1}(\mathbf{a}_1 \times \mathbf{a}_2) \tag{61}$$

Here $\{\mathbf{a}_i\}$ and $\{\mathbf{b}_i\}$ are called *reciprocal bases* to each other. These are very useful in the study of periodic structures and Floquet space harmonics.

Application to Planar Periodic Arrays

Fig. 8a shows a broadside planar array with elements at the square grid intersection points in the $x_1 x_2$ plane and spacing d (in wavelengths). Its pattern function in the $u_1 u_2$ space is also periodic as shown in Fig. 8b, where the small circles indicate the locations of the grating lobes (including the main beam). If the array is of infinite extent, each circle corresponds to a space harmonic. The spacing between two circles along the u_1 or u_2 axis is $1/d$. The visible region is bounded by a unit circle and a typical point in this $u_1 u_2$ plane has the polar coordinates $(\sin \theta, \phi)$. Thus a given point (u_1, u_2) determines uniquely (θ, ϕ) in the physical region $0 \leqq \theta \leqq \pi$, $0 \leqq \phi < 2\pi$, and vice versa.

For a cophasal array with main beam at (θ_0, ϕ_0)

$$J(\mathbf{x}) = |J(\mathbf{x})| \exp(-j2\pi \hat{\mathbf{u}}_0 \cdot \mathbf{x}) \tag{62}$$

where

$$u_{01} = \sin \theta_0 \cos \phi_0, \qquad u_{02} = \sin \theta_0 \sin \phi_0 \tag{63}$$

This additional phase results simply in a translation of the origin of the $u_1 u_2$ space by $(-u_{01}, -u_{02})$ as shown in Fig. 9 with all grating beam circles fixed, or with the origin remains fixed and all grating beam circles translated by (u_{01}, u_{02}). The visible region in this case is given by a unit disc with (u_{01}, u_{02}) as center as shown in Fig. 9a. For a phased array with full scan range, $0 \leqq \theta_0 \leqq \pi$ and $0 \leqq \phi_0 < 2\pi$, the *overall* visible region is bounded by a disc with radius 2 and center at the original origin as shown in Fig. 9b. For a particular scan range, say $0 \leqq \theta_0 \leqq \theta_0''$ and $\phi_0' \leqq \phi_0 \leqq \phi_0''$, the total visible region is bounded by that part of the $u_1 u_2$ plane which is covered by all unit discs with their centers (u_{01}, u_{02}) satisfying the inequalities

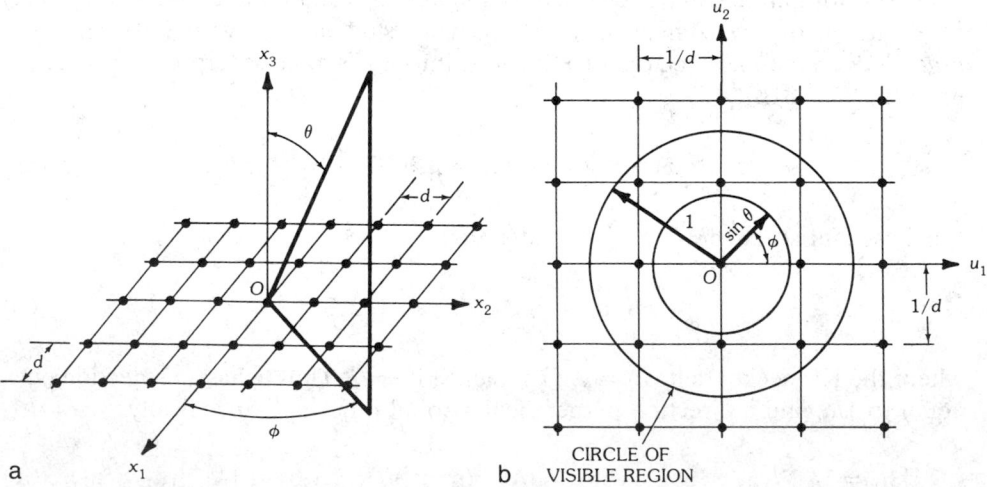

Fig. 8. Broadside planar array and pattern function. (*a*) A periodic planar array with square cells in the $x_1 x_2$ plane. (*b*) Periodic structure of the pattern function of the array (*a*) in the $u_1 u_2$ plane. (*After Lo and Lee [64], © 1965 IEEE*)

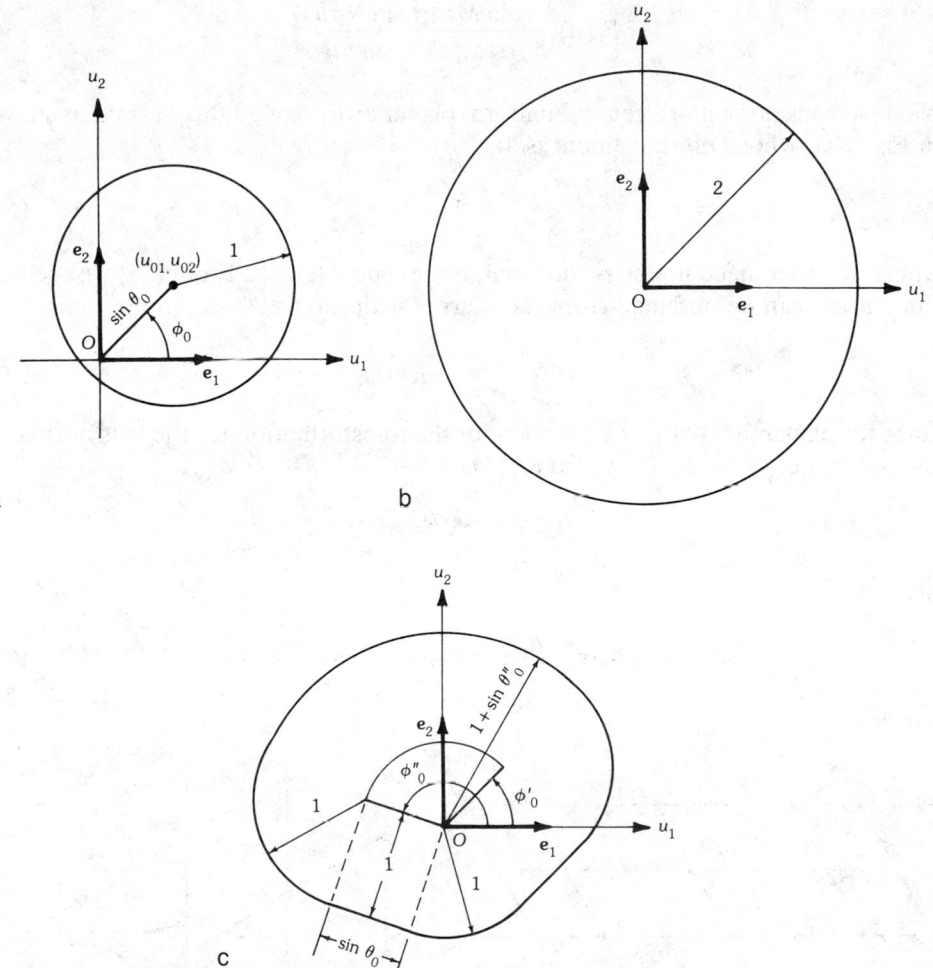

Fig. 9. Visible region in the $u_1 u_2$ plane. (*a*) As the beam is scanned to (θ_0, ϕ_0). (*b*) For a full scan: $0 \leqq \theta_0 \leqq \pi$ and $0 \leqq \phi_0 \leqq 2\pi$. (*c*) For a limited scan: $\phi'_0 \leqq \phi_0 \leqq \phi''_0$ and $0 \leqq \theta_0 \leqq \theta''_0$. (*After Lo and Lee [64], © 1965 IEEE*)

$$0 \leqq (u_{01}^2 + u_{02}^2)^{1/2} \leqq \sin \theta''_0 \quad \text{and} \quad \phi'_0 \leqq \tan^{-1}\left(\frac{u_{02}}{u_{01}}\right) \leqq \phi''_0 \tag{64}$$

as illustrated in Fig. 9c.

An Example—Consider a uniform array (called the *x* array) with square grid size $d = 1$ as shown in Fig. 8a. Then the *mn*th element position is

$$\mathbf{x}_{mn} = m\hat{\mathbf{e}}_1 + n\hat{\mathbf{e}}_2, \qquad 1 \leqq m \leqq M, \qquad 1 \leqq n \leqq N \tag{65}$$

Its pattern is

$$F_x(\mathbf{u}) = \frac{\sin M\pi u_1}{\sin \pi u_1} \frac{\sin N\pi u_2}{\sin \pi u_2} \tag{66}$$

Next we consider a more general uniform planar array, called the y array, as shown in Fig. 10a, whose mnth element is at

$$\mathbf{y}_{mn} = m\mathbf{a}_1 + n\mathbf{a}_2, \qquad 1 \leqq m \leqq M, \qquad 1 \leqq n \leqq N \tag{67}$$

where $\mathbf{a}_1$ and $\mathbf{a}_2$ need not be orthogonal, nor of equal length. From (58) we see that the y array can be obtained from the x array with the transformation

$$\mathbf{A} = (\mathbf{a}_1, \mathbf{a}_2, \hat{\mathbf{e}}_3) \tag{68}$$

since for planar arrays there is no need for the transformation for the axis normal to the array, i.e., $\mathbf{a}_3 = \hat{\mathbf{e}}_3$. Now, from (55),

$$\mathbf{U} = \mathbf{B}^{-1}\mathbf{V} = \mathbf{A}'\mathbf{V}$$

or

$$\mathbf{u}_1 = \mathbf{a}_1 \cdot \mathbf{v}, \qquad \mathbf{u}_2 = \mathbf{a}_2 \cdot \mathbf{v}, \qquad \mathbf{u}_3 = \hat{\mathbf{e}}_3 \cdot \mathbf{v} \tag{69}$$

where

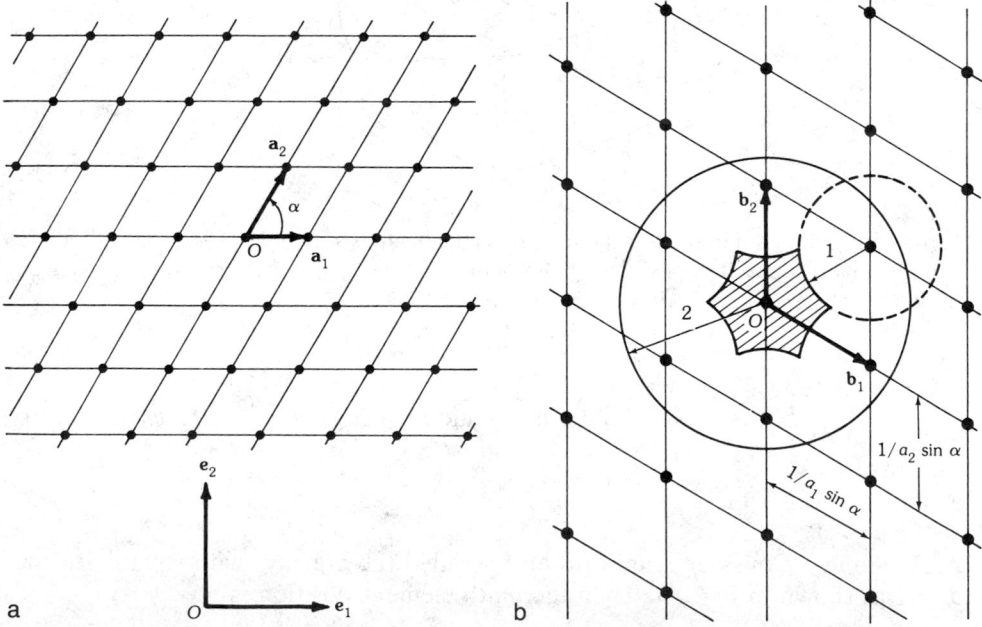

Fig. 10. Periodic planar array and pattern function. (*a*) Periodic planar array with parallelogram cells defined by vectors $\mathbf{a}_1$ and $\mathbf{a}_2$ in y space. (*b*) Periodic pattern function structure in the $v_1 v_2$ plane. (*After Lo and Lee [64],* © *1965 IEEE*)

$$\mathbf{v} = \sin\theta\cos\phi\hat{\mathbf{e}}_1 + \sin\theta\sin\phi\hat{\mathbf{e}}_2 + \cos\theta\hat{\mathbf{e}}_3 \tag{70}$$

Therefore from (66) and (69) the pattern of the y array is simply

$$F_y(\mathbf{v}) = \frac{\sin(M\pi\mathbf{a}_1\cdot\mathbf{v})}{\sin(\pi\mathbf{a}_1\cdot\mathbf{v})}\frac{\sin(N\pi\mathbf{a}_2\cdot\mathbf{v})}{\sin(\pi\mathbf{a}_2\cdot\mathbf{v})} \tag{71}$$

Since $F_x(\mathbf{u})$ has grating lobes at u_1 and u_2 equal to $0, \pm 1, \pm 2, \ldots$, or

$$\mathbf{u} = m\hat{\mathbf{e}}_1 + n\hat{\mathbf{e}}_2, \qquad m, n = 0, \pm 1, \pm 2, \ldots, \tag{72}$$

$F_y(\mathbf{v})$ will have grating lobes at $(\mathbf{a}_1\cdot\mathbf{v})$ and $(\mathbf{a}_2\cdot\mathbf{v})$ equal to $0, \pm 1, \pm 2, \ldots$, or

$$\mathbf{v} = m\mathbf{b}_1 + n\mathbf{b}_2 \tag{73}$$

where $\mathbf{b}_1$ and $\mathbf{b}_2$ are determined from $\mathbf{a}_1$ and $\mathbf{a}_2$ from (61) as shown in Fig. 10b. Note that $\mathbf{b}_1$ and $\mathbf{b}_2$ are perpendicular to $\mathbf{a}_1$ and $\mathbf{a}_2$, respectively, and that the lengths of $\mathbf{b}_1$ and $\mathbf{b}_2$ are equal to the reciprocals of $a_1\sin\alpha$ and $a_2\sin\alpha$, respectively. If $a_1 = a_2$ and the angle between $\mathbf{a}_1$ and $\mathbf{a}_2$, namely α, is $60°$, the y array becomes an equilateral-triangular grid. It is very simple to determine the range of scan for which no grating lobe will appear. This region is bounded by the arcs of unit circles with centers at all adjacent grating lobe locations shown as the shaded parts in Fig. 10b. To translate the boundary of this region into one in the $\theta\phi$ plane, one needs to recall that any point in the v_1v_2 plane has the polar coordinates $(\sin\theta, \phi)$. Thus by placing a transparent overlay with ordinary polar coordinates (except that the concentric circles are marked with θ according to the value of $\sin\theta$ as their radii) on the top of Fig. 10b, one can determine immediately the scanning region in θ and ϕ (see Fig. 14). This method is not only simpler but also more illuminating than others [88].

Nonuniform Excitation and Relation between Aperture Antenna and Discrete Array

Often an array with discrete elements can be considered as a sampled set of an aperture antenna with a distribution function $f_0(\mathbf{y})$. A question of interest will be how close the pattern of the sampling array is to that of the aperture antenna. To see this, let $f_0(\mathbf{y})$ be the given excitation for $\mathbf{y}$ inside the aperture and zero outside the aperture. Then for the array $\{y\}$ one may have

$$f(\mathbf{y}) = f_0(\mathbf{y})\sum_{m,n}\delta(\mathbf{y} - \mathbf{y}_{mn}) \tag{74}$$

where $\delta(\mathbf{y} - \mathbf{y}_{mn})$ is the two-dimensional delta function and the summation is taken over all the integers for both m and n. Now applying the transformation $\mathbf{A}$, one obtains a corresponding array $\{x\}$ with excitation

$$f(\mathbf{AX}) = f_0(\mathbf{AX})\sum_{m,n}\delta(\mathbf{AX} - \mathbf{AX}_{mn}) \tag{75}$$

By substituting (75) into (52) and making use of the convolution integral theorem, one has

$$F_x(\mathbf{u}) = F_{0x}(\mathbf{u}) * \sum_{m,n} \delta(\mathbf{U} - \mathbf{X}_{mn}) |\mathbf{A}|^{-1} \tag{76}$$

where the asterisk denotes the convolution integration and $F_{0x}(\mathbf{u})$ is the Fourier tranform or the pattern function of an aperture antenna with excitation $f_0(\mathbf{AX})$. As in (67), if $\mathbf{y}_{mn} = m\mathbf{a}_1 + n\mathbf{a}_2$, then $\mathbf{x}_{mn} = m\hat{\mathbf{e}}_1 + n\hat{\mathbf{e}}_2$ and $F_x(\mathbf{u})$ is periodic in $\mathbf{u}_1$ and $\mathbf{u}_2$ of unity period. Equation 76 states that $F_x(\mathbf{u})$ is the sum of infinitely many $F_{0x}(\mathbf{u})$s, each displaced from the other by a unit along the u_1 and u_2 axes. Since $f_0(\mathbf{AX})$ is an aperture limited function, $F_{0x}(\mathbf{u})$ becomes negligible for large u. Therefore, for (u_1, u_2) in the neighborhood of (m, n), $F_x(\mathbf{u})$ is essentially equal to $F_{0x}(\mathbf{u})$. From (76) the pattern function of array $\{y\}$ is given immediately by*

$$F_y(\mathbf{v}) = F_x(\mathbf{B}^{-1}\mathbf{V}) = |\mathbf{B}| F_{0x}(\mathbf{B}^{-1}\mathbf{V}) * \sum_{m,n} \delta(\mathbf{V} - \mathbf{BX}_{mn})$$

$$= |\mathbf{B}| F_{0x}(\mathbf{B}^{-1}\mathbf{V}) * \sum_{m,n} \delta(\mathbf{v} - m\mathbf{b}_1 - n\mathbf{b}_2) \tag{77}$$

As in Fig. 10b the pattern function of array $\{y\}$ is also a periodic function with periods inversely proportional to $a_1 \sin \alpha$ and $a_2 \sin \alpha$ along $\mathbf{b}_1$ and $\mathbf{b}_2$, respectively, and it is thus a linearly distorted pattern of $F_x(\mathbf{u})$. Obviously (76) and (77) also show the exact difference between the pattern function for continuous (or sectionally continuous) excitation and that for discrete excitation with periodically spaced elements.

Hexagonal Arrays

Periodic arrays with unit cells other than parallelograms may also be analyzed in a similar manner. As an example, consider a hexagonal array of uniform excitation, as shown in Fig. 11a, where the element positions are marked by crosses. The array may be regarded simply as a superposition of two parallelogram arrays with equal excitations but opposite in sign. The sides of the two parallelograms are defined by

$$\mathbf{a}_1 = \hat{\mathbf{e}}_1, \qquad \mathbf{a}_2 = \cos(60°)\,\hat{\mathbf{e}}_1 + \sin(60°)\,\hat{\mathbf{e}}_2$$

$$\mathbf{a}_1' = 3\hat{\mathbf{e}}_1, \qquad \mathbf{a}_2' = \sqrt{3}\cos(30°)\,\hat{\mathbf{e}}_1 + \sqrt{3}\sin(30°)\,\hat{\mathbf{e}}_2 \tag{78}$$

From the relation $\mathbf{B} = (\mathbf{A}')^{-1}$, one immediately obtains

---

*It is understood that the convolution integral in (77) refers to the variable **v** whereas in (76) it refers to the variable **u**.

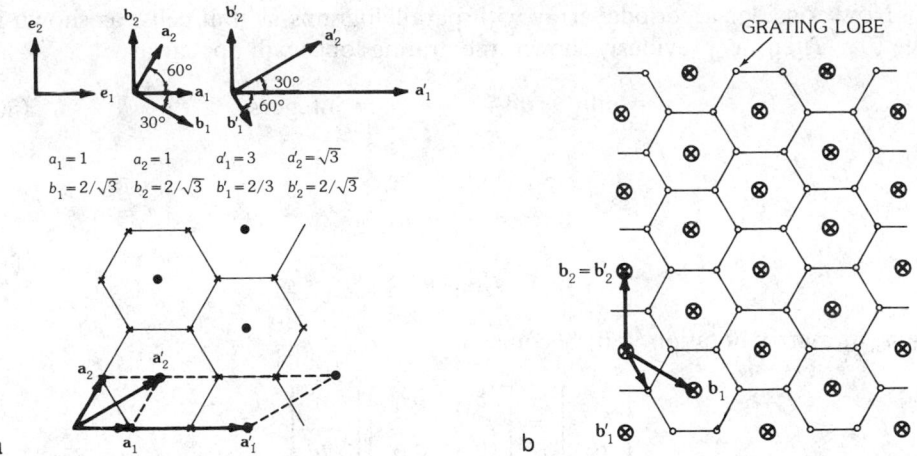

Fig. 11. Periodic array with hexagonal cells and hexagonal pattern function structure. (*a*) Array with cells in *y* space. (*b*) Pattern function structure in *v* space. (*After Lo and Lee [64], © 1965 IEEE*)

$$\mathbf{b}_1 = \frac{2}{\sqrt{3}}\cos(-30°)\,\hat{\mathbf{e}}_1 + \frac{2}{\sqrt{3}}\sin(-30°)\,\hat{\mathbf{e}}_2, \qquad \mathbf{b}_2 = \frac{2}{\sqrt{3}}\hat{\mathbf{e}}_2$$

$$\mathbf{b}_1' = \frac{2}{3}\cos(-60°)\,\hat{\mathbf{e}}_1 + \frac{2}{3}\sin(-60°)\,\hat{\mathbf{e}}_2, \qquad \mathbf{b}_2' = \frac{2}{\sqrt{3}}\hat{\mathbf{e}}_2$$

(79)

as shown in Fig. 11b. It is seen that the grating lobes, denoted by the small circles without crosses, also form a hexagonal periodic structure.

Periodic Arrays with Minimum Number of Elements

In practice it is often desirable to have only a single main beam in the visible region. This condition sets an upper limit on the element spacing of a period array. For the array with elements at the vertices of square cells, as shown in Fig. 8a, it is well known that for the case of full scan the element spacing must be smaller than half wavelength, as seen from Fig. 8b. As a result this sets a lower limit on the required number of elements. In the following we shall show that, for some other periodical arrays, this lower limit can be reduced.

First, it should be understood that the previous statement about a single main beam in the visible region, in fact, has different meanings, depending on the types of arrays under discussion. For example, for a linear array, due to symmetry, a single main beam may actually be a disklike beam when it is broadside, a cone-shaped beam when it is fired at an angle, and a unidirectional beam when it is end-fired. On the other hand, for a planar array, whenever there is beam firing above the plane of the array, there is always one below the plane, due to symmetry, except for the limiting case of end-fire. Customarily, one may regard all of these as being degenerated cases of a three-dimensional array which, in general, can possess a truly single beam in the visible range.

Now consider a periodic array with parallelograms as unit cells, as shown in Fig. 10a. Then, as previously shown, the grating lobes will appear at

$$\mathbf{u} = m\mathbf{b}_1 + n\mathbf{b}_2, \qquad m, n = \text{integers} \tag{80}$$

Let

$$\mathbf{b}_1 = b_{11}\hat{\mathbf{e}}_1 + b_{12}\hat{\mathbf{e}}_2$$

$$\mathbf{b}_2 = b_{21}\hat{\mathbf{e}}_1 + b_{22}\hat{\mathbf{e}}_2 \tag{81}$$

Then, in matrix notation, (80) becomes

$$\begin{bmatrix} u_1 \\ u_2 \end{bmatrix} = \begin{bmatrix} b_{11} & b_{21} \\ b_{12} & b_{22} \end{bmatrix} \begin{bmatrix} m \\ n \end{bmatrix}$$

The objective is to determine all bs under the condition that no grating lobes appear in the visible region* and then determine all as which fix the array arrangement. In order to make the number of elements minimum for a given aperture one must add a condition that the area of each unit cell, namely $|\mathbf{A}|$, should be made as large as possible, or $|\mathbf{B}|$ as small as possible.

To give an example, first consider the case of *full* scan. Then it is required that:

(1) for all grating lobes outside the *full* scan circle in u space: $(b_{11}m + b_{21}n)^2 + (b_{12}m + b_{22}n)^2 \geqq 4$, for m, n, equal to all integers except $m = n = 0$;
(2) for minimum number of elements:
  $|\mathbf{B}| = b_{11}b_{22} - b_{12}b_{21}$ to be minimum

The solution is

$$\mathbf{a}_1 = \frac{1}{\sqrt{3}}\hat{\mathbf{e}}_1 \quad \text{and} \quad \mathbf{a}_2 = \frac{1}{\sqrt{3}}[\cos(60°)\,\hat{\mathbf{e}}_1 + \sin(60°)\,\hat{\mathbf{e}}_2] \tag{82}$$

This is the well-known equilateral-triangular grid structure.

If the scan range is limited to

$$0 \leqq \theta_0 \leqq \theta_0'' \quad \text{and} \quad 0 \leqq \phi_0 \leqq 2\pi$$

the solution can be determined similarly as

$$\mathbf{a}_1 = \frac{2}{\sqrt{3}(1 + \sin\theta_0'')}\hat{\mathbf{e}}_1$$

$$\mathbf{a}_2 = \frac{2}{\sqrt{3}(1 + \sin\theta_0'')}(\hat{\mathbf{e}}_1 \cos 60° + \hat{\mathbf{e}}_2 \sin 60°) \tag{83}$$

---

*This condition is regarded as one where the grating lobe begins to show up in the visible region at the extreme scanning angle. This determines the upper limit of the element spacings. Actually, to avoid the grating lobe *completely*, the spacings should be somewhat *smaller*, depending on the beam shape, width, or the aperture and the excitation function.

Now comparing this array with a conventional one with square cells, as shown in Fig. 8a, the number of elements is reduced approximately by

$$\frac{|\mathbf{B}_{sq}| - |\mathbf{B}|}{|\mathbf{B}_{sq}|} = \frac{4 - 2\sqrt{3}}{4} = 13.4 \text{ percent}$$

where $|\mathbf{B}_{sq}|^{-1}$, $|\mathbf{B}|^{-1}$ are, respectively, the areas of a square cell and a parallelogram given by (83). It can be easily shown that this percentage is dependent on ϕ_0'' only, and the previous value applies when ϕ_0 covers the whole range $(0, 2\pi)$. If ϕ_0 covers only part of this range, condition 1 should be revised. For the special case, $0 \leqq \theta_0 < 45°$, $0 \leqq \phi_0 < 2\pi$, $a_1 = a_2 = 0.676$.

It is easy to extend the previous results to a three-dimensional periodic array with elements over vertices of parallelopiped cells. In this case, for *full* scan, $\mathbf{b}_1$, $\mathbf{b}_2$, and $\mathbf{b}_3$ are found to form an oblique coordinate system with 60° between any two vectors, and each with length 2. With respect to the orthonormal basis $\{\hat{\mathbf{e}}_1, \hat{\mathbf{e}}_2, \hat{\mathbf{e}}_3\}$, they are

$$\mathbf{b}_1 = 2\hat{\mathbf{e}}_1, \qquad \mathbf{b}_2 = \hat{\mathbf{e}}_1 + \sqrt{3}\hat{\mathbf{e}}_2, \qquad \mathbf{b}_3 = \hat{\mathbf{e}}_1 + \frac{1}{\sqrt{3}}\hat{\mathbf{e}}_2 + \frac{2\sqrt{2}}{\sqrt{3}}\hat{\mathbf{e}}_2 \qquad (84)$$

From the relation between as and bs, one obtains

$$\mathbf{a}_1 = \frac{1}{2}\hat{\mathbf{e}}_1 - \frac{1}{2\sqrt{3}}\hat{\mathbf{e}}_2 - \frac{1}{2\sqrt{6}}\hat{\mathbf{e}}_3, \qquad \mathbf{a}_2 = \frac{1}{\sqrt{3}}\hat{\mathbf{e}}_2 - \frac{1}{2\sqrt{6}}\hat{\mathbf{e}}_3, \qquad \mathbf{a}_3 = \frac{\sqrt{3}}{2\sqrt{2}}\hat{\mathbf{e}}_3 \qquad (85)$$

It is readily verified that all as have a length equal to $\sqrt{3}/2\sqrt{2}$ and make an angle of $\cos^{-1}(-1/3) = 109.5°$ between any two of them. In comparison with a conventional array with cubic cells, the number of elements is reduced approximately by 29.3 percent.

Transformation between Circular and Elliptical Arrays

The pattern of a circular array or aperture antenna can be expressed in terms of Bessel functions (see later in this section, under "Circular Arrays"). In a similar manner the pattern of an elliptical array or aperture antenna can be expressed in terms of Mathieu functions. But the numerical computation of the latter is much more difficult than that of the former. By using linear transformation, this difficulty can be alleviated.

Let the Cartesian coordinates of the nth element with excitation I_n of an elliptical array, called the x array, be $\mathbf{x}_n = (x_{n_1}, x_{n_2})$ in the $\hat{\mathbf{e}}_1\hat{\mathbf{e}}_2$ plane; then

$$(x_{n_1}/a)^2 + (x_{n_2}/b)^2 = 1 \qquad (86)$$

where $2a$ and $2b$ are the major and minor axes, respectively. The pattern function of the x array is

$$F_x(\mathbf{u}) = \sum_n I_n e^{j2\pi\mathbf{u}\cdot\mathbf{x}_n} \qquad (87)$$

Now, applying the transformation

$$\mathbf{A} = \begin{bmatrix} 1 & 0 \\ 0 & \tau \end{bmatrix} \quad \text{and} \quad \mathbf{B} = (\mathbf{A}^{-1})^t = \begin{bmatrix} 1 & 0 \\ 0 & t \end{bmatrix} \tag{88}$$

where $\tau = 1/t = a/b =$ axial ratio, one obtains

$$F_x(\mathbf{u}) = \sum_n I_n e^{j2\pi\mathbf{u}\cdot\mathbf{x}_n} = \sum_n I_n e^{j2\pi\mathbf{v}\cdot\mathbf{y}_n} = F_y(\mathbf{v}) \tag{89}$$

where

$$\mathbf{y}_n = \begin{bmatrix} y_{n_1} \\ y_{n_2} \end{bmatrix} = \mathbf{AX} = \begin{bmatrix} 1 & 0 \\ 0 & \tau \end{bmatrix}\begin{bmatrix} x_{n_1} \\ x_{n_2} \end{bmatrix} = \begin{bmatrix} x_{n_1} \\ \tau x_{n_2} \end{bmatrix} \tag{90}$$

and

$$\mathbf{V} = \begin{bmatrix} v_1 \\ v_2 \end{bmatrix} = \mathbf{BU} = \begin{bmatrix} 1 & 0 \\ 0 & t \end{bmatrix}\begin{bmatrix} u_1 \\ u_2 \end{bmatrix} = \begin{bmatrix} u_1 \\ t u_2 \end{bmatrix} \tag{91}$$

From (86) and (90)

$$y_{n_1}{}^2 + y_{n_2}{}^2 = a^2 \tag{92}$$

Thus the y array is a circular array whose nth element excitation is still I_n. Similarly, since

$$u_1{}^2 + u_2{}^2 = \sin^2\theta^{(u)}\cos^2\phi^{(u)} + \sin^2\theta^{(u)}\sin^2\phi^{(u)} = \sin^2\theta^{(u)} \tag{93}$$

one obtains from (91)

$$v_1{}^2 + (\tau v_2)^2 = \sin^2\theta^{(v)}\cos^2\phi^{(v)} + \tau^2\sin^2\phi^{(v)}\sin^2\phi^{(v)} = \sin^2\theta^{(u)} \tag{94}$$

where the superscript u is used to denote the values of θ and ϕ associated with a certain point in the u space and the superscript v is used to denote the values of θ and ϕ for its transformed point in the v space. All the above results can be summarized as in the following:

> "The pattern of an elliptical array, with excitation $\{I_n\}$, major axis $2a$, and axial ratio τ when computed over a circle of radius $\sin\theta^{(u)}$ and $0 \leq \phi^{(u)} < 2\pi$ in the u plane is the same as that of a circular array, with excitations $\{I_n\}$ and radius a when computed over an ellipse with major axis $2\sin\theta^{(u)}$ and axial ratio τ in the v plane, as shown in Fig. 12."

Sometimes it is more convenient to specify element position and observation point by azimuthal angles as shown in Fig. 12. One finds that the element located at angle

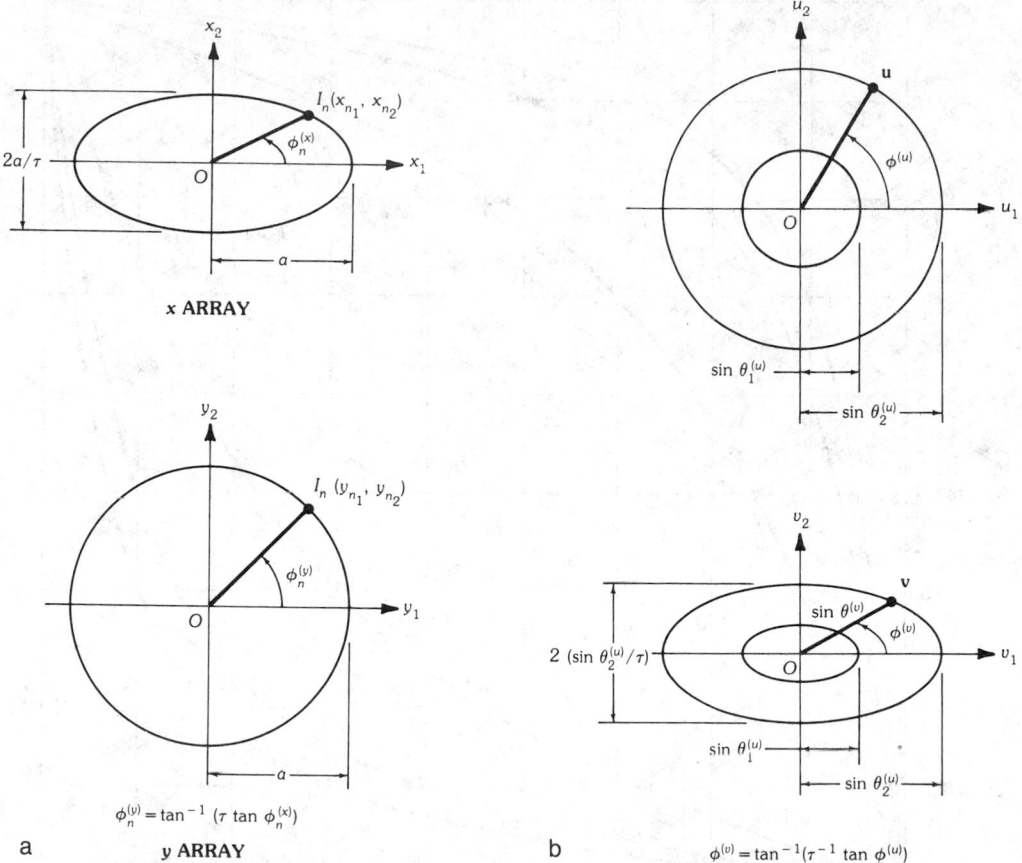

Fig. 12. Far field at u with polar coordinates $(\sin\theta^{(u)}, \phi^{(u)})$ in the $u_1 u_2$ plane due to an elliptical array x is the same as that at v with polar coordinates $(\sin\theta^{(v)}, \phi^{(v)})$ in the $v_1 v_2$ plane due to a circular array y. (a) Array plane. (b) Field plane. (*After Lo and Hsuan [66], © 1965 IEEE*)

$\phi_n^{(x)}$ of the x array will be located for the y array at

$$\phi_n^{(y)} = \tan^{-1}(\tau\tan\phi_n^{(x)}) \tag{95}$$

Similarly, an observation at $\phi^{(u)}$ on the circle of radius $\sin\theta^{(u)}$ in the u plane will be on the ellipse with major axis $2\sin\theta^{(u)}$ and axial ratio τ in the v plane at angle

$$\phi^{(v)} = \tan^{-1}(\tau^{-1}\tan\phi^{(u)}) \tag{96}$$

All of these are shown in Fig. 12, and also Fig. 13 for $\tau = 3$, 7, and 10. A few examples will be given in later in this section, under "Circular Arrays." This technique can also be applied to aperture antennas. In particular the Taylor distribution for a circular aperture [99, 43, 45] can be extended to that for an elliptical aperture [66].

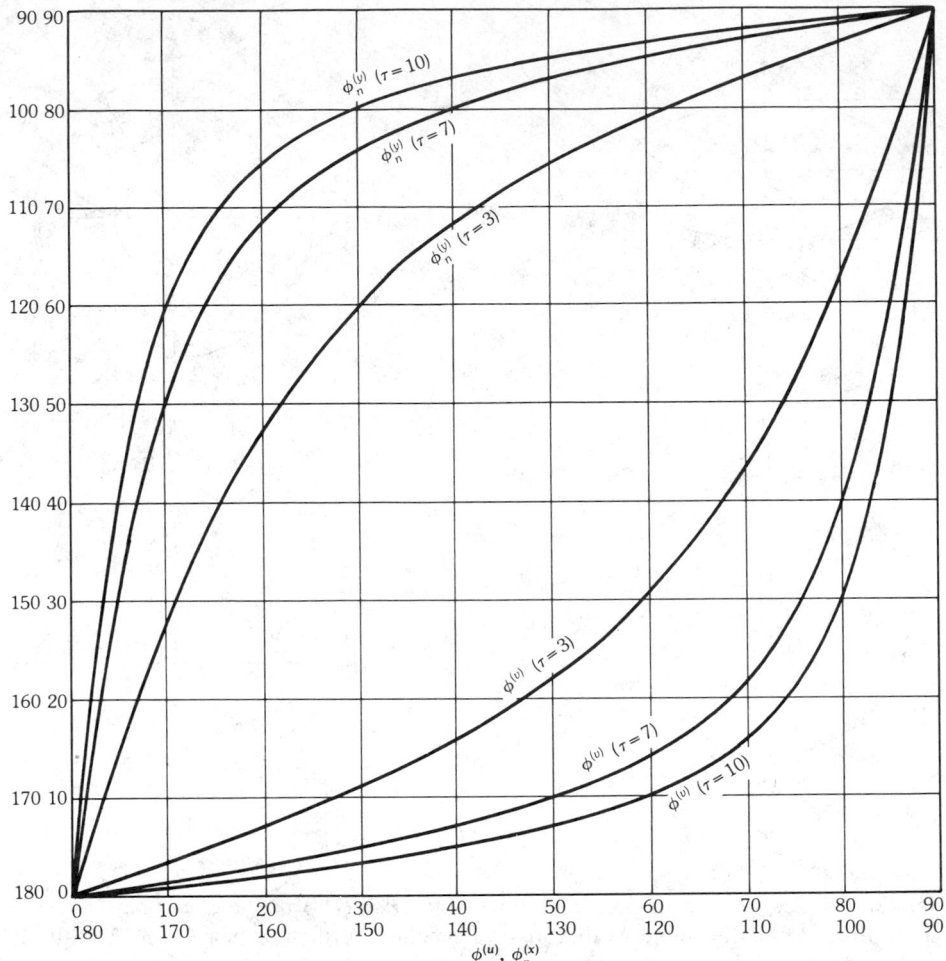

Fig. 13. Transformation of the angular position $\phi_n^{(x)}$ of the nth element of the x array to that of the y array, $\phi_n^{(y)}$, for various axial ratios τ; also, transformations of the observation angle $\phi^{(u)}$ in the u plane to that in the v plane, $\phi^{(v)}$, for various values of τ. (*After Lo and Hsuan* [66], © 1965 IEEE)

Beam and Pattern Distortion Due to Scanning

It is well known that the beam broadens and the pattern distorts in the $\theta\phi$ space as the beam scans away from the broadside direction [28]. However, if the pattern is presented in the u space, the beam scanning due to progressive phase shift as in (62) results only in a translation of the origin of the u space, leaving the beam and the pattern unchanged. The so-called distortion is really a result of the nonlinear-functional relationship between (θ, ϕ) and $\mathbf{u} - \mathbf{u}_0$. An understanding of this relationship is therefore important for determining the distortion, no matter which array is considered.

Consider a hemisphere of unit radius as shown in Fig. 14. A point on the sphere

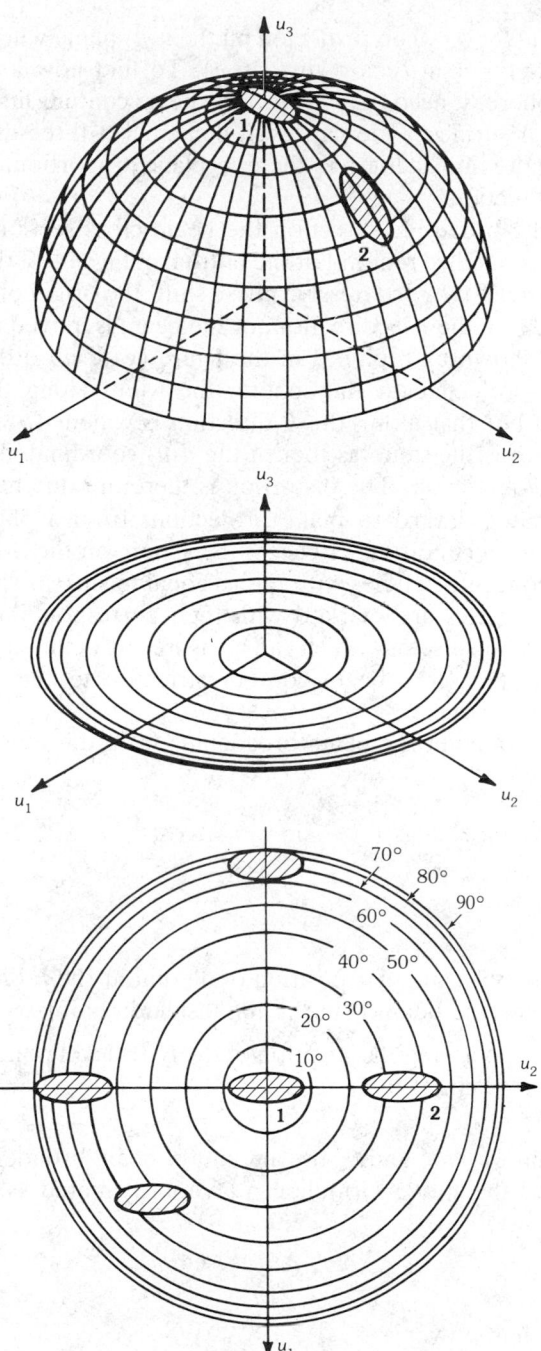

Fig. 14. Beam and pattern distortion as the beam scans away from the polar axis, a result from the nonlinear projection of the u_1u_2 plane on the sphere (the beam shape in the u_1u_2 plane remains unchanged as the beam scans).

with coordinates $(1, \theta, \phi)$ when projected on the $u_1 u_2$ plane will have the Cartesian coordinates $(u_1, u_2) = (\sin \theta \cos \phi, \sin \theta \sin \phi)$. To find how a beam (or pattern) appears on the sphere, it need only project the beam contour lines in the $u_1 u_2$ plane onto the spherical surface. One can even construct a three-dimensional pattern model by making the radial length in the $u_1 u_2$ plane proportional to the level of the contour in that direction.

Suppose in the broadside direction the beam cross section at a certain level looks like the cross-hatched region 1 at the bottom of Fig. 14. If this beam is scanned to $\theta = 33°$ and $\phi = 90°$ by a progressive phase shift, the origin of the $u_1 u_2$ plane will be moved by $\sin 33°$ in the $-u_2$ direction, or the beam is moved by $\sin 33°$ along the $+u_2$ direction as shown by region 2 in the figure (with *no* distortion). However, when beam 2 is projected on the sphere, the width along the ϕ direction remains unchanged but that along the θ direction is widened. Since the azimuthal angle of the sphere is the same as that of the $u_1 u_2$ coordinate system, there is no distortion in the ϕ direction. The distortion is therefore due to $\sin \theta$ only.

It is somewhat awkward to make projections from a plane onto a sphere. Instead, one can project latitude circles of the sphere on the $u_1 u_2$ plane, or simply draw a series of concentric circles with radius equal to $\sin \theta$ as shown at the bottom of Fig. 14. If the circles are marked with their corresponding values of θ, one obtains a set of nonlinear scales which can measure any beam distortion due to scan simply by moving the same beam contour structure to the desired scan angle (θ_0, ϕ_0) in this set of scales.

One can also compute the beam broadening from the following formula:

$$\sin \Delta\theta_p = \sin(\theta_0 + \Delta\theta_p') - \sin \theta_0 = 2 \cos(\theta_0 + \Delta\theta_p'/2) \sin(\Delta\theta_p'/2) \qquad (97)$$

where

$\Delta\theta_p$ = the polar-angular position of a point at the p-dB level of the beam when the beam is in the broadside direction

$\Delta\theta_p' + \theta_0$ = the polar-angular position of the p-dB level point of the beam when it is scanned to θ_0

Once $\Delta\theta_p'$ is found for $\Delta\theta_p$ and θ_0 the amount of beam broadening is determined. For *narrow* beams the above formula can be approximated as

$$\Delta\theta_p'/\Delta\theta_p \cong \sec \theta_0$$

in particular, for half-power

$$\Delta\theta_{\text{HP}}' = \Delta\theta_{\text{HP}}/\cos \theta_0 \qquad (98)$$

which is the well-known result; namely, the half-power beamwidth broadens according to the cosine of the scan angle from the broadside direction.

Linear Transformations on Excitations

For simplicity, consider a linear array of N elements at $z = z_1, z_2, \ldots, z_N$ (in wavelengths) and with excitations $I_1, I_2, \ldots, I_N$, respectively. Consider further a nonsingular matrix $\mathbf{A}$ and let

$$\mathbf{I}' = \mathbf{A}^{-1}\mathbf{I} \quad \text{or} \quad \mathbf{I} = \mathbf{A}\mathbf{I}' \tag{99}$$

where $\mathbf{I}$ is the column vector with elements $I_1, I_2, \ldots, I_N$ and $\mathbf{I}'$ is another column vector, with elements $I'_1, I'_2, \ldots, I'_N$. Then the pattern of the array can be written as

$$
\begin{aligned}
f(\theta) &= \sum_n^N I_n e^{j\psi_n} = \mathbf{V}'\mathbf{I} = \mathbf{V}'\mathbf{A}\mathbf{I}' \\
&= \mathbf{V}^t(\mathbf{A}_1, \mathbf{A}_2, \ldots, \mathbf{A}_N)\,\mathbf{I}' \\
&= \mathbf{V}^t\mathbf{A}_1 I'_1 + \mathbf{V}^t\mathbf{A}_2 I'_2 + \cdots + \mathbf{V}^t\mathbf{A}_N I'_N \\
&= F_1(\theta)\,I'_1 + F_2(\theta)\,I'_2 + \cdots + F_N(\theta)\,I'_N
\end{aligned}
\tag{100}
$$

where the superscript t denotes the transpose, and

$$\mathbf{V}^t = (e^{j\psi_1}, e^{j\psi_2}, \ldots, e^{j\psi_N}) \tag{101}$$

$$\psi_n = 2\pi z_n \cos\theta \tag{102}$$

$\mathbf{A}_1, \mathbf{A}_2, \ldots, \mathbf{A}_N$ are the first, second, $\ldots$, and Nth *columns* of matrix $\mathbf{A}$

$$F_n(\theta) = \mathbf{V}^t\mathbf{A}_n = \sum_m^N A_{mn} e^{j\psi_m} \tag{103}$$

Equation 103 states that $F_n(\theta)$ is the pattern of an array with excitations $A_{1n}, A_{2n}, \ldots, A_{Nn}$, and (100) states that the pattern of the array with excitations $I_1, I_2, \ldots, I_N$ is the sum of $F_1(\theta), F_2(\theta), \ldots, F_N(\theta)$ weighed by $I'_1, I'_2, \ldots, I'_N$, respectively. In other words, the pattern of an array with excitations $\{I_n\}$ can be expressed in terms of the patterns associated with some other, say canonical, excitations. An interesting application is given below.

Consider a linear uniformly spaced array of N elements with interelement spacing d (in wavelengths) and excitations $I_1, I_2, \ldots, I_N$. Let

$$
\mathbf{A} =
\begin{bmatrix}
1 & 1 & 1 & 1 & \cdots & 1 \\
1 & e^{-j2\pi/N} & e^{-j4\pi/N} & e^{-j6\pi/N} & \cdots & e^{-j2\pi(N-1)/N} \\
1 & e^{-j4\pi/N} & e^{-j8\pi/N} & e^{-j12\pi/N} & \cdots & e^{-j4\pi(N-1)/N} \\
\vdots & \vdots & \vdots & \vdots & & \vdots \\
1 & e^{-j2\pi(N-1)/N} & e^{-j4\pi(N-1)/N} & e^{-j6\pi(N-1)/N} & \cdots & e^{-j2\pi(N-1)^2/N}
\end{bmatrix}
\tag{104}
$$

Then

$$F(\theta) = \frac{1}{N}\sum_{n}^{N} I_n e^{j\psi_n} = \mathbf{I}_1' F_1(\theta) + \mathbf{I}_2' F_2(\theta) + \cdots + \mathbf{I}_N' F_N(\theta) \qquad (105)$$

where the factor $1/N$ has been added for convenience, and

$$F_1(\theta) = \mathbf{V}^t \mathbf{A}_1 = \frac{1}{N}\sum_{n=0}^{N-1} e^{j2\pi nd\cos\theta} = \frac{1}{N}\frac{\sin(N\psi/2)}{\sin(\psi/2)} \qquad (106)$$

$$F_2(\theta) = \mathbf{V}^t \mathbf{A}_2 = \frac{1}{N}\sum_{n=0}^{N-1} e^{j2\pi nd(\cos\theta - 1/Nd)} = \frac{1}{N}\frac{\sin[N(\psi - \psi^{(1)})/2]}{\sin[(\psi - \psi^{(1)})/2]} \qquad (107)$$

$$F_3(\theta) = \mathbf{V}^t \mathbf{A}_3 = \frac{1}{N}\sum_{n=0}^{N-1} e^{j2\pi nd(\cos\theta - 2/Nd)} = \frac{1}{N}\frac{\sin[N(\psi - \psi^{(2)})/2]}{\sin[(\psi - \psi^{(2)})/2]} \qquad (108)$$

$$F_N(\theta) = \mathbf{V}^t \mathbf{A}_N = \frac{1}{N}\sum_{n=0}^{N-1} e^{j2\pi nd[\cos\theta - (N-1)/Nd]} = \frac{1}{N}\frac{\sin[N(\psi - \psi^{(N-1)})/2]}{\sin[(\psi - \psi^{(N-1)})/2]} \qquad (109)$$

$$\psi = 2\pi d\cos\theta$$
$$\psi^{(m)} = 2\pi m/N$$

Equations 106–109 are simply uniform array patterns with a progressive phase shift such that the main beam appears at

$$\psi = \psi^{(m)} \quad \text{or} \quad \theta = \cos^{-1}(m/Nd) \qquad (110)$$

The set of functions $\{F_n(\theta)\}$ has the following interesting properties:
  (a)   All their nulls coincide.
  (b)   Let the beam maximum of $F_n(\theta)$ be at $\theta_{max}^{(n)}$. Then at this angle all other members of $\{F_n(\theta)\}$ vanish, i.e., $F_p(\theta_{max}^{(n)}) = 1$ as $p = n$, and 0 as $p \neq n$. This result has been used for pattern synthesis [117, 118]. From (105)

$$F(\theta_{max}^{(n)}) = I_n', \qquad n = 1, 2, \ldots, N \qquad (111)$$

Thus let the desired pattern be sampled at equispaced points in $\psi_{max}^{(n)} = 2\pi d\cos\theta_{max}^{(n)}$, for $n = 1, 2, \ldots, N$; then (111) states that these samples are simply $\{I_n'\}$ and from (99) one determines the required excitations:

$$\mathbf{I} = \mathbf{A}\mathbf{I}' = \mathbf{A}F(\theta_{max}^{(n)}) \qquad (112)$$

Of course, the synthesizing pattern agrees with the desired pattern exactly *only* at this set of points. Since in general there is some arbitrariness in selecting this set (such as the starting point, sampling rate, and sampling numbers—see "An Example," below), the solution is not unique. One should in general try a few possibilities, bearing in mind that the sampling rate depends on d, sampling number is the number of elements N, and array size is Nd (in wavelengths). For close agreement with the desired pattern, a high sampling rate and a large sampling number are required, resulting in closely spaced elements. In that case impractical unstable solutions, as for the so-called superdirectivity problem, may occur (see Section 7 for more discussion).

To recapitulate, the philosophy of this method can be viewed as a linear transformation from the N degrees of freedom of $\{I_n\}$ to another N degrees of freedom of $\{I'_n\}$ as shown in (99). Therefore its application can be very general. For example, $F(\theta)$ can also be expressed in terms of a *set* of patterns, each associated with a *uniform aperture* excitation function $f_n(z)$ along z:

$$I'_n f_n(z) = I'_n e^{j2\pi z \cos\theta_n}, \qquad n = 1, 2, \ldots, N \qquad (113)$$

The array so synthesized will be an aperture antenna. Mathematically speaking, the difference between this and the discrete array is rather trivial, mainly in the choice of "basis" functions only.

An Example—Fig. 15 shows an example considered by Balanis [4] in which the desired pattern is

$$F_d(\theta) = \begin{cases} 1 & \text{if } \pi/4 \leqq \theta \leqq 3\pi/4 \\ 0 & \text{elsewhere} \end{cases}$$

It is synthesized first by a linear array with 11 elements and $d = 1/2$, then by a continuous aperture distribution of length 5λ, both sampled at 11 points as shown in Table 2. Fig. 15b shows seven nonzero composing functions $\{F_n(\theta)\}$ and their sum.

Another Example—Fig. 16 shows two among many possible solutions for synthesizing a $\csc\theta$ pattern with a linear array of 21 elements and spacing $d = 1/2$. The first solution is obtained by sampling $\csc\theta$ such that the synthesizing array pattern is zero at $\theta = 0°$. The second solution is obtained by choosing the sampling points such that the synthesizing pattern is 0.45 at $\theta = 0°$. It is seen that the latter gives a closer solution to the desired pattern $\csc\theta$ over the useful region.

Circular Arrays

First consider a circular array of N elements uniformly excited and uniformly distributed over a circle of radius a (in wavelengths). Then the nth element will be located at

$$\mathbf{r}_n = \hat{x}a\cos\phi_n + \hat{y}a\sin\phi_n \qquad (114)$$

where

$$\phi_n = 2\pi n/N, \qquad n = 1, 2, \ldots, N \qquad (115)$$

Then the pattern function is

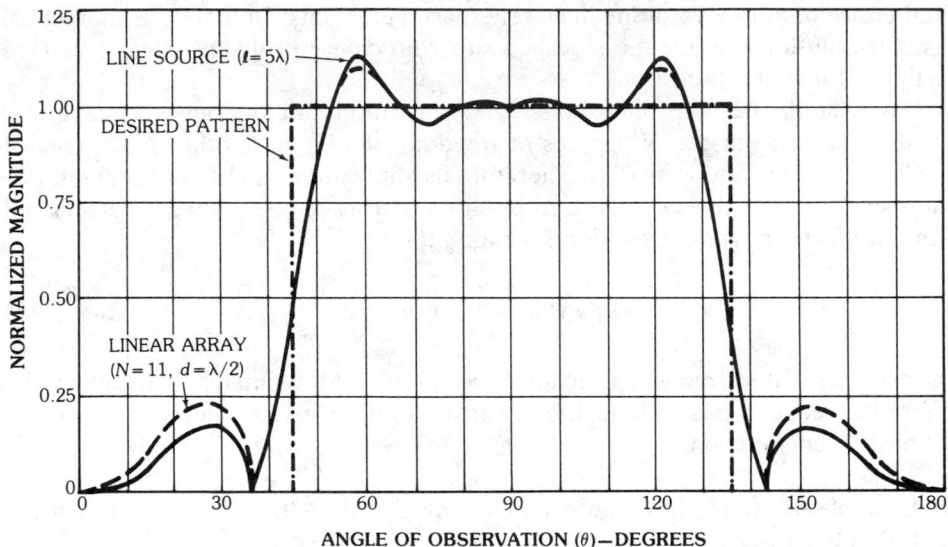

a

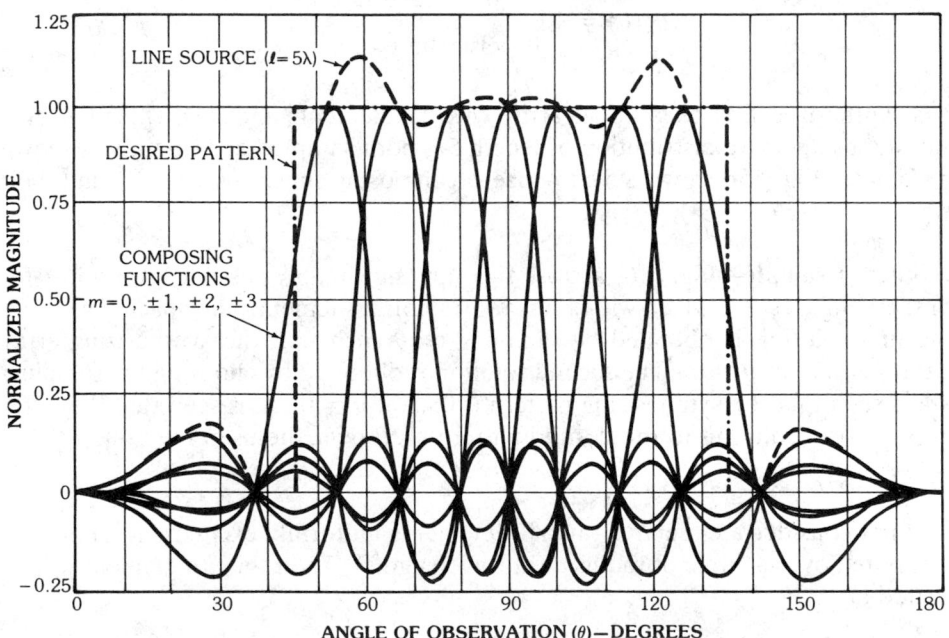

b

Fig. 15. Desired and synthesized patterns and composing functions, using Woodward's method [117, 4] for a linear array with 11 elements at $\lambda/2$ spacing and a line aperture 5 wavelengths long. (*a*) Normalized amplitude patterns. (*b*) Composing functions for line source (*l* = 5λ). (*After Balanis [4], © 1982, Harper & Row, Publishers, Inc.; reprinted by permission of the publisher*)

Table 2. Pattern Synthesized at $\{\theta_m\}$

m	θ_m	$F(\theta_m)$	m	θ_m	$F(\theta_m)$
0	90°	1			
1	78.46°	1	-1	101.54°	1
2	66.42°	1	-2	113.58°	1
3	53.13°	1	-3	126.87°	1
4	36.87°	0	-4	143.13°	0
5	0°	0	-5	180°	0

(After Balanis [4], © 1982 Harper & Row, Publishers, Inc., reprinted by permission of the publisher)

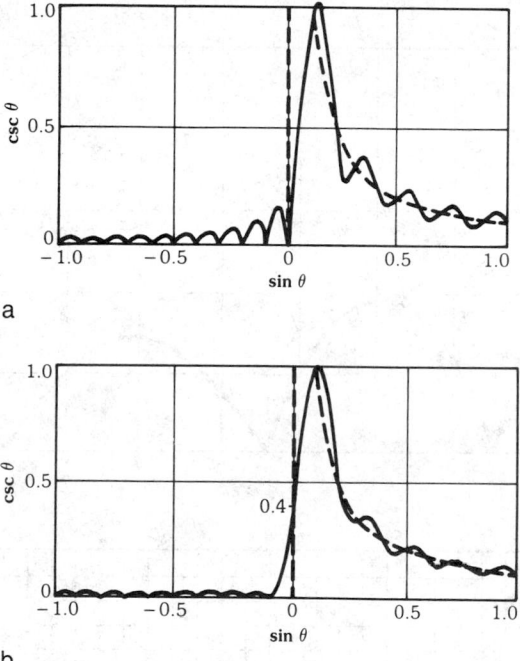

a

b

Fig. 16. Two possible synthesized patterns for the csc θ in dashed line. (*a*) With a value of 0 at $\theta = 0°$. (*b*) With a value of 0.45 at $\theta = 0°$. (*After Woodward and Lawson [119]*, © *British Crown Copyright 1948*)

$$F(\theta, \phi) = \frac{1}{N} \sum_{n=1}^{N} e^{j2\pi \mathbf{r}_n \cdot \mathbf{u}} = \frac{1}{N} \sum_{n=1}^{N} e^{j2\pi a \sin\theta \cos(\phi - \phi_n)}$$

$$= \sum_{m=-\infty}^{\infty} J_{mN}(2\pi a \sin\theta) e^{jmN(\pi/2 - \phi)} \tag{116}$$

$$\cong J_0(2\pi a \sin\theta) \qquad \text{if } 2\pi a/N \ll 1 \tag{117}$$

where use has been made of the expansion

$$e^{jz\cos\phi} = \sum_{m=-\infty}^{\infty} (j)^m J_m(z) e^{jm\phi} \tag{118}$$

and

$$\frac{1}{N}\sum_{n=1}^{N} e^{j2\pi mn/N} = \begin{cases} 1 & \text{if } m = Np,\ p = 0, \pm 1, \pm 2, \ldots \\ 0 & \text{otherwise} \end{cases} \tag{119}$$

The last approximation in (117) applies if $N \gg 2\pi a$, i.e., if the circumferential distance between any two adjacent elements, $2\pi a/N$, is sufficiently less than λ. In that case the pattern varies with $\sin\theta$ according to the $J_0(\cdot)$ function and independent of ϕ, as shown in Fig. 17a, where $|J_0(\cdot)|$ is plotted. When the above

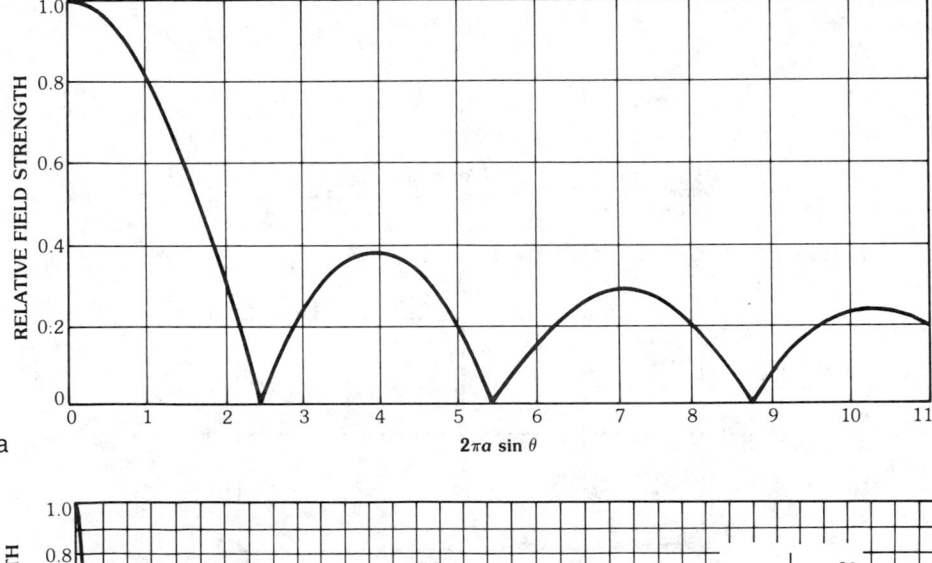

a

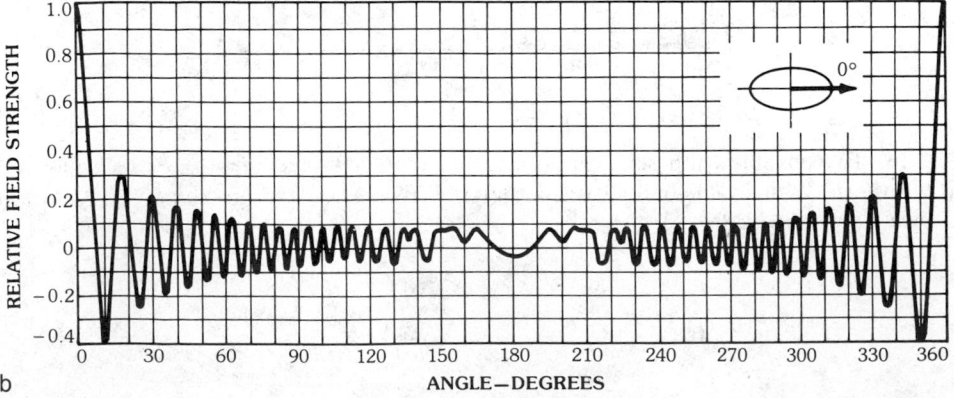

b

Fig. 17. Patterns of uniformly excited circular and elliptical arrays. (*a*) Pattern of a circular array with radius a and small element spacing. (*b*) Pattern of a cophasal elliptical array with major axis $2a = 20\lambda$, axial ratio $\tau = 3$, and main beam at $\phi_0 = 0°$. (*c*) Same as (*b*) but with $\phi_0 = 30°$. (*d*) Same as (*b*) but with $\phi_0 = 60°$. (*e*) Same as (*b*) but with $\phi_0 = 90°$. (*After Lo and Hsuan [66], © 1965 IEEE*)

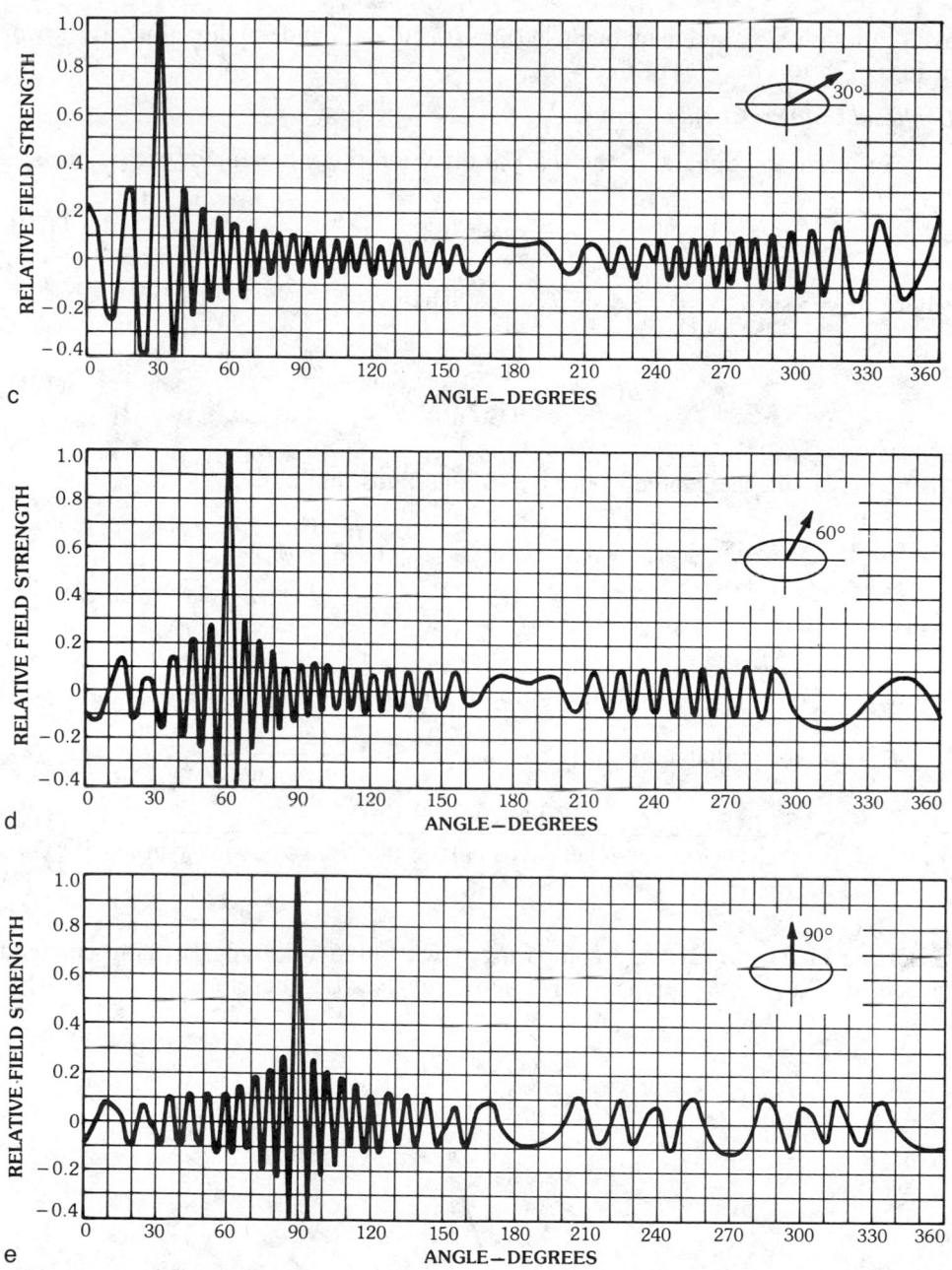

Fig. 17, *continued.*

condition is not satisfied, higher-order terms become significant and the pattern varies with ϕ. Using the transform discussed earlier in this section, under "Transformation between Circular and Elliptical Arrays," one can readily find the patterns of a uniformly excited cophasal elliptical array with major axis $2a = 20\lambda$, axial ratio $= \tau = 3$, and main beam in $\phi_0 = 0°$, $30°$, $60°$, and $90°$ directions as shown in Figs. 17b to 17e.

Cophasal Uniform Circular Arrays

For an array cophasal in the (θ_0, ϕ_0) direction the nth-element excitation is

$$I_n = e^{j2\pi \mathbf{r}_n \cdot \mathbf{u}_0} \tag{120}$$

where $\mathbf{u}_0 = \hat{\mathbf{x}} \sin\theta_0 \cos\phi_0 + \hat{\mathbf{y}} \sin\theta_0 \sin\phi_0$. Then

$$F(\theta, \phi) = \frac{1}{N} \sum_{n=1}^{N} e^{j2\pi \mathbf{r}_n \cdot (\hat{\mathbf{u}} - \mathbf{u}_0)} \tag{121}$$

Using (114) and the geometry in Fig. 18, one finds that

$$\mathbf{r}_n \cdot (\hat{\mathbf{u}} - \mathbf{u}_0) = a|\mathbf{u}_{xy} - \hat{\mathbf{u}}_0| \cos(\xi - \phi_n) \tag{122}$$

where

$\mathbf{u}_{xy} = \hat{\mathbf{u}} \cdot (\hat{\mathbf{x}}\hat{\mathbf{x}} + \hat{\mathbf{y}}\hat{\mathbf{y}}) = $ projection of $\hat{\mathbf{u}}$ on the xy plane

$\xi = $ the azimuthal angle of $(\mathbf{u}_{xy} - \mathbf{u}_0) = \cos^{-1} \dfrac{\hat{\mathbf{x}} \cdot (\hat{\mathbf{u}} - \mathbf{u}_0)}{|\mathbf{u}_{xy} - \mathbf{u}_0|}$

$$= \cos^{-1} \left\{ \frac{\sin\theta \cos\phi - \sin\theta_0 \cos\phi_0}{[(\sin\theta \cos\phi - \sin\theta_0 \cos\phi_0)^2 + (\sin\theta \sin\phi - \sin\theta_0 \sin\phi_0)^2]^{1/2}} \right\} \tag{123}$$

Inserting (122) in (121) and comparing it with (116), nothing is really changed except that $|\mathbf{u}_{xy} - \mathbf{u}_0|$ replaces $\sin\theta$, and ξ replaces ϕ. Therefore

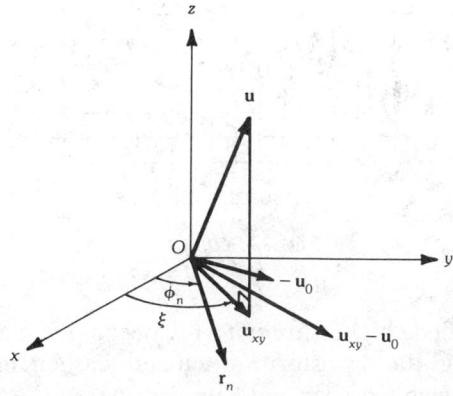

Fig. 18. Geometrical relation for cophasal circular array formulas (123)–(125).

$$F(\theta, \phi) = \sum_{m=-\infty}^{\infty} J_{mN}(2\pi a |\mathbf{u}_{xy} - \mathbf{u}_0|) e^{jmN(\pi/2-\xi)}$$
$$\cong J_0(2\pi a |\mathbf{u}_{xy} - \mathbf{u}_0|), \quad \text{if } 2\pi a/N \ll 1 \tag{124}$$

where ξ is given by (123) and

$$|\mathbf{u}_{xy} - \mathbf{u}_0| = [(\sin\theta\cos\phi - \sin\theta_0\cos\phi_0)^2$$
$$+ (\sin\theta\sin\phi - \sin\theta_0\sin\phi_0)^2]^{1/2} \tag{125}$$

Nonuniformly Excited Circular Arrays

A nonuniformly excited circular array can be analyzed in terms of uniformly excited circular arrays by applying a linear transformation on excitation. For example let $\{I_n\}$ be any given excitation. Now applying the transformation

$$\mathbf{I}' = \mathbf{A}^{-1}\mathbf{I} \quad \text{or} \quad \mathbf{I} = \mathbf{A}\mathbf{I}' \tag{126}$$

where

$$\mathbf{A} = \begin{bmatrix} 1 & e^{j2\pi/N} & e^{j4\pi/N} & \cdots & e^{j2\pi(N-1)/N} \\ 1 & e^{j4\pi/N} & e^{j8\pi/N} & \cdots & e^{j4\pi(N-1)/N} \\ \cdot & \cdot & \cdot & \cdots & \cdot \\ 1 & e^{j2\pi N/N} & e^{j4\pi N/N} & \cdots & e^{j2\pi(N-1)N/N} \end{bmatrix}$$

Then, similar to (105) through (109),

$$F(\theta, \phi) = \frac{1}{N} \sum_{n=1}^{N} I_n e^{j2\pi \mathbf{r}_n \cdot \mathbf{\hat{u}}}$$
$$= I_1' F_1(\theta, \phi) + I_2' F_2(\theta, \phi) + \cdots + I_N' F_N(\theta, \phi) \tag{127}$$

where

$$F_1(\theta, \phi) = \mathbf{V}^t \mathbf{A}_1 = \frac{1}{N} \sum_{n=1}^{N} e^{j2\pi a \sin\theta \cos(\phi - \phi_n)}$$
$$= \sum_{m=-\infty}^{\infty} J_{mN}(2\pi a \sin\theta) e^{jmN(\pi/2-\phi)} \cong J_0(2\pi a \sin\theta) \tag{128}$$

$$F_2(\theta, \phi) = \mathbf{V}^t \mathbf{A}_2 = \frac{1}{N} \sum_{n=1}^{N} e^{j2\pi[a \sin\theta \cos(\phi - \phi_n) + (n/N)]}$$
$$= \sum_{m=-\infty}^{\infty} J_{mN+1}(2\pi a \sin\theta) e^{j(mN+1)(\pi/2+\phi)}$$
$$\cong J_1(2\pi a \sin\theta) e^{j(\pi/2+\phi)} \tag{129}$$

$$F_3(\theta, \phi) = \mathbf{V}^t \mathbf{A}_3 = \frac{1}{N} \sum_{n=1}^{N} e^{j2\pi[a\sin\theta\cos(\phi-\phi_m)+(2n/N)]}$$

$$= \sum_{m=-\infty}^{\infty} J_{mN+2}(2\pi a\sin\theta)\, e^{j(mN+2)(\pi/2+\phi)}\, e^{j4\pi/N}$$

$$\cong J_2(2\pi a\sin\theta)\, e^{j2(\pi/2+\phi)}$$

$$F_N(\theta, \phi) = \mathbf{V}^t \mathbf{A}_N = \sum_{m=-\infty}^{\infty} J_{mN+N-1}(2\pi a\sin\theta)\, e^{j(mN+N-1)(\pi/2+\phi)}$$

$$\cong J_{N-1}(2\pi a\sin\theta)\, e^{j(N-1)(\pi/2+\phi)} \tag{130}$$

As before, all the approximate solutions are valid only if $2\pi a/N \ll 1$. Similar to the linear array in (105), the nonuniform excitation is now expressed in terms of the uniform excitation given by (126). This transformation is known as the method of symmetrical components in power engineering for analyzing an unbalanced polyphase system in terms of the balanced ones and also multiple-arm spiral antennas [122].

Elliptical Arrays With Nonuniform Excitations

By applying the linear transformations (95) and (96) to each of the $\{F_n(\theta, \phi)\}$ of (127) one can then analyze an elliptical array with nonuniform excitation in terms of that with *uniformly* excited *circular* arrays. This case serves as an interesting example for applying simultaneously both transformations, one for the geometry and the other for the excitations. If the minor axis of the ellipse approaches zero, the transformation will give a closed-form solution for an *unequally* spaced *nonuniformly* excited linear array.

5. Planar Arrays

When a planar array has a separable excitation, say in x and y, an analysis can be made simply by regarding each row (or column) subarray as a single element and then considering all rows (or columns) to form a "linear" column (or row) array. In so doing, the theory for linear arrays applies. For example, the pattern function of a uniform rectangular array cophasal in the (θ_0, ϕ_0) direction is given by

$$F(\theta, \phi) = \frac{1}{M} \frac{\sin(M\psi_x/2)}{\sin(\psi_x/2)} \frac{1}{N} \frac{\sin(N\psi_y/2)}{\sin(\psi_y/2)} \tag{131}$$

where M and N are the number of elements along the x and y axes, respectively, and

$$\psi_x = 2\pi d_x(\sin\theta\cos\phi - \sin\theta_0\cos\phi_0)$$
$$\psi_y = 2\pi d_y(\sin\theta\sin\phi - \sin\theta_0\sin\phi_0)$$

and d_x and d_y are interelement spacings (in wavelengths) along the x and y axes, respectively.

A three-dimensional pattern for a 5×5 element array with $d_x = d_y = \frac{1}{2}$ and $\theta_0 = 0°$ is shown in Fig. 19. The first side lobe level in the $\phi = 0°$ and $90°$ planes is approximately -12 dB (as compared with -13.2 dB for a uniformly excited linear aperture) while that in the $\phi = 45°$ and $135°$ planes is -24 dB. Fig. 20 shows the pattern for the same array but with $d_x = d_y = 1$. In this case four grating lobes appear at $\phi = 0°$, $90°$, $180°$, and $270°$ in the $\theta = 90°$ plane.

For nonseparable excitations the pattern function in general cannot be expressed in terms of a polynomial, and the pattern may have to be evaluated numerically.

Two-Dimensional Dolph-Chebyshev Arrays

One of the interesting problems is how to design a planar array which will have side lobes of equal level in the three-dimensional pattern. Evidently this cannot be achieved with two separable Dolph-Chebyshev excitations along the x and y directions as in the method discussed above, because in any plane other than $\phi = 0°$ and $90°$, the side lobes will not have the same level. Baklanov [123] and Tseng and Cheng [107] have given a simple but interesting solution to this problem. In essence, they introduced a transformation which generates from a linear Dolph-Chebyshev array a planar array with a pattern having equal side lobe levels

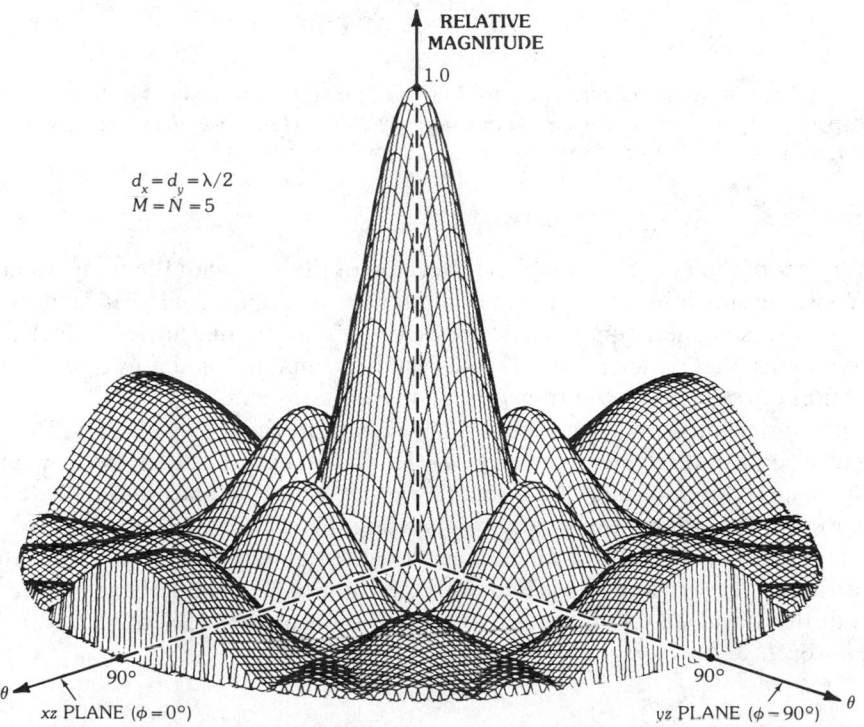

Fig. 19. Three-dimensional antenna pattern of a uniform planar array of isotropic elements with spacing $d_x = d_y = \lambda/2$. (*After Balanis [4], © 1982, Harper & Row, Publishers, Inc.; reprinted by permission of the publisher*)

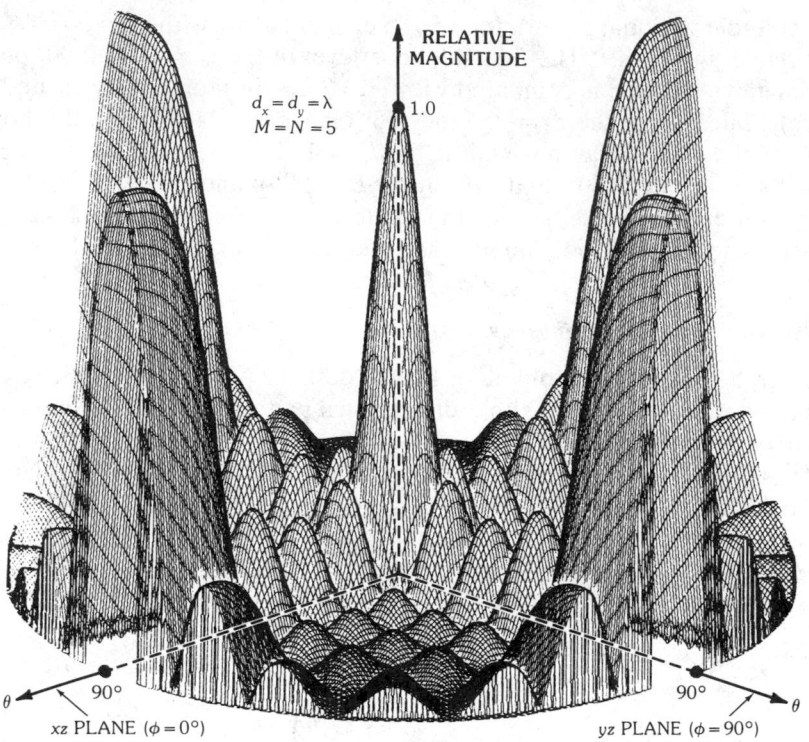

Fig. 20. Three-dimensional antenna pattern of a uniform planar array of isotropic elements with spacing $d_x = d_y = \lambda$. (*After Balanis [4], © 1982, Harper & Row, Publishers, Inc.; reprinted by permission of the publisher*)

in *every* cut of the ϕ = constant plane. Because of the nature of the transformation, however, the side lobe peaks do not occur at the same polar angle θ in all cuts, thus forming the so-called ringlike side lobes (i.e., not in concentric circles) in the three-dimensional patterns [28]. This, as will be seen later, is simply a result of the distortion introduced by the transformation in the u plane.

First, for simplicity, consider a square array in the *xy* plane with $2N \times 2N$ elements and interelement spacing d (in wavelengths) along both *x* and *y* directions. (If the unit cell is not a square, i.e., $d_x \neq d_y$, the pattern can be deduced from that of the square one simply by changing the scale of one of the *u* axes, not the magnitude of the pattern function as discussed in Section 4.) Let the center of the array be the origin and assume the excitation to be symmetrical with respect to both the *x* and *y* axes. Further let the excitation of the element at $x = md$ and $y = nd$ be I_{mn}, for $m, n = -N, \ldots, N$. Then

$$I_{mn} = I_{-m,n} = I_{m,-n} = I_{-m,-n} \tag{132}$$

and the pattern function

$$F(\theta, \phi) = \sum_{m,n=1}^{N} I_{mn} \cos[(2m - 1) \cos^{-1}\alpha] \cos[(2n - 1) \cos^{-1}\beta]$$

$$= \sum_{m,n=1}^{N} I_{mn} T_{2m-1}(\alpha) T_{2n-1}(\beta) \tag{133}$$

where

$$\cos^{-1}\alpha = u_1 = \pi d \sin\theta \cos\phi \tag{134}$$

$$\cos^{-1}\beta = u_2 = \pi d \sin\theta \sin\phi \tag{135}$$

$$T_m(\cdot) = \text{the Chebyshev polynomial of degree } m$$

Equation 133 states that $F(\theta, \phi)$ is a polynomial in α and β; thus one can write

$$F(\theta, \phi) = \sum_{m,n=1}^{N} A_{mn} \alpha^{2m-1} \beta^{2n-1} \tag{136}$$

If I_{mn} can be chosen such that

$$A_{mn} = \delta_{mn} A_{mm} \tag{137}$$

i.e., all $A_{mn} = 0$ if $m \neq n$, then

$$F(\theta, \phi) = \sum_{m=1}^{N} A_{mm}(\alpha\beta)^{2m-1} \tag{138}$$

which is a polynomial of degree $2N - 1$ in $\alpha\beta$. Let us equate this polynomial to the Chebyshev polynomial

$$F(\theta, \phi) = \sum_{m=1}^{N} A_{mm}(\alpha\beta)^{2m-1} = T_{2N-1}(w_0\alpha\beta) \tag{139}$$

with

$$T_{2N-1}(w_0) = R \tag{140}$$

Then, for every plane cut $\phi = $ constant, $\alpha\beta$ is a function of θ only; thus $F(\theta, \phi)$ follows a Chebyshev pattern in θ. As $\theta = 0$, $\alpha\beta = 1$, and, therefore,

$$F(0, \phi) = T_{2N-1}(w_0) = R \tag{141}$$

which is the main-beam magnitude with the side lobe level being 1. To recapitulate, the key point in this method is to reduce (136) to (139) by imposing (137). The former is a general planar array pattern while the latter is that of a linear array. One may thus consider this process as one which generates a two-dimensional

Dolph-Chebyshev pattern from a one-dimensional one, $T_{2n-1}(z)$, by using the transformation

$$z = w_0 \alpha \beta = w_0 \cos u_1 \cos u_2 \qquad (142)$$

Beam Scanning—From the discussion for the one-dimensional case, if the main beam is scanned to (θ_0, ϕ_0), it is sufficient only to replace the definitions of u_1 and u_2 of (134) and (135) in the above analysis by

$$\cos^{-1}\alpha = u_1 - u_{01} = \pi d (\sin\theta\cos\phi - \sin\theta_0\cos\phi_0)$$
$$\cos^{-1}\beta = u_2 - u_{01} = \pi d (\sin\theta\sin\phi - \sin\theta_0\cos\phi_0)$$

Odd Number of Elements—From the basic principle given above, it is also evident that the method can also be applied to an array with $(2N + 1) \times (2N + 1)$ elements.

Symmetry—As a result of (137) one will find an additional symmetry of the excitation to those shown in (132):

$$I_{mn} = I_{nm} \qquad (143)$$

Thus for such an array one needs to determine only $N(N + 1)/2$ excitation currents for an array with $2N \times 2N$ elements and $(N + 1)(N + 2)/2$ excitation currents for an array with $(2N + 1) \times (2N + 1)$ elements.

A Few Major Results*

(a) For an array with $L \times L$ elements, the excitation for the mnth element as determined from (133), (138), and (139) can be reduced to the following form:

$$I_{mn} = \left(\frac{4}{L}\right)^2 \sum_{p=1}^{N} \sum_{q=1}^{N} T_{L-1}\left(w_0 \cos p - \frac{1}{2}\frac{\pi}{L}\cos q - \frac{1}{2}\frac{\pi}{L}\right)$$
$$\times \cos\left(\frac{2\pi}{L}m - \frac{1}{2}p - \frac{1}{2}\right)\cos\left(\frac{2\pi}{L}n - \frac{1}{2}q - \frac{1}{2}\right), \quad \text{for } L = 2N \qquad (144)$$

and

$$I_{mn} = \left(\frac{2}{L}\right)^2 \sum_{p=1}^{N+1} \sum_{q=1}^{N+1} \varepsilon_p \varepsilon_q T_{L-1}\{w_0 \cos[(p-1)\pi/L]\cos[(q-1)\pi/L]\}$$
$$\times \cos\left[\frac{2\pi}{L}(m-1)(p-1)\right]\cos\left[\frac{2\pi}{L}(n-1)(q-1)\right], \quad \text{for } L = 2N+1 \qquad (145)$$

---

*These are taken from Tseng and Cheng [107].

where $w_0 - \cosh\left(\dfrac{1}{L-1}\cosh^{-1}R\right)$.

(b) Beamwidth $\Delta\theta_c$ at level c/R, or $20\log_{10}(R/c)$ dB below the main-beam maximum at (θ_0, ϕ_0). Let θ_c be the solution to the equation

$$\cosh\left(\frac{1}{L-1}\cosh^{-1}c\right) = w_c = w_0 \cos u_{1c} \cos u_{2c}$$

where

$$u_{1c} = \pi d(\sin\theta_c - \sin\theta_0)\cos\phi_0$$
$$u_{2c} = \pi d(\sin\theta_c - \sin\theta_0)\sin\phi_0$$

Because of the nonlinearity in transformation (i.e., nonlinear distortion from the u plane to the $\theta\phi$ plane) the beamwidth at level c/R, as determined by θ_c in the above equations, varies with ϕ_0. In particular, if $\phi_0 = 0$,

$$\Delta\theta_c = \sin^{-1}\left[\sin\theta_0 + \frac{1}{\pi d}\cos^{-1}\left(\frac{w_c}{w_0}\right)\right]$$
$$- \sin^{-1}\left[\sin\theta_0 - \frac{1}{\pi d\cos\theta_0}\cos^{-1}\left(\frac{w_c}{w_0}\right)\right]$$
$$\cong 2\sin^{-1}\left[\frac{1}{\pi d\cos\theta_0}\cos^{-1}\left(\frac{w_c}{w_0}\right)\right], \qquad \text{when } L \text{ is large} \qquad (146)$$

If $\phi_0 = \pi/4$,

$$\Delta\theta_c = \sin^{-1}\left[\sin\theta_0 + \frac{\sqrt{2}}{\pi d}\cos^{-1}\left(\frac{w_c}{w_0}\right)\right] - \sin^{-1}\left[\sin\theta_0 - \frac{\sqrt{2}}{\pi d}\cos^{-1}\left(\frac{w_c}{w_0}\right)\right]$$
$$\cong 2\sin^{-1}\left(\frac{\sqrt{2}}{\pi d\cos\theta_0}\right)\cos^{-1}\left(\frac{w_c}{w_0}\right), \qquad \text{when } L \text{ is large} \qquad (147)$$

(c) Minimum number of elements. This is determined by the maximum permissible spacing d for a given array area so that no grating lobe will appear in a given scanning range. As discussed in Section 4, under "Beam and Pattern Distortion Due to Scanning," since there is no distortion in azimuthal angles in the transformation from the (θ, ϕ) plane to the u plane, one may expect that the maximum value of d depends only on the maximum scan angle of θ_0, designated by θ_M. Then

$$d \leqq \frac{1 - (1/\pi)\cos^{-1}(1/w_0)}{1 + \sin\theta_M} \qquad (148)$$

(d) Some numerical results for (c). Figs. 21 and 22 show the largest beamwidth $(\Delta\theta_c)_M$ at the 30-dB level below the main beam for maximum scanning angles $\theta_M =$

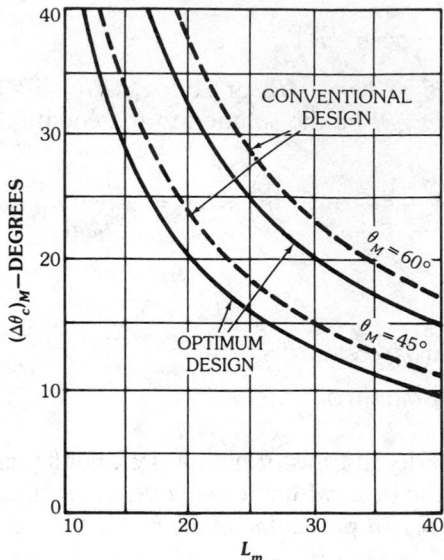

Fig. 21. Beamwidth versus minimum element number L_m on each side of a square array ($c/R = 0.03$, 30-dB side lobe level). (*After Tseng and Cheng [107], © 1968 IEEE*)

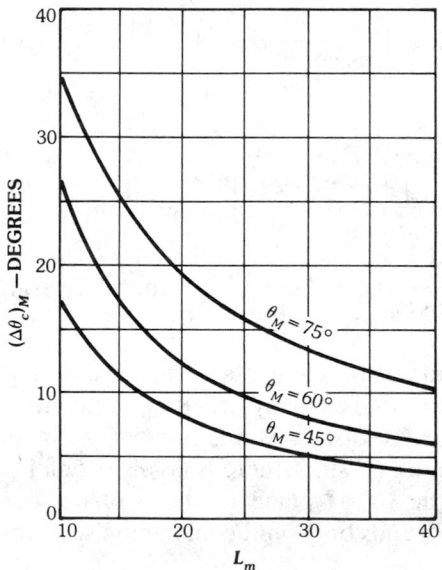

Fig. 22. Beamwidth versus minimum element number L_m on each side of a square array (optimum design: $c/R = 0.707$, 30-dB side lobe level). (*After Tseng and Cheng [107], © 1968 IEEE*)

60° and 45° and an array with minimum number of elements $L_m \times L_m$, or largest permissible spacing as given by (148). The design is for an equal side lobe level of −30 dB in all planes cutting through the main-beam maximum. Also shown for comparison is the beamwidth (in dashed curves) under the same condition but based on the conventional separable product of two Dolph-Chebyshev linear array designs. The beamwidth of the Baklanov-Tseng-Cheng design is always narrower than that of the conventional design, but the side lobes of the latter in planes other than the principal ones are lower. Fig. 22 shows the half-power beamwidth versus L_m for the same array condition as in Fig. 21. Fig. 23 shows the largest element spacing d_M permissible versus L for various scanning ranges $\theta_M = 45°$, 60°, and 75° and for a −30-dB side lobe level. Directivity of this design may be higher or lower than that of the conventional design.

A Numerical Example—An array with 10 × 10 elements, spacing $d_x = 1/2$, $d_y = 3/4$, −20-dB side lobe level, and $\theta_0 = 0°$ has been considered by Elliott [28]. Table 3 shows the relative currents in all elements, and Figs. 24a through 24d show the patterns in four different planes $\phi = 0°$, 30°, 60°, and 90°. It is seen that all side lobes have equal levels of −20 dB but are displaced along the θ axis. Since for this example $d_x \neq d_y$, the displacements are different in these planes. In fact the beamwidth is narrower in the $\phi = 90°$ plane than that in the $\phi = 0°$ plane because $d_y > d_x$. If $d_x = d_y$, the patterns in these two planes would be identical and the beamwidth would become the widest in the $\phi = 45°$ plane.

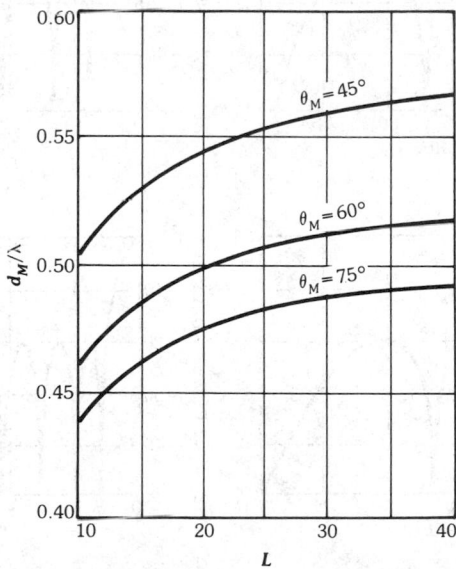

Fig. 23. Largest element spacing d_M (in wavelengths) versus L, the element number on each side of a square array, for nonappearance of grating lobes for a 30-dB side lobe level. (*After Tseng and Cheng [107], © 1968 IEEE*)

Table 3. Current Distributions for 10 × 10 Array to Give 20-dB Tseng-Cheng Pattern [28]

I_{mn}	20-dB Tseng-Cheng
I_{11}	0.773
$I_{21} = I_{12}$	0.569
$I_{31} = I_{13}$	0.796
$I_{41} = I_{14}$	0.029
$I_{51} = I_{15}$	1.000
I_{22}	0.946
$I_{32} = I_{23}$	0.119
$I_{42} = I_{24}$	0.618
$I_{52} = I_{25}$	0.667
I_{33}	0.486
$I_{43} = I_{34}$	0.777
$I_{53} = I_{35}$	0.286
I_{44}	0.387
$I_{54} = I_{45}$	0.071
I_{55}	0.008

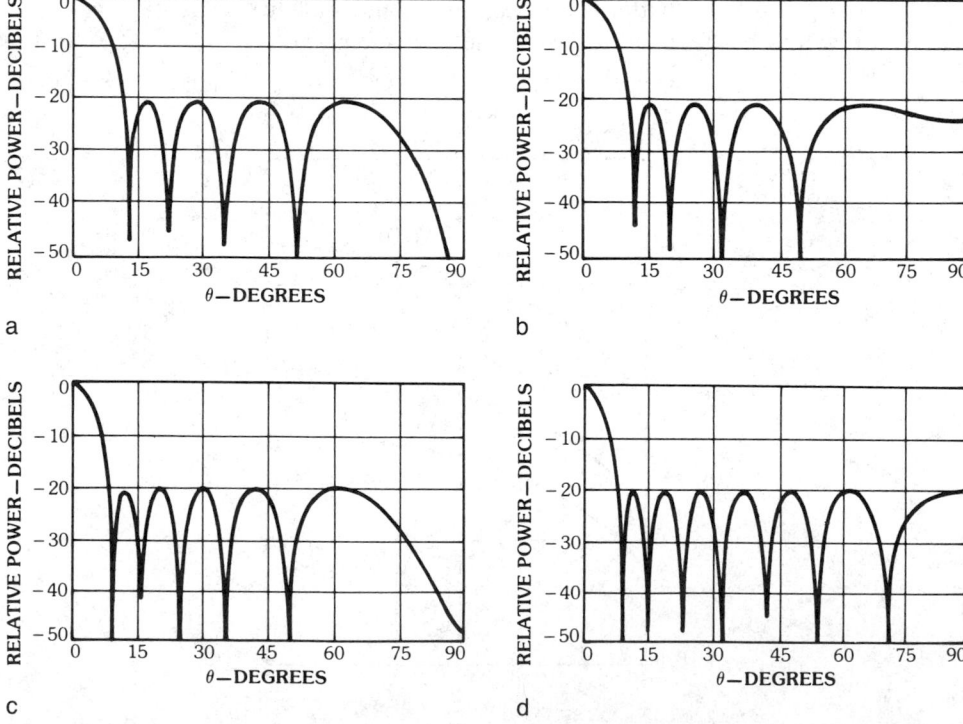

Fig. 24. Patterns in four ϕ cuts for a 10 × 10 rectangular Baklanov-Tseng-Cheng array with $d_x = 0.5\lambda$, $d_y = 0.75\lambda$, and side lobe level of 20 dB. (*a*) For $\phi = 0°$. (*b*) For $\phi = 30°$. (*c*) For $\phi = 60°$. (*d*) For $\phi = 90°$. (*After Tseng and Cheng,* Radio Science, *vol. 12, pp. 653–57,* © *American Geophysical Union*)

General Discussion of the Transformation

To see the distortion introduced by the transformation (142), a set of universal contour lines defined by

$$\cos u_1 \cos u_2 = c \quad \text{(constant)}$$

for various values of c from 0.01 to 1.00 are plotted in Fig. 25, with u_1 and u_2 defined by (134) and (135). Along the u_1 axis (i.e., $\phi = 0$) the pattern, as given by (139), is exactly that of a linear Dolph-Chebyshev array $T_{2N-1}(w_0\alpha)$ with $\alpha = \cos(\pi d_x \sin\theta)$. Along the u_2 axis the pattern is $T_{2N-1}(w_0\beta)$ with $\beta = \cos(\pi d_y \sin\theta)$. They are identical if $d_x = d_y$. (If $d_x \neq d_y$, only a linear change of the scale in u is needed.) For any $\phi = $ constant plane, the pattern is given by $T_{2N-1}(w_0 \cos u_1 \cos u_2)$, which is again the same as that of a linear Dolph-Chebyshev array except that the scale in $\sin\theta$ is changed, depending on

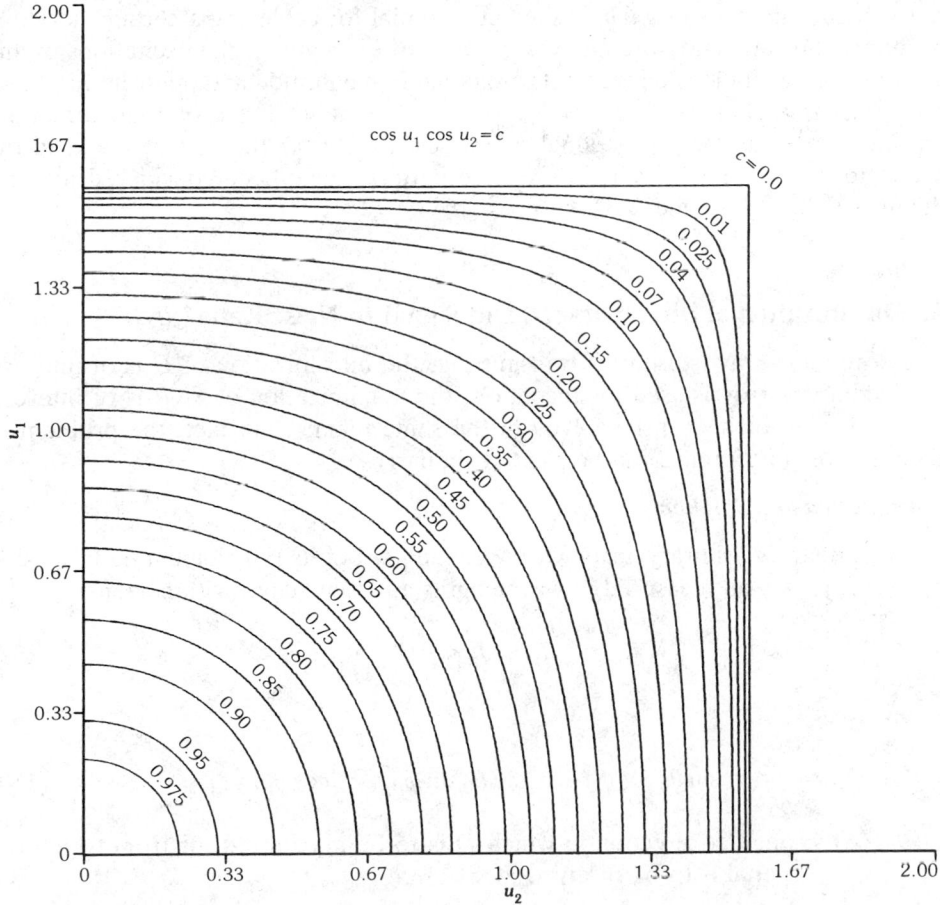

Fig. 25. Plots of $\cos u_1 \cos u_2 = c$ in the $u_1 u_2$ plane for various values of c. (*Prepared by Michael L. Oberhart*)

the value of ϕ. Fig. 25 shows exactly the dependence of this scale change on ϕ. (Note that the polar angle of any point (u_1, u_2) in the plane is simply ϕ.) As shown, the scale stretches to the largest extent at $\phi = 45°$. To find how this change is reflected in the $\theta\phi$ scale, one needs only to place an overlay on this figure as shown in Fig. 14.

As discussed earlier, a polynomial $P(z)$ can always be factorized in terms of its roots z_k:

$$P(z) = A \prod_k (z - z_k)$$

The pattern magnitude is thus proportional to the product of all the phasor magnitudes $|z - z_k|$. When z passes through a certain z_k to z_{k+1} one obtains either a main beam or a side lobe. In the latter case the locations of z_k and z_{k+1} will generally have a major influence on the height of that side lobe level, since all other roots are presumably farther away from them. Using this fact Elliott [28] is able to perturb certain roots of a Chebyshev polynomial for achieving a certain side lobe topology. For an array with an even number of elements, if the excitations of any two symmetrically located elements are equal in magnitude and opposite in phase, one obtains a sharp null, instead of a main beam, in the broadside direction. Optimum designs for nearly equal side lobes, except for the first two next to the broadside null of the so-called difference pattern, have been considered in great detail by Elliott [28] and Bayliss [6].

6. Optimization of Directivity (*D*) and Signal-to-Noise Ratio (*SNR*)

When an array is used for transmitting, the maximization of D is of interest, and when an array is used for reception, the maximization of *SNR* is of interest. These two problems can be solved in the same manner. In fact, the problem of jamming or interference can be treated similarly.

Formulation and Solution

Consider an arbitrary array with N elements. Let its nth element be located at $\mathbf{r}_n = (x_n, y_n, z_n)$ all measured in wavelengths and be excited with current

$$I_n = J_n e^{-j\mathbf{r}_n \cdot \mathbf{u}_0} \tag{149}$$

where

$$\mathbf{r}_n \cdot \mathbf{u}_0 = 2\pi(x_n \sin\theta_0 \cos\phi_0 + y_n \sin\theta_0 \sin\phi_0 + z_n \cos\theta_0) \tag{150}$$

(θ_0, ϕ_0) = angular direction in which D is to be maximized, or from which the signal is to be received

In (149), $\{J_n\}$ can be complex and will be determined. If $\{J_n\}$ is real, $\{I_n\}$ will be cophasal excitation.

For convenience, we shall use *bra-ket* notation as defined below:

$$\langle J = \text{a } row \text{ vector} = (J_1, J_2, \ldots, J_N) \tag{151}$$

$$\langle V = \text{a } row \text{ vector} = (\exp[j(\psi_1^0 - \psi_1)], \exp[j(\psi_2^0 - \psi_2)], \ldots,$$
$$\exp[j(\psi_N^0 - \psi_N)]) \tag{152}$$

$J\rangle$ and $V\rangle$ are transposes of $\langle J$ and $\langle V$, respectively, and are therefore column vectors

where

$$\psi_n = 2\pi(x_n \sin\theta \cos\phi + y_n \sin\theta \sin\phi + z_n \cos\theta) \tag{153}$$

$$\psi_n^0 = \psi_n\big|_{\theta-\theta_0, \phi=\phi_0} \tag{154}$$

Then the pattern function is given by

$$F(\theta, \phi) = \sum_n^N J_n \exp[j(\psi_n - \psi_n^0)] = \langle JV^* \rangle \tag{155}$$

The power density, or the received signal in the direction (θ_0, ϕ_0), is proportional to

$$|F(\theta, \phi)|^2 = \langle JV_1^* \rangle \langle V_1 J^* \rangle = \langle JCJ^* \rangle \tag{156}$$

where

$$V_1\rangle = V\rangle\big|_{\theta=\theta_0, \phi=\phi_0}$$

$\mathbf{C}$ = a dyad, or a special $N \times N$ matrix:

$$\mathbf{C} = V_1^* \rangle \langle V_1 = V^* \rangle \langle V \big|_{\phi=\theta_0, \phi=\phi_0} = \begin{bmatrix} 1 & 1 & \cdots & 1 \\ 1 & 1 & \cdots & 1 \\ \cdot & \cdot & \cdots & \cdot \\ 1 & 1 & \cdots & 1 \end{bmatrix} \tag{157}$$

If the array is in a noisy environment with temperature distribution $T(\theta, \phi)$, the total received noise power is proportional to

$$P_n = \langle JAJ^* \rangle \tag{158}$$

where $\mathbf{A}$ is an $N \times N$ matrix given by

$$\mathbf{A} = \frac{1}{4\pi} \int_{4\pi} V^* \rangle \langle V T(\theta, \phi) \, d\Omega \tag{159}$$

Therefore the signal-to-noise ratio is proportional to

$$SNR = \frac{\langle JCJ^* \rangle}{\langle JAJ^* \rangle} \tag{160}$$

Similarly, the average of total radiated power is proportional to

$$P_{av} = \frac{1}{4\pi} \int_{4\pi} \langle JV^* \rangle \langle VJ^* \rangle \, d\Omega = \langle JBJ^* \rangle \tag{161}$$

where **B** is an $N \times N$ matrix given by

$$\mathbf{B} = \frac{1}{4\pi} \int_{4\pi} V^* \rangle \langle V d\Omega = \mathbf{A} \big|_{T=1} \tag{162}$$

Therefore the directivity is given by

$$D = \frac{\langle JCJ^* \rangle}{\langle JBJ^* \rangle} \tag{163}$$

The objective is to maximize D for directivity, or to maximize the SNR for reception in the given noise environment $T(\theta, \phi)$, with the N degrees of freedom $J_1, J_2, \ldots, J_N$. Before proceeding, it may be noted that

$$SNR = \frac{\langle JCJ^* \rangle}{\langle JBJ^* \rangle} \bigg/ \frac{\langle JAJ^* \rangle}{\langle JBJ^* \rangle} = D/T_{av} \tag{164}$$

where

$$T_{av} = \langle J(1/4\pi) \int_{4\pi} V^* \rangle \langle VT(\theta, \phi) \, d\Omega J^* \rangle \Big/ \langle J(1/4\pi) \int_{4\pi} V^* \rangle \langle V \, d\Omega J^* \rangle$$

    = temperature of the received noise power if it were distributed uniformly in space

    = equivalent uniform temperature distribution for the same noise power received by the array   (165)

It may also be noted that for interference, or jamming signal, with power density distribution, $I(\theta, \phi)$, uncorrelated in space, the signal-to-interference (S/I) ratio is also given by (160) with $T(\theta, \phi)$ replaced by $I(\theta, \phi)$.

Therefore (*a*) the problem of maximizing D/T_{av} is exactly the same as that of maximizing the SNR, (*b*) the problem of maximizing D is the same as that of maximizing the SNR with $T = 1$, i.e., uniform temperature distribution in space, and (*c*) the problem of maximizing S/I is the same as that of maximizing the SNR. Hence, for all these problems, one needs only to consider the maximization of the SNR; but the optimum solution for $\{I_n\}$ depends on the matrix **A**, or **B**, or both.

Solutions to all these problems may lead to unrealistic results which are generally referred to as the "supergain," "superdirectivity," "ill-conditioned," or "improperly posed" problems [9, 65, 120]. To make the solutions more physically meaningful, a constraint may be imposed on the so-called array Q factor

$$Q = \frac{\langle JJ^* \rangle}{\langle JVJ^* \rangle} \qquad (166)$$

which follows from Taylor's definition of *supergain ratio* for an aperture antenna [100]. Except for a constant and the quality factor of a single element, the above definition of the array Q factor has the usual meaning of 2π times the total stored energy divided by the energy radiated per cycle if the mutual impedance effect can be neglected.

The most general optimization problem can thus be stated as follows:

Given: Array geometry, direction of desired signal (θ_0, ϕ_0), $T(\theta, \phi)$, and constraint on the Q factor.

To find: $\langle J$ (or $\langle I$) such that the *SNR* is maximized.

Solution: Using the fact that matrices **A**, **B**, and **C** are Hermitian, the solutions can be found for various cases as summarized in Table 4 [65]. The optimum $\langle J$ as listed in the last row in Table 4 is for maximum *SNR* with a prescribed Q factor. When there is a constraint on the Q factor, one needs first to solve for p in the following eigenvalue equation:

$$\det(V_1\rangle, W_2\rangle, \ldots, W_N\rangle) = 0 \qquad (167)$$

where

$$\mathbf{I} = \text{identity matrix}$$
$$V_1\rangle = (1, 1, \ldots, 1)^t$$
$$W_n\rangle = p^2 (Q\mathbf{B} - \mathbf{I}) V_n^*\rangle + 2p\mathbf{A} V_n^*\rangle + \mathbf{A}(Q\mathbf{B} - \mathbf{I})^{-1}\mathbf{A} V_n^*\rangle,$$
$$\text{for } n = 2, 3, \ldots, N$$
$$V_2\rangle = (-1, 1, 0, 0, \ldots, 0)^t$$
$$\cdots \qquad \cdots$$
$$V_N\rangle = (-1, 0, 0, \ldots, 1)^t$$

In words, $V_n\rangle$, for $n = 2, \ldots, N$, is a column vector with -1 as its first element, $+1$ as its nth element, and 0 for all other elements. Once p is found, one can compute the matrix

$$\mathbf{K} = \mathbf{A} + p(Q\mathbf{B} - \mathbf{I}) \qquad (168)$$

and the optimum $J\rangle$ as shown in the last row in Table 4:

$$J\rangle = \mathbf{K}^{-1*} V_1\rangle \qquad (169)$$

Table 4. Formulas for Optimum Gain and SNR of an Arbitrary Array

	Current	D	SNR	Q Factor
Definition	$J\rangle$	$\dfrac{|\langle JV_1^*\rangle|^2}{\langle JBJ^*\rangle}$	$\dfrac{|\langle JV_1^*\rangle|^2}{\langle JAJ^*\rangle}$	$\dfrac{\langle JJ^*\rangle}{\langle JBJ^*\rangle}$
Uniform current excitation	$V_1^*\rangle$	$\dfrac{N^2}{\langle V_1BV_1^*\rangle}$	$\dfrac{N^2}{\langle V_1AV_1^*\rangle}$	$\dfrac{N}{\langle V_1BV_1^*\rangle}$
Optimum D without constraint on Q	$\mathbf{B}^{*-1}V_1^*\rangle$	$\langle V_1\mathbf{B}^{-1}V_1^*\rangle$	$\dfrac{|\langle V_1\mathbf{B}^{-1}V_1^*\rangle|^2}{\langle V_1\mathbf{B}^{-1}A\mathbf{B}^{-1}V_1^*\rangle}$	$\dfrac{\langle V_1\mathbf{B}^{-2}V_1^*\rangle}{\langle V_1\mathbf{B}^{-1}V_1^*\rangle}$
Optimum D with a prescribed Q	$\mathbf{F}^{*-1}V_1^*\rangle$	$\dfrac{|\langle V_1\mathbf{F}^{-1}V_1^*\rangle|^2}{\langle V_1\mathbf{F}^{-1}B\mathbf{F}^{-1}V_1^*\rangle}$	$\dfrac{|\langle V_1\mathbf{F}^{-1}V_1^*\rangle|^2}{\langle V_1\mathbf{F}^{-1}A\mathbf{F}^{-1}V_1^*\rangle}$	A given constant
Optimum SNR without constraint on Q	$\mathbf{A}^{*-1}V_1^*\rangle$	$\dfrac{|\langle V_1\mathbf{A}^{-1}V_1^*\rangle|^2}{\langle V_1\mathbf{A}^{-1}B\mathbf{A}^{-1}V_1^*\rangle}$	$\langle V_1\mathbf{A}^{-1}V_1^*\rangle$	$\dfrac{\langle V_1\mathbf{A}^{-2}V_1^*\rangle}{\langle V_1\mathbf{A}^{-1}B\mathbf{A}^{-1}V_1^*\rangle}$
Optimum SNR with a prescribed Q	$\mathbf{K}^{*-1}V_1^*\rangle$	$\dfrac{|\langle V_1\mathbf{K}^{-1}V_1^*\rangle|^2}{\langle V_1\mathbf{K}^{-1}B\mathbf{K}^{-1}V_1^*\rangle}$	$\dfrac{|\langle V_1\mathbf{K}^{-1}V_1^*\rangle|^2}{\langle V_1\mathbf{K}^{-1}A\mathbf{K}^{-1}V_1^*\rangle}$	A given constant

Symbolism
Actual current in the nth element $= J_n e^{-j\psi_n^0}$
$\langle J = (J_1, J_2, \cdots, J_N), \qquad \langle V_1 = (1, 1, \cdots, 1)$
$\psi_n = 2\pi(x_n \sin\theta\cos\phi + y_n \sin\theta\sin\phi + z_n\cos\theta)$
$\psi_n^0 = \psi_n\big|_{\theta=\theta_0,\,\phi=\phi_0}$
$\mathbf{K} = \mathbf{A} + p(Q\mathbf{B} - \mathbf{I}), \qquad\qquad \mathbf{F} = \mathbf{K}\big|_{A=B}$
$\mathbf{A} = (1/4\pi)\int_{4\pi} V^*\rangle\langle VT(\theta,\phi)\,d\Omega, \qquad \mathbf{B} = \mathbf{A}\big|_{T=1}$

(After Lo, Lee, and Lee [65], © 1966 IEEE)

From (168) one can compute its associated directivity

$$D = \frac{|\langle V_1\mathbf{K}^{-1}V_1^*\rangle|^2}{\langle V_1\mathbf{K}^{-1}B\mathbf{K}^{-1}V_1^*\rangle} \tag{170}$$

and its associated SNR

$$SNR = \frac{|\langle V_1\mathbf{K}^{-1}V_1^*\rangle|^2}{\langle V_1\mathbf{K}^{-1}A\mathbf{K}^{-1}V_1^*\rangle} \tag{171}$$

One of the eigenvalues p of (167) which gives the largest value of SNR is the optimum solution. Since the solution $J\rangle$ is determined for maximum SNR, D as computed from (170) using that $J\rangle$ is not necessarily maximum.

● For maximum *SNR* without constraint on Q the solution is given in row 5 of Table 4. For this case, $p = 0$ and no eigenvalue needs to be computed. Therefore from (168) $\mathbf{K} = \mathbf{A}$ and

$$J\rangle = \mathbf{A}^{-1*} V_1 \rangle \tag{172}$$

● For maximum D with and without constraint on the Q factor, the solutions are listed in rows 4 and 3 of Table 4, respectively. The solutions are exactly the same as those for maximum *SNR* except that $T(\theta, \phi) = 1$ and $\mathbf{A} = \mathbf{B}$. The corresponding $\mathbf{K}$ matrix is denoted by

$$\mathbf{F} = \mathbf{B} + p(Q\mathbf{B} - \mathbf{I}), \quad \text{for a constraint on the } Q \text{ factor} \tag{173}$$

$$\mathbf{F} = \mathbf{B}, \quad \text{for no constraint} \tag{174}$$

● Since $V_1 \rangle = (1, 1, \ldots, 1)^t$, the optimum solution for J_n is simply the *sum* of all the elements in the *n*th row of $\mathbf{B}^{*-1}$, $\mathbf{F}^{-1*}$, $\mathbf{A}^{-1*}$, and $\mathbf{K}^{-1*}$, respectively, for each case. The actual current for the *n*th element, by definition, is given by (149).

Planar Array with Isotropic Elements or Vertical Dipoles in the (x, y) *Plane*

Referring to Fig. 26, the elements of matrix $\mathbf{B}$ for this array can be integrated in closed form:

$$b_{nm} = b_{mn}^* = e^{-j\psi_{nm}^0}\left[\frac{\sin 2\pi \varrho_{nm}}{2\pi \varrho_{mn}} + q\frac{\cos 2\pi \varrho_{nm}}{(2\pi \varrho_{nm})^2} - q\frac{\sin 2\pi \varrho_{nm}}{(2\pi \varrho_{nm})^2}\right],$$
$$\text{for } n < m \tag{175}$$

$$b_{nn} = 1 - q/3 \tag{176}$$

where

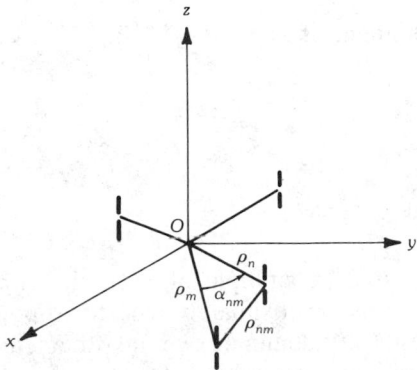

Fig. 26. A planar array with elements (vertical dipoles or isotropic) in the *xy* plane.

$$\psi_{mn}^0 = 2\pi\varrho_{nm} \sin\theta_0 \cos(\phi_0 - \alpha_{nm})$$

$$\varrho_{nm} = [(x_n - x_m)^2 + (y_n - y_m)^2]^{1/2}$$

$$= \text{distance between the } n\text{th and } m\text{th elements}$$

$$(x_n, y_n) = \text{coordinates of the } n\text{th element}$$

$$\alpha_{nm} = \tan^{-1}[(y_n - y_m)|(x_n - x_m)], \text{ where } 0 \leqq \alpha_{nm} < \pi$$

$$q = \begin{cases} 0 & \text{for isotropic elements} \\ 1 & \text{for vertical dipoles} \end{cases}$$

From the above a few interesting results can be obtained:

(*a*) In particular for a linear array, say along the *x* axis, with *N* isotropic elements,

$$b_{nm} = b_{mn}^* = e^{-j\psi_{nm}^0} \frac{\sin 2\pi\varrho_{nm}}{2\pi\varrho_{mn}}, \qquad n < m \tag{177}$$

$$b_{nn} = 1 \tag{178}$$

where

$$\varrho_{nm} = x_n - x_m$$

$$\psi_{nm}^0 = 2\pi\varrho_{mn} \sin\theta_0 \cos\phi_0$$

For uniform interelement spacing equal to $\lambda/2$ or its multiples,

$$b_{nm} = \begin{cases} 0 & \text{for } n \neq m \\ 1 & \text{for } n = m \end{cases}$$

Thus

$$\mathbf{B} = \mathbf{I} = \text{identity matrix}$$

$$Q = 1, \text{ independent of } J\rangle$$

$$\max D = N \tag{179}$$

and

$$\text{optimum } \langle J = (1, 1, \ldots, 1) \tag{180}$$

In other words, if $d = m/2$ (a multiple of $\lambda/2$), the *cophasal uniform array is optimum*. This conclusion is approximately true for dipole arrays since, for $d = m/2$, the diagonal elements of **B** dominate over all others, and **B** is approximately **I**.

(*b*) For broadside planar array $\psi_{nm}^0 = 0$. Thus the **B** matrix is real and the optimum excitation is real. In other words, for broadside arrays *optimum* *excitations are either in phase or antiphase.* A few examples are shown in Figs. 27 through 31. In Figs. 30 and 31 the term "cophasal" implies that only the excitation amplitudes are optimized while the phases are kept cophasal in the end-fire direction.

(*c*) For planar arrays with large interelement spacings, say much greater than a wavelength, **B** $\cong$ **I**. *The uniform cophasal excitation is nearly optimum, max* D $\cong$ N *and* Q $\cong$ *1.* Some workers have sought optimum spacings and excitations for a thinned array; this result shows that the uniform cophasal excitation is the solution no matter what element spacings are as long as they are sufficiently large as in a thinned array.

(*d*) From the definition of the array Q factor, the value of $1/Q$ is bounded between the smallest and the largest eigenvalue of the **B** matrix. For equally spaced linear arrays with isotropic elements, $Q = 1$ as $d = m/2, m = 1, 2, \ldots$, and $Q \cong 1$ as $d > 1$, no matter what $\langle J$ is.

(*e*) From the example to be discussed next, it will be seen that the optimum excitation for a linear end-fire array with uniform spacings is nearly *antiphase for* d $\leqq$ *1/2 and nearly cophasal for* d $\geqq$ *1/2.*

(*f*) Gilbert and Morgan [124] showed that the *average* of *maximum* directivity over all directions is equal to N, the number of isotropic elements, i.e.,

$$\frac{1}{4\pi} \int_{4\pi} \max D(\theta_0, \phi_0)\, d\Omega = N \qquad (181)$$

Thus if max D is greater than N in some direction, it must be smaller than N in some other direction.

(*g*) In general, ohmic losses in most antennas with low Q are of little importance in antenna efficiency. However, for antennas with strong local fields and large circulating currents, such as in superdirective arrays, the efficiency for radiation in (θ_0, ϕ_0) may be defined as

$$\eta = \frac{\text{radiated power density in } (\theta_0, \phi_0)}{\text{radiated power density in } (\theta_0, \phi_0) + \text{ohmic power loss}} = \frac{1}{1 + rS} \quad (182)$$

where

r = ohmic resistance of each element,

$$S = \frac{Q}{D} = \frac{\langle JJ^* \rangle}{\langle JCJ^* \rangle} \qquad (183)$$

The S parameter, called the *sensitivity factor* [109], is a measure of the mean-square variation of the maximum field with respect to the mean-square deviation of the excitation. Thus large Q results in not only low efficiency but also high sensitivity, as will be seen in the following example, Section 7, and [8, 9, 53, 120].

A Typical Example for Maximum Directivity

Consider an equally spaced linear end-fire array with 10 isotropic elements along the *x* axis. Using the results in row 3 of Table 4 the maximum directivity can be

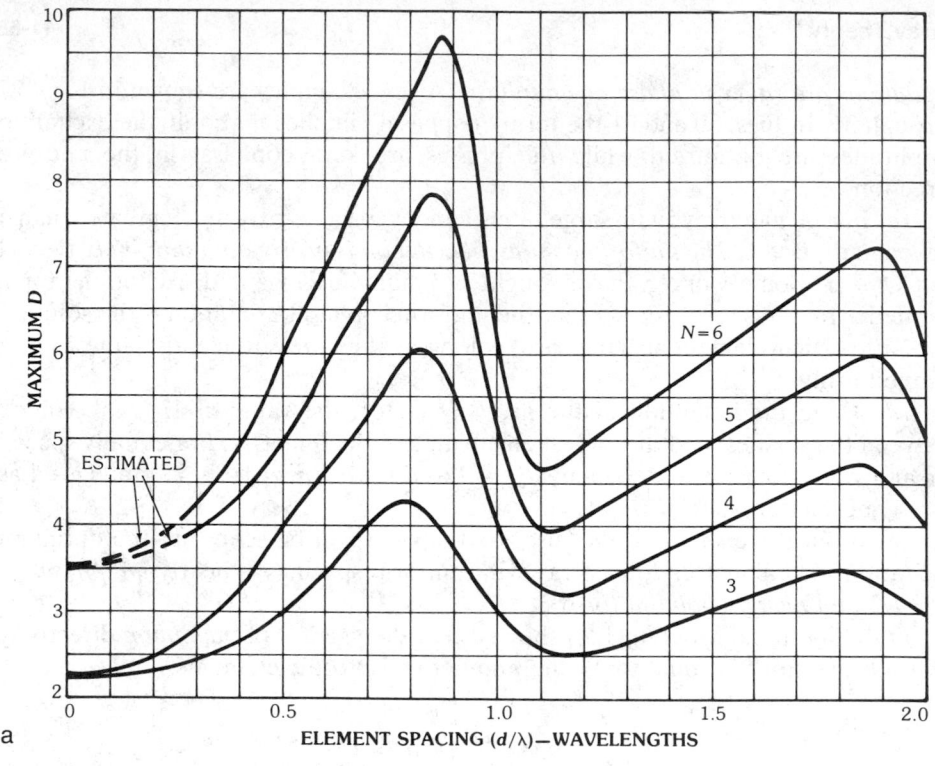

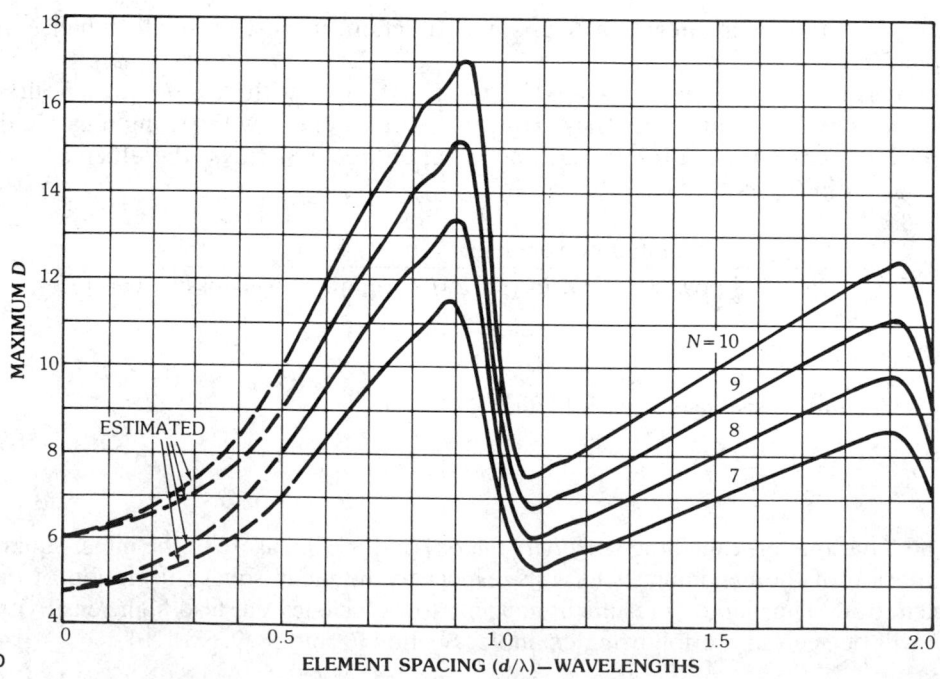

Fig. 27. Maximum directivity D of a linear broadside array of N isotropic elements versus element spacing d in wavelengths. (*a*) For $N = 3$ through 6. (*b*) For $N = 7$ through 11. (*Courtesy C. T. Tai*)

11-66

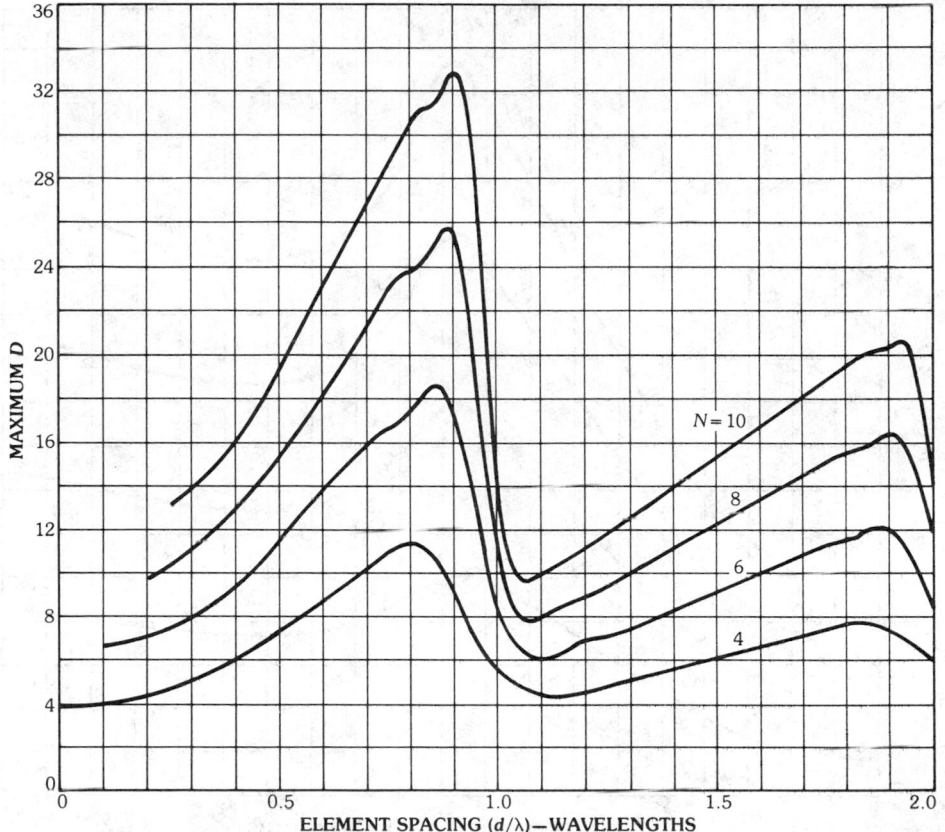

Fig. 28. Maximum directivity of a linear broadside array of N parallel dipoles versus element spacing d (in wavelengths) for $N = 4, 6, 8,$ and 10. (*Courtesy C. T. Tai*)

computed for various values of spacing d as shown by the curve $D(O)$ in Fig. 32. As expected, max $D = N = 10$ at $d = m/2$, $m = 1, 2, \ldots$. As d decreases below $\frac{1}{2}$, D increases rapidly and approaches $N^2 = 100$ as d approaches zero. For comparison a few other cases are also shown in the figure:

$D(\text{OC})$ = maximum directivity with only the excitation *magnitudes* subject to optimization and with the phases confined to the cophasal condition in the end-fire direction

$D(\text{U})$ = directivity for a uniform cophasal excitation

$D(\text{HW})$ = directivity for the Hansen-Woodyard (HW) excitation (see Section 3, under "Uniform Arrays With N Elements")

Fig. 33 shows the Q factor versus d for all the cases just stated. From these two figures the following remarks can be made:

(*a*) $D(O)$ is considerably higher than $D(U)$ only when $d \leqq 0.4$. A moderate

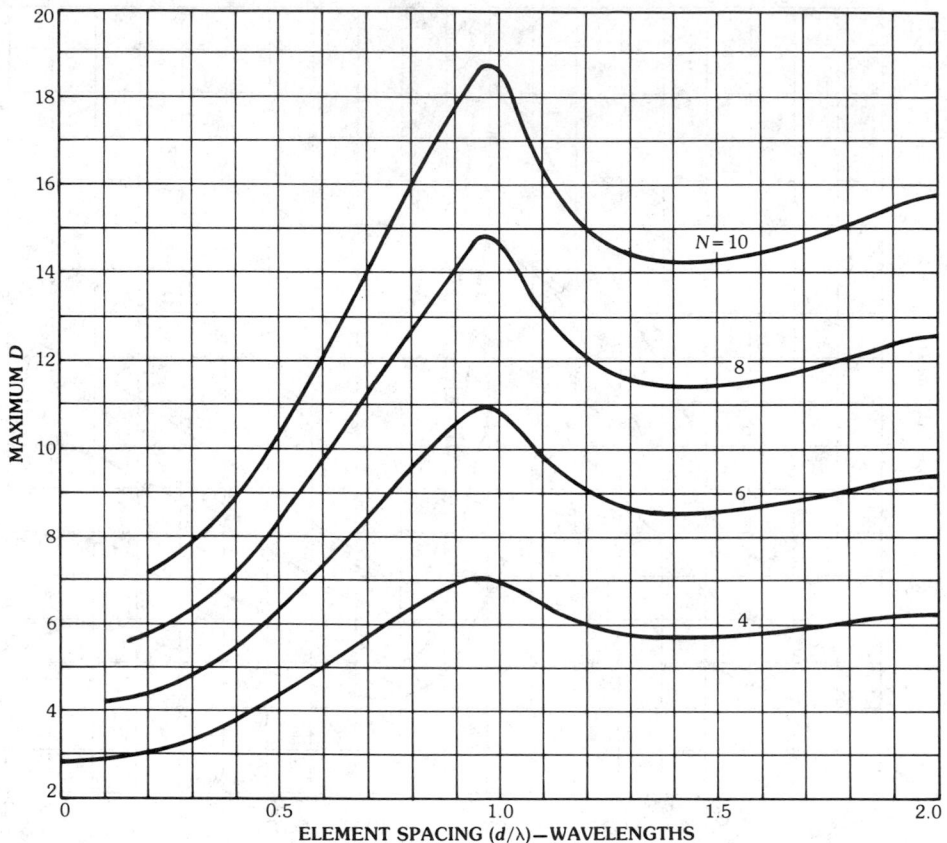

Fig. 29. Maximum directivity of a linear broadside array of N collinear dipoles versus element spacing d (in wavelengths) for $N = 4, 6, 8,$ and 10. (*Courtesy C. T. Tai*)

improvement of $D(\mathrm{O})$ over $D(\mathrm{U})$ can be obtained only at the price of astronomically large Q, which implies extremely high local field, dissipation loss, sensitivity, and extremely low efficiency and narrow bandwidth. Therefore a superdirective array is more a fantasy than reality.

(*b*) Hansen-Woodyard excitation is not truly optimum; in fact its directivity is much lower than $D(\mathrm{U})$ for $d > 1/2$.

(*c*) $D(\mathrm{U}) \to 1$ as $d \to 0$, and $D(\mathrm{U}) = N$ as $d = m/2, m = 1, 2, \ldots$.

(*d*) $D(\mathrm{O}) \cong D(\mathrm{U})$ as $d \geqq 1/2$. In view of the simplicity of uniform excitation the uniform array is indeed an excellent practical antenna so far as the directivity is concerned.

(*e*) A superdirective antenna, according to the IEEE Standard, is one whose directivity is "significantly" higher than that of a uniform excitation. A top-loaded monopole, for example, is not a superdirective antenna since the top loading is in effect to make the excitation more uniform. The radiation leak from an open-circuit, balanced, two-wire transmission line with small spacing can be considered as a superdirective antenna, but obviously it is a very inefficient poor

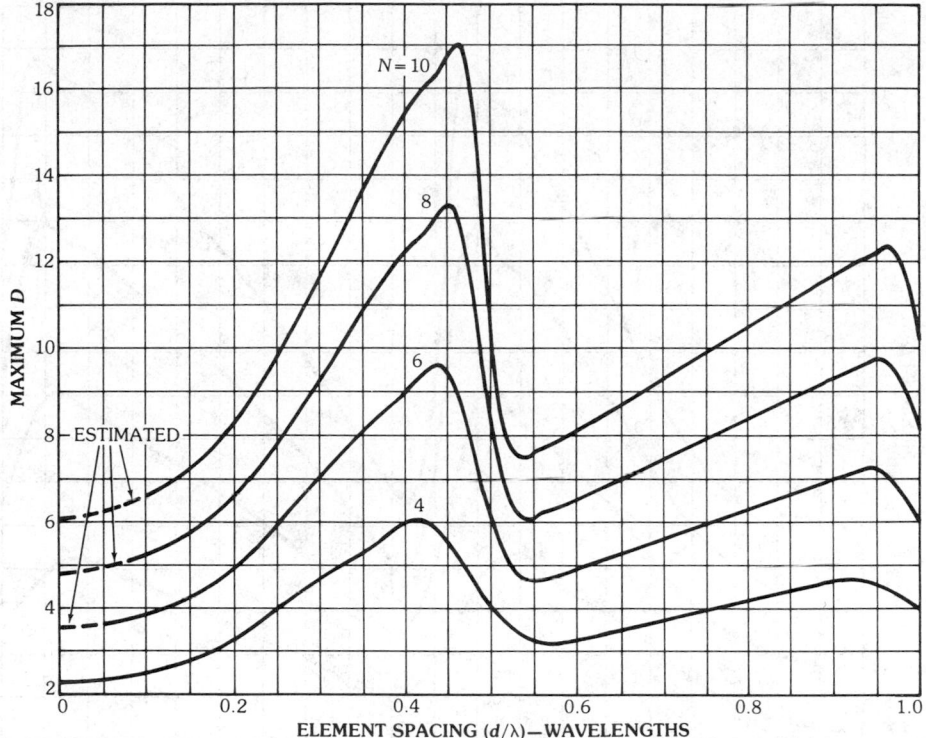

Fig. 30. Maximum directivity of a linear cophasal end-fire array of N isotropic elements versus element spacing d (in wavelengths) for $N = 4$, 6, 8, and 10. (*Courtesy C. T. Tai*)

antenna. So are some other small antennas whose largest dimensions are smaller than a few thousandths of a wavelength.

For academic interest the optimum excitation magnitude and phase for the above example are shown in Figs. 34 and 35. In both figures the curves for $d = 0.2\lambda$ should not be taken seriously, and in Fig. 35 the phases of the excitations of elements 9 and 10 are omitted for brevity, but they can be obtained by extrapolation from the curves shown. From these two figures the following remarks can be made:

(*a*) The optimum excitation for small spacing is highly tapered toward the ends of the array and approximately uniform for spacing equal to or larger than $\lambda/2$. This is consistent with the directivity characteristics stated above, namely, for $d \geqq 1/2$ the maximum directivity is nearly that of a uniform array.

(*b*) The optimum phase is nearly cophasal for $d \geqq 1/2$, again in agreement with the directivity characteristics stated above. However, for $d < 1/2$ the optimum phase distribution is nearly antiphase, about 170°, which results in large local field, low efficiency, etc.

(*c*) For $d = 0.4$ wavelength, where the strong superdirectivity begins to show $[D(O) - D(U) \cong 3.5 \text{ dB}]$, the optimum pattern is shown in Fig. 36a. For com-

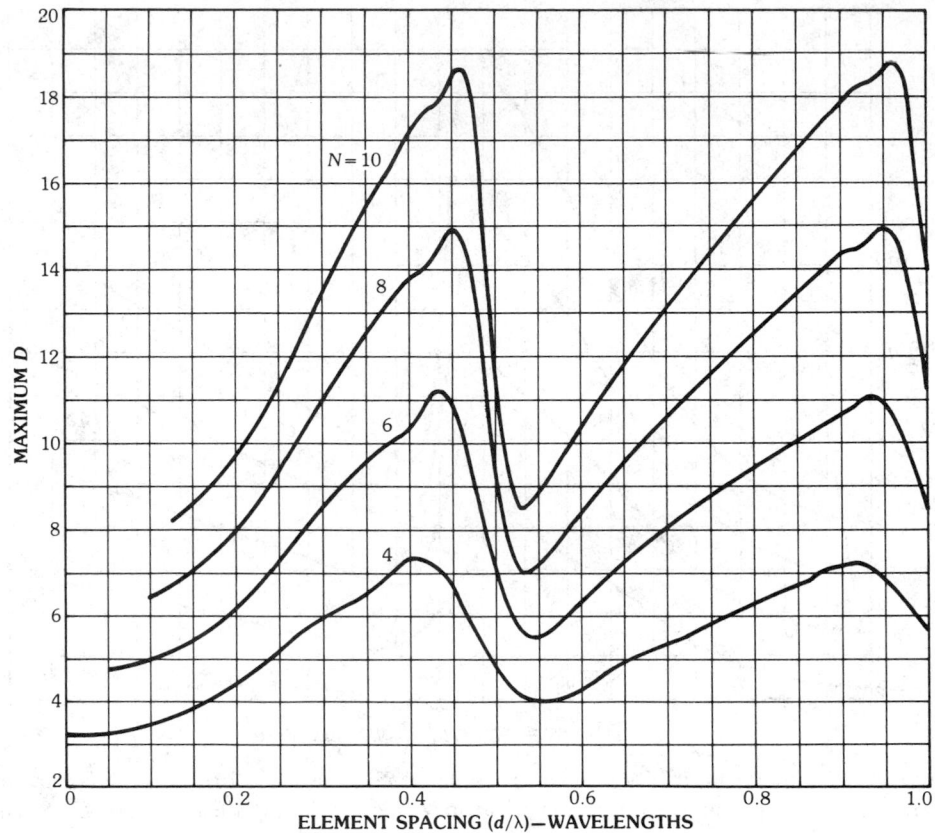

Fig. 31. Maximum directivity of a linear cophasal end-fire array of N parallel dipoles versus element spacing d (in wavelengths) for $N = 4$, 6, 8, and 10. (*Courtesy C. T. Tai*)

parison, the pattern for the same array but with Hansen-Woodyard excitation is shown in Fig. 36b. It is seen that the latter has a much broader beam.

(*d*) Fig. 37 shows how the maximum D varies with the angle for which D is maximized for two different cases: $D(O)$ with both excitation magnitudes and phases subject to optimization and $D(OC)$ with only the magnitudes subject to optimization and cophasal phase distribution. It is seen that for both cases D is larger than N, namely 10 here, for small angles (i.e., near end-fire) and smaller than N, namely 10, in some other directions, as expected from (181).

An Example for Maximum SNR

Superdirective arrays are impractical as stated above. However, for reception, it is the *SNR*, not D, that is of concern. In particular, one is interested in finding out whether a significant improvement of *SNR* over that of a uniform array is possible without paying a high price on the Q factor. To show that this is possible a simple semicircular array, consisting of nine uniformly distributed isotropic elements in the xz plane is considered. Let thc signal come from the z axis and $T(\theta, \phi) =$ constant for $z < 0$ and 0 for $z > 0$ as shown in Fig. 38. Using the formulas in Table 4

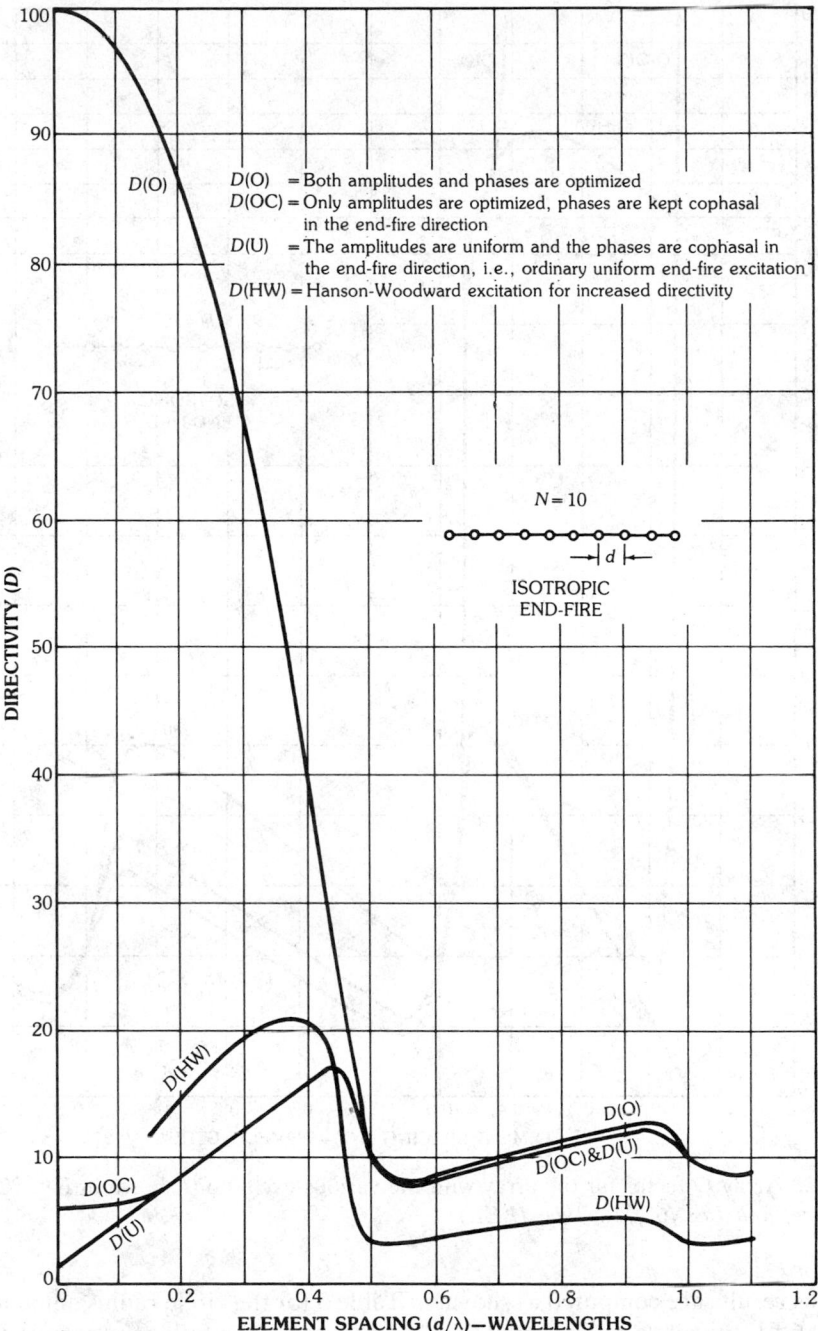

Fig. 32. Directivities of a linear end-fire array of 10 isotropic elements with various excitations. (*After Lo, Lee, and Lee [65], © 1966 IEEE*)

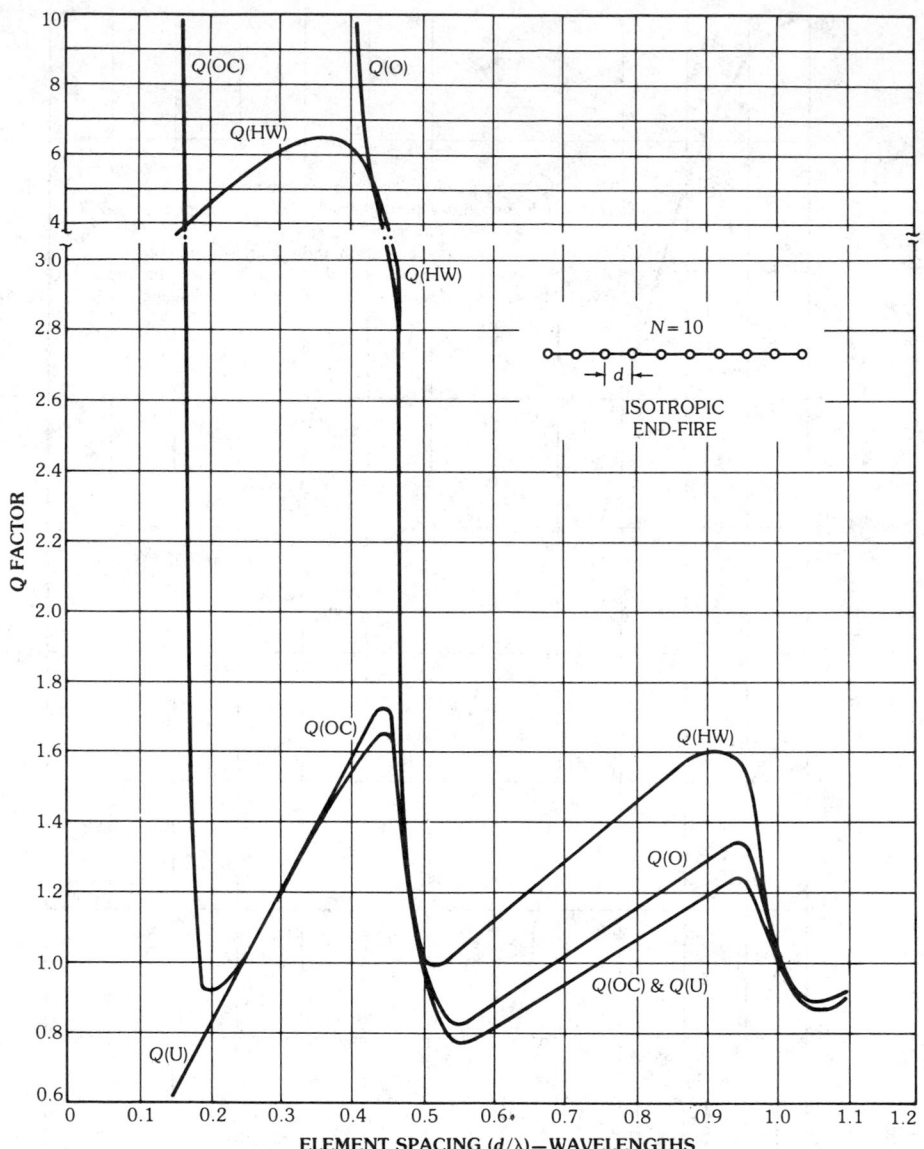

Fig. 33. Array Q factor for the array with the various excitations shown in Fig. 32. (*After Lo, Lee, and Lee [65], © 1966 IEEE*)

various results are computed as shown in Table 5 for the circle radius equal to λ and Table 6 for the circle radius equal to $\lambda/4$. From these the following remarks can be made:

(*a*) From Table 5 the highest *SNR* of 81.6 with $Q = 1.14$ is obtained, as compared with 35.5 with $Q = 0.916$ for a uniform excitation. Thus an improvement of 3.6 dB is obtained without paying a high price on the Q factor.

(*b*) From Table 6 where the elements are closely spaced, the highest *SNR* of

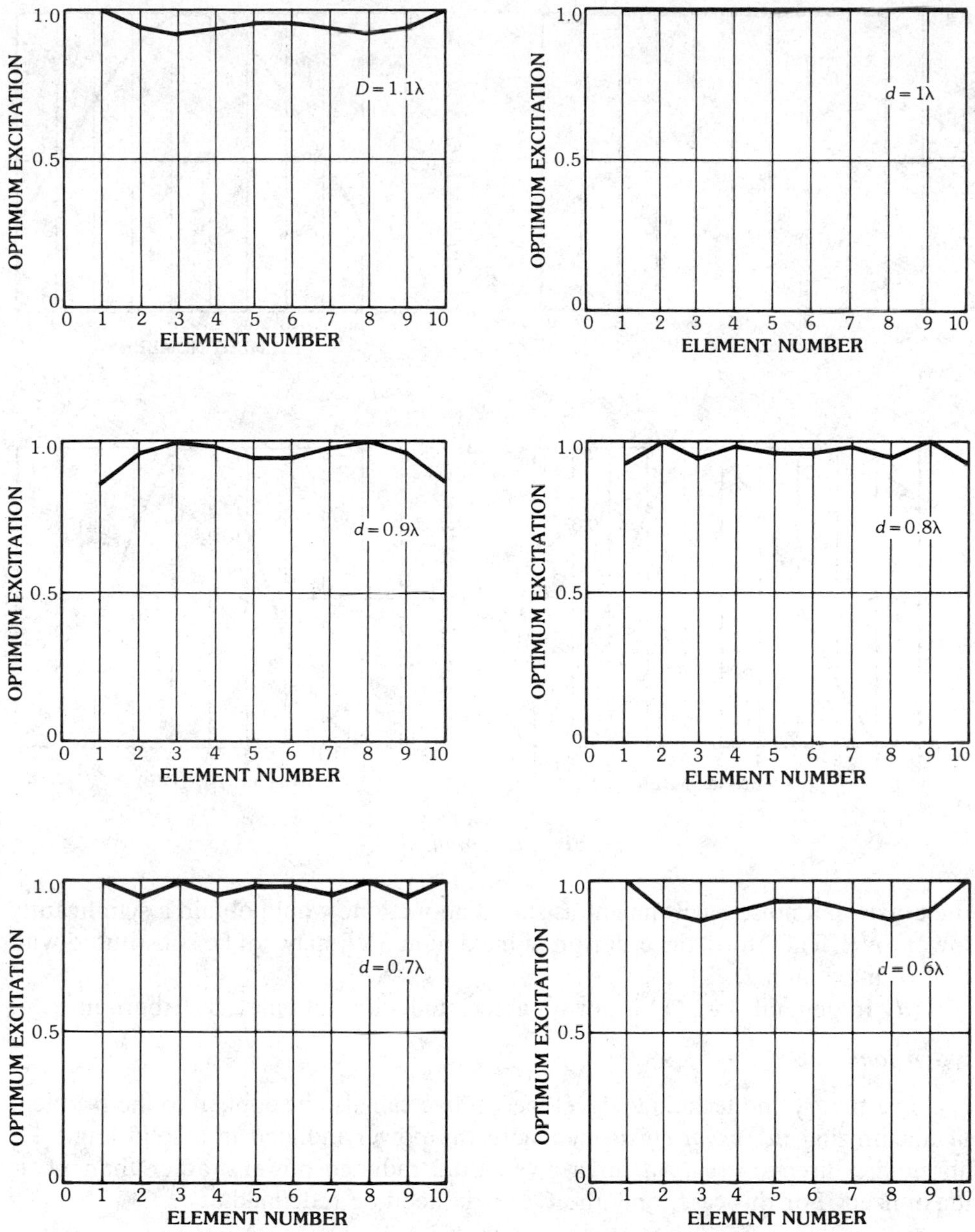

Fig. 34. Relative amplitudes of the optimum excitations for $D(O)$ in Fig. 32 for various element spacings d. (*After Lo, Lee, and Lee [65], © 1966 IEEE*)

47.1 is obtained at the price of $Q = 3.26 \times 10^3$. But with a prescribed value of 20 for the Q factor, $SNR = 21.8$, which is about 5.2 dB higher than that of a uniform excitation.

(*c*) If D, rather than the SNR, is optimized for the array to receive a signal from

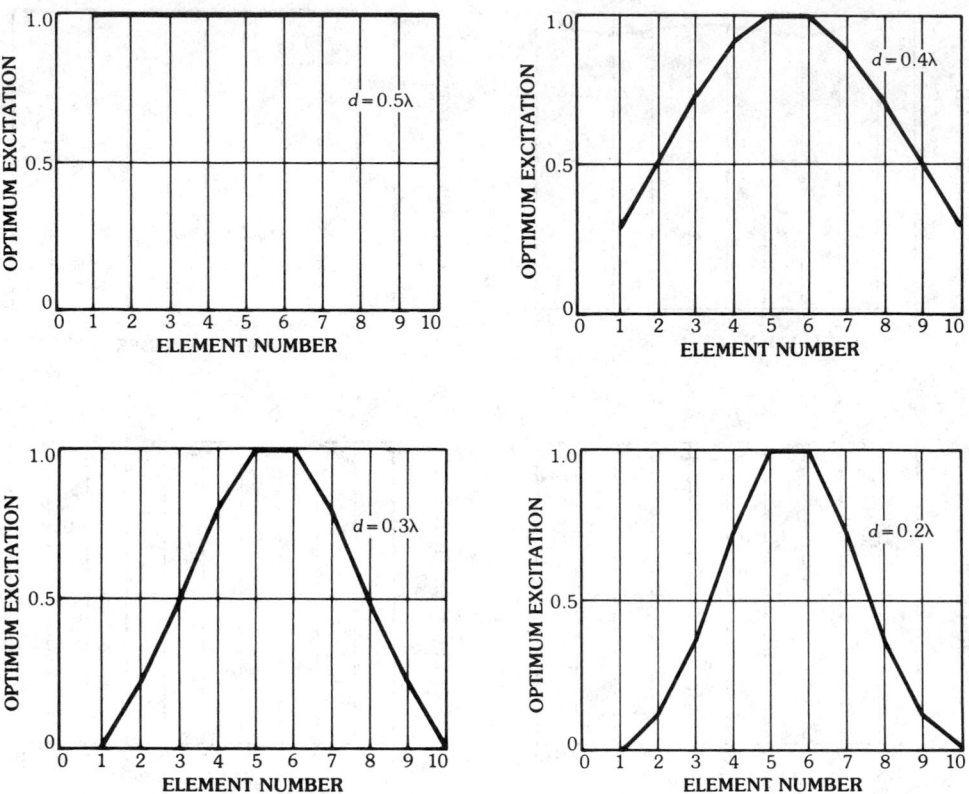

Fig. 34, *continued.*

the z axis in a noise environment as stated above, one would obtain a significantly lower *SNR*. Therefore, the criterion of maximum *SNR* may not be substituted with that of maximum D.

(d) In general, large Q is always associated with antiphase distribution.

Extensions

The theory and technique developed above can also be applied to the problem of maximizing the *beam efficiency* where the power radiated in a solid angle Ω_0 about the direction (θ_0, ϕ_0) for a given total radiated power, rather than D, is maximized. For this case, only the **C** matrix need be redefined as

$$\mathbf{C} = \int_{\Omega_0} V^* \rangle \langle V d\Omega \tag{184}$$

This solution is given by solving the eigenvalue problem $(\mathbf{C} - \lambda\mathbf{B})(J) = 0$. The optimization of an aperture antenna can also be solved in a similar manner except that a set of basis functions, or modal functions, over the aperture, instead of elements, should be considered [65].

The technique can also be applied for pattern synthesis. In that case the

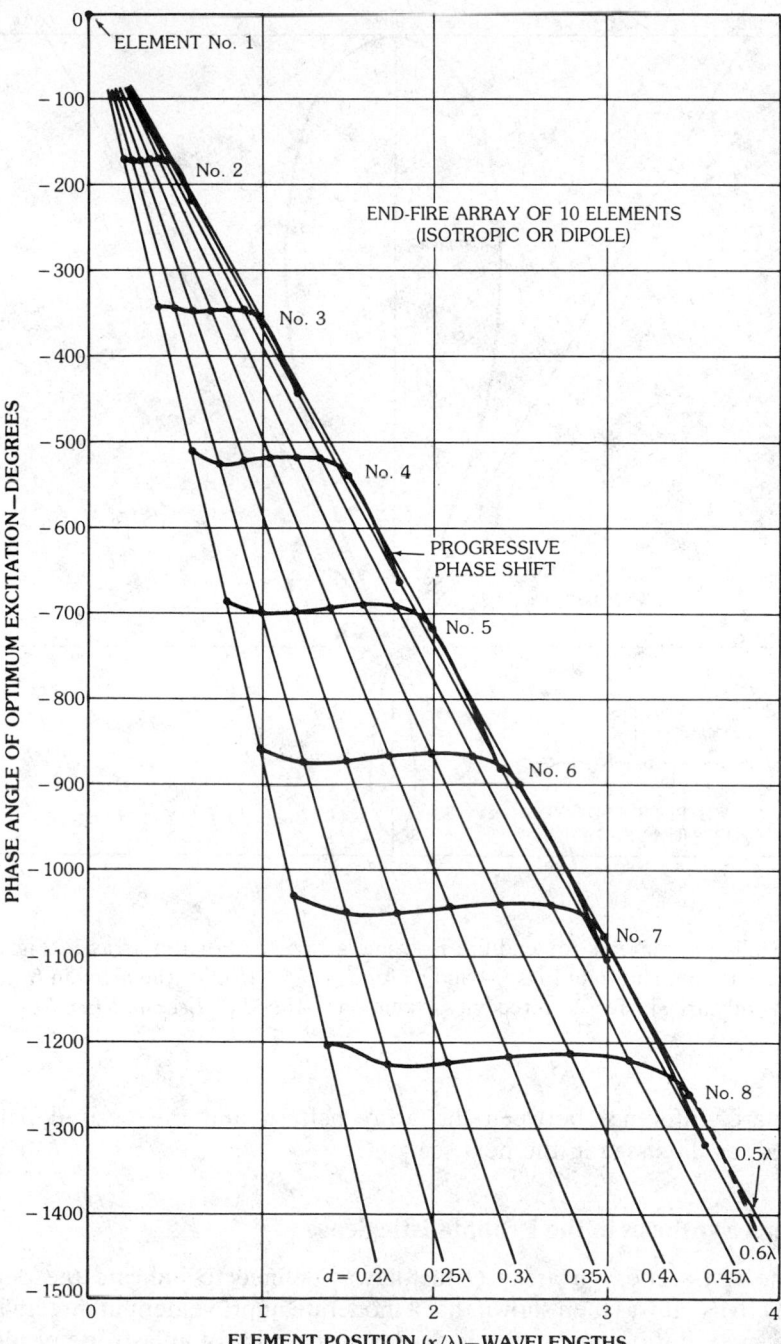

Fig. 35. Relative phases of the optimum excitations for $D(O)$ in Fig. 32 for various element spacings d, where only the phases for the first eight elements are shown.

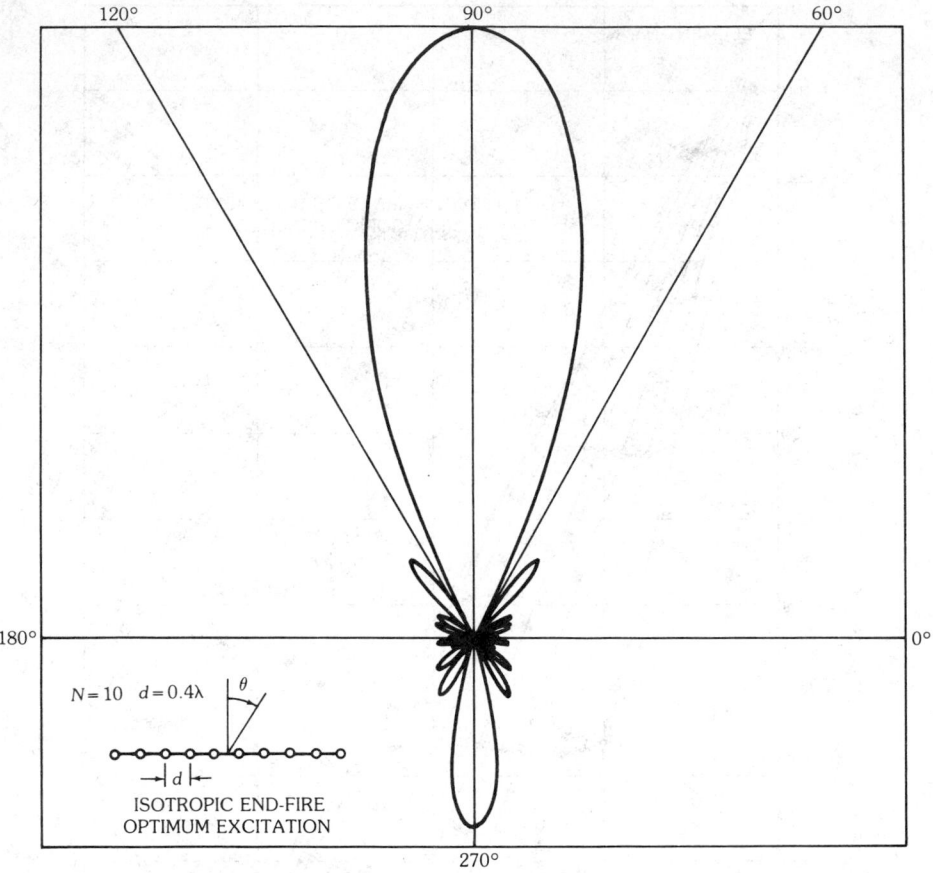

a

Fig. 36. Radiation patterns for end-fire isotropic array. (*a*) For the array in Fig. 32 with optimum excitations shown in Figs. 34 and 35 for $d = 0.4\lambda$. (*b*) For the array in (*a*) but with Hansen-Woodyard's increased directivity excitation. (*After Lo, Lee, and Lee [65], © 1966 IEEE*)

mean-square difference between the array pattern and the desired pattern is minimized, as discussed in the next section.

7. Pattern Synthesis in the Probabilistic Sense

In the last section an array Q factor was defined to indicate the degree of superdirectivity. It was then shown that a moderate improvement in directivity over the uniform excitation can be obtained only at the price of an astronomically large Q factor. To make the design practical a constraint on the Q factor should be imposed. However, the question of how large a value of the Q factor can be considered practical is left unanswered. One possible approach to this problem is to take the parametric uncertainty into consideration. To illustrate this, a pattern synthesis problem is discussed. Let

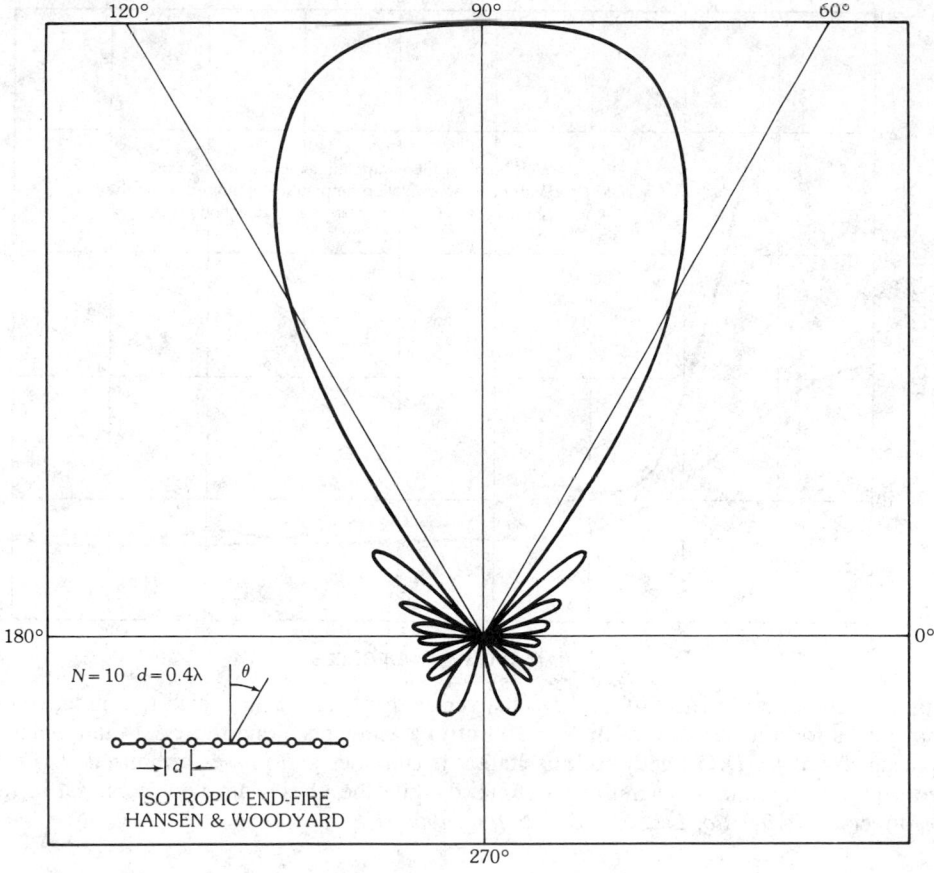

Fig. 36, *continued.*

$$f_d(\theta, \phi) = \text{desired pattern function}$$
$$f(\theta, \phi) = \text{actually realizable pattern of the array} = \langle JV^* \rangle \qquad (185)$$

where

$\langle J = \text{a row vector} = (J_1, J_2, \ldots, J_N)$

$J_n = \text{excitation current for the } n\text{th element}$

$V \rangle = \text{a column vector} = (V_1, V_2, \ldots, V_N)^t$

$V_n^* = \text{pattern function of the } n\text{th element}$

Then, for least-square optimization, the following norm in the L_2 space is to be minimized:

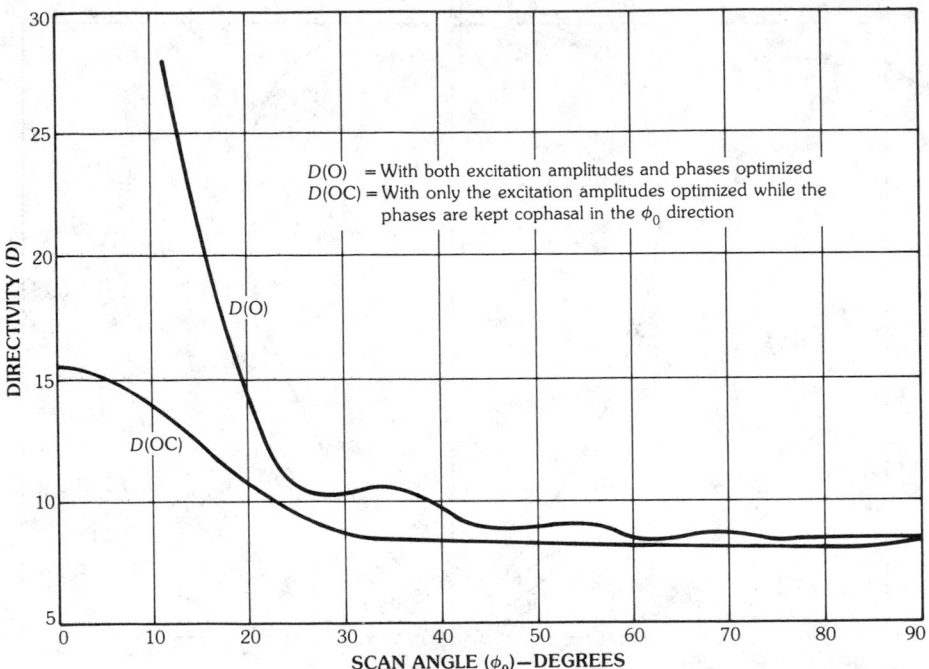

$D(O)$ = With both excitation amplitudes and phases optimized
$D(OC)$ = With only the excitation amplitudes optimized while the
phases are kept cophasal in the ϕ_0 direction

Fig. 37. The directivities $D(O)$ and $D(OC)$ versus scan angle ϕ_0 in which the directivity is maximized for a linear array with $N = 10$ isotropic elements along the x axis and element spacing $d = 0.4\lambda$. $D(O)$: with both excitation magnitudes and phases optimized. $D(OC)$: with only the excitation magnitudes optimized while the phases are kept cophasal in the ϕ_0 direction. (*After Lo, Lee, and Lee [65], © 1966 IEEE*)

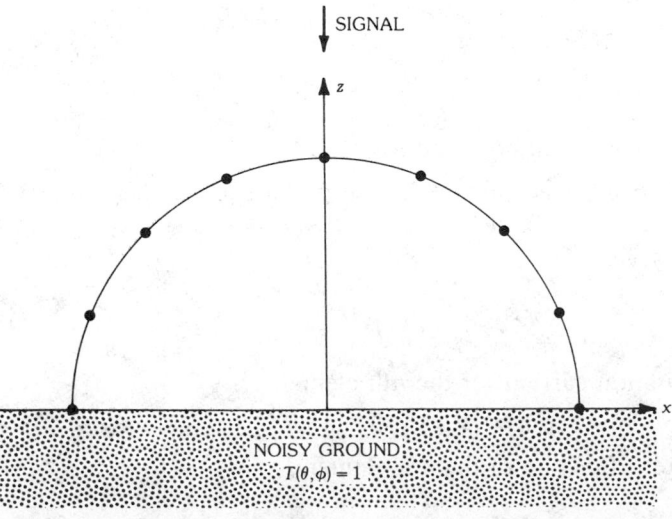

Fig. 38. Geometry of a semicircular array above a noisy ground to receive a signal from the z axis. (*After Lo, Lee, and Lee [65], © 1966 IEEE*)

Table 5. Optimum Semicircular Array ($r = \lambda$) of Nine Cophasally Excited Isotropic Elements

	Current	D	SNR	Q Factor
Uniform excitation	$J_1 = J_2 = J_3 = J_4 = J_5 =$ $J_6 = J_7 = J_8 = J_9 = 1$	8.24	35.5	0.916
Optimum D without constraint on Q factor	$J_1 = J_9 = 1.123$ $J_2 = J_8 = 1.29$ $J_3 = J_7 = 0.881$ $J_4 = J_6 = 0.757$ $J_5 = 0.600$	8.71	55.0	1.03
Optimum D with a prescribed Q factor	$J_1 = J_9 = 1.082$ $J_2 = J_8 = 1.218$ $J_8 = J_7 = 0.898$ $J_4 = J_6 = 0.816$ $J_5 = 0.659$	8.67	50.5	1.0 (prescribed)
Optimum SNR without constraint on Q factor	$J_1 = J_9 = 11.436$ $J_2 = J_8 = 15.396$ $J_3 = J_7 = 10.446$ $J_4 = J_6 = 3.746$ $J_5 = -0.421$	7.76	81.6	1.14
Optimum SNR with a prescribed Q factor	$J_1 = J_9 = 5.835$ $J_2 = J_8 = 7.719$ $J_3 = J_7 = 7.451$ $J_4 = J_6 = 5.223$ $J_5 = 2.664$	8.44	55.1	1.0 (prescribed)

Element positions (in wavelengths): $x_1 = -x_9 = 1.000$ $z_1 = z_9 = 0.0$
$x_2 = -x_8 = 0.924$ $z_2 = z_8 = 0.383$
$x_3 = -x_7 = 0.707$ $z_3 = z_7 = 0.707$
$x_4 = -x_6 = 0.383$ $z_4 = z_6 = 0.924$
$x_5 = 0.0$ $z_5 = 1.0$

Main beam: $\theta_0 = 0$

Thermal noise distribution: $T(\theta, \phi) = \begin{cases} 1 & \text{for } \pi/2 < \theta \leq \pi \\ 0 & \text{otherwise} \end{cases}$

(After Lo, Lee, and Lee [65], © 1966 IEEE)

$$\varepsilon = \| f_d - f \|^2 = \int_{4\pi} (f_d - f)(f_d - f)^* \, w \, d\Omega$$

$$= \| f_d \|^2 + \int_{4\pi} |f|^2 \, w \, d\Omega - 2 \operatorname{Re} \left\{ \int f f_d^* \, w \, d\Omega \right\}$$

$$= \| f_d \|^2 + \left\langle J \int_{4\pi} V^* \right\rangle \langle V_d \, w \, d\Omega \, J^* \rangle - 2 \operatorname{Re} \left\{ \left\langle J \int_{4\pi} V^* \right\rangle f_d^* \, w \, d\Omega \right\}$$

$$= \| f_d \|^2 + \langle J G J^* \rangle - 2 \operatorname{Re} \{ \langle J C \rangle \} \tag{186}$$

where

Table 6. Optimum Semicircular Array ($r = 0.25\lambda$) of Nine Cophasally Excited Isotropic Elements

	Current	D	SNR	Q Factor
Uniform excitation	$J_1 = J_2 = J_3 = J_4 = J_5 =$ $J_6 = J_7 = J_8 = J_9 = 1$	2.19	6.63	0.244
Optimum D without constraint on Q factor	$J_1 = J_9 = 5.23$ $J_2 = J_8 = -15.74$ $J_3 = J_7 = 34.81$ $J_4 = J_6 = -55.83$ $J_5 = 66.69$	3.63	37.8	3.76×10^3
Optimum D with a prescribed Q factor	$J_1 = J_9 = 2.24$ $J_2 = J_8 = -2.92$ $J_3 = J_7 = 3.35$ $J_4 = J_6 = -2.23$ $J_5 = 2.37$	3.25	20.2	20.0 (prescribed)
Optimum SNR without constraint on Q factor	$J_1 = J_9 = 58.86$ $J_2 = J_8 = -179.6$ $J_3 = J_7 = 412.72$ $J_4 = J_6 = -686.83$ $J_5 = 836.80$	3.52	47.1	3.26×10^3
Optimum SNR with a prescribed Q factor	$J_1 = J_9 = 12.80$ $J_2 = J_8 = -15.58$ $J_3 = J_7 = 19.70$ $J_4 = J_6 = -18.96$ $J_5 = 25.87$	3.19	21.8	20.0 (prescribed)

Element positions (in wavelengths):
$$x_1 = -x_9 = r \qquad z_1 = z_9 = 0$$
$$x_2 = -x_8 = 0.9239r \qquad z_2 = z_8 = 0.3827r$$
$$x_3 = -x_7 = 0.7071r \qquad z_3 = z_7 = 0.7071r$$
$$x_4 = -x_6 = 0.3827r \qquad z_4 = z_6 = 0.9239r$$
$$x_5 = 0 \qquad z_5 = r$$

Main beam: $\theta_0 = 0$

Thermal noise distribution: $T(\theta, \phi) = \begin{cases} 1 & \text{for } \pi/2 < \theta \leq \pi \\ 0 & \text{otherwise} \end{cases}$

(After Lo, Lee, and Lee [65], © 1966 IEEE)

w = a possible weighting function that stresses a closer approximation of f to f_d for some angular regions

$$\mathbf{G} = \int_{4\pi} V^*\rangle\langle V\, w\, d\Omega \quad \text{(a positive-definite Hermitian matrix)} \tag{187}$$

$$\mathbf{C} = \int_{4\pi} V^*\rangle f_d^*\, w\, d\Omega \tag{188}$$

Using the variational method, the solution of $J\rangle$ which minimizes ε is found to be

$$J_0\rangle = (\mathbf{G}^{-1}\mathbf{C}\rangle)^* = U\,\mathrm{diag}[(\lambda_1)^{-1}, (\lambda_2)^{-1}, \ldots, (\lambda_N)^{-1}]\,U^\dagger\mathbf{C}^*\rangle \qquad (189)$$

where

$$\lambda_1, \lambda_2, \ldots, \lambda_N = \text{the eigenvalues, in descent order, of } \mathbf{G}^*$$
$$U = \text{a unitary matrix which diagonalizes } \mathbf{G}^*$$
$$U^\dagger = \text{the complex conjugate transpose of } U$$

Since $\mathbf{G}$ depends on the array geometry, its largest to smallest eigenvalue ratio, λ_1/λ_N, can increase very rapidly to an extremely large value as the array element spacing decreases. This finally leads to an ill-conditioned $\mathbf{G}$ and a very unstable solution $J_0\rangle$. Mathematically, this phenomenon is identical with the super-directivity. Such a solution is not only impractical, but is also difficult to compute accurately. Therefore, the so-called optimum solution (189) in such a case should not be taken seriously.

Because of the physical limitation, instrumental error, environmental variation, etc., the excitation $J\rangle$ can only be adjusted in practice within an uncertain random error, say $\delta\tilde{J}\rangle$. Therefore for a more practically meaningful design, this error should be taken into consideration. Assume that the error $\delta\tilde{J}\rangle$ is, as in most practical cases, only *relative* to $J\rangle$:

$$\delta\tilde{J}\rangle = \tilde{\mathbf{A}}J\rangle \qquad (190)$$

where $\tilde{\mathbf{A}}$ is a matrix with stochastic elements, which may result from, for example, some uncertainties in the feeding network. Thus the pattern function due to the actual current $J\rangle + \delta\tilde{J}\rangle$ will be a random function,

$$\tilde{f}(\theta, \phi) = \langle (J + \delta\tilde{J}), V^*(\theta, \phi)\rangle \qquad (191)$$

and so must be the difference between $\tilde{f}(\theta, \phi)$ and $f_d(\theta, \phi)$. Following (186), let

$$\tilde{\varepsilon} = \| f_d - \tilde{f} \|^2 \qquad (192)$$

Then for a given $J\rangle$ and a given realization of the random error $\delta\tilde{J}\rangle$, there is a realization of $\tilde{f}$ and also $\tilde{\varepsilon}$. Hence, for the totality of all these realizations there is a probability distribution $F(\varepsilon; J\rangle)$ which defines the probability for $\tilde{\varepsilon} < \varepsilon$ for the given $J\rangle$. With this distribution function one can define the probability mean of a quantity in the following formula:

$$E\{\cdot\} = \int_{\substack{\text{entire prob.}\\ \text{space}}} (\cdot)\,dF(\varepsilon; J\rangle) \qquad (193)$$

With this preparation one can define the optimization problem in many different ways. Among them we may state three theoretically possible situations as shown in Fig. 39, each giving an optimum solution different from the others,

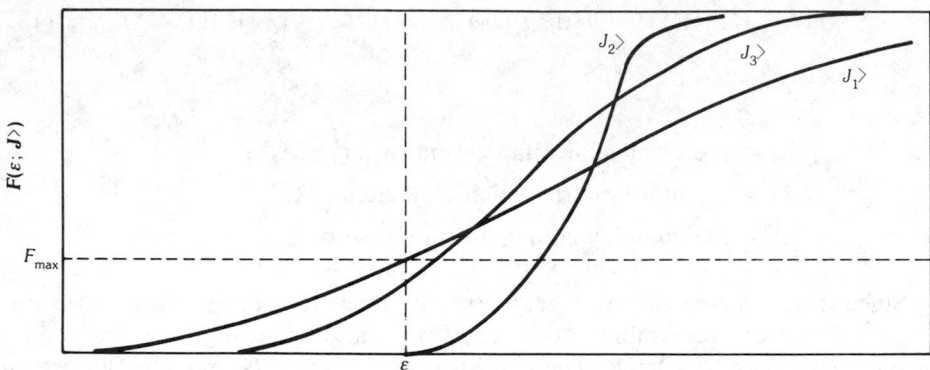

a

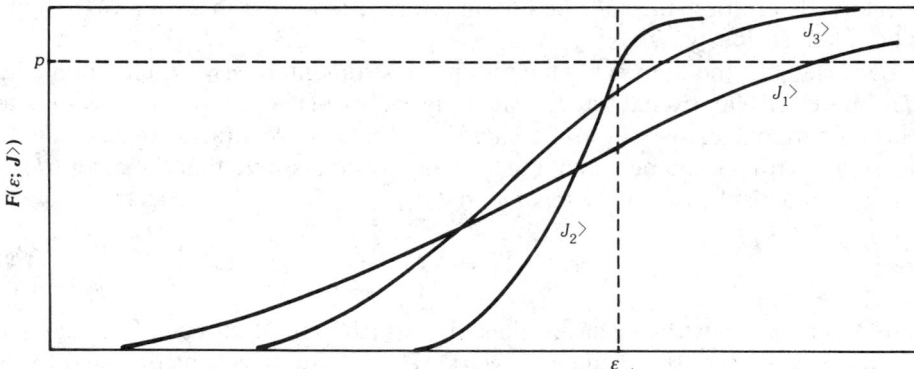

b

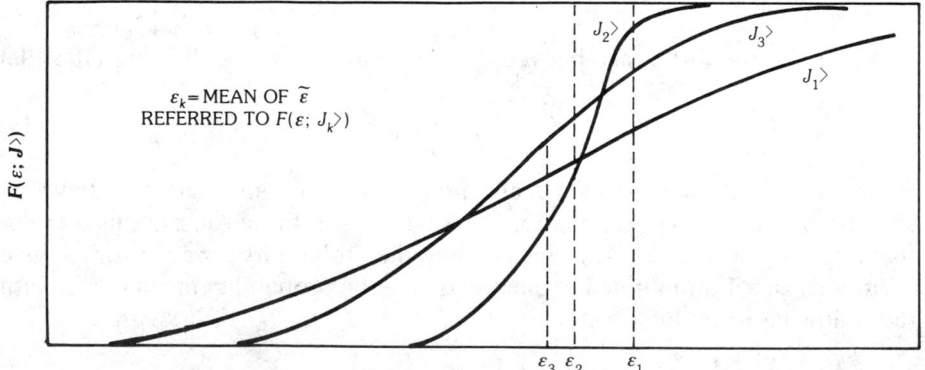

c

Fig. 39. Three possible philosophies of optimization. (*a*) Vertical optimization: $J_1\rangle$ is the optimum. (*b*) Horizontal optimization: $J_2\rangle$ is the optimum. (*c*) Optimization of $E\{\tilde{\varepsilon}\}$: $J_3\rangle$ is the optimum. (*After Richards and Lo [80], © 1975 IEEE*)

depending on the sense of optimization. The first scenario shown in Fig. 39a, which may be called the *vertical optimization*, illustrates that for a *given* tolerable error ε, $J\rangle$ is to be determined under the condition $F(\varepsilon; J\rangle)$ is the maximum, i.e., the largest probability for having error less than the given tolerance ε. Comparing the three possible distribution curves, $J_1\rangle$ is apparently the best solution since it gives the largest value of $F(\varepsilon; J\rangle)$ for the given ε.

The second possible optimization, which may be called the *horizontal optimization*, is defined by seeking $J\rangle$ such that for a given probability, say p, the error ε is the minimum as shown in Fig. 39b. It is seen that for this scenario $J_2\rangle$ is the best solution. The third possible optimization is to seek $J\rangle$ which minimizes the mean value of ε, namely $\bar{\varepsilon} = E\{\bar{\varepsilon}\}$. As can be seen from Fig. 39c, $J_3\rangle$ is the best solution. As stated, these three distribution curves are hypothetical and whether they would occur in a certain problem is not clear. For simplicity, in what follows the last sense of optimization is considered and it is found later that the solution so obtained also leads to approximately the optimum solutions in the other two senses.

Without loss of generality one may assume that

$$E\{\delta\tilde{J}\rangle\} = 0 \tag{194}$$

Since now the excitation is $J\rangle + \delta\tilde{J}\rangle$, similar to (186), the random error between the realized pattern f and desired pattern f_d is simply

$$\bar{\varepsilon} = \|f_d\|^2 + \langle(J + \delta\tilde{J})\,\mathbf{G}(J + \delta\tilde{J})^*\rangle - 2\,\mathrm{Re}\{\langle(J + \delta\tilde{J})\,C\rangle\} \tag{195}$$

Since $\delta\tilde{J}\rangle = \tilde{A}J\rangle$,

$$\bar{\varepsilon} = E\{\bar{\varepsilon}\} = \|f_d\|^2 + \langle J\mathbf{G}J^*\rangle - 2\,\mathrm{Re}\{\langle JC\rangle\} + E\{\langle\delta\tilde{J}\,\mathbf{G}\,\delta\tilde{J}^*\rangle\} \tag{196}$$

where

$$E\{\langle\delta\tilde{J}\,\mathbf{G}\,\delta\tilde{J}^*\rangle\} = \langle J\mathbf{K}_2 J^*\rangle \quad \text{and} \quad \mathbf{K}_2 = E\{\tilde{A}^t\mathbf{G}\tilde{A}^*\} \tag{197}$$

The *mn*th element of matrix $\mathbf{K}_2$ is given by

$$(\mathbf{K}_2)_{mn} = \sum_{i,j} G_{ij} E\{\tilde{A}_{jm}\tilde{A}_{in}^*\} \tag{198}$$

where G_{ij} is the *ij*th element of $\mathbf{G}$, and $E\{\tilde{A}_{jm}\tilde{A}_{in}^*\}$ is the covariance (or joint moment) of elements (jm) and (in) of $\tilde{A}$. Using (196)–(198) the optimum solution of $J\rangle$ which minimizes $\bar{\varepsilon}$ for a given probabilistic property of $\delta\tilde{J}$ (or $\tilde{A}$) is given by

$$J\rangle = [(\mathbf{G} + \mathbf{K}_2)^{-1}C\rangle]^* \tag{199}$$

For a typical simple case where all array elements are identical, $\delta\tilde{J}\rangle$ has uncorrelated elements and identical variance σ^2, the matrix $\mathbf{K}_2$ becomes diagonal:

$$\mathbf{K}_2 = \sigma^2 G_{11}\mathbf{I} = \alpha\mathbf{I} \tag{200}$$

where $\mathbf{I}$ is an identity matrix and α is a scalar equal to $\sigma^2 G_{11}$. Therefore, following (189), the optimum solution for this case is simply

$$J\rangle = \mathbf{U}\,\mathrm{Diag}[(\lambda_1 + \alpha)^{-1},(\lambda_2 + \alpha)^{-1},\ldots,(\lambda_N + \alpha)^{-1}]\,\mathbf{U}^\dagger C^*\rangle \tag{201}$$

where $\lambda_1,\ldots,\lambda_N$ are, as before, the eigenvalues of $\mathbf{G}$. It is interesting to compare (201) with (189) and note that the two solutions are identical except for the constant α. Now even if $\lambda_1,\ldots,\lambda_N$ march off to zero as the element spacing decreases, the solution $J\rangle$ is no longer unstable and dominated by those small λ_ns. Tihonov [125] introduced the so-called regularization method for ill-conditioned problems such as the one in (189). Cabayan's group [126], in making use of that method for the pattern synthesis problem, proposed to minimize $\| f_d - f \|^2 + \alpha\langle JJ^*\rangle$, instead of $\| f_d - f \|^2$. It is clear, however, that α must assume a proper value because if too large the minimization will be applied mainly for $\langle JJ^*\rangle$ and, if too small, the solution may be again unstable. For a given probability distribution of $\delta\tilde{J}\rangle$, Cabayan used the Monte Carlo method to determine the proper value of α numerically. The above analysis shows that the proper value is simply $\sigma^2 G_{11}$.

An Example—Richards and Lo [80] considered a planar array with 10 concentric rings as shown in Fig. 40. The radius of the kth ring is

$$\varrho_k = 2\pi\left[\frac{1}{4}\left(k - \frac{2}{\pi}\right) - \frac{1}{2\pi}\left(\frac{1}{2}\right)^{10-k}\right], \qquad k = 1,2,\ldots,10$$

A total of 200 vertical dipoles are placed over the intersecting points between the rings and 20 rays from the origin uniformly spaced in angles over 360° as illustrated in the figure. The objective is to determine the optimum excitation which will produce a pattern "closest" to a secant pattern:

$$f_d(\theta) = \begin{cases} \sec\theta & \text{for} \quad 0 < \theta \leq 70° \\ \sec 70° & \text{for } 70° \leq \theta \leq 90° \end{cases}$$

The solid curve in Fig. 41a is the nominal elevation pattern for $\sigma = 0$ using the solution given by (189) for no tolerance. The solid curve in Fig. 41b is the nominal pattern for $\sigma = 5$ percent using the solution given by (201). It is seen that the former is closer than the latter to the desired pattern f_d which is also shown in both figures. However, if the actual error $\delta\tilde{J}\rangle$ is added to $J\rangle$, the results are completely different. Using the Monte Carlo method, five sets of sample errors of $\delta\tilde{J}\rangle$ with zero mean and 5 percent standard deviation are generated. From these five sets five sample patterns are computed for both cases, shown as the dotted curves in Fig. 41a and dots only in Fig. 41b. It is clearly seen that in the former case, where the unavoidable error is ignored in the optimization, none of the five realizations even remotely resembles the desired, or the nominal, pattern. As for the latter case, although the nominal pattern is less close to f_d than that of the first case, with all five sets of real errors included, the five realizations of the pattern are so close to

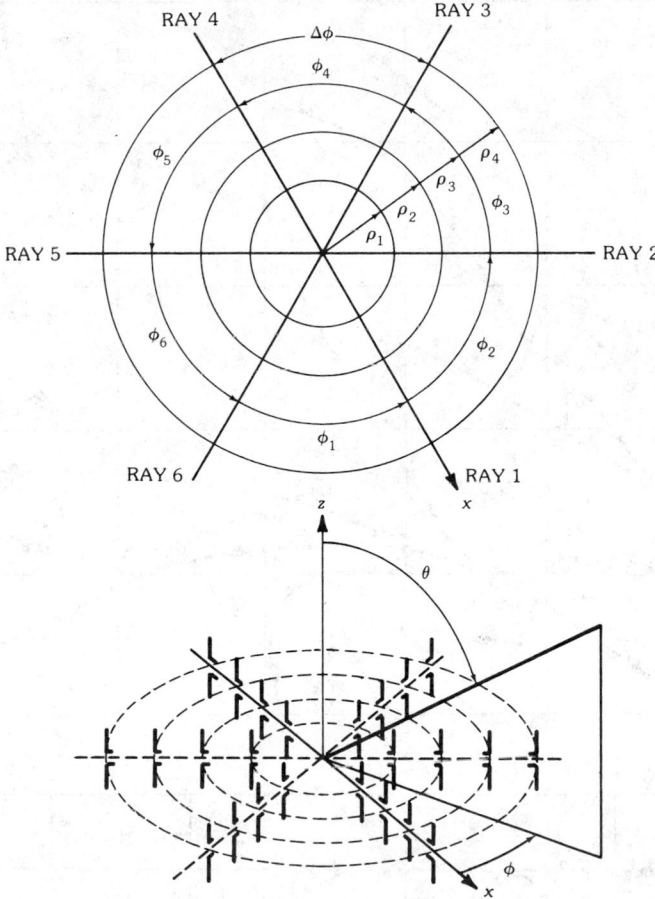

Fig. 40. Geometry of a planar array of vertical dipoles (or monopoles above the ground). (*After Richards and Lo [80],* © *1975 IEEE*)

the nominal pattern that, to avoid confusion, no attempt was made to connect the dots for each realization. This result vividly confirms that in some design problems, where the stability is questionable, to demand less may *actually* obtain a better optimum solution.

For the example discussed above, Richards and Lo [80] has derived an Edgeworth series representation for the distribution function $F(\varepsilon; J\rangle)$ and also showed the validity of this solution with 2500 samples using the Monte Carlo simulation with a computer. They also verified the theoretical solution for one value of σ while using Monte Carlo simulations for a wide range of σ values. For the example studied, the sample distribution curves for $\sigma = 3$, 5, and 10 percent are very close to each other, indicating that a precise information of the statistics of $\delta \bar{J}$ may not be very important for determining the optimum solution $J\rangle$. Another interesting result shows that $\| J \|$ for $\sigma = 0$ is several orders of magnitude larger than that for $\sigma = 5$ percent, implying very high Q, low efficiency, and unstable solution for the former, in agreement with previous discussion for maximum D.

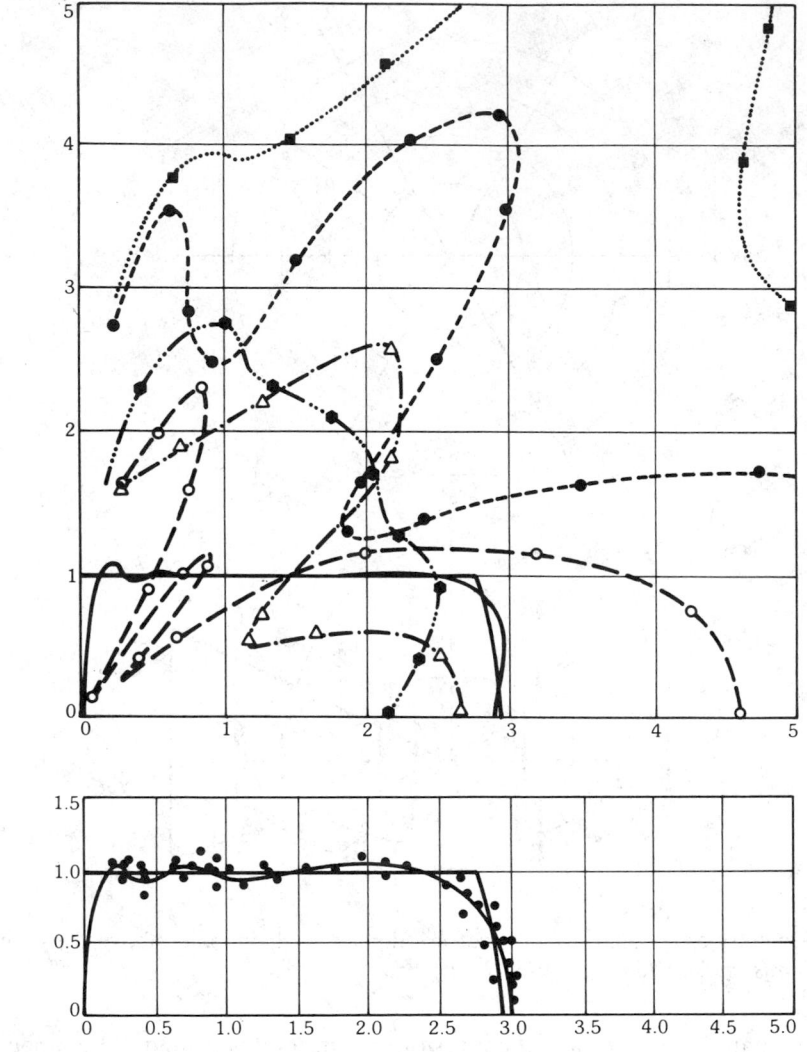

Fig. 41. Synthesis for a desired pattern function: $f_d = \sec\theta$ for $0° < \theta \leqq 70°$ and $\sec 70°$ for $70° \leqq \theta \leqq 90°$. (*a*) Nominal pattern (solid curve), in polar plot, corresponding to the deterministic optimum solution with zero tolerance and five actual sample patterns (dotted) for 5% standard deviation of the tolerance, generated by the Monte Carlo method. (*b*) Nominal pattern (solid curve), in polar plot, corresponding to the probabilistic optimum with the 5% tolerance taken into consideration and five actual sample patterns (dots only), generated as in (*a*) by the Monte Carlo method. (*After Richards and Lo [80], © 1975 IEEE*)

8. References

[1] J. L. Allen et al., "Phased array radar studies," *Tech. Rep. 238*, MIT Lincoln Labs, August 1960.

[2] N. Amitay et al., *Theory and Analysis of Phased Array Antennas*, New York: John Wiley & Sons, 1972.

[3] H. Bach and J. E. Hansen, "Uniformly spaced arrays," chapter 5 in *Antenna Theory,*

Part 1, ed. by R. E. Collin and F. J. Zucker, New York: McGraw-Hill Book Co., 1969.

[4] C. A. Balanis, *Antenna Theory: Analysis and Design*, New York: Harper & Row, 1982.

[5] D. Barbiere, "A method for calculating the current distribution of Tschebyscheff arrays," *Proc. IRE*, vol. 40, pp. 78–82, 1952.

[6] E. T. Bayliss, "Design of monopulse antenna difference patterns with low side lobes," *Bell Syst. Tech. J.*, vol. 47, pp. 623–640, 1968.

[7] A. Bloch, R. G. Medhurst, and S. D. Pool, "A new approach to the design of super-directive aerial arrays," *IEE Proc.*, vol. 100, pt. III, pp. 303–314, 1953.

[8] A. Bloch, R. G. Medhurst, and S. D. Pool, "Superdirectivity," *Proc. IRE*, vol. 48, p. 1164, 1960.

[9] C. J. Bouwkamp and N. G. De Bruijn, "The problem of optimum antenna current distribution," *Philips Res. Rep.*, vol. 1, p. 135, 1946.

[10] P. A. Bricout, "Pattern synthesis using weighted functions," *IRE Trans. Antennas Propag.*, vol. AP-8, pp. 441–444, 1960.

[11] J. L. Brown, "A simplified derivation of the Fourier coefficients for Chebyshev patterns," *Proc. IEE*, vol. 105C, pp. 167–168, 1957.

[12] J. L. Brown, "On the determination of excitation coefficients for a Tchebycheff pattern," *IRE Trans. Antennas Propag.*, vol. AP-10, pp. 215–216, 1962.

[13] L. B. Brown and G. A. Scharp, "Tschebyscheff antenna distribution, beam-width, and gain tables," *NOLC Rep. 383*, February 1958.

[14] D. K. Cheng, "Optimization techniques for antenna arrays," *Proc. IEEE*, vol. 59, pp. 1664–1674, 1971.

[15] D. K. Cheng and M. T. Ma, "A new mathematical approach for linear array analysis," *IRE Trans. Antennas Propag.*, vol. AP-8, pp. 255–259, 1960.

[16] L. J. Chu, "Physical limitations of omnidirectional antennas," *J. Appl. Phys.*, vol. 19, pp. 1163–1175, 1948.

[17] H. P. Coleman, "An iterative technique for reducing side lobes of circular arrays," *IEEE Trans. Antennas Propag.*, vol. AP-18, pp. 566–567, July 1970.

[18] R. E. Collin and S. Rothschild, "Evaluation of antenna Q," *IEEE Trans. Antennas Propag.*, vol. AP-12, pp. 23–27, January 1964.

[19] A. G. Dermerud and J. J. Gustincic, "The interpolation of general active array impedance from multielement simulators," *IEEE Trans. Antennas Propag.*, vol. AP-27, pp. 68–71, 1979.

[20] B. L. Diamond, "A generalized approach to the analysis of infinite planar array antennas," *Proc. IEEE*, vol. 56, pp. 1837–1851, 1968.

[21] B. L. Diamond and G. H. Knittel, "A new procedure for the design of a waveguide element for a phased-array antenna," *Phased Array Antennas*, ed. by A. A. Oliner and G. H. Knittel, Dedham, Mass.: Artech House, 1972, pp. 149–156.

[22] C. L. Dolph, "A current distribution for broadside arrays which optimizes the relationship between beam width and side lobe level," *Proc. IRE*, vol. 34, pp. 335–348, 1946.

[23] C. J. Drane, "Derivation of excitation coefficients for Chebyshev arrays," *Proc. IEE*, vol. 110, pp. 1755–1758, 1963.

[24] C. J. Drane, "Dolph-Chebyshev excitation coefficient approximation," *IEEE Trans. Antennas Propag.*, vol. AP-12, pp. 781–782, 1964.

[25] C. J. Drane, "Useful approximations for the directivity and beamwidth of large scanning Dolph-Chebyshev arrays," *Proc. IEEE*, vol. 56, pp. 1779–1787, 1968.

[26] R. H. DuHamel, "Optimum patterns for end-fire arrays," *Proc. IRE*, vol. 41, pp. 652–659, 1953.

[27] R. H. DuHamel, "Pattern synthesis for antenna arrays on circular, elliptical, and spherical surfaces," *Tech. Rep. no. 16*, E. E. Research Lab, Univ. of Illinois, Urbana, 1952.

[28] R. S. Elliott, *Antenna Theory and Design*, Englewood Cliffs: Prentice-Hall, 1981.

[29] R. S. Elliott, "Mechanical and electrical tolerances for two-dimensional scanning

antenna arrays," *IRE Trans. Antennas Propag.*, vol. AP-6, pp. 114–120, 1958.

[30] R. S. Elliott, "The theory of antenna arrays," in chapter 1 of *Microwave Scanning Antennas, Volume II*, ed. by R. C. Hansen, New York: Academic Press, 1966.

[31] R. S. Elliott, "Design of line-source antennas for sum patterns with side lobes of individually arbitrary heights," *IEEE Trans. Antennas Propag.*, vol. AP-24, pp. 76–83, 1976.

[32] R. S. Elliott, "On discretizing continuous aperture distributions," *IEEE Trans. Antennas Propag.*, vol. AP-25, pp. 617–621, 1977.

[33] R. S. Elliott and R. M. Johnson, "Experimental results on a linear array designed for asymmetric side lobes," *IEEE Trans. Antennas Propag.*, vol. AP-26, pp. 351–352, 1978.

[34] J. E. Evans, "Synthesis of equiripple sector antenna patterns," *IEEE Trans. Antennas Propag.*, vol. AP-24, pp. 347–353, 1976.

[35] R. J. Evans and T. E. Fortmann, "Design of optimal line-source antennas," *IEEE Trans. Antennas Propag.*, vol. AP-23, pp. 342–347, 1975.

[36] R. L. Fante, "Optimum distribution over a circular aperture for best mean-square approximation to a given radiation pattern," *IEEE Trans. Antennas Propag.*, vol. AP-18, pp. 177–181, 1970.

[37] G. R. Forbes, "An end-fire array continuously proximity-coupled to a two-wire line," *IEEE Trans. Antennas Propag.*, vol. AP-8, pp. 518–519, 1960.

[38] A. C. Gately, Jr., et al., "A network description for antenna problems," *Proc. IEEE*, vol. 56, pp. 1181–1193, 1968.

[39] W. S. Gregorwich et al., "A waveguide simulator for the determination of a phased-array resonance," *IEEE G-AP Intl. Symp. Dig.*, pp. 134–141, 1968.

[40] J. J. Gustincic, "The determination of active array impedance with multielement waveguide simulators," *IEEE Trans. Antennas Propag.*, vol. AP-20, pp. 589–595, 1972.

[41] P. W. Hannan and M. A. Balfour, "Simulation of a phased-array antenna in waveguide," *IEEE Trans. Antennas Propag.*, vol. AP-13, pp. 342–353, 1965.

[42] P. W. Hannan, "Discovery of an array surface wave in a simulator," *IEEE Trans. Antennas Propag.*, vol. AP-15, pp. 574–576, 1967.

[43] R. C. Hansen, "Tables of Taylor distributions for circular aperture antennas," *IRE Trans. Antennas Propag.*, vol. AP-8, pp. 23–26, 1960.

[44] R. C. Hansen, "Comparison of square array directivity formulas," *IEEE Trans. Antennas Propag.*, vol. AP-20, pp. 100–102, 1972.

[45] R. C. Hansen, "A one-parameter circular aperture distribution with narrow beamwidth and low sidelobes," *IEEE Trans. Antennas Propag.*, vol. AP-24, pp. 477–480, 1976.

[46] R. C. Hansen, "Gain limitations of large antennas," *IRE Trans. Antennas Propag.*, vol. AP-8, pp. 490–495, 1970; see also correction in vol. AP-13, p. 997, 1965.

[47] R. C. Hansen, *Microwave Scanning Antennas, Volume 1*, chapter 1, New York: Academic Press, 1964.

[48] R. C. Hansen (ed.), Special issue on "Electronic scanning," *Proc. IEEE*, vol. 56, pp. 1761–2038, 1968.

[49] R. C. Hansen, *Significant Phased Array Papers*, Dedham: Artech House, 1973.

[50] W. W. Hansen and J. R. Woodyard, "A new principle in directional antenna design," *Proc. IRE*, vol. 26, pp. 333–345, 1938.

[51] R. F. Harrington, "Effect of antenna size on gain, bandwidth, and efficiency," *J. Res. Natl. Bur. Stand.* (US), vol. 64D, pp. 1–12, 1960.

[52] W. G. Jaeckle, "Antenna synthesis by weighted Fourier coefficients," *IEEE Trans. Antennas Propag.*, vol. AP-12, pp. 369–370, 1964.

[53] E. C. Jordan and K. G. Balmain, "Electromagnetic waves and radiating systems," section 14.08, Englewood Cliffs: Prentice-Hall, 1968.

[54] W. K. Kahn, "Ideal efficiency of a radiating element in an infinite array," *IEEE Trans. Antennas Propag.*, vol. AP-15, pp. 534–538, 1967.

[55] W. K. Kahn, "Impedance-match and element-pattern constraints for finite arrays," *IEEE Trans. Antennas Propag.*, vol. AP-25, pp. 747–755, 1977.

[56] H. E. King, "Directivity of a broadside array of isotropic radiator," *IRE Trans. Antennas Propag.*, vol. AP-7, pp. 197–198, 1959.

[57] M. J. King and R. K. Thomas, "Gain of large scanned arrays," *IRE Trans. Antennas Propag.*, vol. AP-8, pp. 635–636, 1960.

[58] R. W. P. King, *The Theory of Linear Antennas*, Cambridge: Harvard Univ. Press, 1956.

[59] H. L. Knudsen, "Radiation from ring quasi-arrays," *IRE Trans. Antennas Propag.*, vol. AP-4, pp. 452–472, 1956.

[60] J. D. Kraus, *Antennas*, New York: McGraw-Hill Book Co., 1950.

[61] H. N. Kritikos, "Optimal signal-to-noise ratio for linear arrays by the Schwartz inequality," *J. Franklin Inst.*, vol. 276, pp. 195–304, 1963.

[62] A. Ksienski, "Maximally flat and quasi-smooth sector beams," *IRE Trans. Antennas Propag.*, vol. AP-8, pp. 476–484, 1960.

[63] L. Lapaz and G. A. Miller, "Optimum current distributions on vertical antennas," *Proc. IRE*, vol. 31, pp. 214–232, 1943.

[64] Y. T. Lo and S. W. Lee, "Affine transformation and its application to antenna arrays," *IEEE Trans. Antennas Propag.*, vol. AP-13, pp. 890–896, 1965.

[65] Y. T. Lo, S. W. Lee, and Q. H. Lee, "Optimization of directivity and signal-to-noise ratio of an arbitrary antenna array," *Proc. IEEE*, vol. 54, pp. 1033–1045, 1966.

[66] Y. T. Lo and H. C. Hsuan, "An equivalence theory between elliptical and circular arrays," *IEEE Trans. Antennas Propag.*, vol. AP-13, pp. 247–256, 1965.

[67] M. T. Ma, *Theory and Application of Antenna Arrays*, New York: John Wiley & Sons, 1974.

[68] M. T. Ma, "A new mathematical approach for linear array analysis and synthesis," PhD dissertation, Syracuse University, 1961.

[69] M. T. Ma and L. C. Walters, "Synthesis of concentric ring antenna arrays yielding approximately equal sidelobes," *Radio Sci.*, vol. 3, pp. 465–470, May 1968.

[70] R. B. Mack, "A study of circular arrays," technical report, Cruft Lab, Harvard University, May 1963.

[71] B. A. Munk et al., "Scan independent phased arrays," *Radio Sci.*, vol. 14, pp. 979–990, 1979.

[72] E. A. Nelson, "Quantization sidelobes of a phased array with a triangular element arrangement," *IEEE Trans. Antennas Propag.*, vol. AP-17, pp. 363–365, 1969.

[73] E. H. Newman et al., "Superdirective receiving arrays," *IEEE Trans. Antennas Propag.*, vol. AP-26, pp. 629–635, 1978.

[74] R. L. Pritchard, "Optimum directivity patterns for linear point arrays," *J. Acoust. Soc. Am.*, vol. 25, pp. 879–891, 1953.

[75] D. R. Rhodes, "The optimum line source of the best mean-square approximation to a given radiation pattern," *IEEE Trans. Antennas Propag.*, vol. AP-11, pp. 440–446, 1963.

[76] D. R. Rhodes, "On an optimum line source for maximum directivity," *IEEE Trans. Antennas Propag.*, vol. AP-17, pp. 485–492, 1969.

[77] D. R. Rhodes, "On a fundamental principle in the theory of planar antennas," *Proc. IEEE*, vol. 52, pp. 1013–1021, 1965.

[78] D. R. Rhodes, *Synthesis of Planar Antenna Sources*, London: Clarendon Press, 1974.

[79] H. J. Riblet, "Discussion on 'A current distribution for broadside arrays which optimizes the relationship between beam width and side-lobe level,'" *Proc. IRE*, vol. 35, pp. 489–492, 1947.

[80] W. F. Richards and Y. T. Lo, "Antenna pattern synthesis based on optimization in a probabilistic sense," *IEEE Trans. Antennas Propag.*, vol. AP-23, pp. 165–172, 1975.

[81] G. M. Royer, "Directive gain and impedance of a ring array of antennas," *IEEE Trans. Antennas Propag.*, vol. AP-14, pp. 566–573, 1966.

[82] L. A. Rondinelli, "Effects of random errors on the performance of antenna arrays of many elements," *IRE Natl. Conv. Rec.*, pt. 1, pp. 174–189, 1959.

[83] J. Ruze, "Antenna tolerance theory—a review," *Proc. IEEE*, vol. 54, pp. 633–640, 1966.

[84] H. E. Salzer, "Calculating Fourier coefficients for Chebyshev patterns," *Proc.*

IEEE, vol. 63, pp. 195–197, 1975.

[85] S. A. Schelkunoff, "A mathematical theory of linear arrays," *Bell Syst. Tech. J.*, vol. 22, pp. 80–107, 1943.

[86] S. A. Schelkunoff and H. T. Friis, *Antenna Theory and Practice*, New York: John Wiley & Sons, 1942, pp. 368 and 401.

[87] A. C. Schell and A. Ishimaru, "Antenna pattern synthesis," chapter 7. in *Antenna Theory*, ed. by R. E. Collin and F. J. Zucker, New York: McGraw-Hill Book Co., 1969.

[88] E. D. Sharp, "A triangular arrangement of planar-array elements that reduces the number needed," *IRE Trans. Antennas Propag.*, vol. AP-9, pp. 126–129, 1961.

[89] J. C. Simon, "Application of periodic functions approximation to antenna pattern synthesis and circuit theory," *IRE Trans. Antennas Propag.*, vol. AP-4, pp. 429–440, 1956.

[90] G. C. Southworth, "Arrays of linear elements," in *Antenna Engineering Handbook*, ed. by H. Jasik, New York: McGraw-Hill Book Co., 1961.

[91] R. J. Spellmire, "Tables of Taylor aperture distributions," *Rep. TM 581*, Hughes Aircraft Co., Culver City, CA, 1958.

[92] L. Stark, "Radiation impedance of a dipole in an infinite planar phased array," *Radio Sci.*, vol. 1, pp. 361–377, 1966.

[93] R. J. Stegen, "Excitation coefficients and beamwidths of Tschebyscheff arrays," *Proc. IRE*, vol. 41, pp. 1671–1674, 1953.

[94] R. J. Stegen, "Gain of Tchebyscheff arrays," *IRE Trans. Antennas Propag.*, vol. AP-8, pp. 629–631, 1960.

[95] W. L. Stutzman, "Synthesis of shaped-beam radiation patterns using the iterative sampling method," *IEEE Trans. Antennas Propag.*, vol. AP-19, pp. 36–41, 1971.

[96] W. L. Stutzman, "Side lobe control of antenna patterns," *IEEE Trans. Antennas Propag.*, vol. AP-20, pp. 102–104, 1972.

[97] W. L. Stutzman and G. A. Thiele, *Antenna Theory and Design*, New York: John Wiley & Sons, 1981.

[98] C. T. Tai, "The optimum directivity of uniformly spaced broadside arrays of dipoles," *IEEE Trans. Antennas Propag.*, vol. AP-12, pp. 447–454, 1964.

[99] T. T. Taylor, "Design of circular apertures of narrow beamwidth and low side lobes," *IRE Trans. Antennas Propag.*, vol. AP-8, pp. 17–22, 1960.

[100] T. T. Taylor, "Design of line-source antennas for narrow beamwidth and low side lobes," *IRE Trans. Antennas Propag.*, vol. AP-3, pp. 16–28, 1955.

[101] T. T. Taylor, "A synthesis method for circular and cylindrical antennas composed of discrete elements," *IRE Trans. Antennas Propag.*, vol. AP-1, pp. 251–261, 1952.

[102] J. D. Tillman, Jr., "The theory and design of circular antenna arrays," Univ. of Tennessee, Eng. Exp. Station, Knoxville, 1966.

[103] G. Toraldo Di Francia, "Directivity, super-gain, and information," *IRE Trans.*, 1956, pp. 473–478; also "Theory of antenna arrays," special issue, *Radio Sci.*, vols. 3–5, 1968.

[104] G. N. Tsandoulas, "Tolerance control in an array antenna," *Microwave J.*, vol. 20, pp. 24–30, 1977.

[105] F. I. Tseng, "Design of array and line-source antennas for Taylor patterns with a null," *IEEE Trans. Antennas Propag.*, vol. AP-27, pp. 474–479, 1979.

[106] F. I. Tseng and D. K. Cheng, "Antenna pattern response to arbitrary time signals," *Can. J. Phys.*, vol. 42, pp. 1358–1368, 1964.

[107] F. I. Tseng and D. K. Cheng, "Optimum scannable planar arrays with an invariant side lobe level," *Proc. IEEE*, vol. 56, pp. 1771–1778, 1968.

[108] A. I. Uzkov, "An approach to the problem of optimum directive antenna design," *C. R. Acad. Sci. URSS*, vol. 53, p. 35, 1946.

[109] M. Uzsoky and L. Solymar, "Theory of super-directive linear arrays," *Acta Phys. Hung.*, vol. 6, pp. 185–205, 1956.

[110] G. J. van der Maas, "A simplified calculation for Dolph-Tchebycheff arrays," *J. Appl. Phys.*, vol. 25, pp. 121–124, 1954.

[111] R. C. Voges and J. K. Butler, "Phase optimization of antenna array gain with constrained amplitude excitation," *IEEE Trans. Antennas Propag.*, vol. AP-20, pp. 432–436, 1972.

[112] W. H. von Aulock, "Properties of phased arrays," *Proc. IRE*, vol. 48, pp. 1715–1727, 1960.

[113] W. Wasylkiwskyj and W. J. Kahn, "Element pattern bounds in uniform phased arrays," *IEEE Trans. Antennas Propag.*, vol. AP-25, pp. 597–604, 1977.

[114] W. Wasylkiwskyj and W. K. Kahn, "Element patterns and active reflection coefficient in uniform phased arrays," *IEEE Trans. Antennas Propag.*, vol. AP-22, pp. 207–212, 1974.

[115] H. A. Wheeler, "A survey of the simulator technique for designing a radiating element," in *Phased Array Antennas*, ed. by A. A. Oliner and G. H. Knittel, pp. 132–148, Dedham: Artech House, 1972.

[116] L. P. Winkler and M. Schwartz, "A fast numerical method for determining the optimum snr of an array subject to a Q factor constraint," *IEEE Trans. Antennas Propag.*, vol. AP-20, pp. 503–505, 1972.

[117] P. M. Woodward, "A method for calculating the field over a plane aperture required to produce a given polar diagram," *Proc. IEE*, vol. 93, pt. IIIA, pp. 1554–1558, 1946.

[118] P. M. Woodward, "A method of calculating the field over a plane aperture required to produce a given polar diagram," *Proc. IEE*, vol. 93, pt. III, pp. 1554–1558, 1947.

[119] P. M. Woodward and J. D. Lawson, "The theoretical precision with which an arbitrary radiation pattern may be obtained from a source of finite size," *Proc. IEE*, vol. 95, pt. III, pp. 363–370, 1948.

[120] N. Yaru, "A note on supergain antenna arrays," *Proc. IRE*, vol. 39, pp. 1081–1085, 1951.

[121] R. S. Elliott, "Beamwidth and directivity of large scanning arrays," first of two parts, *Microwave J.*, December 1963.

[122] C. S. Liang and Y. T. Lo, "A multipole-field study for the multiarm log-spiral antennas," *IEEE Trans. Antennas Propag.*, vol. AP-16, pp. 656–664, 1968.

[123] Ye. V. Baklanov, "Chebyshev distribution of currents for a plane array of radiators," *Radio Eng. Electron. Phys.*, vol. 11, pp. 640–642, April 1966 (English translation from Russian).

[124] E. N. Gilbert and S. P. Morgan, "Optimum design of directive antenna arrays subject to random variations," *Bell Syst. Tech. J.*, vol. 34, pp. 637–661, 1955.

[125] A. H. Tihonov, "Solution of incorrectly formulated problems and the regularization method," *Sov. Math.*, vol. 4, pp. 1035–1038, 1964.

[126] H. S. Cabayan, P. E. Mayes, and G. A. Deschamps, "Techniques for computation and realization of stable solutions for synthesis of antenna patterns," *Antenna Lab Rep. 70-13*, Univ. of Illinois at Urbana, October 1970.

Chapter 12

The Design of Waveguide-Fed Slot Arrays

Robert S. Elliott
University of California at Los Angeles

CONTENTS

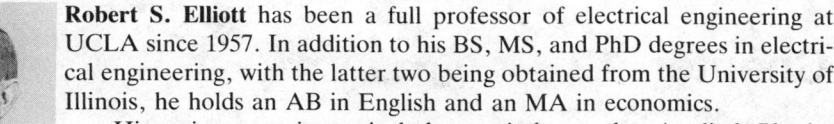

Robert S. Elliott has been a full professor of electrical engineering at UCLA since 1957. In addition to his BS, MS, and PhD degrees in electrical engineering, with the latter two being obtained from the University of Illinois, he holds an AB in English and an MA in economics.

His prior experience includes periods at the Applied Physics Laboratory of Johns Hopkins University and at the Hughes Research Laboratories, where he headed antenna research. Dr. Elliott has also been on the faculty of the University of Illinois and was a founder of Rantec Corporation, serving as its first vice president and technical director. He has recently completed a second two-year stint as a distinguished lecturer for the IEEE and was chairman of the coordinating committee for the 1981 IEEE Symposium, held in Los Angeles. In addition to being a Fellow of the IEEE he is also a member of Sigma Xi, Tau Beta Pi, and the New York Academy of Sciences. Dr. Elliott is the author of 60 journal papers and two textbooks. He is the recipient of two Best Paper Awards from the IEEE.

1. Introduction

Many different types of wave-guiding structures have found practical application in a variety of microwave devices. These guiding structures include the coaxial line, rectangular and circular waveguides, stripline, and microstrip. A characteristic of each of these structures is that it will support a set of modes, usually described as consisting of two subsets: TE (transverse electric) and TM (transverse magnetic). For the multiple-conductor structures (coaxial line, stripline, and microstrip) a TEM mode is also possible—indeed, is the dominant mode.

For simple geometries the space-time description of the electric and magnetic fields which comprise a particular mode is obtained readily from Maxwell's equations. When the guiding structure is composed of good conductor, as is usually the case, the lineal current density $\mathbf{K}$ in the conductor surface can be found from $\mathbf{K} = \mathbf{n} \times \mathbf{H}$, where $\mathbf{n}$ is a unit vector normal to the conductor surface and $\mathbf{H}$ is the magnetic field for the particular mode, evaluated at the surface of the conductor.

If a hole is cut in the conductor surface such as to interrupt $\mathbf{K}$, and if this hole is open to outer space, radiation can occur. A family of such holes constitutes an antenna array. Judicious choice of the shape, orientation, and relative placement of these holes can produce a variety of useful antenna patterns. Efficient extraction of power from the mode and its transformation to radiation must also be considered. These design questions are the subject of the present chapter.

Most applications of the foregoing idea involve the use of rectangular waveguide as the guiding structure, with transverse dimensions chosen so that only the TE_{10} mode can propagate. The wall currents associated with this mode are sketched in Fig. 1a. The most commonly used slots are shown in Figs. 1b through 1d. Slot A, which interrupts negligible current, is essentially nonradiative and is thereby ideal for use with an inserted movable probe, thus effecting a measurement of the fields within the guide (vswr indicator).

Slot B, which is offset, primarily interrupts transverse current; the induced $\mathbf{E}$ field in slot B is increased by increasing the offset; the polarity of the induced $\mathbf{E}$ field is reversed by reversing the direction of offset.

Slot C, which is tilted, primarily interrupts longitudinal current; the induced $\mathbf{E}$ field in slot C is increased by increasing the tilt; the polarity of the induced $\mathbf{E}$ field is reversed by reversing the direction of the tilt.

The inclination of slot D causes interruption of the transverse current in the narrow wall; the induced $\mathbf{E}$ field in slot D is increased by increasing the inclination and is reversed by reversing the direction of inclination.

The three slot types B, C, and D can all radiate, and one can see from the foregoing that the intensity of radiation as well as its polarity is controllable by the amount and direction of offset or tilt. Thus, if λ_g is the guide wavelength for the TE_{10} mode, and if slots are placed $\lambda_g/2$ apart and alternately offset or tilted,

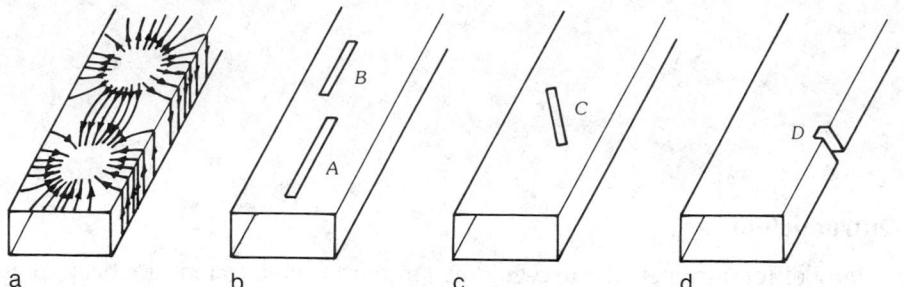

Fig. 1. Rectangular waveguide, TE$_{10}$ mode. (*a*) Wall current distribution. (*b*) Longitudinal broad wall slots. (*c*) Centered, inclined broad wall slot. (*d*) Inclined narrow wall slot.

as shown in Fig. 2, the result is a linear array for which the induced **E** fields in the various slots can have a common phase and an arbitrary amplitude distribution. Linear arrays of this type produce a useful antenna pattern consisting of a main beam and side lobes of governed heights.

If linear arrays of the types shown in Fig. 2a or 2b or 2c are placed side by side, the result is a planar slot array. An example is shown in Fig. 3. By controlling the relative excitation of the individual linear arrays, one is able to achieve a governable two-dimensional aperture distribution and thereby produce a desired radiation pattern.

When the slots are spaced $\lambda_g/2$ apart in a common waveguide the array is said to be *standing-wave fed*. But they need not be $\lambda_g/2$ apart, and many applications

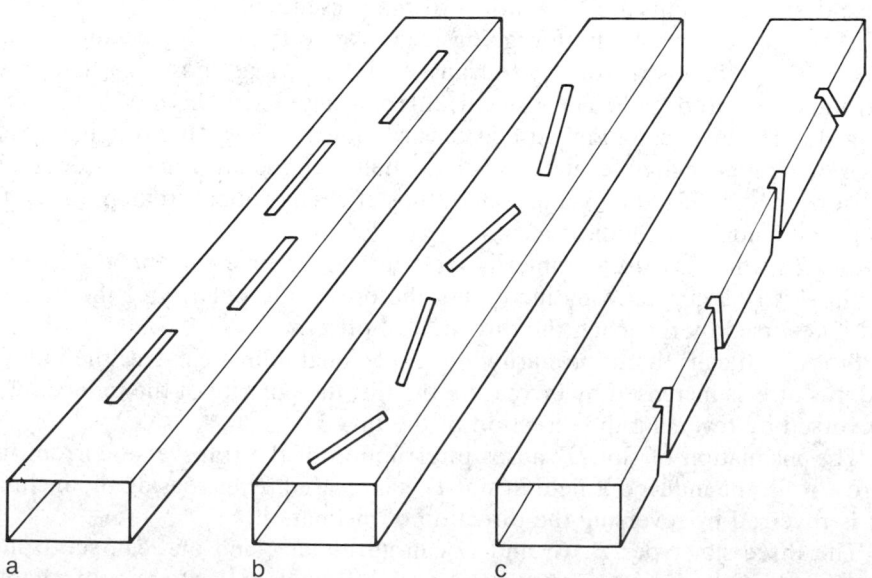

Fig. 2. Linear slot arrays. (*a*) Longitudinal broad wall slots. (*b*) Centered, inclined broad wall slots. (*c*) Inclined narrow wall slots.

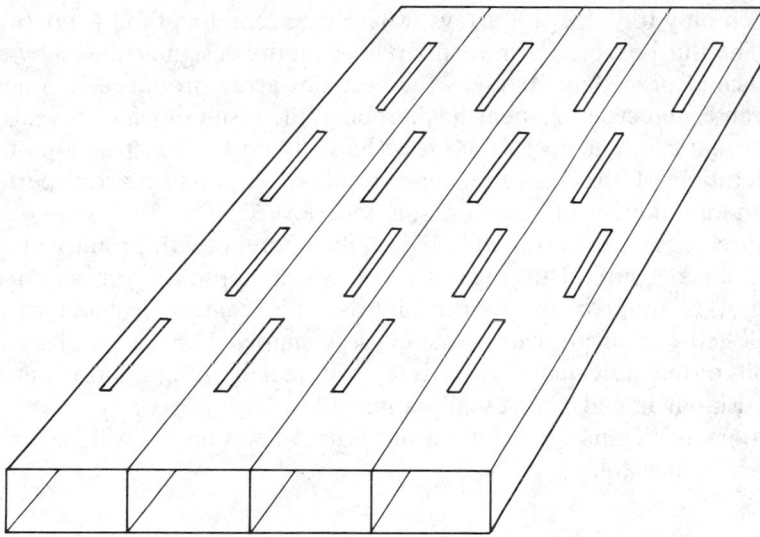

Fig. 3. Planar array of longitudinal slots.

require that they have a spacing other than this. Such arrays are said to be *traveling-wave fed*. Both types will be discussed in this chapter.

Pattern requirements, such as side lobe level in a sum or difference pattern, or ripple in a shaped-beam pattern, are usually sufficiently stringent that external mutual coupling between slots in an array cannot be ignored. There are basically three design procedures currently in vogue which account for mutual coupling:

(1) The first procedure employs an infinite array assumption and posits that the mutual coupling experienced by every slot in the array is the same, being what that slot would experience if it were imbedded in an infinite array, with the array excited in a uniform-amplitude/uniform-progressive-phase distribution. Mutual coupling effects are usually accounted for by an experimental measurement of active impedance or active admittance, using large sample arrays. This approach is acceptable when the array is large, when the actual amplitude distribution is gently tapered, and when the side lobe level requirement is modest (say −20 dB to 30 dB). It is not applicable when the side lobe level is low and/or asymmetric, nor is it applicable when the needed excitation exhibits considerable variability, such as for shaped-beam patterns.

(2) The second procedure relies on a theoretical formulation and calculation of the external mutual coupling between any two slots in the array, with these effects summed for each slot. This approach is, at the present time, fully developed only for cases in which the slots lie in a ground plane sufficiently large to be modeled by an infinite ground plane, or when they lie in a conducting surface of large curvature. Thus the method focuses on broad wall slots. It uses a computer program which is moderately difficult to construct and render error-free. But once that is done, the design costs are modest when compared with the "infinite array" procedure, even for arrays containing several thousand slots, and the inherent accuracy is much higher. Because of this the "infinite array" approach is usually

undertaken only for edge slot arrays, where the second method is not applicable.

(3) The third procedure in reality refines on the design achieved by either the first or second procedure. It takes the best slot array produced by one of those methods and subjects it to a near-field probing, the result of which reveals errors in excitation that can be corrected, at least partially, in the construction of a second array. Iteration of this procedure can result in impressive array performance, notably the attainment of very low side lobe levels.

All three design methods will be described herein, with primary emphasis on the second procedure. But regardless of which approach, or combination of approaches, is adopted in a particular case, the central problem in designing waveguide-fed slot arrays can be stated very simply. How does one choose the dimensions of the individual slots so as to produce a desired radiation pattern and a prescribed input impedance (usually a match)?

Solutions to this problem for various types of slot arrays will be presented in the sections which follow.

2. The E-Field Distribution in a Longitudinal Slot

Longitudinal broad wall slot arrays of the types shown in Figs. 2a and 3 are attractive because of their flush-mounting capability. Additionally, the sturdy box-beam construction evidenced by Fig. 3 is a desirable feature. For these reasons they are widely used in fuze antenna, seeker antenna, and various other radar applications. The design of such arrays begins with an understanding of the E-field distribution in the aperture of a single longitudinal slot when it is excited by a TE_{10} mode.

The slot shown in Fig. 4 is assumed to be followed by a matched load and to be fed by a matched generator which causes a TE_{10} mode, traveling in the $+z$ direction, to be incident on the slot. The slot has a length 2ℓ, a width w, and is offset a distance x_0 from the side wall. It is cut in the upper broad wall of the rectangular waveguide; this wall is assumed to be imbedded in a large ground plane of good conductivity. Under these circumstances, what E-field distribution is induced in the slot and how does it depend on the slot's offset and length?

When the waveguide walls are very thin (a modern trend) this question can be answered by modeling the actual situation with a "zero" wall thickness replica and matching internal and external expressions for tangential **H** at all points P in the slot aperture. In mathematical terms, one imposes the boundary condition

$$\mathbf{1}_y \times [\mathbf{H}_{int}^{inc}(P) + \mathbf{H}_{int}^{scat}(P)] = \mathbf{1}_y \times \mathbf{H}^{ext}(P) \qquad (1)$$

where $\mathbf{1}_y$ is a unit vector in the y direction, $\mathbf{H}_{int}^{inc}$ is the field of the incident TE_{10} mode, $\mathbf{H}_{int}^{scat}$ is the field scattered by the slot into the waveguide, and $\mathbf{H}^{ext}$ is the field scattered by the slot into the outer half-space.

Equation 1 can be written in the component form

$$H_x^{ext}(P) - H_x^{scat}(P) = H_x^{inc}(P) \qquad (2)$$

$$H_z^{ext}(P) - H_z^{scat}(P) = H_z^{inc}(P)$$

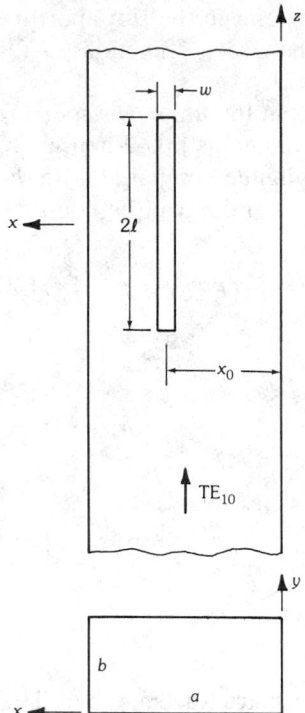

Fig. 4. A single longitudinal slot excited by a TE_{10} mode.

The incident field serves as the driving function in (2) and (3) and is given by

$$H_x^{\text{inc}} = -\frac{\beta_{10}}{\pi/a} A_{10} \sin\left(\frac{\pi x}{a}\right) e^{-j\beta_{10}z} \tag{4}$$

$$H_z^{\text{inc}} = jA_{10}\cos\left(\frac{\pi x}{a}\right) e^{-j\beta_{10}z} \tag{5}$$

with $\beta_{10} = [k^2 - (\pi/a)^2]^{1/2}$ the propagation constant of the TE_{10} mode, where $k = \omega(\mu_0\epsilon)^{1/2}$ is the wave number and ϵ is the permittivity of the dielectric filling the waveguide.

The internal scattered field which appears in (1)–(3) can be expressed in terms of Stevenson's Green's functions [1]. Similarly, the external field can be expressed in terms of the Green's function for a half-space [2]. These introductions permit (2) and (3) to be rewritten in the forms

$$\int_{\text{slot}} [G_{xx}(P, P')E_x(P') + G_{xz}(P, P')E_z(P')]dS' = -\frac{\beta_{10}}{\pi/a} A_{10}\sin\left(\frac{\pi x}{a}\right) e^{-j\beta_{10}z} \tag{6}$$

$$\int_{\text{slot}} [G_{zx}(P, P')E_x(P') + G_{zz}(P, P')E_z(P')]dS' = jA_{10}\cos\left(\frac{\pi x}{a}\right) e^{-j\beta_{10}z} \tag{7}$$

in which $P(x, z)$ is any field point in the slot aperture and $P'(x', z')$ is any source point in the slot aperture. The Green's functions $G_{ij}(P, P')$ which appear in (6) and (7) are known functions [1, 2].

These integral equations in the unknown aperture field $E_x(P')$, $E_z(P')$ can be solved using the method of moments [3]. When $w \ll 2\ell$, which is the usual case, one finds that $E_z(P')$ is negligible compared with $E_x(P')$.* The problem is then considerably simplified and attention can be confined to (7) in the reduced form

$$\int_{\text{slot}} G_{zx}(P, P')E_x(P')dx'dz' = jA_{10}\cos\left(\frac{\pi x}{a}\right)e^{-jB_{10}z} \tag{8}$$

where

$$G_{zx}(P, P') = \frac{1}{2\pi j\omega\mu_0}\left(\frac{\partial^2}{\partial z'^2} + k_o^2\right)\frac{e^{-jk_oR}}{R}$$

$$+ \frac{2}{j\omega\mu_0 ab}\sum_{m=0}^{\infty}\sum_{n=0}^{\infty}\frac{\varepsilon_{mn}^2}{\gamma_{mn}}\cos\left(\frac{m\pi x}{a}\right)\cos\left(\frac{m\pi x'}{a}\right)$$

$$\times \left(\frac{\partial^2}{\partial z'^2} + k^2\right)e^{-\gamma_{mn}|z-z'|} \tag{9}$$

with $R = \overline{PP'}$ and k_o the free-space wave number. The propagation constant γ_{mn} of the mnth mode is given by $\gamma_{mn} = [(m\pi/a)^2 + (n\pi/b)^2 - k^2]^{1/2}$ and the Neumann numbers ε_{mn}^2 are such that $\varepsilon_{00}^2 = 1/4$, $\varepsilon_{0n}^2 = \varepsilon_{m0}^2 = 1/2$, $\varepsilon_{mn}^2 = 1$ otherwise.

Various method of moment types of solution of (8) give results typified by the curves shown in Fig. 5, where E_x along the center line of a longitudinal slot is plotted as a function of z'/ℓ. It can be observed that the magnitude of E_x is an almost-symmetric function, peaking near the center of the slot, and that the phase of E_x is nearly constant over the length of the slot, these observations applying for slot lengths around the first resonance ($2\ell_{\text{res}} \cong \lambda_o/2$ for air-filled guide, less for dielectric-filled guide). The peak amplitude is greatest at resonance and the phase distribution is most nearly constant at resonance.

The three amplitude distributions shown in Fig. 5 (for $\ell = 0.95\ell_{\text{res}}$, $1.00\ell_{\text{res}}$, $1.05\ell_{\text{res}}$) are redrawn in Fig. 6 to a common normalized scale. It can be seen that the curves differ slightly over the half of the slot closer to the generator but are indistinguishable over the other half. Also shown in Fig. 6 is a half-cosinusoid distribution. The normalized results from a method of moments calculation can be characterized as slightly bulged out when compared to a half-cosinusoid, with the bulge slightly asymmetric. Investigation has disclosed that this bulge is inconsequential[†] in determining the dimensions of a typical slot array [4]. Thus the important conclusion is reached that the electric-field distribution along the center line of a narrow longitudinal slot, when the slot is excited by a TE_{10} mode, can be represented by the function

---

*This is not true for tilted slots.
†It is *not* inconsequential in determining the equivalent circuit of an isolated slot.

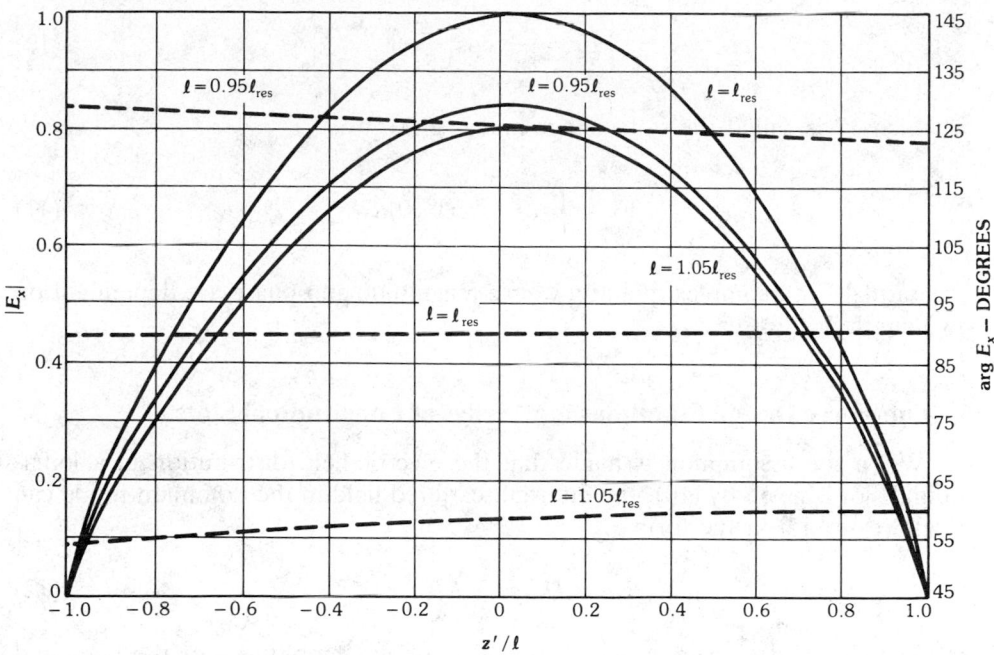

Fig. 5. Amplitude and phase distribution of **E** field in longitudinal slot versus its length. (*After Elliott [5]*)

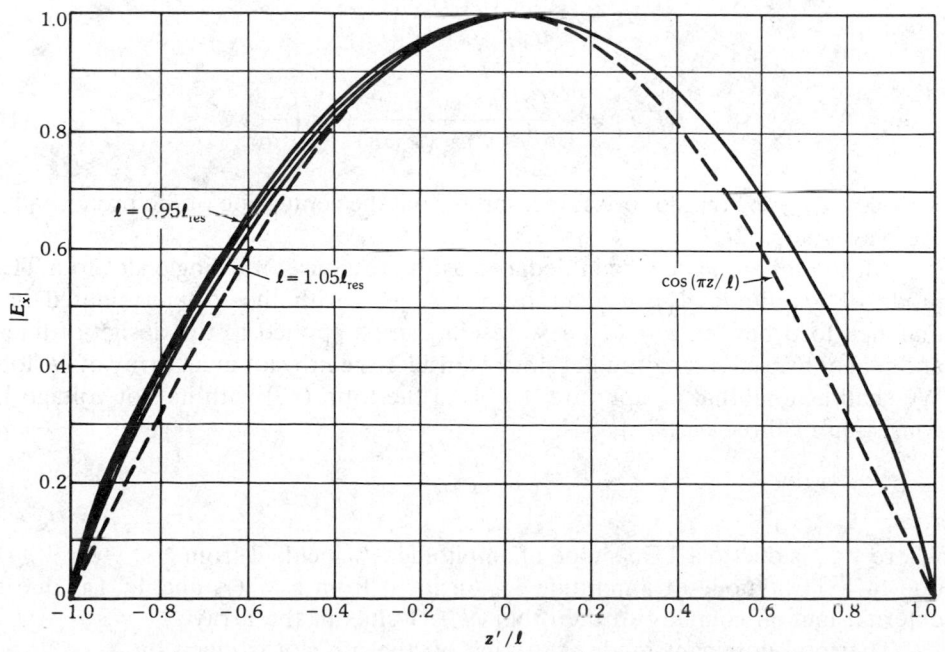

Fig. 6. Normalized **E**-field amplitude distribution in longitudinal slot compared with half-cosinusoid.

$$E_x = \frac{V^s}{w} \cos\left(\frac{\pi z'}{2\ell}\right) \tag{10}$$

in which V^s is called the slot voltage and is given by

$$V^s = \int_{x_o - w/2}^{x_o + w/2} E_x(x', 0)dx' \tag{11}$$

The term V^s is a complex quantity whose magnitude and phase are dependent on slot length and offset.

3. The Three Design Equations for Arrays of Longitudinal Slots

When the assumption is made that the electric-field distribution in a longitudinal slot is given by (10), the internal scattered field in the dominant mode can be expressed [5] in the form

$$B_{10} = C_{10} = -Kf(x, \ell)V^s \tag{12}$$

in which B_{10} and C_{10} are, respectively, the complex amplitudes of the back and forward scattered TE_{10} modes. The constant K and the function $f(x, \ell)$ are given by

$$K = \frac{2(\pi/a)^2}{j\omega\mu_0(\beta_{10}/k)(ka)(kb)} \tag{13}$$

$$f(x, \ell) = \frac{(\pi/2k\ell)\cos\beta_{10}\ell}{(\pi/2k\ell)^2 - (\beta_{10}/k)^2} \sin\left(\frac{\pi x}{a}\right) \tag{14}$$

with $x = x_o - a/2$ the slot offset measured from the center line of the broad wall of the waveguide.

All the foregoing has been deduced as the response of a single slot to a TE_{10} mode of amplitude A_{10} incident from $z < -\ell$, with the slot terminated in a matched load beyond $z = \ell$. These results can be applied to the design of linear and planar arrays of longitudinal slots. Consider the nth slot in an array of N slots. We shall assume that its aperture field is in the form (10) with its slot voltage V_n^s consisting of three parts, that is,

$$V_n^s = V_{n,1}^s + V_{n,2}^s + V_{n,3}^s \tag{15}$$

where $V_{n,1}^s$ is due to a TE_{10} mode of amplitude A_{10}^n incident from $z < -\ell n$, $V_{n,2}^s$ is due to a TE_{10} mode of amplitude D_{10}^n incident from $z > \ell_n$, and $V_{n,3}^s$ is due to external mutual coupling to the other $N - 1$ slots in the array.

The total dominant mode scattering off the nth slot is given by

$$B_{10}^n = C_{10}^n = -Kf(x_n, \ell_n)V_n^s \tag{16}$$

This symmetrical scattering can be modeled by a shunt admittance on an equivalent lossless transmission line [6] for which

$$B_n = C_n = -\frac{1}{2}\frac{Y_n^a}{G_o}V_n \tag{17}$$

with Y_n^a the *active* admittance of the nth slot (including external mutual coupling). The term V_n is the mode voltage on the equivalent transmission line at the site of Y_n^a; G_o is the characteristic conductance of the equivalent transmission line.

Since B_{10}^n and B_n should be proportional, the combination of (16) and (17) yields

$$\frac{Y_n^a}{G_o} = K_1 f(x_n, \ell_n)\frac{V_n^s}{V_n} \tag{18}$$

where K_1 is a proportionality constant. Equation 18 is the first of three design equations needed to determine the lengths and offsets of all slots in the array.

For an isolated slot

$$\frac{Y}{G_o}(x, \ell) = -\frac{2B_{10}/A_{10}}{1 + (B_{10}/A_{10})} \tag{19}$$

which means that

$$B_{10}^{n,1} = -\frac{Y(x_n, \ell_n)/G_o}{2 + Y(x_n, \ell_n)/G_o}A_{10}^n \tag{20}$$

with $B_{10}^{n,1}$ that part of B_{10}^n due to $V_{n,1}^s$. Since $B_{10}^{n,1} = -Kf(x_n, \ell_n)V_{n,1}^s$, it follows that

$$V_{n,1}^s = \frac{1}{Kf(x_n, \ell_n)}\frac{Y(x_n, \ell_n)/G_o}{2 + Y(x_n, \ell_n)/G_o}A_{10}^n \tag{21}$$

Similarly,

$$V_{n,2}^s = \frac{1}{Kf(x_n, \ell_n)}\frac{Y(x_n, \ell_n)/G_o}{2 + Y(x_n, \ell_n)/G_o}D_{10}^n \tag{22}$$

A development [5] which entails use of the reciprocity theorem reveals that the third component of slot voltage is given by

$$V_{n,3}^s = -j(\beta_{10}/k)(k_o b)(a/\lambda)^3\frac{1}{f^2(x_n, \ell_n)}\frac{Y(x_n, \ell_n)/G_o}{2 + Y(x_n, \ell_n)/G_o}$$

$$\times \sum_{m=1}^{N}{}' V_m^s g_{mn}(x_m, \ell_m, x_n, \ell_n) \tag{23}$$

where

$$g_{mn} = \int_{-k_o\ell_m}^{k_o\ell_m} \cos\left(\frac{u'_m}{4\ell_m/\lambda_o}\right)\left\{\frac{1}{(4\ell_n/\lambda_o)}\left[\frac{e^{-jk_oR_1}}{k_o R_1} + \frac{e^{-jk_oR_2}}{k_oR_2}\right]\right.$$
$$\left. + \left[1 - \frac{1}{(4\ell_n/\lambda_o)^2}\right]\int_{-k_o\ell_n}^{k_o\ell_n} \cos\left(\frac{u'_n}{4\ell_n/\lambda_o}\right)\frac{e^{-jk_oR}}{k_oR}\,du'_n\right\}du'_m \tag{24}$$

In (23) the prime on the summation sign means that the term $m = n$ is to be excluded. In (24) the surrogate variable $u' = k_o z'$ has been introduced. The variable distances R, R_1, and R_2 are defined in Fig. 7. Since generally

$$\frac{Y_n^a}{G_o} = -\frac{2B_{10}^n}{A_{10}^n + D_{10}^n + B_{10}^n} \tag{25}$$

where $B_{10}^n = B_{10}^{n,1} + B_{10}^{n,2} + B_{10}^{n,3}$, one can relate all quantities which appear on the right side of (25) to $V_{n,1}^s$, $V_{n,2}^s$, and $V_{n,3}^s$. When this is done, the second design equation results, viz.,

$$\frac{Y_n^a}{G_o} = \frac{2f^2(x_n, \ell_n)}{\dfrac{2f^2(x_n, \ell_n)}{Y(x_n, \ell_n)/G_o} + j\dfrac{\beta_{10}}{k}(k_ob)\left(\dfrac{a}{\lambda}\right)^3 \displaystyle\sum_{m=1}^{N}{}' \dfrac{V_m^s}{V_n^s}g_{mn}} \tag{26}$$

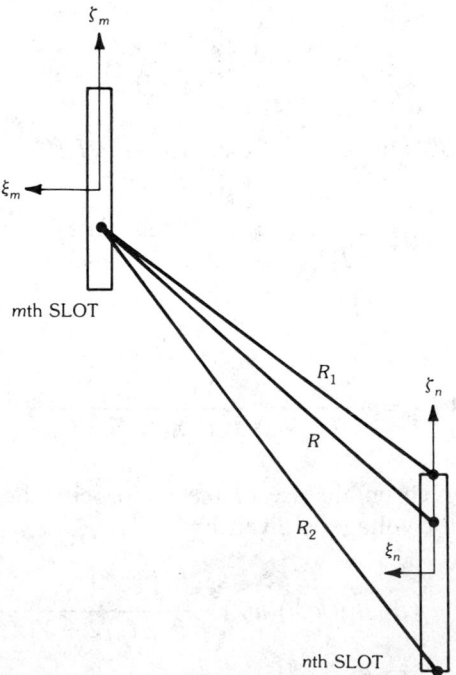

Fig. 7. Geometry for mutual coupling calculation.

If there are P slots in the waveguide containing the pth slot, the third design equation is simply

$$\sum_{p=1}^{P} \frac{Y_p^a}{G_o} = \text{specified constant} \qquad (27)$$

For a linear array (single waveguide) the specified constant will usually be unity, indicating that an input match is desired. But in a planar array the individual branch line waveguides need not be matched since the sum of the active admittances is transformed down to an equivalent load on the main line waveguide; it is the sum of these equivalent loads that is often specified to be a match.

4. The Design of a Linear Array of Resonantly Spaced Longitudinal Slots (Standing-Wave Feed)

When all of the slots are in a common waveguide and resonantly spaced ($\lambda_g/2$) on centers, the mode voltage V_n is common to all slots except for an alternation in sign. This alternation is compensated by an alternation in the direction of slot offset, which causes $\sin(\pi x_n/a)$ to change sign at successive slots. In this case the first design equation (18) can be written in ratio form

$$\frac{Y_n^a}{G_o} \bigg/ \frac{Y_m^a}{G_o} = \frac{|f(x_n, \ell_n)| V_n^s}{|f(x_m, \ell_m)| V_m^s} \qquad (28)$$

with m and n any two slots in the linear array and $f(x, \ell)$ given by (14).

To use the second design equation one needs first to express in functional form the input data that has been accumulated on the isolated self-admittance Y/G_o versus slot length and offset. That data can be obtained in either of two ways:

(1) Experimentally, by constructing a family of waveguides, each with a single slot but with different offsets and shorter-than-resonant lengths. After Y/G_o has been measured for each member of the family, the slots can be lengthened and the measurements repeated, etc. For most applications it is sufficient to obtain data in the range $0.95 \ell_{res} \leq \ell \leq 1.05 \ell_{res}$ and seven well-spaced data points per family member are usually sufficient.

(2) Computationally, by solving (8) via the method of moments to find $E_x(P')$ as a function of slot offset and length. The knowledge thus gained can be used to calculate B_{10}/A_{10} and then, using (19), to deduce $Y(x, \ell)/G_o$. If need be, a correction for wall thickness can be applied to these calculations [7].

Whichever of these two approaches is used, one should find that the input data can be normalized in the following way. Let

$$g(x) = \frac{G_{res}}{G_o}(x, \ell_{res}) \qquad (29)$$

$$v(x) = k \ell_{res}(x) \qquad (30)$$

$$h(y) = h_1(y) + j h_2(y) \tag{31}$$

$$h_1(y) = \frac{G(x,y)/G_o}{g(x)} \tag{32}$$

$$h_2(y) = \frac{B(x,y)/G_o}{g(x)} \tag{33}$$

$$y = \ell/\ell_{\mathrm{res}} \tag{34}$$

The function $g(x)$ gives the conductance of a resonant-length slot versus offset and can be fitted with a few terms of a polynomial (polyfitted) or often better fitted by a transcendental function. The function $v(x)$ is seen to be the resonant length versus offset multiplied by the wave number. The function $h(y)$ is a complex function such that

$$Y(x,y)/G_o = g(x)\, h(y) \tag{35}$$

The two components of $h(y)$ are real functions, with $h_1(y)$ representing $G(x,y)/G_o$ in ratio to $g(x)$ and $h_2(y)$ similarly representing $B(x,y)/G_o$ in ratio to $g(x)$. The virtue of assembling the data in this form, with Y/G_o normalized to $g(x)$, is that when the data are plotted versus $y = \ell/\ell_{\mathrm{res}}$, the result is found to be essentially independent of offset. Since $h_1(y)$ and $h_2(y)$ are easily polyfitted, the representation of Y/G_o given in the form (35) is helpful in the ensuing calculations.

With Y/G_o now a known input function, the next step is to find *starting* values for the lengths and offsets of all the slots. Since the design procedure converges so rapidly, it would be sufficient to start with all the slots on the center line and resonant-length. However, it is prudent, and a good benchmark, to begin instead by finding the slot dimensions which would prevail were there no mutual coupling. To do this, one need work only with (28), which takes on the special form*

$$\frac{g(x_n)}{g(x_m)} = \frac{|f(x_n, \ell_n^{\mathrm{res}})|V_n^s}{|f(x_m, \ell_m^{\mathrm{res}})|V_m^s} \tag{36}$$

The desired slot voltage ratio V_n^s/V_m^s is known from pattern specifications. If we let $m = 1$, and choose x_1 arbitrarily, the only unknown in (36) is x_n, which can be obtained by a search routine. In this fashion a family of slot offsets $(x_1, x_2, \ldots, x_n)$ is determined. The corresponding resonant conductances can be found from $g(x)$ and the input conductance from

$$\frac{G_{\mathrm{in}}}{G_o} = \sum_{n=1}^{N} g(x_n) \tag{37}$$

---

*It is being assumed here, because this is a resonantly spaced array, that the aperture distribution is equiphase.

If this sum, which is a use of the third design equation, does not satisfy the input admittance specification (usually a match), x_1 can be adjusted upward or downward as appropriate, and the process repeated. A relatively few iterations will usually produce a set of offsets $(\mathring{x}_1, \mathring{x}_2, \ldots, \mathring{x}_N)$ which do cause agreement with the input admittance specification. Then the corresponding resonant lengths $(\mathring{\ell}_1^{\text{res}}, \mathring{\ell}_2^{\text{res}}, \ldots, \mathring{\ell}_N^{\text{res}})$ can be found by using the known function $v(x)$.

The next step is to compute g_{mn} from (24) for $m = 1, 2, \ldots, N$ and $n = 1, 2, \ldots, N$ but $m \neq n$, with the starting values $(\mathring{x}_n, \mathring{\ell}_n^{\text{res}})$ and $(\mathring{x}_m, \mathring{\ell}_m^{\text{res}})$ used in the determination of R, R_1, and R_2. The values of g_{mn} should be stored and retrieved to compute the mutual coupling terms

$$MC_n = j(\beta_{10}/k)(k_o b)(a/\lambda)^3 \sum_{m=1}^{N}{}' \frac{V_m^s}{V_n^s} g_{mn} \tag{38}$$

Equation 37 can be recognized as the second term in the denominator of the second design equation (26). Absent this term, (26) reduces to the simple identity that the active admittance of the nth slot equals its self-admittance. Thus MC_n, accounts for external mutual coupling and is seen to contain the sum of the g_{mn} values weighted by the aperture distribution. In calculating MC_n, one should use the desired ratio V_m^s/V_n^s.

After these starting values of MC_n are known for each slot in the array, a search needs to be undertaken to find a couplet (x_n, ℓ_n) which will cause

$$\mathscr{I}m\left\{ \frac{2f^2(x_n, \ell_n)}{Y(x_n, \ell_n)/G_o} \right\} = -\mathscr{I}m\{MC_n\} \tag{39}$$

Satisfaction of (39) will make Y_n^a/G_o in (26) be pure real, resulting in what one might call active resonance.[*]

It will be found that a continuum of couplets (x_n, ℓ_n) satisfies (39). Similarly, a continuum of couplets (x_m, ℓ_m) will satisfy (39) when the mth slot is considered in place of the nth slot. But for a given choice (x_n', ℓ_n'), there is only one couplet (x_m', ℓ_m') which also satisfies the first design equation in the form (28). Using this criterion, one can assemble a compatible family of slot offsets and lengths (x_1', ℓ_1'), $(x_2', \ell_2'), \ldots, (x_N', \ell_N')$.

But was the right family selected? This question can be answered by seeing if the sum of the active admittances satisfies the specification implicit in the third design equation (27). If not, a new family needs to be chosen by starting with a different couplet (x_n', ℓ_n').

After the right family of couplets has been found, the process starting with (37) must be iterated, because now it is possible to improve on the calculation of g_{mn}, using $(x_n', \ell_n', x_m', \ell_m')$ in place of $(\mathring{x}_n, \mathring{\ell}_n, \mathring{x}_m, \mathring{\ell}_m)$. The iterations should be continued until the latest set of slot dimensions differs from the penultimate set by

---

[*]It is somewhat arbitrary to require that Y_n^a/G_o be pure real for all n. Doing so ensures that the aperture distribution V_n^s/V_m^s will be equiphase. But that result could be accomplished by requiring that all the active admittances have any common phase. However, demanding that their common phase be zero degrees helps to optimize the bandwidth of the array.

less than the achievable construction tolerances.* At this point the design of the linear array of longitudinal slots is completed.

A simple example which illustrates how the input data on the admittance of a single slot can be functionally fitted to give $g(x)$, $v(x)$, and $h_1(y) + jh_2(y)$, and how the three design equations can be used to determine the slot dimensions for a two-by-four array, can be found in [4].

5. The Design of a Linear Array of Nonresonantly Spaced Longitudinal Slots (Traveling-Wave Feed)

When all of the slots are in a common waveguide and equispaced a distance d, but with $d \neq \lambda_g/2$, and when the last slot is terminated by a matched load, the array is said to be traveling-wave fed. Here, in contrast to the situation considered in the previous section, it is not correct to say that the mode voltage V_n is common to all slots except for an alternation in sign. This introduces an additional complexity into the design.

The equivalent circuit of a linear traveling-wave array is shown in Fig. 8. With reference to that figure, let Y_n be the total admittance seen looking into the nth junction toward the matched load. Then

$$\frac{Y_n}{G_o} = \frac{Y_n^a}{G_o} + \frac{(Y_{n-1}/G_o)\cos\beta_{10}d + j\sin\beta_{10}d}{\cos\beta_{10}d + j(Y_{n-1}/G_o)\sin\beta_{10}d} \tag{40}$$

The mode voltages at successive junctions are connected by the relation

$$V_n = V_{n-1}[\cos\beta_{10}d + j(Y_{n-1}/G_o)\sin\beta_{10}d] \tag{41}$$

When the first design equation (18) is written in ratio form for two successive slots and (41) is employed, one finds that

$$\frac{Y_n^a/G_o}{f(x_n,\ell_n)} = \frac{Y_{n-1}^a/G_o}{f(x_{n-1},\ell_{n-1})} \cdot \frac{V_n^s/V_{n-1}^s}{\cos\beta_{10}d + j(Y_{n-1}/G_o)\sin\beta_{10}d} \tag{42}$$

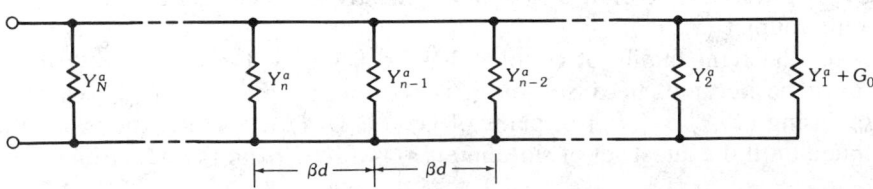

Fig. 8. Equivalent circuit of a linear array of longitudinal slots, traveling-wave fed.

*Usually only a few iterations are needed to achieve this.

Equation 42 is a recurrence relation which, in conjunction with the second design equation (26), permits determination of the length and offset of the nth slot, once the length and offset of the $(n - 1)$st slot are known.*

As in the case of the standing-wave array discussed in the previous section, the first step in the design is to find starting values for the slot lengths and offsets. Once again it is advisable to do this by ignoring (temporarily) all external mutual coupling. Then, if self-resonant slots are assumed initially, (42) becomes

$$\frac{g(\mathring{x}_n)}{f(\mathring{x}_n, \mathring{\ell}_n^{res})} = \frac{g(\mathring{x}_{n-1})}{f(\mathring{x}_{n-1}, \mathring{\ell}_{n-1}^{res})} \cdot \frac{V_n^s/V_{n-1}^s}{\cos\beta_{10}d + j(Y_{n-1}/G_o)\sin\beta_{10}d}$$

It is necessary to digress at this point in order to discuss an important consideration which impacts on the design process. Most practical applications involving traveling-wave linear slot arrays are such that the number of slots N is large (say 30 or more) and the fraction of the power absorbed in the terminating load is modest (say 5 to 10 percent). With 90 to 95 percent of the power being radiated by N slots, and with a typical aperture distribution designed to give a sum pattern with low side lobes, the slot closest to the matched load will be called on to radiate less than 1 percent of the power. Thus $g(x_1)$ will typically be an order of magnitude less than unity and Y_1/G_o will be close to unity. As one considers the total admittance at junctions farther and farther removed from the matched load, since $g(x_n)$ is small compared to unity for all n, it follows that $Y_{n-1}/G_o \cong 1 + j0$, for all n. Thus (42) is given to good approximation by

$$\frac{g(\mathring{x}_n)}{f(\mathring{x}_n, \mathring{\ell}_n^{res})} = \frac{g(\mathring{x}_{n-1})}{f(\mathring{x}_{n-1}, \mathring{\ell}_{n-1}^{res})} \cdot \frac{V_n^s/V_{n-1}^s}{e^{j\beta_{10}d}} \tag{43}$$

This equation puts a limitation on the *phase progression* one can specify for V_n^s/V_{n-1}^s. Three cases need to be examined:

Case 1. All Slots on the Same Side of the Center Line

When this is so, $f(\mathring{x}_n, \mathring{\ell}_n^{res})$ and $f(\mathring{x}_{n-1}, \mathring{\ell}_{n-1}^{res})$ have the same sign and a solution of (43) is only possible if V_n^s/V_{n-1}^s has a phase progression of $-\beta_{10}d$ radians. Since the array factor for a sum pattern is given by

$$S(\theta) = \sum_{n=1}^{N} (V_n^s/V_1^s)e^{-jnk_od\cos\theta}$$

with $\theta = 0°$ directed along the array toward the matched load, we get for this case

$$S(\theta) = \sum_{n=1}^{N} |V_n^s/V_1^s|e^{-jnd(k_0\cos\theta - \beta_{10})}$$

---

*It is assumed here that an input match is desired; thus by implication the third design equation is also being taken into account.

and the main beam points at an angle θ_o given by

$$\theta_o = \cos^{-1}(\beta_{10}/k_o), \tag{44}$$

a result which is seen to be independent of the spacing d. For economy reasons, d should be chosen as large as possible without permitting an extra main beam to occur [8].

Case 2. Slots Alternately Displaced, $\beta_{10}d < \pi$

When this is so, (43) becomes

$$\frac{g(\mathring{x}_n)}{|f(\mathring{x}_n, \mathring{\ell}_n^{res})|} = \frac{g(\mathring{x}_{n-1})}{|f(\mathring{x}_{n-1}, \mathring{\ell}_{n-1}^{res})|} \cdot \frac{V_n^s/V_{n-1}^s}{e^{j(\beta_{10}d - \pi)}}$$

The phase progression of V_n^s/V_{n-1}^s must be $\beta_{10}d - \pi$ radians and the main beam points at an angle θ_o, which can be determined from

$$\theta_o = \cos^{-1}\left(\frac{\beta_{10}d - \pi}{k_o d}\right) \tag{45}$$

With $\beta_{10}d < \pi$, angle θ_o lies somewhere between broadside and reverse end-fire. In this case, θ_o is not independent of d. However, there is a practical lower limit on d because the slots should not overlap, otherwise internal higher-order-mode coupling becomes nonnegligible. Similarly, there is an upper limit on d because, as $\beta_{10}d - \pi \rightarrow 0$, scattering off the different slots tends to add up in phase at the input, causing an unsatisfactory mismatch.

Case 3. Slots Alternately Displaced, $\beta_{10}d > \pi$

Here the analysis is the same as in case 2, and equation (46) still applies but now, with $\beta_{10}d > \pi$, θ_o lies somewhere between broadside and forward end-fire. Once again, θ_o is not independent of d. However, d has a lower bound because $\beta_{10}d - \pi$ should not be so close to zero as to cause the scattering from the various slots to add up at the input. The term d also has an upper bound since multiple main beams usually should be avoided.

The discussion of these three cases has been based on (42), which ignores mutual coupling. However, it applies equally well to (41). As will be seen shortly, when mutual coupling is taken into account the individual slots are detuned to make the appropriate adjustments so that the desired aperture distribution V_n^s/V_{n-1}^s is achieved. But if what is desired does not fit one of the three natural cases just described, the range of adjustment possible in the slot lengths and offsets is not adequate to accomplish the task.

This digression can be summarized by saying that the desired aperture distribution must be restricted to be in the form

$$\frac{V_n^s}{V_{n-1}^s} = \left| \frac{V_n^s}{V_{n-1}^s} \right| e^{j(\beta_{10}d - p\pi)} \tag{46}$$

where, if $p = 0$, the main beam position θ_o points somewhere between broadside and forward end-fire and is given by (44). If $p = 1$, θ_o points somewhere between broadside and reverse (forward) end-fire if $\beta_{10}d$ is less (greater) than π. No other values of p are allowed. If $p = 0$, θ_o is independent of d; if $p = 1$, θ_o is dependent on d. More will need to be said about the choice of the value of d shortly.

With these restrictions on the allowable desired aperture distributions, (43) reduces to

$$\frac{g(\mathring{x}_n)}{f(\mathring{x}_n, \mathring{\ell}_n^{res})} = \frac{g(\mathring{x}_{n-1})}{f(\mathring{x}_{n-1}, \mathring{\ell}_{n-1}^{res})} \cdot \left| \frac{V_n^s}{V_{n-1}^s} \right| \tag{47}$$

with $(\mathring{x}_n, \mathring{\ell}_n^{res})$ the sought-for starting couplet for the nth slot. These starting values can be found by the following procedure:

First, *assume* a value for $\mathring{x}_1$, that is, for the offset of the slot closest to the matched load. Then find $v(\mathring{x}_1)$ from (30). This yields $\mathring{\ell}_1^{res}$. Next find $f(\mathring{x}_1, \mathring{\ell}_1^{res})$ from (14). At this point the right side of (47) can be computed for the case $n = 2$, using the desired slot voltage ratio $|V_2^s/V_1^s|$. One now searches for the couplet $(\mathring{x}_2, \mathring{\ell}_2^{res})$ which brings the left side of (47) into harmony with the right side. This process is then repeated for $n = 3$, etc., until values have been obtained for all slot offsets and lengths.

This set of couplets needs to be examined for appropriateness in the following sense: If $\mathring{x}_1$ has been chosen too low, the maximum offset, say $\mathring{x}_m$, is below the upper limit for which the input data on $g(x)$ and $v(x)$ are considered reliable. In this case the fraction of the input power that is dissipated in the load is unnecessarily high. Conversely, if $\mathring{x}_1$ has been chosen too high, the maximum offset, say $\mathring{x}_m$, is beyond the upper limit for which the input data on $g(x)$ and $v(x)$ are trusted. In this case the power dissipated in the load is too small. In either event the direction to be taken for the next guess for the value of $\mathring{x}_1$ is indicated. Usually a few such trials are sufficient to yield the proper set of starting slot lengths and offsets.

With the starting couplets ascertained, attention can be focused on the second design equation. The quantities g_{mn} should be calculated for $m = 1, 2, \ldots, N$ and $n = 1, 2, \ldots, N$, but $m \neq n$. (Before performing these computations one should be sure to input whether or not the slots are alternately offset. See the earlier discussion in this section of the three cases.) In order to find g_{mn}, one must first decide on the value of d. For case 1, d should be as large as possible, consistent with the avoidance of extra main beams. For cases 2 and 3, d should be chosen so as to satisfy (45), with beam placement at the desired angle θ_o.

Knowledge of the g_{mn} values permits calculation of the mutual coupling terms MC_n, given by (38). With the offset of the first slot left unchanged, one should search for a new slot length ℓ_1 such that

$$\mathcal{I}m\left\{ \frac{2f^2(\mathring{x}_1, \ell_1)}{Y(\mathring{x}_1, \ell_1)/G_o} \right\} = -\mathcal{I}m\{MC_1\} \tag{48}$$

With this length known, Y_1^a/G_o can be found from (26) and Y_1/G_o from (40). The next step is to find a new couplet (x_2, ℓ_2) which simultaneously satisfies (39) and (42) with $n = 2$. This process is then repeated to find a new couplet (x_3, ℓ_3), etc., until finally new values exist for all the couplets and the input admittance Y_N^a/G_o is known.

At this point it is wise to pause and take stock of several factors. Is Y_N^a/G_o close to $1 + j0$? If not, there is probably an error in the computer program. Is the largest slot offset comfortably close to the upper limit of reliable self-admittance input data? If not, there is probably too much (or too little) power going into the matched load. This can be checked easily by using the formula

$$\frac{P_{\text{load}}}{P_{\text{in}}} = \frac{V_1 V_1^*}{V_N V_N^* G_N^a/G_o} \tag{49}$$

with the ratio V_1/V_N found through recursive use of (41).

One is now ready to iterate the design since the new couplet values (x_n, ℓ_n) allow an improved calculation of g_{mn}. This time it may be desirable to adjust x_1 up or down, depending on what was found in taking stock. The entire process should be iterated as many times as necessary to reach the point at which the ultimate set of slot offsets and lengths differs from the penultimate set by less than construction tolerances. Experience has shown that usually this requires only a few iterations.

It may prove desirable, as a final polish on the design, to make a minor adjustment in the value of d in order to optimize the input admittance Y_N^a/G_o. It can be shown [9] that an input match is achieved when

$$\sum_{n=1}^{N} \frac{Y_n^a}{G_o} e^{j2n\beta_{10}d} = 0 \tag{50}$$

If the ultimate values of Y_n^a/G_o are used in (50), a value of d can be found very close to the original value, which should improve on the input admittance. Indeed, one could choose to change d at every iteration, using (50), thus converging on a final value of d in concert with converging on final values of the couplets (x_n, ℓ_n).

An actual example of the use of this design technique can be found in [9].

6. The Design of a Planar Array of Longitudinal Slots

Two-dimensional arrays of slots of the type shown in Fig. 3 can be designed by a simple extension of the procedures detailed in the previous two sections. The analysis to be presented here will be limited to situations in which all waveguides containing radiating slots have the same a and b dimensions and a common vanishingly thin wall thickness. Adjacent waveguides share a common narrow wall, giving a box-beam construction. All slots have a rectangular periphery and a common width w and all are longitudinal, with individual offsets and lengths (x_{mn}, ℓ_{mn}). There is a common longitudinal spacing d_z and a common waveguide-to-waveguide spacing d_x.

The new ingredient, not present in the earlier designs of linear arrays, is the need to couple to the different waveguides in the correct ratios. A common coupling mechanism is shown in Fig. 9. It can be seen from a study of this figure that a single waveguide (hereafter called the "main line guide") has been placed behind the array and transverse to it. It shares a broad wall sequentially with all the radiating waveguides (hereafter called the "branch line guides"). In that shared wall a sequence of centered-inclined slots has been cut. These serve to couple power from the main line guide to the branch line guides. The relative amounts of coupled powers are governed by the tilts θ_n and lengths ℓ_n of the coupling slots. It is customary (and desirable for bandwidth purposes) to make the coupling slots resonant-length. The scattering matrix of the nth coupling junction is simply [10]

$$\mathbf{S} = \begin{bmatrix} s_{11} & (1-s_{11}) & s_{13} & -s_{13} \\ (1-s_{11}) & s_{11} & -s_{13} & s_{13} \\ s_{13} & -s_{13} & (1-s_{11}) & s_{11} \\ -s_{13} & s_{13} & s_{11} & (1-s_{11}) \end{bmatrix} \qquad (51)$$

in which $s_{11}(\theta_n)$ can be deduced by measurement or calculation, and in which $s_{13} = [s_{11}(1 - s_{11})]^{1/2}$. The back-scattering coefficient $s_{11}(\theta_n)$ is obtained by placing matched loads in a branch line devoid of radiating slots, as well as placing a matched load in the main line guide just beyond the nth coupling slot, and measuring or calculating B_1/A_1, with A_1 the voltage wave leaving the matched generator and B_1 the voltage wave returning to it, both referenced at the slot center. For a given tilt θ_n, as the coupling slot length is varied, resonance is said to occur when s_{11} is pure real.

Since

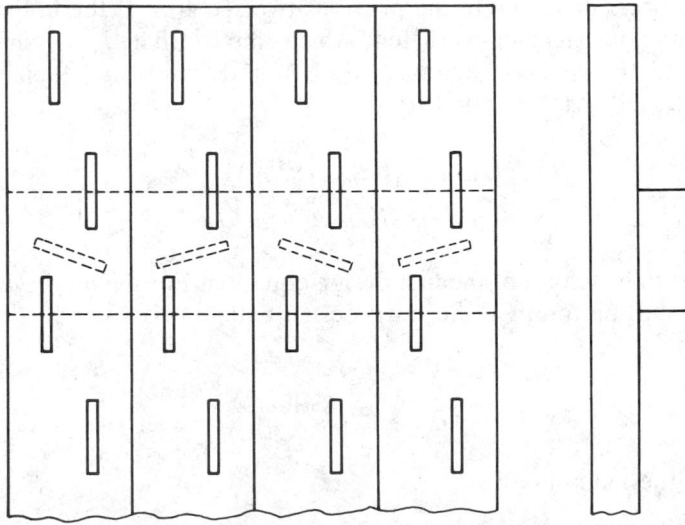

Fig. 9. Planar array of longitudinal broad wall slots fed from below by a transverse waveguide containing centered, inclined coupling slots.

$$[B] = [S][A] \tag{52}$$

with $[A]$ an arbitrary set of voltage waves incident on the four ports and $[B]$ the corresponding set of back-scattered voltage waves, if a total active admittance

$$\frac{Y_n^a}{G_o} = \sum_{m=1}^{M(n)} \frac{Y_{mn}^a}{G_o} \tag{53}$$

occurs at a cross section in the nth radiating branch line a distance $\lambda_g/4$ from the coupling slot center, it is a simple exercise to show that this total active admittance is transformed to appear as a series load in the main line, the relation being

$$\frac{Z_n^a}{R_o} = \varkappa_n^2 \frac{Y_n^a}{G_o} \tag{54}$$

where $\varkappa_n^2 = s_{11}(\theta_n)/[1 - s_{11}(\theta_n)]$. Because equivalent expressions for complex power flow must equate, it follows that $I_n I_n^* Z_n^a/R_o = V_n V_n^* Y_n^a/G_o$ and thus

$$V_n = \varkappa_n I_n \tag{55}$$

with I_n the mode current in the main line at the nth junction and V_n the mode voltage in the nth branch line at a cross section $\lambda_g/4$ from the junction center. Equation 55 is a key result which must be included in the design of a planar array. Obtained here for the case of a centered-inclined slot, it is equally valid for other coupling mechanisms, such as iris-coupled slots on the broad wall center line, and corporate feeds.

The new ingredient (the set of coupling junctions) does not fundamentally alter the design process described in the previous two sections. If the branch lines and main line are both standing-wave fed, which implies that $|I_n|$ is common to all junctions, and that $|V_n|$ is common to all M radiating slots, then the starting couplets $(\mathring{x}_{mn}, \mathring{\ell}_{mn}^{\text{res}})$ can be found from

$$\frac{g(\mathring{x}_{mn})}{g(\mathring{x}_{pq})} = \frac{|f(\mathring{x}_{mn}, \mathring{\ell}_{mn}^{\text{res}})|}{|f(\mathring{x}_{pq}, \mathring{\ell}_{pq}^{\text{res}})|} \frac{V_{mn}^s}{V_{pq}^s} \frac{\varkappa_n}{\varkappa_q} \tag{56}$$

which is seen to be a form of the first design equation. For the nth branch line, the appropriate starting form for the third design equation is

$$\frac{G_{\text{in}}^n}{G_o} = \sum_{m=1}^{M(n)} g_{mn}(\mathring{x}_{mn}) \tag{57}$$

whereas for the main line it is

$$\frac{R_{\text{in}}}{R_o} = \sum_{n=1}^{N} \frac{R_n}{R_o} = \sum_{n=1}^{N} \varkappa_n^2 \sum_{m=1}^{M(n)} g(\mathring{x}_{mn}) \tag{58}$$

One can find the starting offsets by considering (56)–(58) in concert, but it is important to note that there is not a unique solution to this set of equations. It is recommended that the slot which is to have the greatest slot voltage be singled out and assigned initially an offset near the upper limit of reliable input data. Equation 56 can then be used to obtain initial offsets for all the other slots in the same branch line.

At this point it is necessary to guess a set of values for the coupling coefficients $\varkappa_n$. Having done this, one can then use (56) to obtain initial offsets for all the other slots in all the other branch lines.

Equation 58 can be examined for an initial estimate of the input resistance to the main line. If this is too high (low), the coupling coefficients $\varkappa_n$ can be collectively adjusted down (up). Further, if the slots in a particular branch line are too little (much) offset, that particular coupling coefficient can be decreased (increased). A certain amount of adjusting of the various coupling coefficients is usually required before all slots have offsets in the most reliable range* and the desired main line input resistance is achieved, but this is an important step in the design process. On the other hand, too much accuracy is not called for since these are only starting values and they will need to be modified to include the effects of external mutual coupling.

With $\mathring{x}_{mn}$ selected for every slot, the corresponding resonant lengths can be determined from $v(\mathring{x}_{mn})$. The couplets $(\mathring{x}_{mn}, \ell_{mn}^{res})$ and $(\mathring{x}_{pq}, \ell_{pq}^{res})$ can then be used in the calculation of g_{mnpq}, using (24) but with double subscript notation. A starting value for the mutual coupling term MC_{mn} can next be calculated from

$$MC_{mn} = j(\beta_{10}/k)(k_o b)(a/\lambda)^3 \sum_{p=1}^{M(q)} \sum_{q=1}^{N} \frac{V_{pq}^s}{V_{mn}^s} g_{mnpq} \qquad (59)$$

for every slot in the planar array. Equation 59 is seen to be a restatement of (38) in double subscript notation, and is the second term in the denominator of the second design equation.

Equation 39 is used to find suitable couplets (x_{mn}, ℓ_{mn}) and then acceptable families of couplets are assembled by satisfying

$$\frac{G_{mn}^a(x_{mn}, \ell_{mn})/G_o}{G_{pq}^a(x_{pq}, \ell_{pq})/G_o} = \frac{|f(x_{mn}, \ell_{mn})|}{|f(x_{pq}, \ell_{pq})|} \cdot \frac{V_{mn}^s}{V_{pq}^s} \cdot \frac{\varkappa_n}{\varkappa_q} \qquad (60)$$

The proper family is the one which satisfies (58) with G_{mn}^a/G_o replacing $g(\mathring{x}_{mn})$. The procedure will need to be iterated since improved values of g_{mnpq} are now possible. Some adjustment of the coupling coefficients $\varkappa_n$ may prove necessary as the iterations progress.

If the branch lines are traveling-wave fed, the procedure outlined in Section 5 needs to be folded into the design process. If the main line is traveling-wave fed, that procedure can be applied to the main line as well.

---

*An alternate criterion for determining the coupling coefficients involves choosing the admittance levels of the branch lines so as to optimize input vswr and pattern side lobe level over a specified frequency range.

In many planar array applications the aperture is divided into four quadrants for the purpose of providing sum and difference patterns. In such cases, if the sum pattern is to have quadrant symmetry, the slot offsets and lengths display quadrant I/quadrant III and quadrant II/quadrant IV symmetry, and one needs to find the offsets and lengths of only half the slots. However, external mutual coupling with *all* the other slots must still be included.

7. The Achieved Aperture Distribution of Arrays of Longitudinal Slots

Given the lengths and offsets of all slots in a linear or planar array it is possible to find the resulting aperture distribution. This is accomplished by equating the right sides of the first and second design equations which, after rearrangement, gives

$$
\frac{2f^2(x_{mn}, \ell_{mn})}{Y(x_{mn}, \ell_{mn})/G_o} V^s_{mn} + j(\beta_{10}/k)(k_ob)(a/\lambda)^3 \sum_{p=1}^{M(q)} \sum_{q=1}^{N}{}' g_{mnpq} V^s_{pq}
$$
$$
= V_{mn} f(x_{mn}, \ell_{mn}) \tag{61}
$$

(The immaterial constant $2/K_1$ has been suppressed on the right side of this equation.)

Equation 61 is given in double subscript notation but it applies equally well for linear arrays. With the slot lengths and offsets known, (61) is seen to be a set of simultaneous linear equations in the unknown slot voltages V^s_{mn}. Matrix inversion gives the aperture distribution.

This equation is useful as a design check, but in the case of planar arrays with quadrantal symmetry it has further utility. Suppose such an array has been designed to produce a specified sum pattern and that (61) has provided a gratifying check on the aperture distribution. If half of the mode voltages are reversed in sign, the *E*-plane or *H*-plane difference pattern can be computed by first using (61) to deduce the relevant aperture distribution. In a similar vein, if the array has been designed to produce a specified *E*-plane (*H*-plane) difference pattern, reversing half of the mode voltages permits calculation of the aperture distributions, and subsequently the patterns, for the sum mode and *H*-plane (*E*-plane) difference mode.

8. The Design of Arrays of Centered-Inclined Broad Wall Slots

All of the analysis presented in Sections 4 through 7 for longitudinal slots can be duplicated for arrays which use centered-inclined slots in the broad wall. This has been done by Orefice in an earlier version of the theory [11]. The use of centered-inclined slots is less common because, unless the array is large, the tilt angles are big enough to cause a sizable cross-polarized radiation pattern, a side effect which is often not desirable.

Cross polarization can be eliminated by leaving the slots on the center line but untilted. (Such slots can also be viewed as longitudinal but not offset.) In this case the slots are excited by irises [12]. The theory described in earlier sections is then applicable with Y/G_o represented in terms of slot length and iris dimensions.

9. The Design of Arrays of Inclined Narrow Wall Slots

Linear arrays of edge slots of the type shown in Fig. 2c find application as the line source feeds for cylindrical reflector antennas. *Planar* arrays of edge slots are attractive when beam scanning is desired in the *H*-plane. The reason for this is that adjacent waveguides can be placed close together ($\lambda_o/2$ spacing), permitting wide-angle scanning without the intrusion of extra main beams.

Unfortunately, unlike broad wall slots, inclined narrow wall slots cannot be assumed to lie in an infinite ground plane, because they actually "wrap around" and invade both broad walls in order to achieve resonant length. For this reason adjacent waveguides in planar arrays of edge slots need to be spaced apart, with metallic barriers often placed in the interstices, as suggested in Fig. 10. Although this results in a conducting surface which, viewed from the outer half-space, is closed except for the slots, that surface is far from a plane.* Thus the theoretical computation of external mutual coupling embodied in formulas such as (24) is not valid for planar arrays of narrow wall slots. Even in the case of a linear array, embedment in a ground plane is not feasible.

One is forced to seek an *experimental* estimation of mutual coupling in arrays of this type. Depending on the accuracy desired, the complexity of the measurements can vary. Consider first a standing-wave-fed linear array of inclined narrow wall slots. Because these slots are almost parallel, mutual coupling is significant. If the array is long, and the desired aperture distribution is equiphase and not extreme in its amplitude variation, one can argue that each slot has immediate neighbors whose excitations are essentially the same as its own. If one further assumes that most of the mutual coupling is due to nearest neighbors, then it follows that the active admittance of any edge slot in the linear array is practically the same as if it were part of an infinitely long array of identically excited slots.

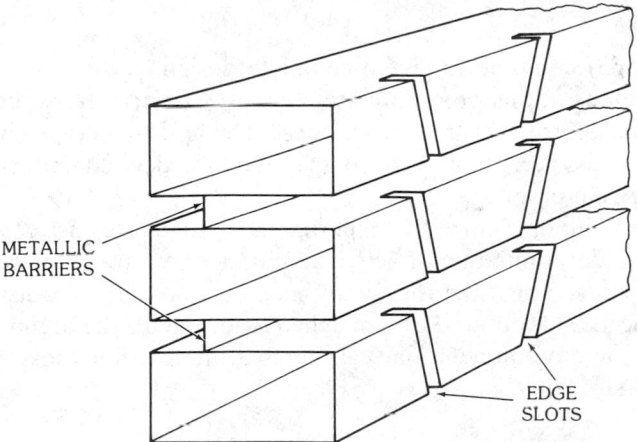

METALLIC
BARRIERS

EDGE
SLOTS

Fig. 10. Planar array of edge slots with baffles placed in the interstices.

---

*To complicate matters still further, in some applications baffles or wire grids are placed in front of the array to cut down further on cross polarization.

This argument paves the way for an experimental procedure to determine $Y^a(\theta, \ell)/G_o$ for an edge slot. One constructs a family of linear arrays of edge slots. Each member of the family is characterized by having N slots of common length and common but alternating tilt angle θ. The number N should be large ($N \gtreqqless 10$). The slots are spaced $\lambda_g/2$ apart and the last one is followed by a short-circuit $\lambda_g/4$ beyond. With the slot closest to the generator covered over with conducting tape, the normalized input admittance is measured for each family member. Then with the first slot uncovered, the measurement is repeated. Arguably, the difference in the two readings is $Y^a(\theta, \ell)G_o$ for a central slot in the array. Alternatively, with N slots in the array, one can assume that $Y^a(\theta, \ell)/G_o = N^{-1} Y_{in}/G_o$.

After this, ℓ is lengthened and the measurements repeated on all members of the family. In this manner a mosaic of active admittance values can be assembled with θ and ℓ the independent variables.

Usually one is interested in extracting from this mosaic the resonant-length information,* that is, the function $G^a_{res}(\theta, \ell_{res})/G_o$. Since $VV^*G^a_{res}(\theta_n, \ell_n^{res})/G_o$ measures the power radiated by the nth slot, with V the common mode voltage, and since this power is approximately proportional to $(V_n^s)^2$, it follows that

$$\frac{G^a_{res}(\theta_n, \ell_n^{res})/G_o}{G^a_{res}(\theta_m, \ell_m^{res})/G_o} = \left(\frac{V_n^s}{V_m^s}\right)^2 \tag{62}$$

With the desired aperture distribution known, (62) can be used to determine all the slot tilt angles and the corresponding slot lengths once these quantities are known for a reference slot. A unique solution results when the input admittance is specified. This is usually a match; in that case

$$\sum_{n=1}^{N} G^a_{res}(\theta_n, \ell_n^{res})/G_o = 1 \tag{63}$$

and (62) and (63) combine to give a complete design.

The limitations of this procedure are obvious. The array must be long and the variations in the aperture distribution modest. The approximations inherent in this method become less acceptable as the demands on the desired pattern (such as very low side lobes) increase.

With only a minor change the experimental procedure just described can be extended to the determination of active admittance in a traveling-wave-fed linear array of edge slots. This time the family members contain N slots that are non-resonantly spaced and followed by a matched load. Under the argument that every slot sees the same environment, one can say that the fractional loss in power is the same at every slot, that is,

$$\frac{P_{1N} - P_N}{P_{1N}} = \frac{P_N - P_{N-1}}{P_N} = \cdots = \frac{P_n - P_{n-1}}{P_n} = \cdots = \frac{P_1 - P_{LOAD}}{P_1} \tag{64}$$

---

*For this reason the procedure just described is generally referred to as the incremental conductance technique.

Solution of this train of equations gives

$$\frac{P_{1N}}{P_{\text{LOAD}}} = \left(\frac{P_1}{P_{\text{LOAD}}}\right)^{N+1} \tag{65}$$

But the power radiated by the slot closest to the load is $1/2\,\text{Re}\{V_1 V_1^* Y_1^a\} = P_1 - P_{\text{LOAD}}$, whereas the power absorbed in the load is $1/2\,V_1 V_1^* G_o = P_{\text{LOAD}}$. Therefore

$$\text{Re}\left\{\frac{Y_1^a}{G_o}\right\} = \frac{P_1 - P_{\text{LOAD}}}{P_{\text{LOAD}}} \tag{66}$$

The combination of (65) and (66) yields

$$\text{Re}\left\{\frac{Y_1^a}{G_o}\right\} = \left(\frac{P_{1N}}{P_{\text{LOAD}}}\right)^{1/(N+1)} - 1 \tag{67}$$

If one measures the ratio P_{LOAD}/P_{1N}, (67) can be used to measure the normalized active conductance of a slot in the test array. This can be repeated as the lengths of all the slots in the test array are changed to a common longer value. Resonance can be defined as occurring for that length which results in minimum power entering the load (maximum power being radiated). This is the condition usually desired, for which

$$\frac{G_{\text{res}}^a(\theta, \ell^{\text{res}})}{G_o} = \left(\frac{P_{1N}}{P_{\text{LOAD}}^{\text{MIN}}}\right)^{1/(N+1)} - 1 \tag{68}$$

With data assembled on G_{res}^a/G_o versus tilt angle one can proceed to the design of a traveling-wave-fed linear array. Fig. 8 and (40) and (41) are applicable, but now (62) is replaced by

$$\frac{V_n V_n^* G_{\text{res}}^a(\theta_n, \ell_n^{\text{res}})/G_o}{V_m V_m^* G_{\text{res}}^a(\theta_m, \ell_m^{\text{res}})/G_o} = \left|\frac{V_n^s}{V_m^s}\right|^2 \tag{69}$$

Implicit in (69) is the assumption that a traveling-wave aperture distribution is desired (arbitrary amplitude variation, but uniform progressive phase).

One begins by assuming a tilt angle for the slot closest to the load. With $m = n - 1$, (41) is used to determine V_n/V_{n-1} and then (69) is used to determine the tilt angle of the second slot from the load. This process is repeated until the inclinations of all the slots are known. At this point the fraction of the input power going into the load can be calculated and the maximum tilt angle noted. A change in the guess for θ_1 might be indicated, in which case the process can be repeated.

For planar arrays of edge slots, if one assumes that external mutual coupling between waveguides can be ignored, all of the discussion just concluded about linear arrays can be carried over intact.* If it is felt that a more exact accounting for

---

*An additional ingredient is that the optimum set of coupling coefficients will also need to be determined. See Section 6.

mutual coupling is needed, small planar arrays (say 10 by 10) can be constructed, each containing slots with common but alternating tilt. The active resonant conductance of a central slot can be measured, with all waveguides equally excited, using the incremental conductance technique described earlier in this section. The resulting information can be used in (62) and/or (69) to design the array, given the desired aperture distribution. Here again, the applicability is limited to situations in which the array is large and variations in the aperture distribution are not extreme.

What does one do when these conditions are not met? One solution is offered in Section 12.

10. Difficulties in the Design of Large Arrays

A study of the second design equation for longitudinal slot arrays (26), plus an appreciation of the design procedure described in Sections 4 through 6, reveals that the calculation of mutual coupling involves a computation of the double integral g_{mn} for all values* of m and n. This can produce a sizable computer cost for large arrays.[†]

A possible way around this difficulty exists if the mutual coupling experienced by a slot is mostly due to its nearest neighbors, for then the calculation of g_{mn} could be truncated for some maximum value of m, given n. For a central element in a large planar array one could consider square rings of surrounding slots, as suggested in Fig. 11, and ask how many rings would need to be included in the calculation of MC_n before the value of MC_n stabilized. It can be anticipated that convergence would be slow because the magnitude of g_{mn} is roughly inversely proportional to R, whereas the number of slots in a ring is proportional to R.

If the special case of uniform excitation is considered, the question of convergence reduces to a study of the summation

$$S = \frac{MC_n}{j(\beta_{10}/k)(k_o b)(a/\lambda)^3} = \sum_m{}' g_{mn} \qquad (70)$$

The term S has been evaluated for the case of longitudinal slots in air-filled waveguide with a common length $\lambda_o/2$ and a common spacing $0.7\lambda_o$. (Offsets were ignored.) The results are shown in Fig. 12 for a central element, an end element, and a penultimate element in a linear array. Fig. 13 shows equivalent results for a central element, a corner element, and middle-of-a-side elements in a planar array.

One can observe from Figs. 12a and 12b that convergence for a linear array is fairly rapid in the H-plane but less so in the E-plane. A frequently posed question is: Does the second element in from the end of a linear array see a mutual coupling environment similar to that seen by a central element? A study of Figs. 12c and 12d indicates that the answer is clearly no for short arrays but is approximately yes for

---

*Symmetries can reduce the number of calculations, but not significantly.
[†]However, this cost is not prohibitive. As an example, for a slot array of one thousand elements the entire design procedure only entails a computer cost of about one thousand US dollars.

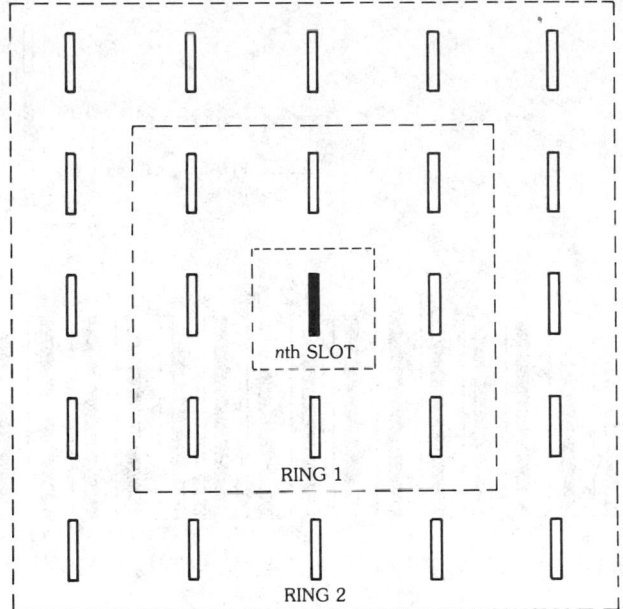

Fig. 11. A longitudinal slot surrounded by rings of similar slots.

longitudinal arrays of eight or more elements and for transverse arrays of fourteen or more elements.

Fig. 13a shows that the variability in mutual coupling for a central element in a planar array is still significant even after 24 rings are included. This is also true for a corner element, as Fig. 13b attests. Also it is interesting to note that a corner element has mutual coupling that is about half that of a central element, not one-quarter. Surprisingly, Fig. 13c reveals that a middle-of-a-side element, with the slots perpendicular to that side, has mutual coupling comparable to that of a central element. For a middle-of-a-side element in a side parallel to the slots, the mutual coupling is at about the level of a corner element, that is, half that of a central element, as indicated by Fig. 13d.

The results given in Figs. 12 and 13 are for a uniform distribution. When a tapered distribution is assumed, convergence is affected slightly [13]. It is clear from this study that truncation must be used with caution if one wishes to obtain accurate values for mutual coupling.

If the procedure outlined in Sections 4 through 6 is to be used for the design of large arrays, it is recommended that a routine be set up for the calculation of MC_n that involves the use of rings of neighboring elements, but that the rings be rectangular,* rather than square as in Fig. 11. The desired slot voltage distribution should be used as a weighting function on g_{mn}. More specifically, the full expression (38) should be used, with truncation occurring when the variation in MC_n has become less than a prescribed value.

---

*This will accommodate the fact that E-plane mutual coupling is stronger than H-plane mutual coupling, resulting in slower convergence in the E-plane direction.

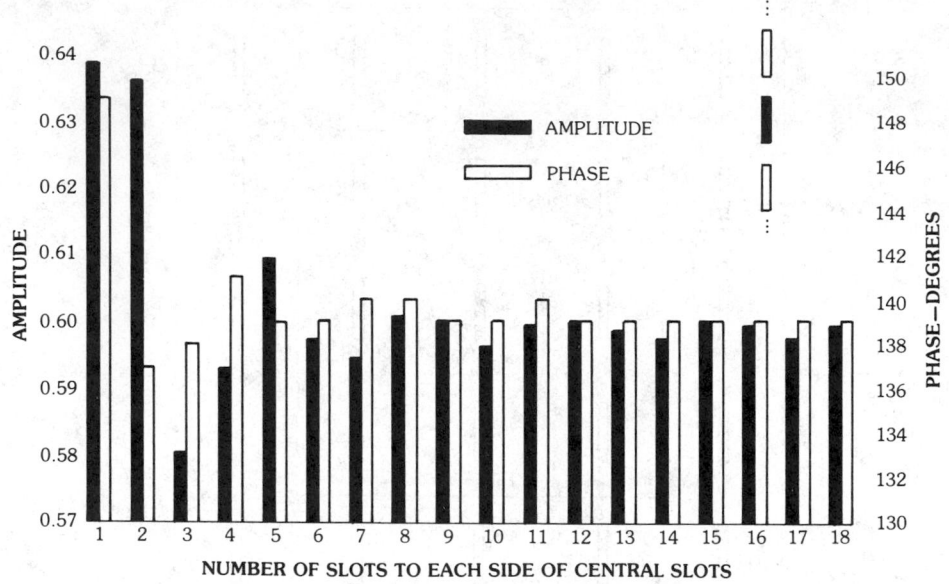

a

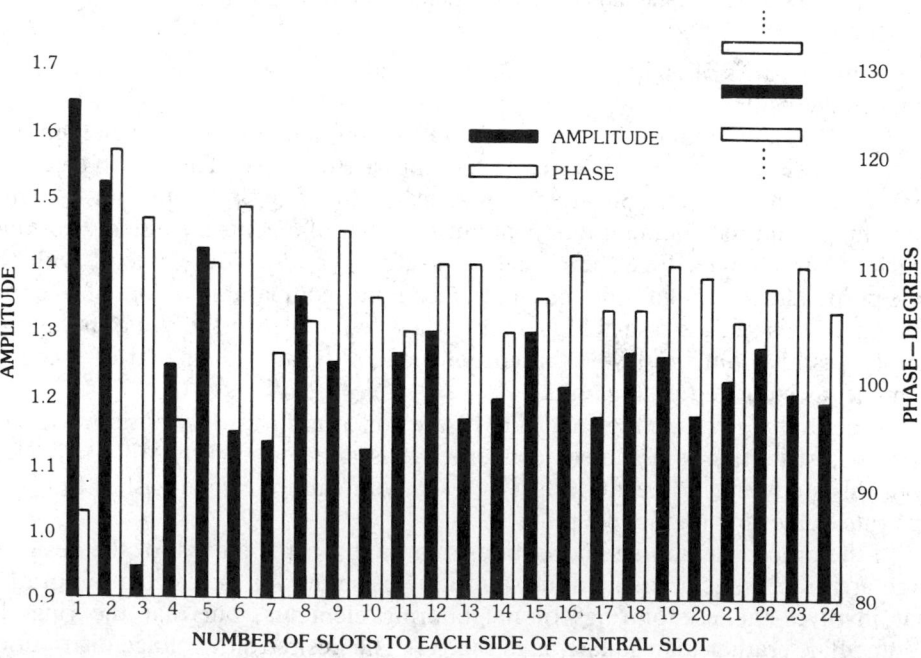

b

Fig. 12. Mutual coupling in a linear array, with $2\ell = \lambda_o/2$, $d = 0.7\lambda_o$. (*a*) Longitudinal slots, central element (divide by 2 for end element). (*b*) Transverse slots, central element (divide by 2 for end element). (*c*) Longitudinal slots, penultimate element. (*d*) Transverse slots, penultimate element.

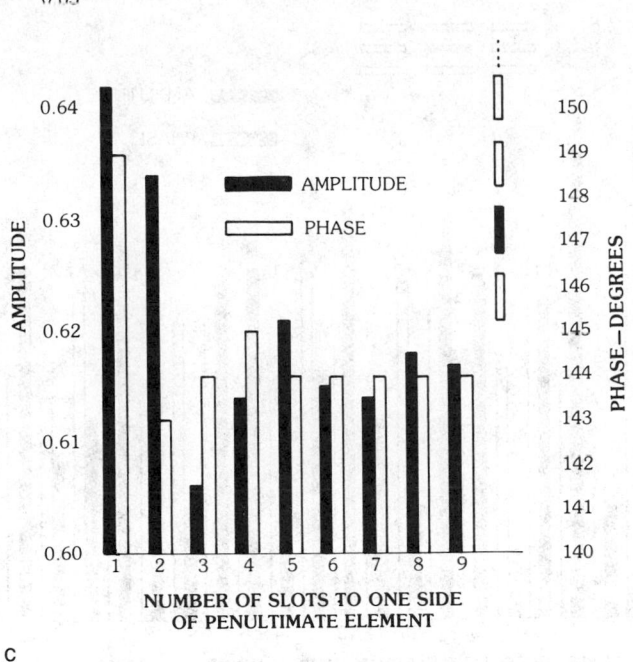

c

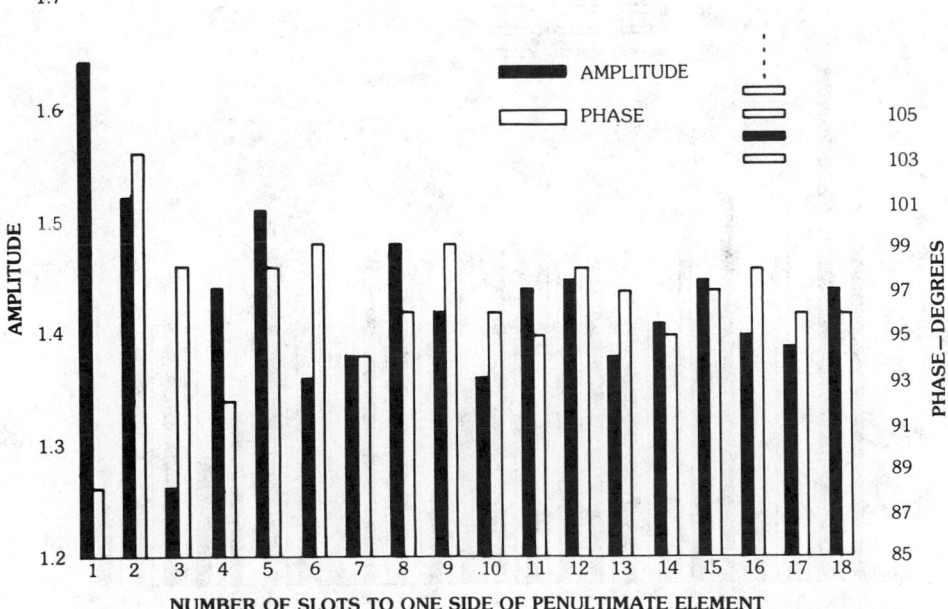

d

Fig. 12, *continued*

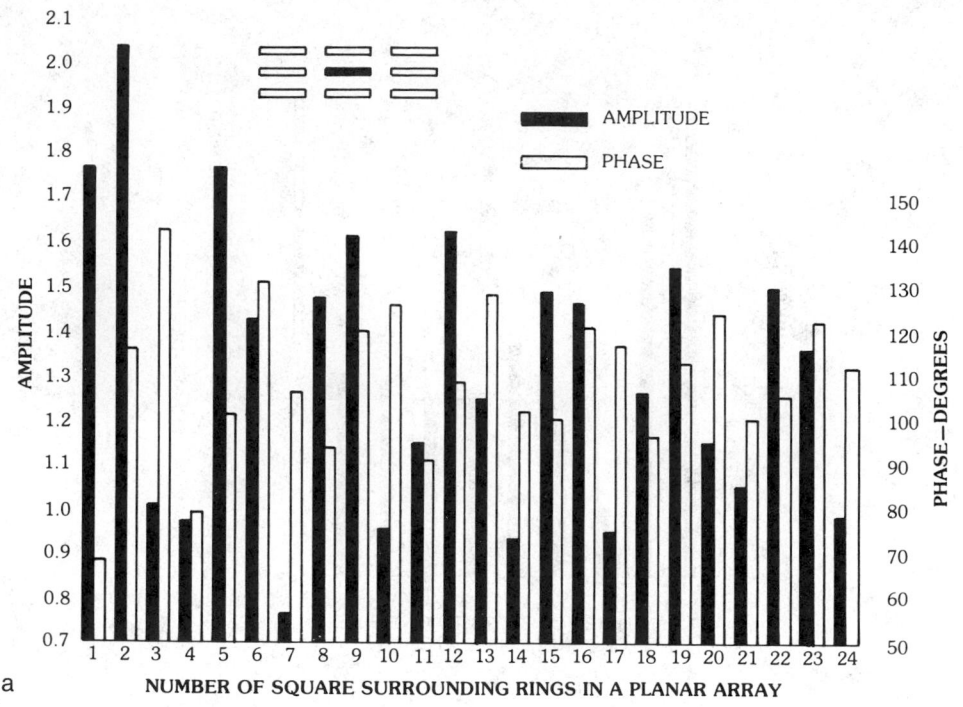

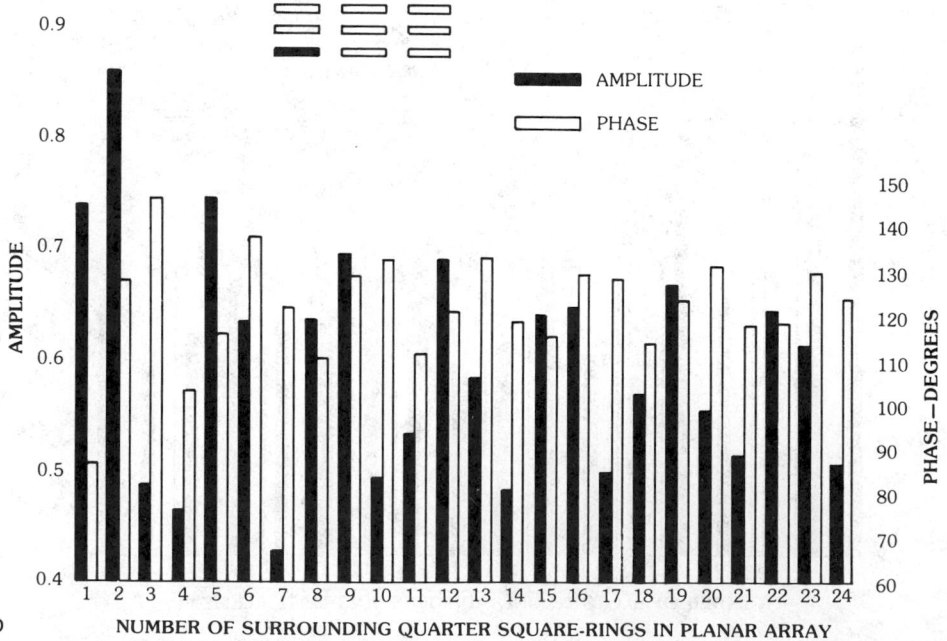

Fig. 13. Mutual coupling in a planar array, with $2\ell = \lambda_o/2$, $d = 0.7\lambda_o$. (a) Central element. (b) Corner element. (c) Element in middle of longitudinal side. (d) Element in middle of transverse side.

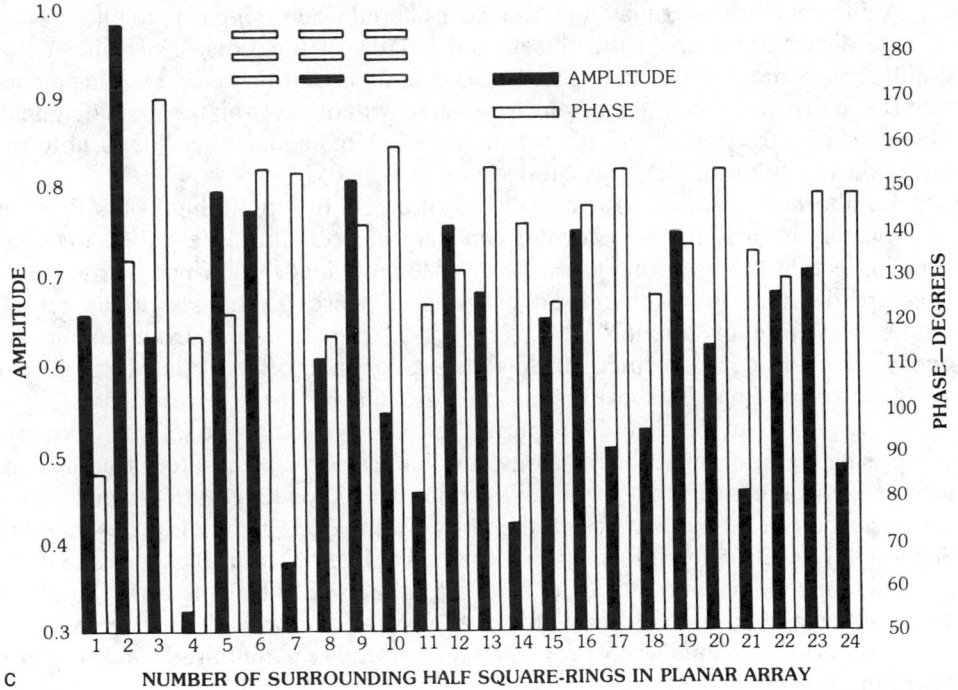

c NUMBER OF SURROUNDING HALF SQUARE-RINGS IN PLANAR ARRAY

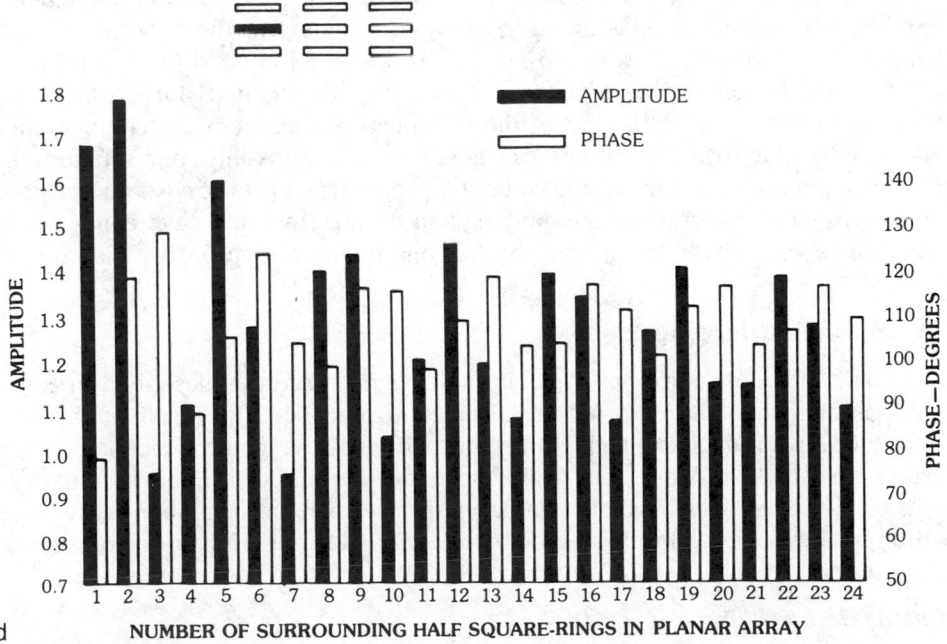

d NUMBER OF SURROUNDING HALF SQUARE-RINGS IN PLANAR ARRAY

Fig. 13, *continued.*

A different type of truncation can be considered when using (61) to find the slot voltage distribution, given the offsets and lengths of the slots. The value g_{mnpq} diminishes as the distance between the mnth and pqth slots increases. This means that the matrix in (61) has terms which decrease with offset from the main diagonal. One can set a fraction of the far-out matrix elements equal to zero and note the effect on the solution $[V_{mn}^s]$ as the fraction is varied.

An alternative approach to the design of arrays of longitudinal slots that has been widely used is the incremental conductance technique, described for edge slots in Section 9. For a long linear array of longitudinal slots which is to have an aperture distribution with uniform progressive phase (perhaps equiphase) and modest variation in amplitude, $G_{res}^a(x, \ell_{res})/G_o$ can be found from families of linear arrays, in each of which all slots have a common offset. For a large planar array of longitudinal slots, with similar restrictions on the aperture distribution, $G_{res}^a(x, \ell_{res})/G_o$ can be found by putting baffles on each side of linear arrays comprising the aforementioned family; the baffles serve to create an equivalent infinite environment in the E-plane direction. This knowledge of resonant active conductance can be used to design the arrays in the same manner as described in Section 9. However, the information contained in Figs. 12 and 13 cautions that the incremental conductance technique should not be used unless the arrays are very large and even then is not accurate for slots near the end of a linear array nor near the periphery of a planar array, nor when the variability in amplitude and/or phase distribution is significant.

From this discussion it can be seen that errors in slot lengths and offsets should be expected if truncation is used in calculating MC_n or if the incremental conductance technique is assumed to apply for all slots in the array. A method which will reduce these errors substantially is described in Section 12.

This discussion applies equally well to centered, inclined slots in the broad wall, and limitations on the use of the incremental conductance technique apply also to edge slot arrays. In situations where a choice is possible, one should weigh the costs and errors of a truncated computer program against the costs and errors of an experimental program which gathers data on incremental active conductance. The choice will usually tip in favor of the computational approach.

11. Second-Order Effects

Some of the assumptions made in the design procedures described in earlier sections of this chapter are not strictly met in practice. The result of this is to introduce errors which can be identified but not easily overcome. Fortunately these errors are small and are thus called second-order effects. Even second-order effects, however, can become significant if very high performance (such as an ultralow side lobe level) is demanded of the array. A listing of the principal second-order effects follows.

Infinite Ground Plane

This assumption is made in deriving a formula for g_{mn} for broad wall slots of either the longitudinal or centered, inclined type. For linear arrays not embedded in ground planes it is most questionable in the transverse direction. For planar arrays not embedded in a ground plane (a frequently encountered situation) it has the least validity for peripheral slots.

In some applications, e.g., fuze antennas, a linear broad wall slot array is embedded in a conducting cylinder of small enough radius that the infinite ground plane assumption is inadequate. Lee and Safari-Naini [14] have derived an appropriate equivalent expression for g_{mn} for such cases.

Wall Thickness

The theory presented in Section 2 assumes that the upper broad wall, in which the slots are cut, is vanishingly thin. Finite wall thickness can be taken into account [15, 7] but the additional complexity adds to the computer cost of finding $Y(x, \ell)/G_0$ computationally for a single longitudinal slot, or $Z(\theta, \ell)/R_0$ computationally for a single centered, inclined slot, and the result is approximate. Of course, if these quantities are measured, the effect of wall thickness is automatically included. However, measurement of the input data is inherently less accurate and more costly than a theoretical determination. Either way a small error in $Y(x, \ell)/G_0$ or $Z(\theta, \ell)/R_0$ creeps into the design procedure. Since the modern trend is toward thinner and thinner walls, the computational approach should gain in favor.

E-Field Distribution in the Slot Aperture

Figs. 5 and 6 show the transverse **E** field along the center line of a longitudinal slot in amplitude and phase as a function of the longitudinal coordinate. This **E** field is caused by an incident TE_{10} mode and is seen to be almost equiphase and almost symmetrical in amplitude. The asymmetries are small and are ignored in deriving a design procedure based on (10), as is the symmetrical bulging of the amplitude distribution relative to a half-cosinusoid, as is the transverse variation of transverse **E**, as is the small longitudinal component of the **E** field in the slot aperture.

These small ignored quantities, if included, would combine to cause $B_{10} \neq C_{10}$ for a longitudinal broad wall slot. Similarly, they have the effect of causing $B_{10} \neq -C_{10}$ for a centered, inclined broad wall slot. This means that the simple equivalent circuits which represent the longitudinal slot as a shunt element and the centered-inclined slot as a series element are not strictly correct.

A more rigorous formulation requires T or π networks. That representation is so cumbersome that it is then preferable to design the arrays in terms of scattered waves rather than equivalent admittances and impedances. As pattern and input match requirements become more demanding, it will become increasingly important to take these scattering asymmetries into account.

Internal Higher-Order-Mode Mutual Coupling between Radiating Slots

All of the design procedures presented in this chapter contain the underlying premise that only TE_{10} modes matter—that higher-order-mode scattering off any radiating slot has died out to a negligible level before reaching a neighboring slot.*

---

*This premise implies that the a dimension is chosen so that all modes but TE_{10} are cut off.

This is not strictly true; particularly, neglect of TE_{20} scattering introduces an error which is more serious as the b dimension lessens. This is because the resonant length of a longitudinal slot increases as b decreases, thus putting the tips of adjacent slots closer together. The TE_{20} mode effects can be included at some cost in complexity of the design procedures. They affect array performance in two ways: (1) TE_{20} mode scattering off one slot induces an additional electric field in its two neighboring slots in the same guide. That affects the pattern. (2) This additional electric field in turn causes a change in the TE_{10} scattering off the neighboring slots. That affects the overall slot voltage and the input impedance. When high performance is required, these effects cannot be ignored.

Higher-Order-Mode Coupling in Junctions

This only occurs in planar arrays, but it can be the most serious of all the second-order effects. A classic example occurs in slot arrays of the type shown in Fig. 9. The tilted coupling slot is seen to be only $\lambda_g/4$ from the two radiating slots which straddle it. Compared to the problem of higher-order-mode internal coupling between neighboring radiating slots, this situation is more severe because the cutoff modes have only half the distance in which to die out. Further complicating the situation, the coupling slot and the two radiating slots are essentially transverse, so that more than the TE_{20} mode is involved in the supplemental coupling mechanism. If one ignores this effect, the results are that the straddling slots are improperly excited, thereby degrading the pattern, and the main line match is not achieved.

12. Far-Field and Near-Field Diagnostics as Design Tools

Sections 10 and 11 chronicled some of the difficulties in the design of linear and planar slot arrays—difficulties due to imperfect assumptions in theory and experiment, and difficulties due to prohibitive costs in computation or in the amassing of extensive experimental input data. The result, particularly for high-performance arrays, is that actual pattern and input impedance fall somewhat short of the design goals. When this is the case, far- and near-field measurements can be used to improve on the design procedures.

As an example, a seeker antenna is often fed by four main lines, one per quadrant, so that sum and difference patterns can be produced. Each main line contains a sequence of coupling junctions, one per branch line. The radiating slots which straddle a junction have their slot voltages perturbed by higher-order-mode scattering (see also Section 11). These perturbations occur at regularly spaced positions in the aperture and are the cause of grating lobes. In some applications they can be seen unambiguously in the far-field pattern. Their levels and positions can be used to deduce the errors in slot voltages of the straddling slots and these errors can be corrected by small changes in the offsets and lengths of the affected slots.

More potent is the use of a near-field probe facility to determine the amplitude and phase of each slot in the array. An illustration of the way in which near-field data can improve on the design is the case of a traveling-wave linear array of edge slots. Section 9 contains a description of how the real part of the normalized active admittance, $\text{Re}\{Y^a(\theta, \ell)/G_o\}$, is obtained using a family of test arrays, each with N slots of alternating tilt, but with a different θ value for each family member. If one takes near-field data on each test array as ℓ is altered, one finds that the uniform progressive phase changes with ℓ. If the uniform progressive phase corresponding to resonant length (minimum power into the load) is taken as reference, the differential phase can be related to the phase of $Y^a(\theta, \ell)/G_o$. Thus one possesses more information than just $\text{Re}\{Y^a(\theta, \ell)/G_o\}$. If the actual array is now designed according to the procedure described in Section 9, and if its performance falls short of expectation, one can then probe the near field of the actual array and determine the errors in the various slot voltages. With the foreknowledge of $Y^a(\theta, \ell)/G_o$, the error in slot voltage for the slot closest to the load can be compensated. One then moves back to the next slot and institutes another compensation, proceeding in this way to the slot farthest from the load. Experience has shown that a second array, built to incorporate these compensations, has an improved performance over the first array.

Acknowledgments

The author wishes to thank M. Armstrong, L. Josefsson, L. Kurtz, R. Mailloux, and R. Shavit for criticizing all or portions of the manuscript. Discussions with W. Lange and J. Thomas were helpful and appreciated. Phyllis Parris typed the manuscript with her customary skill and her cheerfulness is warmly acknowledged.

13. References

[1] A. F. Stevenson, "Theory of slots in rectangular waveguides," *J. Appl. Phys.*, vol. 19, pp. 24–38, 1948.

[2] T. V. Khac, "A study of some slot discontinuities in rectangular waveguides," PhD dissertation, Monash University, Australia, November 1974.

[3] R. F. Harrington, *Field Computation by Moment Methods*, New York: Macmillan Co., 1968.

[4] Y. U. Kim, "Electric field distribution in a longitudinal slot and its effect on the design of slot array antennas," MS thesis, University of California, Los Angeles, June 1983.

[5] R. S. Elliott, "An improved design procedure for small arrays of shunt slots," *IEEE Trans. Antennas Propag.*, vol. AP-31, pp. 48–54, January 1983.

[6] R. S. Elliott, *Antenna Theory and Design*, Englewood Cliffs: Prentice-Hall, 1981, pp. 95–96.

[7] H. Y. Yee, "Impedance of a narrow longitudinal shunt slot in a slotted waveguide array," *IEEE Trans. Antennas Propag.*, vol. AP-22, pp. 589–592, July 1974.

[8] See, for example, Reference 6, pp. 134–136.

[9] See Reference 6, p. 471.

[10] R. S. Elliott, "Dominant mode analysis of coupling junctions for flat-plate slot array antennas," *Report No. IDC 56 B1. 30/68*, Hughes Missile Systems Group, Canoga Park, California, July 1983.

[11] M. Orefice and R. S. Elliott, "Design of waveguide-fed series slot arrays," *IEE Proc.*, pt. H, vol. 129, pp. 165–169, August 1982.

[12] R. Tang, "A slot with variable coupling and its application to a linear array," *IRE Trans. Antennas Propag.*, vol. AP-8, pp. 97–101, January 1960.

[13] Y. U. Kim and R. S. Elliott, "External mutual coupling in large arrays of longitudinal

slots," *Report No. AP-201*, Dept. of Electr. Eng., University of California, Los Angeles, July 15, 1983.

[14] S. W. Lee and S. Safari-Naini, "Approximate asymptotic solution of surface field due to magnetic dipole on a cylinder," *IEEE Trans. Antennas Propag.*, vol. AP-26, pp. 593–598, July 1978.

[15] A. A. Oliner, "The impedance properties of narrow radiating slots in the broad face of rectangular waveguide," *IRE Trans. Antennas Propag.*, vol. AP-5, pp. 1–20, January 1957.

Chapter 13

Periodic Arrays

R. J. Mailloux
Rome Air Development Center, Electromagnetic Sciences Division

CONTENTS

 **Robert J. Mailloux** was born in Lynn, Massachusetts. He received the BS degree in electrical engineering from Northeastern University, Boston, Massachusetts, in 1961, and the SM and PhD degrees from Harvard University, Cambridge, Massachusetts, in 1962 and 1965, respectively.

He was with the NASA Electronics Research Center, in Cambridge, from 1965 to 1970, and with the Air Force Cambridge Research Laboratories from 1970 to 1976. He is presently acting chief of the Antennas and Components Division, Rome Air Development Center, Electromagnetics Directorate. His research interests are in the area of periodic structures and antenna arrays. He has published numerous papers on antennas and arrays, and book chapters on antenna research topics, on hybrid systems of arrays and reflectors or lenses, and on conformal arrays. He was elected to the grade of Fellow of the IEEE in 1978.

Dr. Mailloux is a member of Tau Beta Pi, Eta Kappa Nu, Sigma Xi, and Commission B of URSI. He was Technical Activities Chairman for Commission B of URSI from 1979 to 1982, and President of the Hanscom Chapter, Sigma Xi, in 1980. He was President of the Antenna and Propagation Society in 1983 and has previously been AP-S Distinguished Lecturer, AdCom Member, and Meetings Chairman.

1. Introduction

Most scanning array antennas are composed of rows and columns of periodically spaced antenna elements. Periodic arrays can be designed to provide extremely low side lobes and high gain when element spacing is kept relatively small, and they are chosen in preference to arrays with aperiodic lattices for many radar and communications applications. In addition, when hundreds or thousands of closely spaced elements are required, it is also simpler and cheaper to construct periodic than aperiodic arrays. Aperiodic arrays, treated in the next chapter, have advantages in high-resolution thinned configurations, and when used to achieve equivalent amplitude taper without the use of a complex feed network.

Chapter 11 deals with a number of theoretical aspects of phased arrays and establishes a basis in mathematics by which one can compute or synthesize detailed radiation patterns. This chapter will provide some background in theory to increase comprehension, but is intended mainly to present fundamental engineering data to aid in antenna design and evaluation.

Although phase-scanned arrays are the main topic of the chapter, some consideration is given to time-delay scanned arrays. There are discussions of pattern distortions resulting from grating lobes and array errors and a brief treatment of mutual coupling effects. Pattern synthesis is covered in more detail in Chapter 11, so the treatment in this chapter is restricted to giving engineering data and some examples of the effects of discreting line source data. The results of organizing an array into subarrays are presented and several approaches described that achieve wide instantaneous bandwidth.

The discussion of specific devices for arrays is restricted to passive components (phase shifters, power dividers, array elements, etc.) and does not include amplifiers or any of a wide variety of array control systems, mixer steering, or signal processing means of pattern control.

Pattern and Excitation

The most general form of radiation pattern used in this section is given below. Each element at position (x_i, y_i, z_i) of Fig. 1 is excited by a complex weighting a_i and radiates with a vector element pattern $\mathbf{f}_i(\theta, \phi)$ so that the total far-field radiated field is given by

$$\mathbf{E}(\mathbf{r}) = \frac{e^{-jkR_0}}{R_0} \sum_i a_i \mathbf{f}_i(\theta, \phi)\, e^{+jk\mathbf{r}_i \cdot \hat{\varrho}} \tag{1}$$

where $k = 2\pi/\lambda$, the vectors $\mathbf{r}_i$ define the locations of array elements relative to the element with index zero, and the unit vector $\hat{\varrho}$ is the position vector locating the observation point P a distance R_0 from the origin at the zeroth element:

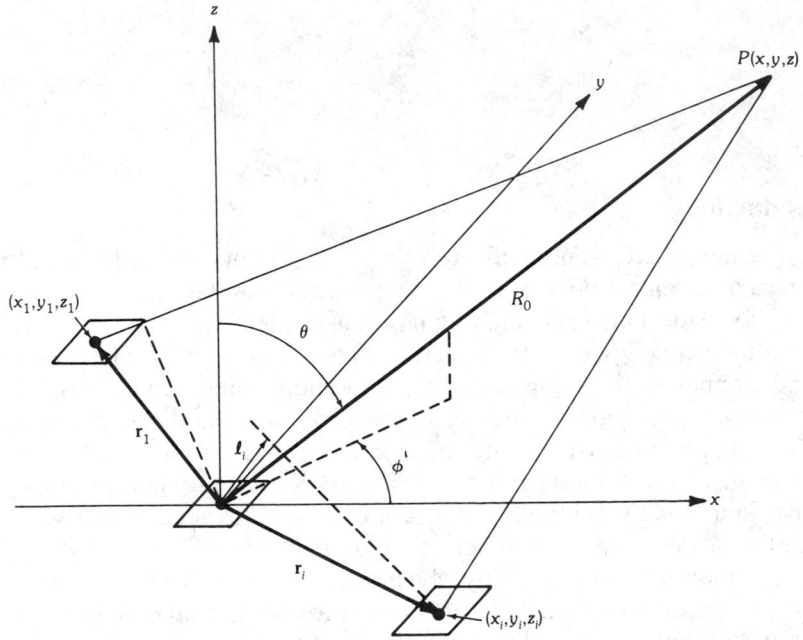

Fig. 1. Generalized array geometry.

$$\hat{\varrho} = \hat{x}u + \hat{y}v + \hat{z}\cos\theta$$
$$\mathbf{r}_i = \hat{x}x_i + \hat{y}y_i + \hat{z}z_i \qquad (2)$$
$$\mathbf{R}_0 = |r - r_0| = \sqrt{x^2 + y^2 + z^2}$$

The direction cosines u and v are

$$u = \sin\theta\cos\phi$$
$$v = \sin\theta\sin\phi$$

with $u^2 + v^2 \leqq 1$.

In this generalized form the weighting a_i is the *applied* excitation, an incident mode amplitude for a transmission-line or waveguide-fed array, or an *applied* voltage for an array of wire elements. The feeding sources are assumed matched to their characteristic impedance. The variables $\mathbf{f}_i(\theta, \phi)$ are the radiation patterns of each ith element, in the presence of all other elements, and so are called *element patterns* (sometimes termed "active element patterns"). The vector element patterns $\mathbf{f}_i$ are in general different, even in an array of like elements. The difference results from the interaction between individual elements and proximity to the array edge. In the formal treatment of antenna radiation the element patterns are unknowns, obtained by solving an electromagnetic boundary value problem.

Before leaving the topic of interpreting (1), it should be noted that an alternative but fully as general formalism considers the a_i to be the actual (unknown) currents for wire elements, or electric fields for slots or apertures. From this perspective, when mutual coupling is considered, the $a_i f_i(\theta, \phi)$ are replaced by the radiation pattern of the unknown distribution of current or field on the ith element: a distribution that is found from the solution of a boundary value problem, and will not, in general, be directly proportional to the applied excitations. In the absence of mutual coupling, however, the two formalisms are identical.

Throughout this section the a_i will be the applied excitation, as in the element pattern formalism described earlier. Furthermore, throughout Sections 2, 3, and 5, the element patterns will be assumed equal and isotropic over the hemisphere $z \geqq 0$, but element pattern distortion will be described in Section 4, in connection with array mutual coupling. In addition, it is assumed that all expressions relate to the axis of principal polarization, and so the patterns are written as scalar functions.

The usual purpose of an array is to form a beam at some specific angle in space (θ_0, ϕ_0). In the absence of mutual coupling this can be done at all frequencies by choosing the excitation

$$a_i = |a_i| e^{-jk\mathbf{r}_i \cdot \hat{\varrho}_0} = |a_i| e^{-jk\ell_i} \tag{3}$$

with ϱ_0 given by (2) using θ_0 and ϕ_0 in the direction cosine expressions. At such a point the fully collimated beam field strength is a simple vector summation of the element patterns weighted by the amplitudes $|a_i|$

$$\mathbf{E}_0(\mathbf{R}_0) = \frac{e^{-jkR_0}}{R_0} \sum_i \mathbf{f}_i(\theta_0, \phi_0)|a_i| \tag{4}$$

and if the element patterns $\mathbf{f}_i$ are equal and isotropic, $(|\mathbf{f}_i| = 1)$; this is the largest possible value of the field $\mathbf{E}(\mathbf{r})$ for any given R_0 in the far field. Selection of the excitations of (3) is understood intuitively by considering that the projected distance to the observer at (R_0, θ_0, ϕ_0) is different for each array element by the length ℓ_i in Fig. 1. Removal of this path length difference will cause the contributions from each element to add in phase in the far field. The envelope of coefficients $|a_i|$ is the array illumination and is the primary determinant of the radiation side lobe levels, just as it is for aperture antennas.

Time Delay and Phase Steering

Applying signals of the form of (3) is called *time-delay steering* because the phase of the excitation signals exactly compensates for the time delay of a signal traveling in the projected distances ℓ_i. Time-delay steering results in a fully collimated beam at all frequencies, but is extremely expensive, bulky, and lossy since it depends on switching relatively long delay lines. For this reason true time delay is not often used at the array element level but more commonly incorporated into the feed circuits of arrays divided into subarrays. Examples of subarray excitation are described in Section 3.

Alternatively, at some fixed frequency f_0, with wavelength λ_0 and wave number $k_0 = 2\pi/\lambda_0$, phase weighting can be substituted for the time-delay steering. In such a case the weighting factors a_i are

$$a_i = |a_i| e^{-jk_0\mathbf{r}_i \cdot \hat{\varrho}_0} \tag{5}$$

and

$$\mathbf{E}_0(\mathbf{r}) = \frac{e^{-jkR_0}}{R_0} \sum_i \mathbf{f}_i(\theta, \phi)|a_i|e^{+j\mathbf{r}_i \cdot (k\hat{\varrho} - k_0\hat{\varrho}_0)}$$

which represents exact collimation only at fixed frequencies $\lambda = \lambda_0$ and is called *phase steering*. Most arrays are phase steered, but when wide operating bandwidths are required it may be necessary to investigate options for time delay. Section 3 describes several broadbanding approaches.

Examples of Array Collimation

Several examples of array collimation are given below for the arrays of Fig. 2. Note m and n are half-integers, $\pm\frac{1}{2}, \ldots$, to $\pm(N_y - 1)/2$ or $\pm(N_x - 1)/2$ for arrays with even numbers of elements, or integers, $0, \pm1, \ldots, (N_y - 1)/2$, etc.

Periodic Column Array in One Plane

$$\mathbf{r}_m = \hat{\mathbf{x}}x_m = \hat{\mathbf{x}}md_x, \qquad u_0 = \cos\theta_0 \tag{6}$$

Steering excitation:

$$a_m = |a_m|e^{-jk_0md_xu_0}$$

Radiation pattern:

$$\mathbf{E}(\mathbf{r}) = \frac{e^{-jkR_0}}{R_0} \sum_m \mathbf{f}_m(\theta, \phi)|a_m|e^{jmd_x(ku - k_0u_0)}$$

Periodic Two-Dimensional Array (Rectangular Lattice)

$$\mathbf{r}_{mn} = \hat{\mathbf{x}}md_x + \hat{\mathbf{y}}nd_y, \qquad u_0 = \sin\theta_0\cos\phi_0, \qquad v_0 = \sin\theta_0\sin\phi_0$$

Steering excitation:

$$a_{mn} = |a_{mn}|e^{-jk_0(md_xu_0 + nd_yv_0)} \tag{7}$$

Radiation pattern:

$$\mathbf{E}(\mathbf{r}) = \frac{e^{-jkR_0}}{R_0} \sum_{m,n} \mathbf{f}_{mn}(\theta, \phi)|a_{mn}|e^{j[md_x(ku - k_0u_0) + nd_y(kv - k_0v_0)]}$$

Circular Array Section

The circular array section of Fig. 2c is another characteristic array shape that requires a simple regular excitation vector to form a beam in the principal plane $(\theta, 0)$:

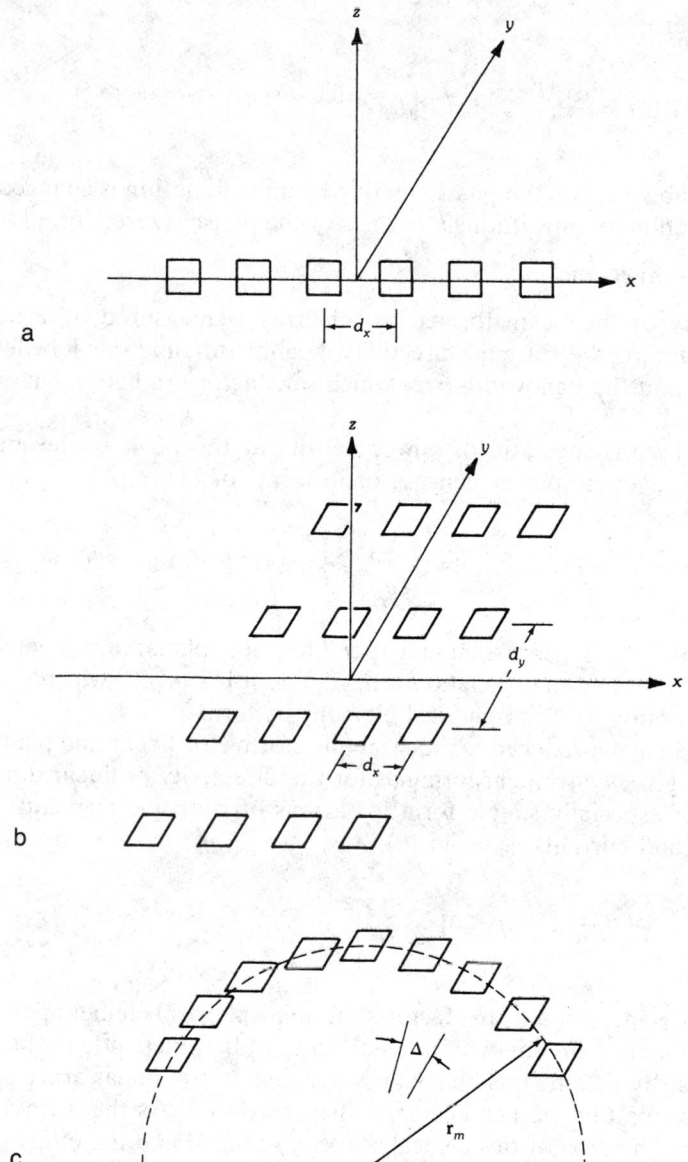

Fig. 2. Common array orientations. (*a*) One-dimensional array. (*b*) Two-dimensional rectangular grid array. (*c*) Sector of circular array.

$$\mathbf{r}_m = \hat{\mathbf{x}}a\cos\theta_m + \hat{\mathbf{y}}a\sin\theta_m, \qquad \hat{\boldsymbol{\varrho}} = \hat{\mathbf{x}}\cos\theta + \hat{\mathbf{y}}\sin\theta \tag{8}$$

Steering excitation:

$$a_m = |a_m|e^{-jk_0 a\cos(\theta_0 - \theta_m)}, \qquad \theta_m = m\Delta$$

Radiation pattern:

$$\mathbf{E}(\mathbf{r}) = \frac{e^{-jkR_0}}{R_0} \sum_m \mathbf{f}_m(\theta)|a_m| e^{ja[k\cos(\theta-\theta_m)-k_0\cos(\theta_0-\theta_m)]}$$

In each of the above cases the phase steering beam collimation is changed to time-delayed collimation by substituting $k = 2\pi/\lambda$ for the phase steered term $k_0 = 2\pi/\lambda_0$.

Quality of the Array Beam

The quality of the beam formed by an array is measured by a number of factors. Chief among these are the directivity, beamwidth, and side lobe level of the array pattern, and the bandwidth over which satisfactory radiation characteristics can be obtained.

The *directivity* is the ratio of power density at the peak of the main beam ($\mathbf{r} = \mathbf{r}_0$) to the average power density, or in terms of (1):

$$D = E(\mathbf{r}_0) E^*(\mathbf{r}_0) \bigg/ (4\pi)^{-2} \int_0^{2\pi} \int_0^{\pi} E(\mathbf{r}) E^*(\mathbf{r}) \sin\theta \, d\theta \, d\phi \qquad (9)$$

The integral over θ is often carried only to $\pi/2$ for most planar arrays with a ground screen, as it is assumed that radiation is negligible for $\theta > \pi/2$, which is consistent with the assumption of hemispherical element patterns.

Equation 9 can be reduced to much simpler forms for linear and planar arrays.

Elliott [1] gives convenient formulas for the directivity of linear dipole arrays and derives an especially simple form for arrays of isotropic elements with half-wave spacing and currents a_m as in (6).

$$D = \left(\sum a_m \right)^2 \bigg/ \sum a_m^2 \qquad (10)$$

The directivity is increased by the factor 2 for hemispherical element patterns. This expression shows the directivity of a linear array to be independent of scan angle. As pointed out by Elliott, this behavior is peculiar to the linear array and results from the broad pattern perpendicular to the array axis. As the array is scanned toward end-fire the area of this conical shape is reduced and the effect offsets the beam broadening in the plane of scan that tends to reduce gain.

Fig. 3 shows the broadside directivity (dashed curve) of a linear array of eight isotropic elements with uniform excitation for various array element spacings [2]. It indicates that directivity is a severe function of element spacing. The solid curve shown in the figure is the maximum directivity for the array, as derived by Tai, and serves to indicate that the directivity of a broadside array is very nearly optimum except in the supergain region ($d/\lambda < 0.5$). The reduced directivity occurring near $d/\lambda = 1$ results directly from the radiation of an additional set of primary lobes, called *grating lobes*, which will be discussed in detail later. Since the presence of grating lobes is unacceptable for most applications, the directivity of a uniformly

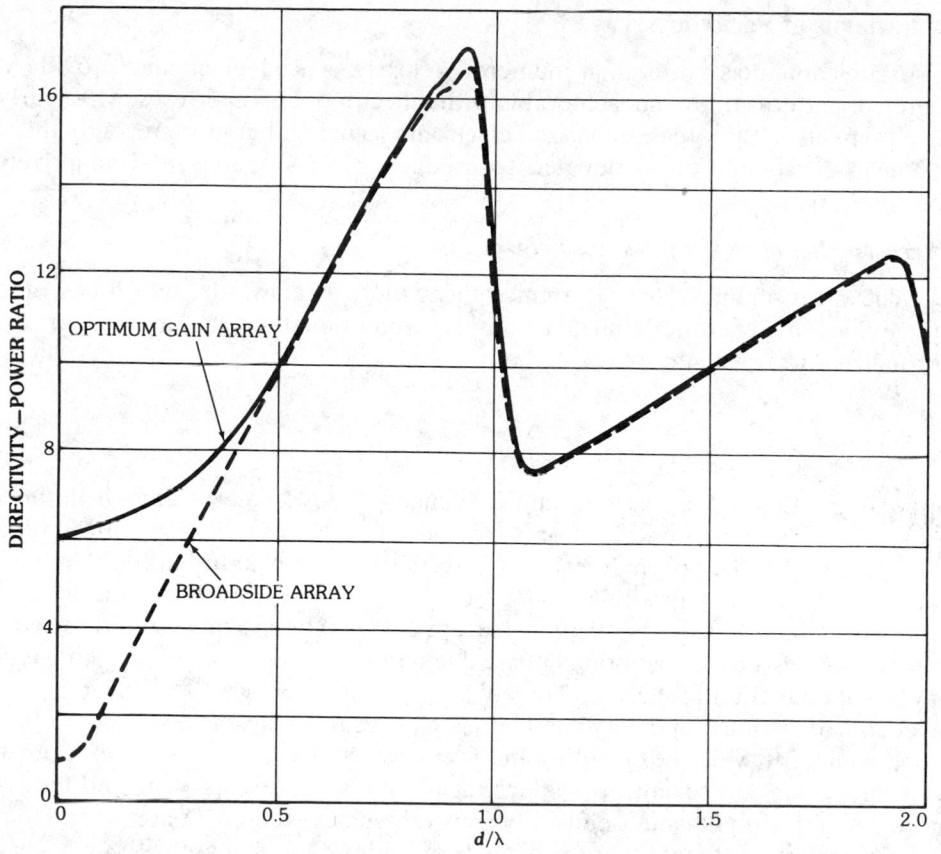

Fig. 3. Array gain for eight isotropic elements versus element spacing. (*After Tai [2],* © *1964 IEEE*)

illuminated roadside array is well approximated by the following expression throughout the linear (useful) portion of the curve shown in Fig. 3:

$$D \cong 2Nd/\lambda \tag{11}$$

The directivity of an ideal planar array (one that is perfectly matched and has $\cos\theta$ hemispherical element patterns) is given approximately by

$$D = 32\,400/B \tag{12}$$

where B is the 3-dB beamwidth product for scan angle θ

$$B = \theta_{x_3}\theta_{y_3}\sec\theta \tag{13}$$

with beamwidths given in degrees.

2. Patterns of Periodic Arrays

Array antennas, with their numerous closely spaced elements, provide a degree of pattern control not achievable with reflector or lens apertures. Most early developments in this area emphasized electronic scanning, but more recently there has been substantial effort devoted to producing low side lobe and adaptively controlled patterns.

Characteristics of an Array Scanned in One Plane

The array illumination determines the pattern beamwidth, side lobes, and directivity. In general, the beamwidth is proportional to the inverse of the normalized array length, or

$$\Delta\theta = K\lambda/L \tag{14}$$

with $\Delta\theta$ in radians, K a constant, and L defined to be Nd, for N elements in the θ plane. This equation gives the half-power beamwidth of this radiation pattern for a linear array or in the principal planes of a rectangular array at broadside. Uniform illumination ($|a_i| = 1$), with $K = 0.886$, produces the highest directivity and narrowest beam of any illumination (except for certain special "superdirective" illuminations associated with rapid phase fluctuations and closely spaced elements), but this illumination produces relatively high side lobe patterns (about -13 dB). Selection of various tapered illuminations can result in much lower side lobes, accompanied by wider beamwidths and lower directivity. As the array is scanned from broadside the beamwidth widens, again like $\sec\theta_0$ except near end-fire. A more general formula and a plot of beamwidth versus scan angle are given later, but it is important to note that the array beamwidth does increase with scan when observed in the (θ, ϕ) space. The solid-line curves of Fig. 4 show the radiation pattern of an eight-element, one-dimensional array at broadside and scanned to 45° ($u = 0.7071$). The element spacing is 0.5λ. The comparison shows that the array pattern is invariant in the parameter $(d/\lambda)(u - u_0)$, so that no beam broadening or other pattern change is evident when plotted in direction cosine space. The observed $\sec\theta$ beam broadening factor for large arrays is thus the result of mapping the uniform beamwidth in u space onto the θ plane. The advantage of plotting antenna patterns in u ($\sin\theta$) space is that as far as the pattern function is concerned, there is no need to recompute patterns for any other scan angle. Later we shall see that mutual coupling introduces angle-dependent effects that can drastically alter the radiation characteristics.

Similarly, the form of the pattern is not dependent on the spacing d/λ, except that its scale in $u - u_0$ space expands or contracts with the choice of spacing d/λ. This subject is addressed in the next sections.

Array Lattice Spacings and Comparison With Continuous Line Source—The 0.5λ spacing used in the array of Fig. 4 was chosen to minimize the effects of periodicity and makes that array pattern very little different from that of a continuous aperture. Normalized broadside radiation patterns for an N-element array [of actual length $(N - 1)d_x$] and a line source of length L are given as follows:

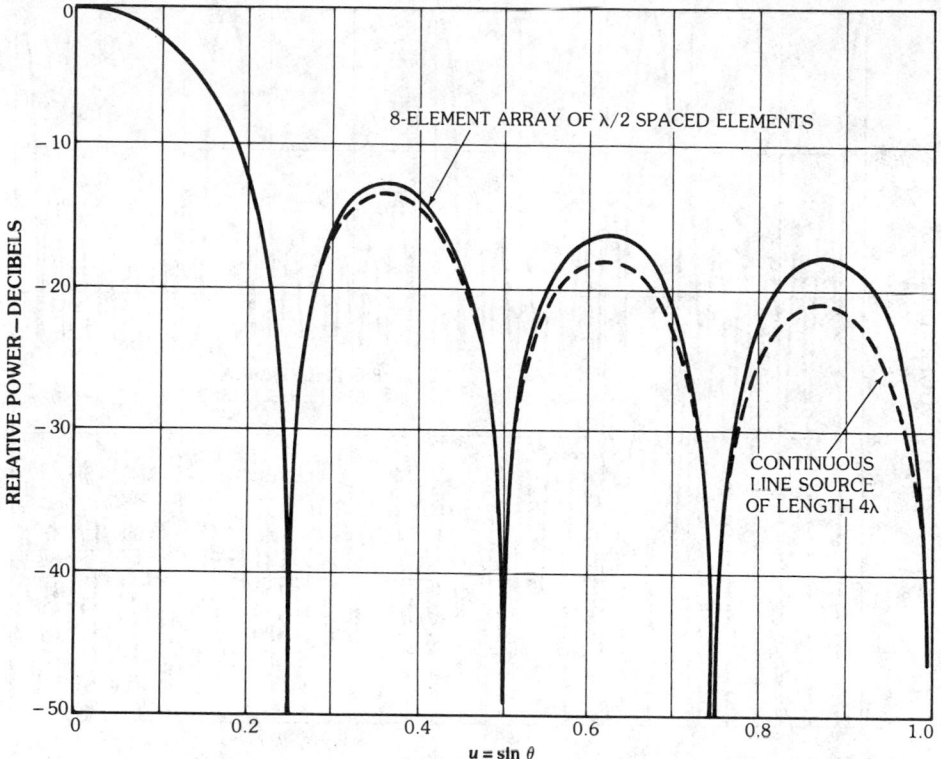

Fig. 4. Patterns of uniformly illuminated array and line source.

Linear array:

$$f(\theta) = \frac{\sin(N\pi d_x u/\lambda)}{N\sin \pi d_x u/\lambda} \tag{15}$$

Line source:

$$f(\theta) = \frac{\sin(L\pi u/\lambda)}{L\pi u/\lambda} \tag{16}$$

The radiation patterns for a continuous line source of length 4λ and an eight-element array of $\lambda/2$ spaced elements with uniform excitation are shown in Fig. 4. The line source pattern differs very little from the array pattern up to the second side lobe, and the null positions are unchanged, since these are determined from the numerators of the two expressions in (15) and (16). The similarity of these expressions makes it convenient to define the length parameter $L = Nd$ for arrays.

Fig. 5 shows the pattern (solid) of an eight-element array with 4λ spacings, as compared with the (dashed) pattern of a continuous line source with $L = 32\lambda$. Here the dramatic difference is the occurrence of lobes in the periodic array pattern at the zeroes of the denominator of (15). These lobes, which are called *grating*

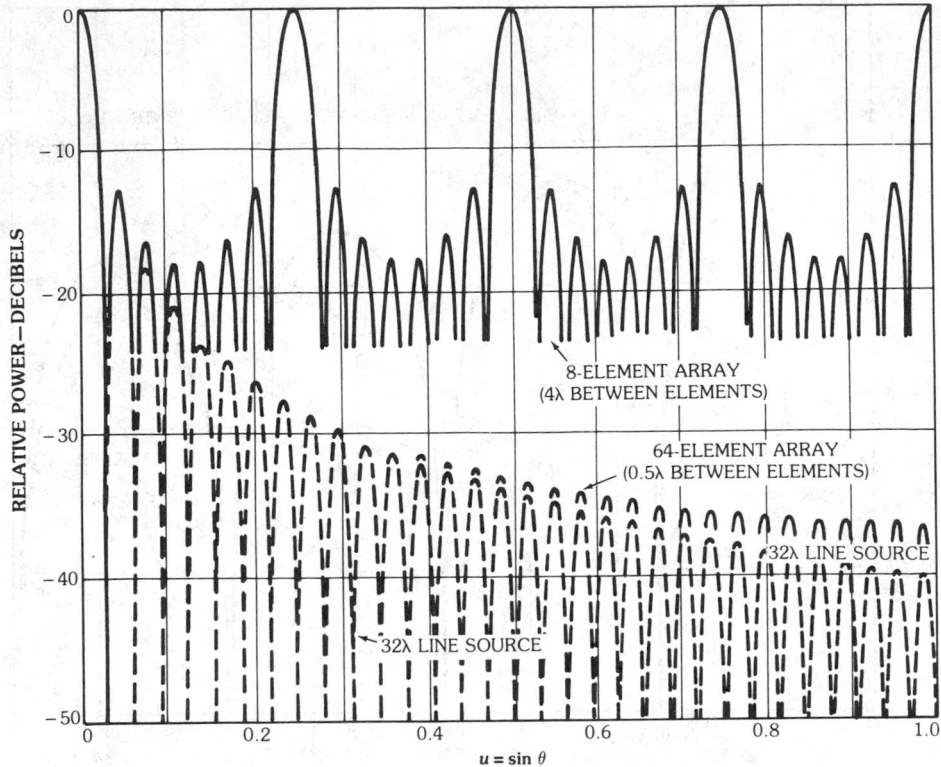

Fig. 5. Patterns of uniformly illuminated 64-element array with 0.5λ spacing, 32λ line source, and 8-element array with 0.5λ spacing.

lobes, have the same peak value as the main beam and are located at distances $p\lambda/d_x$ from the main beam in u space, for all integers p that define angles in real space ($|u| \leqq 1$). Grating lobes are a direct consequence of the periodicity and occur independently of any chosen amplitude distribution across the array. When the array is scanned to u_0 they occur at angles

$$u_p = u_0 + p\frac{\lambda}{d_x}, \qquad \text{for } p = \pm 1, \pm 2, \cdots \qquad (17)$$

subject to $-1 \leqq u_p \leqq +1$.

The array factor is thus completely periodic in u space (for $|u| \leqq 1$), with period equal to grating lobe separation (λ/d_x). The region $|u| \leqq 1$ is called *real space* because $\sin\theta$ has a geometric interpretation in this regime. Before leaving the comparison of continuous apertures and discrete arrays it should be noted that the similarity is maintained almost halfway between the main beam and the nearest grating lobe, because it is on this scale that the pattern of the periodic array is repeated. Fig. 5 also shows (solid) the side lobe peaks of the 64-element array with 0.5λ spacing.

Fig. 6 further emphasizes the pattern invariance in the scale parameters

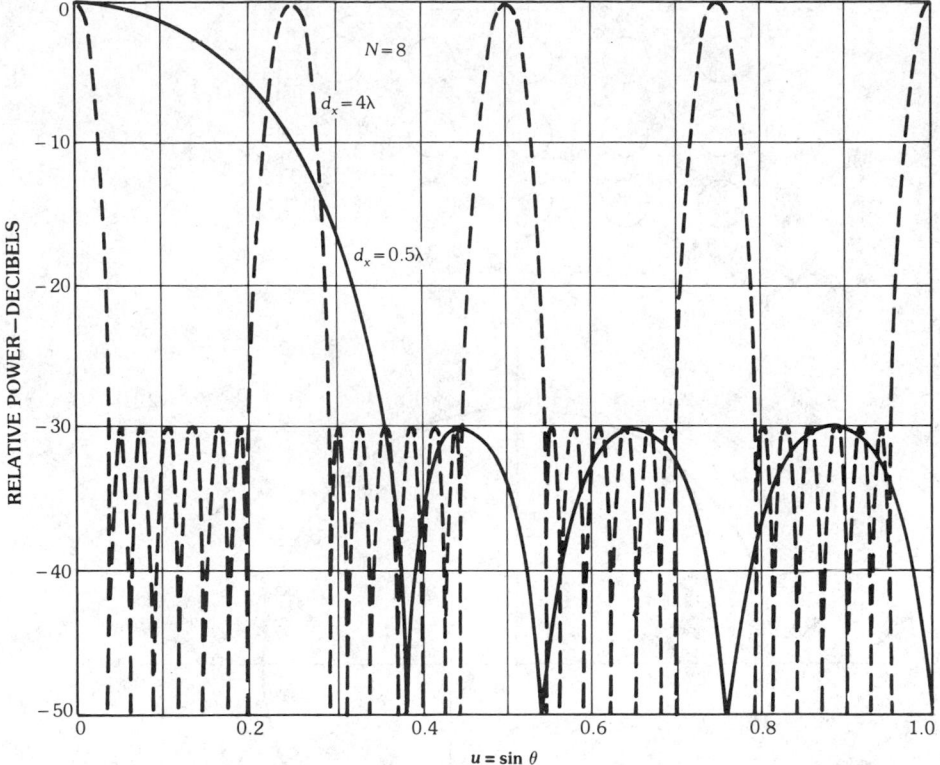

Fig. 6. Patterns of eight-element arrays with −30-dB Chebyshev illuminations: 0.5λ spacing, and 4λ spacing.

$d_x(u - u_0)/\lambda$ by showing the patterns of two eight-element arrays with one of the low side lobe excitations that will be described in a later section, a −30-dB Chebyshev taper. The two patterns are for arrays with $d_x = 0.5\lambda$ and 4λ. Increasing d_x shrinks the scale of the pattern and so brings additional grating lobes into real space but produces no other change in the pattern.

Scanning in Two Planes

Equation 7 gives the pattern of a rectangular planar array like that of Fig. 7a scanned in two planes. The occurrence of grating lobes in rectangular-grid two-dimensional arrays is directly apparent from (7), where the substitutions

$$u_p = u_0 + p\frac{\lambda}{d_x}, \qquad v_q = v_0 + q\frac{\lambda}{d_y} \qquad (18)$$

leave these expressions unchanged. Not all values of q and p correspond to allowed angles of radiation, however, for the direction θ of radiation measured from the array normal is given by

$$\cos\theta_{pq} = \sqrt{1 - u_p^2 - v_q^2} \qquad (19)$$

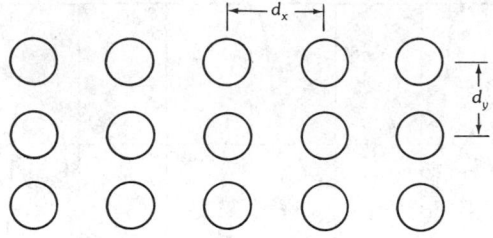

Rectangular Grid Array

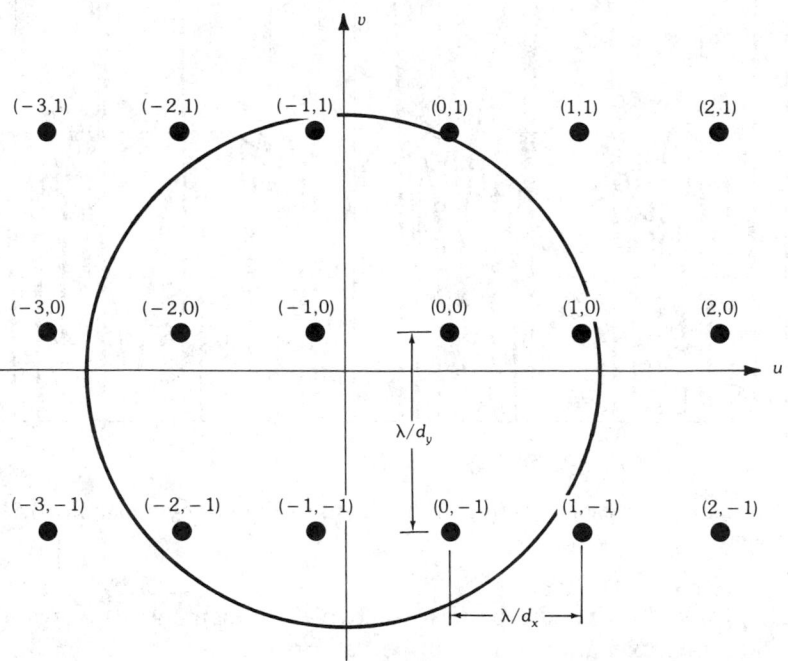

a

Fig. 7. Rectangular and triangular grid planar arrays and their grating lobe lattices. (*a*) Grating lobe lattice for rectangular grid. (*b*) Grating lobe lattice for triangular grid.

and so real angles of radiation θ_{pq} require that the allowed values of u_p and v_q are constrained by the condition

$$u_p^2 + v_q^2 < 1 \qquad\qquad (20)$$

These points are shown in (u, v) space as a regularly spaced grating lobe lattice about the main beam location (u_0, v_0) in Fig. 7a. The circle with unity radius represents the bounds of the above inequality; all grating lobes within the circle represent those radiating into real space, and those outside do not radiate.

Fig. 7b shows a triangular array lattice and pertinent grating lobe locations for that lattice. In this case, (5) is still valid and there are still grating lobes, but the nearest lobes in the azimuth scan plane are removed. The grating lobe locations shown in Fig. 7b are given by

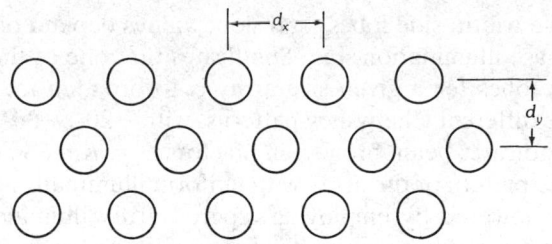

Triangular Grid Array

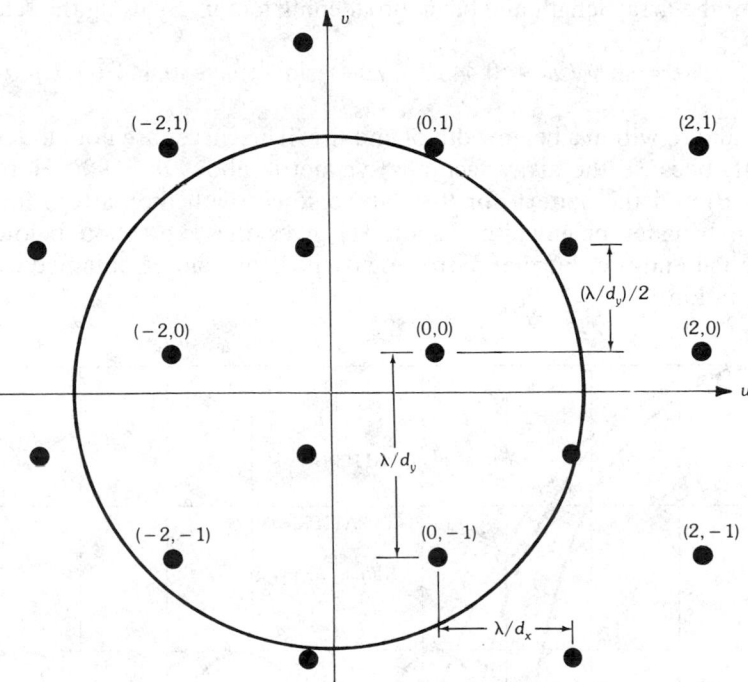

b

Fig. 7, *continued.*

$$u_p = u_0 + p\frac{\lambda}{d_x}$$

$$v_q = \begin{cases} v_0 + q\dfrac{\lambda}{d_y} & \text{for } p = 0, \pm 2, \cdots \\[2ex] v_0 + \left(q - \dfrac{1}{2}\right)\dfrac{\lambda}{d_y} & \text{for } p = \pm 1, \pm 3, \cdots \end{cases} \tag{21}$$

Pattern Shape and Beam Broadening

In the previous section it was pointed out that for a periodic array the pattern shape does not change with scan angle if it is plotted in *u* space, and changes in array spacing alter only the scale (width) of the pattern. The actual shape of the pattern, its relative width, side lobes, and slope values depend only on the applied amplitude and phase illumination. Fig. 8 indicates that one of the major results of reducing the side lobes for a given size array is to broaden the beamwidth. The figure shows three different Chebyshev patterns, with -20-, -30-, and -40-dB side lobe levels, and indicates beam broadening factors $B = K/0.886$ of 1.12, 1.29, and 1.43 relative to the pattern of the array with uniform illumination. In general, then, side lobes can be lowered by employing tapered array illuminations, but this is achieved at the expense of broadening the beamwidth and reducing the array gain. Fig. 9 gives the beamwidth in degrees for a scanned array with arbitrary taper as a function of the array length and beam broadening factor. By using the relationship

$$\Delta\theta = \sin^{-1}(u_0 + 0.443B\lambda_0/L) - \sin^{-1}(u_0 - 0.443B\lambda_0/L) \qquad (22)$$

which is valid to within a beamwidth of end-fire. The curves are not plotted beyond that point, because the array factor is symmetric about $\theta = 90°$. Here $F(\theta) = F(180° - \theta)$ and the pattern for $\theta < 90°$ coalesces with the pattern for $\theta > 90°$ to form a broader beamwidth. Elliott [1] gives the expression below for the beamwidth at end-fire. Further narrowing of the beam can be obtained using slow-wave excitation [3]:

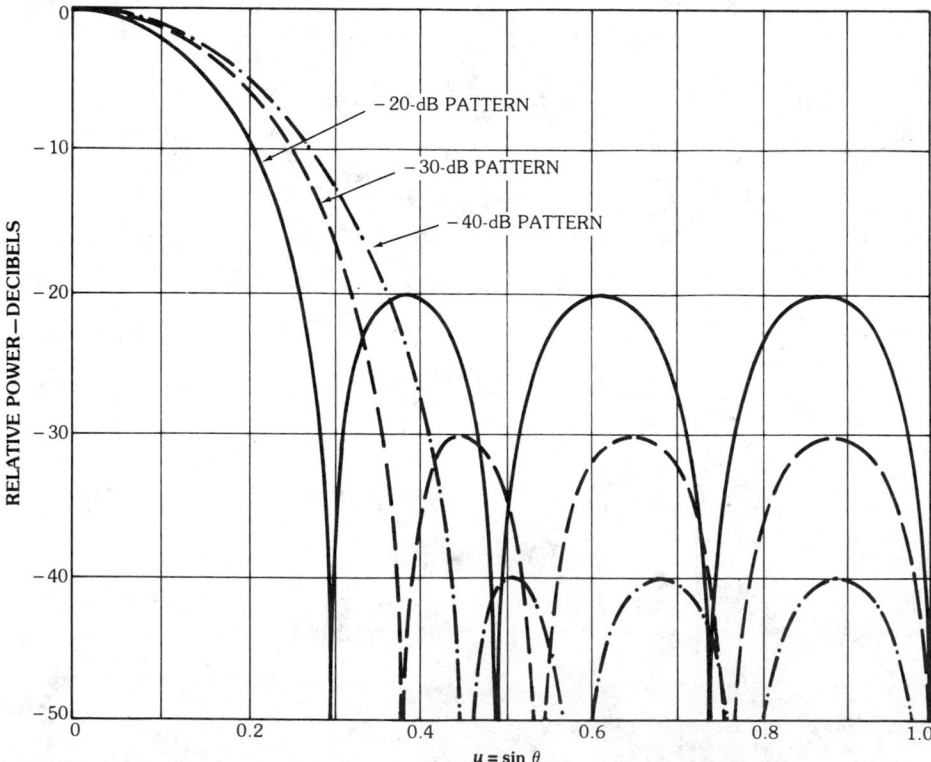

Fig. 8. Chebyshev patterns for an 8-element array ($\lambda/2$ spacing) with illuminations for -20 dB, -30 dB, and -40 dB.

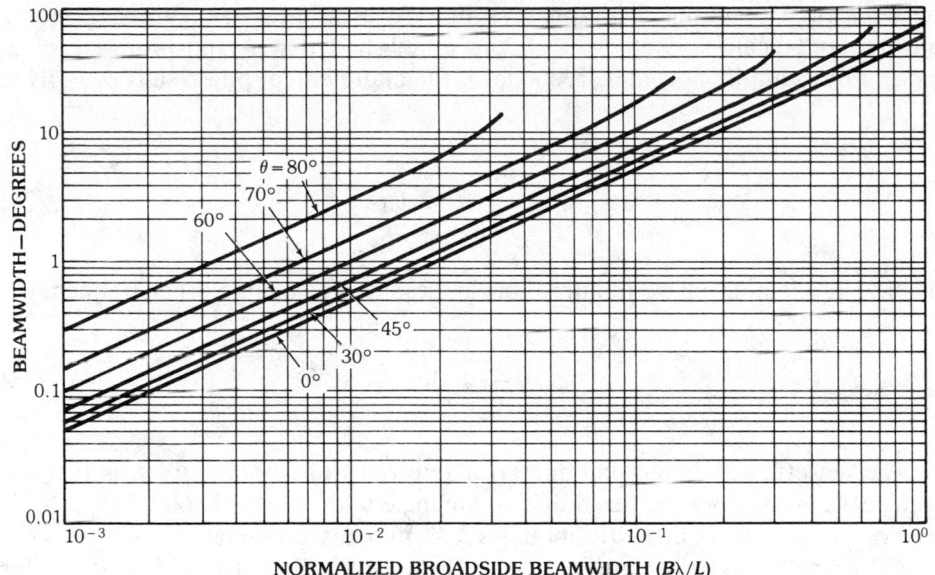

Fig. 9. Beamwidth versus normalized broadside beamwidth for a scanned array.

$$\Delta\theta = 2\cos^{-1}(1 - 0.443B\lambda_0/L) \tag{23}$$

Phased Array Bandwidth

As indicated in Section 1, most arrays are designed using phase shifters, not time-delay units. A phase-steered array establishes a progressive phase front to match a wave at a single frequency. At a different frequency the progressive phase front corresponds to a wave at a different angle. This effect is shown schematically in Fig. 10.

Equations 3 and 5 give array excitation coefficients for time-delay and phase-

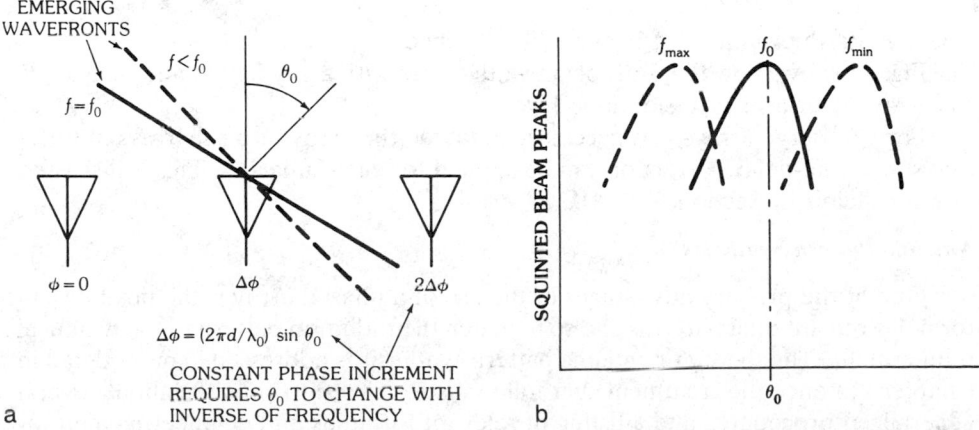

Fig. 10. Beam squint for a phase-scanned array. (*a*) Wavefronts. (*b*) Beam-pointing errors.

steered arrays. Scanning with phase shifters causes the peak gain angle of a phase-steered beam to not occur at the established (θ_0, ϕ_0) position except at $\lambda = \lambda_0$. For example, at center frequency f_0 the interelement phase shift ϕ_0 is given by

$$\phi_0 = \frac{2\pi}{\lambda_0} d_x \sin \theta_0 \tag{24}$$

but the interelement phase required for this scan angle at some other frequency is

$$\phi = \frac{2\pi}{\lambda} d_x \sin \theta_0 \tag{25}$$

The net effect, a beam-pointing error referred to as *beam squint*, is that the beam scans away from the desired θ_0 to an angle with $\sin \theta = (\lambda/\lambda_0) \sin \theta_0$; this is the most significant bandwidth limiting effect in array antennas.

Assuming an approximate half-power beamwidth $K\lambda/L$ (note $K = 0.886$ for uniform illumination) one can solve for the array bandwidth under the assumption that the gain at each frequency limit is reduced to half-power (that the squint is equal to a half-beamwidth at each limit). The resulting fractional bandwidth is given by

$$\frac{\Delta f}{f_0} = 0.886B \left(\frac{\lambda_0}{L} \right) \frac{1}{\sin \theta} \tag{26}$$

and for small scan angles

$$\frac{\Delta f}{f_0} \cong \frac{1}{n_B} \tag{27}$$

where n_B is the number of beamwidths scanned.

Fig. 11 shows how the 3-dB beamwidth varies with array length and scan angle for arrays with arbitrary side lobe levels.

For very large arrays it is necessary to divide the array into subarrays in order that some time-delay correction can be applied for each subarray. This is discussed in more detail in Section 3.

Antenna Pattern Synthesis

One of the primary advantages of the use of a phased array is the flexibility to form desired antenna patterns and so to match the radiation pattern to the technical requirement. The theory of antenna pattern synthesis is addressed in more detail in Chapter 2, hence the treatment that follows is restricted to remarks about several generalized procedures and a listing of relevant formulas and engineering data for the special case of low side lobe methods.

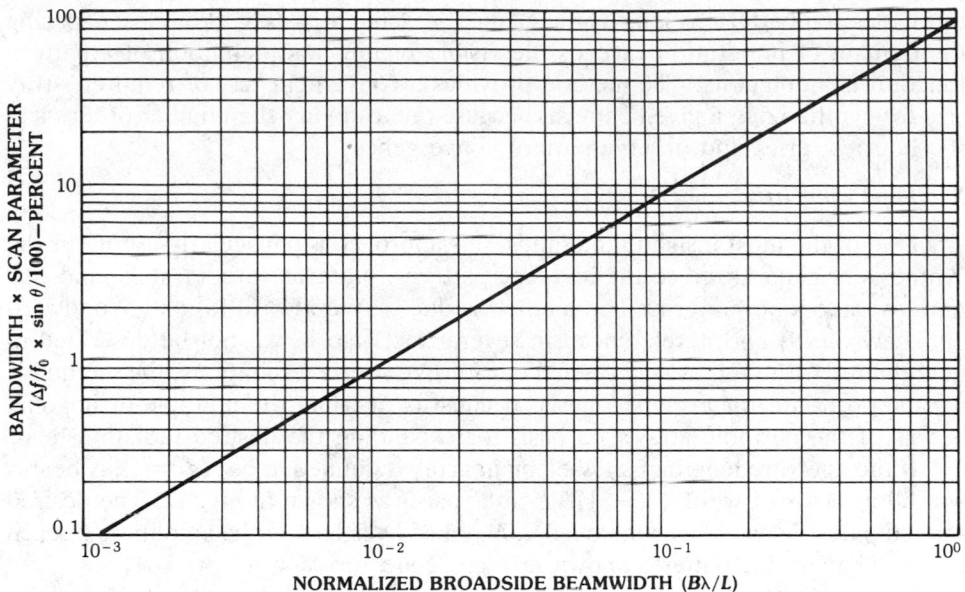

Fig. 11. Bandwidth scan product versus normalized broadside beamwidth for a scanned array.

General Procedures

Schelkunoff's method [4], introduced in Chapter 3, is based on the manipulation of zeros of the pattern function in the complex plane. The procedure is very useful of itself, and in addition is the foundation of computer-based methods with great power and flexibility. Elliott [5, 6] has described a generalized method based on a null perturbation approach to modify the side lobe structure of Taylor and Bayliss line source patterns. The procedure develops simultaneous linear equations solved by matrix inversion for a perturbed pattern and is repeated iteratively to converge to the desired result.

Fourier series methods can be readily applied to array synthesis problems by recognizing that the pattern function

$$E(u) = \sum_{m=-(M-1)/2}^{(M-1)/2} a_m e^{jkmud_x} \qquad m = \begin{cases} \pm\frac{1}{2}, \pm\frac{3}{2}, \ldots, & \text{for } M \text{ even} \\ 0, \pm 1, \ldots, & \text{for } M \text{ odd} \end{cases} \qquad (28)$$

is a finite Fourier series and is periodic in u space with the interval of the grating lobe distance λ/d_x. Thus, given a desired field distribution $E(u)$, one can obtain an expression for the currents from orthogonality, with

$$a_m = \frac{d_x}{\lambda} \int_{-\lambda/2d_x}^{\lambda/2d_x} e^{-j(2\pi/\lambda)ud_x m} E(u)\, du \qquad (29)$$

Stutzman and Thiele [7] show the application of this procedure to synthesis of a sector pattern. The method provides the least mean squared error approximation to the desired pattern as long as the element spacing $d_x \geqq 0.5\lambda$. For closer spacing the domain of integration exceeds the visible region, and definition of a pattern function is ambiguous. The method provides a convenient test of required array size for synthesizing a given pattern, because one can vary the number of terms in the Fourier series and observe pattern convergence.

Synthesis with Orthogonal Beams

One of the most insightful methods, this approach is particularly useful for the synthesis of generalized sector coverage patterns, patterns without nulls, and certain low side lobe patterns. The method is due to Woodward [8] and Woodward and Lawson [9] and is referenced in several texts, so it will not be described in detail here. Although usually carried out for line source excitations, the method is very appropriate for array synthesis. It consists of using the multiple orthogonal beams of the periodic array as a basis for expanding the desired radiation field.

If the aperture length of an N-element array is defined to be $Nd_x = L$, N beams will fill a sector of width $(N - 1)\lambda/L$ in u space, as shown in Fig. 12. The desired shaped pattern can then be matched at N points by selecting the amplitude of each beam. The specific patterns shown in Fig. 12 are for $N = 8$.

To excite the ith beam the elements are excited by progressive phase distribution

$$a_n = e^{-j(2\pi/\lambda)d_x u_i n} \tag{30}$$

where

$$u_i = (\lambda/L)i = (1/N)(\lambda/d_x)i \quad \text{and} \quad i, n = \pm\tfrac{1}{2}, \cdots, \pm(N - 1)/2$$

The ith beam is given by

$$g_i(u) = \sum_{n=-(N-1)/2}^{(N-1)/2} e^{j(2\pi/\lambda)x_n(u-u_i)}$$

$$= N\left\{\frac{\sin\left[(N\pi d_x/\lambda)(u - u_i)\right]}{N\sin\left[(\pi d_x/\lambda)(u - u_i)\right]}\right\} \tag{31}$$

The set of beams $g_i(u)$ is also orthogonal and occupies the beam positions shown in Fig. 12 (the example was done for $N = 8$). Fig. 12b shows two of the normalized orthogonal beams ($i = -7/2$ and $i = +1/2$) and clearly indicates that the domain of pattern synthesis must be restricted to $|Nd_x \sin\theta| \leqq 0.5(N - 1)$, for beyond that the grating lobes of the outermost beams present an ambiguity that leads to significant pattern distortion. One of the most significant results of synthesis with orthogonal beams is that the resultant feed networks can be lossless, since the progressive phase sequences can be formed by lossless Butler matrices or orthogonal beam lenses.

Section 3 shows an example of the use of orthogonal beams in a subarraying feed configuration to synthesize a pulse-type subarray pattern.

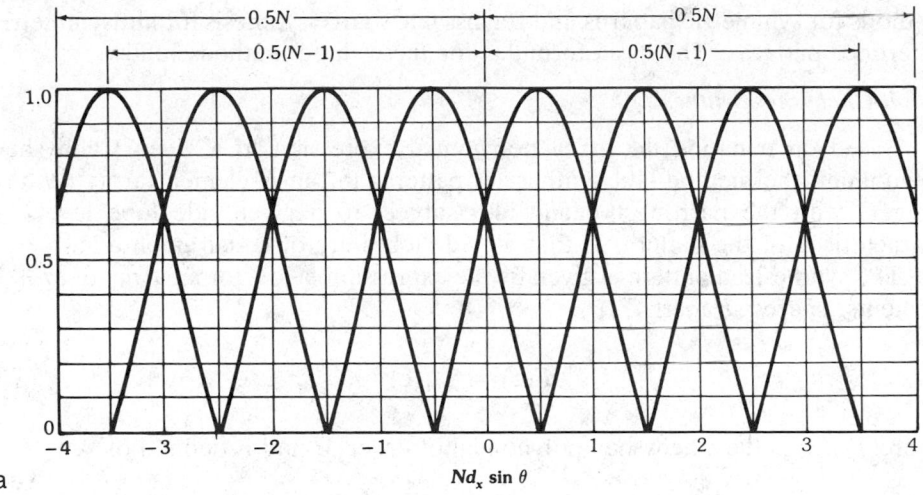

a

$Nd_x \sin \theta$

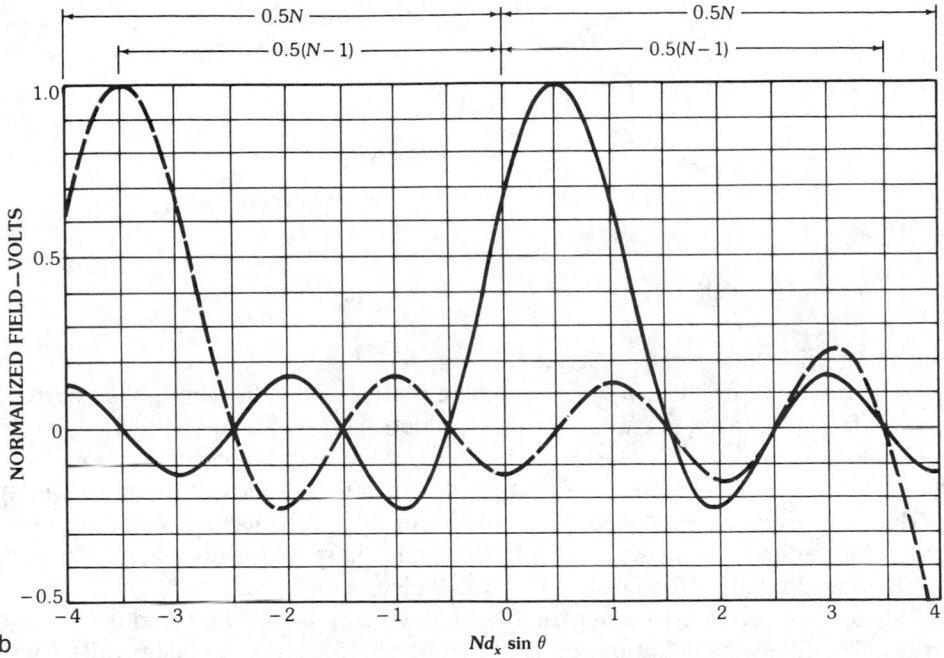

b

$Nd_x \sin \theta$

Fig. 12. Orthogonal beams for an N-element array. (a) Complete set of beams (plotted to first zeros). (b) Two orthogonal beams (plotted over domain of orthogonality) with $i = 1/2$ (solid) and $i = -7/2$ (dashed).

Low Side Lobe Solutions: Basic Formulas and Engineering Data

Among the most important and useful synthesis procedures are those that result in minimum beamwidth consistent with very low side lobes. From this class of methods three of the most important are Chebyshev and Taylor line source methods for symmetric patterns and Bayliss line source synthesis for antisymmetric difference patterns. The basic formulas for these three methods follow.

Dolph-Chebyshev Synthesis

Based on equating the array polynomial expression to a given Chebyshev polynomial, the method [10] synthesizes patterns for an N-element array (N any integer) with the narrowest beamwidth subject to a given side lobe level. A characteristic of the patterns is that all side lobes are of equal amplitude.

The synthesized pattern is given by the expression below for an array of $M + 1$ elements, spaced d apart:

$$F(z) = T_M(z) \tag{32}$$

where $T_M(z)$ is the Chebyshev polynomial of order M and is defined by

$$\begin{aligned} T_M(z) &= \cos(M\cos^{-1}z) && \text{for } |z| \leqq 1 \\ &= \cosh(M\cosh^{-1}z) && \text{for } |z| \geqq 1 \end{aligned} \tag{33}$$

The parameter z is given by

$$z = z_0\cos[(\pi d/\lambda)\sin\theta] \tag{34}$$

where

$$z_0 = \cosh(M^{-1}\cosh^{-1}R)$$

for voltage side lobe level R ($\mathrm{SL_{dB}} = 20\log_{10}R$).

The original formulation of this problem is attributable to Dolph, who derived results for $\lambda/2$-spaced elements. Later Riblett [11] extended the analysis to elements greater than $\lambda/2$. Barbiere [12] derived the expression for z_0 above, and derived convenient relations for the currents. Stegen [13] derived the most widely used exact expressions for the required current distribution, and extensive tabulations of his results are given by Brown and Scharp [14], who also include gain and beamwidth values for arrays of up to 40 elements.

Stegen also gives a formula for Chebyshev array beamwidth, valid for large arrays. The following equation, derived by Drane [15], gives the beamwidth (here converted to degrees) for an array of length L with side lobe level $|\mathrm{SL}|$(dB)

$$\mathrm{BW} = \left(\frac{\lambda}{L}\right)(10.314)(|\mathrm{SL}| + 4.52)^{1/2} \tag{35}$$

Fig. 13 shows the normalized beamwidth parameter $(L/\lambda)\mathrm{BW}$ (in degrees) as a function of side lobe level R as computed from the above equation. Drane also gives an equation for directivity for array spacings greater or less than a half-wavelength. For spacings greater than $\lambda/2$ the directivity D is [15]

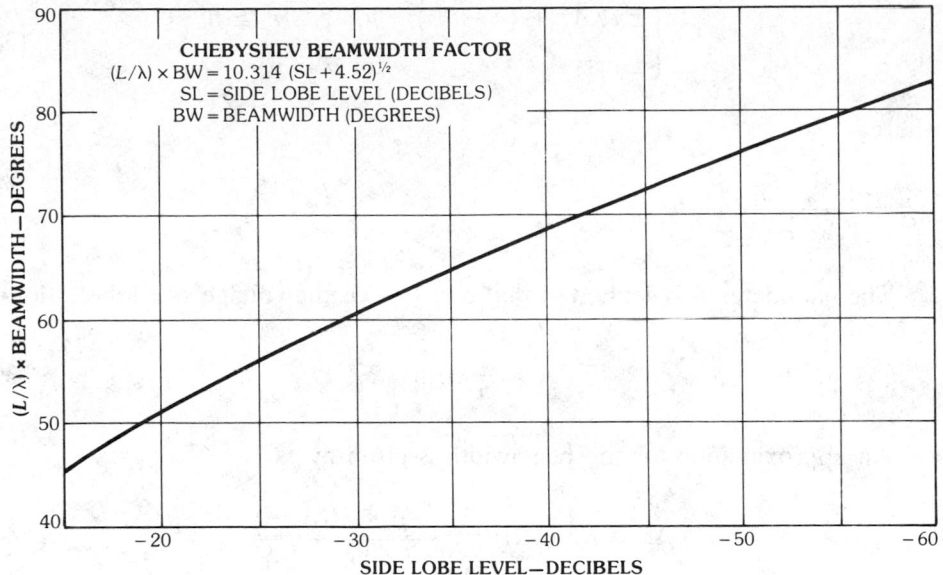

Fig. 13. Normalized beamwidth versus side lobe level for Chebyshev arrays. (*After Drane [15], © 1968 IEEE*)

$$D \cong \frac{2R^2}{1 + (\lambda/L)R^2\sqrt{(\ln 2R)/\pi}} \tag{36}$$

This result is in close agreement with the equation of Elliott [1]. For large arrays this directivity reaches the maximum value $2R^2$, or 3 dB greater than the specific side lobe level.

Chebyshev patterns have been used for purposes of illustration throughout this chapter, and so no additional patterns will be shown here.

Taylor Line Source Synthesis

The classic paper by Taylor [16] investigated the synthesis of equal side lobe patterns with continuous line source excitations and satisfying the same criteria as do the Chebyshev linear array patterns. He showed that the idealized pattern is the solution to this problem is not physically realizable but could be approximated arbitrarily closely by a function of two parameters, A and $\bar{n}$. The family of patterns derived by Taylor has the first $\bar{n}$ side lobes at the desired level, and all side lobes beyond the $\bar{n}$th fall off as $(\sin \pi z)/\pi z$ (for $z = Lu/\lambda$) for a line source of length L.

The synthesized pattern, normalized to unity, is given by

$$F(z, A, \bar{n}) = \frac{\sin \pi z}{\pi z} \prod_{n=1}^{\bar{n}-1} \frac{1 - z^2/z_n^2}{1 - z^2/n^2} \tag{37}$$

for $z = (L/\lambda)u$.

The zeros of the function are at

$$z_n = \begin{cases} \pm\sigma\sqrt{A^2 + (n - \frac{1}{2})^2} & \text{for } 1 \leq n \leq \bar{n} \\ \pm n & \text{for } \bar{n} \leq n < \infty \end{cases} \qquad (38)$$

where

$$\beta = \frac{\bar{n}}{\sqrt{A^2 + (\bar{n} - \frac{1}{2})^2}}$$

The parameter A is defined so that $\cosh(\pi A)$ is the voltage side lobe ratio, or

$$A = \frac{1}{\pi}\cosh^{-1}R \qquad (39)$$

An approximation for the beamwidth is given by

$$\Delta\theta \cong \sigma\beta_0(\lambda/L) = \frac{\beta_0\bar{n}(\lambda/L)}{\sqrt{A^2 + (\bar{n} - \frac{1}{2})^2}} \qquad (40)$$

where

$$\beta_0 = (2/\pi)\sqrt{(\cosh^{-1}R)^2 - (\cosh^{-1}R/\sqrt{2})^2}$$

The current or aperture field distribution necessary to produce this pattern family is

$$g(x) \doteq F(0, A, \bar{n}) + 2\sum_{m=1}^{\bar{n}-1} F(m, A, \bar{n})\cos(2m\pi x/L) \qquad \text{for } -L/2 \leq x \leq L/2 \quad (41)$$

where the coefficients $F(z, A, \bar{n})$ are the pattern values of (37) at the integers $(z = m \leq \bar{n})$. In abbreviated form these are

$$F(m, A, \bar{n}) = \frac{[(\bar{n} - 1)!]^2}{(\bar{n} - 1 + m)!(\bar{n} - 1 - m)!}\prod_{n=1}^{\bar{n}-1}(1 - m^2/z_n^2) \qquad (42)$$

These coefficients $F(m, A, \bar{n})$ are tabulated in Hansen [17] for various n for side lobe levels between -30 and -40 dB in 5-dB increments.

Fig. 14 shows two Taylor patterns. The solid one is the line source pattern computed from (37), while the dashed curve is the radiation pattern of the $g(x)$ distribution sampled at the points: $(L/N)i$ for $\pm i = \frac{1}{2}, \frac{3}{2}, \ldots, (N-1)/2$ for an array of N elements using the tabulated values of F. The patterns are for a -30-dB Taylor distribution with $\bar{n} = 4$ and $N = 16$. The figure indicates that the sampled line source distribution is a very good approximation of the line source pattern, even for an array with only 16 elements.

The practical selection of the $\bar{n}$ parameter is discussed by Taylor. This parameter must be large enough or the beam will be broader than necessary; $\bar{n}$

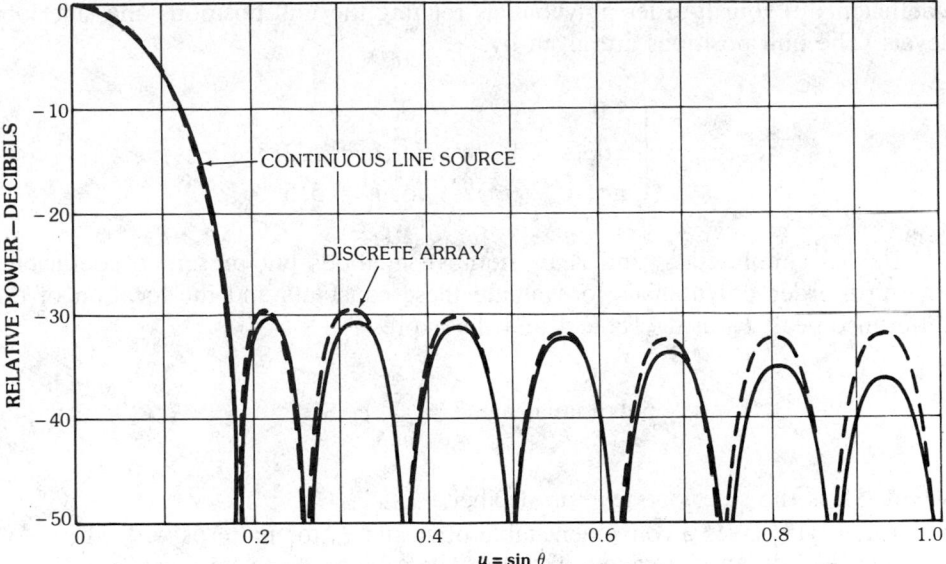

Fig. 14. Taylor $\bar{n} = 4$ line source patterns: continuous line source and discrete array pattern ($N = 16$) using sampled Taylor illumination.

should be at least 3 for −25-dB side lobes, and at least 6 for −40 dB. Increasing $\bar{n}$ further narrows the main beam, but supergaining results if it is made too large.

Bayliss Line Source Synthesis

E. T. Bayliss [18] has developed a method of synthesizing line source difference patterns that parallel the essential features of Taylor line source patterns. The difference pattern is fully described by two parameters, A and $\bar{n}$, which completely control the side lobe level and decay behavior. Bayliss' method results in a pattern of the following form:

$$F(z) = \pi z \cos \pi z \prod_{n=1}^{\bar{n}-1} [1 - (z/\sigma z_n)^2] \Big/ \prod_{n=0}^{\bar{n}-1} \{1 - [z/(n + \frac{1}{2})]\} \qquad (43)$$

where

$$z = (L/\lambda)u$$
$$\sigma = \frac{\bar{n} + \frac{1}{2}}{z_{\bar{n}}}$$
$$z_{\bar{n}} = (A^2 + \bar{n}^2)^{1/2}$$

Here, as in the Taylor method, the first few side lobes are at the design level, and beyond about $z = \bar{n}$ the side lobes decay as $z^{-3/2}$. Unlike the Taylor line source method, the z_n terms are not available in closed form but are given in terms of the coefficients of fourth-order polynomials relating the null positions and side lobe levels. The null positions are given by:

$$\sigma z_n = \begin{cases} 0 & \text{for } n = 0 \\ \pm\sigma\xi_n & \text{for } n = 1, 2, 3, 4 \\ \pm\sigma(A^2 + n^2)^{1/2} & \text{for } n = 5, 6, \ldots \end{cases} \tag{44}$$

Bayliss computed ξ_n and A by iterative methods but presented coefficients of fourth-order polynomials to evaluate these constants and the location of the difference peak P_0 using Table 1 and the expression

$$\text{polynomial name} = \sum_{n=0}^{4} c_n (\text{SL})^n \tag{45}$$

where SL is the side lobe level in decibels.

Elliott [19] gives a convenient table of A and ξ_n for patterns with SL = -15 through -40 dB in increments of 5 dB. The table is given here as Table 2.

The line source excitation required to produce this pattern is given by

$$g(x) = \sum_{n=0}^{\bar{n}-1} B_n \sin[(2\pi x/L)(n + \frac{1}{2})], \quad -L/2 \leq x \leq L/2 \tag{46}$$

with Fourier coefficients

Table 1. Polynomial Coefficients

Polynomial Name	c_0	c_1	c_2	c_3	c_4
A	0.303 875 30	$-0.050 429 22$	$-0.000 279 89$	$-0.000 003 43$	$-0.000 000 02$
ξ_1	0.985 830 20	$-0.033 388 50$	0.000 140 64	0.000 001 90	0.000 000 01
ξ_2	2.003 374 87	$-0.011 415 48$	0.000 415 90	0.000 003 73	0.000 000 01
ξ_3	3.006 363 21	$-0.006 833 94$	0.000 292 81	0.000 001 61	0.000 000 00
ξ_4	4.005 184 23	$-0.005 017 95$	0.000 217 35	0.000 000 88	0.000 000 00
p_0	0.479 721 20	$-0.014 566 92$	$-0.000 187 39$	$-0.000 002 18$	$-0.000 000 01$

Table 2. Values of A and ξ_n

Polynomial Name	Side Lobe Level (dB)					
	15	20	25	30	35	40
A	1.0079	1.2247	1.4355	1.6413	1.8431	2.0415
ξ_1	1.5124	1.6962	1.8826	2.0708	2.2602	2.4504
ξ_2	2.2561	2.3698	2.4943	2.6275	2.7675	2.9123
ξ_3	3.1693	3.2473	3.3351	3.4314	3.5352	3.6452
ξ_4	4.1264	4.1854	4.2527	4.3276	4.4093	4.4973

$$B_m = \frac{1}{2j}(-1)^m(m + \tfrac{1}{2})^2 \prod_{n=1}^{\bar{n}-1} \{1 - [(m + \tfrac{1}{2})/\sigma z_n]^2\} \Big/$$

$$\prod_{\substack{n=0 \\ n \neq m}}^{\bar{n}-1} \{1 - [(m + \tfrac{1}{2})/(n + \tfrac{1}{2})]^2\} \tag{47}$$

for $m = 0, 1, \cdots, \bar{n} - 1$. Also

$$B_m = 0, \qquad \text{for } m \geqq \bar{n}$$

Fig. 15 shows two Bayliss patterns. The solid one is computed directly from (43) and is the line source pattern, while the dashed one is the radiation pattern of the $g(x)$ distribution sampled at the points (L/Ni) for $\pm i = \tfrac{1}{2}, \tfrac{3}{2}, \ldots, (N-1)/2$. The patterns are for a -30-dB illumination with $\bar{n} = 4$ and $N = 16$.

Although the discretized pattern is a good approximation to the line source pattern, even for as few as 16 elements, in some cases it may be important to improve the discretized patterns. Elliott has applied perturbation methods for this purpose and derived a set of linear equations from the perturbations of peak side lobe values. The procedure is iterated until convergence is adequate.

The above methods have been extended to treat circular apertures by Taylor [20] and Bayliss [18], and in addition Tseng and Cheng [21] have synthesized a class of circularly rectangular arrays with rectangular lattices and circularly symmetric patterns.

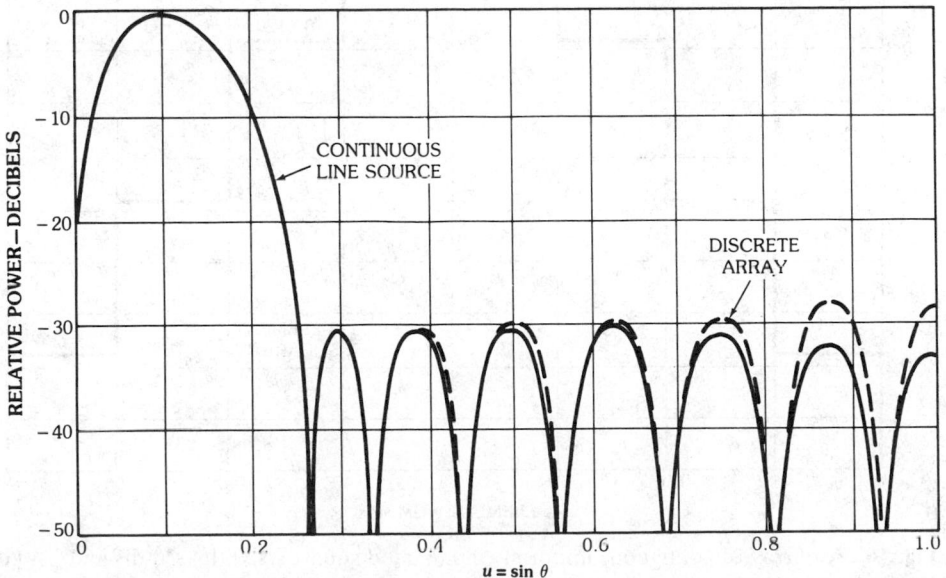

Fig. 15. Bayliss $\bar{n} = 10$ line source patterns: continuous line source and discrete array pattern ($N = 16$) using sampled Bayliss illumination.

3. Array Organization: Subarrays and Broadband Feeds

Often it is convenient to treat a large array as an array of smaller arrays. This is done to simplify power distribution networks, to incorporate low-power, light-weight, or compact circuitry that may have high losses, to integrate amplifier stages into the feed network, or to introduce time-delay networks to improve the broadband properties of the array.

Grouping the array into subarrays may be quite advantageous, but for several reasons it tends to increase the array side lobe level. In Section 4 it is shown that random errors in subarray excitation result in larger side lobes than errors at the array elements because there are so few subarrays. Furthermore, the periodic phase errors that occur in arrays of equal-size time-delayed subarrays produce grating lobes in the array factor. Techniques for producing wide-band behavior with an array of phase-steered elements are described in this section.

Aperture Illumination Control at Subarray Input Ports

One example of an application of subarrays is shown in Fig. 16. An array of $m \times M$ elements is divided into m arrays of M elements each. For simplicity the identical M-way power dividers provide in-phase, equal-amplitude output signals. The m-way beamformer provides feed coefficients a_i applied at the subarray ports. Since the subarrays are formed by equal-amplitude power dividers the array illumination has a staircase appearance shown in the figure, with steps at the a_i level. The normalized radiation pattern at center frequency, unscanned, is

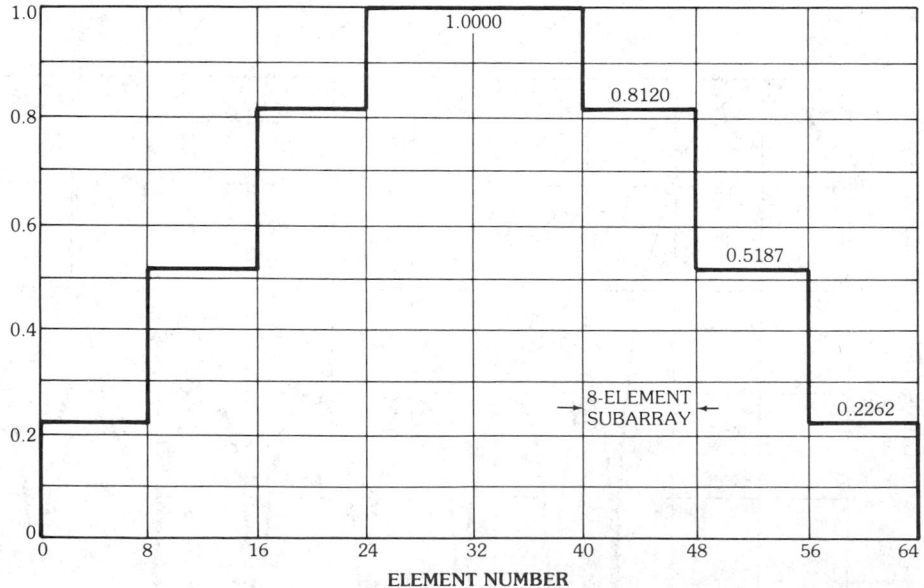

Fig. 16. An array of contiguous uniformly illuminated subarrays with −30-dB and −40-dB Chebyshev array illuminations. (*a*) Array illumination (−30-dB case) for eight subarrays with eight elements each. (*b*) Subarray pattern and array factor (for −30-dB case). (*c*) Total array pattern for −30-dB case (−40-dB case shown partially).

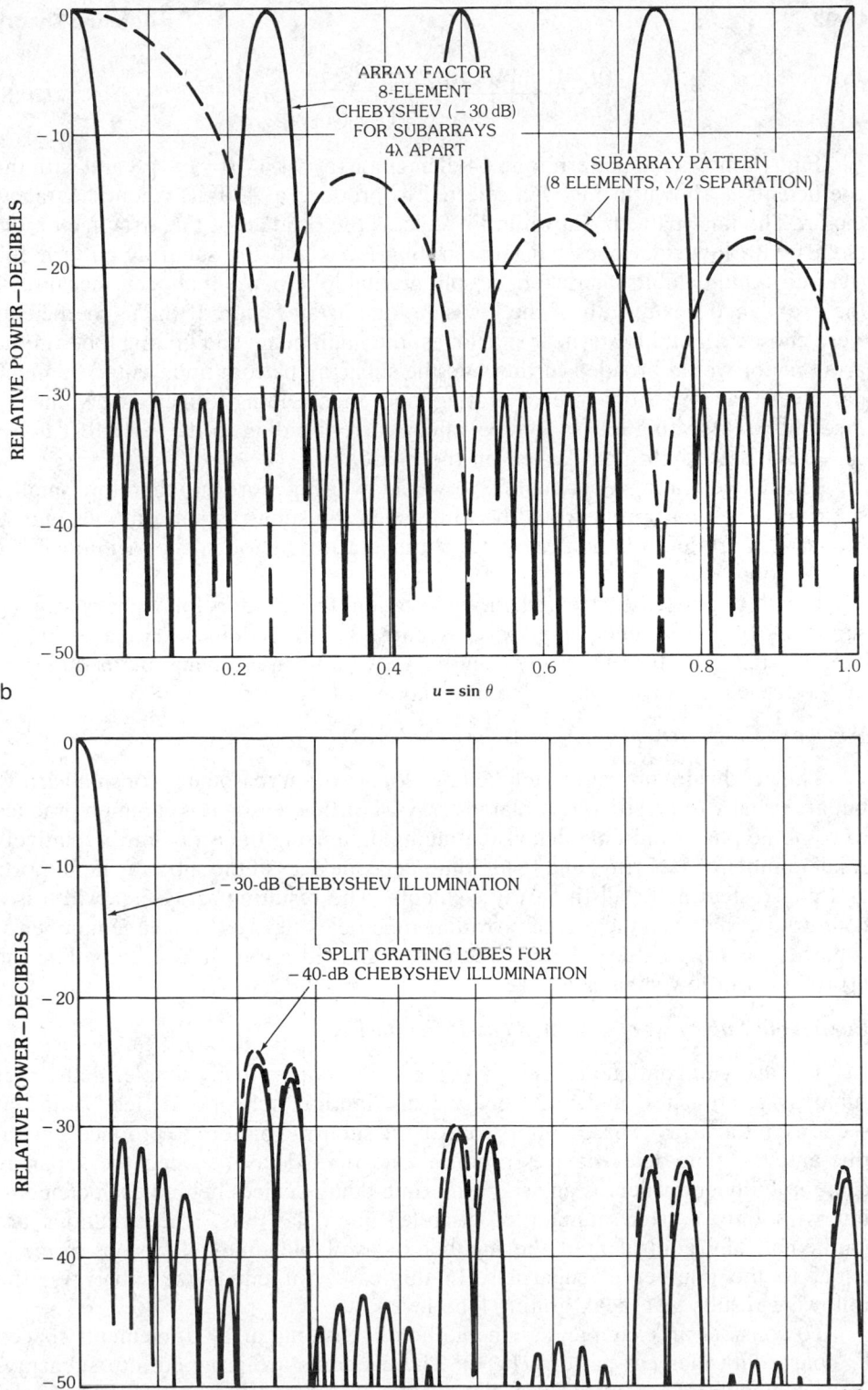

b

ARRAY FACTOR
8-ELEMENT
CHEBYSHEV (−30 dB)
FOR SUBARRAYS
4λ APART

SUBARRAY PATTERN
(8 ELEMENTS, λ/2 SEPARATION)

−30-dB CHEBYSHEV ILLUMINATION

SPLIT GRATING LOBES FOR
−40-dB CHEBYSHEV ILLUMINATION

c

Fig. 16, *continued.*

x

13-31

$$E(r) = \frac{e(u,v)}{m} \frac{\sin[Mu\pi(d_x/\lambda)]}{M\sin[\pi u(d_x/\lambda)]} \sum_{i=-(m-1)/2}^{(m-1)/2} a_i e^{ji(2\pi/\lambda)uD_x} \qquad (48)$$

Fig. 16 shows the pattern for a 64-element array with $M = m = 8$ and with the coefficients a_i shown in Fig. 16a selected to produce a -30-dB Chebyshev array factor. The final pattern, shown in Fig. 16c, is the product of the array factor 16b (solid) with low side lobes but large grating lobes and the subarray pattern 16b (dashed), and exhibits characteristic split grating lobe peaks that occur because of the broadened beamwidth of the low side lobe array factor. If the a_i coefficients were chosen for still lower near side lobes, the main beam and grating lobes in the array factor would broaden further and the subarray pattern nulls would be much narrower than the width of each grating lobe, thus leading to increased values of the split-peak grating lobes. Sample values corresponding to the -40-dB Chebyshev results are indicated dashed on the figure.

The above grating lobes could be lowered by using more and therefore smaller subarrays. The end result would be to broaden the subarray pattern nulls and so reduce the product of subarray pattern times array factor in the vicinity of the grating lobes.

In all of the cases treated in this section the grating lobe peaks can be suppressed by using unequally spaced subarrays. This modification leaves higher average side lobes than the array would have without subarraying, but is often the method selected for introducing time delay in a large array.

Wideband Characteristics of Time-Delayed Subarrays

The bandwidth limitations implied by (26) are often reasonable for small arrays but are usually too restrictive for large arrays. For this reason it is common practice to combine phase and time-delay steering by organizing the array into a relatively small number of subarrays and to use time-delay devices at the subarray input ports and phase steering at all the array elements. The resulting array bandwidth is a compromise between the cost of providing time-delay devices for a large number of subarrays and the pattern deterioration and bandwidth limitations of dividing the array into too few subarrays.

Contiguous Subarrays of Discrete Time-Delay Devices

The array of contiguous subarrays (Fig. 17) is conceptually simpler than other subarray approaches, and uses separate distribution networks to feed adjacent sections of the array. Phase shifters control the subarray pattern to produce a beam tilt, and the time-delay devices produce true time delay between the subarray centers. Fully equivalent is an array with time-delay devices behind each element, but with only a fixed number of discrete time-delay bits. The situations are mathematically equivalent if the number of available time-delay steps is made equal to the number of subarrays. In this case, for equal-size subarrays, the following results give peak grating lobe levels.

To consider an example, assume a one-dimensional array of elements spaced d_x apart, with element pattern $e(u,v)$. The elements are grouped into subarrays of M elements. The entire array has m equally spaced subarrays. Each of the subarrays has a subarray pattern that is the same as the middle term in (48), and

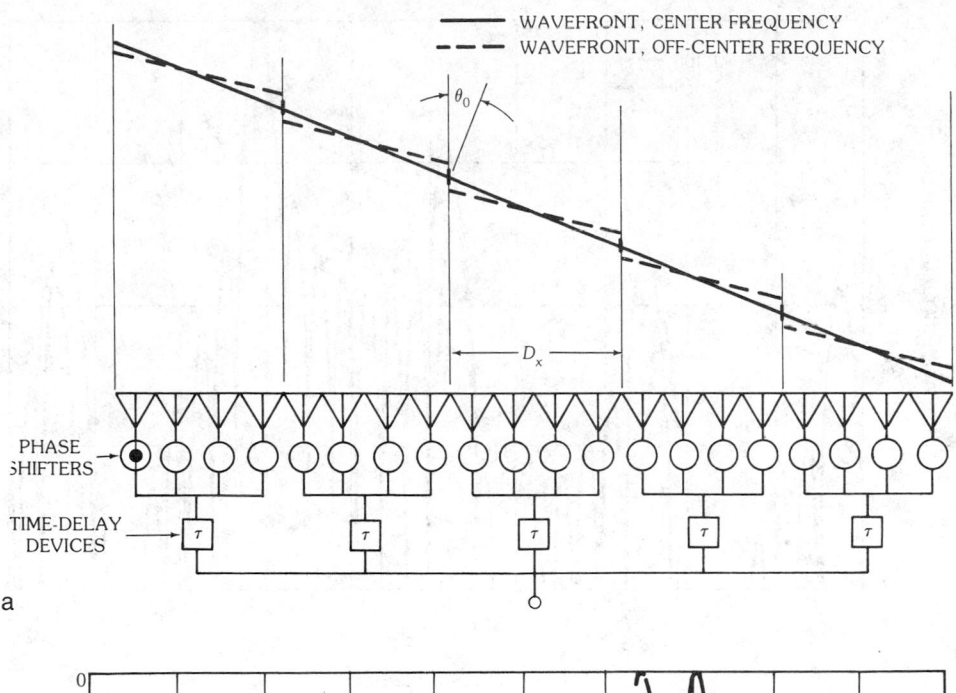

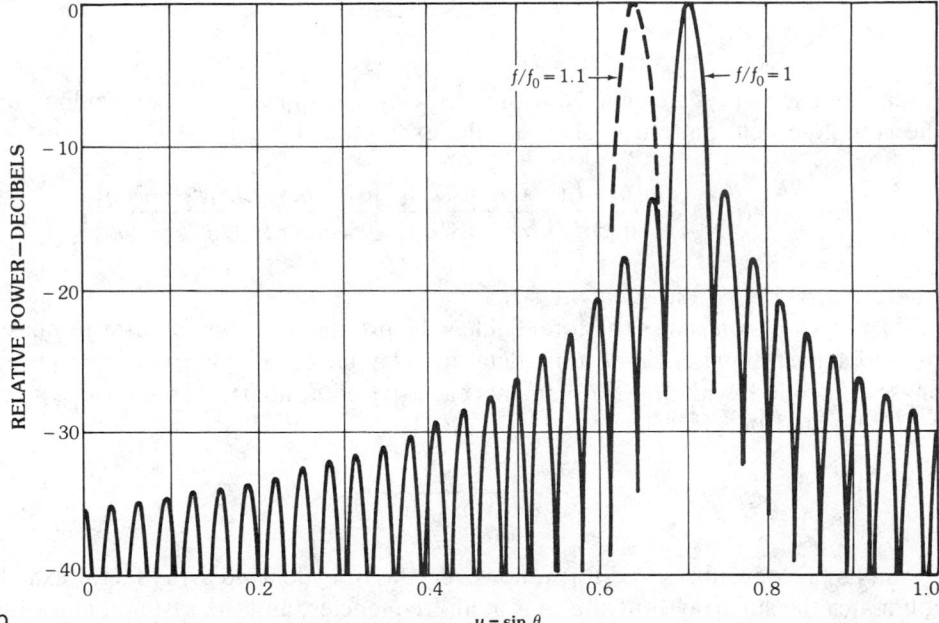

Fig. 17. Broadband characteristics of an array with time delay at subarray level, and contiguous phase-steered subarrays. (*a*) Array geometry, showing wavefront at center and off-center frequencies. (*b*) Patterns of 64-element array with phase shift steering. (*c*) Patterns of array organized with eight time-delayed subarrays.

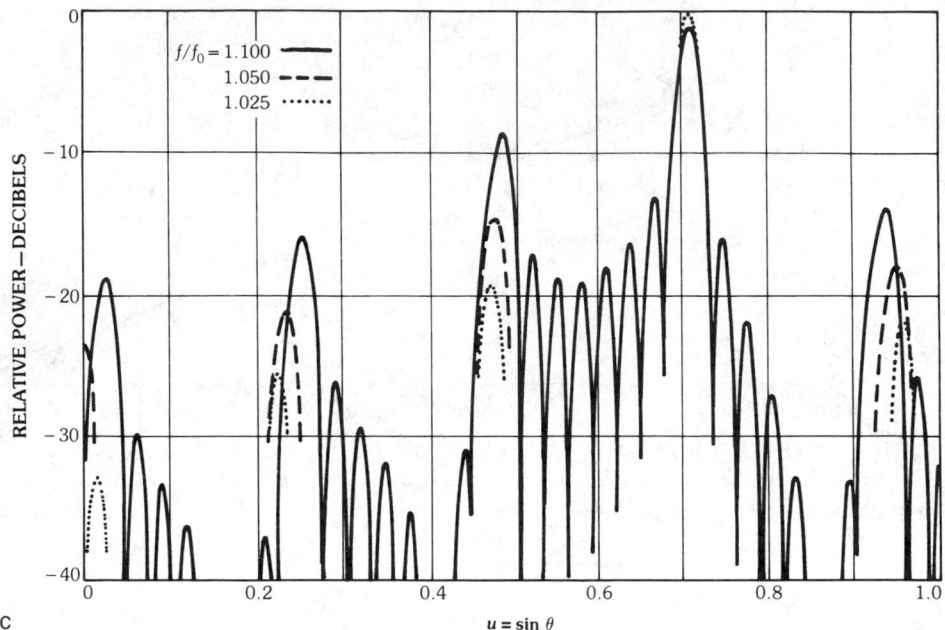

Fig. 17, *continued.*

when these subarrays are arrayed with time delay appropriate for beam collimation the complete field pattern is given by the expression

$$E(r) = e(u, v) \left\{ \frac{\sin[M\pi d_x(u/\lambda - u_0/\lambda_0)]}{M\sin[\pi d_x(u/\lambda - u_0/\lambda_0)]} \right\} \left\{ \frac{\sin[(m\pi D_x/\lambda)(u - u_0)]}{m\sin[(\pi D_x/\lambda)(u - u_0)]} \right\} \quad (49)$$

where $D_x = Md_x$ is the subarray size.

This expression shows the total field as the product of element pattern, phase-steered subarray pattern, and time-delayed array factor. If the array were purely phase controlled, with $M \times M$ elements each spaced d_x apart, its bandwidth given by (26) would be

$$\frac{\Delta f}{f_0} = \frac{K\lambda_0}{Mmd_x \sin \theta_0} \quad (50)$$

In its present subarrayed form, however, the time-delayed array factor exactly collimates the subarray contributions at all frequencies, and the system bandwidth is essentially the same as the subarray bandwidth:

$$\frac{\Delta f}{f_0} = \frac{K\lambda_0}{Md_x \sin \theta_0} \quad (51)$$

For example, an array of ten subarrays of ten elements each has approximately ten times the bandwidth of the phase-steered array of one hundred elements.

The above description emphasizes bandwidth based on gain, but in fact subarraying can introduce severe pattern degradation in the form of grating lobes that arise as frequency is changed. Grating lobes exist in this case, even though the subarray phase centers are appropriately delayed to form a beam at θ_0. This is because each subarray has a phase squint that causes the peak of the subarray pattern to move off the position θ_0 and the subarray pattern nulls to move so that they do not suppress the array pattern grating lobes. Fig. 17 shows the pattern of a uniformly illuminated array of 64 elements, arranged in subarrays of 8 elements, with each element 0.5λ apart. The array is scanned to 45°. In Fig. 17b, the array is steered by phase controls alone, and its main beam squints from the desired 45° to 40° for $f/f_0 = 1.10$. Fig. 17c shows that the same array with time delay at the sub-array input ports exhibits no beam squint, but that large grating lobes (about 8 dB below the main beam) seriously distort the pattern and cause a loss of gain at $f/f_0 = 1.1$. Grating lobes at $f/f_0 = 1.05$ and 1.025 are shown dashed and dotted, respectively. Clearly, the use of contiguous time-delayed subarrays leads to intolerable pattern deterioration for all but extremely small fractional bandwidths.

Overlapped Subarrays

A technique for implementing time-delay steering at the subarray level without the occurrence of large grating lobes involves the synthesis of subarray illumina-tions that are not merely contiguous but actually overlap. By using an aperture illumination wider than the intersubarray period it is possible to produce subarray patterns that have flat tops and are narrow enough to suppress the array pattern grating lobes [22]. This synthesis is achieved using two back-to-back transform networks in order to form a number of flat-topped subarray patterns, using the orthogonal beams as in a Woodward-type synthesis. The transform networks could be Butler matrices, as described here, or confocal lenses (Fig. 18), or reflectors, or some combination of these.

Fig. 18a shows the basic configuration of two Butler matrixes back to back used to excite an array that has phase shifters at each array element. The phase shifters are controlled in accordance with (24). A signal applied to the ith input port of the matrix at right (the $M \times M$ matrix) produces a progressive set of phases at the N array elements and radiates with the pattern

$$g_i(u) = Nf^e(u) \frac{\sin[(N\pi d_x/\lambda_0)(fu/f_0 - u_i)]}{N\sin[(\pi d_x/\lambda_0)(fu/f_0 - u_i)]} \tag{52}$$

where $f^e(u)$ is the array element pattern (assumed equal for all elements) and

$$u_i = i\frac{\lambda_0}{Nd_x} + u_0 \tag{53}$$

Each of the orthogonal beams is displaced from the angle of its peak radiation with all phase shifters set to zero by the amount u_0. When the matrix at left is used to provide the signals at the input to the $M \times N$ matrix at right, each input J_m excites a whole set of signals I_{im}, given by

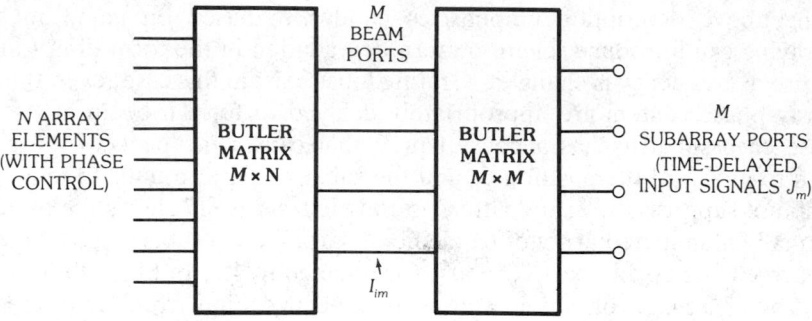

a

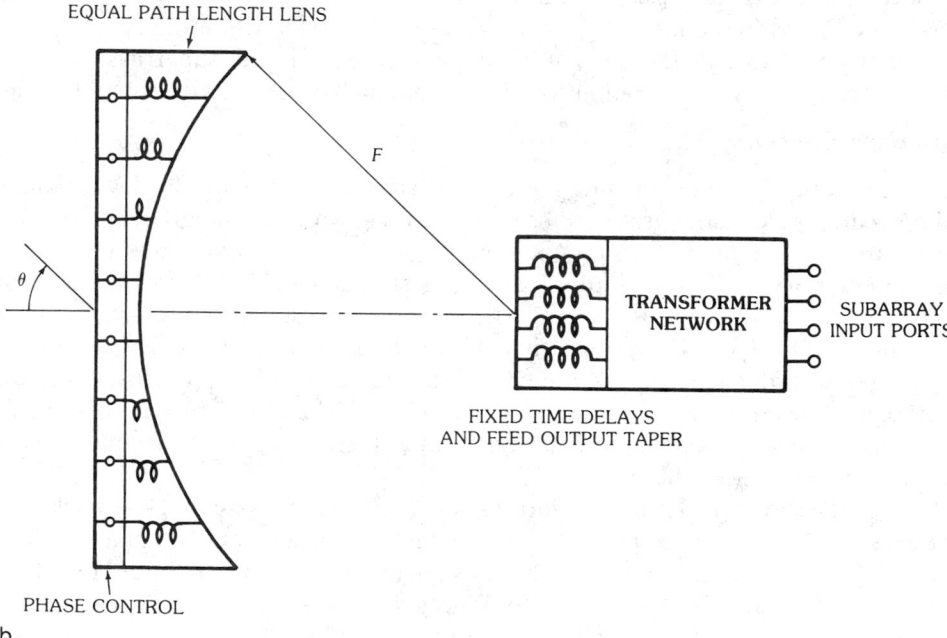

b

Fig. 18. Constrained and space-fed systems for overlapped subarray formation. (*a*) Constrained network. (*b*) Space-fed network.

$$I_{im} = \frac{J_m}{\sqrt{M}} e^{j2\pi(m/M)i}, \qquad -\left(\frac{M-1}{2}\right) \leqq i \leqq \frac{M-1}{2} \qquad (54)$$

The aperture illumination (phase scanned to u_0) corresponding to the mth subarray is, for each nth element of the N-element array,

$$i_m^{(n)} = e^{-j(2\pi/\lambda_0)u_0} \frac{1}{\sqrt{N}} \sum_{i=-(M-1)/2}^{(M-1)/2} I_{im} e^{-j2\pi(n/N)i}$$

$$= \frac{NJ_m e^{-j(2\pi/\lambda_0)u_0}}{\sqrt{NM}} \frac{\sin M\pi[(mN-nM)/MN]}{M\sin\pi[(mN-nM)/MN]} \qquad (55)$$

This illumination has a maximum at the element with index $n = m(N/M)$, and overlaps all the elements of the array. An example of one such subarray illumination is the dashed curve of Fig. 19a for the subarray ($m = 4$) of an array of 64 elements ($N = 64$) with $\lambda/2$ separation. The array has eight subarrays ($M = 8$).

The radiated subarray patterns are given by

$$f_m(u) = \sum_i I_{im} g_i(u)$$

$$= \frac{NJ_m f^e(u)}{\sqrt{MN}} \sum_{i=-(M-1)/2}^{(M-1)/2} e^{j2\pi(m/M)i} \left(\frac{\sin[(N\pi d_x/\lambda_0)(fu/f_0 - u_i)]}{N\sin[(\pi d_x/\lambda_0)(fu/f_0 - u_i)]} \right) \quad (56)$$

This expression is a sum of M orthogonal pencil beams arranged to fill the sector, and taken together to form a flat-topped pattern for the mth subarray, which is shifted in angle so that its center is at u_0 at center frequency.

Fig. 19b shows two subarray patterns at center frequency for the same 64-element array. The selected subarrays are an edge ($m = -7/2$) and one of the two central subarrays ($m = 1/2$). The edge subarray has higher side lobes and a highly rippled pass region because its illumination is truncated at the array edge.

The array excitation with all subarrays excited is

$$\mathcal{I}_n = \sum_{m=-(M-1)/2}^{(M-1)/2} J_m i_n^{(m)} \quad (57)$$

where the $i_n^{(m)}$ are given in (55). An example of such a composite excitation is shown in the solid curve of Fig. 19a, for all subarrays weighted with a -30-dB Chebyshev illumination. This excitation is much smoother than the illumination

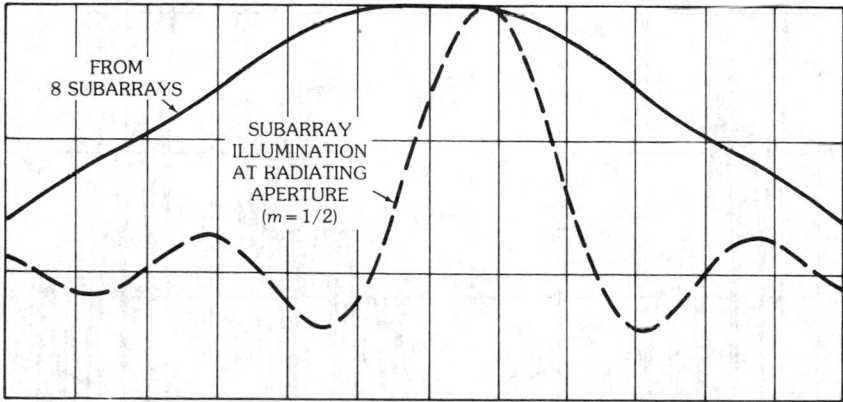

a

Fig. 19. Broadband characteristics of an array with time delay at the subarray level and completely overlapped phase-steered subarrays. (a) Typical subarray illumination at radiating aperture and total illumination from eight subarrays with -30-dB Chebyshev weighting. (b) Radiated subarray patterns near array edge and array center. (c) Array radiation pattern at broadside and 45° scan ($f = f_0$) and 45° scan ($f = 1.1f_0$).

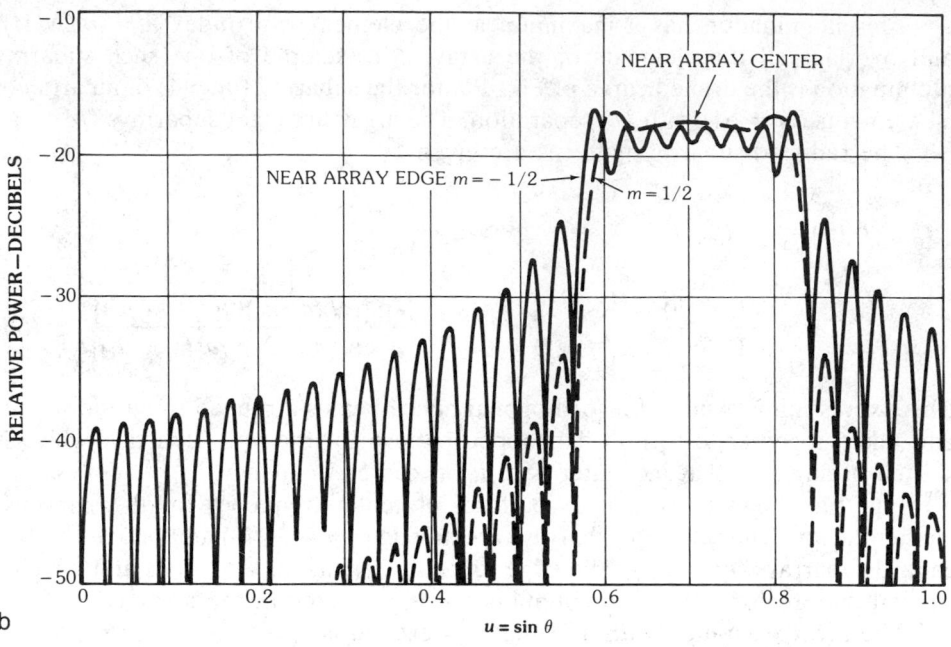

b

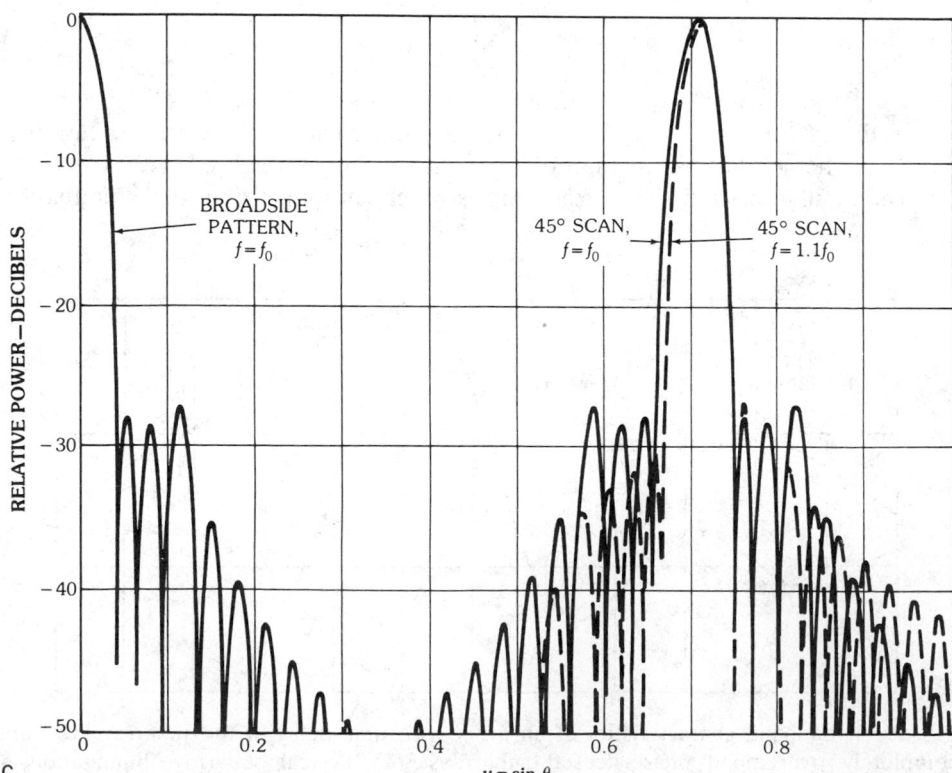

c

Fig. 19, *continued.*

shown in Fig. 16a, which shows the same weightings applied to an array of contiguous subarrays. So it can be expected that the side lobes would be lower for the overlapped subarray case.

The radiated array pattern is given by

$$F(u) = \sum_{M=-(M-1)/2}^{(M-1)/2} J_m f_m(u) \tag{58}$$

To scan the array beam to u_0 with time delay, one applies the signals

$$J_m = |J_m| e^{-j(2\pi/\lambda_0)D_x(f/f_0-1)u_0} \tag{59}$$

where $D_x = (N/M)d_x$ is the distance between subarray centers at the array face.

The absolute values of the input signals $|J_m|$ are weighted directly to provide the appropriate near side lobe distribution.

Figs. 19c and 19d show several array patterns for the array with the -30-dB illumination. As shown in the figures the side lobe levels exceed -30 dB because of the rippled subarray patterns. At center frequency ($f/f_0 = 1$) the pattern scanned to $u = 0.707$ (45°) has the same form as the broadside pattern, and even at 10 percent above center frequency ($f/f_0 = 1.1$) the main beam is not altered in location or gain, though the side lobe structure is.

The bandwidth of such a subarraying is on the order of

$$\frac{\Delta f}{f_0} \cong \frac{(M-1)}{N} \frac{\lambda}{d_x} \frac{1}{\sin\theta_0} \tag{60}$$

Overlapped subarray systems have been implemented using multiple-beam lens systems [22, 23, 24] and Butler matrix [25] networks. With the emergence of digital beam-forming technology it is likely that it will be convenient to form subarrayed patterns digitally for future system applications.

The networks described above produce completely overlapped subarrays; each subarray extends over the whole array. However, convenient networks have also been developed to overlap small groups of elements. Such techniques form approximations of the ideal flat-topped pattern and are useful for limited scan applications [26, 27].

Broadband Array Feeds with Time-Delayed Offset Beams

Equation 26 gives the fractional bandwidth of a phase-steered array as a function of its beamwidth and the maximum scan angle θ_0. The bandwidth can be relatively large if the array scan remains small, so that the product $(L/\lambda_0)\sin\theta_0$ does not become a large number. Similarly, if an array is fed by a system that produces a time-delayed beam at some angle θ_T, and the beam is phase steered to an angle θ_0 by phase shifters, then the bandwidth is given by

$$\frac{\Delta f}{f_0} = \frac{0.866 B\lambda_0}{L|\sin\theta_T - \sin\theta_0|} \tag{61}$$

This equation shows that by using a feed system that provides a number of fixed-offset time-delayed beams it is possible to scan those beams over the limited angular regions between the beams, and so operate over substantially increased instantaneous bandwidth. One such implementation, suggested by Rotman and Franchi [28], is indicated in Fig. 20a, which shows an active lens with four feed horns equally spaced along the focal arc of a two-dimensional microwave-cons-

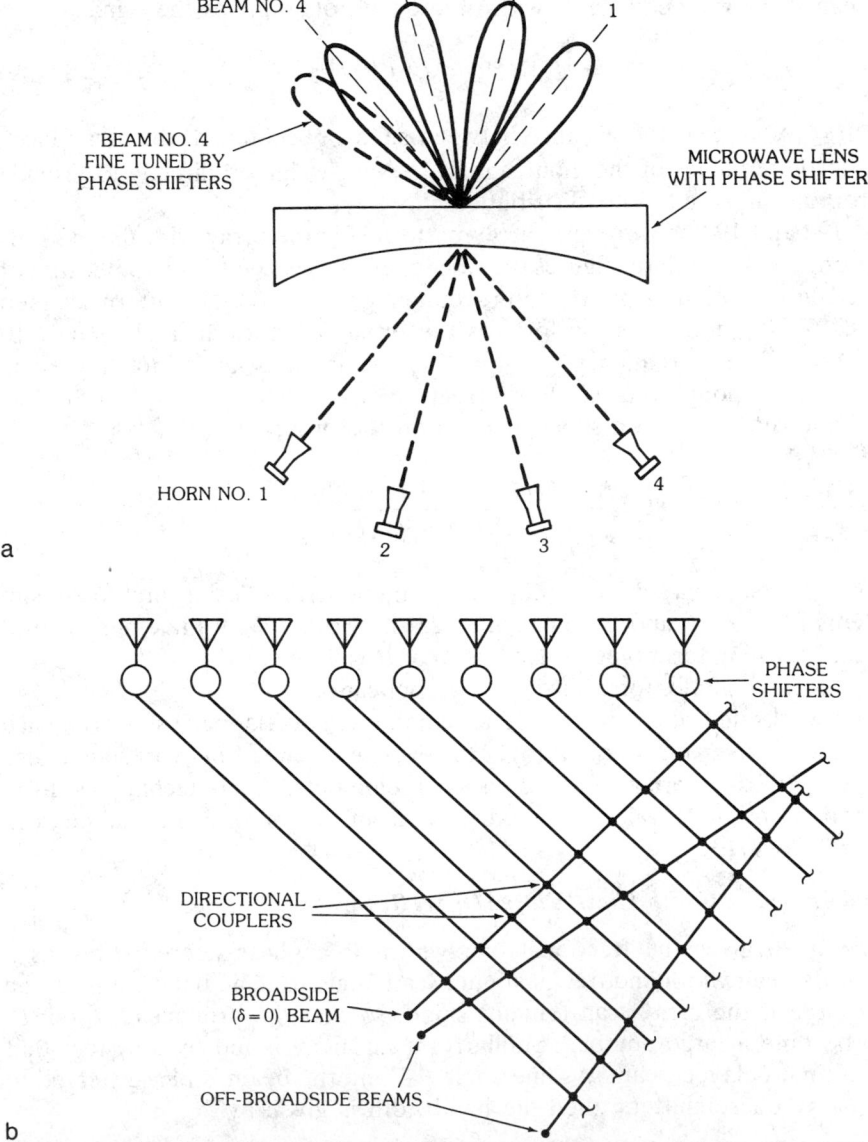

Fig. 20. Wideband scanning using phase shifter and fixed time-delayed beams. (*a*) Microwave lens configuration. (*b*) True time-delay matrix.

trained lens. Energy from a transmitter can be directed to any one of the horns by means of the switching tree. Each horn, in turn, will form a beam in a different azimuth direction for the zero phase shifter setting. A typical beam for the Mth horn is sketched as the solid curve (Mth beam). As the phase of the lens illumination is changed by the phase shifters the beam scans to either side of its no-phase-shift position. This effect is illustrated by the dotted curve (phase-scanned Mth beam). The bandwidth limitation imposed by this phase scanning is given by the following equation, with θ_0 the maximum scan angle and N the number of beam positions (4 for the example in the figure). The system bandwidth, given below and readily derived from (26), is wide because of the limited scan angle:

$$\frac{\Delta f}{f_0} \cong \frac{0.886 B N}{(L/\lambda_0)\sin\theta_0} \tag{62}$$

Other means for achieving fixed offset beams include the use of constrained multiple-beam systems, such as the true time-delay matrix of Fig. 20b, or precut switched time delays at each array element. In addition to providing wideband gain, systems that use offset beams have no phase discontinuities and therefore no grating lobes. In principle they can have very low side lobes.

4. Practical Arrays

The previous sections on periodic arrays treat the array in an idealized case, with perfectly regular lattices, prescribed exact phase controls, half-space isotropic element patterns, and, most importantly, with all element patterns equal.

This section deals with a number of problems that confront array designers. Array mutual coupling leads to unequal element patterns and to a need to solve coupled integral equations before applying any of the synthesis methods mentioned earlier. Conformal nonplanar arrays have lattices that are, at most, periodic in one plane, and so present special problems in synthesis and pattern control. Finally, the section addresses array component errors of several types, and it reviews components used to distribute power and scan the beam of a phased array.

Mutual Coupling and Element Patterns

One of the most important and complex aspects of modern array design is that element excitation coefficients are not proportional to applied sources (voltages or currents) and that the element patterns are nonisotropic and not equal to the pattern of an isolated element. These phenomena occur because each of the elements couples through radiation to all of the others, and hence the relationship between applied sources and element excitation must be expressed in terms of a complex matrix. This phenomenon is called *mutual coupling*.

A detailed treatment of mutual coupling is beyond the scope or intent of this chapter. The solution of wire antenna problems, such as the dipole array of Fig. 21a, is carried out by satisfying a boundary condition (usually that the tangential electric field is zero) at the surface of the wire and equal but opposite to the applied field at the source point. For the dipole array of Fig. 21a, with dipole axes along the $\hat{y}$ direction and their centers located at $(x_m, y_m, c/2)$ with and without a ground

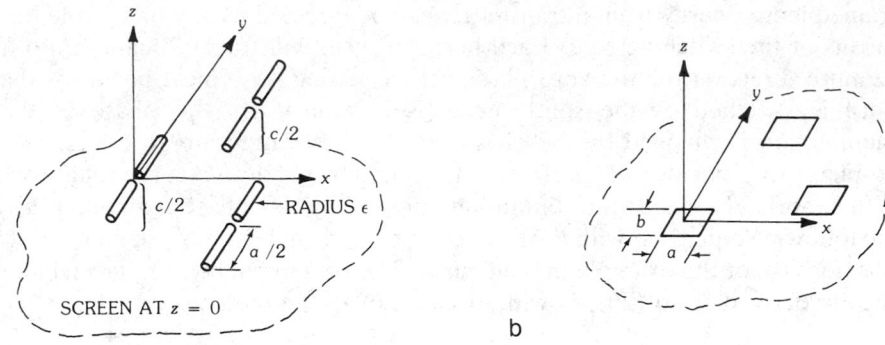

Fig. 21. Elements for scanning arrays. (*a*) Dipoles over ground screen. (*b*) Waveguide apertures in ground screen.

plane, it is convenient to introduce a single component of the vector potential defined at $\mathbf{r} = \hat{\mathbf{x}}x + \hat{\mathbf{y}}y + \hat{\mathbf{z}}z$ as

$$A_y(x, y, z) = \frac{\mu_0}{4\pi} \sum_m \int_{V_m} I_m(y') G(r, r') dy' \qquad (63)$$

where

$$G(r, r') = \frac{e^{-jk_0|\mathbf{r}-\mathbf{r}_m|}}{|\mathbf{r}-\mathbf{r}_m|}$$
$$|\mathbf{r} - \mathbf{r}_m| = \sqrt{(x - x_m)^2 + (y - y')^2 + (z - c/2)^2}$$

for the array without a ground plane, and

$$G(r, r') = \frac{e^{-jk_0|\mathbf{r}-\mathbf{r}_m|}}{|\mathbf{r}-\mathbf{r}_m|} - \frac{e^{-jk_0|\mathbf{r}-\mathbf{r}_m^{(i)}|}}{|\mathbf{r}-\mathbf{r}_m^{(i)}|}$$
$$|\mathbf{r} - \mathbf{r}_m^{(i)}| = \sqrt{(x - x_m)^2 + (y - y')^2 + (z - c/2)^2}$$

for the array with ground plane at $z = 0$.

The second expression accounts for the image dipole at $z = -c/2$.

The set of integral equations equating the tangential **E** field to zero at each dipole radius is written at the nth dipole

$$E_y(x_n, y, c/2) = -V^{(n)}\delta(y - y_n) = -j\frac{\omega}{k_0^2}\left[\frac{\partial^2 Ay}{\partial y^2} + k^2 Ay\right],$$
$$\text{for } -a/2 \leqq y - y_n \leqq a/2 \quad (64)$$

where $V^{(n)}$ is the potential across the source at the nth antenna. The source is assumed to be a delta function of the electric field located at the center of each dipole, and the equation sets the induced field equal to the negative of the source field at this point.

In this form the integral equation is often called *Pocklington's equation* and is frequently chosen for digital computer solution of dipole and wire antenna problems. In this procedure each dipole is considered as made up of a number of connected segments, with each segment radiating and coupling to all other segments in all of the radiators. For detailed descriptions of this procedure the reader is referred to [29] and [30].

Another traditional form of equation for dipole arrays is obtained by constructing a solution to the above differential form to obtain a generalized form of Hallen's equation [31]:

$$
\left.
\begin{aligned}
A_y(x_n, y, z_n) &= c_1^{(n)}\cos ky + c_2^{(n)}\sin ky - j\frac{k}{2}\frac{V^{(n)}}{\omega}\sin ky, && 0 \leq y - y_n \leq a/2 \\
&= c_3^{(n)}\cos ky + c_4^{(n)}\sin ky + j\frac{k}{2}\frac{V^{(n)}}{\omega}\sin ky, && -a/2 \leq y - y_n \leq 0
\end{aligned}
\right\}
\quad (65)
$$

The potential A_y on the left side of the equation is evaluated at the center line of each dipole (using $|\mathbf{r} - \mathbf{r}_m| = [\varepsilon^2 + (y - y')^2]^{1/2}$ for $n = m$). The evaluation of the potential function in this manner is valid for thin dipoles (with ε much less all other dimensions, including wavelength) and assumes the potential outside of and even at the surface of the dipole is the same as if there were a filament of current at the dipole axis.

The constant c_4 is equal to c_2.

The above is a set of M integral equations, written along each dipole, and consisting of M unknown current functions and $3M$ constants $(c_1^{(n)}, c_2^{(n)}, c_3^{(n)})$ that arose from the integration of the integrodifferential equations. The solution of the above equations is necessarily approximate and there have been numerous forms of published solutions, each with different degrees of validity and complexity. The simplest approximation is one that assumes that the form of the current distribution is the same on all dipoles,

$$
I_m(y) = I(m)f(y) \quad (66)
$$

This approach is most appropriate for resonant dipoles with their source points all in the same plane ($y_n = 0$). Examples of this sort of solution exist in the literature and have been carried out by assuming a form $f(y)$ and solving the resulting simultaneous equations.

The end result after elimination of constants is an impedance matrix

$$
\mathbf{V} = \mathbf{ZI} \quad (67)
$$

with column vectors $\mathbf{I}$ and $\mathbf{V}$ and square impedance matrix $\mathbf{Z}$ that one can use for computation of currents, given applied voltages, or, in the case of synthesizing required antenna patterns, can be used to compute the required voltages to provide desired current terms. Commonly used single-mode impedance formulas are given by Brown and King [32], Carter [33], and Tai [34].

King and his colleagues have employed several higher-order solutions that are more realistic for computing near-field effects or coupling between column arrays of dipoles with parallel axes. The most comprehensive of these is the five-term theory [35, 36] that includes asymmetric current terms for evaluation of the radiation properties of arrays scanned in two dimensions.

Of these two basic methods, the expansion of the current by a finite number of functions that span the entire dipole or by a piecewise approximation in sections across the dipole, solutions based on higher-order current expansions have to date seen far wider application to large arrays than the multisegment solutions, because the former involve the inversion of much smaller matrices.

Finite waveguide arrays can be treated in a similar manner. In this case there are no electric current sources in the half-space $z \geqq 0$, and the only sources are the magnetic current sources as represented by the tangential aperture fields. In this case there is no single vector component that serves to completely represent the fields except for special two-dimensional cases. In general, however, for a finite waveguide aperture the solution is vector and is formulated by expanding the aperture field in a set of functions and matching fields in the waveguides and free space. For open-ended waveguides it is convenient to choose as basis functions the waveguide normal-mode fields, and for unloaded rectangular waveguides one can choose the orthogonally polarized transverse electric fields. The transverse electric field for the waveguide at the origin of the coordinate system of Fig. 21 is

$$\mathbf{E}^T = \mathbf{e}_0(x, y)e^{-jk_z(0)z} + \sum_n V_n \mathbf{e}_n(x, y)e^{+jk_z(n)z} \tag{68}$$

where the $\mathbf{e}_n(x, y)$ are the transverse-mode functions for the two possible polarizations, the $k_z(n)$ are the modal propagation constants, and V_n are undetermined modal amplitude coefficients. This expression represents a single incident mode in the waveguide and an infinite series of reflected modes. Typically all but the $k_z(0)$ propagation constants are imaginary, indicating that those are beyond waveguide cutoff, but they enter into the solution to match boundary conditions. The solution proceeds by expanding the transverse magnetic waveguide fields in terms of these and writing the half-space fields ($z > 0$) as the aperture field. Construction of the free-space Green's function ensures that the tangential $\mathbf{E}$ field is continuous, and imposed continuity of the magnetic field components results in a vector integrodifferential equation

$$\hat{\mathbf{z}} \times \mathbf{B}(z = 0^-) = \hat{\mathbf{z}} \times \mathbf{B}(z = 0^+)$$
$$= j2\omega\varepsilon\hat{\mathbf{z}} \times \sum_s \bar{\bar{\Gamma}}^0(r, r') \cdot (\hat{\mathbf{z}} \times \mathbf{E})ds' \tag{69}$$

where the free-space dyadic Green's function is given by

$$\bar{\bar{\Gamma}}^0(\mathbf{r}, \mathbf{r}') = \left(\bar{\bar{\mathbf{U}}} + \frac{1}{k^2}\nabla\nabla\right)G(\mathbf{r}, \mathbf{r}')$$

where

$$G(\mathbf{r}, \mathbf{r}') = \frac{e^{-jk|\mathbf{r}-\mathbf{r}'|}}{4\pi|\mathbf{r} - \mathbf{r}'|}$$

$$\mathbf{r} - \mathbf{r}' = \sqrt{(x - x')^2 + (y - y')^2 + z^2}$$

is the scalar Green's function. The unit dyad $\bar{\bar{\mathbf{U}}}$ is defined in rectangular coordinates as

$$\bar{\bar{\mathbf{U}}} = \hat{x}\hat{x} + \hat{y}\hat{y} + \hat{z}\hat{z}$$

The magnetic field **B** for $z = 0$ can be obtained from the expansion (68) using waveguide modal admittances. By following Galerkin's procedure the above can then be reduced to a matrix equation and solved for aperture fields. The details of this procedure will not be carried further, but are described in a number of available references [37, 38].

Waveguide array solutions using single-mode approximations in each aperture have much more limited applicability than the simple theories for dipole arrays. Single-mode waveguide coupling solutions fail to predict blindness effects (see next section) but can be used successfully for small arrays and for relatively closely spaced elements at scan angles far from the grating lobe onset. Single-mode solutions have also been used by Golden [39], Steyskal [40], and others for elements conformal to curved surfaces.

Fig. 22 [41] shows several of the most significant effects due to mutual coupling. Fig. 22a shows the element pattern of the central element in an array of N parallel plane elements with the incident waveguide fields in the plane of scan. The presence of multiple ripples in the element pattern is due to reflections from the array edge, as indicated by the higher angular frequency for increased N. The infinite array results, also shown in the figure, demonstrate element pattern narrowing due to mutual coupling that forces the pattern to be zero at the horizon. Fig. 22b shows the associated reflection coefficient for the central element and again evidences the rippling effect for finite arrays and the unity reflection coefficient for the infinite array model at end-fire (about $\psi = 140°$) and throughout the slow-wave region $\psi > 140°$.

Array Blindness

In certain circumstances it is possible to have mutual coupling effects that actually create a null in the array element patterns so that the array cannot transmit energy in given directions.

Fig. 23 shows the basic phenomena as described by Farrell and Kuhn [42, 43] in the first published analytical work on the subject. The figure shows a measured deep null in an element pattern of a waveguide array and compares the data with results computed using a single-mode grating lobe series for an infinite array and a full modal array solution. The null is due to the cumulative effects of mutual coupling and can be related to surface wave type behavior at the array face. In many cases the existence of the null is understood as a cancellation process involving waveguide higher-order modes, and this is why the single-mode grating lobe solution bears little correlation to the data in Fig. 23. In the years since this initial discovery these blindnesses have been found in most waveguide array

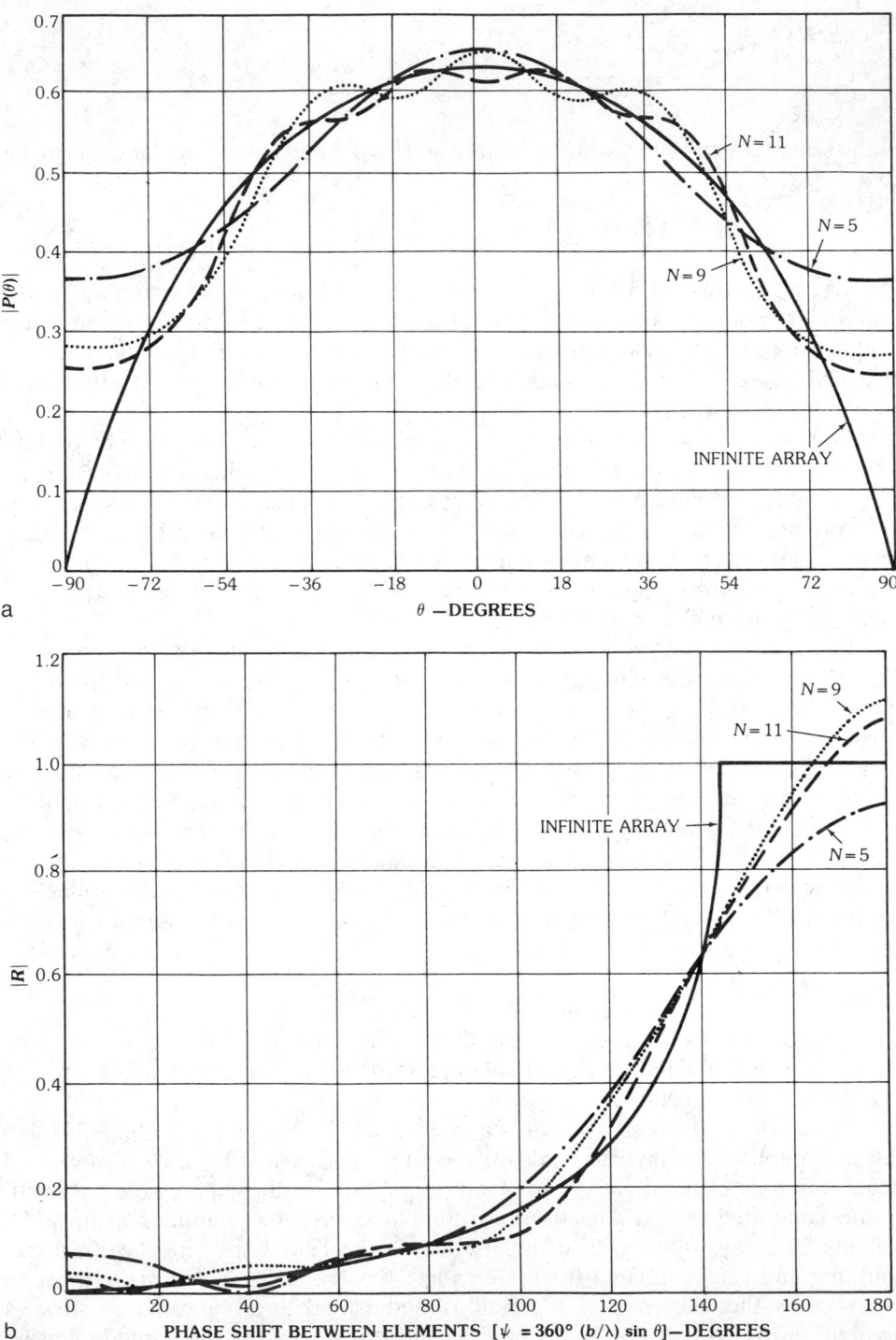

Fig. 22. Element patterns $P(\theta)$ and reflection coefficients R of center elements in unloaded waveguide array ($b/\lambda = a/\lambda = 0.4$). (*a*) Radiation patterns. (*b*) Reflection coefficients. (*After Wu [41], © 1970 IEEE*)

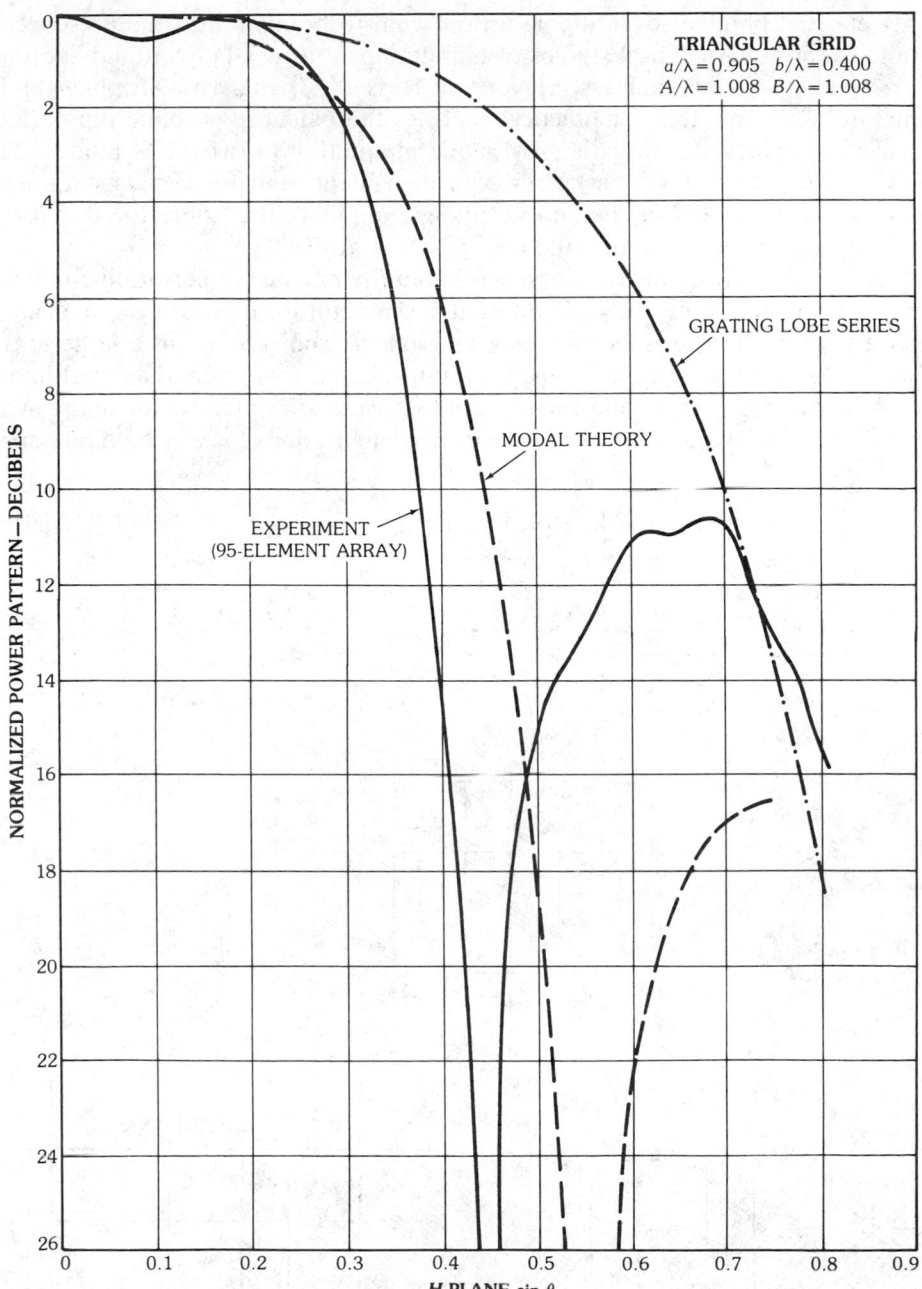

Fig. 23. Array element power pattern showing array blindness. (*After Farrell and Kuhn [42, 43], © 1968 IEEE*)

configurations and in some dipole and stripline arrays [44]. The problem can usually be reduced or eliminated by keeping the element lattice dimensions d_x and d_y small enough so that grating lobes are well beyond the maximum scan angle throughout the operating frequency range [45].

Reported blindnesses in dipole arrays seem to be related to the presence of dipole supports. Analytical studies of infinite dipole arrays [46] without supports do not exhibit array blindness. Mayer and Hessel [47] analyze a stripline dipole structure and show that for practical spacings the balanced stripline dipole feed structure supports a propagating TM mode in addition to two TEM modes. The TEM mode propagation constant is scan dependent, and for certain parameter selections it occurs before the onset of the grating lobe. It is conjectured that this mode might be the cause of blindness in dipole arrays.

Experience with array blindness has led to the practice of performing infinite-array studies, measuring the array element in simulator, or fabricating a small array for element pattern tests before embarking on the construction of a large array. Multimodal infinite array solutions are far simpler to obtain than multimode solutions for large arrays and such solutions have been obtained for many array types. Fig. 24 shows a few of the basic array configurations for which infinite array

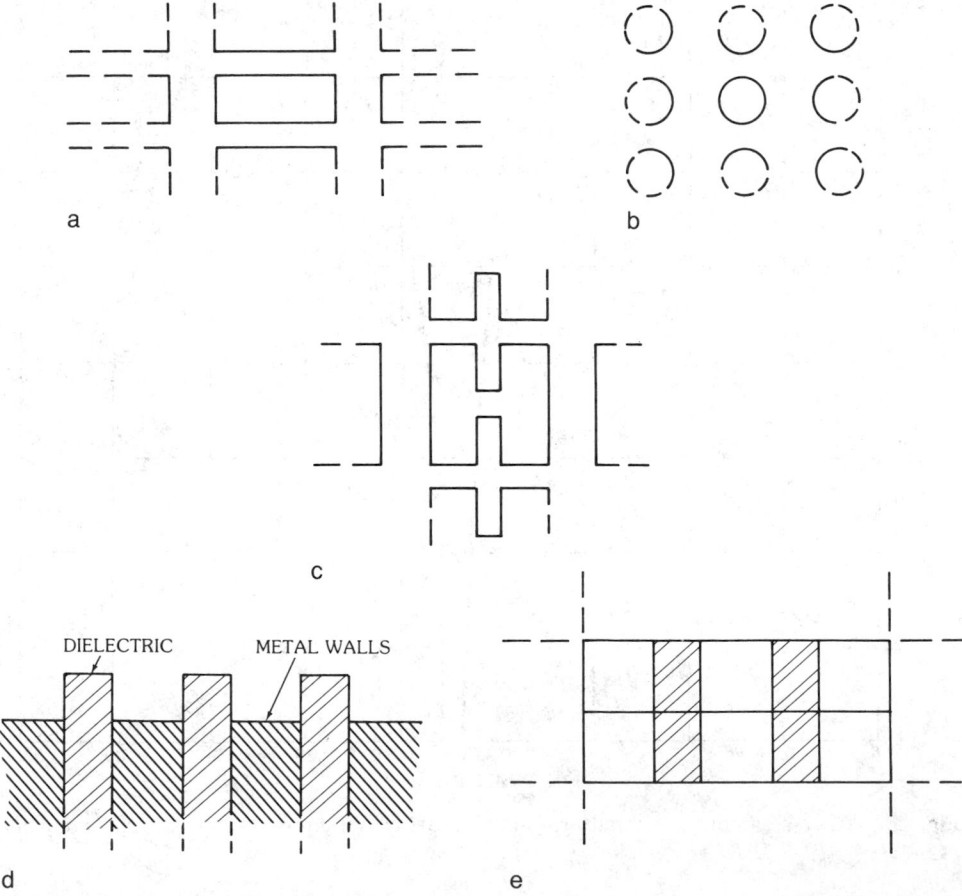

Fig. 24. Several configurations with existing infinite array solutions. (*a*) Rectangular waveguide array. (*b*) Circular waveguide array. (*c*) Ridge loaded waveguide array. (*d*) Protruding dielectric waveguide array. (*e*) Dual-frequency array.

solutions are published. Included in the figure are flush mounted arrays of rectangular [48] and circular [49] elements, ridge loaded elements [50, 51, 52], protruding dielectric (TEM solution) [53], and a dual-frequency dielectric loaded configuration [54]. Among other published solutions are numerous interlaced multiple-frequency configurations [55], examples of dielectric loading [56], iris loading [57] and fence [58], and corrugated plate [59] loadings for impedance match, as well as several very wide band configurations for waveguide [60] and stripline [61]. In addition there have been a number of infinite dipole array solutions published [62].

Conformal Arrays

The need for conformal or low profile arrays for aircraft and missile applications, and for ground-based arrays with 360° azimuth coverage or hemispherical coverage, has grown continually with requirements that emphasize maximum utilization of available space and minimum cost. The earliest and continuing stimulus for cylindrical and circular array development is the need for inexpensive systems with mechanical or electronic scanning with constant gain throughout the 360° coverage sector. There are also a number of spacecraft and aircraft applications requiring low profile or conformal arrays. Fig. 25 shows a possible configuration of an airborne array for satellite communication.

Array elements on curved bodies point in different directions, and so it is usually necessary to turn off those elements that radiate primarily away from the desired direction of radiation. For this reason also, one cannot factor an element pattern out of the total radiation pattern and therefore conformal array synthesis is very difficult. In addition, mutual coupling problems can be severe and difficult

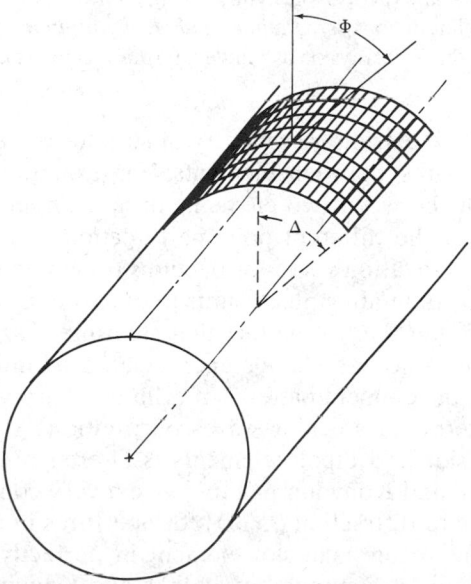

Fig. 25. Conformal array geometry for 2Δ arc array on cylinder.

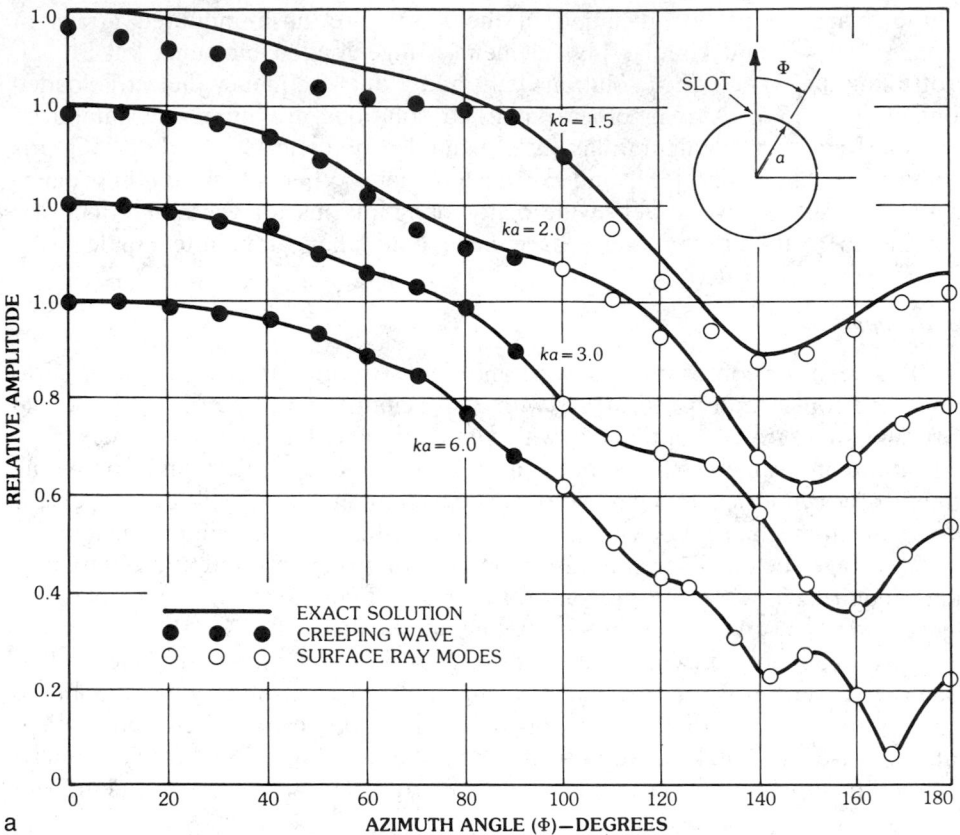

Fig. 26. Patterns of slots and arrays on cylinders. (*a*) Patterns of a thin axial slot in a perfectly conducting cylinder. (*After Pathak and Kouyoumjian [63], © 1974 IEEE*) (*b*) Element patterns for dipole arrays on a cylinder. (*After Herper et al. [68], © 1980 IEEE*)

to analyze because of the extreme asymmetry of structures like cones and because of multiple coupling paths between elements (for example, the clockwise and counterclockwise paths between two elements on a cylinder). Cross-polarization effects arise because of the different pointing directions for elements on curved surfaces causing the polarization vector projections to be nonaligned. There is also a need to use different collimating phase shifts in the azimuth plane of a cylindrical array scanned in elevation due to the fact that steering in azimuth and elevation planes is not separate. Another phenomenon related to mutual coupling is the evidence of ripples on the element patterns of cylindrical arrays. This phenomenon can be explained in terms of creeping-wave contributions.

The behavior of slot and dipole elements is altered by the presence of the curved surface. Pathak and Kouyoumjian [63] give a very convenient extension of the geometrical theory of diffraction (GTD) for apertures in curved surfaces. Fig. 26a shows the patterns of an axial slot element in perfectly conducting circular cylinders of various radii, as computed by Pathak and Kouyoumjian. The pattern compares the exact solution with that obtained using the appropriate GTD

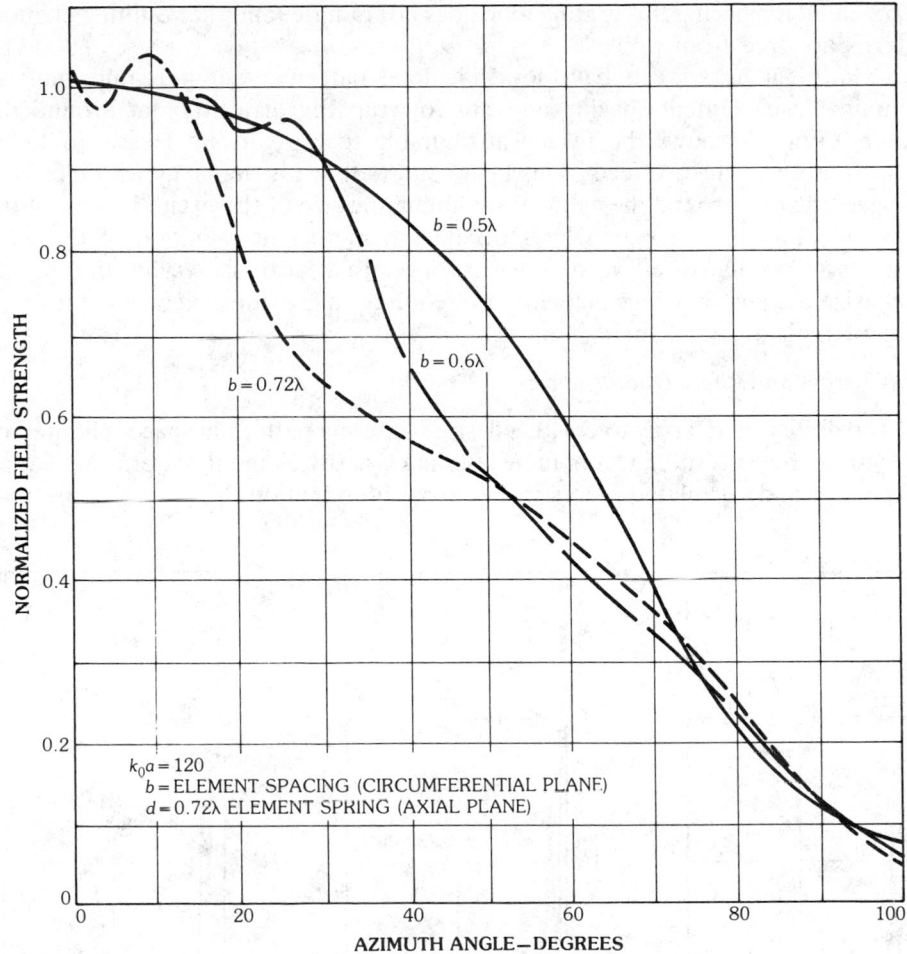

b

Fig. 26, *continued.*

expressions in several regions of space, and shows the GTD formalism to be highly accurate, even for relatively small cylinders. The figure also emphasizes the way in which the finite cylinder alters slot element patterns, for a similar slot in a flat ground plane would have a constant radiation pattern from $\phi = 0°$ to $180°$. A number of other references give the radiation patterns of slots and dipole elements on a variety of generalized surfaces [64, 65, 66].

Arrays of slots or dipoles on curved surfaces also behave differently from those on plane surfaces, and often have highly rippled element patterns [67] that make low side lobe synthesis impossible. Fig. 26b, however, shows that the rippled element pattern characteristic shown for a dipole array over a cylinder does not occur if the element spacing is restricted to about a half-wavelength [67]. The unwelcome rippled effect shown for larger spacings has been attributed [67] to interference between the grating lobe of the fast creeping wave and the direct ray.

At $\lambda/2$ element spacing this grating lobe does not radiate, and the resulting element patterns are free from ripples.

Cylindrical arrays can have low side lobe patterns, but it is important to maintain close element spacing and not to wrap the array too far around the cylinder. Fig. 27 shows the results attributable to Sureau and Hessel [67] that illustrate both of these effects. This figure shows that for the array with elements wrapped entirely around the cylinder, doubling the size of the excited sector of the array only increases the gain by 1 dB due to inefficient radiation of the edge elements. The narrowed array element patterns, also shown in the figure, emphasize the fact that edge elements are required to provide coverage at the outer limits of their active element patterns.

Array Errors and Phase Quantization

The ability of an array to create a desired antenna pattern in space is limited by diffraction effects resulting from finite antenna size, by element pattern ripple, and by random and correlated errors in the array illumination.

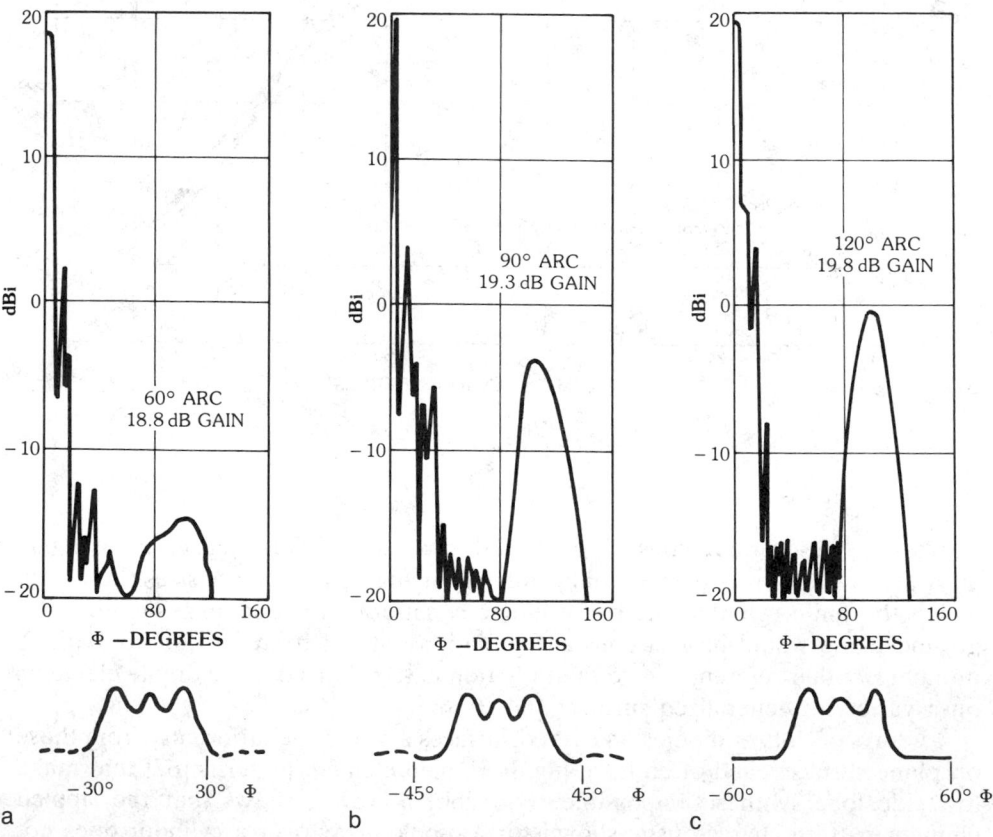

Fig. 27. Array gain (in dBi = dB relative to isotropic) and radiation patterns for 60°, 90°, and 120° arc arrays on a cylinder, where the lower curves show angular extent of element patterns used for given arc. (*After Sureau and Hessel [67], © 1972 Artech House*)

In the case of an array with random phase and amplitude errors, and including randomly failed elements, the average side lobe level far from the beam peak is given by [69, 70]

$$\bar{\sigma}^2 = \frac{\bar{\varepsilon}^2}{P_e N \eta} = \frac{[(1 - P_e) + \bar{\Delta}^2 + P_e \bar{\delta}^2]}{P_e N \eta} \tag{70}$$

where

$\bar{\Delta}^2 =$ the amplitude error variance normalized to unity

$\bar{\delta}^2 =$ the phase error variance

$\bar{\varepsilon}^2 =$ the error variance

$P_e =$ the probability of survival for any element in the array

$\eta =$ array efficiency

$N =$ total number of elements

This equation gives the normalized side lobe level relative to the average array gain. The failed elements in the array are assumed to be randomly located, and the average value of the phase and amplitude errors is assumed to be zero. The side lobe level above should be considered the average of a number of antenna patterns, not the average level of any one antenna.

If the broadside, no-error gain of an array with elements $\lambda/2$ apart is $\pi P_e N \eta$, the side lobe level is given from (70) as

$$\bar{\sigma}^2 = \pi \bar{\varepsilon}^2 / G \tag{71}$$

in terms of the gain G and the error variance.

Peak side lobe levels are also given in the literature. A convenient result is obtained when the errors are sufficiently large compared to side lobes or null depths that structured minor-lobe radiation is negligible and the statistics of the field intensity pattern are described by a Rayleigh density function. In this case the probability $P(v > v_0)$ that a particular side lobe level v_0^2 is exceeded at any point is [70]

$$P(v > v_0) = e^{-v_0^2/\bar{\sigma}^2} \tag{72}$$

where $\bar{\sigma}^2$ is the average side lobe level of (70).

Starting with the expression above, valid at a particular point, Allen [69] derives the following rule of thumb for the error $\bar{\varepsilon}^2$ allowable for an array with gain G, in terms of the far side lobe level $1/R$, assuming 99-percent probability that all side lobes are below the given level

$$\bar{\varepsilon}^2 \leqq \frac{1}{10\pi} \frac{G}{R} \tag{73}$$

which results in an allowable phase error of about $10°$ when the side lobe level is numerically equal to the gain. This important relationship explains why it is not difficult to design arrays with side lobes at the isotropic level $(G = R)$, but to maintain side lobes of 20 dB below the isotropic level would require $1°$ phase error, an extremely difficult goal and one barely within the present state of the art.

In this expression, and in all equations given in this section, when the parameter N is the number of elements in the array, the side lobe levels are those distributed throughout all real space; but when the errors are correlated in one plane, as they would be for power divider or phase shifter errors in the plane of scan for an array of columns, then N is the number of columns, and the side lobe level given by these equations is in the principal scan plane.

For the case of an array of columns of $\lambda/4$-spaced elements the gain of a column is equal to $\sqrt{\pi}\eta N P_e$, and (70) becomes

$$\bar{\sigma}^2 = \frac{\sqrt{\pi}\,\bar{\varepsilon}^2}{G_1} \tag{74}$$

where G_1 is the column gain, $\bar{\varepsilon}^2$ is the error variance between columns, and $\bar{\sigma}^2$ the average side lobe level in the plane perpendicular to the columns. For a square array this reduces to

$$\bar{\sigma}^2 = \frac{\sqrt{\pi}\,\bar{\varepsilon}^2}{\sqrt{G}} \tag{75}$$

where now G is the total array gain.

Fig. 28 shows the average side lobe level, (71), for a square array with random phase shift errors, and a square array of columns (75) with errors completely

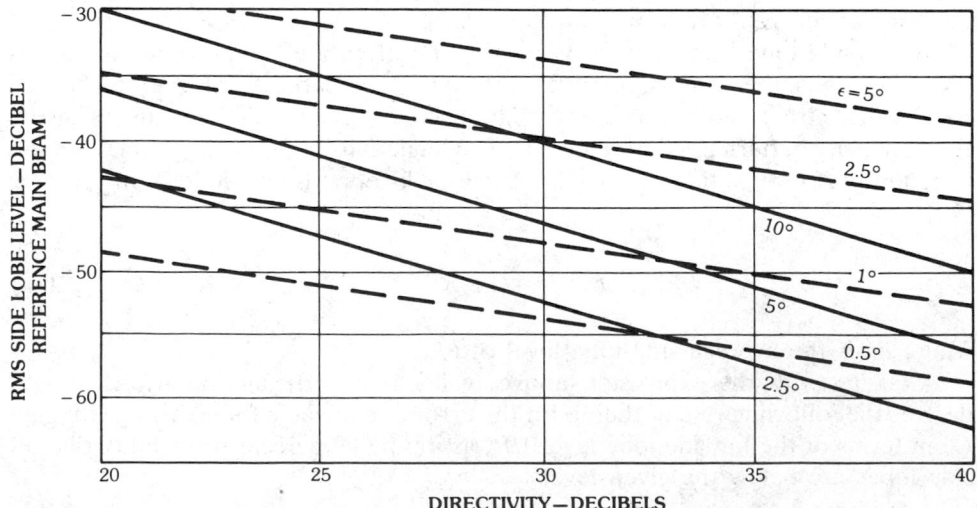

Fig. 28. Root-mean-square side lobe level versus directivity for square array with phase errors at elements (solid lines) or at columns (dashed lines, in plane of arrayed columns).

correlated in the plane of the columns, and the phase error between columns randomly distributed in the scan plane. This figure illustrates that extreme precision is required for arrays organized into columns when low principal-plane side lobes are required.

The reduction in directivity due to these errors is given by

$$\frac{D}{D_0} = \frac{P_e}{1 + \overline{\Delta}^2 + \overline{\delta}^2} \tag{76}$$

In the case of digitally controlled phase shifts, a p bit phase shifter has 2^p phase states separated by phase steps of $2\pi/(2^p)$. Miller [71] has analyzed the resulting peak and rms side lobe levels for this staircase approximation to the desired linear phase progression and has shown that the loss in array gain due to the triangular error distribution is

$$\Delta G = \frac{1}{3} \frac{\pi^2}{2^{2p}} \tag{77}$$

which is on the order of 0.23 for a 3-bit phase shifter and 0.06 for a 4-bit phase shifter. More significant are the average side lobe levels which, based on an average array loss of 2 dB to account for illumination taper and scan degradation, are

$$\text{rms side lobes} = \frac{5}{N \times 2^{2p}} \tag{78}$$

where N is the number of elements in the array. For a one-dimensionally scanned array, N is the number of phase controls, and the rms side lobe level above is measured in the plane of scan. The net result, as before, is to require extreme precision for unidimensional scanned arrays. Fig. 29 shows the side lobe level for various phase shifter bits p and N up to 10 000 elements. For −50-dB rms side lobes an array of 1000 elements requires 5-bit phase shifters, but an array of 10 000 elements can maintain 50-dB side lobes with only 3 phase bits.

Of greater significance to antenna design is that the phase errors have a periodic variation across the array and tend to collimate as individual side lobes, called *phase-quantization side lobes*, that are much larger than the rms levels. A detailed discussion of this phenomenon is given by Miller along with simple formulas for evaluating the resulting lobes. In the case of a perfectly triangular quantization error the quantization lobe level is $1/2^p$, which gives −30 dB for 5-bit phase shifters. Cheston and Frank [72] show that for discrete phase shifters the error is not triangular and that the maximum quantization lobe can be substantially larger.

One solution to the peak quantization lobe problem is to decorrelate the phase shifter errors. Decorrelation occurs naturally in space-fed arrays, where the phase shifters collimate the beam as well as steer it. In such arrays the phase error is distorted from the triangular shape and the quantization lobe is substantially reduced. Alternatively, in an array with in-phase power division one can introduce

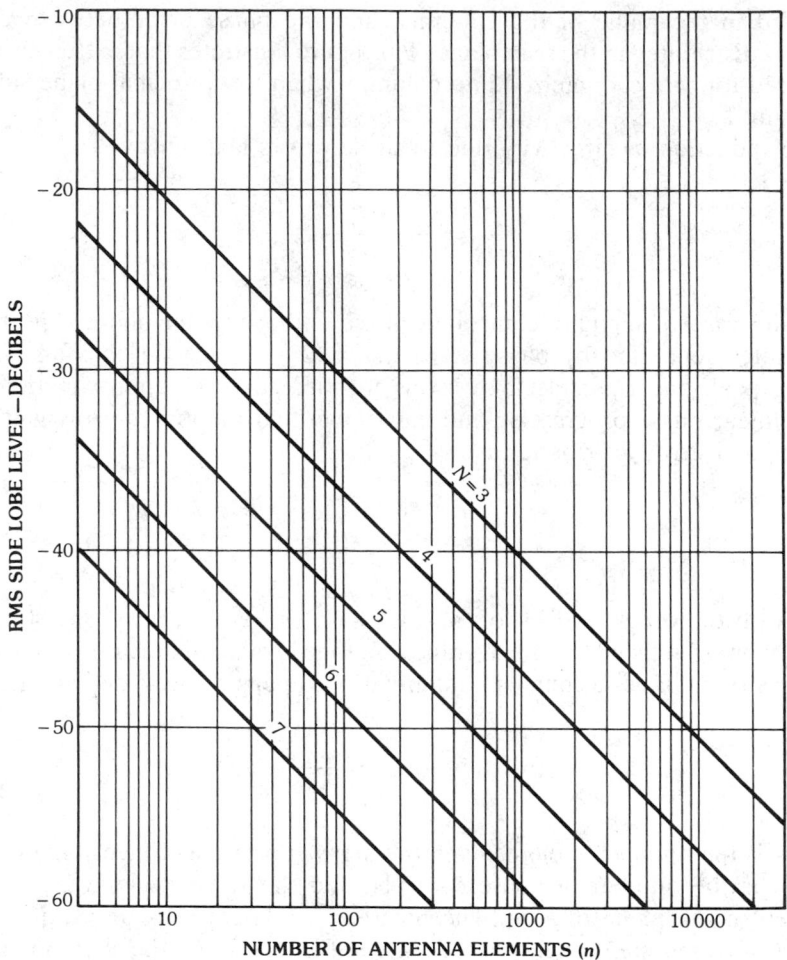

Fig. 29. Root-mean-square side lobes due to quantization, where N is the number of phase shifter bits. (*After Miller [71]*)

a phase error into each path and then program the phase shifter to remove the error in addition to steering the beam. Optimizing this error can reduce the peak side lobes very close to the rms side lobe level, but this is a consideration that must be carefully accounted for in the array design.

An entirely different solution to the quantization lobe problem is often achieved at the system level by recycling all the phase shifters between consecutive radar pulses or between transmit and receive. This process, called *beam dithering* [73], consists of adding a fixed phase shift to the phase command and recomputing phase shifts. The net result is to change all the phase states so that the quantization is made differently for each pulse (or between transmit and receive). If this procedure is compatible with other radar processing, one can use simple row-column steering but introduce randomness into the quantization steps to reduce the peak quantization lobes.

Array Elements

Array elements are usually some form of dipole or slot excited by a waveguide or other transmission line. Waveguide arrays, though heavy, tend to have low loss, good bandwidth, and relatively graceful scan degradation. They also have been the subject of numerous design studies, and so their behavior is well documented and predictable. Early examples of specific waveguide element designs are the studies of Wheeler [74, 75] in which matching networks were derived using waveguide transmission circuits like that shown in Fig. 30 consisting of dielectric slabs mounted in and above the waveguide. McGill and Wheeler [76] introduced the use of a dielectric sheet, often called a WAIM (wide-angle impedance-matching) sheet, to produce a susceptance variation with scan angle that partially cancels the scan mismatch of the array face. A significant development in impedance matching of waveguide arrays is the synthesis of double-tuned response characteristics achieved using dielectric loading and a cutoff waveguide section [77]. Fig. 31 shows a loaded rectangular waveguide (phase shifter), a transformer to circular guide, two dielectric disks, and an unloaded section of guide that is below cutoff at the operating frequencies.

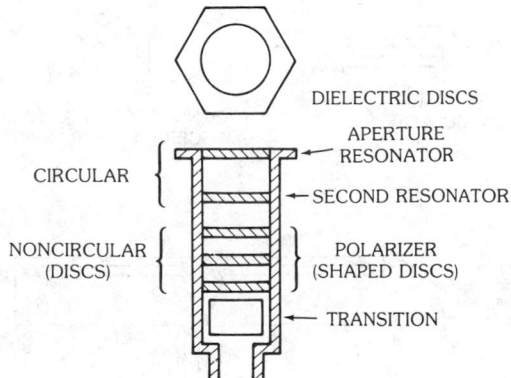

Fig. 30. Circular element for triangular grid array. (*After Wheeler [74, 75], © 1968 IEEE*)

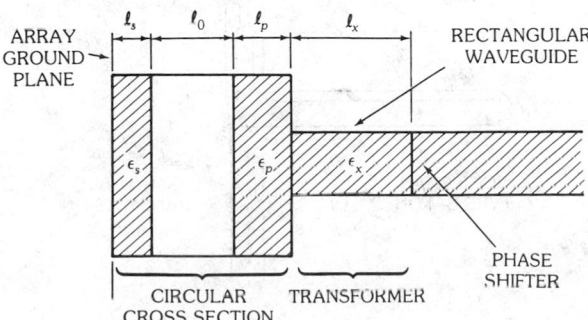

Fig. 31. Doubly tuned waveguide array element. (*After Lewis, Kaplan, and Hanfling [77], © 1974 IEEE*)

Dipole arrays have also received substantial attention and have generally graceful scan properties when properly designed. Fig. 32 shows several common varieties of stripline printed dipoles. One design, attributable to Wilkinson [78], uses metallization on two sides of a microstrip line to produce a complete dipole fed by a two-wire line in the plane of the dipoles. This dipole and printed circuit distribution network is fabricated by two photographic exposures using a two-sided printed-circuit board, and so is an example of low-cost technology. The array was mounted a quarter wavelength above a ground plane and uniformally illuminated by a reactive power divider to form a pencil beam. Another convenient circuit for dipole design, shown in Fig. 32, is described in a report by Hanley and Perini [79] and is a printed-stripline folded dipole with a Schiffman balun. One major advantage of this element is that it is printed in a single process, all on one side of a circuit board and so is relatively inexpensive to produce.

Fig. 33 shows one means of exciting flush-mounted stripline slot antennas. Most often these elements are isolated from the rest of the stripline medium by the

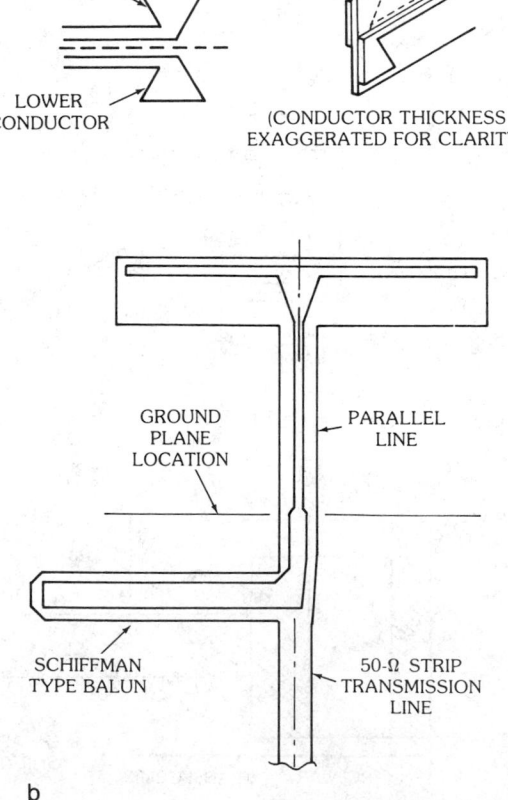

Fig. 32. Printed-circuit dipole configurations. (*a*) Conductors and dielectric. (*b*) Printed-stripline folded dipole and balun.

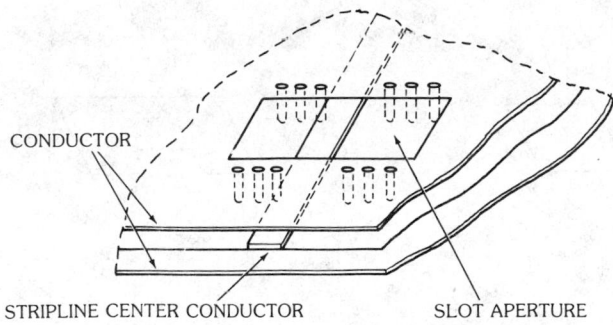

Fig. 33. Stripline slot radiator.

use of plated through holes or rivets that form a cavity as shown in the figure, which also serves to suppress higher-order modes.

Several very broadband elements have been described in the literature. Among these the broadest-band element is the flared-notch antenna (Fig. 34) studied by Lewis and others [61], which exhibits up to an octave band when used in an array, but needs careful design for any given array configuration because of the possibility of array blindness effects for critical frequencies and scan angles.

The microstrip patch element and its variations are inexpensive and lightweight, and have found increasing use in a variety of array applications. The basic path element (Fig. 35) is narrow-band, with percentage bandwidth given approximately by

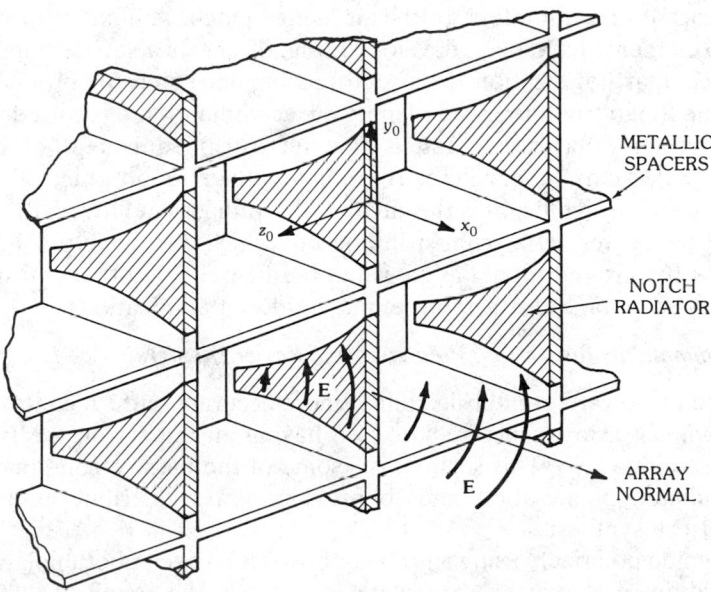

Fig. 34. Flared-notch array elements. (*After Lewis, Kaplan, and Hanfling [77], © 1974 IEEE*)

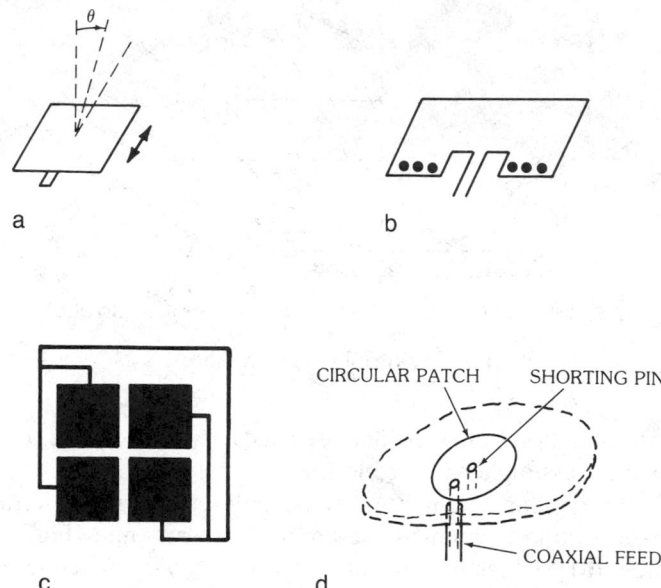

Fig. 35. Useful microstrip radiator types. (*a*) Patch antenna elements. (*b*) Shorted patch. (*c*) Crossed slot of four shorted patches. (*d*) Circular disk.

$$100 \frac{\Delta f}{f_0} = 5tf_0 \qquad\qquad (79)$$

for a thickness t in centimeters for the air-loaded patch. A number of very creative microstrip elements have been developed. Fig. 35 shows a square patch design for radiating circular polarization, a shorted rectangular patch for producing a much wider element pattern in the scan plane, and a combination of shorted patches that radiates circular polarization and is the microstrip equivalent of crossed slot radiations [80]. Many other microstrip radiators have found practical application. Of these, the most significant is the circular disk radiator of Howell [81] (Fig. 35d), which can be excited by a microstrip line but which also is very suitably excited from below the ground plane, as shown in the figure. Still other transmission line media offer advantages for array use in a variety of applications.

Passive Components for Arrays: Polarizers and Power Dividers

The trends toward greater flexibility, more accurate pattern control, and lightweight, compact array structures are also having an impact on the type of components used in arrays. This section lists some of the passive components used to control radiated polarization and to provide power distribution networks for precise pattern synthesis.

Waveguide polarizers using a probe or obstacle to excite both polarizations and some variation of a quarter-wave plate to produce the requisite ±90° delay for circular polarization have long been used as phased array components.

Waveguide polarizers in current use include the tapered septum polarizer [82] for converting linear to circular polarization, which operates over about a 2- to 5-percent bandwidth, and the stepped-septum polarizer of Chen and Tsandoulas, a three-port device that allows polarization agility over at least 20-percent bandwidth, with isolation greater than 26 dB [83].

Often it is less expensive to insert a polarizer in front of the whole array. The earliest polarizer of that type is the use of a grid of quarter-wave plates, but a much more popular recent solution that is compatible with wide-angle electronic scanning is the use of meander line polarizers following the work of Young's group [84].

Power divider networks for array feeds need to be extremely precise to synthesize low side lobe radiation patterns. Other requirements often impose extreme high-power specifications and still others demand very lightweight or compact construction practices. Waveguide and coaxial-line corporate feed networks are most often used for high-power arrays but the increasing ease and quality of stripline construction has made that the medium of choice for many new array developments. Often it is convenient to produce hybrid combinations of waveguide, coaxial line, and stripline to take advantage of inexpensive stripline network techniques for lower-power sections of the array while using waveguide or coaxial line at the high-power regions of the feed network.

Precise feed synthesis requires both equal and unequal power dividers. Among the various developments in stripline components the most commonly used are reactive tee power dividers, branch line couplers, and parallel coupled line and in-line power dividers. Although they are most simple and inexpensive to construct, reactive tee power divider networks have no isolated port and hence offer serious mismatch and isolation problems when used to feed mismatched elements. Reactive corporate-feed networks are therefore useful mainly for fixed beam arrays or for power division in the unscanned plane of arrays with one plane of scan. Single-section [85] branch line couplers occupy an area approximately $\lambda/4$ square and are most useful for coupling ranges from 3 to 9 dB. These couplers are easily fabricated using a conventional stripline by machining or etching the center conductor. Parallel coupled stripline power dividers for loose coupling (greater than 10 dB) can also be designed from conventional stripline using side coupled parallel lines, but tighter coupling requires the use of three-layer stripline for broadside coupled lines (3 to 6 dB coupling) or variable overlap couplers for intermediate values. Single-section parallel coupled power dividers are $\lambda/4$ long but occupy less area than branch line hybrids. Another likely choice for array feed networks is the Wilkinson [86] in-line power divider or its impedance-compensated derivative split-tee power dividers [87].

Single-section Wilkinson power dividers are $\lambda/4$ long for equal power division and $\lambda/2$ long for unequal power division. Split-tee power dividers have an extra stage of impedance matching and so are longer by approximately $\lambda/4$, although they have the advantage of wider bandwidth.

In either case the in-line power dividers have excellent broadband characteristics in comparison with branch and coupled line hybrids because the coupling ratio is determined by relative impedance ratios, not line length. Similarly, in-line hybrids are in-phase power dividers and so there is little phase error introduced with frequency change. The output ports of branch and coupled line hybrids have substantially different phases ($\pi/2$ for equal power division) and, although this can be compensated at center frequency, networks of these hybrids tend to be very narrow band relative to in-line hybrids. Typical bandwidths for individual in-line hybrids can reach an octave. The selection of power division networks is critical to array design and the choice can vary substantially with the application.

Array Phase Control

A discussion of time-delay devices for arrays is omitted because the primary components used to date are switched transmission lines and are governed by the same critical components as phase shifters.

Phasing networks are most often implemented at the rf operating frequency because this is usually the most efficient process. Notable exceptions use phase shifters at intermediate frequencies and up-convert to rf with amplifiers to improve efficiency. In addition, many ingenious intermediate-frequency phase scanning systems have used harmonically derived phased shifts or frequency displaced signals across an array to produce time-varying beam positions. Systems of this type are described in the literature and their operation is beyond the scope of this chapter. In general, diode phase shifters dominate the frequency range below 2 GHz, and ferrites are usually selected above 5 GHz for high-power applications. Diode phase shifters are compact and very lightweight compared with ferrite devices and so are gaining popularity in lightweight array configurations through 40 GHz, often in combination with low-cost monolithic microstrip antenna circuits. The rapid development of solid-state amplifier modules for arrays also favors the use of diode phase shifters at all frequencies.

At present the most popular types of diode phase shifter designs for arrays are the hybrid coupled, switched-line, and loaded-line phase shifters using pin diodes. These three fundamental networks are shown in Fig. 36. Hybrid and loaded-line (transmission) phase shifters require two diodes per bit, and switched-line phase shifters require four per bit. Switched-line hybrids also have greater insertion loss and an undesirable phase dependence with frequency that usually makes them unsuitable for low side lobe array control. They do possess distinct advantages in weight and compactness, however, and have been used successfully in monolithic microstrip antennas for many years. Hybrid and transmission phase shifters have lower loss and better bandwidth performance. An S-band stripline hybrid phase shifter is reported by White [88] to have an average phase loss of only 3° over 20 percent bandwidth. This device had about 0.8 dB average loss from 3.0 to 3.5 GHz and was tested to high-power burnout at 4 kW peak with 0.1-ms pulses and 0.05 duty cycle. Insertion loss for X-band and K_u-band phase shifters was about 2 and 3 dB.

Broadband low–side-lobe array designs are possible using Schiffman phase shifters because this phase shifter produces a nearly constant phase shift over extremely wide frequency ranges. White gives data for a 90° bit over a frequency ratio of 2.27:1.

A precise low–side-lobe array developed by Tsandoulas [89] used six-bit diode phase shifters with phase tolerance limits of less than 0.0° rms for the 90° and 180° bits, 0.4° for 45°, 22.5°, and 11.25° bits, and 0.2° for the 5.625°. These remarkable results were achieved in a practical testbed array described in the reference.

Ferrite phase shifters have been built to operate up to 60 GHz and possess

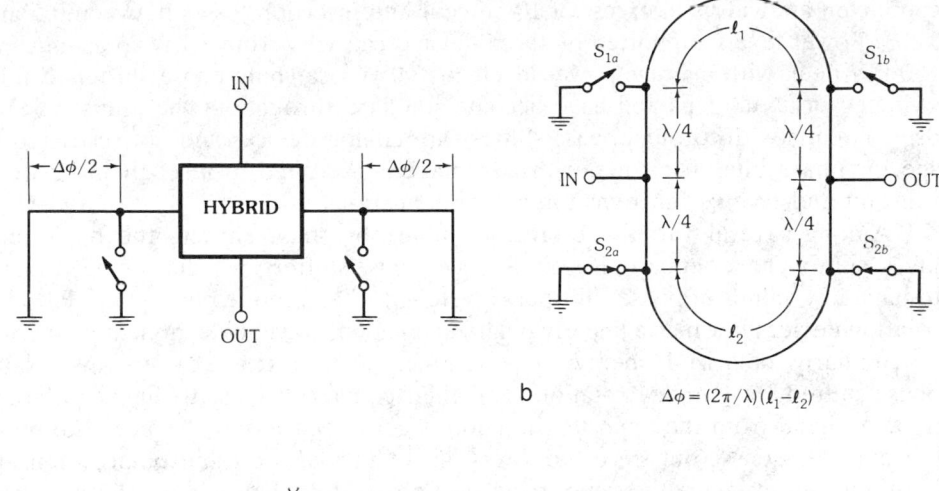

a   b $\Delta\phi = (2\pi/\lambda)(\ell_1 - \ell_2)$

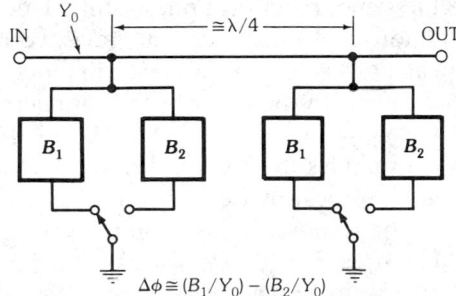

c $\Delta\phi \cong (B_1/Y_0) - (B_2/Y_0)$

Fig. 36. Diode phase shifter circuits. (*a*) Hybrid coupled reflection phase shifter. (*b*) Switched-line phase shifter. (*c*) Loaded-line phase shifter.

excellent characteristics for many phased array applications. Several recent survey articles and an annotated bibliography [90, 91, 92] summarize progress in this field and list numerous references to devices and to the fundamental theory of ferrite phasor operation. Nonreciprocal ferrite phase shifters include early twin-slab designs (Fig. 37) that require a transverse-switched external magnetic field and the well-known toroid designs that use a longitudinal wire to drive the ferrite magnetization to saturation as in a latching phase shifter, or to various points on the magnetization curve with flux drive circuitry. Typical digital latching phase shifters

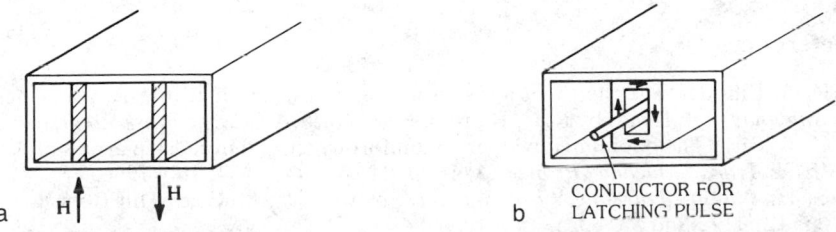

a   H   H   b   CONDUCTOR FOR LATCHING PULSE

Fig. 37. Dual-slab and toroid ferrite phase shifters. (*a*) Dual-slab. (*b*) Toroid.

can have bandwidths in excess of 10 percent and insertion losses between 0.5 and 1 dB. Power levels supported by these devices can vary from 1 kW to as much as 150 kW peak with average power levels to 400 W. Latching phase shifters can be switched in about 1 μs and have become standard throughout the industry. Flux drive circuits with toroidal phase shifters are analog devices and not restricted to specific phase bits. Their major disadvantage is the need to reset them between transmit and receive functions for radar applications.

Among several varieties of reciprocal ferrite phase shifters the dual-mode phase shifters have replaced Reggia-Spencer phase shifters, which have been found to have low values of phase shift per wavelength. Dual-mode phasors are Faraday rotation devices in which a linearly polarized incident waveguide mode is converted to circularly polarized energy by a nonreciprocal ferrite quarter-wave plate, phase-shifted by Faraday rotation, and then converted back to linear polarization. A signal from the opposite direction is converted to circular polarization of the opposite sense, but since the directions of propagation and polarization are both opposite it incurs the same, reciprocal phase shift. Dual-mode phase shifters are very competitive with toroid phase shifters and have average power levels up to 1.5 kW at S-band and peak powers to 150 kW. Insertion loss can be 0.6 dB through X-band. Switching speed can be on the order of tens of microseconds for a latched design, depending on the application. The reference by Ince and Temme [90] compares specific phasor examples for S-band through K_a-band.

One other important reciprocal phase shifter is the analog rotary-field phase shifter of Boyd [93]. This phase shifter is based on the principle of the commercial rotary-vane phase shifters after Fox [94] and uses a ferrite rod of circular cross section fitted with a slotted stator in which are wound two sets of coils each generating a four-pole field. In comparison with the well-known rotating half-wave plate of the Fox phase shifter, the stator windings produce a rotating four-pole field distribution with the orientation of the principal axes proportional to the coil driving current values. The dc distribution in the ferrite serves to rotate a virtual half-wave plate that is converted to a phase shift just as is the mechanical half-wave plate rotation of the Fox phase shifter. This circuit has the disadvantages of requiring substantial drive power and having relatively long switching times (200 to 500 μs). Its advantages for many applications far outweigh these disadvantages because it has nearly dispersionless phase shift that can be maintained within a degree or two over substantial bandwidths, has insertion loss well under 1 dB, and can handle very high peak and average power levels. An S-band model operates at 90 kW peak, 3 kW average power, 0.5 dB insertion loss, and phase tracking within ±1.5° over about 9 percent bandwidth.

5. References

[1] R. S. Elliott, "The theory of antenna arrays," in Chapter I, in *Microwave Scanning Antennas*, vol. II, ed. by R. C. Hansen, New York: Academic Press, 1966, pp. 29–31.

[2] C. T. Tai, "The optimum directivity of uniformly spaced broadside arrays of dipoles," *IEEE Trans. Antennas Propag.*, vol. AP-12, pp. 447–454, July 1964.

[3] C. H. Walter, *Traveling-Wave Antennas*, New York: McGraw-Hill Book Co., 1965, pp. 121–122 and 322–325.

[4] S. A. Schelkunoff, "A mathematical theory of linear arrays," *Bell Syst. Tech. J.*,

vol. 22, pp. 80–107, 1943.

[5] R. S. Elliott, "Design of line source antennas for sum patterns with side lobes of individually arbitrary heights," *IEEE Trans. Antennas Propag.*, vol. AP-24, pp. 76–83, June 1976.

[6] R. S. Elliott, "Design of line source antennas for difference patterns with side lobes of individually arbitrary heights," *IEEE Trans. Antennas Propag.*, vol. AP-24, pp. 310–316, June 1976.

[7] W. L. Stutzman and G. A. Thiele, *Antenna Theory and Design*, New York: John Wiley and Sons, 1966.

[8] P. M. Woodward, "A method of calculating the field over a plane aperture required to produce a given polar diagram," *J. IEE* (London), pt. IIIA, vol. 93, pp. 1554–1558, 1947.

[9] P. M. Woodward and J. P. Lawson, "The theoretical precision with which an arbitrary radiation pattern may be obtained from a source of finite size," *J. AIEE*, vol. 95, P1, pp. 362–370, September 1948.

[10] C. L. Dolph, "A current distribution for broadside arrays which optimizes the relationship between beamwidth and side lobe level," *Proc. IRE*, vol. 34, pp. 335–345, June 1946.

[11] H. J. Riblett, "Discussion of Dolph's paper," *Proc. IRE*, vol. 35, pp. 489–492, May 1947.

[12] D. Barbiere, "A method for calculating the current distribution of Tschebyscheff arrays," *Proc. IRE*, vol. 40, pp. 78–82, January 1952.

[13] R. J. Stegen, "Excitation coefficients and beamwidths of Tschebyscheff arrays," *Proc. IRE*, vol. 41, pp. 1671–1674, November 1953.

[14] L. B. Brown and G. A. Scharp, "Tschebyscheff antenna distribution, beamwidth, and gain tables," *NAVORD Rep. 4629 (NOLC Rep. 383)*, Naval Ordnance Lab., Corona, California, February 1958.

[15] C. J. Drane, Jr., "Useful approximations for the directivity and beamwidth of large scanning Dolph-Chebyshev arrays," *Proc. IEEE*, vol. 56, pp. 1779–1787, November 1968.

[16] T. T. Taylor, "Design of line source antennas for narrow beamwidth and low side lobes," *IEEE Trans. Antennas Propag.*, vol. AP-3, pp. 16–28, January 1955.

[17] R. C. Hansen, ed., *Microwave Scanning Antennas, Volume 1*, app. 1, New York: Academic Press, 1966.

[18] E. T. Bayliss, "Design of monopulse antenna difference patterns with low side lobes," *Bell Syst. Tech. J.*, vol. 47, pp. 623–640, 1968.

[19] R. S. Elliott, "Design of line source antennas for difference patterns with side lobes of individually arbitrary heights," *IEEE Trans. Antennas Propag.*, vol. AP-24, no. 3, pp. 310–316, May 1976.

[20] T. T. Taylor, "Design of circular apertures for narrow beamwidth and low side lobes," *IRE Trans. Antennas Propag.*, vol. AP-8, pp. 17–22, January 1960.

[21] F. I. Tseng and D. K. Cheng, "Optimum scannable planar arrays with an invariant side lobe level," *Proc. IEEE*, vol. 46, pp. 1771–1778, 1968.

[22] R. Tang, "Survey of time-delay beam-steering techniques," *Phased Array Antennas: Proceedings of the 1970 Phased Array Antenna Symposium*, Dedham: Artech House, pp. 254–260, 1972.

[23] V. Borgiotti, "An antenna for limited scan in one plane: design criteria and numerical simulation," *IEEE Trans. Antennas Propag.*, vol. AP-25, pp. 232–243, March 1977.

[24] R. L. Fante, "Systems study of overlapped subarrayed scanning antennas," *IEEE Trans. Antennas Propag.*, vol. AP-28, pp. 668–679, September 1980.

[25] P. D. Hrycak, "The theoretical and experimental investigation of a constrained-feed totally overlapped subarray antenna system," *IEEE 1982 AP-S Symp. Dig.*, pp. 695–698, May 24–28, 1982.

[26] R. J. Mailloux, "An overlapped subarray for limited scan application," *IEEE Trans. Antennas Propag.*, vol. AP-22, pp. 487–489, May 1974.

[27] E. C. Dufort, "Constrained feeds for limited scan arrays," *IEEE Trans. Antennas*

Propag., vol. AP-26, pp. 407–413, May 1978.

[28]   W. Rotman and P. Franchi, "Cylindrical microwave lens antenna for wideband scanning application," *IEEE Intl. Symp. Dig.*, vol. AP-S, pp. 564–567, June 1980.

[29]   J. H. Richmond, "Digital computer solutions of the rigorous equations for scattering problems," *Proc. IEEE*, vol. 53, pp. 796–804, August 1965.

[30]   R. F. Harrington, *Field Computation by Moment Methods*, New York: Macmillan Co., 1968.

[31]   R. W. P. King, *The Theory of Linear Antennas*, Cambridge: Harvard University Press, 1956, p. 79.

[32]   G. H. Brown and R. King, "High-frequency models in antenna investigations," *Proc. IRE*, vol. 22, pp. 457–480, April 1934.

[33]   P. S. Carter, "Circuit relations in radiating systems and applications to antenna problems," *Proc. IRE*, vol. 20, pp. 1004, 1041, June 1932.

[34]   C. T. Tai, "Coupled antennas," *Proc. IRE*, vol. 36, pp. 487–500, April 1948.

[35]   R. W. P. King, R. B. Mack, and S. S. Sandler, *Arrays of Cylindrical Dipoles*, London: Cambridge University Press, 1968, pp. 282–283.

[36]   V. W. H. Chang and R. W. P. King, "Theoretical study of dipole arrays of N parallel elements," *Radio Sci.*, vol. 3 (New Series), no. 5, September–October 1968.

[37]   N. Amitay, V. Galindo, and C. P. Wu, *Theory and Analysis of Phased Array Antennas*, New York: Wiley Interscience, 1972.

[38]   R. J. Mailloux, "First-order solutions for mutual coupling between waveguides which propagate two orthogonal modes," *IEEE Trans. Antennas Propag.*, vol. AP-17, no. 6, pp. 740–746, November 1969.

[39]   K. E. Golden et al., "Approximation techniques for the mutual admittance of slot antennas in metallic cones," *IEEE Trans. Antennas Propag.*, vol. AP-22, pp. 44, 48, 1979.

[40]   H. Steyskal, "Analysis of circular waveguide arrays on cylinders," *IEEE Trans. Antennas Propag.*, vol. AP-25, pp. 610–616, 1977.

[41]   C. P. Wu, "Analysis of finite parallel-plate waveguide arrays," *IEEE Trans. Antennas Propag.*, vol. AP-18, no. 3, pp. 328–334, May 1970.

[42]   G. F. Farrell, Jr., and D. H. Kuhn, "Mutual coupling effects of triangular grid arrays by modal analysis," *IEEE Trans. Antennas Propag.*, vol. AP-14, pp. 652–654, September 1966.

[43]   G. F. Farrell, Jr., and D. H. Kuhn, "Mutual coupling in infinite planar arrays of rectangular waveguide horns," *IEEE Trans. Antennas Propag.*, vol. AP-16, pp. 405–414, July 1968.

[44]   J. C. Herper, C. J. Esposito, C. Rottenberg, and A. Hessel, "Surface resonances in a radome covered dipole array," *1977 IEEE AP-S Intl. Symp. Dig.*, pp. 198–201.

[45]   G. H. Knittel, A. Hessel, and A. A. Oliner, "Element pattern nulls in phased arrays and their relation to guided waves," *Proc. IEEE*, vol. 56, no. 11, pp. 1822–1836, November 1968.

[46]   V. W. H. Chang, "Infinite phased dipole array," *Proc. IEEE*, vol. 56, no. 11, pp. 1892–1900, November 1968.

[47]   E. Mayer and A. Hessel, "Feed region modes in dipole phased arrays," *IEEE Trans. Antennas Propag.*, to be published.

[48]   G. N. Tsandoulas and G. H. Knittel, "The analysis and design of dual-polarization square waveguide phased arrays," *IEEE Trans. Antennas Propag.*, vol. AP-21, pp. 796–808, November 1973.

[49]   N. Amitay and V. Galindo, "The analysis of circular waveguide phased arrays," *Bell Syst. Tech. J.*, vol. 47, pp. 1903–1931, November 1968.

[50]   M. H. Chen and G. N. Tsandoulas, "Bandwidth properties of quadruple ridge circular and square waveguide radiators," *IEEE AP-S Intl. Symp. Rec.*, pp. 391–394, June 1973.

[51]   J. P. Montgomery, "Ridged waveguide phased array elements," *IEEE Trans. Antennas Propag.*, vol. AP-24, no. 1, pp. 46–53, January 1976.

[52]   S. S. Wang and A. Hessell, "Aperture performance of a double-ridge rectangular

waveguide in a phased array," *IEEE Trans. Antennas Propag.*, vol. AP-26, pp. 204–214, March 1978.

[53] L. R. Lewis, A. Hessell, and G. H. Knittel, "Performance of a protruding-dielectric waveguide element in a phased array," *IEEE Trans. Antennas Propag.*, vol. AP-20, pp. 712–722, November 1972.

[54] R. J. Mailloux and H. Steyskal, "Analysis of a dual frequency array technique," *IEEE Trans. Antennas Propag.*, vol. AP-27, no. 2, pp. 130–134, March 1979.

[55] J. K. Hsiao, "Computer aided impedance matching of an interleaved waveguide phased array," *IEEE Trans. Antennas Propag.*, vol. AP-20, pp. 505–506, July 1972.

[56] V. Galindo and C. P. Wu, "Dielectric loaded and covered rectangular waveguide phased arrays," *Bell Syst. Tech. J.*, vol. 47, pp. 93–116, January 1978.

[57] S. W. Lee and W. R. Jones, "On the suppression of radiation nulls and broadband impedance matching of rectangular waveguide phased arrays," *IEEE Trans. Antennas Propag.*, vol. AP-19, pp. 41–51, May 1971.

[58] R. J. Mailloux, "Surface waves and anomalous wave radiation null phased arrays of TEM waveguides with fences," *IEEE Trans. Antennas Propag.*, vol. AP-20, pp. 160–166, January 1972.

[59] E. C. Dufort, "Design of corrugated plates for phased array matching," *IEEE Trans. Antennas Propag.*, vol. AP-16, pp. 37–46, January 1968.

[60] C. C. Chen, "Octave band waveguide radiators for wave-angle scan phased arrays," *IEEE AP-S Intl. Symp. Rec.*, pp. 376–377, June 1972.

[61] L. R. Lewis, M. Fassett, and J. Hunt, "A broadband stripline array element," *IEEE AP-S Intl. Symp. Dig.*, pp. 335–337, June 1974.

[62] W. H. Chang, "Infinite phased dipole array," *Proc. IEEE*, vol. 56, pp. 1892–1900, 1968.

[63] B. H. Pathak and R. G. Kouyoumjian, "An analysis of the radiation from apertures in curved surfaces by the geometrical theory of diffraction," *Proc. IEEE*, vol. 62, no. 11, pp. 1433–1447, November 1974.

[64] W. D. Burnside, R. J. Marhefka, and C. L. Yu, "Roll-plane analysis of on-aircraft antennas," *IEEE Trans. Antennas Propag.*, vol. AP-21, no. 6, pp. 780–786, November 1973.

[65] W. D. Burnside, M. C. Gilreath, R. J. Marhefka, and C. L. Yu, "A study of KC-135 aircraft antenna patterns," *IEEE Trans. Antennas Propag.*, vol. AP-23, no. 3, pp. 309–316, May 1975.

[66] B. H. Pathak, N. Wang, W. D. Burnside, and R. G. Kouyoumjian, "A uniform GTD solution for the radiation from sources on a convex surface," *IEEE Trans. Antennas Propag.*, vol. AP-29, pp. 609–622, July 1981.

[67] J. C. Sureau and A. S. Hessel, "Realized gain function for a cylindrical array of open-ended waveguides," *1970 Proc. Phased Array Antennas*, ed. by A. A. Oliner and G. H. Knittel, Dedham: Artech House, 1972, p. 283.

[68] J. C. Herper, C. Mandarino, R. Hessel, and B. Tomasic, "Performance of a dipole element in a cylindrical array—a modal approach," *IEEE AP-S Intl. Symp.*, pp. 162, 165, 1980.

[69] J. L. Allen, "The theory of array antennas," *Tech Rep. 323*, MIT Lincoln Labs, July 1963.

[70] R. E. Collin and F. J. Zucker, eds., *Antenna Theory*, New York: McGraw-Hill Book Co., 1969.

[71] C. J. Miller, "Minimizing the effects of phase quantization errors in an electronically scanned array," *Proc. 1964 Symp. on Electron. Scanned Array Tech. Appl.*, RADC TR-64-225, vol. 1, pp. 17–38, RADC, GAFB, NY.

[72] T. C. Cheston and J. Frank, *Radar Handbook*, Chapter 11, ed. by M. I. Skolnik, New York: McGraw-Hill Book Co., 1970, pp. 11–42.

[73] E. Brookner, ed., *Radar Technology*, Dedham: Artech House, 1977.

[74] H. A. Wheeler, "A systematic approach to the design of a radiator element for a phased array antenna," *Proc. IEEE*, vol. 56, pp. 1940–1951, November 1968.

[75] H. A. Wheeler, "A survey of the simulator techniques for designing a radiating

element," *Array Antennas: Proceedings of the 1970 Phased Array Antenna Symposium*, ed. by A. A. Oliner and G. H. Knittel, Dedham: Artech House, 1972.

[76] E. G. McGill and H. A. Wheeler, "Wide-angle impedance matching of a planar array antenna by a dielectric sheet," *IEEE Trans. Antennas Propag.*, vol. AP-14, no. 1, pp. 49–53, January 1966.

[77] L. R. Lewis, L. J. Kaplan, and J. L. D. Hanfling, "Synthesis of a waveguide phased array element," *IEEE Trans. Antennas Propag.*, vol. AP-22, no. 4, pp. 536–540, July 1974.

[78] W. C. Wilkinson, "A class of printed circuit antennas," *IEEE AP-S Intl. Symp. Dig.*, pp. 270–273, 1974.

[79] G. R. Hanley and H. R. Perini, "Column network study for a planar array used with an unattended radar," *RADC TR-80*, final report, RADC, GAFB, NY, March 1980.

[80] G. Sanford and L. Klein, "Increasing the beamwidth of a microstrip radiating element," *Intl. Symp. Dig. of Antennas Propag. Soc.*, Univ. of Washington, pp. 126–129, June 1979.

[81] J. Q. Howell, "Microstrip antennas," *IEEE Trans. Antennas Propag.*, vol. AP-23, pp. 90–93, January 1975.

[82] D. Davis, O. J. Digiandomenico, and J. A. Kempic, "A new type of circularly polarized antenna element," *IEEE G-AP Symp. Dig.*, pp. 2–23, 1967.

[83] M. H. Chen and G. N. Tsandoulas, "A wideband square waveguide array polarizer," *IEEE Trans. Antennas Propag.*, vol. AP-21, pp. 389–391, May 1973.

[84] L. Young, L. A. Robinson, and C. A. Hacking, "Meander line polarizer," *IEEE Trans. Antennas Propag.*, vol. AP-21, pp. 376–378, May 1973.

[85] H. Howe, *Stripline Circuit Design*, Dedham: Artech House, 1974.

[86] E. Wilkinson, "An *n*-way hybrid power divider," *IEEE Trans. Microwave Theory Tech.*, vol. MTT-8, no. 1, pp. 116–118, January 1960.

[87] L. I. Parad and R. L. Moynihan, "Split-tee power divider," *IEEE Trans. Microwave Theory Tech.*, vol. MTT-13, pp. 91–95, January 1965.

[88] J. F. White, *Semiconductor Control*, Dedham: Artech House, 1977.

[89] G. N. Tsandoulas, "Unidimensionally scanned phased arrays," *IEEE Trans. Antennas Propag.*, vol. AP-28, no. 1, pp. 86–98, January 1980.

[90] W. J. Ince and D. H. Temme, "Phasors and time-delay elements," *Advances in Microwaves*, vol. 4, pp. 2–183, New York: Academic Press, 1969.

[91] L. R. Whicker and C. W. Young, "The evolution of ferrite control components," *Microwave J.*, vol. 21, pp. 33–37, November 1978.

[92] L. R. Whicker and D. M. Bolle, "Annotated literature survey of microwave ferrite control components and materials for 1968–1974," *IEEE Trans. Microwave Theory Tech.*, vol. MTT-23, no. 11, pp. 908–918, November 1975.

[93] C. R. Boyd, "Analog rotary-field ferrite phase shifters," *Microwave J.*, vol. 20, pp. 41–43, December 1977.

[94] A. G. Fox, "An adjustable waveguide phase changer," *Proc. IRE*, vol. 35, pp. 1489–1498, December 1947.

Chapter 14

Aperiodic Arrays

Y. T. Lo
University of Illinois

CONTENTS

Yuen T. Lo is a professor and the director of the Electromagnetics Laboratory (formerly the Antenna Laboratory) in the Electrical and Computer Engineering Department, University of Illinois at Urbana-Champaign. He is a member of the National Academy of Engineering and a Fellow of IEEE, and a member of the International Union of Radio Science. He received the 1964 IEEE AP-S Bolljahn Memorial Award, the 1964 IEEE AP-S Best Paper Award, the 1979 IEEE AP-S Best Paper Award, the IEEE Centennial Medal, and the Halliburton Education Leadership award. He served as an AP-S AdCom member, the Chairman of the AP-S Education and Tutorial Papers Committee, and twice (1979–1982 and 1984–1987) as IEEE AP-S National Distinguished Lecturer. Dr. Lo is an honorary professor of the Northwest Telecommunication Engineering Institute and also the Northwestern Polytechnical University, both at Xian, China. He has published over a hundred technical articles in refereed journals covering a wide spectrum, from theoretical to experimental works. His works include large antenna arrays, radio telescopes, multiple-beam antennas, multiple scattering, antenna synthesis, antijamming antennas, antenna in plasmas, corrugated guides and horns, artificial dielectrics, and microstrip antennas. He designed the University of Illinois Radio Telescope, considered to be the world's largest antenna in the early 1960s.

1. Introduction

Arrays with uniformly spaced elements (often called *uniformly spaced arrays*) have been widely studied, mainly for two reasons: mathematical tractability in many cases and simplicity in fabrication. As discussed in Chapter 11, the pattern function of a uniformly spaced array, no matter how it is excited, is always periodic in the wave-vector **k**-space, or sometimes expressed in the so-called **u**-space. Thus a beam in the pattern function will repeat itself an infinite number of times. However, the visible region corresponds only to a finite region of this space. To ensure that only a single beam appears in the visible region (i.e., no grating lobes), adjacent element spacing must be kept sufficiently small, and as a result a large number of elements must be used to fill a given aperture, which is determined by the desired beamwidth and, sometimes, directivity also.

In contrast, arrays with incommensurable element spacings have aperiodic pattern functions, and are thus called *aperiodic arrays*. In general, they have no grating lobes, and, as a result, the required number of elements is not directly related to the grating lobe condition. Some nonuniformly spaced arrays, even with commensurable element spacings, may have very large periods in the pattern functions as compared with that of a uniformly spaced array and thus exhibit characteristics similar to those of aperiodic arrays. They may be regarded as pseudo-aperiodic arrays or, for all practical purposes, simply aperiodic arrays. This situation occurs actually in *all* practical arrays since all element spacings must be rounded up numerically to a finite number of the digits and are therefore divisible by a common unit.

Aperiodic arrays may be used in many different ways, but the most interesting applications are in array thinning, beamwidth narrowing, and element-interaction reduction. In regard to directivity it is shown in Chapter 11 that for element spacing greater than a half-wavelength the directivity for a uniform excitation is nearly maximum, approximately equal to the number of elements. This conclusion can also be deduced from the fact that, in general, the *mutual* radiation resistance becomes less significant in comparison with the self-radiation impedance. As an example, for half-wavelength dipoles, as the element spacing increases beyond a half-wavelength the total power radiated depends mainly on the self-radiation resistance and thus reaches approximately a constant value. From this point of view one can expect that the same conclusion can be drawn even for nonuniformly spaced arrays, as most element spacings, if not all, are greater than $\lambda/2$. On the other hand, if element spacings are allowed to assume any value, the result in Chapter 11 shows that maximum directivity can be reached when element spacings become vanishingly small. However, this is the so-called superdirectivity and cannot be realized in practice. From this argument one can therefore conclude that the optimization of element spacings for maximum directivity is a problem of little practical significance. Thus in this chapter we shall direct our attention to other aspects of the array.

One of the interesting array synthesis problems is to find an optimum set of element spacings and excitations that would minimize the highest side lobe level in the entire visible region. Many attempts have been made, including some analytic methods (using the Poisson summation formula), numerical methods (using the quadrature approximation and the so-called dynamic programming), and statistical and probabilistic approaches [1–21]. It is shown [22] that none of these methods can yield a solution even close to the true optimum. The difficulty of this problem may be attributed to the fact that the side lobe level depends on the element spacings in a highly nonlinear manner, and that, in general, there is no known analytical method to determine the highest side lobe level, or the angular direction where the highest side lobe may occur, even with all element positions given. Except for small arrays a numerical search with a modern computer is considered impractical. The reason is that the highest side lobe position does not change *continuously* with the element positions; thus there is no simple way to keep track of the highest side lobe as the element position changes unless an entire three-dimensional pattern is computed. With this understanding it is not surprising to find that so far all the attempts are not successful. In the following we shall make a brief review of these works and finally focus our discussion on the probabilistic approach because, first, the theory is more complete, second, it has been supported by various experiments, and, last, it can provide a useful practical solution which, though not optimum in the ordinary sense, is optimum in some probabilistic sense (see below). It will be seen that the array so obtained can have a performance much superior to that of conventional arrays in many respects.

2. A Brief Review

Unz [1] used the Fourier-Bessel expansion to relate the pattern function to the element positions. As pointed out by various authors it is difficult to make use of this expansion to yield useful numerical results. King, Packard, and Thomas [2] computed the pattern functions of a few sample arrays with preassigned spacings. Their computed results reveal some interesting properties of nonuniformly spaced arrays. At about the same time, nonuniformly spaced arrays were studied at the University of Illinois in connection with the feed of a radio telescope [3]. In that paper a certain optimization procedure and the method of relating the element spacings to the excitation function were proposed and applied. Maffett [4] independently proposed the same method of relating the element position function to the excitation function. Andreason [5] suggested a procedure for using a computer to optimize the element positions such that the side lobe level in the visible region is minimized. This method has been independently applied by Lo [3], who also stated that, at best, the solution so obtained is only optimum locally. For large arrays this procedure becomes tedious and time consuming, if not completely impossible. In his article Andreason gave an interesting lower bound of the side lobe level of widely spaced arrays by using the fact that the directivity of these arrays is proportional to the number of elements.

Ishimaru [6], Yen and Chow [7], and Ishimaru and Chen [8], using the Poisson summation formula, reduced the pattern function of a finite sum to an infinite sum. In general, this would make the computation more difficult; however, for small u, it is reasonable to assume that only the first term of the series is important. However, for large u or, equivalently, for very wide average element spacing, which, unfortunately, is the case of greater interest, other terms also become significant. Ishimaru and Chen in particular considered a spacing function of the type $x + (2A_1/\pi) \sin \pi x$ with $A_1 < \frac{1}{2}$, which is a perturbed uniform spacing, particularly when A_1 is small. As $A_1 = \frac{1}{2}$, this spacing function becomes identical with that considered earlier by Lo [3]. However, using this type of spacing function, Ishimaru expressed the pattern function in terms of the Anger function, and, for convenience, he and his associate compiled a table for this function [9]. In principle, for a general spacing function one may need to consider more terms of this type as indicated in their paper, but it is doubtful that the numerical convenience could be retained. Ishimaru and others claimed that the choice of a single term with $A_1 = \frac{1}{6}$ gave a pattern quite close to the optimum. As shown in [22], this is not so.

Skolnik, Nemhauser, and Sherman [11] realized that a straightforward search for an optimum set of element spacings, using a high-speed computer, is next to impossible, even for a moderately large array. As a result they proposed a systematical method using the so-called dynamic programming technique. Unfortunately, this method does not lead to a truly optimal solution since, as noted by the authors, the "principle of optimality" (which is essential in the successful use of the dynamic programming) does not apply to this problem [11]. In particular, the assumption that the optimal position of the first element depends *only* on the position of the second elements or in general that the $(n - 1)$st element depends on the nth element is not valid for large arrays.

In addition, it is perhaps proper to cite the works of Harrington [14] and Baklanov, Pokrovshi, and Surdotovich [15], whose investigations are not concerned with reducing the number of elements but rather with achieving certain pattern characteristics by spacing weighting. Baklanov, Pokrovshi, and Surdotovich, in particular, used some complicated numerical techniques in an attempt to derive a spacing-weighted array from the Dolph-Chebyshev counterpart. From a practical point of view this may appear to have some advantage since it is much easier to achieve spacing weighting than excitation weighting. In doing so, however, some element spacings have to be smaller than usual, and the strong coupling effects between elements may offset this apparent advantage appreciably. Furthermore, the method seems to be too cumbersome to be applicable to large arrays.

Finally, we may mention the probabilistic and statistical approaches to this problem. Rabinowitz and Kolar [16] analyzed the case that the placement of elements over a uniform grid system in an aperture is determined by the outcome of a random experiment. They obtained only the *mean* side lobe level which, for the example studied in their paper, is about 7 dB lower than what was computed actually. Later, Skolnik, Sherman, and Ogg [17] considered the same problem and made a very similar analysis. They too obtained only the mean side lobe level; likewise, their examples showed that the predicted level is about 7 to 8 dB lower than the actually computed result in what they called "principal" planes.

Maher and Cheng [18] studied the problem of random removal of elements in a uniformly spaced array. Their assumption that the removals of elements are statistically independent events may result in the removal of the same element many times.

Starting with the probability distribution function of the elements in the aperture, Lo [12] obtained the distribution function of the antenna response at each observation angle, the half-power beamwidth, and the directivity, etc. He also obtained an approximate distribution of the side lobe level in any range of u. Later, these results have been verified by using the Monte Carlo simulation [13] and actual holey plate experiment [23]. Later Agrawal and Lo [24] refined the analysis, extended the theory for small arrays, and included the effect of element interactions. Clearly this method does not lead to an optimum design in the true sense.

The confusion of various theories finally prompted Lo and Lee [22] to conduct a comparison study with a few small arrays. They found that none of the theories yields the optimum solution; in fact, most are far from it. Similar to the so-called space-tapered arrays, some of these designs can control near-in side lobes only to some extent, and the far-out side lobe level not at all.

3. Spaced-Tapered Arrays

A spaced-tapered array is one in which the interelement spacing varies in some fashion across the array, usually increasing from the center to the edge of an array. This method is motivated by an attempt to remove the following two disadvantages of a conventional uniformly spaced array [3]:

(1) *Inefficiency*, in the sense that, in most low side lobe level designs, a large number of elements in the outer portion of the array serves only the purpose to cancel partially the high side lobes produced by the central portion of the array but contributes very little to the radiated power.

(2) *Feeding difficulty*, in that the design may require the element excitation power to be level over a range of several orders of magnitude.

As discussed in Chapter 11 a discrete array could be regarded as a sampling of an aperture antenna with "continuous" illumination functions. When the sampling is uniform one obtains a uniformly spaced array with element excitation weighted according to the illumination function. If the sampling interval is small enough, one expects a close similarity between the pattern functions of the discrete and the continuous array, such as $\sin(N\psi/2)/\sin(\psi/2)$ for the former against $\sin(L\psi/2)/(\psi/2)$ for the latter in the case of a uniform excitation with aperture length $L \cong Nd/\lambda$. On the other hand, if the sampling is made such that each interval contains the same amount of excitation power, one obtains a space-tapered array with all elements excited equally in power as illustrated in Fig. 1.

Design Procedure for a Symmetrical Space-Tapered Array

(*a*) Choose a proper illumination function $f(x)$ (see comments below).
(*b*) Define and compute

$$g(x) = \int_{-a/2}^{x} f(x)\, dx \qquad (1)$$

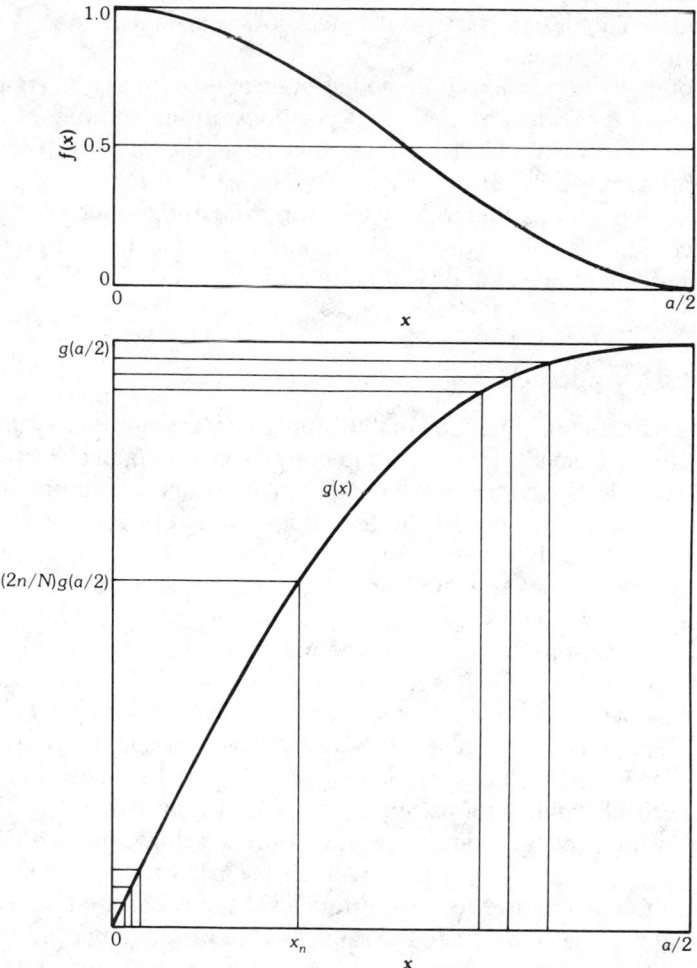

Fig. 1. A graphical method for determining element location of a space-tapered array.

where a = aperture dimension in wavelengths.

(c) Let N be the number of elements, which is approximately equal to the array directivity.

(d) Solve for x_n, the nth element location, in

$$g(x_n) = \frac{2g(a/2)}{N}n, \qquad n = 1, 2, \ldots, N/2 \tag{2}$$

which could easily be solved graphically as illustrated in Fig. 1.

Advantages—Usually a smaller number of elements are needed and all elements excited *equally*.

Disadvantages—With the exception of the region near the main beam, the array pattern may differ greatly from that of the corresponding aperture antenna; usually the side lobe level increases with observation angle.

Comments—The design is useful if only a low near-in side lobe is desired, or directive elements are used.

The set of element positions obtained above can be used as a starting set for an iteration procedure in which the element positions are moved one at a time in a direction where the side level is decreasing. Of course, the final set so determined is at best an optimum locally [3].

The extension of this method to two- or even three-dimensional arrays is obvious, since it needs only to divide the illumination function in equal power per area or volume where an element will be placed.

4. Probabilistic Approach

Since the pattern function is an analytic function it is theoretically impossible to have contributions from all elements completely canceled for all observation angles except for a single direction in which the main beam maximum is intended. Therefore all patterns (except for the trivial case of a very wide beam) have side lobes. Consequently the best design one could hope for in many applications would be one having all side lobes equal in level. When a large number of elements is used the uniformly spaced Chebyshev-Dolph array has already provided the solution. But for economic reasons one may ask whether or not a similar performance could be achieved with substantially fewer elements. As discussed in the review section, so far all attempts have failed. In view of this fact it is natural to resort to a probabilistic approach so that contributions from all elements would be as incoherent as possible for all observation angles except in the main beam maximum direction where all contributions should be completely coherent. This can be achieved by using random element positions (or spacings, but with progressive phase excitation in case of scan). It will be seen later that, in so doing, the probability for the pattern to exceed a certain level at any observation angle outside the main beam region can be made the same as at any other angle. In the language of information theory this is a state of maximum entropy. It is only in this sense that the probabilistic design is optimum; in other words, all side lobes are equal in a probability sense. Therefore this design is, in a sense, a probabilistic approach to the Chebyshev-Dolph array.

For simplicity consider first a linear array, while the extension to the planar array is straightforward as will be shown later. In the following, all length dimensions are understood to be measured in wavelength λ. Let $g(X)$ = probability density function for placing an element at X, with $|X| \leq a/2$ and

$$\int_{-a/2}^{a/2} g(X)\,dX = 1 \tag{3}$$

If there are N equally excited elements each of which is to be placed within the array aperture $(-a/2, a/2)$ according to the same probability function $g(X)$ but independently of each other, then for each set of random samples $\{X\}$: $(X_1, X_2, \ldots, X_N)$ there is associated a sample pattern function

$$F(u) = \frac{1}{N} \sum_{n=1}^{N} \exp[j2\pi(\sin\theta - \sin\alpha)X_n] = \frac{1}{N} \sum_{n=1}^{N} \exp(jux_n) \qquad (4)$$

where the last expression is a normalized pattern function with

θ = any observation angle

α = main beam angle

$x_n = 2X_n/a$ = normalized element position

$u = a\pi(\sin\theta - \sin\alpha)$

As a result of the normalization in X, the aperture becomes $(-1, 1)$ and (3) can be rewritten as

$$g(x) \equiv 0 \quad \text{for } |x| > 1$$

$$\int_{-1}^{1} g(x)\, dx = 1 \qquad (5)$$

The ensemble of all sample sets $\{x\}$ constitutes a corresponding ensemble of $\{F(u)\}$. Our objective is to study the probabilistic properties of $F(u)$, in particular, those related to the array performance. For later discussion it is noted that the array dimension a appears *only* in the parameter u. Thus the pattern of an array with any value a can be obtained simply from that of a normalized array. In other words, if a is increased by a factor of, say, two, one needs only to extend the computation of $F(u)$ for twice the range of u while $F(u)$ in the first half range remains the same except that the scale of u changes by a factor of two. This scaling is not important if one is only interested in the highest side lobe level in the visible region.

Theoretical Results

Let $E\{\cdot\}$ be a probability average operator. A few important results are summarized below:

(*a*) The mean of $F(u)$ is

$$\phi(u) \equiv E\{F(u)\} = \int_{-\infty}^{\infty} g(x)\, e^{jux}\, dx \qquad (6)$$

which is simply the Fourier transform of $g(x)$, or in probability theory the characteristic function of the random variable x, or in antenna theory the pattern function of a "continuous" aperture antenna with illumination function $g(x)$. This implies that by choosing a proper $g(x)$, at least the mean pattern $\phi(u)$ can behave in some desired manner. It may also be noted that since $g(x)$ is an aperture-limited function, $\phi(u)$ is analytic.

(*b*) For *any given* u, the joined probability density function of the real and imaginary parts of $F(u)$, namely $F_1(u)$ and $F_2(u)$, is asymptotically normal:

$$f(F_1, F_2) = \frac{1}{2\pi\sigma_1\sigma_2} \exp\left\{-\left[\frac{1}{2}\frac{(F_1 - \phi)^2}{\sigma_1^2} + \frac{F_2^2}{\sigma_2^2}\right]\right\} \qquad (7)$$

where the independent variable u has been suppressed, and $g(x)$, for simplicity, is assumed to be even, implying

$$\text{Im}\{\phi(u)\} = 0$$
$$\sigma_1^2(u) = \text{variance of } F_1(u) = E\{[F_1(u) - \phi(u)]^2\}$$
$$= \frac{1}{2N}[1 + \phi(2u)] - \frac{1}{N}\phi^2(u) \qquad (8)$$
$$\sigma_2^2(u) = \text{variance of } F_2(u) = E\{F_2^2(u)\}$$
$$= \frac{1}{2N}[1 - \phi(2u)] \qquad (9)$$

(c) Let $\text{Pr}\{\cdot\}$ be the probability measure for the event in the curly braces; then

$$\text{Pr}\{|F(u)| < r\} = \iint_{(F_1^2 + F_2^2) < r^2} f(F_1, F_2)\, dF_1 dF_2 \qquad (10)$$

which is a generalized noncentral chi-square distribution with two degrees of freedom. Tables of percentiles in r for various values of the parameters are available [25]. When r is large compared with σ_1 and σ_2, an asymptotic solution is given by

$$\text{Pr}\{|F| < r\} = \frac{1}{2}\text{erf}(sr)[\text{erf}(r + m) + \text{erf}(r - m)]$$
$$- \frac{1}{4\sqrt{\pi}s^2 r}\exp\left(-\delta^2\left\{1 + \frac{3\delta}{4s^2 r} + \left[\frac{3}{8s^2} + \frac{15(2\delta^2 - 1)}{48s^4}\right]\frac{1}{r^2}\right.\right.$$
$$\left.\left. + 0(s^{-4}r^{-3}) + 0[\sqrt{skr}\exp(-s^2k^2r^2)]\right\}\right) \qquad (11)$$

where $m = \phi/\sigma_1$, $s = \sigma_1/\sigma_2$, $\delta = r - m$, and $0 < k < 1$.

For the special case when $\sigma_1 = \sigma_2$ (i.e., $s = 1$) which occurs for u outside the main beam region, the above expansion reduces to that given by Rice [26], who utilized the asymptotic expansion of the Bessel function in his derivation. For large $|u|$, $\phi(u) \cong 0$, and the distribution becomes simply Rayleigh. For the general case, however, it can approximately be computed by a method due to Patnaik [27] which approximates a noncentral chi-square distribution by a central one with different degrees of freedom. The latter can then be read from an incomplete gamma function table prepared by Pearson [28], which is shown as $I(v, p)$ in Fig. 2 with

$$\text{Pr}\{|F| < r\} \cong I(v, p) \qquad (12)$$

where

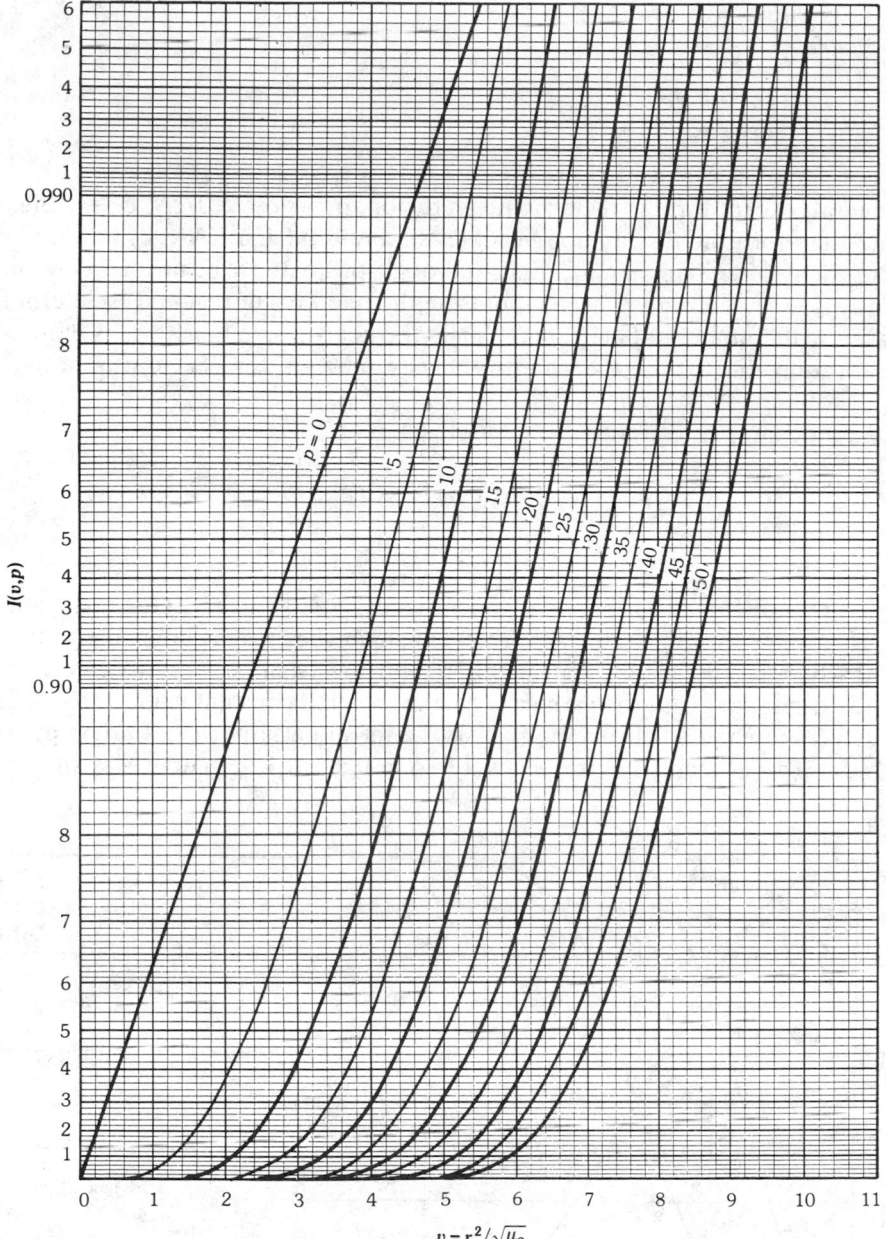

Fig. 2. Incomplete gamma function approximation for noncentral chi-square distribution. The curve for $p = 0$ is the Gausian distribution. (*After Lo [13], © 1964 by the American Geophysical Union*)

$$v = r^2/\sqrt{\mu_2}$$
$$p = \mu_1^2/\mu_2 - 1$$
$$\mu_1 = \sigma_1^2 + \sigma_2^2 + \phi^2$$
$$\mu_2 = 2(\sigma_1^4 + \sigma_2^4 + 2\phi^2\sigma_1^2)$$

In general, except for u in the main beam region, $\sigma_1^2 \cong \sigma_2^2 \cong 1/2N$, $\phi \cong 0$ and $p \cong$ 0, and therefore the distribution becomes simply Rayleigh $I(Nr^2, 0)$ in Fig. 2 and independent of u. This implies that, although the pattern behavior in the main beam region is determined by $g(x)$, outside the main beam region it is determined *only* by N, the number of elements. Unless the near-in side lobe level is of interest it is advantageous to use the uniform density function for $g(x)$ so that a narrow beam is obtained.

As an example, consider

$$g(x) = \begin{cases} \cos^2(\pi x/2) & \text{for } |x| \leqq 1 \\ 0 & \text{for } |x| > 1 \end{cases}$$

Fig. 3 shows the mean pattern $\phi(u)$ and variances $\sigma_1^2(u)$ and $\sigma_2^2(u)$ for each value of u/π. It is seen that in the neighborhood of main beam maximum the variances are nearly zero, indicating that in this region the beam is almost deterministic. Outside the main beam the variances quickly reach a constant value $1/2N$. Figs. 4 through 6 show the level curve as well as the mean pattern $|\phi(u)|$ for comparison, for $N = 10^2$, 10^4, and 10^5, respectively. A *p-percent* level curve is a plot of r_p

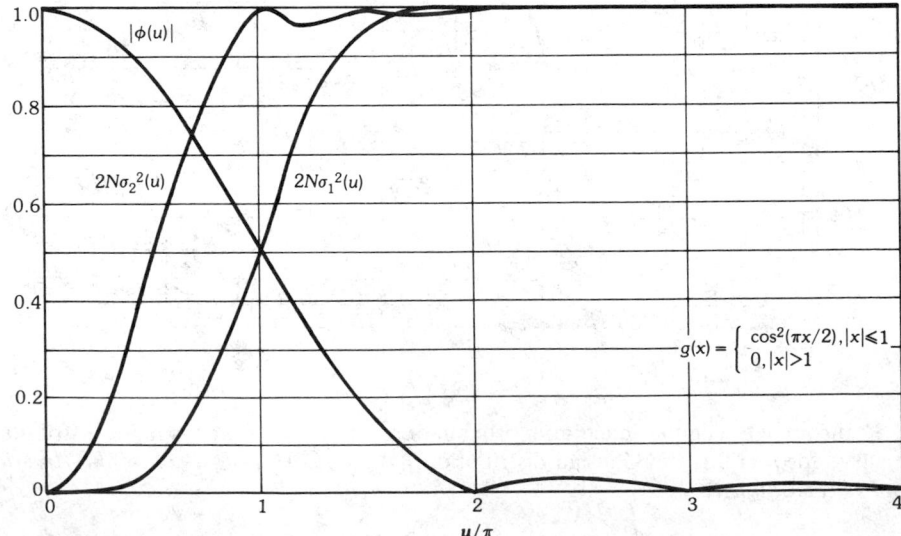

Fig. 3. The mean $\phi(u)$ and variances σ_1^2 and σ_2^2 of the real and imaginary parts of the random pattern function $|F(u)|$ as functions of u, for a cosine-square probability density function and uniform excitation. (*After Lo [12]*, © *1964 IEEE*)

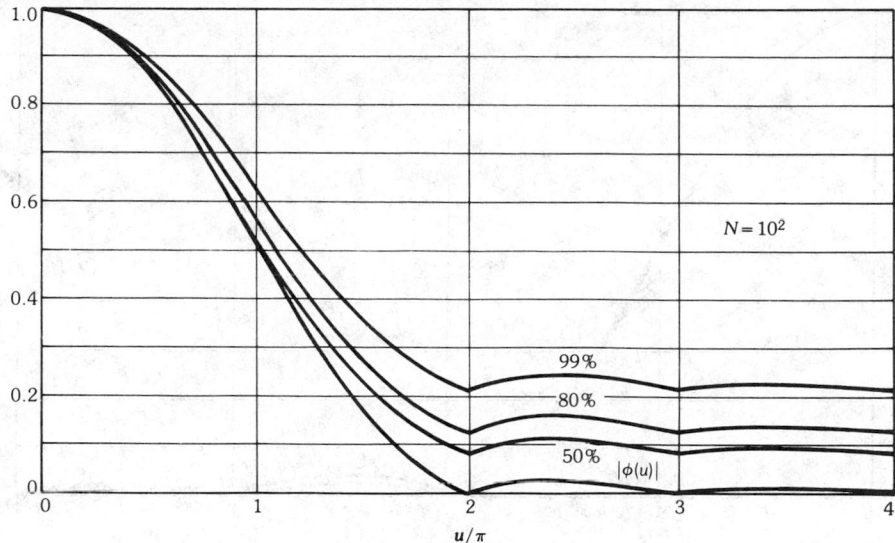

Fig. 4. Level curves of $|F(u)|$ for cumulative probability equal to 99 percent, 80 percent, and 50 percent with a cosine-square probability density function and uniform excitation, for $N = 10^2$. (*After Lo [12], © 1964 IEEE*)

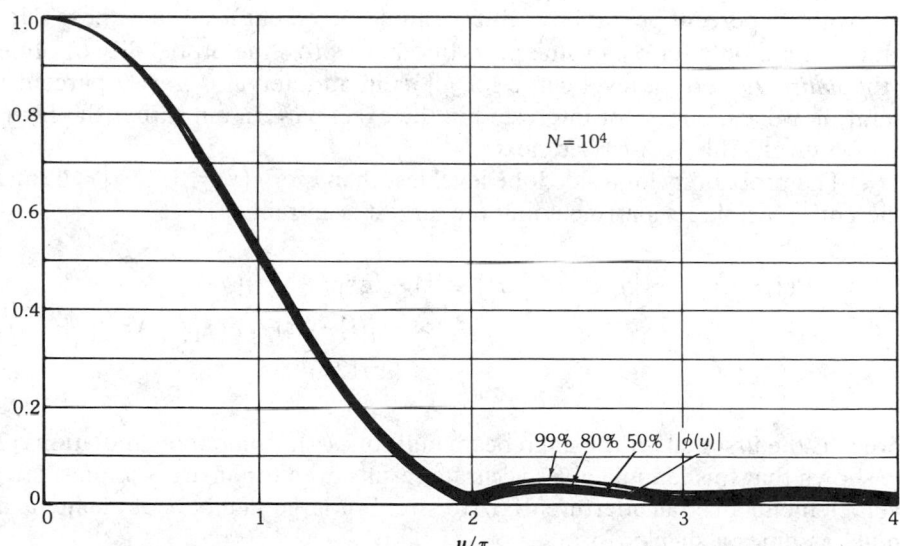

Fig. 5. Level curves of $|F(u)|$ for cumulative probability equal to 99 percent, 80 percent, and 50 percent with a cosine-square probability density function and uniform excitation, for $N = 10^4$. (*After Lo [12], © 1964 IEEE*)

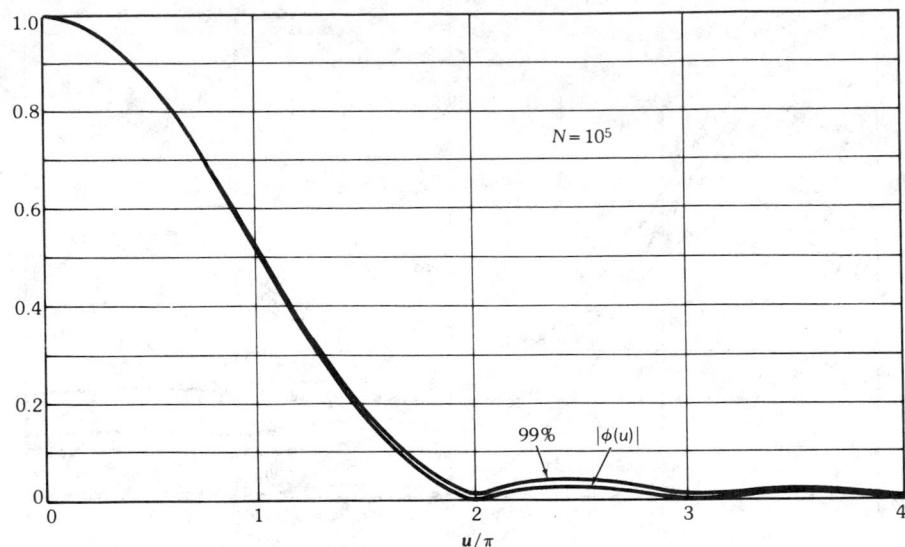

Fig. 6. Level curves of $|F(u)|$ for cumulative probability equal to 99 percent with a cosine-square probability density function and uniform excitation, for $N = 10^5$. *(After Lo [12],* © *1964 IEEE)*

versus u with $\Pr\{|F(u)| < r_p\} = p$ percent. For $N \geqq 10^4$, $|F(u)|$ is almost equal to $|\phi(u)|$ with 99 percent probability. But it should be emphasized that these curves are meaningful only for *each* value of u. In other words, the probability of obtaining the *entire* 99 percent level curve for all *us* in an interval is *not* 99 percent. In general, however, one is not interested in the exact pattern but rather the highest side lobe level. This is discussed next.

(*d*) The probability for a side lobe level less than r for $g(x) = 1/2$, $|x| < 1$, and u in the entire visible region (including the largest scan range) is

$$\Pr\{|F(u)| < r, |u| \in (u_f, 2\pi)\} \cong [1 - \exp(-Nr^2)]$$
$$\times \exp\{[-4\pi\sqrt{N}\,r\exp(-Nr^2)]$$
$$\times (a^2/12\pi)^{1/2}\} \tag{13}$$

where u_f is the first null (or the main beam null) of $\phi(u)$. Computer simulations [24] have shown that this formula gives accurate results even for an array with as few as eleven elements over an aperture of 5λ to 10λ. For large numbers of elements the formula assumes a simpler form:

$$\Pr\{|F(u)| < r, |u| \in (u_f, 2\pi)\} \cong [1 - 10^{-0.4343Nr^2}]^{[4a]} \tag{14}$$

where $[4a]$ is the integer part of $4a$. This formula shows that the probability for achieving a certain side lobe level r is nearly zero as N is below a certain value and increases sharply as N increases to a certain critical value. After that the probability increases very slowly. Thus to ensure a good probability of success a sufficient number of elements must be used for a desired side lobe level. For a 90 percent probability the required number N is plotted in Fig. 7 versus side lobe level in both ratio and decibel scale for various aperture dimensions $a = 10^q \lambda$. In this figure the graphs to the left of the dash-dot line are less accurate; here (13) should be used. It is of interest to see that for a 25-dB side lobe level the array needs about 4700 elements over $10^5 \lambda$ aperture with a half-power beamwidth of about $0.0005°$, or $(5 \times 10^{-4})°$, and average element spacing of 20λ. If these same elements (4700) are spread over an aperture ten times larger, i.e., $10^6 \lambda$, the side lobe level increases only by about 0.5 dB but the beamwidth will be reduced by a factor of 10, i.e., to $0.00005°$, or $(5 \times 10^{-5})°$. Now the average element spacing becomes 200λ, and for this spacing an ordinary equally spaced array would have hundreds of grating lobes. Thus it is not necessary to use a large number of elements in order to achieve a narrow beam. In the case of a planar array the saving in the number of elements is even more dramatic. However, as will be seen later, this does not imply that the directivity can be increased in this manner. The common notion of associating a narrow beamwidth with high directivity is not always correct.

(e) The random variable u_0 defined by the smallest positive root of the following equation determines the half-power beamwidth $2\theta_0$:

$$|F(u_0)| = 1/\sqrt{2} \qquad (15)$$

where $F(u)$ is the random pattern function given by (4) and $u_0 = \pi a(\sin\theta_0 - \sin\alpha)$. Even for the deterministic problem the solution u_0 to the above problem cannot be obtained analytically. Since in practice one is interested only in large arrays with narrow beams an approximate distribution for u_0 can be found from the joint probability of $F(u)$ and $dF(u)/du$ for u in the neighborhood of the mean pattern half-power value u_1, i.e., $|\phi(u_1)| = 1/\sqrt{2}$. Details can be found in Lo [12]. For example, when $g(x) = \cos^2(\pi x/2)$ and $N = 10^4$ the half-power angle θ_0 will fall in $0.7244/a < \sin\theta_0 - \sin\alpha < 0.7377/a$ with 90 percent probability, where α is the scan angle. This implies that the half-power beamwidth deviates from its mean no more than ± 0.91 percent with 90 percent probability; in other words, it is almost certain in practice that the half-power beamwidth is equal to that of the mean pattern. This conclusion is generally true for most functions $g(x)$ of interest and, in fact, verified by Monte Carlo simulations and holey plate experiments to be discussed later.

(f) Since the element locations are drawn from a collection of random numbers, for each sample set of these numbers there is associated a sample radiation pattern and thus a sample directivity. It is of interest to know how the directivity is distributed probabilistically for all possible sets of the element arrangement. Using the Karhunen and Loeve theorem for the expansion of a random function [29], one can obtain the following approximate distribution for the norm $\|F(u) - \phi(u)\|$:

$$\Pr\{\|P(u) - \phi(u)\| < k \|\phi(u)\|\} = \Phi\left[\{(k^2\|g(x)\|^2/d_{av}) - 1\}\sqrt{\frac{a}{2}}\bigg/\|g(x)\|\right] \qquad (16)$$

where

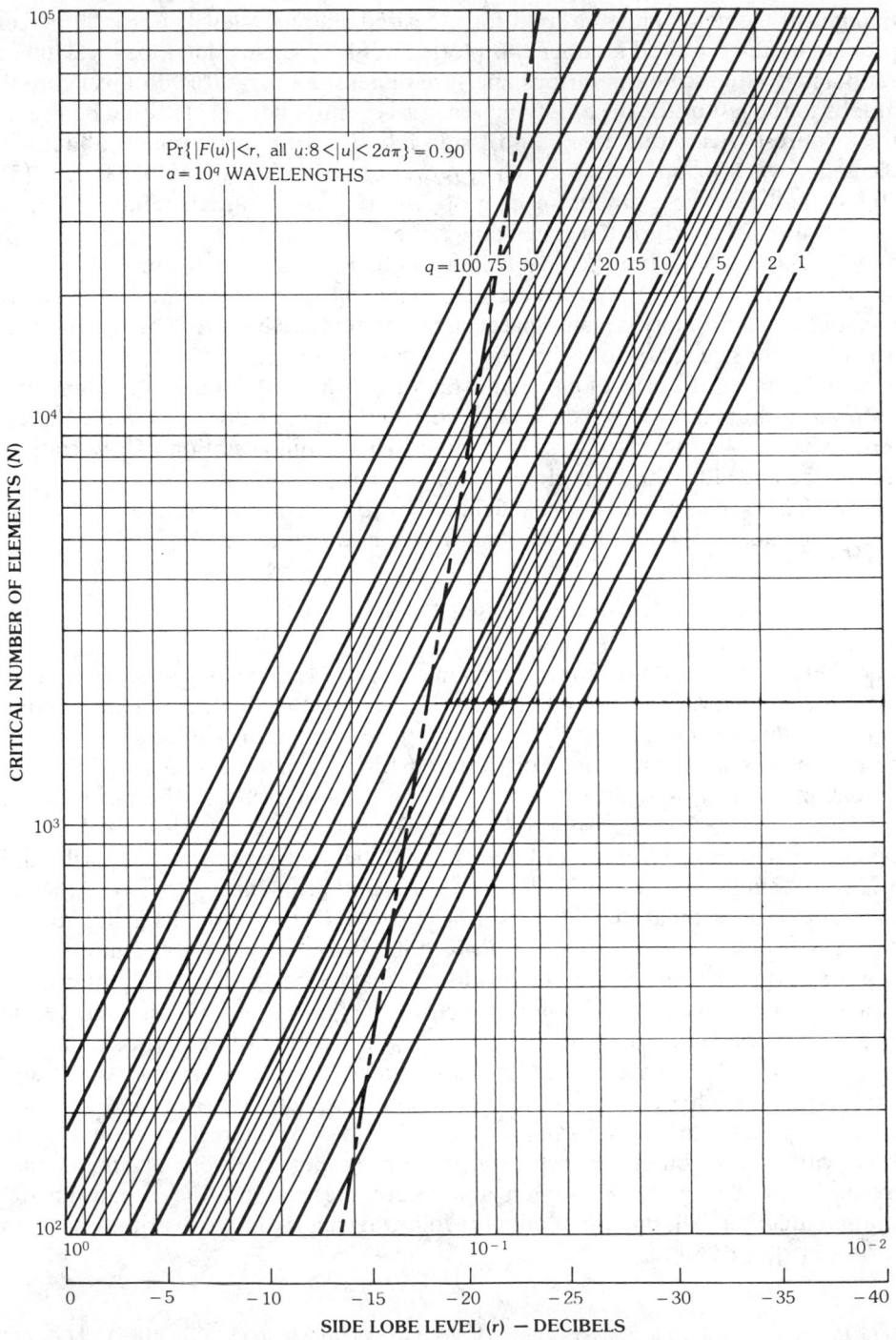

$\Pr\{|F(u)| < r, \text{ all } u : 8 < |u| < 2a\pi\} = 0.90$
$a = 10^q$ WAVELENGTHS

$q = 100 \ 75 \ 50 \qquad 20 \ 15 \ 10 \qquad 5 \qquad 2 \quad 1$

Fig. 7. Number of elements required as a function of the side lobe level for various values of $a = 10^q \lambda$ with a 90 percent probability. (*After Lo [12],* © *1964 IEEE*)

$$\|F(u) - \phi(u)\|^2 = \int_{\substack{\text{visible} \\ \text{region}}} |F(u) - \phi(u)|^2 \, du$$

$$\|g(x)\|^2 = \int_{-1}^{1} |g(x)|^2 \, dx$$

d_{av} = average element spacing $\cong a/N$

k = a positive real number

$\Phi(\cdot)$ = standardized normal distribution function (i.e., one with zero mean and unit variance)

Because generally a is a very large number, if k^2 is slightly greater than $d_{\text{av}}/\|g\|^2$, the above probability is nearly unity, while if k^2 is slightly less than $d_{\text{av}}/\|g\|^2$, the above probability drops sharply to zero. From this result and the definition of directivity D associated with each sample pattern function, and the directivity D_0 associated with the mean pattern function $\phi(u)$, one can conclude that with probability nearly one

$$(D_0 - D)\,\text{dB} \leqq 20\log(1 + \sqrt{d_{\text{av}}}/\|g(x)\|) \tag{17}$$

In other words, it is practically certain that D is lower than D_0 by a quantity no greater than $20\log(1 + \sqrt{d_{\text{av}}}/\|g\|)\,\text{dB}$, which for large average element spacing equals $20\log\sqrt{d_{\text{av}}}/\|g(x)\|$. The interpretation of this result can easily be understood if we consider two arrays of a different number of elements, say, N_1 and N_2. Let their corresponding directivities be D_1 and D_2, respectively; then the above results show that

$$(D_1 - D_2)\,\text{dB} \cong 10\log(N_1/N_2) \tag{18}$$

with a probability nearly equal to 1. In other words it is practically certain that the directivity D is proportional to N, the number of elements. This result is in agreement with that obtained in Chapter 11. The reason why the probability is not exactly 100 percent is that there exist particular element arrangements, but with very small probabilities, for which D is not exactly equal to N, such as superdirectivity, as shown in Chapter 11.

(g) There is little difficulty in extending most of the above results to an array of higher dimensions. For example, consider a rectangular aperture of $ab\lambda^2$ in the xy plane with a probability density function $g(x, y)$ for the element locations which would produce a mean pattern with sufficiently low side lobe level as before. Then the relation between the total number of elements and the side lobe level is still approximately given by (14), except that [4a] is replaced by [16ab]. Assuming that $a = 10^q$ and $b = 10^p$, this relation is again shown by Fig. 7 except that q should be replaced by $(p + q)$, and 90 percent probability by 80 percent (approximately).

To give an example, the Benelux Cross antenna [30] is considered. This antenna consists of two perpendicular linear arrays each having an aperture of roughly $7 \times 10^3\lambda \times 80\lambda$, or $ab \cong 10^6\lambda^2$. From Fig. 7, with a -30-db side lobe level and 90 percent chance of success, each arm requires only about 1.9×10^4 elements. But according to Christiansen and Hogbem [30], each arm would require 2×10^6 uniformly spaced dipoles. This is about 100 times larger than the former estimate. Moreover, if elements were randomly spread over a square aperture $10^4 \lambda \times 10^4 \lambda$, a total of only 2.3×10^4 elements would be needed to achieve a -30-dB side lobe level. This is very significant since the cross type telescope, being a multiplicative antenna system, will have a side lobe level of -15 dB in the two principal planes even if each arm is designed to have a -30-dB side lobe level. Of course, as already noted before, the directive gain of the randomly spaced array is perhaps 10 or 15 dB lower than that of the uniformly spaced array. Since the former has uniform amplitude weighting function while a uniformly spaced array must have a strongly tapered illumination in order to achieve a -30-dB side lobe level, then from a practical point of view (considering the feeding system and the aperture efficiency) the loss in directive gain may not be as much as indicated above.

Illustrative Examples

Suppose that it is desired to design a linear array whose array pattern will have approximately a half-power beamwidth of 1 minute of arc and a side lobe level of -18 dB. If $g(x)$ is chosen to be uniform, the near-in side lobe level would likely be -13 dB. To ensure that it is below -18 dB, one may, although not necessarily, let

$$g(x) = \cos^2(\pi x/2) \qquad (19)$$

For this function the half-power beamwidth in degrees is approximately $83/a$ and, therefore, $a \cong 83 \times 60 \cong 5 \times 10^3(\lambda)$. From Fig. 7, N (the number of elements needed for a 90 percent probability of success) is approximately 800. The next step is to generate 800 random numbers, between $(0, 1)$ under the probability density function $g(x)$ given above, and these numbers, after scaled by the factor a, determine the actual element locations in the aperture. According to the theory this array when excited uniformly will have a 90 percent probability to yield a pattern with 1 minute of arc in half-power beamwidths and -18-dB side lobe level. Of course, before an investment is made for constructing this array it is prudent to first compute the pattern with the given set of element positions just found, since there is still a 10 percent chance of failure. In a sense the theory predicts that in about nine out of ten trials one should obtain the desired pattern. Hundreds of computer simulations have been made for many different arrays, and in almost all cases the desired properties as predicted by the theory were obtained in the very first trial. Since the *same* basic array theory is used for computing this pattern as for any conventional equally spaced array, there is no reason to doubt this array performance cannot be realized. In fact in this case, all elements being widely spaced, the mutual coupling will have a much less deleterious effect on the pattern. After the pattern properties are confirmed by the computations the array can then be constructed. From that step on, the array is as *deterministic* as any conventional array. This will be demonstrated in a holey plate experiment later.

For simplicity in pattern computation a symmetrical array has been considered for the example stated above. In this case the required number of elements becomes about 1000, somewhat more than that shown in Fig. 7, since there are in effect only 500 random numbers [13]. These numbers were taken directly from Owen's table [31]. Since these numbers are uniformly distributed, the probability integral transformation [32] is then used to convert the set of numbers into another set which obeys the cosine-square law prescribed in (19). This is done by solving the following equation for x_n:

$$y_n = \int_0^{x_n} g(x)\,dx \tag{20}$$

where y_n is the uniformly distributed random number between 0 and 1, and x_n is the $g(x)$ distributed random number. This method is similar to that used to convert a set of uniform element spacings to a set of nonuniform element spacings in Fig. 1. With this transformation the distribution of the 500 random numbers taken from Owen's table is shown as dotted curves in Fig. 8. For comparison the exact cosine-square distribution is shown in solid line. The computed pattern for $0 \leq u/\pi \leq 19$ is shown as the solid curve in Fig. 9. Also shown are three other cases with $N = 100$, 300, and 600. It is clear that all four cases have almost identical main beams, but the side lobe level, as expected, increases as N decreases even for the small region of angle of observation shown in the figure. It is impractical to show the pattern for the entire visible range. Instead, a statistical distribution of $|F(u_i)/\sigma|^2$ for roughly 4×10^4 values of u_i uniformly spaced in the visible region is made for the case $N = 1000$ shown as the stepped sample curve in Fig. 10. The theory developed before predicts an χ^2 distribution which is also shown in the figure in a normalized scale. The agreement is nearly perfect. This implies that the theory, although it cannot predict the exact shape of the pattern curve, can predict quite accurately the *distribution* of the nearly 4×10^4 numbers of $\{|F(u_i)|\}$ actually computed. Unfortunately, a very accurate prediction of this distribution tells practically nothing about the side lobe level since if *one* of the numbers in the set $\{|F(u_i)|\}$ is changed to a very large value, it will have no effect on the distribution yet it alone will determine a high side lobe level. It is seen once again that the determination of side lobe level is a much more difficult problem. Using the approximate theory given earlier, however, one can predict the side lobe level as the aperture size a or visible region increases. This is shown in Fig. 11 for the four different cases along with the actually computed values for the sample arrays. In this figure the side lobe level in decibels is plotted against both a in wavelengths and the half-power beamwidth in minutes of arc. It is seen that even with as few as 300 uniformly excited elements one can obtain a half-power beamwidth of about 1 minute with a side lobe level of -13 dB. If the conventional design with a half-wavelength spacing is used, it would require 10^4 elements to produce the same beamwidth and side lobe level. The reduction in the number of elements is at a ratio of 100:3; in other words, 97 percent of the elements are saved. The directivity, however, being roughly proportional to N, cannot be obtained without paying the correct price. On the other hand, for the same number of elements as required for the directivity one can achieve a much narrower beam and lower side lobe level with a simple uniform excitation. As will be seen below, there are other significant advantages (such as absence of blind angles and a smaller deleterious effect due to phase errors in a digitalized phase system) for using this type of array.

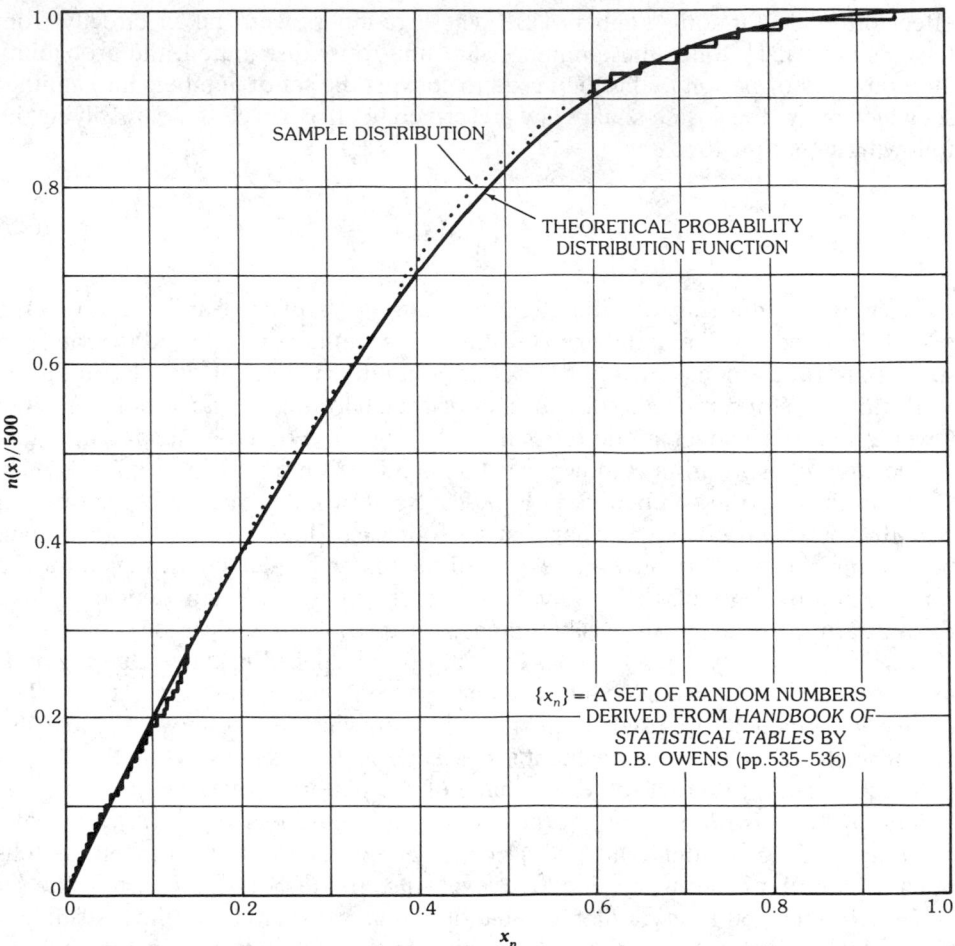

Fig. 8. Distribution of element positions $\{x_n\}$ for an assumed cosine-square density function. (*After Lo [13], © 1964 by the American Geophysical Union*)

The Mutual Coupling Effect and Blind Angles

In general, the mutual coupling in a phased array with randomly spaced elements is not a serious problem since for such an array elements are commonly many wavelengths apart from each other. However, it may be shown that even if array thinning is not the objective and elements are not spaced far apart, random spacings can still be used advantageously in some applications.

Rigorously speaking, the mutual coupling effect in a practical array is a difficult problem to analyze. The approximate methods commonly used, as discussed in other chapters, are two. The first is to consider a practical array as a truncated portion of a corresponding infinite array [33]. Only when the array is a periodic structure is this method applicable. For arrays with unequal or random element spacings one must resort to the second approach in which the mutual impedances between two elements at various spacings are determined first and their effect in an

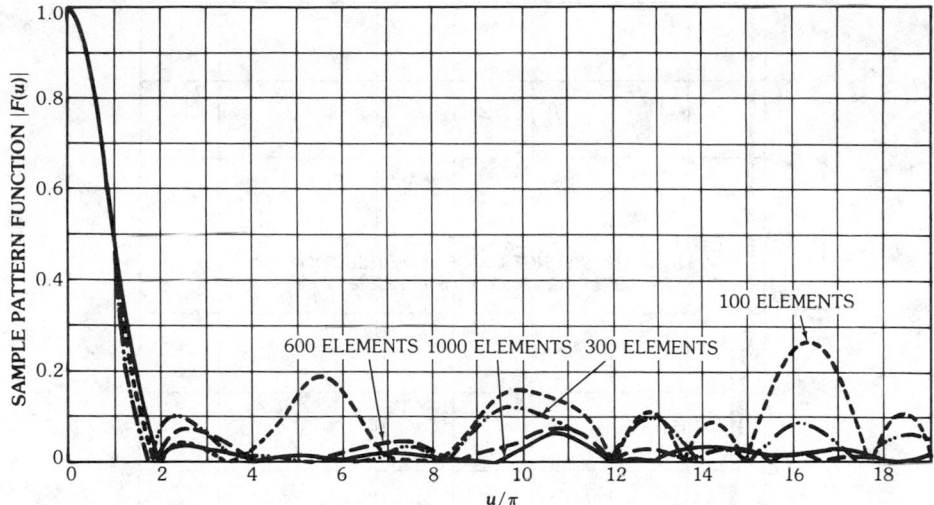

Fig. 9. Sample pattern functions for $0 \leqq u/\pi \leqq 19$ of four symmetrical arrays with $N = 100$, 300, 600, and 1000 elements, respectively. (*After Lo [13], © 1964 by the American Geophysical Union*)

array is then studied with a circuit treatment. In this approach it is assumed that the current or aperture field distributions on the two elements remain unchanged whether or not all other elements are present. This seems generally to be a valid assumption for elements of many types at not too close a spacing.

Let I_n be the actual feed current of the nth element in an array and $I_n^{(0)}$ be the current when all mutual couplings are absent. Then by circuit analysis,

$$I_n = I_n^{(0)} - \sum_{\substack{m=1 \\ m \neq n}}^{N} \frac{Z_{mn}}{Z_{11} + Z_0} I_m \tag{21}$$

where

Z_{mn} = mutual impedance between the mth and nth elements

Z_{11} = self-impedance of the element at the feed terminals

Z_0 = generator or the feed-line impedance

For all the elements, (21) can be written as

$$\mathbf{I} = \mathbf{I}^{(0)} - \mathbf{C}\mathbf{I} \quad \text{or} \quad \mathbf{I} = (\mathbf{E} + \mathbf{C})^{-1}\mathbf{I}^{(0)} \tag{22}$$

where

$$\mathbf{I} = \begin{bmatrix} I_1 \\ \vdots \\ I_N \end{bmatrix}, \qquad \mathbf{I}^{(0)} = \begin{bmatrix} I_1^{(0)} \\ \vdots \\ I_N^{(0)} \end{bmatrix}$$

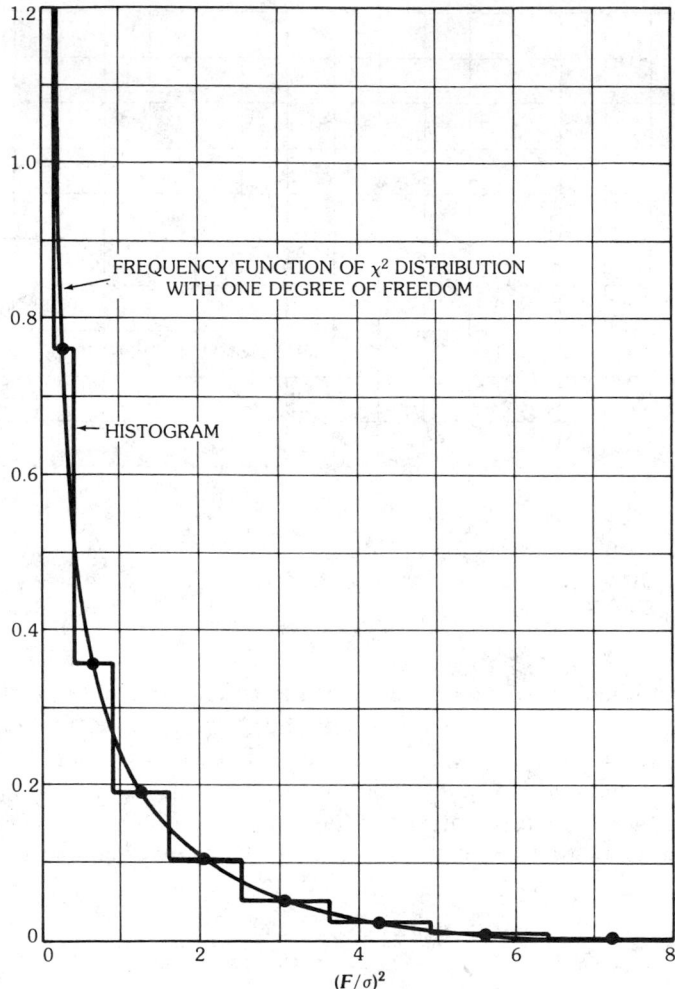

Fig. 10. Sample distribution of 4×10^4 points of the normalized pattern function squared $|F/\sigma|^2$, equally spaced in u/π from 0 to 10^4 (corresponding to $a = 10^4\lambda$) for $N = 1000$ as compared with the theoretical density function, namely the chi-square density function with one degree of freedom. (*After Lo [13], © 1964 by the American Geophysical Union*)

C is an $N \times N$ matrix with elements

$$C_{mn} = \frac{Z_{mn}}{Z_{11} + Z_0} \quad \text{and} \quad C_{mm} = 0$$

E = identity matrix

For uniform cophasal excitation $I_n^{(0)} = \exp(-jvx_n)$, the pattern function

$$F(u, v) = \frac{1}{N} \sum_{n=1}^{N} I_n \exp(jux_n) \tag{23}$$

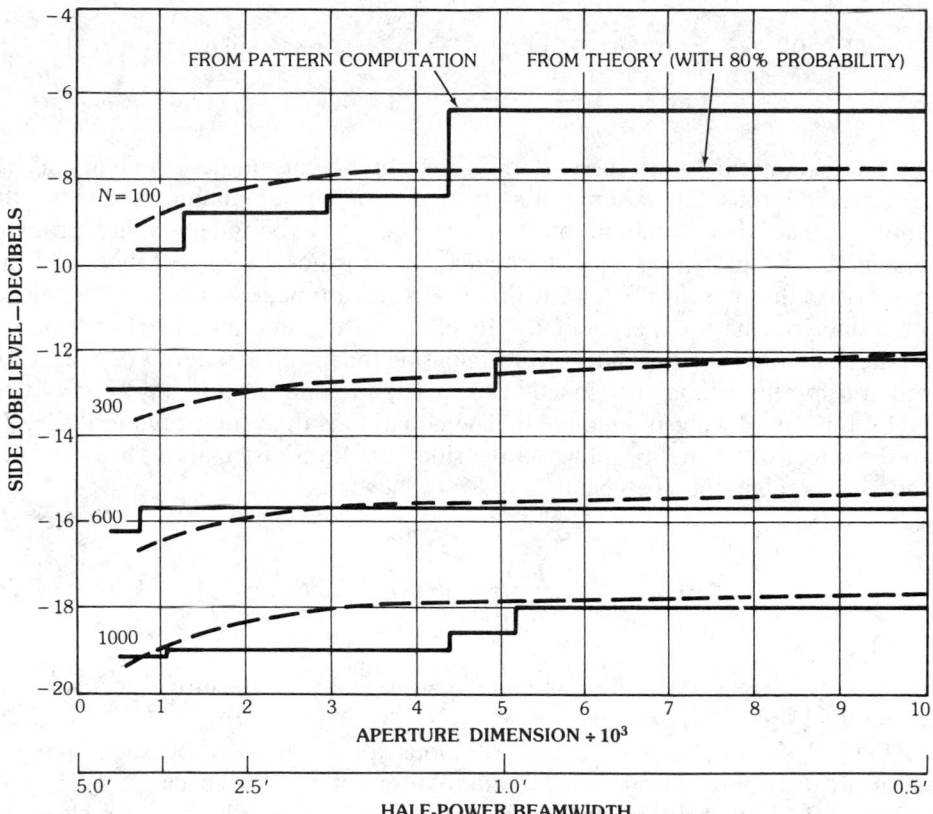

Fig. 11. Side lobe level for four sample symmetrical arrays with $N = 100, 300, 600,$ and 1000 elements as a function of the aperture dimension in wavelengths, or the half-power beamwidth in minutes of arc. (*After Lo [13], © 1964 by the American Geophysical Union*)

where

$v = 2\pi \sin \alpha$

α = beam scan angle

$u = 2\pi \sin \theta$

θ = observation angle

$x_n = n$th element position in wavelengths

Except for the uninteresting case of close element spacing one may use the following approximation for (22),

$$\mathbf{I} \cong (\mathbf{E} - \mathbf{C})\mathbf{I}^{(0)} \qquad (24)$$

Then

$$F(u,v) = \frac{1}{N} \sum_{n=1}^{N} \exp[j(u-v)x_n] - \frac{1}{N} \sum_{n=1}^{N} \sum_{\substack{m=1 \\ m \neq n}}^{N} C_{mn} \exp[j(ux_n - vx_m)] \quad (25)$$

The first summation gives the pattern in the absence of mutual coupling and has been studied thus far, whereas the second, double summation is due to the coupling. The determination of Z_{mn}, or C_{mn}, is a boundary value problem depending on a particular type of antenna, which is not the subject matter of this chapter. At the present it is assumed that Z_{mn}, as a function of spacing, is known either analytically or experimentally. Its effect on the side lobe level distribution can then be found approximately by formulating the problem in terms of a diffusion equation for the probability function of the up-crossing of $|F(u,v)|$ at any given level η for u in any given region [24]. The equation is then solved numerically. To see the effect of mutual coupling on the side lobe level distribution, use has been made of the coupling coefficient

$$C_{mn} = \frac{0.11}{0.22 + |x_{mn}|} \exp(-j2.22\pi|x_{mn}|) \quad (26)$$

where $x_{mn} = x_m - x_n$, as determined experimentally by Lechtreck for two horn elements [34].

Fig. 12 shows the side lobe distributions for an array of 50 such elements uniformly distributed at random with three different average spacings: $d_{av} = 0.5\lambda$, 1.0λ, and 2.5λ. When the coupling is considered, the distributions are shown in solid curves, and when ignored, in dashed curves. As expected the coupling results in a higher side lobe level, but to a lesser degree for larger average spacing. Fig. 13 shows more clearly how the increase in side lobe level depends on the average spacings. In this case, $a = 500\lambda$ and $d_{av} = 0.5\lambda$, 1.0λ, 2.5λ, 5.0λ, and 10λ are considered. It is seen that when $d_{av} \geqq 2.5\lambda$ the coupling may be ignored. It may be noted that for $d_{av} = 2.5\lambda$ and $N = 200$, the increase in side lobe level is less than that shown in Fig. 12 for the same d_{av} and for $N = 50$. Thus even for the same average spacing, the effect of mutual coupling on side lobe level becomes less as N increases. This is also expected as the double summation in (25), consist of terms in random phase, contributes less and less as N increases, for the same reason that the side lobe level decreases when N increases even in the case of no coupling as shown in (14).

If $u = v$ in (25), one obtains the main beam maximum. For most antenna elements, if not spaced very closely, the phase of C_{mn} is nearly linear in spacing as shown in (26). Let

$$C_{mn} = |C_{mn}| \exp(j\beta|x_{mn}|), \qquad x_{mn} = x_m - x_n$$

Then the main beam magnitude as a function of scan angle α, or its corresponding value v, sometimes called the *array scan characteristics*, is

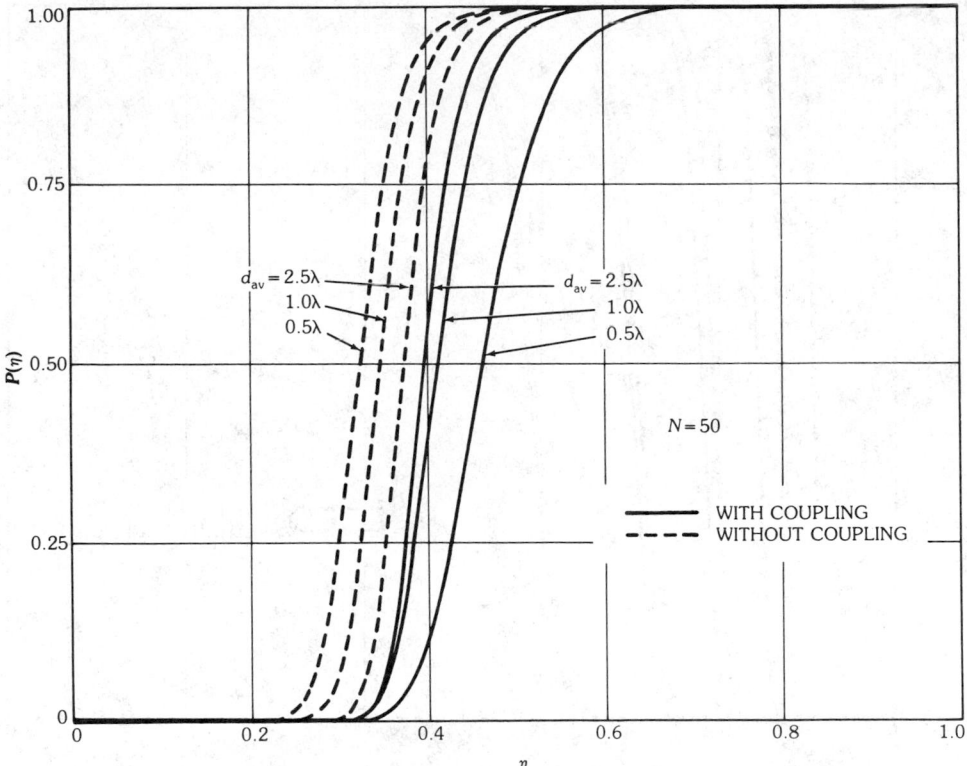

Fig. 12. Distribution of side lobe level for random arrays of 50 elements and average spacing $d_{av} = 0.5\lambda$, 1.0λ, and 2.5λ with and without mutual coupling. (*After Agrawal and Lo [24],* © *1972 IEEE*)

$$F(v,v) = \frac{1}{N} \sum_{n=1}^{N} \left[1 - \sum_{\substack{n=1 \\ n \neq m}}^{N} |C_{mn}| \exp(-jvx_{mn} + j\beta|x_{mn}|) \right] \qquad (27)$$

The quantity in the square brackets is exactly the pattern function of the nth element with all other elements terminated in the match load Z_0. It may be called the *array element pattern*, namely, the pattern of an element in an array environment, which can be vastly different from that of an isolated element. The scan characteristic is therefore the average of all array element patterns. It is interesting to see that if $x_n = nd$ (i.e., uniform spacing), and if

$$(v \pm \beta)d = 2\pi p, \qquad p = 0, 1, 2, \ldots$$

the terms in the second sum of (27) will add up in phase to cause a sharp decrease in $F(v,v)$. For a very large array this sum could be as large as unity, thus resulting in the so-called blind angle phenomenon; namely, in that scan direction, v, the main beam maximum and therefore the total radiated power drop to nearly zero. In other words the array ceases to function as an antenna and all the power sent by

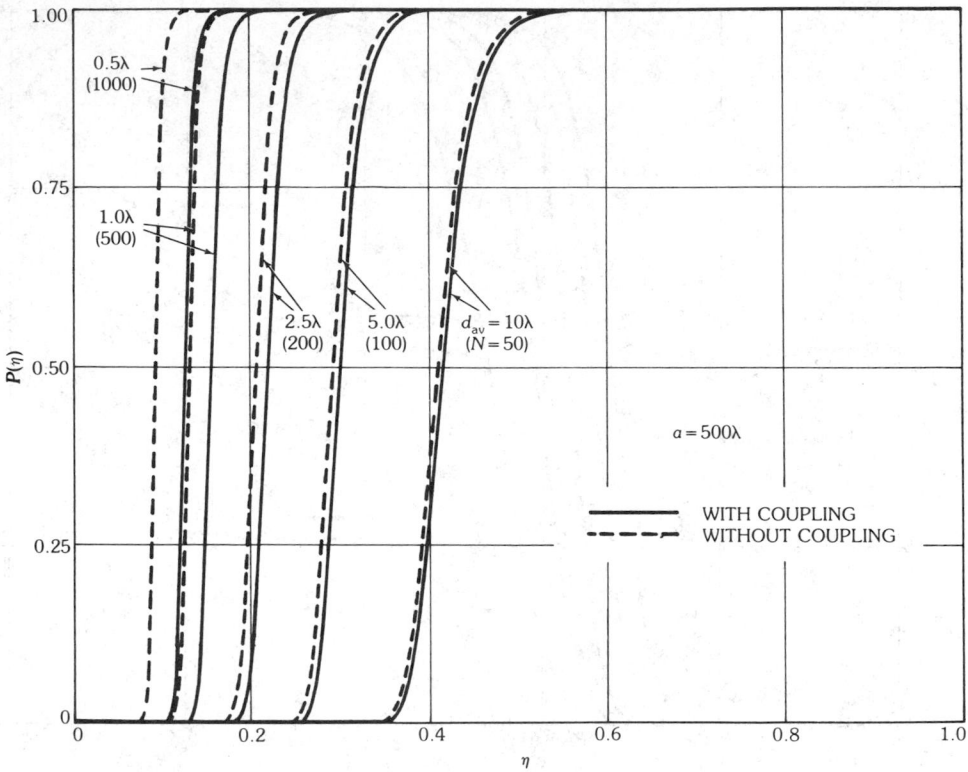

Fig. 13. Distribution of side lobe level for a 500λ aperture and 50, 100, 200, 500, and 1000 elements with and without mutual coupling. (*After Agrawal and Lo [24], © 1972 IEEE*)

the exciting generator will be reflected, or the element impedance in the array becomes reactive. Now if $x_n \neq nd$ and $\{x_n\}$ is a set of random numbers, the in-phase condition for all the terms in the summation will be unlikely to occur. In fact, the larger the array (i.e., more elements), the smaller will be the sum and thus the smaller the drop of main beam magnitude. This conclusion is in exact opposition to that of a uniformly spaced array, a very gratifying result indeed.

The probability distribution of the main beam magnitude $F(v,v)$ can be determined approximately once the coupling coefficient C_{mn} for a given type of element to be used and the probability density function $g(x)$ for placing the elements are known. The readers may refer to Agrawal and Lo [24] for details. In the following only some typical results are given.

Let the *fluctuation* of the main beam magnitude be defined as

$$r = \max_{v \in [0,2\pi]} F(v,v) - \min_{v \in [0,2\pi]} F(v,v) \qquad (28)$$

When $g(x) = 1/2$ for $|x| \leq 1$ and 0 otherwise, C_{mn} is given by (26), $N = 50$, and $d_{av} = 0.5\lambda$, the probability for $r < \xi$ versus ξ can be computed and is plotted in Fig. 14. Also shown is the "experimental" result for 50 random sample arrays simulated by using the Monte Carlo method. The close agreement between the two results

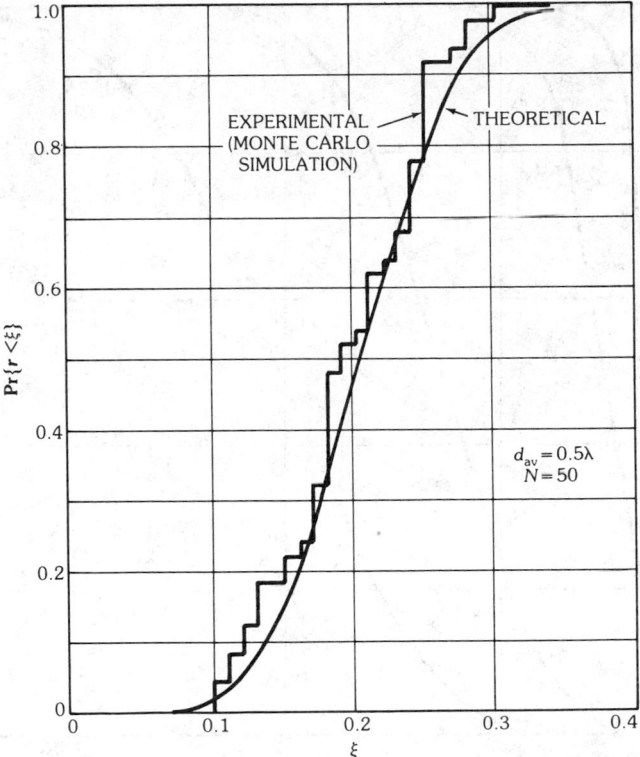

Fig. 14. Distribution of main beam amplitude fluctuation r over the entire scanning range for 50 elements and average spacing 0.5λ. (*After Agrawal and Lo [24], © 1972 IEEE*)

provides some confidence in the approximate theory. Fig. 15 shows the theoretical results for the same array except that $d_{av} = 0.5, 1.0, 2.5, 5$, and 10λ. As expected, the larger the average spacing d_{av}, the smaller the coupling effect and the smaller the fluctuation of main beam magnitude over the scanning range.

The Holey Plate Experiment

Side Lobe Level and Half-Power Beamwidth—A physical experiment for a large array is a rather costly enterprise. Fortunately, the holey plate method as described by Stone at optical frequency [35] and by Skolnik at microwave frequency [36] provides a simple modeling technique. In this method the array is modeled by a large conducting plate perforated with small circular holes, each simulating an antenna element. The "holey" plate is illuminated by an incident plane wave from one side, and the field is measured from the other side for all directions. The pattern so obtained would be nearly the same as that of the actual array. The sketch in Fig. 16 shows the major components used. The holey plate is placed against a microwave lens which converts a spherical wave emitted from a reduced open waveguide at the focus into a plane wave. The assembly of these components is enclosed in a box made of absorbing material, and placed over a turntable.

In the actual setup the holey plate is used as a receiving array. Fig. 17 shows the

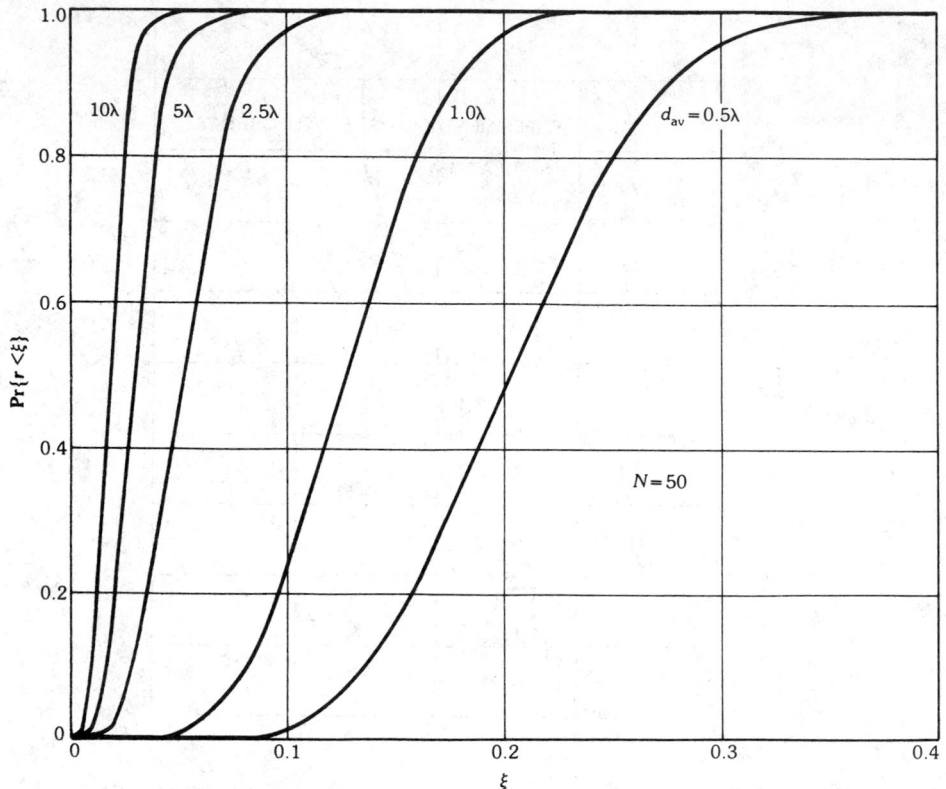

Fig. 15. Distribution of main beam amplitude fluctuation r over the entire scanning range for 50-element arrays with 0.5λ, 1.0λ, 2.5λ, 5λ, and 10λ average spacing. (*After Agrawal and Lo [24], © 1972 IEEE*)

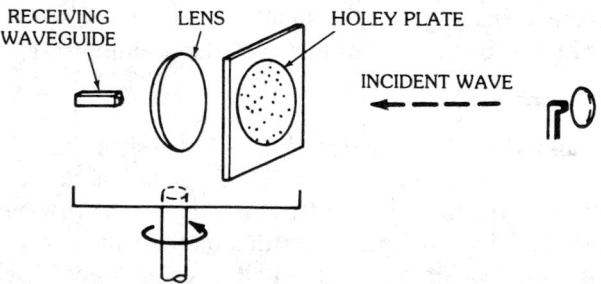

Fig. 16. Major components for the holey plate experiment. (*After Lo and Simcoe [23], © 1967 IEEE*)

physical experimental setup at 4 mm. The paraboloidal reflector at the far right is the transmitting antenna. The incident wave is diffracted by the holey plate and then focused on an open-ended waveguide on the left by a polystyrene lens. The receiving assembly, consisting of the plate, lens, and the open-ended waveguide, is

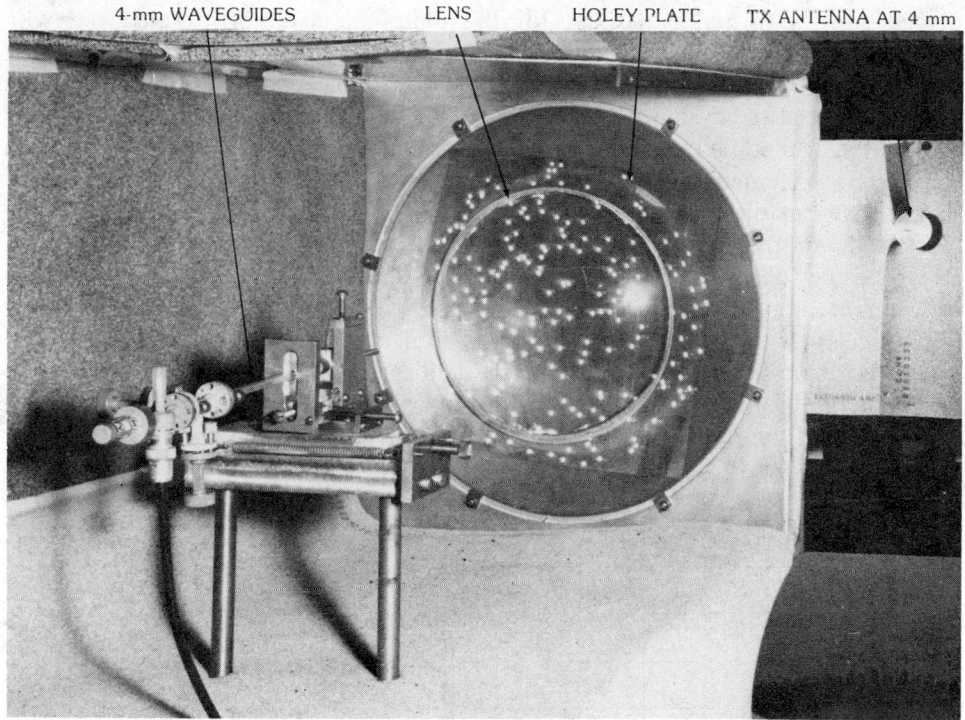

4-mm WAVEGUIDES   LENS   HOLEY PLATE   TX ANTENNA AT 4 mm

Fig. 17. Experimental setup for 4-mm waves. (*After Lo and Simcoe [23], © 1967 IEEE*)

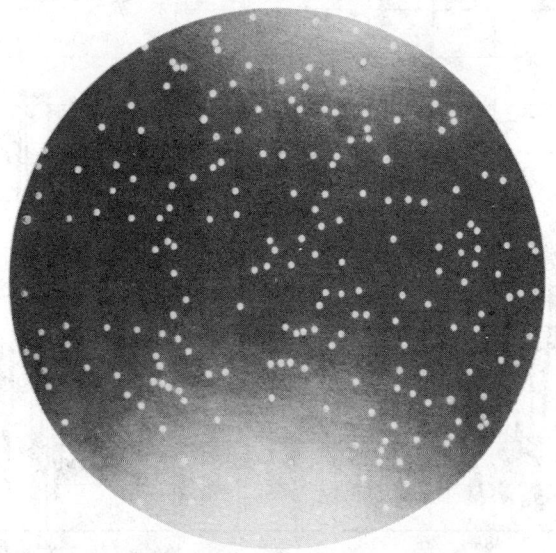

Fig. 18. Photo of the holey plate sample array, where the coordinates of elements are obtained from a set of uniformly distributed random numbers. (*After Lo and Simcoe [23], © 1967 IEEE*)

placed over a turntable. Except for the plate the assembly is enclosed with microwave-absorbing materials which are partially removed in order to show the details inside. In this picture the range has been reduced so that the transmitting antenna can also be shown. The holey plate contains 210 elements uniformly distributed at random over a circular aperture of about 56λ in diameter as shown in Fig. 18. Each element is a hole of about $\lambda/4$ in diameter. As is clear from this figure there exists no plane of symmetry and thus no principal plane. Therefore the measured patterns in a few planes cannot be used to infer the overall performance. It would be desirable to measure a three-dimensional pattern, but in this experiment a total of 90 cuts was taken (roughly twice as many as the highest spatial frequencies). A typical pattern with expanded main beam is shown in Fig. 19. The measured beams in all cuts are almost identical, with a half-power beamwidth ranging from 0.9° to 1.25° and a statistical mean of 1.04°. The difference between this value and the theoretical mean of 1.05° is well within the experimental error.

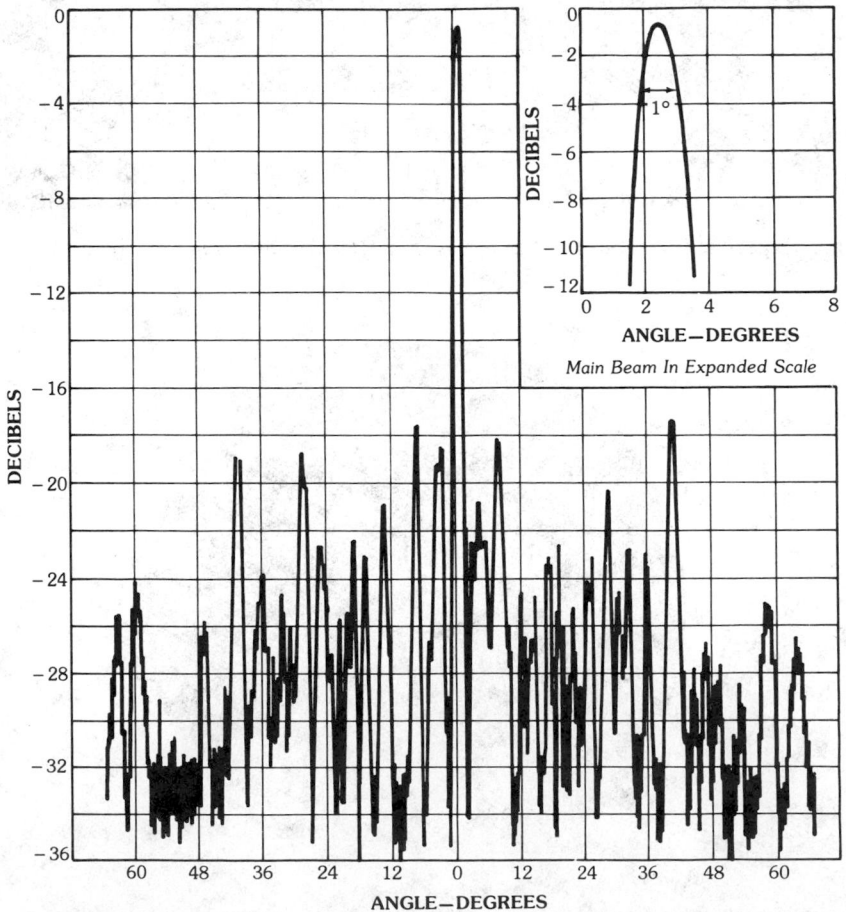

Fig. 19. A typical pattern in a plane cutting through the sample array as shown in Fig. 18. (*After Lo and Simcoe [23], © 1967 IEEE*)

The 90 measured patterns are statistically "identical" in the sense that all have nearly the same beam and a few high side lobes, ranging approximately from −18 dB to −13 dB. However, no two patterns are really identical in details. Fig. 20a shows the measured as well as the theoretically predicted distributions of the highest side lobe levels associated with the 90 patterns. The agreement is remarkable. To safeguard against the luck in this experiment, a second completely independently designed holey plate was also constructed and tested. The same close agreement was obtained as shown in Fig. 20b. Finally, it is worth noting again that if a conventional array design (i.e., using uniformly spaced elements) is used to produce the same beam and side lobe level, 10^4 elements would be required. This number is about 50 times what was used in the experiment. Obviously the

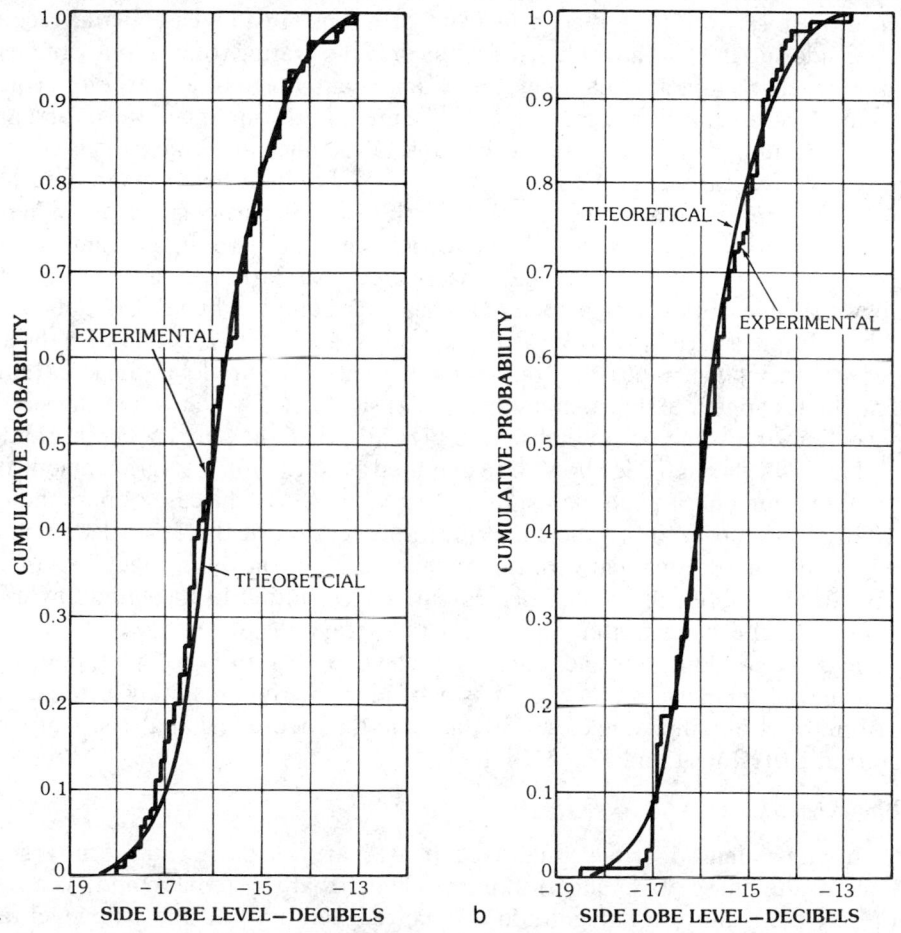

Fig. 20. Comparison of the theoretical and experimental distributions for the side lobe levels of linear arrays. (*a*) Experimental distribution obtained from a set of 90 pseudo-random linear arrays generated by the planar array 1. (*b*) Experimental distribution obtained from a set of 90 pseudo-random linear arrays generated by the planar array 2. (*After Lo and Simcoe [23], © 1967 IEEE*)

difference is by no means insignificant. Of course, the two designs would not have the same directivity. On the other hand, suppose that the desired directivity is less than 10^4, say 10^3. Then the random spacing approach will provide a design which uses not only 1/10 of the elements but also can yield a lower side lobe level and/or narrower beam. In other words, in this random design the required number of elements is determined not directly by the aperture size but rather by the desired directivity and side lobe level.

Scanning Characteristics—It is a simple matter to obtain cophasal excitation in a holey plate experiment. If the holey plate is rotated at an angle α with respect to the lens as shown in Fig. 21, all elements will automatically be excited cophasally to produce a beam at an angle α from the broadside. In Fig. 21 a linear uniformly spaced slot array (or grating), etched from a copper-clad substrate with relative permittivity $\epsilon_r = 4.25$, is placed in front of a lens. In this case if the receiving waveguide and the lens are held fixed while only the grating rotates, one obtains the scanning characteristics, namely, the main beam magnitude variation (or the scanning pattern) with the scan angle α. The measured scanning pattern for a linear array of 61 uniformly spaced slots when **E** is perpendicular to the slots is shown in Fig. 22. It is clearly shown that at about $\pm 18°$ the main beam magnitude drops sharply to -20 dB. These are the blind angles for this particular array. When the same 61 slots were rearranged to have random spacings but the same average spacing $d_{av} = 0.6\lambda$, the blind angles disappeared, as clearly seen in Fig. 22. Fig. 23 shows the superimposed measured main beams for another linear array of 45 slots, both uniformly spaced (Fig. 23a) and randomly spaced (Fig. 23b), all with the same average spacing $d_{av} = 0.6\lambda$. For this case the **E** vector is parallel to the slots and the blind angles appear at approximately $\pm 30°$ as shown in Fig. 23a when the slots are spaced uniformly and become absent in Fig. 23b when they are spaced at random.

Figs. 24a through 24c show the computed patterns for a single element in an array environment of 21 elements, i.e., $F(v, v)$ in (27), for the case of perpendicular polarization and uniform spacing. With the exception of the few edge elements, they all are almost identical with the scanning pattern shown in Fig. 22 so that the sharp drop at about $\pm 18°$ becomes cumulative, resulting in the blind angle [37]. From this, one expects that the larger the array, the deeper the drop. When spacings are random, however, the array element pattern is completely different from one element to another as shown in Fig. 25, thus resulting in a scanning pattern free from blind angles. For details and the theoretical analysis, readers are referred to Agrawal and Lo [24, 37].

Other Remarks

In most phased arrays quantized phases are used. For uniformly spaced elements the phase errors incurred in the elements are systematic and, as a result, not only the main beam magnitude may be reduced but the side lobe level raised also. In case of randomly spaced elements such deleterious effects can be reduced particularly for large arrays. Another method that can be used to achieve this goal in a similar manner is to deliberately add a small amount of random phase to each quantized level. However, this method is effective only when a large number of levels (or bits) is used in the system.

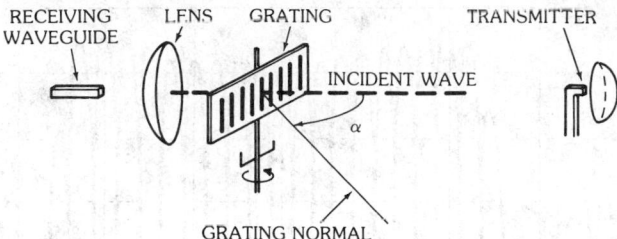

Fig. 21. Sketch of the experimental setup for measuring main beam magnitude fluctuation. (*After Agrawal and Lo [37]*)

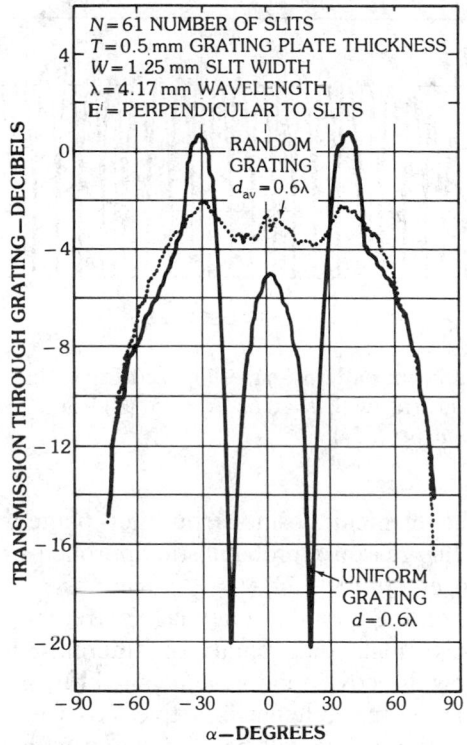

Fig. 22. Measured transmitted power through uniform and random gratings versus orientation angle α (perpendicular polarization). (*After Agrawal and Lo [37]*)

For many conformal arrays, such as cylindrical and spherical arrays, even if the elements are placed in a regular manner (for example, equiangular spacings), their pattern functions cannot be simply expressed in terms of polynomials as for linear arrays. Thus there is, in general, no analytic advantage to considering a particular element arrangement. Commonly the patterns are computed numerically, and clearly for large arrays this brute force method is very costly. However, the contributions from all elements to a pattern in a plane have phases proportional to

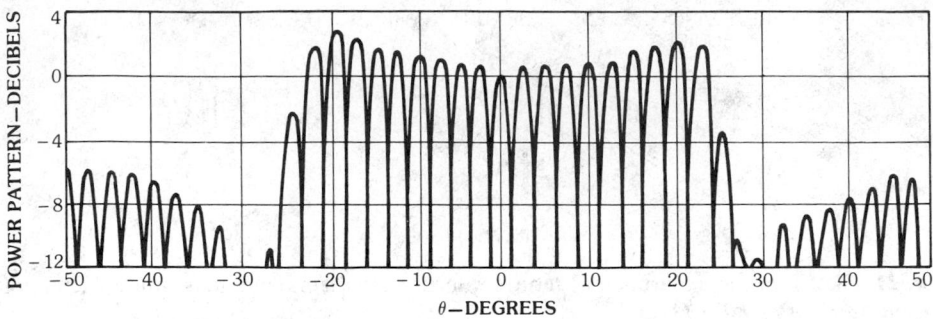

a

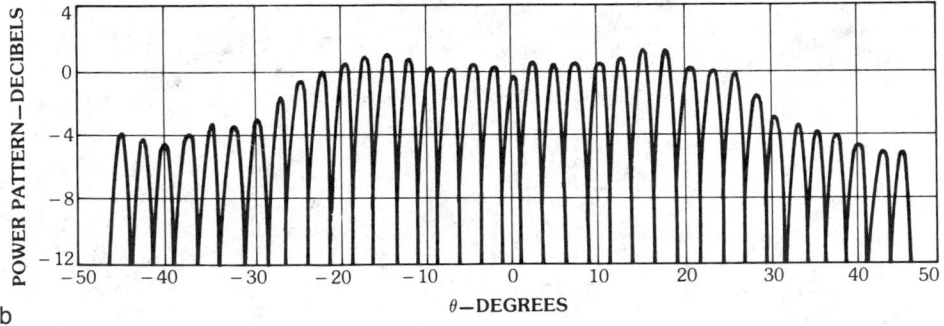

b

Fig. 23. Measured main beam patterns, superimposed for various scan angles α (parallel polarization). (*a*) Uniform array with $d = 0.6\lambda$, $N = 45$. (*b*) Random array with $d_{av} = 0.6\lambda$, $N = 45$. (*After Agrawal and Lo [37]*)

the projections of the element positions on that plane which are in general pseudo-random. For this reason a probabilistic approach to large cylindrical and spherical arrays has been studied [38].

With the advent of log-periodic antennas, extremely wide band antennas become a reality. This outstanding family of antennas, however, has its own limitations, namely, low directivity and wide beamwidth, since for each frequency only a limited portion of the antenna is active. To form an array with these elements for high directivity, one will be confronted with the difficulty that the physical element spacing increases with frequency, causing a serious grating lobe problem. To alleviate this difficulty random spacings of log-periodic elements, particularly for large arrays, should be used since such an array can tolerate large element spacings and is inherently frequency insensitive, as shown in this chapter.

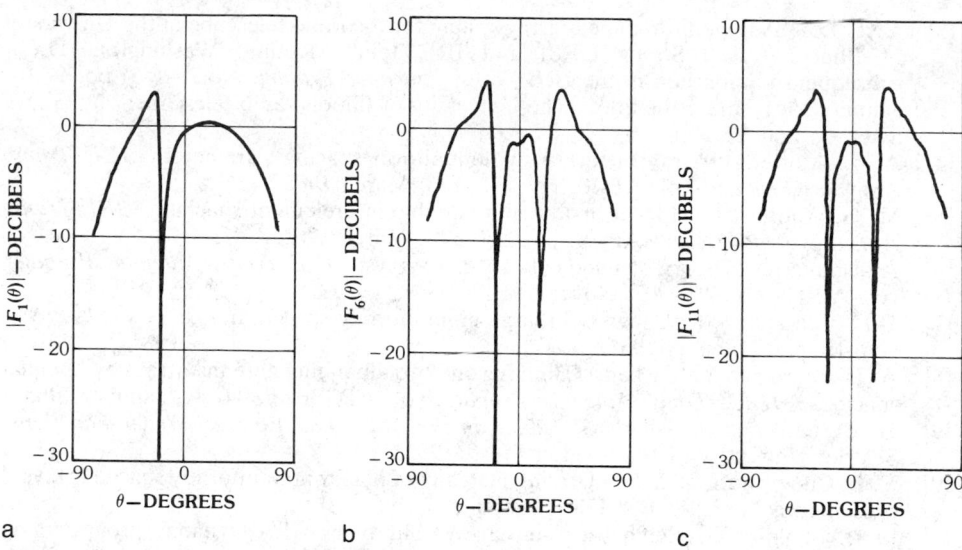

Fig. 24. Computed single-slit pattern $|F_k(\theta)|$ when only the kth element of a uniformly spaced array with 21 elements is excited, where $d = 0.6\lambda$, $\epsilon_r = 4.25$, $T = 0.5$ mm, $W = 1.25$ m, $\lambda = 4.17$ mm, and **E** is perpendicular to slit. (*a*) First-element pattern. (*b*) Sixth-element pattern. (*c*) Central-element pattern. (*After Agrawal and Lo [37]*)

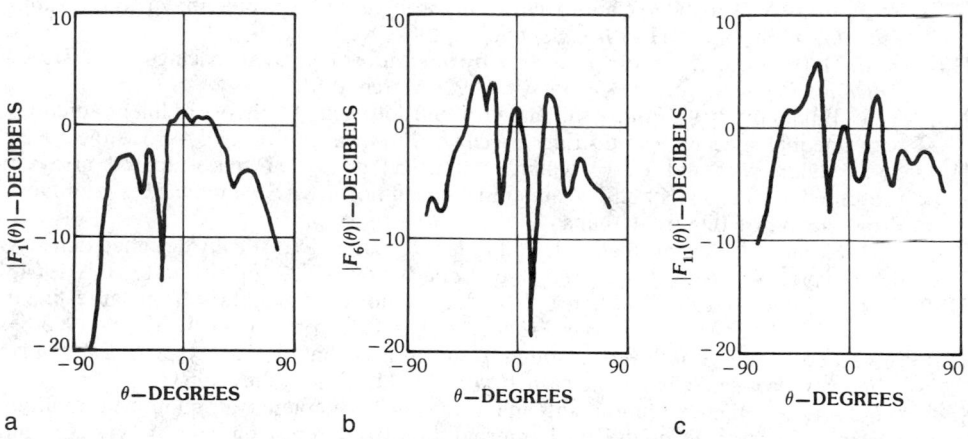

Fig. 25. Same as for Fig. 24 except that the 21 elements are randomly spaced with $d_{av} = 0.6\lambda$. (*a*) First-element pattern. (*b*) Sixth-element pattern. (*c*) Central-element pattern. (*After Agrawal and Lo [37]*)

5. References

[1]   H. Unz, "Linear arrays with arbitrarily distributed elements," *IRE Trans. Antennas Propag.*, vol. AP-8, pp. 222–223, March 1960.

[2]   D. D. King, R. F. Packard, and R. K. Thomas, "Unequally spaced broadband antenna arrays," *IRE Trans. Antennas Propag.*, vol. AP-8, pp. 380–385, July 1960.

[3]   Y. T. Lo, "A nonuniform linear array system for the radio telescope at the University of Illinois," 1960 Spring URSI and IRE Joint Meeting, Washington, D.C., subsequently published in the *IRE Trans. Antennas Propag.*, vol. AP-9, pp. 9–16, January 1961, under the title, "The University of Illinois radio telescope," by G. W. Swenson and Y. T. Lo.

[4]   A. L. Maffett, "Array factors with nonuniform spacing parameter," *IRE Trans. Antennas Propag.*, vol. AP-10, pp. 131–136, March 1962.

[5]   M. G. Andreason, "Linear arrays with variable interelement spacings," *IRE Trans. Antennas Propag.*, vol. AP-10, pp. 137–143, March 1962.

[6]   A. Ishimaru, "Theory of unequally spaced arrays," *IRE Trans. Antennas Propag.*, vol. AP-10, pp. 691–702, November 1962.

[7]   J. L. Yen and Y. L. Chow, "On large nonuniformly spaced arrays," *Can. J. Phys.*, vol. 41, p. 1, January 1963.

[8]   A. Ishimaru and Y. S. Chen, "Thinning and broadbanding antenna arrays by unequal spacings," *IEEE Trans. Antennas Propag.*, vol. AP-13, pp. 34–42, January 1965.

[9]   G. Bernard and A. Ishimaru, *Tables on the Anger and Lommel-Weber Functions*, Seattle: University of Washington Press, 1962.

[10] Y. L. Chow and J. L. Yen, "Grating plateaux of planar nonuniformly spaced arrays," presented at the 1964 Fall URSI Meeting, Urbana, Il.

[11] M. I. Skolnik, G. Nemhauser, and J. W. Sherman, III, "Dynamic programming applied to unequally spaced arrays," *IEEE Trans. Antennas Propag.*, vol. AP-12, pp. 35–43, January 1964.

[12] Y. T. Lo, "On the theory of randomly spaced antenna arrays," *Tech. Rep. 1*, Antenna Lab., Dept. of Electr. Eng., Univ. of Illinois, Urbana, under NSF G 14894, 1962. Part of this report appeared in "A mathematical theory of antenna arrays with randomly spaced elements," *IEEE Trans. Antennas Propag.*, vol. AP-12, pp. 257–268, May 1964.

[13] Y. T. Lo, "A probabilistic approach to the problem of large antenna arrays," *Radio Sci.*, vol. 68D, pp. 1011–1019, September 1964.

[14] R. F. Harrington, "Side lobe reduction by nonuniform element spacing," *IRE Trans. Antennas Propag.*, vol. AP-9, pp. 187–192, March 1961.

[15] Y. V. Baklanov, V. L. Pokrovshi, and G. I. Surdotovich, "A theory of linear antennas with unequal spacing," *Radio Eng. Electron. Phys.*, no. 6, pp. 905–913, June 1962.

[16] S. J. Rabinowitz and R. F. Kolar, "Statistical design of space-tapered arrays," presented at the 1962 Twelfth Annual Symposium on USAF Antenna Res. and Dev. Prog., Univ. of Illinois, Urbana.

[17] M. I. Skolnik, J. W. Sherman, III, and F. C. Ogg, Jr., "Statistically designed density-tapered arrays," *IEEE Trans. Antennas Propag.*, vol. AP-12, pp. 408–417, July 1964.

[18] T. M. Maher and D. K. Cheng, "Random removal of radiators from large linear arrays," *IEEE Trans. Antennas Propag.*, vol. AP-11, pp. 106–111, March 1963.

[19] Y. T. Lo, "Side lobe level in nonuniformly spaced antenna arrays," *IEEE Trans. Antennas Propag. (Communications)*, vol. AP-11, p. 511, July 1963.

[20] Y. T. Lo, "High-resolution antenna arrays with elements at quantized random spacings," presented at the 1964 International Conference on Microwaves, Circuit Theory, and Information Theory, Tokyo.

[21] Y. L. Chow, "On grating plateaux of nonuniformly spaced arrays," *IEEE Trans. Antennas Propag.*, vol. AP-13, pp. 208–215, March 1965.

[22] Y. T. Lo and S. W. Lee, "A study of spaced-tapered arrays," *IEEE Trans. Antennas Propag.*, vol. AP-14, pp. 22–30, January 1966.

[23] Y. T. Lo and R. J. Simcoe, "An experiment on antenna arrays with randomly spaced elements," *IEEE Trans. Antennas Propag.*, vol. AP-15, pp. 231–235, March 1967.

[24] V. D. Agrawal and Y. T. Lo, "Mutual coupling in phased arrays of randomly spaced antennas," *IEEE Trans. Antennas Propag.*, vol. AP-20, pp. 288–295, May 1972.

[25] A. R. DiDonato and M. P. Jarnagin, "Integration of the general bivariate Gaussian distribution over an offset ellipse," *NWL Report 1710*, Naval Weapons Laboratory, Dahlgren, Va., 1960.

[26] S. O. Rice, "Mathematical analysis of random noise," *Bell Syst. Tech. J.*, vol. 24, p. 46, 1945.

[27] P. B. Patnaik, "The noncentral χ^2 and F distributions and their application," *Biometrica*, vol. 36, pp. 202–232, 1949.

[28] K. Pearson, *Tables of Incomplete Gamma Functions*, Department of Scientific and Industrial Research, Cambridge: Cambridge University Press, 1934.

[29] K. Karhunen, "Über lineare methoden in der wahrscheinlichkeits rechnung," *Ann. Acad. Sci. Fennicae*, ser. A, no. I, pp. 37–79, 1947.

[30] W. N. Christiansen and J. H. Hogbem, "A design for the Benelux Cross antenna," *BCAP Tech. Rep. No. 3*, Benelux Cross Antenna Project, Sterrewacht, Leiden, Netherlands.

[31] D. B. Owen, *Handbook of Statistical Tables*, pp. 535–536, Reading: Addison-Wesley Pub. Co., 1962.

[32] E. Parzen, *Modern Probability Theory and Its Applications*, New York: John Wiley & Sons, 1960.

[33] N. Amitay, V. Galindo, and C. P. Wu, *Theory and Analysis of Phased Array Antennas*, New York: Wiley-Interscience, 1972.

[34] L. W. Lechtreck, "Effects of coupling accumulation in antenna arrays," *IEEE Trans. Antennas Propag.*, vol. AP-16, pp. 31–37, January 1968.

[35] J. M. Stone, *Radiation and Optics*, New York: McGraw-Hill Book Co., pp. 146–152, 1963.

[36] M. I. Skolnik, "A method of modeling array antennas," *IEEE Trans. Antennas Propag.*, vol. AP-11, pp. 97–98, January 1963.

[37] V. D. Agrawal and Y. T. Lo, "Anomalies of dielectric-coated gratings," *Appl. Opt.*, vol. 11, pp. 1946–1951, September 1972.

[38] A. R. Panicali and Y. T. Lo, "A probabilistic approach to large circular and spherical arrays," *IEEE Trans. Antennas Propag.*, vol. AP-17, pp. 514–522, July 1969.

Chapter 15

Reflector Antennas

Y. Rahmat-Samii
Jet Propulsion Laboratory

CONTENTS

 **Yahya Rahmat-Samii** received the MS and PhD degrees in electrical engineering in 1972 and 1975, respectively, from the University of Illinois, Champaign-Urbana. From 1975 to 1978 he was a visiting assistant professor at the University of Illinois with research interests in the areas of high-frequency diffraction (GTD), electromagnetic pulse (EMP), and microwave techniques. In 1978 he joined the NASA Jet Propulsion Laboratory/California Institute of Technology, where he is a senior research scientist and has contributed significantly to antenna technology for space programs, large space systems, microwave holography techniques, and mathematical and computer modeling for which he has received many NASA certificates of recognition. He was a guest professor at the Technical University of Denmark in the summer of 1986. He currently also is an Adjunct Professor of Electrical Engineering at the University of California, Los Angeles.

Dr. Rahmat-Samii is an elected fellow of the IEEE (1985) and the 1984 recipient of the Henry George Booker Fellow Award of URSI. He has been appointed an IEEE Antennas and Propagation Society Distinguished Lecturer and has presented lectures internationally. He was an elected AdCom member and Associate Editor of the IEEE Antennas and Propagation Society. He has authored, or coauthored, over 120 technical journal articles and conference papers and has written chapters in five books.

1. Introduction

The fine art of synthesizing, analyzing, and designing reflector antennas of many various geometries did not really advance until the days of World War II, when numerous radar applications evolved to satisfy diverse technical demands. Subsequent demands for reflectors for use in radio astronomy, microwave communications, satellite communications and tracking, and the like have resulted in both the development of sophisticated reflector configurations and analytical and experimental design techniques. Reflector antennas may take many configurations, some of the most popular ones being plane, corner, curved reflectors, and so on. In this chapter only the curved reflectors, such as parabolic and Cassegrain, will be discussed. The reader is referred to references [1, 2, 3, 4] for other shapes.

The reflector antennas (curved) can be classified in a variety of ways, and one of the more recently suggested models shown in Fig. 1 is obtained from reference [5]. Fig. 1 identifies reflectors according to pattern type, reflector type, and feed type. In Fig. 1, pencil-beam reflectors are very popular and are commonly used in point-to-point microwave communications, since their patterns yield the maximum boresight gain and typically their beam directions are fixed at the time of antenna installation. In satellite communication systems the uplink beam of these pencil-beam reflectors may be either fully steerable by reflector movements, as in INTELSAT ground stations, or capable of limited steering, as in the Canadian domestic systems. Many new generations of satellite reflectors have produced other popular types of pattern classifications: contour (shaped) beams and multiple beams. These applications demand reflectors with improved off-axis beam characteristics, which result in more sophisticated configurations as suggested in Fig. 1. Many microwave communication antennas operate with one sense of polarization at a given frequency and require only reasonable discrimination between orthogonal polarizations. However, many of the current generation of microwave communication antennas operate with dual polarizations at the same frequency to enhance their so-called frequency reuse capabilities. Such requirements impose stringent demands on the polarization performance of reflectors and could be used as factors in pattern classification (Fig. 1).

Finally, the performance of reflectors cannot be properly examined without knowledge of the feed configurations of the reflectors. For this reason column three of Fig. 1 suggests a classification based on feed types. Horn or waveguide feeds operating in a single pure mode have been the most popular feeds for reflectors. However, in order to meet radio astronomy, earth station, and satellite antenna requirements considerable efforts have been focused on the development of new feeds to efficiently illuminate either the main reflector or the subreflector of the antenna. For example, hybrid-mode feeds (combining TE and TM fields) are used to match efficiently the feed distribution with the desired focal field distribution of the reflector, which also results in an ideal feed for reducing cross

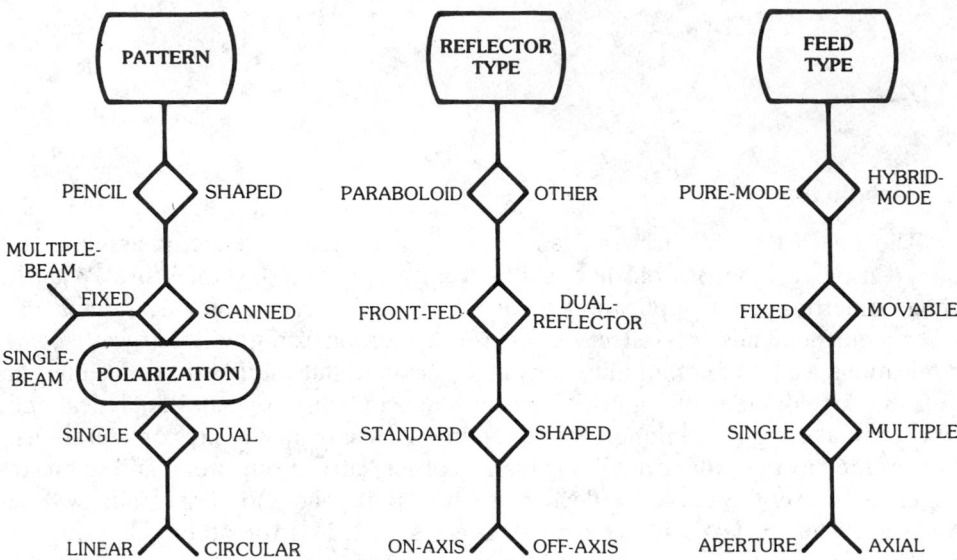

Fig. 1. Classification of reflector antennas based on pattern, reflector, and feed types. (*After Clarricoats and Poulton [5], © 1977 IEEE*)

polarization. Among the hybrid-mode feeds are feeds with corrugated metal walls. Another important consideration is the complexity of the feed system in terms of the number of feed elements used, for instance, to create contour or multiple beams. The proximity of feed elements necessitates an understanding of mutual coupling and the overcoming of difficulties in designing an acceptable beam-forming network. The reader may refer to references [6, 7] for a detailed discussion of the feeds.

The objective of this chapter is to provide a summarized description of the performances of many modern reflector configurations as discussed above and of advanced analytical techniques for analyzing reflectors. The emphasis is to present useful design data that complement information given in other reflector handbooks [1, 3, 7, 8]. In particular a considerable amount of data is provided for offset reflector configurations which are used extensively in today's antenna design. The are a few recently published review papers and collections of papers that the reader is strongly advised to study [1, 5, 9, 10]. The additional references assembled in these papers are not repeated here and only the most recent and relevant ones are referred to in this chapter.

2. Basic Formulations for Reflector Antenna Analysis

Since analytical/numerical techniques are used very extensively in the performance evaluation of reflector antennas and since there are numerous successful checks of the results of these techniques against measured data, this section describes application of some of the most currently applied techniques. Dual-offset reflector antennas are used as the base configuration for the description of these techniques.

The geometry of a dual-offset reflector with a feed arbitrarily positioned is shown in Fig. 2. This is the most general configuration and all other cases, such as single reflectors, symmetric configurations, Cassegrains, etc., can be regarded as special cases. Reflectors with more than one subreflector (beam waveguide systems) may also be handled by replacing the subreflector of Fig. 2 with the appropriate beam waveguide systems. Three coordinate systems are erected to define the main reflector, the subreflector, and the feed position (or array of feeds). The position and field vectors of these coordinate systems can be interrelated using the Eulerian angles construction [11] (see Appendix). For instance, the fields of feed can be expressed in feed coordinates (x_F, y_F, z_F) and then transformed into subreflector coordinates (x_S, y_S, z_S) to determine the scattered field from the subreflector and then transformed again into main reflector coordinates (x_M, y_M, z_M) to finally obtain the radiated field of the main reflector.

There are many different analytical/numerical techniques to determine the radiation characteristics of reflectors. Among them one may refer to physical optics (PO), aperture field (AF), geometrical optics (GO), geometrical theory of diffraction (GTD), method of moments (MOM), or any combination of these. All of these techniques have advantages or disadvantages, depending on the particular reflector configuration, far-field pattern domain, polarization, computation time, accuracy, and so on. For example, MOM gives the most accurate result but it is impossible to use it economically for reflectors larger than approximately 5 wavelengths. The aperture field method is not very accurate for offset configuration with displaced feeds when the edge diffracted rays are not included in the construction of the aperture fields. The GTD method is not easily applied in the caustics regions of pencil beams. The PO method can take an excessive amount of computation time for large reflectors, in particular, in the dual-reflector configuration, etc. Many advances have been reported in the last decade that improve both the accuracy and the computation time of analysis by applying a variety of techniques. One of the most appealing techniques has been the application of GTD for the subreflector and physical optics for the main reflector, in conjunction with

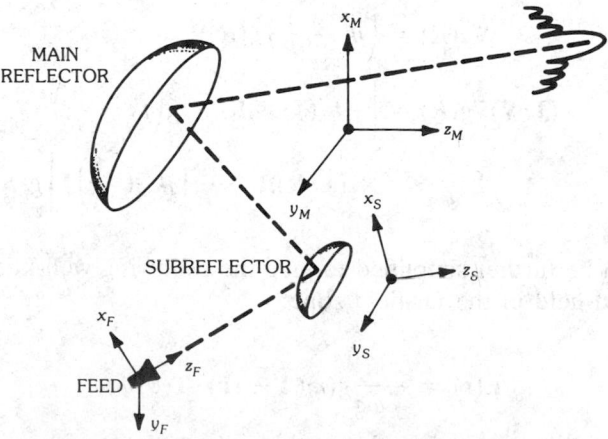

Fig. 2. Geometry of a dual-offset reflector antenna.

efficient expansions, such as Jacobi-Bessel, sampling theorem, etc. It is these last procedures which are summarized in this paper.

Physical Optics Analysis

The foundation of physical optics, or PO, rests on the assumption that the induced current on the reflector surface is given by (for a perfect conductor)

$$
\mathbf{J} = \begin{cases} 2\hat{\mathbf{n}} \times \mathbf{H}^i & \text{illuminated region} \\ 0 & \text{otherwise} \end{cases}
$$

where $\hat{\mathbf{n}}$ is the unit normal to the surface and $\mathbf{H}^i$ is the incident magnetic field. This incident field may emanate directly from the source or be scattered from the subreflector. Although PO current is an approximation for the true current on the reflector surface (obtainable using MOM), it nevertheless gives very accurate results for far fields of reflectors as small as approximately 5 wavelengths in diameter. The PO current displays certain errors, particularly near the edge of the reflector, which can be augmented by the incorporation of the fringe current [12]. Unfortunately the incorporation of the fringe current is not in general an easy task, although it may be required to accurately determine the far fields for a very wide range of observation angles. The radiated $\mathbf{E}$ field can be constructed using

$$
\mathbf{E}(\mathbf{r}) = -\frac{j}{\omega\epsilon} \int_{\Sigma} [(\mathbf{J} \cdot \nabla)\nabla + k^2 \mathbf{J}] g(R)\, dS' \tag{1}
$$

which is applicable for both the near- and far-field zones [4, 13]. The above expression can be easily derived using the concept of vector potentials. In (1), $\mathbf{r}'$, $\mathbf{r}$, and $\mathbf{R}$ are described in Fig. 3, $k = \omega\sqrt{\mu\epsilon}$, and

$$
g(R) = \frac{e^{-jkR}}{4\pi R}
$$

$$
\nabla g(R) = \left(jk + \frac{1}{R} \right) g(R)\hat{\mathbf{R}}
$$

$$
(\mathbf{J} \cdot \nabla)\nabla g(R) = \left[-k^2(\mathbf{J} \cdot \hat{\mathbf{R}})\hat{\mathbf{R}} + \frac{3}{R}\left(jk + \frac{1}{R} \right) \right.
$$

$$
\left. \times (\mathbf{J} \cdot \hat{\mathbf{R}})\hat{\mathbf{R}} - \frac{1}{R}\left(jk + \frac{1}{R} \right)\mathbf{J} \right] g(R) \tag{2}
$$

Equation 1 can be further simplified to give the following well-known expression for the radiated field in the far-field zone:

$$
\mathbf{E}(\mathbf{r}) = -\frac{jk^2}{\omega\epsilon} g(r)(\bar{\bar{\mathbf{I}}} - \hat{\mathbf{r}}\hat{\mathbf{r}}) \cdot \mathbf{T}(\theta, \phi) \tag{3}
$$

where

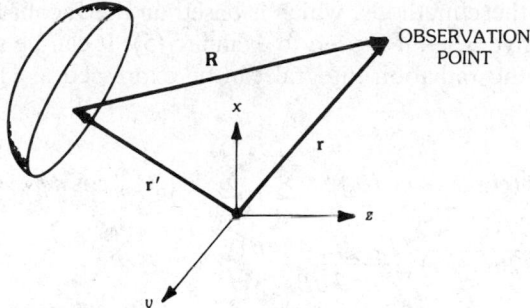

Fig. 3. Reflector geometry and integration and observation parameters for constructing the physical optics (PO) formulation.

$$\mathbf{T} = \int_{\Sigma} \mathbf{J}e^{jk\hat{\mathbf{r}}\cdot\mathbf{r}'}d\Sigma' \tag{4}$$

In deriving the above expression the standard far-field approximations are used, namely,

$$R = |\mathbf{r} - \mathbf{r}'| \cong r - \hat{\mathbf{r}}\cdot\mathbf{r}'$$

A similar but slightly more complicated expression can also be developed for the field in the Fresnel zone [13].

The surface integration in (4) is performed on the curved surface Σ. This integration, however, can be recast in terms of integration variables defined in a planar aperture by properly using the surface projection transformation to obtain

$$\mathbf{T} = \int_{\sigma} \mathbf{J}_{\text{eff}}e^{jk\hat{\mathbf{r}}\cdot\mathbf{r}'}dS' \tag{5}$$

where

$$\mathbf{J}_{\text{eff}} = \mathbf{J}\sqrt{1 + \left(\frac{\partial f}{\partial x'}\right)^2 + \left(\frac{\partial f}{\partial y'}\right)^2}$$

where σ is the projected aperture in this plane (for example, the xy plane), and $z' = f(x', y')$ defines the description of the reflector surface. One can show that this integral may be represented in terms of a summation series of many two-dimensional Fourier integrals [14, 15]. Typically one needs the first few terms of this series [14] to achieve a converging solution.

The PO radiation integral may be evaluated in many different ways and its efficient evaluation has been a challenging problem. A successful numerical method has been used based on the small patch constant phase technique [16] which can be applied to both the subreflector and main reflector geometries. Efficient numerical methods, which are particularly applicable to the main reflector with circular projected apertures, have been developed in [14, 17] and used frequently. One of these methods, which is based on the Jacobi-Bessel expansion, is particularly effective when it is used to evaluate (5). It can be shown that, based on this expansion, the radiation integral can be expressed as [14, 15]

$$T \propto \sum_{p=0}^{P \to \infty} \frac{1}{p} (jk)^p (\cos\theta - \cos\theta_B)^p \sum_{n=0}^{N \to \infty} \sum_{m=0}^{M \to \infty} J^n [_p\mathbf{C}_{nm} \cos n\phi + _p\mathbf{D}_{nm} \sin n\phi]$$

$$\times \sqrt{2(n + 2m + 1)} \left(\frac{J_{n+2m+1}[kaB]}{kaB} \right) \tag{6}$$

In the previous equation, a is the radius of the geometrically projected circular aperture, J is the Bessel function, and

$$B = \sqrt{(\sin\theta \sin\phi - \sin\theta_B \sin\phi_B)^2 + (\sin\theta \cos\phi - \sin\theta_B \cos\phi_B)^2}$$

$$\phi = \tan^{-1} \left(\frac{\sin\theta \sin\phi - \sin\theta_B \sin\phi_B}{\sin\theta \cos\phi - \sin\theta_B \cos\phi_B} \right) \tag{7}$$

where (θ_B, ϕ_B) is the direction of the anticipated beam maximum (or its vicinity). Note that when $B = 0$, i.e., in the boresight direction, only the first term of the series contributes. This first term is the Airy disk function, a typical far-field pattern of a uniform aperture. Another important feature of (6) is the fact that the coefficients $\mathbf{C}$ and $\mathbf{D}$ are independent of the observation points. These coefficients take the following form:

$$\begin{Bmatrix} _p\mathbf{C}_{nm} \\ _p\mathbf{D}_{nm} \end{Bmatrix} = \frac{\varepsilon_n}{2\pi} \int_0^{2\pi} \int_0^1 \mathbf{Q}_p \begin{Bmatrix} \cos n\phi' \\ \sin n\phi' \end{Bmatrix} F_m{}^n(s') s' d\phi' ds' \tag{8}$$

where ε_n is the Neumann constant ($\varepsilon_n = 1$ for $n = 1$ and $\varepsilon_n = 2$ otherwise), $F_m{}^n$ is the modified Jacobi polynomial [15], and $\mathbf{Q}_p$ is directly related to the physical optics induced current. The exact mathematical form of $\mathbf{Q}_p$ can be found in [14, 15].

Another useful representation of the far field is obtained by applying the sampling theorem, which allows one to express the far fields from a knowledge of field at a limited number of sampling points [18], viz.,

$$\mathbf{E} = \sum \sum \mathbf{E}_{nm} \frac{\sin(u - n\pi)}{u - n\pi} \frac{\sin(v - m\pi)}{v - m\pi} \tag{9}$$

where $u = ka' \sin\theta \cos\phi$ and $v = ka' \sin\theta \sin\phi$ (a' is the radius of an enlarged aperture [18] in front of the reflector) and the $\mathbf{E}_{nm}$ are the fields at the sampling points dictated by the Wittaker-Shannon theorem with spacing $\Delta u \leqq \pi/a'$ and $\Delta v \leqq \pi/a'$. This far-field representation is particularly useful when one is evaluating the fields at many observation points. A generalization of this procedure can also be performed using nonuniform sampling techniques [19].

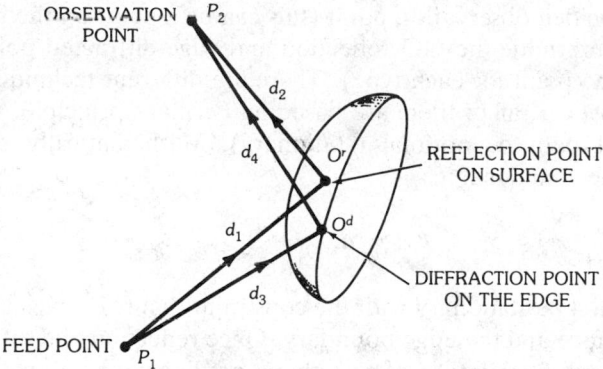

Fig. 4. Reflector geometry and reflection and diffraction parameters for constructing the geometrical theory of diffraction (GTD) formulation.

Geometrical Theory of Diffraction Analysis

Since the fundamentals of the geometrical theory of diffraction, or GTD, are discussed in another chapter, here only its application to reflectors is summarized. The essence of GTD lies in its concepts of localization and geometrical rays satisfying Fermat's principle. These rays may be classified as the geometrical and edge-diffracted rays emanating from surface reflection and edge-diffracted points in accordance with Snell's law and Keller's theory [20]. For the reflector of Fig. 4 the total scattered **H** field can be expressed as

$$\mathbf{H}^t = \theta^i \mathbf{H}^i + \sum_i \theta_i^r \mathbf{H}_i^r + \sum_j \mathbf{H}_j^d \tag{10}$$

where the superscripts i, r, and d designate the incident, reflected, and diffracted fields, respectively, and the subscripts i and j refer to the number of surface reflection and edge diffraction specular points, respectively, for a given feed and observation location. Furthermore, θ^i and θ^r designate the incident and reflected shadow indicators (i.e., $\theta = 0$ if the ray is blocked, and $\theta = 1$ if the ray path is unblocked) [21, 22, 23]. The **E** field can also be constructed in a straightforward manner using the fact that on each ray the **E** and **H** fields assume a plane wave relationship, namely,

$$\mathbf{E} = Z\mathbf{H} \times \hat{\mathbf{r}}$$

where Z is the intrinsic impedance of the medium ($Z = 120\pi$ ohms for the free space) and $\hat{\mathbf{r}}$ is the unit vector in the direction of each ray.

To apply the ray construction it is necessary to associate a ray field with the radiation field of the feed. For a single feed with a well-defined phase center one typically associates rays emanating from the phase center. For more distributed sources (arrays of feeds) one may have to associate many phase centers (one per feed element) and repeat the GTD construction for each of them. For a given feed position and specified observation point (this can be a point on the main reflector) one must first determine the GO reflection and edge-diffracted points (there can be more than one point for each type). There are different techniques for this determination; however, all of them are based on Fermat's principle, which requires that the optical path be optimal (stationary). Mathematically speaking, it is required that the functional

$$d = d_{1(3)} + d_{2(4)} \tag{11}$$

as defined in Fig. 4 be stationary with the constraint that (11) gives a solution(s) on the reflector surface and the edge boundary. Once reflection and diffraction points are obtained from the solution of the above nonlinear equations, one can then implement the field construction of GTD to determine both the reflected and diffracted fields. The details of this construction may be found in [22] and other references [23].

The geometrical optics field (reflected field) transported by the reflected rays can be formulated as

$$\mathbf{H}^r(P_2) = (\mathrm{DF})e^{-jkd_2}\{\mathbf{H}^i(O') - 2[\mathbf{H}^i(O') \cdot \hat{\mathbf{n}}]\hat{\mathbf{n}}\} \tag{12}$$

which is given in terms of the incident field $\mathbf{H}^i$ at the reflection point O', the unit surface normal $\hat{\mathbf{n}}$ at O', and a divergence factor (DF). This factor can be expressed as

$$\mathrm{DF} = \frac{1}{\sqrt{1 + (d_2/R_1^r)}} \frac{1}{\sqrt{1 + (d_2/R_2^r)}} \tag{13}$$

where the square roots take positive real, negative imaginary, or zero values (so that DF is positive real, positive imaginary, or infinite). The terms R_1^r and R_2^r are the radii of principal curvature of the reflected wavefront passing through O'. Their computation can be found in [22] or in the GTD chapter of this handbook.

Corresponding to each diffraction point O^d of Fig. 4 there is a contribution to the diffracted field $\mathbf{H}^d$ in (10). Following the formulation of [22], the diffracted field can be expressed as

$$\mathbf{H}^d(P_2) = \frac{1}{2\sqrt{2\pi d_4}} e^{-j(d_4 + \pi/4)} \frac{1}{\sqrt{1 + (d_4/R_1)}} \frac{1}{\sin \beta} [\boldsymbol{\beta} D^h H_\beta^i + \boldsymbol{\alpha} D^s H_\alpha^i] \tag{14}$$

where

$R_1 = $ the radius of curvature of the diffracted wavefront passing through O^d

$\beta = $ the angle between the tangent to the edge and $\mathbf{d}_4$

$\hat{\boldsymbol{\alpha}}, \hat{\boldsymbol{\beta}} = $ the unit vectors of the diffracted-ray coordinates

$D^h, D^s = $ the soft and hard diffraction coefficients

$H_\alpha^i, H_\beta^i = $ the projections of the $\mathbf{H}^i$ incident field on the ray coordinates at O^d

The detailed mathematical construction of these parameters can be found in [22]. It is worthwhile to mention that Keller's standard diffraction coefficients diverge at the incident and reflected shadow boundaries, which can be remedied by applying a uniform diffraction coefficient. There are different uniform expressions for the diffraction coefficients and the reader may refer to [24, 25, 26, 27] for a detailed comparison.

Aperture Field Method

Another widely used reflector analysis technique is based on the application of the aperture field, or AF, method. In this method first a hypothetical planar aperture is erected in front of the reflector and then the tangential fields are determined in this aperture using the GO and GTD constructions. The aperture is typically truncated to the reflector projected aperture size when one deals with well-focused pencil-beam antennas and only the GO field construction is used. For cases where the feed is defocused or for fields at wide observation angles, however, one must use larger apertures to properly incorporate the contribution of the edge-diffracted fields. (Alternatively one may use an aperture which caps the reflector's rim.) Once the tangential **E** and **H** fields in the aperture are determined, the far fields can then be obtained.

There are many different representations for the construction of far fields which use either the aperture tangential **E** or **H** fields separately or a combination of them. These different representations are customarily referred to as different Kirchhoff approximations [2]. One of the representations which uses both the aperture tangential **E** and **H** fields and results in the far-field radiated **E** field can be formulated as

$$\mathbf{E} = \frac{-jk^2}{\omega\epsilon} g(r) \int_A \left\{ \hat{\mathbf{n}} \times \mathbf{H}^A - [(\hat{\mathbf{n}} \times \mathbf{H}^A) \cdot \hat{\mathbf{r}}]\hat{\mathbf{r}} - \frac{1}{Z}[(\hat{\mathbf{n}} \times \mathbf{E}^A) \times \hat{\mathbf{r}}] \right\} e^{jk\hat{\mathbf{r}}\cdot\mathbf{r}'} dA \quad (15)$$

where $\mathbf{E}^A$ and $\mathbf{H}^A$ are the **E** and **H** fields in the aperture with aperture size A and $\hat{\mathbf{n}}$ is the normal to this planar aperture as shown in Fig. 5. Furthermore, $g(r)$ is defined

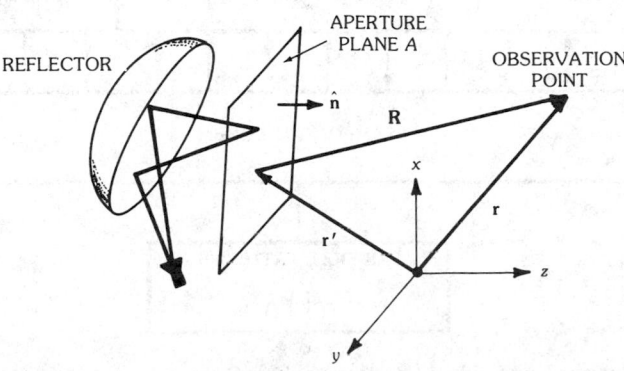

Fig. 5. Reflector geometry and integration and observation parameters for constructing the aperture field (AF) formulation.

in (2). For a planar aperture, (15) is a two-dimensional Fourier transform and can be evaluated in many different ways, one of which includes the use of the fast Fourier transform (FFT) algorithm. The reader is referred to [1, 28, 29] for different computational methods and comparisons among them (see also Fig. 6).

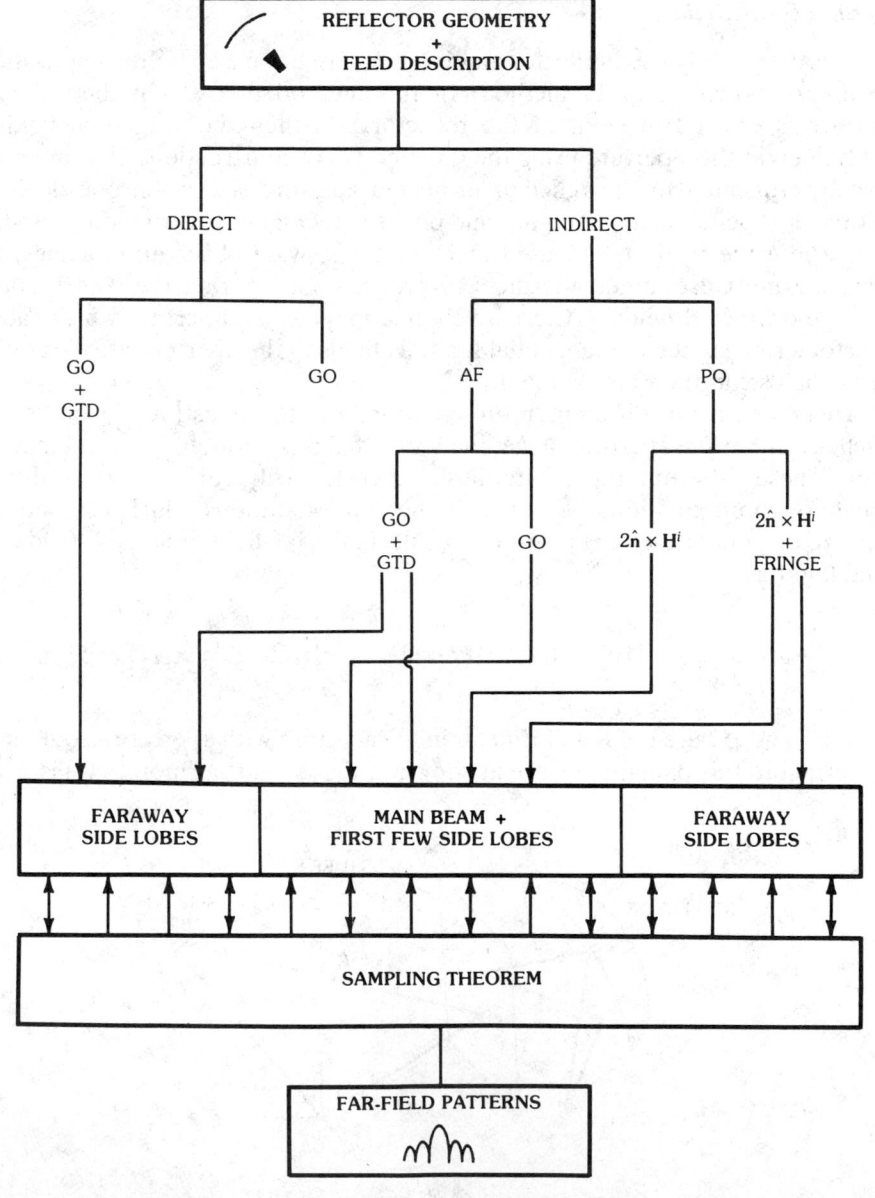

Fig. 6. A summary of available diffraction analysis techniques for focusing reflectors, where PO = physical optics, AF = aperture field, GO = geometrical optics, and GTD = geometrical theory of diffraction. (*After Rahmat-Samii [29], © 1984 IEEE*)

3. Simple Formulas for Far Fields of Tapered-Aperture Distributions

For many applications it is difficult to employ the formulations of PO or GTD (without the use of efficient computer programs) and arrive at simple expressions for the fields or to initiate a preliminary design. For reflectors with pencil-beam radiation characteristics, such as parabolic reflectors, there are simple models which predict their performances. Two such models are discussed in this section.

Two-Parameter Model

A useful and simple analytical model for characterizing the far-field patterns of aperturelike antennas with uniform phase distributions involves the application of the following amplitude distribution across the aperture:

$$Q(\varrho) = C + (1 - C)[1 - (\varrho/a)^2]^P \tag{16}$$

where P and C are parameters used to control the shape of the amplitude distribution and a is the radius of the circular aperture. In particular, C can be related to the aperture edge taper ET by

$$ET = 20 \log C \tag{17}$$

Figs. 7 and 8 show the plot of (16) for values of $P = 1$ and $P = 2$ and different edge tapers. The far-field pattern of this aperture distribution is simply proportional to its Fourier transform. For the aperture of radius a and the blockage radius of a_0, one can arrive at the following integral expression for the far-field pattern:

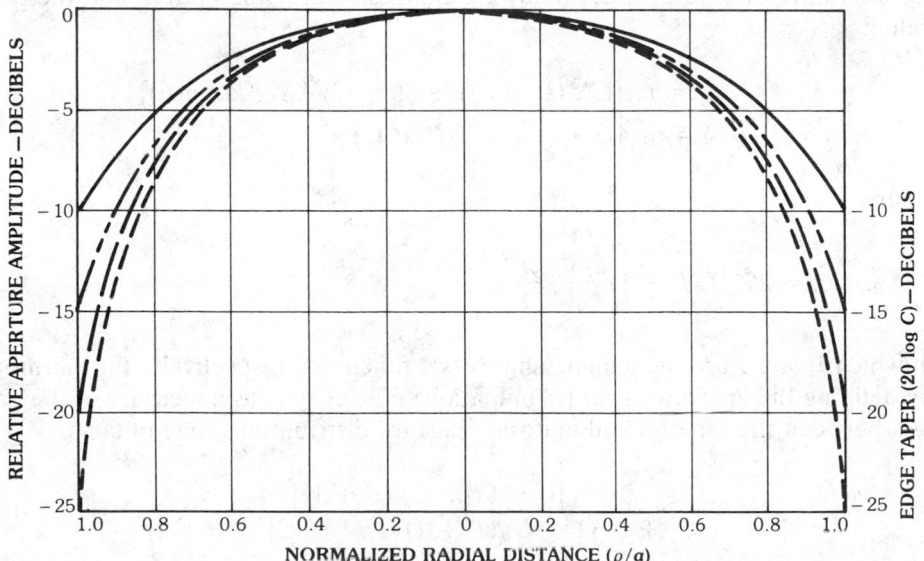

Fig. 7. Aperture distribution based on (16) for different edge tapers (-10, -15, -20, and -25 dB) and $P = 1$.

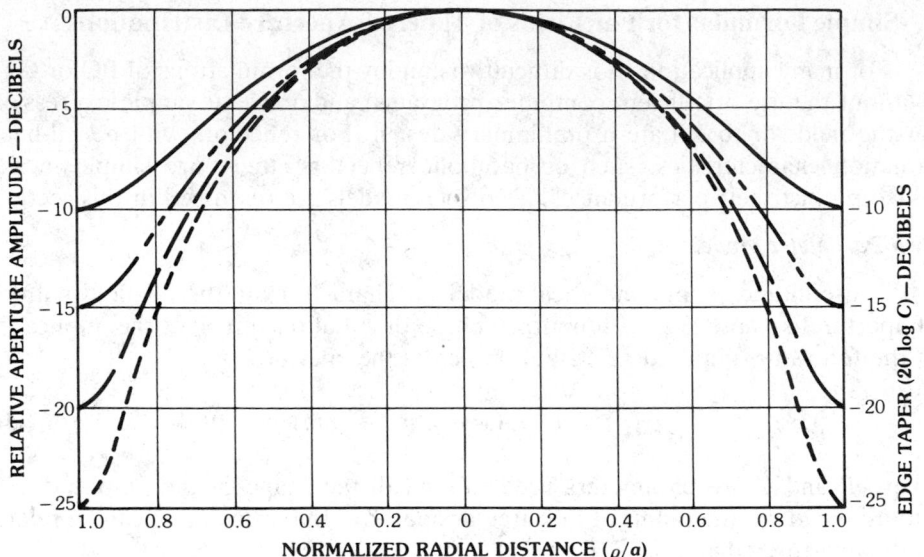

Fig. 8. Aperture distribution based on (16) for different edge tapers (-10, -15, -20, and -25 dB) and $P = 2$.

$$T = \int_{a_0}^{a} \int_{0}^{2\pi} Q(\varrho')e^{jk\varrho'\sin\theta\cos(\phi-\phi')}\varrho'd\varrho'd\phi' \tag{18}$$

For the cases of no blockage, i.e., $a_0 = 0$, (18) can be evaluated in closed forms for all values of P to give the following expressions for the pattern and the peak value:

$$T = \pi a^2[C\Lambda_1(ka\sin\theta) + (1 - C)\Lambda_{P+1}(ka\sin\theta)]$$
$$T(0°) = \pi a^2[C + (1 - C)/(P + 1)] \tag{19}$$

where

$$\Lambda_{P+1}(\zeta) = 2^{P+1}\Gamma(P + 1)\frac{J_{P+1}(\zeta)}{\zeta^{P+1}} \tag{20}$$

in which Γ and J are the gamma and Bessel functions, respectively. Furthermore, by defining the aperture taper (illumination) efficiency η_t as a measure of the gain loss between the tapered and uniform aperture distributions, one obtains

$$\eta_t = \frac{[C + (1 - C)/(P + 1)]^2}{C^2 + 2C(1 - C)/(P + 1) + (1 - C)^2/(2P + 1)} \tag{21}$$

It can readily be observed how the beamwidth and side lobe levels vary as a function of edge tapers and aperture parameter P. For $P = 1$ and 2 cases Table 1 contains the values of pertinent parameters for quick reference.

Table 1. Characteristics of Tapered Circular-Aperture Distribution Based on Two-Parameter Model

Edge Illumination		P = 1			P = 2		
ET (dB)	C	Half-Power Beamwidth (rad)	Side Lobe Level (dB)	η_t	Half-Power Beamwidth (rad)	Side Lobe Level (dB)	η_t
0	1	$1.01\lambda/2a$	−17.6	1	$1.01\lambda/2a$	−17.6	1
−8	0.398	$1.12\lambda/2a$	−21.5	0.942	$1.14\lambda/2a$	−24.7	0.918
−10	0.316	$1.14\lambda/2a$	−22.3	0.917	$1.17\lambda/2a$	−27.0	0.877
−12	0.251	$1.16\lambda/2a$	−22.9	0.893	$1.20\lambda/2a$	−29.5	0.834
−14	0.200	$1.17\lambda/2a$	−23.4	0.871	$1.23\lambda/2a$	−31.7	0.792
−16	0.158	$1.19\lambda/2a$	−23.8	0.850	$1.26\lambda/2a$	−33.5	0.754
−18	0.126	$1.20\lambda/2a$	−24.1	0.833	$1.29\lambda/2a$	−34.5	0.719
−20	0.100	$1.21\lambda/2a$	−24.3	0.817	$1.32\lambda/2a$	−34.7	0.690

For apertures with central blockage ($a_0 \neq 0$), (18) can only be evaluated in closed forms for integer values of P. It can be shown that the following expressions hold [30] for $P = 1$ and $P = 2$:

$$T = T_1 - T_0 \tag{22}$$

for $P = 1$:

$$T_n = \pi a_n \left[C \frac{2}{u_n} J_1(u_n) + (1 - C) \left\{ \frac{2}{u_n} J_1(u_n) \right. \right.$$
$$\left. \left. - \frac{a_n^2}{a^2} \left[\frac{2}{u_n} J_1(u_n) - \left(\frac{2}{u_n}\right)^2 J_2(u_n) \right] \right\} \right] \tag{23}$$

$$T_n(0°) = \pi a_n^2 \left[C + (1 - C)\left(1 - \frac{a_n^2}{2a^2} \right) \right]$$

and for $P = 2$:

$$T_n = \pi a_n^2 \left[C \frac{2}{u_n} J_1(u_n) + (1 - C) \left\{ \frac{2}{u_n} J_1(u_n) \right. \right.$$
$$- \frac{2a_n^2}{a^2} \left[\frac{2}{u_n} J_1(u_n) - \frac{4}{u_n^2} J_2(u_n) \right]$$
$$+ \frac{a_n^4}{a^4} \left[\frac{2}{u_n} J_1(u_n) - 2\left(\frac{2}{u_n}\right)^2 J_2(u_n) \right.$$
$$\left. \left. \left. + 2\left(\frac{2}{u_n}\right)^3 J_3(u_n) \right] \right\} \right] \tag{24}$$

$$T_n(0°) = \pi a_n^2 \left[C + (1 - C)\left(1 - \frac{a_n^2}{a^2} + \frac{a_n^4}{3a^4} \right) \right\} \right]$$

where, in the above equations, the following holds:

$$u_n = ka_n \sin \theta; \qquad a_1 = a, \quad a_0 = a_0 \qquad (25)$$

Obviously, for the cases $a_0 = 0$, (23) and (24) are directly obtainable from (19).

Next, numerical results for apertures with different blockage values are presented. First, let us define the blockage parameter b as the ratio of the blockage diameter to the aperture diameter, viz.,

$$b = \frac{D_0}{D} = \frac{2a_0}{2a}$$

For different values of parameter b, the far-field patterns are shown in Figs. 9 and 10. Again, these far-field patterns are normalized with respect to the no-blockage case. The reader can develop a good understanding by comparing these results with the no-blockage cases, in order to identify the effects of central blockage. Also shown in Figs. 11 and 12 are the boresight gain loss and first side lobe levels as a function of the blockage parameter b for different edge tapers and cases $P = 1$ and $P = 2$.

Plots of Figs. 7 through 12 should provide a good basis for predicting the performance of most aperturelike antennas. The values of C and P should be chosen such that they provide similar edge tapers and amplitude slopes at the edge as compared with the actual feed/reflector configuration. In most cases the desired values of P are $1 \leqq P \leqq 2$.

One-Parameter Model

A one-parameter simple expression is also available for modeling the radiated field of a circular aperture and the reader may refer to [31] for details. In this model, which applies to the no-blockage circular aperture case, the far-field pattern is proportional to

$$T = \frac{2J_1[\sqrt{u^2 - H^2}]}{\sqrt{u^2 - H^2}}, \qquad u \geqq H$$

$$T = \frac{2I_1[\sqrt{H^2 - u^2}]}{\sqrt{H^2 - u^2}}, \qquad H \geqq u \qquad (26)$$

where $u = ka \sin \theta$, J_1 and I_1 are Bessel and modified Bessel functions of the first kind of the first order, and the constant H is the single parameter. The J_1 part of the pattern provides the side lobe structure plus part of the main beam, and the remainder of the main beam is the I_1 part. The transition occurs at $u = H$, when the pattern function is unity. For this model the first side lobe level (SL) is at

$$SL = -17.57 \text{ dB} - 20 \log(2I_1(\pi H)/\pi H) \qquad (27)$$

Furthermore, the inverse Fourier transform of (26) can be obtained analytically [31] to result in the following expression for the aperture distribution, viz.,

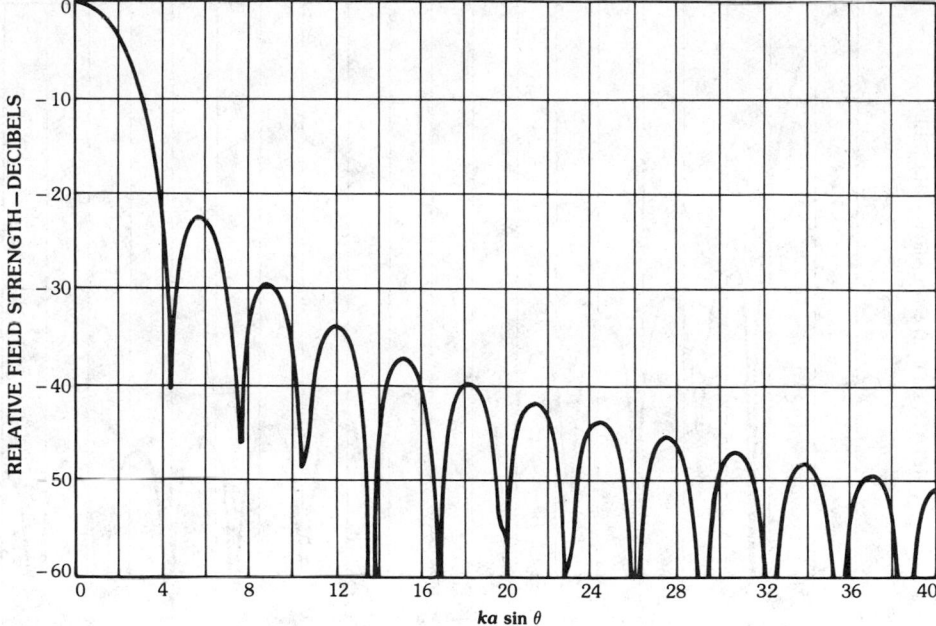

a

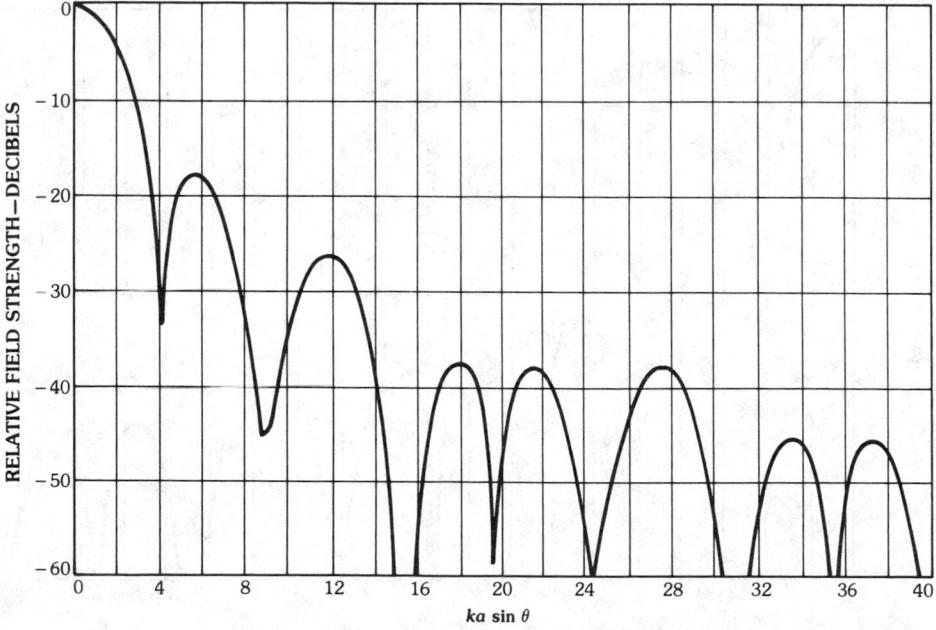

b

Fig. 9. Far-field patterns based on the aperture distribution (16), with ET = −10 dB and $P = 1$, for different values of $b = D_0/D$. (*a*) For central blockage ratio $b = 0$. (*b*) For central blockage ratio $b = 0.2$.

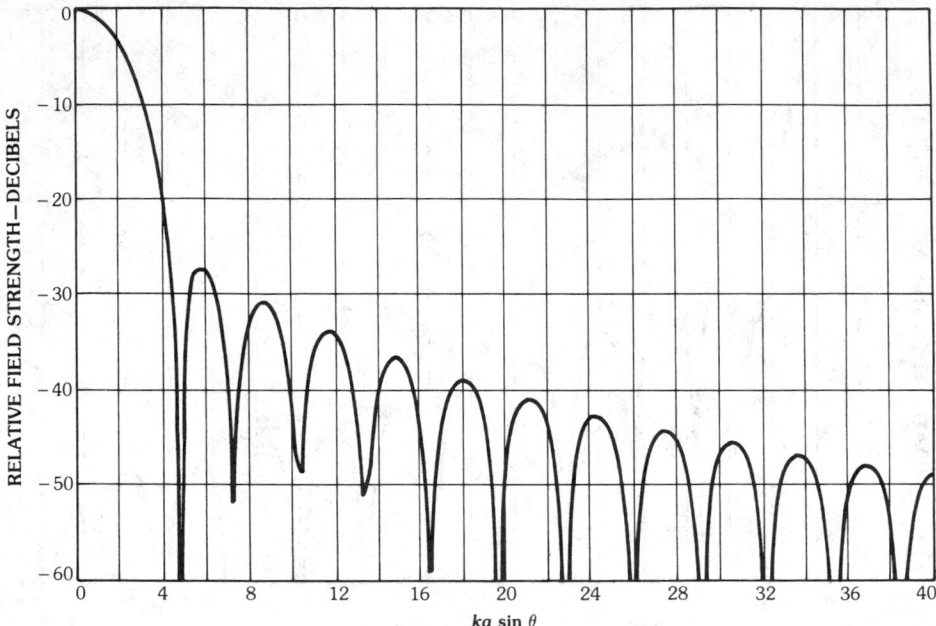

a

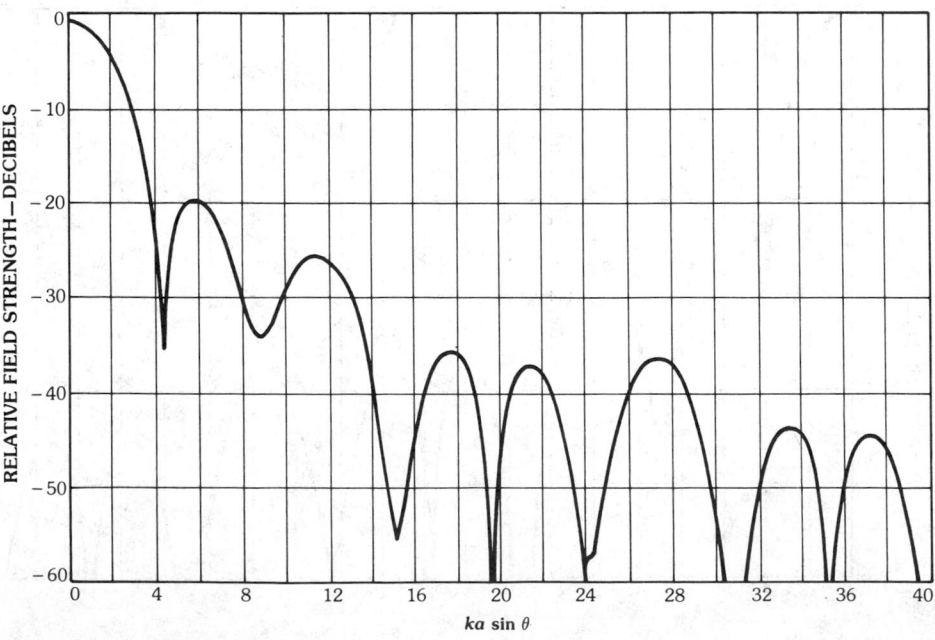

b

Fig. 10. Far-field patterns based on the aperture distribution (16), with ET = −10 dB and $P = 2$, for different values of $b = D_0/D$. (*a*) For central blockage ratio $b = 0$. (*b*) For central blockage ratio $b = 0.2$.

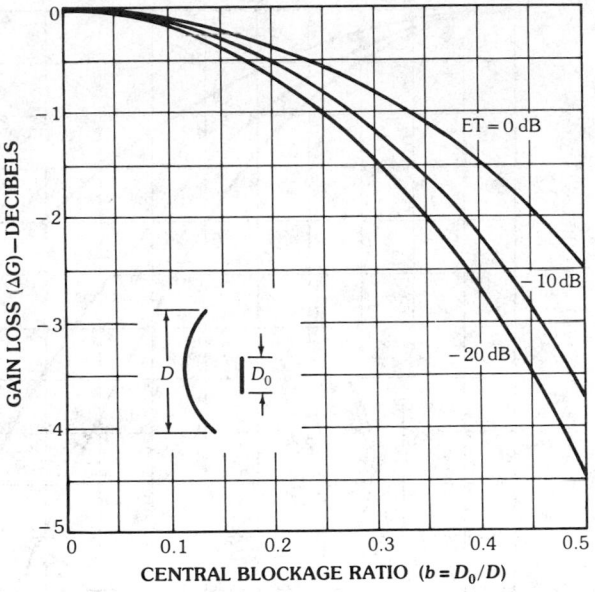

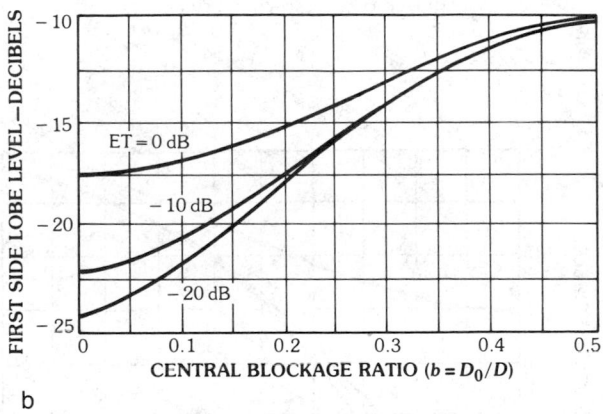

Fig. 11. Gain loss and first side lobe level as a function of $b = D_0/D$ for different edge tapers when (16) with $P = 1$ is the aperture distribution. (*a*) Gain loss. (*b*) First side lobe level.

$$Q(\varrho) = I_0[H\sqrt{1 - (\varrho/a)^2}] \qquad (28)$$

where I_0 is the modified Bessel function of the zeroth order. By using expressions (26)–(28) different characteristics of the pattern can be determined, as detailed in Table 2 [31].

Near, Fresnel, and Far Fields

Before closing this subsection it is worthwhile to present data on the near, Fresnel, and far fields of aperturelike antennas. This information is useful for

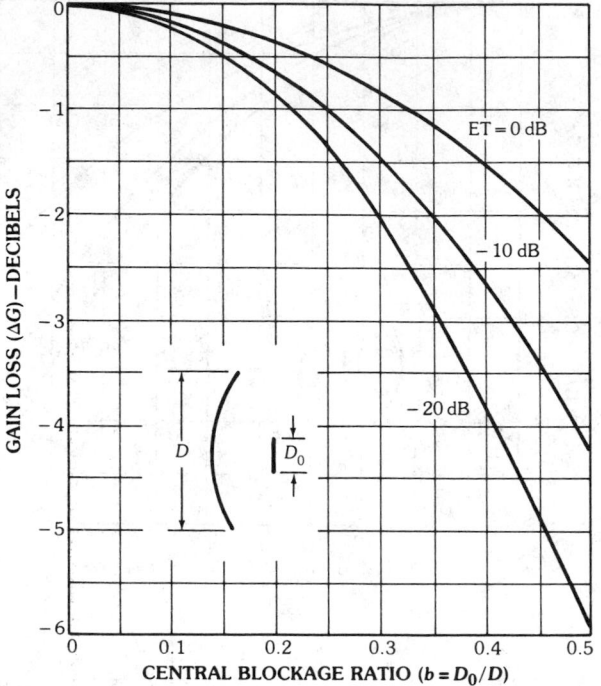

a

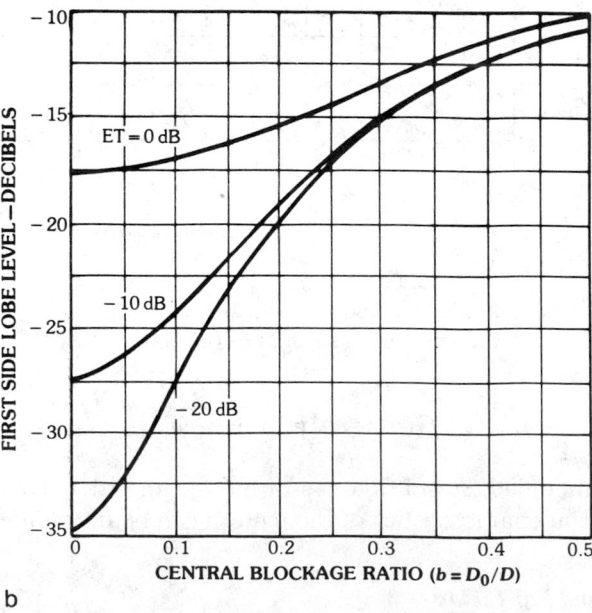

b

Fig. 12. Gain loss and first side lobe level as a function of $b = D_0/D$ for different edge tapers when (16) with $P = 2$ is the aperture distribution (a) Gain loss. (b) First side lobe level.

Table 2. Characteristics of Tapered Circular-Aperture Distribution Based on One-Parameter Model (*After Hansen [31]*, © *1975 IEE*)

First Side Lobe Level	H/π	Edge Taper (dB)	Half-Power Beamwidth*	Efficiency η_t
−17.57	0	0	1.0000	1.0000
−20	0.4872	−4.49	1.0483	0.9786
−25	0.8899	−12.35	1.1408	0.8711
−30	1.1977	−19.29	1.2252	0.7595
−35	1.4708	−25.78	1.3025	0.6683
−40	1.7254	−31.98	1.3741	0.5964

*Relative to a uniform distribution.

developing an understanding of the evolution of radiated fields away from the reflector. For the sake of simplicity, only the uniform aperture distribution is considered and its radiated fields at many different distances from the aperture are constructed by numerically evaluating (1). Results are shown in Fig. 13, where the reader can readily observe how the far-field pattern is formed at distances $R > 2D^2/\lambda$. For very low side-lobe reflectors, distances much greater than $2D^2/\lambda$ are required to obtain an accurate description of inner side lobes.

4. Some Important Geometrical Features of Conic-Section–Generated Reflector Antennas

Both single- and dual-offset reflector antennas are being applied more frequently in the design of low side lobe antenna systems. Due to their very unique optical focusing characteristics, surfaces generated from conic sections are most commonly used in practice. Typically these surfaces are constructed as a result of a translation or rotation of the conic sections. Offset reflectors are carved-out portions of these surfaces, resulting from their intersections with cylinders or cones. The cylinders have their axes parallel to the axes of the parent reflector surfaces, and the cones have their tips at one of the foci of the reflectors. In this section geometrical characteristics of offset (symmetric) conic section reflectors are presented in a unified fashion and some of their important geometrical features are described. Only the final results are given here and the reader is referred to [32, 33] for details.

Conic Sections

Conic sections are basically second-degree planar curves which can be generated in many ways, including the intersection of a circular cone with a planar surface. Referring to Fig. 14, with z as the abscissa and x as the ordinate of a plane Cartesian coordinate system, the equation of a conic section can be expressed in the following general form (when the principal axes of a conic section are tilted with respect to the x and z axes a more general quadratic expression results):

$$\frac{(z-c)^2}{(f+c)^2} + \frac{x^2}{(f+c)^2 - c^2} = 1 \tag{29}$$

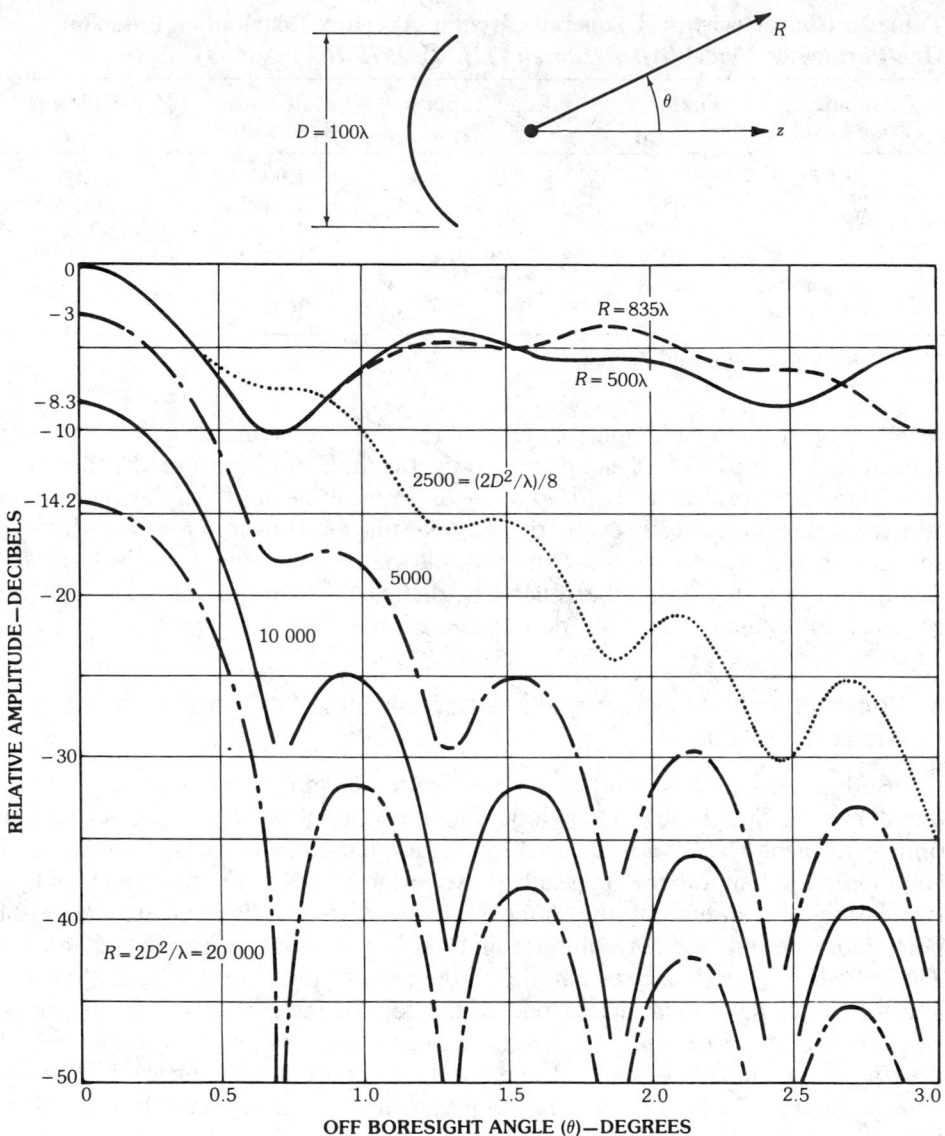

Fig. 13. Reflector antenna patterns as a function of distance R.

In polar coordinates this equation is written as

$$\varrho = \frac{f + 2c}{f + (1 - \cos\phi)c}f \tag{30}$$

where in both of the previous equations the parameters have the following definitions: f is the focal length (distance from a focus to the nearest apex) having a positive or zero value, and $2c = F_1F_2 \cdot \hat{z}$ is the distance between the two foci having

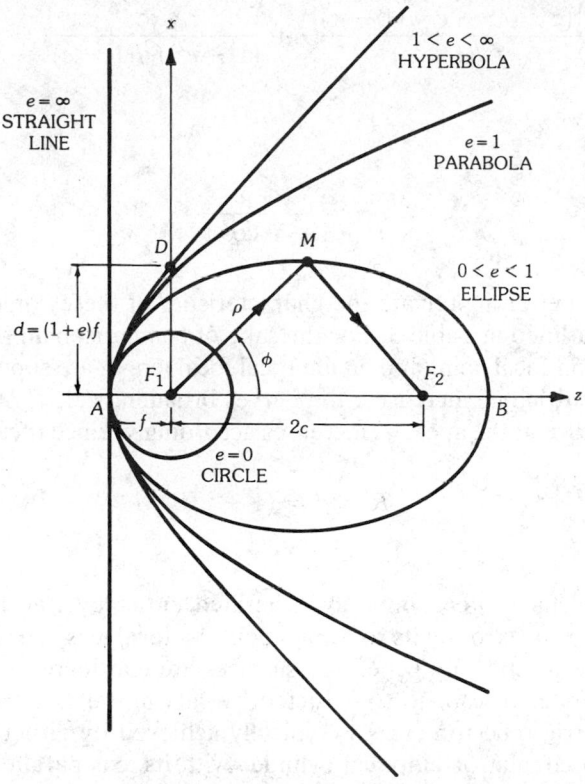

Fig. 14. Geometrical configuration of conic sections as a function of eccentricity e and for focal length f. (*After Jamnejad and Rahmat-Samii [32], © 1980 IEEE*)

an algebraic value which is positive if F_2 is to the right of F_1 (ellipse) and negative otherwise (hyperbola) where $\hat{z}$ is a unit vector along the positive z direction. One of the common properties of the conic sections is that either the sum or difference of distances from any point on the curve to the focal points is constant. Another manifestation of this property from the geometrical optics viewpoint is that any ray passing through one focus and crossing the curve at a given point is reflected along the line connecting the crossing point to the second focus.

Eccentricity e is defined as

$$e = \frac{c}{c + f} = \frac{1}{1 + f/c} \tag{31}$$

which is always a positive number. This parameter is a measure of the off-centeredness of the focal points for a given focal length. Note that the center is located at the midpoint of the foci. Thus, when the foci are at the center (a circle), the eccentricity is zero, and when they are infinitely apart, it is unity (a parabola), and so on.

In terms of this new parameter (29) and (30) can be rewritten as

$$\frac{[(z - ez - ef)/(1 - e)]^2}{f^2/(1 - e)^2} + \frac{x^2}{(1 - e^2)f^2/(1 - e)^2} = 1 \tag{32}$$

and

$$\varrho = \frac{1 + e}{1 - e \cos \phi} f \tag{33}$$

In order to better appreciate the characteristics of these conic section curves, the cases are outlined in Table 3. For the sake of comparison all the conic sections with one common focal point and an identical focal length are superimposed in Fig. 14. As the eccentricity e increases, the curves broaden, i.e., $F_1D = d$ and R, the radius of curvature at the apex A, increases accordingly, since it can be shown that

$$R = d = (1 + e)f \tag{34}$$

Reflector Surfaces

Reflector surfaces are commonly generated either by translation of a conic section along the y axis, or by its rotation about the focal axis z, as in Fig. 15. In this presentation the rotationally generated surfaces are considered.

An offset reflector can be constructed by carving out a portion of the rotationally symmetric reflector. This is typically achieved by either intersecting the reflector with a circular or elliptical cylinder with its axis parallel to the reflector axis, or by intersecting the reflector with a circular or elliptical cone with its tip at the focal point. The following subsections discuss the geometrical characteristics of reflectors thus obtained.

Intersection with a Circular Cone

The cone of rays emanating from a source located at the focal point F_1 intersects the reflector surface as shown in Fig. 15. The equation of the reflector surface, produced by the revolution of a conic section about the z axis, is obtained by simply replacing x^2 by $x^2 + y^2$ in (32) to arrive at

Table 3. Classification of Conic Curves

Foci Separation/2	Eccentricity	Equation	Type
$c = 0$	$e = 0$	$\dfrac{z^2}{f^2} + \dfrac{x^2}{f^2} = 1$	Circle
$0 < c < \infty$	$0 < e < 1$	(32)	Ellipse
$c = \infty$	$e = 1$	$(z + f) - \dfrac{1}{4f} x^2 = 0$	Parabola
$-\infty < c < -2f$	$1 < e < \infty$	(32)	Hyperbola
$c = -f$	$e = \infty$	$z + f = 0$	Straight line

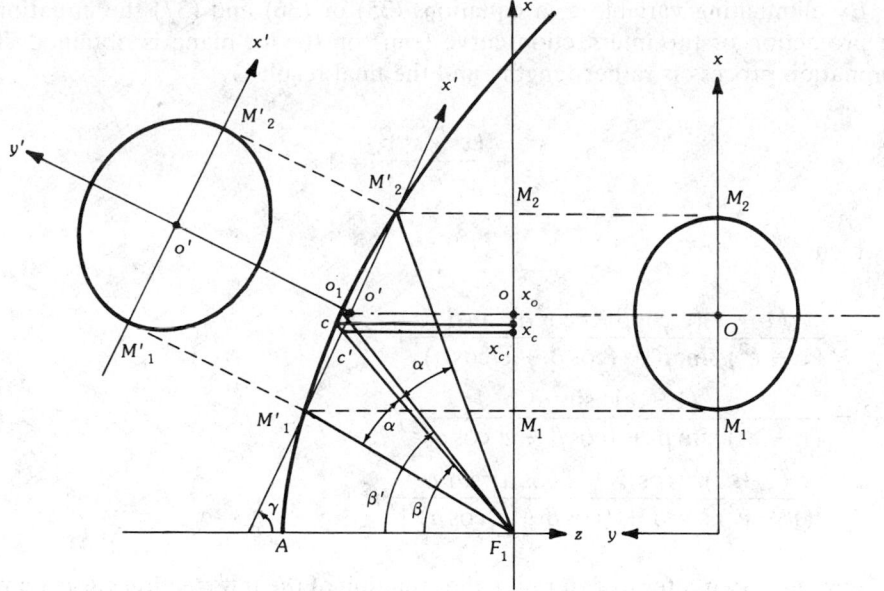

Fig. 15. Geometry of an offset reflector obtained from the intersection of a circular cone and a surface generated by rotation of a conic section about its focal axis. (*After Jamnejad and Rahmat-Samii [32], © 1980 IEEE*)

$$\frac{[(z - ez - ef)/(1 - e)]^2}{f^2/(1 - e)^2} + \frac{x^2 + y^2}{(1 - e^2)f^2/(1 - e)^2} = 1 \tag{35}$$

The equation of a cone whose axis lies in the xz plane making an angle β with the negative z axis and having a half-angle α, as shown in Fig. 15, is

$$(x^2 + y^2 + z^2)\cos^2\alpha = (x \sin\beta - z \cos\beta)^2 \tag{36}$$

The equation of the intersection curve of the surface of revolution and the cone is the solution of (35) and (36). Elimination of the variable y from these equations leads to the projection of the intersection curve on the xz plane with a final result of

$$x = \frac{\cos\beta + e \cos\alpha}{\sin\beta} z + \frac{(1 + e) \cos\alpha}{\sin\beta} f \tag{37}$$

which is the equation of the projection of the intersection on the xz plane. Clearly this projection is a *straight* line, indicating that the intersection curve (rim) lies in a *plane* perpendicular to the xz plane. In other words it is shown that the intersection of a reflector surface, generated from the rotation of a conic section about its focal axis, and a circular cone, with its tip on a focal point of the reflector surface, is always a *planar curve*. In fact it will be shortly demonstrated that this curve is an ellipse.

By eliminating variable z in equations (35) or (36) and (37) the equation of the projection of the intersection curve (rim) on the xz plane is obtained. This elimination process is rather lengthy and the final result is

$$\frac{y^2}{a^2} + \frac{(x - x_o)^2}{b^2} = 1 \tag{38}$$

in which

$$
\begin{aligned}
x_o &= \frac{(1 + e)(e \cos\beta + \cos\alpha) \sin\beta}{(1 - e^2) \sin^2\beta + (\cos\beta + e \cos\alpha)^2} f \\
a^2 &= \frac{(1 + e)^2 \sin^2\alpha}{(1 - e^2) \sin^2\beta + (\cos\beta + e \cos\alpha)^2} f^2 \\
b^2 &= \frac{(1 + e)^2 (\cos\beta + e \cos\alpha)^2 \sin^2\alpha}{[(1 - e^2) \sin^2\beta + (\cos\beta + e \cos\alpha)^2]^2} f^2
\end{aligned}
\tag{39}
$$

It is now an easy matter to determine the equation of the intersection curve (rim) in the plane of intersection. This is done by designating the $x'y'$ plane as the intersection plane, as shown in Fig. 15, and by choosing o' as the origin of this coordinate system to finally arrive at

$$\frac{y'^2}{a'^2} + \frac{x'^2}{b'^2} = 1 \tag{40}$$

where

$$
\begin{aligned}
a'^2 &= a^2 \\
b'^2 &= \frac{(1 + e)^2 [\sin^2\beta + (\cos\beta + e \cos\alpha)^2] \sin^2\alpha}{[(1 - e^2) \sin^2\beta + (\cos\beta + e \cos\alpha)^2]^2}
\end{aligned}
\tag{41}
$$

in which a is given in (39).

It is of some interest to note that the point of intersection of the cone axis with the intersection plane, i.e., point c', does not coincide with the center of the intersection curve, i.e., point o' (Fig. 15). This is easily demonstrated by calculating the ordinate of point c' in the xz plane, which results in

$$x_{c'} = \frac{(1 + e) \cos\alpha \sin\beta}{1 + e \cos\alpha \cos\beta} f \tag{42}$$

which is clearly different from $x_{o'} = x_o$ given in (39). Neither do points c (intersection of the cone axis and the reflector surface) and o_1 (intersection of the central axis of the reflected cylinder, $o'o$, and the reflector surface) coincide. It can be shown that

$$x_c = x_o \Big|_{\alpha=0} = \frac{(1 + e) \sin \beta}{1 + e \cos \beta} f \tag{43}$$

which is different from $x_{o_1} = x_o$.

Special Cases

The equation of the intersection curve (rim) in the $x'y'$ plane and its projection on the xy plane as given by (40) and (38), respectively, and also the ordinate of some points of interest as given in (43), (42), and (39), are simplified in the following cases:

(a) spherical surface ($e = 0$)

$$\begin{aligned} a^2 = f^2 \sin^2\alpha, && a'^2 = a^2 \\ b^2 = a^2 \cos^2\beta, && b'^2 = a^2 \end{aligned} \tag{44}$$

and

$$\begin{aligned} x_o &= f \sin \beta \cos \alpha \\ x_c &= x_o \\ x_{c'} &= f \sin \beta \end{aligned} \tag{45}$$

In this case the intersection curve (rim) is a *circle* and its projection on the xy plane is an *ellipse* with its minor axis along the y axis.

(b) paraboloidal surface ($e = 1$)

$$\begin{aligned} a^2 &= \frac{4 \sin^2\alpha}{(\cos \beta + \cos \alpha)^2} f^2, && a'^2 = a^2 \\ b^2 &= a^2, && b'^2 = a^2 \left[1 + \frac{\sin^2\beta}{(\cos \beta + \cos \alpha)^2} \right] \end{aligned} \tag{46}$$

and

$$\begin{aligned} x_o &= \frac{2 \sin \beta}{\cos \beta + \cos \alpha} f \\ x_c &= \frac{2 \sin \beta}{1 + \cos \beta} f \\ x_{c'} &= \frac{2 \sin \beta \cos \alpha}{1 + \cos \alpha \cos \beta} f \end{aligned} \tag{47}$$

In this case the intersection curve (rim) is an *ellipse* with its major axis along the x' axis, while its projection on the xy plane reduces to a *circle*.

It is also of interest in this case to determine the relationship between the angles β and β' as shown in Fig. 15. After some manipulations one arrives at

$$\tan \beta' = \frac{1}{1 - (\sin^2\alpha)/2(\cos^2\beta + \cos\beta\cos\alpha)} \tan\beta \qquad (48)$$

(c) planar surface ($e = \infty$)

$$a^2 = \frac{\sin^2\alpha}{\cos^2\alpha - \sin^2\beta} f^2, \qquad a'^2 = a^2$$

$$b^2 = a^2\cos^2\alpha, \qquad\qquad b'^2 = b^2 \qquad (49)$$

and

$$x_o = \frac{\cos\beta\sin\beta}{\cos^2\alpha - \sin^2\beta} f$$

$$x_c = f\tan\beta \qquad\qquad\qquad (50)$$

$$x_{c'} = x_c$$

It is obvious that, in this case, the intersection curve (rim) and its projection on the xy plane are the same and display an *ellipse* elongated in the x direction.

As a concluding remark for this subsection one can say that, starting from zero eccentricity $e = 0$ (a sphere), where the intersection curve (rim) is a circle, as eccentricity increases, the rim curve becomes more oblong in the x' direction. On projection onto the xy plane, however, the curve becomes contracted in the x direction to the extent that for $e = 1$ (a paraboloid) the projected curve becomes a circle. For values of $e < 1$ (an ellipsoid) the projected curve has its minor axis in the x direction, while for values of $e > 1$ (a hyperboloid) it has its major axis in that direction.

Intersection with a Circular Cylinder

The equation of a circular cylinder with its axis in the yz plane and parallel to the z axis is given as

$$y^2 + (x - x_o)^2 = R^2 \qquad (51)$$

where x_o and R are the offset height and the radius of the cylinder, respectively. The intersection of this cylinder with the reflector surface can be found from (51) and (22). The projection of the intersection curve (rim) on the yz plane is found by eliminating y in these two equations, which results in

$$(1 - e^2)z^2 - 2e(1 + e)fz + 2xx_o = (1 + e)^2f^2 + x_o^2 - R^2 \qquad (52)$$

The above equation, in general, describes a parabola. It is therefore concluded that the rim lies on a *parabolic cylinder* and is *not* a planar curve. However, in the special case of $e = 1$ (paraboloid surface) the curve given by (52) is a line, which indicates that the rim is planar and, because it is the intersection of a circular cylinder and a plane, is indeed an ellipse.

5. Offset (Symmetric) Parabolic Reflectors

In this section the radiation characteristics of offset (symmetric) parabolic reflector antennas illuminated by a single feed element (fixed phase center) are presented. As is typical in any reflector design, there are too many almost-independent parameters which may be varied to achieve a particular design goal. For instance, to design an offset parabolic reflector one must study the effects of such parameters as offset angles, illumination tapers, F/D ratios, locations and orientations of the feed, polarizations, etc., on such far-field pattern characteristics as scan loss, beamwidth, side lobe level, cross-polarization level, efficiency, and so on. Obviously it is not possible to perform a comprehensive study of all these parameters; rather, attempts are made to present the most important reflector characteristics based on a few key parameters. The results of this section can therefore be used as a guideline for an initial design, which can then be refined into greater detail by using computer programs and measured data.

Geometrical Parameters

Although some of the general geometrical features of conic sections were presented in Section 4, features which are particular to parabolic reflectors are presented in this section. The geometry of an offset parabolic reflector with focal length F, diameter D, and offset height H is shown in Fig. 16. There are also other parameters used to characterize offset parabolic reflectors which are defined below:

F = focal length

D = reflector diameter (diameter of the circular projected aperture)

H = offset height ($H = -D/2$ for symmetric reflectors)

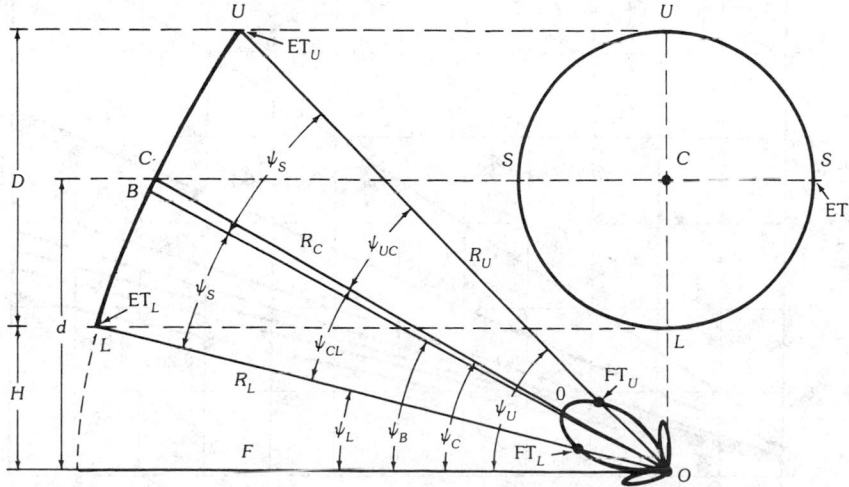

Fig. 16. Geometrical parameters of an offset parabolic reflector (for the symmetric case $H = -D/2$).

d = offset height of the circular projected aperture center
$= D/2 + H$

ψ_U = angle subtended to the upper tip
$= 2 \tan^{-1}[(D + H)/2F]$

ψ_C = angle subtended to the center of the projected aperture
$= 2 \tan^{-1}[D/2 + H)/2F]$

ψ_L = angle subtended to the lower tip
$= 2 \tan^{-1}[H/2F]$

ψ_B = angle subtended in the bisect direction
$= (\psi_U + \psi_L)/2$

ψ_S = half-angle subtended to the upper and lower tips
$= (\psi_U - \psi_L)/2$

ψ_{SC} = half-angle subtended to the right and left sides

$$= \sin^{-1}\left(\frac{D}{2}\middle/\sqrt{\left(H + \frac{D}{2}\right)^2 + \left(\frac{D}{2}\right)^2 + \left[\frac{(D/2)^2 + (H + D/2)^2}{4F} - F\right]^2}\right)$$

D_p = parent parabola diameter
$= 2(D + H)$ for $H \geqq -D/2$

In some cases F, ψ_B, and ψ_S are given, and from them D and H may be constructed. This can be done by using the following expressions:

$$D = 4F \sin \psi_S/(\cos \psi_B + \cos \psi_S)$$
$$H = 2F (\sin \psi_B - \sin \psi_S)/(\cos \psi_B + \cos \psi_S)$$

$$(53)$$

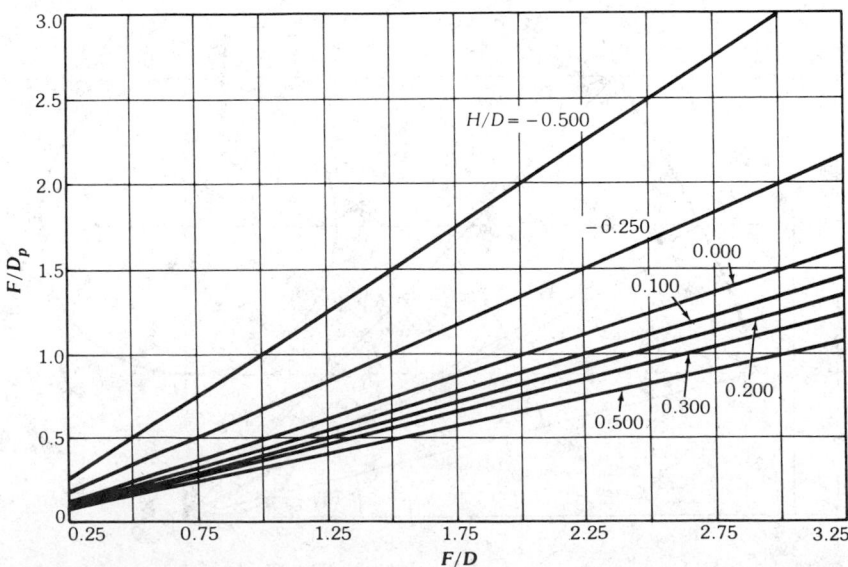

Fig. 17. The ratio F/D_p as a function of F/D and for different values of H/D.

In the design of offset reflectors any combination of the previously mentioned parameters may be given and the others may be constructed. In this subsection the parameters D, F/D, and H/D are used as independent parameters and the others are obtained. For example, Figs. 17 and 18a through 18c show plots of F/D_p, ψ_{UC}, ψ_{CL}, ψ_{CS}, and $20\log(R_C/R_{U,L,S})$ as functions of F/D for different H/D values. It is clear for symmetric reflectors that $F/D_p = F/D$ and $\psi_{UC} = \psi_{CL} = \psi_{CS}$, because $H/D = -0.5$. In the following subsections it will become apparent that these parameters play important roles to the extent that the far-field characteristics of reflectors are concerned. For example, different path losses at different edges (tips) of the reflector can cause different side lobe levels in the far-field pattern cuts. For most offset parabolic antennas, H/D typically varies between $0.1 \leqq H/D \leqq 0.3$ to provide clearance for the feed assemblies.

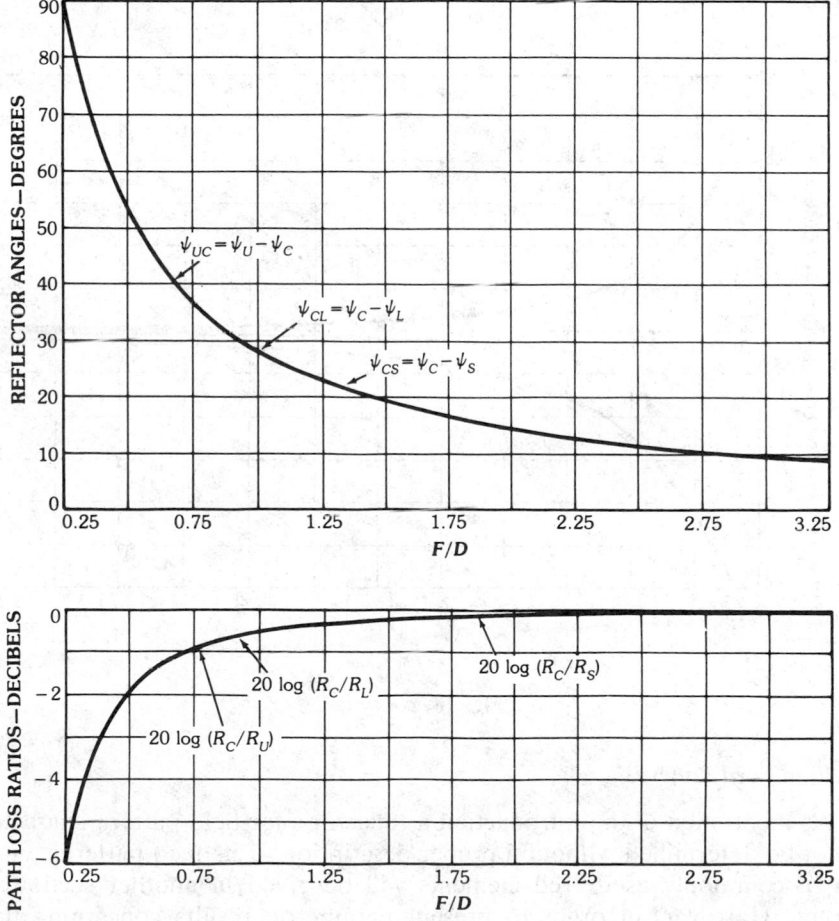

a

Fig. 18. Reflector angles and path loss ratios, as defined in Fig. 16, as a function of F/D. (*a*) For $H/D = -0.5$ (symmetric reflector case). (*b*) For $H/D = 0.10$. (*c*) For $H/D = 0.20$.

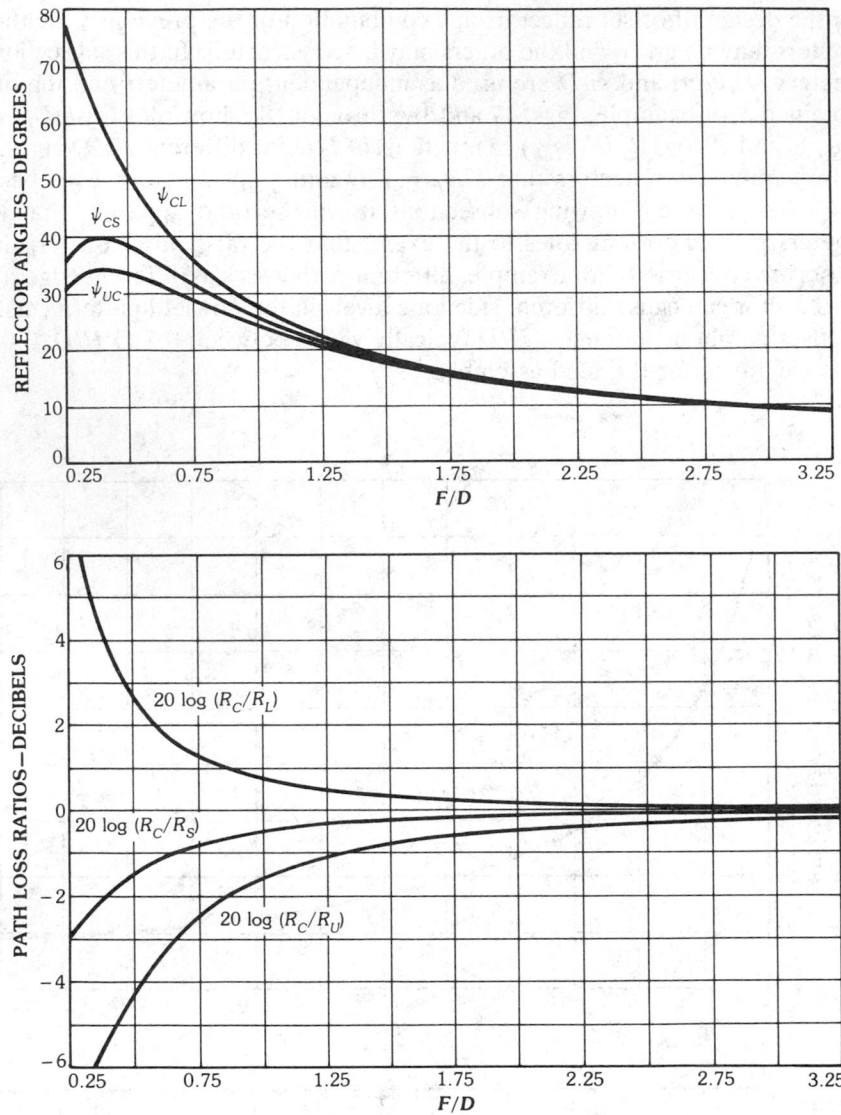

b

Fig. 18, *continued.*

Idealized Feed Patterns

As mentioned in the introduction a reflector's far-field pattern characteristics cannot be determined without a proper description of its feed patterns. A discussion of commonly used feed elements will be given in another section of this chapter. However, in order to present parametric results concerning the performance of parabolic reflectors, idealized feed patterns, which have proven to be very useful models, are used. For an idealized feed with a fixed phase center, its radiation pattern may be described as

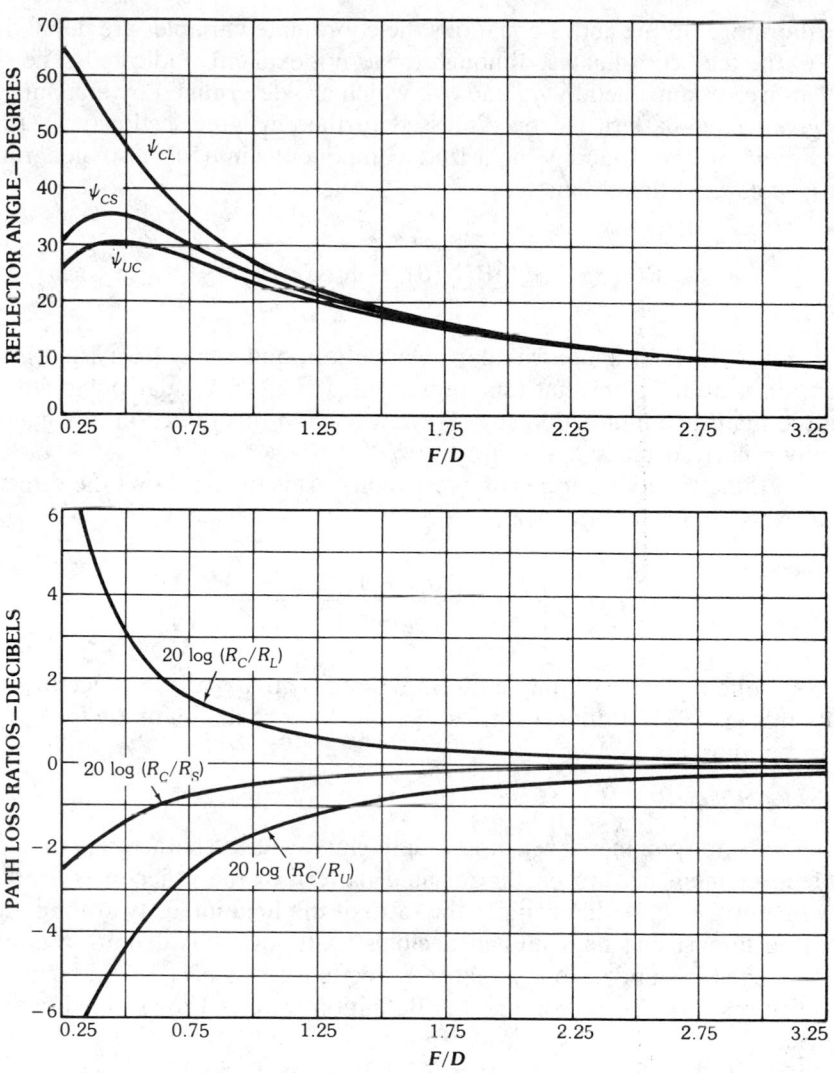

c

Fig. 18, *continued*.

$$\mathbf{E}(\mathbf{r}) = A_0 \begin{cases} \hat{\boldsymbol{\theta}} C_E(\theta) \cos\phi - \hat{\boldsymbol{\phi}} C_H(\theta) \sin\phi \\ \hat{\boldsymbol{\theta}} C_E(\theta) \sin\phi + \hat{\boldsymbol{\phi}} C_H(\theta) \cos\phi \end{cases} \frac{e^{-jkr}}{r}, \quad \begin{array}{l} \text{for } \hat{\mathbf{x}} \text{ polarized} \\ \text{for } \hat{\mathbf{y}} \text{ polarized} \end{array} \quad (54)$$

where A_0 is a complex constant and

$$C_E(\theta) = (\cos\theta)^{q_E} = E\text{-plane pattern}$$
$$C_H(\theta) = (\cos\theta)^{q_H} = H\text{-plane pattern} \quad (55)$$

for $0 \leqq \theta \leqq \pi/2$ and zero otherwise.

Furthermore, in the above equations the coordinate variables are defined with respect to the feed coordinates, although this is not explicitly indicated. The shape of the pattern is controlled by q_E and q_H, which are determined by matching (55) to the given feed pattern (to be discussed further in later sections). A proper superposition of the x and y polarized components in (54) also generates a circularly polarized field, namely,

$$\mathbf{E}(\mathbf{r}) = A_0 e^{j\tau\phi}[\hat{\boldsymbol{\theta}}C_E(\theta) + \hat{\boldsymbol{\phi}}j\tau C_H(\theta)]\frac{e^{-jkr}}{r} \tag{56}$$

where $\tau = +1$ for left-handed circular polarization, and $\tau = -1$ for right-handed circular polarization. Note that (56) represents a perfect circular polarized wave only in the main beam direction ($\theta = 0$). Away from this direction it is generally elliptically polarized unless $q_E = q_H$.

One of the attractive features of expression (54) is that it allows the directivity to be derived in closed form, viz.,

$$D(\theta = 0) = \frac{2(2q_E + 1)(2q_H + 1)}{q_E + q_H + 1} \tag{57}$$

which consequently allows computation of the directivity of the reflector illuminated by this feed [34]. It must be noted that the back radiation of the feed pattern is ignored in deriving (57).

Edge and Feed Tapers

An important parameter which is widely employed to characterize the effects of the feed element pattern on the far-field pattern of the reflector is the "edge taper." In a broad sense this signifies the ratio of the field intensity at the reflector edge to the intensity at its center in decibels. Although this definition is unambiguous when it is applied to symmetric reflectors it can become ambiguous for offset reflectors. For this reason another definition, referred to as "feed taper," is introduced.

Referring to Fig. 16, the feed taper FT_U in the upper-tip direction is defined as

$$FT_U = 20\log\left[\frac{C(\psi_U - \psi_C)}{C(0°)}\right] \tag{58}$$

where $0°$ refers to the central direction (i.e., OC in Fig. 16) and C denotes the feed pattern as defined in (55). Similar definitions for FT_L and FT_S can also be given at the lower and side angles by using $\psi_C - \psi_L$ and ψ_S, respectively. The edge taper ET_U at the upper tip may now be expressed as

$$ET_U = FT_U + 20\log(R_C/R_U) \tag{59}$$

where R_C and R_U are the path lengths from the feed to the center and the upper tip of the reflector, respectively. The second term in (59) is also called the *path loss* term. Similar definitions can also be given for ET_L and ET_S. It is noted that, for symmetric reflectors, ET and FT only need to be defined at one edge (tip). For different values of F/D and H/D the path loss curves are shown in Figs. 18a through 18c. It is worthwhile to add that for most cases of interest ET_U, ET_L, and ET_S have nearly equal values. Since ET directly controls the reflector's aperture amplitude taper it has a more dominant effect on the far-field pattern than EF does. From the results of Section 3 it is noted that it is not only the taper level which controls the reflector pattern but also the overall shape of the illumination distribution. In particular, the slopes of the illumination pattern at the reflector's edge can affect the side lobe levels. The results of this section are primarily based on the $\cos^q(\theta)$ type illumination patterns.

Reflector Pattern Characteristics for On-Focus Feeds

In this section, reflector pattern characteristics are discussed for beams generated by on-focus feeds. Results are shown for beamwidths, side lobe levels, first-null positions, cross-polarization levels, directivity efficiency, etc., as functions of the edge taper ET (or feed taper FT) and the reflector geometries. To simplify the presentation and limit the number of graphs, cases are considered in which the path losses are small (less than 0.5 dB) and, therefore, no substantial differences may be observed for path losses at the different tips of the reflector. For these cases, $\psi_{UC} \cong \psi_{CL} \cong \psi_{CS} \cong \psi_S$ and FT $\cong$ ET. The reader should attempt to properly interpret the results when the path losses are substantial. This effect is clearly demonstrated in Fig. 19, which displays far-field patterns for symmetric reflectors ($H/D = -0.5$) for the cases of $F/D = 0.4$ and $F/D = 2.0$, with an edge taper of ET $= -10$ dB. From Fig. 18a it is concluded that the path losses are -2.86 dB and -0.13 dB for F/Ds of 0.4 and 2.0, respectively. Nevertheless, the patterns are very similar, which indicates that ET $= -10$ dB is the controlling factor. Note that for these values of F/D the feed taper FT takes the values of -7.14 dB and -9.87 dB, respectively. In the latter case, ET $\cong$ FT. Similar results are shown in Fig. 20, where the feed taper FT is kept constant at -10 dB. The patterns for cases in which $F/D = 0.4$ and 2.0 show differences of up to 2.8 dB at the first side lobe level which is obviously a manifestation of the effects of the path loss differences.

For offset parabolic reflectors the far-field patterns in different cuts are, in general, different even when the feed has a symmetric pattern. Also the pattern can be slightly asymmetric in the plane of offset (xz plane of Fig. 16) depending on the F/D ratio (see Fig. 21) [29]. However, for the results shown here, F/D is large in order to reduce the path loss effects and asymmetry of the pattern. Fig. 22 shows the half-beamwidth, first and second null positions, and first side lobe positions as functions of the edge taper ET. It is worthwhile to mention that for edge tapers beyond -20 dB the pattern characteristics depend heavily on the actual feed pattern description and results shown here are for $\cos^q(\theta)$ feed patterns. Also shown are the first side lobe levels and spillover and taper efficiencies as functions of edge tapers in Fig. 23. From Fig. 23 it is readily observed that the resulting efficiency $\eta = \eta_s \eta_t$ is maximized for edge tapers about -11 dB with a value of 81 percent. Some representative far-field patterns are shown in Figs. 24a through 24f for different edge tapers. Again, it is obvious that for edge tapers in the neighborhood of ET $= -20$ dB, the first side lobe starts to merge with the main beam,

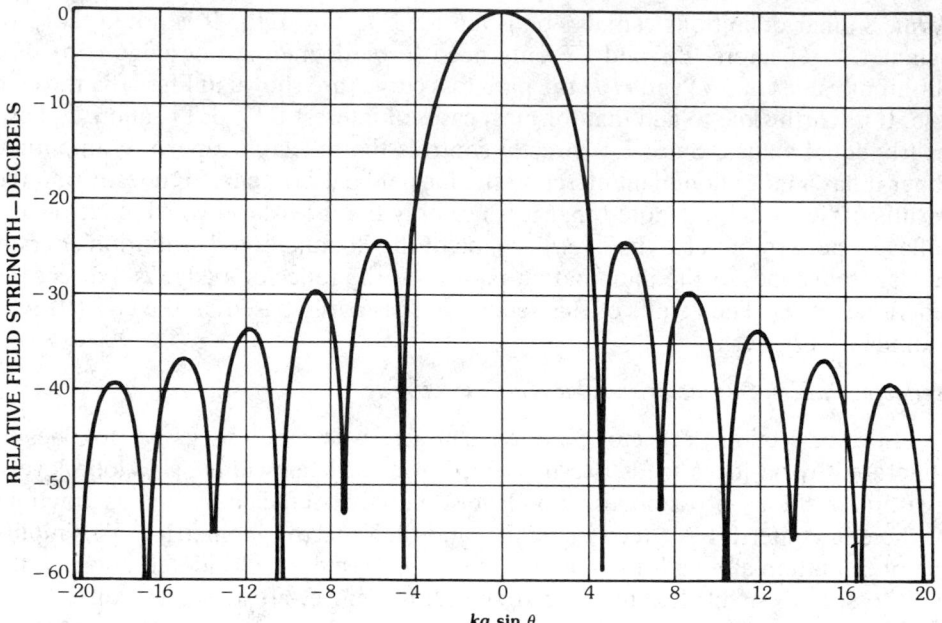

a

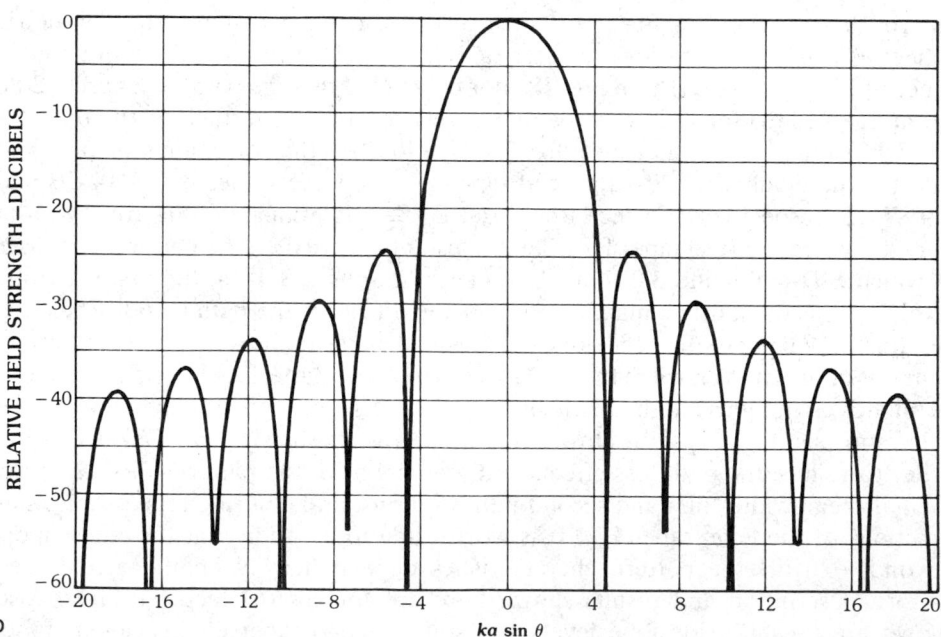

b

Fig. 19. Far-field patterns of a symmetric reflector antenna with an edge taper illumination of ET = −10 dB and different path losses (different F/D values). (*a*) For F/D = 0.4. (*b*) For F/D = 2.0.

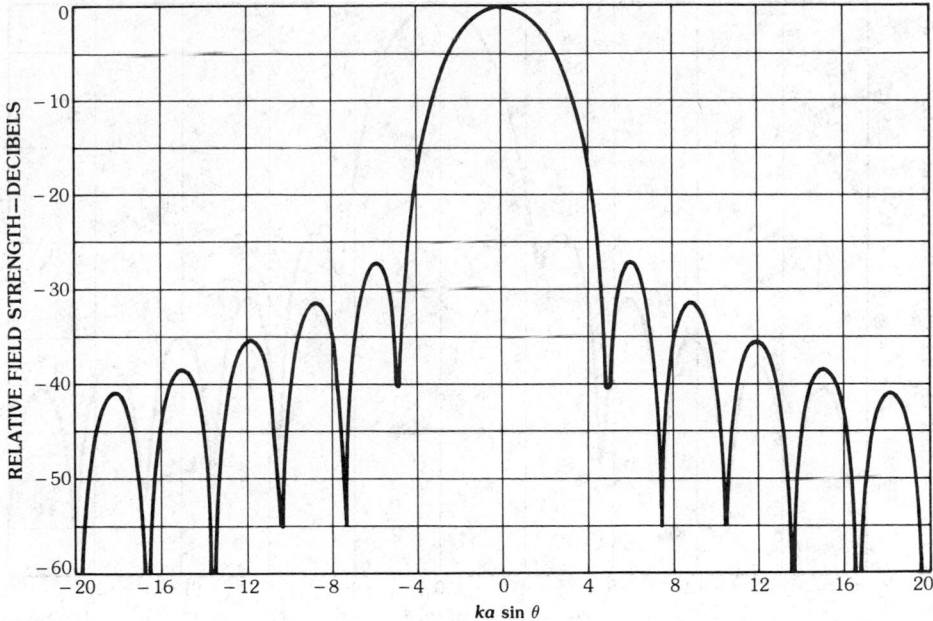

a

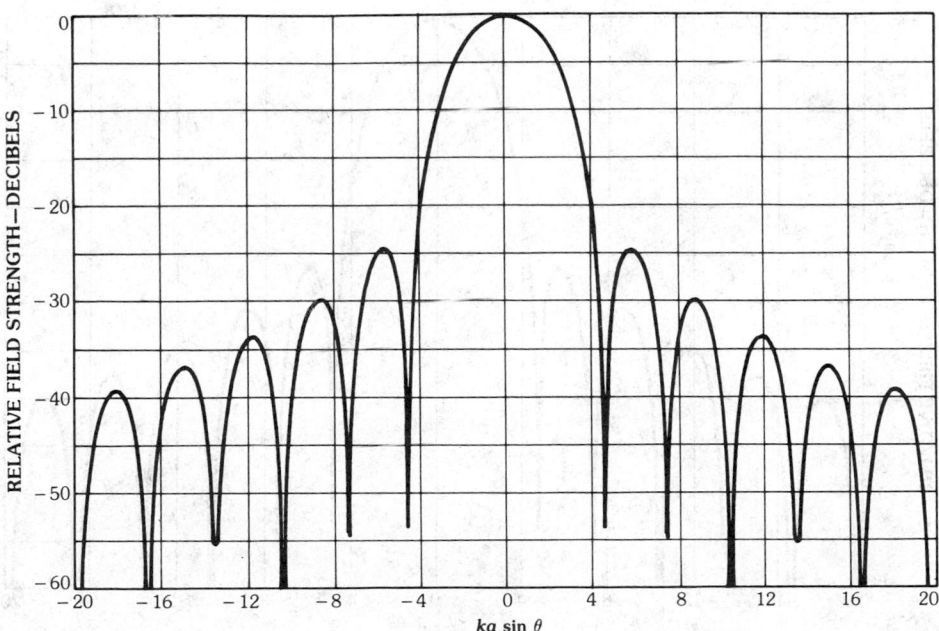

b

Fig. 20. Far-field patterns of a symmetric reflector antenna with a feed taper illumination of FT = −10 dB and different path losses (different F/D values). (*a*) For F/D = 0.4. (*b*) For F/D = 2.0.

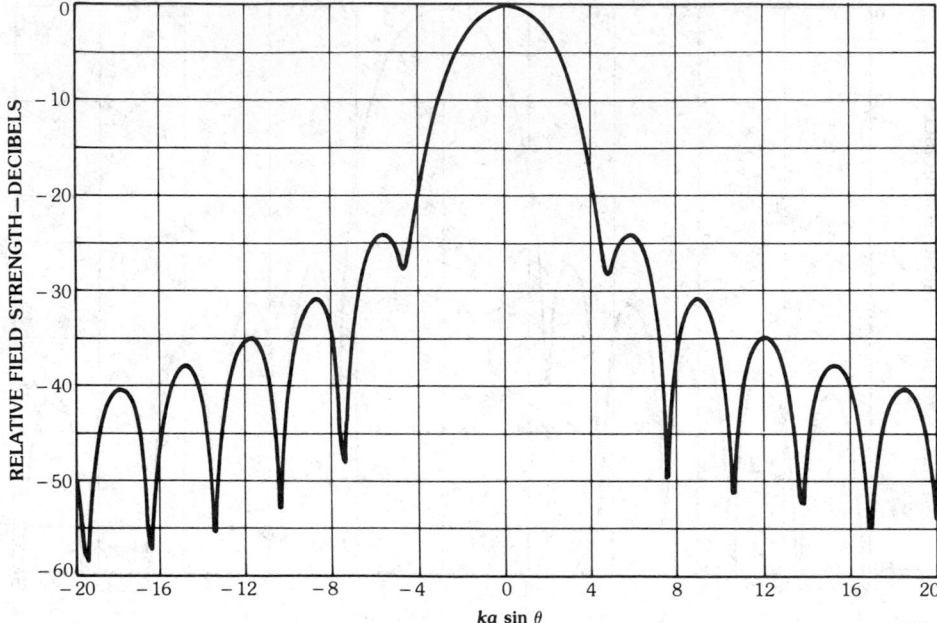

a

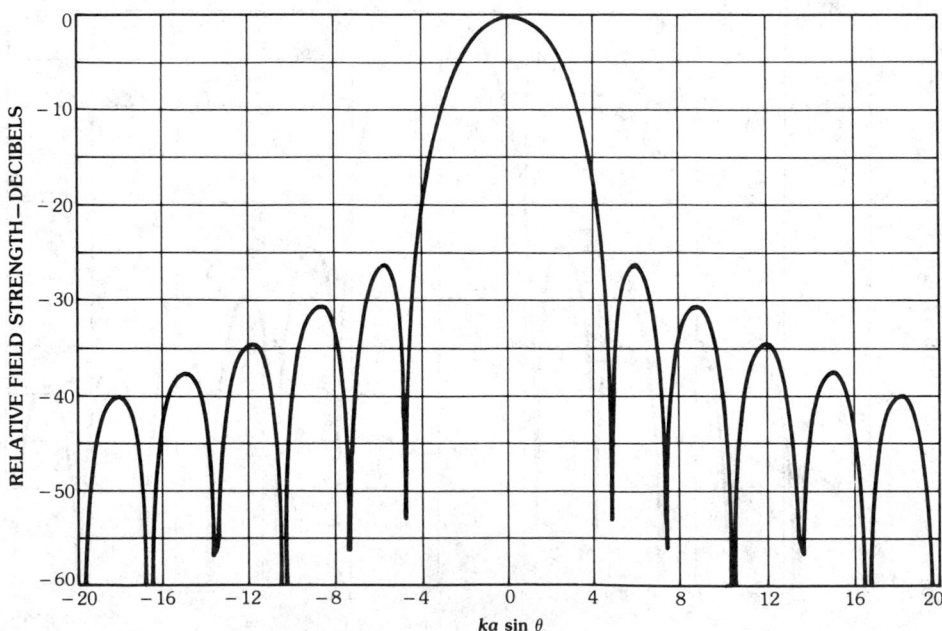

b

Fig. 21. Patterns of an offset reflector with $H/D = 0.5$ ($F/D_p = 0.25$) illuminated by a symmetric feed with a feed taper of FT = -10 dB. (a) For pattern in plane $\phi = 0°$ (plane of offset). (b) For pattern in plane $\phi = 90°$.

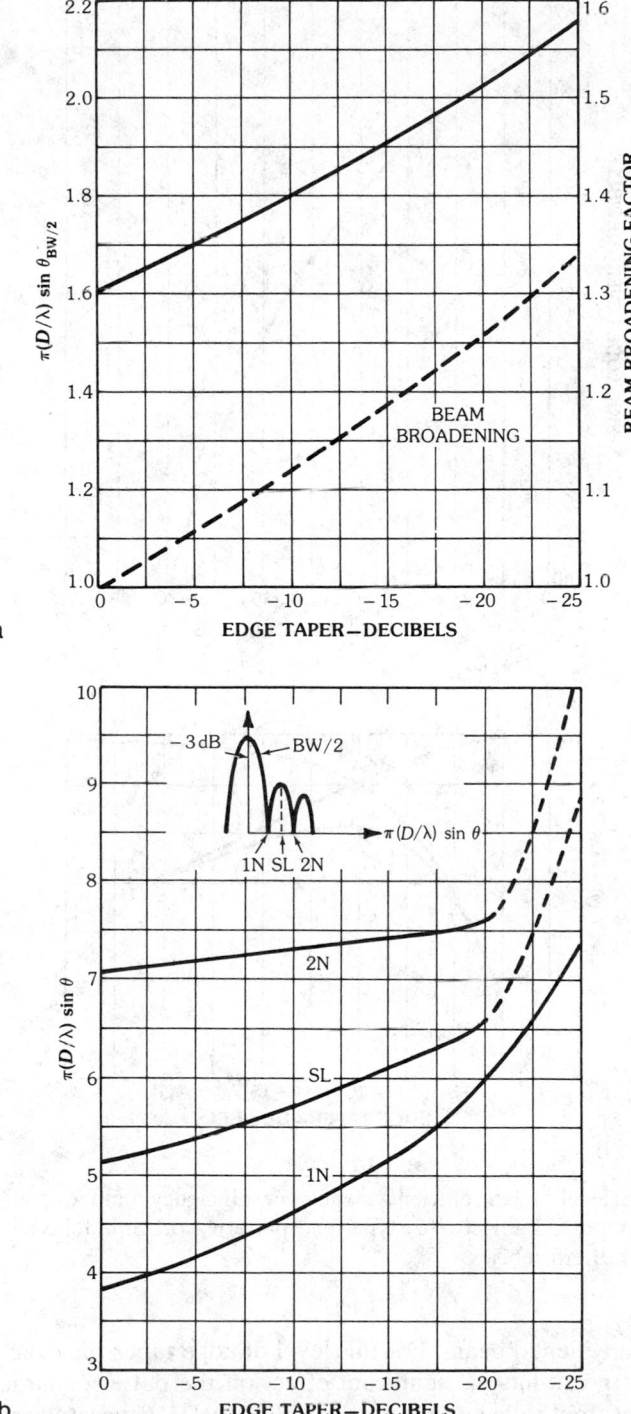

Fig. 22. Half-power beamwidth, first-null, second-null, and side lobe positions as functions of the edge taper, where D/λ is the diameter of the reflector in terms of the wavelength. (*a*) Half-power beamwidth. (*b*) First and second nulls and side lobe positions.

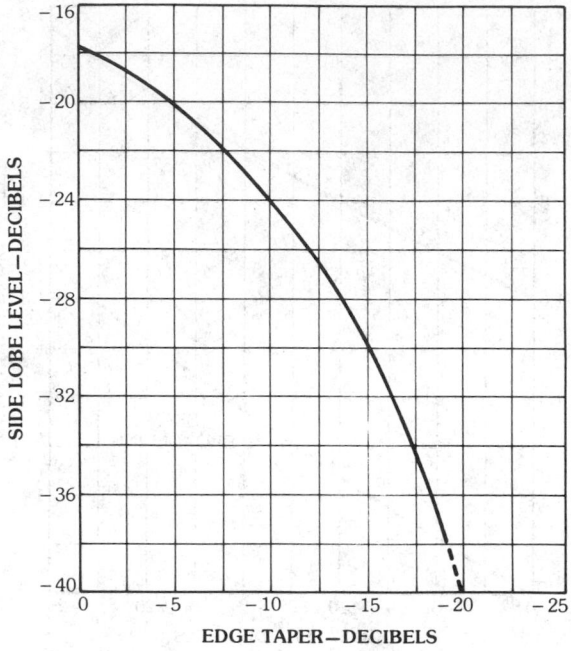

a

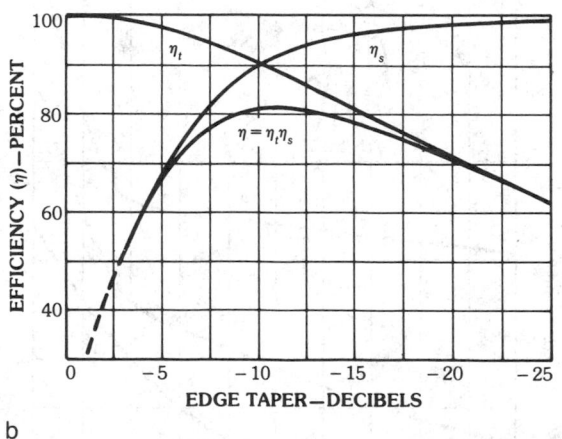

b

Fig. 23. Side lobe level, taper efficiency, spillover efficiency, and overall efficiency as functions of edge taper ET for $\cos^q(\theta)$ type feed patterns. (a) Side lobe level. (b) Taper, spillover, and overall efficiencies.

which results in a widened beam. For this level of edge taper the exact distribution of the feed pattern can have a significant effect on the pattern characteristics.

Another important reflector parameter is the level of generated cross-polarized field. This topic has been addressed in [1, 35, 36] for both symmetric and offset reflectors. Here, some representative cases are presented. There are many dif-

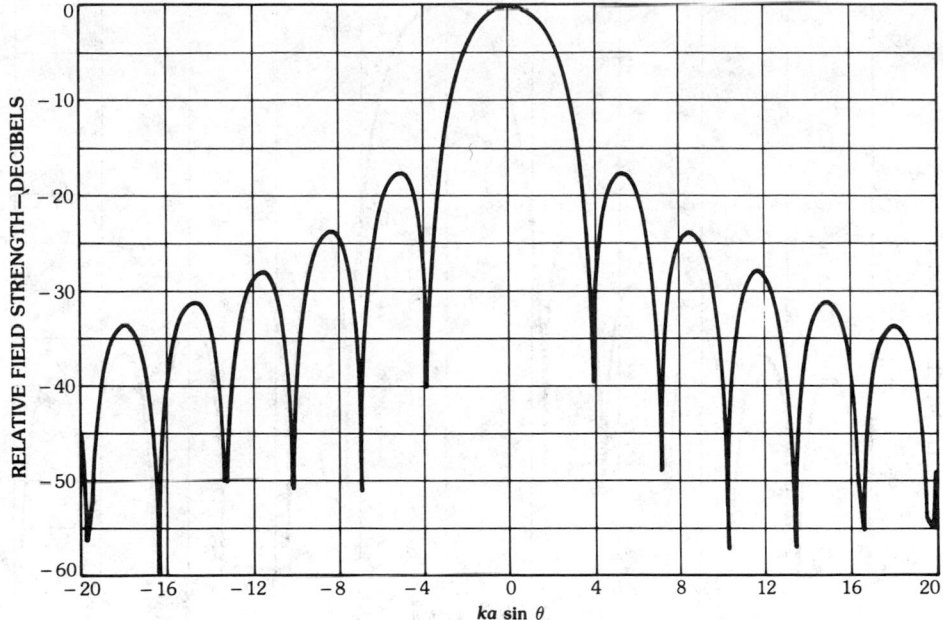

a

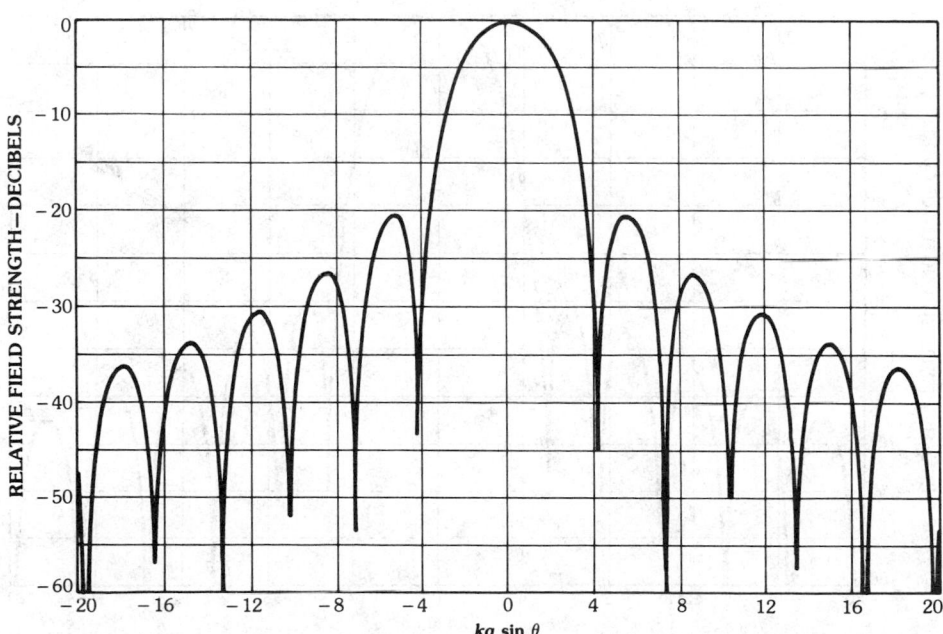

b

Fig. 24. Reflector far-field patterns for different edge tapers. (*a*) ET = 0 dB. (*b*) ET = −5 dB. (*c*) ET = −10 dB. (*d*) ET = −15 dB. (*e*) ET = −20 dB. (*f*) ET = −25 dB.

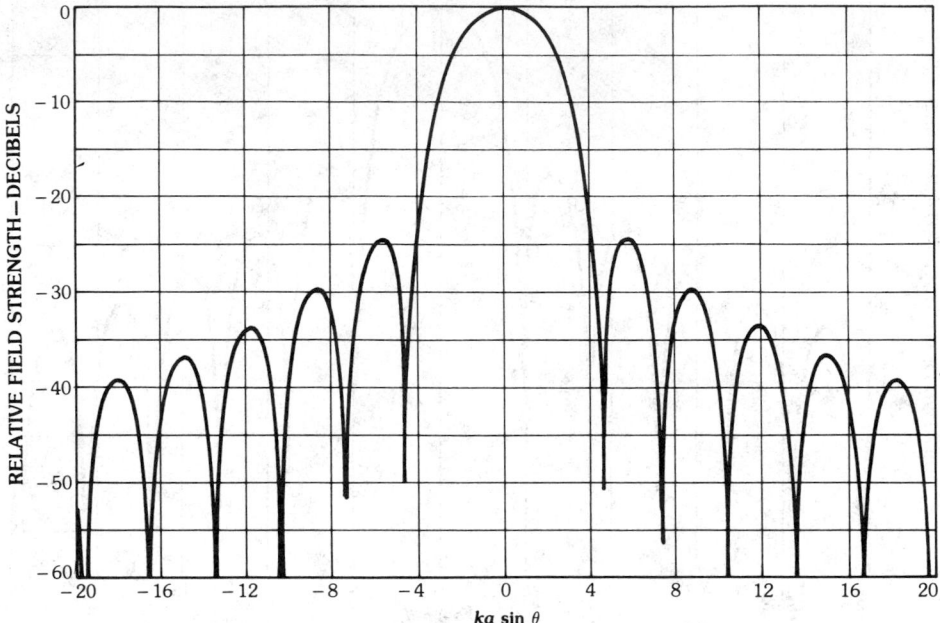

c

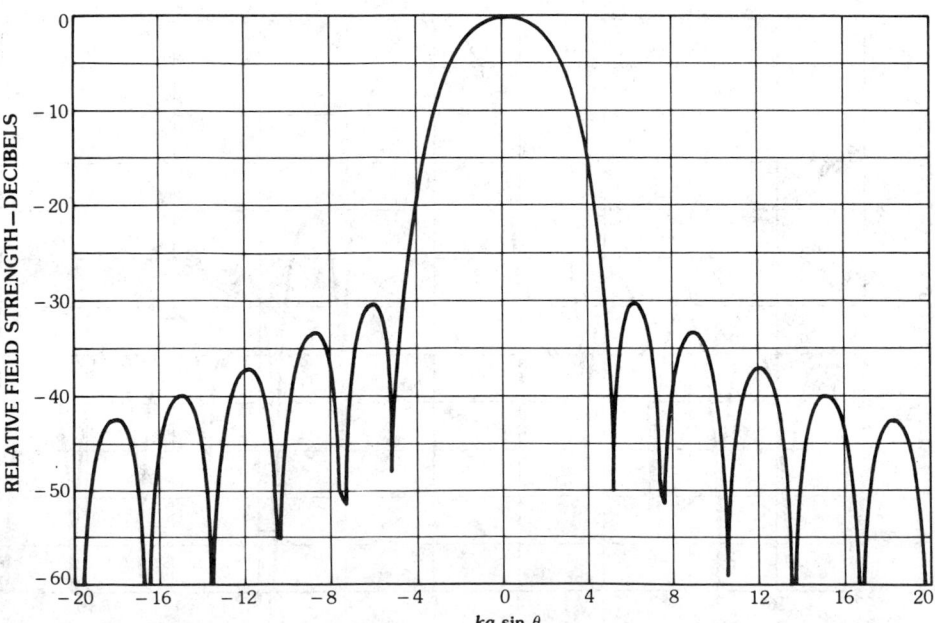

d

Fig. 24, *continued.*

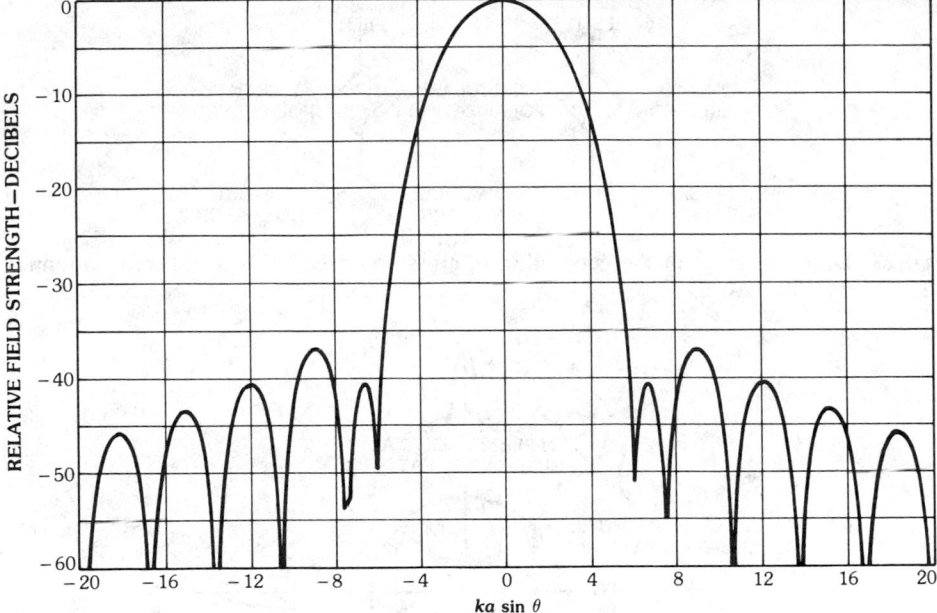

e

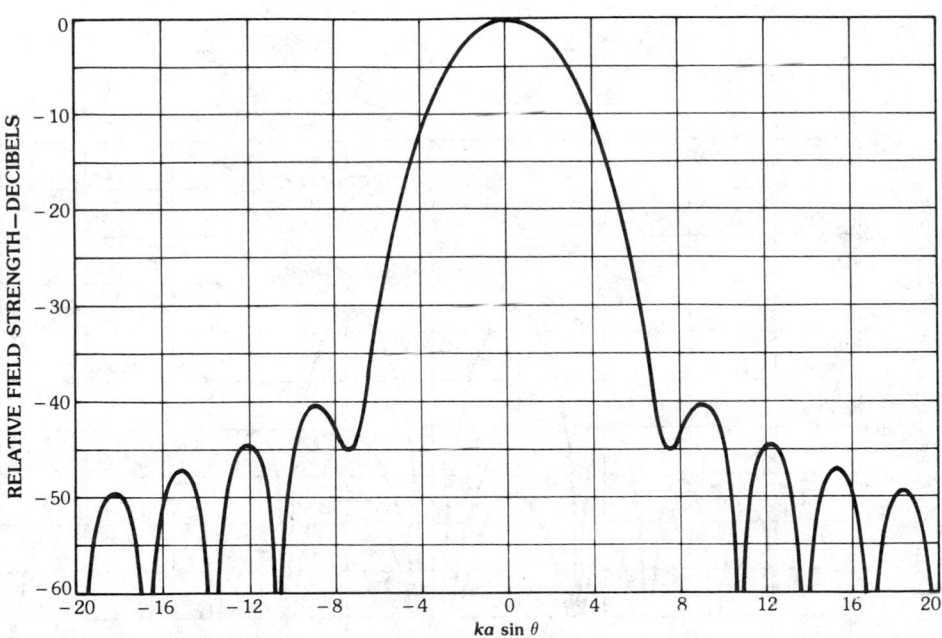

f

Fig. 24, *continued.*

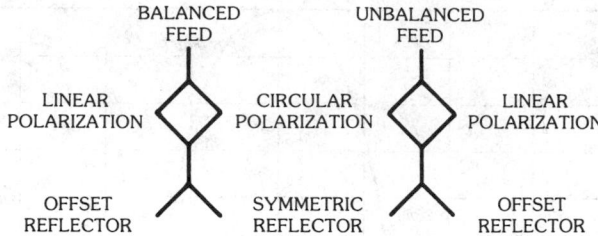

Fig. 25. Different cases of the generation of cross-polarized fields in reflector antennas.

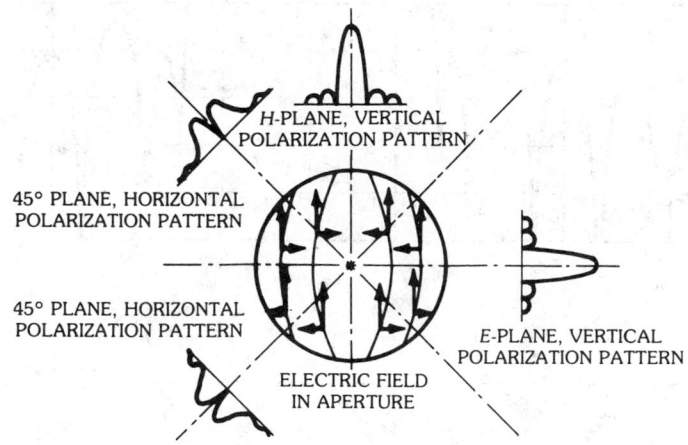

a

b

Fig. 26. The generation of cross-polarized fields and some typical patterns of symmetric reflectors. (*a*) Mechanism of generation of cross-polarized fields for symmetric parabolic reflectors. (*b*) Cross-polarized field (normalized with respect to the peak) of a symmetric reflector ($F/D = 0.3$) illuminated by a linearly polarized feed with unbalanced *E*- and *H*-patterns.

ferent cases which must be studied, summarized in Fig. 25. In this figure the unbalanced feed refers to a feed with different *E*- and *H*-plane patterns. For the symmetric reflector the mechanism of generation of the cross-polarized field is shown in Fig. 26a and some typical patterns are given in Fig. 26b. It should be noted that for a symmetric parabolic reflector illuminated by a linearly polarized feed the maximum of cross-polarized fields occurs in the plane $\phi = 45°$. As the feed becomes more balanced, the level of the cross-polarized field decreases substantially, as demonstrated by the results in Fig. 26. It is worth mentioning that these results are dependent on the *F/D* ratio and the edge taper.

Next, the cross-polarization characteristics of offset parabolic reflectors are considered. In contrast to symmetric reflectors, which have very low levels of cross-polarized fields for balanced and linearly polarized feeds, offset parabolic reflectors can have high levels of cross polarization. Even for balanced feeds located at the focal point, levels of cross-polarized fields can be high, depending on the tilt angle of the feed axis with respect to the reflector axis for linearly polarized feeds. For example, Fig. 27 shows the generation of cross-polarized fields for various feed-axis tilt angles for a fixed offset reflector configuration. In this example the feed pattern is chosen to be isotropic in order to show clearly the generation of the cross-polarized fields. It should be mentioned that for offset parabolic reflectors the cross-polarized field is predominantly observed in the plane $\phi = 90°$ (normal to the plane of the offset). Clearly, in practice, in order to reduce spillover the feed axis is always tilted toward the center of the reflector; hence a high level of cross-polarized field should be expected. The levels of cross-polarized fields for different values of

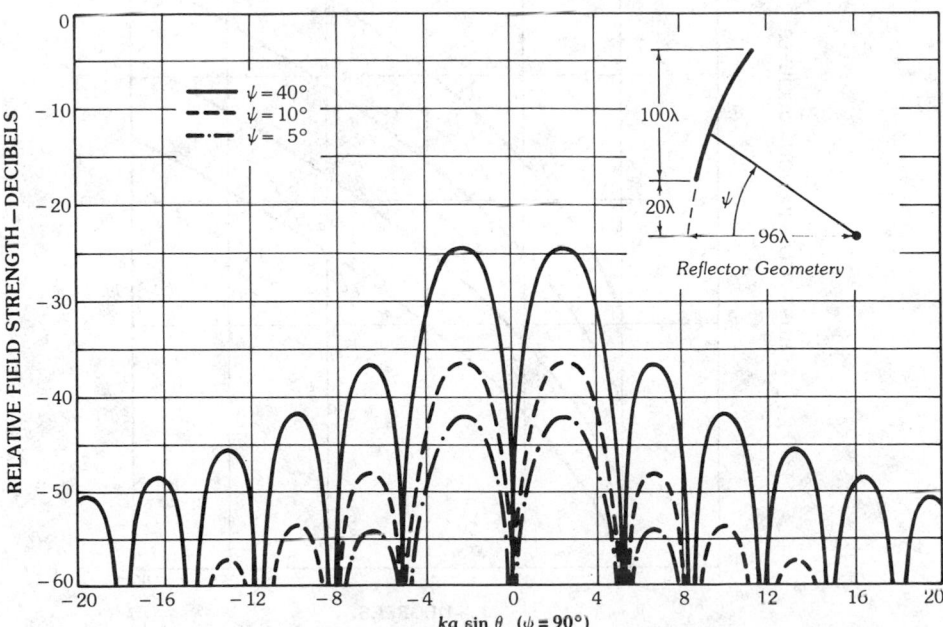

Fig. 27. Cross-polarized fields in the plane $\phi = 90°$ (normal to the offset plane) for an offset parabolic reflector for various feed-axis tilt angles ψ and illuminated by a balanced linearly polarized feed.

bisect angle ψ_B and the half-angle ψ_S (see Fig. 16) are shown in Fig. 28 [35].

When symmetric or offset parabolic reflectors are illuminated with balanced circularly polarized feeds very low levels of cross-polarized field result. For offset parabolic reflectors illuminated by circularly polarized feeds, however, an additional feature is observed, which is referred to as the *beam squint*. This means that the beam peak is shifted from the axis on the plane normal to the plane of offset (*yz* plane in Fig. 16). The amount of squint depends on the tilt angle of the feed axis and the reflector geometry. It can be shown that the following expression is a good approximation:

$$\sin \theta_s = \mp \frac{\sin \psi_B}{4\pi(F/\lambda)} \tag{60}$$

where $\mp$ signs are for right and left circularly polarized cases and θ_s is the amount of squint. As an example, Fig. 29 shows the squinted patterns for right and left circularly polarized feeds located at the focal point. (Angles ψ_B and ψ_S are defined in Fig. 16.) The amount of squint obtained from experimental data and diffraction

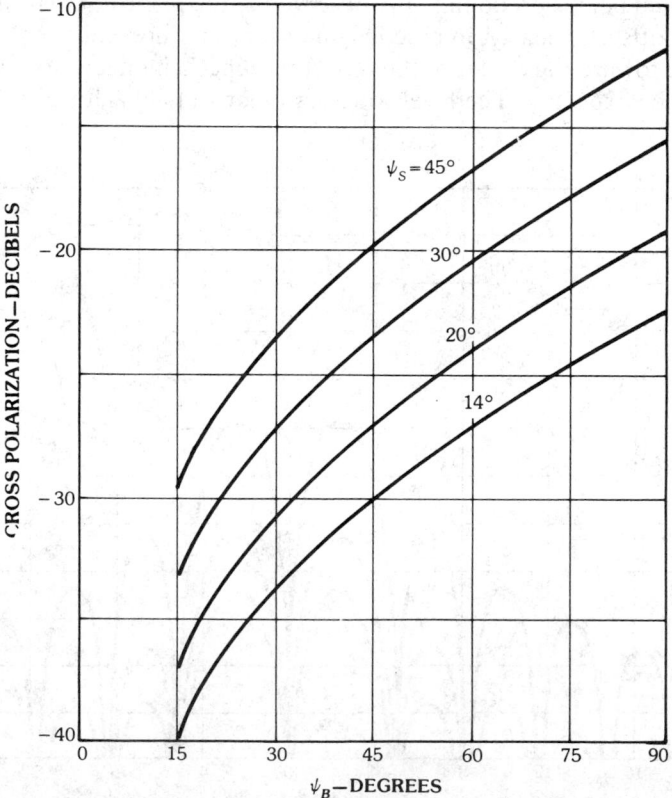

Fig. 28. Maximum cross-polarization level of an offset parabolic reflector illuminated by a balanced linearly polarized feed. (*After Chu and Turrin [35], © 1973 IEEE*))

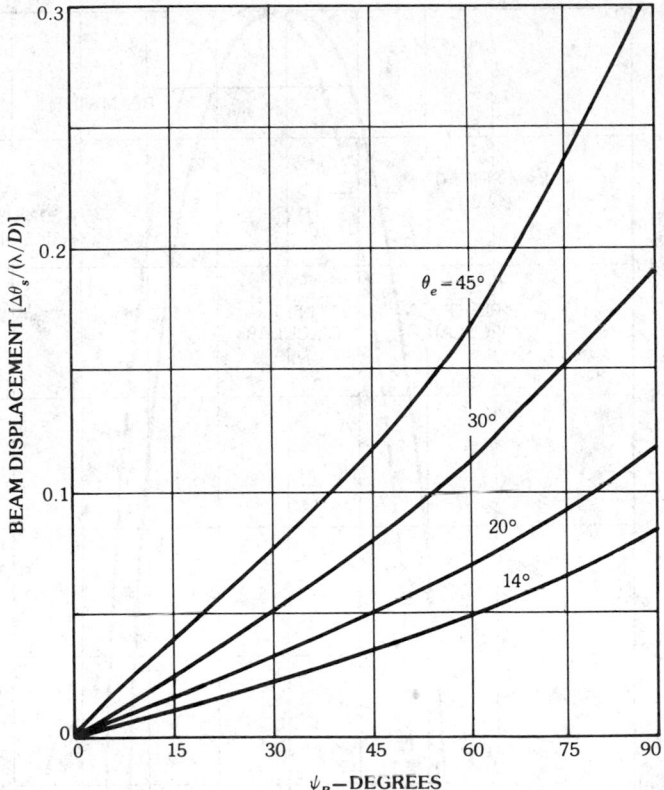

Fig. 29. Normalized beam squint $\theta_S/(\lambda/D)$ of an offset parabolic reflector illuminated by a circularly polarized feed located at the focal point. (*After Chu and Turrin [35], © 1973 IEEE*))

analysis [35] agrees well with the approximate formula. Additional results are shown in Fig. 30 [35], where again ψ_B and ψ_S are used to characterize the reflector.

Reflector Pattern Characteristics for Off-Focus Feeds

In many applications, such as the design of multiple and contour beam reflectors, it becomes necessary to illuminate the reflector with feeds positioned away from the reflector focal point. This feed displacement introduces phase aberration, which results in pattern distortion in terms of gain loss, side lobe degradations, etc. In this subsection some of the key distortion characteristics of reflectors are presented for both the symmetric and offset parabolic reflectors and for both the axial and lateral feed defocusings.

Axial Displacements—For the symmetric reflector the results of feed axial defocusing for gain loss and patterns are shown in Figs. 31 through 33. It is assumed that the feed provides a −10-dB edge taper when it is located at the focal point. Results are demonstrated for feed displacements toward and away from the reflector. A small asymmetry can be observed. These results are dependent on the

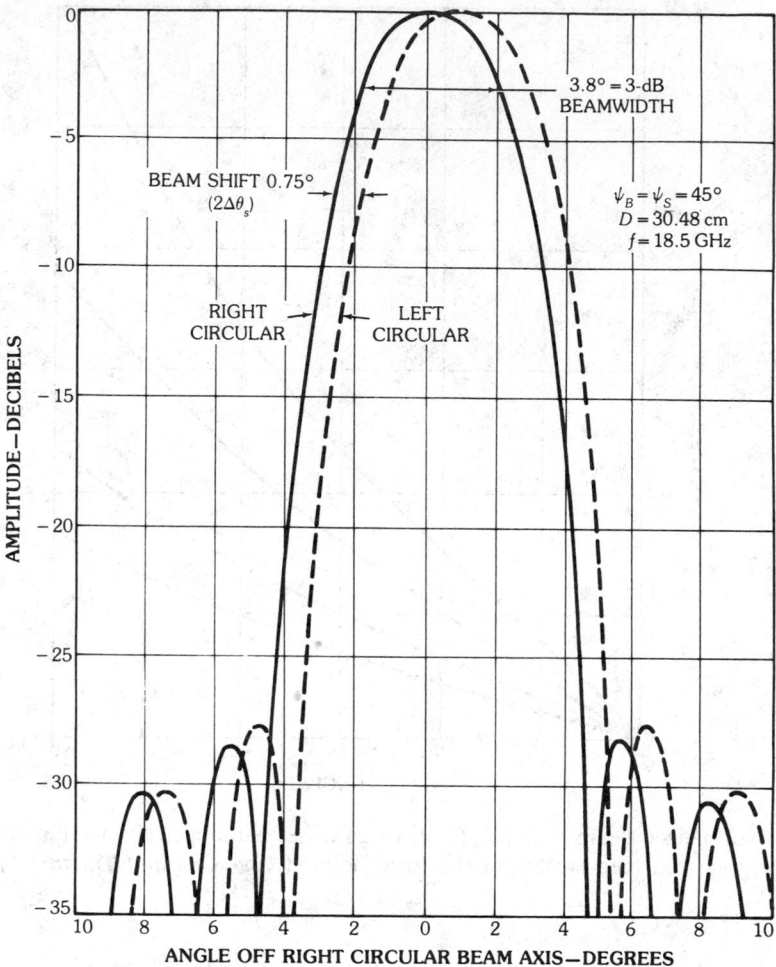

Fig. 30. Squinted far-field patterns in the plane $\phi = 90°$ of an offset parabolic reflector illuminated by right and left circularly polarized feeds located at its focal point. (*After Chu and Turrin [35], © 1973 IEEE*)

edge taper as discussed in [37]. Furthermore, for the cases where $\delta \ll F$ and the feed is an infinitesimal dipole with moment **p** polarized in the $\hat{y}$ direction, one can obtain the following approximate expression [37] for the field on axis, namely,

$$\mathbf{E}(R,0,0) = -\frac{jk^2 p}{4\pi\epsilon}(2kF)\,\hat{y}\,\frac{\exp(-jkR)}{R}\,\frac{1}{(4F/D)^2 + 1}$$

$$\exp\left[-jk\delta\,\frac{(4F/D)^2}{(4F/D)^2 + 1}\right]\frac{\sin\xi}{\xi} \qquad (61)$$

where

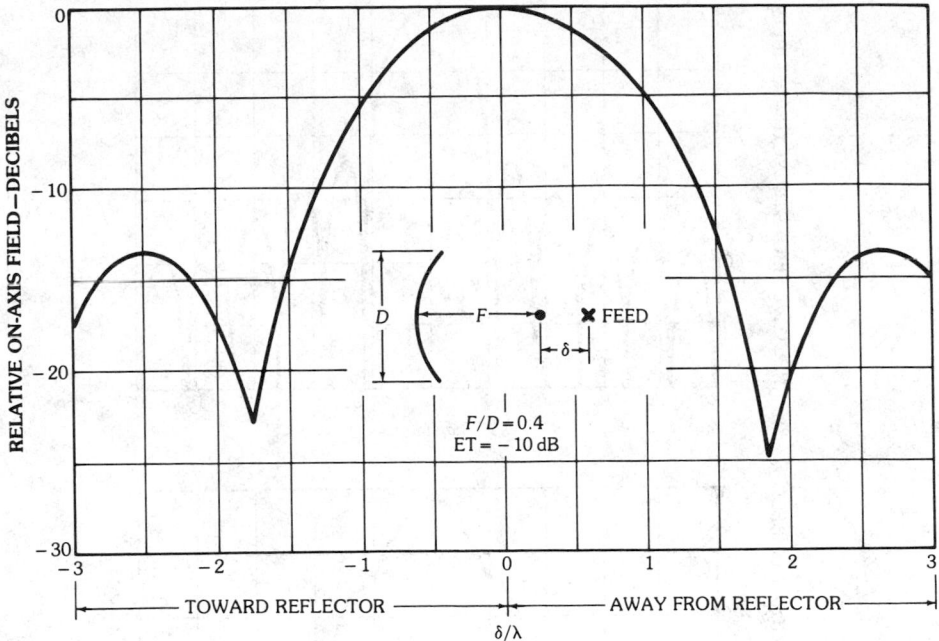

Fig. 31. Relative on-axis field as a function of axial feed displacement.

$$\xi = \frac{2\pi(\delta/\lambda)}{(4F/D)^2 + 1} \tag{62}$$

The dominant effect of the axial defocusing is the generation of the quadratic phase error across the reflector aperture. The axial field (61) becomes virtually zero for values of $\zeta = \pm\pi, \pm 2\pi, \pm 3\pi$, etc., which results in

$$\frac{\delta}{\lambda} = \pm\frac{n}{2}[(4F/D)^2 + 1], \qquad n = 1, 2, 3, \ldots \tag{63}$$

For these values the beam widens considerably and may also be bifurcated (see Figs. 32 and 33). The reader may have noticed a resemblance between these patterns and those resultant from the field of an aperture in the Fresnel zone.

Lateral Displacements—For a parabolic reflector, lateral feed displacements result in scanned beams. It is well known that these reflectors have limited scan capability which strongly depends on F/D and F/D_p ratios for symmetric and offset reflectors, respectively. First, results are presented for symmetric reflectors. An important parameter in dealing with scanned beams is the beam deviation factor (BDF) which is defined as

$$\text{BDF} = \frac{\theta_B}{\theta_F} \tag{64}$$

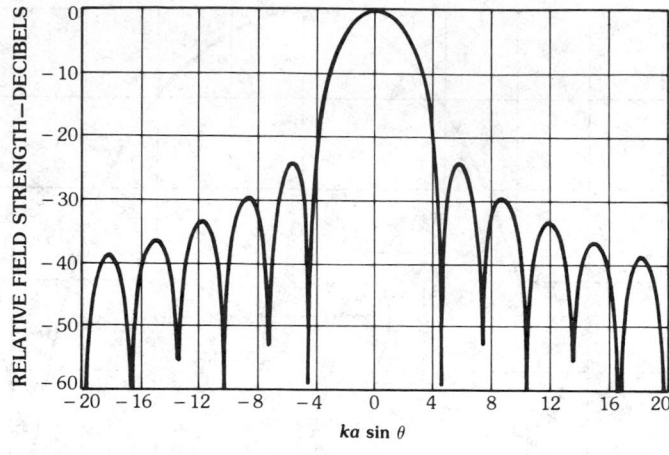

a

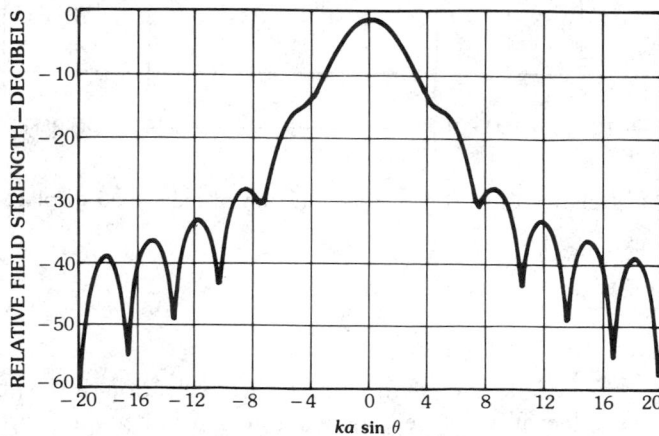

b

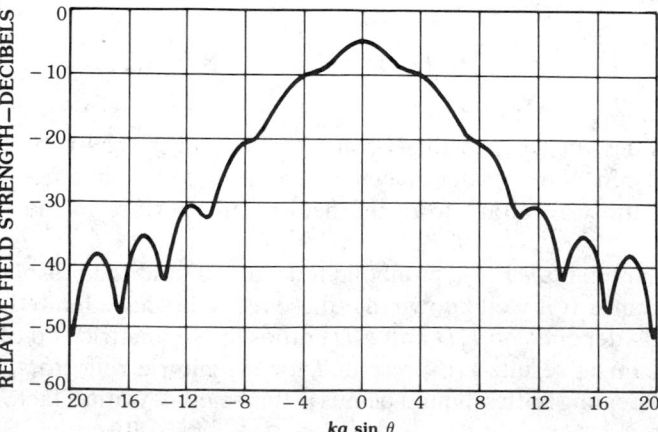

c

Fig. 32. Reflector far-field patterns for $\phi = 0°$ as a function of axial feed displacement away from the reflector (see Fig. 31). (*a*) For $\delta/\lambda = 0$. (*b*) For $\delta/\lambda = 0.5$. (*c*) For $\delta/\lambda = 1.0$. (*d*) For $\delta/\lambda = 2.0$. (*e*) For $\delta/\lambda = 3.0$.

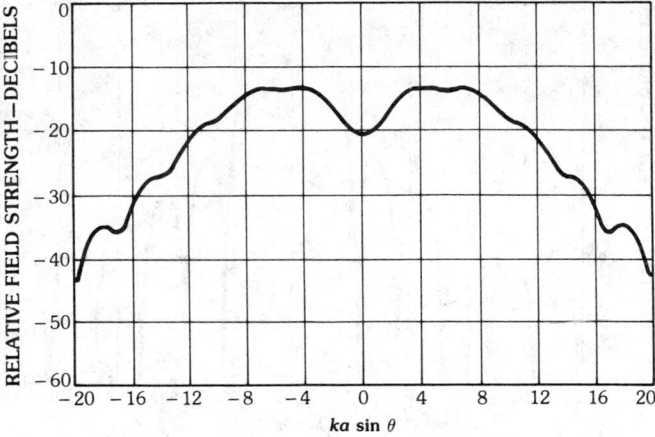

d

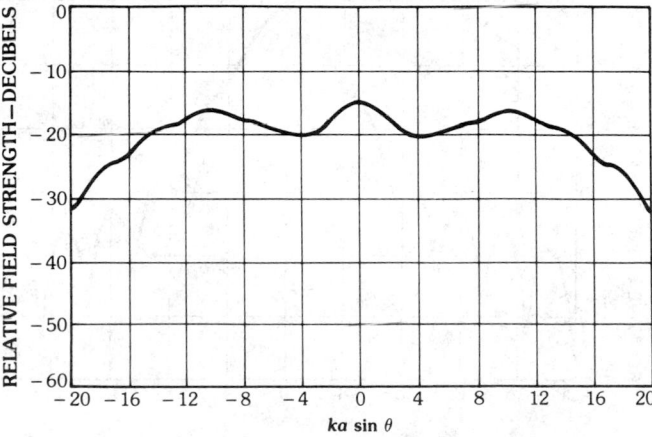

e

Fig. 32, *continued.*

where θ_B and θ_F are the beam scan angle and feed tilt angle, respectively, as shown in Fig. 34. The BDF has a strong dependence on F/D and weakly depends on the edge taper and δ/F (δ is the lateral feed displacement). A good approximation for BDF is [38]

$$\text{BDF} = \frac{\sin^{-1}(\delta/F)\{[1 + k(D/4F)^2]/[1 + (D/4F)^2]\}}{\tan^{-1}(\delta/F)} \tag{65}$$

where $0 < k < 1$ and $k = 0.36$ provides very accurate results when the BDF is compared with experimental and diffraction analysis data. For small δ/F, (65) may be simplified to give

$$\text{BDF}_0 = \frac{1 + k(D/4F)^2}{1 + (D/4F)^2} \tag{66}$$

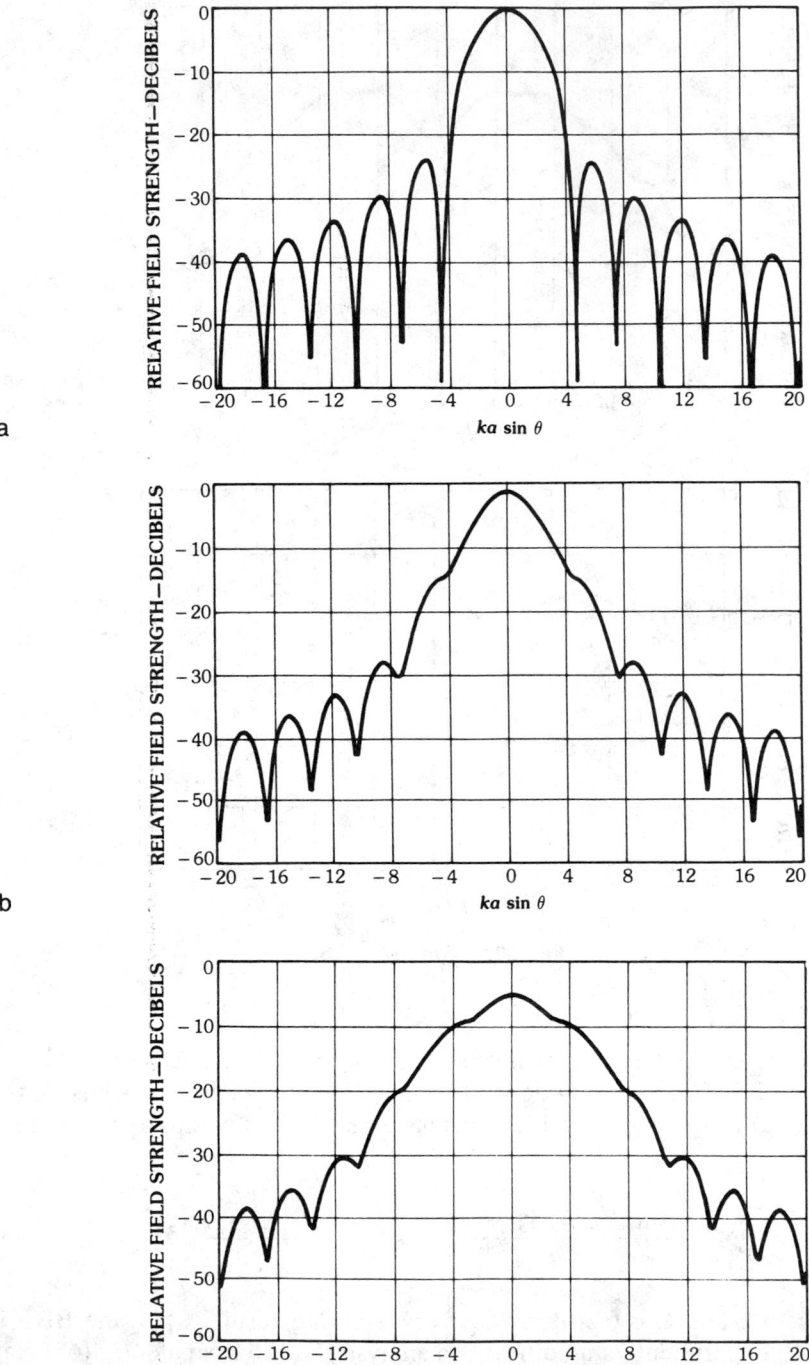

Fig. 33. Reflector far-field patterns for $\phi = 0°$ as a function of axial feed displacement toward the reflector (see Fig. 31). (*a*) For $\delta/\lambda = 0$. (*b*) For $\delta/\lambda = -0.5$. (*c*) For $\delta/\lambda = -1.0$. (*d*) For $\delta/\lambda = -2.0$. (*e*) For $\delta/\lambda = -3.0$.

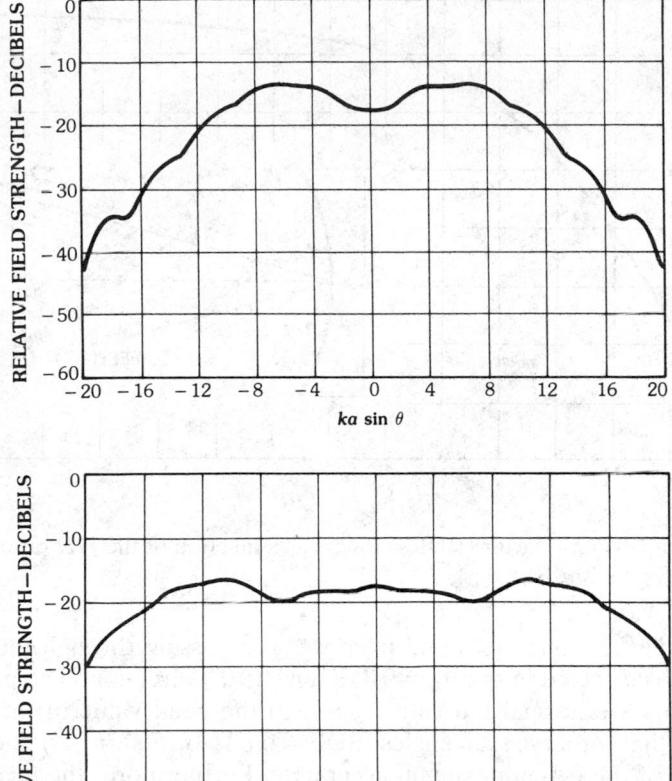

Fig. 33, *continued.*

which is independent of δ/F. Fig. 34 shows the BDF as a function of F/D and for $k = 0.36$. For large feed displacements one may have to resort to diffraction analysis results in order to obtain more accurate values for the BDF.

Due to the phase aberration introduced by defocused feeds the far-field pattern can be substantially degraded. This degradation depends very strongly on the F/D ratio and the angle of scan in terms of number of beamwidths scanned (bmws). There are some approximate formulas available for predicting the peak gain loss as a function of beamwidths scanned [39]. In many applications, however, the knowledge of peak gain loss is not sufficient and one must know the overall degradation effects on the far-field pattern. For this reason many selective but representative cases are given here to provide a clear picture of the pattern degradation. First, an $F/D = 0.4$ symmetric parabolic reflector illuminated by a $\cos^q(\theta)$ type feed with ET $= -10$ dB is considered. Far-field patterns for different numbers of beamwidths scanned are shown in Figs. 35b through 35e. It is important to note that the patterns are plotted versus the universal parameter $ka \sin \theta_p = \pi(D/\lambda) \sin \theta_p$

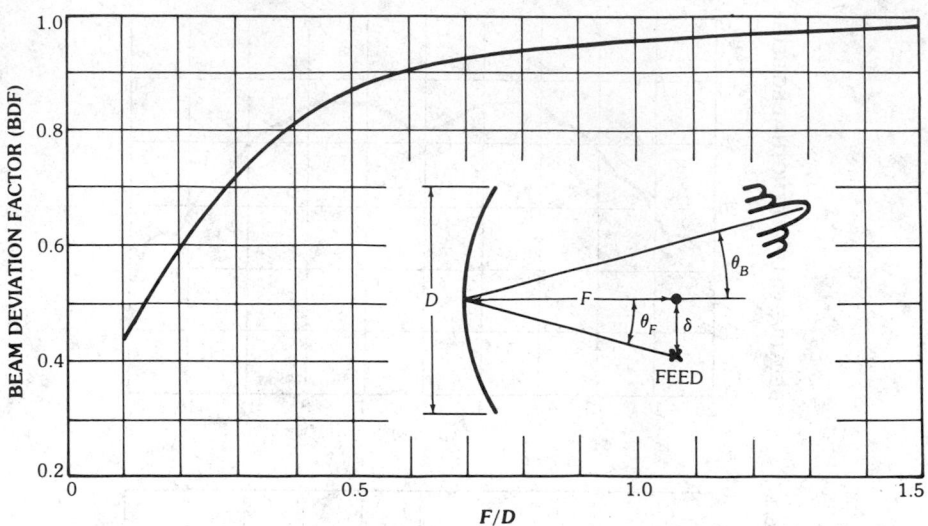

Fig. 34. Beam deviation factor (BDF $= \theta_B/\theta_F$) as a function of the F/D ratio for symmetric reflectors (ET $= -10$ dB).

in which θ_p is the angle measured from the axis passing through the peak of the beam which is directed in the $\theta_B =$ BDF $\tan^{-1}(\delta/F)$ direction. The peak values of these patterns are normalized with respect to the peak value of the nonscanned beam. Note that for large scan angles, BDF $= 0.82$ (for $F/D = 0.4$) does not predict the location of the beam maximum accurately. Furthermore, the feed axis is kept parallel to the reflector axis and has not been tilted toward the reflector center. This condition is more practical when feeds are used in a planar array in contour and multiple beam applications. For directive feeds with large displacements, however, it may be necessary to tilt the feed toward the reflector center in order to reduce the amount of spillover. In this case the feeds may be arrayed on the spherical surface rather than the planar surface. Graphs of Figs. 35b through 35e clearly demonstrate how rapidly the reflector pattern can degrade for scan angles beyond two beamwidths scanned when $F/D = 0.4$. One simple way to improve this very limited scan capability is to increase the F/D ratio. Figs. 36a through 36d show the far-field patterns for $F/D = 1.0$ as a function of the beamwidths scanned. Note that for this value of F/D, beams with much larger scan angles can be generated with adequate characteristics. The prime drawbacks of employing large F/D values are the larger structural size and the need for more directive feeds to provide the required edge taper. Finally, Fig. 37 shows the gain loss curves for different values of F/D as a function of the beamwidths scanned (bmws). Although most of the results are shown for the case of ET $= -10$ dB, similar observations can be made for other edge tapers. For example, Fig. 38 shows the results of experimental data [40]. Excellent agreement has been observed as far as the effects of the feed displacements are concerned. There are, however, differences in the third side lobe levels due to the fact that the experimental feed patterns are not exactly modeled by the $\cos^q(\theta)$ type pattern.

The previously mentioned results for symmetric reflectors only demonstrate

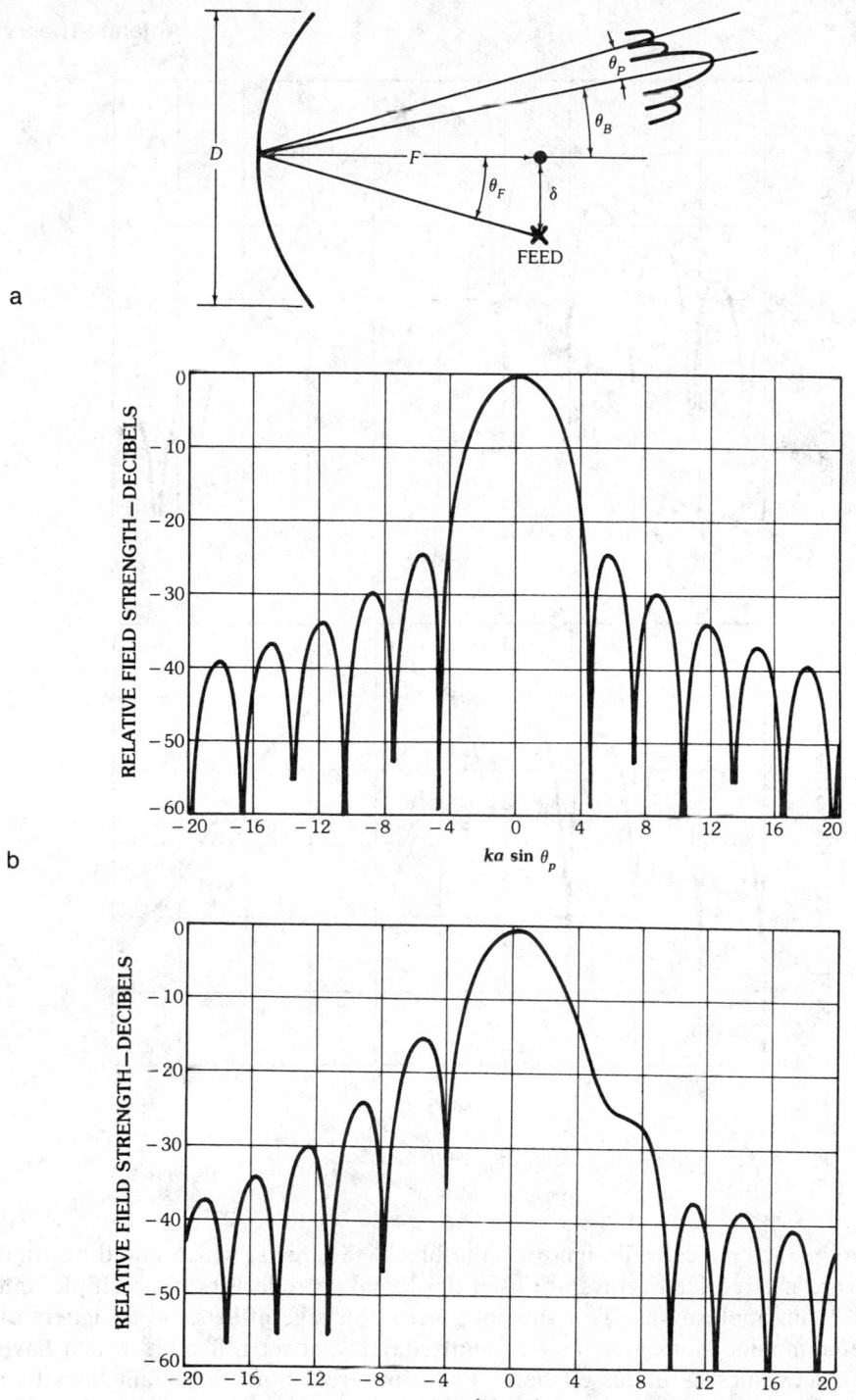

Fig. 35. Symmetric reflector far-field patterns for $\phi = 0°$ as a function of lateral feed displacement in terms of the number of beamwidths scanned ($F/D = 0.4$, ET $= -10$ dB, BDF $= 0.82$). (*a*) Reflector parameters. (*b*) 0 bmws. (*c*) 2 bmws. (*d*) 6 bmws. (*e*) 10 bmws.

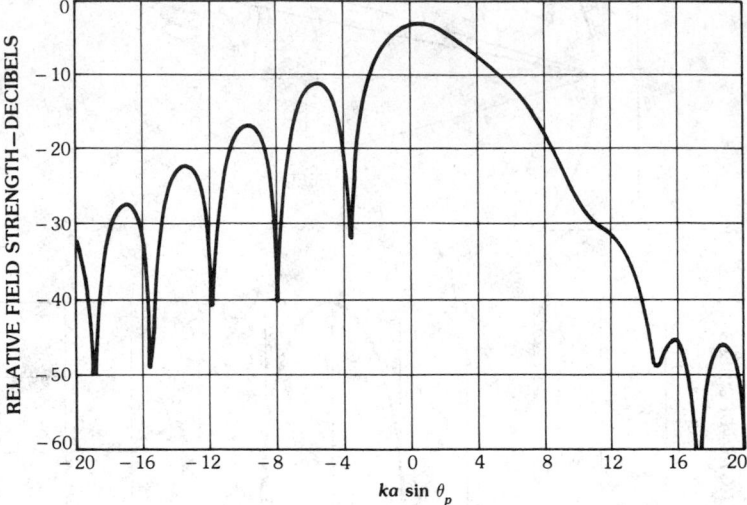

d

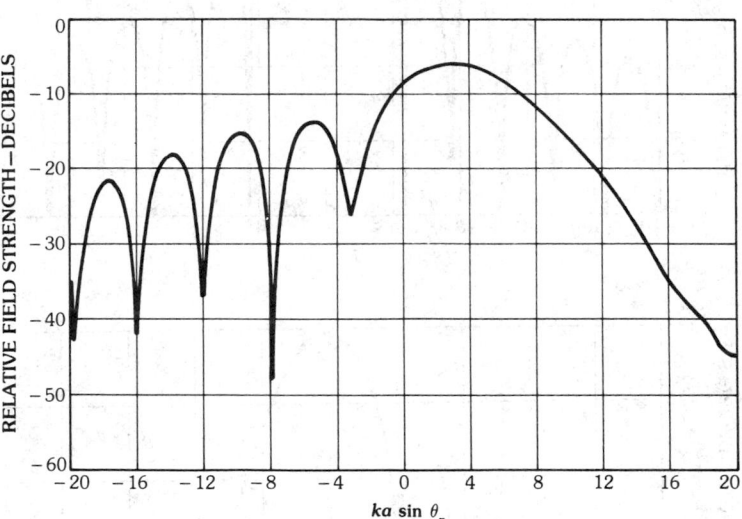

e

Fig. 35, *continued.*

the scan characteristics while ignoring the blockage effects, which could become very severe if large feed arrays are used to illuminate reflectors for multiple and contour beam applications. To overcome these blockage effects the designers of satellite communications systems frequently employ offset reflectors which have scan characteristics as discussed here. For offset reflectors the beam deviation factor is computed as shown in Fig. 39. Many studies have shown that the F/D_p ratio characterizes the offset reflector patterns better than the F/D ratio. Far-field patterns as a function of the number of beamwidths scanned are shown in Figs. 40a through 40d for $F/D_p = 0.4$ and ET = -10 dB. For this case $F/D = 0.96$ ($F = 96\lambda$, $D = 100\lambda$, $H/D = 0.2$), which is almost two and a half times larger than F/D_p.

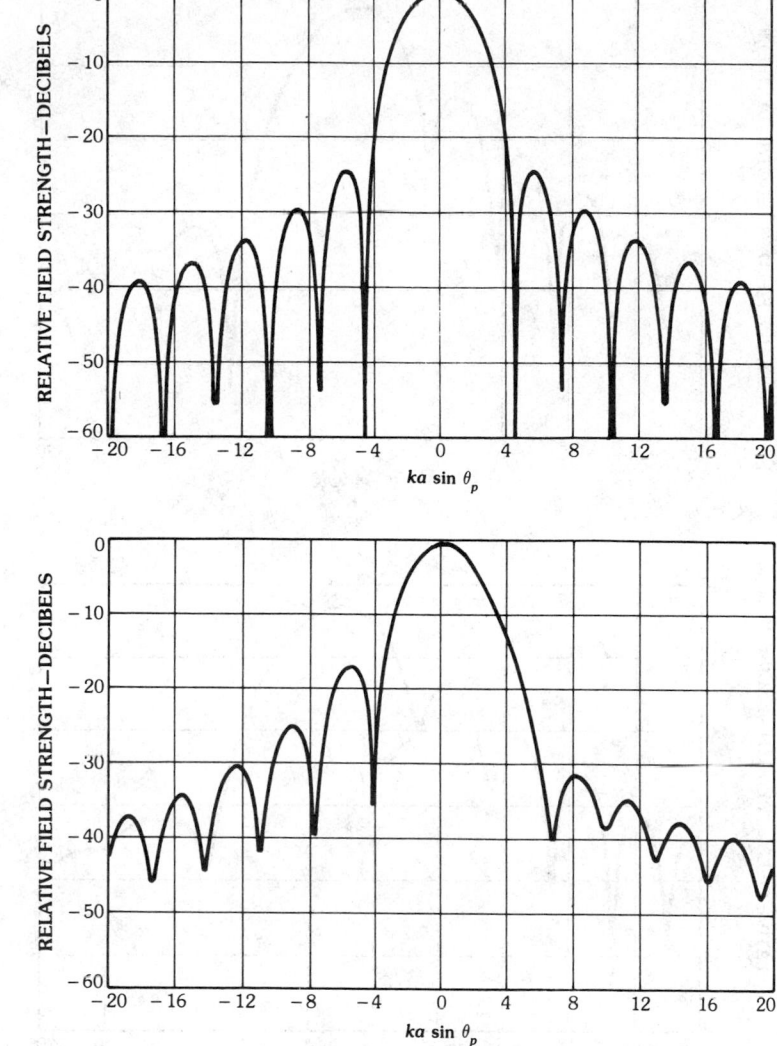

Fig. 36. Symmetric reflector far-field patterns for $\phi = 0°$ as a function of lateral feed displacement in terms of the number of beamwidths scanned ($F/D = 1.0$, ET $= -10$ dB, BDF $= 0.966$). (*a*) 0 bmws. (*b*) 8 bmws. (*c*) 16 bmws. (*d*) 20 bmws.

Notice the similarity between these patterns and those of the symmetric reflector with $F/D = 0.4$ (Fig. 35). Similar results are also shown for an offset reflector with $F/D_p = 1.0$ and $F/D = 2.4$. Figs. 41a through 41c show the patterns for the cases in which the feed direction is kept fixed as the feed is displaced in the plane orthogonal to the line joining the focal point with the reflector center. For this large value of F/D_p, the feed is very directive in order to provide the -10-dB edge taper and, therefore, for large scans the reflector would be very poorly illuminated. Results for the cases in which the feed is tilted toward the reflector center while it

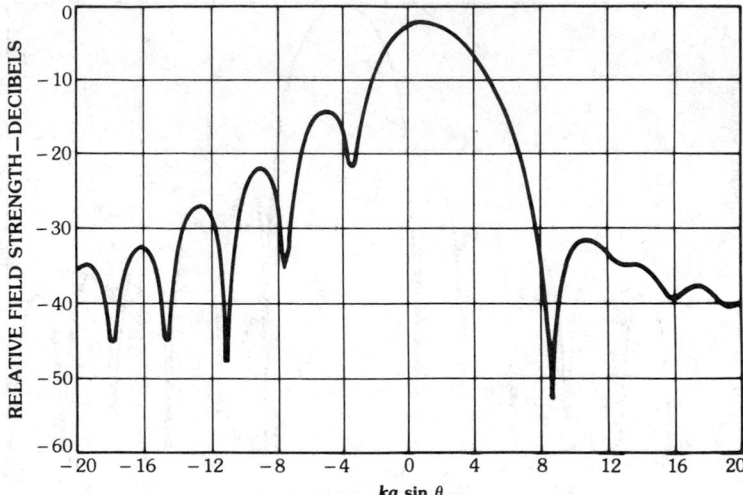

c

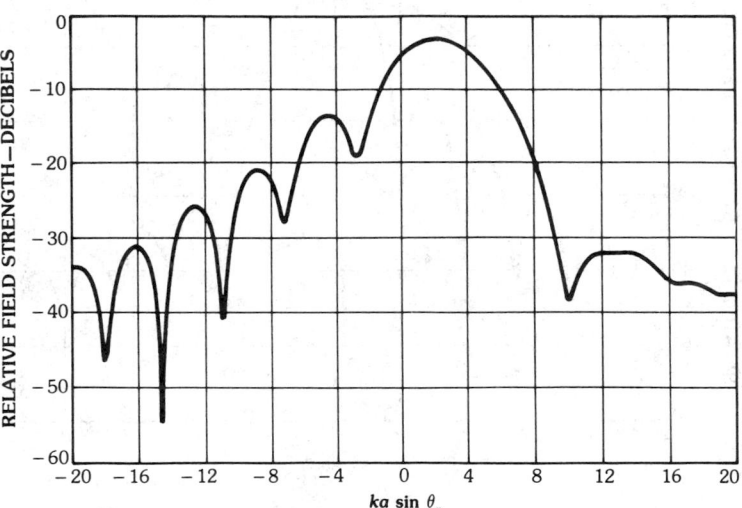

d

Fig. 36, *continued.*

is displaced in the focal plane are shown in Figs. 42a through 42f. Notice that a marked improvement can be observed in the patterns. Finally, the peak gain losses as a function of beamwidths scanned and for different values of F/D_p are plotted in Fig. 43. It is worthwhile to mention that there are a considerable number of ongoing attempts to improve the scan performance of single-reflector antennas by employing the concept of conjugate matched focal-plane feed arrays. This approach may also be utilized to overcome the deterministic reflector surface distortions.

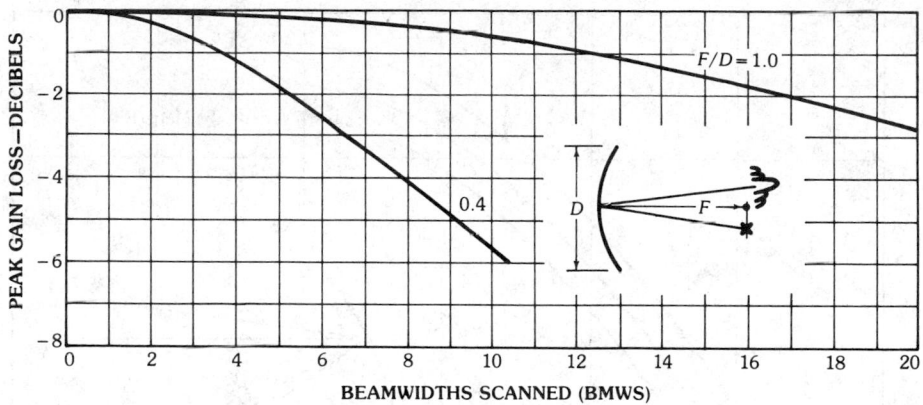

Fig. 37. Peak gain loss of symmetric parabolic reflectors as a function of beamwidths scanned for different F/D ratios.

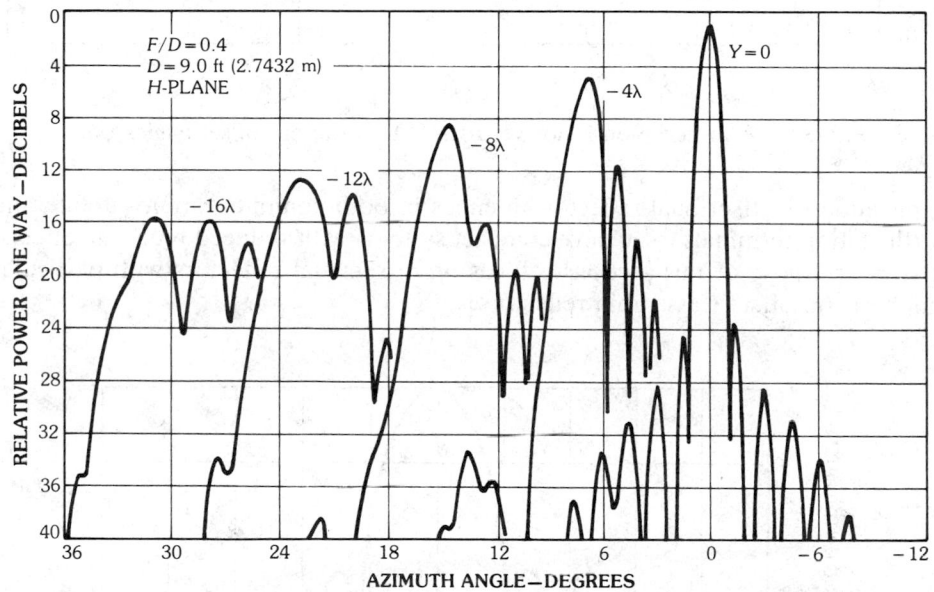

Fig. 38. Measured secondary patterns as a function of lateral primary feed displacement for a symmetric parabolic reflector. (*After Imbriale, Ingerson, and Wong [40],* © *1974 IEEE*)

6. Dual-Reflector Antennas

The application of dual reflectors at microwave frequencies has evolved from their counterparts used in optical telescopes. Originally, most of the designs were symmetric configurations with large main reflectors to reduce the blockage effects of the subreflector. These designs include Cassegrain, Gregorian, and shaped. The reader may refer to references [1, 7, 33, 39, 41, 42], which discuss many aspects of these reflectors. Recent stringent performance requirements have increased the

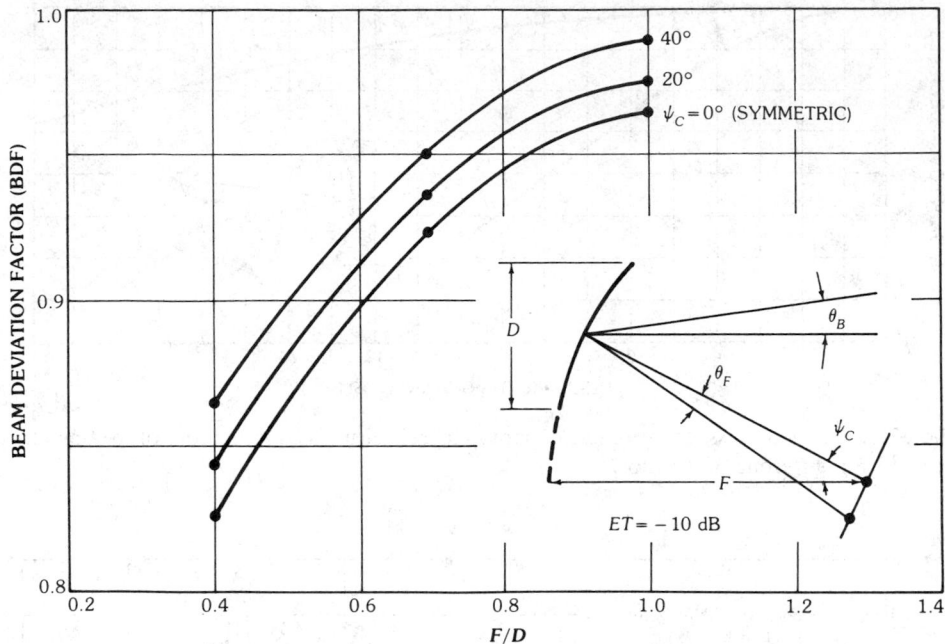

Fig. 39. Beam deviation factor versus F/D for different offset angles ψ_C.

application of offset dual-reflector antennas in both communication satellite and earth station terminals to achieve compact systems with reduced blockage effects. It is the purpose of this section to focus on offset dual reflectors with particular emphasis on offset Cassegrain reflectors.

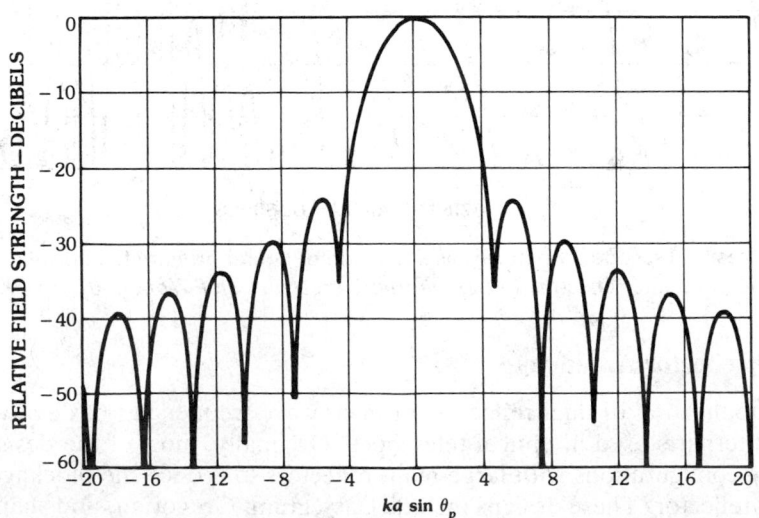

a

Fig. 40. Offset reflector far-field patterns as a function of feed displacement in terms of the number of beamwidths scanned ($F/D_p = 0.4$, $F/D = 0.96$, ET = −10 dB, BDF = 0.983, $\phi = 0°$). (a) 0 bmws. (b) 4 bmws. (c) 6 bmws. (d) 10 bmws.

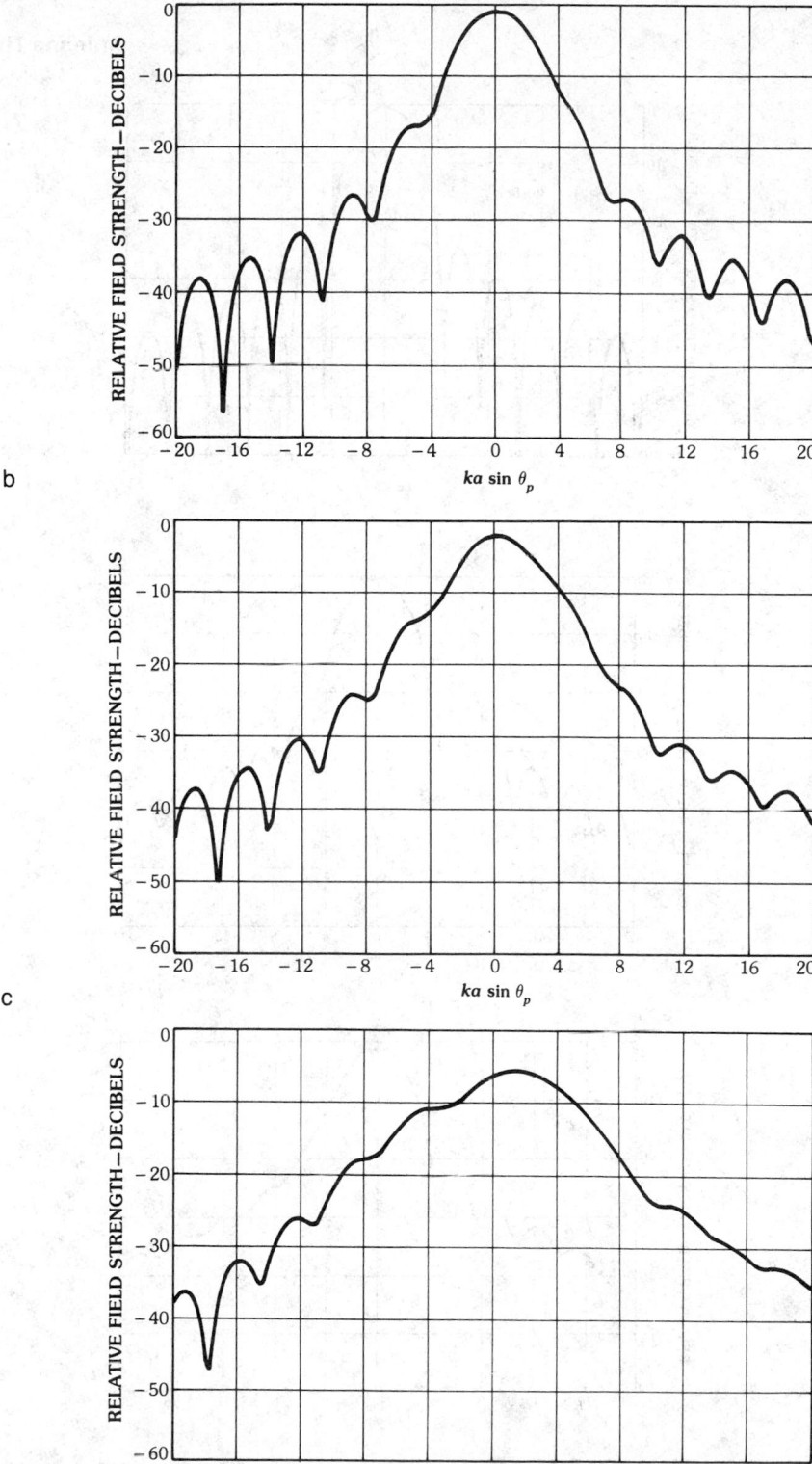

Fig. 40, *continued.*

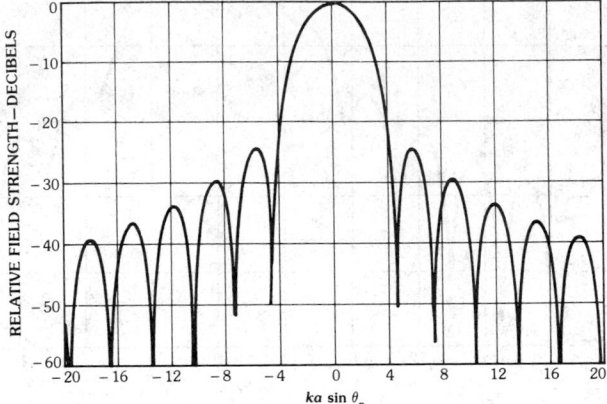

a

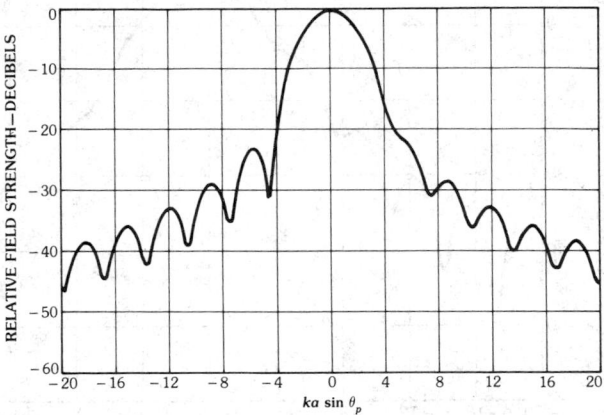

b

15-41B

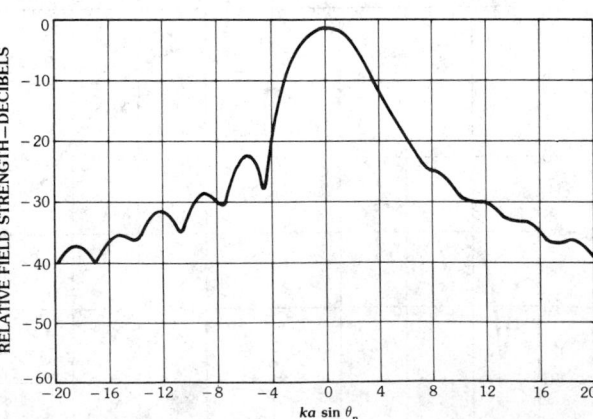

c

Fig. 41. Offset reflector far-field patterns as a function of feed displacement in terms of the number of beamwidths scanned ($F/D_p = 1.0$, $F/D = 2.4$, ET $= -10$ dB, BDF $= 0.998$, $\phi = 0°$). (*a*) 0 bmws. (*b*) 4 bmws. (*c*) 8 bmws.

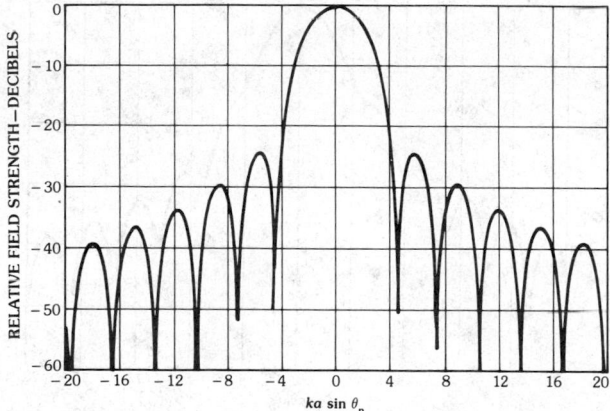

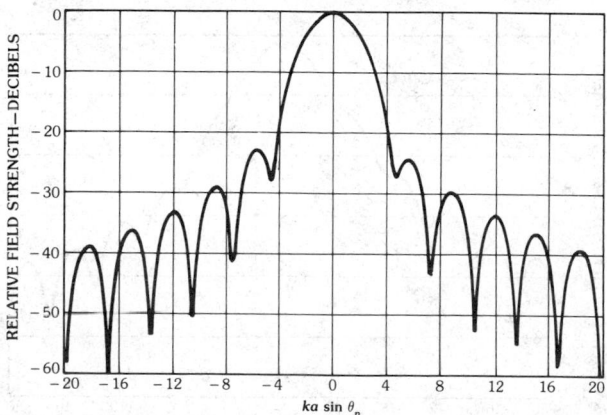

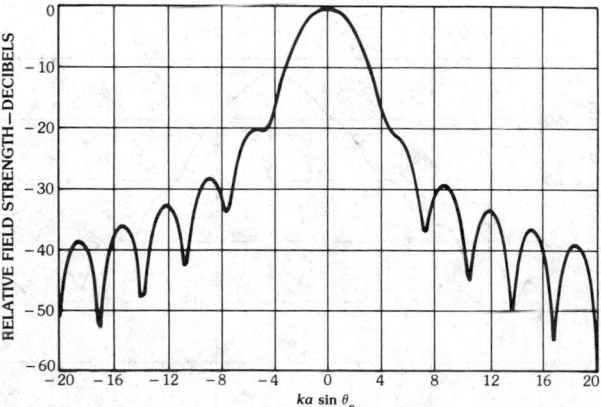

Fig. 42. Offset reflector far-field patterns for tilted feed as a function of feed displacement in terms of the number of beamwidths scanned (F/D_p = 1.0, F/D = 2.4, ET = −10 dB, BDF = 0.998, ϕ = 0°). (*a*) 0 bmws. (*b*) 4 bmws. (*c*) 8 bmws. (*d*) 12 bmws. (*e*) 16 bmws. (*f*) 20 bmws.

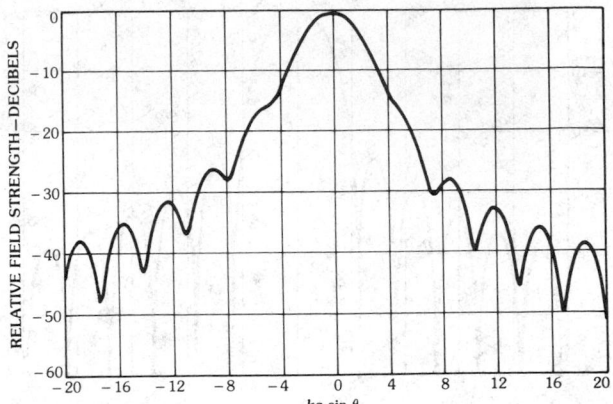

d

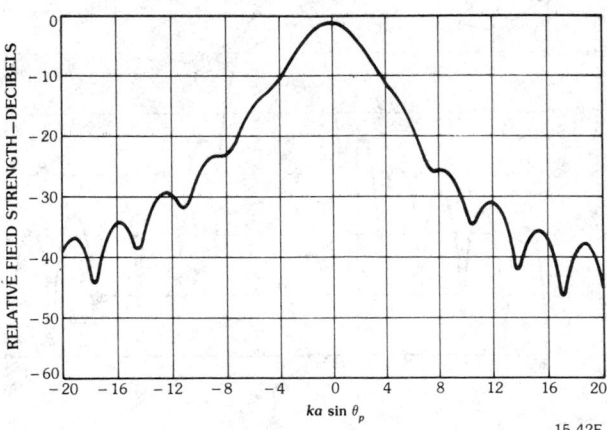

e

15-42E

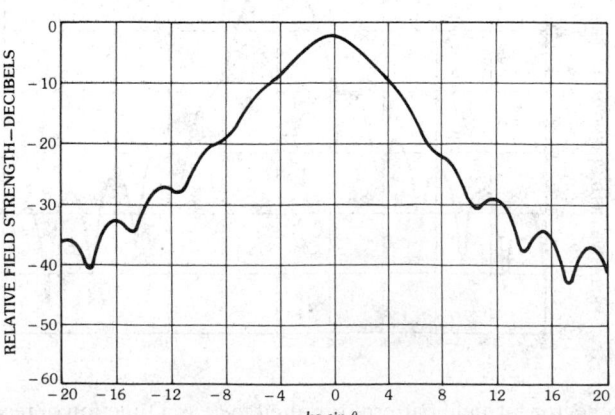

f

Fig. 42, *continued.*

General Parameters

Geometries of an offset Cassegrain antenna and its equivalent paraboloid are shown in Fig. 44, and some of their important geometrical relationships among different parameters are listed below.

Main reflector:

$$\theta^B = \frac{\theta^U + \theta^L}{2}$$

$$\theta^C = \frac{\theta^U - \theta^L}{2}$$

$$D = \frac{4F \sin \theta^C}{\cos \theta^B + \cos \theta^C} \tag{67}$$

$$d = \frac{2F \sin \theta^B}{\cos \theta^B + \cos \theta^C}$$

Subreflector:

$$e = \frac{\sin (\theta^U + \theta_s^U)/2}{\sin (\theta^U - \theta_s^U)/2}$$

$$M = \frac{e + 1}{e - 1}, \qquad e = \frac{M + 1}{M - 1}$$

$$\frac{z_0}{D'_s/2} = \frac{1}{\tan \theta^U} + \frac{1}{\tan \theta_s^U}$$

$$\frac{z_0}{D'_s/2 - D_s} = \frac{1}{\tan \theta^L} + \frac{1}{\tan \theta_s^L} \tag{68}$$

$$\frac{2L}{z_0} = 1 - \frac{\sin (\theta^U - \theta_s^U)/2}{\sin (\theta^U + \theta_s^U)/2} = \frac{2}{1 + M}$$

$$\tan (\theta_s/2) = M^{-1} \tan (\theta/2)$$

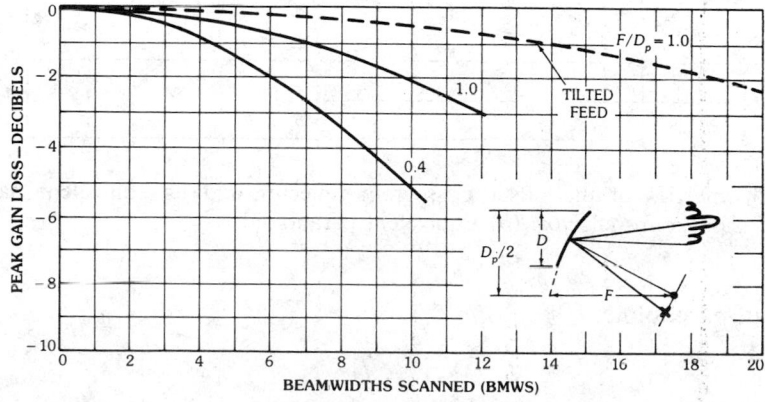

Fig. 43. Peak gain loss of offset parabolic reflectors as a function of the number of beamwidths scanned for different F/D_p ratios.

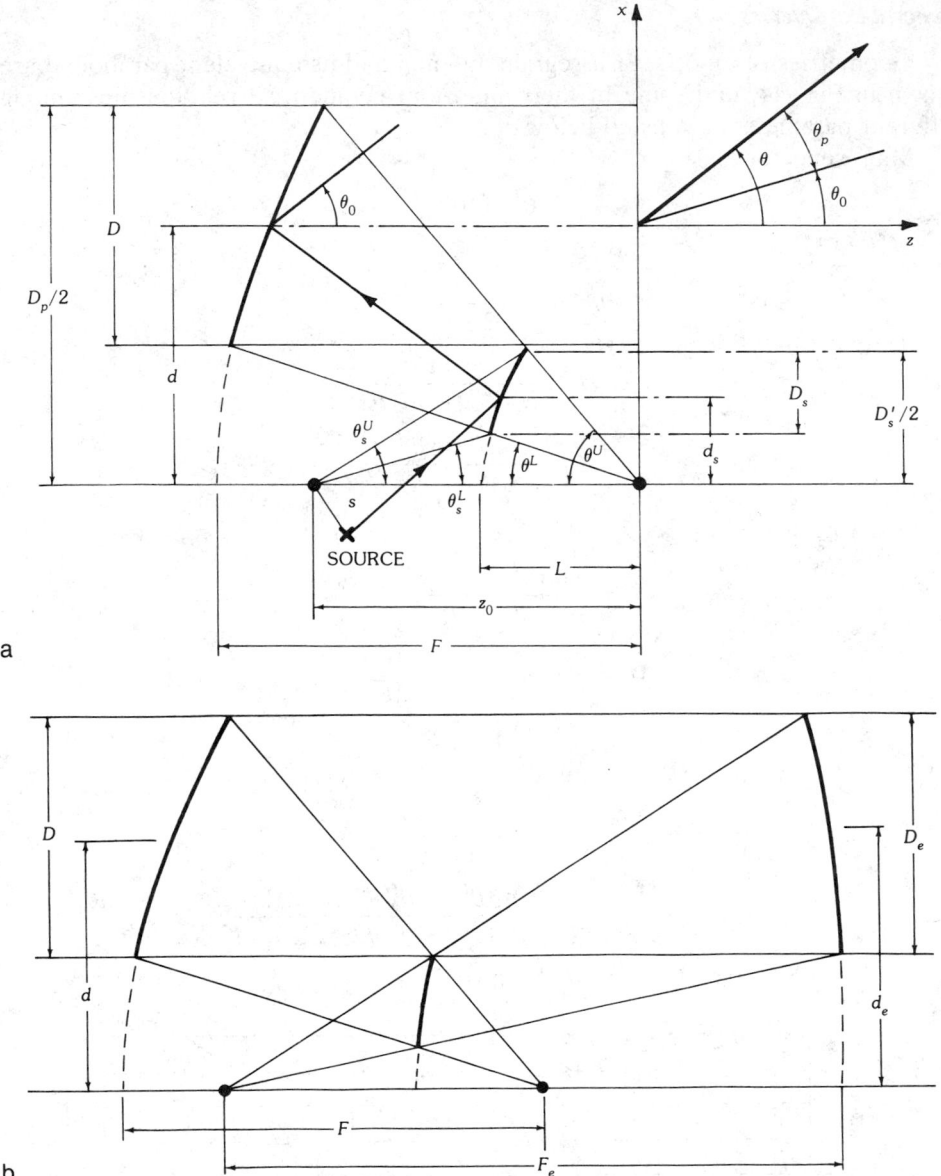

Fig. 44. Geometries of the offset Cassegrain reflector and its equivalent paraboloid. (*a*) Offset Cassegrain reflector. (*b*) Equivalent paraboloid.

Equivalent paraboloid:

$$D_e = D$$
$$d_e = d$$
$$F_e = M \cdot F$$

(69)

Performance Evaluation

Due to the appearance of the subreflector the analysis of dual reflectors is more complicated than for the single reflector. There are many different methods available based on techniques discussed in Section 2 and a comparative study has been done in [43]. Among the various techniques the applications of GTD from the subreflector and PO from the main reflector (GTD/PO), and PO from both the subreflector and the main reflector (PO/PO), have been most popular. Although in a majority of cases close agreement between the two approaches has been observed, there are situations, such as in small subreflectors (less than 10λ), where the results of GTD/PO do not agree with PO/PO, in particular, in the prediction of the cross-polarization level. As a matter of fact, in these cases PO/PO typically predicts a higher level of cross-polarized fields. It has been conjectured that for small subreflectors PO/PO provides a more accurate result.

For large subreflectors, however, PO/PO can become very time consuming and, for this reason, GTD/PO is used more frequently [1, 43]. One of the advantages of using this technique is that the effects of edge diffraction from the subreflector can be identified, when comparison is made between GO/PO and GTD/PO. In particular, it is known that, once the subreflector boundary is extended 1 to 2 wavelengths beyond the geometrical optical boundary, the diffraction effects can be significantly reduced.

In what follows, a few representative examples are shown. In all cases the offset Cassegrain of Fig. 44 is used, which is illuminated with a $\cos^q(\theta)$ type feed with different edge tapers. The results of the far-field patterns, each normalized to its own peak, for both GO/PO and GTD/PO are shown in Figs. 45 and 46. From these figures one can readily identify the contribution of edge diffracted rays. It is worthwhile to mention that for these cases subreflectors with optical extensions 0.5λ and 1λ beyond the optical extensions are used. Also shown in Fig. 47 is the variation of aperture efficiency as a function of feed edge taper for the cases of GO and GTD. It is interesting to note that for the optical extension case GTD predicts almost 8 percent less efficiency than GO due to the edge diffraction. However, in accordance with Fig. 47 considerable improvement can be achieved by extending the subreflector as small as 1λ beyond its optical edge.

Cross-Polarization Reduction

One of the important features of any dual-reflector system is to provide an additional degree of freedom to optimize a particular characteristic of these reflectors. For example, in a Cassegrain or Gregorian reflector system one can reduce considerably the level of cross-polarized field by simply tilting the axis of the subreflector with respect to the axis of the main reflector. Using the geometrical optics construction one can determine the amount of tilt to be [10, 44]

$$\tan(\alpha/2) = \frac{1}{M} \tan(\zeta/2) \tag{70}$$

where M is the subreflector magnification, and angles ζ and α are angles shown in Fig. 48. As an example, for the offset Cassegrain reflector with dimensions as

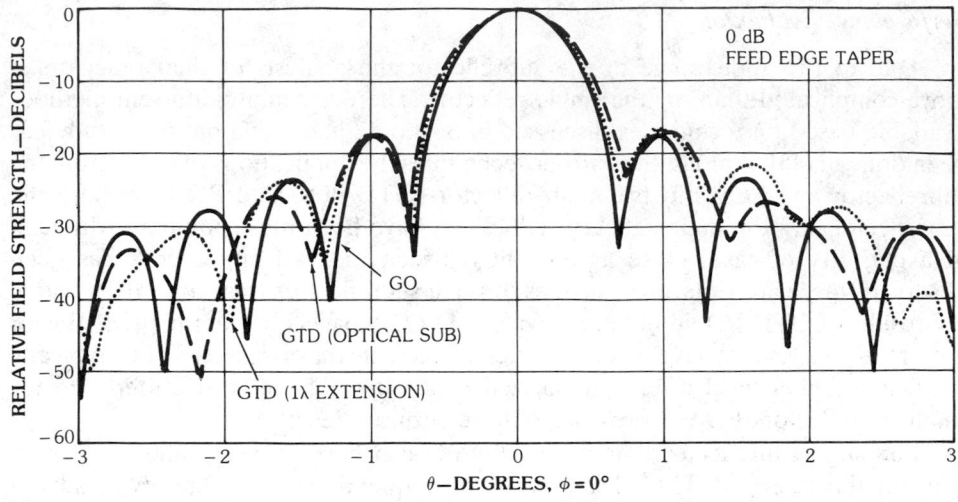

a

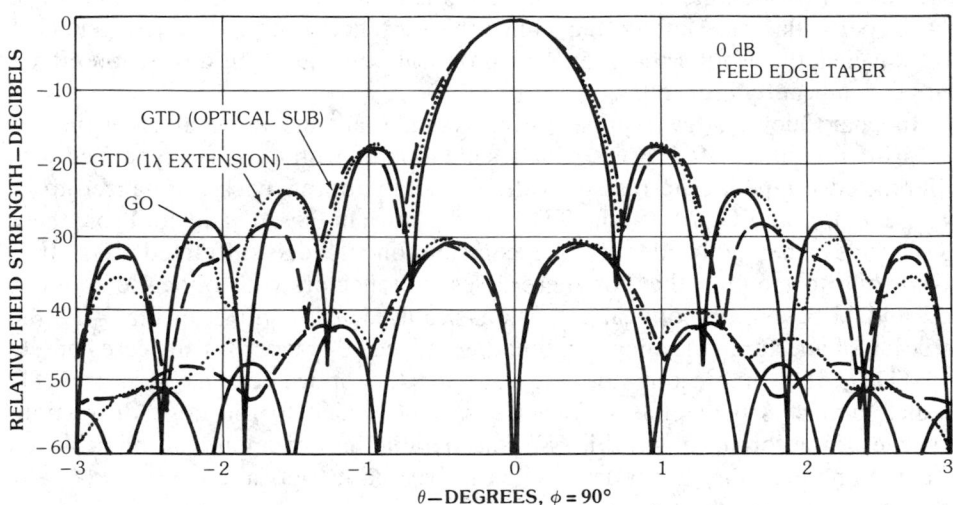

b

Fig. 45. A comparison between GO and GTD for a 0-dB tapered feed illuminating an offset Cassegrain antenna ($D = 100\lambda$, $d = 70\lambda$, $F = 96\lambda$, $Z_0 = 25\lambda$, $M = 1.5$). (*a*) $\phi = 0°$ cut (*xz* plane). (*b*) $\phi = 90°$ cut (*yz* plane). (*After Rahmat-Samii, "Subreflector extension for improved efficiencies in Cassegrain antennas—GTD/PO analysis,"* IEEE Trans. Antennas Propag., *vol. AP-34, pp. 1266–1269, October 1986,* © *1986 IEEE*)

shown in Fig. 49, the optimal tilt angle is $\zeta = 9.26°$. For two cases of no tilt $\zeta = 0°$ and optimal tilt $\zeta = 9.26°$ the contour of the cross-polarized fields based on diffraction analysis is shown in Fig. 50, which clearly demonstrates how effectively the cross-polarization level can be reduced. Although the level of cross polarization can be reduced drastically for the feed at the focal point, the amount of reduction is not as drastic as for the case with a feed off the focal point.

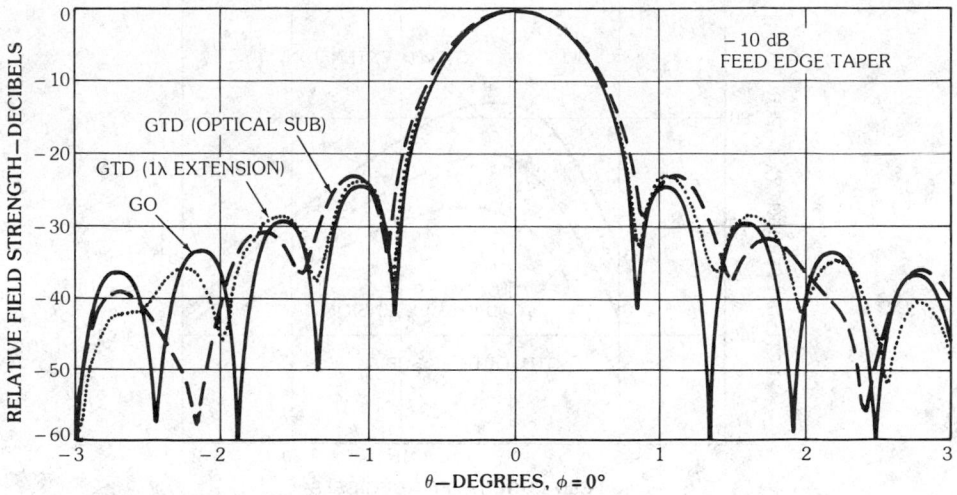

a

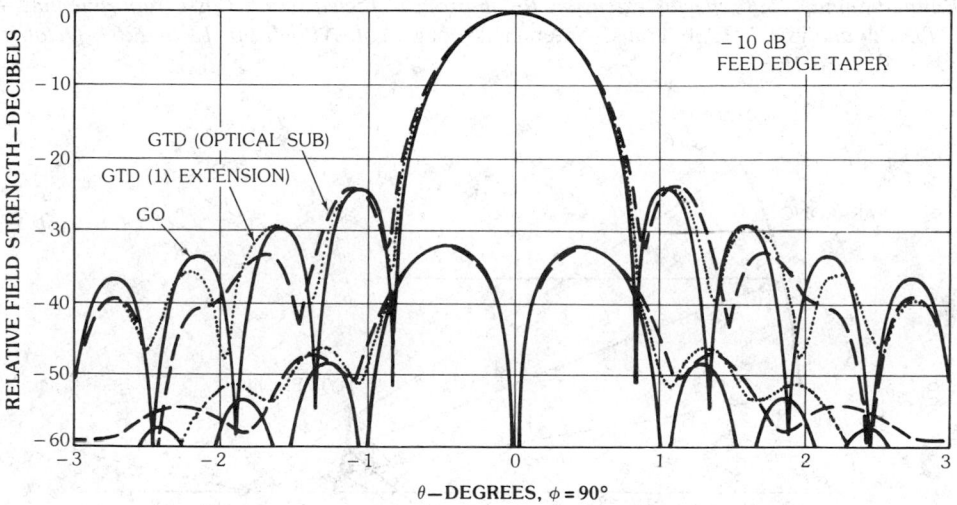

b

Fig. 46. A comparison between GO and GTD for a −10-dB tapered feed illuminating an offset Cassegrain antenna. (*a*) $\phi = 0°$ cut (*xz* plane). (*b*) $\phi = 90°$ cut (*yz* plane). (*After Rahmat-Samii, "Subreflector extension for improved efficiencies in Cassegrain antennas—GTD/PO analysis," IEEE Trans. Antennas Propag., vol. AP-34, pp. 1266–1269, October 1986,* © *1986 IEEE*)

Scan Performance

Due to the fact that the equivalent paraboloid of a Cassegrain reflector has a longer focal length than the main reflector, it is expected to demonstrate a better scan performance than the single reflector. Earlier work reported in [33] on symmetric Cassegrain reflectors suggested that the scan performance of these

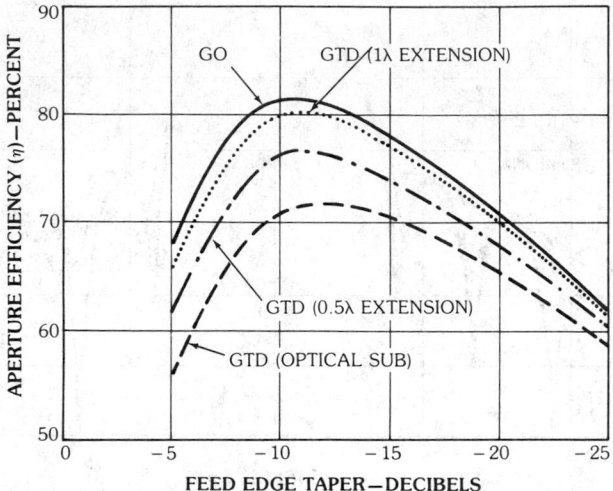

Fig. 47. A comparison between GO and GTD for reflector aperture efficiency. (*After Rahmat-Samii, "Subreflector extension for improved efficiencies in Cassegrain antennas—GTD/PO analysis," IEEE Trans. Antennas Propag., vol. AP-34, pp. 1266–1269, October 1986, © 1986 IEEE*)

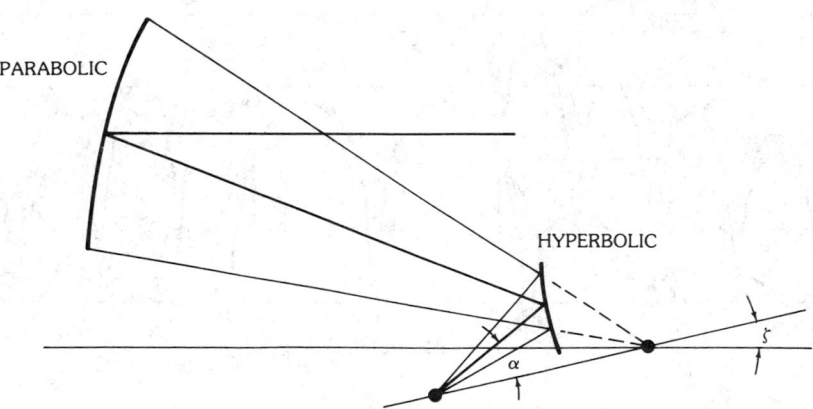

Fig. 48. A tilted subreflector for cross-polarization reduction.

reflectors may be predicted from their equivalent paraboloid counterparts. To this effect, results were presented in [45] based on the PO/PO diffraction analysis, which demonstrated close agreement between the Cassegrain and equivalent paraboloid. Fig. 51 shows the result for the scan performance. Similar results were also obtained using GTD/PO as shown in Fig. 52. More recent investigations have demonstrated that the concept of the equivalent paraboloid is only applicable for small scans and should be employed with care [43]. For example, for an offset Cassegrain with parameters as given in Fig. 53, comparisons are made between the main reflector, Cassegrain, and its equivalent paraboloid in Figs. 53–55. These

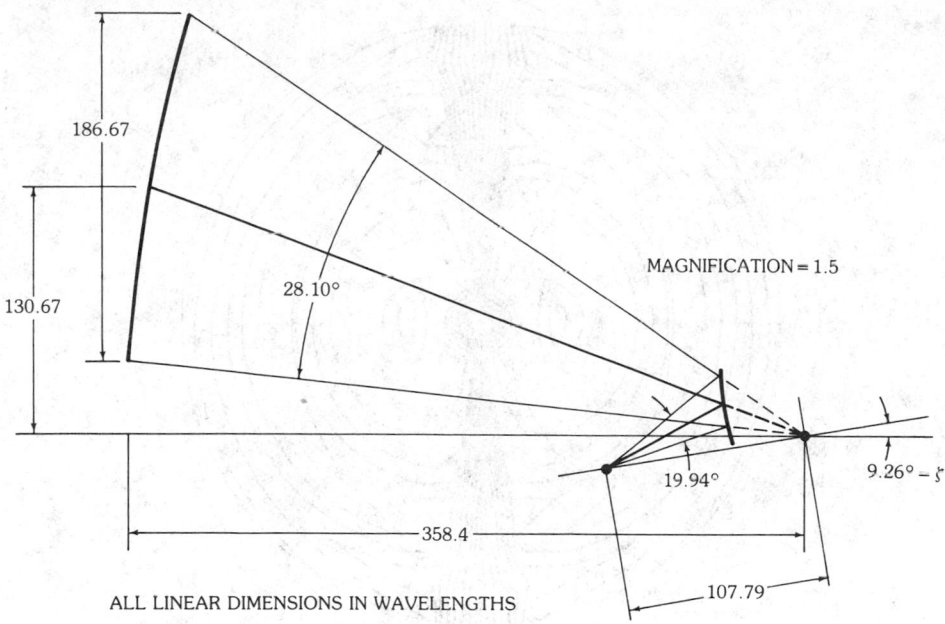

MAGNIFICATION = 1.5

ALL LINEAR DIMENSIONS IN WAVELENGTHS

Fig. 49. An offset Cassegrain reflector with tilted subreflector.

results indicate that for small scans the performance of the Cassegrain is similar to the equivalent paraboloid, whereas for the large scan it deviates substantially. In all these cases an oversized subreflector was used for better illumination of the main reflector for large scans. Fig. 56 shows the gain loss as a function of the beam-widths scanned for all three cases. Additional improvement can be expected by locating the feeds on the optimal focal surface rather than the optimal focal plane tangent to this surface [46]. Similar observations can also be made for Gregorian reflectors [47].

Shaped Reflectors

An important class of dual reflectors is the shaped reflectors, with properly shaped subreflector and main reflector, which provide desired aperture amplitude and phase distributions. These reflectors are used particularly in the design of high-gain antennas and also extensively in the design of many ground and telemetry antennas. The foundations of these reflectors can be found in [1, 48, 49, 50, 51], where most of the designs have been for symmetric configurations. Recently, however, in order to further improve the overall efficiency, offset shaped reflectors have also been considered [52, 53, 54, 55, 56]. In particular, a recent evaluation [54] of this type of antenna has been reported which gives an overall efficiency of nearly 85 percent! A brief description of the concept behind these reflectors is given below [55].

With some exceptions reflector antennas are synthesized on the basis of geo-metrical optics (GO) with the assumptions that the wavelength is short relative to the overall dimensions and that the reflective surfaces have large radii of curvature.

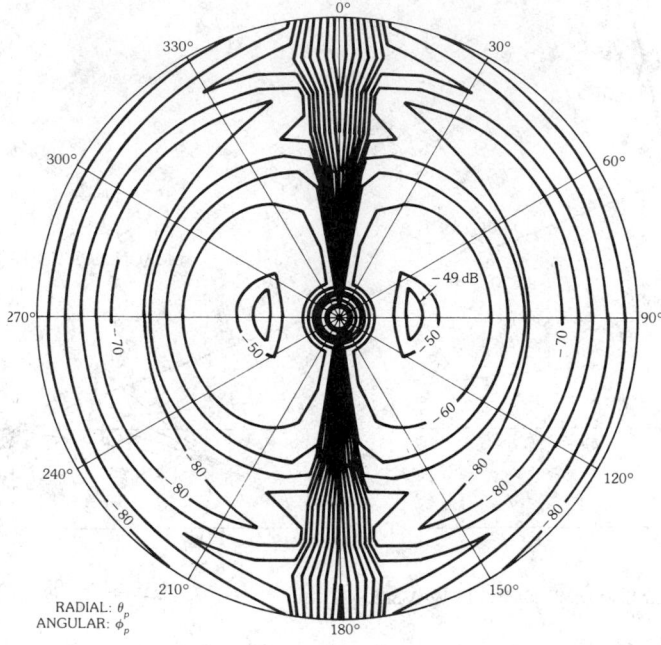

a

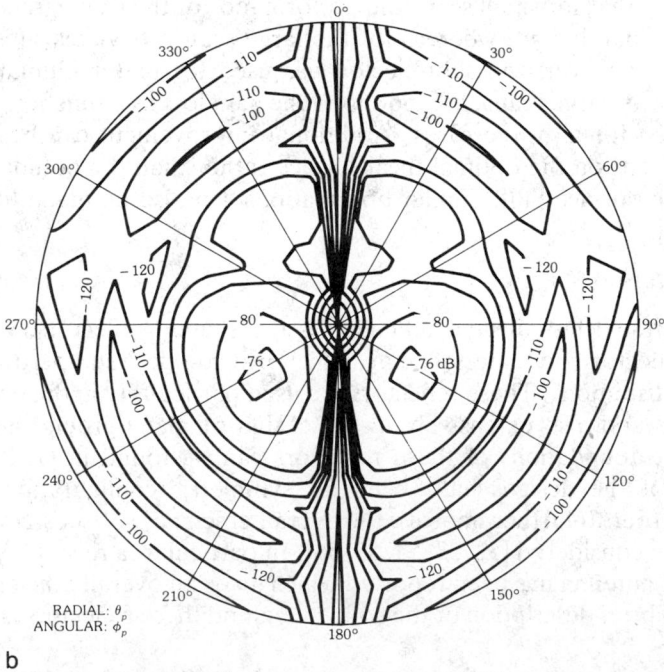

b

Fig. 50. A contour plot of the cross-polarized field for the Cassegrain reflector of Fig. 49. (a) With $\zeta = 0°$. (b) With $\zeta = 9.24°$.

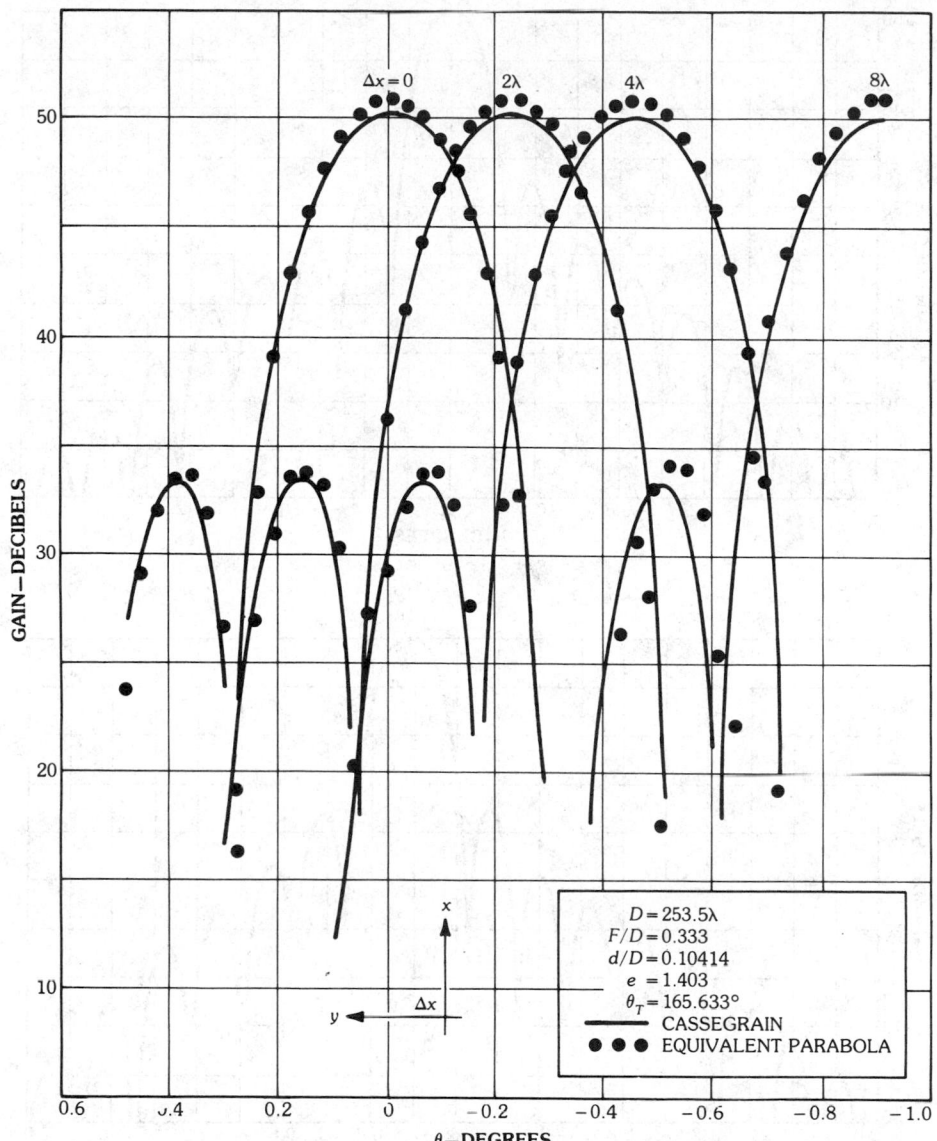

Fig. 51. A comparison between a symmetric Cassegrain reflector and its equivalent paraboloid as a function of feed displacement using PO/PO analysis. (*After Wong [45],* © *1975 IEEE*)

Classically the paraboloid converts a spherical wave emanated from the source (feed) into a plane wave; hence one may state that the reflector transforms the shape of the *phase front*. Some single reflectors are also designed to convert the feed pattern into a given optical *energy* distribution, e.g., into a $\csc^2(\theta)$ type pattern [57]. It is interesting that a doubly curved noncylindrical reflector of this type has no exact solution. Dual reflectors have also been used to convert a spheri-

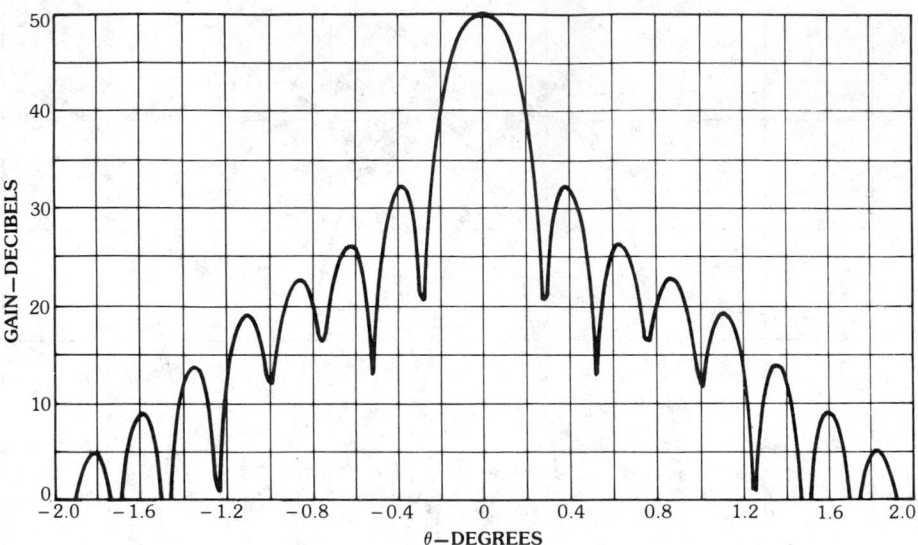

a

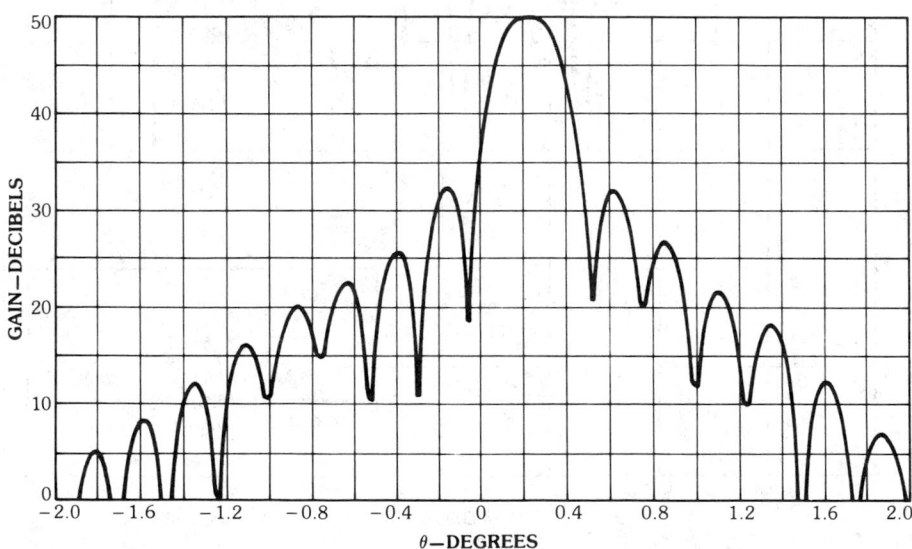

b

Fig. 52. Same as Fig. 51 except that GTD/PO diffraction analysis is used. (*a*) $\Delta = 0$. (*b*) $\Delta = -2\lambda$. (*c*) $\Delta = -4\lambda$. (*d*) $\Delta = -8\lambda$.

cal phase front into a plane phase front; for example, one may refer to the Cassegrain (paraboloid-hyperboloid) and Gregorian (paraboloid-ellipsoid) reflectors [33].

When two sequential reflections from the feed to the aperture are used for aperture magnification, it is possible to exercise control over both the resultant

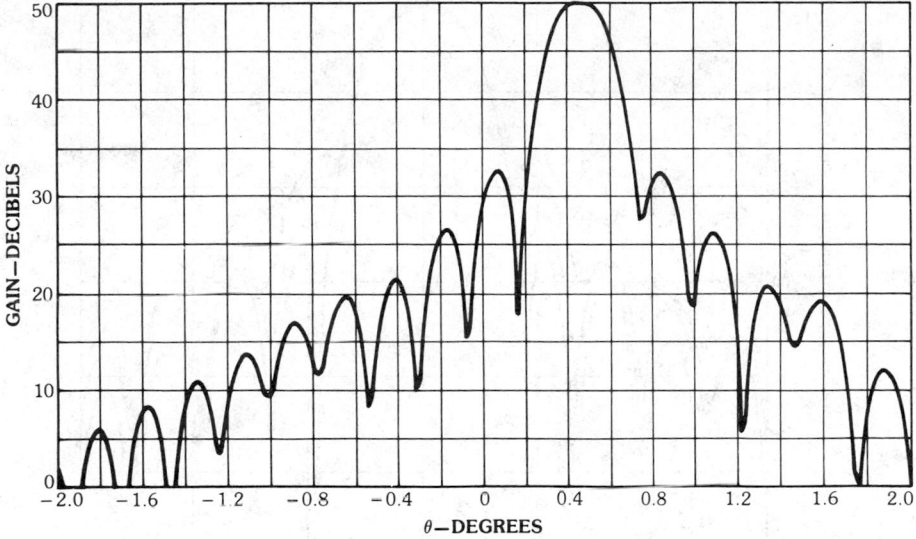

c

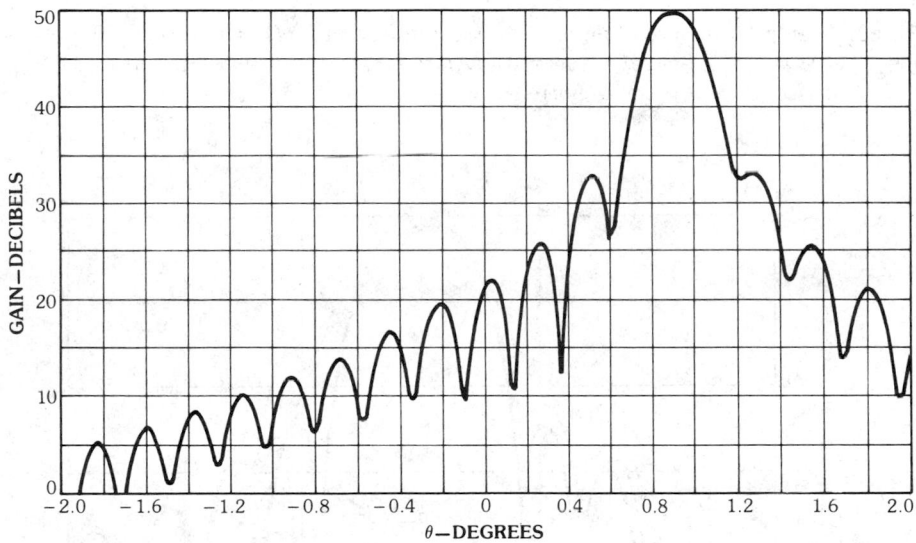

d

Fig. 52, *continued.*

aperture phase-front shape and the aperture energy distribution by appropriate shaping of the reflector surfaces. The basic objective of the GO dual-shaped synthesis is illustrated in Fig. 57. A bundle of rays radiated by the feed is a representation of the feed pattern in both amplitude and phase. This bundle, which has a well-defined periphery, is intercepted by the subreflector and then by the main reflector. The output bundle of rays, after two reflections, is required to have: (*a*)

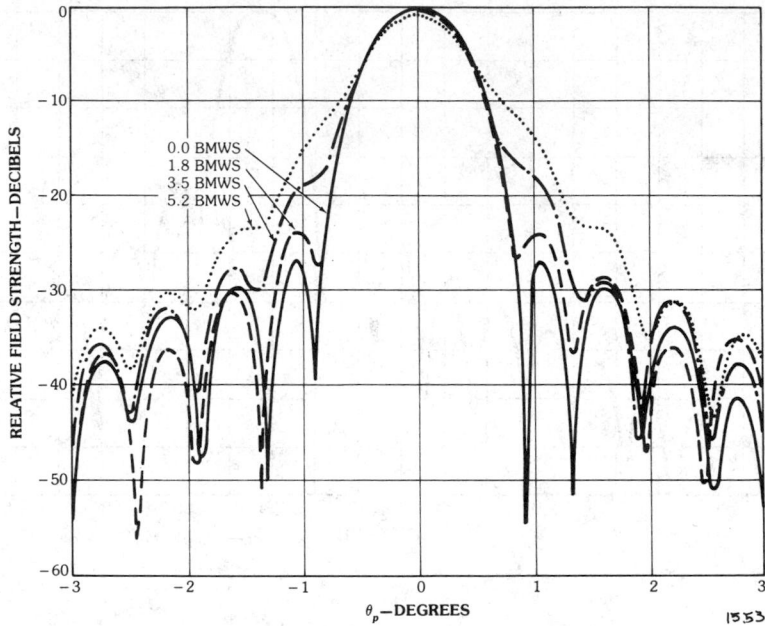

Fig. 53. Scan performance of an offset Cassegrain reflector ($D = 100\lambda$, $d = 70\lambda$, $F = 96\lambda$, $z_0 = 25\lambda$, $M = 1.5$, $D_s = 26\lambda$, oversized). (*After Rahmat-Samii and Galindo-Israel [43]*, © *1981, American Geophysical Union*)

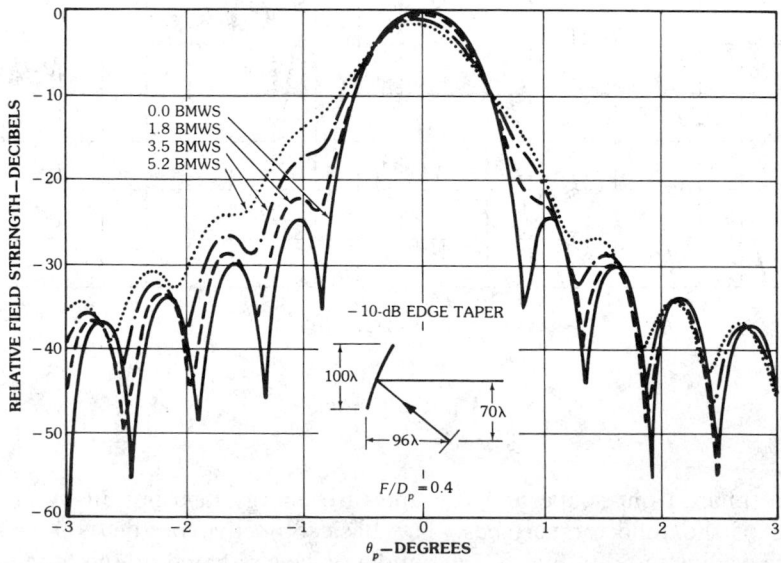

Fig. 54. Scan performance of an offset parabolic reflector with $F/D_p = 0.4$ illuminated by a linearly polarized incident field with -10-dB edge taper. (*After Rahmat-Samii and Galindo-Israel [43]*, © *1981, American Geophysical Union*)

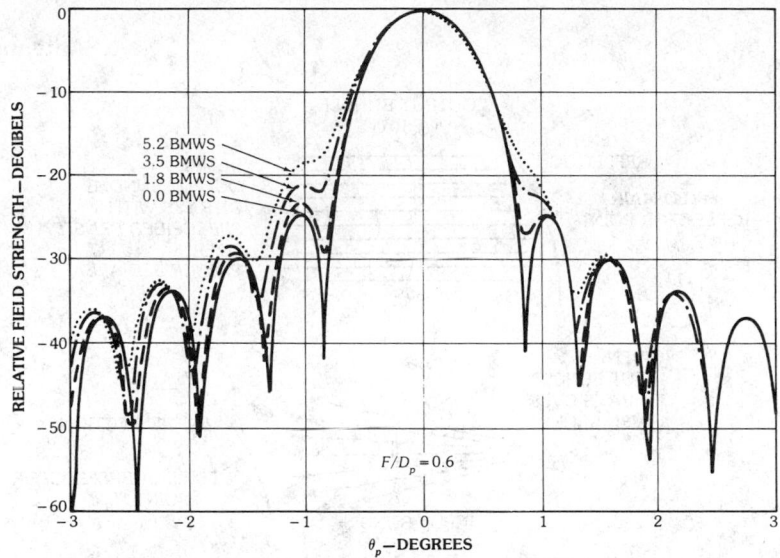

Fig. 55. Same as Fig. 54 except that $F = 144\lambda$ and $F/D_p = 0.6$. (*After Rahmat-Samii and Galindo-Israel [43], © 1981, American Geophysical Union*)

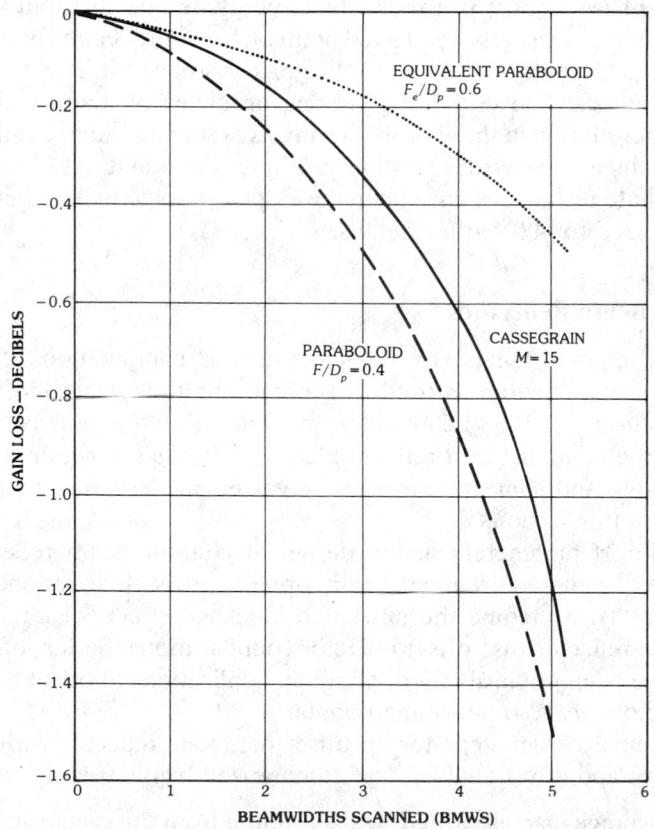

Fig. 56. Cassegrain gain loss performance with a comparison between the main-reflector paraboloid ($F/D_p = 0.4$) and its equivalent paraboloid ($F_e/D_p = 0.6$). (*After Rahmat-Samii and Galindo-Israel [43], © 1981, American Geophysical Union*)

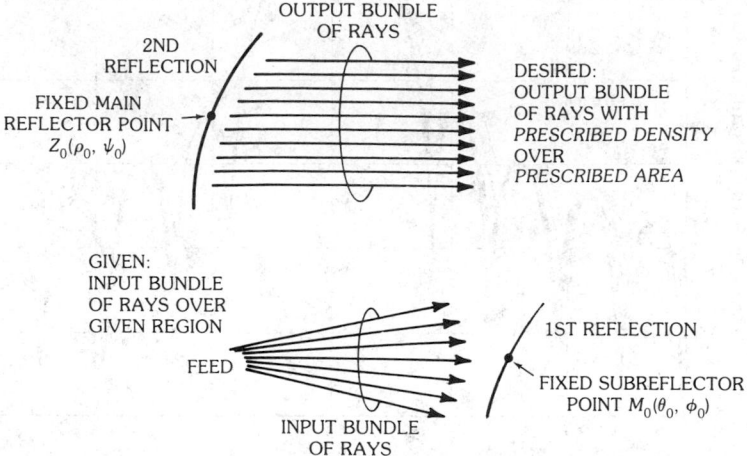

Fig. 57. Geometrical optics synthesis of a dual-shaped reflector—a statement of the problem.

a prescribed phase front, (*b*) a prescribed energy density distribution, (*c*) a prescribed periphery, (*d*) a prescribed fixed point on the subreflector (or a line), and (*e*) a prescribed fixed point on the main reflector (or a line) for a given feed pattern. These constraints define useful engineering problems of many types. It can be shown that the solution to these problems involves solving (numerically) two simultaneous, nonlinear, first-order, ordinary differential equations. Many numerical and approximate techniques have been developed to solve these equations and the reader is referred to [48–59] for details.

7. Contour Beam Reflectors

Another important class of reflectors is the combination of arrays with reflectors. This combination is used to generate multiple-beam [60, 61] improved scan performance [7, 62] and contour beam [7, 61, 63, 64] antennas. Among these applications the contour beam reflectors are widely used in communication satellite applications and some of the steps involved in designing them are briefly summarized in this section.

The major rf parameters in the design of contour beam reflectors are the coverage area, gain requirements, gain ripples, cross-polarization levels, overall losses, etc. To overcome the unwanted blockage effects of large arrays most contour beam reflectors use offset reflector configurations. So far, offset parabolic reflectors have been primarily used, although applications of offset Cassegrain and shaped reflectors are also becoming popular.

The principal design steps for an offset parabolic reflector with diameter D, focal length F, and offset angle ψ_c are summarized below [61]:

1. Plot a coverage map as viewed by the satellite from the synchronous orbit, or a composite coverage map including the pointing error and views from different orbit positions. Fig. 58 shows the typical azimuth and elevation coordinates

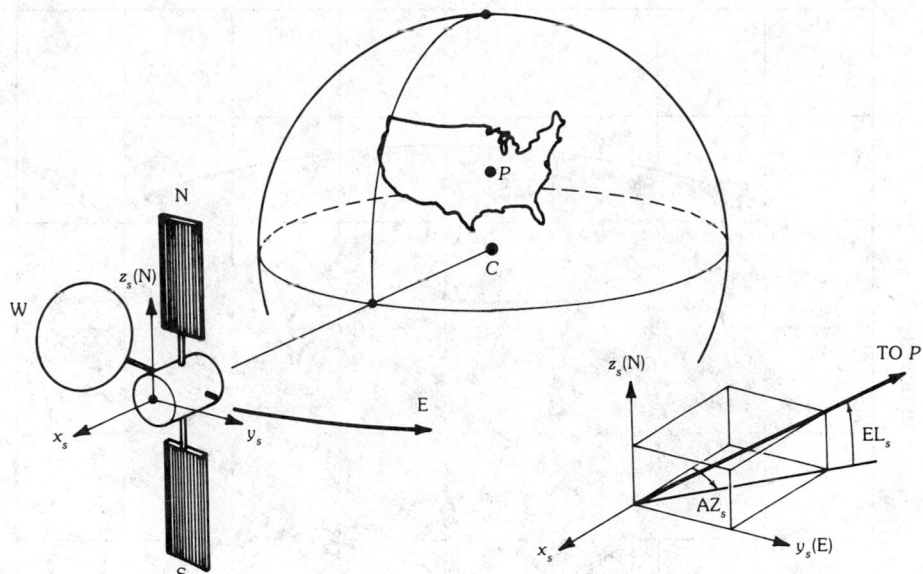

Fig. 58. An observation point P on earth as described by azimuth angle AZ_s and elevation angle EL_s in satellite coordinates.

used to define the coverage map and Fig. 59 depicts the United States from a synchronous satellite at 98° west longitude [61].

2. Select a set of contiguous (sometimes noncontiguous may also be used) squares or circles which represent the −3- to −4-dB beamwidth of the reflector illuminated by a single feed element. It takes a bit of experience to find the best arrangement. However, by setting the center of the outer elements close to the boundary of the desired coverage area, satisfactory results can be expected. Fig. 60 shows one such arrangement to cover the Eastern Time Zone (ETZ) of the United States.

3. Select an appropriate location for the antenna boresight and determine the proper scale factor between the lattice element centers and actual feed location in the feed plane. Beware of the beam deviation factor. For example, it has been found that the following expression provides a satisfactory estimate of the feed size:

$$\Delta = \tan\left(\theta_{\text{bmw}}/\text{BDF}\right) \frac{2F}{1 + \cos\psi_c} \tag{71}$$

where ψ_c is the offset angle of the feed array, BDF is the beam deviation factor that can be approximated by

$$\text{BDF} \cong \frac{1 + k(D/4F)^2}{1 + (D/4F)^2}, \qquad k = 0.36, \tag{72}$$

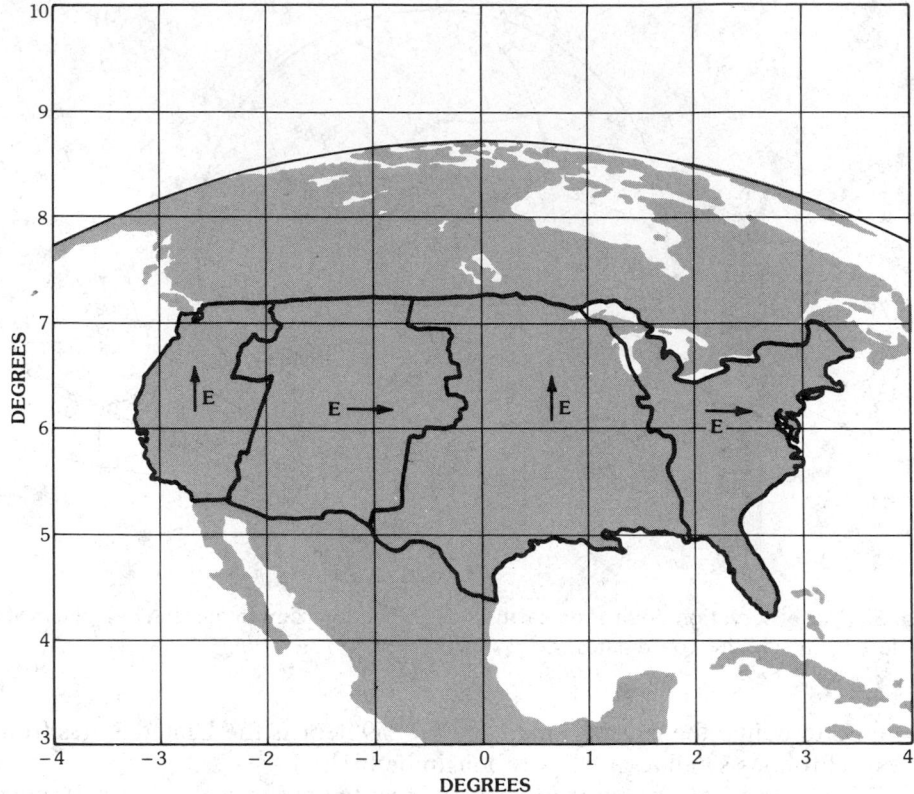

Fig. 59. Continental U.S. time zone map as viewed from a geosynchronous satellite at 98° west longitude. (*After Chen and Franklin [61]*)

and θ_{bmw} is the -3-dB beamwidth of the reflector with diameter D/λ ($\lambda =$ wavelength) illuminated by a single feed, viz.,

$$\theta_{bmw} \cong 2 \sin^{-1}\left(\frac{1.6 - 0.02ET}{\pi D/\lambda}\right) \tag{73}$$

In the above equation, ET is a single-feed edge taper which typically takes the value ET = -4 to -5 dB. The previous expression can be further approximated to give [61]

$$\Delta = \frac{1.06}{D/\lambda} \frac{2F}{1 + \cos \psi_c} \tag{74}$$

4. Select sample points on the map for gain optimization. Fig. 61 shows an example of the selection of sample points on ETZ. Typically the gain variations at a subset of these points are specified by the system requirements.
5. Determine the reflector far field due to each feed with unity excitation.
6. Determine the optimum amplitude and phase of each feed by numerical

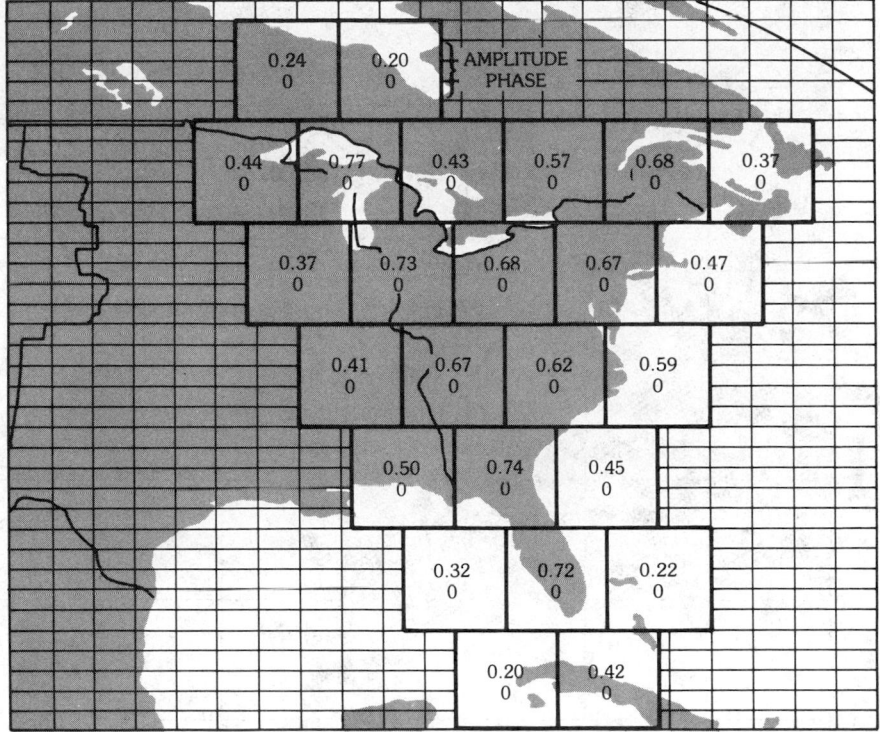

Fig. 60. Reflector image of ETZ feed cluster and its amplitude and phase distribution overlaid on the ETZ map. (*After Chen and Franklin [61]*)

optimization techniques. The proper selection of the optimization algorithm is a very crucial step in obtaining contours which satisfy prescribed constraints. Some algorithms are discussed in [7, 61]. For example, Fig. 62 shows a block diagram of the steps involved in an optimization. In this figure, P designates the total power radiated by the array for the directivity computation [34].

7. Construct the contour pattern of the reflector illuminated by the feed array with optimized coefficients. For example, the excitation coefficients of an array for the ETZ coverage is shown in Fig. 60 and the reflector contour map is depicted in Fig. 63 [61].

8. Finally, verify the results of the computer simulation with the experimental data. One such comparison is reported in [61] and the results are depicted in Fig. 64.

One of the difficult steps in accurately predicting the performance of a reflector illuminated by a densely packed feed array is the inclusion of mutual coupling effects. So far, a simple approach has not yet been developed for incorporating these coupling effects. This is typically done either by using the measured pattern of the feed elements in the array environment or by using numerical methods for simple feed array configurations [64, 65]. In a recent work [66], the planar near-field probing technique has provided a means to significantly reduce the effects of

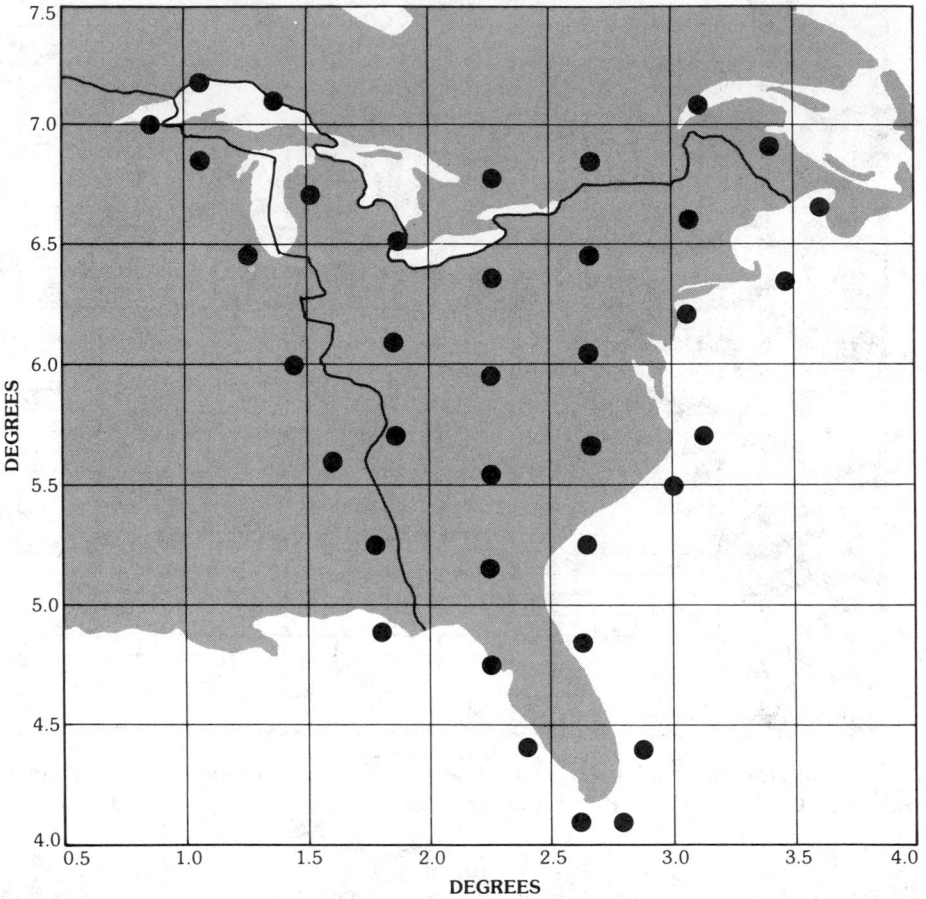

Fig. 61. Selected sample points for gain optimization, with satellite at 98° west longitude. (*After Chen and Franklin [61]*)

mutual coupling by the insertion of additional tuning into the feed lines of appropriate elements.

8. Feeds for Reflectors

Antenna feeds share a major role in the successful and efficient operation of any reflector system, independently of its optical setup. Unfortunately there is no such thing as an all-purpose universal feed, as in practice for a given reflector configuration and performance requirements the feed must be properly tailored. In the previous sections, independently of the actual feed elements, the feed patterns were approximated by $\cos^q(\theta)$ type patterns to allow for detailed parametric studies. In this section the characteristics of the most commonly used reflector feed elements are presented. These elements may take a variety of configurations including dipoles, log spirals, open-ended rectangular and circular waveguides, rectangular and circular horns, corrugated horns, dual- and hybrid-mode horns,

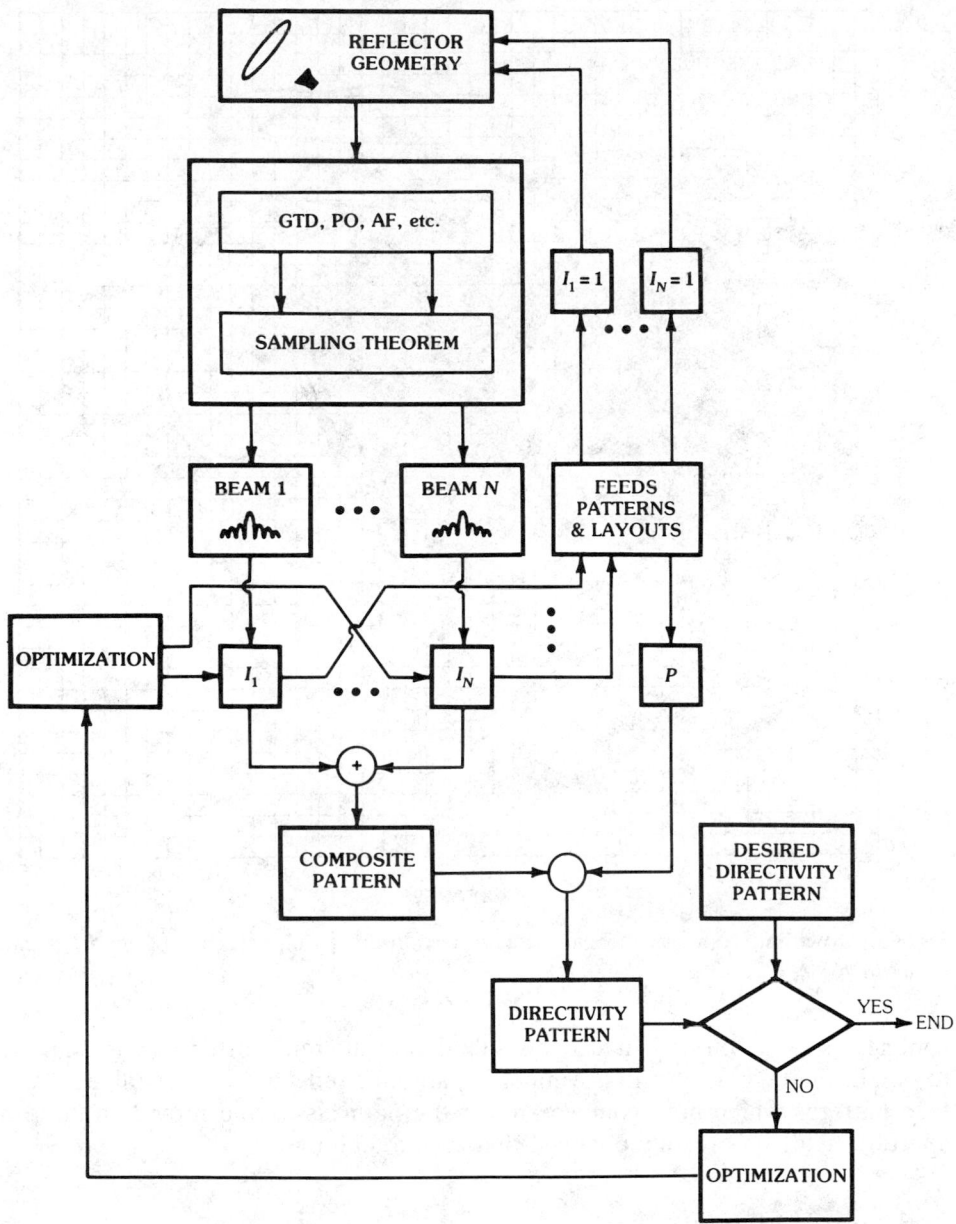

Fig. 62. Basic steps of a contour beam design.

microstrips, etc., or arrays of these elements. In this section only a few are considered and the reader may refer to [6, 7, 9, 67] for a more comprehensive treatment of the subject.

In the ideal situation, in which the reflector optics possess symmetric path losses (such as symmetric reflectors), the ideal feed element (although unachievable in reality) is one that has symmetric patterns, has a unique phase center, and

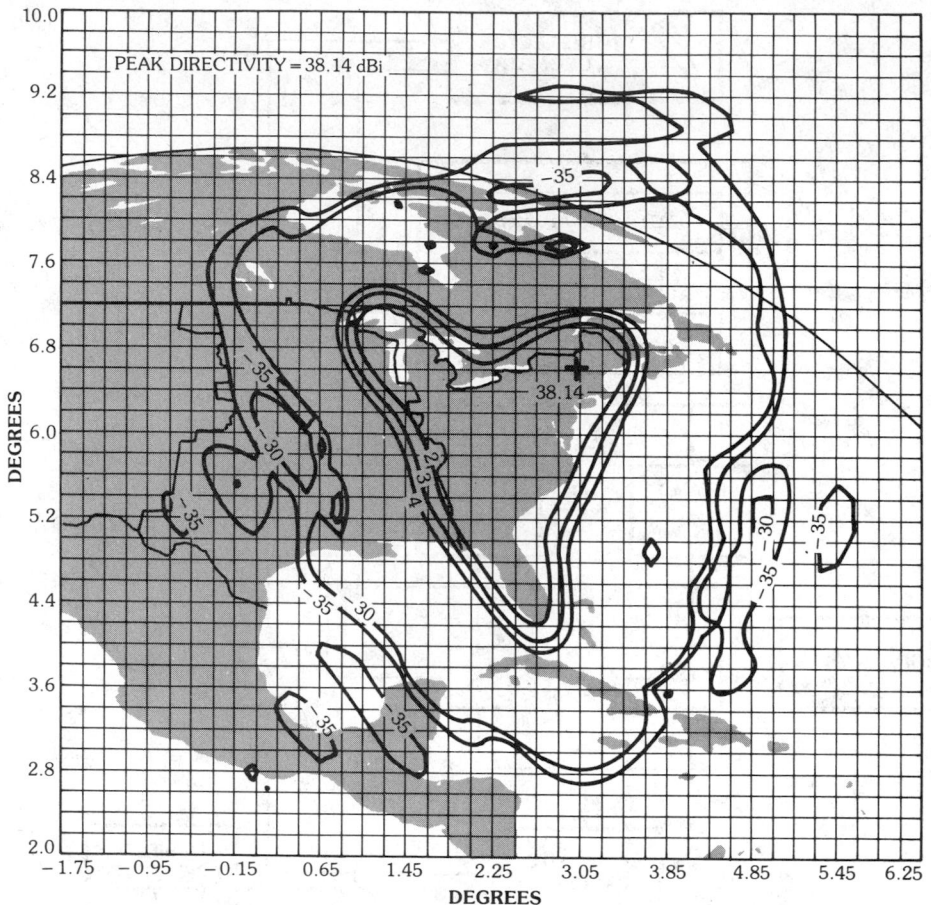

Fig. 63. Downlink copolarized beam isolation contours at 11.95 GHz. (*After Chen and Franklin [61]*)

confines its radiation only in a cone subtended to the reflector edge. For example, to obtain the highest gain in a symmetric parabolic reflector it is desirable to have feed patterns which both compensate for the path losses and provide a uniform aperture field. It is readily obtained that such a feed pattern is

$$C(\theta) = \sec^2(\theta/2) \tag{75}$$

where θ is the spherical angle from the feed axis which coincides with the reflector axis (see Section 5, under "Idealized Feed Patterns"). Note that, in terms of the power pattern, $\sec^4(\theta/2)$ will be used.

In the spherical coordinates of the feed as shown in Fig. 65, the radiated field may, in general, be presented as

$$\mathbf{E}(\mathbf{r}) = A_0[\hat{\boldsymbol{\theta}}C_\theta(\theta, \phi) + \hat{\boldsymbol{\phi}}C_\phi(\theta, \phi)]\frac{e^{-jkr}}{r} \tag{76}$$

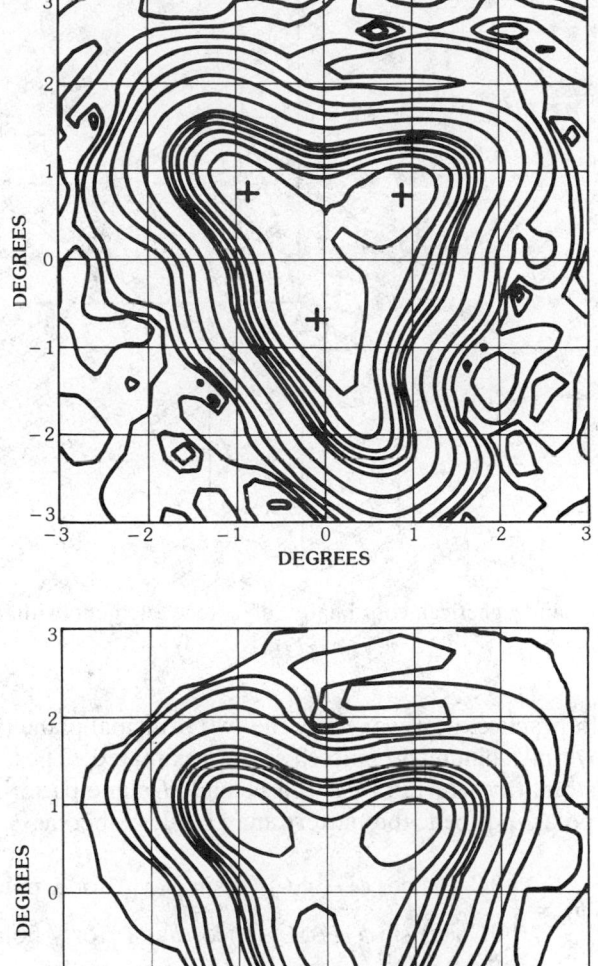

Fig. 64. Calculated and measured far-field gain contours. (*a*) Calculated gain contours from measured near-field feed data. (*b*) Measured far-field gain contours. (*After Chen and Franklin [61]*)

where A_0 is a complex constant. For a specified feed element its pattern can be measured in many $\phi = $ constant planes and then interpolated or expanded in terms of spherical harmonics to give the overall volumetric pattern. Or it may be numerically computed if the feed element possesses a mathematically tractable

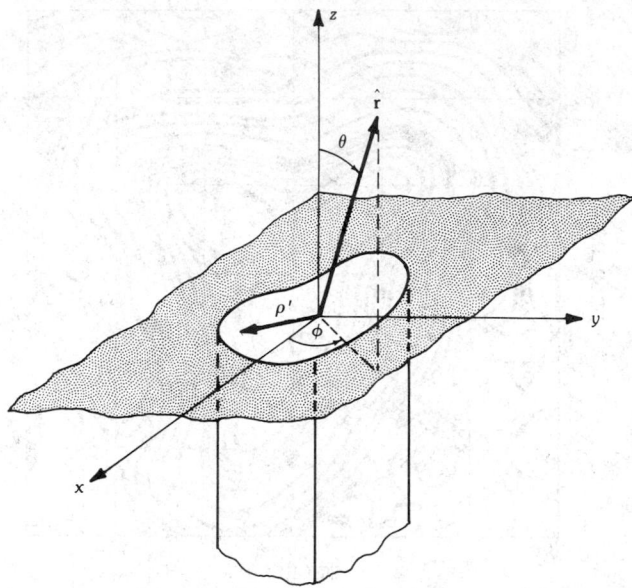

Fig. 65. Cartesian and spherical coordinates of a feed element with a planar aperture opening.

configuration. In practice, however, only the two principal plane patterns are used and then the overall volumetric pattern is approximated. These principal plane patterns are customarily referred to as the *E*- and *H*-plane patterns. For example, for an $\hat{x}$ and $\hat{y}$ polarized feed, the pattern may be described as

$$\mathbf{C}(\theta, \phi) = \begin{cases} \hat{\boldsymbol{\theta}}C_E(\theta) \cos\phi - \hat{\boldsymbol{\phi}}C_H(\theta)\sin\phi, & \text{for } \hat{x} \text{ polarized} \\ \hat{\boldsymbol{\theta}}C_H(\theta) \sin\phi + \hat{\boldsymbol{\phi}}C_E(\theta)\cos\phi, & \text{for } \hat{y} \text{ polarized} \end{cases} \quad (77)$$

For the case of a feed element with the above pattern illuminating a parabolic reflector (or equivalent parabola in dual reflectors) it is possible to construct the aperture field in a straightforward fashion, resulting in

$$\mathbf{E}_A = A_0[-\hat{x}(C_\theta \cos\phi - C_\phi \sin\phi) + \hat{y}(C_\theta \sin\phi + C_\phi \cos\phi)]\frac{e^{-j2kF}}{r} \quad (78)$$

where $r = \sec^2(\theta/2)$ and it is assumed that the z axis of the feed and reflector coordinates coincide (but lie in opposite directions) and that the aperture plane passes through the focal point with a focal length of F. The above expression for the aperture field is correct as long as it is further assumed that the incident and reflected fields have spherical and planar wavefronts, respectively. More complicated expressions result from general cases. It is worthwhile to emphasize that the reader should pay careful attention so as to clearly differentiate between feed and reflector coordinates, even if similar notations are used to describe them. For instance, in (78) $\hat{x}$ and $\hat{y}$ are referred to as the antenna coordinates, whereas in (77) they are referred to as feed coordinates. These coordinates can, in general, be related using Eulerian angles as described in the Appendix. Expression (78) allows one to clearly identify the generation of copolarized and cross-polarized fields in the aperture.

Radiation Patterns of Simple Feeds

There are different analytical approaches available to determine radiated fields of simple feeds. Among these methods the aperture E-field model has been used very successfully because of both its simplicity and the good comparison of its results with measured data. This model assumes that the tangential electric field is known in the aperture and is zero outside the aperture, which is the same as assuming that the aperture is in a ground plane. Denoting the feed's tangential electric field by $E^a(\varrho')$, one obtains the following Cartesian components for the radiated field:

$$C_x(\theta,\phi)\hat{\mathbf{x}} + C_y(\theta,\phi)\hat{\mathbf{y}} = 2\iint_{\substack{\text{feed} \\ \text{aperture}}} E^a(\varrho')\, e^{jk\hat{\mathbf{r}}\cdot\varrho'}\,dS' \tag{79}$$

where $\hat{\mathbf{r}}$ and ϱ' are shown in Fig. 65. The far-field spherical components are (using the vector potential approach)

$$\begin{bmatrix} C_\theta \\ C_\phi \end{bmatrix} = \begin{bmatrix} \cos\phi & \sin\phi \\ -\cos\theta\,\sin\phi & \cos\theta\,\cos\phi \end{bmatrix}\begin{bmatrix} C_x \\ C_y \end{bmatrix} \tag{80}$$

If Ludwig's third definition [36] of cross polarization is used, the copolar C_p and cross-polar C_q components of the radiated field are

$$\begin{bmatrix} C_p \\ C_q \end{bmatrix} = \begin{bmatrix} \sin\phi & \cos\theta \\ \cos\phi & -\sin\phi \end{bmatrix}\begin{bmatrix} C_\theta \\ C_\phi \end{bmatrix} \tag{81}$$

Note that this assumes that the feed principal polarization lies along the y axis. If the principal polarization lies along the x axis, C_p and C_q must be interchanged. The above expression may further be combined with (80) to result in

$$\begin{bmatrix} C_p \\ C_q \end{bmatrix} = \cos^2(\theta/2)\begin{bmatrix} \sin 2\phi\,\tan^2(\theta/2) & 1-\tan^2(\theta/2)\cos 2\phi \\ 1+\tan^2(\theta/2)\cos 2\phi & \sin 2\phi\,\tan^2(\theta/2) \end{bmatrix}\begin{bmatrix} C_x \\ C_y \end{bmatrix} \tag{82}$$

which is a simple form for determining the copolar and cross-polar fields in planes $\phi = 0°$, $45°$, and $90°$.

Open-Ended Rectangular Waveguides—For an open-ended rectangular waveguide propagating the TE_{10} mode (y polarized) and with sides $x = 2a$ and $y = 2b$, the aperture field is (ignoring the end effects)

$$\mathbf{E}^a(x', y') = E_0 \cos(\pi x'/2a)\hat{\mathbf{y}} \tag{83}$$

The radiated field from (79) and (80) is

$$C_\theta(\theta, \phi) = 4\pi a b E_0 \sin\phi \frac{\cos U}{(\pi/2)^2 - U^2} \frac{\sin V}{V}$$

$$C_\phi(\theta, \phi) = 4\pi a b E_0 \cos\theta \cos\phi \frac{\cos U}{(\pi/2)^2 - U^2} \frac{\sin V}{V} \tag{84}$$

where

$$U = ka \sin\theta \cos\phi$$
$$V = kb \sin\theta \sin\phi \tag{85}$$

Figs. 66 and 67 show some important characteristics of the open-ended rectangular waveguides [68].

Open-Ended Circular Waveguides—For an open-ended circular waveguide propagating the TE_{11} mode (y polarized) and with radius a, the aperture field may be approximated by

$$\mathbf{E}^a(\varrho', \phi') = E_0 \frac{1}{\varrho'} J_1\left(\frac{\chi}{a}\varrho'\right) \sin\phi \,\hat{\mathbf{\varrho}} + E_0 \frac{\partial}{\partial\varrho'}\left[J_1\left(\frac{\chi}{a}\varrho'\right)\right] \cos\phi' \,\mathbf{\phi} \tag{86}$$

where $\chi = 1.8411$. The radiated field from (79) and (80) is

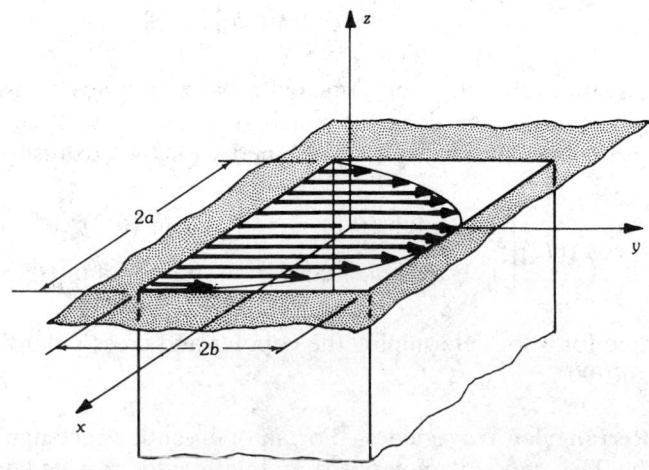

Fig. 66. A TE_{10}-mode open-ended rectangular waveguide feed with dimensions $2a$ and $2b$.

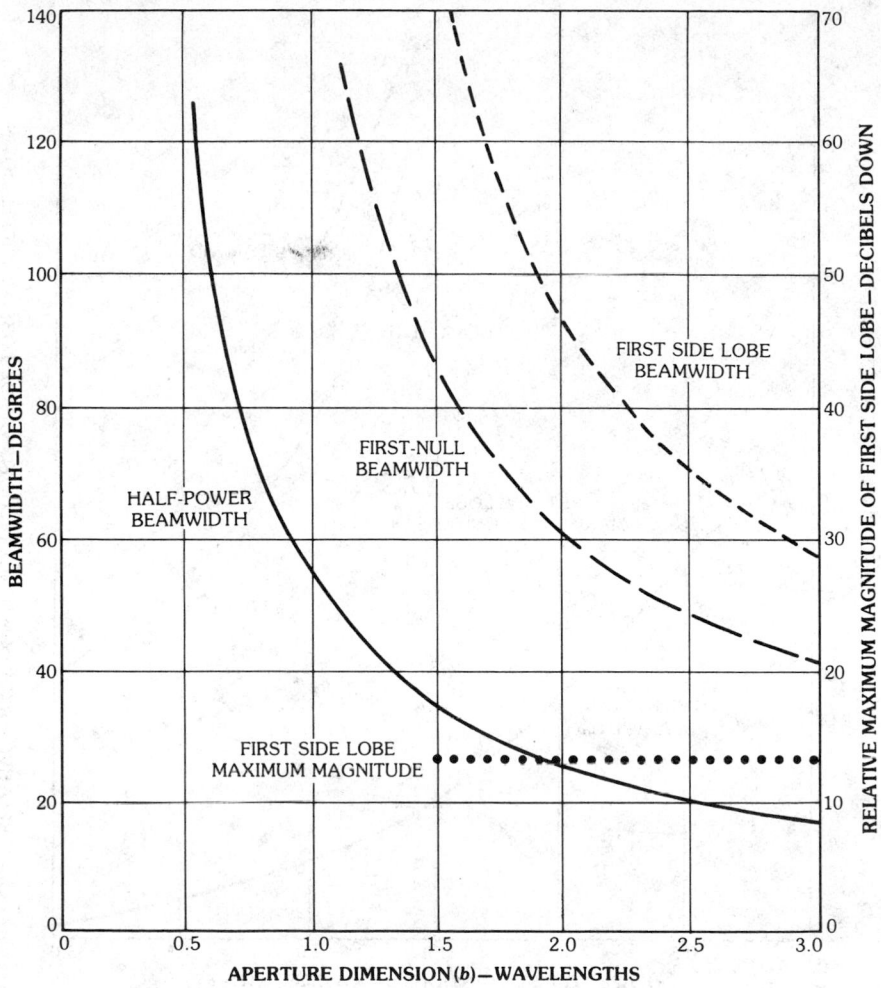

a

Fig. 67. Different beamwidths and first side lobe level for TE_{10}-mode open-ended rectangular waveguide on ground plane. (*a*) *E*-plane. (*b*) *H*-plane. (*After Balanis [68],* © *1982 Harper & Row, Publishers, Inc.; reprinted with permission of the publishers*)

$$C_\theta(\theta, \phi) = 4\pi a E_0 J_1(\chi) \sin\theta \, \frac{J_1(U)}{U}$$

$$C_\phi(\theta, \phi) = 4\pi a E_0 J_1(\chi) \cos\theta \, \cos\phi \, \frac{J_1'(U)}{1 - (U/\chi)^2} \qquad (87)$$

where

$$U = ka \sin\theta$$
$$J_1'(U) = J_0(U) - J_1(U)/U \qquad (88)$$

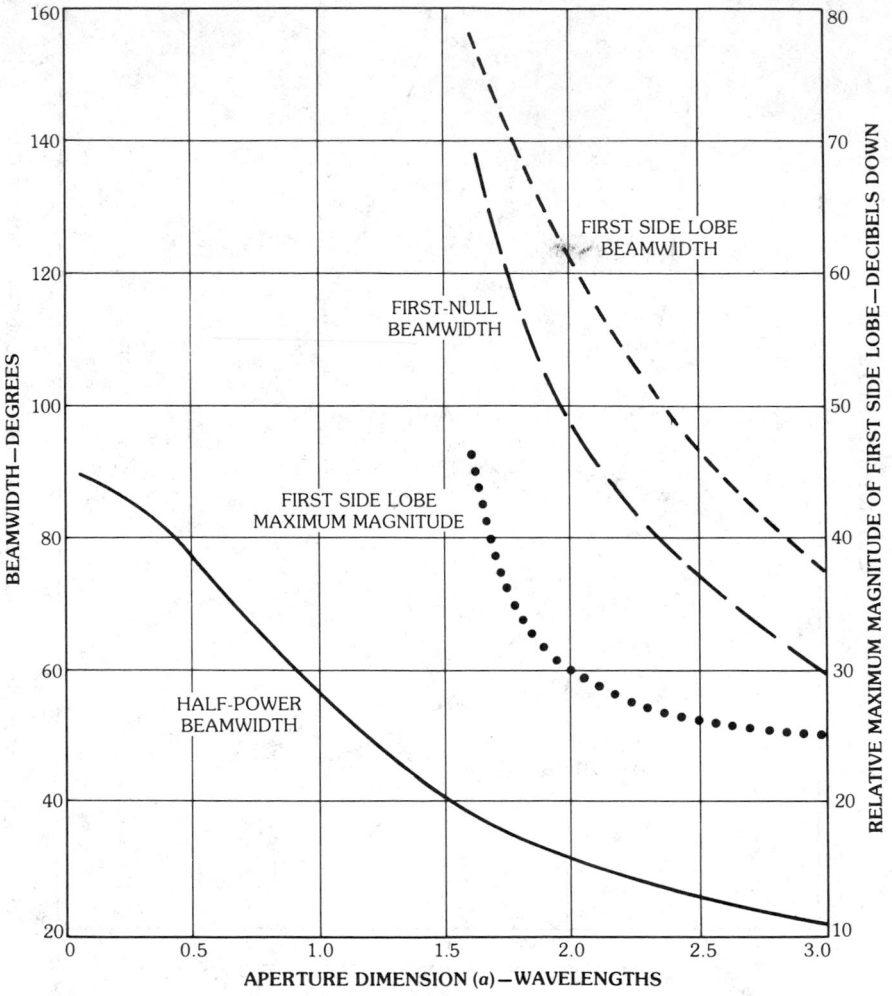

Fig. 67, *continued.*

Figs. 68 and 69 show some important characteristics of the open-ended circular waveguides [68].

Pyramidal Horns—Pyramidal horns may be classified as *E*-plane sectoral, *H*-plane sectoral, or pyramidal horns depending on to what plane its opening tapers. The type, direction, and amount of taper can have a significant effect on the overall performance of these horns. The radiation characteristics of these horns can be determined using the aperture field method or GTD construction [68, 69, 70]. The latter also provides the near-field characteristics of these horns. There are considerable amounts of design data available on these horns and the reader is referred to the just-mentioned references.

To apply the aperture field method the horn's aperture field is approximated by

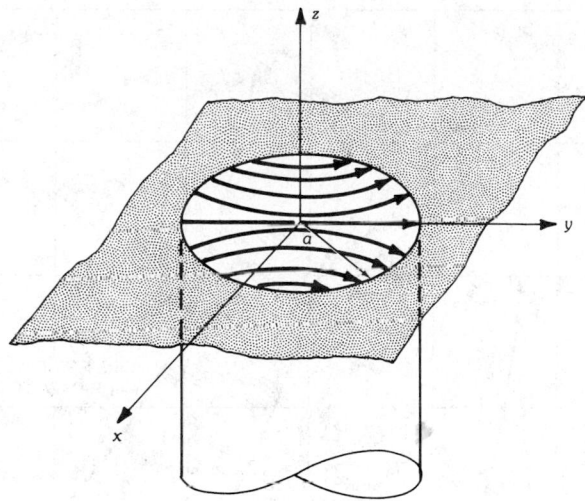

Fig. 68. A TE_{11}-mode circular waveguide feed with radius a.

$$\mathbf{E}^a(x', y') = E_0 \cos\left(\frac{\pi x'}{2a_1}\right) \exp\left\{-jk\left[\frac{(x')^2}{2L_x} + \frac{(y')^2}{2L_y}\right]\right\}\hat{\mathbf{y}} \qquad (89)$$

where L_x and L_y are the flare lengths for a pyramidal horn as shown in Fig. 70c. For E- and H-plane sectoral horns, only one of the terms in the exponent is used. The need for the quadratic phase term stems from the fact that the phase in the horn's aperture is not constant as it propagates from the throat to the opening. By substituting (89) into (79), the radiated field can finally be expressed in terms of the well-known sine and cosine Fresnel integrals [68]. Here only representative results are shown.

Figs. 71 and 72 are the plots of universal patterns for E- and H-plane sectoral horns as a function of their geometrical parameters. For pyramidal horns the geometrical parameters may be chosen to achieve the so-called optimum horn. Fig. 73 shows the relationship between the parameters to design an optimum horn for a specified gain (the overall efficiency of these horns is typically about 50 percent). Fig. 74 is the plot of the E- and H-plane patterns of such horns.

Conical Horns—The geometry of a circular horn is shown in Fig. 75. In contrast to pyramidal horns which are typically fed by rectangular waveguides, the circular horn is usually fed by circular waveguides. The aperture field of these horns can be constructed in a fashion similar to that of pyramidal horns by simply multiplying the aperture field of the circular waveguide by a quadratic phase term. The resulting integral for the computation of the radiated field can then be evaluated numerically. Fig. 75 gives the proper horn dimensions for constructing an "optimum" horn for a specified gain. These horns typically possess more symmetric E- and H-plane patterns than do their pyramidal horn counterparts. They can also be used more effectively to create circularly polarized fields.

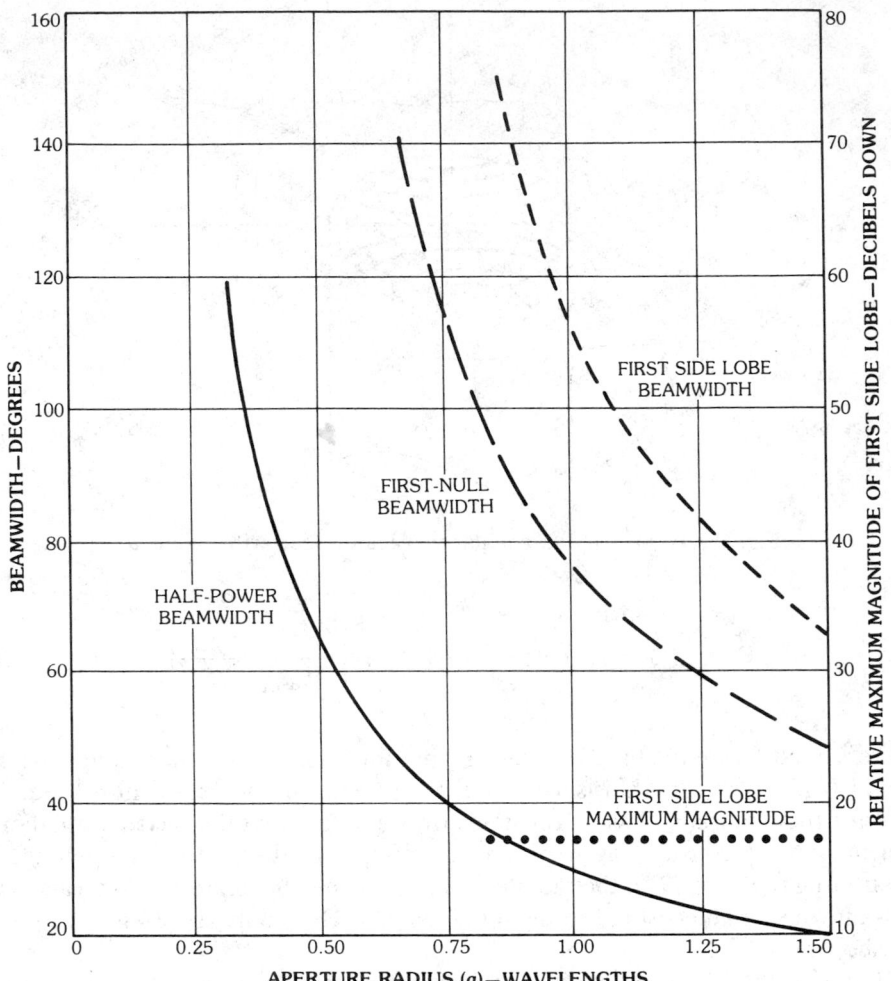

Fig. 69. Different beamwidths and first side lobe level for a TE_{11}-mode open-ended circular waveguide on ground plane. (*a*) *E*-plane. (*b*) *H*-plane. (*After Balanis [68],* © *1982 Harper & Row, Publishers, Inc.; reprinted with permission of the publishers*)

Complex Feeds

So far the feed discussion has been concentrated around the single dominant-mode TE_{10} and TE_{11} structures. These feeds in general do not possess the needed ideal pattern symmetry in the subtended angle of the reflector, nor do they have good cross-polarization characteristics. Nevertheless it is possible to achieve an acceptable symmetry down to levels of 10 to 12 dB by properly adjusting the aperture dimensions. The prime advantage of using these single-mode feeds is their compactness which makes them in particular very attractive for contour and multiple-beam feed array designs.

The requirements for an optimum compromise between illumination efficiency, spillover, and cross polarization for a wide range of *F/D* values and frequencies

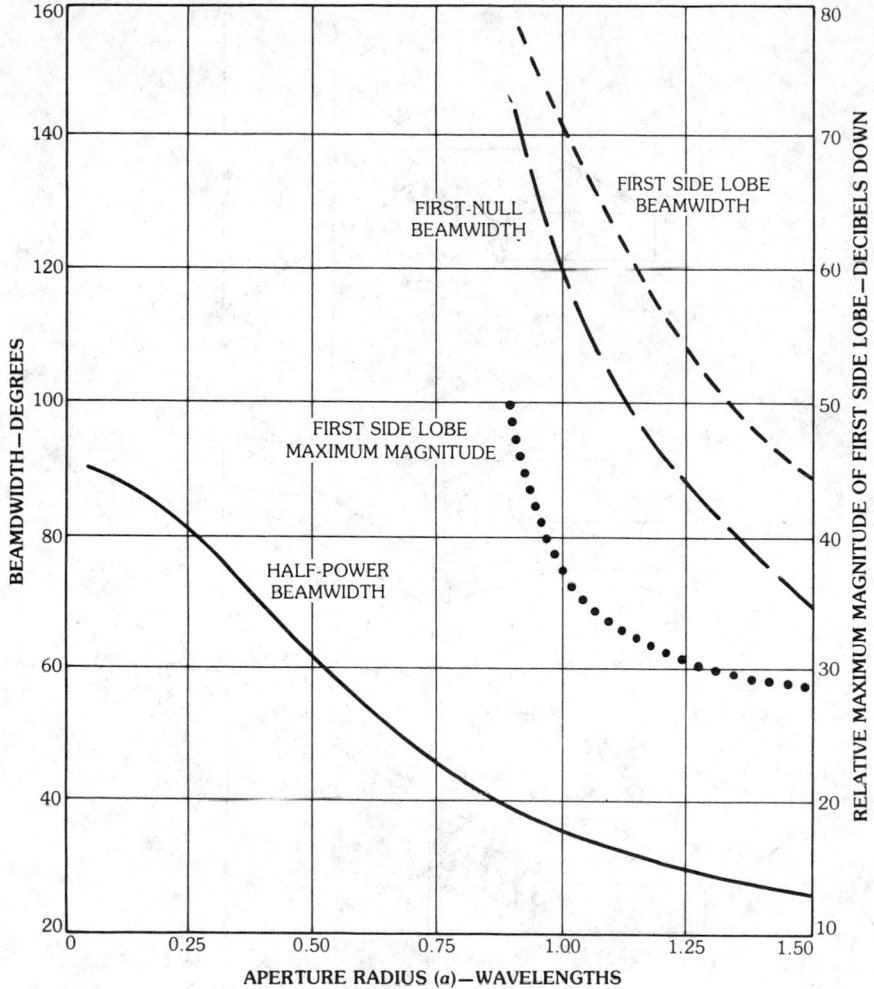

b

Fig. 69, *continued.*

have resulted in the generation of a new class of feeds. These feeds in particular are used in large reflectors for radioastronomy and tracking applications and offset reflectors. Among them one may refer to multimode, corrugated (hybrid-mode), matched, etc., feed horns [6,7].

Multimode horns were invented to equalize the pattern asymmetry of single-mode horns. For instance, in the Potter horn [71] the TM_{11} mode is generated along with the dominant TE_{11} mode of a circular horn. Although this new TM_{11} mode does not have any appreciable effect on the H-plane radiation pattern, when it is properly phased and combined with the TE_{11} mode it can effectively alter the E-plane aperture distribution, which results in a symmetric radiation pattern. All these favorable features, however, are not properly realized until the feed aperture diameter exceeds about 1.3λ. For example, Fig. 76 shows the radiation pattern of these dual-mode horns as reported in [71]. As shown in this figure a partial con-

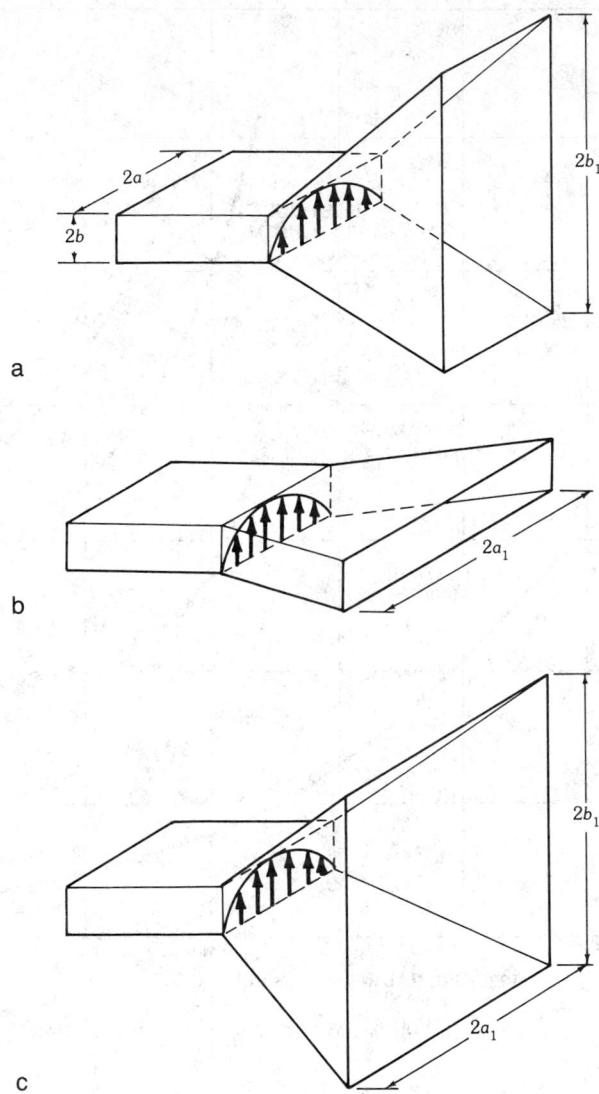

Fig. 70. Typical horn feeds. (*a*) *E*-plane sectoral horn. (*b*) *H*-plane sectoral horn. (*c*) Pyramidal horn.

version of the TE_{11} mode energy to a TM_{11} mode happens in the flared section of the horn while the straight section of length ℓ enforces the condition that both modes have the proper phase relationship at the aperture which can be maintained over a bandwidth of less than 10 percent. In general these feeds are very well suited for large *F/D* reflectors. There are also available other types of multimode horns, which result from a combination of modes such as TE_{10}, TE_{12}, and TM_{12} in a square-aperture pyramidal horn.

Corrugated horns are capable of creating similar boundary conditions at all polarizations which result in similar tapers in the aperture field distribution in all

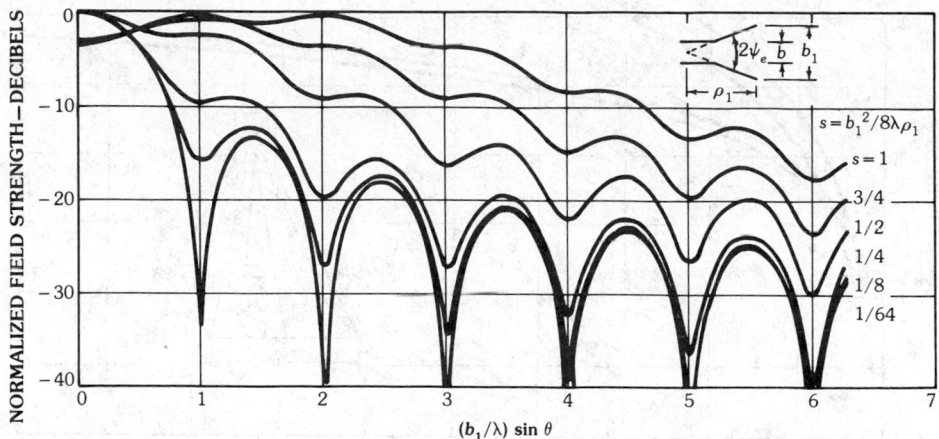

Fig. 71. *E*-plane universal patterns for *E*-plane sectoral and pyramidal horns. (*After Balanis [68], © 1982 Harper & Row, Publishers, Inc.; reprinted with permission of the publishers*)

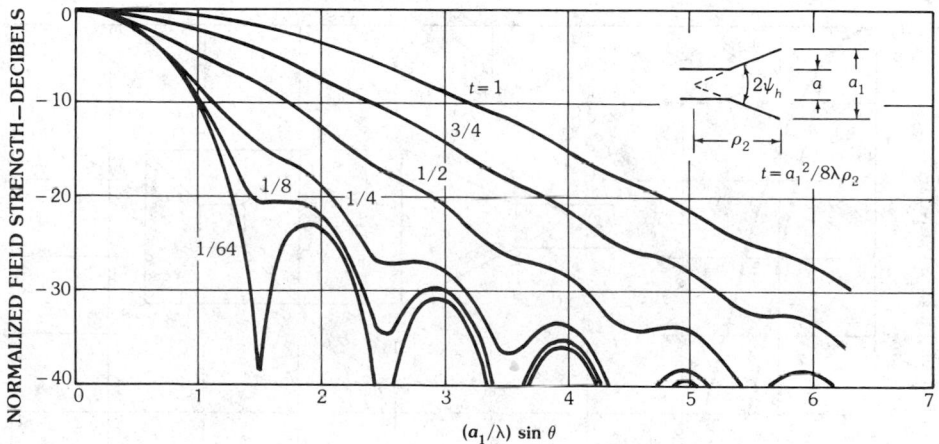

Fig. 72. *H*-plane universal patterns for *H*-plane sectoral and pyramidal horns. (*After Balanis [68], © 1982 Harper & Row, Publishers, Inc.; reprinted with permission of the publishers*)

planes. Due to these boundary conditions symmetric radiation patterns can be obtained at levels as low as −25 dB in both the *E*- and *H*-planes. A corrugated horn can be realized by grooving the *E*-plane of a pyramidal horn or the entire wall of a circular horn with, typically, ten or more slots (corrugations) per wavelength. For circular corrugated horns Fig. 77 shows the plots of the pattern widths at different levels as a function of the opening angle. The appearance of the corrugations, especially near the waveguide-horn junction, affects the vswr of the horn. The usual practice is to begin the corrugations at a small distance from the junction. These horns are also classified as hybrid-mode horns because they support modes in

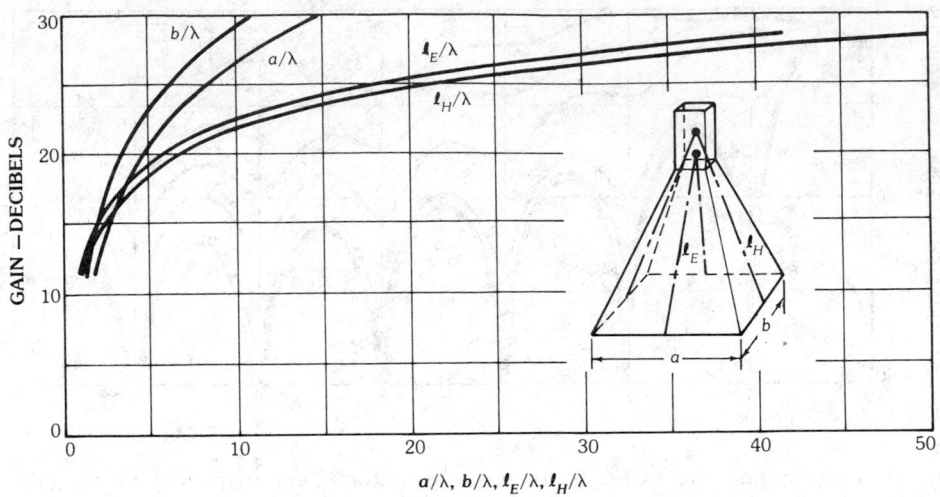

Fig. 73. Gain characteristics of pyramidal horns of optimum design.

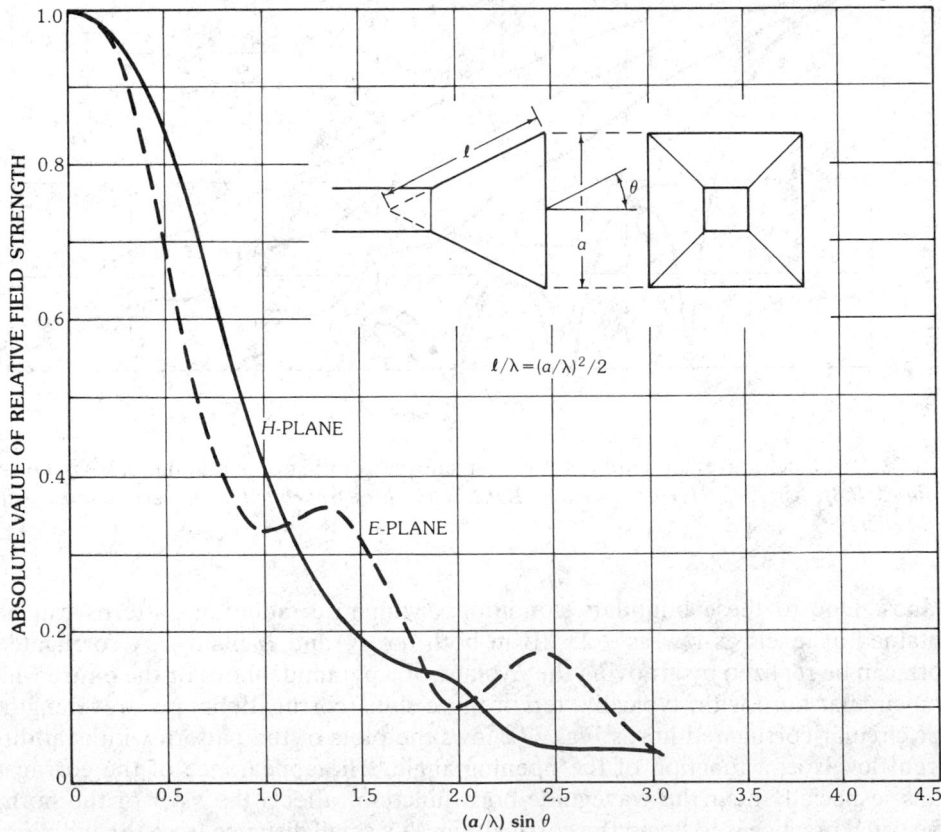

Fig. 74. Universal patterns of pyramidal horns of optimum design.

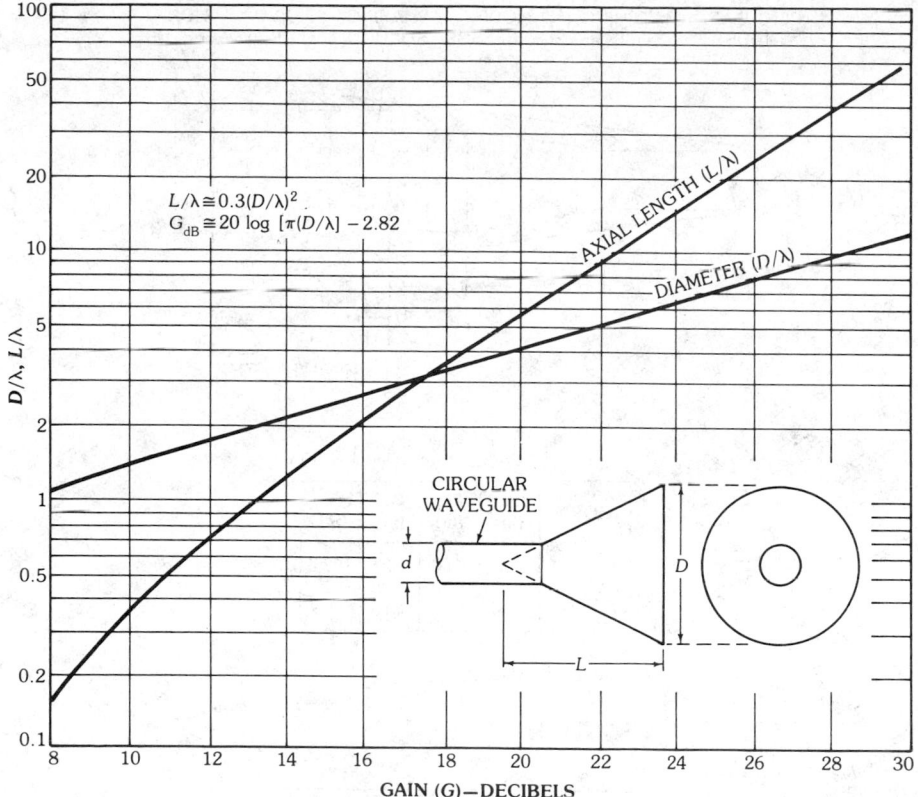

Fig. 75. Gain characteristics of conical horns of optimum design.

which both longitudinal **E**- and **H**-field components are present. In a circular corrugated horn a natural mixture of TE_{11} and TM_{11} results in the generation of a hybrid-mode HE_{11}. In contrast to dual-mode horns there is no need for a mode converter, and therefore the hybrid-mode horns have typically wider bandwidths than the dual-mode horns. In particular, one version of these hybrid horns (scalar horns), which has a large flare angle with a relatively short horn, has radiation characteristics with little dependence on frequency.

There are other classes of feeds which are properly tailored to specifically overcome some undesirable characteristics of reflectors. For example, matched feeds [72, 73] are used to significantly reduce the generation of unwanted cross-polarized fields in an offset parabolic reflector illuminated with a tilted feed from its axis. The basic concept behind these feeds is to match the feed aperture field distribution with the receiving focal plane distribution of the reflector as closely as possible.

cos^q (θ) Type Patterns

In the previous sections, independently of the actual feed elements, the feed patterns were approximated by $cos^q(\theta)$ type patterns to allow for detailed parametric studies of reflector characteristics. Some simple expressions are

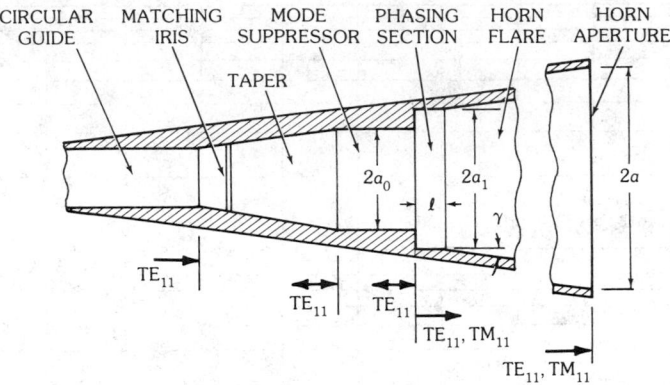

CIRCULAR GUIDE · MATCHING IRIS · MODE SUPPRESSOR · PHASING SECTION · HORN FLARE · HORN APERTURE · TAPER

$2a_0$ $2a_1$ ℓ γ $2a$

TE_{11} TE_{11} TE_{11} TE_{11}, TM_{11} TE_{11}, TM_{11}

a

HORN	$2a$ (in)	$2a_1$ (in)	$2a_0$ (in)	ℓ (in)
1	5.74	1.60	1.25	0.25
2	5.27	1.60	1.25	0.40
3	4.80	1.60	1.25	0.54

$\gamma = 6.25°$
Frequency = 9600 MHz

b

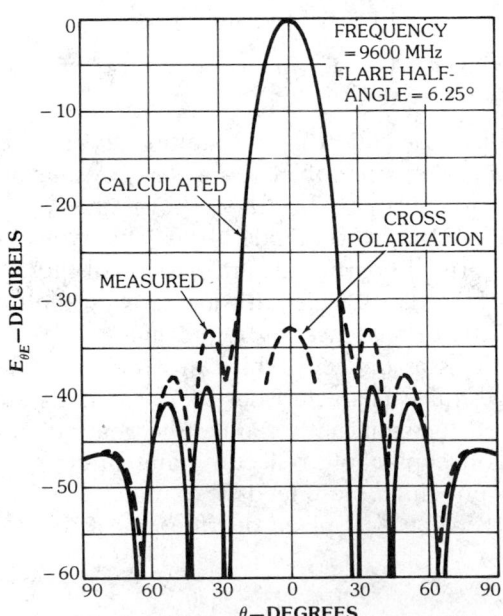

FREQUENCY = 9600 MHz
FLARE HALF-ANGLE = 6.25°

CALCULATED

CROSS POLARIZATION

MEASURED

$E_{\theta E}$ —DECIBELS

θ —DEGREES

c

Fig. 76. Dual-mode conical horn of Potter [71]. (*a*) Horn structure. (*b*) Experimental data. (*c*) *E*-plane. (*d*) 45° plane. (*e*) *H*-plane. (*Reprinted with permission of* Microwave Journal, *from June 1963 issue,* © *1963 Horizon House–Microwave, Inc.*)

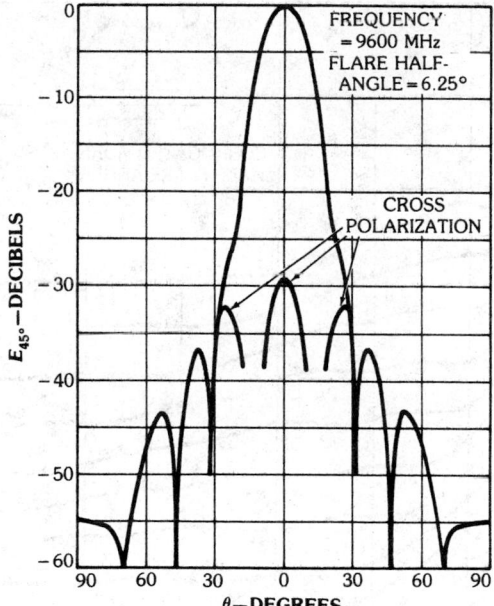

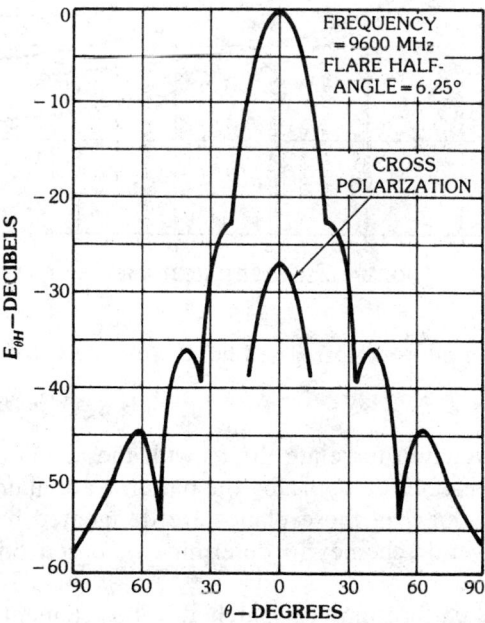

Fig. 76, *continued.*

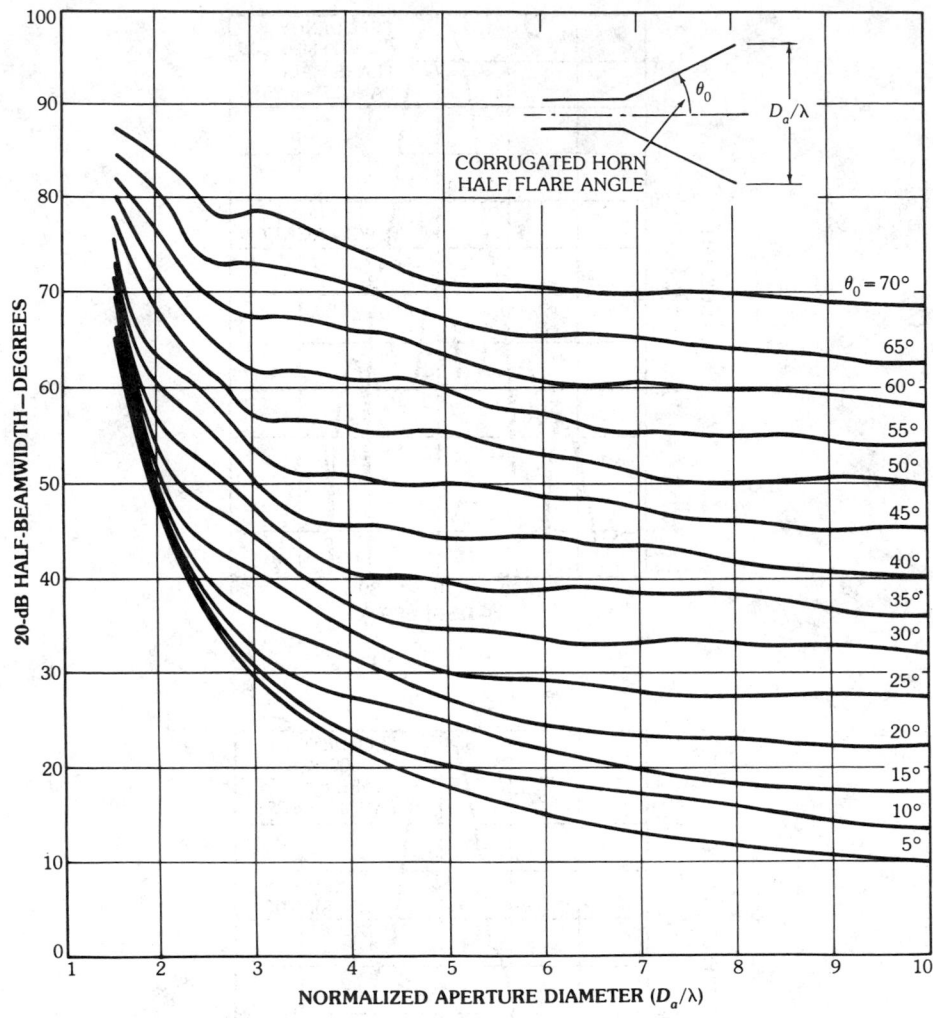

Fig. 77. Half-beamwidths of corrugated horns. (*a*) 20-dB. (*b*) 10-dB. (*c*) 3-dB.

presented in this subsection to relate the *q*s with the actual patterns of the most commonly used feed elements. Typically the patterns are matched in the principal planes and the values of *q* in these planes are designated by q_1 and q_2, respectively. There are several schemes to determine *q*, based on different matching requirements.

The $\cos^q(\theta)$ type pattern may be matched to the element pattern in the principal planes at both the peak and at a specified off-axis angle, which may coincide with the subtended angle of the reflector in these planes. In this case one obtains the same feed taper as the actual feed element at the reflector edge. Denoting the principal plane patterns by $C_E(\theta)$ and $C_H(\theta)$, and the subtended angles by θ_1 and θ_2, one obtains

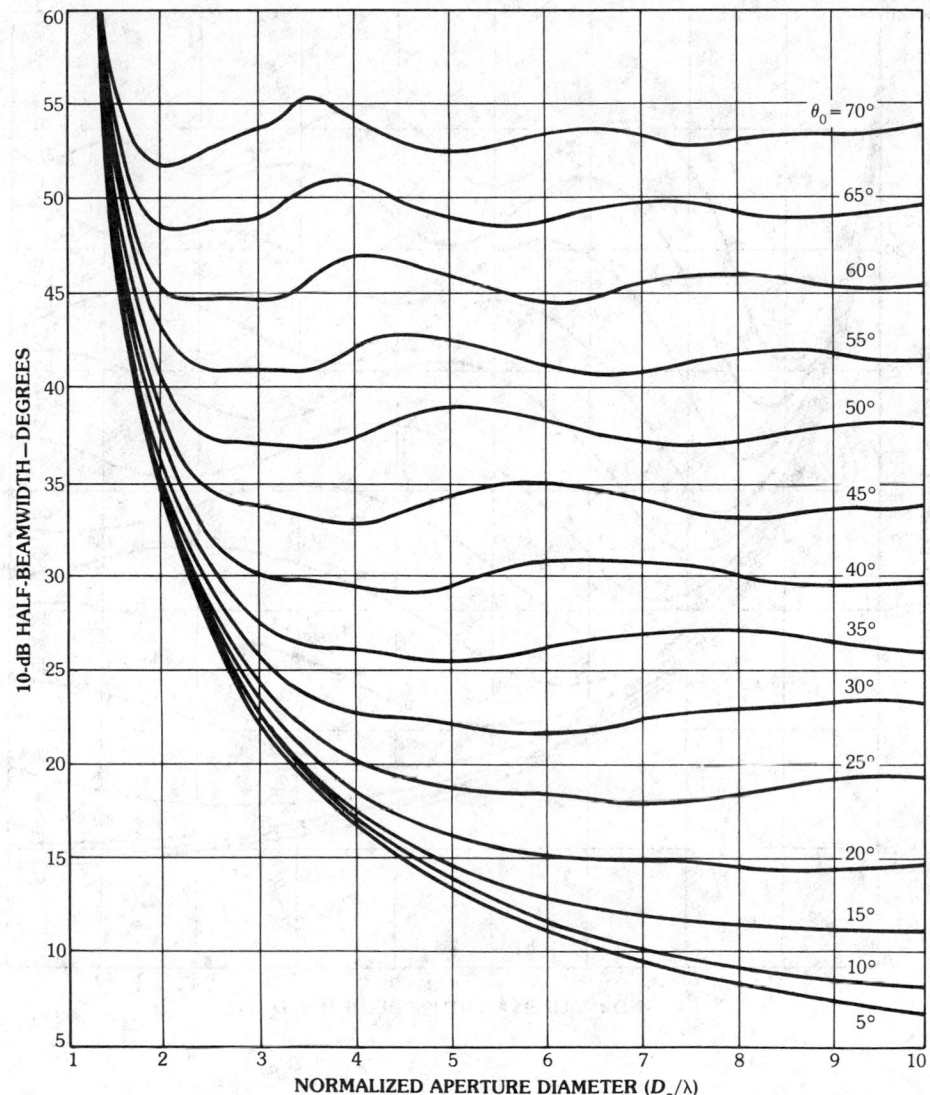

b

Fig. 77, *continued.*

$$q_{1,2} = \frac{\log C_{E,H}(\theta_{1,2})}{\log \cos \theta_{1,2}} \tag{90}$$

Patterns C_E and C_H may be given in either a closed form or in terms of measured data.

The match in (90) can be imposed at the 3-dB beamwidth level in the E- and H-planes. In this case, q is determined as

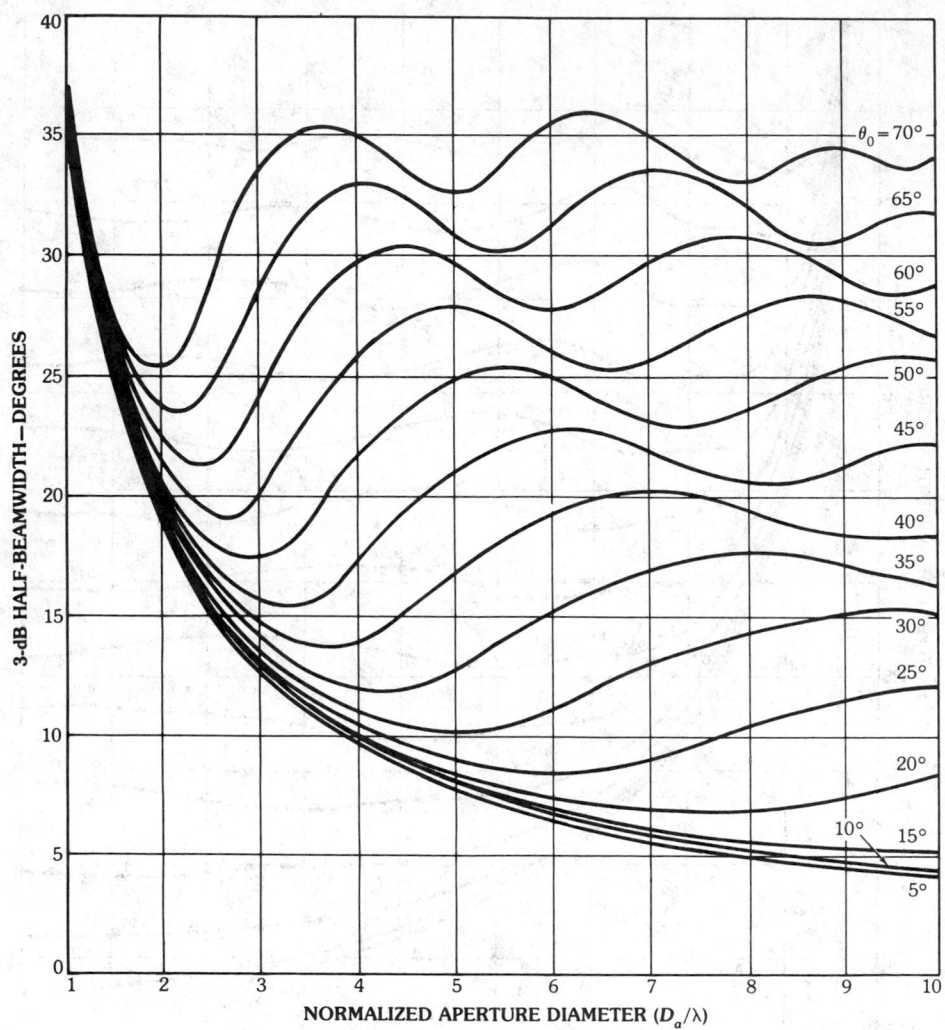

C

Fig. 77, *continued.*

$$q = \frac{-0.15}{\log \cos \theta_{1/2}} \tag{91}$$

where $\theta_{1/2}$ is one-half of the half-power beamwidths in each of the principal planes. For instance, for an open-ended rectangular waveguide feed with sides $2a$ (*H*-plane) and $2b$ (*E*-plane) the following is obtained:

$$2\theta_{1/2} = \begin{cases} \dfrac{50.6°}{2b/\lambda}, & \text{for } E\text{-plane} \\[2ex] \dfrac{68.8°}{2a/\lambda}, & \text{for } H\text{-plane} \end{cases} \tag{92}$$

Similarly, the following expressions are derived for the beamwidths of an open-ended circular waveguide feed with radius a:

$$2\theta_{1/2} = \begin{cases} \dfrac{29.2°}{a/\lambda} \\ \dfrac{37.0°}{a/\lambda} \end{cases} \tag{93}$$

Finally, values of q may be determined from directivity considerations. In this case the condition is imposed that the directivity of the feed element be the same as the directivity of the feed with $\cos^q(\theta)$ patterns. For the latter it has been shown that

$$D = \frac{2(2q_1 + 1)(2q_2 + 1)}{q_1 + q_2 + 1} \tag{94}$$

Furthermore, if it is assumed that the E- and H-plane patterns are the same, the result is

$$D = 2(2q + 1) \tag{95}$$

From (95), q may be obtained by making D equal to the directivity of the actual feed element. For example, for an open-ended rectangular waveguide the following is obtained:

$$D = \frac{32}{\pi} \frac{(2a)(2b)}{\lambda^2} \tag{96}$$

Similarly, for an open-ended circular waveguide the following expression is obtained:

$$D = 10.5\pi(a/\lambda)^2 \tag{97}$$

In some cases the value of q is approximated as an average value of all the qs obtained using the three previously mentioned methods.

9. Effects of Random Surface Errors

Due to imperfections of the manufacturing process, reflectors can only be constructed within tolerance ranges, which results in reflector surfaces that are randomly distorted from the ideal paraboloidal configuration. Random distortions cause radiation patterns to deteriorate by reducing the reflector aperture efficiency and increasing its side lobe envelope. A successful model has been suggested in [30, 74, 75] for the determination of these distortion effects. If it is assumed that the reflector's aperture phase error at a given point in the aperture has a zero mean with a Gaussian distribution and correlation radius c, it can then be shown that the distorted pattern is [74]

$$G(\theta,\phi) = G_0(\theta,\phi)e^{-\sigma^2} + \left(\frac{\pi D}{\lambda}\right)^2 e^{-\sigma^2}\left(\frac{2c}{D}\right)^2 f(\sigma, c\sin\theta) \tag{98}$$

where the "diffuse term" f is

$$f(\sigma, c\sin\theta) = \sum_{n=1}^{\infty} \frac{(\sigma^2)^n}{n!n} e^{-[(\pi c\sin\theta)/\lambda]^2/n} \tag{99}$$

In the above equations G_0 is the nondistorted pattern, D is the diameter of the circular aperture, and

$$\sigma \cong 2\varkappa(2\pi/\lambda)\epsilon_{rms} \tag{100}$$

where ϵ_{rms} is the rms surface error, the factor of 2 accounts for the two-way path incurred by the reflected ray, $\varkappa \lesssim 1$ depends on the reflector geometry (F/D ratio), and $\varkappa = 1$ for shallow reflectors. As a rule of thumb the rms surface error is approximately one-fourth to one-third that of the surface peak error.

From (98) the peak gain reduction at $\theta = 0°$ becomes

$$G(0,0) = \eta\left(\frac{\pi D}{\lambda}\right)^2 e^{-\sigma^2}\left[1 + \frac{1}{\eta}\left(\frac{2c}{D}\right)^2 \sum_{n=1}^{\infty} \frac{(\sigma^2)^n}{n!n}\right] \tag{101}$$

where η is the aperture efficiency. For small correlation intervals one finally obtains the following well-known result:

$$G(0,0) \cong \eta\left(\frac{\pi D}{\lambda}\right)^2 e^{-(4\pi\varkappa\epsilon_{rms}/\lambda)^2} \tag{102}$$

where values of $\varkappa$ may be estimated from [74]

$$\varkappa = \frac{4F}{D}\sqrt{\ln[1 + 1/(4F/D)^2]} \tag{103}$$

or can more accurately be determined [30] from Fig. 78.

From (102) and with the assumption of constant aperture efficiency, the peak gain first increases with the square of the frequency until the tolerance effect becomes significant. The shortest wavelength at which the gain is maximized can then be obtained by differentiating (102), which results in

$$\lambda_{shortest} = 4\pi\varkappa\epsilon_{rms} \tag{104}$$

At this wavelength the reflector gain is 4.3 dB (equal to $10 \log e$) below the error-free gain. Fig. 79 shows the antenna gain as a function of frequency for different surface roughness values [76]. In this figure $\varkappa = 1$ is used. It is worthwhile to observe that at $\lambda_{shortest}$ ($\epsilon_{rms}/\lambda \cong 1/16$), the gain is maximized; nevertheless the higher-order side lobes can be unacceptably distorted, as will be shown shortly.

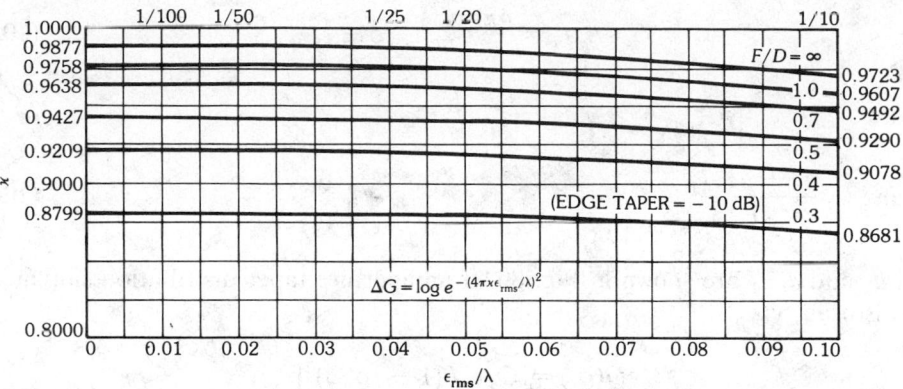

Fig. 78. The correction factor $\varkappa$ as a function of ϵ_{rms}/λ for different F/D values. (*After Rahmat-Samii [30], © 1983 IEEE*)

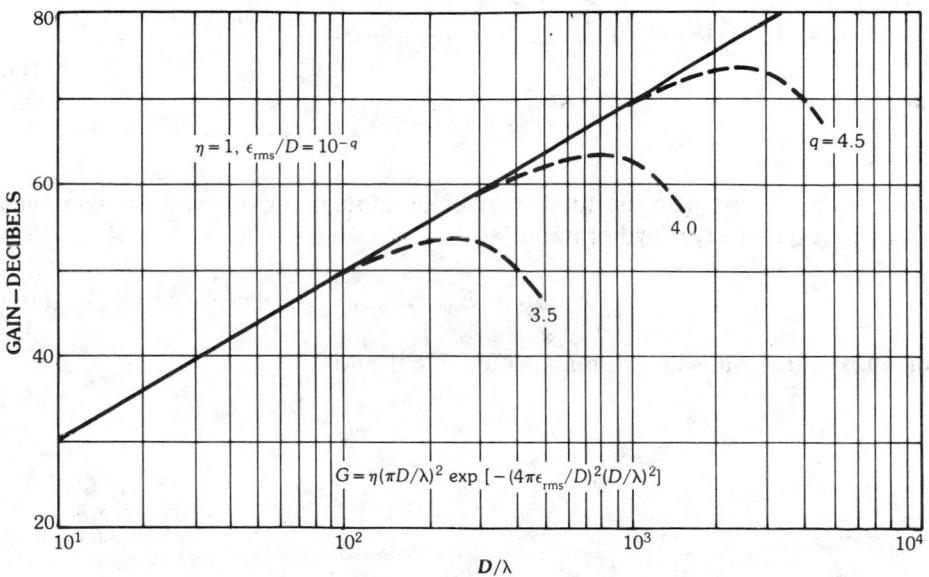

Fig. 79. The effect of reflector roughness on antenna gain. (*After Love [76], © 1976 American Geophysical Union*)

Although (98) determines the effects of surface tolerances on the reflector pattern it does not properly account for the tapered aperture distributions and nonuniform rms distortions. A simplified model has been developed in [30], which allows for the incorporation of the just-mentioned factors. The model assumes that in each prescribed annular region of the antenna the geometrical rms surface value is known and is designated by $\epsilon_{(rms)n}$ in the nth annular region. The rms aperture phase error can then be defined as

$$\sigma_n = \frac{4\pi}{\lambda}\,\epsilon_{(\text{rms})n}\cos\xi \tag{105}$$

where

$$\tan\xi = \frac{a_n + a_{n-1}}{4F} \tag{106}$$

and a_n and a_{n-1} are shown in Fig. 80. If an aperture taper distribution similar to (16) is used, viz.,

$$Q(\varrho') = C + B[1 - (\varrho'/a)^2]^P$$
$$B + C = 1 \tag{107}$$

one can then arrive at the following expression for the average power pattern [30]:

$$G(\theta,\phi) = \sum_{n=1}^{N}\sum_{m=1}^{N} E_{n,n-1}\,E^*_{m,m-1}\,e^{-0.5(\sigma_n^2 + \sigma_m^2)}$$
$$+ \sum_{n=1}^{N} E_{n,n-1}\,E^*_{n,n-1}\left(1 - e^{-\sigma_n^2}\right) \tag{108}$$

where N and $*$ designate the total number of annular regions and the conjugate operator, respectively. Furthermore,

$$E_{n,n-1} = E_n - E_{n-1} \tag{109}$$

which takes the following expressions for $P = 1$ and 2.

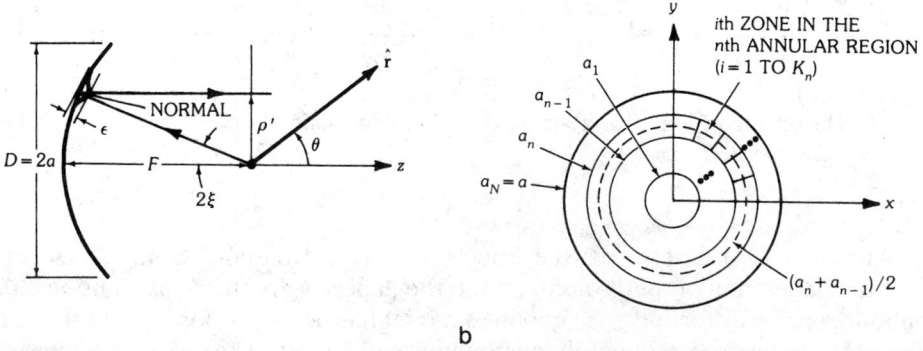

a   b

Fig. 80. Treatment of random surface errors. (*a*) Parabolic reflector with random surface errors. (*b*) Reflector aperture divided into N annular regions, which are further divided into K_n zones in the nth region ($n = 1$ to N).

$P = 1$:

$$E_n = \pi a_n^2 \left[C \frac{2}{u_n} J_1(u_n) + B \left\{ \frac{2}{u_n} J_1(u_n) - \frac{a_n^2}{a^2} \left[\frac{2}{u_n} J_1(u_n) \right. \right. \right.$$

$$\left. \left. \left. - \left(\frac{2}{u_n} \right)^2 J_2(u_n) \right] \right\} \right] \tag{110}$$

$$E_n(0°) = \pi a_n^2 \left[C + B \left(1 - \frac{a_n^2}{2a^2} \right) \right]$$

$P = 2$:

$$E_n = \pi a_n^2 \left[C \frac{2}{u_n} J_1(u_n) + B \left\{ \frac{2}{u_n} J_1(u_n) - \frac{2a_n^2}{a^2} \left[\frac{2}{u_n} J_1(u_n) - \frac{4}{u_n^2} J_2(u_n) \right] \right. \right.$$

$$\left. \left. + \frac{a_n^4}{a^4} \left[\frac{2}{u_n} J_1(u_n) - 2 \left(\frac{2}{u_n} \right)^2 J_2(u_n) + 2 \left(\frac{2}{u_n} \right)^3 J_3(u_n) \right] \right\} \right] \tag{111}$$

$$E_n(0°) = \pi a_n^2 \left[C + B \left(1 - \frac{a_n^2}{a^2} + \frac{a_n^4}{3a^4} \right) \right]$$

In the previous equations,

$$u_n = ka_n \sin \theta$$

For the special case of the uniform aperture phase distortion

$$\sigma_n = \sigma_m = \sigma$$

(109) can be further simplified as

$$G(\theta, \phi) = e^{-\sigma^2} G_0(\theta) + (1 - e^{-\sigma^2}) \sum_{n=1}^{N} E_{n,n-1} E_{n,n-1}^* \tag{112}$$

where the undistorted gain function is proportional to

$$G_0(\theta) = \pi^2 a^4 \left[C \frac{2}{u} J_1(u) + B \frac{2^{P+1} P!}{u^{P+1}} J_{P+1}(u) \right]^2 \tag{113}$$

in which $u = ka \sin \theta$. Typically, for values of $N > 30$, insignificant changes are observed in the numerical results of (108).

Numerical results based on (108) for different values of edge tapers and $\epsilon_{\text{rms}}/\lambda$ are shown in Figs. 81 and 82. From these results one can estimate how the average power pattern becomes degraded. In particular, the variation of the peak and first three side lobes for the ET $= -10$-dB case are shown in Fig. 83. For these values of $P = 2$ and $F/D = 0.7$ the first, second, and third side lobe levels for the undistorted case are at -27.06, -30.83, and -33.85 dB, respectively.

The preceding results were constructed for a parabolic reflector. Similar results can also be derived for dual reflectors [77]. For example, the peak gain of a dual

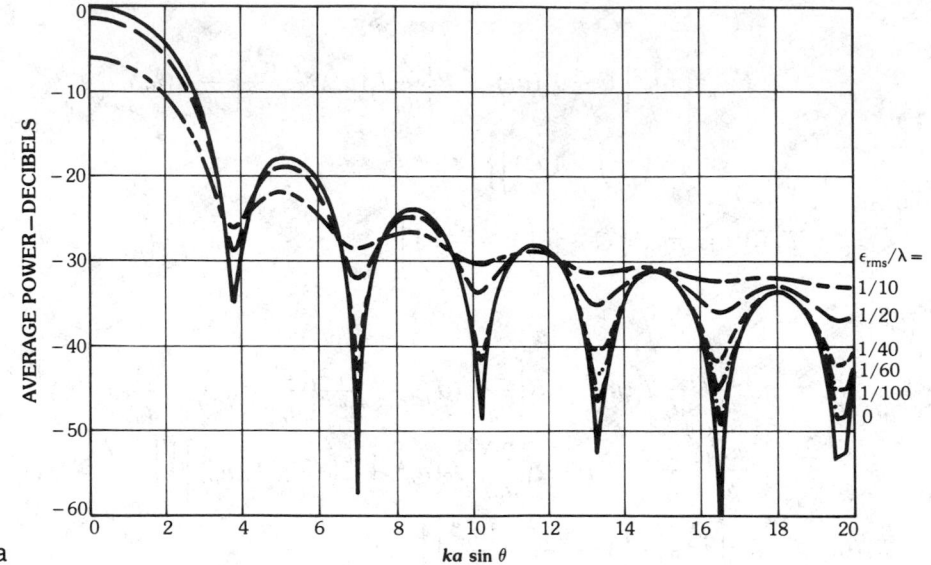

a

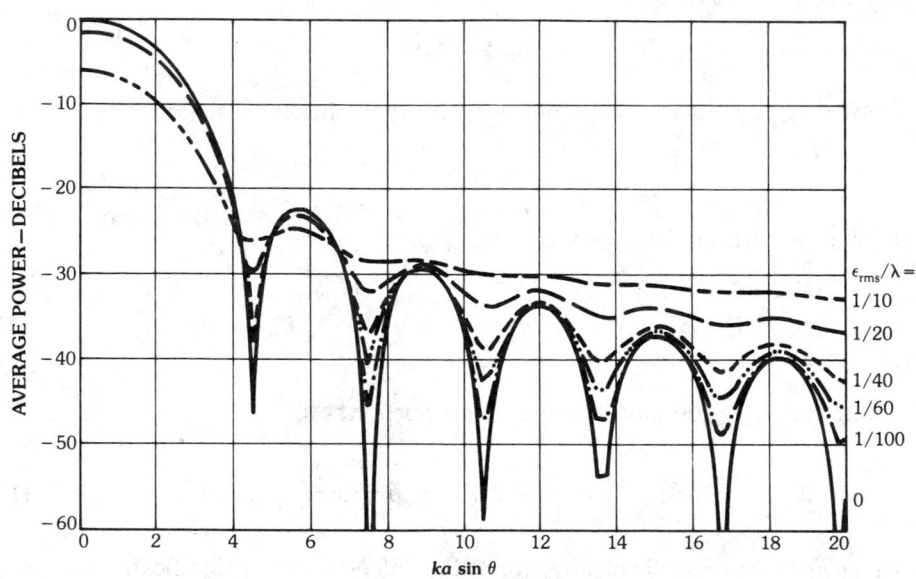

b

Fig. 81. Far-field patterns for various surface errors, with $P = 1$. (*a*) For ET = 0 dB and F/D = 0.7. (*b*) For ET = −10 dB and F/D = 0.7. (*c*) For ET = 0 dB and F/D = 0.3. (*d*) For ET = −10 dB and F/D = 0.3. (*After Rahmat-Samii [30], © 1983 IEEE*)

reflector may be estimated from

$$G(0,0) \cong \eta \left(\frac{\pi D}{\lambda} \right)^2 e^{-(4\pi\varkappa\epsilon_{\mathrm{rms}}/\lambda)^2} e^{-(4\pi\varkappa'\epsilon'_{\mathrm{rms}}/\lambda)^2} \tag{114}$$

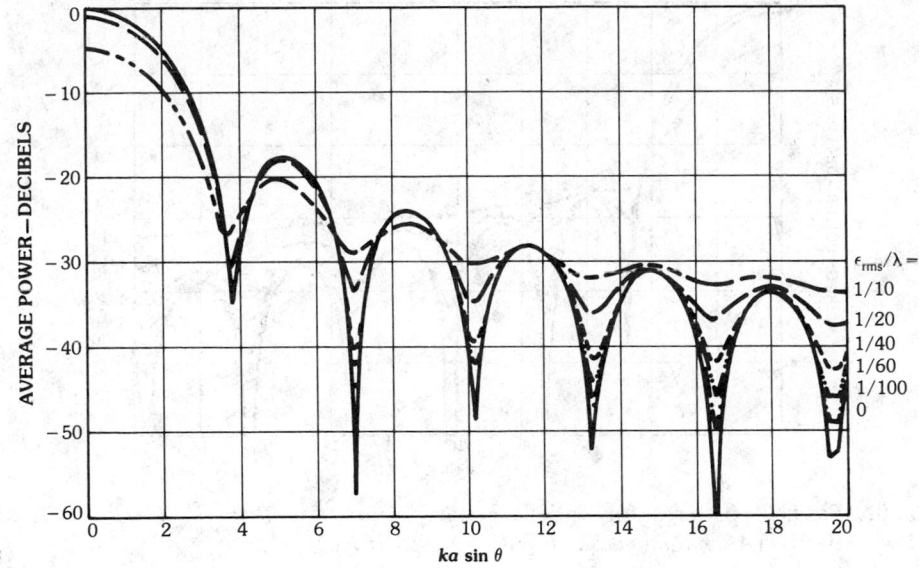

c

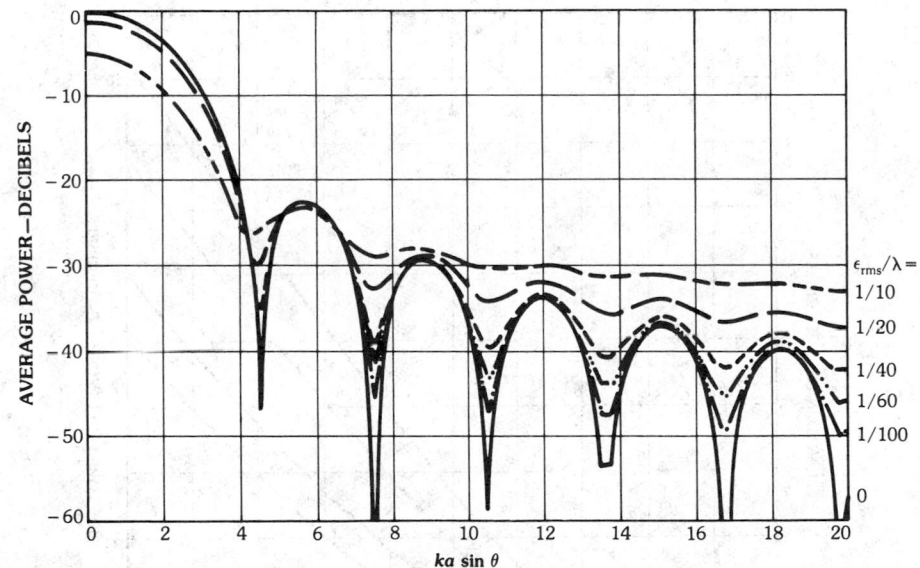

d

Fig. 81, *continued.*

where η is the aperture efficiency, D is the main parabolic reflector diameter, ϵ_{rms} and ϵ'_{rms} are, respectively, the main and subreflector surface rms, and $\varkappa$ and $\varkappa'$ are the correction factors for the main and subreflector, respectively, as shown in Figs. 78 and 84. The results of Fig. 84 are for uniform aperture illumination and do not vary substantially for other tapers. In this figure $\varkappa'^2$ is shown in terms of the paraboloid F/D ratio and the subreflector's eccentricity.

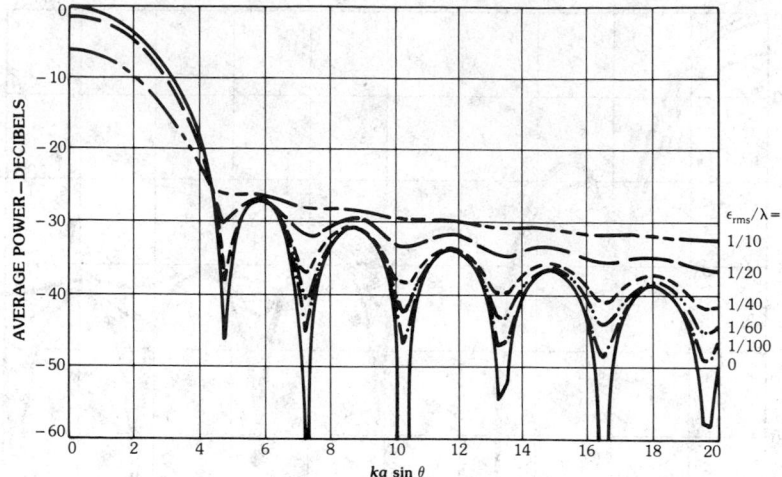

Fig. 82. Far-field patterns for various surface errors, for $P = 2$, $F/D = 0.7$, and ET $= -10$ dB. (*After Rahmat-Samii [30], © 1983 IEEE*)

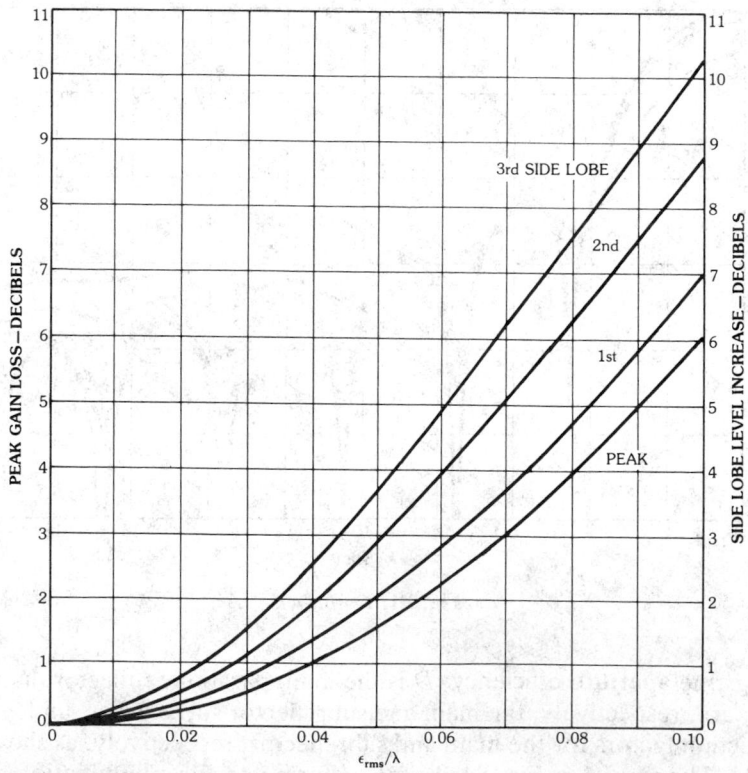

Fig. 83. Peak gain loss and side lobe level increase as a function of surface errors, with $P = 2$, $F/D = 0.7$, and ET $= -10$ dB. (*After Rahmat-Samii [30], © 1983 IEEE*)

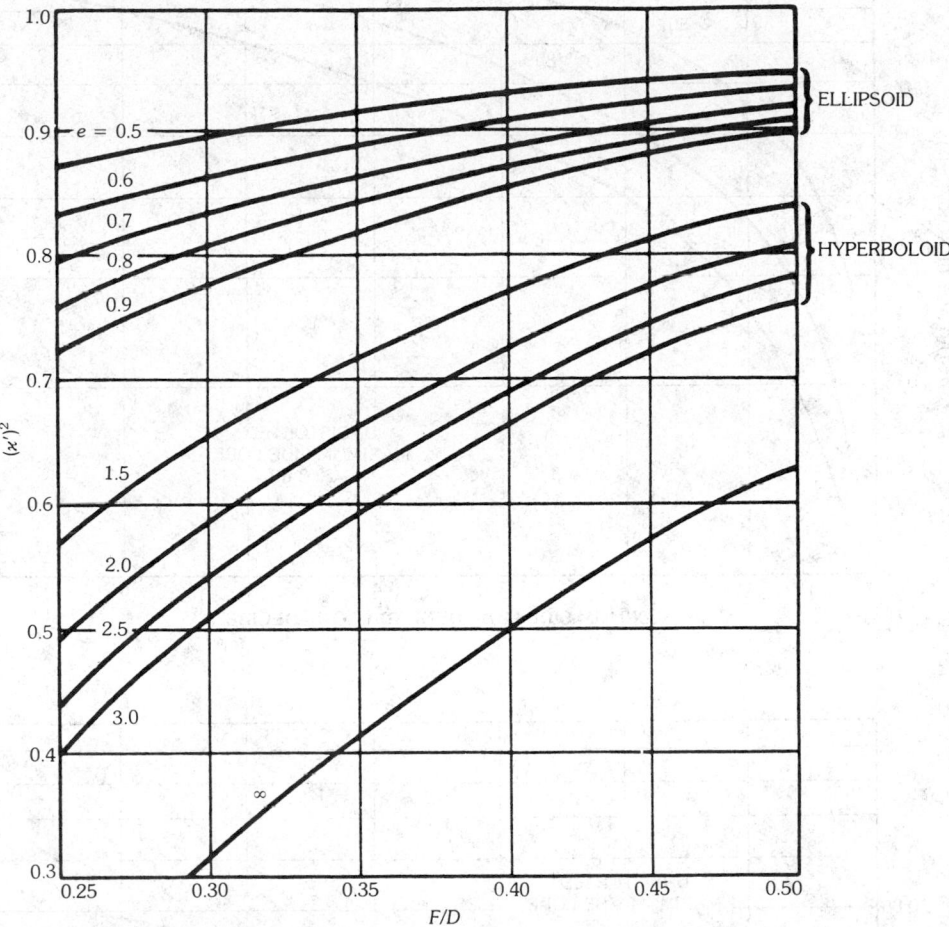

Fig. 84. The correction factor $(\varkappa')^2$ for subreflectors. (*After Rusch and Wolleben [77],* © *1982 IEEE*)

So far results have only been shown for average power patterns. However, it is more meaningful to determine the probabilistic relationship between random reflector surface errors and side lobe levels [78]. For example, Figs. 85a and 85b provide the surface tolerance versus degradation of maximum side lobe level for different values of the probability of occurrence of the specified degradation. Clearly, as the required side lobe levels become lower, more stringent surface rms values would be needed for a desired probability. These kind of data should be of interest to system engineers in assessing the reliability of a particular design.

Additionally, one should incorporate the effects of other degradations, such as systematic surface distortions, mesh surface effects, strut blockages, etc., in the evaluation of the performance of reflectors. These evaluations arc, in particular, important for the currently designed high-performance satellite and ground reflector antennas. The reader may refer to [79, 80] for many representative test cases.

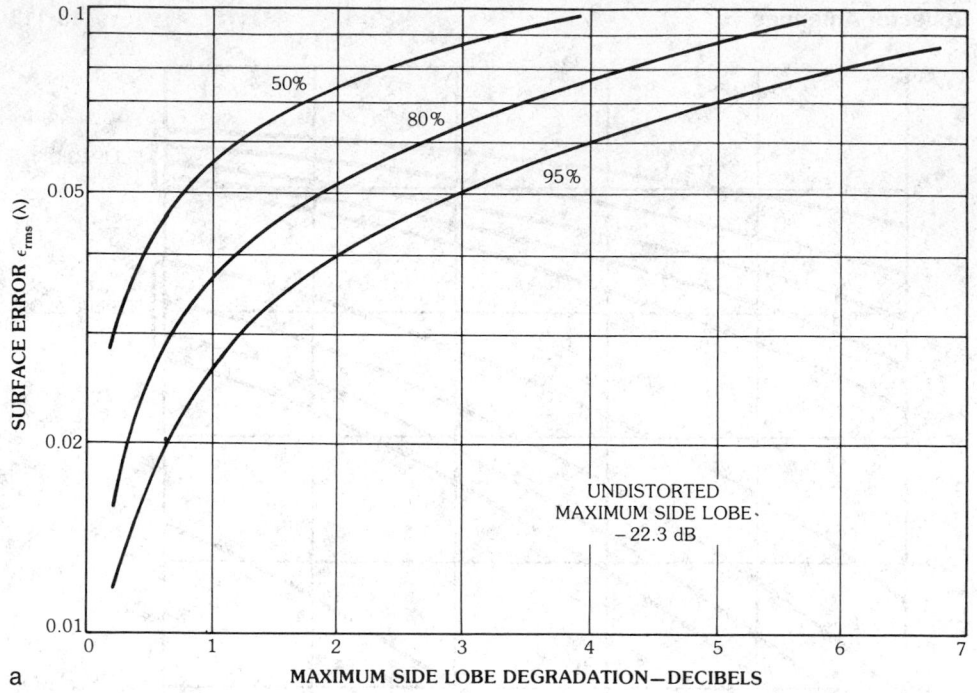

a

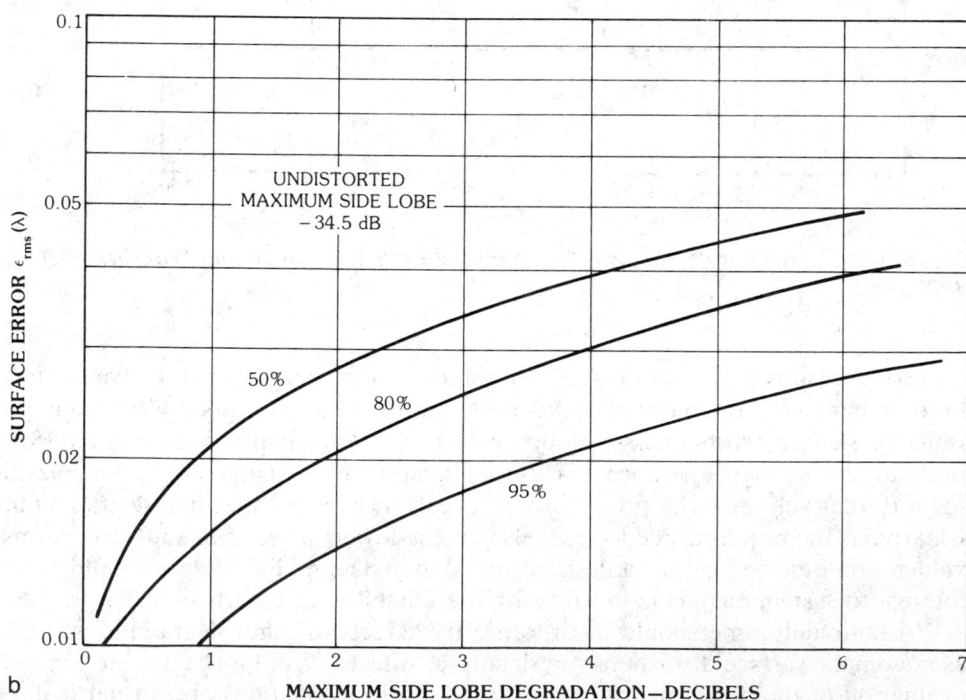

b

Fig. 85. Surface tolerance versus degradation of maximum side lobe level. (*a*) For a reflector with design side lobe level of −22.3 dB. (*b*) For a reflector with design side lobe level of −34.5 dB.

10. Appendix: Coordinate Transformations for Antenna Applications

Feed coordinates, far-field pattern coordinates, and reflector coordinates do not, in general, coincide. In many practical design and measurement applications it is necessary to have a general coordinate transformation which relates the Cartesian or spherical components of one coordinate system to the components of another system. For example, when a reflector antenna is illuminated by an array of feed horns one is required to find the illuminating field on the reflector in the reflector coordinates. The radiated field of each horn is known in its own system of coordinates which in general do not coincide with the reflector coordinates; therefore a coordinate transformation is needed to relate the two systems. Also, in many cases one does not want to compute the far-field pattern in the spherical coordinates of the antenna. Instead, it is desirable to compute them in the spherical coordinates which coincide with the main beam direction. In this appendix some basic transformations which are frequently applied to relate the Cartesian and spherical components of one coordinate system with those of another system are presented in a coherent and an easily usable format. In general, these two systems can be constructed with the use of some translations and rotations. For details the reader is referred to [11].

Cartesian and Spherical Components

Fig. A-1 illustrates Cartesian and spherical coordinate systems for defining points or vectors. A vector field **H** can be expressed by its Cartesian or spherical components as follows:

$$\mathbf{H} = \sum_{i=1}^{3} H_i^c \hat{\mathbf{c}}_i = \sum_{i=1}^{3} H_i^s \hat{\mathbf{s}}_i \tag{115}$$

where $\hat{\mathbf{c}}_i$ and $\hat{\mathbf{s}}_i$ are Cartesian ($\hat{\mathbf{c}}_1 = \hat{\mathbf{x}}$, $\hat{\mathbf{c}}_2 = \hat{\mathbf{y}}$, $\hat{\mathbf{c}}_3 = \hat{\mathbf{z}}$) and spherical ($\hat{\mathbf{s}}_1 = \hat{\mathbf{r}}$, $\hat{\mathbf{s}}_2 = \hat{\boldsymbol{\theta}}$, $\hat{\mathbf{s}}_3 = \hat{\boldsymbol{\phi}}$) unit vectors, respectively. The objective is to relate the $\{\hat{\mathbf{s}}\}$ to the $\{\hat{\mathbf{c}}\}$ and the $\{H^s\}$ to the $\{H^c\}$, where $\{\hat{\mathbf{s}}\} = \{\hat{\mathbf{s}}_1, \hat{\mathbf{s}}_2, \hat{\mathbf{s}}_3\}$. Note that $H_1 = H_r$, $H_2 = H_\theta$, and $H_3 = H_\phi$ and similarly for $\{H^c\}$. The desired connection can best be made by introducing the transformation matrix $[^s T^c]$ defined below:

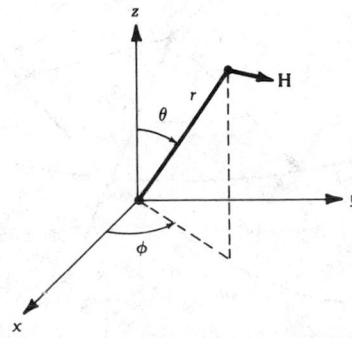

Fig. A-1. Cartesian and spherical coordinates. (*After Rahmat-Samii [11],* © *1979 IEEE*)

$$[{}^{s}T^{c}] = \begin{bmatrix} \sin\theta\cos\phi & \sin\theta\sin\phi & \cos\theta \\ \cos\theta\cos\phi & \cos\theta\sin\phi & -\sin\theta \\ -\sin\phi & \cos\phi & 0 \end{bmatrix} \qquad (116)$$

where superscripts s and c are used to denote transformations from Cartesian to spherical components. One can then easily demonstrate that for $(\hat{s}) = (\hat{s}_1, \hat{s}_2, \hat{s}_3)'$,

$$(\hat{s}) = [{}^{s}T^{c}](\hat{c}), \qquad (H^{c}), = [{}^{s}T^{c}](H^{c}) \qquad (117)$$

i.e., the same transformation holds for both the unit vectors and the vector components. Furthermore, it can be shown that

$$[{}^{c}T^{s}] = [{}^{s}T^{c}]^{-1} = [{}^{s}T^{c}]^{t} \qquad (118)$$

where, in this case, $[{}^{c}T^{s}]$ defines a transformation from spherical to Cartesian components.

Eulerian Angles

Let us consider two Cartesian coordinate systems $\{\hat{c}\}$ and $\{\hat{c}'\}$ as shown in Fig. A-2. In the most general case these Cartesian systems can be aligned via three rotations. The angles of these rotations are known as *Eulerian angles*. Although one can find different definitions for these angles in the literature, the definition used in [11] is used. As displayed in Fig. A-2, angles α, β, and γ are employed to define the Eulerian angles. Angle α describes a counterclockwise rotation about

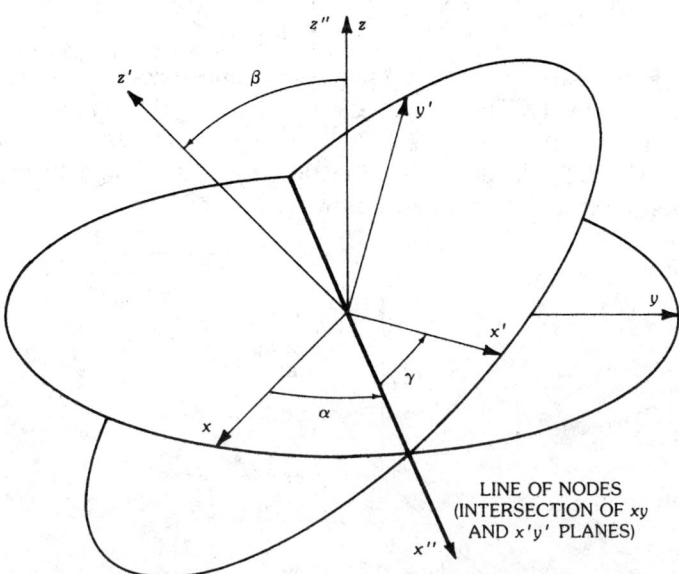

Fig. A-2. Eulerian angles relating primed and unprinted coordinates. (*After Rahmat-Samii [11], © 1979 IEEE*)

the z axis which brings the x axis to the x'' axis aligned with the line of nodes (line of intersection between xy and $x'y'$ planes), angle β defines a rotation about the line of nodes in a counterclockwise sense as indicated, bringing the z axis to z', and finally angle γ is another rotation about the z' axis which aligns the x'' axis with the x' axis in a counterclockwise sense.

By successively performing these three rotations one arrives at the following transformation equations among unit Cartesian components and Cartesian components of vectors, viz., for column vectors $(\hat{e}')$, $(\hat{e})$, $(H^{c'})$, and (H^c):

$$(\hat{e}') = [^{c'}A^c](\hat{e}), \qquad (H^{c'}) = [^{c'}A^c](H^c) \tag{119}$$

where $[^{c'}A^c]$ is the transformation matrix from Cartesian coordinates $(\hat{e})$ to $(\hat{e}')$ defined as

$$[^{c'}A^c] = \begin{bmatrix} \cos\gamma & \sin\gamma & 0 \\ -\sin\gamma & \cos\gamma & 0 \\ 0 & 0 & 1 \end{bmatrix} \begin{bmatrix} 1 & 0 & 0 \\ 0 & \cos\beta & \sin\beta \\ 0 & -\sin\beta & \cos\beta \end{bmatrix}$$

$$\times \begin{bmatrix} \cos\alpha & \sin\alpha & 0 \\ -\sin\alpha & \cos\alpha & 0 \\ 0 & 0 & 1 \end{bmatrix} \tag{120}$$

with components

$$A_{11} = \cos\gamma\cos\alpha - \sin\gamma\cos\beta\sin\alpha$$
$$A_{12} = \cos\gamma\sin\alpha + \sin\gamma\cos\beta\cos\alpha$$
$$A_{13} = \sin\gamma\sin\beta$$
$$A_{21} = -\sin\gamma\cos\alpha - \cos\gamma\cos\beta\sin\alpha$$
$$A_{22} = -\sin\gamma\sin\alpha + \cos\gamma\cos\beta\cos\alpha \tag{121}$$
$$A_{23} = \cos\gamma\sin\beta$$
$$A_{31} = \sin\beta\sin\alpha$$
$$A_{32} = -\sin\beta\cos\alpha$$
$$A_{33} = \cos\beta$$

The superscripts c and c' are deleted in the above expressions for simplicity, and the first and second subscripts, respectively, denote rows and columns of a 3×3 matrix. In addition, one can easily establish the following relations:

$$[^cA^{c'}] = [^{c'}A^c]^{-1} = [^{c'}A^c]^t \tag{122}$$

Feed and Reflector Coordinates

Having derived the transformation matrices (116) and (120) one can now solve an important antenna design and analysis problem. As depicted in Fig. A-3, the

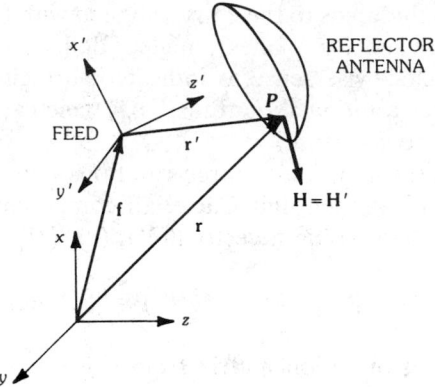

Fig. A-3. Feed and antenna coordinate systems. (*After Rahmat-Samii [11], 1979 IEEE*)

reflector coordinate system is $\{\hat{\mathbf{c}}\}$ and the feed coordinate system is $\{\hat{\mathbf{c}}'\}$. The radiated field of the feed is given in $\{\hat{\mathbf{c}}'\}$ in terms of its spherical components. The aim is to determine this radiated field at a point on the reflector antenna in the reflector coordinate system $\{\hat{\mathbf{c}}\}$ in terms of both the Cartesian and spherical components. Coordinates $\{\hat{\mathbf{c}}\}$ and $\{\hat{\mathbf{c}}'\}$ are, in general, related through both a translation and three rotations using Eulerian angles. This problem can be solved in two steps. First, the spherical coordinates of a point given on the reflector surface are determined in terms of the feed coordinate system $\{\hat{\mathbf{c}}'\}$. Second, the field expression of the radiated field of the feed is derived in the reflector coordinate system $\{\hat{\mathbf{c}}\}$.

The vectors $\mathbf{r}$ and $\mathbf{r}'$ are designated as the position vectors for point P on the antenna. Furthermore, as shown in Fig. A-3, the vector $\mathbf{f}$ is used to define the location of the feed in the reflector or unprimed system $\{\hat{\mathbf{c}}\}$. Clearly, the following holds:

$$\mathbf{r}' = \mathbf{r} - \mathbf{f}$$

With $\mathbf{r}$ and $\mathbf{f}$ given, one first determines the Cartesian components of $\mathbf{r}'$ in $\{\hat{\mathbf{c}}\}$ and then employs the transformation matrix to find its Cartesian components in $\{\hat{\mathbf{c}}'\}$. Cartesian components of $\mathbf{r}'$ in $\{\hat{\mathbf{c}}\}$ can be written as

$$\mathbf{r}'^c = (r'^c) = \{r\sin\theta\cos\phi - f_1,\ r\sin\theta\sin\phi - f_2,\ r\cos\theta - f_3\}^t \qquad (123)$$

where $f_1, f_2,$ and f_3 are the Cartesian components of $\mathbf{f}$ in $\{\hat{\mathbf{c}}\}$ and t is the transpose operator. Using the second equation in (119), one can determine the Cartesian components of $\mathbf{r}'$ in $\{\hat{\mathbf{c}}'\}$, namely

$$\mathbf{r}'^{c'} = (r'^{c'}) = [{}^{c'}A^c](r'^c) \qquad (124)$$

Since

$$\mathbf{r}'^{c'} = (r'\sin\theta'\cos\phi',\ r'\sin\theta'\sin\phi',\ r'\cos\theta')^t \qquad (125)$$

one can then easily determine r', θ', and ϕ' in terms of r, θ, and ψ via the application of (123) and (124).

Once the coordinates of reflector point P in $\{\hat{c}'\}$ are obtained, one next designates the spherical components of the radiated field of the feed at this point by $(H'^{s'})$. The task is now to describe the Cartesian and spherical components of $\mathbf{H}' = \mathbf{H}$, namely $\mathbf{H}^c$ and $\mathbf{H}^s$, in $\{\hat{c}\}$. Using (118) in the primed coordinates one can find the Cartesian components of $\mathbf{H}'^{s'}$ from

$$\mathbf{H}' = (H'^{c'}) = [^{c'}T^{s'}](H'^{s'}) \tag{126}$$

The Cartesian components of this vector field in $\{\hat{c}\}$ can then be found from (122), namely,

$$\mathbf{H} = (H^c) = [^cA^{c'}](H'^{c'}) = [^cA^{c'}][^{c'}T^{s'}](H'^{s'}) \tag{127}$$

It is worth mentioning that both $\mathbf{H}'$ and $\mathbf{H}$ represent the same vector and that the only difference lies in their functional dependence on primed and unprimed coordinates. With the use of (117), one can next derive the spherical components of $\mathbf{H}$ from (127) to arrive at

$$\mathbf{H} = (H^s) = [^sT^c](H^c) = [^sT^c][^cA^{c'}][^{c'}T^{s'}](H'^{s'}) \tag{128}$$

In summary, expressions (124) and (128) may finally be written more explicitly as follows:

$$\begin{bmatrix} r'\sin\theta'\cos\phi' \\ r'\sin\theta'\sin\phi' \\ r'\cos\theta' \end{bmatrix} = \begin{bmatrix} A_{11} & A_{12} & A_{13} \\ A_{21} & A_{22} & A_{23} \\ A_{31} & A_{32} & A_{33} \end{bmatrix} \begin{bmatrix} r\sin\theta\cos\phi - f_1 \\ r\sin\theta\sin\phi - f_2 \\ r\cos\theta \qquad - f_3 \end{bmatrix} \tag{129}$$

and

$$\begin{bmatrix} H_r(r,\theta,\phi) \\ H_\theta(r,\theta,\phi) \\ H_\phi(r,\theta,\phi) \end{bmatrix} = \begin{bmatrix} \sin\theta\cos\phi & \sin\theta\sin\phi & \cos\theta \\ \cos\theta\cos\phi & \cos\theta\sin\phi & -\sin\theta \\ -\sin\phi & \cos\phi & 0 \end{bmatrix}$$

$$\times \begin{bmatrix} A_{11} & A_{21} & A_{31} \\ A_{12} & A_{22} & A_{32} \\ A_{13} & A_{23} & A_{33} \end{bmatrix}$$

$$\times \begin{bmatrix} \sin\theta'\cos\phi' & \cos\theta'\cos\phi' & -\sin\phi' \\ \sin\theta'\sin\phi' & \cos\theta'\sin\phi' & \cos\phi' \\ \cos\theta' & -\sin\theta' & 0 \end{bmatrix}$$

$$\times \begin{bmatrix} H'_{r'}(r',\theta',\phi') \\ H'_{\theta'}(r',\theta',\phi') \\ H'_{\phi'}(r',\theta',\phi') \end{bmatrix} \tag{130}$$

In most cases one deals with the far field of the feed and therefore H_r is zero in (130). Expressions 129 and 130 are used in constructing the physical optics radiation integral for a reflector illuminated by an arbitrarily positioned feed source as described in Section 2.

Determination of Eulerian Angles

Eulerian angles can be determined by properly rotating the (x, y, z) system and aligning it with the (x', y', z') system, as shown in Fig. A-2, and then calculating the rotation angles. In many cases, however, it is desirable to have simple formulas which provide these angles. From Fig. A-2 the Eulerian angles α, β, and γ can be obtained by first assuming that the direction $\hat{z}'$ is known, which then allows the construction of x'' as

$$\hat{z}' = \hat{k}' = k_x'\hat{x} + k_y'\hat{y} + k_z'\hat{z}$$
$$\hat{x}'' = \hat{z} \times \hat{z}'/\Delta, \qquad \Delta^2 = k_x'^2 + k_y'^2 = 1 - k_z'^2 \tag{131}$$

Furthermore, by using the above expressions, α, β, and γ can be determined from the following equations:

$$\sin\alpha = (\hat{x} \times \hat{x}')\cdot\hat{z} = \hat{x}\cdot\hat{z}'/\Delta = k_x'/\Delta$$
$$\cos\alpha = \hat{x}\cdot\hat{x}'' = -\hat{y}\cdot\hat{z}'/\Delta = -k_y'/\Delta \tag{132}$$

$$\sin\beta = (\hat{z} \times \hat{z}')\cdot\hat{x}'' = \Delta$$
$$\cos\beta = \hat{z}\cdot\hat{z}' = k_z' \tag{133}$$

$$\sin\gamma = (\hat{x}'' \times \hat{x}')\cdot\hat{z}' = \hat{z}\cdot\hat{x}'/\Delta$$
$$\cos\gamma = \hat{x}''\cdot\hat{x}' = \hat{z}\cdot\hat{y}'/\Delta \tag{134}$$

It is worthwhile to mention that α and β are determined once the direction of $\hat{z}'$ is identified. Angle γ defines a rotation about $\hat{z}'$, which is not fixed, but can be chosen, based on such considerations as polarization and the like. For example, once the $\hat{z}'$ axis of the feed coordinates in Fig. A-3 is known, the α and β angles can be constructed from (132) and (133), respectively. Then γ can be used, for instance, to adjust the polarization of the feed pattern with respect to the antenna coordinates.

11. References

[1]   A. W. Love, ed., *Reflector Antennas*, New York: IEEE Press, 1978.
[2]   E. C. Collin and F. J. Zucker, eds., *Antenna Theory, Parts 1 and 2*, New York: McGraw-Hill Book Co., 1969.
[3]   H. Jasik, *Antenna Engineering Handbook*, New York: McGraw-Hill Book Co., 1961.
[4]   S. Silver, ed., *Microwave Antenna Theory and Design*, chapter 12, New York: McGraw-Hill Book Co., 1949.
[5]   P. J. B. Clarricoats and G. T. Poulton, "High-efficiency microwave reflector antennas —a review," *Proc. IEEE*, vol. 65, no. 10, pp. 1470–1504, October 1977.
[6]   A. W. Love, *Electromagnetic Horn Antennas*, New York: IEEE Press, 1976.
[7]   A. W. Rudge et al., eds., *The Handbook of Antenna Design*, *Volumes I and II*, IEE

Electromagnetic Waves Series 15, London: Peter Peregrinus Ltd., 1982.

[8] M. I. Skolnik, ed., *Radar Handbook*, chapter 9, New York: McGraw-Hill Book Co., 1970.

[9] P. J. B. Clarricoats, "Feeds for reflector antennas—a review," Part 1: Antennas, *Second Intl. Conf. Antennas Propag.*, April 13–16, 1981, York, England, pp. 309–317.

[10] A. W. Rudge and N. A. Adatia, "Offset-parabolic-reflector antennas: a review," *Proc. IEEE*, vol. 66, no. 12, pp. 1592–1618, December 1978.

[11] Y. Rahmat-Samii, "Useful coordinate transformations for antenna applications," *IEEE Trans. Antennas Propag.*, vol. AP-27, pp. 571–574, July 1979.

[12] Y. Hwang, C. H. Tsao, and C. C. Han, "Uniform analysis of reflector antenna for satellite application," *IEEE/AP-S Intl. Symp. Natl. Radio Sci. Mtg.*, Houston, Texas, May 23–26, 1983.

[13] V. Galindo-Israel and Y. Rahmat-Samii, "A new look at a Fresnel field computation using the Jacobi-Bessel series," *IEEE Trans. Antennas Propag.*, vol. AP-29, pp. 885–898, November 1981.

[14] Y. Rahmat-Samii and V. Galindo-Israel, "Shaped reflector antenna analysis using the Jacobi-Bessel series," *IEEE Trans. Antennas Propag.*, vol. AP-28, no. 4, pp. 425–435, July 1980.

[15] Y. Rahmat-Samii, R. Mittra, and V. Galindo-Israel, "Computation of Fresnel and Fraunhofer fields of planar apertures and reflector antennas by Jacobi-Bessel series—a review," *J. Electromagnetics*, vol. 1, no. 2, pp. 155–185, April–June 1981.

[16] A. C. Ludwig, "Calculation of scattered patterns from symmetrical reflectors," *Tech. Rep. No. 32-1430*, Jet Propulsion Laboratory, California Institute of Technology, Pasadena, February 1970.

[17] V. Galindo-Israel and R. Mittra, "A new series representation for the radiation integral with application to reflector antennas," *IEEE Trans. Antennas Propag.*, vol. AP-25, pp. 631–641, September 1977.

[18] G. Franceschetti, "Analytical techniques for reduction of computational effort in reflector antenna analysis," *IEE Intl. Symp. Antennas Propag. Dig.*, York, England, pp. 457–465, April 1981.

[19] Y. Rahmat-Samii and R. Chueng, "Nonuniform sampling techniques for antenna applications," *IEEE Trans. Antennas Propag.*, vol. AP-35, pp. 268–279, March 1987.

[20] J. B. Keller, "Geometrical theory of diffraction," *J. Opt. Soc. Am.*, vol. 52, pp. 116–130, 1962.

[21] S. W. Lee, "Uniform asymptotic theory of electromagnetic edge diffraction: a review," in *Electromagnetic Scattering*, New York: Academic Press, 1978, pp. 67–119.

[22] S. W. Lee, P. Cramer, Jr., K. Woo, and Y. Rahmat-Samii, "Diffraction by an arbitrary subreflector: GTD solution," *IEEE Trans. Antennas Propag.*, vol. AP-27, pp. 305–316, May 1979.

[23] R. G. Kouyoumjian, "The geometrical theory of diffraction and its application," in *Topics in Applied Physics, Volume 3: Numerical and Asymptotic Techniques in Electromagnetics*, Heidelberg: Springer-Verlag, 1975, pp. 165–215.

[24] S. W. Lee and G. A. Deschamps, "A uniform asymptotic theory of electromagnetic diffraction by a curved wedge," *IEEE Trans. Antennas Propag.*, vol. AP-24, no. 1, pp. 25–34, January 1976.

[25] R. G. Kouyoumjian and P. H. Pathak, "A uniform geometrical theory of diffraction for an edge in a perfectly conducting surface," *Proc. IEEE*, vol. 62, no. 11, pp. 1448–1461, November 1974.

[26] Y. Rahmat-Samii and R. Mittra, "Spectral analysis of high-frequency diffraction of an arbitrary incident field by a half-plane—comparison with four asymptotic techniques," *Radio Sci.*, vol. 13, pp. 31–48, January–February 1978.

[27] J. Boersma and Y. Rahmat-Samii, "Comparison of two leading uniform theories of edge diffraction with the exact uniform asymptotic solution," *Radio Sci.*, vol. 15, pp. 1179–1194, November–December 1980.

[28] J. F. Kauffman, W. F. Croswell, and L. J. Jowers, "Analysis of the radiation patterns

of reflector antennas," *IEEE Trans. Antennas Propag.*, vol. AP-24, pp. 53–65, January 1976.

[29] Y. Rahmat-Samii, "A comparison between GO/aperture field and physical optics methods for offset reflectors," *IEEE Trans. Antennas Propag.*, vol. AP-32, pp. 301–306, March 1984.

[30] Y. Rahmat-Samii, "An efficient computational method for characterizing the effects of random surface errors on the average power pattern of reflectors," *IEEE Trans. Antennas Propag.*, vol. AP-31, pp. 92–98, January 1983.

[31] R. C. Hansen, "Circular aperture distribution with one parameter," *Electron. Lett.*, vol. II, p. 184, April 1975.

[32] V. Jamnejad and Y. Rahmat-Samii, "Some important geometrical features of conic-section–generated offset reflector antennas," *IEEE Trans. Antennas Propag.*, vol. AP-28, pp. 952–957, November 1980.

[33] P. W. Hannan, "Microwave antennas derived from the Cassegrain telescope," *IRE Trans. Antennas Propag.*, vol. AP-9, pp. 140–153, March 1961.

[34] Y. Rahmat-Samii and S. W. Lee, "Directivity of planar array feeds for satellite reflector applications," *IEEE Trans. Antennas Propag.*, vol. AP-31, pp. 463–470, May 1983.

[35] T. S. Chu and R. H. Turrin, "Depolarization properties of offset reflector antennas," *IEEE Trans. Antennas Propag.*, vol. AP-21, pp. 339–345, May 1973.

[36] A. C. Ludwig, "The definition of cross polarization," *IEEE Trans. Antennas Propag.*, vol. AP-21, no. 1, pp. 116–119, January 1973.

[37] P. G. Ingerson and W. V. T. Rusch, "Radiation from a paraboloid with an axially defocused feed," *IEEE Trans. Antennas Propag.*, vol. AP-21, no. 1, pp. 104–106, January 1973.

[38] Y. T. Lo, "On the beam deviation factor of a parabolic reflector," *IRE Trans. Antennas Propag.*, vol. AP-8, pp. 347–349, May 1960.

[39] J. Ruze, "Lateral-feed displacement in a paraboloid," *IEEE Trans. Antennas Propag.*, vol. AP-13, pp. 660–665, September 1965.

[40] W. A. Imbriale, P. G. Ingerson, and W. C. Wong, "Large lateral feed displacements in a parabolic reflector," *IEEE Trans. Antennas Propag.*, vol. AP-22, pp. 742–745, November 1974.

[41] P. J. Woods, *Reflector Antenna Analysis and Design*, London: Peter Peregrinus, 1980.

[42] W. V. T. Rusch and P. D. Potter, *Analysis of Reflector Antennas*, New York: Academic Press, 1970.

[43] Y. Rahmat-Samii and V. Galindo-Israel, "Scan performance of dual offset reflector antennas for satellite communications," *Radio Sci.*, vol. 16, no. 6, pp. 1093–1099, November–December 1981.

[44] Y. Mitzugutch, M. Akagawa, and H. Yokoi, "Offset dual reflector antenna," *IEEE Intl. Symp.*, pp. 2–5, Amherst, Massachusetts, October 11–15, 1976.

[45] W. C. Wong, "On the equivalent parabola technique to predict the performance characteristics of a Cassegrainian system with an offset feed," *IEEE Trans. Antennas Propag.*, vol. AP-21, no. 3, pp. 335–339, May 1973.

[46] V. Krichevsky and D. F. DiFonzo, "Optimum feed locus for beam scanning in offset Cassegrain antennas," IEEE/AP-S International Symposium, Quebec, Canada, June 1980.

[47] V. Krichevsky, and D. F. DiFonzo, "Beam scanning in the offset Gregorian antenna," *Comsat Tech. Rev.*, vol. 12, pp. 251–269, 1982.

[48] B. Y. Kinber, "On two reflector antennas," *Radio Eng. Electron. Phys.*, vol. 6, June 1962.

[49] V. Galindo, "Design of dual reflector antennas with arbitrary phase and amplitude distribution," *IEEE Trans. Antennas Propag.*, vol. AP-12, pp. 403–408, July 1964.

[50] W. F. Williams, "High-efficiency antenna reflector," *Microwave J.*, vol. 8, p. 79, 1965.

[51] P. J. Woods, "Reflector profiles for the pencil-beam Cassegrain antenna," *Marconi Rev.*, pp. 121–138, 2nd quarter, 1972.

[52] V. Galindo-Israel, R. Mittra, and A. Cha, "Aperture amplitude and phase control of

offset dual reflectors," *IEEE Trans. Antennas Propag.*, vol. AP-27, no. 2, pp. 154–164, March 1979.

[53] J. J. Lee, L. I. Parad, and R. S. Chu, "A shaped offset-fed dual-reflector antenna," *IEEE Trans. Antennas Propag.*, vol. AP-27, no. 2, pp. 165–171, March 1979.

[54] A. Cha and D. A. Bathker, "Preliminary announcement of an 85-percent efficient reflector antenna," *IEEE Trans. Antennas Propag.*, vol. AP-31, no. 2, pp. 341–342, March 1983.

[55] V. Galindo-Israel, Y. Rahmat-Samii, W. Imbriale, and R. Mittra, "Recent advances in electromagnetic synthesis and analysis of dual-shaped reflector antennas," *SPIE*, vol. 294, pp. 98–112, 1981.

[56] P. Balling, "Wavefront synthesis of contoured beam antennas with low cross-polarization," *Rep. S-92-01*, under Grant 1112/69–78, from the Danish Space Board, TICRA, ApS, 1979.

[57] R. G. Spencer, *Microwave Antenna Theory*, Section 13.8, Radiation Lab Series No. 12, ed. by S. Silver, New York: McGraw-Hill Book Co., 1949, pp. 502–509.

[58] C. A. Smith, "A review of the state of the art in large spaceborne antenna technology," *Pub. 78–88*, Jet Propulsion Lab., Pasadena, California, November 15, 1978.

[59] R. S. Elliott, *Antenna Theory and Design*, Englewood Cliffs: Prentice-Hall, 1981.

[60] Y. Rahmat-Samii, "Chapter 3 of 30/20-GHz Lewis antenna support," *Final Report*, *NASA No. 643-10-01-04-00*, February 1981.

[61] C. C. Chen and C. F. Franklin, "*Ku*-band multiple-beam antenna," NASA *Contract Rep. 154364* for contract no. NAS 1-14814, NASA Langley Research Center, Hampton, Virginia 23665, December 1980.

[62] A. V. Mrstik, "Scan limits of off-axis fed parabolic reflectors," *IEEE Trans. Antennas Propag.*, vol. AP-27, no. 5, pp. 647–651, September 1979.

[63] D. T. Nakatani, F. A. Taormina, G. G. Kuhn, and D. K. McCarty, "Design aspects of commercial satellite antennas," Communications Satellite Antenna Technology Short Course, UCLA, March 1976.

[64] C. C. Han, A. E. Smoll, II. W. Bilenko, C. A. Chuang, and C. A. Klein, "A general beam shaping technique—multiple-feed offset reflector antenna system," AIAA Sixth Communications Satellite Systems Conference, Montreal, Canada, April 5–8, 1976.

[65] W. A. Imbriale, "Applications of the method of moments to thin-wire elements and Arrays," chapter 7 in *Topics in Applied Physics*, *Volume 3: Numerical and Asymptotic Techniques in Electromagnetics*, Heidelberg: Springer-Verlag, 1975.

[66] S. J. Hamada, P. G. Ingerson, and W. V. Rusch, "Reflector radiation from planar near-field measurements of array feed," *IEEE Trans. Antennas Propag.*, vol. AP-28, no. 4, pp. 436–442, July 1980.

[67] Y. Rahmat-Samii, P. Cramer, Jr., K. Woo, and S. W. Lee, "Realizable feed-element patterns for multibeam reflector antenna analysis," *IEEE Trans. Antennas Propag.*, vol. AP-29, pp. 961–963, November 1981.

[68] C. A. Balanis, *Antenna Theory Analysis and Design*, New York: Harper & Row, 1982.

[69] J. Huang, Y. Rahmat-Samii, and K. Woo, "A GTD study of pyramidal horns for offset reflector antenna applications," *IEEE Trans. Antennas Propag.*, vol. AP-31, no. 2, pp. 305–309, March 1983.

[70] W. L. Stutzman and G. A. Thiele, *Antenna Theory and Design*, New York: John Wiley & Sons, 1981.

[71] P. D. Potter, "A new horn antenna with suppressed side lobes and equal beam-widths," *Microwave J.*, vol. 6, pp. 71–78, June 1963.

[72] A. W. Rudge and N. A. Adatia, "Matched feeds for offset parabolic reflector antennas," *Proc. of the Sixth Eur. Microwave Conf.*, Rome, Italy, pp. 1–5, September 1976.

[73] A. W. Rudge and N. A. Adatia, "New class of primary-feed antennas for use with offset parabolic reflector antennas," *Electron. Lett.*, vol. 11, pp. 597–599, November 27, 1975.

[74] J. Ruze, "Antenna tolerance theory—a review," *Proc. IEEE*, vol. 54, no. 4, pp. 633–640, April 1966.

[75] T. B. Vu, "The effect of aperture errors on the antenna radiation pattern," *Proc. Inst. of Electr. Eng. (IEE)*, vol. 116, pp. 195–202, 1969.

[76] A. W. Love, "Some highlights in reflector antenna development," *Radio Sci.*, vol. 11, pp. 671–684, August–September 1976.

[77] W. V. T. Rusch and R. Wohlleben, "Surface tolerance loss for dual-reflector antennas," *IEEE Trans. Antennas Propag.*, vol. AP-30, no. 4, pp. 784–785, July 1982.

[78] H. Ling, Y. T. Lo, and Y. Rahmat-Samii, "Reflector sidelobe degradation due to random surface errors," *IEEE Trans. Antennas Propag.*, vol. AP-34, January 1986.

[79] Y. Rahmat-Samii and S. W. Lee, "Vector diffraction analysis of reflector antennas with mesh surfaces," *IEEE Trans. Antennas Propag.*, vol. AP-33, pp. 76–90, January 1985.

[80] Y. Rahmat-Samii, "Effects of deterministic surface distortions on reflector antenna performance," *Annales des Telecommunications*, France, vol. 40, pp. 350–360, August 1985.

Chapter 16

Lens Antennas

J. J. Lee
Hughes Aircraft Company

CONTENTS

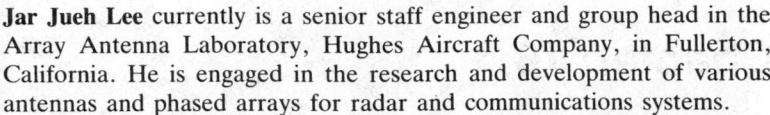

Jar Jueh Lee currently is a senior staff engineer and group head in the Array Antenna Laboratory, Hughes Aircraft Company, in Fullerton, California. He is engaged in the research and development of various antennas and phased arrays for radar and communications systems.

From 1977 to 1982 he worked for Rockwell International, in Anaheim, California, where he was responsible for the design of antennas and adaptive phased arrays for satellite communications. Prior to that he was with GTE Sylvania, in Needham, Massachusetts, involved in the study of missile-site communications systems, ground-wave propagation, electromagnetic-pulse effects, and high-voltage simulation tests. From 1973 to 1974 he was associated with Cornell University in the Nuclear and Plasma Physics Laboratory.

Dr. Lee received his BSEE from National Taiwan University, R.O.C., in 1967 and his PhD in electrical engineering from the Case Institute of Technology, Cleveland, Ohio, in 1973. He is a member of the IEEE Antennas and Propagation Society. He has authored and coauthored more than 20 papers in various technical journals, and served as a reviewer for the *IEEE Transactions on Antennas and Propagation.*

1. Introduction

The recent advent of millimeter-wave technology and the increasing demand for higher frequencies have stimulated a renewed interest in the design of lens antennas for radar and communication systems. For frequencies above K_u-band the penalty in weight and size associated with a lens is relatively small. In fact, for many ehf applications, lenses are selected for reasons of economy, performance, and reliability.

In general, lens antennas are easy to design and construct. They are mechanically rigid and are relatively tolerant of surface imperfections and load distortions. Compared with a dual-reflector design the lens antenna has a more desirable form factor due to its feedthrough characteristics. This unique feature eliminates the blockage problem and simplifies the mechanical design in the control of spillover with absorbing materials. It also offers greater design flexibility to meet various stringent requirements, such as low side lobes and wide scan coverage, by incorporating phase shifters or other phase compensation devices in the lens system.

Lenses, however, are not without drawbacks and problems. For high-gain antennas, especially those for spaceborne applications, the required aperture size, physical weight, and the transmission loss may prohibit the use of lenses. Edge diffraction, shadowing effect, and bandwidth limitation associated with lens zoning are also some of the common problems encountered in a lens design. In addition, impedance mismatch at the surface boundaries is always an important issue when very low side lobes are required.

Depending on their design, lenses can be broadly classified into two categories: dielectric lenses and constrained lenses. *Dielectric lens* refers to any focusing device of dielectric material which may or may not be homogeneous in its index of refraction. A *constrained lens* is defined as any optical transforming device in which the rays are guided and constrained to follow discrete paths that may have different propagation characteristics. The path lengths and the geometries of these guiding elements, which constitute the lens, are so designed that the exit rays produce the desired phase and amplitude distributions across the aperture. Constrained lenses include metal plate lenses, waveguide lenses, and other microwave lenses that consist of receiving and radiating elements with active or passive components for phase adjustments.

With computer controlled machines as readily available as they are today, a dielectric lens can be fabricated at very low cost. It can be easily contoured to any reasonable shape to produce a specified beam pattern. Recent advances in the development of low-loss dielectric materials make these lenses even more attractive to many antenna designers. Constrained lenses, on the other hand, are more expensive to construct. But normally they are lighter and more radiation resistant and therefore more suitable for space applications. Furthermore, they can be designed to meet very stringent requirements which cannot be fulfilled by dielectric lenses.

The design principles of lens antennas are very well known. Geometric optics is all that is required to formulate the problem. The lens can be considered as an optical transformer which, in most cases, transforms a spherical wavefront from a point source into a plane wave. In addition to this fundamental requirement the lens may be required to satisfy other conditions, depending on the application.

Sometimes one of the two surfaces is chosen to be flat for simplicity. For multibeam systems the lens may be designed to be coma free for a wide scan. For other applications the lens may be required to produce a special beam pattern and side lobe structure. Thus actual lens profiles depend on the specific conditions imposed in the design.

For uniform dielectric lenses Fermat's principle implies that Snell's law is automatically satisfied at each surface boundary of the lens. For constrained lenses, however, Snell's law is not necessarily satisfied at the boundary. Since Snell's law is not imposed at the boundaries of a constrained lens, other conditions must be specified to define the lens profiles in addition to the path length constraint. This provides much freedom in the design and for this reason many unique features can be achieved with a constrained lens.

Despite all the geometric differences, a dielectric lens is analogous to a dual-reflector system in many ways in an optical sense. The pickup surface of the lens (the one facing the feed element) is equivalent to the subreflector, which primarily controls the power distribution across the aperture. The second surface is equivalent to the main reflector, which chiefly corrects the phase. The space between the two reflectors can be viewed as a uniform dielectric medium which has a refraction index of -1. In this case, refraction is replaced by reflection at the boundary as a special case of Snell's law.

Since the lens design is based on geometric optics the aperture size of the lens must be sufficiently large, preferably more than 20 wavelengths, to make the geometric optics assumption valid. This consideration also limits the maximum number of zones that can be used for a given aperture size. Each zone must not be so small as to violate the geometric optics assumption. After the lens is designed by ray optics actual antenna characteristics must be determined by physical optics with diffraction taken into account. The far-field pattern is computed by standard methods, such as surface integration over the lens aperture as described in many antenna books.

The subject of lens antenna design has been treated in the past by many authors in considerable detail. Risser has summarized the early lens developments in Silver's book [1]. Brown also provides an excellent review on many lens designs in [2] and [3]. Detailed lens formulas with practical design considerations are given by Cohn [4] and Sengupta and Hiatt [5]. It is not the intent of this chapter to duplicate these works, but rather to present a complementary and systematic overview on various lenses as a design guide, with emphasis placed on more recently developed lens systems.

In the following sections, design formulas and performance characteristics of various dielectric and constrained lenses are presented. For brevity, discussions on theory and mathematical derivations are kept to a minimum. System applications are not stressed in this chapter, as these topics will be treated in more detail in Part C.

2. Design Principles of the Dielectric Lens

Fig. 1 shows a dielectric lens with its general contours S_1 and S_2 represented by (x_1, y_1) and (x_2, y_2), respectively. It is assumed that the lens is rotationally symmetric, and therefore only the cross section of the lens is of interest. Consider a lens of dielectric constant ϵ_r illuminated by a feed pattern emanating from a phase center located at the origin of the coordinates. The distance between the origin and the first surface of the lens is F. Let the central thickness of the lens be T and the diameter of the lens aperture be D.

Rectangular Coordinates

The most important condition to be imposed in the derivation of the lens profiles is the path length constraint. For phase reference an aperture plane is established at $x = S$, to the right of the lens for clarity. In general, the aperture phase distribution can be any rotationally symmetric function specified by $\varrho(\phi)$, where ϱ is the distance from the axis of the lens. For a nonuniform aperture phase front the lens profiles will be a function of S, but the dependence is minor since the

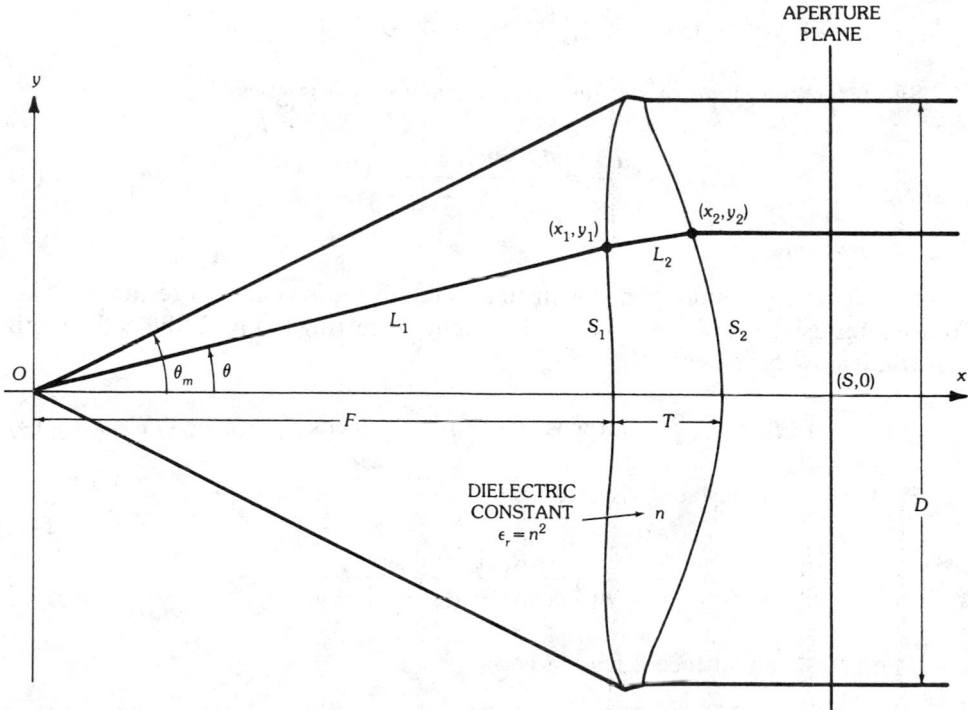

Fig. 1. Geometry for dielectric lens design.

aperture plane is usually very close to the lens. Examples of nonuniform phase distributions will be discussed in more detail later, but for the moment the aperture phase is assumed to be uniform. Mathematically the path length condition is given by

$$(x_1^2 + y_1^2)^{1/2} + n[(x_2 - x_1)^2 + (y_2 - y_1)^2]^{1/2} - x_2 = (n - 1)T \tag{1}$$

where the central ray has been used as a path length reference and $n = \sqrt{\epsilon_r}$ is the refractive index of the lens, which is greater than unity for real dielectrics. It should be noted that the parameter S for the reference plane has been canceled out of both sides of (1), and the focal length F is implicitly imposed as an initial condition for x_1.

The second condition to be imposed is Snell's law derived from Fermat's principle. By taking the differential of y_1 with respect to x_1 in (1), the slope of the lens at (x_1, y_1) is obtained:

$$\frac{dy_1}{dx_1} = \frac{nL_1(x_2 - x_1) - L_2 x_1}{L_2 y_1 - nL_1(y_2 - y_1)} \tag{2}$$

where

$$L_1 = (x_1^2 + y_1^2)^{1/2}$$
$$L_2 = [(x_2 - x_1)^2 + (y_2 - y_1)^2]^{1/2}$$

Similarly, the slope of the lens contour at (x_2, y_2) is given by

$$\frac{dy_2}{dx_2} = \frac{L_2 - n(x_2 - x_1)}{n(y_2 - y_1)} \tag{3}$$

Polar Coordinates

Sometimes it is more convenient to formulate the problem in terms of polar coordinates using the phase center as the origin. Referring to Fig. 2, the path length constraint now becomes

$$r + n[R^2 + r^2 - 2Rr\cos(\theta - \phi)]^{1/2} - R\cos\phi = (n - 1)T \tag{4}$$

Snell's law on surface 1 is given by

$$\frac{dr}{d\theta} = \frac{nRr\sin(\theta - \phi)}{n[R\cos(\theta - \phi) - r] - L_2} \tag{5}$$

and Snell's law on surface 2 leads to

$$\frac{dR}{d\phi} = \frac{nRr\sin(\theta - \phi) - L_2 R\sin\phi}{n[R - r\cos(\theta - \phi)] - L_2\cos\phi} \tag{6}$$

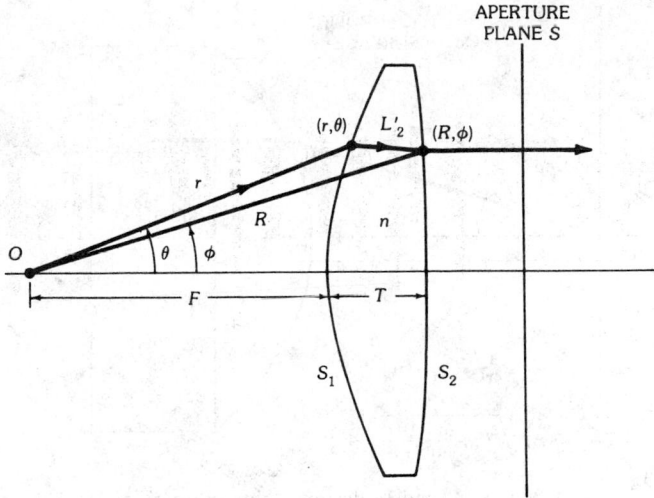

Fig. 2. Dielectric lens formulated in polar coordinates.

where

$$L_2 = n^{-1}[(n - 1)T + R\cos\phi - r] = [R^2 + r^2 - 2Rr\cos(\theta - \phi)]^{1/2}$$

In the lens formulation there are four variables in (1)–(3): $x_1, y_1, x_2,$ and y_2. With one of these quantities considered to be the independent variable three independent equations are required to uniquely determine the lens contours. The details of how to apply and solve these equations are discussed in the following sections, where different types of lenses are presented.

3. Simple Lenses with Analytic Surfaces

A *simple lens* is defined to be a lens which can be described by an analytic expression for one or both surfaces. To be examined in this section are four commonly used designs.

Lens with a Flat Surface on S_2

The simplest case is the one that has a flat surface on S_2 as shown in Fig. 3. This lens transforms a spherical wavefront into a plane wave, or conversely focuses a beam from infinity onto one point. The conditions imposed to derive S_1 are $x_2 = T$, the slope on S_2 being infinity, and the equal path length constraint. It can be readily shown that S_1 is a hyperbolic surface given by [1]

$$y_1 = [(n^2 - 1)(x_1 - F)^2 + 2(n - 1)F(x_1 - F)]^{1/2} \tag{7}$$

or equivalently

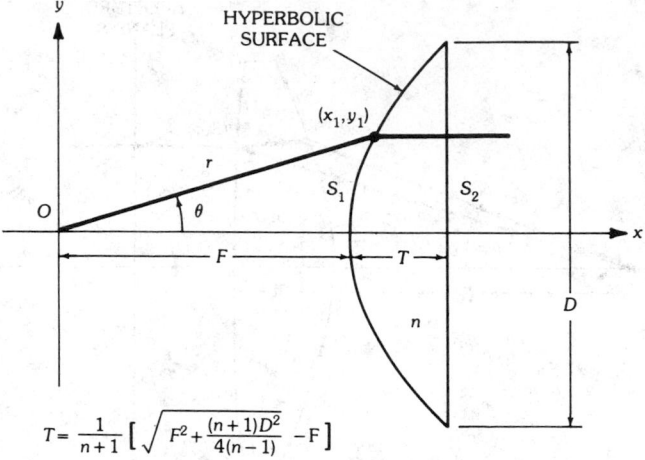

$$T = \frac{1}{n+1}\left[\sqrt{F^2 + \frac{(n+1)D^2}{4(n-1)}} - F\right]$$

Fig. 3. Hyperbolic lens, planar on S_2.

$$r = \frac{(n-1)F}{n\cos\theta - 1} \qquad (8)$$

For a given diameter D, the central thickness of the lens is

$$T = \frac{1}{n+1}\left[\sqrt{F^2 + \frac{(n+1)D^2}{4(n-1)}} - F\right] \qquad (9)$$

Lens with a Flat Surface on S_1

Fig. 4 is a lens with a flat surface on S_1. By setting $x_1 = F$ and the slope on S_1 equal to infinity, a parametric solution for S_2 can be found from the path length constraint

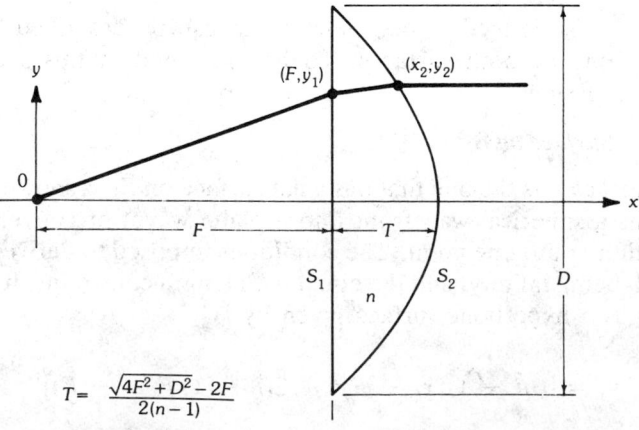

$$T = \frac{\sqrt{4F^2 + D^2} - 2F}{2(n-1)}$$

Fig. 4. Convex lens with planar surface on S_1.

$$x_2 = \frac{[(n-1)T - \sqrt{F^2 + y_1^2}]\sqrt{(n^2-1)y_1^2 + n^2F^2} + n^2F\sqrt{F^2 + y_1^2}}{n^2\sqrt{F^2 + y_1^2} - \sqrt{(n^2-1)y_1^2 + n^2F^2}} \quad (10)$$

$$y_2 = y_1\left[1 + \frac{x_2 - F}{\sqrt{(n^2-1)y_1^2 + n^2F^2}}\right] \quad (11)$$

In this case the central thickness is

$$T = \frac{\sqrt{4F^2 + D^2} - 2F}{2(n-1)} \quad (12)$$

Lens with a Spherical Surface on S_1

If S_1 is spherical, as shown in Fig. 5, the outer surface S_2 is an ellipse specified by

$$R = \frac{(n-1)(F+T)}{n - \cos\phi} \quad (13)$$

This is obtained by setting $\theta = \phi$ in (6) and integrating the differential equation. In rectangular coordinates the solution is

$$y_2 = \left[\left[\frac{x_2 + (n-1)(F+T)}{n}\right]^2 - x_2^2\right]^{1/2} \quad (14)$$

In this case the central thickness is

$$T = \frac{2F - \sqrt{4F^2 - D^2}}{2(n-1)} \quad (15)$$

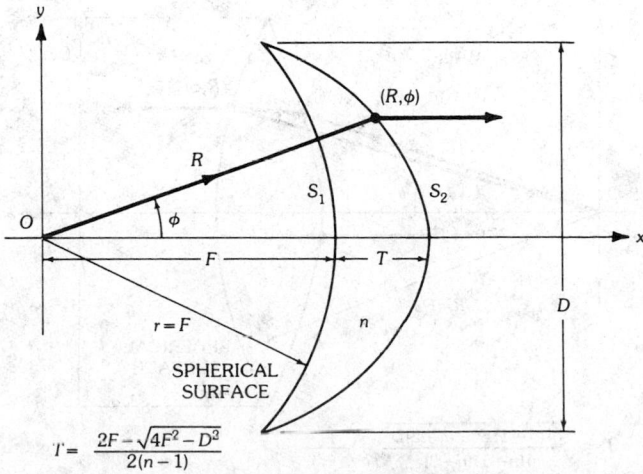

Fig. 5. Lens with spherical surface on S_1.

Lens with a Spherical Surface on S_2

Conversely, if S_2 is specified to be spherical, as shown in Fig. 6, the inner surface (r, θ) is defined by the following equations:

$$n^2[R^2 + r^2 - 2Rr\cos(\theta - \phi)] = [(n - 1)T + R\cos\phi - r]^2 \qquad (16)$$

$$n^2 r\sin(\theta - \phi) = (\sin\phi)[(n - 1)T + R\cos\phi - r] \qquad (17)$$

where $R = F + T$, and ϕ is the parametric variable. Equation 16 is derived from (4) and (17) is obtained by setting $dR/d\phi = 0$ in (6). For this lens the central thickness is related to D by

$$T = \sqrt{\frac{4(n - 1)F^2 - (n - 3)D^2}{4(n - 1)(n - 3)^2}} + \frac{F}{n - 3} \qquad (18)$$

Shown in Fig. 7 are plots of the central thickness versus dielectric constant for these four designs with F/D as a parameter. It can be seen that for small F/D the lens with spherical surface on S_1 is much thicker than the other three designs. For larger F/D, however, all these lenses have about the same thickness for a given dielectric constant.

4. Lens Aberrations and Tolerance Criteria

Before proceeding to the discussion of other lens designs a brief review of various lens aberrations may be in order. The term "aberration" is primarily used by optical designers referring to any imperfection caused by a lens in reproducing the image of an object. For antenna engineers the more commonly used nomenclature is *phase error*. Phase error is defined as the deviation in phase of the

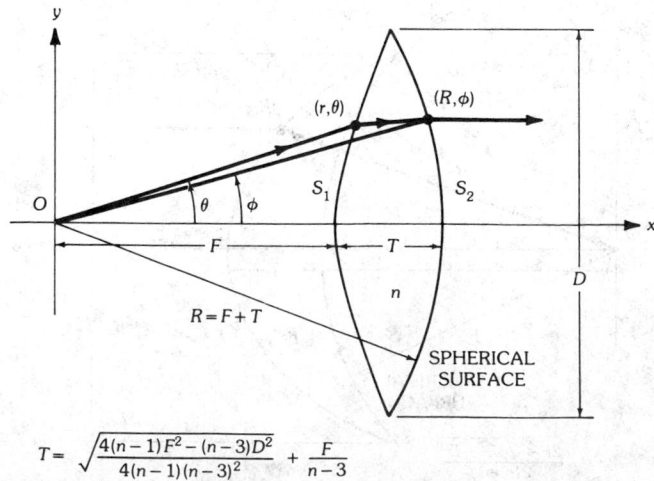

$$T = \sqrt{\frac{4(n-1)F^2 - (n-3)D^2}{4(n-1)(n-3)^2}} + \frac{F}{n-3}$$

Fig. 6. Lens with spherical surface on S_2.

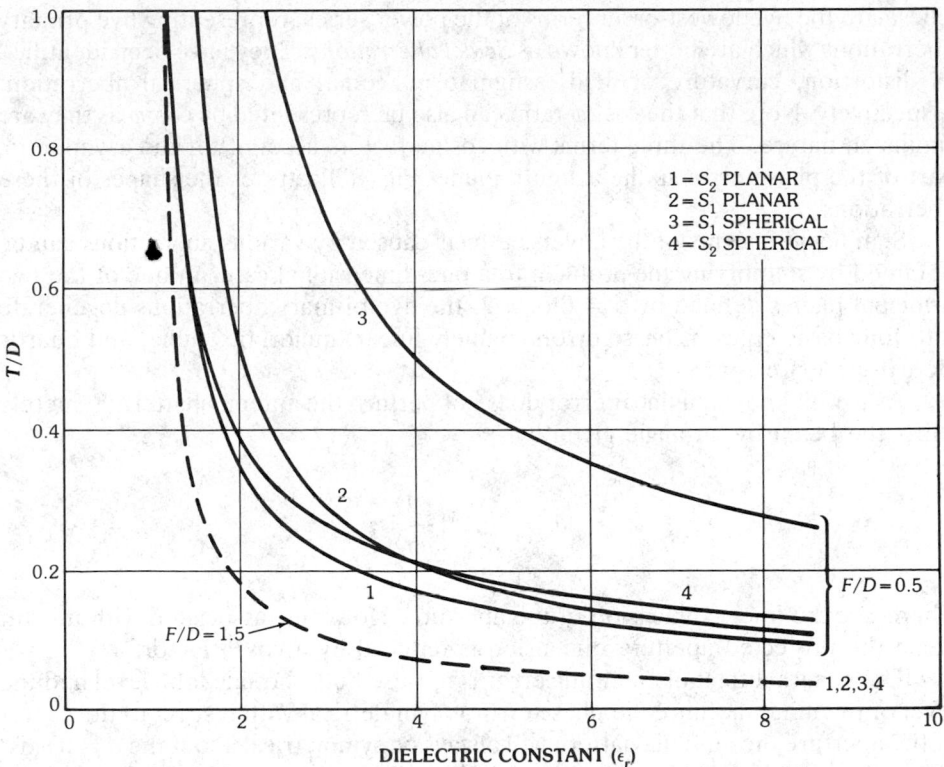

Fig. 7. Center thickness versus dielectric constant for different lenses.

wavefront from a prescribed reference, usually the aperture plane in front of the lens. Since amplitude errors generally do not occur in the lens design, they are therefore ignored here.

Phase errors may result from errors in the illumination pattern, displacement of the source from the focal point, or from the intrinsic errors of the lens. The phase error is normally expressed in terms of path length difference between a general ray and the reference central ray as a function of the exit point through the aperture. This aberration function, as defined in optics, can be represented as a surface above the aperture plane, defined by (ϱ, ϕ) in this case.

In general, degradations of the far-field pattern caused by a two-dimensional phase error distribution are difficult to visualize. However, the subject can be better understood if the error distribution is expanded into a power series in ϱ and $\cos\phi$, or equivalently in terms of the so-called Zernike cylindrical polynomials, as discussed in Born and Wolf [6]. The Zernike polynomials form an orthogonal set, which simplifies the evaluation of the diffraction integral over a circular aperture in the study of various aberration effects.

It has been shown that any aberration function of small amplitude can be approximated by the following series:

$$\Delta L(\varrho, \phi) = \alpha\varrho\cos\phi + \beta\varrho^2 + \gamma\varrho^2\cos 2\phi + \delta\varrho^3\cos\phi + \varepsilon\varrho^4 \tag{19}$$

These are the five lowest-order terms of the power series, representing five primary aberrations which are better known as *Seidel aberrations*. They have been identified as distortion, curvature of field, astigmatism, coma, and spherical aberration, respectively. Note that the $\cos 2\phi$ term can also be represented by $\cos^2\phi$ as they are similar in nature. The three terms with cosine factors account for the asymmetric part of the phase error in the azimuth plane. Fig. 8 illustrates the shapes of these aberrations.

Significant insights of the adverse effects caused by various aberrations can be obtained by simplifying the problem to a one-dimensional case. In one of the two principal planes defined by $\phi = 0$ or $\pi/2$, the five primary aberrations degenerate into four basic types of phase errors, namely linear, quadratic, cubic, and quartic (fourth-order) errors.

As is well known, a linear error does not perturb the antenna pattern; it merely shifts the beam by an angle given by

$$\theta_s = \sin^{-1}\left(\frac{\alpha\lambda}{a\pi}\right)$$

where a is the linear dimension of the aperture. However, associated with this tilt angle the projected aperture dimension is reduced by a $\cos\theta_s$ factor.

The general effect of quadratic error is to raise both the side lobe level and the level of the minima. Since the phase error is symmetrical with respect to the center of the aperture the antenna pattern will always be symmetrical about the $\theta = 0$ axis. To gain further insights Silver's analysis [7] is followed. Let the far-field pattern be denoted by

$$g(u) = \frac{a}{2}\int_{-1}^{1} f(x)\, e^{j[ux - \beta x^2]}\, dx \tag{20}$$

where $u = (\pi a/\lambda)\sin\theta$, βx^2 stands for the quadratic phase error, and $f(x)$ is assumed to be an even function for the amplitude distribution over a normalized aperture. Then the power pattern for small β is given by

$$P(u) \cong \frac{a^2}{4}\{g_0^2(u) + \beta^2[g_0''(u)]^2\} \tag{21}$$

where $g_0(u)$ is the pattern in the absence of phase error ($\beta = 0$), and $g_0''(u)$ is the second derivative of $g_0(u)$.

Fig. 9 shows far-field patterns of a linear aperture with uniform amplitude distribution and quadratic phase error. It has been found that when β gets sufficiently large the main lobe becomes bifurcated, with maxima appearing on either side of the $\theta = 0$ axis. Fig. 10 exemplifies the effect of quadratic error on gain.

The cubic phase errors (δx^3) can be treated by the same technique. In this case the power pattern is

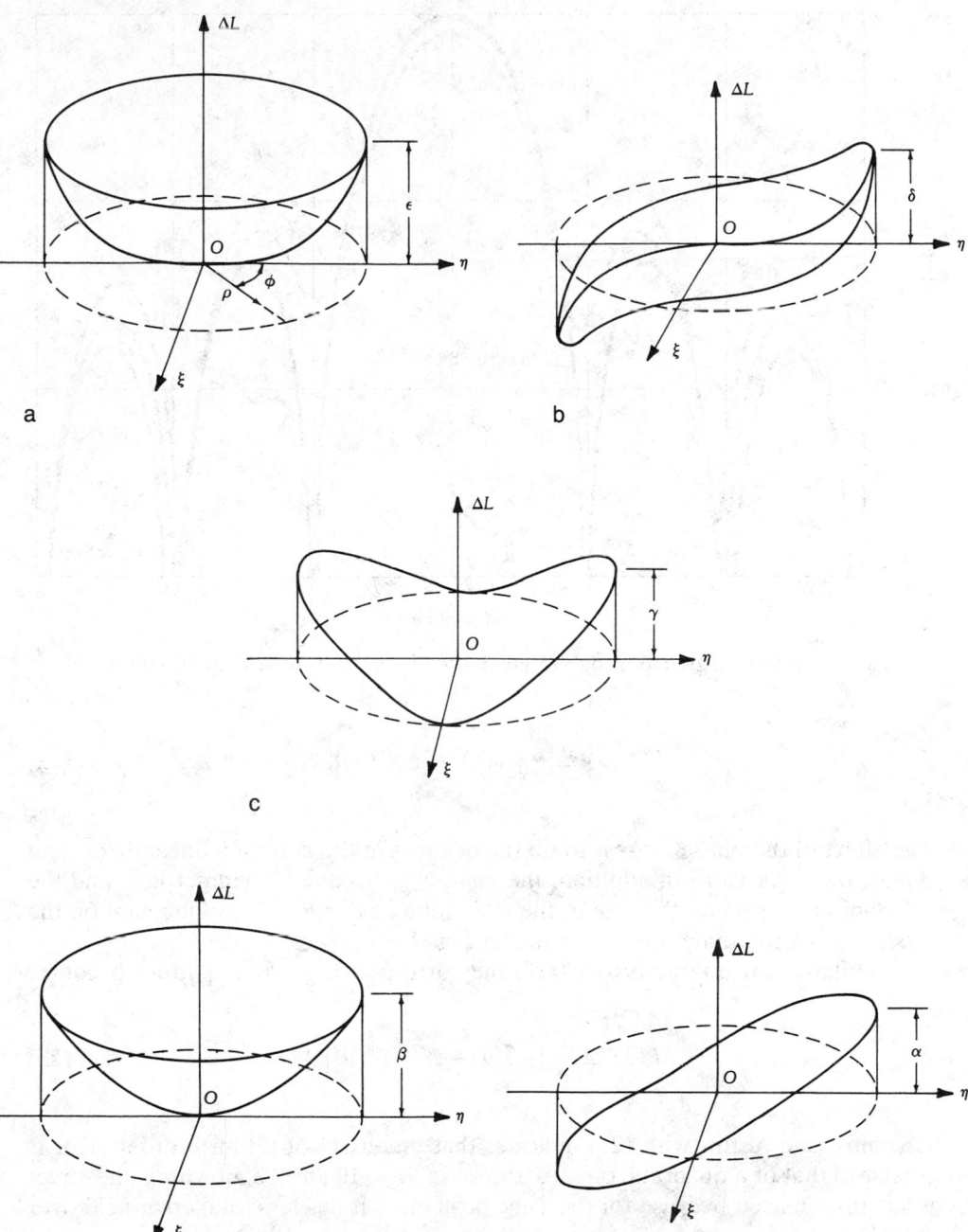

Fig. 8. Primary aberrations of a lens. (*a*) Spherical aberration: $\Delta L = \epsilon \varrho^4$. (*b*) Coma: $\Delta L = \delta \varrho^3 \cos \phi$. (*c*) Astigmatism: $\Delta L = \gamma \varrho^2 \cos^2 \phi$ (similar to $\varrho^2 \cos 2\phi$ in nature). (*d*) Curvature of field: $\Delta L = \beta \varrho^2$. (*e*) Distortion: $\Delta L = \alpha \varrho \cos \phi$. (*After Born and Wolf [6]; reproduced by permission of M. Born and E. Wolf*, Principles of Optics, *Third Edition, New York and London: Pergamon Press, 1965*)

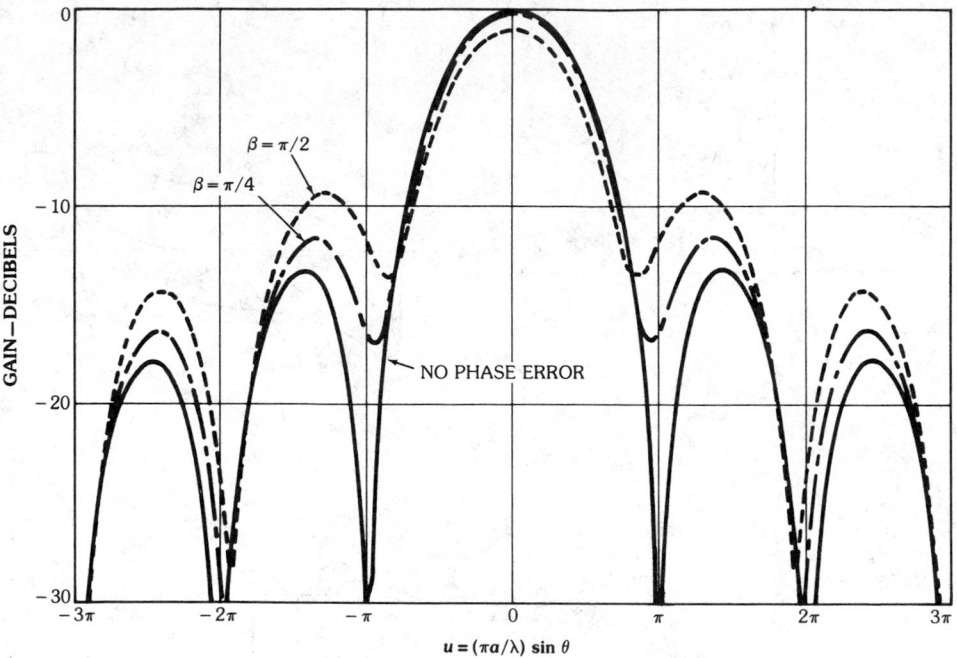

Fig. 9. Pattern degradation due to quadratic phase error (one-dimensional case).

$$P(u) \cong \frac{a^2}{4} [g_0(u) + \delta g_0'''(u)]^2 \tag{22}$$

The effect of the phase error is to tilt the beam as in the case of a linear error, and reduce the peak gain. In addition, the main lobe becomes asymmetrical, and the side lobes decrease on the side of the main lobe nearer $\theta = 0$ and increase on the other side of the main lobe as shown in Fig. 11.

Similarly, for quartic errors (εx^4) the corresponding power pattern becomes

$$P(u) \cong \frac{a^2}{4} \{g_0^2(u) + \varepsilon^2 [g_0^{(4)}(u)]^2\} \tag{23}$$

A comparison of this with (21) indicates that the effect of a fourth-order error is similar to that of a quadratic one. As shown in Figs. 10 and 12, however, the effect is less pronounced because for the same peak error it has less total error field over the aperture.

Based on the simple analysis given above, tolerance criteria of various aberrations can be discussed. If gain degradation is a major concern in the design, a maximum allowable loss in the peak gain must be specified to establish a tolerance level for a particular phase error. For example, if a 1-dB drop in gain is allowed, a maximum deviation of $\lambda/4$ is tolerable for the quadratic error in a one-dimensional case. For a fourth-order error the tolerance level is relatively higher, but for cubic

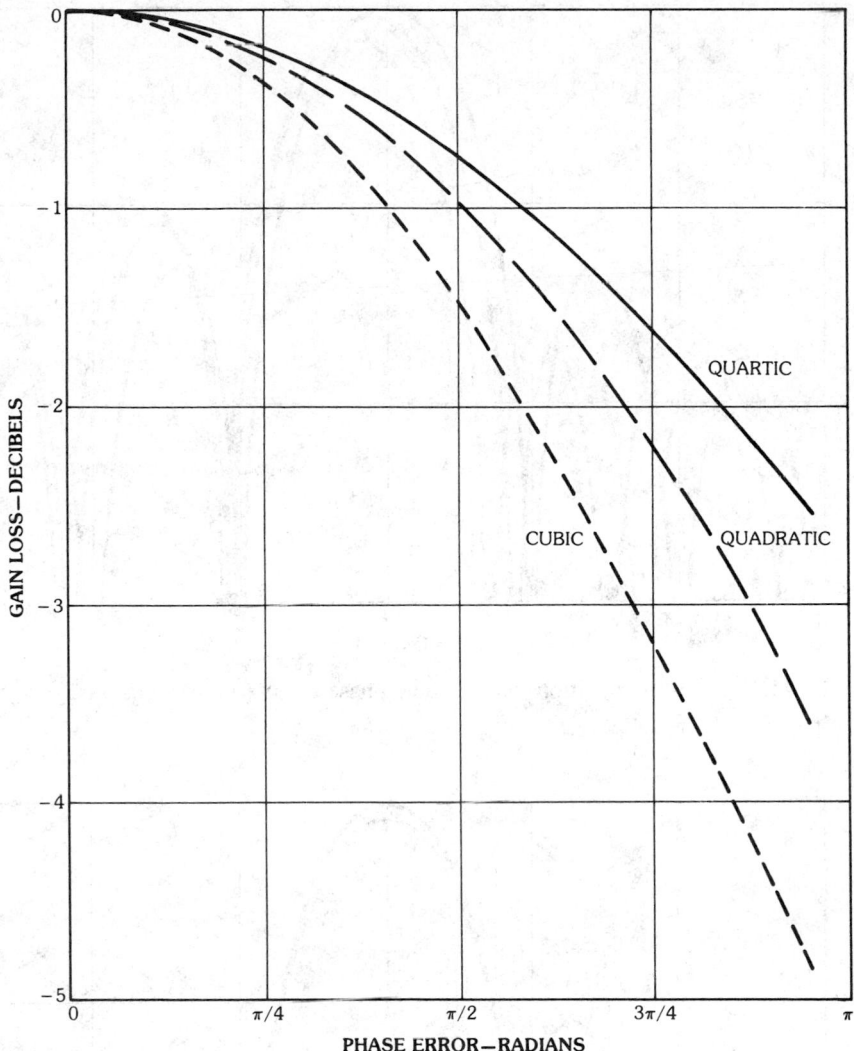

Fig. 10. Gain loss due to various phase errors (one-dimensional case).

error the tolerance for the same boresight gain loss is much tighter, as illustrated in Fig. 10.

On the other hand, if side lobe degradation is more of a concern, the criterion is taken to be the maximum error allowed for a peak side lobe level. Among the various aberrations the cubic error, which causes asymmetrical side lobe distortion, is the most undesirable one. In two-dimensional cases a contour plot of the diffraction pattern of a circular aperture with a cubic phase error (astigmatism) will show that concentric rings of equal intensity in the main lobe are shifted to one side and distorted into pear-shaped contours. This is commonly described as coma distortion in optics.

Usually quadratic errors can be corrected by displacing the feed from the focus

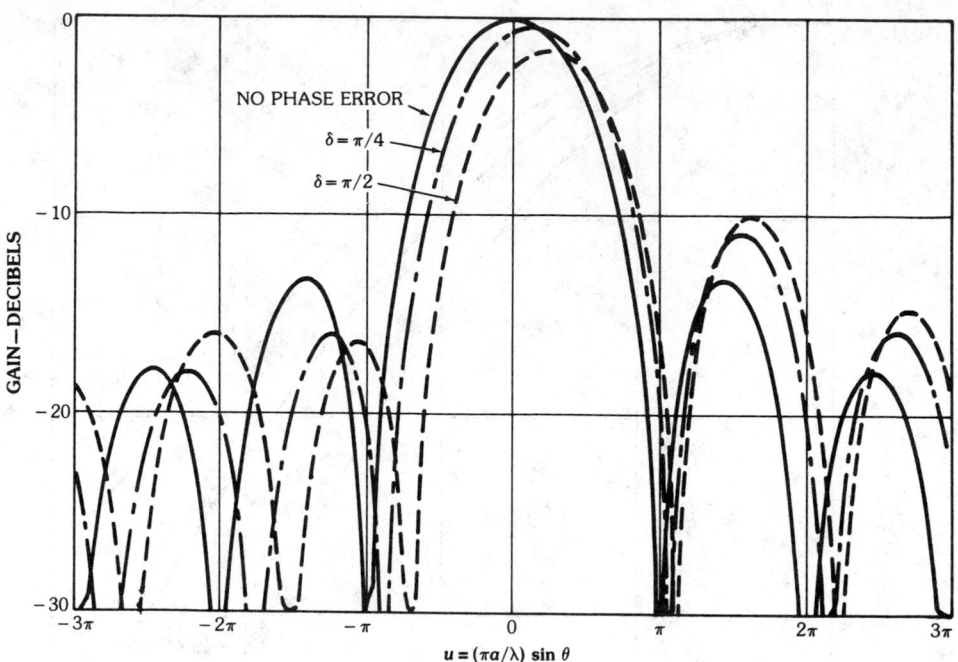

Fig. 11. Pattern degradation due to cubic phase error (one-dimensional case).

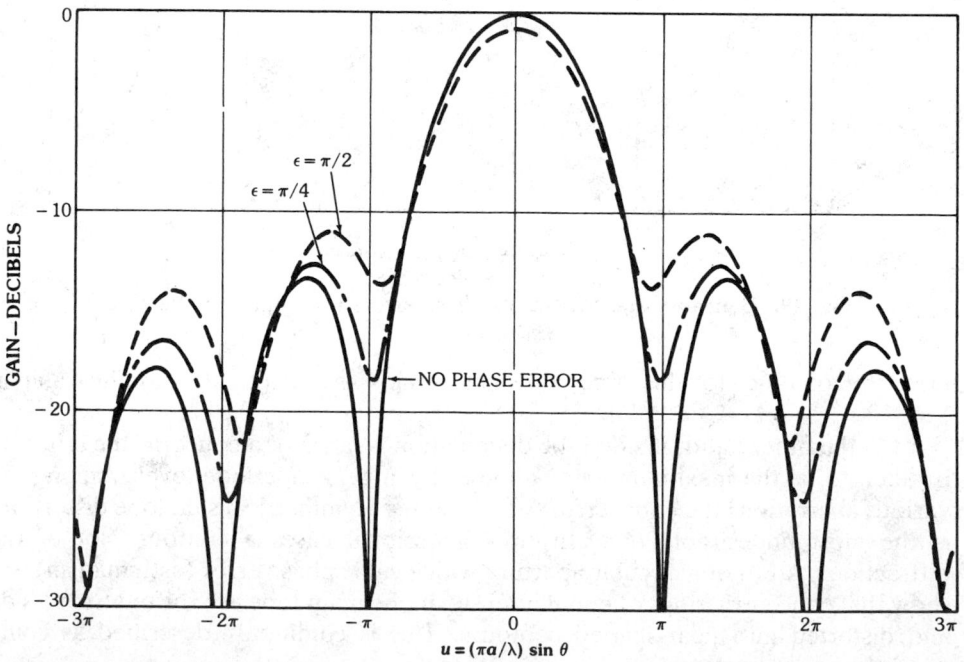

Fig. 12. Pattern degradation due to quartic phase error (one-dimensional case).

slightly along the axis. This defocusing technique cannot, however, compensate for the cubic or fourth-order errors. To minimize the coma aberration the system must be designed to satisfy the so-called Abbe sine condition, which will be discussed in the next section. For some applications the lens system is optimized to produce a minimum mean error instead of a particular type of error distribution. Thus the tolerance criterion is subject to system requirements and applications.

Although tolerance criteria for two-dimensional aberrations are more complex to analyze, the adverse effects are characterized by similar features derived in the one-dimensional case. If $\Delta L(\varrho, \phi) = a\varrho^n \cos m\phi$ is the distribution of the phase error, and $(1 - \varrho^2)^P$ is the amplitude taper over a normalized circular aperture, then the aperture efficiency (gain factor) is computed by

$$
\eta = \frac{1}{\pi} \frac{\left[\int_0^1 \int_0^{2\pi} (1 - \varrho^2)^P e^{ja\varrho^n \cos m\phi} \varrho \, d\varrho \, d\phi \right]^2}{\int_0^1 \int_0^{2\pi} (1 - \varrho^2)^{2P} \varrho \, d\varrho \, d\phi}
$$

$$
= 2(4P + 2) \left[\int_0^1 (1 - \varrho^2)^P J_0(a\varrho^n) \varrho \, d\varrho \right]^2
$$

Based on this formula the gain drop due to the phase error can be calculated. If gain loss is taken as the criterion for tolerance condition, then in general the tolerance of phase error for a circular aperture is not as tight as that for a one-dimensional case, and the tolerance increases with the power of ϱ. Shown in Table 1 are some numerical examples of gain loss as a function of the phase errors. Using 1-dB loss as a criterion, one can specify tolerance conditions for various errors as given in Table 2.

The phase errors discussed so far are strictly systematic errors. Random errors are not considered here. For a more detailed treatment on two-dimensional lens aberrations and random errors, readers are referred to Born and Wolf [6] and the collection of papers in [8] by Love.

5. Wide-Angle Dielectric Lenses

For multibeam or scanning applications a lens must be designed for wide-angle coverage. As the phase center is displaced from the axis to produce an off-axis beam, various lens aberrations will be generated. Among these aberrations the cubic phase error is the most undesirable effect as far as antenna performance is concerned. The resultant beam degradations are manifested by gain loss and coma lobe distortions [7]. Thus to maintain satisfactory scanning characteristics the cubic phase error due to an offset feed must be minimized. In this section several techniques to achieve relatively wide-angle coverage are introduced.

Abbe Sine Condition

As is well known in optics, a collimating lens can be designed to be coma free for a limited scan by imposing the Abbe sine condition [9]. Mathematically the Abbe sine condition requires that

Table 1. Boresight Gain Drop As a Function of Phase Error for Aperture Distribution $E(\varrho, \phi) = (1 - \varrho^2)^P \exp(ja\varrho^n \cos m\phi)$

	Gain Drop (dB)							
	$P = 0$				$P = 1$			
Peak Error	$n = 1$	$n = 2$	$n = 3$	$n = 4$	$n = 1$	$n = 2$	$n = 3$	$n = 4$
20°	0.13	0.09	0.07	0.05	0.09	0.04	0.02	0.02
40°	0.54	0.35	0.26	0.21	0.36	0.18	0.10	0.07
60°	1.22	0.80	0.59	0.47	0.80	0.40	0.23	0.15
80°	2.21	1.42	1.05	0.83	1.44	0.70	0.41	0.27
100°	3.55	2.23	1.63	1.28	2.28	1.09	0.64	0.42

Table 2. Primary Aberrations and Tolerance Conditions of a Circular

		Tolerance	
Type of Aberration	Representation	$P = 0$ (approximation)	$P = 1$
Spherical aberration	$\epsilon \varrho^4$	$\epsilon \lesssim 0.25\lambda$	$\epsilon \leq 0.43\lambda$
Coma	$\delta \varrho^3 \cos \phi$	$\delta \lesssim 0.22\lambda$	$\delta \leq 0.35\lambda$
Astigmatism	$\gamma \varrho^2 \cos 2\phi$	$\gamma \lesssim 0.19\lambda$	$\gamma \leq 0.27\lambda$
Curvature of field	$\beta \varrho^2$	$\beta \lesssim 0.19\lambda$	$\beta \leq 0.27\lambda$
Distortion	$\alpha \varrho \cos \phi$	$\alpha \lesssim 0.15\lambda$	$\alpha \leq 0.19\lambda$

$$y = F_e \sin \theta \tag{24}$$

This condition is automatically fulfilled if the inner surface of a conventional waveguide lens is spherical. For a thin dielectric lens it is sufficient if the average shape of the lens is spherical [10]. The interpretation of this condition for a thick lens is that the initial and the final ray, when extended, intersect inside the lens on a circle of radius F_e. This is illustrated in Fig. 13, where F_e is the effective focal length, which is different from the F parameter defined previously.

Since the derivation of this condition is based on the approximation of paraxial rays, care should be taken not to apply the Abbe sine condition to systems with subtended feed angle much larger than 20°.

The first wide-angle dielectric lens based on the Abbe sine condition was developed by Friedlander [11] in 1946. He discovered that for a given focal length and thickness there is a family of lenses that satisfies the coma-free condition. But among these there is just one for which the aperture size is a maximum, characterized by the fact that the surfaces of the lenses meet at the edge. Furthermore, it is pointed out that if the dielectric constant is close to 2.6, the lens can be made to nearly satisfy the Abbe sine condition even with a flat inner surface. As reported in [11] a plano-convex lens of 50λ in diameter had been constructed

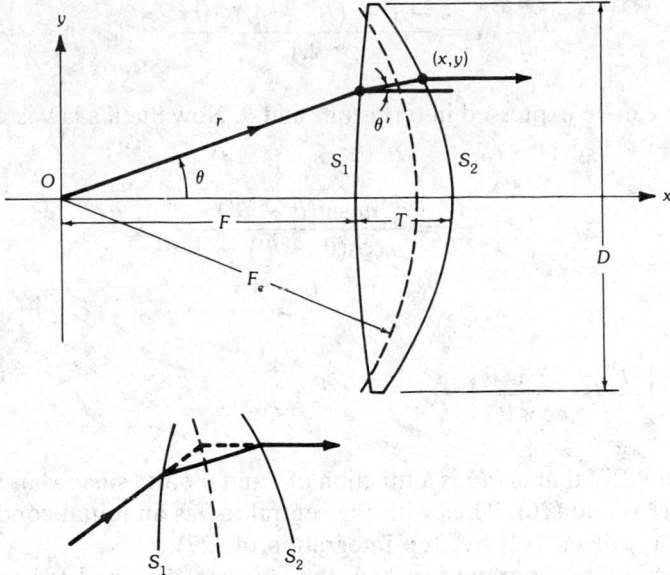

Fig. 13. Dielectric lens constrained by Abbe sine condition.

which produced a beam free of coma distortion over a scan coverage up to eight beamwidths.

In general, numerical integration is required to derive a coma-free lens. The design procedures are summarized in the following. When appropriate boundary conditions are specified, condition (24) and the phase constraint completely determine the lens contours. However, this does not imply that the solution can be analytically derived. It only means that a numerical solution can be obtained by step integration of the governing equations. To see this, first note that in reference to Fig. 13 the phase constraint can be written as

$$r + n[(y - r\sin\theta)^2 + (x - r\cos\theta)^2]^{1/2} - x = K \tag{25}$$

where $K = (n - 1)T$ is a constant determined by the central ray as a boundary condition. Next, substitute (24) into (25) to eliminate y. After some manipulation a quadratic equation in x can be deduced:

$$Ax^2 + Bx + C = 0 \tag{26}$$

where

$$A = \epsilon_r - 1$$
$$B = 2(r - K) - 2\epsilon_r r\cos\theta$$
$$C = \epsilon_r r^2 \cos^2\theta + \epsilon_r(F_e - r)^2 \sin^2\theta - (r - K)^2$$

The solution for x is simply

$$x = \frac{-B + (B^2 - 4AC)^{1/2}}{2A}$$

Thus x and y can be expressed in terms of r and θ. Now Snell's law is applied to get [12]

$$\frac{dr}{d\theta} = \frac{nr\sin(\theta - \theta')}{n\cos(\theta - \theta') - 1} \tag{27}$$

where

$$\theta' = \tan^{-1}\left[\frac{(F_e - r)\sin\theta}{x - r\cos\theta}\right]$$

It is clear from (27) that $dr/d\theta$ is a function of r and θ only, since x and y have been replaced by (24) and (26). Thus with the central ray as an initial condition, S_1 can be numerically solved [12] by step integration of (27).

Fig. 14 shows an example in which the lens was designed for satellite multibeam applications at 44 GHz. The Rexolite material has a dielectric constant of 2.54. The step size was 0.1° in θ. Other input parameters for the computer program include the aperture diameter D, the central thickness T, the distance F from the phase center to the vertex of the lens, and the effective focal length F_e.

The measured far-field patterns of the unmatched lens are shown in Fig. 15. The feed pattern is basically a $(\sin u)/u$ function of a square horn with a 20-dB edge taper. In this example the beam was only required to scan up to three beamwidths. Within this limited scan, as can be seen, there is virtually no coma distortion in the patterns. The first side lobe level of the lens is 23 dB below the peak. If the surface impedance is well matched [13], the side lobe level can be reduced by a few decibels.

It should be noted that not any combination of the input parameters can yield a

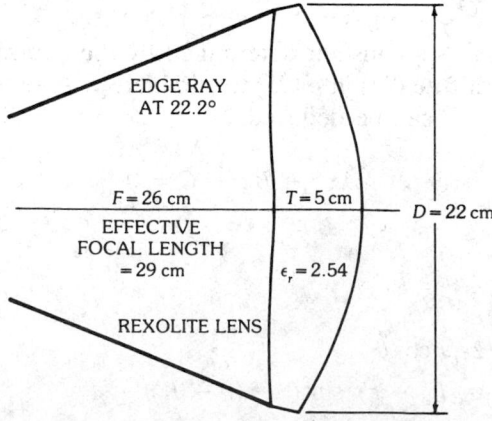

Fig. 14. Cross section of a coma-free dielectric lens.

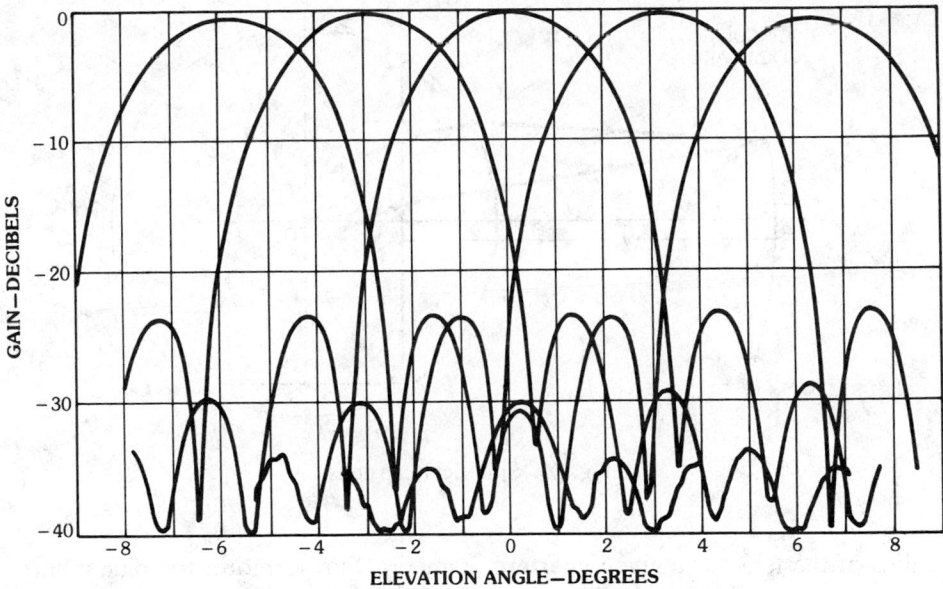

Fig. 15. Far-field patterns of the coma-free lens shown in Fig. 14.

complete lens. Some iterations are required to optimize a particular design. The numerical solution so derived for a given aperture size is certainly not unique. Any variation in the step size or the geometric parameters will lead to a different solution for the lens. Despite the fact that this is only an approximate solution, for all practical applications this numerical method provides versatile options and trade-offs for optimizing the performance.

It should be remarked that when the Abbe sine condition is imposed, the aperture power distribution can no longer be independently specified. In this case, as will be discussed in a later section, the aperture taper is mainly determined by the feed pattern. Hence a coma-free lens cannot provide very low side lobes if the feed pattern does not have enough illumination taper to begin with.

Schmidt Corrector

It is well known that a Luneburg lens is a truly wide-angle lens, because its radiation pattern is independent of the scan angle due to its inherent spherical symmetry and the property of possessing a perfect focus. Despite these unique features, however, the Luneburg lens has two basic drawbacks. One is that the peak side lobes cannot be held lower than about -17 dB in practice due to the inverse amplitude taper introduced by the lens [14]. The second disadvantage lies in the fact that the dielectric constant of the lens must vary as a function of radius, determined by $2 - (r/R)^2$, which makes it very difficult to fabricate. For these reasons a hemispherical or truncated spherical cap lens of uniform dielectric material, as shown in Fig. 16, has been used as an alternative for wide-scan applications.

A spherical cap lens does not have a well-defined focus, but the spherical aberrations can be minimized by properly adjusting the ratio of focal length to the

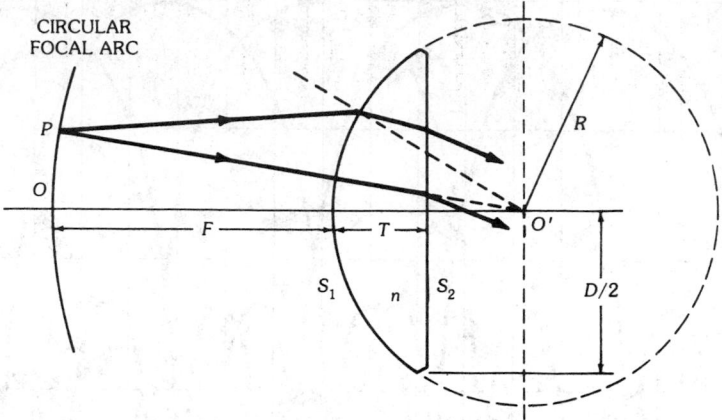

Fig. 16. Spherical cap lens.

radius of the lens for a given aperture diameter. Furthermore, the planar surface can be easily corrected to achieve a perfect focus if the aberration is small. Such corrected spherical cap lenses have been found to have excellent scanning characteristics. Other desirable features of these cap lenses are light weight, low volume, and simplicity in fabrication compared with the Luneburg lens. The only disadvantage of the cap lens is that the projected aperture decreases with scan angle as opposed to a constant aperture for a fully spherical lens.

Shown in Fig. 17 are some typical examples of the phase errors plotted across the aperture at various scan angles for two spherical cap lenses with different radii. It is seen that the phase error is almost symmetrical with respect to the central ray for scans up to $\pm30°$. The maximum error is a function of the dielectric constant, radius of the inner surface, and the F/D ratio. If maximally flat phase over the central region is required, the F/R ratio [14] is $(n - 1)^{-1}$. But the optimum condition is the one that produces a minimum average phase error over the whole aperture, similar to that shown in Fig. 17a. This kind of quadratic error can be easily corrected to yield perfect phase for the beam at boresight. Furthermore, the residual phase errors for other off-axis beams can be kept within an acceptable level.

A corrected spherical cap lens can be considered as a superposition of a cap lens and a Schmidt type corrector, which is commonly used in the optical design of a spherical mirror where spherical aberration is corrected.

To determine the corrector profile, basic lens equations presented in Section 2 are applied. Using x_1 as a running variable, the spherical pickup surface is defined by

$$y_1 = [R^2 - (x_1 - F - R)^2]^{1/2} \tag{28}$$

With this given, the second surface can be found by solving (1) and (2). It can be shown that in terms of x_1 and y_1, the variable x_2 is given by

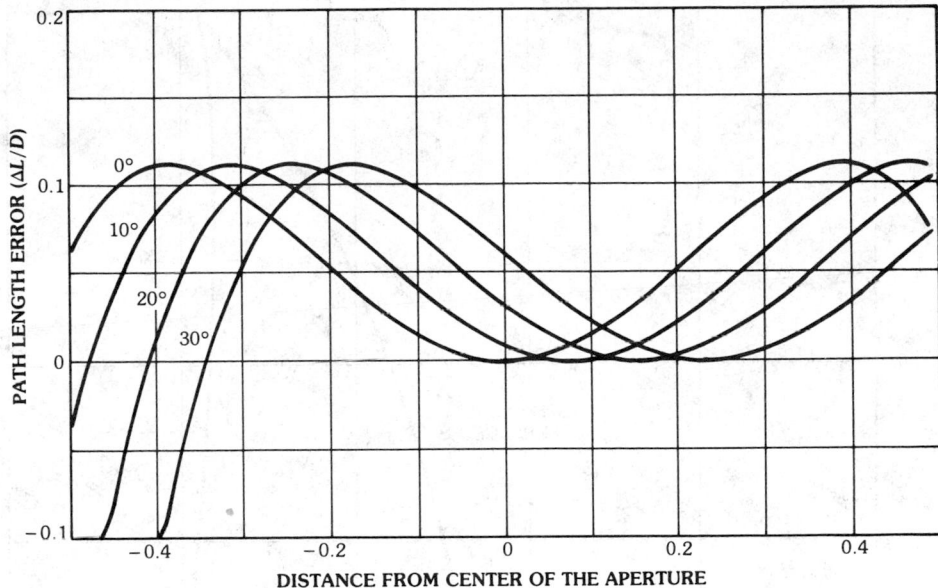

a

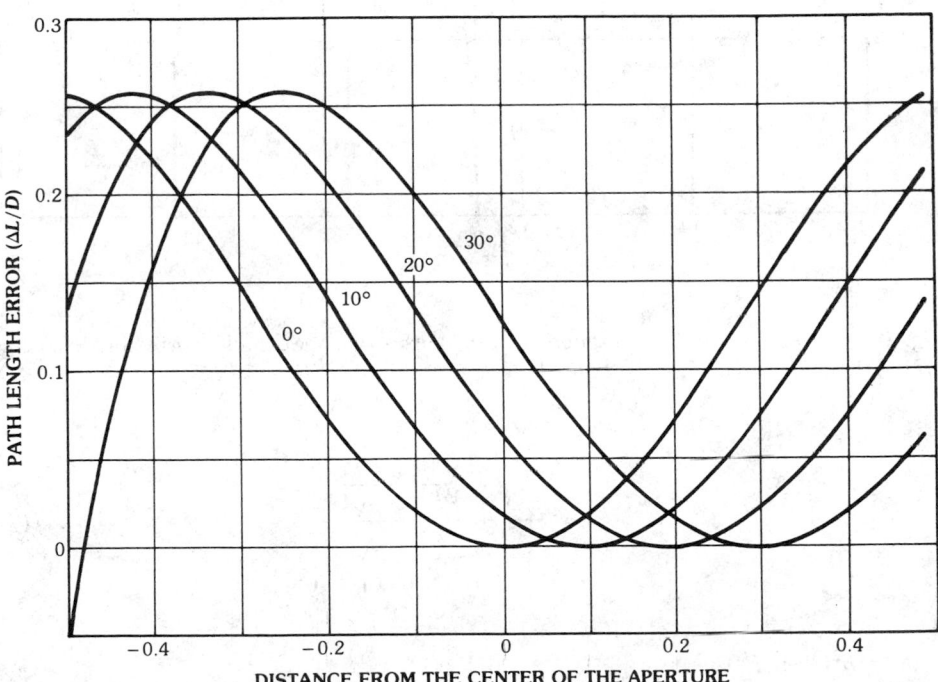

b

Fig. 17. Path length error for two spherical cap lenses across the aperture for different scan angles. (*a*) With *F/D* = 1.0, *R/D* = 0.85, ϵ_r = 2.54. (*b*) With *F/D* = 1.0, *R/D* = 1.0, ϵ_r = 2.54.

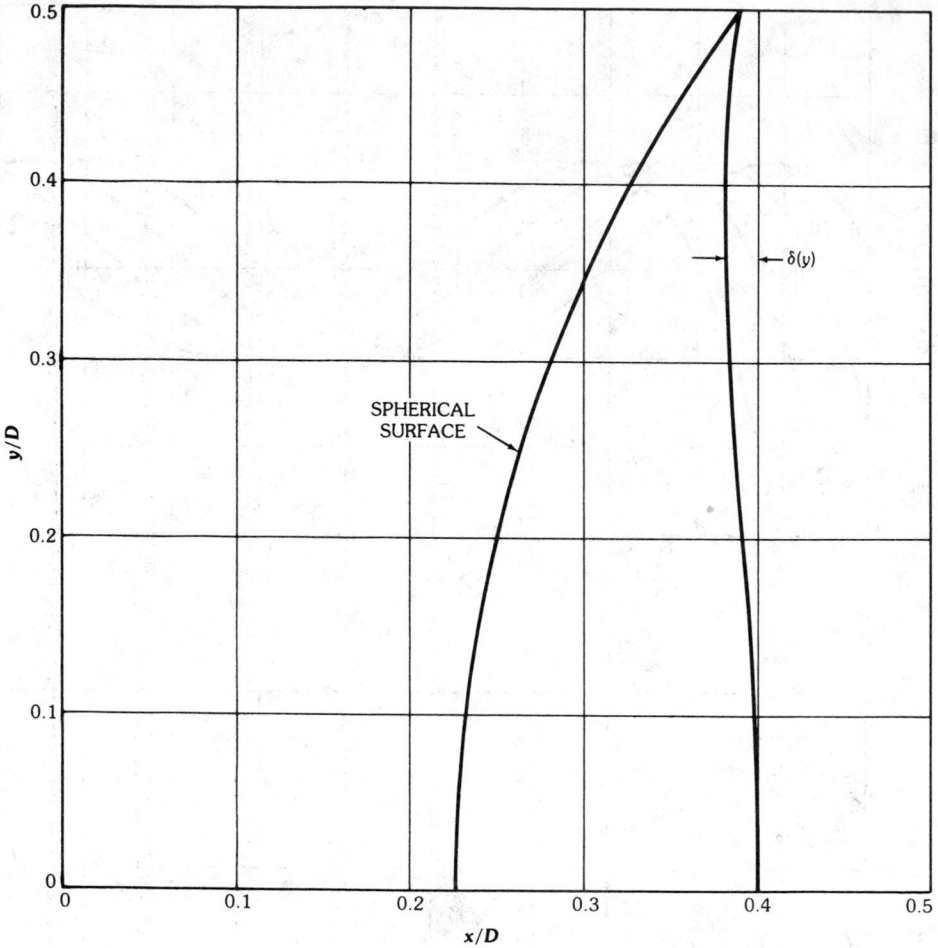

Fig. 18. Corrections for two spherical cap lenses with Schmidt correctors. (*a*) For $F/D = 1.0$, $R/D = 0.85$, $\epsilon_r = 2.54$. (*b*) For $F/D = 1.0$, $R/D = 1.0$, $\epsilon_r = 2.54$.

$$x_2 = \frac{-B + \sqrt{B^2 - 4AC}}{2A} \qquad (29)$$

where

$$A = n^2(1 + P^2) - 1$$
$$B = 2n^2 P(Q - y_1) + 2\sqrt{x_1^2 + y_1^2} - 2n^2 x_1 - 2(n - 1)T$$
$$C = n^2(Q^2 - 2Qy_1 + x_1^2 + y_1^2) - [(n - 1)T - \sqrt{x_1^2 + y_1^2}]^2$$

and

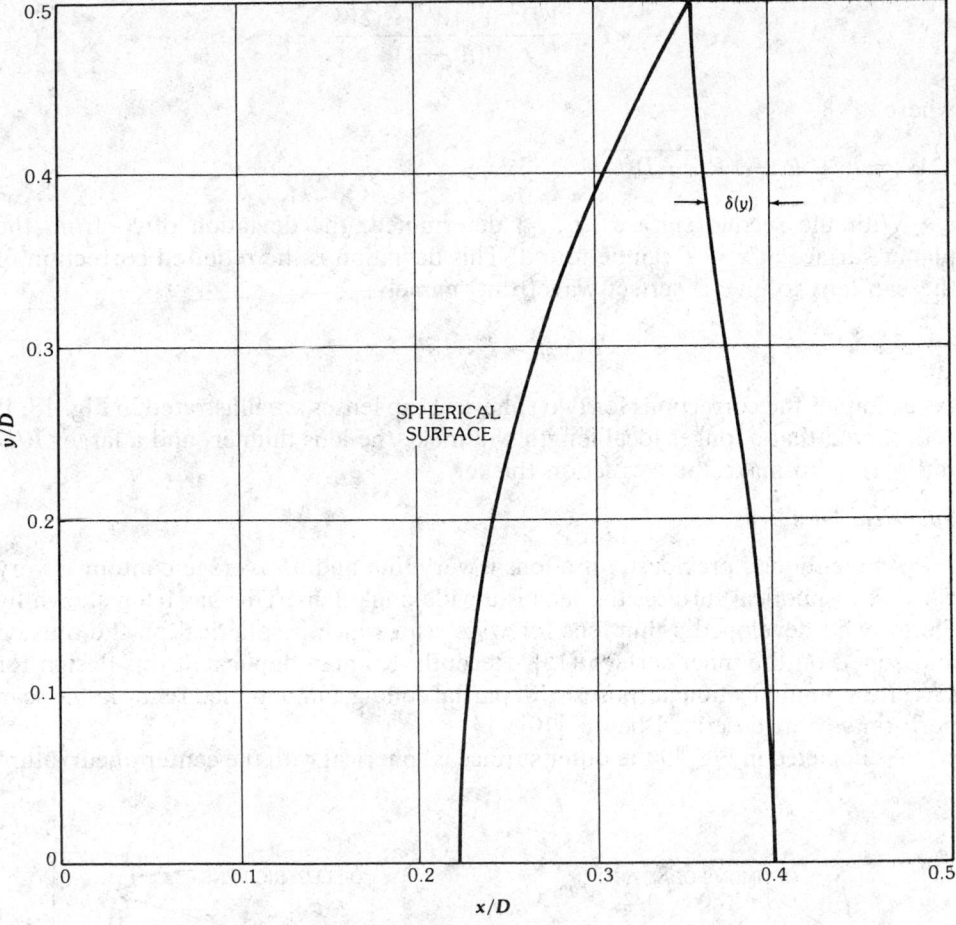

Fig. 18, *continued.*

b

$$P = \frac{F + R - n^2\sqrt{x_1^2 + y_1^2}}{n^2(F + R - x_1)} \frac{y_1}{\sqrt{x_1^2 + y_1^2}}$$

$$Q = (F + R) \frac{(n^2 - 1)\sqrt{x_1^2 + y_1^2} + (n - 1)T}{n^2(F + R - x_1)} \frac{y_1}{\sqrt{x_1^2 + y_1^2}}$$

The corresponding y_2 is computed by

$$y_2 = Px_2 + Q \tag{30}$$

If the lens is designed to have zero thickness at the edge, the central thickness must be chosen to be

$$T = \frac{\sqrt{D^2 + 4W^2} - 2W}{2(n-1)} \tag{31}$$

where

$$W = F + R - \sqrt{R^2 - D^2/4}$$

With the second surface (x_2, y_2) determined, the deviation of S_2 from the planar surface at $x = T$ can be found. This deviation is the required correction of the cap lens to give a perfect wavefront, namely,

$$\delta(y_2) = F + T - x_2 \tag{32}$$

As examples the corrections for two spherical cap lenses are illustrated in Fig. 18. It is observed that a longer focal length will make the lens thinner, and a larger R/D ratio tends to make the correction thicker.

Spherical Thin Lens

As mentioned previously, if a lens is very thin and its average contour is very close to a spherical surface, the lens is a wide-angle lens. This has been shown by Shinn, who developed a thin lens for wide scan which is spherical on the outside and zoned on the inner surface [15]. Recently Rotman duplicated this design for satellite communication purposes with partial zoning and obtained remarkable scan performance as described below [10].

As depicted in Fig. 19 the outer surface is spherical with the center of curvature

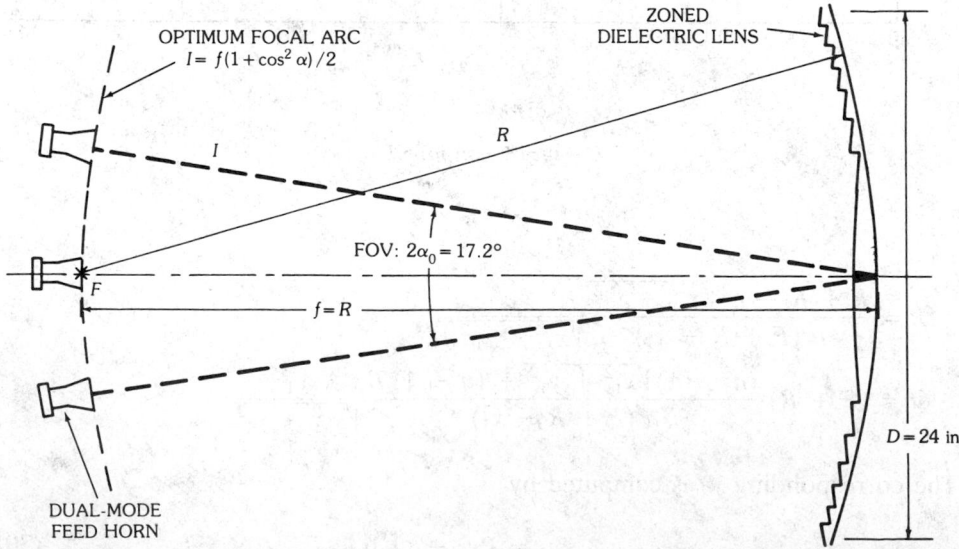

Fig. 19. Ehf aplanatic dielectric lens with multiple feeds. (*After Rotman [10]; reprinted with permission of Lincoln Laboratory, Massachusetts Institute of Technology, Lexington, Massachusetts*)

located at the focal point. The inner surface is determined by solving (16) and (17). Starting from the center the lens is stepped whenever the thickness reduces to a certain minimum value. For machining, the central and first outer zones of the inner surface of the lens are represented by a fourth-degree polynomial, while the outer five zones are approximated by straight line segments, determined by the points of discontinuity and corners on the contour. These approximations specify the inner surface of the lens to within 2 mils of the theoretical values. The lens was machined on a numerically controlled lathe in accordance with the numerical inputs.

As the beam is scanned and as the frequency changes, phase errors will occur across the radiating aperture. Since this lens is very thin, with its front surface radius R equal to its focal length, it obeys the Abbe sine condition and hence has minimum coma distortions. The only remaining significant phase error is the spherical aberration which, according to Shinn, is determined by the scanning locus (focal arc) and is independent of the shape of the lens.

The spherical aberration, measured as the path length error with respect to the central ray, is given by

$$\delta = \frac{1}{2}\frac{y^2}{f}\left(\frac{f}{\ell}\cos^2\alpha - 1\right) + \frac{1}{2}\frac{x^2}{f}\left(\frac{f}{\ell} - 1\right) \tag{33}$$

where the parameters are defined in Fig. 20. A compromised locus for the focal points, which balances the aberrations in both the yz and xz planes, is chosen to be [10]

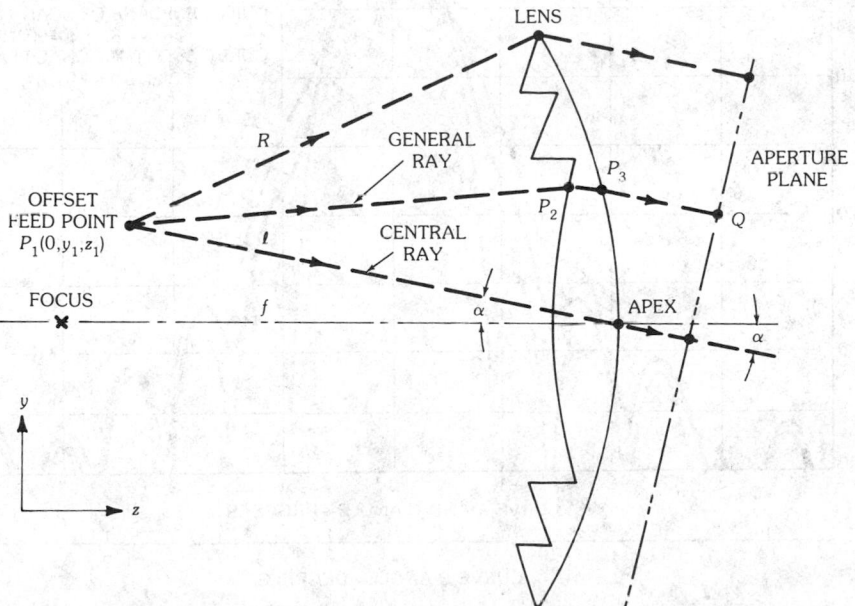

Fig. 20. Ray-trace analysis of zoned lens. (*After Rotman [10]; reprinted with permission of Lincoln Laboratory, Massachusetts Institute of Technology, Lexington, Massachusetts*)

$$\ell = \frac{1}{2}f(1 + \cos^2\alpha) \tag{34}$$

An experimental 90λ-diameter zoned lens operating at 44 GHz has been designed which generates a beam of 0.7° half-power beamwidth and 47 dBi gain, measured at the subsatellite point, over a 5-percent bandwidth. The beam can be steered over a total scan angle of 18° for full earth coverage with a scanning loss of less than 1 dB. Typical measured radiation patterns of the zoned lens are shown in Fig. 21.

Bifocal Lenses

Wide-scan capabilities can also be achieved by using bifocal systems, which are designed to have two perfect foci in the principal plane for two off-axis beams symmetrically displaced with respect to the axis. The aberrations of other beams that lie in between the limiting scans are relatively small compared with the cases where the system is designed for only one focal point on axis. Bifocal systems have been employed in both reflector [16, 17] and dielectric lens antennas [18]. It is, however, much more complex to design the bifocal dielectric lens because the algebra is complicated. For simplicity, only the design procedures are summarized below.

As shown in Fig. 22, let $(0, a)$ and $(0, -a)$ be the conjugate focal points of the

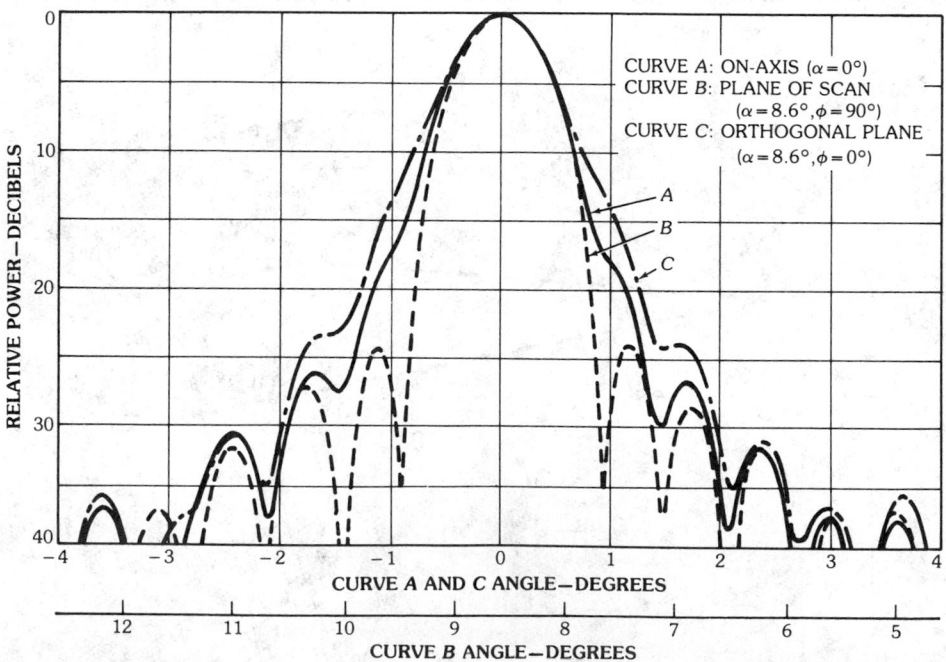

Fig. 21. Measured radiation patterns of zoned spherical thin lens (45.5 GHz). (*After Rotman [10]; reprinted with permission of Lincoln Laboratory, Massachusetts Institute of Technology, Lexington, Massachusetts*)

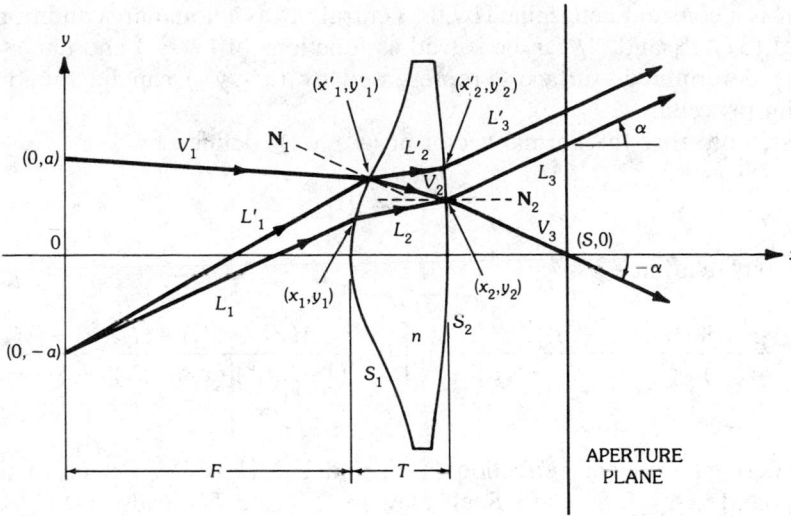

Fig. 22. Bifocal lens design.

lens for the two limiting scans at angles α and $-\alpha$ with respect to the boresight. The normal vector at (x_2, y_2) on S_2 can be denoted by

$$\mathbf{N_2} = \hat{\mathbf{i}} + D_2\hat{\mathbf{j}}$$

Then Snell's law requires that

$$n^2\left[1 - \frac{[x_2 - x_1 + D_2(y_2 - y_1)]^2}{(1 + D_2^2)[(x_2 - x_1)^2 + (y_2 - y_1)^2]}\right] = 1 - \frac{(1 + D_2\tan\alpha)^2}{(\sec^2\alpha)(1 + D_2^2)} \quad (35)$$

which is derived from the angular relation between $\mathbf{L_2}$, $\mathbf{L_3}$, and $\mathbf{N_2}$ by taking the dot product of the vectors. From this equation D_2 can be solved as a function of (x_1, y_1) and (x_2, y_2), which are either given as initial conditions or previously determined in the step computations. Now the task is to compute (x'_1, y'_1) in terms of (x_1, y_1) and (x_2, y_2), where (x'_1, y'_1) is the point of incidence of the ray from the upper focal point which emerges at (x_2, y_2) with scan angle $-\alpha$. To this end, apply Snell's law at (x_2, y_2) for the ray with scan angle $-\alpha$, which leads to

$$n^2\left[1 - \frac{[x_2 - x'_1 + D_2(y_2 - y'_1)]^2}{(1 + D_2^2)[(x_2 - x'_1)^2 + (y_2 - y'_1)^2]}\right] = 1 - \frac{(1 - D_2\tan\alpha)^2}{(\sec^2\alpha)(1 + D_2^2)} \quad (36)$$

This is defined by $\mathbf{V_2}$, $\mathbf{V_3}$, and $\mathbf{N_2}$. Next, impose the path length constraint for the ray along $\mathbf{V_1}$, $\mathbf{V_2}$, and $\mathbf{V_3}$,

$$[x'_1{}^2 + (y'_1 - a)^2]^{1/2} + n[(x_2 - x'_1)^2 + (y_2 - y'_1)^2]^{1/2}$$
$$+ (S - x_2)\sec\alpha + [y_2 - (S - x_2)\tan\alpha]\sin\alpha = K \quad (37)$$

where K is a constant determined by the central ray as a boundary condition. From (36) and (37) x'_1 and y'_1 can be solved as functions of (x_1, y_1) and (x_2, y_2). With (x'_1, y'_1) determined, surface 2 represented by (x'_2, y'_2) can be found by the following procedures.

First, note that the normal vector at (x'_1, y'_1), defined by

$$\mathbf{N}_1 = \hat{\mathbf{i}} + D_1\hat{\mathbf{j}}$$

satisfies the relation

$$1 - \frac{[x'_1 + D_1(y'_1 - a)]^2}{(1 + D_1{}^2)[(x'_1{}^2 + (y'_1 - a)^2]} = n^2\left\{1 - \frac{[(x_2 - x'_1) + D_1(y_2 - y'_1)]^2}{(1 + D_1{}^2)[(x_2 - x'_1)^2 + (y_2 - y'_1)^2]}\right\} \tag{38}$$

This is derived from the refraction of V_1 and V_2. Thus D_1 is given in terms of (x'_1, y'_1) and (x_2, y_2). Similarly Snell's law for L'_1 and L'_2 leads to

$$1 - \frac{[x'_1 + D_1(y'_1 + a)]^2}{(1 + D_1^2)[x'_1{}^2 + (y'_1 + a)^2]} = n^2\left\{1 - \frac{[(x'_2 - x'_1) + D_1(y'_2 - y'_1)]^2}{(1 + D_1^2)[(x'_2 - x'_1)^2 + (y'_2 - y'_1)^2]}\right\} \tag{39}$$

This is one of the two equations to be solved for x'_2 and y'_2. The other equation is obtained by imposing the path length constraint along $\mathbf{L}'_1$, $\mathbf{L}'_2$, and $\mathbf{L}'_3$, which is

$$[x'_1{}^2 + (y'_1 + a)^2]^{1/2} + n[(x'_2 - x'_1)^2 + (y'_2 - y'_1)^2]^{1/2}$$
$$+ (S - x'_2)\sec\alpha - [y'_2 + (S - x'_2)\tan\alpha]\sin\alpha = K \tag{40}$$

From (39) and (40) the two unknowns x'_2 and y'_2 can be uniquely determined. Repeating the process until the lens is completely shaped, a bifocal system is derived numerically. Clearly, the formulation is complex and extensive programming efforts are necessary; nevertheless, the design procedures are actually straightforward. Similar techniques have been applied to reflector systems with excellent results [16, 19].

As input parameters the focal points $(0, \pm a)$, the focal length F, the central thickness T, and the limiting scan angle α must be specified. The shaping starts with the central ray through P_1 and P_2 from the lower focal point. The exit point P_2 is given by $(F + T, b)$ where b is given by

$$\frac{a}{\sqrt{a^2 + F^2}} = n\frac{b}{\sqrt{b^2 + T^2}}$$

The step computations continue until a desirable aperture size is achieved. Due to the inherent symmetry of the lens in the principal plane, only one-half of the lens cross section is required in the computations. A complete lens is obtained by taking a figure of revolution of the contour so generated. This implies that a ring focus is obtained.

The shaping technique discussed in this section for dielectric lenses with bifocal points is different from those presented previously in that no step integration is involved and the step increments are relatively large. To completely define the surface points in between, a smoothing process of curve fitting is necessary. Due to the symmetry, only even power terms are needed. For most applications a fourth-order polynomial is sufficient. If, however, the geometry is such that the resultant step size is too large to warrant a smooth lens, this bifocal approach may not be acceptable. The other imperfection of this design is that there is a small amount of quadratic phase error in the orthogonal plane for any scan in the principal plane. This is due to the fact that the design is based on a two-dimensional analysis, whereas the actual lens is a figure of revolution of the contour generated.

6. Taper-Control Lenses

So far the treatment of lens design has been limited to the subjects of phase correction and wide-angle scanning. In addition to these lens there is another class of dielectric lens that is shaped to control the aperture taper for a special beam pattern and side lobe structure. The design principle for this type of lens is to explicitly impose the power conservation law in place of one of the constraints applied in the shaping formulation.

The significance of imposing the power law in the design can be illustrated by examining the aperture taper of a wide-angle lens constrained by the Abbe sine condition as discussed previously. Recall that when the Abbe sine condition is imposed on the lens design the aperture power distribution no longer can be independently specified. In this case the aperture power distribution is almost identical with that of the feed pattern. In other words, the lens performs no special transformation in the power distribution, a function that often is required in other applications. An examination of the power conservation law associated with the wide-angle lens verifies this last statement.

In differential form the power law requires that

$$f(\theta) \sin \theta \, d\theta = P(y) y \, dy$$

where $f(\theta)$ is the feed power pattern (in watts per unit solid angle) and $P(y)$ is the aperture power distribution (in watts per unit area). Using (24)

$$P(y) = \frac{f(\theta)}{F_e^2 \cos \theta}$$

For a feed angle (half-cone) less than 40°, the difference in edge taper between the feed pattern and the aperture distribution is less than 1 dB. Therefore, if taper control is desired, the power conservation law which specifies the aperture distribution must be imposed.

Referring to Fig. 23, consider a feed horn illuminating a dielectric lens with rotational symmetry. A reference plane for the aperture is taken at $x = S$. For nonuniform phase distribution, such as the one required for a special beam pattern [20], the lens shape is a function of the location of the reference plane, but the

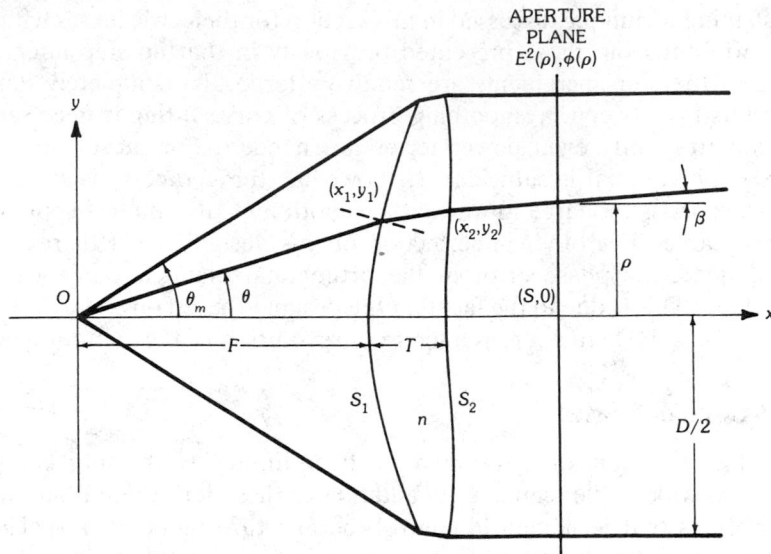

Fig. 23. Taper-control lens.

effect is inconsequential as long as the reference plane is not too far from the lens. Let the half-angle of the feed pattern subtended by the lens be θ_m. Then the power conservation law can be expressed as

$$\frac{\int_0^\theta g^2(\theta)\sin\theta\,d\theta}{\int_0^{\theta_m} g^2(\theta)\sin\theta\,d\theta} = \frac{\int_0^\varrho E^2(\varrho)\varrho\left[1 - \left(\frac{1}{k}\frac{d\phi}{d\varrho}\right)^2\right]^{1/2}d\varrho}{\int_0^{D/2} E^2(\varrho)\varrho\left[1 - \left(\frac{1}{k}\frac{d\phi}{d\varrho}\right)^2\right]^{1/2}d\varrho} \tag{41}$$

where $g^2(\theta)$ is the given feed pattern, watts per unit solid angle, and $E(\varrho)$ and $\phi(\varrho)$ are the specified aperture amplitude and phase distribution, respectively, all referred to the plane at $x = S$.

In (41), $k = 2\pi/\lambda$ is the wave number in free space, and the expression within the square brackets accounts for the cosine projection of the Poynting vector on the lens axis for a nonuniform wavefront. For the cases where a uniform phase is assumed, this factor reduces to unity. With $g(\theta)$, $E(\varrho)$, and $\phi(\varrho)$ given, a mapping relationship between θ and ϱ is established through (41).

The second condition to be imposed is Snell's law on surface 1, which was given in (2):

$$\frac{dy_1}{dx_1} = \frac{nL_1(x_2 - x_1) - L_2 x_1}{L_2 y_1 - nL_1(y_2 - y_1)} \tag{42}$$

The last constraint to be imposed is the path length condition specified by the phase distribution $\phi(\varrho)$, i.e.,

$$L_1 + nL_2 + [(S - x_2)^2 + (\varrho - y_2)^2]^{1/2} - \phi(\varrho)/k = K' \tag{43}$$

where $K' = F + S + (n - 1)T$.

To proceed, note that from the slope of the exit ray, y_2 is given by

$$y_2 = \varrho - (S - x_2)\tan\beta \tag{44}$$

The goal now is to solve (x_2, y_2) in terms of (x_1, y_1) and other related points on the same ray trajectory. This is accomplished by substituting (44) in (43). After a lengthy but straightforward algebraic manipulation one can show that [21]

$$x_2 = \frac{-B + (B^2 - 4AC)^{1/2}}{2A} \tag{45}$$

where

$$A = \frac{n^2 - 1}{n^2 \cos^2\beta}$$

$$B = 2\left(P\tan\beta - x_1 - \frac{Q}{n^2 \cos\beta}\right)$$

$$C = x_1{}^2 + P^2 - Q^2/n^2$$

and

$$\beta = \sin^{-1}\left(\frac{1}{k}\frac{d\phi(\varrho)}{d\varrho}\right)$$

$$P = \varrho - S\tan\beta - y_1$$

$$Q = K' - (x_1{}^2 + y_1{}^2)^{1/2} - S\sec\beta + \phi(\varrho)/k$$

To derive the contours of the lens the aperture radius is first divided into a large number of equal steps $\Delta\varrho$. The mapping then generates the corresponding delta increment $\Delta\theta$ for each $\Delta\varrho$. Using the central ray as an initial condition one can readily compute the slope at the starting point of surface 1. From (42), or by symmetry, the slope at the vertex is infinity. The intersection of this vertical tangent and the first emerging ray from the focal point with an angle $\Delta\theta$ is the first point of (x_1, y_1). The corresponding (x_2, y_2) on surface 2 can be found from (44) and (45). By repeating the same procedures until the whole aperture is covered in ϱ, a complete lens is shaped.

Geometrically (41) maps a cone of half-angle θ into a circle of radius ϱ on the aperture, a priori, without advance knowledge of the actual profiles of the lens. The design objective is to shape the lens to serve as an optical transformer, which transforms the given feed pattern into the desired aperture distribution.

In the step computation the process of generating the lookup table for ϱ and θ can be carried out in either direction. However, for those cases where the lens is shaped to converge the power density toward the axis for low side lobes, tracing backward to find $\Delta\theta$ with equal steps $\Delta\varrho$ is more convenient. The step size, $\Delta\varrho$, plays an important role in the accuracy of the numerical solution. Normally 200 points for a lens of 20λ in radius are sufficient. The accuracy is judged by the asymptotic behavior of the solution after iterations and how close Snell's law is satisfied on surface 2 as a check. For very rigorous solutions, of course, a higher-order approximation, such as the Runge-Kutta method, may be employed.

For most applications the specified aperture phase distribution is uniform. In this case the preceding formulation can be much simplified, because $\phi(\varrho)$ and β are set to zero. Also it should be remarked that for the computer program the feed pattern and the aperture distribution need not be given in closed analytic form. The (ϱ, θ) mapping can be generated with discrete input data points.

Based on the formulation just described, a general program can be developed. The inputs to the program are listed in the following:

feed location from the lens	F
feed power pattern (assumed axisymmetric)	$g^2(\theta)$
maximum feed angle	θ_m
aperture amplitude distribution	$E(\varrho)$
aperture phase distribution	$\phi(\varrho)$
aperture size	D
dielectric constant	ϵ_r
central thickness	T
reference plane for aperture	S

The outputs of the program specify the coordinates of the lens profiles for both sides, (x_1, y_1) and (x_2, y_2). The design procedure starts with a proper choice of the focal length F for the lens. An initial estimate of F can be made by finding the location of the lens at which the cross section of the feed cone is about the size of the aperture. For a given dielectric constant ϵ_r, the central thickness T must be large enough to warrant a complete solution. Too small a T sometimes does not give enough dielectric medium at the outer rim of the lens for the rays to converge and perform the power transformation. Some optimization efforts are required in the design.

Since no restriction on the shape of the aperture taper has been imposed in the formulation, any reasonable power distribution with rotational symmetry can be specified to suit a particular application. It may be a standard $(1 - \varrho^2)^P$ taper; or a special function, such as a truncated Bessel type distribution discussed by Love [22], where the antenna pattern has a dip at the center for better earth coverage from a geostationary satellite orbit. For satellite multibeam antennas it may be desirable for each beam to have a very broad beamwidth with a fast rolloff and low side lobes, so that a more uniform earth coverage can be achieved. To produce such shaped beams the generalized Taylor distributions with complex zeros [20, 23] are more appropriate.

As an example, Fig. 24 shows a lens shaped for an aperture distribution $E(\varrho) = [1 - (\varrho/1.05)^2]^3$ with a uniform phase designed for -35-dB side lobes. The feed pattern is a standard E-plane pattern of a square horn. The horn size was chosen to have an edge taper of 20 dB at $\theta_m = 20°$. The dielectric constant is 2.54 for Rexolite

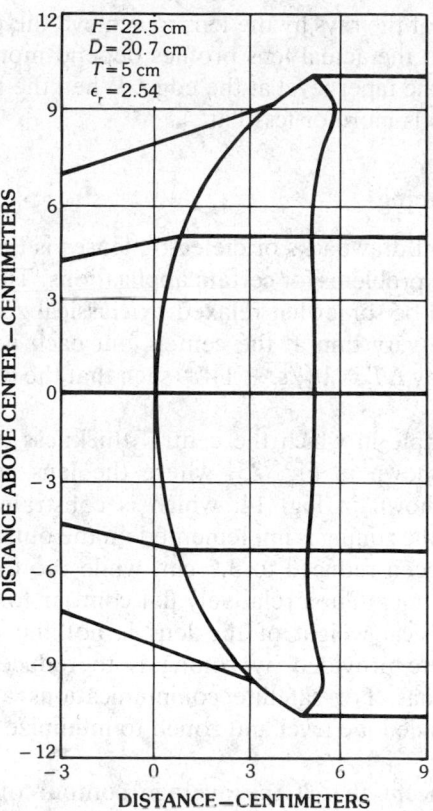

Fig. 24. A Rexolite lens with an aperture amplitude distribution of $[1 - (\varrho/1.05)^2]^3$.

material. The central thickness is 5 cm and the focal length is 22.5 cm. The aperture efficiency of this lens is 48 percent. The computed far-field pattern has a 3.2° beamwidth with a directivity of 36.4 dB for $D = 30.3\lambda$ (20.7 cm) at 44 GHz.

The transformation between the feed pattern and the aperture distribution is primarily accomplished by the first surface through refraction. The second surface is contoured mainly to satisfy the phase constraint. In general the solution of the lens is sensitive to the choice of F and T. This is usually true when the aperture distribution differs substantially from the feed pattern. When the required transformation is not too drastic, solutions exist over a wider range of F and T.

It should be noted that many lens solutions exist for the same power transformation if different input parameters, such as focal lengths, central thickness, etc., are specified. For example, a thinner lens can perform the same power transformation if a material of higher dielectric constant is used for the lens. Naturally, as the dielectric constant increases and the lens becomes thinner, surface mismatch becomes a limiting factor in choosing a very high dielectric constant. The F/D ratio is another factor that can be traded off with other design parameters. For a given D, a smaller F will make the subtended feed angle θ_m larger. If the horn size remains fixed, a larger θ_m means smaller spillover. In this case the lens will be less concave, since the illumination taper more closely matches the aperture taper, thus requiring less bending of the rays by the lens to achieve the power transformation. It should be noted that the actual lens profiles depend more on the shape of the feed pattern than just the taper level at the edge. When the input and output taper match closely, the lens is more or less flat.

7. Dielectric Lens Zoning

One of the inherent drawbacks of dielectric lenses is the physical volume and weight which may pose problems for certain applications. The penalty in weight for having a thick lens can be somewhat relaxed by classical zoning. That is, the lens can be designed to be very thin at the center, but each new zone is allowed to increase the thickness by $\Delta T = \lambda/(\sqrt{\epsilon_r} - 1)^{1/2}$, such that the overall phase difference is 2π.

Fig. 19 is an example in which the central thickness is kept to a minimum. Another example is shown in Fig. 25, where the lens accomplishes the same function as the one shown in Fig. 14, which is constrained by the Abbe sine condition. In this case the zoning is implemented on the outside of the lens, and the central thickness has been reduced to 3.5 cm, while the focal length F has been adjusted to 26.7 cm to maintain a relatively flat contour for the lens.

Reducing the physical weight of the lens is not the only merit of zoning. Another unique feature provided by zoning is to reduce coma aberrations in multibeam lens antennas. For satellite communications a lens antenna can be shaped to control the side lobe level and zoned to minimize the cubic phase errors for off-axis beams [21].

Based on the concept that if the average contour of a collimating lens is spherical the Abbe sine condition is approximately satisfied [24, 25], a special zoning technique for coma correction has been developed. This is accomplished by making the inner surface of the lens follow a circular arc on average. As shown in Fig. 26, when the inner surface of the shaped lens deviates from the circle by more than, in this case, 1λ, the lens is zoned to have the surface moved back by 2λ along

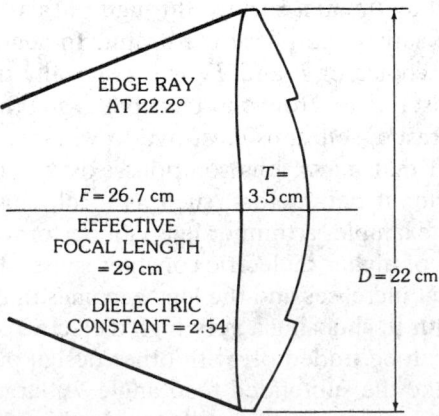

Fig. 25. Cross section of a coma-free zoned lens.

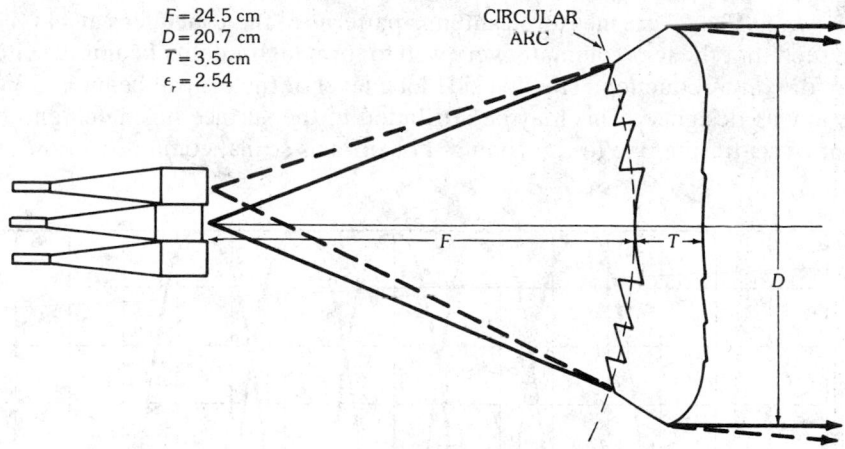

$F = 24.5$ cm
$D = 20.7$ cm
$T = 3.5$ cm
$\epsilon_r = 2.54$

CIRCULAR ARC

Fig. 26. Coma-corrected zoned lens for multibeam applications.

the ray path. Note that the zoning does not have to be 2λ per step. It can be any step size as long as the overall path length at the aperture satisfies the phase constraint. A smaller step, however, will result in a larger number of zones to conform to the spherical shape, which will cause more edge diffractions for off-axis beams and reduce the bandwidth to some extent.

The absolute phase, of course, can differ by any integral number of 2λ across each zone. In fact, to keep the second surface reasonably smooth in this example an additional wavelength is added to the total path length for each new zone to offset the effect of zoning on surface 1.

Shown in Fig. 26 is a zoned lens of Rexolite designed for the same aperture distribution and feed pattern as discussed in Fig. 24. The focal length is 24.5 cm with $T = 3.5$ cm and $D = 20.7$ cm.

It should be remarked that the zoning process does not significantly perturb the aperture distribution or the illumination function compared with the original design. The constraint of power conservation is unchanged; the additional step required in the shaping program is to give new boundary conditions as inputs whenever the inner surface deviates from the reference circle by a preset value. Effectively the lens is made up of zones of different lenses with different initial central thicknesses and focal lengths, but each zone is constrained to produce the corresponding portion of the same specified aperture distribution. Naturally, over an infinitesimal interval across the junction between zones the ray trajectory is not defined due to the discontinuity.

Since the edge of each zone is cut along the incident ray direction, very little scattering is expected for the beam on axis. For off-axis scans, however, a small amount of scattering and shadowing effect due to the steps occurs. The zoning technique presented here is similar to that proposed by Ronchi and Toraldo Di Francia [26] to correct the coma of a reflector, which was subsequently demonstrated by Provencher [27]. In 1961 Dasgupta and Lo presented a more detailed analysis on the same subject based on the diffraction theory [28]. Although this analysis does not apply to the case of lenses the approach is very similar.

Shown in Fig. 27 are measured antenna patterns of the zoned lens at 44 GHz. It can be seen that the lens collimates very well to form high-quality beams despite the zoning for coma reduction. The first side lobe level of the central beam is -30 dB, not as low as designed. This may be attributed to the surface mismatch and finite amount of scattering due to the zoning. For off-axis scans, coma distortion of the

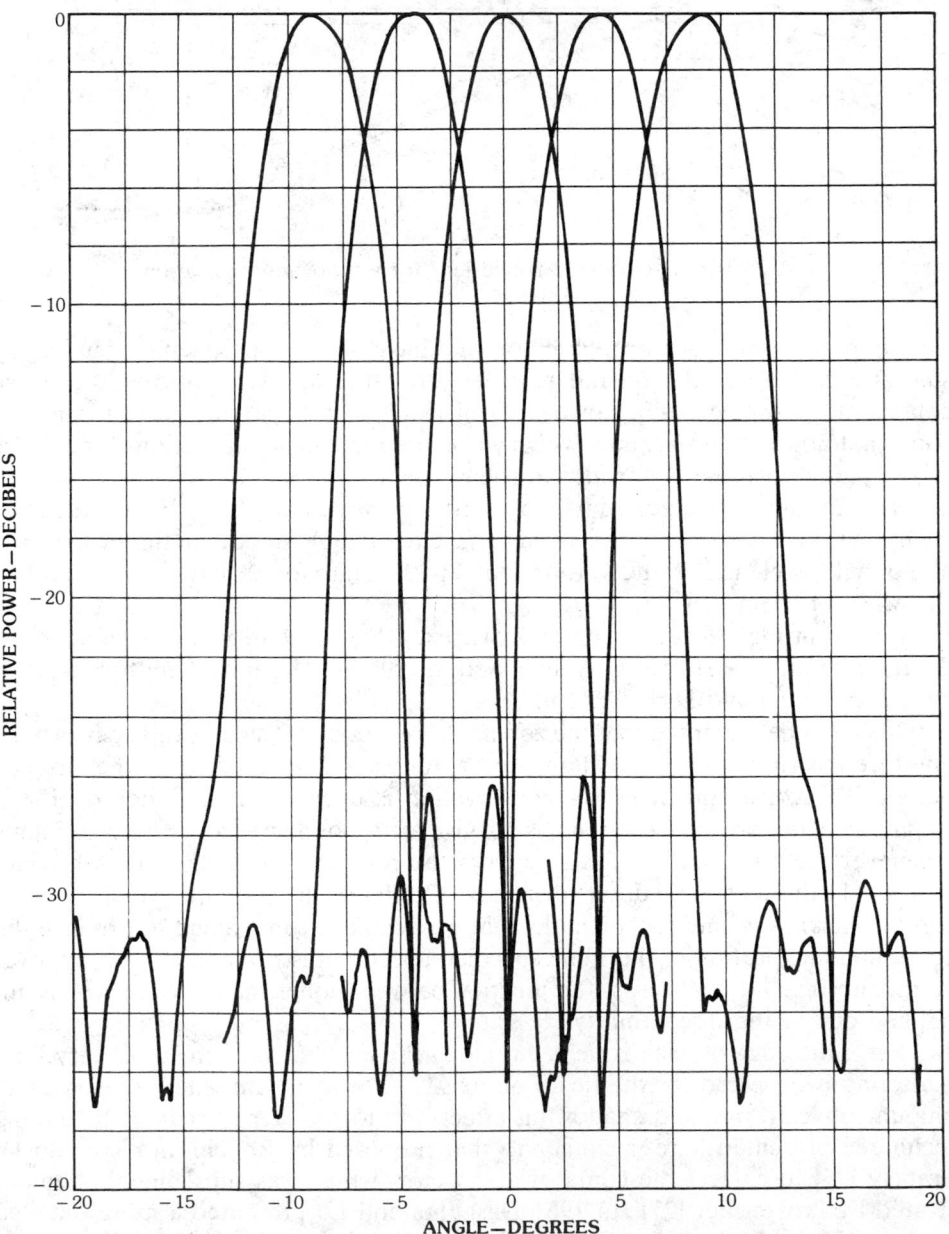

Fig. 27. Measured patterns of the coma-corrected lens.

main beam is negligible and the side lobe degradation is small. Thus it has been demonstrated that zoning and shaping can be combined to design a low side lobe, coma-corrected lens antenna.

8. Constrained Lenses

A constrained lens usually consists of an equal number of receiving and radiating elements on each surface of the lens, with corresponding elements connected by a transmission line for phase adjustment on a one-to-one basis. The lens transforms a divergent wavefront from a focal point into a collimated beam by providing the correct line lengths to compensate for the path differences. The term "transmission line" is used in a very broad sense here. Actually it can be any guiding structure such as waveguide, microstrip, coaxial line, etc., that serves as a phase or, ideally, time-delay medium.

The design principle of a constrained lens is quite simple. The only condition imposed in the design is the fundamental equal path length condition. As illustrated in Fig. 28, for a two-dimensional case this condition can be expressed as

$$[(x_1 - a)^2 + (y_1 - b)^2]^{1/2} + w + (T - x_2)\cos\alpha + y_2\sin\alpha = F + w_0 \qquad (46)$$

in which there are five variables, (x_1, y_1), (x_2, y_2), and w, that can be constrained. To uniquely determine the contour of either surface of a rotationally symmetric lens three more conditions must be specified. This freedom of choice in selecting additional constraints provides numerous options in design. For example, one of the two surfaces may be chosen to be planar or spherical. For some cases multiple focal points may be desired to enhance the scanning capabilities. Also, the question may be asked as to whether equal line lengths or unequal line lengths are to be used. Different considerations lead to different lens designs.

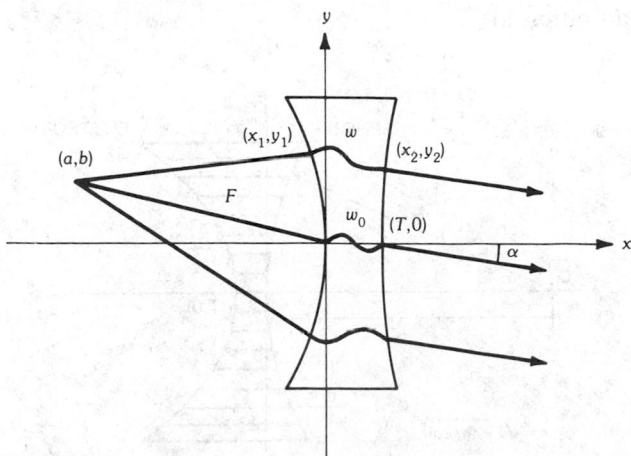

Fig. 28. Constrained lens design.

Waveguide Lens and Zoning

The use of waveguide lenses for microwave antenna applications has been reported by various authors [29, 30]. The waveguide lens is a special case of a constrained lens where straight sections of open-end waveguides are used as radiating elements. The radial positions of the elements on one side are identical with that of the corresponding ones on the opposite side of the lens. The simplest design is the one that has rotational symmetry with a single focal point on axis. The inner surface is a sphere of revolution of radius F centered at $(-F, 0)$, as shown in Fig. 29, while the outer surface is an ellipsoid centered at $[nF/(1 - n), 0]$ given by

$$\frac{y_2^2}{F^2} + \left[\frac{x_2 - T - Fn/(1 - n)}{Fn/(1 - n)}\right]^2 = 1 \tag{47}$$

where T is the central thickness and n is the index of refraction defined by λ/λ_g, with λ_g being the waveguide wavelength [31].

A two-dimensional waveguide lens exhibiting two focal points may also be designed. Following the treatment of Dion and Ricardi [32], let $(F \sin \alpha, -F \cos \alpha)$ be one of the two conjugate focal points of the lens in the principal plane, where F is the focal length and α the scan angle. Then the cross-section profile of the inner surface is

$$\frac{y_1^2}{F^2} + \frac{(x_1 + F \cos \alpha)^2}{F^2 \cos^2\alpha} = 1 \tag{48}$$

i.e., an ellipse with foci at the two point sources, while the equation of the outer surface is given by

$$\frac{y_2^2}{F^2} + \left[\frac{x_2 - T - (nF \cos \alpha)/(\cos \alpha - n)}{(nF \cos \alpha)/(\cos \alpha - n)}\right]^2 = 1 \tag{49}$$

which is also an ellipsoid.

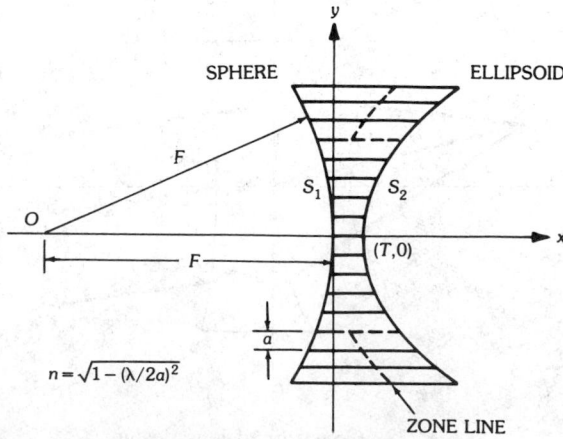

Fig. 29. Waveguide lens.

The thickness of the lens depends on the index of refraction; consequently the inside width of each waveguide determines the number of waveguides required for a given aperture size and the thickness of the lens, because n is a function of λ_g. For circular polarization, using square waveguides, the waveguide dimension a must be large enough to permit propagation of the dominant TE_{10} and TE_{01} modes and sufficiently small to suppress the higher-order modes. The latter requires that $a <$ 0.7λ and determines a maximum value of $n = 0.7$; usual values for n lie between 0.58 and 0.68. The lower limit is determined by the need to prevent an intolerable reflection loss at the surfaces. The F/D ratio determines the number of zones in the lens; increasing F/D decreases the number of zones, resulting in an increase in antenna efficiency.

In general it is necessary to zone the lens to reduce its thickness and to increase its bandwidth. Zoning a waveguide lens gives increased bandwidth because it reduces the dispersion in the waveguide. The discontinuity in the step is an integral multiple of $\lambda/(1 - n)$ so that the overall path length is increased by one or more wavelengths. The lens may be stepped on either or both of its surfaces. However, steps on the surface opposite the feed appear preferable in order to reduce shadowing effects.

As described in [33] by Dion, a number of zoning techniques can be applied to increase the bandwidth of a waveguide lens. One way to achieve this goal is to zone the lens for minimum thickness. In this design, as illustrated in Fig. 30, the location of the zones is obtained by making d equal to $d_0 + md_\lambda$, where $d_\lambda = \lambda/(1 - n)$, yielding $z_m = m\lambda$. At the start of each zone the waveguide length is the same as that of the central element, i.e., $d = d_0$ at $z = m\lambda$.

The phase error of a zoned lens across the aperture is given by

$$\Delta\phi = \frac{2\pi z}{\lambda} \frac{1 + n}{n} \left[1 - \frac{m\lambda}{(1 + n)z} \right] \frac{\Delta f}{f} \qquad (50)$$

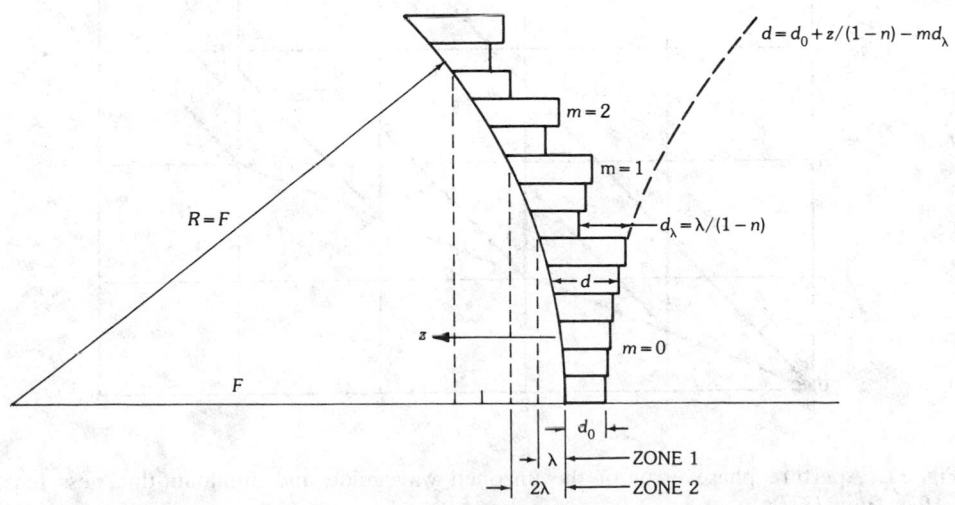

Fig. 30. Waveguide lens zoned for minimum thickness. (*After Dion [33]*)

where z is related to the radial distance r by $z = F - (F^2 - r^2)^{1/2}$, and $z \cong r^2/2F$ for large F/D. For the case of minimum-thickness lens the normalized phase error is a sawtooth curve superimposed with a linear slope, as plotted in Fig. 31 for $n = 0.6$.

On the other hand, a waveguide lens can be zoned to achieve minimum phase error at the expense of increased thickness at the center, as first reported by Colborn [34]. The basic concept is to ensure equal time delay at discrete points in addition to the constraint of equal phase delay for all rays. These discrete points correspond to the step locations of new zones. They can be found by first requiring that the time delay, based on group velocity, of a general ray from the focal point to an aperture plane is equal to that of the central ray. This condition leads to

$$(1 - n)(d - d_0) + nz = 0 \tag{51}$$

Next impose the equal phase constraint for a zoned lens to obtain

$$d = d_0 + \frac{z - m\lambda}{1 - n} \tag{52}$$

From (51) and (52) the step locations can be determined to be

$$z_m = \frac{m\lambda}{1 + n}, \qquad m = 1, 2, \ldots \tag{53}$$

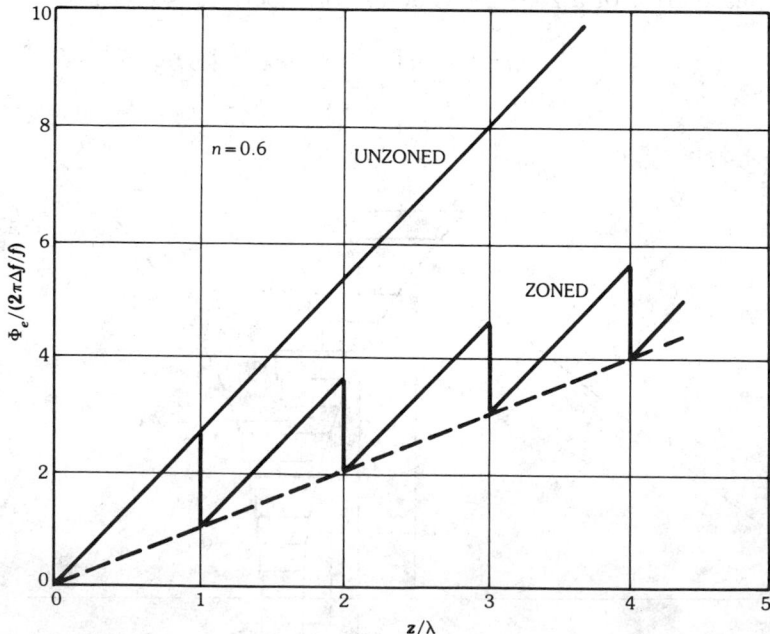

Fig. 31. Aperture phase error of the unzoned waveguide and minimum-thickness lens. (*After Dion [33]*)

This happens to be the condition that will reduce the phase error of the zoned lens to zero at the step locations, as can be proved from (50). A minimum phase error lens is illustrated in Fig. 32, and the resultant phase error distribution is shown in Fig. 33.

Equal Group Delay Lens

Waveguide lenses discussed so far are basically conventional designs that are based on the concept of equal phase delay. These lenses are intrinsically narrow-banded, because there is a large difference in time delay between the central and edge rays. Recently, Ajioka and Ramsey [35] developed an equal group delay waveguide lens which improves the bandwidth significantly. In this design all rays from the focal point to the aperture plane have equal time delay at the center frequency. The equality of time delay does not ensure equality of phase, however. The phase is then made equal by proper adjustment of a half-wave-plate phase

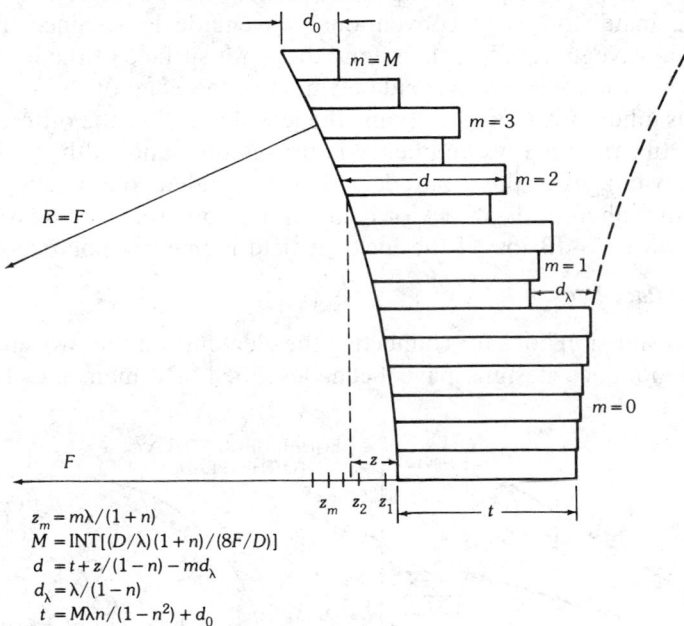

$$z_m = m\lambda/(1+n)$$
$$M = \text{INT}[(D/\lambda)(1+n)/(8F/D)]$$
$$d = t + z/(1-n) - md_\lambda$$
$$d_\lambda = \lambda/(1-n)$$
$$t = M\lambda n/(1-n^2) + d_0$$

Fig. 32. Minimum phase error lens. (*After Dion [33]*)

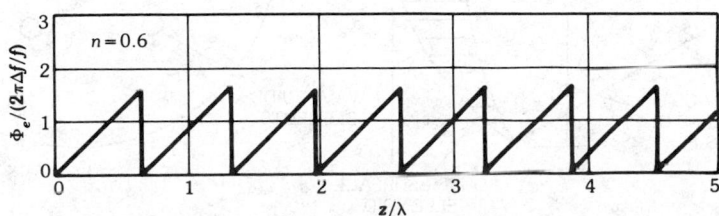

Fig. 33. Aperture phase error of the minimum phase error lens. (*After Dion [33]*)

shifter in each element. The half-wave plates add a constant length to all elements producing no phase error.

The design results in an aperture phase distribution which remains essentially constant over a much greater bandwidth than in the other lenses. The bandwidth of this group delay lens is only limited by the impedance mismatch between the lens radiating elements and free space, a limitation shared by all lenses, and the bandwidth of the half-wave plates.

As shown in Fig. 34 this lens consists of an array of waveguide elements that are uniformly spaced with various lengths. Each waveguide element contains a half-wave-plate phase shifter which orients in such a way that the phase of a circularly polarized wave incident on it is shifted by the proper amount to produce the correct phase at the aperture. The inner surface of the lens is spherical, for the benefit of the Abbe sine condition which ensures wide-angle performance as discussed previously. The outer lens surface is determined by the constraint of equal group delay for all rays. It can be shown that the outer surface is an ellipsoid.

The principle of the equal group delay lens also explains clearly why zoning enhances the bandwidth of a conventional waveguide lens. Since the index of refraction of a waveguide is less than unity the outer surface of the lens is concave in contour, with the longest waveguide element at the edge of the lens where the group delay is inherently longest. Zoning the lens diminishes the difference in time delay among the various rays and hence improves the bandwidth.

The disadvantages of the group delay lens are added complexity of the phase shifters and the inherent drawback of being in only one sense circularly polarized [35]. There will be 3-dB loss if the incident field is linearly polarized.

Multifocal Bootlace Lens

Using transmission lines for connecting the elements on the two surfaces of the lens, several bootlace designs have been described. Rotman and Turner have

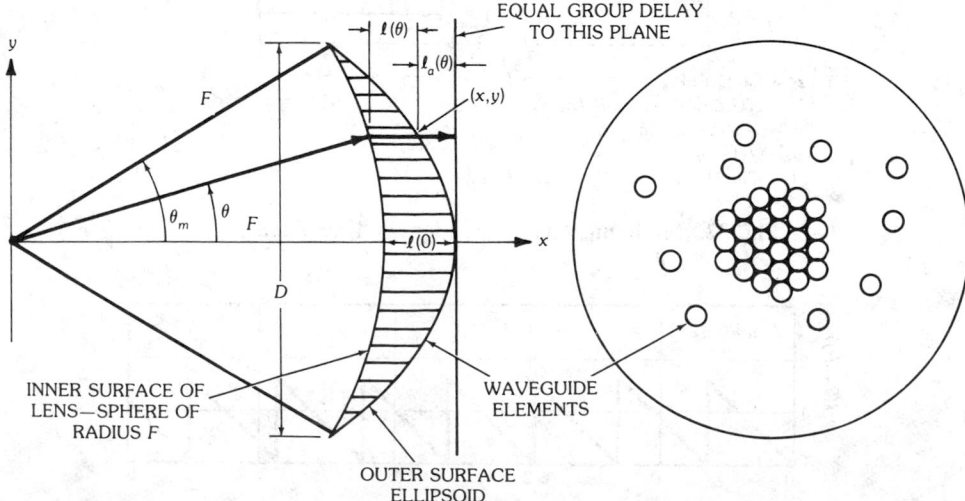

Fig. 34. Equal group delay lens. (*After Ajioka and Ramsey [35], © 1978 IEEE*)

derived a trifocal lens with a circular focal arc and a straight outer edge for a linear array [36]. Cornbleet also presents design details of quadrifocal lenses in [37]. Recently Rao [38] has investigated several bootlace lenses with simpler geometry as described in the following. As shown in Fig. 35 the bootlace lens in this case has a planar outer surface with focal points located on a vertical line. The shape of its inner surface is dependent on the condition of whether equal line length or unequal line length is imposed. If equal line length is used, a bifocal design can be deduced with the inner surface specified by

$$y_1^2 \cos^2\alpha + (x_1 + F_0)^2 = F_0^2 \tag{54}$$

and $y_1 = y_2$, $w = w_0$.

If the condition of unequal line length is selected, a trifocal system with the third focal point located on axis is available, where the inner surface is given by

$$y_1^2 + (x_1 + F_0)^2 = (F_0 - w + w_0)^2 \tag{55}$$

with

$$y_1 = y_2[1 - (w - w_0)(\cos\alpha)/F_0] \tag{56}$$

and

$$w - w_0 = (y_2^2/F_0)(\cos\alpha)\cos^2(\alpha/2) \tag{57}$$

Rao has also studied the phase error across the aperture when the feed is displaced from a focal point to scan the beam to other angles. With the feed confined to a straight line passing through the focal points, the maximum path length error versus scan angle for three multifocal systems plus the single focus case, with $F/D = 1$, is plotted in Fig. 36. It may be noted that, for a maximum path

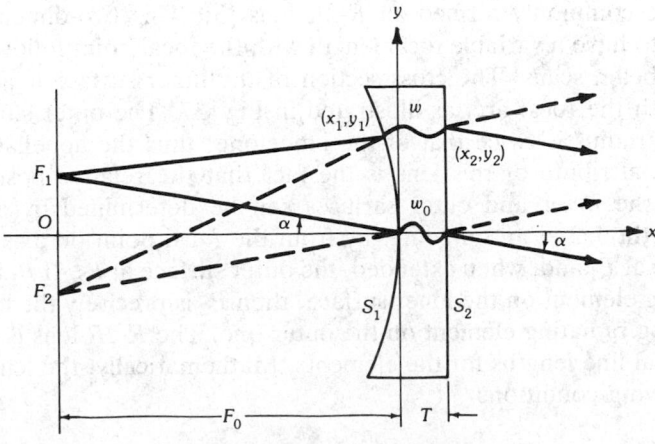

Fig. 35. Multifocal bootlace lens.

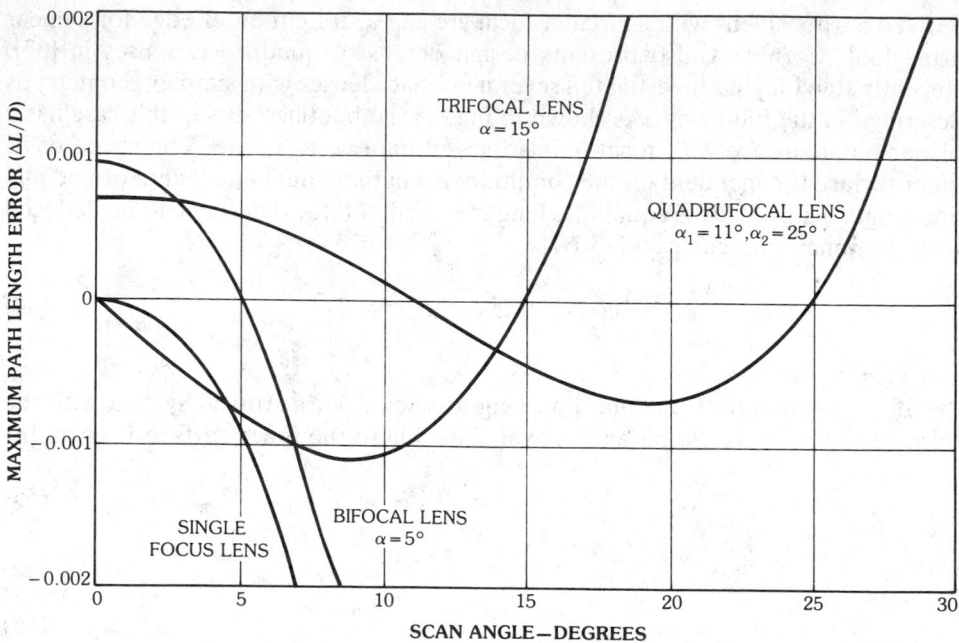

Fig. 36. Path length error versus scan angle for $F/D = 1$. (*After Rao [38], © 1979 IEEE*)

length error less than $0.001D$, a bifocal lens designed for $\alpha = 5°$ can be scanned up to $\pm7°$, whereas a trifocal lens with $\alpha = 15°$ can be scanned up to $\pm16.5°$. The comparison also suggests that if extremely wide scanning is required, a quadrifocal lens discussed by Cornbleet [37] is a better choice, since this lens minimizes the phase error over a much wider range in scan.

R-2R Lens

Another interesting design which is characteristized by perfect focusing at any scan angle is commonly termed an *R-2R* lens [3]. This two-dimensional lens is constrained to have a variable focal length with the focal point following a circular path as the beam scans. The cross section of the inner surface is a circle, which coincides with the focal arc, as illustrated in Fig. 37. The outer surface is also a circle whose radius is twice that of the inner one, thus the appellation *R-2R*.

Another attribute of this lens is the fact that the relative positions of each element on the inner and outer surfaces can be determined by a very simple graphical method. Let an outgoing ray from the focal point on axis intercept the inner surface at P_1 and, when extended, the outer surface at P_2. If P_1 is the location of a receiving element on the inner surface, then P_2 is precisely the location of the corresponding radiating element on the outer one. The *R-2R* lens is characterized by using equal line lengths for the elements. Mathematically, the lens is described by the following conditions:

$$y_2^2 = F_0^2 - x_2^2 \tag{58}$$

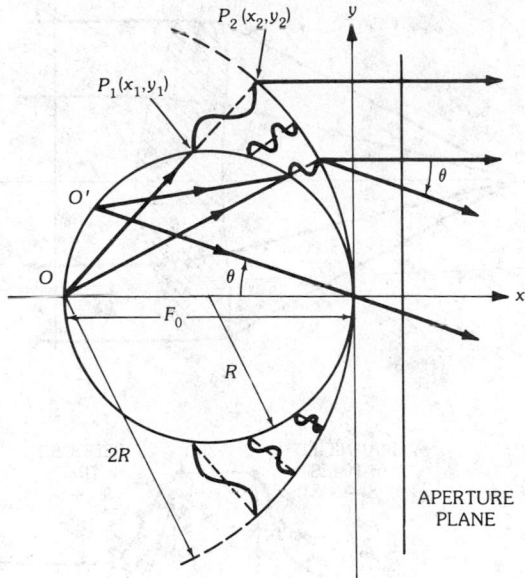

Fig. 37. *R-2R* lens.

$$x_1 = F_0 - y_2^2/F_0 \tag{59}$$

$$y_1^2 = y_2^2 - y_2^4/F_0^2 \tag{60}$$

and $F = F_0 \cos \theta$.

In principle, there is no aberration with this lens at any scan angle. However, scanning is restricted to 60° in practice due to the fact that the effective aperture, and hence the peak gain, will drop significantly beyond that limit.

Constrained Analog of Dielectric Lens [39]

It is interesting to note that for any dielectric lens there is an equivalent constrained lens system that performs the same optical transformation as the dielectric lens. The converse, however, is not necessarily true. In other words, a dielectric lens can be replaced by a constrained lens system operating in free space without the dielectric medium, but a constrained lens cannot be substituted by a dielectric lens in general. Shown in Fig. 38 is a dielectric lens and its constrained analog that consists of two constrained lenses separated by a region of free space. The two lenses are identified as lens 1 and lens 2. Lens 1 is formed by two surfaces that are different from each other only by a scale factor. On the left the surface S_1 is identical with the inner surface of the dielectric lens. On the right the surface S_1' is an exact replica of S_1 except expanded by a factor n, where n is the refractive index of the dielectric lens. The lens has an equal number of radiating elements on each surface, with corresponding elements connected by equal line lengths on a one-to-one basis. The element spacing on S_1' is n times that on S_1 in all directions.

For lens 2 the above process is reversed. The surface on the right, S_2, is

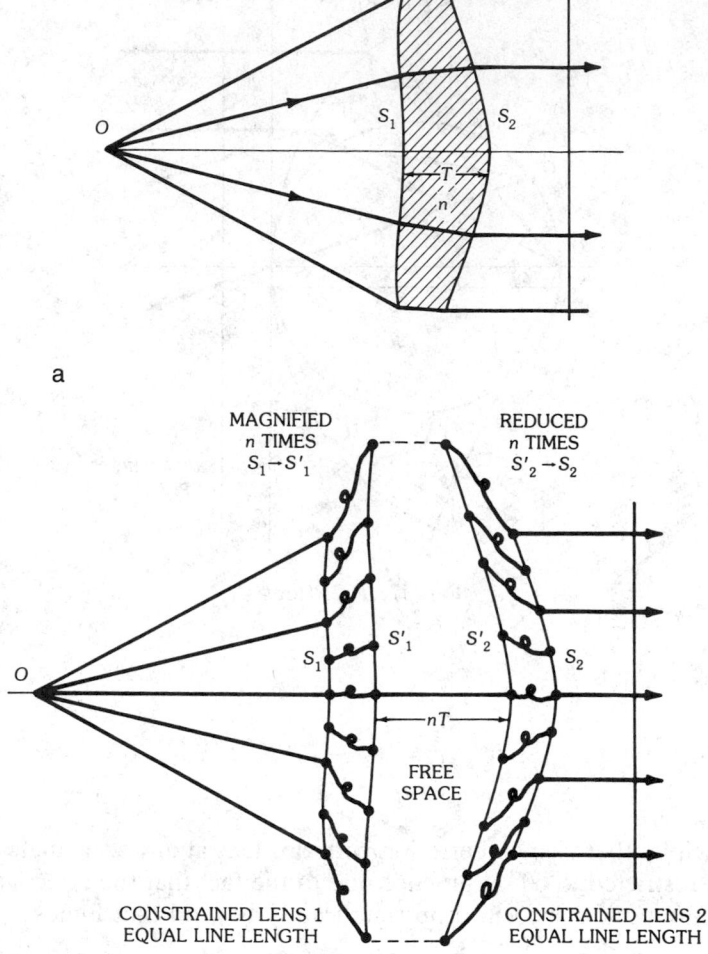

Fig. 38. Constrained analog of dielectric lens. (*a*) Dielectric lens. (*b*) Constrained analog of (*a*).

identical with the outer surface of the dielectric lens, whereas the surface on the left, S_2', is an expanded version of S_2 with a scale factor of n. Similarly, equal line lengths are used to connect the corresponding elements on lens 2. The separation between S_1' and S_2' is n times the central thickness of the dielectric lens. Thus the region of free space confined by S_1' and S_2' has the same proportion as the dielectric lens in all directions.

To show that the constrained system accomplishes the same function as the dielectric lens the phase relationship illustrated in Fig. 39 is examined. Assume that the incidence angle of a ray at S_1 with respect to the local normal vector is θ_i and the exit angle at S_1' is θ_r. Then the differential phase delay between two adjacent elements at S_1 is $kd\sin\theta_i$, and the corresponding phase difference at S_1' is $nkd\sin\theta_r$.

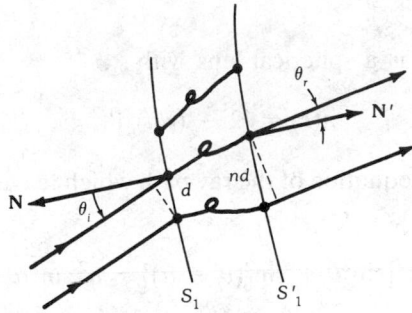

Fig. 39. Refraction through a constrained lens.

Since equal line lengths are used the phase distribution is preserved from S_1 to S_1'. This implies that

$$kd \sin \theta_i = nkd \sin \theta_r$$

or

$$\theta_r = \sin^{-1}\left(\frac{\sin \theta_i}{n}\right)$$

It can be recognized that this is Snell's law of refraction. Since S_1 and S_1' share the same form of surface contour, the normal vectors at the corresponding points on S_1 and S_2' are parallel to each other. Thus the ray refracts in the same way as it would across the inner surface of the dielectric lens. Likewise, the same kind of refraction takes place from S_2' to S_2. Furthermore, the time delay of each ray in the free space between S_1' and S_2' will be the same as in the dielectric lens, because the region is n times larger than the dielectric lens. It is evident that this constrained lens system behaves the same as the dielectric lens, and therefore the analogy between them is proved.

With this one-to-one correspondence established, a dielectric lens can be realized by a constrained lens, thereby eliminating such drawbacks of the dielectric lens as physical weight and complexity in machining. Moreover, all the design formulas presented in the sections on dielectric lenses can be applied to develop constrained lenses to meet special requirements.

9. Inhomogeneous Lenses

So far the refractive index of the lenses discussed has been assumed to be uniform. If, however, the lens medium is allowed to be inhomogeneous—as a function of radius, for example—a new class of lenses can be designed. These lenses have certain unique features that cannot be achieved with uniform lenses. The Luneburg lens and Maxwell fish-eye lens are two classical examples of such lenses.

Luneburg Lens

The Luneburg lens is a spherical lens with

$$n(r) = [2 - (r/a)^2]^{1/2} \tag{61}$$

It can be shown that the equation of the ray path which leaves the feed at an angle α to the lens axis is [3]

$$r^2[\sin^2\theta + \sin^2(\theta - \alpha)] = a^2\sin^2\alpha \tag{62}$$

and all rays leaving a point P_1 on the surface emerge at the opposite end as a parallel beam, as shown in Fig. 40. Thus scanning is accomplished simply by moving the feed point around the surface. Due to the spherical symmetry the scanning performance is independent of the beam direction. The power distribution across the aperture is related to the feed power pattern by $P(y) = P(\theta)\sec\theta$, where $y = a\sin\theta$.

The Luneburg lens is difficult to construct due to the fact that its refractive index must vary continuously from $\sqrt{2}$ at the center to unity at the surface. However, satisfactory performance can be obtained by using a number of spherical shells of uniform dielectric constant to approximate a continuous gradient. Another drawback of this lens is the special feed required. It is difficult to design a useful feed which has a phase center that can be placed directly on the surface of the lens. This problem can be solved by moving the phase center away from the lens and requiring the lens to have a modified refractive index distribution. This has been investigated by Wolff [40] and Elliott [41] where numerical solutions of $n(r)$ are derived.

Maxwell Fish-Eye Lens

The Maxwell fish-eye lens is also a spherical lens which is characterized by

$$n(r) = \frac{n_0}{1 + (r/a)^2} \tag{63}$$

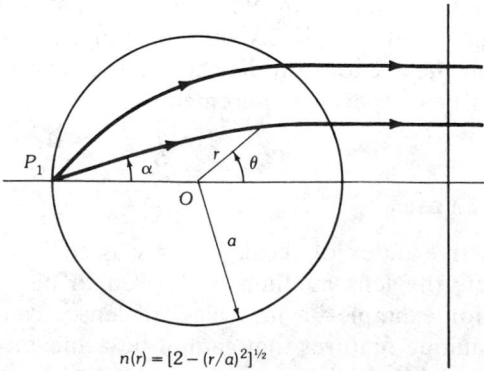

$$n(r) = [2 - (r/a)^2]^{1/2}$$

Fig. 40. Luneburg lens.

In this case, rays leaving point P_1 converge at P_2 diametrically opposite the point P_1, as shown in Fig. 41. The ray trajectory is given by

$$\sin\theta = \frac{C}{\sqrt{a^2 n_0{}^2 - 4C^2}} \frac{a^2 - r^2}{ar} \tag{64}$$

where C is a parametric constant determined by the product of $n(r)$ and r at Q, with the ray path normal to the line OQ. The ray paths are segments of circular arcs which form figures shaped like the eye of a fish.

To form a collimated beam the sphere in Fig. 41 has to be cut in half along OQ so that the lens becomes a hemisphere. This, however, may have a drastic effect on its scanning properties.

Hyperbolic Cosine Lens

A rectangular slab of dielectric material with index of refraction given by

$$n(y) = n_0 \operatorname{sech}(\pi y/2F) \tag{65}$$

also exhibits focusing effect. All rays leaving O, as in Fig. 42, emerge parallel to the x axis. If a line source at O radiates uniformly into each angular sector $d\theta$, the field intensity along the front surface follows the distribution $\cosh(\pi y/2F)$. The scanning performance of this lens is not known at this time.

Recently Brown has reviewed the subject and technology of gradient index optics [42]. A survey is given on the types of gradient available and the manufacturing processes being developed. When production techniques for inhomogeneous dielectrics become mature enough and cost effective, various inhomogeneous lenses such as those just discussed can be developed for special applications.

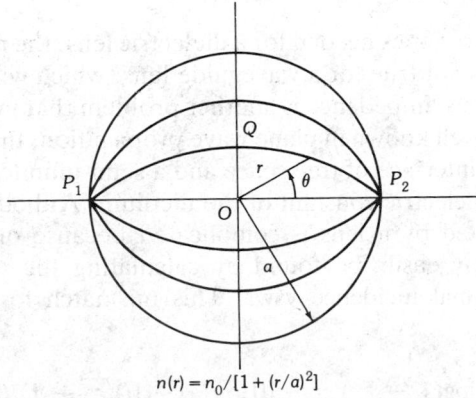

$$n(r) = n_0/[1 + (r/a)^2]$$

Fig. 41. Maxwell fish-eye lens.

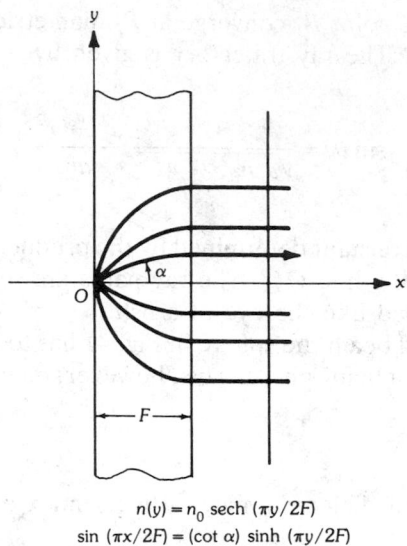

$$n(y) = n_0 \text{ sech } (\pi y/2F)$$
$$\sin (\pi x/2F) = (\cot \alpha) \sinh (\pi y/2F)$$

Fig. 42. Lens with hyperbolic cosine distribution.

10. Bandwidth Limitation and Surface Mismatch

A dielectric lens is a time-delay optical device which has virtually unlimited bandwidth for microwave frequencies if no zoning is introduced. However, the bandwidth of the dielectric lens is restricted once zoning is employed. Since zoning is implemented by reducing the lens thickness in steps by the amount of $\lambda/(n-1)$ with respect to the center frequency, error in the path length occurs as the frequency changes. If a total path length error of one-eighth wavelength from center to edge is tolerated, the bandwidth of a dielectric lens with M zones is given by [41]

$$B = \frac{25}{M-1} \text{ percent} \tag{66}$$

It is clear that the more zones needed for a dielectric lens, the more narrowband the lens becomes. This is not true for a waveguide lens, which will be discussed later.

Mismatch in surface impedance is another problem that must be dealt with in a dielectric lens. As is well known in plane wave propagation, the vswr for a normally incident wave at the interface of free space and a semi-infinite dielectric medium is $\sqrt{\epsilon_r}$, where ϵ_r is the dielectric constant of the medium. Although the exact calculation of power reflected by a lens is complicated because of the lens curvature, a rough estimate may easily be found by calculating the mismatch loss corresponding to the normal incidence vswr. This mismatch loss is simply given in decibels by

$$\text{loss} = -10 \log(1 - \Gamma^2) = -10 \log\{1 - [(\sqrt{\epsilon_r} - 1)/(\sqrt{\epsilon_r} + 1)]^2\}$$
$$= 10 \log[(\sqrt{\epsilon_r} + 1)^2/4\sqrt{\epsilon_r}]$$

It is therefore good practice to match the lens surface if the dielectric constant is large. A standard technique is to use a quarter-wave layer of intermediate refractive index. Other techniques involve cutting grooves or grids on the surface to simulate a region of intermediate dielectric constant. A detailed treatment on this subject has been given by Cohn in [4]. At high frequencies such as K_u-band or above, however, it becomes extremely difficult to implement these matching layers due to practical engineering problems. Thus a performance margin must be allowed to account for the gain drop and side lobe degradation in the design of a dielectric lens at high frequencies.

For constrained lenses the bandwidth is determined by many factors. The properties of the radiating elements, the connecting transmission lines, the array geometry, and the mutual coupling effects all play an important role in the bandwidth calculation. Many constrained lenses make use of phase shifting elements; therefore their frequency response affects the bandwidth as well. As a matter of fact, the criterion of bandwidth for a given lens must be carefully defined in terms of array performance. The definition of bandwidth from different points of view has been examined and summarized in [43] by Frank, in which the effects of aperture, feed, fill-up time, etc., on the bandwidth of a phased array are also discussed.

The bandwidth of a metal plate or waveguide lens is limited due to its dispersive characteristics. As discussed in [1], the bandwidth of an unzoned lens is

$$\text{bw} \cong 8.3 \frac{\lambda_0}{(1 - n_0) T} \text{ percent} \tag{67}$$

where λ_0 and n_0 are respectively the wavelength and refractive index of the lens at center frequency, and T is the difference between the thickness of the lens at the edge and at the center. Usually the bandwidth is only a few percent of the operating frequency. Unlike dielectric lenses, zoning actually improves the bandwidth of a waveguide lens because it reduces the dispersion of the waveguide. According to Risser in [1], the bandwidth of a zoned lens is

$$\text{bw} \cong 25 \frac{n_0}{1 + K n_0} \text{ percent} \tag{68}$$

where K is the number of zones introduced. A comparison of equivalent zoned and unzoned lenses indicates that the bandwidth of a zoned lens is two to three times better than that of an unzoned one.

For a waveguide lens zoned for minimum thickness as discussed in [33], the bandwidth of the lens with a phase error less than $\pm\lambda/16$ is

$$\text{bw} \cong \frac{200n(F/D)}{(1 + n)(D/\lambda)} \frac{1}{1 - 8K(F/D)/[(1 + n)(D/\lambda)]} \text{ percent}$$

where K is the number of zones. If the lens is zoned for minimum phase error, the bandwidth is approximately equal to $25n$ percent.

Matching the surface impedance of a constrained lens is probably the most difficult part of the lens design. The inner surface facing the feed should be spherical, if possible, primarily for better impedance match due to normal incidence from a point source. A spherical surface also offers the desirable wide-scan features according to the Abbe sine condition. If planar surface is considered, the F/D ratio should be chosen as large as practical.

For wide-band and wide-scan applications internal matching of each element may be necessary. For waveguide elements this may involve impedance matching with use of dielectric plugs, stub chokes, and reactive irises at the apertures of the radiators. A dielectric sheet placed in front of the aperture is also one of the most common techniques used to achieve wide-angle impedance matching. The matching process can be best described by the equivalent circuit shown in Fig. 43. The dielectric sheet acts as an impedance transformer whose characteristic impedance and propagation constant change with scan [44]. For a thin sheet it behaves approximately as a variable shunt susceptance. By optimizing the dielec-

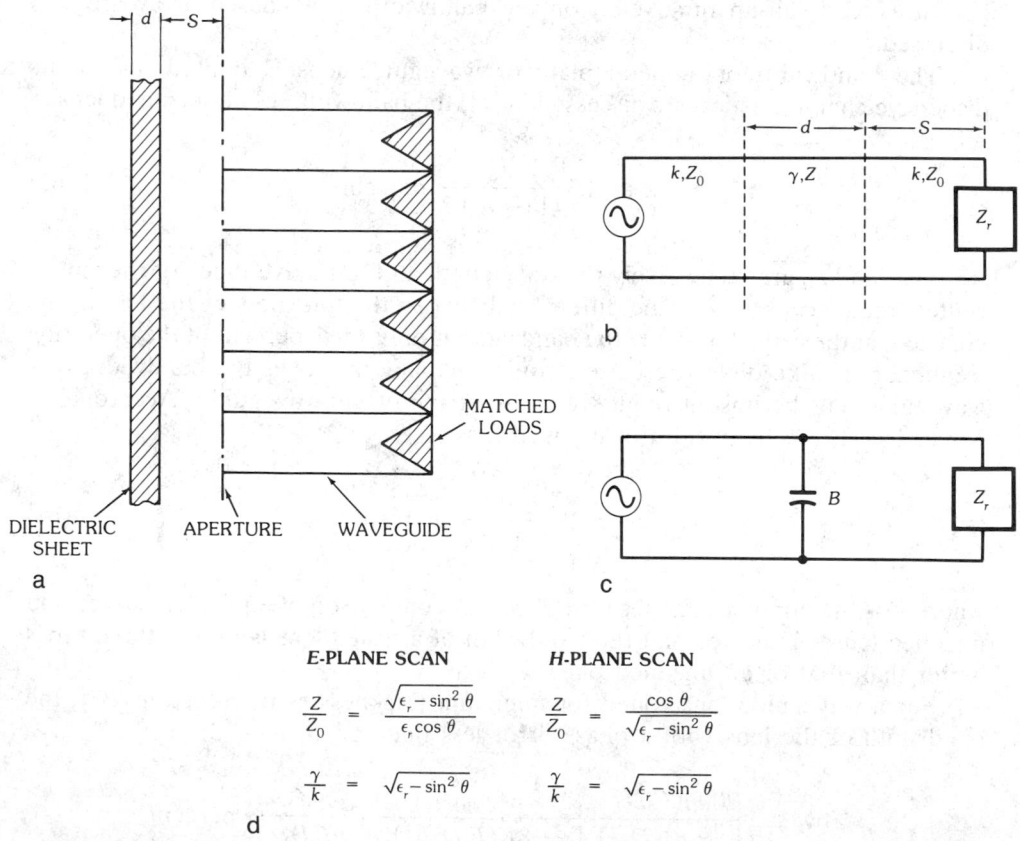

E-PLANE SCAN

$$\frac{Z}{Z_0} = \frac{\sqrt{\epsilon_r - \sin^2 \theta}}{\epsilon_r \cos \theta}$$

$$\frac{\gamma}{k} = \sqrt{\epsilon_r - \sin^2 \theta}$$

H-PLANE SCAN

$$\frac{Z}{Z_0} = \frac{\cos \theta}{\sqrt{\epsilon_r - \sin^2 \theta}}$$

$$\frac{\gamma}{k} = \sqrt{\epsilon_r - \sin^2 \theta}$$

Fig. 43. Equivalent circuit of dielectric matching sheet. (*a*) Dielectric matching sheet. (*b*) Equivalent circuit. (*c*) Simplified equivalent circuit for thin sheet. (*d*) Mathematical relationships.

tric constant of the sheet and its thickness and distance from the elements through iterations, improved performance in matching can be obtained [45, 46].

11. Summary

Lenses have found many applications not only in optics but also in microwave antenna systems. The subject of lens design has been treated by numerous authors, and this field is relatively familiar to many antenna designers. In this chapter, only a very brief review has been offered due to space limitations. Care has been taken not to reproduce the design details that can be readily found in the existing reference literature. Emphasis has been placed on the more recently developed systems. For those who would like to explore more details of a particular design, the references listed below should be consulted.

Acknowledgments

Dr. Allan Love and Leni Parad's enlightening comments and corrections in proofreading this chapter are highly appreciated. Ms. Lynda Schenet's excellent typing of the manuscript is also gratefully acknowledged. Her patience and efficiency made writing this chapter a very pleasant task.

12. References

[1] S. Silver, ed., *Microwave Antenna Theory and Design*, chapter 11, New York: McGraw-Hill Book Co., 1949.

[2] J. Brown, *Microwave Lenses*, London: Methuen & Co., 1953.

[3] R. E. Collin and F. J. Zucker, eds., *Antenna Theory*, chapter 18, New York: McGraw-Hill Book Co., 1969.

[4] S. B. Cohn, "Lens-type radiators," chapter 14 in *Antenna Engineering Handbook*, ed. by H. Jasik, New York: McGraw-Hill Book Co., 1961.

[5] D. L. Sengupta and R. E. Hiatt, "Reflectors and lenses," chapter 10 in *Radar Handbook*, ed. by M. Skolnik, New York: McGraw-Hill Book Co., 1970.

[6] M. Born and E. Wolf, *Principles of Optics*, 3rd ed., New York: Pergamon Press, 1965, pp. 211–218 and pp. 468–484.

[7] S. Silver, ed., *Microwave Antenna Theory and Design*, section 6.7, New York: McGraw-Hill Book Co., 1949, p. 186.

[8] A. W. Love, *Reflector Antennas*, IEEE Press, 1978, pp. 291–336.

[9] F. A. Jenkins and H. E. White, *Fundamentals of Optics*, 4th ed., chapter 9, p. 173, New York: McGraw-Hill Book Co., 1973.

[10] W. Rotman, "Ehf dielectric lens antenna for satellite communication systems," *Report 620*, ESD-TR-82-102, Lincoln Labs, MIT, 1983.

[11] F. G. Friedlander, "A dielectric lens aerial for wide-angle beam scanning," *J. IEE*, vol. 93, pt. 3A, pp. 658–662, 1946.

[12] J. J. Lee, "Numerical methods make lens antennas practical," *Microwave*, pp. 81–84, September 1982.

[13] S. B. Cohn, "Lens-type radiators," chapter 14 in *Antenna Engineering Handbook*, ed. by H. Jasik, New York: McGraw-Hill Book Co., 1961.

[14] J. McFarland, D. Wesley, and R. Dell-Imagine, Private communication.

[15] D. H. Shinn, "The design of a zoned dielectric lens for wide-angle scanning," *Marconi Rev.*, no. 117, p. 37, 1955.

[16] J. B. L. Rao, "Bifocal dual-reflector antennas," *IEEE Trans. Antennas Propag.*, vol.

AP-22, pp. 711–714, 1974.

[17] K. C. Lang et al., "Gain, side lobe and cross-polarization performance of dual offset reflector antennas," *Proc. IEEE AP-S Symp.*, p. 269, 1982.

[18] R. M. Brown, "Dielectric bifocal lenses," *IRE Natl. Conv. Rec.*, pp. 180–187, 1956.

[19] H. Kumazawa and M. Kariokomi, "Multiple-beam antenna for domestic communication satellites," *IEEE Trans. Antennas Propag.*, vol. 21, p. 876, 1973.

[20] J. J. Lee, "A dielectric lens shaped for a generalized Taylor distribution," *Proc. IEEE AP-S Symp. vol. AP-5*, p. 124, 1982.

[21] J. J. Lee, "Dielectric lens shaping and coma correction zoning," *IEEE Trans. Antennas Propag.*, vol. AP-31, p. 211, 1983.

[22] A. W. Love, "Shaped-beam antennas for Earth coverage," Rockwell Space Div. internal report, *SD 74-SA-0074*, 1974.

[23] D. K. Waineo, "Low side lobe sector patterns for a circular aperture," *IEEE Antennas Propag. Symp. Dig.*, p. 527, 1977.

[24] K. S. Kelleher, "Microwave optics at NRL," *Proc. McGill Symp. Microwave Opt.*, AFCRC-TR59-118(I), p. 5, 1959.

[25] T. C. Cheston, "Microwave lenses," *Proc. McGill Symp. Microwave Opt.*, AFCRC-TR-59-118(I), p. 8, 1959.

[26] L. Ronchi and G. Toraldo Di Francia, "An application of parageometric optics to the design of microwave mirrors," *IRE Trans. Antennas Propag.*, vol. AP-6, p. 129, 1958.

[27] J. H. Provencher, "Experimental study of a diffraction reflector," *IRE Trans. Antennas Propag.*, vol. AP-8, p. 331, 1960.

[28] S. Dasgupta and Y. T. Lo, "A study of the coma-corrected zoned mirror by diffraction theory," *IRE Trans. Antennas Propag.*, vol. AP-9, p. 130, 1961.

[29] J. Ruze, "Wide-angle metal plate optics," *Proc. IRE*, vol. 38, pp. 53–59.

[30] S. Silver, ed., *Microwave Antenna Theory and Design*, chapter 11, New York: McGraw-Hill Book Co., 1949, pp. 401–412.

[31] R. W. Major and J. M. Devan, "Ehf multiple-beam antennas," *NOSC TR730*, San Diego 92152: Naval Ocean Systems Center, 1981.

[32] A. R. Dion and L. J. Ricardi, "A variable-coverage satellite antenna system," *Proc. IEEE*, vol. 59, p. 252, 1971.

[33] A. R. Dion, *Handbook of Antenna Design*, chapter 3.7, ed. by A. W. Rudge et al., Stevenage, Herts., UK: Peter Peregrinus, 1982, p. 293.

[34] C. B. Colborn, Jr., "Increased bandwidth waveguide lens antenna," *Aerospace Corp., Rep. No. TOR-0076(6403-01)-3*, December 8, 1975.

[35] J. S. Ajioka and V. W. Ramsey, "An equal group delay waveguide lens," *IEEE Trans. Antennas Propag.*, vol. AP-26, p. 519, 1978.

[36] W. Rotman and R. F. Turner, "Wide-angle microwave lens for line source applications," *IEEE Trans. Antennas Propag.*, vol. AP-11, pp. 623–632, 1963.

[37] S. Cornbleet, *Microwave Optics*, chapter 1, New York: Academic Press, 1976, pp. 73–77.

[38] J. B. L. Rao, "Multifocal three-dimensional bootlace lenses," *Proc. IEEE AP-S Symp.*, vol. AP-5(a)-2, p. 332, 1979.

[39] This analogy was pointed out by the late J. L. McFarland in private discussions.

[40] E. A. Wolff, *Antenna Analysis*, New York: John Wiley & Sons, 1966, p. 496.

[41] R. S. Elliott, *Antenna Theory and Design*, Englewood Cliffs: Prentice-Hall, 1981, p. 545.

[42] S. T. S. Brown, "An introduction to gradient index optics," *Marconi Rev.*, first quarter, p. 3, 1981.

[43] J. Frank, "Bandwidth criterion for phased array antennas," *Proc. 1970 Phased Array Antenna Symp.*, Dedham: Artech House, 1972.

[44] R. E. Collin, *Field Theory of Guided Waves*, New York: McGraw-Hill Book Co., 1960, pp. 87–94.

[45] S. W. Lee and W. R. Jones, "On the suppression of radiation nulls and broadband impedance matching of rectangular waveguide phased arrays," *IEEE Trans. Antennas Propag.*, vol. AP-19, pp. 41–51, 1971.

[46] C. C. Chen, "Wideband wide-angle impedance matching and polarization charac-
       teristics of circular waveguide phased arrays," *IEEE Trans. Antennas Propag.*, vol.
       AP-22, pp. 414–418, 1974.

PART C

Applications

Chapter 17

Millimeter-Wave Antennas

F. Schwering
US Army CECOM

A. A. Oliner
Polytechnic University

CONTENTS

Felix Schwering was born on June 4, 1930, in Cologne, Germany. He received the Diplom-Ingenieur degree in electrical engineering and the PhD degree from the Technical University of Aachen, West Germany, in 1954 and 1957, respectively.

From 1956 to 1958 he was an assistant professor at the Technical University of Aachen. In 1958 he joined the US Army Research and Development Laboratory at Fort Monmouth, New Jersey, where he performed basic research in free space and guided propagation of electromagnetic waves. From 1961 to 1964 he worked as a member of the Research Staff of the Telefunken Company, Ulm, West Germany, on radar propagation and missile electronics. In 1964 he returned to the US Army Communication Electronics Command (CECOM), Fort Monmouth, and has since been active in electromagnetic-wave propagation, diffraction and scatter theory, theoretical optics, and antenna theory. Recently he has been interested in particular in millimeter-wave antennas and propagation. At present he is also a visiting professor at Rutgers University and at New Jersey Institute of Technology.

Arthur A. Oliner was born in Shanghai, China, on March 5, 1921. He received the PhD in physics from Cornell University in 1946. He joined the Microwave Research Institute of the Polytechnic Institute of Brooklyn in 1946 and was made professor in 1957. He served as Department Head from 1966 through 1974 and Director of the Microwave Research Institute for fifteen years, from 1967 through 1982. Dr. Oliner is the author of over 150 papers and coauthor or coeditor of three books. Two of his papers have earned prizes: the IEEE Microwave Prize in 1967, and the Institution Premium, the highest award of the British IEE, in 1964. In 1982 he received the Microwave Career Award, the highest award of the IEEE Microwave Theory and Techniques Society, and he is one of only six Honorary Life Members of that society.

Dr. Oliner's research in microwaves includes network representations of microwave structures, precision measurement methods, guided-wave theory, traveling-wave antennas, plasmas, periodic-structure theory, and phased arrays. More recently he has been interested in guiding and radiating structures for the millimeter- and near–millimeter-wave ranges.

1. Introduction

The millimeter-wave region of the electromagnetic spectrum is commonly defined as the 30- to 300-GHz frequency band or the 1-cm to 1-mm wavelength range. Utilization of this frequency band for the design of data transmission and sensing systems has a number of advantages:

1. The very large bandwidth resolves the spectrum crowding problem and permits communication at very high data rates.
2. The short wavelength allows the design of antennas of high directivity but reasonable size, so that high-resolution radar and radiometric systems and very compact guidance systems become feasible.
3. Millimeter waves can travel through fog, snow, and dust much more readily than infrared or optical waves.
4. Finally, millimeter-wave transmitters and receivers lend themselves to integrated and, eventually, monolithic design approaches, resulting in rf heads which are rugged, compact, and inexpensive.

Propagation effects have a strong influence on the design and performance of millimeter-wave systems, and for this reason are briefly reviewed here. As a general rule, millimeter-wave transmission requires unobstructed line-of-sight paths, but propagation into shadow zones is possible by edge diffraction and scatter, though at a reduced signal level. Recent propagation experiments in woods and forests have shown, moreover, that under favorable conditions (trunk region with little underbrush), transmission ranges of several hundred meters can be achieved in vegetated areas.

Amplitude and angle of arrival scintillations caused by atmospheric turbulence are usually small in the millimeter region. For path lengths in the order of a few kilometers, the interesting range, they are of no consequence for most applications.* But atmospheric absorption can be pronounced. Fig. 1 shows the frequency dependence of millimeter-wave attenuation by atmospheric oxygen and water vapor, by rain, and by fog or clouds. Snow absorption is negligible at frequencies below 100 GHz but can be substantial above 140 GHz, even at moderate snowfalls.**

Absorption by rain and atmospheric gases is the dominant effect and it is evident that the choice of the operating frequency of a millimeter-wave system will depend strongly on the desired transmission range. Large transmission distances can be obtained in the low-attenuation windows at 35, 94, 140, 220, and 340 GHz.

---

*High-resolution radar systems are an exception.
**Applies to dry snow. Data on millimeter-wave attenuation by wet snow are not yet available.

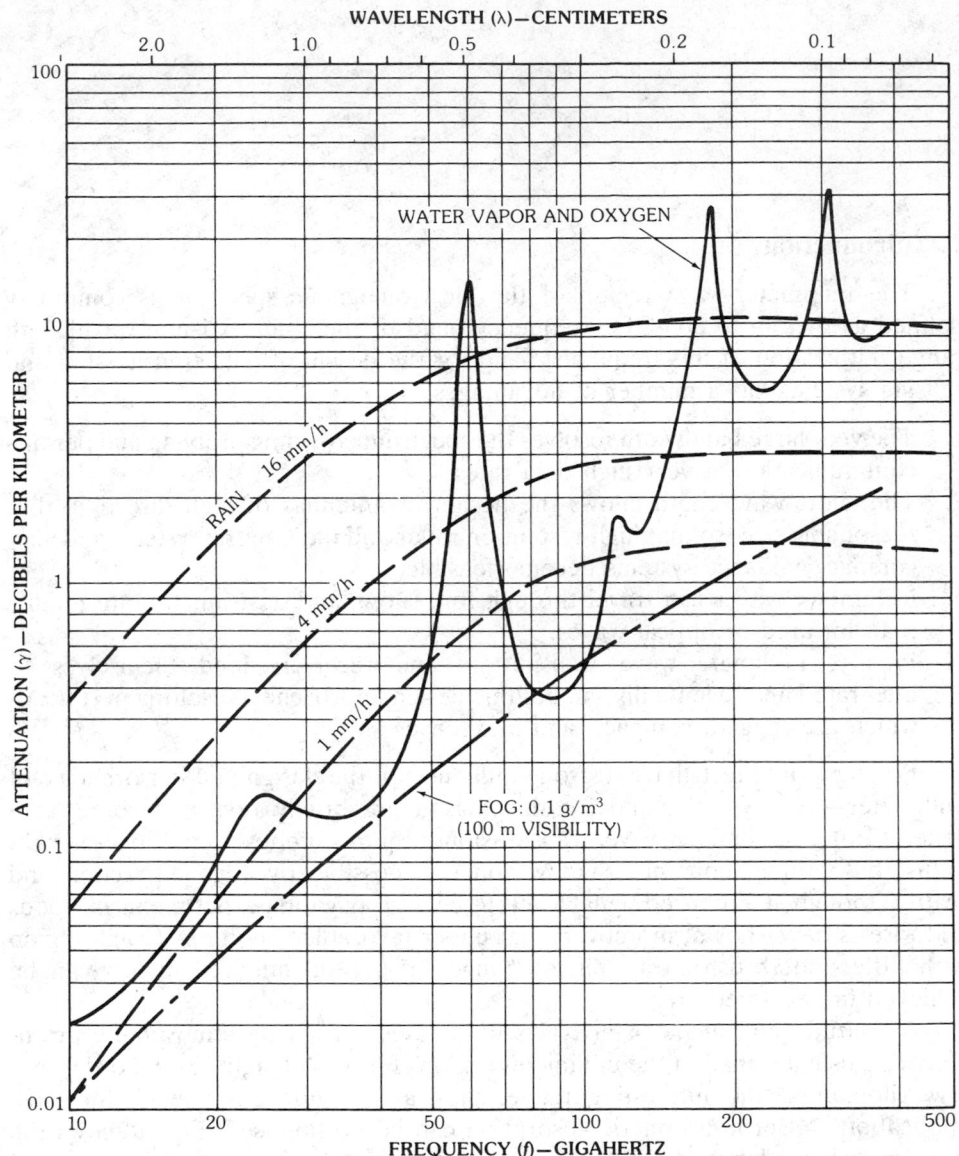

WAVELENGTH (λ)—CENTIMETERS

Fig. 1. Attenuation of millimeter waves by atmospheric effects.

On the other hand, operation near one of the steep absorption lines, for example the O_2 line at 60 GHz, will allow one to control the transmission distance by adjusting the frequency. In this way "overshoot" can be minimized and hence the probability of detection and interference by an unfriendly observer. Since the millimeter range, moreover, permits the design of antennas of high directivity and very small beamwidth a system of optimum transmission security can be realized.

This security aspect, in connection with the other advantages of millimeter-wave systems, including the fact that millimeter waves travel through dust and battlefield smoke with little attenuation, makes this frequency range particularly attractive for military applications, including communication, guidance, fuzing, and radar. Also, numerous commercial and scientific applications are developing, including millimeter-wave radiometry, remote sensing and mapping, and satellite communication at 30/20 GHz, which means 30 GHz uplink, 20 GHz downlink. (The frequencies of military satellites are 44/20 GHz.) Millimeter-wave telescopes for radioastronomy have been constructed since the mid-sixties and millimeter-wave imaging seems to be an interesting possibility for the upper millimeter-wave region.

Most millimeter-wave systems require antennas of high directivity gain and these antennas have been predominantly investigated. As mentioned previously, it is one of the advantages of the millimeter-wave region that such antennas have very reasonable dimensions. Fig. 2 shows the antenna aperture area needed to achieve a given directivity. But certain applications in millimeter-wave guidance and radar also require fan-shaped radiation characteristics, and vehicular millimeter-wave communication implies the use of omnidirectional antennas providing a circular symmetric pattern in the azimuth plane and moderate directivity in the elevation plane. Such antennas are included in this chapter.

For the purpose of this discussion millimeter-wave antennas are subdivided into four classes: antennas of conventional configuration, such as horn, lens, and reflector antennas; surface-wave and leaky-wave antennas based on open wave-guide technology; microstrip resonator and printed-circuit antennas; and some newer developments using integrated and monolithic design approaches. The design principles of the first class of antennas are well established at microwave frequencies, and scaling into the millimeter-wave region is straightforward in most cases. These antennas are particularly suited for frequencies below 100 GHz, although the usefulness of lens and reflector antennas extends well into the infrared and optical regions. The millimeter-wave aspects of these antennas will be summarized here. The second and third classes of antennas are of interest for both the lower and upper millimeter-wave regions. Most of these antennas have microwave counterparts, but scaling is usually not straightforward, while some are novel with no microwave heritage. These antennas will be discussed in some detail. The fourth class of antennas is still in the research stage; systematic design information is not as yet available but an attempt is made to indicate trends.

Millimeter-wave antennas are a fast-developing area, and any discussion of them must be dated. This chapter was written in 1983–84 and, except for a few later revisions, reflects the state of development at that time.

2. Antennas of Conventional Configuration

A large class of millimeter-wave antennas is obtained by wavelength scaling of well-established antenna configurations developed for the microwave and lower frequency bands [1]. This class of antennas includes reflector, lens, horn, and array antennas as radiating structures of high directivity gain, spiral antennas as broadband radiating elements of medium gain, and pillbox, geodesic, biconical, and linear antennas for fan-beam and omnidirectional applications. The design principles and performance characteristics of these antennas are discussed in pre-

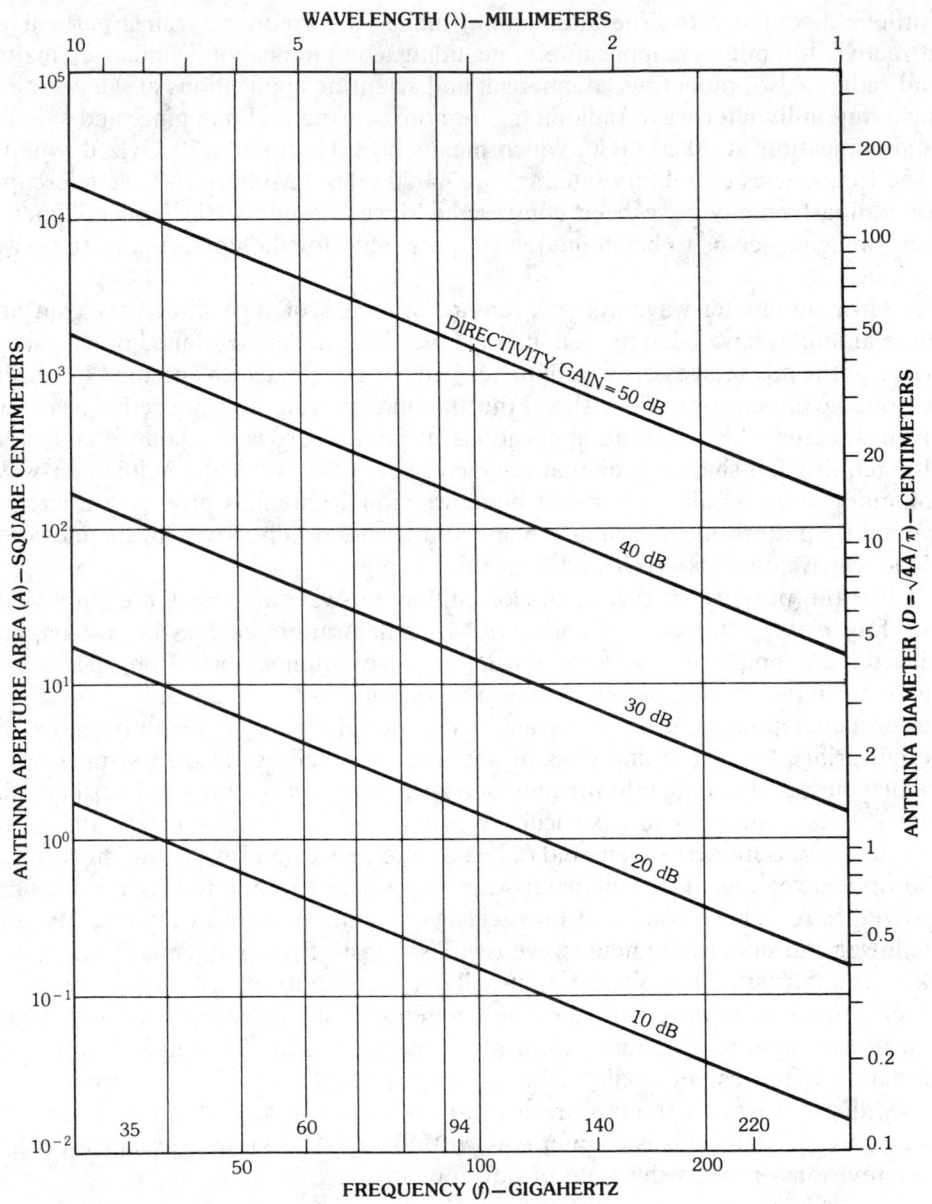

Fig. 2. Aperture area of millimeter-wave antennas as a function of directivity gain and frequency assuming an aperture efficiency of 50 percent.

ceding chapters. Scaling is, to a large degree, straightforward and the design information presented can be extended readily into the millimeter-wave region.

In practical realizations, however, conventional millimeter-wave antennas are not necessarily exact replicas of their microwave counterparts. The smaller physical size and tighter fabrication tolerances suggest modifications in their design. Structural simplicity, in particular, becomes an important design goal, while increased dimensions (as counted in wavelengths) usually do not cause problems but facilitate fabrication. The question of which type of antenna should be used for a given application may be answered differently for the microwave and millimeter-wave regions.

In the following the millimeter-wave aspects of the various conventional antenna configurations are reviewed. There is no need for a further discussion on the antenna fundamentals, which can be found in other chapters; the purpose of the review is rather to point out design options and performance limitations peculiar to the millimeter-wave region. In the process some interesting features emerge. For example, because of the much smaller size and weight of millimeter-wave antennas rapid mechanical scanning now becomes a practical option, in particular for reflector antennas. For the same reason of physical smallness, antennas made from dielectric material are much more useful in the millimeter-wave region; new types of dielectric lenses have been proposed, and dielectric horns find increasing attention.

High- and Medium-Gain Antennas

Reflector Antennas—Reflector antennas are widely used in the millimeter-wave region [2, 3]. They are typically single-beam antennas of moderate or high directivity gain for communication, radar, and sensing, and monopulse antennas for tracking and guidance. Multibeam antennas are under development for millimeter-wave satellite communication [4, 5].

Offset feed arrangements eliminate aperture blockage problems and, as in the microwave region, can be used to achieve low side lobes. For large antennas Cassegrain feed systems are particularly attractive from an efficiency viewpoint, since the length of waveguide runs is minimized, and from a fabrication viewpoint, since the feed horn can have a comparatively large aperture. Dual-reflector antennas with an offset feed geometry combine the advantages of both approaches and, in addition, can be designed for high aperture efficiency. Examples of such antennas are available with aperture efficiencies close to 70 percent and peak side lobes down by 30 dB from the main beam level [6]. Furthermore, depolarization effects caused by the asymmetry of the main reflector can be compensated in dual-reflector antennas, and a cross-polarization level as low as −40 dB can be achieved [6]. In most cases such antennas can be operated with conventional feed horns.

Because of their small size and weight millimeter-wave reflector antennas are suitable for *rapid mechanical scanning*, and antennas of this type may be designed for functions for which frequency-scanned arrays or phased arrays would be considered at microwave frequencies. The complexity and high cost of electronically scanned millimeter-wave arrays provide a further argument in favor of mechanically scanned antennas. Antenna configurations where the reflector is moved while the feed system remains stationary are of particular interest since no rotary joints are needed in the feeding waveguide.

Figs. 3a and 3b illustrate two different design approaches for such antennas. The design shown in Fig. 3a uses a cavity-backed spiral antenna as an offset feed to illuminate a shaped reflector [7]. The feed is stationary and provides a practically frequency-independent feed pattern of axial symmetry and circular polarization.

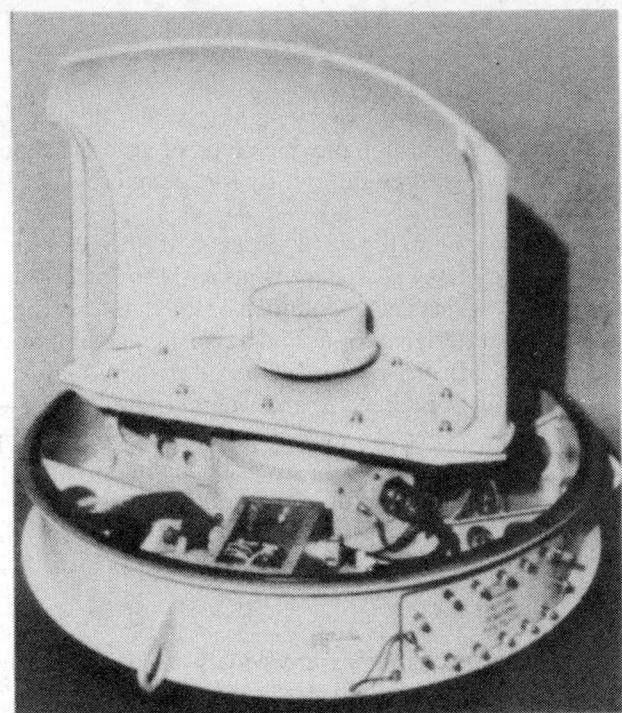

a

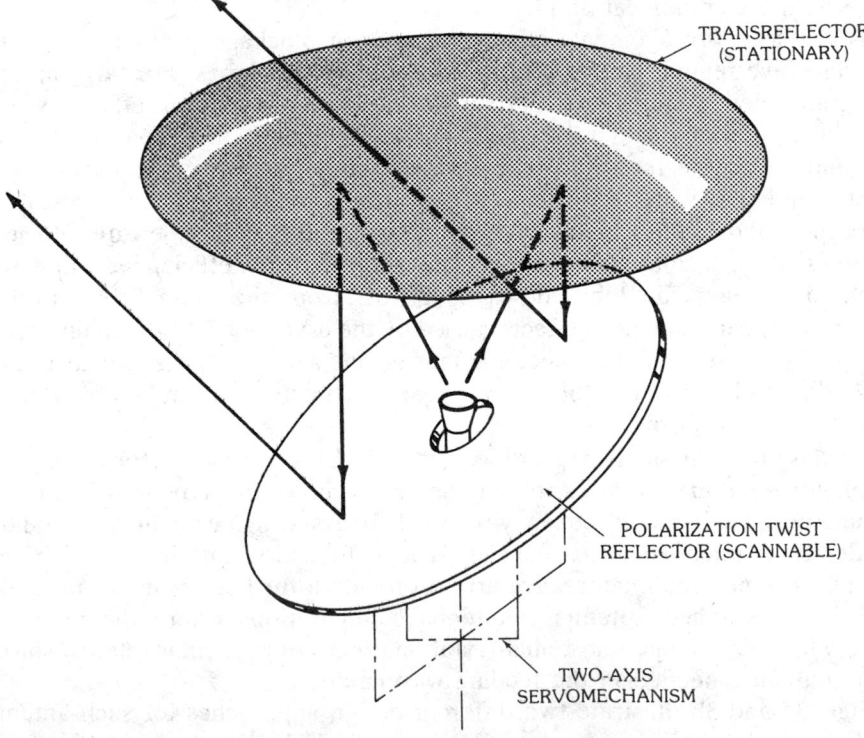

TRANSREFLECTOR
(STATIONARY)

POLARIZATION TWIST
REFLECTOR (SCANNABLE)

TWO-AXIS
SERVOMECHANISM

b

Fig. 3. Mechanically scanned millimeter-wave antennas. (*a*) Single-reflector antenna. (*Reprinted with permission of publisher from* Microwave System News, *November 1981; all rights reserved*) (*b*) Dual-reflector antenna. (*After Waineo and Konieczny [8],* © *1979 IEEE*)

The reflector rotates about the feed spiral axis; its shape is synthesized to yield a narrow beamwidth in the order of a few degrees in the azimuth plane and a wider beamwidth of approximately 20° in the elevation plane. Designed to operate over the entire 1- to 40-GHz band, the antenna has a height of 53 cm and a diameter of 61 cm. These rather large dimensions are determined by the lower frequency limit and could be reduced significantly for an antenna whose band of operation does not extend below the millimeter-wave region. The reflector of the present antenna rotates with a speed of up to 300 rpm. The primary application of antennas of this type is in broadband signal acquisition and direction finding [7].

The antenna depicted in Fig. 3b operates with two reflectors, a stationary parallel-wire transreflector and a scanning polarization-twist reflector [8,9]. Pointing of the twist reflector is typically adjusted with the help of a two-axis servomechanism. The beam radiated from the feed horn is polarized to be reflected at the stationary mirror, which has a parabolic shape and serves as a highly directive antenna. The beam is reflected a second time at the polarization-twist reflector, a planar mirror, which by its orientation determines the final beam direction. At the same time, it turns the plane of polarization of the reflected beam by 90° so that the outgoing beam is now transmitted through the stationary mirror without reflection. Antennas of this type are useful as radar tracking antennas; with a quadrature feed horn they may be used as monopulse radar. Dimensions depend on the desired gain of the sum channel and, to a lesser degree, on the required null depth of the difference channel and the side lobe level [8,9].

The twist reflector, the only moving part of the antenna, can be designed as a lightweight printed-circuit plate. Since, moreover, tilting of the reflector by $\Delta\theta$ causes beam deflection by $2\Delta\theta$, these antennas permit high-speed beam scanning. A 94-GHz model [8] had an overall diameter of 18 cm and allowed for beam acceleration in excess of $20\,000°/s^2$.

Solid-state power sources have considerable size, weight, and cost advantages over electron tubes and are preferred for many applications. Their output power, however, is limited in the millimeter-wave region, and several such sources have to be combined when a comparatively large transmitter power is required as in short-range millimeter-wave radars and in satellite communication. Reflector antennas permit *space-combination*, i.e., a cluster of several closely spaced feed horns would be used, individually driven by solid-state amplifiers phase-locked to a common source [10]. This approach eliminates most of the rf losses associated with circuit techniques for power combining. Since waveguide losses increase with frequency this is an important design consideration in the millimeter-wave region. In addition, the method provides for graceful performance degradation if some of the amplifiers should fail. If, furthermore, each amplifier is coupled with a phase shifter, a very versatile feed system is obtained, permitting dynamic control of beam pointing and illumination contour. One obtains a so-called hybrid antenna, i.e., a reflector (or lens) antenna fed by a phased array. Offset dual-reflector antennas with such feed arrays are, at present, under investigation by NASA [5] for possible use as high-gain multibeam antennas for advanced future communication satellites operating in the 30/20-GHz bands. These antennas will have to provide up to twenty fixed spot beams and up to six scannable spot beams with a gain of 50 to 53 dB and a 3-dB beamwidth of 0.3°. A main objective of the NASA investigation is to demonstrate the feasibility of feed arrays where 20-GHz power amplifier and phase shifter modules are designed in monolithic microwave integrated circuit (MMIC) technology and integrated with each radiating element near its aperture. Predictable benefits derived from such feed systems include enhanced reliability inherent in the use of a large number of amplifiers, a further reduction in cost and weight, a high-speed scan capability, and improved beam isolation combined with reduced scan loss through the fine adjustment achievable with a large number of individually controlled phase shifters and amplifiers [5].

The directivity gain of reflector antennas is limited by their *size* and by their *surface errors*. Antenna size is usually not a problem in the millimeter-wave region. Fig. 2 shows, for example, that millimeter-wave antennas with gains as high as 40 or 50 dB still have diameters less than 1 m. Accuracy requirements, on the other hand, are proportional to wavelength (for constant electrical performance) and can become difficult to satisfy in the millimeter-wave region.

Theoretically the tolerance requirements of reflector antennas are well under-stood [11–13]. Ruze has shown that random surface irregularities reduce the directivity gain of a reflector antenna according to the relation

$$G_D = G_0 e^{-\sigma^2} \tag{1a}$$

where G_D is the directivity gain in the presence of surface errors, G_0 is the directivity of the ideal, undistorted antenna, and σ is proportional to the rms surface error ϵ_{rms}:

$$\sigma = 2\pi\Upsilon \frac{\epsilon_{rms}}{\lambda_0} \tag{1b}$$

Here, ϵ_{rms} is measured in the axial direction, λ_0 is the free-space wavelength, and Υ is an adjustment factor which is approximately 2 for a shallow reflector and 1.33 for a uniformly illuminated focal plane paraboloid. Equations 1a and 1b apply to antennas with surface errors of negligibly small correlation distances. More uniform reflector deformations with finite correlation intervals cause less scattering than uncorrelated errors of the same rms value and, as a consequence, lead to a smaller reduction in directivity gain [12]. Equations 1a and 1b can thus be regarded as a worst-case estimate.*

It is instructive to examine the frequency dependence of the directivity gain for a reflector antenna of given diameter D and surface accuracy ϵ_{rms}. The directivity of the perfect, undistorted antenna is determined by the well-known formula $G_0 = \eta(\pi D/\lambda_0)^2$, so that we have, with (1)

$$G_D = \eta \left(\frac{\pi D}{\lambda_0}\right)^2 \exp\left[-\left(\frac{4\pi\epsilon_{rms}}{\lambda_0}\right)^2\right] \propto f^2 e^{-\alpha f^2}$$

---

*The effect of surface irregularities on side lobe levels has been studied by Vu The Bao [12].

where η is the aperture efficiency and γ is assumed to be 2. In the "low" frequency range where λ_0 remains large compared with $4\pi\epsilon_{rms}$, the gain is proportional to the aperture area measured in wavelengths and thus increases with f^2. At higher frequencies, in the region where λ_0 is in the order of $4\pi\epsilon_{rms}$, the directivity reaches a peak value

$$G_{pk} = \frac{\eta}{43}\left(\frac{D}{\epsilon_{rms}}\right)^2$$

which is proportional to the square of the precision of manufacture D/ϵ_{rms}. The peak gain occurs at $\lambda_0 = 4\pi\epsilon_{rms}$ and remains 4.3 dB below the gain G_0 of the perfect antenna. If the frequency is further increased into the range where $\lambda_0 \ll 4\pi\epsilon_{rms}$, scattering at the surface irregularities intensifies and the power of the main beam decreases exponentially with f^2. The power lost from the principal pattern reappears in the scatter pattern of the antenna. This scatter pattern is centered about the antenna axis and can be rather narrow; its beamwidth (in radians) and gain depend on the correlation distance d_c of the surface errors and for $\lambda_0 \ll 4\pi\epsilon_{rms}$ are given approximately by

$$\Delta\theta_s = \frac{7\epsilon_{rms}}{d_c} \quad \text{and} \quad G_s = \left(\frac{d_c}{2\epsilon_{rms}}\right)^2$$

respectively [11]. For reflectors with large correlation distances $d_c \gg \epsilon_{rms}$, the gain G_s can be rather high so that the main beam of the principal pattern becomes obscured by the scatter pattern.

The useful frequency band of a reflector antenna is the region below the peak gain, where the gain reduction due to scattering remains small. If this frequency band is to include the millimeter-wave region, high surface accuracy must be achieved and tolerance requirements become important for moderate-size reflectors and critical for very large reflectors. According to (1), the gain reduction will remain below 0.1 dB provided a surface accuracy of $0.012\lambda_0$ or better is maintained, i.e.,

$$\epsilon_{rms} < 0.12 \quad \text{mm at} \quad 30 \text{ GHz}$$
$$\epsilon_{rms} < 0.036 \text{ mm at } 100 \text{ GHz}$$

If a gain reduction of 1 dB is acceptable, the surface tolerances are relaxed to

$$\epsilon_{rms} < 0.38 \quad \text{mm at} \quad 30 \text{ GHz}$$
$$\epsilon_{rms} < 0.115 \text{ mm at } 100 \text{ GHz}$$

For Cassegrain antennas these tolerances must be shared between the two reflectors; a subdivision proportional to the reflector diameters is an appropriate choice [1]. On the other hand, Cassegrain antennas permit compensation of profile deformations of the primary reflector by appropriate shaping of the subreflector [14, 15].

Most millimeter-wave antennas are designed for directivity gains of less than 50 dB and diameters of less than 1 m. For reflectors of this size the tolerance requirements usually do not constitute a critical problem, but they do require precise fabrication techniques [2, 6, 16] and, in the case of high-gain antennas, a rigid support structure. Reflectors with a surface precision D/ϵ_{rms} of 2×10^4 are commercially available in diameters up to 1.2 m.

Examples of millimeter-wave systems which operate with larger antennas, having dimensions in the order of several meters, are ground stations for satellite communication and compact antenna test ranges [17]. (Compact millimeter-wave ranges are of interest not only for antenna measurements but also for scaled experiments on the microwave radar cross section of large objects.) The tolerance requirements for test ranges are particularly stringent. In order to achieve the desired plane wave distribution in the test zone with an amplitude and phase accuracy of better than 0.5 dB and 10°, respectively, a surface accuracy of $0.007\lambda_0$ at the highest frequency is required [17]. Such accuracies can be achieved by constructing the antennas from highly precise surface panels whose positions and orientations can be adjusted individually; this technique is generally applied in the design of very large reflector antennas.

The largest millimeter-wave antennas currently in operation are *radio-astronomy telescopes*. An overview with respect to the telescopes in existence in the United States and abroad in 1970 can be found in the literature [18]. The antennas typically use Cassegrain feeds and have main reflector diameters in the range from 5 m to 20 m. Receiver quality in the millimeter-wave region is by an order of magnitude poorer than in the microwave range, and large and accurate telescopes are needed to achieve high system performance. Recently, several new antennas have been designed, among them a 25-m antenna planned for the National Radio Astronomy Observatory but subsequently canceled, a 30-m antenna [19] constructed under Franco-German cooperation (IRAM) at Pico Veleta in Spain, and a 45-m antenna built by Tokyo Observatory at Nobeyama, Japan. The highest operating frequencies of these antennas are 400 GHz, 300 GHz, and 150 GHz, respectively. The configurations of the first two antennas are shown in Fig. 4; their beamwidths at their highest frequencies are 7.5 and 8.2 arc seconds. The requirements on the mechanical accuracy ($\epsilon_{rms} < 0.1$ mm) and pointing/tracking accuracy (approximately 1 arc sec) of such antennas in the presence of gravitational deformations, temperature gradients, and wind load push current technology to its limits and call for new design and fabrication methods for large reflectors [19]. One such method, termed *homology*, allows for large gravitational reflector deformations when the antenna is turned in the elevation plane but imposes the constraint that the reflector surface maintains a parabolic shape [20]. Adaptive position adjustment of the feed system ensures that the antenna remains focused at all times.

Lens Antennas—Lens antennas have found numerous applications in the millimeter-wave region [2, 3]. Together with reflector and horn antennas they are the most widely used millimeter-wave antennas.

Tolerance requirements are less stringent than for reflector antennas since lenses are usually made from materials of relatively low refractive index to minimize reflection at the lens surface. Thus the antennas are easy to machine and comparatively inexpensive. Lenses have very large bandwidth and excellent wide-angle scanning properties. Furthermore, there is no aperture blockage by the feed system, and lens antennas can be designed to have low side lobes (less than −35 dB) and a high front-to-back ratio. In the case of zoned lenses, however, the side lobe level becomes frequency sensitive.

Lens antennas are *more* attractive in the millimeter-wave region than at lower frequencies because their *dielectric weight is significantly reduced*. For given electrical performance the volume and weight of a lens decrease with λ_0^3 and are usually very reasonable at millimeter wavelengths. An example may illustrate this point: projected military satellite communication systems operating at 44/20 GHz will require multiple-beam antennas capable of directing spot beams and clusters of beams from a synchronous satellite to any point on the earth's surface. A recent design study [21] has shown that the rather stringent performance requirements on these antennas can be satisfied by an appropriately shaped zoned lens with an aperture diameter of 60 cm. This antenna would provide beams with a gain of 48 dB at 44.5 GHz, the center frequency of the uplink; the first side lobes would be more than 20 dB below the beam maximum; and pattern degradation for off-axis beams would remain small across the entire field-of-view scan range, which for a synchronous satellite has a width of ±8.6°. The weight of a polystyrene (Rexolite) antenna of this type would be less than 10 lb (4.5 kg) [21].

Because of the reduction in dielectric weight mentioned above, *spherically symmetric lenses* also become practical in the millimeter-wave region [22–30]. Luneburg lenses which focus an incident plane wave into a single point on the lens surface would provide excellent performance, but the required gradation of the refractive index profile of these lenses would probably be difficult to realize in the millimeter-wave region. Moreover, much simpler lenses, for example, homogeneous spheres, provide already good performance, and two-layer lenses consisting of a homogeneous spherical core surrounded by a concentric shell of different refractive index can be sufficiently well corrected to satisfy almost all practical requirements [28].

In addition to their simple configuration and their good electrical performance, such spherically symmetric lenses permit wide-angle beam scanning without pattern degradation. If the antennas are operated with an array of identical feed horns, beam scanning can be implemented digitally by electronic switching from feed horn to feed horn. With these advantages a further discussion of spherically symmetric lenses appears justified here, in particular since they are not widely used in the microwave region (where their weight would be excessive).

A theory of *two-layer lenses* based on ray tracing* is available in the literature and has been confirmed by experiments [28]. The theory includes homogeneous lenses, the so-called constant-K lenses, as a special case. These latter lenses [26] should be made from materials with dielectric constants in the range

$$2 \leqq \epsilon_r \leqq 3.5$$

---

*Wave theory approaches to the analysis of spherically symmetric lenses (based on Mie-series expansions) may be found in [29] and [30].

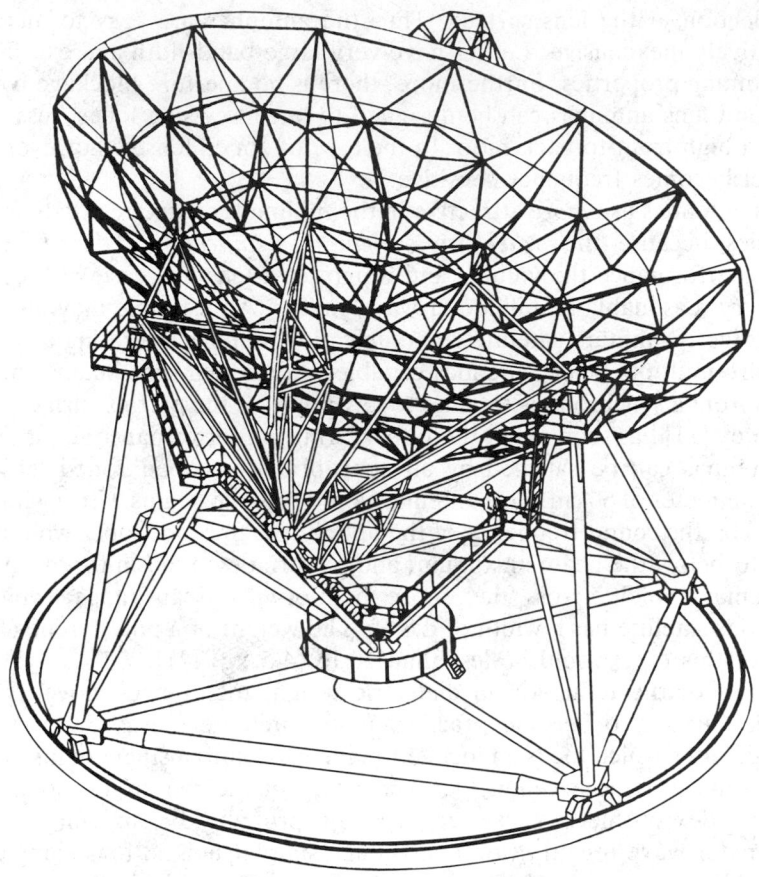

a

Fig. 4. New millimeter-wave radiotelescopes. (*a*) The 25-m antenna designed for National Radio Astronomy Observatory. (*b*) The 30-m antenna constructed at Pico Veleta, Spain. (*After Baars [19]*)

For $\epsilon_r > 3.5$ the focal point would be located in the interior of the lens, which is undesirable, and for $\epsilon_r < 2$ problems associated with large focal lengths may arise. The distance R' of the focal point from the center of a lens of radius R is approximately given by

$$R' = R \frac{\epsilon_r}{\sqrt{2}(\epsilon_r - 1)}$$

which defines the position of the feed horn. Particularly for lenses with permittivity values close to the upper limit of 3.5, spherical aberrations are small and such lenses can be designed for diameters of up to $15\lambda_0$ and a gain of up to 30 dB while maintaining good pattern quality, including a side lobe level below −20 dB.

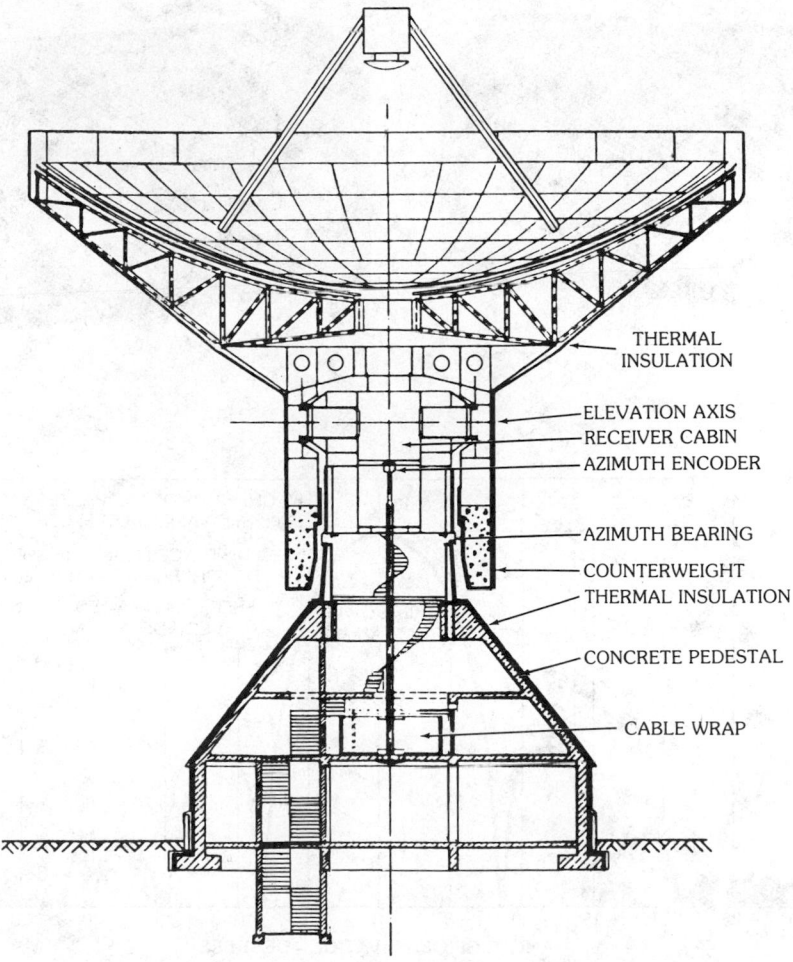

THERMAL INSULATION

ELEVATION AXIS
RECEIVER CABIN
AZIMUTH ENCODER

AZIMUTH BEARING

COUNTERWEIGHT
THERMAL INSULATION

CONCRETE PEDESTAL

CABLE WRAP

b

Fig. 4, *continued.*

As an example, Fig. 5a shows an experimental 60-GHz radio which uses two homogeneous Teflon spheres as antennas [31]. The lenses, which are fed by small waveguide horns, have a diameter $2R = 10\lambda_0 = 5$ cm. Fig. 5b shows the radiation pattern. The gain of these lenses is 26 dB, the main beam pattern is nearly circularly symmetric, and first side lobes are down by -20 dB.

The parameters of a two-layer lens can, in principle, be determined by imposing suitably defined optimum performance criteria. In the practical implementation the dielectric constants of core and shell are restricted by the limited number of available low-loss millimeter-wave materials. A lens which has shown good performance consists of a polystyrene core ($\epsilon_r = 2.46$) and a quartz shell ($\epsilon_r = 3.80$) with an outer-to-inner diameter ratio of 1.63 [28]. An experimental lens of this type with a diameter of 6.8 cm, which corresponds to approximately $16\lambda_0$ at the test frequency of 70 GHz, has yielded a gain of 32 dB and side lobes below -27 dB,

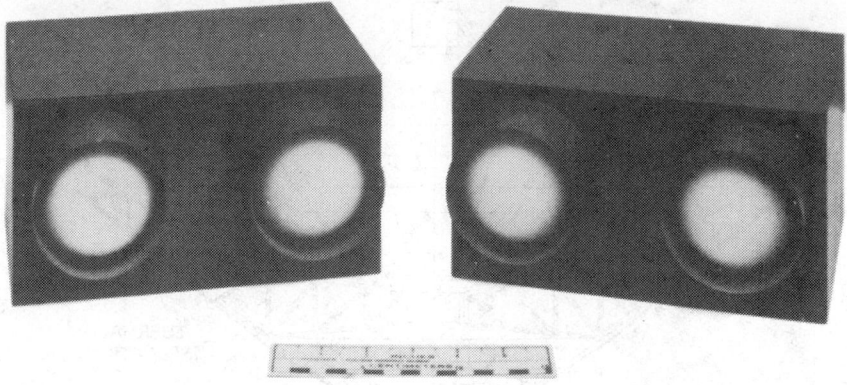

a

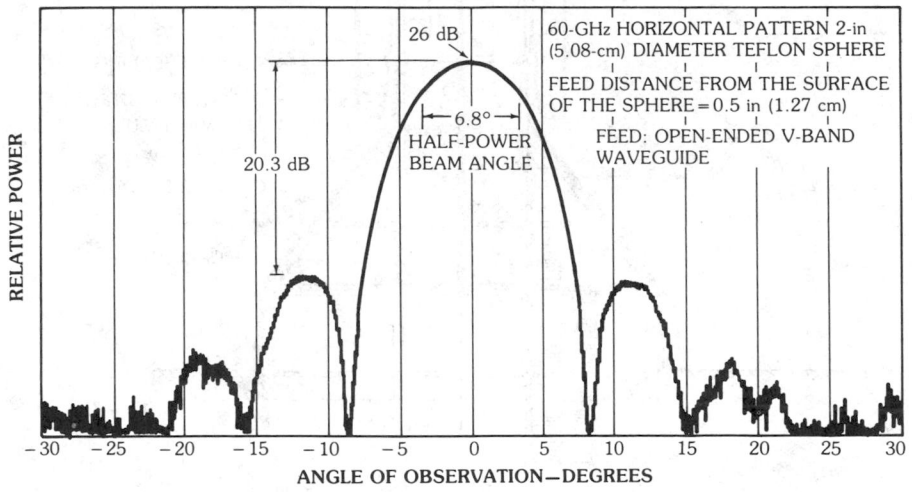

b

Fig. 5. Teflon-sphere lens antennas. (*a*) Experimental 60-GHz radios with lens antennas. (*b*) Horizontal radiation pattern at 60 GHz. (*Courtesy Chang, Paul, and Ngan [31]*)

in very good agreement with theoretical predictions.* The theory [28] shows, moreover, that lenses with ten times this diameter could be designed providing substantially increased directivity gain (51 dB) while maintaining good pattern quality. Dielectric spheres with such diameters would be large and bulky, even at

---

*It is interesting to note that for the two-layer lenses considered here, the permittivity of the shell is larger than that of the core, while in Luneburg lenses the refractive index decreases with distance from the center. In other words, these lenses should not be viewed as first-order approximations to a Luneburg lens. Their performance can rather be explained in terms of two counteracting aberrations which are caused by the core and shell, respectively [24]. If the lens parameters are chosen appropriately, these aberrations will compensate each other to a large extent, thus leading to enhanced lens performance.

millimeter wavelength, and such limiting factors as fabrication tolerances, homogeneity requirements, and dielectric losses would most likely make them impractical. But the point is that spherically symmetric lenses of simple structure are capable of providing very good electrical performance [28].

Practical requirements on spherically symmetric lenses include a very uniform permittivity throughout the lens and correctness in shape to avoid beam tilting and pattern deformations [26]. Dielectric losses in constant-K lenses can be estimated from the formula

$$L_{\text{diel}} = 36 \frac{R}{\lambda_0} [\epsilon_r^{3/2} - (\epsilon_r - 1)^{3/2}] \tan \delta$$

where $\tan \delta$ is the loss factor of the lens material and L is in decibels. For a quartz lens with a diameter of $15\lambda_0$ and $\tan \delta = 0.00025$, the loss would be 0.18 dB. About the same loss would occur in a $10\lambda_0$ Teflon lens (Fig. 5) with $\tan \delta = 0.0006$. In the case of a lens consisting of core and shell, the step in refractive index at the outer surface is, in general, large enough to require an antireflection layer. A single matching layer can be effective up to large angles of incidence [28]. The step between core and shell is usually sufficiently small so that reflection is negligible, and no matching layer is needed at the inner lens surface.

Shaping techniques for millimeter-wave lens antennas have recently been developed by Lee [32, 33]. Similar to reflector shaping methods, these techniques are based on ray tracing approaches and are analytically and computationally involved. Speaking in general terms, the design procedure imposes the power conservation law to control the aperture taper, Snell's law to define the first surface of the lens (as seen from the feed point), and the path length condition to determine the second surface. The path length condition is specified by the desired phase distribution across the aperture. The radiation pattern of the lens is calculated in the usual way with scatter effects at the zoning edges neglected.

Note, however, [32] that these lens shaping methods differ from the well-established design techniques for optical lenses. Optical correction techniques are concerned primarily with spherical lenses and the compensation of their phase errors (aberrations), while the objective of millimeter-band shaping techniques is the realization of a desired phase *and* amplitude distribution across the lens aperture.

For multibeam antennas, furthermore, Lee [32, 33] has introduced a new method, termed *coma-correction zoning*, which permits one to minimize distortions of off-axis beams caused by cubic phase errors. The basic idea is that coma aberrations of a thin lens are significantly reduced provided the inner surface of the lens, i.e., the surface facing the feed horn, is spherical on the average. This condition can be satisfied by zoning. Lenses which combine contour shaping with coma-correction zoning are of particular interest for use in millimeter-wave satellite communication systems. This application requires spaceborne multibeam antennas capable of radiating a large number of closely spaced, highly directive beams of high pattern quality, which implies minimum pattern degradation for off-axis beams and low side lobes (to allow frequency reuse). In addition, a high crossover level of 4 dB is required for adjacent beams in order to ensure continuous ground coverage.*

Fig. 6 shows the cross-section profile of a shaped, coma-reduced lens designed by Lee [32]. The lens has a focal length of 24.5 cm, an aperture diameter of 20.7 cm = $30\lambda_0$, and a center thickness of 3.5 cm = $5\lambda_0$ at the design frequency of 44 GHz. The feed pattern is assumed to be the standard E-plane pattern of a square horn, i.e.,

$$g(\theta) = (1 + \cos\theta)\frac{\sin[(\pi d/\lambda_0)\sin\theta]}{(\pi d/\lambda_0)\sin\theta}$$

where θ is the angle counted from the horn axis. The horn size d is chosen to provide an edge taper of 30 dB at $\theta = 20°$, the aperture half-angle of the lens as seen from the feed location. The lens is designed to produce an aperture distribution $E(r) = [1 - (r/1.05)^2]^3$ with a uniform phase. Made from Rexolite with $\epsilon_r = 2.54$, the lens has a weight of 0.5 kg.

An experimental model of the lens has yielded a gain of 33 dB, a beamwidth of 3.3°, and a side lobe level of −30 dB for individual beams [33]. These beams could be scanned up to ±12° without significant coma degradation. For multi-beam operation the lens can be fed by a cluster of small single-beam feed horns generating closely spaced beams with the desired high crossover level of 4 dB. The aperture efficiency of the antenna is 48 percent. Total losses amount to 2 to 3 dB.

Satellite communication systems will require larger antennas of much higher directivity in many cases. But the experimental antenna demonstrates the excellent performance that can be obtained through the use of lens shaping methods.

Another type of lens antenna which should be useful for millimeter-wave applications is the *Fresnel zone-plate lens*. This device was proposed many years ago in optics as a flat version of a lens. It differs from the usual lens in that it focuses by employing diffraction instead of refraction. The zone-plate lens is based on the fact that ray bundles traced from the source point to the image point via alternate Fresnel zones of the lens plane differ in path length by a half-wavelength. The Fresnel zones form concentric rings about the optical axis; the radii of these rings are

$$r_n = \left[(n + 1/2)\lambda_0\frac{d_1 d_2}{d_1 + d_2}\right]^{1/2}, \qquad n = 1, 2\ldots$$

---

*If a conventional, spherical lens antenna is used, the two requirements of high crossover level and low side lobes can lead to a packaging problem of the feed horn assembly [33]. A low side lobe level implies the use of feed horns of comparatively large aperture (to achieve an appropriately tapered illumination across the lens). But horns of large aperture dimensions cannot be spaced sufficiently close to obtain a high crossover level between adjacent beams. The problem is commonly solved by using an assembly of horns of small aperture and synthesizing the desired feed pattern by exciting a subset of adjacent horns for each beam [33]. Since each beam is generated by several horns (and some horns contribute to several beams) the method requires a large, complex feed system which in the millimeter-wave region would be rather lossy. In the case of a shaped lens, on the other hand, the beam forming and side lobe control functions are shifted from the feed horns to the lens. Thus, single-beam horns of small aperture can be used and the packaging problem finds a straightforward solution.

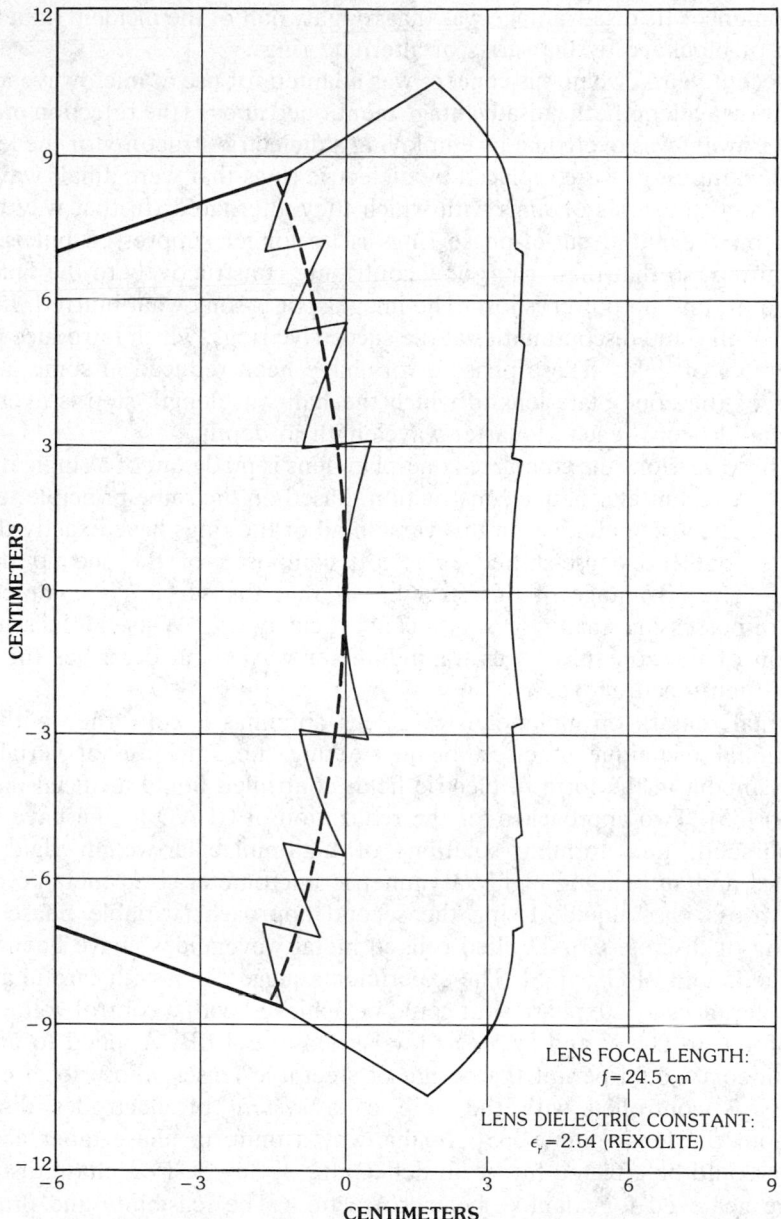

Fig. 6. Cross-section profile of coma-reduced, shaped lens for multibeam operation at 44 GHz. (*After Lee [32], © 1983 IEEE*)

where λ_0 is the free-space wavelength and d_1 and d_2 are the distances of the source point and image point from the lens plane. By blocking out each second ring the lens thus permits only those rays to pass which add constructively at the image point. Its important advantage was that the lens is flat, thereby offering an element of convenience. Its disadvantage was that roughly half of the incident light was lost because of blockage by the series of alternate rings.

In recent years, when this concept was adapted for use at microwave and then millimeter wavelength, the disadvantage mentioned above (the rejection of half the incident power) was overcome by employing a dielectric structure for the lens. The metal blocking rings were replaced by dielectric rings that were a half-wavelength thinner than the series of rings with which they alternated. In that way the field incident on the initial out-of-phase rings is no longer suppressed but is instead phase shifted, so that these rings now contribute constructively to the field at the image point, and no power is lost. The image spot is somewhat blurred, however, because of the step discontinuities at the successive rings, which introduce periodic phase errors of ±90°. These phase errors have been reduced in some advanced versions of the zone-plate lens in which the half-wavelength step is averaged in three smaller steps, each a quarter-wavelength in depth.

In these versions the complete zone-plate lens is made out of a single dielectric material. A recent alternative construction, based on the same principle, uses two different dielectric materials. In this version all of the rings have exactly the same thickness, but the phase-shifted rings are composed of the second dielectric material. The advantage of this structure is that the surface is completely flat instead of possessing small ridges. A recent article by J. C. Wiltse [34] discusses the operation of the zone-plate lens for millimeter waves and describes the various versions mentioned above.

A final remark on millimeter-wave lens antennas is concerned with an unconventional technique to effect beam steering, i.e., the use of variable permittivity media in the form of electric field–controlled liquid artificial dielectrics (CLAD) [35]. Two approaches for the realization of CLAD media have recently been pursued, i.e., forming solutions of high-molecular-weight rigid macromolecules and suspending highly asymmetric micrometer-sized metallic particles in a low-loss base liquid. Using the second approach, variable phase shifters consisting of discrete CLAD filled cells in metal waveguides* have been demonstrated at 35 and 94 GHz [35]. The experiments suggest that with careful attention to cell tolerances a 360° phase shift could be achieved with a control voltage in the order of 100 to 200 V and losses as low as 1.8 to 2.4 dB. Applied to antennas, CLAD media would permit the design of steerable lenses whose local electrical thickness is controlled with the help of a system of electrodes distributed throughout the lens volume [35]. In this way a uniform phase taper across the aperture could be effected for beam deflection, or any desired phase distribution could be achieved for adaptive beam correction. The feasibility and practicality of liquid dielectric lenses as millimeter-wave antennas remains to be established. But the number of useful CLAD media is probably very large, and only very few have been studied [35]. Some of those explored so far were both flammable and toxic.

---

*To ensure a uniform particle density throughout the CLAD medium, the liquid was kept in a continuous motion during the experiments, i.e., it was passed through the phase shifter cell at a constant flux rate. (The specific weight of the suspended metal particles is larger than that of the base liquid.)

Horn Antennas—Horn antennas for the millimeter-wave region are well developed [3, 36]. In design and performance there is little difference between these antennas and their microwave counterparts. Horns with large flare angles and lens correction provide gain levels of up to 30 dB in compact designs. Corrugated horns can be designed for very low side lobe and back lobe levels, for low cross polarization, and a practically circular symmetric radiation pattern [37–40]. For square-pyramidal and conical horns in particular, the *E*- and *H*-plane patterns are very similar, and these antennas are well suited not only for radiation at linear polarization but at circular polarization as well [41]. High mechanical precision can be achieved by using electroforming techniques. Standard gain horns are available over the complete millimeter-wave band.

As an example of a high-performance millimeter-wave horn antenna, Fig. 7a shows a *corrugated conical horn* designed for the 33-GHz band [42]. The antenna has a half-power beamwidth of 7° and a very broad bandwidth. The flare angle is 10°, and a moderate groove spacing (two grooves per wavelength) permits compact and easy construction. The choke grooves on the rim face suppress potential back lobe radiation. The *H*- and *E*-plane radiation patterns, Figs. 7b and 7c, show an extremely low side lobe level, which in the far side lobe region is more than −75 dB below the main peak. (Measurement at this level required a specially designed test range.) The antenna was originally designed for satellite-based measurements of the properties of the cosmic background radiation but it should also be very well suited as a feed horn for large parabolic reflectors.

A different type of horn antenna, i.e., *dielectric horns*, are indicated in Figs. 8a and 8b. The first figure shows a rhombic dielectric plate antenna (sectoral horn) fed by a rectangular waveguide [43], and the second figure illustrates a conical dielectric horn protruding from a circular waveguide [44]. When fabricated from materials of comparatively low dielectric constant in the range of 1.5 to 2.5, such antennas are well suited as array elements and as primary feeds for large reflectors or lenses. Dielectric horns have been found to have good bandwidth of approximately 20 percent, and a beamwidth smaller than that of metallic horns of the same flare angle and axial length. If used as a primary feed, they cause little aperture blockage. A further advantage is that dielectric horns have less critical tolerances than metal horns and are easier to machine. Furthermore, the cross-polarized component of conical dielectric horns has been shown to be very small [44]. Dielectric horns have also been utilized as Gaussian beammode launchers for quasi-optical devices operating in the upper millimeter-wave region [45].

Array Antennas—Array antennas of conventional design have not been widely used up to now in the millimeter-wave region. The technology for fixed-beam and frequency-scanned arrays is available in principle, while the utilization of phased arrays will depend on the development of suitable millimeter-wave phase shifters [2].* All of these antennas, and in particular large phased arrays, require highly

---

*Ferrite phase shifters for the 35-GHz band with characteristics suitable for systems applications have recently beome available [77].

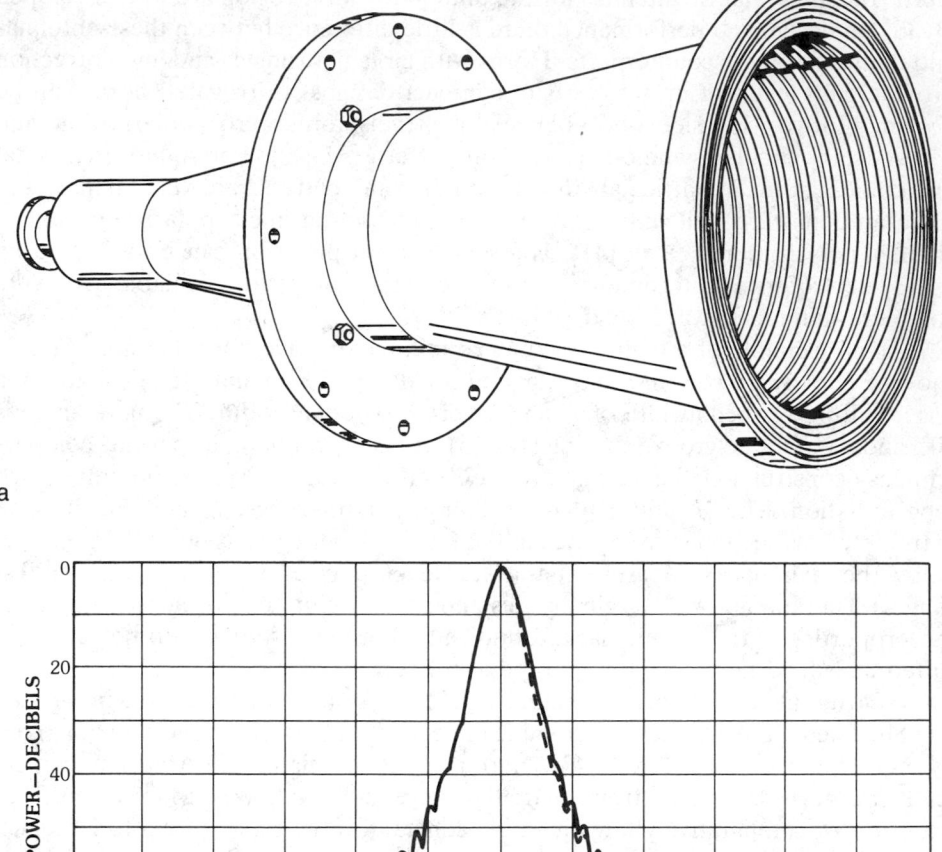

a

b

Fig. 7. High-performance horn antenna for the 33-GHz band. (*a*) Horn configuration. (*b*) *H*-plane pattern. (*c*) *E*-plane pattern. (*After Jansen et al. [42], © 1979 IEEE*)

precise fabrication techniques and are likely to be expensive. Now and for the immediate future, mechanically scanned fixed-beam antennas appear to be the best choice for many millimeter-wave radar, sensing, and communication uses [2]. A contributing factor is the compactness of millimeter-wave antennas which facilitates rapid mechanical scanning [7–9]. In the more distant future it is likely that the

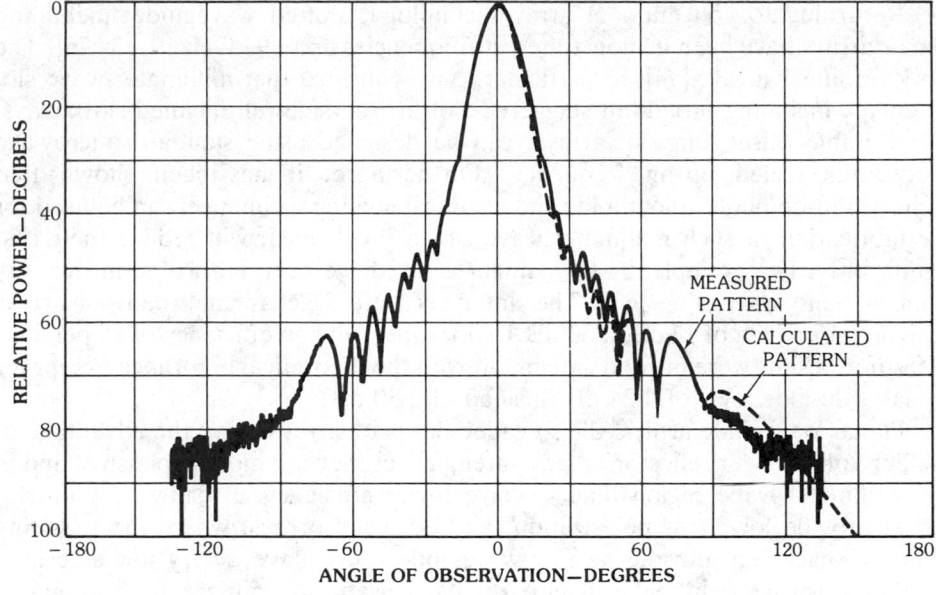

c

Fig. 7, *continued.*

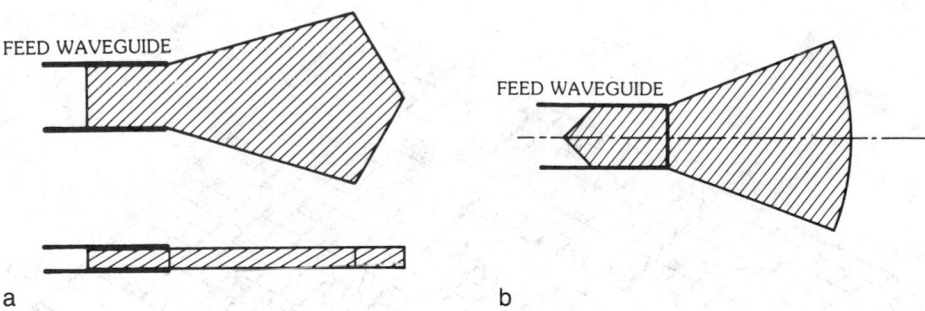

a   b

Fig. 8. Dielectric horn antennas. (*a*) Rhombic dielectric plate. (*After Ohtera and Ujiie* *[43], © 1981 IEEE*) (*b*) Conical dielectric horn. (*After Hombach [44]*)

design of millimeter-wave arrays will be dominated by the utilization of printed-circuit antenna technology and microstrip antenna techniques in particular. This holds for fixed-beam arrays as well as for frequency-scanned and phased arrays. Microstrip antennas and printed antennas in general can be conveniently and precisely fabricated by the use of etching techniques, and integrated designs are possible where feed lines, phase shifters, and eventually power amplifiers and low-noise detectors are integrated with the radiating elements on the same substrate (though possibly on different substrate layers). Advanced work in this area is in progress [8,9,46–72]. Recent results and future trends are discussed in Sections 4 and 5.

Returning to conventional array technology, slotted waveguide linear and planar arrays have been demonstrated at frequencies up to 94 GHz [9, 73–76]. The work on linear arrays [74], in particular, has confirmed that millimeter-wave slot antennas, including broadwall staggered-slot arrays, sidewall inclined-slot arrays, and variable offset long-slot arrays, can be designed using standard microwave procedures scaled up in frequency. Furthermore, it has been shown that high-resolution photolithographic and chemical etching techniques can be used for the fabrication of such millimeter-wave slot arrays, which will reduce their cost significantly. For example, 94-GHz linear arrays have been fabricated in this way from *W*-band waveguide [73]. The antennas were 21-element broadwall arrays designed for a gain of 17 dB, a 30-dB Taylor taper, and an efficiency of 85 percent. Measured results were in good agreement with the theoretical perorance except for a peak side lobe level of −24 dB (instead of −30 dB).

Planar waveguide arrays, the so-called flat plate arrays, have the advantage of smaller volume over reflector or lens antennas but they are more expensive, and in the millimeter-wave region tend to have lower efficiency. Broadwall shunt-slot arrays provide low cross polarization, and edge-slot arrays, where the radiating slots are machined into the narrow waveguide walls, have a very low side lobe capability which would make them particularly useful for military applications.

As an example, Fig. 9 shows a frequency-scanned 400-element *K*-band

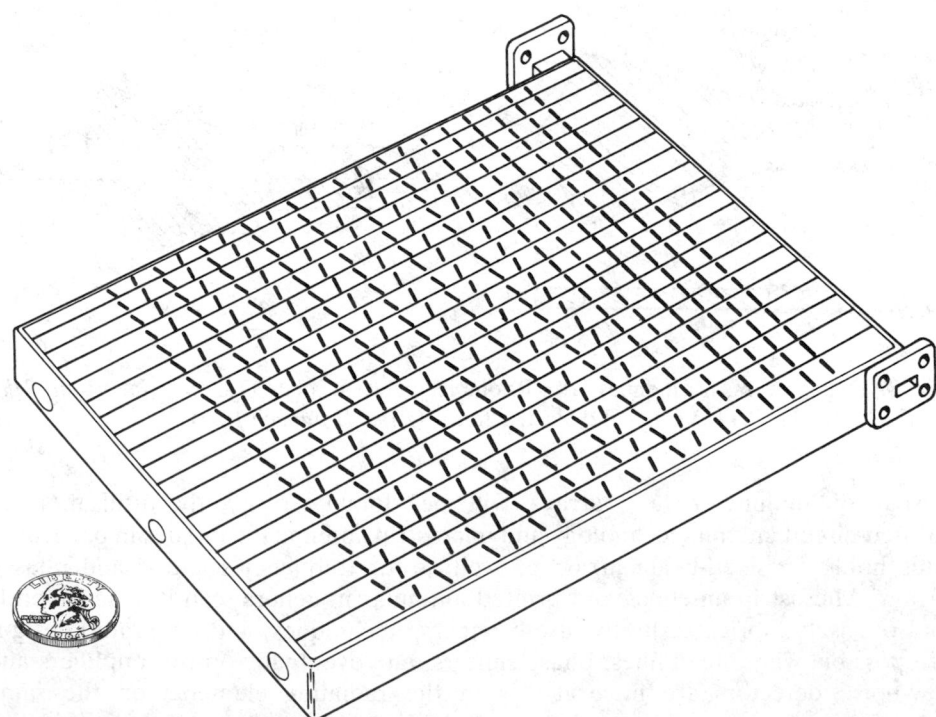

Fig. 9. A *K*-band frequency-scanned edge-slot array with 400 radiating elements. (*Courtesy Hilburn and Prestwood [78], © 1974 IEEE*)

edge-slot array in RG-96/U waveguide technology [78]. The array, designed for a 30-dB Taylor distribution with three equal side lobes and an overall efficiency of 65 percent, exhibits a gain of 23, 25.5, and 28.5 dB at frequencies of 26.5, 28.5, and 30 GHz, respectively. The corresponding pointing angles are 23°, 13°, and 6°, so that a scan range of 17° is obtained at a scan rate of approximately 15° per 10-percent frequency change. No side lobes or cross-polarized components were observed down to −30 dB below the main beam level. Machining tolerances for slot depth and slot-to-slot spacing (±0.025 mm and ±0.05 mm, respectively) did not present great difficulties, but precision requirements would be more difficult to achieve at higher frequencies [78]. Note that frequency scaling of such arrays is not entirely straightforward in that the wall thickness of metal waveguides, as measured in wavelengths, is larger in the millimeter-wave region than it is at microwave frequencies.

Purcell-type arrays [79, 80], though known for many years, have not found much attention up to now. They are mentioned here since their mechanical aspects, i.e., substantially increased slot dimensions and relaxed fabrication tolerances, should make them well suited for millimeter-wave applications. In these arrays a comparatively large wall thickness is used for the waveguide wall carrying the radiating slots; see Fig. 10, which shows a sectoral horn excited by a Purcell-type slotted waveguide. The slots themselves, in this case, are regarded as waveguide sections coupling the interior region of the feeding waveguide to the outside region. If the wall thickness is chosen equal to an electrical quarter-wavelength of these waveguide sections (or an odd multiple of $\lambda_g/4$), an impedance transformation will occur which in effect nearly short-circuits the slots as seen from the interior of the feeding waveguide. Hence little energy leakage will take place from each slot and the cross-sectional slot dimensions can be made comparatively large. The quarter-wavelength condition on the wall thickness limits the bandwidth of these arrays.

Spiral Antennas and Fan-Shaped Beam Antennas

Spiral Antennas—These antennas are well known as "frequency independent" antennas which provide essentially constant directivity gain, beamwidth, and impedance over broad frequency ranges. In addition, it is advantageous for many applications that the antennas operate at circular polarization.

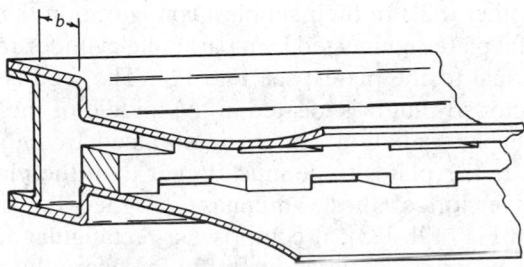

Fig. 10. A Purcell-type slot array shown feeding an *E*-plane sectoral horn. (*After Silver [80], © 1949; reprinted with permission of McGraw-Hill Book Company*)

a

Fig. 11. Millimeter-wave spiral antenna. (*a*) Antenna configuration. (*b*) Radiation pattern at 18 GHz. (*c*) Radiation pattern at 40 GHz. (*After E. M. Systems [81]; reprinted with permission of* Microwave Journal, *from the December 1975 issue,* © *1975 Horizon House–Microwave, Inc.*)

Logarithmic spiral antennas have been designed for the lower millimeter-wave region up to 60 GHz [7, 81]. Fig. 11a is a photograph of an antenna covering the 18- to 40-GHz band [81]; the spiral diameter is approximately 1.9 cm. Figs. 11b and 11c show the radiation patterns at 18 and 40 GHz. The two curves in each graph refer to horizontal and vertical polarization. The beamwidth is 70° and the axial ratio remains below 2 dB across the band.

Antennas with these characteristics are useful in particular for broadband direction-finding systems [81]. They may also be used as reflector feeds or as array elements in highly directive communication and sensing systems when a broad bandwidth is required.

Pillbox and Geodesic Lens Antennas—Pillbox antennas provide fan-shaped radiation patterns of broad beamwidth in one principal plane and narrow beamwidth in the other [82]. In their simplest configuration these antennas consist of two parallel metal plates connected by a parabolic-cylinder reflector whose focal line is directed normal to the plates; see Fig. 12a. The feed is placed in the focal region of the reflector so that it is located in the middle of the radiating aperture. The desired fan-shaped radiation pattern is obtained by choosing the aperture dimension parallel to the plates to be much larger than the plate spacing.

An improved version of these antennas, designed for the millimeter-wave region, is shown in Fig. 12b [83]; it comprises a rectangular feed horn, a doubly folded pillbox, and a corrugated flare section. In the figure the structure is bisected vertically and only one-half is shown. The doubly folded geometry permits feeding the antennas from the back. Hence aperture blockage is eliminated, resulting in

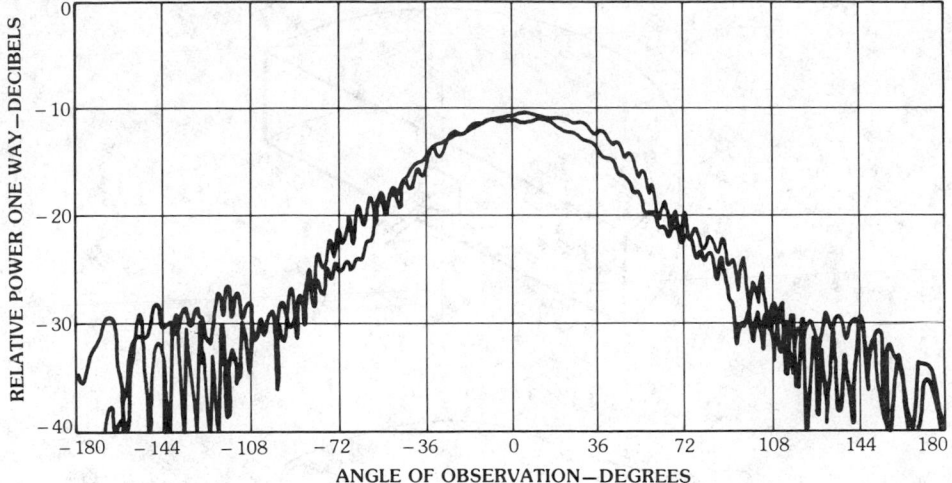

b

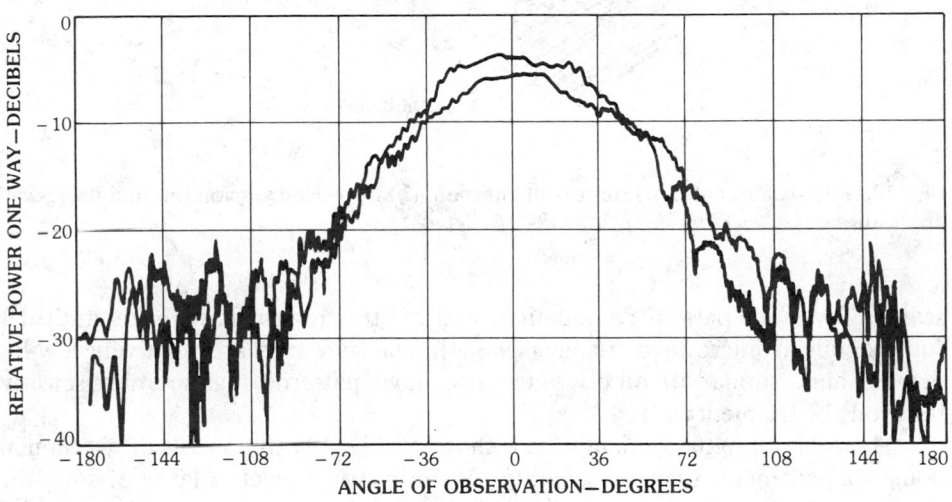

c

Fig. 11, *continued.*

improved side lobe performance, and the length of feed waveguide runs is reduced, leading to enhanced efficiency. An experimental model of the antenna designed for the 60-GHz band had aperture plane dimensions of 20 cm by 1.75 cm and provided a gain of 27 dB corresponding to a beamwidth of 2° in the *H*-plane and 30° in the *E*-plane. *H*-plane side lobes were −26 dB below the main beam, and it appears that by careful design a side lobe level as low as −35 dB can be achieved both in the *E*- and *H*-planes [83]. The bandwidth of the antenna is greater than 30 percent.

Geodesic lenses are a second type of antenna that can be used to produce fan-shaped radiation patterns. In addition, these antennas permit wide-angle beam

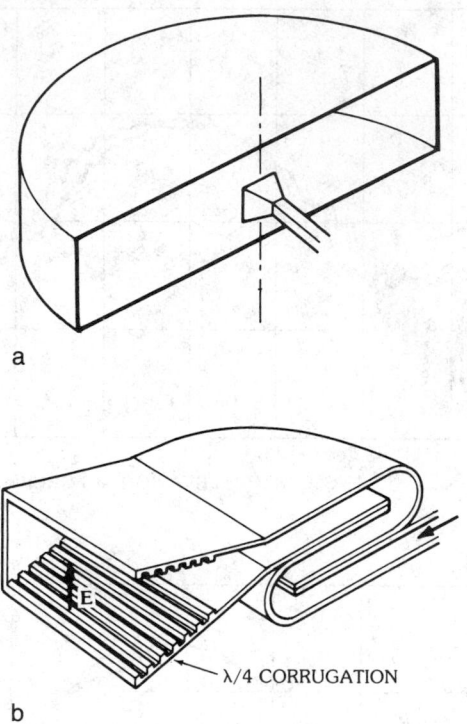

a

b

λ/4 CORRUGATION

Fig. 12. Pillbox antennas. (*a*) Basic configuration. (*b*) Improved version for millimeter-wave applications. (*After Chen [83],* © *1981 IEEE*)

scanning without pattern degradation, and for this reason have been studied in some detail at microwave frequencies [84]. The antennas are relatively low-loss devices and, similar to pillbox antennas, have patterns that are not sensitive functions of frequency [3].

The basic antenna configuration is shown in Fig. 13a. It consists of two similar, dome-shaped metal surfaces separated by an air or dielectric layer of small but constant thickness. The antenna is fed by a small horn, usually in the form of an open waveguide, placed at the edge opening between the two surfaces. The EM waves or rays emanating from the horn are guided between the metal surfaces toward the radiating aperture on the diametrically opposite side of the edge slit. The ray paths are determined by Fermat's principle of minimum path length and follow geodesic lines. By adjusting the shape of the metal domes, the ray paths and thus the amplitude and phase distribution in the radiating aperture can be controlled. Usually these antennas are designed to transform a point source at the input into a straight, collimated line source at the output. The size of the aperture is limited in one dimension (*E*-plane) by the spacing of the metal surfaces but can be very wide in the other dimension (*H*-plane), so that the antennas are well suited to generate fan-shaped beams. Furthermore, if the metal domes are surfaces of revolution about the vertical axis, the beam direction can be scanned in the *H*-plane over a wide angular range without pattern degradation simply by moving the feed

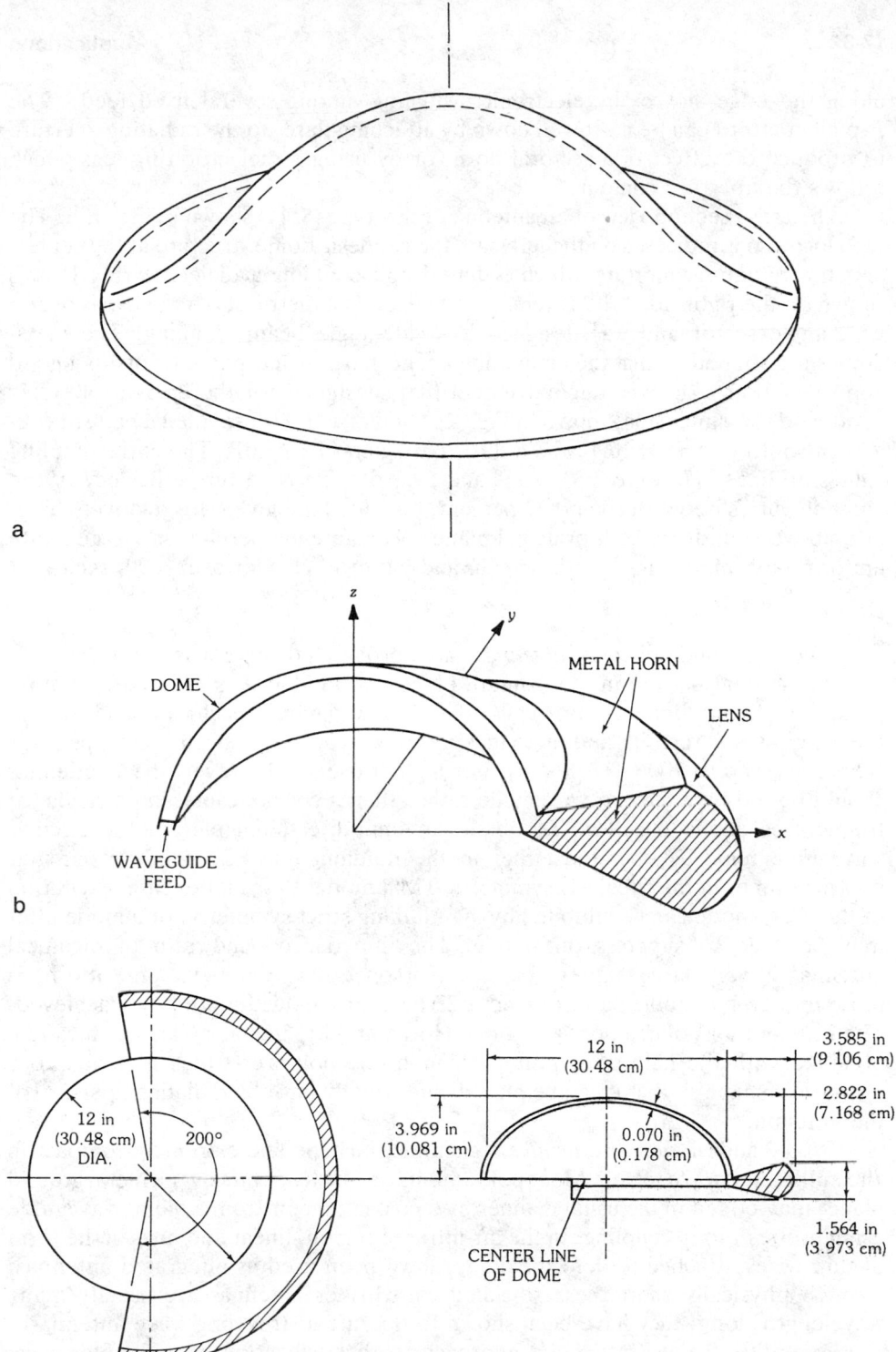

Fig. 13. Geodesic lens antennas. (*a*) Basic configuration. (*b*) Cross section of antenna with metal flares and dielectric ring lens attached to output edge. (*c*) Dimensions of experimental model for 26.5- to 40-GHz band. (*b* and *c*: *After Dufort and Uyeda [85]*, © *1981 IEEE*)

along the edge slot or by electronic switching among several fixed feeds. The *E*-plane pattern can be narrowed down by attaching flares to the radiating aperture to produce the effect of a sectoral horn, or by using a dielectric ring lens which follows the aperture contour.

The cross-section view of an antenna of this type [85] is shown in Fig. 13b. The feed location is indicated on the left side, the geodesic dome structure at the center, and the radiating aperture which is flared and contains a dielectric ring lens is shown on the right side. The antenna is circularly symmetric about the *z* axis over a wide-angle sector and was designed for wide-angle beam scanning. The metal domes are shaped so that the emanating beams have planar phase fronts to a good approximation. An experimental model [85], designed for the 26.5- to 40-GHz band, had the dimensions shown in Fig. 13c, and at 40 GHz provided a pattern with a beamwidth of $1.7° \times 10.7°$ and a directivity gain of 31.5 dB. The corresponding values at 26.5 GHz were $2.5° \times 15°$ and 29.5 dB. The aperture efficiency of the antenna varies between 60 and 72 percent, but side lobe levels are relatively high, i.e., above -20 dB in both principal planes. The antenna permits one to generate uniform multiple beams of similar gain and pattern behavior over a 90° sector.

Omnidirectional Antennas

Biconical and *discone antennas* are broadband antennas providing an omnidirectional pattern in the azimuth plane and moderate gain in the elevation plane. For the millimeter-wave region the same design approach can be used as in the microwave region, and antennas of this type have been developed for frequencies up to 100 GHz [86, 87]. For high efficiency, however, these antennas should be fed by a circular waveguide rather than a coaxial cable, in particular at frequencies above 60 GHz. To obtain an omnidirectional pattern the circular waveguide must be operated either in the fundamental TE_{11} mode at circular polarization or in the circularly symmetric TM_{01} mode. In the latter case, excitation of the TE_{11} mode can be inhibited by maintaining strict symmetry, or a mode filter may be used to suppress this mode. The impedance bandwidth of biconical antennas is very large; pattern bandwidth (concerning the elevation pattern) is more limited but stable patterns over a 25 percent bandwidth are easily achieved. The direction of polarization is vertical. Horizontal or circular polarization can be obtained with the help of a printed meanderline polarizer which is bent into a cylindrical shape so that it can be placed conformally into the radiating aperture of the antenna.

Dipole and *monopole antennas* are simple in shape and extremely compact in the millimeter-wave region. Monopole antennas, which operate over a metal ground plane, may be fed in the usual manner by a coaxial line or from a metal waveguide using probe or loop coupling. In the far-infrared region, linear antennas in the form of thin wires attached to detector diodes have been used as integrated antennas. Though physically short these so-called cat-whisker antennas are usually many wavelengths long; they have been shown to operate as traveling-wave antennas.

Recently, a novel integrated approach to the realization of millimeter-wave and submillimeter-wave dipole antennas has been suggested where these antennas would take the form of semiconducting strips or films grown on insulating or semiconducting substrates [88]. The primary advantage of such antennas is that their conductivity profile can be adjusted by modulating the carrier concentration along the antenna. This could be achieved, for example, through the utilization of surface field effects, biased pn junctions, or electro-optical methods [88]. In this way the current distribution of the antenna and hence its input impedance and radiation pattern can be controlled electronically. An additional advantage, which these antennas share with printed-circuit antennas, is that they could be integrated readily with other solid-state components created on the same substrate. At present, however, such semiconducting film dipoles are a concept only; an experimental model does not seem to be available as yet.

Linear arrays can be realized in the form of slotted waveguides. The slots, which are operated at resonance, are equivalent to magnetic dipoles placed tangentially on the (closed) waveguide surface. A waveguide with an axial sequence of longitudinal slots represents a collinear array, and such arrays are useful as omnidirectional antennas. For this application the waveguide axis will usually be oriented vertically, and it is desirable that maximum radiation occurs in the horizontal plane. This condition can be satisfied by operating the waveguide as a resonant array where all slots are excited in equal phase.

Because of the presence of the waveguide, however, the array configuration is not strictly symmetric about the vertical axis, and the azimuth pattern will show a certain amount of directivity. Since the waveguide has limited width this directivity will not be prohibitive, but achieving a truly omnidirectional pattern requires special care. The antenna radiates horizontal polarization.

As an example, Fig. 14 shows a broadwall, longitudinal-shunt-slot array antenna which is operated as a resonant array in the 50-GHz band [89]. In the usual manner, adjacent slots are spaced one-half guide wavelength apart and are offset by equal and opposite amounts from the center line to obtain excitation in equal phase. Eight radiating slots are provided both in the front and rear walls of the waveguide in identical arrangements. The fins attached to the narrow walls assist in achieving the desired omnidirectional pattern in the azimuth plane within a ± 2-dB ripple. The beamwidth in the elevation plane is approximately $10°$, corresponding to a gain of 8 dB. The bandwidth of the antenna is in the order of a few percent. Since the antenna is comparatively short the bandwidth is limited by the frequency response of the individual slot elements. For long arrays with many slots, on the other hand, the array resonance condition would be the limiting factor, restricting

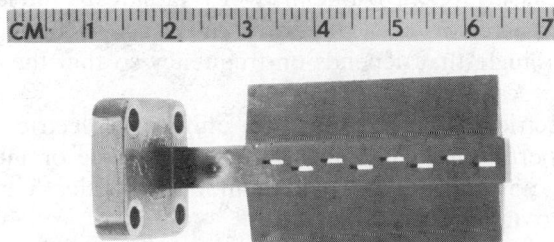

Fig. 14. Slotted-waveguide linear array antenna for 50-GHz band. (*Courtesy U.S. Army CECOM, Ft. Monmouth, NJ, and Norden Systems, Norwalk, CT*)

the bandwidth to approximately $\pm 50/n$ percent, where n is the number of array elements [90].

3. Surface-Wave and Leaky-Wave Antennas Based on Open Millimeter Waveguides

Most millimeter waveguides are open guiding structures; to minimize conduction losses and simplify fabrication they do not enclose the guided fields on all sides by metal walls. Hence, if these guides are not excited in the appropriate mode or if the uniformity of the guiding structure is perturbed, energy leakage will occur and part of the guided energy will be radiated into the surrounding medium. In a waveguide, of course, radiation losses are undesirable and the design problem is to minimize energy leakage. But the leakage effect may be utilized to advantage for the design of antennas, by modifying an open guide in such a way that radiation occurs in a controlled manner. Antennas of this type will have the same advantages as the guides from which they are derived, i.e., a low profile and structural simplicity. In addition, they are directly compatible with these guides and will be well suited for integrated designs.

Considering the numerous of open waveguides which have been suggested for use in the millimeter-wave region and the various ways in which these guides may be modified to obtain radiation, it is evident that a wide variety of millimeter-wave antennas can be constructed by using this general approach. But few such antennas had been studied until recently.

Two antenna structures which have been investigated in some depth are tapered dielectric-rod antennas and periodic dielectric antennas. Although both of these antennas, which are derived from dielectric waveguides, have been known for many years, the latter antenna was not studied extensively until relatively recently and in fact in the context of millimeter waves. Several new traveling-wave antennas which are uniform rather than periodic have been proposed recently; they are based on groove guide, microstrip line, and NRD (nonradiative dielectric) guide, which is a recently described variant of H guide. All of these antennas show significant differences in their operating principles and each may be regarded as typical for a class of antennas.

In tapered-rod antennas the cross section of the dielectric waveguide is decreased monotonically in the forward direction so that a wave traveling along the guide is gradually transformed from a bounded wave into a free-space wave. The structure is a typical surface-wave antenna which radiates in the forward direction.

The remaining antennas are leaky-wave antennas. A common feature is that they radiate at an angle that depends on frequency so that the antennas may be used for frequency scanning.

Periodic dielectric antennas consist of a uniform dielectric waveguide with a periodic surface perturbation in the form of a dielectric or metal grating. Diffraction at this grating transforms a guided mode into a leaky mode and, hence, the waveguide into an antenna.

The new uniform leaky-wave antennas are based on low-loss millimeter waveguides (except for the microstrip antenna), and the leakage of radiation can be caused in two different ways. One way is to perturb the structure longitudinally (along the guiding direction) in an asymmetric but uniform fashion. The difference here from the grating dielectric antennas is that there the perturbation is periodic but here it is uniform, so that these antennas have the outward appearance of a uniform open waveguide. An advantage of the uniform perturbation is its simplicity, permitting easier fabrication at the shorter millimeter wavelengths. These uniform leaky-wave antennas are less versatile than the periodic ones, however, since they can radiate only into the forward quadrant, whereas the periodic leaky-wave antennas can radiate into the backward quadrant as well.

The second way in which radiation can be produced on uniform open waveguides is to employ a higher mode. The fundamental mode on these guides is always purely bound (if the guiding structure is unperturbed), but higher modes can leak, although sometimes only over a restricted frequency range.

Examples of uniformly perturbed millimeter waveguides that give rise to leaky-wave antennas are the groove guide and nonradiative dielectric guide. Examples of leaky higher modes which can be employed for antenna purposes occur on groove guide and on microstrip. There, the dimensions of the guide are chosen such that it supports the first higher mode, and an appropriate feed arrangement ensures that only this mode is excited.

In the following, tapered dielectric-rod antennas and periodic dielectric antennas are discussed in some detail. These antennas are already of substantial practical interest since dielectric waveguides will be employed in many applications throughout the millimeter-wave band and, in some form, are likely to be used as the basic transmission medium for integrated millimeter-wave devices operating above 100 GHz. The antennas are obtained simply by letting a dielectric waveguide terminate into a tapered section or a section with a periodic surface corrugation whose function it is to radiate out the guided energy.

The new uniform leaky-wave antennas will be reviewed with respect to their operating principles and applicational aspects in and near the millimeter-wave region. These antennas are still under study and available design information was limited when this chapter was first written. Latest available information can be found in a comprehensive report [221] that was prepared very recently. The report also contains the results of several studies concerned with linear *arrays* of such leaky-wave line-source antennas, where the individual antennas in the array are fed in parallel from one end in phased-array fashion, permitting phase-shift scanning in the cross plane.

A concluding remark is concerned with dielectric resonator antennas, a different type of millimeter-wave antenna which has been examined recently [91]. These antennas are mentioned here since, in a broad sense, they are also derived from open millimeter waveguides. Typically, they take the form of short, resonating sections of rectangular or circular dielectric waveguide placed vertically on a metal ground plane. Since these dielectric resonators are not enclosed within metal walls, all energy delivered by the feed system will ultimately be radiated out (except for dielectric and metal losses, which may be significant) and the resonators will act as antennas with relatively narrow bandwidth. If the antennas are made from a material of sufficiently high permittivity ($\epsilon_r > 5$), the dielectric cavity will support well-defined resonant modes and the radiation patterns will be determined primarily by the shape of the resonator. Hence, these antennas may be well suited for beam shaping, although the relationship between resonator configuration and mode pattern is not a simple one. The antennas should be particularly useful for the upper millimeter-wave region; at present they are studied in scaled microwave experiments.

Table 1 lists materials suitable for millimeter-wave dielectric antennas and microstrip antennas [92, 93]. The constitutive parameters shown in the table apply to the (upper) microwave range. In the millimeter-wave region the permittivity values should be approximately the same, but the loss factors may be higher. Additional values, applying to the near-millimeter-wave/far-infrared region, may be found in [94].

Tapered Dielectric-Rod Antennas

Dielectric-rod antennas have been known for many years [95–98], but they have not been widely used as microwave antennas. In the millimeter-wave region these antennas have very reasonable size and weight, and many applications can be expected because of their compatibility with dielectric waveguides.

Tapered-rod antennas can take any of the configurations shown in Fig. 15. Their cross section may be rectangular or circular, where, in the rectangular case, the taper may occur in one cross-sectional dimension or, symmetrically, in both. Millimeter-wave antennas will usually have a rectangular cross section; a linear taper, as indicated in Fig. 15a, is preferable for reasons of simplicity. Such antennas have been designed for wavelengths down to 0.12 mm, i.e., well into the submillimeter-wave region [99, 100].

In most experimental investigations reported in the literature the antennas are fed by metal waveguides, as indicated for the shaped antenna of Fig. 15c. In this case, radiation from the feed aperture can have a strong effect on pattern shape and side lobe level, and an appropriate feed arrangement must be used to minimize transition effects [101]. If the antennas are fed by dielectric waveguides made from

Table 1. Millimeter-Wave Materials

	Relative Dielectric Constant (ϵ_r)	Loss Factor ($\tan \delta$)
Teflon	2.08	0.0006
PTFE-glass	2.17	0.0009
	2.33	0.0015
Polyethylene	2.26	0.0006
Duroid 5880	2.20	0.0006
Duroid 5870	2.33	0.0005
Quartz Teflon	2.47	0.0006
Polystyrene	2.54	0.0012
Fused quartz	3.78	0.00025
Boron nitride	4.40	0.0003
Sapphire	9.0	0.0001
Alumina	9.8	0.0001
Silicon	11.8	
Gallium arsenide	13.2	0.001
Magnesium titanate	16.1	0.0002

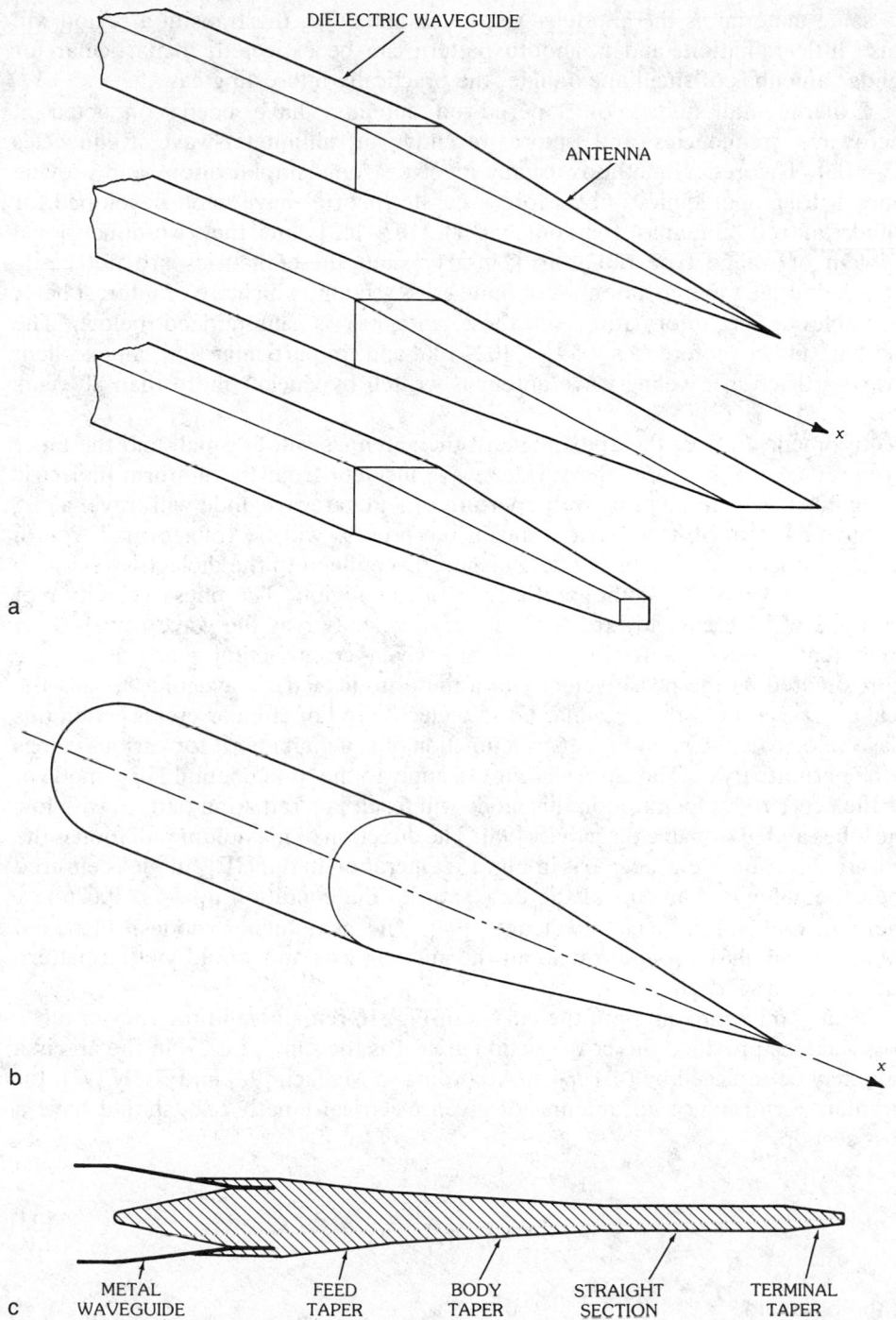

Fig. 15. Tapered dielectric-rod antennas. (*a*) Rectangular-cross-section antennas with linear taper. (*b*) Circular-cross-section antenna with linear taper. (*c*) Shaped antenna. (*After Zucker [98]*)

the same material as the antennas (Figs. 15a and 15b), the transition region will cause little radiation, and a smooth pattern can be expected, in particular for slender antennas of small apex angle, the practically interesting case.

Experimental studies on tapered-rod antennas have been conducted at microwave frequencies and, more recently, at millimeter-wave frequencies (95–106). Theoretical methods usually involve strong simplifications and provide general design guidelines only. More accurate theories have been developed for cylindrical rod antennas (without taper) [107–113] and the two-dimensional problem of wedge type radiators [114–117], but these theories are not easily extended to the tapered antennas of finite cross section, which are of interest here. Available design information on these antennas is summarized below. The summary utilizes references [95–97, 102–106] and, in particular [98], an excellent review article on traveling-wave antennas written by Zucker more than 20 years ago.

In practical cases, the apex angle of the antennas will be small and the taper very gradual. In this case, the surface wave incident from the uniform dielectric waveguide or excited by the horn aperture of a metal waveguide will travel along the antenna with little reflection and in the process will be transformed from a strongly bounded wave whose energy is mostly confined to the dielectric region to a free-space wave propagating entirely in the air region. The phase velocity v of this wave will increase toward the free-space velocity c as the wave travels from the antenna base toward its tip. In any given cross-section plane it can be approximated by the phase velocity of a uniform dielectric waveguide having the local cross section of the antenna. For a dielectric rod of circular cross section this phase velocity is shown in Fig. 16 as a function of rod diameter d for various values of the permittivity ϵ_r. The curves of Fig. 16 apply to the fundamental HE_{11} mode of the dielectric rod. Operation in this mode will result in a radiation pattern with low side lobes and reasonable directivity [98]. The direction of maximum radiation is the end-fire direction, i.e., the x axis in Fig. 15. Operation in the HE_{11} mode is ensured when the antenna diameter at the base satisfies the condition $d_B/\lambda_0 < 0.626/\sqrt{\epsilon_r}$, where λ_0 is the free-space wavelength [98]. The next higher modes, TE_{01} and TM_{01}, are circularly symmetric about the antenna axis and would yield a pattern null in the forward direction.

To a good approximation the curves of Fig. 16 remain valid for rods of other cross sections provided the cross-section area A is the same, i.e., d in the abscissa scale may be replaced by $(4A/\pi)^{1/2}$. According to Mallach [96] and Kiely [97], for optimum performance an antenna of given electrical length L/λ_0 should have a cross section

$$A_{max} = \frac{\lambda_0^2}{4(\epsilon_r - 1)} \tag{2a}$$

at the base and

$$A_{min} = \frac{\lambda_0^2}{10(\epsilon_r - 1)} \tag{2b}$$

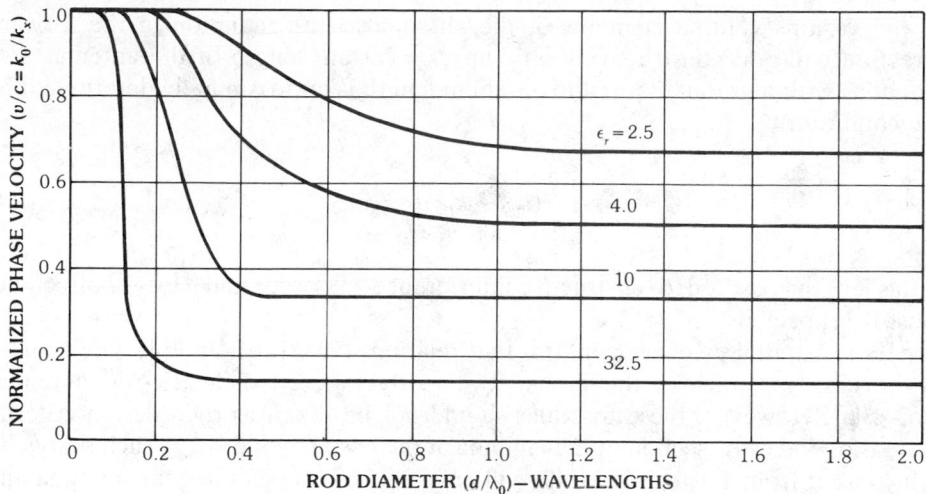

Fig. 16. Dispersion curves of fundamental HE$_{11}$ mode of cylindrical dielectric rod. (*After Mueller and Tyrell [95]; reprinted with permission from the* AT&T Technical Journals, © *1947 AT&T*)

at the termination. For permittivity values in the range $2.5 < \epsilon_r < 20$, which covers most millimeter-wave materials, this rule corresponds roughly to letting $c/v = k_x/k_0$ be 1.1 at the feed and very close to 1.0 at the termination, where k_x is the (local) phase constant of the guided wave and k_0 is the free-space wave number.

From (2a) and (2b) it is evident that the higher the dielectric constant of the antenna, the smaller its cross section and, hence, as a general rule, its weight. On the other hand, Fig. 16 shows that the dispersion curves become steeper with increasing ϵ_r so that parameter dependences will be more sensitive and bandwidth will be reduced [98].

For tapered rods satisfying (2a) and (2b), the directivity gain is determined primarily by the antenna length L. The gain initially increases with L according to the relation [98]

$$G = p \left(\frac{L}{\lambda_0} \right) \tag{3}$$

where p is approximately 7 for long antennas having a linear taper, while p can be as large as 10 for the maximum-gain antennas discussed below.* The largest attainable gain, however, is limited to 18 to 20 dB, which corresponds to a half-power beamwidth of 17° to 20°. This gain limitation is easy to understand on the basis of the equivalence principle. The polarization current in each volume element of the antenna radiates as a Hertz dipole. The phases of these dipoles are determined by the phase velocity, of the surface wave guided by the antenna. Since this phase velocity remains below the free-space wave velocity, the contributions

---

*The value $p = 10$ applies to maximum-gain antennas of moderate length L between $3\lambda_0$ and $8\lambda_0$.

of the various volume elements of the antenna to the radiation in the forward direction will add constructively only up to a certain length of the antenna. For antennas with a gradual taper the optimum length is approximately determined by the condition

$$\int_0^L (k_x - k_0)\, dx = \pi \tag{4}$$

If this length is exceeded, destructive interference will occur and a loss of directivity must be expected.*

Experimental evidence confirms that relations (2a), (2b), and (4) may be used as working formulas in the design of linearly tapered dielectric-rod antennas [102, 106]. However, these formulas should not be taken as rigid design criteria. Increasing the cross section of the antenna at the base beyond A_{max}, such that k_x/k_0 is increased from 1.1 to 1.25–1.40, will enlarge the bandwidth of the antenna and can also be used as a method for reducing the side lobe level (at the expense of a moderate gain reduction) [98]. Furthermore, for long antennas with a gradual taper it should be of little consequence if the antenna termination at the far end is blunt as suggested by (2b) or if the antenna is physically continued to its apex. The field strength within this tip section is small since most of the energy is now traveling in the air region; and the phase velocity is practically that of free space so that the left side of (4) will not noticeably be affected. For short antennas with a length in the order of a few wavelengths, on the other hand, tapering to a sharp tip has the advantage of a reduced reflection coefficient, a feature which will be discussed later in this section.

Zucker has described a procedure based on physical reasoning and experimental evidence for the design of surface-wave antennas of maximum-gain [98]. In the case of tapered-rod antennas the basic configuration of a maximum-gain antenna takes the form shown in Fig. 15c. It includes a feed taper, a body taper, a straight section, and a terminal taper. The feed taper establishes a surface wave which is assumed to radiate continuously as it travels along the body taper and the straight section. The terminal taper reduces reflections which would be caused by an abrupt discontinuity. Details of the design procedure are given in [98]; the main results are summarized in Fig. 17.** The solid curves show the gain and beamwidth of maximum-gain antennas as a function of antenna length L/λ_0 according to experimental data reported in the literature. The half-power beamwidth, which is approximately given in degrees by

$$\Delta \theta_{HP} = 55 \sqrt{\frac{\lambda_0}{L}}$$

is an average value; the beamwidth is usually slightly smaller in the *E*-plane and slightly larger in the *H*-plane [98].

---

*A formulation of condition (4) appropriate for maximum-gain antennas has been given by Zucker [98].
**Fig. 17 applies not only to tapered-rod antennas but to surface-wave antennas in general.

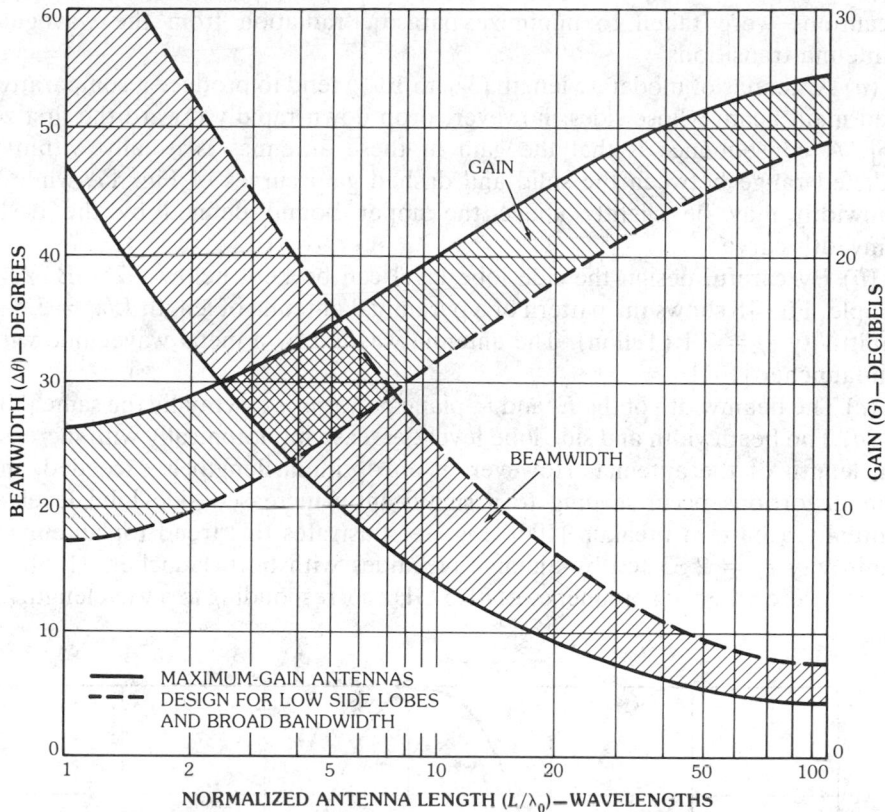

Fig. 17. Gain and beamwidth of surface-wave antennas as a function of normalized antenna length. (*After Zucker [98]*)

A problem with maximum-gain antennas is that they have comparatively small pattern bandwidth (± 10 percent) and high side lobes no more than -6 dB to -10 dB down from the main peak. Appropriate design of the body taper will solve these problems, though at the expense of a reduced gain [98]. The dashed curves in Fig. 17 characterize surface-wave antennas designed for broad bandwidth and low side lobe levels. Most tapered-rod antennas can be expected to have gain and beamwidth values in the shaded region between the dashed and solid curves. Experimental evidence [105, 106] shows, in particular, that antennas with a linear taper can be designed to have a gain not significantly lower than that of maximum-gain antennas while their pattern quality is substantially better (side lobes less than -20 dB). Since these antennas have the additional advantage of structural simplicity they should be of more interest for millimeter-wave applications than maximum-gain antennas.

With regard to the radiation patterns of linearly tapered dielectric-rod antennas the following general observations can be derived from experimental results reported in the literature [98, 102–104], in particular, by Shiau [105] and by Kobayashi, Mittra, and Lampe [106]. In the experiments of interest here, precautions were taken to minimize parasitic radiation from the waveguide-to-antenna transitions.

(*a*) Antennas of moderate length ($3\lambda_0$ to $10\lambda_0$) tend to produce a comparatively broad main beam whose sides, however, drop down rapidly toward the first zero [106]. A consequence is that the gain of these antennas is usually within the predicted range between the solid and dashed gain curves of Fig. 17, while the beamwidth may be slightly above the upper bound defined by the dashed beamwidth curve.

(*b*) By careful design the side lobe level can be kept below -25 dB. As an example, Fig. 18 shows the pattern of a pyramidal antenna of length $L/\lambda_0 = 5.5$ and permittivity $\epsilon_r = 2.1$ (Teflon). The antenna was fed by a metal waveguide with a horn launcher [101].

(*c*) The beamwidth of the *E*- and *H*-plane patterns is essentially the same [105].

(*d*) The beamwidth and side lobe level decrease monotonically with increasing axial length of the antenna. However, if an optimum length is exceeded, main beam distortions occur leading to a significantly increased side lobe level and eventually a pattern breakup [106]. Fig. 19 illustrates this trend for antennas of permittivity $\epsilon_r = 2.33$ fed by metal waveguides with horn launchers [106]. The patterns were taken at a frequency of 81.5 GHz corresponding to a wavelength $\lambda_0 =$

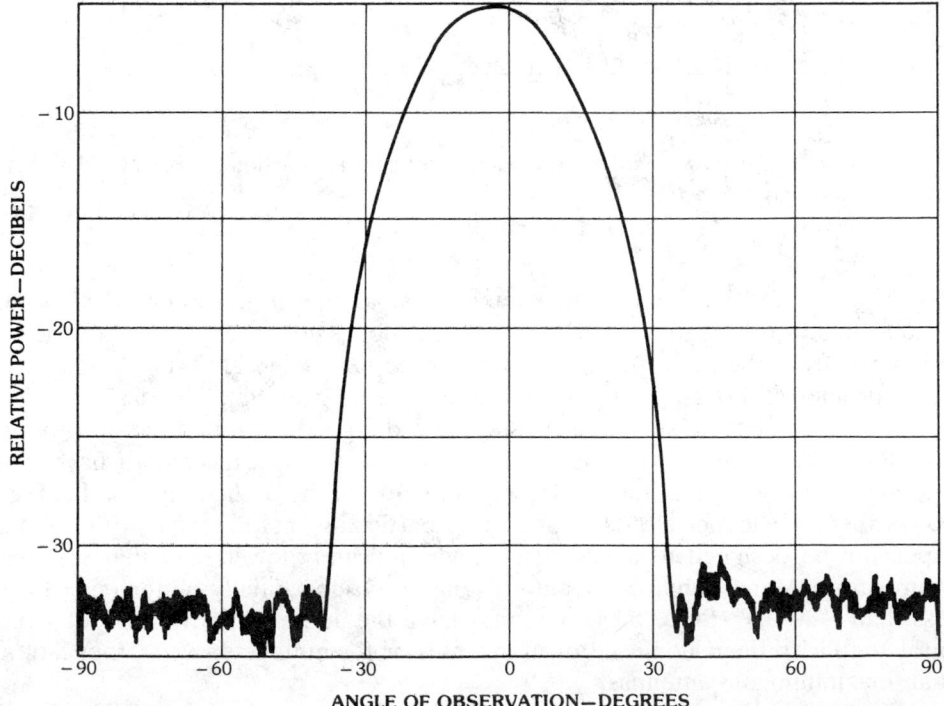

Fig. 18. *H*-plane pattern of pyramidal dielectric-rod antenna fed by metal waveguide with horn launcher. (*After Kobayashi, Mittra, and Lampe [106], © 1982 IEEE*)

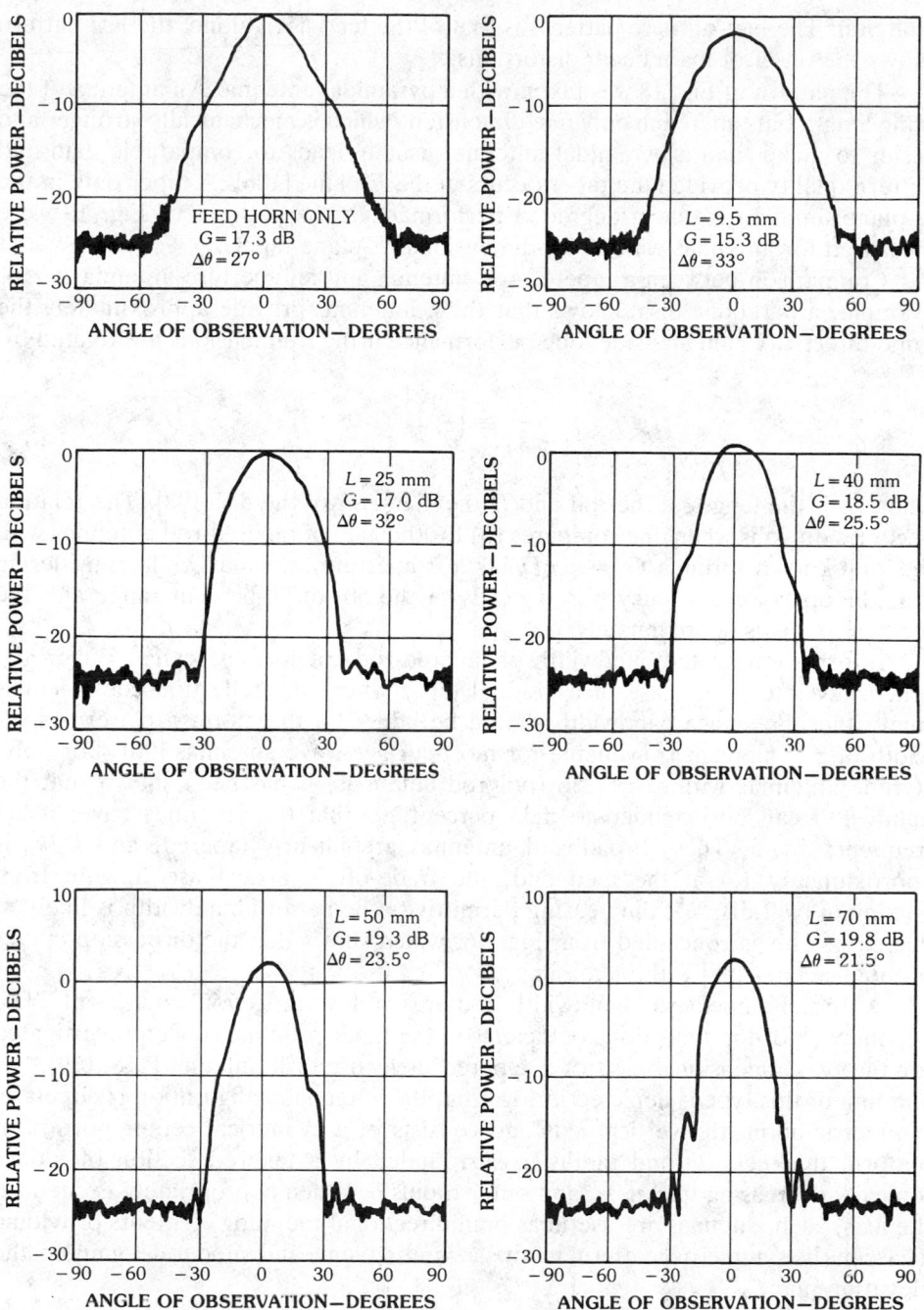

Fig. 19. Dependence of radiation pattern of tapered dielectric-rod antennas on axial length L. (*After Kobayashi, Mittra, and Lampe [106], © 1982 IEEE*)

3.68 mm. The first of these patterns is that of the feed horn alone; the last pattern shows the onset of main beam distortions.*

The pattern of Fig. 18 was taken with a pyramidal antenna. An antenna of the same length but tapered in only one dimension, which is mechanically stronger and easier to make than a pyramidal antenna, usually leads to comparable gain and pattern quality provided the taper occurs in the E-plane [106]. A taper only in the H-plane tends to result in degraded performance. The patterns of Fig. 19 were measured for antennas with a one-dimensional E-plane taper.

Comparison between a tapered-rod antenna and an aperture antenna, as for example, a parabolic dish, shows that these antennas provide approximately the same directivity gain and side lobe performance if their dimensions are related by

$$\frac{L}{\lambda_0} \cong \left(\frac{D}{\lambda_0}\right)^2$$

where L is the length of the rod and D the diameter of the dish [98]. The relation becomes obvious when one compares (3) for the gain of tapered-rod antennas with the well-known formula $G = \eta(\pi D/\lambda_0)^2$ for aperture antennas while considering that the aperture efficiency η is typically in the 50- to 70-percent range and the factor p in (3) is approximately 7.

Information on the bandwidth of tapered-rod antennas is scarce. But in the practically interesting case of a gradual taper where the reflection coefficient is small, the impedance bandwidth should be large. Furthermore, a pattern bandwidth of ± 15 percent is available for most surface-wave antennas [98]. For polystyrene antennas with $\epsilon_r = 2.56$ (polyrod antennas) it has been shown that the bandwidth can be extended to ± 33 percent, so that the antennas cover a 2:1 frequency band. These broadband antennas are linearly tapered, and k_x/k_0 is approximately 1.4 at the feed end; the trade-off is a decrease in gain from maximum by 2 dB. With increasing permittivity the pattern bandwidth is likely to decrease. This is concluded from Fig. 16, which shows that the dispersion curves become steeper [98] with increasing ϵ_r.

A broad impedance bandwidth requires a low reflection coefficient. The dependence of the reflection coefficient on the taper profile has been investigated for the two-dimensional case of a tapered dielectric disk antenna [118, 119]. An antenna of this type is depicted in Fig. 20a; the antenna configuration is circularly symmetric about the vertical axis and consists of a cylindrical center portion of uniform thickness $2h$ and radius a, surrounded by a tapered section of monotonically decreasing thickness, and outer radius b. When appropriately excited (at the axis) such antennas are useful as omnidirectional radiating elements providing a circularly symmetric pattern in the azimuth plane and moderate gain in the elevation plane.

---

*On the other hand, for very long tapered-rod antennas (fed by a dielectric waveguide of the same permittivity), the transformation of a surface wave traveling along the structure from a strongly bounded wave into a free-space wave will proceed very gradually and little reflection will occur. One may conclude that such antennas will radiate a smooth, highly directive pattern. But no theoretical or experimental data to confirm or refute this view seem to be available.

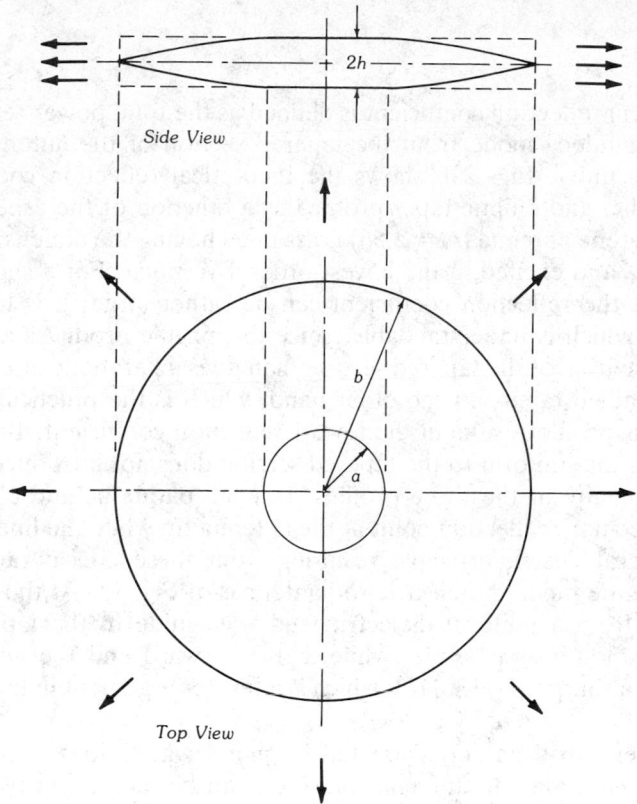

a

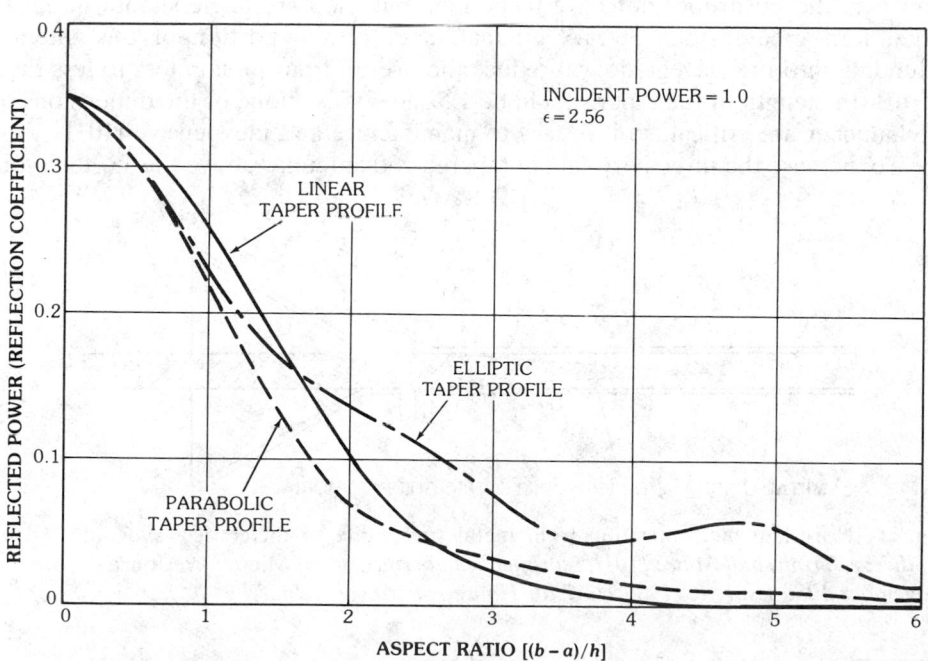

b

Fig. 20. Reflection coefficients of a tapered dielectric disk antenna. (*a*) Antenna configuration. (*b*) Reflection coefficients.

The (power) reflection coefficient is defined as the total power returning in the fundamental (guided) mode from the tapered section of the antenna, when the input power is unity. Fig. 20b shows the theoretical reflection coefficient for a linear, parabolic, and elliptic taper profile* as a function of the aspect ratio $(b - a)/h$. A polystyrene antenna ($\epsilon_r = 2.56$) is assumed having the dimensions $a = 0.2\lambda_0$ and $2h = 0.4\lambda_0$ and excited in the lowest-order TM mode. For a short taper with $(b - a)/h < 2$, the reflection coefficient can be rather large; it is largest for the linear profile, which is understandable, since this profile produces a more abrupt transition at the base of the tapered section than does a parabolic or elliptic profile. For a long, gradual taper, on the other hand, which is the practically interesting case, the linear profile results in the lowest reflection coefficient. In this case the transition from the uniform to the tapered section does not introduce a significant perturbance for any of the three profiles. But the parabolic and elliptic profiles produce a "specular" reflection point at the antenna tip while the linear taper does not. The general design principle resulting from these considerations may be applied also to the tapered dielectric-rod antennas of Fig. 15: At the antenna base the transition from a uniform dielectric-feed waveguide to the tapered antenna should be as smooth as possible, while at the forward end the antenna should terminate into a sharp tip [118, 119], which is a further argument in favor of linearly tapered antennas.

When tapered-rod antennas are fed by metal waveguides, unwanted radiation from the waveguide-to-antenna transition can be minimized by the use of a surface-wave launcher in the form of a small pyramidal horn [101] as indicated in Fig. 21. According to Malherbe and others [101] who have studied this type of launcher, the horn does not have to be long but the flare angle should be fairly large, i.e., greater than 30°. A gradual taper of the portion of the antenna extending into the waveguide will reduce the overall transmission loss to less than 0.5 dB; the length of the taper should be $1.5\lambda_0$ to $2.5\lambda_0$. None of the dimensions of the launcher are critical, so it is easy to manufacture and inexpensive [101].

To increase the directivity gain of tapered-rod antennas above the 18- to 20-dB

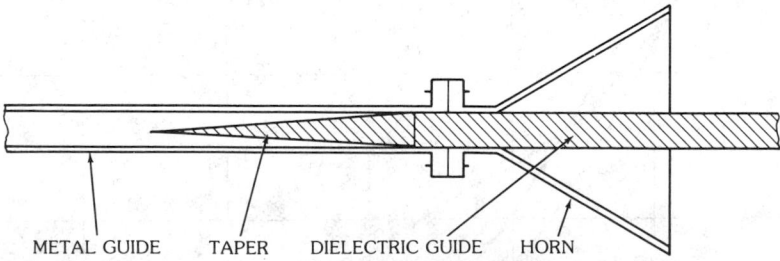

METAL GUIDE   TAPER   DIELECTRIC GUIDE   HORN

Fig. 21. Horn-launcher transition from metal waveguide to dielectric waveguide. (*After Malherbe, Trinh, and Mittra [101]; reprinted with permission of* Microwave Journal, *from the November 1980 issue,* © *1980 Horizon House–Microwave, Inc.*)

---

*For the numerical analysis the actual taper profile was replaced by a staircase approximation with ten steps.

limit, *arrays* of these antennas can be used. Since the individual array elements have already a substantial directivity, element spacing can be chosen comparatively large so that interelement coupling is minimized. According to Zucker [98], the spacing, b, should be in the range

$$0.5 \sqrt{\frac{L}{\lambda_0}} \leq \frac{b}{\lambda_0} \leq \sqrt{\frac{L}{\lambda_0}}$$

where the lower bound formulates the condition that interelement coupling should be negligible while the upper bound is determined by the requirement that the first grating lobe should be substantially reduced by multiplication with the element pattern. Ideally, the first minimum of the element pattern should coincide with the first grating lobe of the array pattern. For scanned arrays the element spacing b must be chosen smaller, in correspondence with the desired scan range [98].

Under these circumstances the array gain and the beamwidth of the main beam of a linear array of n equally spaced elements are

$$G_{\text{array}} = nG_{\text{element}}$$

and

$$\Delta\theta_{\text{array}} = \frac{\lambda_0}{nb} \frac{180}{\pi}$$

respectively where $\Delta\theta_{\text{array}}$ is the 3-dB beamwidth in degrees in the plane of the array. Side lobes can be controlled by tapering the excitation of the elements. The bandwidth of the array will be smaller than that of the individual elements, probably by 5 percent [98].

The array gain, of course, can be further increased by the use of planar arrays (volume arrays) of tapered-rod antennas. It should be noted, however, that the attainable gain is determined by the overall array aperture and, therefore, is not much higher than in the case of a dipole array of the same aperture size. The difference is that highly directive end-fire elements can be spaced much farther apart than dipole elements so that the number of array elements is significantly reduced.

An interesting feed arrangement for a linear array of tapered-rod antennas is shown in Fig. 22. Suggested by Williams and others [121], the feed system is designed in insular guide technology and is shown here supporting a frequency-scanned array of eight elements operating in the 30-GHz band. Alternate array elements are fed from the main insular guide via proximity couplers. The feed circuit is repeated on the reverse side of the ground plane to feed the second half of the array. An *E*-plane power splitter is used to excite each array half with the same amplitude. The arrangement permits the use of couplers whose radii of curvature are sufficiently large to prevent excessive radiation from the curved sections but maintains the close element spacing necessary for wide-angle scanning [120]. A uniform scan performance over an angular range of more than 40° was obtained at a

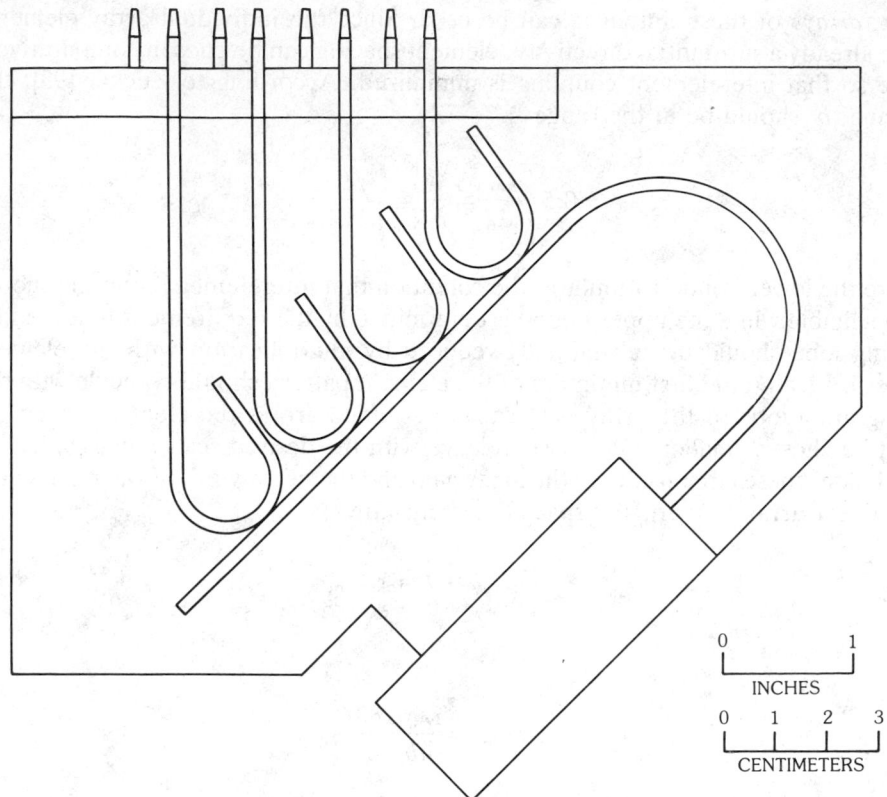

0   1
INCHES

0   1   2   3
CENTIMETERS

Fig. 22. Feed arrangement for eight-element frequency-scanned linear array of tapered dielectric-rod antennas. (*After Williams, Rudge, and Gibbs [121],* © *1977 IEEE*)

frequency variation of 10 percent. The material from which the waveguides and antennas are made has a dielectric constant of approximately 10.

Periodic Dielectric Antennas

Periodic dielectric antennas [122–137] consist of a uniform dielectric waveguide with a periodic surface perturbation. The uniform waveguide supports a traveling wave; the surface perturbation acts as a grating that radiates out the guided energy. As is true for dielectric-rod antennas, these grating antennas are directly compatible with dielectric waveguides. Additional advantages include ruggedness and a low profile—the antennas can be installed conformally with a (planar) metal surface—and a capability for beam scanning by frequency variation.

Periodic antennas can have various configurations; recently investigated structures are shown in Fig. 23. The first of these antennas, Fig. 23a, operates with a dielectric grating in the form of a periodic surface corrugation [125–128]. The geometry of the teeth and grooves can take a variety of forms. For example, when the antenna is not backed by a metal plane, the corrugation profile can be designed to prevent radiation in the downward direction so that all radiated power is con-

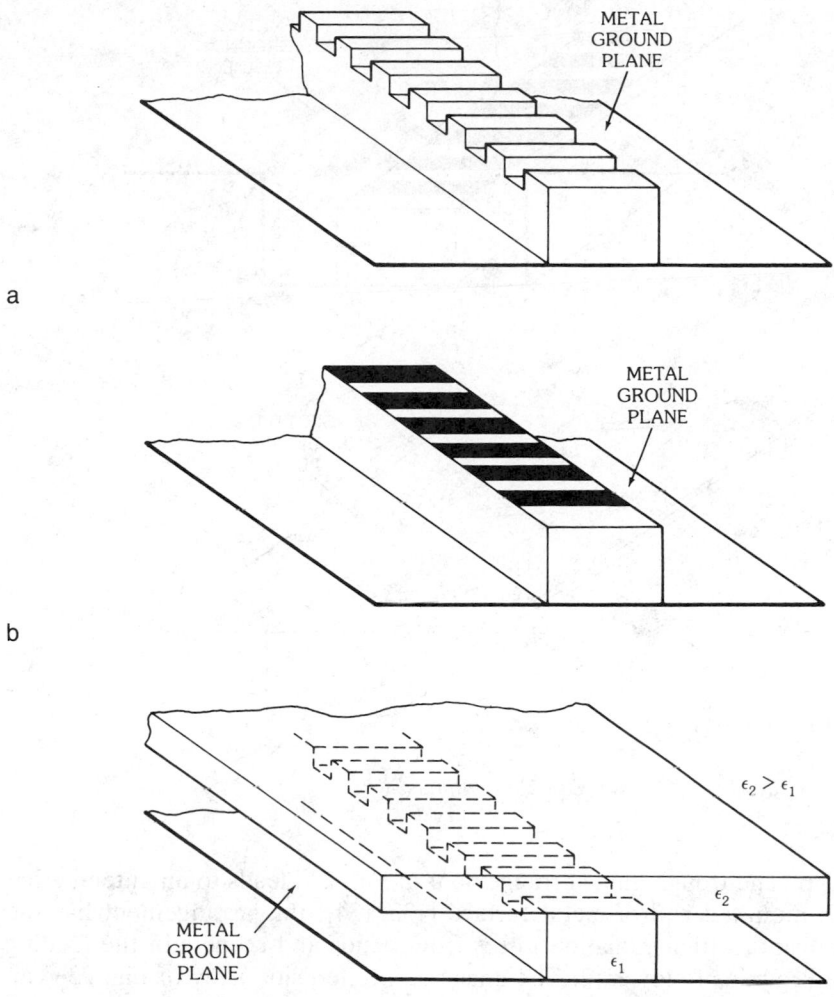

Fig. 23. Periodic dielectric antennas. (*a*) Antenna with dielectric grating. (*b*) Antenna with metal grating. (*c*) Inverted dielectric strip guide antenna. (*d*) Trapped image-guide antenna. (*e*) Image-guide–fed slot array.

centrated into a single upward-directed beam [138–141]. In the presence of a ground plane, shaping of grooves and teeth is not necessary and a rectangular profile is preferred since it is easiest to machine. A grating of metal strips, shown in Fig. 23b, is an alternative to a dielectric grating; the metal strips tend to increase the leakage constant, i.e., the radiation rate per period [122–124, 129–131]. Inverted-strip dielectric waveguide antennas, as shown in Fig. 23c, provide a very rugged design which integrates a planar radome with the antenna [132], but results in a lower leakage constant. The reason for the lower leakage constant is that the permittivity of the top plate is greater than that of the dielectric strip so that the guided energy travels primarily in the top plate in the region just above the

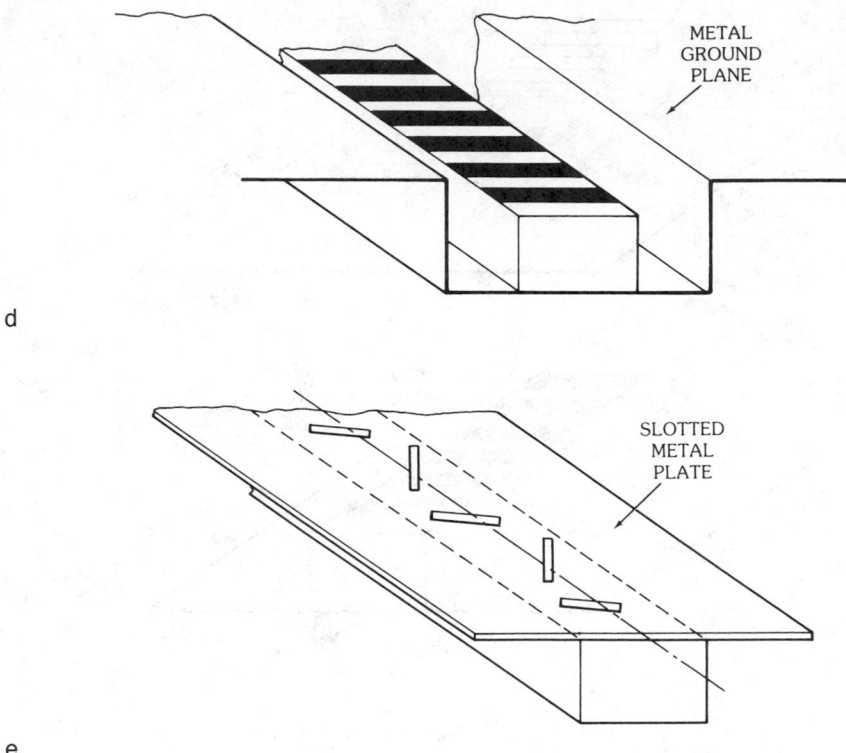

d

e

Fig. 23, *continued.*

dielectric strip. The trough guide arrangement of Fig. 23d leads to an antenna that can be flush mounted with a metal surface [133, 134]; this arrangement has the additional advantage that radiation losses from bends and corners in the feeding dielectric waveguide are lowered. The image-guide–fed slot array of Fig. 23e can be designed to provide a circularly polarized radiation pattern [142].

Review of Fundamentals—The theory of dielectric-grating antennas can be regarded as well developed due primarily to the work of Peng [125–127, 143–146]. This theory is based on a space harmonics representation of the fields guided by the antenna and should be regarded as comprising two parts. The first part applies to antennas of large width [125–127], and the second introduces modifications that allow the theory to be extended to narrower antennas [144–146], which are more likely to be employed in practice. The theory for the wide antennas is closely related to the analysis of optical grating couplers [147–152] and, to a certain extent, can be adapted from this analysis. Modifications, however, are required to accommodate the much smaller lengths of millimeter-wave devices, as expressed in wavelengths, and the effect of the much higher permittivity of millimeter-wave materials, such as silicon and alumina. Numerical evaluations have been performed primarily for antennas with dielectric gratings of rectangular profile [125–127], and design curve for these antennas are available.

For dielectric antennas with *metal* gratings, an alternative formulation in terms of an eigenvalue problem in the Fournier transform domain has been suggested by Mittra and Kastner [153], but a complete solution has not been provided. Very recently, an analysis that is complete, containing all the information required for practical design, was presented [219]. That analysis differs from the analysis below of corrugated dielectric antennas, but all of the fundamentals are similar.

For a detailed description of the theory of periodic leaky-wave antennas one may refer to the literature cited above. Here we restrict ourselves to the presentation of some of the pertinent results in order to provide a basis for the discussion on design guidelines to follow later. The present remarks apply to any of the periodic structures shown in Fig. 23, but for convenience we shall refer to the antenna of Fig. 24.

The waves supported by periodic dielectric antennas radiate as they travel along the antennas; they are leaky modes which decay exponentially in the forward direction. The lowest leaky mode is of primary interest for antenna applications. Higher leaky modes can be suppressed by choosing the antenna dimensions below the cutoff conditions of these modes or by using an appropriate feed arrangement.

A leaky mode has a phase constant β and an attenuation, or leakage, constant α, the latter being a measure of the power leaking away per unit length along the antenna. The periodicity produces an infinity of space harmonics associated with that leaky mode; the phase constants β_n of the various space harmonics are related to the phase constant β of the basic wave by

$$\beta_n = \beta + \frac{2\pi n}{d}, \qquad -\infty \leqq n \leqq +\infty$$

where n is the order of a given space harmonic, $n = 0$ corresponds to the basic wave, and d is the period of the modulated waveguiding structure.

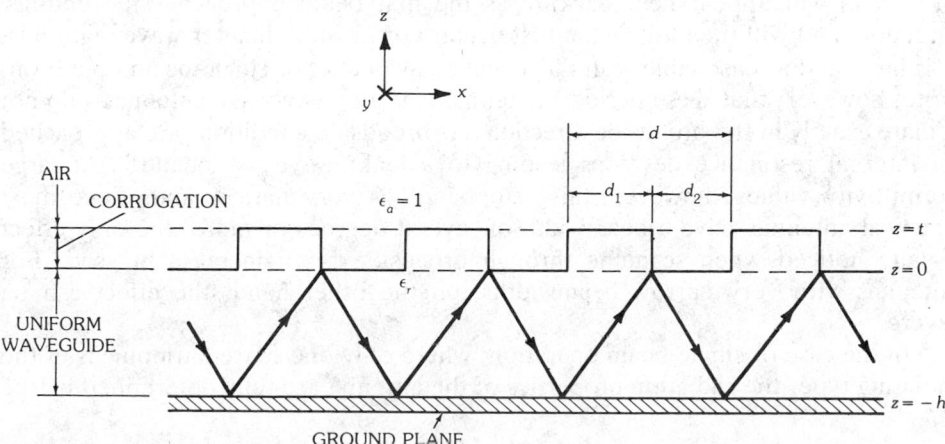

Fig. 24. Dielectric grating antenna: geometry and coordinate system. (*After Schwering and Penn [127], © 1983 IEEE*)

If a space harmonic is slow along the antenna surface, it is purely bound; if it is fast, it will radiate power away at an angle, given by

$$\phi_n = \sin^{-1}\left(\frac{\beta_n}{k_0}\right) = \sin^{-1}\left(\frac{\beta}{k_0} + n\frac{\lambda_0}{d}\right) \tag{5}$$

where λ_0 is the free-space wavelength, $k_0 = 2\pi/\lambda_0$ is the free-space wave number, and ϕ_n is the radiation angle measured from the positive z axis, i.e., from the normal to the antenna surface; see Fig. 24. The nonradiating space harmonics are evanescent waves that decay exponentially in the z direction. The number of these waves is infinite. The number of the radiating space harmonics, on the other hand, depends on d/λ_0 and is always finite. If d/λ_0 is large, several space harmonics are of the propagating type and the antenna will radiate in several directions simultaneously. Most antenna applications, however, require single-beam operation, which leads to the following condition* [125] on d:

$$\frac{\lambda_0}{\beta/k_0 + 1} \leqq d \leqq \frac{\lambda_0}{\beta/k_0 - 1}, \qquad \text{for } \beta/k_0 > 3$$

$$\frac{\lambda_0}{\beta/k_0 + 1} \leqq d \leqq \frac{2\lambda_0}{\beta/k_0 + 1}, \qquad \text{for } \beta/k_0 < 3 \tag{6}$$

If this condition is satisfied, only the space harmonic of order $n = -1$ will radiate. As the frequency is increased, the radiation direction

$$\phi_{-1} = \sin^{-1}\left(\frac{\beta}{k_0} - \frac{\lambda_0}{d}\right) \tag{7}$$

will scan from backfire over broadside to end-fire. If $\beta/k_0 > 3$, the full scan range is in principle available without a second beam appearing. If $\beta/k_0 < 3$, a second beam ($n = -2$) will appear near backfire as the first beam approaches the end-fire direction, and will then follow the first beam. For most millimeter-wave materials, ϵ_r is large and a reasonable scan range will be available for single-beam operation. Note, however, that these periodic antennas—as all leaky-wave antennas—do not radiate exactly in the broadside direction. As broadside conditions are approached an internal resonance develops leading to a leaky-wave "stopband." At large permittivity values, however, this "stopband" is very narrow and, since most practical antennas have a beamwidth of several degrees or more, the only effect usually noticed when scanning through broadside is an increase in vswr. For antennas with very narrow beamwidths, on the other hand, the effect can be severe.

In the case of single-beam operation, where only one space harmonic is of the radiating type, the radiation properties of the antenna are fully determined by the

---

*A more stringent condition taking into account the finite beamwidth of the antenna has been formulated by Kobayashi et al. [129].

complex propagation constant $k_x = \beta - j\alpha$. Hence β and α are the key parameters to be calculated in the theory of periodic dielectric antennas.

Phase Constant β and Leakage Constant α—The leaky modes supported by dielectric grating antennas of finite width cannot be described in terms of TE or TM waves alone. They are hybrid modes involving TE-TM coupling, and a rigorous determination of their complex propagation constants would require the solution of a vector boundary value problem [143, 144]. But approximate methods are available which provide results of good accuracy and reduce the computational burden significantly.

In the discussion below, we first present results for *wide antennas*. The phase constant β will be calculated using the so-called effective dielectric constant (EDC) method [154], which is a simple procedure that yields very accurate results for β for a large range of parameter values [125]. Furthermore, for antennas with large lateral width $w > \lambda_0/\sqrt{\epsilon_{\text{eff}} - 1}$ the leakage constant α is, to a good approximation, the same as that for an antenna of infinite width [125–127]. For this limiting case the leaky modes become TE or TM modes and the vector boundary value problem is reduced to a much simpler scalar problem. (For a precise definition of ϵ_{eff}, see below.)

The condition $w > \lambda_0/\sqrt{\epsilon_{\text{eff}} - 1}$ will not always be satisfied in practical cases. To prevent the occurrence of higher-order leaky modes, w will be chosen significantly smaller than $\lambda_0/\sqrt{\epsilon_{\text{eff}} - 1}$ in applications where a large H-plane beamwidth is acceptable, and it is noted that most experimental studies reported in the literature have used antennas of small w. In the second part of the discussion below, modifications of the theory will be introduced which allow the solution methods for wide antennas to be extended to *much narrower* antennas with $w < \lambda_0/\sqrt{\epsilon_{\text{eff}} - 1}$ [144–146]. In both parts it will be assumed that the antennas operate over a metal ground plane, that a dielectric grating is used with a rectangular profile of teeth and grooves, as indicated in Fig. 24, and that only the lowest-order leaky mode is present. (The principal field-strength components of this mode are E_z and H_y.) Since typical millimeter-wave materials, like silicon and gallium arsenide, have a dielectric constant of $\epsilon_r \cong 12$, this permittivity is assumed in most of the design curves presented here.

In applying the EDC method to antennas of large width the corrugation region is replaced by an equivalent layer of uniform dielectric constant. This dielectric constant is chosen to be the volume average permittivity

$$\epsilon_{\text{avg}} = \frac{\epsilon_1 d_1 + \epsilon_2 d_2}{d}$$

$$= 1 + (\epsilon_r - 1)\frac{d_1}{d}, \qquad \text{for } \epsilon_1 = \epsilon_r, \epsilon_2 = 1 \qquad (8)$$

where ϵ_1, d_1 and ϵ_2, d_2 are the permittivity and width of the teeth and grooves, respectively, as shown in Fig. 24. The second expression holds if the permittivity of the teeth is that of the uniform guide and the permittivity of the grooves is that of air. This will be assumed in the following.

The EDC method [154] uses a two-step approach to derive a characteristic equation for the phase constant of the modes guided by dielectric waveguides of rectangular cross section. In the first step an effective dielectric constant, ϵ_{eff}, is defined by formulating the transverse resonance condition for the z direction, while temporarily assuming that the guide is infinitely extended in the y direction. In the second step a slab guide of permittivity ϵ_{eff} is considered which has finite width in the y direction but infinite width in the z direction. Formulation of the transverse resonance condition (in the y direction) for this guide yields the desired characteristic equation for β. The approximation consists in assuming that the geometrical discontinuities occurring at the sides of the waveguide are neglectable, in the sense that all higher transverse modes can be ignored and only the dominant vertical and horizontal transverse modes are considered. As a result one can formulate the transverse resonance conditions for the y and z directions separately. The condition for the z direction leads to the following characteristic equation for ϵ_{eff}:

$$\frac{\epsilon_{\text{avg}}}{\epsilon_r} \frac{k_z^{(\epsilon)}}{k_z^{(m)}} \tan(k_z^{(\epsilon)}h) = \frac{\epsilon_{\text{avg}} |k_z^{(o)}|/k_z^{(m)} - \tan(k_z^{(m)}t)}{1 + \epsilon_{\text{avg}}(|k_z^{(o)}|/k_z^{(m)}) \tan(k_z^{(m)}t)} \tag{9a}$$

with

$$|k_z^{(o)}| = k_0(\epsilon_{\text{eff}} - 1)^{1/2}$$
$$k_z^{(m)} = k_0(\epsilon_{\text{avg}} - \epsilon_{\text{eff}})^{1/2}$$
$$k_z^{(\epsilon)} = k_0(\epsilon_r - \epsilon_{\text{eff}})^{1/2}$$

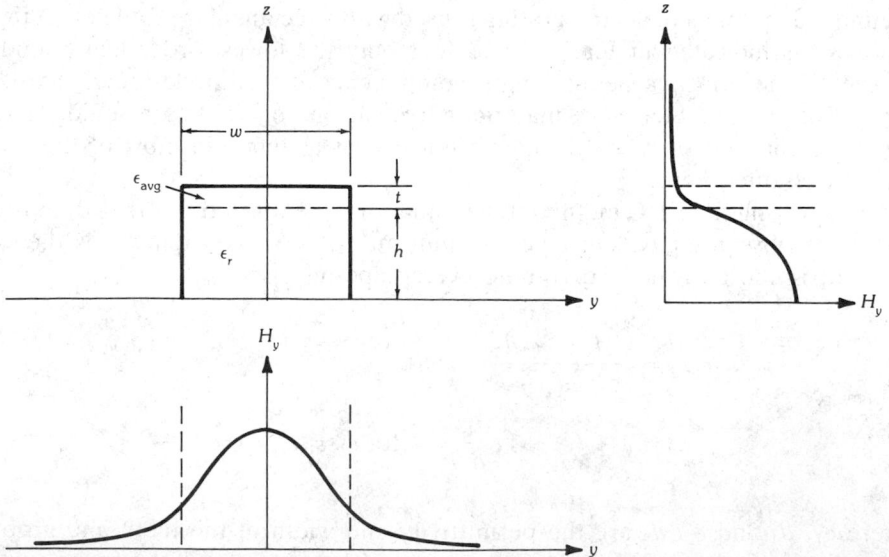

Fig. 25. Field distribution (H_y) of fundamental quasi-TM mode supported by dielectric antenna of width $w > \lambda_0\sqrt{\epsilon_{\text{eff}} - 1}$. (*After Schwering and Penn [127]*, © *1983 IEEE*)

and the condition for the y direction results in a second characteristic equation determining the normalized phase constant $\epsilon_{ant} = \beta^2/k_0^2$:

$$\frac{k_y^{(\epsilon)}}{|k_y^{(o)}|} \tan(k_y^{(\epsilon)} w/2) = 1 \tag{9b}$$

with

$$|k_y^{(o)}| = k_0(\epsilon_{ant} - 1)^{1/2}$$
$$k_y^{(\epsilon)} = k_0(\epsilon_{eff} - \epsilon_{ant})^{1/2}$$

The square roots in these equations are all positive real, with the exception of $k_z^{(m)}$, which is either positive real or negative imaginary (depending on whether the field in the corrugation region is of the propagating or evanescent type). The superscripts (o), (m), and (ϵ) indicate the air region, modulation region, and dielectric region, respectively. The fundamental mode of the antenna* is associated with the lowest-order solutions of (9a) and (9b). With these solutions, ϵ_{eff} and ϵ_{ant}, determined, the phase constant of the fundamental mode is obtained as

$$\beta = k_0 \sqrt{\epsilon_{ant}} \tag{9c}$$

The fundamental mode does not have a cutoff frequency; its cross-sectional field distribution is indicated in Fig. 25, with the corrugation region of the antenna replaced by a layer of uniform permittivity ϵ_{avg}. The *effective dielectric constant* for this mode is shown in Fig. 26 as a function of the *corrugation depth t* for various values of the height h of the uniform portion of the antenna [126, 127]. The teeth and grooves of the corrugation are assumed to have equal width $d_1 = d_2 = d/2$, resulting in an average permittivity of 6.5 for the corrugation layer. Fig. 27** shows the normalized phase constant $\epsilon_{ant} = (\beta/k_0)^2$ as a function of w. Evidently, when w is large, ϵ_{ant} does not deviate significantly from ϵ_{eff} and the width w of the antenna has little effect on the phase constant β, which can be approximated by $k_0 \sqrt{\epsilon_{eff}}$.

Returning to Fig. 26 it is seen now that the phase constant β of a wide antenna does not significantly vary with the corrugation depth t, provided t and h are

---

*In the form in which the characteristic equations (9a) and (9b) are presented here, they apply to the fundamental mode of the antenna and to those higher modes whose polarization and symmetry properties are similar to those of the fundamental mode. The principal field-strength components of this mode are E_z and H_y, and these components are of even symmetry with regard to the planes $y = 0$ and $z = -h$ in Fig. 24. (We assume here that the plane $y = 0$ bisects the antenna which, thus, extends laterally by $w/2$ to either side of this plane.)

**As explained above, the curves of Figs. 26 and 27 are the lowest-order (fundamental mode) solutions of the characteristic equations (9a) and (9b) which formulate the transverse resonance conditions for the z and y directions, respectively. In formulating the former condition, the antenna is temporarily extended to infinity in the y direction (see insert in Fig. 26) and the fundamental mode, whose principal field components are E_z and H_y, becomes a pure TM wave. In formulating the latter condition, the antenna is infinitely extended in the z direction (see insert in Fig. 27) and the fundamental mode becomes a pure TE wave. The purpose of the legends "TM mode" in Fig. 26 and "TE mode" in Fig. 27 is to express this situation.

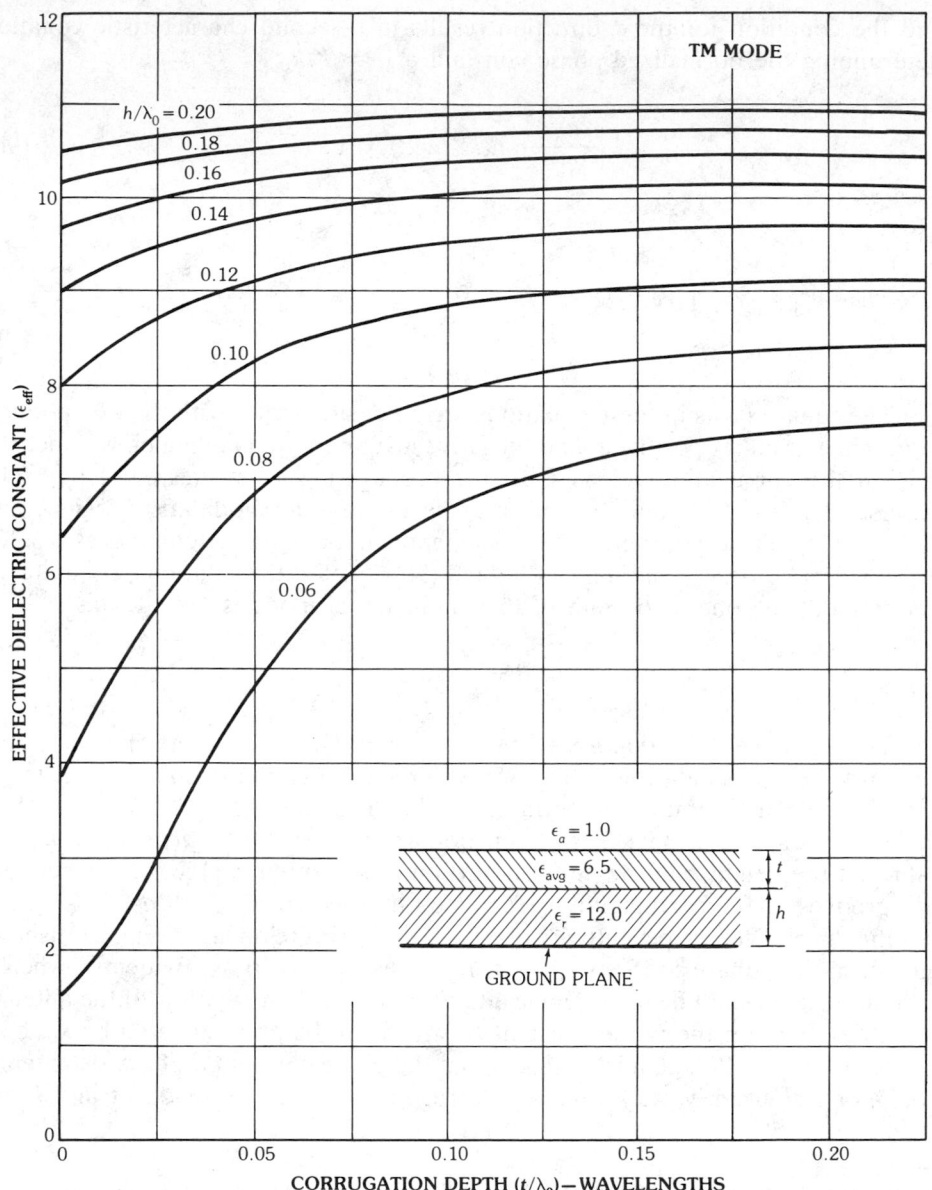

Fig. 26. Effective dielectric constant as a function of corrugation depth and substrate height for $d_1/d = 0.5$. (*After Schwering and Penn [127]*, © *1983 IEEE*)

sufficiently large, i.e., $t > 0.05\lambda_0$, $h > 0.1\lambda_0$. If the surface corrugation is sufficiently deep, the actual value of t is not critical.

The *grating period d* has little influence on β, but it has a determining influence on the *radiation angle* ϕ_{-1}. The dependence of ϕ_{-1} on both d and h is shown in Fig. 28 for $t = 0.05\lambda_0$ and $d_1 = d_2$. The requirement of single-beam operation is satisfied in the range below the dashed curve $\phi_{-2} = -90°$. Above this curve the $n = -2$

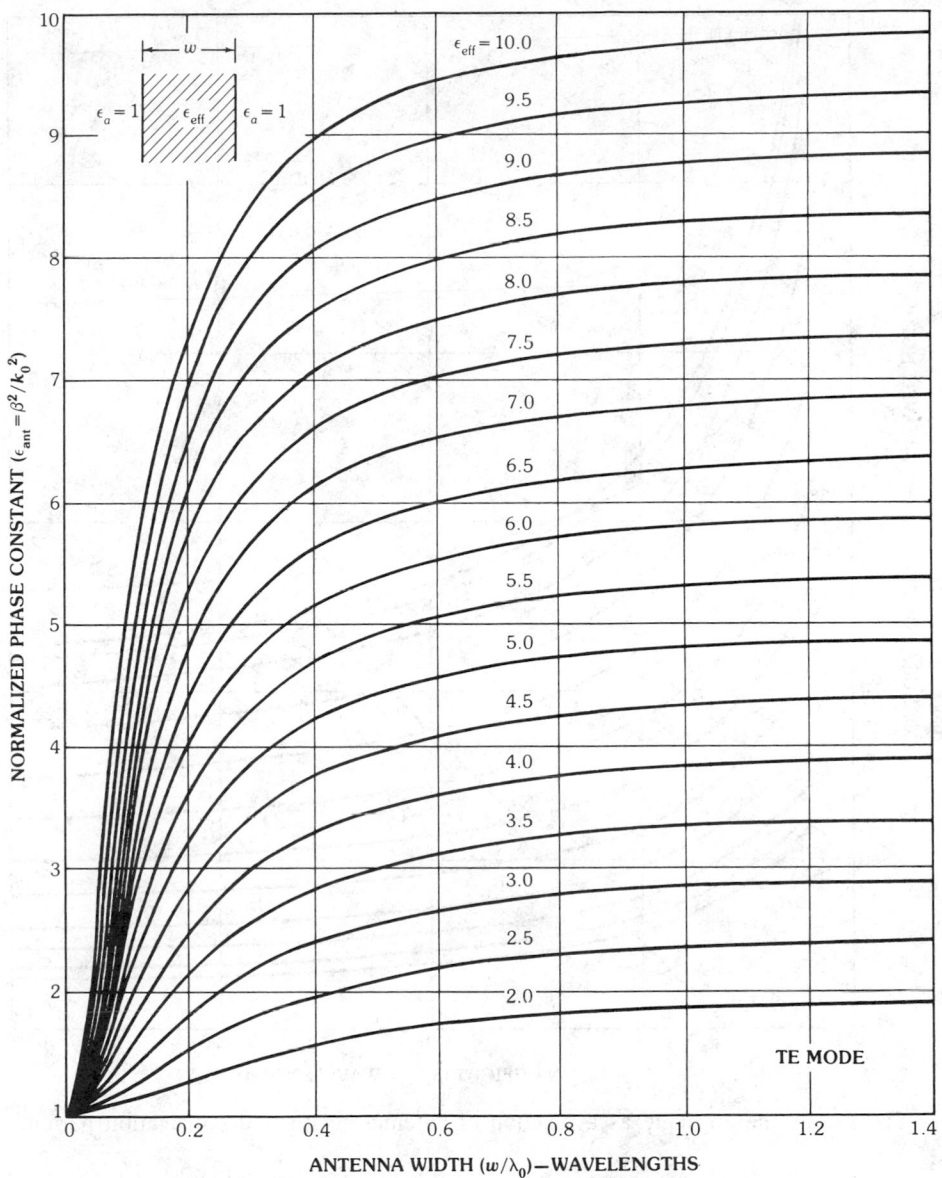

Fig. 27. Normalized phase constant as a function of antenna width and effective dielectric constant. (*After Schwering and Penn [127],* © *1983 IEEE*)

space harmonic is transformed from an evanescent wave into a propagating wave, and the antenna will radiate two and eventually three and more beams simultaneously. In the range below the (solid) curve $\phi_{-1} = -90°$, where the $n = -1$ space harmonic becomes evanescent, the antenna ceases to radiate. An antenna of given dimensions h and d is characterized in Fig. 28 by a straight line through the origin. The short-dashed line, for example, represents an antenna with $d = 2.5h$,

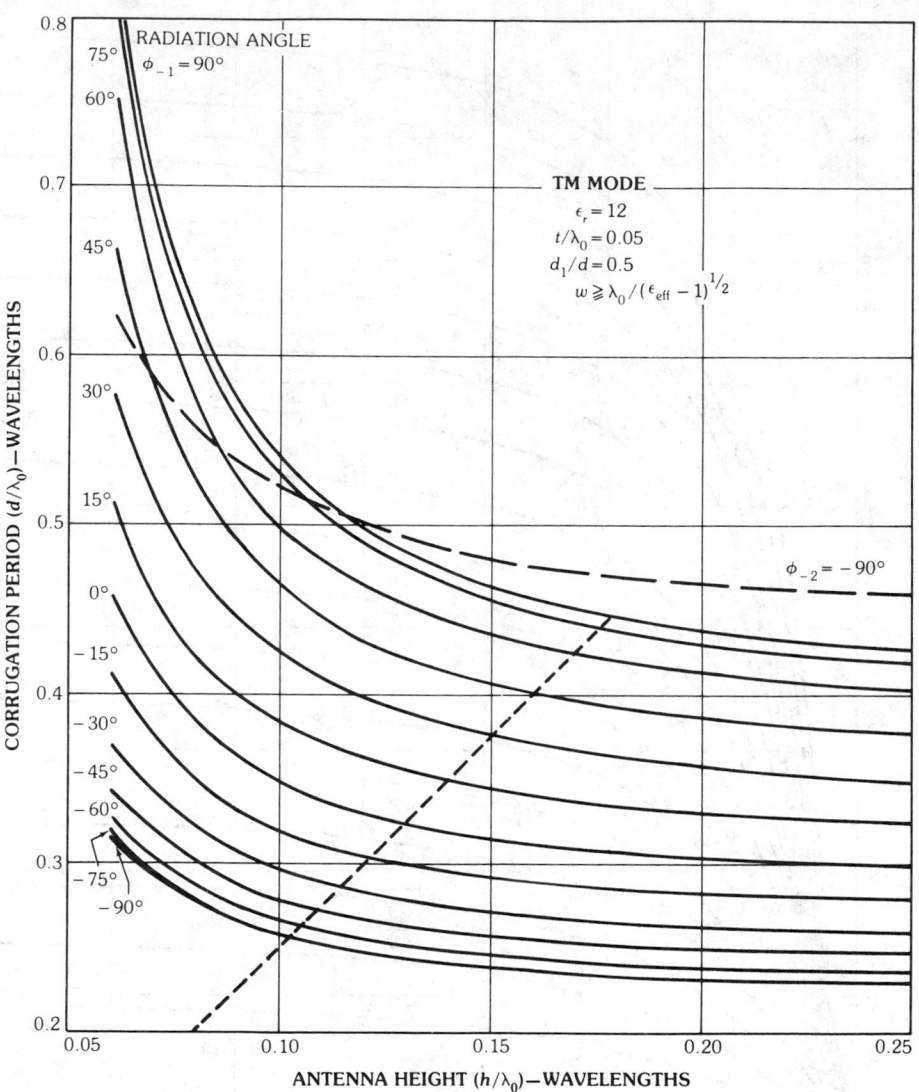

Fig. 28. Radiation angle as a function of antenna height and corrugation period.

and the frequencies associated with given scan angles ϕ_{-1} can be read conveniently from either the abscissa or the ordinate scale so that the scan characteristic of the antenna is readily established.

The *leakage constant* α of wide antennas is found by solution of the (scalar) boundary value problem for infinite width, mentioned above. The leakage constant of the fundamental leaky mode and its dependence on the antenna parameters h, t, d, and d_1/d is shown in Figs. 29–32 [126–127]. For completeness, the radiation angle ϕ_{-1} is indicated in these figures by a curve or else numerically at a few discrete points. The graphs are discussed below.

Approximate closed-form expressions for $\alpha\lambda_0$ have been derived by Tamir and

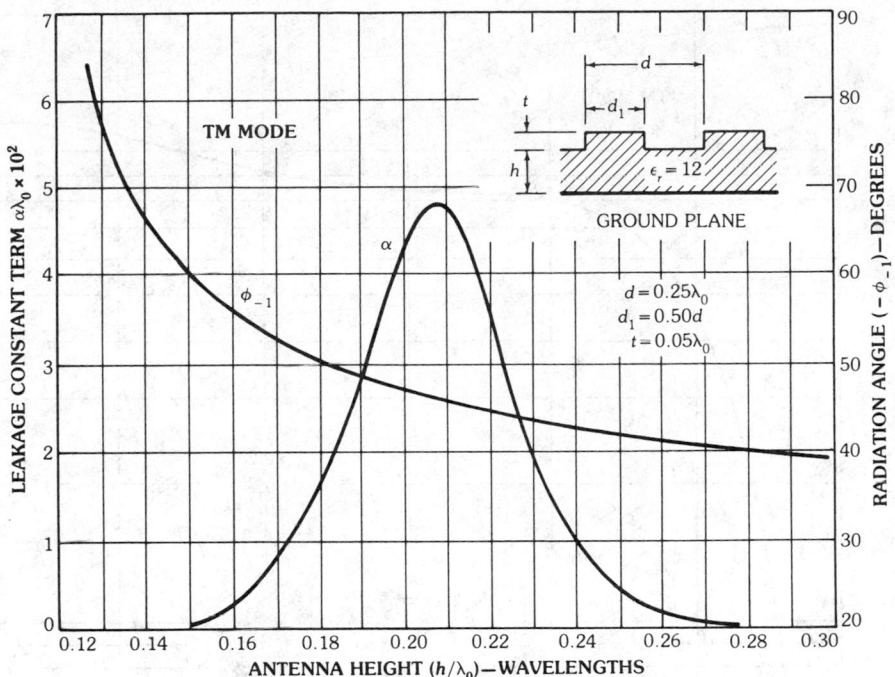

Fig. 29. Dependence of leakage constant α on antenna height. (*After Schwering and Penn [127], © 1983 IEEE*)

Peng [152]. These expressions apply to antennas with a rectangular grating profile whose teeth and grooves have equal width $d_1 = d_2 = d/2$. It is assumed, furthermore, that $\epsilon_{\text{eff}} > \epsilon_{\text{avg}}$, so that the leaky mode guided by the antenna is evanescent in the corrugation region; for a grating profile with the assumed aspect ratio of $d_1/d = 0.5$ this is usually the case. The expression for α takes a different form for small corrugation depths, where α increases proportional to t^2, and for large corrugation depths, where α reaches a saturation level and (apart from a superimposed oscillation of modest amplitude) is practically independent of t.

For $2\pi(\epsilon_{\text{eff}} - \epsilon_{\text{avg}})^{1/2}\left(\dfrac{t}{\lambda_0}\right) \lesssim 1$ in the region of small t, we have

$$\alpha\lambda_0 = \frac{2}{\sqrt{\epsilon_{\text{eff}}}}(\epsilon_r - \epsilon_{\text{eff}})(\epsilon_r - 1)\left(\frac{Q(1)}{\Gamma\tau(1)}\right)\left(\frac{t^2}{h\lambda_0}\right)$$

and for $2\pi(\epsilon_{\text{eff}} - \epsilon_{\text{avg}})^{1/2}\left(\dfrac{t}{\lambda_0}\right) \gtrsim 1$ in the region of large t, we have

$$\alpha\lambda_0 = \frac{1}{2\pi^2\sqrt{\epsilon_{\text{eff}}}}\frac{(\epsilon_r - \epsilon_{\text{eff}})(\epsilon_r - 1)^2}{(\epsilon_r - \epsilon_{\text{avg}})(2\sqrt{\epsilon_{\text{eff}}} - \lambda_0/d)}\left[\frac{Q(\epsilon_{\text{avg}})}{\Gamma\tau(\epsilon_{\text{avg}})}\right]\left(\frac{d}{h}\right)$$

where

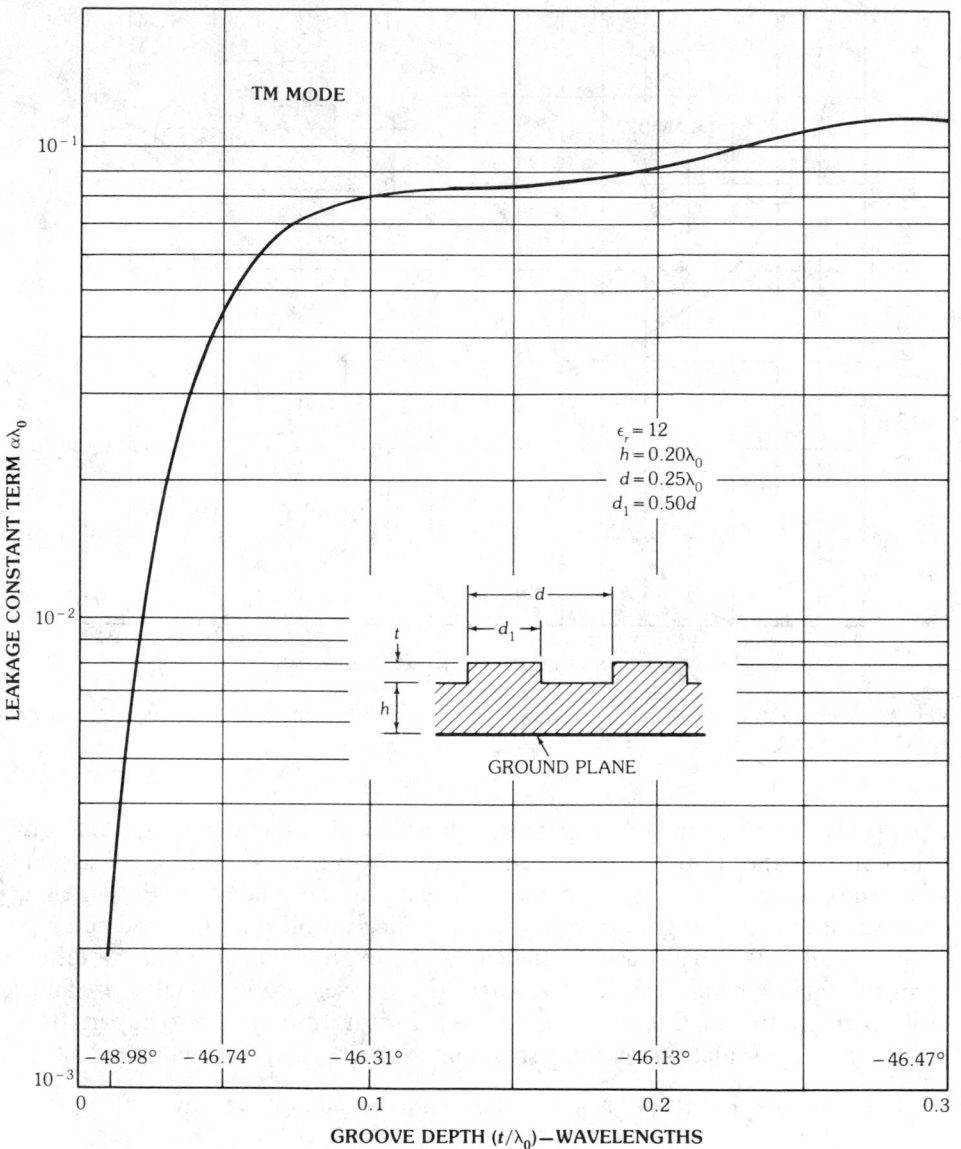

a

Fig. 30. Dependence of leakage constant α on groove depth. (*a*) For *h* constant. (*b*) For *h + t* constant. (*After Schwering and Penn [127], © 1983 IEEE*)

$$\Gamma = [\epsilon_{\text{avg}} - (\sqrt{\epsilon_{\text{eff}}} - \lambda_0/d)^2]^{1/2}$$

$$\tau(\epsilon_{\text{avg}}) = \begin{cases} 1 + \dfrac{1}{k_0 h(\epsilon_{\text{eff}} - \epsilon_{\text{avg}})^{1/2}} & \text{for TE modes} \\[4mm] 1 + \dfrac{1}{k_0 h(\epsilon_{\text{eff}} - \epsilon_{\text{avg}})^{1/2}} \cdot \dfrac{1}{(\epsilon_{\text{eff}}/\epsilon_r \epsilon_{\text{avg}})(\epsilon_r - \epsilon_{\text{avg}}) - 1} & \text{for TM modes} \end{cases}$$

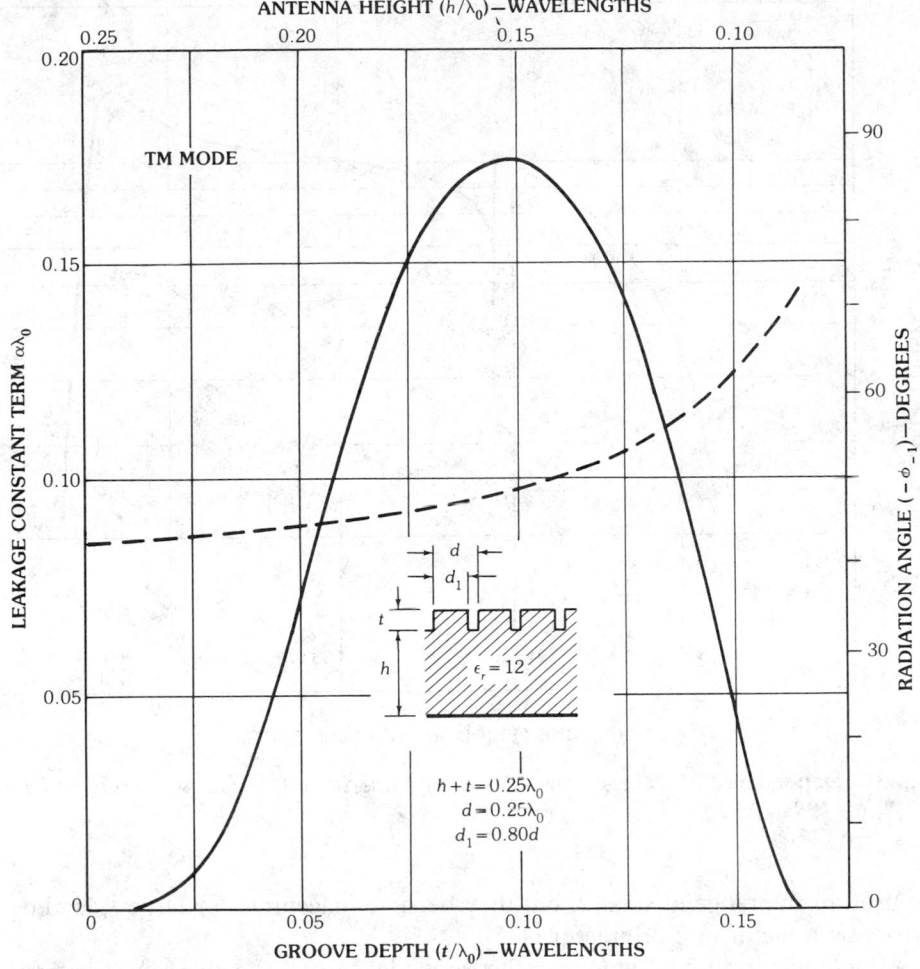

Fig. 30, *continued.*

$$Q(\epsilon_{\text{avg}}) = \begin{cases} 1 & \text{for TE modes} \\[2mm] \dfrac{\epsilon_r - \epsilon_{\text{avg}}}{\epsilon_r \epsilon_{\text{avg}}} \cdot \dfrac{\Gamma^2(\epsilon_r - \epsilon_{\text{avg}}) + \epsilon_{\text{eff}}(\sqrt{\epsilon_{\text{eff}}} - \lambda_0/d)^2}{(\epsilon_r - \epsilon_{\text{avg}}) + (\epsilon_{\text{avg}}/\epsilon_r)^2(\epsilon_r - \epsilon_{\text{eff}})} & \text{for TM modes} \end{cases}$$

The expression for Γ holds in both regions. The expressions for τ and Q apply to the formula for $\alpha\lambda_0$ for large t. In the formula for small t, ϵ_{avg} in τ and Q is replaced by unity (except where it occurs in Γ).

Comparison with values for $\alpha\lambda_0$ obtained by numerical solution of the scalar boundary value problem mentioned earlier has shown that the approximate formulas yield results of good accuracy for antennas of low permittivity in the order of 2 to 4. For high-permittivity antennas with $\epsilon_r = 10$ to 12, the formulas may still

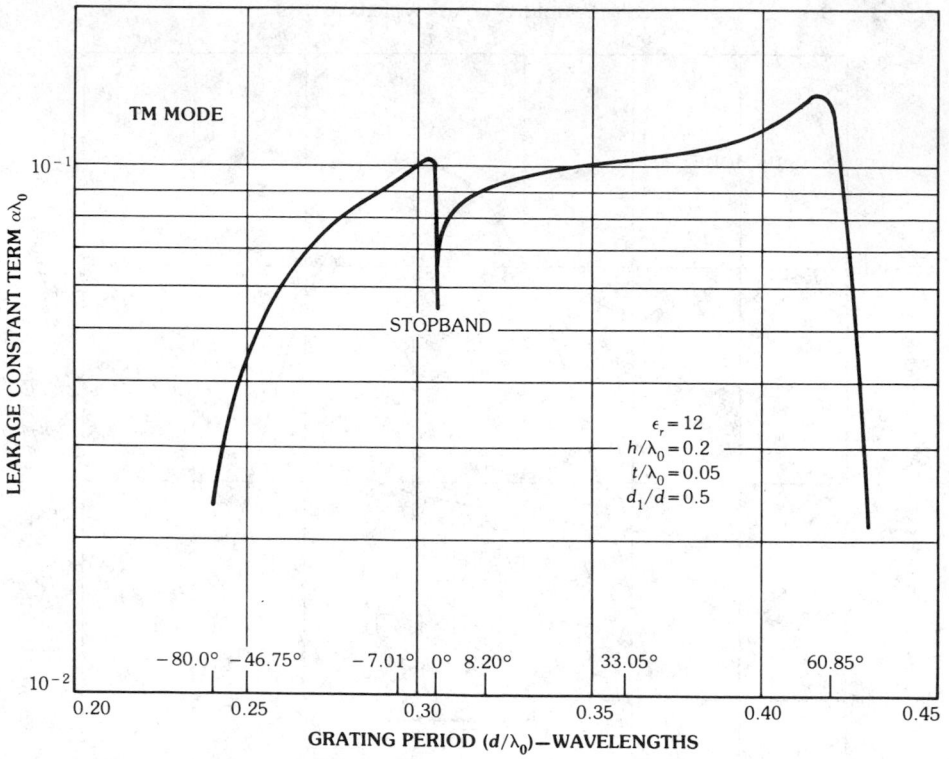

Fig. 31. Dependence of leakage constant α on grating period. (*After Schwering and Penn* [127], © 1983 IEEE)

be used in the range of small t, but they become inaccurate for large t, leading to errors by a factor of 2 or greater [125].

For low-permittivity antennas the dependence of α on the aspect ratio d_1/d approaches a sine-squared relationship with the maximum occurring at $d_1/d = 0.5$, the value assumed in the above formulas. We have in good approximation [152]

$$\alpha\left(\frac{d_1}{d}\right) = \alpha\left(\frac{1}{2}\right)\sin^2\left(\pi\frac{d_1}{d}\right)$$

For antennas of large permittivity and large corrugation depths the sine-squared relationship is no longer valid; the design curves [126, 127] shown below indicate that for $\epsilon_r = 10$ to 12 the maximum of α is shifted to $d_1/d = 0.7$ to 0.8.

Design Guidelines for Wide Dielectric Grating Antennas—An efficient grating antenna must have an axial length L sufficiently large so that a major portion of its input power is radiated before it reaches the antenna termination at the far end, where the remaining power must be absorbed to avoid pattern distortions. To obtain high efficiency at reasonable antenna length the leakage constant must be sufficiently large. This, in turn, implies the use of high-permittivity materials, such

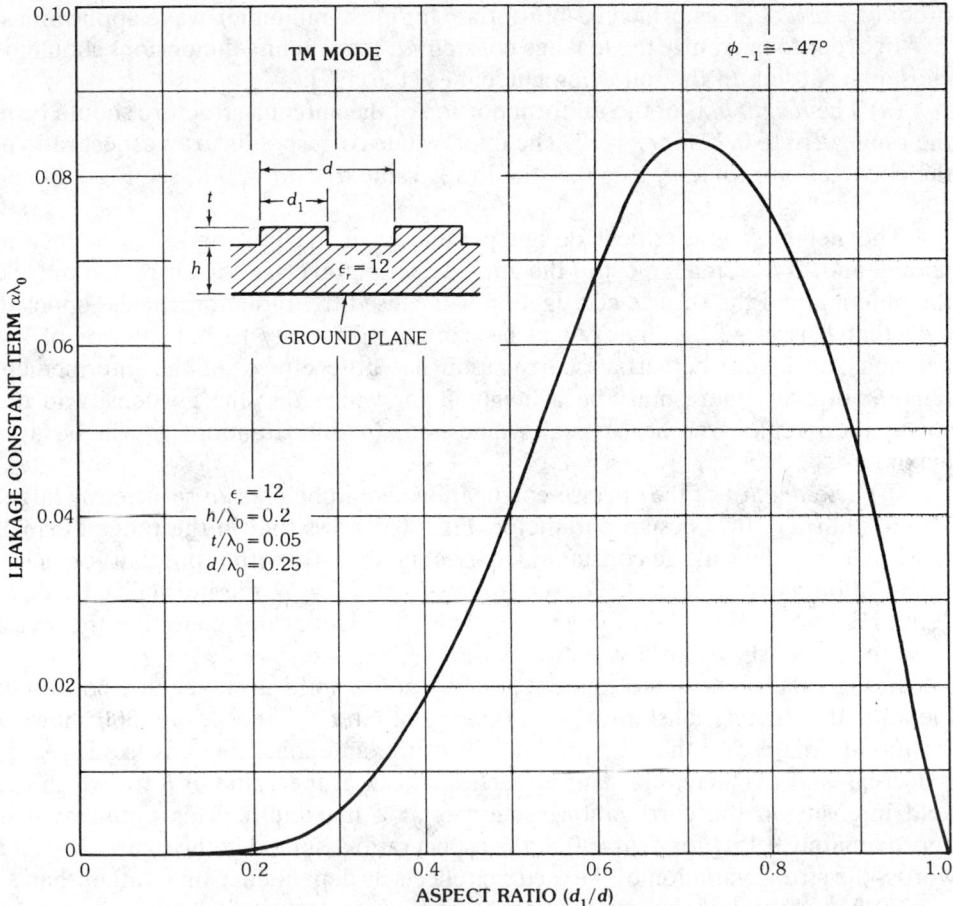

Fig. 32. Dependence of leakage constant α on aspect ratio of surface corrugation. (*After Schwering and Penn [127], © 1983 IEEE*)

as silicon or gallium arsenide [125]. Under optimized conditions, $\alpha\lambda_0$ is typically in the order of 0.1 in this case, while for lower-permittivity materials, such as boron nitride and polystyrene, $\alpha\lambda_0$ would be an order of magnitude smaller.

Neglecting dielectric losses, the antenna efficiency is given by

$$\eta = 1 - e^{-2\alpha L} \qquad (10)$$

If $\alpha\lambda_0 = 0.1$, then

$$\frac{L}{\lambda_0} = \begin{cases} 11.5 & \text{for } \eta = 90 \text{ percent} \\ 23 & \text{for } \eta = 99 \text{ percent} \end{cases}$$

which leads to reasonable antenna dimensions even in the lower millimeter-wave region near $\lambda_0 = 1$ cm. The corresponding E-plane beamwidth (xz plane) is in the order of a few degrees, which is appropriate for most millimeter-wave applications.

In order to maximize the leakage constant α, the antenna dimensions should be chosen according to the following guidelines [126, 127]:

(a) The *height* h/λ_0 of the uniform portion of the antenna structure should be in the range 0.15 to 0.21 for $\epsilon_r = 12$. The upper value corresponds to an aspect ratio of the corrugation profile $d_1/d = 0.5$, the lower value to $d_1/d = 0.8$; see Figs. 29 and 30b.

The height h is a critical design parameter in maximizing α, as is easy to understand. If h is small, most of the guided energy travels in the air region outside the antenna and the surface corrugation will cause little radiation. In the opposite case that h is large the energy will be confined primarily to the interior of the antenna and again the surface corrugation has little effect. In the intermediate range, however, there must be a height h for which the energy density in the corrugation region reaches a peak value and, in this situation, α will be at a maximum.

(b) The *depth* t of the surface corrugation should be chosen sufficiently large, but it is not a critical design parameter. Fig. 30a shows that in the range of small groove depths the leakage constant increases significantly with t, but that it reaches a saturation value at large t. For an antenna with $\epsilon_r = 12$ the normalized groove depth t/λ_0 should be equal to or greater than 0.05. Under this condition the phase constant β also shows little variation with t; see Fig. 26.

In Fig. 30a it is assumed that the height h of the uniform waveguide portion of the antenna remains constant while t varies. For later reference, Fig. 30b shows α for the alternate case that the *total* height of the antenna, $h + t$, is fixed. Hence h decreases as t is increased, and α reaches a peak at the value of h for which the field intensity in the corrugation section is at a maximum. This optimum h is approximately $0.15\lambda_0$ for $d_1/d = 0.8$, the aspect ratio assumed in the figure. In other words, the strong variation of the α curve reflects its dependence on h rather than t.

(c) The *period* of the surface corrugation, d, is determined primarily by the desired radiation angle and the condition (6) for single-beam operation, but it is not a parameter available for maximizing α. Fig. 31 shows, moreover, that over much of the angular range, α does not vary strongly with d/λ_0, and is fairly constant over a wide range of d values. The stopband which occurs when broadside conditions are approached is very narrow.

(d) The *aspect ratio* d_1/d of the surface corrugation should be chosen close to 0.7, i.e., the width of the grooves should be significantly smaller than that of the teeth; see Fig. 32. This is a consequence of the high permittivity considered here. For low-permittivity antennas, maximum energy leakage would occur near $d_1/d = 0.5$.

(e) As pointed out before, the effect of the *lateral width* w of the antenna on α is small for $w > \lambda_0/\sqrt{\epsilon_{\text{eff}} - 1}$, the case considered here.

Design Procedure for Narrow Dielectric Grating Antennas—In the above discussion on periodic dielectric antennas of *large width* $(w > \lambda_0/\sqrt{\epsilon_{\text{eff}} - 1})$ it was pointed out that the propagation constant of these antennas can be obtained simply and accurately by the use of the EDC method, and that their leakage constant is practically the same as that of an antenna of infinite width (with the leaky modes traveling in the direction normal to the surface grooves). This equivalence reduces the computational burden substantially. Except for $w \to \infty$, the infinite antenna has the same geometrical parameters and the same permittivity as the finite antenna. The discussion has shown, furthermore, that α depends strongly on the fraction of the guided energy traveling within the corrugation region of the antenna.

For antennas of *narrow width* ($w < \lambda_0/\sqrt{\epsilon_{\text{eff}} - 1}$), the propagation constant can still be obtained in good accuracy by the EDC procedure, but the leakage constant can no longer be approximated by that of an infinitely wide antenna with the same permittivity as the finite antenna. With decreasing w, the phase velocity of the leaky modes guided by the actual antenna increases; furthermore, an increasing portion of the guided energy will now travel in the air region on both sides of the antenna, thus reducing the energy density within the corrugation region. Both effects combine to decrease the leakage constant of the antenna of narrow width.

However, an equivalent antenna of infinite width yielding approximately the same α as the narrow antenna can still be defined with the help of the EDC procedure [146] which allows one to replace an antenna of finite width by an equivalent antenna of infinite width having an effective dielectric constant which is smaller than the permittivity of the original antenna.* The decreased permittivity reduces the perturbation of the guided wave caused by the surface modulation and hence the leakage constant of the infinite antenna. The geometrical parameters of this equivalent antenna are the same as those of the original antenna (apart, of course, from w). There is a great advantage in the availability of an equivalent antenna of infinite width. As pointed out before, for $w \to \infty$, the leaky modes become pure TE or TM modes, depending on their polarization, so that the boundary value problem, from which α is determined, is reduced from a vector problem to a much simpler scalar problem. Since, in addition, the equivalent antenna has the same height and corrugation profile as the original antenna, the formulas and graphs for α and β, already available and discussed before, can be readily utilized.

In order to determine the equivalent structure of infinite width we use the EDC procedure to carry out a transverse resonance in the lateral (y) direction [146]. In doing so, we temporarily extend the structure in the vertical direction to positive and negative infinity, producing a *uniform* dielectric layer, as shown in the inset of Fig. 33a. The transverse resonance then results in the dispersion relation (9b), which yields dispersion curves of the form shown in Fig. 33a as a function of the width w of the fictitious uniform layer and therefore of the original antenna.** The

---

*Note, however, that the effective dielectric constant is determined from the condition that both antennas have the same *phase* constant.

**Note the difference in the way in which the effective dielectric constant is determined here and in the section on the antennas of large width. In that section the effective permittivity is calculated from the transverse resonance condition in the z direction, while here it is determined from the resonance condition for the y direction. To signify this difference the symbol $\bar{\epsilon}_{\text{eff}}$ is used here instead of ϵ_{eff}. This also means that in calculating $\bar{\epsilon}_{\text{eff}}$ from (9b), the quantities $k_y^{(o)}$ and $k_y^{(e)}$, which appear on the left side of this equation, must be rewritten as

$$|k_y^{(o)}| = k_0(\bar{\epsilon}_{\text{eff}} - 1)^{1/2} \quad \text{and} \quad k_y^{(e)} = k_0(\epsilon_r - \bar{\epsilon}_{\text{eff}})^{1/2}$$

The equation then applies to TE polarization, i.e., the polarization associated with the fundamental mode. The dispersion relation for TM polarization is obtained if, in addition, the left side of (9b) is multiplied by $1/\epsilon_r$.

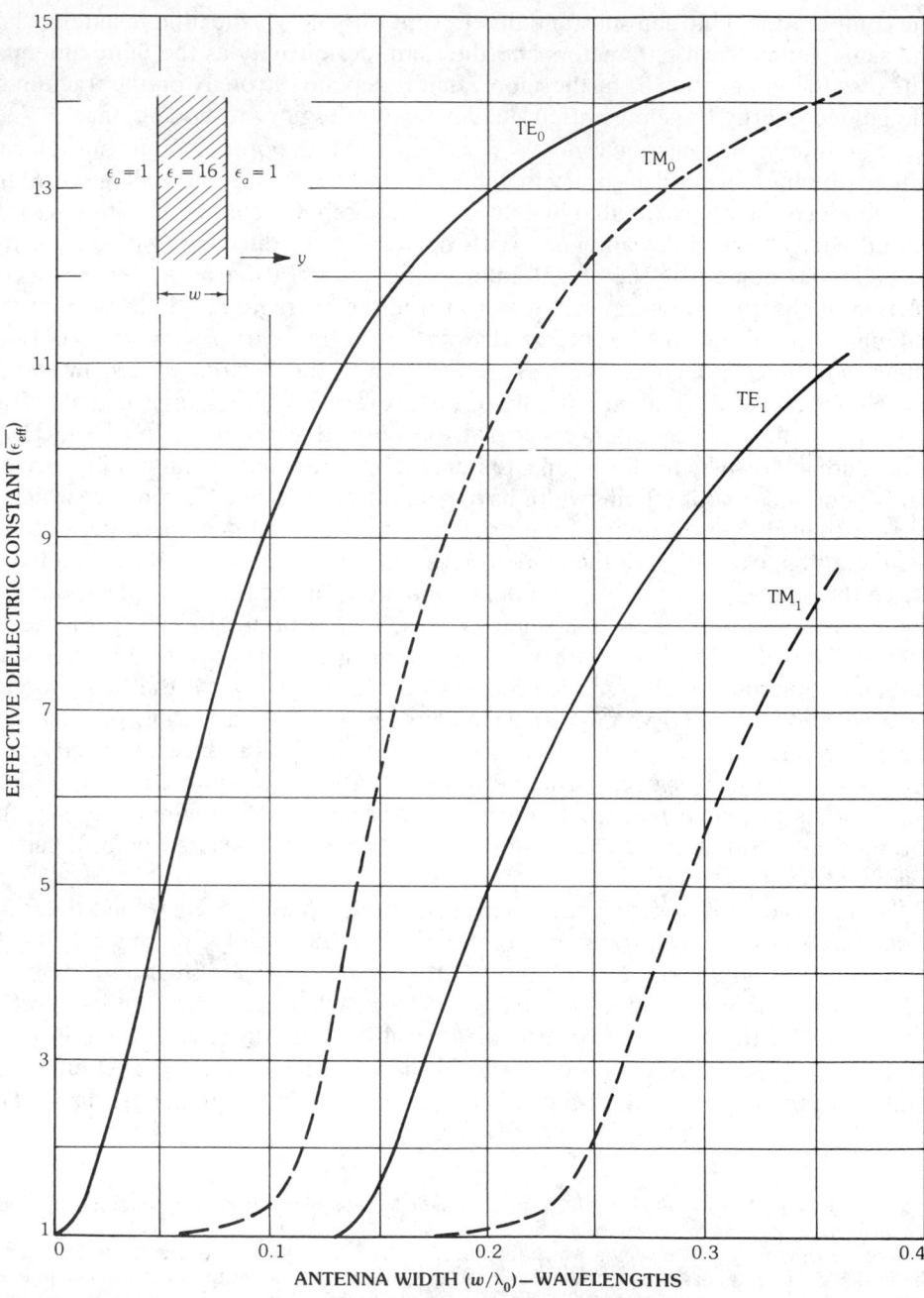

a

Fig. 33. Approximate procedure for calculation of the leakage constant of dielectric grating antennas of narrow width. (*a*) Dispersion curves for uniform dielectric slab, to be used in obtaining the equivalent antenna of infinite width. (*b*) Comparison between theoretical values and experimental results for the leakage constant of a 30-GHz antenna of $w = 1.3$ mm and $\epsilon_r = 16$. (*After Peng et al. [146], © 1984 IEEE*)

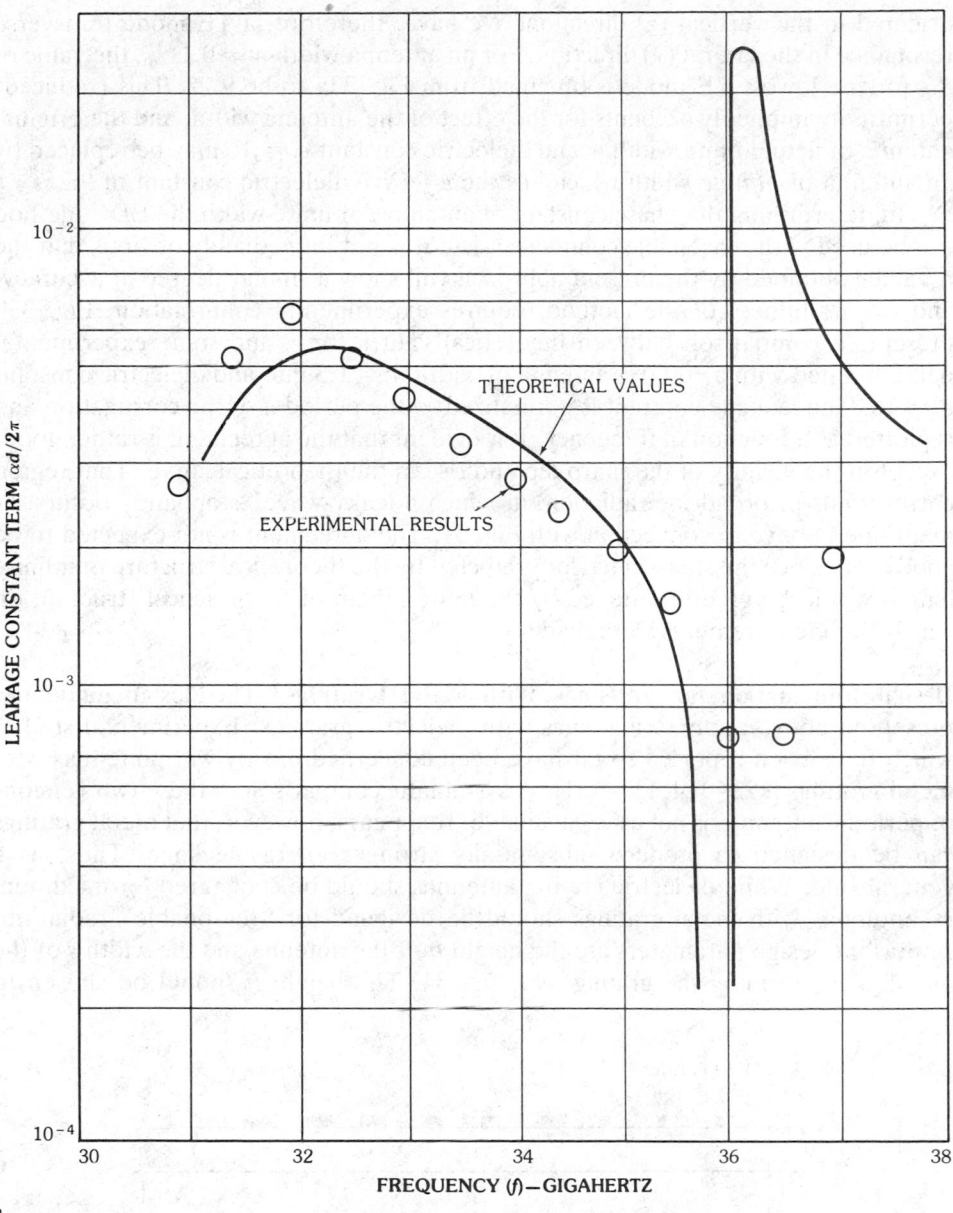

Fig. 33, *continued.*

actual curves appearing in Fig. 33a correspond to a dielectric constant of 16 for the layer, for both TE polarization (solid lines) and TM polarization (dashed lines). The ordinate is $\bar{\epsilon}_{\text{eff}}$, the effective dielectric constant of the equivalent antenna of infinite width.

To show how the curves are to be used in the present context, consider as an example the fundamental leaky mode for which the main electric field component is oriented in the vertical (z) direction. We have, therefore, a TE mode transverse resonance in the lateral (y) direction. For an antenna width $w = 0.11\lambda_0$, the value of $\bar{\epsilon}_{\text{eff}}$ for the lowest TE mode is obtained from Fig. 33a to be 9.25. This (reduced) permittivity implicitly accounts for the effect of the antenna width, and the original antenna of actual finite width w and dielectric constant $\epsilon_r = 16$ may be replaced by an antenna of *infinite* width which has the *effective* dielectric constant of 9.25.

In determining the phase constant of antennas of finite width the EDC method can be used with reasonable confidence. But it is not immediately obvious that the α values obtained by the present approach will show a similar degree of accuracy, and the usefulness of the method requires experimental confirmation. Fig. 33b presents a comparison between theoretical values for α and some experimental data obtained with a 30-GHz antenna of width $w = 1.3$ mm and dielectric constant $\epsilon_r = 16$. The leakage constant is normalized to the period d of the corrugation and is plotted as a function of frequency. It is evident that the agreement is rather good, except in the vicinity of the sharp dip and rise in the theoretical curve. That region corresponds to broadside radiation, at which a leaky-wave "stopband" occurs, as mentioned above in connection with Fig. 31. The agreement is not expected to be good there since the sharp behavior predicted by the theoretical structure of infinite length would be greatly softened by the finite length of the practical structure on which the measurements were made.

Design Information for Antennas With Metal Gratings—The design guidelines presented above apply to antennas with dielectric gratings. Experimental studies which have been reported so far have been concerned mostly with antennas with metal gratings [122–124, 129–131]. A systematic comparison of these two versions of periodic antennas is not as yet available. It appears, however, that metal gratings can be designed to produce substantially stronger energy leakage. Thus, as a general rule: While dielectric grating antennas should be configured for maximum α, antennas with metal gratings should be designed for "reasonable" radiation. Important design parameters are the height h of the antenna and the width s of the metal strips forming the grating; see Fig. 34. The height h should be chosen to

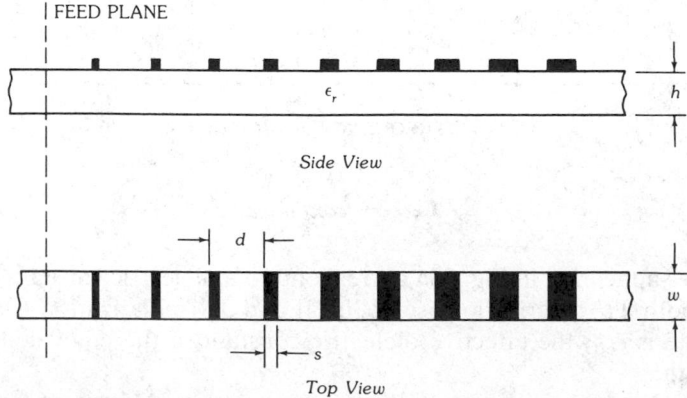

Fig. 34. Antenna with metal grating of variable strip width s.

ensure that a large field intensity exists on the grating surface, and the strip width s can then be adjusted to obtain a desired radiation rate. The optimum height is, thus, roughly the same as in the case of an antenna with a dielectric grating, though it should be taken into account that a metal grating tends to have a stronger influence on the field distribution in the antenna than a surface corrugation, in particular if the metal strip width s is large.

Trinh and others have shown in a recent experimental study [130, 131] that the optimum strip width s is approximately $0.4\lambda_g$, where $\lambda_g = \lambda_0(k_0/\beta)$ is the guide wavelength. If s is much smaller, i.e., less than $0.2\lambda_g$, a noticeable amount of residual radiation in the end-fire direction has always been observed, even for long antennas (50 strips). If the metal strips are too wide, $s > 0.5\lambda_g$, a major portion of the guided power is radiated by the first few strips and the effective antenna aperture is very small, leading not only to a large beamwidth but to high side lobes, which are probably due to a large mismatch at the waveguide-to-antenna transition [130].

Furthermore, it was found that linear tapering of s from period to period, i.e., starting with a small strip width at the feed point and gradually increasing s until it reaches the final value of $0.4\lambda_g$, substantially reduces the side lobes to a level less than -25 dB below the main beam [129]. An optimization study [130, 131] has resulted in the following empirical relation for the width of the nth strip:

$$s_n = \begin{cases} [0.145 + 0.015(n-1)]\lambda_g & \text{for } n \leqq 18 \\ 0.4\lambda_g & \text{for } n > 18 \end{cases}$$

The first few strips in this case produce little radiation since their function is apparently limited to providing an impedance match in the transition region.

Similar improvements in pattern quality can be expected for antennas with dielectric gratings, when the groove width or depth are appropriately varied from period to period. Design procedures for such gratings are discussed under the heading "Pattern Shaping."

Beam Scanning—The scan performance of antennas with dielectric gratings may be seen from Fig. 28. It is discussed here in more detail for an antenna with the specific dimensions indicated in Fig. 35 [125]. These dimensions apply to beam scanning in the 3-mm band (94 GHz) but can be scaled, of course, to any other frequency region. The figure shows the radiation angle ϕ_{-1} and the leakage constant α for two antennas, one designed for radiation in the forward direction ($d = 1.05$ mm), the other for radiation in the backward direction ($d = 0.81$ mm). For a 10-percent variation in wavelength, say from $\lambda_0 = 2.83$ mm to 3.15 mm, the main beam direction sweeps almost linearly over an angular range of about 25° for backward radiation and 20° for forward radiation. In both cases the leakage constant reaches a maximum near $\lambda_0 = 3$ mm and decreases almost symmetrically when λ_0 is varied in either direction. This is a very desirable characteristic. If one designates $\lambda_0 = 3$ mm as the center wavelength of a frequency scan, the radiation constant and hence the beamwidth will change comparatively little as the antenna is scanned and a stable radiation pattern with little beam degradation can be expected [125].

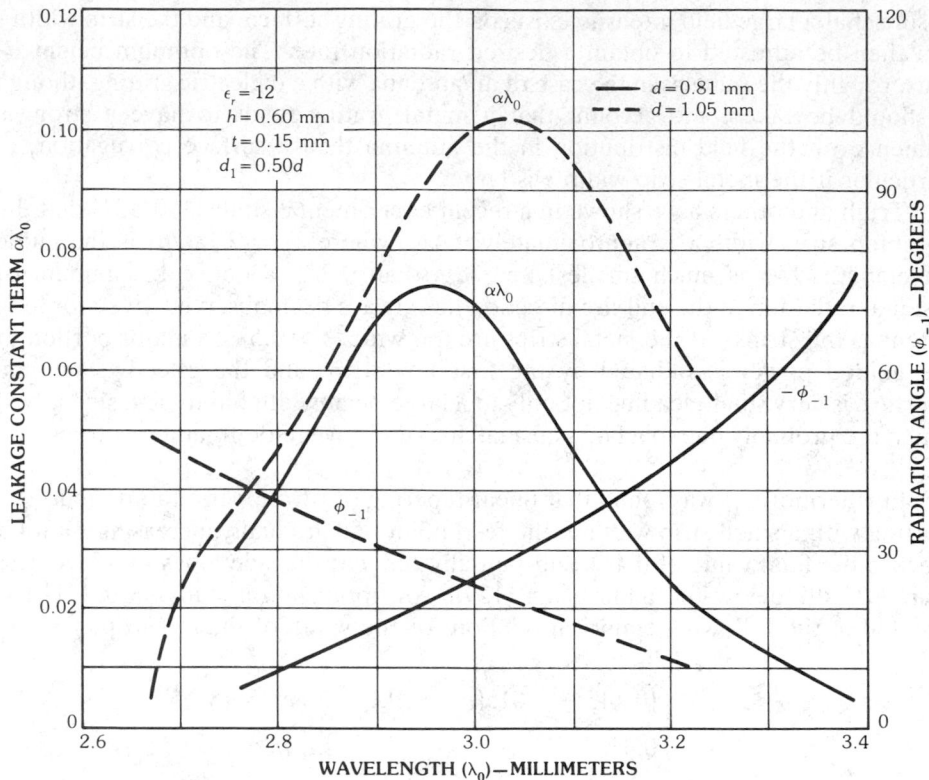

Fig. 35. Scan performance of dielectric grating antennas: variation of radiation angle ϕ_{-1} and leakage constant α with wavelength.

Recent experiments indicate that the theoretical scan rate of 20° to 25° per 10-percent change in frequency is conservative. For grating antennas made from silicon ($\epsilon_r = 12$) and magnesium titanate ($\epsilon_r = 16$), a scan rate of approximately 40° per 10-percent frequency variation was measured [122, 128]. The silicon antenna used a metal grating while the magnesium titanate antenna operated with a dielectric grating with a rectangular profile of large aspect ratio d_1/d.

Under the heading "Review of Fundamentals" it was pointed out that periodic dielectric antennas do not radiate in the broadside direction; as broadside conditions are approached, an internal resonance develops which prohibits radiation. According to a method proposed by James and Hall [155] this problem can be alleviated, however, by the use of a surface modulation whose elements are *pairs* of grooves or metal strips rather than *single* scatter elements. The spacing between the elements of each pair is equal to a quarter guide wavelength at the design frequency. In this case the wave reflected by the first element of each pair will nearly be canceled by the wave reflected from the second element, so that the total reflected power is greatly reduced. The pair-to-pair spacing (i.e., the overall period of the surface modulation), of course, will be chosen at one guide wavelength at the design frequency in order to obtain broadside radiation.

The validity of the method has been confirmed experimentally by Solbach and

Adelseck [156], who used an image-guide antenna made from Duroid, which was designed to operate as a frequency-scanned antenna over the 60- to 75-GHz band. A periodic surface perturbation was provided by a sequence of 27 pairs of small, circular metal disks placed on the upper surface of the guide. The measured reflection coefficient remained below −20 dB over most of the band, but showed an increase to about −12 dB in the range of broadside radiation near 70 GHz. While this increase still indicates the existence of a "stopband," the absolute value of the reflection coefficient near 70 GHz remained by an order of magnitude below that measured for a similar antenna using a sequence of single disks. Furthermore, the gain of the new antenna showed a smooth frequency dependence and did not exhibit a noticeable dip near broadside.

For many applications an antenna would be desirable which permits electronic beam scanning at constant frequency, i.e., by other means than frequency variation. A promising approach, suggested by Jacobs, is indicated in Fig. 36. The basic idea is to vary the phase constant β of the antenna, and thus its radiation angle, by electronically controlling the antenna cross section. This is accomplished with the help of semiconducting fins attached to the antenna on one or both sides. In a recent experimental study [123, 124], these fins were made from silicon (same ϵ_r as antenna) which was processed so that the fins could be operated as distributed pin diodes. Through biasing currents applied with the help of electrodes placed on the top and bottom of the diodes one can control the conductivity of the fins and hence the depth to which the fields guided by the antenna extend into the fin region. In this way one can adjust the effective cross section of the antennas and, in turn, its phase constant and radiation angle.

Conduction losses suffered by the fields extending into the fin region reduce antenna efficiency. But these losses are small in the two limiting cases of very high and very low conductivity (i.e., full bias and off bias) so that the method appears particularly suited for digital beam scanning.

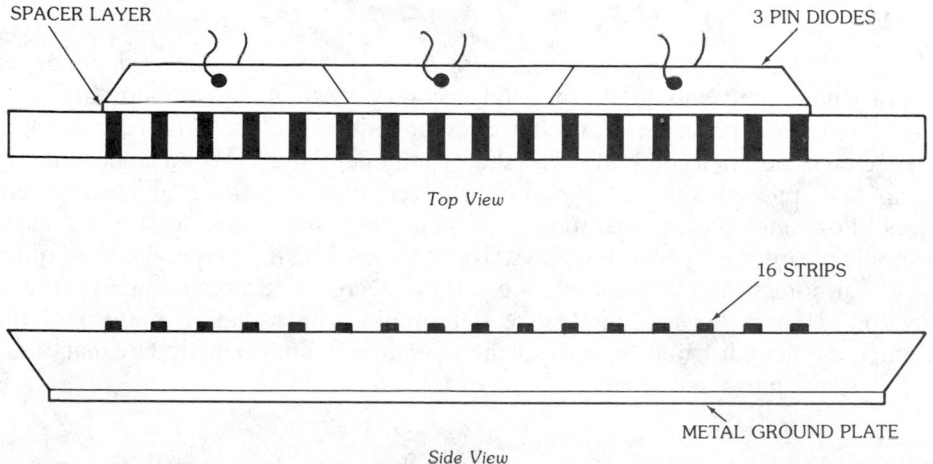

Top View

Side View

Fig. 36. Antenna with metal grating and distributed pin diodes for electronic beam scanning at constant frequency. (*After Horn et al. [123], © 1980 IEEE*)

The experimental antenna [123, 124] was designed for the 60-GHz band and consisted of a 1-mm × 1-mm silicon rod with 16 metal strips. With three pin diodes attached to one side, a scan range of 8° to 10° was obtained at low power consumption (1 V × 100 mA per diode). In some cases the beam shift was as large as 15°. To minimize losses the diodes were separated from the antenna by a thin layer of a low-permittivity insulating material.

An alternate approach to beam steering at constant frequency has been suggested by Bahl and Bhartia [157]. This approach would use a periodic antenna structure containing a liquid artificial dielectric medium whose permittivity can be controlled by a biasing electric field. The practicality of this approach remains to be established, but a theoretical study using data for existing artificial dielectrics has predicted good performance for the structure under investigation, i.e., a large scan range of 40° to 50° for a moderate change in permittivity, and a practically constant beamwidth.

Radiation Pattern—The main-beam direction of a dielectric-grating antenna is found from (7) once the phase constant β is known. Calculation of the radiation pattern is a more difficult task. An approximate method is described in the following that should yield fairly accurate results for the main beam and near-in side lobe regions. Neglected effects, however, may strongly influence the radiation pattern in the far side lobe region.

Fig. 37 shows a grating antenna of length L and width w, where it is assumed that $L \gg \lambda_0$ and $w > \lambda_0/\sqrt{\epsilon_{\text{eff}} - 1}$. The terms θ and ϕ are the angular coordinates of a spherical coordinate system whose axis coincides with the y axis. The tangential magnetic field distribution of the fundamental leaky mode in the antenna aperture, i.e., in the portion of the plane $z = t$ just above the antenna, can be expressed in terms of the space-harmonics representation of this mode. We have

$$H_y(x, y, t) = e^{-\alpha x} \sum_{n=-\infty}^{+\infty} I_n e^{-j\beta_n x} \cos(\pi y/w) \qquad (11)$$

for $0 \leqq x \leqq L$, $-w/2 \leqq y \leqq +w/2$, with $\beta_n = \beta + 2\pi n/d$. The y dependence is approximate but should be of good accuracy since it is assumed that $w > \lambda_0/\sqrt{\epsilon_{\text{eff}} - 1}$. For the same reason the x component of the magnetic field strength is expected to be small compared with the y component and has been neglected.* In the air half-space the space harmonics are decoupled and travel as independent waves. For single-beam operation, the practically interesting case, the space harmonic of order $n = -1$ is the only wave of the propagating type, while all other space harmonics are evanescent, i.e., they decrease exponentially in the z direction. Hence, for the purpose of determining the radiation pattern of the antenna, the field distribution in the plane $z = t$ may be approximated by that of the $n = -1$ space harmonic alone:

---

*The x component of the aperture illumination, $H_x(x, y, t)$, determines the cross-polarized part of the radiation pattern. It is antisymmetric in y so that the cross-polarized radiation vanishes in the E-plane, but it is different from zero elsewhere. For antennas of narrow width the cross-polarized radiation would have to be taken into account since H_x can be significant when $w < \lambda_0/\sqrt{\epsilon_{\text{eff}} - 1}$.

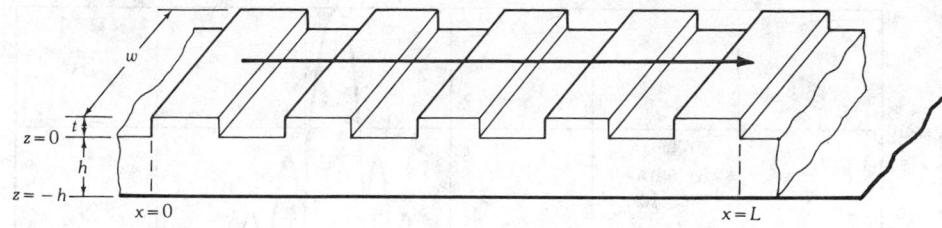

a

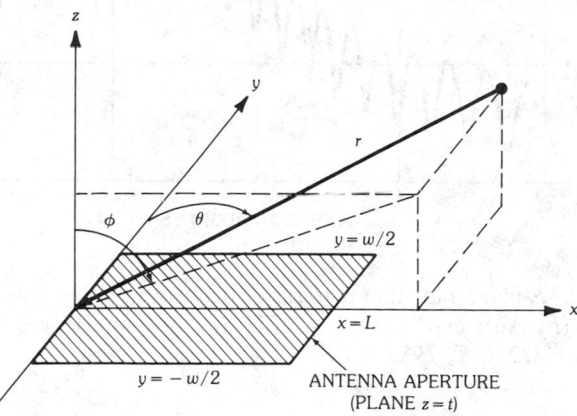

b

Fig. 37. Radiation pattern calculation. (*a*) Antenna geometry. (*b*) Antenna aperture and spherical coordinate system. (*After Schwering and Penn [127],* © *1983 IEEE*)

$$H_y(x,y,t) = \begin{cases} I_{-1}e^{-\alpha x}\, e^{-jk_0 x \sin\phi_{-1}} \cos(\pi y/w) & \text{for } 0 \leqq x \leqq L, \quad |y| \leqq w/2 \\ 0 & \text{elsewhere} \end{cases} \quad (12)$$

Equation 7 was used here to replace $\beta_{-1} = \beta - 2\pi/d$ by $k_0 \sin\phi_{-1}$. For a discussion of the approximations involved in (12) we refer to the literature [127].

The power radiation pattern associated with aperture distribution (12) is given by

$$G(\theta,\phi) = G_D S(\cos\theta)\, T(\sin\phi\sin\theta)$$

where

$$G_D = \frac{16}{\pi^3} k_0^2 \frac{w}{\alpha} \tanh\left(\frac{\alpha L}{2}\right) \cos\phi_{-1} \quad (13a)$$

is the directivity gain of the antenna,

$$S(\cos\theta) = \left(\frac{\pi}{2}\right)^4 \frac{\cos^2[(k_0 w/2)\cos\theta]}{\{(\pi/2)^2 - [(k_0 w/2)\cos\theta]^2\}^2} \sin^2\theta \quad (13b)$$

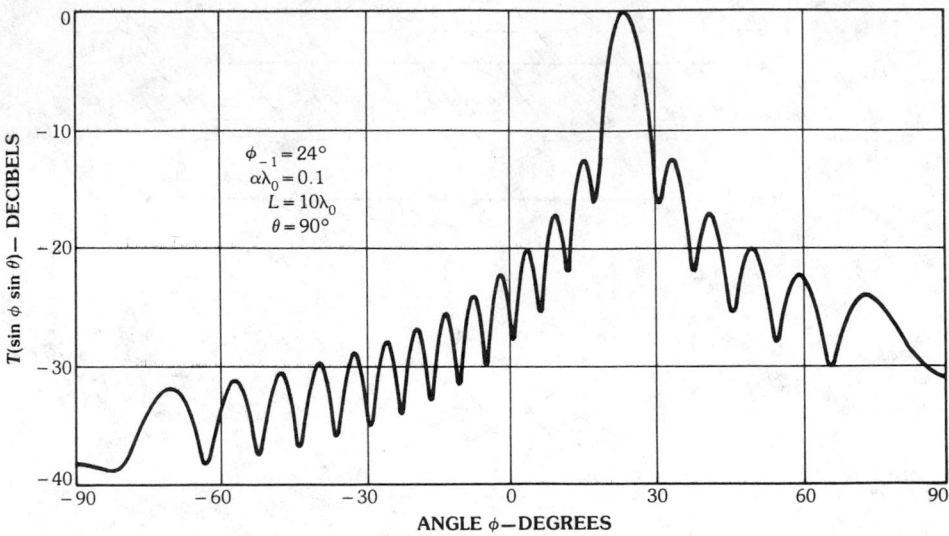

$\phi_{-1} = 24°$
$\alpha\lambda_0 = 0.1$
$L = 10\lambda_0$
$\theta = 90°$

Fig. 38. Theoretical *E*-plane radiation patterns of dielectric grating antennas. (*a*) Antenna length $L = 10\lambda_0$. (*b*) Antenna length $L = 20\lambda_0$. (*c*) Antenna length $L = 150\lambda_0$. (*After Schwering and Penn [127]*, © *1983 IEEE*)

is the *H*-plane pattern, and

$$T(\sin \phi \sin \theta) = \left(\frac{\alpha L}{1 - e^{-\alpha L}}\right)^2 \frac{1 - 2e^{-\alpha L}\cos[k_0 L(\sin \phi \sin \theta - \sin \phi_{-1})] + e^{-2\alpha L}}{(\alpha L)^2 + (k_0 L)^2(\sin \phi \sin \theta - \sin \phi_{-1})^2}$$

(13c)

determines the *E*-plane pattern. Both *S* and *T* are normalized to unity in the main beam direction $\theta = 90°$, $\phi = \phi_{-1}$.

The directivity gain is proportional to the antenna aperture area *wL* in the case that $\alpha L \ll 1$, where the aperture illumination is practically uniform (and the antenna efficiency is low). In the opposite case that $\alpha L \gg 1$, the effective aperture area is determined by the decay constant α and independent of *L*; we have $G_D \propto w/\alpha$.

The H-plane pattern is the well-known radiation characteristic of a cosine-tapered aperture distribution. The 3-dB beamwidth is $\Delta\theta = 1.2\lambda_0/w$ (in radians) and first side lobes are down by -23 dB. The *E*-plane pattern is that of an exponentially tapered source distribution. Figs. 38a through 38c show this pattern for $\alpha\lambda_0 = 0.1$ and $L/\lambda_0 = 10$, 20, and 150. The corresponding antenna efficiencies are $\eta = 86.5$, 98.2, and approximately 100 percent; see (10). The beamwidths of these patterns are in the order of a few degrees, which is appropriate for most millimeter-wave applications. A smaller beamwidth can be realized by designing the antenna to have a smaller leakage constant α and by correspondingly increasing the antenna length *L*. Designing for a broader beamwidth would be more of a problem. Either α would have to be raised, which is not easily done since the assumed value

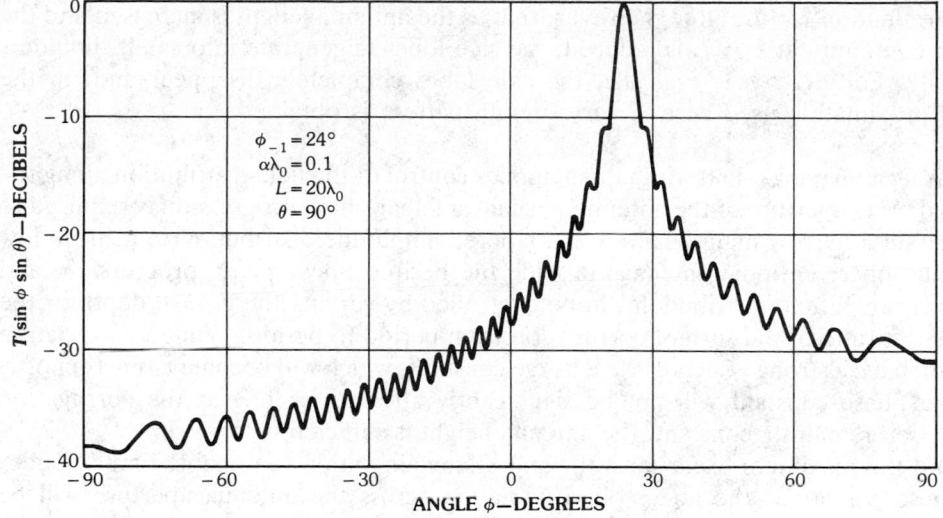

b

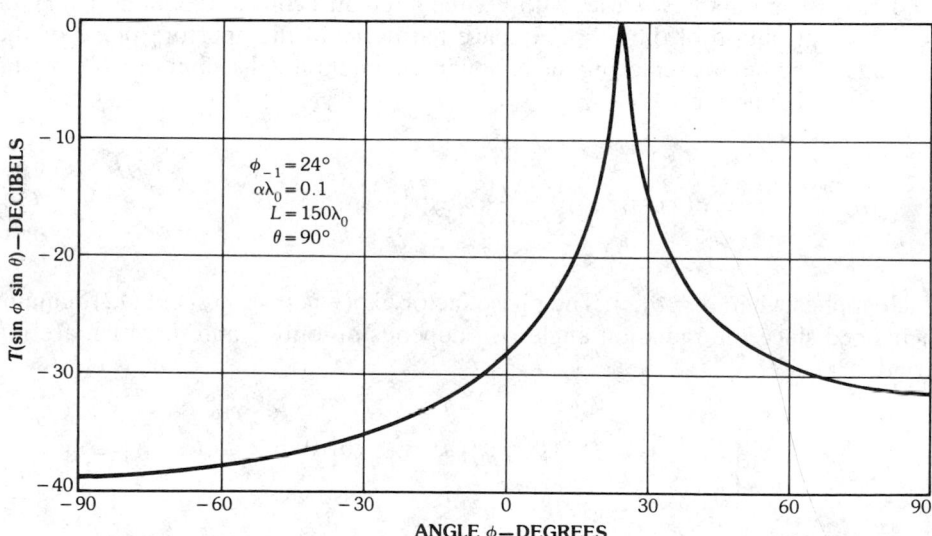

c

Fig. 38, *continued.*

of $\alpha\lambda_0 = 0.1$ is already close to maximum for antennas with $\epsilon_r = 12$, or the antenna length would have to be reduced at the expense of a decreased efficiency. The problem can be solved by the use of a metal grating instead of a dielectric grating. It was mentioned earlier that a metal grating will produce a strong leakage constant when the width of the metal strips is chosen sufficiently large.

The first side lobes of the E-plane pattern are down by only -13 dB. The high side lobe level is caused by the discontinuity in the aperture distribution (12) at the termination $x = L$. Note, however, that as the antenna length is increased and the discontinuity at $x = L$ is reduced, the side lobes degenerate into small shoulders only. For $\alpha L \gg 1$, Fig. 38c, the side lobes completely disappear and, in the approximation used here, a very smooth pattern is obtained.

Pattern Shaping—Pattern shaping implies control of the field distribution along the radiating aperture of the antenna. Usually a linear phase progression is required to obtain a narrow main beam, and a tapered amplitude distribution to achieve low side lobes. Periodic antennas provide the desired linear phase progression, and their amplitude distribution can be controlled by varying the groove depth or the aspect ratio of the surface corrugation from period to period. While this variation will have a strong effect on the leakage constant, which will become a function of x, the phase constant will not be significantly affected as long as the corrugation period remains a constant, the antenna height is sufficiently large ($h + t > 0.2\lambda_0$), and the maximum excursions of t and d_1 remain within reasonable limits. Under these conditions the linear phase variation across the antenna aperture will be preserved while tapering of the leakage constant will permit beam shaping through modification of the amplitude distribution.

Thus, assuming that α varies with x while β remains constant, equation (12) for the field distribution of the $n = -1$ space harmonic in the aperture plane of the antenna is modified by replacing the exponential amplitude distribution $e^{-\alpha x}$ by the more general function

$$F(x) = F(0)\left\{\frac{a(x)}{a(0)}\right\}^{1/2} \exp\left(-\int_0^x a(x)\,dx\right) \tag{14}$$

which applies when $\alpha = \alpha(x)$. The phase factor $\exp(-jk_0 x \sin\phi_{-1})$ in (12) remains unchanged since the radiation angle ϕ_{-1} depends on only β and d, which are not varied. Hence

$$H_y(x,y,t) = \begin{cases} I_{-1}F(x)\,e^{-jk_0x\sin\phi_{-1}}\cos(\pi y/w) & \text{for } 0 \leq x \leq L, \quad |y| \leq w/2 \\ 0 & \text{elsewhere} \end{cases} \tag{15}$$

Assuming that the antenna is sufficiently long and the variation of α from period to period is sufficiently gradual, the leakage constant is treated here as a continuous function rather than a step function of x. The directivity gain and the E-plane radiation pattern associated with aperture illumination (15) are obtained from conventional antenna theory as

$$G_D = \left(\frac{2}{\pi}\right)^3 k_0^2 w \, \frac{\left(\int_0^L F(x)\,dx\right)^2}{\int_0^L F(x)^2 dx} \cos\phi_{-1} \tag{16a}$$

$$T(\sin\phi\sin\theta) = \left| \frac{\displaystyle\int_0^L F(x)\exp\{jk_0 x(\sin\phi\sin\theta - \sin\phi_{-1})\}\, dx}{\displaystyle\int_0^L F(x)\, dx} \right|^2 \tag{16b}$$

The H-plane pattern is independent of $F(x)$ and, as before, determined by (13b).

When the E-plane pattern $T(\sin\phi\sin\theta)$ is prescribed, the aperture illumination $F(x)$ is, in principle, a known function of x, which can be determined from (16b), to good accuracy, by standard methods. The distribution of the leakage constant $\alpha = \alpha(x)$, which will produce a given $F(x)$, is found by inversion of (14). It is not difficult to show that

$$\alpha(x) = \frac{1}{2}\frac{F^2(x)}{F(0)^2/2\alpha(0) - \displaystyle\int_0^x F^2(x)\, dx} = \frac{1}{2}\frac{F^2(x)}{(1/\eta)\displaystyle\int_0^L F^2(x)\, dx - \displaystyle\int_0^x F^2(x)\, dx} \tag{17}$$

where η is the antenna efficiency, i.e., the ratio of the radiated power to the input power of the antenna. The variation of groove depth or aspect ratio from period to period, which will yield a given α profile, can be read from the curves $\alpha = \alpha(t/\lambda_0)$ and $\alpha = \alpha(d_1/d)$ as they are shown in Figs. 30 and 32 for the example of an antenna with $\epsilon_r = 12$ and $d = 0.25\lambda_0$. Together with (17) these curves provide the design information necessary for pattern shaping of dielectric grating antennas.

Fig. 39 shows an example. An aperture illumination with a cosine-squared amplitude taper

$$F(x) = \cos^2\left[\frac{\pi}{2}\left(\frac{2x}{L} - 1\right)\right], \qquad 0 \le x \le L \tag{18}$$

is a desirable aperture distribution; it produces a radiation pattern with side lobes below -30 dB. The function $F(x)$ is indicated in Fig. 39a. Corresponding α profiles for antennas with efficiencies of 75 and 95 percent are shown by the solid curves in Figs. 39b and 39c, respectively. Note that these curves do not show the same symmetry as the aperture illumination $F(x)$. The energy density of the field traveling along the antenna decreases with x due to radiation, and the maximum of the α profile must be shifted to the right to compensate for the decreasing field amplitude and, thus, to produce the desired symmetric aperture illumination.

Assuming that the α profile of Fig. 39b is realized by varying the aspect ratio of the surface corrugation, and the α profile of Fig. 39c is obtained by tapering the groove depth, the dashed curves in these figures show the corresponding d_1 and t variations along the antenna aperture. The curves were obtained from Figs. 30b and 32, and they apply to a silicon antenna having a total height $h + t = 0.25\lambda_0$ and a length $L = 25\lambda_0 = 100d$. The main beam direction ϕ_{-1} is $-47°$. As in the case of α, the d_1 and t profiles are shown as continuous rather than as step functions of x. Fig. 39c suggests that a simple triangular groove depth distribution has the capability of providing very low side lobes.

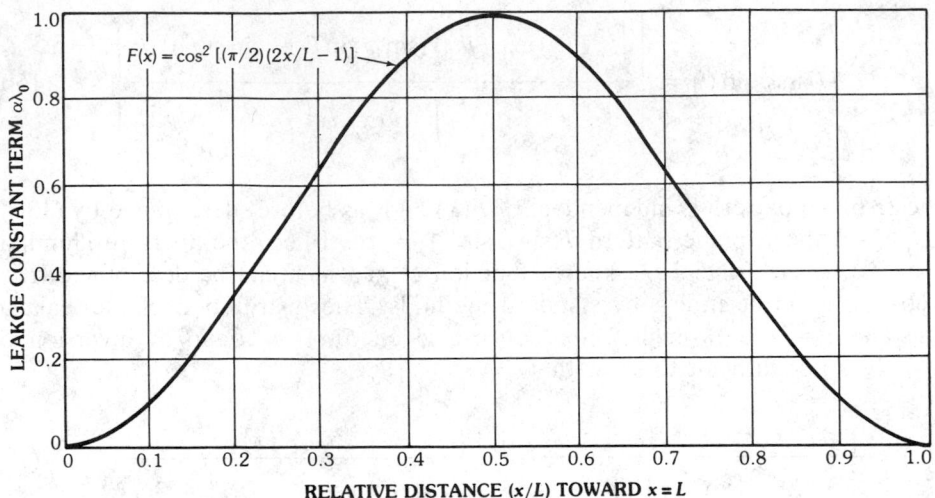

a

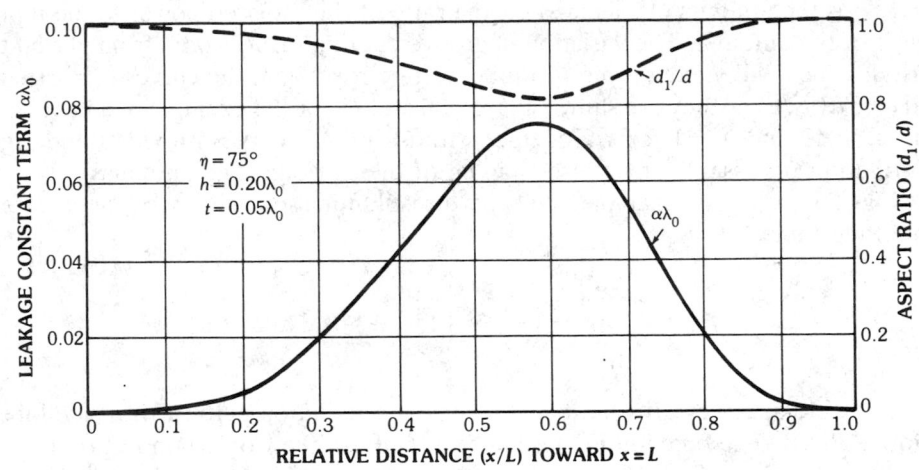

b

Fig. 39. Control of amplitude distribution in aperture plane of periodic dielectric antenna. (*a*) Cosine-squared amplitude distribution. (*b*) Leakage constant α and aspect ratio d_1/d versus x for antenna of efficiency $\eta = 75$ percent. (*c*) Leakage constant α and groove depth t/λ_0 versus x for antenna of efficiency $\eta = 95$ percent.

Close inspection shows that the radiation angle ϕ_{-1} changes slightly with d_1 and t. These deviations could, in principle, be compensated by introducing an appropriate moderate variation in the period of the surface corrugation as well. But in view of the other approximations involved in the pattern calculations, it is not clear how much improvement in pattern quality could be gained by such refinements.

While grating antennas produce *E*-plane patterns of narrow beamwidth, their *H*-plane patterns are usually not very directive since the width *w* of the antennas is

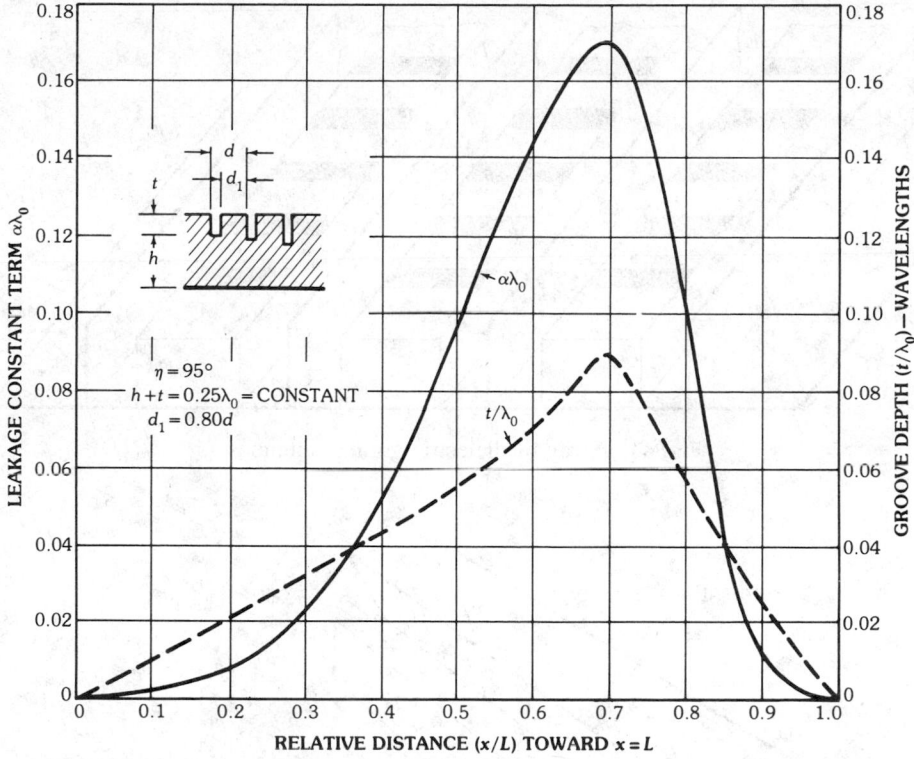

RELATIVE DISTANCE (x/L) **TOWARD** $x = L$

c

Fig. 39. *Continued*

limited to prevent the occurrence of higher-order modes. A small beamwidth in the *H*-plane can be realized by using an array of several antennas in parallel as indicated in Fig. 40. Separating the antennas by metal fins will minimize mutual coupling. The use of arrays has the additional advantage that the excitation of the individual antennas can be controlled independently, thus permitting scanning of the array pattern both in the *E*-plane (by frequency variation) and in the *H*-plane (by phasing).

An alternate method for enhancing *H*-plane directivity has been suggested by Trinh and others [130, 131]. A single grating antenna is embedded in a rectangular metal trough with flares attached to each side. The resulting *H*-plane sectoral horn configuration is shown in Fig. 41; the dimensions apply to *E*-band (81.5 GHz) where this concept was tested. Directivity can be maximized by appropriate choice of the flare angle γ, which should be approximately 15° in the case of near broad-side radiation. The experimental antenna used a dielectric guide of permittivity $\epsilon_r = 2.47$ with a grating of 32 metal strips and a guide wavelength of 2.7 mm. The gain was measured to be 29 dB and side lobes were at least -25 dB below the main beam. The *E*- and *H*-plane half-power beamwidths were 4° and 13°, respectively.

A closely related approach which permits one to reduce the *H*-plane beamwidth even further would use a periodic antenna of small width as a feed for a

Fig. 40. Array of dielectric grating antennas.

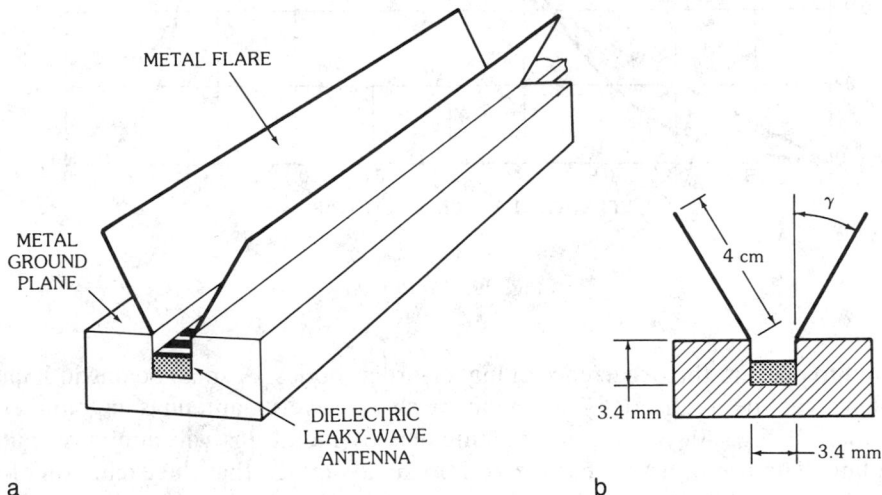

a

b

Fig. 41. Dielectric grating antenna of enhanced *H*-plane directivity. (*After Trinh, Mittra, and Paleta [130], © 1981 IEEE*)

reflector antenna. The reflector in this case would be a parabolic cylinder with the dielectric antenna placed along its focal line. Antennas of this general type have been investigated by Ore [158, 159].

Omnidirectional Dielectric Grating Antennas—Omnidirectional versions of periodic dielectric antennas are shown in Fig. 42 [119, 160, 161]. The antennas possess rotational symmetry about the vertical axis and are assumed to be fed by uniform dielectric waveguides of circular cross section. Fig. 42a shows an antenna with a dielectric grating in the form of a periodic surface corrugation, and Fig. 42b an antenna with a metal grating consisting of a sequence of equally spaced metal rings. Design versatility can be enhanced by combining both options, i.e., by the

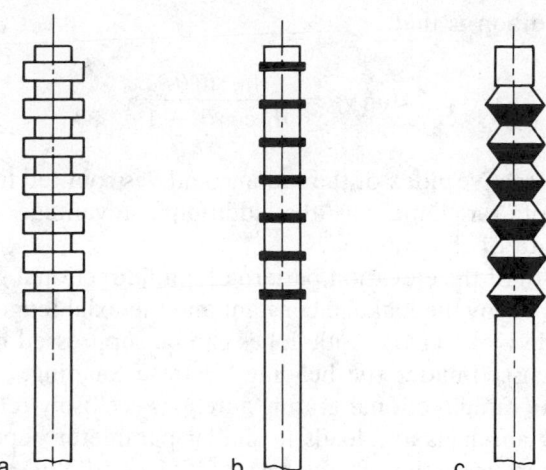

Fig. 42. Omnidirectional dielectric grating antennas. (*a*) Dielectric grating. (*b*) Metal grating. (*c*) Partially metallized dielectric grating. (*After Ore [159], © 1972 IEEE*)

use of an antenna with a partially metallized dielectric grating as indicated in Fig. 42c. Such antennas have been studied by Ore [158, 159].

An omnidirectional radiation pattern is obtained when the antennas are excited in the circular symmetric TM_{01} or TE_{01} modes of the feeding waveguide. Mode TM_{01} operation leads to vertical polarization, TE_{01} operation to horizontal polarization of the radiated field. As an alternative the antennas may be excited in the fundamental HE_{11} mode, provided this mode is circularly polarized. The radiation field will then show elliptical polarization.

The main beam of these antennas has a conical shape. The axis of the radiation cone coincides with the antenna axis, and the apex angle, being determined by (7), varies with frequency; scanning from backfire over broadside to end-fire is, in principle, possible. As with all grating antennas these omnidirectional antennas develop a "stopband" when broadside conditions are approached, i.e., in theory they do not radiate in the azimuth plane and in practice an increased vswr will result.

When a pattern maximum in the horizontal plane is desired, which is a common requirement for omnidirectional antennas, this deficiency can be corrected by the technique discussed under the heading "Beam Scanning," which utilizes a surface modulation consisting of *pairs* of grooves or metal rings, spaced $\lambda_g/4$ apart, to reduce the reflection coefficient of the antenna. An alternate solution would employ a conically shaped radome in which the antenna is embedded such that its axis coincides with the cone axis. It is assumed here that the radome consists of a solid cone of refractive material and that the apex of the cone is at the upper end, its base at the lower end (ground plane) of the antenna. The antenna may then be designed for radiation under an oblique angle (within the radome); but refraction at the radome-to-air interface will redirect the emanating beam into the horizontal direction, provided the aperture half-angle γ_R of the radome is appropriately related to the radiation angle θ of the antenna field within the radome. The condition is that

$$\tan \gamma_R = \frac{n_R \sin \theta}{n_R \cos \theta - 1}$$

where n_R is the refractive index of the radome and θ is counted from the horizontal plane. The use of a radome has the additional advantage of enhancing the mechanical strength of the antenna.

The beamwidth of the elevation pattern of omnidirectional grating antennas is determined primarily by the leakage constant and the axial length of the antennas, and it can be made very narrow. Side lobes can be suppressed by pattern shaping techniques as discussed under the heading "Pattern Shaping."

The theory of omnidirectional grating antennas is closely related to the theory of planar periodic antennas and leads to similar parameter dependencies and the same general design principles. In the case of TM_{01} and TE_{01} excitation the field distribution of the antennas can be described in terms of TM or TE waves alone, which simplifies the analysis considerably.

Uniform-Waveguide Leaky-Wave Antennas

As is the case for tapered-rod and periodically perturbed dielectric antennas, the radiating structures discussed here are derived from open millimeter waveguides. They differ from the above-mentioned antennas in that these guiding structures are maintained *uniform* in the longitudinal direction. It is true that for many applications the aperture dimensions of these antennas will be tapered slightly in order to control the radiation side lobes, but we must understand that it is not necessary to taper these structures or to modulate them periodically in order to produce the radiation.

The radiation is the result of leakage from these open millimeter waveguides; depending on the guiding structure this leakage can be produced in two basic ways. One way is to *perturb* the structure *longitudinally* in a uniform fashion, thereby transforming an initially purely bound mode into a leaky mode. The second way is to employ a *higher mode* on the open waveguide. The dominant mode is generally purely bound, but for certain guides, higher modes, including the lowest mode of the opposite polarization, may leak [162, 163, 172].

An important advantage associated with these uniform leaky-wave antennas is their *simplicity*, permitting easier fabrication at the higher millimeter-wave frequencies where the antenna dimensions become very small. In fact, these antennas have the outward appearance of a uniform open waveguide (except for the slight taper which controls the side lobe level and distribution). On the other hand, these uniform leaky-wave antennas are less versatile than the periodic leaky-wave antennas, since they can radiate into the forward quadrant only, whereas the periodic structures can radiate into the backward quadrant as well and even radiate near the broadside direction.

In principle, uniform leaky-wave antennas may be based on any one of a variety of open waveguides. So far, however, only three waveguides have been considered for this purpose: groove guide, NRD guide, and microstrip line. The first two waveguides are *low-loss* guides specially designed for millimeter wavelengths; they are therefore particularly suitable for the shorter millimeter wavelengths, where the waveguide material losses may otherwise be so great as to compete with the leakage loss, thereby distorting the antenna performance and reducing antenna efficiency. We should recall that moderately lossy waveguides are still acceptable for waveguide circuit components which are only a wavelength or so long, but they may not be suitable for leaky-wave antennas, which may typically be $20\lambda_0$ to $100\lambda_0$ long, where λ_0 is the free-space wavelength. The third waveguide, microstrip line, is not a low-loss guide, but the leaky-wave antenna based on it is so simple that its possible use is worth further consideration.

Each of these leaky-wave antennas may be used as a line-source element in a linear array of such elements arranged in parallel, providing a two-dimensional scanning antenna where the scanning in one direction (the leaky-wave direction) can be controlled electronically or by changing the frequency, and in the orthogonal direction by using phased array methods. This type of phase/frequency two-dimensional scanning antenna was also discussed briefly in connection with the periodically grooved dielectric waveguide treated in the preceding subsection. Such two-dimensional arrays for millimeter waves have not yet been built or tested, and are mentioned here only as a concept likely to be useful.

In view of the small size of waveguiding structures at millimeter wavelengths, and the fabrication difficulties at these wavelengths associated with more conventional line source antennas, such as slot arrays, we believe that leaky-wave antennas, of both the uniform and periodic types, form a *natural* class of antennas for millimeter waves.

Our discussion of uniform leaky-wave antennas follows the waveguides on which the antennas are based. First, we treat the *groove guide*, where leakage can be produced in both of the ways mentioned above, namely, by a uniform longitudinal perturbation and by the use of a higher mode. Second, we consider the *NRD* (nonradiative dielectric) *guide*, where the leakage is obtained by two basically different perturbation methods. Third, we describe an antenna based on *microstrip line*, where the leakage is produced when the *first higher mode* operates in a narrow frequency range near cutoff.

Leaky-Wave Antennas Based on the Groove Guide—Groove guide is one of several waveguiding structures proposed for millimeter wavelengths about 20 years ago in order to overcome the higher attenuation occurring at these higher frequencies.

The cross section of the groove waveguide is shown in Fig. 43, together with an indication of the electric field lines present; the field is in effect vertically polarized. One should note that the structure resembles that of rectangular waveguide with most of its top and bottom walls removed. Since the currents in these walls contribute significantly to the waveguide losses, the overall attenuation of groove waveguide is less than that for rectangular waveguide. The reduced attenuation loss will therefore interfere negligibly with the leakage loss of any leaky-wave antenna based on groove guide.

The greater width in the middle, or central, region was shown by T. Nakahara [164, 165] to serve as the mechanism that confines the field in the vertical direction, much as the dielectric central region does in H guide. The field thus decays

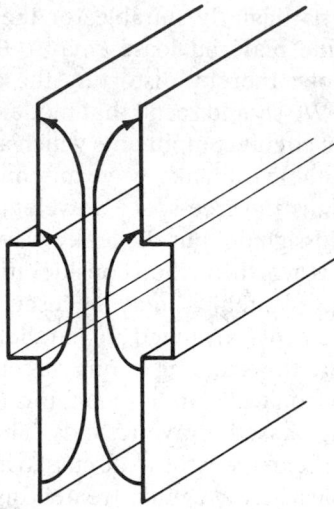

Fig. 43. Cross section of symmetrical, nonradiating groove waveguide. (*After Lamporiello and Oliner [170],* © *1985 IEEE*)

exponentially away from the central region in the narrower regions above and below. If the narrower regions are sufficiently long, it does not matter if they remain open or are closed off at the ends.

Work on the groove guide progressed in Japan and in the United States until the middle 1960s, but then stopped until it was revived and developed further by D. J. Harris and his colleagues [166, 167] in Wales. The first contribution to *antennas* based on the groove waveguide was made by Oliner and Lampariello [168, 169], who proposed and analyzed the first of the leaky-wave antennas to be described. A pair of papers [170, 171] that present the theory and design considerations in detail have also appeared.

(*a*) *Antenna employing an asymmetric strip*—The first of these leaky-wave antennas is shown in Fig. 44. The basic difference between the structures in Figs. 43 and 44 is that in Fig. 44 a *continuous metal strip* of narrow width has been added to the guide in *asymmetrical* fashion. Without that strip the field of the basic mode of the symmetrical groove waveguide is evanescent vertically, so that the field has decayed to negligible values as it reaches the open upper end. The function of the asymmetrically placed metal strip is to produce some amount of net *horizontal* electric field, which in turn sets up a mode akin to a TEM mode between parallel plates. The field of that mode propagates all the way to the top of the waveguide, where it leaks away. It is now necessary to close up the bottom of the waveguide, as seen in Fig. 44, to prevent radiation from the bottom, and (nonelectrically) to hold the structure together. Of course, the upper walls could end as shown in Fig. 44 or they could attach to metal flares or a horizontal ground plane.

We now have available a leaky-wave line-source antenna of simple construction. The value of the phase constant β of the leaky wave is governed primarily by the properties of the original unperturbed groove guide, and the value of the leakage constant α is controlled by the width and location of the perturbing strip.

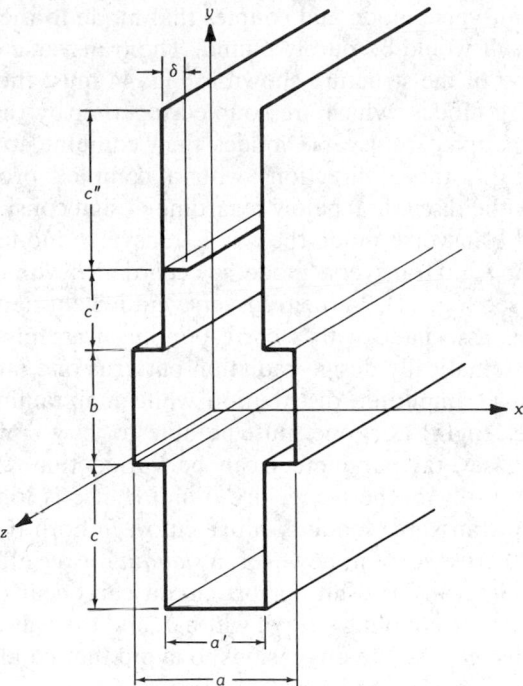

Fig. 44. Leaky-wave antenna derived from groove guide by the introduction of an asymmetric, continuous metal strip. (*After Lamporiello and Oliner [170], © 1985 IEEE*)

As with any leaky-wave antenna one can, by suitably changing the dimensions and the frequency, obtain a variety of scan angles and leakage constants. Let us choose a *typical case* where $a'/a = 0.70$, $b/a = 0.80$, $c/a = 1.215$, $c'/a = 0.145$, $c''/a = 1.50$, $\delta/a = 0.21$, $\lambda_0/a = 1.20$. For this particular set of dimensions, calculations [169] yield $\beta/k_0 = 0.749$ and $\alpha/k_0 = 6.24 \times 10^{-3}$. This value of leakage constant α yields a leakage rate of about 0.34 dB per wavelength, resulting in an antenna about $30\lambda_0$ long if, as is customary, 90 percent of the power is to be radiated, with the remaining 10 percent dumped into a load. The resulting beamwidth of the radiation is approximately 2.9°, and the beam radiates at an angle of about 49° to the normal. At a frequency of 50 GHz, for example, dimension a would be 0.50 cm and the antenna would be 18 cm long.

The leaky-wave antenna has been analyzed [170] by deducing the proper transverse equivalent network, deriving simple closed-form expressions for the various parameters of this network, and then obtaining the dispersion relation for the complex propagation constant of the leaky mode from the lowest resonance of the transverse equivalent network. The resulting expression for the complex propagation constant is also in closed form. From this expression, numerical calculations were made [171] of the antenna's performance characteristics, and their dependence on the various dimensional parameters of the antenna.

As indicated above, it is the added continuous strip of width δ that introduces *asymmetry* into the basic groove guide and creates the leakage. The strip gives rise to an additional transverse mode and couples that mode to the original transverse mode, which by itself would be purely bound. The transverse equivalent network for the cross section of the structure shown in Fig. 44 must therefore be based on these two transverse modes, which are coupled together by the narrow asymmetrical strip. These coupled transverse modes then combine to produce a net TE longitudinal mode (in the z direction) with a complex propagation constant, $\beta - j\alpha$. To assist in the discussion below regarding design considerations, let us call the *original* bound transverse mode the $i = 1$ transverse mode and the *additional* transverse mode the $i = 0$ transverse mode, in accordance with their field variations in the x direction (see Fig. 44). In the transverse equivalent network, each of these transverse modes is associated with a corresponding transmission line.

In order to systematically design radiation patterns one must be able to taper the antenna aperture amplitude distribution while maintaining the phase linear along the aperture length, i.e., one must be able to vary α while keeping β the same. Fortunately, several parameters can be varied that will change α while affecting β hardly at all; the best ones are δ and c, if c is long enough.

Since the $i = 0$ transverse mode is *above* cutoff in both the central and outer regions of the guide cross section, however, a *standing-wave* effect is present in the $i = 0$ transmission line. As a result a short circuit can occur in that transmission line at the position of the coupling strip of width δ, and the value of α then becomes zero. Hence we must choose the dimensions to avoid that condition, and in fact to optimize the value of α.

In the design one first chooses the width a and adjusts a' and b to achieve the desired value of β/k_0, which is determined essentially by the $i = 1$ transverse mode. That value of β/k_0 immediately specifies the angle of the radiated beam. It is then recognized that the value of α can be increased if the coupling strip width δ is increased, or if the distance c' between the step junction and the coupling strip is decreased, since the coupling strip is excited by the $i = 1$ transverse mode, which is evanescent away from the step junction in the outer regions. After those dimensions are chosen, the length c must be determined such that the standing-wave effect mentioned above optimizes the value of α. If c is sufficiently long, it will affect only the $i = 0$ transmission line and influence β negligibly. The length c'' also affects α strongly and β weakly, and it also must be optimized because another, although milder, standing wave exists between the coupling strip and the radiating open end.

We present here, in Figs. 45a and 45b respectively, a curve of α/k_0 as a function of c', and the value of $c + c'$ that must be selected so as to achieve the value of α determined from Fig. 45a. In effect, therefore, Fig. 45b indicates the value of c required once α (via c') is specified. It is interesting to note that $c + c'$ is almost constant for optimization. The curves in Fig. 45 apply to an antenna with $b/a = 0.8$, $a'/a = 0.7$, $c''/a = 1.5$, $\delta/a = 0.21$, and $\lambda_0/a = 1.2$.

It is important to realize that the dimensions for optimization are *independent of frequency*, since the transverse wave numbers are all frequency independent. Of course, when the frequency is altered the values of β and α will change, but the dimensional optimization is undisturbed. In fact, for the dimensions discussed above, the radiated beam can be scanned with frequency from about 15° to nearly 60° from the normal before the next mode begins to propagate.

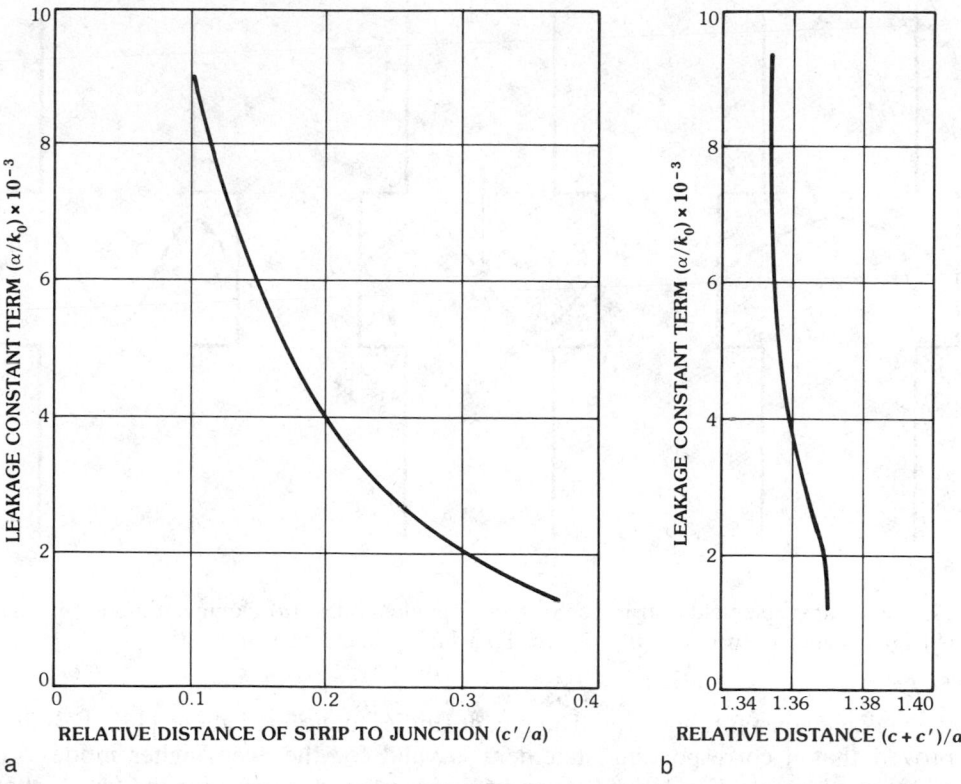

a   b

Fig. 45. Design of groove guide leaky-wave antenna employing asymmetric metal strip. (a) Leakage constant α/k_0 (optimum value) as function of distance c'/a of perturbing strip from the step junction. (b) Value of $(c' + c)/a$ required to achieve the optimum value of α/k_0. (*After Lamporiello and Oliner [170], © 1985 IEEE*)

The leaky-wave antenna discussed above is, therefore, straightforward to understand and amenable to systematic design. It is also sufficiently flexible; a reasonably wide range of pointing angles and beamwidths can be achieved by appropriate adjustment of its dimensional parameters.

(b) *Antenna based on the use of leaky higher modes*—Recent studies by Lampariello and Oliner [172, 173] have shown that only the dominant mode of groove guide is purely bound, and that all of the higher modes of that guide are *leaky*. These higher modes thus provide the basis for new leaky-wave antennas.

In particular, interesting results were obtained [172, 173] for the first higher even mode and the first higher odd mode. These modes are distinguished from the dominant mode by their transverse variations (in the x direction in Fig. 46). The dominant ($n = 1$), first higher even ($n = 2$), and first higher odd ($n = 3$) modes possess, respectively, a half sine wave, a full sine wave, and a three-halves sine wave variation across the width a.

An earlier paper by Nakahara and Kurauchi [165] examined the simple relations among the wave numbers for the odd higher modes and demonstrated that these modes must be leaky. These authors did not indicate the magnitude of

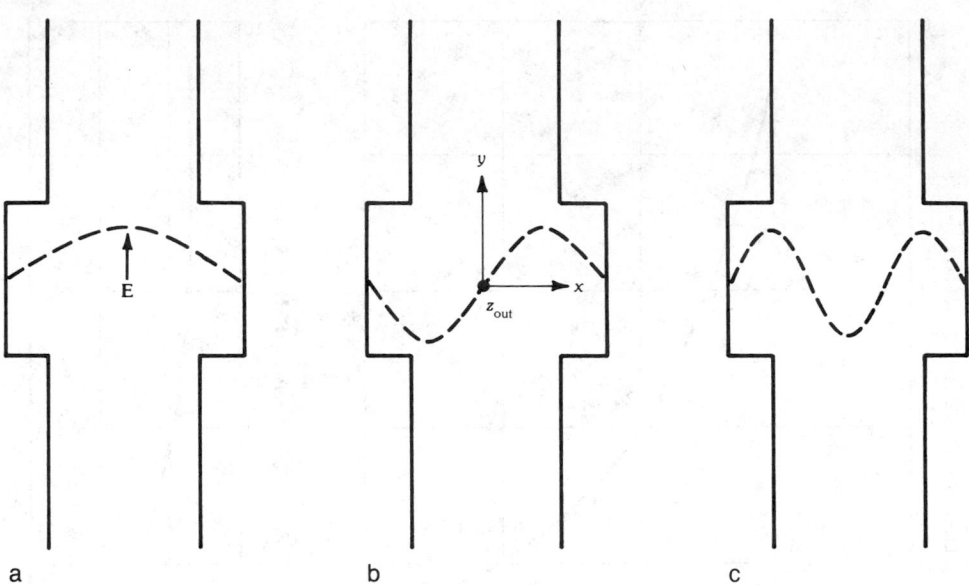

Fig. 46. Transverse field distribution of groove guide modes. (*a*) Dominant mode (*n* = 1). (*b*) First higher even mode (*n* = 2). (*c*) First higher odd mode (*n* = 3).

the leakage constant, however. The recent studies mentioned above [172, 173] first proved that a corresponding statement is valid for the even higher modes. In addition, these studies derived transverse equivalent networks for the first higher odd mode (*n* = 3) and the first higher even mode (*n* = 2). From these transverse equivalent networks dispersion relations were derived in the form of transcendental relations all of whose constituents are in a simple closed form. The leakage constants for each higher mode can be found readily from these dispersion relations.

Let us first consider the propagation behavior of the first higher *odd* mode. To understand that behavior we must recognize that the cross-section dimensions must be large enough to permit both the dominant and the first higher odd longitudinal modes to propagate. Looking in the *y* direction in Fig. 46, we then see that the guide can be initially excited such that the incident power is basically either in the *i* = 1 transverse mode, for which the *x* dependence is a half sine wave, or in the *i* = 3 transverse mode, for which the *x* dependence contains three half sine waves. These excitations result in the *n* = 1 and *n* = 3 longitudinal modes, respectively. In either case, we observe that at the step junction an *i* = 1 or *i* = 3 transverse mode will couple to all other transverse modes of the same symmetry; for example, the *i* = 3 mode will couple to all of the *i* = 1, 5, 7, ... transverse modes. In the transverse equivalent network, the *i* = 1 and *i* = 3 transverse modes are separated out, and separate transmission lines are furnished for each of them. The coupling susceptance between the transmission lines was derived using small obstacle theory in a multimode context.

When groove guide is excited in *dominant mode* fashion, the *i* = 1 transmission line is propagating transversely in the central region of width *a*, but evanescent in the outer narrower regions of width a'. Also, it can be shown that the $i = 3$ transmission line is below cutoff everywhere, so that the dominant longitudinal mode is purely bound. On the other hand, *when the groove guide is excited in the first higher odd longitudinal mode*, corresponding to the $i = 3$ transverse mode, the $i = 3$ transmission line is propagating in the central region but evanescent in the outer regions. But, the $i = 1$ transmission line can now be shown to be propagating in *both* the central and the outer regions. The result is that the first higher odd mode is *leaky*, but with the interesting feature that the energy that leaks has the variation in x of the *dominant* mode, not of the first higher mode. These coupled transverse modes combine to produce a net TE longitudinal mode (in the z direction) with a complex propagation constant, $\beta - j\alpha$.

Some numerical results are presented next for the behavior of the phase constant β and the attenuation constant α of the leaky mode that results when the groove guide is excited in the first higher odd longitudinal mode (the $i = 3$ transverse mode). In Figs. 47a and 47b we note the variation of β and α as a function of frequency. We see from Fig. 47a that β is almost linear with frequency at the higher frequencies, but shows substantial curvature near cutoff. The variation of α with frequency, in Fig. 47b, is seen to be almost hyperbolic at the higher frequencies; nearer to cutoff, α is seen to rise substantially.

These variations follow directly from the wave-number relationship

$$k^2 = k_x^2 + k_y^2 + k_z^2$$

where k_x and k_y are determined by the waveguide dimensions and $k_z = \beta - j\alpha$. By taking the real part, and noting that the transverse wave numbers are independent of frequency, we find that β is approximately linearly proportional to the frequency

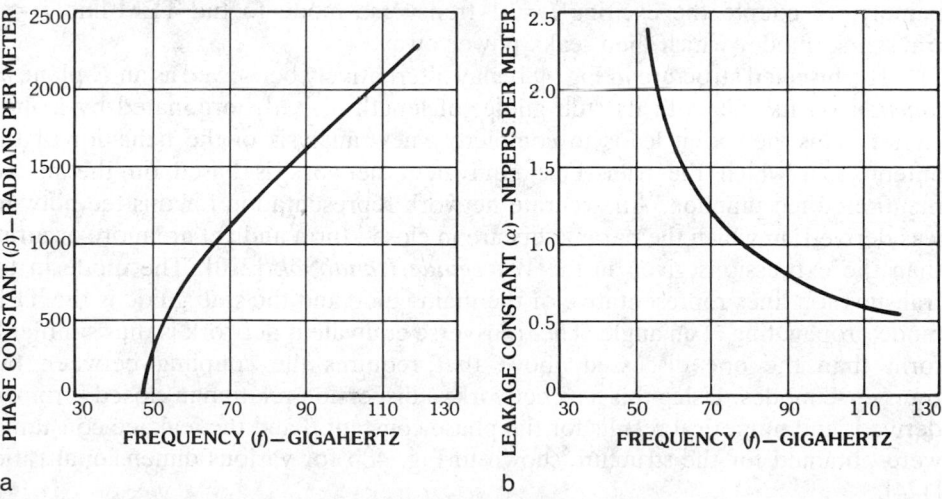

a   b

Fig. 47. Variations with frequency of phase constant and leakage constant of groove guide excited in first higher odd longitudinal mode. (*a*) Phase constant β. (*b*) Leakage constant α. (*After Lamporiello and Oliner [173], © 1983 IEEE*)

when α is small, which occurs for the higher frequencies. Such behavior is in agreement with that in Fig. 47a. When we take the imaginary part, we find that the product $\alpha\beta$ should remain independent of frequency. In the frequency range for which β is proportional to frequency, we thus find that α must vary as the reciprocal of the frequency, in agreement with Fig. 47b. For Fig. 47 the guide dimensions are $a = 1.0$ cm, $a'/a = 0.7$, and $b/a = 0.4$.

The behavior described above applies to the spectrum of *odd* higher modes. Qualitatively similar leakage behavior is found for the *even* higher modes, although certain important differences occur. The step junction then couples only the even transverse modes, of course; in particular, for the first higher even ($n = 2$) mode, the $i = 2$ transverse mode will couple at the step junction to the $i = 0$ transverse mode, which is a TEM-like mode. Thus, when we excite the $n = 2$ mode in the groove guide it will leak, but the transverse form of the leakage is *TEM-like*, and not of the form of the exciting second mode. In a sense this leakage behavior is similar to that found for the first higher odd ($n = 3$) mode, but the polarization of the electric field of the leakage energy is *horizontal* here whereas it was *vertical* there.

This difference in the polarization of the leakage energy is very interesting because it can form the basis for two new types of leaky-wave antenna, with vertical and horizontal polarization, respectively.

An additional interesting observation can be made with respect to the even higher mode case. As shown in Fig. 48a, the vertical midplane is an electric wall, or short-circuit wall, by virtue of the symmetry of the excitation. The waveguide can therefore be bisected, and a metal wall can be placed along the electric-wall midplane without affecting the electric field distribution in either bisected half. The resulting structure, of half the width, as shown in Fig. 48b, may now be viewed as an *asymmetric* groove guide of normal width. That asymmetry serves the same function as the continuous asymmetric strip in the antenna shown in Fig. 44, namely, to couple the exciting $i = 1$ transverse mode to the TEM-line $i = 0$ transverse mode, which then leaks power away.

The bisected structure in Fig. 48b may alternatively be viewed as an *E*-plane *tee junction* on its side with its stub guide, of length $a - a'$, terminated by a short circuit. This viewpoint leads to completely new analysis of the behavior of this antenna, in which the transverse equivalent network is based on the above-mentioned tee junction. An accurate network representation for this tee network was derived, in which the parameters are in closed form and yet are more accurate than the expressions given in the *Waveguide Handbook* [220]. The mode in the transmission lines representative of the main guide and the stub guide is the TEM mode propagating at an angle. The transverse equivalent network is thus simpler in form than the one discussed above that requires the coupling between two transverse modes. Using this new network, a dispersion relation in closed form was derived, and numerical results for the phase constant β and the leakage constant α were obtained for the structure shown in Fig. 48b for various dimensional ratios [174].

It was found that the values of α for this structure are somewhat larger than one would like, with the result that narrow beams are not readily obtained. To overcome this limitation, two additional modifications were introduced, as

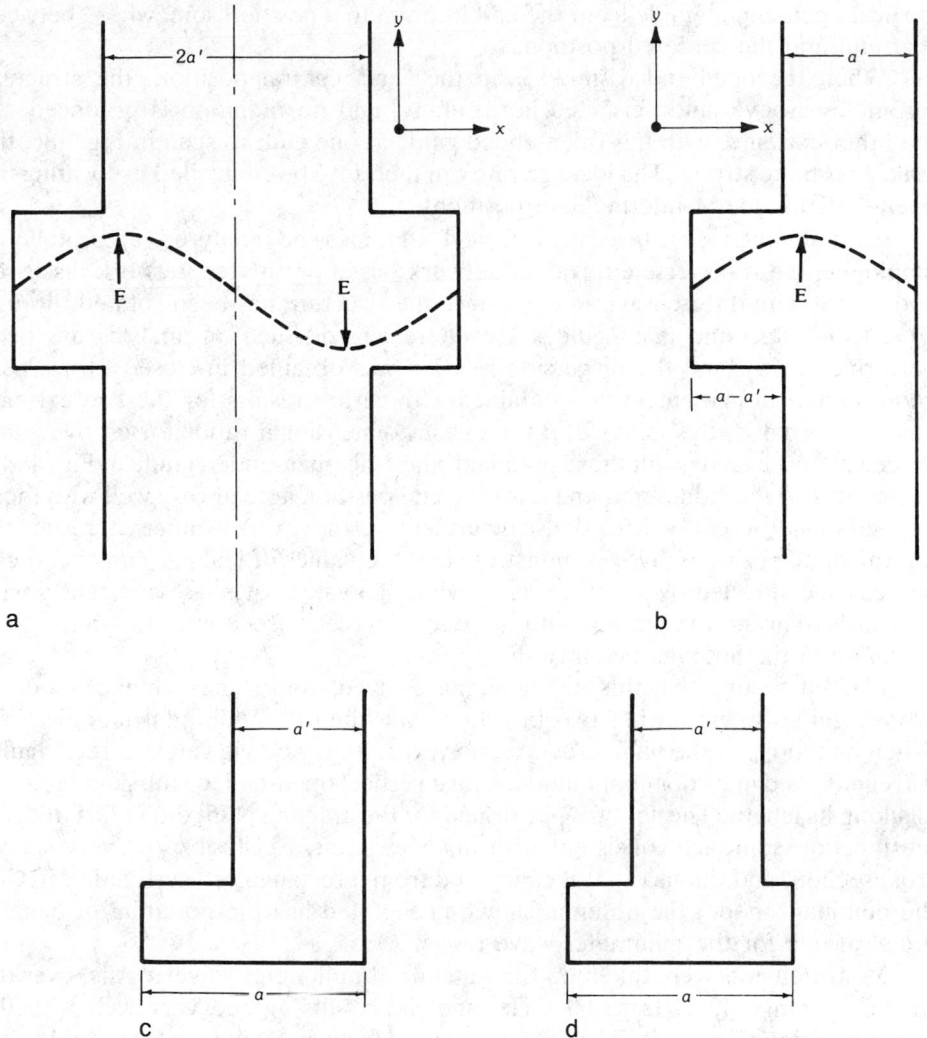

Fig. 48. Groove guide excited in the first higher even ($n = 2$) mode. (*a*) Full guide, showing electric wall symmetry for the vertical midplane. (*b*) Bisected guide, corresponding to an asymmetric groove guide excited in the dominant mode. (*c*) Further bisection of the structure in (*b*), using the fact that the horizontal bisection plane is an electric wall. (*d*) Modification of the structure in (*c*) in which the open stub guide is moved nearer to the central position.

shown in Figs. 48c and 48d. The first of these modifications, in Fig. 48c, produces a structural change but no change in the electrical performance except for some small additional metal loss; it is based on the recognition that, because of symmetry, the horizontal line bisecting the stub guide in Fig. 48b is an electric wall. This structure now resembles an L where the vertical portion of the L is an open-ended guide that permits the leakage of radiation. The second change, shown in Fig. 48d, moves the vertical open-ended guide from the end location to a position somewhere between that end and the centered position.

When the open-ended guide is in the exact central position, the structure resembles groove guide bisected horizontally, and no radiation is produced. At the other extreme, with this open-ended guide at one end, as seen in Fig. 48c, the leakage is quite strong. The leakage rate can therefore be controlled by locating the open-ended guide at intermediate positions.

The modified structure in Fig. 48d can also be analyzed accurately by employing the transverse equivalent network based on the tee network discussed above, but with different parameters since the structure has been rotated through 90° so that stub and main guides are interchanged. Such an analysis has been performed [175], and the dispersion relation was obtained in closed form. Very good numerical agreement was obtained with earlier results for the two extreme positions. That is, the values of β for various dimensional ratios for groove guide agreed almost exactly with those obtained when the open-ended guide in Fig. 48d is centered, and the values of β and α for the end position agreed very well with those derived in a different way for the structure in Fig. 48c (or b). Numerical results for intermediate positions [175] demonstrate that the values of leakage constant α can indeed be controlled over a wide range while the values of phase constant β vary very little. This new structure, with its added degree of freedom, thus yields great *flexibility* in the antenna beamwidth.

It is interesting that this new antenna evolved from a leaky higher mode in groove guide by employing two bisections and then an additional modification. When one looks at the final structure, however, it may be viewed as a rectangular waveguide, fed in its dominant mode, with a vertical open-ended stub guide present all along its length. The analysis performed for the antenna [175] did in fact utilize a tee junction approach consistent with this viewpoint. Because of its very simple cross section, and the fact that it can be fed from a rectangular waveguide carrying the dominant mode, the antenna shown in Fig. 48d has the potential of being a practical one for the millimeter-wave range.

Measurements were taken on this antenna at millimeter wavelengths, over the frequency range 40 GHz to 60 GHz, and the results agreed very well with the above-mentioned analysis. The antenna was fed from a rectangular waveguide, and measurements were made first of the phase and leakage constants and then of the radiation patterns, with the results verifying the theory well in all cases [221].

Leaky-Wave Antennas Based on the Nonradiative Dielectric (NRD) Guide—Two papers appeared recently [176, 177] which proposed a new type of waveguide for millimeter waves, and showed that various components based on it can be readily designed and fabricated. By a seemingly trivial modification, the authors, T. Yoneyama and S. Nishida, transformed the old well-known H guide, which had languished for the past decade and appeared to have no future, into a practical waveguide with attractive features. The old H guide stressed its potential for low-loss long runs of waveguide by making the spacing between the metal plates large, certainly greater than half a wavelength; as a result, the waveguide had lower loss, but discontinuities or bends would produce leakage of power away from the guide. Yoneyama and Nishida simply observed that when the spacing is reduced to less than half a wavelength all the bends and discontinuities become purely reactive; they therefore call their guide "nonradiative dielectric waveguide," or NRD guide. As a result of this modification many components can be constructed easily, and in an integrated circuit fashion, and these authors proceeded to demonstrate how to fabricate some of them, such as feeds, terminations, ring resonators, and filters.

These papers [176, 177] treat only reactive circuit components, and no mention is made of how this type of low-loss waveguide can be used in conjunction with antennas. In this section, two new types of leaky-wave antenna are described which can be readily fabricated with NRD guide and, in fact, can be directly connected to NRD guide circuits in integrated circuit fashion, if desired. The first of these two antenna types was discussed recently by Sanchez and Oliner [178, 179], and the second in two different talks by Oliner, Peng, and Sheng [180] and by Shigesawa, Tsuji, and Oliner [181]. The leakage mechanisms in these two antenna types are different; in the first antenna type the leakage is produced by foreshortening the metal plates on one side of the guide, whereas in the second the leakage is caused by introducing asymmetry in the dielectric strip cross section. In addition, the polarizations of the radiation are also different, being vertical in the first and horizontal in the second. In both types, however, the basic guided wave must be operated in the fast-wave range.

(*a*) *Antenna produced by foreshortening one side of the guide*—The new waveguide, shown in Fig. 49a, looks like the old H guide except that the spacing between plates is less than half a wavelength to ensure the nonradiative feature. In the vertical (*y*) direction, the field is of standing-wave form in the dielectric region and is exponentially decaying in the air regions above and below. The guided wave propagates in the *z* direction. The leaky-wave antenna based on this waveguide is shown in Fig. 49b; the antenna is created simply by decreasing the distance *d* between the dielectric strip and the top of the metal plates. When the distance *d* is small, the fields have not yet decayed to negligible values at the upper open end, and therefore some power *leaks away*. The upper open end forms the antenna

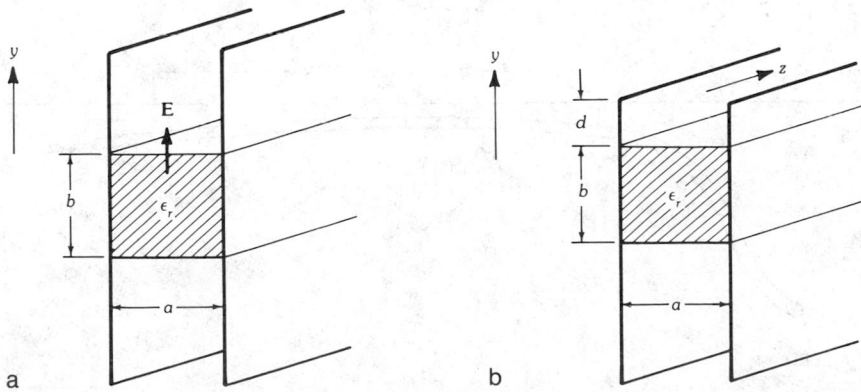

Fig. 49. Leaky-wave antenna derived from nonradiative dielectric (NRD) guide by foreshortening one side of the guide. (*a*) NRD guide. (*b*) Antenna. (*After Sanchez and Oliner [178],* © *1984 American Geophysical Union*)

aperture. The amplitude distribution in this aperture can be controlled by varying the distance d as a function of the longitudinal coordinate z. This may be accomplished either by appropriately shaping the upper edge of the metal plates, or by slightly curving the dielectric strip waveguide so that its distance from the edge of the plates varies in a prescribed fashion with z; the edge in this case would be straight. This leads to a very simple and easy-to-build antenna configuration which is indicated in Fig. 50. Furthermore, the antenna is directly compatible with transmit and receive circuits designed in NRD guide technology. The antenna radiates with vertical polarization.

A leaky-wave antenna *for H guide* of the type shown in Fig. 49b was proposed some years ago by Shigesawa and Takiyama [182, 183]. The principle of operation is identical with that described here; however, the structure presented by them was symmetrical, so that it radiated from both sides, although that was not a necessary feature. On the other hand, their structure would not permit a bending of the dielectric strip to taper the amplitude distribution, as in Fig. 50, because additional radiation would be produced due to the bend. Amplitude tapering could be achieved by keeping the dielectric strip straight and cutting back the open end, however. Additional leakage would occur at the feed and load junctions, in any case, because of the large spacing between the plates. The utilization of the NRD guide instead of the H guide makes this antenna type much more practical.

The NRD guide antenna, in the form shown in Fig. 49b, was analyzed [178, 179] as a leaky waveguide that possesses a complex propagation constant $\beta - j\alpha$, where β is the phase constant and α is the attenuation or leakage constant. An accurate transverse equivalent network for the cross section of the antenna was established, and the dispersion relation for the β and α values was obtained from the resonance of this network. This dispersion relation contains elements all of which are in closed form, thus permitting easy calculation.

The various parametric dependences of α and β on the dimensions and on the dielectric constant have been obtained [184] in order to clarify design information. Here we discuss only a single typical case, for which the parameter values are indicated in Fig. 51. The figure shows the behavior of β and α as a function of the

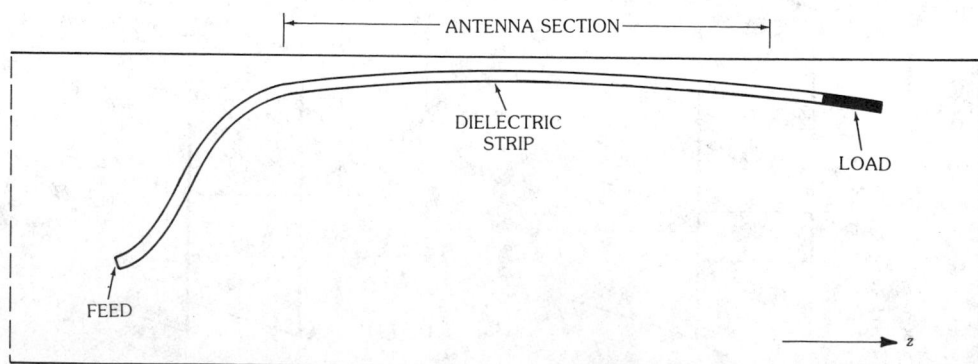

Fig. 50. Side view of NRD waveguide antenna. (*After Sanchez and Oliner [178],* © *1984 American Geophysical Union*)

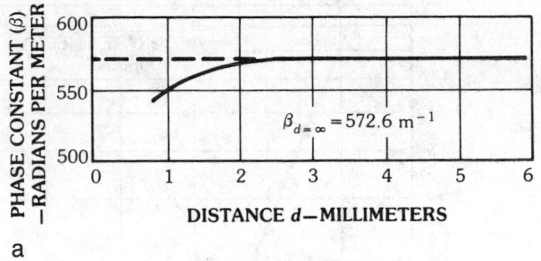

a

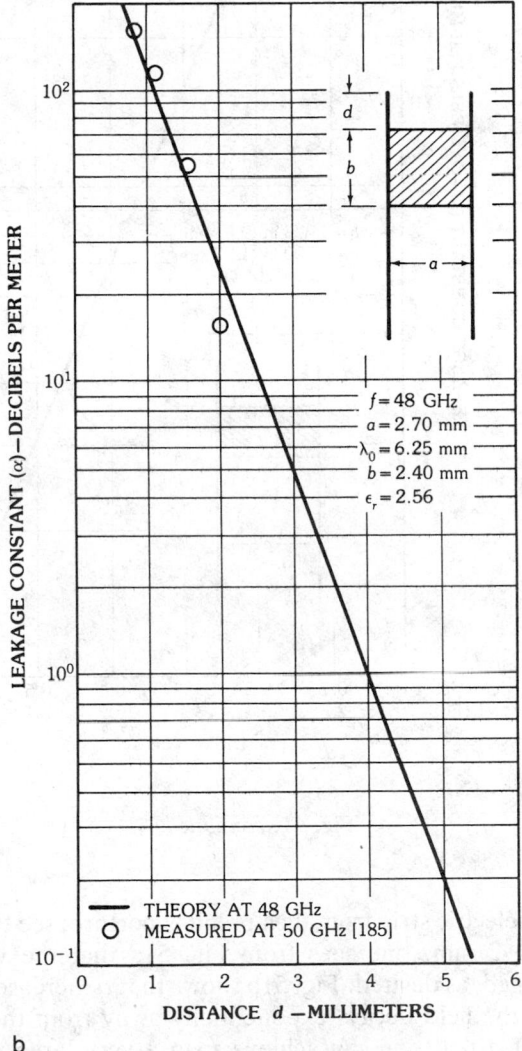

b

Fig. 51. Phase constant and leakage constant of NRD guide antenna as functions of distance of dielectric strip from antenna aperture. (*a*) Phase constant. (*b*) Leakage constant. (*After Sanchez and Oliner [178], © 1984 American Geophysical Union*) (*c*) Comparison with measurements by Han et al. [186]. (*After Han et al. [186], © 1987 IEEE*)

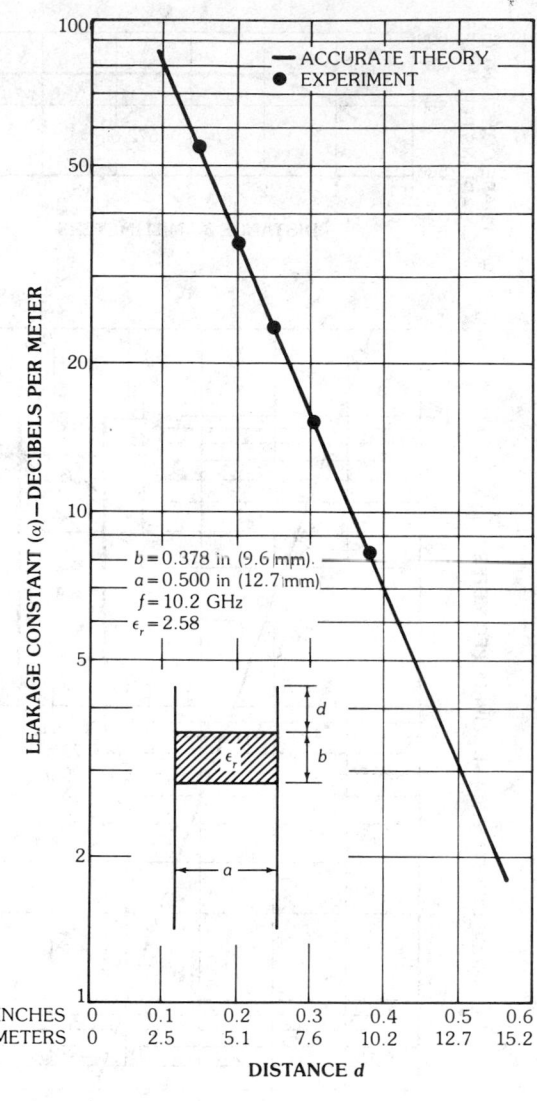

Fig. 51, *continued.*

distance d of the dielectric strip from the antenna aperture; see the inset in Fig. 51b. For distances $d > 2$ mm, one sees from Fig. 51a that the value of β remains essentially unchanged, as desired. Fig. 51b shows that α increases as d is shortened, as expected since the field decays exponentially away from the dielectric region. Thus the values of α that one can achieve span a very large range.

Leaky-wave antennas are usually designed so that 90 percent of the incident power is radiated, and the remaining 10 percent is dumped into a load. Following this criterion, if one selects $d = 2.0$ mm for this geometry, the length of the antenna will be about 40 cm, and the beam will radiate at an angle of about 35° from the normal, with vertical electric field polarization, and with a beamwidth of approximately 1°. A larger (or smaller) value of d will result in a narrower (or wider) beam whose width can be calculated from the curve of α in Fig. 51b.

Measurements of the leakage constant α as a function of distance d have been taken by Yoneyama [185] at a frequency of 50 GHz and by Han, Sanchez, and Oliner [186] at frequencies in the vicinity of 10 GHz on a scaled structure. All of the measurements agree very well with theoretical values. Some of the measurements taken by Yoneyama are superimposed as points on the theoretical values which are represented by the solid line in Fig. 51b. Although the theoretical curve corresponds to a frequency of 48 GHz and the measured points to a frequency of 50 GHz, these frequencies are sufficiently close to permit a comparison between measurement and theory. The scaled measurements by Han and colleagues [186], on the other hand, permitted greater precision and demonstrated excellent agreement with the theory, as seen in Fig. 51c, where the measured points fall almost directly on the theoretical solid curve over a rather large range of values of leakage constant α. Such agreement provides confidence that the theoretical results are reliable for design purposes.

(b) *Antenna produced by asymmetry*—Two antenna structures, based on NRD guide, that leak because of the introduction of asymmetry are shown in Figs. 52a and 52b. The physical mechanism that produces the leakage is the same as that present in the antenna of Fig. 44, where an asymmetric strip was introduced to perturb the symmetry of the open waveguide. In all of these antennas, the asymmetry causes mode conversion to an additional transverse mode of the TEM type that propagates at an angle in the parallel plate region of the cross section, thereby creating radiation polarized with the electric field parallel to the aperture. For the orientation of the structures in Fig. 52 the polarization is horizontal; in contrast, the radiation from the antenna configuration in Fig. 49b is vertically polarized.

The structure in Fig. 52a has been analyzed [180,181] by means of mode

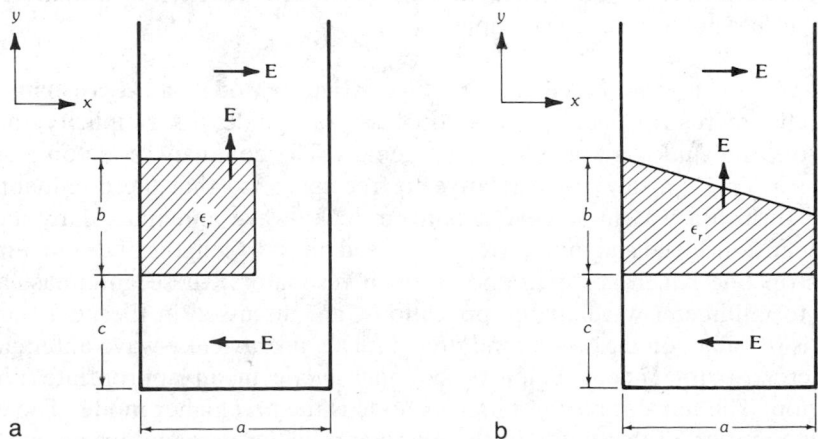

Fig. 52. Two uniform leaky-wave antennas derived from NRD guide by the introduction of asymmetry. (*a*) Antenna with dielectric strip of rectangular cross section and air gap. (*b*) Antenna with dielectric strip of trapezoidal cross section.

matching at the air-dielectric interfaces, and numerical results were obtained as a function of the geometrical parameters. When the NRD guide is operated as a nonradiating waveguide, as in Fig. 49a, the dielectric strip fills the waveguide cross section uniformly. The introduction of a small air gap produces the asymmetry that results in mode conversion to a TEM mode that propagates away at an angle in the outside air-filled parallel plate regions. The air gap does not have to be very large to produce significant leakage; the gap shown in Fig. 52a would yield a large value of α, with a consequent wide beam in the radiation pattern. The geometry can thus be controlled easily to achieve a large range of values for the leakage constant α, and therefore a large range of desired beamwidths.

For certain values of length c in Fig. 52a, additional complications can arise in the radiation behavior. These problems can be avoided by taking care in the selection of c, but most easily by making $c = 0$ and halving the value of b, because the metal plane at $c = 0$ then acts as an imaging mirror. This last modification has the added virtue of a simpler (and perhaps more rugged) structure. In addition, the finite length of open-ended parallel-plate guide on the upper side can cause the introduction of another leaky wave, a modification of the well-known channel guide mode. As a result, interesting coupling effects have been found [181, 221] to occur between the NRD guide leaky mode and this channel guide mode, but the antenna can be designed to avoid such coupling effects so that they need not present a practical difficulty.

In summary, care must be taken in the practical design of this antenna type, and it appears that the simplest structural form is the easiest and safest to design. Accurate theoretical results [180, 181] have been obtained for the phase and leakage behavior, so that such antennas can be designed reliably. By modifying the width of the air gap, a large range of beamwidths can be achieved. No measurements have been taken as yet on this antenna type.

The structure in Fig. 52b, on the other hand, has not been analyzed, but some preliminary measurements [187] have been made on it. The measurements show that radiation indeed occurs, and that good patterns result, but parameter optimization has not yet been accomplished.

Leaky-Wave Antenna Based on the First Higher Mode on Microstrip Line— Although microstrip line is not a low-loss waveguide, its simplicity makes it attractive as a guide on which to base antennas. The dominant mode on a uniform microstrip line is a slow wave relative to free space, so that the dominant mode cannot furnish a way to achieve a uniform leaky-wave antenna. Many antennas have been conceived and built, however, based on short lengths of dominant-mode microstrip line which are operated as open resonators; these antennas, as they relate to millimeter-wavelength applications, are discussed in the next section.

It is possible, on the other hand, to create a *uniform* leaky-wave antenna based on microstrip line if one employs a *higher mode* in an appropriate range of operation. The most convenient higher mode is the *first* higher mode. The electric field distribution of that mode in the microstrip line cross section is shown in Fig. 53. In contrast to the dominant mode the first higher mode has a nonzero cutoff frequency, which depends on the guide width.

Ermert [188, 189] conducted a careful numerical study of the propagation

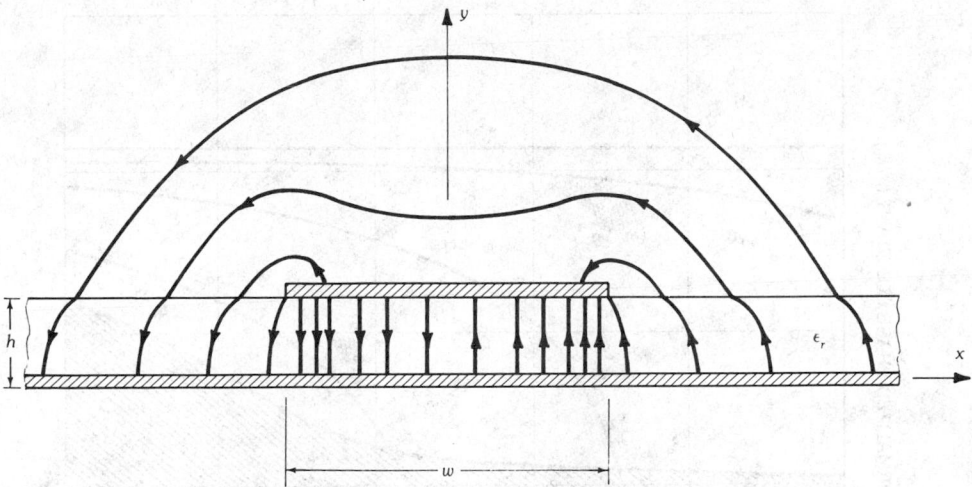

Fig. 53. Cross section of microstrip line, showing electric-field lines for the first higher mode.

characteristics of the dominant and first two higher modes of microstrip line employing an accurate mode-matching procedure, but the microstrip had a top cover which permitted him to employ discrete higher modes in the transverse representation. He found that in a range close to cutoff for the higher modes it was not possible to obtain real values for the propagation wave number; he therefore termed this range the "radiation" range, but he did not interpret it any further except to indicate that this "radiation region" was characterized by a continuous spectrum. One of the figures presented by Ermert [188, 189] is reproduced here, with modifications, as Fig. 54. (For this figure the microstrip line dimensions are: strip width, 3.00 mm; dielectric layer thickness, 0.635 mm; $\epsilon_r = 9.80$; and the height of the top cover is five times the dielectric layer thickness.) Ermert's curves are the solid ones, shown for the normalized propagation wave number β/k_0 for the lowest mode and the first two higher modes of covered microstrip line. All of his wave-number values are real, meaning that the modes are purely bound in those ranges. In the region shown lined, which he called the "radiation region," no real solutions exist. The dashed lines in Fig. 54 do not appear in his figure.

Oliner and Lee [190, 191] added these dashed lines and pointed out that they correspond to complex solutions (of course, only the real part is plotted) which signify physically that the modes have become *leaky* in this "radiation region." Ermert selects a spectral distribution for the modes of microstrip, and in his second paper [189] he rejects any inclusion of leaky-modes since they are nonspectral. Although that statement is a mathematically correct one, his rejection of leaky modes was unfortunate because it prevents one from understanding certain practical consequences. An alternative representation in terms of leaky modes can be obtained from his continuous spectrum by employing the steepest descent representation, which is not a spectral one. Such an investigation [190, 191] shows that the continuous spectrum in Ermert's radiation region is indeed characterized by essentially a *single leaky mode*. The dashed lines in Fig. 54 show the results for

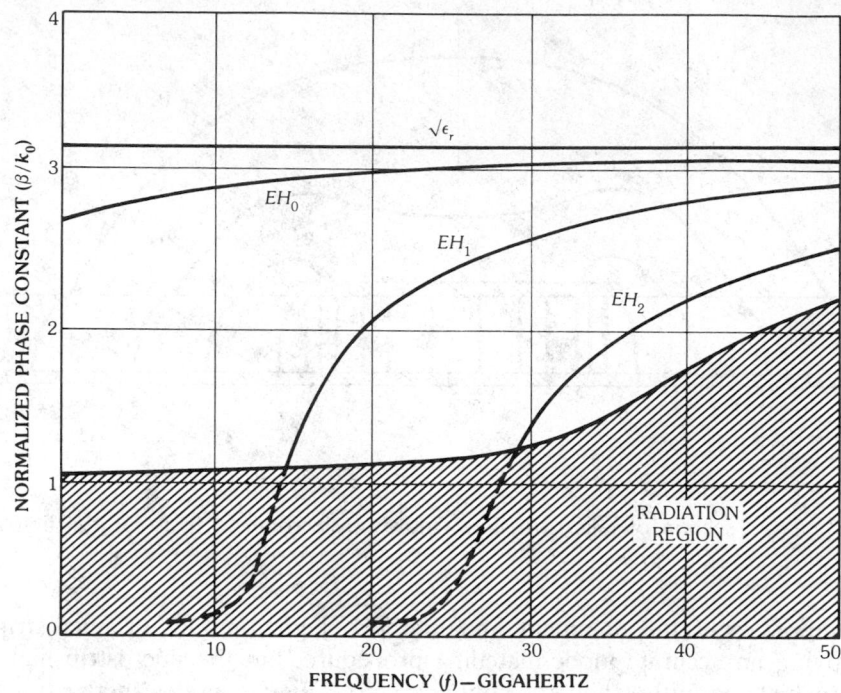

Fig. 54. Dispersion curves for the lowest mode and the first two higher modes in microstrip line with a top cover. (*After Oliner and Lee [190],* © *1986 IEEE*)

the appropriate leaky-mode solution corresponding to each of the two higher modes.

Oliner and Lee [190] have also shown that leakage can occur in two forms: a surface wave and a space wave. Furthermore, the onset of leakage for each form is given by simple conditions.

Let us envision a top view of the strip and the dielectric region around it. With this picture, we examine the case of leakage away from the strip in the form of a *surface wave* on the dielectric layer outside of the strip region. When there is leakage into the surface wave, the modal field propagates axially (in the z direction) with phase constant β, and the surface wave propagates away (on both sides) at some angle with phase constant k_s. The surface-wave wave number k_s has components k_z and k_x in the z and x directions, respectively, where k_z must be equal to β, since all field constituents are part of the same leaky modal field. We may therefore write

$$k_x^2 = k_s^2 - \beta^2$$

For actual leakage, k_x must be real, so that the condition for leakage is $k_x^2 > 0$. Applying this condition, we find that for leakage,

$$\beta < k_s$$

This condition defines the lined region in Fig. 54; the upper boundary of that region is actually the dispersion curve for the surface wave, of wave number k_s, that can be supported by the dielectric layer on a ground plane, if the microstrip line is open above, or by the dielectric layer between parallel plates, if there is a metal top cover. At the onset of the surface wave, it emerges essentially parallel to the strip axis, consistent with the condition $\beta = k_s$.

As the frequency is lowered so that β is decreased below the value of k_s, power leaks away in the form of a surface wave, as discussed above. As β is decreased further, power is then *also* leaked away in another form, the *space wave*. If the microstrip line is *open above*, this space wave actually corresponds to radiation at some angle, the value of this angle changing with the frequency. At the onset of this space wave, the wave emerges essentially parallel to the strip axis, so that $\beta = k_0$ then, where $k_0 = (2\pi/\lambda_0)$ is the free-space wave number. This boundary corresponds to the horizontal line $\beta/k_0 = 1$ in Fig. 54. For values of

$$\beta < k_0$$

power will leak into a space wave in addition to the surface wave.

What happens when the microstrip line has a *top cover*, of height H? If $H < \lambda_0/2$, approximately, such that only the surface wave can propagate in the dielectric-loaded parallel-plate region, then all the other modes are below cutoff, and power can leak away *only* in surface-wave form. If the plate spacing is increased, then some of the non–surface-wave modes are above cutoff, and these modes can also carry away power. The "space wave" then corresponds to the sum of those modes.

A separate study has shown that as the height H of the top cover increases, the proportion of power in the "space wave" also increases. When the top cover is removed far away and the dielectric substrate is thin, the percentage of power going into the surface wave becomes very small indeed. When there is no top cover present, therefore, we should expect that, for frequencies for which $\beta < k_0$, almost all of the power is radiated away in the form of a space wave. A leaky-wave antenna based on a microstrip higher mode would therefore be very efficient, and very little power would be lost to surface waves.

We next recognize that a traveling-wave antenna based on the first higher mode on microstrip line has actually been built and tested by Menzel [192]. It could have been operated as a leaky-wave antenna, but it was instead made very short (a typical length being $2.23\lambda_0$) as a competitor to microstrip resonator antennas. This antenna is therefore not characterizable as a true leaky-wave antenna, but it is novel and interesting, nevertheless, and represents a pioneering structure in this category.

In order to support the first higher mode above cutoff, the strip upper conductor is made somewhat wider than a half-wavelength in the substrate material. The dominant mode, which can be present simultaneously, of course, is suppressed, so that the traveling wave field corresponds only to that of the first higher mode. Based on this traveling wave field, but not taking any attenuation due to leakage into account, Menzel [192] also performed some elementary calculations with respect to the radiation pattern and beamwidth. He also conducted some measurements on typical structures.

In his approach, Menzel assumed that the propagation wave number of the first higher mode was real in the very region where Ermert said no such solutions exist; since his guided wave, with a real wave number, was fast in that frequency range, Menzel presumed that it should radiate. His approximate analysis and his physical reasoning were therefore incomplete, but his proposed antenna was valid and his measurements demonstrated reasonably successful performance. Once we recognize the relevance of leaky modes to the "radiation" region of microstrip line higher modes, it becomes clear that Menzel's antenna is a leaky-wave antenna in principle, even though he did not recognize this fact and did not discuss the antenna's design or behavior in those terms.

Oliner and Lee [190, 191] derived an accurate transverse resonance formulation for the propagation characteristics of the higher modes, both in the purely bound range (real wave numbers) and the "radiation" range (complex wave numbers). In this derivation, they employed a rigorous (Wiener-Hopf) solution derived by D. C. Chang and E. F. Kuester [193] for the reflection from one side of that microstrip line. They made a parametric dependence study of the leakage and phase constants of the first three higher modes in the "radiation" range; they found that the leakage rate α grows rapidly as the mode approaches "cutoff," as expected, but that the phase constant β, after approaching zero, slowly increased again and continued to increase as the frequency was lowered further. Such behavior modifies the usual understandings of the nature of "cutoff" for microstrip line higher modes in open regions.

What is puzzling at first with regard to Menzel's antenna, however, is why the antenna radiated so well in traveling-wave fashion even though it was so short ($2.23\lambda_0$); leaky-wave antennas are usually much longer. To answer this question, Oliner and Lee [191] employed the accurate expression mentioned above for the propagation characteristics of microstrip line higher modes, and obtained the variations of phase constant β and leakage constant α shown in Figs. 55a and 55b for Menzel's structure as a function of frequency.

On use of the relation $\sin \theta_m \cong \beta/k_0$, where θ_m is the angle of the beam maximum measured from the broadside direction, we see from Fig. 55a that after the onset of leakage the beam moves from end-fire toward the broadside direction. From Fig. 55b we note that as the frequency is lowered (and the beam swings up from end-fire) the value of α increases rather substantially, so that the beam width will increase strongly as the beam swings up.

The large value of α/k_0 ($= 0.0378$) explains why Menzel's antenna radiated so well despite its short length ($2.23\lambda_0$); a quick calculation shows that over this short length about 65 percent of the power was actually radiated. The remaining 35 percent would be largely reflected from the end, producing a large back lobe at the same angle from broadside. A look at Menzel's experimental pattern (his Fig. 11 in [192]) indeed verifies that such a back lobe is present with an amplitude about 0.4 of that of the main beam.

Since leaky-wave antennas are designed to radiate 90 percent or so of the incident power, one simply increases the length of the strip appropriately to accomplish that end. For the same cross section as that of Menzel's antenna, the strip length L would be increased to 21.7 cm from 10.0 cm for 90-percent power radiation. In wavelengths, L changes from $2.23\lambda_0$ to $4.85\lambda_0$; the half-power beamwidth then reduces from 26° to about 14°. It is interesting that in this case one needs only to slightly more than double the strip length to substantially improve the efficiency, essentially eliminate the back lobe, and reduce the beamwidth to a more practical value.

The discussion above applies to the radiation pattern in the H-plane. The E-plane pattern is very wide, of course, being dependent on the microstrip line strip width, but it can be narrowed down substantially, if desired, by using an array of parallel microstrip lines etched on the same substrate and excited with equal phase. Varying the phase would permit beam scanning in the E-plane.

A practical antenna would require the suppression of the dominant mode. That suppression can be accomplished by exciting the upper conductor of the microstrip line directly in antisymmetric fashion, using a pair of stubs fed from below in $\pm$ fashion, for example. Another way to produce suppression is to use an asymmetric feed arrangement, as shown in Fig. 56, together with a sequence of transverse slits on the center line of the antenna to inhibit the propagation of the dominant mode [192], which possesses a strong longitudinal current in the center portion of the upper conductor (whereas the longitudinal current there would be zero for the first higher mode). A quarter-wavelength transformer can be used to match the impedance of the antenna to that of the much thinner feed line. In this way, an impedance bandwidth of 10 percent can be obtained within a vswr of 2:1 [192].

Menzel states in his paper [192] that he proposed this antenna because it could yield improved bandwidth relative to microstrip resonator antennas. That is why it is made so short; as a result, as mentioned above, the antenna is not efficient, and a substantial reflected wave from the end is found experimentally, manifesting itself in a large back lobe. If the same structure were made longer and designed as a leaky-wave antenna, one should expect to obtain a uniform leaky-wave antenna of *simple configuration* that would yield a narrower beamwidth and be efficient.

4. Microstrip Resonator Antennas and Other Printed-Circuit Antennas

This section deals with printed-circuit antennas and their use in the millimeter-wave region. The best-known antennas of this type are microstrip antennas, which consist of metal patches or dipoles printed on a single, thin dielectric substrate backed by a ground plane. These antennas are operated *at resonance*, which distinguishes them from the traveling-wave microstrip antennas reviewed in the preceding section. The basic microstrip geometry is the same for both the resonant and the traveling-wave microstrip antennas, of course, but they differ in their design principles and modes of operation, so that treatment in different sections is appropriate.

While microstrip traveling-wave antennas are a comparatively new development, resonant microstrip antennas have been investigated extensively during the past decade at microwave frequencies. In particular, microstrip patch resonator antennas can be regarded as well understood by now although papers on this subject are still appearing quite frequently in the literature. This technique has

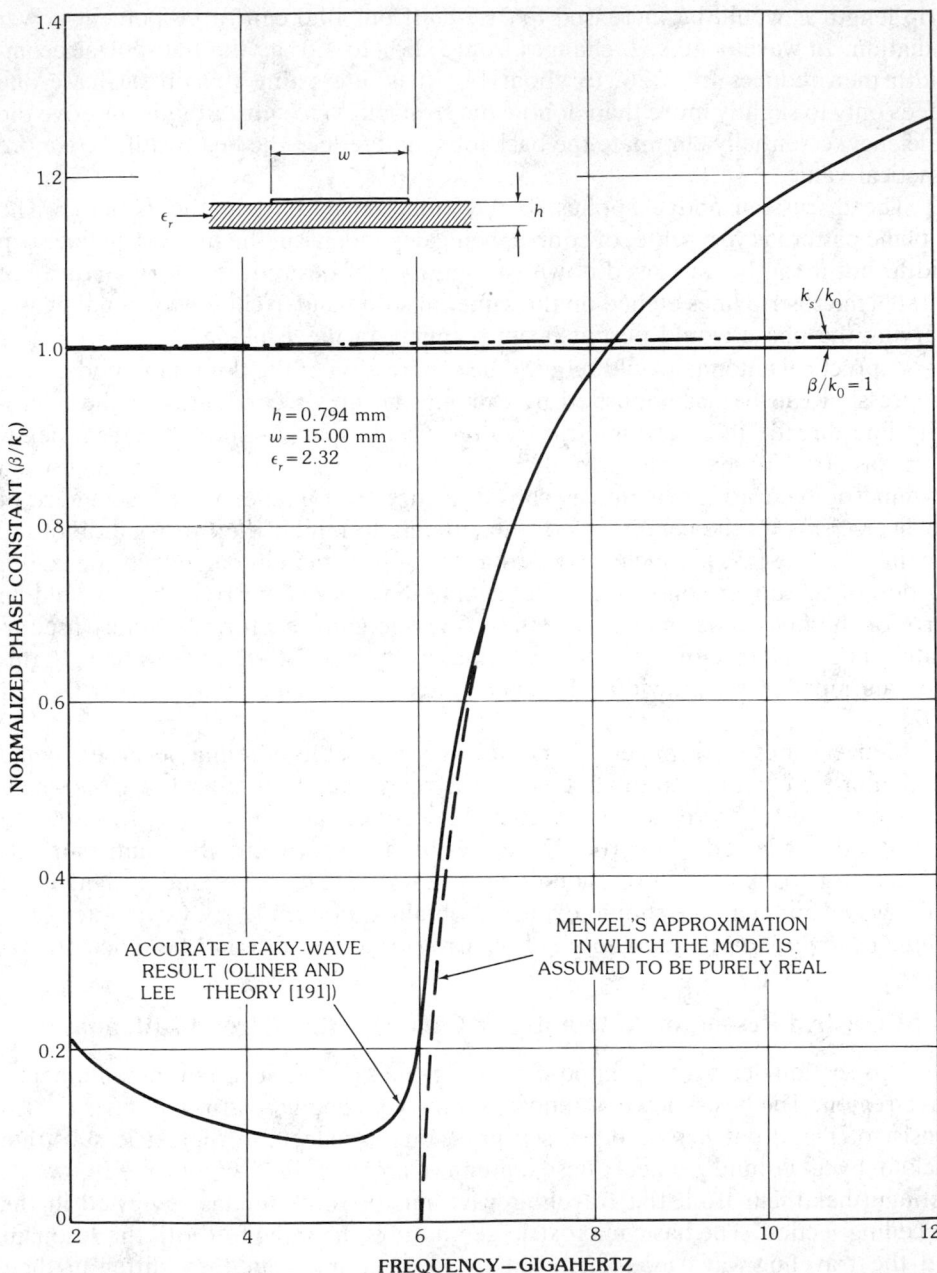

a

Fig. 55. Normalized phase constant β/k_0 and leakage constant α/k_0 as functions of frequency for Menzel's antenna structure [192]. (a) Phase constant β/k_0. (b) Leakage constant α/k_0. (*After Oliner and Lee [191], © 1986 IEEE*)

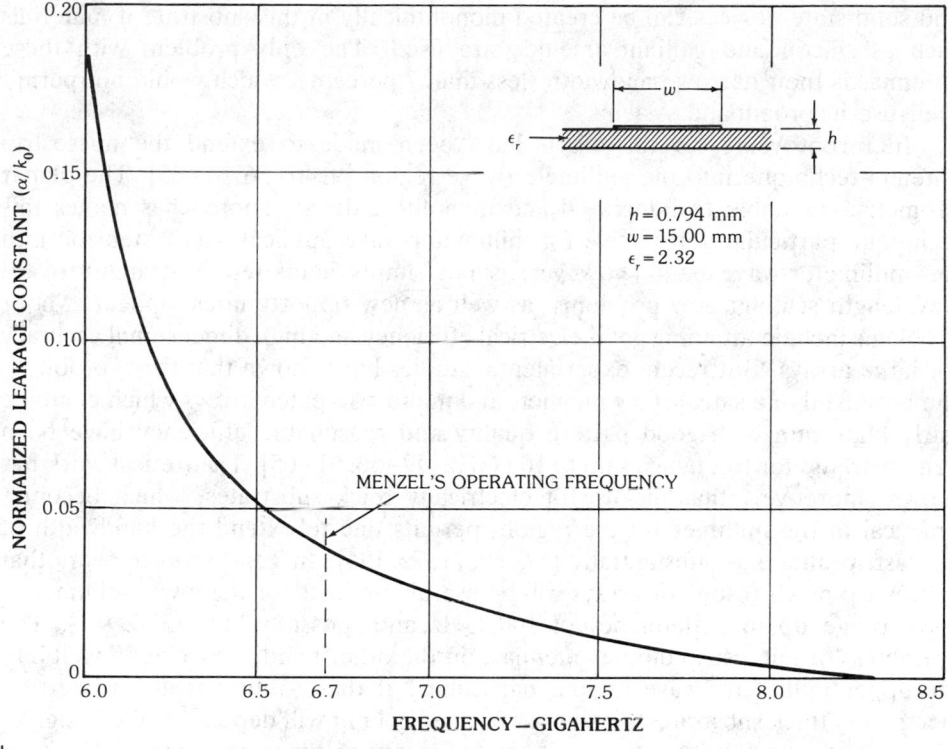

Fig. 55, *continued.*

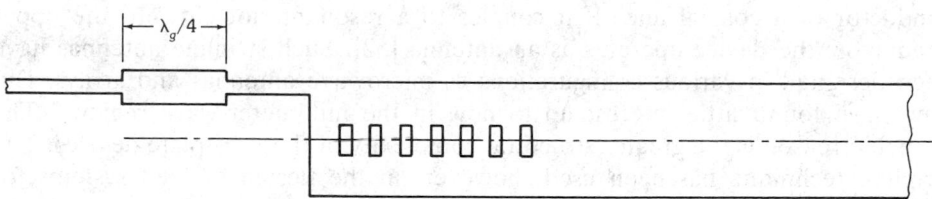

Fig. 56. Asymmetrically excited microstrip traveling-wave antenna. (*After Menzel [192]*)

been used very successfully for the design of single antennas and for a wide variety of antenna arrays for the uhf through microwave regions. The technique is described in detail in Chapter 12; it is this technique which is commonly associated with the term microstrip antennas.

Microstrip patch and dipole antennas possess numerous advantages, including a very low profile, suitability for conformal installation, convenient fabrication by etching techniques, a high degree of reproducibility, ruggedness, low cost, and light weight. Moreover, the microstrip technique is very well suited for the design of integrated antennas. Transmission lines, phase shifter circuits, hybrid couplers, mixer circuits, and filters may be printed with the antennas on the same substrate, and solid-state devices can be created monolithically in this substrate if materials such as silicon and gallium arsenide are used. The only problem with these antennas is their narrow bandwidth (less than 7 percent), which would not permit their use in broadband systems.

In recent years several efforts have been made to extend the microstrip antenna technique into the millimeter-wave region [9, 46–56, 61–65]. The planar geometry amenable to integrated and monolithic design approaches makes this technique particularly attractive for millimeter-wave applications. Extension into the millimeter-wave band, however, is not simply a matter of straightforward wavelength scaling; new problems, as well as new opportunities, appear. Major problems include attaining good electrical efficiency and high dimensional accuracy for large arrays. But recent experimental studies have shown that these problems can be solved in a satisfactory manner, and microstrip patch arrays which combine fairly high gain with good pattern quality and reasonable efficiency have been demonstrated for frequencies up to 100 GHz [49–56, 61–65]. Theoretical work has shown, moreover, that the use of electrically thick substrates, which becomes practical in the millimeter-wave region, permits one to extend the bandwidth of microstrip antennas substantially [57–60, 71, 72, 194]. In general, it appears that microstrip patch resonator arrays will be useful antennas for the lower millimeter-wave range up to a frequency of 100 GHz and, possibly, 140 GHz [62]. The usefulness of microstrip dipole antennas, on the other hand, may extend well into the upper millimeter-wave region, particularly if these antennas are printed on electrically thick substrates. The upper frequency limit will depend on the design of efficient, easy-to-fabricate feed systems for arrays of these antennas.

Stripline antennas, a second class of printed-circuit antennas, consist of three, rather than two, metal layers separated by thin dielectric substrates. The center conductor usually has small width, and its function is similar to that of the center conductor of a coaxial line. If it couples to a resonant slot cut into the upper conductor, the device operates as an antenna [52]. Such stripline antennas have been designed in various configurations as microwave antennas and arrays. But they have found little interest up to now in the millimeter-wave region.* The probable reason is the greater structural complexity of these tri-plate devices. The stripline technique has been used, however, in the design of feed systems for microstrip millimeter-wave arrays [52, 53]. The tri-plate approach permits one to minimize energy leakage, which can be a problem particularly near the array feed point where power levels on the transmission lines are high.

A third type of printed-circuit antenna, i.e., holographic antennas [195], which are rather different from the others, is reviewed briefly. Similar to stripline antennas, these antennas have been studied, up to now, at microwave frequencies. But they appear well suited for millimeter-wave applications and are, therefore, included in the discussion.

Permittivity and loss tangent values for typical substrate materials are listed in Table 1.

---

*There are exceptions, however. For example, Sedivec and Rubin have designed a 4 × 4 element stripline array for the 44-GHz band [196]. The array has the advantage of a large bandwidth.

Microstrip Antennas with Electrically Thin Substrates

Small Arrays of Microstrip Patch Antennas: Feed Arrangements and Tolerance Problem—Fig. 57 shows the metallization pattern of a typical 4 × 4 element patch resonator array including a microstrip feed system. The wide portions of the feed lines are quarter-wavelength impedance transformers matching the element impedances of approximately 120 Ω (for square patches at resonance) to the 50-Ω input port of the overall antenna.

The length dimension b of the patches is determined by the resonance condition which requires that b (with an appropriate edge effect correction) is equal to a half-wavelength in the substrate material. The width w determines the input admittance of the patch resonators and can be utilized to control the excitation amplitudes of the array elements on an individual basis. The substrate thickness h is typically in the order of 5 to 10 mils for antennas printed on low-loss, low-permittivity substrates with $\epsilon_r = 2$ to 4 and designed for operation in the 30- to 100-GHz band. The choice of h involves a trade-off. Greater substrate height will increase bandwidth and reduce feed line losses; smaller height will help to minimize feed system radiation, improve pattern quality, and reduce cross-polarized

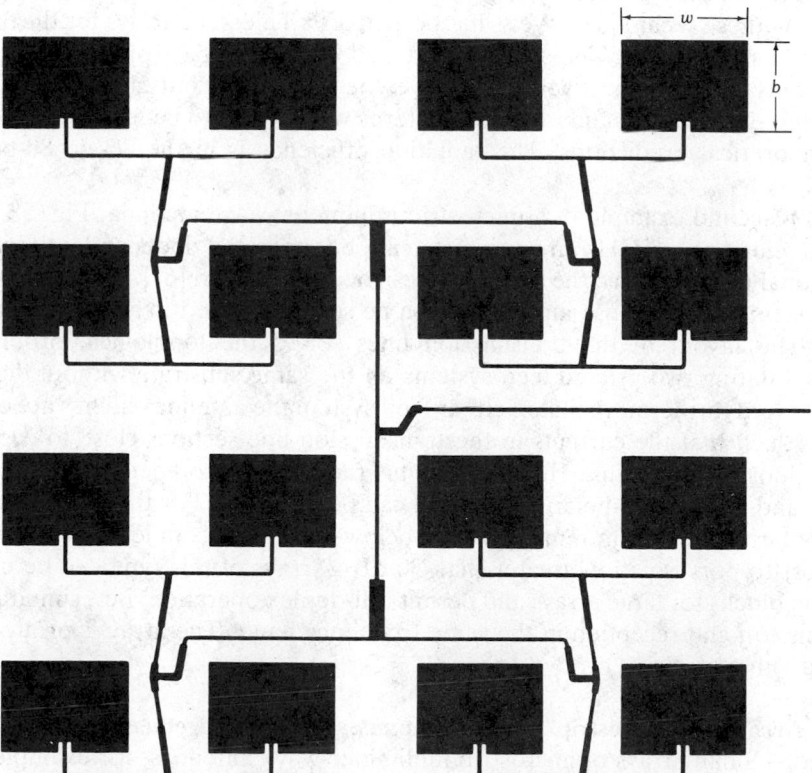

Fig. 57. Metallization pattern of microstrip resonator array with 4 × 4 elements. (*After Weiss and Cassell [49]*)

radiation. Note, however, that the radiation resistance of the antenna patches is practically independent of h as long as the substrate remains electrically thin.

The patch dimensions b and w are very small in the millimeter-wave region, and the width of the microstrip transmission lines feeding the patch elements is even smaller; it is typically in the order of 0.1 mm in the 30- to 100-GHz band and very accurate etching techniques are required for the fabrication of such arrays. Weiss has shown that close tolerances can be achieved through the use of glass-plate negatives and by choosing the thinnest possible copper cladding for the substrate [48–50]. Glass plates provide greater stability and higher resolution than the film negatives typically used in etching microwave antennas; and for typical substrate materials such as Duroid, Polyguide, and quartz the use of ¼ oz (7.09 g, 0.4 mil) copper cladding—instead of the more conventional 1 oz (28.35 g, 1.4 mils) cladding—resolves the "undercutting" problem at the edges of the microstrip patches and feed lines and thus permits maintaining close tolerances. As a general rule, devising precise fabrication techniques should be regarded as an integral part of the design of microstrip millimeter-wave antennas. It is advisable, also, to fabricate and optimize these antennas first at a lower scale frequency, which allows a much greater sensitivity in refining antenna performance [49]. Reduction of the final production mask to the appropriate millimeter-wave dimensions is easily and accurately done by photography.

Using these techniques, Weiss has designed 4×4 element arrays for the 35- and 60-GHz bands [48, 50]. The arrays are etched on Duroid 5880 substrates, 10 mils and 5 mils thick, respectively, and have the general layout shown in Fig. 57. Measured gain (17 dB) and radiation patterns were found to be in good agreement with theoretical predictions. The radiation efficiency is in the 75- to 80-percent region.

As a second example of a microstrip millimeter-wave antenna, Fig. 58 shows a 4×8 element 35-GHz array which can be operated independently in two orthogonal polarizations; the array was designed by Lalezari [61]. A dual corporate feed system is used whose input ports can be seen near the top and bottom of the array. The layout of the transmission lines solves the topological problem of accommodating two printed feed systems on the same substrate without "crossed wires." Note furthermore that in either feed system the antenna patches are excited in pairs such that the currents in the transmission line sections close to each pair have opposite directions. Hence feed-line radiation is compensated to a large degree and a low cross-polarization level can be achieved. For the array of Fig. 58 cross-polarized radiation remains 28 dB below the main beam level, which results in a port-to-port isolation greater than 30 dB. Arrays of this type can be used as building blocks for large arrays and permit full-duplex operation, i.e., simultaneous transmission and reception in the same frequency band. The array is printed on a Duroid substrate.

Large Arrays of Microstrip Patch Antennas: Feed Arrangements and Efficiency Problem—Small arrays of microstrip millimeter-wave antennas, for example, 4×4 or 8×8 element arrays, can be designed to have good efficiency and pattern quality. But in large arrays, which are needed to obtain high directivity gain, losses can be substantial. Most of these losses occur in the feed system of the array. Microstrip lines are not low-loss lines and power dissipation can be large, in particular in a

Fig. 58. Dual-polarization microstrip array for 35-GHz band. (*Courtesy Ball Aerospace Systems Division*)

printed corporate feed system consisting of long runs of microstrip lines. On the other hand, losses in the radiating elements of a microstrip array usually are negligible in comparison to feed system losses. In the lower millimeter-wave region, the radiation efficiency of microstrip patch resonators typically exceeds 80 percent. A second problem associated with microstrip arrays with long feed lines is that inhomogenieties in the substrate material can lead to phase errors in the excitation of the array elements ("detuning" of feed system) and, thus, to pattern distortions.

Accurate formulas for the calculation of the propagation and attenuation constants of microstrip lines are available in the literature. These formulas are lengthy, but they are shown here since in large microstrip arrays, as in all large arrays, the design of the feed system is a critical problem. The equations given below were taken from the book by Gupta, Garg, and Chadha [197], where also references to the earlier literature on this subject can be found. The characteristic impedance Z_0 of a microstrip line is determined by the permittivity ϵ_r and height h of the substrate, and the width w and thickness t of the metal strip.* According to [197],

$$
Z_0 = \begin{cases} \dfrac{1}{2\pi} \dfrac{\eta}{\sqrt{\epsilon_{\text{eff}}}} \ln\left(\dfrac{8h}{w_e} + 0.25\dfrac{w_e}{h}\right) & \text{for } w/h \leqq 1 \\[3mm] \dfrac{\eta}{\sqrt{\epsilon_{\text{eff}}}}\left[\dfrac{w_e}{h} + 1.393 + 0.667\ln\left(\dfrac{w_e}{h} + 1.444\right)\right]^{-1} & \text{for } w/h \geqq 1 \end{cases} \qquad (19a)
$$

*In the microwave region the thickness of the metal strip is usually very small compared with the substrate height, and the dependence of the line parameters on t amounts to a small correction only. In the millimeter-wave region, t/h may be as large as 0.1 to 0.2 and strip thickness effects cannot be neglected. Equation 19 for Z_0 and the subsequent equations for α_{cond} account for these effects approximately; expressions which would yield the t dependence precisely do not seem to be available as yet in the literature.

where

$$\epsilon_{\text{eff}} = \frac{\epsilon_r + 1}{2} + \frac{\epsilon_r - 1}{2}\left[\left(1 + 10\frac{h}{w}\right)^{-1/2} - \frac{1}{2.3}\frac{t}{\sqrt{wh}}\right] \tag{19b}$$

$$\frac{w_e}{h} = \frac{w}{h}\left[1 + \frac{1.25}{\pi}\frac{t}{w}\left(1 + \ln\left(\frac{2B}{t}\right)\right)\right] \tag{19c}$$

$$B = \begin{cases} 2\pi w & \text{for } w/h \leqq 1/2\pi \\ h & \text{for } w/h \geqq 1/2\pi \end{cases} \tag{19d}$$

$$\eta = 120\pi \text{ ohms}$$

Usually, the substrate parameters ϵ_r and h and the thickness t of the metal cladding are given, and Z_0 is prescribed. Equation 19 can then be regarded as implicit relations for the determination of w. With w determined, the propagation constant β is found from

$$\beta = k_0\sqrt{\epsilon_{\text{eff}}}$$

where ϵ_{eff}, the effective dielectric constant of the feed lines, is given by (19b). The attenuation constant α consists of two terms

$$\alpha = \alpha_{\text{cond}} + \alpha_{\text{diel}}$$

where α_{cond} accounts for the conduction loss and α_{diel} for the dielectric loss. We have

$$\alpha_{\text{cond}} = \begin{cases} 1.38A\dfrac{R_s}{Z_0}\dfrac{\lambda_0}{h}\dfrac{32 - (w_e/h)^2}{32 + (w_e/h)^2} \text{ dB}/\lambda_0 & \text{for } w/h \leqq 1 \\[3mm] 6.1 \times 10^{-5}A\,R_s Z_0 \epsilon_{\text{eff}}\dfrac{\lambda_0}{h}\left[\dfrac{w_e}{h} + \dfrac{0.667w_e/h}{w_e/h + 1.444}\right] \text{ dB}/\lambda_0 & \text{for } w/h \geqq 1 \end{cases}$$

$$\alpha_{\text{diel}} = 27.3\frac{\epsilon_r}{\epsilon_r - 1}\frac{\epsilon_{\text{eff}} - 1}{\sqrt{\epsilon_{\text{eff}}}}\tan\delta \text{ dB}/\lambda_0$$

In these expressions,

$$A = 1 + \frac{h}{w_e}\left[1 + \frac{1}{\pi}\ln\left(\frac{2B}{t}\right)\right]$$

$$R_s = \left(\frac{\omega\mu_0}{2\sigma}\right)^{1/2}$$

Characteristic impedance Z_0 and surface resistance R_s are in ohms, σ is the conductivity of the strip, $\tan\delta$ is the loss factor of the substrate, and ϵ_{eff}, w_e/h, and

B are given by (19b), (19c), and (19d), respectively. It is convenient to express α in decibels per free-space wavelength.

Typically, a well-designed 50-Ω feed line on a 5-mil substrate will have an attenuation coefficient [63]

$$\alpha = \begin{cases} 0.12 \text{ dB}/\lambda_0 & \text{for Duroid} \\ 0.14 \text{ dB}/\lambda_0 & \text{for quartz} \\ 0.28 \text{ dB}/\lambda_0 & \text{for alumina} \end{cases}$$

These values apply to the 30- to 100-GHz band and include both conduction and dielectric losses, with the former losses yielding the dominant contribution. Microstrip losses tend to increase with line impedance, and for a given impedance the conduction loss increases with decreasing substrate height, while the dielectric loss remains approximately constant. It is interesting to note, however, that in monolithic circuits where microstrip lines are created on silicon or gallium arsenide substrates, the dielectric loss is much higher than the conduction loss [198].

James and Hall [62] have pointed out that surface roughness of the microstrip metal surfaces can have a substantial effect on the conduction loss. In the presence of surface roughness, α_{cond} in the above expression for α must be replaced by

$$\alpha'_{\text{cond}} = \alpha_{\text{cond}} \left\{ 1 + \frac{2}{\pi} \tan^{-1} \left[1.4 \left(\frac{\Delta}{\delta_s} \right)^2 \right] \right\}$$

where Δ is the rms surface roughness and δ_s the skin depth. Apparently, when Δ significantly exceeds the skin depth the conduction loss will be increased by a factor of 2. The expression for α'_{cond} was derived for a triangular roughness profile but should yield good results also for other profiles. Estimates for Δ were obtained by James and Hall for commercially available Duroid substrates with 1- and ½-oz (28.35-g and 14.17-g) copper claddings [62]. These estimates were made from electron microscope photographs of the surfaces and, for electrodeposited copper, yielded Δ values of 2.5 μm and 1.25 μm for the 1- and ½-oz claddings, respectively. Corresponding values for rolled copper were close to 0.42 μm for both copper weights. Comparison of these values with the skin depth of copper, which is 0.35 μm at 30 GHz and 0.21 μm at 100 GHz, indicates that α'_{cond} will not be far away from the upper limit of $2\alpha_{\text{cond}}$. The numerical values for α given above, therefore, may be somewhat optimistic. For polished copper, on the other hand, Δ will be much smaller than δ_s and surface roughness effects should be insignificant [62].

Fig. 59 shows a 32 × 32 element patch resonator array for the 38-GHz band which was designed for broadside radiation with a directivity of 35 dB and a beam-width of approximately 3° [49]. The array is etched on a 5-mil-thick Duroid substrate and uses a corporate feed system. The problems associated with long runs of microstrip feed lines were minimized, in this case, by subdividing the array into four equal subarrays of 16×16 elements each, and by using a four-way power divider in metal waveguide to feed the subarrays in parallel. This technique also eliminates parasitic radiation from bends and corners of the feed system near the input terminal where in large arrays the power level is high.

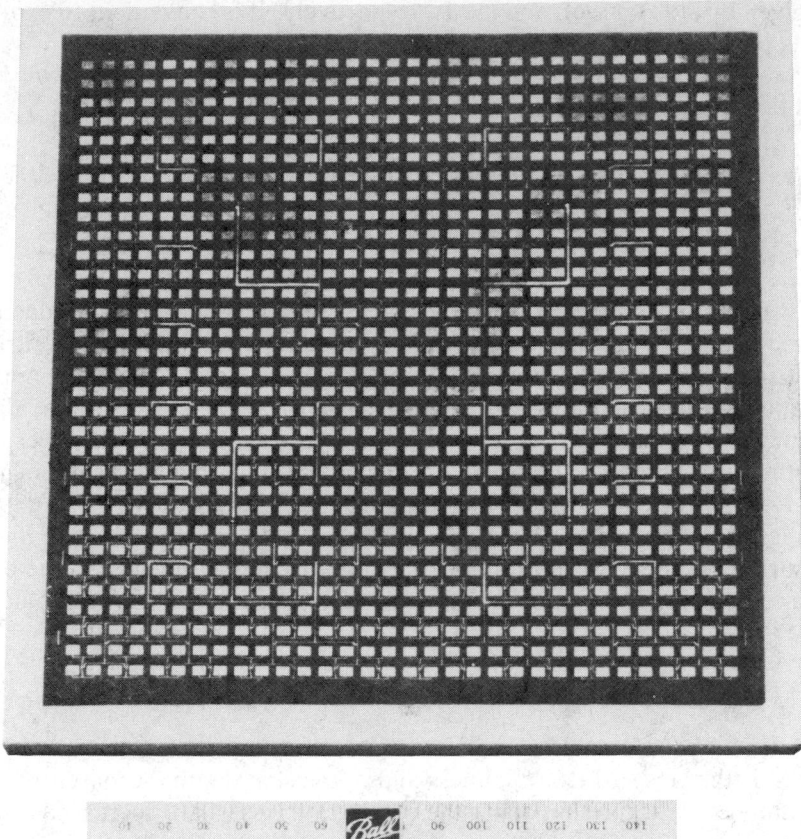

a

Fig. 59. A 32 × 32 element microstrip antenna array for 38-GHz band. (*a*) Front view. (*b*) Back view. (*Courtesy Weiss and Cassell [49]*)

The power divider is milled into the aluminum base plate whose thickness of 6.5 mm is appropriate to accommodate the waveguide. At 38 GHz, losses in the waveguide legs of the power divider (the small screws in Fig. 59b outline its contour) are smaller than 0.008 dB/cm as compared with a loss of approximately 0.24 dB/cm in the microstrip feed lines.* An additional advantage of this solution is that the increased thickness of the base plate enhances the mechanical strength of the antenna.

Experiments with a first model of the array indicated efficiency and pattern quality problems, but subsequent investigations have shown that arrays of this type, when carefully designed, are capable of providing good pattern quality, low vswr over a bandwidth in the order of a few percent, and acceptable efficiency. The

---

*The array shown in Fig. 59 was a first model. By careful design, losses in the microstrip lines could probably be reduced by a factor of 2.

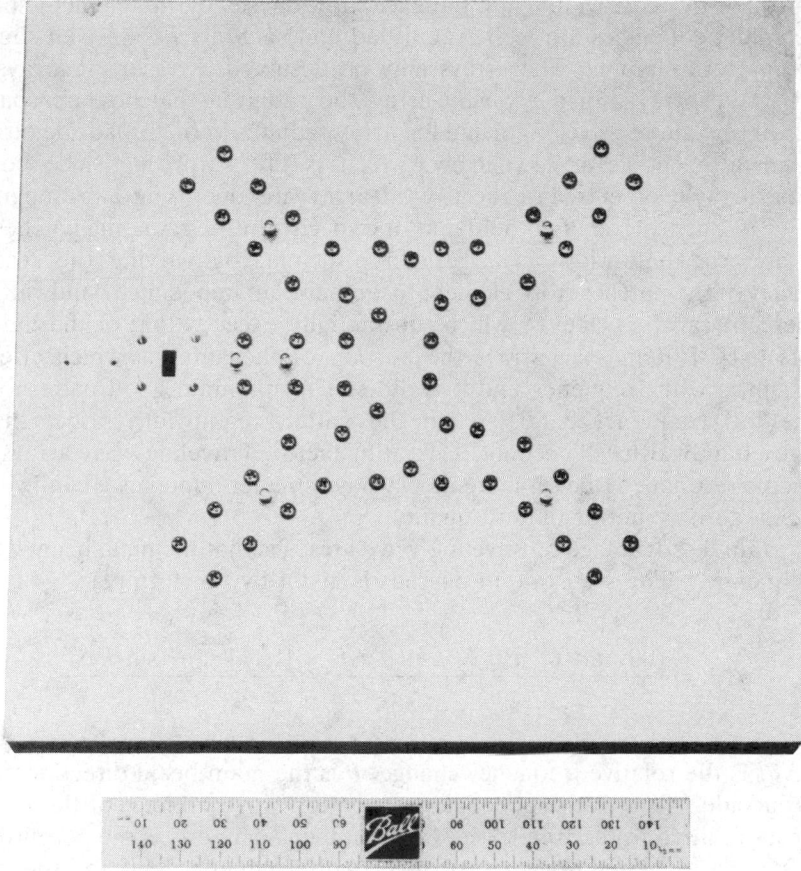

b

Fig. 59, *continued.*

efficiency could be further improved by the use of a higher-order eight- or sixteen-way power divider, i.e., by further extending the metal waveguide portion of the feed system at the expense of the microstrip portion.

An alternative approach to enhancing antenna efficiency is the use of *series-fed* arrays where the length of feed-line runs is substantially reduced in comparison to a corporate feed system [51, 63–65]. Fig. 60 shows the metallization pattern of a linear array of this type. The excitation amplitudes of the various array elements

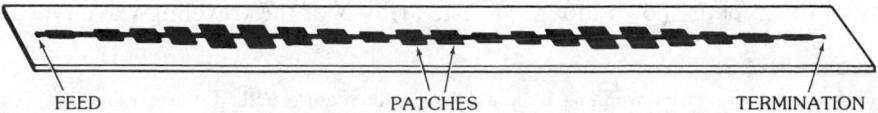

Fig. 60. Series-fed microstrip patch array. (*After Jones, Chow, and Seeto [65],* © *1982 IEEE*)

are controlled by their widths and the excitation phases by the element spacings. Parameter dependencies are well established and such arrays are well suited for accurate pattern shaping. The arrays may be designed as resonant arrays or as traveling-wave arrays. In a *resonant* array the radiating patches are spaced at intervals of one guide wavelength and the array radiates in the broadside direction. The bandwidth, however, is narrow, which is not surprising since both the individual array elements and the overall array are operating at resonance. In *traveling-wave* arrays, on the other hand, the spacing of the radiating patches is not critical and the bandwidth is larger. In principle, by varying the resonance frequencies of the patches from element to element, an impedance bandwidth of 10 to 20 percent can be achieved which substantially exceeds that of the individual elements [64]. But in such arrays the portion of the antenna which effectively radiates shifts with frequency and it is difficult to maintain good pattern quality over the full frequency band so that the pattern bandwidth rather than the impedance bandwidth will become the limiting factor. Traveling-wave arrays where all elements resonate at the same frequency have smaller impedance bandwidth but the advantage of superior pattern quality.

A certain disadvantage of traveling-wave arrays is that the main beam will scan with frequency.* The scan rate in degrees is given by the formula

$$\Delta\theta = \frac{180°}{\pi} \frac{d/df(f \cdot n_{\text{eff}}) - \sin\theta}{\cos\theta} \frac{\Delta f}{f} \cong \frac{180°}{\pi} \frac{\sqrt{\epsilon_r} - \sin\theta}{\cos\theta} \frac{\Delta f}{f} \qquad (20)$$

where $\Delta f/f$ is the relative frequency change, θ is the main beam direction counted from broadside,** n_{eff} is the effective propagation wave number of the traveling-wave array relative to the free-space wave number k_0, and ϵ_r is the permittivity of the substrate; n_{eff} does not vary strongly with frequency and for the present purpose may be approximated by $\sqrt{\epsilon_r}$. A typical microstrip array will be printed on a low-loss substrate of permittivity ϵ_r between 2 and 3 and will be designed to produce a main beam not too far away from broadside. Under these conditions, (20) predicts a scan rate of approximately 1° per 1 percent of frequency change, which is in good agreement with experimental evidence [66]. For high-permittivity substrates, such as alumina and silicon, the scan rate would be about twice as large. The beam-scanning effect may appreciably narrow down the usable frequency band of a high-gain traveling-wave array with a beamwidth in the order of a few degrees, but it is not likely to noticeably affect the performance of a broad-beam, low-gain antenna.

A planar array of series-fed microstrip antennas can be obtained by printing several linear arrays in parallel on the same substrate and by connecting their feed points to a corporate feed or, simply, a common microstrip feed line. A center-fed array of this type is shown in Fig. 61. The array is of the traveling-wave type, with

---

*Applications where beam scanning is desirable usually require a larger scan range than can be accommodated within the limited bandwidth of microstrip antennas. These antennas appear better suited for fixed-beam operation.
**The sign of θ is positive (negative) in the range of forward (backward) directions.

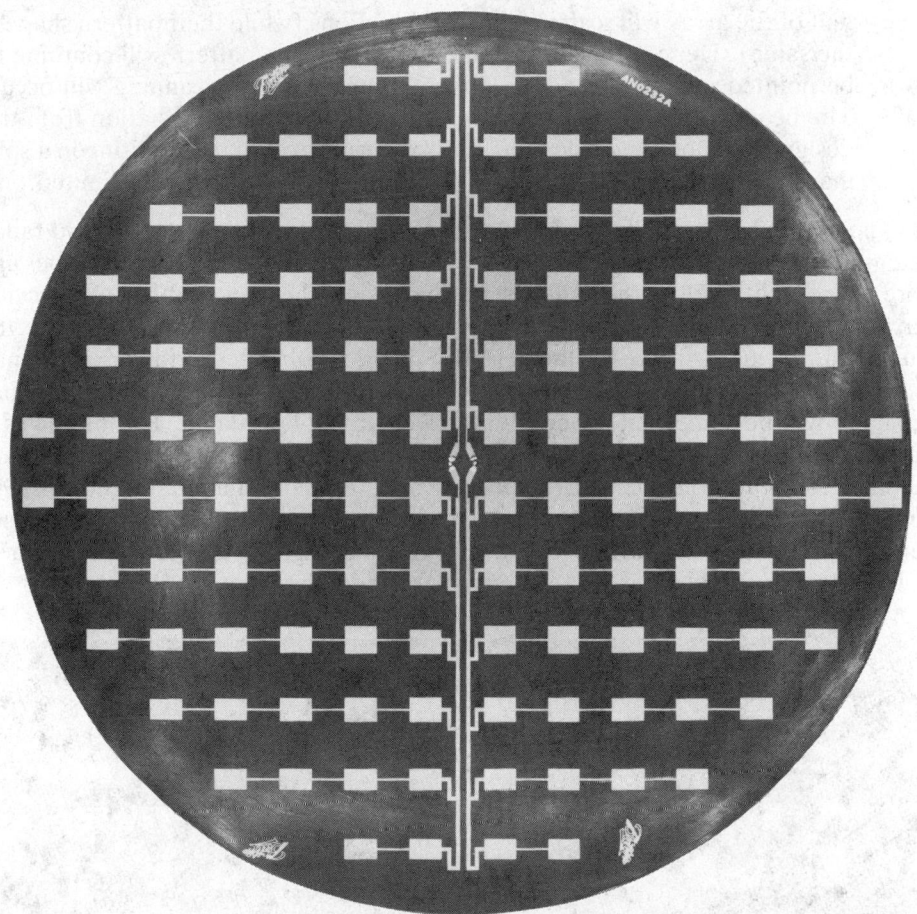

Fig. 61. Series-fed microstrip array for broadside radiation. (*Courtesy Ball Aerospace Systems Division*)

all elements resonating at the same frequency. The antenna is designed for radiation in the broadside direction and is interesting for several reasons:

(a) The short transmission line runs of the series feed-system result in low losses and good antenna efficiency.

(b) The widths of the radiating elements and hence their excitation amplitudes decrease away from the array center in a roughly circular symmetric fashion. The effect is reinforced by the power reduction from element to element due to radiation. Thus a tapered aperture illumination is obtained, resulting in low side lobes.

(c) The array elements are spaced to be excited in equal phase at the design frequency so that the array radiates in the broadside direction. Away from the design frequency the excitation phase will vary from element to element in a linear fashion. But it is easily seen from the symmetry of the array configuration that the two partial beams produced by the right and the left half of the array will scan in opposite directions (while their pattern shape is the same). Hence the main beam of the composite pattern will continue to be pointed into the broadside direction and no beam scanning will occur. The beamwidth, however, will increase with increasing deviation from the design frequency and side lobe levels will rise, eventually resulting in a split main beam. The pattern bandwidth of the array is therefore limited.

The particular array shown in Fig. 61 was designed for use in an X-band radar system. The gain of the array is 27 dB, the efficiency is 75 percent, the side lobes are more than 20 dB below the main beam level, and the bandwidth is 1 percent. Maximum power is 1 kW and the weight is 100 g. The same design principle, of course, can be used in the millimeter-wave band to obtain fixed-beam antennas of high gain and good efficiency, although the efficiency and power-handling capability will be somewhat lower, and the weight will be smaller. Fig. 62 shows a second example of a series-fed traveling-wave type array. Laid out in a four-quadrant configuration, the array operates as a monopulse antenna in the 35-GHz band. The monopulse comparator visible at the center is integrated with the array elements on the same substrate. The arrays shown in Figs. 61 and 62 were designed by Lalezari. Both antennas are etched on Duroid substrates.

Fig. 62. A 35-GHz microstrip monopulse array with integrated comparator network. (*Courtesy Ball Aerospace Systems Division*)

A completely different technique for reducing feed system losses has recently been suggested by James and others [62, 67, 68]. As illustrated by Fig. 63a, microstrip feed lines are eliminated; they are replaced by a low-loss open dielectric waveguide (in the form of an insular guide) whose fringing fields excite the microstrip antenna patches. By varying the distances of the microstrip patches from the feed guide, as indicated in Fig. 63b, the excitation amplitudes of the array elements can be adjusted and a degree of pattern control exercised. Improvements in feed system efficiency that can be achieved by this approach may be seen from Table 2 [69], which applies to the specific example of an alumina feed guide placed on a quartz substrate, where it replaces a microstrip feed line with a characteristic impedance of 50 Ω. The table compares the theoretical attenuation coefficients of the two guides for three frequencies in the lower millimeter-wave band. The attenuation coefficient of silver-plated metal waveguide is included in the table for comparison. While metal waveguide has the lowest attenuation (and the largest size and production costs), insular guide has a substantial efficiency advantage over microstrip line and these improvements tend to increase with frequency. A second advantage of this feed method is its simplicity; the dielectric guide has a reasonable cross section and is easy to fabricate, and no contact has to be established between radiating elements and feed line [67].

Experiments with a 2×40-element array of this type have confirmed the usefulness of the concept [62]. Etched on a Duroid substrate, the array was dimensioned for broadside radiation at 90 GHz. An alumina feed guide was used with a cross section of 0.35×1.10 mm^2. The guide was operated in the fundamental HE$_{11}$ mode, and the spacing of the radiating elements from the feed guide was tapered for uniform aperture illumination. Feed line losses and cross-polarization levels were low. But launching losses and side lobe levels remained a problem requiring further improvements.

The next logical step would be the use of insular guide not only for the feed line of the array but for the radiating elements as well, i.e., the microstrip patches in

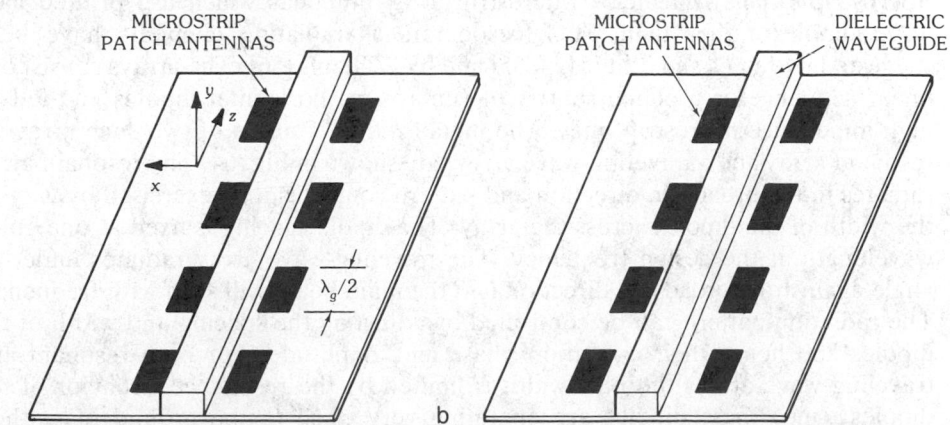

Fig. 63. Microstrip patch resonator arrays fed by dielectric image line. (*a*) Basic configuration. (*b*) Configuration permitting pattern control. (*After James and Henderson* [68])

Table 2. Theoretical Attenuation Coefficient of Microstrip Line, Insular Guide, and Metal Waveguide in the Lower Millimeter-Wave Region (*After Knox* [69], © *1976 IEEE*)

Feed System		Frequency (GHz)	h (cm)	a (cm)	α (dB/cm)
Microstrip line (gold on fused quartz: 50 Ω)		30	0.027	0.054	0.056
		60	0.014	0.027	0.154
		90	0.009	0.018	0.280
Insular waveguide (alumina on fused quartz)		30	0.027	0.268	0.022
		60	0.013	0.134	0.055
		90	0.009	0.090	0.096
Metal waveguide (silver plated)		30		1.067	0.007
		60		0.376	0.016
		90		0.254	0.030

Figs. 63a and 63b would be replaced by dielectric resonators. These resonators may take the form of small rectangular or cylindrical dielectric blocks placed on the substrate. Arrays of this type have been studied by Birand and Gelsthorpe [70].

Microstrip Dipole Antennas—Microstrip array antennas which use printed half-wave dipoles rather than patch resonators as radiating elements have been suggested by James and others [54, 55] and by Williams [56]. The arrays consist of a linear sequence or a planar matrix of (unbroken) horizontal dipoles end-fed by high-impedance microstrip lines. The metallization patterns of two such arrays, a resonant array and a traveling-wave array, are shown in Fig. 64. The resonant array radiates in the broadside direction and pattern control can be exercised by varying the width of the dipoles across the array. The dipole spacing is fixed at one guide wavelength at the design frequency. The traveling-wave array radiates under an angle against the broadside direction and the main beam will scan with frequency. The radiation pattern can be controlled by adjusting the spacing and width of the dipoles and hence their excitation phase and amplitude. For both resonant and traveling-wave arrays the bandwidth is limited by the resonance behavior of the dipoles; since these dipoles are operating very close to the ground plate their impedance bandwidth is small. Measurements at 36 and 70 GHz have shown that such arrays are capable of providing good performance in the lower millimeter-wave region, possibly up to 94 GHz. But further study is needed to improve

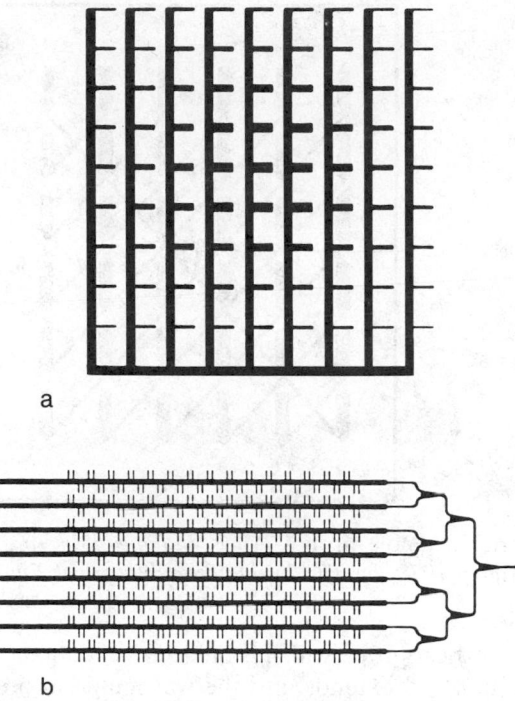

Fig. 64. Metallization patterns of printed horizontal half-wave dipole arrays fed by microstrip lines. (*a*) Resonant array. (*b*) Traveling-wave array. (*After James et al. [55]*)

pattern quality at 70 GHz and above [53]. In particular, radiation from the corporate feed of these arrays seems to be a problem necessitating an enclosed, tri-plate layout of the feed system near the array feed point, where power levels are high.

Fig. 65 shows a center-fed array with a diagonal layout of the feed system [56]. The characteristics of this array are similar to those of the patch array shown in Fig. 61: The use of a series feed-system results in short transmission line runs and good efficiency; power division at the feed-line branching points and radiation losses at the dipoles lead to a tapered amplitude distribution across the array aperture and low side lobes; and a symmetric distribution of the excitation phase about the array center ensures that the main beam remains pointed in the broadside direction when the frequency is shifted away from the design frequency. But the beamwidth increases with Δf and the pattern bandwidth is limited. An experimental 16×16 element array for the 36-GHz band printed on a polythene substrate of 0.8-mm thickness has shown a gain of 25 dB and side lobe and cross-polarization levels below -20 dB [56]. The array size was 6.8×6.8 cm^2. A rather good radiation efficiency of 60 percent was measured at the center frequency, but the bandwidth (corresponding to a 3-dB reduction in gain) was limited to 700 MHz.

Antenna-to-Waveguide Coupling—An effective technique for coupling a microstrip millimeter-wave antenna to a metal waveguide (mounted on the back of the base

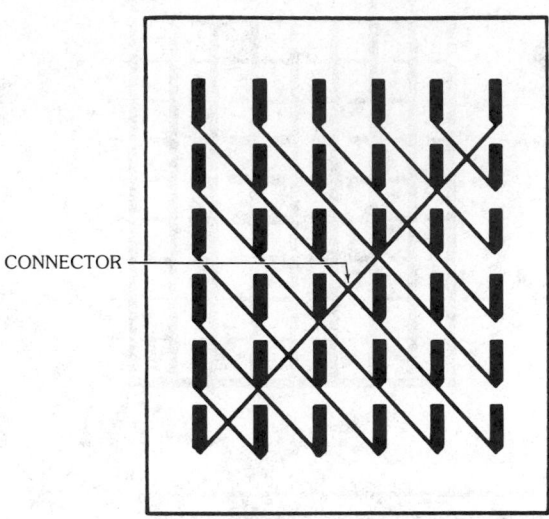

CONNECTOR

Fig. 65. Center-fed printed dipole array. (*After Williams [56], previously published in* Electronics Letters, *March 1978, by The Institution of Electrical Engineers*)

plate) is the use of a vertical probe which is in electrical contact with the microstrip feed line of the antenna and extends into the waveguide at a quarter-wavelength distance from the short-circuited end of this guide [48, 50]; see Fig. 66. With appropriate design, a vswr better than 2:1 can be achieved easily. The bandwidth of this coupling technique exceeds that of the antenna. Coupling to a coaxial line is straightforward and also leads to good results when an arrangement is used where the line feeds the antenna from the back [48]. Both methods can be combined by feeding a microstrip antenna from a metal waveguide via a microstrip-to-coax-to-waveguide transition. The transition would include a short section of coaxial line only so that the high losses normally associated with these lines at millimeter-wave frequencies are avoided. A transition of this type is often used by necessity since a probe which extends from the microstrip feed line into a metal waveguide (Fig. 66) has to pass through a small aperture in the base plate of the antenna. The thickness of the base plate is easily in the order of a quarter-wavelength or more in the millimeter-wave region so that the probe section in the aperture will act as a short piece of coaxial line [63]. Efficient coupling of microstrip antennas to other millimeter-wave guides remains to be studied. Work in this area is in progress and includes the examination of alternative arrangements for feeding these antennas from metal waveguides and coaxial lines [62, 199, 200].

Computer-Aided Design (CAD)—In many cases the design of microstrip array antennas to specifications can be handled efficiently by computer. In particular, the pattern synthesis problem can be solved in this way. The parameter dependencies of microstrip antennas are well understood; their radiation and impedance characteristics depend sensitively on their dimensions; and the overall array configuration (including substrate parameters) is usually specified beforehand so that the pattern synthesis problem reduces to the constrained adjustment of a well-

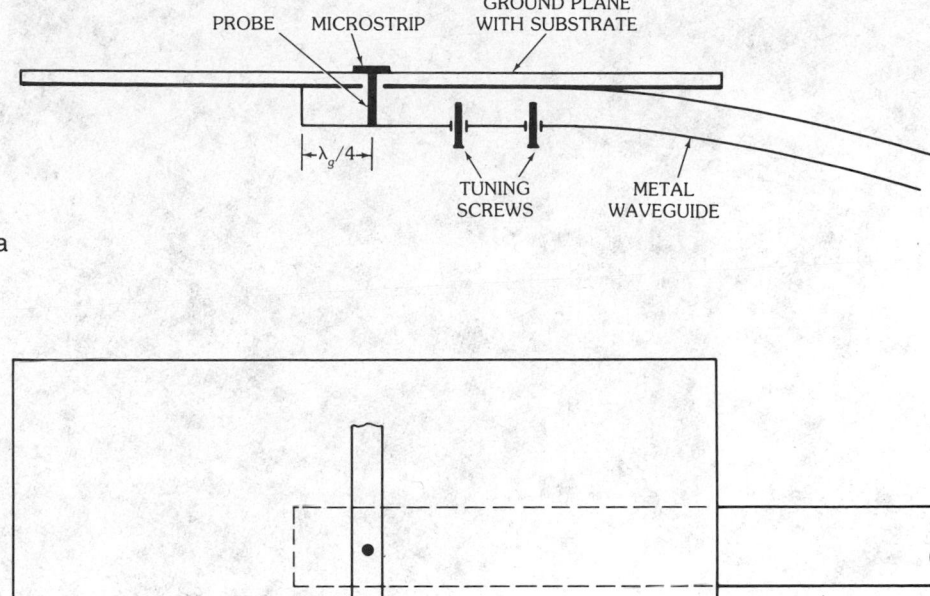

Fig. 66. Microstrip to metal waveguide (probe) transition. (*a*) Side view. (*b*) Top view. (*After Weiss [48]*)

defined set of geometrical parameters. Computer design may be supplemented by computer-controlled fabrication of the array where the fabrication control data are generated as part of the CAD procedure. In this case there is no need for drawings or layout artwork and the engineering process is considerably simplified [64].

As an example a computer code devised by Campi is mentioned here which permits the design and the generation of fabrication control data for series-fed linear arrays of rectangular microstrip patch antennas [64]. The code generates an equivalent network representation of the array, from which the radiation pattern can be calculated. Using an optimization subroutine the code then continuously varies the conductance values of the array elements until the computed radiation pattern approximates in the least-square sense the desired pattern specified by its directivity gain, main beam direction, and side lobe structure, which may be characterized by a Tschebychev, Taylor, binominal, or other distribution function. The code includes provisions for parametric and sensitivity studies of the pattern, a feature useful for engineering and optimization analyses. Measurements on antennas produced with the help of this code have demonstrated the usefulness of the CAD procedure for the 1- to 100-GHz region [64]. Fig. 67 shows a 28-element array designed for the 94-GHz band.

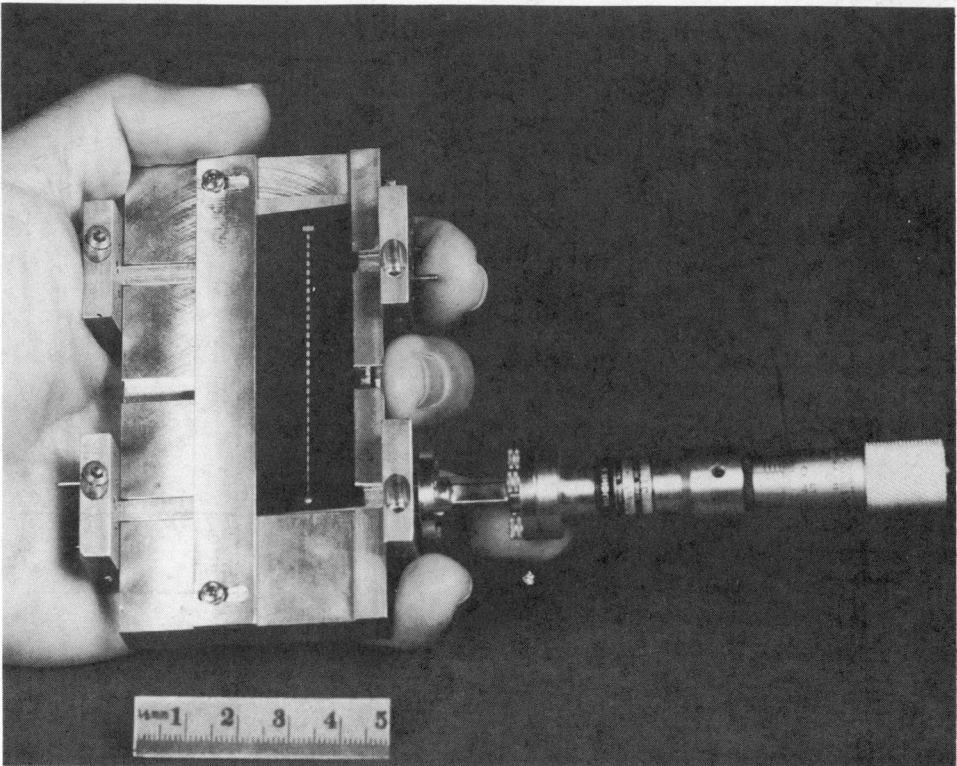

Fig. 67. Computer-designed 94-GHz microstrip antenna array. (*Courtesy Campi [64]*)

Microstrip Antennas with Electrically Thick Substrates

In the microwave region the thickness of typical substrates used in the design of printed antennas is a small fraction of a wavelength only. Thin substrates imply the use of resonator antennas of high Q in order to raise the radiation resistance to reasonable values. This, in turn, leads to a narrow bandwidth. In the millimeter-wave region, on the other hand, a substrate of low geometrical height may have an electrical thickness in the order of a quarter-wavelength to several wavelengths. Hence there is no need for using high-Q antennas, and other antenna configurations can be employed which provide substantially larger bandwidth. An example would be a printed dipole or loop antenna of large width w; Fig. 68a shows a dipole antenna of this type. The use of thick substrates has the additional advantage that fabrication tolerances become less critical.

A problem associated with substrates of large electrical thickness is the generation of *surface waves*. These surface waves are guided away by the substrate even though they do not radiate;* their power must therefore be counted as antenna loss. The design problem is to obtain a large bandwidth while maintaining good efficiency.

---

*Except for producing uncontrolled radiation at substrate edges or at other geometrical discontinuities.

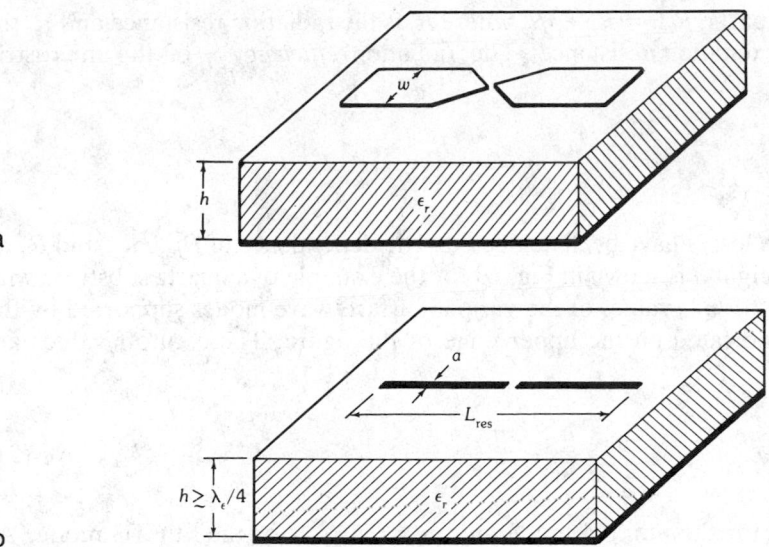

Fig. 68. Printed dipole antennas on electrically thick substrates. (*a*) Dipole of large width *w*. (*b*) Thin-wire dipole.

The theory of microstrip antennas on thick substrates has recently been studied by Alexopoulos, Katehi, and Rutledge [58, 59], who have established the basic parameter dependencies and provided systematic design information for these antennas. The results obtained so far apply to the case of a single printed dipole of small width *w* (wire dipole).* For most millimeter-wave applications high-gain antennas of reasonably broad bandwidth are required, and an array of dipoles of large width would be of interest. A theory of such arrays, however, is not as yet available and the results of the single-dipole study are discussed here in some detail since they are not only interesting in themselves, but provide conditions for substrate optimization which should apply to arrays as well. In addition, the study yields at least a worst-case estimate (thin dipole case) on attainable bandwidth. Alexopoulos [58] has pointed out that the analytical expressions for the field radiated by a microstrip dipole are obtained as products of a substrate factor which is independent of the antenna dimension and an antenna factor which is independent of the substrate parameters. This suggests that substrate optimization arguments will hold for all printed antennas regardless of their shape [58]. An independent study by Pozar [194] has yielded theoretical results in substantial agreement with [58, 59].

The following discussion draws freely from the cited papers by Alexopoulos and colleagues. The antenna configuration considered is shown in Fig. 68b. A center-fed wire dipole is placed on a substrate of permittivity ϵ_r and height *h*; the substrate is backed by a metal ground plane. The dipole is assumed to have the resonance length corresponding to this situation. Its resonance impedance R_{res}

---

*Pairs of dipoles and their mutual interaction effects have been studied in a separate investigation by Alexopoulos and Rana [201].

consists of two parts, $R_{res} = R_r + R_s$, where R_r is the radiation resistance and R_s the surface-wave excitation resistance. The radiation *efficiency* η of the microstrip dipole is

$$\eta = \frac{R_r}{R_r + R_s}$$

where substrate losses have been neglected. The dependence of R_{res}, R_r, and R_s on the substrate height h is shown in Fig. 69 for the example of a quartz substrate with $\epsilon_r = 4$. The cutoff h/λ_0 values of the various surface-wave modes supported by the substrate are indicated on the upper edge of the figure. These cutoff values are given by

$$\frac{h}{\lambda_0} = \frac{m}{4} (\epsilon_r - 1)^{-1/2} \tag{21}$$

where $m = 2n$ for the nth TM mode and $m = 2n + 1$ for the nth TE mode. As before, λ_0 is the free-space wavelength.

The important result illustrated by Fig. 69 is that the radiation efficiency reaches a maximum just below the cutoff thickness of the lowest-order TE surface-wave mode;* careful numerical studies [58] have shown that this result holds not only for quartz but also for materials of any (reasonable) permittivity. From the physical viewpoint this is easy to understand. The lowest-order surface-wave mode, TM_0, does not have a cutoff frequency and is always present, but for substrates of low height it is only weakly excited by a horizontal dipole antenna, while the radiated power increases with h/λ_0. The TE_0 surface-wave mode, on the other hand, would be strongly excited by the antenna. Hence maximum η occurs just before this mode can exist.

With (21), the optimum substrate thickness h_{opt} satisfies

$$h_{opt} \lesssim \frac{\lambda_0}{4\sqrt{\epsilon_r - 1}} \tag{22}$$

The optimum thickness and the corresponding maximum radiation efficiency η_{max} are shown in Fig. 70 as functions of substrate permittivity ϵ_r. For completeness the resonance length L_{res} of the dipole is included in the figure. The graph shows that the requirement of high radiation efficiency implies the use of a substrate of relatively low permittivity. For η_{max} to exceed 50 percent, ϵ_r must be smaller than 4.5. Hence quartz would be a useful substrate material while such high-permittivity materials as silicon and gallium arsenide would not be suitable for this application. The optimum substrate thickness h_{opt} is moderately larger than a quarter-wavelength (in the substrate material) and remains within reasonable limits even in the lower millimeter-wave region. For example, at 38 GHz, h_{opt} would be approximately 1.2 mm for quartz and approximately 1.8 mm for Duroid ($\epsilon_r = 2.20$).

---

*The radiation efficiency approaches 100 percent for $h \rightarrow 0$. But the radiation resistance vanishes at this point so that this first maximum of η is meaningless.

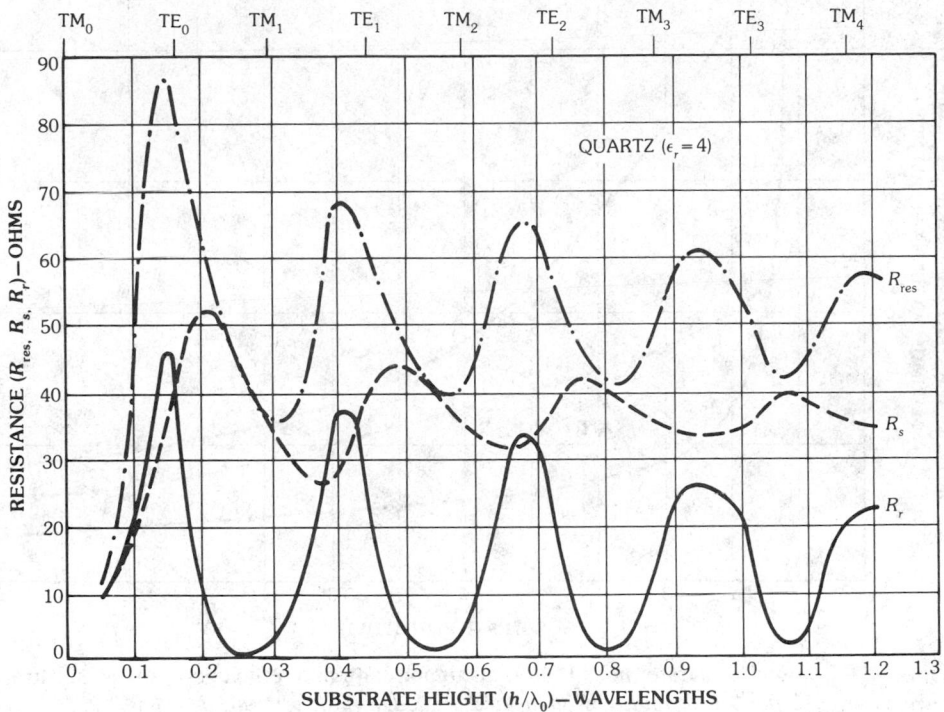

Fig. 69. Resonance resistance of microstrip dipole versus substrate height for quartz. (*After Alexopoulos, Katehi, and Rutledge [58], © 1983 IEEE*)

The impedance *bandwidth* (bw) of a dipole antenna near resonance can be expressed as

$$\text{bw} = \frac{\Delta f}{f_0} = \frac{2R_{\text{res}}}{f_0(dX/df)_{\text{res}}} \tag{23}$$

where $(dX/df)_{\text{res}}$ is the derivative of the input reactance of the antenna at resonance.[*] The dependence of $\Delta f/f_0$ on substrate thickness h/λ_0 is shown in Fig. 71 for $\epsilon_r = 4$. A dipole of radius $a/\lambda_0 = 10^{-4}$ is assumed. Maximum bandwidth, similar to maximum efficiency, occurs just below the cutoff thickness of the TE_0 surface-wave mode [58]. This is typical for low-permittivity substrates, with $\epsilon_r < 5$. For large ϵ_r, on the other hand, maximum bandwidth would be obtained at a substrate thickness noticeably greater than h_{opt}. This is illustrated by Fig. 72. The solid curve shows the maximum attainable bandwidth as a function of substrate permittivity; the dashed curve shows the bandwidth at $h = h_{\text{opt}}$, i.e., under the

---

[*]In computing the curves of Figs. 71 and 72, Alexopoulos and colleagues [58] have used a modified version of (23), which expresses the bandwidth in terms of the parameters discussed above, i.e., $\Delta f/f_0 = 2R_{\text{res}}/[L_\lambda(dX/dL_\lambda)]_{\text{res}}$, where $L_\lambda = L/\lambda_\epsilon$ is the antenna length normalized to wavelength in the dielectric, and the term in square brackets is taken at resonance. This modified definition of bandwidth utilizes the fact that the antenna reactance near resonance changes more rapidly with antenna length than with substrate height.

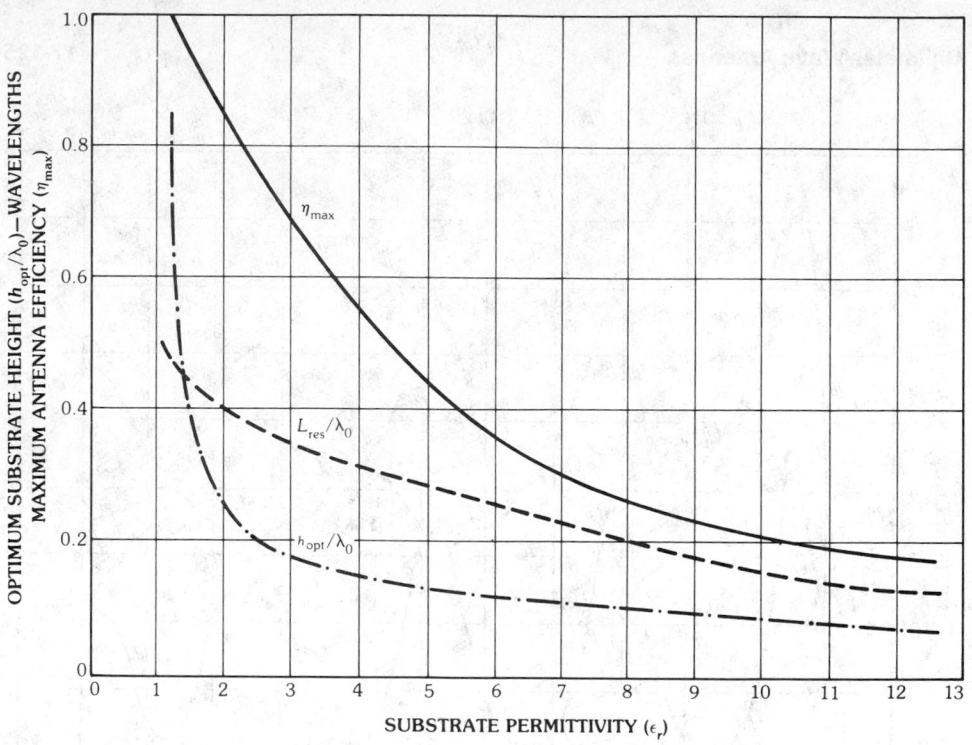

Fig. 70. Optimum substrate height and maximum antenna efficiency versus substrate permittivity. (*After Alexopoulos, Katehi, and Rutledge [58], © 1983 IEEE*)

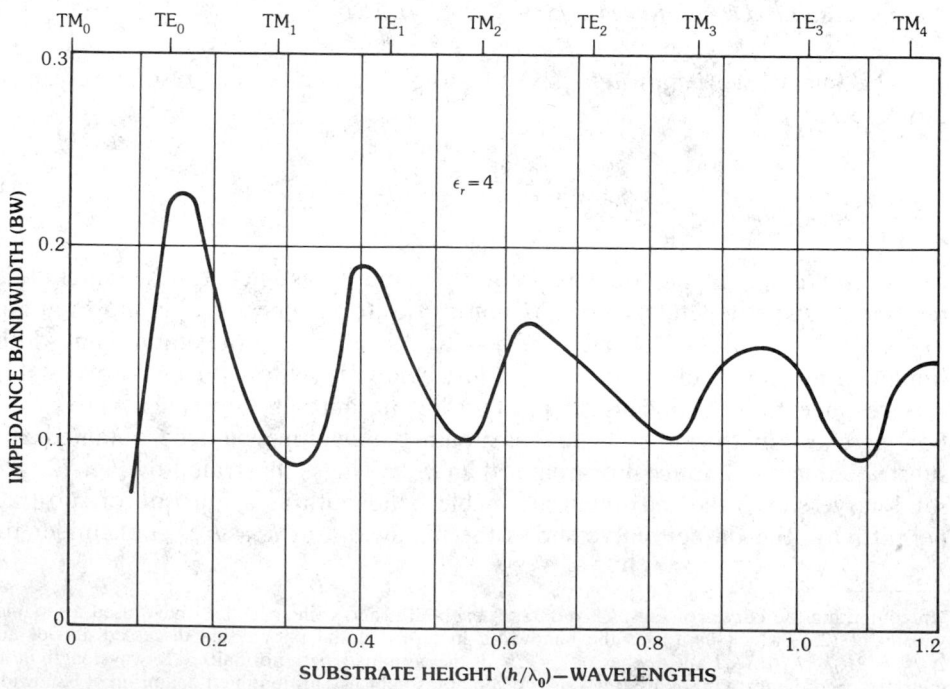

Fig. 71. Bandwidth of thin printed dipole versus substrate height for $\epsilon_r = 4$. (*After Alexopoulos, Katehi, and Rutledge [58], © 1983 IEEE*)

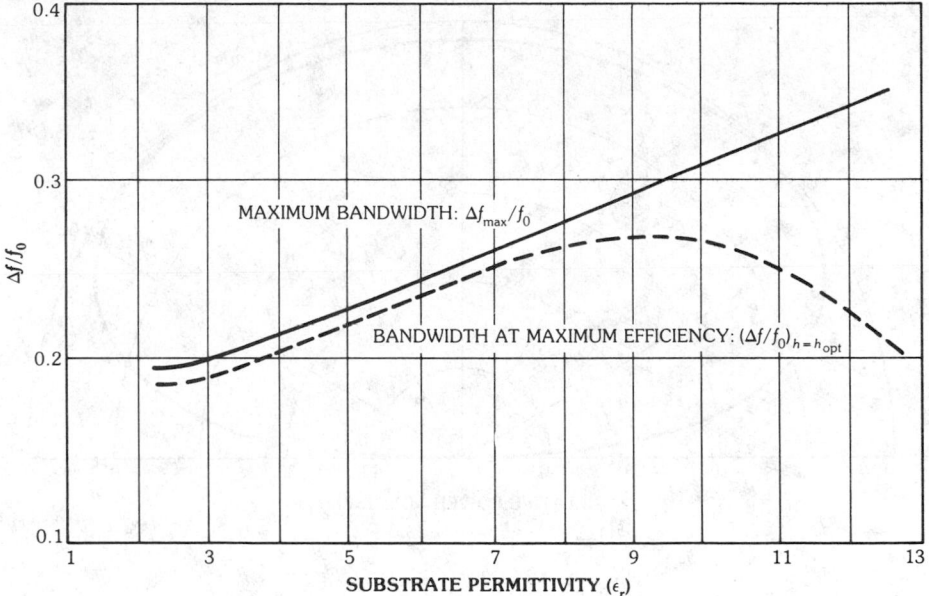

Fig. 72. Bandwidth as a function of substrate permittivity. (*After Alexopoulos, Katehi, and Rutledge [58], © 1983 IEEE*)

maximum efficiency constraint. In this case, $\Delta f/f_0$ reaches a maximum of 26 percent at $\epsilon_r = 9.4$. In the range of low permittivities (less than 4.5), where high radiation efficiencies can be obtained, the bandwidth is limited to approximately 20 percent. This, however, is a substantial improvement over the bandwidth of microstrip patch antennas on thin substrates, which is on the order of a few percent only. Note, furthermore, that this bandwidth applies to wire dipoles of very small radii. Printed dipoles of large width can be expected to have substantially greater bandwidth.

The theoretical radiation pattern of a dipole of length $L = 0.3\lambda_0$ placed on a quartz substrate of thickness $h = 0.2\lambda_0$ is shown in Fig. 73. The pattern is essentially hemispherical, but both the E- and H-plane patterns have a null in the horizontal plane $\theta = \pm 90°$. Evidently any radiation in this plane occurs in the form of surface waves.

Latest theoretical results by Alexopoulos and Jackson [71,72] show that the performance of microstrip dipole antennas can be significantly improved by the use of a low-loss superstrate whose permittivity (or permeability) exceeds that of the substrate.* It appears that by careful selection of the substrate and superstrate parameters substantial improvements in gain and radiation resistance can be realized over a relatively large bandwidth. Furthermore, these parameters can be adjusted such that a nearly omnidirectional E- or H-plane pattern is achieved. Alternatively, a very high radiation efficiency—close to 100 percent with no surface

---

*Both infinitesimal dipoles [71] and dipoles of finite length and width [72] have been considered in this study.

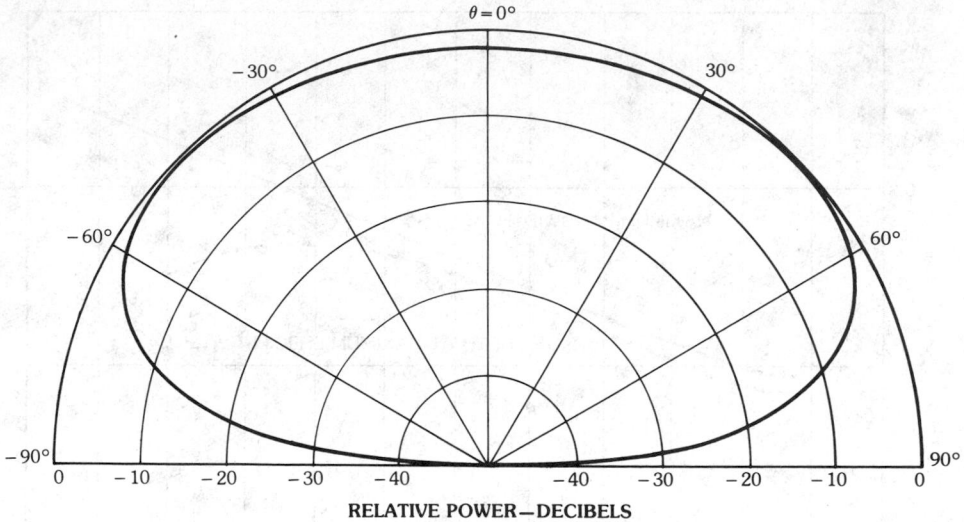

a

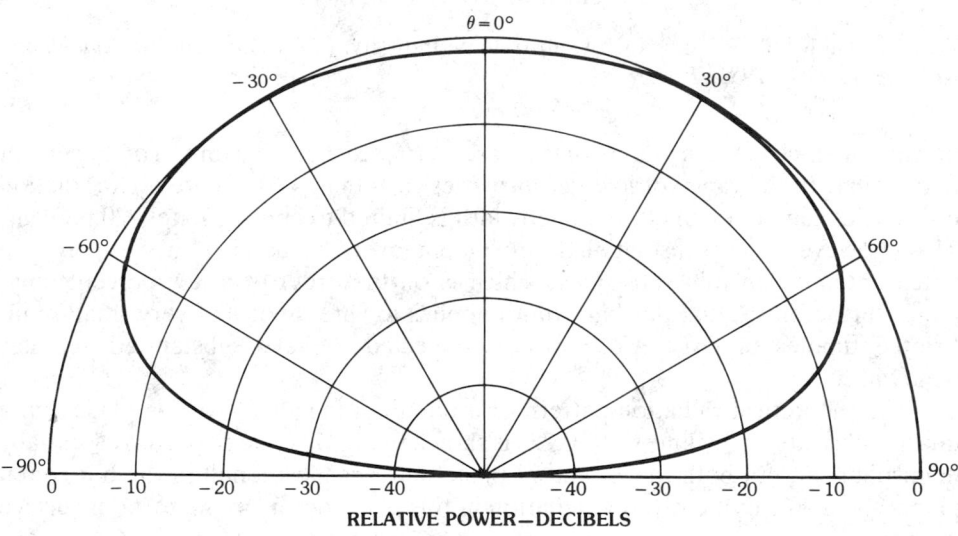

b

Fig. 73. Power radiation pattern of microstrip dipole on quartz substrate, with $h/\lambda_0 = 0.2$, $L/\lambda_0 = 0.3$, and $\epsilon_r = 4$. (a) E-plane pattern. (b) H-plane pattern. (*After Alexopoulos, Katehi, and Rutledge [58], © 1983 IEEE*)

wave excited—can be obtained over a limited bandwidth. An additional very practical advantage is that a superstrate will act as a protective cover for the antenna.

Millimeter-wave antennas are usually designed for high directivity gain, which implies the use of arrays of printed dipoles. This leads to the additional question of mutual coupling between closely spaced printed dipoles—surface waves supported by the substrate are likely to increase this coupling—and to the problem of devising efficient feed systems for such arrays. Ideally, i.e., for simplicity and ease of fabrication, the feed system should be printed with the dipoles on the same substrate. But a balanced, two-wire feed system would be complicated and not advantageous for integrated designs, while an unbalanced single-wire system would lead to feed-line leakage, multimoding, and impedance matching problems because of the large substrate height. The use of a dual substrate appears more promising. In this case a single-line microstrip feed system would be printed on the lower portion of the substrate while the (unbroken) dipoles would be etched on the upper portion, and coupling would be effected electromagnetically, i.e., without conductive contact between feed lines and antennas. In the microwave region, where the substrate thickness is only a small fraction of a wavelength, this method has been used successfully [202–204]. Its feasibility for feeding antennas on fairly thick substrates has been shown recently by Katehi and Alexopoulos for the single-dipole case [205]. But for large microstrip millimeter-wave arrays on thick substrates, the design of efficient, easy-to-fabricate feed systems remains up to now an open question requiring further study. (The Alexopoulos group is investigating this problem at present.) A similar question remains for the problem of mutual coupling in such arrays. A theoretical study on the interaction between two printed dipoles has confirmed that mutual coupling is a significant effect which cannot be neglected [201]. Suppression of surface waves through the use of a superstrate of high ϵ_r or μ_r [71, 72] seems to be a promising approach to this problem deserving further attention.

Holographic Antennas

When a hologram made in accordance with the interference pattern of two propagating waves is illuminated by one of these waves, the amplitude and phase distribution of the other wave is reconstituted on the surface of the hologram. A holographic plate may thus be used as a beam-shaping antenna; it permits one to derive a desired aperture distribution (across the hologram) from a given illumination incident from a primary feed. In the microwave and millimeter-wave regions a hologram can be approximated by an appropriately shaped metallization pattern etched on a printed-circuit board. The use of a metal pattern will, in general, permit one to closely approximate the desired phase distribution across the holographic aperture, whereas the amplitude distribution (which is less critical) may not be entirely independent of that of the illuminating field.

Iizuka and coworkers have described an antenna of this type; the following discussion is based on a paper by these authors [195]. The antenna, depicted in Fig. 74, consists of a holographic plate with a metallization pattern made by photoetching a printed-circuit board in accordance with the interference pattern of a spherical wave originating from the horn aperture C and a plane wave incident normal to the surface of the printed-circuit board. The holographic pattern, in this case, consists of a set of concentric circular rings spaced one wavelength apart. When the plate is illuminated by a spherical wave radiated from the horn aperture the waves scattered from the metal rings have equal phase, and two beams emanate from the plate in opposite directions normal to the plate surface. In holographic terminology the true image and the conjugate image are reconstructed [195].

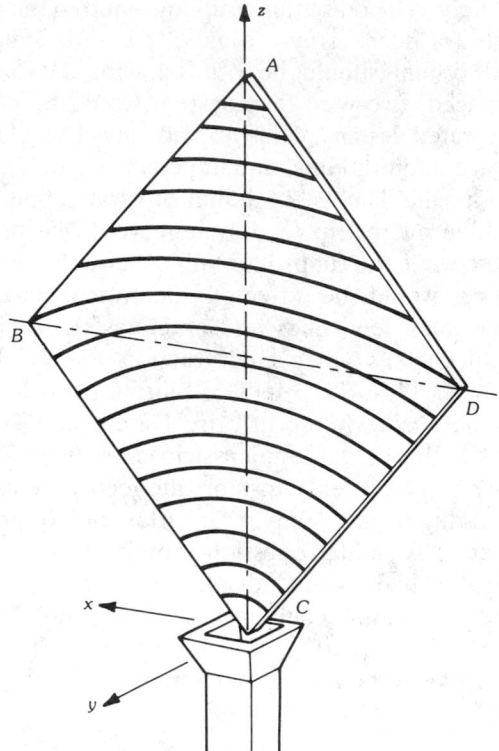

Fig. 74. Holographic antenna. (*After Iizuka et al. [195], © 1975 IEEE*)

If two holographic plates are used in a parallel arrangement, the conjugate beam traveling in the backward direction can be suppressed while the main beam in the forward direction is reinforced [195]. The condition is that the two plates are spaced a quarter-wavelength apart and that the metal rings on the second plate have radii a quarter-wavelength larger than those on the first plate. Figs. 75a and 75b illustrate these conditions. Fig. 75c shows, moreover, that in the presence of the second plate the wave reflected toward the feed horn (downward direction) is canceled. Hence the impedance characteristic of the antenna is improved, resulting in increased bandwidth. Radiation in the upward direction can be reduced to a tolerable level by adjusting the position of the plates relative to the feed horn [195]. Beam shaping is possible by modifying the metallization pattern of the plates.

Fig. 76 compares the measured E- and H-plane patterns of a one-plate and a two-plate antenna. The plates had a size of $12\lambda_0 \times 12\lambda_0$ with sixteen metal rings printed on each board by photoetching. A pyramidal horn with aperture size $0.56\lambda_0 \times 0.64\lambda_0$ was placed such that its apex was located near the common center point of the metal rings of each plate. The dual-plate antenna has a unidirectional pattern with a front-to-back ratio of 20 dB and a main lobe level 7 dB higher than that of the single-plate antenna. Back lobe radiation can be reduced further by increasing the number of holographic plates [195].

By replacing the feed horn with a printed antenna etched with the holographic

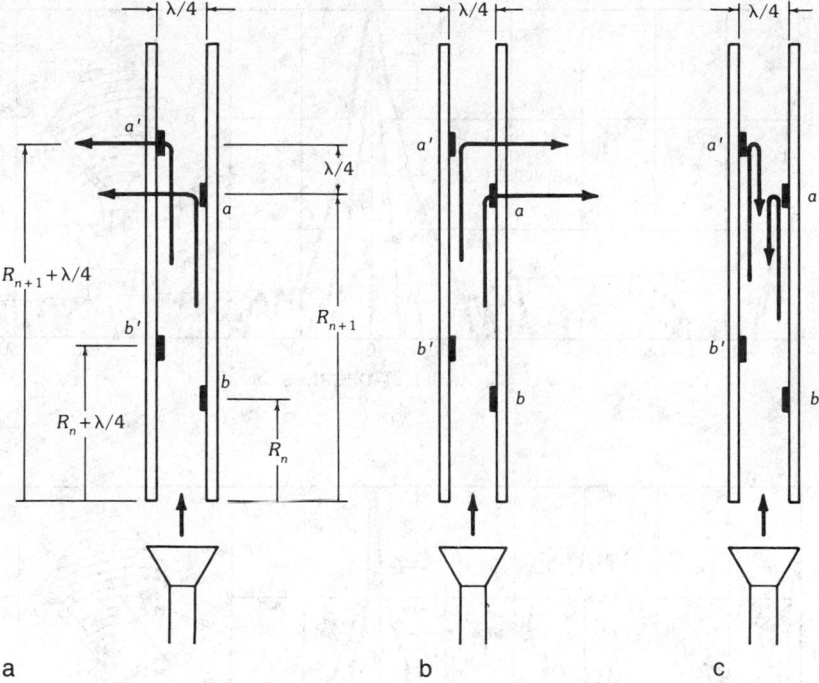

Fig. 75. Dual-plate holographic antenna. The spherical wave emanating from the feed horn is scattered at the metal rings of the two plates. (*a*) In the forward direction the scattered fields of the two plates have the same path lengths and add in phase. (*b*) In the backward direction the path lengths differ by $\lambda_0/2$; the scattered fields are in antiphase and cancel each other. (*c*) In the downward direction the path lengths also differ by $\lambda_0/2$ and the wave reflected toward the feed horn is eliminated. (*After Iizuka et al. [195], © 1975 IEEE*)

pattern on the same circuit board, a fully integrated antenna would be realized [195]. This possibility appears particularly attractive for millimeter-wave applications. The experiments were conducted at 19 GHz.

5. Integrated Antennas

Fundamental contributions to the state of the art in millimeter-wave antennas are currently being made by the development of integrated antennas where printed circuits and active solid-state components, usually in monolithic form, become an integral part of the radiating structure. Such antennas serve not only as electromagnetic radiators, but include additional functions (e.g., mixing, amplification, or phase shifting) in a highly structured but very compact device. Integrated antennas are of interest not only for the lower millimeter-wave region but are particularly well suited for the so-called near–millimeter-wave range above 100 GHz [88, 99, 100, 206–213]; and it can be expected that eventually they will be extended into the sub-millimeter-wave region as well. The small size of radiating elements, components, and circuits in these frequency bands strongly suggests the use of integrated planar design approaches.

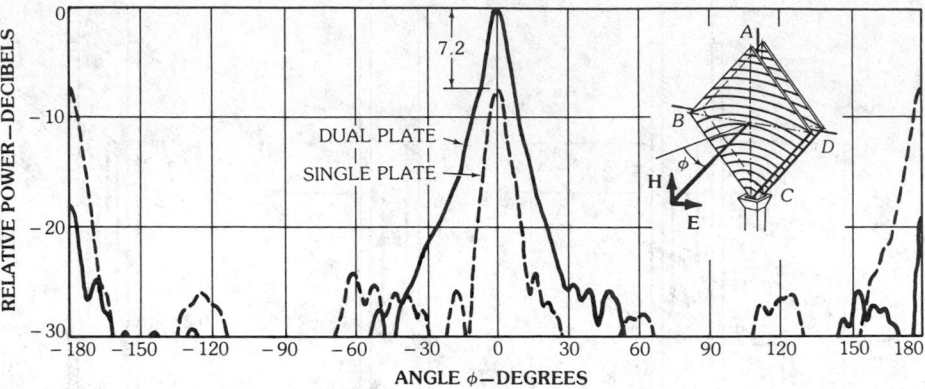

a

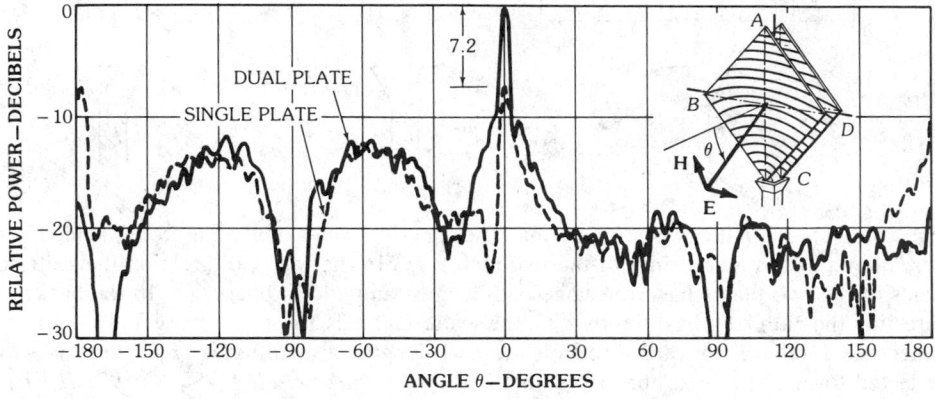

b

Fig. 76. Radiation patterns of single-plate and dual-plate holographic antennas. (*a*) *E*-plane pattern. (*b*) *H*-plane pattern. (*After Iizuka et al. [195], © 1975 IEEE*)

The microstrip antenna technique lends itself to the design of array antennas where transmission lines, phase shifter circuits, and receiver or transmitter circuits are integrated with the radiating elements on the same substrate (which may consist of several layers when circuits for more complex functions have to be accommodated). Both dielectric antennas and printed-circuit antennas are suited for monolithic designs using silicon or gallium arsenide (GaAs) technology. While silicon has been used extensively in the past, it appears that future monolithic designs will prefer gallium arsenide, in particular at millimeter-wave frequencies. Silicon is mechanically stronger and easier to work with. The primary advantage of gallium arsenide is its high electron mobility, which is critical for the performance of monolithic active components at higher frequencies.

Devising precise fabrication techniques is an essential step in the design of integrated antennas and must be regarded as an integral part of the design procedure. It is evident that the feasibility of these small but highly structured devices depends on the availability of appropriate fabrication methods and that the quality and performance of the finished device are ultimately determined by the accuracy of the manufacturing process.

The crucial problem is the design and fabrication of the integrated active components. In comparison to a monolithic mixing or switching diode, for example, the antenna elements themselves are large in size and simple in structure. Practical problems include heat sinking for silicon components, and mechanical stress when gallium arsenide, a mechanically fragile material, is used. Also, available space for integrated components may be limited in the case of printed-circuit antennas since the antenna patches, though physically small, are large from an IC viewpoint and together with their feed lines take up a substantial portion of the substrate surface. In addition, a certain clearance is required between the radiating elements and feed lines on the one hand and any integrated circuits and components on the other in order to minimize undesired near-field coupling into these circuits and avoid pattern distortions. Because of these and similar constraints the electrical performance of integrated antennas may remain somewhat below the state of the art. But this will be traded against substantial advantages in compactness, ruggedness, and cost [100].

Current studies are usually directed at developing a proof-of-concept model to demonstrate a specific point with potentially important future implications, but an overall research pattern is not easily identified. Instead of trying to present a general overview over the field, an attempt is made here to indicate trends by discussing three typical examples: a dielectric antenna with a monolithically integrated silicon mixer diode useful for short-distance, line-of-sight communication at high data rates [209]; a steerable-beam monolithic microstrip antenna array with integrated phase shifters for radar and guidance applications [210]; and an integrated microstrip dipole array for millimeter-wave imaging [212]. The discussion draws freely from the cited publications.

Tapered Dielectric-Rod Antenna with Integrated Mixer Diode

The antenna was designed and investigated by Yao, Schwarz, and Blumenstock [209]. An earlier version, designed for the far-infrared region and using microbolometers as detectors, was developed by Rutledge and others [99, 100]. The configuration of the present antenna is shown in Fig. 77. It consists of a monolithic assembly of a dielectric antenna, a dielectric waveguide, a coupler, and a

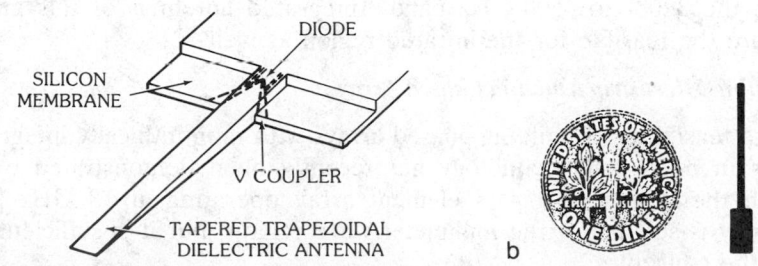

Fig. 77. Dielectric antenna with monolithically integrated Schottky mixer diode. (*a*) Configuration. (*b*) Size comparison with US dime. (*After Yao, Schwarz, and Blumenstock [209], © 1982 IEEE*)

mixer diode. The antenna is a tapered dielectric rod, etched together with the dielectric waveguide from a silicon wafer. A V-shaped metallic coupler printed on the backside of the waveguide serves to concentrate the received electromagnetic energy toward the mixer diode located at the apex. No external metal structure is needed to achieve this coupling; the V-shaped coupler is the only metal part of the device. The mixer diode, which is the critical component of the integrated antenna, is a planar Schottky diode. The diode configuration is indicated in Fig. 78; details on its fabrication process may be found in [209]. The finished antenna, designed for the 100-GHz band, is 17 mm long, 0.21 mm high, and 1.1 mm wide at the position of the coupler. The membrane thickness is 0.03 mm. The polarization of the incident field is assumed to be parallel to the plane of the coupler.

The device was tested at 85 GHz. A 3-dB beamwidth of 49° in the *E*-plane and 56° in the *H*-plane was measured, which corresponds to a directivity gain of 10 dB. The bandwidth of the device is very broad. The lower frequency limit is primarily determined by the antenna and waveguide dimensions, the upper limit by the rolloff of the responsivity of the Schottky diode above a cutoff frequency that depends on the details of its fabrication. Designed for operation in the 100-GHz band, the diode in the experimental antenna leads to a measured maximum responsivity of 35 V/W at the test frequency of 85 GHz.

In its present form, the device has poor conversion efficiency; an overall conversion loss* of 35 dB was measured. But the investigators expect that substantial improvements can be achieved by the use of epitaxial silicon or gallium arsenide. From the available data for these materials they predict a conversion loss of 16 dB for epitaxial silicon and a loss as low as 10 dB for epitaxial gallium arsenide [209].

The device should be regarded as a first step toward the development of fully integrated receiver front ends where, in addition to a mixer, a hybrid coupler, a filter, an if amplifier, and possibly a local oscillator would be monolithically integrated with a dielectric antenna and waveguide. The most likely application of an integrated antenna of this type would be in short-haul terrestrial communication and data transmission in the 100- to 300-GHz band [209]. The device would be used in this case as a feed system for a highly directive lens or mirror.

The diode used in the present antenna has a relatively low cutoff frequency. But Yao and colleagues [209] point out that planar surface oriented diodes have been demonstrated with a zero bias cutoff frequency several hundred times higher, i.e., in the 3200- to 4500-GHz band. Integrated antennas of this kind should, therefore, be feasible for the infrared region as well.

Monolithic Microstrip Antenna Phased Array

The feasibility of building phased arrays with monolithically integrated phase shifters in microstrip technology has recently been demonstrated by Stockton through the design of a 4×4 element array operating at 18 GHz [210]. This frequency is still below the millimeter-wave region, but it is sufficiently close to prove the concept.

---

*Overall conversion loss equals the power delivered to the if amplifier divided by the received radiative power.

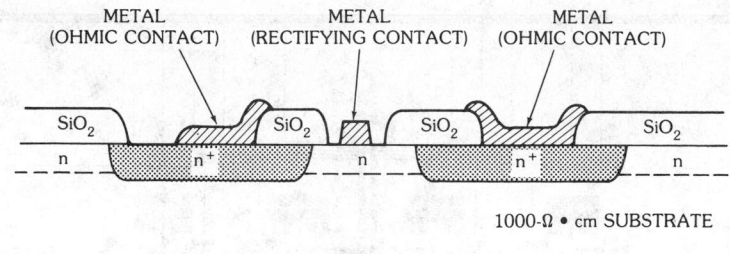

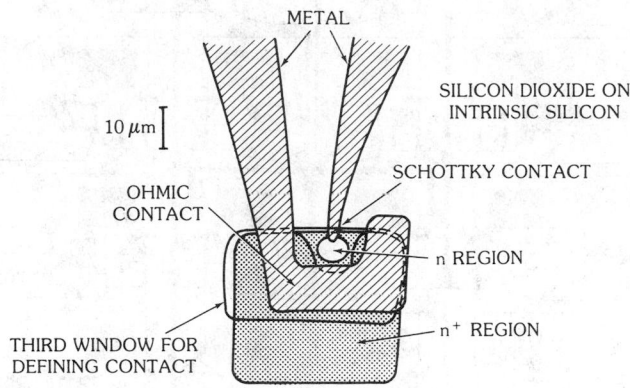

Fig. 78. Surface-oriented Schottky diode. (*a*) Diagram. (*b*) Top view. (*After Yao, Schwarz, and Blumenstock [209], © 1982 IEEE*)

Fig. 79 shows the metallization pattern of the array. The antenna is printed on a 5-cm–diameter silicon wafer of high resistivity (equaling or greater than 5000 Ω-cm) and includes 96 planar beam lead pin diodes which are fabricated in situ in this wafer. The diodes serve as rf switching devices in the integrated phase shifters of the array. The phase shifters are 3-bit devices with two diodes required per bit; there is one phase shifter for each array element. (In the usual manner the first, second, and third stages of each phase shifter provide phase changes of 180°, 90°, and 45°, respectively, thus permitting phasing in increments of 45°.) After fabrication of the monolithic diodes a thin metal layer is deposited by thin film sputtering on the wafer, and the radiating elements, feed networks, and phase shifter circuits of the array are created by photoetching of this layer. All metal parts are fabricated in a single metallization and etching step which also produces the diode interconnections and dc bias lines. Fig. 79 shows the photo mask for the etching process; the dark areas and gray lines define the metal pattern of the array remaining on the wafer after etching.

Array efficiency is estimated at 25 percent, determined by a loss of 3 dB in the complete, 3-bit phase shifters and a loss of 0.35 dB/cm in the 50-Ω microstrip feed lines (based on a postprocessing substrate resistivity of 2500 Ω-cm). The directivity gain of the array is 17.5 dB. Experiments have shown good pattern quality in the *H*-plane. In the *E*-plane certain pattern distortions occurred when the antenna was scanned. These distortions are under study at present.

This investigation leads the way toward the development of large monolithic

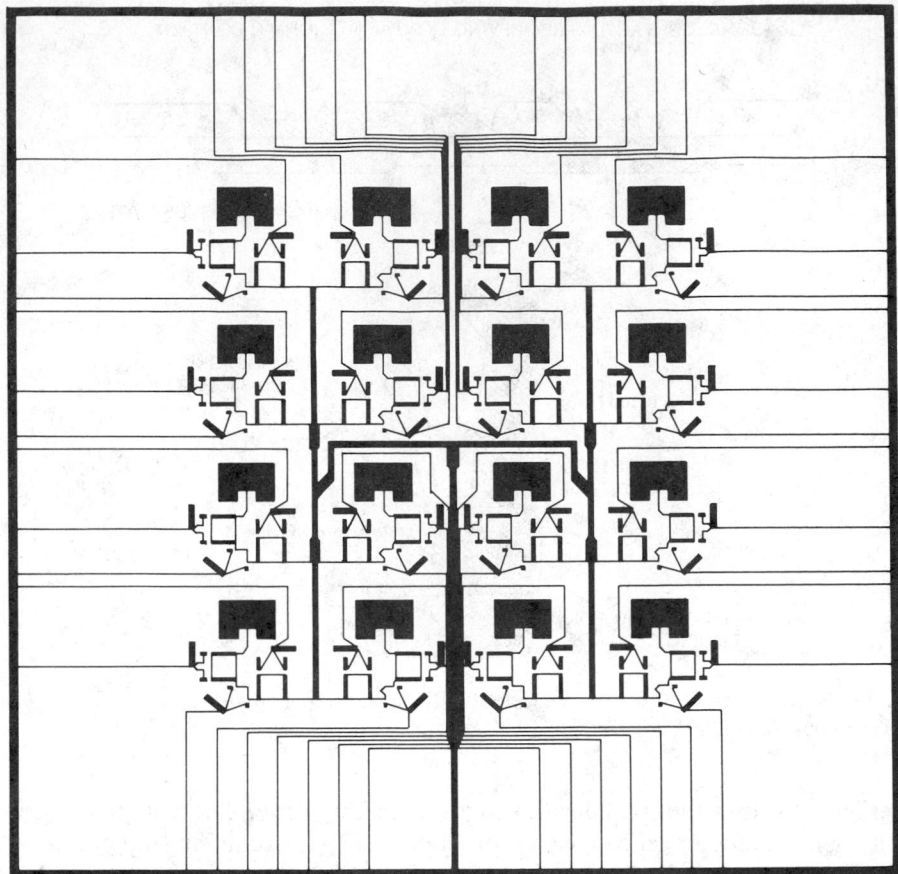

Fig. 79. Photo mask defining metallization pattern for 18-GHz, sixteen-element monolithic phased array in microstrip technology. (*Courtesy Ball Aerospace Systems Division*)

phased arrays of high directivity gain and an operating frequency in the lower millimeter-wave region. The size of the present array is limited by the diameter (approximately 5 cm) of currently available silicon wafers. At millimeter-wave frequencies gallium arsenide provides advantages over silicon in the monolithic fabrication of active devices and is likely to be used as the substrate material.

Eventually, in the design of future high-gain, multielement phased arrays, it will become desirable to integrate not only phase shifters but power amplifiers and low-noise detectors with the radiating elements or else at a suitably chosen subarray level, in order to ensure high efficiency in the transmit mode and high signal-to-noise ratio in the receive mode. Monolithic integration of these devices as well will be the ultimate goal. The use of distributed, individually controlled amplifiers and detectors in phased arrays would have the additional advantages that beam shaping becomes possible, that low side lobes can be achieved over a broad scan range and a sizable frequency band, and that graceful performance degradation will occur if some of the active components should fail. Work on monolithic millimeter-wave transmit and receive modules for array applications is in progress [214, 215]. A main objective is to demonstrate that gallium-arsenide FET technology is capable of meeting the requirements for the 30- to 100-GHz band.

Integrated Near–Millimeter-Wave Imaging Array

The Rutledge group [212] has recently demonstrated an integrated millimeter-wave imaging antenna in the form of a linear array of eight parallel dipoles, each with its own detector, monolithically fabricated on a planar substrate of fused quartz ($\epsilon_r = 3.78$). The array is designed for the 1.2-mm band and, in its present form, provides one-dimensional imaging. The multidetector approach overcomes the limited sensitivity and speed of more conventional imaging systems using a single, scanned detector.

The operating principle is illustrated in Fig. 80. The objective lens projects the object plane onto the substrate surface (image plane), where the receiving array is located. Plotting the output of the antenna detectors produces the received image. In addition to serving as base plate for the array, the planar quartz substrate has the beneficial effect of directing the radiation patterns of the array elements into the direction of the incoming waves* [58, 208, 216]. But the surface waves supported by the (planar) substrate would counteract this effect by reducing the effective receiving cross section of the array elements and by increasing mutual coupling. The primary purpose of the substrate lens is to minimize these ill effects by eliminating the conditions under which surface waves can be trapped in the substrate [216]. Absorption and reflection losses of the substrate lens are minimized by choosing a low-loss dielectric (the lens is made from the same material as the substrate) and by using a quarter-wavelength antireflection coating.

In the practical realization the antennas are bow-tie dipoles spaced 310 μm apart (a half-wavelength in quartz) so that the eight-element array has a width of 2.5 mm; see Fig. 81. The dipoles are fabricated from evaporated silver by photography and lift-off. The detectors are bismuth bolometers located at the antenna feed points, where they connect to the two arms of the bow-ties. The

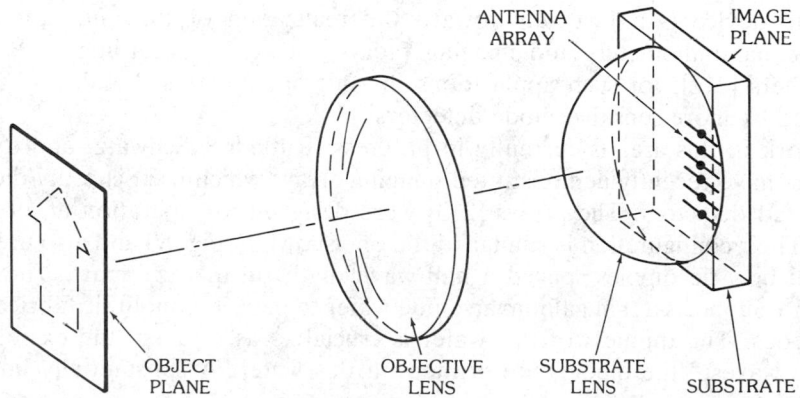

Fig. 80. Operating principle of near–millimeter-wave imagining array. (*After Neikirk et al. [212]*)

---

*The array is located on the *far* side of the substrate.

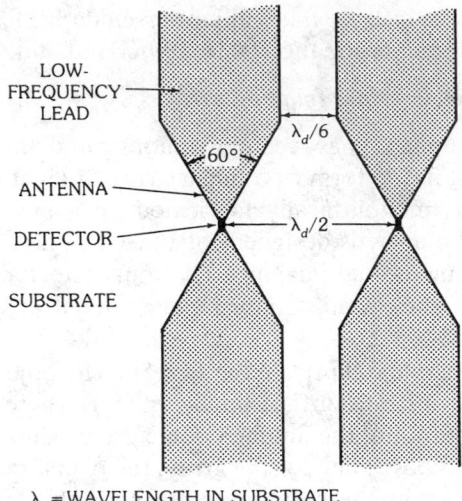

Fig. 81. Bow-tie antennas for imaging array. (*After Rutledge and Muha [208],* © *1982 IEEE*)

dimensions of these microbolometers are 2 μm by 5 μm. The bow-tie angle of 60° results in an antenna impedance of 150 Ω, matching that of the detectors. A special feature of the antenna is that the extension arms of the bow-tie dipoles serve as low-frequency leads with no additional rf isolation required. The substrate lens had a radius of 6.6 mm and the lens-substrate combination a thickness of 10 mm. In the presence of the lens about 50 percent of the available power is coupled into the array.

The element responsivity was measured at 1 to 2 V/W, but with recent improvements a system responsivity of 10 V/W can be expected as a reasonable estimate [217]. The experiments have confirmed, moreover, that the array resolution approaches the diffraction limit.

The device is a first step toward the realization of fully integrated two-dimensional radiometers and imaging radars. However, according to Rutledge and others [212], for such applications the microbolometers should probably be replaced by more sensitive diode detectors.

Work in this area is currently in progress. Rutledge, Schwarz, and their co-workers have recently demonstrated imaging arrays which use Schottky diodes as integrated detectors. The arrays [213] were designed for operation at 69 and 94 GHz. Their configuration is similar to the one shown in Fig. 80 and also uses eight parallel bow-tie dipoles spaced a half-wavelength (in quartz) apart. The quartz substrate supports a thin gallium arsenide wafer to permit monolithic fabrication of the diodes. The thinness of this wafer is crucial so as to avoid the excitation of surface waves;* the dipoles are printed on this wafer. A substantially improved

---

*Due to the asymmetry that the gallium-arsenide wafer is bordered by quartz on one side and by air on the other, *all* surface waves supported by this wafer have cutoff frequencies and can be suppressed [218]. The substrate lens, however, is still neede to eliminate surface waves in the (much thicker) quartz substrate.

system responsivity was obtained; the highest measured value was 330 V/W.

Fig. 82 shows the diode configuration and a micrograph of a diode integrated with one of the dipoles of the array. The fabrication process is described in [213]. The cutoff frequency of the diodes is estimated at 500 to 700 GHz. Fig. 83 is a photograph of a recent version of the 94-GHz array. The device shown here operates with nine dipoles and incorporates some further improvements in design which have resulted in a still higher system responsivity of 600 V/W [218]. The

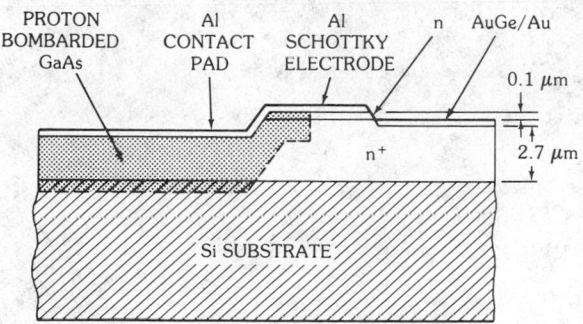

a

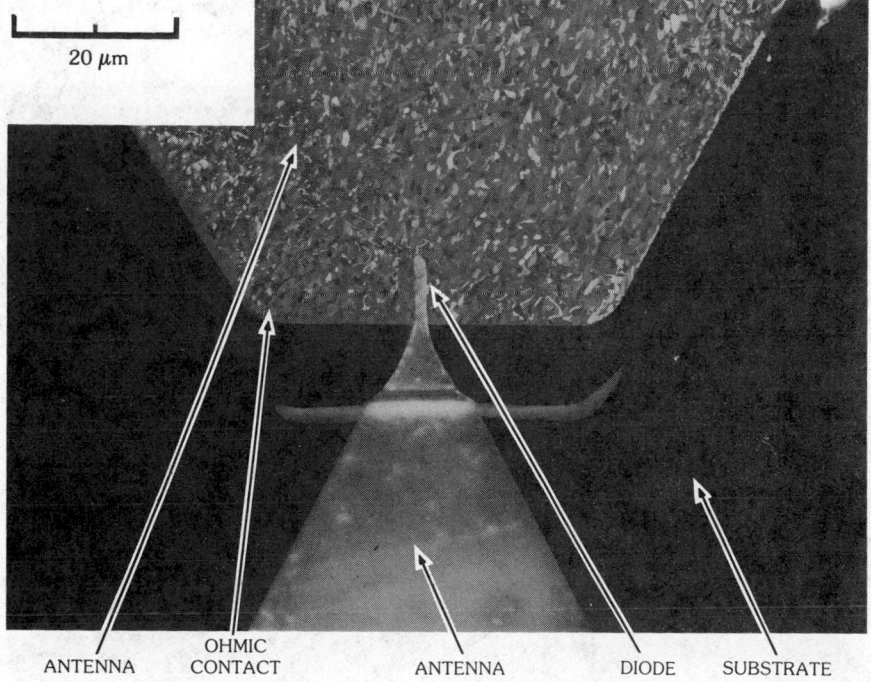

b

Fig. 82. Monolithic Schottky diode for imaging array. (*a*) Diode configuration. (*b*) Diode integrated with bow-tie dipole. (*Courtesy Rutledge, after Rav-Noy et al. [213],* © *1983 IEEE*)

a

b

Fig. 83. Imaging array for 94-GHz band. (*a*) Monolithic 9-element array on gallium arsenide wafer. (*b*) Finished device with substrate lens. (*Courtesy, Rutledge, after Zah et al. [218],* © *1984 IEEE*)

width of the nine-element array is about 2 wavelengths in quartz, i.e., approximately 7 mm.

While the use of diode detectors leads to superior system responsivity, the investigators still expect that microbolometers will provide a better noise performance. The peculiar advantage of diode detectors is that they will permit heterodyning [217].

6. References

[1] A. F. Kay, "Millimeter-wave antennas," *Proc. IEEE*, vol. 54, pp. 641–647, April 1966.

[2] R. B. Dybdal, "Millimeter-wave antenna development," *Proc. 1982 Antenna Appl. Symp.*, University of Illinois, September 22–24, 1982.

[3] E. K. Reedy and G. W. Ewell, "Millimeter-wave radar, in infrared and millimeter waves," in *Infrared and Millimeter Waves, Vol. 4,* ed. by K. J. Button and J. C. Wiltse, New York: Academic Press, 1981, pp. 23–24.

[4] R. W. Myhre, "Advanced 30/20-GHz multiple-beam antennas for communications satellites," *Proc. 1982 Antenna Appl. Symp.*, University of Illinois, September 22–24, 1982.

[5] J. Smetana, "Application of MMIC modules in future multibeam satellite antenna systems," *Proc. 1982 Antenna Appl. Symp.*, University of Illinois, September 22–24, 1982.

[6] N. Williams and N. A. Adatia, "Millimeter-wave antennas," *Proc. Mil. Microwave Conf.*, London, UK, October 22–24, 1980.

[7] R. L. Powers, K. D. Arkind, and D. G. LaRochelle, "Extended design yields compact 18–40 GHz antenna," *Microwave Systems News*, pp. 89–96, November 1981.

[8] D. K. Waineo and J. F. Konieczny, "Millimeter-wave monopulse antennas with rapid scan capability," *IEEE/AP-S 1979 Intl. Symp. Dig.*, pp. 477–480, Seattle, June 18–22, 1979.

[9] O. B. Kessler and J. George, "94-GHz antenna techniques," *Tech. Rep. AFWAL-TR-80-1222*, Texas Instruments, February 1981.

[10] L. M. Schwab, A. R. Dion, and D. L. Washington, "Space-fed, offset, plane-wave Cassegrainian system for ehf applications," *Proc. 1982 Antenna Appl. Symp.*, University of Illinois, September 22–24, 1982.

[11] J. Ruze, "Antenna tolerance theory—a review," *Proc. IEEE*, vol. 54, pp. 633–640, April 1966.

[12] B. E. Vu The Bao, "Influence of correlation interval and illumination taper in antenna tolerance theory," *Proc. Inst. Electr. Eng.* (London), vol. 166, pp. 195–202, 1969.

[13] H. Zucker, "Gain of antennas with random surface deviations," *Bell Syst. Tech. J.*, vol. 47, pp. 1637–1651, 1968.

[14] P. R. Cowles and E. A. Parker, "Reflector surface error compensation in Cassegrainian antennas," *IEEE Trans. Antennas Propag.*, vol. AP-23, pp. 323–328, May 1975.

[15] E. N. Davies, "Proposals for electronic compensation of surface profile errors in large reflectors, design and construction of large steerable aerials," *IEE Conf. Pub. 21*, pp. 80–83, 1966.

[16] R. A. Semplak and R. H. Turrin, "Pressure formed parabolic reflectors for millimeter wavelengths," *IEEE Trans. Antennas Propag.*, vol. AP-16, pp. 762–764, November 1968.

[17] A. G. Repjar and D. P. Kremer, "Accurate evaluation of a millimeter-wave compact range using planar near-field scanning," *IEEE Trans. Antennas Propag.*, vol. AP-30, pp. 419–425, May 1982.

[18] J. R. Cogdell et al., "High-resolution millimeter reflector antennas," *IEEE Trans. Antennas Propag.*, vol. AP-18, pp. 515–529, July 1970.

[19] J. W. M. Baars, "Design of large millimeter-wave radio telescopes," *Proc. 1980 Intl. URSI Symp.*, Munich, Germany, pp. 143A/1–4, August 26–29, 1980.

[20] S. V. Hoerner and W.-Y. Wong, "Gravitational deformation and astigmatism of tiltable radio telescopes," *IEEE Trans. Antennas Propag.*, vol. AP-23, pp. 689–695, September 1975.

[21] W. Rotman, "Ehf dielectric lens antenna for multiple beam MILSATCOM applications," *1982 Intl. IEEE/AP-S Symp. Dig.*, Albuquerque, pp. 132–135, June 1982.

[22] J. T. Mayhan and A. J. Simmons, "A low side lobe K_a-band antenna-radome study," *IEEE Trans. Antennas Propag.*, vol. AP-23, pp. 569–572, July 1975.

[23] G. Bekefi and G. W. Farnell, "A homogeneous dielectric sphere as a microwave lens," *Can. J. Phys.*, vol. 34, 1956.

[24] G. Toraldo diFrancia, "New stigmatic system of the concentric type," *J. Opt. Soc. Am.*, vol. 47, p. 566, June 1957.

[25] G. Toraldo diFrancia, "Spherical lenses for infrared and microwaves," *J. Appl. Phys.*, vol. 32, p. 2051, 1961.

[26] T. C. Cheston and E. J. Luoma, "Constant-K lenses," *APL Tech. Dig.*, April 1963.

[27] S. Cornbleet, "A simple spherical lens with external foci," *Microwave J.*, vol. 8, p. 65, May 1965.

[28] T. L. ApRhys, "The design of radially symmetric lenses," *IEEE Trans. Antennas Propag.*, vol. AP-18, pp. 497–506, July 1970.

[29] H. Mieras, "Radiation pattern computation of a spherical lens using Mie series," *IEEE Trans. Antennas Propag.*, vol. AP-30, pp. 1221–1224, November 1982.

[30] M. S. Narasimhan and S. Ravishankar, "Radiation from aperture antennas radiating in the presence of a dielectric sphere," *IEEE Trans. Antennas Propag.*, vol. AP-30, pp. 1237–1240, November 1982.

[31] Y. W. Chang, J. A. Paul, and Y. C. Ngan, "Millimeter-wave integrated circuit modules for communication interconnect systems," *US Army R&D Tech. Rep. ECOM76-1353-2*, November 1977.

[32] J. J. Lee, "Dielectric lens shaping and coma-correction zoning, Part I: Analysis," *IEEE Trans. Antennas Propag.*, vol. AP-31, pp. 211–216, January 1983.

[33] J. J. Lee and R. L. Carlise, "A coma-corrected multibeam shaped lens antenna, Part II: Experiments," *IEEE Trans. Antennas Propag.*, vol. AP-31, pp. 216–220, January 1983.

[34] J. C. Wiltse, "Fresnel zone-plate lenses," *SPIE Proc.*, vol. 544, Millimeter-Wave Technology III, July 1985.

[35] H. T. Buscher, "Electrically controllable liquid artificial dielectric media," *IEEE Trans. Microwave Theory Tech.*, vol. MTT-27, pp. 540–545, May 1979.

[36] R. Blundell and M. C. Carter, "Millimeter-wave aerials for full illumination radars," *Proc. Mil. Microwaves*, London, UK, October 22–24, 1980.

[37] R. Baldwin and P. A. McInnes, "A rectangular corrugated feedhorn," *IEEE Trans. Antennas Propag.*, vol. AP-23, pp. 814–817, November 1975.

[38] H. P. Coleman, R. M. Brown, and B. D. Wright, "Parabolic reflector offset fed with a corrugated conical horn," *IEEE Trans. Antennas Propag.*, vol. AP-23, pp. 817–819, November 1975.

[39] T.-S. Chu and W. E. Legg, "Gain of corrugated conical horns," *IEEE Trans. Antennas Propag.*, vol. AP-30, pp. 698–703, July 1982.

[40] B. M. Thomas and K. J. Greene, "A curved aperture corrugated horn having very low cross-polar performance," *IEEE Trans. Antennas Propag.*, vol. AP-30, pp. 1068–1072, November 1982.

[41] C. A. Mentzer and L. Peters, "Pattern analysis of corrugated horn antennas," *IEEE Trans. Antennas Propag.*, vol. AP-24, pp. 304–309, 1976.

[42] M. A. Jansen, S. M. Bednarczyk, S. Gulkis, H. W. Marlin, and G. F. Smoot, "Pattern measurement of a low-sidelobe horn antenna," *IEEE Trans. Antennas*

Propag., vol. AP-27, pp. 551–555, July 1979.

[43] T. Ohtera and H. Ujiie, "Radiation performance of a modified rhombic dielectric plate antenna," *IEEE Trans. Antennas Propag.*, vol. AP-29, pp. 660–662, July 1981.

[44] A. Hombach, "Dielectric feeds with low cross polarization," *Proc. 1980 Intl. URSI Symp.*, Munich, Germany, August 26–29, 1980.

[45] N. Nakajima and R. Wantanabe, "A quasioptical circuit technology for short millimeter-wavelength multiplexers," *IEEE Trans. Microwave Theory Tech.*, vol. MTT-29, pp. 897–905, September 1981.

[46] R. J. Eckstein et al., "35-GHz active aperture," *Tech. Rep. AFWAL-TR-81-1079*, Motorola, June 1981.

[47] M. F. Durkin, R. J. Eckstein, M. D. Mills, M. S. Stringfellow, and R. A. Neidhard, "35-GHz active aperture," *1981 IEEE MTT-S Intl. Microwave Symp. Dig.*, Los Angeles, June 1981.

[48] M. A. Weiss, "Microstrip antennas for millimeter waves," *R&D Tech. Rep. ECOM-76-0110-F*, October 1977.

[49] M. A. Weiss and R. B. Cassell, "Microstrip millimeter-wave antenna study," *R&D Tech. Rep. CORADCOM-77-0158-F*, April 1979.

[50] M. A. Weiss, "Microstrip antennas for millimeter waves," *IEEE Trans. Antennas Propag.*, vol. AP-29, pp. 171–174, January 1981.

[51] T. Metzler, "Microstrip series arrays," *Proc. Workshop on Printed-Circuit Antenna Technology*, New Mexico State University, Las Cruces, NM, pp. 21-1–16, October 17–19, 1979.

[52] K. R. Carver and J. W. Mink, "Microstrip antenna technology," *IEEE Trans. Antennas Propag.*, vol. AP-29, pp. 2–24, January 1981.

[53] R. J. Mailloux, J. F. McIlvenna, and N. P. Kernweis, "Microstrip array technology," *IEEE Trans. Antennas Propag.*, vol. AP-29, pp. 25–37, January 1981.

[54] J. R. James, P. S. Hall, C. Wood, and A. Henderson, "Some recent developments in microstrip antenna design," *IEEE Trans. Antennas Propag.*, vol. AP-29, pp. 124–128, January 1981.

[55] J. R. James, P. S. Hall, C. Wood, and A. Henderson, "Microstrip antenna research at the Royal Military College of Sciences," *Proc. Workshop on Printed Circuit Antenna Technology*, New Mexico State University, Las Cruces, pp. 1-1–10, October 17–19, 1979.

[56] J. C. Williams, "A 36-GHz printed planar array," *Electron. Lett.*, vol. 14, pp. 136–137, March 1978.

[57] I. E. Rana and N. G. Alexopoulos, "Current distribution and input impedance of printed dipoles," *IEEE Trans. Antennas Propag.*, vol. AP-29, pp. 99–105, January 1981.

[58] N. G. Alexopoulos, P. B. Katehi, and D. B. Rutledge, "Substrate optimization for integrated-circuit antennas," *IEEE Trans. Microwave Theory and Tech.*, vol. MTT-31, pp. 550–557, July 1983.

[59] P. B. Katehi and N. G. Alexopoulos, "On the effect of substrate thickness and permittivity on printed-circuit dipole properties," *IEEE Trans. Antennas Propag.*, vol. AP-31, pp. 34–39, January 1983.

[60] W.-C. Chew and J.-A. Kong, "Analysis of circular microstrip disk antenna with a thick dielectric substrate," *IEEE Trans. Antennas Propag.*, vol. AP-29, pp. 68–76, January 1981.

[61] F. Lalezari, "Dual-polarized high-efficiency microstrip antenna," U.S. patent no. 322930, November 1981.

[62] J. R. James and C. M. Hall, "Investigation of new concepts for designing millimeter-wave antennas," final technical report on Contract DAJA37-80-C-0183, US Army European Research Office, September 1983.

[63] F. Lalezari and T. Pett, "Millimeter microstrip antennas for use in mil-spec environment," final report to Battelle Columbus Labs/US Army CECOM, November 21, 1983.

[64] M. Campi, "Design of microstrip linear array antennas by computer," 1982 Army

Science Conference, West Point, New York, June 1982.

[65] B. B. Jones, F. Y. M. Chow, and A. W. Seeto, "The synthesis of shaped patterns with series-fed microstrip patch arrays," *IEEE Trans. Antennas Propag.*, vol. AP-30, pp. 1206–1212, November 1982.

[66] F. Lalezari, private communication.

[67] A. Henderson, A. E. England, and J. R. James, "New low-loss millimeter-wave hybrid microstrip antenna array," Eleventh European Microwave Conference, Amsterdam, The Netherlands, September 1981.

[68] J. R. James and A. Henderson, "A critical review of millimeter planar arrays for military applications," Military Microwave Conference, London, England, October 20–22, 1982.

[69] R. M. Knox, "Dielectric waveguide microwave integrated circuits—an overview," *IEEE Trans. Microwave Theory Tech.*, vol. MTT-24, pp. 806–814, November 1976.

[70] M. T. Birand and R. V. Gelsthorpe, "Experimental millimetric array using dielectric radiators fed by means of dielectric waveguide," *Electron. Lett.*, vol. 17, no. 18, pp. 633–635, September 1981.

[71] N. G. Alexopoulos and D. R. Jackson, "Fundamental superstrate effects (cover) on printed-circuit antennas," *Integrated Electromagnetics Lab Rep.*, No. 10, UCLA Rep. no. ENG-83-50, October 14, 1983.

[72] D. R. Jackson and N. G. Alexopoulos, "Superstrate (cover) effects on printed-circuit antennas," *Dig. 1984 Intl. IEEE-APS/URSI Symp.*, pp. 563–565, Boston, June 1984.

[73] B. R. Rao, "94-gigahertz slotted waveguide array fabricated by photolithographic techniques," *Dig. 1983 Intl. IEEE-APS/URSI Meeting*, pp. 688–689, Houston, May 23–26, 1983.

[74] F. G. Farrar, "Millimeter-wave W-band slotted waveguide antennas," *1981 IEEE/AP-S Intl. Symp. Dig.*, Los Angeles, pp. 436–439, June 16–19, 1981.

[75] M. C. Carter and E. R. Cashen, "Linear arrays for centimetric and millimetric wavelengths," *Proc. Mil. Microwave Conf.*, London, UK, October 22–24, 1980.

[76] C. A. Strider, "Millimeter-wave planar arrays," *1974 Millimeter-Wave Techniques Conf.*, NELC, San Diego, pp. B6/1-11, March 26–28, 1974.

[77] C. A. Boyd, Jr., "Practical millimeter-wave ferrite phase shifters," *Microwave J.*, vol. 25, pp. 105–108, December 1982.

[78] J. L. Hilburn and F. H. Prestwood, "K-band frequency scanned waveguide array," *IEEE Trans. Antennas Propag.*, vol. AP-22, pp. 340–342, March 1974.

[79] R. C. Honey, "Line source and linear arrays for millimeter wavelengths," *Proc. Symp. Millimeter Waves*, Polytechnic Institute of Brooklyn, NY, March 31–April 2, 1959.

[80] S. Silver, ed., *Microwave Antenna Theory and Design*, New York: McGraw-Hill Book Co., 1949.

[81] E. M. Systems, Inc., "Millimeter-wave spiral antenna," *Microwave J.*, vol. 18, p. 28, December 1975.

[82] K. S. Kelleher, "High-gain reflector-type antennas," chapter 12 in *Antenna Engineering Handbook*, ed. by H. J. Jasik, New York: McGraw-Hill Book Co., 1961.

[83] C. C. Chen, "High-efficiency V-band fan beam antenna," *1981 Intl. Symp. Dig.*, Los Angeles: IEEE Antennas and Propagation Society, pp. 124–126, June 16–19, 1981.

[84] K. S. Kelleher, "Scanning antennas," chapter 15 in *Antenna Engineering Handbook*, ed. by H. J. Jasik, New York: McGraw-Hill Book Co., 1961.

[85] E. C. Dufort and H. Uyeda, "A wide-angle scanning optical antenna," *IEEE Trans. Antennas Propag.*, vol. AP-31, pp. 60–67, January 1983.

[86] F. E. Ore, "The modified biconical horn millimeter-wave antenna," *Tech. Rep. AFAL-TR-65-156*, Univ. of Illinois, Urbana, July 1965.

[87] A. D. Munger and J. H. Greg, "Design and development of biconical horn antennas for 18–100 GHz," *NOSC Tech. Note 580*, San Diego, December 1978.

[88] F. C. Jain, R. Bansal, and C. V. Valerio, Jr., "Semiconductor antenna: a new device in millimeter- and submillimeter-wave integrated circuits," *IEEE Trans. Microwave*

Theory Tech., vol. MTT-32, pp. 204–207, February 1984.

[89] "Final technical report for 54.5-GIIz omni radio system," *Rep. No. 1301 R 0001*, US Army CORADCOM Contract No. DAAK80-79-C-0765, Norden Systems, October 1980.

[90] M. J. Ehrlich, "Slot antenna arrays," chapter 9 in *Antenna Engineering Handbook*, ed. by H. J. Jasik, New York: McGraw-Hill Book Co., 1961.

[91] S. A. Long, M. V. McAllister, and L. C. Shen, "The resonant cylindrical dielectric cavity antenna," *IEEE Trans. Antennas Propag.*, vol. AP-31, pp. 406–412, May 1983.

[92] T. E. Nowicki, "Microwave substrates, present and future," *Proc. Workshop on Printed Circuit Antenna Technology*, New Mexico State University, Los Cruces, NM, October 17–19, 1979.

[93] H. Howe, Jr., "Dielectric material development," *Microwave J.*, vol. 21, pp. 39–40, November 1978.

[94] P. F. Goldsmith, "Quasi-optical techniques," Chapter 5 in *Infrared and Millimeter Waves, Vol. 6*, ed. by K. J. Button, p. 335, New York: Academic Press, 1982.

[95] G. E. Mueller and W. A. Tyrell, "Polyrod antennas," *Bell Syst. Tech. J.*, vol. 26, pp. 837–851, October 1947.

[96] P. Mallach, "Notes from unpublished German documents," Central Radio Bureau Library, London.

[97] D. G. Kiely, *Dielectric Aerials*, London: Methuen & Co., 1952.

[98] F. J. Zucker, "Surface- and leaky-wave antennas," chapter 16 in *Antenna Engineering Handbook*, ed. by H. J. Jasik, New York: McGraw-Hill Book Co., 1961.

[99] D. B. Rutledge et al., "Antennas and waveguides for far-infrared integrated circuits," *IEEE J. Quantum Electron.*, vol. QE-16, pp. 508–516, May 1980.

[100] S. E. Schwarz and D. B. Rutledge, "Moving toward near mm-wave integrated circuits," *Microwave J.*, vol. 23, pp. 47–67, June 1980.

[101] J. A. G. Malherbe, T. N. Trinh, and R. Mittra, "Transition from metal to dielectric waveguide," *Microwave J.*, vol. 23, pp. 71–74, November 1980.

[102] B. J. Levine and J. E. Kietzer, "Hybrid millimeter-wave integrated circuits," *US Army R&D Tech. Rep.* ECOM-74-0577-F, October 1975.

[103] H. S. Jones, Jr., "Conformal and small antenna designs," *Tech. Rep. HDL-TR-1952*, US Army ERADCOM, Adelphi, Maryland, April 1981.

[104] D. C. Chang and R. Mittra, "Workshop report on modern millimeter-wave systems," *Scientific Rep. 63*, University of Colorado, Boulder, May 1981.

[105] Y. Shiau, "Dielectric-rod antennas for millimeter-wave integrated circuits," *IEEE Trans. Microwave Theory Tech.*, vol. MTT-24, pp. 869–872, November 1976.

[106] S. Kobayashi, R. Mittra, and R. Lampe, "Dielectric tapered-rod antennas for millimeter-wave applications," *IEEE Trans. Antennas Propag.*, vol. AP-30, pp. 54–58, January 1982.

[107] S. P. Schlesinger and A. Vigants, "HE_{11} excited dielectric surface-wave radiators," *Tech. Rep. AFCRC-TN-59-573*, Columbia University, June 1959.

[108] C. M. Angulo and W. S. C. Chang, "A variational expression for the terminal admittance of a semi-infinite dielectric rod," *IEEE Trans. Antennas Propag.*, vol. AP-7, p. 207, July 1959.

[109] F. J. Zucker, "Electromagnetic boundary waves," *Tech. Rep. AFCRL-63-165*, AF Cambridge Research Laboratories, June 1963.

[110] J. R. James, "Theoretical investigation of cylindrical dielectric-rod antennas," *Proc. IEE* (London), vol. 114, pp. 309–319, March 1967.

[111] A. D. Yaghjian and E. D. Kornhauser, "A modal analysis of dielectric-rod antennas excited in the HE_{11} mode," *IEEE Trans. Antennas Propag.*, vol. AP-20, pp. 122–128, January 1972.

[112] J. R. Blakey, "A scattering theory approach to the prediction of dielectric-rod antenna radiation patterns: the TM_{01} mode," *IEEE Trans. Antennas Propag.*, vol. AP-23, pp. 577–579, July 1975.

[113] F. J. Zucker, "Surface-wave antennas," chapter 21 in *Antenna Theory, Part II*, ed. by R. E. Collin and F. J. Zucker, New York: McGraw-Hill Book Co., 1969.

[114] L. B. Felsen, "Radiation from a tapered surface-wave antenna," *IRE Trans. Antennas Propag.*, vol. AP-8, pp. 577–586, November 1960.

[115] P. Balling, "Radiation from the dielectric wedge," Lic. Tech. Dissertation, Technical University of Denmark, December 1971.

[116] P. Balling, "Surface fields on the source-excited dielectric wedge," *IEEE Trans. Antennas Propag.*, vol. AP-21, pp. 113–115, January 1973.

[117] P. Balling, "On the role of lateral waves in the radiation from the dielectric wedge," *IEEE Trans. Antennas Propag.*, vol. AP-21, pp. 247–248, March 1973.

[118] S. T. Peng and F. Schwering, "Effect of taper profile on performance of dielectric taper antennas," *Dig. 1979 Natl. Radio Sci. Mtg. and Intl. IEEE-APS Symp.*, p. 96, Seattle, June 18–22, 1979.

[119] S. T. Peng and F. Schwering, "Omni-directional dielectric antennas," 9th DARPA/Tri-Service Millimeter-Wave Conference, Huntsville, Alabama, October 20–22, 1981.

[120] T. Itoh, "Dielectric waveguide type millimeter-wave integrated circuits," in *Infrared and Millimeter Waves, Vol. 4*, ed. by K. J. Button, pp. 199–273, New York: Academic Press, 1981.

[121] N. Williams, A. W. Rudge, and S. E. Gibbs, *Proc. IEEE MTT-S Intl. Microwave Symp.*, pp. 542–544, San Diego, 1977.

[122] K. L. Klohn, R. E. Horn, H. Jacobs, and E. Freibergs, "Silicon waveguide frequency scanning linear array antenna," *IEEE Trans. Microwave Theory Tech.*, vol. MTT-26, pp. 764–773, October 1978.

[123] R. E. Horn, H. Jacobs, E. Freibergs, and K. L. Klohn, "Electronic modulated beam-steering silicon waveguide array antenna," *IEEE Trans. Microwave Theory Tech.*, vol. MTT-28, pp. 647–653, June 1980.

[124] R. E. Horn, H. Jacobs, K. L. Klohn, and E. Freibergs, "Single-frequency electronic-modulated analog line scanning using a dielectric antenna," *IEEE Trans. Microwave Theory Tech.*, vol. MTT-30, pp. 816–820, May 1982.

[125] S. T. Peng and F. Schwering, "Dielectric grating antennas," R&D technical report, CORADCOM-78-3, Fort Monmouth, July 1978.

[126] F. Schwering and S. T. Peng, "Design of periodically corrugated dielectric antennas for millimeter-wave applications," *Proc. 1982 Antenna Appl. Symp.*, Univ. of Illinois, Urbana, September 22–24, 1982.

[127] F. Schwering and S. T. Peng, "Design of dielectric grating antennas for millimeter-wave applications," *IEEE Trans. Microwave Theory Tech.*, vol. MTT-31, pp. 199–209, February 1983.

[128] J. Borowick, W. Bayha, R. A. Stern, and R. W. Babbitt, "Dielectric waveguide antennas," 1982 Army Science Conference, West Point, New York, June 1982.

[129] S. Kobayashi, R. Lampe, R. Mittra, and S. Ray, "Dielectric-rod leaky-wave antennas for millimeter-wave applications," *IEEE Trans. Antennas Propag.*, vol. AP-29, pp. 822–824, September 1981.

[130] T. N. Trinh, R. Mittra, and R. J. Paleta, "Horn image-guide leaky-wave antenna," *1981 IEEE-MTT-S Intl. Microwave Symp. Dig.*, Los Angeles, June 1981.

[131] T. N. Trinh, R. Mittra, and R. J. Paleta, "Horn image-guide leaky-wave antenna," *IEEE Trans. Microwave Theory Tech.*, vol. MTT-29, pp. 1310–1314, December 1981.

[132] T. Itoh, "Application of gratings in dielectric waveguides for leaky-wave antennas and band-reject filters," *IEEE Trans. Microwave Theory Tech.*, vol. MTT-25, pp. 1134–1138, December 1977.

[133] T. Itoh and B. Adelseck, "Trapped image guide for millimeter-wave circuits," *IEEE Trans. Microwave Theory Tech.*, vol. MTT-28, pp. 1433–1436, December 1980.

[134] T. Itoh and B. Adelseck, "Trapped image-guide leaky-wave antennas for millimeter-wave applications," *IEEE Trans. Antennas Propag.*, vol. AP-30, pp. 505–509, May 1982.

[135] K. Solbach, "*E*-band leaky-wave antenna using dielectric image line with etched radiating elements," *1979 MTT-S Intl. Microwave Symp. Dig.*, pp. 214–216, April 1979.

[136] K. Solbach, "Slots in dielectric image line as mode launchers and circuit elements," *IEEE Trans. Microwave Theory Tech.*, vol. MTT-29, pp. 10–16, January 1981.

[137] W. V. McLevige, T. Itoh, and R. Mittra, "New waveguide structures for millimeter-wave and optical integrated circuits," *IEEE Trans. Microwave Theory Tech.*, vol. MTT-23, pp. 788–794, October 1975.

[138] S. T. Peng and T. Tamir, "Effects of groove profile on the performance of dielectric grating couplers," *Proc. Symp. Opt. Acoust. Micro-Electron.*, Polytechnic Press, Brooklyn, 1974.

[139] D. Marcuse, "Exact theory of TE-wave scattering from blazed dielectric gratings," *Bell Syst. Tech. J.*, vol. 55, pp. 1295–1317, 1976.

[140] K. C. Chang and T. Tamir, "Simplified approach to surface-wave scattering by blazed dielectric gratings," *Appl. Opt.*, vol. 19, pp. 282–288, 1980.

[141] A. Gruss, K. T. Tam, and T. Tamir, "Blazed dielectric gratings with high beam-coupling efficiencies," *Appl. Phys. Lett.*, vol. 36, pp. 523–526, 1980.

[142] T. Hori and T. Itanami, "Circularly polarized linear array antenna using a dielectric image guide," *IEEE Trans. Microwave Theory Tech.*, vol. MTT-29, pp. 967–970, September 1981.

[143] S. T. Peng, "Oblique guidance of surface waves on corrugated dielectric layers," *Proc. URSI Symp. Electromag. Waves*, Paper No. 341B, Munich, Germany, August 1980.

[144] S. T. Peng, A. A. Oliner, and F. Schwering, "Theory of dielectric grating antennas of finite width," *IEEE AP-S Intl. Symp. Dig.*, pp. 529–532, Los Angeles, June 1981.

[145] M. J. Shiau, S. T. Peng, and A. A. Oliner, "Simple and accurate perturbation procedure for millimeter-wave dielectric grating antennas of finite width," *1982 IEEE/AP-S Symp. Dig.*, pp. 648–651, Albuquerque, May 24–28, 1982.

[146] S. T. Peng, M. J. Shiau, A. A. Oliner, J. Borowick, W. Bayha, and F. Schwering, "A simple analysis procedure for dielectric grating antennas of finite width," 1984 IEEE-AP-S Symposium, Boston, June 25–28, 1984.

[147] T. Tamir, *Integrated Optics*, New York: Springer-Verlag, 1975.

[148] S. T. Peng, T. Tamir, and H. L. Bertoni, "Theory of periodic dielectric waveguides," *IEEE Trans. Microwave Theory Tech.*, vol. MTT-23, p. 123, 1975.

[149] M. Neviere, R. Petit, and M. Cadilhac, "About the theory of optical grating coupler-waveguide systems," *Opt. Commun.*, vol. 8, pp. 113–117, 1973.

[150] K. Honda, S. T. Peng, and T. Tamir, "Improved perturbation analysis of dielectric gratings," *Appl. Phys.*, vol. 5, p. 325, 1975.

[151] S. T. Peng and T. Tamir, "TM mode perturbation analysis of dielectric gratings," *Appl. Phys.*, vol. 7, p. 35, 1975.

[152] T. Tamir and S. T. Peng, "Analysis and design of grating couplers," *Appl. Phys.*, vol. 14, pp. 235–254, 1977.

[153] R. Mittra and R. Kastner, "A spectral domain approach for computing the radiation characteristics of a leaky-wave antenna for millimeter waves," *IEEE Trans. Antennas Propag.*, vol. AP-29, pp. 654–656, July 1981.

[154] R. M. Knox and P. P. Toulios, "Integrated circuits for the millimeter through optical frequency range," *Proc. Symp. Millimeter Waves*, Polytechnic Institute of Brooklyn, March 31–April 2, 1970.

[155] J. R. James and P. S. Hall, "Microstrip antennas and arrays, part 2: new array-design technique," *IEE J. Microwaves, Optics and Antennas*, no. 1, pp. 175–181, 1977.

[156] K. Solbach and B. Adelseck, "Dielectric image line leaky-wave antennas for broadside radiation," *Electron. Lett.*, vol. 19, pp. 640–644, August 1983.

[157] I. J. Bahl and P. Bhartia, "Leaky-wave antennas using artificial dielectrics at millimeter-wave frequencies," *IEEE Trans. Microwave Theory Tech.*, vol. MTT-28, pp. 1205–1212, November 1980.

[158] F. R. Ore, "A millimeter-wave receiving antenna with an omnidirectional or

directional scannable azimuthal pattern and a directional vertical pattern," *Tech. Rep. AFAL-TR-72-282*, Univ. of Illinois, September 1971.

[159] F. R. Ore, "A millimeter-wave receiving antenna with an omnidirectional or directional scannable azimuthal pattern and a directional vertical pattern," *IEEE Trans. Antennas Propag.*, vol. AP-20, pp. 481–482, July 1972.

[160] G. E. Mueller, "A broadside dielectric antenna," *Proc. IRE*, vol. 40, pp. 71–75, July 1952.

[161] S. T. Peng, "Omnidirectional dielectric antennas," CECOM R&D report in preparation.

[162] S. T. Peng and A. A. Oliner, "Guidance and leakage properties of a class of open dielectric waveguides, part I: mathematical formulations," *IEEE Trans. Microwave Theory Tech.*, vol. MTT-29, pp. 843–855, September 1981.

[163] A. A. Oliner, S. T. Peng, T. I. Hsu, and A. Sanchez, "Guidance and leakage properties of a class of open dielectric waveguides, part II: new physical effects," *IEEE Trans. Microwave Theory Tech.*, vol. MTT-29, pp. 855–869, September 1981.

[164] Polytechnic Institute of Brooklyn, Microwave Research Institute, *Monthly Performance Summary, Rep. PIBMRI-875*, pp. 17–61, 1961.

[165] T. Nakahara and N. Kurauchi, "Transmission modes in the grooved guide," *J. Inst. Electron. Commun. Eng. Japan*, vol. 47, no. 7, pp. 43–51, July 1964. Also in *Sumitomo Electr. Tech. Rev.*, no. 5, pp. 65–71, January 1965.

[166] D. J. Harris and K. W. Lee, "Groove guide as a low-loss transmission system for short millimetric waves," *Electron. Lett.*, vol. 13, no. 25, pp. 775–776, December 8, 1977. Professor Harris and his colleagues have published many papers on this topic, of which this is one of the first.

[167] D. J. Harris and S. Mak, "Groove-guide microwave detector for 100-GHz operation," *Electron. Lett.*, vol. 17, no. 15, pp. 516–517, July 23, 1981.

[168] A. A. Oliner and P. Lampariello, "Novel leaky-wave antenna for millimeter waves based on groove guide," *Electron. Lett.*, vol. 18, pp. 1105–1106, December 1982.

[169] P. Lampariello and A. A. Oliner, "Theory and design considerations for a new millimeter-wave leaky groove-guide antenna," *Electron. Lett.*, vol. 19, pp. 18–20, January 1983.

[170] P. Lampariello and A. A. Oliner, "A new leaky wave antenna for millimeter waves using an asymmetric strip in groove guide, part I: theory," *IEEE Trans. Antennas Propag.*, vol. AP-33, pp. 1285–1294, December 1985.

[171] P. Lampariello and A. A. Oliner, "A new leaky wave antenna for millimeter waves using an asymmetric strip in groove guide, part II: design considerations," *IEEE Trans. Antennas Propag.*, vol. AP-33, pp. 1295–1303, December 1985.

[172] P. Lampariello and A. A. Oliner, "Bound and leaky modes in symmetrical open groove guide," *Alta Frequenza*, vol. LII, no. 3, pp. 164–166, 1983.

[173] P. Lampariello and A. A. Oliner, "Leaky modes of symmetrical groove guide," *Dig. IEEE/MTT-S Intl. Microwave Symp.*, pp. 390–392, Boston, May 30–June 3, 1983.

[174] A. A. Oliner and P. Lampariello, "A new simple leaky wave antenna for millimeter waves," *Dig. 1985 North American Radio Sci. Meeting*, p. 57, Vancouver, Canada, June 17–21, 1985.

[175] A. A. Oliner and P. Lampariello, "A simple leaky wave antenna that permits flexibility in beam width," *Dig. Natl. Radio Sci. Meeting*, p. 26, Philadelphia, June 9–13, 1986.

[176] T. Yoneyama and S. Nishida, "Nonradiative dielectric waveguide for millimeter-wave integrated circuits," *IEEE Trans. Microwave Theory Tech.*, vol. MTT-29, no. 11, pp. 1188–1192, November 1981.

[177] T. Yoneyama and S. Nishida, "Nonradiative dielectric waveguide circuit components," International Conference on Infrared and Millimeter Waves, Miami, December 1981.

[178] A. Sanchez and A. A. Oliner, "Accurate theory for a new leaky-wave antenna for millimeter waves using nonradiative dielectric waveguide," *Radio Sci.*, vol. 19, no. 5, pp. 1225–1228, September–October 1984.

[179] A. Sanchez and A. A. Oliner, "Microwave network analysis of a leaky-wave

structure in nonradiative dielectric waveguide," *Dig. IEEE MTT-S Intl. Microwave Symp.*, pp. 118–120, San Francisco, May 30–June 1, 1984.

[180] A. A. Oliner, S. T. Peng, and K. M. Sheng, "Leakage from a gap in NRD guide," *Dig. 1985 IEEE Intl. Microwave Symp.*, pp. 619–622, St. Louis, June 3–7, 1985.

[181] H. Shigesawa, M. Tsuji, and A. A. Oliner, "Coupling effects in an NRD guide leaky wave antenna," *Dig. Natl. Radio Sci. Meeting*, p. 27, Philadelphia, June 9–13, 1986.

[182] H. Shigesawa and K. Takiyama, "Study of leaky H-guide," *Paper No. M1-7*, International Conference on Microwaves, Circuit Theory and Information, Tokyo, 1964. More complete version in K. Takiyama and H. Shigesawa, "On the study of a leaky H-guide," *Sci. Eng. Rev. Doshisha University*, vol. 7, no. 4, pp. 203–225, March 1967 (in English).

[183] K. Takiyama and H. Shigesawa, "The radiation characteristics of a leaky H-guide," *J. Inst. Electr. Commun. Eng. Japan* (J.I.E.C.E.), vol. 50, no. 2, pp. 181–188, February 1967 (in Japanese).

[184] A. Sanchez and A. A. Oliner, "A new leaky waveguide for millimeter waves using nonradioactive dielectric (NRD) waveguide, part I: accurate theory," *IEEE Trans. Microwave Theory Tech*, vol. MTT-35, pp. 737–747, August 1987.

[185] Y. Yoneyama, letter to A. A. Oliner, July 4, 1983.

[186] Q. Han, A. A. Oliner, and A. Sanchez, "A new leaky waveguide for millimeter waves using nonradioactive dielectric (NRD) waveguide, part II: comparison with experiments," *IEEE Trans. Microwave Theory Tech.*, vol. MTT-35, pp. 748–752, August 1987.

[187] T. Yoneyama, T. Kuwahara, and S. Nishida, "Experimental study of nonradiative dielectric waveguide leaky wave antenna," *Proc. 1985 Intl. Symp. Antennas Propag. (ISAP)*, Kyoto, August 1985.

[188] H. Ermert, "Guided modes and radiation characteristics of covered microstrip lines," *Archiv für Electronik und Übertragungstechnik*, vol. 30, pp. 65–70, February 1976.

[189] H. Ermert, "Guiding and radiation characteristics of planar waveguides," *Microwaves, Optics and Acoustics*, vol. 3, pp. 59–62, March 1979.

[190] A. A. Oliner and K. S. Lee, "The nature of the leakage from higher modes on microstrip line," *Dig. 1986 IEEE Intl. Microwave Symp.*, pp. 57–60, Baltimore, June 2–4, 1986.

[191] A. A. Oliner and K. S. Lee, "Microstrip leaky wave strip antennas," *Dig. 1986 IEEE Intl. Antennas Propag. Symp.*, pp. 443–446, Philadelphia, June 8–13, 1986.

[192] W. Menzel, "A new traveling-wave antenna in microstrip," *Archiv für Electronik und Übertragungstechnik*, vol. 33, pp. 137–140, April 1979.

[193] D. C. Chang and E. F. Kuester, "Total and partial reflection from the end of a parallel-plate waveguide with an extended dielectric loading," *Radio Sci.*, vol. 16, pp. 1–13, January–February 1981.

[194] D. M. Pozar, "Considerations for millimeter-wave printed antennas," *IEEE Trans. Antennas Propag.*, vol. AP-31, pp. 740–747, September 1983.

[195] K. Iizuka, M. Mizusawa, S. Urasaki, and H. Ushigome, "Volume-type holographic antenna," *IEEE Trans. Antennas Propag.*, vol. AP-23, pp. 807–810, November 1975.

[196] D. F. Sedivec and B. H. Rubin, "A wideband 44-GHz printed-circuit array antenna," *Tech. Rep. RADC-TR-83-198*, Rome Air Development Center, Rome, N.Y., October 1983.

[197] K. C. Gupta, R. Garg, and R. Chadha, *Computer-Aided Design of Microwave Circuits*, Dedham: Artech House, 1981.

[198] T. Itoh and J. Rivera, "A comparative study of millimeter-wave transmission lines," chapter 2, vol. 9, of *Infrared and Millimeter Waves*, ed. by K. J. Button, New York: Academic Press, 1983.

[199] J.-F. Miao and T. Itoh, "Coupling between microstrip line and image guide through small apertures in the common ground plane," *IEEE Trans. Microwave Theory Tech.*, vol. MTT-31, pp. 361–363, April 1983.

[200] R. E. Neidert, "Waveguide-to-coax-to-microstrip transitions for millimeter-wave

monolithic circuits," *Microwave J.*, vol. 27, pp. 93–101, June 1983.

[201] N. G. Alexopoulos and I. E. Rana, "Mutual impedance computation between printed dipoles," *IEEE Trans. Antennas Propag.*, vol. AP-29, pp. 106–111, January 1981.

[202] H. G. Oltman and D. A. Huebner, "Electromagnetically coupled microstrip dipole arrays," *IEEE Trans. Antennas Propag.*, vol. AP-29, pp. 151–157, January 1981.

[203] R. S. Elliott and G. J. Stern, "The design of microstrip dipole arrays including mutual coupling," part I: theory; part II: experiments," *IEEE Trans. Antennas Propag.*, vol. AP-29, pp. 757–765, September 1981.

[204] A. Sabban, "A new broadband stacked two-layer microstrip antenna," *Dig. 1983 Intl. IEEE Symp. Antennas Propag.*, pp. 63–66, University of Houston, May 1983.

[205] P. B. Katehi and N. G. Alexopoulos, "A generalized solution to a class of printed-circuit antennas," *Dig. 1984 Intl. IEEE-APS/URSI Symp.*, pp. 566–568, Boston, June 1984.

[206] K. Mizuno, Y. Daiku, and S. Ono, "Design of printed resonant antennas for monolithic diode detectors," *IEEE Trans. Microwave Theory Tech.*, vol. MTT-25, pp. 470–472, June 1977.

[207] D. B. Rutledge and S. E. Schwarz, "Planar multimode detector arrays for infrared and millimeter-wave applications," *IEEE J. Quantum Electron.*, vol. QE-17, pp. 407–414, March 1981.

[208] D. B. Rutledge and M. S. Muha, "Imaging antenna arrays," *IEEE Trans. Antennas Propag.*, vol. AP-30, pp. 535–540, July 1982.

[209] C. Yao, S. E. Schwarz, and B. J. Blumenstock, "Monolithic integration of a dielectric millimeter-wave antenna and a mixer diode: an embryonic millimeter-wave IC," *IEEE Trans. Microwave Theory Tech.*, vol. MTT-30, pp. 1241–1247, August 1982.

[210] R. J. Stockton, "A monolithic phased array at *K*-band," EHF SATCOM Technology Workshop, San Diego, August 1981.

[211] R. J. Stockton, "Monolithic integrated antenna system—a new trend," Microwave and Millimeter-Wave Monolithic Circuit Symposium, Dallas, June 1982.

[212] D. P. Neikirk, D. B. Rutledge, M. S. Muha, H. Park, and C.-X. Yu, "Far-infrared imaging antenna arrays," *Appl. Phys. Lett.*, vol. 40, pp. 203–205, 1982.

[213] Z. Rav-Noy, C. Zah, U. Schreter, D. B. Rutledge, T. C. Wand, S. E. Schwarz, and T. F. Kuech, "Monolithic Schottky diode imaging arrays at 94 GHz," *Dig. Infrared and Millimeter-Wave Conf.*, Miami Beach, December 1983.

[214] T. A. Midford, M. Feng, R. Hackett, J. M. Schellenberg, E. Watkins, and H. Yamasaki, "Advanced GaAs FET technology for ehf monolithic arrays," *NOSC Contractor Rep. 225*, February 1984.

[215] C. R. Seashore and D. R. Singh, "Millimeter-wave ICs for precision guided weapons," *Microwave J.*, vol. 26, pp. 51–65, June 1983.

[216] C. Zah, R. C. Compton, and D. B. Rutledge, "Efficiencies of elementary integrated-circuit feed antennas," *Electromagnetics*, special issue on printed-circuit antennas and devices, pp. 239–254, March 1983.

[217] D. B. Rutledge, private communication.

[218] C. Zah, W. Lam, J. S. Smith, Z. Rav-Noy, and D. B. Rutledge, "Progress in monolithic Schottky-diode imaging arrays," Ninth International Conference on Infrared and Millimeter Waves, Osaka, November 1984.

[219] M. Guglielmi and A. A. Oliner, "A practical theory for image guide leaky-wave antennas loaded by periodic metal strips," *Proc. 17th European Microwave Conference*, pp. 549–554, Rome, Italy, September 7–11, 1987.

[220] N. Marcuvitz, *Waveguide Handbook*, *Vol. 10*, MIT Radiation Laboratory Series, Sec. 6.1, McGraw-Hill Book Co., New York, 1951.

[221] A. A. Oliner, "Scannable Millimeter Wave Arrays," Final Report on Contract No. F19628-84-K-0025, Rome Air Development Center, Hanscom Field, MA, December 1, 1987.

Chapter 18

Practical Aspects of Phased Array Design

Raymond Tang
Hughes Aircraft Company

CONTENTS

Raymond Tang is the manager of the Antenna Array Laboratory, Electromagnetic Laboratories, Surveillance and Sensor Systems Division, of Hughes Aircraft Company. He received the BSEE degree from the Polytechnic Institute of Brooklyn, and the MSEE degree from the University of Southern California.

During his 30 years at Hughes Aircraft Company Mr. Tang has been concerned with microwave antennas and components. He has participated in the design and development of various types of electronic scanning antennas, such as frequency-scan, phase-scan, and optically scanning lenses. In recent years he has been actively involved in the development of wideband phased arrays and limited-scan phased arrays. He has also been engaged in the development of diode and ferrite phase shifters, switches, and solid-state transmit/receive modules.

Mr. Tang holds 12 patents and has authored or coauthored 15 technical publications. He is a member of the IEEE, PGAP, PGMTT, PGED, and Eta Kappa Nu fraternity.

1. Introduction

The intent of this chapter is to provide the reader with a basic understanding of the practical aspects of phased array antenna design. The theory of phased arrays has been covered in the preceding chapters. In this chapter a treatment of the various design considerations and trade-offs is given, so that the antenna designer can arrive at an optimum antenna configuration in order to meet a given set of radar system requirements. The topics that will be covered are

(a) design specifications and procedure for phased array antennas
(b) selection criteria for array components
(c) effects of component errors on array performance

To establish a common basis of understanding, let us first define the basic components in a phased array antenna. As shown in Fig. 1, the phased array antenna consists of an array of radiating elements with each radiating element connected to a phase shifter. The phase shifters control the phase of the radiated signals at each element to form a beam at the desired direction θ_0. A beam-forming network, commonly called a *feed network*, is used to distribute the output signal from the transmitter to the radiating elements and to provide the required aperture distribution for beam shape and side lobe control. Phase shifter drivers provide the required control/bias currents and voltages for each phase shifter for steering the beam to the desired scan angle θ_0. The control signals (or phase words) for the drivers are calculated by the beam-steering computer and stored in the serial shift registers. When the beam is ready to be scanned, the beam-forming trigger signal causes the stored phase words in the serial shift registers to dump into the parallel latching registers, which in turn set the drivers and phase shifter for the desired scan angle. Using this type of phased array the radar is capable of performing the following functions:

1. Rapid and accurate beam scanning; typically the beam-switching time is 10 to 40 μs
2. Search and automatic target tracking over a hemispherical scan coverage by using four planar array faces
3. Perform multiple functions such as surveillance, multiple target tracking, target illumination and missile guidance, terrain following and avoidance, ground mapping, etc.
4. Pulse-to-pulse frequency and/or beam agility
5. Beam shaping and/or polarization flexibility
6. Low peak and average side lobe levels, typically −40 dB peak and −55 dB average

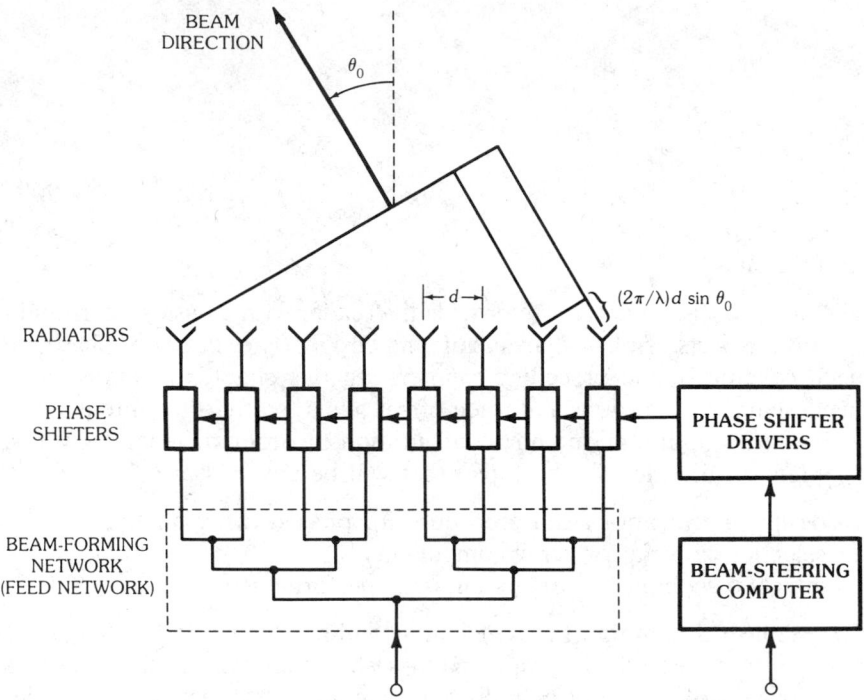

Fig. 1. Basic components of a phased array antenna.

7. High power transmission with multiple distributed transmitters
8. Electronic beam stabilization on moving platforms

With this simplified description of the basic functions of a phased array antenna, let us now proceed to the discussion of the design specifications of phased arrays.

2. Design Specification and Procedure of Phased Array Antennas

The design specifications of a phased array antenna are usually determined by the overall radar system requirements. These requirements are given in terms of radar performance requirements, physical requirements, operating environmental requirements, producibility, maintainability, and reliability requirements. A detailed listing of all the requirements, showing the breakdown of each category, is given in Chart 1.

The antenna engineers, working together with the system and mechanical design engineers, must perform a trade-off study of these requirements (as shown in Chart 1) in order to establish a set of antenna specifications that would satisfy all the radar system requirements at minimum cost. In other words, the antenna requirements must not be overspecified so that the cost would not be affordable. For example, the number of simultaneous beams, beam-switching speed, beamwidth, and total transmitted power can be traded against each other to simplify the

Chart 1. Design Requirements of Phased Array Antennas

1. Performance Requirements
   spatial scan coverage
   tunable and instantaneous bandwidth
   beamwidth
   peak and average side lobe level
   antenna gain
   polarization
   peak and average power
   beam-switching speed
   prime power
   number of simultaneous beams
   beam shape

2. Physical Requirements
   size
   weight
   transportability
   mobility—setup time and march time

3. Environmental Requirements
   operating temperature range
   shock and vibration loads
   humidity, salt, fog, and fungus
   overpressure

4. Producibility, Maintainability, and Reliability

5. Cost (Affordability)

antenna requirements without sacrificing the required power aperture product for a given radar search mode. Once the antenna requirements are specified, the next step is to formulate and draw an overall antenna schematic diagram showing all the functional subassemblies of the antenna system. The subsequent step is to select the array components for the various functional subassemblies that would best meet the system requirements in terms of performance and cost. Following the component selection, the next step is to perform an error analysis to determine the allowable tolerances of these components that would meet the antenna performance requirements. These tolerance requirements are then used as the performance and physical specifications for the design, development, fabrication, and acceptance testing of these components. Detailed discussions on the selection criteria of array components and error analysis follow in Sections 3 and 4.

3. Selection Criteria of Array Components

The three major components of a phased array antenna are the radiators, phase shifters, and beam-forming feed network. Some of the commonly used criteria in selecting these components for a given phased array application are shown in Chart 2.

The requirements in Chart 2 influence the selection of array components in different ways. For example, the selection of the type of radiator is mainly

Chart 2. Selection Criteria of Array Components

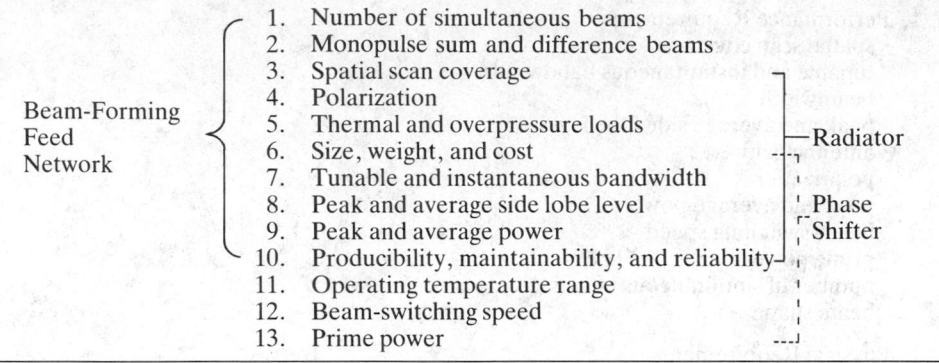

Beam-Forming Feed Network
1. Number of simultaneous beams
2. Monopulse sum and difference beams
3. Spatial scan coverage
4. Polarization
5. Thermal and overpressure loads
6. Size, weight, and cost
7. Tunable and instantaneous bandwidth
8. Peak and average side lobe level
9. Peak and average power
10. Producibility, maintainability, and reliability
11. Operating temperature range
12. Beam-switching speed
13. Prime power

Radiator
Phase Shifter

determined by the requirements of items 3 through 10, whereas the selection of the type of phase shifter is mainly determined by the requirements of items 6 through 13, and the selection of feed network is determined by items 1 through 10. As shown in Chart 2, some of the requirements are unique to only one of the components while the others affect two or all three components. For example, the requirements of beam-switching speed, operating temperature range, and prime power affect only the phase shifter selection, whereas the frequency bandwidth, side lobe, and power level requirements affect all three components. In general, there are many combinations of antenna component specifications that can result in the same overall system performance. These nonunique component requirements allow the antenna designer to perform trade-offs in selecting the component to arrive at an optimum design in terms of meeting performance requirements at minimum cost. A detailed discussion of the various trade-offs available in component selection is given in the following sections.

Radiator Selection

Types of Radiators—Before we begin the discussion on radiator selection, let us first review the various types of radiators that are commonly used in phased arrays. The basic types of phased array radiators are listed below:

1. Open-ended waveguide radiators
2. Dipole radiators
3. Waveguide slot radiators
4. Disk and patch radiators

The open-ended waveguide radiator comes in two basic forms, namely, the rectangular and the circular waveguide radiators. The rectangular waveguide radiator is usually used for linear polarization applications, whereas the circular waveguide radiator is frequently used for dual linear or circular polarization applications. The rectangular waveguide radiator can be packaged into very close spacing in the E-plane of the waveguide (less than 0.5λ) by using reduced height waveguide, thus allowing a very large scan coverage in the E-plane. The H-plane scan is restricted by the width of the guide as determined by the desired ratio of the operating frequency to the cutoff frequency. The rectangular waveguide radiator, however, can be packaged into an equilateral triangular element lattice arrangement in order to provide conical scan coverage as in the case of the circular waveguide radiators, thus allowing an increase in *H*-plane scan coverage. The various possible lattice arrangements for both the rectangular and circular waveguide radiators are shown in Fig. 2. The dimension *d* in Fig. 2 is usually in the order of one-half wavelength at the high end of the operating frequency band. For all cases of lattice arrangements the spacing between the elements can be reduced by dielectrically loading and/or ridge loading the waveguides.

The two common types of dipole radiators are the microstrip dipole [1, 2] and the coaxial dipole as shown in Fig. 3. In the case of the microstrip dipole the dipole wings are either etched or printed on a dielectric substrate, such as a copper-clad Teflon-fiberglass board, or on an alumina (Al_2O_3) substrate. These dipole wings

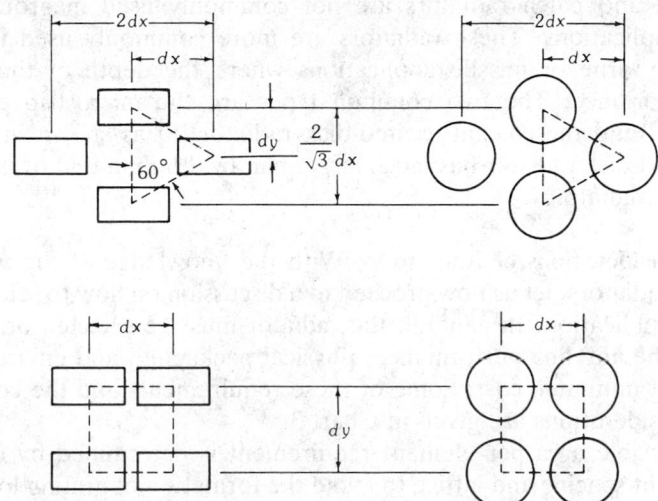

Fig. 2. Lattice arrangements of waveguide radiators.

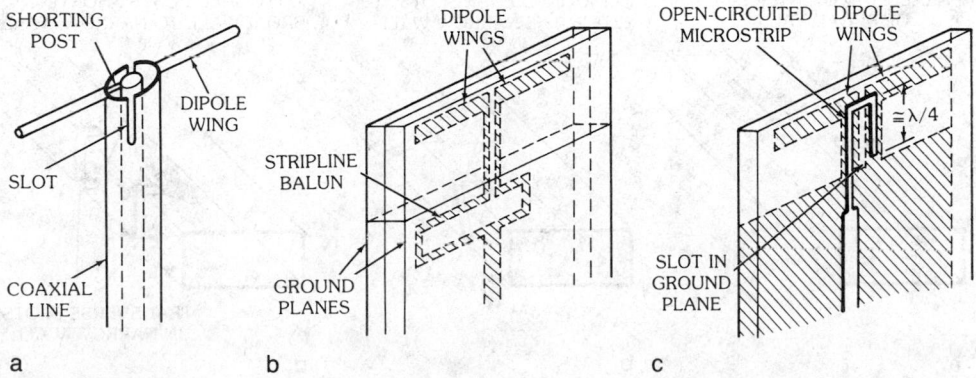

Fig. 3. Common types of dipole radiators. (*a*) Coaxial dipole. (*b*) Stripline dipole. (*c*) Microstrip dipole.

are excited by means of a microstrip balun. In the case of the coaxial dipole the dipole wings are excited by means of a slotted coaxial transmission line as shown in Fig. 3. These dipole radiators can be packaged in a large two-dimensional array just as in the case of the rectangular waveguide radiators.

There are many different types of slot radiators. In the case of slots excited by waveguide transmission lines, there are shunt slots [3] which are cut along and parallel to the centerline of the broadwall of the waveguide and series inclined [4] as well as noninclined slots [5] which are cut along the sidewall of the waveguide (see Fig. 4). The noninclined slot, which is a magnetically coupled transverse slot [5], does not have cross polarization as in the case of the inclined slot. There are also slot radiators which are etched on the ground plane side of a microstrip transmission line. The operating bandwidth of the slot radiators, as in the case of the dipole radiators, is less than that of the open-ended waveguide radiators.

The disk and patch radiators are not commonly used in ground-based or shipboard applications. These radiators are more commonly used in conformal arrays for airborne or missile applications where the depth of the array is of cardinal importance. The two common types are the microstrip excited-patch radiator [6, 7] and the coaxial excited-disk radiator [8] as shown in Fig. 5. The coaxial excited-disk radiator has much larger bandwidth than that of the microstrip excited-patch radiator.

Selection Considerations of Radiators—With the knowledge of the various types of available radiators, let us now proceed to a discussion on how to select a radiator for a given application. In general, the radiator must be selected on the basis of meeting all the antenna performance, physical packaging, and environmental requirements at minimum cost. Some of these requirements and the corresponding selection considerations are given in Chart 3.

The allowable area per element requirement is determined by choosing the proper element spacing and lattice to avoid the formation of grating lobes over the entire volumetric scan coverage. The element lattice is usually chosen to maximize

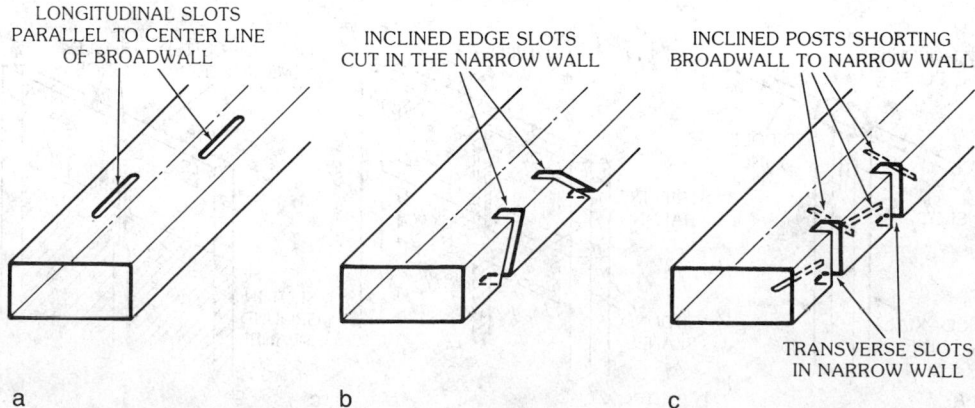

Fig. 4. Various types of slot radiators. (*a*) Broadwall shunt slots. (*b*) Inclined edge slots. (*c*) Magnetically coupled transverse slots.

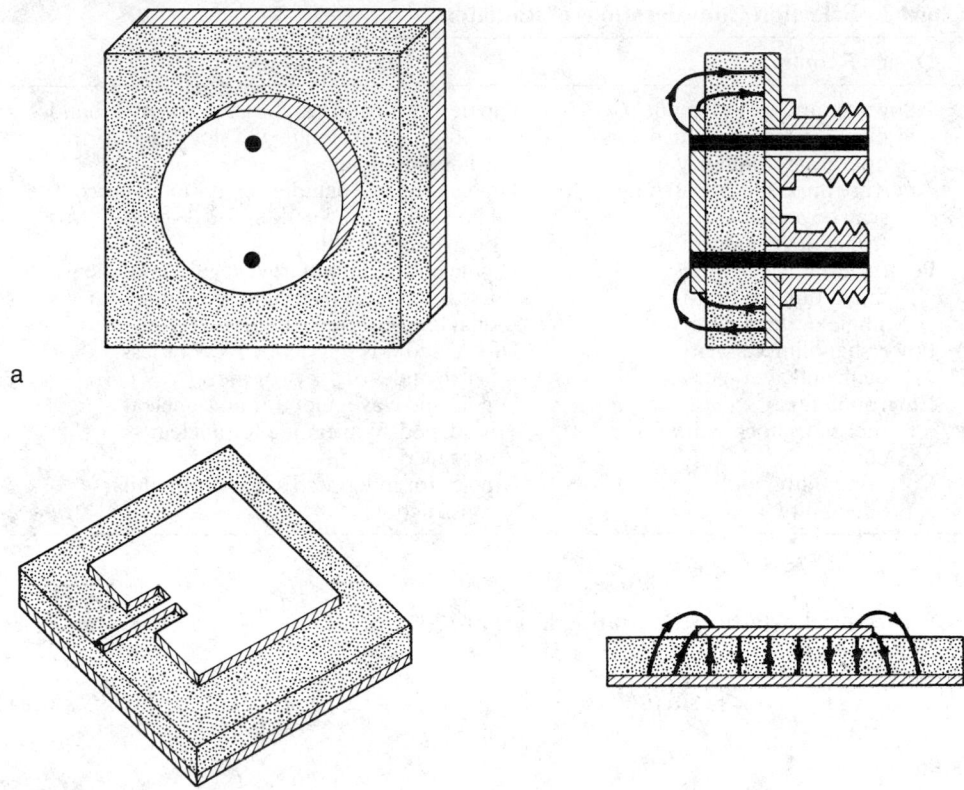

Fig. 5. Microstrip radiators. (*a*) Coaxial excited-disk radiator. (*After Byron [8]; reprinted by permission, © 1972 Artech House, Inc.*) (*b*) Microstrip excited-patch radiator. (*After Carver and Mink [1], © 1960 IEEE*)

the allowable area per element corresponding to the required scan coverage. For example, a triangular lattice should be used for a conical scan coverage and a rectangular lattice for a rectangular scan coverage. The grating lobe locations for a rectangular lattice are given by

$$\sin\theta\cos\phi - \sin\theta_0\cos\phi_0 = \pm\frac{\lambda}{d_x}p \qquad (1a)$$

$$\sin\theta\sin\phi - \sin\theta_0\sin\phi_0 = \pm\frac{\lambda}{d_y}q \qquad (1b)$$

where

θ_0, ϕ_0 = the scan direction of the main beam

d_x, d_y = element spacing along the x and y axis

$p, q = 0, 1, 2, \ldots$

Chart 3. Selection Considerations of Radiators

Design Requirements	Selection Considerations
Allowable area per element element spacing lattice operating frequency	Dipoles, disk, and patch radiators for X-band or lower. Waveguide and slot radiators for S-band or higher
Aperture impedance matching scan coverage frequency bandwidth	Dipoles and waveguides for 10- to 25-percent bandwidth. Waveguides for 10-percent to octave bandwidth
Polarization linear or circular single or dual	Dipole and rectangular waveguide for linear. Cross dipole and circular waveguide for dual linear or dual circular
Power-handling capacity peak and average	Dipole and waveguide for 1 kW or less. Waveguide for 1 kW or more
Environmental thermal, shock, vibration, etc.	Dipole and waveguide for non-nuclear hardened. Waveguide for nuclear hardened
Cost, reliability, and producibility	Dipoles for integrated subarray modular construction

For a triangular lattice the grating lobes are located at

$$\sin \theta \cos \phi - \sin \theta_0 \cos \phi_0 = \pm \frac{\lambda}{2d_x} p \tag{2a}$$

$$\sin \theta \sin \phi - \sin \theta_0 \sin \phi_0 = \pm \frac{\lambda}{2d_y} q \tag{2b}$$

where $p + q$ is even.

In order to prevent the grating lobe formation the maximum projected element spacing d along a given scan plane must satisfy the following formula:

$$\frac{d}{\lambda} \leq \frac{1}{1 + \sin \theta_{max}} \tag{3}$$

For a conical scan coverage of a 60° half-angle cone the maximum allowable area per element is approximately $0.3\lambda^2$ as shown in Fig. 2. Once the area per element is established, a radiator must be selected to fit within that area. Since the element spacing is directly proportional to the wavelength of the operating frequency, the dipole radiators are usually used for X-band or lower frequencies and waveguide/slot radiators are used for S-band or higher frequencies. Another requirement for radiator selection is aperture impedance matching over the required scan coverage and operating frequency bandwidth. For a 60° half-angle cone coverage the dipole and waveguide radiators can be reasonably well matched over a 10-percent bandwidth, whereas waveguide radiators [9] can be matched over almost an octave bandwidth. In general, some of the other considerations in radiator selection for meeting the requirements of polarization, power-handling capacity, environment, cost, etc., are given in Chart 3.

Development Procedure of Phased Array Radiators—Once the radiator selection is made in accordance with the requirements of Chart 3, the usual step-by-step procedure to develop this radiator in a two-dimensional array environment is given below:

Step 1

Select an element spacing and lattice that does not formulate grating lobes or surface-wave resonances over the required scan coverage and the operating frequency band.

Step 2

Formulate an analytical model of the radiating aperture and optimize the performance by varying the design parameters, such as element dimensions.

Step 3

Perform aperture matching, such as by using an inductive iris or a dielectric plug in the opening of the waveguide radiator, metallic fences around dipole radiators, or using dielectric sheets in front of the radiating aperture.

Step 4

Using the analytical results from steps 2 and 3, fabricate waveguide simulators to measure and verify the radiation impedance at discrete scan angles.

Step 5

Fabricate a small test array (usually 9×9 elements) and measure the active element pattern of the central element over the required frequency band. This active element pattern is defined as the pattern of an element with all the neighboring elements terminated into matched loads. It describes the variation in array gain G (including aperture mismatch) as a function of beam scan angle:

$$G = \frac{4\pi A}{\lambda^2} \, \eta \cos \theta \, (1 - |\Gamma(\theta)|^2) \tag{4}$$

where

$\quad A$ = area of element multiplied by the total number of elements in the array antenna

$\quad \eta$ = aperture efficiency corresponding to the amplitude distribution across the array aperture, with $\eta = 1$ for uniform amplitude distribution

$\quad \theta$ = beam scan angle

$\quad |\Gamma(\theta)|$ = magnitude of reflection coefficient at scan angle θ

The net gain of the antenna is given by the above gain minus all the other ohmic losses and mismatch losses, such as the phase shifter loss, beam-forming feed network loss, etc.

Step 6

Establish the performance characteristics of the final radiator design by combining the measured impedances of step 4 with the measured active element patterns of step 5.

Phase Shifter Selection

The selection criteria of a phase shifter for a particular phased array application are listed below:

1. Operating frequency and bandwidth (tunable and instantaneous)
2. Peak and average rf power
3. Insertion loss
4. Switching time (reciprocal and nonreciprocal)
5. Drive power
6. Size and weight
7. Phase quantization error
8. Cost and producibility

Some of these criteria are established by radar requirements, while the others are used to compare the relative merits of various phase shifter types. Among all the above criteria the six most often used criteria for selecting a particular phase shifter are the operating frequency, peak and average rf power, switching time, size, weight, and cost. The cost is an important consideration since the phase shifters contribute to one-third of the cost of most phased arrays. The other two-thirds of the total cost are contributed almost equally by the phase shifter drivers and the beam-forming feed network plus the radiators. A more detailed discussion on the selection criteria of the phase shifters will be given after the various available phase shifter types are described.

Phase Shifter Types—The two types of commonly used phase shifters are the semiconductor diode phase shifters [10, 11, 12] and the ferrite phase shifters [10, 11, 12]. It is not the intention here to describe the theory of operation of these phase shifters since this subject is well covered elsewhere. This section will, however, describe in detail the performance and physical characteristics of these phase shifters so that a comparison can be made to select the proper phase shifter for a particular radar application. The diode phase shifters are generally digital phase shifters, i.e., the phase states of the phase shifter are in discrete binary phase increments. Fig. 6 shows the binary phase states of a 4-bit diode phase shifter. The

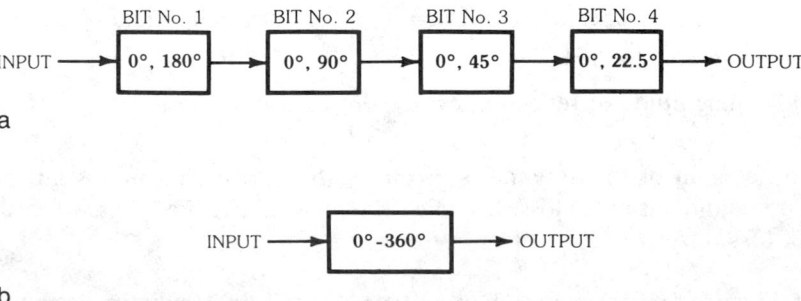

Fig. 6. Basic types of electronic phase shifters. (*a*) Semiconductor diode phase shifter (digital). (*b*) Ferrite phase shifter (digital or analog).

diode phase shifter uses pin diodes to provide phase shifting either by switching in different line length across a transmission line or by changing from inductive to capacitive loading across the transmission line. Typical phase shifter circuits using pin diodes are shown in Fig. 7. There are analog diode phase shifters [13, 14] containing either varactor diodes or pin diodes. However, these analog diode phase

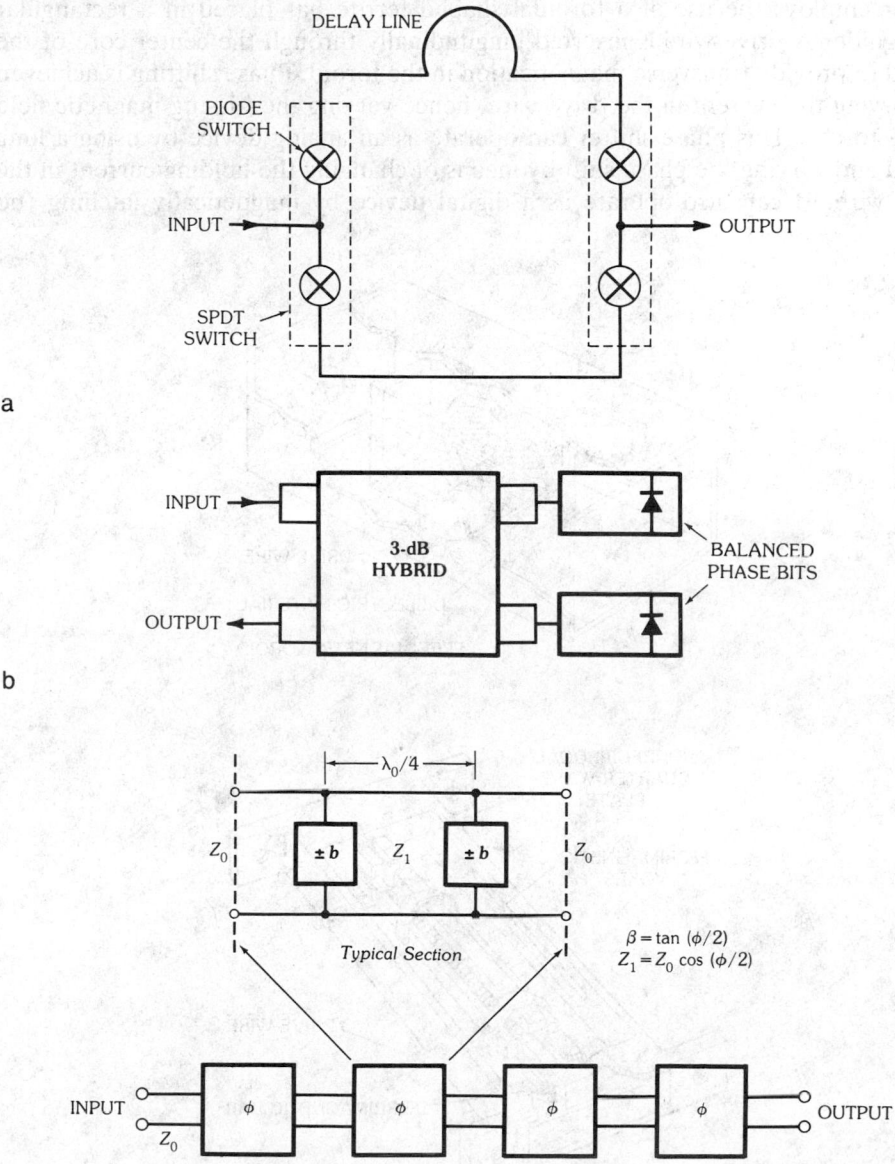

Fig. 7. Circuit designs for diode phase shifter. (*a*) Switched-line phase bit. (*b*) Hybrid-coupled phase bit. (*c*) Loaded-line phase bit. (*After Stark [10], © 1984; reprinted with permission of McGraw-Hill Book Company*)

shifters are extremely low power devices, typically in the milliwatts range. On the other hand, the ferrite phase shifters can be either digital or analog devices as illustrated in Fig. 8. In general, the two types of commonly used ferrite phase shifters are the nonreciprocal, toroidal ferrite phase shifter [15, 16] and the reciprocal, dual-mode ferrite phase shifter [17, 18]. Fig. 8 shows the basic configurations of these two types of ferrite phase shifters. The toroidal ferrite phase shifter employs the use of a toroidal shaped ferrite bar placed in a rectangular waveguide. A drive wire is inserted longitudinally through the center core of the toroid to provide transverse magnetization in the toroid. Phase shifting is achieved by varying the current in the drive wire, hence varying the biasing magnetic field in the toroid. This phase shifter can operate as an analog device by using a long toroid and varying the phase shift by means of changing the holding current in the drive wire. It can also operate as a digital device by magnetically latching (no

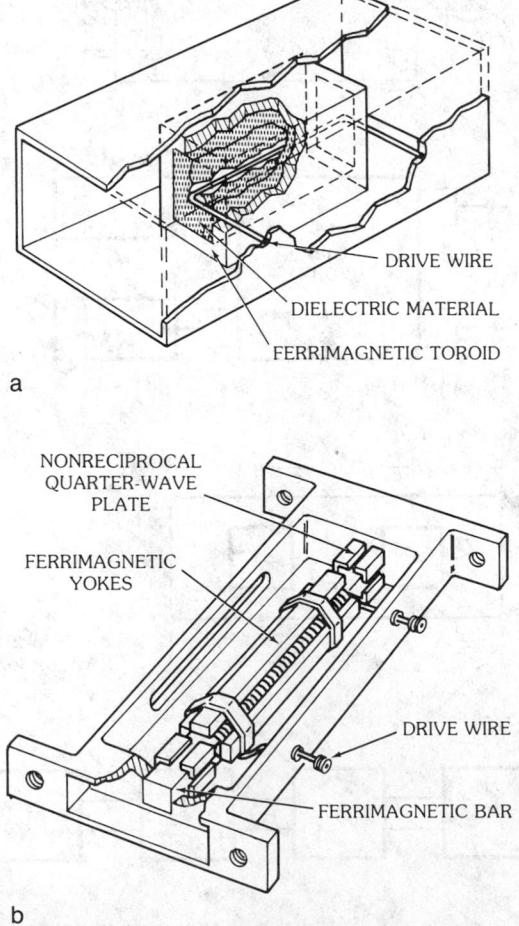

Fig. 8. Basic ferrite phase shifters. (*a*) Nonreciprocal twin-slab toroidal type. (*b*) Reciprocal dual-mode type. (*After Tang and Burns [11]*)

holding current) the toroid to the various minor hysteresis loops. The amount of phase shift is determined by the product of the magnitude and the time duration of the voltage pulse. In applications where fast switching speed is required, the long toroid can be split into smaller sections with the length of each section corresponding to a binary phase bit. In this case each section is magnetized into saturation and is quickly switchable since the volume of ferrite material for each section is significantly smaller. This type of phase shifter is nonreciprocal; hence the phase shift must be reset between the transmit and receive modes of the radar. The dual-mode ferrite phase shifter, however, is reciprocal, and it does not require resetting between transmit and receive. This phase shifter consists of a long metallized ferrite bar in between two nonreciprocal quarter-wave plates. The quarter-wave plate at each end of the ferrite bar converts the incident linearly polarized electric field into a circularly polarized field. This circularly polarized field interacts with the longitudinally magnetized biasing field in the ferrite bar to produce Faraday rotation, resulting in a net phase shift of the wave propagating through the ferrite bar. The amount of phase shift is controlled by the magnitude of the biasing field in the ferrite bar. This device can be used as a digital or analog phase shifter as in the case of the nonreciprocal toroidal phase shifter. However, the switching time of this phase shifter is much longer (50 µs compared with 10 µs for the toroidal phase shifter) due to the eddy current effects of the metal wall around the ferrite bar. There are other types of reciprocal ferrite phase shifters such as the Reggia-Spencer phase shifter [19] and Fox phase shifter [20]. The details of these phase shifters can be found in the references.

Performance Characteristics of Phase Shifters—A comparison of the general characteristics of diode and ferrite phase shifters is shown in Table 1. As shown in this table, most of the diode phase shifters that have been built operate over the frequency range of uhf to X-band, whereas the ferrite phase shifters have been built to operate over the frequency range of S-band to W-band. Ferrite phase shifters of the reciprocal, dual-mode type have been built at 95 GHz with insertion loss of approximately 2.5 dB for 360° of phase shift. The insertion loss of the diode phase shifter varies typically from 0.5 dB at L-band to 1.4 dB at X-band. On the other hand, the insertion loss of the ferrite phase shifter varies typically from 0.6 dB at S-band to approximately 1.0 dB at X-band. The insertion loss of the

Table 1. General Characteristics of Diode and Ferrite Phase Shifters

Parameter	Diode (Digital)	Remanent Ferrite (Analog and Digital)
Frequency	Uhf to X-band	S- to W-band
Insertion loss/2π	0.5 to 1.4 dB	0.6 to 2.5 dB
Temperature sensitivity	Negligible	<0.03 to $0.3\%/°C$
Peak power	$\leqq 8$ kW	$\leqq 100$ kW
Average power	$\leqq 300$ W	$\leqq 800$ W
Bandwidth	10% to 25%	10% to octave
Switching speed	10 ns to 30 µs	1 to 50 µs
Control power	0.1 W to 0.5 W	20–800 µJ

ferrite phase shifter goes up to approximately 1.4 dB at 30 GHz. A comparison of the insertion loss of the diode and ferrite phase shifters as a function of frequency is shown in Fig. 9. At *L*-band frequencies or below, the diode phase shifters are used because of their low insertion loss and lower cost. At *S*-band frequencies the insertion loss of the diode phase shifter is quite comparable to that of the ferrite phase shifter. For example, the insertion loss of the ferrite phase shifter is approximately 0.2 to 0.3 dB lower. Above *S*-band, however, the disparity in insertion loss becomes significantly more in favor of the ferrite phase shifter. The temperature sensitivity of the diode phase shifter is nil compared to 0.03%/°C to 0.3%/°C for the ferrite phase shifter, depending on the type of ferrite material used. The peak and average power handling capacity of most of the diode phase shifters is approximately 8 kW and 300 W, respectively, compared with 100 kW and 800 W, respectively, for the ferrite phase shifter. The bandwidth of most of the diode phase shifters is typically 10 percent. However, an octave bandwidth can be achieved by using Schiffman coupled circuits [21]. In the case of the ferrite phase shifter an octave bandwidth can also be achieved by using multiple stages of ridge-loaded waveguide transformers. The switching speed of the diode phase shifter varies from 0.5 ns to approximately 30 μs depending on the capacitance of the diode and the type of driver used to switch the diode from forward bias to reverse bias. Typically, the switching speed is approximately 30 μs, using a simple resistive pull-up type of driver circuit. However, the switching speed can be reduced to 10 μs or less when an active pull-up circuit is used. The switching speed for the ferrite phase shifter varies from 1 μs for the individual toroidal bit type to about 50 μs for the long Faraday rotator type. The required control power for the diode phase shifter varies from 0.1 to 0.5 W depending on the amount of forward bias current needed to set the quiescent state of each bit to the constant resistance region of diode *I-V* characteristics. Forward bias currents can be reduced from the above settings at the expense of a slight increase in insertion loss. The required control energy for the ferrite phase shifter varies from 20 to 800 μJ, depending on the speed with which the phase shifter has to be switched.

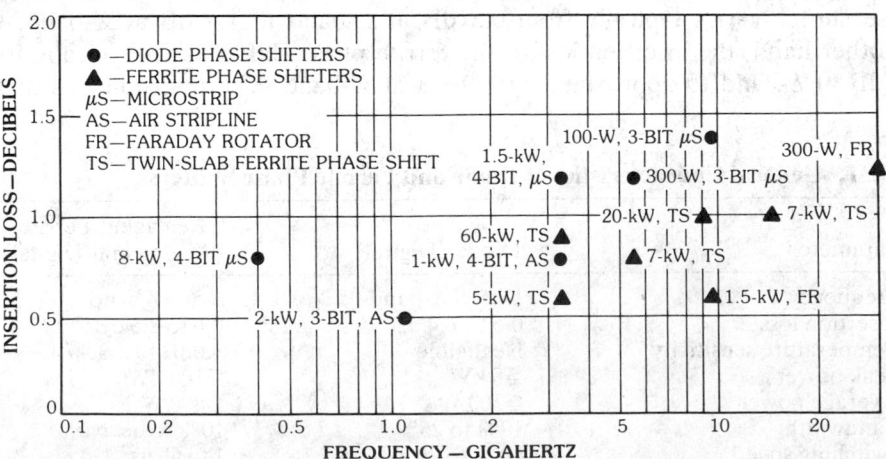

Fig. 9. Insertion loss of diode and ferrite phase shifters. (*After Tang and Burns [11]*)

General Observations in Selecting Phase Shifter Type—Above *S*-band and at peak power level of 2 kW or more per element, ferrite phase shifters are preferred due to their lower insertion loss. However, in situations where the required rf power is low (less than 2 kW) and size and weight constraints dictate a compact and lightweight package, diode phase shifters are preferred due to their simplicity in construction and lower production cost. For example, cost can be minimized by combining several dipole radiators, diode phase shifters, feed networks, drivers, and logic circuits on a common alumina substrate as an integral subarray module.

Beam-Forming Feed Network Selection

Among all the selection criteria, as stated in Chart 2, the most influential criteria in the selection of a beam-forming feed network for a particular radar application are the following:

1. Number of simultaneous beams
2. Monopulse sum and difference beams
3. Peak and average side lobe level
4. Tunable and instantaneous bandwidth
5. Peak and average power
6. Size, weight, and cost

For most applications the above criteria can be narrowed down from the multitude of possible design choices to a few practical design selections. For example, if a large number of multiple simultaneous beams are required, the possible design choices for the beam-forming feed network would be either a multiple-beam optical feed (such as the circular pillbox feed, Rotman lens feed, etc.) or a constrained matrix feed (such as the Butler or Blass matrix feed). Once the choices are narrowed down to only a few possible design approaches, a comparison of these approaches can then be made in terms of side lobe performance, bandwidth, size, weight, and cost in order to determine the optimum design approach. In order to select the optimum feed for a particular application it is necessary for the antenna designer to have a broad and comprehensive knowledge of the various types of available beam-forming feed networks. This subsection summarizes some of the basic types of beam-forming feed networks and their performance capabilities and limitations. A detailed discussion of the various feed network designs is given in Chapter 19, on beam-forming feed networks. In general, all the beam-forming feed networks can be classified into the following three basic categories:

Category 1—Space feeds:
transmission type
reflection type
Category 2—Constrained feeds:
series feed
parallel feed
Category 3—Hybrid feeds (a combination of space and constrained feeds)

Space Feeds—In the category of space feeds there are two basic types, namely, the transmission type [22] and the reflection type [23]. In the case of the transmission type the array elements and phase shifters are connected to an array of pickup elements, which are in turn illuminated by a feed horn located at a given focal distance away from the aperture of the pickup array (see Fig. 10). The ratio of the focal distance to the diameter of the radiating aperture varies nominally from 1/2 to 1. The array of pickup elements in conjunction with the phase shifters and the radiating elements forms an electronic scanning feedthrough lens (transmission lens). The phase shifters are set to provide the required phase increments between the radiating elements for beam scanning and for correcting the phase error introduced by the spherical phase front from the feed horn. When digital phase shifters (constant phase with frequency type) are used in the lens this spherical phase front correction approach can suppress the peak error side lobe introduced by the phase quantization error of the digital phase shifter. A detailed discussion of the effect of phase quantization error on array performance is given by Miller [24]. The tunable and instantaneous bandwidths of this space-fed antenna are typically 10 percent and 40 MHz, respectively. When the time-delay type of phase shifters are used, this antenna system has extremely wide instantaneous bandwidth limited only by the performance of the components. An instantaneous bandwidth of 1000 MHz is achievable. Monopulse sum and difference beams can be formed by using a cluster of feed horns at the focal point. For example, a cluster of 2×2 feed horns combined with magic tees can provide a sum beam, an elevation difference beam, and an azimuth difference beam as shown in Fig. 10. Multiple simultaneous beams can also be formed by using feed horns displaced from the focal point. The number of multiple beams that can be formed, however, is limited by the spherical aberration effects. Various methods of correcting the spherical aberration are treated by Rotman and Turner [25]. Since the signal distribution from the feed horns to the elements is through free space, this space feed is probably the simplest

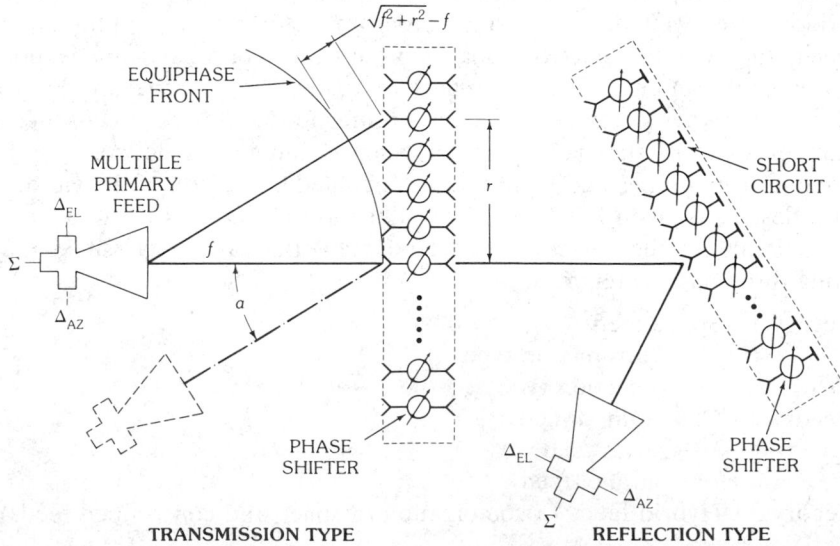

Fig. 10. Space feed systems. (*After Cheston and Frank*, Array Antennas *[22] TG-956, JHU/APL, Laurel, MD, March 1968*)

way to form multiple simultaneous beams. In the case of the reflection type of space feed (see Fig. 10), the concept is the same as that of the transmission type except a short is placed behind each phase shifter so that the signals are reflected and reradiated from the pickup elements. Since the signal travels back and forth through the phase shifters twice, the amount of required phase shift at each element is half of that of the transmission case for the same scan angle. However, the peak power requirement for the phase shifter is quadrupled because of the standing wave. In order to minimize the blockage effect of the feed horn, the feed horn has to be offset with respect to the axis of the reflective lens.

Space feeds have several disadvantages. One of them is the large physical volume required by the space feed. Another is the multiple reflection effects between the mismatches of the two lens apertures as a function of beam scan angle. This effect can be minimized by performing the best impedance matching possible for the two apertures over the required scan coverage and frequency band. A less obvious problem with space feeds is the difficulty of connecting control wires of the phase shifters in the lens because of the lack of access from the front or the back of the lens. The only access is through the peripheral edge of the lens.

Constrained Feeds—In the category of the constrained feeds there are two basic types, namely, the series feeds and the parallel feeds. By definition the radiating elements are fed serially in a series feed, while they are fed in parallel in a parallel feed. Typical examples of the series feed are shown in Fig. 11. The basic form of a series feed is shown in Fig. 11a. The input signal is fed from one end of the feed and the other end is terminated into a matched load. The input signal is then coupled serially through directional couplers to the phase shifters and radiating elements. Since the transmission line length increases from the input to the following radiating elements, there is a progressive phase change between the radiating elements with frequency variations; thus the beam scans with frequency change. The amount of beam squint with frequency is given by

$$\text{beam squint} = \frac{\Delta f}{f} \frac{1}{\cos \theta_0} \tag{5}$$

where Δf is the amount of frequency change and θ_0 is the nominal scan angle. A detailed discussion of the bandwidth limitations of series and parallel feeds is given by Frank [26]. The beam squint, however, can be brought back to the original position by resetting the phase shifters at each radiating element. One of the problems with this type of series feed is that the mismatches from the radiating elements, phase shifters, and couplers can all add up in phase at the frequencies when the path length between the elements is a multiple of a half-wavelength. In order to provide sum and difference beams for monopulse tracking the feed input is moved from the end to the center of the series feed as shown in Fig. 11b. The two halves of the feed are fed by a magic tee so that the sum port of the magic tee provides in-phase excitation of the two halves and the difference port provides out-of-phase excitation to generate a difference beam. Since the same coupling coefficients of the couplers are used for both the in-phase and the out-of-phase excitations, the side lobes for the sum and difference beams cannot be op-

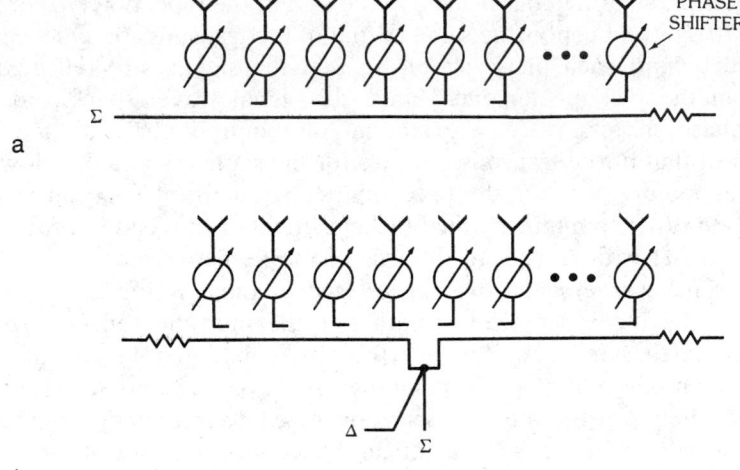

Fig. 11. Constrained feeds (series feed networks). (*a*) End feed. (*b*) Center feed. (*c*) Center fed with separately optimized sum and difference channels. (*d*) Equal path length feed. (*e*) Series phase shifters. (*After Cheston and Frank,* Array Antennas *[22], TG-956, JHU/APL, Laurel, MD, March 1968*)

timized simultaneously. For low side lobes the sum beam requires an even distribution, such as the Taylor distribution [27], and the difference beam requires an odd distribution, such as the Bayliss distribution [28]. In order to achieve low side lobes for both sum and difference beams a center-fed dual series feed (see Fig. 11c) is used. The excitations of the two parallel series feeds are adjusted to optimize the side lobes for both the sum and the difference beams. A detailed description of the dual series feed is given by Lopez [29]. At broadside beam position the two halves of the center-fed series feed scan in opposite directions with frequency. This results in a beam broadening with no change in direction. If the two halves scan too

Fig. 11, *continued*.

far apart, it could even result in a splitting of the beam. Therefore the bandwidth of the center-fed series feed is significantly worse than that of the parallel feed. However, at large scan angles such as 60° from broadside the bandwidth of the center-fed series feed is quite comparable to that of a parallel feed (see Frank [26]). In order to broaden the bandwidth at broadside for a center-fed array the path length from the feed input to the radiating elements can be made equal as shown in Fig. 11d. In this case the bandwidth is only limited by the in-phase addition of the mismatches of the couplers. For a series-fed array the phase shifters can be also inserted serially between the couplers along the series feed line as shown in Fig. 11e. In this design the amount of required phase shift for each phase shifter is greatly reduced compared with that of the phase shifters at the radiating elements. However, the insertion losses of these serial phase shifters are additive, resulting in a reduction in array gain and an increase in side lobe level from the asymmetrical amplitude distribution.

Typical examples of the parallel feeds are shown in Fig. 12. The basic form of a parallel feed is shown in Fig. 12a. In this basic form the input signal is divided in a corporate tree fashion to all the radiating elements. The path lengths from the input to each output are made equal. The bandwidth at broadside is ideally infinite, except for the practical limitations of such components as the couplers, phase shifters, and radiators. When the beam is scanned away from broadside the beam scans with frequency. The amount of beam squint with frequency is given by

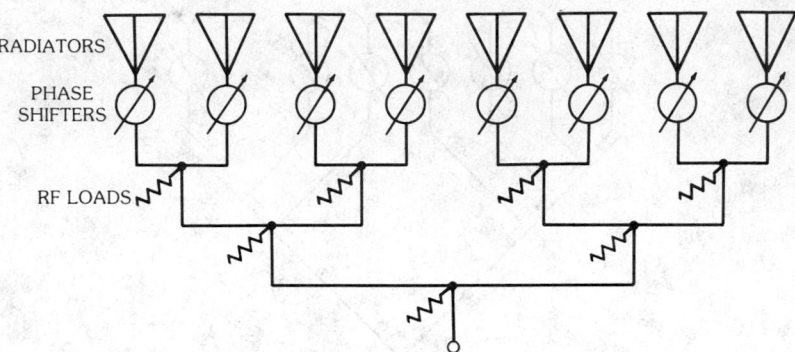

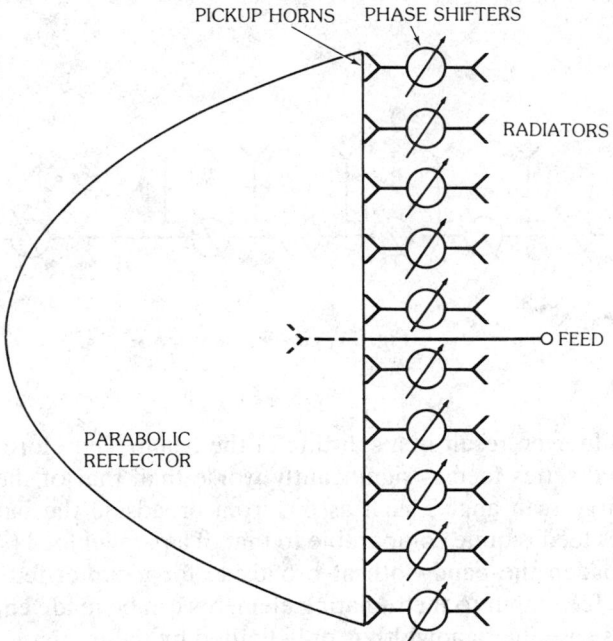

Fig. 12. Constrained feeds (parallel feed networks). (*a*) Corporate feed. (*b*) Pillbox feed. (*After Rotman [34]*, © *1958 IEEE*) (*c*) Butler matrix feed (multiple beams). (*After Butler and Lowe [30]*) (*d*) Time-delay matrix feed (multiple beams). (*After Blass [31]*, © *1960 IEEE*)

$$\text{beam squint} = \frac{\Delta f}{f} \tan \theta_0 \tag{6}$$

The bandwidth at 60° scan (see Frank [26]) is given by

$$\text{percent bandwidth} = \text{beamwidth in degrees}$$

When magic tees or hybrid couplers are used at each level of the corporate feed the mismatches from the radiating elements are reasonably well isolated from each other. Hence the parallel corporate feed does not have the additive effect as in the

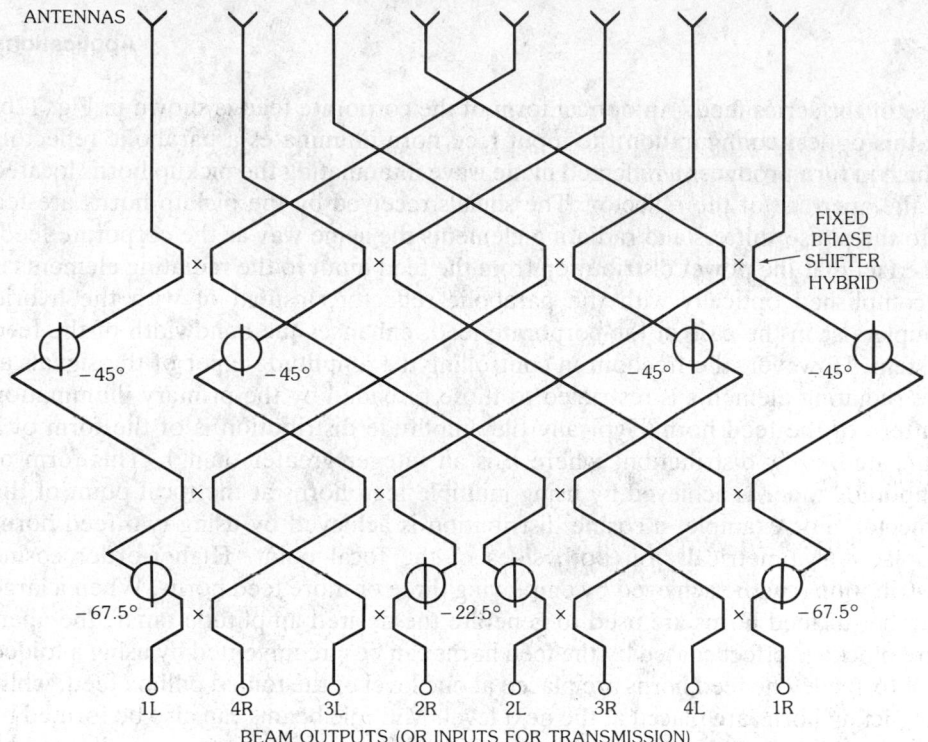

ANTENNAS

FIXED PHASE SHIFTER HYBRID

−45°   −45°   −45°   −45°

−67.5°   −22.5°   −67.5°

1L   4R   3L   2R   2L   3R   4L   1R

c   BEAM OUTPUTS (OR INPUTS FOR TRANSMISSION)

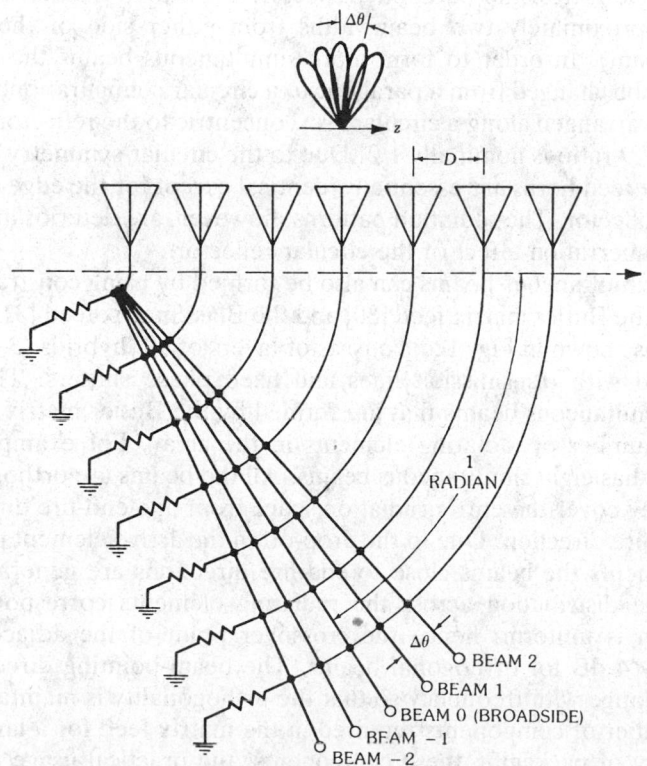

1 RADIAN

Δθ

BEAM 2

BEAM 1

BEAM 0 (BROADSIDE)

BEAM −1

BEAM −2

d

Fig. 12, *continued*.

case of the series feed. An optical form of the corporate feed is shown in Fig. 12b. In this optical configuration the input feed horn illuminates a parabolic reflector, which in turn produces a reflected plane wave illuminating the pickup horns located at the aperture of the reflector. The signals received by the pickup horns are fed into the phase shifters and radiating elements the same way as the corporate feed. The fact that the power distribution from the feed input to the radiating elements is accomplished optically with the parabolic reflector, instead of with the hybrid couplers as in the case of the corporate feed, enhances the bandwidth of the feed system. However, the freedom in controlling the amplitude taper of the signals at the radiating elements is restricted to those provided by the primary illumination pattern of the feed horn. Typically the amplitude distribution is of the form of a truncated $\cos^n x$ distribution, where n is an integer greater than 1. This form of amplitude taper is achieved by using multiple feed horns at the focal point of the reflector. For example, a cosine distribution is achieved by using two feed horns displaced symmetrically on both sides of the focal point. Higher-order cosine distributions can be achieved by employing three or more feed horns. When a large number of feed horns are used to generate the desired amplitude taper, the aperture blockage effect caused by the feed horns can be circumvented by using a folded pillbox feed. The feed horns are placed at one level of the folded pillbox feed, while the pickup horns are placed at the next level. Multiple beams can also be formed by using multiple-feed horns. The number of multiple beams, however, is limited by the defocusing effect of the parabolic reflector. Typically, scanned beams can be formed to approximately two beamwidths from either side of the focal beam (broadside beam). In order to form more simultaneous beams the shape of the reflector must be changed from a parabolic to a circular configuration, and the feed horns must be arranged along a circular arc, concentric to the reflector surface. For this case the f/D ratio is nominally 1/2. Due to the circular symmetry the radiation patterns of the feed horns are essentially identical except for the edge or truncation effect of the reflector. The radiation patterns, however, are deteriorated slightly by the spherical aberration effect of the circular reflector.

Multiple simultaneous beams can also be formed by using constrained parallel feeds such as the Butler matrix feed [30] and the Blass matrix feed [31]. The Butler matrix feed, as shown in Fig. 12c, consists of layers of 90° hybrids (3-dB couplers) interconnected with transmission lines and fixed phase shifters. The maximum number of simultaneous beams that are formed by the Butler matrix feed is equal to the total number of radiating elements in the array. For example, an eight-element array has eight simultaneous beams. All the beams are orthogonal to each other, and they cover the entire radiation space from one end-fire direction to the opposite end-fire direction. Due to the drop-off in the active element pattern of the radiating elements the beams close to end-fire directions are generally not used. The amplitude distribution across the radiating elements corresponding to any beam position is uniform; hence the crossover point of the adjacent beams is approximately 4 dB for orthogonal beams. The beam-pointing direction and the beamwidth change with frequency so that the orthogonality is maintained. Due to the large number of components required in the matrix feed for a large array and the complexity of packaging these components, the practical usage of the Butler matrix feed is generally limited to a sixteen-element array.

In order to prevent the beam from scanning with frequency a time-delay matrix feed of the form shown in Fig. 12d can be used. This time-delay matrix feed is a special form of the Blass matrix feed [31]. For the broadside beam (beam 0 in Fig. 12d) all the path lengths from the input to the radiating elements are equal. For all the scanned beams the difference in path lengths between two adjacent feed lines from the beam input terminal to the radiating elements is exactly equal to the required time delay between these elements for that particular scan angle. Hence this time-delay matrix feed has extremely wide bandwidth, and it is only limited by the bandwidth of the couplers in the feed. One of the practical design problems of this time-delay matrix feed is the coupling between the feed lines. The amount of coupling is determined by the crossover level of the corresponding beams and the directivity of the couplers. High crossover level between beams (higher than the level corresponding to orthogonal beams) appears as a cross-coupling loss which in turn degrades the antenna gain. Poor directivity in the couplers causes circulating power between feed lines, which produces amplitude and phase errors at the radiating elements. In order to minimize amplitude and phase errors it is imperative that the beams are spaced as closely to the orthogonal condition as possible, and the directivity of the couplers is as high as possible.

Independently controlled sum and difference beams for monopulse tracking can be formed in a parallel constrained feed as shown in Fig. 13. The signals from symmetrical pairs of radiating elements located diametrically opposite from the centerline are combined in 180° hybrids (magic tees) to form in-phase and out-of-phase signals. The in-phase signals are combined in a feed network with the proper amplitude weighting to form a sum beam. The out-of-phase signals from the magic

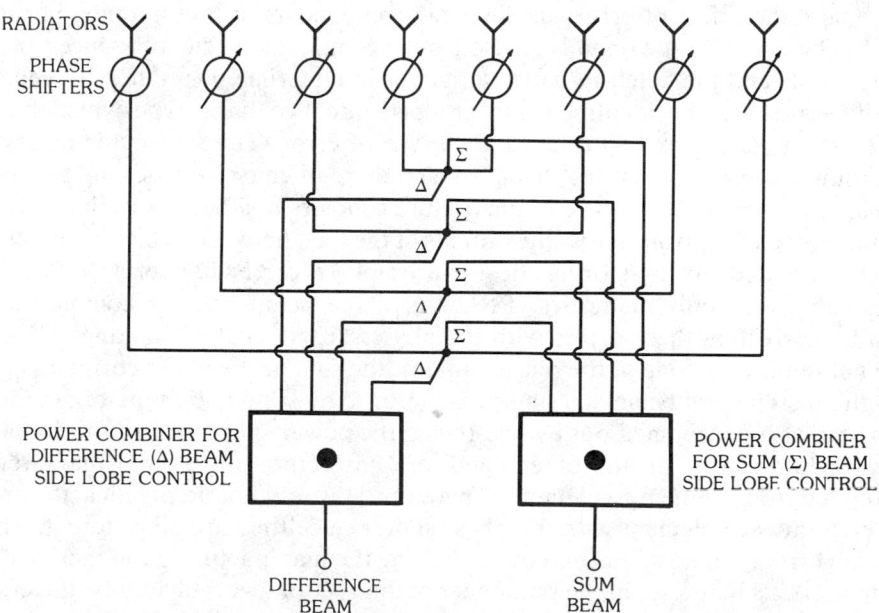

Fig. 13. Constrained monopulse feed with independently controllable sum and difference beam side lobes. (*After Tang and Burns [11]*)

tees are combined in a separate feed network to form a difference beam. Different amplitude distributions, such as the Taylor distribution for the sum beam feed network and the Bayliss distribution for the difference beam feed network, can be used to independently control the side lobes of both beams. When the lengths of the transmission lines from the input to the radiating elements are made equal, the bandwidth of this monopulse feed network is the same as that of a corporate feed.

General Observations in Selecting Beam-Forming Feeds—In general, the series feeds are more frequency sensitive than the parallel feeds; hence the series feeds are more bandwidth limited. Furthermore, the mismatches from the couplers in a series feed are additive when the transmission lines between the couplers are integer multiples of a half-wavelength. However, the lengths of transmission lines between the couplers in a parallel feed are nonuniform so that the mismatches do not add in phase. Also, the mismatches from the couplers can be isolated by the magic tees in order to prevent the reflected signal from circulating in the feed network and, as a result, cause high-power breakdown problems. In order to avoid the resonance problem in the series feed the length of transmission line in the series feed is chosen to avoid resonance over the operating band. One advantage of the series feed is that it is physically more compact than the parallel feed; hence it is more useful for applications where antenna volume is of prime importance.

4. Effect of Component Errors on Array Performance

One of the most important considerations in the design of a phased array antenna is the effect of errors in the array components on the antenna performance. These errors are mainly caused by the manufacturing tolerances of the components and the batch-to-batch variations in material consistency. In general, all the errors can be identified and grouped into two basic types: namely, the systematic type of error and the random type of error. The systematic errors are deterministic errors resulting from some inherent characteristics of the components. For example, the use of quadrature couplers in a beam-forming network results in fixed 90° errors across the outputs of the feed network. These phase errors can be trimmed out by storing the required phase correction terms in the programmable read-only memories (PROMs) of the beam-steering computer, and then by correcting these errors with the phase shifters at the radiating elements. The net required setting of the phase shifter is the sum of the phase correction term and the incremental beam-steering phase term. Any systematic type of amplitude error can also be trimmed out by fine-tuning the power split ratios of the couplers. The random errors, on the other hand, are not deterministic, and they are not correlated from element to element. Therefore, it would not be practical to correct the error at each element with the phase shifter since they are all different. These random errors can be minimized by controlling the manufacturing tolerances of the components. Therefore, in the remainder of this section, we will address the effects of random errors on antenna performance.

Before we can analyze the effect of random errors on array performance the following assumptions relating to the characteristics of these errors are made:

1. Errors in each radiating element are statistically independent from those in all the other elements.
2. Errors in all elements possess the same statistic.

Based on the above assumptions the array performance can be calculated in terms of the net rms amplitude and phase errors at each element of a large two-dimensional array. The net rms error at each element is determined by statistically summing all the contributions from the various components to that element using the central limit theorem. Detailed treatments of the error effects on array performance were done by Ruze [32], Allen [33], Miller [24], etc. They have expressed the array performance in terms of the rms error as follows:

1. The rms side lobe power, $\bar{p}$, is

$$\bar{p} = \frac{\sigma^2}{\eta N} \tag{7}$$

where

$\sigma^2 = \sigma_a^2 + \sigma_p^2$

σ_a = rms amplitude error

σ_p = rms phase error in radians

η = aperture efficiency, which is less than or equal to 1

N = total number of elements

2. The peak side lobe level,* p_{pk}, is

$$p_{pk} = |S_0| + 2\sigma/\sqrt{\eta N} \tag{8}$$

where S_0 is the amplitude of the design side lobe without errors.

3. The reduction in antenna gain is

$$\frac{G}{G_0} = \frac{1}{1 + (3\pi/4)(d/\lambda)^2\sigma^2} \tag{9}$$

where

d/λ = element spacing in wavelengths

G_0 = antenna gain with the absence of errors

4. The beam-pointing error is

$$\frac{\delta_\psi}{\Delta\theta_{rad}} = \sqrt{\frac{3}{N}}\frac{\sigma}{0.88\pi} \tag{10}$$

---

*Estimate is based on power not to be exceeded with 98-percent probability.

where

δ_ψ = rms beam-pointing error in radians

$\Delta\theta_{rad}$ = beamwidth in radians

The above formulas can be used in conjunction with actual radiation pattern calculations to estimate the allowable error budgets for the array components in meeting a given set of array performance specifications. Conversely, if the amplitude and phase errors of the components are known, then the array performance can be estimated by the above formulas. An example which illustrates the effects of errors on the array performance is given below. This example is for the case of an array of 2000 elements with element spacing d of 0.5λ. The error-free side lobe level $(p_{pk})_0$ is asumed to be -40 dB; correspondingly, the aperture efficiency η is equal to 0.64. Assuming the composite rms amplitude error σ_a and the rms phase error σ_p are 0.5 dB and 4°, respectively, then the net rms error σ is given by

$$\sigma^2 = \sigma_a^2 + \sigma_p^2 = 0.0035 + 0.0049 = 0.0084$$

The average side lobe level of the array with errors is given by

$$\bar{p} = \frac{\sigma^2}{\eta N} = 6.6 \times 10^{-6} \quad \text{or} \quad -51.8 \text{ dB}$$

The peak side lobe level p_{pk} is obtained from $p_{pk} = |S_{pk}|^2$, where

$$S_{pk} = S_0 + \frac{2\sigma}{\sqrt{\eta N}} = 0.01 + 0.005 = 0.015$$

so that

$$p_{pk} = -36.5 \text{ dB}$$

The gain reduction is given by

$$\frac{G}{G_0} = \frac{1}{1 + (3\pi/4)(d/\lambda)^2\sigma^2} = 0.995 \quad \text{or} \quad -0.02 \text{ dB}$$

The beam-pointing error is given by

$$\frac{\delta_\psi}{\Delta\theta_{rad}} = \sqrt{\frac{3}{N}}\frac{\sigma}{0.88\pi} = 0.001\,28 \quad \text{or} \quad \delta_\psi = \Delta\theta_{rad}/800$$

5. References

[1] K. R. Carver and J. W. Mink, "Microstrip antenna technology," *IEEE Trans. Antennas Propag.*, vol. AP-29, no. 1, pp. 2–24, January 1981.

[2] R. J. Mailloux, J. F. McIlvenna, and N. P. Kernweis, "Microstrip array technology," *IEEE Trans. Antennas Propag.*, vol. AP-29, no. 1, pp. 25–37, January 1981.

[3] R. J. Stegen, "Longitudinal shunt slot characteristics," *Tech. Memo. No. 261*, Hughes Aircraft Company, 1951.

[4] A. F. Stevenson, "Theory of slots in rectangular waveguide," *J. Appl. Phys.*, no. 19, pp. 24–38, 1948.

[5] J. S. Ajioka, "Frequency-scan antennas," *Antenna Engineering Handbook*, 2nd ed., ed. by R. C. Johnson, New York: McGraw-Hill Book Co., 1984, pp. 19-1–19-30.

[6] R. S. Munson, "Microstrip antennas," *Antenna Engineering Handbook*, 2nd ed., ed. by R. C. Johnson, New York: McGraw-Hill Book Co., 1984, pp. 7-1–7-28.

[7] J. Q. Howell, "Microstrip antennas," in *Dig. Intl. Symp. Antennas Propag. Soc.*, pp. 177–180, Williamsburg, Virginia, December 1972.

[8] E. V. Byron, "A new flush-mounted antenna element for phased array application," in *Proc. 1970 Phased Array Antenna Symp.*, pp. 187–192, Polytechnic Institute of Brooklyn, June 1970. Reprinted in *Phased Array Antennas*, ed. by A. A. Oliner and G. H. Knittel, Dedham: Artech House, 1972.

[9] C. C. Chen, "Broadband impedance matching of rectangular waveguide phased arrays," *IEEE Trans. Antennas Propag.*, vol. AP-21, pp. 298–302, May 1973.

[10] L. Stark, R. W. Burns, and W. P. Clark, "Phase shifters for arrays," *Radar Handbook*, ed. by M. I. Skolnick, New York: McGraw-Hill Book Co., 1970, pp. 12-1–12-65.

[11] R. Tang and R. W. Burns, "Phased arrays," *Antenna Engineering Handbook*, 2nd ed., ed. by R. C. Johnson, New York: McGraw-Hill Book Co., 1984, pp. 20-1–20-67.

[12] D. H. Temme, "Diode and ferrite phaser technology," in *Proc. Phased Array Antenna Symp.*, 1970, pp. 212–218, reprinted in *Phased Array Antennas*, ed. by A. A. Oliner and G. H. Knittel, Dedham: Artech House, 1972.

[13] C. A. Liecht and G. W. Epprechi, "Controlled wideband differential phase shifters using varactor diodes," *IEEE Trans. Microwave Theory Tech.*, vol. MTT-15, pp. 586–589, October 1967.

[14] J. F. White, "Figure of merit for varactor reflection type phase shifters," *NEREM Rec.*, pp. 206–207, November 1965.

[15] W. J. Ince et al., "The use of manganese-doped iron garnets and high dielectric constant loading for microwave latching ferrite phasers," *G-MTT Dig.*, pp. 327–331, 1970.

[16] D. H. Temme et al., "A low cost latching ferrite phaser fabrication technique," *G-MTT Dig.*, pp. 88–96, 1969.

[17] R. G. Roberts, "An *X*-band reciprocal latching Faraday rotator phase shifter," *G-MTT Dig.*, pp. 341–345, 1970.

[18] C. R. Boyd, Jr., "A dual-mode latching reciprocal ferrite phase shifter," *G-MTT Dig.*, pp. 337–340, 1970.

[19] F. Reggia and E. G. Spencer, "A new technique in ferrite phase shifting for beam scanning of microwave antennas," *Proc. IRE*, vol. 45, pp. 1510–1517, November 1957.

[20] A. G. Fox, "An adjustable waveguide phase changer," *Proc. IRE*, vol. 35, pp. 1489–1498, December 1947.

[21] B. M. Schiffman, "A new class of broadband microwave 90-degree phase shifters," *PGMTT-MTT-6*, pp. 232–237, 1958.

[22] T. C. Cheston and J. Frank, "Array antennas," *Tech. Memo. TG-956*, The Johns Hopkins University Applied Physics Laboratory, March 1968.

[23] R. Tang, R. W. Burns, and N. S. Wong, "Phased array antenna for airborne application," *Microwave J.*, vol. 14, no. 1, pp. 31–38, January 1971.

[24] C. J. Miller, "Minimizing the effects of phase quantization errors in an electronically

scanned array," *Proc. 1964 Symp. Electronically Scanned Array Techniques and Applications*, RADC-TDR-64-225, vol. 1, pp. 17–38.

[25] W. Rotman and R. F. Turner, "Wide angle microwave lens for line source," *IEEE Trans. Antennas Propag.*, pp. 623–632, November 1963.

[26] J. Frank, "Bandwidth criteria for phased array antennas," *Proc. Phased Array Antenna Symp.*, pp. 243–253, Polytechnic Institute of Brooklyn, June 1970. Reprinted in *Phased Array Antennas*, ed. by A. A. Oliner and G. H. Knittel, Dedham: Artech House, 1972.

[27] T. T. Taylor, "Design of line-source antennas for narrow beamwidth and low side-lobes," *IRE Trans. Antennas Propag.*, vol. AP-3, pp. 16–28, 1955.

[28] E. T. Bayliss, "Design of monopulse antenna difference patterns with low sidelobes," *Bell Syst. Tech. J.*, pp. 623–650, May–June 1968.

[29] A. R. Lopez, "Monopulse networks for series feeding an antenna," *IEEE Trans. Antennas Propag.*, vol. AP-16, pp. 436–440, June 1968.

[30] J. Butler and R. Lowe, "Beamforming matrix simplifies design of electronically scanned antennas," *Electron. Des.*, vol. 9, no. 7, pp. 170–173, April 1961.

[31] J. Blass, "The multi-directional antennas: a new approach to stacked beams," *IRE Conv. Proc.*, vol. 8, pt. 1, pp. 48–51, 1960.

[32] J. Ruze, "Physical limitations on antennas," *Tech. Rep. No. 248*, Massachusetts Institute of Technology, October 30, 1952.

[33] J. L. Allen, "The theory of array antennas," *Tech. Rep. No. 323*, Lincoln Lab, Massachusetts Institute of Technology, July 25, 1963.

[34] W. Rotman, "Wide-angle scanning with microwave double-layer pillboxes," *IRE Trans. Antennas Propag.*, vol. AP-6, pp. 96–105, January 1958.

Chapter 19

Beam-Forming Feeds

J. S. Ajioka
Hughes Aircraft Company

J. L. McFarland
(Late) Hughes Aircraft Company

CONTENTS

James S. Ajioka was born in Thornton, Idaho. He received the BS and MS degrees in electrical engineering from the University of Utah in 1949 and 1951, respectively.

From 1949 to 1955 he was a group leader and project engineer in the Antenna Design Section of the Navy Electronics Laboratory. In 1955 he joined Hughes Aircraft Company, where he is manager of the Electromagnetics Laboratories, Hughes Ground Systems Group, in Fullerton, California. He has been active in the design and development of electronically scanning arrays, waveguide slot arrays, geodesic antennas, and wide-angle multiple-beam optical antenna systems, and also low-noise high-efficiency feeds, broadband multioctave multiplexing feeds for reflectors and lenses, and millimeter-wave phased arrays.

Mr. Ajioka has more than 20 patents awarded and several are pending. He has presented and published numerous papers on antenna and microwave systems and has contributed to several books. He has also won awards for outstanding achievements in antenna engineering. Mr. Ajioka is a Fellow of the IEEE.

Jerry L. McFarland was born in Iraan, Texas. He received the BS and MS degrees in electrical engineering, with the latter being from the University of Southern California, Los Angeles.

From 1958 to 1964 he was a staff engineer with Hughes Aircraft Company involved with the development of frequency scanning antennas and research on multiple-beam geodesic lenses. From 1964 to 1972 he was engaged in similar work with Autonetics, a division of Rockwell International. In 1972 he became the technical director of EMP, Chatsworth, California, developing antennas and single-axis trackers. In 1977 he joined Lockheed Missiles and Space Company, where he was concerned with adaptive arrays and multiple-beam antennas. He rejoined Hughes Aircraft Company, Fullerton, California, in 1982 as a senior scientist and was most recently involved in antenna development and investigation of nonconventional slot radiators.

Mr. McFarland's untimely death is a great loss to his profession and to all his friends and associates.

1. Introduction

When phased arrays were relatively simple, antenna subassemblies were easy to identify as feed networks, phasors, and radiating aperture. Modern phased arrays, however, have become quite complex, with a wide variety of designs and physical implementations depending on the particular application. With simple phased arrays the feed network was a passive network of branching transmission lines to distribute the power from a single transmitter to each of the radiating elements in the array via the phasors, and, conversely on receive, it combined the power received by each of the radiating elements to the input to a single receiver. Modern phased arrays may have multiple distributed transmitters, multiple distributed preamplifiers, multiple duplexing switches, and multiple simultaneous beam ports, each with its own final receiver. In addition, adaptive arrays may have adaptive control loops distributed throughout the feeding network with a significant amount of signal processing done within the antenna. For these reasons, general categories and general definitions become somewhat ambiguous. However, since generality is necessary to discuss phased arrays in general, an attempt is made to organize feed systems into general categories, and the reader should be aware of the shortcomings.

2. Constrained Feeds (Transmission-Line Networks)

The simplest method of feeding an array is to use simple passive transmission-line networks which take the transmitter power from a single source, or possibly multiple sources, and to distribute the power to each radiating element via transmission lines and associated passive microwave devices, such as hybrids, magic-Ts, or directional couplers, etc. The network itself is usually a combination of directional couplers, hybrids, and/or Ts in waveguide, coax, stripline, or microstrip. Printed-circuit techniques are popular for some of the approaches to be discussed. Fig. 1 is a simplified diagram of a constrained feed showing the basic subassemblies. In the case of multiple beams, there would be more than one input and more than one beam. In the following sections the series-fed, parallel-fed, time-delay-fed, and multiple-beam matrix-fed approaches are discussed.

Series Feed Networks

The simplest power distribution feed is an end-fed transmission line in which power is coupled off at (usually) periodic intervals to the radiating elements as shown schematically in Fig. 2. Although the series feed is simple, low cost, and packages easily, it has a drawback in that it "frequency scans" because there is an interelement progressive phase delay equal to $k_g d$, which is proportional to frequency, where $k_g = 2\pi/\lambda_g = \omega/v_g$ is the wave number of the feeding transmission line and d is the line length between branch line couplers. Although the phase

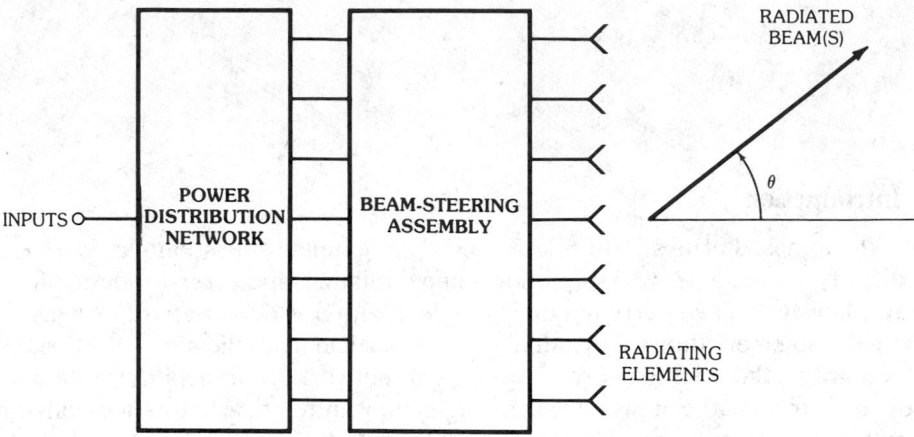

Fig. 1. Constrained feed approach.

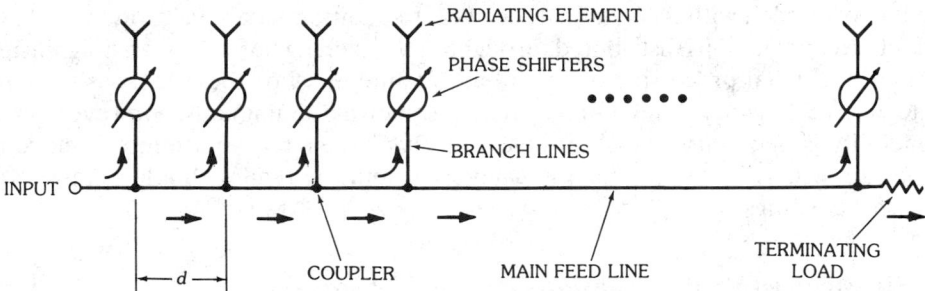

Fig. 2. Schematic diagram of a series feed.

shifters are usually reset for each frequency change, the instantaneous or signal bandwidth is degraded because the antenna beams corresponding to each of the frequency components in a narrow pulse will be pointing in slightly different directions in space. Thus the composite antenna beam is "smeared out" in angular space, thereby broadening the effective beamwidth to reduce the antenna gain and angular resolution.

There are many coupler designs that can couple the desired amount of power from the main feed line to the branch lines. The couplers can be nondirective three-port reactive couplers, such as pure series or pure shunt couplers, nondirective matched couplers, or directive four-port couplers. Nondirective couplers are simple and inexpensive to fabricate. Pure series or shunt couplers are the simplest but are inherently mismatched. For arrays with a large number of elements (on the order of 50 or more elements) the individual reflections due to coupler mismatch are quite small because the coupling is very loose. For large arrays the maximum voltage coupling coefficient is on the order of 0.1. These coupler mismatch reflections, for the most part, add randomly in phase at the input to give a low feed input vswr except at or very near the resonant frequency, that is, the frequency corresponding to a broadside beam. At the resonant frequency the couplers are an integral number of wavelengths apart (or half-wavelengths if the phase of the couplers is alternately 0° and 180°) and all the reflections add in phase to result in a very high input vswr. For purely series couplers the impedances add, and for purely shunt couplers the admittances add. The input vswr versus frequency curve shows high resonance for large arrays, and the width of the vswr versus frequency curve is inversely proportional to the number of elements in the array. In short, the input vswr versus frequency curve resembles an antenna pattern in all respects except the independent variable is frequency instead of spatial angle. That is, tapered aperture distributions result in lower side lobes and broader beamwidth in antenna patterns, and tapered coupling coefficients result in lower side lobes and broader resonance bandwidths in the vswr curve.

Because of the high vswr at resonance the resonant frequency is usually designed to be out of the operating frequency range when simple nondirective couplers are used. Matched nondirective couplers have been designed in waveguide that allow operation at the resonant frequency [1]. Although they are more complex, more costly, and, in general, more difficult to package, directional couplers have superior performance because they are inherently matched at all four ports.

In designs using nondirective couplers the reflections from the couplers can couple into the preceding branch arms and cause spurious side lobes in the antenna pattern; hence they are called *reflection lobes*. In particular, a reflection from a poorly matched terminating load will cause a reflected beam in the conjugate direction from broadside as the desired beam. Reflections from mismatches in the branch lines, such as reflections from the input end of the phase shifters or connectors on the feed side of the phase shifter, will usually reflect specularly as a whole and the reflected wave reflects back into the feed main line and will be phased such that it will be dissipated in the terminating load of the series feed.

Parallel Feed Networks

The simplest parallel feed network consisting of branching transmission lines is commonly known as a corporate feed (see Fig. 3). Since the path lengths from the feed point to the radiating elements are equal, there is no progressive phase delay between radiating elements, and hence no frequency scan, so that the instantaneous bandwidth is much greater. If the corporate feed were made up nondispersive TEM transmission line, the feed itself would be true time delay and would, in principle, have unlimited bandwidth.

The power dividing branch points of the feed are usually matched four-port hybrid junctions instead of reactive three-port T junctions. Since a matched four-port junction is impedance matched in all four ports, spurious reflections from connectors, phase shifters, radiating apertures, etc., will not be scattered from the various junctions back into the radiating aperture to cause undesired side lobes and other antenna pattern degradation. Instead, these spurious reflections are absorbed in the terminating loads of the hybrids. Also, multiple reflections among the junctions can cause high resonance effects that can lead to high-power breakdown in the feed network. The use of directional couplers helps alleviate this potential problem.

As with the use of directional couplers in the series feed, the unused terminal

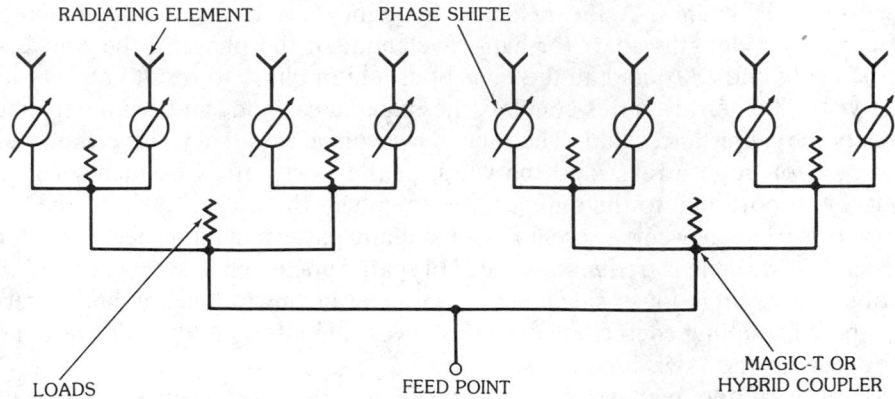

RADIATING ELEMENT   PHASE SHIFTE

LOADS   FEED POINT   MAGIC-T OR
                                   HYBRID COUPLER

Fig. 3. Diagram of a corporate feed, phase shifters, and radiating elements.

(the port that is normally terminated in a matched load) can be used to form a multitude of auxiliary antennas such as side lobe blanking and coherent side lobe cancellation antennas [2].

Again, as in the series feed, when a plane wave is incident on the antenna from the direction of the main beam, most of the receive power arrives at the input port of the feed and very little power ends up in the terminating loads. In an ideally lossless feed network, that portion of the power in the incident plane wave which ends up in the terminating loads accounts for the fact that the aperture efficiency η is less than unity for aperture distributions other than uniform. For uniform aperture distributions none of the received power is dissipated in the terminating loads. For tapered aperture distributions most of the received power that does not arrive at the feed input is dissipated in the terminating loads of the junctions that have the highest power split ratios which are near the edges of the aperture. With reactive three-port power splitters the receive power that is not received at the input port of the feed is reflected back into free space and can result in a high radar cross section. In any case in which matched four-port junctions are used in the feed, the radar cross section of the ideal impedance-matched aperture antenna is zero. Of course it is assumed that the incident radar wave has the same polarization and frequency as the antenna. That is, all the energy incident in the direction of the main beam of the antenna is absorbed at the antenna input (receiver or duplexer load)—none of it is reflected back into space. For plane waves incident on the antenna from any other direction than in the main beam, the incident power nearly all (except for the small amount of energy associated with side lobes of the antenna pattern) ends up in the terminating or "unused" ports of the hybrid junctions. That small amount of incident power (if there is any) associated with the side lobe in that direction ends up at the input port of the antenna feed. Hence the unused ports of the hybrid junctions can be combined in various ways to create a large variety of antenna pattern shapes outside the region of the main beam of the main antenna. Such patterns can be used as auxiliary antennas such as for coherent side lobe cancellation or for communications outside the region of the main beam of the main antenna. All these auxiliary antenna patterns would have essentially a null in the direction of the main beam of the main antenna [2].

True Time-Delay Feeds

A true time-delay antenna is in which that the time delay from an incoming wavefront to a feed point is the same for every path from the wavefront via each of the radiating elements, phase shifters, etc., to the feed point. This is illustrated in Fig. 4. The equal time delay ensures that signals via all paths add in phase at the feed point for every frequency component in the pulse. In a non-true-time-delay antenna the time delays for the various paths may differ by an integral number of rf cycles of the center frequency in the pulse. In general, other frequencies in the pulse do not add exactly in phase at the feed point. Hence there is a reduction in the peak of the received pulse and the pulse is smeared out or broadened in time. From the transmit point of view the interelement phase shift ψ is correct only at the center frequency to have its main beam point in the desired direction in space. Antenna patterns corresponding to other frequencies in the pulse will be scanned off from the desired direction according to the expression $\sin\theta = c\psi/\omega d$, which shows that the beam-pointing angle θ depends on the frequency ω since ψ is constant. The antenna is a bandpass filter in that the frequency spectrum of the

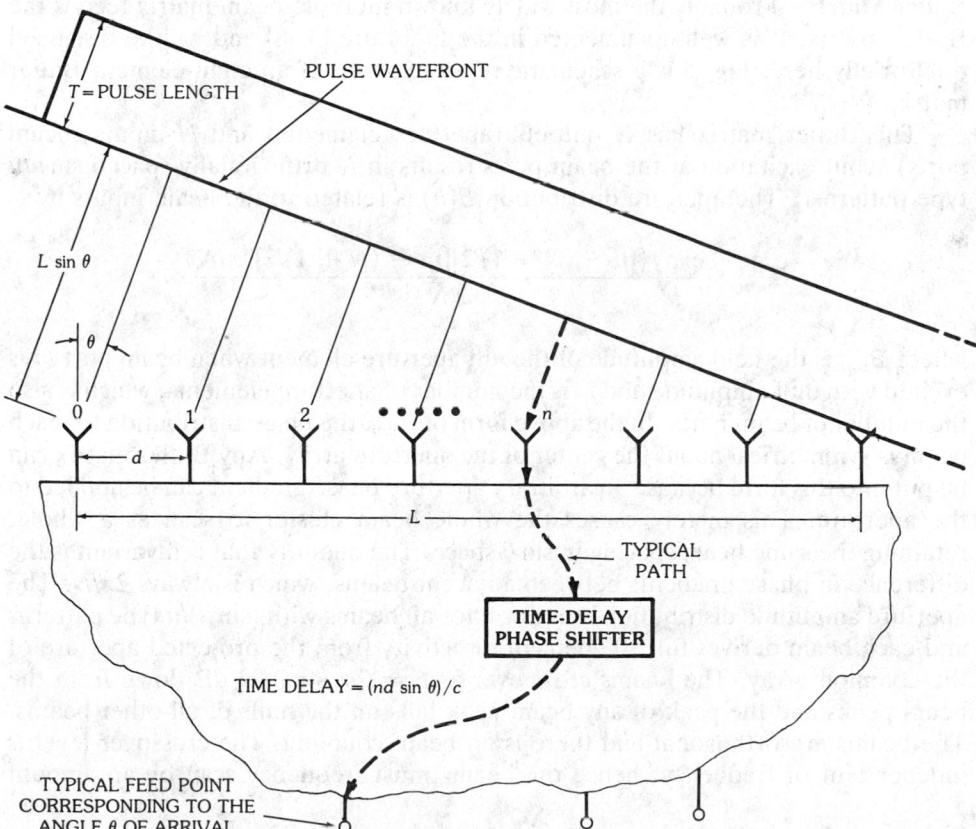

Fig. 4. Schematic diagram of true-time-delay antenna using time-delay phase shifters. (*Courtesy Hughes Aircraft Co., Fullerton, Calif.*)

pulse radiating toward a target is modified. The frequency component corresponding to the center frequency of the pulse is weighted most heavily because its corresponding antenna pattern is, by design, pointing in the direction of the target, whereas the other frequency components in the short pulse have their antenna patterns scanned off from the direction of the target; hence their weighting is reduced by the amount that their antenna patterns are scanned off. By the principle of reciprocity this spectral transformation on receive is the same as that previously discussed for transmit.

In a true time-delay array the interelement phase shift ψ must not be constant but must be proportional to frequency so that all frequency components in the pulse will have their corresponding antenna patterns all pointing in the same direction.

Examples of true time-delay feeding techniques include the use of the circular folded pillbox covered in Section 3, under "Pillbox," and the Meyer geodesic lens covered in Section 3, under "Meyer Lens," and in Section 6, under "Hughes Matrix-Fed Meyer Geodesic Lens," among others.

Multiple-Beam Matrix Feeds

Butler Matrix—Probably the most widely known multiple-beam matrix feed is the Butler matrix. It is well documented in the literature [3–8] and will be discussed only briefly here. Fig. 5 is a schematic representation of an eight-element Butler matrix.

The Butler matrix has N outputs (aperture elements) and N inputs (beam ports). Unit excitation at the beam ports results in N orthogonally spaced $\sin u/u$ type patterns.* The aperture distribution $B(n)$ is related to the beam inputs by

$$B_{nm} = \frac{\exp j\{[n - (N + 1)/2][m - (N + 1)/2]2\pi/N\}}{\sqrt{N}}$$

where B_{nm} is the field amplitude of the nth aperture element when beam port m is excited with unit amplitude and N is the number of aperture elements, which is also the number of beam ports. In the above form of B_{nm}, the phase distribution for each beam is symmetrical about the center of the aperture array. Any Butler matrix can be put into this form because an arbitrary aperture phase gradient can be applied to the aperture. This merely causes the whole beam cluster to scan as a whole, retaining the same beam spacing in $\sin \theta$ space. The quantity that is invariant is the difference in phase gradients between adjacent beams, which is always $2\pi/N$. The aperture amplitude distribution is uniform for all beams with $(\sin u)/u$ type patterns and each beam derives full 100-percent directivity from the projected aperture of the common array. The beams cross over at $E = 2/\pi$ or 3.92 dB down from the beam peaks and the peak of any beam peak falls on the nulls of all other beams. The beams are orthogonal and there is no beam coupling. The crossover level is independent of frequency; hence the beams must frequency scan by an amount

---

*More accurately, for discrete arrays it is $(\sin n\psi)/n \sin \psi$, where $\psi = (\pi d/\lambda)\sin \theta$, which is the counterpart for $(\sin u)/u$ for continuous distributions.

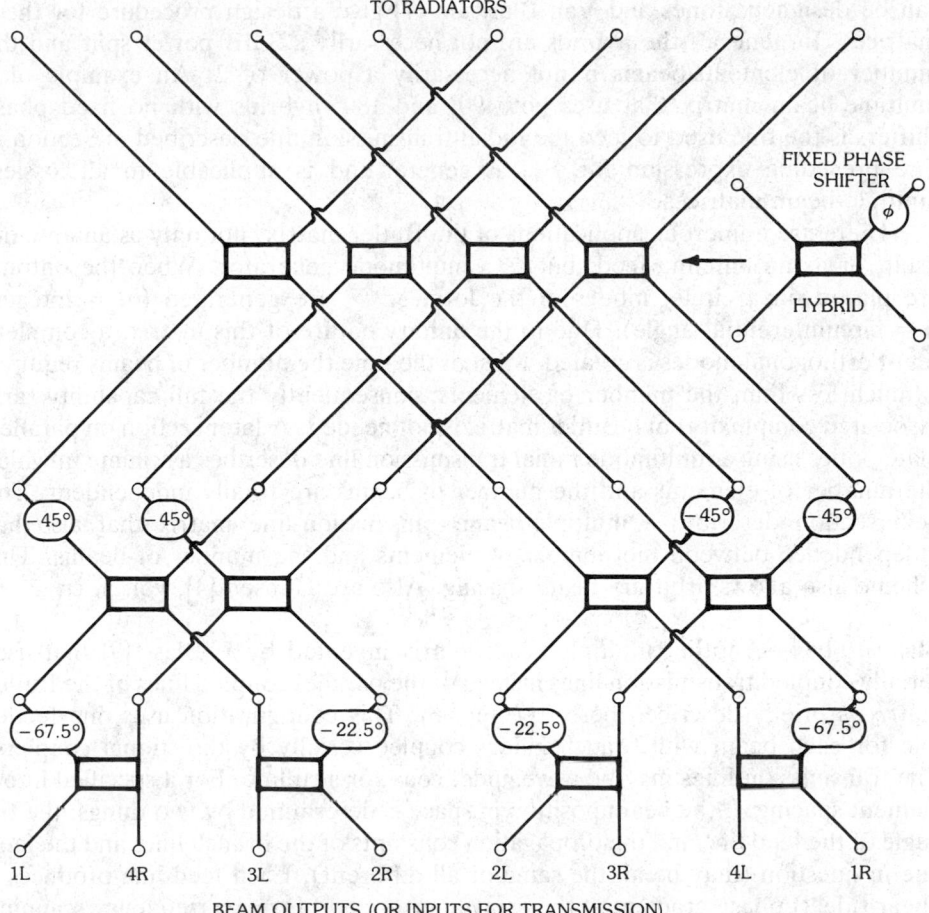

Fig. 5. Eight-port Butler matrix. (*Courtesy Hughes Aircraft Co., Fullerton, Calif.*)

proportional to their separation from broadside in order to retain orthogonality and a fixed crossover level. For large arrays the beams span an angular coverage of $2\sin^{-1}(\lambda/2d)$, where d is the element spacing and λ is the free-space wavelength. For $d > \lambda/2$, grating lobes will fill in the region near end-fire, and for $d < \lambda/2$, some beams will be in imaginary space. In all cases all of visible space is covered.

As exemplified by Fig. 5, numerous hybrid junctions and fixed phase shifters must be cascaded to generate a Butler matrix. Specifically [5], the number of hybrids is equal to $(N/2)\log_2 N$ and the number of fixed phase shifters is equal to $(N/2)\log_2(N - 1)$. Moody [8] gives a systematic design procedure for generating such matrices.

It should be mentioned that the multiple-beam matrices of the type described in [5] (and most others in the literature) use all 90° 3-dB or all 180° 3-dB couplers. In these matrices the number of elements/beams is a power of 2 and requires fixed phase shifts. However, since any fixed phase shift can be generated by a combination of 90° and 180° hybrids, matrices without separate fixed phase shifters can be designed. Jones and Van Blaricum [7] give a design procedure for these matrices. In general, the hybrids are not necessarily a 3-dB power split and the number of elements/beams is not necessarily a power of 2. An example of a multiple-beam matrix that uses both 90° and 180° hybrids with no fixed phase shifters is the one used to feed the radial transmission line described in Section 3. The preceding expression for B_{nm} is general and is applicable to all lossless multiple-beam matrices.

There are numerous applications of the Butler matrix, not only as an antenna itself, or as an antenna feed, but as a multimode generator. When the outputs are placed on a circle, modes of the form $e^{\pm jm\phi}$ are generated (m = integer, ϕ = circumferential angle). Due to the unitary nature of this matrix, a complete set of orthogonal modes is created. Most of the time the number of beams required is much less than the number of elements; consequently the full capability (and associated complexity) of a Butler matrix is not needed. A later section on parallel-plate optics using a multimode radial transmission line describes a scheme in which the number of elements and the number of beams are totally independent. The next section describes a multiple-beam transmission-line matrix that also has independence between the number of elements and the number of beams. This scheme also allows arbitrary beam spacing. Also see Hansen [4], vol. 3, ch. 3.

Blass Matrix—Another multiple-beam matrix invented by J. Blass [9] that uses serially coupled transmission lines instead of the parallel coupled lines of the Butler matrix is briefly described here (see Fig. 6). This configuration uses one feeder line for each beam with branch guides coupled serially by directional couplers. The transmission lines may be waveguide, coax, or stripline. For a specified inter-element spacing d, the beam position in space is determined by two things: the tilt angle of the feed line and the propagation constants of the branch lines and the feed line in question (may be all the same or all different). Each feed line produces a linear (ideal) phase gradient at the array aperture and forms a frequency scanning beam in space. By proper choice of the feed line tilt angles, propagation constants, and coupling distributions, multiple beams can be generated in space having whatever crossover value the designer chooses. These multiple beams, in general, may not be precisely orthogonal over a large frequency band but can be designed to be nearly so from a loss point of view. All the beams frequency scan together. For a small number of beams (two or so) this array works quite well. For a large number of beams, or for very low side lobes, care must be taken to account for the phase and amplitude changes that occur in the transmission coefficient as the wave passes by a multiplicity of nonideal directional couplers. The larger the number of beams, the more difficult this becomes. As can be seen from Fig. 6, feed line 1 produces a branch guide field that does not have to pass by any other couplers, while beam M must pass by $M - 1$ other couplers before entering free space; hence perturbations are present in feed line M that are not there for feed line 1. These perturbations will manifest themselves to some degree in the form of radiation pattern degradation and/or excessive power dissipated in the terminating loads. These properties are manifestations of the fact that ideal beam orthogonality is not guaranteed as in the Butler matrix or radial line approaches. It is conceivable that orthogonality could be forced by a design procedure similar to the Gram-Schmidt procedure [10], but in

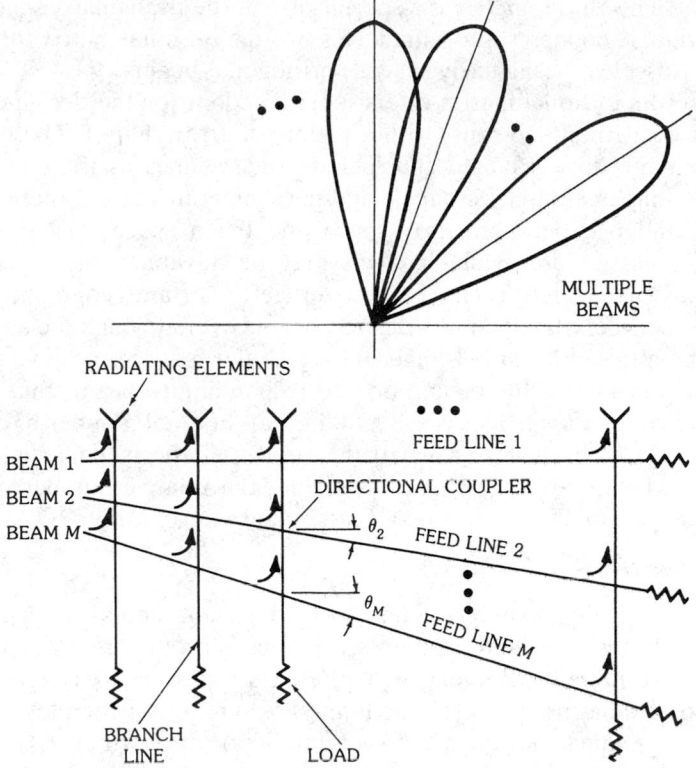

Fig. 6. Multiple-beam Blass matrix. (*After Butler and Lowe [3], © 1961; reprinted with permission of Hayden Publishing Co.*)

doing so, some of the advantages of the Blass matrix, such as simplicity and arbitrary beam spacing, could be restricted. However, this allows arbitrary beam spacing.

Various derivatives of the Blass matrix are also given in Hansen [4], including a true time-delay version of this matrix. The reader is referred to this reference for more details.

Two-Dimensional, Isosceles, Triangularly Spaced, Multiple-Beam Matrix—This class of two-dimensional multiple-beam matrices introduced by Chadwick and Glass has been extensively investigated by Chadwick, McFarland, Charitat, Gee, and Hung of the Lockheed Missile and Space Company [11, 12]. The analysis of this relatively new class of two-dimensional multiple-beam matrices is quite involved and, compared with other topics of this handbook, the required length of a self-contained explanation would not be justified. For this reason, only an acknowledgment that such an antenna technique does exist and a discussion of its possible advantage over the more established techniques are given. For further information the reader is referred to the references cited.

The Butler matrix multiple-beam antenna is one-dimensional and applies to multiple-beam linear arrays. A two-dimensional array is made by an array of these linear arrays. The Butler matrix is a special case of the triangularly spaced array in which the triangle collapses into a line. This two-dimensional matrix forms a family of lossless, isosceles, triangularly spaced, orthogonal beams.

This two-dimensional matrix offers more freedom for the designer to choose the number of elements (beams) with a variety of array shapes. These shapes are derived from the basic triangle and include such shapes as irregular hexagons, V shapes, Z shapes, hourglass shapes, and parallelograms. Since there is a one-to-one correspondence between array space and beam space, a variety of beam coverage geometries are available. This may be advantageous where irregular beam coverage is desired, such as the coverage of certain geographical areas on earth from a spacecraft antenna. A desired beam coverage might be approximated by one of the allowed beam coverage shapes. Butler matrices achieve an irregular coverage by terminating the beam ports corresponding to beams that fall outside the desired coverage area. This is wasteful in that the total number of components in the matrix is greater than for a matrix that utilizes all the available beams. Due to the variety of beam coverage shapes available, the relative number of "wasted" beams should be smaller for this two-dimensional matrix array.

Multimode Element Array Technique

For electronically scanning antennas with requirements of high directivity and high gain but with limited field of view (typically on the order of 10°), it is very wasteful to have radiating elements with phase shifters at half-wavelength intervals over the entire radiating aperture. The limited scan requirement allows a reduction in these devices (which may include power amplifiers and low-noise amplifiers in an active array) in proportion to the limited solid angle of scan. Limited-scan antennas are designed to minimize the number of these costly devices with minimal degradation of antenna performance. Constrained feeding techniques using large directive elements spaced much greater than at $\lambda/2$ intervals are discussed in this section. Limited-scan techniques using optical type devices are described in Section 4 on unconstrained optical feeds and in Section 5 on optical transform feeds.

The use of large directive elements (e.g., waveguide horns) in a phased array usually results in high grating lobe levels as scanning is performed; however, for limited scan, Mailloux and Forbes [13, 14] have found that by properly exciting waveguide flared horns with not only the dominant LSE_{10} mode but with controlled higher-order odd modes as well, the grating lobe that would ordinarily be the worst can be suppressed. For example, consider first that array scanning is to be performed in the E-plane only, with elements as depicted by Fig. 7. The array factor will be scanned by controlling the phase shifter labeled η, while the odd-mode amplitude control will utilize the phase shifter labeled $\eta + \Delta$. An example of far-field radiation patterns for the LSE_{10} and LSE_{11} modes is shown in Fig. 8. This particular example is predicated on forming a null at $\eta = -0.75$. This null location would correspond to a grating lobe position for a positive main beam scan angle. The null location can be controlled by an appropriate linear combination of these two modes. By linearly combining the two modes, a scanned element pattern that suppresses the worst grating lobe can be achieved. An example of this is given by Fig. 9 for waveguide horns that are 2.9λ by 2.9λ on a side. Also shown, for reference purposes, is the element pattern produced by the dominant LSE_{10} mode only.

Fig. 10 is a photograph of an experimental eight-element array taken from

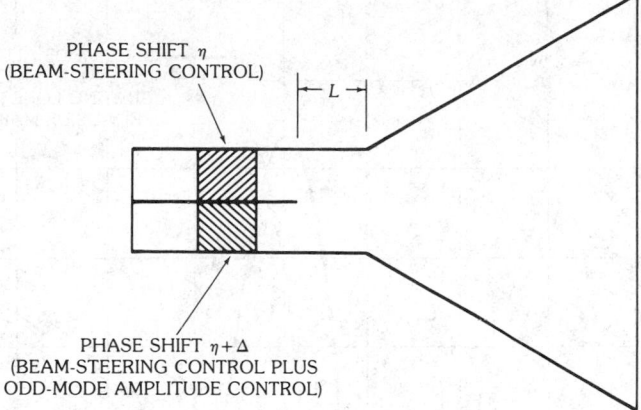

Fig. 7. Even/odd-mode power divider circuit for *E*-plane scanning. (*After Mailloux [13]*)

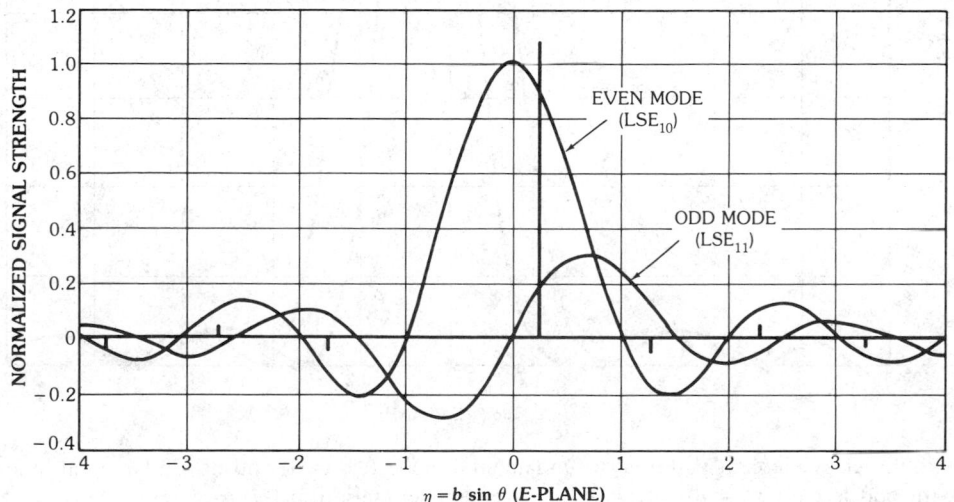

Fig. 8. *E*-plane field patterns for LSE_{10} and LSE_{11} waveguide modes. (*After Mailloux [13]*)

Mailloux [13]. It is designed for *E*-plane scan of $\pm 12°$. The amplitude distribution is such that the center four are uniform, the second element in from each end is -3 dB, and the outer elements are -6 dB. Examples of the measured far-field patterns are given in Figs. 11 and 12, where for comparison the latter also shows the calculated pattern without odd-mode control. Without odd mode control, the highest grating lobe for 12° scan of the main beam would be about 4 dB higher than the main beam. With odd-mode control, however, this grating lobe is suppressed by about 20 dB for $\pm 12°$ scan.

For odd-mode control in both planes the configuration shown by Fig. 13 can be used [14]. This configuration would allow large directive elements to be used in both dimensions, while suppressing the offending grating lobe ordinarily encountered. For more details the reader is referred to Mailloux and Forbes [13, 14].

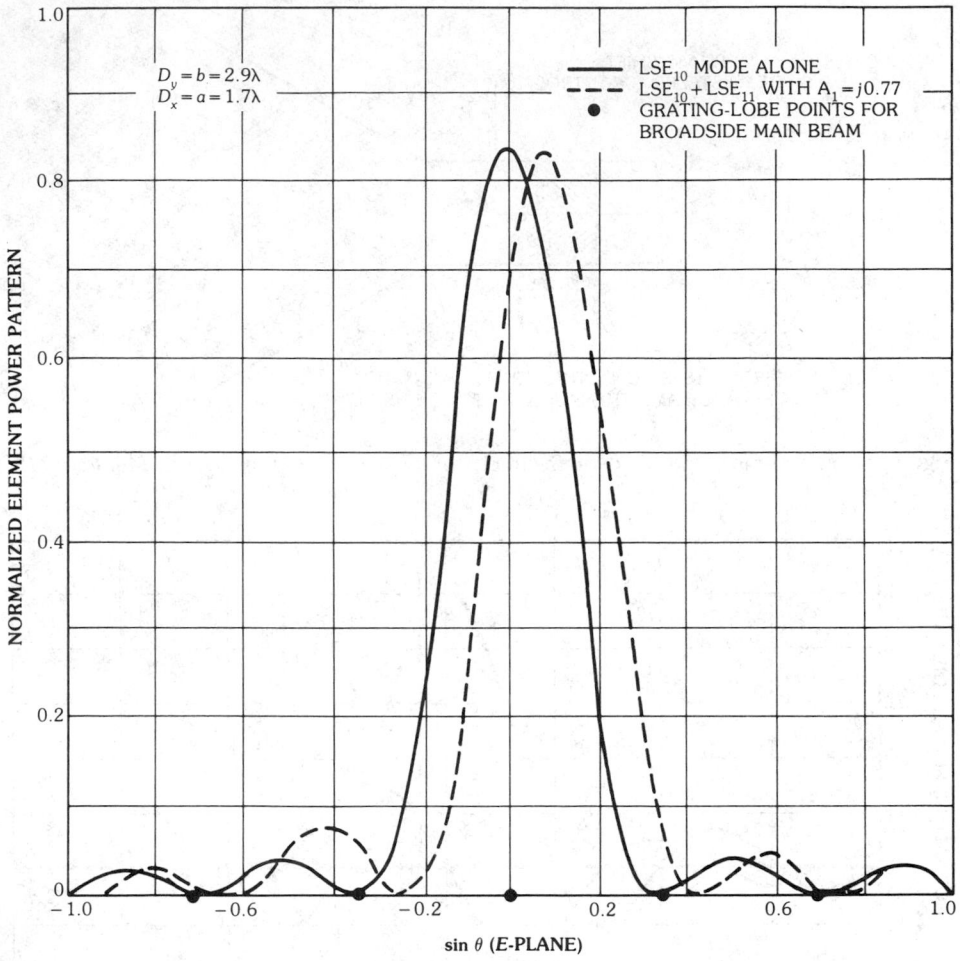

Fig. 9. Array element patterns of fundamental mode ($A_1 = 0$) and of fundamental mode with odd mode ($A_1 = j0.77$), with 2.9λ element separation in the *E*-plane.

3. Semiconstrained Feeds (Parallel-Plate Optics)

Feed systems that are constrained in one dimension but are not constrained in the other dimension are used because of simplicity, low cost, low loss, and high power handling capability. The volume occupied, however, is generally greater than that for fully constrained feeds. These feeds utilize propagation in the quasi-TEM mode between closely spaced (less than $\lambda/2$) parallel metallic surfaces. The surfaces need not be planar but can also be singly or doubly curved as long as the spacing between the surfaces is small in terms of wavelength and the radii of curvature are large with respect to the wavelength. Under these conditions the wave is considered to be propagating on the mean surface. In the dimension of the mean surface the wave is essentially unbounded and is not constrained by guiding structures. The rf power "radiates" from the primary feed along the mean surface to a pickup array that, in turn, feeds the radiating elements of the antenna. Hence

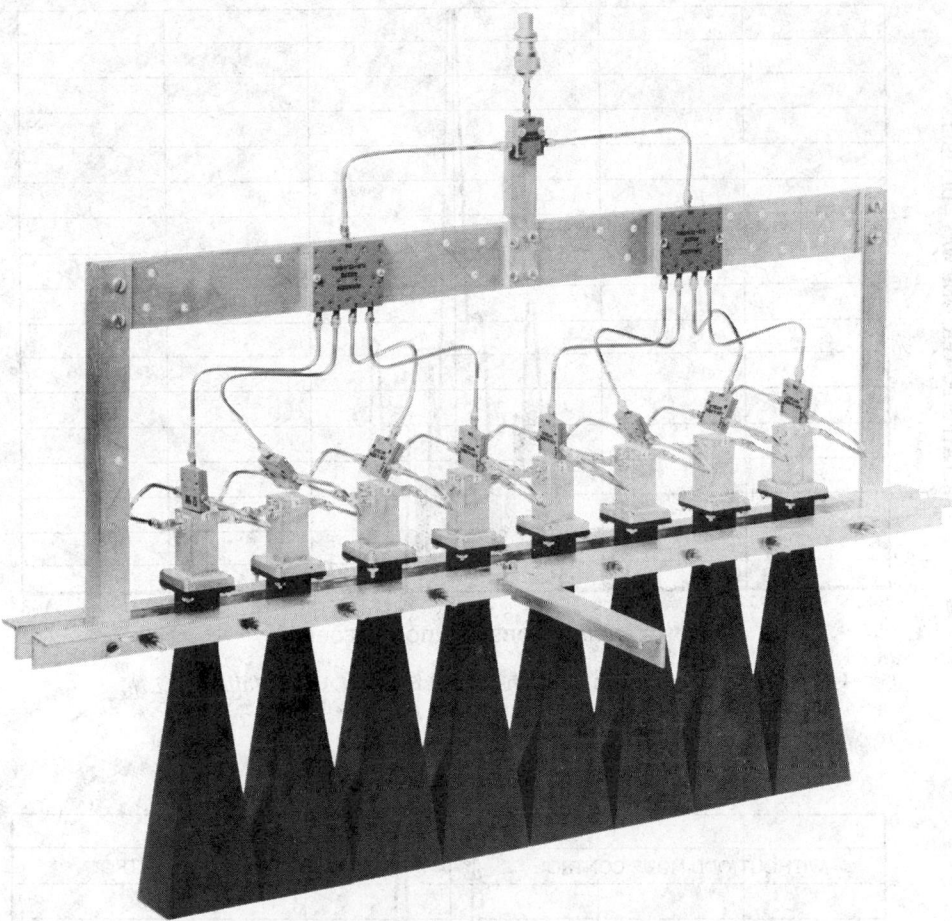

Fig. 10. Prototype array for *E*-plane scan (*W* = 3.01). (*After Mailloux [13]*)

the aperture distribution of the antenna array is determined by the radiation pattern of the primary feed and the geometric properties of the mean surface. If the mean surface is planar, the aperture illumination is determined by the feed pattern in the usual sense. However, if the mean surface is doubly curved or if the index of refraction is a function of position, the feed pattern is modified, as will be discussed later.

Under the restrictions of closely spaced surfaces and other dimensions large compared with the wavelength, the laws of geometric optics are quite valid on the mean surface. That is, Fermat's principle stating that the path of a ray in geometric optics is stationary applies. Stated mathematically,

$$\delta \int_{\text{ray path}} n\, ds = 0$$

where *n* is the index of refraction of the medium between the surfaces. The index of refraction may vary as a function of position on the mean surface. This is the same

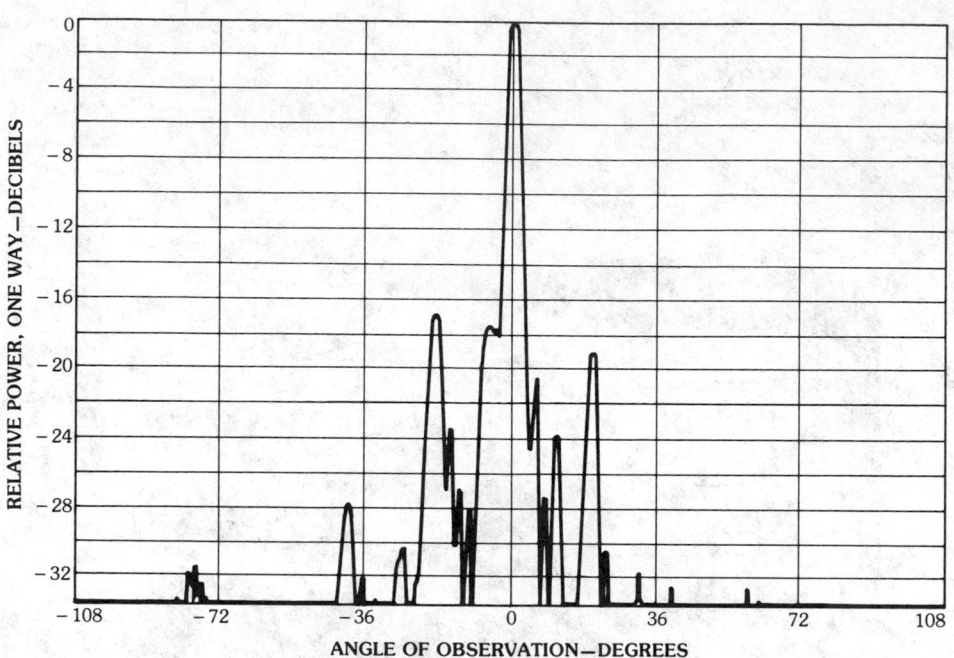

Fig. 11. Array radiation patterns (broadside). (*After Mailloux [13]*)

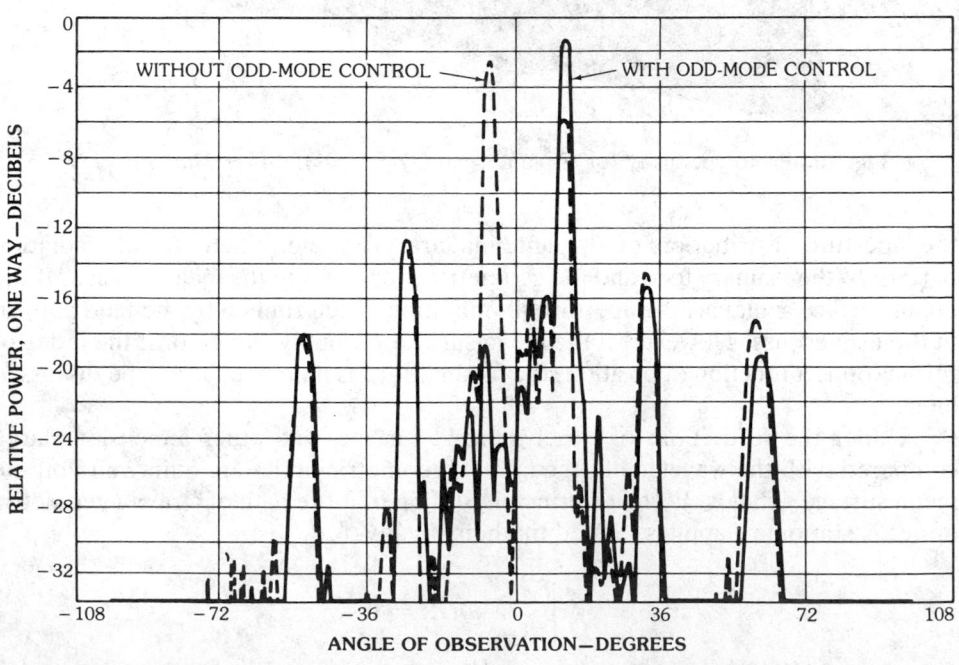

Fig. 12. Array radiation patterns (12° scan). (*After Mailloux [13]*)

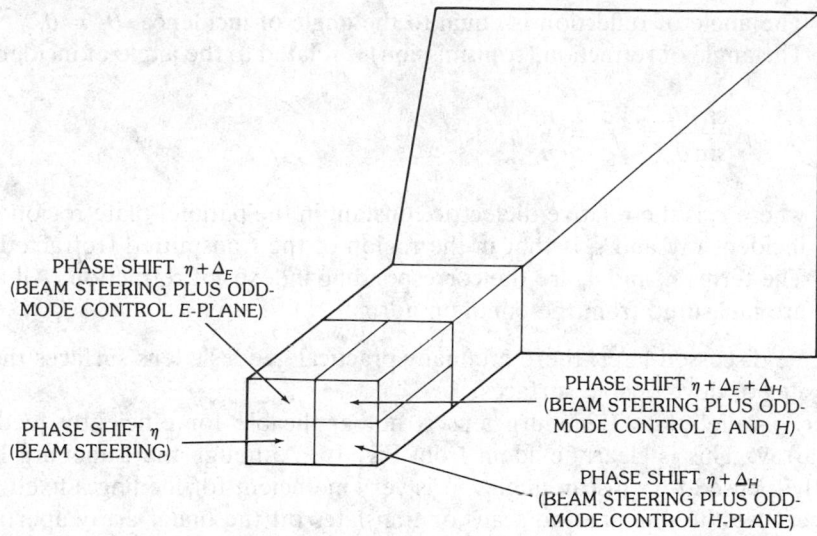

PHASE SHIFT $\eta + \Delta_E$
(BEAM STEERING PLUS ODD-
MODE CONTROL *E*-PLANE)

PHASE SHIFT $\eta + \Delta_E + \Delta_H$
(BEAM STEERING PLUS ODD-
MODE CONTROL *E* AND *H*)

PHASE SHIFT η
(BEAM STEERING)

PHASE SHIFT $\eta + \Delta_H$
(BEAM STEERING PLUS ODD-
MODE CONTROL *H*-PLANE)

Fig. 13. Four-mode horn and power divider circuit. (*After Mailloux [13]*)

as saying that the time of transit of a ray between two points on the mean surface is an extremum (usually a minimum) and the rays are geodesics [15]. For this reason such feeds are generally called *geodesic feeds* or *lenses*.

The simplest and probably the earliest geodesic feed for phased arrays is the folded or two-layered pillbox [16–19]. It consists of two parallel-plate regions connected by a 180° bend that is parabolic in the plane of the plates. A point source feed is located in one layer. A cylindrical wave radiated from the feed goes around the 180° bend to enter the second layer, where it is collimated into a plane wave by the parabolic shape of the bend. The bend may be abrupt [16, 17], circular, or mitered [19].

The mean surface of geodesic lenses may be planar, singly curved, doubly curved, or composite. For planar surfaces the geodesics are straight lines. For singly curved (cylindrical) surfaces the geodesics are straight lines on the developed (flattened into a plane without stretching or tearing) surface. Doubly curved surfaces cannot be flattened into a plane without stretching or tearing the surface. The geodesics on doubly curved surfaces are the curves that a tightly stretched string would take with the string constrained to stay on the surface everywhere. Mathematically the geodesic is an extremum, which is the shortest or the longest path length on the surface between two points. As a simple example the shortest arc of a great circle through two points on a sphere is a geodesic, as is the longest arc of the same great circle through the points. Usually the geodesics of interest are the shortest paths between two points.

If two surfaces intersect* along a line, a geodesic (ray) across the intersection obeys Snell's laws of reflection and transmission, namely,

---

*Of course, any arbitrary "cut" in a geodesic surface can be considered as an intersection but most often the intersection is between two portions that are each developable.

(1) The angle of reflection is equal to the angle of incidence, $\theta_r = \theta_i$.
(2) The angle of refraction (transmission) is related to the angle of incidence by

$$\frac{\sin \theta_t}{\sin \theta_i} = \frac{\sqrt{\epsilon_i}}{\sqrt{\epsilon_t}} = \frac{n_i}{n_t}$$

where ϵ_i is the relative dielectric constant in the parallel-plate region of the incident ray and ϵ_t is that in the region of the transmitted (refracted) ray. The terms n_i and n_t are the corresponding indexes of refraction. All angles are measured from the common normal.

As will be discussed later, there are many practical geodesic lens surfaces that are composites of developable surfaces.

The parallel-plate Luneburg lens is not applicable for efficiently feeding a linear array. This is clearly evident from Fig. 14. Although the Luneburg lens is perfectly focused for all scan angles, it is very inefficient for feeding a fixed linear array because the illumination scans or translates off the linear array aperture as

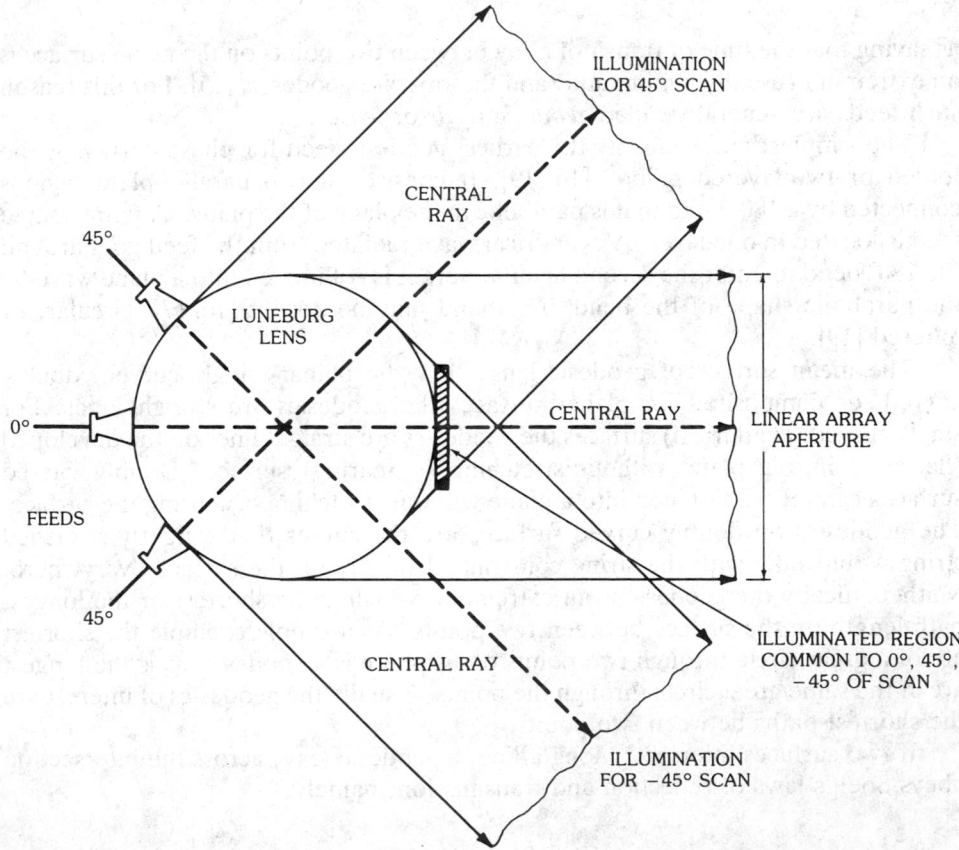

Fig. 14. Luneburg lens illuminating a linear array. (*Courtesy Hughes Aircraft Co., Fullerton, Calif.*)

the beam is scanned. The darkened portion of the linear array is all that is common for angles of scan out to ±45°. The amount of feed spillover and illumination asymmetry that results with beam scan is intolerable. Note that the central ray does not pass through the center of the linear array for scanned off beams. This results in asymmetric array illumination for tapered distributions. The circular pillbox and Meyer lens (these lenses are discussed later) do not suffer this shortcoming, as can be seen in Fig. 15, which shows that for the Meyer lens (the same is true for the circular folded pillbox) the central ray for all beams passes through the center of the linear array, i.e., the illumination does not translate across the linear array aperture. It should be mentioned that since the projected aperture of the linear array decreases as the cosine of the angle of scan, to keep the illumination efficiency from degrading, the feed illumination pattern should be made more directive according to the secant of the scan angles. This requires larger feeds as the scan increases. Fortunately, there is just enough space to do this (in terms of beamwidths of scan) because the radiated beamwidth also increases as the secant of the scan angle. This effect is true for *all* optical type feeding techniques for linear or planar arrays.

Pillbox

Fig. 16 shows a sketch of a parabolic folded pillbox [16–19]. This device has been used since the early 1950s for IFF and air traffic control antennas and as a feed for phased arrays.

The input primary feed is usually an open-ended waveguide, or two of them side by side for monopulse operation. Sometimes electric probes fed by coaxial lines backed by a quarter-wave short or cavity are used for the primary feed. The primary feed radiates in the region between a pair of parallel plates, as depicted by Fig. 16. The phase center of the primary feed is located at the focus of the parabola. The field then propagates in the TEM parallel-plate mode to the 180° bend. The bend is designed such that the "reflected" (or more accurately, transmitted around

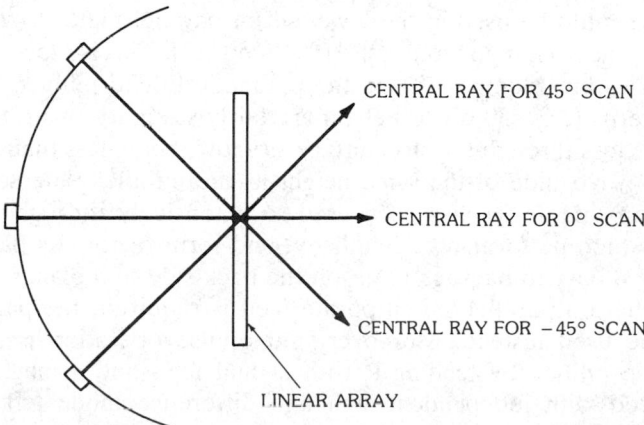

Fig. 15. Meyer lens illuminating a linear array. (*Courtesy Hughes Aircraft Co., Fullerton, Calif.*)

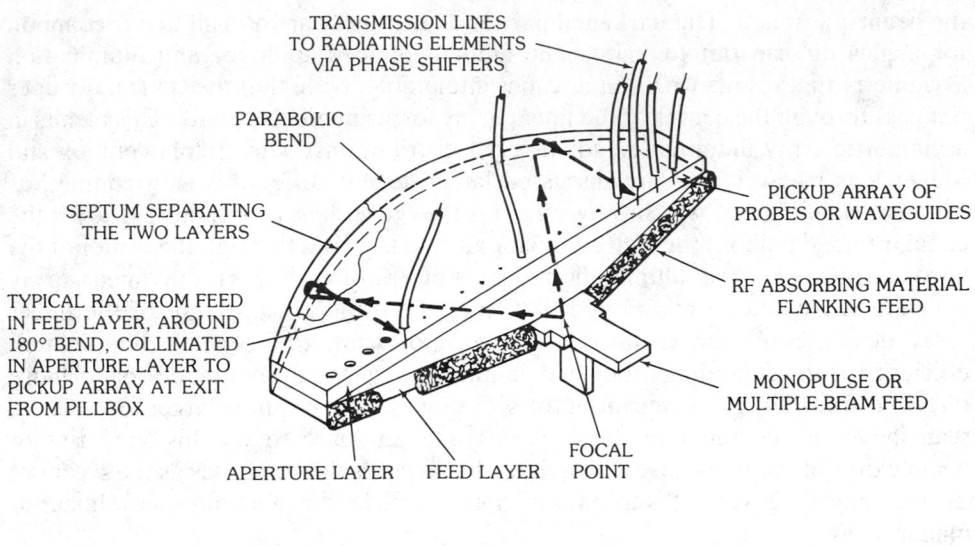

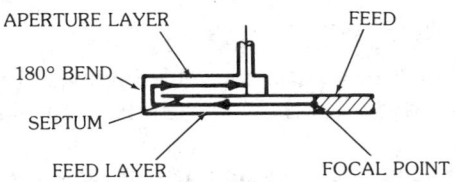

Fig. 16. Folded pillbox feed for phased array. (*Courtesy Hughes Aircraft Co., Fullerton, Calif.*)

the bend) field emerges between the upper set (ideally) of TEM parallel plates, and is collimated because of the parabolic shape. The output is located in the upper set of parallel plates at any convenient terminal plane. The output is transitioned to coax waveguide or stripline, thus forming the pickup array. Radio-frequency absorbing material should be used in the lower set of parallel plates to absorb any reflected power that did not go around the 180° bend [18]. This is especially important if low side lobes are required. Using the parabolic folded pillbox, low side lobe radiation patterns (e.g., 30 dB or better) are fairly straightforward to obtain [18]. Moreover, the loss through the structure is very low, being less than an equal path length run of waveguide of the same height as the parallel-plate separation.

The reason this device has been used so extensively throughout industry is because of its high performance, simplicity, and form factor. Its flat, thin profile usually makes it easy to package (e.g., on the back side of a planar array). In any application where a parallel-fed corporate feed is required, the parabolic folded pillbox can be used instead. Moreover, monopulse operation is simple to implement in the pillbox by feeding it with a dual feed and a magic-T or with a multimode feed with independent sum and difference modes. It is much less complicated (mechanically) than the corporate feed and does not require the multiplicity of components (hybrid junctions or directional couplers, bends, etc.) inherent in the corporate feed.

If wide-angle-coverage multiple beams are desired, the circular folded pillbox [19] can be used instead of the parabolic folded pillbox. Fig. 17 is a photograph of a 29-beam circular folded pillbox at *X*-band. Its principle of operation is similar to that of the parabolic folded pillbox, except that a circular arc is used at the 180° bend rather than a parabolic arc, and the feed locus is a circle whose radius is slightly greater than half the radius of the circular reflector (at the 180° bend). Without dielectric loading, the usable portion of the output aperture is restricted to about one-half the diameter, or slightly greater. The reason for this is that spherical aberration produces phase errors that limit the usable portion of the aperture. Most of the phase errors at the aperture can be corrected by trimming the line lengths at the output aperture, similar to a Schmidt correction in optics. The spherical aberration for the scanned beams resembles the spherical aberration for the on-axis beam; hence the spherical aberration that is common to all beams can be removed at the output aperture by line length adjustment.

For example, consider a design that forms multiple beams over a 90° sector (±45°). Fig. 18 shows the calculated spherical aberration for scanned beams over ±45°. The curves all resemble each other, grossly speaking; consequently the mean value of the spherical aberration can be removed. The long and short dashed curve is the mean value, or compensation curve. The negative of this curve is inserted at the aperture by line length adjustments in the pickup array. The maximum phase error after compensation is shown dashed. The spherical aberration has been reduced by a factor of about 5 or 6. This factor depends on the total field of view over which multiple beams are formed, and the allowable degree of spherical aberration (after compensation) that can be tolerated.

With partial dielectric loading of the entrance layer the optics can be modified such that nearly 80 percent of the output diameter is usable aperture. For example, the pillbox in Fig. 17 uses Rexolite 1422 loading ($\epsilon_r = 2.56$) between the feed arc and $0.8R$ in the entrance layer and gives respectable phase error over 80 percent of *D*. The extreme scanned beams suffered from what appeared to be residual spherical aberration phase error, indicating that the use of 80 percent of *D* is

Fig. 17. Circular, folded, multiple-beam pillbox, shown with 12-in (30.48-cm) ruler. (*Courtesy North American Rockwell*)

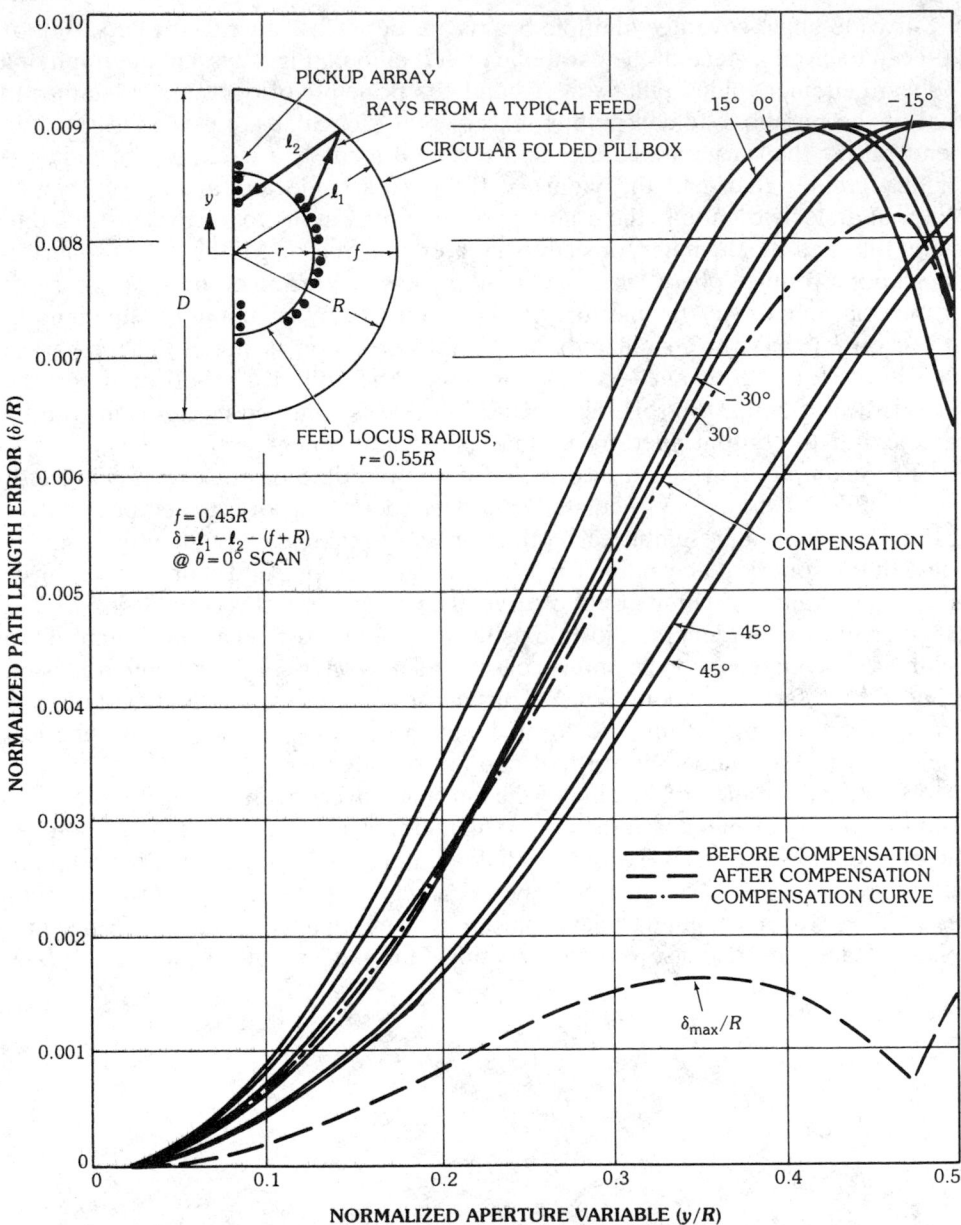

Fig. 18. Path length error versus aperture variable before compensation, and maximum path length error after compensation. (*Courtesy Hughes Aircraft Co., Fullerton, Calif.*)

probably a little too much. There was no mechanism for focusing (i.e., feed probe positions were not adjustable). In this design electric probes are used as feeds, which yields a very nearly uniform amplitude distribution at the aperture, with beam crossover levels of about 4 dB and side lobe levels consistent with a uniform amplitude distribution (13 dB) for most of the beams. The crossover level varies with frequency consistent with the beamwidths' dependence on frequency. The 29-beam peak positions remain fixed in space independent of frequency. If waveguides were used as feeds with focusing (radial feed position adjustability) capability, improved performance would undoubtedly be realized.

Radial Transmission Line

This section describes a simple, inexpensive, and ideally lossless multiple-beam-forming device whose cost and complexity do not increase rapidly with the number of radiating elements [20–22].

In physical appearance it resembles a constrained lens using parallel-plate optics. From the microwave circuit point of view it is similar to the Butler hybrid matrix but it differs in that the number of beams does not have a definite mathematical relationship with the number of radiating elements. The number of elements is arbitrary, in contrast with the Butler matrix in which the number of beams equals the number of elements. In practice the number of beams is quite limited in the radial transmission-line scheme but the number of radiating elements can be increased arbitrarily with little extra complexity and at a cost that varies only linearly with the number of elements.

Consider a parallel-plate radial transmission line terminated on its periphery by an array of "pickup" probes connected to the radiating elements of a linear array with equal lengths of transmission line as shown in Fig. 19. Suppose that besides the TEM mode, higher-order cylindrical modes with circumferential phase variation could be excited in the radial line. The circumferential variation of these modes can be characterized by $A_m \exp(jm\phi)$, where A_m is the amplitude of the mth mode and m is an integer (positive or negative). Because of the orthogonality properties of the modes they do not couple. By virtue of equal line length connection between the circumferentially dispersed pickup probes and the elements of the linear array,

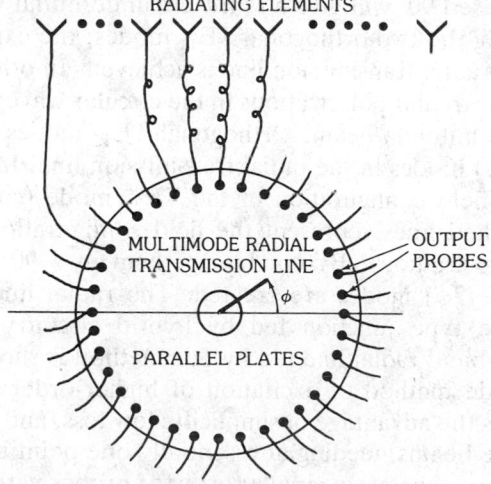

Fig. 19. Radial-transmission-line multiple-beam-forming network. (*Courtesy Hughes Aircraft Co., Fullerton, Calif.*)

the circumferential phase variation $\exp(jm\phi)$ is transformed to a linear progressive phase $\exp(j2m\pi x/L)$, where L is the length of the array and x is the aperture variable ($\phi = 0$ corresponds to $x = 0$ and $\phi = 2\pi$ corresponds to $x = L$). Thus there is a beam for each value of m. Positive and negative values of m correspond to left and right beams, respectively. For $m = 0$, there is a beam at broadside. The amplitude distribution for each of the beams is uniform and the far-field pattern is given by

$$E(\psi) = \frac{\sin(N\psi/2)}{N\sin(\psi/2)}$$

where

N = number of elements in the array

$\psi = 2\pi d/\lambda \sin(\theta - \alpha)$

θ = angle of beam from broadside

$d = L/(N - 1)$, interelement spacing

$\alpha = 2m\pi/N$, interelement phase shift for the mth beam

This is exactly the same as a Butler matrix array of N elements with m beams being used.

The $m = 0$ mode can be excited at the center of the radial transmission line with a circular waveguide operating in the TM_{01} or with a coaxial TEM mode as shown in Fig. 20. The method of Fig. 20 also generates the $m = \pm 1$ modes. The peak power-handling capability is limited by the E-plane magic-T. Fig. 21 shows a very high power transition using a multimode turnstile junction to excite these modes. The TE_{11} excitation results in a $\cos\phi$ circumferential variation. An orthogonal TE_{11} mode phased 90° will give a $j\sin\phi$ circumferential variation. Hence, by adding or subtracting the two orthogonal TE_{11} modes, the $\exp(\pm j\phi)$ circumferential variation in the radial transmission line is achieved. In other words, an excitation of right and left circular polarizations in the circular waveguide feed will result in a right and a left antenna beam. Orthogonal TE_{12} modes as shown in Fig. 22 will excite $\exp(\pm 2j\phi)$ modes in the radial transmission line. In this figure the solid lines represent the field configuration of the TE_{21} mode ($\cos 2\phi$ circumferential variation). The dashed lines represent the field configuration of the orthogonal TE_{21} mode ($\sin 2\phi$ variation). By combining them in a 90° phase relationship, $\exp(j2\phi)$ and $\exp(-j2\phi)$ modes are created. The radial line can be fed with a multimode turnstile type junction fed by hybrid circuitry as in Fig. 21a. A photograph of an S-band radial line fed by this method is shown in Fig. 21b. This hybrid-fed waveguide method of excitation of higher-order modes in the radial transmission line has the advantage of simplicity, low loss, and high power handling capability. For more beams, feeding at essentially one point becomes impractical. For a greater number of beams a circular array of probes with less than half-wave spacing fed with a Butler matrix would probably be the most efficient. The obvious question may be raised as to why not dispose of the radial line and just use the

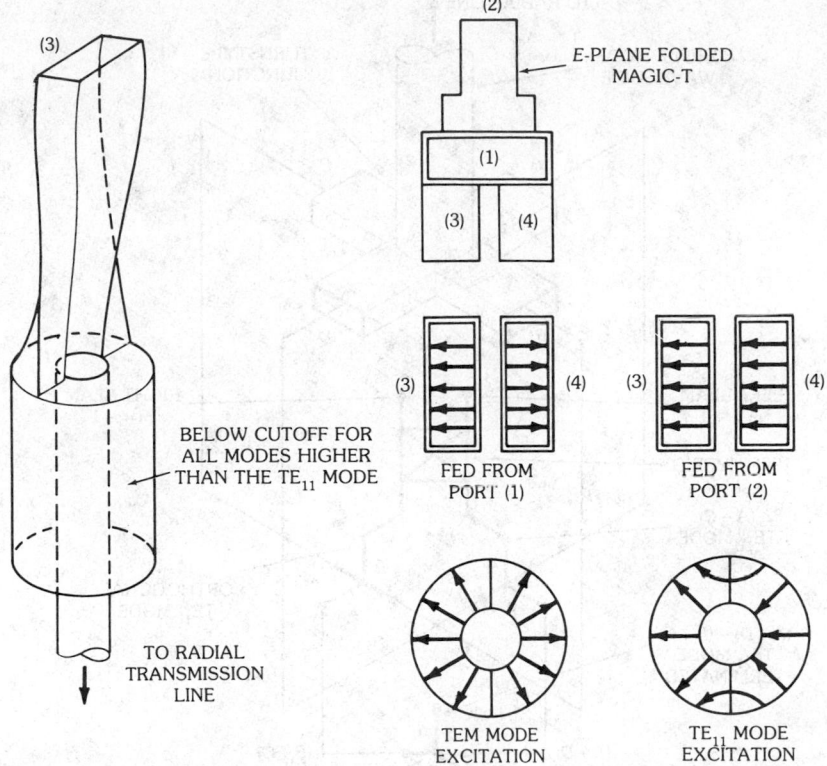

Fig. 20. Operation of high-power, dual-mode feed for radial transmission line. (*Courtesy Hughes Aircraft Co., Fullerton, Calif.*)

Butler matrix [3, 5]. The advantage of using the radial line is that regardless of the number of beams the number of outputs from the radial line can be increased to any arbitrary number by merely enlarging the diameter of the radial line to accommodate a larger number of output elements. In many systems applications requiring large arrays the number of simultaneous beams required is often much less than the number of radiating elements. For example, many phased-array systems may require only two beams for monopulse capability with beam steering achieved with ferrite or diode phase shifters. Obviously in this extreme case a radial line feed would be much simpler and much less expensive than a Butler matrix. In other applications a relatively small multiple-beam cluster that can be scanned in synchronism may satisfy the system requirements.

As an example of a possible application of the multimode radial line beam-forming network, suppose that on transmit a uniform distribution is desired and on receive a 30-dB side lobe tapered aperture distribution is desired with monopulse capability. A feeding arrangement that is still different from those of Figs. 20 and 21 is shown in Fig. 23 and is used for this example. The radial transmission line has a coaxial input (excites $m = 0$ mode) and a dual orthogonal mode (TE_{11} mode and orthogonal TE_{11} mode) waveguide input which can excite the $m = \pm 1$ mode. On transmit the coaxial line input only is used, resulting in a uniform aperture

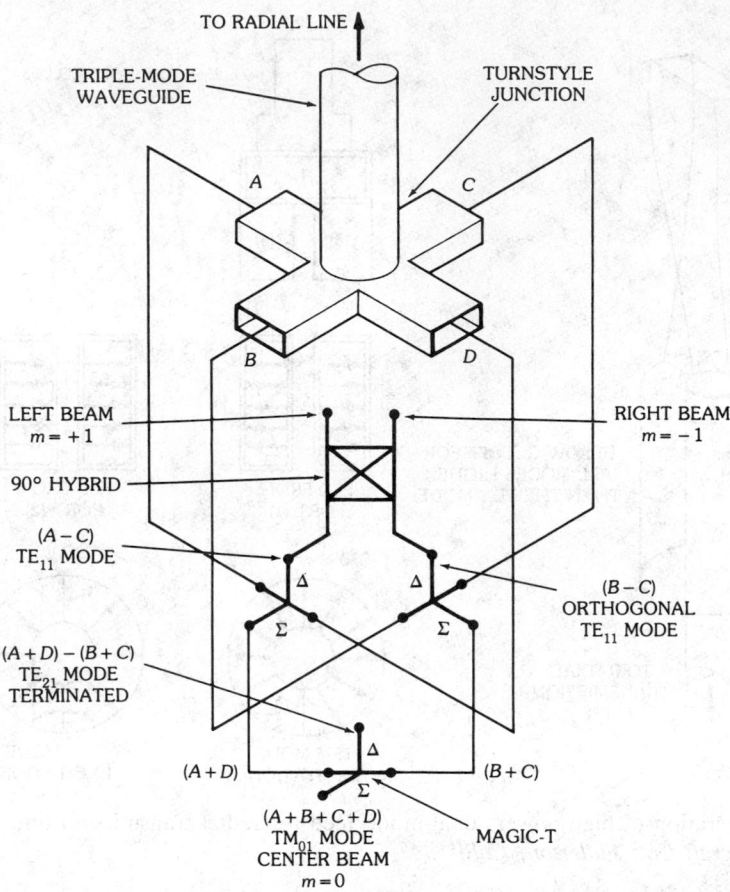

TO RADIAL LINE

TRIPLE-MODE
WAVEGUIDE

TURNSTYLE
JUNCTION

A   C

B   D

LEFT BEAM
$m = +1$

RIGHT BEAM
$m = -1$

90° HYBRID

$(A - C)$
TE_{11} MODE

$(B - C)$
ORTHOGONAL
TE_{11} MODE

$(A + D) - (B + C)$
TE_{21} MODE
TERMINATED

Δ Δ

Σ Σ

Δ

$(A + D)$ $(B + C)$

Σ

$(A + B + C + D)$
TM_{01} MODE
CENTER BEAM
$m = 0$

MAGIC-T

a

Fig. 21. Turnstyle junction and antenna. (*a*) High-power turnstyle junction to generate three beams. (*b*) Multiple-beam antenna using a multimode radial transmission line fed by a hybrid turnstyle junction. (*Courtesy Hughes Aircraft Co., Fullerton, Calif.*)

distribution on transmit. On receive, the $m = 0$ (uniform) mode and the $m = 1$ ($\cos\phi$) mode are combined in the power ratio of 2 to 1 to form a voltage distribution of $1 + \cos\phi$, which is equal to $2\cos^2(\phi/2)$, which is theoretically a −32-dB side lobe level distribution. This creates the low side lobe "sum" monopulse pattern. By changing the power division ratio a more efficient and lower side lobe cosine squared on a pedestal distribution is easily achieved. The "difference" monopulse pattern is taken from the orthogonal TE_{11} mode port, which gives a $\sin\phi$ aperture distribution which results in a difference pattern with good side lobe characteristics because there is no abrupt discontinuity in the aperture distribution as in the case of split aperture phase monopulse schemes.

A technique for a larger number of beams uses a small circular array of probes concentric with the center of the radial line, which are properly phased to generate the different modes. A seven-beam system was designed and tested [22]. This feed system utilized a hybrid network of 90° and 180° 3-dB couplers. A Butler matrix of

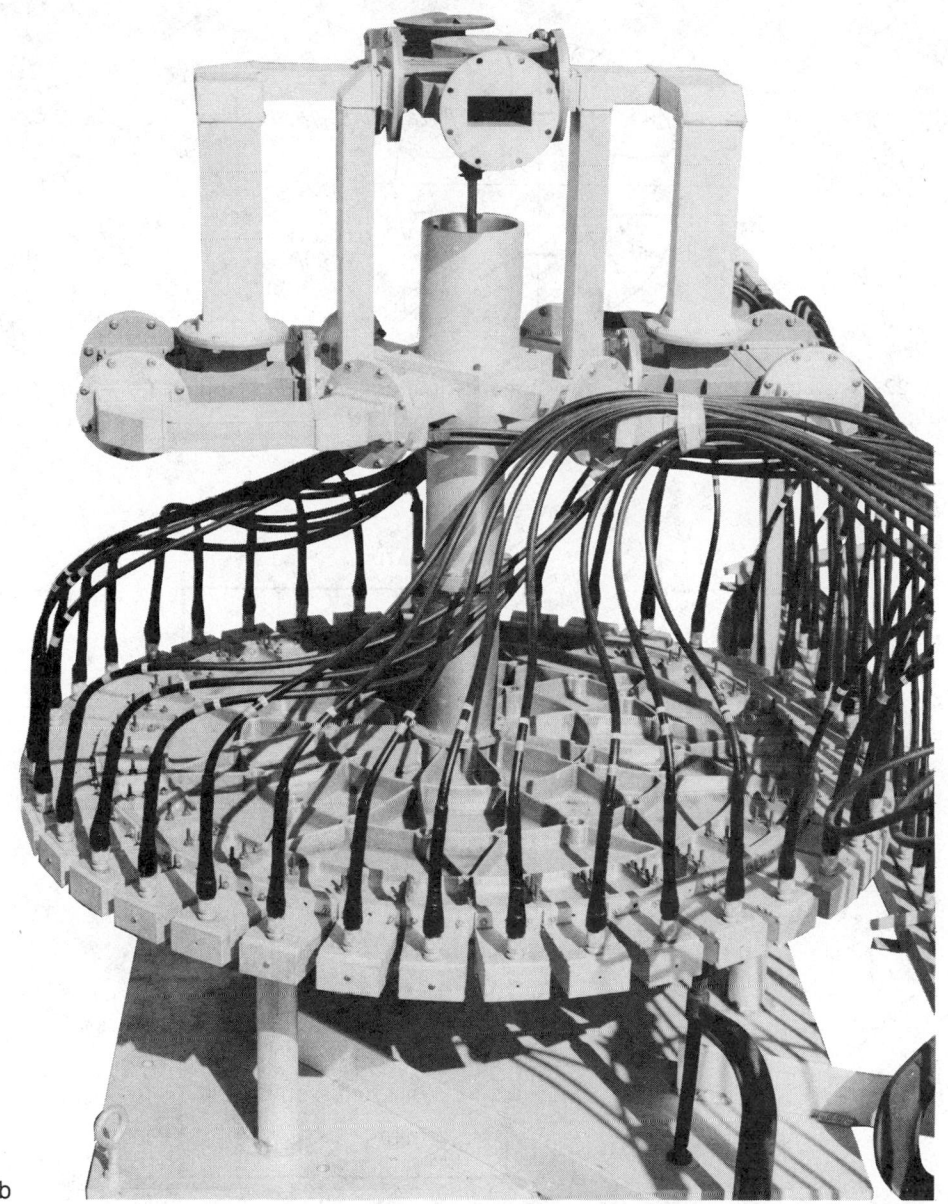

b

Fig. 21, *continued.*

the type shown in Fig. 5 could have been used just as well. It is interesting to note that, as mentioned in Section 2, the matrix in Fig. 24 does not require fixed phase shifters, in contrast to that of Fig. 5. Fig. 24 is a schematic diagram of the network; the 180° (magic-T) is designated by the letter T and the 90° coupler by the letter H. The phase progression around the circular array for each antenna

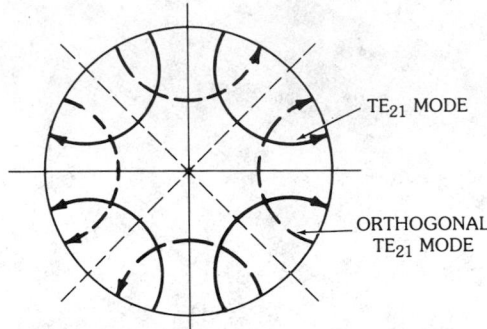

Fig. 22. Waveguide excitation of $m = 2$ modes. (*Courtesy Hughes Aircraft Co., Fullerton, Calif.*)

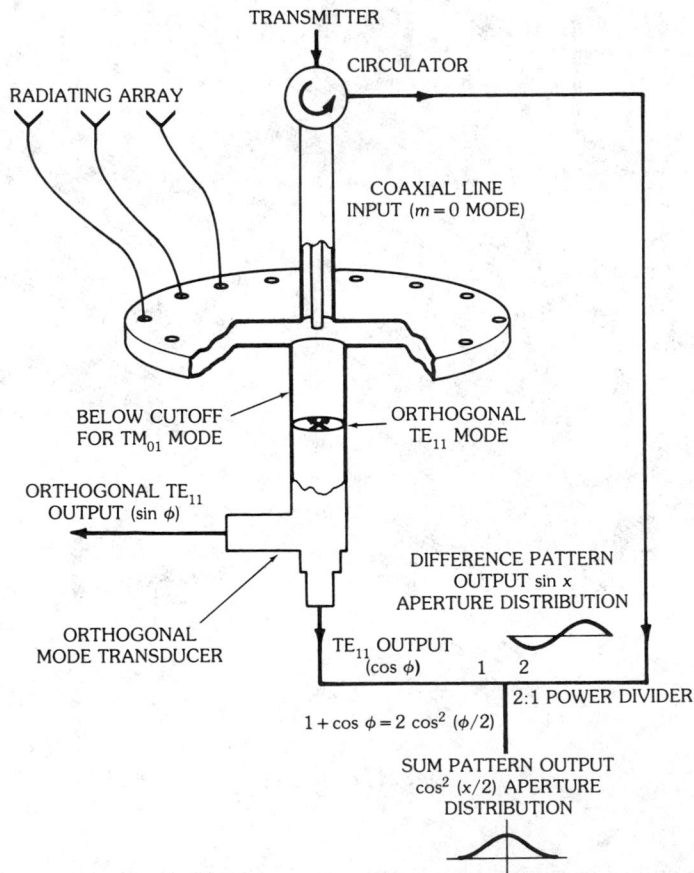

Fig. 23. Arrangement to achieve uniform aperture distribution on transmit and cosine-squared distribution on receive with monopulse capability. (*Courtesy Hughes Aircraft Co., Fullerton, Calif.*)

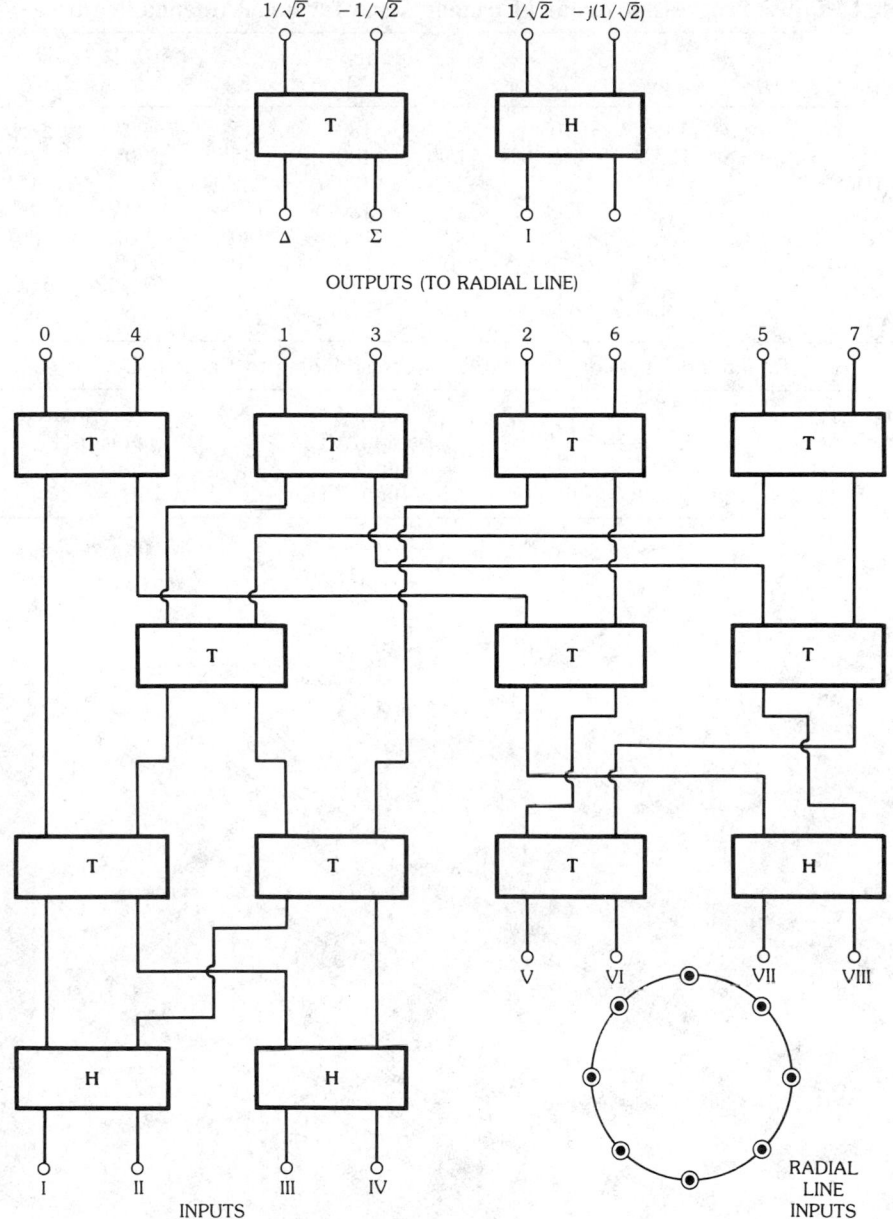

Fig. 24. Block diagram of feed circuitry. (*Courtesy Hughes Aircraft Co., Fullerton, Calif.*)

beam (mode) is given in Table 1. Note that input V, which corresponds to output phases that are alternately 0° and 180°, gives rise to a split end-fire beam that results in two actual opposing end-fire beams. This input is not used and is terminated in a matched load. Fig. 25 is a photograph of the stripline network. Because of the finite number of probes the $\exp(\pm jm\phi)$ variation is only discretely approximated, and if the probe separation is too great, undesirable higher-order modes are also

Table 1. Phase Progression Around Circular Array for Each Antenna Beam

				Outputs				
Input	0	1	2	3	4	5	6	7
I	0°	135°	270°	45°	180°	315°	90°	225°
II	0°	225°	90°	315°	180°	45°	270°	135°
III	0°	45°	90°	135°	180°	225°	270°	315°
IV	0°	315°	270°	225°	180°	135°	90°	45°
V	0°	180°	0°	180°	0°	180°	0°	180°
VI	0°	0°	0°	0°	0°	0°	0°	0°
VII	0°	90°	180°	270°	0°	90°	180°	270°
VIII	0°	270°	180°	90°	0°	270°	180°	90°

Radial Transmission-Line Modes Corresponding to Each Input

Input I	$m = +3$	Input V	(terminated)
Input II	$m = -3$	Input VI	$m = 0$
Input III	$m = +1$	Input VII	$m = +2$
Input IV	$m = -1$	Input VIII	$m = -2$

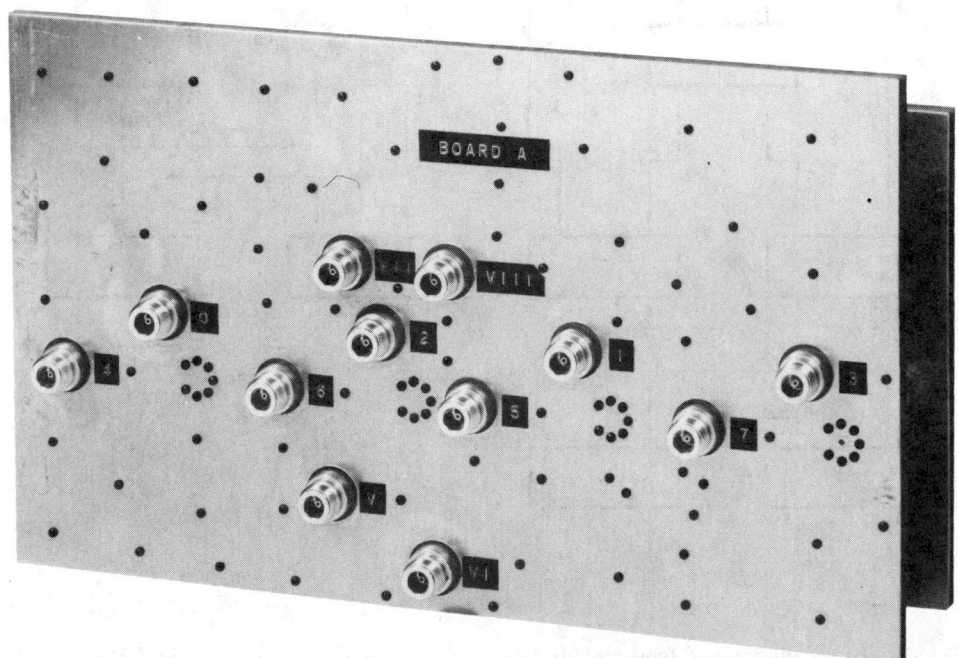

Fig. 25. Stripline hybrid-feed network. (*Courtesy Hughes Aircraft Co., Fullerton, Calif.*)

generated. However, if the feed probe spacing is somewhat closer than $\lambda/2$, the undesirable higher-order modes are "below cutoff." This corresponds to $n > kR_0$ in the cylindrical modal expansion, where n is any undesired higher-order mode number ($n > m$), R_0 is the radius of the circular array of probes, and k is $2\pi/\lambda$. In this case the higher-order modes are "below" cutoff and do not radiate from the feed circle. The undesired higher-order modes, if they did radiate, would result in extraneous beams in directions corresponding to these modes which can also be identified as grating lobes from the linear array due to the periodic amplitude ripple in the aperture distribution. With a $\lambda/2$ probe spacing, the undesired higher-order modes were not sufficiently suppressed; hence a $3\lambda/8$ spacing was used in the experimental model. It should be emphasized that by higher-order modes we mean modes that are higher than the desired modes intentionally generated by the feed. A photograph of the experimental model including the hybrid feed network, the radial line, and the linear array is shown in Fig. 26. The radial line outputs are waveguides and the linear array consists of open-ended waveguide elements with a common horn.

Fig. 27 gives the measured patterns of the seven beams superposed. The patterns were taken without individual tuning or gain adjustment of the different beams. Hence the patterns as recorded indicate the relative gains of the separate beams including impedance-mismatch, circuit, and scan losses. As with the waveguide method of excitation the patterns agree quite well with the theoretical patterns. The vswr was less than 1.3 for all beams. To verify the beam-combining technique to produce tapered aperture distributions, three beams were combined to give a cosine-squared function on a pedestal aperture distribution. Fig. 28 is a typical antenna pattern which shows that low side lobe distributions are achievable. Also, as with other multiple-beam antennas, sector beams and other shaped

Fig. 26. Experimental model of seven-beam antenna. (*Courtesy Hughes Aircraft Co., Fullerton, Calif.*)

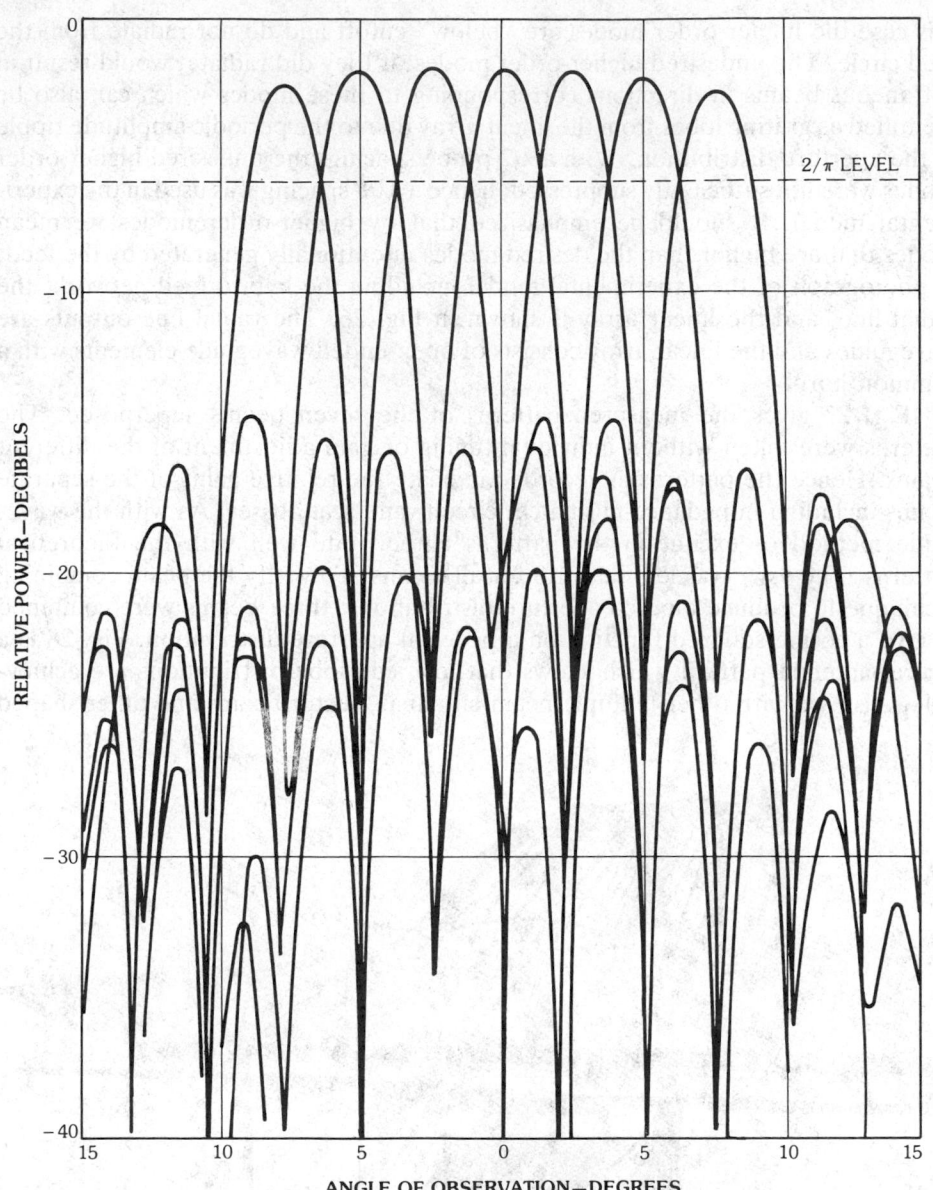

Fig. 27. Measured patterns from the seven-beam antenna system. (*Courtesy Hughes Aircraft Co., Fullerton, Calif.*)

patterns are achievable by combination of beams in the proper relative amplitudes and phases.

Meyer Lens

The principle of the Meyer lens is easily seen by considering a point-source feed radiating in the region between parallel plates as shown in Fig. 29a. The phase

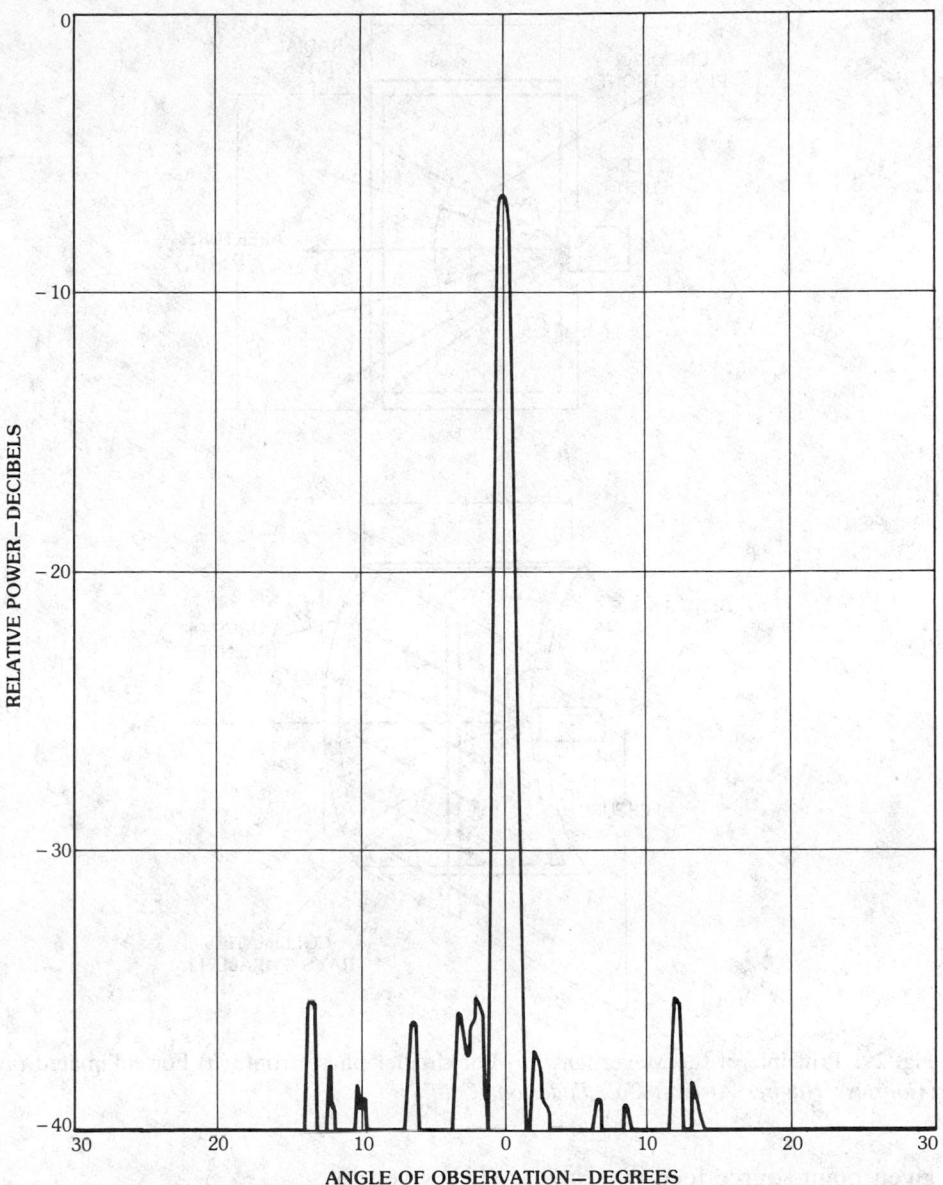

Fig. 28. Measured pattern of cosine-squared function on a pedestal distribution. (*Courtesy Hughes Aircraft Co., Fullerton, Calif.*)

front in the parallel plates is circular and the rays are radial from the feed. The circular phase front can be made linear to collimate the rays by simply curving the parallel plates in the form of a cylinder and putting a 90° bend at the base of the cylinder to direct the collimated rays normal to the axis of the cylinder as shown in Fig. 29b. The Meyer concept is more general in that the bend angle need not be 90° but is arbitrary. The shape of the curve of the cylinder for perfect focus from a

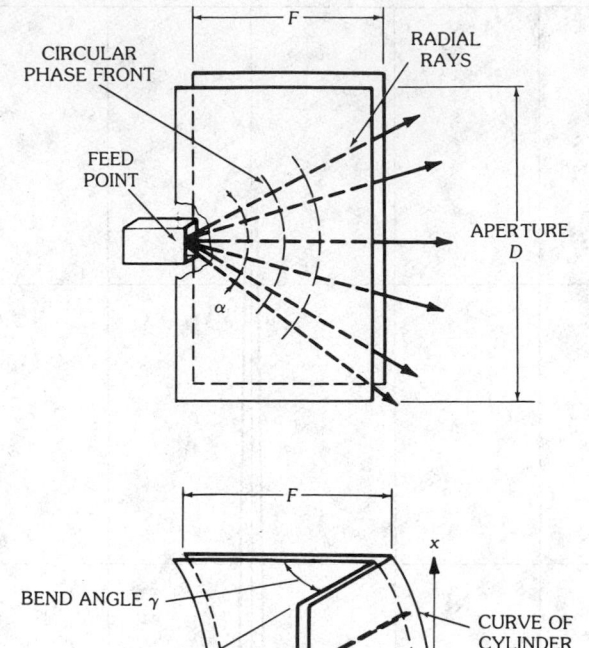

Fig. 29. Principle of the Meyer lens. (*a*) For circular phase front. (*b*) For collimated rays. (*Courtesy Hughes Aircraft Co., Fullerton, Calif.*)

given point-source feed is a catenary [23] given by

$$\frac{y}{\delta} = \cosh\left(\frac{x}{\delta}\right) - 1, \qquad \delta = f(\sec\phi + \tan\phi), \qquad \gamma = \frac{\pi}{2} - \phi$$

where γ is the angle of the bend. If the feed is moved laterally and rotated about its phase center to maintain the proper illumination of the common aperture, the beam will scan from that of the original beam by an amount proportional to lateral displacement, just as with a parabolic antenna. In general, as with all optical devices, the wide-angle scan capability improves with larger F/D, or more precisely, the smaller the feed subtending angle α. By making the bend angle γ larger, keeping the aperture size D the same, the F/D ratio is made larger, the feed angle α is smaller, and the wide-angle scannability is improved. As with all optical devices, however, the feed must be larger and the whole structure less compact by virtue of larger F.

It is interesting to note that if the bend angle γ is made $0°$, the catenary degenerates into a parabola and the Meyer lens becomes the familiar folded pillbox described in Section 3, under the heading "Pillbox." For bend angles on the order of $90°$ or greater the catenary curve is a much better fit to a circle than is a parabola; hence the Meyer lens designed with a nonperfectly focusing circular cylinder has better aperture phase characteristics than a circular pillbox with the same F/D. Due to their circular geometry both have unlimited scannability. A modified $360°$ Meyer lens with dielectric loading can be used to feed a cylindrical array. This is discussed in the last part of Section 6.

The circular Meyer lens [24] is depicted by Figs. 30 and 31 in its undeveloped and developed states, respectively.

A feed may be placed and properly oriented at any point along the feed circle, and the beam radiated from the lens will point in the direction of a principal ray from the feed point through the center of the lens. The analysis may be more easily understood by developing the lens onto a flat surface, as in Fig. 31. In the developed lens all the geodesic paths become straight lines, making analysis very simple.

Two developed* surfaces can be connected together by joining corresponding points of the two surfaces with equal lengths of TEM cable. In this case Snell's laws become

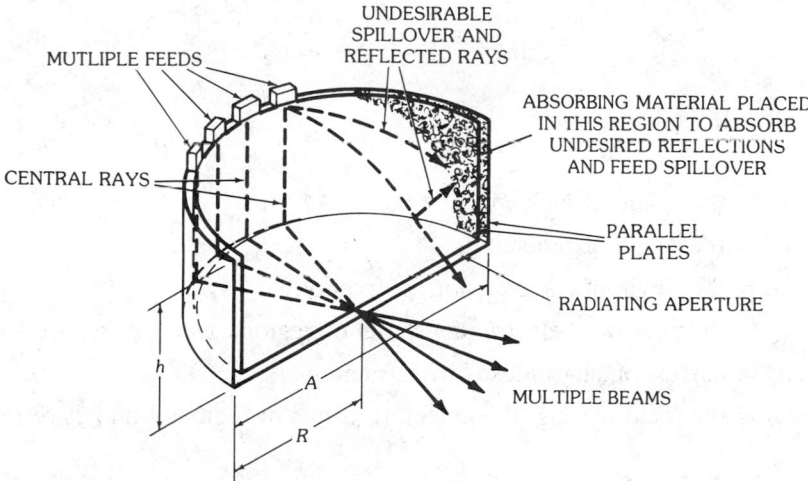

Fig. 30. Schematic diagram of undeveloped multiple-beam geodesic line source. (*Courtesy Hughes Aircraft Co., Fullerton, Calif.*)

---

*The joining surfaces need not, in general, be developable or even be physically joinable but in most practical cases the parts are separated at an intersection in the undeveloped configuration.

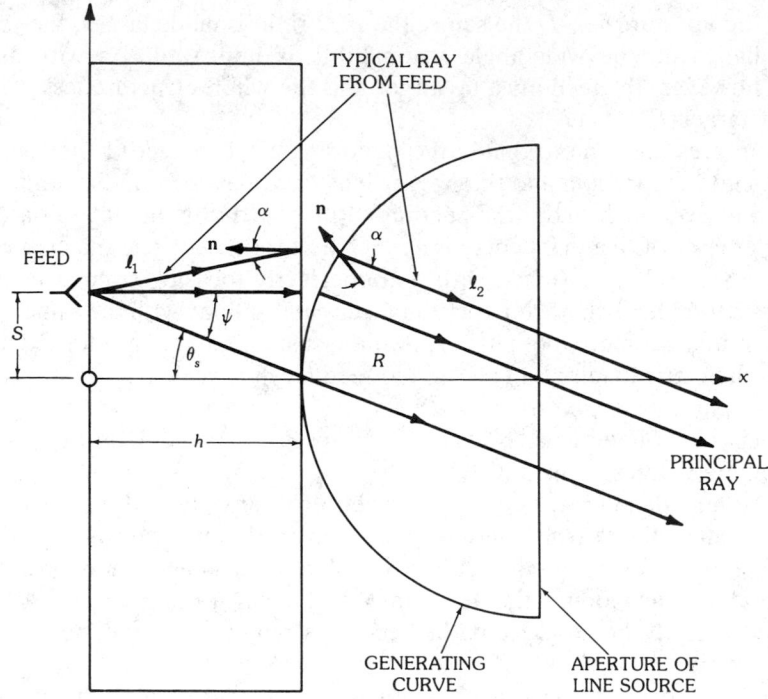

Fig. 31. Developed geodesic lens. (*Courtesy Hughes Aircraft Co., Fullerton, Calif.*)

$$\alpha_i = \alpha_r$$

$$\frac{\sin \alpha_t}{\sin \alpha_i} = \frac{\sqrt{\epsilon_1}\,ds_1}{\sqrt{\epsilon_2}\,ds_2} = \frac{n_1 ds_1}{n_2 ds_2}$$

where

α_i = the angle of incidence

α_r = the angle of reflection

α_t = the angle of transmission (refraction)

ϵ_1, ϵ_2 = the relative dielectric constants of regions 1 and 2, respectively

n_1, n_2 = corresponding indexes of refraction

ds_1, ds_2 = the local spacing of connecting points of regions 1 and 2, respectively

It can be seen from these relationships that the relative spacings ds_1 and ds_2 of the connecting points have the same effects as the relative index of refraction, n_1 and n_2.

An obvious further generalization can be made to make the cable lengths not necessarily equal. Then the differential line length $d\ell$ between adjacent lines times the index of refraction $\sqrt{\epsilon_3}$ in the cables must be added to give the relation

$$\sqrt{\epsilon_1}\, ds_1 \sin \alpha_1 = \sqrt{\epsilon_2}\, ds_2 \sin \alpha_2 + \sqrt{\epsilon_2}\, d\ell$$

The law of reflection remains essentially the same with the assumption that the major portion of the reflected energy is due to the parallel plate–to-cable transition mismatch.

Rotman and Turner [25] used these concepts to design wide-angle multiple-beam lenses using planar parallel plates interconnected by cables. Consequently, these types of lenses are known as *Rotman lenses*. Excerpts from a paper by Rotman and Turner are given in the next section.

In Fig. 31 the beam direction is determined by the principal ray which passes through the center of the aperture. The phase across the aperture may be found by comparing the geodesic path length $\ell = \ell_1 + \ell_2$ for any ray to the principal ray of path length $h + R$.

This type of lens is not perfectly focusing, that is, it will not provide a perfectly plane phase front over the entire aperture. It does, however, provide a reasonably flat phase front over approximately 70 to 80 percent of the aperture. The spherical aberration of this lens is similar to that of the circular folded pillbox covered earlier, but smaller as explained before, allowing more usable aperture.

For phasing a linear array, a pickup array is located at the output aperture which could be waveguide, coax, or stripline. The line lengths at the pickup array would be adjusted to remove the portion of the spherical aberration that is common to all beam positions in exactly the same fashion as discussed previously for the circular folded pillbox.

A multiplicity of feeds located on the circular feed arc is used to create multiple simultaneous beams. The pointing angles of the multiple beams are frequency independent, which implies that the crossover level is frequency dependent, since the beamwidths depend on frequency.

The main advantage of this lens over the circular folded pillbox is that it has more usable aperture. Its main disadvantage is its form factor (in its undeveloped form), which could represent a packaging problem for some applications. However, the circular portion of this lens could be folded, thereby halving its height. It could also be folded more than once to decrease its height even more, but there is certainly a point of diminishing returns.

Radio-frequency absorbing material should be placed in the curved plates at places that do not interfere with the principal optical paths of any feed. This is to absorb any reflected power that does not go around the 90° bend. Materials such as synthane (linen base phenolic) can be used as structural members to hold the plate separation fixed while also acting as rf absorbers. The loss through the Meyer lens is very low, similar to that of the folded pillbox.

Rotman and Turner Line Source Microwave Lens

The Rotman and Turner lens [25] is a parallel-plate constrained lens consisting of a focal arc on which multiple feeds are placed (see Figs. 32 and 33), a set of parallel plates whose plate separation is less than $\lambda/2$ into which the feeds radiate, a pickup array along a surface designated Σ_1 in Fig. 32, a set of interconnecting cables of variable line lengths, and a radiating array designated Σ_2 along a straight line. Fig. 33 illustrates the physical configuration. Four independent conditions are

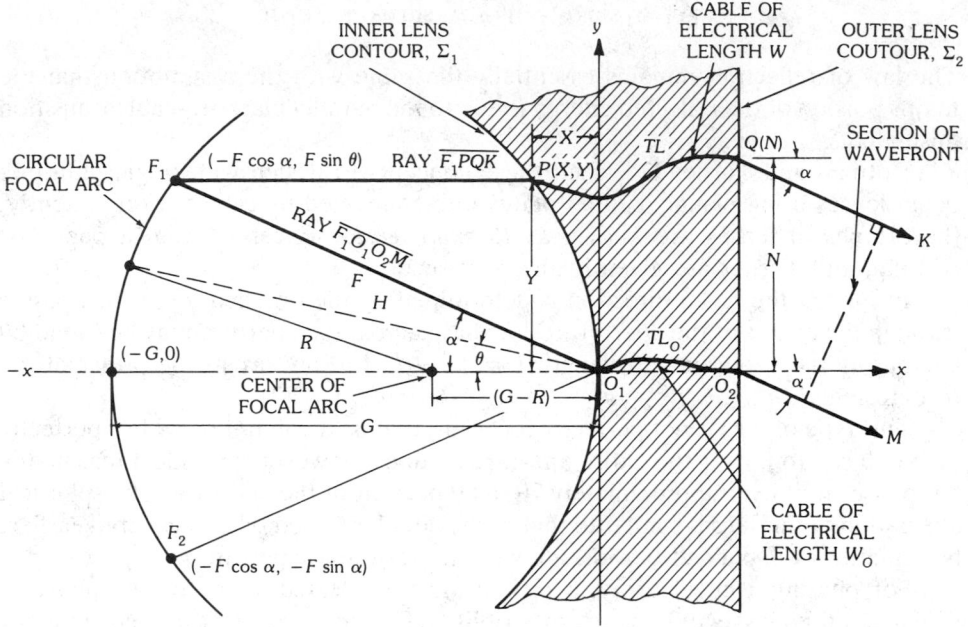

Fig. 32. Microwave lens parameters. (*After Rotman and Turner [25], © 1963 IEEE*)

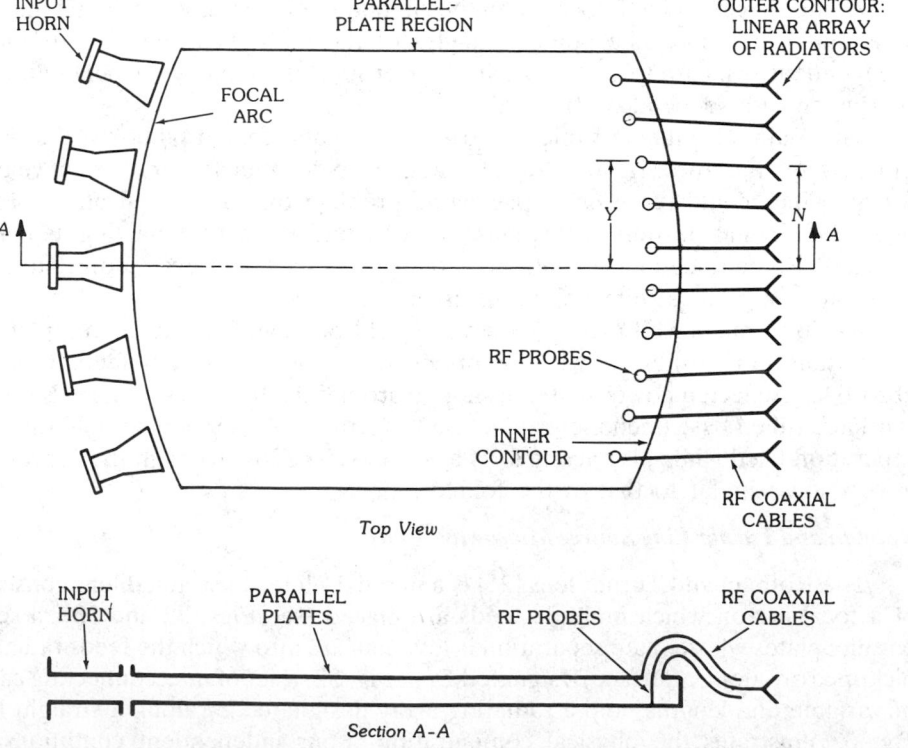

Fig. 33. Parallel-plate microwave lens. (*After Rotman and Turner [25], © 1963 IEEE*)

imposed on the lens system to uniquely determine its configuration: the four conditions imposed are a straight front face, two symmetrical off-axis focal points, and an on-axis focal point. This lens is similar to Ruze's [26] lens except that an additional degree of freedom is available (that is, $y \neq N$, see Fig. 33) which will manifest itself in improved performance for very large apertures. In Ruze's design the lens has two perfect off-axis symmetrical focal points and an on-axis focal point for which the second, but not higher-order, phase deviation is zero. One hundred beamwidths of scan are possible. For these no-second-order lenses, both the front and back lens contours are curved. Ruze [26] also discusses a straight front-face design that has two perfect symmetrical off-axis foci and one highly corrected on-axis focal point. For very large apertures there is some degree of residual higher-order coma aberration in the design.

Rotman and Turner use the generalized lens design principles developed by Gent [27] and others and impose the four previously mentioned conditions to arrive at their configuration.

The following three equations condition perfect focusing at the three* foci using a straight radiating face (see Fig. 32):

$$\overline{(F_1P)} + W + N \sin \alpha = F + W_0 \tag{1}$$

$$\overline{(F_2P)} + W - N \sin \alpha = F + W_0 \tag{2}$$

and

$$\overline{(GP)} + W = G + W_0 \tag{3}$$

where

$$\overline{(F_1P)}^2 = F^2 + X^2 + Y^2 + 2FX \cos \alpha - 2FY \sin \alpha, \tag{4}$$

$$\overline{(F_2P)}^2 = F^2 + X^2 + Y^2 + 2FX \cos \alpha + 2FY \sin \alpha \tag{5}$$

and

$$\overline{(GP)}^2 = (G + X)^2 + Y^2 \tag{6}$$

A set of parameters is normalized relative to the focal length F:

$$\eta = N/F, \qquad x = X/F, \qquad y = Y/F \tag{7}$$

$$\omega = \frac{W - W_0}{F}, \qquad g = G/F \tag{8}$$

Also,

---

*Very recently a quadrufocal bootlace lens has been reported by Rappaport and Zaghloul [28].

$$a_0 = \cos \alpha \qquad b_0 = \sin \alpha \tag{9}$$

After normalizing and using definition (9), Equations 4, 5, and 6 can be combined with (7):

$$\frac{\overline{(F_1P)}^2}{F^2} = 1 + x^2 + y^2 + 2a_0x - 2b_0y \tag{10}$$

$$\frac{\overline{(F_2P)}^2}{F^2} = 1 + x^2 + y^2 + 2a_0x + 2b_0y \tag{11}$$

$$\frac{\overline{(GP)}^2}{F^2} = (g + x)^2 + y^2 \tag{12}$$

The normalized forms of (1) and (10) are combined to yield

$$\frac{\overline{(F_1P)}^2}{F^2} = (1 - \omega - b_0\eta)^2$$
$$= 1 + \omega^2 + b_0^2\eta^2 + 2b_0\omega\eta - 2\omega - 2b_0\eta$$
$$= 1 + x^2 + y^2 + 2a_0x - 2b_0y \tag{13}$$

Since the off-axis focal points are located symmetrically about the center axis, the lens contours must also be symmetrical. Therefore, (13) remains unchanged and can be separated into two independent equations if η is replaced by $-\eta$ and y by $-y$. One equation contains only odd powers of y and η while the other contains the even powers. Thus,

$$-2b_0\eta + 2b_0\omega\eta = -2b_0y \tag{14}$$

or

$$y = \eta(1 - \omega) \tag{15}$$

Also,

$$x^2 + y^2 + 2a_0x = \omega^2 + b_0^2\eta^2 - 2\omega \tag{16}$$

Equations 3 and 6 relating to on-axis focus together with definitions (8) and (9) are similarly combined to give

$$\frac{\overline{(GP)}^2}{F^2} = (g - \omega)^2 = (g + x)^2 + y^2 \tag{17}$$

or

$$x^2 + y^2 + 2gx = \omega^2 - 2g\omega \tag{18}$$

After algebraic manipulation, (15), (16), and (18) give the following relations between ω and η:

$$a\omega^2 + b\omega + c = 0 \tag{19}$$

where

$$a = \left[1 - \eta^2 - \left(\frac{g - 1}{g - a_0}\right)^2\right] \tag{20}$$

$$b = \left[2g\left(\frac{g - 1}{g - a_0}\right) - \frac{(g - 1)}{(g - a_0)^2}b_0^2\eta^2 + 2\eta^2\right] \tag{21}$$

and

$$c = \left[\frac{gb_0^2\eta^2}{g - a_0} - \frac{b_0^4\eta^4}{4(g - a_0)^2} - \eta^2\right] \tag{22}$$

For a fixed set of values of the design parameters α and g, ω can be computed as a function of η from (19). These values of ω and η are then substituted into (14) and (18) to give x and y, which completes the solution of the lens design.

Choosing $\alpha = 30°$, Figs. 34a, 34b, and 34c show the computed lens shape and mapping function $\eta(y)$, as well as the feed locus, for values of $g = 1.10, 1, 1.137$, respectively. The corresponding path length errors $\Delta\ell$ for scan angles of 5°, 10°, 20°, 30° are given in Fig. 35, illustrating the extremely small degree of error over these scan angles.

Fig. 36 illustrates the hardware implementation of this technique, while Figs. 37a, 37b, and 37c illustrate the measured scanned radiation patterns.

According to Rotman and Turner's calculations this type of antenna design is capable of scanning 800 beamwidths of scan without appreciable degradation. For further details, the reader is referred to Rotman and Turner [25].*

Rinehart-Luneburg Lens

Up to this point, this section has dealt with the feeding of linear or planar phased arrays. Although the Rinehart-Luneburg lens is not particularly applicable to the feeding of linear or planar phased arrays, which is the subject of Section 2, under "Multiple-Beam Matrix Feeds," it is introduced here because it is a parallel-plate device. It is applicable for feeding circular or cylindrical arrays and is discussed more fully for that application in Section 6.

The Rinehart [29] geodesic lens is an analog of the nonuniform index of

---

*The reader should be warned that there are some typographical errors in some of the equations in [25]. The most important error is in the expressions for b (Equation 21 of this chapter). The term $-2g$ is missing in the corresponding equation (12) in [25].

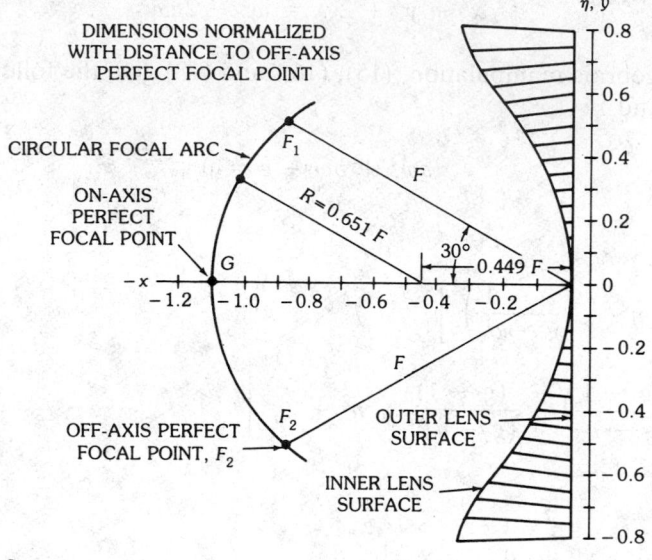

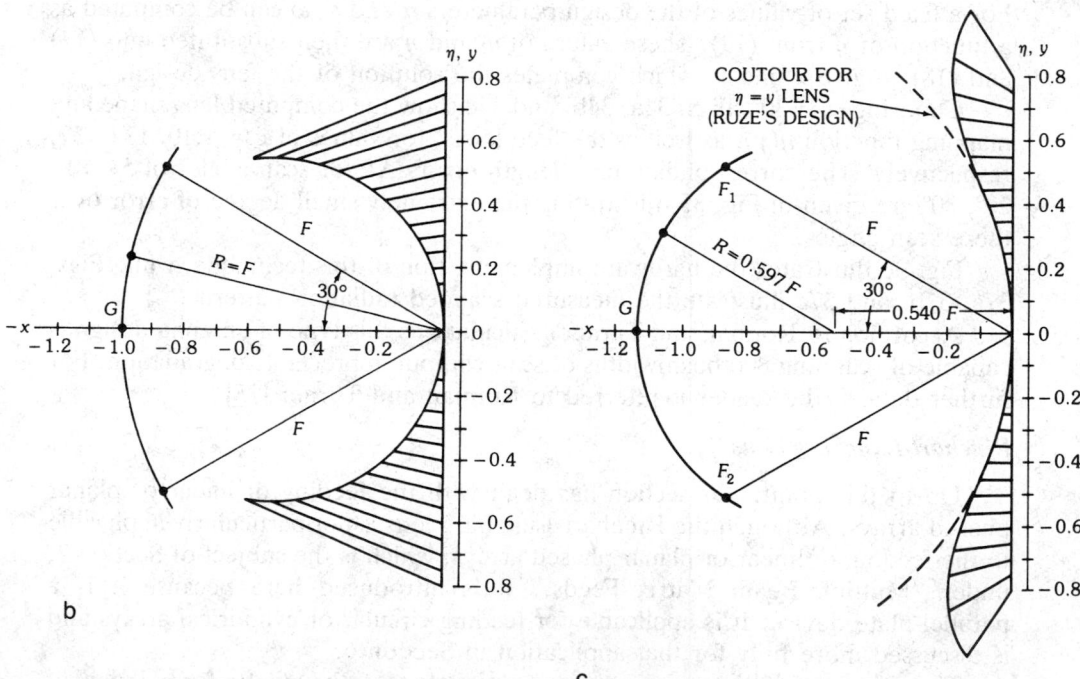

Fig. 34. Microwave lens contours. (*a*) With *g* = 1.10. (*b*) With *g* = 1. (*c*) With *g* = 1.137. (*After Rotman and Turner [25], © 1963 IEEE*)

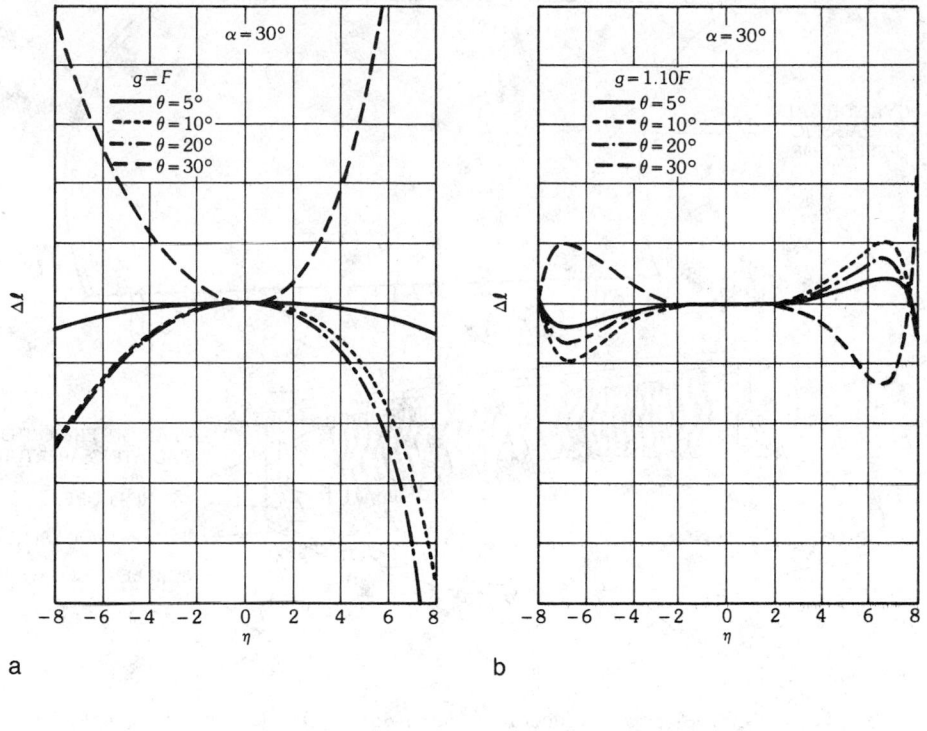

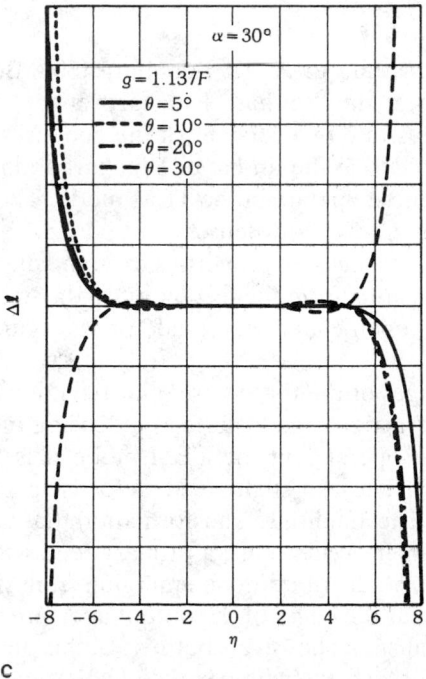

Fig. 35. Path length errors in a microwave lens. (*a*) With $g = 1.00$. (*b*) With $g = 1.10$. (*c*) With $g = 1.137$. (*After Rotman and Turner [25], © 1963 IEEE*)

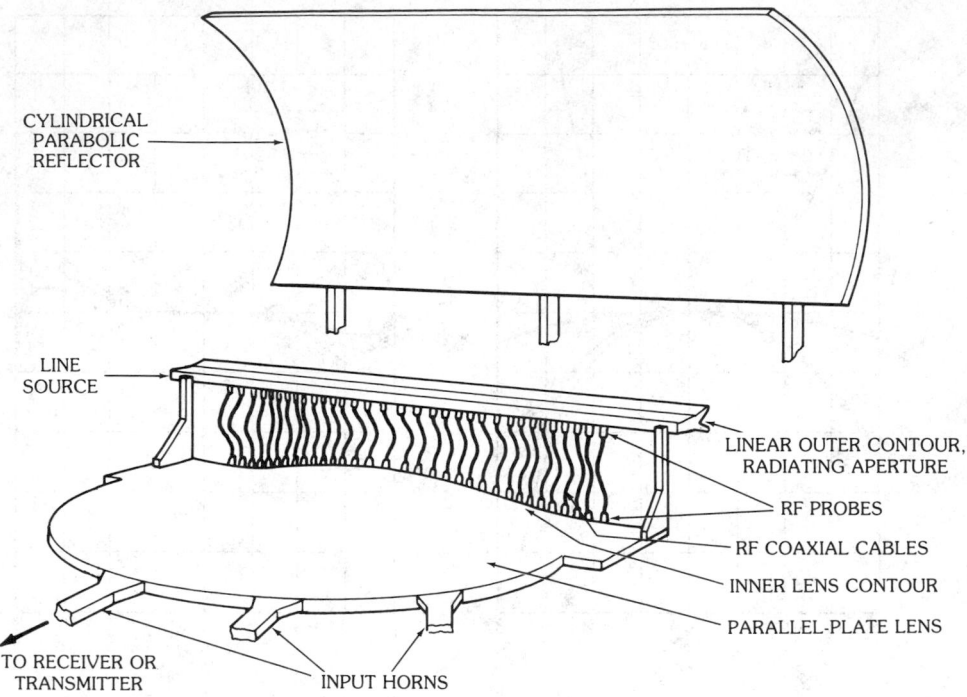

CYLINDRICAL
PARABOLIC
REFLECTOR

LINE
SOURCE

LINEAR OUTER CONTOUR,
RADIATING APERTURE

RF PROBES

RF COAXIAL CABLES

INNER LENS CONTOUR

PARALLEL-PLATE LENS

TO RECEIVER OR
TRANSMITTER

INPUT HORNS

Fig. 36. Parallel-plate lens and reflector. (*After Rotman and Turner [25], © 1963 IEEE*)

refraction spherical lens due to R. K. Luneburg [30]. Because of its shape it is commonly referred to as the "tin-hat" Rinehart lens.

Figs. 38 and 39 illustrate two versions of the geodesic analog of the Luneburg lens, commonly referred to as the tin-hat or Rinehart and flat-plate Luneburg lens, respectively. Both of these operate in the TEM mode. The electrical parallel-plate separation is less than $\lambda/2$ for both lenses.

The tin-hat derives its focusing properties from the physical length that the rays must traverse in propagating from the feed point to the linear aperture as depicted. The feed is usually an open-ended waveguide or waveguide horn, although other types may be used.

The main advantages of the tin-hat are that it contains no dielectric materials, is very simple in construction, and is perfectly focusing over 360°. Moreover, it is very low loss. Since it operates in the TEM mode it is fairly forgiving of small perturbations in shape because slight feed defocusing can partially correct for slowly varying errors. The amplitude transformation that occurs through this lens is such that an inverse taper occurs, e.g., a primary feed with a $\cos\theta$ power pattern transforms into a uniform distribution on emerging from the lens, as illustrated by the bunching of rays near the edges of the output aperture in Fig. 38. Consequently a high gain factor is realizable; however, because of this phenomenon, ultralow side lobes may be difficult to achieve with a simple feed, and a highly directive feed is required to produce very low side lobes.

The bandwidth capability of the tin-hat and flat-plate Luneburg lens is

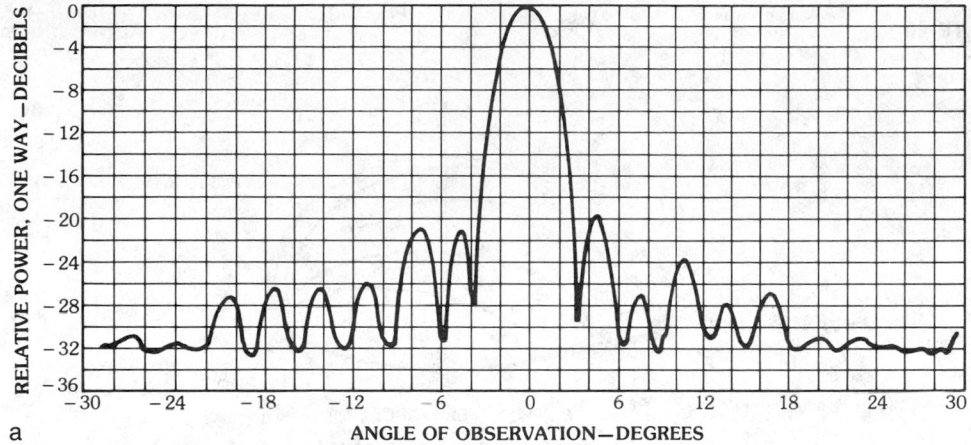

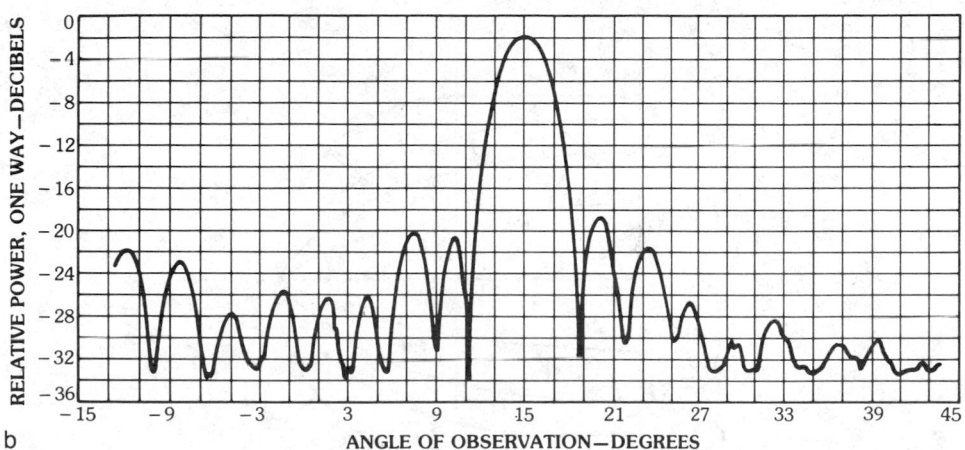

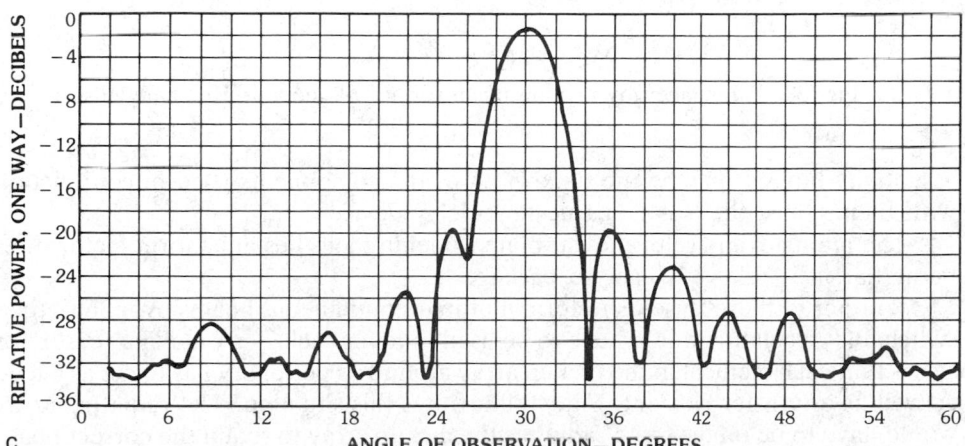

Fig. 37. Radiation patterns of microwave lens antenna. (*a*) For $\theta = 0°$ (on axis). (*b*) For $\theta = 15°$. (*c*) For $\theta = 30°$. (*After Rotman and Turner [25]*, © *1963 IEEE*)

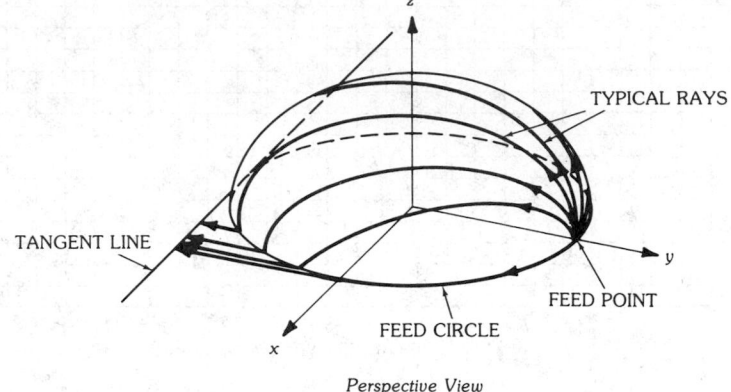

Perspective View

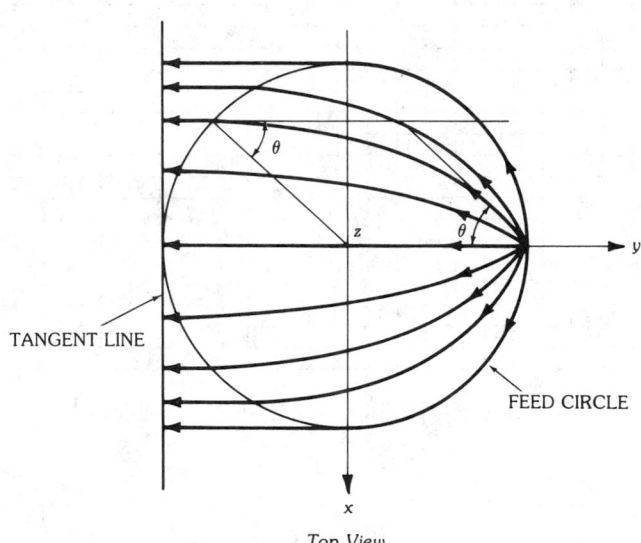

Top View

Fig. 38. Paths taken by rays on mean surface of geodesic Rinehart lens.

essentially limited only by the bandwidth of the components used in conjunction with them, since the lens is a true time-delay device.

The main advantage of the flat-plate Luneburg lens lies in its form factor, i.e., being flat, it is relatively easy to package.

If either of these lenses is to form multiple simultaneous beams over 360°, then switchable circulators would have to be used, one for each element. The reason for this is that each element in the pickup array around its periphery has to act as a feed as well as a transfer element. Fig. 40 illustrates this situation. The radiating array would have to be on the same radius as the pickup array to retain the correct phase distribution. The two arrays are depicted as lying on different radii for clarity only.

If 360° coverage is needed, but not simultaneously, then diode switches could be used rather than circulators. The reason for this is because, at any given

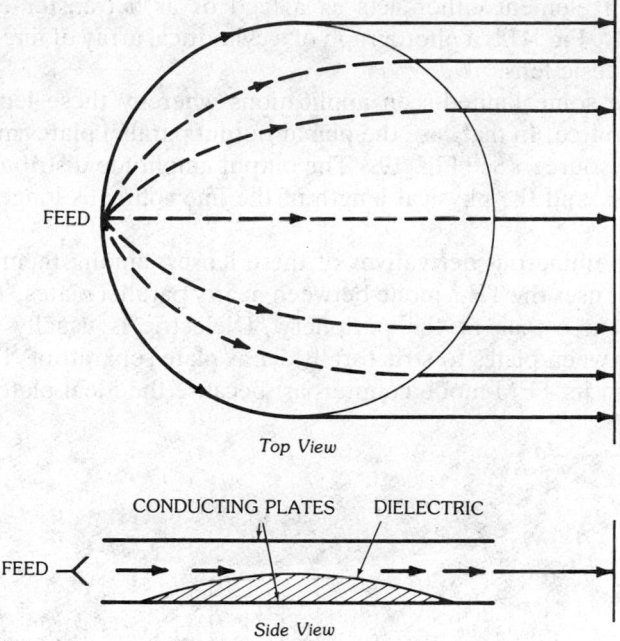

Fig. 39. Flat-plate Luneburg TEM-mode lens.

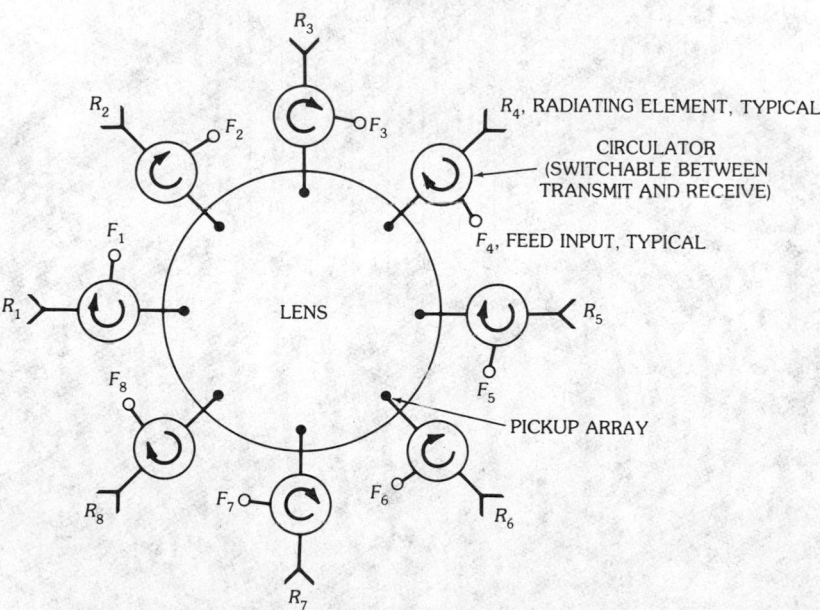

Fig. 40. Flat-plate Luneburg or Rinehart lens for simultaneous beams over 360°. (*Courtesy Hughes Aircraft Co., Fullerton, Calif.*)

moment, each element either acts as a feed or as a transfer element but not simultaneously. Fig. 41 is a photograph of a cylindrical array of line sources fed by a Rinehart geodesic lens.

There are some limited-scan applications whereby these lenses are used to phase a line source. In that case the planar output parallel plates must be extended to form a line source. (See Fig. 19.) The output amplitude distribution moves with scan, however, and the physical length of the line source is longer than its active length.

There are numerous derivatives of these lenses, among them the TE_{10} mode Luneburg that uses the TE_{10} mode between nearly parallel plates. The central plate spacing is greater than at the periphery. Dielectric is usually used to fill the separation between plates to structurally act as plate separators. The bandwidth is much less than its TEM mode counterpart because the ideal plate spacing for the

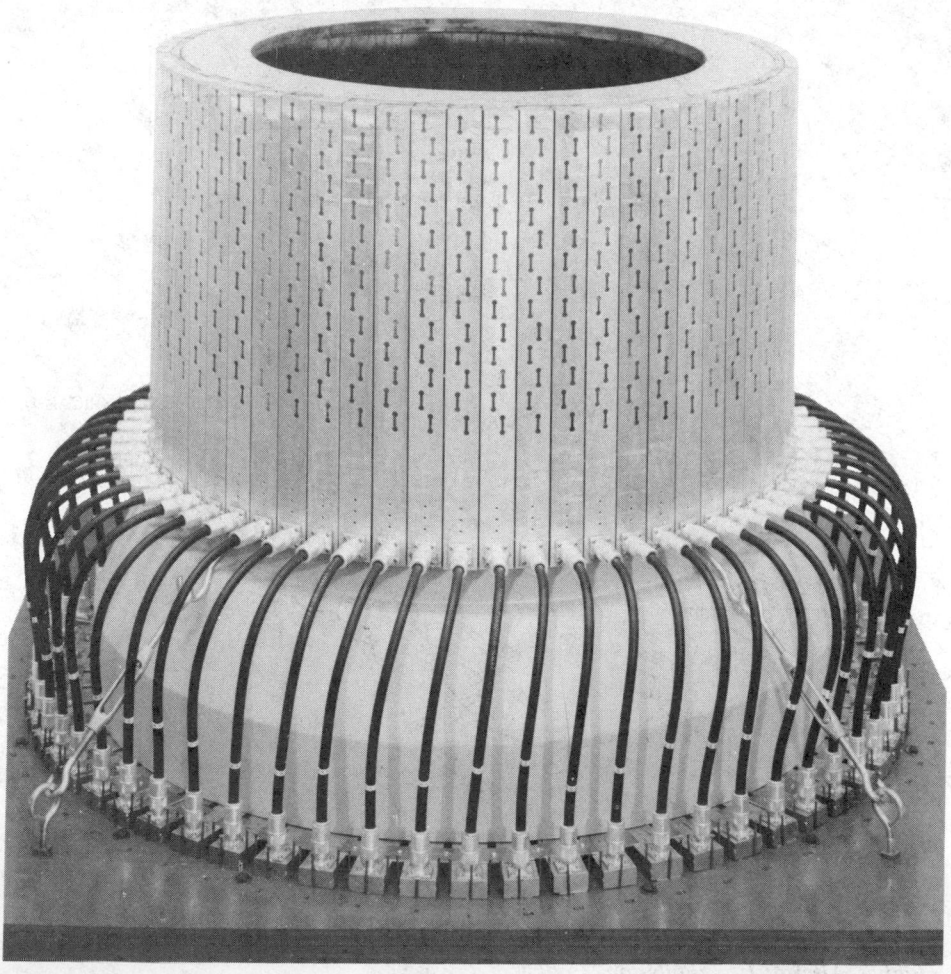

Fig. 41. Cylindrical array fed by a Rinehart geodesic lens. (*Courtesy Hughes Aircraft Co., Fullerton, Calif.*)

TE$_{10}$ mode lens depends on wavelength. Since the correct spacing can be specified only at center frequency, it results in a design that is correct only at the center frequency. The actual bandwidth is a function of the size.

The folded Luneburg lens is, in principle, geodesically the same as the nonfolded one. It resembles a convex dome with a concave dent in the top. From a practical point of view the problem that arises is that of reflections occurring at the points where the slope changes. A mitered bend could probably be used to impedance-match the rays for certain angles of incidence; however, it is doubtful that such a design would work well for all the angles of incidence that the lens requires (up to grazing angles of incidence). Using a smooth bend in the region where the slope changes could be attempted but this changes the geodesics. Little or no information is available to draw definite conclusions regarding the folded Luneburg.

DuFort-Uyeda Lens

DuFort and Uyeda [31] reported a modified Rinehart metal dome antenna with a dielectric ring lens at the output for decreased beamwidth in the elevation plane. The ring lens provides beam collimation in the plane perpendicular to the azimuth scan plane and results in a much smaller antenna than a conventional Rinehart lens fitted with an *E*-plane flared horn.

Even for moderate directivity in the plane perpendicular to the plane of the multiple beams, the length of a low-phase-error, low-flare-angle horn is quite an appreciable fraction of the radius of the dome. This added radius does not contribute to the effective aperture in the azimuth plane which, of course, is equal to the diameter of the dome only. If a shorter horn with a larger flare angle were used, a dielectric lens would be required to collimate the circular phase front of the horn. With dielectric in the toroidal horn it becomes clear by tracing a few rays that the Rinehart lens will no longer focus in the azimuthal plane. The DuFort-Uyeda lens focuses perfectly in the azimuthal plane with an effective aperture equal to the total diameter of the dome and dielectric ring-horn while achieving directivity in the perpendicular plane.

An experimental model was built and tested at K_a-band (26.5 to 40 GHz). It provided a 1.7° by 10.7° beam at 40 GHz. Pattern and gain data show excellent performance.

Fig. 42 shows the top view and cross section of the experimental model. Fig. 43 is a photograph of the experimental model. Fig. 44 shows a typical beam measured at 32 GHz using reduced-height WR28 waveguides for feeds. Fig. 45 depicts an *H*-plane multiple-beam overlay of five contiguous beams measured at 32 GHz. Fig. 46 shows a lower side lobe pattern measured at 32 GHz using a 3/4-inch (1.905-cm) *H*-plane flared horn for the feed.

Aperture efficiencies of between 60 and 72 percent were obtained over the 26.5- to 40-GHz band. Uniform multiple beams of similar gain and pattern behavior can be generated over a 90° sector [31].

4. Unconstrained (Optical) Feeds

An *unconstrained feed* is one in which free space exists between the feed(s) and the radiating aperture. The rf power distribution from feed to aperture is achieved

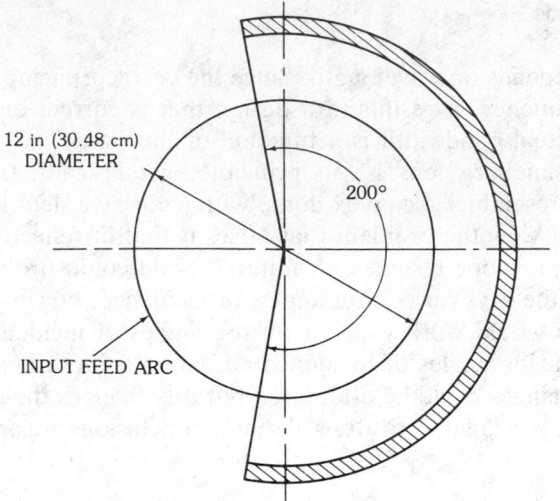

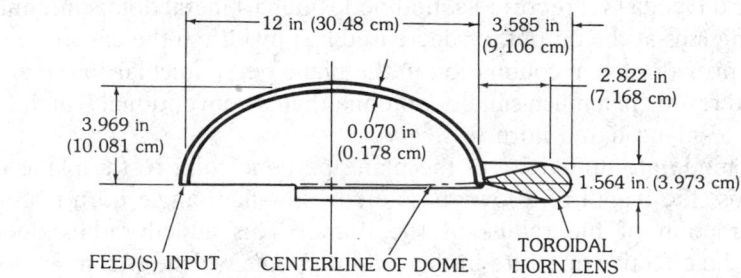

Fig. 42. Design for K_a-band geodesic dome and lens aperture. (*a*) Top view. (*b*) Cross section. (*Courtesy Hughes Aircraft Co., Fullerton, Calif.*)

Fig. 43. Experimental model DuFort-Uyeda lens. (*Courtesy Hughes Aircraft Co., Fullerton, Calif.*)

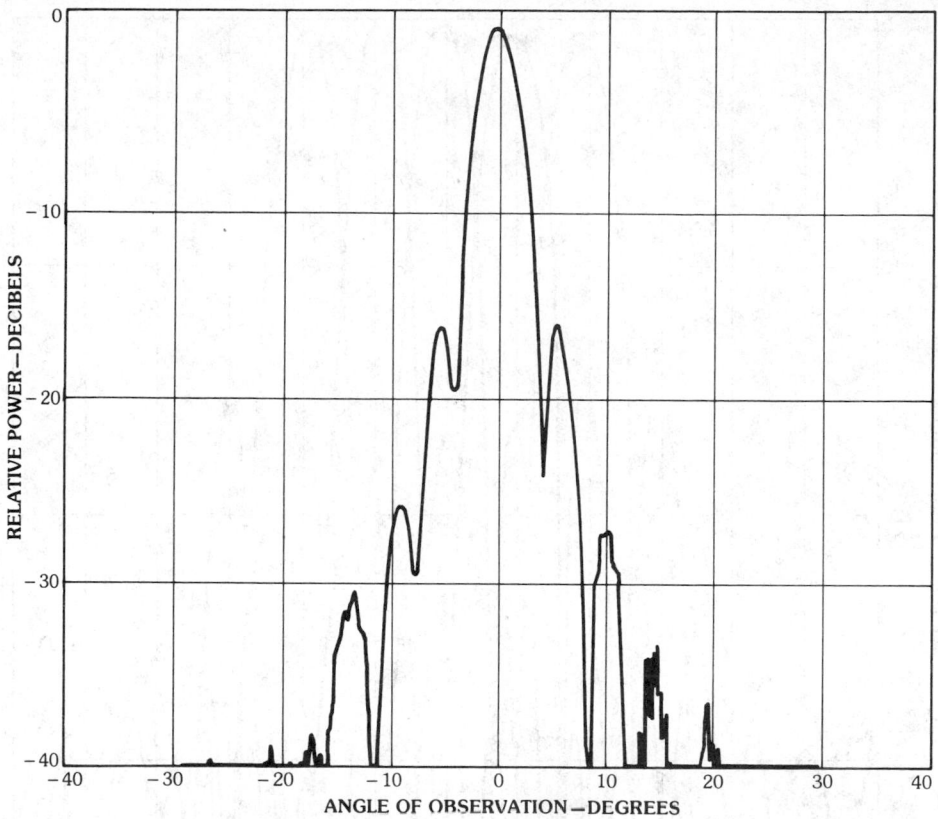

Fig. 44. *E*-plane pattern measurement with WR28 reduced-height waveguide feed at 32 GHz. (*Courtesy Hughes Aircraft Co., Fullerton, Calif.*)

by unconstrained radiation from the feed to the aperture. Hence the aperture distribution is essentially determined by the radiation pattern of the feed. Fig. 47 illustrates such a feed in its most general form. For example, the feed could be a one- or two-dimensional array of radiating elements while the beamformer which collimates the beams could be a reflector(s) or lens which could limit the FOV (field of view) to a relatively narrow sector of space because of off-axis aberrations. As another example the feed could be a simple monopulse feed and the beamformer could be a two-dimensional, nominally half-wavelength-spaced pickup array and radiating array with a phase shifter between a pickup element and a radiating element. In the latter case the FOV is limited only by the wide angle of aperture matching and grating lobe formation (see Chapter 13). A FOV of, say, 90° to 120° cone or greater can be achieved with proper aperture design. For a specified gain (or beamwidth) one would probably use entirely different techniques for achieving a large FOV as opposed to a narrow FOV, or limited scan. In general, the larger the FOV, the greater the complexity and cost of the antenna system. Also, the greater the instantaneous bandwidth, the greater the complexity and cost. Trade-offs can be made between FOV and instantaneous bandwidths.

Any attempt to categorize the various unconstrained-feed approaches is

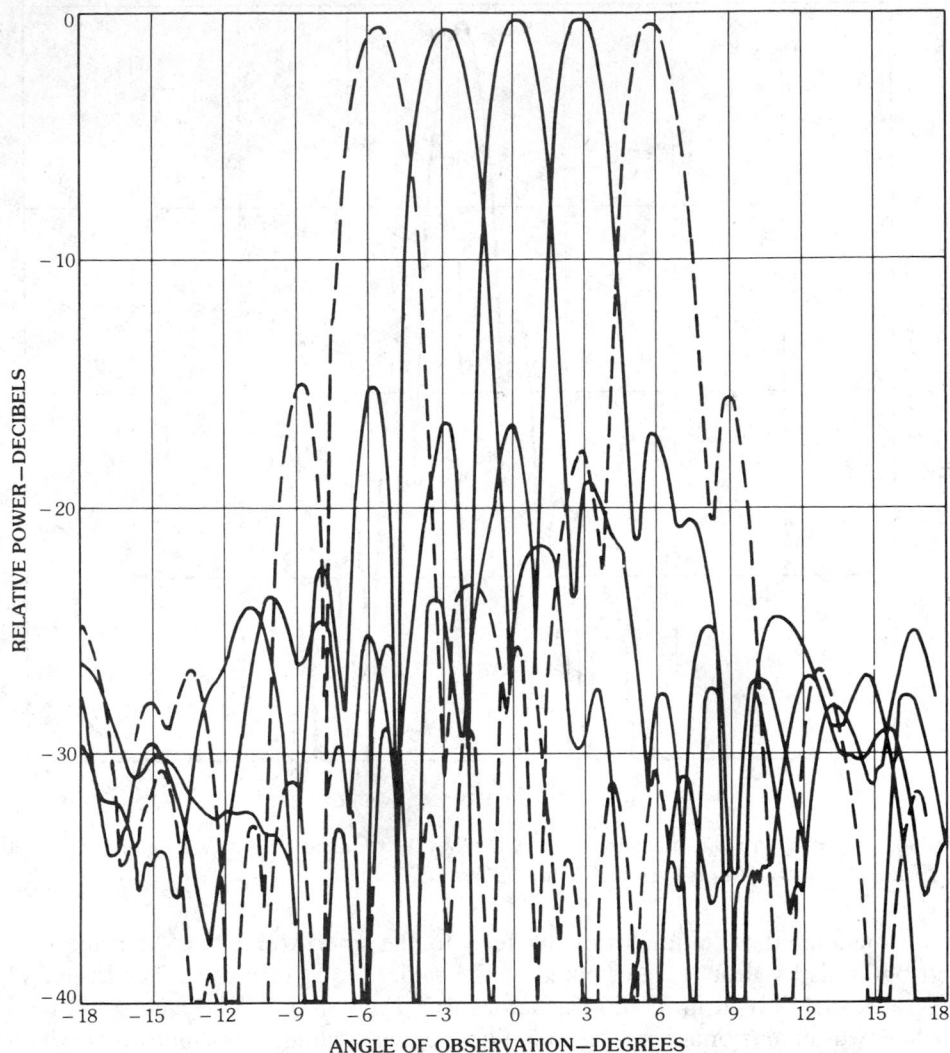

Fig. 45. *H*-plane multiple-beam overlay of five contiguous beams measured at 32 GHz with WR28 reduced-height feeds. (*Courtesy Hughes Aircraft Co., Fullerton, Calif.*)

difficult because there are overlaps in the categories. Nevertheless, a review of the various techniques suggests four categories for the purpose of discussion. These are the following:

1. Wide FOV (0° to 360° phase shifters, nominally half-wavelength-spaced elements
2. Limited scan (0° to 360° phase shifters and optics with aperture magnification)
3. Subarray or partial time delay using overlapping or non-overlapping subarrays, 0° to 360° phase shifters, and time delay at the subarray level. *Note*: Totally constrained limited-scan techniques are discussed in Section 3 of Chapter 13,

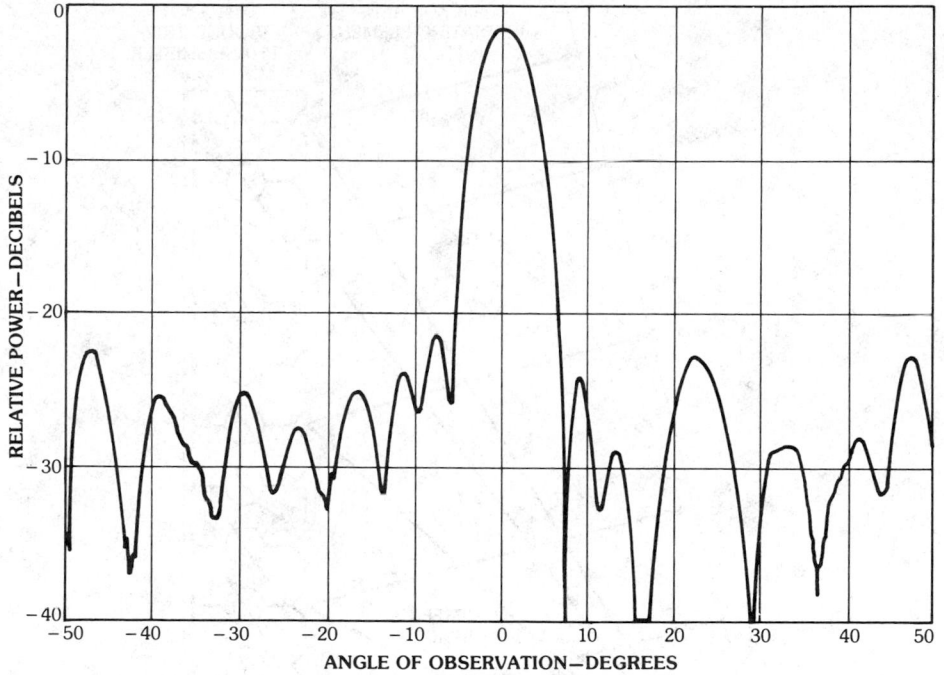

Fig. 46. Low side lobe *H*-plane pattern measured at 32 GHz with 0.75-in (1.905-cm) *H*-plane flared horn. (*Courtesy Hughes Aircraft Co., Fullerton, Calif.*)

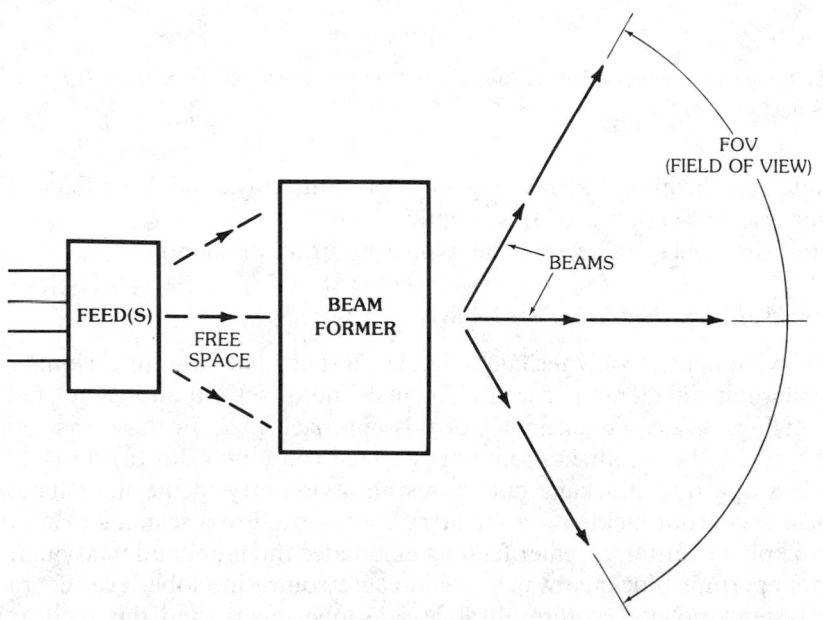

Fig. 47. Unconstrained feed. (*Courtesy Hughes Aircraft Co., Fullerton, Calif.*)

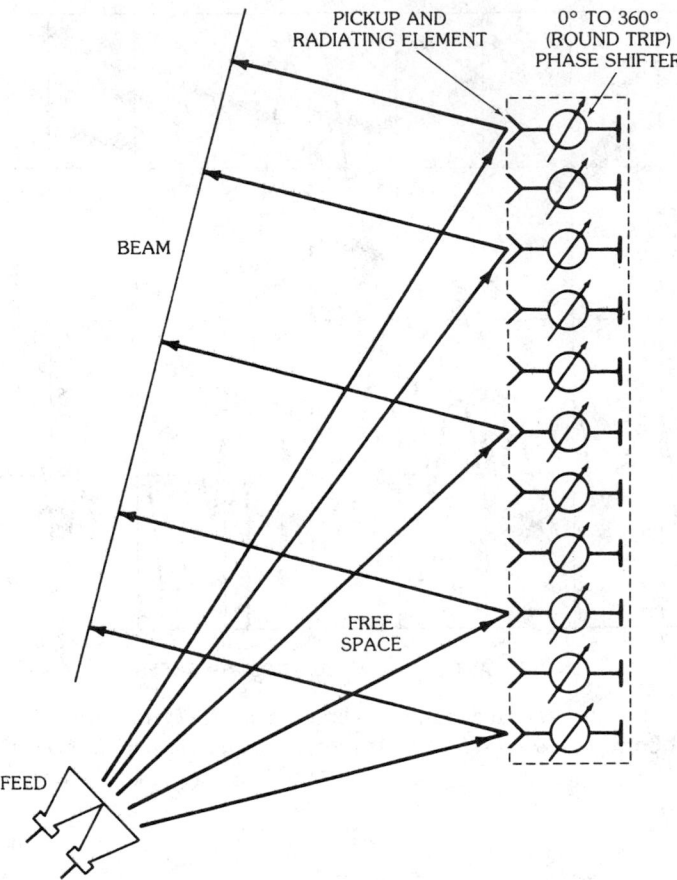

PICKUP AND
RADIATING ELEMENT

0° TO 360°
(ROUND TRIP)
PHASE SHIFTER

BEAM

FREE
SPACE

FEED

a

Fig. 48. Space-fed reflectarray. (*a*) Offset-fed. (*b*) Center-fed. (*Courtesy Hughes Aircraft Co., Fullerton, Calif.*)

under the heading "Array Organization: Subarrays and Broadband Feeds," and also in Section 2 of this chapter.

4. True time delay (true time-delay phasing of all elements)

Wide Field of View (Nontrue Time Delay)

The wide field of view technique implies a space-fed two-dimensional array of nominally half-wavelength-spaced elements. The offset-fed and center-fed reflectarray of Fig. 48 depicts examples of this approach [32]. In these cases the feed would typically be a four-horn monopulse feed (only two shown). Offset feeding offers less aperture blockage but causes an asymmetry in the illumination. The spherical wavefront incident on the array is converted to a scanned beam by 0° to 360° type phase shifters. Center feeding eliminates the illumination asymmetry but presents aperture blockage which would cause some side lobe level degradation. Many systems do not require ultralow side lobe levels, and this feed approach offers a straightforward solution. The Raytheon MTR Radar and Radome Antenna

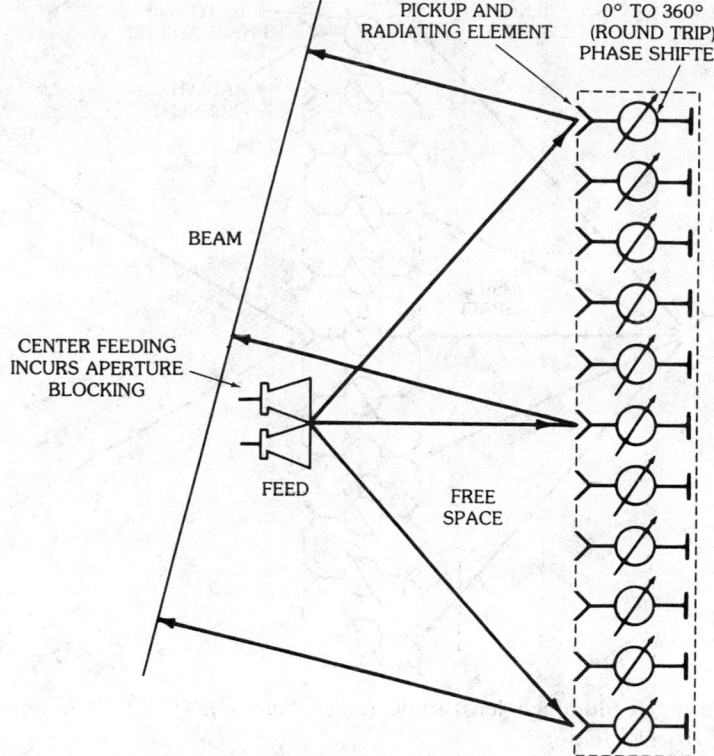

PICKUP AND
RADIATING ELEMENT

0° TO 360°
(ROUND TRIP)
PHASE SHIFTER

BEAM

CENTER FEEDING
INCURS APERTURE
BLOCKING

FEED

FREE
SPACE

b

Fig. 48, *continued.*

and RF Circuitry (RARF) array are examples of the center-fed reflectarray [32]. In the case of the RARF array six waveguides (monopulse sum and two difference channels for two orthogonal polarizations) are routed to the feed in order to achieve polarization agility. Other examples are the NRL *S*-band Reflect Array Radar (RAR) developed by Blass Antenna Electronics Company [32] and the Hughes Aircraft Company Electronically Scanning Airborne Intercept Radar Antenna (ESAIRA).

Aperture blockage is eliminated entirely using a space-fed feed through lens array as depicted in Fig. 49. Since no aperture blockage is presented, the feed could be as large and complex as desired, without degraded side lobe level effects. There are numerous advantages to this feeding scheme. The Raytheon Sam-D Radar uses this approach [32]. It enables a space-duplexing feature to be employed; that is, on transmit the phase shifters are set to focus on the transmit feed, and on receive the phase shifters are set to focus on a separate receive feed. Moreover, two transmitter feeds could be employed using two transmitters for redundancy, where different phase shifter settings are used, depending on which transmitter is being used [32]. (The offset-fed reflectarray could also have this feature without increased aperture blockage.) Since there is room available, optimized sum and difference aperture illuminations could be realized by a sophisticated monopulse feed design. A multiple stack of vertical feed horns could be combined for elevation beam

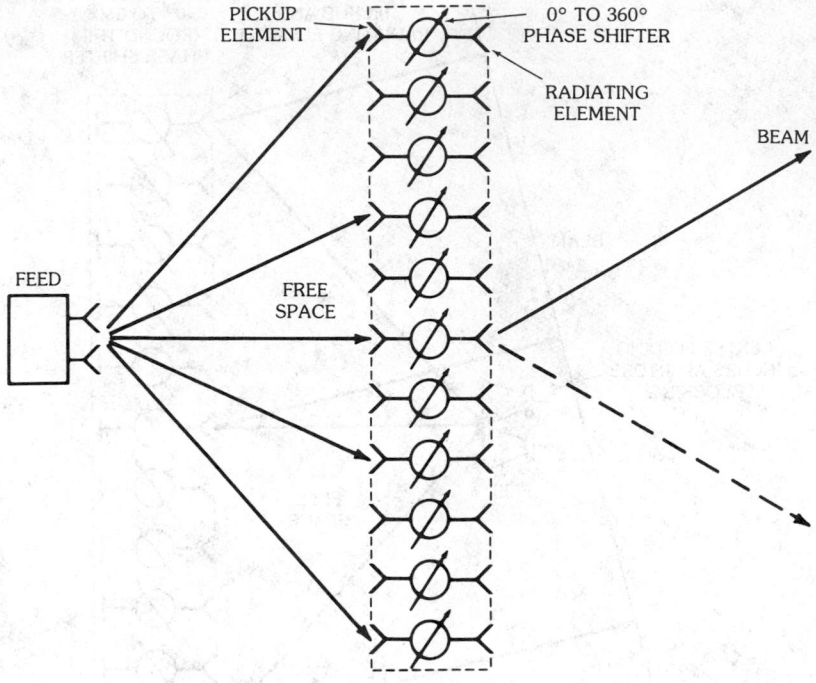

Fig. 49. Space-fed feedthrough lens array. (*After Patton [34], © 1972; reprinted with permission of Artech House*)

shaping, and azimuth patterns could be generated by using multimode sum and difference extraction of each horn in the azimuth plane [32].

Although Fig. 49 depicts the space-fed feed through array with both faces flat, they need not be.

The sperry HAPDAR radar uses a curved pickup face and flat radiating face, with phase shifters in the transmission lines between faces. This structure is called TACOL, for thinned aperture computed lens [32, 33]. The feed is a five-horn cluster that provides independent control over the sum and difference illuminations, using an *F/D* ratio of 1. All five horns are used for the sum pattern but only the outer four are used for the difference patterns. In this way, low side lobe sum and acceptable difference patterns are realized.

Limited Scan

High-gain antennas with wide field of view and/or wide instantaneous bandwidths are very costly due to the large number of active devices. For phased arrays a phase shifter and driver are required for each radiating element, which may total to many thousands for a high-gain array. For multiple-beam antennas with beam switching, a similar number of switching elements and drivers are required. In many radar applications wide field of view is not required although high angular resolution and high gain are. Examples of such limited-scan applications are the airport precision-approach radar, weapon-locating radars, and earth-coverage antennas for synchronous satellites.

It is the objective of limited-scan antenna design to take advantage of this limited field of view to design an antenna with the fewest number of active devices but retain the narrow beamwidth and high gain. To achieve high gain and narrow beamwidth requires a large antenna aperture, and a large field of view requires a phase shifter and drivers for every nominally half-wavelength-spaced radiating element. Hence high gain and large FOV result in an extremely costly antenna system. By restricting the FOV the cost can be greatly reduced.

Limited-scan antennas can employ optical (unconstrained) feeds, totally constrained feeds, or combinations of these. In this section on unconstrained feeds, only those antennas employing optical techniques are discussed. Limited-scan techniques using totally constrained feeds in which elements of an array are grouped together to form identical subarrays with a phase shifter per subarray are discussed in Chapter 13, Section 3, on array organization.

Optically Fed Aperiodic Array—An array of large equal-size directive elements (e.g., a nonscanning subarray) whose element pattern is tailored to the required FOV could be considered, using a phase shifter for each large element or subarray. For regular element spacing, however, this approach produces excessively high grating lobes as the beam is scanned, which would not meet most system requirements. The grating lobes result from the fact that radiation from periodic widely spaced elements add in phase at angles in space other than the desired direction of the main beam. If the periodicity is broken up, the energy concentrated in the grating lobes is smeared out "randomly" in all directions, resulting in lower peak values. The overall antenna gain, however, is not improved since it is only a redistribution of the grating lobe energy. In most cases there is a slight though usually negligible reduction in gain.

One method of minimizing the grating lobe problem, while still using large directive elements, is given by Patton [13, 34]. In this approach he uses concentric rings of equal-area subarray elements, as shown in Fig. 50 (depicted for only a few subarrays). Fig. 51 depicts a large number of subarrays of equal area. The subarrays are dual-polarized dipoles above a ground plane. Each polarization is summed (in phase) within each subarray. Thus each subarray has two outputs—one for each polarization. Fig. 52 illustrates how the concentric ring array is fed. For each subarray pair of cables there is a corresponding pair of terminals (one for each polarization) located on the surface of a partial sphere. The cables must track in phase over the frequency band of the system. The feed is located at the center of the spherical cap. The output final aperture distribution is thus determined by the primary Σ and Δ patterns produced by the primary feed. Since dual-polarized elements are used, the final radiated polarization is determined solely by the polarization of the primary feed. There are phase shifters in every semirigid coaxial line between the concentric ring array and the spherical cap.

By using this technique the periodicity of the usual array factor (for regular spacing) is broken up; consequently, grating lobes are suppressed. The degree of suppression is determined by the total number of subarrays used. Patton [34] shows calculated data for 10 ft and 30 ft diameter (3.0 and 9.1 m) arrays at C-band using a $\pm5°$ FOV. For the 30 ft diameter case, using 1000 subarrays, the vestigial grating lobe is suppressed to about -21 dB for maximum scan of $\pm5°$. The 10 ft diameter case (100 subarrays) corresponds to -15-dB suppression at $\pm5°$ scan.

Predicted and measured loss, including illumination taper (10-dB tapered

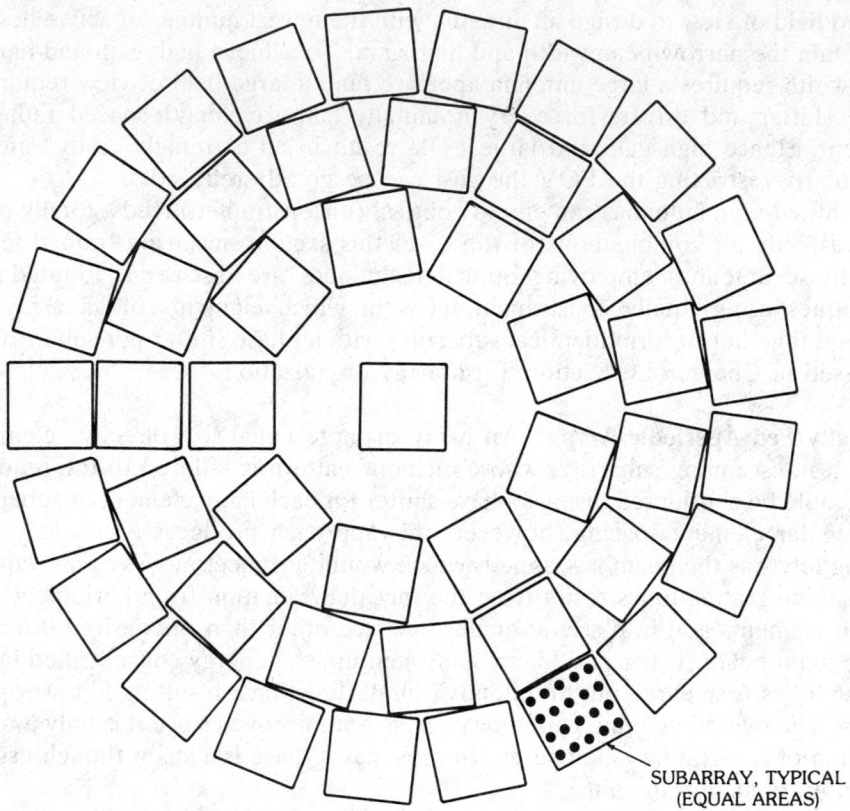

SUBARRAY, TYPICAL
(EQUAL AREAS)

Fig. 50. Concentric ring array of equal-area subarrays. (*After Patton [34], © 1972; reprinted with permission of Artech House*)

Gaussian) for the 10 ft diameter model are 5.94 dB and 5.58 dB, respectively, at midband. Measured first side lobes were about −15 dB, due to effective aperture blockage. The gain decreased with scan by about 2.5 dB at ±5° from broadside.

The primary feed is a multimode monopulse feed, dual polarized for the sum channel, and horizontally polarized only for the difference. Conventional reflector feed techniques for achieving dual monopulse, dual polarization, for both Σ and Δ channels are applicable here; moreover, techniques for achieving independent control over the Σ and Δ patterns can also be used. Since there is room available, a more complex feeding structure could be considered for any additional advantages that may be realized.

Phased Array with Paraboloid—Fig. 53 depicts a paraboloidal reflector fed by a phased reflectarray, after Winter [35]. The reflectarray radiating face is not located at the focus, but is shifted toward the vertex of the paraboloid. Conceptually viewed on receive, the idea is to place the reflectarray aperture in the region forward of the focus, pick up the converging field, and phase-shift it to refocus on the primary feed(s). Geometrical optics was used to determine the phase shifter values, as a function of scan angle θ_s. As θ_s varies, the converging field

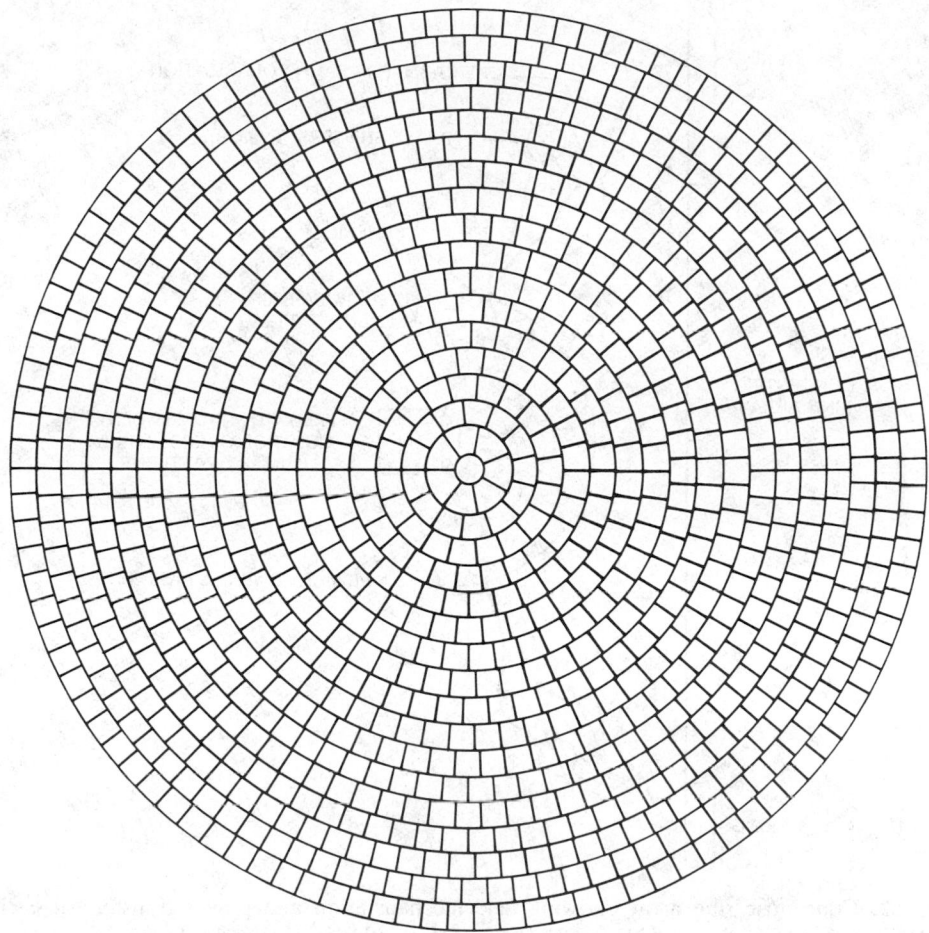

Fig. 51. Aperiodic array for limited scan. (*After Patton [34], © 1972; reprinted with permission of Artech House*)

moves, but for small enough θ_s, most of the reflected power will still be incident on the reflectarray face. Thus one can still change the phase of the converging field to refocus on the primary feed. Note that only the phase is controllable, not amplitude. As θ_s gets larger, only part of the reflected incident field is intercepted by the radiating aperture of the reflectarray antenna; hence, spillover occurs (on receive), causing irretrievable loss. Furthermore, for large θ_s, the effects of aberrations are to cause the reflectarray amplitude distribution to become skewed, introducing both an even and an odd part. The simple primary feed horn couples only to the even part, and the energy in the odd part is lost. Ideally, a horn containing an odd-mode component as well as the even mode, or a small array with the proper amounts of even and odd parts, could be employed to regain that lost energy. For maximum gain, one would use a simple feed designed to produce an illumination across the reflectarray face that best fits the even part of the received converging illumination across the reflectarray face over some specified range of

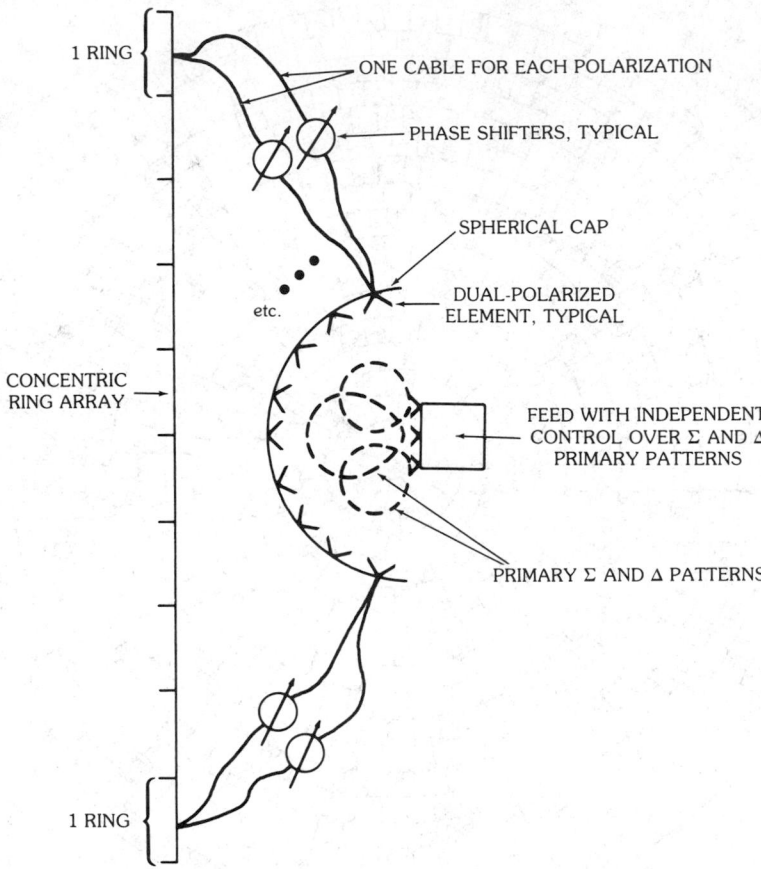

Fig. 52. Concentric ring array showing feed mechanization and power transfer for each polarization. (*After Patton [34],* © *1972; reprinted with permission of Artech House*)

scan angles $\pm\theta_s$. This would produce the best amplitude fit that is achievable using a simple primary feed.

Viewing the process on transmit, the primary feed and reflectarray could be designed to illuminate the entire reflector* for a broadside beam; however, as scanning is performed, because of the optics of the system, only a portion of the reflector can be illuminated correctly to form a beam in an arbitrary given direction. Thus the illuminated portion of the reflector moves with scan, which diminishes the efficiency of the antenna system. The aperture blockage presented by the relatively large reflectarray causes a further reduction in efficiency, as well as increased side lobe levels. One of Winter's [35] experimental models used a 988-element reflectarray and simple feed horn and achieved eight beamwidths of scan with 15-dB side lobes in the *H*-plane and 10 dB in the *E*-plane. The reflector was

---

*In Winter's experiment 80 inches (203 cm) of the 96-inch (244-cm) diameter was used for a broadside beam.

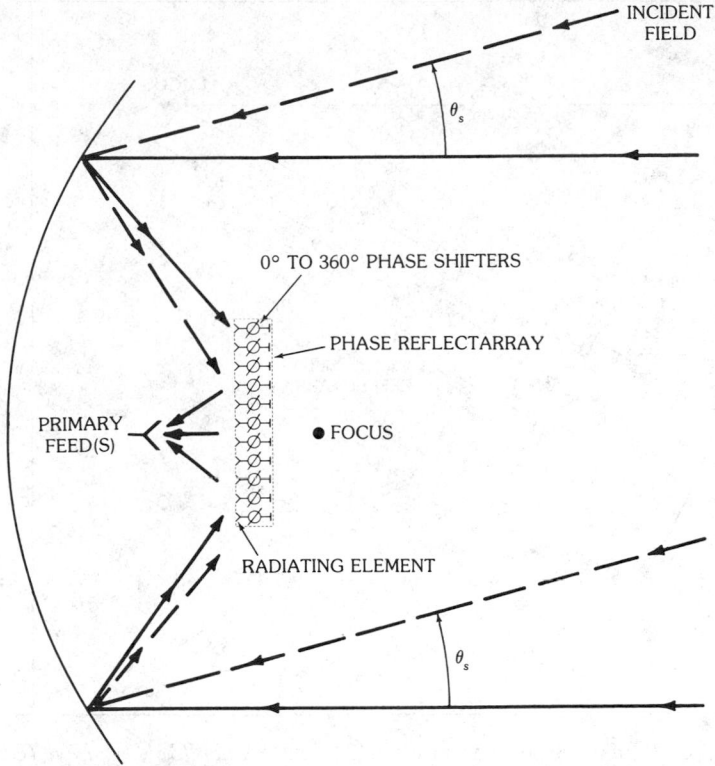

Fig. 53. Winter's approach, using phased array with paraboloid. (*After Winter [35], © 1968 IEEE*)

8 ft (244 cm) in diameter with a 51-inch (130-cm) focal length, offset fed as shown in Fig. 54. His results show that it is feasible to utilize a phased array in conjunction with a reflector to achieve a limited degree of scan. The high blockage presented by the phased array could be circumvented by using the phased array in conjunction with a transmission type feedthrough lens, or by offset-feeding to a greater degree. By using a more complex primary feeding arrangement, such as a small active array that is commanded to change with scan, more control over the amplitude would be possible. Thus the radiation characteristics would be expected to improve, provided the design is not blockage limited, as appears to be the case in Winter's experiment.

Offset Phased Array with Hyperbolic Reflector—By moving the phased array out of the FOV of the reflector the aperture blockage would be eliminated and improved radiation characteristics obtained. The GCA System TPN-19 uses such an approach [36], utilizing 824 phase shifters in conjunction with a 9- × 11½-ft (2.74- × 3.5-m) hyperbolic main reflector. Improved side lobe levels of 22 to 24 dB were obtained. The aperture efficiency was 30 percent at broadside and dropped 3½ dB over ±10 beamwidths (elevation) by ±7.15 (azimuth) beamwidths of scan. Azimuth and elevation beamwidths are 1.4° by 0.75°, respectively. The illumination

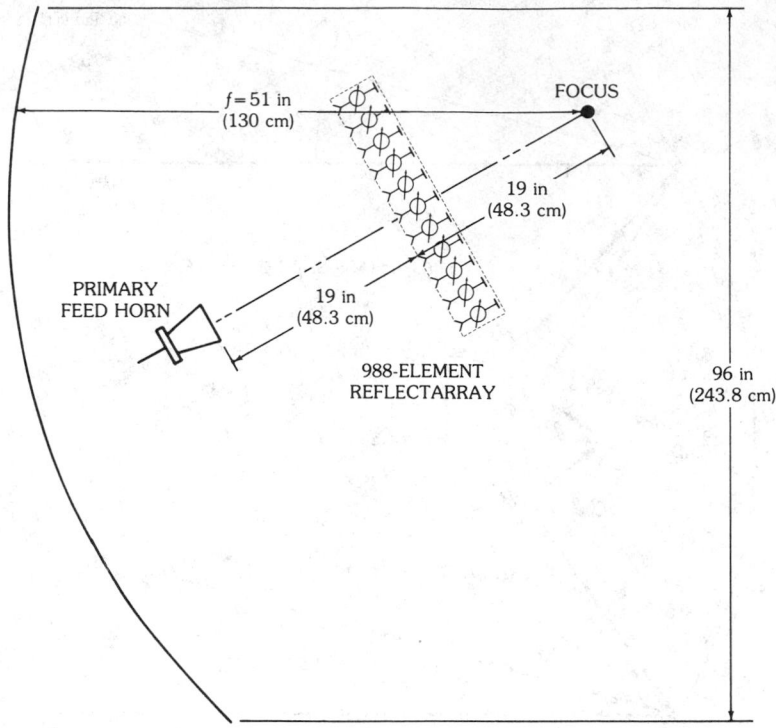

f = 51 in
(130 cm)

FOCUS

19 in
(48.3 cm)

PRIMARY
FEED HORN

19 in
(48.3 cm)

988-ELEMENT
REFLECTARRAY

96 in
(243.8 cm)

Fig. 54. Winter's experimental model. (*After Winter [35], © 1968 IEEE*)

moves over the surface of the reflector as scanning is performed; hence the aperture efficiency is low at the extreme scanned beam positions. The system operates at *X*-band over a 2-percent bandwidth, with an on-axis gain of 42.5 dB.

Near-Field Cassegrain System—Fig. 55 depicts the near-field Cassegrain antenna system. A typical application of this system would have a limited field of view (LFOV) of 5° to 10°, and a very large electrical aperture, e.g., $D/\lambda > 250$. It is composed of two confocal paraboloids with $f/d = F/D$, and is fed by a planar two-dimensional phased array of 0° to 360° type phase shifters. Since the subreflector is in the near field of the phased array, the rays remain collimated from the phased array to the subreflector such that there is no space attenuation. The optics of this system are such that the amplitude distribution emerging from the main reflector is the same as the amplitude distribution across the phased-array feed. In particular, a uniform amplitude distribution over the phased-array feed transforms into a uniform amplitude distribution emerging from the main dish. There is, however, a relatively large blockage by the subreflector or by the effective blockage (d_e) presented by the phased-array feed. Minimum blockage occurs when these two blockages are equal. A typical design [36] would have d/D ratios between 0.25 for $D/\lambda = 400$ and 0.35 for $D/\lambda = 250$. (Also see reference 37.)

The near-field Cassegrain system can be considered as the limiting case of the standard Cassegrain system as the feed point focus of the hyperboloid recedes to

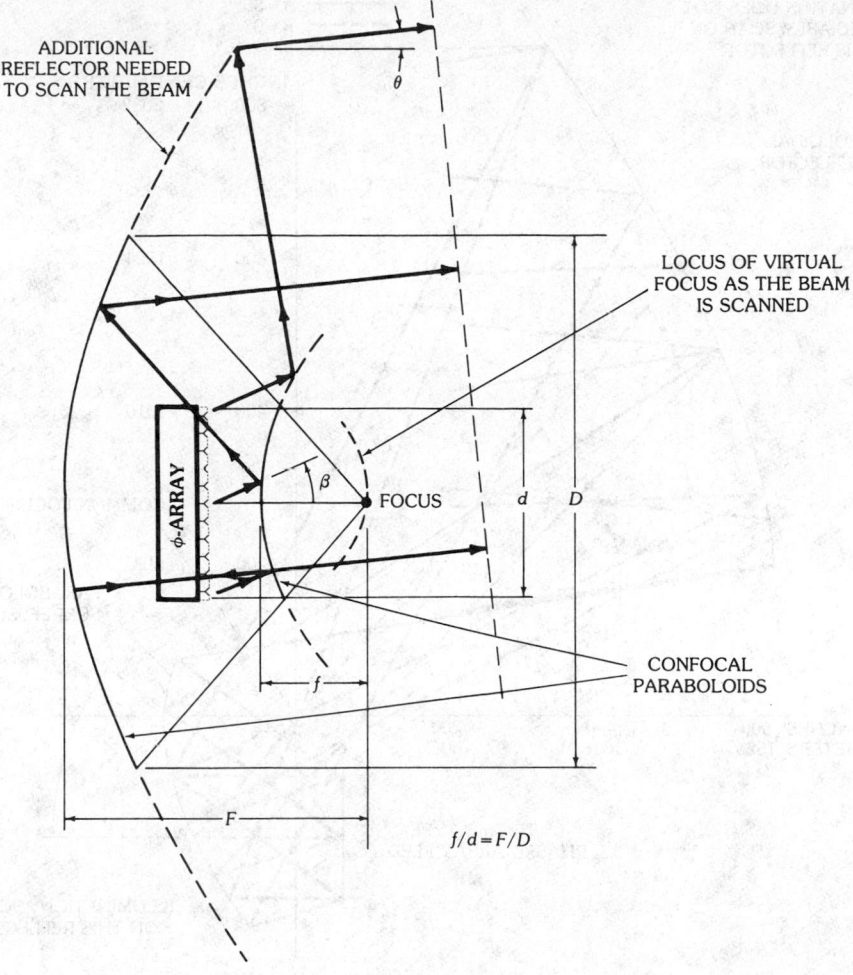

Fig. 55. Near-field Cassegrain system. (*Courtesy Hughes Aircraft Co., Fullerton, Calif.*)

infinity, when the hyperboloid subreflector becomes paraboloidal. The main-beam scan angle θ is related to the angle β by $\tan\beta = (D/d)\tan\theta$. The remotely located primary feed (at infinity) is replaced by the phased-array feed near the vertex of the main reflector with a linear phase gradient to produce rays making an angle β with respect to the z axis, which scans the main beam by the angle θ. Only linear phase gradients are required to scan the main beam; thus one has the advantage of row and column beam steering. A major disadvantage of this scheme is that the illuminated portion of the large reflector moves with beam scan, which causes inefficient usage of the large reflector aperture. This is illustrated in Fig. 55.

Lincoln Lab's Offset-Fed Gregorian System—Fig. 56 depicts a cross-sectional view of the Lincoln Lab's offset-fed Gregorian antenna system. The main reflector is a paraboloid and the subreflector is also a paraboloid with a common focus. Fig. 57

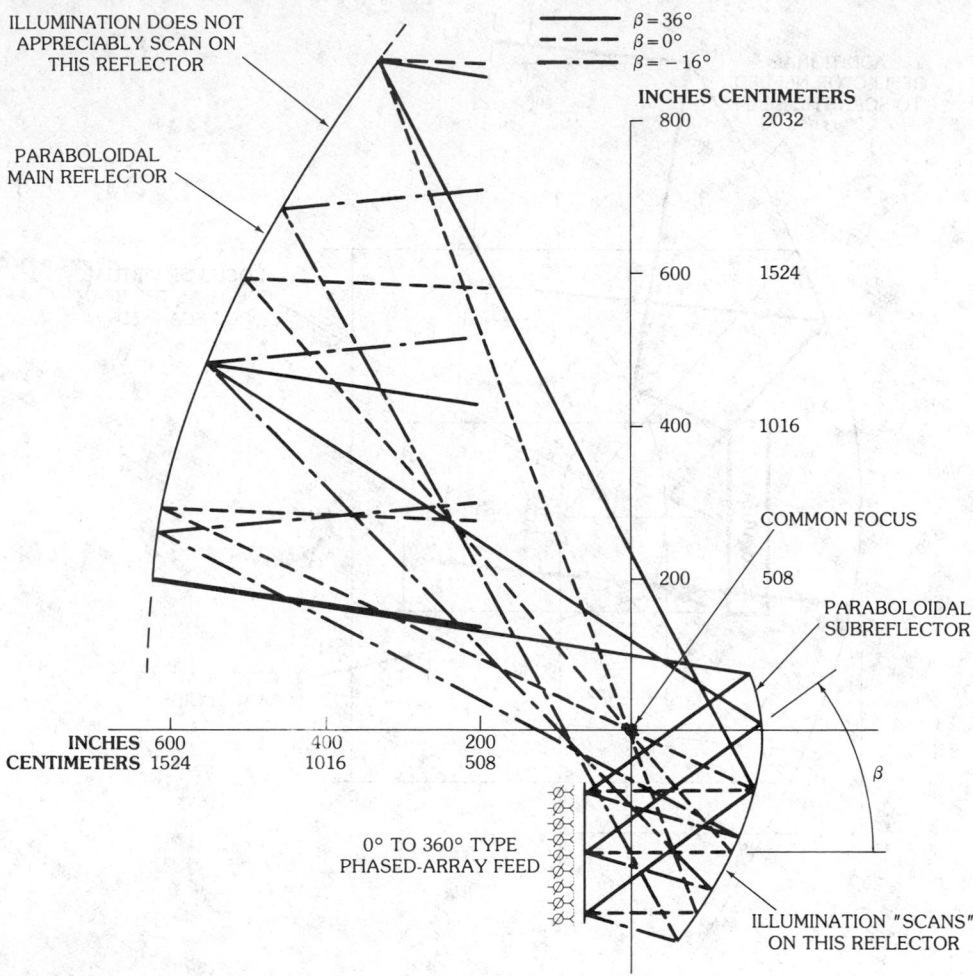

Fig. 56. Preliminary design of an offset-fed Gregorian system and ray tracing for three scan angles. (*After Interim Report [38]*)

shows an artist's concept of such a system. The primary feed for this system is a planar two-dimensional phased array using 0° to 360° type phase shifters. The basic idea here is to use two offset-fed paraboloids in such a way that the off-axis aberrations tend to cancel each other. If this could be achieved, then a small, truncated plane wave leaving the relatively small phased-array feed would be converted to a large, truncated plane wave emerging from the large main reflector. Fig. 56 illustrates ray tracings [38] for three angles β of scan for the phased array. (Also see references 39 and 40.) Although not perfectly parallel, the corresponding rays leaving the main reflector remain nearly collimated, where the scan in terms of beamwidths for the system as a whole would be about the same as the scan in terms of beamwidths for the phased-array feed alone. As can be seen from the ray tracings most of the main reflector is utilized as aperture for all β scan angles; consequently, the efficiency of this system is good for all scan angles. Mailloux and

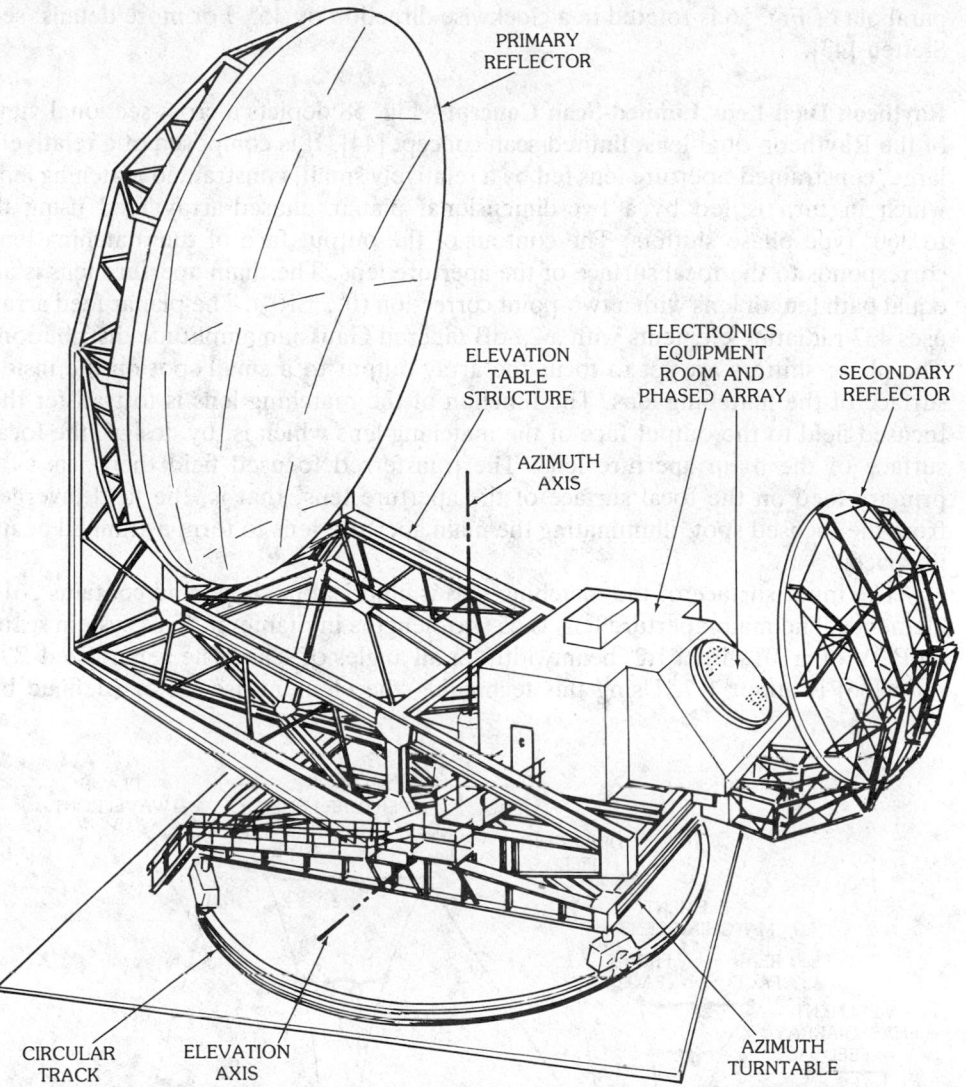

PRIMARY
REFLECTOR

ELEVATION
TABLE
STRUCTURE

ELECTRONICS
EQUIPMENT
ROOM AND
PHASED ARRAY

SECONDARY
REFLECTOR

AZIMUTH
AXIS

CIRCULAR
TRACK

ELEVATION
AXIS

AZIMUTH
TURNTABLE

Fig. 57. Physical configuration of the antenna. (*Courtesy Hughes Aircraft Co., Fullerton, Calif.*)

Blacksmith [36] cite an example of a ½°-beamwidth system being scanned 14 beamwidths using a $45\lambda \times 45\lambda$ phased-array feed. Side lobe levels were −15 to −17 dB for all scan angles. The generalized [36] F/D ratio was 1.5.

A study [41, 42] to develop a computer program that determines the optimum main reflector and subreflector contours for maximum gain for a given scan range showed that appreciable improvement over the confocal paraboloidal system is possible. For details the reader is referred to references 41 and 42.

Improved performance can be obtained by matching the caustics of the main reflector and the subreflector, arriving at a configuration in which the smaller parabola of Fig. 56 is rotated in a clockwise direction by 45°. For more details, see Sletten [43].

Raytheon Dual-Lens Limited-Scan Concept—Fig. 58 depicts a cross-sectional view of the Raytheon dual-lens, limited-scan concept [44]. It is composed of a relatively large, constrained-aperture lens fed by a relatively small, constrained matching lens which in turn is fed by a two-dimensional planar phased-array feed using 0° to 360° type phase shifters. The contour of the output face of the matching lens corresponds to the focal surface of the aperture lens. The main aperture lens is an equal path length lens with a two-point correction (0°, ±10°). The planar feed array uses 437 radiating elements with a 23-dB tapered Gaussian amplitude distribution. The phase shifters are set to focus the array output to a small spot on the inside surface of the matching lens. The function of the matching lens is to transfer the focused field to the output face of the matching lens which is, by design, the focal surface of the main aperture lens. The transferred focused field then acts as a primary feed on the focal surface of the aperture lens, that is, the field diverges from the focused spot, illuminating the main aperture lens to form a scanned beam in space.

The inner surface of the matching lens is elliptical in shape and contains 2617 elements. The main aperture lens is 65 wavelengths in diameter. This system scans an FOV of ±10° with a 1.2° beamwidth for all angles of scan. The generalized *F/D* ratio [36] is about 1.7. Using this technique, the element-use factor (defined by

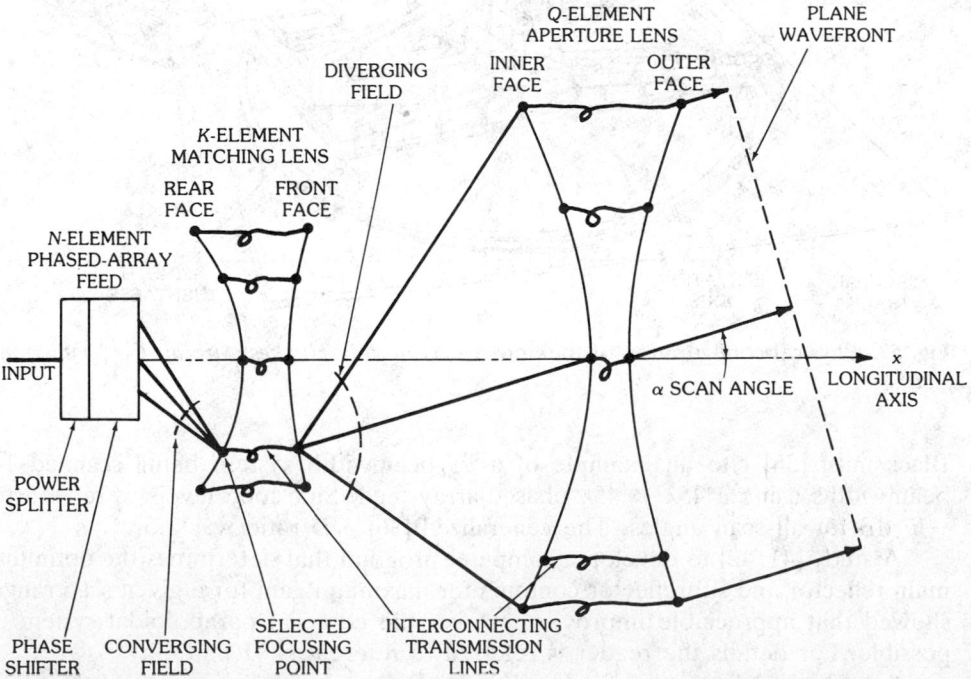

Fig. 58. Dual-lens, limited-scan concept. (*After C. H. Tang, Raytheon Co.*)

Patton [34]) is very good (about 1.5) because of the highly efficient use of the main aperture lens.

Hughes Reflector-Lens Limited-Scan Concept—Fig. 59 shows a multiple-beam constrained lens feeding an offset parabolic reflector. Ideally, points from the feed array are mapped to points on the aperture of the reflector. These points are conjugate focii of the overall optical system as shown by ray tracing, say, from feed point A to aperture point A'. Other points, such as B to B', are likewise mapped. This mapping is not perfect due to the optical aberrations of the system but is adequate for $\pm 10°$ field of view. It is well known [45] that the focal spot of con-

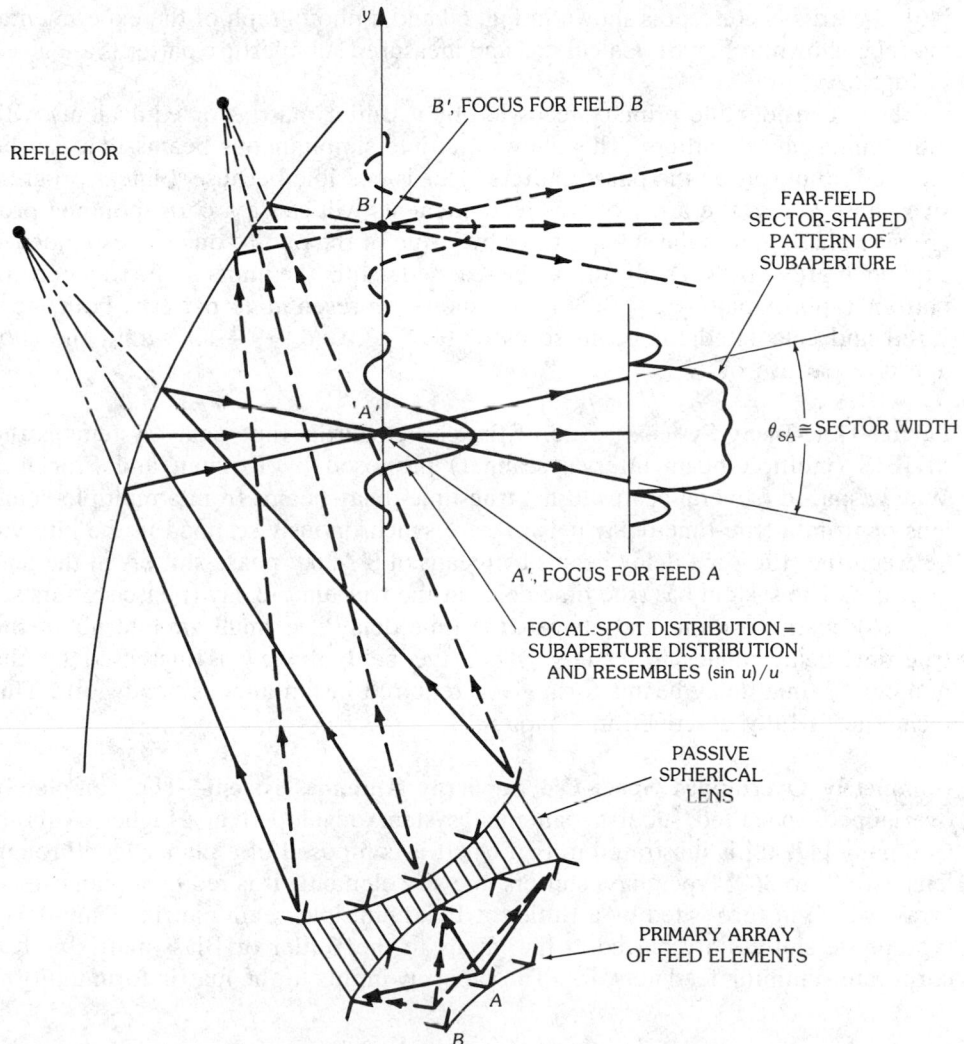

Fig. 59. Idealized geometrical mapping of rays from feed to final aperture. (*Courtesy Hughes Aircraft Co., Fullerton, Calif.*)

verging or diverging waves uniform over a sector has a focal plane distribution like $J_1(u)/u$ for spherical waves and like $(\sin u)/u$ for cylindrical waves. These distributions form overlapping $(\sin u)/u$-like subapertures that can be made orthogonal, or nearly so, by the proper amount of overlap, which is controlled by primary feed spacing along the feed arc. To be truly orthogonal, $(\sin x)/x$ distributions must be all of equal width and crossover such that the peak of any one distribution occurs at the nulls of all others. Proper radial positioning of the feed is made so that the conjugate focal point falls on the aperture plane. The $(\sin u)/u$-like subaperture distributions give rise to sector-shaped far-field patterns whose width is approximately that of the geometrical angle subtended by the rays focused at the aperture, as shown in Fig. 59.

An experimental two-dimensional parallel-plate model was built and tested [46]. An artist's concept is shown in Fig. 60 and a photograph of the experimental model is shown in Fig. 61. Calculated and measured subaperture patterns are given in Fig. 62.

Now consider the primary feeds fed by a Butler matrix (or equivalent) with intervening phase shifters. This allows multiple simultaneous beams that can be scanned in unison by the phase shifters. This is possible because a linear progressive phase across the array of lens feed elements will cause a corresponding progressive phase across the subapertures by virtue of the one-to-one correspondence explained previously. The beam can be scanned within the limits of the subaperture pattern (approximately $\pm 10°$). Fig. 63 shows representative patterns both measured and calculated for beam scans of $0°$, $-5°$, and $-9°$—all within the subaperture pattern of $10°$.

Partial-Time-Delay Systems—One of the earlier partial-time-delay systems is the MUBIS (multiple-beam interval scanner) proposed by Rotman and Franchi.* Widely spaced (several beamwidths) true-time-delay beams from a multiple-beam lens or from a true-time-delay network are synchronously scanned in the interval between the true-time-delay beams by means of $0°-360°$ phase shifters in the lens aperture. The system has true time delay at the true-time-delay (phase shifters set to zero) beam positions and almost true time delay if a small amount about the true-time-delay angles is scanned. Thus the field of view is increased by the number of time-delay beams for a given required instantaneous bandwidth. This technique is fully described in Chapter 13.

Completely Overlapped Space-Fed Subarray Antenna System—The completely overlapped space-fed subarray antenna system developed at Hughes Aircraft Company [47, 48] is illustrated in Fig. 64. It is composed of a planar feedthrough lens with $0°$ to $360°$ type phase shifters in every element. It is fed by a planar-feed array, which in turn is fed by a Butler or Blass multiple-beam matrix. Time-delay type phase shifters are placed at the inputs to the Butler or Blass matrix with a corporate summing feed network. The subarray inputs to the matrix form uniform

---

*See Chapter 13, Section 3, under the heading "Broadband Array Feeds With Time-Delayed Offset Beams."

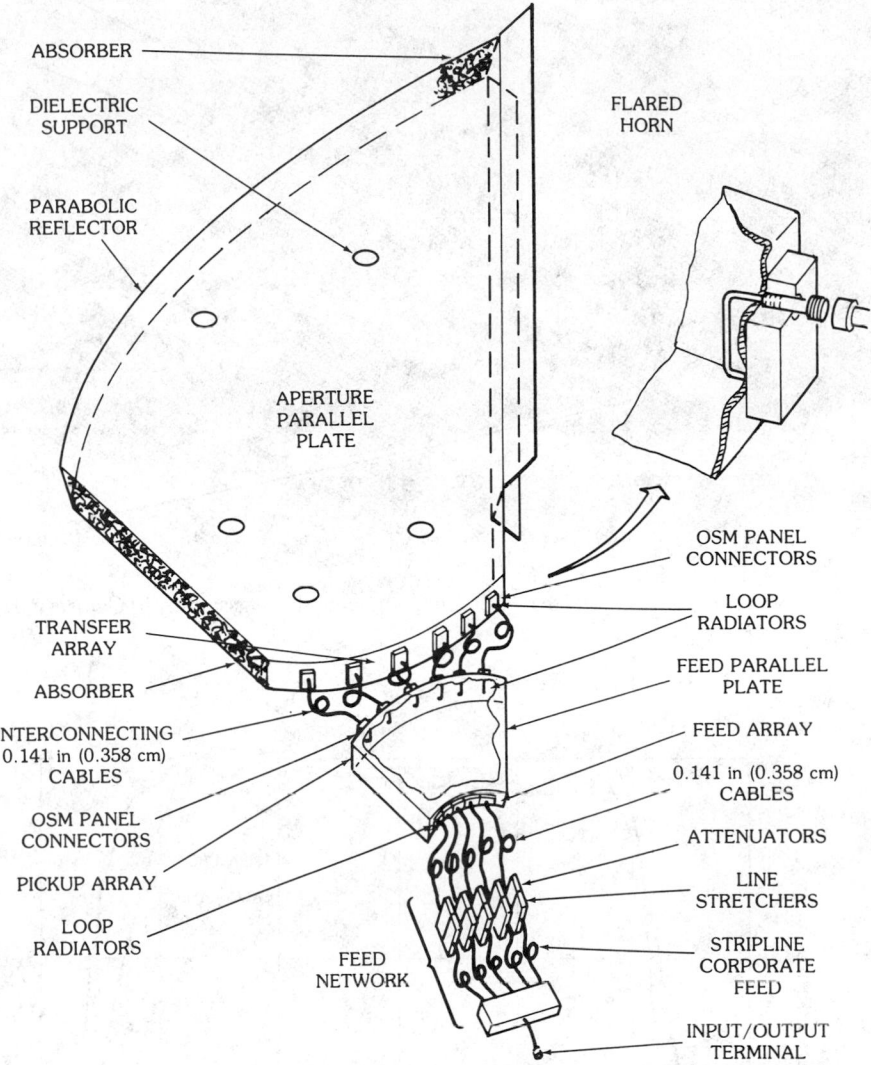

Fig. 60. Artist's concept of the experimental reflector-lens, limited-scan antenna.

distributions at the output of the feed array, which in turn form an orthogonal set of $(\sin u)/u$ type primary radiation patterns that illuminate the pickup array. Except for the slight amount of inverse distance-squared power decay, the emerging distributions from the radiating array form a near-orthogonal set of amplitude subarray distributions which completely overlap each other over the entire radiating aperture face as shown in Fig. 64. The subarray far-field radiation patterns form near-rectangular patterns superimposed in space but whose phase centers are displaced across the face of the radiating aperture. In order to minimize gain loss as a function of scan and to apply grating lobes over a wide range of scan angles and wide bandwidth, the ideal subarray pattern would be rectangular in shape but narrower than the angular separation between the first grating lobes [47].

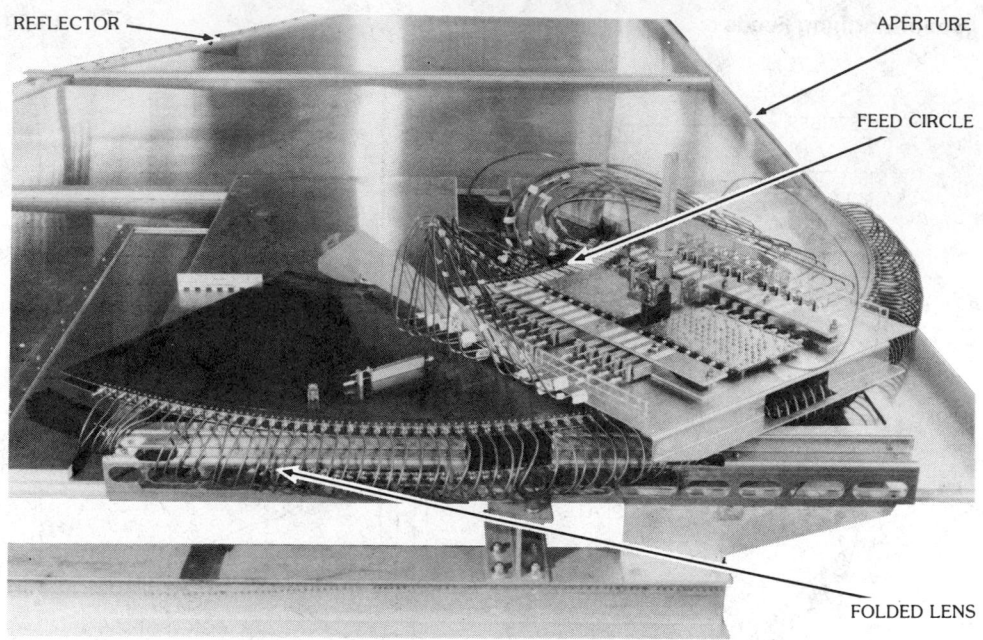

Fig. 61. Reflector-lens, limited-scan antenna, showing lens assembly. (*Courtesy Hughes Aircraft Co., Fullerton, Calif.*)

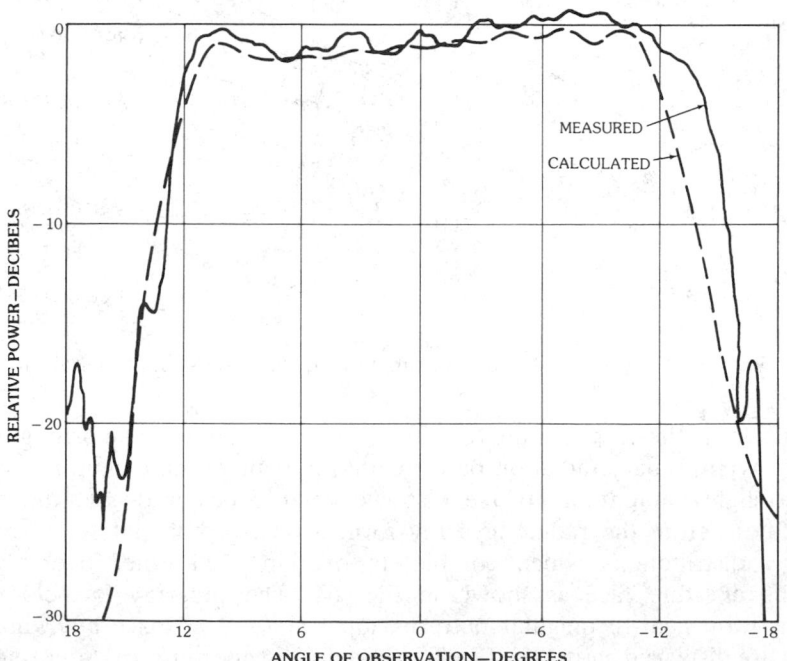

Fig. 62. Subaperture far-field patterns. (*a*) Subaperture centered at $y = 0$. (*b*) Subaperture centered at $y = -11.5\lambda$. (*c*) Subaperture centered at $y = -23\lambda$. (*Courtesy Hughes Aircraft Co., Fullerton, Calif.*)

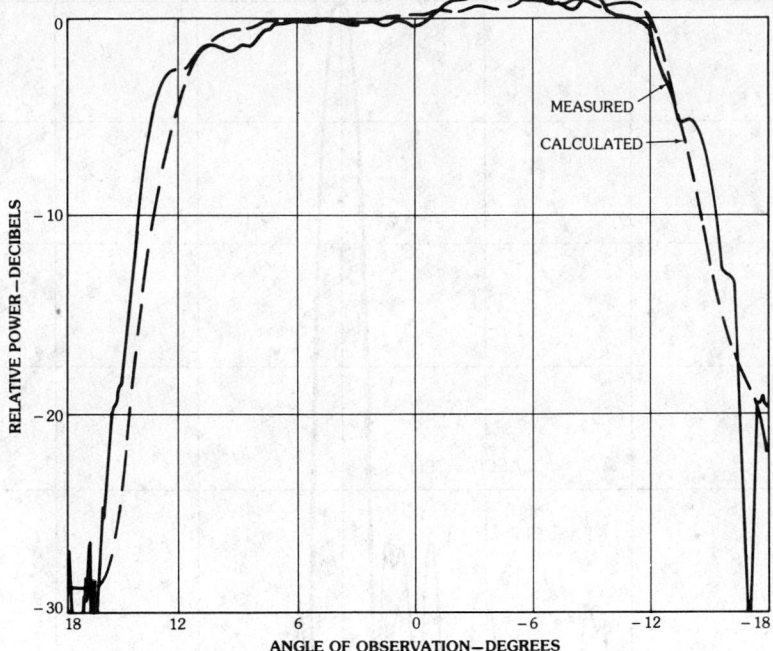

b

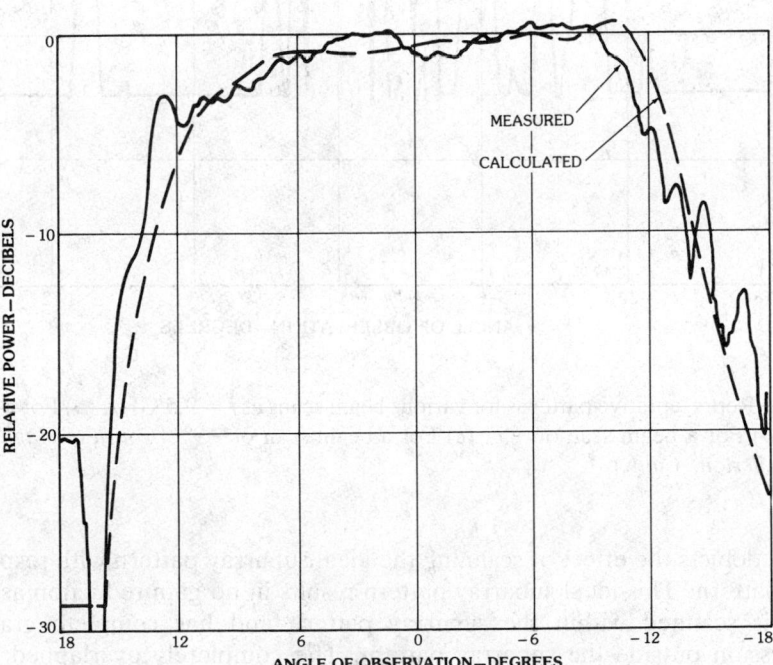

c

Fig. 62, *continued.*

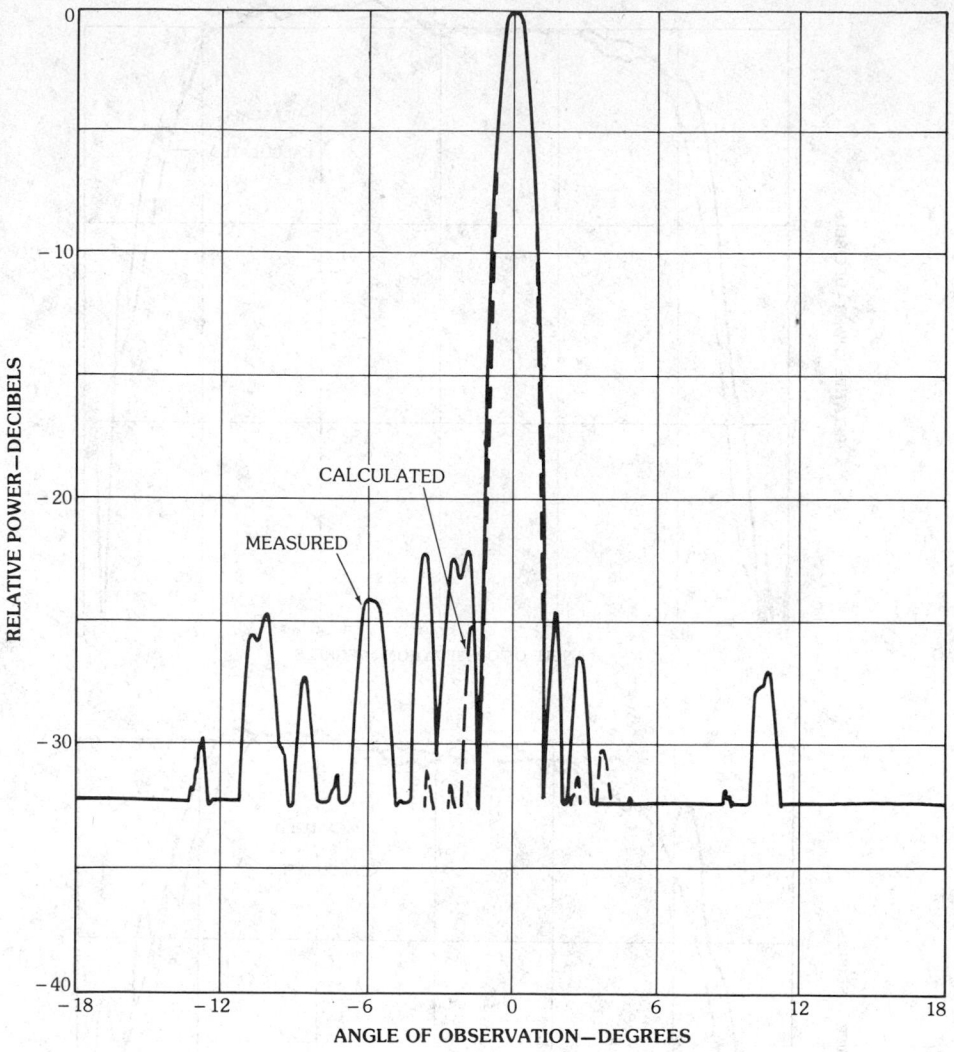

a

Fig. 63. Representative patterns for various beam scans at $f = 9.5$ GHz. (*a*) For a beam scan of $0°$. (*b*) For a beam scan of $-5°$. (*c*) For a beam scan of $-9°$. (*Courtesy Hughes Aircraft Co., Fullerton, Calif.*)

Fig. 65 depicts the effect of scanning the ideal subarray pattern with respect to the array pattern. This ideal subarray pattern results in no gain reduction as the main beam is scanned within the subarray pattern and has complete grating lobe suppression outside the subarray pattern. The completely overlapped space-fed subarray technique achieves a close approximation to the ideal subarray pattern. The amplitude distribution over the radiating aperture for each subarray input terminal is a discrete truncated $(\sin x)/x$ function which gives an approximate rectangular subarray pattern.

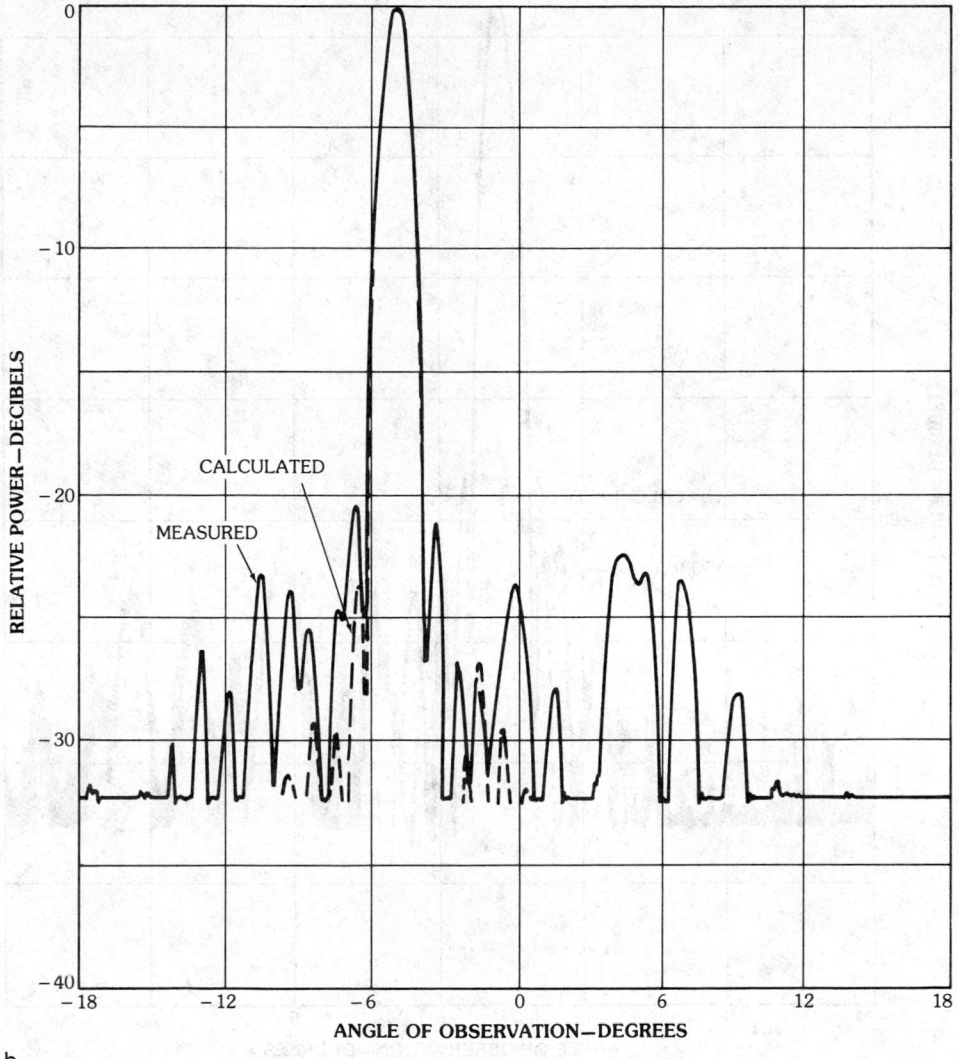

b

Fig. 63, *continued.*

Completely overlapped subarrays using transform networks are described in Chapter 13.

DuFort's Optical Technique—DuFort's optical technique [49] for broadbanding a phased array is illustrated in Fig. 66. It is composed of a relatively large feed-through aperture lens with 0° to 360° type phase shifters, an intermediate passive lens with fixed phase shifters and fixed time delayers, a small feed array with variable phase shifters and variable time delayers, and a summing network.

The aperture lens and the feed lens have a common focus, with a magnification $m = F/f$.

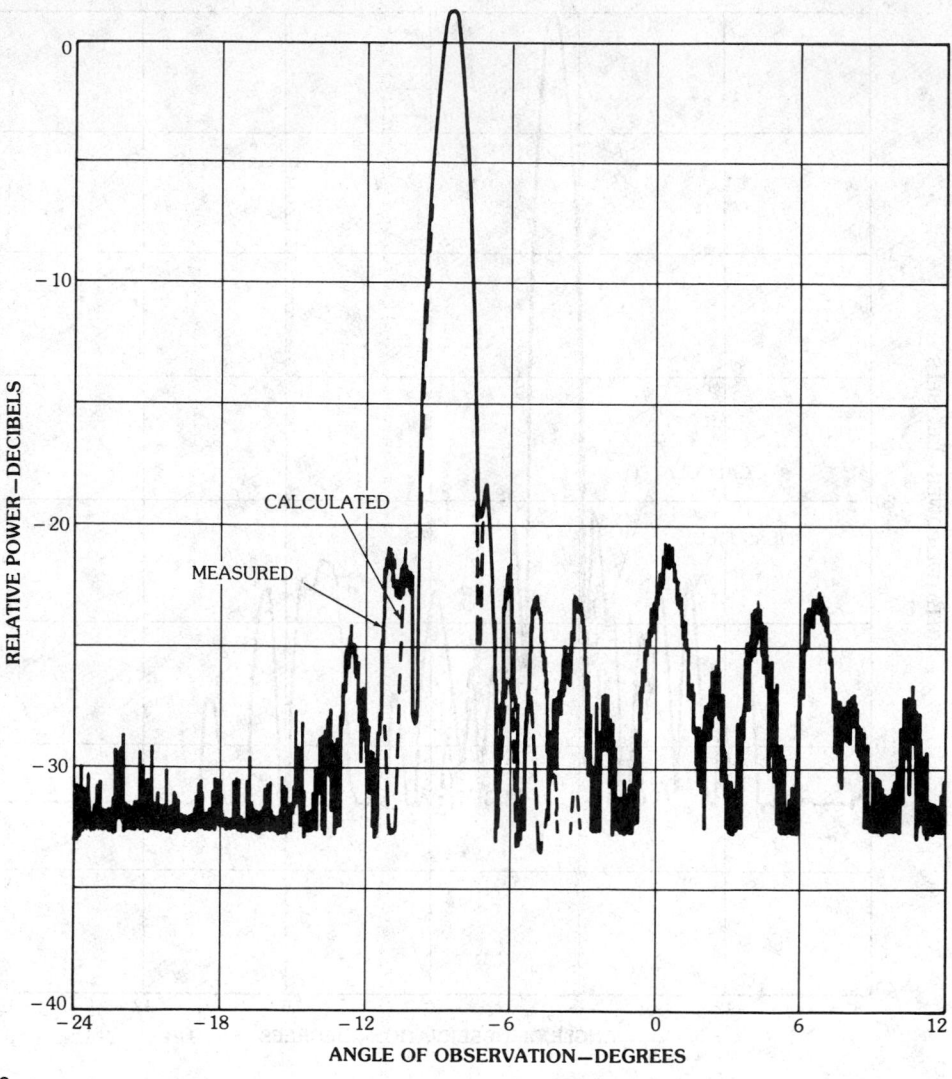

Fig. 63, *continued.*

The feed array is located at a distance *h* behind the confocal feed lens as shown.

The operation of this lens technique is as follows: A plane wave is incident on the lens aperture* x_1, making some general angle θ_0 with respect to the aperture lens normal. The 0° to 360° type phase shifters in the aperture lens are set such that at center frequency the received rays are focused to the center-frequency focal point as shown. These focused rays then diverge and are incident as a spherical

*Note that x_1 is an axis or aperture, not a specific point.

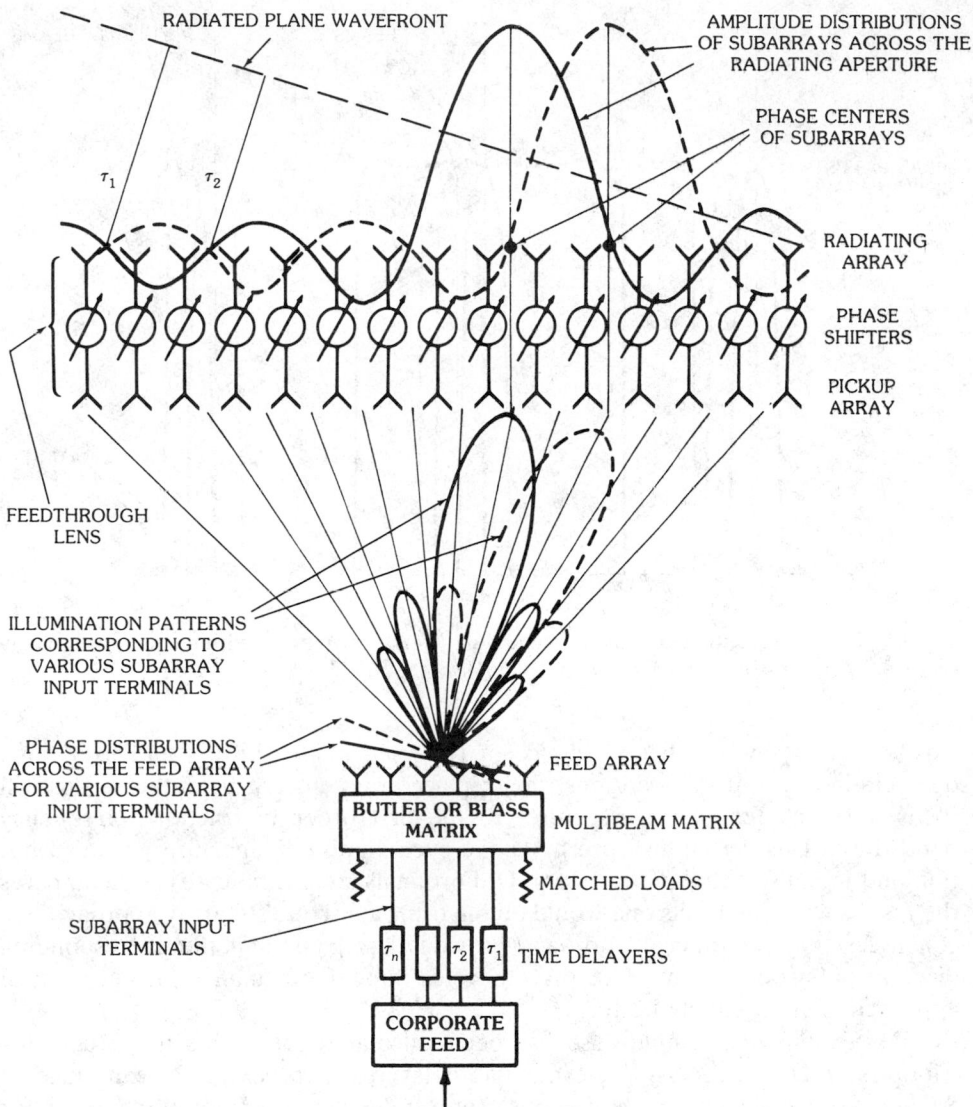

Fig. 64. Completely overlapped subarray pattern antenna system. (*After Tang [47]*)

phase front on the confocal feed lens. The fixed phase shifters and fixed time delayers in the confocal feed lens are such that the divergent incident rays are collimated and normal to the feed lens aperture or the x_2 axis. The feed array then receives these normally incident rays and coherently sums them through the appropriate settings of variable phase shifters and variable time delayers. Off the center frequency the aperture-lens phase shifter settings cause the received plane wave to be defocused from the center-frequency focal point and shifted laterally to an off-frequency focal region.

The divergent rays incident on the confocal feed lens are displaced along the x_2

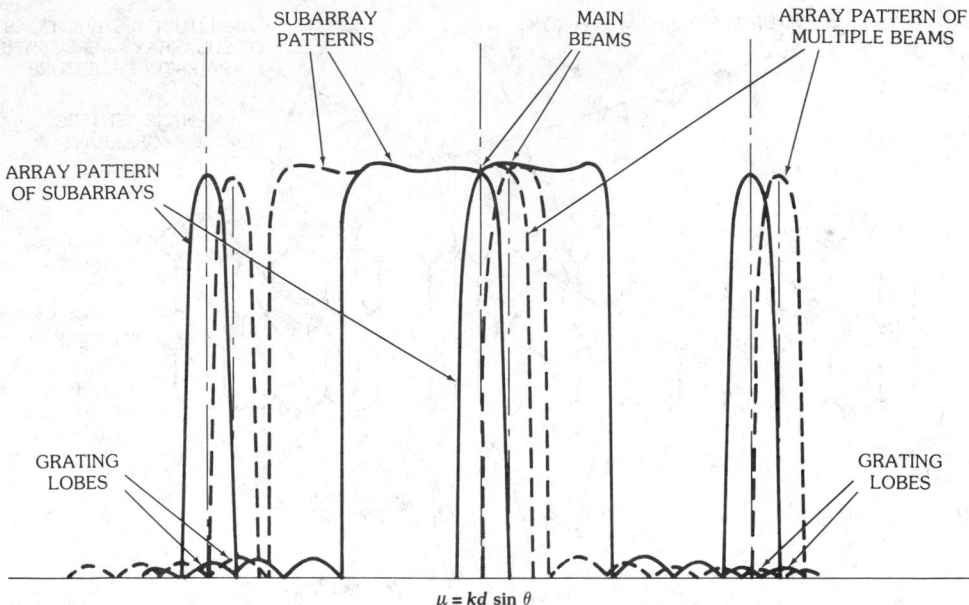

Fig. 65. Effect of scanning the ideal "rectangular" subarray pattern with respect to the array pattern. (*Courtesy Hughes Aircraft Co., Fullerton, Calif.*)

axis; however, by a judicious choice of the fixed phase shifters and fixed time delayers in the confocal feed lens, the displaced rays emerge from the x_2 axis still focused to the feed array aperture, to the first order in fractional frequency bandwidth. This design thus preserves the even-amplitude aperture lens distribution and minimizes the spillover loss. DuFort analyzes several cases and compares the results with the results one would obtain using a discrete subarray approach. He concludes that this optical approach provides less gain than the corresponding discrete subarray system, but that the side lobe performance of the optical approach is substantially better.

For example, a sample case has been calculated for cosine amplitude distribution, $F/D = 1/\sqrt{2}$, using eleven time delayers, at the extreme scan angle of 50° and at a frequency 5 percent above center frequency. The radiation pattern for the optical case is given by Fig. 67. The normalized gain is down about 1.7 dB from the maximum value of 0.816 due to cosine tapering. The far-out side lobe level is −18.5 dB, with other side lobes mostly below −25 dB. Spillover loss is only 1.1 percent. The gain loss is due to off-axis phase error that causes beam broadening. The corresponding case for the discrete subarray is shown in Fig. 68. The main beam is narrower than the optical case and the gain is down only 0.7 dB, but the grating lobe level is up to −9.9 dB, with several above −20 dB.

For more details the reader is referred to DuFort [49].

Wide Field of View True-Time-Delay Antenna Systems

Most unconstrained feeds that offer true-time-delay performance are optical in nature. The transit time between the incident wavefront, from any angle within the

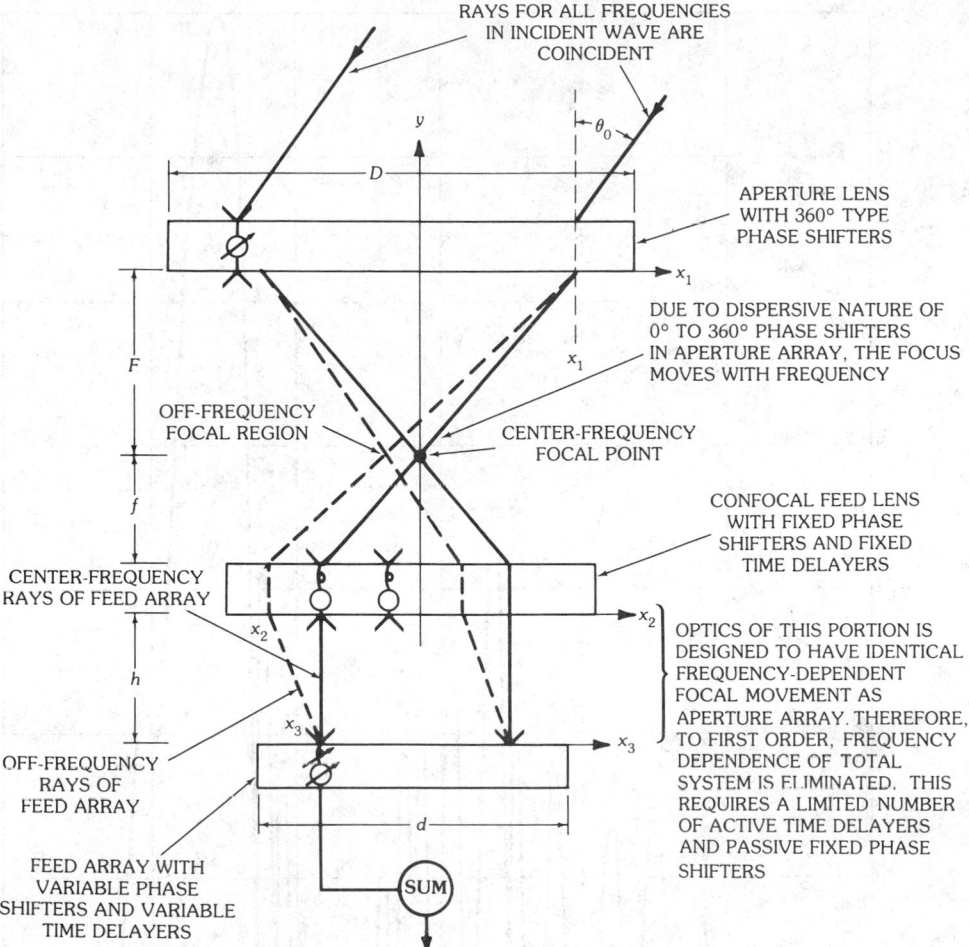

RAYS FOR ALL FREQUENCIES
IN INCIDENT WAVE ARE
COINCIDENT

θ_0

D

y

APERTURE LENS
WITH 360° TYPE
PHASE SHIFTERS

x_1

DUE TO DISPERSIVE NATURE OF
0° TO 360° PHASE SHIFTERS
IN APERTURE ARRAY, THE FOCUS
MOVES WITH FREQUENCY

F

x_1

OFF-FREQUENCY
FOCAL REGION

CENTER-FREQUENCY
FOCAL POINT

f

CONFOCAL FEED LENS
WITH FIXED PHASE
SHIFTERS AND FIXED
TIME DELAYERS

CENTER-FREQUENCY
RAYS OF FEED ARRAY

x_2

x_2

h

OPTICS OF THIS PORTION IS
DESIGNED TO HAVE IDENTICAL
FREQUENCY-DEPENDENT
FOCAL MOVEMENT AS
APERTURE ARRAY. THEREFORE,
TO FIRST ORDER, FREQUENCY
DEPENDENCE OF TOTAL
SYSTEM IS ELIMINATED. THIS
REQUIRES A LIMITED NUMBER
OF ACTIVE TIME DELAYERS
AND PASSIVE FIXED PHASE
SHIFTERS

x_3

x_3

OFF-FREQUENCY
RAYS OF
FEED ARRAY

d

SUM

FEED ARRAY WITH
VARIABLE PHASE
SHIFTERS AND VARIABLE
TIME DELAYERS

Fig. 66. Components of optical broadband phased array. (*Courtesy Hughes Aircraft Co., Fullerton, Calif.*)

FOV, to the corresponding collecting feed is a constant, regardless of the actual path that is taken for any given ray. Consequently, an extremely short time pulse, approaching a delta function, could, in principle, be transmitted or received by such an antenna system without distortion or spreading in the time domain. In practice, however, the instantaneous bandwidth is determined by the components that are used in devising such a system. If waveguides are used, or other types of elements, they will not possess infinite bandwidth when impedance matching or grating lobe phenomena are addressed. If a switching matrix is used to switch between beams, it will function only over some finite bandwidth. Thus the instantaneous bandwidth limitation of a true-time-delay system will not be established by the optics of the system but rather by the bandwidth of the components used in implementing it. For example, all parabolic reflector systems and dielectric lenses that are not zoned are true-time-delay antenna systems. Their

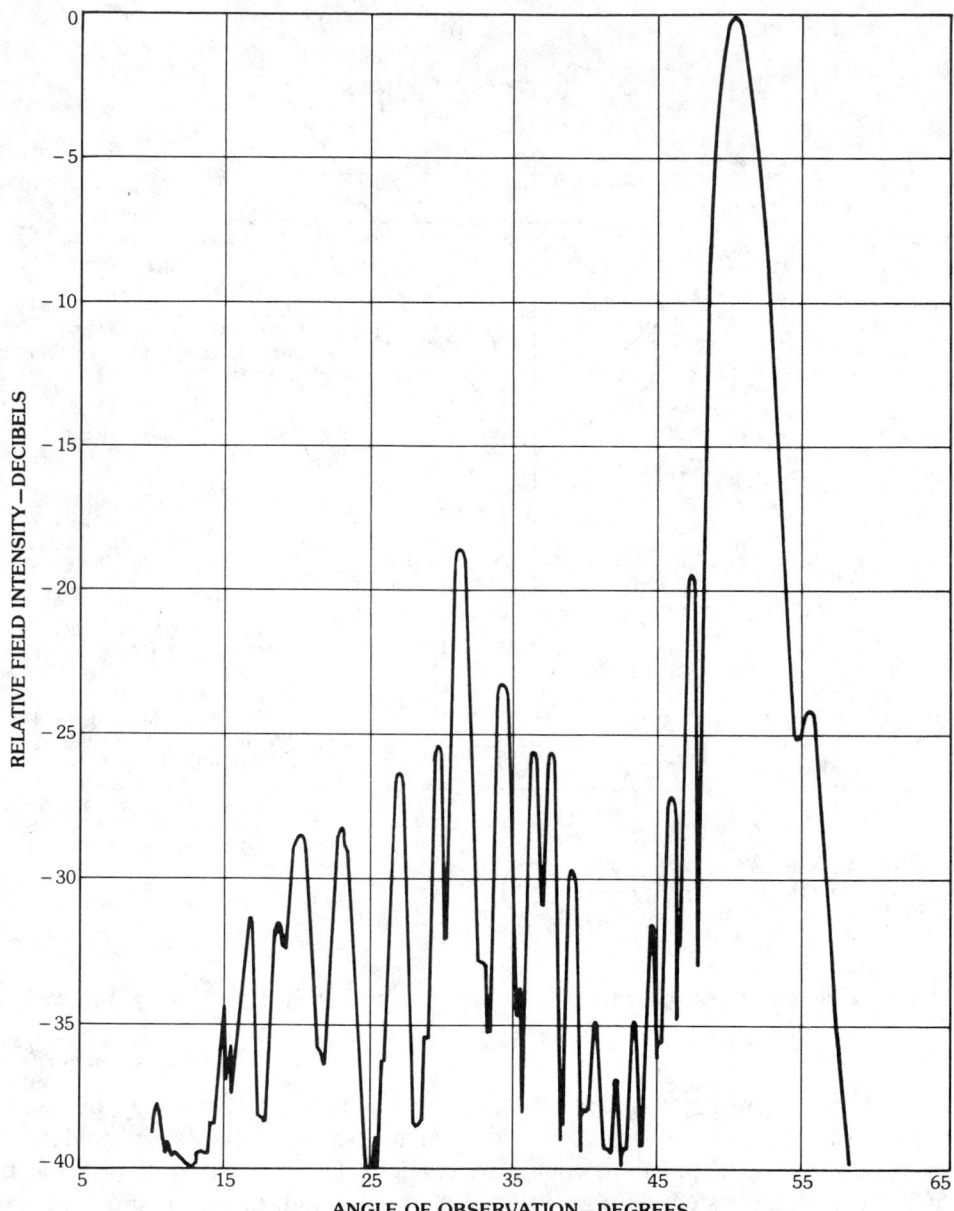

Fig. 67. Optical system radiation pattern. (*Courtesy Hughes Aircraft Co., Fullerton, Calif.*)

bandwidth, both instantaneous and tunable, is determined by the bandwidth of the feed system, or, perhaps to some extent in the case of dielectric lenses, the bandwidth may be determined by the impedance-matching transformers on the lens surfaces, as well as the primary feed(s).

Conventional monopulse feeds for conventional reflectors and lenses are well documented in the literature and will not be covered here, since most of these

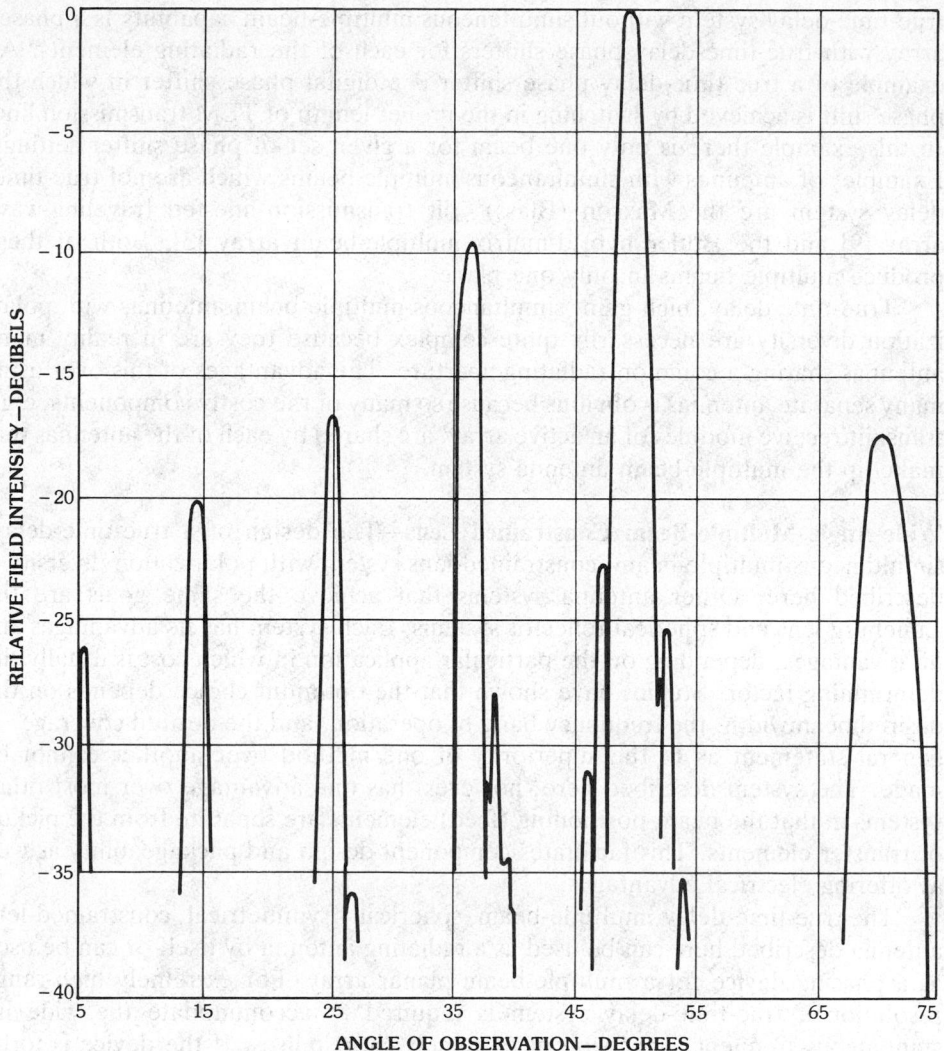

Fig. 68. Discrete subarray radiation pattern. (*Courtesy Hughes Aircraft Co., Fullerton, Calif.*)

antenna systems do not offer wide-angle performance. The purpose of this subsection is to consider and illustrate some true-time-delay approaches that can be used to feed and phase a planar array over a wide angular FOV, e.g., a 90° to 120° conical FOV. Since this section is on unconstrained feeds it may seem out of place to discuss systems that are partially constrained; however, if free space exists between the feeds and the final aperture, then they are considered unconstrained in this context.

True-time-delay antennas can be further subdivided into those that have simultaneous multiple-beam capability and those that do not. Simultaneous multiple-beam capability is important for high-data-rate systems. An example of a true-time-delay system without simultaneous multiple-beam capability is a phased array with true-time-delay phase shifters for each of the radiating elements. An example of a true-time-delay phase shifter is a digital phase shifter in which the phase shift is achieved by switching in the proper length of TEM transmission line. In this example there is only one beam for a given set of phase shifter settings. Examples of antennas with simultaneous multiple beams which are not true-time-delay system are the Maxson (Blass)* tilt transmission-line-fed traveling-wave array [9] and the Butler hybrid-matrix multiple-beam array [3]. Both of these produce multiple beams in only one plane.

True-time-delay, high-gain, simultaneous-multiple-beam antennas with polarization diversity are necessarily quite complex because they are in reality many antennas sharing a common radiating aperture. The advantages of this over using many separate antennas is obvious because so many of the costly components, e.g., transmit/receive modules of an active array, are shared by each of the antennas that make up the multiple-beam antenna system.

Wide-Angle Multiple-Beam Constrained Lens—The design of a true-time-delay, simultaneous-multiple-beam, constrained-lens system with polarization diversity is described here. Other antenna systems that achieve the same goals are the Luneburg lens and spherical reflector systems. Each system has its advantages and disadvantages, depending on the particular application in which cost is usually the determining factor. Studies have shown that the optimum choice depends on the desired beamwidth, the frequency band of operation, and the desired coverage. A general statement as to the superiority of one method over another cannot be made. The system described here, however, has one advantage over most other systems in that the beam-positioning (feed) elements are separate from the pickup or transfer elements. This facilitates component design and packageability as well as offering electrical advantages.

The true-time-delay, multiple-beam, spherically symmetrical, constrained-lens antenna described here can be used as a radiating antenna by itself or can be used as a phasing device for a multiple-beam planar array. For extremely high-range resolution a true-time-delay system is required to accommodate the wide instantaneous frequency spectrum of extremely short pulses. If the device is to be used as a phasing device for a planar array as shown in Fig. 69, it should illuminate the array aperture with the same amplitude distribution for all beam positions. That is, the illuminated portion of the aperture should not scan as the beam is scanned. Concentric (spherical) lenses or reflector systems, such as the hemispherical reflector system shown in Fig. 69, automatically satisfy the nonscanning-aperture requirement because the central ray for every beam passes through the center of the aperture.

Concentric systems have inherent wide-scan capability because they are spherically symmetrical, and hence the beam-forming device, by itself, has phase errors independent of scan angle. However, when the latter is used as a phasing

---

*The Maxson or Blass array can be made true time delay but in doing so, much of its appealing simplicity, low cost, and compact packaging is sacrificed. See Hansen [4], vol. 3, p. 254.

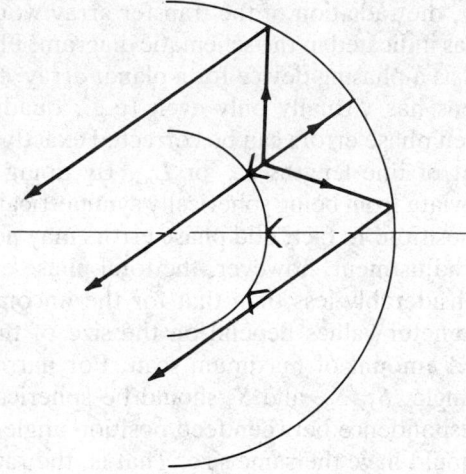

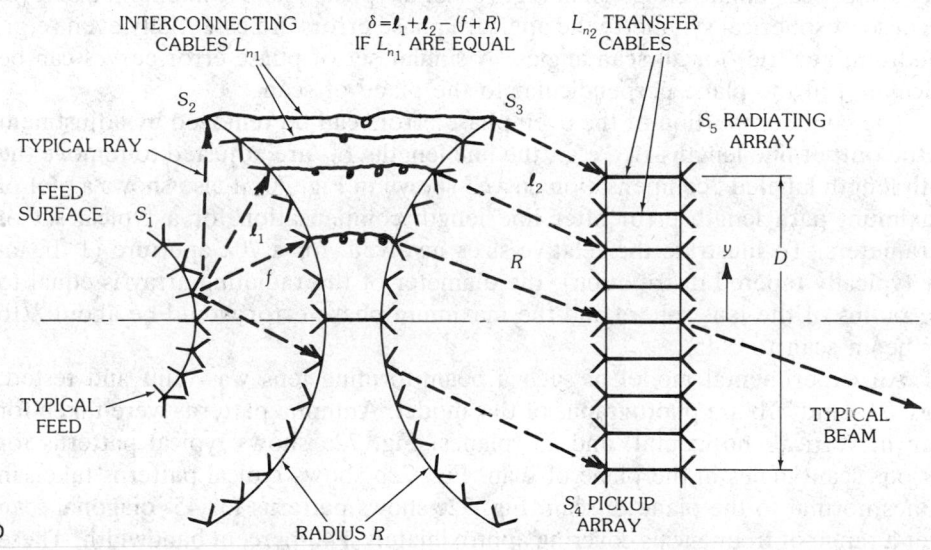

Fig. 69. Hemispherical reflector and lens counterpart. (*a*) Reflector. (*b*) Lens. (*Courtesy Hughes Aircraft Co., Fullerton, Calif.*)

device for a planar array the phase errors may actually decrease as the beam is scanned off broadside. The reason for this is that only the central portion of the phase front is intercepted by the planar array. This is true to the extent that phase errors at one plane can be projected to phase errors at another plane.

A hemsipherical lens [50] developed at Huges Aircraft Company, Fullerton, California, is the transmission (lens) counterpart to a hemispherical reflector system, as shown in Fig. 69a. If the current distribution on the reflector of the hemisphere were transferred via pickup elements with equal line lengths to an identical transfer array, the radiation of the transfer array would be identical with that from the reflector as indicated in the schematic diagram, Fig. 69b. The diagram shows such a lens used as a phasing device for a planar array. Because of spherical symmetry the basic lens has virtually only even (e.g., quadratic, quadric, etc.) phase errors. These even phase errors can be corrected exactly for any single beam position by adjustment of line lengths L_{n_1} or L_{n_2}. By doing this, however, one causes the system to deviate from being spherically symmetrical and the phase error now depends on scan position. In fact, odd phase errors may now arise for scanned beams. By line length adjustment, however, the total phase errors, both odd and even, can be made considerably less than that for the uncompensated lens.

The optimum parameter values depend on the size of the antenna in wavelengths and the desired amount of maximum scan. For narrow beamwidths and large maximum scan angle, S_1, S_2, and S_3 should be spherical.

If an angular correspondence between feed position angle and beam direction is desired, S_4 and S_5 should have the same size. That is, the ratio of the beam scan angle to feed position angle is the ratio of the linear dimensions of S_4 and S_5. With no line length phase compensation the path length error curves in the plane of scan for various scan angles are given in Fig. 70 (for a typical f/R). As mentioned before, because of spherical symmetry the aperture phase errors are essentially even (e.g., quadratic quadric) for all scan angles. A similar set of phase error curves can be calculated in the plane perpendicular to the plane of scan.

The common portion of the even phase errors can be removed by adjustment of the output line lengths L_{n_2}, e.g., the line lengths L_{n_2} are adjusted to remove the path length labeled "compensation curve" shown in Fig. 70; it also shows a plot of maximum path length error after line length compensation for a typical set of parameters. To illustrate the relative sizes involved, for a 70λ aperture ($1°$ beam for typically tapered distribution), the diameter of the radiating array is equal to the radius of the lens sphere and the maximum phase error would be about $\lambda/16$ for beam scan to $\pm 45°$.

An experimental model of such a beam-forming lens was built and tested. Figs. 71a and 71b are photographs of this model. Antenna patterns were taken for scan in vertical, horizontal, and $45°$ planes. Fig. 72a shows typical patterns for various scan angles in the plane of scan. Fig. 72b shows typical patterns taken in planes normal to the plane of scan. Fig. 72c shows patterns for $45°$ diagonal scan over a range of frequencies covering approximately a 40-percent bandwidth. These test results establish the feasibility of this type of three-dimensional, multiple-beam-forming device.

Below is a summary of the properties of the lens:

1. Wide angular coverage
2. Simultaneous multiple true-time-delay beams
3. Nonscanning aperture, hence efficient as phasing device for planar array
4. Polarization diversity capability. With dual-polarized lens elements the polarization is determined solely by the polarization of the feed
5. The number of beams and beam positions are independent of the number and locations of the transfer or pickup elements of the lens since these elements are not used for a dual purpose of feed and transfer or pickup elements

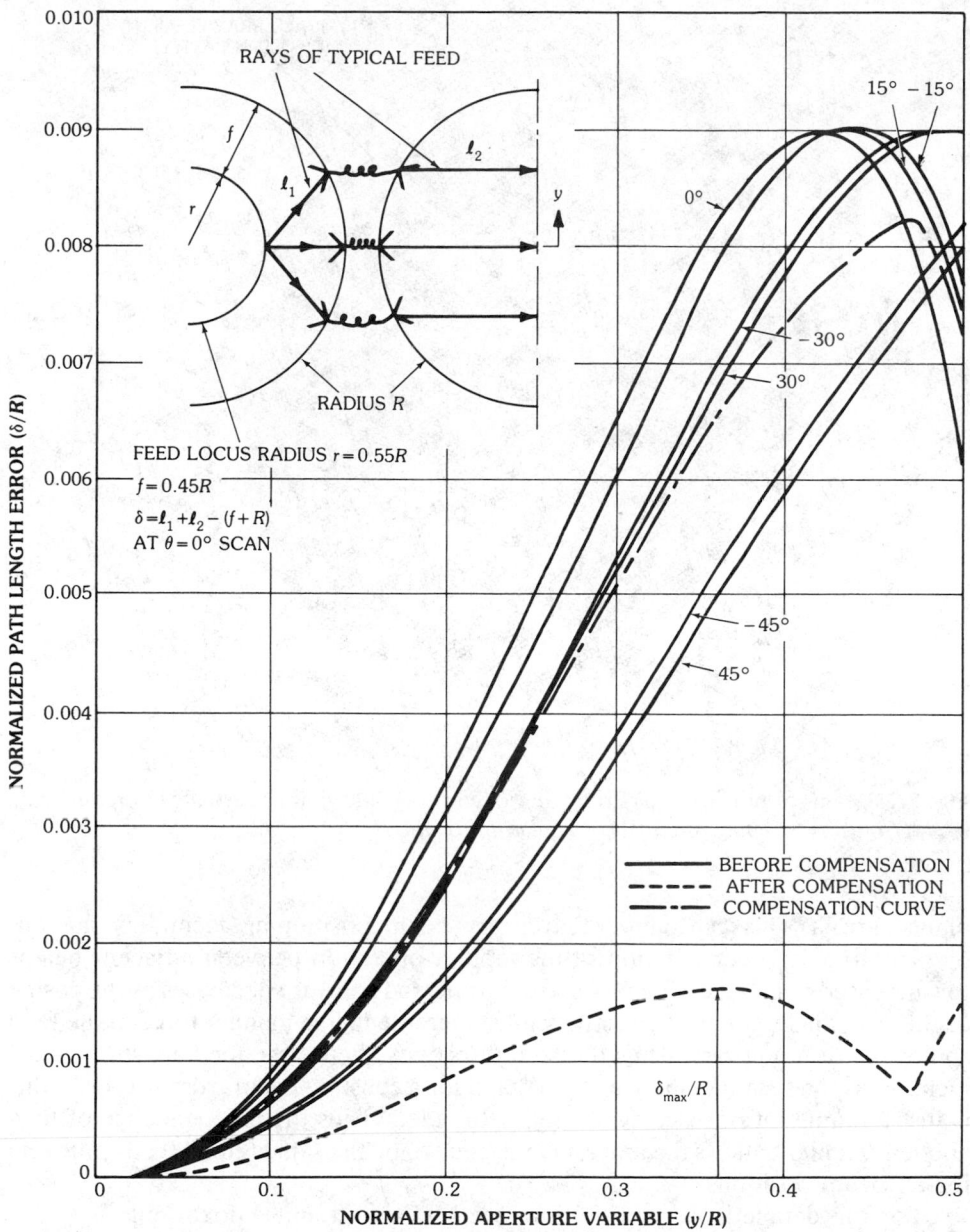

Fig. 70. Path length error versus aperture variable before compensation, and maximum path length error after compensation. (*Courtesy Hughes Aircraft Co., Fullerton, Calif.*)

A consequence of the property stated above is that the beams can be placed in any desired manner in space by proper placement of feeds on surface S_1 of Fig. 69b. For example, if the device is used to phase a planar array, the feeds can and should have variable spacing to keep a constant "crossover level" between adjacent beams since the radiated beam broadens as the beam is scanned off broadside to the

TYPICAL ARRAY ELEMENT
(DIPOLE IN CAVITY)

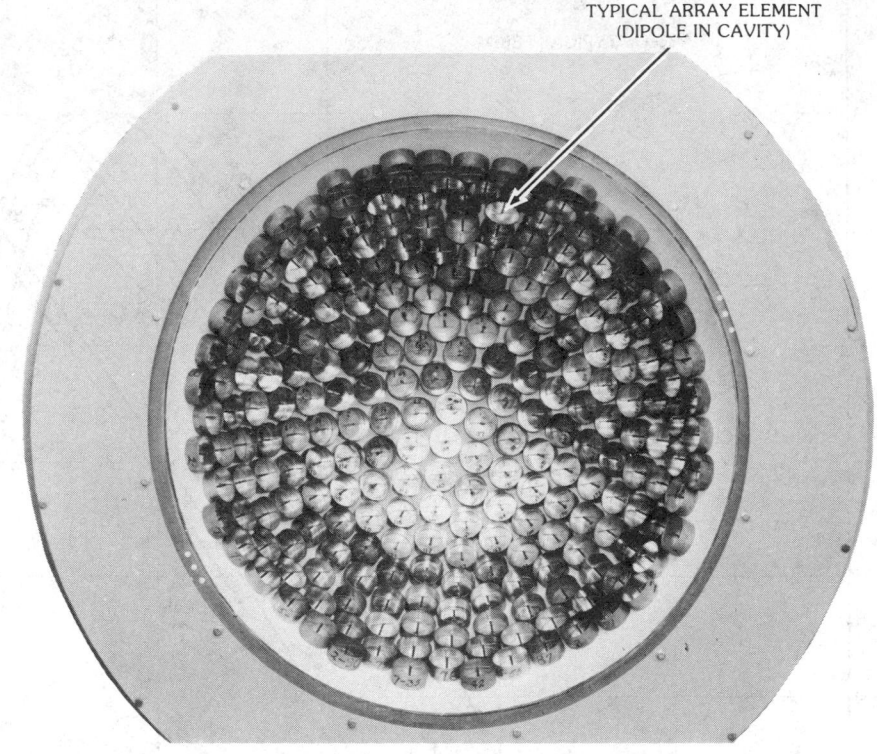

a

Fig. 71. Beam-forming lens. (*a*) Front view of lens. (*b*) Side view of partially interconnected lens. (*Courtesy Hughes Aircraft Co., Fullerton, Calif.*)

planar array. This capability of arbitrary beam positioning facilitates the implementation and computation of monopulse operation between adjacent beams for increased angular accuracy. Since the projected aperture decreases as the cosine of the scan angle, the feed directivity must increase in like manner to keep the feed spillover loss a constant. Thus the feed directivity (hence the feed aperture) must increase as the secant of the scan angle, and for a constant beam crossover level this is just the amount of space available for the feed. Thus the independence of feed and lens element allows the antenna efficiency to be the same (equal feed spillover) for all beam positions.

For the detailed design procedure see McFarland and Ajioka [50].

Modifications of the Basic Lens—Two modifications of the basic lens have been studied in an attempt to improve the aperture efficiency or to increase the usable portion of the approximate plane wave. The first scheme is similar to a Cassegrain technique, as illustrated in Fig. 73a.

This scheme slightly increases the usable portion D of the S_3 surface diameter $2R$, as is shown by Fig. 73b, where it is compared with the basic lens. However, it has the disadvantage of using the radiators covering surface S_2 for the dual purpose of acting as feeds and transfer radiators. To accomplish this function would require the appropriate circuitry in the interconnecting cables L_{n_1}.

PICKUP HEMISPHERE S_2 INTERCONNECTING LINES L_{n_1} TRANSFER HEMISPHERE S_3

b

Fig. 71, *continued.*

The second scheme is similar to a Mangin Mirror, as shown in Fig. 74. It uses a constant concentric dielectric correcting lens at the feed side of the basic lens.

This scheme appreciably increases the usable portion D of the S_3 surface diameter $2R$, as is shown by Fig. 73b, where it is compared with the basic lens.

This modification retains all of the merits of the basic lens while considerably improving the output phase.

A similar design procedure can be followed in designing the modified versions of the lens as has been used for the basic lens.

High-Resolution Hemispherical Reflector Antenna (HIHAT)—The HIHAT multiple-beam antenna system was invented [51] at Hughes Aircraft Company in the early 1960s by Louis Stark of the Ground Systems Group. Fig. 75 illustrates the

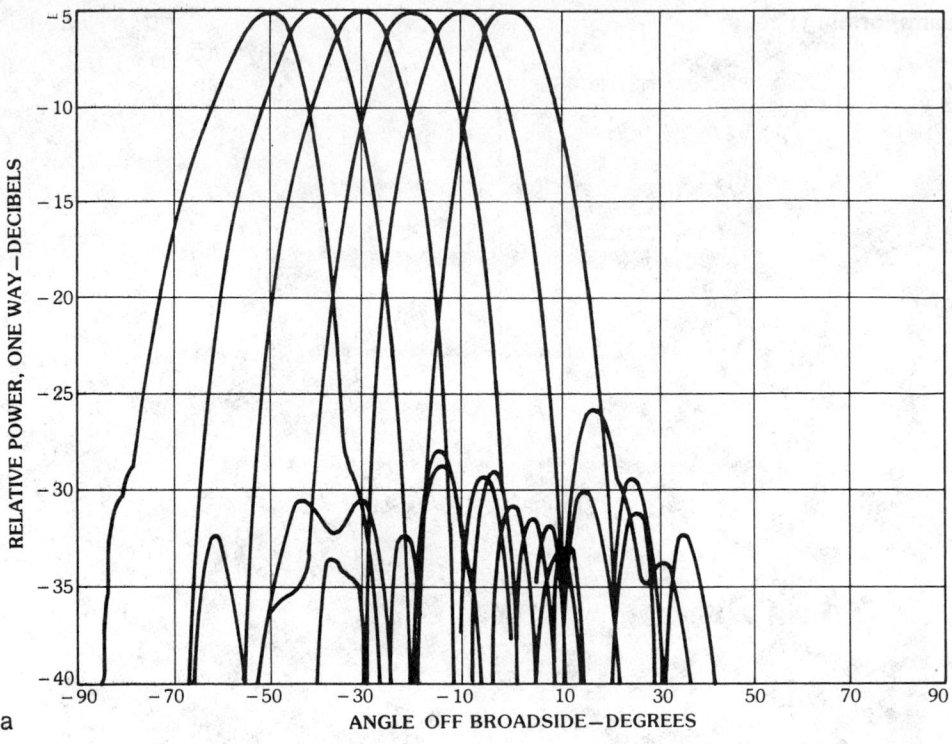

a

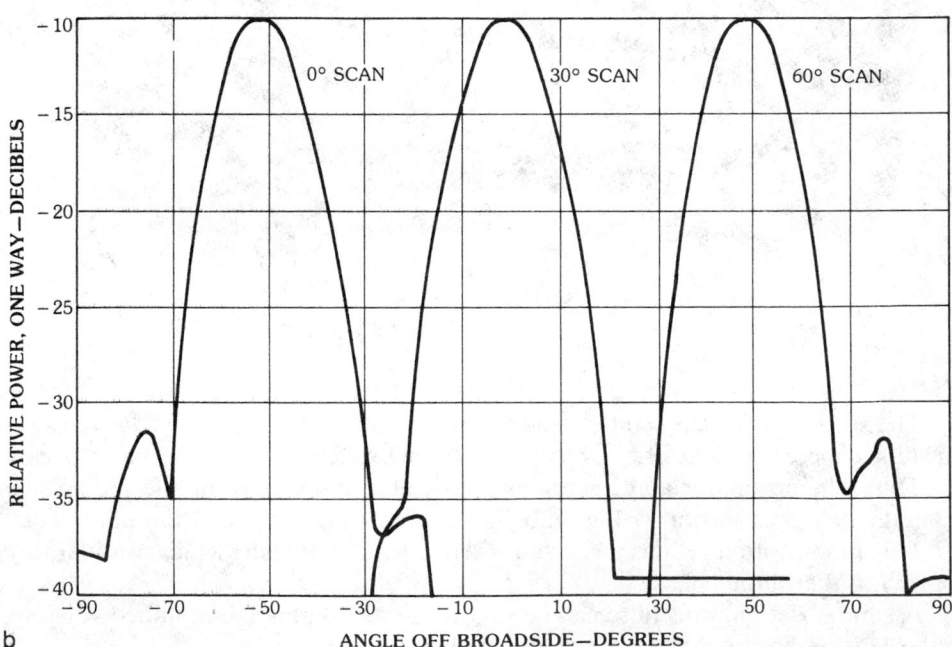

b

Fig. 72. Typical beam-forming lens antenna patterns. (*a*) In plane of scan, showing wide-scan capability. (*b*) Normal to plane of scan for various scan angles. (*c*) Over frequency range (beam scanned 45° in diagonal plane). (*Courtesy Hughes Aircraft Co., Fullerton, Calif.*)

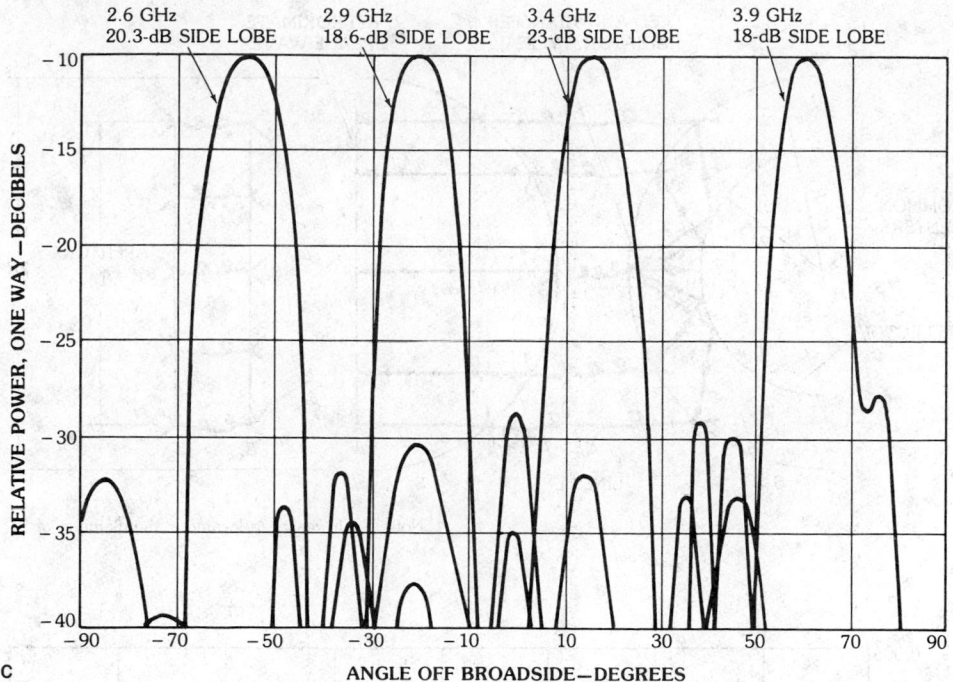

Fig. 72, continued.

basic concept. A spherical reflector is fed by a concentric array of feeds located on a sphere of slightly greater than half that of the spherical reflector. This concentric array of duplexing feeds is dual circularly polarized. This novel feature enables the aperture blockage that would ordinarily be presented by the feed array to become essentially "invisible"; thus a given feed or feed cluster is selected by a beam-switching matrix to be excited as a circularly polarized primary feed, e.g., right circular. The primary feed(s) illuminates a portion of the spherical reflector and becomes nearly collimated with the opposite sense circular polarization (e.g., left circular) after reflection off the spherical reflector. The dualpolarized duplexing feed array picks up the nearly collimated rays in the left circularly polarized channel (for this example) and transfers the nearly collimated wavefront to a spherical transfer array. The spherical transfer array is designed to have its radiators' phase centers lie on the same radius sphere as for the duplexing array. Equal line lengths are used for this transfer. From the spherical transfer array, radiation into free space could take place, or it could be used to phase a planar array, as shown in Fig. 76. In the latter case the common portion of the spherical aberration could be removed prior to radiation into free space, thereby improving the flatness of the radiated phase front. This technique was discussed in the previous section and is applicable here. Excellent radiation patterns are obtainable since the central portion of the reflected phase front is highly correctable by using variable line lengths between the pickup and radiating planar array. Fig. 77 shows typical measured Σ and Δ patterns using two adjacent feeds for primary illumination.

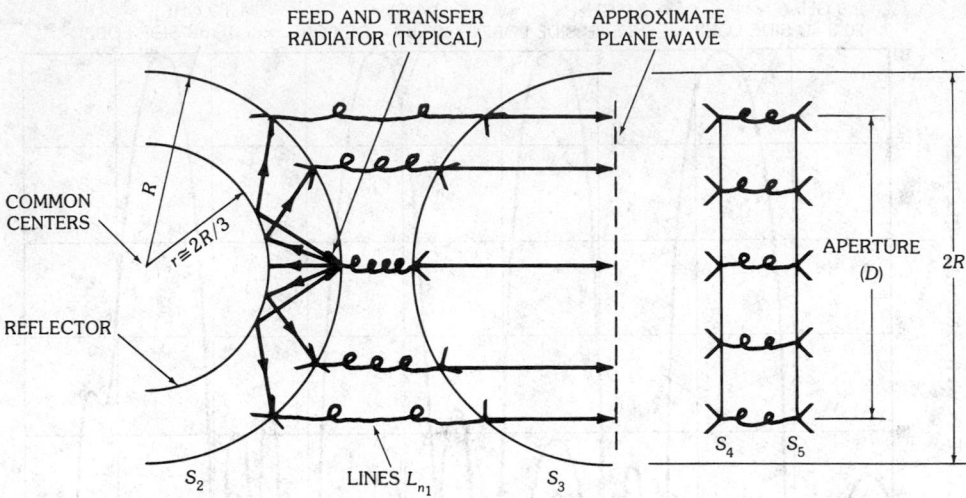

FEED AND TRANSFER
RADIATOR (TYPICAL)

APPROXIMATE
PLANE WAVE

COMMON
CENTERS

REFLECTOR

R

$r \cong 2R/3$

APERTURE
(D)

$2R$

S_2 LINES L_{n_1} S_3 S_4 S_5

Note: S_4, in reality, is located at the diameter of S_3

a

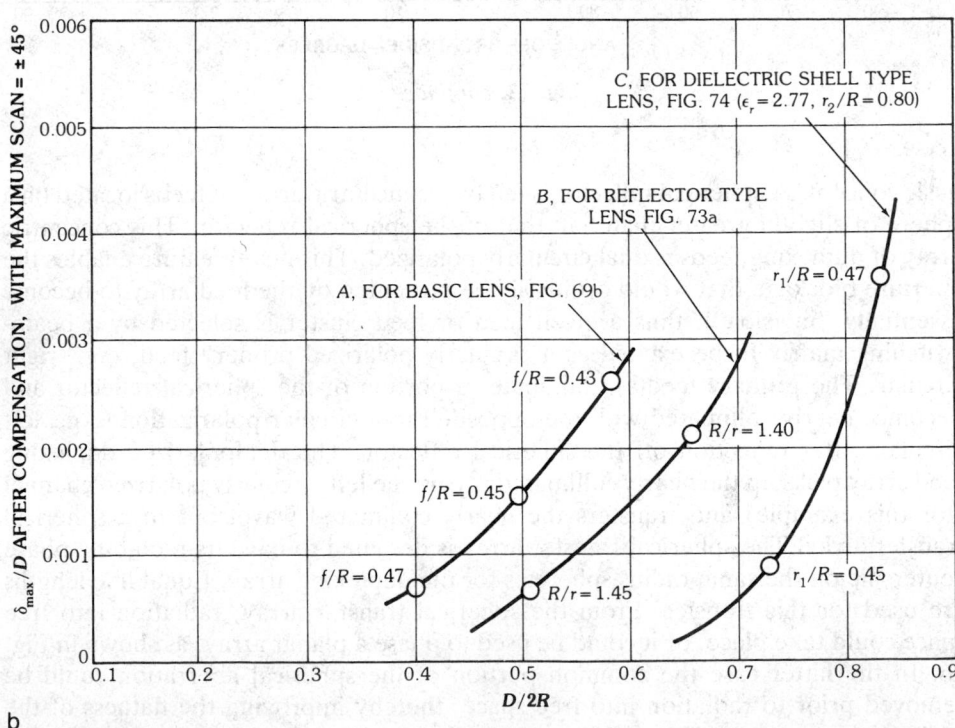

C, FOR DIELECTRIC SHELL TYPE
LENS, FIG. 74 ($\epsilon_r = 2.77$, $r_2/R = 0.80$)

B, FOR REFLECTOR TYPE
LENS FIG. 73a

$r_1/R = 0.47$

A, FOR BASIC LENS, FIG. 69b

$f/R = 0.43$

$R/r = 1.40$

$f/R = 0.45$

$f/R = 0.47$ $R/r = 1.45$ $r_1/R = 0.45$

b

Fig. 73. Cassegrain modification of basic lens. (*a*) Cassegrain reflector type variation of basic lens. (*b*) Normalized maximum path length error, after compensation, versus usable portion of the output hemisphere diameter as radiating aperture. (*Courtesy Hughes Aircraft Co., Fullerton, Calif.*)

R
r2
r1
COMMON CENTERS
FEED
DIELECTRIC, ϵr
APPROXIMATE PLANE WAVE RADIATED

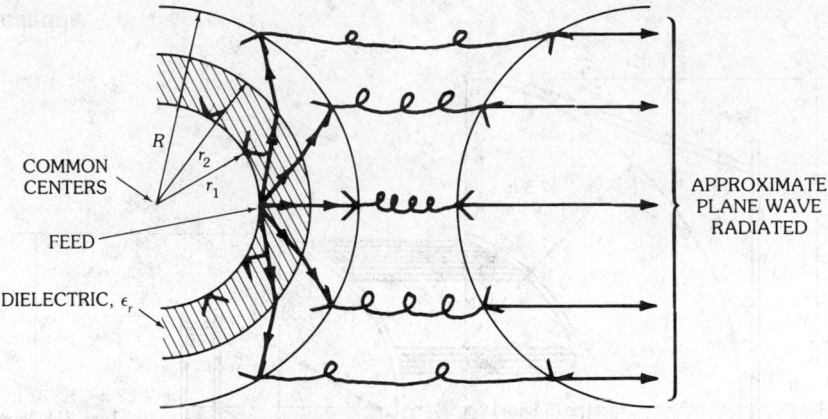

Fig. 74. Dielectric-loaded variation of basic lens. (*Courtesy Hughes Aircraft Co., Fullerton, Calif.*)

SPHERICAL REFLECTOR
SPHERICAL DUPLEXING ARRAY
DUAL CIRCULARLY POLARIZED ELEMENTS, TYPICAL
LCP
RCP
RCP
LCP
TO BEAM-SWITCHING MATRIX (BULTER MATRIX)
θ
r
R
BEAM AXIS
SPHERICAL TRANSFER ARRAY

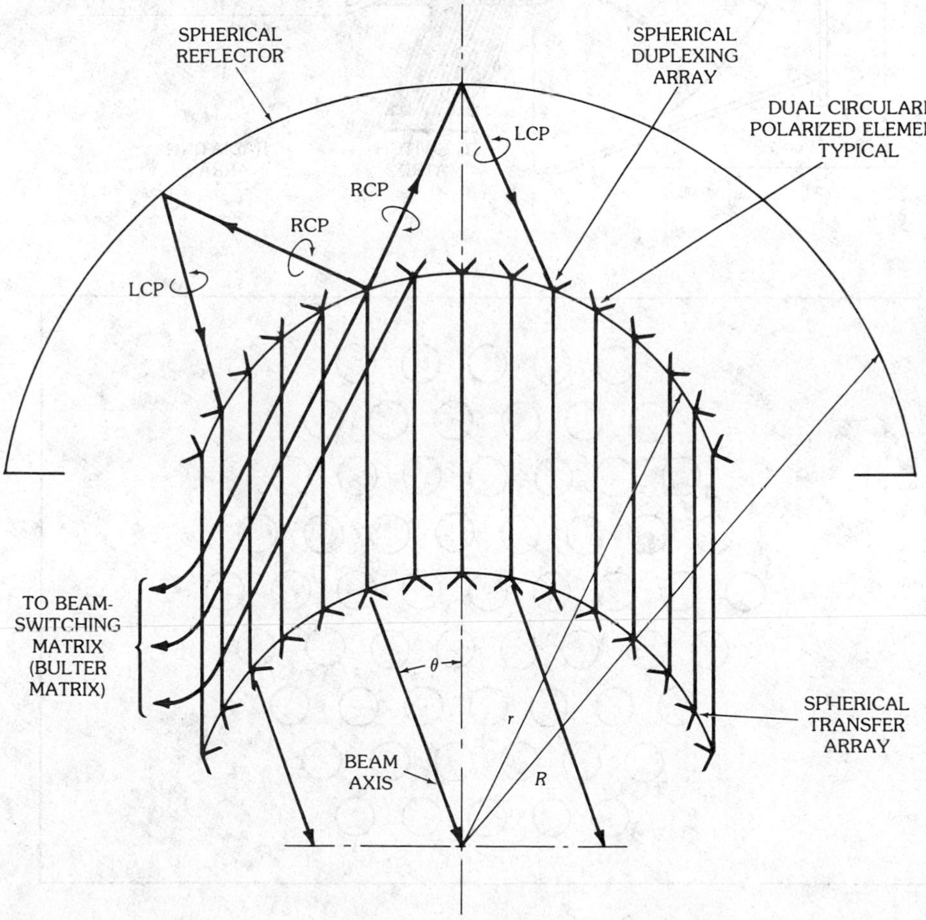

Fig. 75. The HIHAT antenna as a multiple-beam device. (*Courtesy Hughes Aircraft Co., Fullerton, Calif.*)

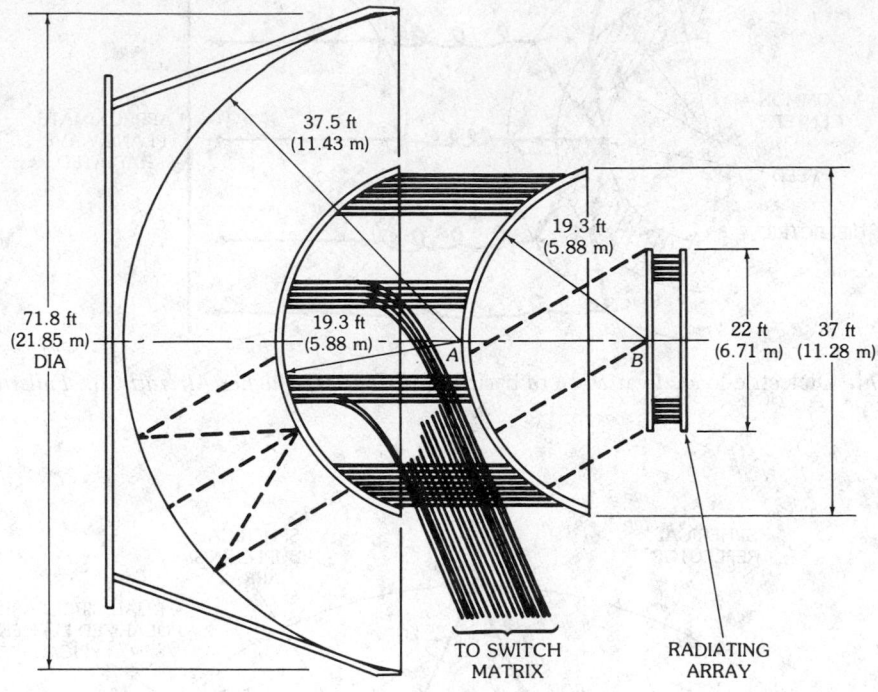

a

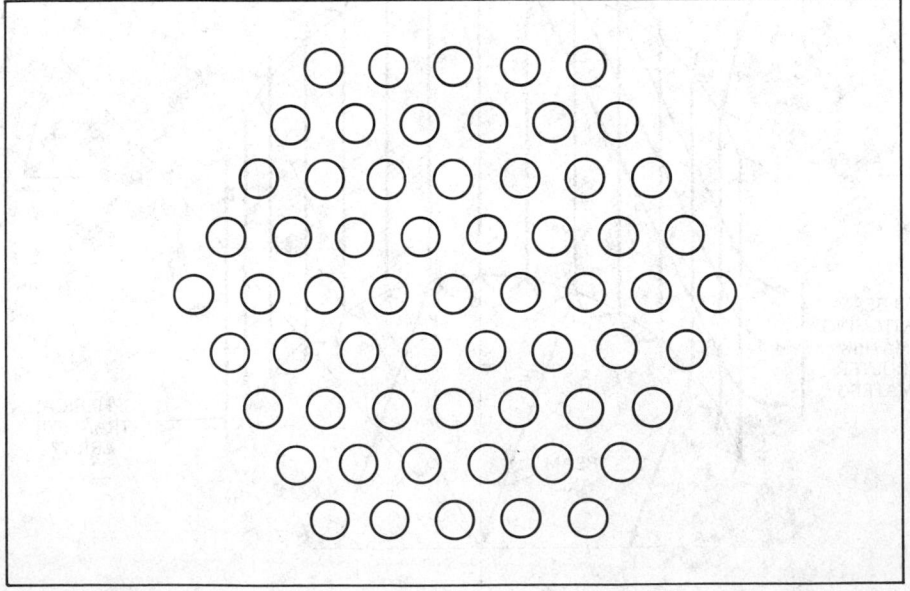

b

Fig. 76. The HIHAT used to phase a planar array. (*a*) System setup. (*b*) Radiating array. (*Courtesy Hughes Aircraft Co., Fullerton, Calif.*)

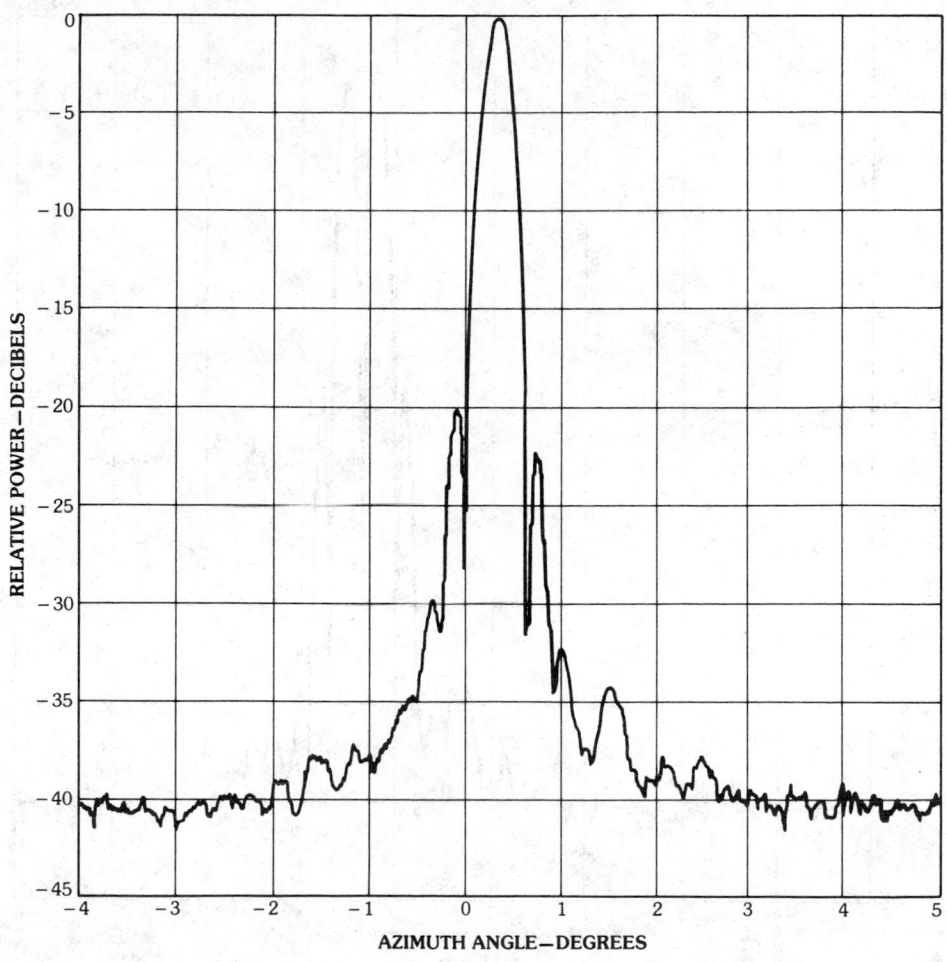

a

Fig. 77. Typical measured Σ and Δ patterns using two adjacent feeds (in the plane of scan). (*a*) Σ pattern. (*b*) Δ pattern. (*Courtesy Hughes Aircraft Co., Fullerton, Calif.*)

5. Optical Transform Feeds

In certain feed systems the input to the feed and the resulting aperture distribution of the array are related by one or more Fourier transforms. This has been evident in many of the limited-scan techniques that were described in Section 4. These feeds are called *transform feeds* [47, 49, 52]. As stated earlier it is prohibitively expensive to have active elements, such as phase shifters,* power amplifiers, low-noise amplifiers, or adaptive control loops, for every radiating element in a large phased array. But if we back down to having active elements only

---

*For broad instantaneous bandwidths, true-time-delay phase shifters may be required.

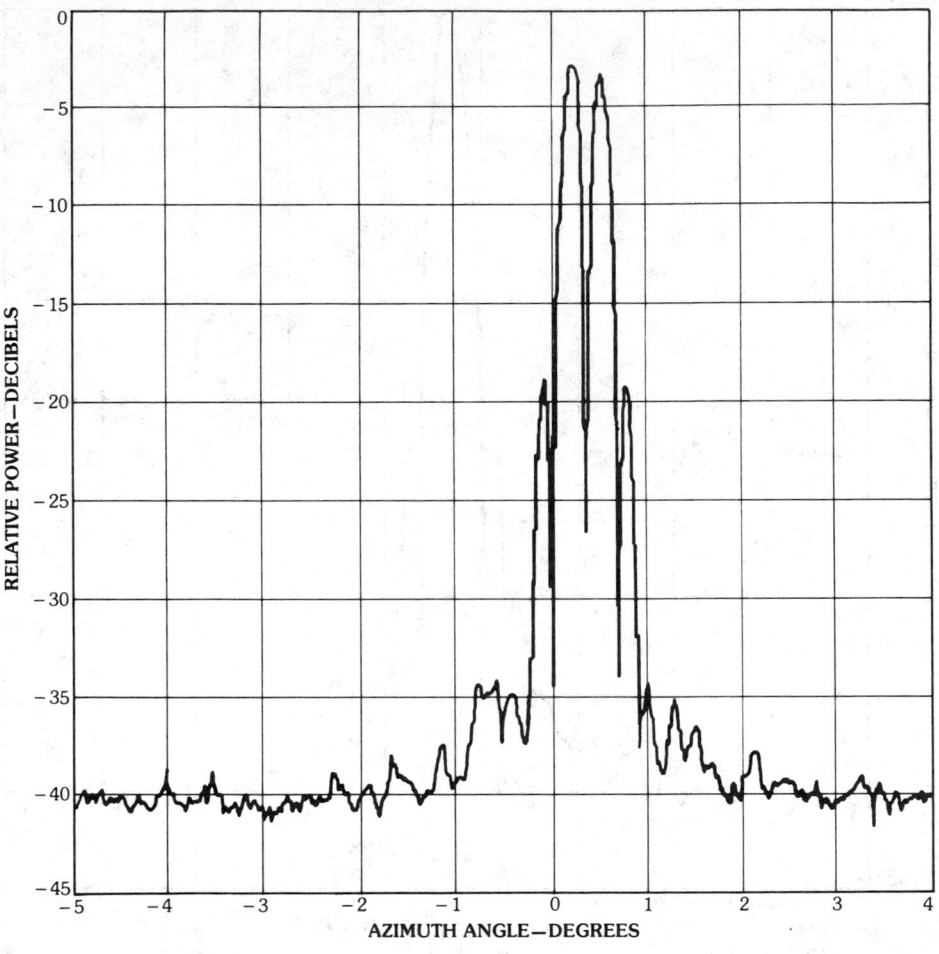

b

Fig. 77, *continued.*

at one feed point of the antenna, we do not have an electronically steerable antenna at all. It is therefore necessary for us to back down to an intermediate level where the number of devices becomes affordable. In doing so, a certain amount of performance must be sacrificed, i.e., the amount of scan coverage is reduced by the ratio of the number of active elements used to the number of active elements required to scan the larger field of view.

Two basic concepts of transform feeds that involve several Fourier transforms are depicted in Figs. 78 and 79. The first concept is a fully constrained feed that uses a small Butler matrix to feed a large Butler matrix whose outputs feed the radiating elements of the antenna. The second concept is an example of optical Fourier transformers in which a large lens is fed by a small lens. The active elements (e.g., phase shifters) are placed in a small phased array that feeds or illuminates the small

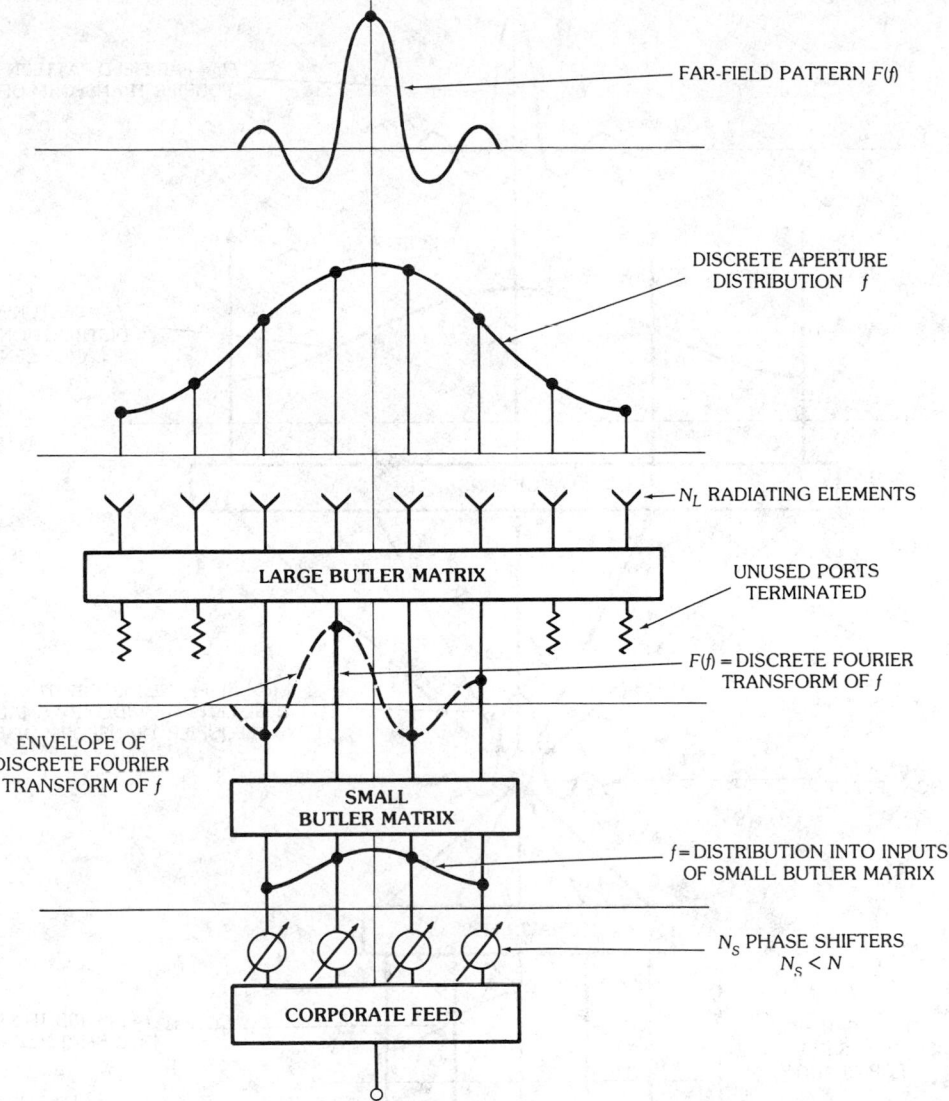

FAR-FIELD PATTERN $F(f)$

DISCRETE APERTURE DISTRIBUTION f

N_L RADIATING ELEMENTS

LARGE BUTLER MATRIX

UNUSED PORTS TERMINATED

$F(f)$ = DISCRETE FOURIER TRANSFORM OF f

ENVELOPE OF DISCRETE FOURIER TRANSFORM OF f

SMALL BUTLER MATRIX

f = DISTRIBUTION INTO INPUTS OF SMALL BUTLER MATRIX

N_S PHASE SHIFTERS $N_S < N$

CORPORATE FEED

Fig. 78. Example of constrained transform feed. (*Courtesy Hughes Aircraft Co., Fullerton, Calif.*)

matrix or optical device. Figs. 78 and 79 are conceptual schemes only and may not represent practical antenna systems. For practical transform antennas refer to Section 4 and the quoted literature.

The ideal Butler matrix is a perfect discrete Fourier transformer. A reflector or lens that focuses a plane wavefront to a point is also a Fourier transformer, but not perfect. The Butler matrix will be briefly discussed, while the optical transformers will be discussed in greater length because of their imperfections.

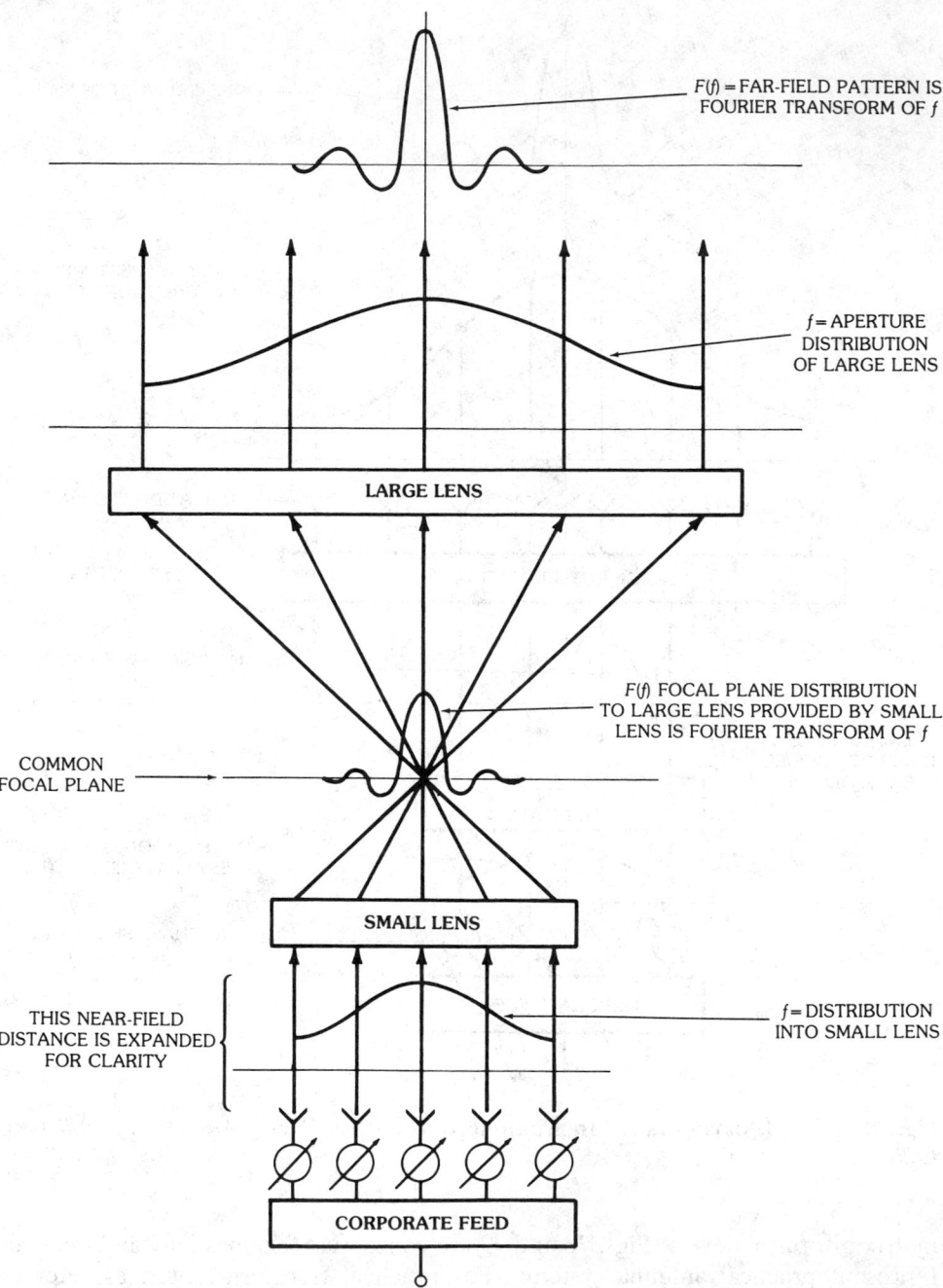

$F(f)$ = FAR-FIELD PATTERN IS FOURIER TRANSFORM OF f

f = APERTURE DISTRIBUTION OF LARGE LENS

LARGE LENS

$F(f)$ FOCAL PLANE DISTRIBUTION TO LARGE LENS PROVIDED BY SMALL LENS IS FOURIER TRANSFORM OF f

COMMON FOCAL PLANE

SMALL LENS

THIS NEAR-FIELD DISTANCE IS EXPANDED FOR CLARITY

f = DISTRIBUTION INTO SMALL LENS

CORPORATE FEED

Fig. 79. Example of optical transform feed. (*Courtesy Hughes Aircraft Co., Fullerton, Calif.*)

Butler Matrix as a Fourier Transformer

The inputs and outputs of an N-element Butler matrix are related by (Section 2)

$$g_{nm} = \frac{\exp j\{(2\pi/N)[n - (N + 1)/2][m - (N + 1)/2]\}}{\sqrt{N}}$$

referenced to the center of the array, $(N + 1)/2$ spaces, where g_{nm} is the field amplitude at the nth output element due to unit excitation at the mth input or beam port. That is, a delta-function input (each port is a discrete delta function) results in a plane wave (linear progressive phase) over all the outputs. By the principle of superposition an arbitrary input distribution $f(m)$ will result in a superposition of discrete plane waves weighted by $f(m)$, resulting in the Fourier transform

$$g(n) = \frac{1}{\sqrt{N}} \sum_{m}^{N} f(m) \exp j\{(2\pi/N)[n - (N + 1)/2][m - (N + 1)/2]\}$$

Fig. 78 illustrates the use of Butler matrices in a transform feed application. The figure is self-explanatory.

Optical Devices as Fourier Transformers

Fig. 79 is an idealized optical analog to the matrix method of Fig. 78. With idealized assumptions to be discussed shortly, Fig. 79 is self-explanatory. Refer to Fig. 80 for the following discussion.

For purposes of discussion consider a wide-angle lens or reflector that will focus over a relatively large region of the focal plane (see Section 4, under "Wide-Angle Multiple-Beam Constrained Lens"). A point-source feed at some point on the focal plane will radiate a spherical wave that will be collimated into a continuous plane wave over the aperture of the lens. Then, just as in the Butler matrix, a point-feed (delta function) input will result in a plane wave output. Again, by the principle of superposition, any focal plane distribution can be generated by a superposition of point feeds over the usable portion of the focal plane.* Again, we have an expansion into plane waves radiating in different directions in space. In general, any device that can focus electromagnetic energy from one point to another is a Fourier transformer. Usually one of the points is infinity.

It should be mentioned that the focal region field is planar on the focal plane [45]. A beam waveguide is an example where both focii are finite. A Gaussian distribution where f and $F(f)$ are identical is used in this application.

Roughly speaking, the primary feed pattern that illuminates a lens or reflector is the Fourier transform of the feed aperture distribution. After collimation this primary feed pattern becomes the aperture distribution of the lens or reflector. Now the Fourier transform of this aperture distribution is the far-field pattern of the lens or reflector. Thus, for real, even feed distributions, the far-field pattern should be identical in shape with the aperture distribution of the primary feed. It is

---

*The term "usable portion" means that off-focal point aberrations are tolerable over that region.

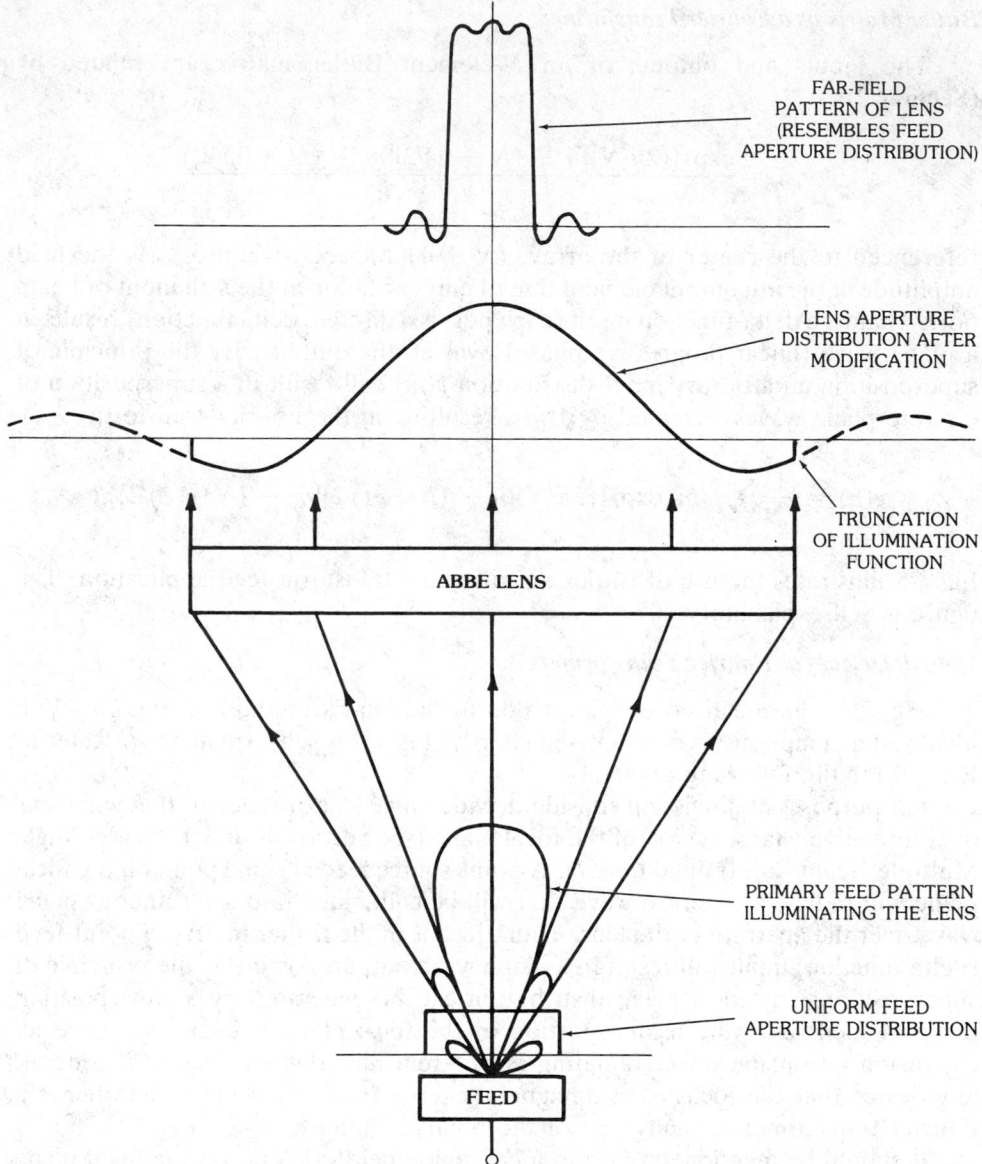

Fig. 80. Lens configuration used for discussion. (*Courtesy Hughes Aircraft Co., Fullerton, Calif.*)

assumed that the lens or reflector is in the far field of the feed, the feed spillover effects are negligible, and the amplitude modification due to the $1/r^2$ spreading of the rays and the modification of the amplitude due to the lens configuration are also negligible. The latter two assumptions are the most significant. Since all the radiation from the primary feed is not intercepted by the finite lens, the primary illumination function is truncated. The effect of this truncation is discussed later.

The other major effect is the $1/r^2$ effect and the modification of amplitude of the illumination function by the lens design. Depending on the lens configuration the net result may be an additional taper to the incident illumination, or it may add an additional inverse taper. For example, a parabolic reflector or a hyperbolic dielectric lens with one flat face will give an additional tapering effect, while an elliptical dielectric lens with a spherical inner surface results in an inverse tapering effect. Lenses obeying the Abbe sine condition have an inverse tapering effect in power that varies as the secant of the angle between the axis of the lens and a general ray from the focal point. The Abbe sine condition states that the distance of a collimated exit ray from the axis of the lens is proportional to the sine of the angle θ that the corresponding ray from the focal point makes with the lens axis. This results in a modifying factor $k\sqrt{\sec\theta}$ to the illumination function.

To help one get a "feel" for the aforementioned effects a special case of a uniform feed distribution is calculated. This is a relatively stringent test because highly tapered distributions would be less affected by feed pattern truncation due to the low amplitude at the point of truncation. Aside from these effects the uniform feed distribution should give a uniform (sector-shaped) far-field pattern from the lens. Fig. 81 shows a calculated far-field pattern from a lens that satisfies the Abbe

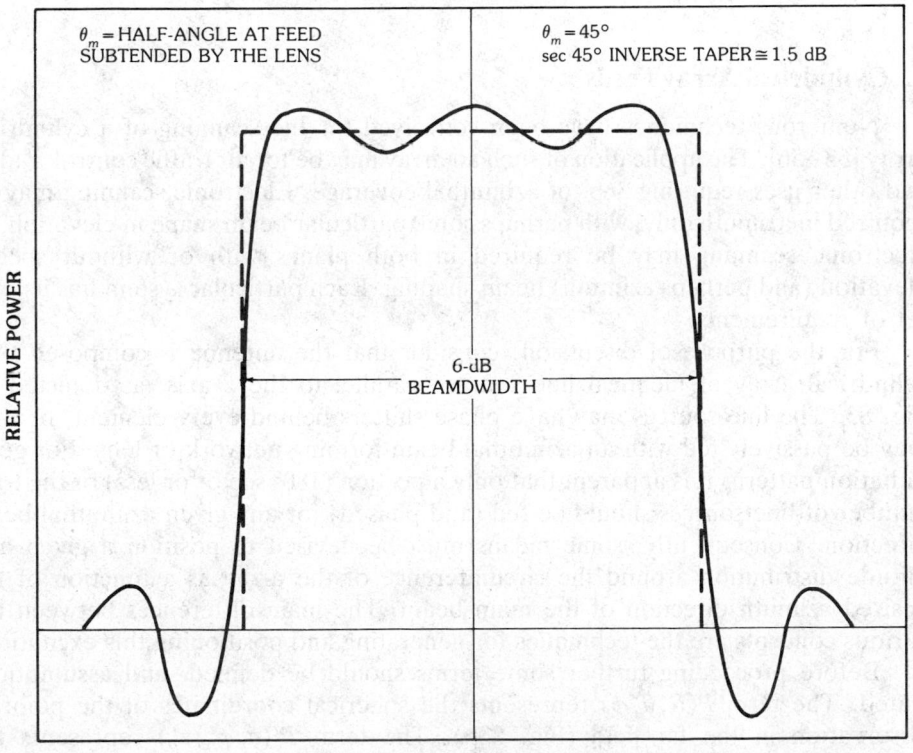

Fig. 81. Far-field pattern of a lens obeying the Abbe sine condition when the primary feed pattern is truncated to the peaks of the first side lobes of the primary pattern. (*Courtesy Hughes Aircraft Co., Fullerton, Calif.*)

sine condition. The lens truncated the $(\sin u)/u$ type primary feed pattern at the peaks of the first side lobes. This is a rather severe truncation. One would expect that at least several side lobes of the illuminating function would have to be included. However, it is seen that, even so, the resulting far-field pattern does resemble a sector-shape pattern with small ripples. In this case the inverse tapering effect of the Abbe lens was 1.5 dB. It has also been shown that the sector shape and width are nearly frequency independent. This is also a consequence of the Fourier transform. At a higher frequency the primary pattern is more directive, which effectively illuminates less of the lens aperture, thereby causing the final far-field pattern to broaden to compensate for the narrowing of the pattern due to the increase in frequency.

A three-dimensional case was both measured and calculated. A circular aperture feed with uniform aperture distribution was used to feed a lens. The far-field pattern of the lens was measured and calculated by the Fourier transform methods. A comparison of the measured and calculated far-field patterns is shown in Fig. 82. It can be seen that the expected sector shape was obtained and the measured and calculated patterns show excellent agreement.

This example, both measured and calculated parts, lends justification to the concept of an optical type antenna as a Fourier transformer.

6. Cylindrical Array Feeds

Numerous techniques have been conceived for the scanning of a cylindrical array [53–56]. The application of such an array may be for air traffic control, radar, and other uses requiring 360° of azimuthal coverage. Electronic scanning may be required in azimuth only, with perhaps some particular beam shape in elevation, or electronic scanning may be required in both planes with or without special elevation (and perhaps azimuth) beam shaping. Each particular system has its own set of requirements.

For the purpose of discussion, consider that the antenna is composed of a cylindrical array of identical line sources parallel to the z axis, as depicted by Fig. 83. The line sources may have phase shifters behind every element, or they may be passively fed with an azimuthal beam-forming network or lens. For good radiation patterns it is apparent that only a portion (180° sector or less) of the total number of line sources should be fed (and phased) for any given azimuthal beam direction. Consequently, some means must be devised to position a given amplitude distribution around the circumference of the array as a function of the desired azimuth direction of the main beam. The main differences between the various concepts are the techniques for generating and positioning this excitation.

Before proceeding further some terms should be defined, and assumptions stated. The term $P(R, \theta, \phi)$ represents the spherical coordinates of the point of observation in the far field (Fig. 83b). The term $P'(a, \phi'_n, z'_p)$ represents the cylindrical coordinates on a cylinder of radius a of a typical radiating element. The terms n and p are integers associated with (ϕ'_n, z'_p), using the notation of [53], except for the angle θ. It is assumed that the element is impedance matched for wide-angle scanning and the interelement spacing in both directions is close enough

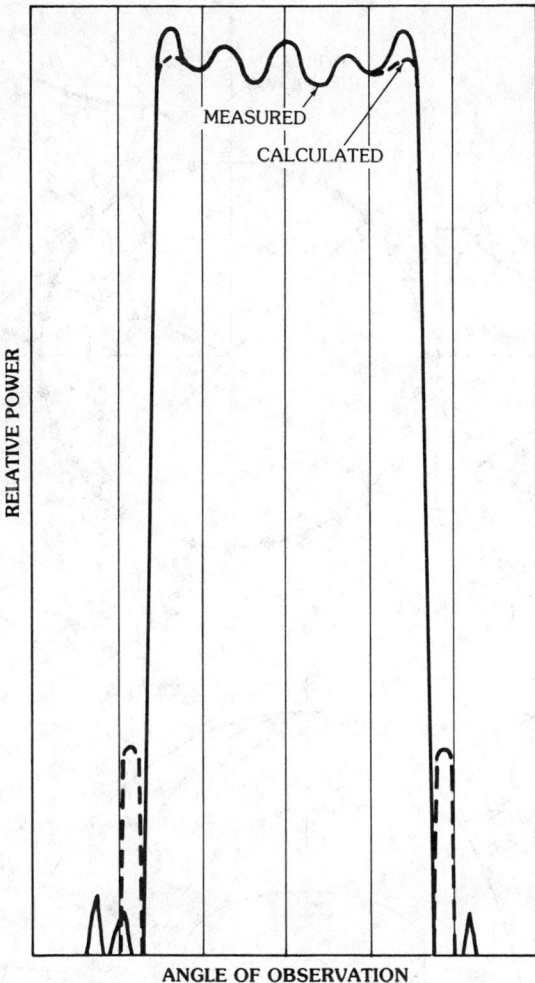

Fig. 82. Measured and calculated far-field patterns from a lens with a circular aperture feed with a uniform distribution. (*Courtesy Hughes Aircraft Co., Fullerton, Calif.*)

(near $\lambda/2$) to produce an ideal element factor $\sqrt{\cos\gamma}$, where γ is the angle between the normal to the array and the point of observation.

The far-field pattern is related to the surface illumination by

$$E(\theta,\phi) = \iint \varepsilon(\phi',z') \frac{e^{-jkR}}{R} \sqrt{\cos\gamma}\, d\phi'\, dz' \tag{23}$$

Let the complex array illumination function be

$$\varepsilon(\phi_n,z_p') = \ell(\phi_n')f(z_p') \tag{24}$$

Then the far field for the array is given [54] by

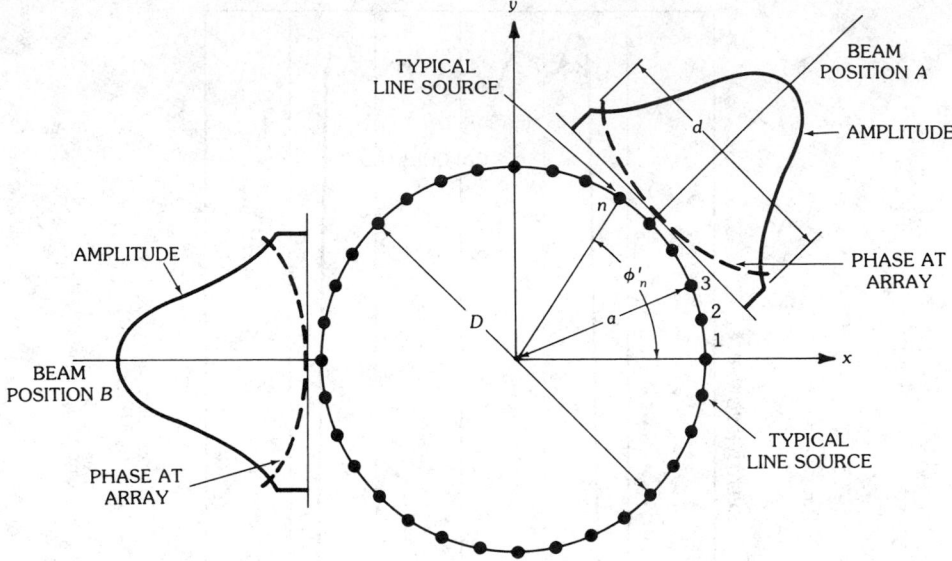

a

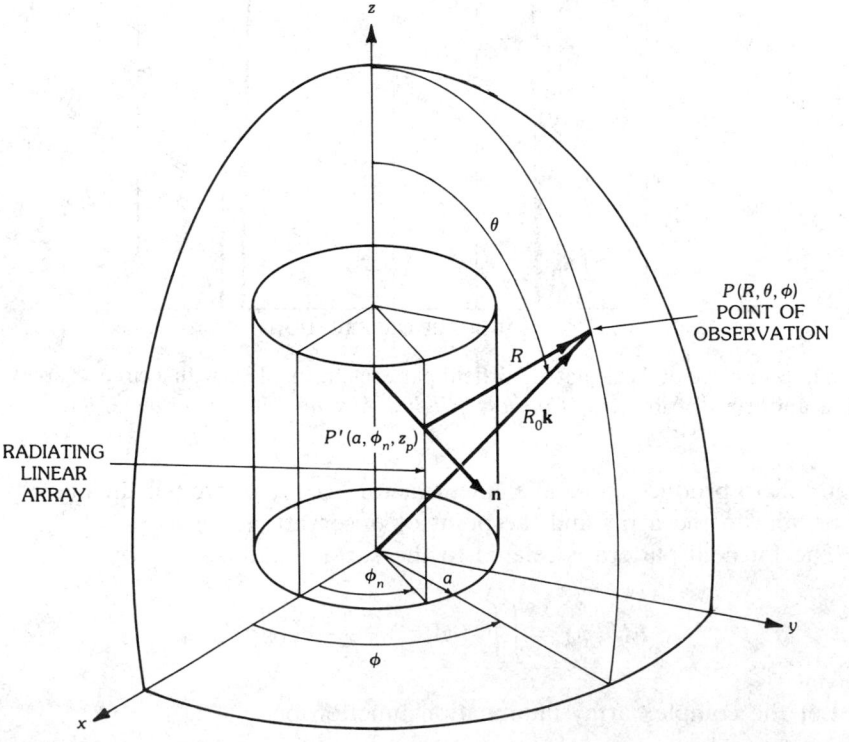

b

Fig. 83. Antenna as a cylindrical array of line sources. (*a*) Cylindrical array plan view. (*b*) Cylindrical array oblique view. (*Courtesy Hughes Aircraft Co., Fullerton, Calif.*)

$$E(\theta) = \sqrt{\cos\theta} \sum_p f(z'_p) \, e^{jkz'_p\cos\theta} \sum_n \ell(\phi'_n) \sqrt{\cos(\phi - \phi'_n)} \, e^{jka\sin\theta\sqrt{\cos(\phi-\phi'_n)}} \quad (25)$$

or

$$E(\theta, \phi) = L(\theta) R(\theta, \phi) \quad (26)$$

where

$$L(\theta) = \sqrt{\cos\theta} \sum_p f(z'_p) \, e^{jkz'_p\cos\theta} \quad (27)$$

and

$$R(\theta, \phi) = \sum_n \ell(\phi'_n) \sqrt{\cos(\phi - \phi'_n)} \, e^{jka\sin\theta\cos(\phi-\phi'_n)} \quad (28)$$

Thus the required array excitation to generate a main beam at (θ_m, ϕ_m) is given by

$$|f(z'_p)| |\ell(\phi'_n)| \, e^{-jk[z'_p\cos\theta_m + a\sin\theta_m\cos(\phi_m-\phi'_n)]} \quad (29)$$

Various techniques for generating this cylindrical array excitation will be discussed briefly. For purposes of clarity some of the diagrams to follow show the use of only eight-array elements.

Matrix-Fed Cylindrical Arrays

The various matrix-fed arrays can be readily understood by use of the following tutorial model.

Consider two back-to-back Butler matrices with controllable phase shifters in between them as shown in Fig. 84a. If an input port a_n of Butler matrix A is excited with any amplitude and phase, a corresponding port b_n of Butler matrix B will be excited with equal amplitude and phase when the phase shifters are all set equal. If the a_ns are weighted in amplitude and phase to any desired distribution, a corresponding distribution is achieved at the outputs b_n. Now if a linear phase distribution with total phase of $m2\pi$ (m an integer) is applied across the array of phase shifters, the entire distribution over b_n will translate across the output in increments of one element spacing for each successive value of m. The translation of the output distribution by an increment less than one element spacing requires a total phase that is a nonintegral multiple K of 2π. That is, $|m| < |K| < |m| + 1$. In these cases there is no longer a one-to-one correspondence between an input port excitation and an output port excitation. Instead, if one input port is excited, not one but all output ports are excited to some extent. However, as would be expected the two adjacent ports corresponding to phases of $m2\pi$ and $(m + 1)2\pi$ are excited most strongly since they straddle the "phantom" element corresponding to the phase of $K2\pi$. In fact, the output distribution is a discrete sampling of a sharp $(\sin x)/x$ distribution as shown in Fig. 84b. The width of the "main lobe" of the

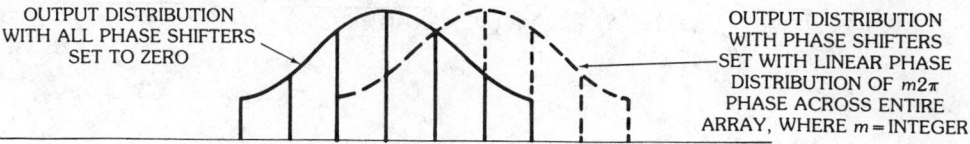

OUTPUT DISTRIBUTION
WITH ALL PHASE SHIFTERS
SET TO ZERO

OUTPUT DISTRIBUTION
WITH PHASE SHIFTERS
SET WITH LINEAR PHASE
DISTRIBUTION OF $m2\pi$
PHASE ACROSS ENTIRE
ARRAY, WHERE m = INTEGER

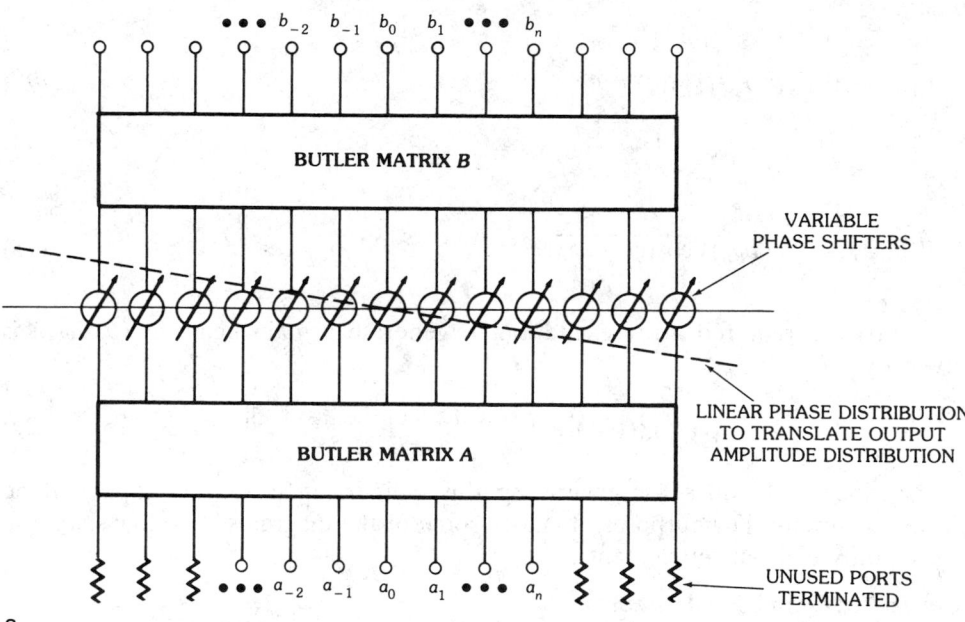

$\bullet\bullet\bullet$ b_{-2} b_{-1} b_0 b_1 $\bullet\bullet\bullet$ b_n

BUTLER MATRIX B

VARIABLE
PHASE SHIFTERS

LINEAR PHASE DISTRIBUTION
TO TRANSLATE OUTPUT
AMPLITUDE DISTRIBUTION

BUTLER MATRIX A

UNUSED PORTS
TERMINATED

$\bullet\bullet\bullet$ a_{-2} a_{-1} a_0 a_1 $\bullet\bullet\bullet$ a_n

a

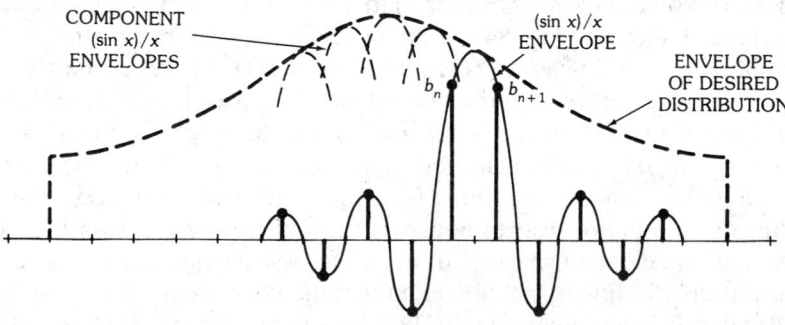

COMPONENT
$(\sin x)/x$
ENVELOPES

$(\sin x)/x$
ENVELOPE

ENVELOPE
OF DESIRED
DISTRIBUTION

b_n b_{n+1}

b

Fig. 84. Back-to-back Butler matrix feeding technique. (*a*) Tutorial model of matrix feeding. (*b*) Output excitation when only one input is excited and linear phase across bank of phase shifters is not an integral multiple of 2π. (*Courtesy Hughes Aircraft Co., Fullerton, Calif.*)

envelope $(\sin x)/x$ is two elements and the side lobe width is one element. The output distribution corresponding to phases of integral multiples of 2π consists of discrete values identical with those at the input to the first Butler matrix A because of the one-to-one correspondence between a_ns and b_ns. This is consistent with the sampled $(\sin x)/x$ output in these special cases; the sampling points are all at the nulls of the $(\sin x)/x$ function except at the peak of the main lobe, hence only that element is excited. For intermediate cases corresponding to nonintegral multiples of 2π, the output distribution is a superposition of sampled $(\sin x)/x$ functions of the kind just described. The net result is that the envelope of this output distribution is almost identical with that of the envelope of the designed output distribution, which is identical with the input distribution of matrix A.

Now, if the output ports b_n are connected by equal line lengths to a similar number of elements arranged uniformly along a circle of radius R, linear phase shifter adjustments will cause the distribution over the b_ns to move along the periphery of the circular array. If the phase distribution is such as to radiate a plane wavefront from the circular array, a directive beam is formed from the circular array that can be steered by means of changing the linear progressive phase of the phase shifters. It is obvious that this peripherally scanned distribution should not extend more than a semicircle because elements to the rear should not be excited. The input ports corresponding to these rearward-looking elements are shown terminated in Fig. 84a. In practice, less than 180° of the circular (or cylindrical) array is excited for one beam direction because of active element patterns, local grating lobe formation, and aperture matching problems associated with the edge elements that are locally phased for near end-fire. Also, for intermediate phasing, the elements near the edge of the active portion of the cylinder may have their corresponding sampled $(\sin x)/x$ lobes to the rear that are greater than desired for low side lobe and rear lobe patterns.

Collimation of the beam from the circular array requires a phase distribution of $\phi_n = kR(1 - \cos n\Delta\theta)$, where $\Delta\theta$ is the angle subtended by one-element spacing, R is the radius of the cylinder, and $k = 2\pi/\lambda$. This phase is easily incorporated in the weighting of the a_ns.

It is interesting to note that the amount of beam scan in terms of antenna beamwidths for each 2π of total phase change is equal to one beamwidth if the element spacing Δs is $\lambda/2$:

$$\Delta s = R\,\Delta\theta = \frac{\lambda}{2}, \qquad \Delta\theta = \frac{\lambda}{2R}$$

This is also the beamwidth for a uniform array of aperture $2R$.

The foregoing serves as a tutorial basis for describing some of the feeding techniques for cylindrical arrays to be discussed.

Sheleg Method—In the preceding discussion it is readily seen that the output of the first Butler matrix A (input to the bank of phase shifters) has an amplitude and phase distribution that is fixed (independent of beam steering); therefore it can be replaced by a passive feed network that produces that same distribution. This reduces to the Sheleg approach. A simplified schematic diagram of the approach for an eight-element array is shown in Fig. 85. Sheleg arrived at this solution from a different point of view. He expressed the far-field pattern in cylindrical modes in the form

$$E(\phi) = \sum_{-N}^{N} C_m e^{jm\phi}$$

where the number of elements is 2N, and

$$C_m = 2\pi K j^m I_m J_m(ka)$$

$$K = \text{constant}$$

Sheleg made the observation that it is not necessary or even desirable to use all the *m* modes of the Butler matrix. This concurs with the foregoing tutorial discussion in that modes that excite elements to the rearward direction of the cylinder should not be used. As stated earlier, in practice less than half the number of available modes should be used to avoid local end-fire radiation at the extremes of the active semicircle. The Sheleg scheme is depicted for an eight-element array in Fig. 85.

This technique can be used to produce uniform or tapered illuminations for the cylindrical array. Sheleg gives computed and measured results for uniform, cosine, and cosine-squared array distributions with quite close correlation between

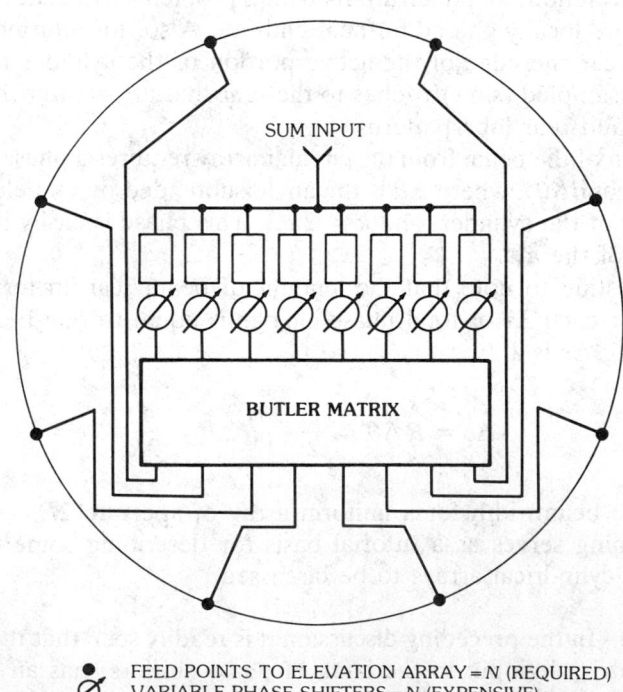

FEED POINTS TO ELEVATION ARRAY = N (REQUIRED)
VARIABLE PHASE SHIFTERS = N (EXPENSIVE)
FIXED PHASE SHIFTERS = N (INEXPENSIVE)

Fig. 85. Sheleg's approach. (*Courtesy Hughes Aircraft Co., Fullerton, Calif.*)

measurement and theory except in the remote side lobe region. He shows the patterns as a function of the number of modes used. Up to a point the patterns improve as more modes are used; then, beyond this point, the radiation pattern degrades somewhat. This has been explained by use of the tutorial model previously discussed.

For uniform distribution the azimuth pattern shape as a function of azimuth scanning of the main beam is shown to be reasonably invariant, i.e., it makes little difference whether the peak of the array distribution falls on an element, midway between two elements, or any fractional part thereof. This is also explained by use of the tutorial model. For more details the reader is referred to Sheleg's paper [56].

Wheeler Lab Approach—Fig. 86 depicts the Wheeler Lab approach [53, 54] to feeding a cylindrical array. It consists of a power divider followed by fixed phase shifts, variable phase shifters, a sequencing switch, an array of diode switches, and the radiating array.

This technique utilizes a *switching network* to step the amplitude distribution around the cylindrical array in coarse steps equal to the interelement angular spacing. The phase and amplitude distribution are established by the power divider and a set of fixed and variable phase shifters. *The variable phase shifters provide fine steering.*

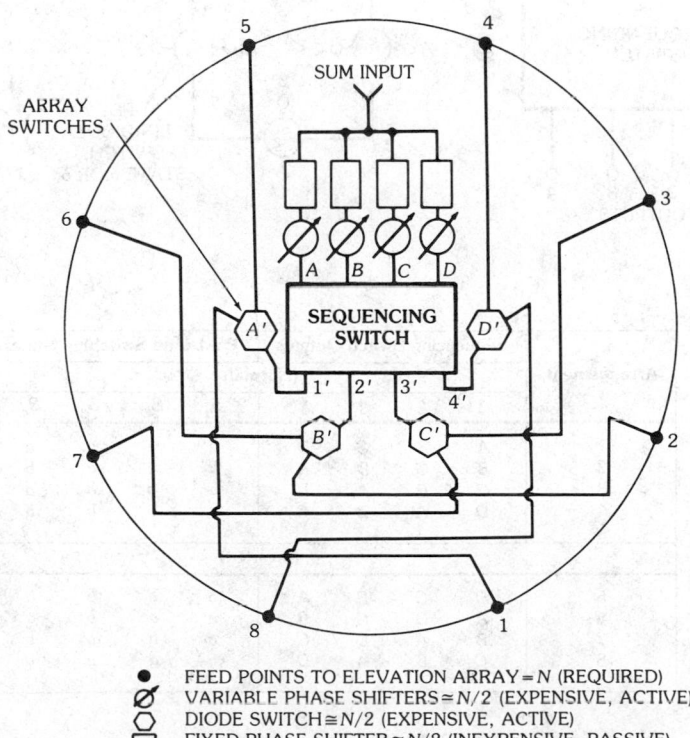

Fig. 86. Wheeler Lab approach. (*Courtesy Hughes Aircraft Co., Fullerton, Calif.*)

In this technique the ratio of the total number of azimuth elements to the number of active elements used at any beam position must be an integer.

In Fig. 86 the number of elements used at a given time is four since the corporate feed is shown as having four outputs. The desired output amplitude and phase distribution is designated A, B, C, D. For a given state of the sequencing switch, for example, elements $1, 2, 3, 4$ are excited with the complex amplitude A, B, C, D, respectively. Now suppose it is desired to move the distribution by one element, say, elements $2, 3, 4, 5$. With no sequencing switch the output amplitude distribution would be B, C, D, A, which is not the desired distribution. The sequencing switch reswitches the distribution back to A, B, C, D, the desired output. The same technique will allow the desired distribution, A, B, C, D, to be positioned all the way around the array in steps of one element at a time.

For the example shown, the sequencing switch would consist of four dpdt pin diode switches, as depicted by Fig. 87a. The complex inputs to the sequencing switch are labeled A, B, C, D. The outputs of the sequencing switch are labeled $1', 2', 3', 4'$. Each of the four pin diode switches has two possible states, labeled a or

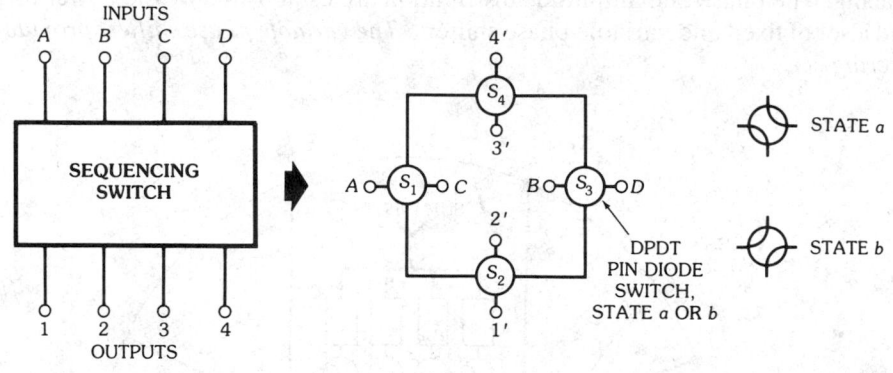

a

Arrangement	Sequencing Switch Outputs				Pin Diode Switching States			
	(Normal)							
	1'	**2'**	**3'**	**4'**	S_1	S_2	S_3	S_4
1	A	B	C	D	a	a	a	a
2	B	C	D	A	b	b	a	b
3	C	D	A	B	b	a	b	a
4	D	A	B	C	a	b	b	b
	(Inverted)							
5	D	C	B	A	b	b	b	b
6	A	D	C	B	a	a	b	a
7	B	A	D	C	a	b	a	b
8	C	B	A	D	b	a	a	a

b

Fig. 87. Sequencing switch circuit and logic. (*a*) Sequencing switch circuit. (*Courtesy Hughes Aircraft Co., Fullerton, Calif.*) (*b*) Logic of switch circuit.

b. The table of Fig. 87b shows four normal combinations and four inverted combinations of complex distributions emerging from the output of the sequencing switch, along with their corresponding switching states for the four dpdt switches labeled S_1, S_2, S_3, S_4. By using the inverted as well as the normal arrangements, beam positions mirror-imaged about the center line of the active cylindrical array normal are generated; consequently the number of phase states required by the phase shifters is halved. The switches labeled array switches in Fig. 86 have a single input with one of two possible outputs, depending on which radiating elements are to be excited.

For fine beam steering, for beam positions corresponding to that between elements for example, the variable phase shifters are used while the elements excited remain fixed at the same position as for a beam position corresponding to an angle through an element.

The power divider as depicted in Fig. 86 shows only a sum input. In actuality, both a sum and a difference port would typically be available by appropriate design of the power divider network.

Hughes Phased-Lens–Fed Approach—This technique utilizes a rotationally symmetric lens such as a Luneburg or Rinehart lens to perform the azimuth beam forming, a switching system to select lens feed points for coarse steering, and phase shifters at each lens output to the cylindrical array to do the fine steering. Fig. 88 schematically depicts this approach.

The sum (and difference, not shown) input is power-divided by the number of lens probes to be fed simultaneously, forming the lens primary feed illumination, two for the example of Fig. 88. Suppose lens probes 1 and 2 are chosen to form a beam diametrically opposite to the position lying between probes 1 and 2. Then switch *A* would select probe position 1 while switch *B* would select position 2. All the probes dispersed over the arc labeled 4, 5, 6, 7 would then pick up the focused field and switches labeled *D, C, J, I* would switch the fields to the elements labeled 4, 5, 6, 7 via phase shifters behind each element. The coarse beam position is determined by choosing which probes to excite. The fine beam steering is accomplished by the phase shifters behind each element.

The number of actively excited probes for a given beam position must divide integrally into the total number of elements; that is, the ratio of the total number of azimuth elements to the number of simultaneously excited lens probes must be an integer.

For example, if three probes needed to be excited simultaneously to achieve the desired primary lens feed distribution, then the total number of elements must be a multiple of three. If nine elements were chosen to satisfy this requirement, then the sum (and difference) input would have to be divided into three outputs followed by three switches each with three positions, rather than four as depicted in Fig. 88.

The lens probes are seen to have to perform the dual function of not only acting as primary feed elements but as transfer elements as well.

Matrix-Fed Conventional Lens Approach—For coarse beam stepping this technique is identical with the phased lens–fed approach described previously. For

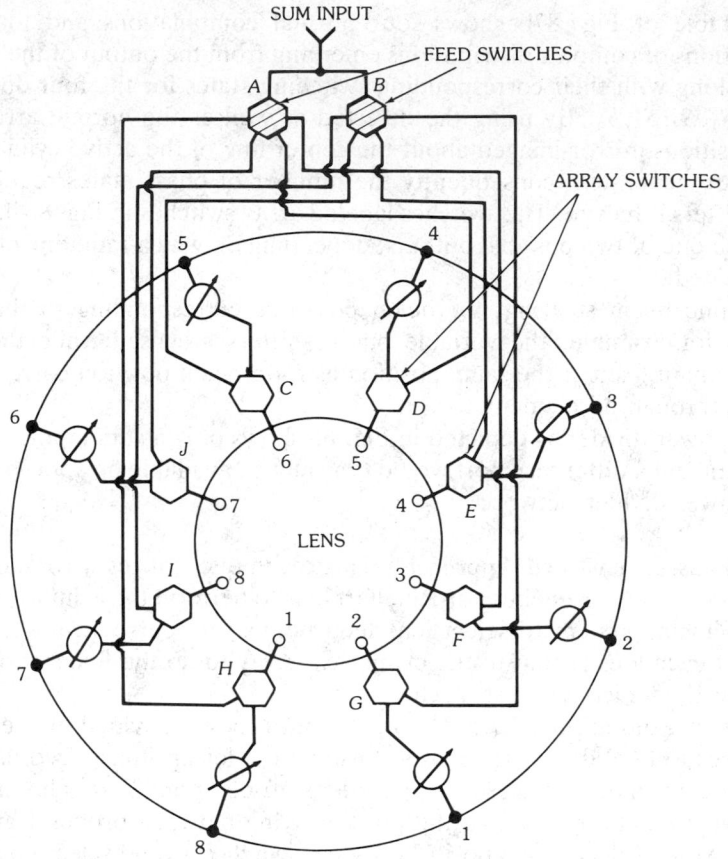

Fig. 88. Phased-lens–fed approach. (*Courtesy Hughes Aircraft Co., Fullerton, Calif.*)

fine steering a "Sheleg" technique is used on a relatively small feed array using an *F*-element Butler matrix and *F* variable phase shifters, where *F* is the small number of lens elements used to form the desired feed pattern. The feed distribution is translated as described before in Fig. 84a, which translates the phase center of the feed for fine steering.

Hughes Matrix-Fed Meyer Geodesic Lens—This technique [54, 57] is a variation of the matrix-fed conventional lens approach discussed in the previous section but utilizes a modified Meyer geodesic lens (figure of revolution) for the beam forming to eliminate active switches and uses passive diplexers or circulators instead. The

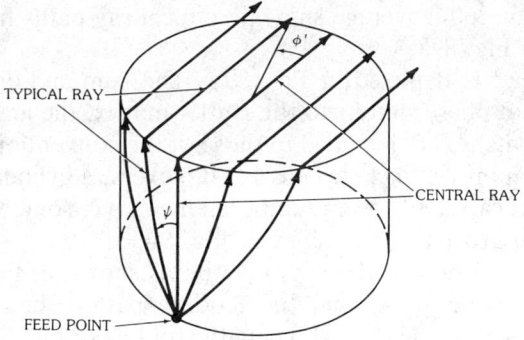

a

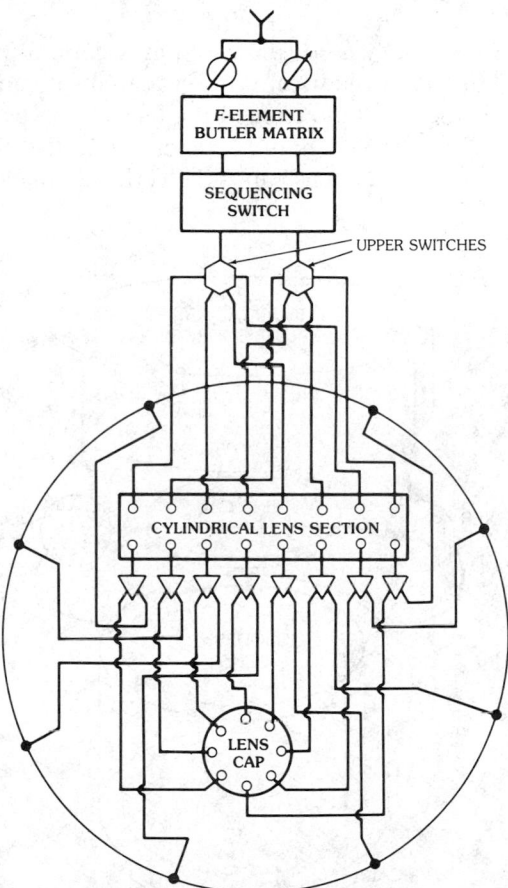

b

Fig. 89. Meyer geodesic lens. (*a*) Basic Meyer lens. (*b*) Matrix feed. (*Courtesy Hughes Aircraft Co., Fullerton, Calif.*)

basic Meyer lens for 360° coverage showing typical ray paths from a feed point is given pictorially in Fig. 89a.

The matrix feed is depicted in Fig. 89b. The sum-and-difference corporate power divider, input phase shifters to the Butler matrix, the amplitude sequencing switch, and feed switches are identical to those of the conventional lens above. The remainder of this approach includes passive diplexers, a cylindrical portion of the Meyer lens, a lens cap, and the radiating elements. A somewhat more physical depiction of this approach is illustrated in Fig. 90.

The cylindrical portion of the lens may be as shown in Fig. 90 or it may be folded, since folding does not change the geodesic paths. The rf paths on transmit and receive are depicted in Fig. 91, showing the folding of the cylindrical portion of the lens. A schematic of the diplexer/circulator is given in Fig. 92. A functional diagram showing transmit and receive signal flow in the diplexer/circulator is shown in Fig. 93.

The Meyer lens, as usually used (discussed in Section 3) phases a line source; however, by completing the cylindrical portion to 360°, it can be used to phase a cylindrical array. It is then possible to feed the lens at any point on the 360° feed circle, and the output of the lens may be taken along the circumference of the circular cap. Furthermore, if the lens cap is dielectrically loaded, it can be smaller

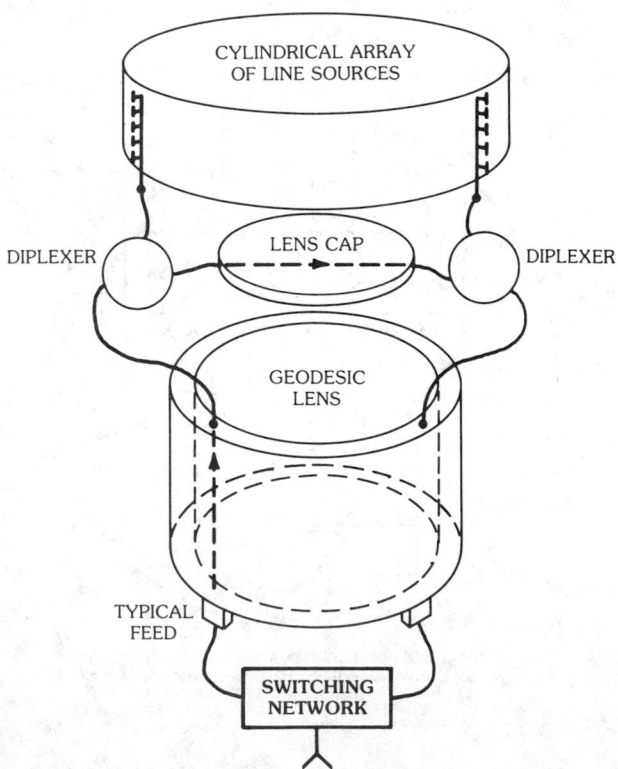

Fig. 90. Meyer system configuration. (*Courtesy Hughes Aircraft Co., Fullerton, Calif.*)

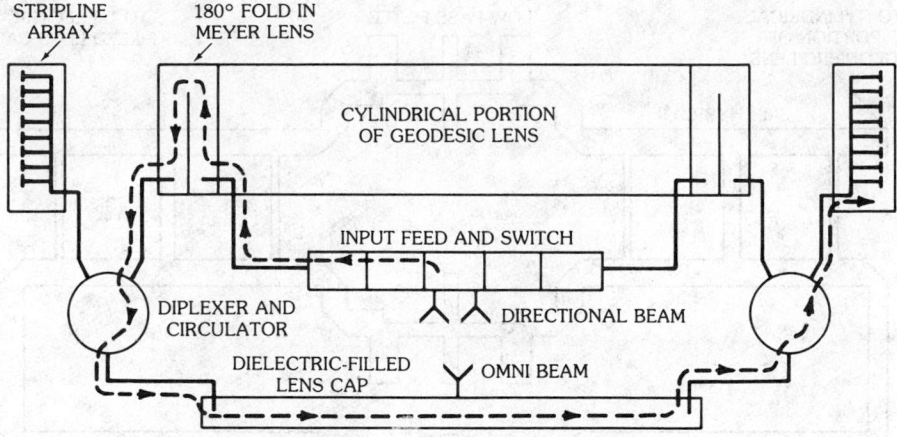

a

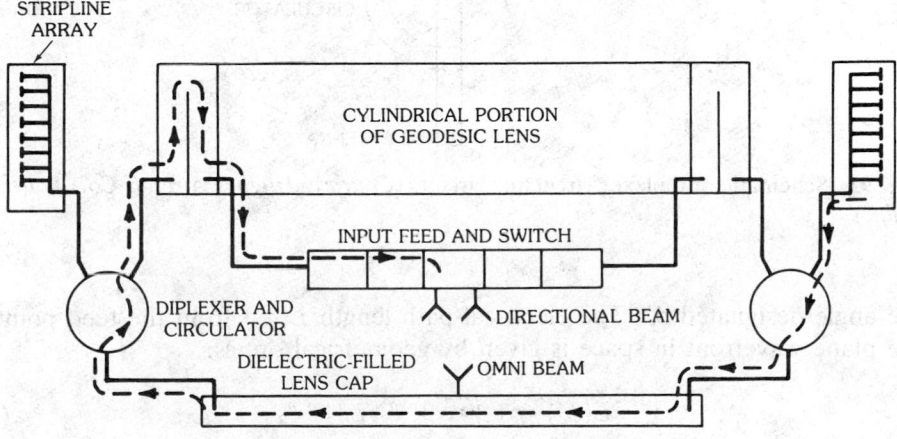

b

Fig. 91. Functional block diagram showing rf paths of Meyer lens. (*a*) Directional-beam signal flow. (*b*) Receive signal flow. (*Courtesy Hughes Aircraft Co., Fullerton, Calif.*)

by the square root of the relative dielectric constant. This allows for easier packaging of the associated components.

A brief analysis of the dielectrically loaded lens is given and experimental results are presented. Since the cylinder is a singly curved surface, the cylinder and flat circular portion can be developed, or broken apart and flattened for the sake of easy analysis. Refer to Fig. 94. The geodesics all become straight lines in the developed case. As in Fig. 94 equal lengths of transmission line are used for the rf connection between the developed cylinder and the cap, where the arc length s is preserved, or is linearly mapped onto the lens cap through some proportionality constant α. The choice of α determines the physical size of the lens cap and the necessary dielectric loading ϵ_2 to correctly phase the radiating elements. The same is true between the cap and the radiating elements, but the mapping here preserves

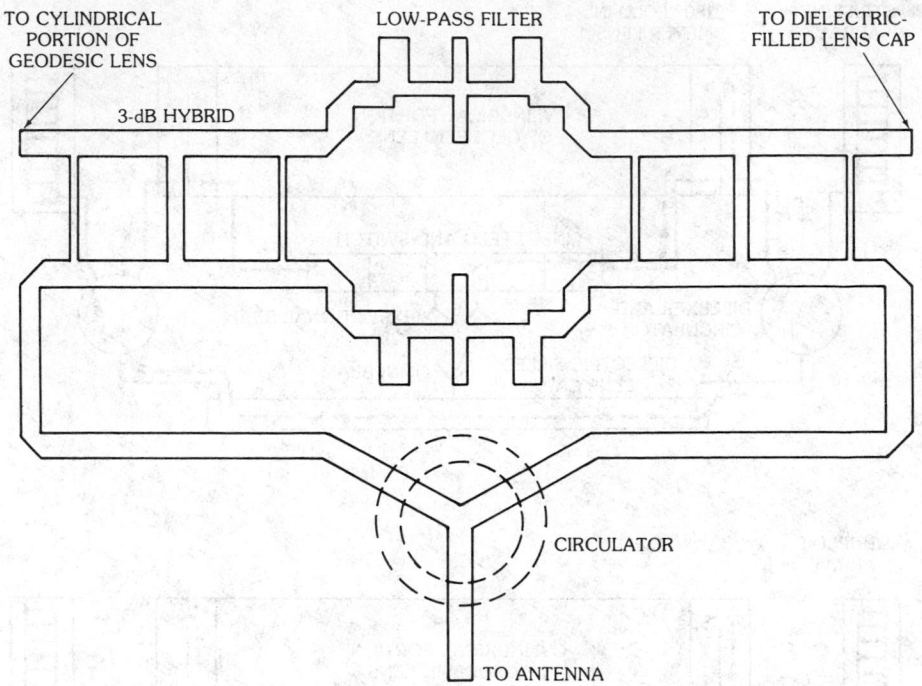

Fig. 92. Schematic of diplexer/circulator circuit. (*Courtesy Hughes Aircraft Co., Fullerton, Calif.*)

the angle designated ϕ'. The electrical path length $L(\phi')$ from the feed point to the plane wavefront in space is given by geometrical optics:

$$L(\phi') = \ell_1\sqrt{\epsilon_1} + \ell_2\sqrt{\epsilon_2} + \ell_3 \tag{30}$$

The reference path length through the center is ℓ_0, given by

$$\ell_0 = h\sqrt{\epsilon_1} + 2b\sqrt{\epsilon_2} \tag{31}$$

Thus the electrical path length error δ is given by

$$\delta = L(\phi') - \ell_0 \tag{32}$$

The output phase error $\Phi(\phi')$ is then

$$\Phi(\phi') = k\delta(\phi') \tag{33}$$

Snell's law is obeyed because of equal line lengths; thus

$$\frac{\sin\gamma}{\sin\psi} = \frac{n_1}{an_2} \tag{34}$$

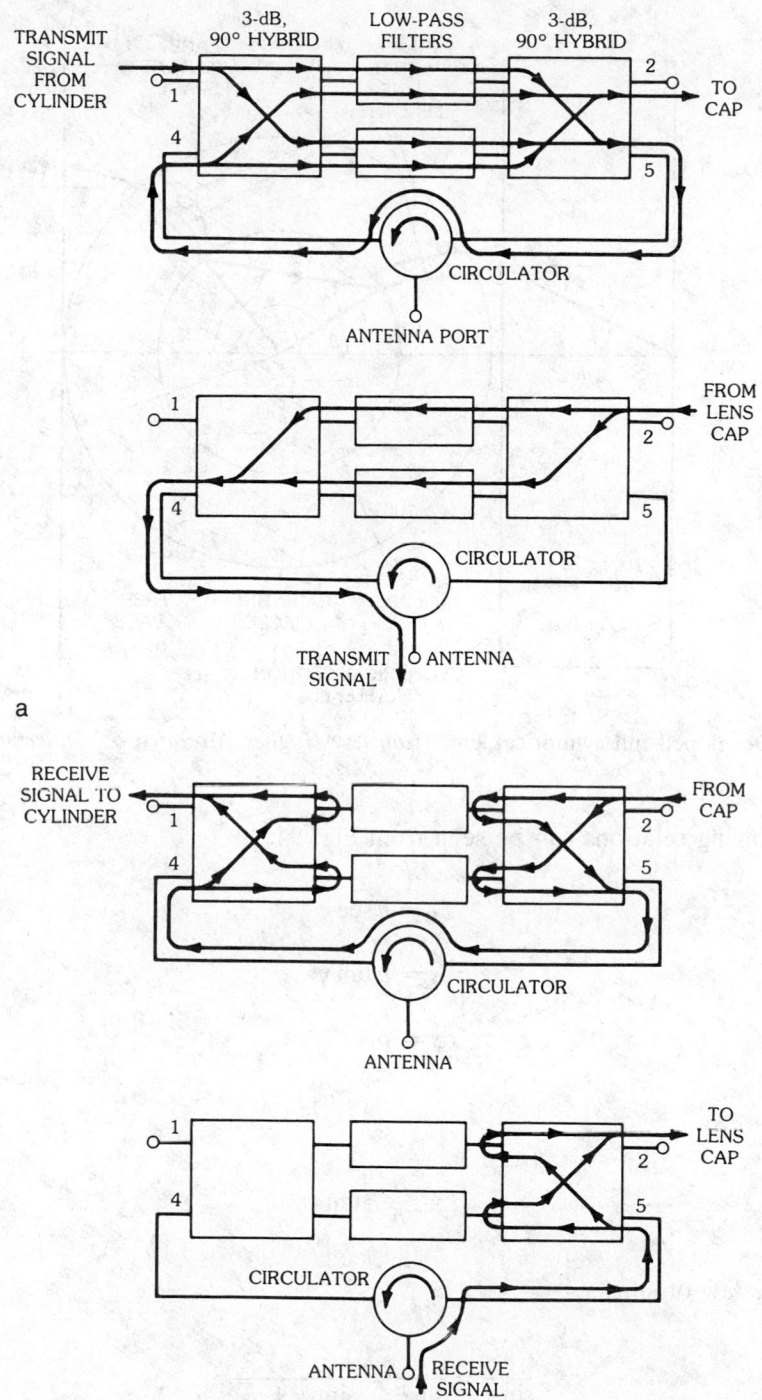

Fig. 93. Functional diagram of diplexer/circulator, showing signal flow. (*a*) Transmit signal paths. (*b*) Receive signal paths. (*Courtesy Hughes Aircraft Co., Fullerton, Calif.*)

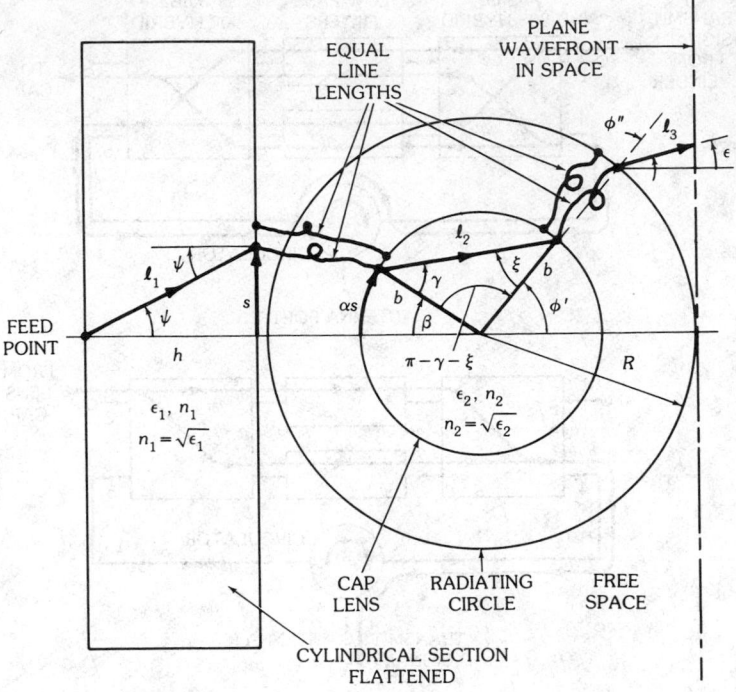

Fig. 94. Developed full cylindrical lens. (*Courtesy Hughes Aircraft Co., Fullerton, Calif.*)

The following relations can be seen from Fig. 94:

$$\ell_1 = h \sec \psi \tag{35}$$

$$s = h \tan \psi \tag{36}$$

$$\alpha s = b\beta \tag{37}$$

Thus

$$\beta = \frac{\alpha}{b} h \tan \psi \tag{38}$$

From the law of sines,

$$\frac{b}{\sin \gamma} = \frac{b}{\sin \xi} = \frac{\ell_2}{\sin(\gamma + \xi)} \tag{39}$$

Thus

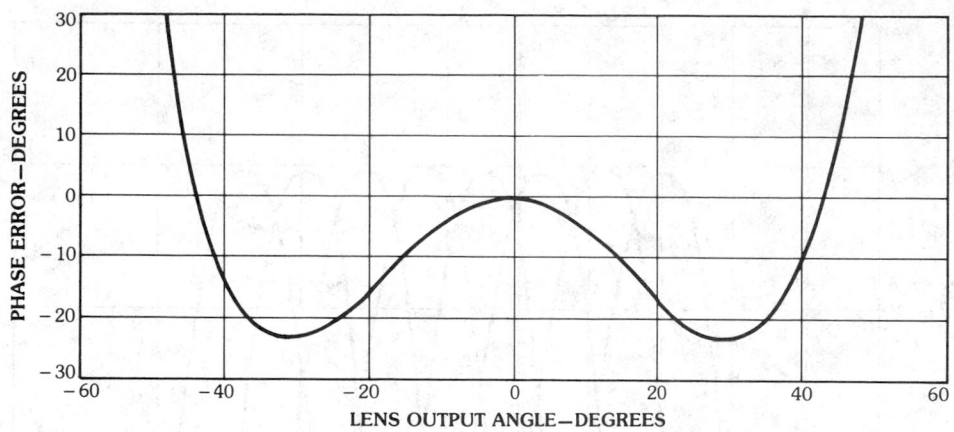

Fig. 95. Phase error at output of geodesic lens. (*Courtesy Hughes Aircraft Co., Fullerton, Calif.*)

a

Fig. 96. Geodesic lens and patterns. (*a*) Modified Meyer lens. (*b*) Pattern (eight adjacent beams). (*c*) Sum pattern. (*d*) Expanded sum and difference pattern. (*Courtesy Hughes Aircraft Co., Fullerton, Calif.*)

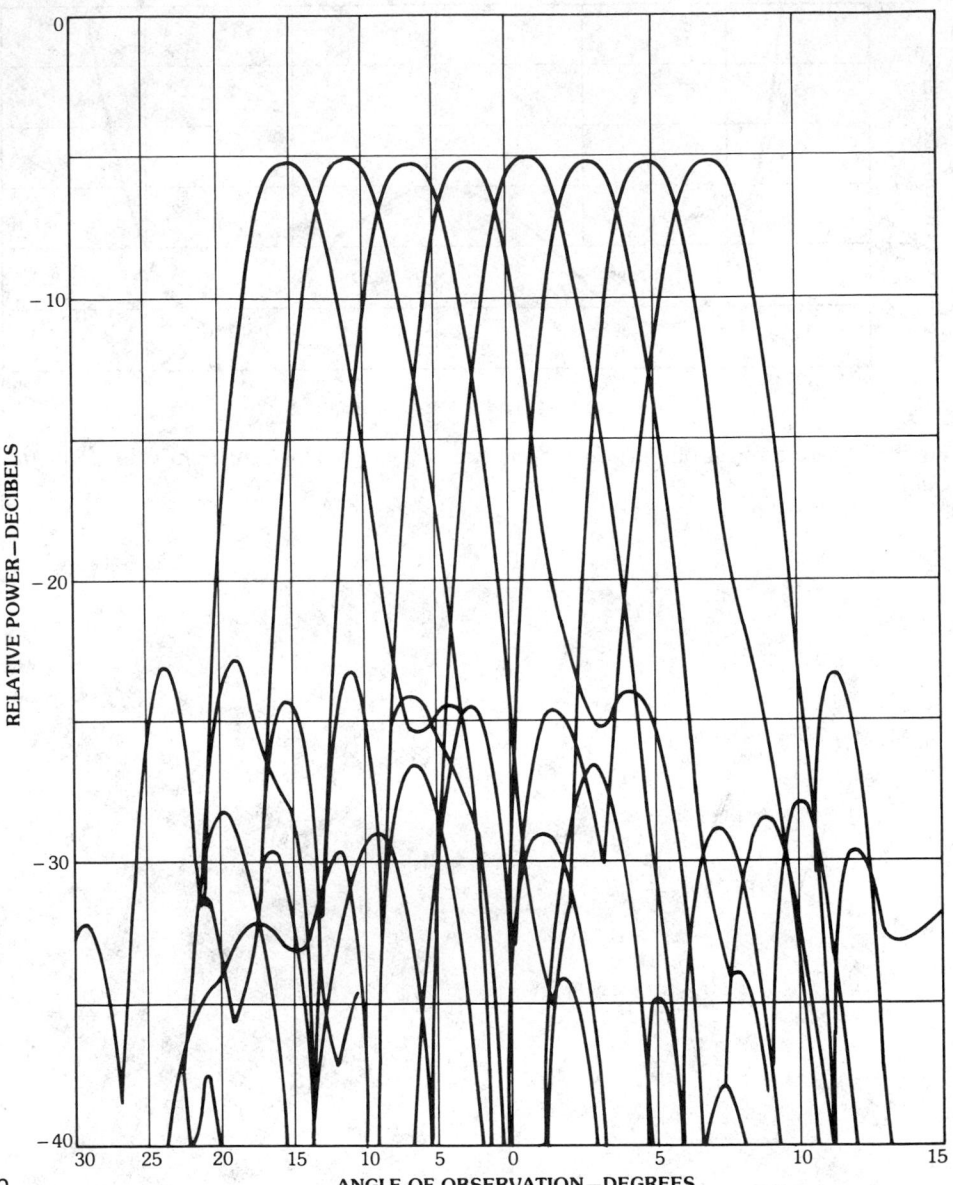

Fig. 96, *continued.*

$$\xi = \gamma \tag{40}$$

$$\ell_2 = \frac{b \sin 2\gamma}{\sin \gamma} = 2b \cos \gamma \tag{41}$$

From geometry it follows that

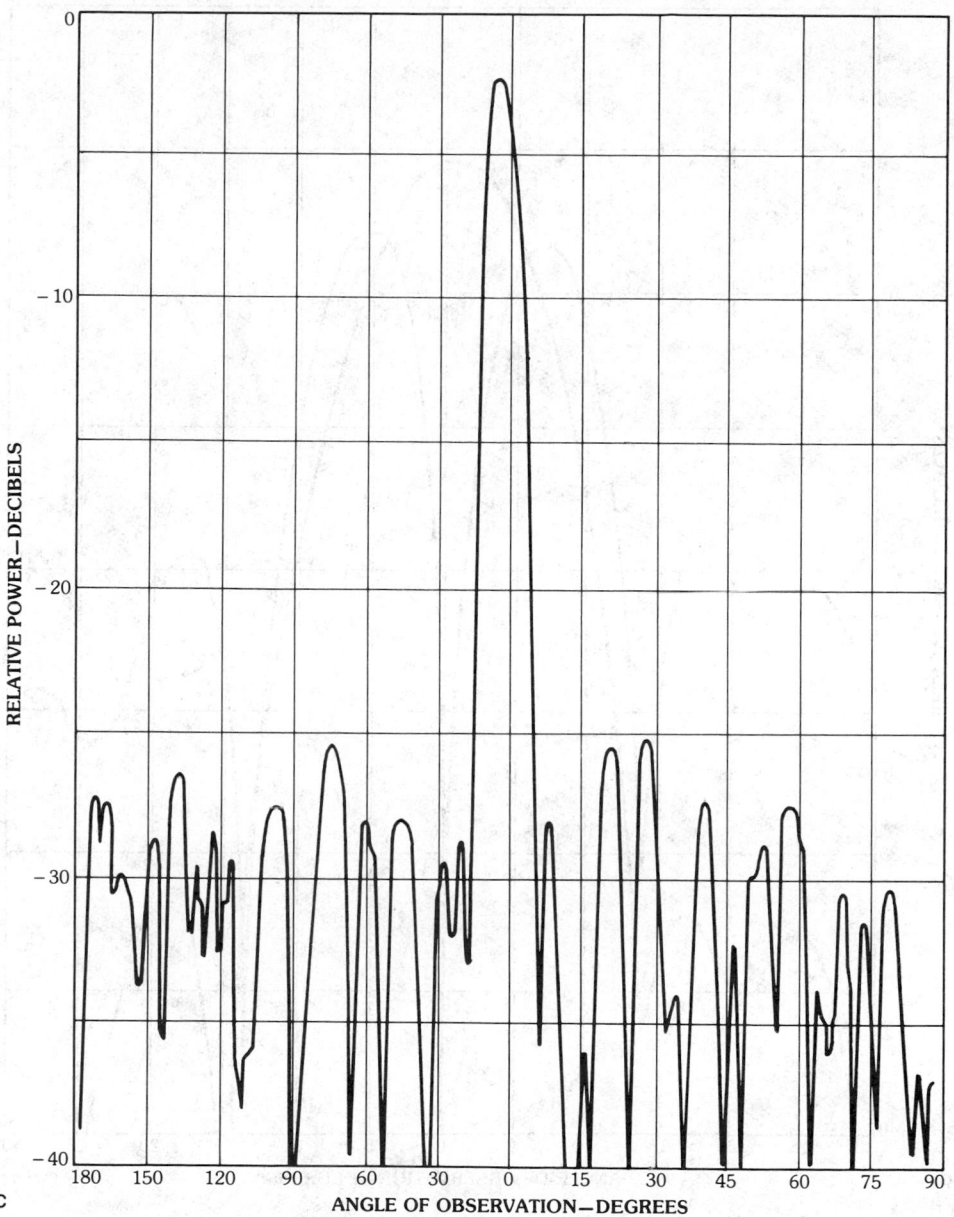

Fig. 96, *continued.*

$$\phi' = 2\gamma - \beta \tag{42}$$

Again using Snell's law, it follows that

$$\frac{\sin \phi''}{\sin \xi} = \frac{bn_2}{R} \tag{43}$$

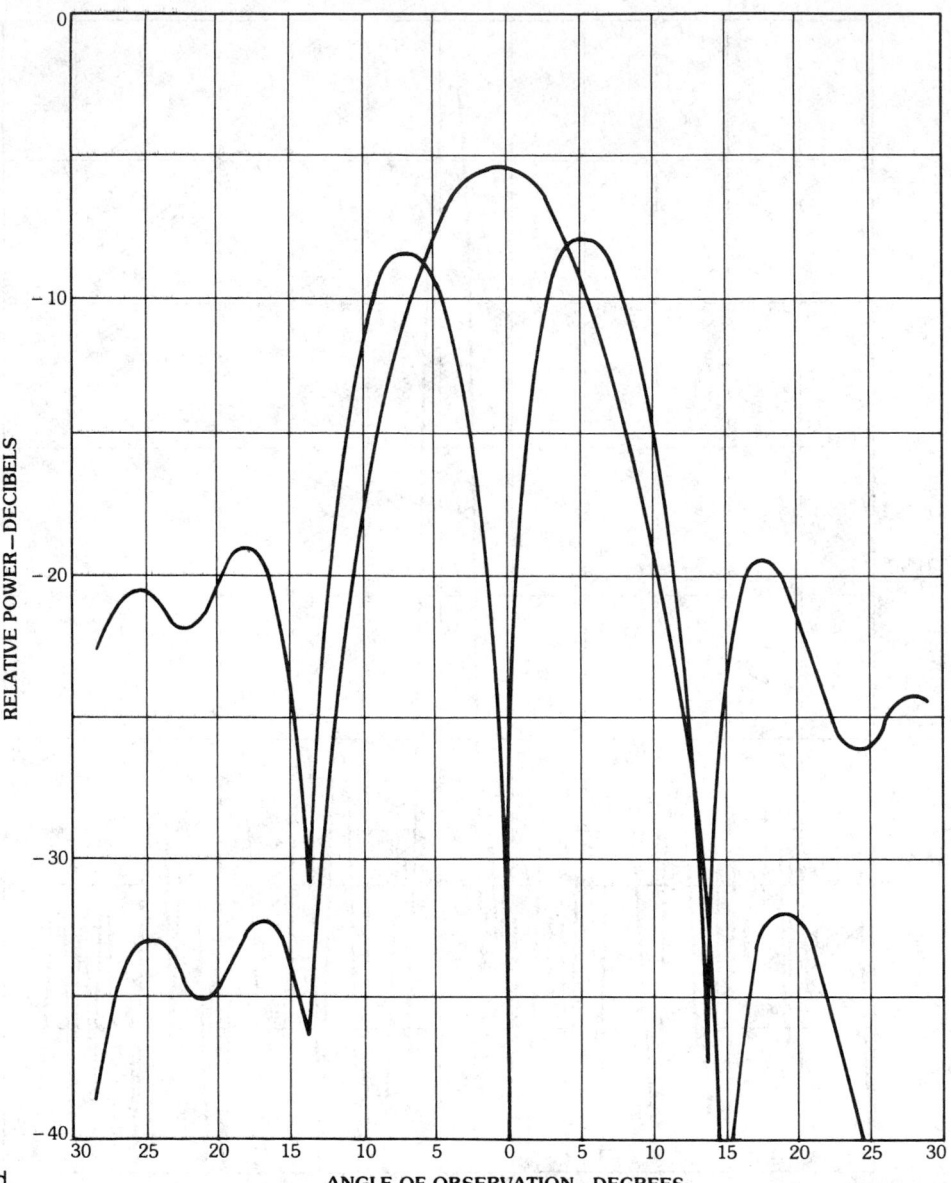

d

Fig. 96, *continued.*

From geometry,

$$\epsilon = \phi'' - \phi' \qquad (44)$$

$$\ell_3 = R(1 - \cos\phi')\sec\epsilon \qquad (45)$$

$$y = R\sin\phi' + R(1 - \cos\phi')\tan\epsilon \qquad (46)$$

Relations 34 through 46 then enable one to compute the path length error for all the rays, evaluated on the hypothetical plane wavefront in space. The parameters h/b, α, b/R, n_1, and n_2 are selected to minimize the overall path length error, as well as to establish physical sizes.

For example, Fig. 95 is the calculated phase error for a 256-element array, illustrating that the spherical aberration phase error is less than $\pm 20°$ over a $\pm 45°$ sector. Thus over 70 percent of the cylindrical array diameter is nearly perfectly phased. Moreover, the associated amplitude distribution falls off rapidly beyond $\pm 45°$. Consequently, a low-azimuth side lobe distribution is realizable using this technique. A high degree of amplitude distribution control is inherent, for both the sum and difference distributions independently, since an eight-element array of probes was used, each of which is amplitude controllable.

To demonstrate the feasibility of this approach, a 128-element array was constructed and tested at X-band. The beamwidth of the model was 5°. Fig. 96a shows the actual model. In the interest of time and funding, the lens was fabricated in one piece rather than connecting the cap and cylinder with cables; however, the array was coupled to the lens by coaxial probes.

The measured patterns of this model are shown in Figs. 96b through 96d. Fig. 96b shows several adjacent beams superposed. Fig. 96c shows a typical sum pattern, and Fig. 96d shows expanded sum and difference patterns. For economy of cost and time the feed used was a simple dual open-ended waveguide. Time did not allow an array-type feed which would have given more precise illumination with independent control over the sum and difference illuminations. Had an array-type feed been used, better patterns would have resulted, especially for the difference pattern.

Acknowledgments

The authors of this chapter gratefully acknowledge their colleagues in the Electromagnetics Laboratories of Hughes Aircraft Company, Fullerton, California, for their consultation and constructive criticism and especially Mrs. Lynda Schenet for typing, revising, and editing many cycles of the manuscript.

7. References

[1] L. A. Gustafson, "S-band two-dimensional slot array," *Tech. Memo 462*, Hughes Aircraft Company, Culver City, California, March 1957.

[2] R. M. Brown, "Performance of an antenna sharing the aperture of a frequency scanned array," *Rep. 8226*, Naval Research Lab, Washington, DC, May 1978.

[3] J. Butler and R. Lowe, "Beam forming matrix simplifies design of electronically scanned antenna," *Electron. Design*, vol. 9, no. 8, pp. 170–173, April 12, 1961.

[4] R. C. Hansen, *Microwave Scanning Antennas*, vols. 1, 2, 3, Array Systems, New York: Academic Press, 1966.

[5] W. P. Delaney, "An rf multiple beam forming technique," *IRE Trans. Mil. Electron.*, pp. 179–186, April 1962.

[6] J. P. Shelton and K. S. Kelleher, "Multiple beams from linear arrays," *IRE Trans.*, vol. AP-9, pp. 154–161, March 1961.

[7] W. R. Jones and G. F. Van Blaricum, "Multiple-beam forming hybrid networks,"

Hughes Aircraft Co., Ground Systems Group, *Tech. Memo. No. TP71-14-2*, April 7, 1970.

[8] H. J. Moody, "The systematic design of the Butler matrix," *IEEE Trans. Antennas Propag.*, vol. AP-12, pp. 786–788, November 1964.

[9] J. Blass, "The multidirectional antenna: a new approach to stacked beams," *1960 IRE Conv. Rec.*, pt. 1, pp. 48–50, 1960.

[10] J. Matthews and R. L. Walker, *Mathematical Methods of Physics*, 2nd ed., Menlo Park: Benjamin-Cummings, pp. 152–153, 1970.

[11] G. G. Chadwick, W. Gee, P. T. Lam, and J. L. McFarland, "An algebraic synthesis method for RN^2 multi-beam matrix network," internal publication of Lockheed Missile and Space Co., Sunnyvale, Calif., 94086. Also in *Proc. 1981 Antenna Appl. Symp.*, Allerton Park, Univ. of Illinois, September 23, 1981.

[12] J. L. McFarland, "The RN^2 multiple beam array family and beam forming matrix," *Proc. 1981 Antenna Appl. Symp.*, Allerton Park, Univ. of Illinois, September 23–25, 1981.

[13] R. J. Mailloux, "Array techniques for limited-scan applications," *Phys. Sci. Res. Papers, No. 503*, AFCRL-72-0421, Air Force Cambridge Research Laboratories, July 19, 1972.

[14] R. J. Mailloux and G. R. Forbes, "Experimental studies of a multiple mode array technique for limited-scan applications," *Phys. Sci. Res. Papers, No. 575*, AFCRL-TR-73-0686, Air Force Cambridge Research Laboratories, November 6, 1973.

[15] G. T. DiFrancia, "A family of perfect configuration lenses of revolution," *Optica Acta*, vol. 1, no. 4, pp. 157–163, February 1955.

[16] L. J. Chu and M. A. Taggart, "Pillbox antenna," US Patent No. 2,688,546, May 12, 1945.

[17] M. A. Taggart and E. C. Fine, "Parallel-plate bends," *MIT Radiation Lab Rep. 760*, September 5, 1945.

[18] J. S. Ajioka, "The development of an integral mkx iff antenna for low-frequency radar," *Rep. No. 534*, US Navy Electronic Lab, San Diego, January 1955.

[19] W. Rotman, "A study of microwave double-layer pillboxes," *AFCRC-TR-54-102*, Air Force Cambridge Research Center, Cambridge, Massachusetts, July 1954.

[20] J. S. Ajioka, "A multiple beam forming network using a multimode radial transmission line," *1963 NEREM Conv. Rec.*

[21] J. S. Ajioka, "A multiple beam forming antenna apparatus," US Patent No. 3,290,682, December 6, 1966.

[22] J. S. Ajioka and H. A. Uyeda, "Experimental performance of a multimode radial transmission line beam-forming network," *Microwave J.*, pp. 53–56, December 1968.

[23] J. L. McFarland, "Catenary geodesic lens antenna," US Patent No. 3,383,691, assigned to Hughes Aircraft Co., May 14, 1968.

[24] S. B. Meyer, "Parallel-plate optics for rapid scanning," *J. Appl. Phys.*, vol. 18, pp. 221–229, 1947.

[25] W. Rotman and R. F. Turner, "Wide-angle microwave lens for line source applications," *IEEE Trans. Antennas Propag.*, pp. 623–632, November 1963.

[26] J. Ruze, "Wide-angle metal plate optics," *Proc. IRE*, vol. 38, pp. 53–58, January 1950.

[27] H. Gent, "The bootlace aerial," *Royal Radar Establishment J.*, pp. 47–57, October 1957.

[28] C. M. Rappaport and A. I. Zaghoul, "Optimized three-dimensional lenses for two-dimensional scanning," *IEEE AP-S 1984 Symp. Dig.*, vol. II, June 25–29, 1984.

[29] R. F. Rinehart, "A solution of the problem of rapid scanning for radar antennae," *J. Appl. Phys.*, vol. 19, September 1948.

[30] R. K. Luneburg, *Mathematical Theory of Optics*, Providence: Brown University, pp. 189–213, 1944.

[31] E. C. DuFort and H. A. Uyeda, "A wide-angle scanning optical antenna," *IEEE Trans. Antennas Propag.*, vol. AP-31, no. 1, p. 60, January 1983.

[32] R. T. Hill, "Phased array feed systems," in *Phased Array Antennas*, cd. by A. A.

Oliner and G. H. Knittel, Dedham: Artech House, pp. 197–211, April 1972.

[33] J. R. Kahrilas, "HAPDAR—an operational phased array radar," *Proc. IEEE*, November 1968.

[34] W. T. Patton, "Limited-scan arrays," *Proc. 1970 Phased Array Antenna Symp.*, ed. by A. A. Oliner and G. H. Knittel, Dedham: Artech House, pp. 332–343, 1970.

[35] C. Winter, "Phase-scanning experiments with two reflector antenna systems," *Proc. IEEE*, vol. 56, no. 11, pp. 1984–1999, 1968.

[36] R. J. Mailloux and P. Blacksmith, "Array and reflector techniques for precision approach radars," *AGARD Conf. Proc. No. 139 on Antennas for Avionics*, NATO 26–30, November 1973.

[37] W. D. Fitzgerald, "Limited electronic scanning with a near-field Cassegrainian system," *Tech. Rep. 484*, ESD-TR-71-271, Lincoln Lab, 1971.

[38] Hughes Aircraft Company, Ground Systems Group, "Tradex S-band phased array design study," *Interim Rep. FP71-14-3*, Fullerton, California, May 27, 1971.

[39] W. D. Fitzgerald, "Limited electronic scanning with an offset-feed near-field Gregorian system," *Tech. Rep. 486*, ESD-TR-272, Lincoln Lab, MIT, 1971.

[40] Lincoln Lab, "A KREMS phased array radar," report dated April 26, 1971.

[41] C. J. Miller and D. Davis, "LFOV optimization study, final report," prepared by Westinghouse Defense and Electronics Systems Center, Systems Development Division, for MIT Lincoln Lab, Contract No. F19628-70-C-0230, May 1, 1972.

[42] Hughes Aircraft Company Ground Systems Group, "Study program for reflector surface optimization LFOV system (TRADEX/KREMS)," *FP No. 71-14-138*, Fullerton, California, August 19, 1971.

[43] C. J. Sletten, "Caustic matching: a new technique for improving limited-scan antennas," US Air Force Cambridge Research Laboratories, Hanscom Air Force Base, Massachusetts, December 17, 1974.

[44] C. H. Tang and C. F. Winter, final report on Contract No. AF19628-72-C-0213.

[45] M. Born and E. Wolf, *Principles of Optics*, New York: Pergamon Press, 1959.

[46] Hughes Aircraft Company Ground Systems Group, "Limited-scan antenna techniques study," final report on Contract No. F1962B-73-C-0129, Fullerton, California 92634, August 14, 1975.

[47] R. Tang, "Survey of time delay beam steering techniques," *Proc. 1970 Phased Array Antenna Symp.*, ed. by A. A. Oliner and G. H. Knittel, Dedham: Artech House, pp. 254–260, 1972.

[48] N. S. Wong, R. Tang, and E. E. Barber, "A multielement high-power monopulse feed for low side lobes and high aperture efficiency," *IEEE Trans. Antennas Propag.*, vol. AP-22, pp. 402–407, May 1974.

[49] E. C. DuFort, "Optical technique for broadbanding phased arrays," *IEEE Trans. Antennas Propag.*, vol. AP-23, pp. 516–523, July 1975.

[50] J. L. McFarland and J. S. Ajioka, "Multiple beam constrained lens design," *NEREM Rec.*, November 1962.

[51] L. Stark, "High-resolution hemispherical reflector antenna," US Patent No. 3,852,748, assigned to Hughes Aircraft Company, December 3, 1974.

[52] G. V. Borgiotti, "An antenna for limited scan in one plane: design criteria and numerical simulation," *IEEE Trans. Antennas Propag.*, vol. AP-25, no. 2, March 1977.

[53] R. J. Giannini, "An electronically scanned cylindrical array based on a switching and phasing technique," *1969 G-AP Intl. Symp. Program Dig.*, pp. 199–204, 1969.

[54] A. E. Holley, E. C. DuFort, and R. A. Dell-Imagine, "An electronically scanned beacon antenna," *IEEE Trans. Antennas Propag.*, vol. AP-22, no. 1, pp. 3–12, January 1974.

[55] J. E. Boyns, C. W. Gorham, A. D. Munger, J. H. Provencher, J. Reindel, and B. I. Small, "Step-scanned circular array antenna," *IEEE Trans. Antennas Propag.*, vol. AP-18, no. 5, pp. 590–595, September 1970.

[56] P. Sheleg, "A matrix-fed circular array for continuous scanning," *Proc. IEEE*, vol. 56, no. 11, November 1968.

[57] "Proposal for air traffic control radar beacon system (ATCRBS) electronic scan antenna," *RFP No. WA5R-1-0059*, submitted by Hughes Aircraft Company, Ground Systems Group, Fullerton, California, March 19, 1971.

Chapter 20

Antennas on Aircraft, Ships, or Any Large, Complex Environment

W. D. Burnside
The Ohio State University ElectroScience Laboratory

R. J. Marhefka
The Ohio State University ElectroScience Laboratory

CONTENTS

Walter D. Burnside is a professor in the Electrical Engineering Department at The Ohio State University (OSU), from which he received his BSEE, MS (1968), and PhD (1972) in electrical engineering. He graduated valedictorian of his undergraduate class and received from the IEEE (Institute of Electrical and Electronic Engineers) Antennas and Propagation Society the R. W. P. King Award (1975), the Best Application Paper Award (1975), and the Best Paper Award (1980). From OSU, he received the Teacher of the Year Award in Electrical Engineering (1980) and the College of Engineering Harrison Award for Outstanding Young Faculty Member (1983). Dr. Burnside is a Fellow of the IEEE (1985); is a member of Sigma Xi, Eta Kappa Nu, Phi Eta Sigma, and Tau Beta Pi; and was Associate Editor of the *IEEE Transactions on Antennas and Propagation* for five years. He has engaged in a wide variety of electromagnetic studies, which include both theoretical and experimental efforts. His main contributions have been in the application of electromagnetic analysis or measurement techniques to solve complex real-world problems; and, more recently, he has been very active in the analysis, measurement, and control of scattering from complex targets.

Ronald Joseph Marhefka was born in Cleveland, Ohio, on June 2, 1947. He received the BSEE degree from Ohio University, Athens, in 1969, and the MS and PhD degrees in electrical engineering from The Ohio State University, Columbus, in 1971 and 1976, respectively.

Since 1969, he has been with The Ohio State University Electro-Science Laboratory, where he currently is a research scientist. His research interests are in the areas of developing and applying high-frequency asymptotic solutions, such as the uniform geometrical theory of diffraction, hybrid solutions, and other scattering techniques. He has applied these methods to numerous practical antenna and scattering problems, including airborne, spacecraft, and shipboard antenna analysis and radar cross-section prediction.

He is the author of the user-oriented computer code, the NEC-Basic Scattering Code. He has written over 70 technical reports and papers. In 1975, he coauthored a paper that won the IEEE Antennas and Propagation Society's Best Application Paper and one that won the R. W. P. King Award.

Dr. Marhefka is a member of IEEE, Tau Beta Pi, Eta Kappa Nu, Phi Kappa Phi, and Sigma Xi. He has served as Secretary-Treasurer, Vice Chairman, and Chairman of the Columbus joint chapter of the IEEE Antennas and Propagation and Microwave Theory and Techniques societies during 1977–1980, respectively.

1. Introduction

The overall capability of an electromagnetic radiating system is dependent on its ability to operate effectively in a complex environment, in that its pattern performance can be adversely limited by pattern distortion effects, such as blockage and structural scattering. In many cases these detrimental effects can be minimized by judiciously locating the antennas. This task is complicated by the large number of systems that are competing for prime locations on, for example, a modern military ship. Without an efficient means to position such systems one normally attempts to use locations similar to previous designs, which may be inexpensive but are certainly not optimum. As a result there is a great need for electromagnetic tools that can efficiently evaluate the pattern performance of radiating systems in their proposed environment.

As with most engineering performance questions the antenna engineer can resort to experimental as well as theoretical solutions. Before applying either approach the engineer needs to examine its properties so that he can most effectively determine and evaluate his design. Due to the size and complexity of most structures it is rather obvious that experimental results will be expensive; yet they are very pleasing in that they potentially provide real-world results. For simple structures, such as a horn antenna radiating in free space, one can gain some insight into modifications necessary to improve its performance. For a horn antenna radiating on a military aircraft, however, the radiation patterns are normally far too complex to allow one to relate structural effects to corresponding measured results. Therefore experimental results are not very useful as a diagnostic tool but are most appropriate for evaluating the performance of various competing systems. Theoretical solutions, on the contrary, are used to numerically simulate the real-world situation and provide insight into the scattering mechanisms creating the resulting pattern performance. From the previous comments it is apparent that, whenever possible, it is most appropriate to use both a theoretical and an experimental approach to design a radiating system for application in a complex environment. The theoretical solution would be used to efficiently evaluate various alternatives and suggest a few candidate systems. These systems could then be evaluated in detail using experimental measurements. Of course, this combined approach sounds very attractive; however, it is predicated on the availability of a theoretical solution. The development of such numerical solutions is the main theme of this chapter.

Before the advent of digital computers, numerical electromagnetic solutions were limited to more classical geometries such as cylinders, cones, spheres, etc. With the ready access to powerful computer systems a designer is now able to construct numerical electromagnetic solutions for much more complex structures. These computer solutions have also required the development of new techniques which are more accurate and efficient for numerical calculations. Two major numerical solutions which have found great success in such applications are the method of moments and the uniform geometrical theory of diffraction (UTD). These two theories complement each other nicely in that the former is used to solve structures which are small in terms of the wavelength, whereas the latter is applicable for electrically large geometries. This chapter is devoted to electrically large structures which can be solved using the UTD; however, many of the general statements also apply to the method of moments solutions.

The uniform geometrical theory of diffraction (UTD) is used here based on its ability to simulate many complex structures [1–8] as well as being numerically amenable, accurate, and efficient. It is based on a ray optical format which is used to determine components of the field incident on and diffracted by the various structures. Components of the diffracted field are found using the UTD solutions in terms of individual rays which are summed with the geometrical optics terms in the far field of each scattering center. Note that one can be in the far field of all the scattering centers yet in the near field of the structure. The rays from a given scatterer tend to interact with other structures, causing various higher-order terms. These various possible combinations of rays that interact between scatterers can be traced out with only the dominant ones included in the final solution. Thus one need not be concerned with all other higher-order terms. This method normally leads to accurate and efficient computer codes that can be systematically written and tested, in that complex problems can be built up from simpler pieces.

The limitations associated with the UTD numerical solutions result mainly from the basic nature of the theoretical analysis. This point is discussed further in Chapter 4; in addition, the limitations associated with the application of the UTD for practical results are presented throughout this chapter.

The UTD simulation models are based on representing the actual structure in terms of a collection of much simpler parts, such as flat plates and elliptic cylinders. The use and validity of these simulation models are presented and verified through numerous applications and comparisons with experimental results. The theory only limits the structures to be locally defined by simple canonical shapes; however, the simple structures mentioned lead to less complex computer codes. The Airborne Code and the NEC-Basic Scattering Code (BSC) are used to demonstrate the usefulness and accuracy of UTD user-oriented numerical solutions.

In order to represent a large variety of antennas it is most appropriate to develop a Green's function type solution for infinitesimal current radiators. If the current distribution of the antenna is known, such as from a moment method analysis, an iteration scheme, or other means, the field from the various weighted current segments can be summed to give the total field. If the scattering centers are in the far zone of the antenna as a whole, the antenna can be represented by its pattern factor or, in the case of an array of antennas, by its array factor. This can save a large amount of computation time by eliminating multiple calculations for each infinitesimal current element.

The antenna can be mounted either on or off the various structures; however, if the antenna is mounted on a curved surface, then one must use the UTD radiation solutions. On the other hand, if the antenna is not located on or near a curved surface, then the UTD scattering solutions are applicable. Consequently these two problems require two different analyses and result in distinct computer codes as will be more apparent later.

The UTD theoretical solutions simply require the receiver to be in the far field of the individual scattering centers. This is extremely significant in that one can develop a UTD numerical solution in terms of a general receiver location, or a solution which does not change as the receiver moves from the near to the far field. This allows one to put a patch type antenna, for example, on an actual aircraft and measure any convenient pattern in the near field of the structure. Then one could construct a simulation structure, verify the model, and compute the desired far-field pattern. This could also greatly reduce the cost associated with obtaining experimental results.

Once the UTD numerical solutions are constructed for general structures one could envision an antenna designer sitting at a graphics display terminal, trying various configurations and determining some candidate antennas and sites for the desired application. These various candidates could then be evaluated using appropriate near-field experimental results. This approach should allow an antenna designer to integrate his antenna efficiently into a complex environment.

2. Numerical Simulation of the Antenna

The general-purpose solutions discussed here can be used with a wide range of antennas. They are basically a Green's function solution; that is, the solution has been developed for infinitesimally small current elements. Any antenna can be modeled, therefore, as a array of fundamental radiators if its current distribution is known. This can be quite a challenge in practice, since many of the antennas used are heuristically designed and are of a very complex physical nature. The current distribution of an antenna can be obtained in many different ways, such as a method of moments analysis [9–11], a synthesis scheme [12, 13], known aperture distributions, measurements, or other means.

The degree to which the current and/or radiation pattern of the antenna under investigation needs to be known depends on the problem being solved. The impedance of an antenna is highly dependent on the exact current distribution on the antenna, since the impedance is proportional to the voltage divided by the current at the terminals. If the impedance of an antenna in the presence of the complex environment is desired, it is necessary to use a method of moments technique or a hybrid method of moments and UTD approach [14] from which the currents can be calculated. The radiation pattern of an antenna, however, is not as sensitive to the current distribution on the antenna. This is because the radiation pattern is an end result of an integration of the currents, which averages out small variations in the distribution. The question is then: given a complex antenna, what is the best method to reproduce a good facsimile of the radiation pattern in the most efficient way possible? One assumption that can be made here is that the radial component of the elemental radiators is not needed. If the near field of an antenna is desired, it is necessary to use infinitesimal elements and integrate them numerically, reproducing the near field for the antenna as a whole.

If the scattering centers are in the far zone of the antenna as a whole, the antenna can most efficiently be represented by its pattern factor or, in the case of an array of antennas, by its array factor [15, 16]. This can save a large amount of computation time by eliminating multiple calculations for each infinitesimal current element. The general form of the transmitted electric field **E** from an antenna at a distance r using pattern factors is given by

$$\mathbf{E} = j Z_0 I_m k \, \mathbf{h} \frac{e^{-jkr}}{4\pi r} \tag{1}$$

for an electric source type and

$$\mathbf{E} = -j K_m k \, \mathbf{h} \frac{e^{-jkr}}{4\pi r} \tag{2}$$

for a magnetic source type, where k is the wave number and Z_0 is the wave impedance of free space. The function **h** is the vector effective height of the antenna, which can be given as

$$I_m \mathbf{h} = \hat{\boldsymbol{\theta}} \, L w J_m \, \sin \theta \, F_x(\theta, \phi) \, F_z(\theta) \, F_a(\theta, \phi) \tag{3}$$

for an electric source type. The magnetic source type is the same except $\hat{\boldsymbol{\phi}}$ is substituted for $\hat{\boldsymbol{\theta}}$, K_m for I_m, and M_m for J_m. The length of the source is given as L and the width as w. The $\sin \theta$ term comes from the pattern factor of a elemental radiator. The function F_z is due to the length, F_x is due to the width, and F_a is an array pattern factor. For a line source $w \to 0$; therefore $w J_m \to I_m$ and $F_x = 1$. The form of the pattern factors depends on the current distribution [15, 16] and will not be given here.

The array pattern factor is a useful concept if the scattering centers are relatively far away from the elements of an array or the infinitesimal elements of a discretized current distribution. A number of the elements can be clustered together with a common phase center, hence a single ray origin. Weaker elements in an array can be clustered more easily since their overall effect on the pattern is less critical.

Many times the pattern of the antenna in free space or on a ground plane is the only information that is known about an antenna. It is possible to use a synthesis scheme to find an equivalent distribution across the aperture of an antenna that will reproduce the antenna pattern reasonably well. The synthesis scheme could be found from the far-zone power pattern of the antenna [12] or more accurately from near-zone measurements, which include phase information [13]. It should be noted that it is not necessary that the calculated currents be the exact currents on the antenna if all that is desired is a facsimile of the pattern.

Interpolation of the measured pattern data is another possible scheme that can be used if that is all that is available. This method is not as desirable as some of the other methods because usually only the principal-plane patterns have been measured. These patterns cannot always be used in a manner to reproduce the patterns well in the off principal planes.

The spatial coupling between antennas can be obtained approximately by using the vector antenna height concept. This can be derived from the generalized reaction principle for antennas in a scattering environment. The total open-circuit voltage at the terminals of the receiver can be obtained by adding the dot product of the **E** fields from the various rays incident on the receiving antenna by the vector effective height of the receiving antenna in that direction, that is,

$$V_{\text{oc}} = \mathbf{h}(\theta, \phi) \cdot \mathbf{E}^i \tag{4}$$

This quantity can be normalized to a more useful term, such as power out over power in, by the following equation [7]:

$$\frac{P_{\text{out}}}{P_{\text{in}}} = \frac{|1/2\, I_m\, V_{\text{oc}}|^2}{P_{rt}\, P_{rr}} \left[\frac{R_r R_L}{|Z_r + Z_L|^2} \right] \tag{5}$$

where P_{rt} is the power radiated by the transmitter and P_{rr} is the power radiated by the receiver if it were a transmitter. The current weight at the receiving terminals is I_m, the radiation impedance is Z_r, and the load impedance is Z_L. The Rs are the resistive parts of the impedances. Note that the power radiated terms can be obtained by a method of moments code or they can be approximated by integrating the volumetric radiation pattern of the antenna in free space or on a ground plane if necessary. This coupling formula is a more general statement of Friis' transmission formula that takes into account the scattering of the environment around the antennas. It is a useful approximation if the antennas are sufficiently far apart from each other or are not near large scattering structures, that is, so that the currents on the antennas are not significantly modified by each other or the environment. This is reasonably accurate as long as the antennas are more than a half-wavelength apart and if they are about a quarter-wavelength from an edge [14]. Of course, if an antenna is over a large ground plane or some other large reflecting structure, the method applied to find the currents can use image theory to include the ground effect.

3. Numerical Simulation of the Environment

There are many questions that must be addressed before one embarks on the development of a numerical electromagnetic solution for a complex structure. The first problem considered here is usually the last question addressed by the electromagnetic analyst; however, it is of great concern to the antenna designer. Using a numerical solution the designer has to specify the geometry to the computer, which implies by necessity that the numerical model simply simulates the actual antenna environment. How well, then, does the computer model actually represent the real world? The most obvious simulation model would include all features associated with the actual structure; in addition it would represent each feature to a high degree of accuracy.

If one attempts to define the geometry associated with a shipboard antenna, it would quickly become very clear that the complete structure can not be simulated to a high degree without a great deal of effort. Even if one could describe such a geometry to the computer the electromagnetic analysis needed to solve such a configuration would be overwhelming to develop, very inefficient to run, and most likely would be limited to specific configurations. Since measurements can ultimately provide the real-world results, in most cases the efficiency of a numerical solution is its most outstanding attribute. With an efficient solution an antenna designer can very quickly evaluate various configurations and narrow design choices down to a few practical solutions.

Since the efficiency of the numerical solution plays such a dominant role in its usefulness, let us examine the features of the solution which dictate its execution time. It has already been mentioned that the geometry used in the analysis must be of such a form that it can be easily defined by the user; however, it must be general enough to describe a complex antenna environment, such as an aircraft. In addition it is most appropriate if the computer model represents the basic shapes being analyzed by the code. This is very useful in terms of evaluating the accuracy of the resulting patterns. With these concepts in mind it is advantageous to divide a complex structure into simpler parts or substructures. These substructures should then be the basic theoretical geometries analyzed by the code so that they can be very efficiently calculated as well as being easily input to the computer.

The evaluation of the substructures used in the simulation model revolves around the electromagnetic analysis needed to solve the class of problems being studied. Since this chapter is devoted to electrically large structures the UTD as described in Chapter 4 will be exclusively applied here. For the sake of brevity the actual equations used to analyze the various structures will not, in general, be given, except to introduce the basic concepts of the UTD approach.

In order to demonstrate the implementation of a UTD solution let us consider the radiation pattern of a short dipole in the presence of a finite flat plate as shown in Fig. 1. Note that the flat-plate geometry is considered here because it can be easily input to the computer and can be efficiently analyzed using the UTD. The total UTD solution consists to the first order of the superposition of the incident, reflected, edge-diffracted, and corner-diffracted fields as described in the following.

The incident field is basically the line-of-sight signal going directly from the dipole to the receiver. Using the UTD methodology the incident field is set to zero if the ray from the dipole intercepts the plate before reaching the receiver, as shown in Fig. 2. Thus the incident field is discontinuous at the incident shadow boundaries, where the field abruptly goes to zero.

The reflected field from the finite flat plate is determined using image theory. To begin the solution the receiver image position is found as illustrated in Fig. 3. The source solution is then used to compute the electric field $\mathbf{E}^S$ at the image position. This field would exist at that point if the plate were not present. The reflected field $\mathbf{E}^r$ is then given by

$$\begin{bmatrix} E^r_x \\ E^r_y \\ E^r_z \end{bmatrix} = \begin{bmatrix} T_{xx} & T_{xy} & T_{xz} \\ T_{yx} & T_{yy} & T_{yz} \\ T_{zx} & T_{zy} & T_{zz} \end{bmatrix} \begin{bmatrix} E^S_x \\ E^S_y \\ E^S_z \end{bmatrix} \tag{6}$$

where $[T]$ represents the reflected field polarization transformation matrix which satisfies the boundary conditions on the plate. This matrix is determined using the normal to the plate such that

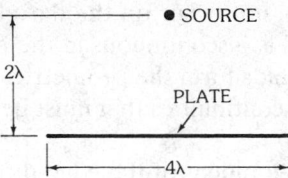

Fig. 1. Geometry for a source in the presence of a plate.

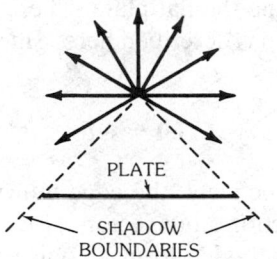

Fig. 2. Illustration of source-ray paths.

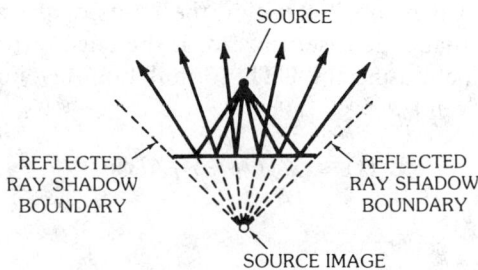

Fig. 3. Illustration of plate-reflected ray paths.

$$\mathbf{E}^r = (\hat{\mathbf{n}} \cdot \mathbf{E}^S)\hat{\mathbf{n}} - (\hat{\mathbf{t}} \cdot \mathbf{E}^S)\hat{\mathbf{t}}$$
$$(\hat{\mathbf{t}} \cdot \mathbf{E}^S)\hat{\mathbf{t}} = \mathbf{E}^S - (\mathbf{E}^S \cdot \hat{\mathbf{n}})\hat{\mathbf{n}} \quad (7)$$
$$\mathbf{E}^r = 2(\hat{\mathbf{n}} \cdot \mathbf{E}^S)\hat{\mathbf{n}} - \mathbf{E}^S$$

or

$$T_{xx}\hat{\mathbf{x}} + T_{yx}\hat{\mathbf{y}} + T_{zx}\hat{\mathbf{z}} = 2(\hat{\mathbf{n}} \cdot \hat{\mathbf{x}})\hat{\mathbf{n}} - \hat{\mathbf{x}}$$
$$T_{xy}\hat{\mathbf{x}} + T_{yy}\hat{\mathbf{y}} + T_{zy}\hat{\mathbf{z}} = 2(\hat{\mathbf{n}} \cdot \hat{\mathbf{y}})\hat{\mathbf{n}} - \hat{\mathbf{y}} \quad (8)$$
$$T_{xz}\hat{\mathbf{x}} + T_{yz}\hat{\mathbf{y}} + T_{zz}\hat{\mathbf{z}} = 2(\hat{\mathbf{n}} \cdot \hat{\mathbf{z}})\hat{\mathbf{n}} - \hat{\mathbf{z}}$$

where the wedge (^) denotes a unit vector. Since the **T** matrix is independent of the receiver location it is stored in order to optimize the computational eficiency of the numerical solution. The total reflected field is then computed using (6), unless the reflection point does not occur on the flat plate, in which case it is set to zero. Thus the reflected field is discontinuous at the reflection shadow boundaries. The incident plus reflected fields form the geometrical optics (GO) solution, which is inherently composed of discontinuities that must be corrected before the solution is complete.

The first step in the development of the edge-diffraction solution is to find the edge-diffraction point. It is known that for a given receiver location there is only one point along an infinitely long straight edge at which the diffracted field can emanate for a near-zone source and/or receiver. Thus this point must be found for each of the edges that describe the flat plate. There are many ways of finding this diffraction point, one of which is described here. Since it is known that $\beta_0 = \beta_0'$ (see Fig. 4), it is obvious that

$$\hat{\mathbf{e}} \cdot \hat{\mathbf{i}} = \hat{\mathbf{e}} \cdot \hat{\mathbf{d}} \tag{9}$$

where ($\hat{\mathbf{e}}$, $\hat{\mathbf{i}}$, and $\hat{\mathbf{d}}$) are, respectively, the edge unit vector, incident direction unit vector, and diffraction direction unit vector. Note that the edge-diffracted rays emanating from a point on an edge form a conical surface making the half-cone angle β_0. In order to solve for the unique edge-diffraction point one merely projects the source and receiver onto the edge in a normal to the edge vector sense. Then by using similar triangles one can determine the isolated diffraction point. If the edge-diffraction point does not fall within the limits of the actual edge, then the diffracted field from that edge is set to zero. If the edge-diffraction point is on the edge, the diffracted field using the UTD formulation developed by Kouyoumjian and Pathak [17] is given by

$$\mathbf{E}^d(s) \cong \mathbf{E}^i(Q_E) \cdot \bar{\bar{\mathbf{D}}}_E A(s) e^{-jks} \tag{10}$$

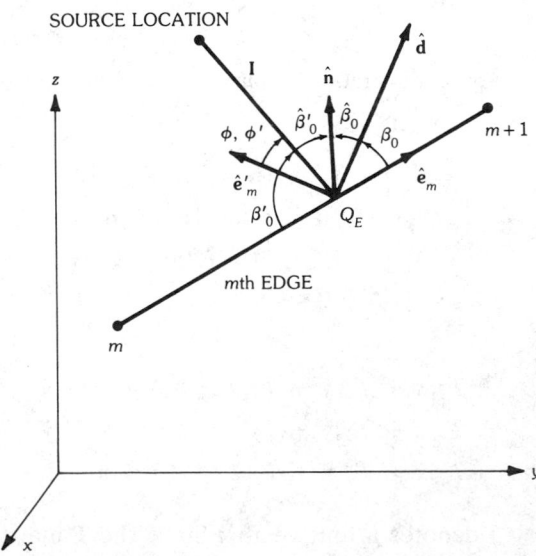

Fig. 4. Edge coordinate system at point of diffraction (Q_E).

where $A(s)$ is the spread factor given by $A(s) = \sqrt{s'/s(s + s')}$ with s and s' the distance from the edge to the receiver and source, respectively. Note that the diffraction coefficient is given by

$$\bar{\bar{\mathbf{D}}}_E = -\hat{\beta}'_0\hat{\beta}_0 D_s - \hat{\phi}'\hat{\phi}D_h \qquad (11)$$

and

$$D_{s,h}(\phi, \phi', \beta_0) = \frac{-e^{-j\pi/4}}{2n\sqrt{2\pi k}\,\sin\beta_0}\left[\cot\left(\frac{\pi + \beta^-}{2n}\right)F[kLa^+(\beta^-)]\right.$$

$$+ \cot\left(\frac{\pi - \beta^-}{2n}\right)F[kLa^-(\beta^-)] \mp \left\{\cot\left(\frac{\pi + \beta^+}{2n}\right)\right.$$

$$\left.\left. \times F[kLa^+(\beta^+)] + \cot\left(\frac{\pi - \beta^+}{2n}\right)F[kLa^-(\beta^+)]\right\}\right] \qquad (12)$$

and

$$F(X) = 2j|\sqrt{X}|e^{jX}\int_{|\sqrt{X}|}^{\infty} e^{-j\tau^2}\,d\tau$$

$$\beta^{\mp} = \phi \mp \phi' \qquad (13)$$

where $F(X)$ is called the *transition function* and the parameters $a^{\pm}(\beta^{\pm})$ are defined in Reference 17 and Chapter 4. The geometry associated with this solution is illustrated in Fig. 5. Note that $\mathbf{E}^i(Q_E)$ is the field incident on the edge at the diffraction point. The complexity of the edge-diffraction solution is lumped into the

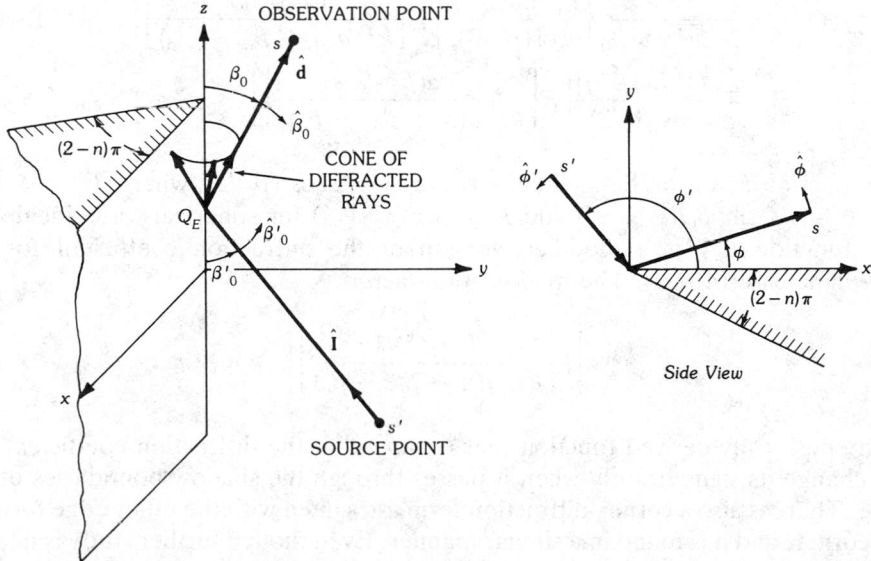

Fig. 5. Geometry for three-dimensional wedge-diffraction problem.

transition function which is illustrated in Fig. 6. Due to the simplicity of this term the edge-diffracted field is rapidly computed, as will be discussed later.

The edge-diffracted field is also abruptly discontinuous at the incident and reflection shadow boundaries in just the precise way to make the total field continuous. On the lit side of the shadow boundary the diffracted field approaches minus one-half the shadowed field, whereas on the shadow side the diffracted field simply changes sign. Thus at the shadow boundary the shadowed plus edge-diffracted field is approximately one-half the shadowed field magnitude.

The first-order UTD solution is not complete until the corner-diffracted field is added. Since the edge-diffracted field is also abruptly discontinuous when the diffraction point falls off the actual edge, the corner term is needed to compensate for this behavior.

The corner-diffraction problem is illustrated in Fig. 7. The corner-diffracted fields associated with one corner and one edge in the near field with spherical-wave incidence are given by [3, 18]

$$
\begin{Bmatrix} E_\parallel^c \\ E_\perp^c \end{Bmatrix} = \begin{Bmatrix} IZ_0 \\ MY_0 \end{Bmatrix} \frac{\sqrt{\sin\beta_c \sin\beta_{0c}}}{(\cos\beta_{0c} - \cos\beta_c)} F[kL_c a(\pi + \beta_{0c} - \beta_c)] \frac{e^{-jks}}{4\pi s} \tag{14}
$$

where

$$
\begin{Bmatrix} I \\ M \end{Bmatrix} = - \begin{Bmatrix} E_\parallel^i(Q_C) \\ E_\perp^i(Q_C) \end{Bmatrix} \begin{Bmatrix} C_s(Q_E)Y_0 \\ C_h(Q_E)Z_0 \end{Bmatrix} \sqrt{\frac{8\pi}{k}} e^{-j(\pi/4)} \tag{15}
$$

and

$$
C_{s,h}(Q_E) = \frac{-e^{-j(\pi/4)}}{2\sqrt{2\pi k}\,\sin\beta_0} \left\{ \frac{F[kLa(\beta^-)]}{\cos(\beta^-/2)} \left| F\left[\frac{La(\beta^-)/\lambda}{kL_c a(\pi + \beta_{0c} - \beta_c)} \right] \right| \right.
$$
$$
\left. \mp \frac{F[kLa(\beta^+)]}{\cos(\beta^+/2)} \left| F\left[\frac{La(\beta^+)/\lambda}{kL_c a(\pi + \beta_{0c} - \beta_c)} \right] \right| \right\} \tag{16}
$$

The function $F(X)$ was defined earlier, $a(\beta^\pm) = 2\cos^2(\beta^\pm/2)$, where $\beta^\pm = \phi \pm \phi'$ and $L = s's''\sin^2\beta_0/(s' + s'')$ and $L_c = s_c s/(s_c + s)$ for spherical-wave incidence. The function $C_{s,h}$ is a modified version of the diffraction coefficient for the half-plane case $(n = 2)$. The modification factor

$$
\left| F\left[\frac{La(\beta^\pm)/\lambda}{kL_c a(\pi + \beta_{0c} - \beta_c)} \right] \right| \tag{17}
$$

is an empirically derived function that ensures that the diffraction coefficient will not change its sign abruptly when it passes through the shadow boundaries of the edge. There is also a corner-diffraction term associated with the other edge forming the corner, and it is found in a similar manner. Even though further study is needed to improve the description of this diffraction mechanism, the writers believe that the benefits obtained from its inclusion warrant its use here.

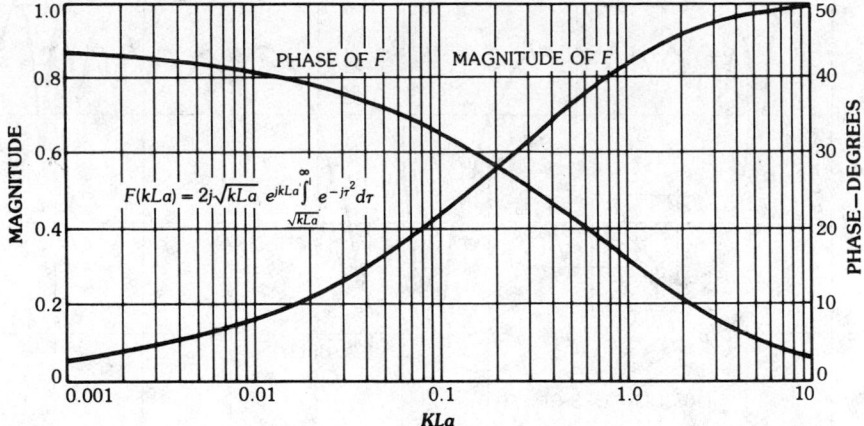

$$F(kLa) = 2j\sqrt{kLa}\ e^{jkLa}\int_{\sqrt{kLa}}^{\infty} e^{-j\tau^2}d\tau$$

Fig. 6. Transition function.

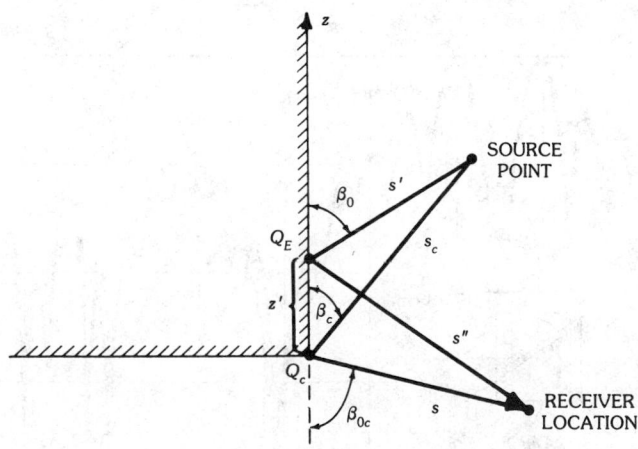

Fig. 7. Geometry for corner-diffraction problem.

This example illustrates very nicely the various attributes associated with UTD numerical solutions. First, the complexity of the solution is lumped into various diffraction coefficients which can be written in subroutine form. Since these coefficients consist of trigonometric and transition function terms, they are very easily and rapidly computed.

This UTD solution for a short dipole radiating in the presence of a finite flat plate can be used to compute the radiation pattern in either the near or far field of the structure. The only limitation is that the receiver not get any closer than a wavelength to an edge or the dipole. The total far-zone *E*- and *H*-plane patterns are shown in Fig. 8. Note the very good agreement obtained between the calculated and measured results in the *H*-plane pattern. In order to illustrate the significance of the various UTD terms they are individually plotted in Fig. 9. The individual terms are shown to the same radiation level so one can acquire a feel for the

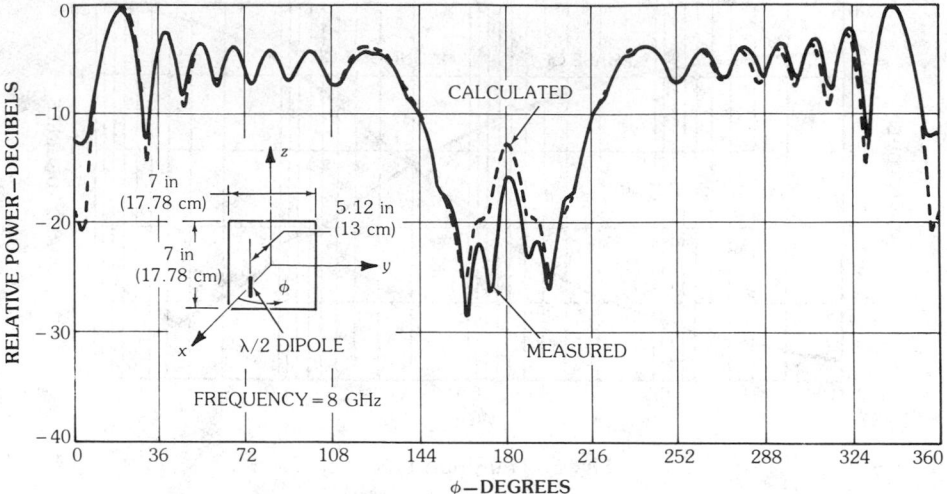

a

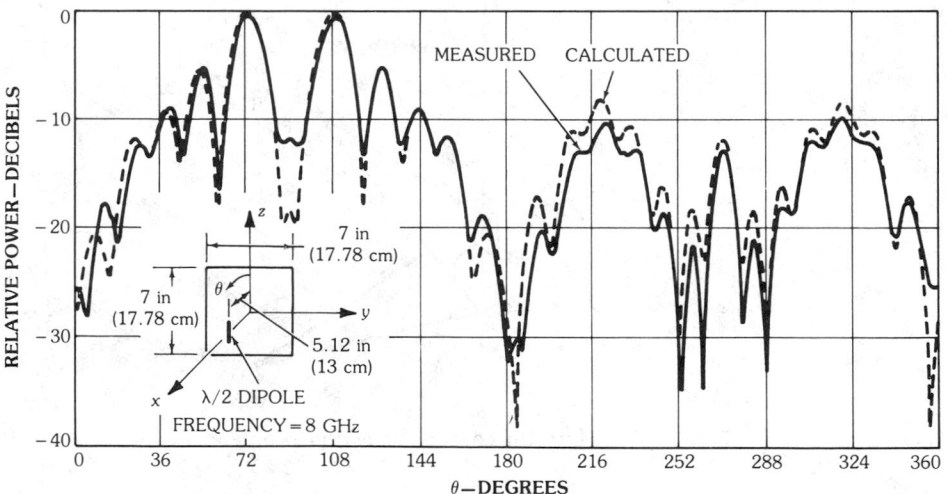

b

Fig. 8. Measured and calculated *H*- and *E*-plane patterns for a half-wave dipole located above a square plate. (*a*) *H*-plane patterns: $E_{\theta p}$. (*b*) *E*-plane patterns: $E_{\phi p}$.

significance of each mechanism. In addition the various discontinuities associated with each term are clearly illustrated.

The total field solution consists of the superposition of a few terms, each of which is associated with an isolated scattering point on the structure. This leads to two useful aspects of the solution: (1) the pattern indicates when the solution fails in that each term introduces a discontinuity which must be corrected in a precise way, and (2) pattern problem sectors can be associated with particular mechanisms which in turn can pinpoint regions on the structure from which the problem term or

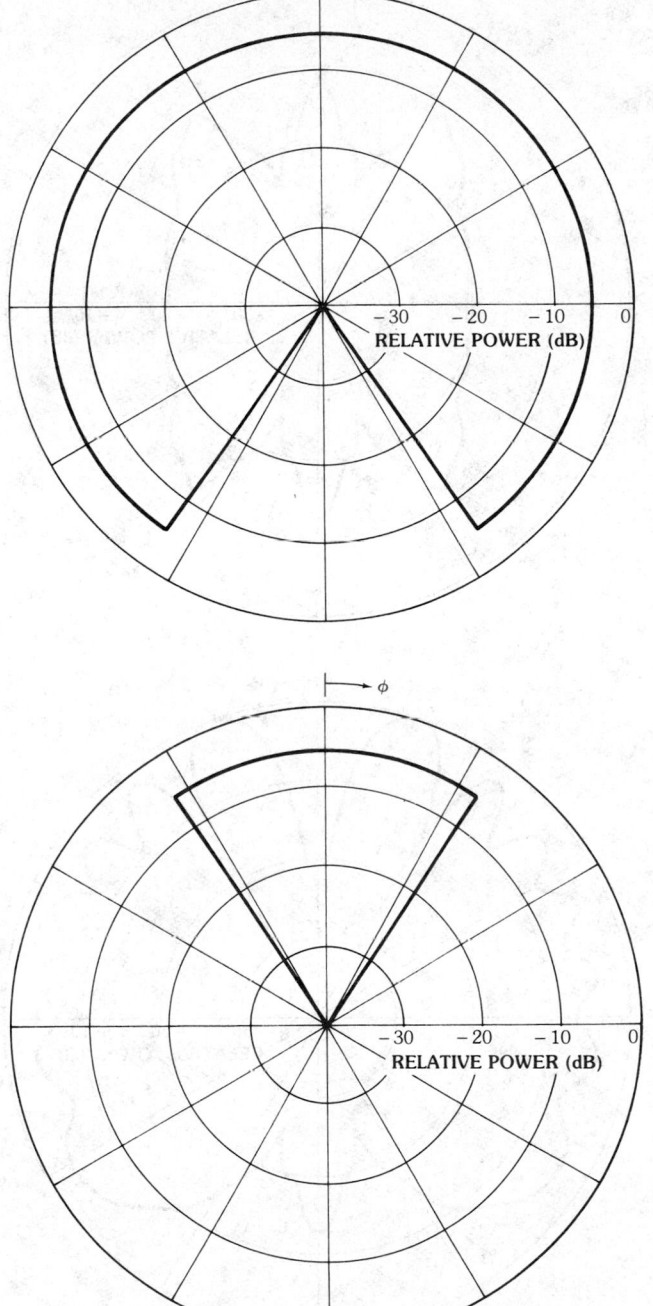

Fig. 9. *H*-plane patterns due to various ray mechanisms for a half-wave dipole located above a square plate. (*a*) Source field: *S*. (*b*) Reflected field: *R*. (*c*) Diffracted field: *D*. (*d*) *S* + *R* + *D*.

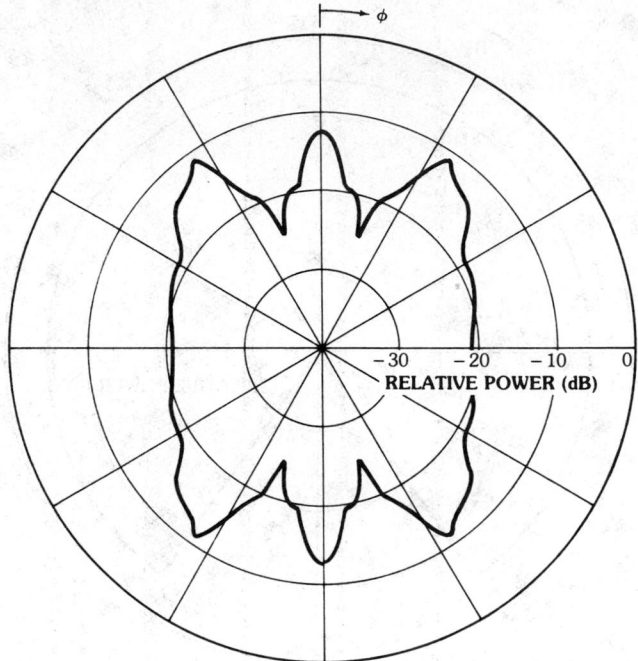

c

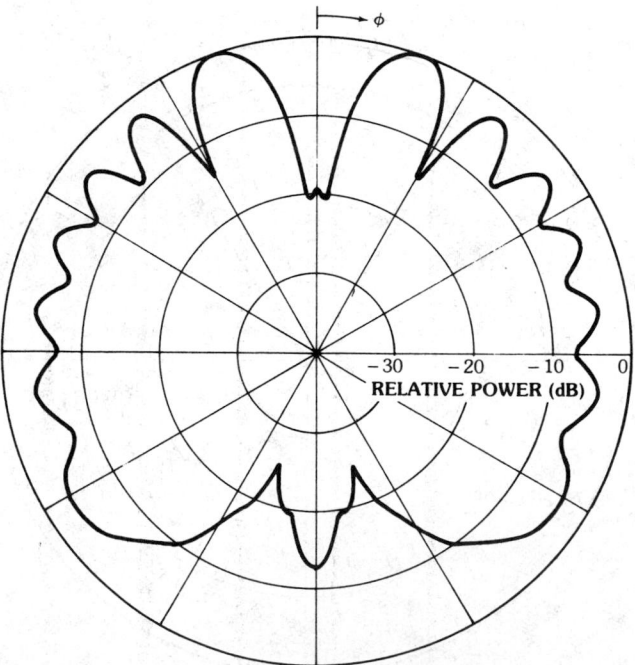

d

Fig. 9, *continued.*

terms emanate. By using the former, one can tell by his pattern calculation if the solution is valid theoretically. By using the latter, one could consider using corrective measures to eliminate various unwanted terms.

The UTD wedge-diffraction solutions have been extended in [19] to treat finite flat plates covered with a thin dielectric or absorber layer. The absorber could be used here to reduce the reflected-field magnitude, which in turn would decrease the strong interference pattern shown in Fig. 10a. A thin ferrite absorber was placed on the flat plate, and the calculated and measured results are shown in Fig. 10b. Again the comparison between results is very encouraging. One should note that with the introduction of the absorber model into the numerical solution he or she can begin

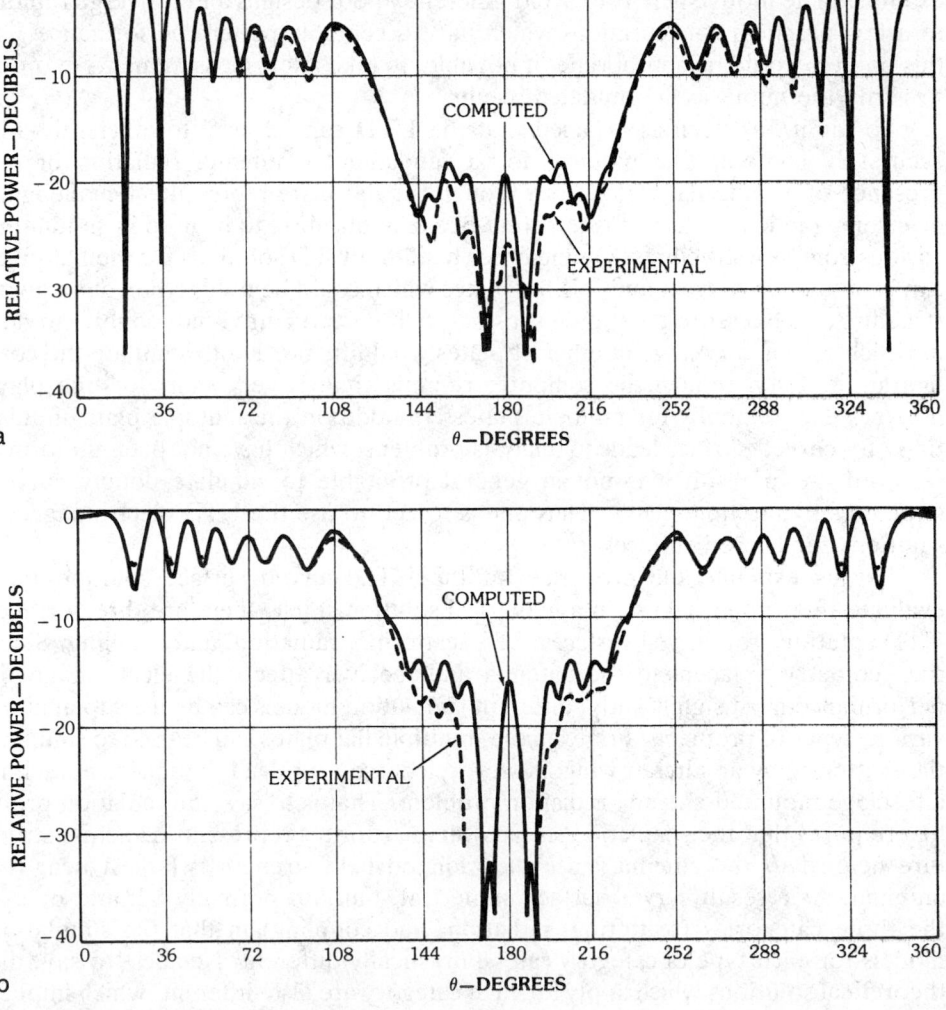

Fig. 10. Measured and calculated $E_{\theta p}$ patterns. (*a*) For a perfectly conducting metallic plate. (*b*) For a thin ferrite absorber on a metallic plate.

to take realistic measures to improve pattern performance by using properly located absorber panels. In addition the absorber model does not add significantly to the computation time, so an antenna designer can quickly examine possible alternatives.

On a standard high-speed digital computer a complete pattern for the previous geometry can be run in less than 10 s. As a result one could envision an antenna designer sitting at a graphics display terminal, which is used to draw the input geometry and resulting patterns, trying various configurations and determining some realistic geometries which provide the desired performance. Since the computation time is so short the antenna designer can observe the cause and effect associated with various changes in the structure. In such a case he or she will be able to quickly eliminate poor designs of the same general class in that they will all have the same characteristic problem, i.e., plate reflections too strong, for example. The motive here is to avoid inherently poor designs and converge rapidly to a few practical configurations which have acceptable pattern performance. At this point it would be appropriate, if possible, to take a few measurements in order to verify the numerically simulated results.

So far it has been ascertained that the UTD can be used to efficiently and accurately compute the patterns for a fundamental antenna radiating in the presence of a finite flat plate. Even though the flat plate represents a challenging electromagnetic problem, it is too great of a simplification to be used in simulating various complex structures. On the other hand the UTD solutions for the flat plate can be extended to treat multiple flat plates which could be used to box out a given structure. Such a representation is satisfactory if the structure is not doubly curved, in which case a very large number of plates would be needed to simulate the configuration. Even though the computer running time is very short for one plate it grows exponentially for multiple plates. In addition the multiple-plate simulation of a curved surface leads to analysis problems which have not been uniformly resolved. As a result it is not in general profitable to simulate doubly curved structures by multiple plates. Thus one is forced to use the UTD electromagnetic solutions for curved surfaces.

Let us examine the properties of the UTD curved surface solutions and evaluate their potential use in a numerical solution. First, there are three distinct UTD solutions for curved surfaces, i.e., scattering, radiation, and coupling. Since the geometry adjacent to the antenna and receiver affects the electromagnetic performance most significantly, different simulation models can be used to analyze various types of problems. For example, multiple flat plates can be used to simulate the scattering by an aircraft wing; however, a curved surface is better for modeling a fuselage mounted antenna radiation problem. That is to say, the radiation problem requires that the simulation surface very accurately represent the actual structure nearest to the antenna since the radiated field strength is largest near the antenna. As a result curved-surface numerical solutions normally fall into one of the three categories (scattering, radiation, and coupling) in that the simulation models for each type of category can be drastically different. Needless to say, the theoretical solutions which apply for each category are also different, which implies that one must associate the appropriate type of problem with an efficient representation of the curved surface.

For the plate geometries examined earlier the ray paths are basically straight lines either going directly from the source to the receiver or going from the source, scattering off the plates, and then arriving at the receiver. As the energy expands outward from the source on a curved surface the rays traverse the surface along *geodesic* paths, a geodesic path being the shortest path between the initial and final points of the trajectory, i.e., it is a great circle on a sphere and a helix on a cylinder. The geodesic paths are defined, for example, by a pair of nonlinear coupled differential equations, and as a result they are very inefficient to track. Because of this inefficiency one is tempted to use as simple a curved surface as possible so that the geodesic paths can be found in some rapid sense. For the coupling problem where both the source and receiver are mounted on the curved surface, one is forced to very accurately represent the complete surface between the two antennas. This requires that one solve the geodesic equations for a very complete representation of the geometry, which implies that coupling solutions are rather inefficient. This is not necessarily the case in that for a coupling problem the source and receiver locations are fixed, so there is basically a small set of geodesic paths needed for the total calculation. Referring now to the radiation problem, one has to find new geodesic paths for each different radiation direction; however, only the source is mounted on the curved surface, so only that portion of the surface needs to be accurately simulated. With this in mind it has been shown in [20, 21] that a perturbation solution can be used to very efficiently determine the geodesics on elongated prolate spheroids and ellipsoids. These solutions are very rapid since they relate the pattern direction to a specific geodesic path; in addition they base the next geodesic path on the last path in that they change very little from one pattern angle to the next. For example, the complete radiation pattern for a short monopole mounted on a prolate spheroid can be run in less than 10 s. The basic curved surface solutions have been treated in Chapter 4, so they will not be reiterated here.

In order to use the UTD to represent complex structures it is necessary to combine the curved-surface- and wedge-diffraction solutions so that both curved surfaces and flat plates can be used in the simulation model. Using these two basic geometries in this analysis let us consider development of the radiation and scattering numerical solutions.

Basic Model Simulations—Antennas Mounted on a Curved Surface

This section describes the development of UTD simulations for antenna problems in which the antenna is mounted on a curved surface. Most of this discussion will involve aircraft shapes since there are many such problems for which this theory can be applied in the development of airborne antenna systems. Although the examples tend to aerodynamic shapes one can use this solution approach to solve a wide variety of problems, as will be noted as the simulation structures evolve. For example, this approach could be used to determine the pattern performance for a mast-mounted antenna radiating in the presence of other ship structures. In another case it could be used to examine the pattern distortion associated with a satellite-mounted antenna.

In order to begin our simulation studies with a simple two-dimensional problem let us consider the antena problem illustrated in Fig. 11. In this example an electric line monopole is mounted on the surface of a perfectly conducting circu-

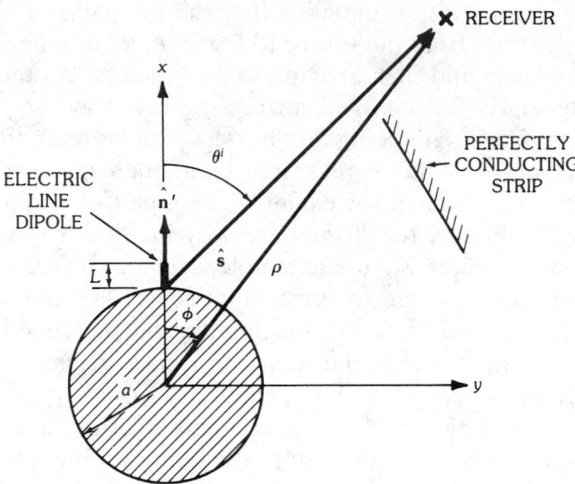

Fig. 11. Electric line dipole mounted on a circular cylinder and radiating in the presence of a perfectly conducting strip.

lar cylinder. In addition a strip is located in close proximity to the cylinder. Before developing the various scattering terms let us review the basic radiation mechanisms associated with antennas mounted on a curved surface. First, the analysis is divided into two parts, these being the lit and shadow region solutions. In the lit region the radiation appears to come directly from the source and has a field value given by

$$H_{\text{lit}}^c(\varrho, \phi) \cong \left\{ \frac{\cos(kL\hat{\mathbf{n}}\cdot\hat{\mathbf{s}}) - \cos kL}{[1 - (\hat{\mathbf{n}}\cdot\hat{\mathbf{s}})^2]\sin kL} \right\}(\sin \theta^i)G(\xi^\ell) \frac{e^{-jks}}{\sqrt{s}} \tag{18}$$

where $G(\xi)$ is known as a Fock function, which is defined in Reference 18 and in Chapter 4. In the shadow region the radiation appears to come from the two effective sources shown in Fig. 12. These two effective sources result from the two creeping waves excited by the source which propagate outward from the source in a clockwise and counterclockwise sense. The shadow region field value for either of the two creeping waves is given by

$$H_{\text{shadow}}^c(\varrho, \phi) \cong \left[\frac{1 - \cos kL}{\sin kL} \right] G(\xi)e^{-jk\ell} \frac{e^{-jks}}{\sqrt{s}} \tag{19}$$

where ℓ is the propagation length of the creeping-wave path. Since the radiated wavefront appears to come from the source or effective sources, these apparent radiation centers can be used to compute the reflected and diffracted fields from the strip. Furthermore, (18) and (19) can be used as the field incident on the strip and its associated edges. Note that the incident field, that is, the field going from the antenna (possibly around the cylinder) to the receiver as given by (18) and (19), is computed provided the ray going from the effective source to the receiver does not

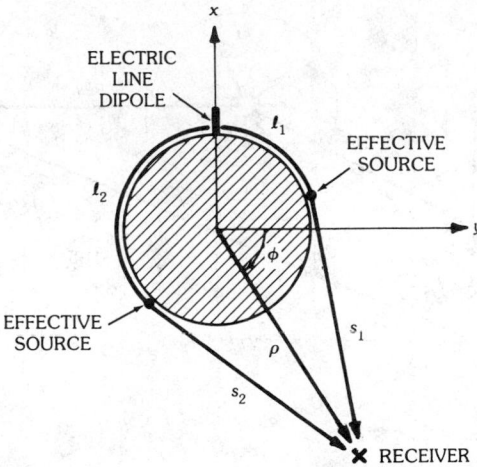

Fig. 12. Radiation geometry with receiver in shadow region.

intersect the strip. From this point forward the effective source terminology is used to indicate the apparent radiation center for a given ray, which implies it is the actual source if the receiver is in the lit region.

The field reflected from the strip is computed using the image position of the receiver in the strip as shown in Fig. 13. The field incident on the strip is computed using (18) and (19) with the observation point located at the image position. For an electric line dipole the field at the image position is identical with the field at the receiver, provided the ray going to the image from the effective source as shown in Fig. 13 intersects the plate. In fact the intersection point is actually the point of reflection. Note that the reflected field will only exist for a small sector of space; however, the strip edge-diffracted fields radiate in all directions, as shown in Fig. 14, and are discussed next.

The edge-diffracted field for the top edge of the strip is considered first because it is illuminated by a lit region field shown in Fig. 14a. In this case the source of the edge-diffracted field is the source itself, and its field value is given by

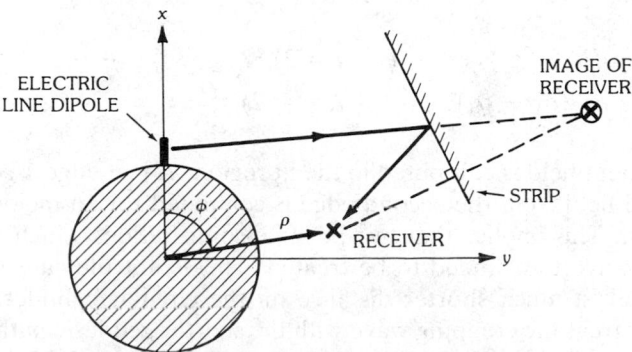

Fig. 13. Strip-reflected field geometry.

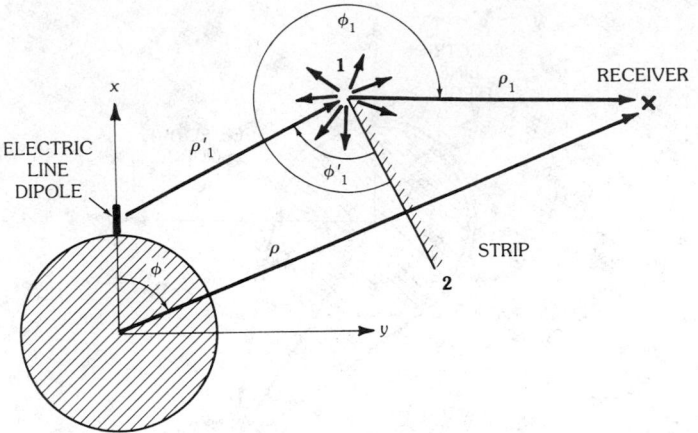

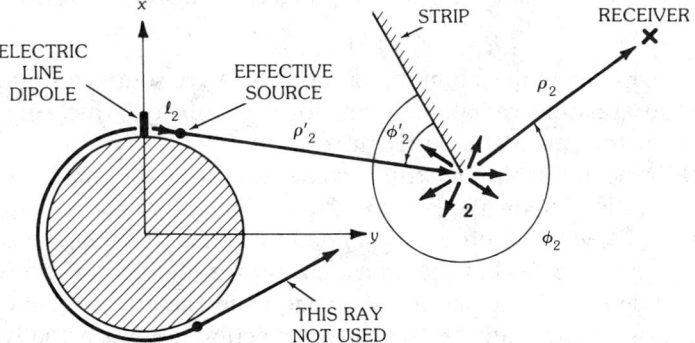

Fig. 14. Diffracted field geometry. (*a*) For edge 1. (*b*) For edge 2.

$$H_1^d = D_h[H_{\text{lit}}^c]_{\text{at edge 1}} \frac{e^{-jk\varrho_1}}{\sqrt{\varrho_1}} \tag{20}$$

where

$$
\begin{aligned}
D_h = &\ D(\varrho_1'\varrho_1/(\varrho_1' + \varrho_1), \phi_1 - \phi_1', n = 2) \\
&+ D(\varrho_1'\varrho_1/(\varrho_1' + \varrho_1), \phi_1 + \phi_1', n = 2)
\end{aligned} \tag{21}
$$

Since the incident field is computed in the lit region, H_{lit}^c at edge 1 is given by (18). The diffracted field from the second edge is computed for an incident field in the shadow region. This implies that two possible contributions which result from the two creeping-wave terms need to be treated. This is not the case in that one ray normally travels a much shorter distance on the circular cylinder; consequently one need only treat the creeping wave with the shorter geodesic path on the curved surface such as shown in Fig. 14b. In this case the diffracted field from edge 2 is given by

$$H_2^d = D_h [H_{\text{shadow}}^c]_{\text{at edge 2}} \frac{e^{-jk\varrho_2}}{\sqrt{\varrho_2}} \qquad (22)$$

where

$$
\begin{aligned}
D_h = \; & D(\varrho_2'\varrho_2/(\varrho_2' + \varrho_2), \; \phi_2 - \phi_2', \; n = 2) \\
& + D(\varrho_2'\varrho_2/(\varrho_2' + \varrho_2), \; \phi_2 + \phi_2', \; n = 2)
\end{aligned}
\qquad (23)
$$

using the geometry illustrated in Fig. 14b. Even though the plate-scattered field can be subsequently scattered by the circular cylinder, this type of interaction is not considered here. It is useful in a practical sense, however, to shadow the plate-scattered field if the rays appear to intersect the cylinder. This is done so that a discontinuity in the resulting pattern can be observed, which is indicative of the strength of the higher-order scattering terms which are missing in the solution. The total solution is then the sum of the incident, strip-reflected, and two edge-diffraction terms which are obtained using (18) through (23).

Before considering the next example, recall that there is the possibility of multiple diffractions between the two edges of the strip. Normally one need only include up to double diffraction, which is significant when the source or receiver and the two edges align since the edge diffraction from the first edge can be nearly as significant as the source field in terms of the second edge illumination. Since double diffraction is normally a secondary contributor it is quite often neglected in the total solution. Again the effect of neglecting double diffraction will be apparent in the computed patterns in that a discontinuity will exist in the directions of the strip orientation, and the magnitude of this jump will indicate its importance.

If the model illustrated in Fig. 11 is to be used to simulate an aircraft structure, it is apparent that the strip has to be attached to the cylinder as shown in Fig. 15. Using this configuration the previous procedure is followed to find the reflected field from the strip and the diffracted field from edge 1. A complication arises in determining the diffracted field from the junction edge formed by the cylinder and strip. Recall that the wedge-diffraction coefficient is not valid for this type of structure in that the field incident on the junction edge results from a creeping wave propagating around the cylinder. In order to develop an approximate solution to this problem the solution can be modified based on the properties associated with this type of structure as shown in Reference 18.

An example is presented which illustrates the different UTD mechanisms included in the solution to this type of problem. For this purpose a single flat plate is attached to a circular cylinder as shown in Fig. 16. The various terms treated using the previous solution are shown in Fig. 17. Note that each pattern is normalized to the same level such that the relative significance of each term can be seen. In order to better illustrate how the total solution evolves, various terms are superimposed in Fig. 18. Note that the double-diffraction term is introduced to the solution to illustrate its significance. An interesting result is shown in Fig. 18a, where the source and reflected fields are superimposed. These two terms form a GO solution which is far from complete, as can be observed from the discon-

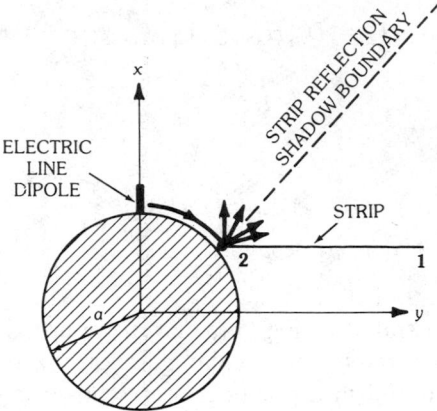

Fig. 15. Geometry of a strip attached to the cylinder.

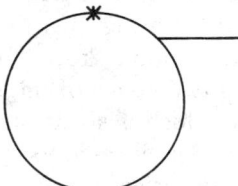

Fig. 16. Geometry of a flat plate attached to a circular cylinder.

tinuities in the pattern. The total UTD solution shown in Fig. 18d is continuous and should agree with experimental results, be presented later.

In order to simulate the wide variety of structures found in practice, more components need to be added into the total UTD solution. Since introducing more perfectly conducting strips into the solution is rather straightforward and efficient to analyze, consider a more general solution in which a cylinder-mounted antenna is radiating in the presence of multiple strips. Using this general approach, more, complex situations can be treated by boxing out the various structures making up the environment surrounding the antenna. The various UTD mechanisms used in this solution are depicted by the ray paths illustrated in Fig. 19. The field solutions used for each of these terms follow the same format applied in the previous cylinder/strip problem.

For completeness consider the analysis used to develop the diffracted-reflected field term illustrated in Fig. 20a. Since a reflection from strip 2 is being considered, the first step in this solution is to find the image of the receiver in strip 2 as shown in Fig. 20b. The diffracted field from the junction edge is then computed to find the field at the image point. That is the same field that would be incident on the receiver provided that the diffracted ray from the junction edge to the image intersects the virtual limits of strip 2. Note that the diffracted field incident on the receiver image position is found with strip 2 removed. The actual diffracted-reflected field is then found using the method described earlier (6). Finally, the

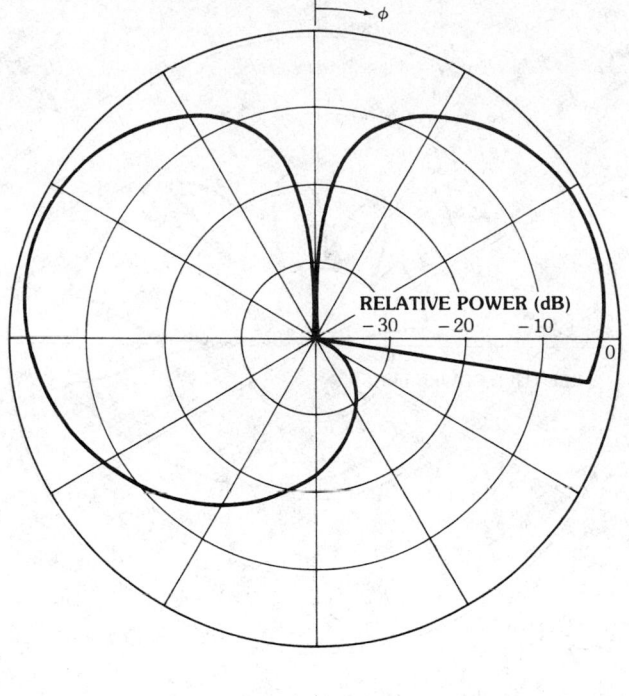

a

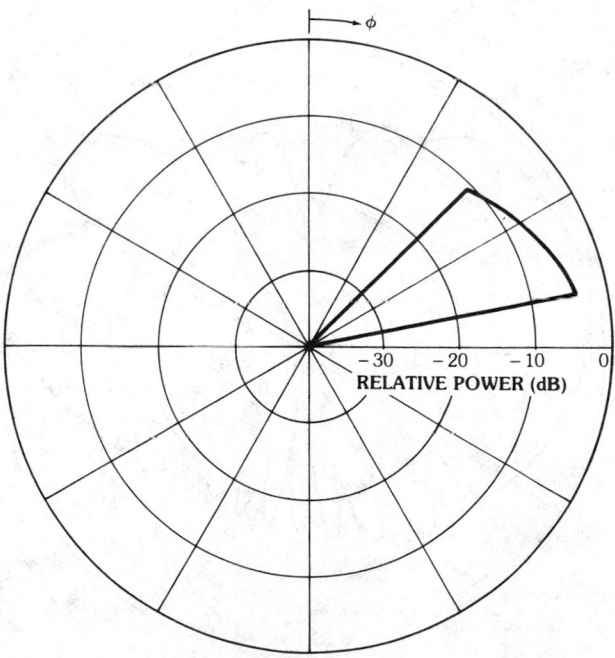

b

Fig. 17. Radiation patterns due to various ray mechanisms. (*a*) Source field: *S*. (*b*) Reflected field: *R*. (*c*) Diffracted field: *D*. (*d*) *S* + *R* + *D*.

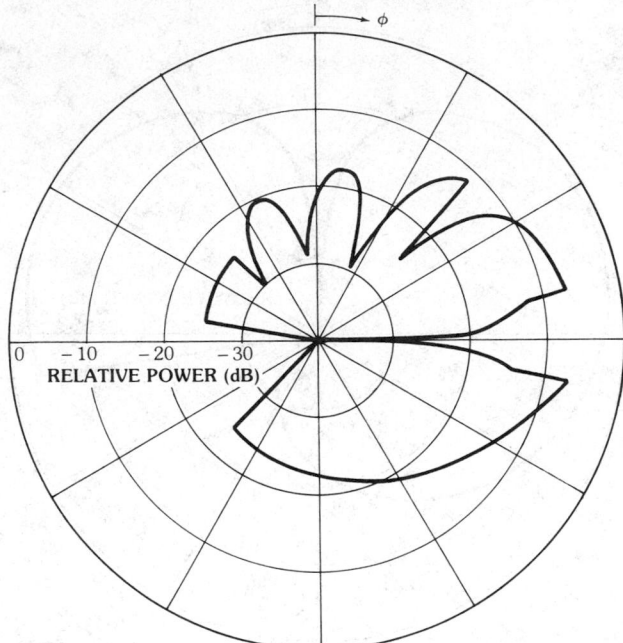

c

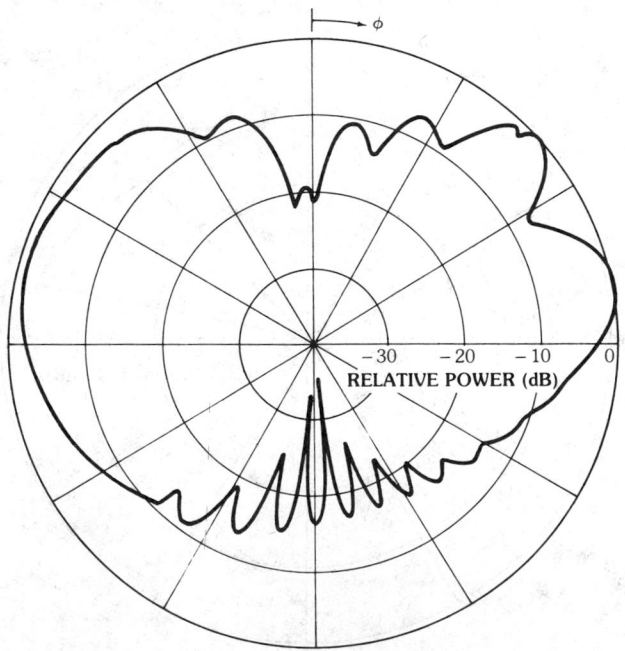

d

Fig. 17, *continued.*

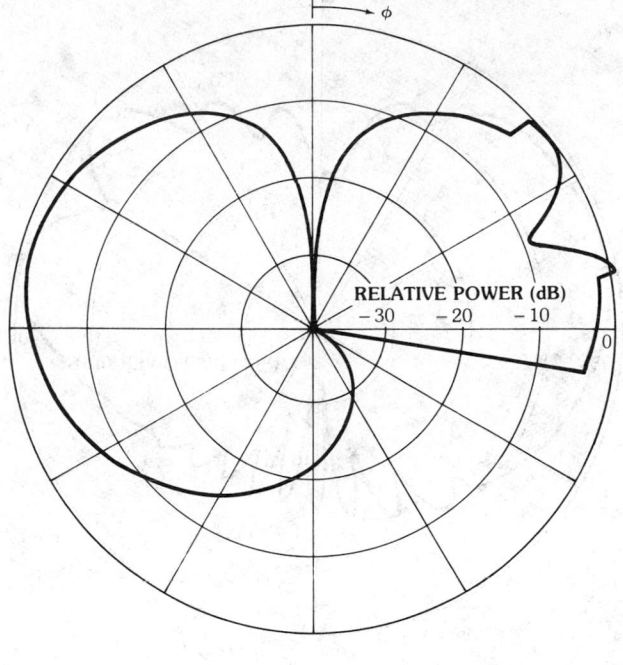

a

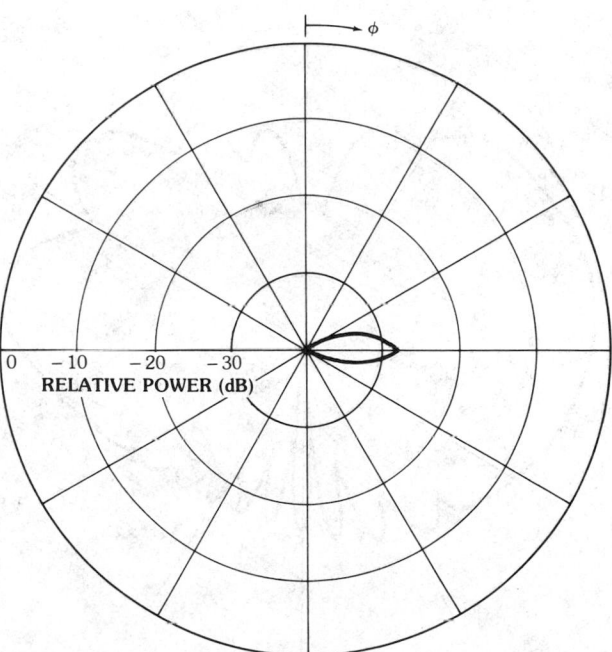

b

Fig. 18. Radiation patterns due to various combinations of the UTD terms. (*a*) *S* + *R*. (*b*) Double diffracted field: *DD*. (*c*) *S* + *R* + *D*. (*d*) *S* + *R* + *D* + *DD*.

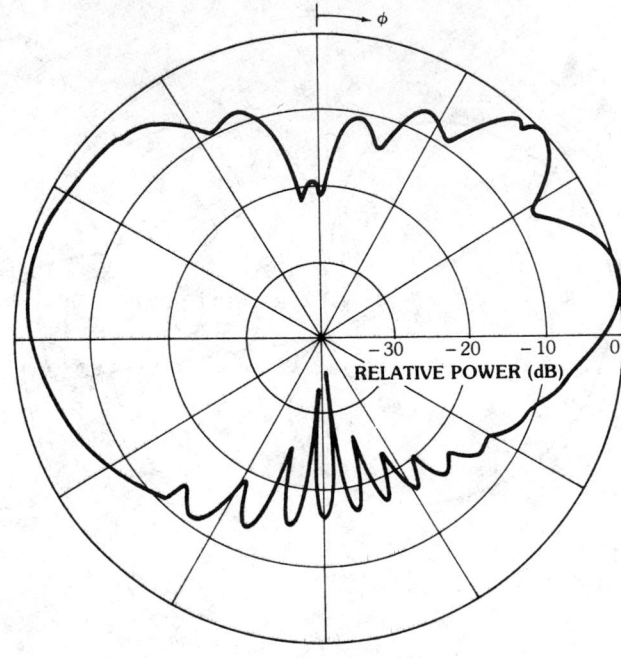

c

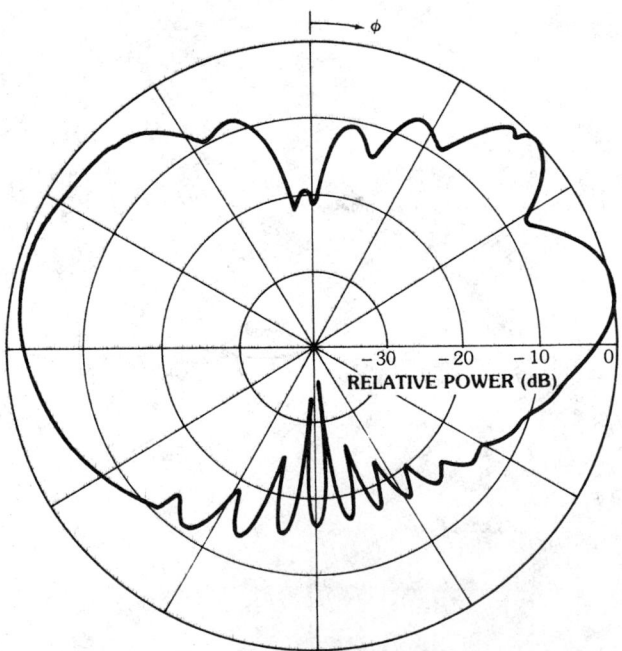

d

Fig. 18, *continued.*

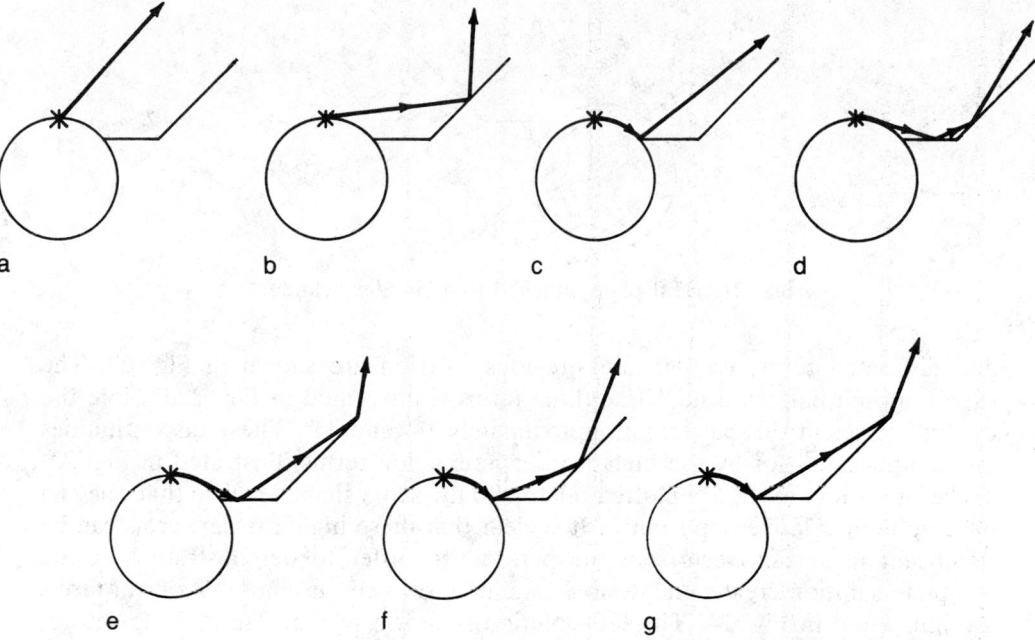

Fig. 19. Various UTD terms. (*a*) Source field. (*b*) Reflected field. (*c*) Diffracted field. (*d*) Reflected-reflected field. (*e*) Reflected-diffracted field. (*f*) Diffracted-reflected field. (*g*) Diffracted-diffracted field. (*After Burnside, Rudduck, and Marhefka [4], © 1980 IEEE*)

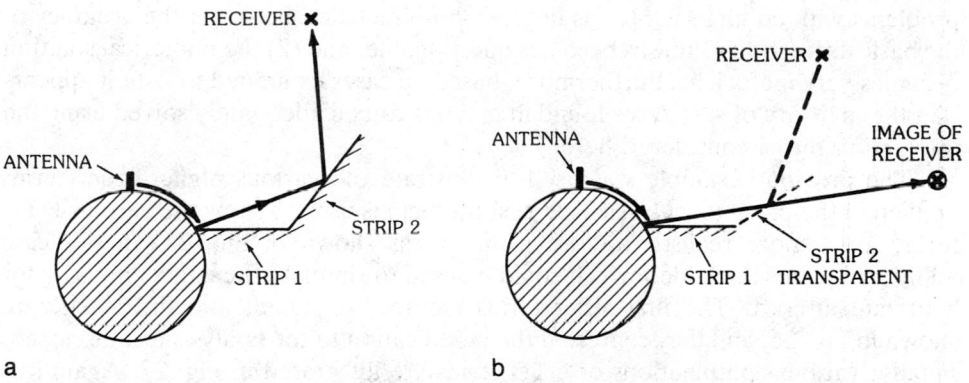

Fig. 20. Analysis for a diffracted-reflected field. (*a*) Diffracted-reflected field component. (*b*) Image receiver in strip 2 to find the diffracted-reflected field.

complete actual ray path must be traced out in order to ensure that some other strip does not block the ray; if so, that field component is set to zero. Note that other field solutions basically follow the same format.

The next example is used to illustrate the additional UTD mechanisms added to treat multiple strips. This geometry is shown in Fig. 21, in which a second plate is added to the model treated in the previous discussion. The source, reflected,

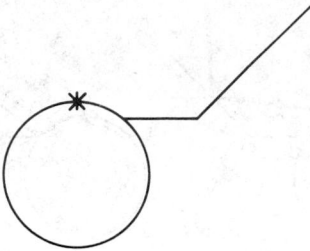

Fig. 21. Bent plate attached to a circular cylinder.

and diffracted terms used in our previous solution are shown in Fig. 22. The superimposed pattern using these three terms is illustrated in Fig. 22d. Note the discontinuities in this pattern at approximately 14° and 45°. These discontinuities are compensated for by the higher-order interaction terms illustrated in Fig. 23. As before, these terms are plotted relative to the same signal level so that one can observe their relative importance. It is clear that these higher-order terms can be significant in certain sectors of the pattern. In order to demonstrate how the complete solution creates the desired total pattern, various combinations of terms are illustrated in Fig. 24. The GO solution is shown in Fig. 24b and, as before, several discontinuities exist in this pattern. The total superposition of the recently introduced higher-order terms is shown in Fig. 24c. It is very interesting to recall that the patterns illustrated in Figs. 22d and 24c superimpose to give the total pattern shown in Fig. 24d. As with any UTD solution for a complex structure, one can compute higher- and higher-order terms. There are, however, two major problems with adding such terms beyond those included here: (1) the accuracy of the basic diffraction solutions becomes questionable, and (2) the numerical solution becomes very inefficient. Furthermore, based on cases examined to date it appears that the majority of structures found in practice can be adequately solved using the interaction terms considered here.

The previous example was used to illustrate the various higher-order terms applied in the present analysis. This next problem is used to show the various UTD terms for a more realistic aircraft geometry as shown in Fig. 25. In this case additional plates are added which could be used to simulate the engine housing for a private aircraft. The first-order terms (source, reflected, and diffracted) are shown in Fig. 26, and they represent the modification to the analysis just discussed. Finally, various combinations of UTD terms are illustrated in Fig. 27. Again it is very interesting how all the different discontinuous terms combine to give the nice, smooth total pattern shown in Fig. 27d.

In order to investigate the accuracy of the present solution a long structure was constructed which included a cylinder, flat-plate wing, and movable engine housing shape. The patterns are then taken in the plane perpendicular to the cylinder axis to simulate the two-dimensional solutions. The first example is used to verify the numerical results generated for the geometry shown in the insert of Fig. 28 along with the calculated and measured roll-plane patterns. Note that a roll-plane pattern is taken in the plane perpendicular to the axis of the aircraft. In the next case the engine plate model is moved to the wing tip, and the measured and calculated roll-

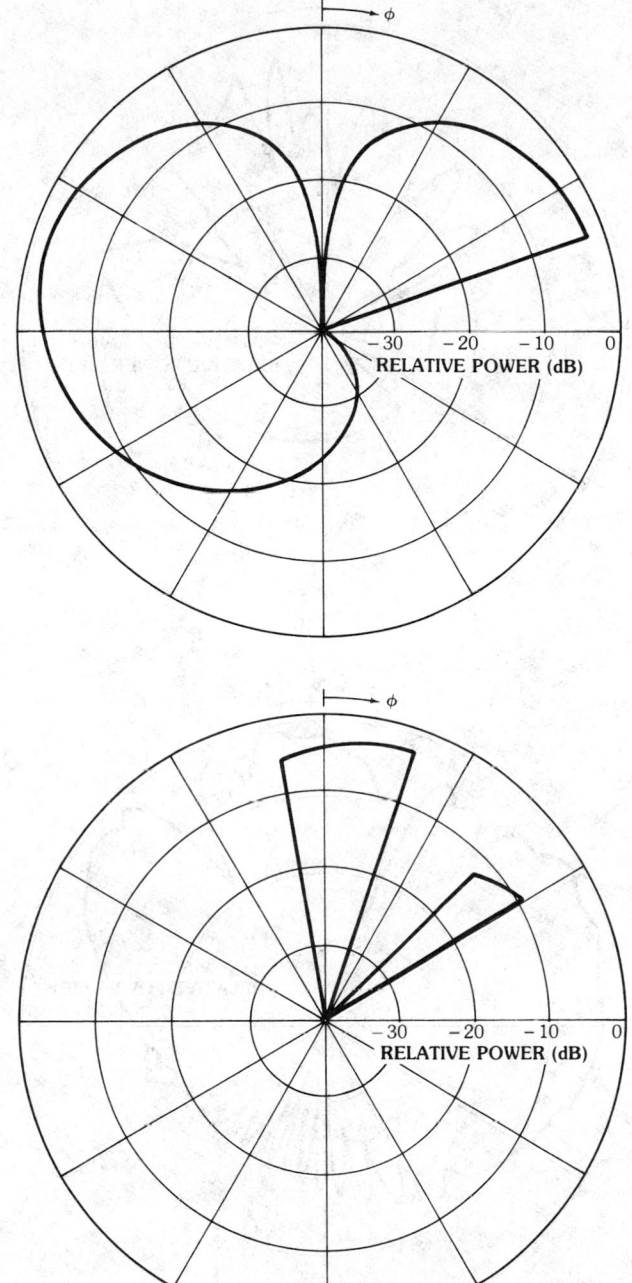

Fig. 22. Radiation patterns due to various UTD terms for Fig. 21. (*a*) Source field: *S*. (*b*) Reflected field: *R*. (*c*) Diffracted field: *D*. (*d*) *S* + *R* + *D*. (*After Burnside, Rudduck, and Marhefka [4],* © *1980 IEEE*)

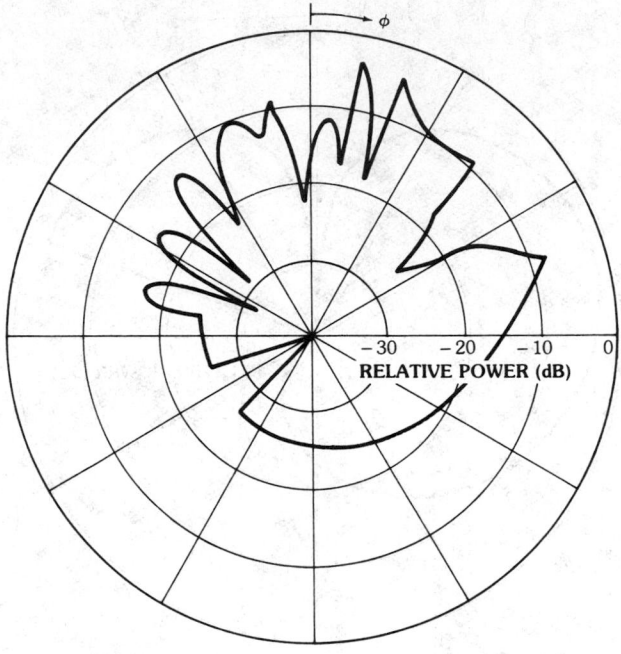

c

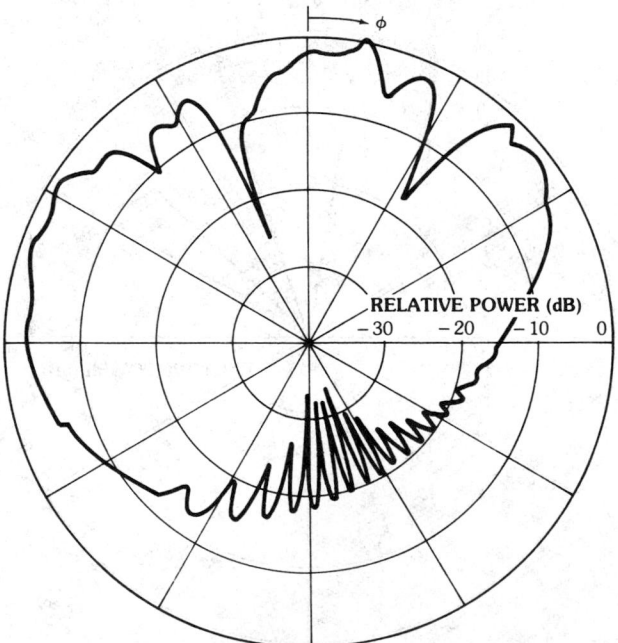

d

Fig. 22, *continued.*

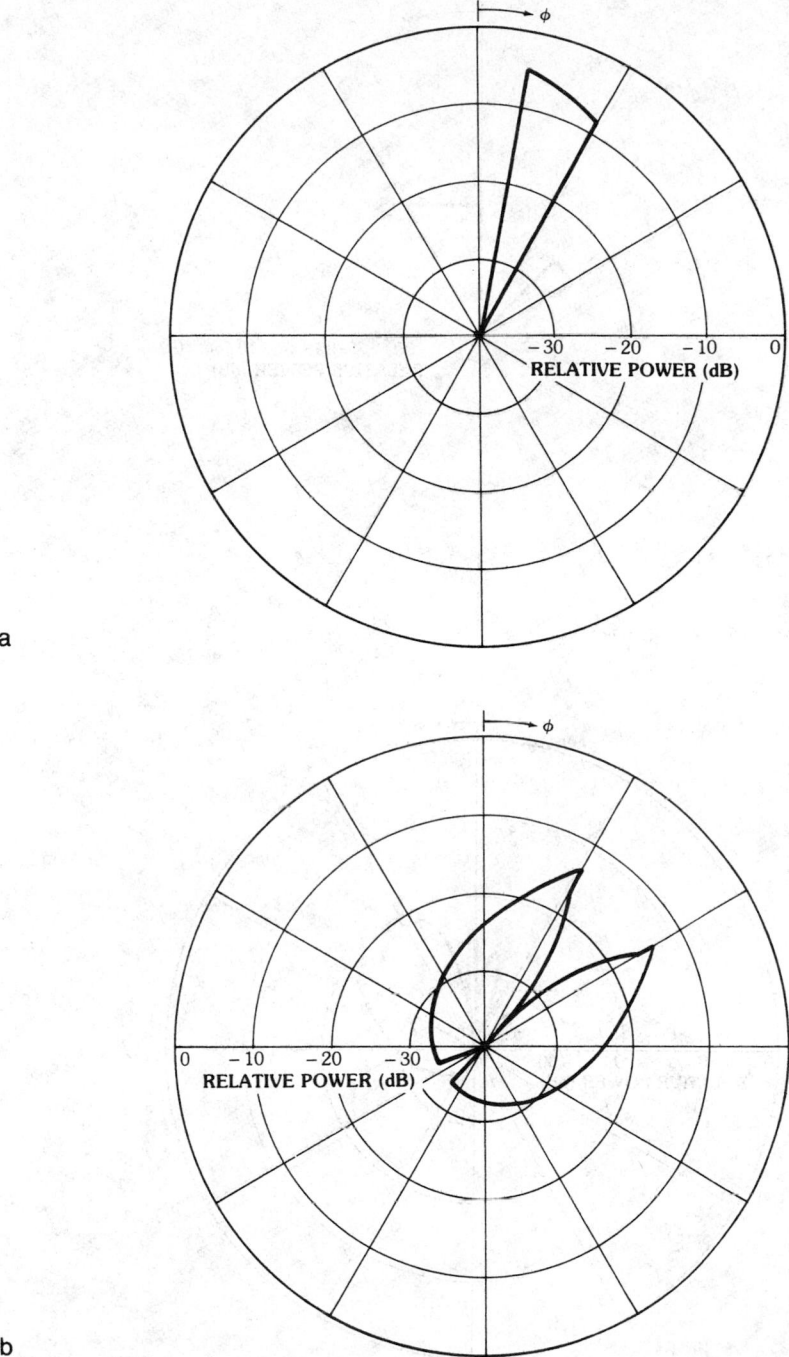

Fig. 23. Radiation patterns due to second-order-interaction UTD terms for Fig. 21. (a) Reflected-reflected field: *RR*. (b) Reflected-diffracted field: *RD*. (c) Diffracted-reflected field: *DR*. (d) Diffracted-diffracted field: *DD*. (*After Burnside, Rudduck, and Marhefka [4],* © *1980 IEEE*)

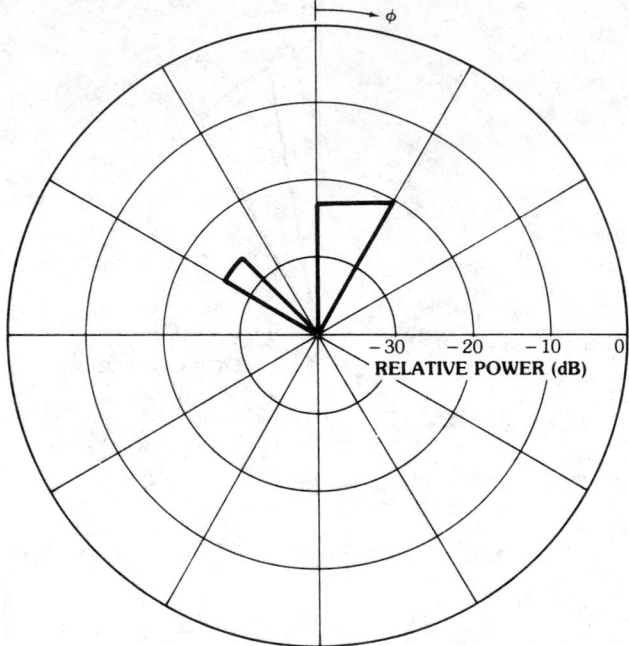

c

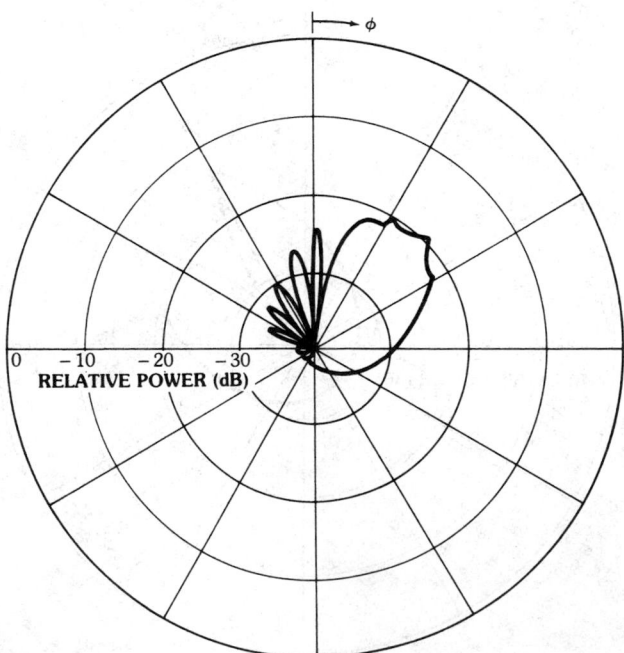

d

Fig. 23, *continued.*

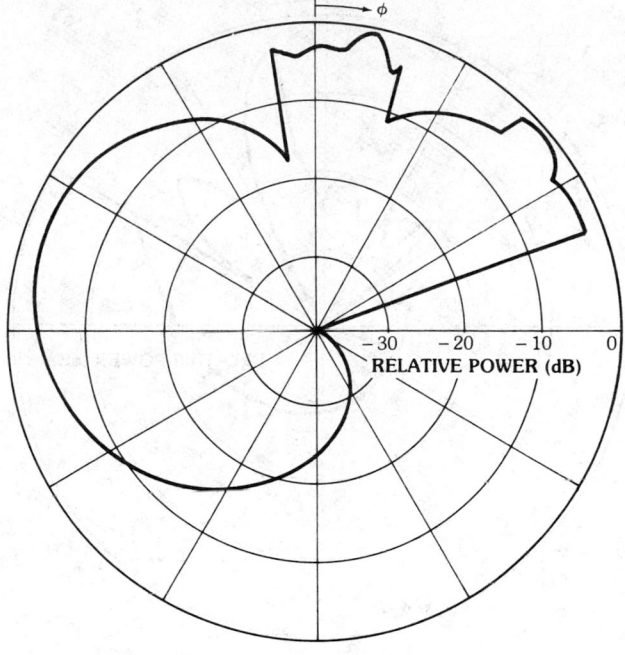

a

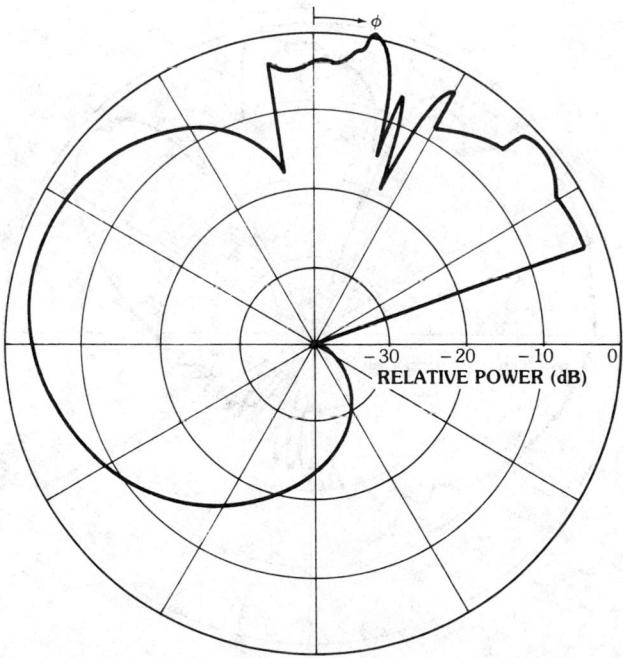

b

Fig. 24. Radiation patterns due to various combinations of UTD terms for Fig. 21. (a) $S + R$. (b) GO solution: $S + R + RR$. (c) Second-order-interaction GTD terms: $RR + RD + DR + DD$. (d) Total solution. (*After Burnside, Ruddeck, and Marhefka [4], 1980 IEEE*)

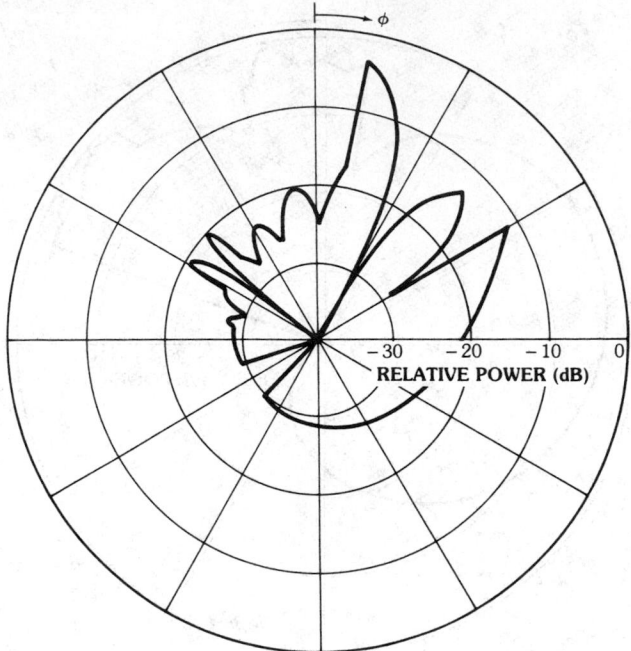

c

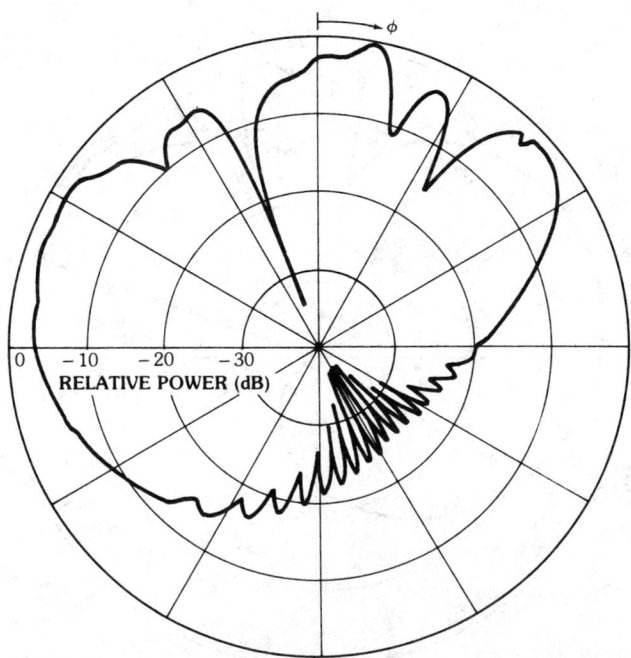

d

Fig. 24, *continued.*

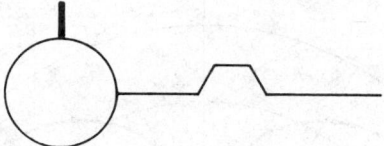

Fig. 25. Flat plate model for the wing and engine.

plane patterns for this geometry are shown in Fig. 29. In each case excellent agreement between measured and calculated results is obtained.

The UTD solution for the three-dimensional solutions follows the same format as for the two-dimensional ones except that the ray paths are not as easily computed. As a result the efficiency of a three-dimensional solution is dictated by the ray trajectory calculations, in which case one is referred to [5, 18, 20, 21, 22].

Basic Model Simulations—Antennas Not Mounted on a Curved Surface

When simulating a scattering structure, such as an aircraft or ship mast with a yardarm, it has been shown that it is very convenient to use both a cylinder and a plate in the model. If the source is not mounted on the curved surface, the field terms differ slightly from those. In the following analysis the various field terms that are needed to find the UTD scattered fields by a two-dimensional cylinder and plate model with the source off the cylinder will be discussed. This is basically an extension of the discussions given in the section on plates. In this two-dimensional analysis the interpretation of the fields and their parameters will be emphasized. The discussion on the means of finding the ray paths in terms of the three-dimensional model can be found in [5]. As before, however, one must keep in mind that any ray must be shadowed that intersects a plate or cylinder that is not a part of the specific mechanism being analyzed. This means that the shadowed ray will be set to zero and will not be included in the sum total of the pattern. As has been demonstrated earlier, the UTD compensates for these discontinuities and produces a continuous field up to the order of the solution.

As a review of the fields that have been discussed previously, the ones needed here will be briefly outlined and others will be introduced. Each of these field terms is illustrated by the example which follows this discussion. The source field is given by

$$u^i = u_0 \frac{e^{-jks^i}}{\sqrt{s^i}} \tag{24}$$

where it can represent either an electric or magnetic line source, and u_0 represents a complex excitation constant. It is used as the direct field from the source to the receiver or as the field incident on a scattering structure producing the fields to follow. Based on various comparisons with experimental results it has been determined that there are five dominant scattered field terms which must be included in a multiple-plate analysis. These are the reflected, edge-diffracted, doubly reflected, reflected-diffracted, and diffracted-reflected fields. The plate-reflected field is given by

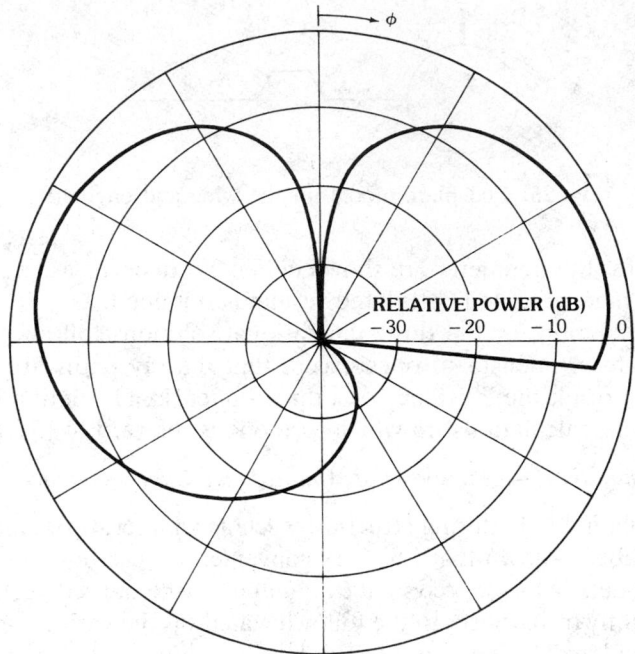

a

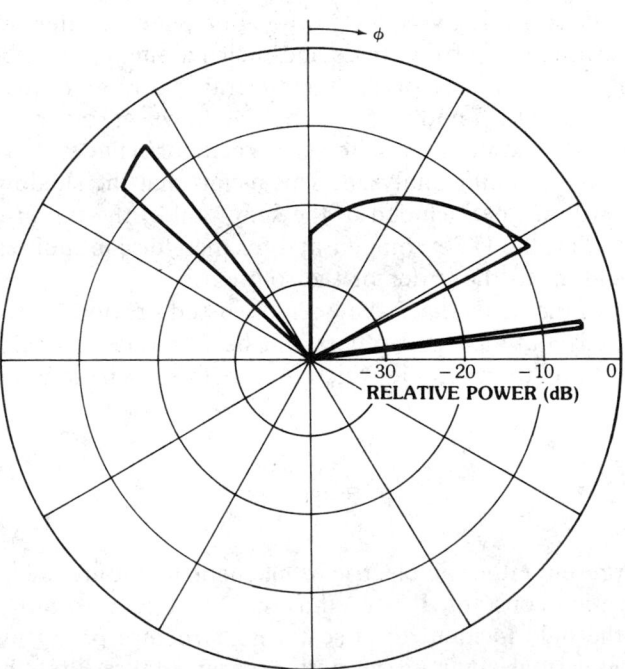

b

Fig. 26. Radiation patterns due to various UTD terms for Fig. 25. (*a*) Source field: *S*. (*b*) Reflected field: *R*. (*c*) Diffracted field: *D*. (*d*) *S* + *R* + *D*.

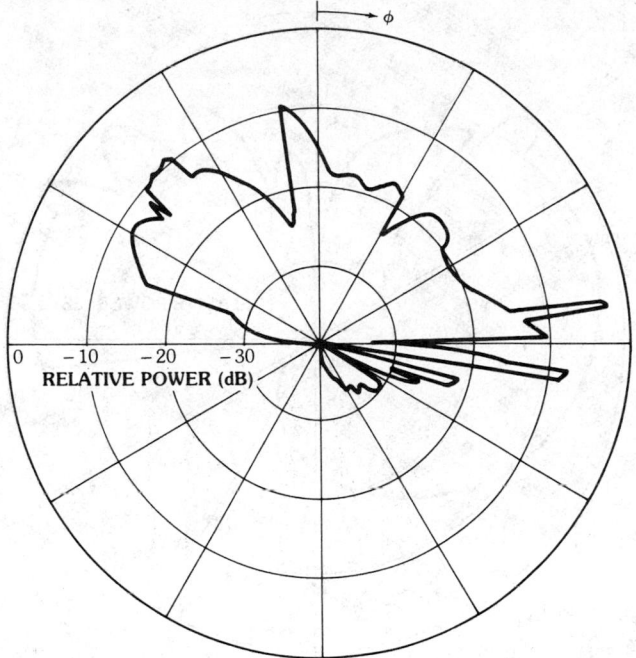

c

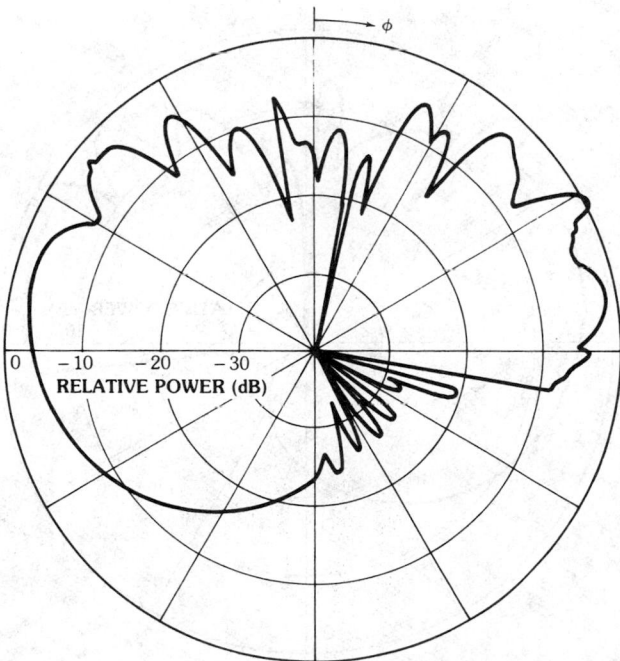

d

Fig. 26, *continued.*

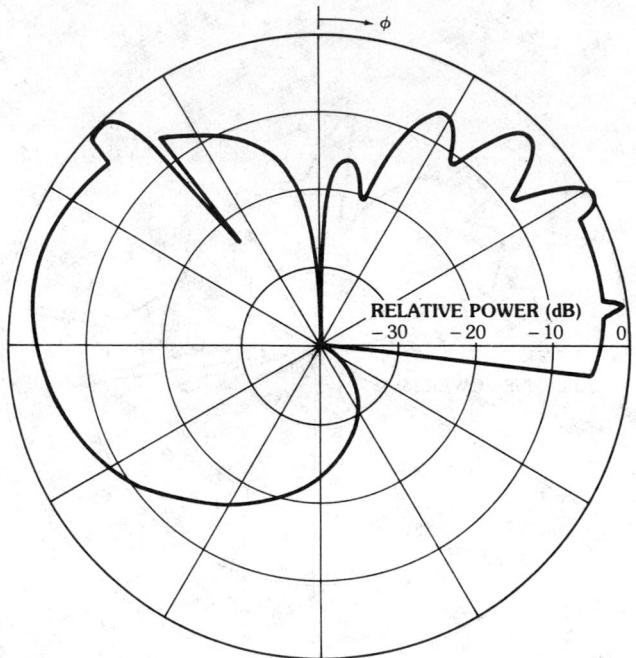

a

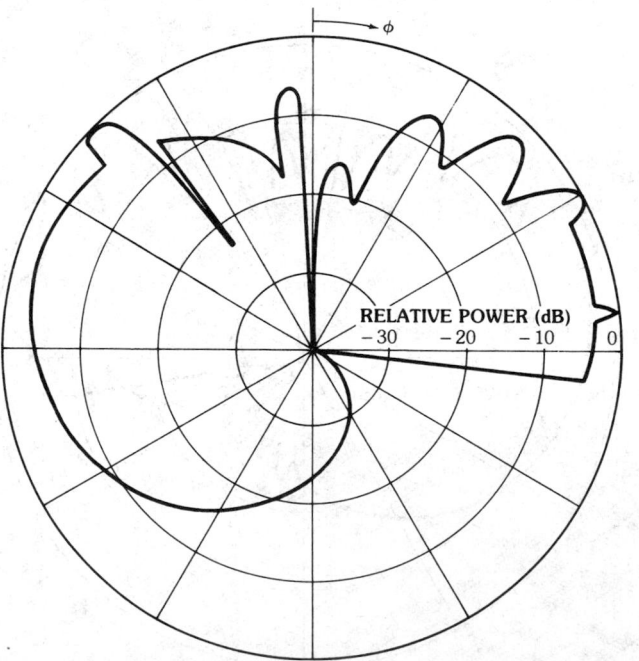

b

Fig. 27. Radiation patterns due to various combinations of the UTD terms for Fig. 25. (*a*) *S* + *R*. (*b*) GO solution: *S* + *R* + *RR*. (*c*) Second-order-interaction GTD terms: *RR* + *RD* + *DR* + *DD*. (*d*) Total solution.

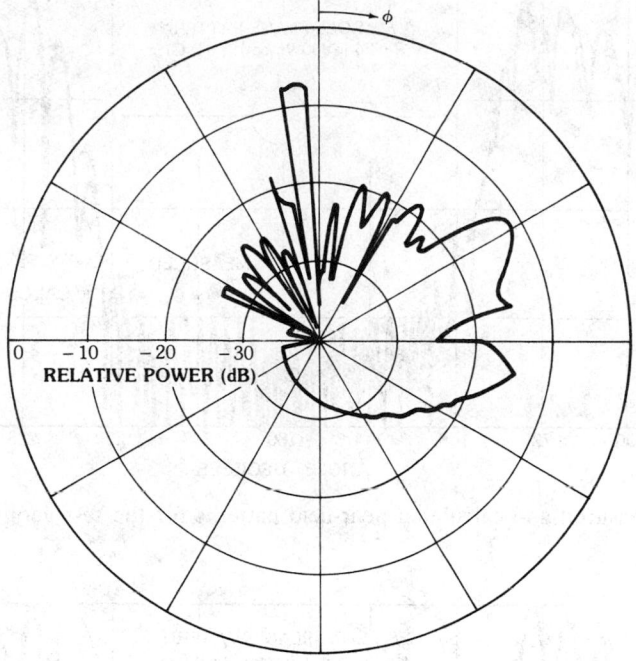

c

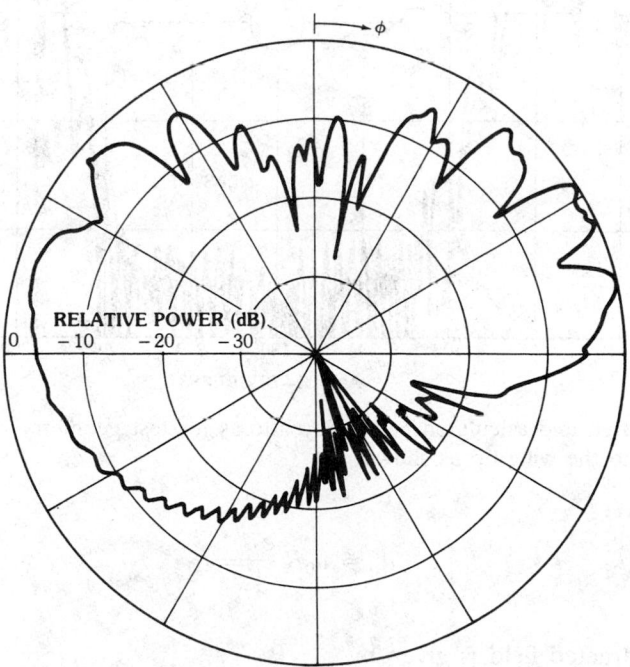

d

Fig. 27, *continued.*

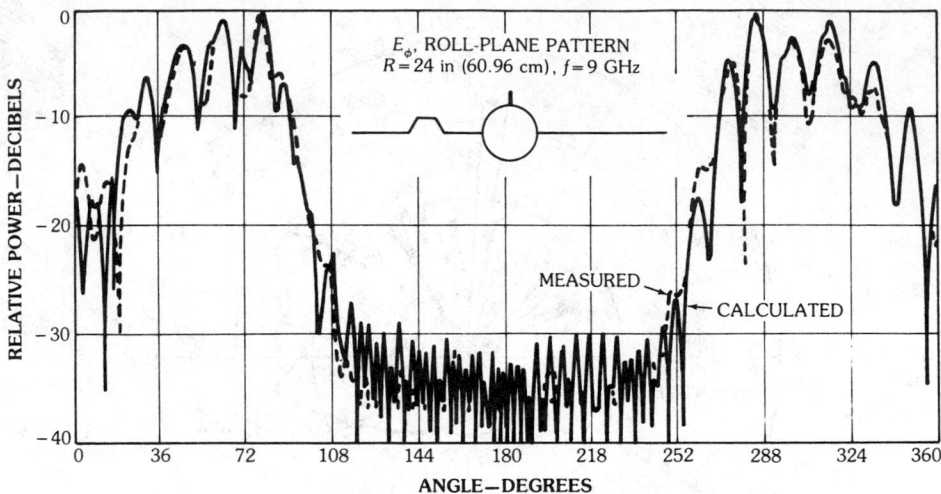

Fig. 28. Measured and calculated near-field patterns for the test geometry shown.

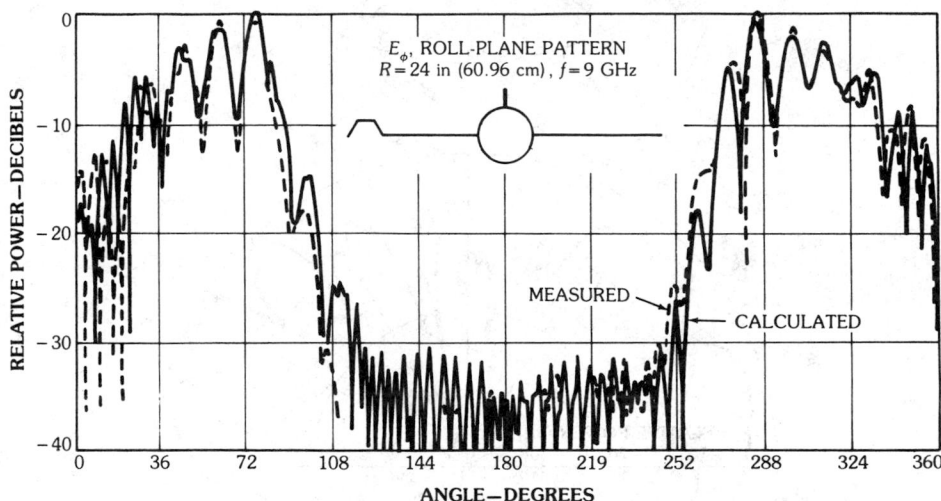

Fig. 29. Measured and calculated near-field patterns for test geometry with engine plate model moved to the wing tip as shown.

$$u_p^r = u_0 R \, \frac{e^{-jks^r}}{\sqrt{s^r}} \tag{25}$$

The plate-diffracted field is given by

$$u_p^d = u^i(Q_d) D \, \frac{e^{-jks^d}}{\sqrt{s^d}} \tag{26a}$$

or

$$u_p^d = u_0 D \frac{e^{-jk(s^i + s^d)}}{\sqrt{s^i s^d}} \tag{26b}$$

The plate doubly reflected field is given by

$$u_{pp}^{rr} = u_0 \frac{e^{-jks^{rr}}}{\sqrt{s^{rr}}} \tag{27}$$

The plate reflected-diffracted field is given by

$$u_{pp}^{rd} = u_p^r(Q_d)D \frac{e^{-jks^d}}{\sqrt{s^d}} \tag{28a}$$

or

$$u_{pp}^{rd} = u_0 RD \frac{e^{-jk(s^r + s^d)}}{\sqrt{s^r s^d}} \tag{28b}$$

The plate diffracted-reflected field is given by

$$u_{pp}^{dr} = u^i(Q_d)D \frac{e^{-jks^d}}{\sqrt{s^d}} R \sqrt{\frac{s^d}{s^d + s^r}} e^{-jks^r} \tag{29a}$$

or

$$u_{pp}^{dr} = u_0 DR \frac{e^{-jk(s^i + s^d + s^r)}}{\sqrt{s^i(s^d + s^r)}} \tag{29b}$$

This is like an imaged wedge, that is

$$u_{pp}^{dr} = u_0 DR \frac{e^{-jk(s^i + s^{dr})}}{\sqrt{s^i s^{dr}}} \tag{29c}$$

where $s^{dr} = s^d + s^r$. Note that R is the soft or hard reflection coefficient, and D is the soft or hard diffraction coefficient. Recall that "soft" or "hard" indicates whether an electric or a magnetic line source is being considered.

The uniform reflected field from the cylinder is given by

$$u_c^r = u^i(Q_R)\mathscr{R} \sqrt{\frac{\varrho^r}{\varrho^r + s_c^r}} e^{-jks_c^r} \tag{30a}$$

or

$$u_c^r = u_0 \mathscr{R} \sqrt{\frac{\varrho^r}{s^i(\varrho^r + s_c^r)}} \, e^{-jk(s^i + s_c^r)} \tag{30b}$$

In addition, the cylinder has a second field component, known as the *uniform diffracted field*, which is given by

$$u_c^d = u^i(Q_1)\mathscr{T} \frac{e^{-jks_c^d}}{\sqrt{s_c^d}} \tag{31a}$$

or

$$u_c^d = u_0 \mathscr{T} \frac{e^{-jk(s^i + s_c^d)}}{\sqrt{s^i s_c^d}} \tag{31b}$$

Note that $\mathscr{R}$ is the modified soft and hard reflection coefficient, and $\mathscr{T}$ is the modified curved surface diffraction coefficient found in Chapter 4. The definition of ϱ^r is given in (37a). It is also possible to have cylinder-to-cylinder interactions.

The combination of plates and cylinders gives rise to fields that interact between them. The fields that are reflected from a plate as one of the interactions are fairly straightforward since image theory can be applied. These terms include the field reflected by the plate and then reflected or diffracted by the cylinder, and the fields reflected or diffracted by the cylinder and then reflected by a plate. The terms that deal with a diffraction off a plate are more complicated, as will be seen below.

The field reflected by a plate and then reflected by a cylinder uses image theory for the reflection off the plate and the uniform reflection coefficient for the reflection off of the cylinder. This field component is given by

$$u_{pc}^{rr} = u_p^r(Q_R)\mathscr{R} \sqrt{\frac{\varrho^r}{\varrho^r + s_c^r}} \, e^{-jks_c^r} \tag{32a}$$

or

$$u_{pc}^{rr} = u_0 R \mathscr{R} \sqrt{\frac{\varrho^r}{s^r(\varrho^r + s_c^r)}} \, e^{-jk(s^r + s_c^r)} \tag{32b}$$

The various parameters are defined in Chapter 4, with the source distance being defined from the source image point. The field reflected by a plate and then diffracted by a cylinder follows in a similar manner as above, that is, the field is given by

$$u_{pc}^{rd} = u_p^r(Q_1)\mathscr{T} \frac{e^{-jks_c^d}}{\sqrt{s_c^d}} \tag{33a}$$

or

$$u_{pc}^{rd} = u_0 R \mathcal{T} \frac{e^{-jk(s'+s_c^d)}}{\sqrt{s' s_c^d}} \tag{33b}$$

The clockwise and counterclockwise terms are included for the creeping-wave terms. The field that uniformly reflects from the cylinder and then reflects from a plate also follows in a straightforward manner. This field is given by

$$u_{cp}^{rr} = u_c^r(Q) R \sqrt{\frac{\varrho^r + s_c^r}{\varrho^r + s_c^r + s^r}} \, e^{-jks^r} \tag{34a}$$

where the square-root term follows from the fact that the source of the plate reflection is the image of the cylinder reflected-field caustic. This can also be written as

$$u_{cp}^{rr} = u_0 \frac{e^{-jks^i}}{\sqrt{s^i}} \mathcal{R} \sqrt{\frac{\varrho^r}{\varrho^r + s_c^r}} \, e^{-jks_c^r} R \sqrt{\frac{\varrho^r + s_c^r}{\varrho^r + s_c^r + s^r}} \, e^{-jks^r} \tag{34b}$$

or

$$u_{cp}^{rr} = u_0 \mathcal{R} R \sqrt{\frac{\varrho^r}{s^i(\varrho^r + s_c^r + s^r)}} \, e^{-jk(s^i + s_c^r + s^r)} \tag{34c}$$

which makes this image relationship quite apparent. Similarly, the field diffracted by the cylinder and then reflected by the plate is given by

$$u_{cp}^{dr} = u_c^d(Q) R \sqrt{\frac{s_c^d}{s_c^d + s^r}} \, e^{-jks^r} \tag{35a}$$

or

$$u_{cp}^{dr} = u_0 \mathcal{T} R \frac{e^{-jk(s^i + s_c^d + s^r)}}{\sqrt{s^i(s_c^d + s^r)}} \tag{35b}$$

Again, the parameters needed for the above fields can be found in Chapter 4, with the scattered distance being replaced with the distance from the scatter point to the image of the observation point.

The field terms dealing with the interaction between a plate edge and a cylinder have an added complication as compared with the terms discussed above. It is assumed in the UTD coefficients that the incident wavefronts are in the form of a ray optical field. This is true for the reflected field off a cylinder and the diffracted field off a plate edge when the fields do not fall within the transition region of the structures. This is one reason why the fields reflected from the cylinder are assumed to be of the geometrical optics type here. The field reflected by a cylinder and diffracted by a plate, therefore, is composed of a geometrical optics field incident on an edge and is given by

$$u_{cp}^{rd} = u_c^r(Q_d)D \frac{e^{jks^d}}{\sqrt{s^d}}$$

(36a)

or

$$u_{cp}^{rd} = u_0 \mathscr{R} D \sqrt{\frac{\varrho_r}{s^i s^d(\varrho^r + s_c^r)}} \, e^{-jk(s^i+s^d+s_c^r)}$$

(36b)

where the reflection coefficient for the cylinder $\mathscr{R}$ approaches ± 1. The reflected-field caustic distance is given by

$$\varrho^r = \frac{s^i R_1 \cos \theta^i}{R_1 \cos \theta^i + 2s^i}$$

(37a)

However, the distance parameter for the wedge-diffraction coefficient must take into account that the incident wavefront looks like it is coming from the caustic of the reflected field off the cylinder. This means that the distance parameter is given by

$$L = \frac{\varrho^i s^d}{\varrho^i + s^d}$$

(37b)

where

$$\varrho^i = \varrho^r + s_c^r$$

(37c)

This will produce a continuous field at the shadow boundary of the wedge. This solution will not be valid, however, when the reflection point is in the transition region of the cylinder, that is, near the shadow boundary of the cylinder.

The field diffracted by a plate and reflected by the cylinder is given by

$$u_{pc}^{dr} = u_p^d(Q_R)\mathscr{R} \sqrt{\frac{\varrho^r}{\varrho^r + s_c^r}} \, e^{-jks_c^r}$$

(38a)

or

$$u_{pc}^{dr} = u_0 \mathscr{R} D \sqrt{\frac{\varrho^r}{s^i s^d(\varrho^r + s_c^r)}} \, e^{-jk(s^i+s^d+s_c^r)}$$

(38b)

The distance parameter for this case is simply given by

$$L = \frac{s^i s^d}{s^i + s^d}$$

(39a)

as before. However, the reflection caustic distance is now given by

$$\varrho^r = \frac{s^d R_1 \cos \theta^i}{R_1 \cos \theta^i + 2s^d} \tag{39b}$$

This assumes that the incident field looks like it is emanating from the plate edge. This will only be true when the diffracted ray is not in the transition region of the edge. In the transition region the true representation is much more complicated and beyond the scope of this discussion. It is assumed that the fields in the transition region using the chosen parameters will give a reasonable engineering representation for the field in any event. In practice these approximations appear to give usable answers to this complicated problem, since the regions in which they are not valid are rather small in extent.

The total field for a problem consisting of plates and cylinders is then composed of all the fields described above from the various scattering parts of the structure. The fields must be shadowed properly even though this may result in a discon-

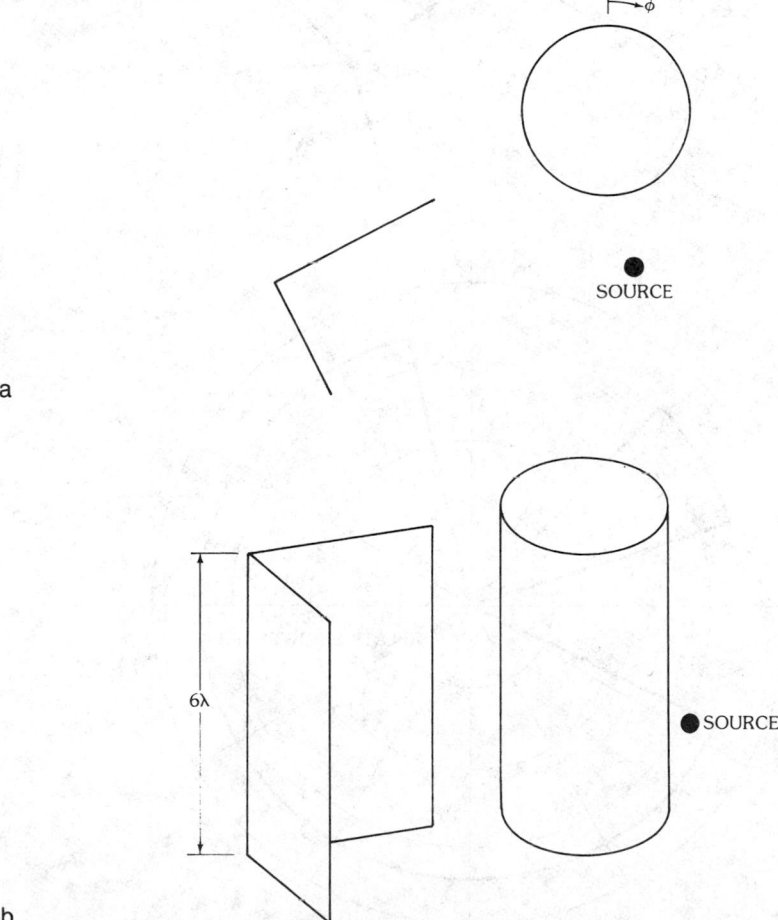

Fig. 30. Illustration of source and scattering elements. (*a*) Top view. (*b*) Side view.

tinuous pattern because all the higher-order interactions have not been included. These discontinuities give a gauge to the accuracy of the solution, which will be seen in the examples that follow.

The following example illustrates how the various field terms previously discussed individually contribute to produce a usable continuous pattern for a complex shape. The example consists of two plates forming a corner reflector and a circular cylinder as shown in Fig. 30. The pattern is taken in the plane normal to the cylinder so it is assumed that a two-dimensional analysis applies here. This is reasonable to assume since in many cases the top and bottom plate edges will not

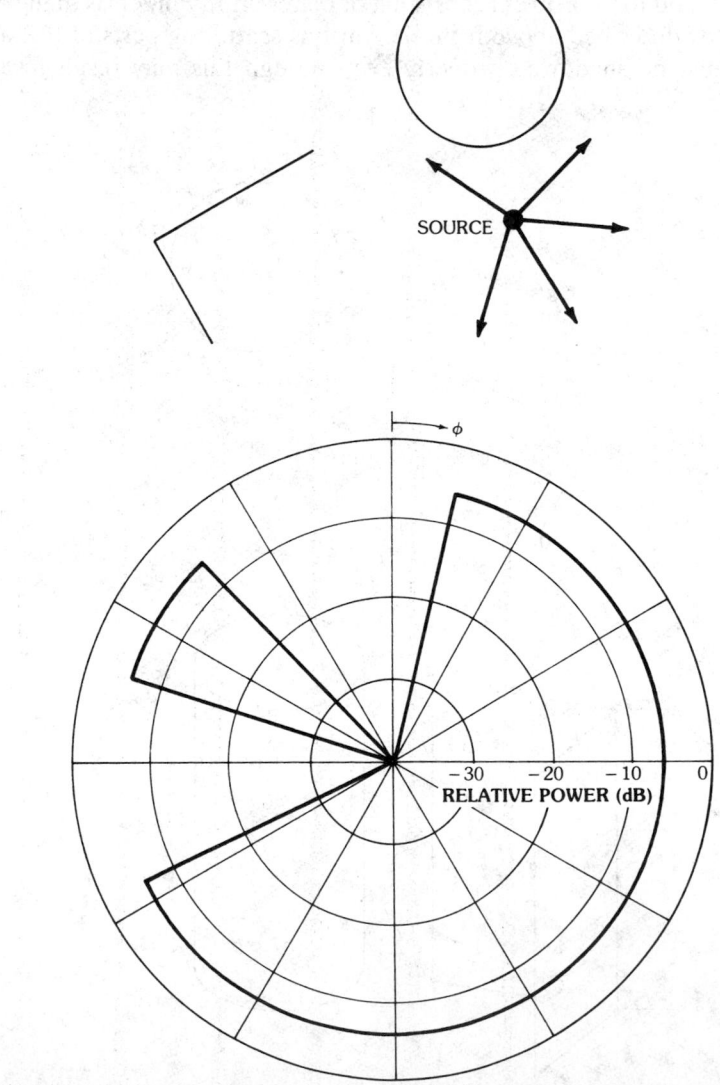

Fig. 31. Source fields. (*a*) Ray paths. (*b*) General pattern.

contribute significantly to the pattern cut illustrated by the top view. The patterns are plotted with a scale from 0 to −40 dB, and they are normalized to the maximum value of the total pattern. The first-order terms associated with the plate (source, reflected, and diffracted fields) are shown in Figs. 31 and 32.

In addition to first-order plate terms there are other first-order terms, such as: (1) the scattered (reflected and diffracted) fields from the cylinders, (2) the field reflected from the end caps, and (3) the fields diffracted by the end cap rims. These are shown in Figs. 33 to 35. Note that in the geometry presented in Fig. 30, end cap reflections will not occur. Therefore a different geometry is shown in Fig. 35 to demonstrate end cap reflections. Note that with more than one body present, individual terms are often shadowed by other bodies in the structure, creating discontinuities as shown in many of the figures (as in Fig. 33 for the cylinder scattered field).

In addition to first-order mechanisms second-order scattering occurs where a ray is scattered by one body and then scattered by a second one. Several different double-scattering (or second-order) terms are computed. Double reflection, where a ray is reflected by one plate and then by another, is shown in Fig. 36.

Another second-order scattering mechanism involving plates is reflection-diffraction, where a ray is reflected from one plate and is then diffracted as shown in Fig. 37. The inverse mechanism, diffraction-reflection, illustrated in Fig. 38, involves fields diffracted from a plate edge and then reflected by another plate.

A number of the scattering mechanisms involve interactions between the cylinder and one of the plates. Two such terms result from scattering of the fields by the cylinder and then reflection by a plate, and vice versa. Fig. 39 illustrates the ray paths and fields of rays which are reflected from a plate and then scattered by the cylinder. Fig. 40 illustrates ray paths and fields resulting from rays scattered by the cylinder and then reflected from a plate.

Another second-order scattering mechanism involves fields reflected by the cylinder and then diffracted by a plate edge. The ray paths and fields for this term

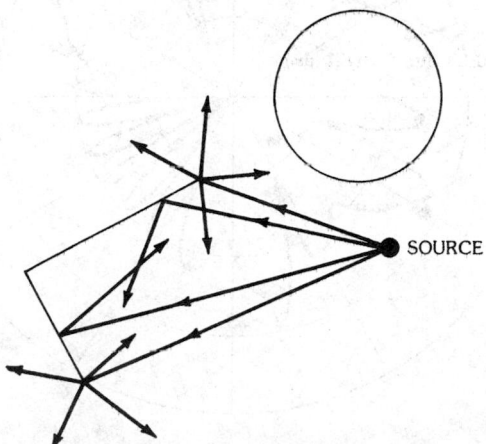

a

Fig. 32. First-order terms associated with the plate. (*a*) Ray paths. (*b*) Field due to plate reflection. (*c*) Field due to plate diffraction.

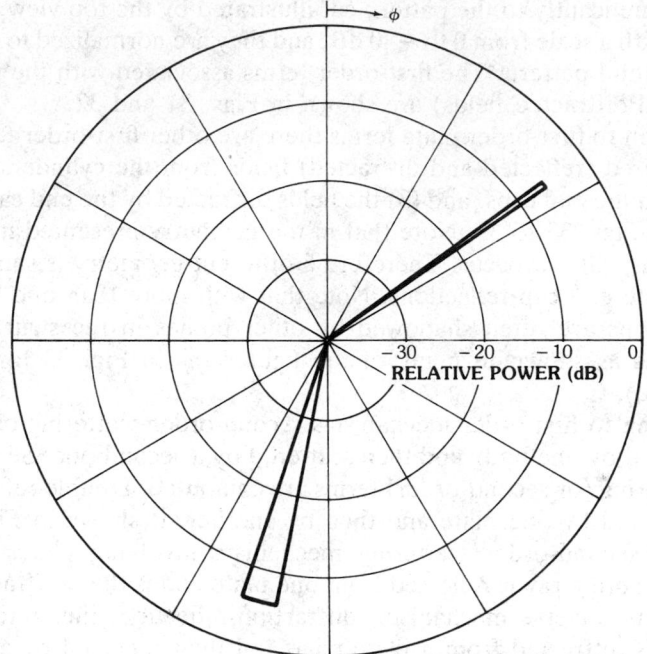

b

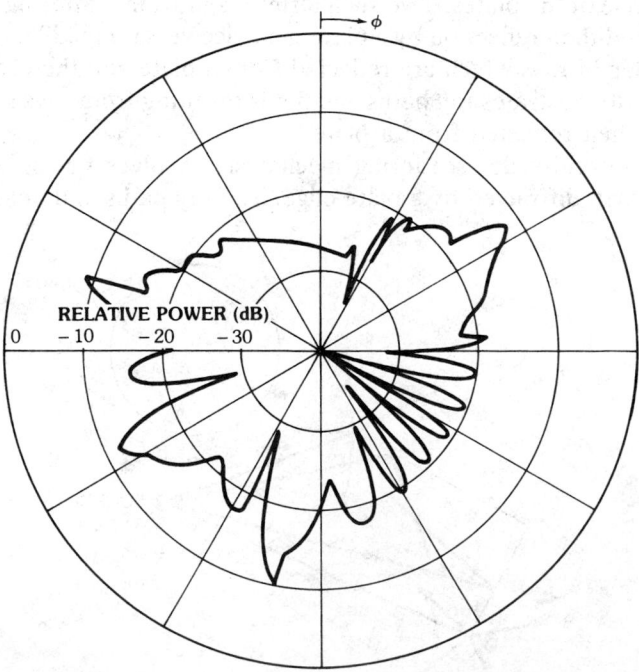

c

Fig. 32, *continued*.

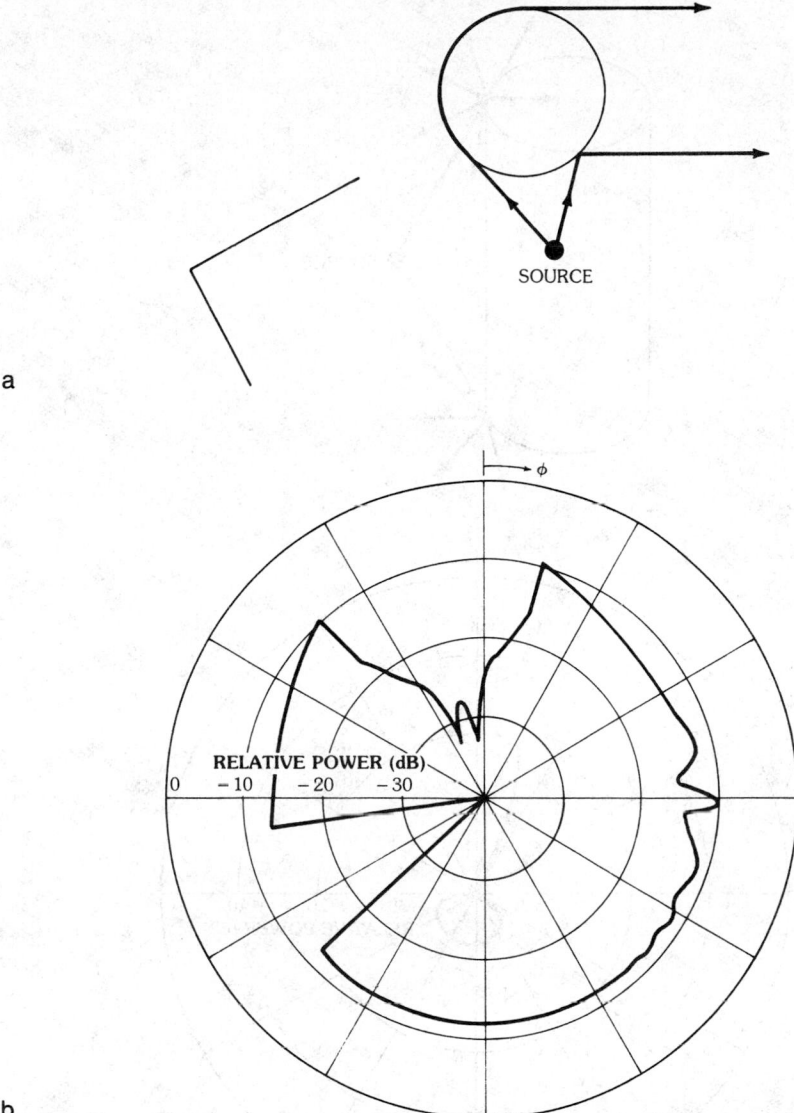

Fig. 33. First-order terms associated with the cylinder. (*a*) Ray paths. (*b*) Scattered field from cylinder's curved surface.

are illustrated in Fig. 41. The reverse of this term is the fields of rays diffracted by a plate edge and then reflected from the cylinder, as shown in Fig. 42.

The total pattern is obtained by summing the field components for the mechanisms mentioned previously. The total field pattern is illustrated in Fig. 43. One should especially note the continuity of the total pattern, which indicates that that is calculated properly and that higher-order terms are not significant for this configuration.

Higher-order scattering terms can also be computed, which will in some cases

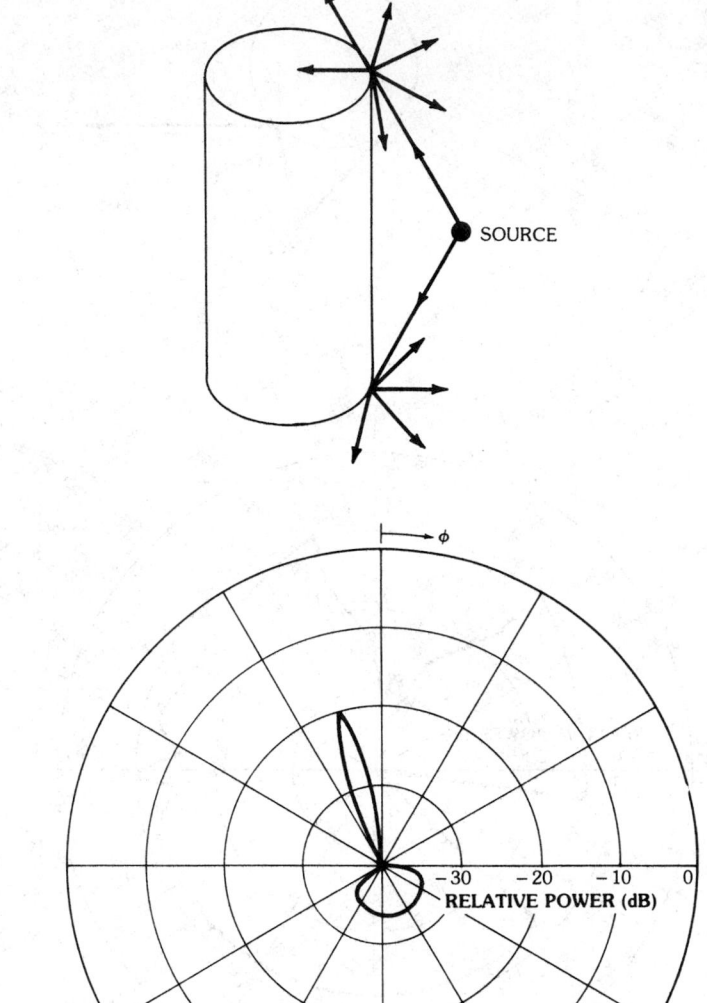

a

b

Fig. 34. First-order terms associated with the end cap diffracted fields. (*a*) Ray paths. (*b*) Fields due to end cap diffraction.

improve the accuracy of the field computations. Generally it is found that such terms are negligible in magnitude as well as being difficult to compute and therefore need not be included. The presence of discontinuities in a final field pattern, however, indicates the presence of regions where higher-order terms are needed. An approximation to the truc pattern can often be obtained by visually averaging

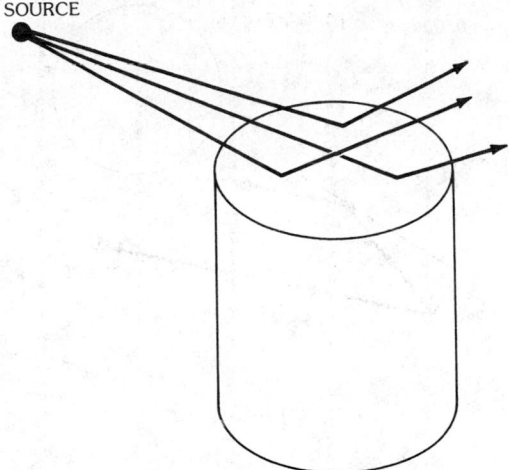

SOURCE

Fig. 35. Illustration of ray paths for end cap reflected fields.

the discontinuities. In any event a pattern generated using UTD gives good visual clues as to its accuracy.

The analysis of three-dimensional models consisting of multiple flat plates and finite elliptic cylinders is a natural extension of these solutions. The analysis has progressed in a systematic fashion by defining the basic elements of the problem and building them into the necessary tools to model real-world scattering structures. The two-dimensional analysis given above is intended to give a feeling for the way the fields of the various terms fit together. More information on the three-dimensional analysis, such as the methods used to find the ray paths, can be found in References 5 and 18.

4. Far Field of Antenna Versus Far Field of Structure

To be in the true far field the receiver has to be an infinite distance away from the transmitting antenna. This implies that the various scattered rays emanating from the antenna and its environment are all parallel. From a practical point of view, a measurement of the far-field pattern of an antenna in its environment must be made at a reasonable distance. The question then arises as to what range is necessary to measure a very good estimate of the true far-field pattern.

Based on a reasonable phase error for nearly parallel rays it is clear that an antenna in free space can be measured at a range given by

$$R = 2D^2/\lambda$$

where D is the maximum extent of the antenna and λ is the wavelength. If the same antenna is placed on or near a significant structure, then one must distinguish the antenna far field from the antenna-and-structure far field. To illustrate this distinction consider a parabolic reflector antenna. The far field of the feed may be a matter of a few feet (or meters), whereas the far field of the feed-and-reflector

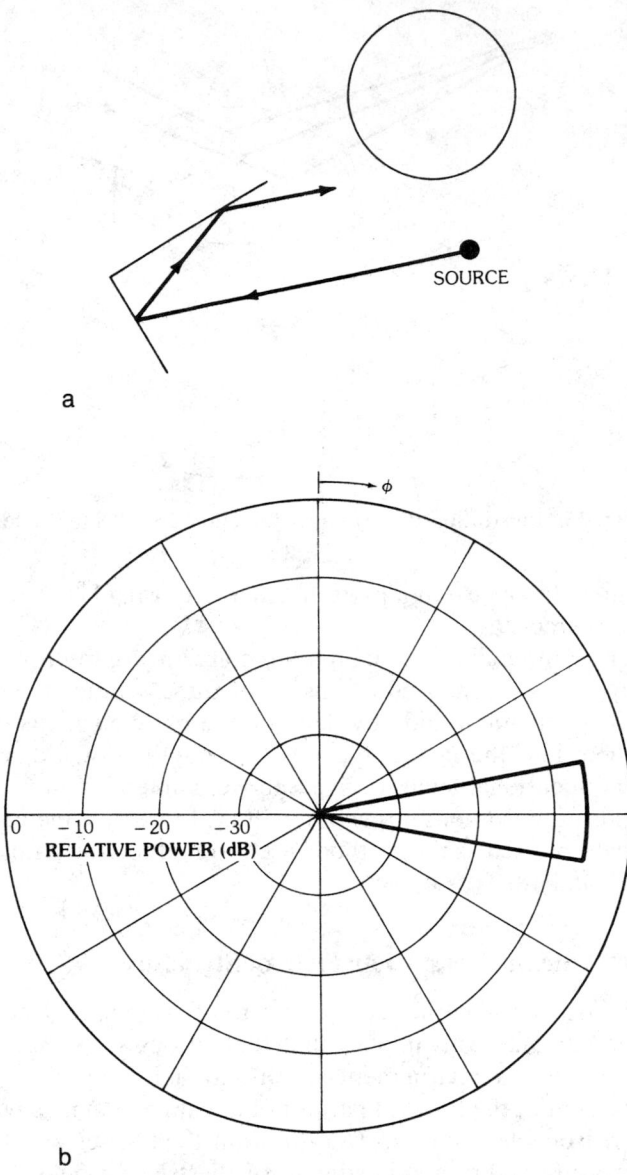

Fig. 36. Illustration of plate double reflection. (*a*) Ray path. (*b*) Fields due to doubly reflected rays.

system may be a mile (or kilometer) or more. In this case the feed is designed to illuminate the reflector in a precise way to generate a narrow-beam far-field pattern. So the reflector is actually an integral part of the antenna system, and the true far-field range is $2D_r^2/\lambda$, where D_r is the diameter of the reflector.

When an antenna radiates in the presence of a large structure that significantly influences the antenna radiation pattern the structure dimensions must be included in the distance (D) used to define the far-field range. In fact the antenna and its

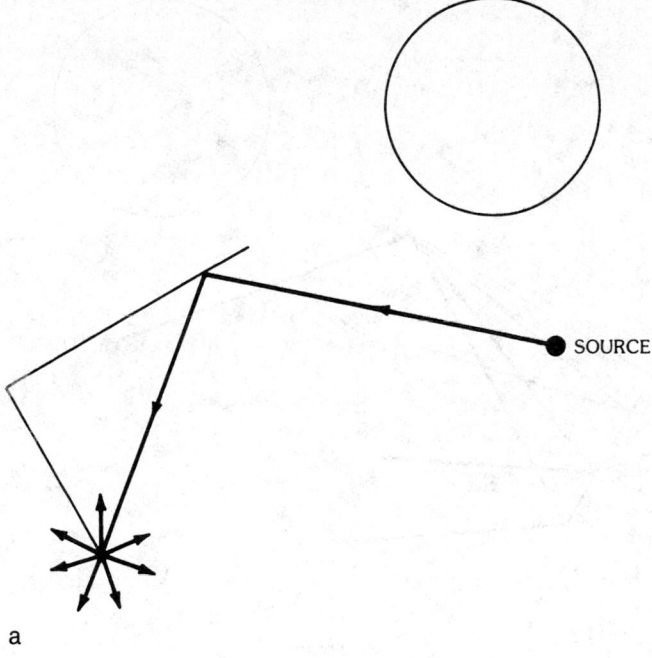

a

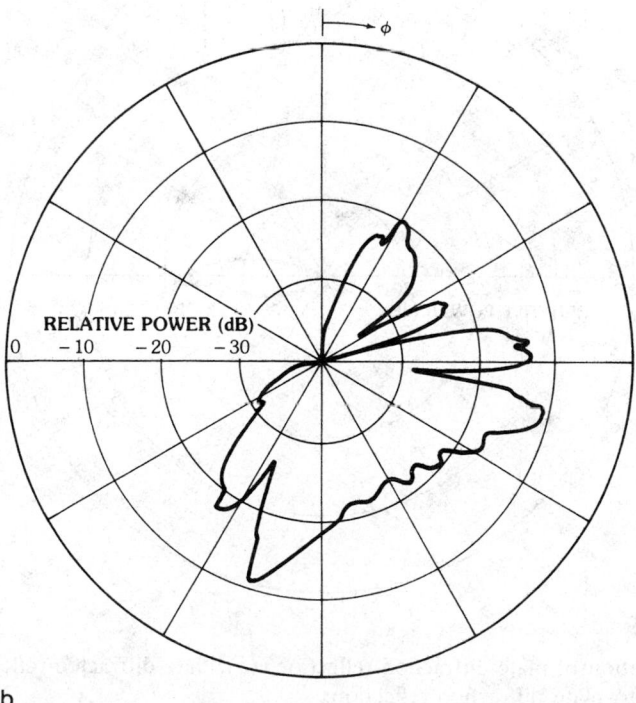

b

Fig. 37. Illustration of plate reflection-diffraction. (*a*) Ray path. (*b*) Fields resulting from plate reflection-diffraction.

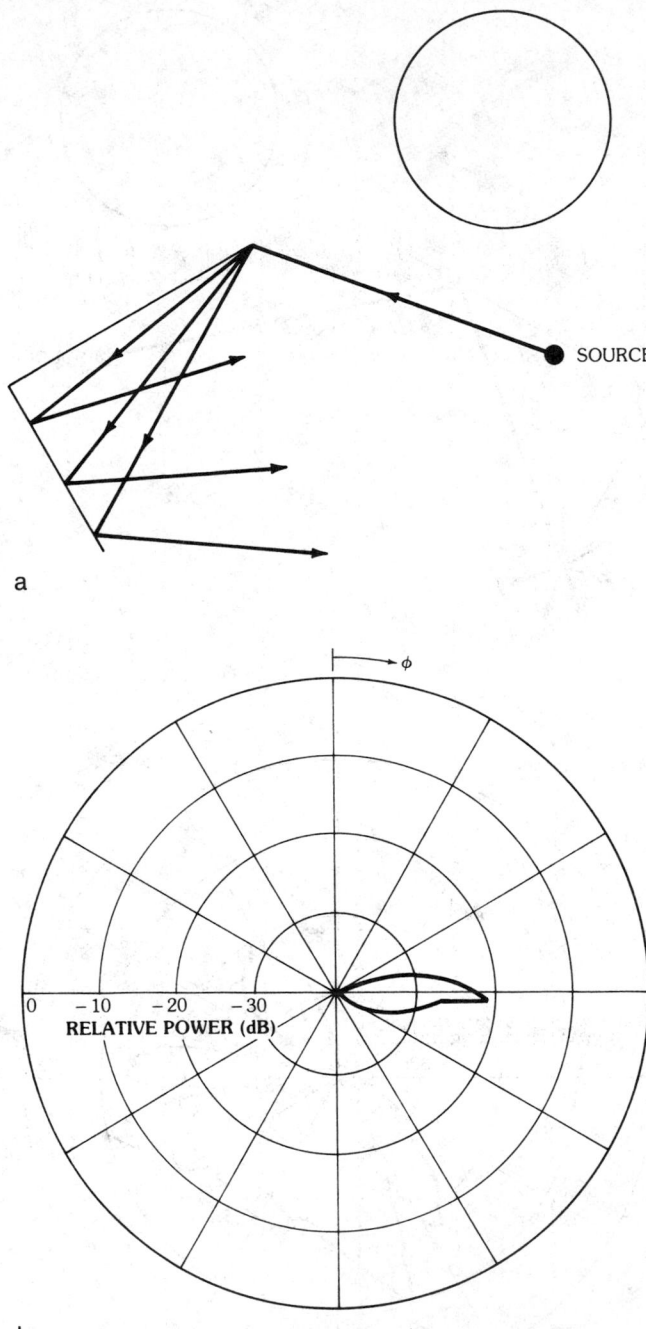

Fig. 38. Illustration of plate diffraction-reflection. (*a*) Plate diffracted-reflected ray paths. (*b*) Fields due to plate diffraction-reflection.

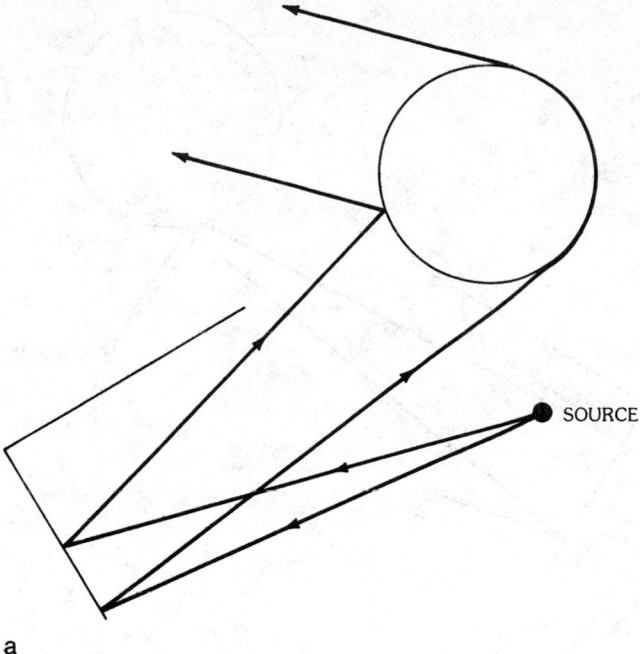

a

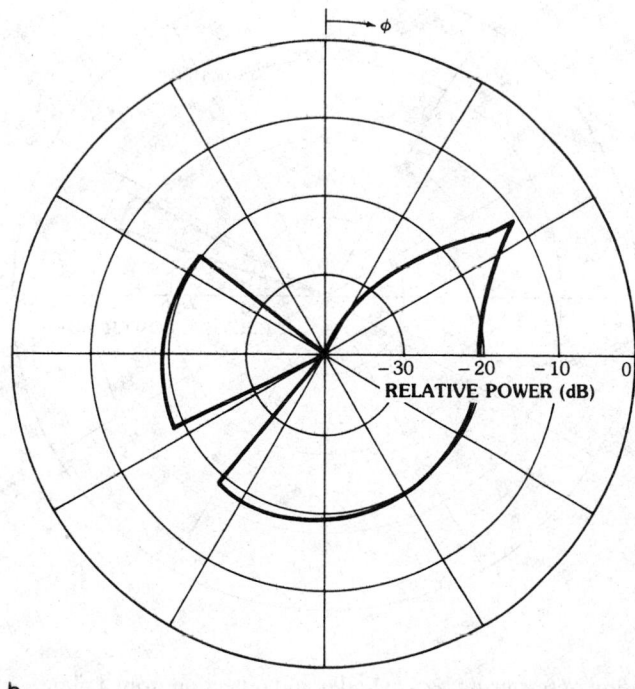

b

Fig. 39. Illustration of reflection from a plate and scattering by a cylinder. (*a*) Ray paths. (*b*) Pattern of fields resulting from (*a*).

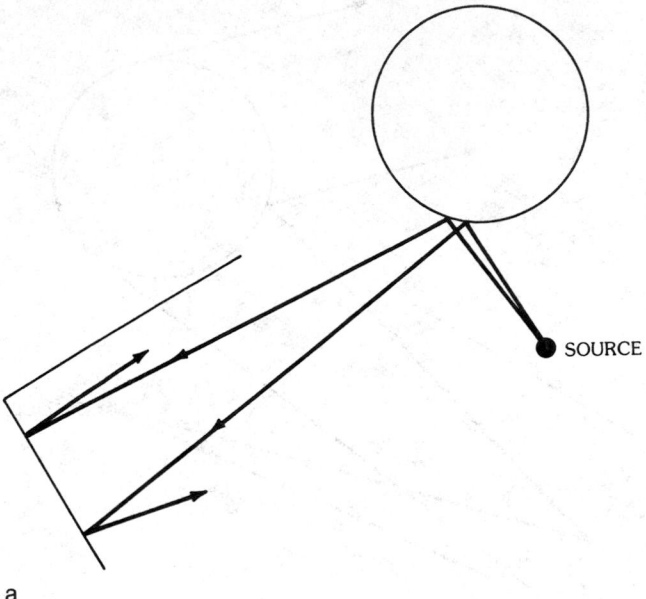

a

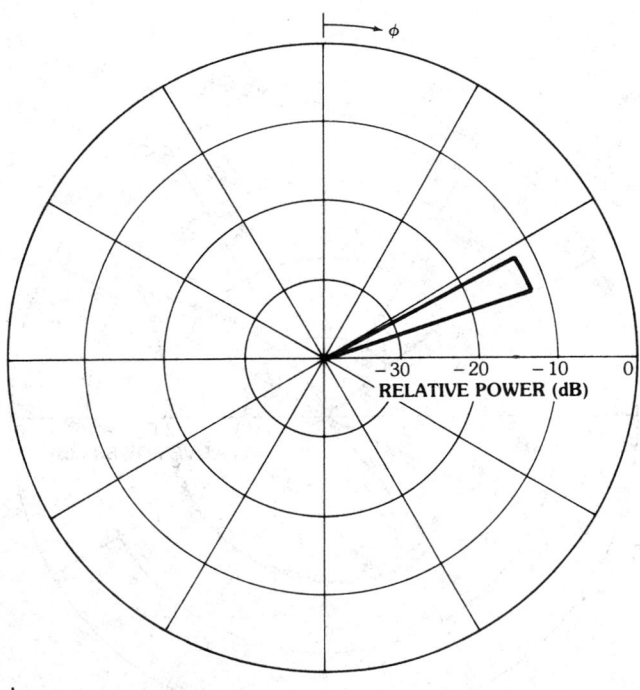

b

Fig. 40. Illustration of scattering by a cylinder and reflection from a plate. (*a*) Ray paths. (*b*) Fields resulting from (*a*).

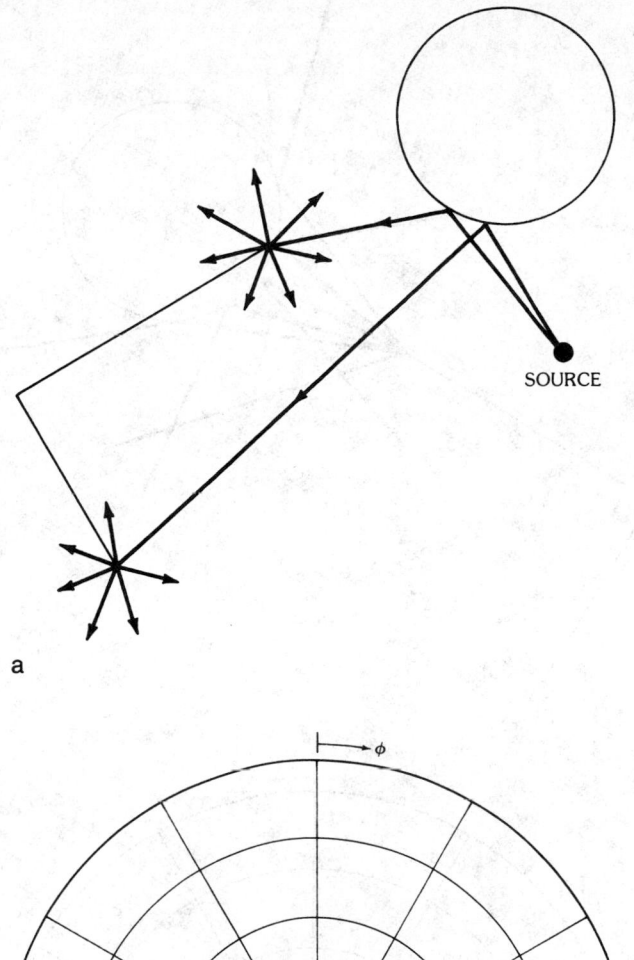

a

b

Fig. 41. Illustration of reflection from a cylinder and diffraction by plate edges. (*a*) Ray path. (*b*) Fields due to reflection from cylinder and diffraction by plate edges.

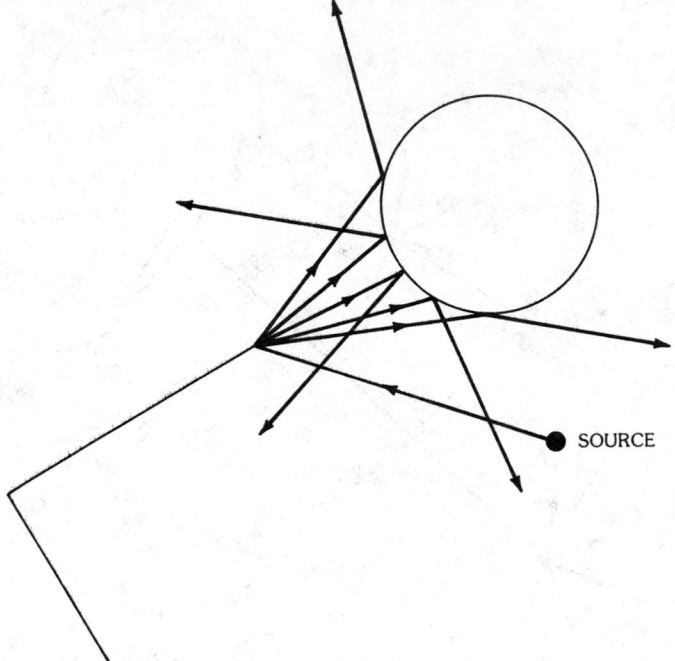

a

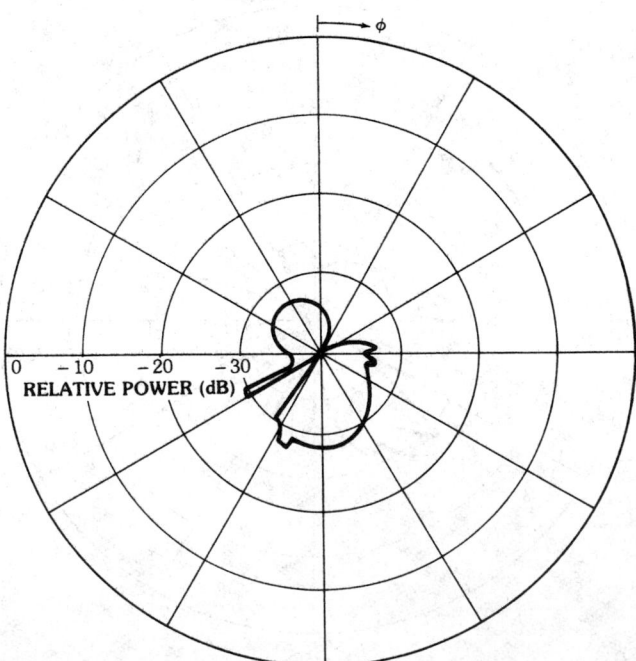

b

Fig. 42. Illustration of diffraction by a plate edge and reflection from a cylinder. (*a*) Ray paths. (*b*) Fields resulting from (*a*).

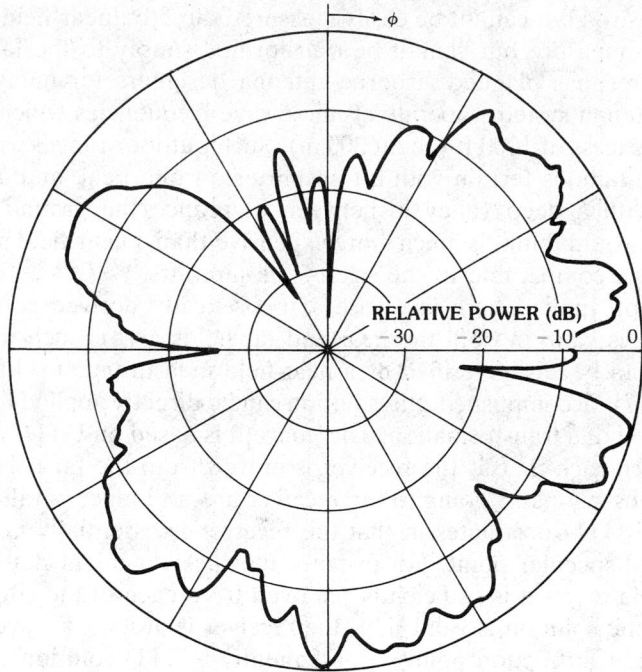

Fig. 43. Total fields of the source in the presence of scattering bodies.

significant environment should be encompassed within a sphere of diameter D; then the far-field range of the antenna-and-structure is given by

$$R = 2D^2/\lambda$$

For a low-gain communications antenna mounted on an aircraft the complete structure is strongly illuminated, and as a result it significantly affects the antenna radiation patterns. Thus the entire aircraft has to be considered as an integral part of the antenna, and the receiver must be moved to the far-field range of the aircraft if the true far field is to be examined. This point will be discussed in more detail later.

As a result of the excessive ranges needed to measure large antenna-and-structure systems, there has been a great deal of interest in determining far-field patterns based on near-field pattern measurements since near-field measurements can be taken efficiently in small anechoic chambers. In order to determine the desired far-field patterns based on these near-field measurements, most of the attention has focused on plane-, cylindrical-, and spherical-wave spectrum approaches. However, each of these spectrum approaches must transform the near-field measured data to the desired far-field result. This transform is basically an integral relationship which in itself can be tedious and expensive. The real solution to this problem lies in a direct approach that simply converts near-field data to the desired far-field results. Thus the following dilemma prevails: far-field patterns are desired but cannot be easily measured directly; near-field patterns are much easier to measure but cannot be transformed simply to the far field.

This dilemma has plagued airborne-antenna designers for many years. The majority of antenna systems operate at microwave frequencies which require far-field ranges in excess of 1000 ft (about 300 m). Such outdoor ranges usually exist in a hilly or mountainous terrain with a transmitter on one peak and a receiver on another and with a deep valley in between to reduce the ground effect. This approach, one would think, is much more expensive than a near-field measurement taken at a range comparable to the size of the aircraft.

The concept applied here is to use a theoretically derived solution to this problem which is valid in both the near and far fields and is such that the near-field solution can be easily verified by a near-field measurement. Once this near-field verification is accomplished, this solution can be directly applied to the far field without the need of a transformation. This concept is based on UTD, which is a ray optical approach such that as the receiver is moved from the far field to the near field, the various ray paths going to the receiver are no longer parallel. This does not violate the UTD postulates in that the receiver is essentially in the far field of each isolated specular point. For instance the receiver might not be in the far field of a flat plate, yet it is sufficiently removed from each of the edge-diffraction points so that the solution is valid (i.e., the receiver is at least a wavelength away from the isolated diffraction point). Consequently a UTD solution can be developed which efficiently solves for near-field patterns. This solution can then be used directly to compute a far-field pattern simply by using the receiver range dictated by the far-field criterion. If one is selective in taking near-field measurements, the ray paths and associated specular points do not deviate greatly from the near field to the far field. This point will be stressed later.

The calculated and measured data presented in Fig. 44 serve as a verification of this near-field analysis, showing results for a test geometry composed of a monopole on a circular cylinder with one attached plate. As indicated, the plane of the plate is tilted 16° with respect to the cylinder axis such that a curved edge exists at the plate-cylinder junction. A 124° conical pattern is presented at a range of 24 in (60.96 cm) from the origin. Note that the receiver moves within about a wavelength of the wing tip for this case. With this in mind the agreement between measured and calculated results is very encouraging.

A practical example demonstrating the improved results obtainable with the near-field analysis is depicted in Fig. 45, which compares measured and calculated elevation plane patterns for a monopole antenna mounted over the cockpit of a Boeing 737 aircraft. The measured data shown were taken in an anechoic chamber using a highly detailed 1:11 scale model aircraft. The receiver range used for the measured data was 26 ft (7.9 m). (This is equivalent to a range of 286 ft or 87.2 m, on a full-scale 737 aircraft.) Two computed patterns are also shown in Fig. 45 for comparison. One of the computed patterns was obtained using the far-field range, while the other was obtained using the near-field analysis. Both of the computed solutions employ the same cylinder and plates model (depicted in Fig. 45a) to simulate the aircraft structure, and comparable scattering terms are included in both solutions. The near-field solution, of course, employs the same receiver range as the measured data. The improved result obtained with the near-field analysis

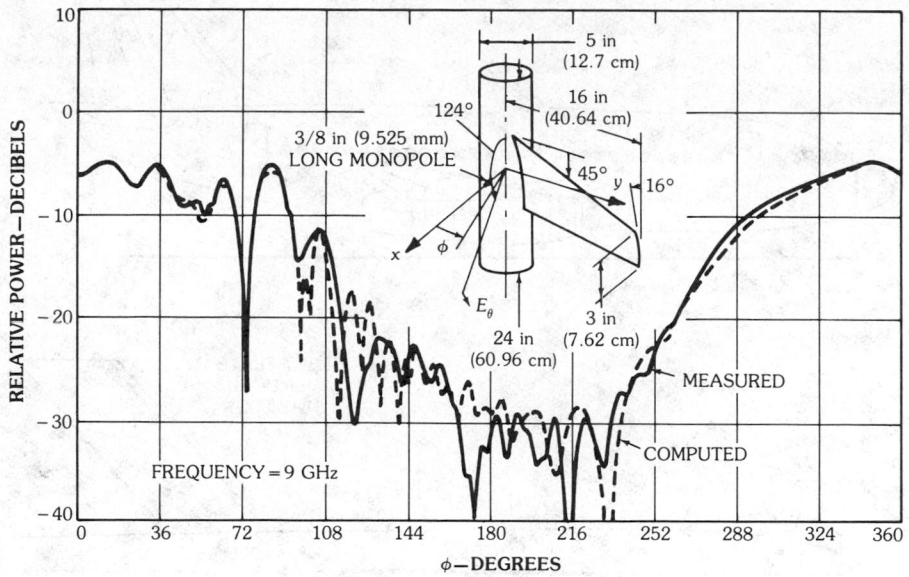

a

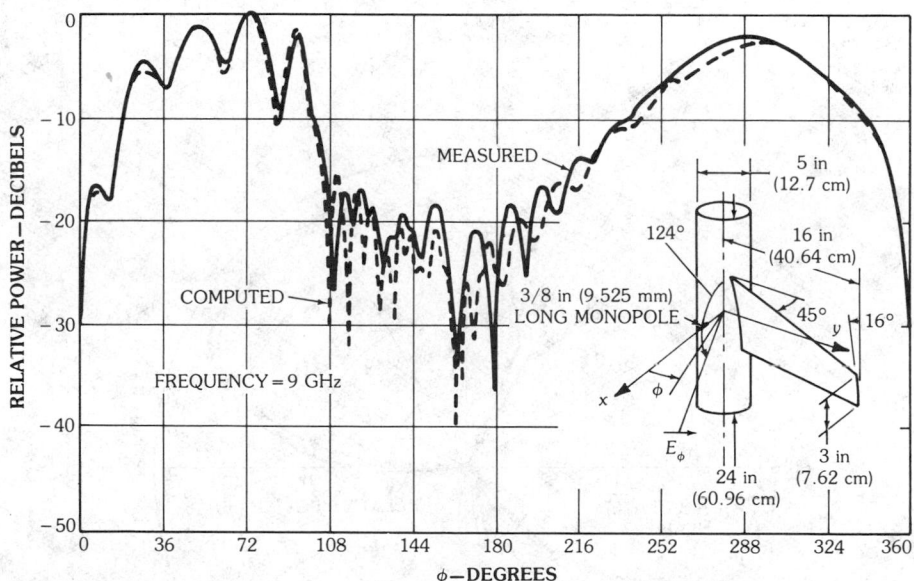

b

Fig. 44. Measured and calculated near-field patterns for the test geometry shown. (*a*) Component E_θ. (*b*) Component E_ϕ.

is especially apparent in the lower left quadrant of the pattern. From the previous results it is clear that the UTD solution can be used to accurately compute the near fields of airborne antennas.

There are many ways to simulate a given aircraft structure using this approach

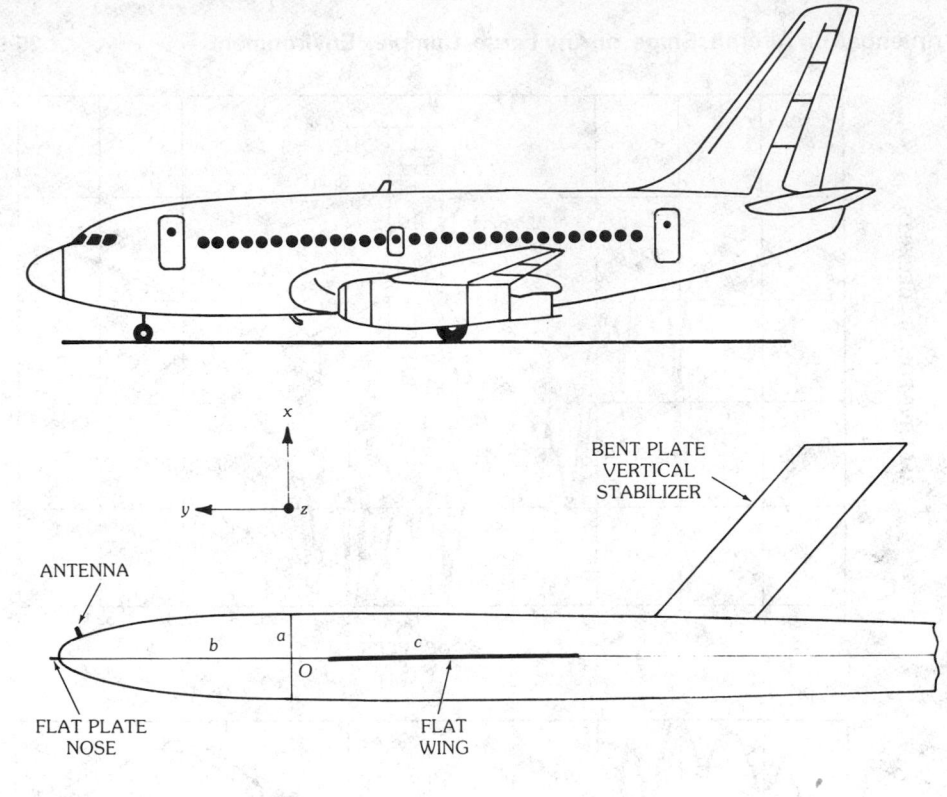

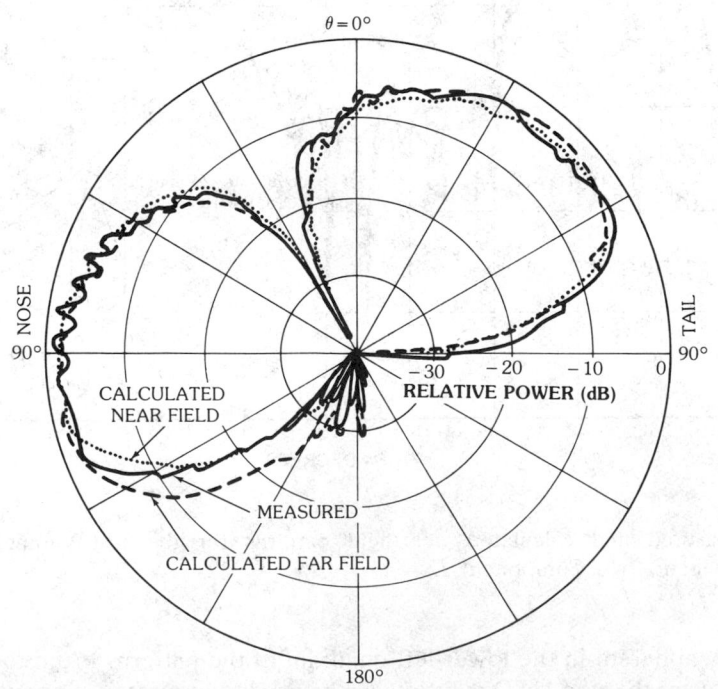

Fig. 45. Boeing 737 aircraft and elevation-plane patterns of a monopole antenna mounted on it. (*a*) Profile views of actual aircraft and model used for analysis. (*b*) Measured and calculated patterns. (*After Burnside, Wang, and Pelton [3], © 1980 IEEE*)

even though specific guidelines are suggested—for example, how much detail of a specific configuration must be included in a simulation model. With this in mind let us assume that an antenna designer has used the code to specify an appropriate antenna and location. At this point it would be satisfying to verify that he or she did achieve the desired far-field pattern performance. Taking the true far-field pattern is out of the question because such a measurement range is not available; however, the designer can simply take some near-field patterns by placing the antenna on the actual aircraft skin. If he or she is going to use this approach, it would be most satisfying to take a near-field pattern whose radiation mechanisms most nearly resemble those of the far field. In order to examine this situation consider the series of elevation-plane patterns taken at a constant range about the three pattern origins shown in Fig. 46. The near-field pattern effects are very noticeable, as shown in Fig. 47. Since the line-of-sight signal from the source to the receiver dominates the significant radiation portion of the pattern, it is not surprising that the near-field pattern should be taken about an origin located at the antenna position. This conclusion is very clearly emphasized by the near-field patterns, where the

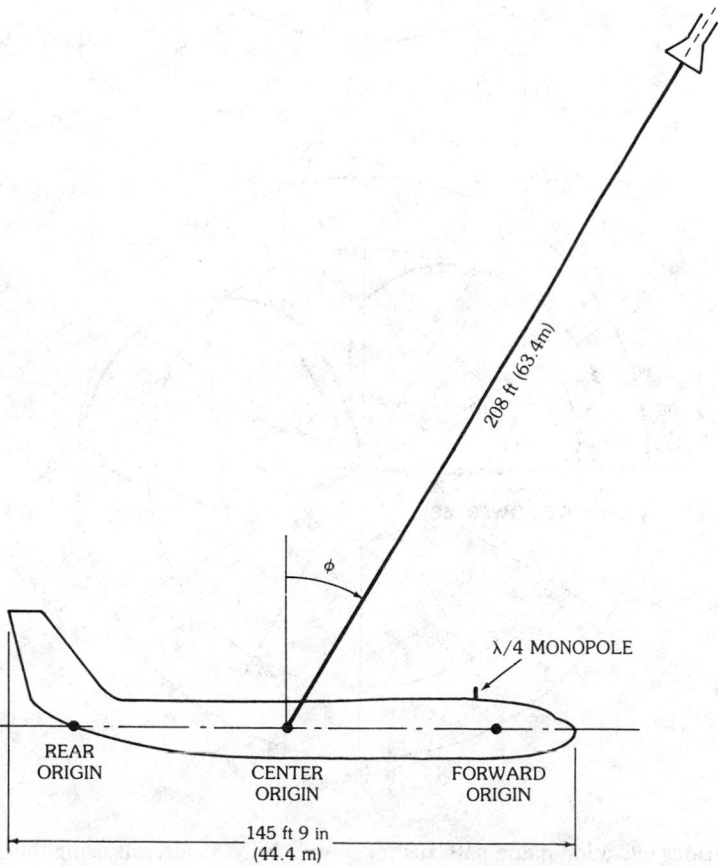

Fig. 46. Geometry associated with near-field pattern computations for the KC-135 aircraft. (*After Burnside, Wang, and Pelton [3], © 1980 IEEE*)

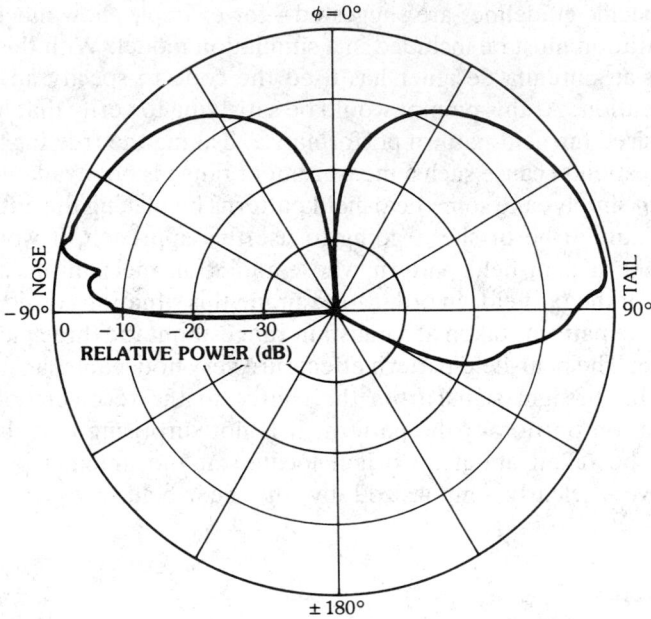

a

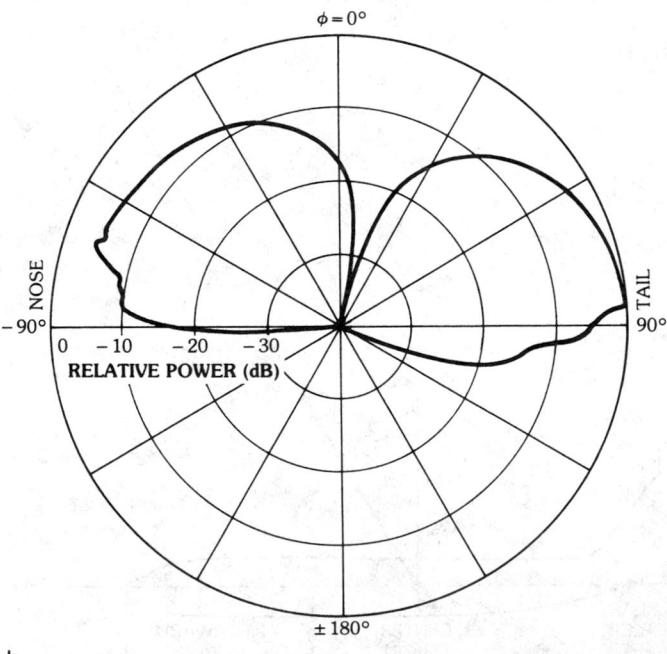

b

Fig. 47. Various elevation-plane patterns taken on a KC-135 aircraft using the geometry in Fig. 46. (*a*) Forward origin. (*b*) Center origin. (*c*) Rear origin. (*d*) Far-field result. (*After Burnside, Wang, and Pelton [3], © 1980 IEEE*)

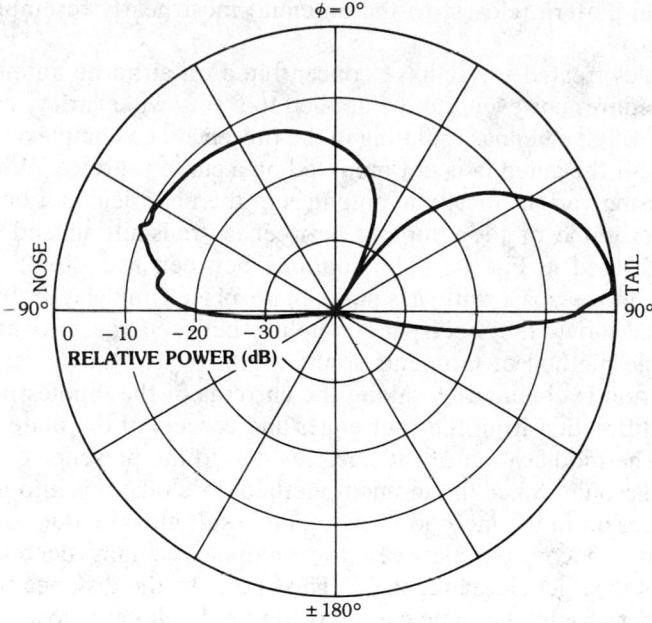

c

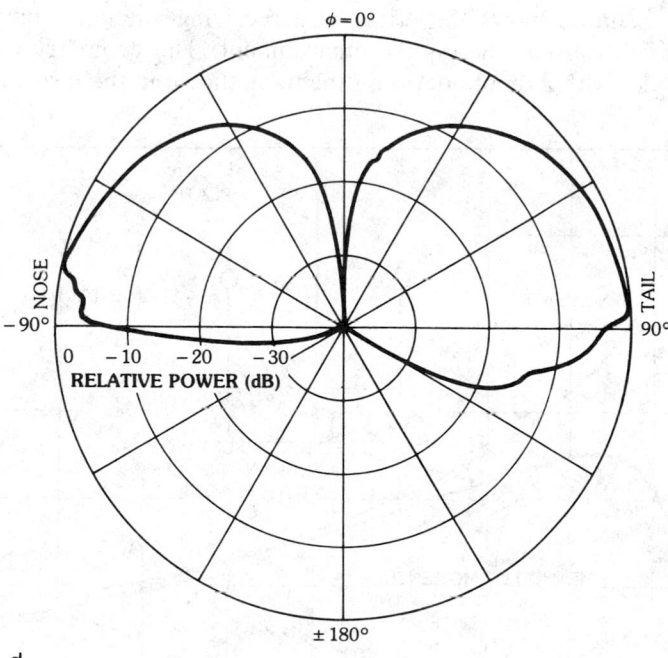

d

Fig. 47, *continued.*

forward-location pattern (closest to the antenna) most nearly resembles the true far-field pattern.

The examples treated so far have concentrated on airborne antenna applications; this procedure, however, can be applied to a very wide variety of problems. Using the NEC BSC, antennas radiating in the presence of a complex environment can be analyzed if the antenna is not mounted on a curved surface. With this code the effects of using various receiving antennas in the near field can be studied.

A simple example of the coupling between a transmitting and a receiving antenna is illustrated in Fig. 48. The coupling between two dipole antennas is shown as they move apart with a square plate placed midway between them. This figure also shows the free-space coupling between the two antennas for comparison. The method of moments solution [10] is also shown on this graph. The UTD solution is obtained by taking the currents of the dipoles in free space and using the diffraction from the four edges and corners of the plate to compute the coupling. The modification of the currents due to the presence of the plate is not taken into account. Since the moment method does take this into account, the ripple that is present in the method of moments result must be due to this effect.

Notice that the coupling between the antennas actually decreases after a certain point as they get closer together. This is due to the distance that the rays must travel to get around the plate even though the dipoles are closer together. In addition the pattern factor of the dipole affects the amplitude of the fields incident on the top and bottom edges as the dipoles get closer.

This next example shows the effect of a receiving antenna's pattern in determining the fields around a shiplike environment. The geometry consists of a plate and cylinder with a dipole between them, as shown in the insert of Fig. 49a.

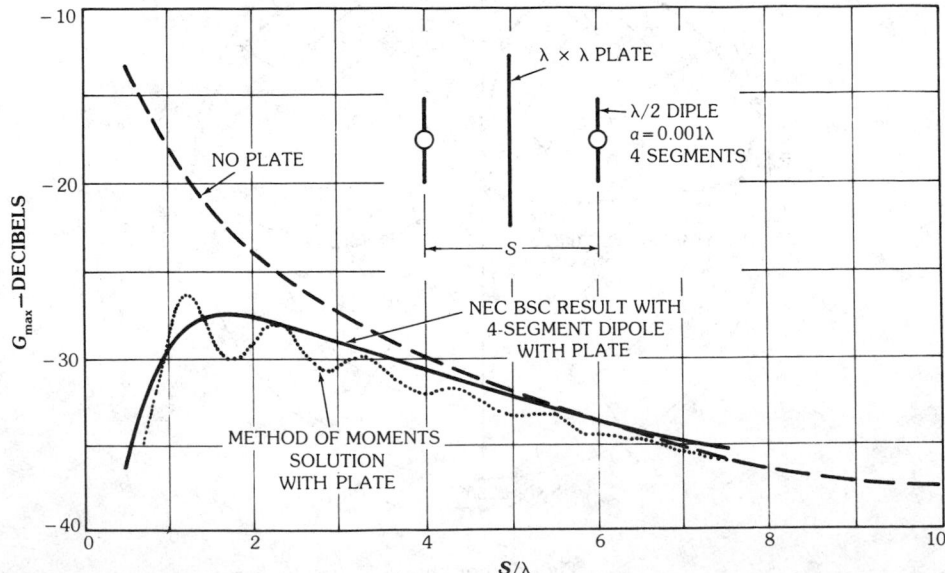

Fig. 48. The coupling between two dipoles in the presence of a plate compared with method of moments results [10].

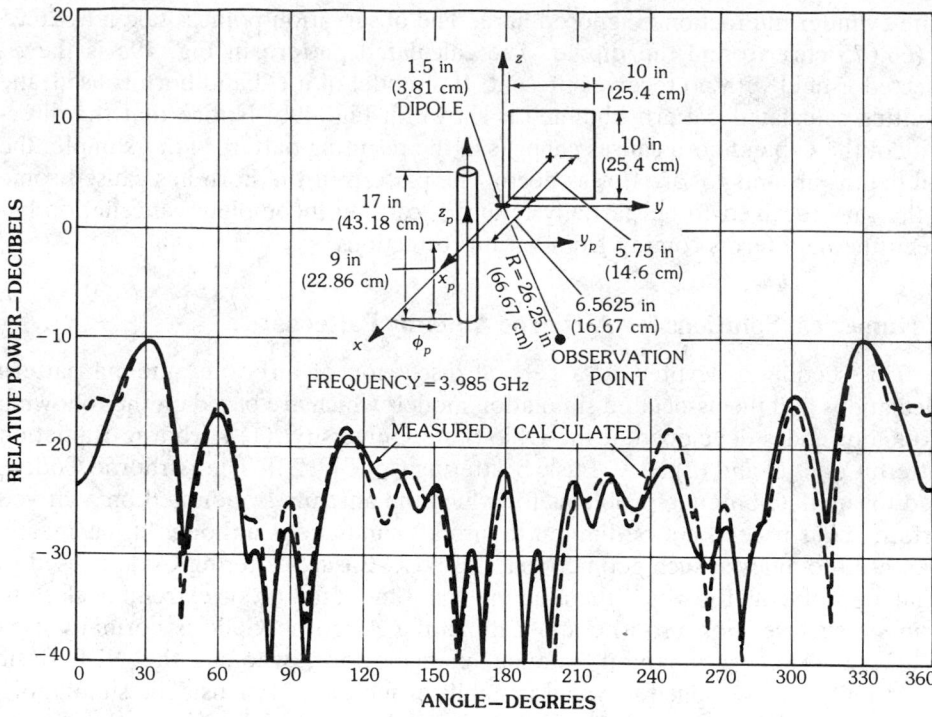

a

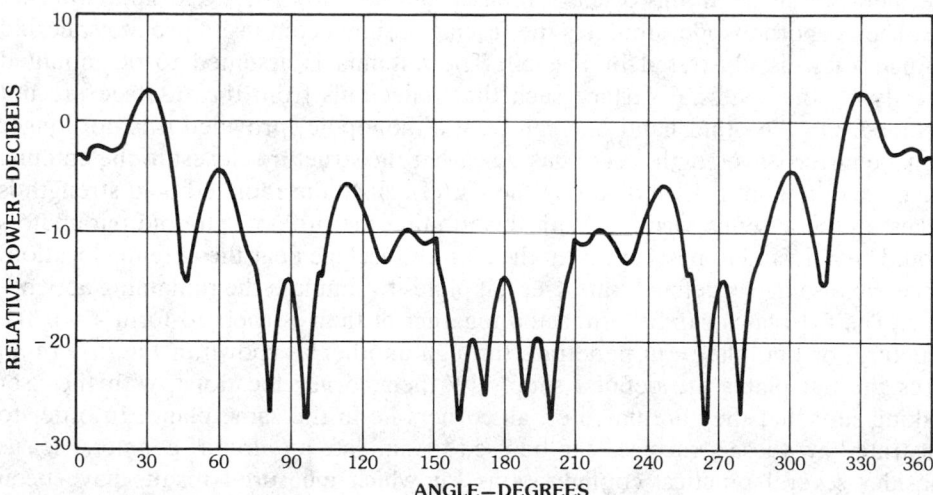

b

Fig. 49. Measured and calculated $E_{\phi p}$ near-zone radiation patterns 26.25 in (66.67 cm) from a dipole in the presence of a square plate and circular cylinder. (*a*) Patterns and test geometry. (*b*) Pattern for the coupling between a *C*-band horn at the observation point and the dipole in the presence of a square plate and a circular cylinder as in (*a*).

Plate-cylinder interaction is ignored here. The observation point is taken as 26.25 in (66.675 cm) around the dipole. The calculated pattern in Fig. 49a is the ϕ-directed **E** field with no receiver present. If a model of a C-band horn is used, the resulting calculated pattern obtained is shown in Fig. 49b. Notice that the directivity of the C-band horn causes changes in the resulting pattern. For example, the null depths around 45° are not as deep. The pattern of the horn has caused some of the rays received to be partially reduced, causing incomplete cancellation between the field terms coming from various directions.

5. Numerical Solutions for Airborne Antenna Patterns

This section is devoted to a general discussion of airborne antenna pattern calculations and the associated simulation models which are based on the following two major codes developed at the Ohio State University: (1) Airborne Radiation Pattern Code [6] and (2) NEC-Basic Scattering Code [7, 23]. The Airborne Code is used to simulate antenna problems for which the antenna is mounted on a curved surface. Thus it need not be limited to aircraft simulations although it has mainly been used to analyze such geometries. The NEC-Basic Scattering Code is used to simulate problems for which the antenna is not mounted on or extremely close to a curved surface. For airborne applications the Airborne Code is normally used to analyze fuselage mounted antenna configurations, whereas the NEC-Basic Scattering Code is applied to wing or stabilizer mounted designs. The simulations used to model a given aircraft are therefore different for the fuselage and wing mounted antenna problems. Since fuselage mounted antennas are much more prevalent, let us begin this discussion based on the Airborne Code simulations.

The Airborne Code simulates the fuselage using a composite prolate spheroid configuration as illustrated in Fig. 50. The antenna is assumed to be mounted directly on the spheroid surface such that reflections from the fuselage are not included. On the other hand one can treat a monopole, provided it is not longer than a quarter-wavelength. As discussed earlier the structure closest to the antenna plays the dominant role in dictating the pattern since the radiated field strength is largest in the antenna vicinity. With this in mind the prolate spheroid parameters should be adjusted to best represent the aircraft fuselage near the antenna location. Once the fuselage is defined, one uses flat plates to simulate the remaining appendages. The flat plates can be connected together at their corners to form a boxlike structure, or one plate can penetrate through another as shown in Fig. 51. In all cases the flat plates are defined simply by their corner locations, with the user making sure that they are flat, i.e., all corners lie in the same plane. In order to illustrate how these structures can be used to simulate an aircraft geometry let us consider several practical configurations for which measured results have been obtained. These results also allow one to evaluate the accuracy of the numerical solutions.

The Boeing 737 commercial jet is studied initially because of its simplicity. The line drawings shown in Fig. 52 are taken from *Jane's All the World's Aircraft* [24] and will be used to generate the simulation model. In the first case, a monopole antenna is mounted on the top center line of the 737 at station 220, which is just above the pilot's head as illustrated in Fig. 53. The prolate spheroid is defined to

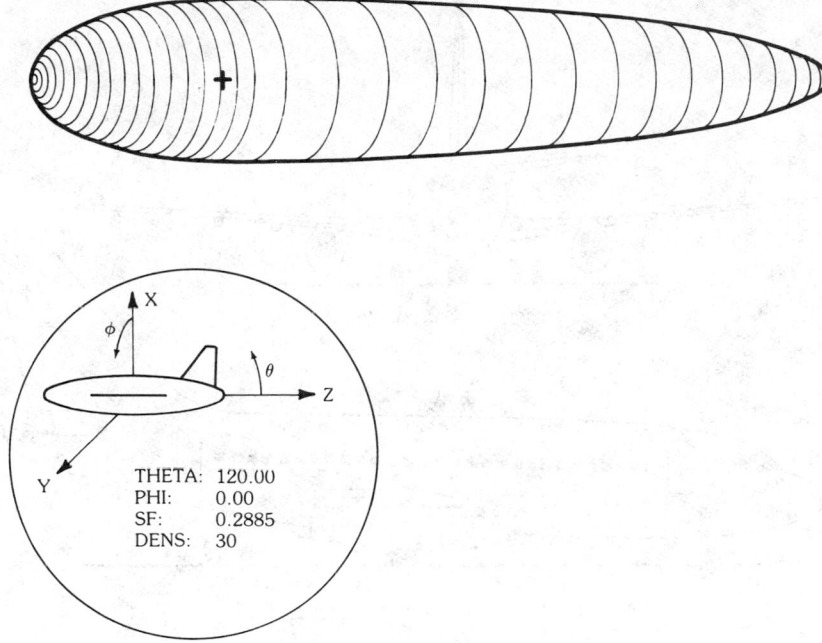

Fig. 50. Composite spheroid fuselage.

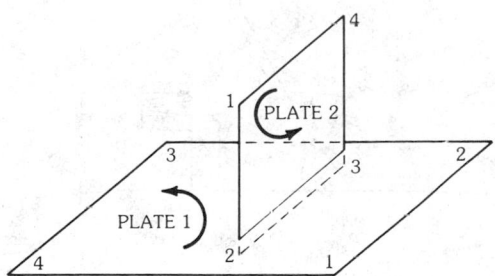

Fig. 51. A flat plate intersecting another flat plate.

represent the nose of the aircraft in a best-fit sense. The wings and horizontal stabilizers are simulated next, using finite flat plates which are attached to the fuselage. Note that one need not actually connect the plates to the fuselage, but simply indicate to the code that they should be attached. The curved junction edge is then generated internal to the code. The front view of the 737 as shown in Fig. 52 is then used to determine the x coordinate values of the simulation plates (for example, $x = 0$ for the wings). Since the horizontal stabilizers are so far away from the antenna and shadowed by the fuselage, they can be neglected. The top view of the 737 as shown in Fig. 52 is used to determine the y and z coordinate values for the wings. With this information the wing simulation plates are defined as shown in Fig. 53.

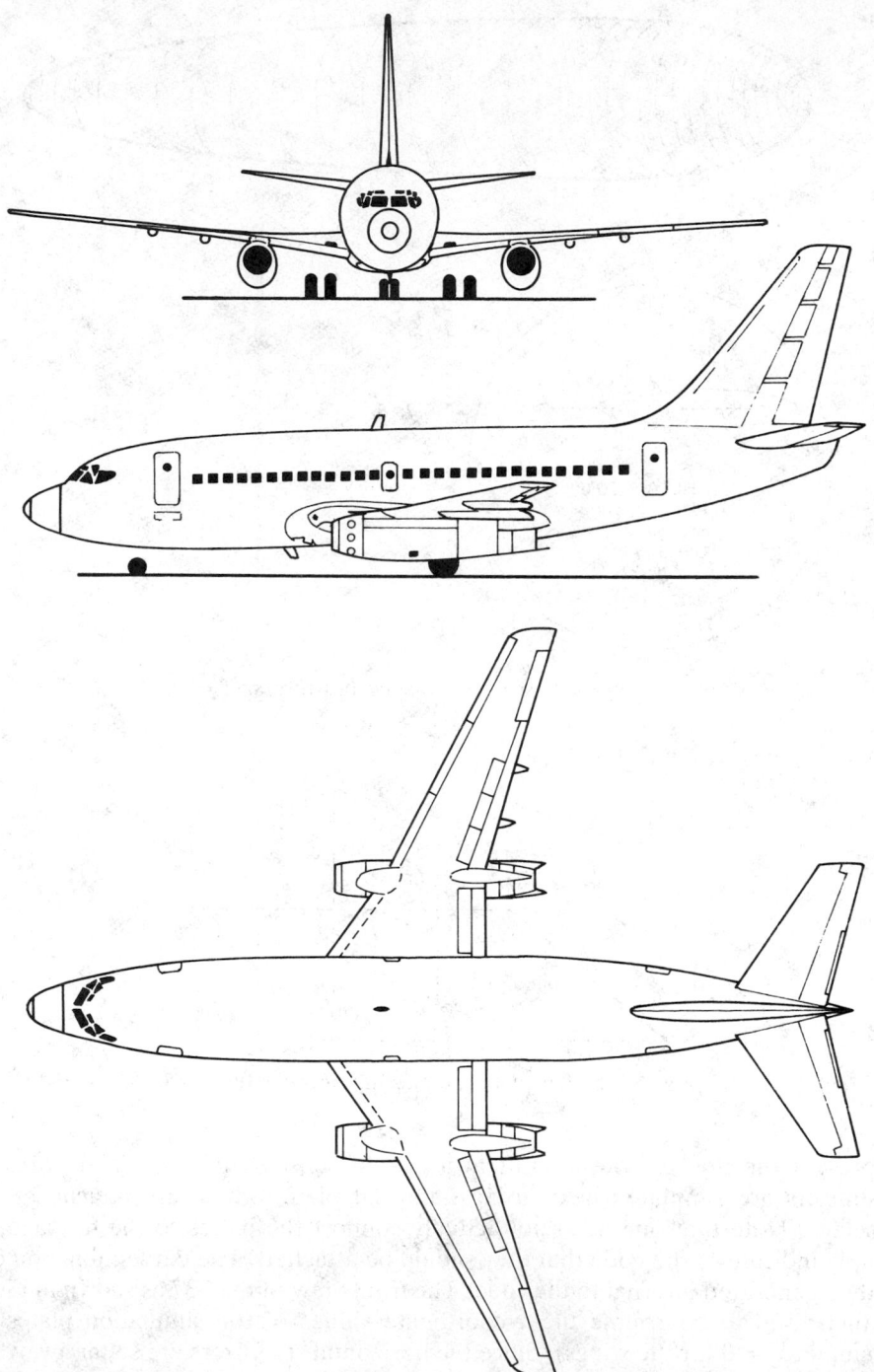

Fig. 52. Line drawing of Boeing 737 aircraft. (*After Taylor [24], © 1974 Jane's Publishing*)

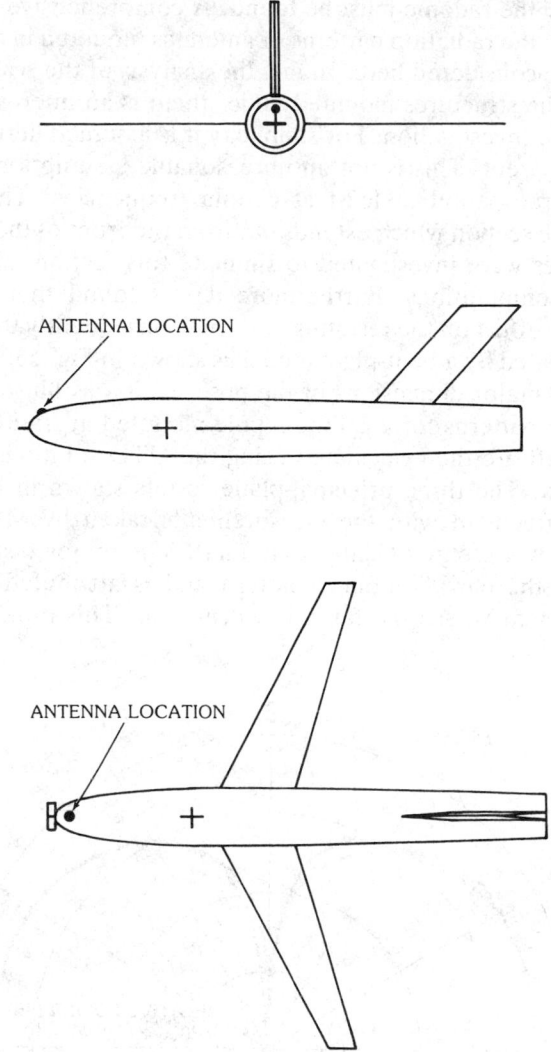

ANTENNA LOCATION

ANTENNA LOCATION

Fig. 53. Simulation model of the Boeing 737 aircraft used for calculations.

Since many airborne antennas are mounted on or near the fuselage center line the vertical stabilizer must be simulated by a structure with finite thickness. This thickness is very significant in that it tends to shadow the direct field from the antenna for aft radiation directions. In order to approximate this effect and maintain the finite flat plate representation, the vertical stabilizer is simulated by a bent plate as shown in Fig. 53. In addition the vertical stabilizer simulation must represent the location and orientation of the actual leading edge. The leading-edge geometry is important because it dictates the edge-diffracted field scattering cone.

For an antenna mounted closed to the nose of the aircraft one must simulate the nose section that extends outward from the base of the windshield. As described in [2], before the nose section can be adequately simulated, a practical representation for the radome must be found. A comprehensive study of radomes and their effect on the radiation patterns of antennas mounted in their vicinity is far too complex to be considered here. In fact the analysis of the scattering properties of radomes and the structures mounted under them is an interesting and relevant problem worthy of investigation. For simplicity it is assumed here that the radome is perfectly transparent. This is not an unreasonable assumption, as radomes are designed to be transparent at least at certain frequencies. This leaves a short blunt-looking nose section which extends out from the front of the aircraft. Various complex structures were investigated to simulate this section, all of which led to very inefficient computations. Furthermore it was found that the nose section normally has little effect on the resulting pattern. Consequently the nose section for simplicity is simulated by a bent-plate model as shown in Fig. 53. These bent plates simply model the major dimensions of the nose section as illustrated in Fig. 52.

The radiation patterns for a $\lambda/4$ monopole mounted at station 220 on top of a Boeing 737 aircraft are then calculated using the UTD solutions described in the simulation section. The three principal-plane results shown in Fig. 54 are found to be in good agreement with the measurements taken by Melvin Gilreath [2] at NASA (Langley Research Center, Virginia). The major discrepancy in these results occurs in the elevation-plane pattern and is attributed to the near-field pattern measurement versus the far-field calculation. This topic was discussed in the previous section.

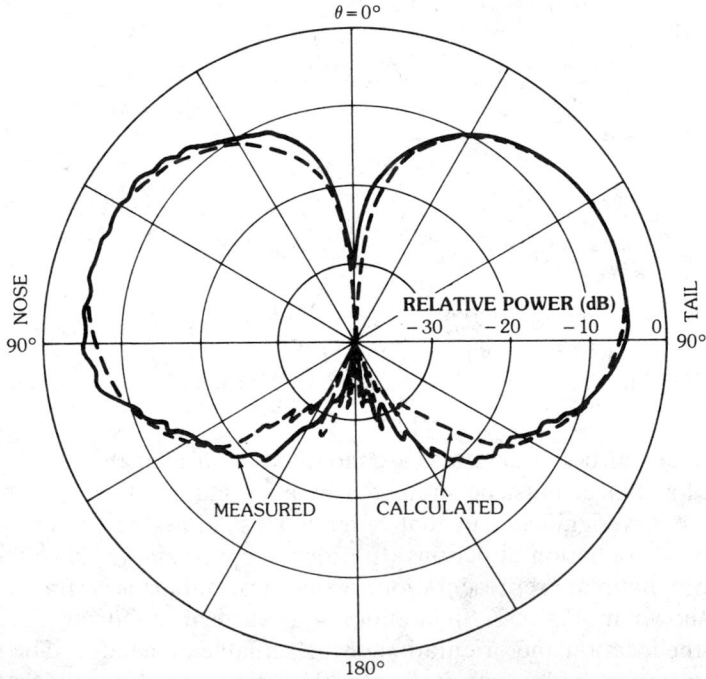

a

Fig. 54. Principal-plane radiation patterns for a $\lambda/4$ monopole at station 220 on top of a Boeing 737 aircraft. (*a*) Roll-plane radiation pattern (E_ϕ). (*b*) Elevation-plane radiation pattern (E_ϕ). (*c*) Azimuth-plane radiation pattern (E_θ).

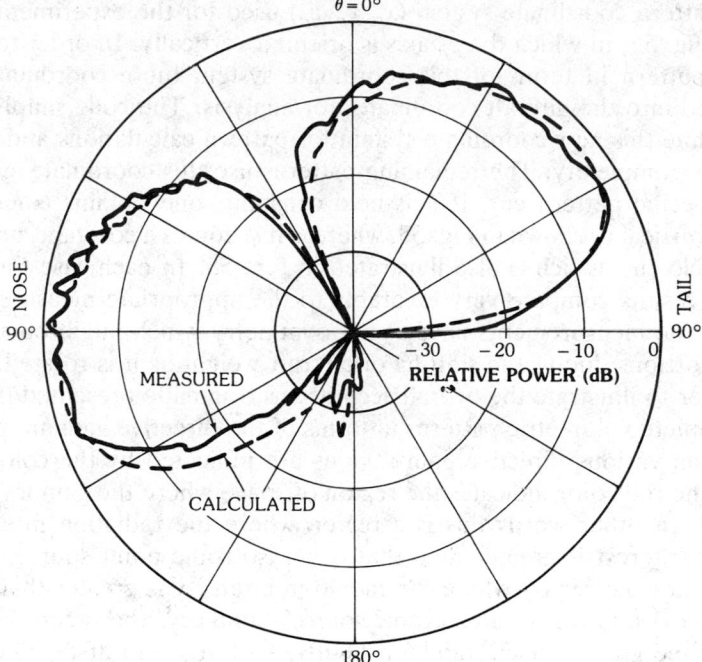

b

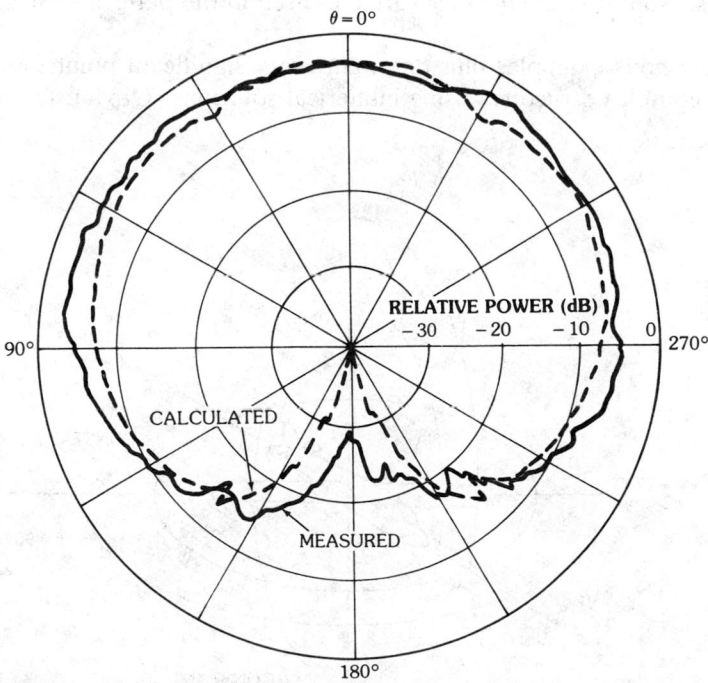

c

Fig. 54, *continued*.

The pattern coordinate system (x_p, y_p, z_p) used for the experimental results is shown in Fig. 55, in which the z_p axis is oriented vertically. In order to calculate a radiation pattern in terms of this coordinate system these coordinates must be transformed into the aircraft coordinates for analysis. The code simply allows the user to define this new coordinate system for pattern calculations and relieves the user of this complexity. The remaining patterns use this coordinate system to define the specific pattern cut. If θ is held constant, one obtains conical patterns about the vertical as shown in Fig. 55, whereas if ϕ equals a constant, one generates a great circle cut, which is also illustrated in Fig. 55. In each case the previously calculated results compare very favorably to the appropriate measurements. It is noted that the measurements have some asymmetry which might be attributed to model distortions due to the shifting of aircraft weight as it is rotated.

In order to illustrate the overall accuracy as originally presented in Reference 2, the complete volumetric patterns in terms of the directive gain are presented in Fig. 56. The various directive gain regions are indicated by the color code. For example, the red color indicates the region of space where the gain level is greater than 0 dB. In other words this is a region where the radiation intensity of the antenna of interest is greater than that of an isotropic point source. The yellow color indicates the region where the radiation intensity is greater than -3 dB but less than 0 dB relative to an isotropic source. Similarly, the green, blue, purple, dark blue, and gray or black stand for -6-dB, -10-dB, -15-dB, -20-dB, and less than -20-dB gain levels, respectively. Good agreement is obtained for each of the gain levels. Note that the θ and ϕ variables used in the pattern plots are defined in Fig. 55.

The previous examples illustrate many very significant points associated with modeling complex structures using numerical solutions. Obviously the simulation

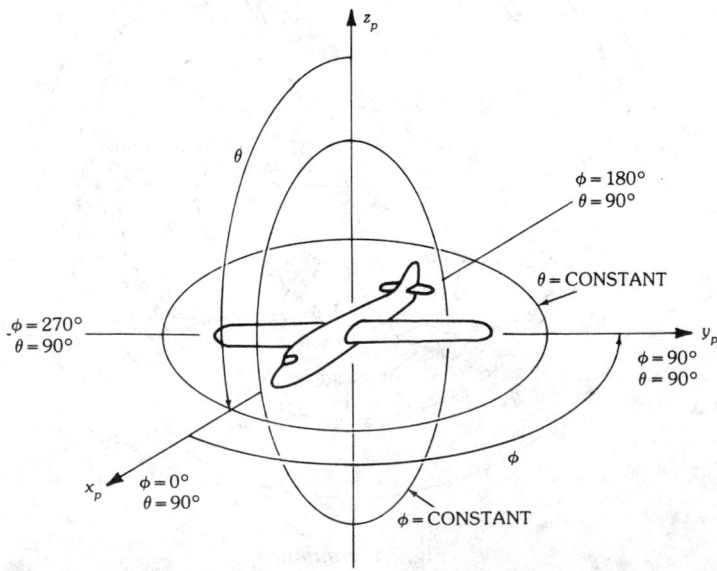

Fig. 55. Coordinate system used for experimental measurements.

a

MAXIMUM DIRECTIVITY = 8.063 dB

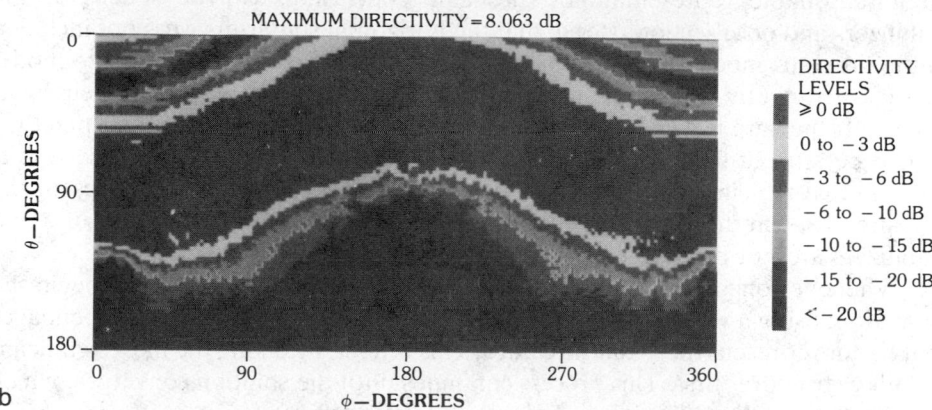

DIRECTIVITY
LEVELS

≥ 0 dB

0 to − 3 dB

− 3 to − 6 dB

− 6 to − 10 dB

− 10 to − 15 dB

− 15 to − 20 dB

< − 20 dB

b

MAXIMUM DIRECTIVITY = 5.747 dB

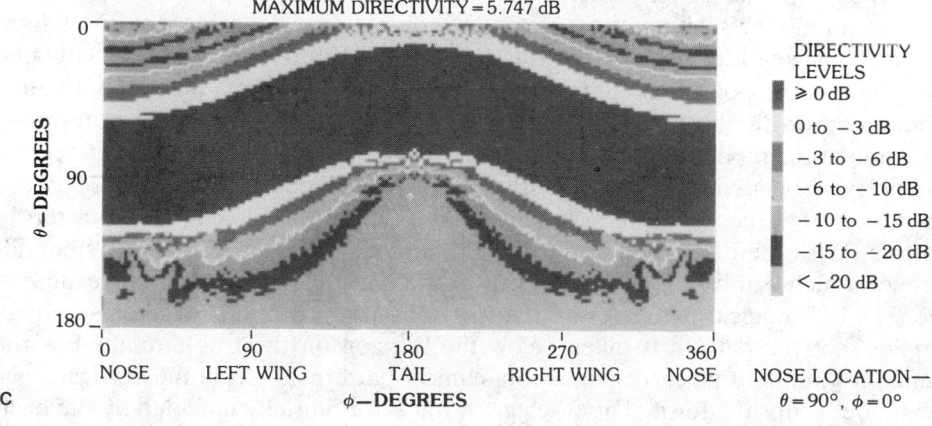

DIRECTIVITY
LEVELS

≥ 0 dB

0 to − 3 dB

− 3 to − 6 dB

− 6 to − 10 dB

− 10 to − 15 dB

− 15 to − 20 dB

< − 20 dB

c

ϕ — DEGREES

NOSE   LEFT WING   TAIL   RIGHT WING   NOSE

NOSE LOCATION—
$\theta = 90°$, $\phi = 0°$

Fig. 56. A Boeing 737 model and volumetric patterns of a monopole at station 220 in terms of directive gain. (*a*) A 1:11 scale model. (*b*) Measured volumetric directive gain pattern. (*c*) Calculated volumetric directive gain pattern. (*After Yu, Burnside, and Gilreath [2], © 1978 IEEE*)

20-77

model represents a greatly simplified version of the actual aircraft. On the other hand the agreement achieved between the calculated and measured results indicates that this simplified version is an adequate representation. How is this possible? To understand this idealization let us examine the scattering from each structure.

The fuselage has the dominant effect on the pattern; more specifically, the fuselage nearest the antenna is most significant. The simulation structure represents this region of the aircraft very nicely, so one might expect good agreement for the scattering associated with the fuselage. When one looks at the aircraft, the wings represent the most significant appendages and must be accurately modeled. The wings affect the pattern in two dominant ways: (1) since they are large and basically flat structures, they create a large reflected field; (2) they shadow the energy flowing from the antenna as the receiver moves under the wings. Since each of these effects can be modeled using a simple flat plate, our wing simulation is not an unreasonable representation. The same conclusions can be stated for the stabilizers and nose section. Recall that the horizontal stabilizers were not included in the previous model for the 737 aircraft. One can investigate the need to model these structures by simply adding them to the model and determining their significance. If they are not needed in the model for a given antenna location, then they should be left out in order to simplify the calculations. In any event these results indicate that the simulation model shown in Fig. 53 does adequately represent the 737 aircraft even though the fuselage is not a prolate spheroid and the wings and stabilizers are not flat plates.

Whenever one attempts to simulate a new configuration one should begin the simulation using a very simplistic model to which can be added various appendages that better represent the actual structure. This is done by taking the most significant or bulky structures first. This process continues until the solution converges, which results in a great simplification of a very complex aircraft.

Since the bulk of airborne antenna designs involve military aircraft which have numerous complex appendages, it is most illuminating to discuss the modeling considerations for such an aircraft. As an example let us consider the F-4 fighter aircraft illustrated in Fig. 57. In this case a uhf blade antenna, which is simulated by a short monopole, is mounted on the belly of the aircraft. Since the antenna is mounted on the bottom of the F-4, one must turn the aircraft over in terms of the simulation model. This requirement is done to ensure that one defines the geometry associated with the belly of the aircraft differently from the top, because most aircraft, especially military ones, have a definite shape change from top to bottom. With the F-4 turned upside down, there will exist a coordinate system difference which can be corrected in terms of the pattern coordinates. For example, the $\theta = 105°$ conical pattern from the vertically upward axis is of interest, which implies that the pattern is taken below the horizon on the true aircraft. For the simulation model this corresponds to a conical pattern 75° from the vertical since the model is upside down. The fuselage of the F-4 is initially modeled by the best-fit composite prolate spheroid to represent the fuselage in the vicinity of the blade antenna. The fuselage-alone pattern is shown in Fig. 58a for a frequency of 375 MHz. Next, the flat-plate wings are added to the model, taking the data directly off the appropriate line drawings. The fuselage-plus-wings pattern, as

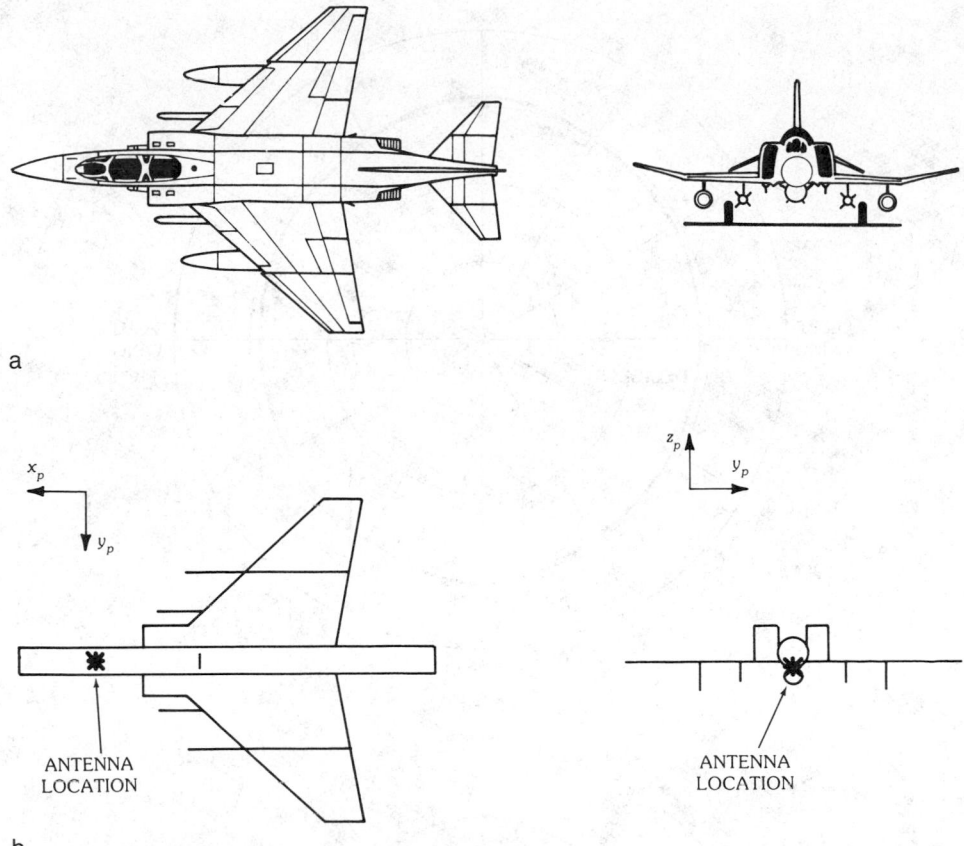

Fig. 57. The F-4 aircraft and computer-simulated model (roll plane only). (*a*) Aircraft. (*After Taylor [24], © 1974 Jane's Publishing*) (*b*) Model.

shown in Fig. 58b, indicates that the wings play very little role in this conical pattern cut. This particular F-4 had a bomb mounted along the fuselage center line, two missile racks on the inboard side of the wings, and two wing fuel tanks as shown in Fig. 57. Using our general modeling approach these structures will be simulated in a very simplistic form. Since their dominant effect on the pattern is shadowing of the incident field let us simulate these structures using single flat plates to represent each one. The plate cross-sectional shapes are chosen to represent the shadowing outline of the given structure. Considering the bomb first, its shadowing outline is basically circular, so the flat-plate simulation is represented by a hexagonal plate which is located at the midpoint of the bomb. The resulting pattern with bomb added is shown in Fig. 58c. Again note that the bomb does not play much of a role for this case. The missile racks are examined next, using a simple flat-plate representation. The composite pattern for this configuration is shown in Fig. 58d. It is apparent from these results that the missile racks play a significant role; thus one should devote additional effort to better represent them. Next, let us consider the outboard fuel tanks which are, also, initially represented

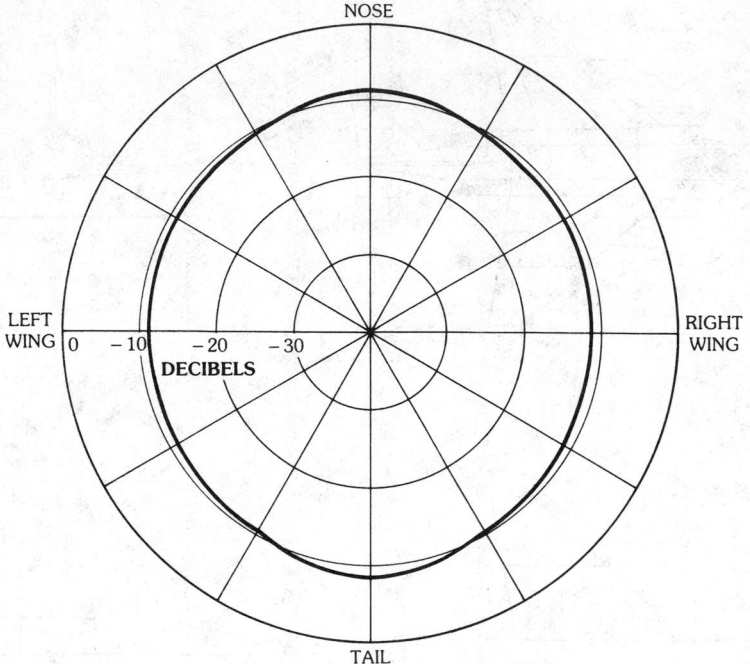

a

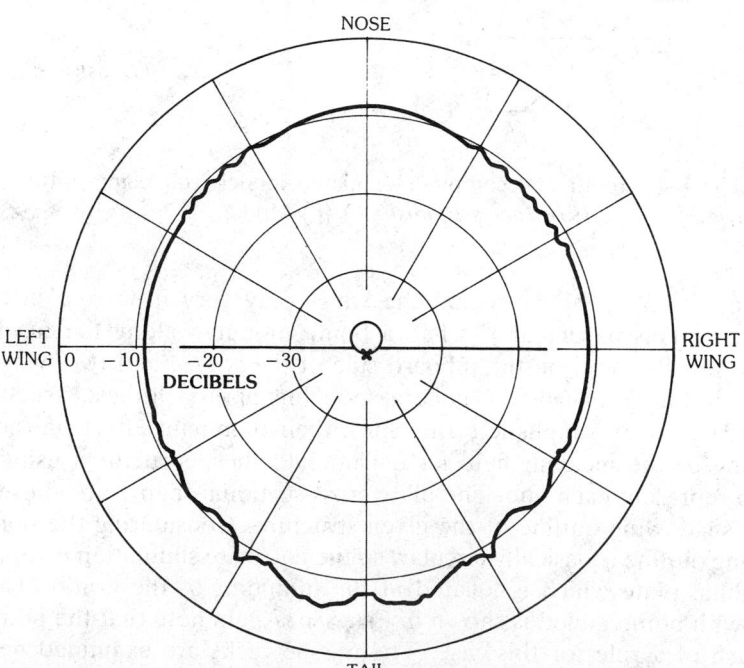

b

Fig. 58. Calculated F-4 azimuthal conical patterns at $\theta_p = 105°$ and $f = 375$ MHz. (*a*) For cylinder-only model. (*b*) For cylinder and two wings model. (*c*) For cylinder, two wings, and blockage under fuselage model. (*d*) For model with missile racks added to model (*c*). (*e*) Measured and calculated final results with outboard fuel tanks added to (*d*). (*Aircraft drawing after Taylor [24], © 1974 Jane's Publishing*)

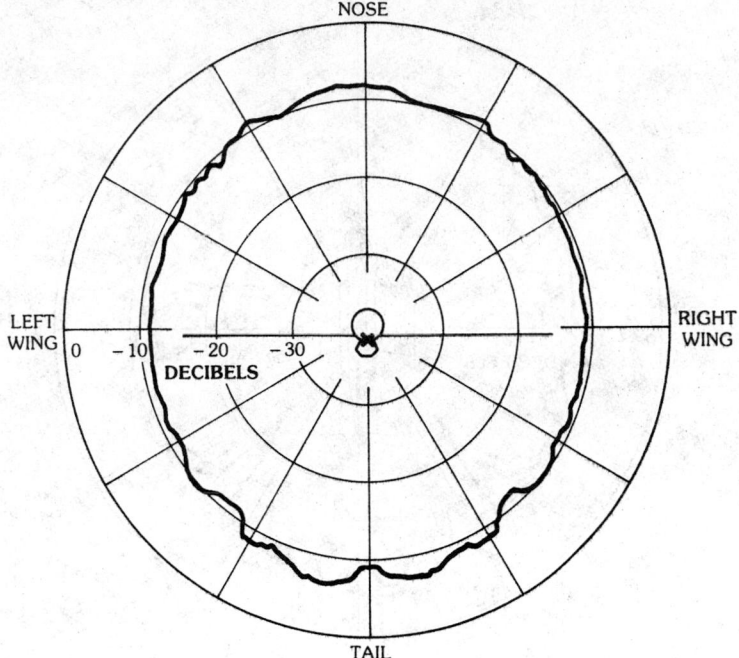

c

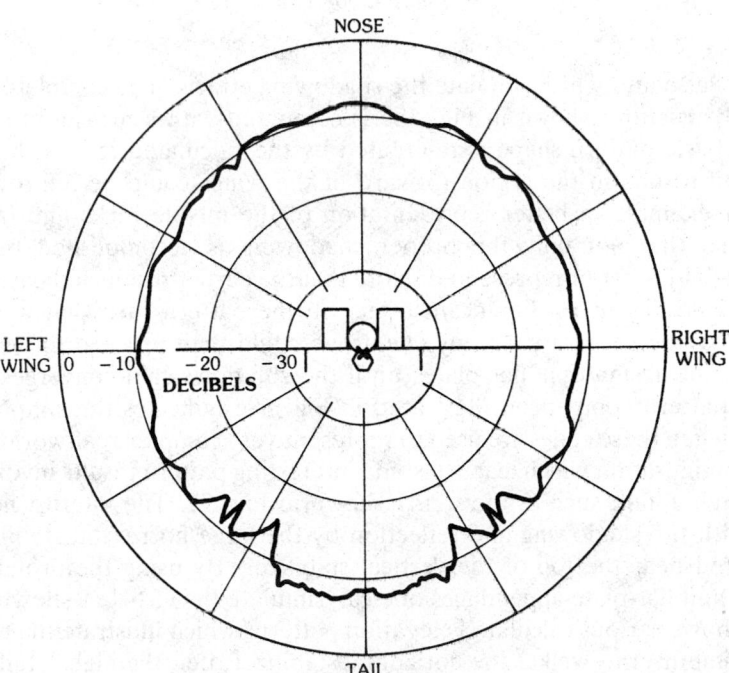

d

Fig. 58, *continued.*

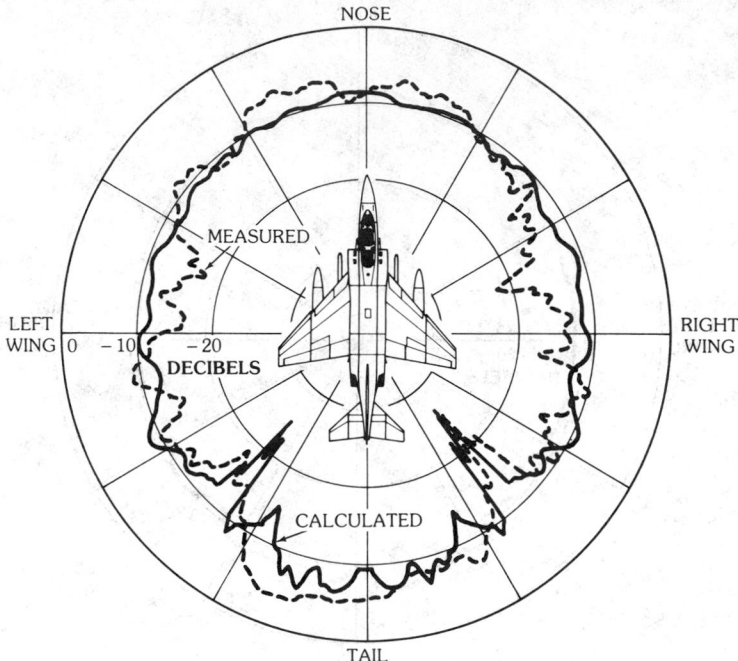

NOSE

MEASURED

LEFT
WING   0   − 10   − 20
       DECIBELS

RIGHT
WING

CALCULATED

e

TAIL

Fig. 58, *continued.*

by finite flat plates which simulate the shadowing effects. The complete pattern for this final structure, shown in Fig. 58e is compared with a measured result. Note that the basic pattern shape is re-created by the calculated results; however, the calculated results in the region forward of the wings could be improved. In this area, for example, a better representation of the missile racks and fuel tanks is needed so that not only the proper shadowing is accomplished but also the reflected field is better represented. Note that the earlier results indicated that both these geometries had a significant effect on the pattern such that more realistic models may be necessary. In any event one could then proceed to box out these structures using multiple flat plates until the solution again converges. Nonetheless the pattern comparison illustrated in Fig. 58e indicates the simplicity that is possible when these codes are used to represent very complex real-world structures.

A configuration which leads to some interesting pattern results involves aircraft with a high T tail, such as the C-141 shown in Fig. 59. The interest here is associated with the shadowing and reflection by the large horizontal stabilizer, which is mounted near the top of the vertical stabilizer. By using the prolate spheroid fuselage and flat-plate appendages one can simulate the C-141 as shown in Fig. 59. Fig. 60 shows various calculated elevation patterns which illustrate the effect of the vertical stabilizer as well as the horizontal stabilizer, i.e., the high T tail. Note that the horizontal stabilizer not only creates a shadow above the horizon but also a large reflected field below the horizon. For various clutter reasons this reflected lobe can be very undesirable in some applications.

As a final example of the Airborne Code the F-16 fighter illustrates the simula-

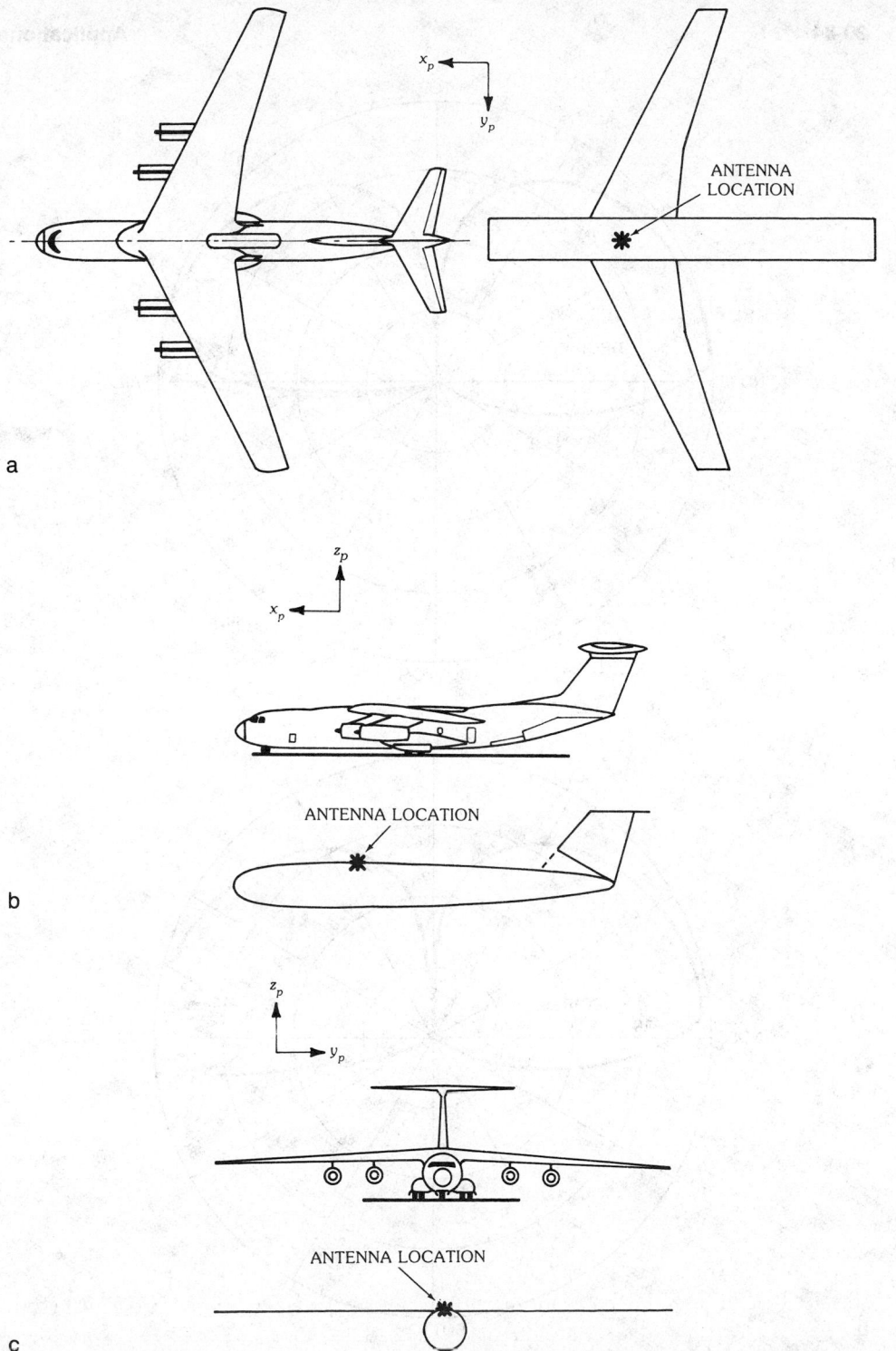

Fig. 59. Computer-simulated model for the C-141 aircraft. (*a*) Roll-plane model, top view. (*b*) Elevation-plane model, side view. (*c*) Roll-plane model, front view. (*Courtesy of USAF*)

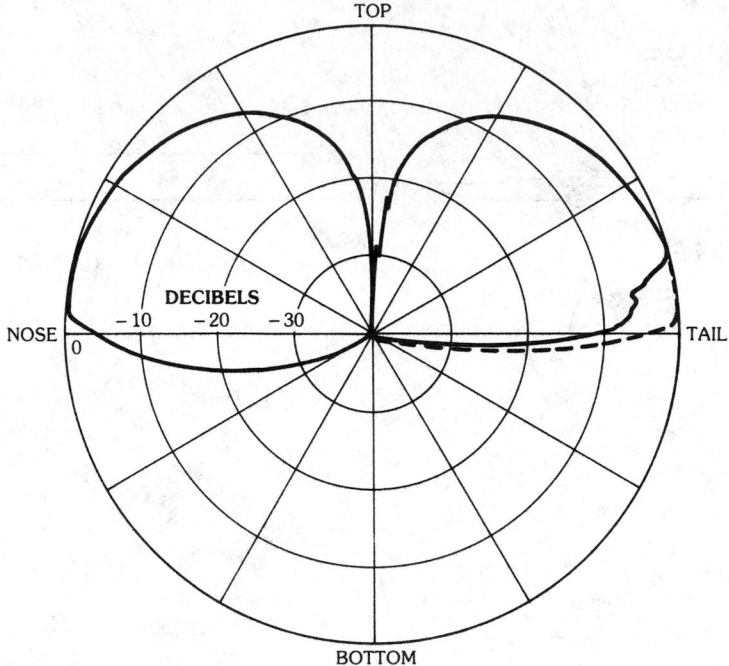

a

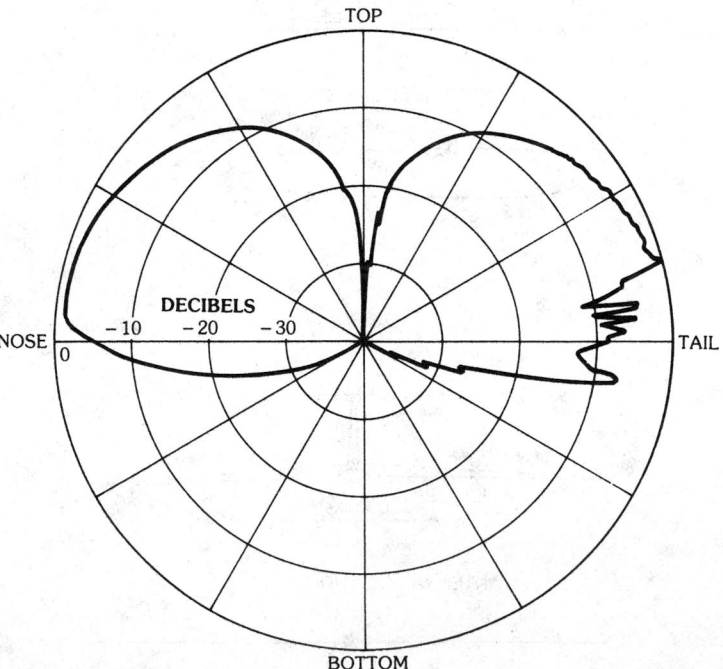

b

Fig. 60. Calculated elevation patterns. (*a*) Without T tail and vertical stabilizer (dashed line) and without T tail but with vertical stabilizer (solid line). (*b*) With T tail and vertical stabilizer.

tion capability in terms of a very complex geometry. The F-16 line drawings are illustrated in Fig. 61, and the simulation model is shown for comparison purposes in Fig. 62. Note that the flat plates are used here to complete the representation of the fuselage as well as simulating the wings, canards, and stabilizers. The three principal-plane patterns for this configuration are shown in Fig. 63 and compared with measured results obtained by General Dynamics (Dallas, Texas).

As mentioned previously, if the antennas are not mounted on the fuselage of the aircraft, the NEC-Basic Scattering Code can be used to predict the antenna patterns. For example, measurements of a slot antenna mounted on the wing of a 1:20 scale model of a Boeing 737 were taken at NASA (Langley Research Center, Virginia). An analytic model composed of a finite elliptic cylinder and flat plates is used to simulate this structure as shown in Fig. 64. In this figure the larger radius along the positive x axis is used for an antenna mounted on top of the wings, and the smaller radius along the negative x axis is used for an antenna mounted on the bottom. Note that the back of the fuselage of the aircraft is simulated by cutting the elliptic cylinder at an angle for these calculations. The curvature at the nose of the scale model has been neglected. The full scale model frequency is 1.75 GHz, which corresponds to 35 GHz for the scale model. The source is a K_a-band wave-

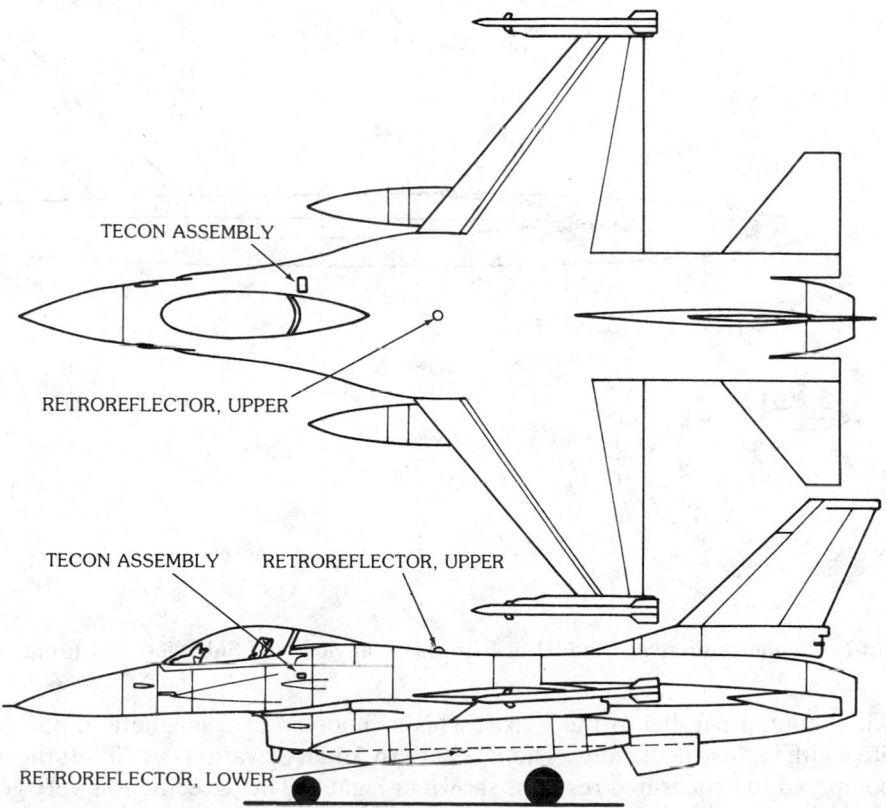

Fig. 61. Line drawing of F-16. (*Courtesy of USAF*)

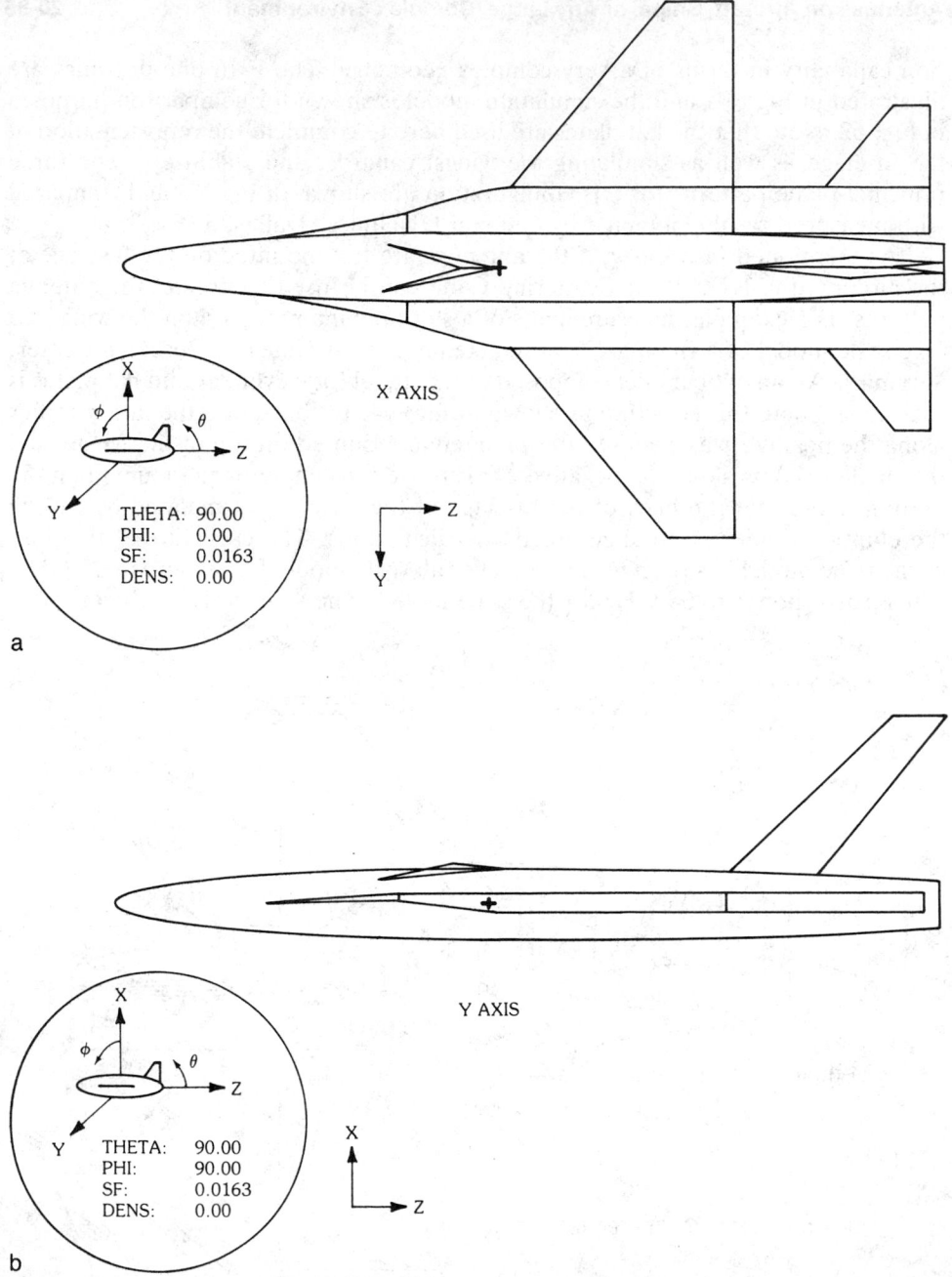

X AXIS

THETA: 90.00
PHI: 0.00
SF: 0.0163
DENS: 0.00

a

Y AXIS

THETA: 90.00
PHI: 90.00
SF: 0.0163
DENS: 0.00

b

Fig. 62. Computer-simulated model of F-16. (*a*) Top view. (*b*) Side view. (*c*) Front view.

guide mounted parallel to the *y* axis. This is modeled as a magnetic dipole with finite width in the calculations. The $E_{\phi p}$ pattern for an elevation cut 30° off the nose is compared to a measured result as shown in Fig. 65. The agreement is very good. The discrepancies that exist can be partially attributed to the simple elliptic cylinder representation of the fuselage in the calculations along with normal experimental

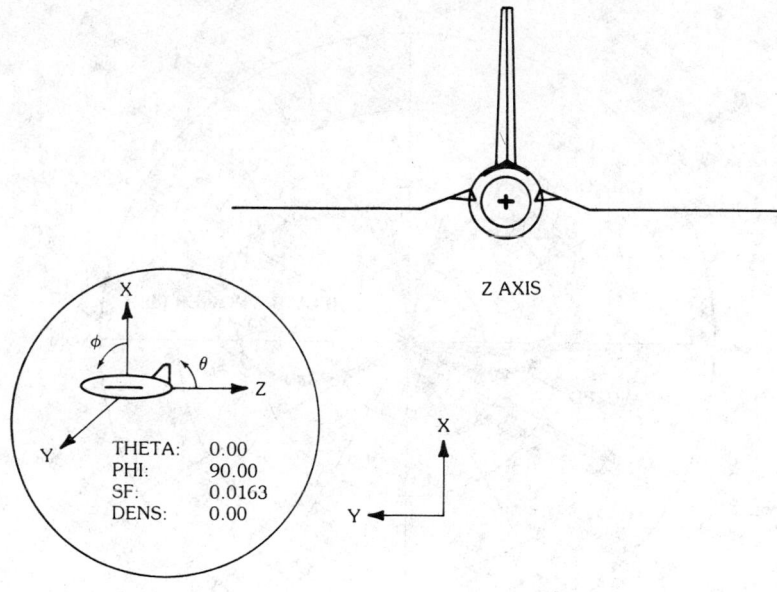

THETA: 0.00
PHI: 90.00
SF: 0.0163
DENS: 0.00

c

Fig. 62, *continued.*

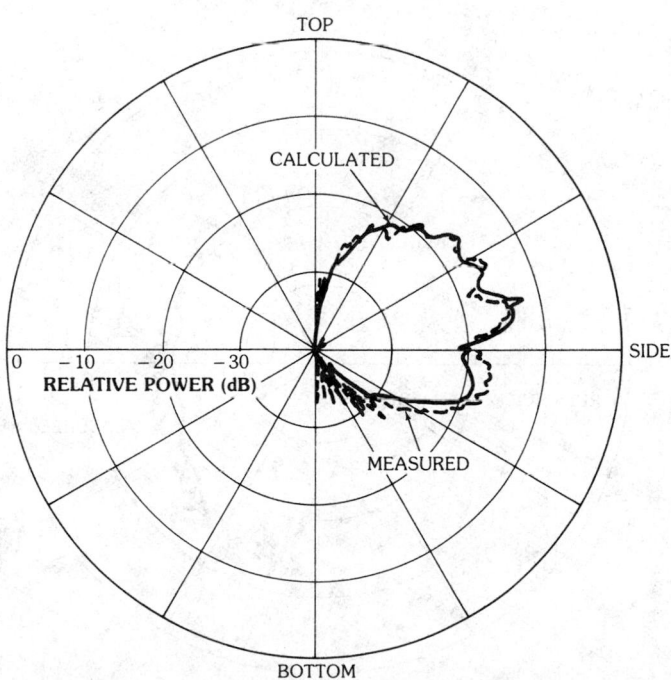

a

Fig. 63. Radiation patterns of a $\lambda/4$ monopole antenna mounted on top of an F-16 aircraft. (*a*) Roll-plane (E_ϕ) pattern. (*b*) Elevation-plane (E_ϕ) pattern. (*c*) Azimuth-plane (E_θ) pattern.

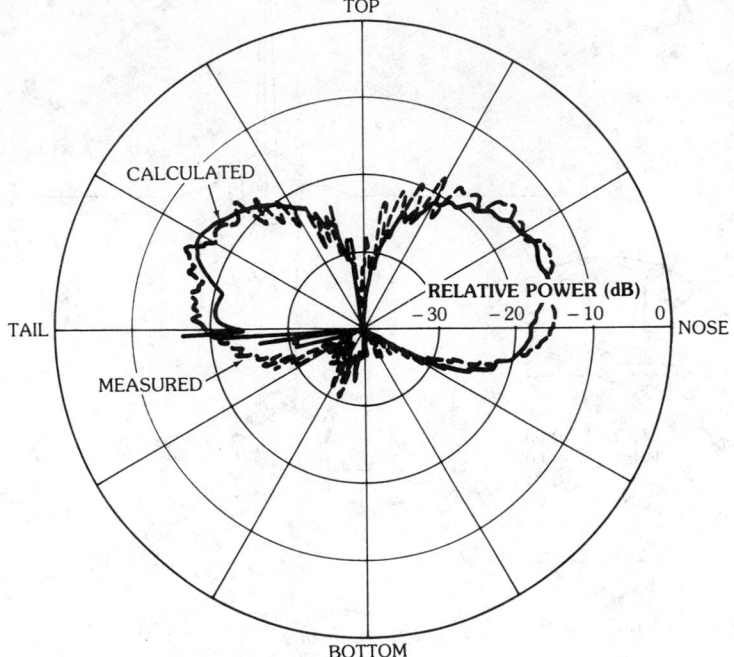

b

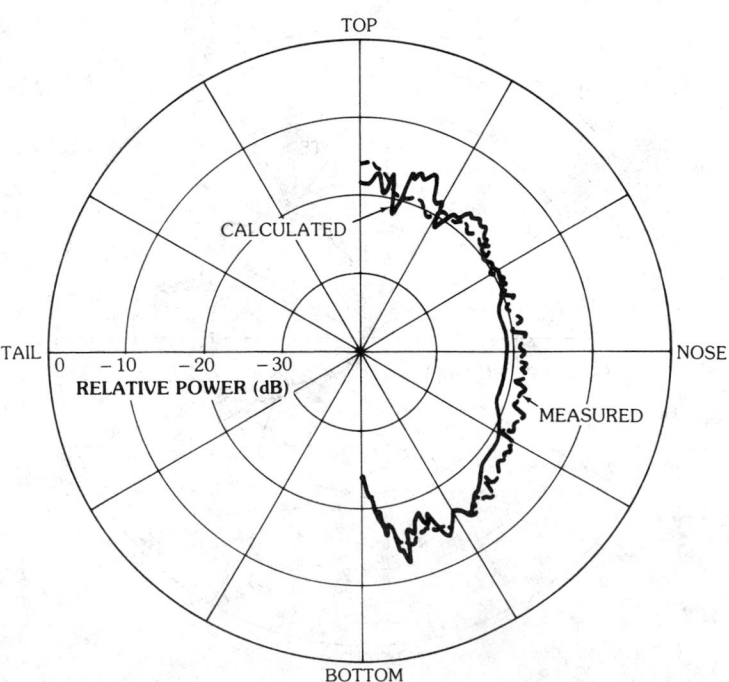

c

Fig. 63, *continued.*

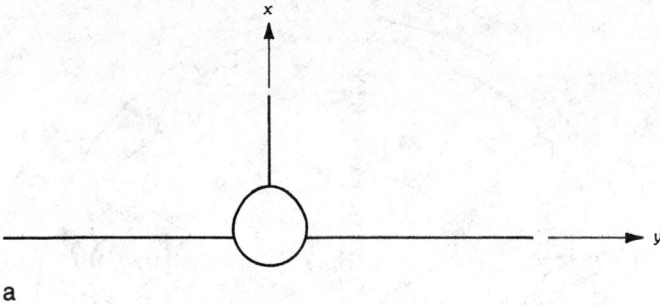

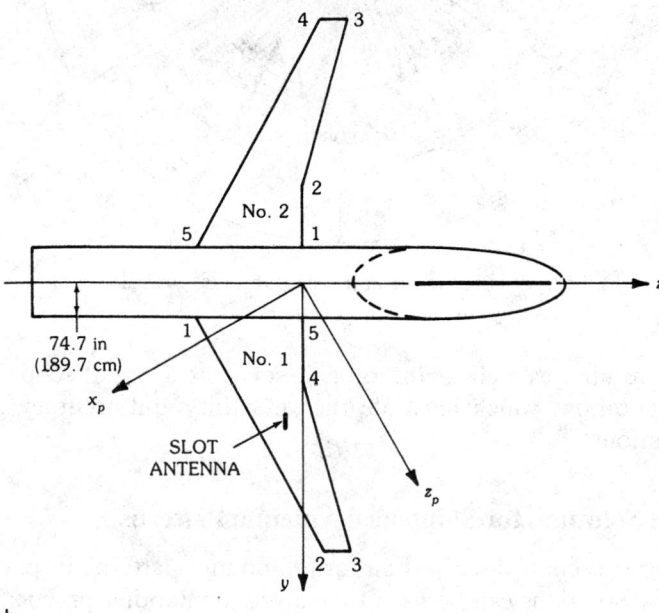

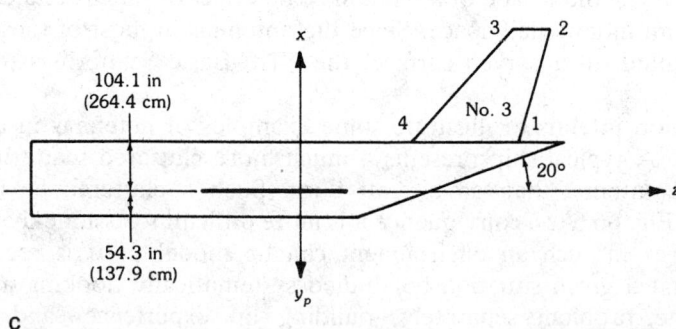

Fig. 64. Geometry of Boeing 737 aircraft model with slot antenna mounted on wing. (*a*) Front view. (*b*) Top view. (*c*) Side view.

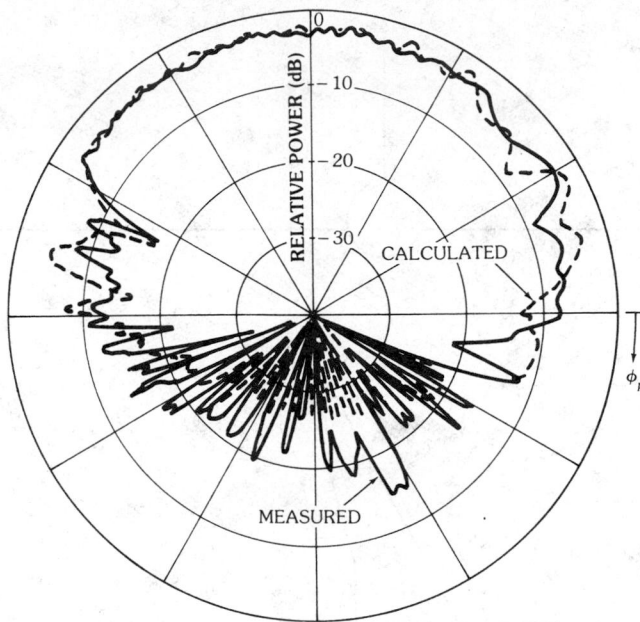

Fig. 65. Measured and calculated $E_{\phi p}$ results.

errors. These results, as well as the others, serve as a broad scope of airborne antenna configurations which illustrate the versatility and accuracy of the UTD numerical solutions.

6. Numerical Solutions for Shipboard Antenna Patterns

The previous section discussed airborne antenna patterns in particular. The UTD numerical solutions can be used to analyze many other practical type structures. For example, the plates can be combined together in such a way as to model the superstructure of a ship, the body of a truck or tank, the body of a satellite with its solar panels, or the facade of a building. The effect of the ground can be simulated by a semi-infinite half-space. Since the antennas in most of these situations are not mounted on a curved surface, the NEC-Basic Scattering Code is most suitable here.

This section intends to illustrate some examples of antennas in a shipboard environment. A typical ship presents a much more cluttered scattering environment for an antenna than an aircraft does. Such a cluttered environment is illustrated in Fig. 66. As a consequence it is more difficult to assume that a majority of the features in such an environment can be modeled. It is recommended, therefore, that a given situation be studied systematically, looking at individual pieces of the problem separately, building up experience, and developing intermediate conclusions. When the designer starts out with a simpler problem he or she can better diagnose the results to make sure that they are reasonable. The code also runs much faster using fewer components. This gives the user the feeling

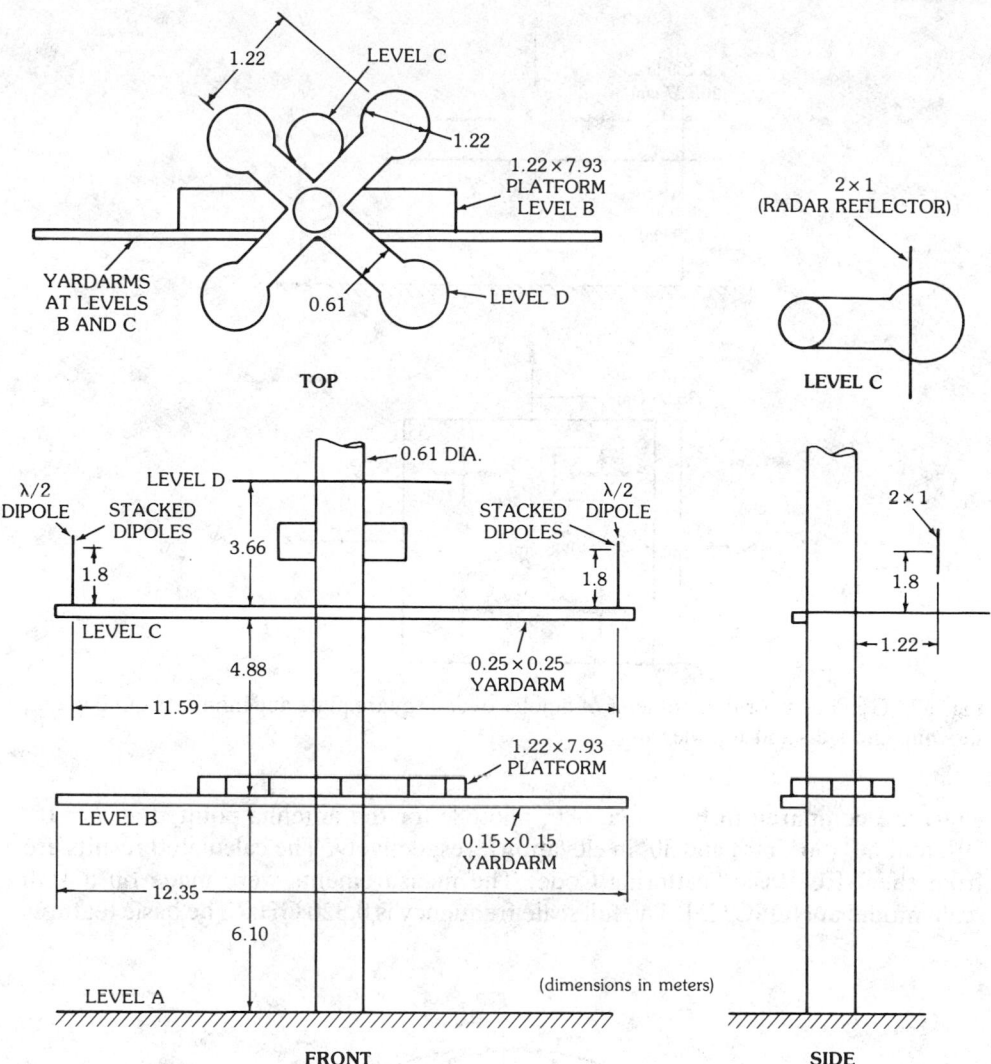

Fig. 66. Geometry of a potential ship environment.

that he or she can try many different situations cost-effectively. As experience is gained for given classes of problems, more details can be added.

The first example illustrates the modeling of an antenna of interest to a shipboard antenna designer. The antenna configuration is shown in Fig. 67. It is composed of four circularly polarized antennas that can be represented as crossed dipoles over a square ground plane. The currents on the antennas were obtained using the NEC-Moment Method Code [11] with six segments per dipole. To save time an analytic representation of the dipoles can be used in this case with equivalent accuracy. The effect of the ship's deck and ocean is modeled as an infinite ground plane here. This model is used to compare against measured results from a 1:10 scale model [25]. The results for the circularly polarized field (relative directive

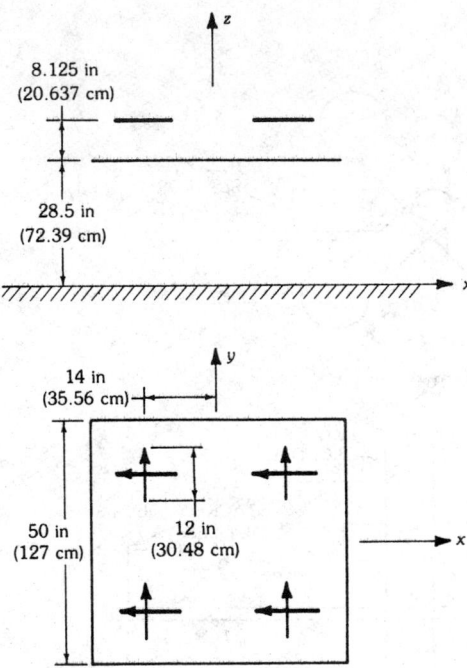

Fig. 67. Geometry for the problems of dipoles over a square plate and infinite ground plane, showing the side and top views.

gain) are compared in Figs. 68a, 68b, and 68c for the antenna pointed at $\theta = 0°$, 30°, and 60° (90°, 60°, and 30° in elevation), respectively. The calculated results are from the NEC-Basic Scattering Code. The measurements were made on a 1:10 scale model at NOSC [25]. The full scale frequency is 0.320 GHz. The basic features

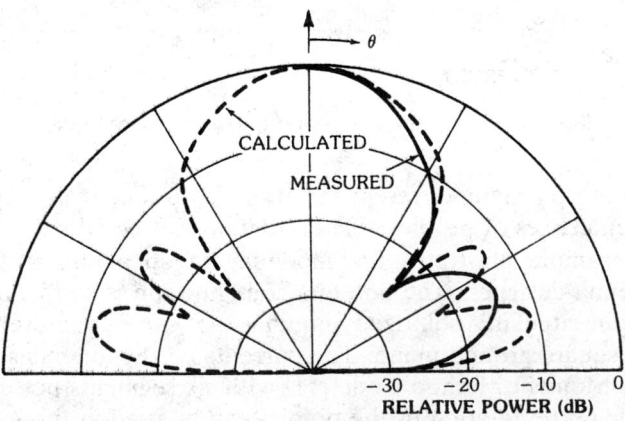

Fig. 68. Measured and calculated circularly polarized patterns for an antenna. (*a*) Antenna pointed at $\theta = 0°$ (90° elevation). (*b*) Antenna pointed at $\theta = 30°$ (60° elevation). (*c*) Antenna pointed at $\theta = 60°$ (30° elevation).

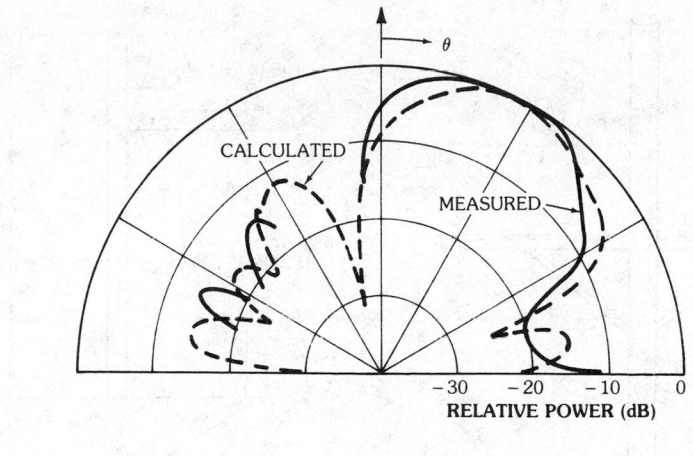

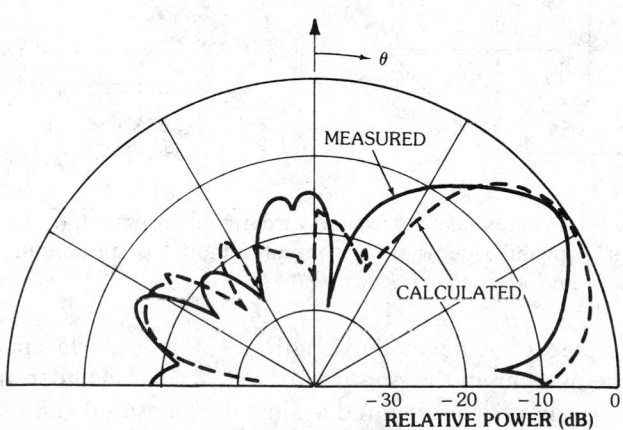

Fig. 68, *continued.*

of these patterns are reconstructed by the calculated results; however, the secondary lobes are off slightly. This is due to the simplified model used to calculate the results along with normal experimental errors. For example, it is difficult to simulate circularly polarized antennas because the low-level fields depend on the precise phasing of the elements. The model used for the computations can be greatly improved if more details are known about the experimental models and setup. For example, the polarization of the pattern could be more closely matched if more details of the receiving antenna were known. Also, the model of the antenna could be improved by adding more detail, such as using a circular cylinder to represent the antenna pedestal and using more plates to model the back plate, and so on.

The next example is for an antenna mounted on the yardarm of a mast as shown in Fig. 69a. Measurements were made for this configuration at NOSC [25] using a 1:10 scale model at a frequency of 4.0 GHz. The geometry used to model this structure for the calculations is illustrated in Fig. 69b. The antenna is simply

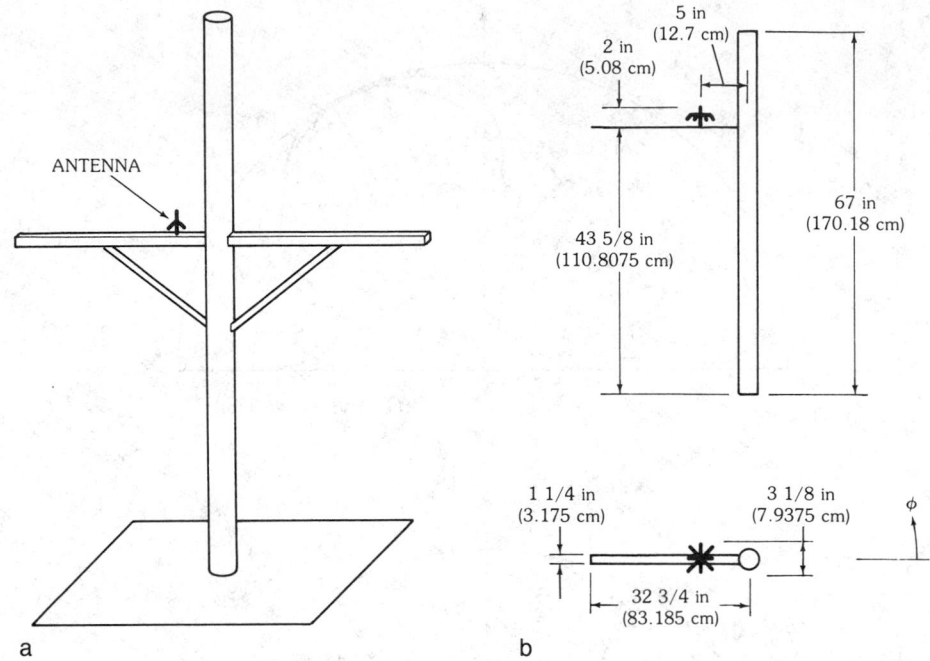

Fig. 69. Example for an antenna in a mast and yardarm environment. (*a*) Geometry of mast and yardarm. (*b*) Top and side view of computer model corresponding to (*a*). (*After Burnside, Rudduck, and Marhefka [4], © 1980 IEEE*)

modeled as a dipole of scale model length 1.925 in (4.8895 cm). The results calculated for the antenna in the presence of the circular cylinder and a flat plate representing the yardarm are compared against the measured results in Fig. 70 for the azimuth plane. The calculated result is from the NEC-Basic Scattering Code including the mast and yardarm [25]. The measurements were made on a 1:10 scale model at NOSC by L. S. Hansen [25]. The result for the cylinder alone gives the basic shape of the pattern. The plate gives the results better agreement in the null depths and in the shape of the lobe at 180°. The lack of a small feature at about 160° might be due to something present in the measurements that is not modeled by the code, or it could be higher interactions between the plate and the cylinder that have not been included. It is also possible that the feature is caused by the currents produced on the antenna due to the reflected field from the cylinder incident on the antenna structure.

The next example illustrates the blockage due to multiple masts [26] as shown in Fig. 71. The antenna system is composed of four broadband dipoles mounted around one of the masts. They are nominally supposed to produce an omnidirectional pattern. The antennas for this calculation are modeled as simple analytically defined dipoles. The frequency for this problem is 0.3125 GHz. Note that this means that the radius of the mast around which the antennas are mounted is only 0.1458 wavelength and that the dipoles are very close to the surface. This is pushing the accuracy of UTD, so the solution should be checked against a known result. This is accomplished by comparing the pattern for one of the dipoles in the

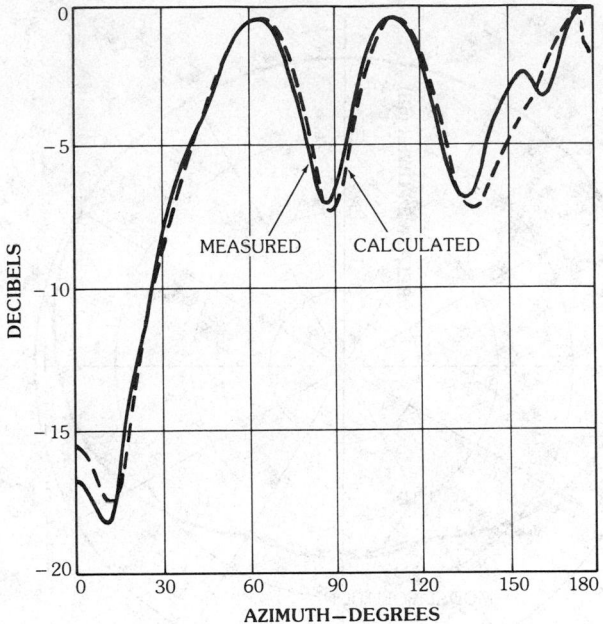

Fig. 70. Measured and calculated azimuth patterns for the mast antenna example. (*After Burnside, Rudduck, and Marhefka [4]*, © *1980 IEEE*)

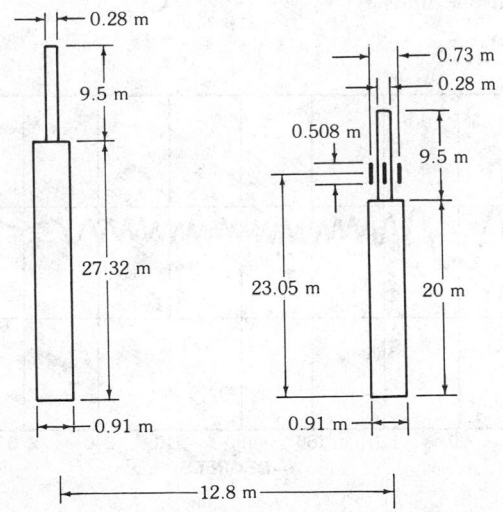

Fig. 71. Geometry used to study the pattern of four antennas in the presence of two masts.

presence of an infinite cylinder, using the exact modal solution against the UTD result for a very long cylinder. The results are compared in Fig. 72, with excellent agreement verifying the validity of the UTD solution. The azimuth-plane pattern for the vertically polarized electric field in the presence of two masts is shown in Fig. 73. This figure indicates a prediction of the amount of ripple that will

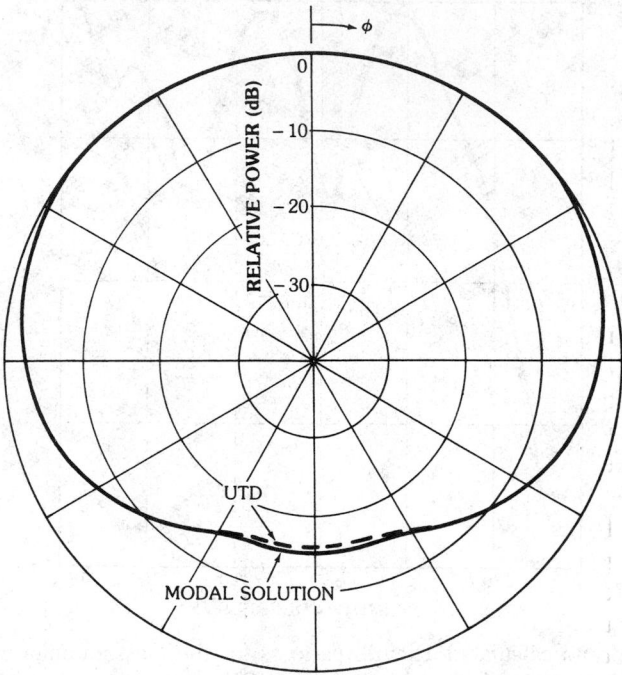

Fig. 72. The exact modal solution and the UTD solution of the E_θ pattern in the azimuth plane for one dipole on one mast.

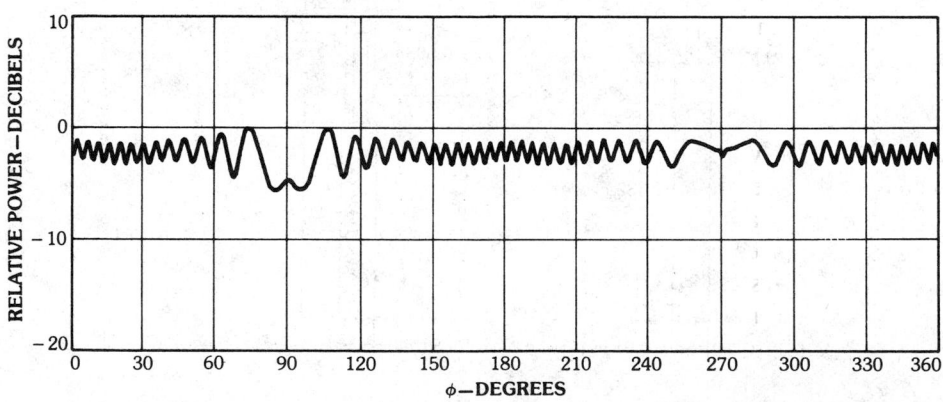

Fig. 73. Calculated azimuth plane pattern at $f = 0.3125$ GHz and a range of 250 m for four dipoles located around a mast in the presence of a second mast with higher-order cylinder interaction terms.

be produced by the presence of a second mast. The pattern for dipoles without the second mast is a straight line that goes through the average of the pattern in Fig. 73; as a result it is not shown here.

7. Summary

An antenna designer often must consider many contradictory situations when he or she is deciding where to place an antenna in a complex environment. First of all, there are usually many antennas competing for the same prime locations, such as the top of a ship's mast or the nose of an aircraft. Obviously this is not possible or even desirable from the standpoint of needed isolation between the various systems.

In many cases the environment can be used in the designer's favor. For example, if it is desired to minimize the coupling between two antennas, Friis' transmission formula says that the further the two antennas are apart, the less coupling there will be between them for the same orientation. If there is an intervening structure between them, it is possible that the minimum coupling will occur when the two antennas are closer together. This can happen when the direct path between the antennas is blocked so that the rays must travel around the structure.

In the case of radiation patterns the structure can also help shape the pattern by directing the gain away from a given direction and into a desired region; however, such terms can produce unwanted ripple. The frequency of the ripple can be minimized by placing the radiating antenna close to the reflecting or diffracting structures. The depth of the nulls in the ripple, however, can be greater because of the increased strength of the interfering fields.

In general it is not possible or even desirable from a time standpoint to try to model the complete scattering environment or even to perfectly define the antenna. It is difficult to make statements that are appropriate all the time since every situation is different; it is possible, however, to make some useful comments. First, when starting to model a new class of problems it is better to begin simple and systematically build the model up. This enables a logical evaluation of the results to make sure that they make sense, and it allows the designer to build up an intuitive understanding of the physical phenomena involved. It is also most cost-effective to start with a simple dipole test antenna or array of simple dipoles. This is usually possible since most small antennas, no matter how complicated they seem physically, radiate a pattern very similar to a dipole. This preliminary design stage often requires many trial-and-error computer runs. After the validity of the computer model is established and the most desirable location is determined, then the antenna and model can be "tweaked" up to confirm the conclusions.

When one is defining the model within the limited bounds of the computer code's capabilities it is often necessary to make compromises on the appearance of the model. The following hints can be of use in this process. The designer should emphasize the structure in the main beam of the antenna, that is, in the directions where the majority of the radiated energy strikes the structure. For low-gain antennas one should emphasize the model closest to the antenna because the power decays as $1/R^2$, where R is the range from the antenna. Always give the most consideration to the parts of the model in the direction of the pattern. It may therefore be necessary to have more than one model for different pattern regions.

As seen in the previous sections it is very useful to study the effects of the different parts of the structures separately. One of the benefits to a UTD computer model is that the major contributors to the pattern can be easily dissected and studied. This gives added insight into the problems involved as well as possible corrective measures. A rule of thumb concerning the higher-order interaction ray terms is that the higher the order of the term, the narrower its pattern effect. The magnitude of the field may be comparably large at the shadow boundaries, but on the whole these terms will be less significant than the first-order effects. This implies that these computationally time-consuming fields can be left out of the analysis until the latter stages of the design phase. Finally, whenever possible, use both numerical and experimental results to get a total picture of the problem. Either method can lead the designer to wrong conclusions if there are errors in the system. If both methods show agreement, or at least the same trends, the designer will be much more confident of his or her conclusions.

The use of high-frequency numerical techniques in solving practical engineering problems has been discussed in this chapter. The validity of these methods has been demonstrated through numerous comparisons with experimental results. The versatility of user-oriented numerical solutions has been illustrated by the wide variety of problems solved using the Airborne and NEC-BSC codes. All of this indicates that high-frequency numerical solutions are a very powerful tool for designing antenna systems located in a complex environment.

8. References

[1]   W. D. Burnside, M. C. Gilreath, R. J. Marhefka, and C. L. Yu, "A study of KC-135 aircraft antenna patterns," *IEEE Trans. Antennas Propag.*, vol. AP-23, pp. 309–316, May 1975.

[2]   C. L. Yu, W. D. Burnside, and M. C. Gilreath, "Volumetric pattern analysis of airborne antennas," *IEEE Trans. Antennas Propag.*, vol. AP-26, pp. 636–641, September 1978.

[3]   W. D. Burnside, N. Wang, and E. L. Pelton, "Near-field pattern analysis of airborne antennas," *IEEE Trans. Antennas Propag.*, vol. AP-28, no. 3, pp. 318–327, May 1980.

[4]   W. D. Burnside, R. C. Rudduck, and R. J. Marhefka, "Summary of GTD computer codes developed at the Ohio State University," *IEEE Trans. on Electromagnetic Compatibility*, vol. EMC-22, no. 4, pp. 238–243, November 1980.

[5]   R. J. Marhefka, "Analysis of aircraft wing-mounted antenna patterns," *Tech. Rep. 2902-25*, June 1976, the Ohio State University ElectroScience Laboratory, Department of Electrical Engineering; prepared under Grant NGL 36-008-138 for NASA, Washington, DC. Also a doctoral dissertation, June 1976, Ohio State University.

[6]   H. H. Chung and W. D. Burnside, "General 3D airborne radiation pattern code user's manual," *Tech. Rep. 711679-10*, July 1982, the Ohio State University ElectroScience Laboratory, Department of Electrical Engineering; prepared under contract F30602-79-C-0068 for Rome Air Development Center.

[7]   R. J. Marhefka and W. D. Burnside, "Numerical electromagnetic code–basic scattering code, NEC-BSC (version 2), part I: user's manual," *Rep. 712242-14*, December 1982, the Ohio State University ElectroScience Laboratory, Department of Electrical Engineering; prepared under Contract No. N00123-79-C-1469 for Naval Regional Contracting Office.

[8]   R. C. Rudduck and Y. C. Chang, "Numerical electromagnetic code (NEC)–reflector antenna code, part I: user's manual (version 2)," *Rep. 712242-16*, December 1982, the Ohio State University ElectroScience Laboratory, Department of Electrical Engineering; prepared under Contract No. N00123-79-C-1469 for Naval Regional

Procurement Office.

[9] J. H. Richmond, "Radiation and scattering by thin-wire structures in a homogeneous conducting medium," *IEEE Trans. Antennas Propag.*, vol. AP-22, no. 2, p. 365, March 1974.

[10] E. H. Newman, "A user's manual for electromagnetic surface patch code (ESP)," *Rep. 713402-1*, July 1981, the Ohio State University ElectroScience Laboratory, Department of Electrical Engineering; prepared under Contract No. DAAG29-81-K-0020 for Army Research Office.

[11] G. J. Burke and A. J. Poggio, "Numerical electromagnetic code (NEC)–method of moments," *NOSC/TD 116*, Naval Ocean Systems Center, San Diego, California 92152, July 1977.

[12] E. L. Pelton, R. J. Marhefka, and W. D. Burnside, "An iterative approach for computing an antenna aperture distribution from given radiation pattern data," *Rep. 784583-6*, June 1978, the Ohio State University ElectroScience Laboratory, Department of Electrical Engineering; prepared under Contract No. N62269-76-C-0554 for Naval Air Development Center.

[13] R. Backhus, "The determination of the aperture distribution of a linear array through near-field measurements," *Rep. 713303-2*, June 1982, the Ohio State University ElectroScience Laboratory, Department of Electrical Engineering; prepared under Contract No. N62269-80-C-0384 for Naval Air Development Center.

[14] G. A. Thiele and T. M. Newhouse, "A hybrid technique for combining moment methods with the geometrical theory of diffraction," *IEEE Trans. Antennas Propag.*, vol. AP-23, no. 1, pp. 62–69, January 1975.

[15] C. H. Walter, *Traveling-Wave Antennas*, New York: Dover Publications, 1979, pp. 15–16.

[16] J. D. Kraus, *Antennas*, New York: McGraw-Hill Book Company, 1950, pp. 464–477.

[17] R. G. Kouyoumjian and P. H. Pathak, "A uniform geometrical theory of diffraction for an edge in a perfectly conducting surface," *Proc. IEEE*, vol. 62, pp. 1448–1461, November 1974.

[18] Short course notes on "The modern geometrical theory of diffraction," vols. 1, 2, and 3, the Ohio State University ElectroScience Laboratory, Department of Electrical Engineering, 1980.

[19] W. D. Burnside and K. W. Burgener, "High-frequency scattering by a thin lossless dielectric slab," *IEEE Trans. Antennas Propag.*, vol. AP-31, pp. 104–110, January 1983.

[20] C. C. Huang, N. Wang, and W. D. Burnside, "The high-frequency radiation patterns of a spheroid-mounted antenna," *Rep. 712527-2*, March 1980, the Ohio State University ElectroScience Laboratory, Department of Electrical Engineering; prepared under Contract N00019-80-C-0050 for Naval Air Systems Command.

[21] J. G. Kim, W. D. Burnside, and N. Wang, "Geodesic solution for an antenna mounted on an ellipsoid," *Rep. 713321-3*, March 1982, the Ohio State University ElectroScience Laboratory, Department of Electrical Engineering; prepared under Contract N00019-80-PR-RJ015 for Naval Air Systems Command.

[22] W. D. Burnside, N. Wang, and E. L. Pelton, "Near-field pattern computations for airborne antennas," *Rep. 784685-4*, June 1978, the Ohio State University ElectroScience Laboratory, Department of Electrical Engineering; prepared under Contract N00019-77-C-0299 for Naval Air Systems Command.

[23] R. J. Marhefka, "Numerical electromagnetic code (NEC)–basic scattering code, part II: code manual (version 2)," *Rep. 712242-15*, December 1982, the Ohio State University ElectroScience Laboratory, Department of Electrical Engineering; prepared under Contract No. N00123-79-C1469 for Naval Regional Procurement Office.

[24] J. W. R. Taylor, *Janes's All the World's Aircraft*, New York: McGraw-Hill Book Co., 1973–74 (published yearly).

[25] R. L. Mather, G. V. Vaughn, and J. C. Logan, "Computer techniques for modeling shipboard communications antennas at UHF and above," *NELC Tech. Note 3313*, February 1977, Naval Ocean Systems Center, San Diego, California 92152.

[26] E. D. Green, "High-frequency scattering from multiple finite elliptic cylinders," *Rep. 712242-7*, June 1981, the Ohio State University ElectroScience Laboratory, Department of Electrical Engineering; prepared under Contract No. N00123-79-1469 for Naval Regional Procurement Office.

Chapter 21

Satellite Antennas

C. C. Han
Equatorial Communications Company

Y. Hwang
Ford Aerospace & Communications Corporation

CONTENTS

Ching Chun Han received the BS, MS, and PhD degrees in electrical engineering, with the doctorate being from Stanford University in 1971.

He joined Equatorial Communications Company in 1985 as Director of Microwave Systems in charge of VSET antenna and microwave converter development. He is also the project manager of Equatorial's K-220 program—the first K_u-band spread-spectrum satellite data communication network in the world.

He was employed by Ford Aerospace from 1967 to 1985, where he was the manager of the Antenna Systems Department. He worked on the INTELSAT-V frequency reuse shaped-beam antenna, the first few INTELSAT standard-A stations, the NATO-III antenna, the NASA 20/30-GHz multiple-beam antenna, low side lobe and low-cost stations, and many components, such as polarizers, OMTs, filters and diplexers, MICs, Gunn oscillators, pin and waveguide switches, VPDs, and stripline devices. He also serves as a part-time senior research associate at Stanford University's Communication Satellite Planning Center.

Yeongming Hwang received the BS in electrical engineering from the National Taiwan University, Taipei, Taiwan, in 1963, the MS in electrical engineering from the Institute of Electronics of National Chiao-tung University, Tsiuchu, Taiwan, in 1965, the masters degree in business administration from Golden Gate University, San Francisco, California, in 1982, and the PhD in electrical engineering from the Ohio State University in 1973.

He was a research associate in 1974 and Assistant Supervisor in 1975 in the Electro-Science Laboratory of the Ohio State University. He joined Ford Aerospace in 1975 and has worked in the satellite and ground antenna systems. At present he is the supervisor of the Antenna Technology Section at Ford Aerospace.

Dr. Hwang is interested in the geometrical theory of diffraction (GTD), numerical techniques, in applying the hybrid method of GTD and numerical techniques to antenna radiation problems, in antenna scattering, EMP problems, multibeam antenna systems, low side lobe ground antennas, beam optimization, frequency-selective surfaces, patched-array antennas, and microwave components for spacecraft antenna applications.

1. Introduction

The design of antennas for satellite applications differs in several respects from other applications. An antenna radiation pattern varies from omnidirectional to highly directional. It can be fixed or changed to accommodate specific needs as they arise. A satellite antenna must be designed to withstand the dynamic mechanical and thermal stresses for the satellite. The design constraints imposed by the satellite on size, shape, and weight are also important factors in design consideration. As the requirements on the side lobes and cross polarization become more stringent, the interference on antenna performance due to the presence of the satellite body, solar cell panel, and other antenna systems cannot be neglected.

The major function of a satellite antenna is communication. A communication satellite is a radio relay in space. A signal is sent from station A on earth to station B via the satellite as shown in Fig. 1. In the synchronous orbit, 22 286 mi (35 865 km) away from the earth, a satellite travels around the earth in exactly the earth's rotation time. Therefore the satellite appears to be stationary over any point on earth. Three satellites in synchronous orbit can cover over 90 percent of all inhabited regions of the earth.

The earth-subtended angle at the synchronous satellite is 17.34°. A satellite antenna can provide an earth coverage area of this size. The antenna radiation pattern can also be shaped to conform to the shape of an intended coverage area in order to improve antenna gain. Most satellite antennas are directional. Except for the tracking, telemetry, and command antenna, which must provide ranging, telemetry, and command operations throughout all mission phases, a nondirectional antenna is required to ensure the continuous reception in every mission operation.

The growing demand for increased communications satellite capacity and the crowding of synchronous orbital slots have led to more stringent demands on satellite antenna designs. The complexity and the size of satellite antennas have increased rapidly. Limited power resources in a satellite place an additional demand for designing more efficient antenna systems. Proper system design requires accurate predictions of designs for an efficient antenna system. As a result, an antenna engineer relies more on accurate theories for analysis and synthesis. Significant strides have been made in developing the analytical and synthesized techniques. Computer-aided design tools and advanced manufacturing techniques are essential not only to provide accurate antenna performance prediction, but also to reduce design cycle times.

This chapter addresses three types of satellite antenna design: communication, earth coverage, and tracking, telemetry, and command (TT&C).

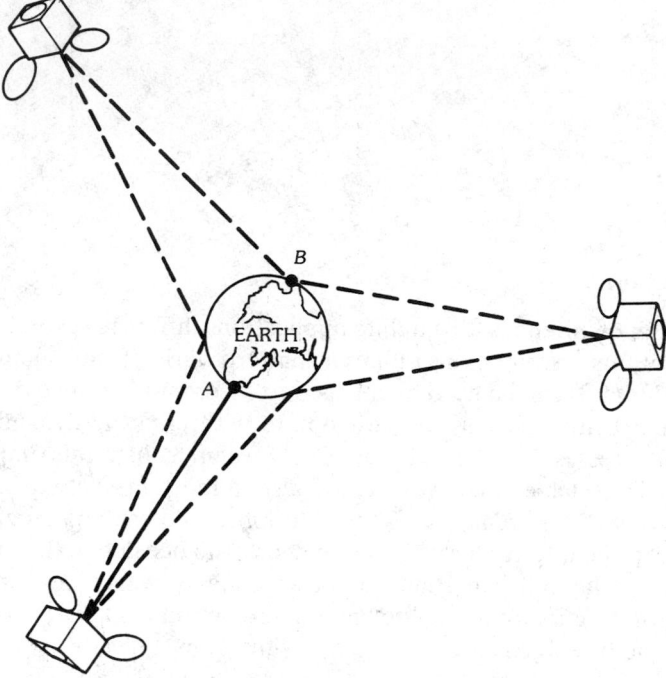

Fig. 1. Communication between two earth stations via satellite.

2. Communication Antennas

A communication satellite functions as a radio relay in space. A satellite antenna can provide the communication link not only between stations on earth but also between antennas from other satellites.

It can be shown [1] that in a communication downlink

$$A_{\text{cov}} \Delta f \cong P_T \frac{A_R}{T_S} \frac{1}{L_i} \left(\frac{C}{N} \right)^{-1} \tag{1}$$

where

A_{cov} = coverage area

Δf = bandwidth

P_T = satellite transmitter power

A_R = earth station antenna effective area

T_S = equivalent system temperature

L_i = incidental loss

C/N = carrier-to-noise power ratio

Increasing the capacity of the communication system can be directly achieved by increasing the bandwidth Δf through frequency reuse. On the other hand, for a fixed C/N and prechosen modulation system, the bandwidth can be increased by reducing the coverage area A_{cov}. The coverage area can be reduced by using a multibeam antenna. The power is divided among the beams and the bandwidth Δf remains constant for each beam. As a result, the total bandwidth available increases by the number of beams.

The demands of the frequency reuse and multiple-beam antenna systems have taken the form of requirements for more antenna pattern control for increased antenna gain, low side lobes, and cross polarizations. Low side lobes and cross polarization are required to prevent excessive interference among beams. The typical isolation requirement for a communication system is 27 dB or higher.

A typical satellite communication antenna consists of an optical aperture, a feed array, and a beam-forming network (BFN). The function of the aperture is to focus the energy from the point source to the desired direction to yield high gain. A reflector, lens, or phased array may be used as an optical aperture. A feed array is located near the focal point to direct the energy to the aperture. A BFN, which is connected to the array, distributes the energy to the proper beam port. The network includes the passive power divider, switch, or variable power divider to control the beam steering and beam shape.

The antenna beams generated can be fixed or scanned over the coverage area. A beam can be scanned mechanically or electrically. A mechanically scanning beam is achieved by gimbaling the antenna system. This approach is adopted in the application where the scanning rate is of no major concern. Certain communication systems using satellite-switched time-division multiple access require a fast scanning rate that can only be achieved by using switches electrically controlled in the BFN.

A beam is generally distorted as scanned from the antenna boresight due to the phase aberration of the antenna optics. As a result it lowers antenna gain and raises the side lobes and cross-polarization levels. In general, the antenna performance is evaluated in terms of gain, side lobes, and cross polarization achieved at the farthest scanning position.

A fixed beam can be a pencil beam or a highly shaped beam. A highly shaped beam is used to improve antenna gain over the prescribed area and to reduce interference outside the coverage area. An ideally shaped beam is one that has the highest possible antenna gain without any side lobes. The radiation pattern turns out to be flat over the coverage area. The theoretically achievable maximum value of the edge-of-coverage gain is given by

$$G = \frac{41\,253}{A_{cov}} \tag{2}$$

where A_{cov} is the coverage area in degrees square. The performance of a shaped-beam antenna is evaluated in terms of beam-shaping efficiency defined as

$$\text{beam-shaping efficiency} = \frac{G_{min} A_{cov}}{41\,253} \tag{3}$$

or the gain-area product $G_{min} A_{cov}$ in which G_{min} is the minimum gain over the coverage area.

The edge rolloff of an optimum elliptical beam [2] is 4.3 dB below the peak. It has a beam-shaping efficiency of 40 to 45 percent. In other words, the gain-area product (GA) is between 16 500 and 18 560. Practically, the gain-area product of a satellite antenna is ranged between 10 000 and 16 000, depending on the size of the antenna aperture and the shape of the coverage.

Antenna System Design

The major parameters involved in the design of a satellite antenna system are listed in Table 1. The requirements of a satellite antenna system are specified in terms of the minimum gain over a coverage area, the frequency band, polarization, and isolation. There is a system trade-off among the transmit power, satellite antenna gain, and the ground terminal antenna gain in a communication link. The frequency is designated from the frequency assignment. Frequency bands for satellite communications and other services have been advocated by both international and national regulatory and policy-making agencies. The range of the frequency may be from 0.1 to 150 GHz. The frequency bands in this range are listed in Table 2. Due to the spectrum crowding in the S- and C-bands, the K_u-band and

Table 1. Design Parameters of a Spacecraft Antenna System

Requirements	Optics	Feed Elements	Beam-Forming Network

Gain
Gain ripple
Gain slope
Coverage area
Frequency
Frequency band
Polarization
Scanning or
  fixed beam
Number of beams
Isolation
	Phased array
	Lens
	Reflector
	Aperture size
	Number of
	  optics
		Type of feed
		  element
		Feed spacing
		Number of feeds
		Polarizer
		Orthomode transducer
			Feed excitation coefficients
			Switch
			Variable phase shifter
			Variable power divider
			Power divider
			Diplexer
			Odd/even-mode converter
			Frequency-selective surface
			Transmission media

Table 2. Frequency Allocation for Satellite Communication

Frequency (GHz)	Frequency (GHz)
1.53–1.599	12.5–12.7
1.6265–1.6605	12.75–13.25
2.5–2.69	14.0–14.5
3.4–4.2	17.7–21.2
4.5–4.8	27.5–31.0
5.85–7.075	40.0–41.0
7.25–7.75	43.5–47.0
7.9–8.4	92.0–95.0
10.9–11.2	102.0–105.0
11.4–12.2	140.0–152.0

higher frequencies are being used for broader bandwidths. Signals may be linearly or circularly polarized. Orthogonal polarization can be reused in the same frequency band to increase satellite capacity. Isolation is specified between beam ports from the system requirements. The isolation requirements are then translated to the required side lobe or cross-polarization levels of the beams.

The design process starts with choosing the size of optical aperture. Theoretically, the larger the aperture size, the higher the antenna gain will be. An ideally shaped beam is generated by an aperture of infinite extent. Practically, the aperture size should be adequately large to meet gain and side lobe requirements. An optimal size is the minimum size that can be placed on the satellite that satisfies the system requirements. The same optical aperture may be used for multifrequency bands, dual polarization, or multibeam at the expense of increasing complexity of the antenna system. The minimum number of apertures required is determined as a result of trade-off of the achievable performance of the system.

Three basic generic types of antennas are: phased array, lens, and reflector. A phased array requires N corporate feed networks for N beams and therefore is generally heavy and complex. A lens is attractive for its good scanning performance and compactness. However, it is also heavy in the low-frequency application. A reflector is widely used for its light weight and structural simplicity.

The choice of feed elements is governed by the achieved gain and cross-polarization level over the required subtended angles, the frequency bandwidth, and the type of optical aperture. Horn is widely used in most applications. A ridged waveguide, helix, turnstile, cross yagi, small reflector, and slot array have also been used. An orthomode junction is required for dual polarization. A polarizer is used to generate circularly polarized waves for the antenna system. A polarization reflector can be used to separate two linearly polarized signals or to provide lower linear cross-polarization signals.

The design of a beam-forming network is heavily influenced by the loss, weight, and size in a satellite antenna system. Most power dividers and transmission lines are waveguides typed for frequencies above K-band because of the low insertion loss. Air stripline is widely used as a transmission medium in the C-band or lower frequency because of its compactness.

There are three main types of switches: mechanical, ferrite control, and diode switches. The choice of the switches is mostly determined by the insertion loss and switching speed. The variable phase shifters and variable power dividers are categorized to be electrical-mechanical, ferrite, or diode variable phase shifter/variable power dividers.

The odd/even converter is basically a two-to-N port network, which is used to combine the odd and even channels of a transponder output multiplexer. In order to alleviate the severe restrictions on the design of a closely separate output multiplexer, odd-numbered channels are often combined to a single output, and even-numbered channels to a second output. These two outputs are then connected to the input ports of an odd/even mode converter.

A diplexer is used to combine two frequency bands in the beam-forming network. It is a two-to-one filter microwave component. A frequency selective surface can be used as a spatial filter to replace the filter type diplexer to alleviate the congestion and complexity of beam-forming networks.

Types of Antennas

Phased array, lens, and reflectors are the three basic antennas. A phased array has a number of advantages over lens or reflector: distribution of power amplification at the elementary radiation level, higher aperture efficiency, no spillover loss, no aperture blockage, and better reliability. Disadvantages are weight, complexity, and relatively high losses in the power distribution system. The advantages of a lens antenna are excellent optics, no feed blockage, and compactness. A lens can be made rotationally symmetrical to preserve good optical characteristics. Disadvantages are heaviness, especially in the low-frequency application, and the mismatch of the lens surface. The lens itself is also a blockage to the feed array. A reflector is the most desirable candidate for spacecraft antennas because of its light weight, structural simplicity, and design maturity. The reflector, however, has to be offset to avoid feed blockage. The offset destroys the rotational symmetry of the surface and limits the range of scan before aberrations seriously degrade the scanning performance. Table 3 summarizes the characteristics of these three antennas.

The Phased Array—A phased array consists of more than one antenna element radiating in phase coherence. Horns, dipoles, helices, spiral antennas, parabolic dishes, and many other types of antennas can be used as radiating elements. The radiation pattern of a radiating element in an array environment is called the *active element pattern*. It may be substantially different from the pattern of an isolated radiating element because of the effect of mutual coupling.

A phased array for satellite applications can be used for

(*a*)   a fixed beam that can be either single or multifixed
(*b*)   electronically steerable scanning beams
(*c*)   a feed array of a lens or reflector antenna system

Table 3. Characteristics of Three Basic Antennas

Antenna	Advantages	Disadvantages
Phased Array	Distribution of power amplification at the elementary radiation levels Reliability No spillover losses No aperture blockage	Complexity Heavy Higher beam-forming network losses
Lens	No feed blockage Better scanning performance	Heavy in low-frequency application Aperture mismatch
Reflector	Simple Lightweight Design maturity	Offset to avoid feed blockage Poor scanning performance

(*d*) an adaptive array that can automatically null a beam in the direction of oncoming jammers

If the mutual coupling among the array elements is neglected, it can be shown [3] that for a planar array of $(2M + 1) \times (2N + 1)$ apertures embedded in an infinite conducting ground plane at $z = 0$ as shown in Fig. 2, the far-field radiation pattern can be expressed as

$$E(r) = f(T_x, T_y)\, S_a(T_x, T_y) \tag{4}$$

where

$$S_a(T_x, T_y) = \sum_{m=-M}^{M} \sum_{n=-N}^{N} V_{mn} e^{j(\mathbf{k}_i \cdot \varrho_{mn})} \tag{5a}$$

and

$$f(T_x, T_y) = \frac{j e^{-jk_0 r}}{2\pi r} (k_0 \sqrt{1 - T_x^2 - T_y^2}) \int_{A_{00}} \mathbf{E}_{00}(x^0, y^0)\, e^{jk_0(T_x x^0 + T_y y^0)}\, dx^0\, dy^0 \tag{5b}$$

in which k_0 is the wave number, $\varrho_{mn} = mb\hat{\mathbf{x}} + nd\hat{\mathbf{y}}$, b and d are the separations between adjacent channels in the x and y directions,

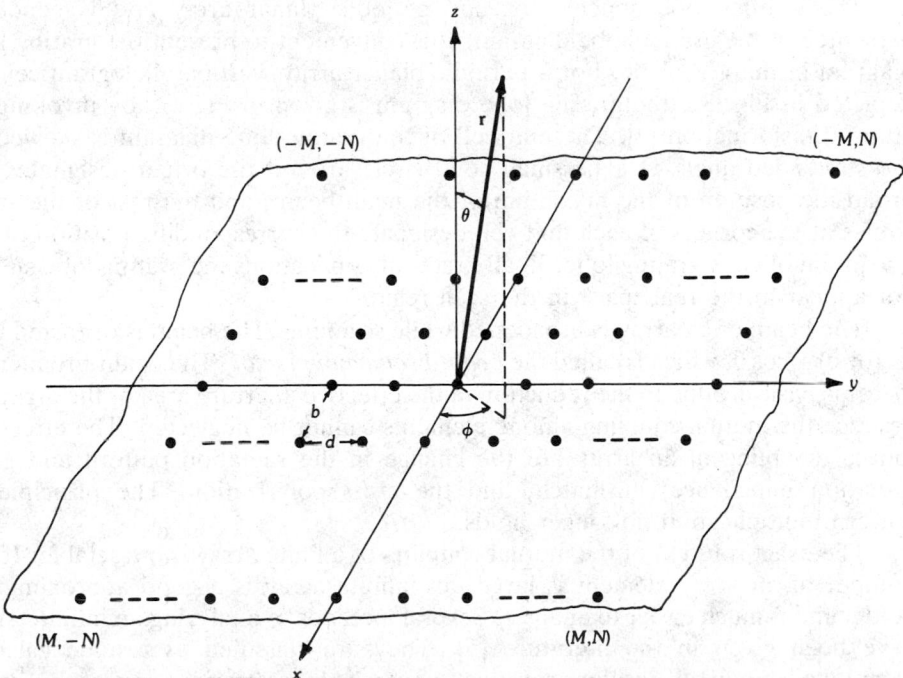

Fig. 2. Planar array in a ground plane. (*After Amitay, Galindo, and Wu [3], © 1972 Bell Telephone Laboratories, Incorporated*)

$$\mathbf{k}'_t = \hat{\mathbf{x}} k'_x + \hat{\mathbf{y}} k'_y$$
$$k'_x = k \sin \theta \cos \phi$$
$$k'_y = k \sin \theta \cos \phi$$

X^0 and Y^0 are the coordinates of the aperture A_{00} over the unit element cell, $E_{00}(x^0, y^0)$ is the field distribution in the aperture A_{00}, V_{mn} is the voltage generator of the (m, n)th element,

$$T_x = \sin \theta \cos \phi$$
$$T_y = \sin \theta \sin \phi \qquad (6)$$
$$T_z = \cos \theta$$

f is the element pattern, and S_a is the array factor. The radiation pattern is the product of the element pattern and the array factor. The element pattern determines the rolloff and the polarization of the far-zone field. It is a slowly varying function in comparison with the array factor. The array factor is a double function of U and V with periods λ/b and λ/d, respectively. Because of the periodicity in the UV plane, the peak of the array factor will be repeated every λ/b and λ/d intervals in the U and V directions, respectively. The peak of the main beam will also be repeated and its peak is tapered off following the rolloff of the element pattern. Such repetition of the main beam is called a *grating lobe*.

The grating lobe appears for any periodic planar array and is generally presented in the grating lobe diagram. It is convenient to present the grating lobe diagram in the UV plane. For a periodic planar array with parallelogram cells as depicted in Fig. 3a, the grating lobe diagram is given in Fig. 3b by invoking an affine transformation [4]. The unit cell of the grating lobe diagram is skewed by the subtended angle Ω. The small solid circle around the origin designates the broadside position of the maximum of the main beam, and the rest of the small circles at the corners of each unit cell designate the corresponding position of the maximum of each grating lobe. In all practical applications the grating lobe should not appear in the real space in the scan region.

The beam of an array is broadened while scanning. The beam is distorted by a factor of $1/\cos \theta$, which is called the *beam-broadening factor*. The beam-broadening factor is basically due to the reduction in the effective aperture area of the array. In practice the mutual coupling among elements cannot be neglected. The effects of mutual coupling in an array are the change in the radiation pattern and gain, radiation impedance, mismatch, and the cross polarization. The principle of pattern multiplication no longer holds.

The exact solution of the mutual coupling of a finite array is intractable. If the number of an array element is large, an infinite array is a good approximation model and is much easier to analyze. Several methods of analyzing an infinite array have been given in the literature [5]. They are classified as residue calculus technique, modified residue calculus, generalized scattering approach, mode matching method, complex power, integral equation, and grating lobe series.

To meet an antenna specification, trade-offs have to be made among the size of

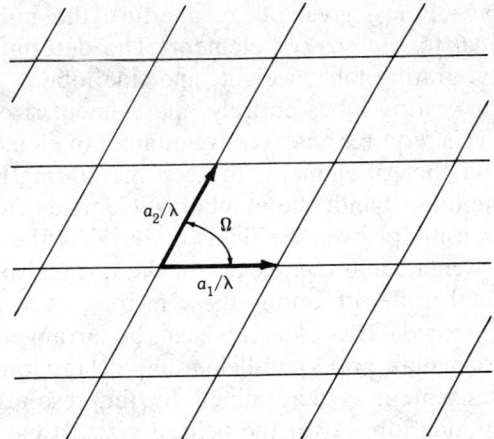

a

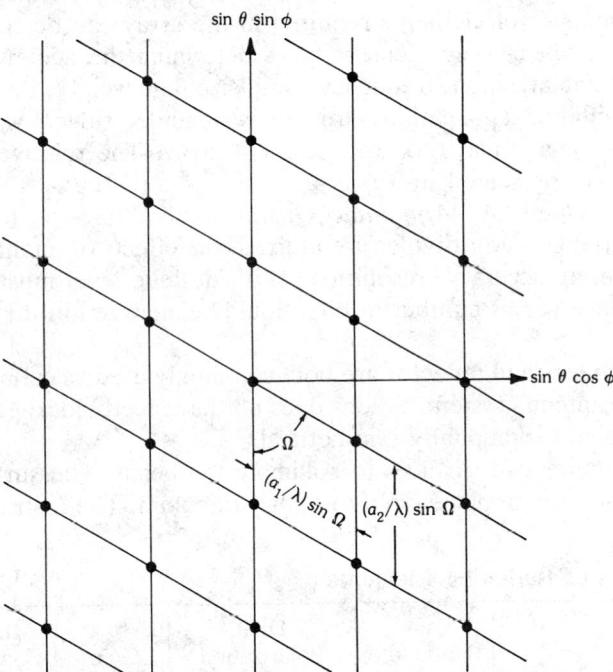

b

Fig. 3. Periodic planar array with parallelogram cells and its grating lobe diagram. (*a*) Periodic planar array. (*b*) Grating lobe diagram. (*After Lo and Lee [4], © 1965 IEEE*)

aperture, number of elements, element arrangement, and the type of elements. Most satellite antenna applications concern not only the peak gain of the beam but also the gain at the edge of coverage while scanning. The critical design parameters are as follows.

Size of the Array—The size of the array is mostly determined by the gain and beamwidth requirements.

Element Number—For a given array aperture the number of elements is inversely proportional to the size of element. The determinant factors are the resolution capability, grating lobe, weight, and side lobe.

To eliminate the grating lobes entirely, the element size should be kept less than a wavelength. This requires an excessive number of elements. Practical design uses the minimum number of elements to keep the grating lobes only out of the field of view. On the other hand, the number of elements cannot be too small to degrade the array sensitivity over the field of view. As the number of elements increases, the cost, weight, and complexity of the array also increase. Therefore there is a fundamental trade-off among these factors.

Element Arrangement—The elements can be arranged in a square grid, rectangular grid, triangular grid, parallelogram, or random arrangement. The choice of each arrangement is determined by the resolution, the number of elements, and the grating lobes over the field of view. It has been shown that by arranging the elements of a beam-scanning planar array in a triangular pattern [6] rather than a rectangular pattern, or in a random pattern [7] rather than a periodic pattern, the number of elements required in the array can be reduced.

Types of Element—The element types determine the achievable array gain, scan loss, cross polarization, frequency band, size, and weight. Possible candidates for satellite antenna applications are the waveguide, ridged waveguide, helix, turnstile, cross yagi, small reflector, and slot array. The relative merits of each element type are presented in Table 4.

Allowable Phase and Amplitude Quantization Errors—If the digital phase shifter and variable power divider are utilized, the effects of quantization steps on the beam-pointing accuracy, resolution, and side lobe level must be considered. The performance versus number of bits should be analyzed in the design process.

The Lens—The lens and reflector are both commonly used as collimating elements in a high-gain antenna system. A lens does not have feed blockage and its surface can be maintained rotationally symmetrical.

A lens requires two surfaces to collimate the beam. The surfaces in general have the simple form of plane, spheroid, or paraboloid. The restriction to surfaces

Table 4. Types of Radiating Elements

Type	Bandwidth	Dual Polarization	Gain	Mutual Coupling	Volume
Waveguide	10%	Yes	Medium	Medium	Medium
Ridged waveguide	Wider	More difficult	Medium	Medium	Medium
Helix	1.67 <35 GHz	One sense only	High	High	Large
Turnstile	10%	Yes	Low	Medium	Small
Crossed yagis	10%	Yes	High	High	Large
Reflector	Wideband	Yes	High	Medium	Large
Waveguide slotted array	Narrow	Difficult	High	High	Small

of simple forms imposes limitations on the ultimate performance that the systems can attain. A shaped surface or aspheric surface is proposed in some systems. The surface can be adequately represented by

$$z = \frac{\gamma \varrho^2}{1 + \sqrt{1 + (1 + K)\gamma^2 \varrho^2}} + A\varrho^4 + B\varrho^6 + C\varrho^8 + D\varrho^{10} \qquad (7)$$

where K is called the *conic coefficient*, ϱ is the distance measured from the axis of the surface of revolution, γ governs the curvature of the surface at the apex, and A, B, C, and D are the coefficients of the polynomial. The first term represents the surface of the form of hyperboloid ($K < -1$), paraboloid ($K = -1$), ellipsoid ($-1 < K < 0$), sphere ($K = 0$), or spheroid ($K > 0$).

The lowest-order aberration function of a rotationally symmetrical optical system is the primary or Seidel aberration [8]. There are five primary aberrations: spherical, coma, astigmatic, field curvature, and distortion as depicted in Fig. 4. Such aberrations, or phase errors, may arise from a displacement of the primary

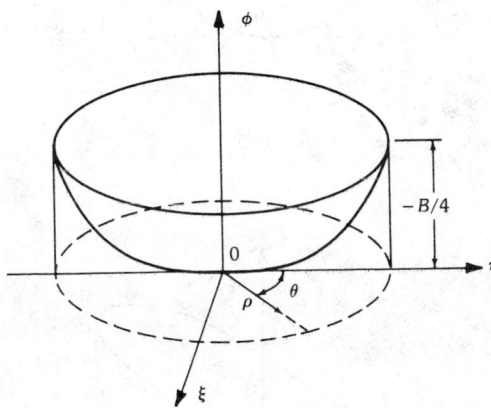

a

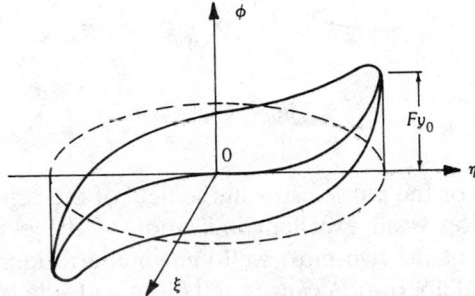

b

Fig. 4. The primary wave aberrations. (*a*) Spherical aberration: $\theta = B\varrho^4/4$. (*b*) Coma: $\phi = Fy_0\varrho^3\cos\theta$. (*c*) Astigmatism: $\phi = -Cy_0^2\varrho^2\cos^3\theta$. (*d*) Curvature of field: $\phi = Dy_0^2\varrho^2/2$. (*e*) Distortion: $\phi = Ey_0^3\varrho\cos\theta$. (*After Born and Wolf [8]; reproduced with permission of Pergamon Press, © 1980*)

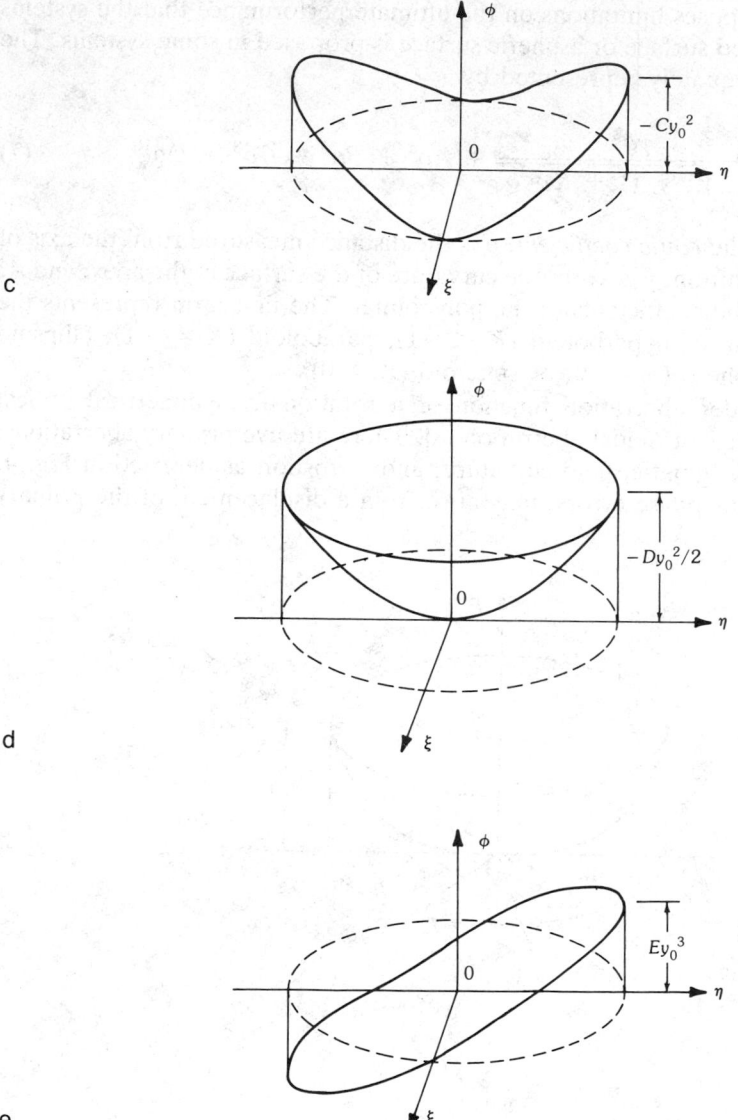

Fig. 4, *continued.*

feed from the focus or the phase error in the field of the primary feed. The degree of primary aberration is an excellent indication of the performance of the lens system. The effects of the two most well-known aberrations, spherical and coma, are that the spherical aberration causes null fill-in and side lobe increases, whereas the coma aberration causes imbalance of the side lobe levels. Fig. 5 presents the effects of the five primary aberrations on the radiation patterns.

The effect of aperture amplitude distribution on the radiation pattern is well studied. It has been shown that side lobes are related to the aperture field dis-

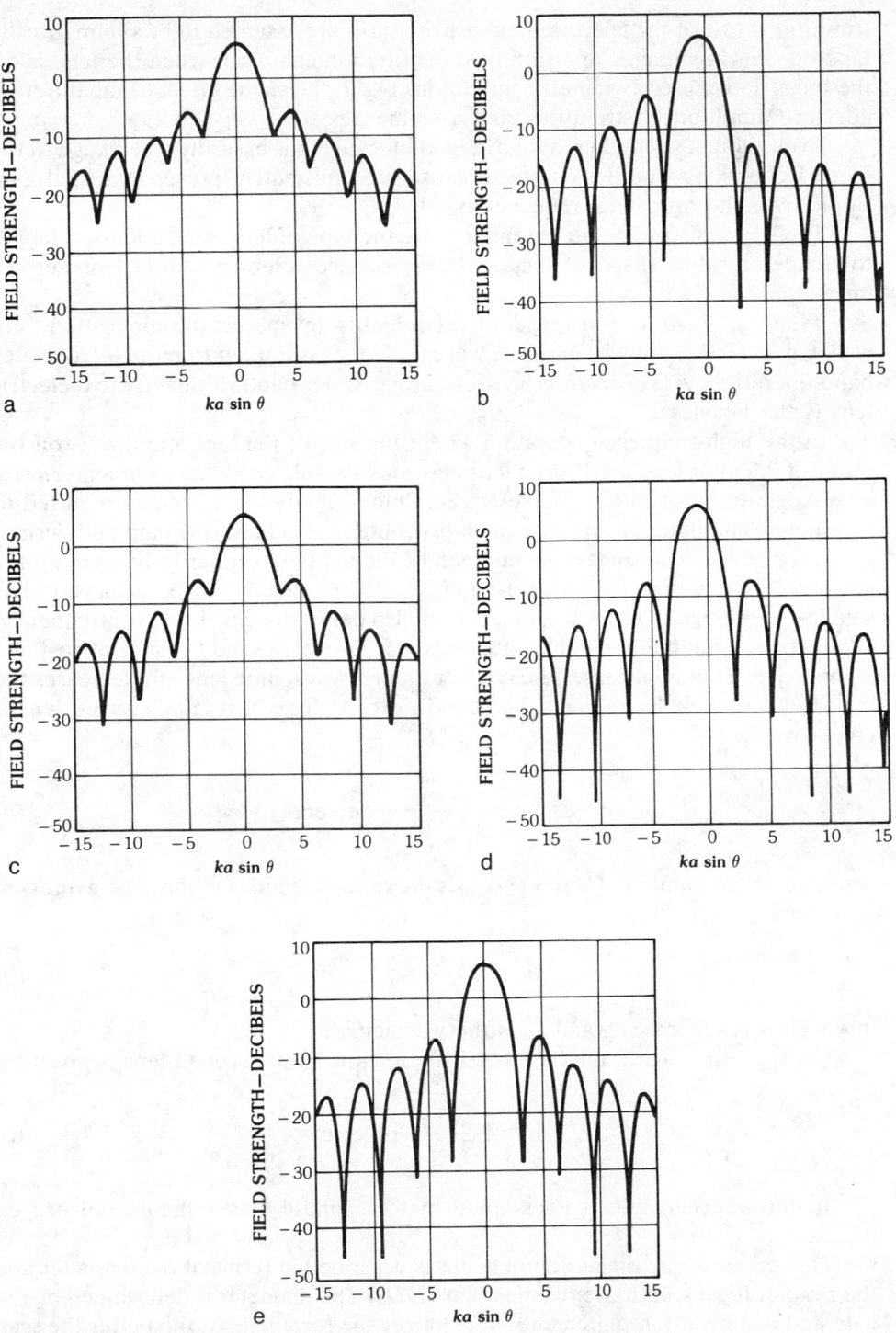

Fig. 5. Antenna pattern degradation due to aberrations. (*a*) $\phi = -\varrho^4$. (*b*) $\phi = -\varrho^3$. (*c*) $\phi = -\varrho^2$. (*d*) $\phi = -\varrho$. (*e*) $\phi = 0$.

tribution. Most of the aperture field distributions are assumed to be symmetrically tapered. An asymmetric aperture field distribution can occur when the feed is off the focus and causes asymmetric null fill-in. Fig. 6 shows the effect of the different aperture amplitude distributions and asymmetries.

Even though a lens does not have feed blockage it is basically a blockage to the feed. Reflections from lens surfaces cause feed mismatch, power loss, and gain ripple over the operating frequency band.

The key design trade-off parameters are the type of lens, size, and focal length to diameter ratio, shape of lens surfaces, surface tolerances, and lens surface mismatch.

Types of Lenses—Most widely used lenses in spacecraft applications are waveguide, TEM, and dielectric. A waveguide lens is limited to narrow frequency band operation. A TEM lens is heavier but has wider bandwidth and the dielectric lens is the heaviest.

In the high-frequency operation where the size of the lens aperture is on the order of 0.9 m or less, a dielectric lens becomes a viable candidate. For a given size of waveguide or TEM lens, there is a minimum number of elements required to adequately simulate the aperture field distribution to achieve the gain and desired side lobe level. The maximum number of elements is primarily limited by the smallest element size for which a good impedance match can be achieved. The smallest element size in a waveguide lens is also determined by its cutoff frequency. In practice, a minimum number of elements is always desired because of cost and complexity. Zoning increases the bandwidth of a waveguide lens but decreases the bandwidth of a dielectric lens. The bandwidth Δf for a waveguide zoned lens is given by [9]

$$\Delta f \cong 25 \frac{n_0}{1 + K_z n_0} \text{ percent} \tag{8}$$

where K_z is the number of zones and n_0 is the refractive index that can be expressed as

$$n_0 = \sqrt{1 - (\lambda_0/2a)^2} \tag{9}$$

in which a is the spacing and λ_0 is the wavelength.

On the other hand, the bandwidth Δf for a dielectric zoned lens is given by

$$\Delta f \cong \frac{25}{K_z - 1} \text{ percent} \tag{10}$$

In the above results it is assumed that the phase errors should not exceed 0.125λ.

Dimension—The dimension of a lens is described in terms of the diameter and the ratio of focal length over its diameter (F/D). The diameter is determined by the gain and beamwidth requirements. The longer the focal length, the better the scan performance. In general, the focal length is determined as a result of compromise between the scan performance and the spacecraft constraint.

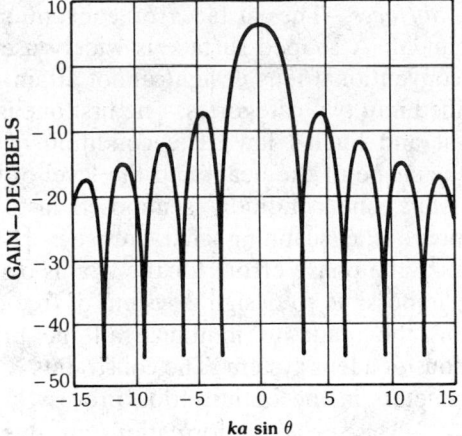

a

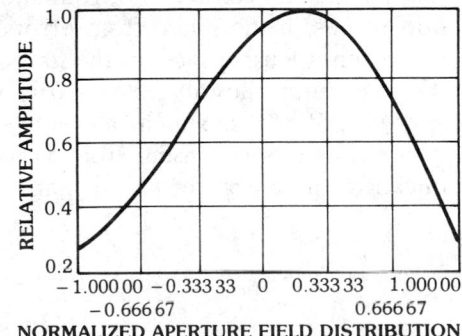

b   d

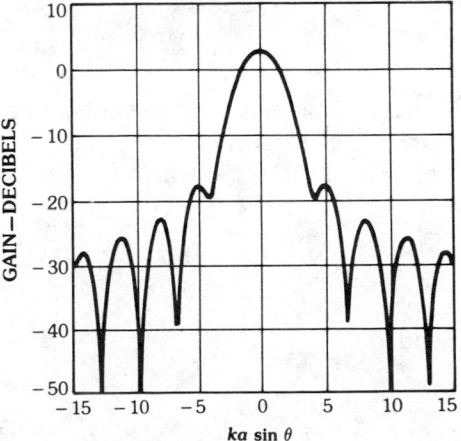

c   e

Fig. 6. Antenna pattern versus amplitude distribution. (*a*) Uniform distribution. (*b*) Tapered distribution. (*c*) Antenna pattern with amplitude distribution (*b*). (*d*) Asymmetric distribution. (*e*) Antenna pattern with amplitude distribution (*d*).

Shapes of Lens Surfaces—The surfaces of conventional lenses are planar, spherical, or paraboloidal. A shaped surface is widely used to obtain the better performance that a conventional lens design cannot attain. The principles of lens shaping can be classified into two categories. The first one is the application of the path-length constraint and Snell's law. The condition of power distribution is imposed to control the shape of the beam and the level of the side lobes [10, 11]. In some cases the Abbe sine condition is imposed instead of invoking power conservation to improve the scanning characteristics [12, 13, 14]. The second approach is to minimize the phase errors for the beams on-focus and off-focus as well. In general, it is impossible to design a system of free-of-phase error. A suitable compromise must be made to minimize all the primary as well as the higher-order aberrations of a lens system. The constraints K, A, B, C, and D in (7) are the variable parameters in the optimization process.

Surface Tolerances—The surface deformations are due to the manufacturing tolerances and thermal effects in space. The deviations of a lens surface from its ideal shape can cause loss of gain and pattern degradation. The surface deformation results in the phase front errors in the aperture field. The tolerance to be placed on a lens surface, in the first-order approximation, can then be related to the maximum allowable irregularity in the equiphase front formed by the lens aperture. If the maximum allowable phase error is taken to be $\lambda/16$ on wavefront irregularities arising from variation in either t or n, where t is the lens thickness, it can be shown [9] that

$$\Delta t \leqq \frac{\lambda}{16(1 - n)} \qquad (11a)$$

and

$$\frac{\Delta a}{a} \leqq \frac{n}{16(n + 1)} \qquad (11b)$$

where Δt and Δa are the tolerance on the waveguide lens thickness and plate spacing a, respectively.

For a dielectric lens,

$$\Delta t \leqq \frac{\lambda}{16(n - 1)} \qquad (12a)$$

and

$$\Delta n \leqq \frac{\lambda}{16t} \qquad (12b)$$

In general, the effect of the surface deformation on the gain, side lobe levels, and cross polarization should be analyzed in a more rigorous way. The radiation pattern can be obtained accurately by integrating the aperture field distribution once the details of the phase front errors are known. The surface tolerance specifications should be examined on a case-by-case basis against system requirements.

Surface Mismatch—In order to design a good lens antenna system it is necessary to match the surface of the lens to reduce effects of direct reflections. Three techniques can be used for lens surface matching:

1. Quarter-wave matching layers
2. Multiple quarter-wavelength impedance-matching transformer [15, 16]
3. Artificial-dielectric quarter-wave plate

In the first technique, the permittivity of the layer should be $\epsilon^{1/2}$ to give a perfect match at a single frequency. If a match is required over a wide bandwidth, then the second or third technique has to be used. The problems with the first and second techniques are the lack of materials and fabrication difficulties. The third technique requires that the dielectric surface be slotted to give the equivalent matching effect. The slots are either parallel or perpendicular to the electric field [17, 18].

The Reflector—The reflector is the most desirable optical candidate in spacecraft antenna systems because of its light weight, structural simplicity, and design maturity. The disadvantage is that the reflector has to be offset to avoid the feed blockage. The offset destroys the rotational symmetry of the optical aperture and limits the range of scan to a few beamwidths before aberrations seriously degrade the performance.

A reflector antenna system consists of one or more reflector surfaces. The surface can take the form of a paraboloid, hyperboloid, spheroid, ellipsoid, or general shape. A single offset parabolic reflector is used most often because it is the simplest in the reflector antenna system.

Geometry—The geometry of a single-offset parabolic reflector antenna is shown in Fig. 7. It can be described by the aperture size, offset distance, and the focal length. The key parameters to determine the characteristics of the system are focal region, scan loss, and cross polarization.

Focal Region—For a center-fed, circularly symmetrical reflector the secondary beam remains symmetrical about the reflector axis when the feed is displaced along the axis. The secondary beam scans when the feed is laterally displaced. For an offset-fed parabolic reflector the offset axis, which is equivalent to the reflector axis of a center-fed reflector, is in the direction of θ_0 [19]:

$$\theta_0 = \frac{1}{2}(\theta_u + \theta_L) \tag{13}$$

where θ_u and θ_L are the angles subtended at the focal point by the upper and lower edges of the reflector in the xz plane. The focal surface is the plane perpendicular to the offset axis.

Polarization—Let an offset reflector be illuminated by a feed. If the feed radiation is circularly polarized everywhere, no cross polarization will appear in the radiation from the reflector. A small beam squint will occur because of variation in the phase shift across the reflector [20]. If the feed radiation is linearly polarized

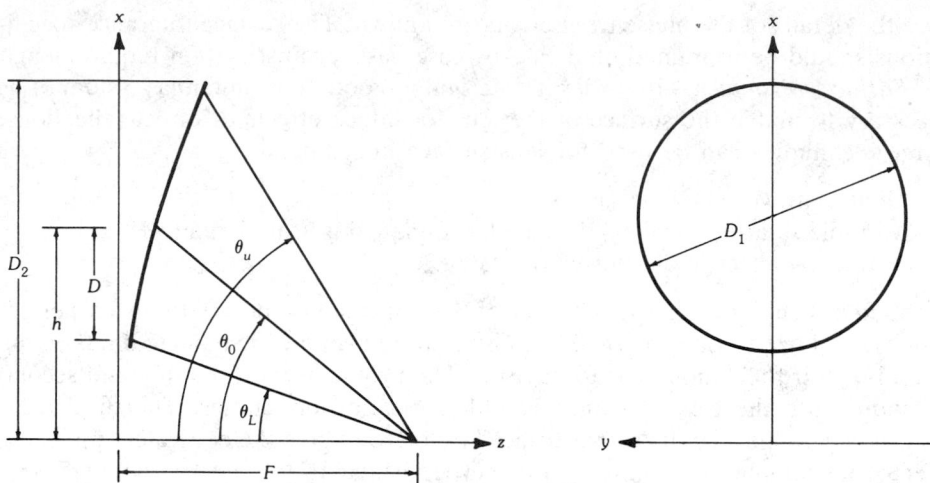

Fig. 7. An offset-fed parabolic reflector antenna system.

everywhere, a small polarization rotation will occur except when the feed is displaced in the plane of asymmetry (*xz* plane). However, there will be a small cross-linear polarization off the beam axis.

Lateral Feed Displacement—The beam generated by the reflector scans when the feed moves laterally from the offset axis. The beam degradation occurs when it scans. The degradation includes beam shape distortion, gain, loss, beamwidth, and higher side lobes. Typical beam shape degradation is depicted in Fig. 8. The gain decreases as the beam scans away the offset axis. The scan loss of a center-fed parabolic reflector antenna can be represented as a function of F/D and the number of half-power beamwidths scanned [21]. The scan loss of an offset-fed parabolic reflector antenna can be represented in a similar way. Fig. 9 shows the scan loss of an offset-fed parabolic reflector antenna for F/D of 1.0 and 1.3.

Design Parameters—The key design parameters are the aperture size, type of reflector, focal length, offset distance, and surface tolerance.

Aperture Size—The reflector aperture size is determined by the gain and the beamwidth required. In a multibeam antenna application the performance for generating the required shaped beam is better if the reflector is larger. The larger reflector allows the radiation energy to concentrate on a small area and provides steeper slope to reduce interference between closely spaced beams and between satellites. Practical aperture size is the smallest aperture that meets the performance requirements.

Type of Reflector—A single reflector is always the first choice in a reflector antenna system. A single offset parabolic reflector is a favorite candidate because of its better performance and design maturity. It is well known that the spherical reflector is a coma- and astigmatism-free optics when the entrance pupil is at the center of curvature. It is therefore suggested that a surface having the form between a paraboloid and a sphere may have better scan performance than a paraboloid. Such a surface can be expressed as

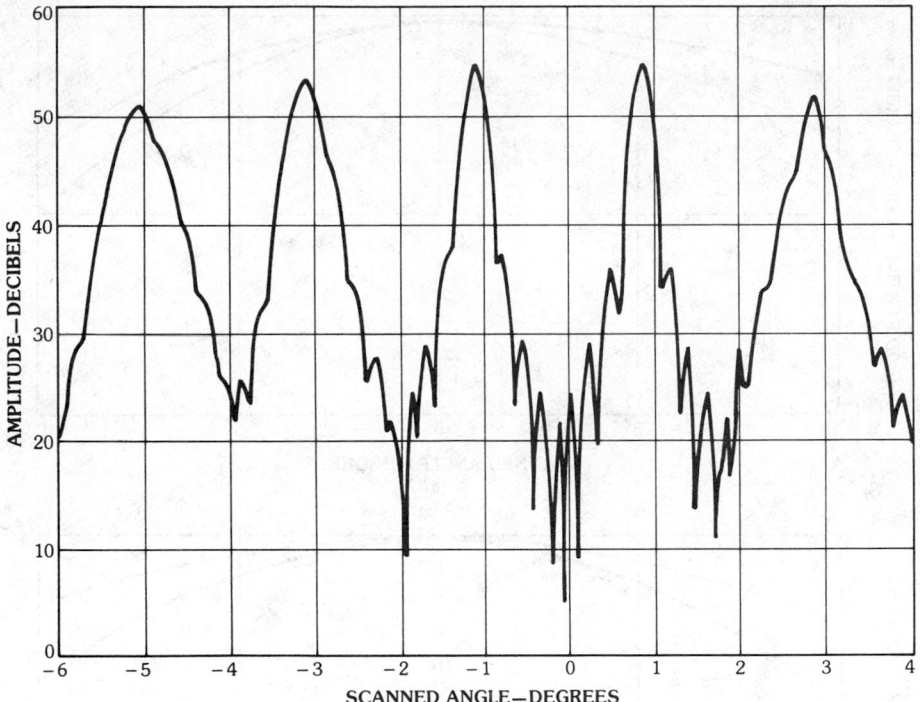

Fig. 8. Antenna radiation pattern versus scan angle (*xz* plane).

$$z = \frac{x^2}{4f} + (1 + b)\frac{x^4}{8(2f)^3} + \cdots \tag{14}$$

in which $b = -1$ for a paraboloid and $b = 0$ for a sphere, and f is a constant.

The reflector surface can be solid for circular polarized waves or gridded for linearly polarized waves. A gridded reflector surface can enhance the purity of linear polarization. Two orthogonal gridded surfaces can also share a common aperture area to form a compact antenna system. The front surface reflects one sense of linearly polarized waves while it allows the other orthogonally polarized wave to pass through with little attenuation. The rear reflector reflects the orthogonally polarized waves and filters the unwanted cross polarization.

Because of the limitation of the scan performance of a single reflector, a dual reflector may be used to replace the single reflector for a better optical system. A dual reflector has one more degree of freedom in the number of surfaces and should be able to reduce the phase aberration for better scan performance. The classical dual-reflector antenna system consists of Cassegrain and Gregorian antenna systems.

Cassegrain antenna reflector surfaces have the form of a paraboloidal primary reflector and a hyperboloidal secondary reflector. The system is electrically equivalent to a parabolic reflector antenna with a longer focal length mF, where m is the magnification factor of the Cassegrain antenna system and F is the focal

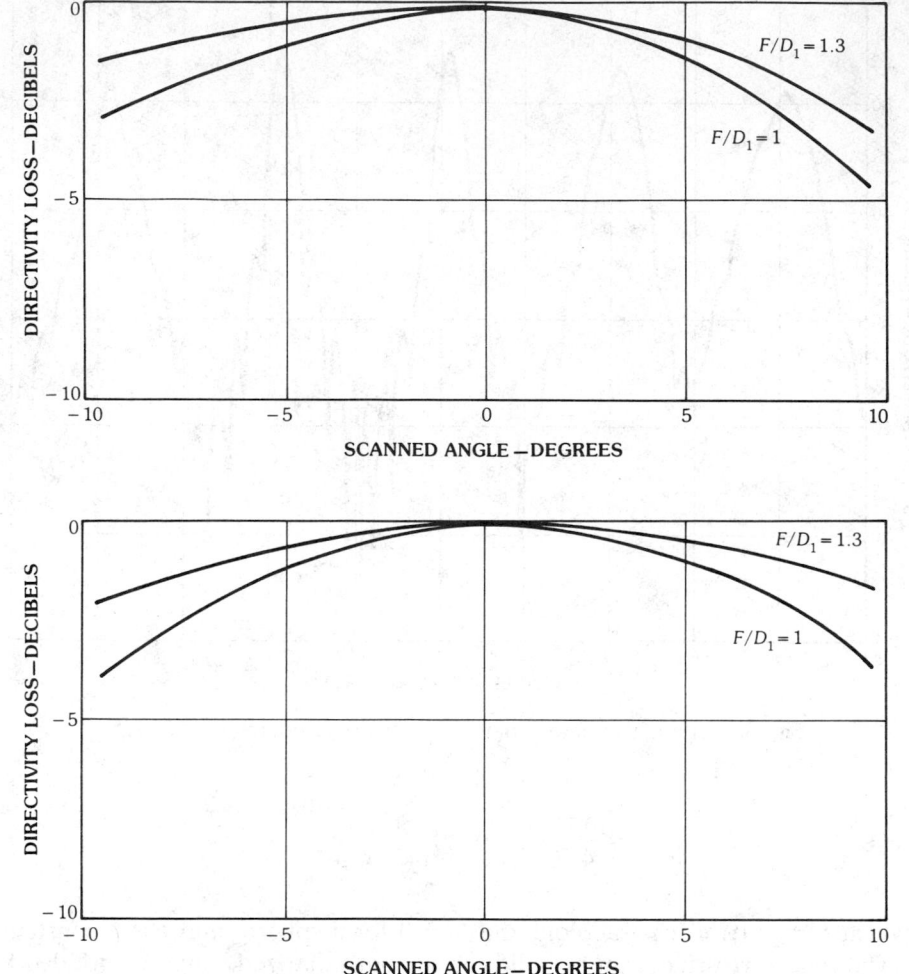

Fig. 9. Scan loss versus scanned angle. (*a*) The *xz* plane scan. (*b*) The *yz* plane scan.

length of the paraboloid. Therefore it can provide a better scan performance than a single parabolic reflector with the focal length *F*. Even though the equivalent principle no longer holds when the feed is not on-focus, a Cassegrain antenna is still a better optical system and compact in spacecraft integration.

A Gregorian antenna reflector surface has the paraboloid form for both the primary and secondary reflectors. Such an optical system forms an image at infinity and requires an additional optical aperture to focus the image at a finite distance. A shaped dual-reflector antenna is also used to improve the scan performance. A semishaped dual-reflector system, such as the surface, has two off-axis foci-bifocal reflectors. A generally shaped dual-reflector antenna is also under study. The idea is to give up the foci in order to reduce the most harmful aberrations for the scanning beams.

Focal Length—One of the most important parameters of a reflector antenna system is the ratio of focal length to parent aperture diameter (F/D), where the parent aperture is the largest circular aperture of the symmetrical paraboloid from which an offset reflector is derived. The larger the F/D is, the better performance for a scanning beam. However, a large F/D results in a small subtended angle for the feed element and the reflector and in turn requires a relatively large feed element. In a multibeam antenna application it implies that the total feed array size is also large. There is a trade-off of the number of feed elements versus the focal length. The aberration, which is worse for a shorter F/D reflector system, can be reduced and compensated by more feed elements. The improvement is due to the fact that more feed elements have more degrees of freedom for beam optimization. From the spacecraft viewpoint a larger F/D means bigger physical size, which is not cost-effective.

Offset Distance—The offset distance, defined as the distance from the reflector axis to the center of the reflector aperture, should be large enough to eliminate the feed blockage. The feed blockage increases the side lobe levels and the cross polarization as well.

Simple approximation formulas have been given [22] for designing an offset parabolic reflector with a circular aperture and feed array configurations. For given SL dB side lobe level of a single pattern, half-power beamwidth $2\theta_0$, maximum scan angle θ_3 with allowable GL dB scan loss, the spacing d between adjacent elements, and the offset distance h, one can determine the aperture taper and efficiency, reflector diameter, focal length, beam deviation factor, and element number in simple formulas. The formulas are given in Table 5 and in Figs. 10 and 11. They are reasonably accurate under the following constraints:

$0 < \Delta < 0.85$
$1 \leqslant GL \leqslant 5$
$F/D \leqslant 1.5$
$\theta_0 < 30°$
θ_3 can be approximated by $\tan \theta_3$ with reasonable accuracy

Surface Tolerance—The reflector surface errors due to the reflector surface distortion can be classified in two categories: random surface errors and deterministic surface errors. Random surface errors are due to manufacturing distortions, treated as a random error because of a lack of precision in constructing the reflector surface. Deterministic surface errors are due to the thermal distortion of a reflector surface and the surface deformation of an unfurlable or mesh reflector.

Classical antenna tolerance theory [23] shows that the effect of random surface distortion results in reduction in the peak gain and increase in side lobe levels. The perturbed radiation field due to the random surface distortion is given as

$$P_s = (2\pi c/\lambda)^2 \bar{\delta}^2 e^{-[(\pi c/\lambda)\sin\theta]^2} \qquad (15)$$

where

$$\bar{\delta}^2 = \left(4\pi a^2 \frac{\Delta z}{\lambda}\right)^2 \qquad (16a)$$

Table 5. Approximated Design Formulas for an Offset Parabolic Reflector

Design Parameters	Formulas	Type A Feed	Type B Feed
Aperture taper (Δ)	$\sum\limits_{n=0}^{3} a_n \left(\dfrac{SL^n}{10}\right)$	$\alpha_0 = -26.55$ $\alpha_1 = 35.17$ $\alpha_2 = -15.59$ $\alpha_3 = 2.37$	$\alpha_0 = -8.87$ $\alpha_1 = 9.32$ $\alpha_2 = -3.0$ $\alpha_3 = 0.32$
Aperture efficiency (η)	$\left(\sum\limits_{n=0}^{3} \beta_n \Delta^n\right) \times 100\%$	$\beta_0 = 1$ $\beta_1 = -0.026$ $\beta_2 = 0.039$ $\beta_3 = -0.263$	Numerical curves
Reflector diameter (D_1/λ)	$\dfrac{1}{\pi \sin \theta_1} \sum\limits_{n=0}^{3} \gamma_n \Delta^n$	$\gamma_0 = 1.609$ $\gamma_1 = 0.245$ $\gamma_2 = -0.259$ $\gamma_3 = 0.396$	$\gamma_0 = 1.61$ $\gamma_1 = 0.57$ $\gamma_2 = -1.43$ $\gamma_3 = 1.47$
Focal length (F)	$\dfrac{\pi(\sin \theta_3/\sin \theta_1)D}{190C \cos^{-1}[1 - (GL/5)]}$	$C = 1 - e^{-0.12\sqrt{D_1/\Lambda}}$	
Beam deviation factor	$\tau(1 - 0.72e^{-3.2(F/\tau D_1)})$	$\tau = \dfrac{\cos \theta_0 + \cos(\theta - \theta_L)}{1 + \cos(\theta_0 - \theta_L)}$ $\tan \theta_0 = \left(\dfrac{h}{F}\right)\left[1 - \dfrac{1}{4}\left(\dfrac{h}{F}\right)^2\right]^{-1}$ $\tan \theta_L = \left(\dfrac{h_1}{F}\right)\left[1 - \dfrac{1}{4}\left(\dfrac{h_1}{F}\right)^2\right]^{-1}$	
Number of feed elements (N)	Nearest integer of ($\theta_3/2\theta_2$)		

$$a = \frac{\tan^{-1}(D/4F)}{D/4F} \tag{16b}$$

and

c = length of correlation interval,

λ = wavelength,

D = antenna diameter,

θ = pattern angle,

$\bar{\delta}^2$ = phase error variance,

Δz = surface tolerance,

F = focal length

An improved and more accurate approach is to assume that the rms surface errors are given in a prescribed annular region. A closed-form expression for the

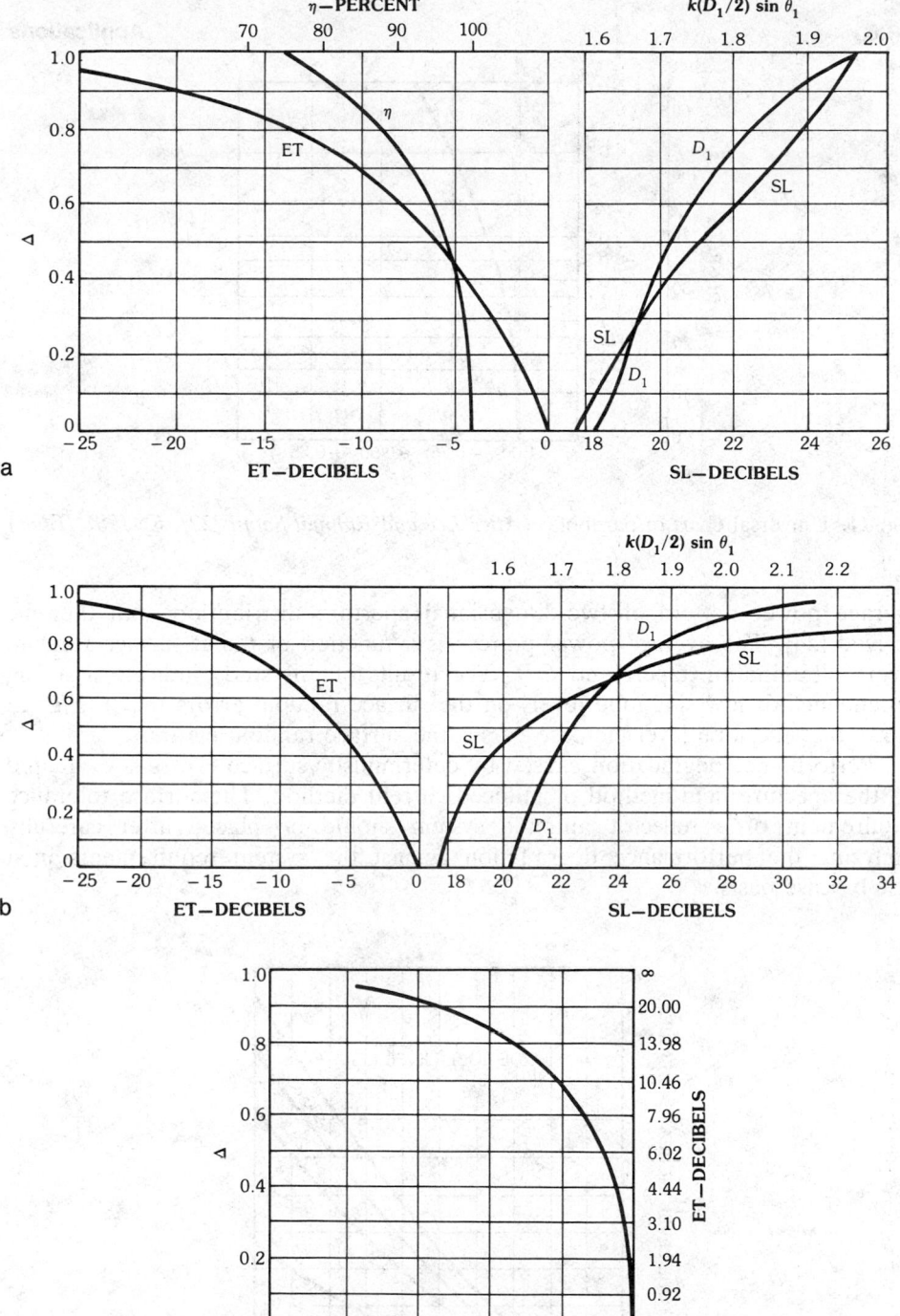

Fig. 10. Universal curves for designing an offset-fed parabolic reflector antenna. (*a*) Relation between Δ and reflector performance parameters when feed element is of type A. (*b*) Relation between Δ and reflector performance parameters when feed element is of type B. (*c*) Aperture efficiency versus edge taper when feed element is of type B. (*After Lee and Rahmat-Samii [22], © 1981 IEEE*)

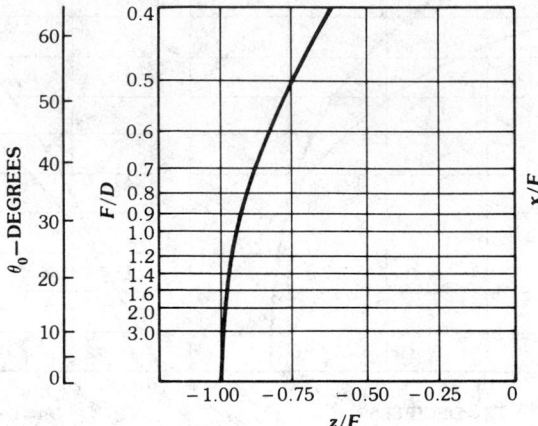

Fig. 11. Universal chart of parabola. (*After Lee and Rahmat-Samii [22], © 1981 IEEE*)

average power pattern of two-dimensional aperture distributions can then be derived [24]. The average power pattern is a function of radial surface random errors, illumination taper, and F/D. The results of this study indicate a strong dependence of low side lobe levels on the surface random errors (ϵ_{rms}). Fig. 12 shows the side lobe level increase versus the surface random errors.

Performance degradation caused by deterministic surface errors is evaluated via the aperture field method or induced current method. The surface tolerance requirement of a reflector antenna system should be placed after carefully analyzing the performance degradation against the system requirements in a case-by-case basis.

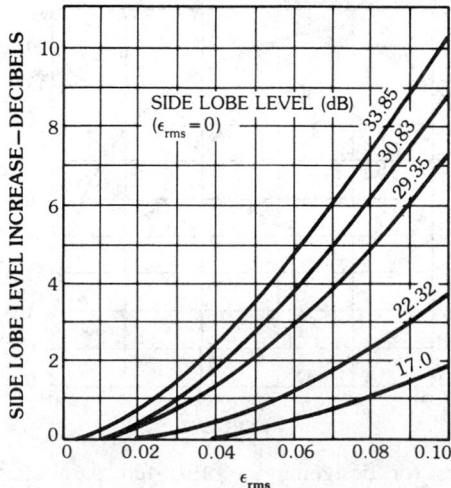

Fig. 12. Side lobe level increase versus random surface errors.

The Feed Array

The design of the feed array differs in scanning beam and fixed beam antenna systems. In a scanning beam antenna system the key design parameters are gain, gain ripple, side lobe, and cross-polarization levels. Gain ripple is due to the crossover of two discrete scanning beams generated by two adjacent feed elements. Gain and side lobe levels are controlled by the optics and the feed illumination taper. In order to improve gain, a larger feed element is preferred to reduce spillover energy. A larger feed element, however, results in a wider feed separation and hence lowers the crossover point of the two adjacent scanning beams. This in turn increases the gain ripple. In order to maintain high gain and reduce gain ripple at the same time, overlapping cluster feed is used. A single beam is generated by a number of feed elements. Some feed elements are used for both of the two adjacent scanning beams. The other approach is to use a continuous scanning system. Variable phase shifters and variable power dividers are used to control the beam scanning continuously from one position to another position. Fig. 13 presents three basic approaches of the scanning beam system.

In a fixed beam antenna system, not only the peak gain, gain ripple, and side lobe level, but also the minimum gain in the coverage area, interference and transmitter power between beams, and the complexity of BFNs have to be considered.

Design Parameters—Key design parameters are the location of feed elements, number of feed array elements, lattice spacing, feed array configuration, radiating elements, and materials.

Location of Feed Elements—From the optical viewpoint it is desirable to place the feed elements on the best focal surface. In general, the best focus surface is not necessarily planar and may be curved. A curved feed array surface introduces design difficulties, such as increased mutual coupling, depolarization, and unavailable design-analytical tools. Trade-offs among the achieved performance, feed array complexity, and design difficulty have to be conducted.

Number of Feed Elements and Lattice Spacing—For a fixed F/D, there is a minimum number of elements required for a given coverage area. The number of elements is inversely proportional to element spacing. The element spacing affects the cross polarization because of mutual coupling. From the beam optimization viewpoint, a greater number of feed elements improves beam shaping and provides higher spatial isolation but results in a more complex beam-forming network.

Grating lobes associated with lattice spacing should be considered. The grating lobes should not fall into the coverage area in a phased array antenna system or fall into the optical aperture to degrade the antenna performance.

Feed Array Configuration—The feed array can be arranged in a periodic or nonperiodic configuration. A nonuniform feed configuration is often used in a multibeam antenna system when the number of feed elements is small. Different feed spacing and different feed size are used for better beam shaping to improve the gain at the edge of coverage.

Radiating Elements—The choice of radiating elements is determined in a

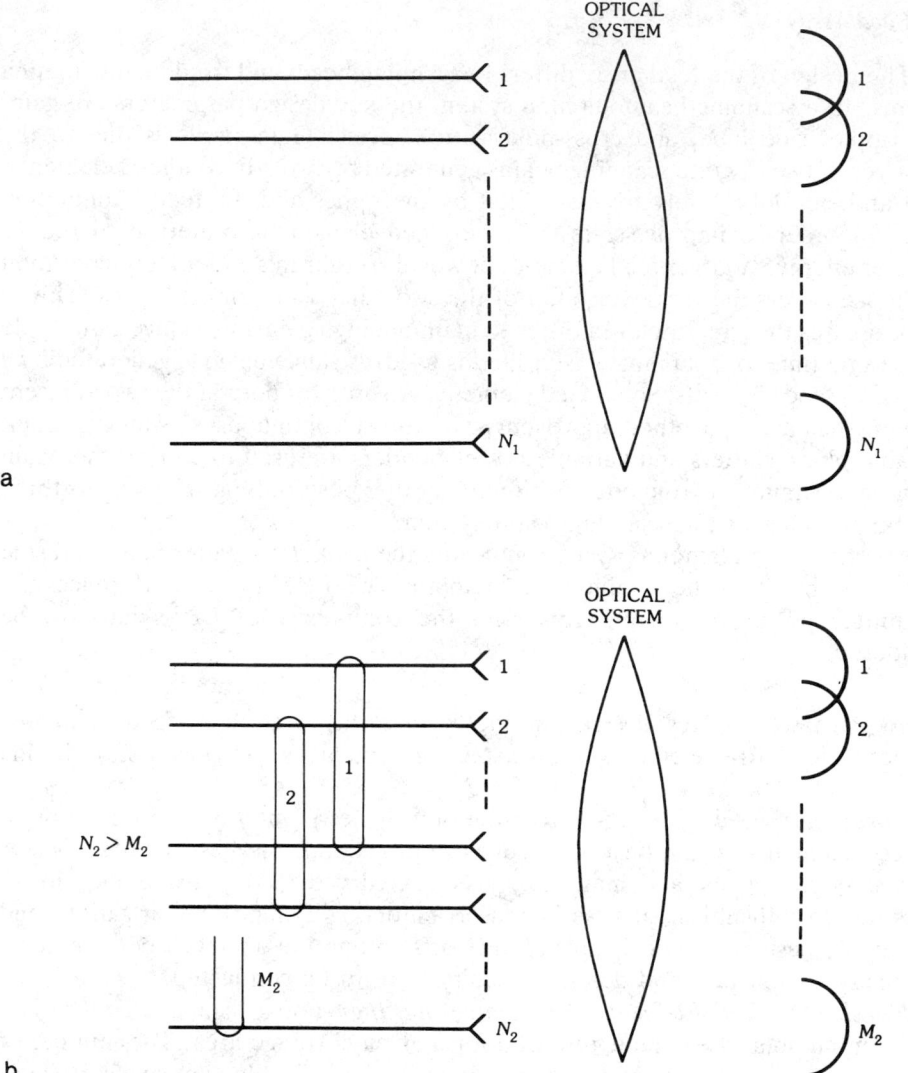

Fig. 13. Three approaches of scanning beam antenna systems. (*a*) Scanning beam feed array. (*b*) Overlapping scanning beam feed array. (*c*) Overlapping scanning beam feed array (variable power dividers).

similar way as for the phased array given under "Types of Element" in the phased array section.

Materials—Feed elements have typically been fabricated using (*a*) cast of machined aluminum alloys, (*b*) electroformed copper or nickel, or (*c*) have been layed up using composite materials such as graphite-epoxy.

Thermal distortion has to be considered in the extreme temperature cycling environment. Among three candidates, copper-clad graphite-cpoxy is the most stable material over a wide temperature range. Weight is also a major design

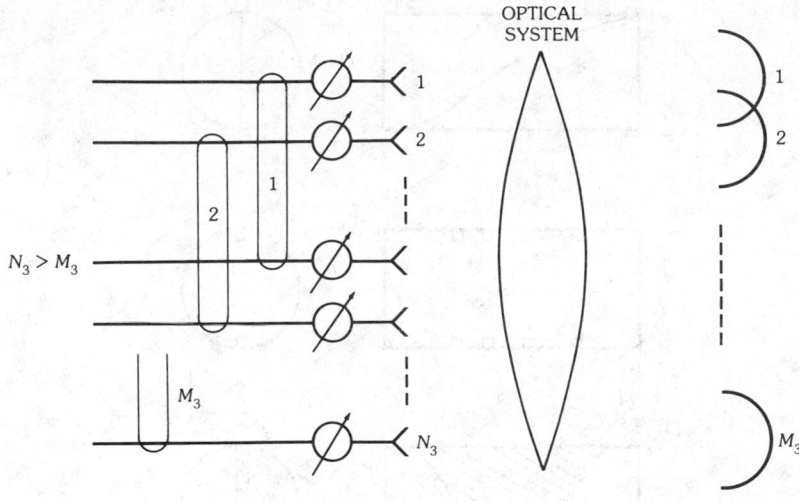

Fig. 13, *continued.*

criterion in feed design. Permanent-mold cast aluminium feed horns have the comparatively thick wall required for manufacturing. Electroformed nickel may provide the required thermal stability and can be accurately manufactured with very thin walls. However, it is heavy because of nickel's high density. The graphite-epoxy feed horn provides the lightest weight.

Polarizer—The next key component of a feed array is a polarizer that is required to provide for circular polarization. The choice of polarizer is decided by performance specifications such as frequency bandwidth, axial ratio, insertion loss, network loss, isolation between ports, and compactness in feed array integration. Fig. 14 shows the most commonly used polarizers: septum, multiprobe launcher, pin, dielectric-slab, pinched-guide, and Meanderline. A Meanderline polarizer uses a square-wave printed-circuit pattern to provide reactive loading to the orthogonal linear component of an electric field. By orienting the Meanderline print at 45° in space with respect to the incident field, one field component, parallel to the printed pattern, is inductively loaded and its orthogonal counterpart is capacitively loaded. This causes a differential electrical phase shift between two orthogonal fields. Multiple sheets are used to provide the required 90° phase differential of two orthogonal incident waves to and from circularly polarized waves. It can be designed to operate in multifrequency bands and be placed in front of a feed array to form a compact system. It has, however, a higher insertion loss and axial ratio than the other types of polarizers.

Beam-Forming Networks

The beam-forming networks (BFNs) are *n*-to-*m* port networks and their purpose is to interconnect the *n* input ports to the individual *m* ports with required amplitudes and phases. The BFNs can be divided into two categories: scanned and fixed.

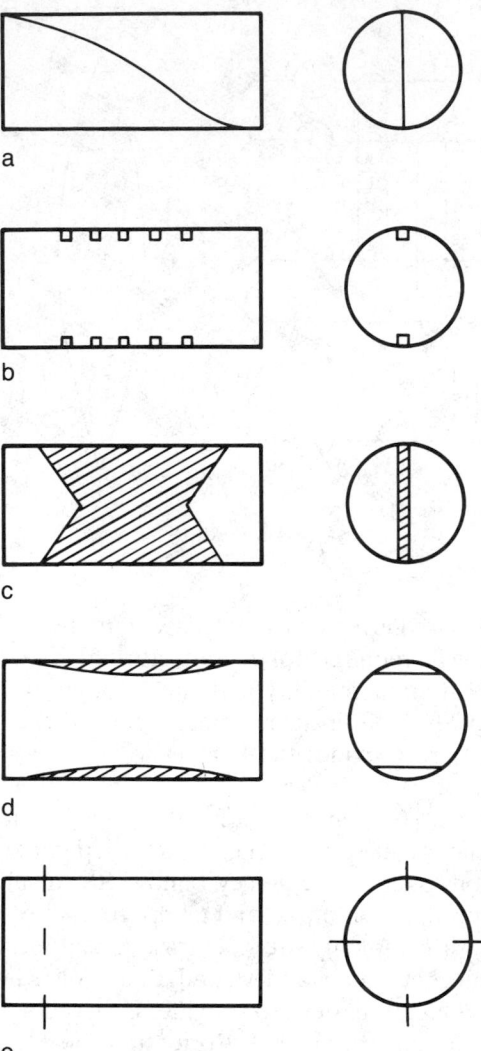

Fig. 14. Types of polarizers. (*a*) Septum polarizer. (*b*) Pin polarizer. (*c*) Dielectric-slab polarizer. (*d*) Pinched-guide polarizer. (*e*) Multiprobe launcher polarizer.

Scanned Beam-Forming Networks—Classical methods of beam steering used either variable phase shifters or interconnecting switches to select a particular individual beam. To these methods can be added scanning with variable amplitude networks, scanning with both variable amplitude and phase networks, and scanning space-fed networks (such as lenses).

Switch Beam Networks—Switch beam networks are composed of many fixed networks, each with an output port connected to a terminal, that represent a unique beam position as shown in Fig. 15.

The BFN for nonoverlapping beam clusters is simple and straightforward. A switch matrix, which allows the selection of one of N available beams formed by

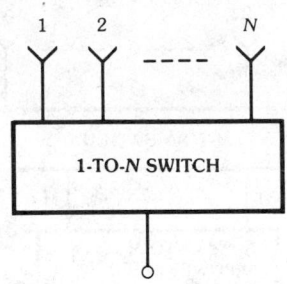

Fig. 15. Nonoverlapping switch beam-forming network.

an array, is shown in Fig. 16. For an overlapping beam the BFNs become more complex since a particular feed element may have to be shared with two or more beams as shown in Fig. 17.

Other variations of switched circuits are Butler and Blass matrices. The Butler matrix (Fig. 18) consists of identical junctions (3-dB hybrids) and fixed phase shifters. This beam-forming technique can be used for planar arrays by combining columns of antenna elements in one set of matrices, and combining the outputs of the column matrices in a group of row matrices. Beams formed by the total matrix may be selected through an auxiliary switching matrix. Such a configuration would be appropriate for feeding a near-field plane wave type of optical system. Characteristics of this configuration are as follows:

1. The number of different beams that can be formed is equal to N, and the number of antenna elements (or clusters) must generally be equal to a power of 2.

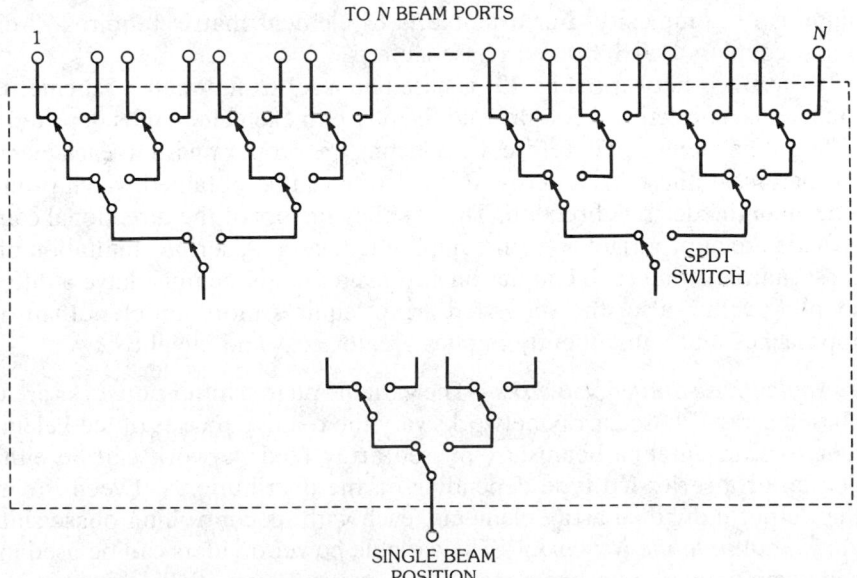

Fig. 16. Multiport switched beam-forming network.

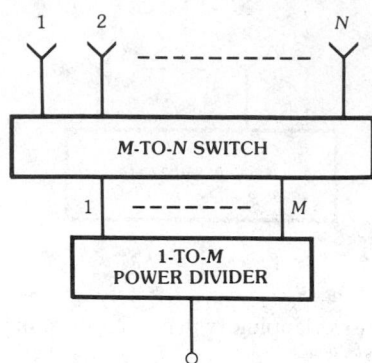

Fig. 17. Overlapping switch beam-forming network.

2. The BFN can be theoretically lossless and matched. The insertion loss of the matrix will be determined by the rf losses of the matrix components.

3. The bandwidth is limited by the components, such as phase shifters and directional couplers. However, the positions of the antenna beam peaks will vary with frequency. The amount of variation with frequency is a function of beam location.

4. The antenna array illumination is uniform. It is possible to achieve $(\text{cosine})^n$ aperture distribution using the weighted addition of selected beams. For example, it is possible to obtain a cosine distribution, but the beam crossover level is 9.5 dB below the peaks. Thus, for lossless beams, the aperture distributions are limited to a uniform distribution that has relatively high first side lobe or tapered distribution that has far-field patterns with undesirably low crossover levels. Another major drawback of this beam-forming technique is its complexity. For example, a 64-element matrix requires 92 directional couplers and 160 fixed phase shifters.

The Blass network (Fig. 19) is based on a series-fed matrix scheme. Each junction is a directional coupler and the phase to each element is depicted by a different line length. Therefore, by selecting the proper phase to each element, a progressive phase front across the aperture can be obtained which produces a beam in the desired direction. The coupling factors of the directional coupler provide the appropriate aperture amplitude tapers. A serious limitation of the Blass matrix is that each coupler on any given radial line must have a different coupling value; also, the series-fed array requires more couplers than other approaches and consequently implies greater cost and weight.

Variable Phase Shifter Networks—The variable phase shifter networks are used in a phased array. Phased array networks vary the relative phases of feed elements in order to scan antenna beams. A phased array feed network can be either a parallel-fed or a series-fed type depending on the distribution between the input and the output individual array elements, each with its controlling phase shifter.

Variable Amplitude Networks—The variable power dividers can be used in the scanned beam networks for beam steering. A switch is essentially operated in one of the special states of a variable power divider in which the power is either on or

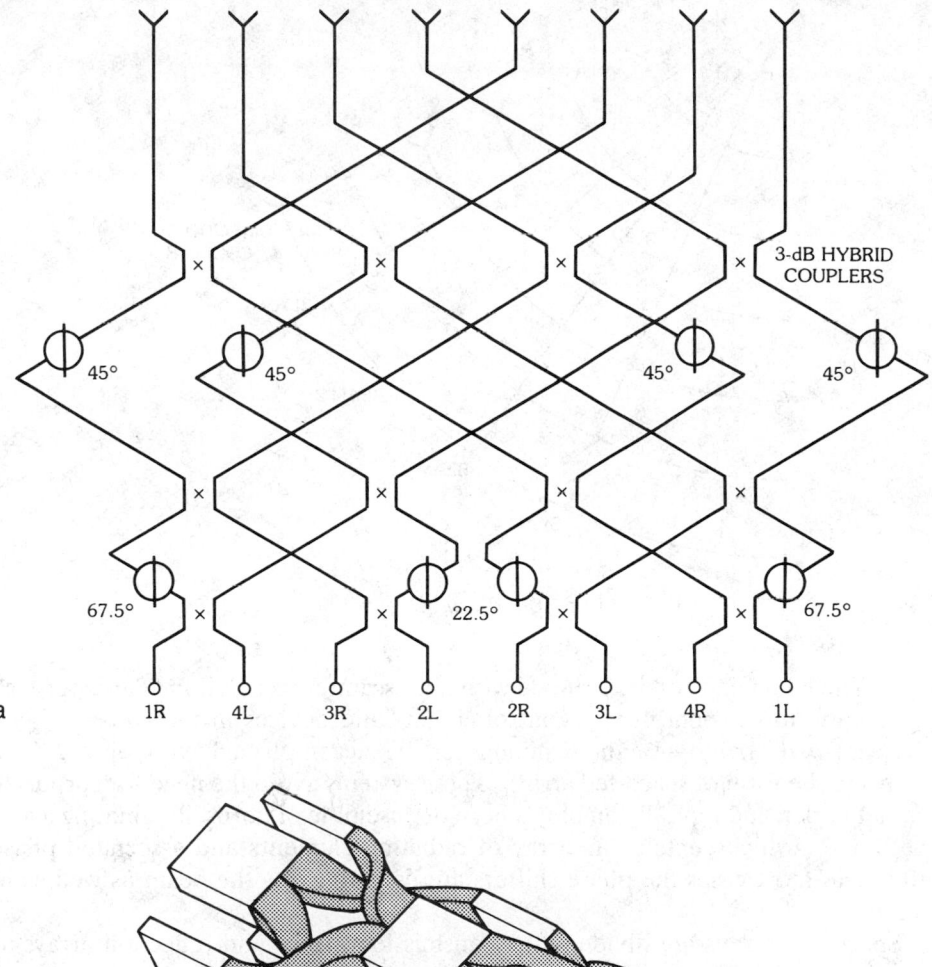

Fig. 18. Butler matrix. (*a*) Eight-element Butler matrix. (*b*) Eight-port Butler matrix in waveguide.

off. Consequently, an antenna excited by a BFN using variable power dividers can produce more flexible coverage than the BFN using switches. However, it requires more sophisticated control circuits than do switch networks.

Variable Phase and Amplitude Networks—The BFNs using both variable phase and amplitude networks provide the maximum degree of beam control. The scanned beam can be optimized at any position by controlling array amplitude and

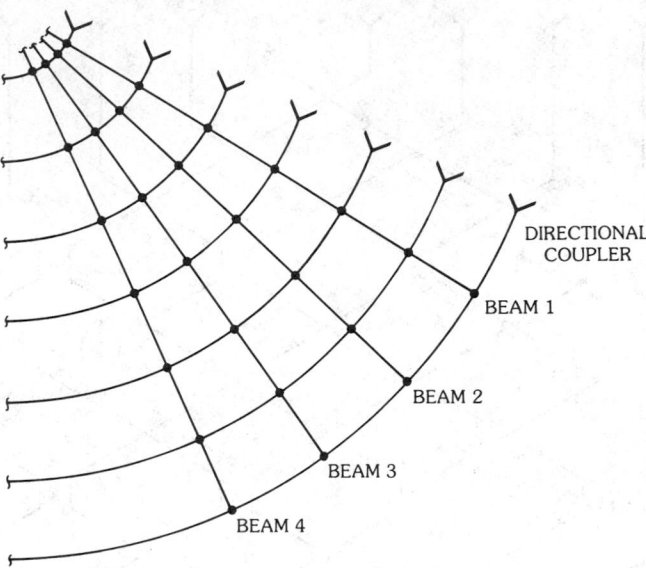

DIRECTIONAL
COUPLER

BEAM 1

BEAM 2

BEAM 3

BEAM 4

Fig. 19. Blass network.

phase. The beam can also be scanned with finer scanning resolution. This approach requires the most sophisticated control circuits and devices in the BFNs.

Space-Fed Arrays—Beam scanning can be accomplished with optical feed systems in the form of space-fed arrays. These systems avoid the need for corporate feed networks, and typically employ a horn or ensemble of horns illuminating a lens type device, which contains an array of radiating elements and associated phase shifters. In most cases the phase shifters are used to focus the beam as well as to steer it.

Space-fed arrays are divided into transmission arrays and reflection arrays as shown in Fig. 20. The transmission and reflection arrays require a combination of lens and array design techniques because they have the properties of conventional lenses, while the secondary radiator functions as a phase-controlled array.

Fixed Beam-Forming Networks—A fixed beam is formed by feeding one or more feed elements, the number required depending on the shape of the beam, gain, and/or side lobe levels. The BFNs can be divided into three categories: constrained, unconstrained, and the hybrid of constrained and unconstrained networks. The constrained BFNs use transmission lines exclusively to transfer energy from input port to output ports. The unconstrained BFNs use free space as a transmission medium. A hybrid of these two types can also be used where energy is constrained in one dimension but not in the other.

The advantages of the constrained BFNs are the accurate controls of amplitude and phase. The physical size can be a small fraction of the overall antenna system. The networks must utilize components which, even though they are separately well matched, are difficult and practically impossible to match to the entire network after assembly. The advantage of the unconstrained BFNs is simple. The physical

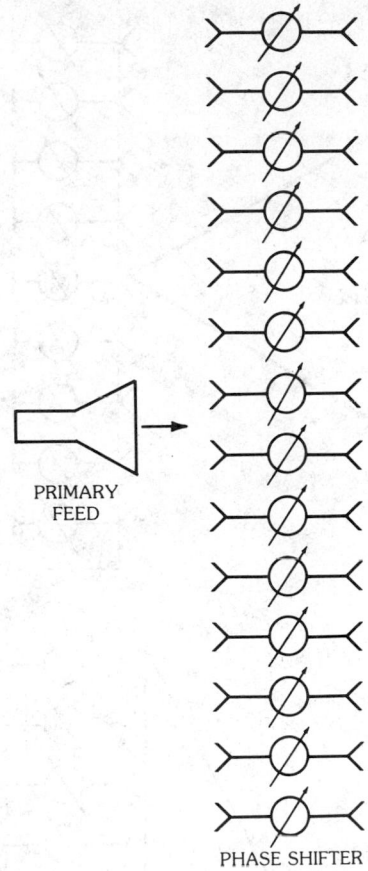

PRIMARY
FEED

PHASE SHIFTER

a

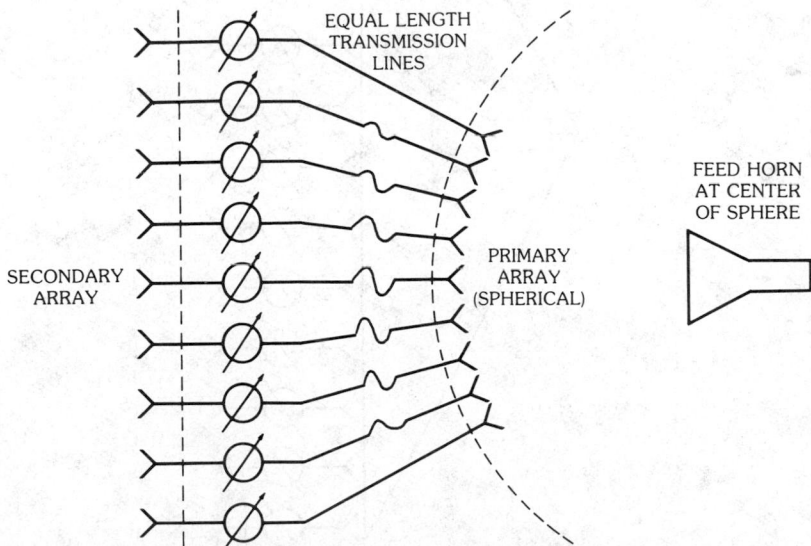

EQUAL LENGTH
TRANSMISSION
LINES

SECONDARY
ARRAY

PRIMARY
ARRAY
(SPHERICAL)

FEED HORN
AT CENTER
OF SPHERE

b

Fig. 20. Types of space-fed beam-forming networks. (*a*) Variable-phase planar-planar transmission array. (*b*) Variable-phase spherical-planar transmission array. (*c*) Variable-phase reflection array. (*d*) Optical parallel-plate phase-scanned array.

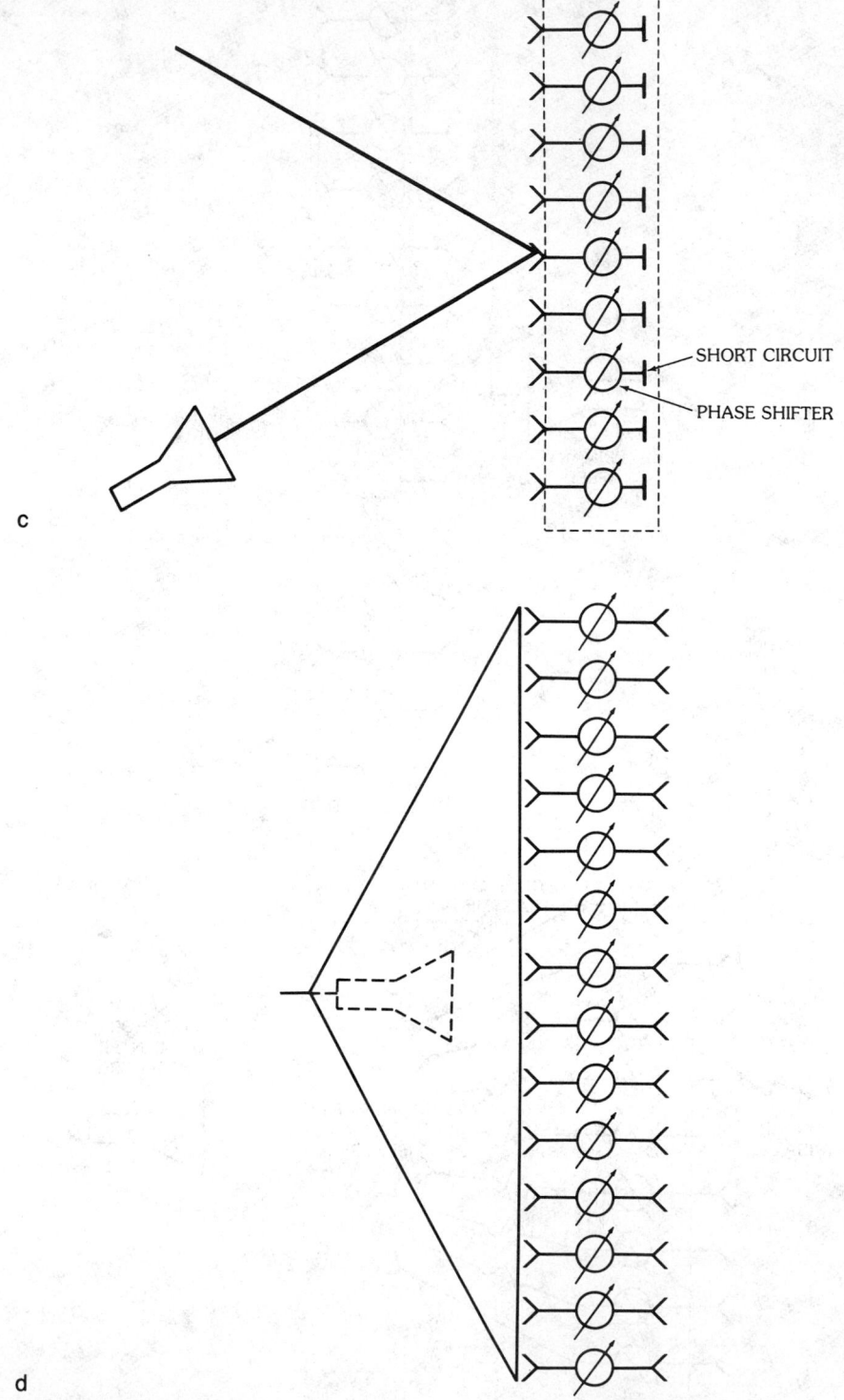

SHORT CIRCUIT

PHASE SHIFTER

c

d

Fig. 20, *continued.*

size is at least half the overall antenna system. However, the networks provide less control of amplitude and phase. The unconstrained feed system may be practical if a large number of components is required in a large array.

Both scanned and fixed BFNs consist of components and feed network media. Components that are generally used include switches, phase shifters, power dividers, hybrids, diplexers, variable phase shifters, variable power dividers, rotary joints, and frequency-selective surfaces. Low insertion loss is extremely important in the BFN for efficient antenna performance. The choice of feed network media is determined by the rf performance requirement, frequency band operation, size, weight, reliability, thermal stability, and manufacturability.

Design Considerations—Selection of an optimum BFN for a given antenna system involves the trade-offs of types of components and feed network media. It also depends on the operating frequency bands. Striplines are often used at C-band or lower frequencies and waveguide transmission lines are used at X-band or higher frequencies. Various components, such as power dividers, switches, variable phase shifters and power dividers, and amplifiers, have to be chosen based on the trade-off of the method of network reconfigurability. The selection of the most suitable feed network transmission media is determined by the insertion loss and the availability of the appropriate components for transmission media.

The following are general trade-off criteria:

Allowable insertion loss
Compactness and weight
Design maturity
Power-handling capacity
Manufacturability
Thermal stability

The first of these criteria is important since the highest achievable performance is the ultimate goal of an efficient antenna system. As a satellite is tightly packed, the BFN must occupy the smallest area as possible. The design must be mature so that essential design information is available. The BFN must also be able to handle the maximum power transferred, must be fabricated with provided tolerances, and must be stable in the thermal environment.

Scanned Beam-Forming Network—The key components are switches, variable phase shifters, and variable power dividers. The design parameters are vswr, insertion loss, isolation, switching time, and size.

1. *Switches*—There are three types of switches: electromechanical, ferrite, and diode. Electromechanical switches offer the lowest insertion loss of the various switch types available with a reasonable degree of reliability. Manufacturers often claim lifetimes of 10^6 actuations or more. Both coaxial and waveguide designs are available, but the latter are usually large and heavy. Actuating mechanisms are usually magnetically controlled, using solenoids to produce mechanical motion of one or more contact bars or levers and permanent magnets to hold them in position to form a latching device. Use of dielectrics to prevent multipacting in many critical

areas is precluded by the mechanical motion. High-power designs are therefore either hermetically sealed or vented to vacuum for space applications.

Single-pole, double-throw (spdt) ferrite switches are available based on the switching circulator principle. The direction of circulation around a circulator structure depends on the direction of the applied magnetic bias field. If this direction is reversed, output power is switched between two alternate output ports. Isolation of 20 to 25 dB is achievable over 10-percent bandwidth, and insertion loss of stripline-type 4- and 6-GHz switches may be as low as 0.15 dB. Cascaded junctions could be used if greater isolation is desired. The switch times are generally under 1 μs. It has been demonstrated that the switch time can be in the range of 200 to 400 ns. Switching energy is in the range of 50 to 100 μJ per switch event. Satisfactory operation over temperature ranges of −40°C to 90°C have been demonstrated.

Microwave diode switches are attractive for their compactness, light weight, moderate control power requirements, and high switching speeds. In addition, multiple-port designs are feasible. Their losses are considerably higher than those of the other two switch candidates, especially at high frequencies. The basic switching diode is a pin junction, which can approximate a short circuit or a small lumped capacitance in the forward or reverse bias states. Such an element can be used either in series or in shunt with a transmission line to form a basic spst switch. Tuning structures can be incorporated to match out diode reactive elements over reasonable bandwidths, and multiple diodes can be incorporated to meet enhanced isolation requirements. High peak and average powers can be handled using diodes with adequate heat dissipation construction and peak inverse voltages in the kilovolt range. Harmonic and intermodulation distortion are possible at higher power levels. Insertion losses are generally 2 dB for 35-dB isolation in the 18-GHz frequency band.

2. *Variable Phase Shifters*—There are three main types of variable phase shifters: ferrite, diode, and electromechanical. Design parameters are: frequency band, insertion loss, phase range, phase steps, phase slope, maximum peak-to-peak phase variation versus frequency band, power-handling capacity, temperature range, size, and weight. The phase shift of a ferrite phase shifter is dependent on the magnetization of the material. Two forms of phase control are analog and digital. Ferrite phase shifters can be divided into the following categories [25]:

Waveguide nonreciprocal
Reciprocal dual-mode
Helical
Reciprocal and nonreciprocal strip transmission line
Microstrip
Latching Reggia-Spencer

Of the types listed, the waveguide nonreciprocal phaser and reciprocal dual-mode phasers have been proved to be electrically superior to the others.

The basic nonreciprocal waveguide phase shifter consists of a toroidal ferrite, rectangular in cross section, located in the center of a rectangular waveguide as shown in Fig. 21. The energy is phase-shifted in the ferrite section. The ferrite

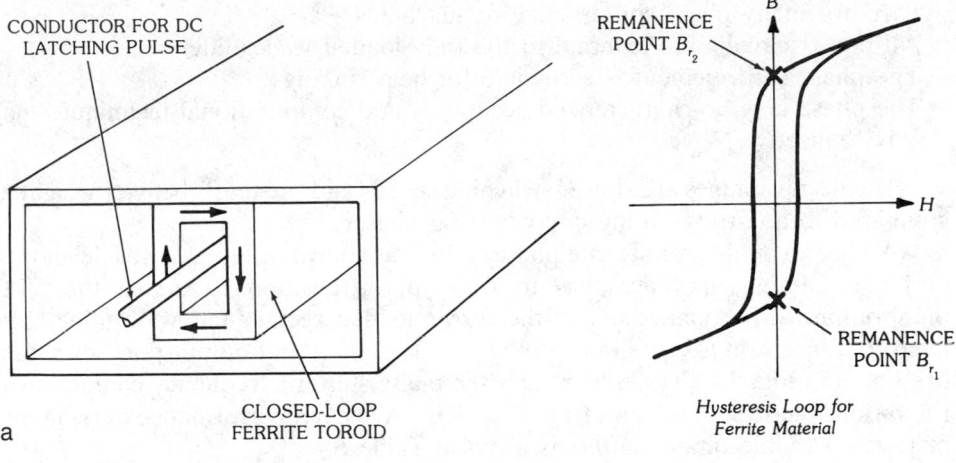

a

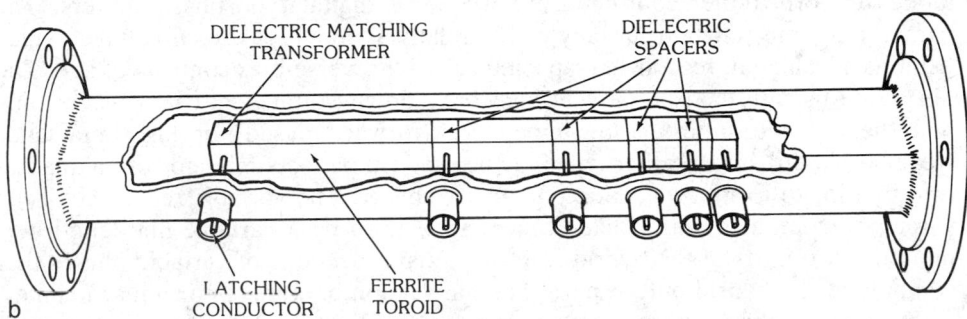

b

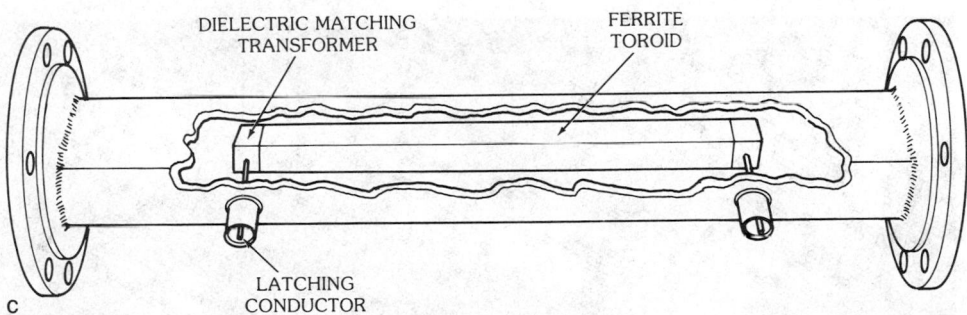

c

Fig. 21. Nonreciprocal toroidal-waveguide phase shifter. (*a*) Cross section and *B*/*H* curve. (*b*) With dielectric spacers. (*c*) Without dielectric spacers. (*After Whicker [25], © 1974; reprinted with permission of Artech House*)

section is a loaded waveguide in which a circular or square ferrimagnetic rod is metallized. Good electrical performance has been observed in the toroidal waveguide phase shifters.

The desirable features of a dual-mode phase shifter are as follows:

Parts for many phase shifters may be machines

All non-rf circuits are external to the fully loaded waveguide

The phase shift element is accessible for heat sinking

The phase may be characterized accurately and computational techniques may be utilized

The disadvantages are slower switching speeds and the unit's heavier weight as compared to a corresponding nonreciprocal phaser.

A *C*-band dual-toroid ferrite phase shifter, as shown in Fig. 22, was developed [26]. The design approach was to develop a transition to match the 50-Ω input/output SMA connection to the ferrite-loaded rectangular waveguide. The measured insertion loss is shown in Fig. 23. The input and output port vswr's are presented in Fig. 24. Fig. 25 illustrates the phase shifter's frequency characteristic for various phase shift settings from 0° to 360°. Achieved performance of two types of ferrite variable phase shifters is given in Table 6.

Diode phase shifters can be constructed using either varactor or pin switching diodes to form either continuously variable or digital type phase shifters. The varactor is a pin junction, usually with a relatively low breakdown voltage, which exhibits a change in reactance (capacitance) with reverse-bias control voltage. The pin diode usually has a relatively high breakdown voltage, but it is useful only in either of two states: a low-impedance (forward-biased) or high-impedance (reverse-biased) state. Either type may be used (in pairs, combined with a quadrature hybrid) to form a reflection type phase shifter. The variable reactance of the varactors, combined with a fixed tuner, appears to be a variable line length terminating two ports of the hybrid, causing equal reflections of variable phases that combine at the hybrid output port. The pin version operates in a similar manner, with the phase states alternating between two fixed values differing approximately

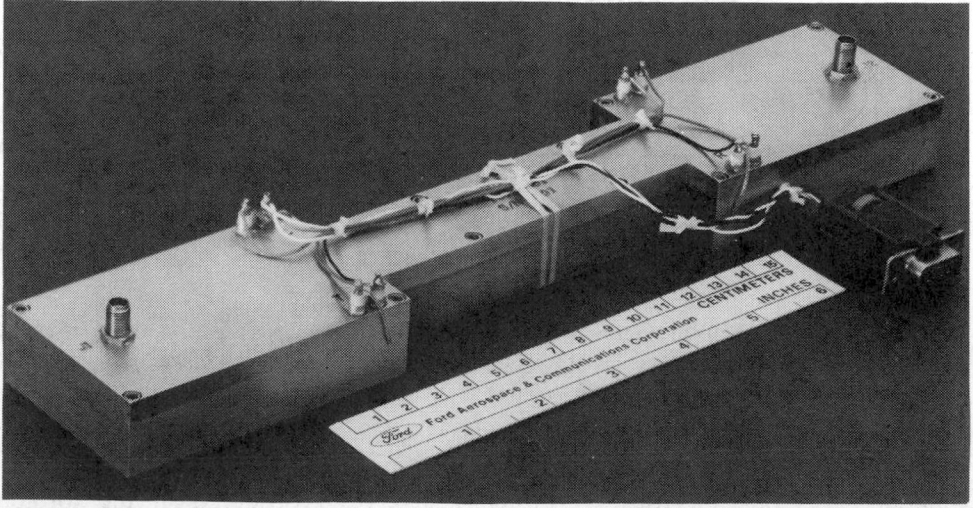

Fig. 22. *C*-band dual-toroid phase shifter. (*Courtesy Ford Aerospace and Communications Corp.*)

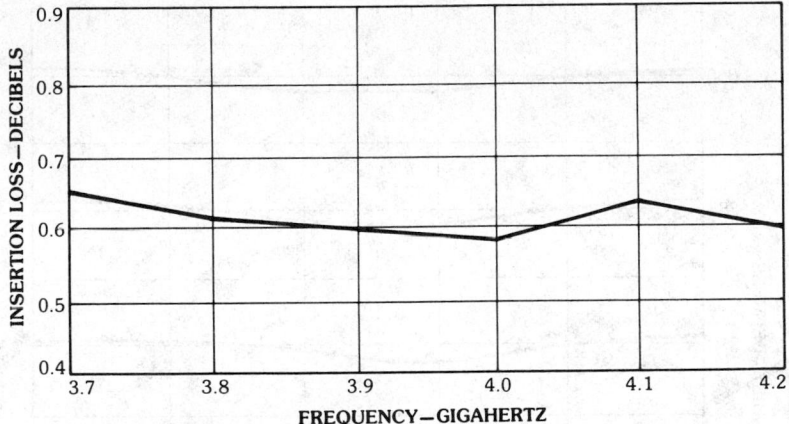

Fig. 23. *C*-band dual-toroid phase-shifter insertion loss versus frequency. (*After Smith, Mathews, and Boyd [26], © 1982 AIAA; reprinted with permission*)

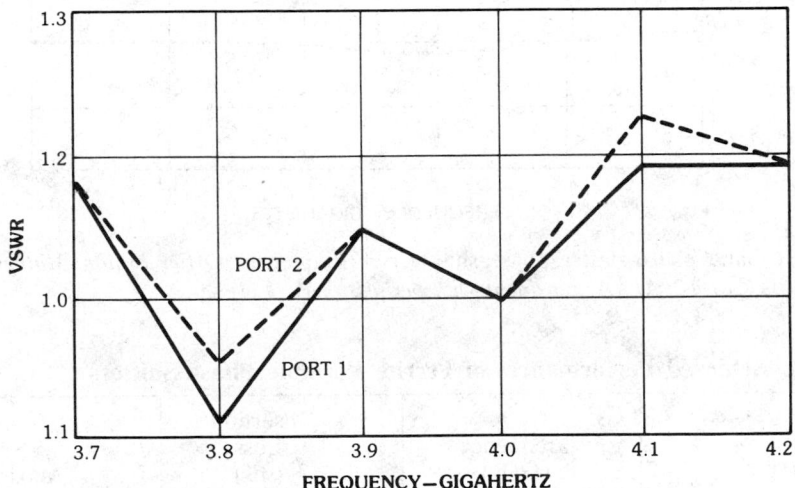

Fig. 24. *C*-band phase-shifter vswr versus frequency. (*After Smith, Mathews, and Boyd [26], © 1982 AIAA; reprinted with permission*)

by twice the line length from the diodes to a pair of fixed short circuits. The pin devices are thus used in combinations of "bits" representing differential phases of 108°, 90°, 45°, etc., depending on the resolution required.

Diode designs are amenable to low-cost quantity production using rf printed circuits and simple transistor drives. The major deficiency of diode phase shifters at microwave frequencies is their relatively high loss compared to other devices. These losses generally increase with the number of bits in the circuit, thus penalizing accuracy of setting. They are attractive, however, in the active antenna system in which the loss can be compensated by the amplifiers.

The phase shift of an electromechanical phase shifter relies on the mechanical

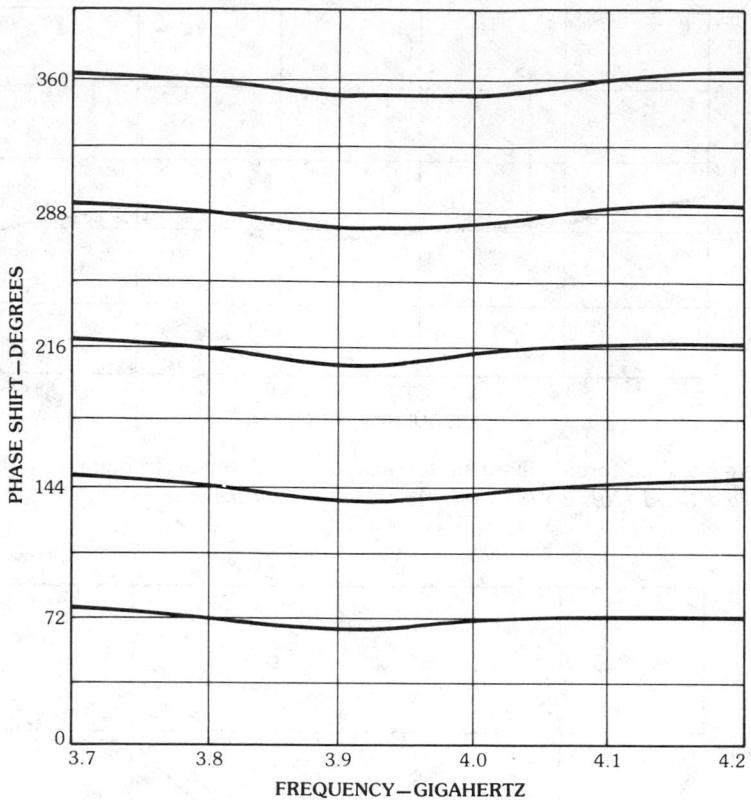

Fig. 25. *C*-band phase-shifter phase shift versus frequency. (*After Smith, Mathews, and Boyd [26],* © *1982 AIAA; reprinted with permission*)

Table 6. Achieved Performances of Ferrite Variable Phase Shifters

Type	Frequency (GHz)	Insertion Loss (dB)	VSWR Maximum
Toroidal	5.4–5.9	0.8	1.2:1
	3.2–4.2	0.7	1.2:1
Dual-mode	8.0–9.0	0.9	1.2:1
	9.0–10.0	0.75	1.2:1
	2.9–3.25	1.0	1.2:1
	5.275–5.725	0.9	1.25:1

movement of a physical component in the device to change the phase. Generally, a line length as in "line stretcher" is changed or a dielectric portion of the rf circuit between regions of high and low electric field strength is moved to change the effective phase velocity of the circuit. All electromechanical phase shifters are bulky and heavy because they require electric drive motors. Many mechanically moving parts severely limit the reliability of the component. Even the low insertion loss cannot compensate for the above disadvantages. Thus, electromechanical phase shifters are not used in satellite applications.

3. *Variable Power Dividers* — There are three main types of variable power dividers: ferrite, diode, and electromechanical. Design parameters are insertion loss, setting accuracy, temperature stability, power handling capacity, isolation to off-port, reset characteristics, vswr, phase characteristics, intermodulation products, gain slope, group delay, size, weight, and reliability.

(*a*) *Ferrite variable power dividers*—The ferrite variable power dividers may be designed for either continuously driven operation or a latched operation using remnant magnetization to maintain set states. The variable power divider may be constructed in waveguide, stripline, or some other transmission medium in the integration of the BFN. The most widely used ferrite variable power dividers are

dual-hybrid,
driven variable polarizer, and
latching Faraday rotator

The basic dual-hybrid variable power divider network, as shown in Fig. 26, consists of a 3-dB 180° hybrid, two variable phase shifters, and a 3-dB 90° hybrid. The output signals at ports 1 and 2 can be shown by

$$V_1 = \cos\left(\frac{\phi_1 - \phi_2}{2} + \frac{\pi}{4}\right) e^{j[(\phi_1 + \phi_2)2 + \phi\pi/4]} \qquad (17a)$$

$$V_2 = \sin\left(\frac{\phi_1 - \phi_2}{2} + \frac{\pi}{4}\right) e^{j[(\phi_1 + \phi_2)/2 + \pi/4]} \qquad (17b)$$

Latching toroidal phase shifters are used to provide the insertion phase shift values ϕ_1 and ϕ_2.

A *C*-band VPD as shown in Fig. 27 was developed [26]. The insertion loss measured for the case of equal power division ratio is shown in Fig. 28. The measured vswr over the 3.7- to 4.2-GHz band is presented in Fig. 29.

The driven-ferrite variable-polarizer variable power divider as shown in Fig. 30 consists of two main sections: a magnetically variable differential phase shift section and a septum polarizer. An incoming signal, which is linearly polarized, is converted into an elliptically polarized wave by the phase shift section. It is then

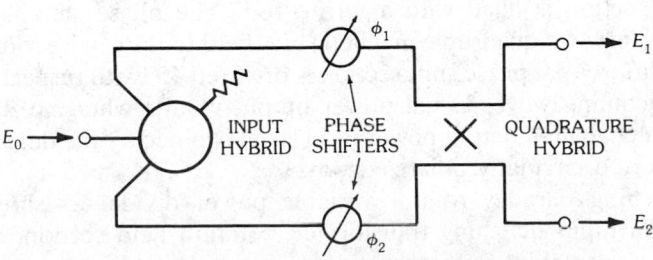

Fig. 26. Dual-hybrid variable power divider circuit.

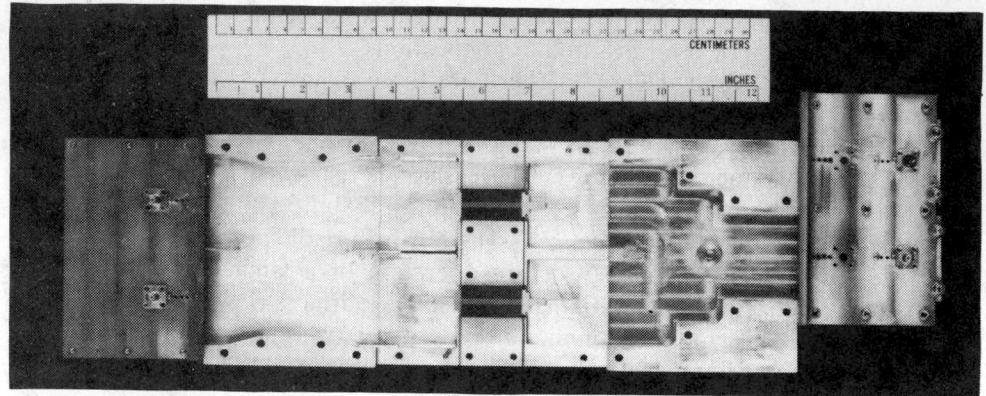

Fig. 27. *C*-band coupler/phase-shifter variable power divider. (*Courtesy Ford Aerospace and Communications Corp.*)

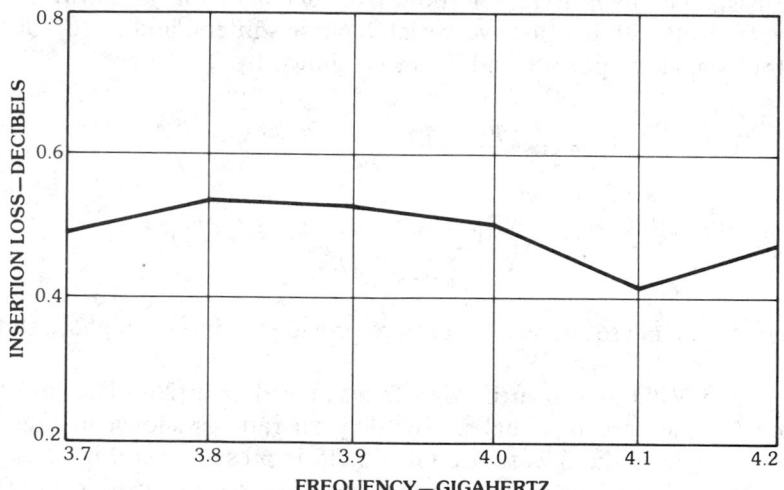

Fig. 28. Coupler/phase-shifter variable power divider insertion loss versus frequency. (*After Smith, Mathews, and Boyd [26], © 1982 AIAA; reprinted with permission*)

converted into two linearly polarized components in separate output ports. The phase shift section is filled with a ferrite rod. The phase shift is controlled by applying a transverse quadruple magnetic bias field to the ferrite rod. The principal axis of the differential phase shift section is oriented 45° with respect to the linearly polarized incoming wave. Equal power output results when no drive current is applied. The change of output power ratio is determined by the degree of ellipticity of the converted circularly polarized waves.

The latching Faraday rotator variable power divider as shown in Fig. 31 consists of an input matching transformer, rotation field section, and an output matching transformer into an orthomode junction as two output ports. The rotation field section is a magnetic yoke that includes a ferrite rod. The input signal is

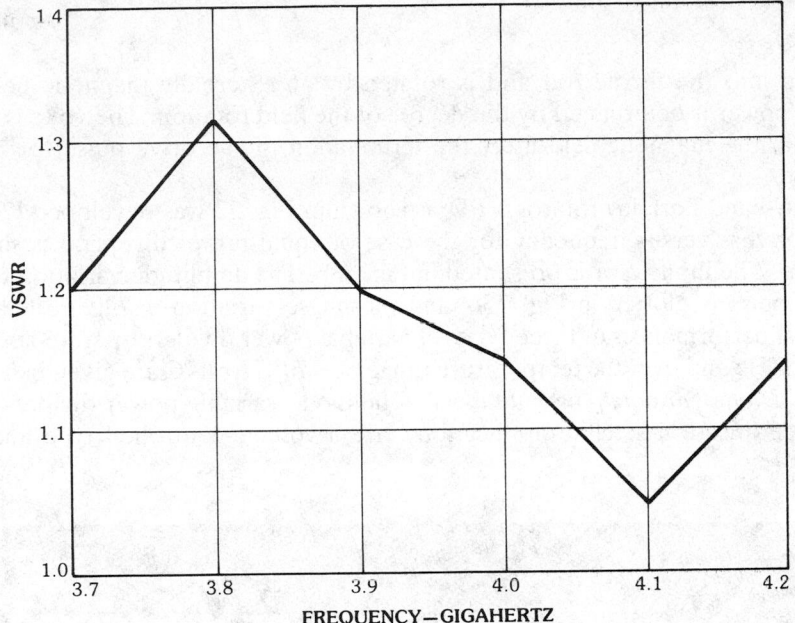

Fig. 29. Coupler/phase-shifter variable power divider vswr versus frequency. (*After Smith, Mathews, and Boyd [26], © 1982 AIAA; reprinted with permission*)

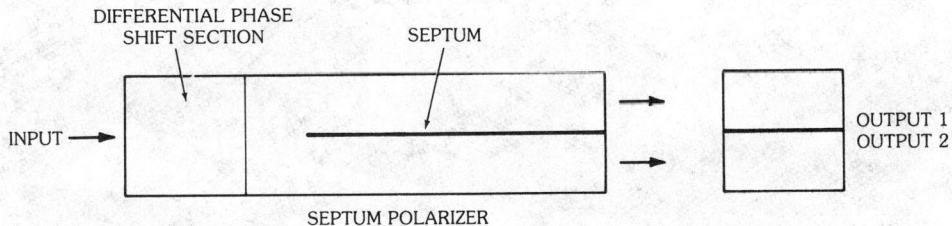

Fig. 30. Driven-ferrite variable-polarizer variable power divider.

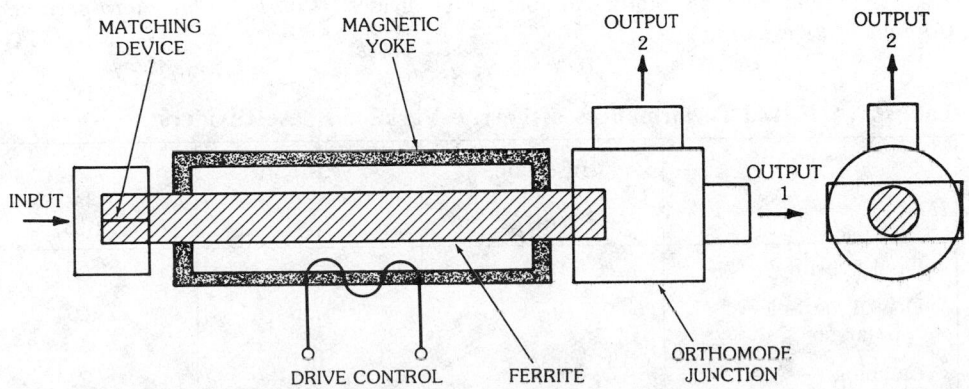

Fig. 31. Latching Faraday rotator variable power divider.

coupled into the ferrite rod and is rotated by an externally magnetic field. The output power is determined by the degree of the field rotation. The yoke is used to maintain the magnetic field after the termination of the drive pulse for latching purpose.

A *C*-band Faraday rotator VPD as shown in Fig. 32 was developed [26]. The insertion loss versus frequency for the case of equal power diversion is shown in Fig. 33a. The input vswr is presented in Fig. 33b. The amplitude variation for both output ports is plotted in Fig. 33c, and the phase variation in Fig. 33d.

Test performances of three types of variable power dividers operated over 7.25 to 7.75 GHz and over the temperature range of $-10°C$ to $40°C$ are given in Table 7.

(*b*) *Diode variable power dividers*—The diode variable power dividers, which have been used in satellite applications, are a voltage-controlled type and a pin

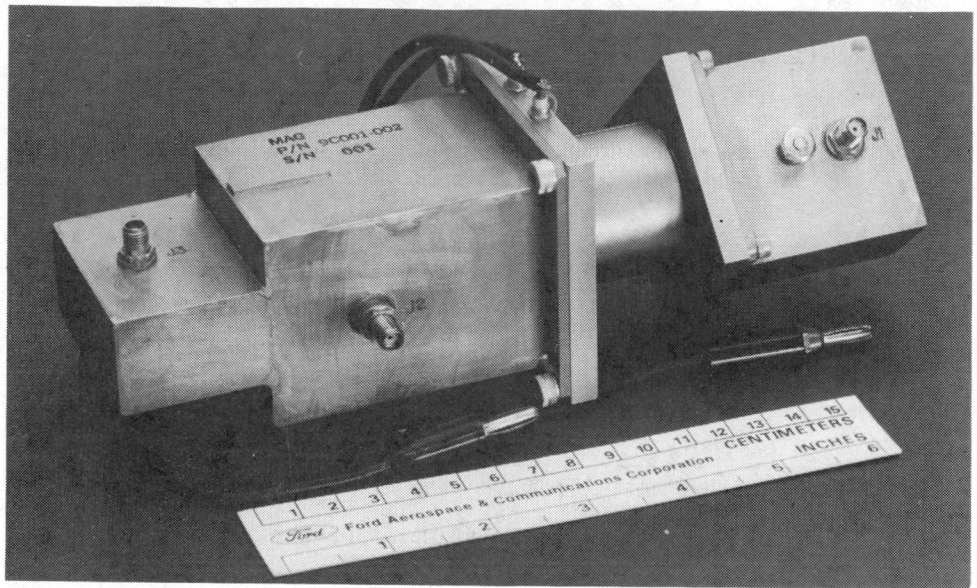

Fig. 32. *C*-band Faraday rotator variable power divider. (*Courtesy Ford Aerospace and Communications Corp.*)

Table 7. Achieved Performances of Ferrite Variable Phase Dividers

Type	Insertion Loss (dB)	VSWR	Isolation (dB)
Dual hybrid	0.34	1.32	30
Driven variable polarizer	0.21	1.56	25
Latching Faraday rotator	0.31	1.91	25

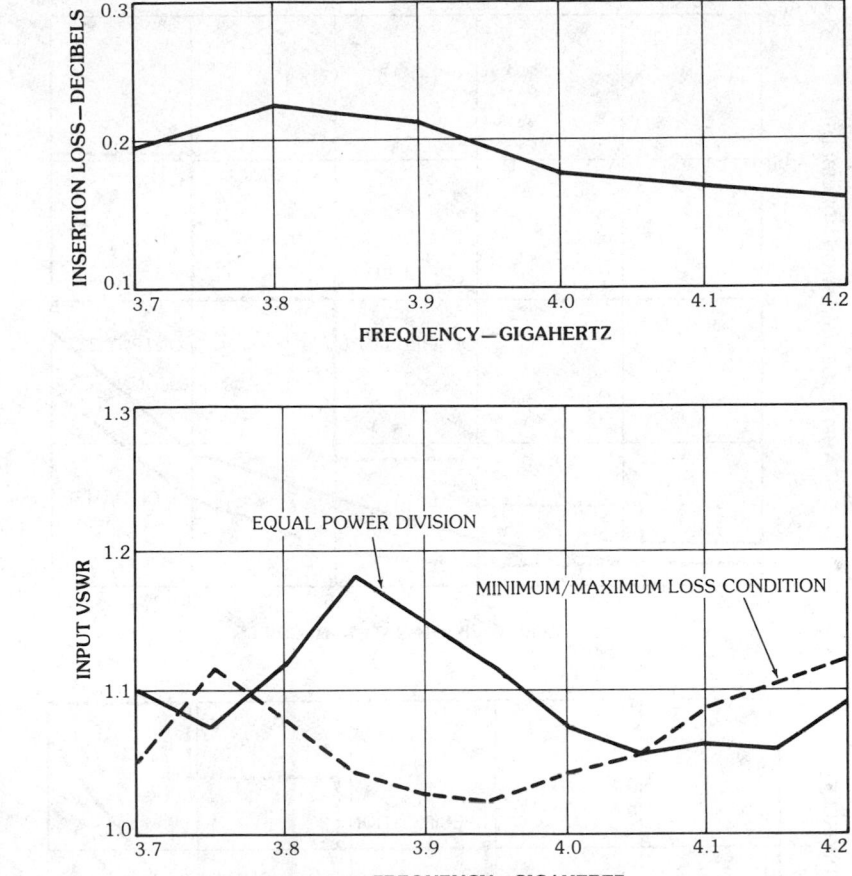

Fig. 33. Faraday rotator variable power divider characteristics. (*a*) Insertion loss versus frequency. (*b*) Input vswr versus frequency. (*c*) Amplitude variation versus power division ratio. (*d*) Phase variation versus power division ratio. (*After Smith, Mathews, and Boyd [26], © 1982 AIAA; reprinted with permission*)

type. Diode variable power dividers operate on the same principle as the dual-hybrid ferrite variable power dividers. The phase shifters are the diode phase shifters, which can be either the continuously variable voltage-controlled varactor type or the pin diode stepped-increment type.

A voltage-controlled diode variable power divider consists of a three-port split-T power divider, two reflection type diode phase shifters, and a 90° hybrid as shown in Fig. 34. A voltage-controlled diode variable power divider without phase shifter has also been designed. It consists of two 3-dB hybrids, two varactor diodes, and two 50-Ω lines one-quarter of a wavelength long as shown in Fig. 35. The input port is port 1 and the two output ports are ports 2 and 3. The varactor diode capacitances control the reflection and transmission of the incident power.

Typical insertion loss for diode variable power dividers is 0.8 dB or higher for *S*-band. The advantage is the fast switching time. As indicated in the application of

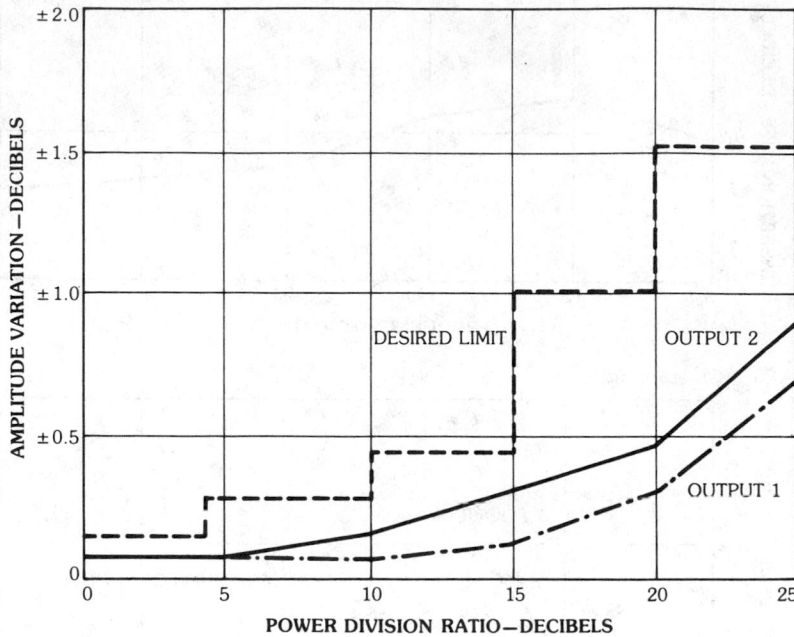

c

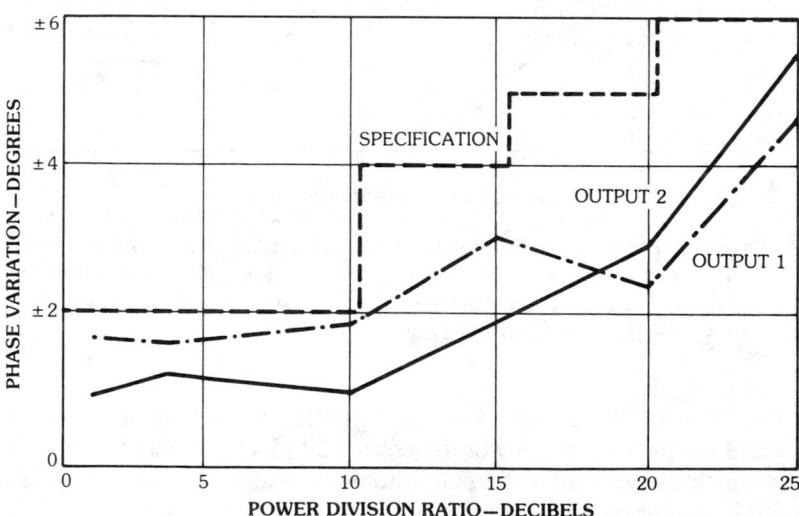

d

Fig. 33, *continued.*

diode variable phase shifter, the diode variable power dividers are lossy. However, they are attractive in an active antenna system. Diode devices which are operated in the higher frequency bands are being developed.

(*c*) *Electromechanical variable power dividers*—Several types of electro-mechanical variable power dividers have been developed. They are not suitable for satellite application for the same reasons given at the end of the variable phase shifter subsection (2).

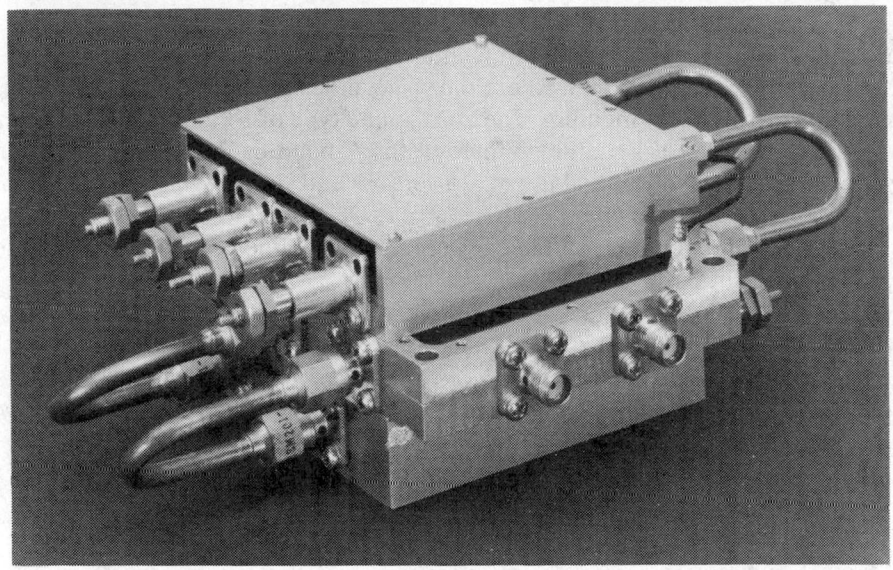

Fig. 34. Voltage-controlled diode variable power divider. (*Courtesy Ford Aerospace and Communications Corp.*)

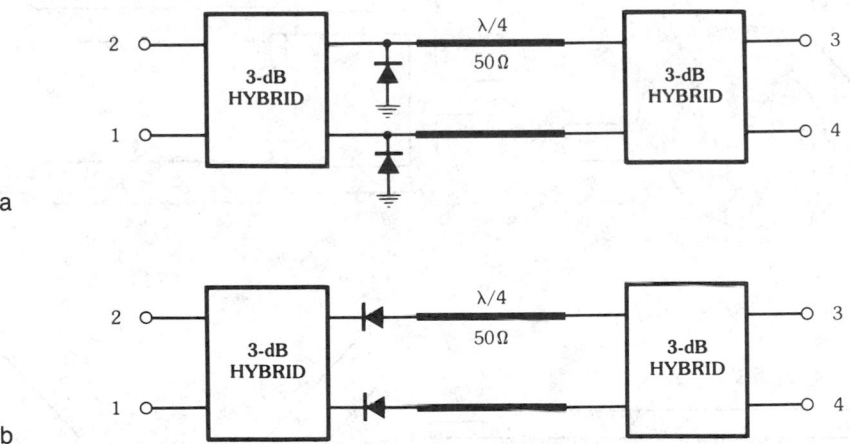

Fig. 35. Voltage-controlled variable power divider without phase shifters. (*a*) With shunt diodes. (*b*) With series diodes.

Fixed Beam-Forming Networks—Fixed beam-forming networks include power division elements, directional couplers, transitions, phase shifters, diplexers, rotary joints, and odd/even mode converters. Most of these components are also used in the scanned beam-forming networks.

1. *Power Division Elements*—The choice of the types of power division elements is based on considerations of the physical layout, manufacturing techniques, the feed transmission medium, and the component insertion losses.

Two main types of power division elements have been widely used: air stripline and waveguide. The air stripline components are used in combination with the air stripline transmission medium, where the waveguide components are used with the waveguide transmission medium. The air stripline type of BFN is compact but more lossy. Its application is generally limited at the C-band or lower-frequency bands. The waveguide BFN is bulky but with lower loss and is mostly used at the K-band or higher-frequency bands.

Several components are suitable for use as air stripline power division elements as depicted in Fig. 36:

Hybrid-ring directional coupler
Split-T power divider
Branchline directional coupler
Symmetrical directional coupler
Semicircular-rod directional coupler
Gysel hybrid

Waveguide power division elements may be the following:

Magic-T junction
Folded-T junction

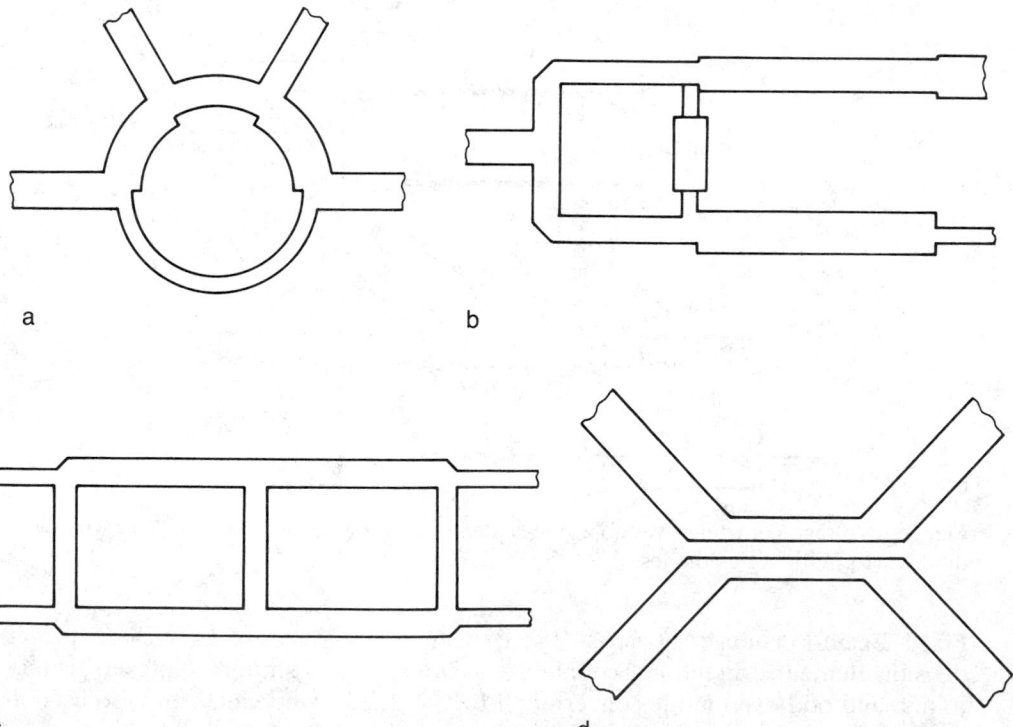

a   b

c   d

Fig. 36. Stripline power division elements. (*a*) Hybrid-ring directional coupler. (*b*) Split-T power divider. (*c*) Branchline directional coupler. (*d*) Symmetrical directional coupler. (*e*) Semicircular-rod directional coupler. (*f*) Gysel hybrid ring.

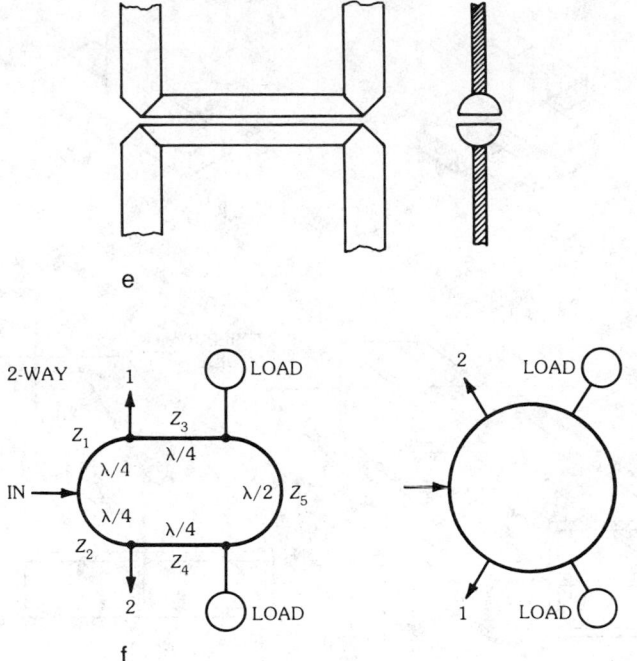

Fig. 36, *continued.*

Top-wall hybrid junction
Narrow-wall coupler
Aperture couplers (single-hole or multihole)
Branch-guide coupler
H-plane septum T
Waveguide hybrid ring

These are depicted in Fig. 37.

Major trade-offs of the types of power division elements are the insertion loss, frequency band, and power split ratios. A combination of types may be required to provide a wide range of power distributions required in a multibeam antenna system. For air stripline division elements, hybrid-ring directional couplers and split-T power dividers are suitable for moderate power split ratios. The Gysel hybrid is more complex but suitable for wider frequency band operation because of the symmetrical treatment of the hybrid. Symmetrical directional couplers are more suitable for the power split ratio that is greater than 10 dB. They can also be cascaded to improve the coupling bandwidth. For waveguide power division elements the magic-T, folded-T, top-wall hybrid junction, and narrow-wall couplers are widely used as 3-dB couplers. The others are suitable for different power split ratios. The hybrid ring's bandwidth is comparatively narrow. Its awkward geometry is not as compact as the corresponding ring-hybrid stripline coupler and does not lend itself very well to compact layouts of BFNs.

Since a wide range of power division ratios is required, a complete design curve

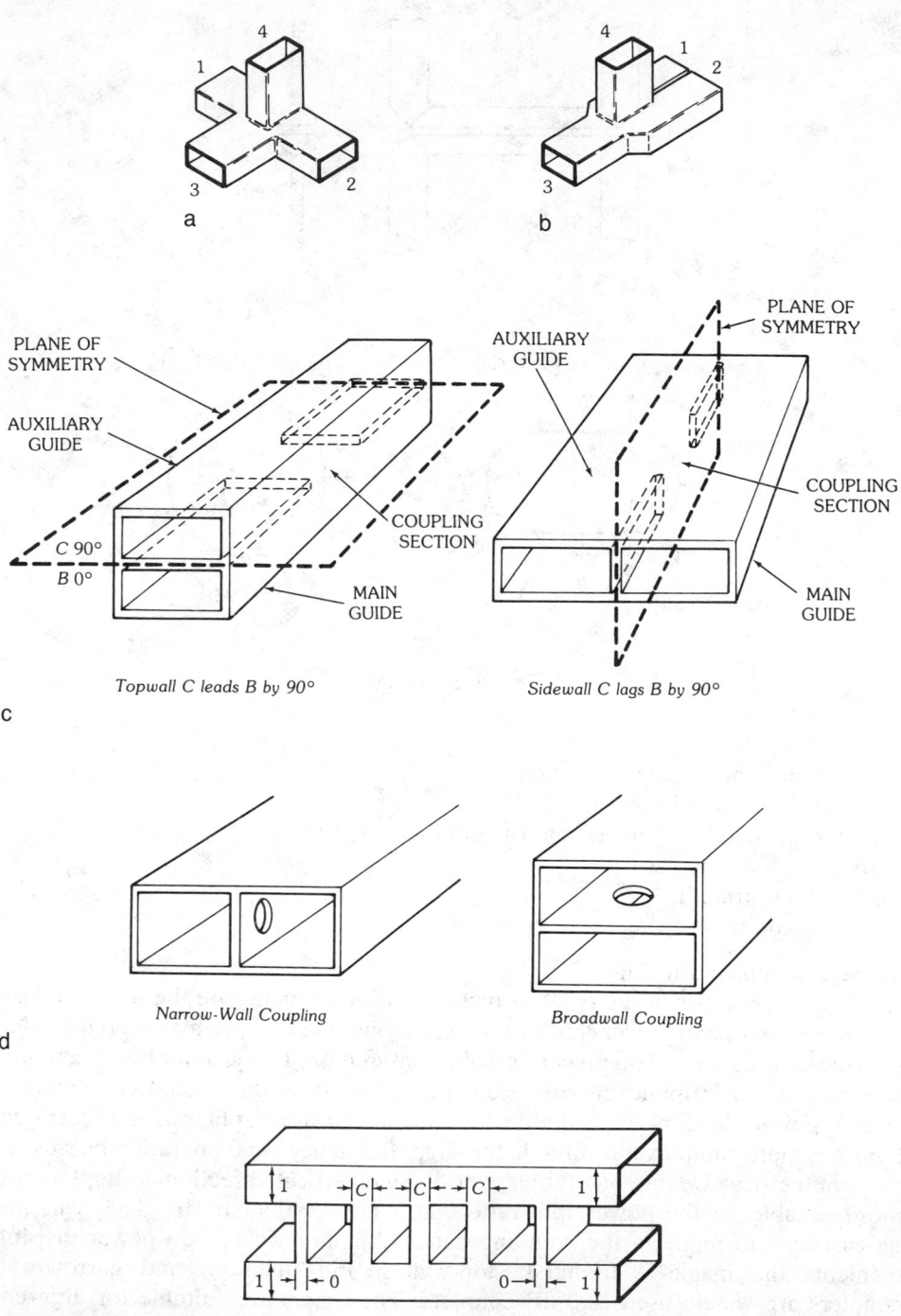

Fig. 37. Waveguide power division elements. (*a*) Conventional magic-T. (*b*) *H*-plane folded T. (*c*) Top-wall hybrid junction. (*d*) Aperture couplers. (*e*) Typical branch-guide coupler with five branches. (*f*) *H*-plane septum T. (*g*) Waveguide hybrid ring.

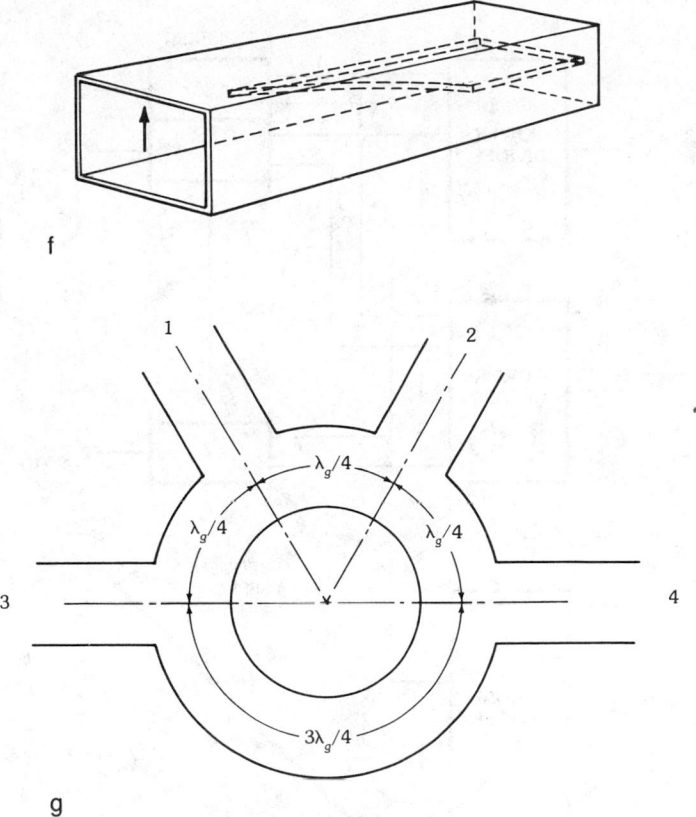

f

g

Fig. 37, *continued.*

is not available. The components are usually designed, fabricated, and tested for a wide range of power division ratios. The design for arbitrary ratios are then extrapolated from the measured data.

2. *Diplexer*—A diplexer can be a filter type or a frequency selective surface. The former approach sometimes imposes severe constraints on the design of the BFN, which is complex and congested in a frequency reuse multibeam antenna system. A frequency selective surface can be designed to be transparent for one frequency band and reflective for the other band, to properly direct and redirect the energy to the same optical aperture as shown in Fig. 38. Trade-offs of the two approaches are the difficulty in the design of the BFN, size of the frequency selective surface, and its thermal stability.

3. *Rotary Joint*—The rotary joint is used to provide a single-channel or multichannel transmission path between a spin spacecraft and its despun antennas. The design trade-off is the choice among waveguide, coaxial, and combined waveguide-coax types with an auxiliary choice of waveguide or coax as an output. Design parameters are insertion loss, performance variation with respect to rotation, size, and weight, and ease of design and construction. Fig. 39 depicts a three-channel rotary joint.

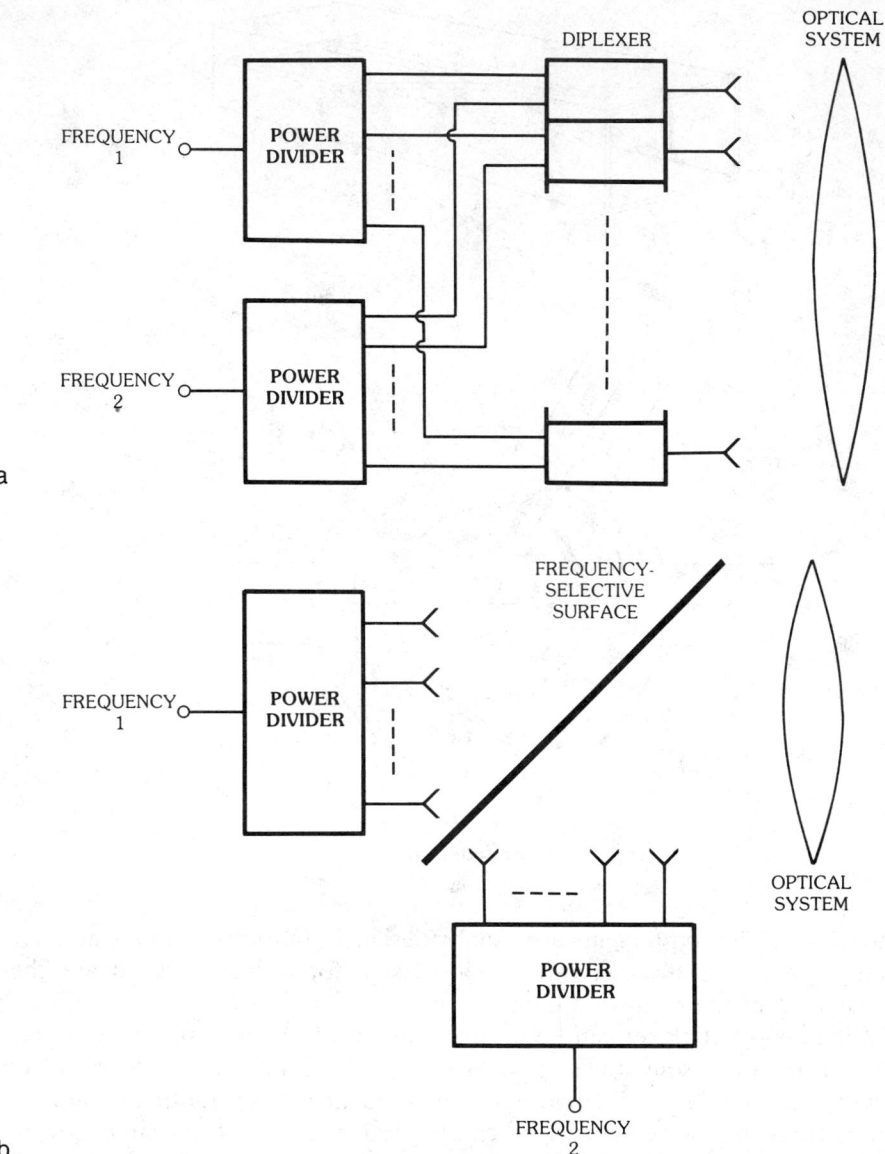

Fig. 38. Types of diplexers. (*a*) Filter-type diplexer. (*b*) Spatial diplexer.

4. *Odd/Even-Mode Converter*—An odd/even-mode converter [27] is used to alleviate the severe design constraint on an output multiplexer. The odd channels of an output multiplexer are combined into a single output, and even channels into a second output. It is generally desirable to use a common optical aperture for both ports. An odd/even-mode converter is a 2-to-N port network to combine the two ports to a feed array that illuminates an optical aperture as depicted in Fig. 40.

Let $[S]$ be the scattering matrix of the network. It can be shown that in a lossless network,

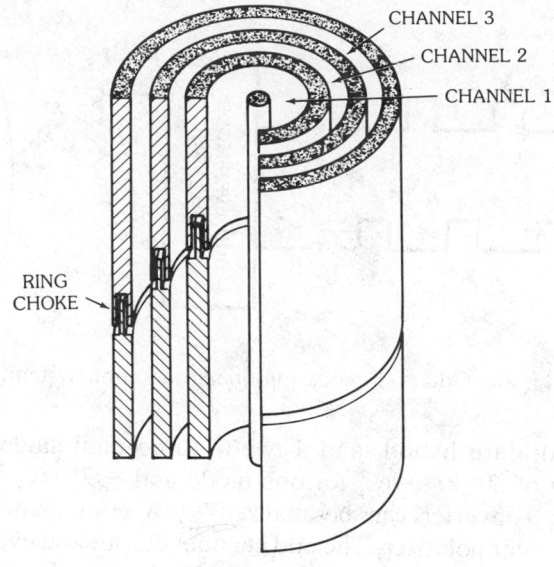

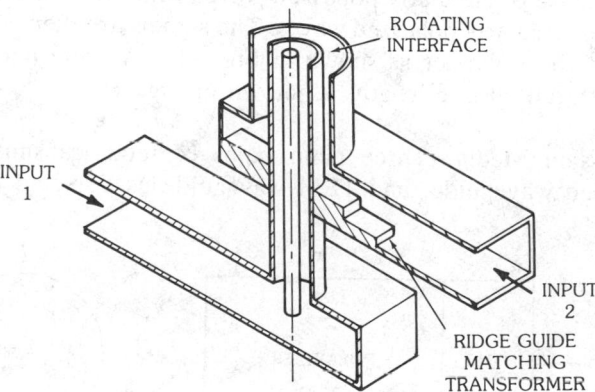

Fig. 39. Three-channel rotary joint. (*a*) Concentric section. (*b*) Transitions, center channel to waveguide.

$$S^*_{11'}S^*_{1'2} + S^*_{12'}S_{2'1} + \cdots + S^*_{in'}S_{n'i} = 0$$

and

$$P_{1'} + P_{2'} + \cdots + P_n = 1$$

where * represents the complex conjugate of the complex number, the prime refers to the input port, and nonprimed numbers the output ports; $P_{i'}$ is the power at the *i*th input port.

For an equal power distributed mode converter in which the output power ratios among the output ports are all equal, a two-to-two dual-mode converter is a

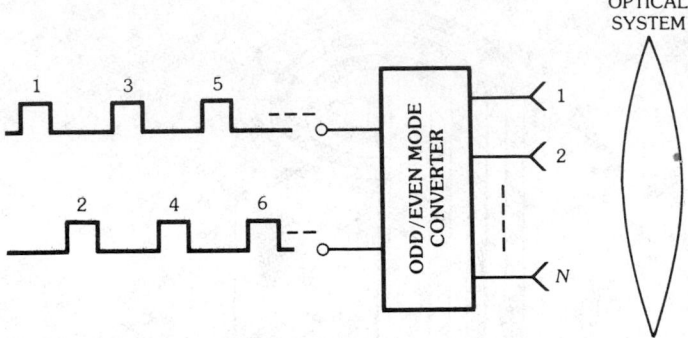

Fig. 40. Odd/even-mode multibeam antenna system.

conventional quadrature hybrid, and a two-to-three dual-mode converter has a phase progression of 60°, 0°, −60° for one mode and −60°, 0°, 60° for the other.

A dual-mode converter can be realized by a combination of orthomode transducer and circular polarizer. The odd channel output and even channel output are connected to the input ports of the orthomode transducer. The odd mode generates one sense of circularly polarized waves while the other mode generates another sense of circularly polarized waves. The signals are then coupled out of the output ports of the polarizer as shown in Fig. 41a. An alternate approach uses hybrids and differential line lengths as shown in Fig. 41b.

Feed Transmission Media—Three main types of feed transmission media are waveguide, ridge waveguide, and TEM waveguide as depicted in Fig. 42a. The

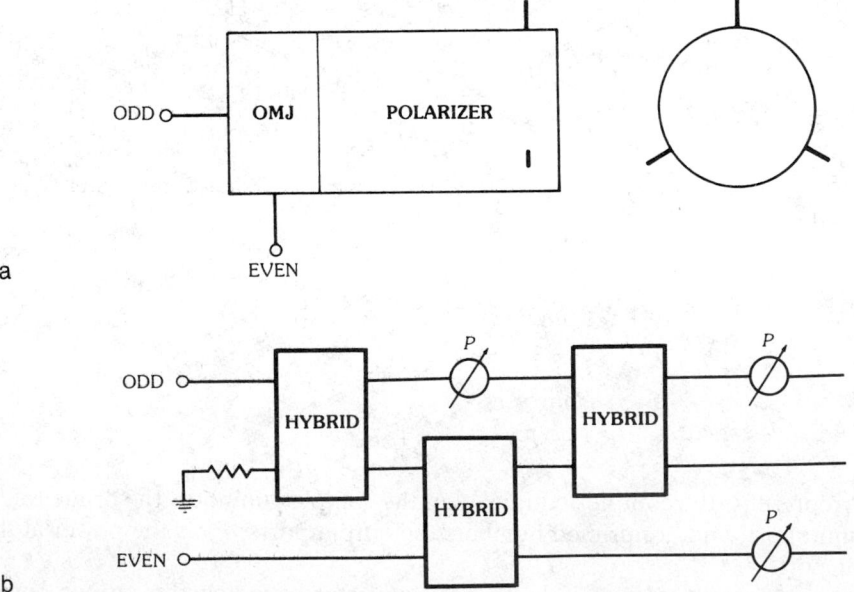

Fig. 41. Odd/even-mode converter. (*a*) Approach 1. (*b*) Approach 2.

waveguides have the advantage of high power-handling capacity. A wide range of commercial waveguide components is also available. The disadvantages are the frequency dispersion of the waveguide, rigidity, and size. The bandwidth capability can be improved by using ridge waveguides. However, ridge waveguide component design is complicated due to the complex structure. The TEM waveguide's bandwidth is much greater than is possible with the other two waveguides since it propagates only the TEM mode. A coaxial line is one of the TEM waveguides. Its modification from a circular to rectangular one yields a planar strip transmission line. The high-Q triplate type of stripline can use printed strips that are compact and easy to construct. As indicated earlier, this air stripline transmission medium is very attractive in the C-band and in lower-frequency–band applications.

The variety of network media spinned off these three main types are shown in Fig. 42b. The relative advantages and disadvantages of each type are summarized in Table 8. Comparisons among the types are given in Table 9.

Beam-Forming Network Topology—The starting point for the circuit layouts are the coordination of the feed array and power division elements.

The initial goal of the network layout is to make all line lengths equal from the common input port to each output element port in order to minimize the effects of frequency dispersion. Lengths can be adjusted slightly to provide the exact phases desired at each element, and to compensate the electrical length of the network paths, such as switches and power divider components. This design of the feasibility layout is an iterative process and is done via a computer graphics software program in the case of designing a complex multibeam antenna system.

Beam-Forming Network Distortion—The distortion of a BFN is due to the manufacturing distortion and thermal distortion. The beam-forming network distortion results in the deviation of amplitude and phase desired for the network. Such deviation yields the performance degradation in an antenna system. The gain will be lower, and side lobe and cross-polarization levels will be higher.

The amplitude and phase variations due to the network distortion can be analyzed. The variations of each component can be measured and used as design information to study the total variations of a beam-forming network. Either the antenna system will be designed with additional margin to compensate the degradation or the BFN will be designed to have smaller tolerance to reduce the distortion in the manufacturing process and thermal environment.

Multibeam Antenna System

A multibeam antenna is the most widely used in satellite antenna systems. It consists of a focusing optics illuminated by an array of feed elements. Each feed element illuminates the optical aperture and generates a constituent beam. Any shaped beam can be formed from a number of these constituent beams by the principle of superposition. Fig. 43 shows a multibeam antenna system. The key features of this antenna system are that it

(*a*) generates a multibeam pattern from one optical aperture,

(*b*) provides pattern shaping and pattern weighting for the radiation pattern, and

(*c*) yields steeper pattern rolloff and results in a higher spatial isolation for the communication system.

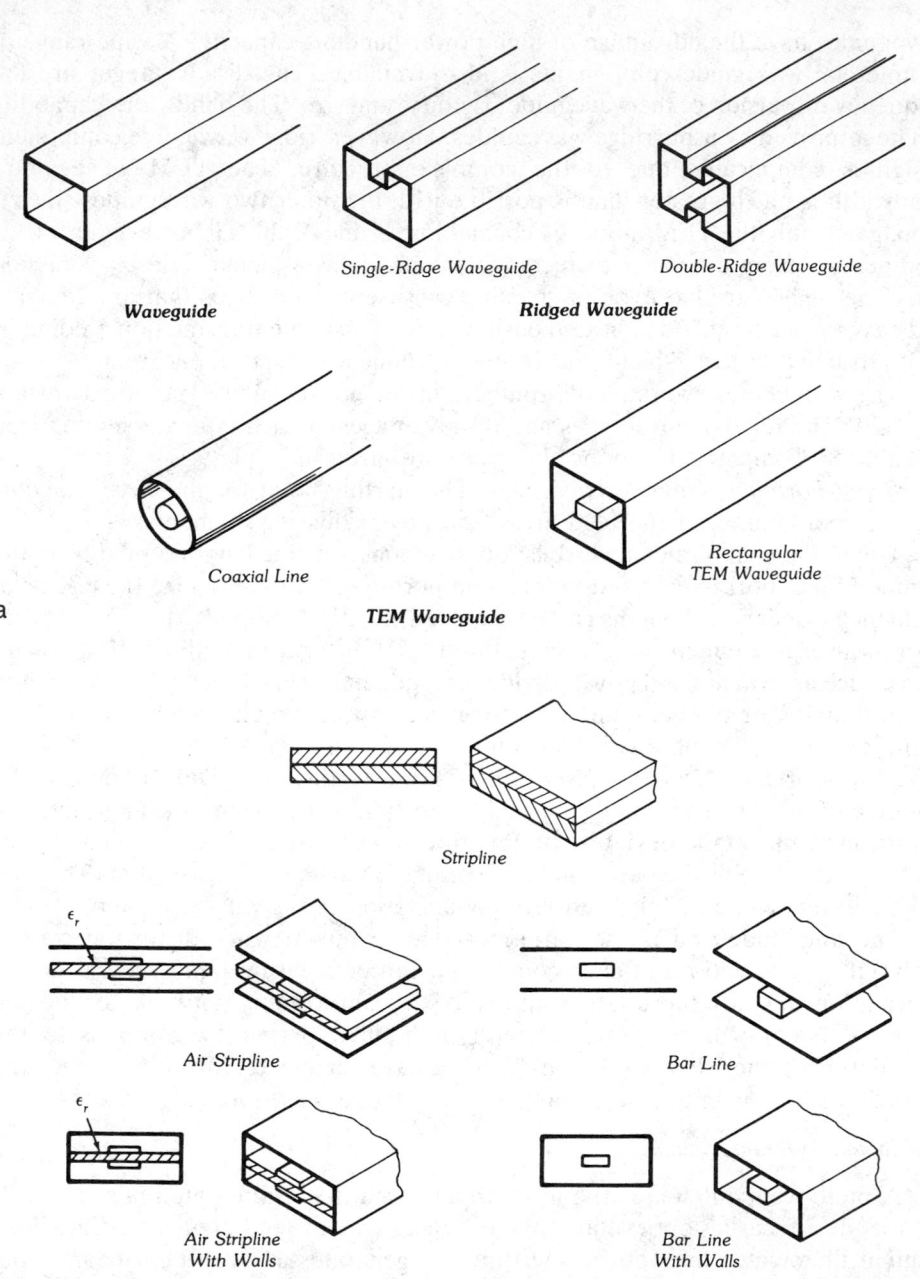

Fig. 42. Feed transmission media. (*a*) Three main types. (*b*) Variety of media.

The beamwidth of a constituent beam is determined by the size of the optical aperture. The position and angular separation of these constituent beams are determined by the feed element separation and the beam deviation factor of an optics. Several possible optical configurations for a multibeam antenna system are

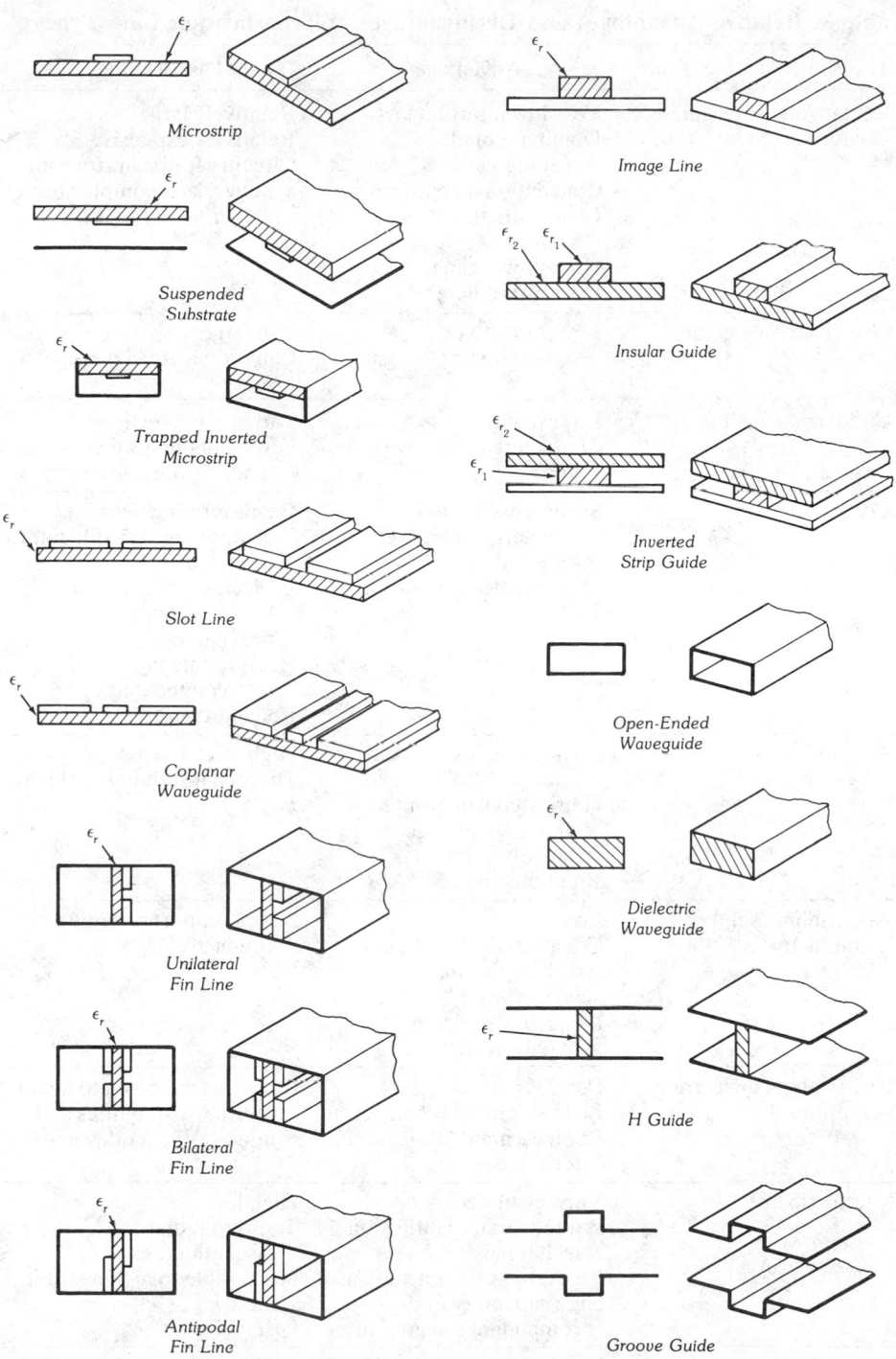

Fig. 42, *continued.*

Table 8. Relative Advantages and Disadvantages of Transmission Line Types

Transmission Line Type	Advantages	Disadvantages
Rectangular waveguide	Very low insertion loss Dominant mode rectangular Conventional approach Characteristics are well known High-power capacity Does not radiate	Relatively large Relatively expensive Circuitry for beam-forming networks is complicated and difficult to fabricate
Oversized waveguide	Lower loss than conventional waveguide	Too large Can propagate higher-order modes
Circular waveguide	Lowest loss Characteristics are well known	Larger cross section Few components can be made in circular guide
Coaxial line	Small cross section Characteristics are well known Moderate loss for air-filled line	Beam-forming network components are difficult to fabricate Dielectric-filled lines are heavy, lossy, and unreliable Requires many interconnections Reliability problems
Stripline	Characteristics are well known Fabricated with printed-circuit techniques Relatively compact Nondispersive	High insertion loss Thermal instability problems
Air stripline, solid conductor	Low loss Compact and lightweight Excellent thermal properties Design information is readily available	Needs support of center conductor
Air stripline, dielectric supported	Low loss Lightweight and compact Printed-circuit fabrication techniques	Care must be taken to avoid thermal instabilities and undesirable modes
Microstrip line	Very small size Printed-circuit fabrication techniques Properties are well known High capability for component integration	High loss Tends to radiate at discontinuities Susceptible to higher-order modes Dispersive
Suspended substrate	Less loss than microstrip Printed-circuit fabrication techniques Tolerances less critical than for microstrip	Suffers same radiation problems as microstrip Dispersive

Table 8, *continued.*

Transmission Line Type	Advantages	Disadvantages
Trapped inverted microstrip	Moderate loss Tolerances not as critical as for microstrip Less radiation at discontinuities than for microstrip Moderate degree of potential for integration Printed-circuit techniques	Somewhat difficult to fabricate circuits
Slot line	Printed-circuit fabrication techniques Compact and lightweight	Very high loss Tends to radiate Not suitable to beam-forming networks
Coplanar waveguide	Printed-circuit fabrication techniques All conducting elements are on the same side of the substrate	Very high loss Tends to radiate Not suitable for beam-forming networks
Fin line	Moderate loss Partially fabricated with printed-circuit techniques	Large cross section Not suitable for beam-forming networks
Dielectric waveguide	Low attenuation loss Relatively simple to fabricate	Radiation occurs at discontinuities, such as bends Difficult to support mechanically Relatively large dimensions Limited bandwidth Not suitable for beam-forming networks
Image guide	Moderately low loss Overcomes support problem of dielectric waveguide	Radiation occurs at discontinuities, such as bends Small range of impedances around 26 Ω available Difficult to achieve zero gap between the ground plane and the dielectric waveguide, causing rf problems Relatively large dimensions Not suitable for beam-forming networks
Insular guide	Moderately low loss Overcomes gap problems of image guide	Radiation problems Relatively large dimensions Not suitable for beam-forming networks

Table 8, *continued*.

Transmission Line Type	Advantages	Disadvantages
Strip dielectric guide	Radiation loss is less than for other dielectric waveguides	Large dimensions Relatively heavy Not suitable for beam-forming networks
Inverted strip guide	Moderate loss Small radiation	Poor guidability at bends Large dimensions Heavy Not suitable for beam-forming networks
Trapped image guide	Does not radiate at bands	Large dimensions Not practical for beam-forming networks
H-guide	Low loss Sections can be joined without connectors	Large cross section dimensions Not practical for beam-forming networks
Groove guide	Low loss Broadband operation	Very large dimensions Little design information available Not suitable for beam-forming networks
Fence guide	Moderately low loss	Little information available Not suitable for beam-forming networks

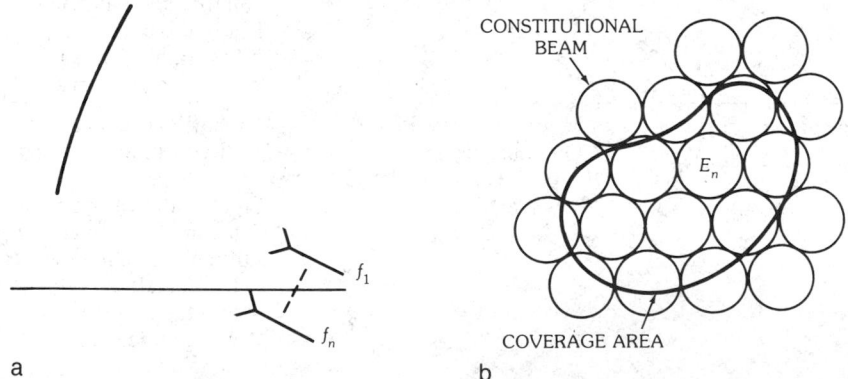

Fig. 43. Multibeam antenna system. (*a*) Feed elements and focusing optics. (*b*) Shaping of beam.

shown in Fig. 44. The lens and reflector are interchangeable in the configurations. The design of the system is the same as those presented previously. The excitation (amplitude and phase) of each beam has to be determined by antenna pattern synthesis.

Table 9. Comparisons of Transmission Lines

Type of Line	Insertion Loss	Impedance Levels (ohms)	Power Handling	Tendency to Radiate	Suitability for Beam-forming Networks	Cross-Sectional Dimensions	Availability of Design Information
Standard waveguide	Very low		Moderate	Zero	Poor	Medium	Excellent
Circular waveguide	Very low		High	Zero	Poor	Large	Excellent
Oversized waveguide	Very low		High	Zero	Poor	Large	Excellent
Coaxial line	Moderate	40–120	Low–mod	Zero	Poor	Small–medium	Excellent
Stripline	High	30–120	Low–mod	Zero	Excellent	Small	Excellent
Air stripline	Low	30–130	Low–mod	Zero	Excellent	Small	Good
Microstrip line	High	20–125	Low	Moderate	Good	Very small	Excellent
Suspended substrate	Moderate	25–130	Low	Moderate	Good	Small	Fair
Trapped inverted microstrip	Moderate	30–140	Low	Small	Poor	Small	Fair
Slot line	Very high	60–200	Low	High	None	Very small	Poor
Coplanar waveguide	Very high	40–150	Low	High	None	Small	Poor
Fin line	Moderate	10–400	Low	Zero	None–poor	Moderate	Poor
Image line	Very low	≅26	Moderate	Mod–high	None–poor	Moderate	Poor
Insular guide	Low		Moderate	Mod–high	None–poor	Moderate	Poor
Inverted strip guide	Moderate		Moderate	Low	None–poor	Moderate	Poor
Dielectric waveguide	Moderate		Moderate	High	None	Moderate	Fair
H-guide	Low		High	Small	None	Large	Fair
Groove guide	Very low		High	Small	None	Large	Poor

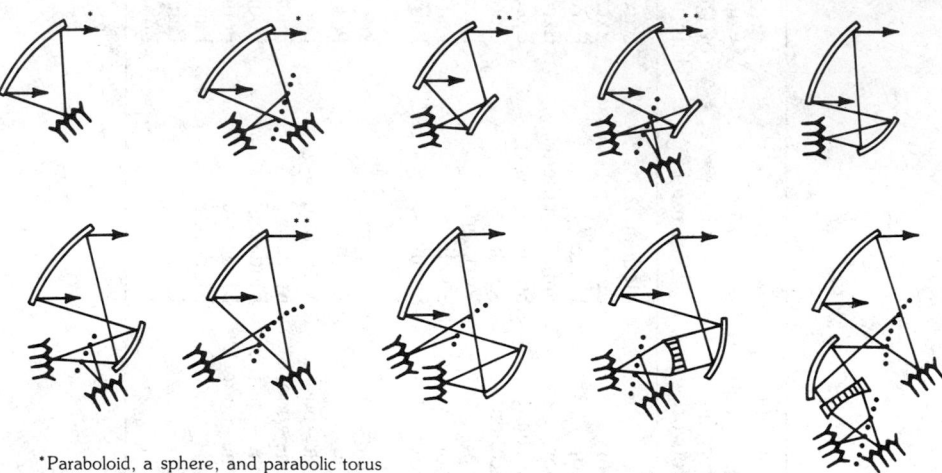

*Paraboloid, a sphere, and parabolic torus
**Classic Cassegrain and Schwarzchild system

Fig. 44. Multibeam antenna system optical configurations.

The effect of weighting the excitations of multifeed elements can be expressed in the power gain calculation. If unit excitation of the ith feed element produces the secondary field $E_n(r, \theta, \phi)$, and voltage coefficients f_n are applied, the power gain is

$$P^2(\theta, \phi) = \frac{4\pi r^2 \left| \sum_{n=1}^{N} f_n E_n(r, \theta, \phi) \right|^2}{n_0 \Sigma |f_n|^2} \tag{18}$$

where $E_n(r, \theta, \phi)$ can be computed via the aperture field method, induced current method, or uniform method [9, 28], and n_0 is the free-space impedance. Equation 18 can be expressed in the following way:

$$[E][f] = [P] \tag{19}$$

in which

$$[f] = \begin{pmatrix} f_1 \\ \vdots \\ f_N \end{pmatrix} \qquad [E] = (E_1, \cdots, E_N), \quad \text{and} \quad P = \begin{pmatrix} P_1 \\ \vdots \\ P_M \end{pmatrix} \tag{20}$$

describes the radiation field sampled at M points. The synthesis problem can be represented by

$$[E][f] \cong [P^d]$$

where $[P^d]$ is the desired pattern vectors. The above equation represents a system of M linear equations with N unknowns. If $[E]$ is a square matrix and nonsingular, the solution is unique. If the rank of $[E]$ is less than N, then more than one solution may exist. If M is greater than N, which is generally the case in a multibeam antenna system, then no exact solution exists but there will be a unique least-squares solution.

Most multibeam antenna systems specify only the power of a required radiation pattern. Let $(P_m^d)^2$ be the desired radiation power at point m; then the deviation from the desired value, which is defined as the *pattern error*, is

$$\epsilon_m = \omega_m \left| \left| \sum_{n=1}^{N} f_n E_{mn} \right| - P_m^d \right|, \qquad m = 1, \ldots, M \tag{21}$$

where ω_m is a weight factor. The sampling points are taken in the coverage and outside the coverage.

The gradient, minimax, and regularization methods of synthesis are used most widely. The *gradient method* minimizes the total errors defined as

$$\epsilon = \sum_{m=1}^{M} |\omega_m (P_m - P_m^d)|^2 \tag{22}$$

by advancing (f_n) along the $-\Delta\epsilon$ direction. The *minimax method* [29, 30] minimizes the maximum errors, which are given in the form of

$$\max \epsilon_m = \omega_m \left| \frac{P_m^d - P_m}{P_m^d} \right| \tag{23}$$

The *regularization method* [31] minimizes the total error

$$\epsilon = \sum_{m=1}^{M} |\omega_m (P_m - P_m^d)|^2 \tag{24a}$$

subject to the constraint of

$$\| f \| \leqslant C \tag{24b}$$

where C is a positive constant. The solution obtained is then used to design the power distribution of the beam-forming network.

The design procedure can be divided into key steps as shown in Fig. 45. The first step is to set requirements. The requirements are given in terms of frequency band and bandwidth, polarization, coverage area, antenna gain, and side lobe and cross-polarization levels. The coverage area is transformed in the antenna coordinate system as viewed from the satellite. The optical aperture size is determined by the resolution of the antenna radiation pattern, gain requirement, and spacecraft constraint. The focal length, as indicated earlier, is determined by the scan loss, side lobe and cross-polarization levels, and the number of feed elements used in the antenna pattern synthesis. The design trade-off parameters of a feed array include the feed array configuration and the shape and size of feed elements. The feed array configuration directly influences the element spacing and the re-

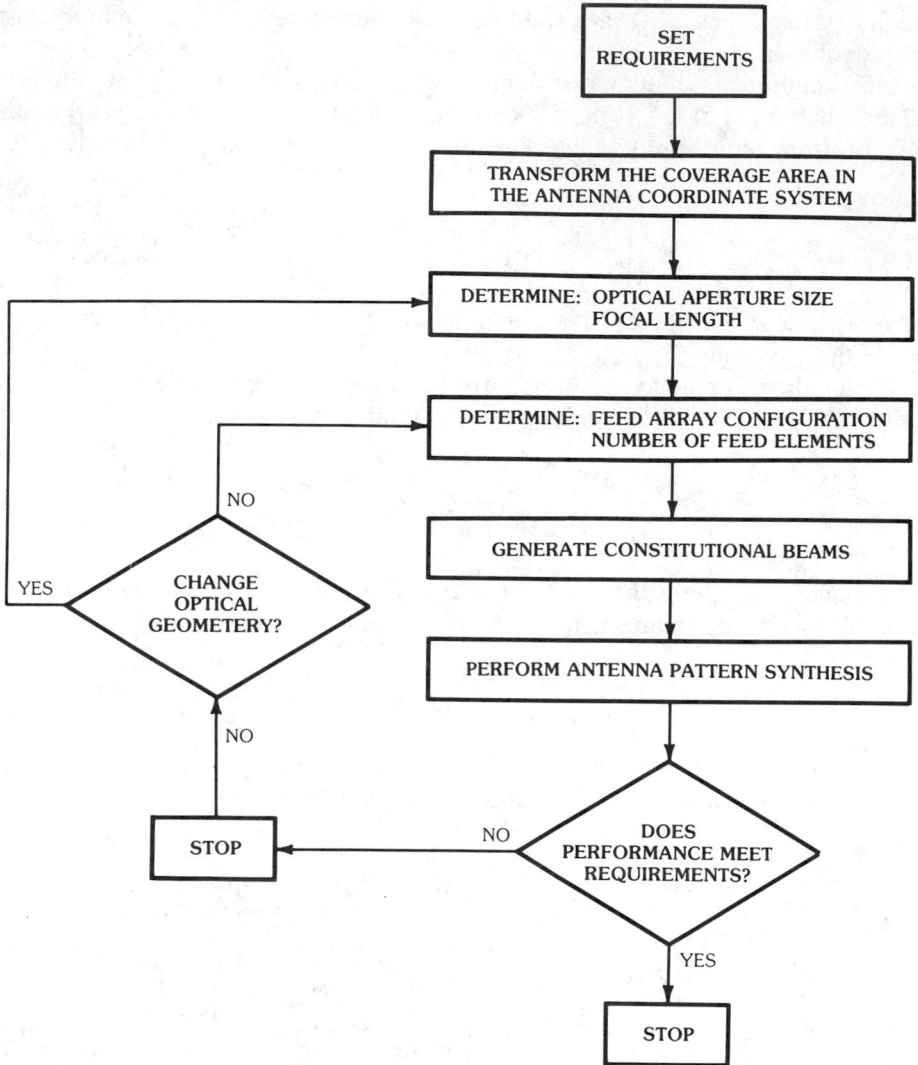

Fig. 45. Flowchart of the design procedure.

quired number of feed elements. It can be periodic or nonperiodic. A periodic configuration is used in a circularly polarized antenna system, while a nonperiodic configuration is sometimes used in a linearly polarized antenna system for pattern control. The constituent beams are then generated. They can be approximated by $(\sin x)/x$ or $J_1(x)/x$, expressed by a Fourier series, computed by rigorous antenna pattern analysis, or measured in a satellite environment. The number of the beams required can be approximately determined by overlaying the beams on the top of the coverage area. Antenna pattern synthesis is applied to obtain the required excitation of each beam. The achieved performance is evaluated against the requirements. The whole procedure is an iterative process.

Design Examples

An *S*-Band Phased Array—An *S*-band multiple-access phased array antenna [32, 33] has been designed for the Tracking and Data Relay Satellite System (TDRSS). The antenna as shown in Fig. 46 consists of a 30-element array with 10 elements commonly used for both transmit and receive frequency bands and 20 elements used only for the receive frequency band. The number of elements is chosen so that minimal numbers are used to reduce weight and simplify the complexity of the beam-forming network.

The requirements on each array element are that each array element must provide 13 dB of gain over ±13.5° field of view. The scan loss of the beam at the edge of field of view is to be 3 to 4 dB. Based on these requirements, three array element configurations have been considered: single element, large subarray of low-gain elements, and a small subarray of relatively high gain elements. The single element is chosen because of the simplicity of the beam-forming network. The reflector, horn helix, and short backfire antenna are considered for an array element. The reflector and horn are precluded because of weight and size. The short backfire antenna is eliminated because of its narrow bandwidth performance. The helix is chosen because of its narrow bandwidth performance, inherent circular polarization, acceptable axial ratio, gain, and simplicity.

The end of the helix is terminated in a cone spiral to improve the axial ratio over the field of view and over the transmit and receive frequency bands. A cup is also used to reduce mutual coupling and suppress the strong normal mode radiation near the base of the helix. Fig. 47 shows the measured active element patterns and the single element pattern. The helix in the array environment behaves almost the same, compared to its single element in free space. The design parameters of the element performance are axial ratio and gain versus array spacing. The measured performances are presented in Fig. 48. Even though the optimum minimum spacing is 3.8 cm, 34.3 cm is chosen for conservative reasons. Fig. 49 depicts the ten-element array. Fig. 50 shows the boresight beam and the scanned beam at $\theta = 15°$, $\phi = 15°$.

Lens Antenna—A waveguide lens has been built and tested [34, 35] for a variable earth coverage *X*-band satellite antenna system. This lens is excited by a variable beam-forming network capable of producing radiation patterns varying from a narrow high-gain beam to the earth coverage beam. The waveguide lens has a 76.2-cm aperture with a 76.2-cm focal length and contains approximately 700 titanium waveguides with a 2.54 × 2.54-cm cross section and a 0.0127-cm wall thickness as shown in Fig. 51 [34]. The lens surface facing the feed is a segment of a sphere centered at the feed and the opposite surface is a segment of a spheroid. This surface is a limiting case of a two-focus design, a design with two focal points. It is expected that the lens has a better scanning performance than those conventional lenses with other different surface shapes. The concave surface facing the feed will improve aperture illumination efficiency. The lens is stepped only on the surface opposite the feed to reduce shadowing effects of the steps.

The waveguides have a square cross section because the circular polarization is required. A 2.54-cm waveguide size is chosen as a compromise of the dominant

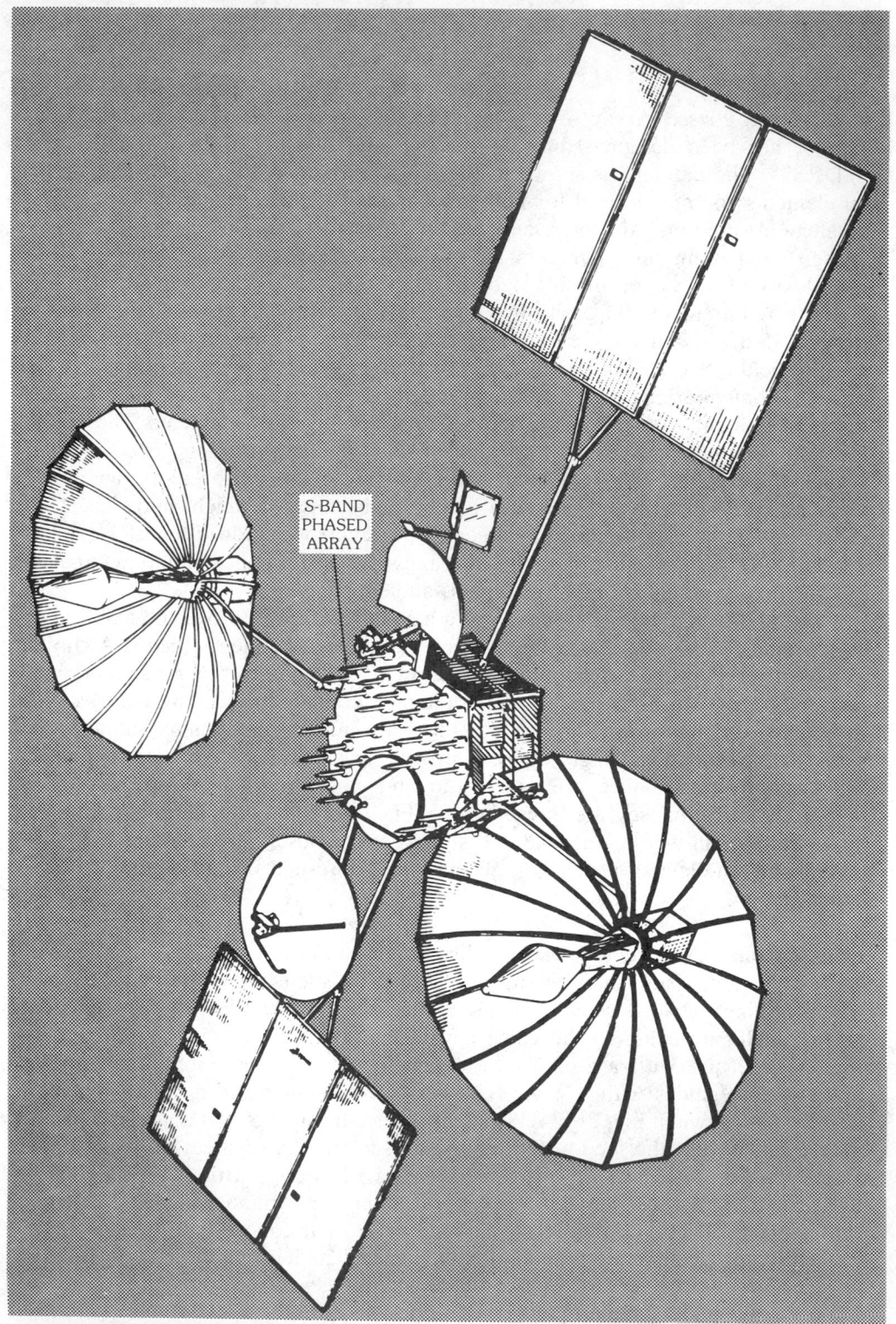

S-BAND
PHASED
ARRAY

Fig. 46. Tracking and data relay satellite system. (*Courtesy C. Donn*)

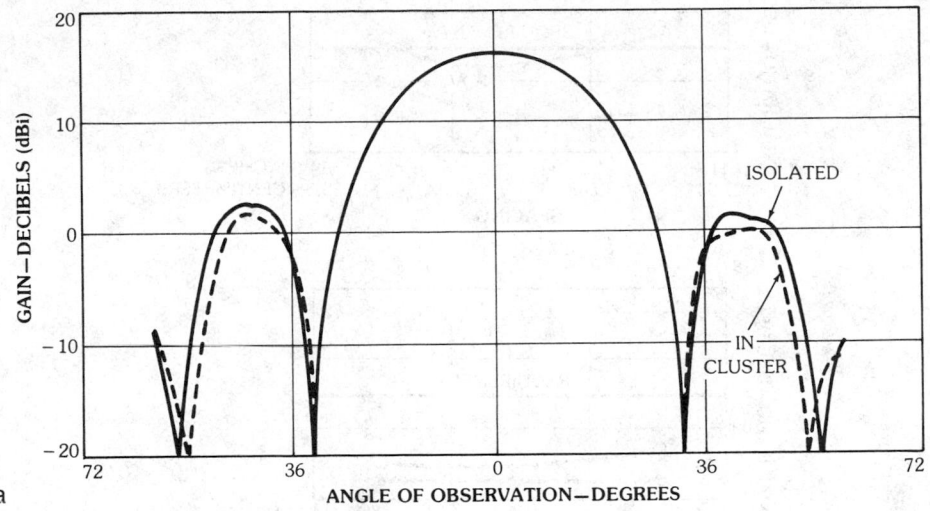

a

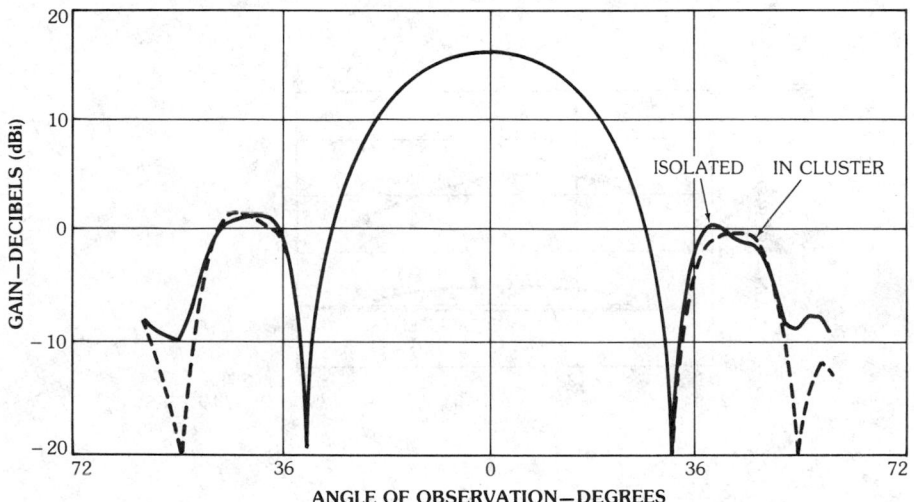

b

Fig. 47. Active element patterns of helix array measured isolated and in seven-element cluster. (*a*) Spacing equal to 13.5 in (34.29 cm). (*b*) Spacing equal to 11.5 in (29.21 cm). (*Courtesy C. Donn*)

mode propagation, higher-order mode suppression, and reflection loss of the lens surface.

Nonoverlapping and overlapping feed clusters are considered in the beam-forming network. The achievable performance of a scanned beam that is generated with either one, two, or three feeds is under study. In order to simplify the beam-forming network, the beam is assumed to scan in a discrete fashion and the feed cluster is excited with equal amplitude and phase. The diameter of the lens and the feed horn spacing are then varied to determine the minimum value of antenna gain anywhere in the field of view. The range of the lens diameter is chosen from 50.8

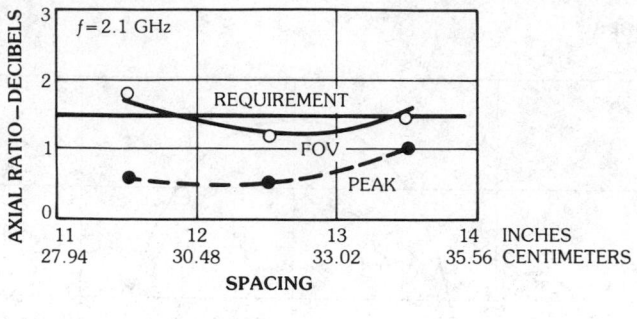

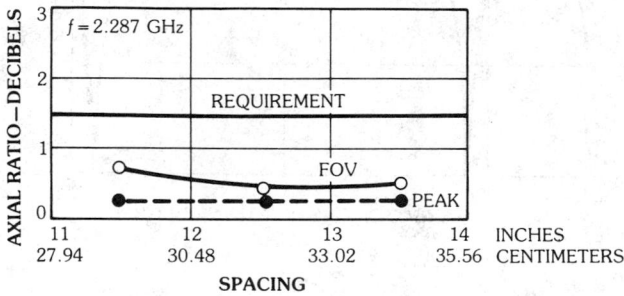

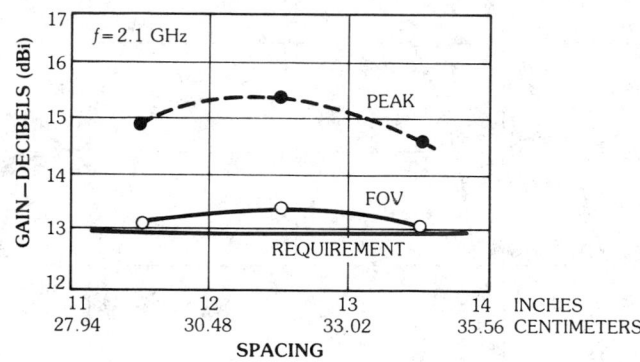

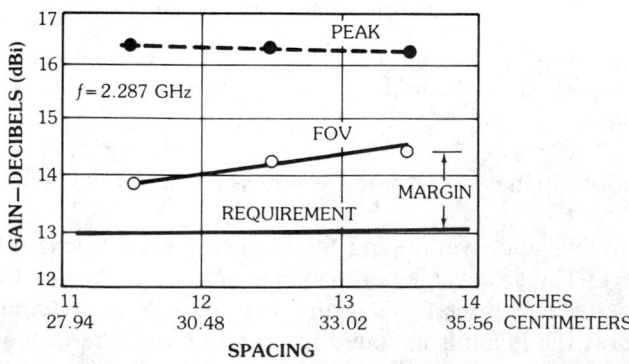

Fig. 48. Design paramenters of helix seven-element cluster. (*a*) Measured axial ratio. (*b*) Measured gain. (*Courtesy C. Donn*)

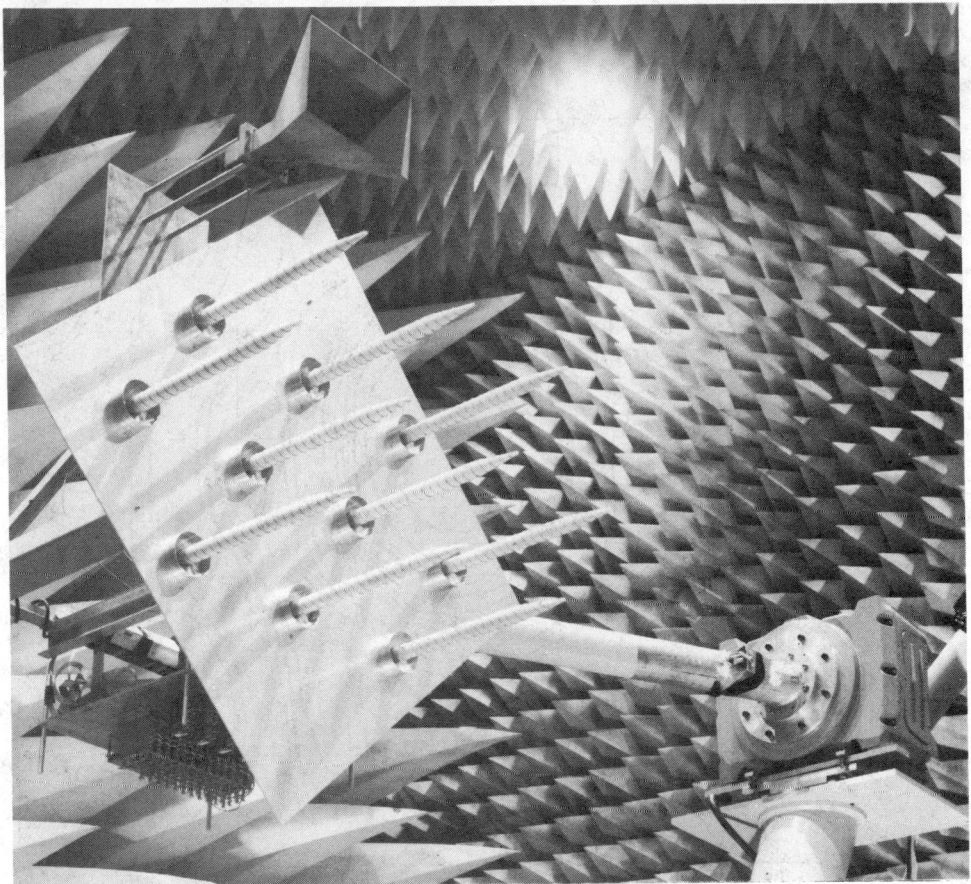

Fig. 49. Breadboard ten-element array. (*After Donn [33], © 1980 IEEE*)

to 86.4 cm. For a given diameter and 19 feed elements, there is unique feed horn spacing that yields a maximum antenna gain achievable over the field of view. The minimum gains achieved over the coverage area versus lens diameters are plotted in Fig. 52. The gain, which is increased with phase correction for the feed cluster, is also given in the figure.

Fig. 53 shows the measured and computed patterns of the on-focused beam while Fig. 54 is for an off-focused beam. The earth coverage radiation pattern obtained by exciting all 19 feed elements with equal amplitude is shown in Fig. 55.

A Reflector Antenna System—A reflector antenna system [36] is designed for Arabsat to provide transmission of C-band signals to operate between 3.2 and 4.2 GHz for fixed satellite service. Frequency reuse is achieved through dual circular polarizations. The antenna system consists of an offset parabolic reflector fed by an array of circular feed elements as depicted in Fig. 56. The reflector has a projected aperture of 1.52 × 1.57 m with focal length of 1.58 m. The feed elements are excited by an air-supported bar-line feed network using a direct plug-in interface as

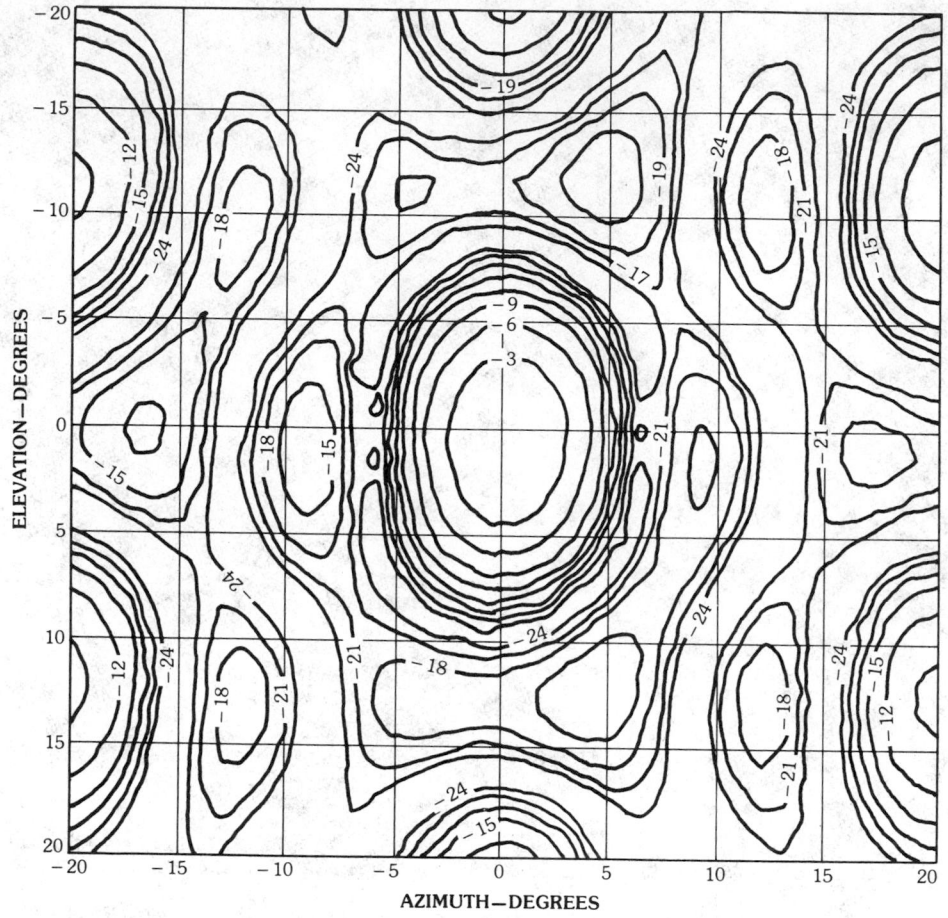

a

Fig. 50. Beam performances. (*a*) Boresight beam. (*b*) Scanned beam. (*After Donn [33],* ©
1980 IEEE)

shown in Fig. 57. The feed network consists of a two-to-four dual-mode converter
and a four-to-thirteen power divider for two circular polarizations. The circuit
diagram of the beam-forming network is presented in Fig. 58.

A circular open-ended waveguide is chosen because of good polarization
performance, wide bandwidth, simplicity of mechanical integration, simple and
mechanically strong structure, and extremely low loss.

Two different sizes of feed spacing are considered: 1.1λ and 1.6λ. The smaller
feed spacing, 9.53 cm, is selected because of better axial ratio performance and
better beam shaping. The septum polarizer is used because of its compactness
to integrate in the feed array and its good performance. The polarizer has a
square-cross-section waveguide at one end and a sloping septum that gradually
divides the square waveguide into a separate rectangular-cross-section waveguide
with a common broadwall at the other end. This septum polarizer also functions as

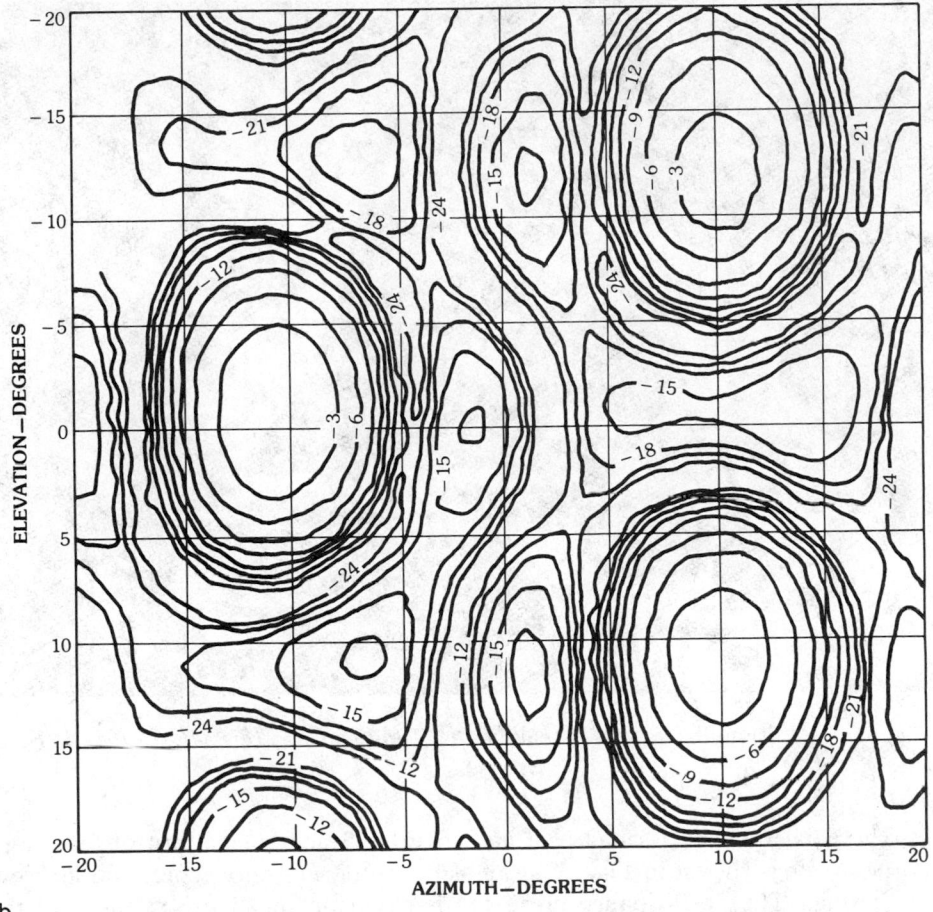

Fig. 50, *continued.*

an orthomode transducer in separating two circularly polarized signals. The achieved performance is summarized in Chart 1. Two types of transition from the septum polarizer to feed horn aperture are considered: stepped and tapered. A stepped transition is physically shorter than the tapered, but it generates both propagating and evanescent higher-order modes. A tapered transition is chosen because it is simpler and will not generate significant higher-order modes that may cause a degradation of the feed patterns.

Chart 1. Measured Performance of the Septum Polarizer

Frequency band	3.7–4.2 GHz
Maximum axial ratio	0.2 dB
Insertion loss	0.1 dB
Return loss	28 dB
Isolation	32 dB

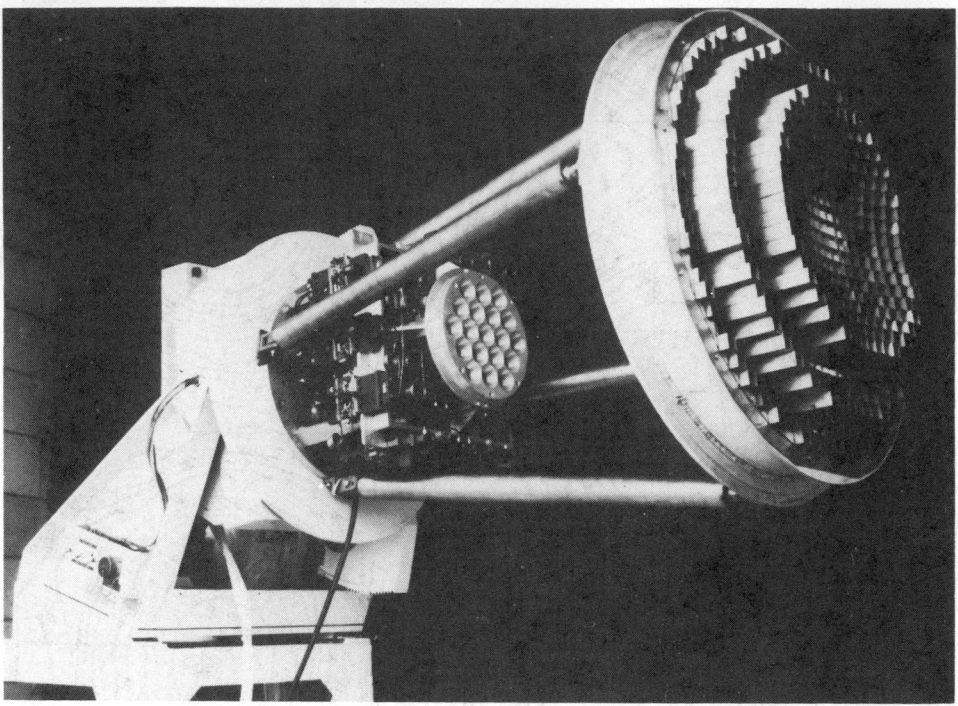

Fig. 51. LES 7 multibeam lens antenna (*After Dion and Ricardi [34], © 1971 IEEE*)

The axial ratio performance of an active feed element pattern of feed horn and polarizer is shown in Fig. 59 at ambient prethermal, hot, cold, and ambient postthermal. The performance presented is based on the figure of merit of the pattern in which the axial ratios are taken at different pattern cuts and then averaged.

Thirteen feed elements are required to cover the Arabic coverage area. They are grouped as shown in Fig. 60 with respect to the four output ports of the odd/even mode converter to provide better beam shaping and to minimize mode shift. A mode shift is due to the fact that the radiation pattern for the odd mode is slightly shifted and different from that for the even mode. The whole feed network consists of four separate circuit segments: RHCP power divider, LHCP power divider, RHCP dual-mode converter, and LHCP dual-mode converter (Fig. 57). The LHCP dual-mode converter, which consists of ring hybrids and differential line lengths, is shown in Fig. 61. The measured and computed patterns are given in Fig. 62.

3. Earth Coverage Antennas

Horns are widely used for earth coverage antennas. The design difficulty lies in the polarization purity requirement over the earth coverage, which is approximately a ±9° circle. This requirement precludes the conventional conical horn

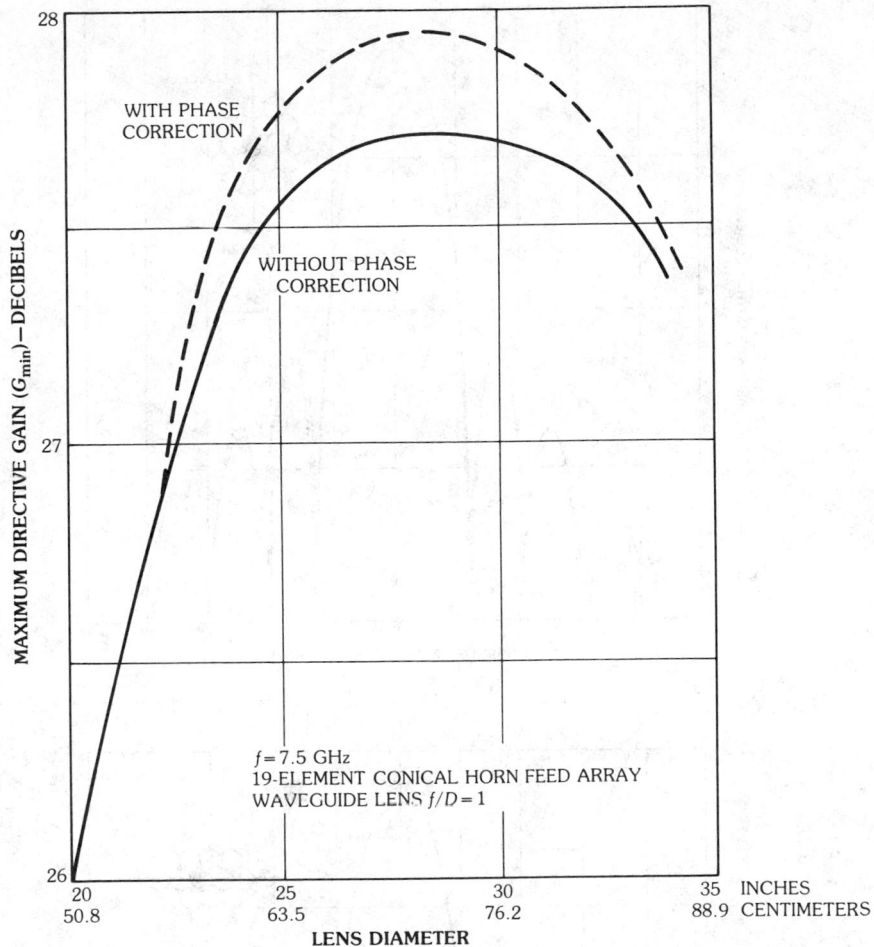

Fig. 52. Minimum directive gain over field of view versus lens diameter.

antenna even though it has the simplest structure. A conical horn has a poor axial ratio at the edge of the beam and relatively high side lobe levels.

In order to meet the polarization purity requirement an earth coverage antenna must have a good polarizer and a rotationally symmetrical beam over at least ±9° for the earth subtended angle over the whole frequency bandwidth. There are two approaches to achieve a rotationally symmetrical beam: corrugation of the surface and generation of a higher-order mode inside the horn. A corrugated horn antenna has wider operating frequency characteristics but it is difficult to fabricate and is relatively heavy. A dual-mode horn is used more frequently because of its simplicity.

A dual-mode horn with an abrupt waveguide discontinuity at the throat [37] is strongly frequency dependent and therefore is a narrow-band radiator. There are two ways to improve the bandwidth of a dual-mode horn: dielectric loading [38] and introduction of dual steps. In the former approach the horn is loaded with a

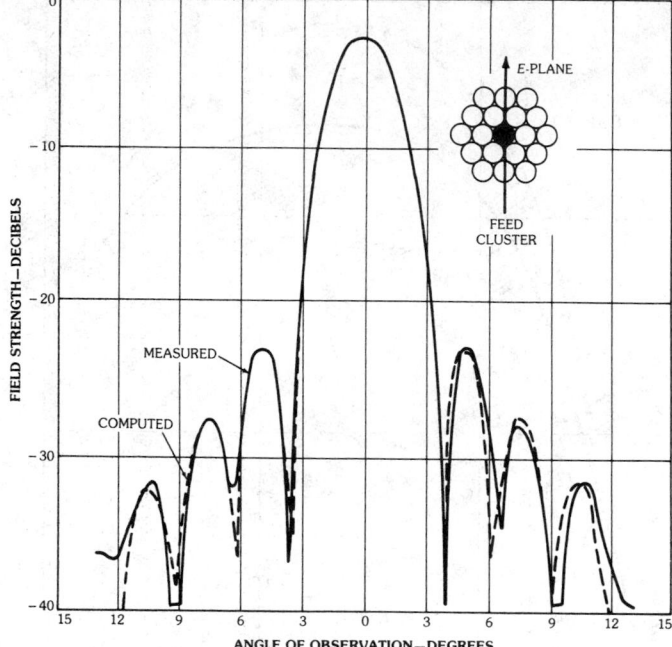

a

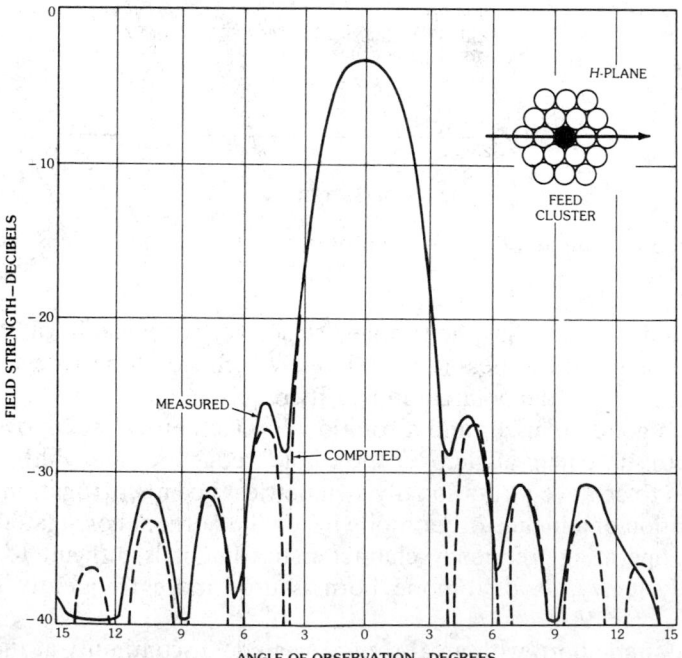

b

Fig. 53. Patterns of center beam. (*a*) *E*-plane. (*b*) *H*-plane. (*After Dion and Ricardi [34],* ©
1971 IEEE)

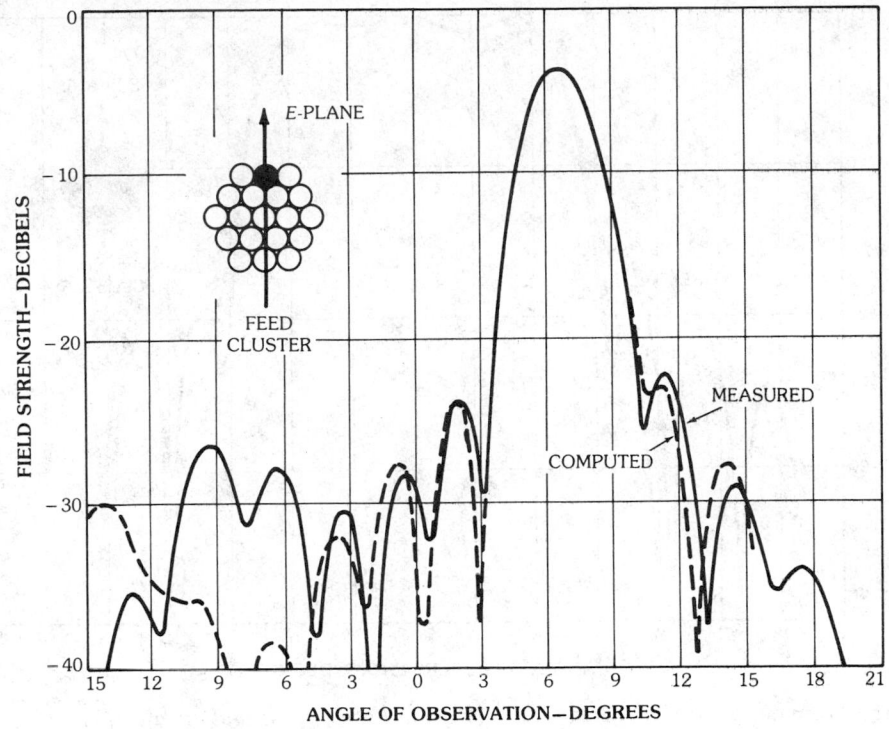

a

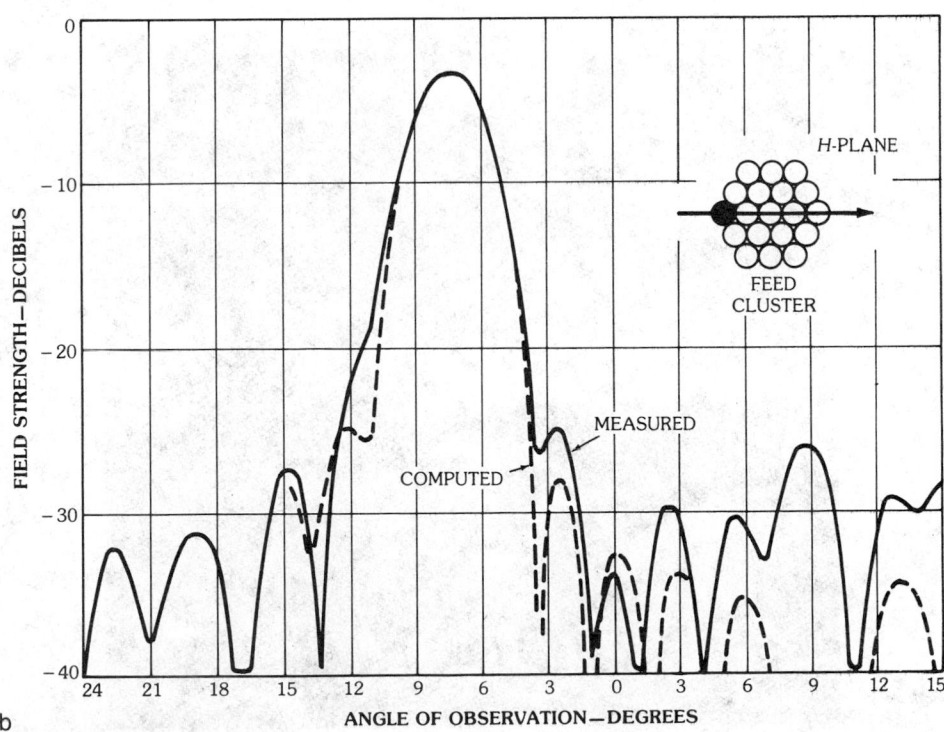

b

Fig. 54. Patterns of off-focused beam. (*a*) *E*-plane. (*b*) *H*-plane. (*After Dion and Ricardi [34], © 1971 IEEE*)

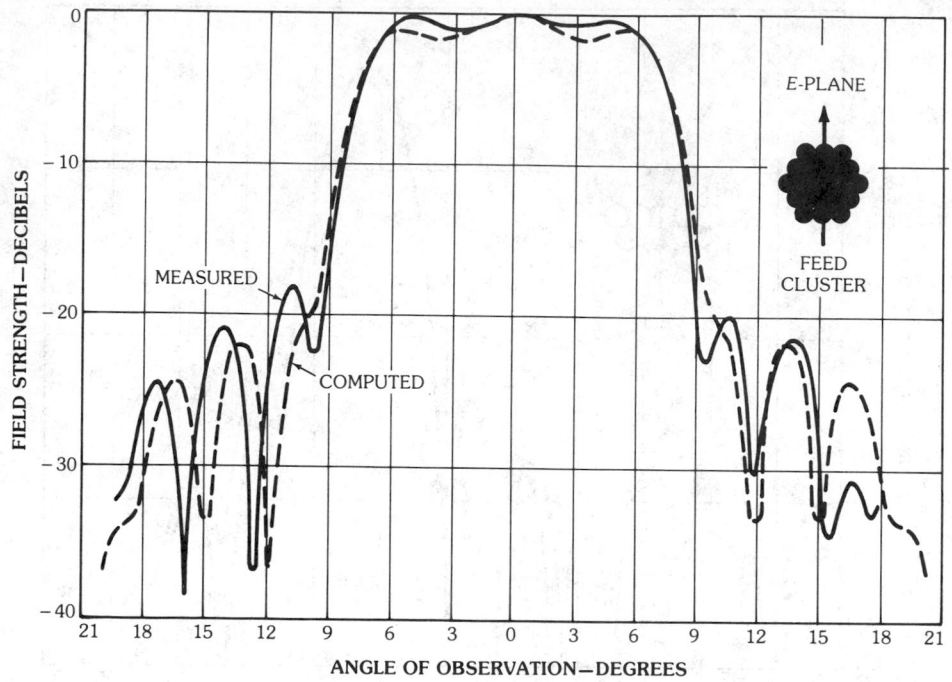

Fig. 55. *E*-plane pattern of earth coverage beam (*After Dion and Ricardi [34], © 1971 IEEE*)

Fig. 56. Arabsat *C*-band transmit antenna—breadboard model. (*Courtesy Ford Aerospace and Communications Corp.*)

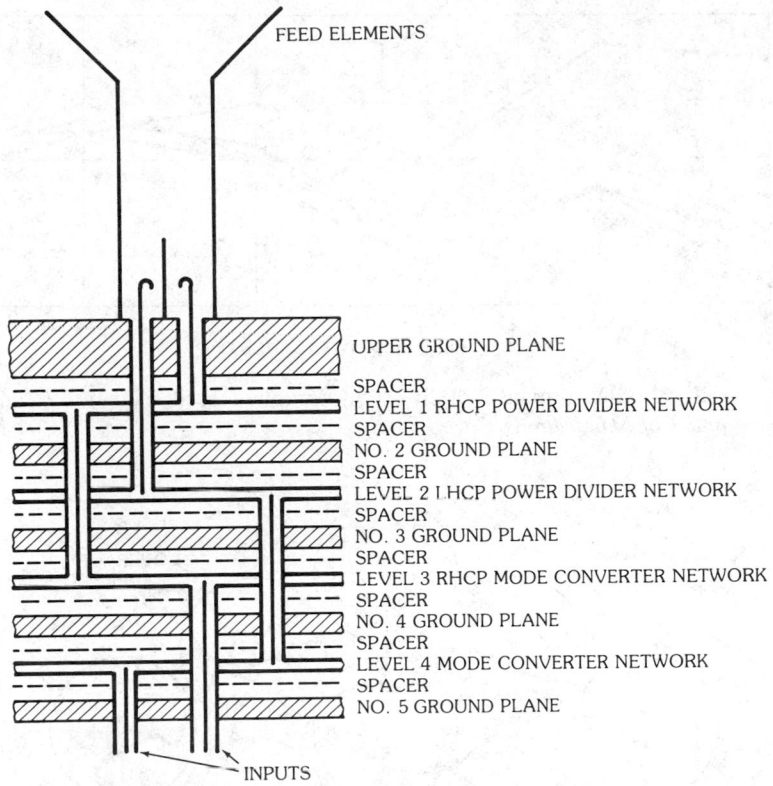

Fig. 57. A 4-GHz transmit feed assembly. (*Courtesy Ford Aerospace and Communications Corp.*)

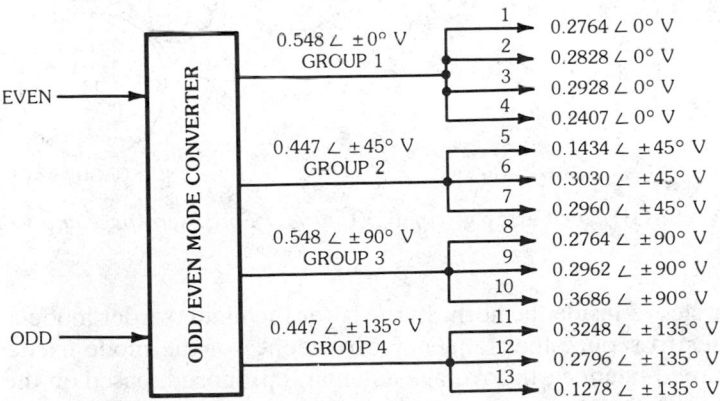

Note: Plus angles are even mode
Minus angles are odd mode

Fig. 58. Diagram of 4-GHz transmit beam-forming network. (*Courtesy Ford Aerospace and Communications Corp.*)

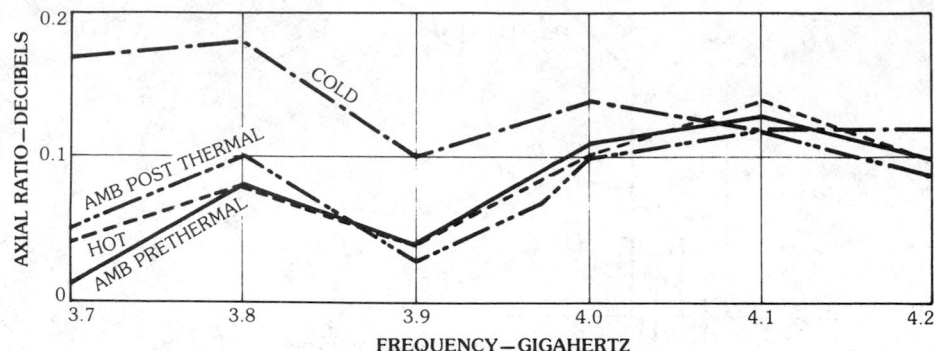

Fig. 59. Axial ratio performance of five-screw graphite feed element. (*Courtesy Ford Aerospace and Communications Corp.*)

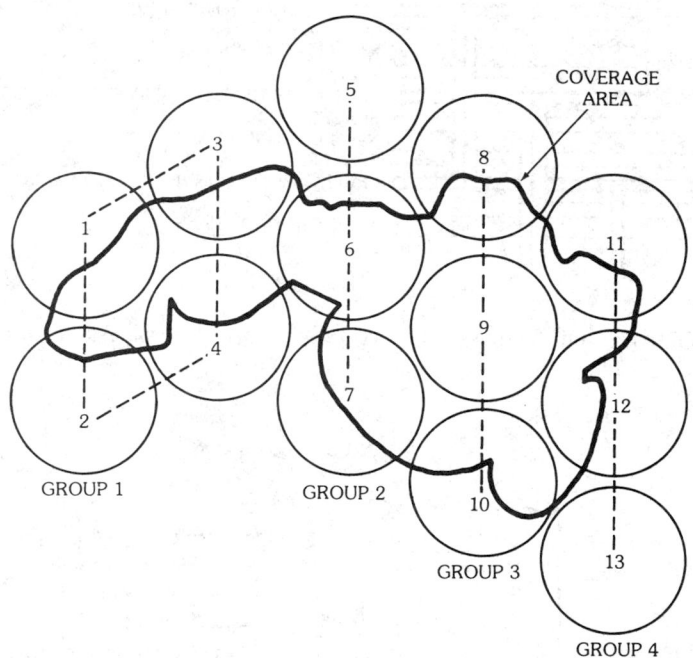

Fig. 60. The 4-GHz feed array grouping. (*Courtesy Ford Aerospace and Communications Corp.*)

dielectric sleeve inside the horn. In the latter the higher-order mode is excited with multi-irises to reduce the frequency dependence of the mode excited.

A shaped-beam earth coverage antenna is proposed, based on the fact that the paths tangential to the earth are longest. As a result the gain should be highest in this region to compensate for the atmospheric attenuation and decrease to a minimum for the path normal to the earth. A nine-horn array [34] has been proposed to achieve the required beam shape. Alternate approaches are the shaped lens and a multibeam reflector antenna system. Table 10 summarizes the advantages and disadvantages of each design.

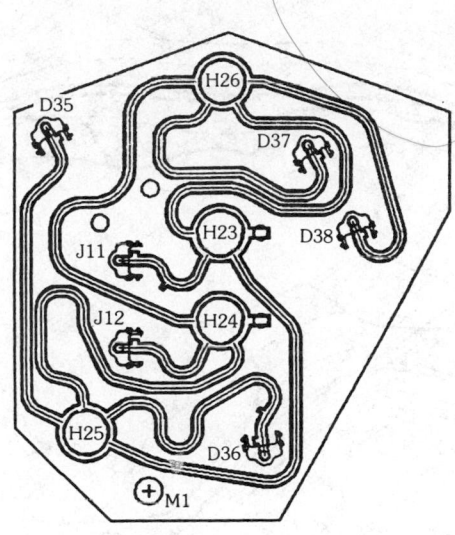

Fig. 61. Level LHCP 2–4 mode converter network. (*Courtesy Ford Aerospace and Communications Corp.*)

Table 10. Alternate Types of Earth Coverage Antennas

Type	Advantages	Disadvantages
Conical horn	Simple Easy to fabricate	Poor axial ratio at the beam edge
Potter horn	Simple Easy to fabricate	Narrow band
Dielectric-loaded horn	Wide band Good axial ratio	
Multistepped dual-mode horn	Wide band Good axial ratio over 11° circle	
Corrugated horn	Wide band Excellent axial ratio	Highest cost Difficult to fabricate Relatively heavy
Shaped-beam nine-horn	Higher gain	Complicated design Heavy

The Stepped Horn

It is well known that for a conical horn, its beamwidths in the E- and H-planes are different when operating in the dominant TE_{11} mode. The beamwidths can be equalized by introducing the higher-order TM_{11} mode. These two modes, when excited in the horn aperture with the appropriate relative amplitude and phase, can affect the beamwidth equalization. A simple step discontinuity used to excite the horn with TE_{11} and TM_{11} modes is proposed [37]. The power conversion coefficients and the launching phases of the TE_{11} and TM_{11} modes are given [39] in Fig. 63. Fig. 64 sketches a K-band earth coverage horn used as a beacon designed

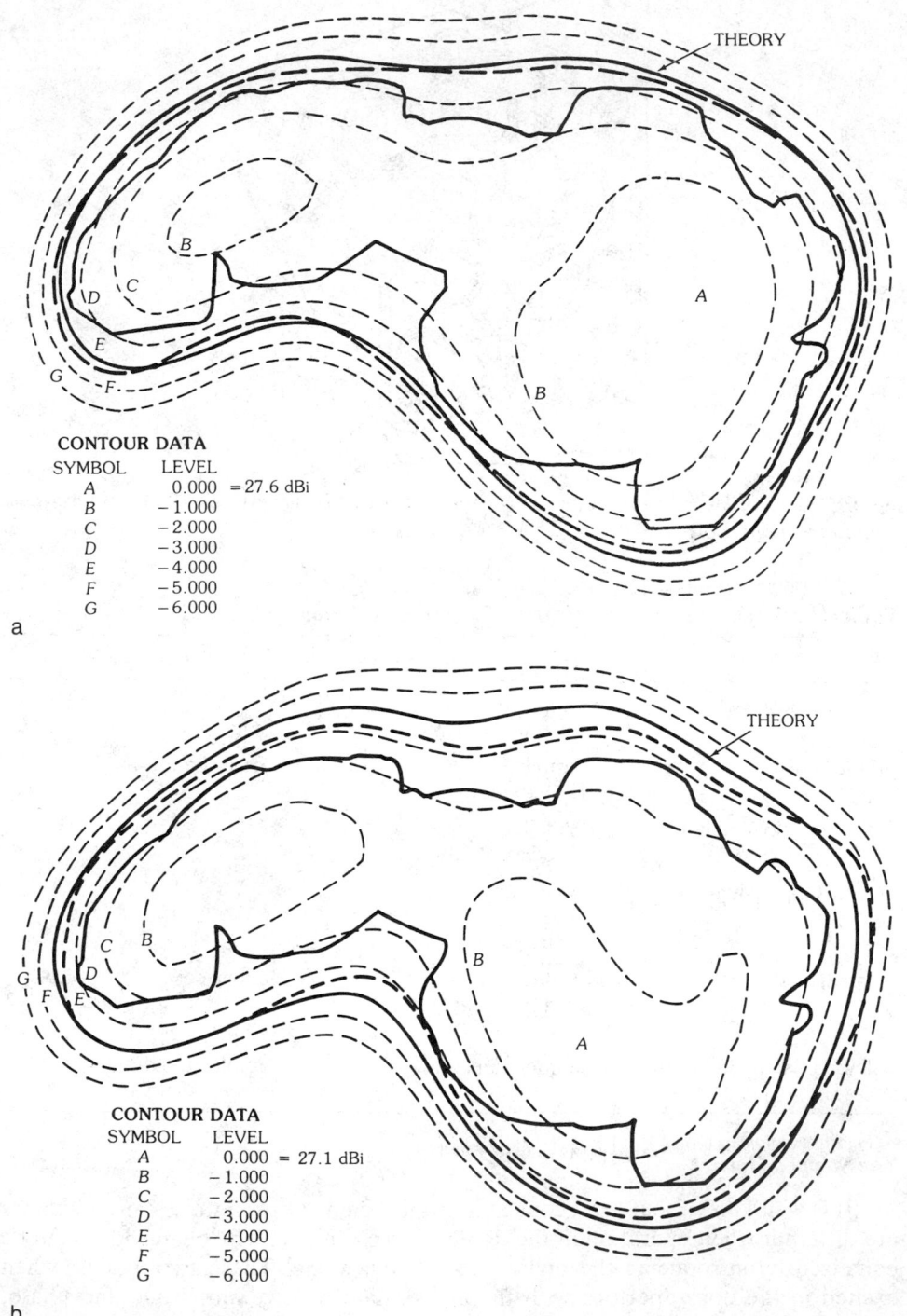

CONTOUR DATA

SYMBOL	LEVEL	
A	0.000	= 27.6 dBi
B	−1.000	
C	−2.000	
D	−3.000	
E	−4.000	
F	−5.000	
G	−6.000	

a

CONTOUR DATA

SYMBOL	LEVEL	
A	0.000	= 27.1 dBi
B	−1.000	
C	−2.000	
D	−3.000	
E	−4.000	
F	−5.000	
G	−6.000	

b

Fig. 62. Theoretical and measured patterns. (*a*) For 3.7-GHz even mode. (*b*) For 3.7-GHz odd mode. (*Courtesy Ford Aerospace and Communications Corp.*)

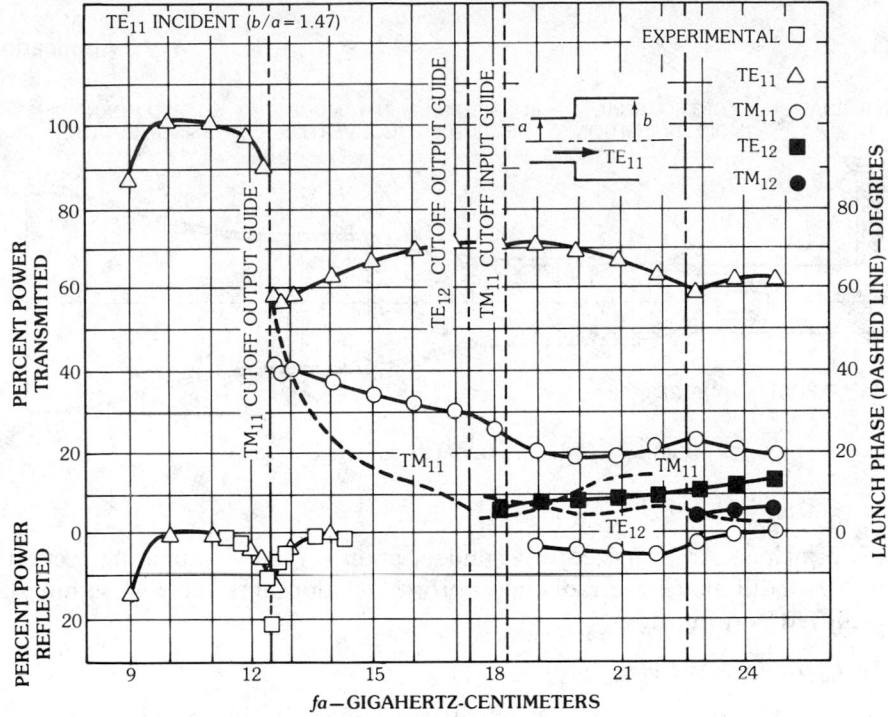

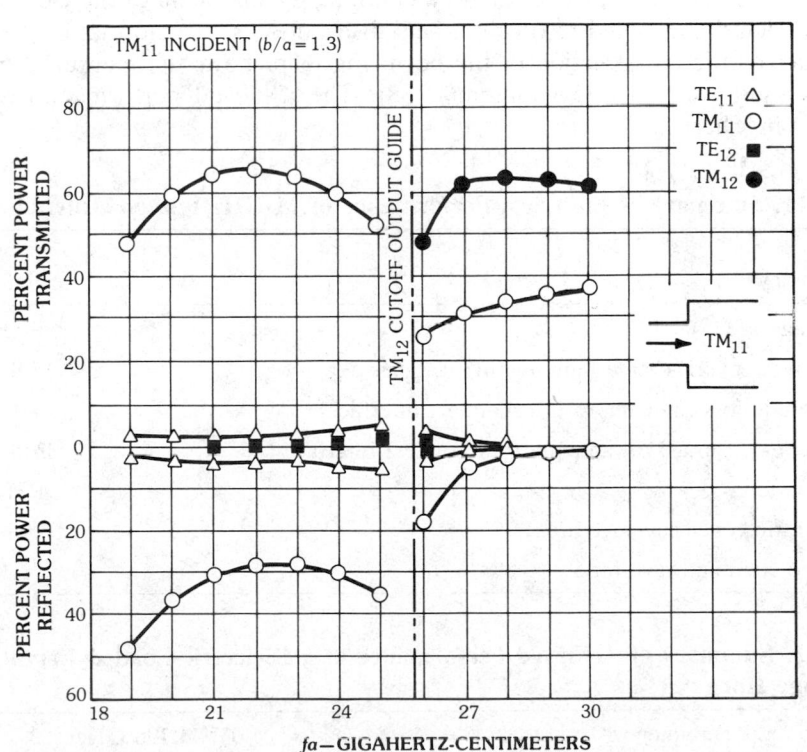

Fig. 63. Mode transducing properties of a symmetric step-discontinuity waveguide junction. (*a*) Illuminated with an incident fundamental TE_{11} mode. (*b*) Illuminated with an incident TM_{11} mode. (*After English [39], © 1973 IEEE*)

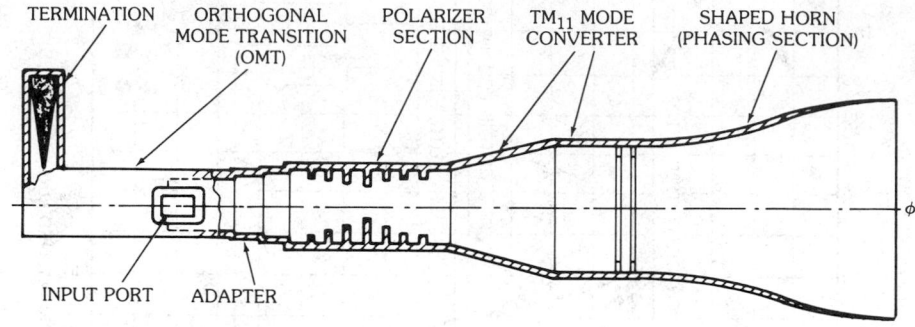

TERMINATION   ORTHOGONAL   POLARIZER   TM$_{11}$ MODE   SHAPED HORN
              MODE TRANSITION   SECTION   CONVERTER   (PHASING SECTION)
              (OMT)

INPUT PORT   ADAPTER

Fig. 64. An 11-GHz earth coverage antenna.

by this principle. An oversized waveguide section is used as a phasing section to ensure that both modes are radiated in proper relationship. Table 11 summarizes the achieved performance.

The Dielectric-Loaded Horn

A dielectric-loaded horn is sketched in Fig. 65a. In this approach the introduction of the dielectric sleeve excites an antiphase component of the electric field in the dielectric band and results in a field distribution similar to that of the TM$_{11}$ mode across the cross section of the horn. Fig. 65b shows the measured pattern of a C-band earth coverage antenna [38]. The achieved performance is summarized in Chart 2.

Table 11. Summary of Electrical Performance of 11-GHz Beacon Antenna

Frequency	11 190 to 11 460 MHz
Coverage area	22° circular
Gain over coverage area (antenna input flange)	15.9 dBi
Waveguide loss (antenna to transponder) interface	1.1 dB
Gain over coverage (at antenna transponder) interface	14.8 dBi
Polarization	RHCP
Axial ratio (over coverage area)	0.98 dB
Voltage standing-wave ratio	1.15

Chart 2. Summary of Achieved Performance of a Dielectric-Loaded Earth Coverage Horn

Frequency	4.037–4.198 GHz
Coverage area	±9°
Antenna gain	17 dBi
Axial ratio	0.4 dB

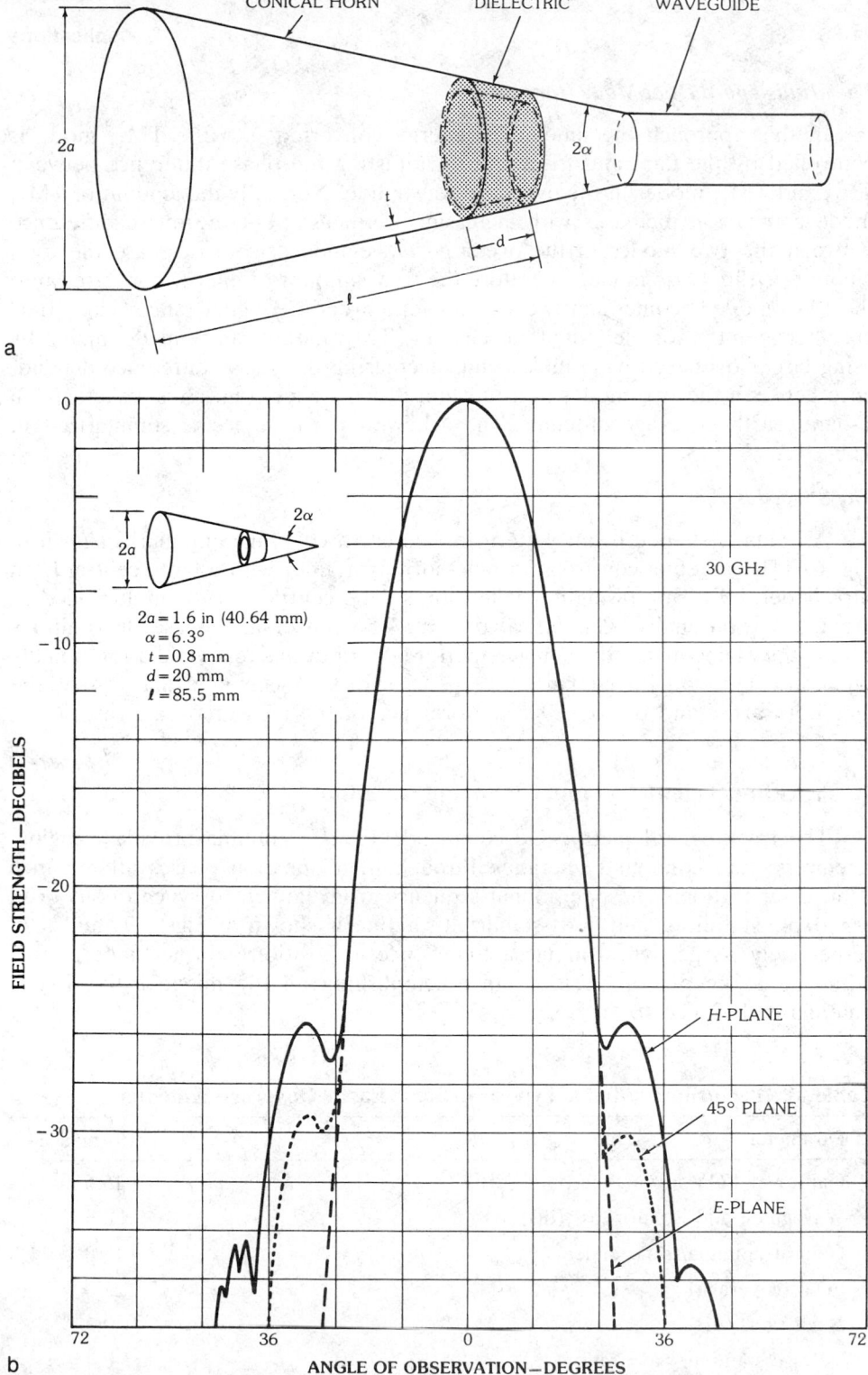

Fig. 65. Dielectric-loaded barn antenna. (*a*) Structure. (*b*) Radiation patterns. (*After Satoh [38],* © *1972 IEEE*)

The Multistepped Dual-Mode Horn

In this approach the amount of energy converted into the TM_{11} mode is controlled by the flare and irises and by adjusting the phase difference between TE_{11} and TM_{11} modes in the oversized waveguide. Normally the amount of TM_{11} mode conversion increases with increasing frequency, but the phase difference between the two modes in the oversized waveguide acts to decrease the total amount of the TM_{11} mode. Therefore the total amount of energy converted into the TM_{11} mode becomes nearly constant over a broad frequency range. The difference between the wavelength of the TE_{11} and TM_{11} modes can be made smaller by using larger oversized waveguides, thus decreasing the phase difference dependence between the two modes as a function of frequency. Fig. 66 is a sketch of a *C*-band earth coverage antenna. The achieved performance is summarized in Table 12.

The Shaped Beam

An ideally shaped beam pattern for the earth coverage antenna is shown in Fig. 67. The antenna consists of a nine-horn array [40] with a large central horn surrounded by a ring of eight smaller horns. The central horn is multimoded to provide a rotationally symmetrical pattern. The power and phase distributions among the elements are to simulate aperture distributions that yield a rotationally symmetrical sector-shaped pattern. Fig. 68 shows a schematic diagram of the nine-horn array and the achieved performance of the measured patterns.

4. Tracking, Telemetry, and Command Antenna

The tracking, telemetry, and command (TT&C) antenna provides ranging, telemetry, and command operation throughout all mission phases after launch vehicle separation. The operational sequence from launch to synchronous orbit for a spin-stabilized and body-stabilized satellite is shown in Figs. 69a and 69b, respectively. A geosynchronous satellite, once in position, still needs occasional adjusting to keep it in position. In accomplishing all the missions the TT&C antenna is designed to

Table 12. Performance for a Typical *C*-Band Earth Coverage Antenna

Parameter	Performance
Gain over FOV at antenna ports (dBi)	15.6
Waveguide and isolator loss (dB)	0.6
Gain at antenna/transponder	15.0
FOV (coverage)	22°
Axial ratio (dB)	0.32
Voltage standing-wave ratio	0.4
Gain variation (dB)	3.0
Isolation, port-to-port (dB)	>40

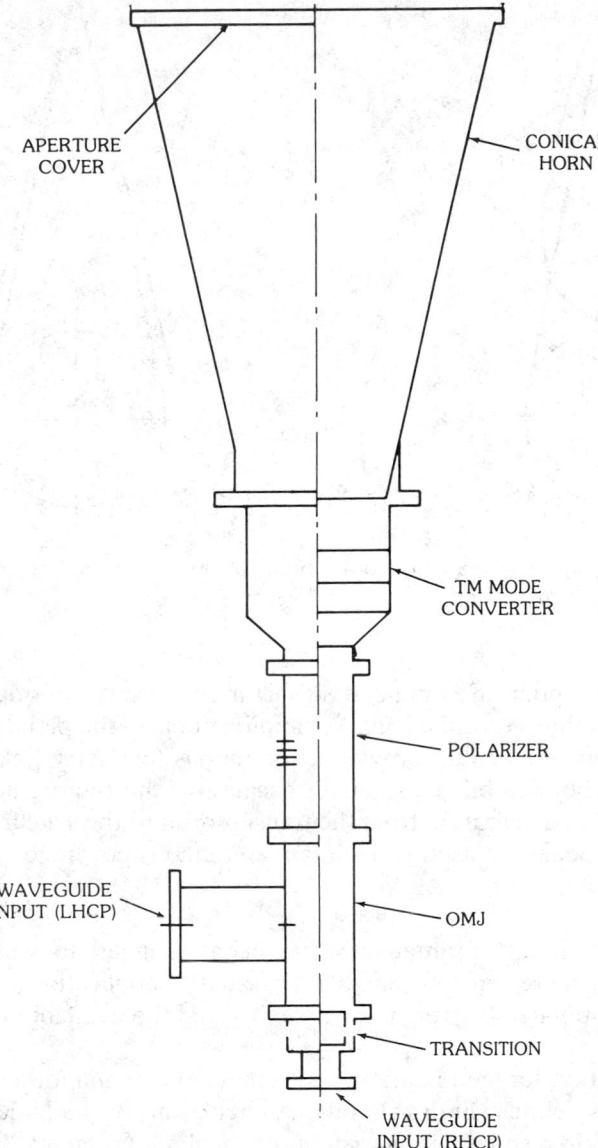

Fig. 66. A 4-GHz-band earth coverage antenna. (*Courtesy Mitsubishi Electric Corp.*)

(*a*)   receive satellite functional commands from ground stations,
(*b*)   transmit satellite functional data (telemetry) to ground stations,
(*c*)   provide a beacon signal to aid ground station acquisition of the satellite, and
(*d*)   receive and retransmit ranging signals.

The typical antenna pattern requirement for the TT&C antenna is illustrated in Fig. 70. A nondirectional and circularly polarized antenna is required to ensure

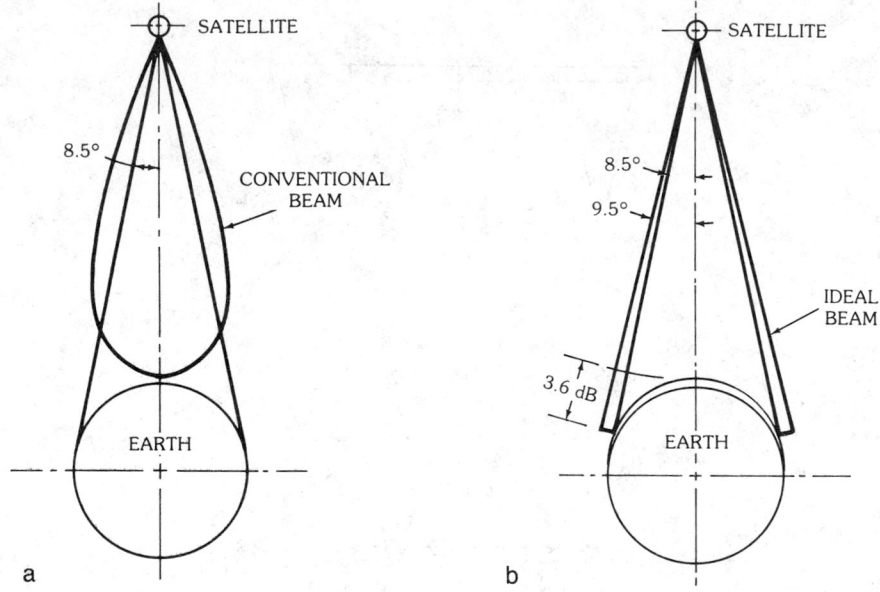

Fig. 67. Earth coverage antennas. (*a*) Conventional beam. (*b*) Shaped beam. (*After Ajioka and Harry [40], © 1970 IEEE*)

the continuous reception of command signals in every mission operation. For a spin-stabilized satellite a toroidal beam (omnidirectional in the plane perpendicular to the satellite spin axis) could provide a continuous telemetry link in almost all directions. For a body-stabilized satellite, because of the requirement of a more complex maneuvering operation from the transfer orbit to the synchronous orbit, a cardioid beam is generally used to maintain continuous coverage.

Types of Antennas

Although a number of antennas have been designed to generate a non-directional beam, there are few suitable for satellite application. The design of the antenna is dominated by weight, complexity, and the available location in the satellite.

An ideal location for the telemetry and command antenna for a spin-stabilized satellite is at the satellite cylindrical body. A circular array is a logical choice as a result of the location restrictions and omnidirectional requirements. Because of the broad pattern of the telemetry and command antenna, interference from the other antenna systems, such as large reflectors for communication antenna systems and the satellite structures, might cause intolerable degradation on the antenna performance. A biconical antenna, which is placed in an optimized location to minimize scattering or reduce blockage of the sight of view, is also used. A cardioid beam is generated by a slotted ring antenna. The pattern shaping is achieved by using a multiring on the cylindrical waveguide or attaching a conical reflector to the waveguide structure. These three types of antennas are described below.

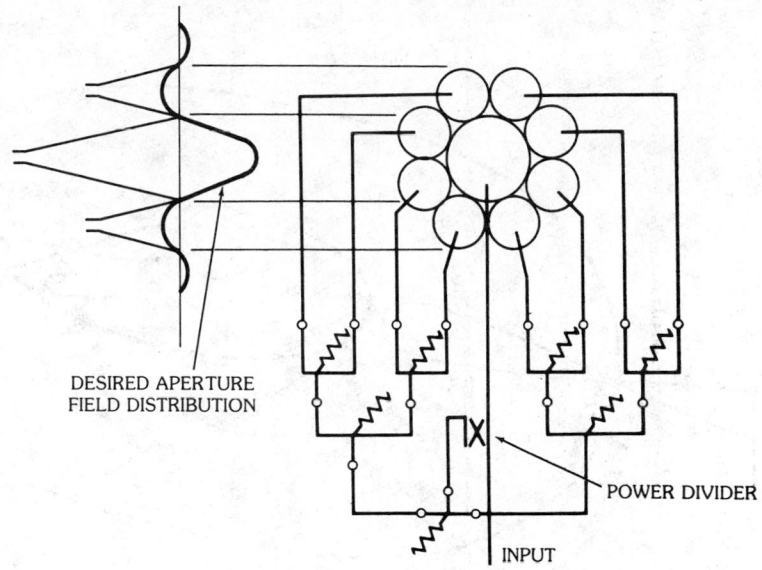

DESIRED APERTURE
FIELD DISTRIBUTION

POWER DIVIDER

INPUT

a

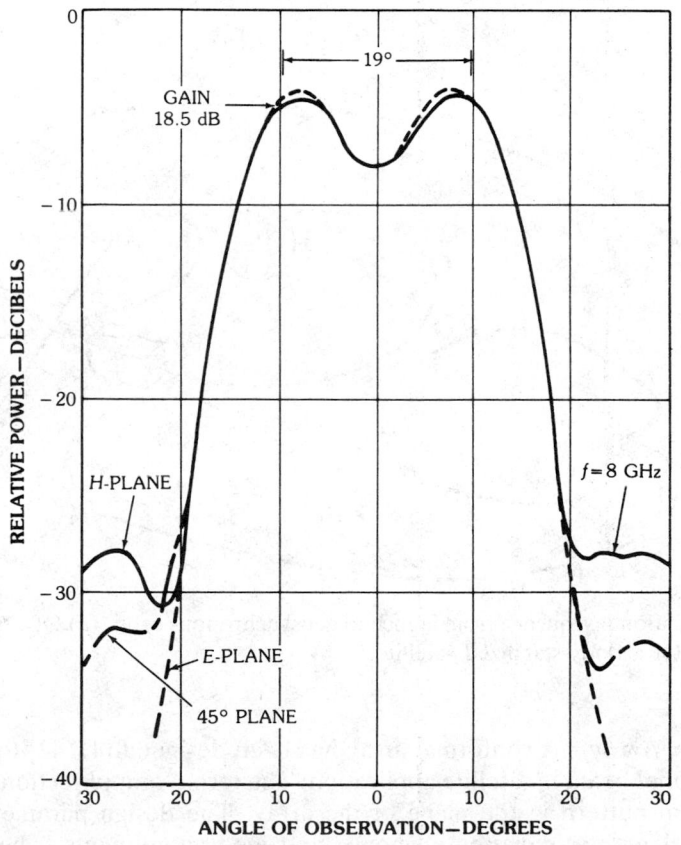

b

Fig. 68. Nine-horn array diagram and pattern. (*a*) Schematic diagram. (*b*) Measured patterns with center horn multimoded. (*After Ajioka and Harry [40], © 1970 IEEE*)

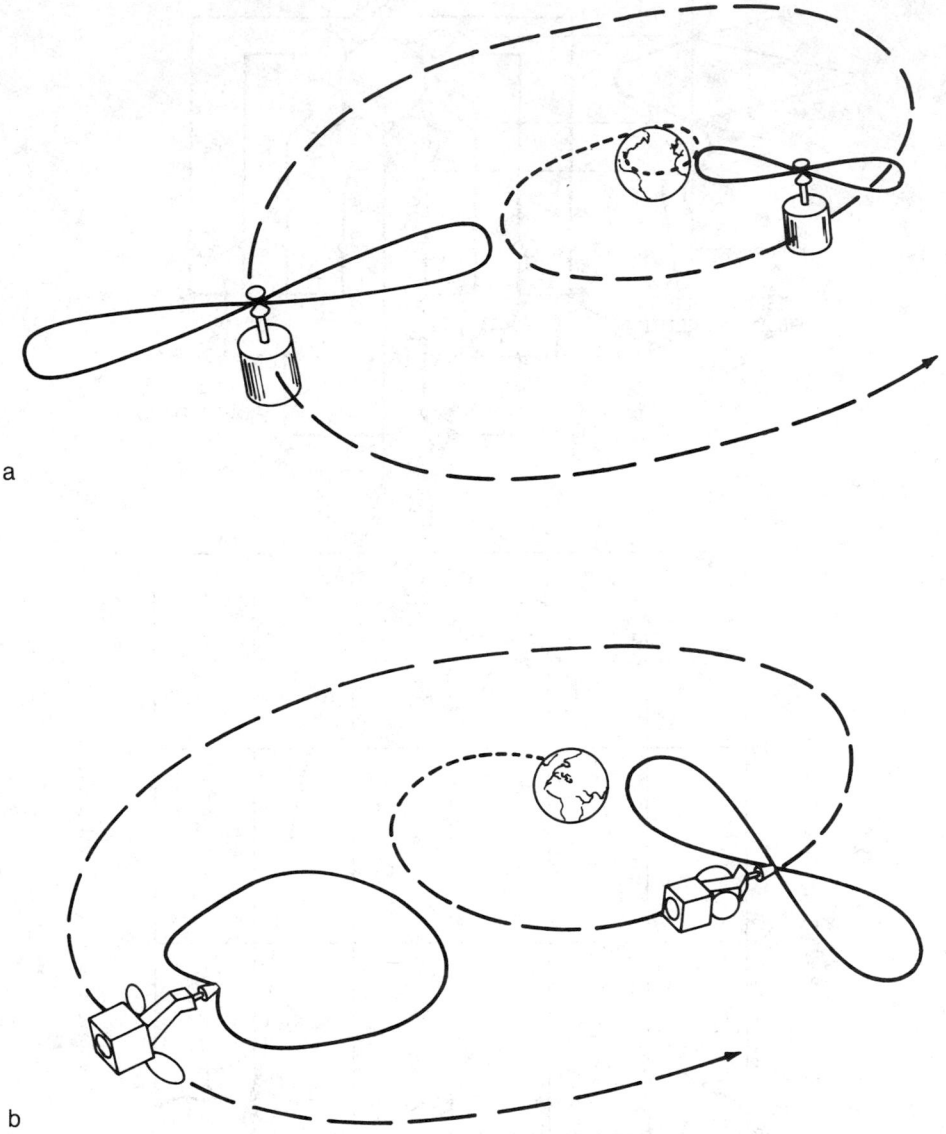

Fig. 69. Operational sequence from launch to geosynchronous orbit. (*a*) Of a spin-stabilized satellite. (*b*) Of a body-stabilized satellite.

The Circular Array—A conformal array has been designed [41, 42] to provide an omnidirectional beam in satellite application. The term "omnidirectional" refers to the azimuthal pattern in the plane of the array. The design parameters are the number of elements, radiating elements, and feed components. The number of elements S is determined to provide a nearly omnidirectional pattern. The minimum number of the elements is decided by the allowed amplitude ripple. The evaluation of amplitude ripple can be given in terms of the fluctuation, which is

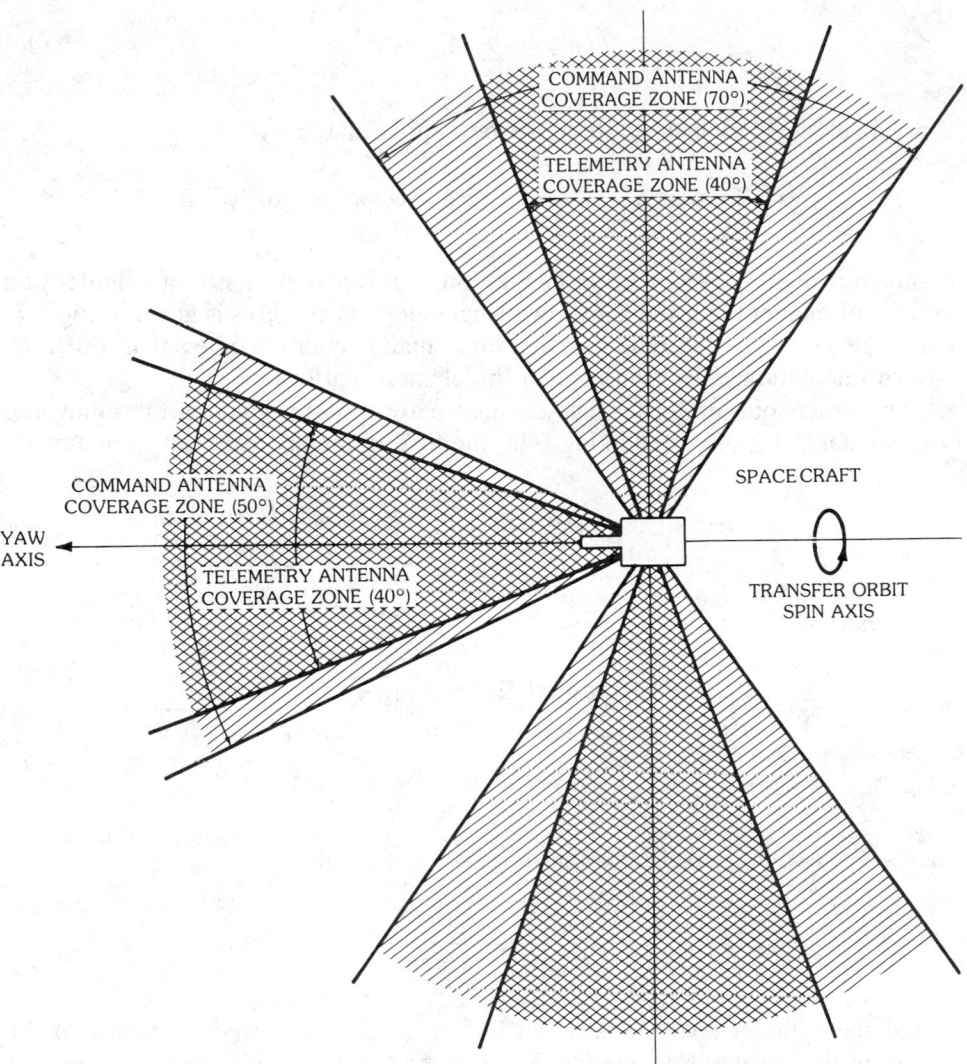

Fig. 70. Telemetry and command coverage requirements.

defined as the ratio of maximum $|\phi|$ to minimum $|\phi|$, where $|\phi|$ is the total far-field pattern of an S-element circular array and is given [41] as

$$\Phi \cong S \sum_{n=0}^{N} A_n(-j)^n \frac{d^n}{dz^n}[J_0(z) + 2(j)^S J_S(z) \cos \phi] \tag{25}$$

where $J_S(z)$ is the Bessel function of the first kind of order S, $z = k_0 a \sin \theta$ in which k_0 is the free-space wave number, and a is the radius of the circular array and is measured from the polar axis. In this derivation it is assumed that the single-element pattern, $F(\phi')$, can be represented by the Fourier cosine series

$$F(\phi') = \sum_{n=0}^{\infty} A_n \cos^n \phi' \qquad (26)$$

A practical single-element pattern can be approximated by

$$(1 + \cos \phi')/2 \quad \text{or} \quad (2 + 3\cos \phi' + \cos 2\phi')/6$$

as shown in Fig. 71. The pattern fluctuation as a function of size of cylinder and number of elements for the above two single-element patterns is given in Figs. 72 and 73, respectively. It can be seen that for a small fluctuation of less than 1 dB, the pattern fluctuation is less sensitive to the element pattern.

The exact solution for a single-element pattern [44] is given in the following. For an axial slot as depicted in Fig. 74a, the field in the aperture can be assumed:

$$E_\phi = \frac{V}{\alpha a} \cos\left(\frac{\pi z}{L}\right), \qquad \begin{cases} -L/2 < z < L/2 \\ -\alpha/2 < \phi < \alpha/2 \end{cases} \qquad (27)$$

The radiation pattern is

$$E_\phi \cong \frac{VL}{\pi^3 a} \frac{e^{-jk_0 r}}{r} \left\{ \frac{\cos(k_0 L/2)\cos\theta}{1 - [(kL/\pi)\cos\theta]^2} \right\} \sum_{n=-\infty}^{\infty} \frac{j^n e^{jn\phi}}{H_n^{(2)\prime}(k_0 a \sin\theta)} \qquad (28)$$

$$E_\theta = 0$$

where $H_n^{(2)}$ is the Hankel function of second kind and

$$\sum_{n=-\infty}^{\infty} \frac{j^n e^{jn\phi}}{H_n^{(2)\prime}(k_0 a \sin\theta)}$$

(called the cylinder space factor) can be regarded as a correction factor to the pattern of the isolated slot antenna. For a circumferential slot as shown in Fig. 74b the field in the aperture can be assumed:

$$E_z = \frac{V}{W} \cos\left(\frac{\pi\phi}{\alpha}\right), \qquad \begin{cases} -W/2 < z < W/2 \\ -\alpha/2 < \phi < \alpha/2 \end{cases} \qquad (29)$$

The radiation pattern is

$$E_\phi = -\frac{V\alpha \cos\theta}{\pi k_0 a \sin\theta} \frac{e^{-jk_0 r}}{r} \sum_{n=-\infty}^{\infty} \frac{nj^n \cos\theta \, (n\alpha/2) e^{jn\phi}}{[\pi^2 - (n\alpha)^2] H_n^{(2)\prime}(k_0 a \sin\theta)} \qquad (30)$$

$$E_\theta = \frac{k_0 V\alpha}{j\pi \sin\theta} \frac{e^{-jk_0 r}}{r} \sum_{n=-\infty}^{\infty} \frac{j^n \cos(n\alpha/2) e^{jn\phi}}{[\pi^2 - (n\alpha)^2] H_n^{(2)}(k_0 a \sin\theta)}$$

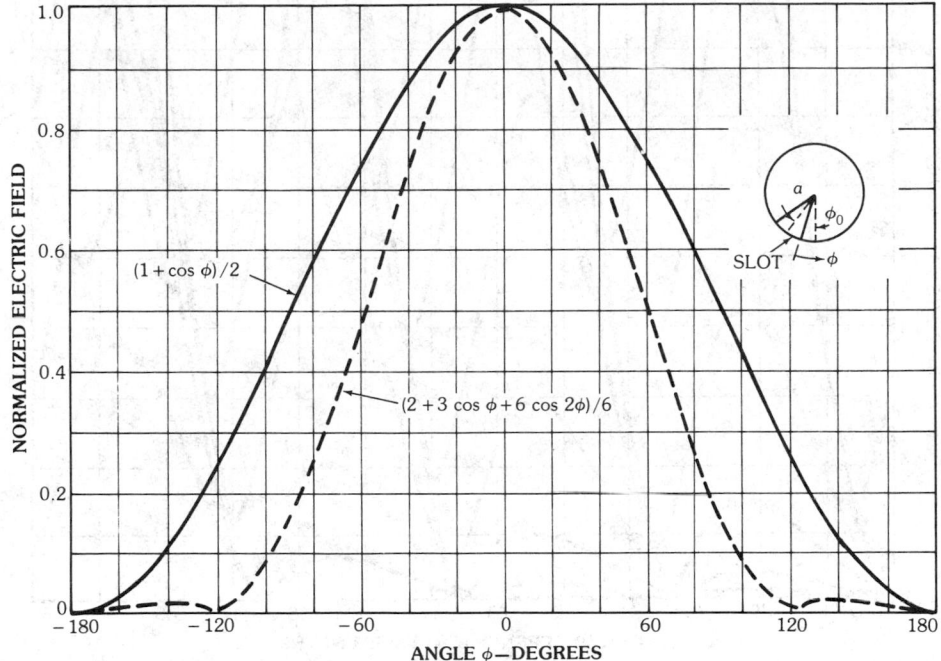

Fig. 71. Approximate electric field patterns.

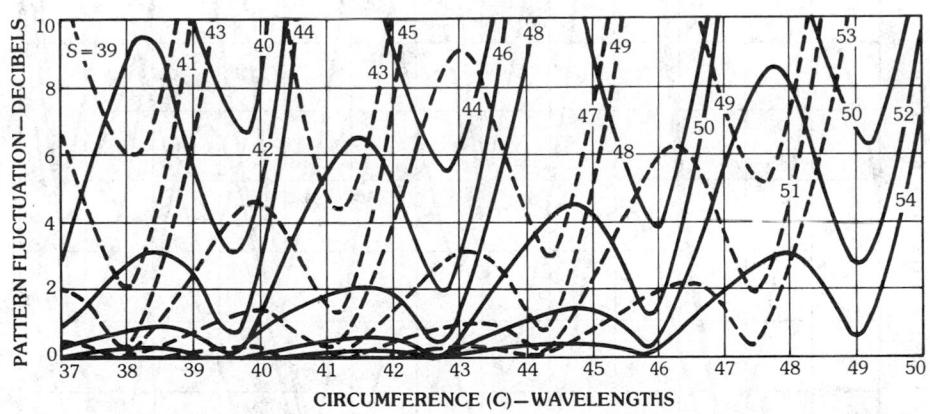

Fig. 72. Pattern fluctuation dependence on number of elements and cylinder size for feed pattern $1 + 1/2 \cos \phi$.

In the azimuthal plane ($\theta = 90°$), the radiation pattern becomes

$$E_\phi = 0$$

$$E_\theta = \frac{k_0 V a}{j\pi} \frac{e^{-jk_0 r}}{r} \sum_{n=-\infty}^{\infty} \frac{j^n \cos(n\alpha/2) e^{jn\phi}}{[\pi^2 - n\alpha]^2 H_n^{(2)}(k_0 a)}$$

$$(31)$$

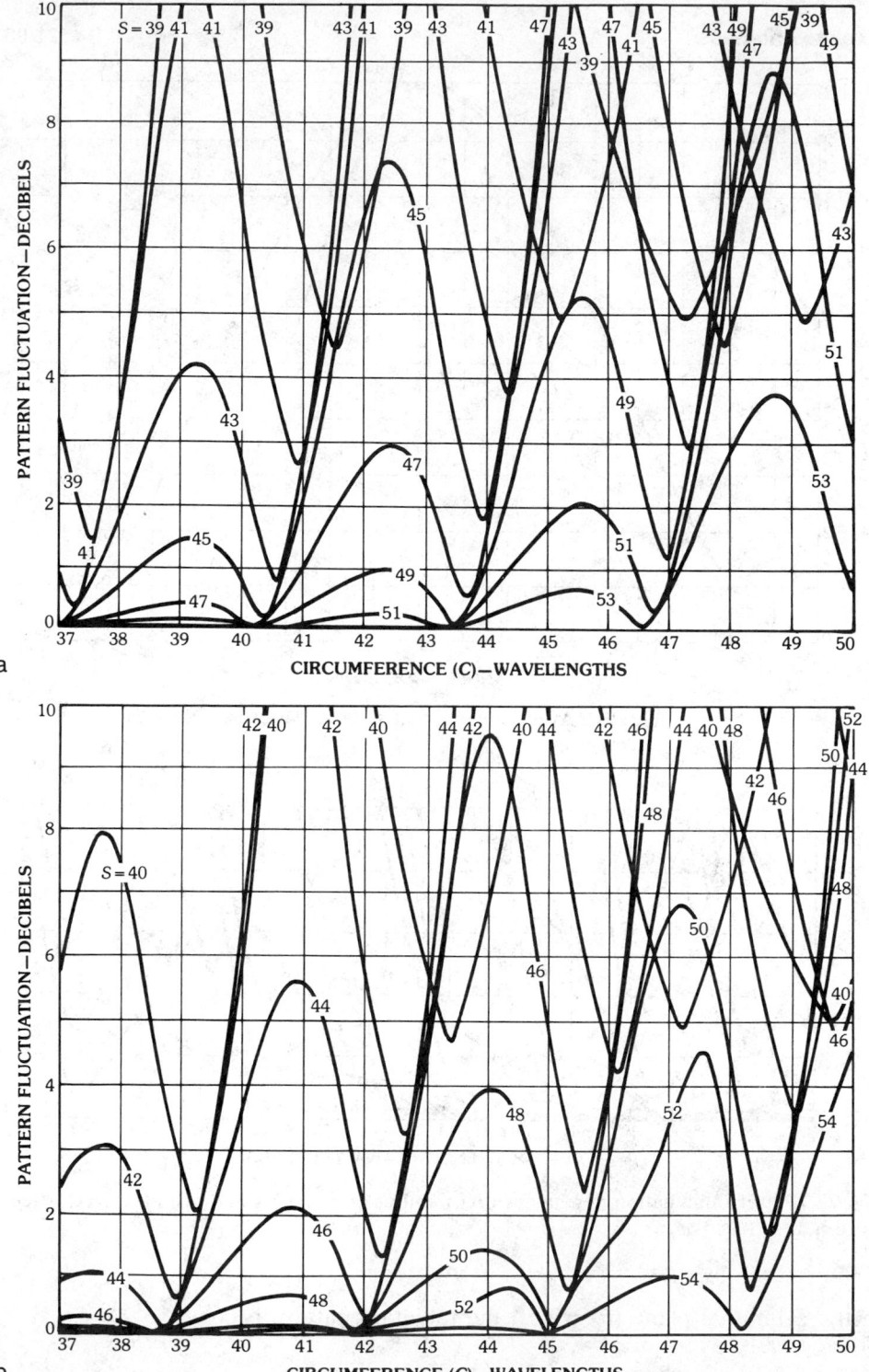

Fig. 73. Pattern fluctuation dependence on number of elements and cylinder size for feed pattern $(2 + 3\cos\phi + \cos 2\phi)/6$. (*a*) For odd numbers of elements. (*b*) For even numbers of elements.

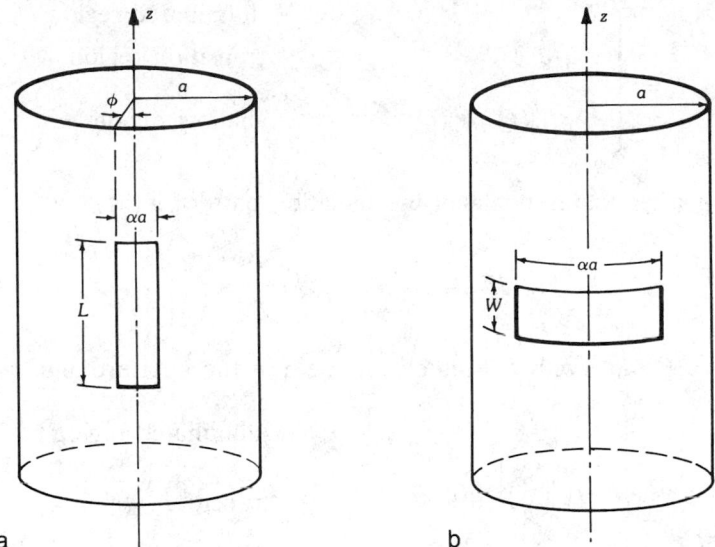

Fig. 74. Conducting cylinder. (*a*) With axial slot. (*b*) With circumferential slot.

The above series is slow convergent when k_0a (circumference of the cylinder in wavelengths) becomes large. The geometrical theory of diffraction [45] is then applied. According to the geometrical theory of diffraction, the far-field region is divided into the illuminated, transition, and shadow regions as shown in Fig. 75. For an axial slot the radiation pattern in the azimuthal plane can be given as

$$dE_\phi = -\frac{jk_0}{4\pi} dP_m F \frac{e^{-jk_0 r}}{r} \tag{32}$$

where dP_m is the infinitesimal magnetic current moment and $dP_m = E_\phi^a\, da'$, where E_ϕ^a is the electric field in the aperture, da' is an area element, and

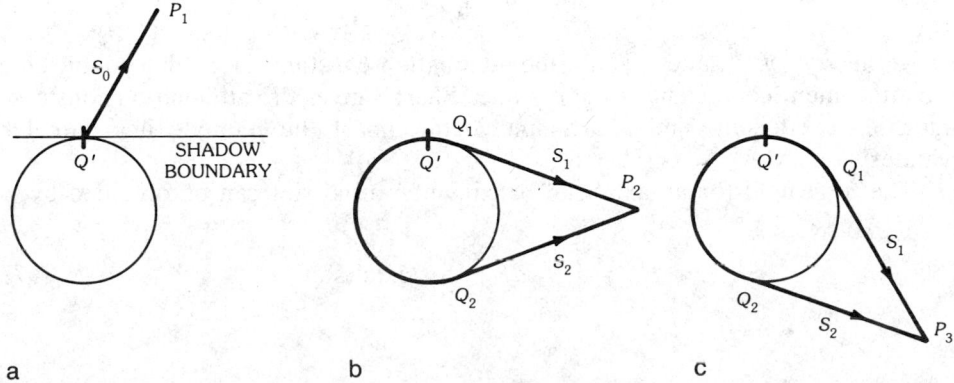

Fig. 75. Rays emanating from a source on a conducting cylinder. (*a*) In the illuminated region. (*b*) In the transition region. (*c*) In the shadow region.

$$
F = \begin{cases} 2 & \text{illuminated region} \\ g(\xi)e^{-jk_0t} & \text{transition region} \\ \displaystyle\sum_{p=1}^{\infty} L_p^h(Q')A_p^h(Q)e^{-(\alpha_p^h+jk_0)t} & \text{shadow region} \end{cases} \tag{33}
$$

Similarly, for a circumferential slot the radiation pattern is

$$
dE_z = -\frac{jk_0}{4\pi}dP_m G\frac{e^{-jk_0r}}{r} \tag{34}
$$

where $dP_m = E_{z'}^a\,da'$, with $E_{z'}^a$ the electric field in the aperture and

$$
G = \begin{cases} 2\cos\theta & \text{illuminated region} \\ -j\left(\dfrac{2}{k_0a}\right)^{1/3}\hat{g}(\xi)e^{-jk_0t} & \text{transition region} \\ \displaystyle\sum_{p=1}^{\infty} L_p^s(Q')A_p^s(Q)e^{-(\alpha_p^s+jk_0)t} & \text{shadow region} \end{cases} \tag{35}
$$

The $g(\xi)$ and $\hat{g}(\xi)$ are the Fock functions that can be expressed as

$$
\begin{aligned}
g(\xi) &= \frac{1}{\sqrt{\pi}}\int_{e^{-j2\pi/3}}^{\infty} \frac{e^{-j\tau\xi}}{W_2'(\tau)}d\tau \\
\hat{g}(\xi) &= \frac{1}{\sqrt{\pi}}\int_{e^{-j2\pi/3}}^{\infty} \frac{e^{-j\tau\xi}}{W_2(\tau)}d\tau
\end{aligned} \tag{36}
$$

in which $W_2(\tau)$ and $W_2'(\tau)$ are the Fock-type Airy function and its derivative, respectively. The term t is the distance parameter as shown in Fig. 76, and

$$
\xi = \frac{1}{a}\left(\frac{k_0a}{2}\right)^{1/3}t
$$

The terms $\alpha_p^{s,h}$, $L_p^{s,h}$, and $A_p^{s,h}$ are the attenuation constant, launching coefficient, and attachment coefficient of the p mode. Chart 3 gives the attenuation constant, launching coefficient, and attachment coefficient of the p mode in a circular cylinder.

The total field for an axial slot or circumferential slot can be obtained by

$$
E_\phi = \int_{\substack{\text{over}\\\text{slot}}} dE_\phi \quad \text{for axial slot} \tag{37}
$$

or

$$
E_z = \int_{\substack{\text{over}\\\text{slot}}} dE_z \quad \text{for circumferential slot} \tag{38}
$$

Chart 3. Launching Coefficient, Attachment Coefficient, and Attenuation Constant of a Conducting Cylinder

Axial slot

$$L_p^h = -j\left(jk_0\frac{\pi}{2}\right)^{1/2} H_{\bar{\nu}_p}^{(2)}(k_0 a)\, D_p^h$$

$$A_p^h(Q) = L_p^h(Q)$$

$$(D_p^h)^2 = \frac{\pi^{-1/2}\, 2^{-5/6}\, a^{1/3}\, e^{-j\pi/12}}{k_0^{1/6}\, \bar{q}_p [A_i(-\bar{q}_p)]^2}\left[1 + \left(\frac{2}{k_0 a}\right)^{2/3}\left(\bar{q}_p/30 - \frac{1}{10\bar{q}_p^2}\right)e^{-j\pi/3}\right]$$

$$\alpha_p^h = \frac{\bar{q}_p}{a}\, e^{j\pi/6}\left(\frac{k_0 a}{2}\right)^{1/3}\left[1 + \left(\frac{2}{k_0 a}\right)^{2/3}\left(\frac{\bar{q}_p}{60} + \frac{1}{10\bar{q}_p^2}\right)e^{-j\pi/3}\right]$$

$$A_i'(-\bar{q}_p) = 0$$
$$\bar{q}_1 = 1.01879$$
$$\bar{q}_2 = 3.2482$$
$$A_i(-\bar{q}_1) = 0.53566$$
$$A_i(-\bar{q}_2) = -0.41902$$

Circumferential slot

$$L_p^s = -\left(jk_0\frac{\pi}{2}\right)^{1/2} H_{\nu_p}^{(2)\prime}(k_0 a)\, D_p^s$$

$$A_p^s(Q) = -L_p^s(Q)$$

$$(D_p^s)^2 = \frac{\pi^{-1/2}\, 2^{-5/6}\, a^{1/3}\, e^{-j\pi/12}}{k^{1/6}[A_i'(-\bar{q}_p)]^2}\left[1 + \left(\frac{2}{k_0 a}\right)^{2/3}\left(\frac{q_p}{30}\right)e^{-j\pi/3}\right]$$

$$\alpha_p^s = \frac{q_p}{a}\, e^{j\pi/6}\left(\frac{k_0 a}{2}\right)^{1/3}\left[1 + \left(\frac{2}{k_0 a}\right)^{2/3}\frac{q_p}{60}e^{-j\pi/3}\right]$$

$$A_i(-q_p) = 0$$
$$q_1 = 2.33811$$
$$q_2 = 4.08795$$
$$A_i'(-q_1) = 0.70121$$
$$A_i' - (-q_2) = -0.80311$$

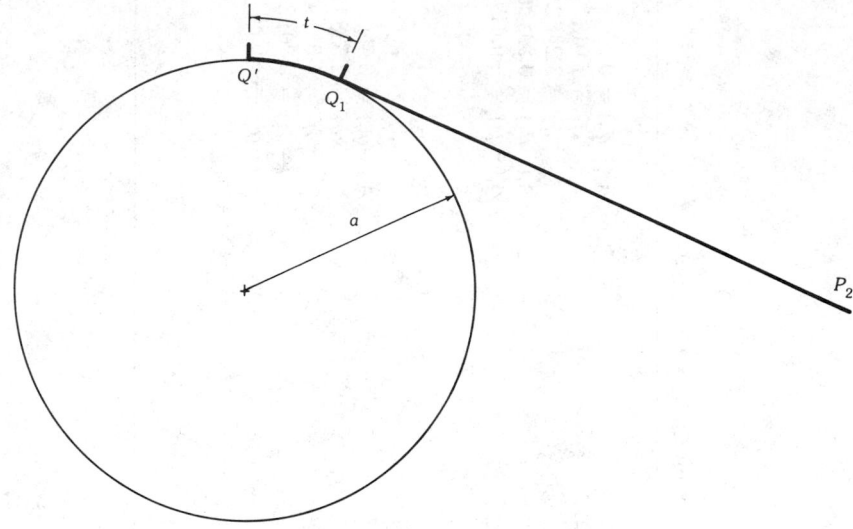

Fig. 76. Diffraction from a source Q'.

The pattern of the cylindrical space factor for a narrow axial slot for different $k_0 a$ is plotted [46] in Fig. 77a, while the pattern for a narrow circumferential slot for different $k_0 a$ is plotted in Fig. 77b.

The total radiation pattern of the array on a cylinder can be obtained by using the known field from the single slot and using superposition directly. The mutual coupling among slots is neglected in the above computation. The mutual coupling among slots on a cylinder has been studied and presented by several authors [47, 48]. The contribution of the mutual coupling to the radiation pattern in this case is not significant.

A typical example of this type of antenna is a 64-element array [49] mounted on a 1.91 m diameter and 1.63 m-long cylindrical satellite design for over the 40 percent bandwidth at S-band. The radiating element is a lightweight cavity-backed turnstile as shown in Fig. 78a. Fig. 78b shows a circumferential section of the conformal array. The measured versus calculated amplitude ripple using (25) in the azimuthal plane is shown in Fig. 78c. The measured elevation radiation pattern (0.7λ interelement spacing) is given in Fig. 78d.

The Biconical Horn—A biconical horn, because of its rotationally symmetrical structure as shown in Fig. 79, could provide an omnidirectional radiation pattern in the azimuthal plane. The two lowest-order modes, the transverse electromagnetic TEM and the transverse electric TE_{01}, are excited to generate vertically and horizontally polarized waves, respectively. The field of the TEM wave is described by [50]

$$E_\theta = -\sqrt{\frac{2k}{\pi}} \frac{1}{r \sin \theta} e^{-jk_0 r}$$

$$H_\phi = \sqrt{\frac{\epsilon_0}{\mu}} E_\theta$$

(39)

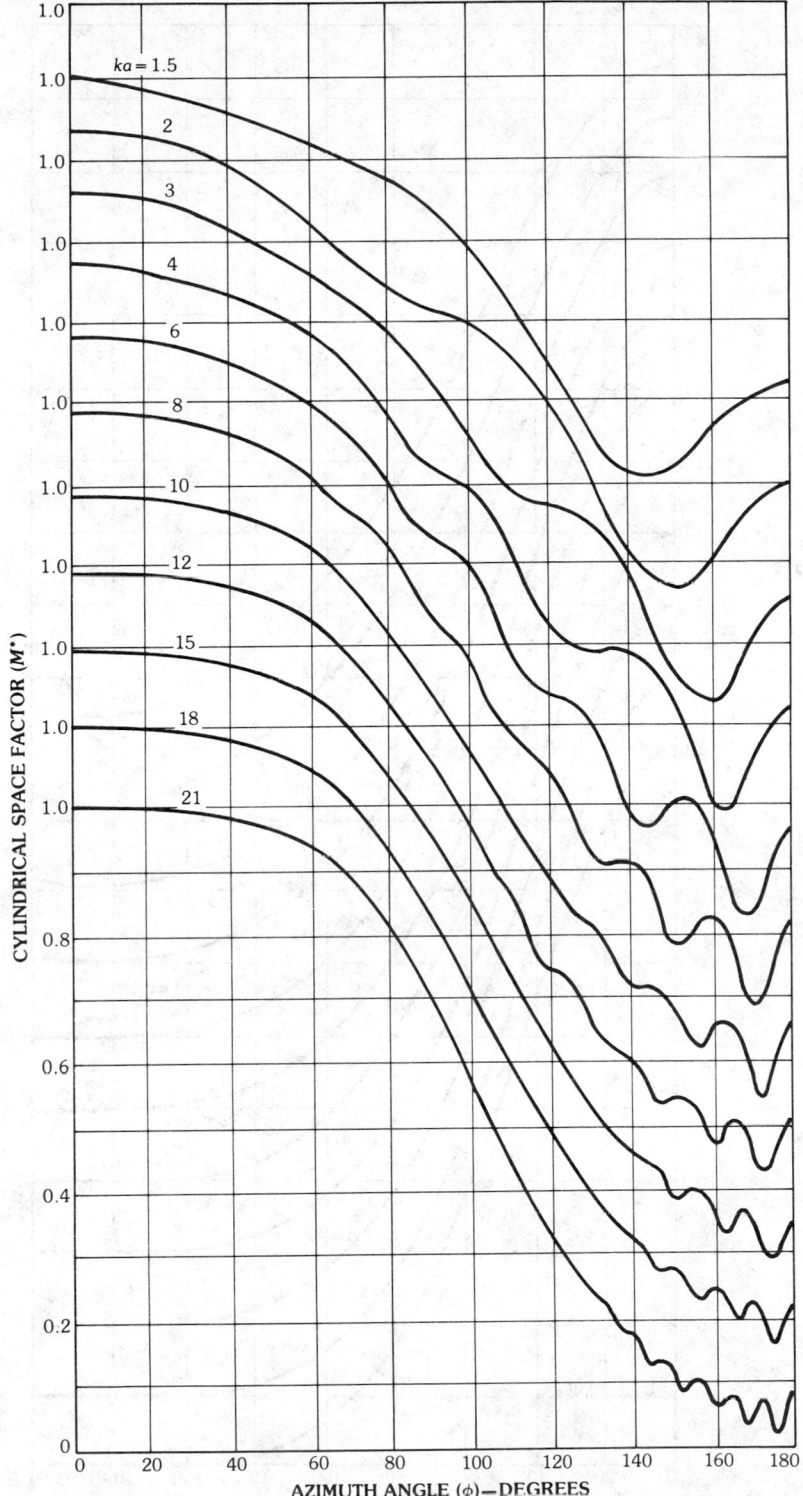

Fig. 77. Cylinder space factors, with vertical scale shifted for each curve by constant amount. (*a*) For narrow axial slot. (*After Wait [46], © 1959 Pergamon Books Ltd.*) (*b*) For narrow circumferential slot.

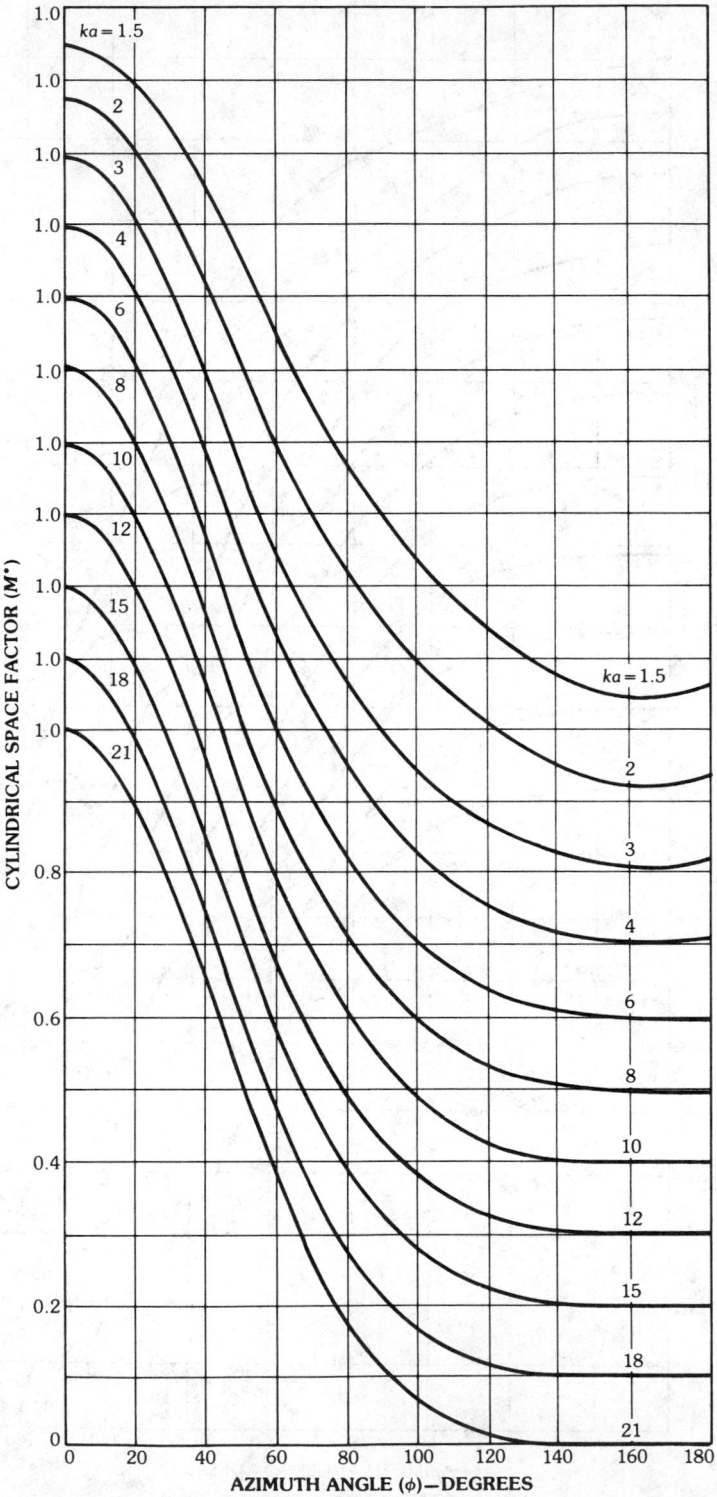

Fig. 77, *continued.*

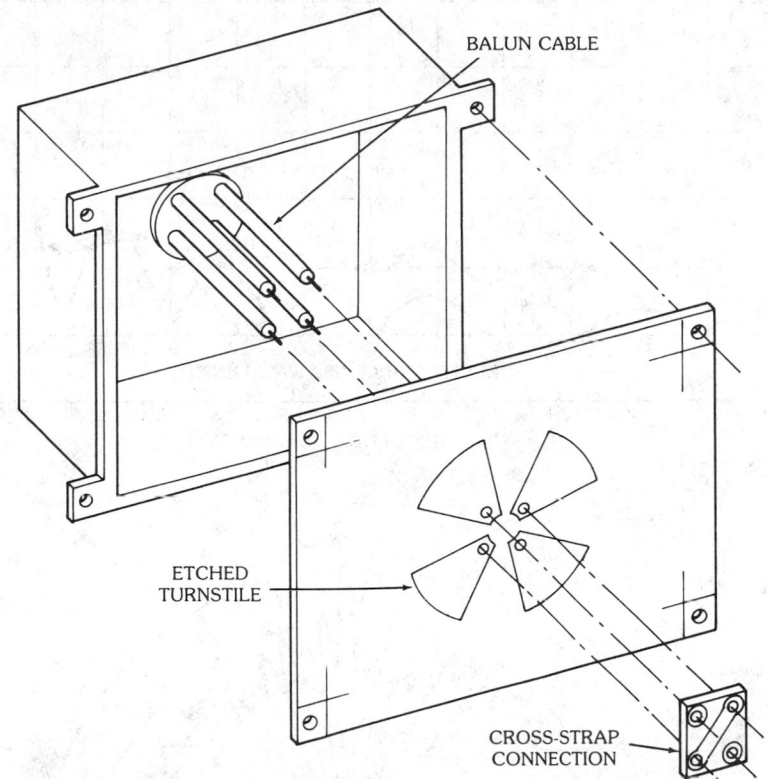

BALUN CABLE

ETCHED
TURNSTILE

CROSS-STRAP
CONNECTION

a

b

Fig. 78. Structure and characteristics of 64-element array. (*a*) Exploded view of array element. (*b*) Circumferential section of conformal array. (*c*) Pattern fluctuation in the ϕ plane. (*d*) Measured elevation (θ-plane) radiation pattern of circular array. (*After Gregorwich [49], © 1979 IEEE*)

while the field of the TE_{01} wave is given by

$$
\begin{aligned}
E_\phi &= -j\omega\mu L'_\ell(\cos\theta)r^{-1/2}H_p^{(2)}(k_0 r) \\
H_r &= L_\ell(\cos\theta)(P^2 - 1/4)r^{-3/2}H_p^{(2)}(k_0 r) \\
H_\theta &= L'_\ell(\cos\theta)[(P^2 - 1/2)r^{-3/2}H_p^{(2)}(k_0 r) - k_0 r^{-1/2}H_{p-1}^{(2)}(k_0 r)]
\end{aligned}
\tag{40}
$$

where

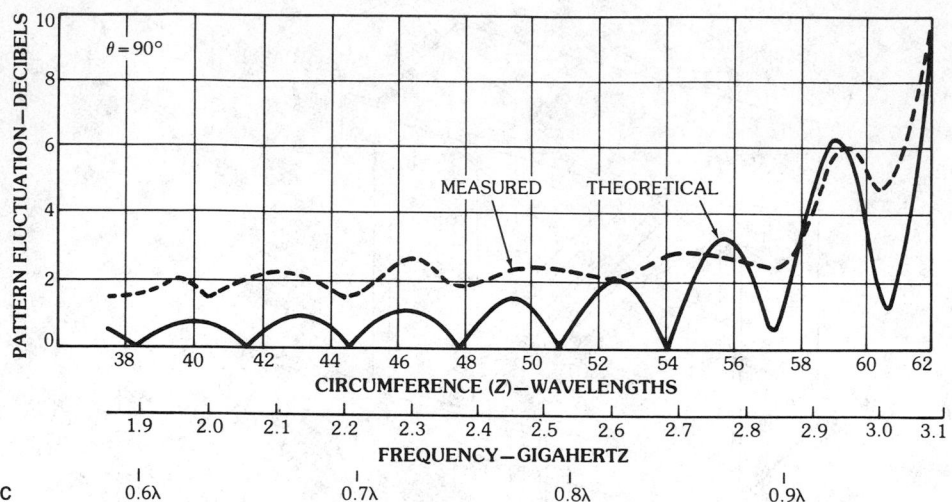

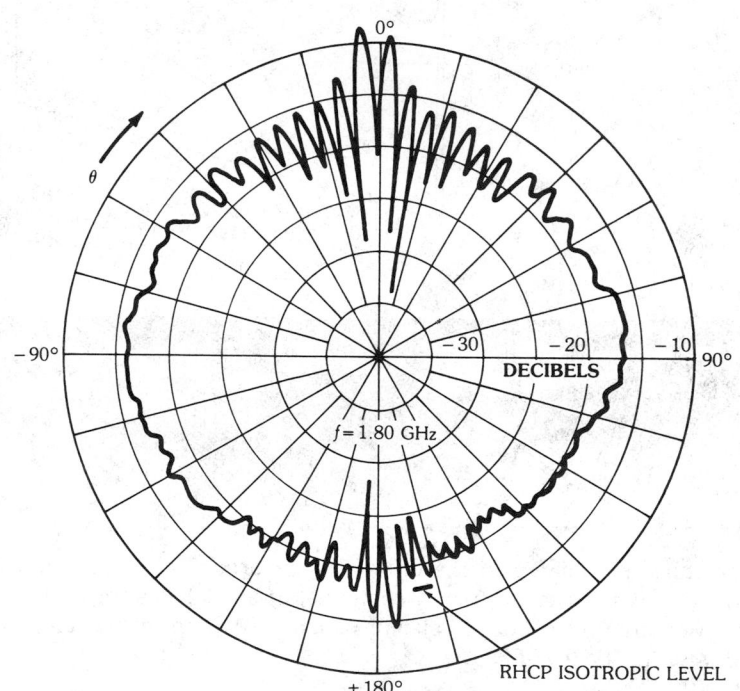

Fig. 78, *continued.*

$$L_\ell(\cos\theta) = AP_\ell(\cos\theta) + BQ_\ell(\cos\theta)$$

P_ℓ = associated Legendre's function of the first kind

Q_ℓ = associated Legendre's function of the second kind

$P = [\ell(\ell + 1) + 1/4]^{1/2}$

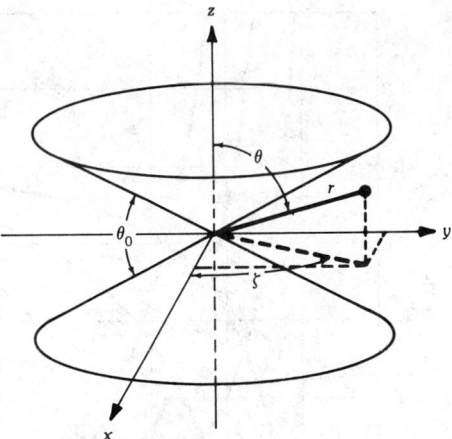

Fig. 79. Biconical horn antenna.

$H_p^{(2)}$ = Hankel function of the second kind

$k = \omega\sqrt{\mu_0\epsilon_0}$

$\mu_0 = 4\pi \times 10^{-7}$ H/m

$\epsilon_0 = 1/(36\pi \times 10^9)$ F/m

The ℓ is determined from the boundary condition that the field E_ϕ must vanish at $\theta = (\pi \pm \theta)/2$, where θ_0 denotes the flat angle between the cones:

$$\frac{\partial}{\partial\theta}L_\ell(\cos\theta) = 0 \quad \text{at } \theta = (\pi \pm \theta_0)/2$$

Let the propagation constant be $\gamma = \alpha + j\beta$. Then

$$\alpha_{\text{TEM}} = \frac{1}{r}$$

$$\beta_{\text{TEM}} = \omega\sqrt{\epsilon_0\mu_0} \tag{41a}$$

and

$$\alpha_{\text{TE}_{01}} = \frac{2P+1}{2r} - \text{Re}\,\{k_0 H_{p-1}^{(2)}(k_0 r)/H_p^{(2)}(k_0 r)\}$$

$$\beta_{\text{TE}_{01}} = -\text{Im}\,\{k_0 H_{p-1}^{(2)}(k_0 r)/H_p^{(2)}(k_0 r)\} \tag{41b}$$

Fig. 80 shows $\alpha_{\text{TE}_{01}}$ and $\beta_{\text{TE}_{01}}$ as functions of r, where the angles are the corresponding flare angles of the horn.

A circular-polarized wave is obtained by using the property of the different phase constants of the two modes so that when the TEM and TE$_{01}$ modes are excited at the throat of the horn, the phase difference between modes will be $\pm 90°$

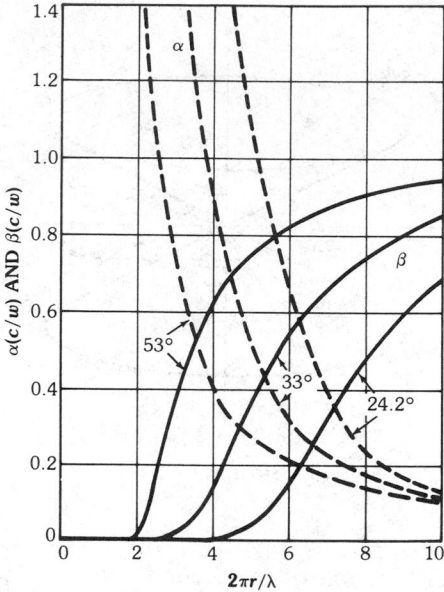

Fig. 80. Variations of attenuation constant α and phase constant β of TE_{01} wave with radial distance. (*After Barrow, Chu, and Jansen [50], © 1939 IRE*)

at the horn aperture. The two modes are generated by the skewed slots on a circular cylinder. The field excited by the slot is then decomposed into vertically polarized and horizontally polarized components. In all practical applications the radius R of the biconical horn is greater than 2 wavelengths. The gain of the horn can be calculated [51] by the following approximations:

$$G_{dB} = 10\log(2a/\lambda) - L_e \quad \text{for TEM mode} \tag{42a}$$

and

$$G_{dB} = 10\log(2a/\lambda) - (L_h + 0.91) \quad \text{for } TE_{01} \text{ mode} \tag{42b}$$

where L_e and L_h are the gain correction factors as shown in Fig. 81. The gain plotted against the flare angle for the TEM and TE_{01} waves is shown in Fig. 82. Values of R/λ and the flare angle exist that yield the maximum gain for a biconical horn. The radiation pattern of the biconical horn is isotropic in the azimuthal plane. In the elevation plane ($\phi = $ constant), the radiation pattern can be given by the universal curves of Fig. 83a for the TEM wave and Fig. 83b for the TE_{01} wave. In the vicinity of the axial direction ($\theta = 0°$ or $180°$), the edge of the horn gives rise to a focusing effect due to the formation of a caustic of the edge-diffracted rays. The equivalent edge currents [52] should be used to evaluate the radiation field. This focused field radiates into the satellite body and radiates into the main beam region. It is very important to reduce the edge currents so that the interference in

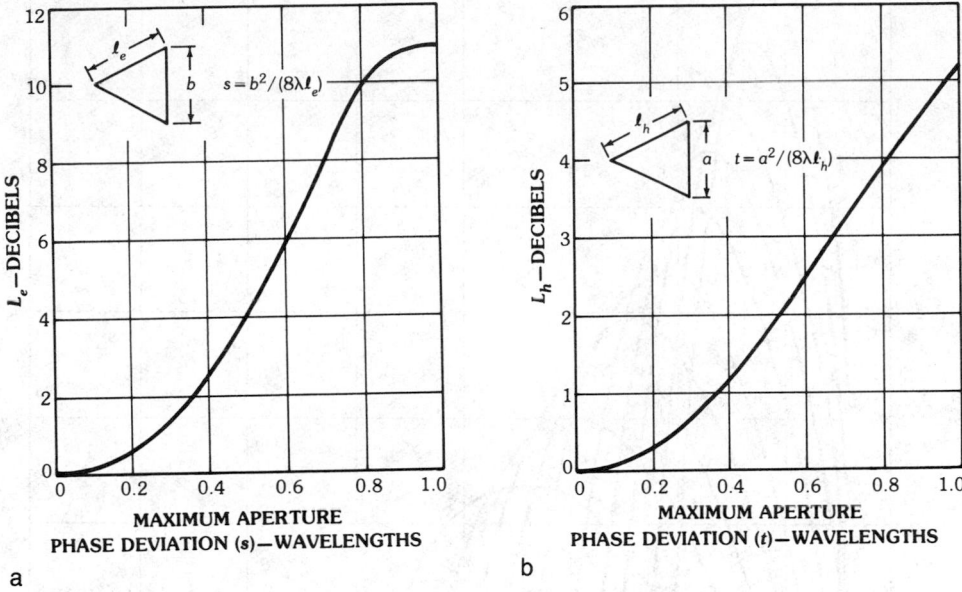

Fig. 81. Gain correction factors. (*a*) For *E*-plane flare. (*b*) For *H*-plane flare. (*After Jasik* [51], © 1961 McGraw-Hill Book Co.)

the main beam region will be minimal. The edge currents could be reduced by curving or serrating the edge. The bottom and the top of the edge can also be designed to radiate out of phase so that their contributions cancel in the axial direction.

The Slotted-Cylinder Reflector Antenna—A slotted-cylinder reflector antenna is used to provide a cardioid shaped pattern. The bottom of the cardioid shaped pattern is facing the satellite body so that the scattering of the satellite can be minimized. The antenna consists of a ring of slots placed near the end of a circular waveguide and a conical reflector as depicted in Fig. 84. The slots are excited by a rotating TE_{11} mode so that the field in the axial direction of the cylinder is circularly polarized.

The pattern shaping in the elevation plane can be achieved by using more than one ring of slot array on the circular waveguide. It is simpler, however, to use a conical reflector attached to the circular waveguide to shape the pattern [53]. The design problem is then divided into two parts: the external, to determine the excitation of slots and the inclined angle of the reflector to yield the desired pattern; and the internal, to decide how to launch the TE_{11} mode in the waveguide and to match the excited mode to the slots to reduce vswr. The radiation pattern of the finite waveguide structure and the conical reflector can be obtained via the method of moments [54]. In this particular case the solution can be simplified by invoking the rotational symmetry of the antenna. Only the fundamental mode is considered in the slot aperture. The rotating TE_{11} mode is launched by two orthogonal probes

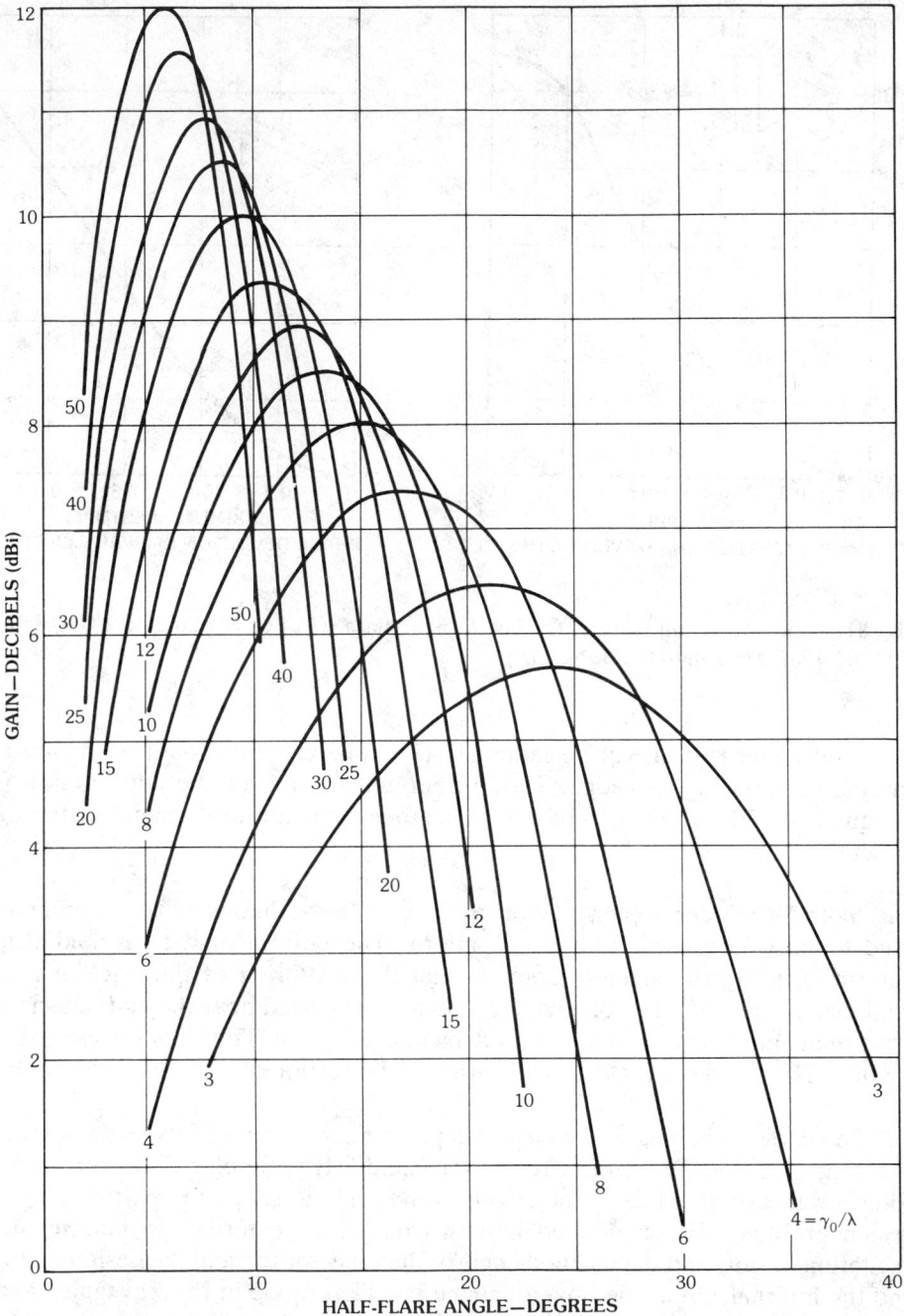

Fig. 82. Biconical horn gain. (*a*) For TEM mode. (*b*) For TE$_{01}$ mode.

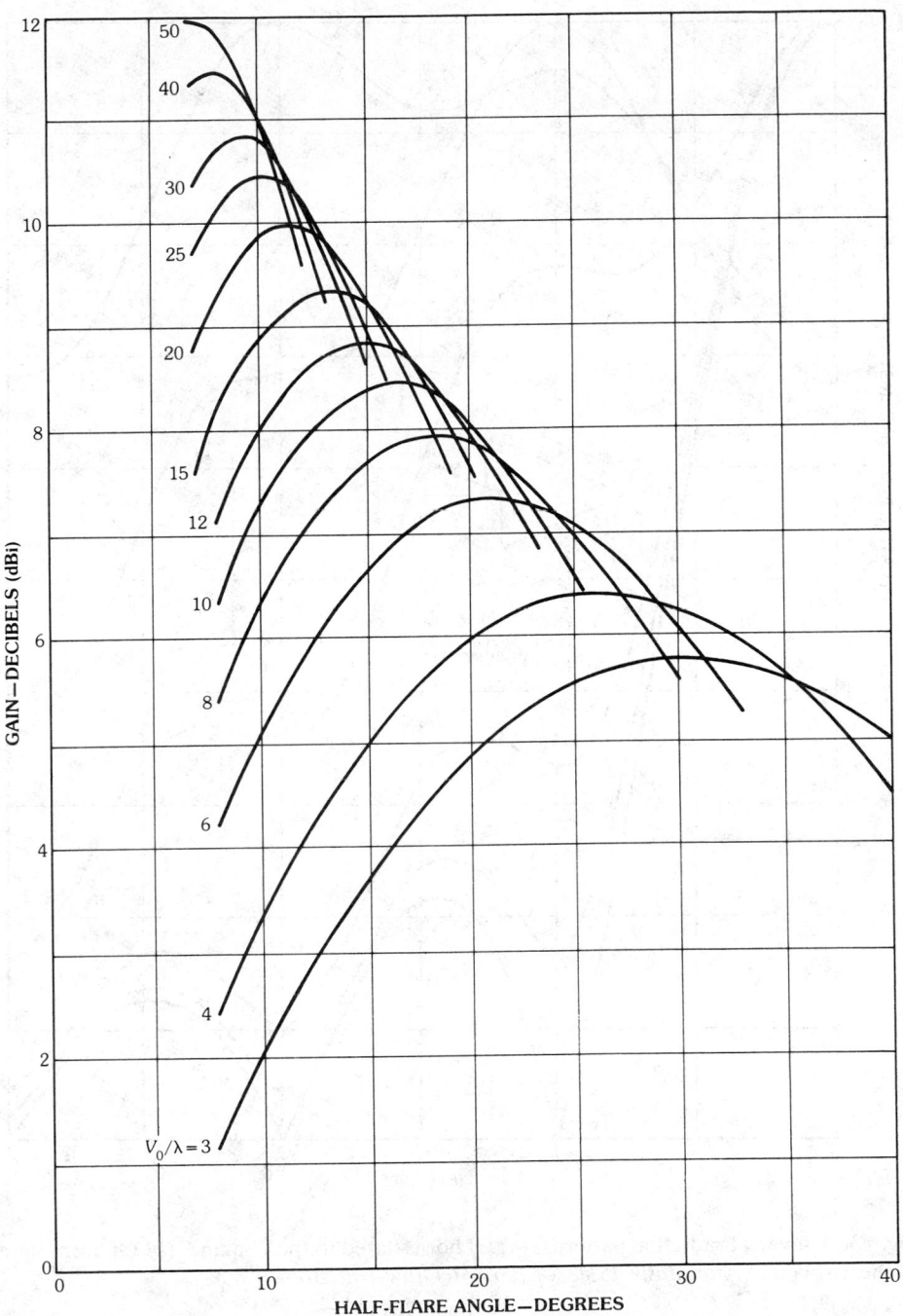

b

Fig. 82, *continued.*

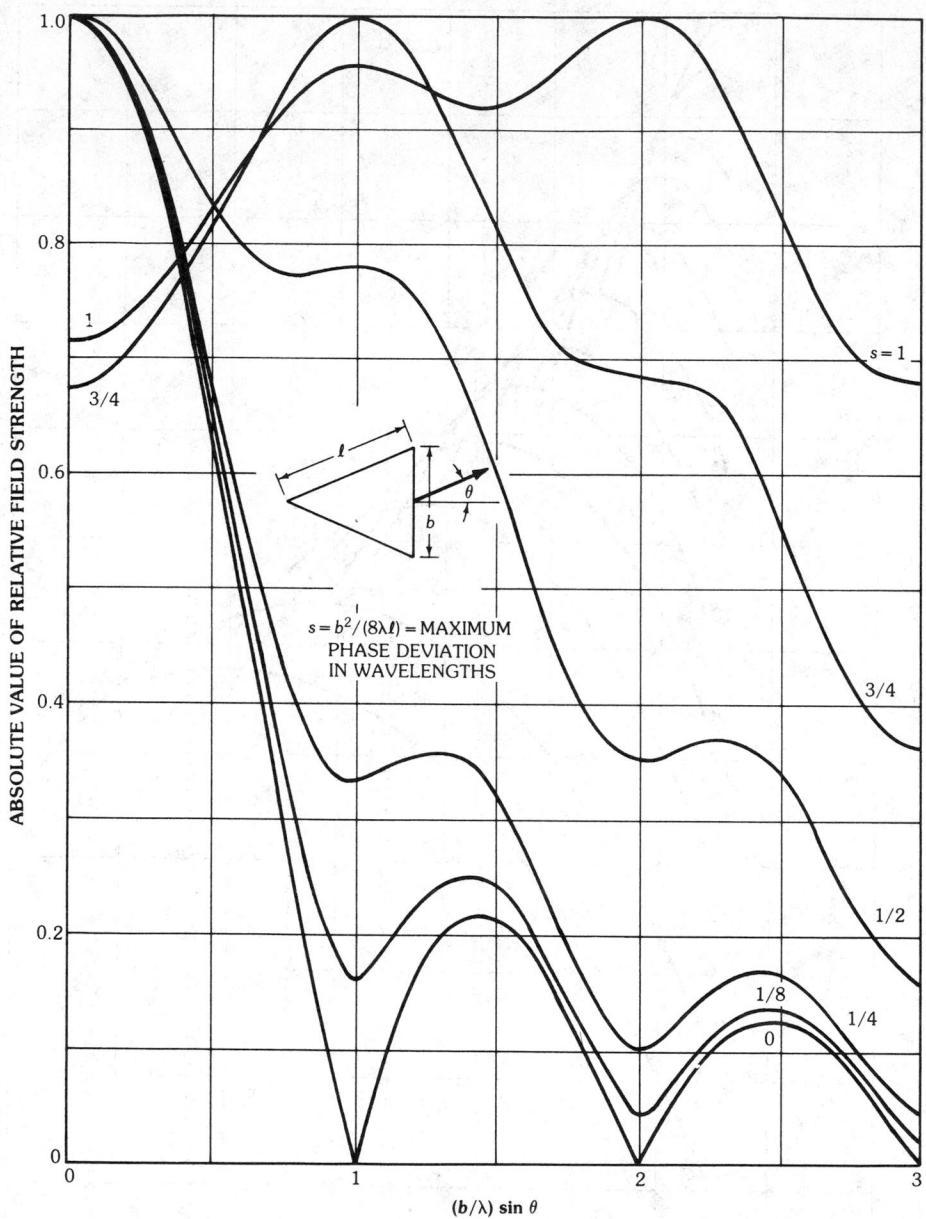

a

Fig. 83. Universal radiation patterns. (*a*) Of horns flared in the *E*-plane. (*b*) Of horns flared in the *H*-plane. (*After Jasik [51], © 1961 McGraw-Hill Book Co.*)

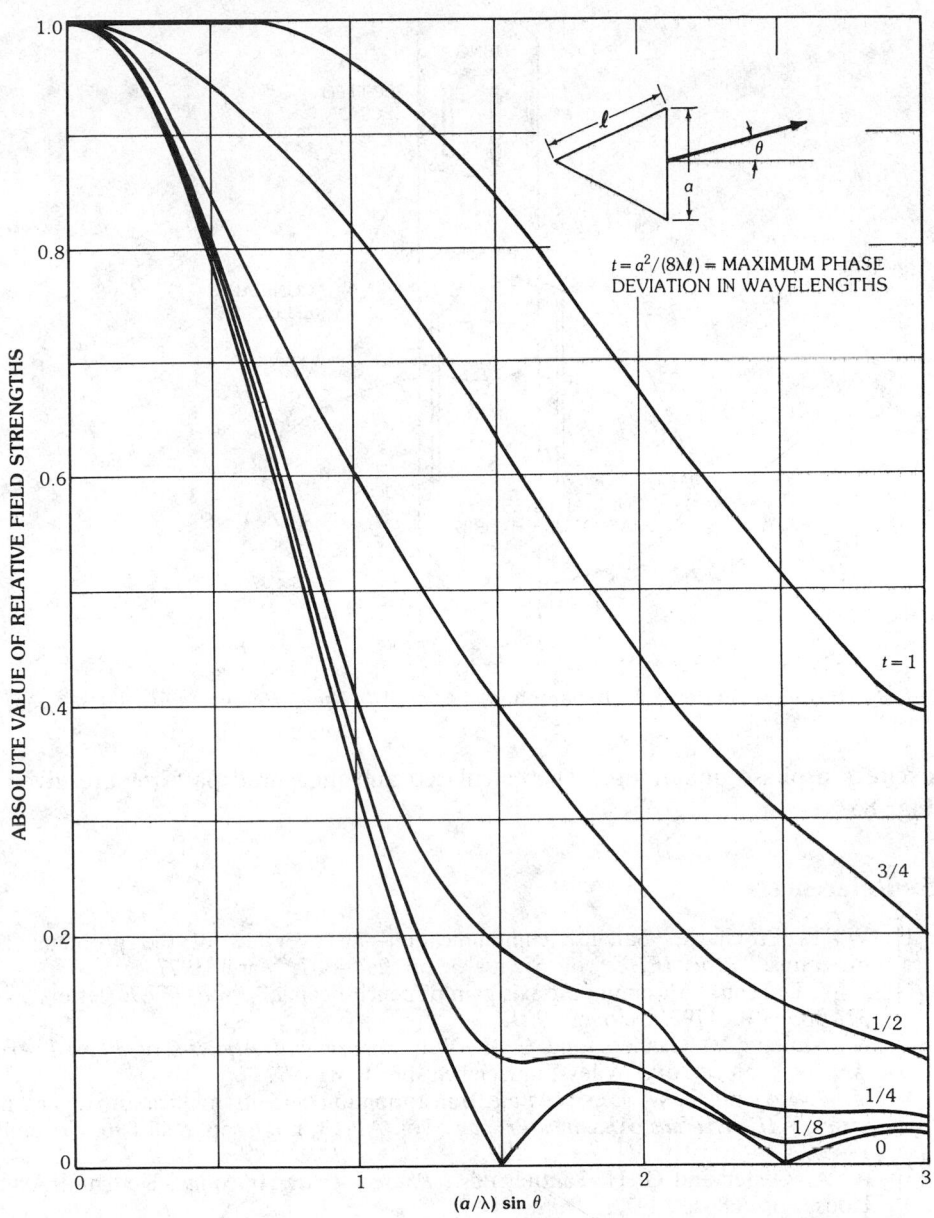

$t = a^2/(8\lambda\ell) = $ MAXIMUM PHASE DEVIATION IN WAVELENGTHS

$t = 1$

$3/4$

$1/2$

$1/4$

$1/8$

0

$(a/\lambda) \sin \theta$

ABSOLUTE VALUE OF RELATIVE FIELD STRENGTHS

b

Fig. 83, *continued.*

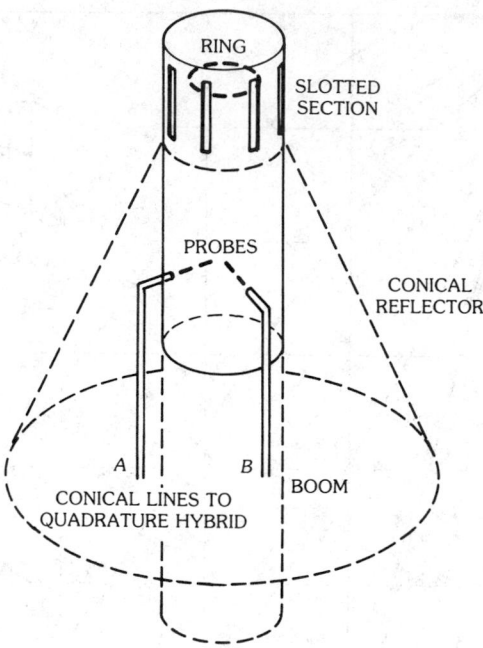

Fig. 84. Slotted-cylinder reflector antenna. (*After Albertsen, Balling, and Laursen [53]*)

excited in phase quadrature. The calculated and measured patterns are given in Fig. 85.

5. References

[1] W. L. Pritchard, "Satellite communication—an overview of the problems and programs," *Proc. IEEE*, vol. 65, no. 3, pp. 294–307, March 1977.

[2] J. W. Duncan, "Maximum off-axis gain of pencil beams," *Proc. IEEE* (letters), vol. 57, pp. 1791–1792, October 1969.

[3] N. Amitay, V. Galindo, and C. T. Wu, *Theory and Analysis of Phased Array Antennas*, New York: Wiley-Interscience, pp. 6–8, 1972.

[4] Y. T. Lo and S. W. Lee, "Affine transformation and its application to antenna arrays," *IEEE Trans. Antennas Propag.*, vol. AP-13, no. 6, pp. 890–896, November 1965.

[5] A. A. Oliner and G. H. Knittel, eds., *Phased Array Antennas*, Dedham: Artech House, pp. 68–82, 1972.

[6] E. D. Sharp, "A triangular arrangement of planar array elements that reduces the number needed," *IEEE Trans. Antennas Propag.*, vol. AP-9, pp. 126–129, March 1961.

[7] Y. T. Lo, "A mathematical theory of antenna arrays with randomly spaced elements," *IEEE Trans. Antennas Propag.*, vol. AP-15, pp. 231–235, March 1967.

[8] M. Born and E. Wolf, *Principles of Optics*, New York: Pergamon Press, 1964, pp. 203–232.

[9] S. Silver, ed., *Microwave Antenna Theory and Design*, New York: McGraw-Hill Book Company, 1949, pp. 389–412.

[10] J. J. Lee, "Numerical methods make lens antennas practical," *Microwaves*, pp. 81–84, September 1982.

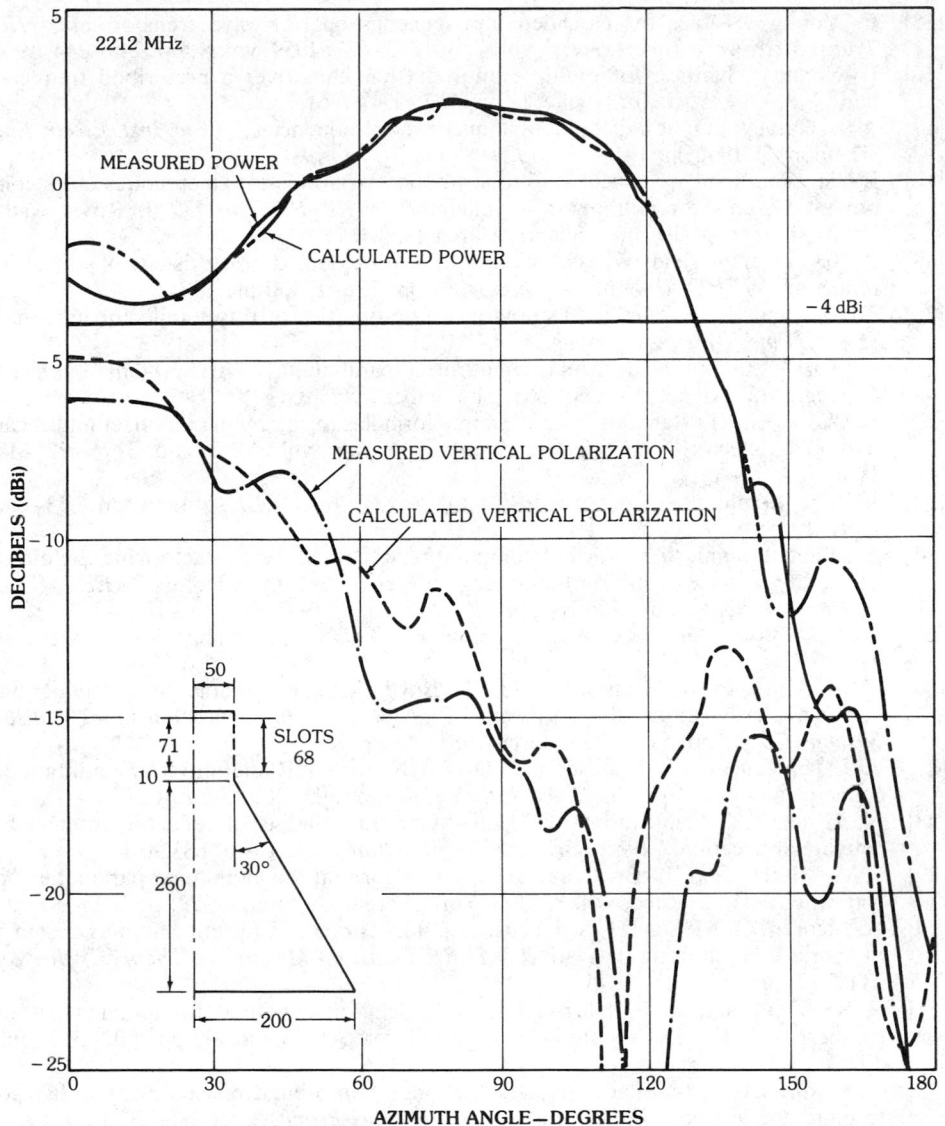

Fig. 85. Measured and computed patterns of slotted-cylinder reflector antenna. (*After Albertsen, Balling, and Laursen [53]*)

[11] D. Waineo, "Lens designed for arbitrary illumination," *Proc. of 1976 IEEE AP-S Symp.*, p. 476, 1976.

[12] F. G. Friedlander, "A dielectric lens aerial for wide-angle beam scanning," *J. Inst. Electr. Eng.*, vol. 93, pt. 3A, pp. 658–662, 1946.

[13] J. A. Jenkins and H. E. White, *Fundamentals of Optics*, chapter 9, New York: McGraw-Hill Book Company, 1957, p. 154.

[14] D. H. Shinn, "The design of a zoned dielectric lens for wide-angle scanning," *Marconi Rev.*, no. 117, p. 37, 1953.

[15] L. Young, "Tables for cascaded homogeneous quarter-wave transformers," *IRE Trans. Microwave Theory Tech.*, vol. 7, pp. 233–237, 1959; vol. 8, pp. 243–244, 1960.

[16] L. Young, "Synthesis of multiple antireflection films over a prescribed frequency band," *J. Opt. Soc. Am.*, vol. 51, pp. 967–974, 1961.

[17] A. F. Harvey, "Optical techniques at microwave frequencies," *Proc. Inst. Electr. Eng.* (London), C106, pp. 141–157, 1959.

[18] R. H. Garnham, "Optical and quasi-optical transmission techniques and component systems for millimeter wavelengths," *RRE Rep. no. 3020*, Royal Radar Establishment, Malvern, England, March 1958.

[19] P. G. Ingerson and W. C. Wong, "Focal region characteristics of offset-fed reflectors," *IEEE AP-S Intl. Symp. Dig.*, pp. 121–123, June 1974.

[20] T. Chu and R. H. Turrin, "Depolarization properties of offset reflector antenna," *IEEE Trans. Antennas Propag.*, vol. AP-21, pp. 339–345, May 1973.

[21] J. Ruze, "Lateral-feed displacement in a paraboloid," *IEEE Trans. Antennas Propag.*, vol. AP-13, pp. 660–665, September 1965.

[22] S. W. Lee and Y. Rahmat-Samii, "Simple formulas for designing an offset multibeam parabolic reflector," *IEEE Trans. Antennas Propag.*, vol. AP-29, no. 3, p. 472, May 1981.

[23] J. Ruze, "Antenna tolerance theory—a review," *Proc. IEEE*, vol. 54, pp. 633–640, April 1966.

[24] Y. Rahmat-Samii, "An efficient computational method for characterizing the effects of random surface errors on the average power pattern of reflectors," *IEEE Trans. Antennas Propag.*, vol. AP-31, pp. 92–98, January 1983.

[25] L. R. Whicker, *Ferrite Control Components, Volume 2*, Dedham: Artech House, 1974.

[26] T. M. Smith, E. W. Mathews, and C. R. Boyd, "*C*-band variable power divider and variable phase shifter development," *Proc. AIAA 9th Communications Satellite Systems Conf.*, pp. 693–697, March 1982.

[27] J. L. Janken, W. J. English, and D. F. DiFonzo, "Radiation from 'multimode' reflector antennas," *IEEE G-AP Intl. Symp.*, pp. 306–309, August 1973.

[28] Y. Hwang, A. Tsao, and C. C. Han, "Uniform analysis of reflector antenna for satellite application," *1983 IEEE AP S Intl. Symp.*, vol. 1, pp. 88–90.

[29] J. E. Heller and J. B. Cruz, Jr., "An algorithm for minimax parameter optimization," *Automatica*, vol. 8, New York: Pergamon Press, 1972, pp. 325–335.

[30] K. Madsen, O. Nielson, H. S. Jacobsen, and L. Thrane, "Efficient Minimax design of networks without using derivatives," *IEEE Trans. on Microwave Theory Tech.*, vol. MTT-23, pp. 507–512, 1975.

[31] J. R. Mantz and R. F. Harrington, "Computational method for antenna pattern synthesis," *IEEE Trans. Antennas Propag.*, vol. AP-23, no. 4, pp. 507–512, July 1975.

[32] C. Donn, W. A. Imbriale, and G. G. Wong, "An *S*-band phased array design for satellite application," *IEEE Intl. Symp. Antennas Propag.*, pp. 60–63, 1977.

[33] C. Donn, "A new helical antenna design for better on-and-off boresight axial ratio performance," *IEEE Trans. Antennas Propag.*, vol. AP-28, no. 2, pp. 264–267, March 1980.

[34] A. R. Dion and L. J. Ricardi, "A variable-coverage satellite antenna system," *Proc. IEEE*, vol. 59, no. 2, pp. 252–262, February 1971.

[35] L. J. Ricardi, A. J. Simmons, A. R. Dion, L. K. DeSize, and B. M. Potts, "Some characteristics of a communication satellite multiple-beam antenna," *MIT Tech. Note 1975-3*, January 1975.

[36] Ford Aerospace & Communications Corporation, Palo Alto, California, "Design for Arabsat *C*-band communication antenna system."

[37] P. D. Potter, "A new horn antenna with suppressed side lobes and equal beamwidths," *Microwave J.*, vol. VI, pp. 71–78, June 1963.

[38] T. Satoh, "Dielectric-loaded horn antenna," *IEEE Trans. Antennas Propag.*, vol. AP-20, pp. 199–201, March 1972.

[39] W. J. English, "The circular waveguide step-discontinuity mode transducer," *IEEE Trans. Microwave Theory Tech.*, vol. MTT-21, pp. 633–636, October 1973.

[40] J. S. Ajioka and H. E. Harry, Jr., "Shaped beam antenna for earth coverage from a stabilized satellite," *IEEE Trans. Antennas Propag.*, vol. AP-18, no. 3, pp. 323–327, May 1970.

[41] T. S. Chu, "On the use of uniform circular arrays to obtain omnidirectional patterns," *IRE Trans. Antennas Propag.*, vol. AP-7, pp. 436–438, October 1959.

[42] V. Galindo and K. Green, "A near-isotropic circular polarized antenna for space vehicles," *IEEE Trans. Antennas Propag.*, vol. AP-13, no. 6, pp. 872–877, November 1965.

[43] W. F. Croswell, C. M. Knop, and D. M. Hatcher, "A dielectric-coated circumferential slot array for omnidirectional coverage at microwave frequencies," *IEEE Trans. Antennas Propag.*, vol. AP-15, no. 6, pp. 722–727, November 1967.

[44] R. F. Harrington, *Time-Harmonic Electromagnetic Fields*, New York: McGraw-Hill Book Company, 1961, pp. 245–250.

[45] P. H. Pathak and R. G. Kouyoumjian, "An analysis of the radiation from apertures in surfaces by the geometrical theory of diffraction," *Proc. IEEE*, vol. 62, pp. 1438–1447, November 1974.

[46] J. R. Wait, *Electromagnetic Radiation From Cylindrical Structures*, New York: Pergamon Press, 1959.

[47] G. E. Stewart and K. E. Gorden, "Mutual admittance for axial rectangular slots in a large conducting cylinder," *IEEE Trans. Antennas Propag.*, vol. AP-19, pp. 120–122, January 1971.

[48] S. W. Lee, "Mutual admittance of slots on a cone: solution by ray techniques," *IEEE Trans. Antennas Propag.*, vol. AP-26, no. 6, pp. 768–773, November 1978.

[49] W. S. Gregorwich, "An electronically despun array flush-mounted on a cylindrical spacecraft," *IEEE Trans. Antennas Propag.*, vol. AP-22, no. 1, January 1974.

[50] W. L. Barrow, L. J. Chu, and J. J. Jansen, "Biconical electromagnetic horns," *Proc. IRE*, pp. 769–779, December 1939.

[51] H. Jasik, ed., *Antenna Engineering Handbook*, New York: McGraw-Hill Book Company, 1961, pp. 10-13 to 10-14.

[52] C. E. Ryan and L. Peters, Jr., "Evaluation of edge-diffracted fields including equivalent currents for the caustic region," *IEEE Trans. Antennas Propag.*, vol. AP-17, no. 3, pp. 292–299, May 1969.

[53] N. C. Albertsen, P. Balling, and F. Laursen, "New low-gain *S*-band satellite antenna with suppressed back radiation," *Sixth Eur. Microwave Conf.*, Rome, Italy, pp. 14–17, September 1976.

[54] R. F. Harrington and J. R. Mantz, "Radiation and scattering from bodies of revolution," *Appl. Sci. Res.*, vol. 20, pp. 405–435, 1969.

Chapter 22

Remote Sensing and Microwave Radiometry

J. C. Shiue
Goddard Space Flight Center

L. R. Dod
Goddard Space Flight Center

CONTENTS

James Chen-Chi Shiue holds a BSEE from Taiwan University, a MSEE from the University of Vermont, an electrical engineer's degree from Stanford University, and a PhD from New York University. He is also a registered professional engineer.

He has more than 20 years of experience in the aerospace industry and in research and academic institutions. Since joining NASA's Goddard Space Flight Center in 1971 he has been working in microwave remote sensing of the earth. He has conducted experiments ranging from testing microwave radiometric and radar techniques for measuring snow depths to mapping soil moisture. At Goddard he is the group leader of the Microwave Sensors Group in the Laboratory for Oceans and is the instrument scientist in charge of developing the advanced microwave sounding unit, a temperature and humidity sounder for the NOAA weather satellites.

Louis R. Dod holds a bachelor of science degree in electrical engineering from Virginia Polytechnic Institute and a master of science degree in electrical engineering from George Washington University in Washington, D.C. He is a member of the IEEE.

He has been employed at the Goddard Space Flight Center of the National Aeronautics and Space Administration for over 25 years. He has been involved in the design and development of antennas and microwave systems for the early Explorer satellites and the space tracking and data acquisition ground antenna network. He was also involved in the initial studies for the tracking and data relay satellites and advanced technology satellites. Recently he worked in microwave remote sensing and has participated in aircraft data flights for microwave imaging of severe storms and sea ice. Currently he is involved in instrumentation of aircraft flights for hurricane measurements in the Caribbean.

1. Basic Principles of Microwave Radiometry

In this introductory section we will consider blackbody microwaves, microwave radiative transfer, and surface emissivity and reflectivity.

Blackbody Radiation at Microwave Frequencies

According to Planck's law of radiation the spectral radiant emittance R_f of a blackbody is isotropic and is given by

$$R_f = \frac{2hf^3}{c^2} \frac{1}{e^{hf/kT} - 1} \tag{1}$$

where

R_f = spectral radiant emittance in (watts)(meter)$^{-2}$(steradian)$^{-1}$(hertz)$^{-1}$ is the power emitted per unit solid angle, per unit area of the emitter, per unit frequency

c = velocity of light, $3 \times 10^8 \, \text{m·s}^{-1}$

h = Planck's constant, $6.63 \times 10^{-34} \, \text{J·s}$

k = Boltzmann's constant, $1.38 \times 10^{-23} \, \text{J·K}^{-1}$

λ = wavelength in meters

f = frequency in hertz

T = temperature in kelvins

When

$$hf \ll kT \tag{2}$$

then

$$R_f \cong \frac{2kT}{\lambda^2} \tag{3}$$

When the condition in (2) is satisfied, then (3) is valid. This is called the *Rayleigh-Jeans approximation*. In the microwave region (1 to 200 GHz) and with the terrestrial temperatures the Rayleigh-Jeans approximation is valid. This approximation makes the radiation power P directly proportional to temperature T. The incremental power received by an antenna from $\Delta\Omega$ is

$$\Delta P = \frac{1}{2} R_f \Delta f \Delta\Omega \, \Delta A_s \tag{4}$$

where

$\Delta\lambda$ = incremental change in wavelength

$\Delta\Omega$ = incremental solid angle subtended by the receiving antenna

ΔA_s = incremental area of the radiation source

The factor ½ in (4) results from the fact that only a single polarization is received.
The receiving antenna effective area is

$$A_R = \frac{\lambda^2}{4\pi} G \tag{5}$$

where G is the antenna gain over isotropic media. Since

$$\Delta\Omega = \frac{A_R}{r^2} = \frac{\lambda^2}{4\pi r^2} G \tag{6}$$

where r is the distance between the emitting surface element and the receiver,

$$\frac{\Delta A_s}{r^2} = \Delta\Omega_s \tag{7}$$

where $\Delta\Omega_s$ is the solid angle subtended at the receiver by ΔA_s. Therefore, by using (3), (6), and (7),

$$\Delta P = \frac{1}{4\pi} kTG \,\Delta\Omega_s \,\Delta f \tag{8}$$

From (8), the total power collected by a receiving antenna, for uniform T, is

$$P = kT\Delta f\left(\frac{1}{4\pi}\int G(\Omega)\,d\Omega\right)$$

or

$$P = kT\Delta f \tag{9}$$

Equation 9 is the Nyquist formula for noise power in bandwidth Δf. As we shall see in a later section, for a nonideal blackbody radiating surface, T should be replaced by $T_b = \epsilon T$, where T_b is the brightness temperature and ϵ the surface emissivity.

Microwave Radiative Transfer

The differential equation governing the transport of radiative energy (or power) in a medium can be expressed in the following form:

$$\frac{dW(x)}{dx} = -\alpha(x)[W(x) - J(x)] \tag{10}$$

where

$W(x)$ = the power density in watts per unit area propagating along path x

$\alpha(x)$ = the attenuation factor per unit distance, which is indicative of the rate at which power is being diminished (by absorption and scattering)

$J(x)$ = the source function (power per unit area) representing the thermal emission of the medium. If the source function $J(x) = 0$, then the solution for the power density $P(x)$ is an exponential decay along the direction x.

The formal solution to (10) is

$$W(x) = W(0)\, e^{-\tau(x,0)} + \int_{u=0}^{u=x} \alpha(u)\, J(u)\, e^{-\tau(x,u)}\, du \tag{11}$$

where

$$\tau(x,x') = \int_{u=x'}^{u=x} \alpha(u)\, du \tag{12}$$

For the special case when $\alpha(x) = \alpha_0$ and $J(x) = J_0$, then

$$W(x) = W(0)\, e^{-\alpha_0 x} + J_0(1 - e^{-\alpha_0 x}) \tag{13}$$

Since the power density W is proportional to temperature T, we can replace W by T in (13) to obtain the relationships between the temperatures.

Similarly, the source function J_0 is replaced by the physical temperature T_p of the emitting medium. Thus

$$T(x) = T(0)\, e^{-\alpha_0 x} + T_p(1 - e^{-\alpha_0 x}) \tag{14}$$

(In the conventional engineering unit of decibels, $L_{\mathrm{dB}} = 10 \log_{10} L$.) Set x equal to d and let L be the loss factor of the path length d, such that

$$L = e^{\alpha_0 x} = e^{\alpha_0 d} \tag{15}$$

Then (14) becomes

$$T(d) = T(0)\frac{1}{L} + \left(1 - \frac{1}{L}\right) T_p \tag{16a}$$

Equation 16 represents the temperature $T(d)$ that results after the energy travels through a medium with loss factor L. This is depicted schematically in Fig. 1. The

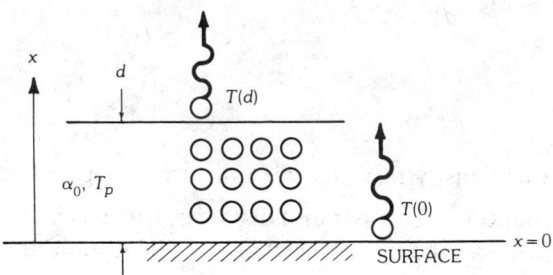

Fig. 1. Radiative transfer model.

brightness temperature $T(0)$ is attenuated by L while the medium is reemitting a component of temperature, $(1 - 1/L)\,T_p$. The radiative transfer theory, as represented by (10), governs only the amplitude, not the phase of a wave propagating in a medium. No coherent phase information of the electromagnetic wave is included in this theory. The radiative transfer theory, however, is an accurate and useful description of many phenomena in microwave radiometry for remote sensing applications.

The following example will demonstrate the use of (16) in the remote sensing of the atmosphere. The presence of high humidity in the atmosphere can be measured by a microwave radiometer looking down at the earth from space. The humid air is detected by the change in the brightness temperature T_b as compared with the adjacent dry air area. As we shall see, it turns out that the change in T_b, or contrast, is larger when detecting the humidity against the ocean background than against the land background. In (16) let $T(d) = T_b$ be the brightness temperature seen by a spaceborne radiometer, and $T(0) = T_s$ be the surface brightness temperature. Further, let $T_p = \overline{T}_A$ be the average temperature of the atmosphere and L be the transmissivity (loss factor) of the atmosphere. Then

$$T_{b_1} = \frac{T_s}{L_{A_1}} + \overline{T}_A\left(1 - \frac{1}{L_{A_1}}\right) \tag{16b}$$

where L_{A_1} is the loss factor of the atmosphere with high humidity. Similarly, the brightness temperature seen by the radiometer over an area with dry air (whose loss
factor is L_{A_2}) is given by

$$T_{b_2} = \frac{T_s}{L_{A_2}} + \overline{T}_A\left(1 - \frac{1}{L_{A_2}}\right) \tag{16c}$$

The change in T_b between the two cases is

$$\Delta T_b = T_{b_1} - T_{b_2} = (T_s - \overline{T}_A)\left(\frac{1}{L_{A_1}} - \frac{1}{L_{A_2}}\right) \tag{16d}$$

Since the humid air has a higher loss factor than dry air, $L_{A_1} > L_{A_2}$, and for land surface $T_s > \overline{T}_A$. Hence $\Delta T_b < 0$. This means that the high-humidity area will appear as a "cooler" spot over land.

To increase the contrast ΔT_b, one can choose the frequency of observation to maximize the second parenthetical term in (16d). For example, the frequency of the water vapor absorption line near 22 GHz is a good frequency for humidity observation. It turns out that because of the difference in brightness temperature between land and air is small the resultant ΔT_b is relatively small (typically 1 to 2 K), making the use of 22 GHz for humidity detection rather difficult over land. Table 1 lists the result of ΔT_b with the assumed values of T_s and $\overline{T}_A$.

High-humidity air against an ocean background produces much larger contrast compared with the land background. Because the surface brightness temperature of the ocean is much *cooler* than the air, the high-humidity area appears as a *warm* spot against an ocean background. In the case of an ocean background, because the difference between $\overline{T}_A$ and T_s is large the contrast is strong. Table 1 shows that the same humidity yields a brightness temperature change greater than 16 K.

Equation 16a can also be used to represent the remote sensing of a snowpack over land. In this case the term T_S stands for the emission from the background soil, and the L of the first term is the extinction factor due primarily to the snow scattering. The second term in (16a) is usually very small for snow. The general principles are similar in both the water vapor and snowpack cases, but the mechanisms contributing to the extinction are quite different. Water vapor absorbs and reradiates the emission from the background surface (whether ocean or land). For the case of detection of humidity over the ocean, where the radiometry technique works most effectively, water vapor absorbs cool ocean background and reemits at its own warmer temperature, thus making the humid air appear as a *warm* spot. In contrast, snow particles (assuming dry snow) absorb very little of the background soil radiation. Their extinction mechanism is primarily due to scattering. For this reason the frequencies suitable for snowpack sensing are usually in the range of 20 to 40 GHz. From (16d) it is clear that in order to infer the property of the intervening media, such as water vapor in the atmosphere or snowpack on the ground, the background radiation T_S must also be determined. This is frequently achieved by multichannel measurements.

Surface Emissivity and Reflectivity

The brightness temperature T_b of an emitting surface is the product of its thermal temperature T_p and the surface emissivity ϵ:

$$T_b = \epsilon T_p \tag{17}$$

The emissivity ϵ is a function of the radiation polarization. The emissivity may be subscripted to account for this polarization dependence:

$$\epsilon_j = 1 - R_j \tag{18}$$

where

Table 1. Brightness Temperatures Expected from a Spaceborne Microwave Radiometer

Condition	T_b (K)	ΔT_b (K)	Assumed Values
Dry air over ocean	187.71		$T_s = 170\,\text{K (ocean)}$
Humid air over ocean	171.54	+16.17	$\overline{T}_A = 280\,\text{K (air)}$
Dry air over land	289.86		$T_s = 290\,\text{K (land)}$
Humid air over land	288.39	−1.47	$L(22\,\text{GHz}) = \begin{cases} 0.06\,\text{dB (dry air)} \\ 0.76\,\text{dB (humid air)} \end{cases}$

$j = v, h$ for vertical or horizontal polarization

$R_j = |\varrho_j|^2$, reflectivity $\qquad\qquad\qquad\qquad\qquad\qquad\qquad$ (19)

ϱ_j = voltage reflection coefficient of the radiation

For a specular surface the voltage reflection coefficient ϱ_j is obtained from the following Fresnel equations [1]:

$$\varrho_h = \frac{\cos\theta_i - \sqrt{(\epsilon_2/\epsilon_1)[1 - (\epsilon_1/\epsilon_2)\sin^2\theta_i]}}{\cos\theta_i + \sqrt{(\epsilon_2/\epsilon_1)[1 - (\epsilon_1/\epsilon_2)\sin^2\theta_i]}} \tag{20}$$

$$\varrho_v = \frac{\cos\theta_i - \sqrt{(\epsilon_1/\epsilon_2)[1 - (\epsilon_1/\epsilon_2)\sin^2\theta_i]}}{\cos\theta_i + \sqrt{(\epsilon_1/\epsilon_2)[1 - (\epsilon_1/\epsilon_2)\sin^2\theta_i]}} \tag{21}$$

where ϵ_1 and ϵ_2 are the dielectric constants of the incident and transmitted media, respectively, and θ_i is the angle of incidence in medium 1. Horizontal polarization is the mode where the electric field vector is perpendicular to the plane of incidence (hence it is tangent to the earth's surface and horizontally oriented). Vertical polarization is the mode in which the electric field vector lies in the plane of incidence. (These definitions of horizontal and vertical polarization will be discussed further in Section 4.) The vertical polarization has total transmission at the Brewster angle in which $\theta_i = \theta_B$, and the emissivity is a maximum:

$$\tan\theta_B = \sqrt{\epsilon_2/\epsilon_1} \tag{22}$$

For nonspecular (rough) surfaces the reflectivity must include not only the scattering in the specular direction but also the power scattered in all other directions by the rough surface. The effective reflection coefficient ϱ is the summation of "differential scattering cross section per unit area" σ^0 over all directions

$$\varrho_j(\theta_i) = \int\int \{[\sigma_{jj}^0(\theta_i, \theta_s, \phi_s)] + [\sigma_{jk}^0(\theta_i, \theta_s, \phi_s)]\} \frac{d\Omega}{4\pi\cos\theta_i} \tag{23}$$

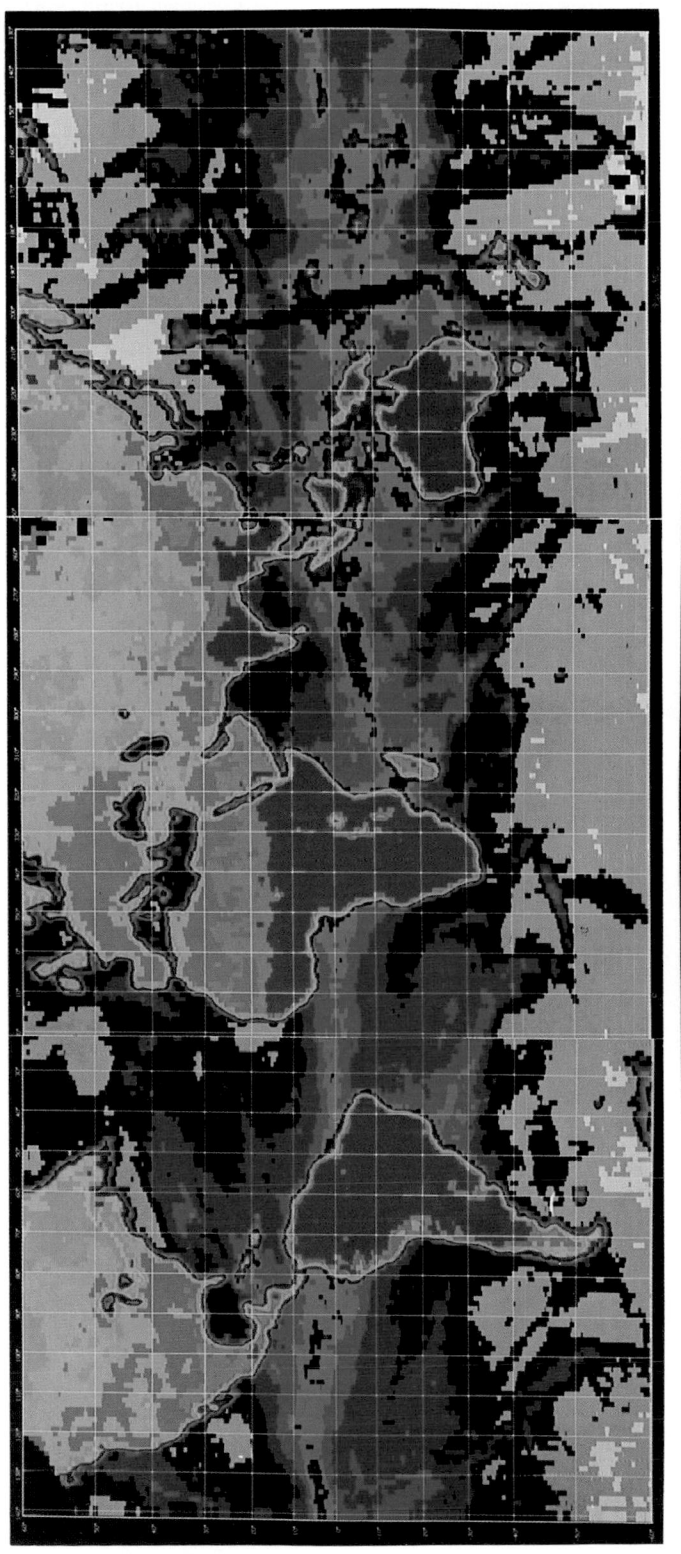

Fig. 22.4. Radio brightness of the world. *(Courtesy NASA)*

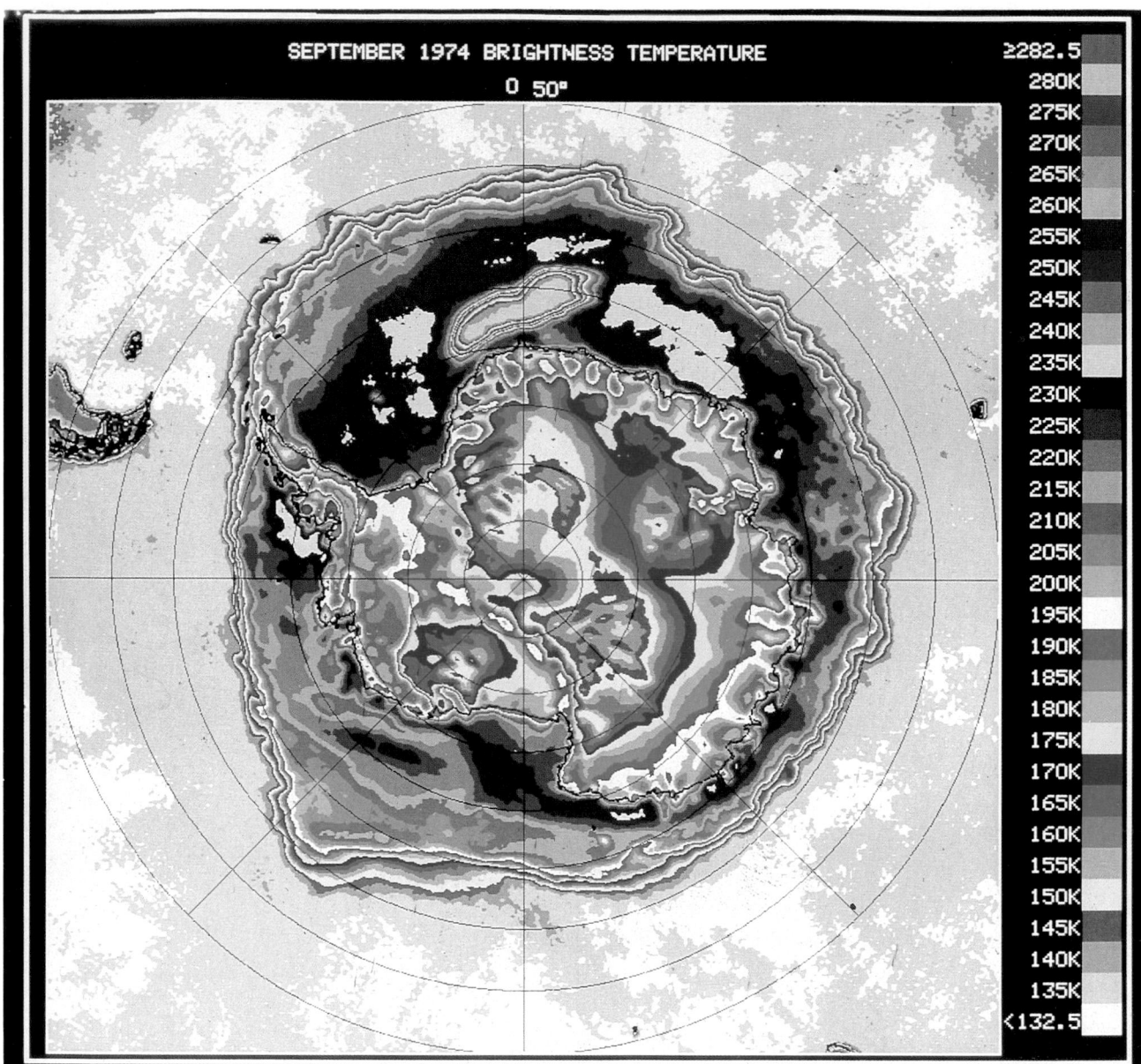

Fig. 22.6. Microwave brightness temperature of Antarctica (August 1974). *(Courtesy NASA)*

where $j = v, h$ for vertical and horizontal polarizations, $d\Omega$ is the differential solid angle, and $\sigma 2_{jk}^0$ is the differential cross section per unit area, from j polarization into k polarization:

$$\sigma_{jk}^0(\theta_i, \theta_s, \phi_s) = 4\pi r_s^2 \frac{\overline{\Delta P_s}}{P_0 \Delta A_0} \qquad (24)$$

where

θ_i = the angle of incidence (elevation)

θ_s = the scattering angle (elevation)

ϕ_s = scattering azimuthal angle

ΔP_s = scattered power density in the direction (θ_s, ϕ_s) in watts per square meter

r_s = distance at which P_s is measured

P_0 = incident power density in watts per square meter

ΔA_0 = differential incident area in square meters

The overbar $(\bar{\ })$ represents the statistical average of many measurements of power density. The subscripts j and k represent the polarizations of incident and scattered waves. The differential scattering cross section σ_{jk}^0 can be obtained by solving the wave equation and matching the boundary conditions incorporating surface roughness characteristics.

Strictly speaking, the formal definition of emissivity is valid only when the temperature of the underlying medium is uniform. If, however, the temperature is uniform to within several skin depths at the measuring frequency, the concept of emissivity can still be applied. An exact approach for the nonuniform temperature case is the solution of the radiative transfer equation for the upwelling energy.

The general emissivity characteristics derived from the Fresnel relations of (20) and (21) have been verified by observations of soil and water surfaces, although the surfaces are modified from the ideal smooth case by nonspecular rough surface effects. Fig. 2 is a plot of the emissivity versus incidence angle for a flat soil surface with two different soil moisture values [2]. Fig. 3 shows the theoretical brightness temperatures of calm sea water [3].

2. Applications of Microwave Radiometry to Remote Sensing

Microwave radiometry is the detection of thermal radiation power at microwave frequencies. A wealth of information can be derived from radiometric observations. In addition to the intensity of the radiation its dependence on frequency, angle of incidence, and polarization can also provide additional information about the source.

The development of microwave radiometry for remote sensing (also called passive microwaves, as compared with active radars) derives its background from radioastronomy. Since the advent of satellites the use of passive microwaves in

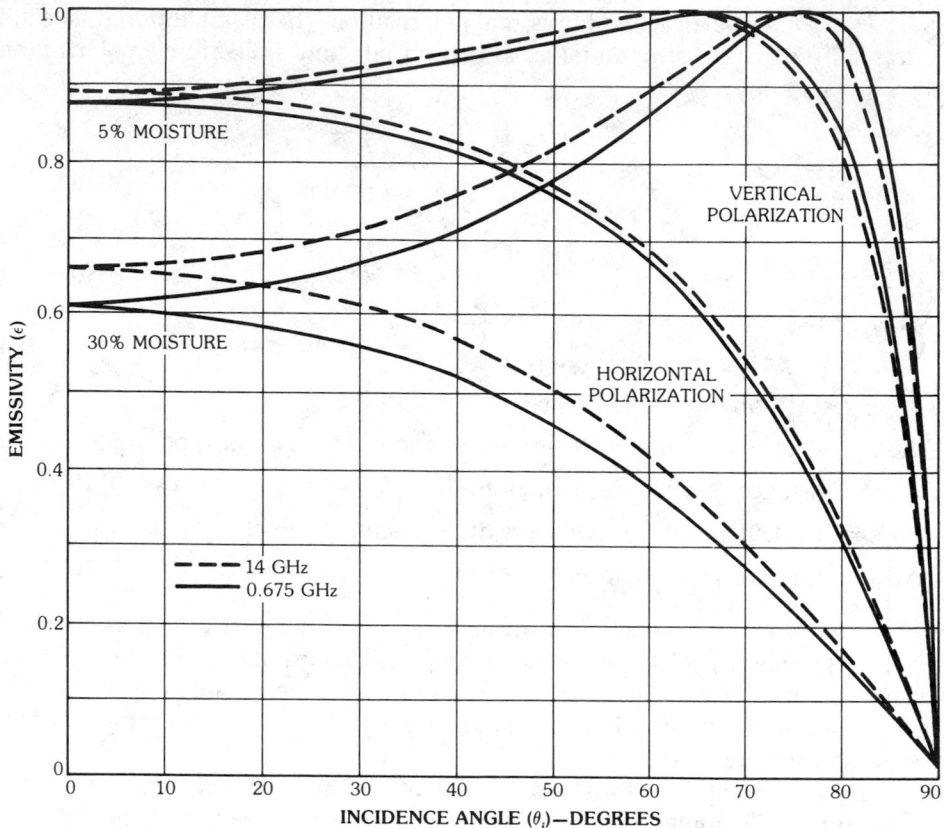

Fig. 2. Theoretical emissivity of bare soil (flat). (*After Njoku and Kong [2], © 1977 American Geophysical Union*)

remote sensing of the earth from space has gone through a rapid evolution in the last two decades, from a laboratory curiosity to daily operational systems.

As we enter the space shuttle era, which will allow one to transport larger and heavier satellites into space at cheaper costs, the use of passive microwaves for earth remote sensing will witness an even faster growth in the future. Fig. 4 is a picture of the microwave brightness temperature of the world obtained from a microwave radiometer [4] called the electrically scanning microwave radiometer (ESMR). From Fig. 4 it is evident that the microwave brightness temperature distribution is quite different from the physical temperature. (See the color insert for Fig. 4.)

Although the physical or thermodynamic temperature (based on the absolute scale, in kelvins) of the earth's surface is fairly uniform, in terms of the microwave brightness temperature scale the land masses stand out much hotter (250 to 280 K) compared with the cool oceans (200 K or less). This is because the land area, in general, has high emissivities, in the range of 0.7 to 0.9. Vegetation and forest covers increase the emissivity over bare soil; surface moisture, on the other hand, reduces the emissivity. A calm ocean surface has a low emissivity, of about 0.3 to

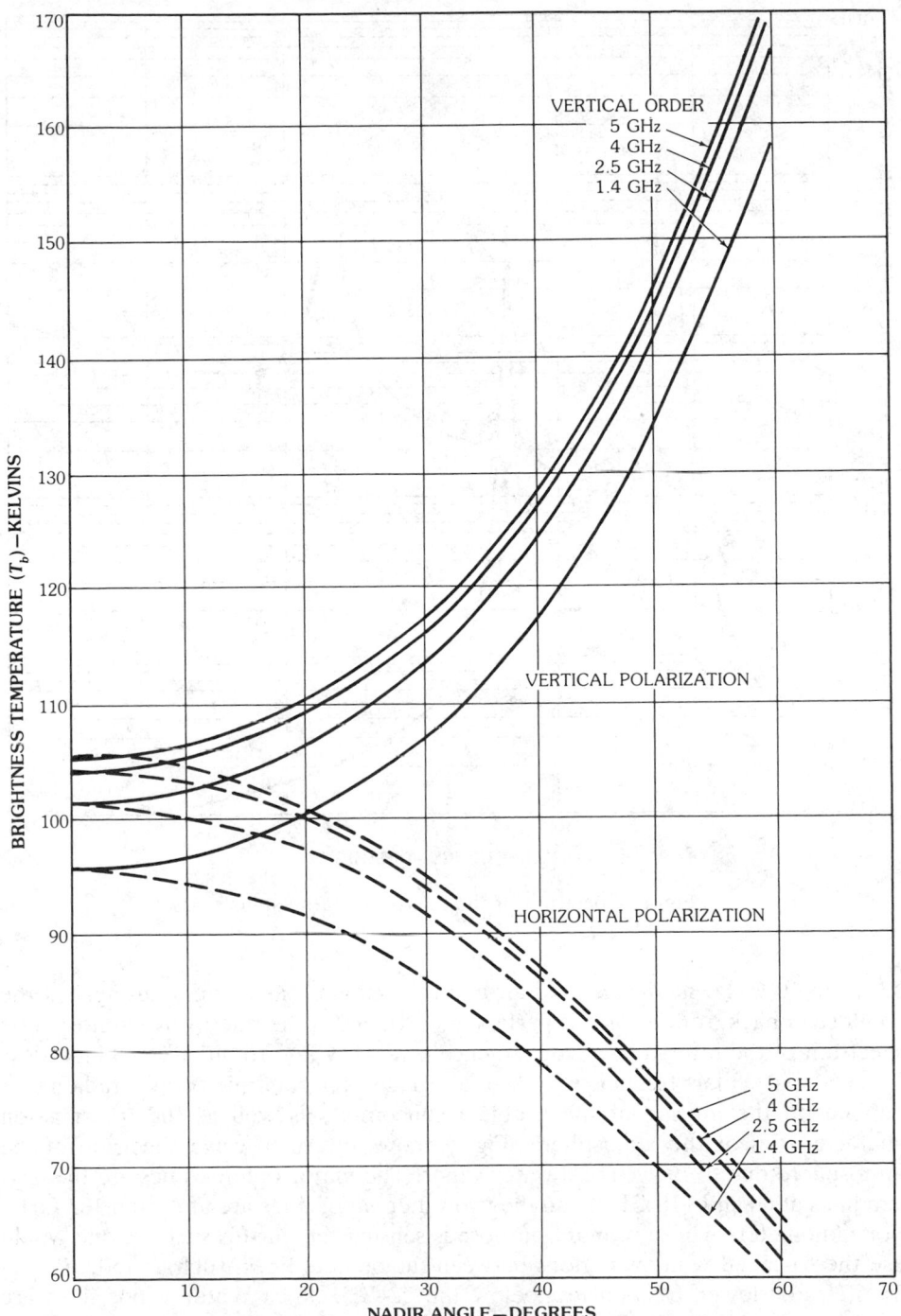

Fig. 3. Specular brightness temperature versus nadir angle of calm sea surface. (*Courtesy NOAA*)

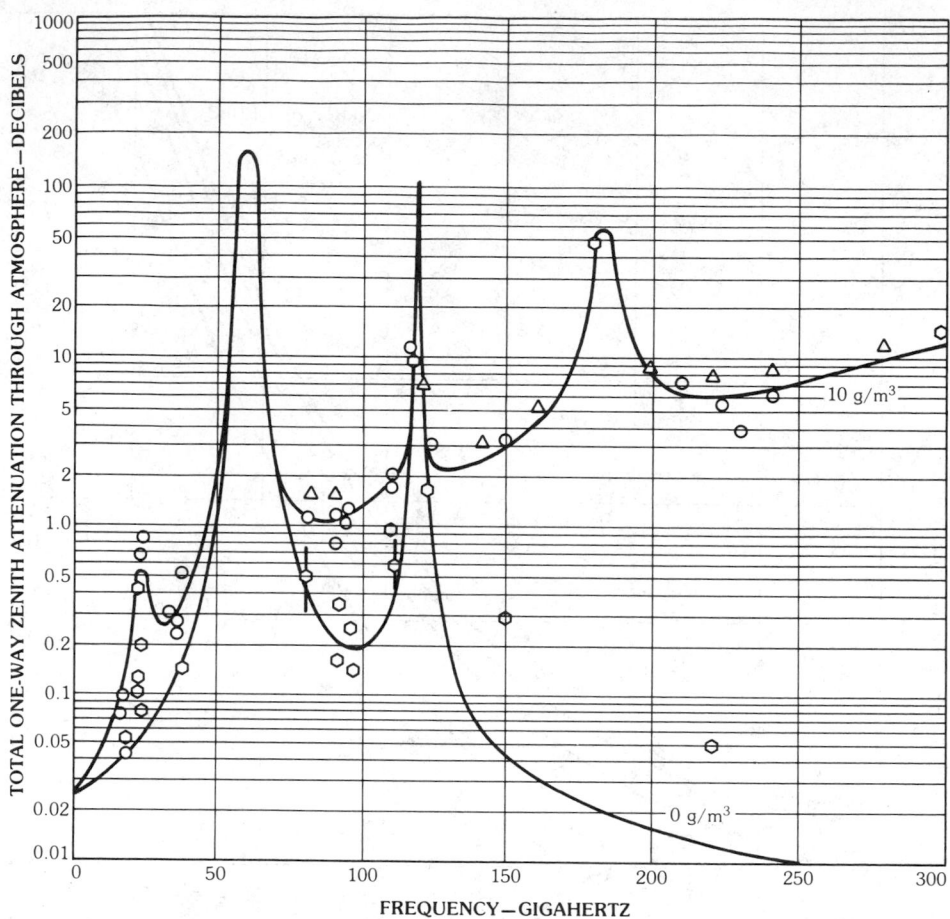

Fig. 5. Spectrum of the microwave atmosphere.

0.6 (1 to 10 GHz, nadir). Humid air and rain show up as warm areas against the cool ocean background. This is particularly evident in the microwave atmospheric spectrum of the intertropical convergence zone (ITCZ) around 10°N, S latitudes.

The particular frequencies used for a spaceborne microwave radiometer depend on the physics of the problem concerned, as well as the transmission characteristics of the atmosphere. Fig. 5 shows the microwave spectrum of the atmosphere due to oxygen and water vapor. The major opaque lines are the oxygen lines at 60 and 118 GHz, and the two water vapor lines are at 22 and 183 GHz. For radiometers whose primary purpose is sensing the earth's surface, one would use the so-called window regions between the opaque lines, such as 150, 90, and 30 GHz, or lower frequencies, below the 22-GHz line. Water vapor lines are used to sense atmospheric humidity and the oxygen lines to sense atmospheric temperature.

Like their infrared counterparts, microwave radiometers operate day or night since they rely on thermal emission rather than sunlight reflections. By virtue of their longer wavelengths, as compared with visible and infrared, microwaves have an important advantage because they can penetrate through the clouds. Therefore they work in nearly all weather conditions. In addition, there are unique features in the microwave spectrum. For example, the presence of water in soil decreases its emissivity. This fact can be used to measure soil moisture. (See Fig. 2.) The longer wavelengths can better penetrate the vegetation coverage to sense the underlying soil moisture. Because of the long wavelengths, however, microwave radiometers need relatively large antennas for good angular resolution.

Furthermore, in order to cover large areas of earth from an orbiting satellite (in a typical low earth polar orbit) the antennas must be able to scan large angular limits (such as $\pm 50°$). Also, remote sensing radiometers usually need several frequencies that are widely separated from one another. To top off these demanding requirements, remote-sensing radiometer antennas must have extremely high beam efficiency and low side lobes. The radiometer must be well calibrated and maintain good stability. In addition, there are the usual spaceflight constraints. It must be lightweight, compact in size, and produce a minimum of heat and mechanical perturbation. In short, the overall performance and engineering requirements of a satellite microwave radiometer antenna for earth remote sensing can be very stringent.

A few of the more important applications of passive microwaves in earth remote sensing are discussed next. For more detailed information, the reader is referred to a recent review article [5].

The Atmosphere

Probably the most useful application of earth satellite remote sensing is the gathering of atmospheric data for meteorological purposes. Because the atmosphere is in a continuous state of change, frequent samples (in both time and space intervals) are needed for weather forecasting. Measurements from orbiting satellites are the most economical methods used to meet this type of data requirement. Microwaves also have an advantage of being able to operate in cloudy regions, where most meteorological actions occur.

Oxygen and water vapor play dominant roles in shaping the absorption spectrum of the atmosphere. Fig. 5 shows a calculated attenuation characteristic for the atmosphere [6, 7]. The lower curve is for dry air (US Standard, 1976) and the upper curve is for humid air with $10 \, g/m^3$ surface water vapor content.

Temperature Sounders—Microwave temperature sounders typically use the 60-GHz oxygen band to measure the atmospheric temperature profiles. This is based on the principle that the oxygen mixing ratio is fairly uniform in space and constant in time; hence the magnitude of the microwave brightness temperature is uniquely related to the atmospheric temperature. By having several frequencies (or channels) spreading down the wings of the 60-GHz oxygen band, each channel will sense a different layer of the atmosphere. The very opaque channels sense only the very top of the atmosphere, as the radiation from the lower layer of the atmosphere is highly attenuated and never arrives at the satellite. The more transparent a channel is, the deeper it will probe into the atmosphere. The complete temperature profile can be retrieved from the brightness temperatures and the associated "weighting functions" of all the channels.

Microwave temperature sounders are in routine use today [8]. For example, the microwave sounder units (MSUs) are aboard the National Aeronautics and Space Administration's (NASA's) TIROS-N, and the US National Oceanic and Atmospheric Administration's (NOAA-series) weather satellites. A similar sounder called the special sensor microwave/temperature (SSM/T) is on the US Air Force's defense meteorological satellites (DMSs). The hardware aspects of these sounders will be discussed in more detail in Section 3. The oxygen line at 118.75 GHz can also be used for temperature sounding [9]. A major advantage of this frequency is that it needs smaller antennas (as compared with the 60-GHz band), which can be very important for geosynchronous satellites. (See Section 6.)

Humidity Sounders—The 22-GHz water vapor line has been used for sensing total water vapor content (humidity) over the ocean [10]. The very opaque water vapor line at 183 GHz can be used to ascertain humidity profiles. In addition to the temperature and humidity profiles, window frequencies can also be used to measure precipitation distributions. The measurement is based on brightness temperature contrast between the precipitation and the "background." For example, at 18 GHz rain cells will appear as warm areas, in contrast to the rather cold ocean, because of their absorption. The rain cells will appear to be colder than the background at high frequencies and high intensities, when scattering is dominant [11]. In addition, satellite radiometers can be used to monitor other species in the atmosphere, such as O_3, CO, and so on [12].

The Ocean Surface

Sea Surface Temperature—The microwave brightness temperature of the ocean is the product of the sea-water thermal temperature and its surface emissivity. The latter is also a function of thermal temperature. The sensitivity of emissivity to thermal temperature change is maximum at 4 to 5 GHz; therefore this is the region of frequency for sea surface temperature measurement [13]. One must also account for emissivity variations caused by other reasons, such as surface roughness (due to waves) and/or atmospheric water. The theoretical sea surface brightness temperature was shown in Fig. 3 (Section 1).

Sea Surface Wind Speed—When the wind disturbs a calm sea the surface emissivity increases from that of a smooth plane surface determined by the Fresnel equations. As the sea becomes very rough, patches of foam begin to form. This also increases the surface emissivity. These relationships can be applied in determining the sea surface wind speed from radiometric observations [14].

Sea Ice—Sea ice has a microwave emissivity value of 0.8 to 0.9 (at nadir), as compared with the value of about 0.3 to 0.6 for calm sea water. Hence, radiometrically, sea ice appears as a warm island against a cold sea-water background. Moreover, microwave radiometers can differentiate between new ice, which is warmer, and the old, multiyear ice, which is comparatively colder. Sea ice concentration maps, derived from orbital microwave radiometer measurements, are very useful in guiding ship routing near the polar regions. Repeated time-series images of the polar region allow one to study and monitor annual polar ice boundary evolution [15]. Fig. 6 is an image of the microwave brightness temperature of Antarctica in the winter of 1974. The data were taken from the ESMR on the Nimbus-5 satellite at a wavelength of 1.55 cm. (See the color insert for Fig. 6.)

Land Applications

There are two major applications of remote sensing with microwave radiometry over land: soil moisture and snow cover.

The need for regional or global data of soil moisture at frequent time intervals can be more efficiently met by remote sensing with microwave radiometry. The conventional point-by-point in situ snow survey is inadequate and expensive.

The uses of soil moisture information are numerous. For example, in hydrology, area-wide soil moisture measurements are needed to assess regional drought conditions. Soil moisture information is also a basis for computing the watershed runoff coefficient, which is used for flood predictions. The evapotranspiration of soil moisture is a part of climate study. The soil moisture, at some critical period of the growth cycle of a plant, determines the yield of that crop at harvest. Timely soil moisture information can be used for irrigation control and yield forecasting.

Timely information on snow-covered areas and water equivalent (amount of water depth or water stored per unit area) in mountain watersheds, such as the western states of the United States, is difficult and expensive to gather during the winter. This snowpack information, however, is important in forecasting the amount of water runoff, which is the basis for the management of this limited and precious natural resource. The passive microwave remote-sensing technique is capable of providing such snowpack information.

Soil Moisture—Water has a much larger dielectric constant than that of dry soil, particularly at lower microwave frequencies, below 5 GHz. The presence of water in the soil increases the dielectric constant of the mixture and consequently lowers its surface emissivity. For example, at 1.4 GHz the emissivity can change from 0.9 for dry soil to 0.6 for wet soil at 30-percent water by dry weight. This range of emissivity variation is the underlying principle [16] for remote sensing of soil moisture with microwave radiometry. A change in emissivity by 0.3 corresponds to a change in microwave brightness temperature of 80 to 90 K. An orbiting microwave radiometer can achieve measurement precision on the order of 1 K. Therefore a microwave radiometer can differentiate not only dry from wet areas, but it is also capable of resolving many levels of soil moisture content.

There are, however, factors that tend to complicate the quantitative determination of moisture content, and research is currently under way to resolve them. The soil dielectric constant is also dependent on the type of soil, because the bounding force between water and the host soil depends on the type of soil. Surface roughness also affects the emissivity. The presence of a vegetation canopy increases the emissivity. The net results are that both vegetation canopy and surface roughness tend to reduce the "sensitivity" of microwave radiometry techniques to measure soil moisture. (Sensitivity is defined as change in microwave brightness temperature per unit change in soil moisture.)

The lower microwave frequencies, from 1 to 5 GHz, are better suited for soil moisture sensing. This is primarily because the difference between dielectric constants of water and dry soil is larger at lower frequencies. Also, longer wavelengths in this range can penetrate deeper into soil and are less vulnerable to the masking effects of vegetation cover and surface roughness. However, a drawback of this low frequency range is that it requires a large antenna for use from satellites. The 1.41-GHz (21-cm wavelength hydrogen line) protected radioastronomy band is a good compromise frequency for the previously mentioned factors.

Snow Hydrology—When a land area is covered by a layer of dry snow its brightness temperature decreases. This is because snow particles scatter background land emission. The snow particles also absorb and reemit the background radiation but this is relatively unimportant; the scattering is the dominant loss for dry snow. The brightness temperature of a snowpack decreases as the snow depth increases. This relationship is used for remote sensing of snow depth by microwave radiometers [17, 18]. A decrease in brightness temperature of 60 K has been measured for a snow depth of 60 cm. When a snowpack begins to melt, the presence of liquid water drastically increases the absorption. As a result the snowpack brightness temperature increases substantially. This fact can be used to monitor the onset of snowpack melting.

As in the case of soil moisture, there are also complicating factors in quantitative determination of snow depth or its water equivalence. The dominant scattering loss implies that the snowpack brightness temperature also depends on snow grain size. Snowpack with smaller grain size scatters less and is, therefore, warmer as compared with larger-grain snowpack. Ice layers embedded in the snow also modify the brightness temperature. Because of these factors multiple frequencies are needed to resolve ambiguities.

3. A Survey of Existing Spaceborne Microwave Radiometer Antennas

Table 2 is a listing of the characteristics of most existing satellite microwave radiometers. Unless mentioned otherwise the country of origin of the satellite is the United States. Most of the radiometers listed are for earth remote sensing, with the exception of Mariner-2. All radiometers are the Dicke switching type.

4. Antenna Requirements for Remote-Sensing Microwave Radiometry

Fig. 5 shows the atmospheric spectrum up to 240 GHz. Most of the present remote-sensing microwave radiometers use frequencies to the left of the 60-GHz oxygen band. This oxygen band has been used for atmospheric temperature sounders because of the well-behaved oxygen mixing ratio in the atmosphere, as mentioned previously in Section 2. The 60-GHz oxygen band in Fig. 5 actually contains a complex band of many individual lines which manifest themselves at higher altitudes. Fig. 7 shows the fine structures of the dominant lines of this band and the frequencies used by many of the microwave temperature sounders. Fig. 7 is presented in a one-way zenith opacity in units of optical depth (OD, where 1 OD = 4.3 dB) versus frequency. These 60-GHz band lines are the rotational lines of the oxygen molecules.

In Fig. 7 the down-pointing arrows indicate the line center frequencies. The numbers with a right superscript (e.g., 9^-, 3^+, etc.) are the line designations. The

Table 2. Existing Satellite Microwave Radiometers

Year of launch	1962	1968	1972
Satellite name	Mariner-2 [19]	Cosmos-243 (USSR) [20]	Cosmos-384 (USSR) [21]
Frequency (GHz)	15.3 22.2	3.5 8.8 22 37	3.5 8.8 22 37
Bandwidth (MHz)			
Antenna type	Parabola, 42-cm diam.	Parabolas	Parabolas
Beamwidth (degree)	2.7 2.0	10 4 4 4	10 4 4 4
Scanning	±60°	Nadir viewing	Nadir viewing
Polarization			
Calibration	Space viewing horns	Spaceview horn and noise generator	Spaceview horn and noise generator
Temperature sensitivity, ΔT, (K)			
Equivalent ΔT at 1 s			
Parameter measured	Limb darkening of planetary emission, temperature of Venus	Atmospheric water vapor and liquid water, sea ice, sea surface temperature	Atmospheric water vapor and liquid water, sea ice, sea surface temperature

Year of launch	1972	1972	1973
Satellite name	Nimbus-5	Nimbus-5	Skylab
Instrument name and acronym	Nimbus-E microwave spectrometer (NEMS) [22]	Electrically scanning microwave radiometer (ESMR) [4]	S-193
Frequency (GHz)	22.235 31.40 53.65 54.90 58.80	19.3	13.9
Bandwidth (MHz)	220 220 220 220 220	240	50
Antenna type	5 lens loaded scalar horns	Phased array, 83 cm × 85 cm	Parabola
Beamwidth (degree)	10 10 10 10 10	1.4° × 1.4°, nadir, 1.4° × 2.2° at ±50° (d) × (c)	1.5
Scanning	Nadir viewing	Electronically scanning ±50°, cross track	Mechanically scanned 0 to ±48° cross- or down-track
Polarization		Horizontal	Horizontal and vertical
Calibration	Microwave load embedded in radiation cooled plate and warm loads	Spaceview horn and hot load	
Temperature sensitivity, ΔT, (K)	0.28 K to 0.56 K (at 2 s)	Better than 1.5 K at 46 ms	1.8 K at 32 ms
Equivalent ΔT at 1 s	0.4 K to 0.8 K	Better than 0.3 K	0.32 K
Parameter measured	Atmospheric temperature profile, total water vapor and liquid water, sea ice	Precipitation, sea ice, and snow cover	Sea surface wind speed
Remarks	Weight = 33 kg, power = 33 W	Weight = 31 kg, power = 41 W	Part of a combined passive/active system

Band 1:

	Skylab / S-194	Nimbus-6 / SCAMS	Meteor (USSR)	Meteor-28	Nimbus-6 / ESMR
Year of launch	1973	1975	1974	1977	1975
Satellite name	Skylab	Nimbus-6 [23]	Meteor (USSR)	Meteor-28	Nimbus-6
Instrument name and acronym	S-194	Scanning microwave spectrometer (SCAMS)			Electrically scanning microwave radiometer (ESMR) [4]
Frequency (GHz)	1.414	22.235 31.60 52.80 53.80 55.40	≅37	≅37	37
Bandwidth (MHz)	27	220 220 220 220 220		1 GHz	220
Antenna type	64-element phased array (102 × 102 cm)	3 hyperbolic reflectors			Phased array, 77 cm × 81 cm
Beamwidth (degree)	15	7.5°		≅1°	1.17° × 0.73° down-track × cross-track
Scanning	Nadir viewing only	±43°, cross track, rotating reflector, fixed feed	35° from nadir ±40°, mechanically scanned		Conical scan, half-cone angle = 45° ±34° azimuth limit
Polarization		V* V* V* V* H**	V&H		V&H
Calibration	Radiation cooled load (200 K) and hot load (370 K)	Reflector views space and an ambient temperature target			Spaceview horn and a hot load
Temperature sensitivity, ΔT, (K)	Better than 0.5 K, at 1 s	0.2 K to 0.5 K at 2 s	0.5 K at 1 s		1 K at 60 ms
Equivalent ΔT at 1 s	Better than 0.5 K	0.28 to 0.71 K	0.5 K		0.25 K
Parameter measured	Soil moisture, sea water salinity	Atmospheric temperature profile liquid water and water vapor	Atmospheric		Atmospheric liquid water, sea ice
Remarks	Weight = 25 kg	Weight = 31 kg, power = 41 W			Weight = 44.6 kg, power = 55 W

Band 2:

	Block 5D / SSM/T	TIROS-N / MSU
Year of launch	1978	1978
Satellite name	Block 5D	TIROS-N
Instrument name and acronym	Special sensor, microwave/temperature (SSM/T) [24]	Microwave sounder unit (MSU) [25]
Frequency (GHz)	50.5 53.2 54.35 54.9 58.4 58.825 59.4	50.30 53.74 54.96 57.9
Bandwidth (MHz)	400 400 400 400 115 400 250	220 220 220 220
Antenna type	1 parabolic reflector	2 hyperbolic reflectors (for upper frequencies and lower halves)
Beamwidth (degree)	14°	7.8×7.1 7.3×7 7.6×7.3 7.3×7
Scanning	±36° cross track, scanning reflector, stationary feed	±47.4° cross track, scanning reflectors, fixed horns
Polarization	V*/H** OMT*** separates lower 3 channels from the rest	V*/H**, OMT*** separates V from H behind each horn
Calibration	Reflector views space and a hot target	Reflector views space and an ambient temperature target
Temperature sensitivity, ΔT, (K)	0.4 to 0.6 K (at 2.7 s)	0.21 0.22 0.18 0.21 (at 1.8 s)
Equivalent ΔT at 1 s	0.66 to 0.99	0.28 0.30 0.24 0.28
Parameter measured	Atmospheric temperature profile	Atmospheric temperature profile
Remarks	Subsequent launches carry the same SSM/T	Weight = 29 kg, power = 30 W, subsequent launches of TIROS-N (renamed NOAA-1, -2, -3, etc.) carry same MSU

	1978	1979	1981
Year of launch	1978	1979	1981
Satellite name	Seasat and Nimbus-7	SEO-1 or Bhaskara-1 (Indian)	SEO-11, Bhaskara-II (Indian)
Instrument name and acronym	Scanning multichannel microwave radiometer (SMMR) [26]	Satellite microwave radiometer (SAMIR) [27]	Microwave radiometer satellite (SAMIR) [28]
Frequency (GHz)	6.60 10.69 18.00 21.00 37.00V 37.00H	19.1 19.6 22.235	19.35 22.235 31
Bandwidth (MHz)	220 220 220 220 220 220	250 250 250	250 250 250
Antenna type	80 cm diam. aperture parabolic reflector with a multifrequency feed horn	Horn Horn	Horn Horn
Beamwidth (degree)	4.0 2.5 1.5 1.2 0.7 0.7	14° 14° 23°	16.5° 26.5° 14°
Scanning	Conical scan, half-cone angle = 42° ±25° azimuthally (Nimbus-7), −3° to +47° (right of track) for Seasat	Spin scan (by spinning satellite)	Spin scan (by spinning satellite)
Polarization	Dual linear	Single linear polarization	Single linear polarization
Calibration	Spaceview horns and hot loads	Views cold space during each spin cycle	Views cold space during each spin cycle
Temperature sensitivity, ΔT, (K)	0.6 0.7 0.8 0.9 1.2 1.3 at 126 ms 62 ms 62 ms 62 ms 30 ms 30 ms	Better than 1 K (350 ms at 19 GHz, 470 ms at 22 GHz)	Better than 1 K (350 ms at 19 GHz, 470 ms at 22 GHz)
Equivalent ΔT at 1 s	0.21 0.17 0.19 0.22 0.21 0.23	Better than 0.55 K	Better than 0.55 K
Parameter measured	Sea surface temperature, sea ice concentration, sea surface wind speed		
Remarks	Weight = 40 kg, power = 55 W		

d = down-track, c = cross-track.
*Vertical.
**Horizontal.
***Orthomode Transducer (OMT).

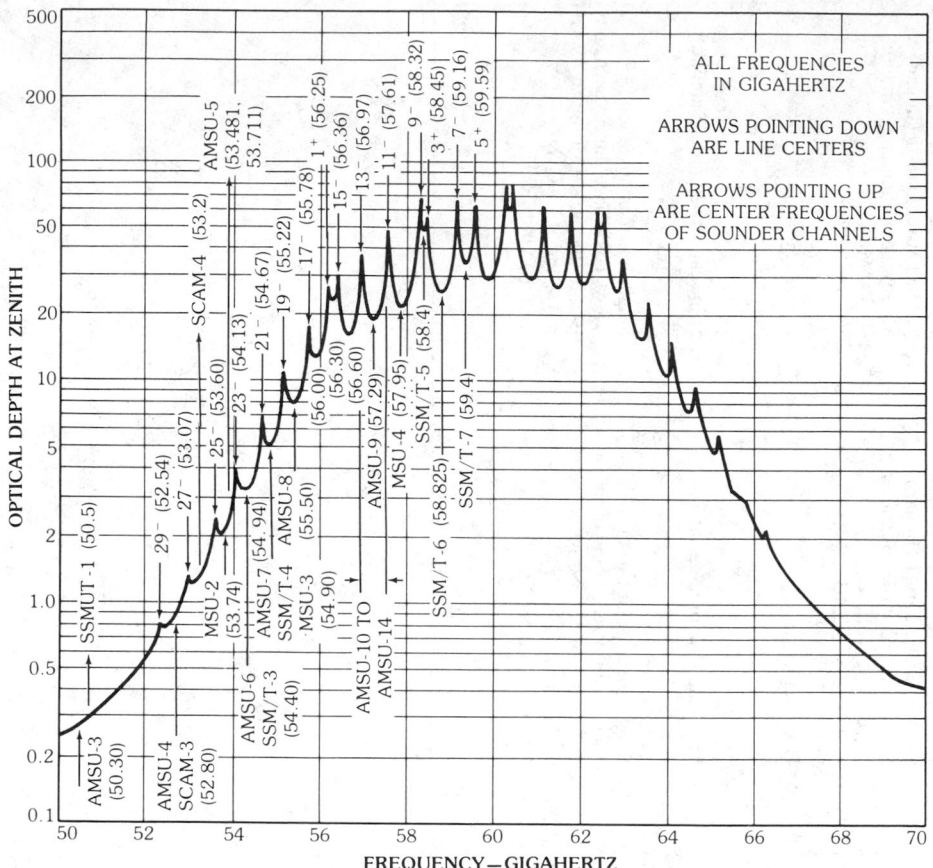

Fig. 7. Oxygen-band spectrum and various microwave sounder frequencies. (*Courtesy NASA*)

base numbers are the rotational angular momentum quantum numbers, and the superscript signs indicate the sign of the total angular momentum changes in a transition. The up-pointing arrows indicate the center frequencies of microwave temperature sounders.

The numbers following each acronym are the channel numbers of that sensor. Generally, the numerical designations of the channels are arranged in the order of ascending sensing height. This feature can also be seen from the opacity curve. The greater the opacity of a channel, the higher it senses above ground.

Note that most of the frequencies are situated at a "valley" between two lines. This is because of two conflicting requirements of a microwave radiometer to be used as a temperature sounder. As we shall see later, the radiometric measurement precision ΔT (generally referred to as the *temperature sensitivity*) is inversely proportional to the square root of the bandwidth. Hence the larger the bandwidth, the better the precision. However, a given point on the opacity curve is related to a particular height of the atmosphere. The higher the opacity, the higher is the height. In order to obtain good vertical resolution in temperature profile it is better for a channel to receive the energy from only a very narrow bandwidth (approaching a point on the spectrum) so that it senses only a very thin layer of the atmosphere (at a chosen height). A too-narrow bandwidth, however, can result in poor (large-value) temperature sensitivity. The valleys are locations where the opacity varies slowly with frequency; hence they allow the use of a maximum of bandwidth to improve the temperature sensitivity but pay little penalty in degrading the vertical resolution. There is also a single oxygen line, 1^-, at 118.75 GHz, which can also be used for temperature-sounding purposes, but this line has not been fully explored yet. Because of its shorter wavelength (by a factor of 2), as compared with 60 GHz, this 118.75-GHz line has the advantage of affording a smaller antenna size. This feature will be a substantial factor in consideration of temperature sounding from a geosynchronous orbit. (See Section 6 for a discussion of a proposed geosynchronous satellite for severe weather monitoring from a geosynchronous orbit.)

Other dominant atmospheric lines are the water vapor rotation lines at 22.2 and 183.3 GHz. Either of these lines can be used to sense the atmospheric humidity. The 22.2-GHz line is a weak line and can only yield the total (integrated) precipitable water. This line (or its vicinity) has been used by many remote sensing radiometers, e.g., Nimbus-7 and Seasat's scanning multichannel microwave radiometer (SMMR). The stronger 183.3-GHz water vapor line can be used to obtain humidity profiles. Because of the state of the art of millimeter-wave technology the use of this line is still in the experimental stage. But it is anticipated that a satellite humidity sounder, based on the 183.3-GHz water vapor line, will be developed soon. (See Section 6.) There are the so-called window regions of lesser absorption around 30, 90, and 150 GHz, with varying degrees of opacity. The 30-GHz window has been a popular one (e.g., Nimbus-6 ESMR and Nimbus-7 SMMR) because the microwave components are readily available at this frequency and reasonably high spatial resolution can be obtained. The 90-GHz window is starting to be used for its high-resolution capability as the microwave components at this frequency become more available. The region around 5 to 6 GHz is an important one, because it is the optimum frequency for sea surface temperature measurement, as mentioned in Section 2. The region between 1 and 2 GHz is sensitive to soil moisture because of the dispersion of water molecules.

Fundamentals of a Microwave Radiometer

A microwave radiometer is similar to a communication receiver except that its main signal is not a coherent carrier signal; its "signal" is the antenna (noise) temperature that a communication receiver is trying to minimize. A radiometer measures the magnitude of the noise power (brightness temperature) radiated by a target or scene.

Referring to the schematic diagram in Fig. 8, a microwave radiometer consists of an antenna for collecting the incoming radiation, whose intensity (in watts per square meter) is represented by the antenna noise temperature T_a (to be called simply "antenna temperature" for short) and a receiver for detecting and determining magnitude of the noise power. The receiver may consist of a preamplifier followed by a detector, or just a detector. It could also be a heterodyne system, in

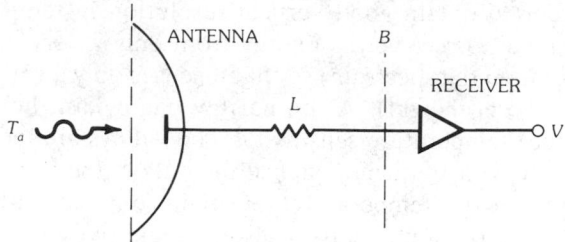

Fig. 8. Total power radiometer.

which the incoming noise power is mixed with a local oscillator and down-shifted to an intermediate frequency before it is detected by a detector.

The output of the receiver (in voltage or digital count) is linearly related to T_a. The ohmic loss of the transmission line between the antenna and receiver is represented by L. More details on radiometer fundamentals can be found in the references [29, 30, 31].

As in any instrument there are errors in a measurement. The errors in a radiometer measurement can be divided into two categories. One is the random, short-term (fast) fluctuations of the output, mainly associated with the noise of the front-end detector of the radiometer system (e.g., the mixer or first-stage amplifier). The other is the systematic errors (which may be slowly varying), resulting from calibration bias and component degradations. The former is commonly known as *temperature sensitivity* (or temperature resolution), and the latter is termed *calibration accuracy*.

Temperature Sensitivity—Temperature sensitivity, commonly represented by the symbol ΔT, is the precision of the radiometer. It is defined as the "minimum detectable change of antenna temperature at the collecting aperture." The "minimum detectable" change is taken to be the standard deviation of the radiometer output when its antenna is viewing a specified constant–brightness-temperature target.

Total Power and Modulating Radiometers—The radiometer, shown schematically in Fig. 8, is a "total power" radiometer. Its temperature sensitivity ΔT, referred to the radiometer input point B, is given by

$$\Delta T = K_1 T_{\text{sys}} \sqrt{\frac{1}{\Delta f t_I} + \left(\frac{\Delta G}{G}\right)^2} \tag{25}$$

where

$$T_{\text{sys}} = T'_a + T_{rn} \tag{26}$$

K_1 = radiometer modulation constant whose value depends on the type of radiometer system, with $K_1 = 1$ for the total power radiometer

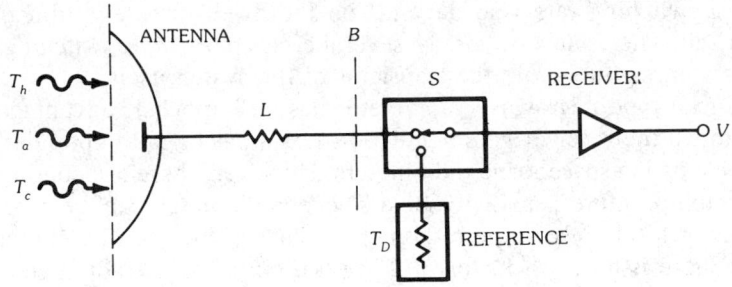

Fig. 9. Modulating radiometer.

T_{sys} = the total system noise temperature referred to the input of the receiver

T_{rn} = the receiver noise temperature referred to its input

T'_a = the effective antenna noise temperature at the input of receiver [see (28)*]

Δf = predetection bandwidth

t_I = integration time of an ideal integrator

G = radiometer gain

ΔG = gain fluctuation

In order to remove the contribution to ΔT due to gain fluctuations, a "modulating" method was introduced by R. M. Dicke and is now commonly known as the *Dicke radiometer*. Fig. 9 is a block diagram of a simplified Dicke radiometer. The receiver is switching between the antenna T'_a and a reference load kept at a known temperature T_D. Since in each half-period of a switching cycle the gain of the radiometer is measured by switching it to the known-temperature T_D load, any gain fluctuation slower than the switching rate is removed. The spectrum of gain fluctuation normally has a $1/f$ dependence on the frequency f, and as long as the switching rate is high enough, the ΔG noise can be removed or substantially reduced. The temperature sensitivity of a balanced Dicke radiometer is

$$\Delta T - \frac{2T_{sys}}{\sqrt{\Delta f t_I}} \qquad (27)$$

where the symbols are as given for (25, 26).

Most spaceborne microwave radiometers are the modulating type, and their modulating frequencies are in the 400- to 1000-Hz range. For example, the ESMR has a modulation frequency of 600 Hz; the SMMR and MSU have 1000 Hz. The

---

*Equation 28 shows that T'_a contains a term T_a, which is the antenna noise temperature received at the aperture. Strictly speaking, ΔT is defined by setting $T_a = 0$ so that it is independent of the scene temperature. This is frequently the case in radioastronomy. For remote-sensing radiometers, however, it is customary to set $T_a \cong 300$ K or some specified scene temperature value.

choice of a switching rate also depends on the dwell or integration time t_I per resolution cell. There must be at least several cycles per t_I. The switching rate must also be slow enough to avoid the transients of the switch itself.

The Dicke type microwave radiometer has an important advantage of being able to remove the receiver gain fluctuation and it has been the popular choice of most of the existing spaceborne radiometers. However, there are conditions under which the temperature sensitivity of a Dicke radiometer can be worse than a corresponding total-power type radiometer. This can be seen by comparing (27) with (25); there is an extra factor of 2 in (27). (In a Dicke radiometer the same available dwell time is split into two halves; only one half is used in viewing the scene, the other half is spent on the reference target. Hence a Dicke radiometer spends only half as long viewing the scene as compared with a total-power radiometer. Since the errors from both the scene and reference half-periods add in a root-square-sum sense, the end result is doubling the temperature sensitivity value in a Dicke radiometer.)

When the receiver gain is stable enough and/or frequent calibrations are possible to remove the gain fluctuation, then the total-power radiometer may yield a better temperature sensitivity than the Dicke type. Two examples of this condition are the special sensor microwave/imager (SSM/I) and the advanced microwave sounding unit (AMSU) (see Section 6). Both are total-power radiometers. The SSM/I is calibrated once every 2 s and the AMSU every 8 and 2.67 s.

Calibration—To remove systematic errors the complete radiometer system shown in Fig. 9 must be calibrated externally from time to time by introducing a known temperature at the antenna aperture. This is needed because during either of the two half-cycles of a modulating period the radiometer output always includes a component which is the receiver noise temperature T_{rn}, which is usually much larger than either T_a or T_D. Therefore, if one were to rely on the gain determination from the reference half-cycle alone, the precise knowledge of T_{rn} and its constancy are essential for accurate calibration. In practice it is sometimes more convenient to devise frequent onboard calibrations which will eliminate the need to know the exact T_{rn} value.

A commonly used calibration method is for the antenna to view two external targets at T_h and T_c (for hot and cold temperatures, respectively) so that the entire radiometer system, including the antenna, can be calibrated. A second, often less satisfying, method is to use two "matched" loads maintained at T_h and T_c, respectively, and connected between the receiver and the antenna as shown in Fig. 10. Switch S oscillates at a typical rate of about 1000 Hz. Switch S_1 is normally connected to the antenna. During calibration S_1 is first switched to T_h, then to T_c. The corresponding output voltages, V_h and V_c, determine the calibration equations. With the incorporation of the two-point calibrations one only needs to relate the amplitude of the demodulated voltage (which is proportional to the difference between T_a and T_D) and T_h and T_c; it is not necessary to know T_{rn} and the gain values explicitly. As long as the radiometer system transfer function is linear, any desired antenna temperature T_a can be interpolated from a calibration curve, as shown in Fig. 11.

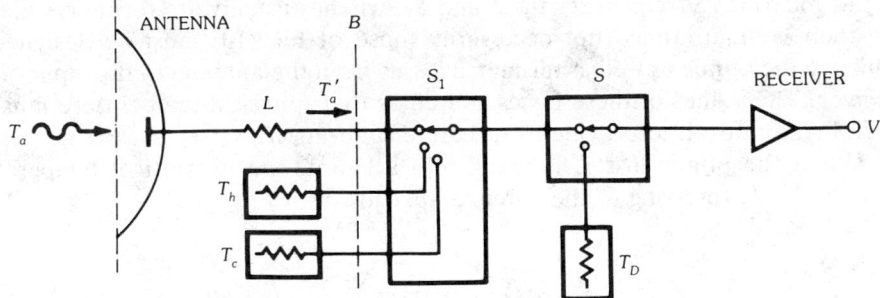

Fig. 10. Modulating radiometer with matched load calibrations.

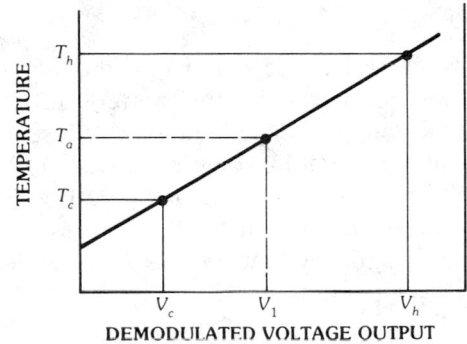

Fig. 11. Linear calibration of a radiometer with two targets: T_h and T_c.

In principle the two-point calibration procedure is needed only once in a pre-launch laboratory test. In practice, however, because of the possibility of slow drift (component degradation) it is a better idea to provide periodic in-orbit calibrations with external calibration targets or matched loads, especially for total power types.

The matched loads calibration method is less desirable because it does not include the antenna characteristics in the procedure, and it also introduces additional losses due to the switch S_1. In satellite operations, however, it is not always feasible to have onboard calibration targets due to their bulkiness. Hence, sometimes the more compact matched loads must be used. For example, in the cases of the Microwave Sounder Unit (MSU) and the scanning microwave spectrometer (SCAMS), onboard reference targets were used for through-the-antenna calibration. In these two cases the sizes of the antennas involved are relatively small, and the required calibration accuracy is high. However, because the antenna aperture sizes were too large (0.8 m in diameter) in the cases of scanning multichannel microwave radiometer (SMMR) and the Electrically Scanning Microwave Radiometer (ESMR), it was impractical to construct external targets, and matched loads were used as hot references. And space-viewing horns were used as cold references for both SMMR and ESMR.

At microwave frequencies the S and S_1 switches usually are ferrite circulators and their configurations (not necessarily those of Fig. 10) must be designed to minimize the ohmic as well as mismatch losses from the antenna to the input of the receiver. The values of these losses as well as their physical temperature must be carefully monitored to provide for accurate calibration.

When the ohmic loss L is taken into account, the theoretical temperature sensitivity ΔT, referring to the antenna aperture, is

$$\Delta T = K_1 \frac{T'_{\text{sys}}}{\sqrt{\Delta f t_I}} \tag{28a}$$

with

$$T'_{\text{sys}} = T_a + (L - 1) T_p + T_{rn} L \tag{28b}$$

where $L > 1$ is the loss factor [see (15)], and T_p is the physical temperature of the lossy element. The loss L represents the sum of total ohmic losses of the transmission line and the antenna lumped into one. It is seen that the ohmic loss L not only produces a noise temperature component $(L - 1) T_p$ due to self-emission, but also increases the effective receiver noise temperature. For radiometer systems with large values of T_{rn}, this latter factor, which causes an increase in T_{rn}, is the dominant term in degrading the system ΔT. As a consequence of the loss L, the effective antenna noise temperature is

$$T'_{\text{ant}} = \frac{T_a}{L} + \left(1 - \frac{1}{L}\right) T_p \tag{29}$$

Similarly, if there are ohmic losses associated with loads T_h and T_c, then their effective noise temperature at S_1 will be

$$T'_h = \frac{T_h}{L_h} + \left(1 - \frac{1}{L_h}\right) T_{ph} \tag{30}$$

$$T'_c = \frac{T_c}{L_c} + \left(1 - \frac{1}{L_c}\right) T_{pc} \tag{31}$$

where L_h and L_c are the losses in the T_h and T_c paths to S_1, respectively, and T_{ph} and T_{pc} are their respective physical temperatures. In practice it is rather difficult to keep track of all the losses and their temperatures. This is why the matched-load type calibration method cannot provide a high degree of accuracy.

For a multiport device, such as switch S_1, the general expression for effective noise temperature at the output of the ith port is

$$T_i = \sum_{j \neq i} \alpha_{ij} T_j + \left(1 - \sum_{j \neq i} \alpha_{ij}\right) T_p \tag{32}$$

where $\alpha_{ij} = 1/L_{ij}$ is the attenuation factor from j to i at a given switch state and T_p is the physical temperature of the switch.*

Special Requirements for Antennas for Remote-Sensing Microwave Radiometers

There are several special features of antennas that are very important to remote-sensing radiometry, although they may not be important in other fields, such as communications. These features are the following:

High beam efficiency
Low ohmic loss
Large scanning angle limits
Provisions for accurate calibration
High polarization purity

Beam Efficiency and Spatial Resolution—Power received by a radiometer antenna is the total sum of power from all directions. Since the power is proportional to temperature we can relate the antenna temperature $T_a(\Omega)$ to the brightness temperature $T_b(\Omega)$ as follows:

$$T_a(\Omega') = \frac{1}{4\pi} \int G(\Omega, \Omega') T_b(\Omega) \, d\Omega \tag{33}$$

where $\Omega' = (\theta', \phi')$ is the direction the antenna main beam is pointing, $d\Omega = \sin\theta \, d\theta \, d\phi$, with the integration being over 4π steradians, and the antenna directive gain over isotropic media is such that

$$\oint G(\Omega) \, d\Omega = 4\pi \tag{34}$$

The *main beam efficiency* (or simply *beam efficiency*) ε_{MB} is the ratio of the power in the main beam to the total power received by the antenna. And the antenna is assumed to be in an isotropic environment (i.e., the brightness temperature is not a function of angular direction). The extent of the main beam has been customarily defined as the null-to-null beamwidths (NNBWs). However, the "2.5 times half-power beamwidth" definition has also been used frequently in place of the null-to-null definition because in practical antennas, due to phase errors, there may not be distinct first nulls. In the following the 2.5 (HPBW) is defined as the main beam width for beam-efficiency computation purposes. The beam efficiency is

$$\varepsilon_{MB} = \frac{1}{4\pi} \int_m G(\Omega) \, d\Omega$$

or $\tag{35}$

---

*For a discussion on the effect due to mismatch reflections, see [31], pp. 404–412.

$$\varepsilon_{MB} = \frac{1}{4\pi} \int_{\phi=0}^{2\pi} \int_{\theta=0}^{\theta_{MB}} G(\theta, \phi) \sin \theta \, d\theta \, d\phi$$

where θ_{MB} is $1.25 \times$ half-power beamwidth.

The *stray efficiency* ε_{ST} is the fraction of power outside of the main beam [29]:

$$\varepsilon_{ST} = \frac{1}{4\pi} \int_{ST} G(\Omega) \, d\Omega \tag{36}$$

where the limit ST is $4\pi - MB$ (i.e., the angles outside the main beam), and

$$\varepsilon_{MB} + \varepsilon_{ST} = 1 \tag{37}$$

From (33), (34), and (35),

$$T_a(\Omega) = \bar{T}_{bMB} \varepsilon_{MB} + \bar{T}_{bSL} \varepsilon_{ST} \tag{38}$$

where

$$\bar{T}_{bMB} = \frac{\int_{MB} G(\Omega, \Omega') \, T(\Omega) \, d\Omega}{\int_{MB} G(\Omega, \Omega') \, d\Omega} \tag{39}$$

So $\bar{T}_{bMB}$ is the "average" brightness temperature within the main beam. Similarly, $\bar{T}_{bSL}$ is the average brightness temperature of the side lobes.

The desired quantity is $\bar{T}_{bMB}$ but the direct radiometer antenna measurement yields only T_a, which includes contributions from side lobes.

In designing the antennas for microwave radiometers, it is important to achieve high main-beam efficiency (frequently 90 to 97 percent is required). In an ideal case, $\varepsilon_{MB} = 1$ and $\varepsilon_{SL} = 0$; then, $T_a = \bar{T}_{bMB}$, the desired brightness temperature of the main-beam area, can be obtained directly from the radiometer measurement.

Usually this is not the case, and the term T_{bMB} must be solved from (38) in terms of T_{bSL} with the main beam and side lobe (stray) efficiencies, ε_{MB} and ε_{SL}, respectively, obtained from careful measurements of the antenna. The side lobe temperature term in (38) must be provided by some other means. For example, if the radiometer antenna scans the complete area cell by cell, then an interactive algorithm can be set up such that the side lobe term of one cell can be computed from the measurements of its immediate neighboring cells.

The beam efficiency as seen in (35) can be further defined for a single polarization. For example, if a vertically polarized beam is desired, then $G(\theta, \phi)$ in (35) can be changed into $G_v(\theta, \phi)$, to signify that only the directive gain of the vertically polarized wave in the main beam is to be counted.

The polarized beam efficiency is slightly lower than the nonpolarized beam efficiency, because there are always some mechanisms which tend to produce cross-polarized components of energy at the expense of the main polarization. For waveguide slotted array antennas the cross-polarized component could result from stray coupling and mechanical imperfections from the slot radiators. For reflector-type antennas the curvature of the reflector will always give rise to some cross polarization even if everything else is perfect. The curvature-related cross polarization decreases with the increase of the focal length to diameter ratio. Imperfections in feed horns and orthomode transducers may also contribute to the cross polarization.

The lowering of beam efficiency due to cross polarization is not a major source of concern in most reflector antenna designs for radiometry because the change caused by it is small. The more detrimental effect is the cross-polarization component of directive gain that will leak some of the orthogonally polarized emission from the earth's surface. This mixing of the wrong polarization can deteriorate the accuracy for some applications. For example, a radiometer antenna designed for the horizontally polarized brightness temperature of a calm sea surface, at an incident angle of 50°, is expected to see a brightness temperature of about 80 kelvins. The brightness temperature of the vertical polarization component at this same angle will be about 150 K. Hence a 2-percent contribution from cross polarization could cause a 1.4-kelvin error, which is appreciable if not accounted for. Of course, leakage in the orthomode transducers following the feed horn or the switches (if used) will result in the same effect.

Both accurate calibration and high beam efficiency are important features of a microwave radiometer to ensure accurate mapping of scene brightness variation. However, each affects the radiometer performance differently. The calibration accuracy affects the bias error of the brightness throughout an entire area containing many spatial resolution cells—all being affected equally. The effect due to low beam efficiency, on the other hand, is to degrade the scene brightness contrast. The effect of high beam efficiency is similar to a low integrated side lobe in a synthetic aperture radar (SAR), which prevents low-contrast targets from being "washed out." For the SAR, only the contrast is of importance in most cases; the absolute radar cross section is not always crucial. In radiometry, however, one needs both the relative contrast and the correct absolute value of the brightness temperatures of individual resolution elements.

Spatial Resolution—Spatial resolution is the "footprint" size, or the diameter of the antenna's main beam projected on the earth's surface. The term *instantaneous field of view* (IFOV) is also commonly used in satellite remote sensing to mean the spatial resolutions. If the antenna beam is a right circular cone of beamwidth θ_b, then the spatial resolution is the diameter of the intersection of the cone and the earth's surface. In general, the intersection is a pear-shaped figure, and its size can be specified by a major and minor diameter, ϱ_M and ϱ_m, respectively. For a scanning antenna the IFOV may vary with scan angle. And the minimum values of ϱ_M and ϱ_m are customarily taken as the spatial resolution values. When the antenna is pointed at nadir, the footprint at nadir is a circle with a diameter $\varrho = h\theta_b$, where h is the satellite orbital height above the earth's surface (see Fig. 12). The beamwidth θ_b is defined as the half-power beamwidth of the antenna main lobe. It is related to the Rayleigh criterion, which states that two point sources are

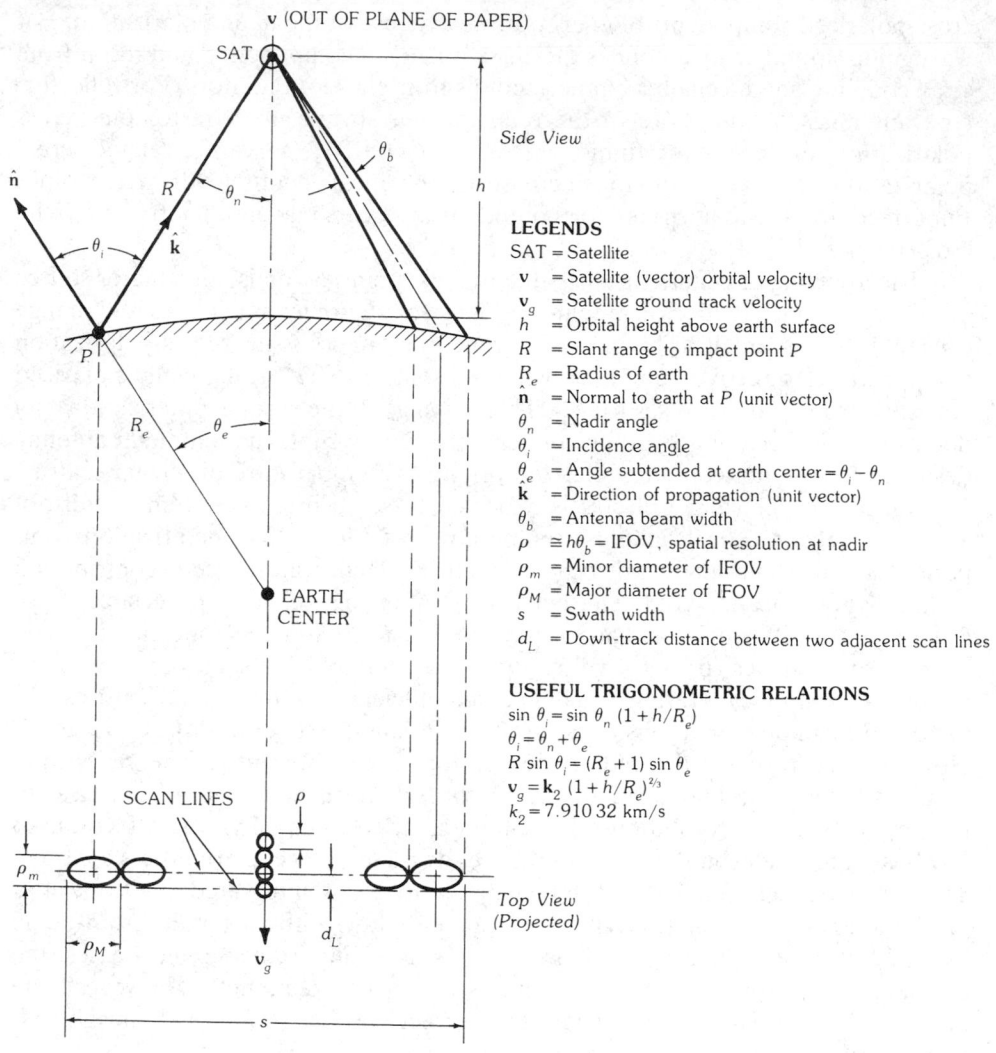

Fig. 12. Planar (cross-track) scan geometry.

resolvable if the angular separation between them is at least such that the first null of one source coincides with the maximum of the other. A uniform aperture distribution for a circular aperture has a Rayleigh resolving power of $1.22\lambda/D$ radians, where λ is the wavelength, and D the aperture diameter. The same uniform circular aperture would have a half-power beamwidth of $1.02\lambda/D$. Hence, for the circular aperture with uniform (nontapered) distribution the resolution as defined by the half-power beamwidth gives a slightly smaller value (i.e., better resolution) than the Rayleigh criterion–defined resolution. The ratio between the half-power beamwidth definition and the Rayleigh criterion is 0.84 for a circular aperture with no taper. Highly tapered antennas tend to increase this ratio as the beam broadens. Spatial resolution determines how small a scale the scene spatial variation can be resolved. But in order to faithfully reproduce scene brightness variation, a radiometer must have a high-beam-efficiency antenna. High resolution (narrow beamwidth) can be achieved with a large-aperture antenna. For a given antenna aperture size, high beam efficiency can be achieved by highly tapered aperture illumination (in addition to other design precautions, such as minimizing phase errors). High taper, however, is an inefficient way to use the aperture and leads to low aperture efficiency and consequently broadened beamwidth. In many microwave radiometer antennas, high beam efficiency is deemed to be of greater importance than narrow beamwidth, and it is often obtained at the expense of lowered aperture efficiency by using highly tapered aperture ilumination. The relationship between the two is shown in Fig. 13 ([29], pp. 219, 221).

Losses—The losses may be categorized as (1) ohmic or (2) scattering. Ohmic loss results from reflector surface resistivity, waveguide feed losses, filter losses, and so on. The scattering losses result from redistribution of energy from the main lobe into other regions of the side lobes and back lobes. The scattered energy may also occur because of undesired cross-polarized energy due to reflector curvature, feed horn cross polarization, reflector surface distortion, and the like.

The ohmic loss degrades the radiometer temperature sensitivity ΔT by increasing the effective system noise temperature as indicated in (28a). Ohmic loss also tends to deteriorate the calibration accuracy of a radiometer due to the self-emission term in (28d) because both the physical temperature of the loss element and the magnitude of the loss contain some uncertainties. Ohmic loss, however, does not affect the beam efficiency ε_{MB} [see (35)] as long as it can be considered a lumped element so that the loss does not depend on direction. Even though ohmic loss diminishes the antenna gain, it does not affect the spatial resolution either, as long as it is not direction (angle) dependent.

Generally the nonohmic losses involve redistribution of energy and may affect any or all of the three radiometer performance parameters: beamwidth, beam efficiency, and temperature sensitivity. Any scattering loss that reduces the energy received by the antenna also degrades the radiometer temperature sensitivity by the same factor. For example, any impedance mismatch causing reflection will lower the energy received, therefore increasing the ΔT value. The mismatch loss L' affects the temperature sensitivity as does the ohmic loss in (28a), except that there is no self-emission term. In other words, one can obtain ΔT from (28b) due to reflection by replacing L with L' and setting $T_p = 0$.

An example of the scattering loss is the antenna reflector surface roughness effect. The roughness produces a random scattering of energy into wide angles (as compared with the coherent main beam) and increases the side lobe envelope. The end result is to reduce the main beam efficiency. The reduction in beam efficiency can be calculated from Ruze's expression for gain reduction [32]:

$$G = G_0 e^{-(4\pi\varepsilon_{\text{rms}}/\lambda)^2} \tag{40}$$

where

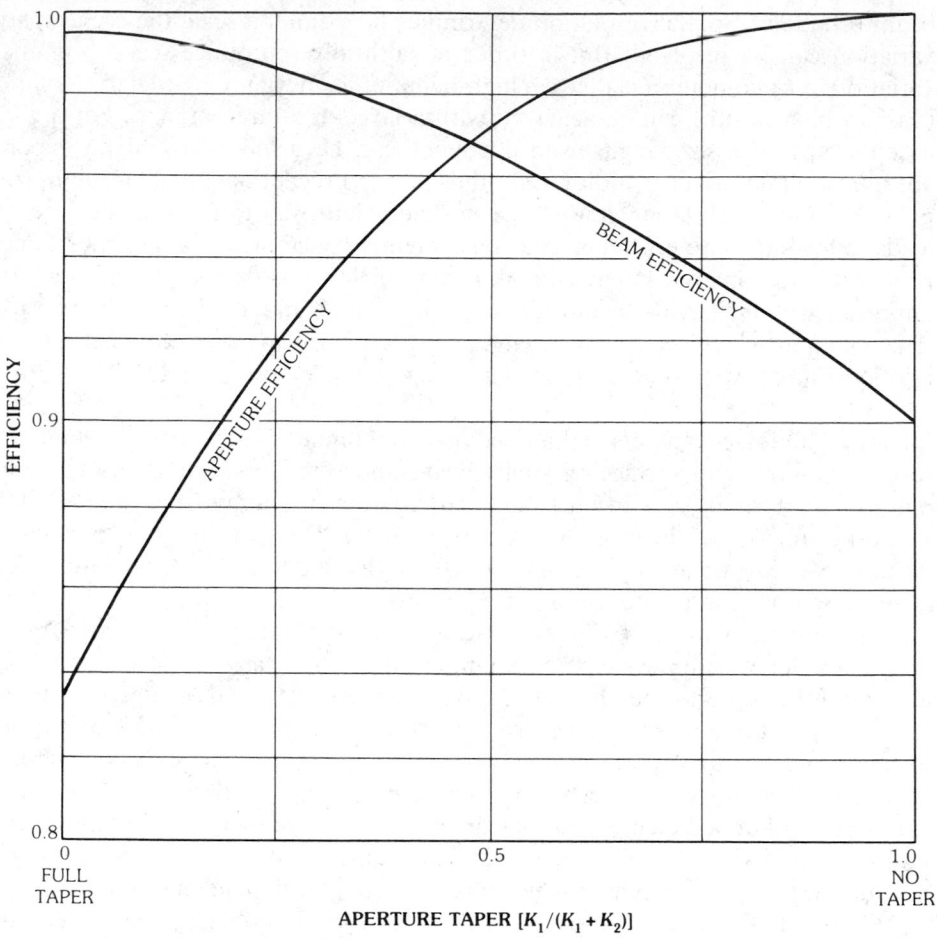

a

Fig. 13. Aperture taper and aperture and beam efficiency. (*a*) For a one-dimensional aperture as a function of taper. (*b*) For a circular aperture as a function of taper and phase error. (*After Kraus [29], from R. T. Nash, Beam efficiency limitations of large antennas*, IEEE Trans. Mil. Electron., *vol. MIL-8, pp. 252–257, July–October 1964; © 1964 IEEE*)

G = antenna gain of antenna with surface roughness

G_0 = antenna gain of a perfect antenna with no surface roughness

ε_{rms} = rms surface roughness

λ = wavelength

Beam Scanning—Most of the remote-sensing microwave radiometer antennas are required to perform scanning of some kind. The purpose of scanning is to produce a two-dimensional image of an area of the earth. Different types of scanning are discussed in the following paragraphs.

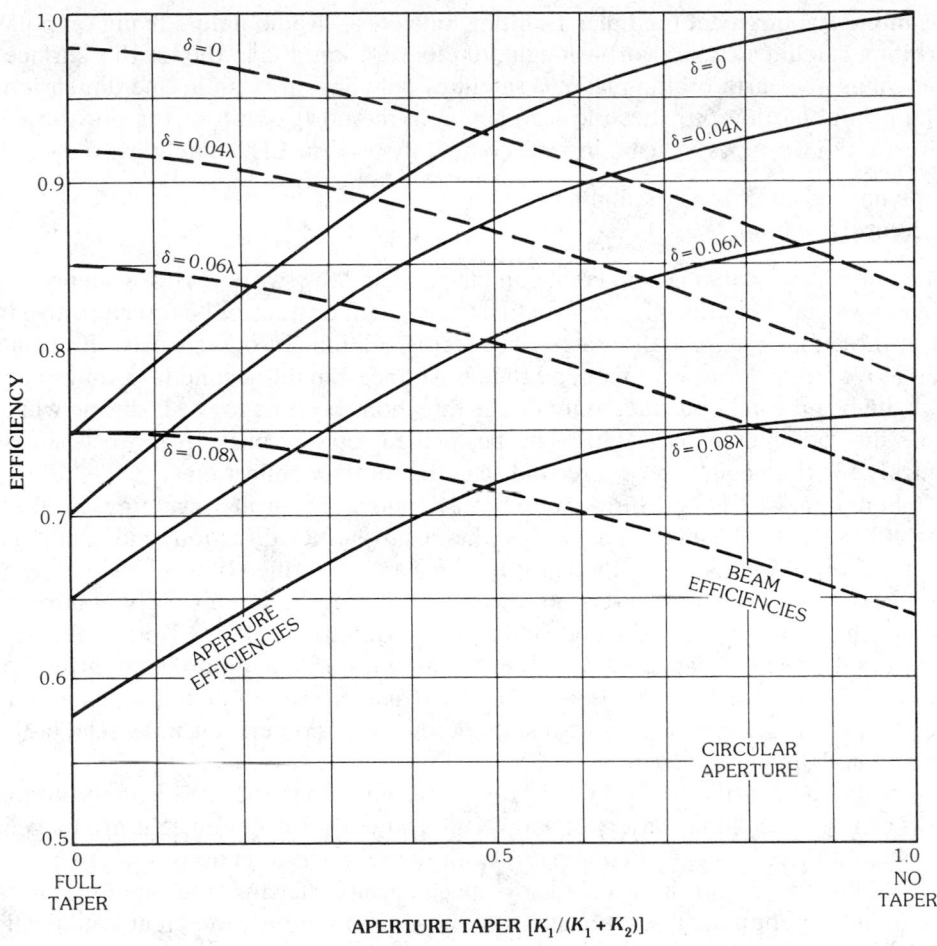

Fig 13, *continued.*

For a spinning satellite, such as the Geostationary Operational Environmental Satellite (GOES), which spins about an axis parallel to the earth's polar axis, the spinning action provides an east-west scan motion and a radiometer on board only has to provide a north-south stepping motion at the end of each scan line.

But the spin-scan is an inefficient scan method in the sense that most of the "available" time is not fully utilized for viewing the earth scene. For example, for each revolution the GOES satellite spins 360° but only a maximum of about 17° (which is the angle the earth subtends from the geosynchronous orbit) can be used for observation. Therefore the "spin-scan" efficiency (which is the ratio of observation time to available time per spin period) can be 4.7 percent, at best. If one only wishes to map a small portion of the earth disk, then the spin-scan efficiency e_{sp} is even less than 4.7 percent. The future GOES satellites are most likely to be of the three-axes stabilized type and the radiometers on board must be able to scan their antenna beam in both E–W and N–S directions.

Scanning Requirement for Polar Orbiting Satellites—Radiometers flying on polar orbiting satellites (typical orbit height 700 to 1000 km above the earth's surface) known as low earth orbiting (LEO) satellites only have to scan in one dimension. The orbital motion provides the scan action in the north-south or the down-track direction. Two types of scanning are commonly used in LEO satellites:

    planar, or cross-track, scanning
    conical scanning

The purpose of scanning is to create an image of an area by successive scan lines of a narrow beam. In principle the scan line (movement or trace of beam) can move in both directions (senses) alternately in a zigzag motion. In other words the beam can move from east to west first and then west to east in the second line, and so on. Or, the beam can be scanned only in one direction, say east to west, during which time the radiometer takes data. The beam then retraces back from west to east quickly for the beginning of a second scan line of data collection.

Since the LEO satellites continuously move in orbit, scanning in both directions (with a beam that moves only in the cross-track direction) will result in a zigzag footprint track on earth. This problem can be rectified by providing a beam motion in the down-track direction to compensate for the satellite orbital motion. While this compensation can be realized with relative ease for optical imagers (e.g., multispectral scanners [MSSs] on Landsats), it is much more complicated to provide the motion compensation in the case of a microwave antenna which usually has a much larger aperture. As a result microwave radiometers usually scan in the cross-track direction only.

Along a scan line (i.e., in the cross-track direction) the adjacent resolution cells can be spaced in a variety of ways. If the adjacent resolution cells are tangent to one another, it is called *contiguous* in the cross-track direction.

When the beam is scanned by mechanically slewing the antenna in a continuous motion the resolution cell also moves continuously, and it automatically results in a contiguous pattern in the cross-track direction. But if the beam is scanned with a stepping motion (i.e., it "dwells" at a resolution cell position for a length of time and then moves quickly to the next resolution cell), the cell spacing in the cross-track direction can be arbitrarily chosen to be either contiguous, overlapping, or undercoverage (leaving gaps between cells).

As we shall see later, the choice of resolution cell spacing is not completely arbitrary. Since the total available time per scan period is fixed, a smaller number of resolution cells per scan line (undersampling) will lead to more integration time per resolution cell, which yields better temperature sensitivity. But spatial undersampling will lose some details of scene spatial variation (aliasing). On the other hand, oversampling (overlap between cells) will result in less integration time per cell and poor temperature sensitivity. The choice of spatial sampling frequency involves a trade-off between spatial resolution and temperature sensitivity of a mapping radiometer. Within a given time t, a scanning radiometer (assuming a single beam for the moment) must cover an area of $A = sv_gt$. (See Fig. 12.) As the sampling frequency increases, individual resolution cell integration time t_I decreases, resulting in poorer temperature sensitivity. The choice of sampling frequency depends on the antenna beamwidth and the degree of cell overlapping.

Let $p = d_L/\varrho$ be the down-track contiguity coefficients; when $p = 1$, there is (down-track) contiguity at nadir. When $p > 1$, there is a gap in the down-track direction, i.e., under coverage at nadir. When $p < 1$, there is some overlap in the down-track direction.

The scan time t_{scan} per line is

$$t_{scan} = d_L/v_g \tag{41}$$

Let $q < 1$ be the scan efficiency, which is the fraction of t_{scan} actually used for taking scene data, and let n be the number of resolution cells per line. Then

$$t_I = \frac{q t_{scan}}{n} = \frac{pq\varrho}{v_g n} \tag{42}$$

For a given total nadir angle scan limit of $2\theta_{nM}$, the number of resolution cells per line is, assuming cross-track continuity,

$$n = \frac{2\theta_{nM}}{\theta_b}, \qquad \varrho = h\theta_b \tag{43}$$

and

$$t_I = \frac{pqh\theta_b^2}{2v_g\theta_{nM}} \tag{44}$$

For earth remote-sensing applications it is usually required that the sensor completely map the earth in a short time period. The consequence is that the scan angle limit θ_{nM} must be as large as practical. For example, in a LEO polar satellite, such as TIROS-N and the NOAA-series weather satellites, it is desired that the onboard sensors map the earth's atmosphere once every 6 to 12 hours in order to update the state of the atmosphere for weather forecasts. This requires two simultaneous satellites (in two polar orbits whose orbital planes are 90° apart), each having to scan to a limit of $\theta_{nM} \cong \pm 50°$. For other earth resource applications, large swath width is also frequently needed in order to completely map the earth once in two to three days.

Equation 44 shows that the integration time t_I increases with the square of the beamwidth θ_b or, equivalently, the square of the spatial resolution ϱ. Since the temperature sensitivity ΔT is inversely proportional to the square root of the integration time t_I, it is therefore inversely proportional to θ_b. In other words, in a mapping radiometer the spatial resolution must be traded off against the temperature sensitivity (unless one can improve the radiometer system noise). Also, the large magnitude of the scan angle limit θ_{nM} makes "small-angle" scan techniques for a reflector antenna (such as mechanical feed displacement in the transverse plane about the focal point) impractical.

Planar scan has its advantages. It is easier to implement, as compared with conical scan. One can design an offset paraboloid reflector geometry in such a manner that the reflector is the only moving part while the feeds and the radiometer receivers are stationary (no rotary joints are required). This scheme has an important advantage because it drastically reduces the mechanical disturbance to the spacecraft, which can generally absorb very little mechanical disturbance. Fig. 14 is a schematic diagram of such a design. An offset paraboloid is driven by a motor to scan about an axis perpendicular to the axis of symmetry of the parent parabola. The planar scan can be easily designed to incorporate onboard calibrations, particularly when the cold space background radiation is used as a reference, simply by rotating the antenna (or reflector) about the same scan axis to view the cold space at T_c (approximately 2.7 K). One disadvantage of a planar scan is that its incidence angle and footprint size both vary with scan angle due to the geometry. Examples of planar scan instruments include MSUs, on TIROS-N, and scanning microwave spectrometers (SCAMSs) and ESMRs, both on Nimbus-5. Both the MSU and the SCAMS have antenna designs similar to that of Fig. 14, except hyperboloids instead of paraboloids were used for reflectors to make the configurations more compact in the SCAMS.

The Nimbus-5 ESMR, on the other hand, is a planar-slotted waveguide array as shown in Fig. 15. A traveling wave is fed into each waveguide "stick." As the wave travels along the guide, its energy is gradually radiated out through the slots. (It is described here as a transmitting antenna.) The beam is scanned by using ferrite phase shifters to change the phase of each waveguide stick with respect to one another.

The advantages of electronic scanning are obvious; it is motionless and agile. It hardly wastes any time for flyback to the starting angle after each scan line. However, there are drawbacks: (*a*) it has larger ohmic losses due to waveguides, terminating resistors, phase shifters, and power dividers as compared with a

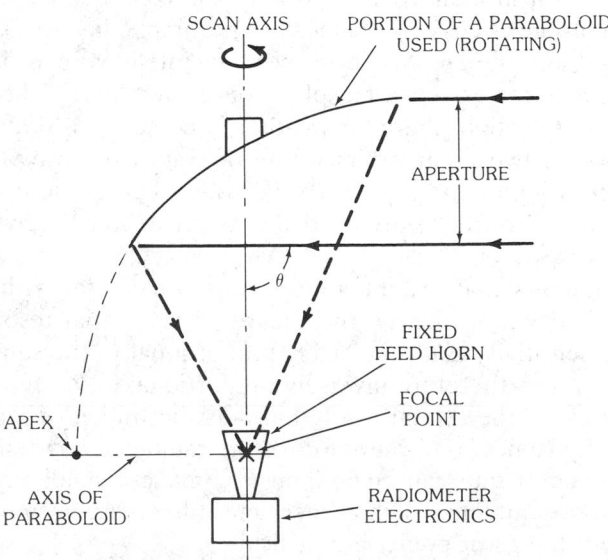

Fig. 14. Schematic of an offset reflector scanned by rotation.

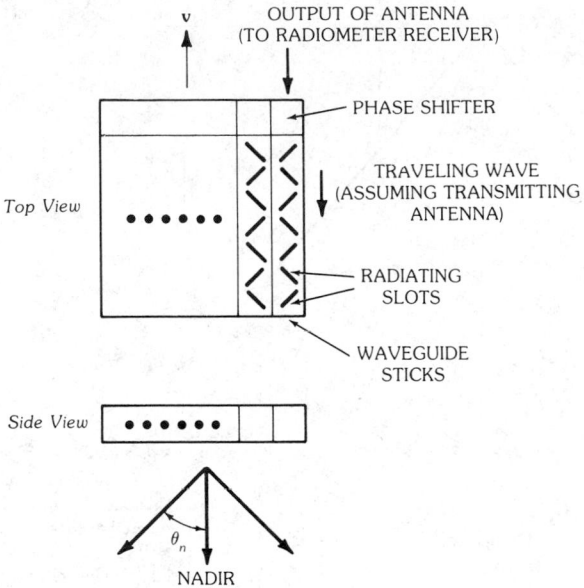

Fig. 15. Schematic of a planar scanning phased array.

reflector, (*b*) the beamwidth broadens as the beam is scanned away from broadside, because the effective aperture is then reduced by a factor of the cosine of the scan angle, and (*c*) it is difficult to share the same aperture in two or more frequencies. In many remote-sensing applications not only are multiple frequencies needed, but both linear polarizations at each frequency are needed as well. This would require many separate phased arrays which would result in heavy weight and large power and require a large earth-viewing area on the satellite. The latter is a premium quantity on a satellite. Reflectors with multiple frequency feeds tend to be the best compromise for many remote-sensing applications.

Conical Scan*—In conical scanning the beam moves on the surface of a cone. Normally, the cone axis is pointed at the earth's center. This results in constant incidence angle θ_i and constant footprint size, both features being advantageous in imaging and data interpretation. Fig. 16 depicts the conical scan geometry.

For a given orbital height h, the ground spatial resolution of a conical scan is not as good as planar scan because the slant range R in the conical system is larger than h. In the conical system, however, the slant range R is constant throughout a scan line. The cross-track contiguity definition is the same as in a planar scan, except that the direction of cross track is really along the scan line (which is not necessarily perpendicular to the ground track). If the scan time t_{scan} is chosen for the nadir resolution cells to be contiguous in the down-track direction between

---

*The term "conical scan" is used here in a different sense from the *IEEE Standard Definition of Terms for Antennas*, no. 145, IEEE, New York, p. 3, 1969.

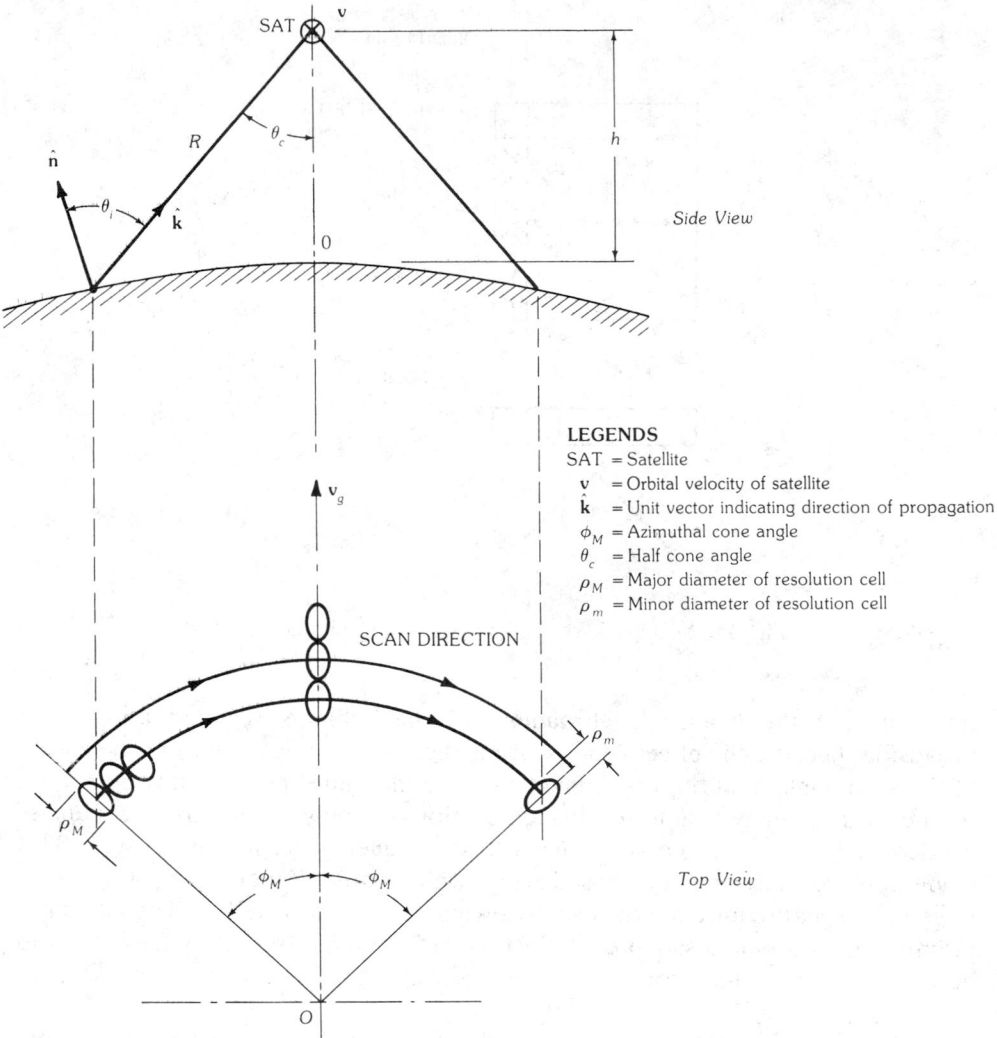

Fig. 16. Conical scanning geometry.

adjacent scan lines (i.e., $p = 1$), then, due to the curvature of the scan line, the swath edge resolution cells will have some overlap, even though the resolutions do not change. The swath width is determined by the scan limit ϕ_M in an azimuthal plane (a tangent plane to the subsatellite point O in Fig. 16 on the earth's surface).

The spatial resolution in a conical scan is represented by the "cross-track" and "down-track" diameters ϱ_m and ϱ_M, respectively.

For the conical scan system the expressions corresponding to the planar scan case are

$$p = d_L/\varrho_M \tag{45}$$

$$t_{\text{scan}} = p\varrho_M/v_g \tag{46}$$

$$\varrho_M \cong R\theta_b/\cos\theta_i \tag{47}$$

$$n = \frac{2\phi_M}{\theta_b \sin\theta_c} \tag{48}$$

and

$$t_I \cong \frac{pqR\theta_b^2 \sin\theta_c}{2v_g\phi_M \cos\theta_i} \tag{49}$$

Equation 49 shows that the individual cell integration time for conical scanning is also proportional to $\theta_b{}^2$, similar to the planar scan case.

As in the case of planar scan the conical scan can also be realized either by mechanically rotating a reflector or by an electronically scanning phased array. The scanning multichannel microwave radiometer (SMMR) is an example of conical scan by mechanical rotation of a reflector, and the Nimbus-6 ESMR is an example of conical scan by a phased array. The latter's geometry is similar to that shown in Fig. 15, except that the waveguide sticks radiate both vertical and horizontal polarizations simultaneously. Unlike Fig. 15, the planar array of the Nimbus-6 ESMR is mounted vertically on the Nimbus-6 satellite's sensory ring and its beam is pointed at θ_c from nadir. The position of the beam is selected by controlling the phasings of each vertical waveguide stick. Each beam is dual polarized. An orthomode transducer separates vertical from horizontal polarization at the output of the array.

The Nimbus-6 ESMR has a spatial resolution on earth of 42 km (down-track) by 20 km (cross-track) from its 1100-km orbit. The scan time per line is $t_{\text{scan}} = 5.33$ s, corresponding to $d_L = 33$ km, hence $p = d_L/\varrho_M = 33/42 = 0.79$, so that there is 21-percent cell overlap in the down-track direction at nadir. The Nimbus-6 ESMR takes data while scanning in only one direction and, because it is electronically scanning, the retrace time needed to swing the beam 70° azimuthally is a very small fraction (a few milliseconds) of the 5.33-s scan time.

The SMMR (on both Nimbus-7 and Seasat) is an example of mechanical implementation of conical scanning. There were many reasons and trade-offs leading to the choice of mechanical scanning of a reflector-type antenna for the SMMR. Chief among them was the fact that the SMMR required five frequencies, ranging from 6.6 GHz to 37 GHz, and each frequency had both linear polarizations. To satisfy this requirement with multiple phase arrays would be impractical for the large earth-viewing areas they would need; heavy weight and high power consumption would also result from multiple-phased array designs.

For a large mechanically scanning antenna such as the SMMR, the "retrace time" can be an appreciable fraction of the scan time and the zigzag shape of the footprint trace on earth becomes a problem. As can be seen in the following description of the SMMR, the overlapping zigzag traces were implemented only in its highest-frequency (37-GHz) channels. For the four lower frequencies, only one direction of the bidirectional scan was used for data taking, thus avoiding the zigzag pattern for these frequencies. Also, because a single reflector is shared by all five diversely different frequencies, the footprint sizes at each frequency are vastly different. In the SMMR design the scan time t_{scan} is chosen to make the higher-frequency (37-GHz) channel contiguous at nadir, hence there are varying degrees of overlapping (down-track) at the lower frequencies. The design of the SMMR is similar to that of Fig. 14, except that the angle θ between the scan axis and the symmetry axis of the parent parabola is no longer 90° but the half-cone angle $\theta_c = 42°$.

Fig. 17 is a sketch of the SMMR, and Fig. 18 is a photograph of the actual hardware.

SMMR design features are described in the following. A stationary five-frequency dual linear polarization feed horn projects the axially symmetric primary radiation patterns vertically upward at an offset paraboloid reflector (assume

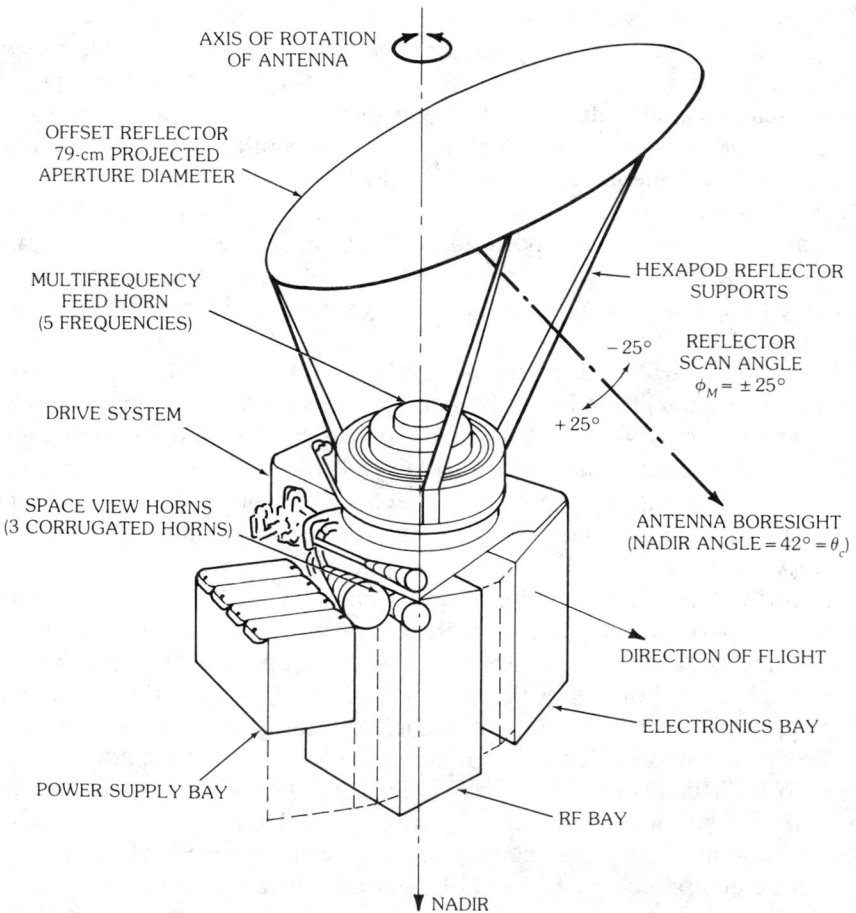

Fig. 17. SMMR configuration.

Fig. 18. Photograph of an SMMR. (*Courtesy NASA*)

transmitting). A beam is formed in the direction of the symmetrical axis of the parent parabola, whose focal point is the phase center of the feed horn. The reflector is rotated in a back-and-forth manner about the nadir axis by a drive motor which also drives a counterrotating mass to compensate the spacecraft for the angular momentum disturbance caused by the oscillating reflector. The azimuthal scan angle limit ϕ_M is $\pm 25°$.

As a result of this design (in which only the reflector moves) the total moving mass (consisting of the graphite epoxy reflector, with major diameter 108 cm by minor diameter 79 cm, ballast and thermal shield) is only 3.4 kg. The total weight of the instrument is 40 kg. The reflector first rotates in a clockwise direction for 2 s and then counterclockwise, also for 2 s. A total scan time period is $t_{scan} = 4$ s (actually 4.096 s). This 4-s scan period is chosen so that the 37-GHz footprints (ϱ_M

= 27 km, ϱ_m = 16 km) are contiguous in the down-track direction (at nadir). The subsatellite speed is v_g = 6.4 km/s, and d_L = 6.4 × 4.096 = 26.2 km, or $p = d_L/\varrho_M$ = 0.97 [see (17)]. Hence it is almost (down-track) contiguous at 37 GHz.

Step Scan Versus Continuous Scan—In a step scan the beam dwells at a given position for a length of time t_I, then moves to the next position and repeats the dwell of t_I seconds. For large antennas this step scan mode becomes impractical if an appreciable amount of time is required to accomplish the stepping motion; or it may consume too much power in order to move the antenna quickly to a new dwell position. An alternative to step scan is to slew the antenna continuously across the total scan angle limit, namely the "continuous scan." Both the SCAMS and the MSU are step scan types, because their antennas are small and their scan periods are relatively long. The SMMR, on the other hand, is a continuous scan type because its antenna is much larger than that of the MSU or the SCAMS, and the scan period at 4 s is much shorter. In a continuous scan the antenna moves (ideally) at a constant angular velocity, and the resolution cell size along the scan direction is determined by the length of integration time. (Sometimes even a constant angular speed is difficult to obtain mechanically, and some kind of velocity variation with time must be accepted. For example, the SMMR antenna actually has a sinusoidal velocity variation with time.) If, in a continuous scan, the integration time is infinitesimally short, then the IFOV along the scan direction is the antenna beamwidth θ_b. For a continuous scan with a finite integration time the effective field of view (EFOV) along the scan direction is larger than the IFOV. In other words the scan motion introduces some smearing effect along the cross-track direction. The EFOV depends on the antenna directive gain pattern $G_D(\phi)$ and the length of time a given point of the scene is viewed by the antenna. Normally the integration time is set so that the antenna moves one beamwidth during the integration time (assuming constant scan velocity).

Scanning Requirements for Geosynchronous Orbiting Microwave Radiometers—Most of the previous discussions on scanning pertain to LEO microwave radiometers. The same radiometer can certainly be used at the geosynchronous orbits. The advantage of this type of orbit is that the satellite appears stationary with respect to earth; it allows one to observe an area continually or repeatedly with high temporal frequency. This could be important for some applications, such as observing severe storms. From the radiometric viewpoint there are two types of geosynchronous earth orbiting (GEO) satellites: the spinning and the three-axis stabilized type. The present GOES satellites are of the spinning type. The other type of GEO satellite is three-axis stabilized, in which the orientation of the satellite with respect to the earth remains unchanged. In this case a sensor must scan in two orthogonal dimensions in order to obtain a map of a given area of the earth.

There is at present no microwave radiometer on a GEO satellite, because the required size of the antennas at a 36 000-km orbit is relatively large; this makes it difficult and expensive for most launch vehicles to carry. With the advent of the space shuttle, however, microwave radiometers for GEO satellites, such as GOES, will soon follow.

Because of the large orbital height, antennas used for GEO satellites will be much larger in order to achieve spatial resolution. Instead of being tens of centimeters in diameter, the antenna diameter will be in meters. There appears to be a very limited utility in having a microwave radiometer on a spinning GEO satellite because of its inherently poor spin-scan efficiency. In addition, large moving antennas may present difficult dynamic problems for the spacecraft attitude control system. For these reasons, microwave radiometers will likely be used on future three-axis stabilized GEO satellites but not on spinning GEO satellites.

Scanning requirements for a microwave radiometer, from a three-axis stabilized geostationary satellite, are quite different from those of polar LEO satellites. The angular scan limits from a GEO satellite are small, since the maximum extent of the full earth disk is only 17° but the pointing accurary must be high. The scan velocity can be much slower, although it depends on the size of a scene area to be covered and the temporal repeat frequency needed. If a 2500-km × 2500-km area near nadir has to be covered in 15 min, with a resolution cell of 42 km, then each cell has about 0.25 s of dwell time. The scan velocity is only 0.26°/s, which is much slower than the 4.3°/s MSU scan speed in low earth orbit.

Polarization—Most microwave radiometers require the reception of linearly polarized waves of either vertical or horizontal polarization, or both. The reason for this is that surface emissivity characteristics of the two modes are distinctly different from each other. Signals from vertical and horizontal channels can be used to delineate the surface from the atmospheric phenomena. For example, the absorption due to the presence of moisture in the atmosphere attenuates both vertical and horizontal polarizations equally, while wind-driven sea surface waves affect vertical and horizontal polarizations differently. Referring to Fig. 19, the vertical polarization is defined as the mode in which the electric field vector $\mathbf{E}$ lies entirely in the plane of incidence (formed by the propagation unit vector $\hat{\mathbf{k}}$ and the normal unit vector $\hat{\mathbf{n}}$). In other words, the magnetic field $\mathbf{H}$ lies transverse to the incidence plane. Hence the term *transverse magnetic (TM) mode* is also used. The horizontal polarization is defined as the mode in which the electric field vector $\mathbf{E}$ is "horizontal" (i.e., transverse to the incidence plane, hence it is also called the TE mode).

To state it more precisely (see Fig. 19), let $\hat{\mathbf{h}}$ and $\hat{\mathbf{v}}$ be the unit vectors representing the directions of horizontal and vertical polarization, respectively. Then

$$\hat{\mathbf{h}} = \frac{\hat{\mathbf{n}} \times \hat{\mathbf{k}}}{|\hat{\mathbf{n}} \times \hat{\mathbf{k}}|} \tag{50}$$

and

$$\hat{\mathbf{v}} = \frac{\hat{\mathbf{h}} \times \hat{\mathbf{k}}}{|\hat{\mathbf{h}} \times \hat{\mathbf{k}}|} \tag{51}$$

Note that while the electric field vector direction in horizontal polarization (i.e., $\hat{\mathbf{n}}$) is truly horizontal, the electric field of the vertical polarization may not be

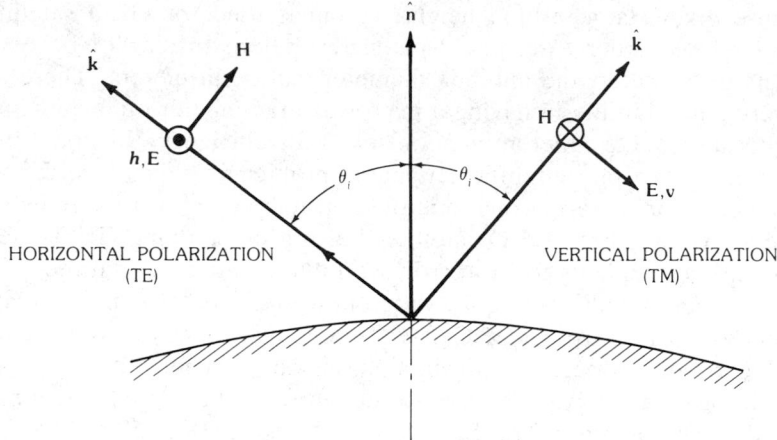

Fig. 19. Horizontal and vertical polarizations.

vertical. In fact, as the incidence angle θ_i approaches zero (approaching normal incidence) the electric field vector in vertical polarization approaches horizontal. At normal incidence there is no distinction between the two polarizations.

For applications where separate vertical and horizontal polarizations are required, the radiometer antenna must be designed for high polarization purity. That is, the amount of orthogonally (cross) polarized component leaking into the main polarization must be kept small. In other words the isolation between the two modes must be good. In general, isolation on the order of 25 dB or better is needed. This isolation is the total amount of energy leaked into vertical polarization from horizontal polarization or vice versa. It includes reflector curvature induced cross polarization (if a reflector type antenna is used), switch and/or orthomode transducer imperfections, and other leakages.

In scanning antennas such as the SMMR, where only the reflector is rotating and the feed is stationary, each of the two orthogonal feeds receives a linear combination of the vertical and horizontal polarizations, as defined in Fig. 19:

$$T_{bf_1} = T_{bv} \cos^2\phi + T_{bh} \sin^2\phi \tag{52}$$

and

$$T_{bf_2} = T_{bv} \sin^2\phi + T_{bh} \cos^2\phi \tag{53}$$

where ϕ is the azimuthal scan angle, T_{bv} and T_{bh} are the vertical and horizontal polarization brightness temperatures, and T_{bf_1} and T_{bf_2} are the output brightness temperatures of the two feeds, respectively. Temperatures T_{bv} and T_{bh} can be computed from the measured T_{bf_1} and T_{bf_2} and scan angle ϕ. This is inconvenient and can also introduce additional errors in retrieving in T_{bv} and T_{bh} individually.

The alternative is to scan the reflector and the feed as a unit, in which case T_{bv} and T_{bh} will be decoupled from each other. Since the front end of a radiometer is normally hardwired to the antenna by waveguides, in scanning with the whole antenna the masses of the radiometer front end must also be carried with the antenna. The penalty in scanning with the whole antenna is that the moving mass is increased and must be compensated for by the spacecraft altitude control system. An example of this type of scan is the SSM/I. (See Section 6.)

For atmospheric sounders most of the channels do not "see" the earth surface; therefore a pure linear polarization for these opaque channels is not essential. Because of this, antenna configurations such as in Fig. 14 (where only the reflector rotates) can be used. Orthomode transducers are often used as a convenient low-loss channel-diplexing technique. This is a particularly useful antenna design technique in the case of sounders where a single antenna must be shared by a large number of channels. For example, the SCAMS, SSM/T, MSU, and AMSU all have this type of design in which only the reflectors are scanning. The feed horns are stationary, with orthomode transducers at the throat of the horn as diplexers to separate signals for different channels. With this type of scan antenna design the output of each port of the orthomode transducer is, in effect, a linear combination of the pure vertical and horizontal polarizations. [See (52) and (53).] To put it in another way, the output of each port represents a rotating polarization. However, this rotation affects only those channels which see the earth's surface (or see substantial effects of the surface). As it turns out in the case of the MSU, the variation in brightness temperature (due to this) for its window channel (at 50.3 GHz) is very small throughout the entire scan angle limit. The smallness in variation is a result of the compensatory nature of the emissivity change and the change in cosine or sine functions. For example, the vertical polarization emissivity ϵ_v increases with the scan angle for angles between zero degrees and the Brewster angle. In the same range of scan angles the cosine-squared function decreases monotonically. Therefore the first term of (52), which is the product of the two, remains nearly constant. Similarly, for the second term, the horizontal emissivity ϵ_h decreases with the angle and is compensated by the increasing sine-squared function.

5. Spacecraft Constraints

Any satellite instrument is subject to the usual constraints of weight, volume (shape), and power limitations as well as thermal and dynamical interactions with the host spacecraft. These constraints vary, depending on the type of satellite and other sensors on board.

Dynamical Interactions

Any momenta produced by scanning motion of an antenna on board a satellite must be compensated. Otherwise, the antenna motion will cause a reaction by the host spacecraft, resulting in a change of its attitude. For small motion disturbances the excess momenta can usually be absorbed by the attitude control system (ACS) gyros of the host spacecraft. For larger antennas, however, the motion may be beyond the capacity of ACS gyros, and momentum compensation devices must be included in the antenna scan system. Both the SMMR and the SSM/I contain momentum compensation devices in the form of counterrotating masses driven by the same scan motor.

Thermal Considerations

The thermal environment affects satellite microwave radiometers in several ways:

1. Temperature gradients across a spacecraft antenna can cause shape distortion, which can lead to antenna performance degradation (reduced gain and beam efficiency); this is an important consideration for larger antennas and higher frequencies.
2. Ambient temperature fluctuation can affect the electronic gain stability of a radiometer (especially important for a total-power type radiometer); it can also introduce errors in the calibration by changing the self-emission part of an ohmic loss element.
3. For a radiometer mounted externally to a satellite, the instrument frequently has to be almost thermally isolated from the host spacecraft. Frequently the radiometer must dissipate its heat with its own radiators into the cold space to maintain a suitable temperature.

Weight, thermal, and mechanical considerations have led to the popular choice of graphite epoxy composite material for many spaceborne microwave radiometer antennas, especially for those with larger apertures (1 m in diameter or larger) and shorter wavelengths (1 cm or smaller). For example, both SMMR and SSM/I radiometers have graphite epoxy antennas.

6. Future Needs and Trends

As remote-sensing microwave radiometry evolves from the research and development stage and becomes mature, we are likely to see some new developments in the following areas of microwave radiometers and their antennas:

1. Low-frequency large antennas (aperture diameters 10 m or larger, *L*-band) for soil moisture measurements from low earth polar orbiting satellites
2. High-frequency radiometers (100 to 300 GHz) for atmospheric temperature, humidity, and precipitation monitoring from both low earth orbiting and geostationary orbiting satellites
3. Multibeam, multifrequency microwave radiometers (with frequencies similar to those of the SMMR and SMM/I) for low earth orbit applications
4. Synthetic aperture antenna radiometry

Future Spaceborne Systems

There are several satellite microwave radiometer systems which already exist in various stages of development. The SSM/I, currently under development, is scheduled to be launched on a defense meteorological satellite in 1988. The advanced microwave sounding unit (AMSU) [33], proposed for the NOAA-series weather satellites beginning in the early 1990s, is also being developed. The microwave scanning radiometer (MSR) [34], proposed for the Japanese marine observation satellite (MOS)-1 [35], and the radiometer for the European Space Agency's (ESA's) remote-sensing satellite (ERS)-1 are both in the planning stage. The SSM/I and AMSU represent two future major spaceborne microwave radiometer systems. A brief description of them is given in the following.

The SSM/I is a four-frequency, dual-polarized microwave radiometer system. It contains seven individual radiometers. Table 3 is a summary of the 54-kg, 33-W SSM/I's characteristics. All four frequencies share a common 66-cm × 61-cm offset parabolic reflector and a multifrequency feed horn. The SSM/I scans by rotating the complete antenna-and-radiometer (the antenna, feed horn, and electronics) about a nadir axis in such a manner that the antenna beams scan conically. The half-cone angle is 45°, and the rotation rate is 31 rpm. By scanning the reflector and its feed horn as one unit, the vertical and horizontal polarizations of each frequency are decoupled. This is one advantage of the SSM/I over the SMMR. The penalty for this advantage is that the moving mass of the SSM/I is much larger than that of the SMMR. Consequently a continuous rotating type scan motion has to be adapted. The zigzag, oscillating type of scan motion, such as that of the SMMR, is impractical for the SSM/I because it would cause too much mechanical disturbance and cannot be compensated easily by the spacecraft attitude control system.

Only 100° of the 360° of each scan (rotation) cycle, however, are utilized for data gathering, resulting in very inefficient use of the available scan time. This makes the integration time per resolution cell, t_I, extremely short. To obtain the required temperature sensitivities, a total-power type radiometer is used for the SSM/I. Since the scan period is only 2 s, calibrations at the end of each scan cycle are used to remove the gain fluctuations.

The advanced microwave sounding unit (AMSU) is a 20-channel microwave radiometer designed for measuring global atmospheric temperature and humidity profiles from the National Oceanic and Atmospheric Administration's NOAA-series polar orbiting weather satellites beginning in the early 1990s.

The AMSU is the next-generation instrument of the current microwave temperature sounder MSU. It, together with an infrared sounder such as the high-resolution infrared sounder (HRIS-2), will form a combined microwave/infrared vertical sounding system for the future NOAA weather satellites.

The AMSU system is still in its design stage at the time of this writing; the following is a brief description of the system.

Table 4 summarizes the essential channel characteristics as specified for the radiometer. The functions of each channel are as follows.

Channels 2, 15, and 16 are the "window" channels. They are relatively trans-

Table 3. Special Sensor Microwave/Imager Characteristics

Frequency (GHz)	Polarization	Beamwidth (°)	Bandwidth (MHz)	Temperature Sensitivity ΔT (K)	Integration Time (ms)
19.35	v & h	1.8	10 to 250	0.4	7.95
22.235	v	1.6	10 to 250	0.7	7.95
37.0	v & h	1.0	100 to 1000	0.4	7.95
85.5	v & h	0.4	100 to 1500	0.7	3.89

Table 4. AMSU Channel Characteristics

Channel Number	Center Frequency	Number of Passbands	Bandwidth (MHz)(+)	Temperature Sensitivity (K) $NE\,\Delta T$	Beam Diameter $\theta_B(°)$	Polarization Polar Angle $\theta_p(°)$
1	23 800 MHz	1	270	0.30	3.3	$90° - \theta**$
2	31 400 MHz	1	180	0.30	3.3	$90° - \theta$
3	50 300 MHz	1	180	0.40	3.3	$90° - \theta$
4	52 800 MHz	1	400	0.25	3.3	$90° - \theta$
5	53 596 MHz $\pm$ 115 MHz	2	170	0.25	3.3	*
6	54 400 MHz	1	400	0.25	3.3	*
7	54 940 MHz	1	400	0.25	3.3	*
8	55 500 MHz	1	330	0.25	3.3	*
9	57 290.344 MHz $\equiv$ LO	1	330	0.25	3.3	*
10	LO $\pm$ 217 MHz	2	78	0.40	3.3	*
11	LO $\pm$ 322.2 $\pm$ 48 MHz	4	36	0.40	3.3	*
12	LO $\pm$ 322.2 $\pm$ 22 MHz	4	16	0.60	3.3	*
13	LO $\pm$ 322.2 $\pm$ 10 MHz	4	8	0.80	3.3	*
14	LO $\pm$ 322.2 $\pm$ 4.5 MHz	4	3	1.20	3.3	*
15	89.0 GHz	1	6000	0.50	3.3	$90° - \theta$
16	89.0 GHz	1	6000	0.60	1.1	$90° - \theta$
17	157.0 GHz	1	4000	0.60	1.1	$90° - \theta$
18	183.31 $\pm$ 1.00 GHz	2	1000	0.80	1.1	*
19	183.31 $\pm$ 3.00 GHz	2	2000	0.80	1.1	*
20	183.31 $\pm$ 7.00 GHz	2	4000	0.80	1.1	*

Notes: (+): Maximum allowable values

θ_p = the angle between (1) the electric-field vector of the incoming radiation and (2) the line which is the intersection of a plane perpendicular to the propagation vector direction and an earth-tangent plane at the resolution cell center

*Unspecified, single polarization

**Scan angle = look angle from nadir

parent as compared with the remaining highly opaque oxygen and water vapor line channels. Channel 1 is on a weak water vapor line. This channel together with the window channels will provide a measurement of liquid, total precipitable water, and other forms of precipitation. They can also be used to map snow coverage and sea-ice distribution.

Channels 4 through 14 use the 5-mm oxygen band for temperature sounding of the atmosphere. The lower seven channels are for the troposphere; the remaining channels are for the stratosphere. Channel 3 is a quasi-surface channel. It provides a direct surface emissivity measurement at a frequency close to the oxygen band. The weighting-functions' peak heights of the 4 through 14 channels are nearly uniformly distributed from surface to about 40 km. Channels 5 through 9 are the "valley" frequencies, i.e., they are at the valley between two oxygen line peaks. The valley channels have the favorable characteristics that permit the use of wider bandwidth and still have narrower weighting-function widths (or sharper vertical resolution) than the nonvalley type. Channels 10 through 14 are for the troposphere temperature. Because narrower bandwidths are required from a given part of a line at these channels in order to avoid too broad weighting functions, energies from two or four similar portions of a line (or two lines) are combined to form a channel. This increases the signal-to-noise ratio by increasing the total bandwidth but not broadening the weighting function. Channels 10 through 14 exploit the symmetry between lines 13^- and 11^-. Channel 10 combines two portions in the valley between the two lines. Each of the channels 11 through 14 combines energies from four passbands, two each from both sides of lines 13^- and 11^- (see Fig. 7). Channels 17 through 20 use the strongly opaque water vapor absorption line at 183.3 GHz for obtaining the humidity profile. Channels 18, 19, and 20 also combine two portions of energy from both sides of the line to enhance the signal-to-noise ratio. Channels 18, 19, and 20 are 1, 3, and 7 GHz, respectively, from the 118.3-GHz line center. Because of its increasing distance from the line center, each succeeding channel has decreasing opacity to the atmospheric water vapor and consequently each is sensing primarily a layer of the water vapor closer to the earth surface. Channel 17 contains a single passband and is located far away from the 183-GHz line. It senses water vapor down to the earth surface. Both channels 15 and 16 serve essentially the same function except that the latter has a 3:1 surface spatial resolution advantage, and hence will be better suited to delineate fine scene features, such as a weather front.

The AMSU is a total-power microwave radiometer system. The high spatial resolution required by the AMSU results in relatively short integration times for each IFOV; they are about 180 ms for channels 1 through 15 and about 18ms for channels 16 to 20.

Onboard calibration targets are provided for periodic calibrations at the end of each antenna scan period of 8 and 2.67 s.

The AMSU is divided into two subsystems. Channels 1 through 15 are called AMSU-A and are primarily used for temperature sounding. Channels 16 to 20 are called AMSU-B, whose primary function is for humidity profiling.

The antenna designs of the AMSU are similar to that shown schematically in Fig. 14. Because of the large spread in the AMSU's frequency range (23 to 183 GHz), four separate reflector type antennas are used, two with 11-in (28-cm) apertures, and two with 6-in (15-cm) apertures.

Channels 1 and 2 share a reflector with about an 11-in (28-cm) aperture diameter. Channels 3, 4, 5, and 8 share another reflector with about a 6-in (15-cm) aperture. Channels 6, 7, and 9 through 15 use a third reflector also with about a 6-in aperture. All of the three reflectors are scanning in synchronous at one revolution per 8 s.

The AMSU-B uses one single reflector of 11-in aperture for channels 16 through 20, which are scanned at three revolutions per 8 s.

The AMSU-A is estimated to weigh 140 lb (63.5 kg) and consumes 115 W of power, while the AMSU-B is estimated to require about 60 lb (27.2 kg) and 60 W of power.

Large Antennas for Low-Frequency Soil Moisture Mapping—As mentioned in Section 1, the soil moisture content can be determined from measurements by a low-frequency microwave radiometer. The frequency range suitable for soil moisture detection is between 1 and 2 GHz, and the most widely used frequency is the 1.41-GHz hydrogen line, which is protected for radioastronomy. At this frequency, a 10 m aperture diameter antenna would produce a 0.6° beam, which corresponds to a 10-km IFOV from a 400-km orbit. To provide a wide swath width suitable for global soil moisture mapping, the antenna must be able to scan through at least ± 40° from nadir. A likely configuration for such an antenna is a phased array similar to that of the ESMR. For antenna apertures larger than 10 m, deployable mesh type reflector antennas with multiple feeds [36] are likely approaches.

Millimeter-Wave Radiometers—As pointed out in Section 4, there is an oxygen line at 118 GHz that can be used for atmospheric temperature sounding, much like the use of 60 GHz in the SSM/T and MSU. There is also a 183-GHz strong water vapor line that is being used by the AMSU for humidity profiling. In addition, there are a host of other absorption lines, at frequencies between 100 and 300 GHz, that can be used to measure other atmospheric parameters and to monitor the abundance of atmospheric constituents. One advantage of the 118-GHz oxygen line and the 183-GHz water vapor line is that their short wavelengths reduce the antenna size needed. This is an important advantage when the radiometer is used from a satellite in a geostationary orbit.

7. References

[1] W. L. Weeks, *Electromagnetic Theory for Engineering Applications*, New York: John Wiley & Sons, 1964, p. 235.

[2] E. G. Njoku and J. A. Kong, "Theory for passive remote sensing of near-surface soil moisture," *J. Geophys. Res.*, vol. 82, no. 20, pp. 3108–3118, 1977.

[3] R. A. Porter and T. J. Wentz, "Microwave radiometric study of sea surface characteristics," *NOAA NESS Report*, no. NOAA 71082701-1, July 1971.

[4] T. T. Wilheit, "The electrically scanning microwave radiometer experiment," *The Nimbus-5 User's Guide*, NASA/Goddard Space Flight Center, Greenbelt, Maryland, pp. 59–105, 1972.

[5] E. G. Njoku, "Passive microwave remote sensing of the earth from space—a review," *Proc. IEEE*, vol. 70, no. 7, pp. 728–750, July 1982.

[6] E. K. Smith, "Centimeter and millimeter wave attenuation and brightness temperature due to atmospheric oxygen water vapor," *Radio Sci.*, vol. 17, pp. 1455–1464, 1982.

[7] International Radio Consultative Committee (CCIR), "Attenuation by atmospheric gases," *Rep. 719*, Proc. CCIR XV Plenary Assembly, Geneva, 1982.

[8] J. W. Waters et al., "Remote sensing of atmospheric temperature profiles with the

Nimbus-5 microwave spectrometer," *J. Atmos. Sci.*, vol. 32, pp. 1935–1969, October 1975.

[9]   A. D. Ali et al., "Atmospheric sounding near 118 GHz," *J. Appl. Meteorol.*, vol. 19, pp. 1234–1238, October 1980.

[10] D. H. Staelin et al., "Remote sensing of atmospheric water vapor with Nimbus-5 microwave spectrometer," *J. Appl. Meteorol.*, vol. 15, pp. 1204–1214, November 1976.

[11] T. T. Wilheit et al., "Monitoring of severe storms," *High-Resolution Microwave Satellites*, ed. by D. H. Staelin and P. W. Rosenkranz, Cambridge: MIT Press, 1978, p. 57.

[12] J. W. Waters and S. C. Wofsy, "Applications of high-resolution passive microwave satellite systems to the stratosphere, mesosphere, and lower thermosphere," *High-Resolution Microwave Satellites*, ed. by D. H. Staelin and P. W. Rosenkranz, Cambridge: MIT Press, 1978, pp. 724–734.

[13] R. L. Berstein et al., "Seasat scanning multichannel microwave radiometry—results of the Gulf of Alaska workshop," *Science*, vol. 204, pp. 1415–1417, June 1979.

[14] J. P. Hollinger, "Passive microwave measurements of sea surface roughness," *IEEE Trans. Geosci. Electron.*, vol. GE-9, p. 165, July 1977.

[15] H. J. Zwally and P. Gloersen, "Passive microwave images of the polar regions and research applications," *Polar Res.*, vol. 18, pp. 116, 431–450, 1977.

[16] T. J. Schmugge, "Microwave approaches in hydrology," *Photogrammetric Engineering and Remote Sensing*, vol. 46, no. 40, pp. 495–507, April 1980.

[17] J. C. Shiue et al., "Remote sensing of snowpack with microwave radiometers for hydrologic applications," *Proc. Twelfth Intl. Symp. Remote Sensing of the Earth Environment*, Environment Research Institute of Michigan, Ann Arbor, pp. 877–886, 1978.

[18] A. Rango, A. T. C. Chang, and J. L. Foster, "The utilization of spaceborne microwave radiometers for monitoring snowpack properties," *Nordic Hydrology*, vol. 10, pp. 25–40, 1979.

[19] F. T. Barath et al., "Mariner-2 microwave radiometer experiment and results," *Astronomical J.*, vol. 69, no. 1, pp. 49–58, 1964.

[20] A. E. Basharinov et al., "Some results of microwave sounding of the atmosphere and ocean from the satellite Cosmos 243," *Space Res. XI*, Berlin: Akademia-Verlag, 1971, pp. 593–600.

[21] A. E. Basharinov et al., "Satellite measurements of microwave and infrared radio brightness temperature of the earth's cover and clouds," *Proc. Eighth Intl. Symp. Remote Sensing of the Earth Environment*, Environment Research Institute of Michigan, Ann Arbor, pp. 291–296, 1971.

[22] J. W. Waters et al., "Remote sensing of atmospheric temperature profiles with the Nimbus-5 microwave spectrometer," *J. Atmos. Sci.*, vol. 32, pp. 1953–1969, October 1975.

[23] D. H. Staelin et al., "Microwave spectroscopic imagery of the earth," *Science*, vol. 197, pp. 991–993, 1977.

[24] H. E. Lauapre and K. A. Paradis, "A multichannel passive microwave atmospheric temperature sounding system," *Proc. Eleventh Intl. Symp. Remote Sensing of the Earth Environment*, Environment Research Institute of Michigan, Ann Arbor, pp. 1212–1228, 1977.

[25] P. N. Swanson et al., "The TIROS-N microwave sounder unit," *Proc. IEEE MTT-S Intl. Microwave Symp.*, IEEE, New York, pp. 123–125, 1980.

[26] P. Gloersen and F. T. Barath, "A multichannel microwave radiometer for Nimbus-C and Seasat-A," *IEEE J. of Ocean Eng.*, vol. OE-2, no. 2, pp. 172–178, 1977.

[27] B. S. Gohil et al., "Remote sensing of atmospheric water content from Bhaskara, Samir data," *Int. J. Remote Sensing*, vol. 3, no. 3, pp. 235–241, 1982.

[28] P. D. Phavsar, G. T. Joseph, and O. P. N. Calls, "Developments of remote-sensing sensors at ISRO," *Proc. and Asian Conf. on Remote Sensing* (ACRS), Beijing, China, pp. 2-2-1 to 2-2-22, 1981.

[29] J. D. Kraus, *Radio Astronomy*, New York: McGraw-Hill Book Co., 1966.

[30] G. Evans and C. W. McLeish, *RF Radiometer Handbook*, Dedham: Artech House, 1977.

[31] F. T. Ulaby, R. T. Moore, and A. K. Fung, *Microwave Remote Sensing, Active and Passive*, vol. 1, Reading, Massachusetts: Addison-Wesley Publishing Co., 1981.

[32] J. Ruze, "Antenna tolerance theory," *Proc. IEEE*, vol. 54, pp. 533–540, 1966.

[33] Y. Ishizawa, "The Japanese MOS and LOS program," *Proc. Fifteenth Intl. Symp. Remote Sensing of the Earth Environment*, Environment Research Institute of Michigan, Ann Arbor, p. 443, 1981.

[34] G. DuChassois and C. Honvault, "The first ESA remote sensing satellite system ERS-1," *Proc. Fifteenth Intl. Symp. Remote Sensing of the Earth Environment*, Environment Research Institute of Michigan, Ann Arbor, pp. 461–477, 1981.

[35] J. C. Shiue, "The next generation microwave sounder for weather satellites," *Proc. Natl. Telesystems Conf.*, IEEE cat. no. 82CH1824-2, pp. C4.4.1–C4.4.7, 1982.

[36] C. E. Farrell and D. A. Strange, "Preliminary analytical results using surface current integration for predicting effects of surface pillows on rf performance," *Space Systems Technology—1981*, NASA Conf. Pub. 2218, pt. 2, pp. 588–690, November 16–19, 1981.

Chapter 23

Antennas for Geophysical Applications

D. A. Hill
National Bureau of Standards

CONTENTS

David A. Hill was born in Cleveland, Ohio, on April 21, 1942. He received the BS and MS degrees from Ohio University, Athens, in 1964 and 1966, and the PhD degree in electrical engineering from Ohio State University, Columbus, in 1970.

Since 1970 he has been a member of the scientific community in Boulder, Colorado. From 1970 to 1971 he was a visiting fellow with the Cooperative Institute for Research in Environmental Sciences, where he worked on pulse propagation. From 1971 to 1982 he was with the Institute for Telecommunication Sciences, where he worked on theoretical problems in antennas and propagation. Since 1982 he has been an electronics engineer in the Electromagnetic Fields Division of the National Bureau of Standards, where he has been working on electromagnetic compatibility and interference problems. He is also a professor adjoint in the Department of Electrical and Computer Engineering of the University of Colorado.

Dr. Hill is a member of URSI Commissions B, E, and F and a Fellow of the IEEE. He has served as a technical editor for the *IEEE Transactions on Geoscience and Remote Sensing* and is now an associate editor for the *IEEE Transactions on Antennas and Propagation*.

1. Introduction

The use of electrical methods in geophysics has expanded greatly in the past two decades, and during the same time period an interest in subsurface communication has developed. The most comprehensive book on electrical methods in geophysics was written in 1966 by Keller and Frischknecht [1], and it is still an excellent reference today. The review article by Murphy and Parkinson [2] is an excellent reference on underground communication. Both geophysical prospecting and underground communication require transmission of signals into the earth and, as a result, the same antenna types are used for both applications.

Because the methods and antennas used in geophysical probing are so varied, it is not possible to attempt a comprehensive discussion in one chapter. However, if we limit the applications to deep, subsurface probing and to through-the-earth communication, then the antennas used are primarily of two types: straight wire antennas which are grounded at the end points and wire loop antennas. Sections 2 and 3 discuss grounded wire antennas for direct current and time-varying excitations, respectively. Section 4 discusses loop antennas. In the analysis and discussion of these antennas some applications in geophysics and underground communication are described for illustrative purposes, but many other applications cannot be mentioned for lack of space. The references should be consulted for a more complete description of the applications. The primary purpose of this chapter is to describe how these antennas perform in the presence of a conducting earth.

In order to penetrate the earth to depths on the order of a hundred meters or more, it is necessary to employ extremely low frequencies (elf) below about 3 kHz. At such frequencies the free-space wavelength is greater than 100 km, and the antennas are electrically small even though they could be physically large (dimensions on the order of a kilometer in some cases). Consequently, the analyses in Sections 2 through 4 utilize the quasi-static assumption that neglects displacement currents in the air. However, no assumption is made regarding the antenna dimensions and separations in terms of the skin depth in the earth, $(2/\omega\mu_0\sigma)^{1/2}$. Here, ω is the angular frequency, σ is the earth conductivity, and the earth permeability is taken to be the free-space value μ_0, which is normally the case. In geophysics it is common to work with the earth resistivity ϱ, which is the reciprocal of the conductivity σ. The dielectric constant of the earth is unimportant at elf because conduction currents are dominant. In the transient results which are given in Section 4 the earth conductivity σ is assumed to be independent of frequency. This is not a good assumption at high frequencies but is a fairly good assumption at elf. The actual value of earth conductivity can commonly vary over a range from about 10^{-1} S/m to 10^{-4} S/m, depending on the moisture content and type of rock [3].

In Section 5, some other antenna types are discussed in much less detail. Many of these antennas are used for shorter ranges and higher frequencies where the quasi-static assumption is not valid.

2. Electrode Arrays for Resistivity Measurements

The most commonly used methods for measuring the direct-current (dc) resistivity of the earth utilize a four-electrode array. An electric current is driven through one pair of electrodes, and the potential established in the earth is measured with the second pair of electrodes. In order to study the variation of the earth resistivity with depth, the spacings between the electrodes are gradually increased. This type of measurement is called *vertical sounding*. In order to study lateral variations of earth conductivity the electrode spacings are kept fixed, and the entire array is moved as a whole along a transverse line. This type of measurement is called *horizontal profiling*.

Theory of Four-Electrode Arrays

We consider a homogeneous half-space with resistivity ϱ as our earth model. If a dc current I is injected at point 1 and removed at point 2, as shown in Fig. 1, then the scalar potential ϕ_P at any point P in the earth is given by

$$\phi_P = \frac{\varrho I}{2\pi}\left(\frac{1}{r_{1,P}} - \frac{1}{r_{2,P}}\right) \tag{1}$$

where $r_{1,P}$ is the separation between points 1 and P and $r_{2,P}$ is the separation between points 2 and P. If we now consider a pair of receiving electrodes 1 and 2, then the voltage difference V between the electrodes is

$$V = \phi_1 - \phi_2 = \frac{\varrho I}{2\pi}\left(\frac{1}{r_{1,1}} - \frac{1}{r_{1,2}} - \frac{1}{r_{2,1}} + \frac{1}{r_{2,2}}\right) \tag{2}$$

where $r_{i,j}$ is the separation between current electrode i and potential electrode j. Normally, the four electrodes are arranged in a straight-line configuration as shown in Fig. 2, but this is not necessary. It is possible to express the resistivity ϱ in (2) in the following form:

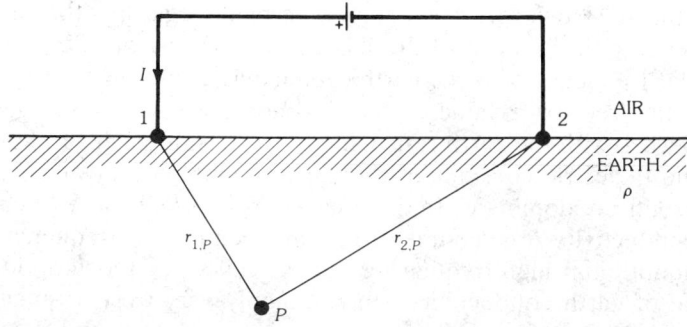

Fig. 1. Direct-current electrodes in a homogeneous earth of resistivity.

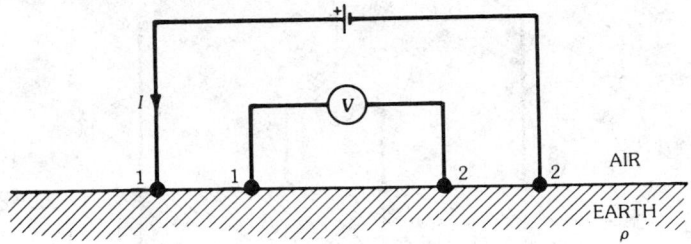

Fig. 2. Four-electrode configuration.

$$\varrho = KV/I \qquad (3)$$

Here K is a strictly geometric factor which depends on the array configuration and, in general, is given by

$$K = \frac{2\pi}{1/r_{1,1} - 1/r_{1,2} - 1/r_{2,1} + 1/r_{2,2}} \qquad (4)$$

The Wenner array is one of the most commonly used arrays for measuring resistivity. It employs an equal spacing, a, between any two electrodes as shown in Fig. 3. The geometric factor K for the Wenner array is

$$K = \frac{2\pi}{1/a - 1/2a - 1/2a + 1/a} = 2\pi a \qquad (5)$$

The Schlumberger array shown in Fig. 4 is also commonly used in measuring earth resistivity. The closely spaced potential electrodes approximately measure the potential gradient at the center of the current array. The geometric factor K for the Schlumberger array is

$$K = \frac{2\pi}{(a - b/2)^{-1} - (a + b/2)^{-1} - (a + b/2)^{-1} + (a - b/2)^{-1}} \qquad (6)$$
$$= \pi\left(\frac{a^2}{b} - \frac{b}{4}\right)$$

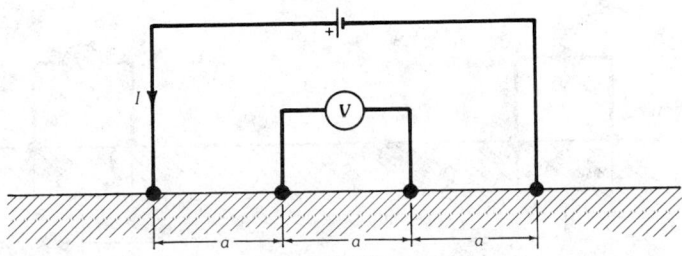

Fig. 3. Wenner array.

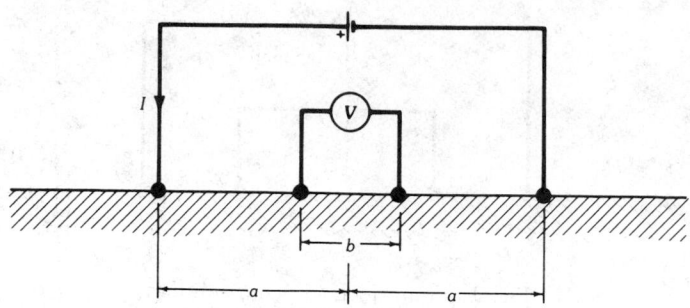

Fig. 4. Schlumberger array.

In the limit of b/a approaching zero, the geometric factor K approaches $\pi a^2/b$, and the potential gradient is measured.

In some cases dipoles (closely spaced electrodes) are used for both the current and potential arrays. In the dipole-dipole array shown in Fig. 5, both b and c are assumed to be small compared with a. The geometric factor K for the dipole-dipole array is

$$K = \frac{2\pi}{(a + b/2 - c/2)^{-1} - (a + b/2 + c/2)^{-1} - (a - b/2 - c/2)^{-1} + (a - b/2 + c/2)^{-1}}$$

$$(7)$$

For the special case of $b = c$, (7) simplifies to

$$K = \pi a \left(\frac{a^2}{b^2} - 1 \right) \tag{8}$$

When the earth is not homogeneous, the situation becomes more complicated and the simple expression (2) for the potential is no longer valid. However, it is still possible to derive an apparent resistivity ϱ_a in terms of the measured current I and potential difference V:

$$\varrho_a = KV/I \tag{9}$$

where K is still given by (4). When the earth is horizontally stratified, as in Fig. 6, it is possible to formulate the apparent resistivity in terms of integrals which must be

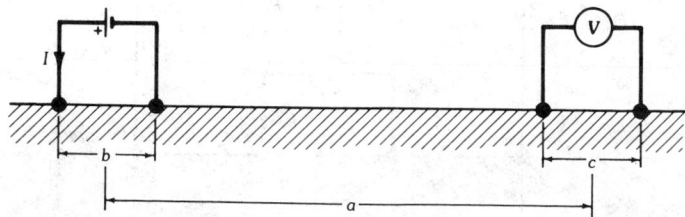

Fig. 5. Dipole-dipole array.

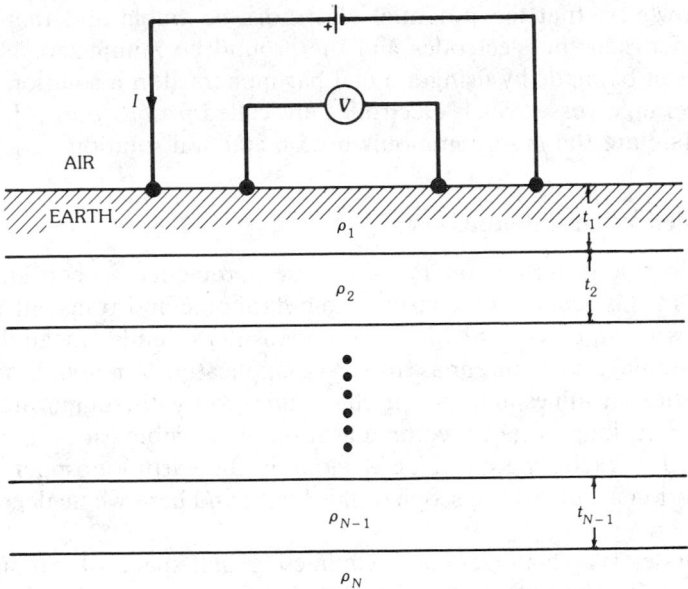

Fig. 6. Four-electrode configuration over a horizontally stratified, N-layer earth.

evaluated numerically. Apparent resistivity curves for Wenner, Schlumberger, and dipole-dipole arrays have been computed for two-, three-, and four-layer earths [4], and a computer code has been published for dipole-dipole arrays for any number of layers up to twelve [5]. The inverse problem of attempting to determine the layer parameters (vertical sounding) from apparent resistivity usually involves some type of curve matching, but that subject is beyond the scope of this chapter.

Instrumentation of Four-Electrode Arrays

The current circuit consists of an insulated wire which is grounded at each end as in Fig. 1. In practice, the wire lies on the ground, and the insulation is important to prevent leakage currents. Grounding can be achieved by driving steel or copper-clad steel stakes into the ground to a depth of several inches or deeper. copper-clad steel stakes into the ground to a depth of several inches or deeper. Spiral-blade electrodes can also be screwed into the ground, and these have the advantage of a larger contact area. The contact resistance can be as low as 10 Ω in moist soil, but can be orders of magnitude larger in dry soils. In dry ground it may be helpful to wet the area around the grounding stake to improve contact between the stake and the ground. It is also possible to employ multiple grounding stakes in parallel to decrease the grounding resistance.

In practice the current is not actually direct current but is usually a low-frequency square wave. A low-frequency sinusoidal current could also be used, but a low-frequency square wave is easier to generate. By avoiding the transient which occurs at the start of each half-cycle of the square wave, the resistivity which is measured is still the desired dc resistivity.

The contact resistance of the potential electrodes is less important because a high-resistance voltmeter is used to measure the potential difference. It is important, however, that the potential electrodes be stable and that polarization potentials between the electrodes and the ground be minimized. Nonpolarizing electrodes can be made by using a metal bar immersed in a solution of one of its salts in a ceramic vessel. Such electrodes are called *porous pots* [6]. Copper and copper sulfate are the most commonly used metal and solution.

3. Grounded Wire Antennas

In the previous section on resistivity measurements we considered only dc excitation. In this section we consider time-harmonic and transient excitation of horizontal wire antennas which are grounded at the ends. In addition to geophysical sounding, such antennas also have application for uplink and downlink communication in mines and for elf communication with submarines [7]. In the application for long-distance communication with submarines the antenna is actually used to excite the quasi-TEM mode in the earth-ionosphere waveguide. This application is beyond the scope of this book, and here we neglect the effect of the ionosphere.

Our model is a homogeneous, conducting half-space of conductivity σ as shown in Fig. 7. The antenna is of length 2ℓ, and the current is assumed to be constant over the length of the antenna. This assumption is valid for insulated wires grounded at the ends when the length of the antenna is much less than a free-space wavelength [8]. The actual input impedance of such antennas is primarily resistive, and the resistance is the sum of the two grounding resistances and the wire resistance. Normally the grounding resistance is dominant.

Time-Harmonic Excitation

In this subsection we consider time-harmonic excitation, and the $\exp(j\omega t)$ time dependence will be suppressed. The subsurface electric and magnetic fields are of interest both in mine communication and in subsurface probing of geophysical

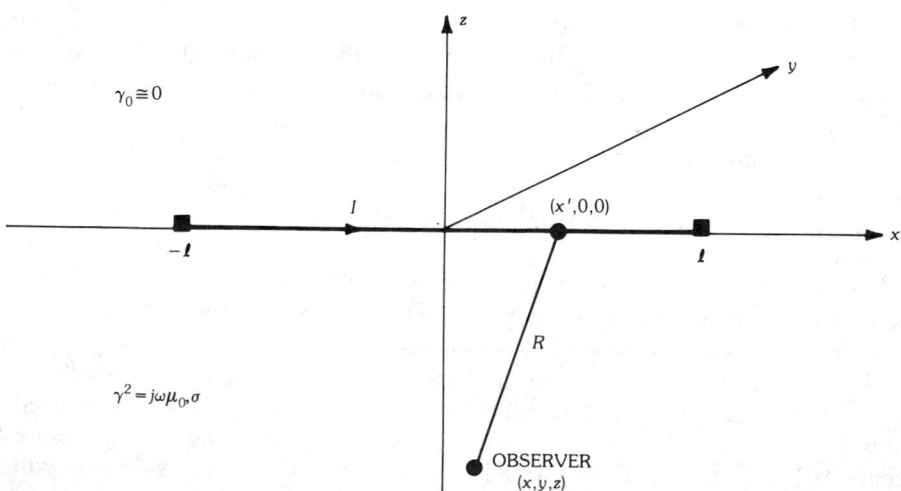

Fig. 7. Finite line source on a homogeneous half-space. (*After Hill and Wait [9]*)

features. In mine communication the subsurface magnetic field is received with a loop antenna, or the subsurface electric field is received with a grounded wire antenna.

We first consider the fields produced by an incremental source of length dx' located at x' as shown in Fig. 7. For the low frequencies of interest here, the free-space propagation constant γ_0 can be set equal to zero and the Sommerfield integral forms for an incremental source of current moment $I\,dx'$ can be greatly simplified [9]. As a result the magnetic field components are

$$dH_x = \frac{I\,dx'}{2\pi\gamma^2}\left(\frac{\partial^4 N}{\partial x\,\partial y\,\partial z^2} - \frac{\partial^3 P}{\partial x\,\partial y\,\partial z}\right) \tag{10a}$$

$$dH_y = \frac{I\,dx'}{2\pi\gamma^2}\left(\frac{\partial^3 P}{\partial z^3} + \frac{\partial^2 P}{\partial x^2\,\partial z} + \frac{\partial^4 N}{\partial z^2\,\partial y^2}\right) \tag{10b}$$

$$dH_z = \frac{I\,dx'}{2\pi\gamma^2}\left(\frac{\partial^4 N}{\partial y\,\partial z^3} - \gamma^2\frac{\partial^2 N}{\partial y\,\partial z} - \frac{\partial^3 P}{\partial y\,\partial z^2}\right) \tag{10c}$$

where

$$N = I_0[(\gamma/2)(R + z)]\,K_0[(\gamma/2)(R - z)] \tag{11a}$$

$$P = R^{-1}\exp(-\gamma R) \tag{11b}$$

$$R = \sqrt{(x - x')^2 + y^2 + z^2} \tag{11c}$$

$$\gamma = \sqrt{j\omega\mu_0\sigma} \tag{11d}$$

The terms I_0 and K_0 are modified Bessel functions of order zero. Similarly, the electric-field components are

$$dE_x = \frac{-I\,dx'}{2\pi\sigma}\left(\frac{\partial^2 P}{\partial z^2} + \frac{\partial^3 N}{\partial y^2\partial z}\right) \tag{12a}$$

$$dE_y = \frac{I\,dx'}{2\pi\sigma}\frac{\partial^3 N}{\partial y\,\partial x\,\partial z} \tag{12b}$$

$$dE_z = \frac{I\,dx'}{2\pi\sigma}\frac{\partial^2 P}{\partial x\,\partial z} \tag{12c}$$

To obtain the fields of a cable of finite length 2ℓ carrying a constant current I, we integrate (10) and (12) over the range of x' from $-\ell$ to ℓ. The dominant components of interest are H_y, H_z, and E_x because all other components vanish in the plane $x = 0$. Also, these are the only nonzero components everywhere if the line source is of infinite length. For normalization purposes it is convenient to write the fields in the following manner:

$$H_y = \frac{I}{2\pi h} A(H, Y, X, L) \tag{13a}$$

$$H_z = \frac{I}{2\pi h} B(H, Y, X, L) \tag{13b}$$

$$E_x = \frac{-j\omega\mu_0 I}{2\pi} F(H, Y, X, L) \tag{13c}$$

where $H = \sqrt{\omega\mu_0\sigma}\, h$, $Y = y/h$, $X = x/h$, $L = \ell/h$, and $h = -z$. Note that Z, B, F, H, Y, X, and L are dimensionless. The specific forms for A, B, and F are

$$A(H, Y, X, L) = \frac{h}{\gamma^2} \int_{-\ell}^{\ell} \left(\frac{\partial^3 P}{\partial z^3} + \frac{\partial^3 P}{\partial x^2 \partial z} + \frac{\partial^4 N}{\partial z^2 \partial y^2} \right) dx' \tag{14a}$$

$$B(H, Y, X, L) = \frac{h}{\gamma^2} \int_{-\ell}^{\ell} \left(\frac{\partial^4 N}{\partial y \, \partial z^3} - \gamma^2 \frac{\partial^2 N}{\partial y \, \partial z} - \frac{\partial^3 P}{\partial y \, \partial z^2} \right) dx' \tag{14b}$$

$$F(H, Y, X, L) = \frac{1}{\gamma^2} \int_{-\ell}^{\ell} \left(\frac{\partial^2 P}{\partial z^2} + \frac{\partial^3 N}{\partial y^2 \, \partial z} \right) dx' \tag{14c}$$

The integral forms in (14) simplify for both the low-frequency (small H) and high-frequency (large H) cases [9] but, in general, numerical integration is required. Typical numerical results are shown for $H = 2$ in Fig. 8. Although A, F, and B are complex for H greater than zero, only the magnitudes are plotted. The phases are relatively constant as a function of L. For very small values of L, the fields are essentially those of a short dipole and are proportional to L as indicated by (10) and (12). For large L, the field components must eventually reach those of an infinite line source. The results in Fig. 8 were found to agree well with earlier results computed for an infinite line source by Wait and Spies [10]. In some geophysical applications it is desirable to make the grounded wire long enough to simulate an infinitely long line source. For practical purposes this is achieved for L greater than about 2. This means that the antenna length 2ℓ should be about 4 times the depth h of interest in the particular geophysical sounding application of interest.

Straight, grounded wire antennas can also be used underground. For example, long wire antennas have been laid out in mine tunnels for uplink transmission as indicated in Fig. 9. The surface fields of such antennas have been computed for the case where the antenna is not parallel to the air-earth interface [11] in order to model cases where either the tunnel or the earth surface is not level. Since the antenna must usually be located at the surface in geophysical applications, we will not go into detail on the subsurface case. The antenna current is normally constant over the length of the antenna, and the surface fields are qualitatively similar to the subsurface fields of surface antennas. Also, the grounding resistance is again normally greater than the wire resistance.

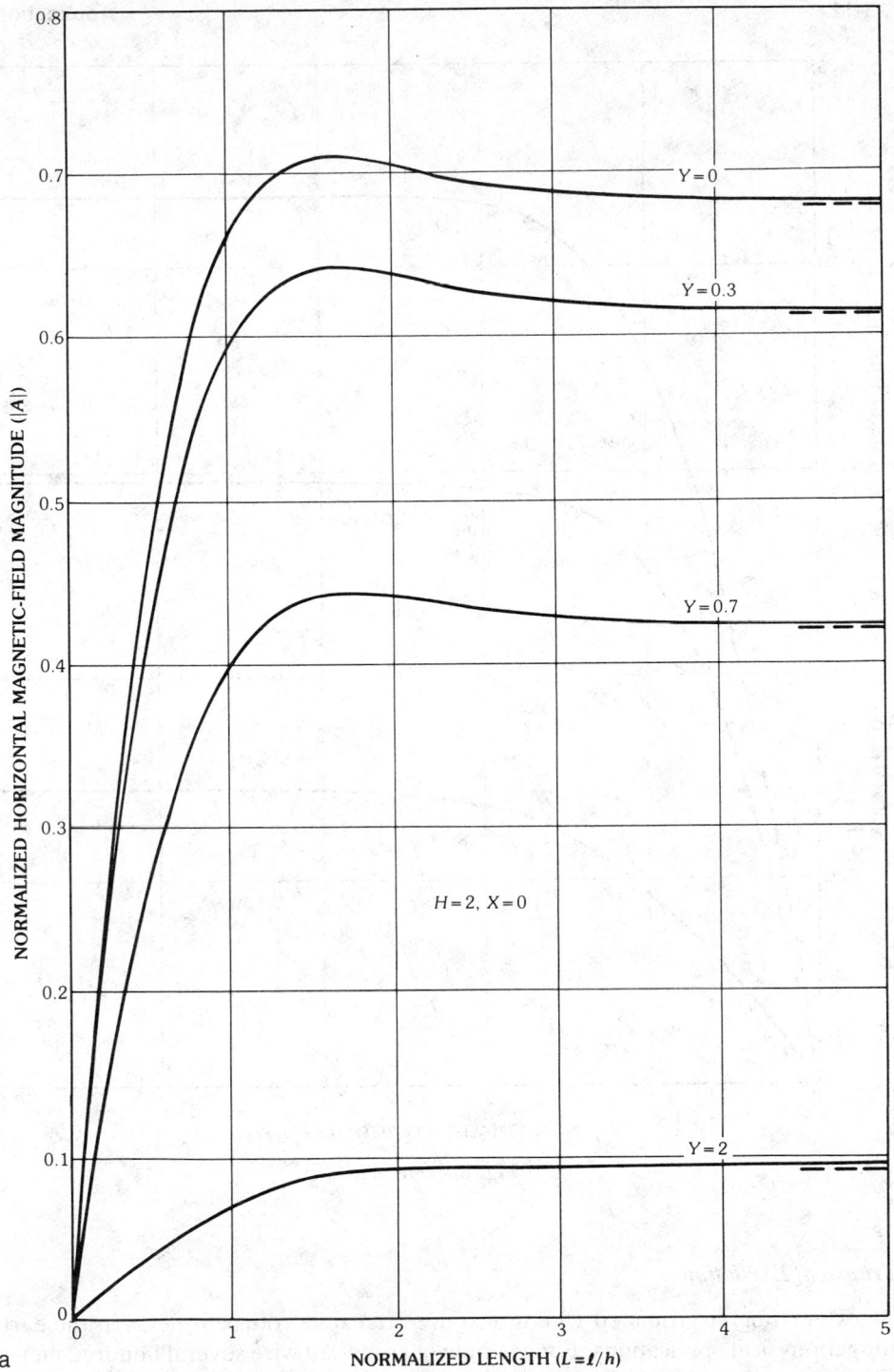

Fig. 8. Magnitudes of the normalized fields as a function of Y and L. (a) Horizontal magnetic field. (b) Vertical magnetic field. (c) Horizontal electric field. (*After Hill and Wait [9]*).

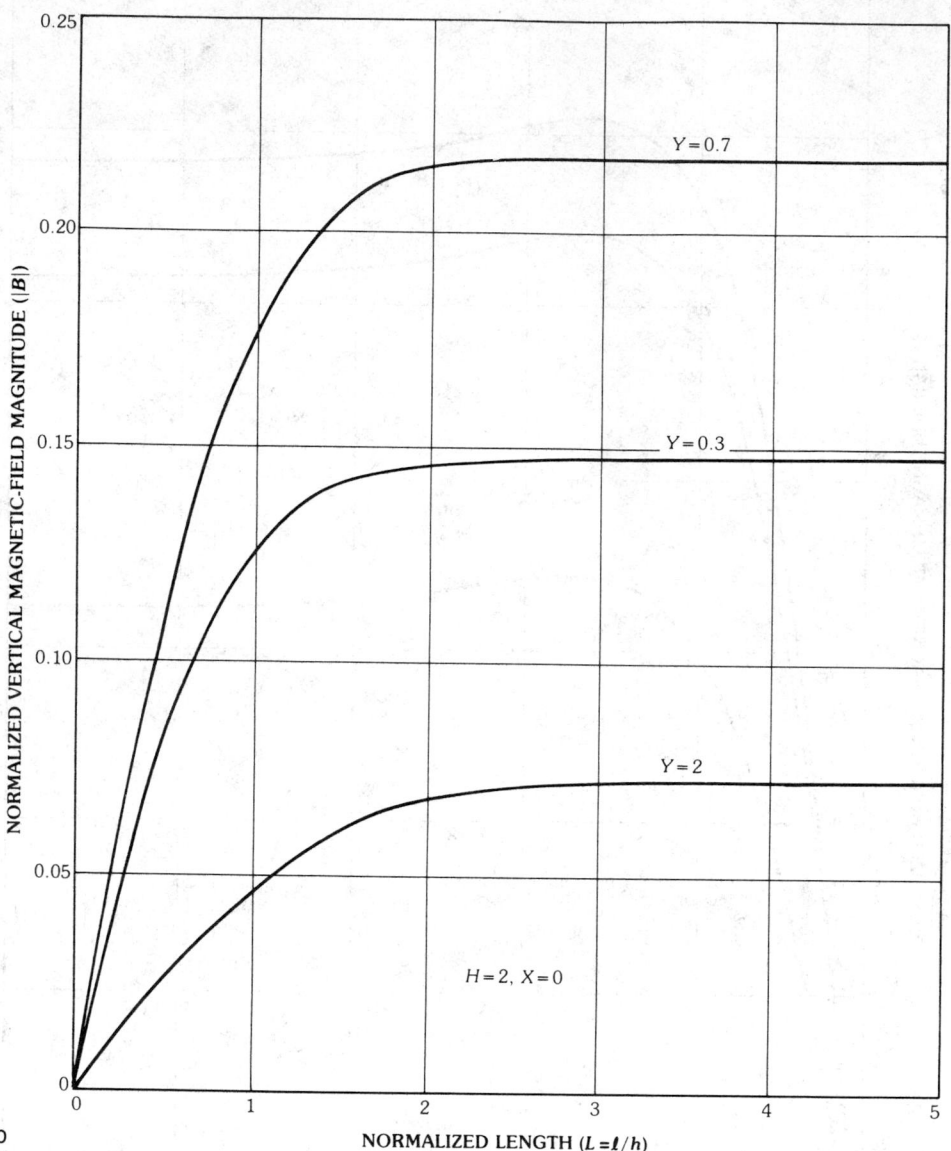

Fig. 8, *continued.*

Transient Excitation

The straight, grounded wire is also used to excite transient fields in the earth for geophysical applications. For example, a grounded wire several hundred meters in length can be excited with a step-function current, and the vertical magnetic field at some remote location can be received with a large horizontal loop antenna. We assume that the transients are sufficiently band limited that the current along the antenna is independent of position.

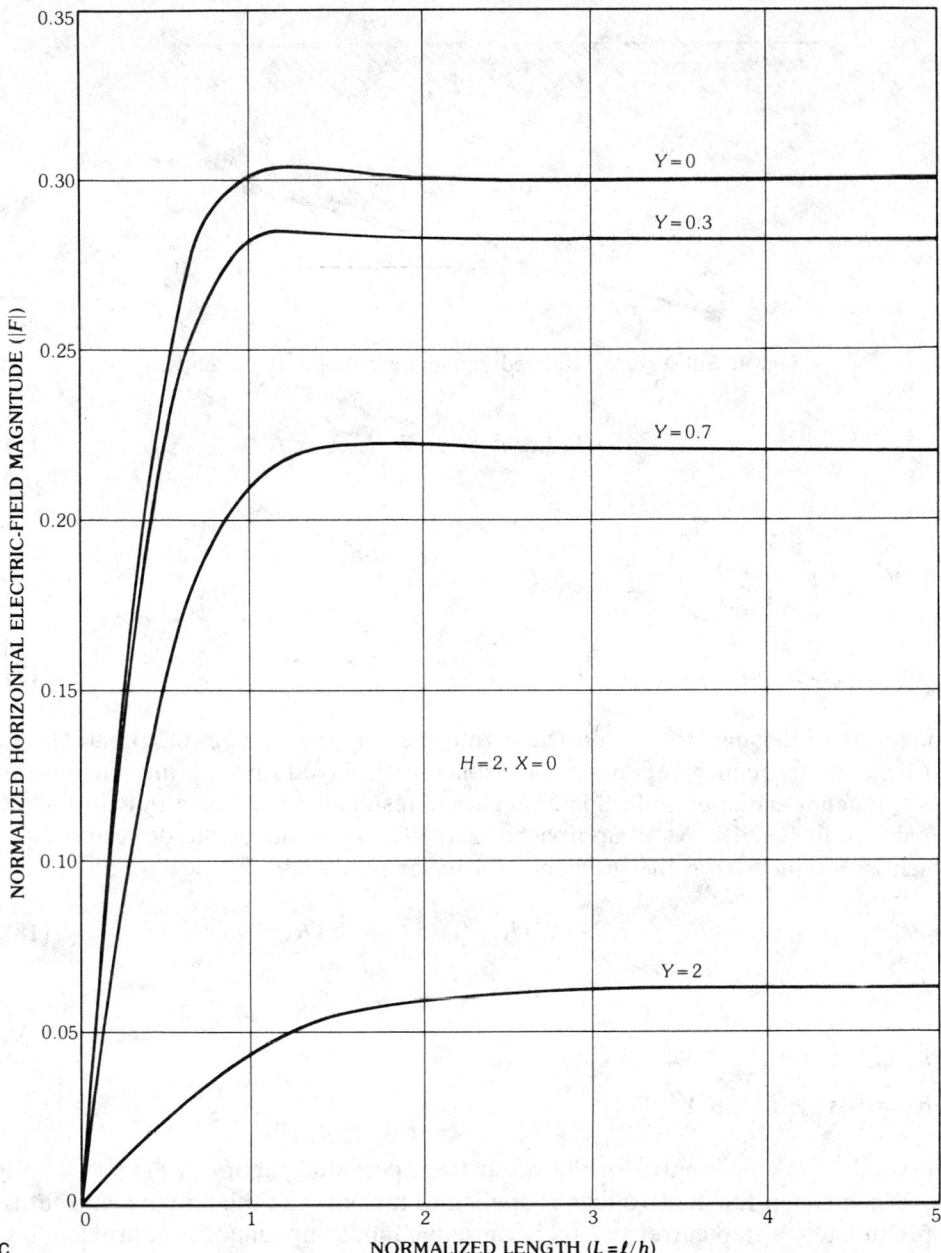

Fig. 8, *continued.*

To illustrate the dispersive nature of the earth we first examine the frequency dependence of the subsurface electric fields. The frequency dependence of all three components has been shown [12], but here we consider only the dominant component E_x:

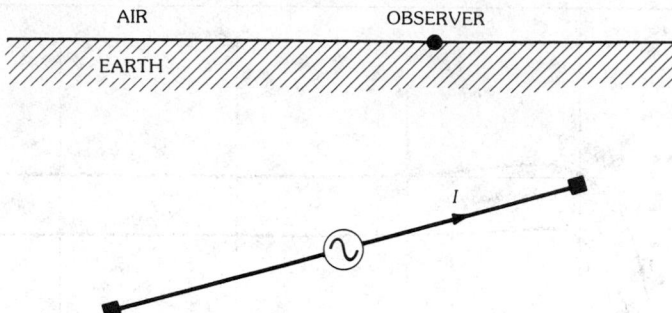

Fig. 9. Subsurface insulated antenna grounded at the ends.

$$E_x = -[I/(2\pi\sigma h^2)]\, E_{xn}(W, L, X, Y) \tag{15}$$

where

$$E_{xn} = h^2 \int_{-\ell}^{\ell} \left(\frac{\partial^2 P}{\partial z^2} + \frac{\partial^3 N}{\partial y^2 \, \partial x} \right) dx' \tag{16}$$

$$W = \omega\mu_0\sigma h^2 \tag{17}$$

Equations 15 through 17 are consistent with the earlier results in (13c) and (14c), but here the frequency dependence is explicitly displayed through the dimensionless frequency parameter W. Some numerical results for $|E_{xn}|$ as a function of W are shown in Fig. 10. As W approaches zero, E_{xn} approaches the dc result, E_{xn}^{dc}, which is obtained from the gradient of a scalar potential:

$$E_{xn}^{dc} = (L - X)\,R_1^{-3} + (L + X)\,R_2^{-3} \tag{18}$$

where

$$R_1 = [(X - L)^2 + Y^2 + 1]^{1/2} \tag{19}$$
$$R_2 = [(X + L)^2 + Y^2 + 1]^{1/2}$$

Equation 18 is consistent with the result from potential theory in (1).

When a step-function voltage is applied to the antenna the antenna current is approximately a step current $I\,U(t)$ because the input impedance is approximately resistive and constant over the frequency range of the band-limited step function. The transient electric field $\tilde{E}_x^s(t)$ is given by the following inverse Fourier transform:

$$\tilde{E}_x^s(t) = \frac{1}{2\pi} \int_{-\infty}^{\infty} \frac{E_x}{j\omega} \exp(j\omega t)\, d\omega \tag{20}$$

It is more convenient to rewrite (20) in the following normalized form:

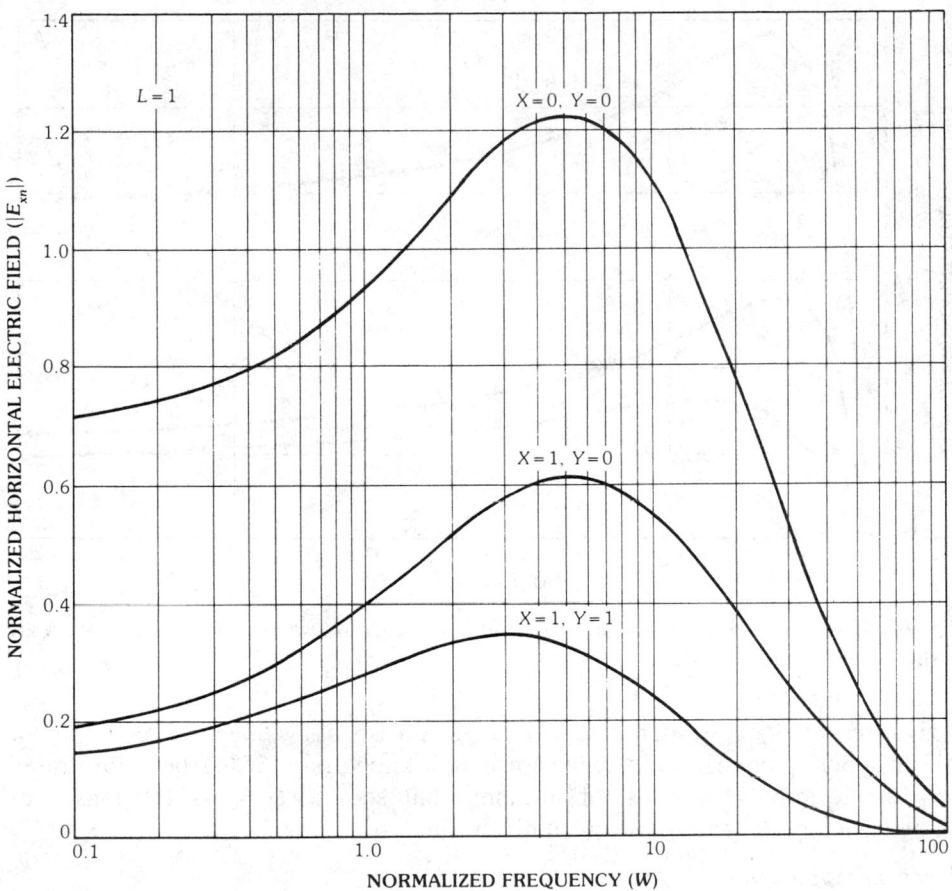

Fig. 10. Magnitude of the normalized horizontal electric field as a function of normalized frequency W. (*After Hill and Wait [12]*)

$$\bar{E}_x^s(t) = -[I/(2\pi\sigma h^2)]\,\bar{E}_{xn}(T, L, X, Y) \tag{21}$$

where

$$\bar{E}_{xn}(T, L, X, Y) = \frac{1}{2\pi}\int_{-\infty}^{\infty} \frac{E_{xn}(W, L, X, Y)}{jW}\exp(jWT)\,dW \tag{22}$$

$$T = t/\sigma\mu_0 h^2 \tag{23}$$

The term T is a normalized time, and $\sigma\mu_0 h^2$ can be thought of as a depth-dependent time constant. The inverse transform in (22) has been carried out numerically, and results for $\bar{E}_{xn}^s$ are shown in Fig. 11.

In all cases the waveforms in Fig. 11 have a fairly rapid rise which overshoots the final value, followed by a slow decay to the dashed final values. The final values are the dc values as given by (18) and (19). The impulse response is the time

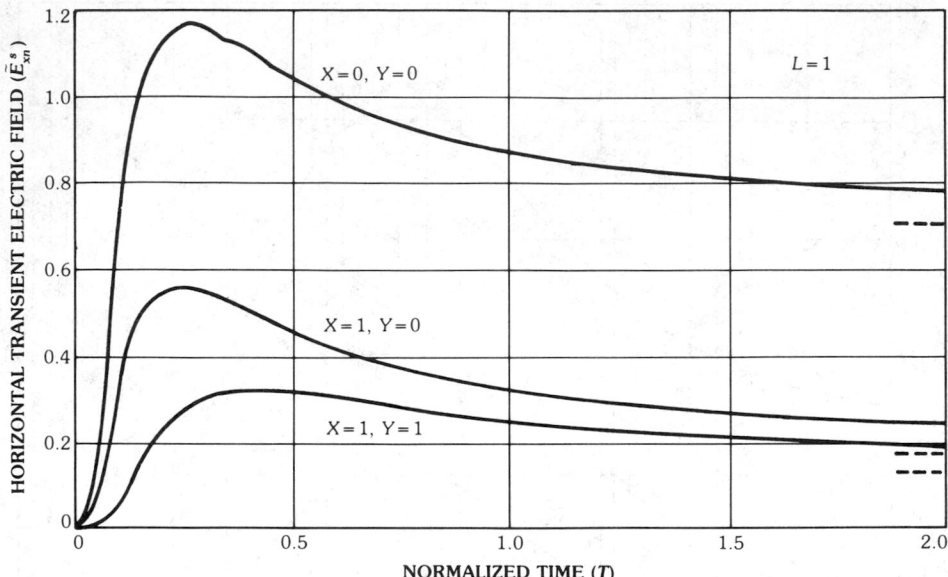

Fig. 11. Horizontal electric field for a step-function current, where the final dc values are dashed. (*After Hill and Wait [12]*)

derivative of the step response in (13), and it consists of a spike followed by a slow tail of opposite polarity. This behavior is well known, and it has been illustrated [13] for the impulse response of an infinite line source ($L = \infty$). Responses for other values of L tend to be qualitatively similar.

Receiving Application

The straight grounded wire antennas shown in Figs. 7 and 9 can also be used as receiving antennas. Typically the subsurface antenna in Fig. 9 would be used to receive a downlink signal as in mine communication or submarine communication. The surface antenna in Fig. 7 could be used to receive an uplink signal, or it could be the receiving antenna in an earth-sounding system. The receiving antennas for dc resistivity in Figs. 2 through 5 are special cases of this receiving application.

Because the current distribution on the antenna is constant over the length, the received voltage is simply the integral of the incident electric field over the length of the antenna. For example, the received voltage v for the antenna in Fig. 7 when used as a receiving antenna is

$$v = \int_{-\ell}^{\ell} E_x(x) \bigg| dx_{z=y=0} \tag{24}$$

The electric field E_x could be either a time-harmonic or a transient field. For example, Wait [14] has used (24) to compute the transient coupling between a pair of grounded wires. In the magneto-telluric method [15], the electric field that is measured is a transient signal produced by lightning or some other natural source.

For the special case of dc fields, (24) simply yields the potential difference between the end points of the antenna as in (2).

4. Loop Antennas

Loop antennas are commonly used in geophysical sounding and subsurface communications, and they have the advantage that no grounding is required. In induction methods, loop antennas transmit a time-varying magnetic field into the earth, and eddy currents are excited in conducting bodies. These eddy currents generate a secondary magnetic field which can be received by a second loop antenna. In mine communications [16], transmitting loops can be used either at or below the earth surface.

Horizontal transmitting loops are typically a large, single turn of insulated wire laid out on the earth. The loop dimensions can be up to a kilometer, and the current can be up to several amperes. Various shapes, such as circular or rectangular, are used depending on the application. When the loop dimensions are small compared with the skin depth in the earth and the observer distance, then the horizontal loop radiates as a vertical magnetic dipole.

Vertical transmitting loops are usually multiple turns of insulated wire on some type of frame with dimensions on the order of meters. Normally such vertical loops radiate as horizontal magnetic dipoles. The inductance of such multiturn loops can be fairly large, and a series capacitor is usually used to tune to the operating frequency.

Time-Harmonic Excitation

In this subsection we consider time-harmonic excitation of a circular loop at the earth surface as shown in Fig. 12. For small loops the results depend only on the magnetic dipole moment IA where I is the loop current and A is the loop area, and the shape is unimportant. Here we wish to consider the effect of finite loop size, and a circular loop is the simplest shape to consider. For a circular loop with constant current I, the nonzero field components are H_z, H_ϱ, and E_ϕ.

The subsurface fields are of interest in mine communications and induction sounding, and here we consider a homogeneous half-space model. The vertical magnetic field is of primary interest for downlink communication between horizontal loops, and it is given by [17]

$$H_z = \frac{-IA}{2\pi h^3} Q \tag{25a}$$

where

$$Q = \int_0^\infty \frac{x^3 e^{-(x^2+H^2)^{1/2}}}{x + (x^2 + iH^2)^{1/2}} J_0(xD) \frac{2\,J_1(xa/h)}{xa/h}\, dx \tag{25b}$$

$$H = \sqrt{\omega\mu_0\sigma}\,h \tag{25c}$$

$$D = d/h \tag{25d}$$

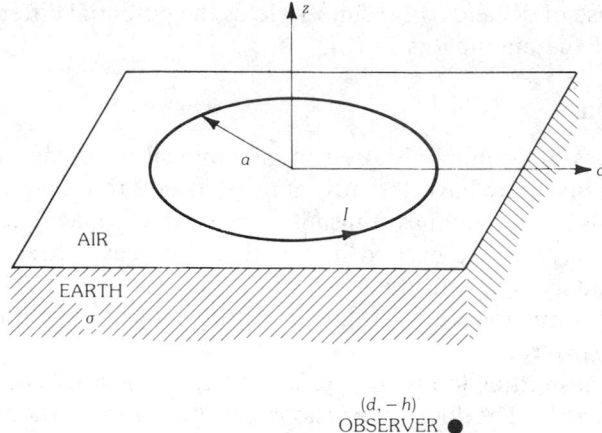

Fig. 12. Circular loop on a conducting half-space with a subsurface observer.

and J_0 and J_1 are the zeroth- and first-order Bessel functions. When a/h is sufficiently small, then the factor $2J_1(xa/h)/(xa/h)$ approaches unity over the important range of x. In that case the dependence on a enters only through the loop area A $(= \pi a^2)$. If the loop is buried at a depth h and the observer is located at the surface (as in uplink communications), then the result for H_z is identical except that the minus sign in (25a) becomes a plus.

For the special case of $H = a/h = 0$, the term Q reduces to the following result for a static magnetic dipole:

$$Q|_{H=a/h=0} = \frac{2 - D^2}{2(1 + D^2)^{5/2}} \tag{26}$$

If, in addition, $D = 0$, then $Q = 1$. Thus Q is the magnetic field normalized to the on-axis field of a static dipole. For $D = 0$, both H_ϱ and E_ϕ are zero.

Numerical results for a magnetic dipole ($a/h = 0$) are shown in Fig. 13. Note that for the static case ($H = 0$), there is a null at $D = \sqrt{2}$ in accordance with (26). For other values of H, the term Q is complex and the null is filled in. The peak value of $|Q|$ always occurs for $D = 0$, and this peak value decreases as H increases. Since H is proportional to the square root of frequency times the depth h, the frequency must be decreased in order to transmit to greater depths. This has been shown experimentally for mine communications. For geophysical probing, vertical sounding [18] is accomplished by varying the frequency, and low frequencies are used to obtain information on the earth conductivity at great depths. The dependence of $|Q|$ on H is shown more explicitly in Fig. 14 for the case of $D = a/h = 0$. For large values of H, the field Q decays exponentially with H just as a plane wave in a lossy medium does. This exponential decay with H will be shown to lengthen the rise time for transient fields of a loop in the following subsection because the high frequencies are severely attenuated.

The field results in Figs. 13 and 14 are actually valid for small loops of any shape which can be represented by a vertical magnetic dipole. When the loop

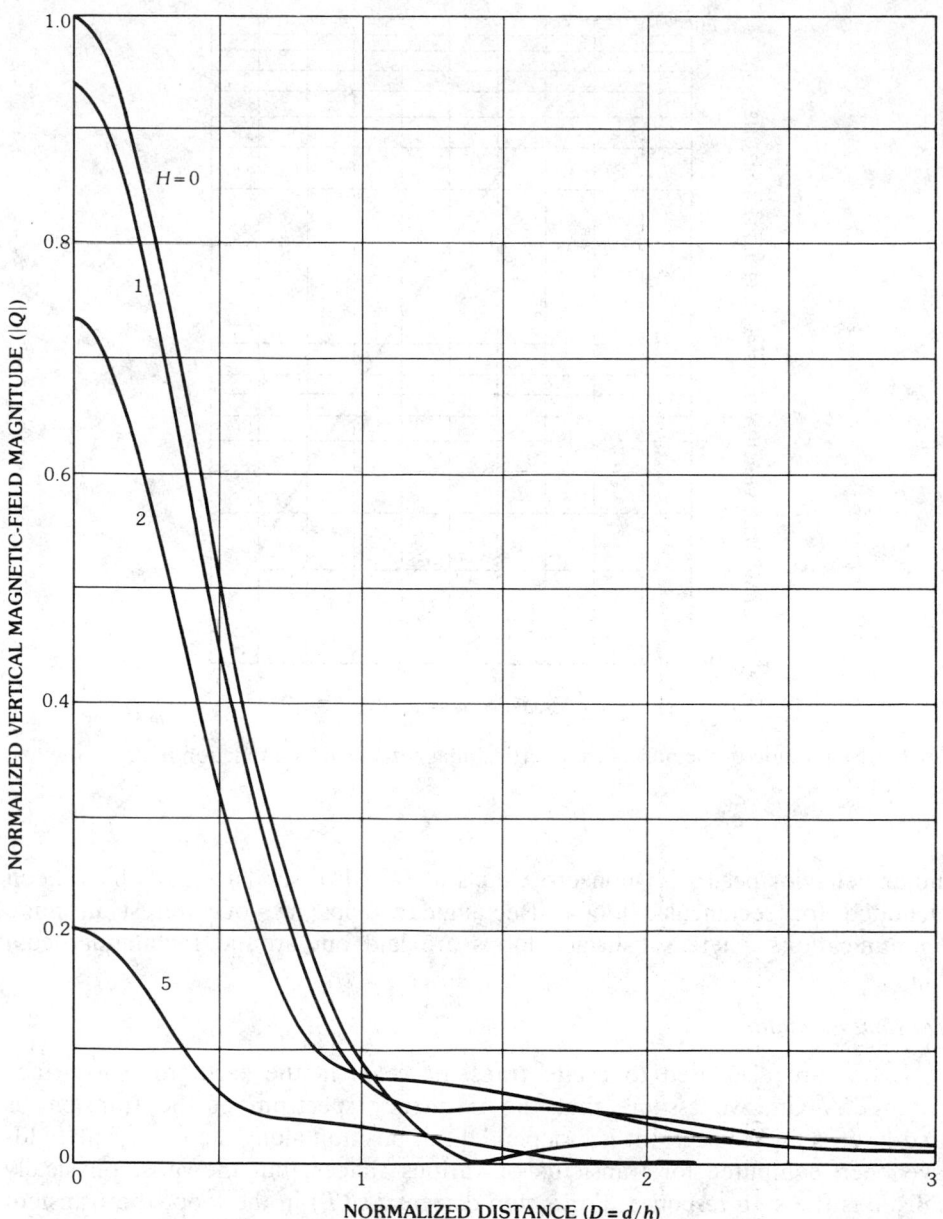

Fig. 13. Magnitude of the normalized vertical magnetic field for a vertical magnetic dipole source ($a/h = 0$). (*After Wait [17]*, © *1971 IEEE*)

dimensions are large, then the field depends strongly on shape, and the theory has been developed for loops of arbitrary shape in a conducting medium which is either homogeneous or layered [19]. Results for a finite circular loop are shown in Fig. 15 for the static case ($H = 0$). As the loop radius is increased, the vertical field is reduced on the axis ($D = 0$), but is increased at the larger horizontal distances. A

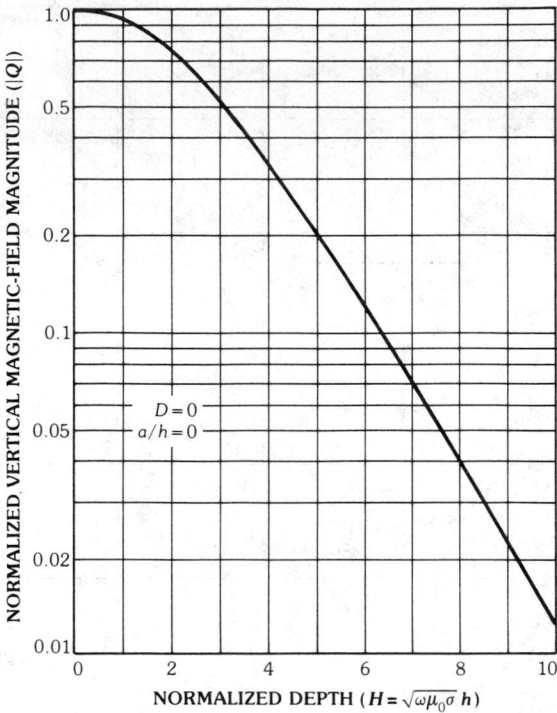

Fig. 14. Magnitude of the normalized vertical magnetic field as a function of H on the axis ($D = 0$).

similar behavior occurs for nonzero values of H. Also, similar results have been calculated for rectangular loops. Rectangular loops are of interest in mine communications where subsurface loops are laid out around rectangular coal pillars.

Transient Excitation

Loops are also used to excite transient fields in the earth for subsurface probing. Again, we assume that the frequency spectrum of the transient is band-limited and the current is independent of position along the loop. The fields have been computed for transients of various shapes, but the most physically realistic is the step response. For a step current $I U(T)$ in the loop, the transient subsurface field $\tilde{H}_z(t)$ can be expressed as [20]

$$\tilde{H}_z(t) = \frac{-IA}{2\pi h^3} Y(D, T) \tag{27a}$$

where

$$T = t/\sigma\mu_0 h^2 \tag{27b}$$

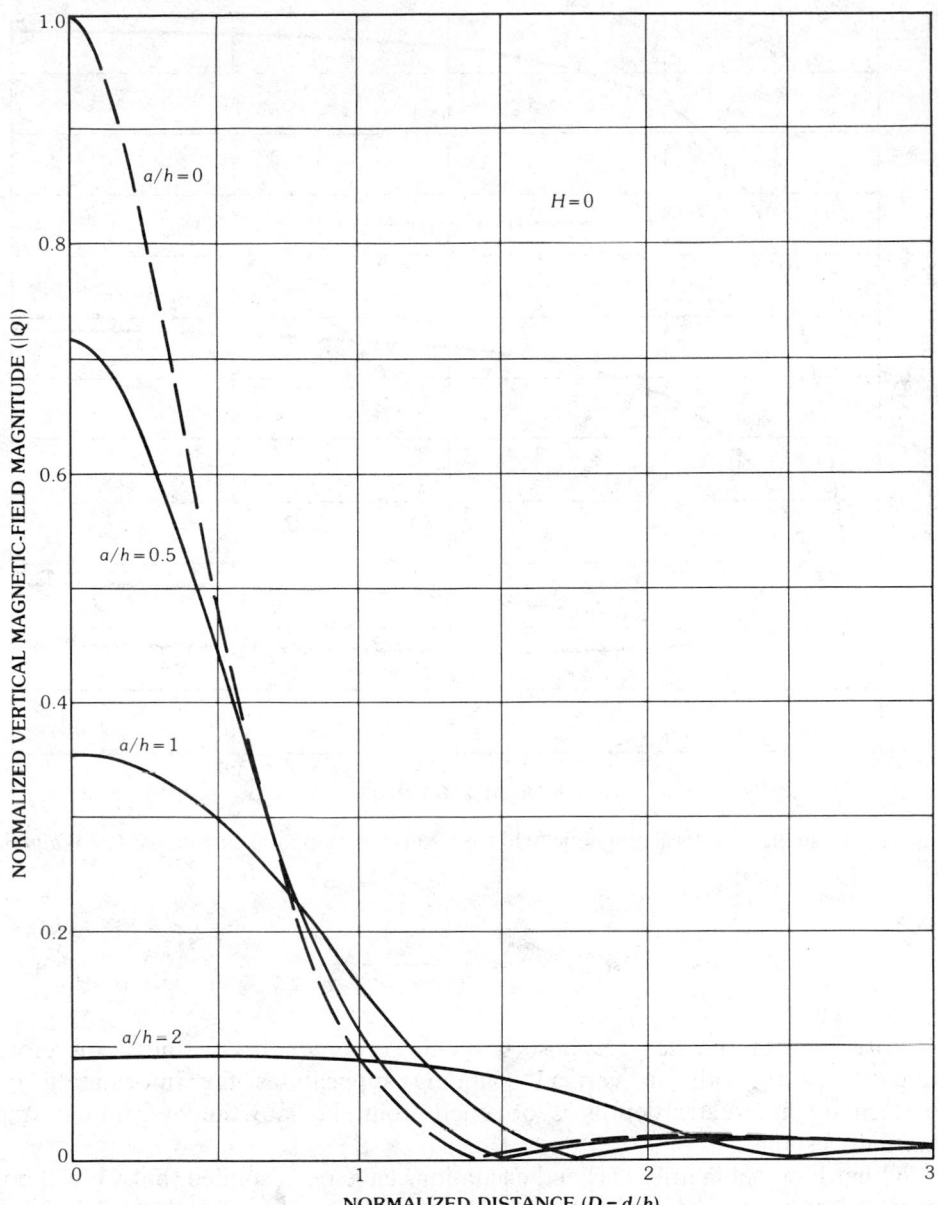

Fig. 15. Magnitude of the normalized vertical magnetic field of a finite circular loop. (*After Wait and Hill [19]*)

Again T is a normalized time, and Y is a normalized step response which can be obtained numerically from the frequency-domain solution in (25).

Numerical results for Y are given in Fig. 16 for the magnetic dipole case ($a/h = 0$). The final values of the curves are obtained from the low-frequency limit of the time-harmonic solution in (26):

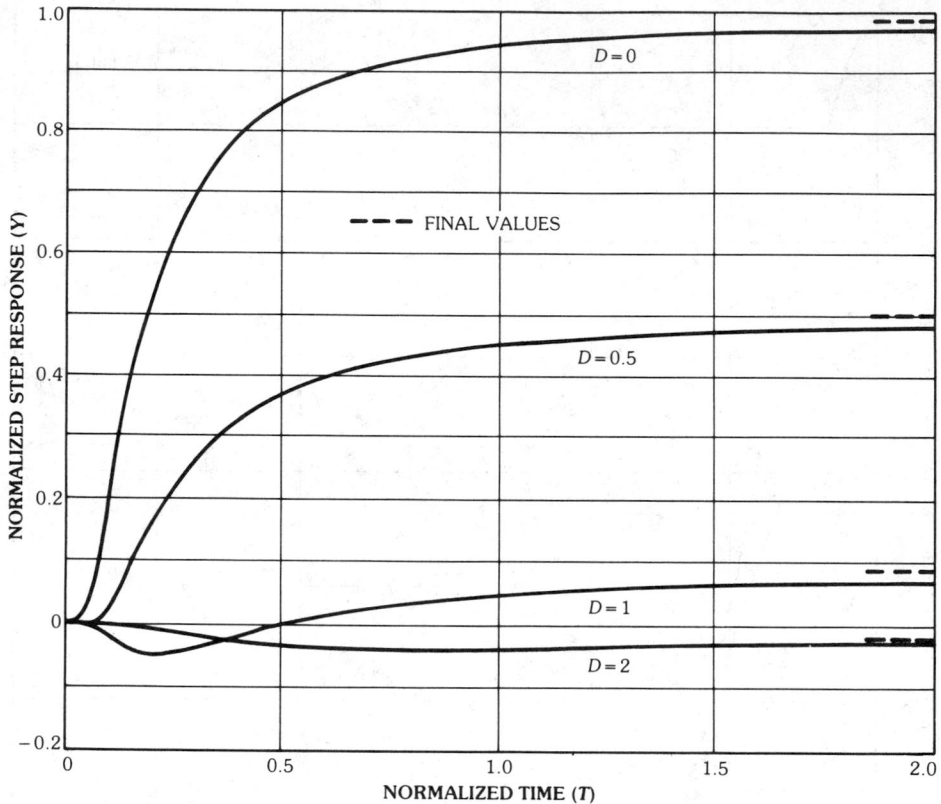

Fig. 16. Normalized vertical magnetic field for a step-function loop current. (*After Wait and Hill [20]*)

$$Y(D, \infty) = \frac{2 - D^2}{2(1 + D^2)^{5/2}} \tag{28}$$

The rise time of the step response is on the order of $\sigma\mu_0 h^2$, and it therefore increases with depth. In vertical sounding applications the information on conductivity at greater depths is obtained from the late time portion of the transient response.

Other loop geometries [21] and excitations have been studied, but we will not cover these here. The excitation by a rectangular pulse of current has been analyzed for both vertical and horizontal loops [22]. This type of pulse is of interest because many transient systems actually employ a periodic square wave.

Receiving Application

Receiving loops are normally small multiturn coils of small wire which can be hand held. Sometimes receiving loops are shielded electrostatically by wrapping them in foil, and the shielding should be grounded at some point to prevent capacitive coupling between the loop and the earth or the operator. Sometimes it is desirable to find the direction of the magnetic field by rotating the loop axis, and then a means of measuring the loop inclination is required. For example, a pair of orthogonal loops is used to measure the tilt of the natural magnetic field in audio-frequency (AFMAG) systems [23].

Since receiving loops are normally small they simply respond to the time derivative of the magnetic flux. For example, the voltage induced in a horizontal loop as in Fig. 12 is given by

$$\tilde{v}(t) = A_R N_R \mu \frac{\partial \tilde{H}_z(t)}{\partial t} \tag{29}$$

where A_R is the receiving loop area, N_R is the number of turns, and $\tilde{H}_z(t)$ is the time-varying vertical magnetic field. If the loop has an air core, then μ is the free-space permeability μ_0; but if the loop has a ferrite core, then μ is some larger effective value. If the loop has some other orientation, then $\tilde{H}_z$ is simply the magnetic field along the loop axis. For example, in the magnetotelluric method the loop is oriented in a vertical plane in order to measure the time-varying horizontal magnetic field.

For time-harmonic fields the received voltage is proportional to $j\omega$ times the axial magnetic field. For a horizontal loop the Fourier transform of (29) yields the following received voltage:

$$v(\omega) = j\omega\mu A_R N_R H_z(\omega) \tag{30}$$

An interesting application of the small receiving loop is the location of a horizontal subsurface transmitting loop in the trapped-miner problem [2]. As indicated in Fig. 17, the horizontal magnetic field has a null and the vertical magnetic field is a maximum directly above the transmitting loop. The transmitting loop is excited with a pulsed cw signal with a carrier frequency of about 1 kHz, and the search coil is equipped with earphones. The null in the horizontal magnetic field can typically be located by the search coil within an accuracy of a few meters for a transmitter depth of a couple hundred meters. The depth of the transmitter is not determined by the null location, but can, in principle, be located from more extensive surface measurements [24].

5. Miscellaneous Antennas

The preceding sections were devoted to the commonly used grounded wire and loop antennas. In this section we discuss a number of antennas and applications, but in much less detail.

Magnetometers are sometimes used in place of loop antennas for receiving magnetic fields. The problem with loop antennas is that the induced voltage is proportional to the time derivative of the magnetic field as shown in (29) or $j\omega$ times the time-harmonic magnetic field as shown in (30). As a result loop antennas are not very sensitive to slowly varying (or low-frequency) fields, and this can be a serious problem in magnetotelluric applications. Several types of magnetometers ([1], pp. 231–243) respond to the magnetic field itself rather than to the time

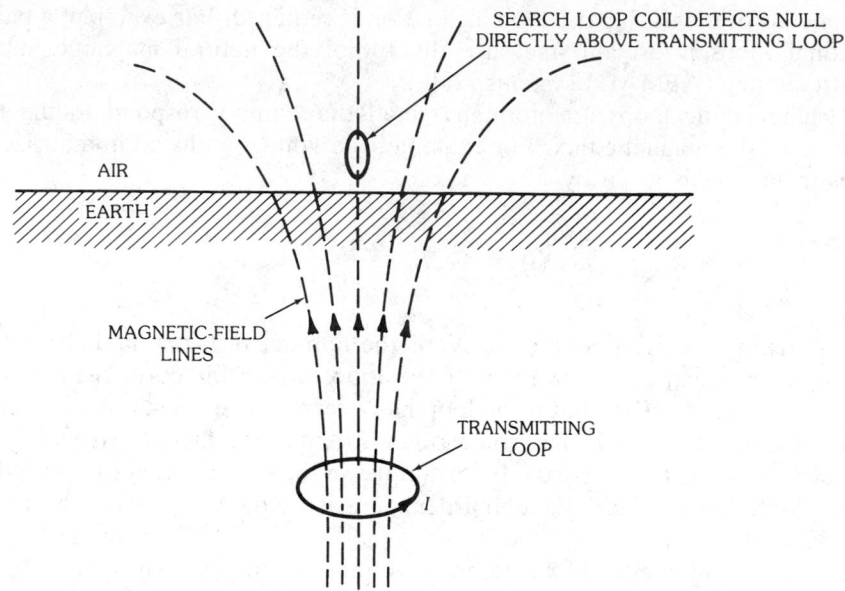

Fig. 17. Geometry for trapped-miner location.

derivative and are useful for measuring slowly varying signals. In the magnetic balance method the torque on a bar magnet due to the magnetic field is balanced against gravity. In the flux-gate magnetometer the nonlinear permeability of a ferromagnetic material is utilized to generate a second harmonic which is proportional to the exciting field. Optical pumping magnetometers utilize the energy level differences within an atom which depend on the ambient magnetic field, and these can have very good sensitivity.

The insulated antenna which is grounded at the end can be assumed to carry a uniform current when the length of the antenna is short compared with the wavelength in the insulation. This is always the case for the elf applications discussed in Section 2, but is not always true for higher frequencies where the current is essentially that of a transmission line whose electrical length is not short. Also, there are cases where the antenna has no insulation, and the current distribution is more complicated. Such cases, where the current distribution on linear antennas in lossy media has to be determined, are discussed thoroughly in the book by King and Smith [25].

Electrical well logging has been used extensively for more precise information on rock conductivity as a function of depth ([1], Chapter II). Single electrodes or various configurations of multiple electrodes are lowered into wells or drill holes on insulated wire, and the apparent resistivity is a function of the material surrounding the electrode or multiple electrodes. In order to have electrical continuity between the electrodes and the surrounding rock, the drill hole must be filled with water or drilling mud. In induction logging, an induction coil excites current in the surrounding rock via a time-varying magnetic field, and electrical contact is not required. Consequently, induction logs can be run in either dry or fluid-filled holes.

The voltage induced in a secondary coil is directly related to the conductivity of the surrounding rock, and the dielectric constant plays no role at the typical frequencies below about 20 kHz. More recently, however, dielectric logging has become quite popular at higher frequencies all the way from high frequencies to microwaves. The crossover frequency, where displacement currents are equal to conduction currents, is determined from

$$\sigma = \omega \epsilon_r \epsilon_0 \tag{31}$$

where ϵ_r is the relative permittivity of the ground and ϵ_0 is the permittivity of free space. For example, for $\sigma = 10$ S/m and $\epsilon_r = 10$, the crossover frequency is approximately 18 MHz. Thus systems operating above high frequency are primarily
sensing the dielectric constant.

When shallow depths on the order of several meters down to a fraction of a meter are of interest, then frequencies all the way up to microwaves can be used because a relatively small skin depth can be tolerated. Frequencies in the range from 1 to 2 GHz have been used in an fm-cw system [26] to sense the thickness of coal remaining on the roof of a mined haulageway for thicknesses up to approximately a half meter. A pair of broadband rectangular-aperture horns which utilize double-ridged waveguide techniques were used for transmission and reception. The same system has been used for probing of soil to shallow depths and for probing the properties of snowpack. A pulsed system with a frequency spectrum from about 100 MHz to 500 MHz has been used for similar applications involving the detection of targets at shallow depths. The antennas used were a pair of broadband crossed dipoles [27] with transmission and reception on orthogonal polarizations. It appears that the use of broadband vhf and microwave systems for high-resolution probing to shallow depths is a promising area which will continue to develop.

6. References

[1]   G. V. Keller and F. C. Frischknecht, *Electrical Methods in Geophysical Prospecting*, Oxford: Pergamon Press, 1966.

[2]   J. N. Murphy and H. E. Parkinson, "Underground mine communication," *Proc. IEEE*, vol. 66, pp. 26–50, 1978.

[3]   E. I. Parkhomenko, *Electrical Properties of Rock*, New York: Plenum Press, 1967.

[4]   P. K. Bhattacharya and H. P. Patra, *Direct Current Geoelectric Sounding*, Amsterdam: Elsevier Publishing Co., 1968.

[5]   E. F. Laine and R. J. Lytle, "A computer program for four probe resistivity measurements in a horizontally layered earth," *IEEE Trans. Geosci. Electron.*, vol. GE-14, no. 4, pp. 232–235, 1976.

[6]   O. Koefoed, *Geosounding Principles, 1*, Amsterdam: Elsevier Publishing Co., 1979.

[7]   M. L. Burrows, *ELF Communications Antennas*, Stevenage, Herts., UK: Peter Peregrinus, 1978.

[8]   E. D. Sunde, *Earth Conduction Effects in Transmission Systems*, Chapter V, New York: D. Van Nostrand Co., 1949.

[9]   D. A. Hill and J. R. Wait, "Subsurface electromagnetic fields of a grounded cable of finite length," *Can. J. Phys.*, vol. 51, pp. 1534–1540, 1973.

[10] J. R. Wait and K. P. Spies, "Subsurface electromagnetic field of a line source on a conducting half-space," *Rad. Sci.*, vol. 6, pp. 781–786, 1971.

[11] D. A. Hill, "Electromagnetic surface fields of an inclined buried cable of finite length," *J. Appl. Phys.*, vol. 44, pp. 5275–5279, 1973.

[12] D. A. Hill and J. R. Wait, "Subsurface electric fields of a ground cable of finite length for both frequency and time domain," *Pure Appl. Geophys.*, vol. 111, pp. 2324–2332, 1973.

[13] D. A. Hill and J. R. Wait, "Diffusion of electromagnetic pulses into the earth from a line source," *IEEE Trans. Antennas Prop.*, vol. AP-22, pp. 145–146, 1974.

[14] J. R. Wait, "Propagation of electromagnetic pulses in a homogeneous conducting earth," *Appl. Sci. Res.*, vol. 8, pp. 213–253, 1960.

[15] A. A. Kaufman and G. V. Keller, *The Magnetotelluric Sounding Method*, Amsterdam: Elsevier Publishing Co., 1981.

[16] D. G. Large, L. Ball, and A. J. Farstad, "Radio transmission to and from underground coal mines—theory and experiment," *IEEE Trans. Commun.*, vol. COM-21, pp. 194–202, 1973.

[17] J. R. Wait, "Electromagnetic induction technique for locating buried source," *IEEE Trans. Geosci. Electron.*, vol. GE-9, pp. 95–98, 1971.

[18] A. A. Kaufman and G. V. Keller, *Frequency and Transient Sounding*, Amsterdam: Elsevier Publishing Co., 1983.

[19] J. R. Wait and D. A. Hill, "Fields of a horizontal loop of arbitrary shape buried in two-layer earth," *Rad. Sci.*, vol. 15, pp. 903–912, 1980.

[20] J. R. Wait and D. A. Hill, "Electromagnetic surface fields produced by a pulse-excited loop buried in the earth," *J. Appl. Phys.*, pp. 3988–3991, 1972.

[21] D. A. Hill, "Transient signals from a buried horizontal magnetic dipole," *Pure and Appl. Geophys.*, vol. III, pp. 2264–2272, 1973.

[22] H. J. Tsaknakis and E. E. Kriezis, "Transient electromagnetic field due to a circular current loop perpendicular or parallel to a conducting half-space," *IEEE Trans. Geosci. Remote Sensing*, vol. GE-20, pp. 122–130, 1982.

[23] S. H. Ward, J. O'Donnell, R. Rivera, G. H. Ware, and D. C. Fraser, "AFMAG—applications and limitations," *Geophys.*, vol. XXXI, pp. 576–605, 1966.

[24] J. R. Wait, "Criteria for locating an oscillating magnetic dipole buried in the earth," *Proc. IEEE*, vol. 59, pp. 1033–1035, 1971.

[25] R. W. P. King and G. S. Smith, *Antennas in Matter*, Cambridge: MIT Press, 1981.

[26] D. A. Ellerbruch and D. R. Belsher, "Fm-cw technique of measuring coal layer thickness," *IEEE Trans. Geosci. Electron.*, vol. GE-16, pp. 126–133, 1978.

[27] L. C. Chan, L. Peters, Jr., and D. L. Moffat, "Improved performance of a subsurface radar target identification system through antenna design," *IEEE Trans. Antennas Propag.*, vol. AP-29, pp. 307–311, 1981.

Chapter 24

Antennas for Medical Applications

C. H. Durney
University of Utah

M. F. Iskander
University of Utah

CONTENTS

Carl H. Durney was born in Blackfoot, Idaho, on April 22, 1931. He received the BS degree in electrical engineering from Utah State University, Logan, in 1958, and the MS and PhD degrees in electrical engineering from the University of Utah, Salt Lake City, in 1961 and 1964, respectively.

From 1958 to 1959 he was employed as an associate research engineer with the Boeing Airplane Company, in Seattle. He has been with the University of Utah since 1963, when he was appointed assistant research professor in electrical engineering.

While on leave in 1971 he studied microwave biological effects at the University of Washington. From 1977 to 1982 he was chairman of the Electrical Engineering Department at the University of Utah, where he is presently a professor of electrical engineering and engaged in teaching and research in electromagnetics, engineering pedagogy, and microwave biological effects. Dr. Durney has received several awards in teaching and research and is a member of many technical societies, editorial boards, and committees.

Magdy F. Iskander was born in Alexandria, Egypt, on August 6, 1946. He received the BSc degree in electrical engineering, University of Alexandria, Egypt, in 1969. From the University of Manitoba, in Winnipeg, Manitoba, Canada, he received the MSc and PhD degrees, both in microwaves, in 1972 and 1976, respectively.

In 1976 Dr. Iskander was awarded a National Research Council of Canada postdoctoral fellowship at the University of Manitoba. Since March 1977 he has been with the Department of Electrical Engineering and the Department of Bioengineering at the University of Utah, in Salt Lake City, where he is currently a professor of electrical engineering.

Dr. Iskander has contributed chapters to four research books. He has also presented and published in technical journals more than 140 papers. He is a member of the editorial boards of the *IEEE Transactions on Microwave Theory and Techniques* and the *Journal of Microwave Power*. His interests include the scattering and diffraction of electromagnetic waves, antenna design, and the biological effects as well as the medical applications of electromagnetic waves.

1. Introduction

Rapid advances in technology have been accompanied by an increasing application of technology to health care. More and more interaction of scientists and engineers with medical people has resulted in significant developments in improved health care using highly sophisticated equipment. Correspondingly, the use of antennas of one kind or another in the practice of medicine has increased. Most of the medical applications of antennas involve coupling electromagnetic energy into the human body or into other biological systems, such as animals used in experimental measurements.

Primarily, these applications can be divided into two categories: therapy and diagnostics. In therapy, antennas have been used mainly for producing hyperthermia for treating cancer. This therapy consists of heating tumors, alone or in combination with either X-ray irradiation or chemotherapy. In diagnostics, electromagnetic energy is coupled into and out of the body to monitor various physiological parameters. Tomographic imaging of the body by electromagnetic measurements has been investigated, but not yet developed to the point of practical
implementation.

The problems encountered in designing antennas for medical applications are somewhat different from those involved in the design of antennas for other applications. For example, in coupling electromagnetic energy into biological systems the near fields are often more important than the far fields, in contrast to other applications. Near-field problems are especially important in the design of hyperthermia applicators. Since biological systems usually have high dielectric constants the antennas for coupling into them must be designed differently from those radiating into free space. It is also very important that antennas used for coupling electromagnetic energy into the human body be designed to produce minimum leakage radiation outside the body. Leakage radiation can sometimes be hazardous to medical personnel, and it can produce artifacts in measurements.

Furthermore, since biological systems are usually quite lossy, there is a fundamental trade-off between depth of penetration and localization of the fields inside the body. In many applications, localization of the internal fields is important. In hyperthermia, for example, it is often desirable to concentrate the internal fields at the tumor, thereby heating it without appreciably affecting the surrounding normal tissue. Concentrating the fields, however, amounts to having narrow-beam radiation, which requires an electrically large aperture. To restrict apertures to physical sizes convenient for use with patients and still get narrow-beam radiation, higher-frequency operation, in the high-megahertz or low-gigahertz range, is required. At these high frequencies, however, the penetration of the electromagnetic radiation is too shallow to be useful. Some insight into this problem can be gained from the data presented in Fig. 1, which shows the plane-wave penetration depth versus frequency in a dielectric half-space with a permittivity equal to the average permittivity of the human body. The power magnitude is normalized to the value at the surface. The profile of electromagnetic field penetration into the body is, of course, a strong function of aperture size. Since the data in Fig. 1 is for an incident plane wave, it corresponds to that of a very large aperture. The curves show that even for a very large aperture the penetration at high frequencies is too shallow to be useful for any application that requires transmission of a signal through the body or heating of tissue in the center of the body. Given the severe attenuation at high frequencies, there is no alternative for some applications except to go to lower frequencies. But at lower frequencies the radiators must often be electrically small, resulting in radiation profiles similar to that of a point source, which means that concentrating the fields in a small region inside the body (such as in a tumor) would be very difficult. Thus there is a fundamental trade-off between depth of penetration and localization. At frequencies high enough that apertures can be electrically large enough to produce good localization of the internal fields, the penetration is too shallow to be useful for many applications. At lower frequencies, while the penetration will be deep enough, it is difficult to make apertures electrically large enough to localize the internal fields to small regions. Electrically small apertures at low frequencies often are described as nonradiating because

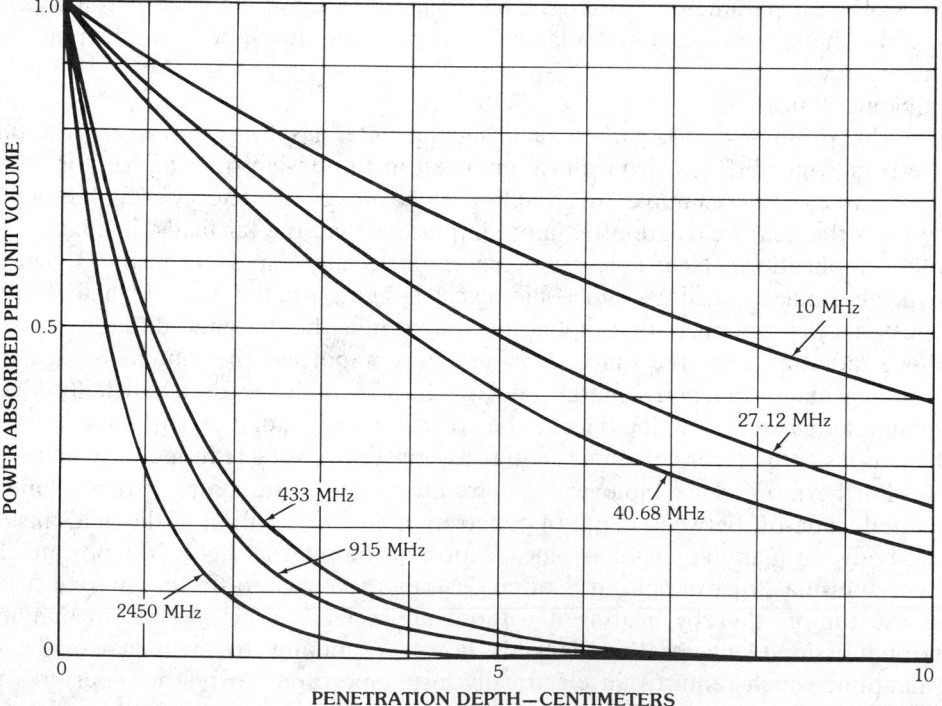

Fig. 1. Relative absorbed power per unit volume versus depth for a plane wave propagating into a dielectric half-space having the properties of tissue. (*After Iskander [1], © 1982 American Institute of Physics*)

their fields are mostly near fields that decay rapidly with distance away from the applicator. These devices are usually thought of as inducing currents, rather than radiating. As explained in the various applications described in this chapter this fundamental problem is encountered over and over again. In producing hyperthermia, for example, the shallow penetration at higher frequencies forces the designer to use low frequencies to get deeper heating; otherwise the surface overheats, even with surface cooling, before sufficient heating occurs deeper inside the body. At lower frequencies, though, the radiation produced by the electrically small apertures is often dominated by near fields, which, because they die away rapidly with distance from the applicator, also cause the surface to overheat. The main problem in hyperthermia applications, therefore, is how to obtain deep heating without overheating the surface.

Another factor that is important in coupling electromagnetic energy into biological systems is the reflection at the interfaces between different kinds of tissues. This reflection depends not only on the specific types of tissues at the interface, but also on the frequency, the polarization, and the angle of incidence of the propagating waves. Fig. 2 illustrates the variation of the magnitude of the reflection coefficient as a function of frequency for various tissue interfaces [1]. These calculations, although they are made for planar dielectric layers exposed to an incident plane wave, nevertheless illustrate what large reflections can occur at

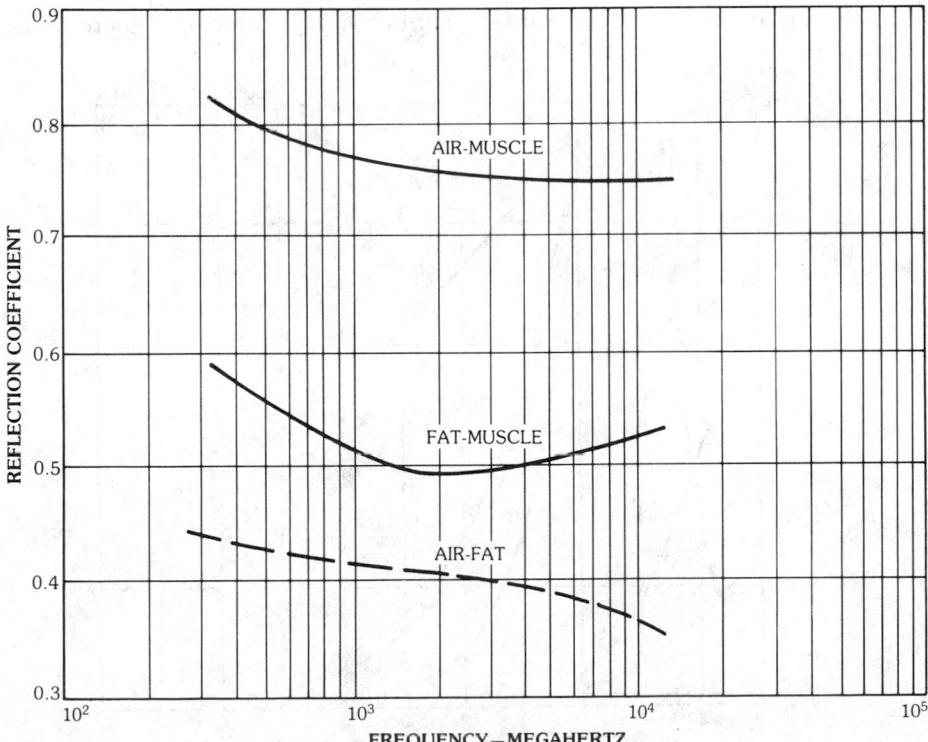

Fig. 2. Plane-wave reflection coefficient for normal incident at the interfaces between different tissue layers. (*After Iskander [1], © 1982 American Institute of Physics*)

the tissue interfaces. The interference between the incident wave and these large reflected waves results in significant standing-wave patterns that may result in excessive heating (hot spots) of certain regions in the body.

Extension calculations of absorbed power in various combinations of tissues have been made by Schwan [2]. Typical results [3] for a planar model of a fat layer in front of a large muscle mass is shown in Fig. 3. Note that the curves are normalized to the value at the fat-muscle interface. Calculations show that the relative absorption is not very sensitive to the thickness of fat, which for subcutaneous fat in people may vary from less than a centimeter to about 2.5 cm. The data in Fig. 3 clearly shows that there is a severe discontinuity in the power absorption at the interface. Furthermore, as the frequency increases, the penetration of the electromagnetic fields into the muscle decreases, and standing waves begin to occur in the fat.

Based on these results, it is clear that significant advantages can be gained by choosing frequencies below 1 GHz for many medical applications. But, once again, the near-field problems as well as the difficulty of localizing the electromagnetic radiation at these lower frequencies often limit the usefulness of devices operating at frequencies much lower than 500 MHz.

This chapter begins with a description of waveguide-type antennas, some used at frequencies as low as 27 MHz. Next, the microstrip and transmission-line kinds of antennas are discussed. Then implantable antennas used in hyperthermia applications and those used to measure the permittivity of tissue *in vivo* are treated.

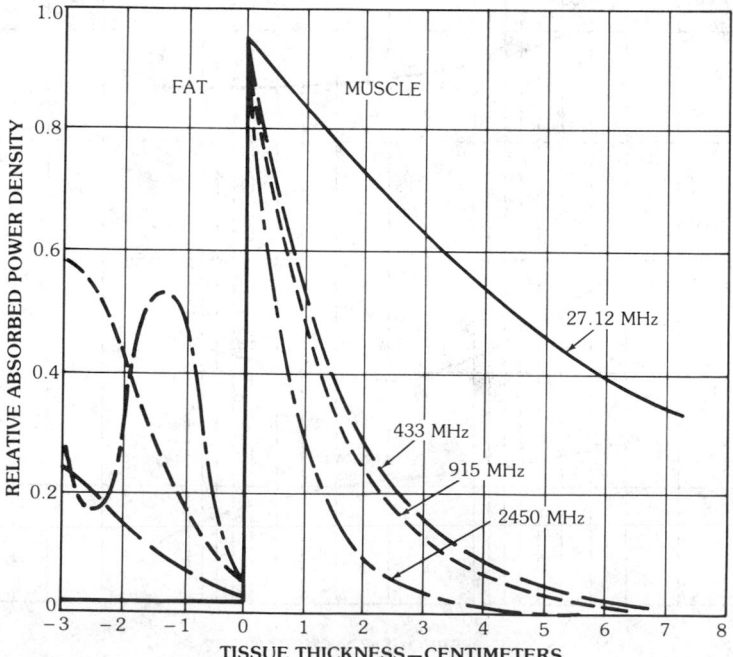

Fig. 3. Relative absorbed power density patterns in planar fat and muscle layers exposed to a plane-wave source. (*After Guy, Lehmann, and Stonebridge [3],* © *1974 IEEE*)

This is followed by a description of antennas used to monitor electromagnetic radiation, primarily in relation to hazard assessment, and a discussion of other antenna applicators used for producing hyperthermia. Finally, some experimental procedures for characterizing antennas used in medical applications are discussed.

2. Waveguide- and Radiation-Type Antennas

One class of commonly used antennas for coupling electromagnetic energy into the human body consists of open-ended waveguides of one kind or another. Such radiators have been used primarily for inducing hyperthermia for cancer therapy, which requires effective heating of muscle tissue without excessive heating of the fat layer that overlies the muscle. Waveguide radiators generally provide good coupling of the electromagnetic energy into the tissue, but they tend to be bulky and awkward. Since the waveguides must be physically very large at low frequencies, they have been used mostly at frequencies of 915 MHz and above, although a ridged waveguide loaded with water has been used at frequencies as low as 27 MHz for inducing hyperthermia. Other radiation-type applicators that are used in electromagnetic hyperthermia applications include various forms of reflector antennas and the annular phased array (APA) applicator.

Direct-Contact Waveguide Applicators

Perhaps the simplest effective waveguide applicator is an open-ended dielectric-loaded waveguide. Loading the waveguide with a lossless dielectric of permittivity approximately equal to that of tissue provides a good impedance match and ensures good transmission into the tissue. The dielectric loading also reduces the size of the waveguide. However, the effective heating of the tissue depends strongly on the size of the aperture and the field distribution in it. In an extensive analysis of the internal fields produced by a TE_{10}-mode rectangular aperture in a two-layer semi-infinite tissue model, Guy [4] calculated internal fields as a function of the size of the aperture and the thickness of the fat layer. He showed that of the two frequencies, 915 MHz and 2450 MHz, that are authorized for medical applications in the higher-frequency ranges in the United States, 915 MHz produces a higher ratio of muscle-to-fat heating, which is desirable for therapeutic applications. At 915 MHz, the optimum aperture size was found to be 13 cm high and between 13 and 26 cm in width. On the basis of these results, a 13-cm by 13-cm aperture applicator that includes provisions for cooling the skin was designed and tested [5]. Details are shown in Fig. 4. Measured heating patterns correlated well with calculated values.

This applicator offers the advantage of providing optimal heating of a fat-muscle configuration at 915 MHz. The disadvantage of the applicator is that it is not suitable for heating muscle at depths of more than about four cm because it operates at 915 MHz.

TEM Waveguide Antenna

Van Koughnett and Wyslouzil [6] suggested the use of an inhomogeneously loaded dielectric waveguide to simulate TEM-mode wave propagation for biological studies. Cheung, Dao, and Robinson [7] used open-ended TEM wave-

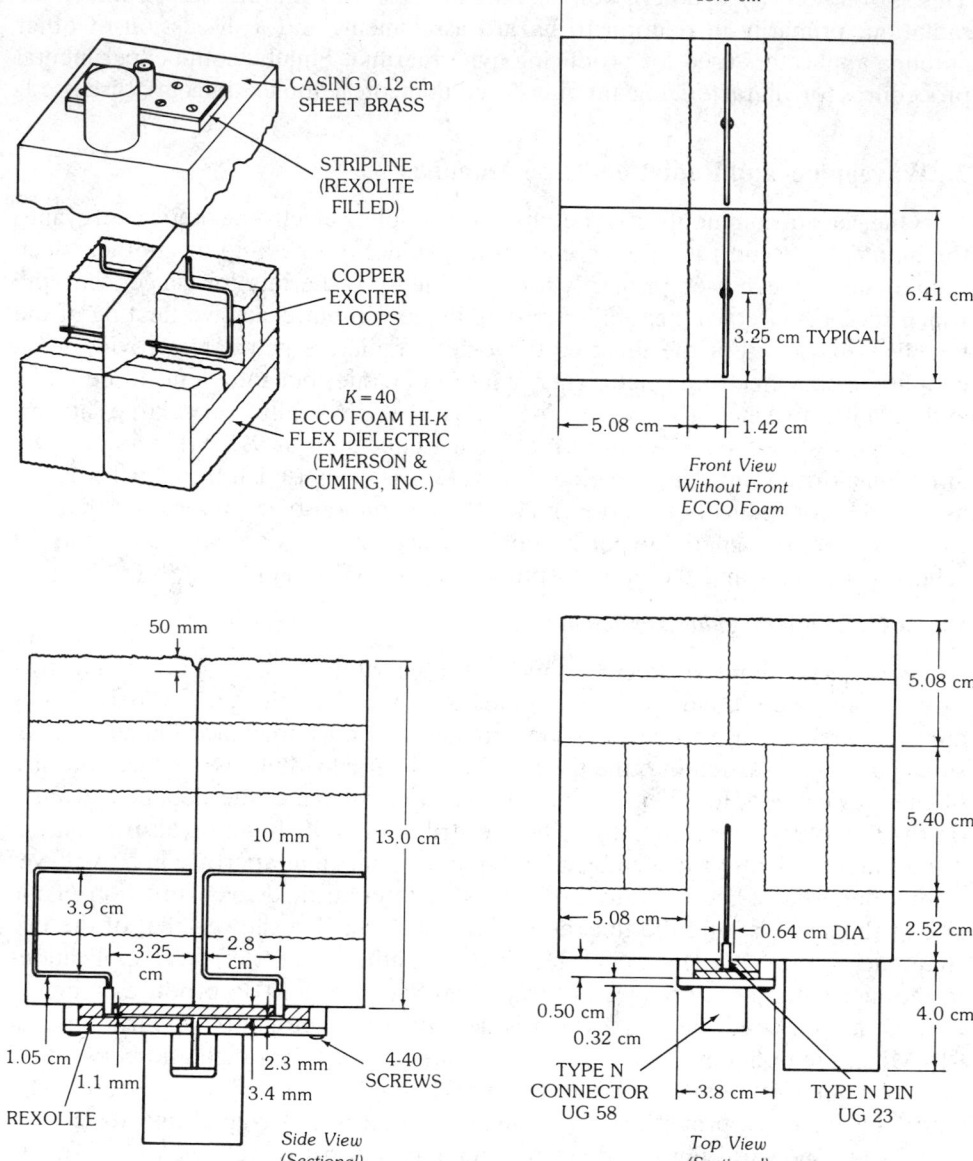

Fig. 4. Details of the air-cooled applicator designed by Guy and others. (*After Guy et al.* [5], © 1978 IEEE)

guide as an applicator for hyperthermia to take advantage of the uniformity of the fields across the aperture of the waveguide. Usually it is desirable to have a power absorption pattern that is uniform in the cross section of the region to be heated. The TEM mode would therefore be more desirable than the TE_{10} waveguide

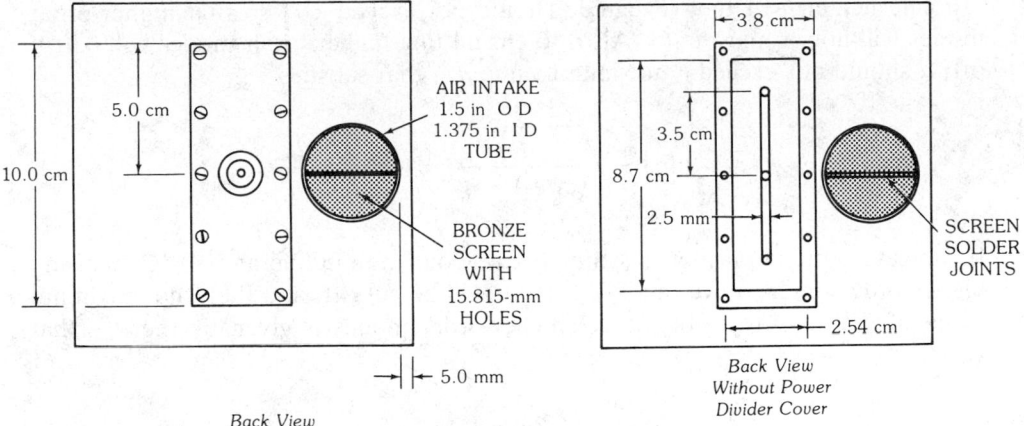

Fig. 4, *continued.*

mode, for example, which has a cosine variation in one direction perpendicular to the direction of propagation.

A diagram of the cross section of the TEM-mode waveguide is shown in Fig. 5. For the field patterns in the central portion of the waveguide to approximate those of the TEM mode, the following conditions must be satisfied [6]:

$$t\sqrt{\epsilon_1/\epsilon_2 - 1} = \lambda_1/4 \tag{1}$$

where λ_1 is the wavelength in an unbounded region of permittivity ϵ_1, and t, ϵ_1, and ϵ_2 are as defined in Fig. 5. This relationship was obtained by writing the field equations in the regions inside the waveguide, matching the boundary conditions, and then choosing the conditions for which the velocity of propagation in the loaded waveguide is the same as the velocity of light in an unbounded region of permittivity ϵ_1, which is one principal characteristic of the TEM mode. The other principal characteristic, that **E** and **H** are uniform, was found to be true in the central region when (1) is satisfied.

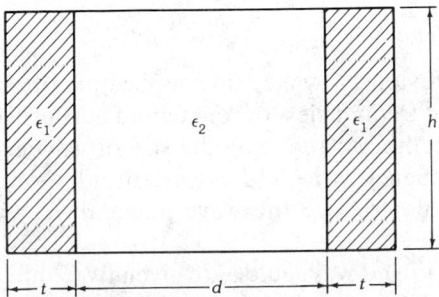

Fig. 5. Cross section of the inhomogeneously dielectric-loaded TEM-mode waveguide.

The height h of the waveguide should not exceed $\lambda_1/2$ so that higher-order modes will not propagate [6]. Also, to ensure that higher-order modes will be cut off, d should not exceed a maximum value d_{m1} that satisfies

$$\sqrt{\epsilon_1/\epsilon_2}\tan\left(\frac{\pi}{2}\sqrt{\frac{\epsilon_1/\epsilon_2}{(\epsilon_1/\epsilon_2)-1}}\right) = -\tan\left(\frac{\pi d_{m1}}{\lambda_1}\right)$$

If, however, the applicator is symmetrical about the midline at $t + d/2$, the first higher-order mode above the TEM mode will be cut off, and then the maximum value of d to ensure cutoff of the higher-order modes is given by the d_{m2} that satisfies

$$\sqrt{\epsilon_1/\epsilon_2}\tan\left(\frac{\pi}{2}\sqrt{\frac{\epsilon_1/\epsilon_2}{(\epsilon_1/\epsilon_2)-1}}\right) = \text{ctn}\left(\frac{\pi d_{m2}}{\lambda_1}\right)$$

In practice, d should be about 10 percent less than d_{m1} or d_{m2} so that the higher-order modes will be sufficiently attenuated.

The impedance of the TEM-mode applicator could be matched to that of tissue by using an ϵ_2 approximately equal to that of tissue. That would cause three problems, however, all related to the rather high permittivity of tissue. First, a high ϵ_2 would probably require a heavy material that would be expensive and would make the applicator very heavy and awkward to use. Secondly, although deionized water could perhaps be used, it is difficult to find a solid material with a high relative permittivity and low enough loss to avoid excessive heating in the large volume occupied by ϵ_2. Thirdly, since ϵ_1 must be greater than ϵ_2, and since the thickness t is related directly to the ratio ϵ_1/ϵ_2, finding a low-loss material with a high enough permittivity for ϵ_1 with a reasonable thickness could be difficult. Cheung, Dao, and Robinson [7] avoided these problems by using air for the central dielectric and using a quarter-wave transformer to match into tissue. Their design for a 2450-MHz applicator is shown in Fig. 6.

The TEM-mode waveguide applicator has the advantage of providing a very uniform field distribution in the cross section. Its disadvantages are that it is relatively narrow-band and, like most other waveguide applicators, is usually practical only at higher frequencies.

Ridged-Waveguide Antennas

As pointed out previously, waveguide applicators often must be too large and bulky to be practical at frequencies low enough to get sufficiently deep penetration into the body. One method of reducing the size of waveguide that will propagate waves at a given frequency is to add ridges. In addition to lowering the cutoff frequency the ridges also reduce the wave impedance, which makes it easier to match into tissue.

Design curves for ridged waveguide are given by Cohn [8], with the dimensions for single- and double-ridged guide denoted as in Fig. 7. Curves for the cutoff frequency for two different values of b_1/a_1 are given in Fig. 8 as a function of a_2/a_1,

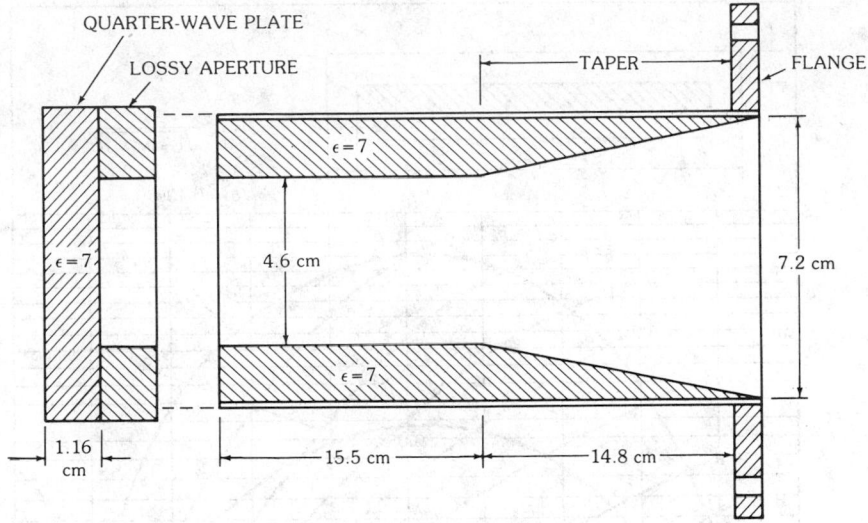

Fig. 6. Top sectional view of the TEM applicator and quarter-wave matching transformer designed by Cheung, Dao, and Robinson. (*After Cheung, Dao, and Robinson [7], © 1977 American Geophysical Union*)

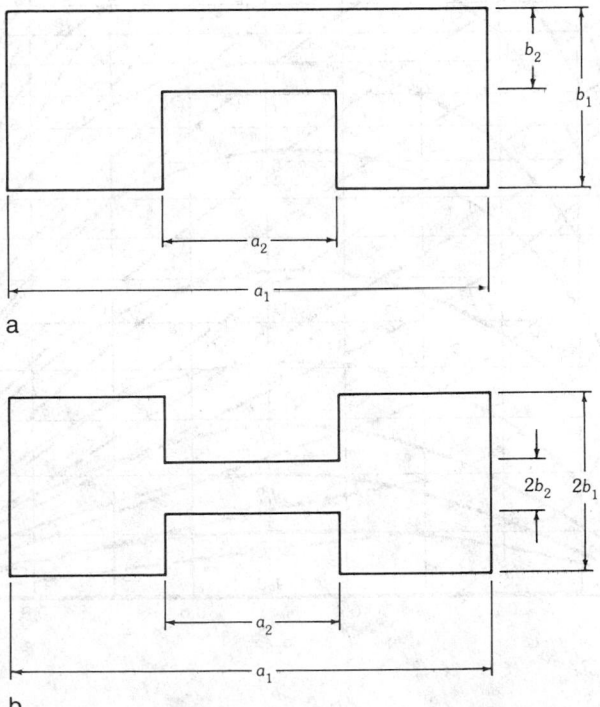

Fig. 7. Dimensions for ridged and double-ridged waveguide. (*a*) For single-ridged waveguide. (*b*) For double-ridged waveguide.

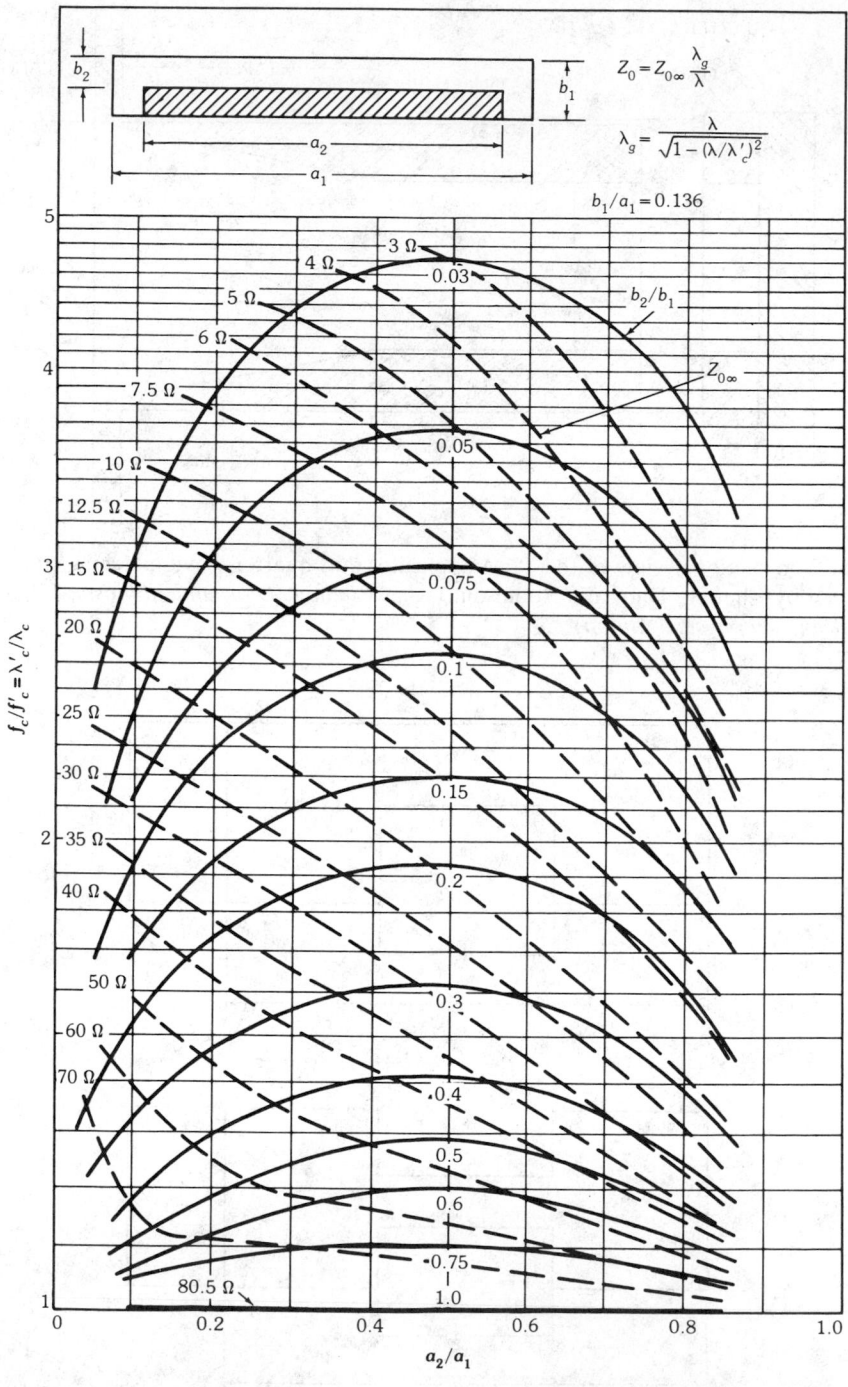

Fig. 8. Characteristic impedance and cutoff wavelength of ridged waveguide. (a) For b_1/a_1 = 0.136. (b) For b_1/a_1 = 0.5. (*After Cohn [8], © 1947 IRE*)

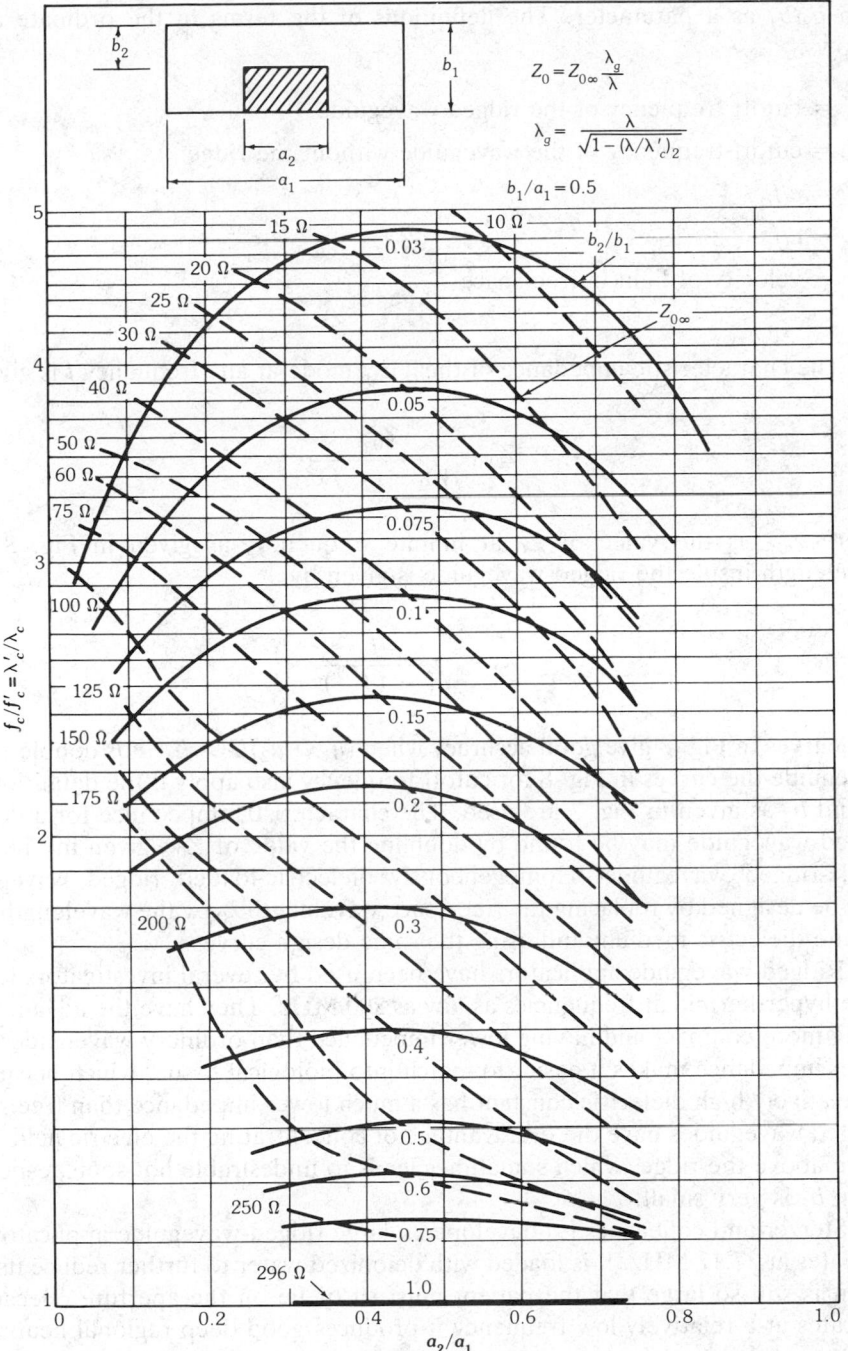

b

Fig. 8, *continued.*

with b_2/b_1 as a parameter. The definitions of the terms in the ordinate are as follows:

f_c' = cutoff frequency of the ridged waveguide

f_c = cutoff frequency of the waveguide without the ridge

$\lambda_c' = c/f_c'$

$\lambda_c = c/f_c$

c = velocity of light in free space

The characteristic impedance of the TE_{10} mode at any frequency f is given by

$$Z_0 = \frac{Z_{0\infty}}{\sqrt{1 - (f_c'/f)^2}}$$

where $Z_{0\infty}$ is the value of Z_0 at infinite frequency, as given in Fig. 8. The wavelength inside the ridged waveguide is given by

$$\lambda_g = \frac{c/f}{\sqrt{1 - (f_c'/f)^2}}$$

The curves in Fig. 8 give good accuracy when $(a_1 - a_2)/2 > b_1$. For double-ridged waveguide the curves in Fig. 8 for cutoff frequency also apply if the definitions for b_1 and b_2 as given in Fig. 7 are used. The characteristic impedance for a double-ridged waveguide may be found by doubling the value of $Z_{0\infty}$ given in Fig. 8 for single-ridged waveguide. Homogeneously dielectric-loaded ridged waveguides may be designed by replacing the free-space wavelength λ_0 by the wavelength in an infinite dielectric medium and using the same design curves.

Ridged-waveguide applicators have been used by several investigators to produce hyperthermia at frequencies as low as 200 MHz. They have the advantage of being more compact and having lower impedance than ordinary waveguides. The lower impedance makes it easier to match into biological tissue, which because of its relatively high dielectric constant has a much lower impedance than free space. Ridged waveguides have the disadvantage of concentrating the electric field in the space above the ridge, which sometimes leads to undesirable hot spots, especially when b_1 is very small.

Sterzer and colleagues [9] developed a large ridged-waveguide applicator that operates at 27.12 MHz. It is loaded with deionized water to further reduce its size, but it is still so large that the patient must sit or lie on the aperture. Because it operates at a relatively low frequency it produces good deep regional heating.

Annular Phased Array Applicator [10]

The annular phased array is an applicator which is specifically designed for hyperthermia applications. It is a large, octagonal, multidirectional applicator with a central opening of 51 cm in diameter. It is designed to treat large body areas, such as the thorax, abdomen, pelvis, and thighs. The basic radiating element in this annular phased array can be any one of many kinds of local applicators, such as those discussed previously. Fig. 9 illustrates the locations of the radiating apertures along the inside surface of the octagonal opening. Fig. 10 shows that the annular phased array applicator actually consists of sixteen applicators arranged in annular form around the patient's body.

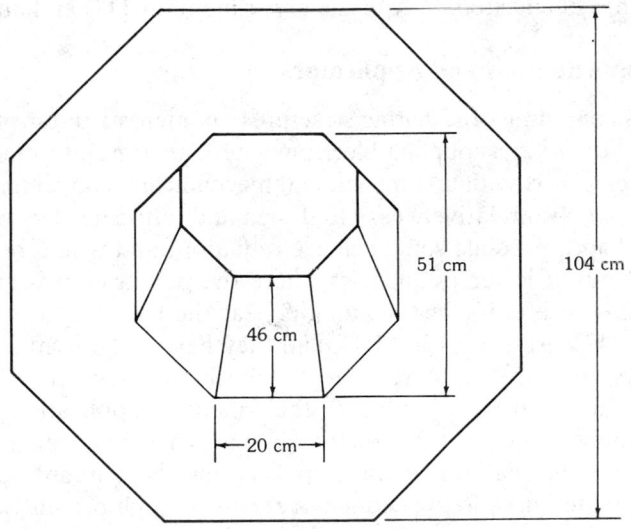

Fig. 9. Annular phased array applicator. (*After Turner [10]*)

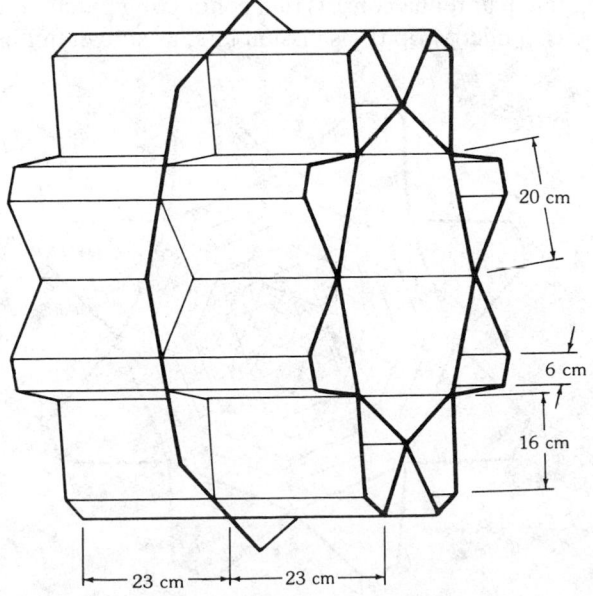

Fig. 10. Annular phased array: view of sixteen apertures. (*After Turner [10]*)

The resultant wavefronts are shown in Fig. 11. The frequencies of operation are usually chosen to be in the range between 50 and 110 MHz. All eight wavefronts operate synchronously in phase with equal amplitudes so that a region of constructive interference occurs at the axis of the array. The array is powered by four coaxial inputs, each activating a quadrant of the array. The apertures are also lined with water-filled pvc plastic bags which are used to direct the energy efficiently into the region of the body to be heated. The heating pattern of the annular phased array has been extensively tested experimentally [11] and numerically [12].

3. Microstrip Antennas and Applicators

Antennas consisting of radiating structures on microstrip transmission line are attractive applicators for coupling electromagnetic energy into biological systems because they can be very flat, compact, and mechanically convenient. These kinds of applicators are also relatively easy to design and fabricate. However, they tend to be narrow-band, produce some leakage radiation, and would be physically too large to be useful at lower frequencies where the penetration is best. They have been found to be useful for heating tumors near the body surface, typically using frequencies of 433, 915, or 2450 MHz, and they have been used for diagnostics.

Since microstrip antennas are treated extensively elsewhere in this book, the discussion in this chapter is limited to their medical applications. The primary difference between designing microstrip antennas for medical applications and other applications is that for medical applications the antenna must usually be designed to radiate into a lossy medium of relatively high permittivity, whereas in other applications the antenna often is designed to radiate into free space.

Microstrip Patch Antennas

A microstrip patch antenna consists of a conducting patch on the side opposite the ground plane of a microstrip transmission line, as shown in Fig. 12a. The patch

Fig. 11. Annular phased array radiated wavefronts. (*After Turner [10]*)

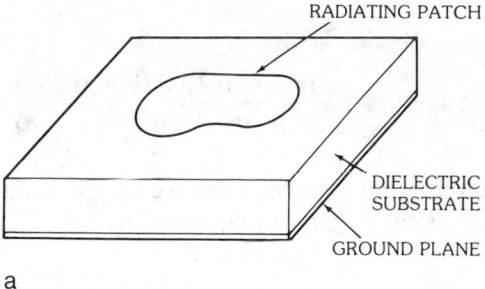

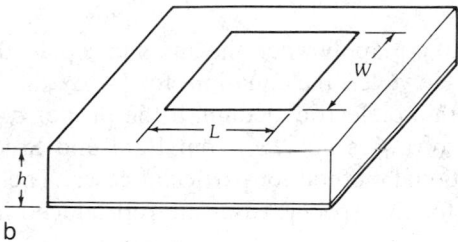

Fig. 12. Microstrip patch antennas. (*a*) Geometry of an arbitrarily shaped radiating patch antenna on microstrip transmission line. (*b*) Rectangular patch antenna geometry and parameters as defined by Bahl and others [13].

can be of arbitrary shape, but since analysis of antenna characteristics is easier for regular shapes, the rectangular, circular, loop, triangular, and elliptical shapes have been used most often. Only the rectangular patch (Fig. 12b) and the loop have been used much in medical applications. The rectangular patch radiator is discussed in this section and the loop radiator is discussed in the next section.

Bahl and his colleagues have developed design procedures for microstrip patch antennas for producing hyperthermia. Their equations for patch antennas radiating into free space are given first, and then the modifications to these equations for radiation into biological tissue are described.

The design procedure outlined by Bahl and Bhartia [13] for microstrip patch antennas to radiate into free space starts with choosing a practical value for the patch width W, which is given by

$$W = \frac{c}{2f_r}\left(\frac{\epsilon_r + 1}{2}\right)^{-1/2} \qquad (2)$$

where f_r is the resonant frequency, c is the velocity of light, and ϵ_r is the relative permittivity of the substrate. For smaller values of W than given by (2), the radiation efficiency is lower, and for larger values, higher-order modes begin to appear. The effective dielectric constant ϵ_{eff} is given by

$$\epsilon_{\text{eff}} = \frac{\epsilon_r + 1}{2} + \left(\frac{\epsilon_r - 1}{2}\right)\left(1 + \frac{12h}{W}\right)^{-1/2} \qquad (3)$$

and $\Delta\ell$ is given by

$$\Delta\ell = 0.412\,h\,\frac{(\epsilon_{\text{eff}} + 0.3)(W/h + 0.264)}{(\epsilon_{\text{eff}} - 0.258)(W/h + 0.8)} \tag{4}$$

where $2\Delta\ell$ is the difference between the effective length and the physical length L of the radiator. The physical length can be found from

$$L = \frac{c}{2f_r\sqrt{\epsilon_{\text{eff}}}} - 2\Delta\ell \tag{5}$$

These equations do not apply when the antenna is placed near to or in contact with biological tissue. However, the equations for W, $\Delta\ell$, and L can be used for an antenna in contact with a dielectric medium if the proper ϵ_{eff} is obtained for that case. The calculation of ϵ_{eff} is not easy, but Bahl and Stuchly [14] give results obtained from a variational method for particular cases. The geometry is shown in Fig. 13. Their results for two specific cases are reproduced in Fig. 14.

Microstrip Loop Radiators

Bahl and colleagues [15] designed and tested microstrip transmission-line loop radiators in the configuration shown in Fig. 15. The ring conductor, which is the radiating element, is fed on one side by a coaxial line protruding through the substrate. A pin shorting the ring to the ground plane through the substrate is located 180° from the feed point. In practice, it was found that best results for heating biological tissue were obtained when a water bolus was placed between the loop radiator and the biological body. Although the relationship beween the size of the ring and the resonant frequency is rather complex, Bahl and colleagues [15] developed a simple empirical formula that applies when the ring is placed on a thick, lossy, dielectric medium. For this case the mean radius R of the ring is given in centimeters by

$$R \cong \frac{6}{f\sqrt{|\epsilon_r|}}$$

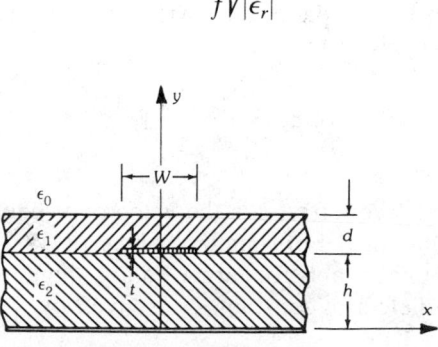

Fig. 13. Geometry of a microstrip covered with dielectric as used by Bahl and Stuchly [14] to calculate the effective dielectric constant. (*After Bahl and Stuchly [14], © 1980 IEEE*)

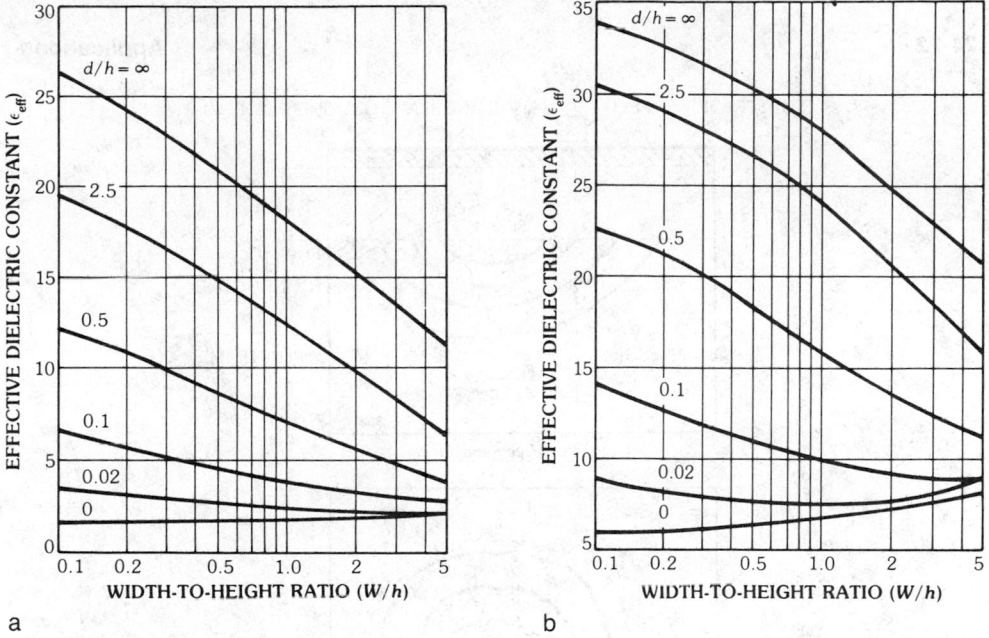

Fig. 14. Effective dielectric constant as a function of W/h for $\epsilon_{r1} = 75.0$ and various values of d/h. (a) For $\epsilon_{r2} = 2.32$. (b) For $\epsilon_{r2} = 10.0$. (*After Bahl and Stuchly [14], © 1980 IEEE*)

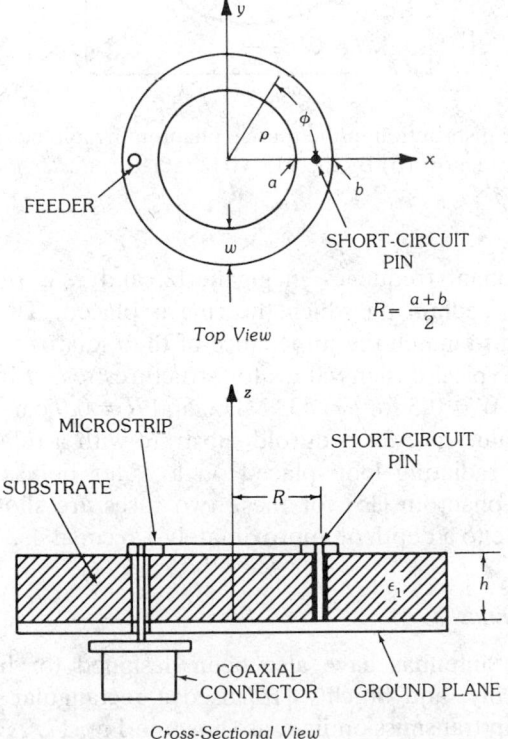

Fig. 15. The microstrip loop radiator designed by Bahl and others. (*After Bahl et al. [15], © 1982 IEEE*)

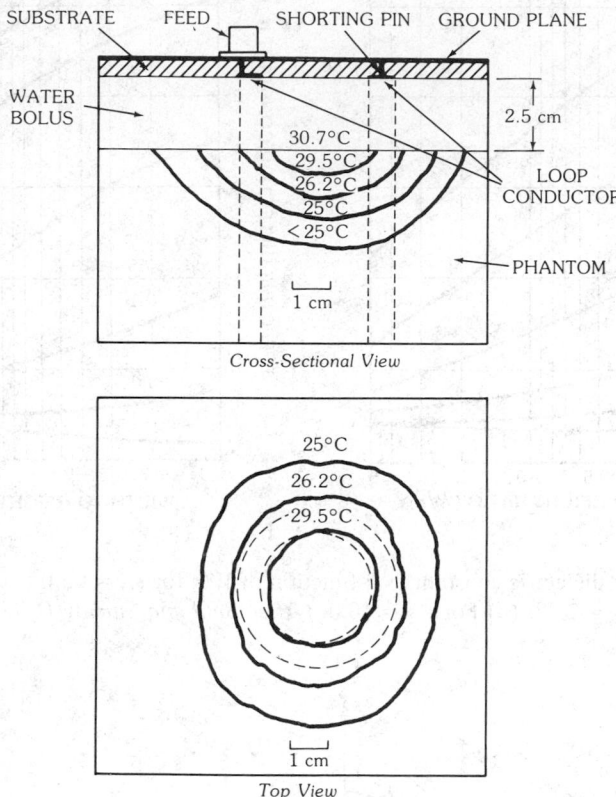

Fig. 16. Temperature distribution in the muscle phantom irradiated through a water bolus. (*a*) By the 433-MHz radiator. (*b*) By the 915-MHz radiator. (*After Bahl et al. [15], © 1982 IEEE*)

where f is the resonant frequency in gigahertz, and ϵ_r is the complex relative permittivity of the medium on which the ring is placed. The width of the ring conductor is chosen to match the impedance of the radiator to that of the coaxial transmission line. Typical design values for structures tested by Bahl's group [15] are $R = 1.6$ cm and $W = 0.5$ for $f = 433$ MHz, and $R = 0.9$ cm, $W = 0.3$ cm for $f = 915$ MHz. These values are for a duroid substrate with a relative permittivity of 2.32 and with the radiating loop placed on a water bolus. Measured heating patterns in muscle phantom [15] for these two cases are shown in Fig. 16. The heating was effective to a depth of approximately 1.6 cm at 433 MHz and 1.1 cm at 915 MHz.

Microstrip Slot Antennas

Microstrip slot antennas have also been designed for hyperthermia applications. Bahl, Stuchly, and Stuchly [16] used a rectangular slot in the ground plane of a microstrip transmission line, as shown in Fig. 17. A metal reflector was used on the opposite side of the ground plane and on the sides of the substrate to restrict the radiation to the half-space above the slot. The distance d was adjusted

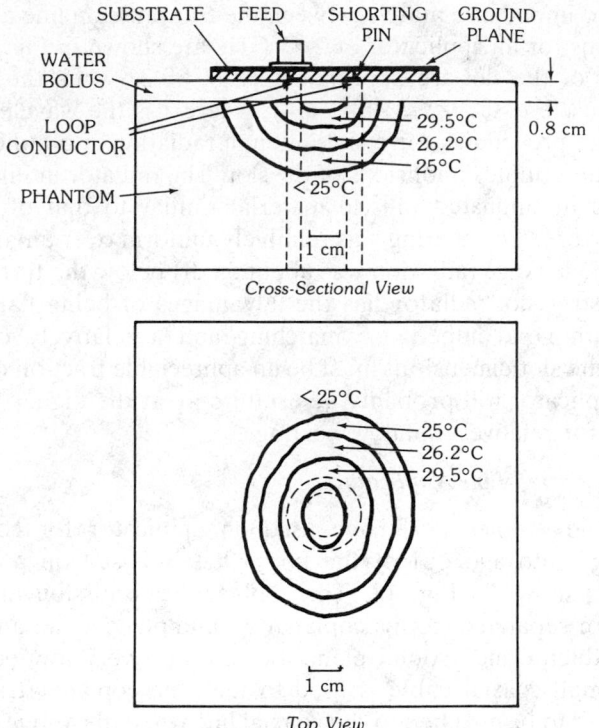

b

Fig. 16, *continued.*

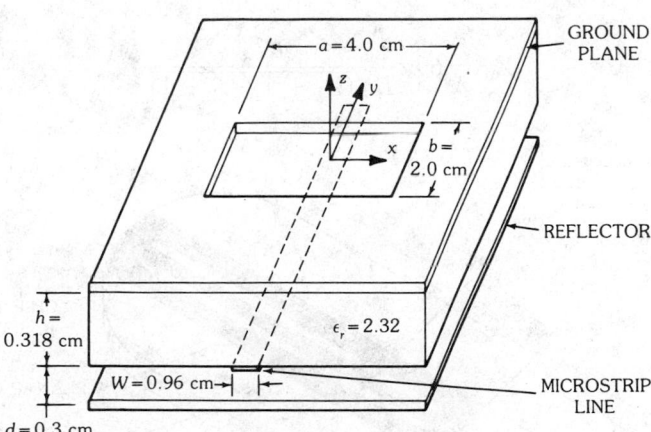

Fig. 17. Geometry of microstrip slot radiator. (*After Bahl, Stuchly, and Stuchly [16],* © *1980 IEEE*)

to obtain a good impedance match between the transmission line and the antenna. Their dimensions for an applicator at 2.45 GHz are shown in Fig. 17. In this case the dimensions of the slot are approximately 1λ by 2λ when the applicator is in contact with muscle tissue ($\epsilon_r \cong 48 - j14$), where λ is the wavelength in muscle. The usual design procedures for a slot antenna radiating into a dielectric medium can be used to determine dimensions of the slot. The radiator in Fig. 17 produced a heating pattern in simulated muscle material similar to that of an open-ended waveguide aperture. The heating was relatively uniform over an area of about 1.5 cm by 1 cm. The leakage radiation was about 35 dB below the transmitted power.

The microstrip slot radiator has the advantages of being flat, compact, and lightweight, with good impedance matching and a relatively uniform heating pattern. Since the slot dimensions must be an appreciable fraction of a wavelength, this kind of applicator will probably be useful only at the higher frequencies and therefore only for relatively shallow heating.

Coplanar Transmission-Line Applicator

Iskander and Durney [17] have used an applicator for coupling electromagnetic energy into and out of the body that is based on a coplanar transmission line, as shown in Fig. 18. The coplanar transmission line consists of a center conductor separated from a coplanar ground plane by an air gap, with both the center conductor and ground plane located on a very low permittivity, thin substrate. A small coaxial cable is used to feed the coplanar transmission line, which is designed to be matched to the coaxial line when the applicator is placed on biological tissue. A resistor equal to the characteristic impedance of the coplanar transmission line when it is on tissue is located at the end of the coplanar line to terminate it.

The design equation for the coplanar transmission line is [18]

$$Z_0 = 60\pi \frac{K'}{K\sqrt{(\epsilon_{r1} + 1)(\epsilon_{r2} + 1)}}$$

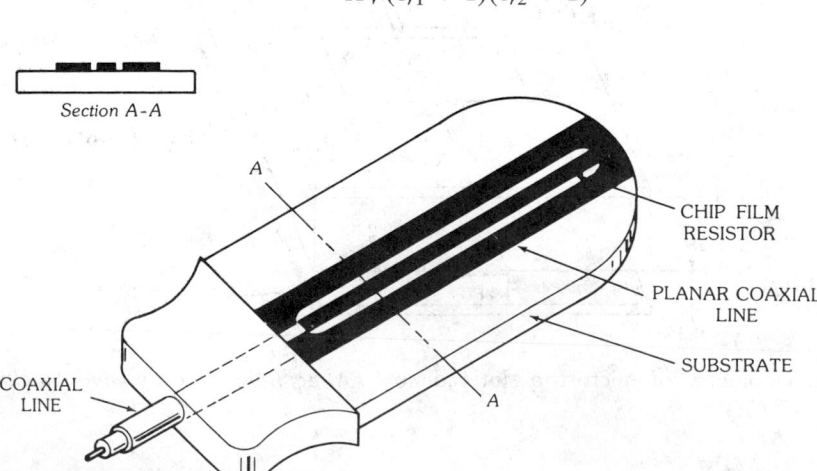

Fig. 18. Geometry of the coplanar transmission-line applicator. (*After Iskander and Durney [17], © 1979 IEEE*)

where

ϵ_{r1} = relative permittivity of region 1

ϵ_{r2} = relative permittivity of region 2

$k = OA/OC$

$k' = \sqrt{1 - k^2}$

K = complete elliptic integral of k

K' = complete elliptic integral of k'

and the dimensions are as given in Fig. 19. To design the transmission line, a convenient width OA is first chosen. Then the gap between the strips is calculated from the design equation. To obtain a good match between the coaxial line and the coplanar line, a taper in the gaps is used, as indicated in Fig. 17, and the transition is adjusted by trial and error using a time-domain reflectometer until the desired match is obtained. In practice, a good match can be obtained without much difficulty.

At 915 MHz this applicator has been found to provide good coupling into the human body, with low leakage radiation. It has the advantages of being light-weight, compact, and easy to tape on the skin. Although it has not been tested as a hyperthermia applicator it would not be expected to provide a very desirable heating pattern because of the nature of the field pattern.

Arrays of Microstrip Antennas

The compactness of microstrip antennas suggests their use in arrays for biological applications. Such arrays can be readily designed using the techniques described elsewhere in this book, but arrays have not been widely used in biological applications because they are practical only at the higher frequencies where penetration of electromagnetic energy into biological tissue is shallow. Sterzer and colleagues [9] have used an array of printed-circuit dipole arrays for heating superficial tumors at 2450 MHz. Other applications, such as receiving antennas for radiometry in biological applications, will probably occur.

4. Implantable Antennas (Radiators) for Localized Cancer Treatment

Heating deep-seated tumors noninvasively using external electromagnetic radiation is extremely difficult because, as explained in the introduction, both the

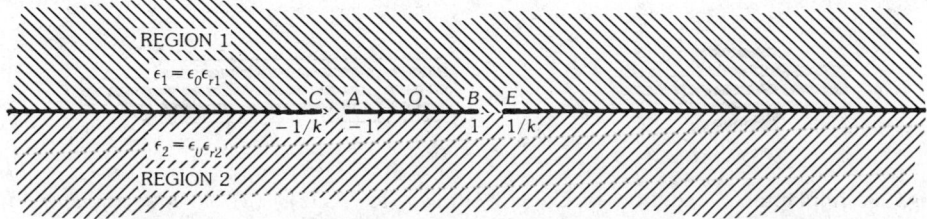

Fig. 19. Dimensions of the coplanar transmission line for use in the design equation. (*After Iskander and Hamid [18], International Microwave Power Institute, Clifton, Virginia*)

shallow penetration of the fields at higher frequencies and the near-field dominance of the radiation from electrically small applicators at lower frequencies tend to overheat the surface before the tumor is heated sufficiently. A large hyperthermia machine, such as the annular phased array system described in Section 2, produces deep regional heating but tends to be expensive and is limited to specialized centers. For these reasons the use of implantable antennas [19] that would produce localized deep heating for cancer treatment without overheating the surface is certainly attractive in situations where the implantation of the antennas is practical.

Several designs of these implantable radiators are available. Fig. 20 shows some antennas suitable for heating tumors in solid organs or in a hollow organ that can be reached through a body orifice [19]. Specifically, Figs. 20a and 20b describe antenna geometries that have been used to heat tumors implanted in solid organs. The radiator in Fig. 20a consists simply of the center conductor extended an appropriate distance from the end of a coaxial cable. The coaxial line may be made of the subminiature semirigid coax that is available in a variety of materials and sizes down to a diameter of 0.2 mm. The extended length of the center conductor may be adjusted to match the impedance of the implanted radiator to the 50-Ω impedance of the feed cable. To some extent the size of the heated area can also be controlled by adjusting the length of the center conductor. The purpose of the dielectric sheathing is to extend the propagation distance along the needle. The radiator shown in Fig. 20b has also been used for invasively heating deep tissues. Fig. 21 shows an alternative structure consisting of a hollow stainless-steel hypodermic needle that serves as a monopole antenna and permits simultaneous chemotherapeutic treatment of tissue [20]. The chemotherapy is accomplished by introducing the chemicals through an inlet port located at the end of the needle. By proper choice of the length-to-radius and the length-to-operating-wavelength ratios of the antenna, both satisfactory radiation patterns and impedance matching can be obtained. In this structure, however, the outer conductor, which is flanged to serve as a ground plane flush to the skin, limits the penetration of the device.

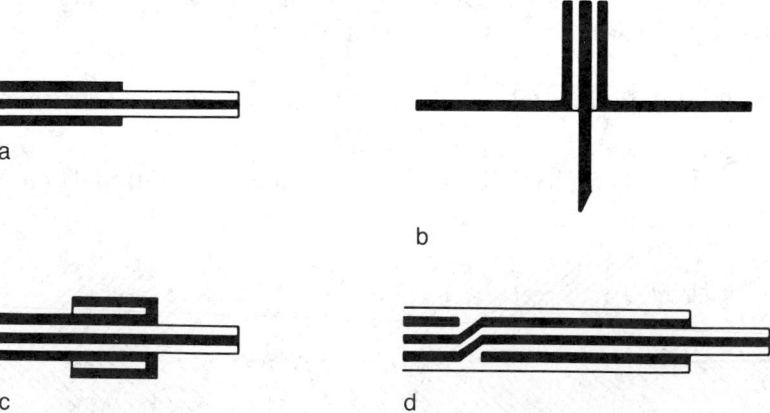

Fig. 20. Implantable radiator types (not drawn to scale). (*a*) Needle radiator. (*b*) Hypodermic monopole and ground-plane flange. (*c*) Sleeve antenna. (*d*) Cross-switch section plus needle. (*After Taylor [19]*, © *1980 IEEE*)

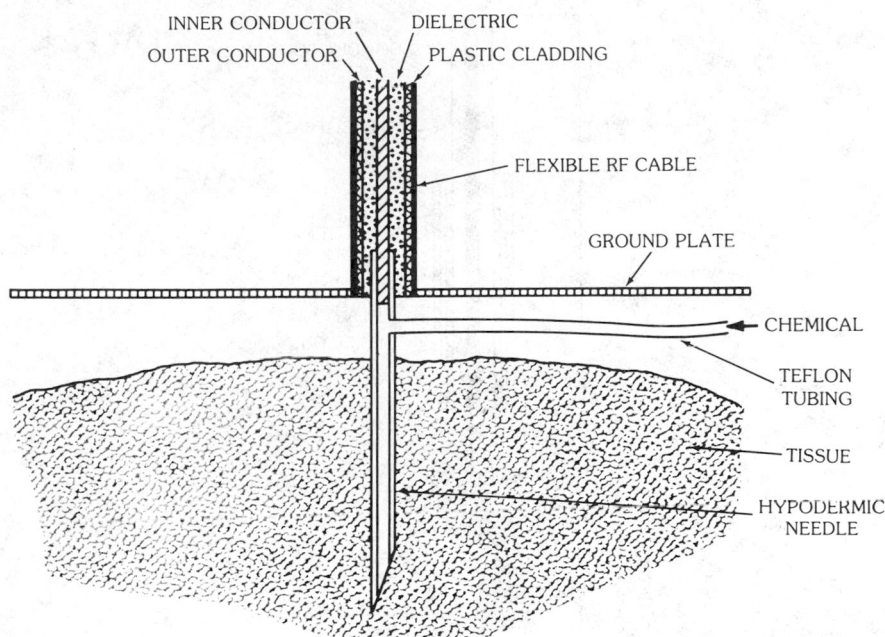

INNER CONDUCTOR
OUTER CONDUCTOR
DIELECTRIC
PLASTIC CLADDING
FLEXIBLE RF CABLE
GROUND PLATE
CHEMICAL
TEFLON TUBING
TISSUE
HYPODERMIC NEEDLE

Fig. 21. Hypodermic needle monopole radiator. (*After Bigu-del-Blanco and Romero-Sierra [20],* © *1977 Pergamon Journals Ltd.*)

In applications where the tumor is in a hollow organ that can be reached through a body orifice, a coaxial radiator located at the center of a body cavity is usually used to heat the organ wall. Fig. 20c illustrates such a radiator, in which the needle radiator configuration is replaced by a coaxial sleeve radiator to terminate transmission back to the generator. The dipole antenna in this case is usually formed of a quarter-wavelength section of the coaxial center conductor with the outer conductor removed and a quarter-wavelength choke (sleeve) placed to minimize the leakage radiation along the outer conductor (see Fig. 22a). Either a metallic semicylindrical reflector (see Fig. 22b) or a polytetrafluoroethylene (ptfe) bulb covering the dipole antenna may be used to direct the energy [21]. Fig. 23 shows a cutaway view of the bulb. This specific bulb was made of ptfe because it has a low dielectric constant. Its cross section is pear shaped to produce the uneven radiation sometimes needed for directional heating [22]. For example, the eccentric position of the antenna within the bulb (see Fig. 23b) places a sector of about 90° of the external surface of the bulb's "active surface" closest to the antenna. In a typical operation the applicator is inserted into a body cavity (such as the rectum) with its active surface facing the site to be treated.

In some clinical situations the geometry of the tumor requires the radiation pattern of the antenna to be cylindrically symmetric and to extend for a length up to 10 cm. A needle radiator (Fig. 20a) is not expected to produce a heating pattern this long, even if a dielectric sheathing is used to extend the propagation distance along the needle. To get a longer pattern it is suggested [19] that the inner and outer conductors of the cable be cross-connected at a point a half-wavelength (in the dielectric) away from the needle radiator (see Fig. 20d). The presence of

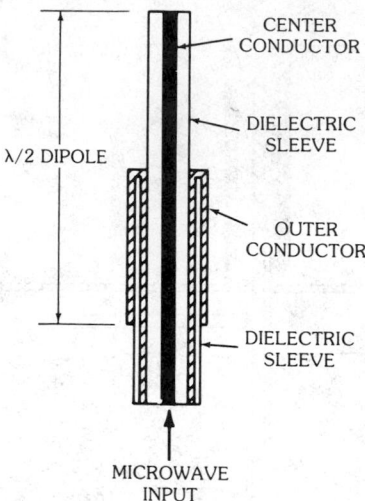

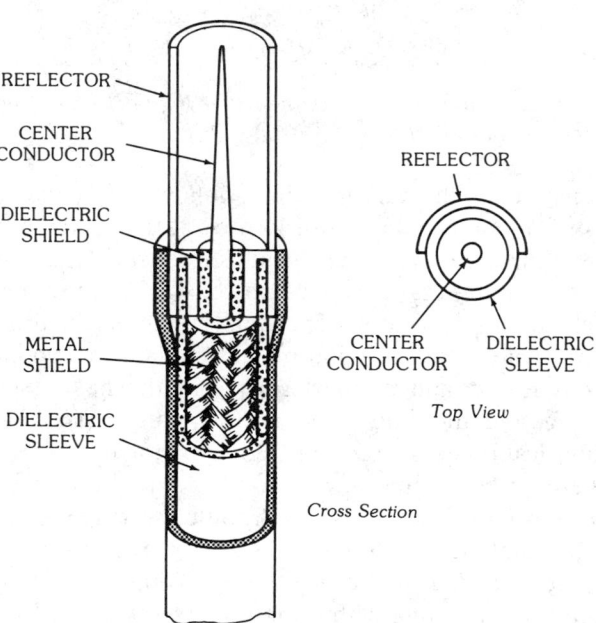

Fig. 22. Dipole antenna and applicator with reflector. (*a*) Longitudinal section of dipole antenna. (*b*) Coaxial applicator with directive reflector. (*After Mendecki et al. [21],* © *1980 Pergamon Journals Ltd.*)

this excited gap extends the current longitudinally on the outer conductor, thus resulting in a longitudinal energy deposition pattern. The geometry of this radiator together with its radiation pattern is shown in Fig. 24. From Fig. 24 it is clear that the radiation pattern displaying two principal peaks with an interference maximum in between is very roughly that of a collinear array of two half-wave dipoles [19].

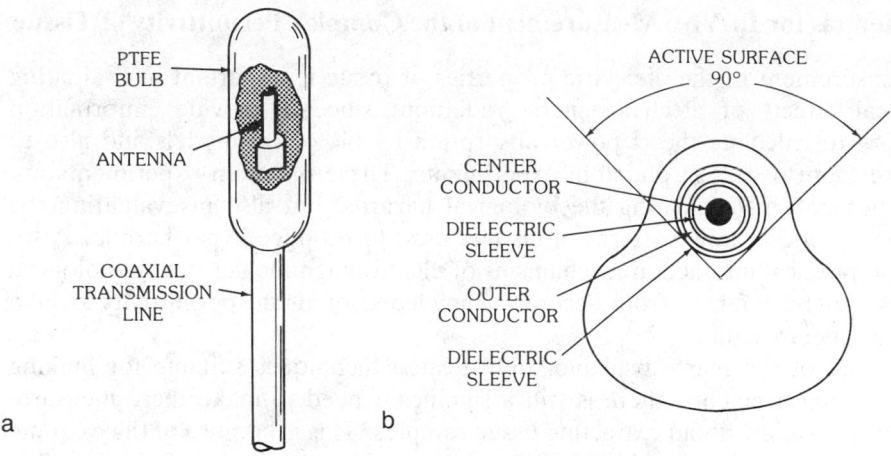

Fig. 23. The ptfe bulb. (*a*) Cutaway view showing the position of the antenna within the bulb. (*b*) Cross section showing the pear-shaped bulb and the eccentric location of the antenna. (*After Mendecki et al. [22], © 1978 Pergamon Journals Ltd.*)

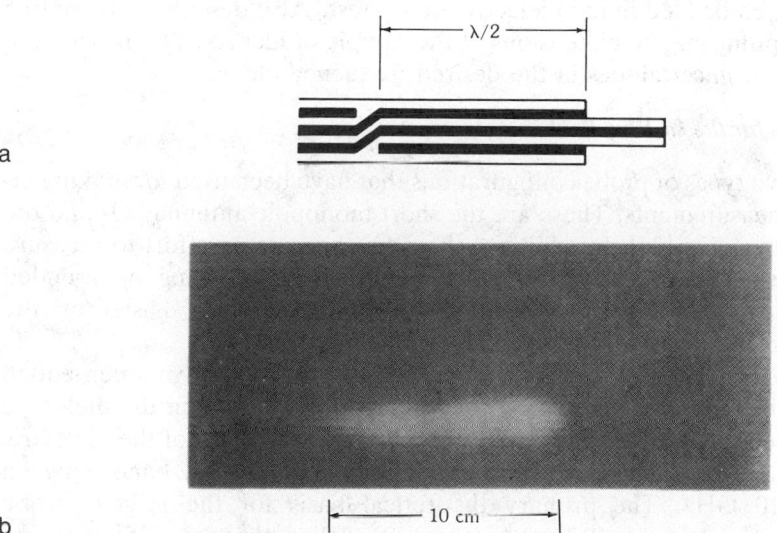

Fig. 24. A cross-switch radiator. (*a*) Diagram of radiator. (*b*) Thermogram of its energy deposition pattern. (*After Taylor [19], © 1980 IEEE*)

The radial pattern is symmetric, which might be required to fit the geometry of the tumor.

Implanted radiators provide the significant advantages of highly localized and highly controlled heating. However, they can be used only in certain cases where the implantation is practical. Detailed results that illustrate the usefulness of these radiators and their ability to produce controllable localized heating of malignant internal structures are described elsewhere.

5. Antennas for In-Vivo Measurement of the Complex Permittivity of Tissue

Measurement of the dielectric properties of tissue is important in evaluating biological effects of electromagnetic radiation, since it provides information necessary to calculate the rf power absorption by biological models and also to simulate these tissues in phantom experiments. These phantom experiments are useful not only in evaluating the biological hazards, but also in evaluating the electromagnetic heating patterns of devices used to produce hyperthermia. Also, many biophysical interaction mechanisms of electromagnetic fields with biological systems can be inferred from accurate knowledge of tissue permittivity over a broad frequency band.

In spite of the many available sophisticated techniques suitable for making dielectric measurements, there is still a significant need to make these measurements *in-vivo* and without extracting tissue samples. This is because of the extreme difficulty in preparing suitable tissue samples and also because of the possible deterioration of the electrical properties of tissue with time in the *in-vivo* measurements. This section describes two of the recently developed procedures for measuring the dielectric properties of tissue. These procedures basically utilize open-ended sections of transmission lines with the extended portion of the center conductor being embedded in the dielectric under test. Also described is a simple procedure for optimizing the dimensions of the sample holder so as to provide the least measurement uncertainties in the desired frequency range.

Theoretical Basis for the In-Vivo Probes

There are two types of probe configurations that have been used for making *in-vivo* dielectric measurements. These are the short monopole antenna [23] and the open-ended coaxial line [24]. In addition, there has been some effort to measure dielectric properties at *X*-band and millimeter wavelengths using open-ended waveguides. The following is a summary of the theoretical basis for the measurement procedure using coaxial sample holders.

The Short Monopole Antenna as a Dielectric Probe [23]—An open-ended coaxial line probe with an extended center conductor embedded in the dielectric under test has been successfully used for the *in-vivo* measurement of the dielectric properties of tissue. Results have been obtained in the frequency band between 10 MHz and 10 GHz. The primary theoretical basis for the *in-vivo* probe measurement concept is found in an antenna modeling theorem [25] that can be made applicable to a short (antenna length much less than 0.1 wavelength) monopole antenna. This theorem simply relates the impedance of a short antenna operating at a frequency ω, and radiating in the material under test, to the impedance value at a frequency $n\omega$ when surrounded by free space. When non-magnetic materials are considered, this theorem has the mathematical form

$$\frac{Z(\omega, \epsilon)}{\eta} = \frac{Z(n\omega, \epsilon_0)}{\eta_0} \tag{6}$$

where

Z = impedance

ϵ = complex permittivity of the material being measured

ϵ_0 = permittivity of free space

$\eta = \sqrt{\mu_0/\epsilon}$ = intrinsic impedance of the material being measured

$\eta_0 = \sqrt{\mu_0/\epsilon_0}$ = intrinsic impedance of free space

$n = \sqrt{\epsilon/\epsilon_0}$ = index of refraction of the material being measured relative to free space

In order for the theorem to be valid, the following conditions must be satisfied:

1. The short antenna or probe must be completely immersed in the material to be measured
2. The material sample size must be larger than the penetration depth ($1/e$) of the field pattern within the material (otherwise, the reaction theorem must be solved for the difference in impedance due to measuring a sample of finite volume)
3. An analytical expression based on the current distribution along the probe must be obtained for the impedance of the probe

When a short monopole antenna is used as the probe the necessary analytical expression for probe impedance has the form

$$Z(n\omega, \epsilon_0) = A\omega^2 + \frac{1}{j\omega C} \tag{7}$$

where A and C are constants determined by the probe's dimensions. This expression is valid so long as the proble length is less than 10 percent of the wavelength in the material being measured.

With a knowledge of the antenna constants A and C as well as the magnitude and phase of the complex impedance $Z(\omega, \epsilon)$ of the short monopole antenna in the material being measured, the values of relative dielectric constant and conductivity can be obtained from (6). In order to characterize the electrical properties of the material in terms of the relative dielectric constant ϵ' and conductivity σ, the complex index of refraction is defined in terms of the loss tangent of the material as

$$n = \sqrt{\epsilon/\epsilon_0} = \sqrt{\epsilon'(1 - j\tan\delta)} \tag{8}$$

where $\tan\delta = \sigma/\omega\epsilon'$. Using the antenna impedance in (7) in the modeling theorem of (6), the following expressions [23] are obtained for the resistance and reactance of the complex impedance $Z(\omega, \epsilon) = R + jX$:

$$R = \frac{\sin 2\delta}{2\epsilon'\omega C} + A\sqrt{\epsilon'}\,\omega^2 \sqrt{\frac{\sec\delta + 1}{2}} \tag{9}$$

and

$$X = \frac{\cos^2\delta}{\epsilon'\omega C} + A\sqrt{\epsilon'}\,\omega^2\sqrt{\frac{\sec\delta - 1}{2}} \tag{10}$$

All the parameters in the above pair of equations except ϵ' and δ are known or can be determined from experimental measurements. Because simultaneous solution of these equations would be quite difficult, an iterative method of solution is usually utilized. The second term in both (9) and (10) is small at low frequencies. When these terms are neglected the following equations result:

$$R = \frac{\sin 2\delta}{2\epsilon'\omega C} \tag{11}$$

and

$$X = \frac{\cos^2\delta}{\epsilon'\omega C} \tag{12}$$

Solutions to these equations are easily obtained by noting that $\tan\delta = R/X$; therefore, by measuring the input impedance of a short monopole antenna inserted into a material, it is possible to calculate both the relative dielectric constant ϵ' and the conductivity σ.

The Open-Ended Coaxial Line as a Dielectric Probe—The other probe that has been used in the *in-vivo* measurements of dielectric properties is in essence a special case of the monopole antenna described previously [24]. An open coaxial line, placed in contact with a test sample, is used as a sensor. The equivalent circuit of the sensor consists of two elements (Fig. 25), a lossy capacitor $C(\epsilon_r)$ and a capacitor C_f that accounts for the fringing field in the Teflon. Here $C(\epsilon_r) = C_0\epsilon_r$, where C_0 is the capacitance when the line is in the air and ϵ_r is the complex relative permittivity. This equivalent circuit is valid only at frequencies for which the dimensions of the line are small compared to a wavelength so that the open end of the line does not radiate. At higher frequencies, increased evanescent TM modes excited at the junction discontinuity cause C_0 to increase with frequency [24]. When the evanescent modes are taken into account, C_0 should be replaced by $C_0 + Af^2$ (see the previous subsection), where A is a constant dependent on the line dimensions. Furthermore, the evanescent modes may become propagating modes when a high dielectric constant material is placed at the interface. For example, the TM_{01} mode propagates when $\lambda_\epsilon < 2.03\,(a - b)$, where λ_ϵ is the wavelength in the external dielectric and a, b are the line dimensions shown in Fig. 25a [24]. To account for the radiation produced by the propagating modes it may be necessary to add a parallel conductivity term to the equivalent circuit of Fig. 25b. When the frequency is low enough that the equivalent circuit of Fig. 25b is valid, the input reflection coefficient at the plane of the discontinuity can be calculated from

$$\hat{\Gamma} = \Gamma e^{j\phi} = \frac{1 - j\omega Z_0[C(\epsilon_r) + C_f]}{1 + j\omega Z_0[C(\epsilon_r) + C_f]} \tag{13}$$

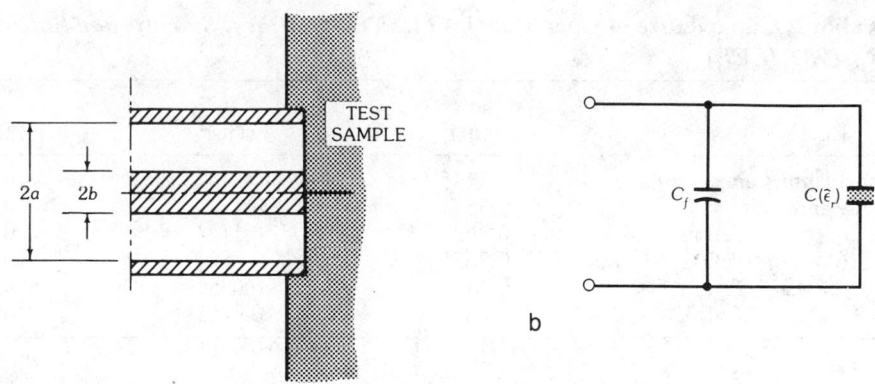

Fig. 25. Open-ended coaxial line sensor. (*a*) Coaxial line terminated with sample. (*b*) Equivalent circuit. (*After Athey, Stuchly, and Stuchly [24],* © *1982 IEEE*)

Solving for the permittivity gives

$$\epsilon_r = \frac{1 - \hat{\Gamma}}{j\omega Z_0 C_0 (1 + \hat{\Gamma})} - \frac{C_f}{C_0} \tag{14}$$

It has been shown [26] that for a given accuracy in measurement of the reflection coefficient, the greatest accuracy in determining the permittivity is obtained when

$$C_0 = \frac{1}{\omega Z_0 \sqrt{\epsilon'^2 + \epsilon''^2}} \tag{15}$$

where $\epsilon_r = \epsilon' - j\epsilon''$. Strictly speaking, this is valid only when the uncertainties in the magnitude and phase of the reflection coefficient are approximately the same, i.e., when $\Delta\phi \cong \Delta\Gamma/\Gamma$. In other cases this is still a good compromise [26].

Table 1 lists approximate values of the optimum capacitance calculated from (15) for several biological substances and reference materials. Table 2 lists the total capacitance C_T for a few commercially available coaxial lines [24].

Table 1. Optimum Capacitance (in Picofarads) for a Probe in Various Materials (*After Athey, Stuchly, and Stuchly [24],* © *1982 IEEE*)

Material	Frequency (GHz)				
	0.01	0.05	0.1	0.05	1.0
Muscle	0.25	0.25	0.25	0.08	0.035
Fat	0.8	0.8	0.8	0.45	0.45
Water, 25°C	4	0.8	0.4	0.08	0.04
0.02 N NaCL, 25°C	0.37	0.37	0.37	0.08	0.04
0.08 N NaCL, 25°C	0.19	0.19	0.19	0.08	0.04
0.25 N NaCL, 25°C	0.08	0.07	0.07	0.055	0.035

Table 2. Capacitance of Open Coaxial Lines (*After Athey, Stuchly, and Stuchly [24],* © *1982 IEEE*)

Line	a (cm)	b (cm)	C_T (pF)
14 mm, air	0.7145	0.3102	0.14
7 mm, air	0.35	0.1520	0.079
8.3 mm, Teflon	0.362	0.1124	0.055
6.4 mm, Teflon	0.2655	0.0824	0.041
3.6 mm, Teflon	0.1499	0.0455	0.027
2.2 mm, Teflon	0.0838	0.0255	0.016

An alternative and more general procedure for determining the optimum value of capacitance for measuring the permittivity of a given dielectric in a specified frequency band is described in the next section.

Detailed Construction of the In-Vivo Probe

A diagram of an *in-vivo* measurement probe [27] is shown in Fig. 26. It can be seen that the probe is essentially a section of open-ended semirigid coaxial cable with a slightly extended center conductor. The small ground plane may or may not be included to minimize fringing effects. The following procedure can be used to construct the *in-vivo* measurement probe:

Step 1

Remove the Teflon dielectric and center conductor from a short piece (approximately 3 cm length) of semirigid coaxial cable.

Step 2

Silver-solder a small circular disk to one end of the tube formed by the now-empty outer conductor. Machine the surface of this disk smooth.

Step 3

Electroplate the resulting item and the center conductor that was previously removed. Use either nickel, gold, or platinum plating.

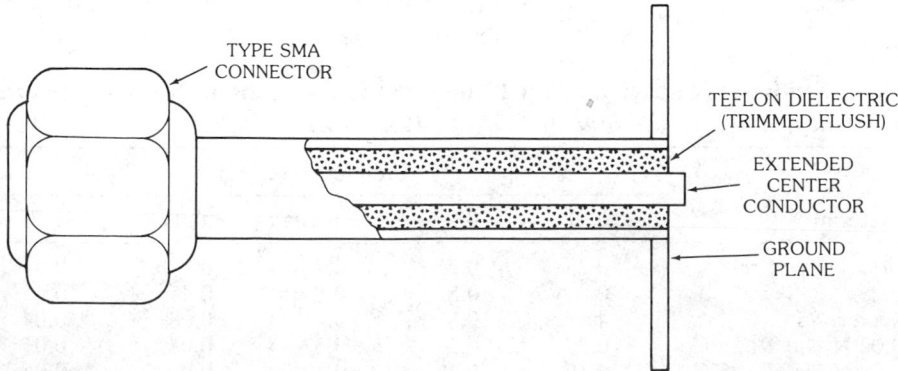

Fig. 26. Diagram of *in-vivo* measurement probe. (*After Toler and Seals [27]*)

Step 4

Solder a connector to the other end of the outer conductor tube.

Step 5

Replace the Teflon dielectric and the plated center conductor, leaving the center conductor slightly extended.

Step 6

Finally, trim any excess Teflon away from the extended center conductor.

Optimizing the Length of the Center Conductor in the Coaxial Sample Holder

In the *in-vivo* measurement procedure involving the open-ended coaxial line with or without the extended portion of the center conductor, the probe can be conveniently represented by a simple shunt capacitor terminating the transmission line at lower frequencies where the dimensions of the transmission line and, in particular, the extended portion of the center conductor are small compared to a wavelength [28]. In using this dielectric probe it is extremely important to determine the optimum value of the capacitance that can be used for measurements over a specified frequency band. There is, however, no easy procedure to determine such a value since the optimum capacitance depends on

(*a*)   the value of the complex permittivity under test, and
(*b*)   the specific frequency band of interest

One of the authors (Iskander) has devised a procedure to calculate such an optimum value of the capacitance that can be used for a given dielectric (i.e., approximate value of the complex permittivity is required) and in a given frequency range [29]. While a detailed description of the method can be obtained from the cited reference, the results generally indicated that larger values of capacitances are required at lower frequencies (which is expected since lumped-circuit elements are usually used at these frequencies), and smaller values of capacitances are preferable at higher frequencies, which is also expected because smaller capacitances in the form of distributed circuit elements are usually used at higher frequencies.

In *in-vivo* dielectric measurements using the open-ended section of the transmission line, smaller values of capacitance are achieved by decreasing the length of the extended portion of the center conductor. The open-ended coaxial transmission line sample holder, therefore, is suitable for making dielectric measurements at higher frequencies and may be used at lower frequencies (e.g., hundreds of megahertz) if dielectrics with high dielectric constants are used. For materials with lower dielectric constants, on the other hand, larger values of capacitances are required. This could be achieved by extending the length of the center conductor further to increase the value of the capacitance of the sample holder. Under this condition, however, the input impedance of the sample holder would not be approximated by a capacitance terminating the transmission line and instead the expression for the input impedance given in (9) and (10) should be used to take into account the radiation resistance of the antenna. These low-frequency extensions, particularly when dielectrics with low complex permittivities are being tested, are still in progress in our laboratories [28].

6. Antennas for Monitoring RF Radiation

In any situation that could involve exposure of people to significant amounts of electromagnetic radiation, monitoring the electromagnetic radiation levels is important. In some cases a device that measures radiation in a given area would be satisfactory. In other cases it would be desirable to have a device that could be worn by a person to monitor the radiation to which that person is exposed. Area radiation monitors are commercially available, but a satisfactory personnel dosimeter is not, probably because of the many difficulties encountered in the design and construction of such a device.

An important aspect of either kind of radiation monitor is what parameter of the electromagnetic fields should be measured. It is generally known that for a simple plane wave any one of a number of field parameters constitutes a valid index and the direction of propagation [30]. These parameters include the magnitudes of the electric field $\mathbf{E}$, of the magnetic field $\mathbf{H}$, and the time-average power density. Provided that the source does not show supergain, plane-wave approximations are appropriate in the practical sense when the distance from the source is large compared to $2D^2/\lambda$, where λ is the wavelength in the medium and D is the largest dimension of the source. Therefore, for quantifying this far-field plane-wave-type radiation, it is sufficient to measure the magnitude of either the electric or magnetic field. Other parameters of interest can easily be computed from these magnitudes without approximation for given properties of the propagation medium. In the near field, however, magnitudes of the electric and magnetic fields are not generally easily related; also, the other radiation parameters cannot simply be calculated from the magnitudes of these fields. For accurate quantification of the electromagnetic radiation in the near field, therefore, the instrumentation suitable for power density measurements (which is reasonably satisfactory in plane-wave radiation) is inadequate and instead independent measurement of the electric and magnetic fields should be performed.

The basic requirements for radiation monitors and some antennas to measure electric and magnetic fields are described in this section.

Basic Requirements for Field Probes

Desirable features in the design of an rf and microwave radiation monitor include the following:

1. The measuring probes should disturb the electromagnetic fields minimally.
2. The probe response should be isotropic and hence independent of the field polarization.
3. It should have a flat frequency response over a broad frequency band. This characteristic is important in making field measurements without precise knowledge of the frequency.
4. It should exhibit linear dynamic response with the field or power quantity being measured.
5. It should be suitable for operation in the near- or far-field regions. This simply requires independent measurement of the electric- and magnetic-field components in three orthogonal directions.

6. The field probes (**E**- and **H**-field sensors) as well as the connecting leads should preferably be made of high-resistance material (carbon-loaded Teflon) to minimize their interference with the fields being measured.

In addition to these basic requirements the probe should also have fast response and be small enough to make spatially resolved measurements in small regions comparable in size to organs of the body.

Detection Methods

In designing a suitable antenna system for detecting the **E** and **H** fields of rf radiation, it is necessary to include in the probe a means for detecting the induced currents. The type of detectors used can generally be segregated into two basic types:

1. Thermocouple type devices that measure the electromagnetic power by sensing the change in temperature between the hot and cold junctions of a thermocouple device. Examples of this thermocouple detecting device include the electromagnetic radiation monitor developed by the Narda Microwave Corporation [31]. In this device the thermocouple materials of the sensor are antimony and bismuth deposited on a substrate of plastic or mica. The sensitivity of the sensor is controlled by variation of the substrate and the dimensions of the hot junction of the thermocouple.
2. Diode detectors, which produce a current or voltage related to the fields of the electromagnetic radiation [32].

The usefulness of thermocouple detection devices is limited by their sensitivity to changes in ambient temperature and other changes in all forms of heat. These devices also have slow response time and may even burn out if overloaded. Their main advantage is linearity of response with intensity of electromagnetic power. Diode detectors are more sensitive than thermocouple detectors, but the diode response is nonlinear with electromagnetic power. For example, if the signal level is such that the diode is operating in the square-law region, the induced current will be proportional to the square of the **E**-field intensity.

E- and H-Field Probes

As indicated in a previous section, a measurement of power density is not sufficient to characterize electromagnetic radiation in the near-field region. The question that is often raised, however, is whether or not we need to measure both the electric- and magnetic-field intensities. Since these fields are indeed related to each other by Maxwell's equations, one can generally measure either one of the fields, say, the electric field, and calculate the other, if necessary, from it. Although the above statements are certainly correct, practical problems in relating these fields make it easier to measure them both. For instance, it is known that the ratio of **E** to **H** in the near field is not the same as the intrinsic wave impedance in free space and that **H** and **E** are not generally perpendicular to each other. In other words, no simple relation exists between these near **E** and **H** fields, and full utilization of Maxwell's equations is required.

Pursuing such an option of fully utilizing Maxwell's equations is not easy, simply because phase information of the measured electric or magnetic fields (in addition to the magnitude) must be obtained. Finally, it is not obvious which fields should be measured, especially at low frequencies where the **E** and **H** fields, according to Maxwell's equations, can be very loosely coupled, and either **E** or **H** can exist almost independently of the other. Therefore, in some applications **E**-field measurements could be easier, more accurate, and more reliable, while in others the **H**-field measurements could be more sensitive and reliable. As an example of the latter case, consider an electric power transformer. It is true that both electric and magnetic fields are present, but the magnetic-field intensity, in this case, is certainly larger than the electric-field intensity and it would be impractical to measure the small **E**-field intensity in the presence of significant errors due to noise and from it find the **H**-field intensity through Maxwell's equations. This is why analyses of transformer action are traditionally based on magnetic flux linkages instead of electric-field quantities. Therefore, it is more advantageous to measure the magnetic-field intensity in this case, while in some other applications it could be advantageous to measure the **E**-field intensity.

In conclusion, it is believed that in order to avoid the prejudgment of what should and should not be measured, an electromagnetic radiation monitoring device should measure both **E**- and **H**-field intensities at the lower frequencies. At higher frequencies (say, above a few hundred megahertz) it is safe to assume that both fields are strongly coupled and the measurement of only one (say, the **E** field since it is usually the easier one to measure) should suffice.

The design of **E**- and **H**-field probes used both for area monitors and for a personnel dosimeter is described below.

Electric-Field Probe

Basic Design and Characteristics—To measure the electric-field intensity small dipole antennas with detecting devices are usually used. The commonly used short electric dipoles are often fabricated from cylindrical wires, with their length limited to a small fraction of the wavelength ($\ell < 0.1\lambda$) to allow a reasonably broad frequency response. For the short electric dipole shown in Fig. 27, with the assumed triangular current distribution I, it can be shown that the open-circuit voltage at the terminals V_{oc} is given by

$$V_{oc} = E_{\parallel} \ell / 2$$

where $E_{\parallel}$ is the component of the electric field parallel to the dipole and ℓ is the dipole length. The output voltage of the antenna can be measured by placing a detector between the terminals of the antenna. As indicated earlier, this detector may be just a thermocouple element with the heat generated in the hot junction being produced by the received rf currents in the dipole circuit [31]. Diode detectors are other possible detecting devices which, when placed between the output terminals of the antenna, would produce a rectified output voltage that can be detected by a dc voltmeter [33]. As mentioned previously, diode detectors are more sensitive than thermocouple detectors, but diodes are nonlinear. Therefore, depending on the intensity of the incident rf signal, the output voltage of the diode

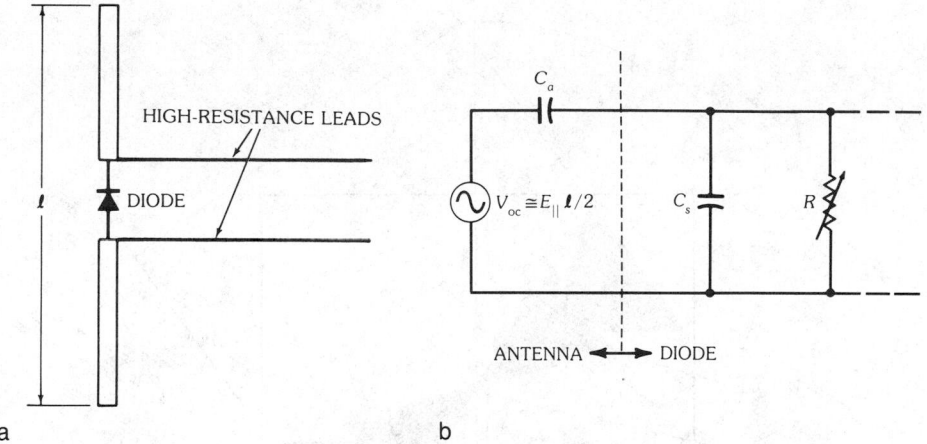

Fig. 27. Diagram of the **E**-field probe and its equivalent circuit. (*a*) The **E**-field probe. (*b*) Approximate equivalent circuit.

detectors may be in the square-law region, thus providing an output voltage that is related to the power density of the incident radiation, or in the linear region, thus providing an output voltage that is linearly proportional to the electric field associated with the incident radiation. From the equivalent circuit in Fig. 27b, it can be shown [34] that if the current in the diode can be assumed to be $I = I_{\text{sat}}(e^{\alpha V} - 1)$, the output voltage from the probe is given by

$$V \cong -(\alpha/2)[V_{\text{oc}}/(1 + C_s/C_a)]^2$$

for $V < 0.01$ V, and

$$V \cong -V_{\text{oc}}/(1 + C_s/C_a)$$

for $V > 1$ V.

The dipole antennas themselves may be of various kinds and shapes, including cylindrical rods, biconical antennas, triangular strips on printed-circuit board, or zigzag antennas. The zigzag dipole has a nearly flat frequency response and good sensitivity. Typical dimensions of a constructed zigzag dipole are given in Fig. 28, which also shows the high-resistance leads used to connect the output of the diode to a dc voltmeter with minimum perturbation of the rf radiation fields.

Isotropic Response—To obtain an isotropic response for the electric-field probe, three orthogonal antennas arranged as shown in Fig. 29 are usually used. A top view of the orthogonal arrangement illustrating the direction of each of the three dipoles is given in Fig. 29b. This specific angle of orientation shown in Fig. 29 coincides with the three orthogonal directions of the three orthogonal diagonals of a cube. Typically the three dipoles are arranged around a ¼-in- (6.35-mm)-diameter cylindrical tube as shown in Fig. 29b. Electrical connections can be made either to measure the sum of the responses of these antennas or to measure these responses individually.

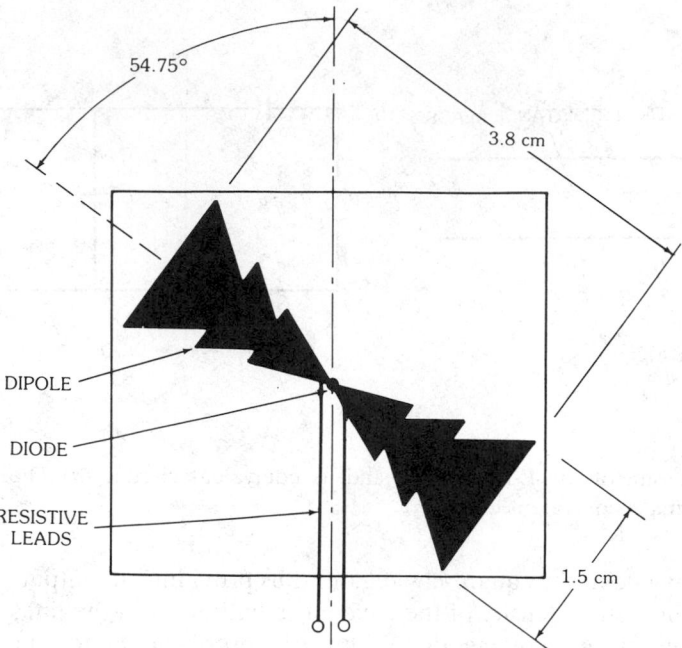

Fig. 28. Diagram illustrating the geometry of the printed-circuit zigzag antenna.

The performance of the isotropic probe can be evaluated experimentally in a Crawford TEM cell (see Section 8). By rotating the three orthogonal dipoles in the cell and measuring the total sum of the output voltages, it was found that the response is omnidirectional within a few percent. In another procedure one dipole is aligned along the direction of the electric field in the calibration (Crawford) cell and the responses of the other dipoles relative to the one in the direction of the **E** field are measured. The relative responses were found to be at least -20 dB, which indicates an adequate orthogonality.

Magnetic-Field Probe

The design of the magnetic-field probe is significantly more involved than that of the electric-field probe. The commonly used diode-loaded circular loops shown in Fig. 30 are generally not adequate for the development of a general-purpose electromagnetic radiation monitor for the following reasons [35]:

1. The frequency response of these loops is linearly proportional to the frequency, which means that the frequency of the incident radiation must be accurately known in order to measure the magnetic field. This characteristic frequency response can be shown as follows. From Faraday's law it is known that

$$\oint_C \mathbf{E} \cdot d\boldsymbol{\ell} = -\frac{\partial}{\partial t} \int_S \mathbf{B} \cdot d\mathbf{S} = -\mu \frac{\partial}{\partial t} \int_S \mathbf{H} \cdot d\mathbf{S}$$

a

b

ANTENNA No. 3

ANTENNA No. 2

ANTENNA No. 1

1/4-in
(6.35-mm)

Fig. 29. Orthogonal arrangement of the **E**-field probes to achieve isotropic receiving characteristics. (*a*) Photograph of **E**-field probe (side view). (*b*) Diagram of **E**-field probe (top view).

where **E** and **H** are the electric and magnetic fields, respectively, and S is any surface bounded by the closed path C. For the sinusoidally time-varying, steady-state case, $\partial/\partial t \rightarrow j\omega$. Hence, by integrating the above equation over the area of the loop (assuming **H** to be uniform over the area of the loop), we have

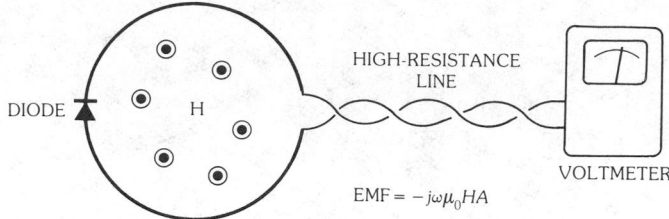

Fig. 30. Diagram illustrating the geometry of the diode-loaded circular loop. (*After Greene [35]*)

$$\oint_C \mathbf{E} \cdot d\ell = \text{emf} = -j\omega\mu HA \qquad (16)$$

where $A = \pi a^2$ is the area of the loop, $\omega = 2\pi f$, f is the frequency, and the emf is the induced voltage across the loop terminals. From (16) it is clear that the induced emf is proportional to the frequency f. Therefore, for measurements in industrial sites where the generation of higher-order harmonics has been observed, the linear frequency response of the loop would be very undesirable.

2. These probes generally respond to the electric field as well as to the magnetic field. This, of course, is undesirable when one is trying to measure the magnetic field. Fig. 31 shows that the loop diameter should not exceed 0.05λ if

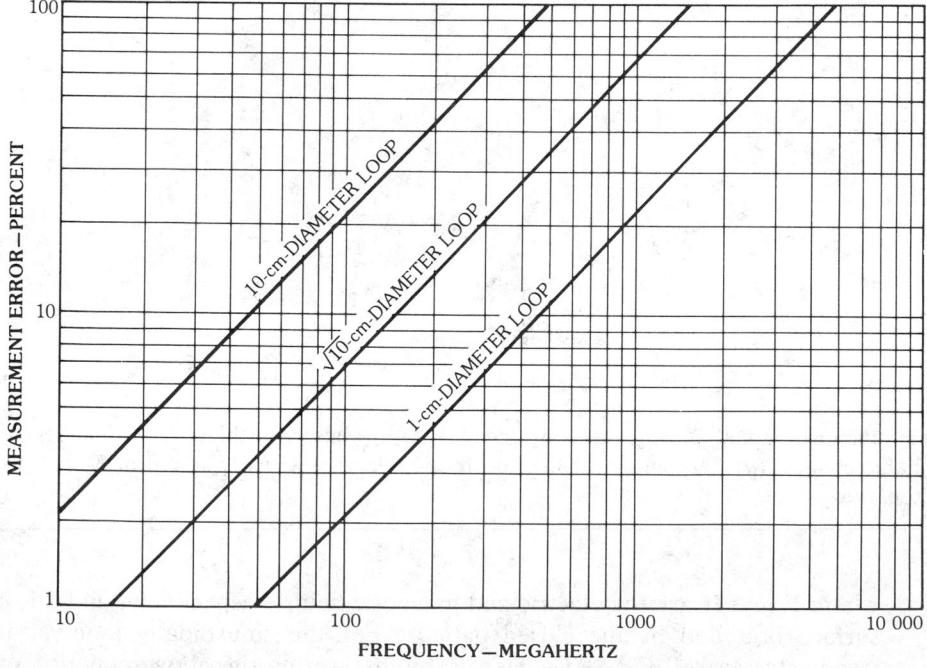

Fig. 31. Worst-case measurement error of the 1-, $\sqrt{10}$-, and 10-cm–diameter loop antennas due to their electric-dipole response. (*After Greene [35]*)

the probe's response to the electric field is to be minimized [35]. A loop of this size, however, is too small to provide the desired sensitivity. Also, at higher frequencies, where the wavelength is shorter, it becomes increasingly difficult to make such small probes sensitive enough.

The frequency response of a loop can be made flatter by making the loop the inductive part of a low-Q resonance circuit [36]. If the resonant frequency of the low-Q circuit is considerably lower than the frequency of operation, the response of the loop in the postresonance frequency range will usually be reasonably flat, but the sensitivity will also be reduced. The reduced sensitivity requires use of zero-bias diodes (HSCH-3486 or HSCH-3171) to rectify the output signal detected by the probe. A schematic diagram illustrating the probe's principle of operation [36] is shown in Fig. 32a. An isotropic **H**-field probe can be constructed by placing three magnetic-field probes on the three orthogonal surfaces of a cube.

Design of an RF Personnel Dosimeter [37]

Although there are several commercially available rf radiation area monitors, there is little available information on the development of a personnel dosimeter. The function of an area monitor is generally to detect electromagnetic radiation and sound an alarm if the level in the working area gets too high. The function of a personnel dosimeter is to measure the electromagnetic radiation to which the person wearing it is exposed, record the amount of radiation, and sound an alarm if excessive exposure occurs. The personnel dosimeter must be able to relate the fields measured on the surface of a worker's body to the free-space radiation, in terms of which the safety standard is written. This is important since this safety standard is the legally binding measurement rather than the significantly distorted fields (due to scattering) measured on the surface of an operator's body.

In summary, an rf personnal dosimeter should consist of the following items:

(a) An antenna system to measure the **E**- and **H**-field intensities associated with the electromagnetic radiation. This antenna system can be basically the same as those described earlier in this section.

(b) A microprocessor system to store the values of the measured fields, calculate the incident fields from the measured fields, compare the calculated values of the incident fields with the safety standard, and sound an alarm if the safe levels are exceeded [38].

It would also be desirable for the dosimeter to require minimal power, to be inexpensive, rugged, and lightweight, and to have overload protection.

7. Other Applicators Used to Produce Hyperthermia

There are some other devices that have been widely used to generate hyperthermia that should be mentioned in this chapter, even though they are not always thought of as antennas because their use is based more on near-field coupling at low frequencies than on traditional antenna radiation characteristics. These applicators are usually thought of as inducing currents in the body instead of producing

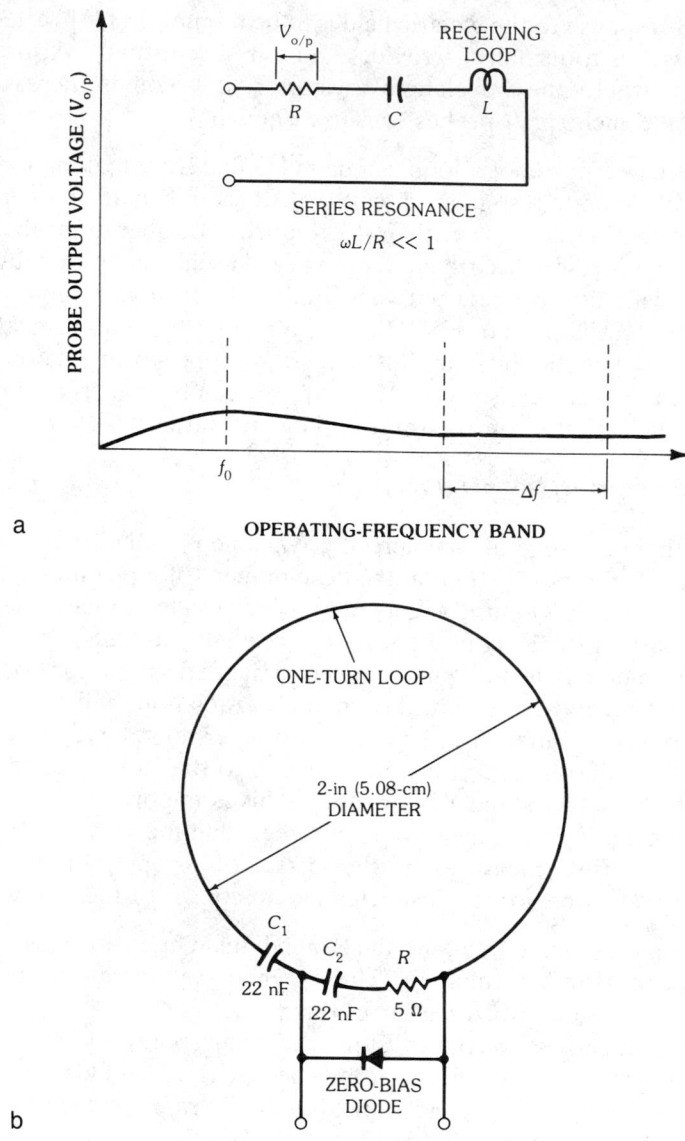

Fig. 32. The low-Q series-resonant circuit magnetic-field probe. (*a*) Principle of operation. (*b*) Design of the loop. (*After Babij and Bassen [36]*)

waves that propagate into the body. For example, one of these applicators is a coil placed coaxially around the body. It induces rf currents that are more conveniently thought of as eddy currents than as propagating waves. Another applicator consists of capacitor plates placed on the body. These plates induce rf currents that are also not conveniently characterized as propagating waves. For this class of applicators the term "depth of penetration" is often used to describe heating patterns, but it is important to note that this term does not have the same meaning as it does for radiating applicators that produce waves that propagate from the surface into the interior of the body. These nonradiating applicators can also overheat the surface, even at low frequencies, because of the spatial variation of the fields that they produce. Thus the fundamental problem in hyperthermia that is described in the introduction to this chapter, how to heat the interior without overheating the surface, is generally not solved by using nonradiating applicators, even at low frequencies. Four couplers of this kind that are described briefly in this section are coaxial current loops, capacitor-plate applicators, pancake coils, and helical coils. The following four figures illustrate the geometries of the four low-frequency couplers and the currents or the fields induced by these couplers inside the body.

Coaxial Current Loops

A simple coaxial current loop placed around a patient is attractive for producing regional hyperthermia because it requires no coupling medium, it is easy to match to a transmitter, and getting a patient into and out of it is easy. (See Fig. 33.) Also, the impedance matching is not very sensitive to patient size and position within the coil. Storm and colleagues [39] have pioneered the use of such an applicator. Their device operates at 13.56 MHz with an input power of 10 to 1000 W.

The coaxial current loop for use at low frequencies is easy to design because the dimensions are not critical if the size of the loop is small compared with a free-space wavelength. The main design task is to construct a matching network of some sort to compensate for the inductive driving-point impedance and to match the coil to the feed line. Standard rf techniques can be used to do this.

Since measurements have shown that the fields inside the coil consist mostly of an axial magnetic field that is nearly uniform in the central portion of the space inside the coil, an approximate expression for the induced **E** field inside a homogeneous body centered within the coil can be obtained from Maxwell's equation for the steady-state sinusoidal case:

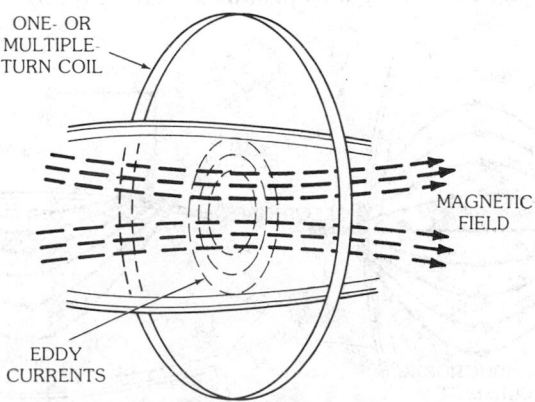

Fig. 33. Magnetic field and eddy currents produced by a coaxial coil "Magnetrode," which surrounds a portion of the patient's body. (*After Iskander [1], © 1982 American Institute of Physics*)

$$\oint_\ell \mathbf{E} \cdot d\boldsymbol{\ell} = -j\omega\mu_0 \int_S \mathbf{H} \cdot d\mathbf{S}$$

where ω is the angular frequency, μ_0 is the permeability of free space, and S is any surface bounded by the closed path ℓ. Assuming circular symmetry and letting ℓ be a circular path inside the body, the equation reduces to

$$2\pi r E_\phi = -j\omega\mu_0 H_z \pi r^2$$

and

$$E_\phi = \frac{-j\omega\mu_0 H_z r}{2}$$

where r is the radius of the path. This expression shows that the induced **E** field, and hence the absorbed power, will be zero at the center and increasing toward the surface. These results thus predict that the current loop will produce primarily surface heating, and if the center is heated, it must be through thermal conduction.

Whether dielectric inhomogeneities would result in better internal heating than that predicted by the homogeneous model is a question that has been widely discussed. Hill, Christensen, and Durney [40] have calculated internal field patterns in a two-dimensional inhomogeneous model, and Paulsen and colleagues [41] combined these calculations with a finite-element calculation of temperature distributions. From their results they concluded that in most cases the current loop does not effectively heat deep-lying tumors. Their conclusions were based on both transient and steady-state temperature calculations, with various patterns of blood flow.

Capacitor-Plate Applicators

At low frequencies capacitive applicators consisting of one plate on one side of a patient and a second plate on the other side have been used for many years in diathermy. (See Fig. 34.) The plates produce a strong electric field perpendicular

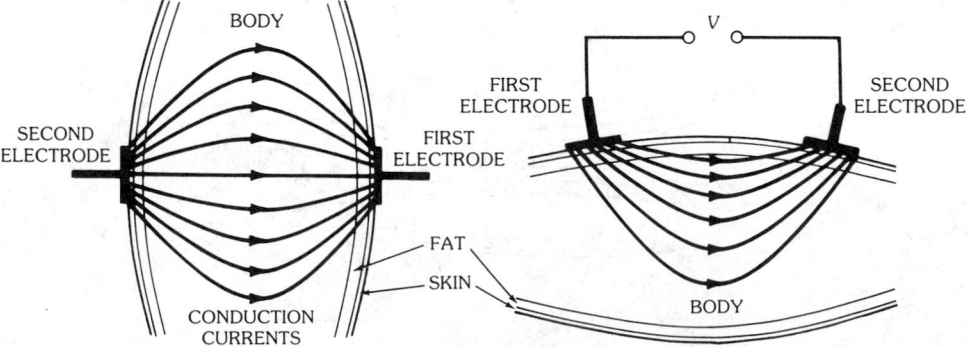

Fig. 34. Two possible arrangements for a regional heating of layered tissue using capacitor-type electrodes for short-wave diathermy. (*After Iskander [1], © 1982 American Institute of Physics*)

to the plates and a negligible magnetic field. The only design procedure required for the applicator is some method of matching its impedance to a transmission feed line.

Although the capacitive applicator can produce good deep heating, it has the serious disadvantage that it tends to overheat the fat [42]. This occurs when the internal **E** field is perpendicular to the muscle-fat interface. At this interface the boundary conditions on the **E** field require that

$$\epsilon_f E_f = \epsilon_m E_m$$

where ϵ_f and ϵ_m are the relative permittivities in the fat and muscle, respectively, and E_f and E_m are the normal components of the **E** field in the fat and muscle, respectively. Thus, at the interface,

$$E_f = \frac{\epsilon_m}{\epsilon_f} E_m$$

and since ϵ_f is considerably smaller than ϵ_m, the field in the fat at the boundary is considerably larger than the field in the muscle. If the fat and muscle layers were approximately planar, the field in the volume of the fat would likewise be considerably stronger than the field in the muscle. Since power varies as E^2, the stronger fields in the fat cause it to overheat, even though the fat is less lossy than the muscle. For this reason capacitive applicators are useful mostly for thin patients.

Pancake Coils

Another commonly used diathermy applicator is the so-called pancake coil, consisting of a planar helical coil placed flat on the surface of the body. (See Fig. 35.) Guy, Lehmann, and Stonebridge [3] have calculated and measured heating

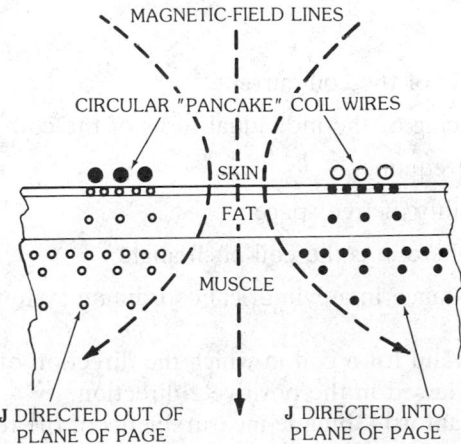

Fig. 35. Magnetically induced current in tissue exposed to short-wave diathermy "pancake" coil. (*After Guy, Lehmann, and Stonebridge [3], © 1974 IEEE*)

patterns of typical coils. Their results show that the pancake coil can heat underlying muscle tissue without overheating the intervening fat. At 27.12 MHz and with a 2-cm fat layer the coil can heat to a depth of about 4 cm (depth where power absorption is e^{-2} of the maximum). The main disadvantage of the pancake coil is the nonuniform heating pattern, which tends to be toroidal in shape.

Helical-Coil Applicators

The helical coil has recently been investigated for use as a hyperthermia applicator [43]. Although enough work has not yet been done to determine its full potential for generating hyperthermia, the helical coil does appear promising. Chute and Vermeulen [44] demonstrated that the axial electric field inside a helical coil is approximately uniform and much stronger than the other electric- and magnetic-field components. (See Fig. 36.) This makes the coil attractive for hyperthermia because the axial field would be mostly parallel to the fat-muscle boundaries in a patient placed inside the coil, and consequently overheating of the fat would not be expected. Since the field is approximately uniform it would also be expected to heat the center of the body much better than a single coaxial current loop.

Chute and Vermeulen [44] showed that when the radius of the helical coil is much smaller than its length and when both dimensions are small compared with a free-space wavelength, the fields inside the coil are given by

$$H_z \cong \frac{I_0}{d} \cos \omega t$$

$$E_\phi \cong \frac{\omega \mu_0 I_0}{2d} \varrho \sin \omega t$$

$$E_z \cong -\frac{\omega \mu_0 a I_0}{2d} \cos \psi \sin \omega t$$

where

I_0 = magnitude of the coil current

d = axial spacing of the individual turns of the coil

ω = angular frequency

μ_0 = permeability of free space

ctn ψ = $2\pi a/d$, where ψ is the coil pitch angle

ϱ = radial distance in a cylindrical coordinate system (ϱ, ϕ, z)

The equations are valid for a coil in which the direction of twist is clockwise with increasing z when viewed in the positive z direction.

Ruggera and Kantor [43] made measurements of the temperature distribution produced by helical coils in arm-sized (8-cm diameter) and thigh-sized (12-cm diameter) fat-muscle cylindrical phantoms. In studies at the FCC-designated Instrument, Scientific, and Medical (ISM) frequencies of 13.56, 27.12, and 40.68

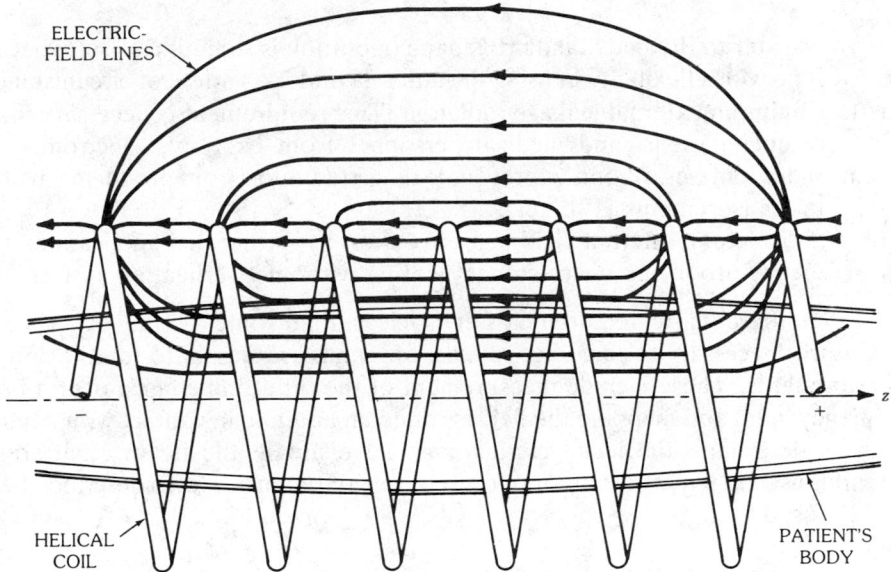

Fig. 36. Helical coil carrying a time-varying current and the electric-field lines associated with it. (*After Chute and Vermeulen [44],* © *1981 IEEE*)

MHz, they found that heating uniform within 10 percent or less over the cross section could be obtained in arm-sized phantoms under the right conditions. Heating uniform within about 20 percent over the cross section could be obtained in thigh-sized phantoms. Their results indicated that two special modes of operation produced the best results, the first when the total wire length in the helix was equal to one-half wavelength, and the second when it was equal to one wavelength. For these two conditions the driving-point impedance was purely resistive. In the full-wavelength operation a coil length-to-diameter ratio of 4 was needed to produce uniform transverse heating, and in half-wavelength operation a ratio of 2 was needed.

Although initial work indicates that the helical coil could be a very effective applicator for hyperthermia generation, much work remains to be done in this area.

8. Experimental Procedures for Characterizing Antennas Used in Medical Applications

This chapter concludes with a brief description of some commonly used techniques for measuring the properties of antenna systems used in medical applications. Since the techniques are designed to measure the most important characteristics of the antennas, it is helpful to summarize those characteristics first. In general, antennas for medical applications should have the following features:

(*a*) Impedance well matched to the human body over a broad frequency band. Good impedance matching is necessary to achieve maximum power transfer to the body, and broadband operation is sometimes useful because it provides flexibility in using the antenna under a variety of circumstances.

(b)   Minimum external leakage radiation. This requirement is necessary first to protect patients and medical personnel from excessive electromagnetic radiation, and second to prevent leakage radiation from producing artifacts in measurements [17].

(c)   Satisfactory internal field patterns. For hyperthermia applications, this means producing deep enough heating without overheating surfaces.

From these basic requirements it is clear that some of the experimental procedures necessary to characterize these antennas are similar to routine antenna measurements. These include measurement of the input impedance over a broad frequency band and mapping the fields around an antenna in contact with a human body to determine the leakage radiation. To evaluate the heating patterns of antennas used in hyperthermia applications, one of the procedures described below may be used.

Evaluation of Heating Patterns of Antennas Used in Hyperthermia

Nonperturbing Implantable Temperature Probes—There are several temperature probes that are especially useful for evaluating heating patterns of electromagnetic antennas because they produce almost no interference with the electromagnetic fields. Among the commercially available and most stable of these probes are those made by Vitek, Inc. [45], and Narda Corporation. Both systems use thermistor detectors and high-resistance leads (carbon-loaded Teflon) to minimize interference with the electromagnetic fields.

Another type of nonperturbing temperature probe was developed by Christensen [46]. It utilizes a small 0.25-mm gallium-arsenide semiconductor sensor at the end of a fiber-optic bundle. The probe relies on the shift in optical absorption at the semiconductor's band edge as a function of temperature. Recent versions of this probe achieved resolutions better than 0.1°C over the temperature range from 37°C to 47°C. This probe is also available commercially from Clini-Therm Corporation. Other fiber-optic probes reported in the literature include a liquid crystal probe [47], a birefringent crystal optical thermometer [48], an etalon fiber-optic probe [49], and a phosphor fiber-optic probe [50]. Another probe called a *viscometric thermometer*, which is based on measurement of the flow resistance of a fluid (a temperature-dependent property) through a small capillary at the tip of the implantable probe, has also provided promising results [51]. In this probe the probe body material and the working fluid can be chosen to match closely the electrical and thermal properties of tissue, thus minimizing perturbation of the electromagnetic and temperature fields in the tissue. This matching of properties is not generally possible in other electrical and optical probes.

In using probe temperature measurements to evaluate heating patterns it is important to keep exposure times to a minimum to minimize the error due to heat conduction and smearing of the heating patterns. With this precaution in mind, this measurement technique is otherwise straightforward and has the advantage of being the most accurate in experiments on phantoms. It also allows temperature regulation to be measured in live animals and is inexpensive. The major disadvantage is the time required to define heating patterns in large or complex geometrical bodies. With the advent of systems having multiple temperature probes [52], this problem has been largely reduced.

Thermographic Cameras—Scanning thermographic cameras can be used to provide rapid acquisition of data on heating patterns in phantoms. The experimental procedure involves using the antenna under test to heat the phantom at high power levels in the shortest possible time. The phantom models are designed to separate along planes perpendicular to tissue interfaces so that cross-sectional relative heating patterns can be measured with a thermograph. After a short exposure the model is quickly disassembled and the temperature pattern over the surface of separation is observed and recorded by means of a thermograph. Details of this technique have been developed at the University of Washington and are described in detail by Guy [53]. The procedure originally involved using a thin sheet of plastic to facilitate separating the two halves of the phantom. Thus the procedure was limited to symmetrical models exposed to a linearly polarized field (**E** field parallel to the interface) to avoid interrupting any induced currents that would normally flow perpendicular to the median plane of separation [53]. For near-field measurements and in evaluating antennas for hyperthermia applicators, however, the procedure was modified by replacing the plastic sheet with a silk screen, thus allowing easy separation without loss of electrical continuity [54]. The major disadvantage of this technique is the high cost of the required equipment.

Field Mapping Using Implantable E-Field Probes

The radiation characteristics of an antenna in tissue can be equivalently evaluated by mapping the electric-field distribution using implantable probes. There are advantages in using these probes for evaluating heating patterns instead of thermographic and thermometric techniques. For example, the latter two techniques determine internal fields only indirectly and under exposure conditions that provide moderate elevation in temperature. The thermoregulatory system of living subjects further complicates thermal dosimetry for localized determination of absorbed energy. For such cases the electric-field measurements provide much more accurate and reliable assessment of the power absorbed by the subject. A comparison of the various dosimetric techniques [55] is presented in Table 3. This comparison simply indicates that the internal-field probe is well suited for *in-vivo* measurement of the absorption characteristics of subjects.

Probes used for measuring electric fields in simulated or actual biological media often use dipole-diode combinations similar to those described in Section 6. The detailed structure of the implantable electric-field probe described by Bassen and colleagues [55] is shown in Fig. 37. The electric-field probe shown in Fig. 37b consists of a single dipole-detector configuration terminated with high-resistance leads that transmit the detected signal to the readout electronics. The use of the three orthogonal dipoles in an I-beam array (Fig. 37a) has been shown to yield a satisfactory isotropic response when the outputs of square-law detectors are summed. Interested readers can find a description of the detailed procedure for fabricating the probe elsewhere [56].

Table 3. Comparison of Internal Dosimetric Techniques (*After Bassen et al. [55],* ©
1977 American Geophysical Union)

	"Ideal" E-Field Probe	Thermometer Probe	Thermography
Directly measured parameter	Electric-field strength	Localized temperature	Spatial distribution of temperatures
Dosimetric approach	Direct, absolute E-field measurement, independent of media. Must know conductivity to compute SAR	Direct thermal measurement. Must know mass density and specific heat to compute SAR	Temperature measured after exposure. Must know mass density, specific heat, and infrared emittance to compute SAR
Sensitivity to external field	$< 10 \, \text{mW/cm}^2$	$> 10 \, \text{mW/cm}^2$	$> 100 \, \text{mW/cm}^2$
Suitable for *in-vivo* E-field dosimetry	Yes	Yes	No
Spatial measurement capability	Point or continuous-line scan. Arbitrary site of measurement	Single-point measurement. Arbitrary site of measurement	Rapid planar scan. Subject must be prepared in advance
Accuracy and limitations	Good absolute accuracy.* Relatively independent of boundaries and ϵ_r	Good absolute thermal accuracy ($\pm 0.05°C$). Measurement of SAR is media dependent	Good relative thermal accuracy ($\pm 0.2°C$) but limited by thermal diffusion. Measurement of SAR is media dependent

*Present prototypes are media dependent (± 2.25 to 4.25 dB).

It is important to note that experimental measurements of internal fields within simulated or actual biological media have shown the response of some probes to be dependent on the dielectric constant of the medium in which the measurements were made, or on the position of the probe with respect to boundaries. Based on experimental and theoretical analyses, sets of principles were formulated for the design of E-field probes whose responses are ideally independent of the media [57].

A typical result obtained by Smith [57] is shown in Fig. 38. In this figure, $V(\epsilon_r)$ is the voltage across the terminals of an electrically short probe which is proportional to the component of the local electric field E_z parallel to the axis of the probe, i.e., $V(\epsilon_r) = K_e E_z$. The proportionality constant K_e depends on the effective height of the probe, its input admittance, and the admittance of the load

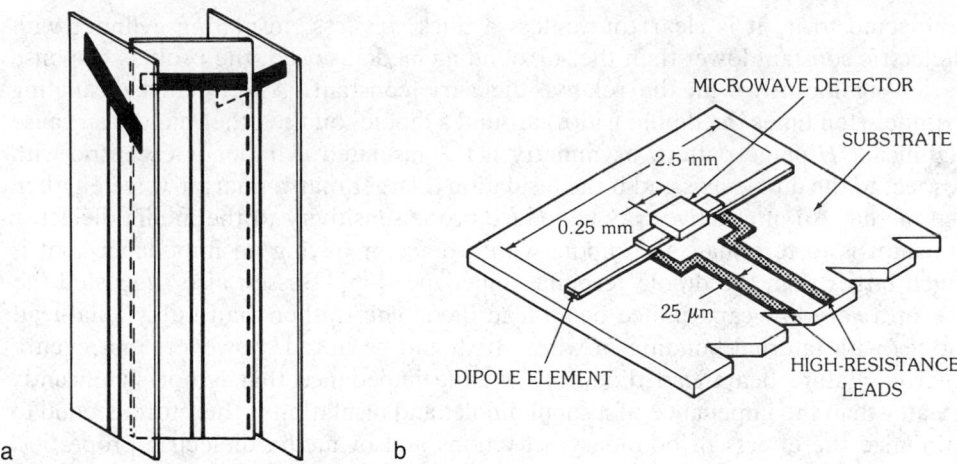

Fig. 37. Structure of implantable electric-field probe. (*a*) Orthogonal detector array. (*After Bassen, Swicord, and Abita [56]*) (*b*) Details of fabrication of probe. (*After Bassen et al. [55], © 1977 American Geophysical Union*)

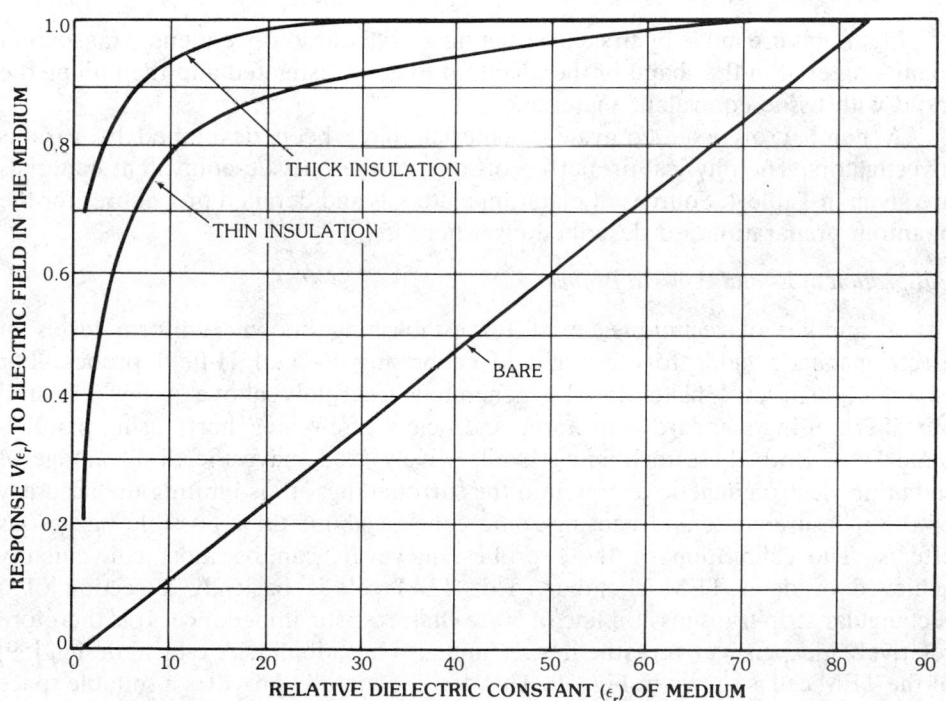

Fig. 38. Response of various probes in media as a function of dielectric constant. (*After Smith [57], © 1975 IEEE*)

connected to it. It is clear that unless a thick, lossless, insulating cylinder with dielectric constant lower than the surrounding media is used, the probe's response varies significantly with the relative dielectric constant. A very thick insulating cylinder (ten times the dipole width) around a dipole, on the other hand, can cause significant *H*-plane pattern asymmetry if the insulated cylinder is eccentric with respect to the dipole axis and if the insulation is larger than a quarter-wavelength in the media. An alternative way to reduce probe sensitivity to the media dielectric constant is to terminate the dipole with a detector having an impedance that is much larger than the dipole terminal impedance [56]. Bassen also suggested the use of a very low capacitance beam-lead diode chip rather than bulky axial-lead diodes with internal bonding of wires. It should be noted, however, that even a microminiature beam-lead diode has a finite impedance that is not significantly greater than the impedance of a small dipole, and insulation is therefore needed to minimize the effects of boundary reflections and of media dielectric properties. Also, at lower frequencies the media dependence of the probe's response increases because of the higher antenna impedance and the higher leakage current of the diode.

Use of Phantoms to Measure Power Deposition Patterns (Heating Patterns) of Antennas

Phantoms are made by first constructing a mold of radio-frequency transparent foam* material in the shape of the phantom to be investigated and then filling the mold with tissue-equivalent material.

A number of tissue-equivalent materials have been developed by various investigators. The physical properties of some of these tissue-equivalent materials are given in Table 4. Sources of phantom materials and detailed procedures for the phantom preparation are described elsewhere [58].

Calibration of E- and H-Field Probes

A number of techniques exist for establishing known, uniform levels of electromagnetic fields for testing and calibrating **E**- and **H**-field probes. For example, high-level fields can be generated accurately above a few hundred megahertz using standard-gain horns and below a few megahertz using parallel-plate lines. Both these techniques, while widely used, have the disadvantage of radiating electromagnetic energy into the surroundings, thus limiting the accuracy of the measurements and causing some concern about their possible hazardous effects. The calibration of these probes, however, can be most conveniently achieved inside a TEM chamber. This TEM cell is basically a section of a rectangular strip transmission line of 50-Ω characteristic impedance. It is therefore relatively inexpensive, versatile in size, and also broadband. A typical design [59] of the TEM cell is shown in Fig. 39. This design generally provides a suitable space where the **E** and **H** fields can be calculated and hence used to calibrate the probes. It should be noted that the dimensions of the cell given in Table 5 can be linearly

---

*Equal proportions of polyol and isocyanate, available at Utah Foam Products, Inc., Salt Lake City, Utah 84119.

Table 4. Properties of Materials for Constructing Phantom Models for Muscle, Brain, Fat, and Bone at Microwave Frequencies [58]

	Physical Properties	
Material	Specific Heat	Specific Density
Muscle	0.84	0.97
Brain	0.83	0.96
Fat and bone	0.29–0.37	1.29–1.38

	Electrical Properties*					
	915 MHz			2450 MHz		
Material	ϵ_r	σ	τ	ϵ	σ	τ
Muscle	50.63	1.355	0.526	49.61	2.286	0.333
Brain	34.37	0.730	0.442	33.56	1.266	0.2713
Fat and bone	5.61	0.0665	0.233	4.51	0.172	0.187

	Composition			
Material	Super Stuff	Polyethylene powder	H_2O	NaCl
Muscle	8.45%	15.20%	75.44%	0.9069%
Brain	7.01%	29.80%	62.61%	0.5823%

	Laminac 4110	Aluminum powder	Acetylene black
Fat and bone**	85.20%	14.5%	0.24%

*Here ϵ_r = relative permittivity, σ = conductivity in siemens per meter, and τ = tan δ (loss tangent).
**For fat and bone add 0.375% of total weight P-102 (60% methyl ethyl ketone peroxide as a catalyst).

scaled with the desired operating frequency band. In constructing a TEM chamber special attention must be paid to the tapered transitions to ensure good impedance matches. Electric and magnetic fields should be mapped inside the chamber to determine their uniformity and ensure the absence of any significant standing waves caused by mismatches at transitions or by the presence of higher-order modes.

When high-order modes and standing waves are not present in the chamber, the electric-field strength can be found from the measurement of the total power transmitted through the cell. It has been shown that [59]

$$E = \sqrt{P/G}/d$$

in volts per meter, and

$$H = E/377$$

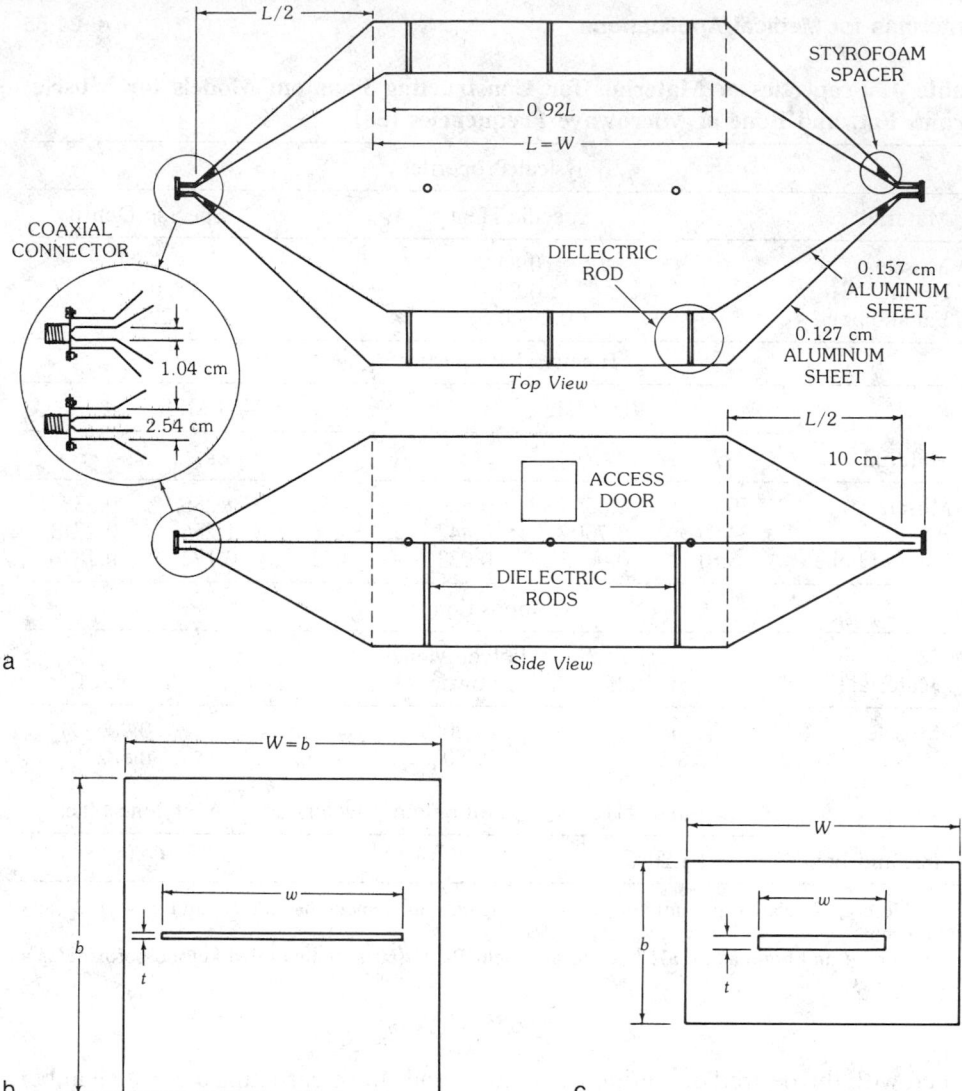

Fig. 39. Typical TEM cell design. (*a*) Design for rectangular TEM transmission cell. (*b*) Cross section of optimum geometry of rectangular transmission line for maximum test area and frequency. (*c*) Cross section of rectangular transmission line with improved **E**-field uniformity. (*After Crawford [59],* © *1974 IEEE*)

Table 5. TEM Cell Dimensions

Cutoff/ Multimode Frequency (MHz)	Square Cell (Fig. 39b)				Rectangular Cell (Fig. 39c)			
	Plate Separation b (cm)	w (cm)	t (cm)	C'_f (pF/cm)	Plate Separation b (cm)	w (cm)	t (cm)	C'_f (pF/cm)
100	150	123.83	0.157	0.087	90	108.15	0.157	0.053
300	50	41.28	0.157	0.087	30	36.05	0.157	0.053
500	30	24.77	0.157	0.087	18	21.83	0.157	0.053

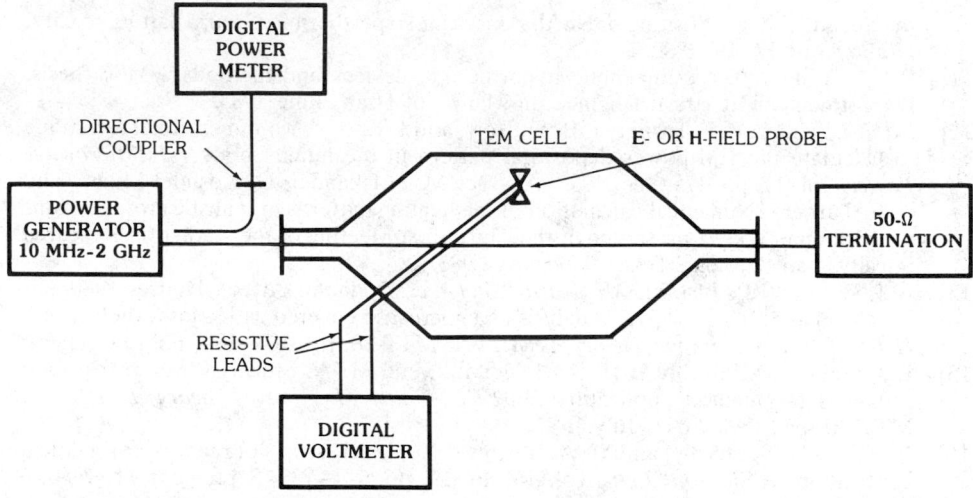

Fig. 40. Diagram of the calibration setup.

in amperes per meter, where P is the net power flowing through the cell, G is the real part (conductance) of the input admittance, and d is the separation distance between the upper and the middle conductors in the cell.

The experimental setup used to calibrate the **E**- and **H**-field probes [60] is shown in Fig. 40. It consists of a microwave power generation unit and a TEM transmission cell terminated by a 50-Ω load.

9. References

[1] M. F. Iskander, "Physical aspects and methods of hyperthermia production by rf currents and microwaves," in *Physical Aspects of Hyperthermia*, ed. by G. H. Nussbaum, New York: American Institute of Physics, 1982.

[2] H. P. Schwan, "Biophysics of diathermy," in *Therapeutic Heat and Cold*, ed. by S. Licht, New Haven: Licht, 1965, sec. 3, pp. 63–125.

[3] A. W. Guy, J. F. Lehmann, and J. B. Stonebridge, "Therapeutic applications of electromagnetic power," *Proc. IEEE*, vol. 62, pp. 55–75, January 1974.

[4] A. W. Guy, "Electromagnetic fields and relative heating patterns due to a rectangular aperture source in direct contact with bilayered biological tissue," *IEEE Trans. Microwave Theory Tech.*, vol. MTT-19, pp. 214–223, February 1971.

[5] A. W. Guy, J. F. Lehmann, J. B. Stonebridge, and C. C. Sorenson, "Development of a 915-MHz direct-contact applicator for therapeutic heating of tissues," *IEEE Trans. Microwave Theory Tech.*, vol. MTT-26, pp. 550–556, August 1978.

[6] A. L. Van Koughnett and W. Wyslouzil, "A waveguide TEM-mode exposure chamber," *J. Microwave Power*, vol. 7, pp. 381–383, December 1972.

[7] A. Y. Cheung, T. Dao, and J. E. Robinson, "Dual-beam TEM applicator for direct-contact heating of dielectrically encapsulated malignant mouse tumor," *Radio Sci.*, vol. 12, pp. 81–85, November–December 1977.

[8] S. B. Cohn, "Properties of ridge waveguide," *Proc. IRE*, vol. 35, pp. 783–788, August 1947.

[9] F. Sterzer, R. W. Paglione, J. Mendecki, E. Friedenthal, and C. Botstein, "RF therapy for malignancy," *IEEE Spectrum*, pp. 32–37, December 1980.

[10] P. F. Turner, "An annual phased array for deep regional hyperthermia," presented at

the Second Annual Meeting of North American Hyperthermia Group, Salt Lake City, Utah, April 17–19, 1982.

[11]   P. F. Turner, "Electromagnetic hyperthermia devices and methods," MSc thesis, Department of Electrical Engineering, Univ. of Utah, June 1983.

[12]   M. F. Iskander, P. F. Turner, J. B. DuBow, and J. Kao, "Two-dimensional technique to calculate the EM power deposition pattern in the human body," *J. Microwave Power*, vol. 17, pp. 175–185, 1982. Also see M. F. Iskander, O. Koshdel-Milani, and P. F. Turner, "Numerical calculation of the heating patterns in realistic cross-sections of the human body," presented at the Thirty-first Meeting of the Radiation Research Society, San Antonio, Texas, February 1983.

[13]   I. J. Bahl and P. Bhartia, *Microstrip Antennas*, Dedham: Artech House, 1980.

[14]   I. J. Bahl and S. S. Stuchly, "Analysis of a microstrip covered with a lossy dielectric," *IEEE Trans. Microwave Theory Tech.*, vol. MTT-28, pp. 104–109, February 1980.

[15]   I. J. Bahl, S. S. Stuchly, J. J. W. Lagendijk, and M. A. Stuchly, "Microstrip loop radiators for medical applications," *IEEE Trans. Microwave Theory Tech.*, vol. MTT-30, pp. 1090–1093, July 1982.

[16]   I. J. Bahl, S. S. Stuchly, and M. A. Stuchly, "New microstrip slot radiator for medical applications," *Electron. Lett.*, vol. 16, no. 19, pp. 731–732, September 11, 1980.

[17]   M. F. Iskander and C. H. Durney, "An electromagnetic energy coupler for medical applications," *Proc. IEEE*, vol. 67, pp. 1463–1465, October 1979.

[18]   M. F. Iskander and M. A. K. Hamid, "A new strip transmission line for moisture content measurements," *J. Microwave Power*, vol. 12, pp. 16–18, 1977.

[19]   L. S. Taylor, "Implantable radiators for cancer therapy by microwave hyperthermia," *Proc. IEEE*, vol. 68, pp. 142–149, January 1980.

[20]   J. Bigu-del-Blanco and C. Romero-Sierra, "The design of a monopole radiator to investigate the effect of microwave radiation in biological systems," *J. Bioengineering*, vol. 1, pp. 181–184, 1977.

[21]   J. Mendecki, E. Friedenthal, C. Botstein, R. Paglione, and F. Sterzer, "Microwave applicators for localized hyperthermia treatment of cancer of the prostate," *Intl. J. Radiation Oncology Biol. Phys.*, vol. 6, pp. 1583–1588, 1980.

[22]   J. Mendecki, E. Friedenthal, C. Botstein, F. Sterzer, R. Paglione, M. Nowogrodzki, and E. Beck, "Microwave-induced hyperthermia in cancer treatment: apparatus and preliminary results," *Intl. J. Radiation Oncology Biol. Phys.*, vol. 4, pp. 1095–1103, 1978.

[23]   E. C. Burdette, F. L. Cain, and J. Seals, "In-vivo probe measurement technique at vhf through microwave frequencies," *IEEE Trans. Microwave Theory Tech.*, vol. MTT-28, pp. 414–427, 1980.

[24]   T. W. Athey, M. A. Stuchly, and S. S. Stuchly, "Measurement of radio frequency permittivity of biological tissues with an open-ended coaxial line: part I," *IEEE Trans. Microwave Theory Tech.*, vol. MTT-30, pp. 82–86, 1982. See also M. A. Stuchly, T. W. Athey, G. M. Samaras, and G. E. Taylor, "Measurement of radio frequency permittivity of biological tissues with an open-ended coaxial line: part II—experimental results," *IEEE Trans. Microwave Theory Tech.*, vol. MTT-30, pp. 87–91, 1982.

[25]   G. A. Deschamps, "Impedance of an antenna in a conducting medium," *IRE Trans. Antennas Propag.*, pp. 648–650, September 1962.

[26]   M. A. Stuchly and S. S. Stuchly, "Coaxial line reflection method for measuring dielectric properties of biological substances at radio and microwave frequencies—a review," *IEEE Trans. Instrum. Meas.*, vol. IM-29, 1980. See also S. S. Stuchly, M. A. Rzepecka, and M. F. Iskander, "Permittivity measurement at microwave frequencies using lumped elements," *IEEE Trans. Instrum. Meas.*, vol. IM-23, pp. 56–62, 1974, and M. A. Rzepecka and S. S. Stuchly, "A lumped capacitance method for the measurement of the permittivity and conductivity in the frequency and time domain—a further analysis," *IEEE Trans. Instrum. Meas.*, vol. IM-24, pp. 27–32, 1975.

[27]   J. Toler and J. Seals, "RF dielectric properties measurement system: human and animal data," NIOSH research report, *HEW (NIOSH) Pub. No. 77-176*, July 1977.

[28] M. F. Iskander and J. B. DuBow, "Time- and frequency-domain techniques for measuring the dielectric properties of rocks: a review," *J. Microwave Power*, special issue on electromagnetics in energy applications, ed. by M. F. Iskander, March 1983. See also S. C. Olson and M. F. Iskander, "A new in-situ procedure for measuring the dielectric properties of low permittivity materials," *IEEE Trans. Instrum. Meas.*, vol. IM-35, pp. 2–7, March 1986.

[29] M. F. Iskander, "Permittivity measurements in time domain," MSc thesis, University of Manitoba, Winnipeg, Manitoba, Canada, 1972.

[30] P. F. Wacker and R. R. Bowman, "Quantifying hazardous electromagnetic fields: scientific basis and practical consideration," *IEEE Trans. Microwave Theory Tech.*, vol. MTT-19, pp. 178–187, 1971.

[31] E. Aslan, "Broadband isotropic electromagnetic radiation monitor," *IEEE Trans. Instrum. Meas.*, vol. 21, pp. 421–424, 1972.

[32] R. L. Moore, S. W. Smith, R. L. Cloke, and D. G. Brown, "Comparison of microwave power density meters," *Non. Ioniz. Radiation*, vol. 2, pp. 15–19, 1971.

[33] F. M. Greene, "A new near-zone electric field strength meter," *NBS Tech. Note 345*, November 1966.

[34] R. R. Bowman, "Some recent developments in the characterization and measurement of hazardous electromagnetic fields," *Proc. Intl. Symp. on Biol. Effects and Health Hazards*, Warsaw, October 15–18, 1973, Warsaw, Polish Medical Publishers, pp. 217–227, 1974.

[35] F. M. Greene, "Development of magnetic near-field probes," NIOSH Technical Information, *HEW (NIOSH) Pub. No. 75-127*, January 1975.

[36] T. M. Babij and H. I. Bassen, "Optimizing frequency response characteristics of an **E/H** probe," presented at the Fourth Annual Bioelectromagnetics Society Meeting, Los Angeles, June 28–July 2, 1982. Also see E. Aslan, "A low-frequency **H**-field radiation monitor," *Selected Papers on Biol. Effects of Electromagnetic Waves*, USNC/URSI Annual Meeting, Boulder, October 20–23, 1975, HEW publication (FDA) 77-8010, December 1976.

[37] M. F. Iskander, H. Massoudi, C. H. Durney, and M. Yafeh, "The development of an rf personal dosimeter," presented at the Fourth Annual Bioelectromagnetics Society Meeting, Los Angeles, June 28–July 2, 1982.

[38] M. F. Iskander, C. H. Durney, and D. L. Jaggard, "The development of a microwave personal dosimeter," presented at the Bioelectromagnetics Society Meeting, San Antonio, September 14–18, 1980.

[39] F. K. Storm, R. S. Elliott, W. H. Harrison, and D. L. Morton, "Clinical rf hyperthermia by magnetic-loop induction: a new approach to human cancer therapy," *IEEE Trans. Microwave Theory Tech.*, vol. MTT-30, pp. 1149–1158, August 1982.

[40] S. C. Hill, D. A. Christensen, and C. H. Durney, "Power deposition patterns in magnetically induced hyperthermia: a two-dimensional quasistatic numerical analysis," *Intl. J. Radiation Oncology Biol. Phys.*, in press.

[41] K. D. Paulsen, J. W. Strohbehn, S. C. Hill, D. R. Lynch, and F. E. Kennedy, "Theoretical temperature profiles for concentric coil induction heating devices in a two-dimensional axi-asymmetric, inhomogeneous patient model," presented at the North American Hyperthermia Group Meeting, San Antonio, February 27, 1983.

[42] D. A. Christensen and C. H. Durney, "Hyperthermia production for cancer therapy: a review of fundamentals and methods," *J. Microwave Power*, vol. 16, pp. 89–105. Also see A. W. Guy, J. F. Lehmann, and J. B. Stonebridge, "Therapeutic applications of electromagnetic power," *Proc. IEEE*, vol. 62, pp. 55–57, January 1974.

[43] P. S. Ruggera and G. Kantor, "Development of a family of rf helical coil applicators which produce transversely uniform, axially disturbed heating in cylindrical fat-muscle phantoms," *IEEE Trans. Biomed. Eng.*, vol. BME-31, pp. 98–106, January 1984.

[44] F. S. Chute and F. E. Vermeulen, "A visual demonstration of the electric field of a coil carrying a time-varying current," *IEEE Trans. Education*, vol. E-24, pp. 278–283, November 1981.

[45] R. R. Bowman, "A probe for measuring temperature in radio frequency heated material," *IEEE Trans. Microwave Theory Tech.*, vol. MTT-24, pp. 43–45, 1976.

[46] D. A. Christensen, "A new nonperturbing temperature probe using semiconductor band edge shift," *J. Bioengineering*, vol. 1, pp. 541–545, 1977.

[47] C. C. Johnson, O. P. Gandhi, and T. C. Rozzell, "A prototype liquid crystal fiberoptic probe for temperature and power measurements in rf fields," *Microwave J.*, vol. 18, pp. 55–59, 1975.

[48] T. C. Cetas, "A birefringent crystal optical thermometer for measurements of electromagnetically induced heating," USNC/URSI 1975 Annual Meeting, Boulder, October 20–23, 1975.

[49] D. A. Christensen, "Temperature measurement using optical etalons," 1975 Annual Meeting of the Optical Society of America, Houston, October 15–18, 1974.

[50] K. A. Wickersheim, R. V. Alves, and J. T. Christol, "Improved fluoroptic thermometry system for hyperthermia," Second Annual Meeting, North American Hyperthermia Group, Salt Lake City, April 17–19, 1982.

[51] M. M. Chen, C. A. Cain, K. L. Lam, and J. Mullin, "The viscometric thermometer: a nonperturbing instrument for measuring temperature in tissues under electromagnetic radiation," *J. Bioengineering*, vol. 1, pp. 547–554, 1977.

[52] D. A. Christensen and R. J. Volz, "A nonperturbing temperature probe system designed for hyperthermia monitoring," URSI Meeting and Bioelectromagnetics Symposium, Seattle, June 18–22, 1979.

[53] A. W. Guy, "Analyses of electromagnetic fields induced in biological tissues by thermographic studies on equivalent phantom models," *IEEE Trans. Microwave Theory Tech.*, vol. MTT-19, pp. 205–214, 1971.

[54] A. W. Guy et al., "A new technique for measuring power deposition patterns in phantoms exposed to em fields of arbitrary polarization—example, the microwave oven," *Proc. Microwave Power Symp.*, University of Waterloo, Ontario, Canada, pp. 36–40, May 1975.

[55] H. Bassen, P. Herchenroeder, A. Cheung, and S. Neuder, "Evaluation of an implantable electric-field probe within finite simulated tissues," *Radio Sci.*, vol. 12, pp. 15–25, 1977.

[56] H. Bassen, M. Swicord, and J. Abita, "A miniature broadband electric-field probe," in *Biological Effects of Nonionizing Radiation*, ed. by P. E. Tyler, Ann. N.Y. Acad. Sci., vol. 247, pp. 481–493, 1975.

[57] G. S. Smith, "A comparison of electrically short bare and insulated probes for measuring the local radio frequency field in biological systems," *IEEE Trans. Biomed. Eng.*, vol. BME-22, pp. 477–483, 1975.

[58] C. H. Durney, H. Massoudi, and M. F. Iskander, *Radiofrequency Radiation Dosimetry Handbook*, 4th ed., Report USAF SAM-TR-85-73, USAF School of Aerospace Medicine, Brooks Air Force Base, Texas 78235, October 1986.

[59] M. L. Crawford, "Generation of standard EM fields using TEM transmission cells," *IEEE Trans. Electromagnetic Compatibility*, vol. EMC-16, pp. 189–195, 1974.

[60] M. F. Iskander, H. Massoudi, and C. H. Durney, "Development of rf personnel dosimeter," final report prepared for R. S. Landauer, Jr., and Co., Department of Electrical Engineering, Univ. of Utah, May 25, 1982.

Chapter 25

Direction-Finding Antennas

R. E. Franks

ESL, a Subsidiary of TRW

CONTENTS

 **Raymond Franks** is an electrical engineer with over 30 years of experience in communication and radar systems, specializing in the antenna and propagation aspects of system analysis and design. He is currently a principal engineer at ESL, a subsidiary of TRW, in Sunnyvale, California. He has been responsible for the antenna and system design of several vhf, uhf, and microwave direction-finding systems from the conceptual stage to field evaluation. The development of antenna systems with extremely broad bandwidths, such as log-periodic arrays and ridged waveguide horns, has occupied much of his career. He has been awarded three patents involving broadband sequential lobing tracking antennas and a dual-polarized ridged waveguide horn design. He has coauthored several technical papers on broadband antennas, polarization coupling, and multiple-signal df processing (MUSIC algorithm).

Prior to joining ESL, he was an engineering specialist at GTE Sylvania Electronic Defense Laboratories, and had previous experience with the General Electric Company. He received the BS degree in electrical engineering from Rice University in 1950.

1. Introduction

The purpose of a radio direction finder (df) is to measure the direction of arrival of a radio signal. Traditionally the azimuth angle of arrival, i.e., the angle of arrival in the horizontal plane, has been considered sufficient to define the direction
of arrival of the signal, but the elevation angle of arrival as measured in a vertical plane is important for some target location schemes and for other purposes, such as propagation research. In any event the elevation angle of arrival of the incoming wave will affect the performance of azimuth-only df systems, even if elevation output is not needed.

Another complication in the use of df systems arises because the signal may propagate via more than one path between transmitting and receiving antennas. This is particularly likely in the case of ionospheric propagation of hf signals. It also occurs to some degree at most ground-based df sites when signals are being received via any mode of propagation because of rf being scattered by objects near the df receiving site. Much of the complexity of df systems and the error in direction of arrival of their output results from the multipath propagation of radio signals.

A direction finder consists of three principal elements: a directional antenna or antenna system to sample the incident wave, a receiver or receivers to limit the signal bandwidth and provide system gain, and a processing and indicating subsystem to provide an output in a useful form. The antenna portion of the df system will be considered in this chapter, with some discussion of the processing and output implementation where necessary to understand the antenna function.

The fact that the signal being received is a traveling wave spreading outward from its source is the basis for all direction-finding antenna techniques. The phase progression of the signal as it travels along its path is the characteristic used to advantage by df antennas. The df antenna is made of two or more elements that are located physically displaced from each other, so that the wave may be sampled at two or more points. The phase differences between these points contain the information on the angle of arrival of the wave. A line perpendicular to the direction of travel of the wave will be a line of constant phase. If the sampling points lie along this line, all of the points will be in phase. This is the condition that is utilized by most df antenna arrays to determine the direction of travel, and thus the angle of arrival of the signal.

The phase differences between the array elements may be directly measured, as in an interferometer; or it may be more convenient to combine the signals from the elements so that the response of the array of elements varies with the direction of arrival of the signal. The directivity of the antenna system can then be used to measure angle of arrival by rotating the antenna pattern and observing how the power out of the antenna changes with angle. Several types of antenna patterns may be generated. A peak in the pattern might indicate the angle of arrival of the signal. Alternately, a null in the pattern could be used. Or the equality of power out of two ports of the array could be used to indicate signal angle of arrival.

2. Rotating Antenna Patterns

When a directive antenna pattern is formed by the df antenna, the angle of arrival of the signal is determined by rotating the pattern and noting the angle at which the distinguishing pattern feature (null or maximum) is seen in the signal output. This angle corresponds to the angle of arrival of the signal. The entire antenna system can be physically rotated about an axis, usually vertical, to rotate the directional pattern. This is usually done for antenna systems that are not large. When the antenna system is physically too large to rotate, the elements may be stationary and the antenna pattern rotated using a goniometer that rotates and couples to the elements in a sequential fashion to produce a rotating pattern.

Physically Rotating Antenna Systems

Fig. 1 shows examples of rotating df antennas. The loop and the Adcock array produce nulls when the incoming wave direction is perpendicular to the plane of

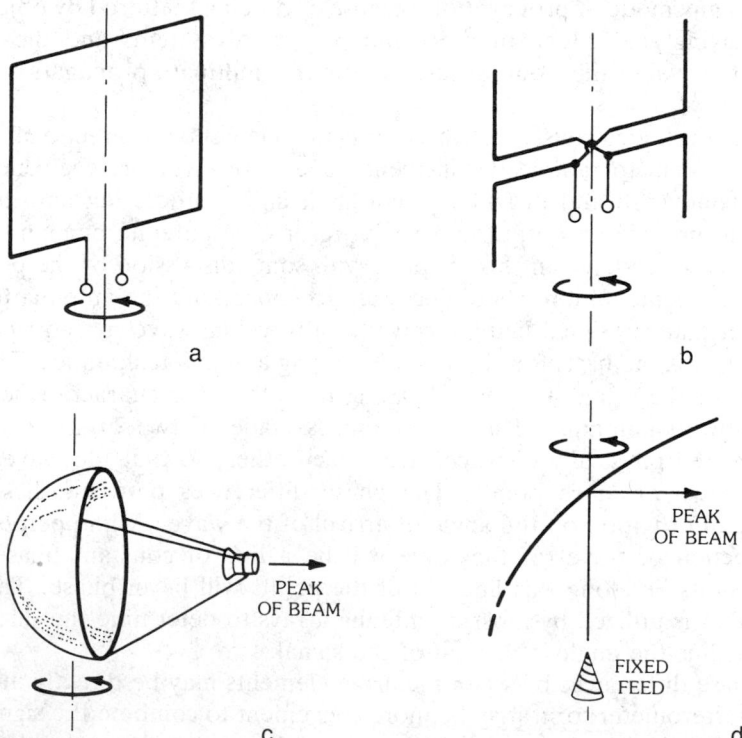

Fig. 1. Rotating df antennas. (*a*) Loop. (*b*) Adcock. (*c*) Reflector with feed. (*d*) Rotating reflector with stationary feed.

the antenna, and they are generally used at longer wavelengths. Rotating reflector antennas with feeds attached and offset-fed rotating reflectors with fixed feeds are typically used at shorter wavelengths where a high-gain pencil beam may be formed in an aperture small enough to rotate. Here the peak of the rotating pattern indicates angle of arrival.

If the antenna is continuously rotated, the instantaneous angular position of the antenna may be read at the antenna and transferred to the deflection circuits of a cathode-ray tube to drive the angular orientation of the display in synchronism with the antenna. Radial amplitude of the display is then driven by the received signal level out of the antenna and the antenna pattern is thus traced out on the display. The angle of the peak for a narrow-beam antenna or the angle of the null for a low-gain element, such as a loop, can be read off an azimuth scale around the face of the tube.

Loop Antenna DF Patterns—The open-circuit voltage of a loop antenna with its axis vertical as shown in Fig. 2 when receiving a vertically polarized signal of wavelength λ arriving in a horizontal plane at an angle θ to the plane of the loop is

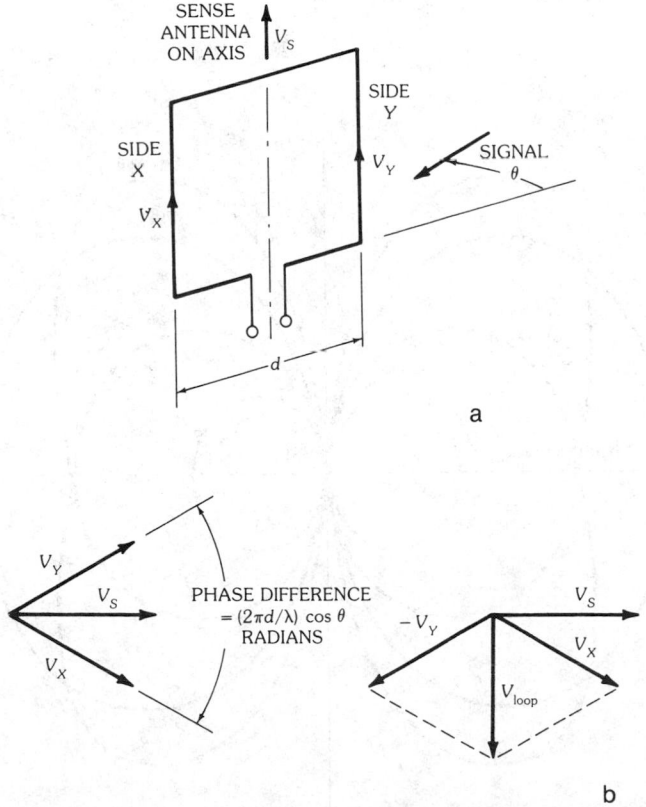

Fig. 2. Loop antenna output voltage. (*a*) Loop geometry. (*b*) Addition of voltages.

$$V_{\text{loop}} = 2V_S \sin\left(\frac{\pi d \cos \theta}{\lambda}\right) \tag{1}$$

where d/λ is the spacing in wavelengths between the sides of the loop and V_S is the voltage induced in one vertical side. This response pattern is plotted in Fig. 3 for several values of d. The nulls at $\theta = 90°$ and $270°$ are the pattern characteristics used for direction finding.

Because there are two nulls, $180°$ apart, the loop presents an ambiguity in its df indication that must be resolved. This can be done by combining the output of the sense antenna shown in Fig. 2, which is a vertical element, with the loop to generate a cardioid pattern. The loop resultant voltage E_R is in quadrature with the voltage induced in the sense antenna as shown in Fig. 2b. If the sense antenna voltage is shifted $90°$ in phase lead and added to the loop as can be done with a $90°$ hybrid combiner, the resultant pattern will be a cardioid if the loop and sense antenna voltages are equal in magnitude. If they are unbalanced, the resulting pattern is distorted as shown in Fig. 4 but will still resolve the ambiguity. In all cases the pattern will produce a sloping response near the two nulls of the loop, and the null that is at the location where the slope increases with increasing angle of arrival of the signal will be the correct null for the signal arriving at $90°$ to the plane of the loop.

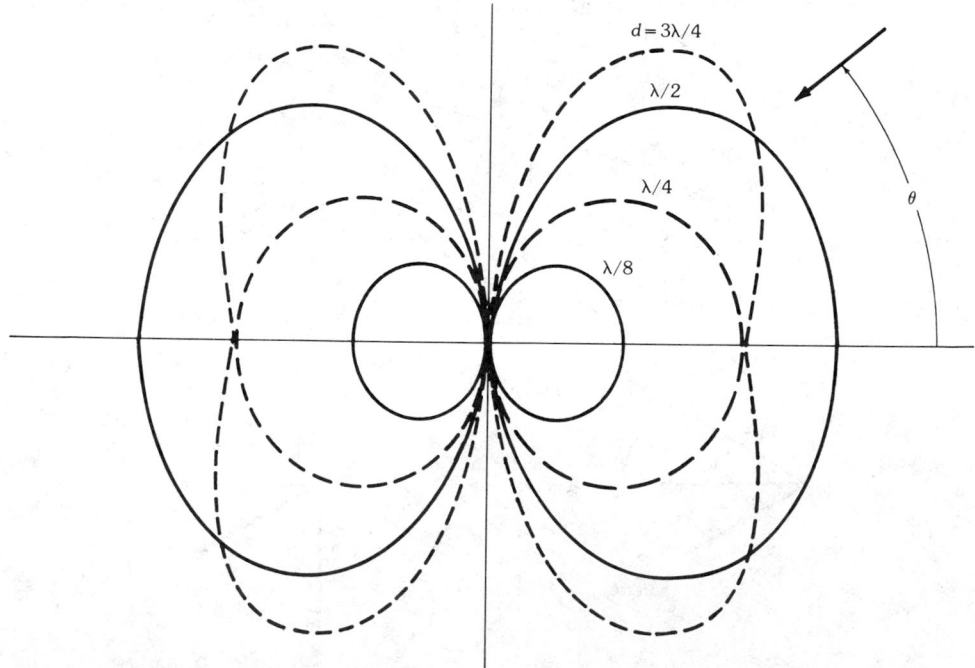

Fig. 3. Loop patterns for several values of spacing between sides.

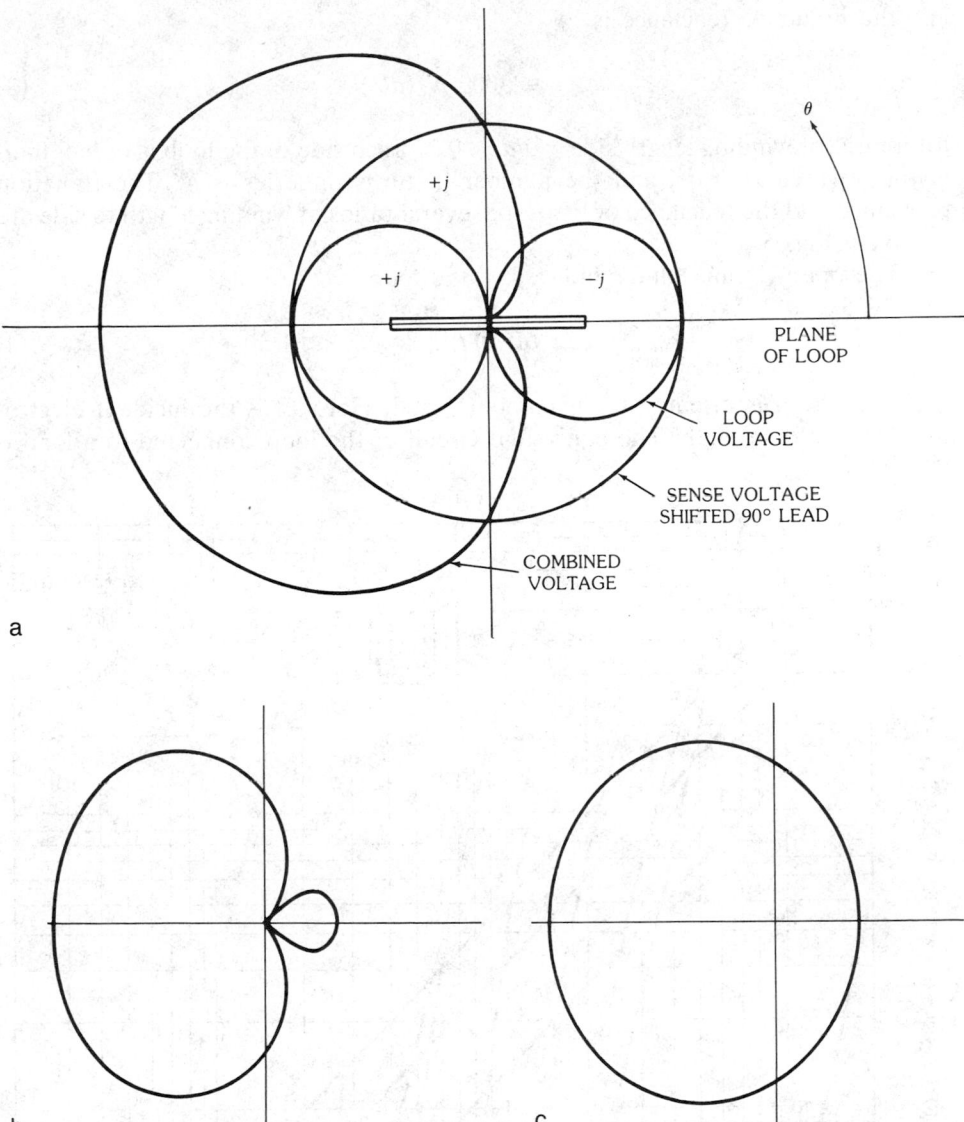

Fig. 4. Combined patterns for loop and sense antenna. (*a*) Cardioid pattern when equal loop and sense antenna voltages are combined. (*b*) Combined pattern with sense antenna voltage of half the loop voltage. (*c*) Combined pattern with sense antenna voltage of twice the loop voltage.

Loop Antenna Impedance and Efficiency—The electrical characteristics of a small, square loop antenna can be expressed as follows. The radiation resistance is

$$R_r \cong 30\,000\ N^2(d/\lambda)^4$$
$$= 30\,000\ (NA/\lambda^2)^2 \tag{2}$$

and the inductive reactance is

$$X_L \cong 3000 \, N^2(d/\lambda) \tag{3}$$

for a ratio of winding length/side $= t/d = 0.2$. Each side of the loop is of length d, giving an area $A = d^2$, and the number of turns in series is N. The radiation resistance and the reactance of loops for several ratios of winding length to side are shown in Fig. 5.

The open-circuit voltage induced in the loop is

$$V_{\text{loop}} = (2\pi N d^2/\lambda) \, E^i, \qquad d/\lambda \ll 1/\pi \tag{4}$$

when the loop is oriented for maximum signal. Here E^i is the incident electro-magnetic-field strength. The equivalent circuit of the loop connected to a load is

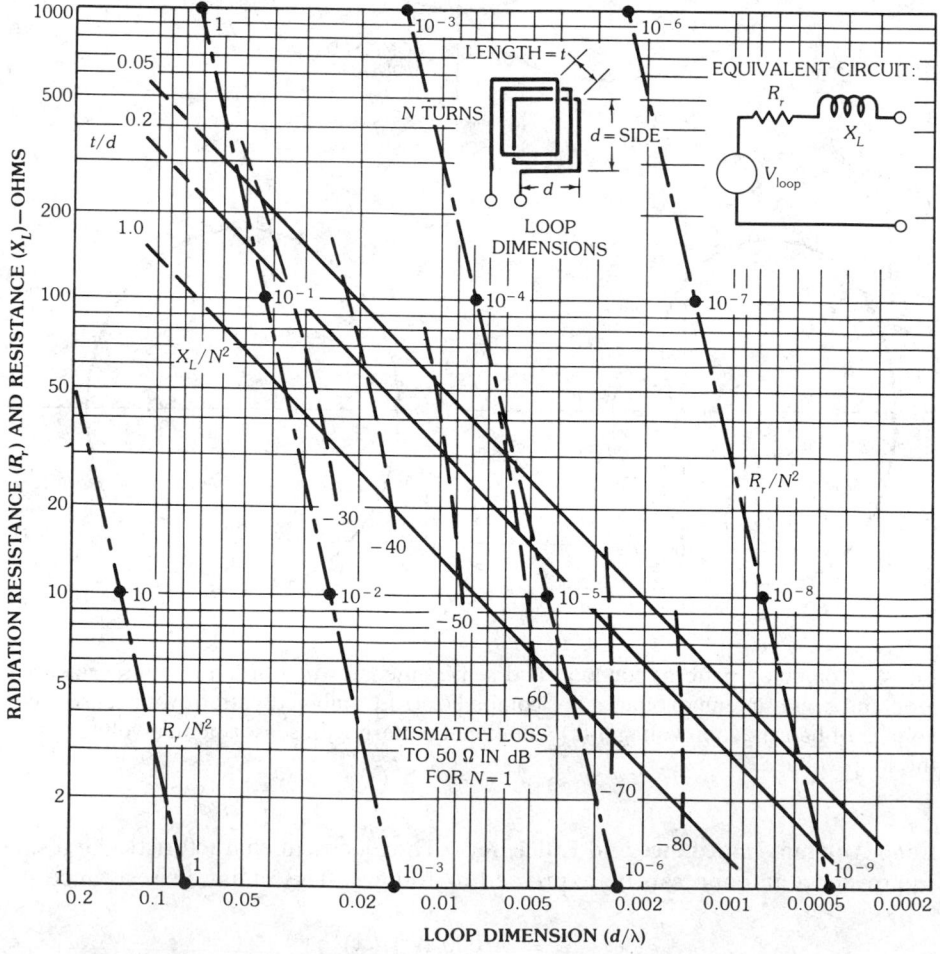

Fig. 5. Radiation resistance and reactance of loop antennas.

given in Fig. 5. The mismatch loss of a single-turn loop directly connected to a 50-Ω transmission line is plotted in the figure. For more information on the use of loops see Chapter 7.

Special Problems With Loops—The environment around the loop can upset the balance between the two vertical sides of the loop and seriously degrade the null. If the capacitance to ground is not the same for both sides, stray voltages may be coupled into the loop and unequal currents may flow into the load. To avoid this condition the loop can be enclosed in a metal shield as in Fig. 6. The shield must have a gap at the center of the loop to prevent it from shorting the loop. The transmission line feeding the loop should be balanced to ground, with a balanced-unbalanced transformer to convert the loop output to single-ended coaxial transmission line for connection to the receiver.

In the hf band the loop is unsatisfactory because of downcoming sky-wave signals. This limitation may also apply in other bands if horizontally polarized waves can fall on the loop.

Elevated H Adcock—An antenna that rejects horizontal polarization is the elevated H Adcock array shown in Fig. 7. This antenna consists of two vertical dipoles supported by a horizontal member which carries balanced shielded transmission lines. The dipoles are connected out of phase to produce nulls broadside to the array. The pattern for vertically polarized signals in the horizontal plane is similar to that given for a loop in Fig. 3. A sense antenna is needed, and this may be a third dipole located at the center of the array and connected to the output of the array in the same manner that the sense antenna is connected to the loop.

The voltage out of the array will be decreased as the cosine squared of the elevation angle because both the phase difference between the dipoles and their induced voltages decrease with elevation.

Spaced Loops—Another type of antenna useful for sky-wave signals is an array

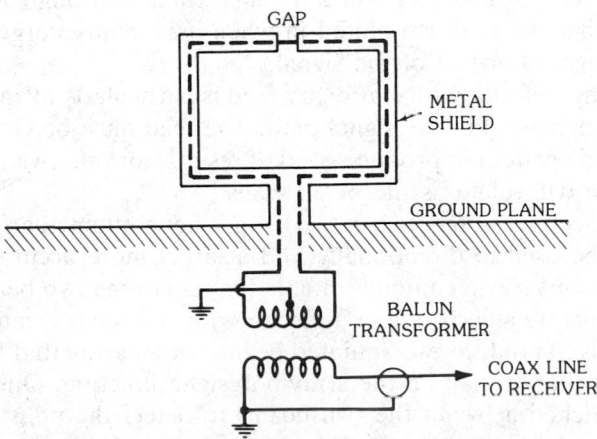

Fig. 6. Balanced shielded loop.

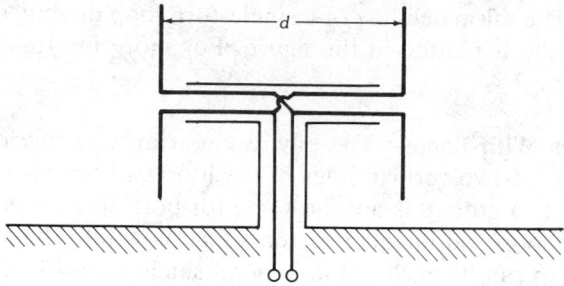

Fig. 7. Elevated H Adcock antenna.

of loops. Three configurations are shown in Fig. 8. The vertical coplanar loops of Fig. 8a generate a pattern similar to that of the single loop, except that the null is deeper because in addition to the array factor forming a null each loop of the array also has a null at 90° and 270° to the plane of the loop. The downcoming horizontally polarized wave produces identical signals in the two loops at the null, and this is canceled by the out-of-phase connection. The induced voltage of downcoming signals is not reduced as it is with Adcock antennas.

A second spaced loop array is the coaxial arrangement of vertical loops shown in Fig. 8b. This reduces the coupling of the loops to the horizontal support structure, but produces a pattern with four nulls per revolution, at 0°, 90°, 180°, and 270°, because now the loop elements generate nulls in the 0° and 180° directions.

Loops that lie in the horizontal plane can be arranged as shown in Fig. 8c. The response of the horizontal coplanar loop array is the same for horizontally polarized signals as that of the elevated H Adcock array for vertically polarized signals.

Rotating High-Gain Antennas—In the frequency range above 1 GHz, a relatively small pencil-beam antenna is feasible. Two types of high-gain aperture antennas were illustrated in Fig. 1. In each of these antennas the antenna pattern will be traced out by receiving a signal as the antenna is rotated through 360° azimuth, and the direction that the peak is pointed in when maximum energy is received will indicate the angle of arrival of the signal.

The rotating reflector with stationary feed is particularly attractive because no rotary rf joint is needed in the signal path. The feed must be circularly polarized with a symmetric pattern to produce good df results, for otherwise the polarization of the df beam will rotate as the beam scans.

Amplitude comparison between two beams that point in nearly the same direction can also be used to directionally find a target more accurately than with a single pencil beam. Any technique suitable for generating two beams can be used, such as two separate reflectors, one reflector with two feeds located on either side of the focal axis to produce two squinted beams, or an array that forms two beams simultaneously that overlap on the array boresight direction. One receiver can be alternately switched between the two beams to detect the point of equal output within the main lobe response of the antenna system [1].

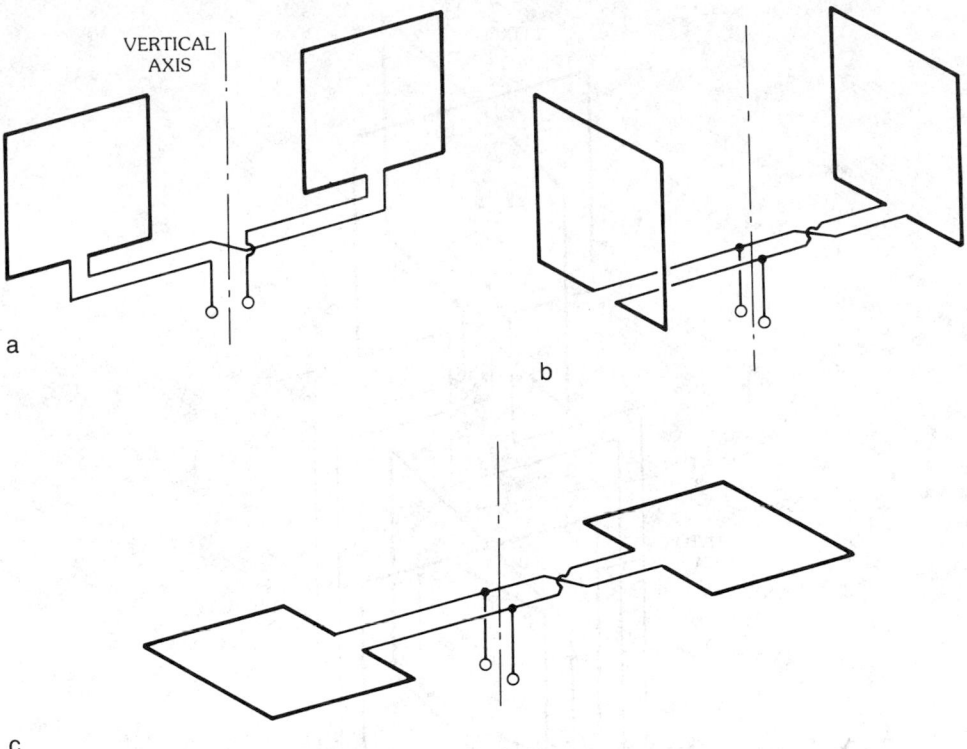

Fig. 8. Spaced loop arrays. (*a*) Vertical coplanar spaced loops. (*b*) Vertical coaxial spaced loops. (*c*) Horizontally polarized spaced loops.

Electrically Rotating Null Patterns

When the antenna array is too large to be physically rotated conveniently, the pattern from stationary elements can be rotated. An example of this using crossed loops is shown in Fig. 9, where the loops feed a goniometer having two stationary coils at right angles and enclosing a smaller rotating coil. The stationary coils simulate the external radio-frequency field within the goniometer so that the rotor behaves as if it were a loop antenna rotating in the external field. Pairs of Adcock elements may be substituted for crossed loops to produce the same performance.

Ideally the external antenna patterns must vary as $\sin \theta$ and $\cos \theta$ with the same maximum response. Coupling between stator and rotor must also vary sinusoidally with rotation. Because of symmetry no error occurs at 0°, 45°, 90°, 135°, etc., but midway between these points, at 22.5°, 67.5°, etc., error builds up as the diagonal of the array (*d* of Figs. 2, 5, or 7) exceeds 0.2λ because the element patterns depart from the ideal. At an array diagonal of 0.5λ, the peak error reaches 7°, and it is 18° at 0.7λ diagonal. If two sets of Adcock elements with an angular separation of 55° between the sets are connected to each stationary coil, the array diagonal can be increased to one wavelength before the error exceeds 3°; however, it builds up rapidly beyond this point. This error decreases with increasing elevation angle of

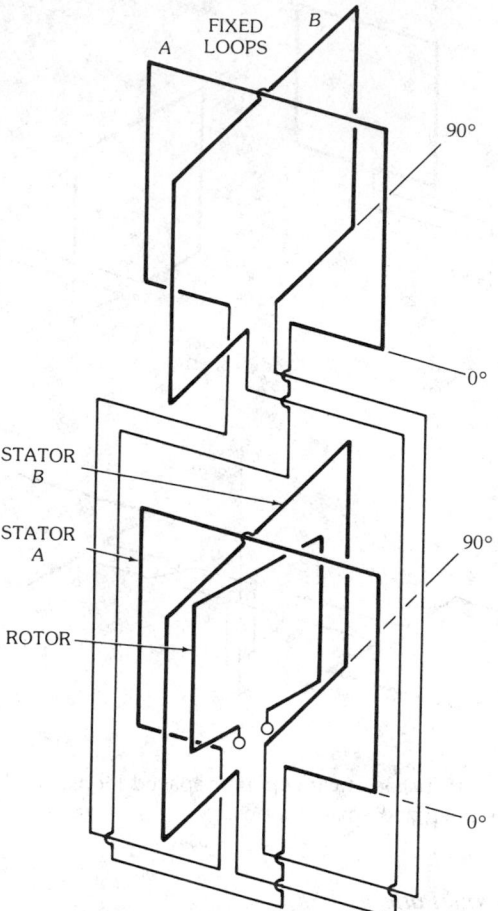

Fig. 9. Electrically rotating loop pattern using a goniometer.

arrival of the wave because the effective antenna diagonal is reduced, causing the antenna patterns to better approximate sinusoidal variation.

Instantaneous df Patterns

The Watson-Watt df system uses two crossed loops or two Adcock arrays at right angles in a manner similar to the rotating goniometer, except that the two rf signals are compared instantaneously to generate angle of arrival. The output of each array pair is amplified by a receiver, usually down-converted to a convenient intermediate frequency, and applied to one pair of deflection plates of a crt. A bi-directional line results, inclined at the signal angle of arrival. The sense antenna may be coupled into the receivers momentarily to resolve the ambiguity. This system is particularly useful for short-duration signals as no pattern scanning is required.

The two coherent receiver channels need to have good phase and amplitude balance to give accurate results. A single-channel system can be developed, but it will no longer generate instantaneous df output. The antenna signals can be combined, as shown in Fig. 10, to produce a composite signal that carries the bearing information in one rf channel on two audio frequencies. Two separate audio tones, one for the north/south (N/S) loop and one for the east/west (E/W) loop, are used to modulate each antenna output as shown in the figure [2].

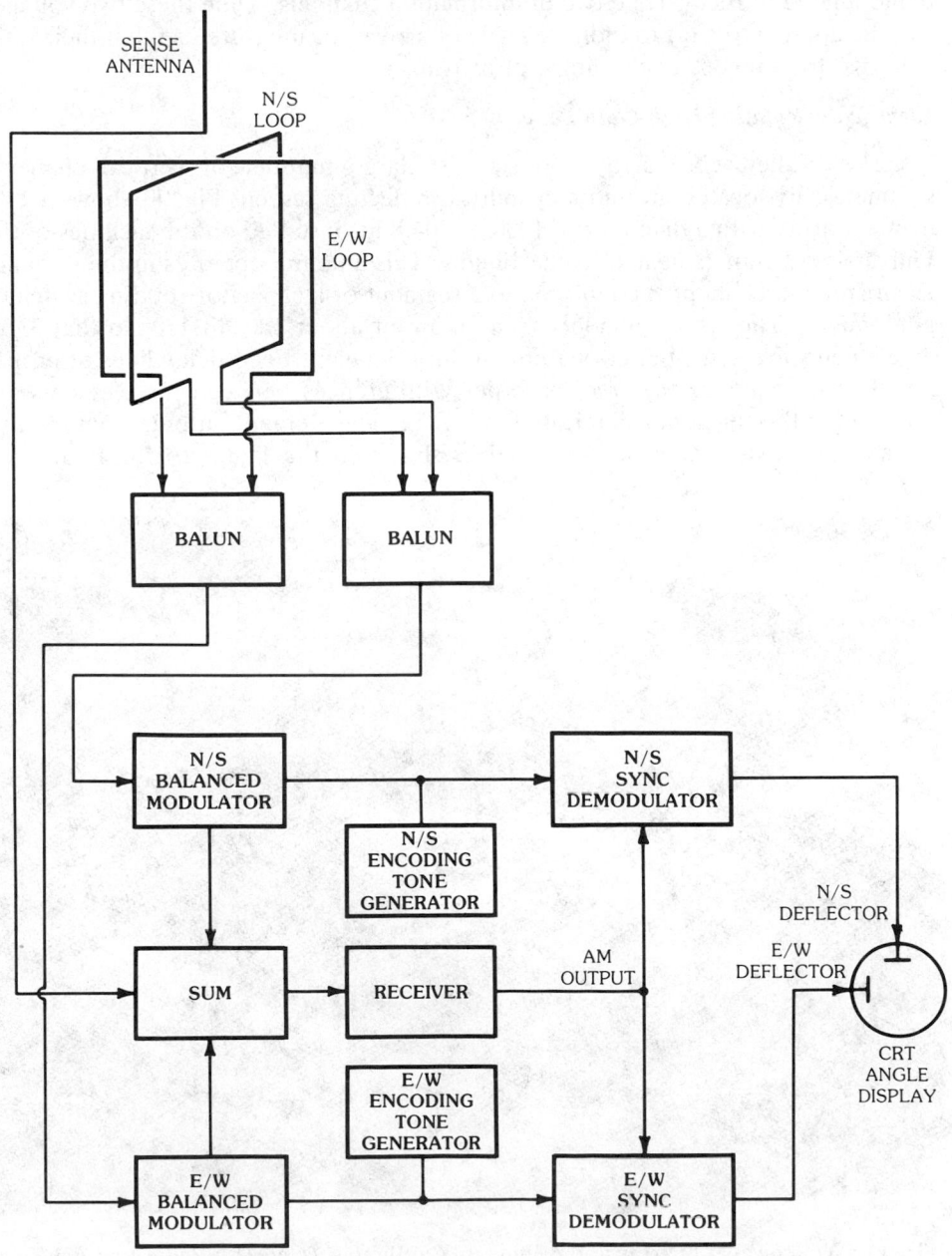

Fig. 10. Single-channel Watson-Watt df system.

The balanced modulators generate double-sideband suppressed-carrier rf outputs which are summed with the sense antenna output. Since this sense signal acts as a reinserted carrier, the composite signal applied to the receiver is an am signal with the two low-frequency audio tones carrying the amplitude and phase information from the df antennas. This modulation is imposed on top of any other modulation that may be on the signal. The receiver am output is synchronously demodulated to recover the two df information channels. Then these two voltages may be applied to a crt to indicate angle of arrival, or measured so that their ratio will give the tangent of the angle of arrival.

Electrically Rotating High-Gain Pattern

The Wullenweber array consists of a large number of vertical elements symmetrically located around a cylindrical reflecting screen. Fig. 11 shows a 120-element array with a diameter of 1000 ft (304.8 m) installed on the facilities of the University of Illinois near Urbana, Illinois. This antenna operates in the hf band. Each antenna element is connected to a segment of a capacitor rotating switch or goniometer. The rotor segments span an arc of about 100° to 120° so that 33 to 40 elements are activated at one time to form a beam. Feed delay lines of proper lengths that equalize the free-space delay differences between the elements are attached to the segment as indicated in Fig. 12. The element outputs from the two equal half-sections may be combined as shown in the figure to form sum and

Fig. 11. Wullenweber hf array at the University of Illinois. (*Courtesy Radio Location Lab., University of Illinois at Urbana*)

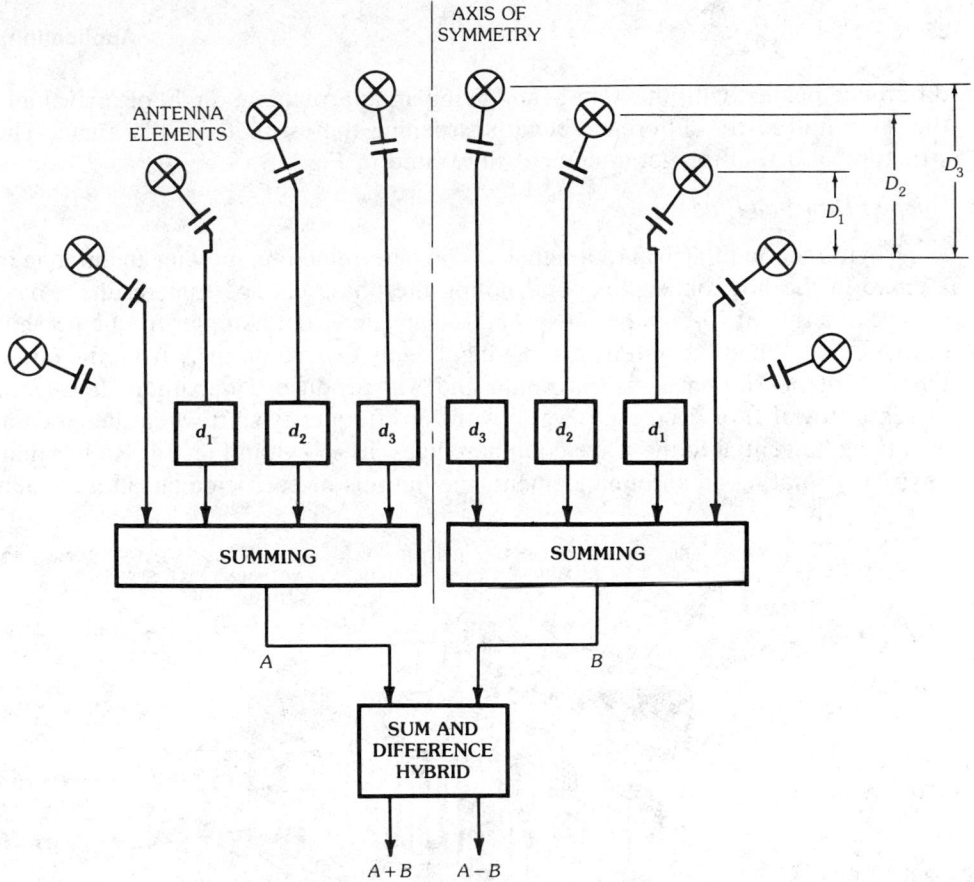

d_1, d_2, d_3 are delay lines with time delays equal to propagation delays D_1, D_2, D_3

a

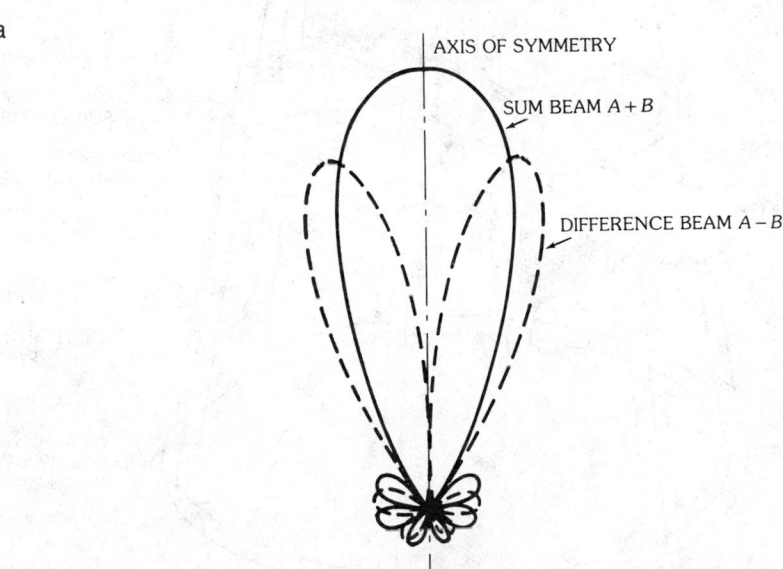

b

Fig. 12. Electrically rotating high-gain df pattern. (*a*) Goniometer circuitry. (*b*) Sum and difference patterns.

difference beams, with the sum beam giving an approximate angle of arrival and the deep null in the difference beam sharpening the estimate of the angle. The structure of a rotating goniometer is illustrated in Fig. 13.

Doppler Direction Finder

The direction of arrival of a signal can be determined by moving an antenna in a circle in the horizontal plane and noting the phase change that results. Phase change at a rate of 2π radians per second is equivalent to a Doppler frequency shift of one hertz. When the antenna is moving directly toward or away from the target the rate of phase change is maximum and will produce a maximum frequency increase (toward) or decrease (away), with zero frequency shift when the antenna is moving tangential to the wave equiphase lines, as shown in Fig. 14. Rather than physically rotating an antenna element, the outputs of each element in a circular

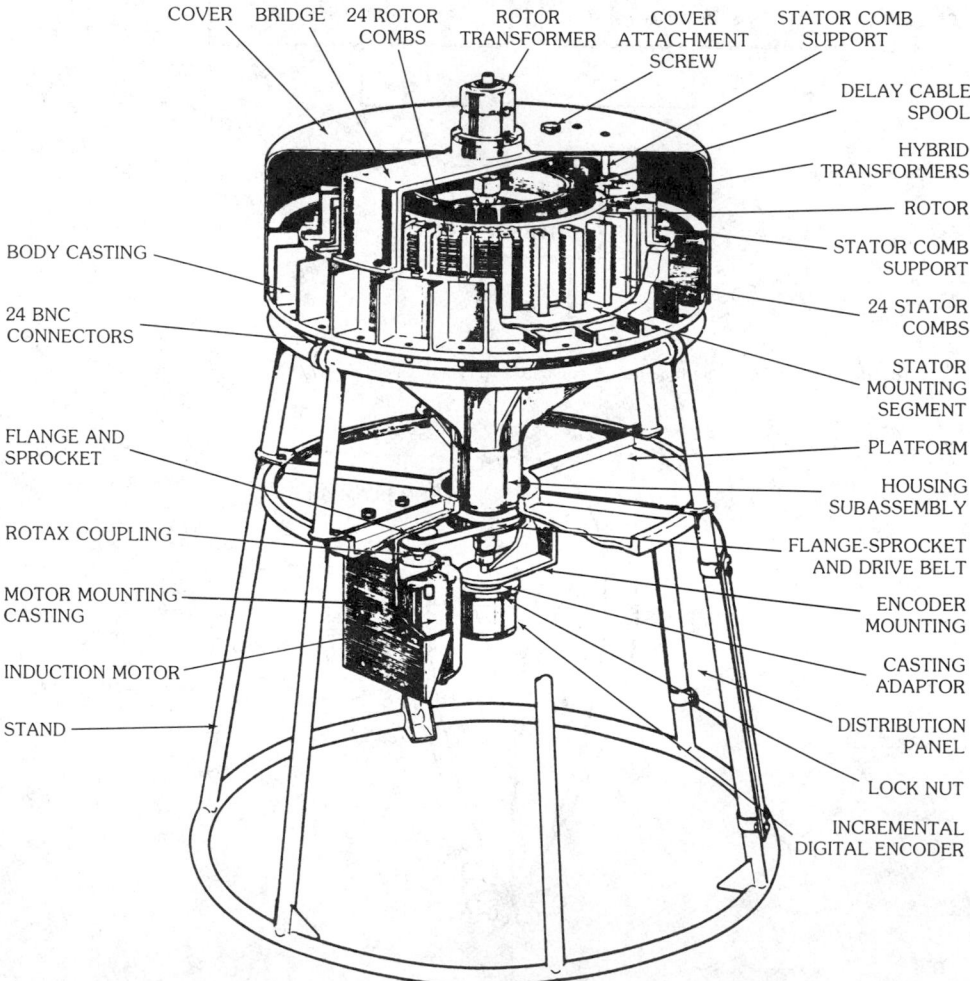

Fig. 13. Mechanical structure of a rotating goniometer. (*After Gething [11]*)

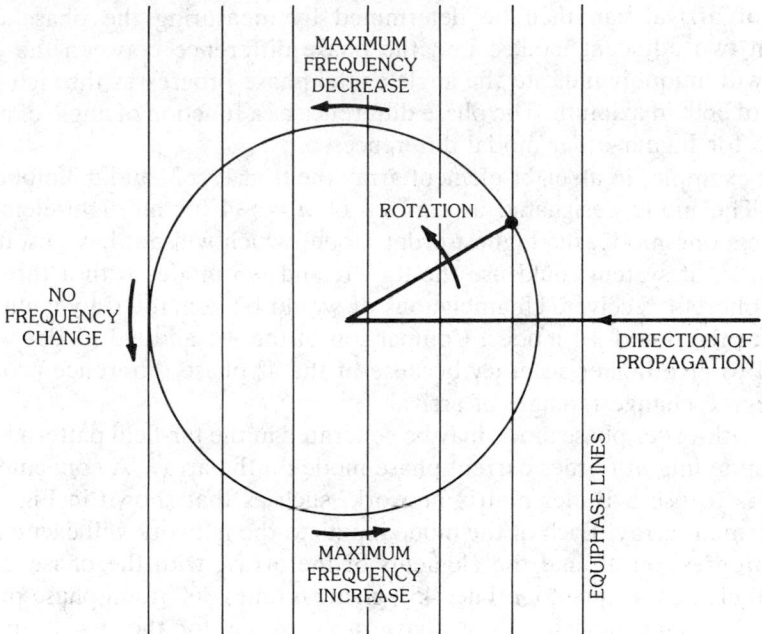

Fig. 14. Doppler shift produced by rotating antenna.

array are sampled sequentially around the circle to generate a phase shift that can be detected to extract the Doppler modulation thus introduced. This modulation will vary at the same rate as the cyclic rate of sampling of the antennas, so if the phase of the modulation is compared with a reference voltage synchronized with the antenna commutation, the angle of arrival of the signal will be indicated. Sampling may be accomplished using a mechanical capacitive goniometer or an electronic commutating switch.

The array elements must be spaced less than one-half wavelength apart around the circle or phase ambiguities will be introduced in the phase modulation. A center antenna may be used as a phase reference so that the phase difference between the rotating samples and the fixed antenna can be detected at the output of a dual-channel phase-coherent receiver.

3. Multimode Circular Arrays

The Wullenweber and the Doppler circular array antennas just described provide direction finding through a scanning process that involves rotating an antenna response through 360° azimuth and deriving angle of arrival from the timing of the antenna output within the scan cycle. Circular arrays can also be used in a nonscanning mode to give an instantaneous df response. This is done by exciting the circular array of elements with a feed network that sets up several far-field phase modes.

The *n*th-order phase mode pattern is defined as having constant amplitude and a phase varying linearly from 0 to $2n\pi$ as the angle of arrival varies through 360°

Angle of arrival can then be determined by measuring the phase difference between two adjacent modes, i.e., the phase difference between the 0 and +1 modes will uniquely indicate the angle as the phase progresses through 360° for a change of 360° in azimuth. The phase difference as a function of angle of arrival will increase for higher-order modal differences.

For example, in an eight-element array the 0, ±1, ±2, and ±3 modes may be useful. The mode designated as $n = +4$ or $n = -4$ for an eight-element array represents one mode, the highest-order mode, which will not have useful characteristics. A df system could use the 0, +1, and −3 modes with a three-channel phase coherent receiver. Unambiguous df would be generated by comparing the phases of the 0 and +1 modes. Comparison of the +1 and −3 modes would also be used to give higher accuracy because of the 4° phase difference produced for each degree change in angle of arrival.

The nth-order phase mode may be generated in the far-field pattern by exciting a corresponding nth-order current phase mode on the array. A convenient way to do this is to use a Butler matrix network, such as that shown in Fig. 15 for an eight-element array. Each of the mode inputs to the network will excite a uniform phase progression around the elements of the array, with the phase differential between elements equal to n times 45°, giving n times 360° total phase progression around the eight elements. The phase progressions for the $n = 0$ and $n = +1$ modes are shown in the figure.

The relative amplitudes of the modes are given by Bessel functions for a continuous circular array of omnidirectional elements excited by a phase progression of order n:

$$\text{relative amplitude of } n\text{th mode} = J_n(\pi D/\lambda) \tag{5}$$

where

J_n = Bessel function of order n

D = diameter of array

An array of discrete omnidirectional elements will produce essentially these same amplitudes as long as the element spacings are well under a half wavelength. In general, the higher the order of the mode and the greater the element spacing, the more the mode phase or amplitude may depart from the ideal.

The Bessel functions that give the mode amplitudes predict nulls for certain combinations of mode number and array diameter, which would seriously limit the bandwidth that could be achieved for a given combination of modes. Fortunately, if directional elements are used in the array with the peak gain of each element pointed away from the center of the array, the mode amplitudes are more constant with array size. The calculated mode amplitudes for omnidirectional elements and for elements with a $1 - \cos\theta$ pattern are shown in Fig. 16 for the 0, 1, and 2 modes. Experimental results have shown that the mode amplitudes do vary smoothly without nulls as the operating frequency varies using directional elements. Dipole elements in front of a reflecting screen or slots on a conducting cylinder would yield element patterns with the desired directivity [3, 4].

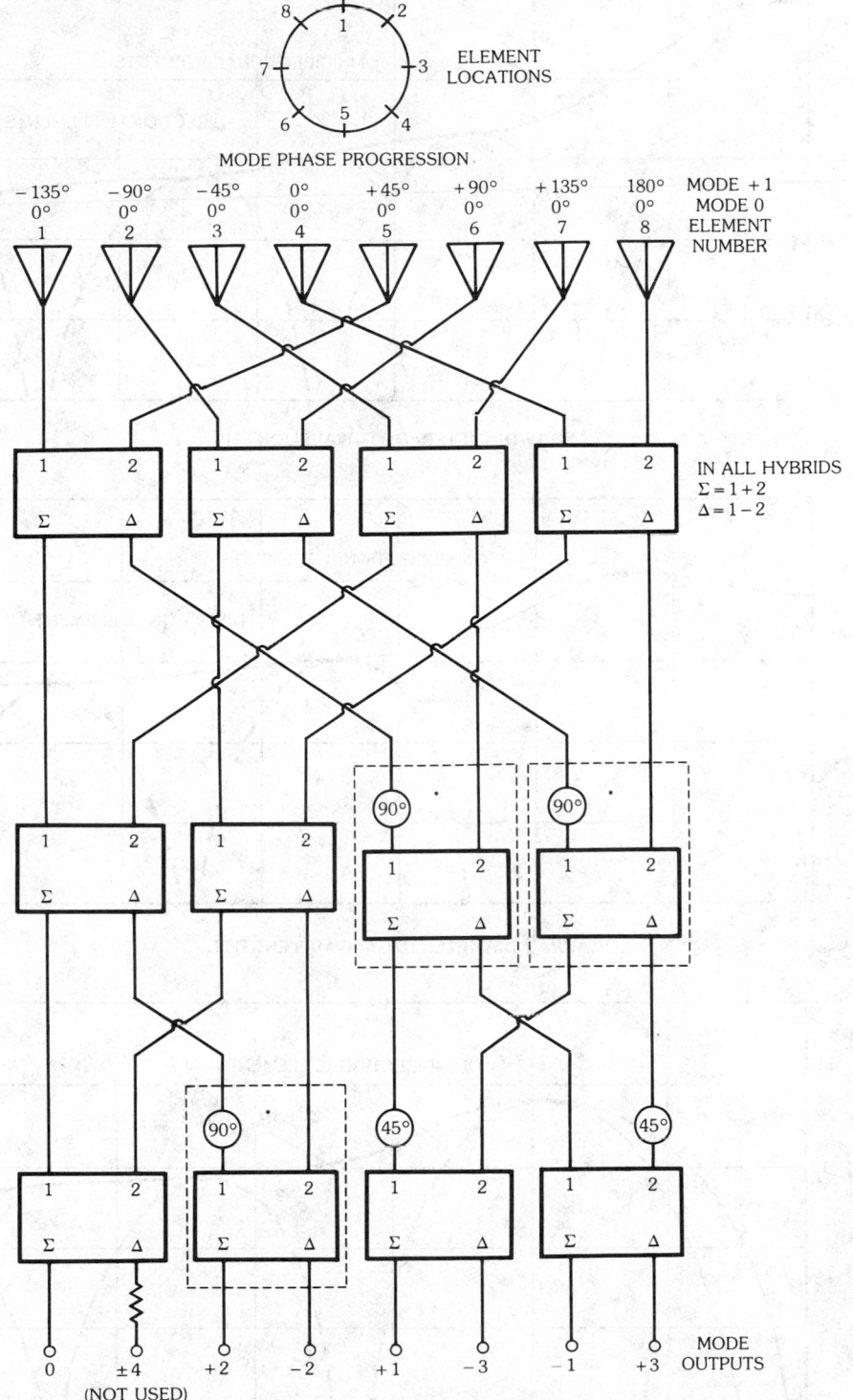

Fig. 15. Butler matrix feed network for a multimode array.

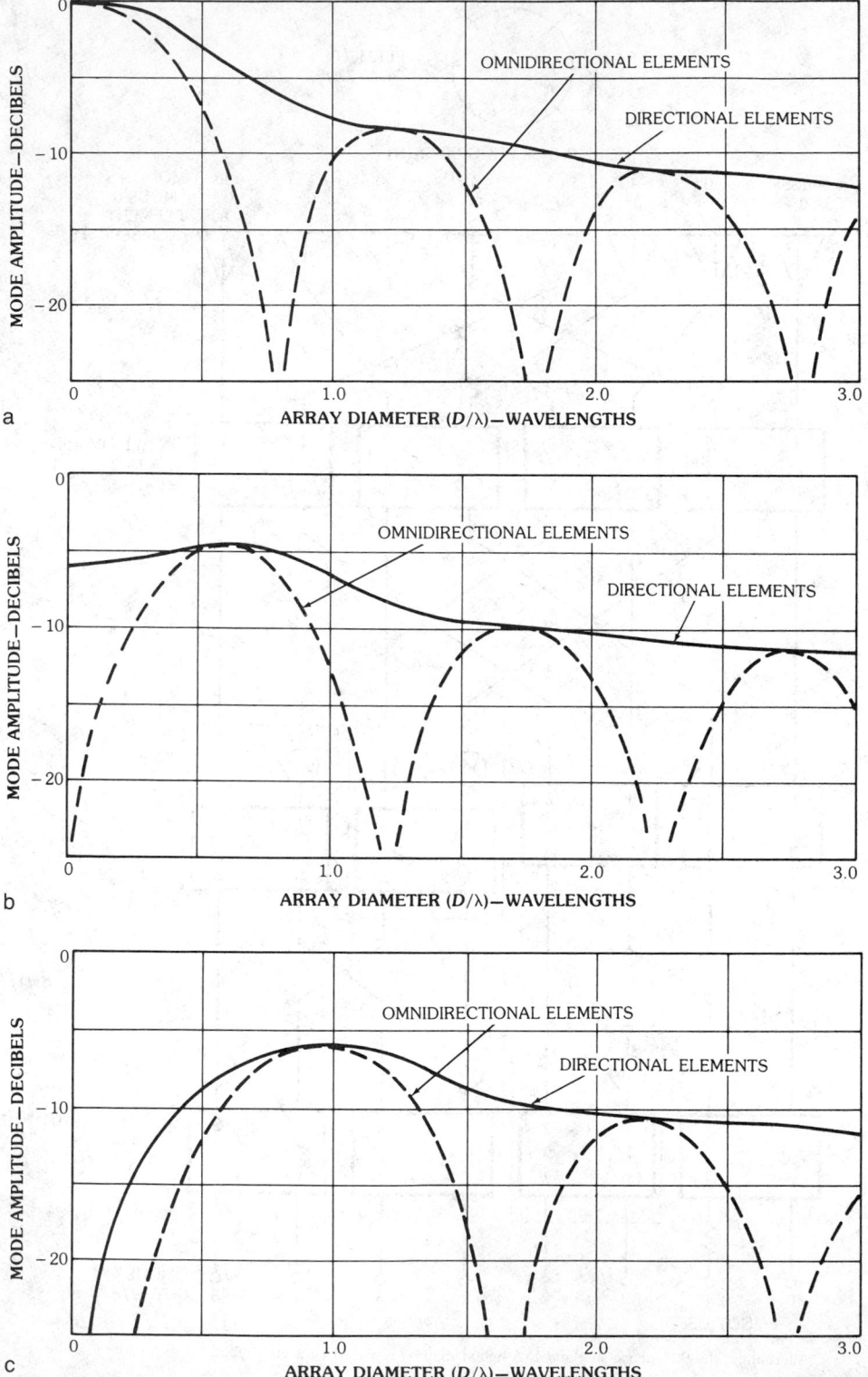

Fig. 16. Mode amplitudes versus array diameter. (*a*) Mode 0. (*b*) Mode 1. (*c*) Mode 2.

4. Interferometers

The df techniques described up to now have employed the amplitude and/or phase pattern characteristics of an antenna array to indicate angle of arrival of a signal. An interferometer df system makes a direct measurement of the phase differences between two or more points on the wavefront and converts the phase information into angle of arrival. Fig. 17 illustrates the relationship between angle of arrival, α, as measured from the normal to the base line and the phase difference, $\Delta\phi$, between two elements spaced along a base line a distance d apart. In Fig. 17a, the output of antenna 2 leads that of antenna 1 by $\Delta\phi = (2\pi d/\lambda) \sin \alpha$ and it can be seen that the phase function has mirror symmetry about the base-line direction, $\alpha = \pm 90°$.

For element spacing less than one-half wavelength (short base lines) each value of phase difference corresponds to two angles of arrival. For element spacings in excess of one-half wavelength additional ambiguities are present because of "foldover" in the phase function when the phase difference exceeds 180°.

The mirror ambiguities for short base lines can be resolved by using two orthogonal base lines and combining the phase differences to get the angle α of arrival as measured from north:

$$\alpha = \tan^{-1}\left(\frac{\Delta\phi \ \text{E/W base line}}{\Delta\phi \ \text{N/S base line}}\right) \tag{6}$$

Mutual coupling between antenna elements will generate an octantal error that is zero for 0°, 45°, 90°, etc., and alternately reaches positive and negative maxima at 22.5°, 67.5°, etc. The magnitude of the error generally increases with element spacing and with element radiation efficiency.

The ambiguities on long base lines can be resolved by using a combination of short and long base lines, or two long base lines whose difference in length is less than one-half wavelength. Angle of arrival measurement errors tend to be reduced in proportion to the increase in base-line length, but the probability of an ambiguity being introduced by a given phase measurement error increases as the base lines are made longer. For example, if $n(\lambda/2)$ and $(n - 1)(\lambda/2)$ are the two base-line lengths, the base lines will both indicate the correct phase values at the true angle of arrival. At the angle corresponding to the nearest ambiguity on the longer base line, the other base line will differ in phase by only $360°/n$ from experiencing an ambiguity also. More antennas and more measurement base lines are needed as the overall length of the interferometer is increased [5, 6].

Antenna Elements

If the interferometer has 360° angular coverage, the antenna elements ideally have omnidirectional patterns. Vertically polarized dipoles, crossed vertical loops fed in quadrature, or monopoles on a ground plane are typically used. Loop antennas in a horizontal plane or a turnstile formed by crossed horizontal dipoles fed in quadrature can give omnidirectional horizontally polarized coverage.

The relative transmission phase stability of the antenna elements must be good to avoid introducing phase errors and resulting angle of arrival errors in the inter-

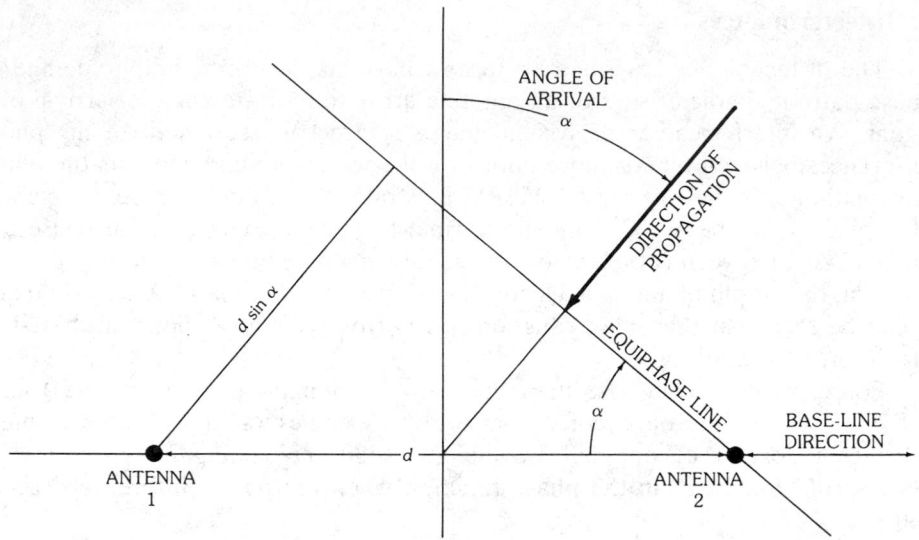

a

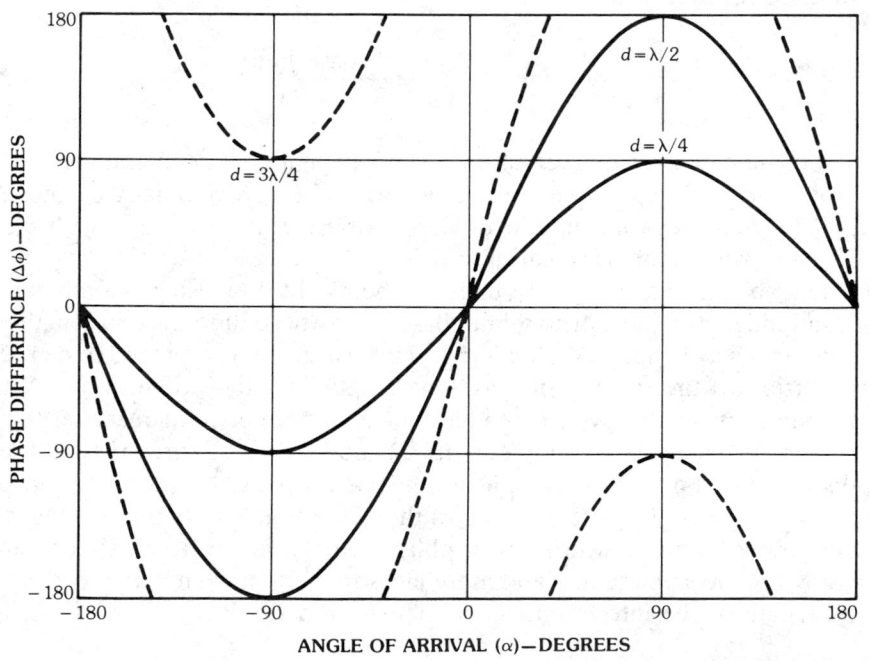

b

Fig. 17. Phase difference versus angle of arrival for a two-element interferometer. (a) Antenna geometry. (b) Phase difference.

ferometer output. If the df system is to operate over a wide frequency bandwidth, which is the usual condition, tuning of the elements to improve efficiency at the operating frequency may not be practical because of the difficulty in making the tuning circuits for all elements track in phase. Broadband impedance matching can be chosen to favor the portion of the band where antenna efficiency is lowest. As an example, electrically small dipoles or monopoles can be connected to a load impedance approximately equal to the magnitude of the antenna reactive impedance at its lowest operating frequency to give a compromise broadband match favoring the lower frequencies.

In the low vhf and the hf bands electrically small antennas may provide sufficient output in spite of their low efficiency. The external noise levels are generally high such that the noise power output from an efficient antenna would be much higher than internal system (receiver) noise. In this situation both signal and external noise will be reduced by the low antenna efficiency but the system signal-to-noise ratio will not be reduced significantly until the external noise power out of the antenna falls to a level near that of the system internal noise level.

At higher frequencies, and particularly into the microwave region, directive antennas may be needed to provide adequate system gain. In this case the angular coverage will be reduced and it may require several interferometer arrays to cover 360° in several sectors. Typically the choice between arrays is made by comparing the antenna amplitude responses to the signal being received to determine from the strongest response which angular sector contains the signal. Then the array covering that sector makes the phase measurements to determine angle of arrival.

Vehicular-Mounted Interferometers

An array of antennas mounted on a ground vehicle or an aircraft will receive energy scattered by the vehicle in addition to the direct signal wavefront. As a result the actual phase differences derived for antenna pairs will usually depart widely from the ideal sinusoidal variation with angle of arrival. Direction-finder outputs based on the antenna geometry alone may not be acceptable. The phase response will be stable, however, and can be measured to generate calibration data for accurate direction finding.

The vehicle with antennas installed may be rotated through 360° while receiving a signal from a known location. The base-line phase responses are measured and stored along with the true angle of arrival. These calibration data can be recalled when a df measurement on an unknown signal is made, and the calibration data searched to find the angle corresponding to the best fit of calibration to measured data. This angle is the best estimate of the signal angle of arrival.

Dipole antennas are preferred over monopoles for vehicle or aircraft df systems. The dipole will usually not generate as much cross-polarized radiation and will generally have a more uniform amplitude pattern. A monopole located on a long structure such as an aircraft fuselage may excite the structure in a long wire mode to radiate a multilobed, cross-polarized pattern. If two monopoles are placed diametrically opposite on the structure and driven out of phase to form a dipole, the excitations from the two monopoles tend to cancel in the structure and suppress long wire patterns.

5. Multiple-Signal Direction Finding

The df techniques described have been applied to measuring the angle of arrival of a single signal when only one wave is incident on the antenna. In general there may be several wavefronts present. These may consist of waves from one source that have traveled via different propagation paths and/or signals from more than one source.

The behavior of the df output under the condition of multiple uncorrelated (independent) signals depends heavily on the processing portions of the df system. A phase interferometer with amplitude limiting in the phase detectors may respond to the strongest signal with little effect from weaker signals. An interferometer that

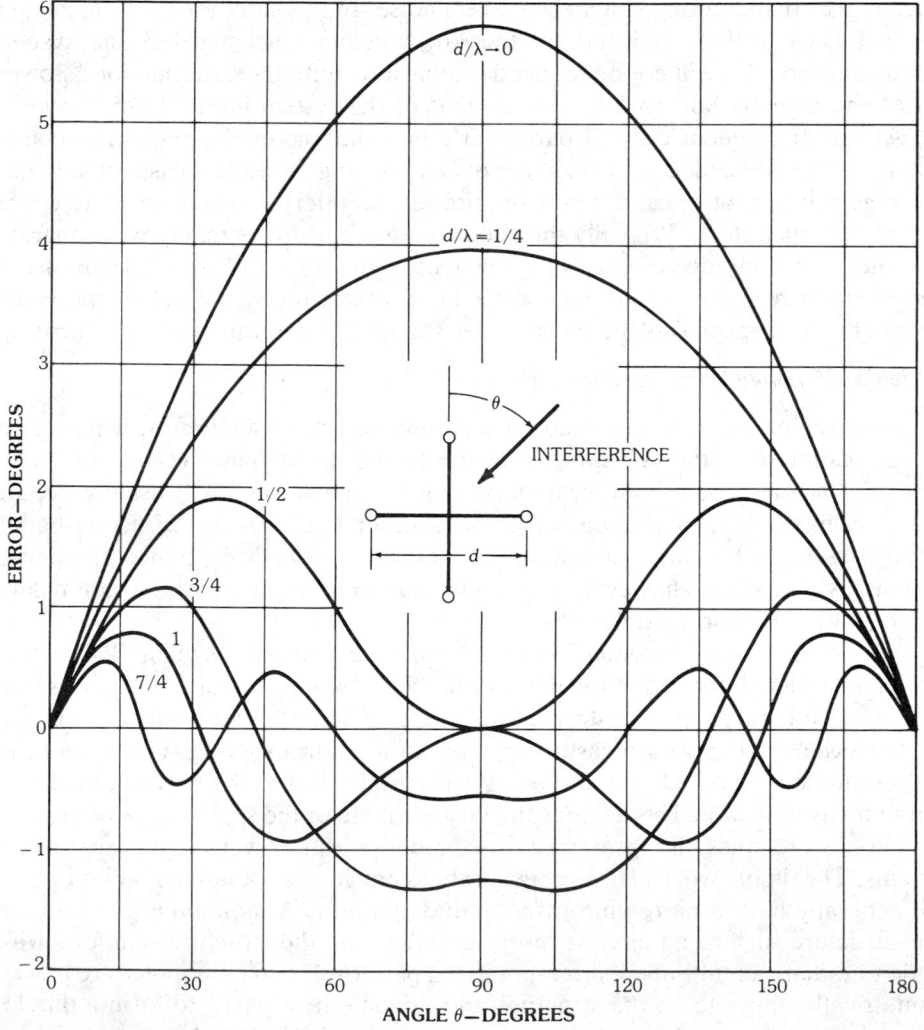

Fig. 18. Direction-finding error in the estimate of a signal at 0° in the presence of interference 10 db below the desired signal.

averages the complex cross-correlation coefficient between antenna outputs to measure phase will tend to produce an output that is the power-weighted average of the two signals; however, for long base lines the error decreases as base-line length increases as shown in Fig. 18 for interference 10 dB below a desired signal. Direction-finding systems with high-gain, narrow-beam antennas can give the angles of arrival of each of several signals if the signals are separated in angle by greater than the beamwidth.

If the signals are correlated, such as by multimode propagation from the same source, the error condition will be similar to that described above for uncorrelated signals except that the error produced in a df system will be related to the voltage and phase of the weaker signal instead of the power ratio. For example, Fig. 18 would apply to the rms error for an interferometer in the presence of multipath 20 dB below the direct path signal as the multipath phase varies over 360°.

To generate angle of arrival on each one of several signals or waves present, high-resolution processing techniques have been developed that use the phase and amplitude data from the df antennas and attempt to model the df system and the signal environment to estimate the angles of arrival in some optimal way. Generally a combination of several signal responses is selected that best coincides with the set of measured outputs from the antenna ports, and the angles of arrival of these selected responses are then an estimate of the angles of arrival of the signals [7, 8, 9, 11].

These system design techniques may impose more stringent requirements on the antenna components. Elements in an array may need to be more alike and their phase and amplitude characteristics better defined. It may be necessary to measure the overall response of the df antennas in their environment, especially where scattering from structures such as an airframe may be significant. Polarization diversity may be required in the array elements to cope with a wide range of signal polarizations [10].

6. References

References 12 through 22 are not indicated in the text of this chapter but are included here as being of general interest.

[1]   R. A. Watson-Watt and J. F. Herd, "An instantaneous direct reading goniometer," *J. IEE*, vol. 64, p. 611, 1926.

[2]   R. Rainer and A. J. Burwasser, "An approach to hf tactical radio direction finding and signal monitoring," *J. Electron. Defense*, vol. 6, no. 10, pp. 173–184, October 1983.

[3]   D. E. N. Davies, "Circular arrays: their properties and potential applications," *Proc. IEE Conf. Antennas Propag.*, York, England, 1981.

[4]   T. Rahim and D. E. N. Davies, "Effect of directional elements on the directional response of circular antenna arrays," *Proc. IEE*, vol. 129, pt. II, February 1982.

[5]   W. M. Sherrill, "Bearing ambiguity and resolution in interference direction finders," *IEEE Trans.*, vol. AES-5, pp. 959–966, November 1969.

[6]   R. L. Goodwin, "Ambiguity-resistant three- and four-channel interferometers," *NRL Report 8005*, Naval Research Laboratory, Washington, D.C. 20375, September 1976.

[7]   W. F. Gabriel, "Spectral analysis and adaptive array superresolution techniques," *Proc. IEEE*, vol. 68, pp. 654–666, June 1980.

[8]   R. Schmidt, "Multiple emitter location and signal parameter estimation," *Proc. RADC Spectrum Estimation Workshop*, RADC-TR-79-63, Rome Air Development

Center, Rome, New York, p. 243. October 1979. Also in *IEEE Trans. Antennas Propag.*, vol. AP-34, pp. 276–280, March 1986.

[9] D. H. Johnson and S. R. DeGraaf, "Improving the resolution of bearing in passive sonar arrays by eigenvalue analysis," *IEEE Trans. Acoust., Speech, Signal Processing*, vol. ASSP-30, pp. 638–647, August 1982.

[10] E. R. Ferrara, Jr., and T. M. Parks, "Direction-finding with an array of antennas having diverse polarizations," *IEEE Trans. Antennas Propag.*, vol. AP-31, pp. 231–236, March 1983.

[11] P. J. D. Gething, *Radio Direction Finding*, Stevenage, Herts., UK: Peter Peregrinus, Ltd., 1978.

[12] "IRE standards on navigation aids: direction finder measurements 1959," *Proc. IRE*, vol. 47, pp. 1349–1371, August 1959.

[13] J. A. Boyd et al., *Electronic Countermeasures*, Los Altos Hills: Peninsula Publishing, 1978, pp. 10-1 to 10-96.

[14] J. E. Browder and V. J. Young, "Design values for loop-antenna input circuits," *Proc. IRE*, vol. 35, pp. 519–526, May 1947.

[15] L. L. Libby, "Special aspects of balanced shielded loops," *Proc. IRE*, vol. 34, pp. 641–646, September 1946.

[16] H. A. Wheeler, "Fundamental limitations of small antennas," *Proc. IRE*, vol. 35, pp. 1479–1484, December 1947.

[17] J. H. Moon, "Design of electromagnetic goniometers for use in medium frequency direction finding," *J. IEE*, vol. 94, p. 69, January 1947.

[18] C. W. Earp and R. M. Godfry, "Radio df by cyclical differential measurement of phase," *J. IEE*, vol. 94, pt. IIIA, pp. 705–721, March 1947.

[19] W. Hausz, "Angular location, monopulse and resolution," *Microwave J.*, vol. 7, no. 2, p. 60, February 1964.

[20] A. D. Bailey and W. C. McClurg, "A sum-and-difference interferometer system for hf radio direction finding," *IEEE Trans. Aerospace and Navigational Electronic*, vol. ANE-10, pp. 65–72, March 1963.

[21] E. C. Hayden, "Propagation studies using direction-finding techniques," *J. Res. Natl. Bur. Stand.*, vol. 65D, pp. 197–212, May 1961.

[22] P. J. D. Gething, "Influence of ionospheric conditions on the accuracy of high-frequency direction finding," *J. Res. Natl. Bur. Stand.*, vol. 65D, pp. 225–228, May 1961.

Chapter 26

Standard AM Antennas

C. E. Smith
Smith Electronics, Inc.

CONTENTS

 **Carl E. Smith** received the BSEE degree from Iowa State University in 1930 and the MSEE and Professional EE degrees from the Ohio State University in 1932 and 1936, respectively. A registered professional engineer, he has gained recognition as an authority in electronics engineering for his achievements in broadcast-station antenna design and in education. In the course of his career he has been responsible for the engineering of scores of am and fm broadcasting stations here and abroad, including antennas, ground systems, and proofs of performance.

Smith Electronics was organized by Mr. Smith to provide research, development, and engineering services to government and industry. It has strong capabilities in antenna research and development, propagation studies, systems engineering, solid-state electronics, and electromagnetic compatibility testing.

The Cleveland Institute of Electronics was founded by Mr. Smith, who authored a major portion of the original advanced engineering course. It is now one of the leading electronics training institutes in this country with a large active student enrollment. Mr. Smith is still active as Chairman of the CIE Educational Committee.

1. Introduction

There is a world-wide medium-frequency broadcast band from 535 to 1605 kHz and a low-frequency broadcast band from 150 to 255 kHz in the Region 1 area.* At these frequencies the normal broadcasting antenna is a vertical conductor, usually a guyed or self-supporting tower. Vertical polarization is used because of its superior ground-wave propagation characteristics and the simplicity of the antenna design.

The purpose of a radio broadcasting station is to transform sound waves into radio waves that can be picked up by radio receiving sets. The utility of this service to the public depends on (1) signal strength, (2) program content, and (3) signal distortion. The broadcast antenna should efficiently radiate the energy supplied to it by the transmitter.

A simple vertical tower radiates the energy quite well equally in all directions along the ground. A secondary purpose of the antenna system may be to concentrate the amount of radiation in the directions that it is wanted and to restrict the radiation in the directions that it is not wanted. A directional antenna system is required if a nondirectional antenna causes interference to other stations or if the signal is not strong enough to adequately serve the populated areas of interest.

2. Standard Reference Antennas

The uniform, omnidirectional, or isotropic radiator in free space is taken as the *standard reference antenna* because it has no directivity. See Fig. 1. Actually such a radiator of radio waves cannot be realized, because all radio antennas have directional properties.

If a uniform radiator is placed at the surface of a perfectly conducting earth, all of the power must be radiated in the hemisphere above the earth as shown in Fig. 2.

Vertical Radiation Characteristics

A nondirectional tower has its own vertical radiation characteristic whether series or shunt fed, sectionalized or nonsectionalized, top loaded or without top loading. The vertical radiation pattern is usually computed using the assumption of a sinusoidal current distributon on the radiating portion of the tower.

The vertical radiation characteristic of a base-fed vertical tower of height H meters is given by

---

*Region 1: Europe, Africa, USSR, Turkey, and Arabia.
 Region 2: The Americas.
 Region 3: All other parts of the world.

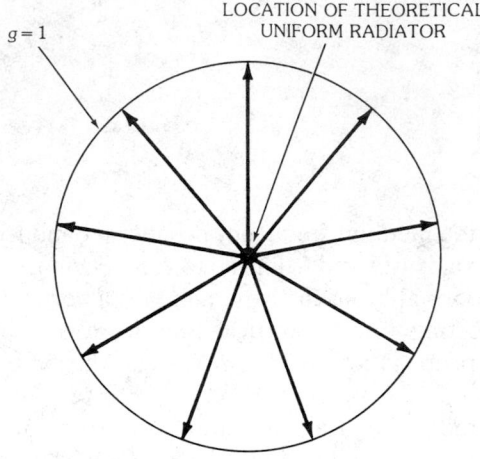

a

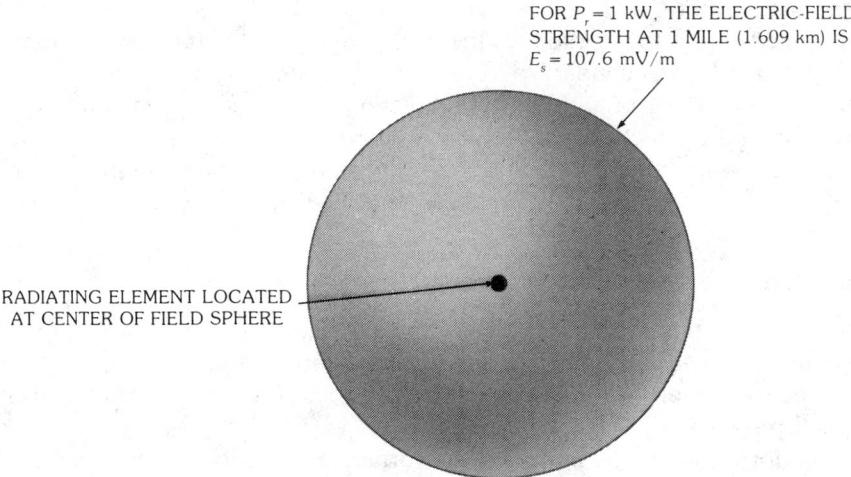

b

Fig. 1. Pattern of a uniform radiator which is the theoretical standard reference antenna. (*a*) Cross section of spherical pattern which has a field-strength gain of 1, a power gain of 1 ($g = 1$), and a decibel gain of 0. (*b*) Spherical radiation pattern surface of a uniform or isotropic radiator with radiated power P_r and electric-field strength E_s.

$$f(\theta) = \frac{\cos(G \sin \theta) - \cos G}{\cos \theta (1 - \cos G)} \tag{1}$$

where

$\quad f(\theta)$ = vertical radiation characteristic

$\quad\quad G$ = electrical height of tower, in degrees

$\quad\quad \theta$ = elevation of observation point, in degrees

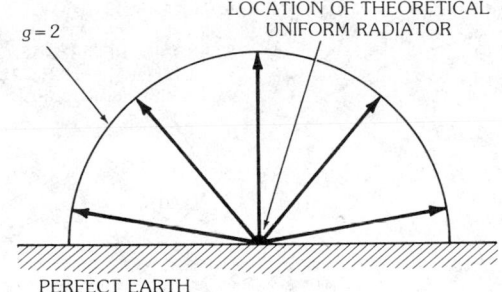

a   PERFECT EARTH

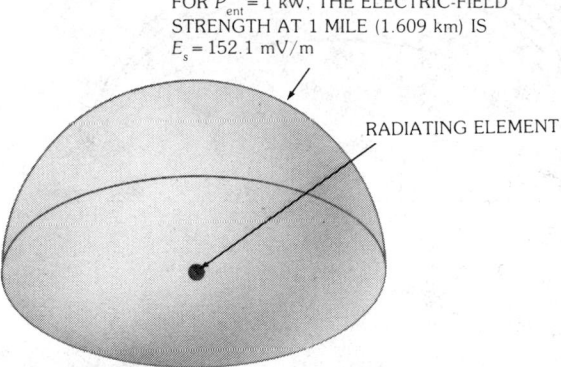

b

Fig. 2. Pattern of a uniform radiator at the surface of a perfect reflecting and conducting earth. (*a*) Cross section of a hemispherical pattern which has a field-strength gain of 1.4, a power gain of $g = 2$, and a decibel gain of 3.01. (*b*) Hemispherical radiation pattern surface of a uniform radiator.

A comparison of the vertical radiation characteristics of several standard reference antennas is shown in Figs. 3 and 4. A summary of several standard reference antennas is given in Table 1. In Fig. 5 the theoretical electric-field strength, loop radiation resistance,* and base radiation resistance are graphed as a function of antenna height.

The vertical radiation characteristic for a base-fed, top-loaded tower is given by

$$f(\theta) = \frac{\cos B \, \cos(A \sin \theta) - \sin \theta \, \sin B \, \sin(A \sin \theta) - \cos G}{\cos \theta (\cos B - \cos G)} \tag{2}$$

where

A = electrical height of the tower, in degrees

B = effective electrical length of top loading, in degrees

$G = A + B$

---

*Radiation resistance at antenna current maximum I_a as shown in Fig. 5.

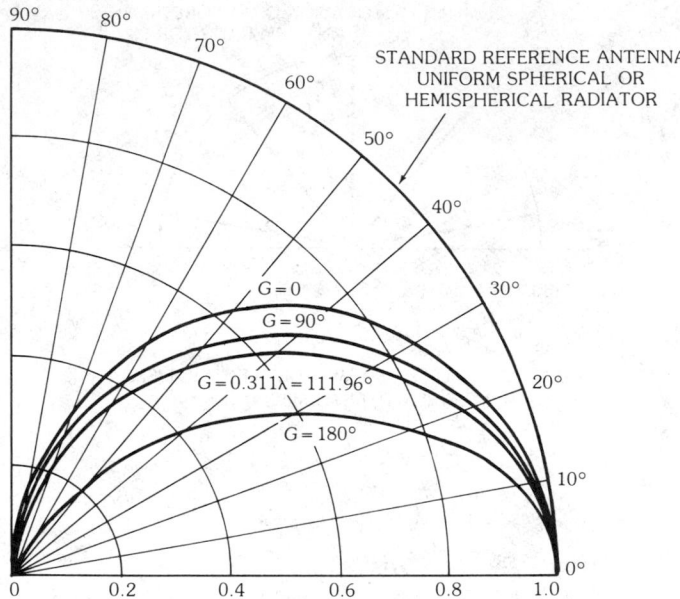

Fig. 3. Comparison of the vertical radiation characteristics for several standard reference antennas.

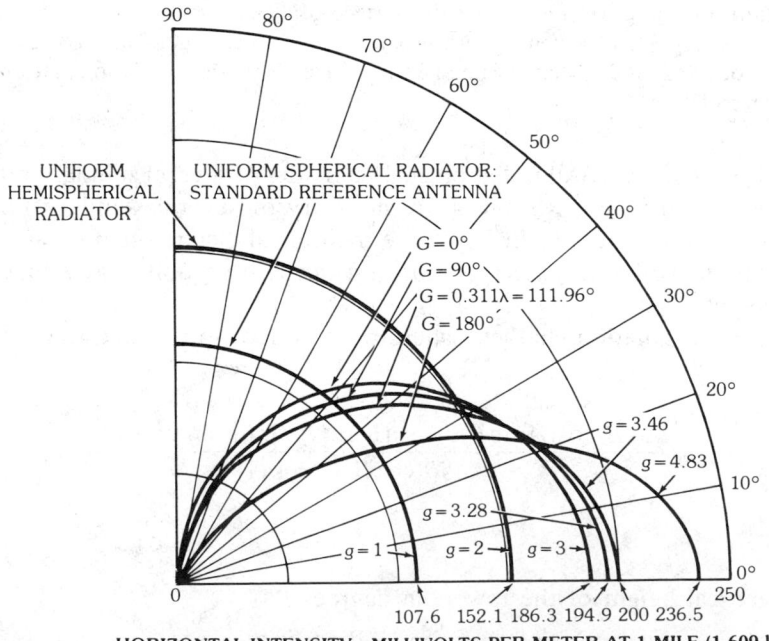

HORIZONTAL INTENSITY—MILLIVOLTS PER METER AT 1 MILE (1.609 km)

Fig. 4. Comparison of vertical radiation patterns of standard reference antennas with a radiated power of 1 kW.

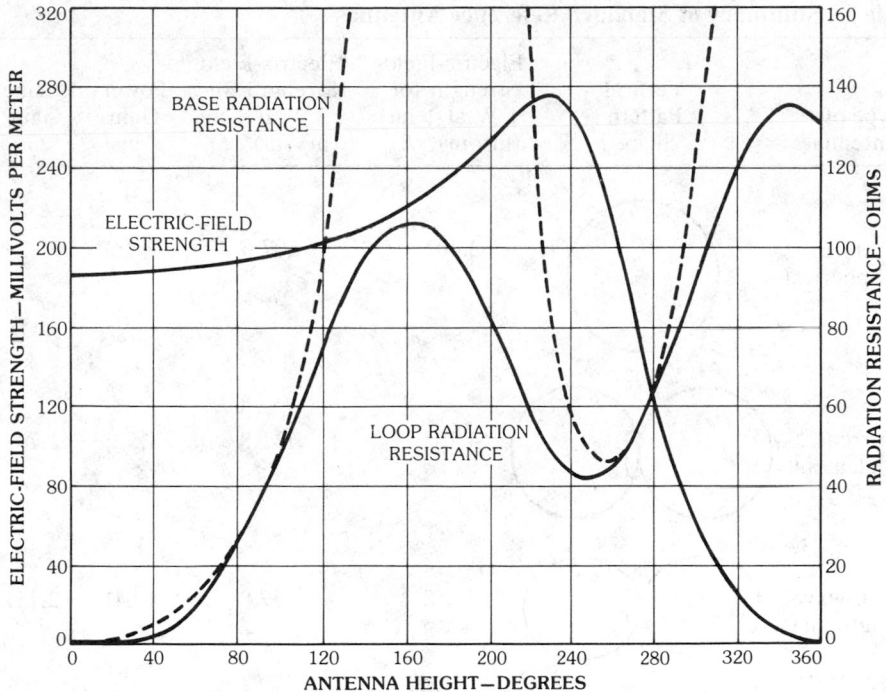

Fig. 5. Electric-field strength at 1 mi (1.609 km) for a radiated power of 1 kW, with loop and base radiation resistance as a function of tower height over a perfectly conducting earth.

The other terms are defined following (1). The theoretical current distribution is shown in Fig. 6. The loop current maximum on the tower is I_a. Equation 2 is valid only if the radiation from the top-loaded section B is negligible.

The vertical radiation characteristic for a base-fed, top-loaded sectionalized tower is given by

$$f(\theta) = \frac{\begin{aligned}\cos B \ \cos(A \sin \theta) - \cos G + \dfrac{\sin B \ \cos(H - C) \cos(C \sin \theta)}{\sin(H - A)} \\[2ex] - \dfrac{\sin B \ \sin \theta \ \sin(H - C) \sin(C \sin \theta)}{\sin(H - A)} - \dfrac{\sin B \ \cos(H - A) \cos(A \sin \theta)}{\sin(H - A)}\end{aligned}}{\cos \theta \{\cos B - \cos G + [\sin B/\sin(H - A)](\cos \overline{H - C} - \cos \overline{H - A})\}}$$

(3)

where

C = electrical height of top section, in degrees

D = electrical length of top loading of top section, in degrees

$H = C + D$

The other terms are defined in (1) and (2). The theoretical current distribution is shown in Fig. 7. The loop current maximum of the top section is I_c.

Table 1. Summary of Standard Reference Antennas

Type of Antenna	Vertical Pattern Shape	Electric-Field Strength for 1 W at 1 mi (mV/m)* E_ϕ	Electric-Field Strength for 1 kW at 1 mi (mV/m)* E_0	Power Gain g	Decibel Gain G
Uniform spherical radiator		3.402	107.6	1	0
Current element		4.167	131.8	1.5	1.761
Half-wave antenna		4.358	137.8	1.641	2.151
0.622λ antenna		4.472	141.4	1.728	2.375
Two ends on half-wave in-phase antenna		5.283	167.1	2.411	3.822

Theoretical Self-Resistance and Field Strength

It is useful to know the theoretical loop and base resistance of a vertical radiator. This information is presented graphically in Fig. 5 along with the theoretical field strength at 1 mi (1.609 km).

Self Base Impedance Characteristics

The loop and base radiation resistances as given in Fig. 5 are not very useful because they are for a very thin conductor and the antenna is not driven at the loop very often. Their primary application is in theoretical calculations. Since most

Table 1, *continued*

Type of Antenna	Vertical Pattern Shape	Electric-Field Strength for 1 W at 1 mi (mV/m)* E_ϕ	Electric-Field Strength for 1 kW at 1 mi (mV/m)* E_0	Power Gain g	Decibel Gain G
Uniform hemi-spherical radiator		4.811	152.1	2	3.010
Vertical current element		5.893	186.3	3	4.771
Quarter-wave vertical antenna		6.163	194.9	1.282	5.161
0.311λ vertical antenna		6.324	200	3.450	5.386
Half-wave vertical antenna		7.471	236.2	4.822	6.832

*E_0 is the electric-field strength on a plane passing through the center and perpendicular to the conductor.

antennas are driven at the base, the self base resistance along with the self base reactance curves are presented in Figs. 8 and 9.

The self-resonant frequency of a tower depends on the cross-sectional size [1]. The first factor to be considered is the average characteristic impedance Z_0 as shown in Figs. 8 and 9. The value of Z_0 depends on the average cross-sectional size, which in ohms for a cylindrical antenna is

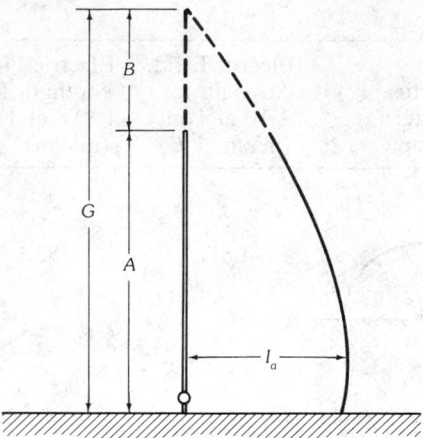

Fig. 6. Theoretical current distribution on top-loaded tower.

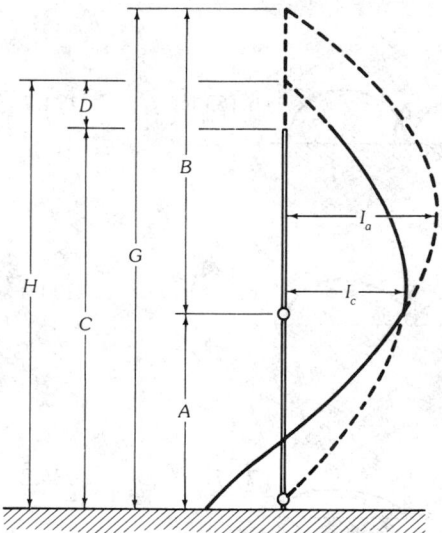

Fig. 7. Theoretical current distribution on top-loaded sectionalized tower.

$$Z_0 = 60[\ln(2G/a) - 1] \tag{4}$$

where

G = antenna height, in degrees

a = antenna radius, in degrees or same units as G if G is not in degrees

Since most antennas have a square or triangular cross section, a good approxima-

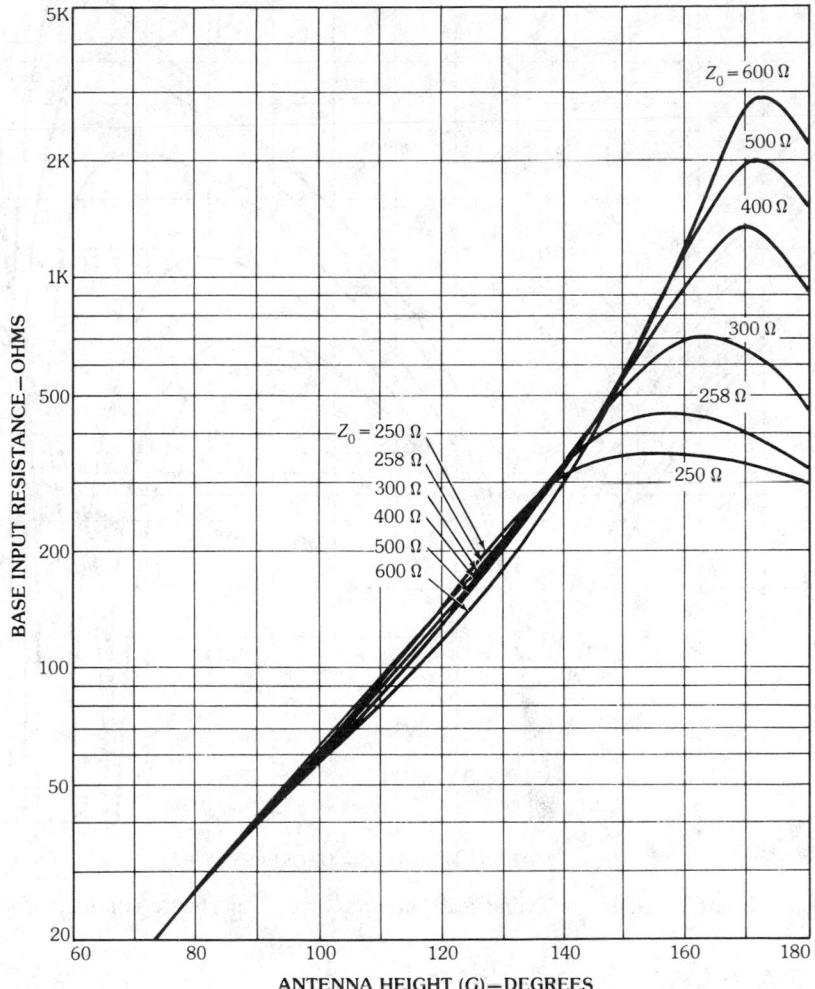

Fig. 8. Base input resistance of cylindrical antennas over a perfectly conducting ground plane.

tion is to determine the cross-sectional area and convert it to a circle with radius *a* to be used in (4). The equivalent radius *a* for the cross section of a regular polygon with *n* sides, each of length *d*, is [2]

n	2	3	4	5	6
a/d	0.25	0.4214	0.590	0.756	0.920

Example 1—Determine the average characteristic impedance of a 400-ft uniform cross-sectional tower that is 6.5 ft square. Then *a* = 0.59 × 6.5 = 3.835 ft.
Solution—Substituting in (4),

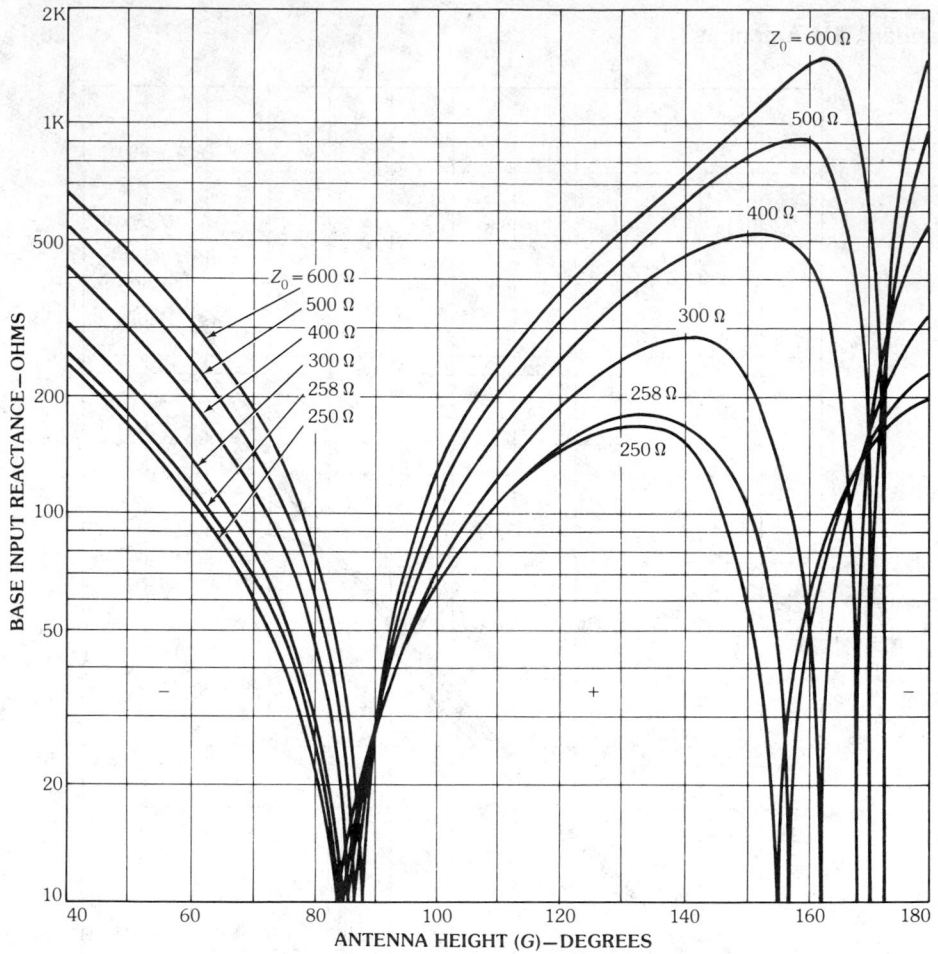

Fig. 9. Base input reactance of cylindrical antennas over a perfectly conducting ground plane.

$$Z_0 = 60 \, [\ln 2(400)/3.835 - 1]$$
$$= 60 \, (\ln 208.6 - 1)$$
$$= 60 \, (5.34 - 1)$$
$$= 260.4 \, \Omega$$

Mutual Base Impedance

The mutual impedance between vertical antennas is needed by the designer of directional antenna feeder systems. Sufficient design equations have been developed and presented in the literature [3, 4] to solve this problem; however, only a few cases of equal-height antennas are presented in Figs. 10 and 11. The loop mutual impedance can be transferred to the base by dividing by the square of the sine of the tower height, $\sin^2 G$.

The general design equation for the mutual base resistance and reactance is given by

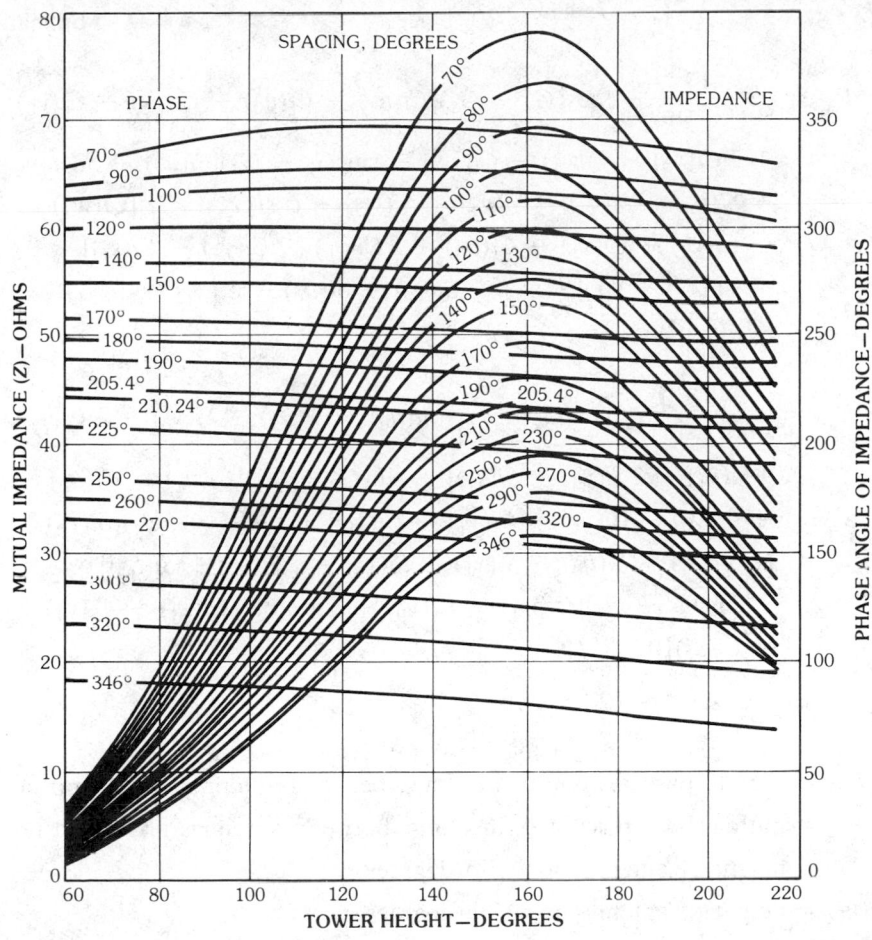

Fig. 10. Loop mutual impedance and phase angle between two towers of equal height.

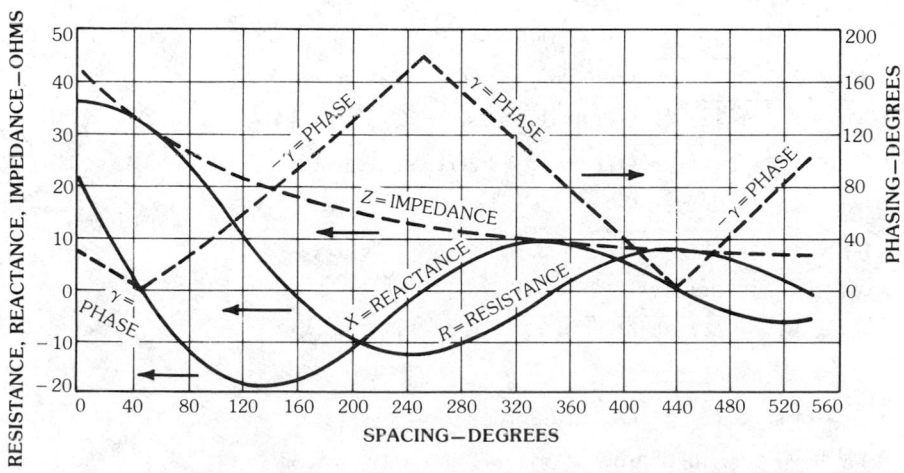

Fig. 11. Loop mutual impedance between quarter-wave vertical towers.

$$R_{12} = \frac{15}{\sin G_1 \sin G_2} \{\cos(G_2 - G_1)[Ci(u_1) - Ci(u_0) + Ci(v_1) - Ci(v_0)$$
$$+ 2\,Ci(y_0) - Ci(y_1) - Ci(s_1)] + \sin(G_2 - G_1)[Si(u_1) - Si(u_0)$$
$$+ Si(v_0) - Si(v_1) - Si(y_1) + Si(s_1)] + \cos(G_2 + G_1)[Ci(w_1)$$
$$- Ci(v_0) + Ci(x_1) - Ci(u_0) + 2\,Ci(y_0) - Ci(y_1) - Ci(s_1)]$$
$$+ \sin(G_2 + G_1)[Si(w_1) - Si(v_0) + Si(u_0) - Si(x_1) - Si(y_1)$$
$$+ Si(s_1)]\} \tag{5}$$

$$X_{12} = \frac{15}{\sin G_1 \sin G_2} \{\cos(G_2 - G_1)[Si(u_0) - Si(u_1) + Si(v_0) - Si(v_1)$$
$$+ Si(y_1) - 2\,Si(y_0) + Si(s_1)] + \sin(G_2 - G_1)[Ci(u_1) - Ci(u_0)$$
$$+ Ci(v_0) - Ci(v_1) - Ci(y_1) + Ci(s_1)] + \cos(G_2 + G_1)[Si(v_0)$$
$$- Si(w_1) + Si(u_0) - Si(x_1) + Si(y_1) - 2\,Si(y_0) + Si(S_1)]$$
$$+ \sin(G_2 + G_1)[Ci(w_1) - Ci(v_0) + Ci(u_0) - Ci(x_1) - Ci(y_1)$$
$$+ Ci(s_1)]\} \tag{6}$$

where

R_{12} = mutual base resistance, in ohms, between antennas no. 1 and no. 2

X_{12} = mutual base reactance, in ohms, between antennas no. 1 and no. 2

G_1 = height of antenna no. 1, in degrees

G_2 = height of antenna no. 2, in degrees

$Si(x)$ = sine integral function of x

$Ci(x)$ = cosine integral function of x

s = spacing between antennas, in degrees

$u_0 = \sqrt{s^2 + G_1^2} - G_1$, in degrees

$u_1 = \sqrt{s^2 + (G_2 - G_1)^2} + G_2 - G_1$, in degrees

$v_0 = \sqrt{s^2 + G_1^2} + G_1$, in degrees

$v_1 = \sqrt{s^2 + (G_2 - G_1)^2} - G_2 + G_1$, in degrees

$w_0 = v_0$, in degrees

$w_1 = \sqrt{s^2 + (G_2 + G_1)^2} + G_2 + G_1$, in degrees

$x_0 = u_0$, in degrees

$x_1 = \sqrt{s^2 + (G_2 + G_1)^2} - G_2 - G_1$, in degrees

$y_0 = s$, in degrees

$y_1 = \sqrt{s^2 + G_2^2} + G_2$, in degrees

$s_0 = y_0 = s$, in degrees

$s_1 = \sqrt{s^2 + G_2^2} - G_2$, in degrees

The desired theoretical mutual base impedance can be obtained by solving (5) and (6). If the antennas are of equal height ($G = G_1 = G_2$), then (5) and (6) reduce to

$$R_{12} = \frac{15}{\sin^2 G} \{4\,Ci(u_1) - 2\,Ci(u_0) - 2\,Ci(v_0) + \cos 2G[Ci(w_1) - 2\,Ci(v_0)$$
$$+ Ci(x_1) - 2\,Ci(u_0) + 2\,Ci(u_1)] + \sin 2G[Si(w_1) - 2\,Si(v_0)$$
$$- Si(x_1) + 2\,Si(u_0)]\} \tag{7}$$

$$X_{12} = \frac{15}{\sin^2 G} \{-4\,Si(u_1) + 2\,Si(u_0) + 2\,Si(v_0) + \cos 2G[-Si(w_1)$$
$$+ 2\,Si(v_0) - Si(x_1) + 2\,Si(u_0) - 2\,Si(u_1)] + \sin 2G[Ci(w_1)$$
$$- 2\,Ci(v_0) - Ci(x_1) + 2\,Ci(u_0)]\} \tag{8}$$

Now, if antenna no. 2 is 90° in height ($G_2 = 90°$), then (5) and (6) reduce to

$$R_{12} = 15\,\{Ci(u_1) + Ci(v_1) - Ci(w_1) - Ci(x_1) + \cot G_1[Si(u_1) - Si(v_1)$$
$$- 2\,Si(y_1) + 2\,Si(s_1) + Si(w_1) - Si(x_1)]\} \tag{9}$$

$$X_{12} = 15\,\{Si(w_1) + Si(x_1) - Si(u_1) - Si(v_1) + \cot G_1[Ci(u_1) - Ci(v_1)$$
$$- 2\,Ci(y_1) + 2\,Ci(s_1) + Ci(w_1) + Ci(x_1)]\} \tag{10}$$

Example 2—Determine the mutual base impedance between two antennas spaced at 160° with electrical heights of 120° and 90°, respectively.
Solution—This is as follows:

$$G_1 = 120°, \; G_2 = 90°, \text{ and } s = 160°$$
$$u_1 = \sqrt{(160)^2 + (30)^2} + 90 - 120 = 132.8°, \text{ or } 2.32 \text{ rad}$$
$$v_1 = \sqrt{(160)^2 + (30)^2} - 90 + 120 = 192.8°, \text{ or } 3.366 \text{ rad}$$
$$w_1 = \sqrt{(160)^2 + (90 + 120)^2} + 90 + 120 = 474°, \text{ or } 8.27 \text{ rad}$$
$$x_1 = \sqrt{(160)^2 + (90 + 120)^2} - 90 - 120 = 54°, \text{ or } 0.943 \text{ radi}$$
$$y_1 = \sqrt{(160)^2 + (90)^2} + 90 = 273.6°, \text{ or } 4.77 \text{ rad}$$
$$s_1 = \sqrt{(160)^2 + (90)^2} - 90 = 93.6°, \text{ or } 1.63 \text{ rad}$$

Substituting these values in (9) and (10):

$$R_{12} = 15\,\{Ci(2.32) + Ci(3.366) - Ci(8.27) - Ci(0.943) + \cot 120°$$
$$\times [Si(2.32) - Si(3.366) - 2\,Si(4.77) + 2\,Si(1.63) + Si(8.27)$$
$$- Si(0.943)]\}$$
$$= -2.935 \; \Omega$$

$$X_{12} = 15\{Si(8.27) + Si(0.943) - Si(2.32) - Si(3.366) + \cot 120°$$
$$\times [Ci(2.32) - Ci(3.366) - 2Ci(4.77) + 2Ci(1.63)$$
$$+ (8.27) - Ci(0.943)]\}$$
$$= -28.85\ \Omega$$

Therefore the mutual base impedance is

$$Z_{12} = -2.935 - j28.85\ \Omega$$

Example 3—Determine the mutual base impedance between two towers of equal height $G = 110°$ and having a spacing of $200°$.

Solution—Since the tables of the sine integral and cosine integral functions are ordinarily tabulated with the arguments in radians, it has been found simpler to convert G and s to radians before substituting in the formulas.

$$G_1 = G_2 = 110° = 1.9199\ \text{rad}$$
$$s = 200° = 3.4907\ \text{rad}$$

Then

$$u_0 = \sqrt{3.4907^2 + 1.9199^2} - 1.9199 = 2.0639$$
$$u_1 = s = 3.4907$$
$$v_0 = \sqrt{3.4907^2 + 1.9199^2} + 1.9199 = 5.9037$$
$$w_1 = \sqrt{3.4907^2 + (2 \times 1.9199)^2} + (2 \times 1.9199)$$
$$= 9.0290$$
$$x_1 = \sqrt{3.4907^2 + (2 \times 1.9199)^2} - (2 \times 1.9199)$$
$$= 1.3496$$
$$\sin^2 G = 0.9397^2 = 0.8830$$

Substituting these values in (7) yields

$$R_{12} = \frac{15}{0.8830}\{4\,Ci(3.4907) - 2\,Ci(2.0639) - 2\,Ci(5.9037) + \cos 220°$$
$$\times [Ci(9.0290) - 2\,Ci(5.9037) + Ci(1.3496) - 2\,Ci(2.0639)$$
$$+ 2\,Ci(3.4907)] + \sin 220°[Si(9.0290) - 2\,Si(5.9037)$$
$$- Si(1.3496) + 2\,Si(2.0639)]\}$$
$$= 16.9875\{-0.1188 - 0.8180 + 0.1666 - 0.7660[0.0524$$
$$+ 0.1666 + 0.4549 - 0.8180 - 0.0594] - 0.6428[1.6663$$
$$-2.8598 - 1.2203 + 3.2672]\}$$
$$= 16.9875\{-0.7702 - 0.7660(-0.2035) - 0.6428(0.8534)\}$$
$$= 16.9875\{-0.7702 + 0.1559 - 0.5486\}$$
$$= 16.9875(-1.1629) = -19.75\ \Omega$$

For the reactance component of the mutual base impedance, substituting in (8) yields

$$
\begin{aligned}
X_{12} &= 16.9875\,\{-4\,Si(3.4907) + 2\,Si(2.0639) + 2\,Si(5.9037) + \cos 220° \\
&\quad \times\,[-Si(9.0290) + 2\,Si(5.9037) - Si(1.3496) + 2\,Si(2.0639) \\
&\quad - 2\,Si(3.4907)] + \sin 220°[Ci(9.0290) - 2\,Ci(5.9037) - Ci(1.3496) \\
&\quad + 2\,Ci(2.0639)]\} \\
&= 16.9875\,\{-7.3360 + 3.2672 + 2.8598 - 0.7660[-1.6663 + 2.8598 \\
&\quad - 1.2203 + 3.2672 - 3.6680] - 0.6428[0.0524 + 0.1666 \\
&\quad - 0.4549 + 0.8180]\} \\
&= -21.33\ \Omega
\end{aligned}
$$

and

$$
Z_{12} = 29.1\ \Omega\ \angle{-132.8°}
$$

It should be pointed out that when any of the towers are in excess of approximately 120° in height, and particularly in the case of self-supporting towers, the values of the base mutual impedance given by these formulas are in considerable error and therefore should not be relied upon. In such cases the use of loop values of mutual impedance and self-impedance will give a better indication of the power division in the directional antenna array.

It is of cardinal importance not to mix measured and theoretical values. If reliable measured values are available, they should be used. It is particularly pertinent not to use measured base self-impedance and theoretical base mutual impedances to determine driving-point base impedance and power division in the directional antenna array. It is usually feasible to measure the self-impedance of each tower with the other towers disabled and then measure the *magnitude* of the mutual impedance between each pair of towers. Since the mutual impedance phase angles are more difficult to measure, it is quite common practice to use the theoretical phase angles since they are usually more reliable than the measured values.

Control of Pattern Shape and Size

There are two basic problems in a directional antenna design. *First*, it is desirable to control the pattern shape to provide the necessary protections to other stations and to serve the community of interest. This is accomplished by proper selection of directional antenna parameters. *Second*, the size of the pattern for a given amount of power is very useful and often necessary in many applications. First the problem of molding the pattern shape will be treated and then methods of determining the size or gain of the directional antenna will be treated.

Space Configuration of the Array

The field strength at any given point in space is a function of the placement of the current elements, along with the magnitude and phase of the field strength produced by each current element. In other words, it is necessary to know the space configuration of the radiating elements and the field-strength magnitude and phase of each radiating element.

It is common practice to treat each tower as a radiating element over a perfectly conducting earth and select the observation point P far enough away from the radiating system to assume that lines joining the radiating elements and the observation point are parallel. This simplifies the mathematics and does not introduce appreciable error except when it is desired to deal with the nearby radiation field. For this case the more general equations must be employed.

In order to establish a system for specifying the location of each antenna in such a manner that the information can be used in the design equations, let us refer to Fig. 12, which is a plan view of the space configuration of the ith antenna. The space reference point is the space origin or point from which all distances are measured. The space reference axis is the reference line from which all azimuth angles are measured. The azimuth angle ϕ_i is measured clockwise in degrees from the space reference axis. The spacing s_i is measured in electrical degrees along the horizontal plane as shown in Fig. 12.

For an observation point P at some azimuth angle ϕ in the horizontal plane, the distance to the ith antenna s_i is shortened by the multiplication factor $\cos(\phi_i - \phi)$ as shown in Fig. 13. This space-phasing quantity is the required difference distance since the observation point P is assumed to be at a great distance; hence the lines from the reference point and the ith antenna are parallel.

When the observation point is at some elevation angle θ, the ith antenna will appear to be closer to the space reference point by the multiplying factor $\cos \theta$. Referring to Fig. 14, observe that a right triangle can be formed by dropping a perpendicular to the line connecting the space reference point and the observation point P.

The ith antenna then appears to have space phasing of $s_i \cos \theta \cos(\phi_i - \phi)$ from

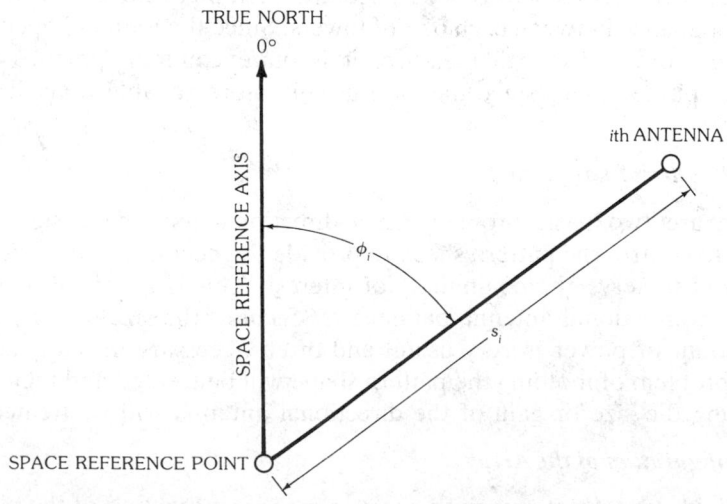

Fig. 12. Plan view of space configuration of the ith antenna.

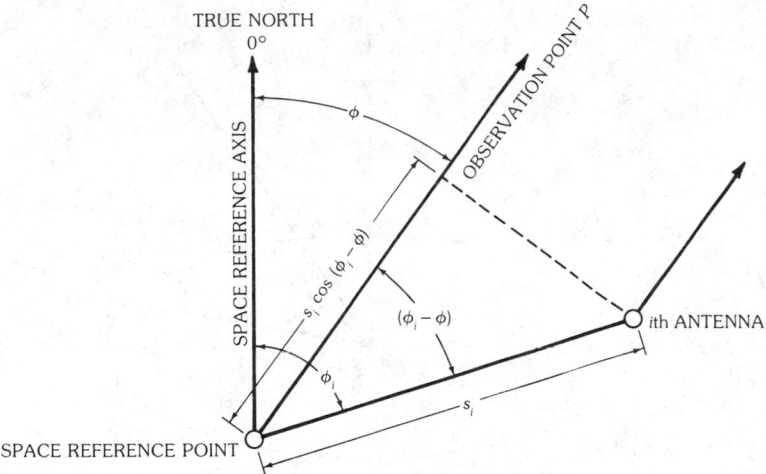

Fig. 13. Plan view of the *i*th antenna showing space phasing in the horizontal plane.

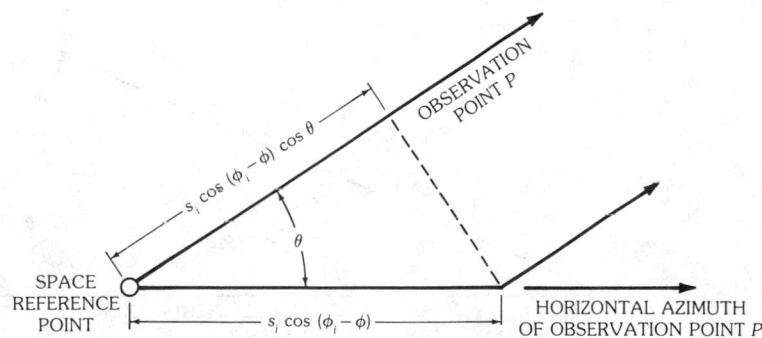

Fig. 14. Elevation angle θ shortens the spacing s_i to the value $s_i \cos \theta$.

the space reference point. This is the complete expression for the space phasing of any antenna in the system and for any observation point P in the hemisphere, as shown in Fig. 15.

Voltage Diagram

For the above generalized space configuration of the *i*th antenna there is a voltage phasor at the point P in space. This voltage phasor for the *i*th antenna depends on E_i, the field strength in the horizontal plane, the vertical radiation characteristic $f(\theta)_i$ as given by (1), (2), or (3), and the time phasing ψ_i.

The electric-field–strength vector for the *i*th antenna makes an angle β_i with respect to the voltage reference phasor axis, which is along the positive x axis, as shown in Fig. 16. The angle β_i is the sum of the space phasing and the time phasing. The space phasing is represented by

$$s_i \cos \theta \, \cos(\phi_i - \phi)$$

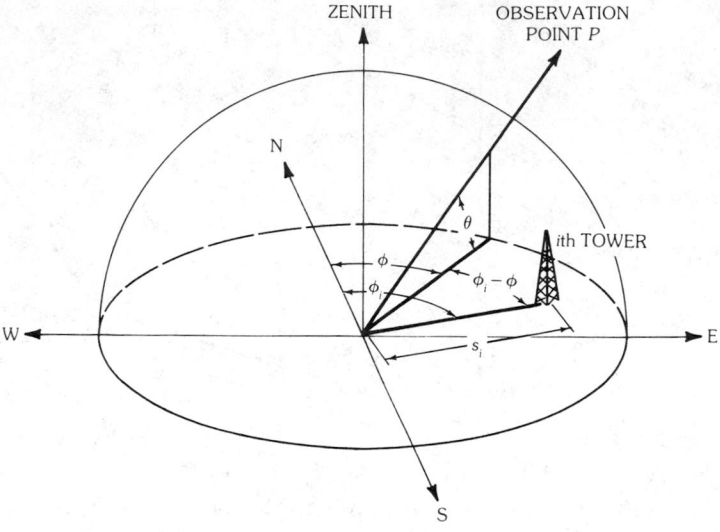

Fig. 15. Space view of observation point P and the ith tower.

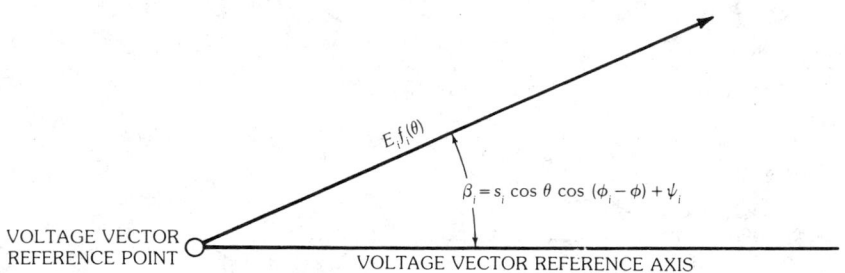

Fig. 16. Voltage vector diagram for the ith antenna.

while the time phase is simply ψ_i.

Generalized Equation

Now, if the electric-field–strength phasors of all the antennas in the directional antenna array are added in series, the resultant field strength at the observation point P can be obtained as shown in Fig. 17. If these phasors for n antennas are added together, the generalized equation can be used to express the hemispherical pattern shape of the directional antenna array with n antennas. This generalized phasor equation can be used to express the pattern shape for any directional antenna array with n antennas. The equation in condensed form is

$$E = \sum_{i=1}^{i=n} E_i f_i(\theta) \angle \beta_i \qquad (11)$$

where

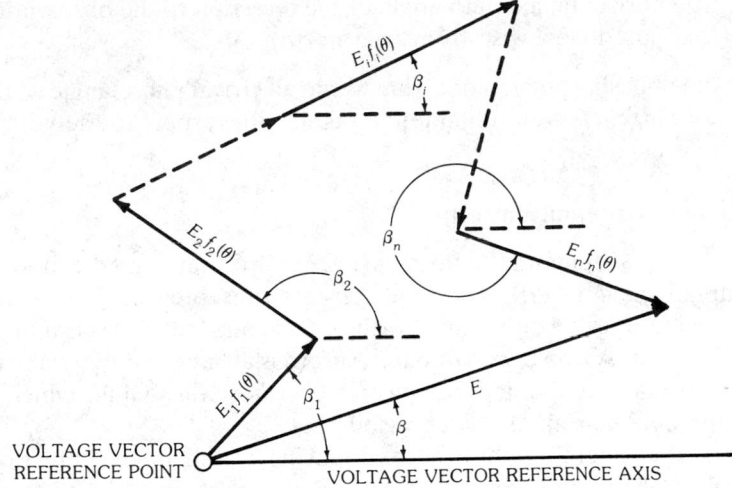

Fig. 17. Summation of electric-field–strength vectors for n antennas in the directional antenna array.

E = the total effective electric-field–strength phasor at unit distance (P) for the antenna array with respect to the voltage phasor reference axis. This resultant phasor makes the angle β with respect to this axis as shown in Fig. 17

i = the ith antenna in the directional antenna system

n = the total number of antennas in the directional antenna array

E_i = the magnitude of the electric-field strength at unit distance in the horizontal plane produced by the ith antenna acting alone

$f_i(\theta)$ = vertical radiation characteristic of the ith antenna as given in (1), (2), or (3)

θ = elevation angle of the observation point P measured up from the horizon in degrees

and

$$\beta_i = s_i \cos \theta \, \cos(\phi_i - \phi) + \psi_i \tag{12}$$

which is phase relation of the electric-field strength at the observation point P for the ith antenna taken with respect to the voltage phasor reference axis, where $s_i \cos \theta \cos(\phi_i - \phi)$ is the space phasing portion of β_i due to the location of the ith antenna, for which

s_i = electrical length of spacing of the ith antenna in the horizontal plane from the space reference point

ϕ_i = true horizontal azimuth, orientation of ith antenna with respect to the space reference axis

ϕ = true horizontal azimuth angle of the direction to the observation point P (measured clockwise from true north)

ψ_i = time phasing portion of β_i due to the electrical phase angle of the voltage (or current) in the ith antenna taken with respect to the voltage phasor reference axis

3. Two-Tower Antenna Patterns

Since two-tower antenna patterns are so useful they have been systematized out to four wavelengths [5]. A few patterns are presented in Fig. 18. If only one null is desired, it must be either due north or due south since this is in line with the antenna of Fig. 18. A two-tower antenna pattern is always symmetrical with respect to the vertical plane containing the towers. Therefore those nulls which are not in line with the antennas appear even in number.

For a more general condition of producing a null in the direction ϕ_n, the azimuth angle of the null from the line of antennas is given by the equation

$$s_2 \cos \phi_n \pm \psi_2 = \pm 180° \tag{13}$$

where

ϕ_n = azimuth angle of the nulls from the line of the antenna, in degrees

= azimuth angle, in degrees, clockwise and counterclockwise from true north to the nulls, if the antennas are on a north-south space reference axis

ψ_2 = total phasing of antenna no. 2 (north antenna) with respect to antenna no. 1 (south antenna), in degrees

s_2 = total spacing between the antennas, in degrees

This equation has been used to prepare the chart in Fig. 19.

4. Power Flow Integration Method to Determine Pattern Size

The total power radiated from a directional antenna array can be computed by integrating the energy flow outward through an imaginary spherical surface surrounding the directional antenna array. This method does not give information regarding the distribution of power to various towers of the directional antenna array. It is, however, very useful for making comparisons of pattern size.

The rate of energy flow in watts per square meter at a given point P in space can be expressed by the Poynting vector, thus

$$\mathbf{P} = \mathbf{E} \times \mathbf{H} \tag{14}$$

where

$\mathbf{P}$ = energy flow in watts per square meter

$\mathbf{E}$ = electric-field strength in volts per meter

$\mathbf{H}$ = magnetic-field intensity in ampere-turns per meter

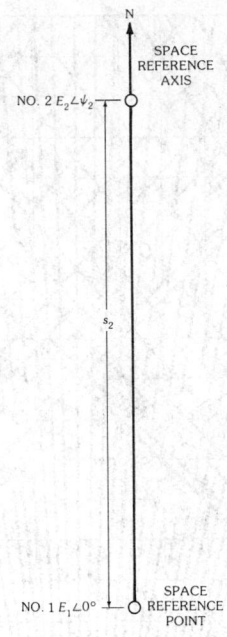

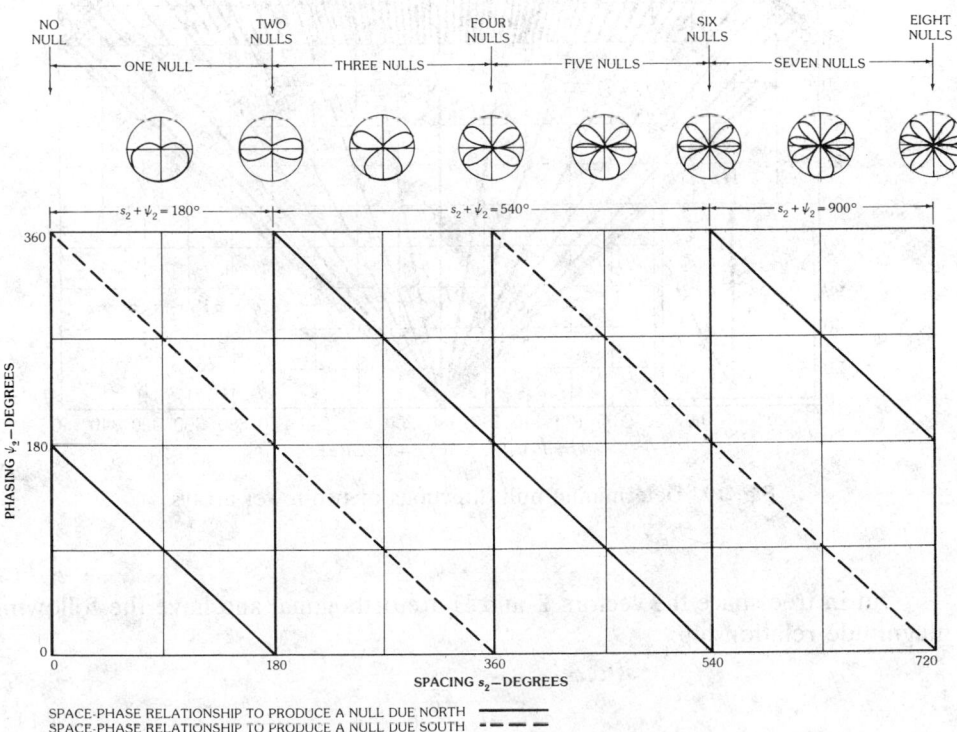

Fig. 18. Two-tower space-phase relationship to produce a null due north.

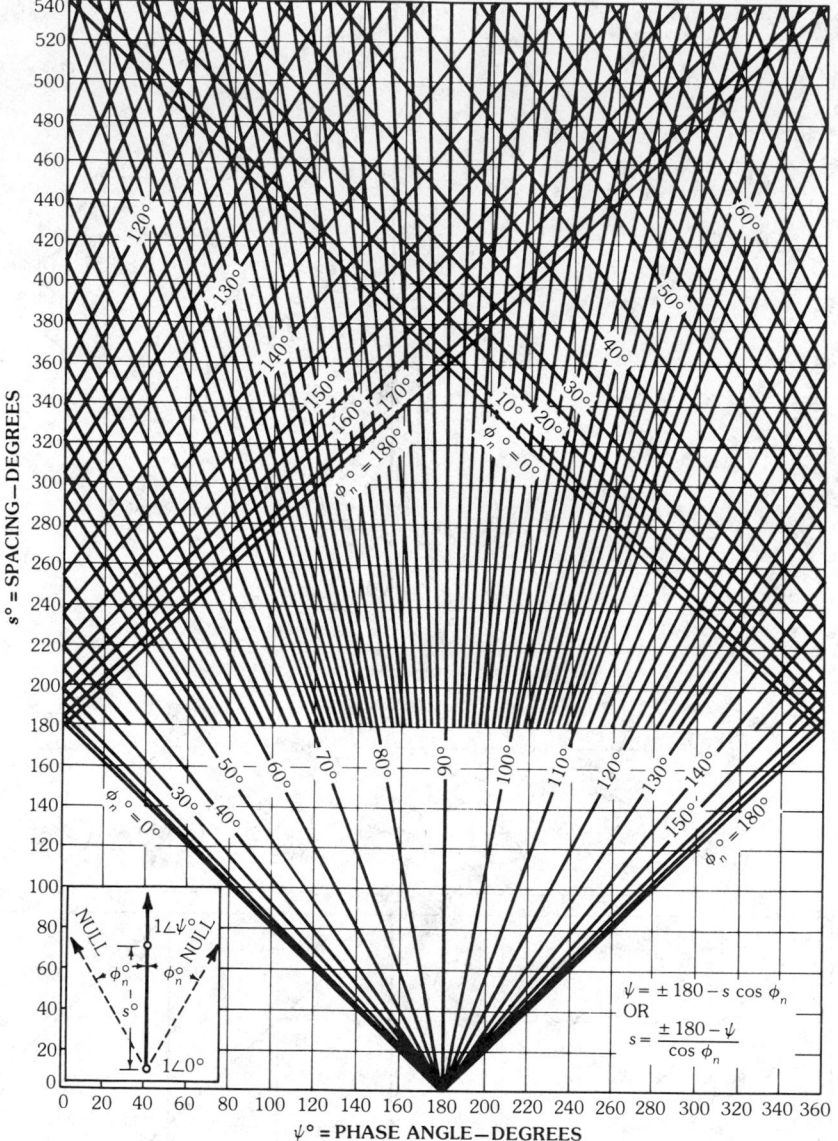

Fig. 19. Determining null directions of two-tower arrays.

Out in free space the vectors **E** and **H** are orthogonal and have the following magnitude relationship:

$$H = \sqrt{\frac{\epsilon_0}{\mu_0}} E \qquad (15)$$

where

μ_0 = the permeability of free space, $4\pi \times 10^{-7}$ henry per meter

ϵ_0 = the permittivity of free space, $1/\mu_0 c^2$ farads per meter

c = the velocity of light, 2.9979×10^8 meters per second

From these values the characteristic resistance of free space can be determined. The characteristic resistance of free space has a value of resistance such that no energy will be reflected. In other words, energy leaving the directional antenna array and flowing into free space will never return. In equation form,

$$R_c = \mu_0 c = \sqrt{\frac{\mu_0}{\epsilon_0}} \qquad (16)$$

where $R_c = 376.710 \ \Omega$, the characteristic resistance of free space.

Substituting (15) and (16) in (14), the power flow can be expressed as

$$P = \frac{E^2}{R_c} \ \Omega \qquad (17)$$

If this power flow is integrated over an imaginary spherical surface surrounding the directional antenna array, the total power radiated is

$$P_r = \frac{1}{R_c} \int_S E^2 \, d\mathbf{S} \qquad (18)$$

where

P_r = total power radiated, in watts

R_c = characteristic resistance of free space, 376.71 Ω

E = total electric-field strength at the surface of the sphere, in volts per meter

$d\mathbf{S}$ = element of area on the surface of the sphere, in square meters

If the energy flow outward through the surface of the sphere is integrated as illustrated in Fig. 20, we can write

$$dS = d^2 \cos \theta \, d\theta \, d\phi \qquad (19)$$

where d is the radius in meters of the spherical surface, and the other values are defined above. Substituting this value of dS in (17) gives

$$P_r = \frac{1}{R_c} \int_0^{2\pi} \int_{-\pi/2}^{+\pi/2} E^2 \, d^2 \cos \theta \, d\theta \, d\phi \qquad (20)$$

as the total power radiated from an antenna system in free space.

If the electric-field strength is the same in all directions, the E in (20) can be

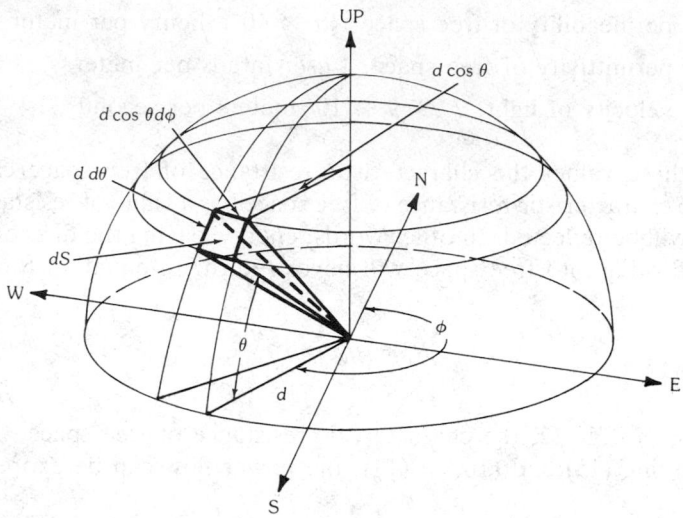

Fig. 20. Spherical surface integration.

replaced by the constant E_s. Moving the constants E_s and d^2 outside of the integral signs and performing the indicated integrations gives

$$P_r = \frac{E_s^2 d^2}{R_c} \int_0^{2\pi} \int_{-\pi/2}^{+\pi/2} \cos\theta \, d\theta \, d\phi$$

$$= \frac{2 E_s^2 d^2}{R_c} \int_0^{2\pi} d\phi = \frac{4\pi E_s^2 d^2}{R_c} \tag{21}$$

Solving this equation for the rms electric-field strength of a uniform spherical radiator gives

$$E_s = \sqrt{\frac{P_r R_c}{4\pi d^2}} \tag{22}$$

Example 4—Determine the electric-field strength at 1 mi (1.609 km) for an antenna power of 1 kW for an isotropic radiator.

Solution—Since there are 1609.347 m in a mile and $R_c = 376.71 \ \Omega$, the characteristic resistance of free space, (22) can be used to obtain

$$E_s = \sqrt{\frac{(1000)(376.710)}{4(3.14159)(1609.347)^2}}$$

$$= 107.584 \ \text{mV/m}$$

This is the electric-field strength at 1 mi for 1 kW from the standard reference spherical radiator.

Example 5—Determine the electric-field strength at 1 mi (1.609 km) for an antenna power of 1 kW for a uniform hemispherical radiator.

Solution—In (20), the integration for θ is from 0 to $\pi/2$. Hence

$$P_r = \frac{1}{R_c} \int_0^{2\pi} \int_0^{\pi/2} E^2 d^2 \cos\theta \, d\theta \, d\phi \tag{23}$$

Substituting the constant E_s for E and solving for E_s similarly to the procedure in (21) and (22) results in

$$E_s = \sqrt{\frac{P_r R_c}{2\pi d^2}} \tag{24}$$

For 1 kW of radiated power the electric-field strength at 1 mi is

$$E_s = \sqrt{\frac{(1000)(376.71)}{2(3.1416)(1609.347)^2}}$$

$$= 152.147 \text{ mV/m}$$

This is the electric-field strength at 1 mi for 1 kW from the standard reference hemispherical radiator. This value can also be obtained by multiplying

$$107.584\sqrt{2} = 152.147$$

The electric-field strength from a center-fed conductor of length $2G$ in free space with a sinusoidal current distribution as shown in Fig. 21 has an electric-field strength of

$$E = E_0 f(\theta) = E_0 \left[\frac{\cos(G \sin\theta) - \cos G}{(1 - \cos G) \cos\theta} \right] \tag{25}$$

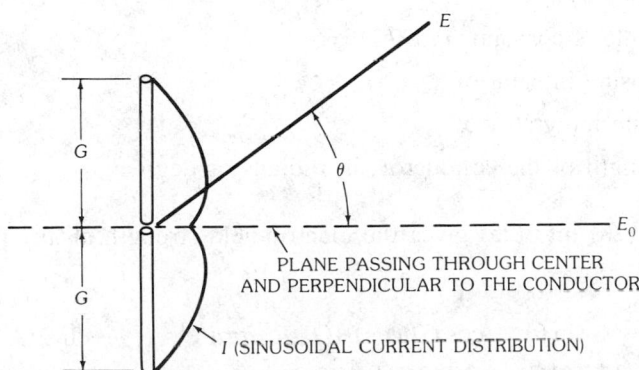

Fig. 21. Determining the electric-field strength of a conductor of length $2G$ in free space.

at any angle θ from a plane passing through the center and perpendicular to the conductor. Substituting this value of E in (20) and integrating with respect to ϕ gives

$$P_r = \frac{2\pi d^2 E_0^2}{R_c(1 - \cos G)^2} \int_{-\pi/2}^{\pi/2} \frac{[\cos(G\sin\theta) - \cos G]^2}{\cos\theta} \, d\theta \tag{26}$$

Performing the indicated integration,

$$P_r = \frac{2\pi d^2 E_0^2}{R_c(1 - \cos G)^2} \{[\gamma + \ln 2G - Ci(2G)]$$
$$+ 1/2 \, (\sin 2G)[Si(4G) - 2Si(2G)]$$
$$+ 1/2 \, (\cos 2G)[\gamma + \ln G - 2Ci(2G) + Ci(4G)]\} \tag{27}$$

Solving this equation for E_0 results in

$$E_0 = \frac{(1 - \cos G)(P_r R_c/\alpha\pi d^2)^{1/2}}{\{[\gamma + \ln 2G - Ci(2G)] + 1/2 \, [Si(4G) - 2Si(2G)]\sin 2G}$$
$$+ 1/2 \, [\gamma + \ln G - 2Ci(2G) + Ci(4G)]\cos 2G\}^{1/2} \tag{28}$$

where

E_0 = the electric-field strength, in volts per meter, on a plane passing through the center and perpendicular to the conductor

P_r = total power radiated, in watts

R_c = 376.710 Ω, the characteristic resistance of free space

α = 2 for a conductor of length $2G$ in free space and $\alpha = 1$ for an antenna of height G over a perfect ground plane

d = distance from the antenna, in meters
   = 1609.347 m in 1 mi

γ = Euler's constant, 0.577 215 66

$Ci(x)$ = cosine integral of x

$Si(x)$ = sine integral of x

$2G$ = length of the conductor, in radians or degrees

Substituting (28) into (25) gives the electric-field strength at any point in space. Thus

$$E = \{[\cos(G\sin\theta) - \cos G]/\cos\theta\}(P_r R_c/\alpha\pi d^2)^{1/2}/\{[\gamma + \ln 2G - Ci(2G)]$$
$$+ 1/2 \, [Si(4G) - 2Si(2G)] \sin 2G$$
$$+ 1/2 \, [\gamma + \ln G - 2Ci(2G) + Ci(4G)] \cos 2G\}^{1/2} \tag{29}$$

where θ is the elevation angle from the plane, and the other terms are defined following (28).

Assuming a sinusoidal current distribution on a vertical antenna of height G over a perfect ground plane as shown in Fig. 22, the integration in (26) is changed to cover the hemisphere above the surface of the earth. This condition also results in $\alpha = 1$ in (28).

Along the horizon, $\theta = 0$ and the theoretical electric-field strength along the ground can be plotted as a function of antenna height G for 1 kW of antenna power at a distance of 1 mi to obtain the curve in Fig. 5.

As a matter of interest the loop radiation resistance for a thin vertical conductor is

$$R_r = 29.99776 \{ [\gamma + \ln 2G - Ci(2G)] + 1/2 [Si(4G) - 2Si(2G)] \sin 2G$$
$$+ 1/2 [\gamma + \ln G - 2Ci(2G) + Ci(4G)] \cos 2G \} \tag{30}$$

where R_r is the loop radiation resistance in ohms of a thin vertical conductor over a perfectly conducting earth. This curve has also been plotted in Fig. 5.

For a vertical quarter-wave antenna the loop radiation resistance is

$$R_r = (29.9978)(1.21884) = 36.5626 \ \Omega$$

The base resistance R_b in ohms can be determined from the approximate equation

$$R_b \cong \frac{R_r}{\sin^2 G} \tag{31}$$

Thus far the power-flow integration method has been applied to the standard reference antennas and to the size of pattern produced by sinusoidal current distribution on a vertical radiator of various heights. The method is not confined to these special cases but can be applied to any directional antenna system so long as the field distribution is known from the individual elements that make up the system.

Considerable space has been devoted to determining the pattern size of a non-directional antenna, that is, an antenna which produces a constant field strength at all azimuth angles in the horizontal plane and for which at any given elevation

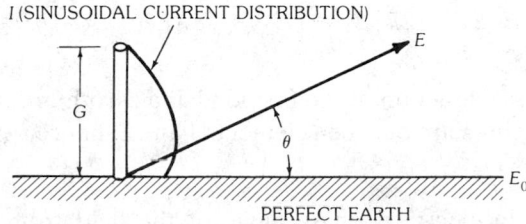

Fig. 22. Determining the electric-field strength of a vertical antenna of height G over a perfect earth.

angle the field strength is constant for all azimuth angles. The reason for this approach is that any directional antenna can be transformed to its equivalent nondirectional antenna pattern. This can be accomplished by determining the root-mean-square valve of the pattern at each elevation angle; then it is only necessary to integrate this equivalent nondirectional pattern with respect to the elevation angle since the determination of the rms values is the result of integrating the pattern with respect to the azimuth angle.

In order to develop this method consider the horizontal electric-field strength from two towers, one at the origin and one at the spacing s_2 due north. The total vector field is then

$$E = E_1 \angle 0 + E_2 \angle s_2 \cos \phi + \psi_2 \tag{32}$$

Changing to rectangular form,

$$E = E_1 + E_2 \cos(s_2 \cos \phi + \psi_2) + JE_2 \sin(s_2 \cos \phi + \psi_2)$$

The magnitude of E^2 is

$$
\begin{aligned}
E^2 &= E_1{}^2 + E_2{}^2 \cos^2(s_2 \cos \phi + \psi_2) + E_2{}^2 \sin^2(s_2 \cos \phi + \psi_2) \\
&\quad + 2E_1 E_2 \cos(s_2 \cos \phi + \psi_2) \\
&= E_1{}^2 + E_2{}^2 + 2E_1 E_2 \cos(s_2 \cos \phi + \psi_2) \tag{33}
\end{aligned}
$$

In performing the azimuth integration in (23) we substitute the value of E^2 from (34) to get

$$
\begin{aligned}
\int_0^{2\pi} E^2 \, d\phi &= \int_0^{2\pi} [E_1{}^2 + E_2{}^2 + 2E_1 E_2 \cos(s_2 \cos \phi + \psi_2)] \, d\phi \\
&= 2\pi \Big\{ E_1{}^2 + E_2{}^2 + 2E_1 E_2 \frac{1}{2\pi} \times \int_0^{2\pi} [\cos(s_2 \cos \phi) \cos \psi_2 \\
&\quad - \sin(s_2 \cos \phi) \sin \psi_2] \, d\phi \Big\}
\end{aligned}
$$

The second term in the integral is zero as can be demonstrated by plotting the function. Therefore,

$$\int_0^{2\pi} E^2 \, d\phi = 2\pi [E_1{}^2 + E_2{}^2 + 2E_1 E_2 (\cos \psi_2) J_0(s_2)]$$

where $J_0(s_2)$ is a Bessel function of the first kind and zeroth order. In this equation the phase ψ and spacing s are between elements 1 and 2; hence a more general form is to write the following:

ψ_{12} = difference in electrical phase angle of the field from the first and the second antenna

s_{12} = electrical length of spacing between the first and second antenna

With this change in nomenclature the above equation becomes

$$\int_0^{2\pi} E^2 \, d\phi = 2\pi [E_1^2 + E_2^2 + 2E_1 E_2 \cos \psi_{12} J_0(s_{12})] \tag{34}$$

In this equation the integration is performed for only two antennas in the horizontal plane. This equation can be generalized to account for any elevation angle θ if the electric-field strengths are multiplied by $f(\theta)$ and the spacing is multiplied by $\cos \theta$. By making these modifications, (34) can be written

$$E(\theta) = \frac{1}{2\pi} \int E^2 \, d\phi = E_1^2 f_1^2(\theta) + E_2^2 f_2^2(\theta)$$
$$+ 2E_1 f_1(\theta) E_2 f_2(\theta) \cos \psi_{12} J_0(s_{12} \cos \theta) \tag{35}$$

where $E(\theta)$ is the rms electric-field strength radiated at the elevation angle θ (in this case for only two antennas). If (35) is generalized for any number of elements in the directional antenna array, the rms electric-field strength at the elevation angle θ can be written

$$E(\theta) = \left[\sum_{p=1}^{p=n} \sum_{q=1}^{q=n} E_p f_p(\theta) E_q f_q(\theta) \cos \psi_{pq} J_0(s_{pq} \cos \theta) \right]^{1/2} \tag{36}$$

where

$\quad E(\theta) =$ rms effective electric-field strength at the elevation angle θ

$\quad\quad\quad p =$ pth antenna in the system

$\quad\quad\quad n =$ number of elements in the complete directional antenna array

$\quad\quad\quad E_p =$ horizontal magnitude of the electric-field strength produced by the pth antenna

$\quad\quad f_p(\theta) =$ vertical radiation characteristic of the pth antenna

$\quad\quad\quad q =$ qth antenna in the system

$\quad\quad\quad E_q =$ horizontal magnitude of the electric-field strength produced by the qth antenna

$\quad\quad f_q(\theta) =$ vertical radiation characteristic of the qth antenna

$\quad\quad \psi_{pq} =$ difference in electrical phase angle of the voltage (or current) between the pth and qth antennas in the directional antenna array

$\quad\quad s_{pq} =$ spacing in degrees or radians between the pth and qth antennas

$J_0(s_{pq} \cos \theta) =$ Bessel function of the first kind and zeroth order of the apparent spacing between the pth and the qth antennas

To clarify the application of (36), consider a three-element directional antenna system. In this case $n = 3$ and the equation can be written

$$E(\theta) = [E_1^2 f_1^2(\theta) + E_2^2 f_2^2(\theta) + E_3^2 f_3^2(\theta) + 2E_1 f_1(\theta) E_2 f_2(\theta) \cos \psi_{12}$$
$$\times J_0(s_{12} \cos \theta) + 2E_1 f_1(\theta) E_3 f_3(\theta) \cos \psi_{13} J_0(s_{13} \cos \theta) + 2E_2 f_2(\theta)$$
$$\times E_3 f_3(\theta) \cos \psi_{23} J_0(s_{23} \cos \theta)]^{1/2} \tag{37}$$

In terms of $E^2(\theta)$ as given in (35), it is possible to rewrite (23) as follows:

$$P_r = \frac{1}{R_c} \int_0^{2\pi} \int_0^{\pi/2} E^2 d^2 \cos \theta \, d\theta \, d\phi$$
$$= \frac{2\pi d^2}{R_c} \int_0^{\pi/2} E^2(\theta) \cos \theta \, d\theta \tag{38}$$

The standard hemispherical electric-field strength produced by the directional antenna system can be obtained by substituting (38) in (24). Thus

$$E_s = \left[\int_0^{\pi/2} E^2(\theta) \cos \theta \, d\theta \right]^{1/2} \tag{39}$$

This is the *exact* formula for determining the size of the directional antenna pattern. To perform the integration for the general case, however, would be quite tedious.

A practical and very useful solution is to determine the value of $E(\theta)$ at a number of elevation angles and to replace the integral with the summation. Thus, for intervals of 10° of elevation the approximate equation can be written, by an application of the trapezoidal rule, as

$$E_s \cong \left\{ \frac{\pi}{18} \left[\frac{E_0^2}{2} + \sum_{n=1}^{n=8} (E_{10n})^2 \cos 10n \right] \right\}^{1/2} \tag{40}$$

where

E_s = the standard hemispherical electric-field strength produced by the directional antenna system (152.1 mV/m for 1 kW at 1 m)

E_0 = rms effective electric-field strength in the horizontal plane

E_{10n} = rms effective electric-field strength at the specified elevation angle

n = integers from 1 to 8, which when multiplied by 10 give the elevation angles θ in degrees

Example 6—A four-tower directional antenna system has the parameters in Table 2.

Determine the values of horizontal electric-field strength from each tower for 5-kW operation, assuming the system loss is 405 W according to FCC standards. What is the value of rms field strength for each 10° elevation angle?

Solution—Initially solving for $E(\theta)$ at every 10° elevation angle by (36) in terms of field ratios F results in column 2 of Table 3.

Table 2. Parameters of Four-Tower System

No.	G (°)	ϕ (°)	s (°)	ψ (°)	F
1	90	0	0	0	1.0
2	90	274.22	176	2	0.786
3	90	302.34	211.3	275	0.841
4	90	358.38	100	260	0.786

Table 3. Values of $E(\theta)$ for Four-Tower System

θ (°)	$E(\theta)$ for $E_1 = 1$	$E(\theta)$ for $E_1 = 275$
0	1.435	395
10	1.432	394
20	1.382	380
30	1.313	361
40	1.212	333
50	1.087	299
60	0.899	247
70	0.657	181
80	0.031	86

Now, by (40) we have

$$E_s = 1.236 \qquad \text{for } E_1 = 1$$

For 5 kW radiated, the standard hemispherical electric-field strength will be

$$E_s = 152.1 \sqrt{5} = 340 \text{ mV/m}$$

Therefore the required values of field strength are as follows:

$$E_1 = 340/1.236 = 275 \text{ mV/m}$$
$$E_2 = 275(0.786) = 216 \text{ mV/m}$$
$$E_3 = 275(0.841) = 231 \text{ mV/m}$$
$$E_4 = 275(0.786) = 216 \text{ mV/m}$$

The final rms electric-field strength for each 10° elevation angle is tabulated in column 3 of Table 3.

The input power for this problem according to FCC standards is 5405 W and the radiated power is 5 kW.

5. RMS Electric-Field Strength in the Horizontal Plane

The rms electric-field strength of a two-tower antenna array can be determined by

$$E_0 = E_1\sqrt{1 + F^2 + (2F\cos\psi)J_0(s)} \qquad (41)$$

where

E_0 = rms electric-field strength at 1 mi (1.609 km), in millivolts per meter

E_1 = rms electric-field strength at 1 mi for reference tower 1 while operating in array, in millivolts per meter

F = ratio of magnitude of electric-field strength from tower 2 divided by that of tower 1

ψ = electrical phase of field from tower 2 with respect to that of tower 1, in degrees

$J_0(s)$ = Bessel function of first kind and zeroth order for tower spacing s

Now, if 90° antennas are used and the field ratio $F = 1$, then (41) can be written

$$E_0 = 195\sqrt{\frac{1 + (\cos\psi)J_0(S)}{1 + \cos(R_{12}/R_{11})}} \qquad (42)$$

where

R_{12} = mutual resistance between tower 1 and tower 2, in ohms

R_{11} = self-resistance of tower 1, 36.6 Ω

The solution of this equation is shown in Fig. 23 for various values of antenna phasing ψ. It gives the theoretical electric-field strength without loss for 1-kW operation.

6. The Theoretical Pattern

The theoretical pattern, as used by the FCC, can be written

$$[E(\phi,\theta)]_{\text{th}} = \left| k \sum_{i=1}^{i=n} F_i f_i(\theta) \angle\beta_i \right| \qquad (43)$$

where

$[E(\phi,\theta)]_{\text{th}}$ = the theoretical pattern inverse-distance electric-field strength at 1 mi (1.609 km), in millivolts per meter

k = multiplying constant which determines the basic pattern size

F_i = field strength ratio of ith element in the array

$E_i = kF_i$ in (11)

and the other terms are defined following (11) and (12).

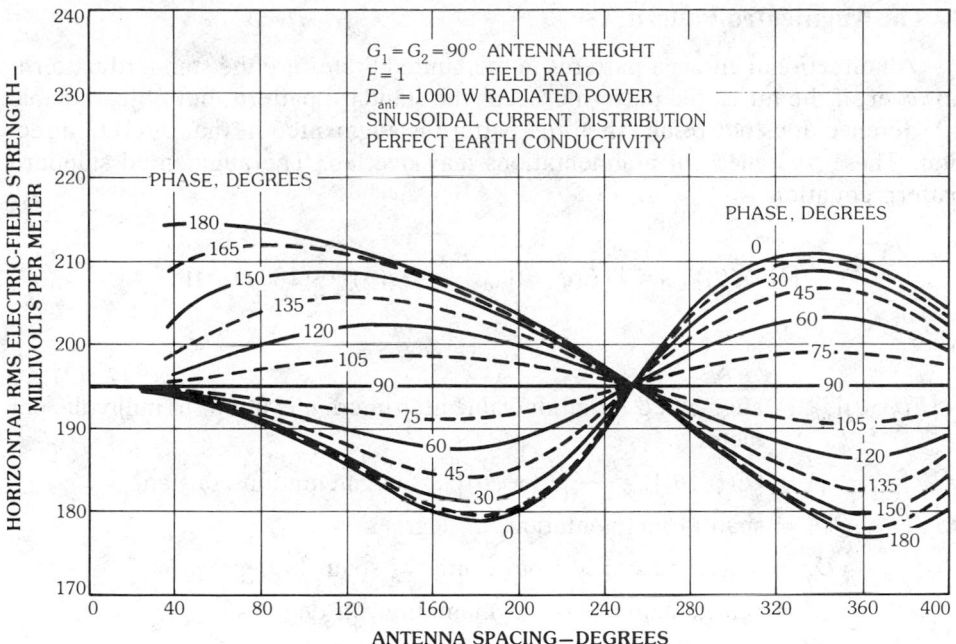

Fig. 23. Horizontal rms electric-field strength of two-antenna directional antenna.

7. The Standard Pattern

The above theoretical pattern in (43) cannot be achieved perfectly in the field, so the standard pattern was introduced by the FCC. To the theoretical value a quadrature term Q is added which primarily fills in the pattern minimums, and the resulting size is increased by 5 percent as shown in the following equation:

$$[E(\phi, \theta)]_{std} = 1.05 \sqrt{[E(\phi, \theta)]_{th}^2 + Q^2} \qquad (44)$$

where

$[E(\phi, \theta)]_{std}$ = the standard pattern inverse-distance electric-field strength at 1 mi (1.609 km), in millivolts per meter

$[E(\phi, \theta)]_{th}$ = the theoretical pattern inverse-distance electric-field strength at 1 mi (1.609 km), in millivolts per meter

Q = is the greater of the following quantities:
$0.025\, g(\theta)\, E_{rss}$ or $6.0\, g(\theta) \sqrt{P_{kW}}$

$$g(\theta) = \sqrt{\frac{f^2(\theta) + 0.0625}{1.030\,776}} \qquad (45)$$

$$E_{rss} = k \sqrt{\sum_{i=1}^{n} F_i^2} \qquad (46)$$

8. The Augmented Pattern

All directional antenna patterns in the United States use the standard pattern; however, if the measured pattern exceeds the standard pattern, but objectionable interference does not result, then the pattern is augmented in that discrete direction. These "patches" or augmentations may overlap. The augmented standard pattern equation is

$$[E(\phi, \theta)]_{\text{aug}} = \sqrt{[E(\phi, \theta)]_{\text{std}}^2 + A[g(\theta)\cos(180\, D_a/s)]^2} \tag{47}$$

where

$[E(\phi, \theta)]_{\text{aug}}$ = augmented radiation value at azimuth/elevation, in millivolts per meter

$A = [E(\phi', \theta)]_{\text{aug}}^2 - [E(\phi', \theta)]_{\text{std}}^2$, augmentation constant

s = span of augmentation, in degrees

D_a = angular distance from center of span, in degrees

ϕ' = central azimuth of augmentation, in degrees

The augmentation term A specifies the magnitude of augmentation at the central angle of augmentation ϕ'. This term A is diminished to zero at the edges of the augmentation span s. An example of theoretical, standard, and augmented patterns is shown in Fig. 24.

9. Power to Provide System Losses

Because of losses in the transmission lines and matching, phasing, and power division networks, plus other losses in the system, such as resistance losses in the tower and ground system and dielectric losses in the insulators, an overfeeding of power is allowed by the FCC at the common point.

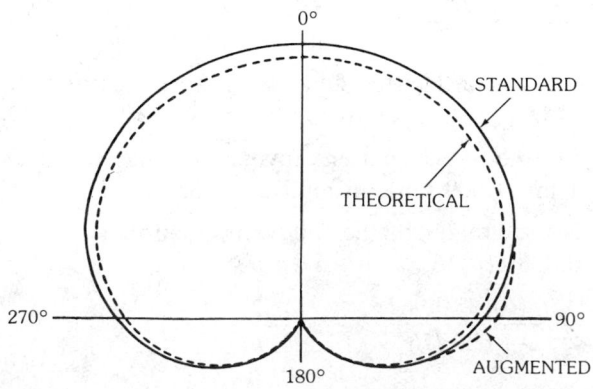

Fig. 24. Theoretical, standard, and augmented patterns.

The calculation of the amount of overfeeding of power at the common point is made as follows.

For stations with directional antennas authorized to radiate 5 kW of power or less, the measured common-point resistance is assumed to be 92.5 percent of its measured value, and for all other directional antennas it is assumed to be 95 percent. This arbitrary reduction in resistance amounts to increasing the current at the common point by $1/\sqrt{0.925} = 1.0398$, or approximately a 4-percent increase for values of transmitter power up to and including 5 kW. For transmitters with powers above 5 kW, the antenna current can be increased $1/\sqrt{0.95} = 1.026$, or approximately 2.6 percent.

Another way of saying this is that for a 1-kW station 1081 W can be fed in at the common point, while for a 50-kW station 52 632 W can be fed in at the common point. For the 1-kW station there are 81 W available for feeder-system loss, and in the 50-kW station there are 2632 W available for feeder-system loss.

10. Ground Systems

Antenna performance is standardized with reference to the ground system being a perfectly conducting flat plane. This assumption serves a useful purpose in designing directional antenna systems; however, in real life the ground-system losses must be considered and reduced to a reasonable value by using a ground system to minimize earth losses.

The **E**-field lines of force extend from the antenna out through the surrounding space to the earth. On entering the earth the current returns to the base of the antenna and produces an **H** field above the ground at right angles to the earth current. Fig. 25 illustrates both the **H**-field vector directions and earth current directions for two antennas with currents in phase. It is interesting to note that the earth currents are not radial and the **H**-field vectors do not follow a circular locus. The criterion is to design the ground system such that the earth currents will be in the copper ground system rather than in lossy earth.

The usual practice is to use no. 10 AWG bare copper wire for the ground system, consisting of 120 radials from each tower buried to a depth of 6 to 10 in (15 to 25 cm). The classic system is composed of radials 90° long, and in many cases a ground screen of expanded copper is installed at the tower base.

11. Directional Antenna Feeder Systems

In general a directional antenna feeder system, as shown in Fig. 26, will consist of a matching network from the transmitter output to the power-divider input, a power-dividing network, phase shifting networks of 0° or ±90°, transmission lines to the towers, and matching networks at the base of each tower.

The tower matching networks transform the driving-point impedance of each tower to the characteristic resistance of the transmission line. In addition, it is also desirable to provide phase shift such that 0° or 90° phase shifters can be used at the input to the transmission lines. The power divider supplies the desired amount of power to the phase shifters. Since the input impedance of the power divider may not match the output of the transmitter, an input matching network is provided.

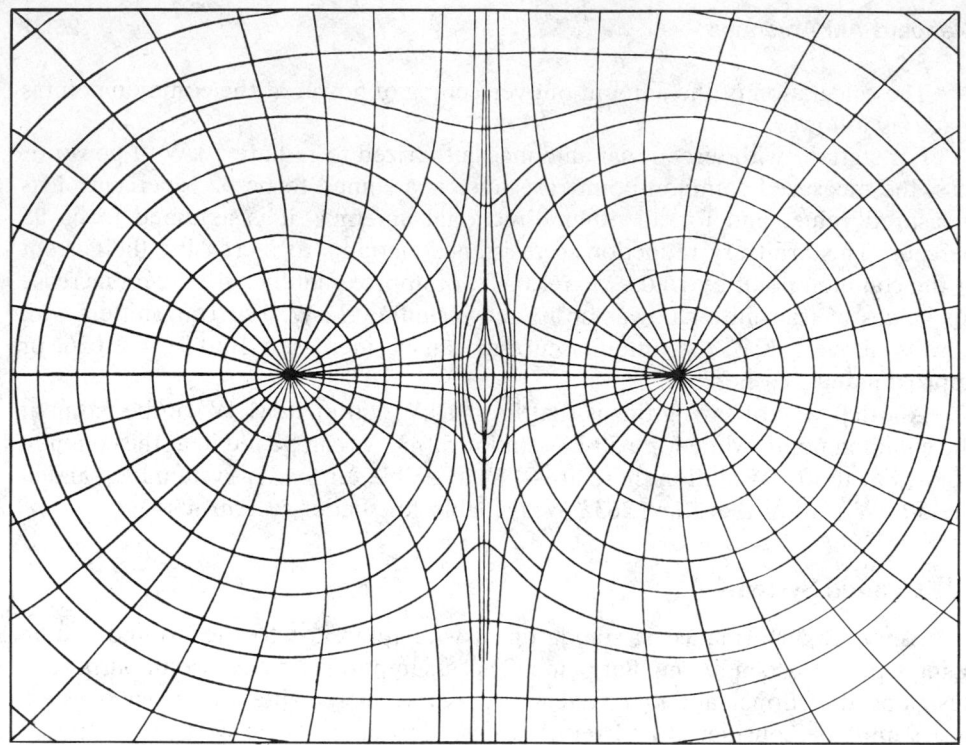

Fig. 25. Typical ground-system current and **H**-field vector direction.

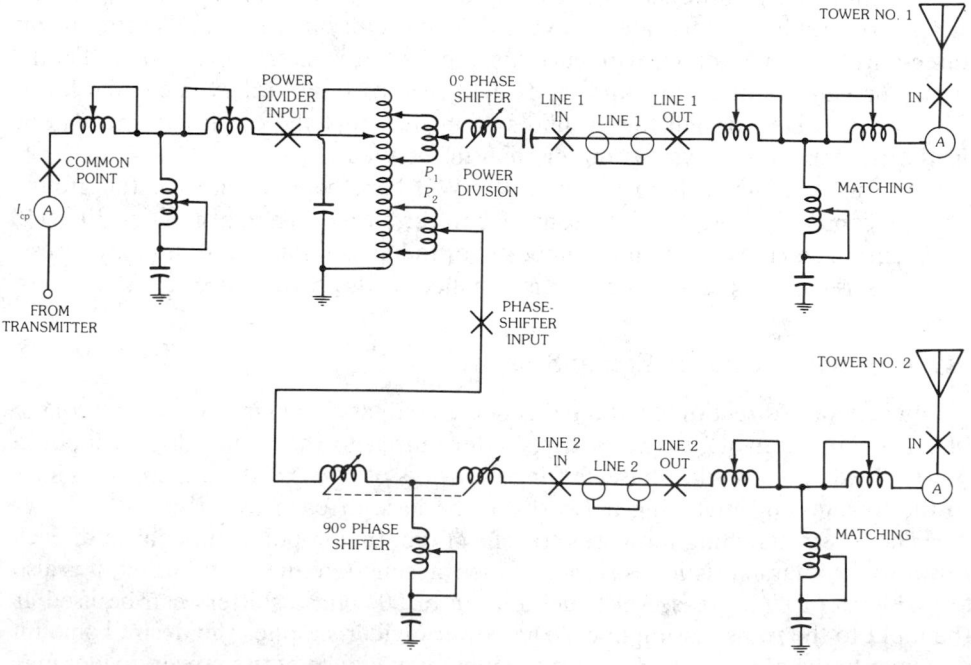

Fig. 26. Directional antenna feeder system.

The shunt coil provides the desired input resistance, while the input series coil provides zero reactance control.

In Fig. 26 there are a number of points marked X in the system. At these points an operating impedance bridge can be inserted to determine the impedance and thus maintain the desired feeder system impedances. By using an rf current meter in series with the operating bridge it is possible to make a useful inventory of the rf power in all parts of the feeder system.

Power Dividing Networks

There are several kinds of power dividing networks as shown in Fig. 27. The push-pull circuit in Fig. 27a produces outputs at terminals 1 and 2 that are 180° out

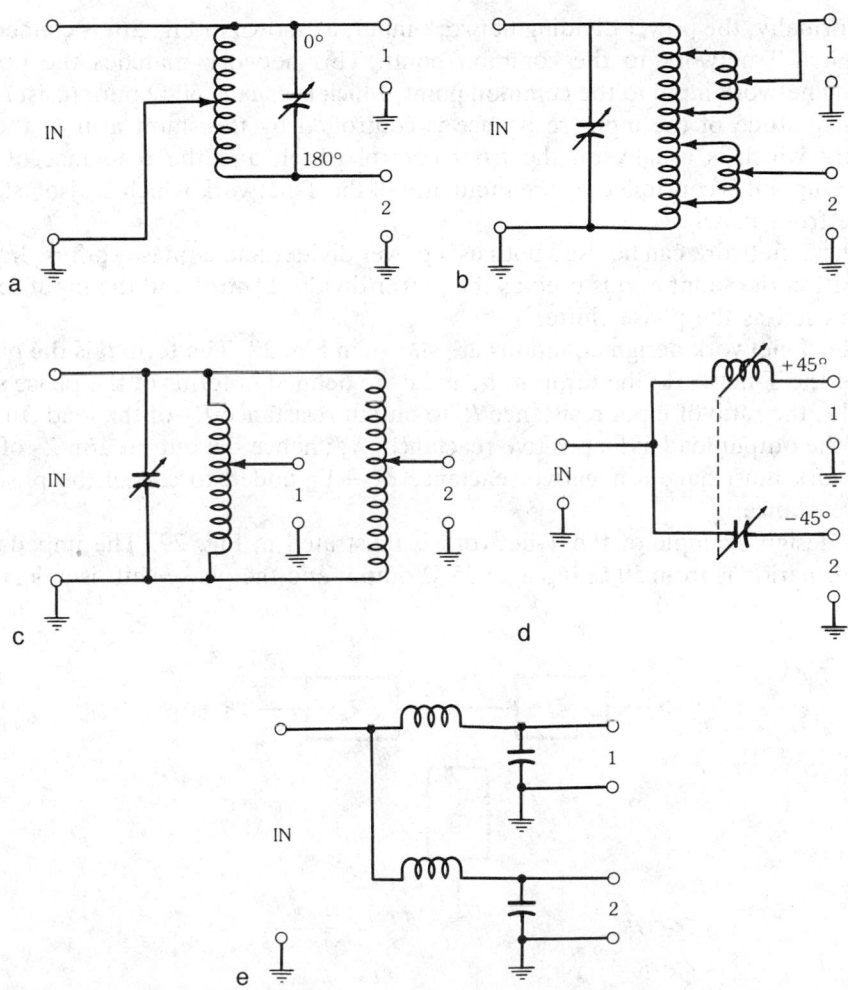

Fig. 27. Typical power-dividing networks. (*a*) Push-pull circuit, 180° feed. (*b*) Series-resonant circuit, 0° feed. (*c*) Parallel-resonant circuit, 0° feed. (*d*) Quadrature circuit, 90° feed. (*e*) Two L-sections in parallel.

of phase. Parallel resonance is achieved by tuning the capacitor while power division is varied by moving the coil tap to ground. The series resonant circuit in Fig. 27b produces output signals that are in phase. The jeep coils are connected by taps on the main coil and the adjustment of the jeep coil is often a front panel control on the phasor cabinet. The parallel variable capacitor tunes the circuit to parallel resonance and increases the input resistance. The parallel resonant circuit in Fig. 27c provides in-phase outputs and variable taps to control the power division. The parallel variable capacitor tunes the circuit to parallel resonance at the input terminal. The quadrature circuit in Fig. 27d provides output signals that are 90° in phase from each other. The loads can be matched to the input terminal in Fig. 27e by using two L-sections.

Impedance-Matching and Phase-Shifting Networks

Normally, the power dividing network input, as shown in Fig. 26, is connected through a T-network to the common point. This network matches the power dividing network input to the common point, which is usually 50 Ω pure resistance. The magnitude of the input resistance is controlled by the shunt arm of the T-network which is usually on the front control panel, and the reactance of the common point is controlled by the input arm of the T-network which is also usually on the front panel.

The T-network can be used both as a power divider and a phase shifter. In this application the shunt arm is used as the power-divider control and the input series arm is used as the phase shifter.

The T-network design equations are shown in Fig. 28. The term β is the phase shift of the T-network, the terms a, b, and c are defined in terms of the phase shift β, and r, the ratio of input resistance R_I to output resistance R_L of the load. In this figure the output load has a positive reactance, X_L; hence the output arm Z_2 of the T-network must have a negative reactance of $-X_L$ added to cancel the positive load reactance.

A design example of the T-network is illustrated in Fig. 29. The impedance transformation is from 50 Ω input to 25 Ω output and the phase shift is a delay of

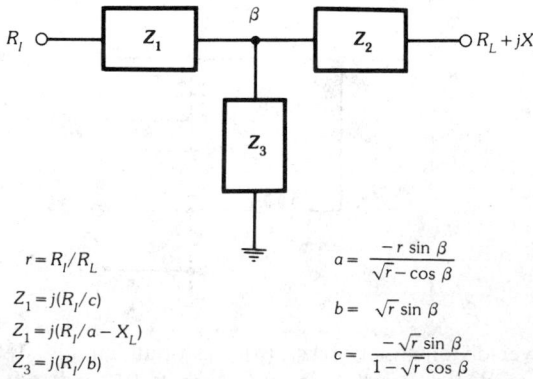

$$r = R_I/R_L$$
$$Z_1 = j(R_I/c)$$
$$Z_1 = j(R_I/a - X_L)$$
$$Z_3 = j(R_I/b)$$

$$a = \frac{-r \sin \beta}{\sqrt{r} - \cos \beta}$$
$$b = \sqrt{r} \sin \beta$$
$$c = \frac{-\sqrt{r} \sin \beta}{1 - \sqrt{r} \cos \beta}$$

Fig. 28. T-network design networks.

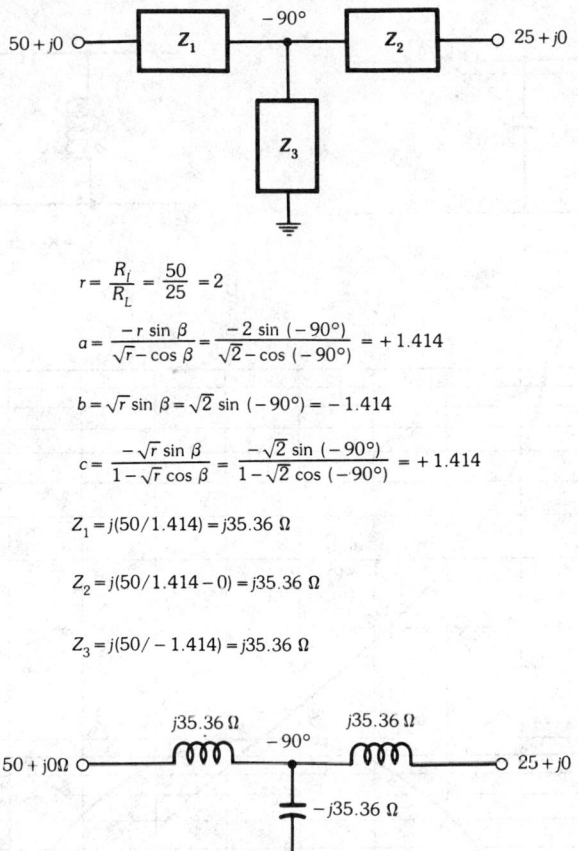

Fig. 29. T-network design examples.

90° from the input to the output. This example shows how the network arms are computed.

The T-network of Fig. 30 is very useful as a 90° ± 15° phase-shifting network. It will be noted that the impedance of the shunt arm X_3 of the network is substantially constant over about 30° of phase shift. The phase shift is accomplished by varying the series arms X_1 and X_2 in unison.

The series circuit of Fig. 31 is called a 0° phase shifter. By varying the reactance X_L from resonance with X_C, the phase can be shifted ±15° without a major disturbance of the network impedance.

Antenna Sampling System

The most useful instrument in adjusting and maintaining proper operation of a directional antenna system is the antenna monitor. An approved antenna monitor system consists of single-turn unshielded loops which are rigid constructed mounted on each tower at the current loop, 90° down from the top of a non–top-loaded tower, or at least 10 ft (3 m) above the ground of short towers, and with

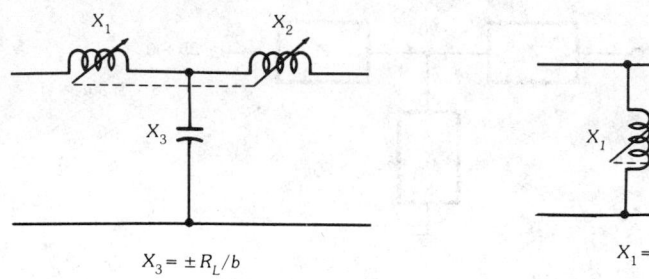

$$X_3 = \pm R_L/b$$

a

$$X_1 = X_2 = \pm R_L/a$$

b

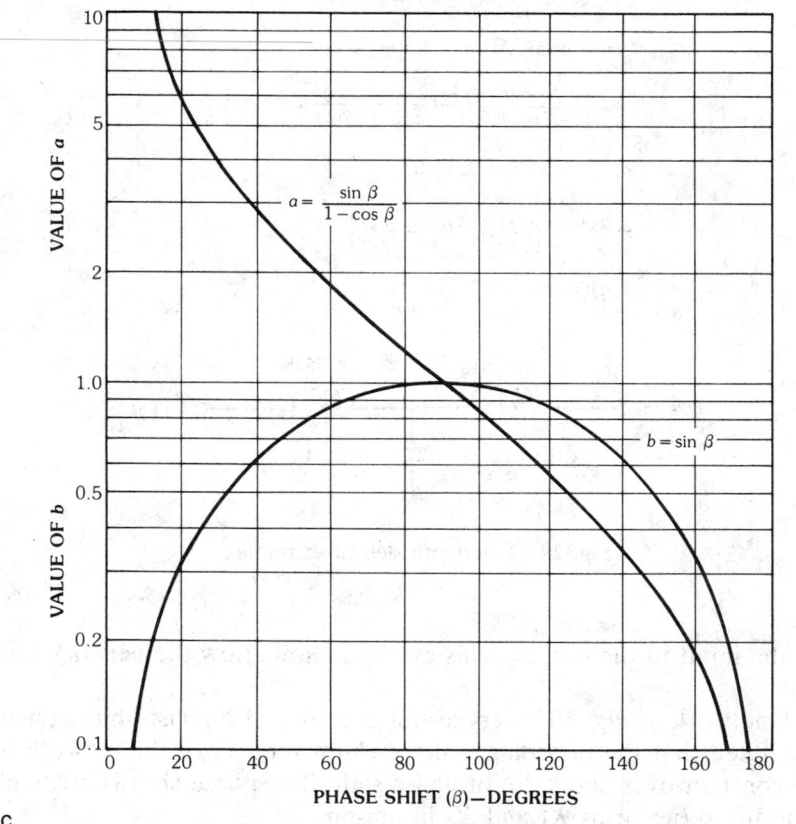

c

Fig. 30. Phase-shifting networks and curves (90° ± 15°). (*a*) Retarding network. (*b*) Advancing network. (*c*) Graphs of *a* and *b*.

equal electrical lengths of rigid outer conductor coaxial cables from each sampling loop to an approved antenna monitor, as shown in Fig. 32.

The antenna monitor provides magnitude and phase values of the current in each tower with respect to the reference tower. This information is very useful in making the initial adjustment and is required information to maintain proper operation of the directional antenna system.

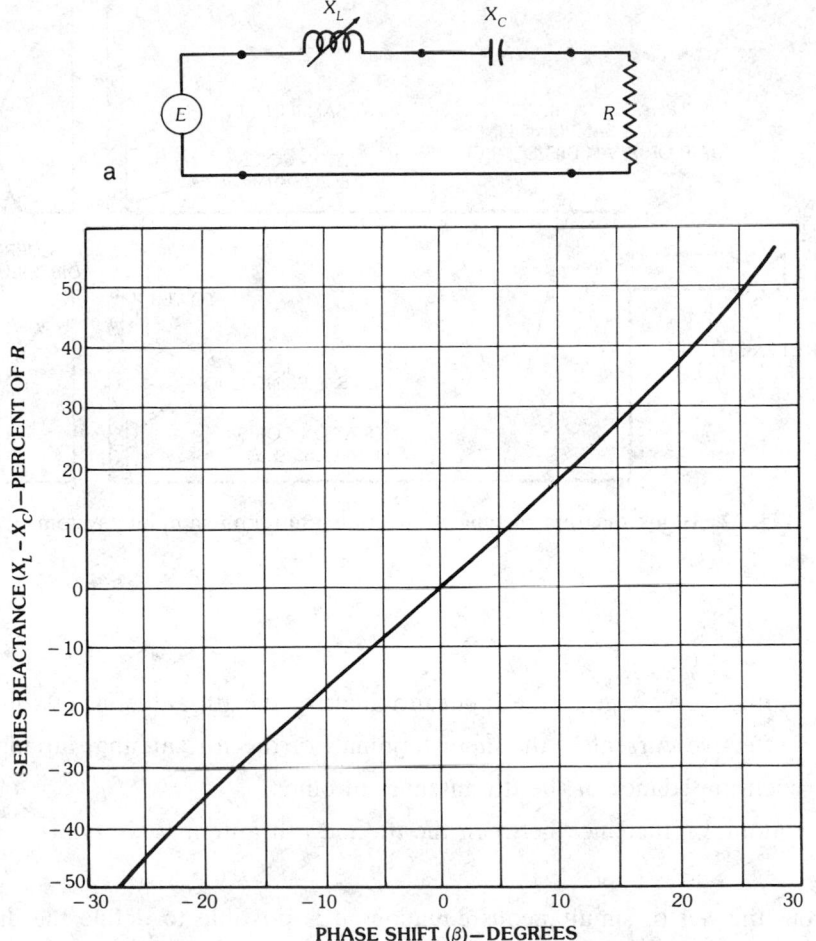

Fig. 31. Phase-shifting network and curve $(0° \pm 15°)$. (*a*) Circuit containing series reactance $X_L - X_C$. (*b*) Graph of series reactance.

Driving-Point Impedance

The driving-point impedance, at the base of each tower, is needed to set up the directional antenna feeder system.

The mesh circuit equations for *n* antennas in the directional antenna system is given by

$$
\begin{aligned}
V_1 &= I_1 Z_{11} + I_2 Z_{12} + \ldots + I_i Z_{1i} + \ldots + I_n Z_{1n} \\
V_2 &= I_1 Z_{21} + I_2 Z_{22} + \ldots + I_i Z_{2i} + \ldots + I_n Z_{2n} \\
&\ \ \vdots \qquad\ \ \vdots \qquad\ \ \vdots \qquad\qquad\ \ \vdots \qquad\qquad\ \ \vdots \\
V_i &= I_1 Z_{i1} + I_2 Z_{i2} + \ldots + I_i Z_{ii} + \ldots + I_n Z_{in} \\
&\ \ \vdots \qquad\ \ \vdots \qquad\ \ \vdots \qquad\qquad\ \ \vdots \qquad\qquad\ \ \vdots \\
V_n &= I_1 Z_{n1} + I_2 Z_{n2} + \ldots + I_i Z_{ni} + \ldots + I_n Z_{nn}
\end{aligned}
\tag{48}
$$

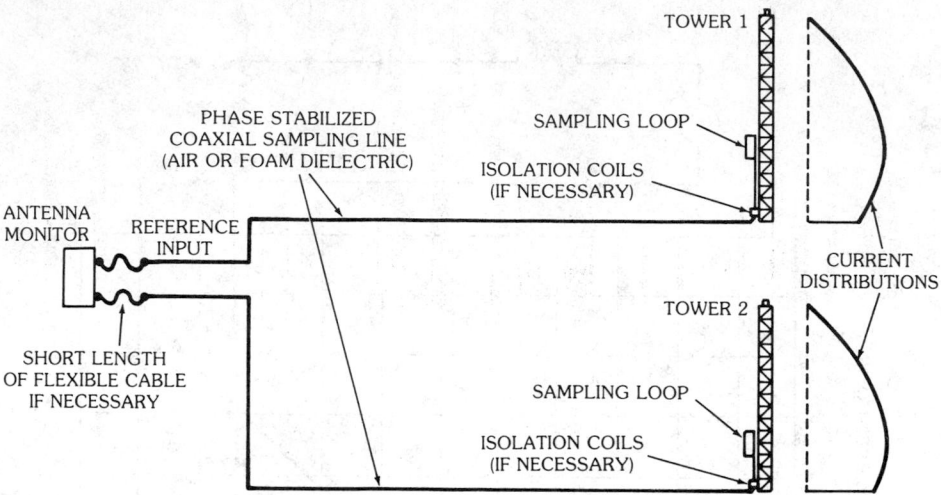

Fig. 32. Block diagram of typical directional antenna sampling system.

where

V_i = effective voltage at the input terminals of the ith antenna, in volts

I_i = effective current at the input terminals of the ith antenna, in amperes

Z_{ii} = self-impedance of the ith antenna, in ohms

Z_{in} = mutual impedance between the ith and nth antennas, in ohms

From this set of simultaneous equations it is possible to define the driving-point impedance of the ith antenna as follows:

$$Z_i = \frac{V_i}{I_i} = \frac{I_1}{I_i} Z_{i1} + \frac{I_2}{I_i} Z_{i2} + \cdots + Z_{ii} + \cdots + \frac{I_n}{I_i} Z_{in} \qquad (49)$$

The resistance component R_i of Z_i is a pure radiation resistance only if there is no loss in the system. The directional antenna system can be calculated on this theoretical basis, and then from a knowledge of the system the losses can be estimated with fair accuracy.

With the desired directional antenna characteristics, one can determine the relative required exciting currents, which in turn determine the respective driving-point impedances from (49). This information is then used to design the impedance-matching networks between the respective transmission lines and towers in the directional antenna array.

To determine the driving-point resistance of the ith antenna, the real part of (49) can be written

$$R_i = \left|\frac{I_1}{I_i}\right| \cdot \left|Z_{1i}\right| \cos(\psi_{1i} + \gamma_{1i}) + \left|\frac{I_2}{I_i}\right| \cdot \left|Z_{2i}\right|$$

$$\times \cos(\psi_{2i} + \gamma_{2i}) + \cdots + \left|Z_{ii}\right| \cos\gamma_{ii} + \cdots$$

$$+ \left|\frac{I_n}{I_i}\right| \cdot \left|Z_{ni}\right| \cos(\psi_{ni} + \gamma_{ni}) \tag{50}$$

and the reactance component is

$$X_i = \left|\frac{I_1}{I_i}\right| \cdot \left|Z_{1i}\right| \sin(\psi_{1i} + \gamma_{1i}) + \left|\frac{I_2}{I_i}\right| \cdot \left|Z_{2i}\right|$$

$$\times \sin(\psi_{2i} + \gamma_{2i}) + \cdots + \left|Z_{ii}\right| \sin\gamma_{ii} + \cdots$$

$$+ \left|\frac{I_n}{I_i}\right| \cdot \left|Z_{ni}\right| \sin(\psi_{ni} + \gamma_{ni}) \tag{51}$$

where

ψ_{ij} = phase angle of I_i minus the phase angle of I_j

γ_{ij} = phase angle of impedance Z_{ij}

Directional Antenna Feeder System Design Example

In order to design a feeder system it is first necessary to have the antenna parameters consisting of current ratios and current phases in each tower. Second, it is necessary to have the driving-point impedances of each tower. Third, the base currents are necessary. Fourth, the power distribution needs to be known. Finally, the transmission-line lengths and their characteristics of impedance and loss are needed.

The directional antenna feeder system can be described by the block diagram of Fig. 33. The various blocks have already been discussed. In this design example the driving-point impedances, powers, base currents, and phases are furnished for each tower.

The first step is to determine the phase shifts desired in each block between the power divider and each tower. Usually the first trial is to start with zero phase of the current in the reference no. 1 tower, with the vector at 0°, as shown in Fig. 33. The coupling network has a delay of 90° and the transmission line has a delay of 244°, resulting in a vector of 334° after going through the 0° phase shifter to the power divider. Assuming that a 90° phase shifter is connected to the power divider and a transmission-line delay of 165° in the transmission line to tower no. 2, the input phasor to the coupling network is 79°. Providing a delay of 59° in this network results in the correct phase of 20° for the current into the driving-point impedance of tower no. 2. This results in the best condition of phase control in the 0° and 90° phase shifters.

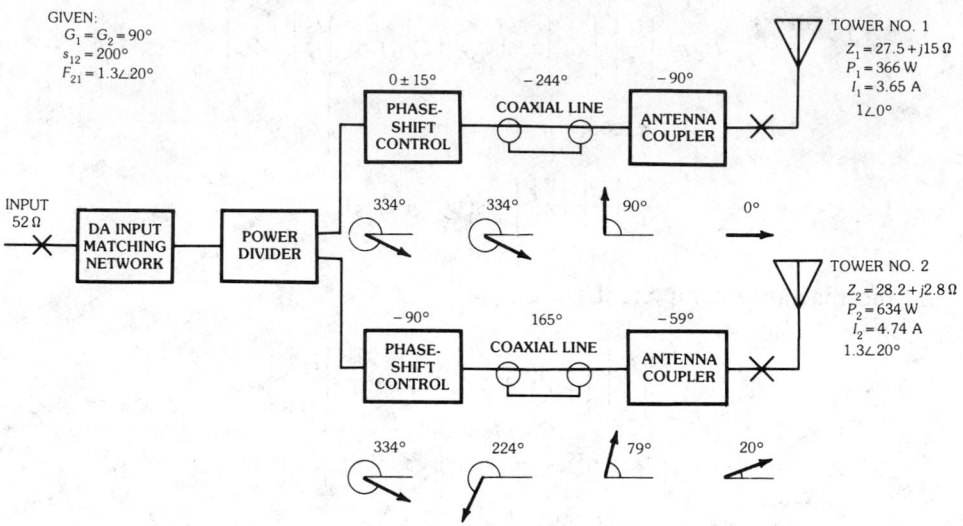

Fig. 33. Block diagram of directional antenna feeder system.

In this example let

$$G_1 = G_2 = 90°$$

$$Z_{11} = Z_{22} = 42.2 \angle 29.8° = 36.6 + j21$$

$$s_{12} = s_{21} = 200°$$

$$Z_{12} = Z_{21} = 14.9 \angle -130° = -9.7 - j11.3$$

$$F_{21} = M_{21} = I_2/I_1 = 1.3 \angle 20°$$

$$\begin{aligned} Z_1 &= Z_{11} + M_{12} Z_{12} \\ &= 42.2 \angle 29.8° + (0.77 \angle -20)(14.9 \angle -130°) \\ &= 27.5 + j15.8 \end{aligned}$$

$$\begin{aligned} Z_2 &= M_{21} Z_{12} + Z_{22} \\ &= (1.3 \angle 20)(14.9 \angle -130°) + 42.2 \angle 29.8° \\ &= 28.2 + j2.8 \end{aligned}$$

The power distribution as shown in Fig. 33 is then

$$P = I_1{}^2 R_1 + I_2{}^2 R_2$$

$$1000 = I_1{}^2(27.5) + (1.3)^2(28.2)$$

$$I_1 = 3.65 \text{ A}$$

$$I_2 = 4.75 \text{ A}$$

$$P_1 = 366 \text{ W}$$

$$P_2 = 634 \text{ W}$$

The next step is to use the design equations in Figs. 28 to 30 to determine the values of reactance in the various impedance-matching and phase-shifting T-

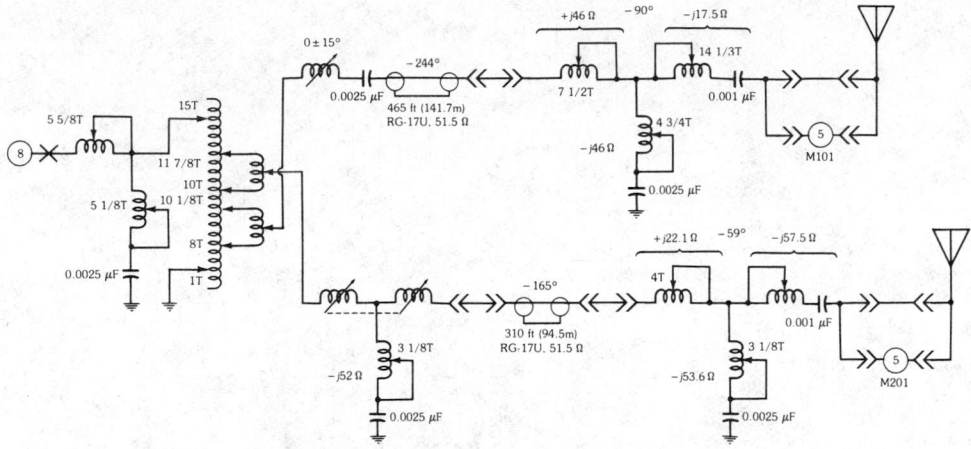

Fig. 34. Schematic diagram of feeder system.

networks. The results of the design are shown in Fig. 34. It will be noted that 0.001-µF capacitors are placed in the series T-network arms toward the antennas to isolate the feeder system from direct currents that are generated on the towers due to static charges. It will also be noted that the antenna meters are replaced with a make-before-break switch with a shorting loop having the same inductance as the meter when it is connected.

In this feeder system the power divider is a coil with jeep coils for fine tuning of the power division. The power divider is then connected to an L-network with a shunt control for the resistance magnitude and a series input arm to series-resonate the circuit, thus presenting a pure-resistance load to the transmitter.

Acknowledgments

The author wants to express his appreciation to many of his associates and seminar students for helpful suggestions. In particular he greatly appreciated the comments and suggestions of Roy E. Christen of the Cleveland Institute of Electronics, and he is indebted to his secretary, Blanche Kotalik, who typed the manuscript.

12. References

[1]   S. A. Schelkunoff, "Theory of antennas of arbitrary size and shape," *Proc. IRE*, vol. 29, pp. 493–521, September 1941.

[2]   Y. T. Lo, "A note on cylindrical antennas of noncircular cross section," *J. Appl. Phys.*, vol. 24, no. 10, pp. 1338–1339, October 1953.

[3]   G. H. Brown, "Directional antennas," *Proc. IRE*, vol. 25, pp. 81–145, January 1937.

[4]   C. R. Cox, "Mutual impedance between vertical antennas of unequal heights," *Proc. IRE*, vol. 35, pp. 1367–1370, November 1947.

[5]   C. E. Smith, "Systematization of two tower patterns," in *Directional Antenna Patterns*, pt. 2, Cleveland: Smith Electronics, 1958, pp. 2.1–2.11.

Chapter 27

TV and FM Broadcast Antennas

G. W. Collins
Harris Corporation

CONTENTS

Gerald W. Collins has over 24 years of antenna and rf engineering experience. Since 1975 he has been with the Harris Broadcast Division, where he has been involved with the design and manufacture of broadcast antennas, including the Harris line of circularly polarized antennas and more recently the waveguide slot antennas. Prior to coming to the Broadcast Division he was with the Harris Government Systems Sector, where he was involved in the design of many types of antennas for military and space applications.

He received a BSEE from the University of Illinois in 1963, an MSEE from the Florida Institute of Technology in 1968, and has taken postgraduate work toward a PhD in electrical engineering through the University of Massachusetts. He is a registered professional engineer in the State of Illinois.

Mr. Collins has published several papers related to antennas and propagation, including various aspects of broadcasting. He has been issued patents for a circularly polarized zigzag antenna and a slotted cylinder antenna. He is a Senior Member of the IEEE.

1. General

Television and fm broadcasting in the United States is assigned to major portions of the vhf and uhf spectrum. Television is assigned to three bands at vhf and one at uhf, namely:

channels 2–4	54–72 MHz
channels 5, 6	76–88 MHz
channels 7–13	174–216 MHz
channels 14–83	470–890 MHz

Frequency-modulation broadcasting is at vhf from 88 to 108 MHz. The channel width for tv is 6 MHz for all channels so that the percentage bandwidth is as high as 10.5 percent at channel 2 to as low as 0.7 percent for channel 83. The fm spectrum is divided into 100 channels of 200 kHz width each, for a percentage bandwidth of about 0.2 percent.

In the design of the antennas it is important that both the pattern and impedance bandwidth meet minimum requirements over these bands. Under normal operating conditions it is generally required that the maximum vswr be less than 1.1:1 over the operating channel although values as high as 1.2:1 are acceptable for fm. For tv, more stringent requirements are placed on the vswr at the visual carrier and color subcarrier frequencies. Generally accepted values are 1.05:1 and 1.08:1, respectively. The visual frequency is 1.25 MHz above the lower band edge of each channel and the color subcarrier frequency is 3.58 MHz above the visual carrier.

The maximum effective radiated power (ERP) of fm and tv stations has been established to permit all stations, no matter the channel assignment, at a given height above average terrain (HAAT) to provide coverage to approximately equal areas. Accordingly, the maximum ERP for tv channels 2 through 6 and all fm channels is 100 kW; for channels 7 through 13 it is 316 kW, and for uhf channels it is 5000 kW. The ERP is specified for the horizontally polarized component at the rms level of an omnidirectional antenna. When circular polarization is used, equal ERP is permitted in the vertical component. The specified ERPs are readily achieved at vhf using relatively low gain antennas, with values of 2 to 6 being common for tv channels 2 through 6 and fm, and values of 6 to 12 being common for tv channels 7 through 13. (These gain values are power ratios relative to a half-wave dipole.) For uhf channels, much higher gains are usually required. This is due to limitations on available transmitter power. To achieve 5000 kW with a 220-kW transmitter usually requires an antenna gain of approximately 30. (Line efficiency is often 70 to 80 percent.)

The actual ERP of a station is dependent on its location and HAAT [1]. The United States and its possessions are divided into three zones as shown in Fig. 1 for

ZONE II
PLUS PUERTO RICO, ALASKA,
HAWAIIAN ISLANDS AND
VIRGIN ISLANDS

ZONE I

ZONE III

Fig. 1. FCC allocations and assignment zones.

tv stations. The maximum ERP for tv stations in each of these zones is a function of the station HAAT as shown in Fig. 2. Since the signal strength at a specific distance for a given ERP increases with increasing HAAT, these curves require a corresponding reduction in ERP for increasing HAAT above 1000 ft (304.8 m) for zone I or 2000 ft (609.6 m) for zones II and III.

For fm stations similar curves apply as shown in Fig. 3. The fm stations are classified as class A, B, B1, C, or C1, depending on the size of community they serve and the zone in which they are located. The zones for fm are somewhat different from those for tv. Zone I is the same as for tv allocations. Zone IA includes Puerto Rico, the Virgin Islands, and the part of California south of the 40th parallel. Zone II includes tv zone III and tv zone II except for those areas in zone IA. Class B and B1 stations may operate in zones I, IA, and II. Class C and C1 stations may operate only in zone II.

A tv station's coverage is stated in terms of distance to the city grade, grade A, and grade B contours. The signal strength at these contours for the various channel groupings is shown in Table 1 and is stated in terms of decibels over a microvolt per meter (dBu) for an ERP of 1 kW (0 dBk). The antenna should be located so that a city grade signal is provided over the principal city of service. The distances to these contours are estimated by the FCC (50/50) curves. Coverage for fm stations is determined by the distance to their 70-dBu and 60-dBu contours. It is essential to realize that signal strength values and contour distances determined from these curves are estimates only, to be expected at 50 percent of locations 50 percent of the time at a height of 30 ft (9.144 m). Actual signal strength at any time or location is dependent on local propagation conditions and receiving antenna factors [2, 3].

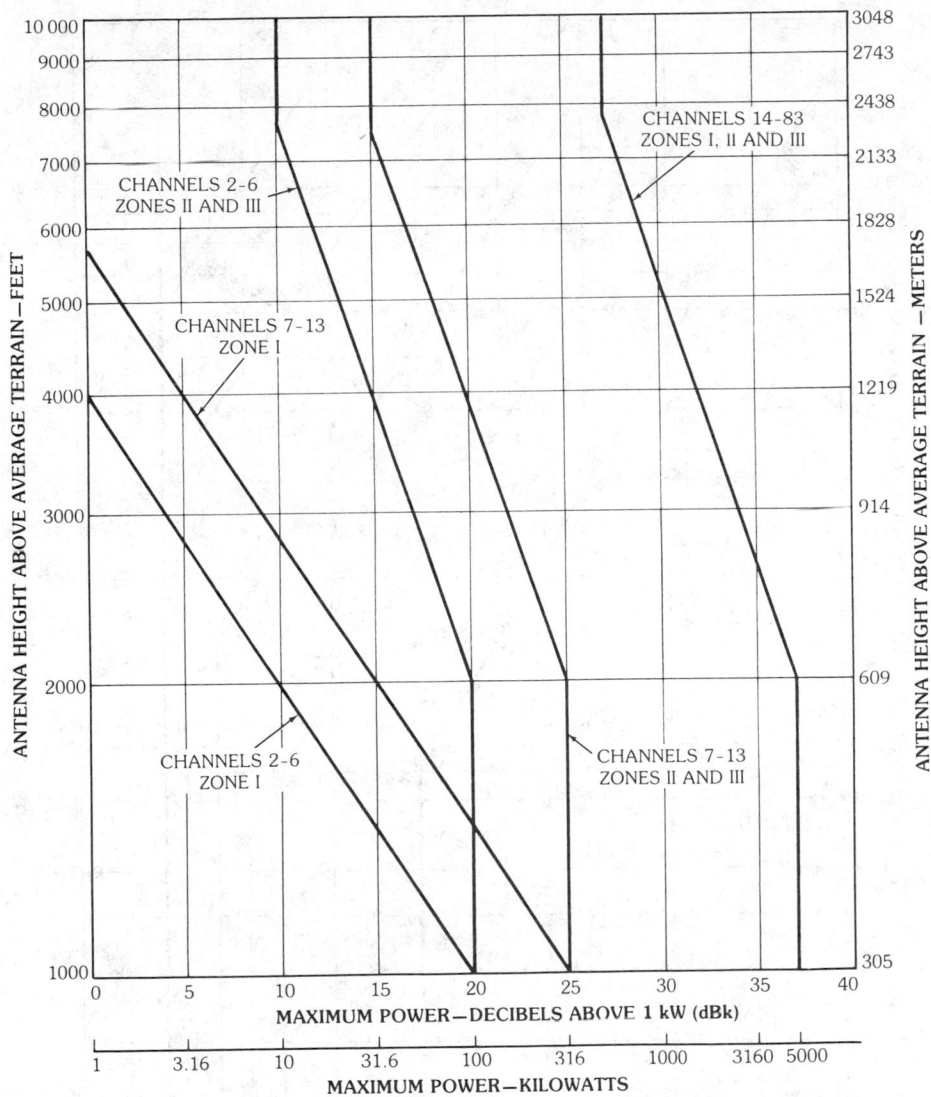

Fig. 2. Maximum effective radiated power versus antenna height for zones I, II, and III.

The usual case is for stations to utilize an omnidirectional pattern to achieve approximately uniform coverage in all directions. In practice, antennas only approximate the ideal pattern. Although there are no hard and fast generally accepted specifications, an antenna is considered to be omnidirectional if the pattern deviates from the rms level by no more than 2 or 3 dB.

It is sometimes necessary for a station to use a directional azimuth pattern. This is particularly important at uhf, where it is often desirable to concentrate the radiated power over a populated area. This permits attainment of a higher ERP over the populated areas while minimizing transmitter power and antenna aperture length. Another reason for the use of directional patterns is to prevent overlapping

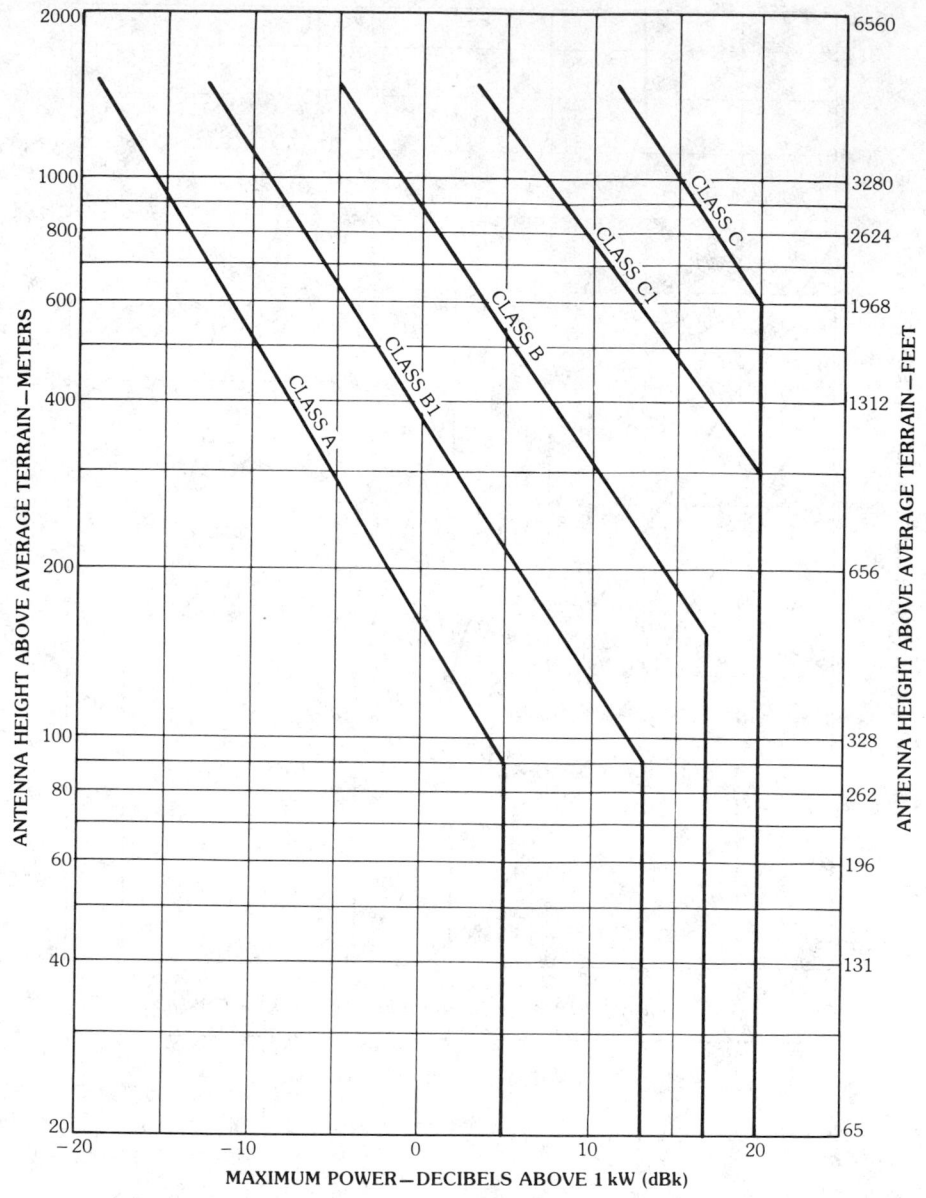

Fig. 3. Maximum effective radiated power versus antenna height for fm stations.

Table 1. Television Station Signal Strength at the Contours

Contour	Channels 2 through 6	Channels 7 through 13	Channels 14 through 83
City grade	74 dBu	77 dBu	80 dBu
Grade A	68 dBu	71 dBu	74 dBu
Grade B	47 dBu	56 dBu	64 dBu

of contours for cochannel stations operating from particular sites. If two stations on the same channel are spaced by less than the minimums specified by the FCC, the ERP along and about the radial connecting the two stations must be reduced to prevent overlapping coverage. The nulls in the pattern must be sufficiently deep to provide the protection. For fm stations the slope in the null must not exceed 2 dB/10°. When a directional antenna is used, the ERP is determined by the ERP at the azimuth of the pattern maximum. The maximum-to-minimum ratio for the azimuth pattern may not exceed 15 dB for fm and uhf tv channels 14 through 69 transmitting more than 1 kW. For tv channels 2 through 13, the maximum-to-minimum ratio is 10 dB.

The beamwidth of the elevation pattern is inversely proportional to the gain. Thus, for very high gain antennas, close attention must be paid to the stability of the beam. This is particularly true for uhf antennas, where high-gain antennas are commonly used, and to a lesser extent for vhf antennas. To ensure beam stability, both mechanical and electrical factors must be considered. The antenna must be stiff enough to minimize wind-induced deflections. In addition, the radiating elements and feed system characteristics must produce a stable beam as a function of frequency. It is also desirable to tilt the beam slightly below the radio horizon for maximum effectiveness. This is of increasing importance as the beamwidth narrows. Fig. 4 includes a plot of the depression angle to the radio horizon as a function of antenna height. A good rule of thumb is to add about 0.2° to the values read from these curves for optimum beam tilt.

Beam tilt is implemented by either electrical or mechanical means. Obviously, the beam is tilted downward in one direction when the antenna is mechanically tilted from the vertical. In the opposite direction the beam is tilted upward, and at orthogonal directions no tilt is introduced. In effect, this process creates a directional azimuth pattern. Electrical beam tilt may be used to tilt the beam downward in all directions. This is accomplished by advancing the phase of the upper portion of the antenna relative to the lower portion. Combinations of electrical and mechanical tilt may be used to increase the tilt in some direction while reducing it in others.

When the antenna is located a substantial distance from populated areas, the elevation pattern nulls need not be filled. However, when the population is close to the antenna or the antenna is unusually high, it is often necessary to fill the first null and in some cases higher-order nulls. Common amounts of first null fill range from 5 to 35 percent. Two methods are used to provide null fill. The first is to feed the upper and lower halves of the antenna with different current amplitudes. This results in incomplete field cancellation in the null regions and filling of the odd-numbered nulls. This has the advantage of minimizing radiation above the horizon and raising the antenna radiation center. The second method is to use a quadrature phase distribution over the vertical aperture. Obviously these methods may also be used together. A more general method is to determine the phase and amplitude of the currents by pattern synthesis and to design the antenna to provide these values.

When determining the power rating of tv antennas, it is necessary to recognize that tv transmitters are rated in terms of peak visual power. The average power in the visual signal with black picture is only 0.6 the peak visual power. The aural power is 10 to 20 percent of the peak visual power. Thus the worst-case average

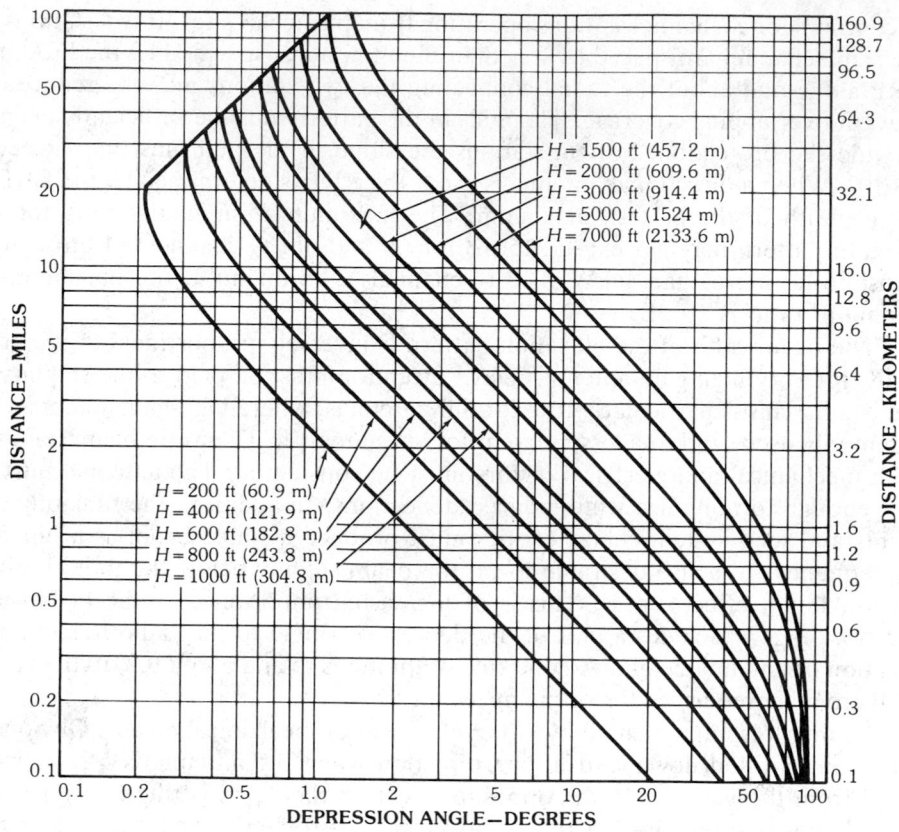

Fig. 4. Depression angle versus distance for various antenna heights.

power is 0.8 of the stated transmitter power. Peak voltage calculation must consider the sum of the voltage peaks for the combined visual and aural signals. For 20 percent aural the peak voltage is 1.477 times that due to the visual signal. Typical transmitter power ratings for tv channels 2 through 6 and fm range from 10 to 60 kW, to 10 to 100 kW for tv channels 7 through 13, and from 30 to 220 kW for uhf. For reliable service it is desirable that the antenna be rated at least as high as the maximum transmitter output.

It is essential that tv and fm antennas be designed to withstand the effects of lightning. Measures used to provide the needed protection include use of lightning rods around the beacon, designing for positive grounding of all parts, and minimizing inductance in radiating elements. To ensure that all parts are grounded, including those at high rf potential, extensive use is made of shorted quarter-wave transmission-line sections in the vicinity of radiator feed points.

2. Circularly Polarized TV Antennas

In 1977 the FCC rules were changed to permit tv stations to transmit circular polarization. These antennas were required to handle up to twice the power of conventional horizontally polarized antennas. This resulted from the FCC rules that permit maintenance of the station's ERP in the horizontally polarized component while adding an equal amount of signal in the vertical component.

In addition to providing beam shapes and impedance characteristics similar to conventional antennas, the CP antennas are required to radiate a right-hand–sense wave with a low axial ratio. It may also be important, in applications where a horizontally polarized antenna is being replaced, that the radiation center of the CP antenna be within 2 m of the antenna being replaced. Other requirements for CP antennas include means of preventing ice accumulation or deicing, lightning protection, provisions for either single or dual inputs, and provisions for either top or side mount.

3. Vee Dipole Array

The vee dipole antenna [4] is used for channels 2 through 13 and is designed for tower top-mounted application. Each bay consists of three crossed vee dipoles mounted at 120° intervals around a vertical mast as shown in Fig. 5. These dipoles are separated by three vertical grids, which isolate the vee dipoles and provide horizontal beam shaping. Each set of cross dipoles is fed in phase quadrature to produce a circularly polarized wave.

The design is rated for power up to 100 kW. A wideband flat dipole is used to handle the required power levels safely. Each dipole is mechanically supported and fed from a balun.

Radomes protect the radiating elements from exposure to moisture, ice, and

Fig. 5. Crossed vee dipoles arranged at 120° intervals.

atmospheric corrosives. Typical elevation patterns for a six-bay low-band antenna are shown in Fig. 6 for the cases of 0- and 15-percent null fill. As the null is raised, the gain is reduced. Typical values are given in Table 2.

For omnidirectional stations the shape of the azimuth pattern varies from circular by less than ±1.5 dB for channels 7 through 13; 2.0 dB for channels 2 through 6. Directional patterns may be obtained by use of nonuniform phase and amplitude around the mast or by the use of parasitic elements on the screens. A typical omnidirectional pattern is shown in Fig. 7. Two examples of directional azimuth patterns are shown in Fig. 8.

The antenna permits the use of either single or dual transmission lines. The low-band antenna is supplied with dual 3⅛-in (79.375-mm), 50-Ω input connec-

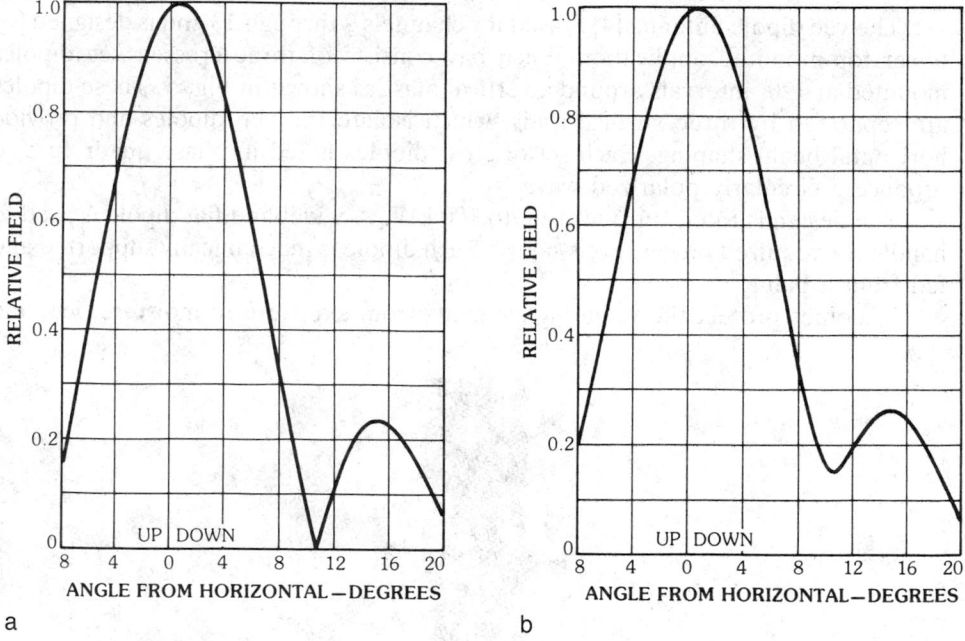

Fig. 6. Typical six-bay vee dipole array elevation patterns. (*a*) 0% null fill, −0.7° beam tilt. (*b*) 15% null fill, −0.7° beam tilt.

Table 2. Six-Bay Vee Dipole Array Gain Versus First Null Fill

Null Fill (%)	Power Gain (per Polarization)
0	2.98
5	2.96
10	2.92
15	2.82
20	2.72
25	2.66
30	2.42

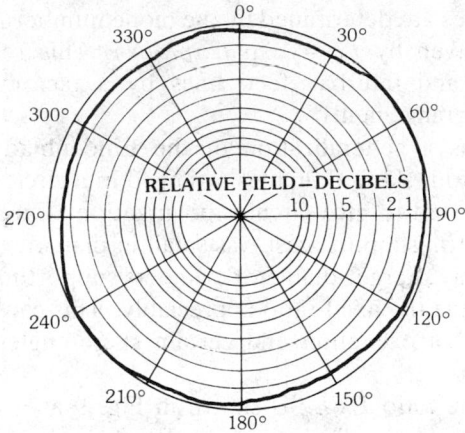

Fig. 7. Typical vee dipole array omnidirectional pattern.

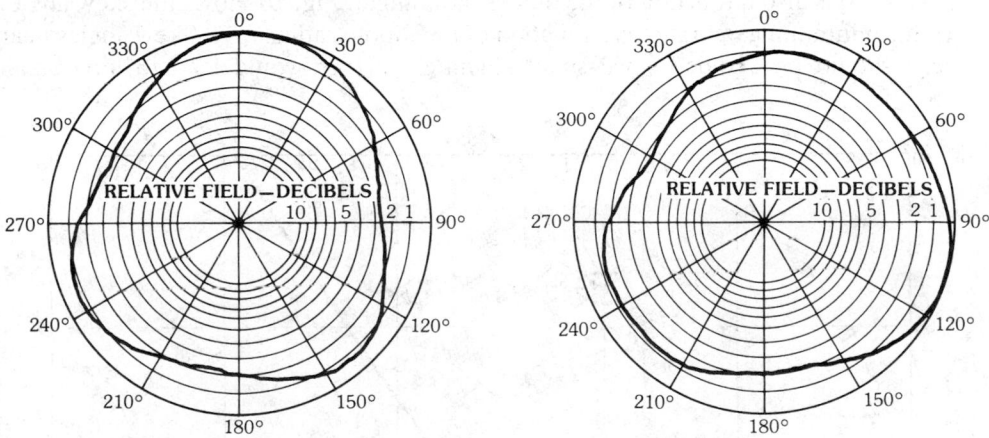

Fig. 8. Typical directional azimuth patterns for vee dipole array.

tions. The upper and lower halves of the antenna are each fed by a separate, independent transmission line to permit using one-half of the antenna in an emergency situation (with appropriate patching). It may also be supplied with a single input connection. The standard high-band antenna is supplied with a single 6⅛-in (155.575-mm), 50-Ω input connection. It may also be supplied with a dual line input which will permit feeding power to one-half of the antenna in an emergency.

4. Skewed-Dipole Antenna

This antenna makes use of circular arrays of skewed dipoles such that each dipole radiates linear polarization but in a configuration so that the complete array radiates circular polarization [5]. This is achieved by exciting N dipoles with equal currents whose phases are determined by the mode number M. That is, the current in each dipole is given by $I_m = \exp(j2\pi nM/N)$. This reduces the number of radiating elements and interbay feed lines by a factor of 2 compared with orthogonal dipole arrangements.

The approach is a generalization of the Lindenblad antenna [6,7]. That antenna consists of four skewed dipoles arranged in a circle and fed in phase with the skew angle adjusted to give circular polarization. Excellent results were obtained since the central support mast was small in diameter and extended only to the center of the array. Larger-diameter cylinders supporting several bays of these skewed arrays degrade the axial ratio appreciably. This can be remedied by using higher-order modes of excitation and certain skew angles, or by adding short vertical elements.

The phase of the ratio E_θ/E_ϕ is shown in Fig. 9 as a function of the radius (ϱ_1) of the supporting mast. The minimum dB of the axial ratio for these phase shifts is also shown on the ordinate. In order to achieve an axial ratio of less than 3 dB it is necessary for the cylinder to have a circumference of less than 0.5λ. The axial ratio is also a function of the dipole skew angle. Fig. 10 shows the skew angle, ψ, for minimum axial ratio as a function of the dipole radius (ϱ_2). Skew angles near zero should be avoided since small alignment errors would lead to large axial

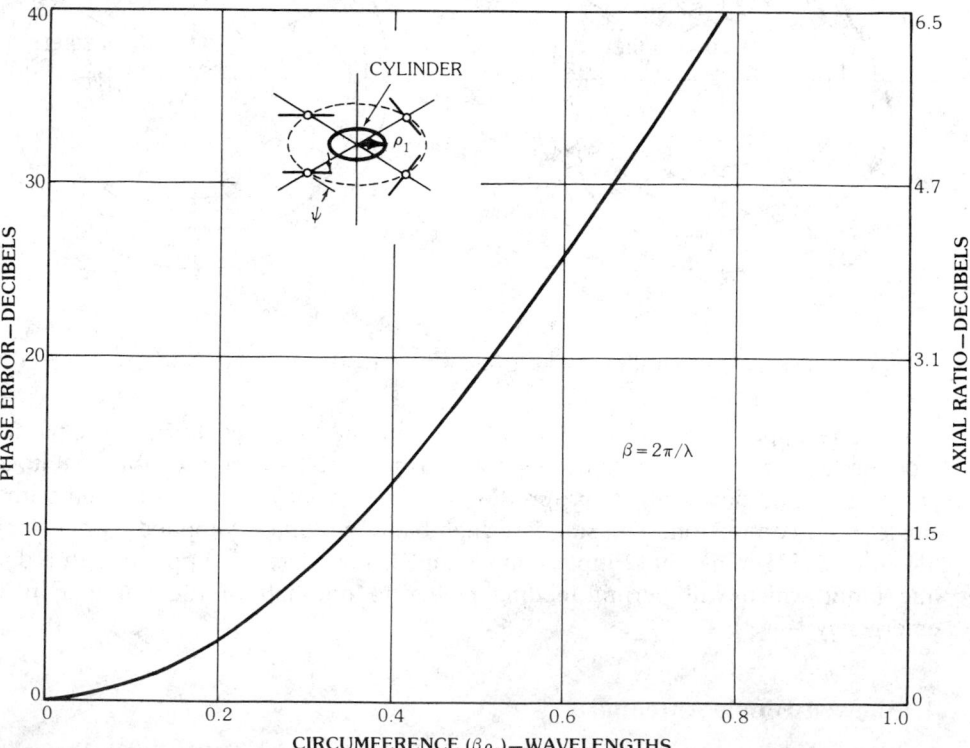

Fig. 9. Phase of E_θ/E_ϕ and axial ratio for $|E_\theta/E_\phi| = 1$ versus cylinder size. (*After DuHamel [5]*)

ratios. This places an upper limit on $\beta\varrho_2$. Fig. 10 is plotted for the case of no supporting mast present and for $\beta\varrho_1$ of 0.5. It is apparent that skew angle is not very dependent on ϱ_1 for reasonable mast diameters. Both Fig. 9 and Fig. 10 are calculated on the basis of short dipoles using the method described by Carter [9]. The results are approximately correct for half-wave dipoles. More accurate results may be obtained by using moment-method modeling of the antenna.

For broadcast applications the antenna consists of one or more bays of these circular arrays of skewed dipoles placed around a conducting cylinder as shown in Fig. 11 [8]. Each layer consists of three radiators fed in phase and mounted symmetrically around the pole. The skew angle is selected to produce equal amounts of horizontal and vertical polarization. When the radiation from all these dipoles is taken into account, an omnidirectional azimuth pattern results for both components (Fig. 12) with a low axial ratio in all azimuthal directions (Fig. 13).

Each radiator is fed by one feed line. A complete seven-layer circularly polarized antenna uses 21 feed lines fed from a branch-type feed system. One junction box feeds the upper four layers and another feeds the lower three layers. A single 3⅛-in (79.375-mm) transmission line is used to feed each junction box. The feed lines are ¾ in (19.05 mm), 50 Ω.

The tee to combine the upper and lower parts of the antenna can be located at the tower top with a single line down the tower (Fig. 14). If standby capability for operating on either the upper four or lower three layers of the antenna is desired, a transmission line from each junction box can be run down the tower and the tee can be mounted in the transmitter building (Fig. 15). The transmission lines from the two junction boxes to the combining tee must be equal in length. Since the two junction boxes are at different elevations on the pole, additional line must be inserted in the line coming from the lower junction box. This line can be inserted at the same location as the combining tee. The combining tee divides the power

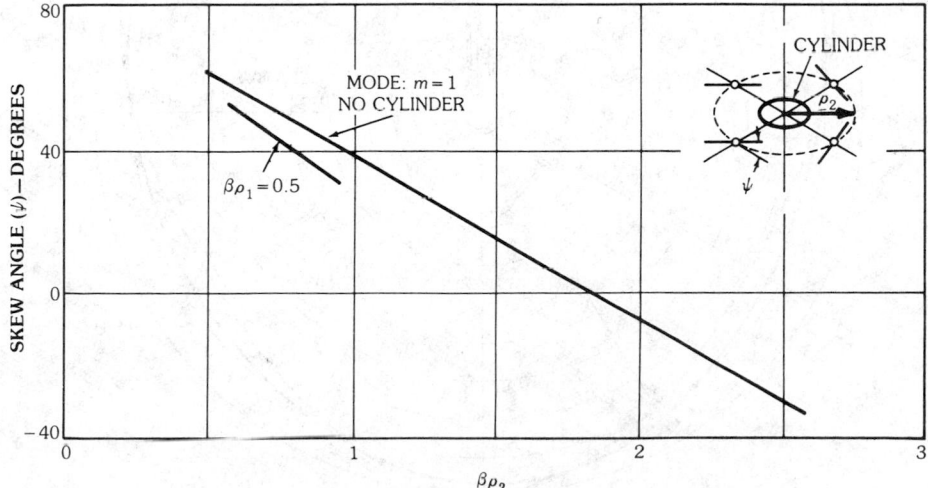

Fig. 10. Skew angle ψ versus coil circumference $\beta\varrho_2$ for mode number $m = 1$: (*After DuHamel [5]*)

Fig. 11. Skewed-dipole circularly polarized antenna. (*Courtesy Dielectric Communications*)

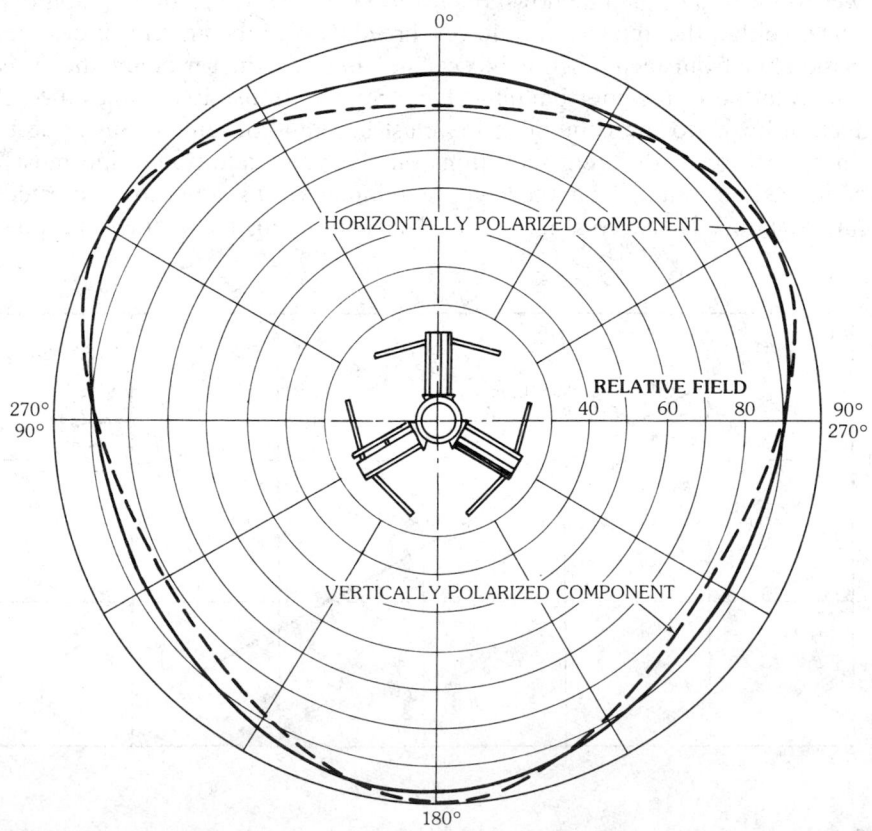

Fig. 12. Measured azimuth pattern of a skewed-dipole antenna. (*Courtesy Dielectric Communications*)

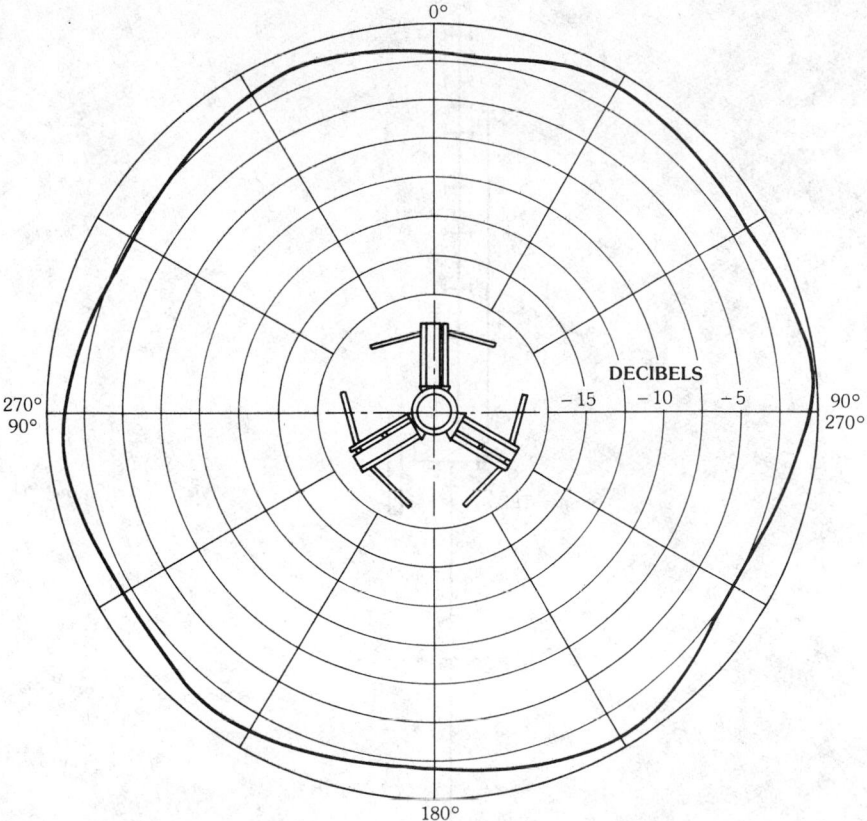

Fig. 13. Measured axial ratio of a skewed-dipole antenna. (*Courtesy Dielectric Communications*)

equally to the two output ports. Therefore each layer of the lower three layers receives more power than each layer of the top four layers. This, in addition to feed-line phasing between the layers, results in 8-percent fill in the first null and 4 percent in the second, as well as 1° beam tilt. A plot of the elevation pattern is shown in Fig. 16. The measured impedance shows a vswr of less than 1.1:1 for a bandwidth of about 12 percent. This is illustrated in Fig. 17.

5. Helical Circularly Polarized Antennas

Top-mounted circularly polarized antennas using multiple helices or spirals have been described [10, 11, 12]. Successful implementation of these using three and four conductors are known to have been made. An example using four conductors is shown in Fig. 18. Each section radiates an omnidirectional pattern broadside to a supporting mast about which four helical conductors are wound. The conductors are spaced from the mast and are equally spaced about the periphery of the mast. The signals are coupled to the conductors so that the phase of the signal coupled to one conductor is 180° out of phase with the signal coupled

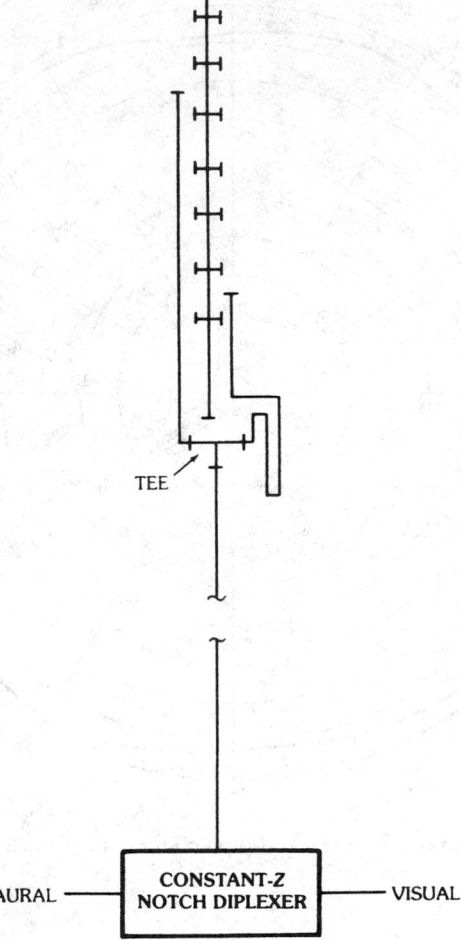

Fig. 14. Single-line feed. (*Courtesy Dielectric Communications*)

to adjacent conductors. The pitch of the helix is selected with the radius so that the radiation is circularly polarized and essentially broadside. This pitch angle ψ to achieve circular polarization and essentially broadside radiation is achieved by satisfying the condition

$$\tan \psi = \frac{J_{m-1}(\beta \varrho)}{2 J_m(\beta \varrho)} \tag{1}$$

where J_{m-1} and J_m are the Bessel functions of order m and $m-1$, respectively (m is the mode number). The term $\beta \varrho$ is the circumference of the coil in wavelengths. Fig. 19 is a plot of the solution of this equation for the pitch angle versus the circumference of the helices for a mode number of 2. It is interesting to compare (1) with the corresponding curve for the skewed-dipole array (Fig. 10) for which the mode number is 1. For small values of $\beta \varrho_2$ the fit is quite good.

Radiating "end loads" are used at the end of each coil to radiate the small

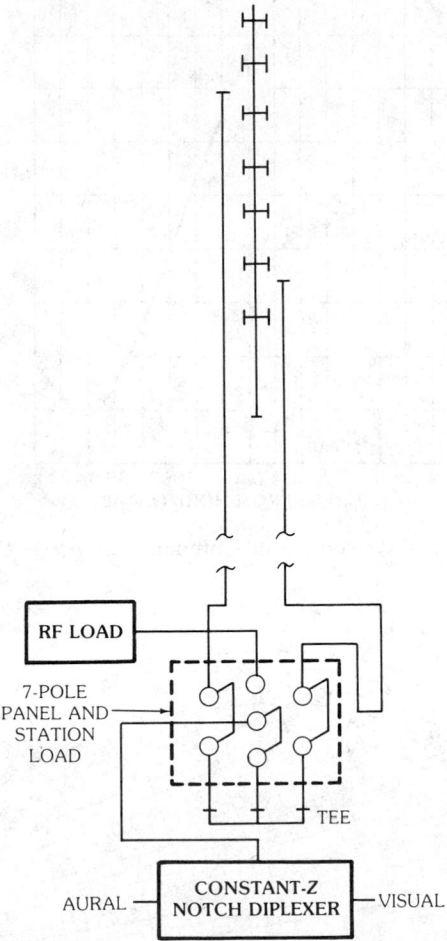

Fig. 15. Dual-line feed. (*Courtesy Dielectric Communications*)

amount of remaining energy from the main radiator. These end loads minimize the reflections of energy from the end toward the input, which otherwise would upset the traveling-wave nature of the antenna, consequently distorting the pattern. Electrical beam tilt is achieved by phasing between the vertical sections. Standard antennas are supplied with a single input of 6⅛ in (15.5 cm), 75-Ω transmission line.

The radiating elements are stainless steel tubing attached to the pole with dielectric supports. The support pole consists of either two or three sections made of galvanized steel. The antenna is designed for flange mount or bury mount at the tower top. Dielectric and steel pole steps are provided for climbing. The stainless steel helical tubes are grounded to the pole for lightning protection.

This type of antenna has also found limited application as a uhf antenna. Although each section is much smaller by virtue of the higher frequencies, more bays are required due to the higher gains usually required for uhf. The fundamental principles remain.

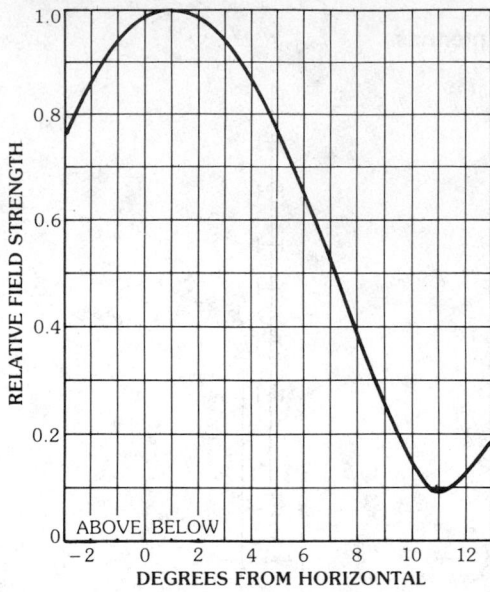

Fig. 16. Elevation pattern of a skewed-dipole antenna. (*Courtesy Dielectric Communications*)

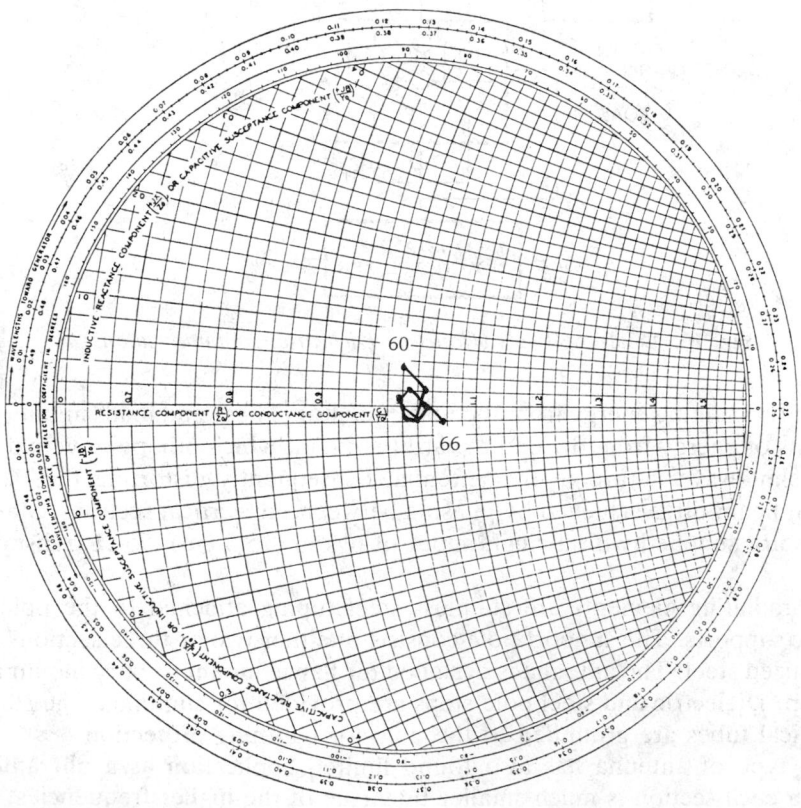

Fig. 17. Measured impedance of a skewed-dipole antenna. (*Courtesy Dielectric Communications*)

Fig. 18. Circularly polarized helical antenna. (*Courtesy Dielectric Communications*)

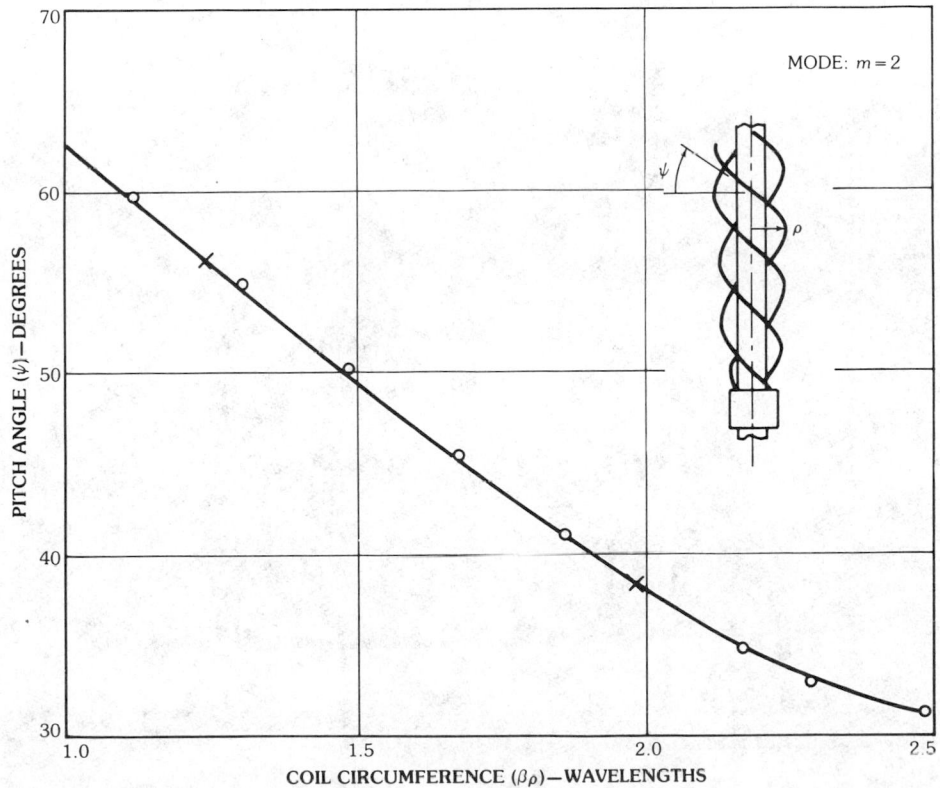

Fig. 19. Pitch angle as a function of the circumference of helices for mode number $m = 2$. (*After Ben Dov [11]*)

6. Side-Mount Circularly Polarized Antennas

When the tower top position is occupied by other antennas, it is often desirable to attach antennas to the side of the tower at lower positions. These are appropriately termed *side mount*.

A design illustrating many of the principles of the side-mount antenna design is the cavity-backed radiator [13], shown in the foreground of Fig. 20. The basic concept is the use of crossed dipoles in the aperture of a cavity. The dipoles are fed with equal currents in quadrature phase to excite the cavity in planes parallel to the dipoles. The beamwidth and radiation patterns are determined primarily by the size of the cavity. The geometry of the dipole controls the antenna impedance. The cavity performs three functions. First, it isolates the antenna from the tower, mounting structure, and adjacent cavities. Secondly, the cavity provides a sharper pattern than is achievable with dipoles alone. Thirdly, the cavity permits equalizing of the beamwidths for horizontal and vertical polarization.

The cavity is circular in shape and uses a grid construction to minimize wind load. Two designs are used, differing principally in electrical size. For a three-around array on a triangular supporting tower, a 120° half-voltage beamwidth is desired. The cavity for this is 0.6 to 0.65λ in diameter. A four-around configuration requires 90° half-voltage beamwidth. This cavity is about 0.8λ to 0.9λ in diameter. The cavities are approximately 0.25λ in depth.

CAVITY

Fig. 20. Broadcast array using cavity-backed radiator elements.

A flat, wideband dipole is used. A flat dipole may be related to an equivalent cylindrical center-fed dipole as shown in Fig. 21. The required bandwidth is achieved with a $1/w$ ratio of about 1.5. Each dipole is supported and fed from a balun. A shunt capacitance near the center of the dipole is used for impedance compensation.

The dipole has separate $1\frac{5}{8}$-in (41.275-mm), 50-Ω, coaxial inputs to the horizontal and vertical elements. The two inputs are fed with a power divider and phasing loop to provide circular polarization. The input vswr of a typical channel 7 through 13 cavity is shown in Fig. 22. The bandwidth is sufficient to permit multiplexing of two channels.

It is often desirable to provide directional azimuth patterns. A typical directional pattern, achieved by reducing power to one cavity in a three-around arrangement, is shown in Fig. 23. The serrations on the pattern are caused by a spinning source dipole to indicate the axial ratio as a function of azimuth.

Another side-mount panel design makes use of two crossed vee dipoles mounted in front of a square screen mesh as shown in Fig. 24. The pairs of dipoles are made unequal in length with the lengths chosen to produce equal quadrature currents in the dipoles. Since the dipoles are also orthogonal, circular polarization results. This design uses a single $\frac{7}{8}$-in (22.2-mm) copper feed line for each panel. The gain per polarization of each layer is slightly less than half that of a half-wave dipole. Beam tilt and null fill may be incorporated into the array pattern with a small reduction in gain.

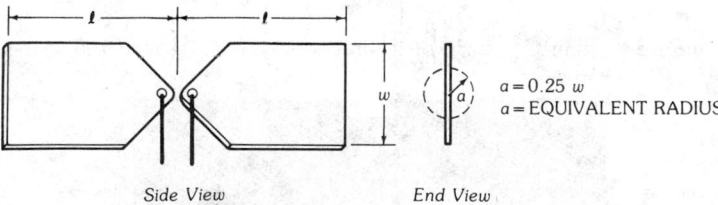

Side View *End View*

Fig. 21. Flat dipole.

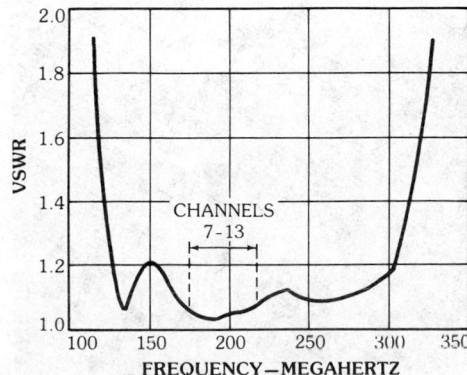

Fig. 22. Measured vswr of a single cavity.

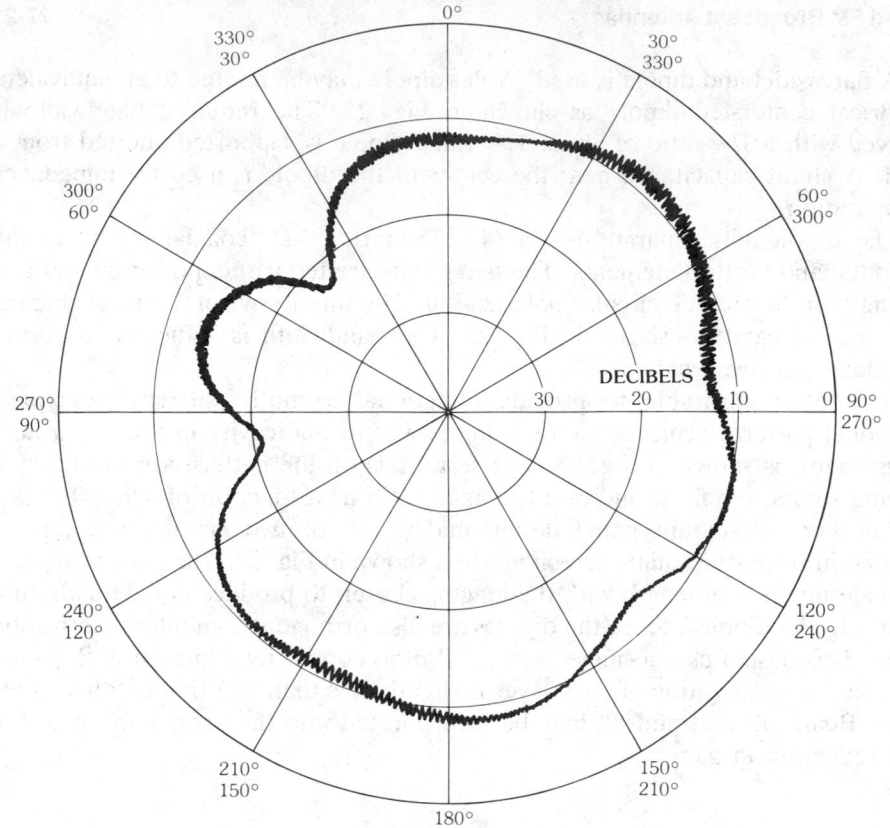

Fig. 23. Measured azimuth pattern of a directional cavity-backed radiator antenna.

Fig. 24. Circularly polarized panel antenna. (*Courtesy Dielectric Communications*)

Phase rotation is employed in the feed system to provide good vswr. This technique requires that each of the panels in a layer be phased 120° apart. The panels are then offset mechanically to restore the phase relationships and maintain good pattern circularity. The phase rotation tends to reduce changes in vswr caused by ice formation.

7. Horizontally Polarized Antennas

Even though the FCC now permits CP transmission, many stations continue to broadcast horizontal polarization (HP). Most stations were on the air prior to 1977, when the installation of HP antennas was all that was permitted. In addition, circular polarization, while improving coverage in many areas, is not a panacea for all propagation-related broadcast problems. Hence, many stations will find it desirable to continue operation in HP. As in the case of CP antennas, both top-mount and side-mount types are used.

8. Batwing Antennas

One of the most widely used designs is the batwing or superturnstile antenna [14]. The simple dipole elements of a turnstile antenna [15] are replaced with an equivalent grid of conductors forming a slot with the supporting mast [16]. (See Fig. 25.)

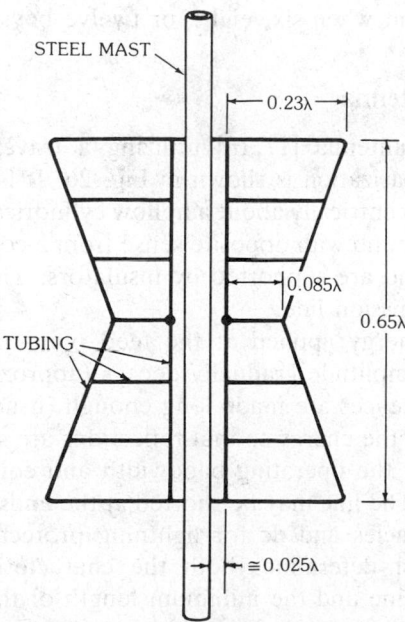

Fig. 25. Single element of batwing antenna with tubing construction, showing method of mounting on mast and typical dimensions. (*After Kraus [16],* © *1950 McGraw-Hill Book Co.*)

The radiators are bolted to the mast at about $\lambda/3$ from the feed point, thus providing structural support and a means of lightning protection. This arrangement provides an swr of less than 1.1:1 over a 35-percent bandwidth, making it convenient for tv service to as low as 47 MHz. There is very little radiation along the mast. Only one bay is required to obtain a gain equivalent to a single $\lambda/2$ dipole. To increase directivity, several bays are stacked with a separation of about λ between centers. The power gain is approximately proportional to the number of bays.

The antenna has an input resistance of about 72 Ω with low reactance over rather wide bandwidth. The bandwidth is sufficient for diplexing channels 4 and 5 or any two channels from 7 to 13. The feed-point impedance is readily matched to $\frac{7}{8}$-in (22.2-mm) coaxial lines without matching networks. Since the feed-point impedance is low, the performance is not severely affected by ordinary icing. Deicers are used where severe icing is expected.

To provide an omnidirectional azimuth pattern, orthogonal pairs of radiators are fed in phase quadrature. The pattern is generally circular and does not deviate from omnidirectional by more than ±1.5 dB.

The antenna can be designed to provide azimuth patterns varying from the normal omnidirectional to a figure-8 shape. This is accomplished by varying the power division and/or phase relationship to the orthogonal pairs of radiators.

Usually, two transmission lines are used up the tower, although a single line feed may be used if the power division between the two halves of the antenna and 90° phasing is placed at tower top. If a single line feed is used, a notch diplexer is required. It is often feasible to feed the upper and lower bays independently. This is particularly convenient when six, eight, or twelve bays are used.

9. Sidefire Helical Antenna

A sidefire helical antenna [17, 18] utilizing a traveling-wave concept and radiating horizontal polarization is shown in Fig. 26. It is composed of a helical conductor mounted concentrically about a hollow cylindrical steel mast. The upper and lower helices are wound with opposite sense from a common feed point at the center of the section and are supported by insulators. The helices and the mast form a radiating transmission line.

Radio-frequency energy applied at the feed point travels along the transmission line, and the amplitude gradually decays (approximately 4 dB per turn) due to radiation. The helices are made long enough (usually six wavelengths) to radiate essentially all of the energy so that reflections are small. The impedance is relatively constant over the operating bandwidth and equal to the characteristic impedance of the line. The line may be shorted at the ends, thereby grounding the radiator at low frequencies and dc for lightning protection. The height of the helices above the mast determines both the characteristic impedance of the radiating transmission line and the minimum length of the helices.

In order to provide an omnidirectional azimuth pattern and broadside radiation, it is necessary that each turn be an integral number of wavelengths long. Two wavelengths per turn is generally used. If a large number of turns are used, this phase relationship is lost for small changes in frequency for turns a long way

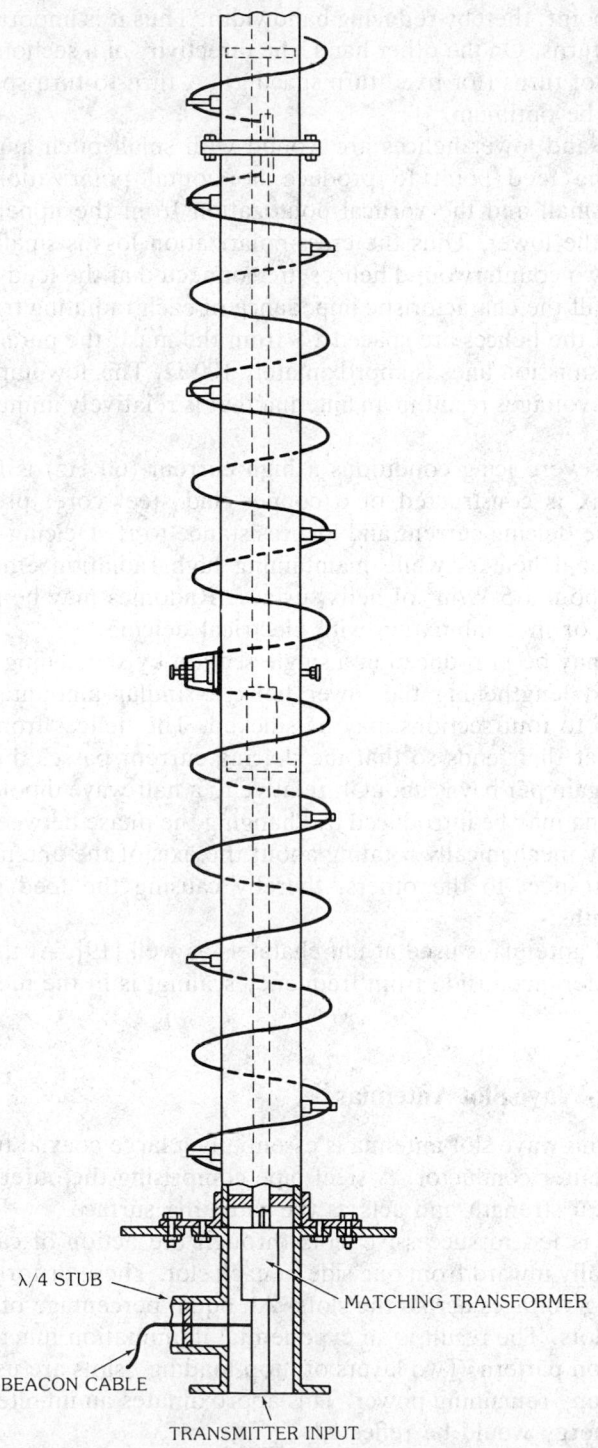

λ/4 STUB

MATCHING TRANSFORMER

BEACON CABLE

TRANSMITTER INPUT

Fig. 26. Single section of sidefire helical antenna. (*After Krause and Smith [17]*)

from the feed point, thereby reducing bandwidth. Thus it is important to minimize the number of turns. On the other hand, the directivity of a section is proportional to the number of turns (for fixed turn spacing). A turn-to-turn spacing of $\lambda/2$ has been found to be optimum.

The upper and lower helices are wound with small pitch angles in opposite senses from the feed point to produce horizontal polarization. The vertical component is small and the vertical polarization from the upper helix tends to cancel that of the lower. Thus the cross-polarization loss is small.

Since the two counterwound helices are connected at the feed point, the input impedance is half the characteristic impedance of each radiating transmission line. For example, if the helices are spaced $\lambda/8$ from the mast, the parallel combination of the two transmission lines is approximately 100 Ω. This low impedance and the resulting low rf voltages result in an antenna that is relatively immune to moderate icing.

For more severe icing conditions a high current (60 Hz) is fed through the helix. The helix is constructed of a copper-clad steel core, presenting a high resistance to the deicing current and low resistance to rf. Deicing is accomplished without additional heaters while maintaining high radiation efficiency. Deicing dissipation is about 1.5 W/in² of helix surface. Radomes may be used to prevent icing in lieu of or in combination with electrical deicing.

Beam tilt may be introduced in a single section by shortening the upper helix fractionally and lengthening the lower helix a similar amount. For additional directivity, two to four sections may be stacked. The helices from adjacent bays are connected at their ends so that the deicing current passes through all of the radiators. The gain per bay is about 4, relative to a half-wave dipole. Beam tilt in a multibay antenna may be introduced by changing the phase between sections. This may be done by mechanically rotating about the axis of the antenna one or more sections with respect to the others, thereby causing the feed points to be in different azimuths.

The helical antenna is used at uhf channels as well [19]. At these frequencies the primary difference, aside from frequency scaling, is in the method of feeding the bays.

10. Traveling-Wave Slot Antennas

The traveling-wave slot antenna is essentially a large coaxial transmission line with a slotted outer conductor. A steel pipe comprising the outer conductor provides mechanical strength and acts as the radiating surface.

The signal is fed to successive slots through the action of capacitive probes projecting radially inward from one side of each slot. These distort the field within the line, placing voltages across the slots. An equal percentage of power is fed to each layer of slots. The result is an exponential illumination function producing a smooth elevation pattern. Two layers of "top-loading" slots are used at the upper end to radiate any remaining power. This approximates an infinitely long antenna in which no energy would be reflected.

The slots are arranged in pairs at each layer with the pairs separated by $\lambda/4$. Successive pairs occupy planes at right angles to each other. The orientation of the capacitive probes in a given plane is alternated so that all slot pairs in a given plane are excited with the correct phase.

The quarter-wave separation of layers and the space quadrature of successive layers produce a turnstile type feed which produces an omnidirectional azimuth pattern. A typical pattern is shown in Fig. 27. In addition, the quarter-wave spacing between slot layers causes reflections from adjacent slot pairs that tend to cancel each other, resulting in an inherently low vswr.

Pattern bandwidth is achieved by introducing the equivalent of a parallel resonant circuit at each slot. This produces compensation which maintains a constant electrical spacing between slot layers for frequencies within a given channel. It is implemented by shaping the slots for the required values of inductance and capacitance and by adjusting slot lengths, the shape of the ends, and the slot width at the center. The elevation beam is tilted downward by selection of the spacing between layers and the design of the capacitive probe compensation between layers.

11. UHF Antennas

Many of the principles of uhf antenna design are the same as for vhf. There are, however, differences due to a variety of factors. Because of the shorter wavelengths, propagation losses are higher, resulting in the need for higher ERP to achieve satisfactory field-strength levels. This generally requires the uhf station to transmit at a higher power level and use an antenna with higher gain. The higher gain is obtained by means of a longer aperture in terms of wavelengths. Directional azimuth patterns are also often used to increase the ERP in desired directions. The longer aperture results in decreased beamwidth so that structural stiffness becomes even more important as it affects beam stability. The stability of beam direction

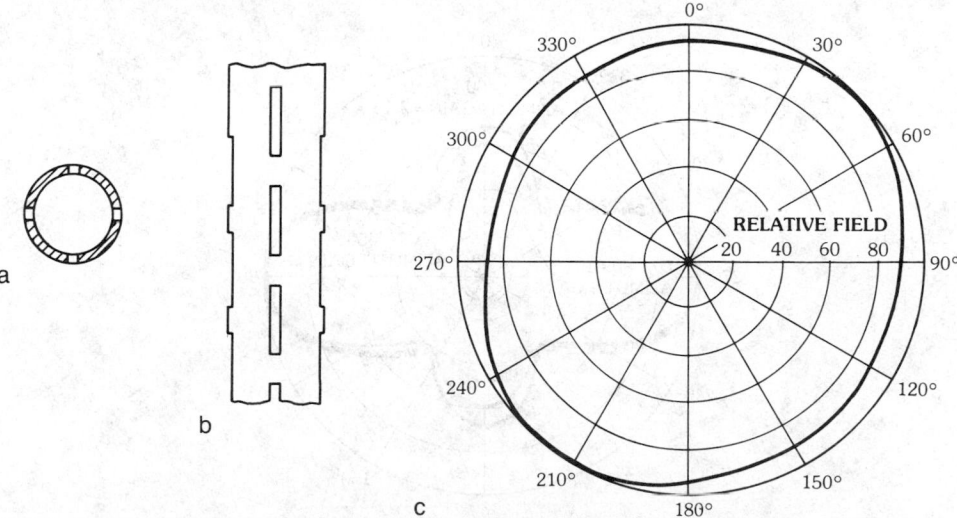

Fig. 27. Traveling-wave slot antenna. (*a*) End view of antenna. (*b*) Side view of antenna. (*c*) Pattern of antenna. (*Courtesy Dielectric Communications*)

and shape as a function of frequency also becomes more important. The increased length also makes simple feed systems more desirable. Finally, the percentage bandwidth of a uhf channel is less than that of the vhf channels, permitting the use of some antenna types (such as resonant slot arrays) that might not otherwise be suitable. Some of the same types of antennas used at vhf are used at uhf, i.e., the sidefire helical antenna. Accordingly, the antennas described most completely are those used primarily or exclusively at uhf.

12. Coaxial Slot Antennas

An antenna of this type is basically a coaxial transmission line with longitudinal radiating slots in the outer conductor fed by bar couplers bolted to the inside edge of each slot. The number of slots per layer is determined by the desired azimuth pattern, such as one slot for a cardioid pattern, two slots for a peanut-shaped pattern, three slots for a trilobe pattern, and four or more slots (depending on cylinder diameter) for an omnidirectional pattern. Typical azimuth patterns are shown in Fig. 28. Omnidirectional patterns are circular within ±1 dB. Parasitic elements have been used in conjunction with a single row of slots to produce omnidirectional patterns.

The layers are spaced at about one wavelength along the length of the antenna with the number of layers determined by the elevation directivity and pattern shape. The amplitude distribution is determined primarily by slot length and coupling bar diameter. Because of the large number of radiating elements this type of antenna lends itself readily to the use of synthesis programs to obtain a wide variety of null fills, beam tilt, and pattern smoothness. Typical elevation patterns are shown in Figs. 29a and 29b. The amount of cross polarization is quite small.

A single feed point is used. High-gain antennas are center fed by means of

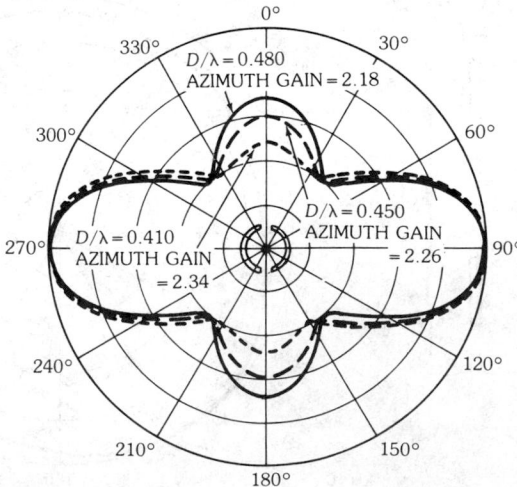

Fig. 28. Typical peanut-shaped patterns, where D is the pole outer diameter and λ is the midchannel wavelength. (*Courtesy Dielectric Communications*)

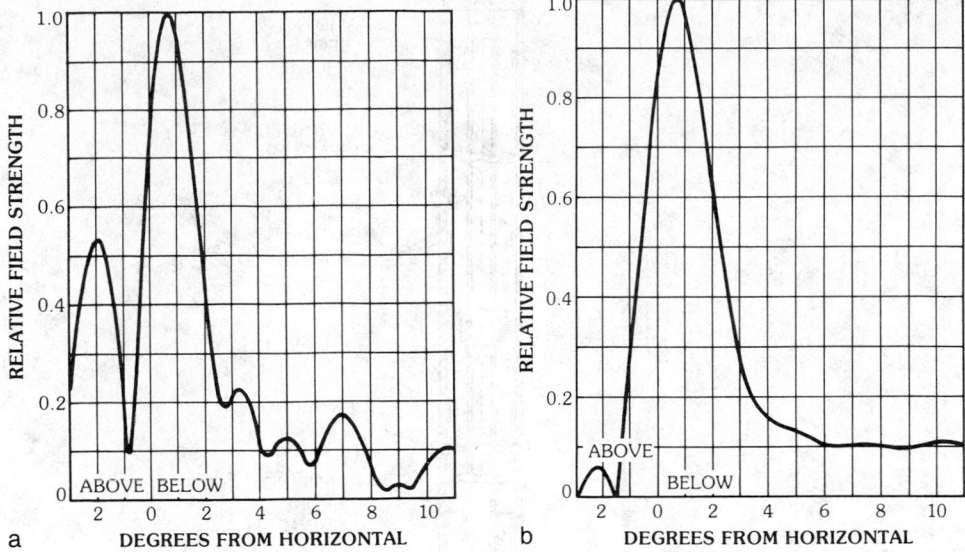

Fig. 29. Typical elevation patterns of a uhf coaxial slot antenna. (*a*) Main lobe power gain 27.5. (*b*) Main lobe power gain 28.0. (*Courtesy Dielectric Communications*)

a coaxial line terminated by a pressure seal at the center of the antenna (Fig. 30). The feed transmission line can be 3⅛ in (7.94 cm) to 9³⁄₁₆ in (23.33 cm) EIA depending on power-handling requirements and/or channel. End-fed antennas are used for moderate gains.

If the antenna is used in areas where icing is likely, it is often equipped with a deicing system. Heaters are usually clamped longitudinally to the outside of the slotted tube and used to prevent or remove ice. An ice detector is often used to detect the presence of ice and to minimize operational costs.

13. Waveguide Slot Antennas

A relatively recent approach to uhf slot antenna design has been the use of waveguide structures. The primary motivation for these designs is to utilize the inherent simplicity and high power-handling capabilities of waveguide. Two waveguide types have been used. One of these makes use of a ridged circular guide operating in the dominant TE_{11} mode. The size and shape of the ridge are selected to minimize the pipe diameter while maintaining λ/λ_g as close as possible to unity. A single row of slots is cut at small angles with respect to the ridge axis. The slot conductance and thus the aperture amplitude distribution are determined solely by this angle. Therefore, no coupling bars are needed to produce radiation from the slots. Since only one row of slots may be cut, this type of waveguide is used only for cardioid pattern requirements. A typical azimuth pattern is shown in Fig. 31. Elevation patterns may be synthesized using well-known techniques [20] to produce smooth contours with beam tilt.

Another waveguide antenna makes use of a hollow circular guide operating in the TM_{01} mode. This mode is similar to the field configuration of the TEM coaxial

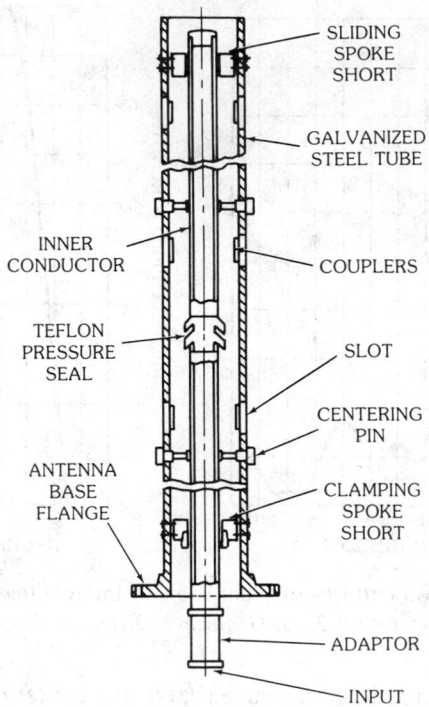

Fig. 30. Cross section of typical center-fed slot antenna. (*Courtesy Dielectric Communications*)

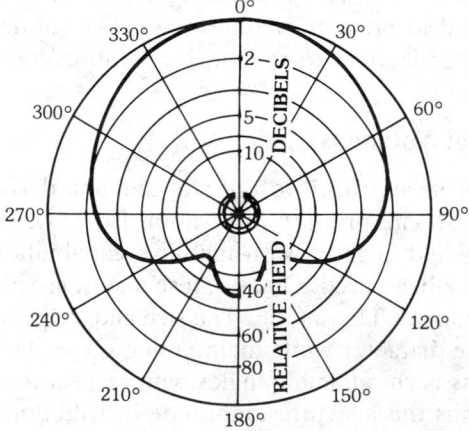

Fig. 31. Typical cardioid azimuth pattern of waveguide slot antenna.

mode. However, no center conductor is used to maintain the mode fields. Coupling devices are required at each slot to produce radiation. A variety of azimuth patterns may be generated by selecting the number and location of slots around the pipe. For an omnidirectional pattern, six equally spaced slots are used. A typical omnidirectional pattern is shown in Fig. 32.

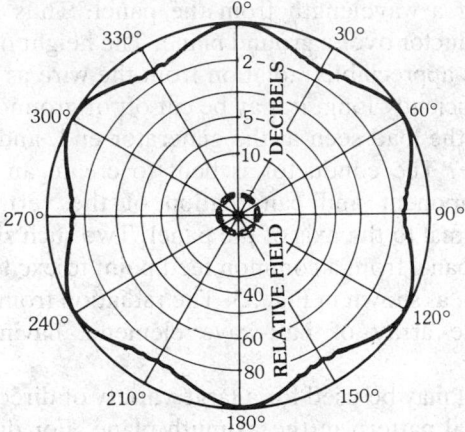

Fig. 32. Typical azimuth pattern of omnidirectional waveguide slot antenna.

The radial symmetry of the TM_{01} mode is essential to achieving a good omnidirectional pattern. The concept of this type of antenna was first described by Riblet [21] but only recently applied to broadcast use.

The construction of these antennas is such as to make approximately the bottom two-thirds a nonresonant traveling-wave array and the top one-third a resonant array. The top of the antenna acts as a matched load for the traveling-wave section while contributing to the desired aperture distribution. Use of the traveling-wave approach for the greater part of the array contributes to bandwidth.

The tendency in an end-fed array is for the beam direction θ to change as a function of frequency by an amount given by

$$\sin \theta = \lambda/\lambda_g - \lambda/2\ell \tag{2}$$

where λ_g is the guide wavelength and ℓ is the slot spacing.

To correct for this, advantage is taken of the nonlinear and opposite phase changes that can result when a controlled standing wave is produced. Capacitive posts are located at selected positions to produce the required correction.

Either waveguide or coaxial line may be used at the input. When a coaxial input is used, the power rating of the antenna is limited to that of the transmission line.

14. Zigzag Antennas

The zigzag antenna [22, 23, 24] is a panel antenna that may be used for omnidirectional or directional applications. The antenna usually consists of four panels per bay mounted on a square mast section. Other structures of three, five, or more sides may also be used. It may be top mounted on a tower or the panels may be side mounted on a tower or other structure.

Like the sidefire helical antenna the zigzag panel uses the traveling-wave principle to excite a large aperture from a single feed point. The panel consists of a rectangular panel with two zigzag conductors mounted on insulators and spaced a fraction of a wavelength from the panel. Thus a transmission line is formed by each conductor over a ground plane. The height of the conductor is 0.1λ to 0.2λ, and there is appreciable radiation from the wire as the current decays. If the conductor is sufficiently long, it may be cut off or grounded at the far end with negligible effect on the load seen at the generator end, and a traveling wave will exist along the wire. The conductor is bent to create an in-phase horizontally polarized field component and cancellation of the vertical component on a horizontal plane normal to the axis of the panel. Two such zigzag radiators are fed at the center of the panel from a common feed point to excite an aperture of seven to eight wavelengths, as shown in Fig. 33. The radiation from the panel is similar to that of a broadside array of half-wave elements having a tapered current distribution.

The zigzag panel may be used for a large variety of directional patterns as well as an omnidirectional pattern in the aximuth plane. For directional patterns the panels facing various directions are fed with unequal amplitudes and phases. The panels may also be used in a tangential fire arrangement on large structures [25, 26, 27].

The zigzag panel has sufficient bandwidth to be used at vhf frequencies. Although the usual panel length is eight wavelengths, the panels may be shortened to about three wavelengths. This has the desirable effect of increasing bandwidth to permit diplexing of two stations [28].

15. FM Antennas

Aside from panel antennas, there are three types of antennas in general use for omnidirectional side-mount fm broadcast applications. Two of these are circularly

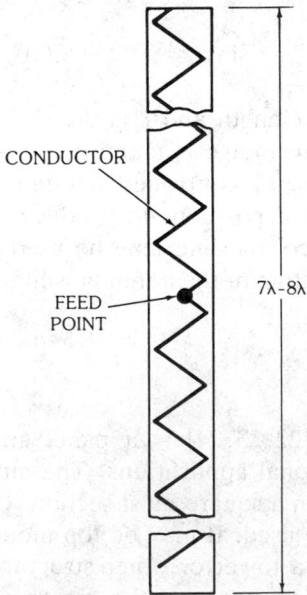

Fig. 33. A zigzag panel.

polarized, the other is horizontally polarized. The horizontally polarized antenna is primarily used for low-power educational stations and consists of one or more bays of circular loops [29] with a spacing of one wavelength between bays (see Fig. 34). They are designed for mounting on a pole having an outside diameter of 2 to 2½ in (5.08 to 6.35 cm). When mounted on such a pole the azimuth pattern is omnidirectional to within ±3 dB. The bandwidth is narrow, being sufficient to cover a single fm frequency ±1.2 MHz with a vswr of less than 1.5:1. The input connection is a uhf connector, and the power rating 800 W for two to four bays.

For circular polarization the most common types may be grouped as (a) loop/dipole combinations and (b) skewed dipoles.

A typical loop/dipole design which radiates vertical and horizontal components is shown in Fig. 35. When the phase of these components is in quadrature, circular polarization results. The rear terminal block is a balun matching the radiator impedance to the transmission line. From 1 to 16 bays may be stacked at nominal one-wavelength intervals. Elevation patterns with beam tilt and null fill are commonly used. Either radomes or electrical heaters may be used for deicing.

The azimuth pattern is omnidirectional for both the horizontal and vertical

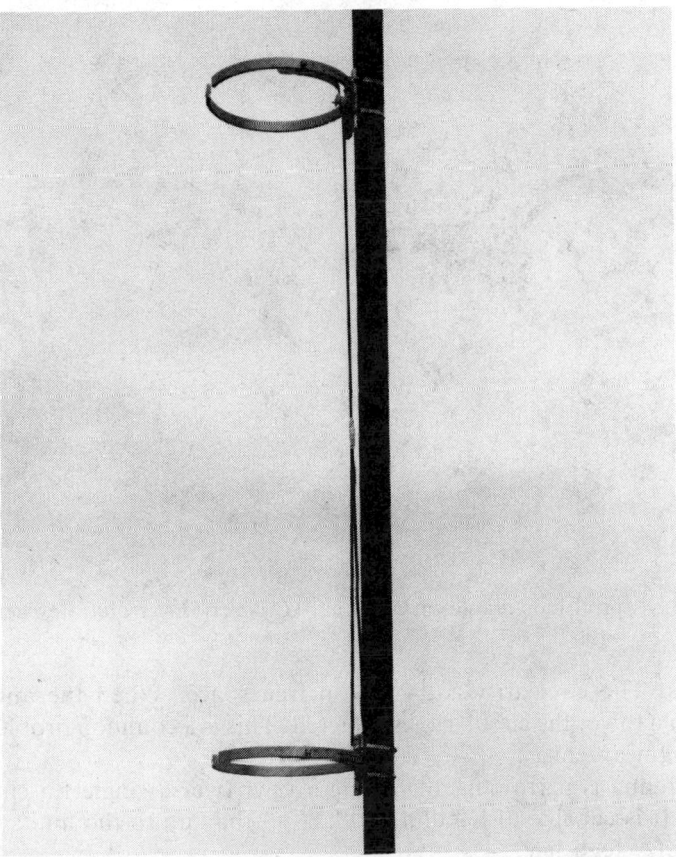

Fig. 34. Horizontally polarized loop fm antenna. (*Courtesy Electronic Research, Inc.*)

a

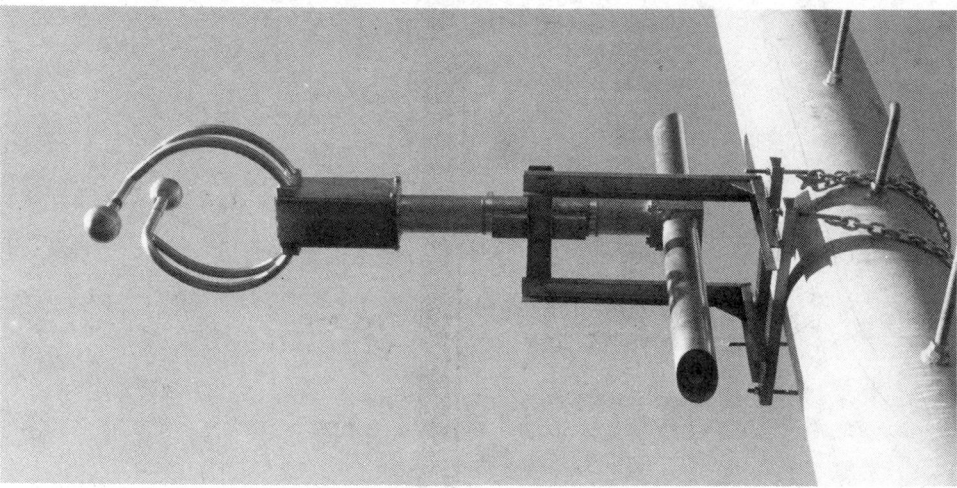

b

Fig. 35. Typical loop/dipole fm antenna. (*Courtesy Electronic Research, Inc.*)

components. The circularity is ±2 dB in free space. When the antenna is side mounted on a tower the circularity is affected. This is a common problem with most side-mount fm antennas.

The antenna is narrow band, having a vswr over a single fm channel of less than 1.1:1. It is capable of handling 10 kW per bay up to the limits of the 3⅛-in (7.94-cm) EIA feed line.

A typical skewed-dipole antenna design is shown in Fig. 36. Two skewed

Fig. 36. A skewed-dipole fm antenna design. (*Courtesy Electronic Research, Inc.*)

dipoles are fed from a common feed line at the middle. The feed point for each dipole is offset from the center. This is a design with very high power-handling capability, radiating up to 40 kW in a single bay. The design is also broadband when compared to many fm designs. Stations having a separation of up to 5 MHz may be diplexed on a single antenna. The vswr for single-station operation is less than 1.07:1 for 200 kHz on either side of the carrier. The azimuth pattern is omnidirectional, having a circularity of less than ±2 dB when mounted on a 14-in (35.56-cm) diameter pole or a 24-in (60.96-cm) triangular tower.

16. Panel FM Antennas

The use of panel antennas has been discussed in detail for vhf tv antennas. The panel antenna is mounted on all sides of the tower, permitting a degree of pattern control not afforded by the ordinary side-mount antenna described previously. Another motivation for the use of panel antennas is the need to multiplex several stations into a common antenna. Panels such as the cavity-backed radiator are quite broadband, exhibiting an impedance bandwidth over the full fm allocation. When high power-handling capability is designed into the radiator and feed system components, it is possible to multiplex several class C stations reliably. For example, a 12-bay antenna [30] operates with 9 class C stations, each using a 25-kW transmitter. Obviously a key component in a system of this type is the multiplexer. This device must combine the outputs of the several transmitters into a common output while preserving audio quality and preventing objectionable intermodulation products.

An alternative to using a single antenna for multiplexing is to use separate antennas for each station with all antennas mounted on a common tower. This is illustrated in Fig. 37 for five stations. Each station uses a single bay of cavity-backed radiator panels to radiate an omnidirectional pattern in azimuth and a

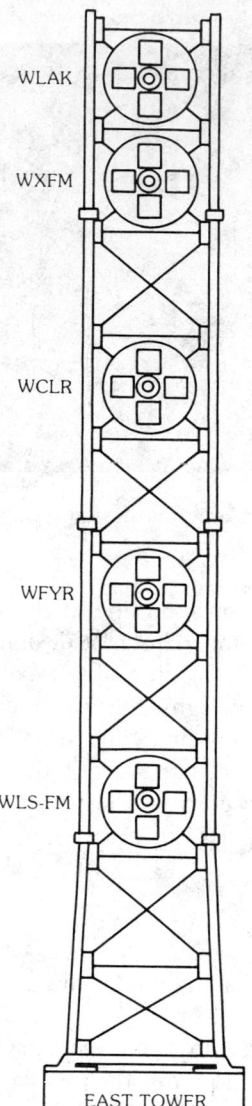

WLAK

WXFM

WCLR

WFYR

WLS-FM

EAST TOWER

Fig. 37. Placement of fm antennas for five stations colocated at the Sears Tower in Chicago.

broad beam in elevation. Isolation is achieved by taking advantage of the natural isolation afforded by the cavities and with the addition of two- or three-section bandpass filters in the lines to each antenna [31]. The stacking plan is selected to optimize the various antenna isolations in accordance with the frequencies involved. This approach is not dependent on the use of a wideband element. Thus, narrow-band elements capable of handling the power may be used. If the pattern circularity is not extremely important, the side-mount antennas described earlier could be used. Care should be taken to provide adequate isolation by proper selection of the mounting arrangement and use of adequate filtering.

17. Multiple-Antenna Installations

The availability of wideband side-mount antennas is just one of the reasons broadcasters in a given market sometimes find it desirable to locate their antennas on a common site and tower. A desirable economic benefit is that only a single piece of favorable real estate is required and a single large tower may be used. If outdoor receiving antennas are in general use, viewers and listeners will point their antennas to the common site, thereby making it possible to receive all signals with approximately equal ease and giving all broadcasters approximately equal coverage.

Furthermore, many communities now consider tower structures to be "unsightly" and therefore prefer to place all such structures in a single locale to minimize the clutter to the landscape.

Several methods are used to accomplish the multistation installation. Two methods have already been described for use with fm stations. Yet another technique for placing multiple antennas on a common site is the use of a tee bar or candelabra at tower top. As in the case of the multiplexing, this technique places all stations at the same HAAT. All of the antennas may be essentially standard single-channel designs and each is afforded a degree of isolation due to the distance separating them. In planning a candelabra or tee-bar installation, several factors, some of which are conflicting, must be considered. These include (*a*) cost, (*b*) pattern circularity, (*c*) video response, (*d*) whether or not any of the antennas will be circularly polarized, (*e*) effects of wind sway, (*f*) isolation, (*g*) echo discernibility, and (*h*) frequency. These factors each influence the selection of the optimum separation.

Cost is a major factor for a candelabra or tee bar. While the antenna wind load and weight and transmission-line loads have some effect on the cost, the major effect is due to the physical separation between antennas. While it is difficult to give exact cost figures, it is generally recognized that, for a fixed number of support points (antennas), the cost increases approximately linearly with increasing spacing. For the antenna to be economically feasible an upper limit on allowable spacing must be set.

Pattern circularity, video response, and the effects of wind sway are all related to the azimuth pattern effects due to scattering from the other antennas. These parameters may be determined theoretically and/or experimentally, depending on the amount of time and effort available. In some cases, such as when the antennas are slotted tube designs, the scattering may be accurately determined using well-known formulas for scattering from round cylinders [32].

For more complex antennas these formulas may be used to approximate the scattered fields with the assumption that the antenna may be represented by an equivalent cylinder [33]. For example, a batwing antenna may be represented by an array of cylinders with radius equal to three-quarters the batwing radius (see Fig. 38). The maximum-to-minimum circularity values as calculated are slightly conservative when compared with measured results.

When more detailed results are required with the more complex antennas, it is necessary to model the actual structure in order to calculate or measure the scattered fields. If the antenna can be modeled as a wire antenna, the scattered

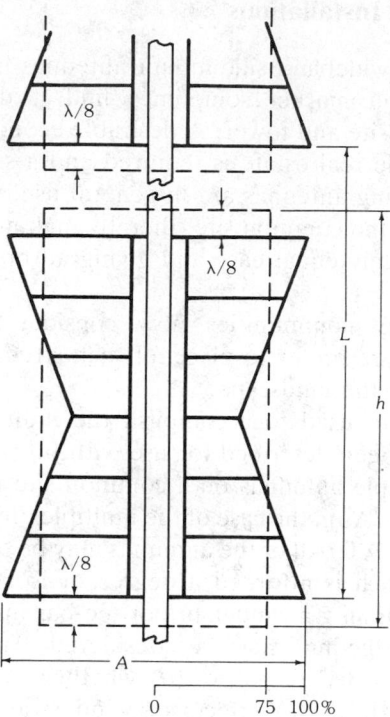

Fig. 38. Equivalent cylinder dimensions of a batwing antenna for horizontally polarized scatter pattern calculations. (*After Siukola [33]*)

fields can be calculated by means of moment methods [34]. For large antennas this approach can become complex. Furthermore, it is important to exercise great care in modeling the junctions in the wire model.

It is often simpler and more useful to build a scale model of the antennas to be used and indirectly measure the scattered fields. This is accomplished by first measuring the total field (amplitude and phase) of the radiating antenna in the presence of the scattering antenna(s) at a typical separation and the field of the radiating antenna by itself. These patterns may be subtracted vectorially to give the scatter pattern of the scattering antenna(s). This scatter pattern may then be used to compute the total fields at other separations.

To compute the circularity of the antennas it is only necessary to determine the total fields as a function of azimuth at the visual carrier frequency. To determine the video response the total fields must be determined as a function of frequency. The variation in the fields at specific azimuths are then used to determine the video response in directions of interest. For small variations the video response may be corrected for by adjustments available to the viewer. The effect of wind sway may be estimated by changing the relative spacing between antennas. The phase of the scattered fields is thereby varied, causing an effect similar to a change in frequency. However, the effect occurs as a function of time and can be observed at a specific location by a viewer.

For fixed type and number of antennas the circularity is generally improved as the separation is increased, although the variation is better described as a slowly damped oscillation as a function of distance. The effect on video response is lessened as the separation is reduced. As the separation increases, the time delay for echo signals from the scattering antennas increases so that for large delays any echos present may be discerned by the viewer. A separation of 50 ft (15.24 m) represents an echo delay of about 0.1 μs. Mitigating this effect is the fact that the magnitude of echos is reduced as separation is increased. Finally, the isolation between antennas increases directly as the spacing increases. Thus it is apparent

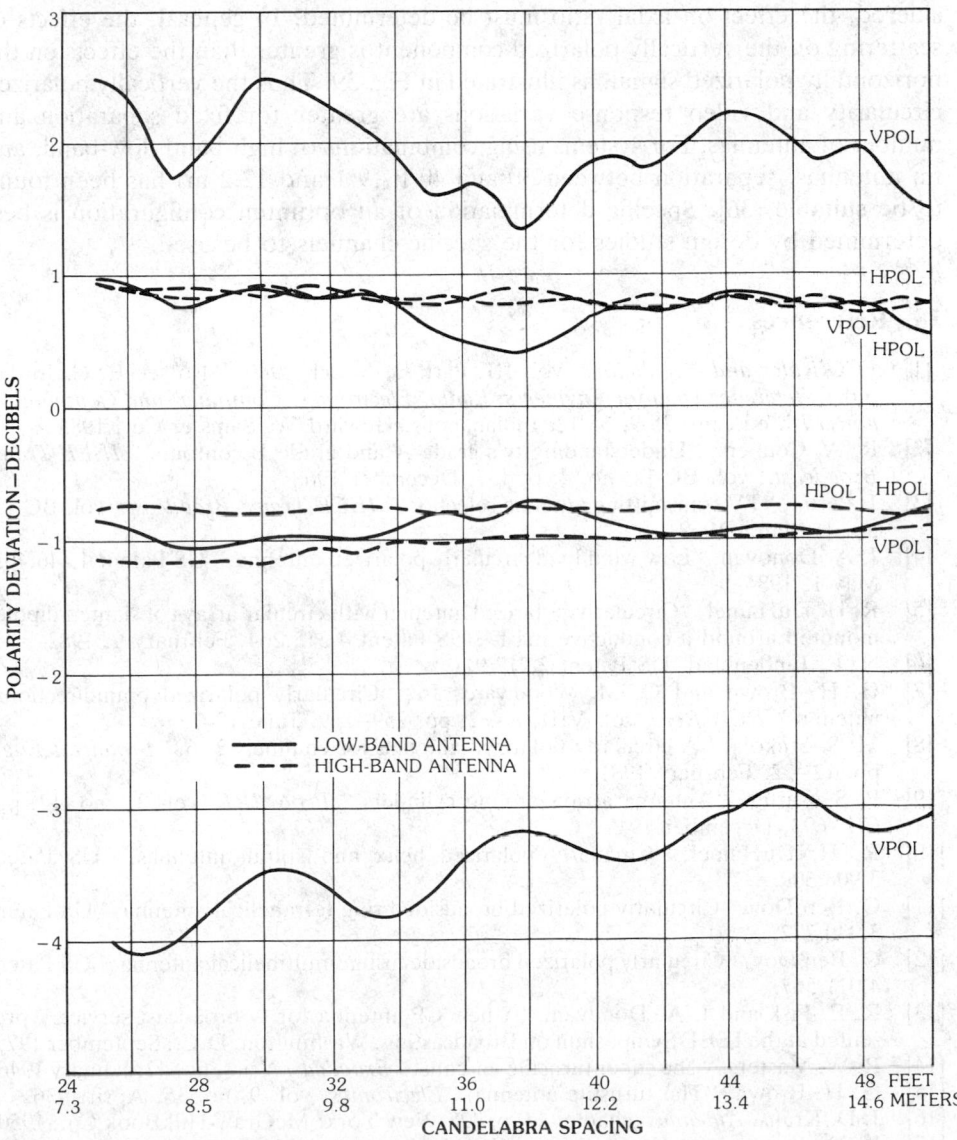

Fig. 39. Circularity deviation from rms versus spacing for three antennas. (*After Johns and Ralston [36], © 1981 IEEE*)

the various parameters impose conflicting requirements on the configuration. In general, spacings between 10 and 100 ft (3 and 30.5 m) have been found to be desirable, depending on the number of antennas, the channels to be used, and polarization [35]. The effect of wind sway can be controlled by keeping the radiation centers as low as possible and by using very stiff antennas. In some cases, such as when stacked antennas are on a candelabra, it may be necessary to guy the antennas.

When circularly polarized antennas are used, the effect on the vertically polarized component becomes important. In addition to the factors already considered, the effect on axial ratio must be determined. In general, the effects of scattering on the vertically polarized component is greater than the effects on the horizontally polarized signal, as illustrated in Fig. 39. Thus the vertically polarized circularity and video response variations are greater for fixed separation and number of antennas. For systems using combinations of high-band, low-band, and fm antennas, separation between 30 and 40 ft (9.1 and 12.2 m) has been found to be suitable [36]. Specific determination of an optimum configuration is best determined by design studies for the specific channels to be used.

18. References

[1]   *FCC Rules and Regulations*, vol. III, Part 73, March 1980. Also see E. C. Jordan (ed.), *Reference Data for Engineers: Radio, Electronics, Computer, and Communications*, 7th ed., pp. 35-9, 35-12, Indianapolis: Howard W. Sams & Co., 1985.

[2]   R. A. Conners, "Understanding tv's grade A and grade B contours," *IEEE Trans. Broadcast.*, vol. BC-14, no. 4, p. 137, December 1968.

[3]   J. Perini, "TV transmitting antenna selection," *IEEE Trans. Broadcast.*, vol. BC-15, no. 1, March 1969.

[4]   J. A. Donovan, "Low wind load circularly polarized antenna," US Patent 4,446,465, May 1, 1984.

[5]   R. H. DuHamel, "Circularly polarized antenna with circular arrays of slanted dipoles mounted around a conductive mast," US Patent 4,315,264, February 9, 1982.

[6]   N. E. Lindenblad, US Patent 2,217,911.

[7]   G. H. Brown and O. M. Woodward, Jr., "Circularly polarized omnidirectional antenna," *RCA Rev.*, vol. VIII, no. 2, pp. 259–269, June 1947.

[8]   M. S. Suikola, "A circularly polarized antenna for channels 2–6," *Broadcast Eng.*, pp. 12–22, February 1981.

[9]   P. S. Carter, "Antenna arrays around cylinders," *Proc. IRE*, vol. 31, no. 12, pp. 671–693, December 1943.

[10] R. H. DuHamel, "Circularly polarized helix and spiral antennas," US Patent 3,906,509.

[11] O. Ben Dov, "Circularly polarized broadside firing tetrahelical antenna," US Patent 3,940,772.

[12] O. Ben Dov, "Circularly polarized broadside firing, multihelical antenna," US Patent 4,011,567.

[13] R. E. Fisk and J. A. Donovan, "A new CP antenna for tv broadcast service," presented at the IEEE Symposium on Broadcasting, Washington, D.C., September 1975.

[14] R. W. Masters, "The super turnstile antenna," *Broadcast News*, no. 42, January 1946.

[15] G. H. Brown, "The turnstile antenna," *Electronics*, vol. 9, no. 15, April 1936.

[16] J. D. Kraus, *Antennas*, chapter 14, p. 428, New York: McGraw-Hill Book Co., 1950.

[17] L. O. Krause and H. G. Smith, "Wound antenna with conductive support," US Patent 2,985,878, May 23, 1961.

[18] R. E. Fisk, "The tv helical antenna adapted to structural tower shapes," *IRE Trans. Broadcast Transmission Systems*, October 1957.

[19] R. E. Fisk, "A simplified 5 megawatt antenna for the uhf broadcaster," presented at the Seventh Annual Symposium of the IRE Professional Group on Broadcast Transmission Systems, Washington, D.C., 1957.

[20] P. C. J. Hall, "Methods for shaping vertical radiation patterns," *Proc. IEEE*, vol. 116, no. 8, p. 1325, August 1969.

[21] H. H. Riblet, "Microwave omnidirectional antennas," *Proc. IRE*, vol. 35, no. 5, pp. 474–478, 1947.

[22] K. B. Hoffman and R. E. Fisk, "Directional zig-zag antenna at KERO, Bakersfield, California," *IEEE Trans. Broadcast.*, p. 12, February 1964.

[23] O. M. Woodward, "Antenna arrays," US Patent 2,759,183.

[24] L. Sin Hoi, "Theory of zig-zag antennas," Univ. of California, March 1969.

[25] J. Perini, "Improvement of pattern circularity of panel antennas mounted on large towers," *IEEE Trans. Broadcast.*, vol. BC-14, no. 1, p. 33, March 1968.

[26] J. Perini, "A method of obtaining a smooth pattern on circular arrays of large diameter," *IEEE Trans. Broadcast.*, vol. BC-14, no. 3, p. 126, September 1968.

[27] R. N. Clark, "The V-Z panels as a side mounted antenna," *IEEE Trans. Broadcast.*, vol. BC-13, no. 1, p. 31, January 1967.

[28] R. E. Fisk, "Empire State antenna installation and performance," presented at the Nineteenth Annual Broadcast Engineering Conference, NAB, Washington, D.C., 1965.

[29] C. A. Balanis, *Antenna Theory*, chapter 5, New York: Harper & Row, 1982.

[30] R. E. Fisk, "Design and application of a multiplex nine station fm antenna for the Senior Road Tower Group," presented at the 1984 NAB Convention, Las Vegas.

[31] W. G. Shulz, "Spurious emission measurements of common site transmitter installations using separate collocated antennas," presented at the 1984 NAB convention, Las Vegas.

[32] R. F. Harrington, *Time-Harmonic Electromagnetic Fields*, New York: McGraw-Hill Book Co., 1961, pp. 223–238.

[33] M. S. Siukola, "Predicting performance of candelabra antennas by mathematical analysis," *RCA Broadcast News*, vol. 97, pp. 63–68.

[34] R. F. Harrington, *Field Computations by Moment Methods*, p. 62. A book available from the author at RD2 Westlake Road, Cazenovia, N.Y. 13035.

[35] M. S. Siukola, "Size and performance trade-offs in multiple arrays of horizontally and circularly polarized tv antennas," *IEEE Trans. Broadcast.*, vol. BC-22, no. 1, pp. 5–12, March 1975.

[36] M. R. Johns and M. A. Ralston, "The first candelabra for CP broadcast antennas," *IEEE Trans. Broadcast*, pp. 77–82, December 1981.

PART D

Related Topics

Chapter 28

Transmission Lines and Waveguides

Y. C. Shih
U.S. Naval Postgraduate School

T. Itoh
The University of Texas

CONTENTS

Yi-Chi Shih received the B Engr degree from National Taiwan University, Taiwan, R.O.C., in 1976, the MSc degree from the University of Ottawa, Ontario, Canada, in 1980, and the PhD degree from the University of Texas at Austin in 1982, all in electrical engineering.

From September 1982 to April 1984 he served as an adjunct professor at the Naval Postgraduate School, Monterey, California, where he offered courses in the areas of electromagnetic field theory, antennas, and propagation. From April 1984 to May 1986 he was with Hughes Aircraft Company as a senior member of technical staff. In May 1986 he joined MM-Wave Technology, Inc., as Technical Director. He is responsible for the development of CAD tools for hybrid and monolithic integrated circuits, including amplifiers, phase shifters, mixers, and filters.

Dr. Shih is a member of the Institute of Electrical and Electronics Engineers. He has published more than 20 papers on the analysis and design of transmission lines, waveguide discontinuities, and millimeter-wave filters.

Tatsuo Itoh received the PhD degree in electrical engineering from the University of Illinois, Urbana, in 1969.

From September 1966 to April 1976 he was with the Electrical Engineering Department, University of Illinois. From April 1976 to August 1977 he was a senior research engineer in the Radio Physics Laboratory, SRI International, Menlo Park, California. From August 1977 to June 1978 he was an associate professor at the University of Kentucky, Lexington. In July 1978 he joined the faculty of the University of Texas at Austin, where he is now a professor of electrical engineering. During the summer of 1979, he was a guest researcher at AEG-Telefunken, Ulm, West Germany. Since September 1983 he has held the Hayden Head Centennial Professorship of Engineering at the University of Texas.

Dr. Itoh is a fellow of the IEEE and a member of the Institute of Electronics and Communication Engineers of Japan, Sigma Xi, and Commission B of USNC/URSI. He serves on the administrative committee of the IEEE Microwave Theory and Techniques Society and is the editor of *IEEE Transactions on Microwave Theory and Techniques*. He is a professional engineer registered in the State of Texas.

1. Introduction

In antenna applications it is necessary to use some form of transmission line to connect the antenna to a transmitter or receiver. The purpose of this chapter is to provide the essential propagation characteristics of the more common forms of transmission lines.

The useful transmission line equations are summarized in Section 2, where most of the quantities are defined and tabulated in tables for convenience. In Sections 3 to 6, four classes of transmission lines are discussed. The classifications, based on the type of the fundamental-mode propagation, are transverse-electromagnetic (TEM) lines, planar quasi-TEM lines, transverse-electric/transverse-magnetic (TE/TM) waveguides, and hybrid-mode waveguides.

A TEM transmission line can support a TEM wave in which both the electric and magnetic field vectors are always perpendicular to the direction of propagation. In the planar quasi-TEM class the transmission lines usually have mixed dielecric boundaries and, therefore, cannot support a pure TEM wave. However, the longitudinal component of the fields is small, so it is neglected under the "quasi-TEM" approximations.

In the transverse-electric (TE) waves, sometimes called H-waves, the electric field vector is always perpendicular to the direction of propagation. In the transverse-magnetic (TM) waves, sometimes called E-waves, the magnetic field vector is always perpendicular to the direction of propagation.

In the case of hybrid-mode waveguides the modal field possesses relatively strong longitudinal components of electric and magnetic fields. Therefore, the quasi-TEM approximation technique cannot be applied.

In the following discussion, quantities are defined in SI units. The formulas are given under the assumption that all the materials are nonmagnetic and loss-free, unless specified.

2. Transmission Line Equations

In conventional analysis a transmission line is often represented by a series of equivalent networks. Fig. 1 shows the equivalent network for a short section Δx. The series impedance and the shunt admittance per unit length are

$$Z = R + j\omega L \tag{1a}$$

$$Y = G + j\omega C \tag{1b}$$

where ω is the angular frequency and R, L, G, and C are distributed resistance, inductance, conductance, and capacitance per unit length, respectively. The volt-

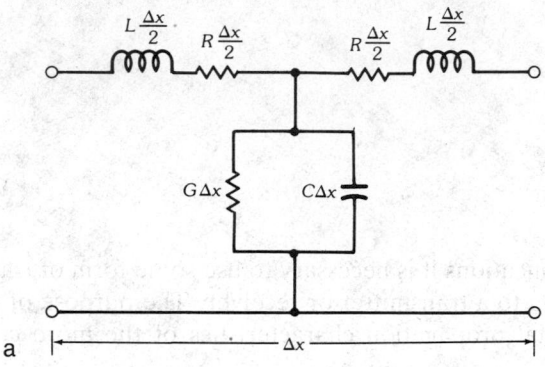

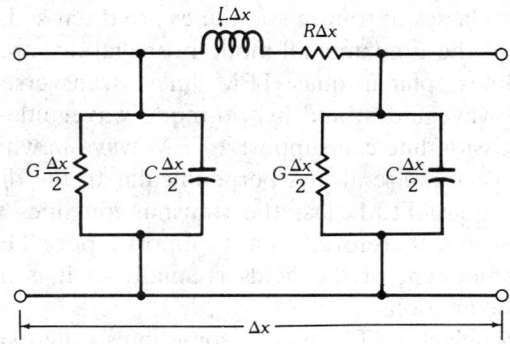

Fig. 1. Equivalent T and π networks of a short section Δx of a transmission line. (*After Jordan [1]*) (*a*) T network. (*b*) π network.

age (V) and current (I) waves along the transmission line are described by the following equations:

$$\frac{dV}{dz} = -ZI = -\gamma Z_0 I \tag{2a}$$

$$\frac{dI}{dz} = -YV = -\gamma Y_0 V \tag{2b}$$

From these equations expressions for important transmission line parameters can be derived and they are summarized in Tables 1 and 2. Fig. 2 defines the reference planes for impedance relations and reflection coefficients. The expressions for voltage, impedance, etc., are generally for the quantities at the input terminal of the line in terms of those at the output terminal. If it is desired to find the quantities at the output in terms of those at the input, it is simply necessary to interchange the subscripts 1 and 2 in the equations and to place a minus sign before ℓ wherever it appears.

The equations summarized here are primarily for transmission lines operating

Table 1. Summary of Transmission Line Equations (*After Ragan [2], © 1948 McGraw-Hill Book Company*)

Item No.	Quantity	General Line	Expression* Ideal Line	Approximation, Low-Loss Line
1	Propagation constant $\gamma = \alpha + j\beta$	$\sqrt{(R+j\omega L)(G+j\omega C)}$	$j\omega\sqrt{LC}$	See items 2 and 3
2	Phase constant β	$\text{Im}\{\gamma\}$	$\omega\sqrt{LC} = \dfrac{\omega}{v_p} = \dfrac{2\pi}{\lambda_g}$	$\beta'\left[1+\dfrac{1}{2}\left(\dfrac{\alpha_c}{\beta'}-\dfrac{\alpha_d}{\beta'}\right)^2\right]$
3	Attenuation constant α	$\text{Re}\{\gamma\} = -\dfrac{1}{2P}\dfrac{dP}{d\ell}$	0	$\alpha_c + \alpha_d = \dfrac{R}{2Z_0'} + \dfrac{G}{2Y_0'}$
4	Characteristic impedance $Z_0 = 1/Y_0$	$\sqrt{\dfrac{R+j\omega L}{G+j\omega C}}$	$\sqrt{\dfrac{L}{C}}$	$Z_0\left[1+\dfrac{1}{2}\left(\dfrac{\alpha_c}{\beta'}-\dfrac{\alpha_d}{\beta'}\right)\left(\dfrac{\alpha_c}{\beta'}+\dfrac{3\alpha_d}{\beta'}\right)\right] - jZ_0'\left(\dfrac{\alpha_c}{\beta'}-\dfrac{\alpha_d}{\beta'}\right)$
5	Input impedance Z_1	$Z_0\dfrac{Z_2+Z_0\tanh\gamma\ell}{Z_0+Z_2\tanh\gamma\ell}$	$Z_0\dfrac{Z_2+jZ_0\tan\beta\ell}{Z_0+jZ_2\tan\beta\ell}$	
6	Impedance of short-circuited line ($Z_2=0$)	$Z_0\tanh\gamma\ell$	$jZ_0\tan\beta\ell$	$Z_0\dfrac{\alpha\ell+j\tan\beta\ell}{1+j\alpha\ell\tan\beta\ell}$
7	Impedance of open-circuited line ($Z_2=\infty$)	$Z_0\coth\gamma\ell$	$-jZ_0\cot\beta\ell$	$Z_0\dfrac{1+j\alpha\ell\tan\beta\ell}{\alpha\ell+j\tan\beta\ell}$
8	Impedance of a line an odd number of quarter-wavelengths long ($\beta\ell = n\pi + \pi/2$)	$Z_0\dfrac{Z_2+Z_0\coth\alpha\ell}{Z_0+Z_2\coth\alpha\ell}$	$\dfrac{Z_0^2}{Z_2}$	$Z_0\dfrac{Z_0+Z_2\alpha\ell}{Z_2+Z_0\alpha\ell}$
9	Impedance of a line an integral number of half-wavelengths long ($\beta\ell = n\pi$)	$Z_0\dfrac{Z_2+Z_0\tanh\alpha\ell}{Z_0+Z_2\tanh\alpha\ell}$	Z_2	$Z_0\dfrac{Z_2+Z_0\alpha\ell}{Z_0+Z_2\alpha\ell}$
10	Voltage V_1 along line	$V_i(1+\Gamma_0 e^{-2\gamma\ell})$	$V_i(1+\Gamma_0 e^{-2j\beta\ell})$	
11	Current I_1 along line	$I_i(1-\Gamma_0 e^{-2\gamma\ell})$	$I_i(1-\Gamma_0 e^{-2j\beta\ell})$	
12	Voltage reflection coefficient at $z=0$, Γ_0	$\dfrac{Z_2-Z_0}{Z_2+Z_0}$	$\dfrac{Z_2-Z_0}{Z_2+Z_0}$	

*The terms not defined above are as follows: $\beta' = \sqrt{LC}$ = phase constant, neglecting losses, $Z_0' = \sqrt{L/C}$ = characteristic impedance, neglecting loss, $Y_0' = 1/Z_0'$, V_i = incident voltage, I_i = incident current, λ_g = wavelength measured along line (guide wavelength), v_p = phase velocity of line, equals velocity of light in dielectric of line for an ideal line, α_c = conductor loss, and α_d = dielectric loss. Other notations refer to Fig. 2.

Table 2. Some Miscellaneous Relations in Low-Loss Transmission Lines
(*After Ragan [2],* © *1948 McGraw-Hill Book Company*)

Item	Equation	Explanation
1a	$r = \dfrac{1 + \lvert\Gamma\rvert}{1 - \lvert\Gamma\rvert}$	r = voltage standing-wave ratio
1b	$\lvert\Gamma\rvert = \dfrac{r - 1}{r + 1}$	$\lvert\Gamma\rvert$ = magnitude of reflection coefficient
2a	$\Gamma = \dfrac{R - Z_0}{R + Z_0}$	Γ = reflection coefficient (real) at a point in a line where the impedance is real (R). Point may be at an actual resistive load or at a voltage max. or min. in standing-wave pattern
2b	$r = \dfrac{R}{Z_0}; \ R = rZ_0$	Conditions for $R > Z_0$, i.e., at voltage maximum
2c	$r = \dfrac{Z_0}{R}; \ R = \dfrac{1}{r}Z_0$	Conditions for $R < Z_0$, i.e., at voltage minimum
3a	$\dfrac{P_r}{P_i} = \lvert\Gamma\rvert^2 = \left(\dfrac{r - 1}{r + 1}\right)^2$	P_r = power in wave reflected by discontinuity or mismatched load
3b	$\dfrac{P_t}{P_i} = 1 - \lvert\Gamma\rvert^2 = \dfrac{4r}{(r + 1)^2}$	P_i = power in incident wave
		P_t = power in transmitted (or absorbed) wave
4	$\dfrac{P_b}{P_m} = \dfrac{1}{r}$	P_b = *net power* transmitted to load at onset of breakdown in a line in which a vswr $= r$ exists
		P_m = same when line is matched, $r = 1$
5	$\dfrac{\alpha_r}{\alpha_m} = \dfrac{1 + \Gamma^2}{1 - \Gamma^2} = \dfrac{r^2 + 1}{2r}$	α_m = ordinary attenuation constant; matched line, $r = 1$
		α_r = attenuation constant allowing for increased *ohmic* loss caused by standing waves (vswr $= r$)
6a	$r_{\max} = r_1 r_2$	Resultant vswr when two separate mismatches combine in worst phase
6b	$r_{\min} = \dfrac{r_2}{r_1}; \ r_1 < r_2$	Resultant when they combine in best phase
6c	$r_{\max} = r_1 r_2 r_3 \cdots r_n$	Resultant for n mismatches, worst phase
6d	$r_{\min} = \dfrac{r_n}{r_1 r_2 \cdots r_{n-1}}$ $r_1 < r_2 < r_2 < \cdots < r_n$	Resultant for n mismatches, best phase. If this gives $r_{\min} < 1$, then $r_{\min} = 1$
7a	$\lvert\Gamma\rvert = \dfrac{\lvert X\rvert}{\sqrt{X^2 + 4}}$	Relations for the case of a normalized reactance X in series with resistance Z_0
7b	$r = \dfrac{\sqrt{X^2 + 4} + \lvert X\rvert}{\sqrt{X^2 + 4} - \lvert X\rvert}$	
7c	$\lvert X\rvert = \dfrac{r - 1}{\sqrt{r}}$	

Table 2, *continued.*

Item	Equation	Explanation
8a	$\|\Gamma\| = \dfrac{\|B\|}{\sqrt{B^2 + 4}}$	Relations for the case of a normalized susceptance B shunting conductance Y_0
8b	$r = \dfrac{\sqrt{B^2 + 4} + \|B\|}{\sqrt{B^2 + 4} - \|B\|}$	
8c	$\|B\| = \dfrac{r - 1}{\sqrt{r}}$	

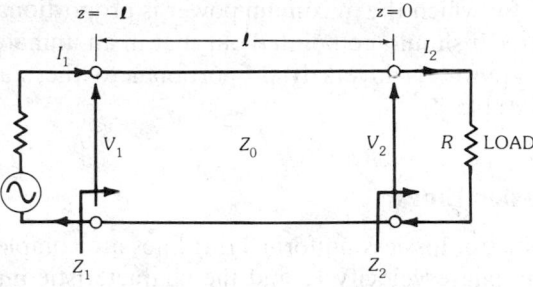

Fig. 2. Transmission line with generator and load.

in the TEM and quasi-TEM modes. The equations are accurate, according to conventional transmission line theory, and are applicable from the lowest power and communication frequencies, from zero frequency up to the frequency where the higher-order mode begins to appear on the line. The information is also valid for non-TEM waveguides under single-mode operation in general, except for those expressions in which the quantities R, L, G, and C are involved.

In addition to the exact equations, low-loss approximation yields simple expressions that are specially adapted for use in radio-frequency problems. In this case the quadratic $(\alpha^2 z^2)$ and higher powers in the expansion of $\exp(\alpha z)$, etc., are neglected. Thus, when $\alpha z = 0.1$ Np, the error in the approximate expressions is on the order of 1 percent.

In practice, the line loss is small and may be neglected in calculation of the characteristic impedance and phase velocity. The attenuation constant, including conductor loss and dielectric loss, may then be obtained by a perturbation method. In all cases it is assumed that the skin depth is small compared with the dimensions of the conductors and that the constructing materials are nonmagnetic. The conductor loss is normally a complicated function of frequency, geometry, and material constants, whereas the dielectric loss in nepers per meter takes the simple form

$$\alpha_d = \frac{\pi}{\lambda_g}(\tan \delta)_{\text{eff}} \tag{3}$$

where the guide wavelength $\lambda_g = 2\pi/\beta'$ is obtained by neglecting the loss in the media and $(\tan \delta)_{eff}$ is the effective loss tangent of the structure. For TEM lines surrounded by a homogeneous dielectric, the effective loss tangent is simply the loss tangent of the dielectric. The conductor loss and the effective loss tangent for non-TEM lines will be given later for each case.

The peak power-handling capacity of a transmission line is determined by the maximum allowable electric field intensity E_b in the dielectric. The maximum E_b, or dielectric strengths, for some commonly used dielectrics are given in Table 3. From this table 3.0 kV/mm is the theoretical maximum for air dielectric at normal temperature and atmosphere pressure, but for proper derating a value of 2.0 kV/mm is more practical. A solid dielectric may be used to obtain higher power level. However, any air pockets within the dielectric may not be able to support the increased **E**-field intensity. A common method of increasing the power limit is to use pressurized air, for which the maximum power is proportional to the square of the absolute pressure. It should be pointed out that in an unmatched transmission line, the maximum power is inversely proportional to the standing-wave ratio (refer to item 4 of Table 2).

3. TEM Transmission Lines

The characteristics of lossless uniform TEM lines are completely described by two parameters, the phase velocity v_p and the characteristic impedance Z_0. The latter is a function of the line cross-sectional dimensions and dielectric constant. The phase velocity in meters per second is given simply by

$$v_p = c/\sqrt{\epsilon_r} \tag{4}$$

where c is the free-space speed of light and ϵ_r is the relative dielectric constant of the filling material. Equation 4 is valid for all TEM structures. Consequently, the characteristic impedance becomes the only parameter necessary to characterize the transmission line.

Table 3. Properties of Commonly Used Dielectric Materials

Material	Dielectric Constant ϵ_r	Loss Tangent at 10 GHz $(\times 10^{-4})$	Thermal Conductivity K (W/m/°C)	Dielectric Strength (kV/mm)
Air	1.0	$\simeq 0$	0.024	3.0
Alumina	9.7	2.0	30.0	400.0
Berylium oxide	6.6	1.0	250.0	
Cuflon	2.1	5.0		78.7
Gallium arsenide	12.9	16.0	30.0	35.0
Polystyrene	2.53	4.7	0.15	28.0
Quartz (fused)	3.78	1.0	1.0	1000.0
RT/Duroid 5880	2.2	10.0	0.26	11.8
Sapphire	11.7	1.0	40.0	400.0
Silicon	11.7	50.0	90.0	30.0

Table 4 is a summary of the impedance formulas for common TEM lines.* They are roughly grouped into three families, namely, open-wire lines, shielded-wire lines, and strip lines. Open-wire lines may consist of either a pair of conductors carrying the going and returning currents (nos. 1 and 2), or of a multiplicity of conductors interconnected in different manners (nos. 3, 4, and 5). The arrangement is sometimes used in conjunction with a ground plate to which the wires are parallel (nos. 6, 7, and 8). Generally, the wires have small dimensions compared with the spacing between the wires and between the wires and the ground plate. Furthermore, the spacing is assumed to be very much less than a wavelength at the operating frequency. This type of transmission line has the advantage of simplicity and economy. However, the open-wire lines become unusable at high frequency because of excessive radiation loss.

Surrounding the wires of a transmission line with a shielding conductor (e.g., nos. 9 through 19) effectively overcomes the radiation problem. The popular coaxial line (no. 16) is essentially a self-shielded transmission, and is widely used for propagation of microwave power.

Strip transmission lines (nos. 20 through 28) are those in which the conductors have the form of flat plates or strips that are amenable to photolithographic techniques for mass production of circuit components. Both open and shielded types exist, but open structures suffer radiation problems at higher frequencies.

For a pair of wires near ground or in a shield, two kinds of modes may be excited:

(a) both "positive" with respect to ground; this is called "even-mode" excitation; or

(b) one "positive" and one "negative" with respect to ground; this is called "odd-mode" excitation.

These modes of excitation are illustrated schematically in Fig. 3, for the case of simple two-wire line near ground. The characteristic impedances, Z_{0e} and Z_{0o}, associated with these modes of excitation are defined as the input impedance of an infinite length of one line, in the presence of the second line, when both are excited in the appropriate manner [(a) or (b) above].

Two-Wire Line

Fig. 4 shows the cross section and field configuration of a two-wire line with wires of diameter d, spaced at a center-to-center distance D. The characteristic impedance in ohms of this line is

$$Z_0 = \frac{119.9}{\sqrt{\epsilon_r}} \cosh^{-1}\left(\frac{D}{d}\right) = \frac{119.9}{\sqrt{\epsilon_r}} \ln\left(\frac{D + \sqrt{D^2 - d^2}}{d}\right) \qquad (5)$$

and is plotted in Fig. 5.

The conductor loss in nepers per meter is

---

*Most of the materials in this table are from [1] and [3]. No. 19 is a quasi-TEM structure; it is included here for convenience of comparison.

Table 4. Impedance Formulas for Common TEM Transmission Lines

No.	Structure	Cross Section	Impedance Formulas*	Range of Validity
1	Open two-wire line		$Z_0 = 119.9 \ln(x + \sqrt{x^2 - 1})$ $x = D/d$	Exact; no limit
2	Balanced two-wire unequal diameters		$Z_0 = 59.95 \ln(x + \sqrt{x^2 - 1})$ $x = \frac{1}{2}\left[\frac{4D^2}{d_1 d_2} - \frac{d_1}{d_2} - \frac{d_2}{d_1}\right]$	Exact; no limit
3	Three-wire line		$Z_0 = 476.6 \ln(1.59\ D/d)$	$d \ll D$
4	Balanced four-wire		$Z_0 = 59.95 \ln\left\{\frac{2D_2}{d}\left[1 + \left(\frac{D_2}{D_1}\right)^2\right]^{-1/2}\right\}$	$d \ll D_1, D_2$

5 — Five-wire line

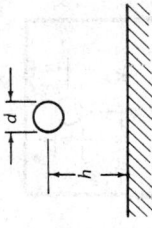

$$Z_0 = 74.95 \ln(D/0.993\,d)$$

$d \ll D$

6 — Single wire near ground

$$Z_0 = 59.95 \ln(x + \sqrt{x^2 - 1})$$
$$x = 2h/d$$

Exact; no limit

7 — Open two-wire near ground

$$Z_{0e} = 29.98 \ln\{(4h/d)[1 + (2h/D)^2]^{1/2}\}$$
$$Z_{0o} = 119.9 \ln\{(2D/d)[1 + (D/2h)^2]^{-1/2}\}$$

$d \ll D, h$

8 — Balanced two-wire near ground

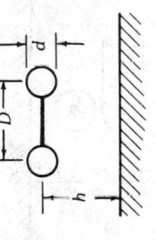

$$Z_0 = 119.9 \ln\left\{\frac{2D}{d}\left[1 + \frac{D^2}{4h_1 h_2}\right]^{-1/2}\right\}$$

$d \ll D, h_1, h_2$

9 — Single wire between two grounded planes

$$Z_0 = 15 \ln[1 + 1.314x + \sqrt{(1.314x)^2 + 2x}]$$
$$x = (1 + 2h/d)^4 - 1$$

Less than 0.1%; no limit

Table 4, *continued.*

No.	Structure	Cross Section	Impedance Formulas*	Range of Validity
10	Two-wire line between grounded planes		$Z_{0e} = 59.95\ln\left\{\coth\left(\frac{\pi D}{2h}\right)\coth\left(\frac{\pi d}{4h}\right)\right\}$ $Z_{0o} = 59.95\ln\left\{\tanh\left(\frac{\pi D}{2h}\right)\coth\left(\frac{\pi d}{4h}\right)\right\}$	$d/h < 0.25$ $D/h > 3d/h$
11	Balanced line between grounded planes		$Z_0 = 119.9\ln(2h/\pi d)$	$d \ll h$
12	Single wire in trough		$Z_0 = 59.95\ln\left\{\tanh\left(\frac{\pi h}{W}\right)\coth\left(\frac{\pi d}{4W}\right)\right\}$	$d/W < 0.25$ $h/W > 3d/2W$
13	Balanced line in trough		$Z_{0o} = 119.9\ln[2W/\pi d(A^{1/2})]$ $A = \csc^2(\pi D/W) + \operatorname{cosech}^2(2\pi h/W)$	$d \ll D, W, h$

14 Single wire in
    square enclosure

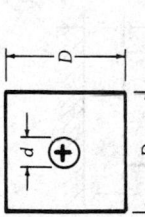

$$Z_0 = 59.95 \ln(1.0787 D/d)$$

Less than 1.5%; $d/D < 0.8$

---

15 Two-wire line
    in rectangular
    enclosure

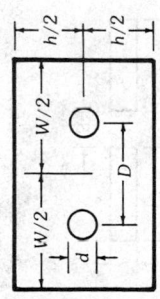

$$Z_{0e} = 119.9 \left[\ln\frac{4h \coth\left(\frac{\pi D}{2h}\right)}{\pi d} + \sum_{m=1} (-1)^m \ln \frac{1 + \dfrac{\cosh^2(\pi D/2h)}{\sinh^2(m\pi W/2h)}}{1 - \dfrac{\cosh^2(\pi D/2h)}{\cosh^2(m\pi W/2h)}} \right]$$

$$Z_{0o} = 119.9 \left[\ln\frac{4h \tanh\left(\frac{\pi D}{2h}\right)}{\pi d} + \sum_{m=1} \ln \frac{1 + \dfrac{\sinh^2(\pi D/2h)}{\cosh^2(m\pi W/2h)}}{1 - \dfrac{\sinh^2(\pi D/2h)}{\sinh^2(m\pi W/2h)}} \right]$$

$d \ll D, W, h$

---

16 Coaxial line

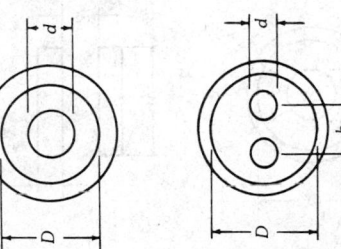

$$Z_0 = 59.95 \ln(D/d)$$

Exact; no limit

---

17 Shielded
    two-wire line

$$Z_{0e} = 29.98 \ln[(v/2\sigma^2)(1 - \sigma^4)]$$
$$Z_{0o} = 119.9 \ln\{2v[(1 - \sigma^2)/(1 + \sigma^2)]\}$$
$$v = h/d; \quad \sigma = h/D$$

$d \ll D, h$

---

18 Eccentric line

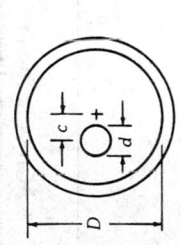

$$Z_0 = 119.9 \ln(x + \sqrt{x^2 - 1})$$
$$x = \frac{1}{2}[(D/d) + (d/D) - (4c^2/dD)]$$

Exact; no limit

Table 4, *continued.*

No.	Structure	Cross Section	Impedance Formulas*	Range of Validity
19	Air coaxial with dielectric supporting wedge		$Z_0 \cong \dfrac{59.95\ln(D/d)}{[1 + (\epsilon_r - 1)(\theta/2\pi)]^{1/2}}$ θ: in rad	—
20	Broadside-coupled stripline		$Z_0 = 2(Z_0 \text{ of No. 22})$	Approx. 0.1%; $W/h < 1000$
21	Edge-coupled stripline		$Z_0 = 376.7\,K(x)^{\dagger}$ $x = S/(S + 2W)$	Exact for $t = 0$
22	Single-dielectric microstrip line		$Z_0 = 59.95\ln\left\{\left[\dfrac{6}{u} + \dfrac{2\pi - 6}{u}\exp\left[-\left(\dfrac{30.666}{u}\right)^{0.7528}\right]\right] + \sqrt{1 + \left(\dfrac{2}{u}\right)^2}\right\}$ $u = \dfrac{W}{h} + \dfrac{t}{\pi h}\ln\left[1 + \dfrac{4e}{\dfrac{t}{h}\coth^2\sqrt{6.517\,W/h}}\right]$	Approx. 0.1%; $W/h < 1000$

23	Triplate stripline	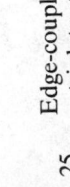	$$Z_0 = 30\ln\left\{1 + \frac{8h}{\pi W'}\left[\frac{16h}{\pi W'} + \sqrt{\left(\frac{16h}{\pi W'}\right)^2 + 6.27}\right]\right\}$$ $$W' = W + \frac{t}{\pi}\ln\left\{e\left[\left(\frac{1}{4h/t+1}\right)^2 + \left(\frac{1/4\pi}{W/t+1.1}\right)^m\right]^{-1/2}\right\}$$ $$m = 2/(1 + t/3h)$$	0.1% or less; $W'/h < 20$
24	Rounded-edge triplate stripline		$$Z_0 = \frac{188.34\ln x}{\pi + 2[(W/D - t/D)/(1 - t/D)]\ln x}$$ $$x = 4/(\pi t/b)$$	Less than 1%; $t/D \leqq 0.5$
25	Edge-coupled triplate stripline		$$Z_{0e} = 29.98\pi K(k_e)$$ $$Z_{0o} = 29.98\pi K(k_0)$$ $$k_e = \tanh\left(\frac{\pi W}{2D}\right)\tanh\left(\frac{\pi}{2}\frac{W+S}{D}\right)$$ $$k_0 = \tanh\left(\frac{\pi W}{2D}\right)\coth\left(\frac{\pi}{2}\frac{W+S}{D}\right)$$	Exact; for $t = 0$
26	Broadside-coupled horizontal stripline		$$Z_{0e} = 59.95\pi/[(W/D)(1 - h/D) + C_e]$$ $$Z_{0o} = 59.95\pi/[(W/D)(1 - h/D) + (W/h) - C_0]$$ $$C_e = 0.4413 - H/\pi; \quad C_0 = HD/h\pi$$ $$H = \ln(1 - h/D) + \ln(h/D)/(D/h - 1)$$	$W/h > 0.35$ $W/D \geqq 0.35(1 - h/D)$

Table 4, *continued.*

No.	Structure	Cross Section	Impedance Formulas*	Range of Validity
27	Square coaxial line		$Z_0 = \dfrac{47.086(1 - W/D)}{0.279 + 0.721\,W/D}$; $W/D > 0.25$ $Z_0 = 59.37\ln(0.9259\,D/W)$; $W/D \leqq 0.5$	Less than 1%; no limit
28	Rectangular coaxial line		$Z_0 = 59.95\ln\left(\dfrac{1 + W'/D}{W/D + t/D}\right)$	10% for $t/D < 0.3$ $W/W' < 0.8$

*Air dielectric is assumed for the formulas; for transmission lines with material of relative dielectric constant ϵ_r, the impedance is $Z_0' = Z_0/\sqrt{\epsilon_r}$.

$^\dagger K(x) = \begin{cases} \pi^{-1}\ln[2(1+\sqrt{x})/(1-\sqrt{x})]; & 1/\sqrt{2} \leqq x \leqq 1 \\ \{\pi/\ln\{2[1 + (1 - x^2)^{1/4}]/[1 - (1 - x^2)^{1/4}]\}; & 0 \leqq x \leqq 1/\sqrt{2} \end{cases}$

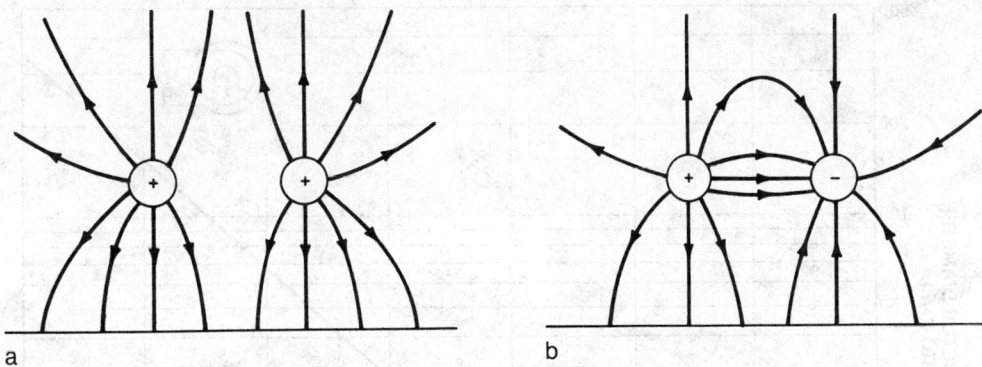

Fig. 3. Schematic representation of even-mode and odd-mode excitation for a typical pair of coupled transmission lines. (*a*) Even-mode excitation. (*b*) Odd-mode excitation.

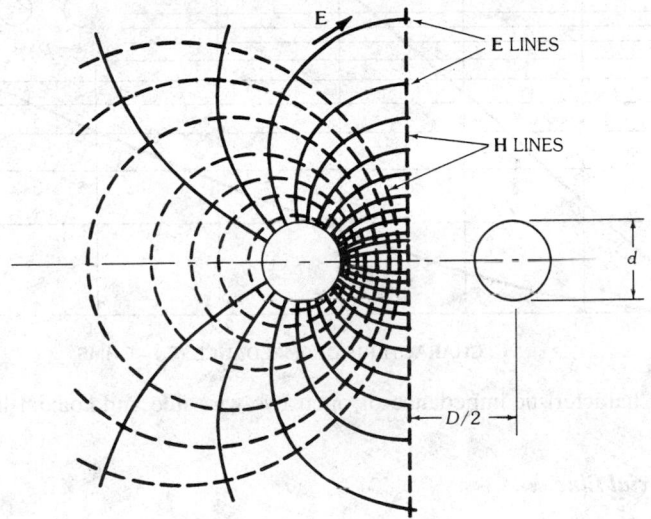

Fig. 4. Cross section and field configuration of a two-wire line.

$$\alpha_c = 5.274 \times 10^{-6} \sqrt{\frac{\epsilon_r f}{\sigma}} \frac{1}{d[\cosh^{-1}(D/d)]} \frac{D}{\sqrt{D^2 - d^2}} \tag{6}$$

where σ is the conductivity of the conductors.

The power-handling capacity in watts is

$$P = \frac{E_b^2 d^2 \sqrt{\epsilon_r}}{239.81} \frac{D - d}{D + d} \cosh^{-1}\left(\frac{D}{d}\right) \tag{7}$$

where E_b, the maximum allowable field intensity, occurs at the innermost surface of each conductor.

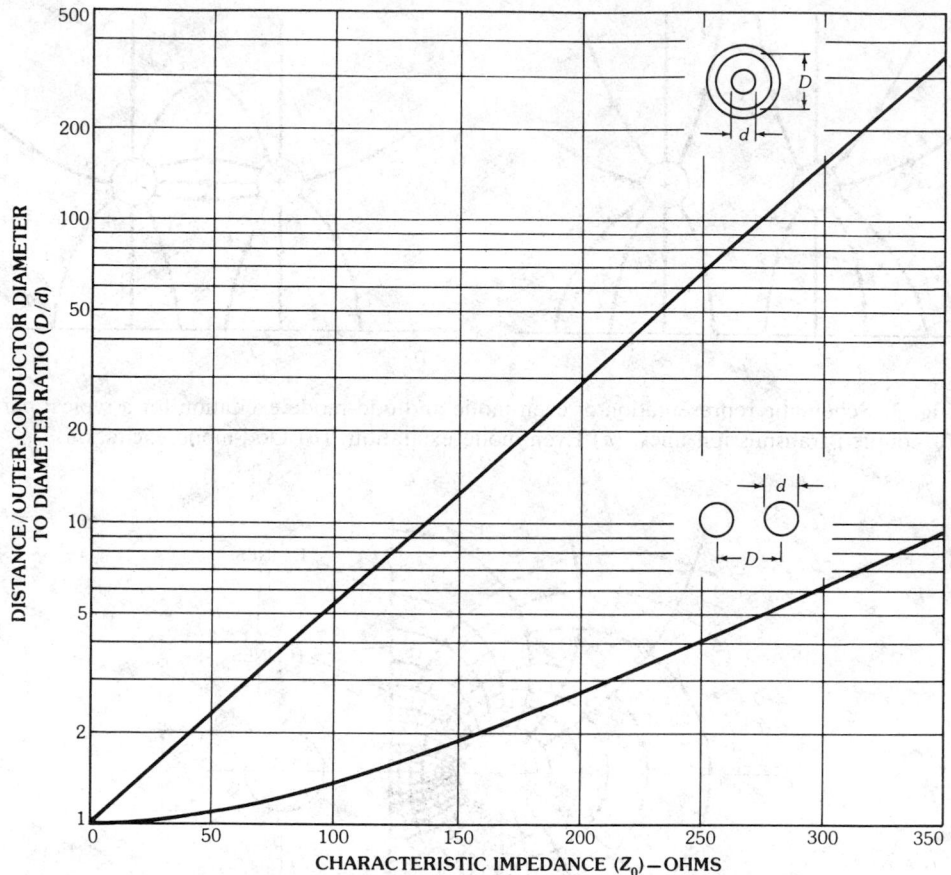

Fig. 5. Characteristic impedance of open two-wire line and coaxial line in air.

Circular Coaxial Line

The circular coaxial line consists of two cylindrical conductors which are located coaxially, one within the other, as shown in Fig. 6. For a line with inner-conductor diameter d and outer-conductor diameter D, the characteristic impedance in ohms is

$$Z_0 = \frac{59.95}{\sqrt{\epsilon_r}} \ln\left(\frac{D}{d}\right) \tag{8}$$

This relation is also plotted in Fig. 5.

The cutoff wavelength in meters of the first higher-order mode is given approximately by

$$\lambda_c = \frac{\pi}{2}\sqrt{\epsilon_r}(D + d) \tag{9}$$

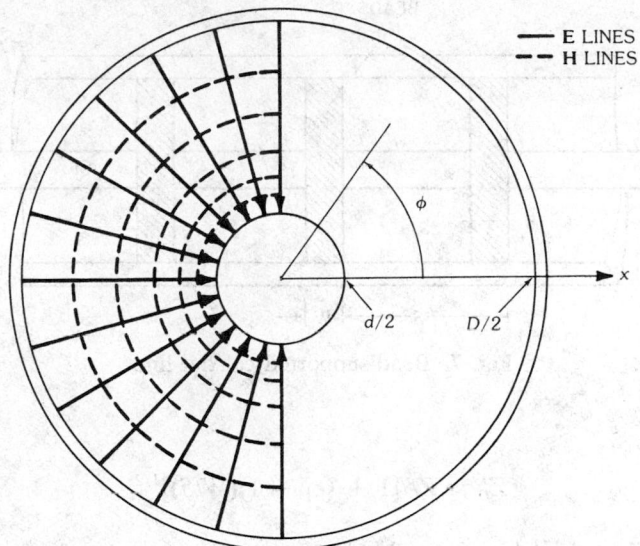

Fig. 6. Cross section and field configuration of a coaxial line.

The conductor loss in nepers per meter for the fundamental mode is

$$\alpha_c = 5.274 \times 10^{-6} \sqrt{\frac{\epsilon_r f}{\sigma}} \left(\frac{1}{D} + \frac{1}{d}\right) \frac{1}{\ln(D/d)} \tag{10}$$

The minimum conductor loss occurs when $D/d = 3.5911$ and is given in nepers per meter by

$$\alpha_{c,\min} = 1.894 \times 10^{-5} \frac{1}{D} \sqrt{\frac{\epsilon_r f}{\sigma}} \tag{11}$$

The power-handling capacity in watts is

$$P = \frac{E_b^2 d^2 \sqrt{\epsilon_r}}{479.6} \ln\left(\frac{D}{d}\right) \tag{12}$$

where E_b is the electric field intensity at the surface of the center conductor. For a fixed frequency the maximum power-handling capacity occurs when $D/d = \sqrt{e} = 1.6487$ and is

$$P_{\max} = 1.043 \times 10^{-3} E_b^2 d^2 \sqrt{\epsilon_r} \tag{13}$$

For air-dielectric coaxial lines, insulating beads are often used to support the center conductor, as shown in Fig. 7. The use of the beads, with relative dielectric constant ϵ_1, changes the line constants. For thin beads at frequent intervals, i.e., $W \ll S \ll \lambda/4$, the new characteristic impedance Z_0' in ohms is

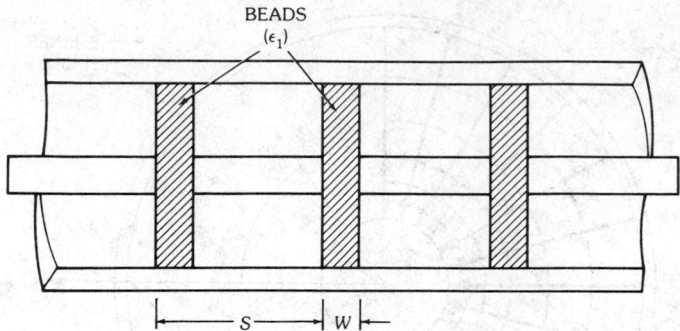

Fig. 7. Bead-supported coaxial line.

$$Z_0' = Z_0/[1 + (\epsilon_1 - 1)(W/S)]^{1/2} \tag{14}$$

"Flexible" coaxial line is probably the most common means of connecting together many separated components of a radio-frequency (rf) system. It consists of a solid or stranded inner conductor, a plastic-dielectric support, and a braided outer conductor. Although the transmission loss is relatively high because of the dielectric loss, the convenience often outweighs this factor in applications where some loss is tolerable. Table 5 summarizes the characteristics of the more common lines.

Figs. 8 and 9 illustrate the approximate attenuation and power-handling capacity of general-purpose rf coaxial cables up to their practical upper frequency limit. For the RG-type cables, only the number is given (for instance, the curve for RG-218/U is labeled 218). The curves for rigid copper coaxial lines are labeled with the diameter of the line only, as $\frac{7}{8}$ in C. Typical curves are also shown for three sizes of 50-Ω semirigid cables labeled by size in inches, as $\frac{7}{8}$ in S. In Fig. 9 the curves are for unity voltage standing-wave ratio. For polyethylene cables an inner-conductor maximum temperature of 80°C is specified. For high-temperature cables (types 211, 228, 225, and 227) the inner-conductor temperature is 250°C. An ambient temperature of 40°C is assumed.

Triplate Stripline

The cross section and field configuration of a triplate stripline is shown in Fig. 10a, where ϵ_r is the relative dielectric constant of the medium filling the entire space between the two ground plates. Practical stripline circuits are frequently constructed from copper-clad printed-circuit boards.

When $t = 0$, the characteristic impedance and a design formula are given by [4]

$$Z_0 = \frac{29.98}{\sqrt{\epsilon_r}} \ln\{1 + \tfrac{1}{2}\chi[\chi + \sqrt{\chi^2 + 6.27}\,]\} \tag{15a}$$

$$\frac{W}{h} = \frac{16}{\pi} \frac{\sqrt{\exp(4\pi r) + 0.568}}{\exp(4\pi r) - 1} \tag{15b}$$

Table 5. List of Radio-Frequency Cables (*Adapted from Jordan [1], © 1985 Howard W. Sams & Company; reprinted with permission*)

Class of Cable		JAN Type RG-	Inner Conductor*	Dielectric Material†	Nominal Diameter of Dielectric (in)	Shielding Braid*	Protective Covering‡	Nominal Overall Diameter (in)	Weight (lb/ft)	Nominal Capacitance (pF/ft)	Maximum Operating Voltage (rms)	Remarks
50 Ω	Single-braid	58C/U	19/0.0071 tc	A	0.116	tc	II	0.195	0.029	28.5	1900	
		213/U	7/0.0296 c	A	0.285	c	II	0.405	0.120	29.5	5000	RG-213/U with armor
		215/U	7/0.0296 c	A	0.285	c	IIa	0.475max	0.160	29.5	5000	Low-attenuation high-power
		218/U	0.195 c	A	0.680	c	II	0.870	0.491	29.5	11000	RG-218/U with armor
		219/U	0.195 c	A	0.680	c	IIa	0.945max	0.603	29.5	11000	Low-attenuation high-power
		220/U	0.260 c	A	0.910	c	II	1.120	0.745	29.5	14000	RG-220/U with armor
		221/U	0.260 c	A	0.910	c	IIa	1.195max	0.925	29.5	14000	Low-attenuation high-power
	Double-braid	55B/U	0.032 sc	A	0.116	sc	III	0.206	0.032	28.5	1900	
		212/U	0.0556 sc	A	0.185	sc	II	0.332	0.093	28.5	3000	
		214/U	7/0.0296 sc	A	0.285	sc	II	0.425	0.158	30.0	5000	
		217/U	0.106 c	A	0.370	c	II	0.545	0.236	29.5	7000	
		223/U	0/035 sc	A	0.116	sc	II	0.216	0.036	28.5	1900	
		224/U	0.106 c	A	0.370	c	IIa	0.615max	0.282	29.5	7000	RG-217/U with armor
75 Ω	Single-braid	11A/U	7/26 AWG tc	A	0.285	c	II	0.412	—	20.5	5000	RG-11A/U with armor
		12A/U	7/26 AWG tc	A	0.285	c	IIa	0.475	—	20.5	5000	High-power low-attenuation
		34B/U	7/0.0249 c	A	0.460	c	II	0.630	0.231	21.5	6500	High-power low-attenuation
		35B/U	0.1045 c	A	0.680	c	IIa	0.945max	0.480	21.5	10000	
		59B/U	0.0230 ccs	A	0.146	c	II	0.242	—	21.0	2300	
		84A/U	9.1045 c	A	0.680	c	II*	1.000	1.325	21.5	10000	*Same as RG-35B/U with lead sheath
		85A/U	0.1045 c	A	0.680	c	II*	1.565max	2.910	21.5	10000	*Same as RG-84A/U, with special armor for underground installations
		164/U	0.1045 c	A	0.680	c	II	0.870	0.490	21.5	10000	RG-35B/U without armor
		307A/U	17/0.0058 sc	A*	0.029	sc	III	0.270	—	20.0	400	*Foamed
	Double-braid	6A/U	21 AWG ccs	A	0.185	In: sc	II	0.332	—	20.0	2700	*Foamed
		216/U	7/0.0159 tc	A	0.285	Out: c	II	0.425	0.121	20.5	5000	

Table 5, *continued.*

Class of Cable	JAN Type RG-	Inner Conductor*	Dielectric Material†	Nominal Diameter of Dielectric (in)	Shielding Braid*	Protective Covering‡	Nominal Overall Diameter (in)	Weight (lb/ft)	Nominal Capacitance (pF/ft)	Maximum Operating Voltage (rms)	Remarks
High temperature Single-braid	144/U	7/0.179 sccs	F1	0.285	sc	X	0.410	0.120	20.5	5000	Z = 75 Ω
	178B/U	7/0.004 sccs	F1	0.034	sc	IX	0.075max	—	29.0	1000	Z = 50 Ω
	179B/U	Same as above	F1	0.063	sc	IX	0.105	—	20.0	1200	Z = 95 Ω
	180B/U	Same as above	F1	0.102	sc	IX	0.145	—	15.5	1500	Z = 95 Ω
	187A/U	7/0.004 ascss	F1	0.060	sc	VII	0.110	—	—	1200	Z = 75 Ω
	195A/U	Same as above	F1	0.102	sc	VII	0.155	—	—	1500	Z = 95 Ω
	196A/U	Same as above	F1	0.034	sc	VII	0.080	—	—	1000	Z = 50 Ω
	211A/U	0.190 c	F1	0.620	c	X	0.730	0.450	29.0	7000	Semiflexible: operating at −55 to +200°C. Z = 50 Ω
	228A/U	0.190 c	F1	0.620	c	Xa	0.795	0.600	29.0	7000	RG-211A/U with armor. Z = 50 Ω
	302/U	0.025 sc	F1	0.146	sc	IX	0.206	—	21.0	2300	Z = 75 Ω
	303/U	0.039 sccs	F1	0.116	sc	IX	0.170	—	28.5	1900	Z = 50 Ω
	304/U	0.059 sccs	F1	0.185	sc	IX	0.280	—	28.5	3000	Z = 50 Ω
	316/U	7/0.0067 ascss	F1	0.060	sc	IX	0.102	—	—	1200	Z = 50 Ω
Double-braid	115/U	7/0.028 sc	F2	0.250	sc	X	0.375	—	29.5	5000	Slightly expandable. Z = 50 Ω
	142B/U	0.039 sccs	F1	0.116	sc	IX	0.195	—	28.5	1900	Z = 50 Ω
	225/U	7/0.0312 sc	F1	0.285	sc	X	0.430	0.176	29.5	5000	Semiflexible; operating at −55 to +200°C. Z = 50 Ω
	226/U	19/0.0254 sc	F2	0.370	c	X	0.500	0.247	29.0	7000	Slightly expandable. Z = 50 Ω
	227/U	7/0.0312 sc	F1	0.285	sc	Xa	0.490	0.224	29.5	5000	RG-225/U with armor. Z = 50 Ω
Pulse Single-braid	26A/U	19/0.0117 tc	E	0.288	tc	IVa	0.505	0.189	50.0	10000	Z = 48 Ω
	27A/U	19/0.0185 tc	D	0.455	tc	IVa	0.670	0.304	50.0	15000 peak	Z = 48 Ω

Category	Type	Conductor	Jacket	Dielectric	Shield					Voltage	Remarks
Double-braid	25A/U	19/0.0117 tc	E	IV	tc	0.288	0.505	0.205	50.0	10000	Z = 48 Ω
	28B/U	19/0.0185 tc	D	IV	In: tc Out: gs	0.455	0.750	0.370	50.0	15000 peak	Z = 48 Ω
	64A/U	19/0.0117 tc	E	IV	tc	0.288	0.475max	0.205	50.0	10000	Z = 48 Ω
	156/U	7/21 AWG tc	AH*	II	In: tc Out: gs	0.285	0.540	0.211	30.0	10000	*1st layer A; 2nd layer H. Z = 50 Ω
	157/U	19/24 AWG tc	HAH*	II	↑	0.455	0.725	0.317	38.0	15000	*1st layer H; 2nd layer A; 3rd layer H. Z = 50 Ω
	158/U	37/21 AWG tc	HAH	II	↑	0.455	0.725	0.380	78.0	15000	Z = 25 Ω
	190/U	19/0.0117 tc	HJH*	VIII	↑	0.380	0.700	0.353	50.0	15000	Z = 50 Ω *1st layer H; 2nd layer J; 3rd layer H.
	191/U	30 AWG c	HJH	VIII	↑	1.065	1.460	1.469	85.0	15000	Z = 25 Ω
Four braids	88/U	19/0.0117 tc	E	III	tc	0.288	0.515	—	50.0	10000	Z = 48 Ω
Low capacitance	Single-braid 62A/U	0.0253 ccs	A	II	c	0.146	0.242	0.382	14.5	750	Z = 93 Ω
	63B/U	0.0253 ccs	A	II	c	0.285	0.405	0.082	10.0	1000	Low-capacitance. Z = 125 Ω
	79B/U	0.0253 ccs	A	IIa	c	0.285	0.475max	0.138	10.0	1000	RG-63B/U with armor. Z = 125 Ω
	Double-braid 71B/U	0.0253 ccs	A	III	tc	0.146	0.250max	—	14.5	750	Low-capacitance. Z = 93 Ω
High attenuation	Single-braid 301/U	7/0.0203 Karma wire	F1	IX	Karma wire	0.185	0.245	—	29.0	3000	High-attenuation. Z = 50 Ω
High delay	Single-braid 65A/U	No. 32 Formex F. Helix diam. 0.128	A	II	c	0.285	0.405	0.096	44.0	1000	High-impedance high-delay line. Z = 950 Ω

*Diameter of strands given in inches (does not apply to AWG types). As, 7/0.0296 = 7 strands, each 0.0296-inch diameter. Conductors: c = copper, gs = galvanized steel, sc = silvered copper, tc = tinned copper, ccs = copper-covered steel, sccs = silvered copper-covered steel, ascss = annealed silvered copper-covered steel.

†Dielectric materials: A = polyethylene, D = layer of synthetic rubber between two layers of conducting rubber, E = layer of conducting rubber plus two layers of synthetic rubber, F1 = solid polytetrafluoroethylene (Teflon), F2 = semisolid or taped polytetrafluoroethylene (Teflon), H = conducting synthetic rubber, and J = insulating butyl rubber.

‡Jacket types: I = polyvinyl chloride (colored black), II = noncontaminating synthetic resin, III = noncontaminating synthetic resin (colored black), IV = chloroprene, V = Fiberglas silicone-impregnated varnish, VII = polytetrafluoroethylene, VIII = polychloroprene, IX = fluorinated ethylene propylene, and X = Teflon-tape moisture seal with double-braid type V jacket; the letter "a" means "with armor," e.g., Ia = I, with armor.

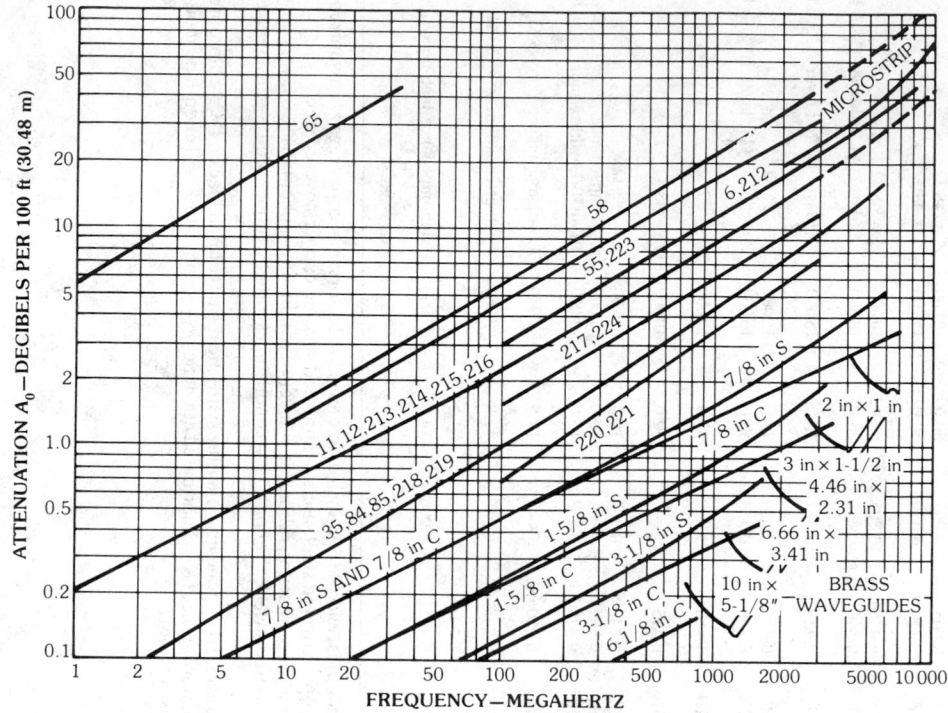

Fig. 8. Attenuation of cables. (*After E. Jordan*, Reference Data for Radio Engineers, *6/e*, © *1965 Howard W. Sams & Company, reproduced with permission*)

where

$$\chi = 16h/\pi W$$
$$r = Z_0 \sqrt{\epsilon_r}/376.7$$

The relative error in Z_0 is less than 0.005 for $W/h < 20$.

When $t > 0$, the thickness effect may be taken into account by replacing the original strip with an equivalent infinitely thin strip of width $W' = W + \Delta W$, where

$$\Delta W = \frac{t}{\pi}\left\{1 - \frac{1}{2}\ln\left[\left(\frac{1}{4h/t + 1}\right)^2 + \left(\frac{1/4\pi}{W/t + 1.1}\right)^m\right]\right\} \tag{16a}$$

or

$$\Delta W = \frac{t}{\pi}\left\{1 - \frac{1}{2}\ln\left[\left(\frac{1}{4h/t + 1}\right)^2 + \left(\frac{1/4\pi}{W'/t - 0.26}\right)^m\right]\right\} \tag{16b}$$

in which

$$m = \frac{6}{3 + t/h}$$

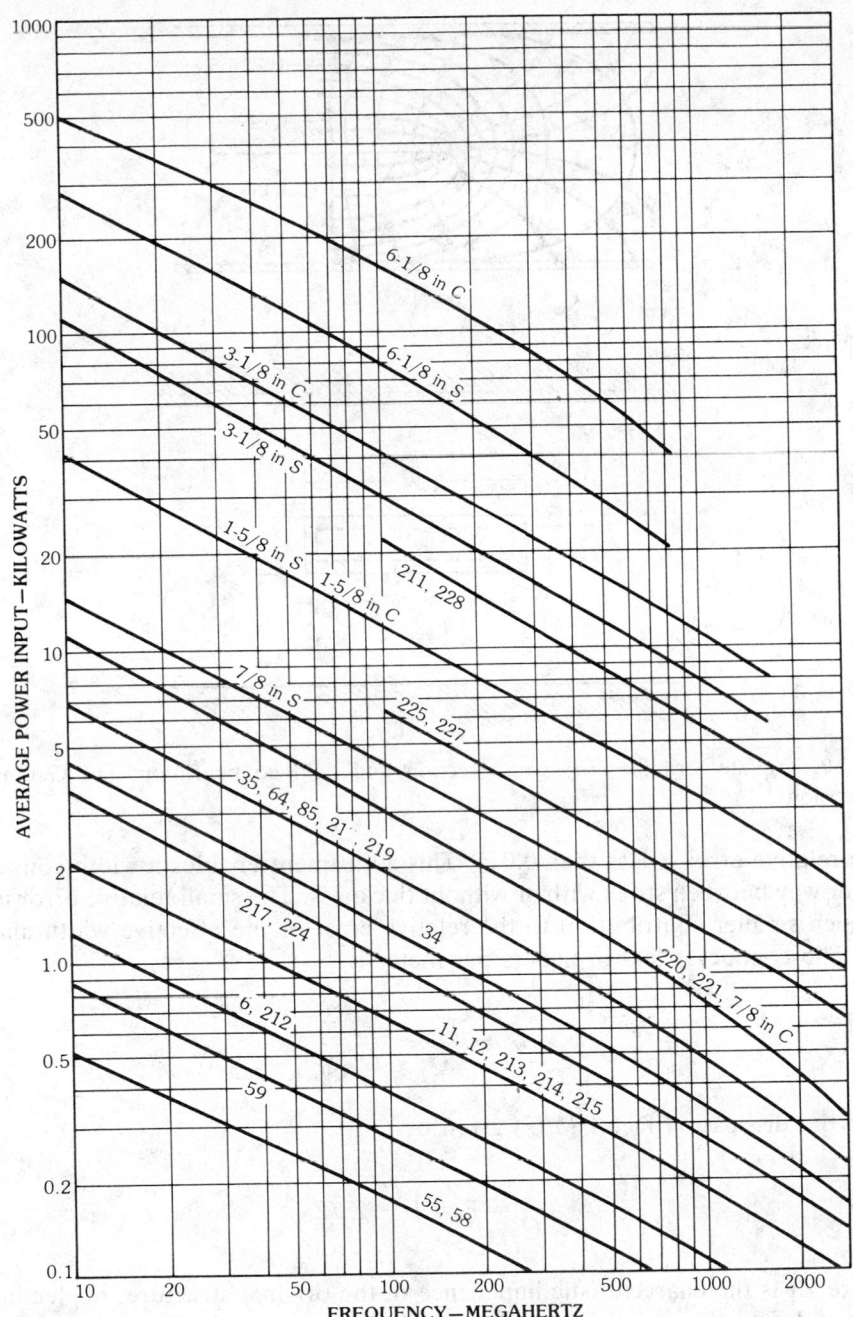

Fig. 9. Power rating of cables. (*After E. Jordan*, Reference Data for Radio Engineers, *6/e*, © *1965 Howard W. Sams & Company, reproduced with permission*)

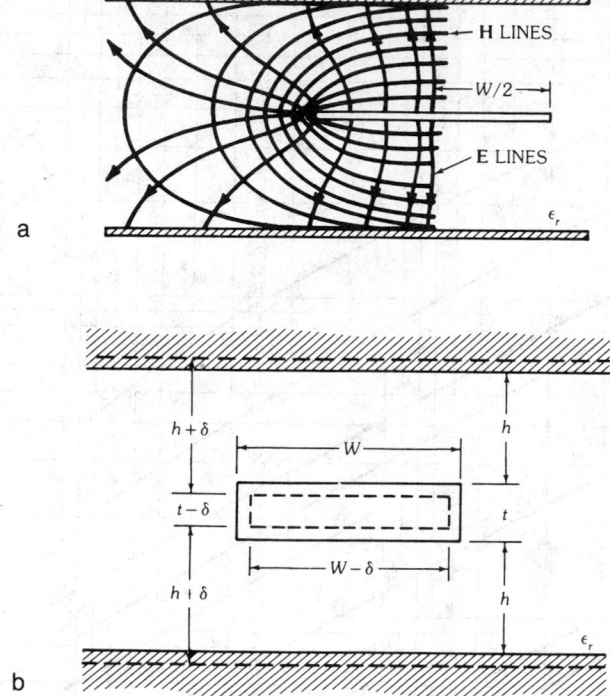

Fig. 10. Triplate stripline. (*a*) Cross section and field configuration. (*b*) Geometrical dimensions.

The relative error is less than 0.015. This adjustment enables a width conversion either way between strips with or without thickness. The small relative error makes a much smaller contribution to the relative error of the effective width and Z_0.

The conductor loss in nepers per meter is

$$\alpha_c = \frac{\beta'}{2Q} = \frac{\sqrt{\epsilon_r}}{Q} \frac{\pi}{\lambda_0} \tag{17}$$

with the dissipation factor $(1/Q)$ given by

$$\frac{1}{Q} = P = 1 - Z_1/Z_\delta \tag{18}$$

where Z_1 is the characteristic impedance of the original structure, neglecting the loss, and Z_δ is the impedance of the modified structure as indicated by the dashed line in Fig. 10b, where δ is the skin depth defined by $\delta = \sqrt{2/\omega\mu_0\sigma}$.

The characteristic impedance and the dissipation factor of the triplate stripline are computed and plotted in Figs. 11 and 12.

The power-handling capacity of stripline is limited by field concentration at sharp corners of the conducting strip. Rounded-edge stripline (e.g., no. 24, Table 4) are necessary for high-power applications.

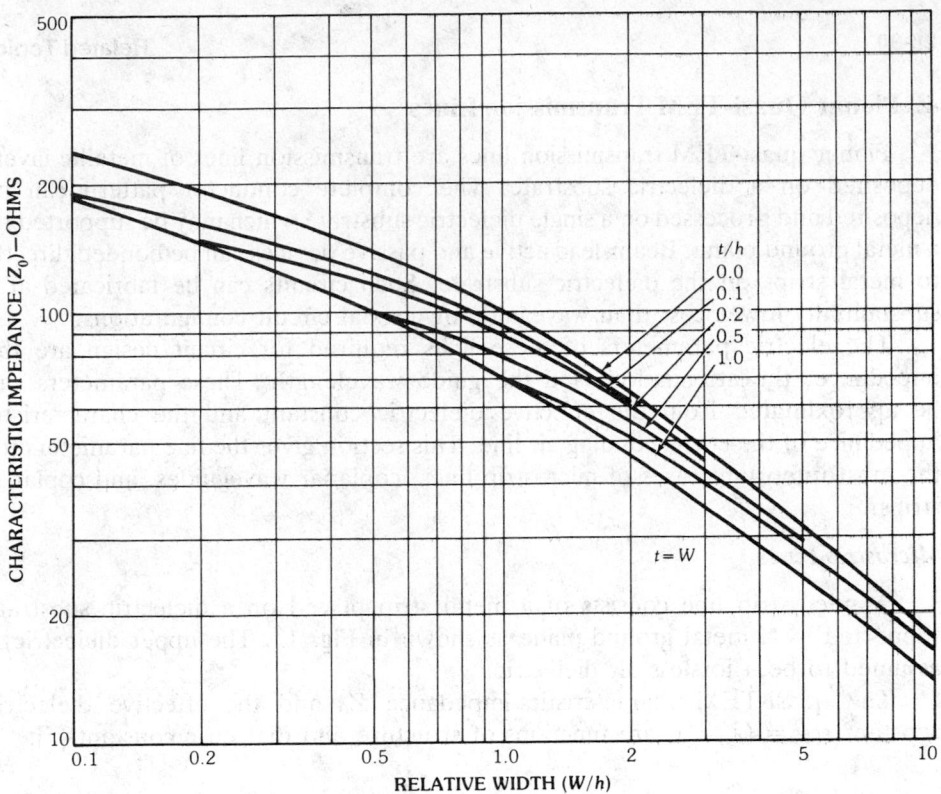

Fig. 11. Characteristic impedance of triplate stripline, showing the effect of thickness.

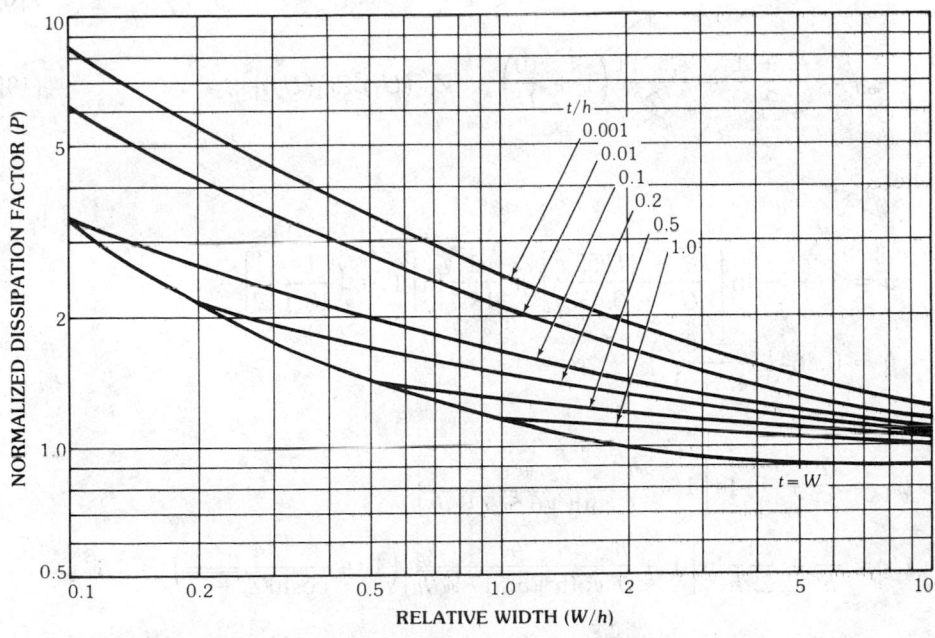

Fig. 12. The normalized dissipation factor of triplate stripline, showing the effect of thickness.

4. Planar Quasi-TEM Transmission Lines

Planar quasi-TEM transmission lines are transmission lines of metallic layers deposited on a dielectric substrate. The complete conductor pattern can be deposited and processed on a single dielectric substrate which may be supported by a metal ground plane. Beam-lead active and passive devices can be bonded directly to metal strips on the dielectric substrate. Such circuits can be fabricated at a substantially lower cost than waveguide or coaxial circuit configurations.

The electric parameters of these lines required for circuit design are the impedance, the attenuation, and the guide wavelength. These parameters can be approximated from the effective dielectric constant and the characteristic impedance of the corresponding air line. This section gives the line parameters for the most important cases of microstrip lines, coplanar waveguides, and coplanar strips.

Microstrip Lines

A microstrip line consists of a metal strip placed on a dielectric substrate supported by a metal ground plane as shown in Fig. 13. The upper dielectric is assumed to be a lossless air dielectric.

The quasi-TEM characteristic impedance Z_0 and the effective dielectric constant $\epsilon_{\text{eff}} = (\lambda_0/\lambda_g)^2$ are functions of structure and dielectric constant [5]:

$$Z_0\sqrt{\epsilon_{\text{eff}}} = Z_{01}(U_1) = 59.95\ln\left\{\frac{6}{U_1} + \frac{2\pi-6}{U_1}\exp\left[-\left(\frac{30.666}{U_1}\right)^{0.7528}\right] + \sqrt{1+\left(\frac{2}{U_1}\right)^2}\right\}$$

(19a)

$$\epsilon_{\text{eff}} = \frac{\epsilon_r + 1}{2} + \frac{\epsilon_r - 1}{2}\left(1 + \frac{10}{U}\right)^{-ab}[Z_{01}(U_1)/Z_{01}(U_2)]^2$$

(19b)

where

$$a = 1 + \frac{1}{49}\ln\left[\frac{U^4 + (U/52)^2}{U^4 + 0.432}\right] + \frac{1}{18.7}\ln\left[1 + \left(\frac{U}{18.1}\right)^3\right]$$

$$b = 0.564\left(\frac{\epsilon_r - 0.9}{\epsilon_r + 3}\right)^{0.053}$$

$$U = W/h$$

$$U_1 = \frac{W}{h} + \frac{t}{\pi h}\ln\left[1 + \frac{h}{t}\frac{4e}{\coth^2\sqrt{6.517\,W/h}}\right]$$

$$U_2 = \frac{W}{h} + \frac{t}{2\pi h}\ln\left[1 + \frac{h}{t}\frac{4e}{\coth^2\sqrt{6.517\,W/h}}\right]\left(1 + \frac{1}{\cosh\sqrt{\epsilon_r - 1}}\right)$$

For $t = 0$, the accuracy of these formulas is on the order of 0.1 percent. The computed data are plotted in Fig. 14.

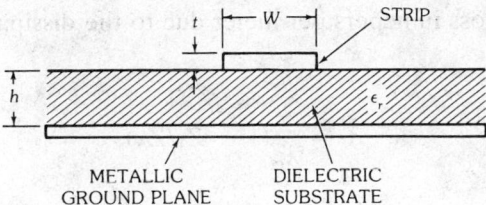

Fig. 13. Cross section of microstrip line. (*After Jordan [1], © 1985 Howard W. Sams & Company; reproduced with permission*)

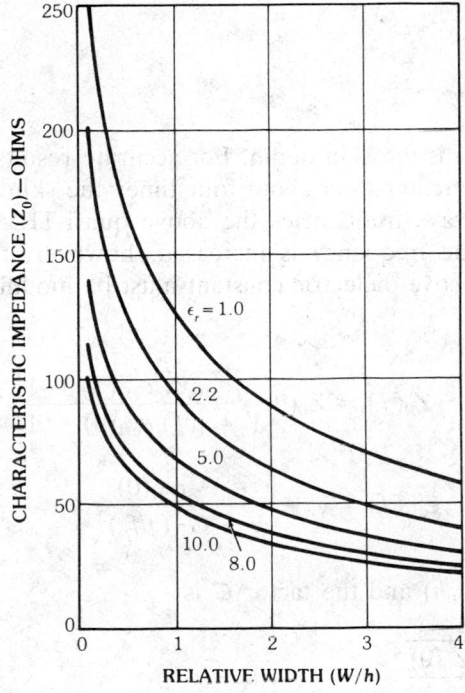

Fig. 14. Characteristic impedance of microstrip line, calculated from quasi-TEM formulas by Hammerstadt and Jansen. (*After Jordan [1], © 1985 Howard W. Sams & Company; reproduced with permission*)

The dielectric loss in a microstrip line in nepers per meter is [6]

$$\alpha_d = \frac{\pi}{\lambda_g}(\tan\delta)_{\text{eff}} = \frac{\pi}{\lambda_g}\left[\tan\delta \Big/ \left(1 + \frac{1}{\epsilon_r}\frac{F-1}{F+1}\right)\right] \qquad (20)$$

where

$$F = \sqrt{1 + 10\,h/W}$$

and the guide wavelength λ_g is calculated from (19b).

The conductor loss in nepers per meter due to the dissipation in the strip and ground plane is

$$\alpha_c = \frac{\pi}{\lambda_g}(1 - Z_1/Z_\delta) \tag{21}$$

where Z_1 is the characteristic impedance of the corresponding air-filled lossless microstrip line obtained by (19), and Z_δ is calculated with the same formula by replacing W, h, and t with W', h', and t' such that

$$\begin{aligned} W' &= W - \delta \\ h' &= h + \delta \\ t' &= t - \delta \end{aligned} \tag{22}$$

where δ $(= \sqrt{2/\omega\mu_0\sigma})$ is the skin depth. For accurate results the thickness of the conductors must be greater than about four times the skin depth.

At lower microwave frequencies the above quasi-TEM approximations are very accurate. As the frequency is increased, however, the characteristic impedance and the effective dielectric constant must be modified to be functions of frequency:

$$Z_0(f) = Z_0(0)\sqrt{\frac{\epsilon_{\text{eff}}(0)}{\epsilon_{\text{eff}}(f)}\frac{\epsilon_{\text{eff}}(f) - 1}{\epsilon_{\text{eff}}(0) - 1}} \tag{23a}$$

$$\epsilon_{\text{eff}}(f) = \epsilon_r - \frac{\epsilon_r - \epsilon_{\text{eff}}(0)}{1 + G \cdot (f/f_p)^2} \tag{23b}$$

where $f_p = Z_0(0)/(2\mu_0 h)$ and the factor G is

$$G = \frac{\pi^2}{12}\frac{\epsilon_r - 1}{\epsilon_{\text{eff}}(0)}\sqrt{\frac{Z_0(0)}{59.95}}$$

In the above, $Z_0(0)$ and $\epsilon_{\text{eff}}(0)$ are dc values obtained from (19).

The power-handling capability of the microstrip line is limited by heating due to ohmic and dielectric losses and by dielectric breakdown. The increase in temperature due to losses limits the average power of the line, while the breakdown between the strip conductor and ground plate limits the peak power. Some theoretical values of average power-handling capability for various substrates at 2 GHz, 10 GHz, and 20 GHz are given in Table 6.

The peak power is limited primarily by the sharp edges of the line due to electric field concentration. To increase the breakdown voltage, one may use thick strip conductors with rounded edges. For a microstrip line composed of a strip 7/32 in (5.55 mm) wide on a Teflon-impregnated Fiberglas base 1/16 in (1.58 mm) thick, corona effects appear at the edge of the strip conductor for pulse power of roughly 10 kW at 9 GHz.

Table 6. Average Power-Handling Capacity for 50-ohm Microstrip Lines

Substrate	Maximum Average Power (kW)*		
	2 GHz	10 GHz	20 GHz
Alumina	12.1	5.2	3.4
Beryllium oxide	174.5	75.7	51.5
Gallium arsenide	3.5	1.5	0.93
Polystyrene	0.32	0.12	0.075
Quartz	1.2	0.52	0.36
RT/Duroid 5880	0.29	0.15	0.049
Sapphire	11.6	5.1	3.5
Silicon	3.2	2.2	1.6

*Maximum temperature = 100°C; ambient temperature = 25°C.

Coplanar Waveguides

Coplanar waveguide consists of a center strip with two ground planes located on the top surface of a dielectric substrate, as shown in Fig. 15. Under the quasi-TEM approximation, the characteristic impedance and the effective dielectric constant are [6]

$$Z_0 = \frac{30\pi}{\sqrt{\epsilon_{\text{eff}}}} \frac{1}{K(k_e)} \tag{24a}$$

$$\epsilon_{\text{eff}} = \left(\frac{\lambda_0}{\lambda_g}\right)^2 = \epsilon_e - \frac{0.7(\epsilon_e - 1)t/W}{K(k) + 0.7\,t/W} \tag{24b}$$

where

$$K(x) = \begin{cases} \dfrac{1}{\pi}\ln\left(2\dfrac{1 + \sqrt{x}}{1 - \sqrt{x}}\right), & \sqrt{1/2} \leqq x \leqq 1 \\[2ex] \pi/\ln\left[2\dfrac{1 + (1 - x^2)^{1/4}}{1 - (1 - x^2)^{1/4}}\right], & 0 \leqq x \leqq \sqrt{1/2} \end{cases} \tag{24c}$$

$$\epsilon_e = \frac{\epsilon_r + 1}{2}\,\{\tanh[0.775\ln(h/W) + 1.75] +$$
$$(kW/h)[0.04 - 0.7k + 0.01(1 - 0.1\epsilon_r)(0.25 + k)]\}$$

$$k_e = \frac{S_e}{S_e + 2W_e} = \frac{S + \Delta}{S + 2W - \Delta}$$

$$k = \frac{S}{S + 2W}$$

$$\Delta = (1.25t/\pi)[1 + \ln(4\pi S/t)]$$

Note that as $t \to 0$, $\Delta \to 0$. For $t = 0$, the accuracy of these expressions is better than 1.5 percent for $h/W \geqq 1$. The characteristic impedance computed from the above formulas is plotted in Fig. 16.

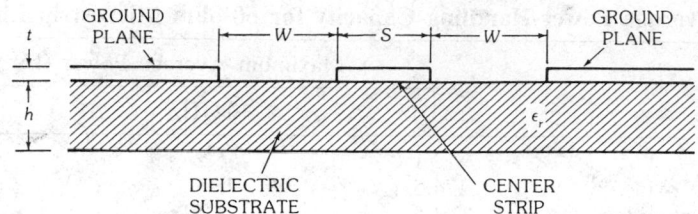

Fig. 15. Cross section of coplanar transmission line. (*After Jordan [1], © 1985 Howard W. Sams & Company; reproduced with permission*)

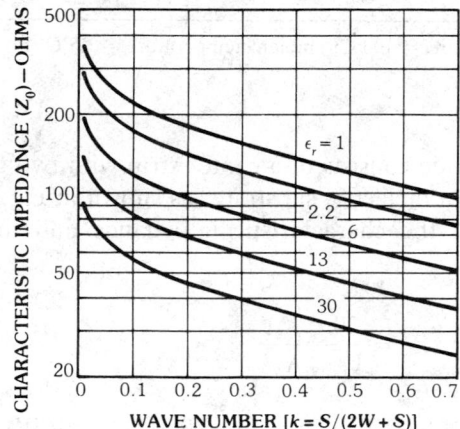

Fig. 16. Characteristic impedance of coplanar transmission line calculated from formula (24) for $t = 0$. (*After Jordan [1], © 1985 Howard W. Sams & Company; reproduced with permission*)

The dielectric loss of this line in nepers per meter is

$$\alpha_d = \frac{\pi \tan \delta}{\lambda_g} \frac{\epsilon_r}{\epsilon_{\text{eff}}} \frac{\epsilon_{\text{eff}} - 1}{\epsilon_r - 1} \tag{25}$$

The conductor loss in nepers per meter is

$$\alpha_c = \frac{\pi}{\lambda_g} (1 - Z_1/Z_\delta) \tag{26}$$

where Z_1 is the characteristic impedance of the line when losses are neglected, Z_δ is the impedance when the dimensions S, W, and t are replaced by S', W', and t' such that

$$S' = S - \delta$$
$$W' = W + \delta \tag{27}$$
$$t' = t - \delta$$

and δ is the skin depth of the conductor.

Coplanar Strips

A configuration of coplanar strips which is complementary to coplanar waveguide is shown in Fig. 17. It consists of two equal-width strips running parallel on the same surface of a dielectric slab.

The characteristic impedance in ohms and the effective dielectric constant of the line are

$$Z_0 = \frac{120\pi}{\sqrt{\epsilon_{\text{eff}}}} K(k_e) \tag{28a}$$

$$\epsilon_{\text{eff}} = \left(\frac{\lambda_0}{\lambda_g}\right)^2 = \epsilon_e - \frac{1.4(\epsilon_e - 1)t/S}{1/K(k) + 1.4t/S} \tag{28b}$$

where $K(x)$, ϵ_e, and k have the same expressions as defined in (24), and

$$k_e = \frac{S - \Delta}{S + 2W + \Delta}$$

$$\Delta = (1.25t/\pi)[1 + \ln(4\pi W/t)]$$

Notice that W is now the strip width and S is the spacing between the strips.

The expressions for the dielectric loss and conductor loss in coplanar strips are given in (25) and (26). But in this case S', W', and t' are

$$S' = S + \delta$$
$$W' = W - \delta \tag{29}$$
$$t' = t - \delta$$

5. TE/TM Waveguides

The most commonly used waveguides operating in TE/TM modes are the hollow-tube waveguides.

For propagation of energy through a hollow metal tube under fixed conditions, two types of waves, i.e., TE and TM waves, are available. (Note: the TEM waves cannot be propagated in a hollow-tube waveguide.) The possible configurations of the fields, commonly termed "modes," are characterized by the introduction of

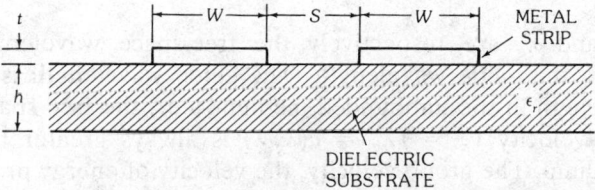

Fig. 17. Cross section of coplanar strips.

integer subscripts m and n, which can take on separate values from 0 or 1 to infinity.

A particular mode can be propagated in a hollow-tube guide only if the frequency is greater than an associated particular value called the cutoff frequency, f_c, for that mode; the corresponding free-space wavelength is called the cutoff wavelength, λ_c. The cutoff frequency is a function of the waveguide dimension, the mode indexes m and n, and the relative dielectric constant ϵ_r of the medium enclosed by the waveguide walls.

As mentioned in Section 2, the transmission line equations are valid for single-mode waveguides, except for those expressions involving the quantities R, L, G, and C. The transmission line description of a waveguide mode is based on the fact that the transverse electric field $\mathbf{E}_t$ and transverse magnetic field $\mathbf{H}_t$ of each mode can be expressed as

$$\mathbf{E}_t(x, y, z) = V(z)\,\mathbf{e}(x, y) \tag{30a}$$

$$\mathbf{H}_t(x, y, z) = I(z)\,\mathbf{h}(x, y) \tag{30b}$$

where $\mathbf{e}(x, y)$ and $\mathbf{h}(x, y)$ are vector functions indicative of the cross-sectional form of the mode field, and $V(z)$ and $I(z)$ are voltage and current functions that measure the rms amplitudes of the transverse electric and magnetic fields at any point z along the direction of propagation. The voltage and current are described by the transmission line equations given in (1), where, for a medium of uniform dielectric constant,

$$\gamma = \alpha + j\beta = \sqrt{k_c^2 - k^2} \tag{31}$$

$$Z_0 = \frac{1}{Y_0} = \begin{cases} \dfrac{j\omega\mu_0}{\gamma} & \text{for TE modes} \\[2mm] \dfrac{\gamma}{j\omega\epsilon} & \text{for TM modes} \end{cases} \tag{32}$$

The parameters k_c, k, and Z_0 are respectively termed the cutoff wave number, the free-space wave number, and the characteristic wave impedance of the mode in question.

When the structure is loss-free, $\alpha = 0$ and the following relationships hold:

$$\begin{aligned} k &= 2\pi/\lambda = 2\pi\sqrt{\epsilon_r}/\lambda_0, & k_c &= 2\pi/\lambda_c \\ \beta &= 2\pi/\lambda_g, & \lambda_g &= \lambda/\sqrt{1 - (\lambda/\lambda_c)^2} \end{aligned} \tag{33}$$

where λ, λ_c, and λ_g are respectively the free-space wavelength, the cutoff wavelength, and the guide wavelength. The guide wavelength is always greater than the wavelength in the unbounded medium, i.e., the free-space wavelength.

The phase velocity $v_p = f\lambda_g = c(\lambda_g/\lambda)$ is always greater than that in an unbounded medium. The group velocity, the velocity of energy propagation down the guide, is less than that in an unbounded medium.

The dielectric loss of the medium in a waveguide may be taken into account by introduction of a complex permittivity

$$\epsilon = \epsilon' - j\epsilon'' = (\epsilon_r' - j\epsilon_r'') \epsilon_0 \tag{34}$$

where ϵ' is the dielectric constant and ϵ'' is the loss factor. For a medium having unity relative permeability, the propagation constant of the waveguide is

$$\gamma = \sqrt{\left(\frac{2\pi}{\lambda_c}\right)^2 - \frac{\epsilon}{\epsilon_0}\left(\frac{2\pi}{\lambda_0}\right)^2} \tag{35}$$

In a waveguide having a cutoff wavelength $\lambda_c > \lambda_0$, the attenuation constant in nepers per meter is

$$\alpha_d = \frac{\pi\lambda_g\epsilon_r''}{\lambda_0^2} = \frac{2\pi}{\lambda_{g0}}\sqrt{\frac{-1 + (1 + x^2)^{1/2}}{2}} = \frac{2\pi}{\lambda_{g0}}\sinh\left(\frac{\sinh^{-1}x}{2}\right) \tag{36a}$$

$$\alpha_d \cong \frac{\pi x}{\lambda_{g0}}\left(1 - \frac{x^2}{8} + \cdots\right), \qquad x \ll 1 \tag{36b}$$

and the phase constant in radians per meter is

$$\beta = \frac{2\pi}{\lambda_g} = \frac{2\pi}{\lambda_{g0}}\sqrt{\frac{1 + (1 + x^2)^{1/2}}{2}} = \frac{2\pi}{\lambda_{g0}}\cosh\left(\frac{\sinh^{-1}x}{2}\right) \tag{37a}$$

$$\beta \cong \frac{2\pi}{\lambda_{g0}}\left(1 + \frac{x^2}{8} + \cdots\right), \qquad x \ll 1 \tag{37b}$$

where $x = \epsilon''(\lambda_{g0}/\lambda_0)^2$ and

$$\lambda_{g0} = \lambda_0/\sqrt{\epsilon_r'' - (\lambda_0/\lambda_c)^2}$$

is the guide wavelength neglecting the loss. The approximations (36b) and (37b) are valid for $\epsilon''/\epsilon' \ll 1$, when λ_0 is not too close to the cutoff wavelength λ_c.

The characteristic wave impedance at a certain point has been defined in (32) as the ratio of the total transverse electric-field strength to the total transverse magnetic-field strength. The wave impedance is constant over the cross section of the guide.

In addition to the wave impedance, the "integrated" characteristic impedance, Z_g, of the line is a quantity of great usefulness in connection with ordinary two-conductor transmission lines. For such lines, Z_g can be defined in terms of the voltage-current ratio or in terms of the power transmitted for a given voltage or a given current. That is, for an infinitely long line (or matched line),

$$Z_{VI} = \frac{V}{I}, \qquad Z_{PI} = \frac{2P}{|I|^2}, \qquad Z_{PV} = \frac{|V|^2}{2P} \tag{38}$$

where V and I are the peak values of voltage and current. For TEM lines these definitions are equivalent, but for non-TEM waveguides they lead to three values that depend on the guide dimensions in the same way, but differ by a constant. Of the three, Z_{PV} is most widely used, but Z_{VI} is found to be more nearly correct in matching a coaxial line to a waveguide.

Rectangular Waveguides

A uniform waveguide of rectangular cross section is described by a Cartesian coordinate system shown in Fig. 18. The inner dimensions of the guide are a and b.

For TE_{mn} modes in rectangular waveguides, m and n may take any integer value from 0 to infinity, except the case $m = n = 0$. For the TM_{mn} modes, m and n may take any value from 1 to infinity. Field patterns for some of the lower-order modes are shown in Fig. 19. The dominant mode in the rectangular waveguide is the TE_{10} mode.

The cutoff wavelength of the TE_{mn} or TM_{mn} mode is

$$\lambda_c = \frac{2\sqrt{ab}}{\sqrt{(b/a)m^2 + (a/b)n^2}} \tag{39}$$

The conductor loss for a propagating TE_{mn} mode due to the dissipative waveguide walls is

$$\alpha_c = \frac{5.274 \times 10^{-6}}{b} \sqrt{\frac{\epsilon_r f}{\sigma}} \left[\frac{\epsilon_n m^2 (b/a) + \epsilon_m n^2}{m^2(b/a) + n^2(a/b)} \sqrt{1 - \left(\frac{\lambda}{\lambda_c}\right)^2} + \frac{[\epsilon_n + \epsilon_m(b/a)](\lambda/\lambda_c)^2}{\sqrt{1 - (\lambda/\lambda_c)^2}} \right] \tag{40}$$

where

$\epsilon_m = 1$ if $m = 0$, $\epsilon_m = 2$ if $m \neq 0$

$\epsilon_n = 1$ if $n = 0$, $\epsilon_n = 2$ if $n \neq 0$

σ = conductivity of conductor walls

ϵ_r = relative permittivity of filling dielectric

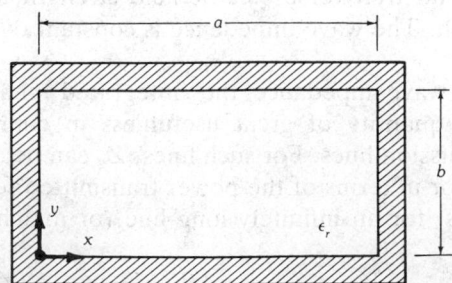

Fig. 18. Rectangular-waveguide cross section.

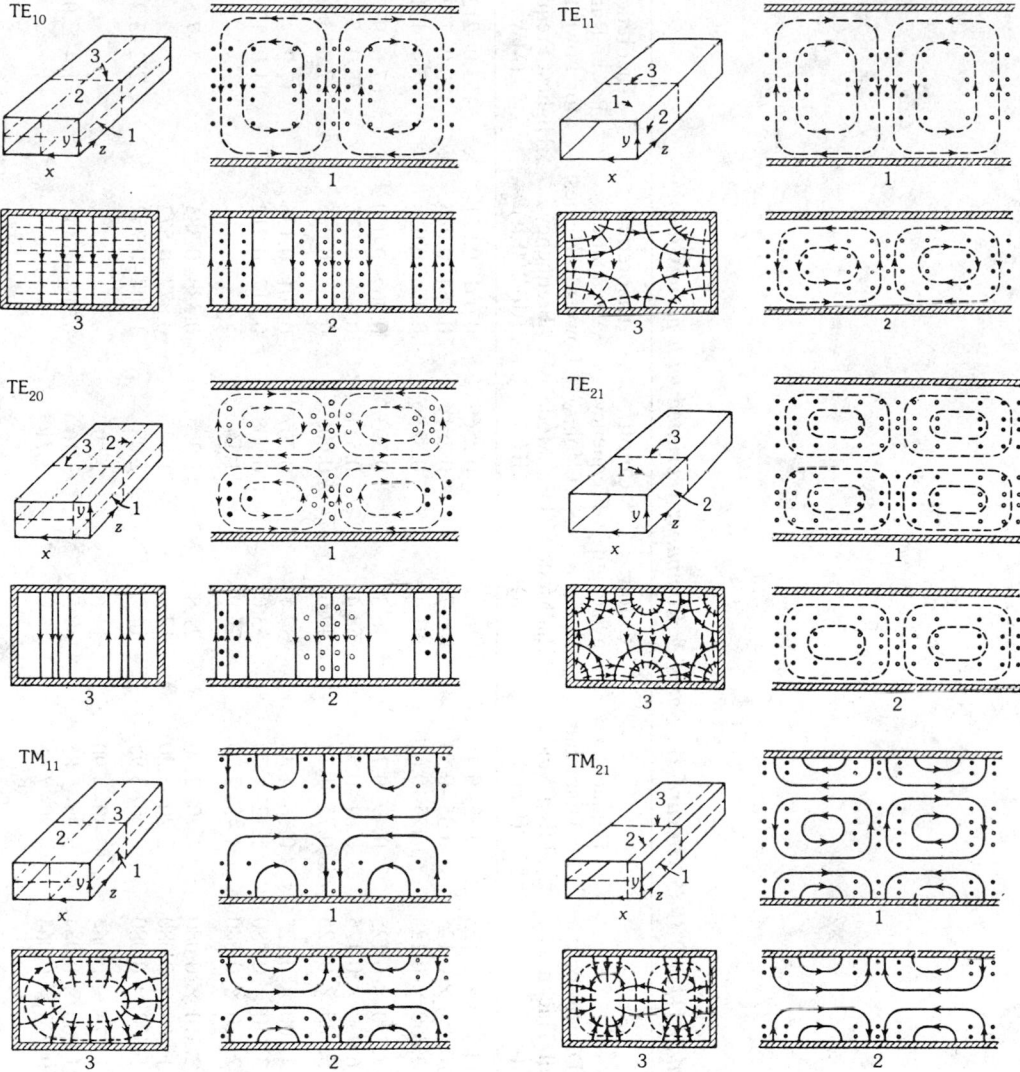

Fig. 19. Field patterns of rectangular-waveguide modes. (*After Blake [7], copyright © 1969 John Wiley & Sons, Inc., reprinted by permission of John Wiley & Sons, Inc.*)

The maximum power-handling capacity in watts for a TE_{mn} mode in a matched nondissipative guide is, for $m \neq 0$, $n \neq 0$,

$$P = 3.318 \times 10^{-4} \sqrt{\epsilon_r} \, ab \left(\frac{mb}{na} + \frac{na}{mb} \right)^2 E_b^{\,2} \sqrt{1 - \left(\frac{\lambda}{\lambda_c} \right)^2} \qquad (41)$$

where E_b is the maximum electric field intensity, which occurs at values of x and y such that

Table 7. Standard Waveguides (*After Jordan [1], © 1985 Howard W. Sams & Company; reprinted with permission*)

EIA Waveguide Designation (Standard RS-261-A)	JAN Waveguide Designation (MIL-HDBK-216, 4 January 1962)	Outer Dimensions and Wall Thickness (in)	Frequency for Dominant (TE$_{10}$) Mode (GHz)	Cutoff Wavelength λ_c for TE$_{10}$ Mode (cm)	Cutoff Frequency f_c for TE$_{10}$ Mode (GHz)	Theoretical Attenuation, Lowest to Highest Frequency (dB/100 ft)	Theoretical Power Rating for Lowest to Highest Frequency* (MW)
WR-2300	RG-290/U†	23.250 × 11.750 × 0.125	0.32–0.49	116.8	0.256	0.051–0.031	153.0–212.0
WR-2100	RG-291/U†	21.250 × 10.750 × 0.125	0.35–0.53	106.7	0.281	0.054–0.034	120.0–173.0
WR-1800	RG-201/U†	18.250 × 9.250 × 0.125	0.425–0.620	91.4	0.328	0.056–0.038	93.4–131.9
WR-1500	RG-202/U†	15.250 × 7.750 × 0.125	0.49–0.740	76.3	0.393	0.069–0.050	67.6–93.3
WR-1150	RG-203/U†	11.750 × 6.000 × 0.125	0.64–0.96	58.4	0.514	0.128–0.075	35.0–53.8
WR-975	RG-204/U†	10.000 × 5.125 × 0.125	0.75–1.12	49.6	0.605	0.137–0.095	27.0–38.5
WR-770	RG-205/U†	7.950 × 4.100 × 0.125	0.96–1.45	39.1	0.767	0.201–0.136	17.2–24.1
WR-650	RG-69/U	6.660 × 3.410 × 0.080	1.12–1.70	33.0	0.908	0.317–0.212	11.9–17.2
WR-510		5.260 × 2.710 × 0.080	1.45–2.20	25.9	1.16		
WR-430	RG-104/U	4.460 × 2.310 × 0.080	1.70–2.60	21.8	1.375	0.588–0.385	5.2–7.5
WR-340	RG-112/U	3.560 × 1.860 × 0.080	2.20–3.30	17.3	1.735	0.877–0.572	
WR-284	RG-48/U	3.000 × 1.500 × 0.080	2.60–3.95	14.2	2.08	1.102–0.752	2.2–3.2
WR-229		2.418 × 1.273 × 0.064	3.30–4.90	11.6	2.59		
WR-187	RG-49/U	2.000 × 1.000 × 0.064	3.95–5.85	9.50	3.16	2.08–1.44	1.4–2.0
WR-159		1.718 × 0.923 × 0.064	4.90–7.05	8.09	3.71		

WR-137	RG-50/U	1.500 × 0.750 × 0.064	6.98	4.29	2.87–2.30	0.56–0.71
WR-112	RG-51/U	1.250 × 0.625 × 0.064	5.70	5.26	4.12–3.21	0.35–0.46
WR-90	RG-52/U	1.000 × 0.500 × 0.050	4.57	6.56	6.45–4.48	0.20–0.29
WR-75	RG-91/U	0.850 × 0.475 × 0.050	3.81	7.88	9.51–8.31	0.12–0.16
WR-62		0.702 × 0.391 × 0.040	3.16	9.49		
WR-51		0.590 × 0.335 × 0.040	2.59	11.6		
WR-42	RG-53/U	0.500 × 0.250 × 0.040	2.13	14.1	20.7–14.8	0.043–0.058
WR-34		0.420 × 0.250 × 0.040	1.73	17.3		
WR-28	RG-96/U‡	0.360 × 0.220 × 0.040	1.42	21.1	21.9–15.0	0.022–0.031
WR-22	RG-97/U‡	0.304 × 0.192 × 0.040	1.14	26.35	31.0–20.9	0.014–0.020
WR-19		0.268 × 0.174 × 0.040	0.955	31.4		
WR-15	RG-98/U‡	0.228 × 0.154 × 0.040	0.753	39.9	52.9–39.1	0.0063–0.0090
WR-12	RG-99/U‡	0.202 × 0.141 × 0.040	0.620	48.4	93.3–52.2	0.0042–0.0060
WR-10		0.180 × 0.130 × 0.040	0.509	59.0		
WR-8	RG-138/U§	0.140 × 0.100 × 0.030	0.406	73.84	152–99	0.0018–0.0026
WR-7	RG-136/U§	0.125 × 0.0925 × 0.030	0.330	90.84	163–137	0.0012–0.0017
WR-5	RG-135/U§	0.111 × 0.0855 × 0.030	0.259	115.75	308–193	0.00071–0.0010
WR-4	RG-137/U§	0.103 × 0.0815 × 0.030	0.218	137.52	384–254	0.00052–0.0007
WR-3	RG-139/U§	0.094 × 0.0770 × 0.030	0.173	173.28	512–348	0.00035–0.0004

*For these computations, the breakdown strength of air was taken as 15 000 V/cm. A safety factor of approximately 2 at sea level has been allowed.

†Aluminum, 2.83×10^{-6} Ω-cm resistivity.

‡Silver, 1.62×10^{-6} Ω-cm resistivity.

§JAN types are silver, with a circular outer diameter of 0.156 in and a rectangular bore matching EIA types. All other types are of a Cu–Zn alloy, 3.9×10^{-6} Ω-cm resistivity.

$$\tan\left(\frac{m\pi}{a}x\right) = \pm\frac{na}{mb}, \qquad \tan\left(\frac{n\pi}{b}y\right) = \pm\frac{mb}{na} \tag{42}$$

For either $m = 0$ or $n = 0$, the maximum power in watts is

$$P = 6.636 \times 10^{-4}\sqrt{\epsilon_r}\,abE_b^{\,2}\sqrt{1 - (\lambda/\lambda_c)^2} \tag{43}$$

where E_b occurs at integer multiples of $x = a/2m$ for $m \neq 0$, $n = 0$ and at integer multiples of $y = b/2n$ for $m = 0$, $n \neq 0$.

The conductor attenuation in nepers per meter for a propagating TM_{mn} mode due to dissipation in the guide walls is

$$\alpha_c = \frac{1.055 \times 10^{-6}}{a}\sqrt{\frac{\epsilon_r f}{\sigma}}\left(\frac{m^2 + n^2a^3/b^3}{m^2 + n^2a^2/b^2}\right)\frac{1}{\sqrt{1 - (\lambda/\lambda_c)^2}} \tag{44}$$

For a matched nondissipative guide, the maximum power-handling capacity of a TM mode is given in watts by

$$P = 3.318 \times 10^{-4}\frac{\sqrt{\epsilon_r}}{\sqrt{1 - (\lambda/\lambda_c)^2}}ab\left(\frac{mb}{na} + \frac{na}{mb}\right)^2 E_b^{\,2} \tag{45}$$

where E_b occurs at values of x and y for which

$$\tan\left(\frac{m\pi}{a}x\right) = \pm\frac{mb}{na}, \qquad \tan\left(\frac{n\pi}{b}y\right) = \pm\frac{na}{mb} \tag{46}$$

Some properties and dimensions of standard rectangular waveguides are listed in Table 7.

Circular Waveguides

A uniform waveguide of circular cross section is most conveniently described by a polar coordinate system as shown in Fig. 20. The cross section has an inner dimension of radius a.

For both TE_{mn} and TM_{mn} modes in circular waveguides, m may take any integer value from 0 to infinity, and n from 1 to infinity. Field patterns for some of the lower-order modes are shown in Fig. 21. The dominant mode in circular waveguide is the TE_{11} mode. Of the circularly symmetrical modes the TM_{01} mode has the lowest cutoff frequency.

The cutoff wavelength of the TE_{mn} mode is

$$\lambda_c = 2\pi a/x'_{mn} \tag{47}$$

where x'_{mn} is the nth-order positive root of the derivative of the mth-order Bessel function. Several of the lower-order roots are given in Table 8.

The TE_{mn} attenuation in nepers per meter due to dissipation in the guide walls is

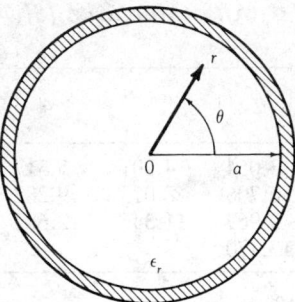

Fig. 20. Circular-waveguide cross section.

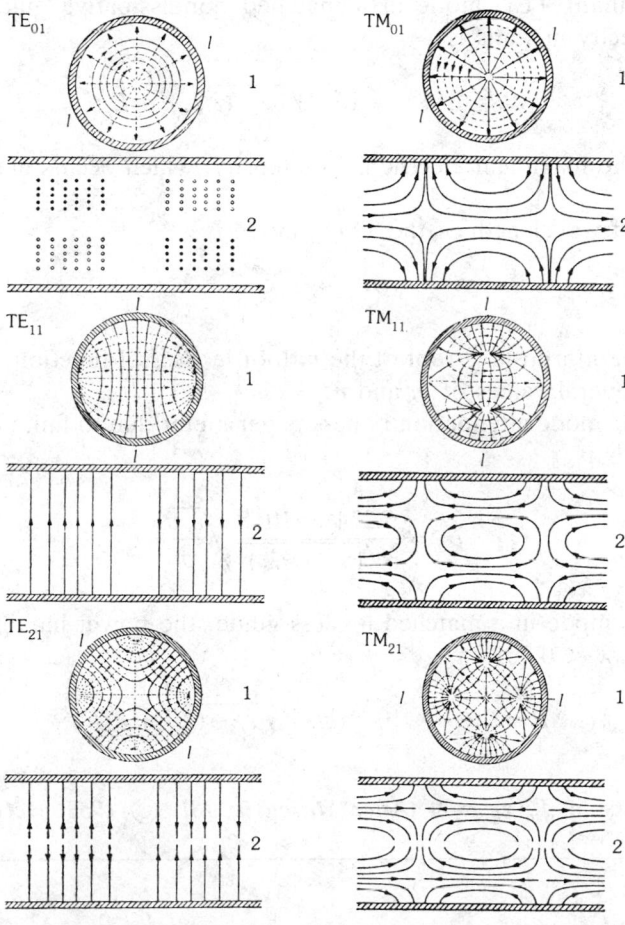

1 Cross-sectional view
2 Longitudinal view through plane *l-l*

Fig. 21. Cross-sectional and longitudinal views of field patterns of six lower-order modes in circular waveguides. (*After Marcuvitz [8], © 1951 McGraw-Hill Book Company*)

Table 8. Roots of $J'_m(x) = 0$ (After Marcuvitz [8], © 1951 McGraw-Hill Book Company)

n \ m	0	1	2	3	4	5	6	7
1	3.832	1.841	3.054	4.201	5.317	6.416	7.501	8.578
2	7.016	5.331	6.706	8.015	9.282	10.520	11.735	12.932
3	10.173	8.536	9.969	11.346	12.682	13.987		
4	13.324	11.706	13.170					

$$\alpha_c = \frac{5.274 \times 10^{-6}}{a\sqrt{1 - (\lambda/\lambda_c)^2}}\sqrt{\frac{\epsilon_r f}{\sigma}}\left[\frac{m^2}{(x'_{mn})^2 - m^2} + \left(\frac{\lambda}{\lambda_c}\right)^2\right] \tag{48}$$

For the dominant TE_{11} mode in a matched nondissipative guide, the power-handling capacity in watts is

$$P = 1.985 \times 10^{-3}\sqrt{1 - (\lambda/\lambda_c)^2}\, a^2 E_b^2 \tag{49}$$

where E_b is the maximum electric field intensity, which occurs at the axis of the guide.

The cutoff wavelength of the TM_{mn} mode is

$$\lambda_c = 2\pi a/x_{mn} \tag{50}$$

where x_{mn}, the nth positive root of the mth-order Bessel function, is tabulated in Table 9 for several values of m and n.

The TM_{mn} mode attenuation in nepers per meter due to finite conductivity of the guide walls is

$$\alpha_c = \frac{5.274 \times 10^{-6}}{a\sqrt{1 - (\lambda/\lambda_c)^2}}\sqrt{\frac{\epsilon_r f}{\sigma}} \tag{51}$$

For the TM_{01} mode in a matched lossless guide, the power-handling capacity in watts is, for $a/\lambda < 0.761$,

$$P = 7.69 \times 10^{-3}(a/\lambda)^2\sqrt{1 - (\lambda/\lambda_c)^2}\, a^2 E_b^2 \tag{52a}$$

Table 9. Roots of $J_m(x) = 0$ (After Marcuvitz [8], © 1951 McGraw-Hill Book Company)

n \ m	0	1	2	3	4	5	6	7
1	2.405	3.832	5.136	6.380	7.588	8.771	9.936	11.086
2	5.520	7.016	8.417	9.761	11.065	12.339	13.589	14.821
3	8.654	10.173	11.620	13.015	14.372			
4	11.792	13.323	14.796					

where E_b occurs at the axis of the guide. For $a/\lambda > 0.761$,

$$P = 3.33 \times 10^{-3} a^2 E_b^2 / \sqrt{1 - (\lambda/\lambda_c)^2} \qquad (52b)$$

In this case the maximum field intensity E_b occurs at $r = 0.765a$ and is independent of the angle.

Ridged Waveguides

A common ridged waveguide is a rectangular waveguide having a rectangular ridge projecting inward from one or both sides, as shown in Fig. 22. The ridge loading in the center of the guide lowers the cutoff frequency of the dominant TE_{10} mode so that a bandwidth ratio of over 4:1 exists between cutoff frequencies for the TE_{20} and TE_{10} modes.

In Fig. 23 the extension factors λ_c/a for the TE_{10} and TE_{20} modes are given as a function of the geometrical parameters. By comparing the factor λ_c/a for the TE_{10} and TE_{20} modes, the useful bandwidth can be easily obtained.

The attenuation in the ridged guides is greater than the equivalent rectangular waveguide of identical cutoff frequency. Furthermore, for ridged guides of identical cutoff frequency, more conductor loss is encountered in the guide with wider operating bandwidth.

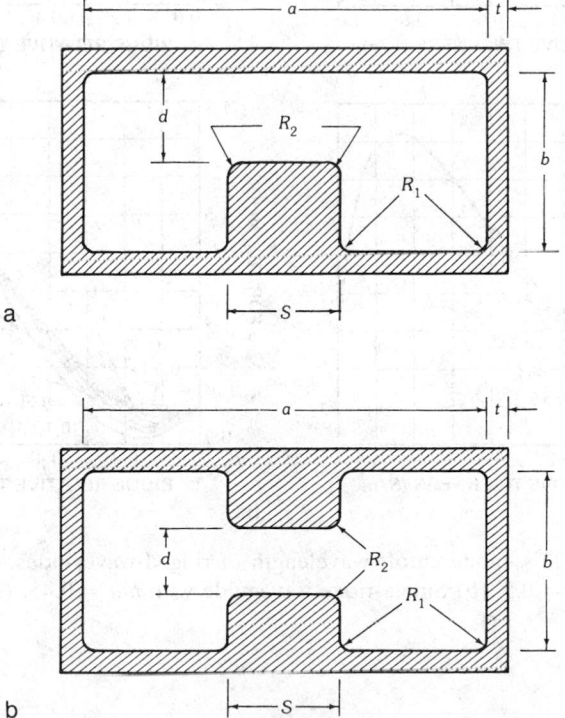

Fig. 22. Cross-sectional views of ridged waveguides. (*a*) Single-ridge waveguide. (*b*) Double-ridge waveguide.

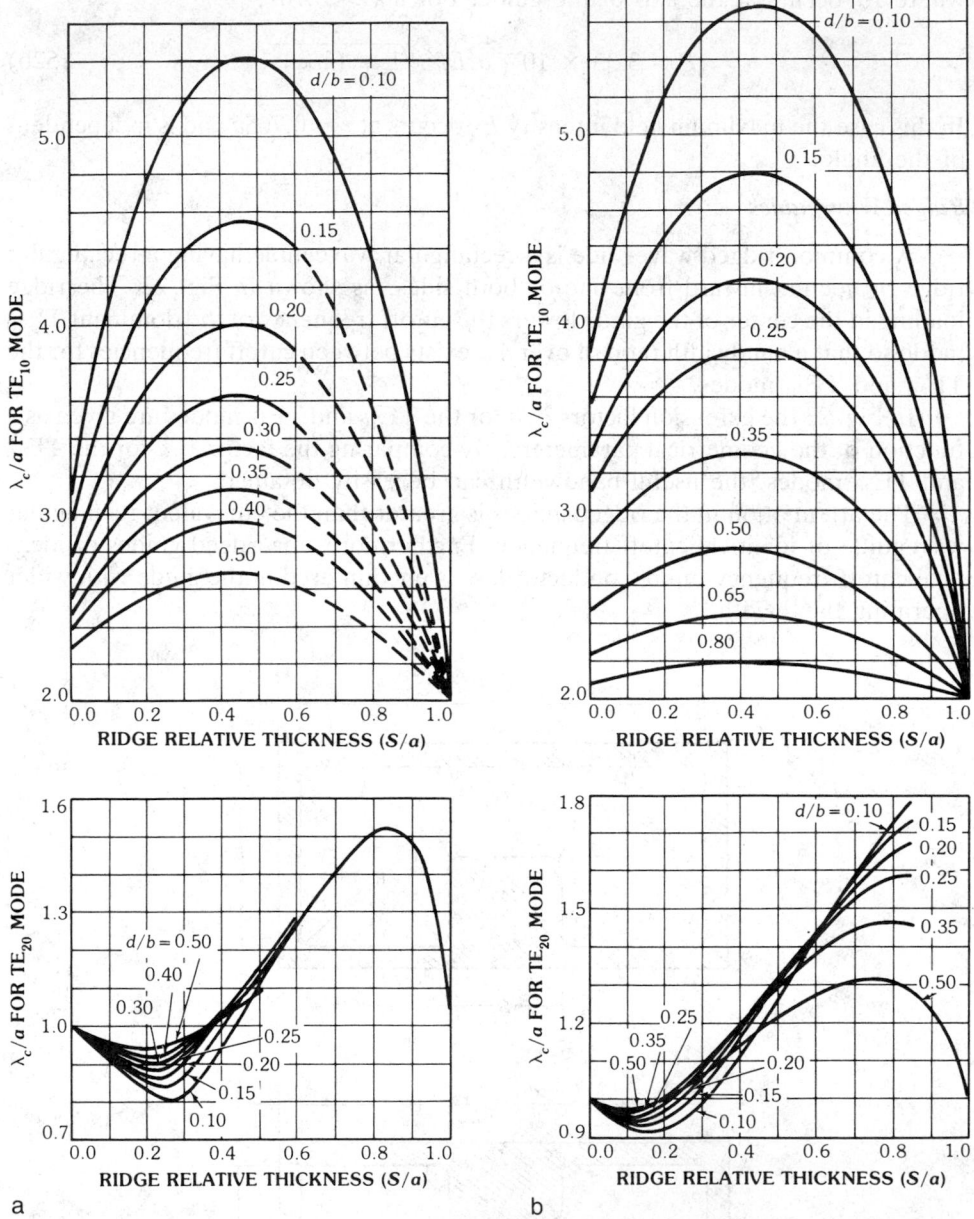

Fig. 23. TE$_{10}$- and TE$_{20}$-mode cutoff wavelength of ridged waveguides. (*a*) Double-ridge waveguide with $b/a = 0.5$. (*b*) Single-ridge waveguide with $b/a = 0.45$. (*After Hopfer [9], © 1955 IEEE*)

Based on a power-voltage definition, the ridge-guide characteristic admittance for the TE_{10} mode is plotted in Fig. 24.

Tables 10 and 11 give the essential characteristics of commonly used single- and double-ridge guides. The parameters are defined in Fig. 22, where R_1 and R_2 are the radii of curvature of the edges.

6. Hybrid-Mode Waveguides

The transmission lines to be discussed in this section include circular and rectangular dielectric waveguides and the corresponding image guides, slot lines, and fin lines. The fundamental mode in these waveguides propagates as a hybrid mode, having longitudinal components of both electric and magnetic fields.

The dielectric waveguides have their major applications in fiber optics and integrated optics, while the slot line is compatible with other planar circuits in microwave frequencies. The image guides and fin lines are proposed for millimeter-wave applications because of their low-loss advantage over the microstrip line and the ease of fabrication compared with the metal-tube waveguides.

The parameters required for circuit design are the guide wavelength, the attenuation, and the impedance. Since the fundamental modes are non-TEM, the definition of the impedance is not unique. The various definitions are given in (32) and (38).

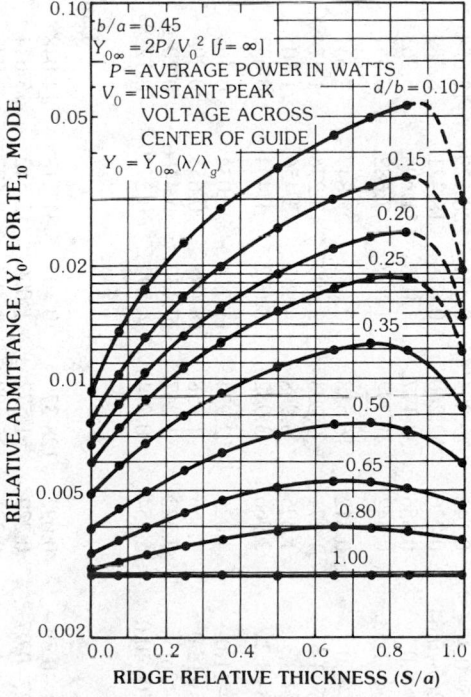

Fig. 24. Admittance of single-ridge waveguide. (*After Hopfer [9], © 1955 IEEE*)

Table 10. Characteristics of Single-Ridge Waveguides *(After Jordan [1], © 1985 Howard W. Sams & Company; reprinted with permission)*

| Frequency Range (GHz) | f_{c10} (GHz) | λ_{c10} (in) | f_{c20} (GHz) | Dimensions in Inches | | | | | | | At $f = (3)^{1/2} f_{c10}$ | |
				a	b	d	s	t	R_1 (max)	R_2	Attenuation* (dB/ft)	Power Rating† (kW)
Bandwidth 2.4:1												
0.175–0.42	0.148	79.803	0.431	28.129	12.658	5.278	4.360	—	—	1.056	0.00024	32870.
0.267–0.64	0.226	52.260	0.658	18.421	8.289	3.457	2.855	—	—	0.691	0.00045	14100.
0.42–1.0	0.356	33.177	1.036	11.695	5.263	2.195	1.813	0.125	0.047	0.439	0.00087	5682.
0.64–1.53	0.542	21.792	1.577	7.682	3.457	1.442	1.191	0.125	0.047	0.288	0.00164	2451.
0.84–2.0	0.712	16.588	2.072	5.847	2.631	1.097	0.906	0.080	0.047	0.219	0.00248	1421.
1.5–3.6	1.271	9.293	3.699	3.276	1.474	0.615	0.508	0.080	0.047	0.123	0.00591	445.8
2.0–4.8	1.695	6.968	4.933	2.456	1.105	0.461	0.381	0.080	0.047	0.092	0.00908	250.6
3.5–8.2	2.966	3.982	8.632	1.404	0.632	0.264	0.218	0.064	0.031	0.053	0.0212	81.87
4.75–11.0	4.025	2.934	11.714	1.034	0.465	0.194	0.160	0.050	0.031	0.039	0.0333	44.43
7.5–18.0‡	6.356	1.858	18.498	0.655	0.295	0.123	0.1015	0.050	0.015	0.025	0.0661	17.82
11.0–26.5‡	9.322	1.267	27.130	0.4466	0.2010	0.0838	0.0692	0.040	0.015	0.017	0.117	8.285
18.0–40.0‡	15.254	0.7743	44.393	0.2729	0.1228	0.0512	0.0423	0.040	0.015	0.010	0.246	3.035
Bandwidth 3.6:1												
0.108–0.39	0.092	128.37	0.404	31.218	14.048	2.402	5.307	—	—	0.480	0.0016	14550.
0.27–0.97	0.229	51.572	1.006	12.542	5.644	0.965	2.132	—	—	0.193	0.0065	2348.
0.39–1.4	0.331	35.680	1.454	8.677	3.905	0.668	1.475	0.125	0.047	0.134	0.0112	1124.
0.97–3.5	0.822	14.367	3.611	3.494	1.572	0.269	0.594	0.080	0.047	0.054	0.0438	182.2
1.4–5.0	1.186	9.958	5.210	2.422	1.090	0.186	0.412	0.080	0.047	0.037	0.0758	87.56
3.5–12.4	2.966	3.982	13.030	0.968	0.436	0.075	0.165	0.050	0.031	0.015	0.300	13.99
5.0–18.0‡	4.237	2.787	18.613	0.678	0.305	0.052	0.115	0.050	0.015	0.010	0.513	6.857
12.4–40.0‡	10.508	1.124	46.162	0.273	0.123	0.021	0.046	0.040	0.015	0.004	2.008	1.115

*Copper.

†Based on breakdown of air—15000 V/cm (safety factor of approximately 2 at sea level). Corner radii considered.

‡Fig. 22a in these frequency ranges only.

Table 11. Characteristics of Double-Ridge Waveguides (*After Jordan* [1], © *1985 Howard W. Sams & Company; reprinted with permission*)

Frequency Range (GHz)	f_{c10} (GHz)	λ_{c10} (in)	f_{c20} (GHz)	a	b	d	s	t	R_1 (max)	R_2	Attenuation* (dB/ft)	Power Rating† (kW)
						Dimensions in Inches					At $f = (3)^{1/2} f_{c10}$	
						Bandwidth 2.4:1						
0.175–0.42	0.092	128.37	0.401	29.667	13.795	5.863	7.417	—	—	1.173	0.0014	28830.
0.267–0.64	0.229	51.572	0.999	19.428	9.034	3.839	4.857	—	—	0.768	0.0055	4653.
0.42–1.0	0.331	35.680	1.444	12.333	5.737	2.437	3.083	0.125	0.050	0.487	0.0097	2227.
0.64–1.53	0.822	14.367	3.587	8.100	3.767	1.601	2.025	0.125	0.050	0.320	0.0378	361.2
0.84–2.0				6.167	2.868	1.219	1.542	0.125	0.050	0.244		
1.5–3.6				3.455	1.607	0.683	0.864	0.080	0.050	0.137		
2.0–4.8				2.590	1.205	0.512	0.648	0.080	0.050	0.102		
3.5–8.2				1.480	0.688	0.292	0.370	0.064	0.030	0.058		
4.75–11.0				1.090	0.506	0.215	0.272	0.050	0.030	0.043		
7.5–18.0				0.691	0.321	0.136	0.173	0.050	0.020	0.027		
11.0–26.5‡				0.471	0.219	0.093	0.118	0.040	0.015	0.019		
18.0–40.0‡				0.288	0.134	0.057	0.072	0.040	0.015	0.011		
						Bandwidth 3.6:1						
0.108–0.39				34.638	14.894	2.904	8.660	—	—	0.581		
0.27–0.97				13.916	5.984	1.167	3.479	—	—	0.233		
0.39–1.4				9.628	4.140	0.807	2.407	0.125	0.050	0.161		
0.97–3.5				3.877	1.667	0.325	0.969	0.080	0.050	0.065		
1.4–5.0	1.186	9.958	5.176	2.687	1.155	0.225	0.672	0.080	0.050	0.045	0.0656	173.5
3.5–12.4	2.966	3.982	12.944	1.074	0.462	0.090	0.269	0.050	0.030	0.018	0.259	27.74
5.0–18.0	4.237	2.787	18.490	0.752	0.323	0.063	0.188	0.050	0.020	0.013	0.443	13.59
12.4–40.0‡	10.508	1.124	45.857	0.303	0.130	0.025	0.076	0.040	0.015	0.005	1.730	2.210

*Copper.

†Based on breakdown of air—15 000 V/cm (safety factor of approximately 2 at sea level). Corner radii considered.

‡Fig. 22b in these frequency ranges only.

Circular Dielectric Waveguides and Image Guides

This type of waveguide has applications in antenna structures, laser devices, fiber optics, and millimeter-wave techniques.

The field structures for the nonradiating modes fall into two classes, circularly symmetric and nonsymmetric modes. The symmetric modes are TE_{0m} and TM_{0m} modes whose cutoff wavelengths λ_c are

$$\lambda_c = \pi d \sqrt{\epsilon_r - 1}/x_{0m} \tag{53}$$

where d is the rod·diameter, ϵ_r the relative dielectric constant, and x_{0m} the mth root of the zeroth-order Bessel function $J_0(x)$.

The nonsymmetric modes are hybrid modes that require the coexistence of an H(TE) wave with an E(TM) wave. These modes are described as HE if the H mode is predominant, and as EH if the E mode is predominant. Among the hybrid modes the HE_{11} mode is the dominant mode of the waveguide, with a zero cutoff frequency. The field pattern of the HE_{11} mode is shown in Fig. 25. Fig. 26 describes the relation between λ_g/λ_0 and d/λ_0 for dielectric guides of different ϵ_r. Fig. 27 is a plot showing the combined effect of ϵ_r on dielectric loss, field spread, and waveguide size, all normalized for frequency and loss tangent.

Since the HE_{11} mode exhibits a plane of symmetry containing the axis of the rod, an image plane may be used to replace half of the rod and surrounding space. This results in an image guide. The image plane reduces the required cross section by one-half, and also largely eliminates the support and shielding problems. In addition, it acts as a polarization anchor and reduces the mode conversion problem.

For the HE_{11} mode the dispersion (λ_0/λ_g) and dielectric attenuation (α_d) in the image guide are the same as those in the corresponding dielectric rod, given in Figs. 26 and 27. The conduction loss due to the dissipation in the image plane remains an order of magnitude less than α_d for reasonable values of $2a/\lambda_0$ and for wavelengths well into the millimeter region [10].

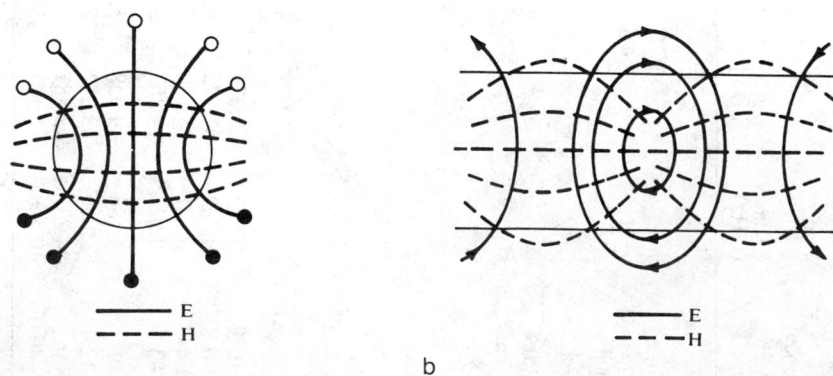

a   b

Fig. 25. Field distribution of the HE_{11} mode. (*a*) Cross section. (*b*) Longitudinal section.

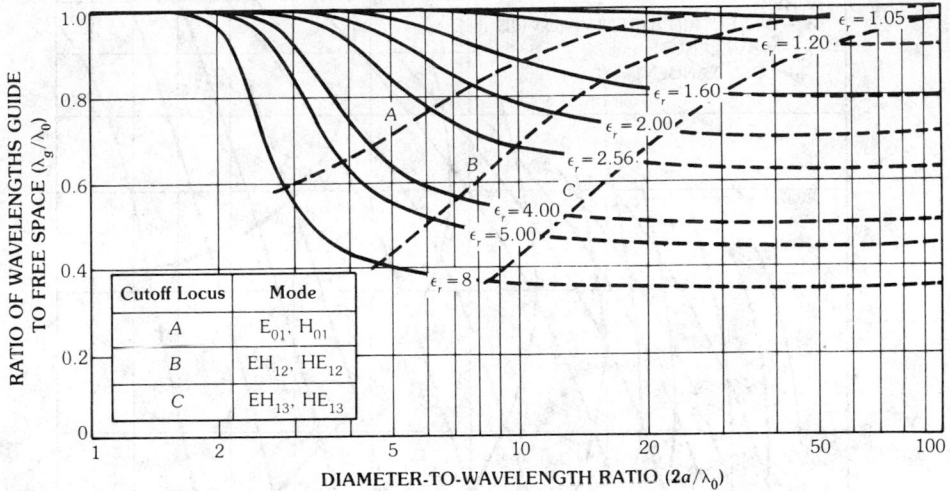

Fig. 26. Effect of dielectric constant on the dispersion characteristics of round rod. (*After Schlesinger and King [10], © 1958 IEEE*)

Rectangular Dielectric Waveguides and Image Guides

Rectangular dielectric waveguides (Fig. 28) find use in integrated optics and in millimeter-wave integrated circuits. In the latter case, image guides (Fig. 29) are more frequently used. Modes are classified into E_{pq}^y and E_{pq}^x. The former has a principal **E** field in the y direction, and the latter in the x direction. The subscripts p and q indicate the number of extrema of electric or magnetic field in the x and y directions.

For a well-guided mode, approximate expressions of phase constant for these modes in dielectric waveguides are [11]

$$\beta = \{\epsilon_1 k_0^2 - (p\pi/a)^2/(1 + 2A_3/\pi a)^2 - (q\pi/b)^2/[1 + (\epsilon_2 A_2 \\ + \epsilon_3 A_3)/(\epsilon_1 \pi b)]^2\}^{1/2} \tag{54a}$$

for the E_{pq}^y mode, with $p, q = 1, 2, \ldots$, and

$$\beta = \{\epsilon_1 k_0^2 - (p\pi/a)^2/(1 + 2\epsilon_3 A_3/\epsilon_1 \pi a)^2 \\ - (q\pi/b)^2/[1 + (A_2 + A_3)/\pi b]^2\}^{1/2} \tag{54b}$$

for the E_{pq}^x mode, with $p, q = 1, 2, \ldots$, where

$$A_2 = \lambda_0/2\sqrt{\epsilon_1 - \epsilon_2}$$
$$A_3 = \lambda_0/2\sqrt{\epsilon_1 - \epsilon_3}$$

in which λ_0 is the free-space wavelength, and ϵ_1, ϵ_2, ϵ_3 are relative dielectric constants of the materials involved.

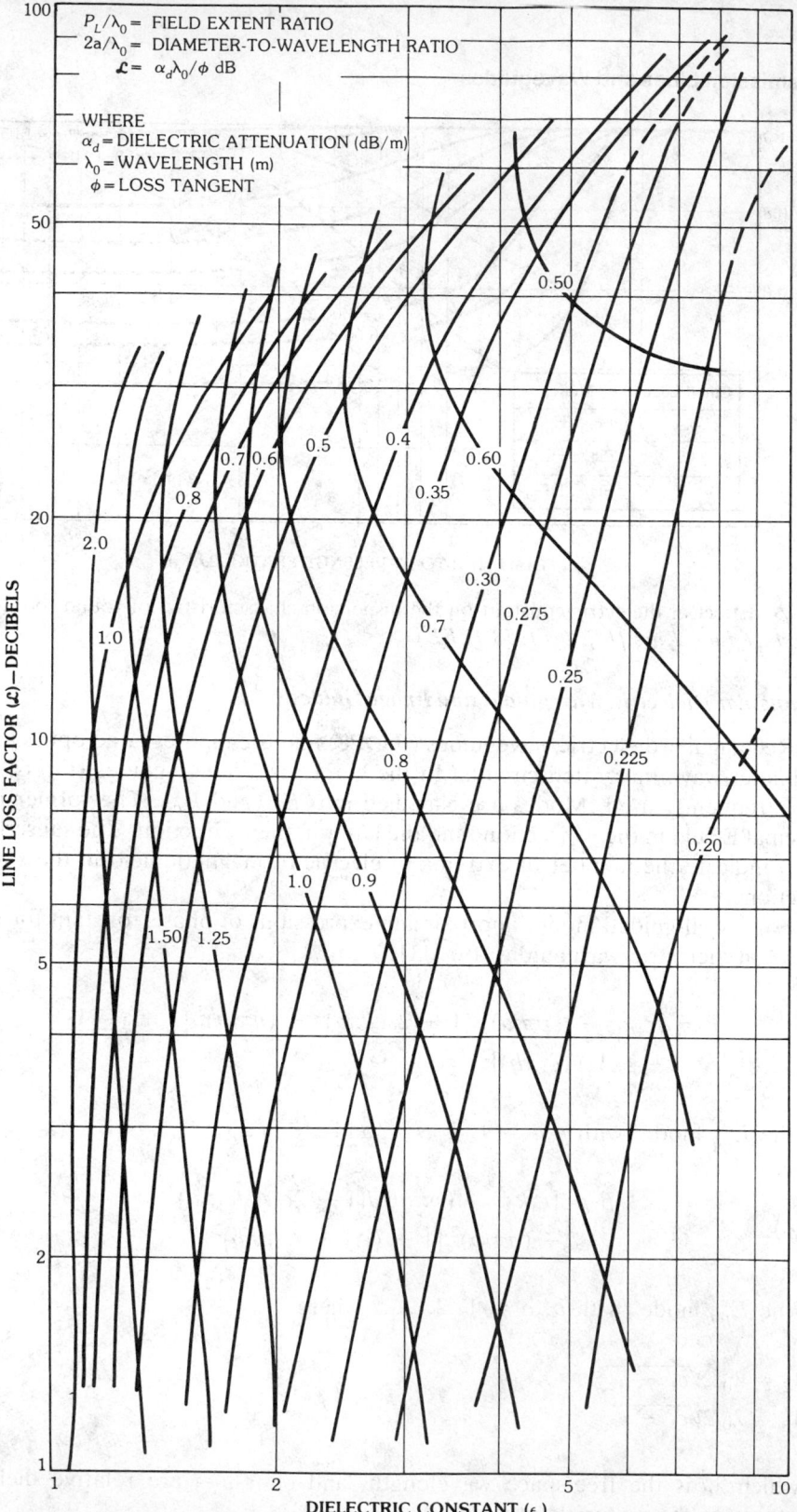

Fig. 27. Normalized loss factor $\mathscr{L}$ as a function of the dielectric constant for constant $2a/\lambda_0$ and ϱ_L/λ_0. (*After Schlesinger and King [10], © 1958 IEEE*)

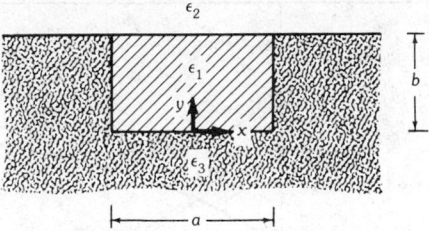

Fig. 28. Cross section of rectangular dielectric waveguide. (*After Jordan [1], © 1985 Howard W. Sams & Company; reprinted with permission*)

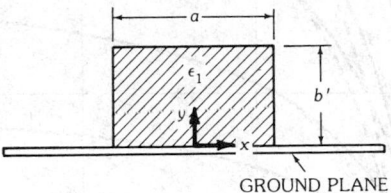

GROUND PLANE

Fig. 29. Cross section of image guide. (*After Jordan [1], © 1985 Howard W. Sams & Company; reprinted with permission*)

The above formulas are not very accurate near the cutoff frequencies. More accurate data are obtainable by using a number of techniques, the simplest of which is the effective dielectric constant (EDC) method [12]. Fig. 30 shows the HE_{11} dispersion characteristics calculated by EDC and by the above formulas for a dielectric waveguide with $a/b = 1$ and $\epsilon_2 = \epsilon_3 = 1$. The dielectric constant ϵ_1 of the guide is varied as a parameter.

As the image guide is usually surrounded by air and is equivalent to a dielectric waveguide of height $2b'$ in free space for the dominant E_{11}^y mode, the approximation formula and the EDC can be used for dispersion characteristics. The results in Fig. 30 are valid for the image guide with the aspect ratio $a/b' = 2$.

Slot Lines

Slot line consists of a narrow gap (or slot) in a conductive coating on a dielectric substrate, as shown in Fig. 31. The geometry is planar and is well suited for its usage in microwave integrated circuits. Normally, the permittivity of the dielectric substrate in a slot line is sufficiently large (e.g., $\epsilon_r = 16$) so that the fields are closely confined to the slot and the radiation loss is minimized. The nature of the slot-mode configuration is such that the electric field extends across the slot while the magnetic field is in a plane perpendicular to the slot and forms closed loops at half-wave intervals. The slot mode is non-TEM and has no low-frequency cutoff.

Since the slot mode is non-TEM, the definition of characteristic impedance is not unique. The power-voltage definition, i.e., $Z_0 = |V|^2/2P$, is normally used.

Over the intervals given by $9.6 < \epsilon_r < 20$, $0.02 < W/H < 2$, and $0.015 < H/\lambda_0 < 0.08$ and $t = 0$, a closed-form approximation to λ_g/λ_0 is expressed as [13]

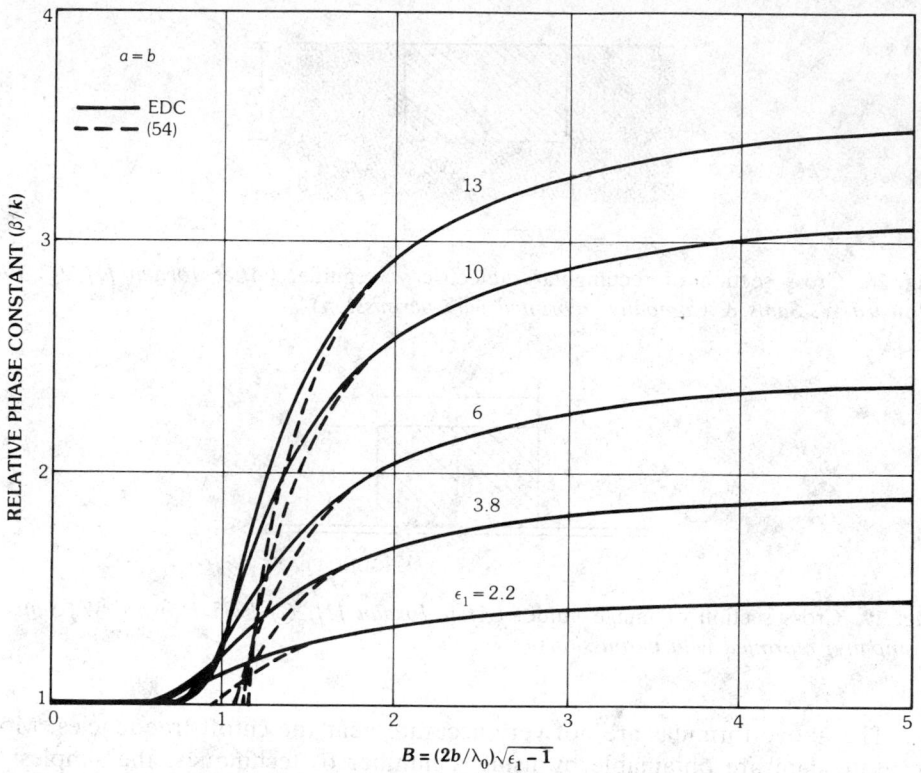

Fig. 30. Dispersion characteristics of rectangular dielectric waveguide.

$$\frac{\lambda_g}{\lambda_0} = f_1(\epsilon_r)\left[f_2\left(\frac{W}{H}\right) + f_3\left(\frac{W}{H}\right)\left(\frac{H}{\lambda_0}\right)^{f_4(W/H)} + f_5\left(\frac{W}{H}\right)\right] \tag{55}$$

where

$$f_1(\epsilon_r) = 3.549\epsilon_r^{-0.56}$$

$$f_2\left(\frac{W}{H}\right) = 0.5632\left(\frac{W}{H}\right)^{0.104(W/H)^{0.266}}$$

$$f_3\left(\frac{W}{H}\right) = -0.877\left(\frac{W}{H}\right)^{0.81} + 0.4233\left(\frac{W}{H}\right) - 0.2492$$

$$f_4\left(\frac{W}{H}\right) = -1.269 \times 10^{-2}\left[\ln\left(50\frac{W}{H}\right)\right]^{1.7} + 0.0674\ln\left(50\frac{W}{H}\right) + 0.20$$

$$f_5\left(\frac{W}{H}\right) = 1.906 \times 10^{-3}\left[\ln\left(50\frac{W}{H}\right)\right]^{2.9} - 7.203 \times 10^3\ln\left(50\frac{W}{H}\right) + 0.1223$$

The accuracy of the expression is less than 3.7 percent.

A closed-form approximation for Z_0 in ohms is

$$Z_0 = \left(\frac{11}{\epsilon_r}\right)^{p(W/H, H/\lambda_0)} g\left(\frac{W}{H}, \frac{H}{\lambda_0}\right) \tag{56}$$

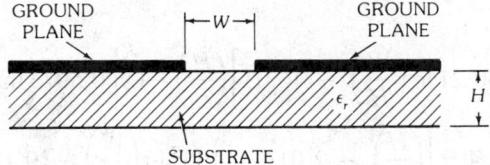

Fig. 31. Cross section of slot line. (*After Jordan [1], © 1985 Howard W. Sams & Company; reprinted with permission*)

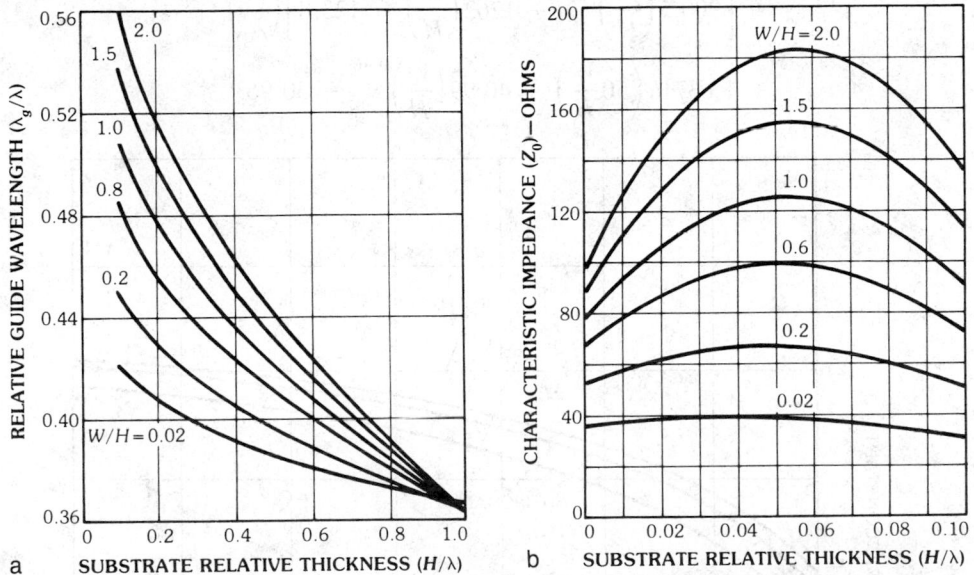

Fig. 32. Characteristics of slot line with $\epsilon_r = 13$. (*a*) Relative guide wavelength. (*b*) Characteristic impedance. (*After Jordan [1], © 1985 Howard W. Sams & Company; reprinted with permission*)

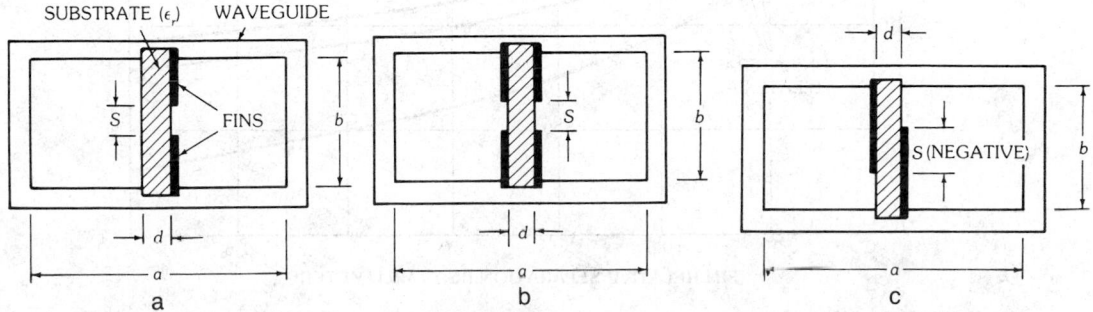

Fig. 33. Cross sections of fin lines. (*a*) Unilateral. (*b*) Bilateral. (*c*) Antipodal. (*After Jordan [1], © 1985 Howard W. Sams & Company; reprinted with permission*)

where

$$p\left(\frac{W}{H}, \frac{H}{\lambda_0}\right) = \left[-30.21 \ln\left(\frac{W}{H}\right) - 46.03\right]\left(\frac{H}{\lambda_0}\right)^2 + \left[0.5073 \ln\left(\frac{W}{H}\right) + 3.358\left(\frac{W}{H}\right)\right.$$
$$\left. + 6.492\right]\left(\frac{H}{\lambda_0}\right) - 2.013 \times 10^{-2} \ln\left(\frac{W}{H}\right) - 0.1374\left(\frac{W}{H}\right) + 0.2365$$

$$g\left(\frac{W}{H}, \frac{H}{\lambda_0}\right) = \left[-1.176 \times 10^4\left(\frac{W}{H}\right)^{0.502} - 6.311 \times 10^3\left(\frac{W}{H}\right) - 162.7\right]\left(\frac{H}{\lambda_0}\right)^2$$
$$+ \left[900.5\left(\frac{W}{H}\right)^{0.28} + 1262\left(\frac{W}{H}\right) - 123.8\right]\left(\frac{H}{\lambda_0}\right)$$
$$+ 1.637 \ln\left(50\frac{W}{H}\right) + 40.99\left(\frac{W}{H}\right)^{0.46} + 30.96$$

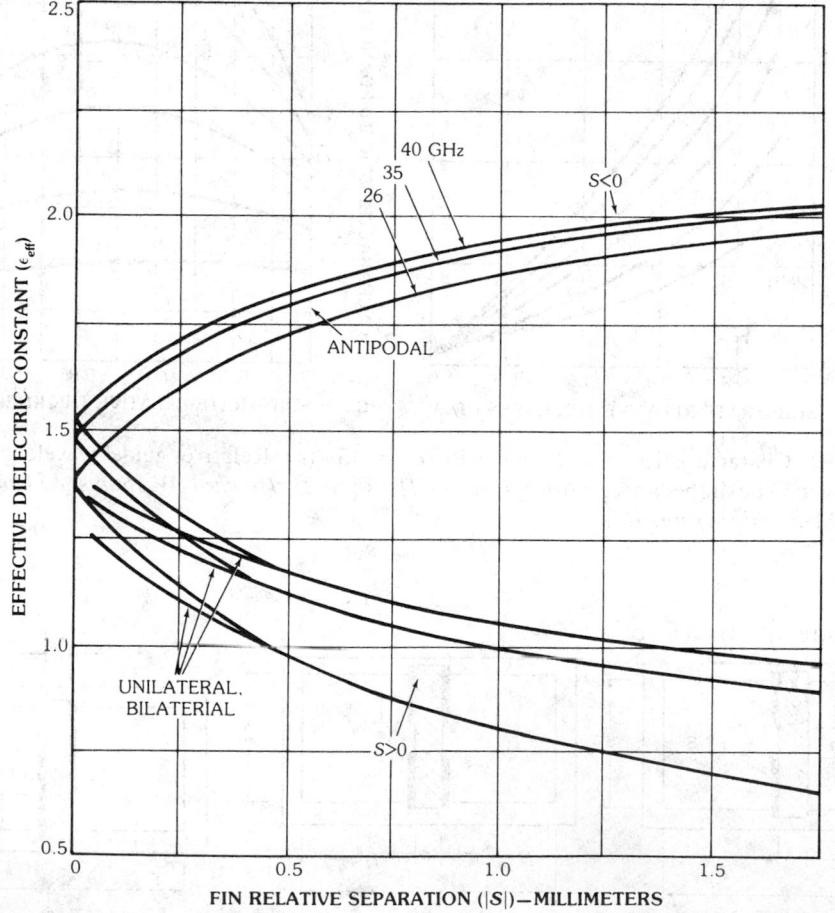

Fig. 34. Effect of fin separation on effective dielectric constant and characteristic impedance of fin lines. (*a*) Effective dielectric constant. (*b*) Characteristic impedance. (*After Hofmann [14], © 1977 IEEE*)

Over the same parameter ranges as above, the accuracy of Z_0 is ±14.5%. For W/H > 0.2, it is better than 4 percent. Data calculated by these formulas are plotted in Fig. 32.

Fin Lines

Fin lines consist of fins separated by a gap printed on one or both sides of a dielectric substrate, which is in turn placed at the center of a rectangular waveguide along its E-plane, as shown in Fig. 33. Therefore, fin lines are considered printed versions of ridged waveguides and have single-mode operating bandwidths wider than the one for the enclosing waveguide itself. Fin lines are widely used for millimeter-wave integrated circuits in the frequency range from 26.5 to 100 GHz and beyond.

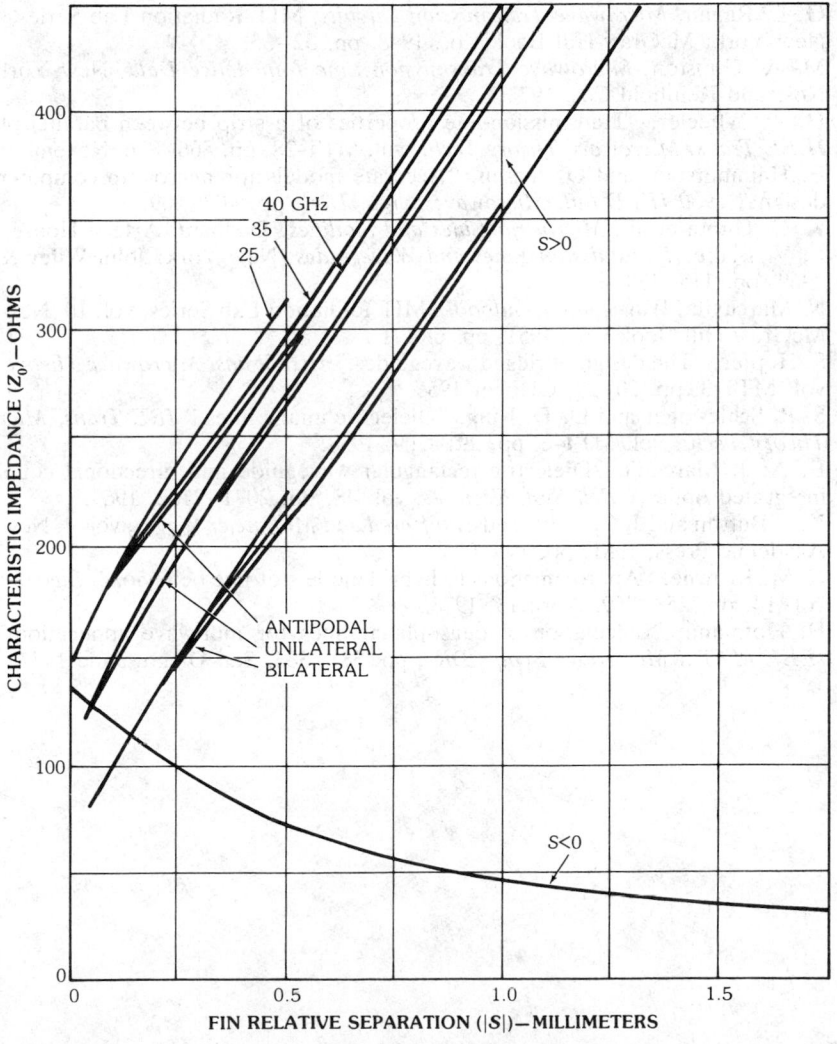

b

Fig. 34, *continued.*

The effective dielectric constant $\epsilon_{\text{eff}} = (\lambda_0/\lambda_g)^2$ and the characteristic impedance Z_0 for fin lines are plotted in Fig. 34 for the K_a-band (26.5 to 40 GHz) applications [14]. The characteristic impedance is defined as the ratio of the voltage across the slot to the current on the fins and is believed useful for small values of substrate thickness and gap width. Negative values of S indicate the overlap of fins in the antipodal fin lines. The waveguides supporting the fins are WR-28 ($a = 7.112$ mm, $b = 3.556$ mm). The substrate used for calculation is RT-Duroid ($\epsilon_r = 2.22$) of thickness $d = 0.254$ mm.

7. References

[1]   E. C. Jordan, ed., *Reference Data for Engineers: Radio, Electronics, Computer, and Communications*, 7th ed., Indianapolis: Howard W. Sams & Co., 1985, pp. 29-3, 29-25, 29-27 to 29-37, 30-10 to 30-12.

[2]   G. L. Ragan, *Microwave Transmission Circuits*, MIT Radiation Lab Series, vol. 8, New York: McGraw-Hill Book Co., 1948, pp. 32–35.

[3]   M. A. Gunston, *Microwave Transmission-Line Impedance Data*, New York: Van Nostrand Reinhold Co., 1972.

[4]   H. A. Wheeler, "Transmission-line properties of a strip between parallel planes," *IEEE Trans. Microwave Theory Tech.*, vol. MTT-28, pp. 866–876, November 1978.

[5]   E. Hammerstadt and O. Jansen, "Accurate models for microstrip computer-aided design," *1980 IEEE Intl. Microwave Symp. Dig.*, pp. 407–409.

[6]   K. C. Gupta et al., *Microstrip Lines and Slotlines*, Dedham: Artech House, 1979.

[7]   L. V. Blake, *Transmission Lines and Waveguides*, New York: John Wiley & Sons, 1969, pp. 115–120.

[8]   N. Marcuvitz, *Waveguide Handbook*, MIT Radiation Lab Series, vol. 10, New York: McGraw-Hill Book Co., 1951, pp. 68–71.

[9]   S. Hopfer, "The design of ridged waveguides," *IRE Trans. Microwave Theory Tech.*, vol. MTT-3, pp. 20–29, October 1955.

[10] S. P. Schlesinger and D. D. King, "Dielectric image lines," *IRE Trans. Microwave Theory Tech.*, vol. MTT-6, pp. 291–299, 1958.

[11] E. A. J. Marcatili, "Dielectric rectangular waveguide and directional coupler for integrated optics," *Bell Syst. Tech. J.*, vol. 48, pp. 2071–2102, 1969.

[12] K. J. Button and J. C. Wiltse, eds., *Infrared and Millimeter Waves*, vol. 4, New York: Academic Press, 1981, pp. 195–273.

[13] C. M. Krowne, "Approximations to hybrid mode slot line behavior," *Electron Lett.*, vol. 14, pp. 258–259, April 13, 1978.

[14] H. Hofmann, "Calculation of quasi-planar lines for mm-wave application," *1977 IEEE MTT-S Microwave Symp. Dig.*, pp. 381–384, San Diego, June 1977.

Chapter 29

Propagation

C. H. Liu
University of Illinois

D. J. Fang
Telectronics International, Inc.

CONTENTS

 **Chao-Han Liu** was born in Kwangsi, China, on January 3, 1939. He received the BS degree in electrical engineering from National Taiwan University in 1960 and the PhD degree in electrical science from Brown University in 1965.

In 1965 he joined the faculty of the Department of Electrical Engineering, University of Illinois at Urbana-Champaign, where he is now a professor. In 1974 and 1977 he was a visiting scientist at the Max Planck Institut Für Aeronomie, Lindau, Germany. In 1981 he was National Science Council Chair Professor at National Taiwan University, Taipei, Taiwan. His research interests are wave propagation, ionospheric and atmospheric physics, and communication. He has published approximately 120 papers and is a coauthor of the book *Theory of Ionospheric Waves*. He served as an associate editor for *IEEE Transactions on Antennas and Propagation* from 1980 to 1987 and currently is the scientific secretary of the Scientific Committee on Solar Terrestrial Physics (SCOSTEP), ICSU.

He is a Fellow of IEEE and a member of the American Physical Society, the American Geophysical Union, US Commissions C, G, and H of the International Scientific Radio Union, Sigma X', and Tau Beta Pi.

 **Dah-Jeng "Dickson" Fang** received his BSEE in 1962 from Taiwan University, Taipei, China, and his MSEE and PhD from Stanford University in 1964 and 1967, respectively. From 1967 to 1974, he was a radio physicist at Stanford Research Institute, studying electromagnetic-wave propagations, radio auroras, and ionospheric irregularities. While on leave from Stanford (1969–1970) he was a visiting professor at Taiwan University, where he published a graduate textbook on electromagnetic theory and five articles on the college education system.

Dr. Fang joined the Propagation Studies Department, COMSAT Laboratories, in 1974 and became manager in 1980. He was involved in both theoretical and experimental studies of ionospheric scintillations, microwave precipitation, multipath effects, and atmospheric optical transmission characterizations. He left COMSAT in 1985 and formed his own consultant company, Telectronics International, Inc. He is the author or coauthor of more than 50 technical papers and reports. He is a member of CCIR (Study Groups 5 and 6) and URSI (Commissions F and G).

1. Introduction

In free space the propagation of radio waves from a transmitter to a receiver is governed by the fundamental equation

$$P_r = \frac{P_t G_t G_r \lambda^2}{(4\pi D)^2} \tag{1}$$

where the subscripts t and r refer to transmitting and receiving antennas, respectively, P is power in watts, G is gain over an isotropic antenna (numerical ratio), λ is the wavelength in meters, and D is distance between transmitter and receiver in meters. If the propagation path is not in free space, the medium's effect can be included by a correction factor F such that

$$P_r = \frac{P_t G_t G_r \lambda^2}{(4\lambda D)^2} |F|^2 \tag{2}$$

Since P is proportional to the square of the electric-field intensity, the factor F is obviously a numerical ratio of actual electric-field intensity E_m in the medium to the free-space electric-field intensity E_0, i.e.,

$$F = \frac{E_m}{E_0} \tag{3}$$

Both E_m and E_0 are in volts per meter.

For practical engineering applications one often writes equation (2) in the form

$$P_r = P_t + G_t + G_r - L_0 - L_p \tag{4}$$

where

P_r = received power in decibels referred to 1 W

P_t = transmitted power, in decibels referred to 1 W

G_t = transmitting antenna gain in decibels

G_r = receiving antenna gain in decibels

L_0 = free-space loss in decibels

L_p = loss due to the medium in decibels

The free-space loss is given by

$$L_0 = 10 \log \left(\frac{4\pi D}{\lambda} \right)^2 \tag{5}$$

If we express D in kilometers and convert λ into the frequency f in megahertz, L_0 becomes

$$L_0 = 32.5 + 20 \log f + 20 \log D \tag{6}$$

The loss L_p due to the medium is related to the F factor by

$$L_p = -10 \log |F|^2 \tag{7}$$

In radio-wave propagations the medium almost invariably behaves as an attenuator, so the factor F is less than unity in magnitude. It follows that L_p is generally a positive number.

The loss L_p due to the medium in general includes the effects due to absorption, scattering, refraction, diffraction, and reflection of the propagation medium as well as from the earth surface. The main text of this chapter is devoted to engineers who require readily useful information on either L_p or F for an assessment of system impact.

Before we move ahead, it is worthwhile to note the utility of the above approach. Generally speaking, in a real-world communications system design problem, the propagation issue is one of the many important issues in considering system constraints and trade-offs. However, the propagation medium is complicated to the extent that usually not a single effect, such as absorption per se, but multiple effects pose as propagation anomalies. System engineers can ill afford the luxury of achieving first an understanding of all the propagation effects before engaging in system design. Therefore, it is believed that the above L_p (or F) approach is a useful one since it essentially treats the propagation medium as a black box. Engineers can apply formulas, curves, or models of L_p or F for system design applications without much detailed knowledge of propagation.

There are, of course, cases where the black box approach is found to be unacceptable. In such cases some knowledge, or an assumption, of the propagation medium has to be provided. One can then proceed with an approach based on principles of physics and calculate the effects of the medium on propagation. For the simplest case, the basic propagation constant for a plane wave in a general isotropic medium is given by

$$k = \omega \sqrt{\mu_0 \epsilon_0}\, n \tag{8}$$

where

k = propagation constant, or wave number in meters^{-1}

$\omega = 2\pi f$, where f is the frequency of the wave in hertz

ϵ_0 = free-space permittivity, 8.854×10^{-12} F/m

μ_0 = free-space permeability, $4\pi \times 10^{-7}$ H/m and n is the refractive index in the medium given by

$$n = \sqrt{\epsilon_r'} = \sqrt{\epsilon_r - j\sigma/\omega\epsilon_0} \qquad (9)$$

in which

ϵ_r = relative permittivity of the medium

σ = conductivity of the medium in siemens per meter

The speed of phase propagation is governed by the real part of k or n while the imaginary part of k or n determines the attenuation of the plane wave. Therefore, to study the propagation of radio waves in a material medium, the first step is to determine the refractive index of the medium for each characteristic mode. In the atmosphere, it turns out, n varies as a function of position as well as time, resulting in complications in propagation characteristics. Radio-wave propagation in such media has been studied by many authors [1–5].

As technology advances, the demands for precision and instantaneity increase. For modern communication and remote-sensing applications, these demands suggest that many of the conventional propagation models that are based on first-order approximations are no longer adequate. Instead of the deterministic models used to describe the media, the complexity and the randomness of the media require a stochastic approach. Phenomena such as scattering by hydrometeors, scintillation, and scattering by rough surfaces are examples in this category. New techniques developed in treating problems of wave propagation in random media are discussed in several books [6, 7]. Specific applications to ionosphere and troposphere problems have been review [8, 9].

In this chapter the authors will attempt to apply the results from the various sources, both theoretical and observational, to estimate the propagation effects for applications in system engineering. Many of the results in terms of empirical equations or graphs are derived by consensus, as documented by empirical modelings, FCC regulations, and/or CCIR documents.* For cases where consensus has not yet been reached, the authors will exercise their best judgment in recommending formulations based on theoretical consideration as well as limited observational data.

The chapter begins with a discussion of satellite-earth propagation with an emphasis on the application to modern satellite communications. Consideration of propagation via the ionosphere follows. Troposphere as well as ground waves are discussed next. A brief discussion of noise and its effect on propagation concludes the chapter.

2. Satellite-Earth Propagation

Previously, satellite communications systems were largely below 10 GHz. Typical examples are the L-band (1.6 GHz) maritime systems and the C-band (4 to

---

*The CCIR documents, which are subjected to periodic reviews and revisions every two years, quoted in this chapter are the output of the XVth Plenary Assembly held in Geneva in 1982, since, at the time of preparing this chapter, formally approved later versions of the documents have not been released.

6 GHz) international communications systems. Due to technology advances and rapid increase of traffic, systems with frequencies higher than 10 GHz have begun to emerge for both fixed services and direct broadcasting services. The K_u-band (14/11 GHz) domestic and international systems are now widely available. The K_a-band (30/20 GHz) systems for the 1990s are topics of intense research.

Propagation anomalies for frequencies above 1 GHz cannot be scaled and estimated from those at lower frequencies. This section provides engineers with a concise summary of major propagation effects experienced at gigahertz frequencies for both satellite-earth and terrestrial paths. These include the gaseous-absorption loss due to atmospheric constituents; path loss due to atmospheric hydrometeors; and ionosphere-induced propagation anomalies.

Free-Space Loss Along a Satellite-Earth Path

A general geometry of wave propagation between two points over a spherical earth is shown in Fig. 1. The slant path distance R_d from an earth station pointing at the satellite with an elevation angle θ_d can be evaluated by the following great-circle equations:

$$\frac{\sin(90° + \theta_d)}{h_2 + a_e} = \frac{\sin \alpha}{h_1 + a_e} = \frac{\sin \Phi}{R_d} \tag{10}$$

where $\Phi = 90° - \theta_d - \alpha$. For a synchronous satellite the altitude h_2 is a constant of 35 860 km and the earth radius a_e is 6376 km.

For example, if the elevation angle θ_d is 10°, pointing to a synchronous satellite, (10) implies

$$h_1 \cong 0, \qquad \alpha = 8.55°$$

$$\Phi = 71.45°, \qquad R_d = 40\,659.5 \text{ km}$$

At a transmission frequency of 4 GHz, the free-space loss L_0, according to (6), will be

$$L_0 = 32.5 + 20 \log 4000 + 20 \log 40\,659.5 = 196.72 \text{ dB} \tag{11}$$

Gaseous Attenuation

Gaseous attenuation is caused by molecular absorption of gaseous constituents in the atmosphere. Absorptions generally have narrow spectrum peaks corresponding to resonance frequencies. Between the peaks, there are "windows," where transmission losses are relatively low. In the lower troposphere the spectral lines are broadened to wide absorption bands. An example of gaseous attenuation at sea level of standard atmosphere is shown in Fig. 2 [10]. To estimate gaseous attenuation along a slant path above the sea level, a CCIR procedure is available as documented in two Study Group 5 documents [11, 12].

For applications in modern satellite communications under 15 GHz, the CCIR procedure can be considerably simplified as below [13]:

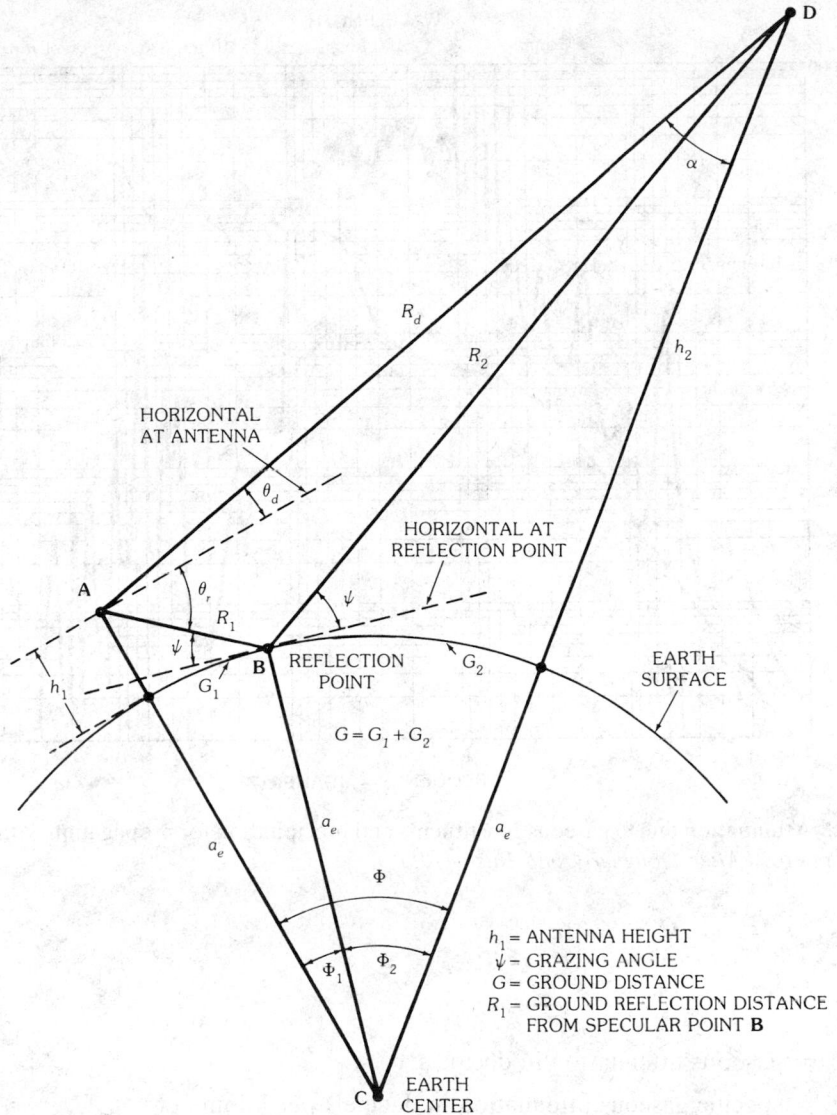

Fig. 1. General geometry of wave propagation between points A and D over a spherical carth. (*After Blake [58]*)

$$L_{\text{gas}} = \gamma_a \ell_a \tag{12}$$

$$\gamma_a = 0.004\,66\,e^{0.1362f} \tag{13}$$

$$\ell_a = \frac{2H_a}{(\sin^2\theta + 2H_a/R_e)^{1/2} + \sin\theta}, \qquad \theta \leqq 10° \tag{14}$$

$$= \frac{H_a}{\sin\theta}, \qquad \theta > 10°$$

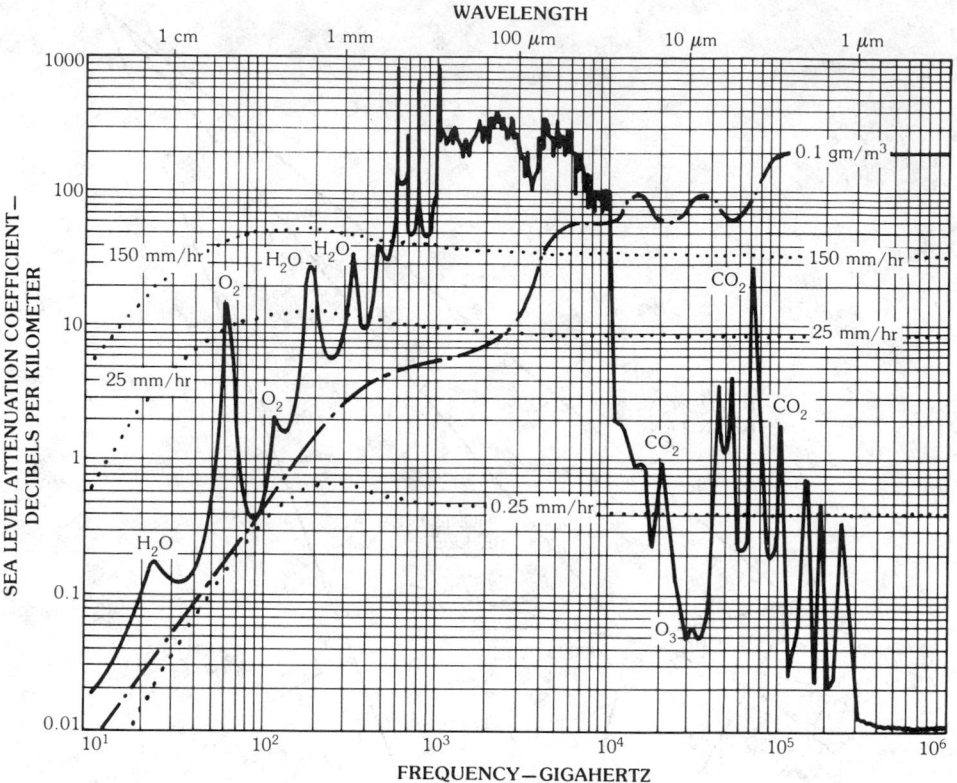

Fig. 2. Attenuation due to gaseous constituents and precipitation for propagation within the atmosphere. (*After Dougherty and Rush [10]*)

$$H_a = 6.01e^{-0.0485f} \tag{15}$$

where

L_{gas} = gaseous attenuation in decibels

γ_a = specific gaseous attenuation in decibels per kilometer

ℓ_a = slant path length in kilometers

f = frequency in gigahertz

H_a = effective height of the absorptive atmosphere in kilometers

R_e = effective earth radius, 8500 km

Rain Attenuation

The effects of rain on radio-wave propagation have been investigated by many authors both theoretically and experimentally [14–18]. These results have been used in constructing practical models for rain effects. In this section we discuss a procedure to compute rain attenuation based on these models.

Since precipitation is a statistical meteorological event, the attenuation due to rain is necessarily evaluated in reference to a percentage of time of an averaged year. The basic equation is still the same as (12), except it includes an additional reduction factor δ, i.e.,

$$L_R(P) = \gamma_R \ell_R \delta \qquad (16)$$

where

L_R = attenuation in decibels at P percent of the time in a year

γ_R = specific rain attenuation in decibels per kilometer

ℓ_R = slant path length in kilometers

δ = reduction factor for P percent of the time

The specific rain attenuation γ_R is related to rain rate at P percent of the time, $R(P)$ in millimeters per hour, by the following power law equation:

$$\gamma_R = K_{H,V} R^{\alpha_{H,V}}(P) \qquad (17)$$

where K and α are power law coefficients. This is based on theoretical scattering computations verified by experimental data. The subscripts H, V refer to horizontal or vertical polarizations of a radio-wave transmission, respectively. Numerical values of K and α as recommended by CCIR [19] are tabulated in Table 1.

To calculate $L_R(P)$, the input parameters are:

$R(P)$: the point rainfall rate in millimeters per hour for the location at a required P percent of the time. Global prediction models based on meteorological observations can be used to estimate this rate [19]

h_0: the height in kilometers above mean sea level of the earth station

θ: the elevation angle in degrees

ϕ: north or south latitude of the earth station, in degrees

According to the CCIR document [20], the calculation of $L_R(P)$ involves the following steps:

Step 1

Evaluate rain height h_R in kilometers:

$$h_R = \varrho_P(\phi)[5.1 - 2.15 \log\{1 + 10^{(\phi-27)/10}\}] \qquad (18)$$

where

$$\varrho_P(\phi) = \begin{cases} 0.6 & \phi \leq 20° \\ 0.6 + 0.02(\phi - 20) & 20° < \phi \leq 40° \\ 1.0 & \phi < 40° \end{cases} \qquad (19)$$

Step 2

Evaluate slant path length ℓ_R in kilometers:

Table 1. Values of Power Law Coefficients $K_{H,V}$ and $\alpha_{H,V}$ to Be Used for Evaluating Specific Rain Attenuation as Given by (17)

Frequency (GHz)	K_H	K_V	α_H	α_V
1	0.000 038 7	0.000 035 2	0.912	0.880
2	0.000 154	0.000 138	0.963	0.923
4	0.000 650	0.000 591	1.12	1.07
6	0.001 75	0.001 55	1.31	1.27
8	0.004 54	0.003 95	1.33	1.31
10	0.010 1	0.008 87	1.28	1.26
12	0.018 8	0.016 8	1.22	1.20
15	0.036 7	0.034 7	1.15	1.13
20	0.075 1	0.069 1	1.10	1.07
25	0.124	0.113	1.06	1.03
30	0.187	0.167	1.02	1.00
35	0.263	0.233	0.979	0.963
40	0.350	0.310	0.939	0.929
45	0.442	0.393	0.903	0.897
50	0.536	0.479	0.873	0.868
60	0.707	0.642	0.826	0.824
70	0.851	0.784	0.793	0.793
80	0.975	0.906	0.769	0.769
90	1.06	0.999	0.753	0.754
100	1.12	1.06	0.743	0.744
120	1.18	1.13	0.731	0.732
150	1.31	1.27	0.710	0.711
200	1.45	1.42	0.689	0.690
300	1.36	1.35	0.688	0.689
400	1.32	1.31	0.683	0.684

$$\ell_R = \frac{2(h_R - h_0)}{[\sin^2\theta + 2(h_R - h_0)/R_e]^{1/2} + \sin\theta}, \quad \theta < 10°$$

$$= \frac{h_R - h_0}{\sin\theta}, \quad \theta > 10° \tag{20}$$

where $R_e = 8500$ km.

Step 3

Evaluate reduction factor δ at P percent of the time.

The reduction factor is due to the fact that during heavier rain the rainfall rate is distributed nonuniformly over the path in the form of rain cells or rain front. The rainfall rate decreases from the center of the rain cell outwards. The rate of decrease is higher for heavier rains. The empirically determined formula for the reduction factor at P percent of the time for the satellite-earth propagation geometry is given by

$$\delta = \frac{90\,C}{90 + 4\ell_R\cos\theta}\left(\frac{P}{0.01}\right)^{-a}$$

where

$$a = 0.33, \quad C = 1; \quad 0.001 \leqq P \leqq 0.01 \tag{21}$$

$$a = 0.41, \quad C = 1; \quad 0.01 < P \leqq 0.1 \tag{22}$$

$$a = 0.50, \quad C = 1.3; \quad 0.1 < P \leqq 1.0 \tag{23}$$

Step 4

Evaluate $L_R(P)$ by (16).

Among more than a dozen methods available, the above method is regarded to be the one that produces least overall errors against currently available global experimental data collected from more than 30 locations [20]. For an individual site, however, the method could make predictions which deviate considerably from experimental results. Typical comparisons between predicted and measured results [21, 22, 23] are provided in Table 2.

In many digital communications applications, rain attenuation statistics for a worst month in a year is required for system design. In the absence of reliable measurement data the statistics can be estimated from the annual statistics by the following conversion relationship:

$$P = 0.29 P_w^{1.15} \quad \text{or} \quad P_w = 2.93 P^{0.87} \tag{24}$$

where P_w is the percentage time in a worst month. It suggests that the worst month $L_R(P_w)$ is the same numeric decibel value as that of $L_R(P)$ in an averaged year.

Table 2. Comparison between Measured Rain Attenuation Data and Predictions

Location (Latitude, h_0)	Frequency (GHz)	Polarization	Elevation Angle (°)	Time (%)	Rain Rate R (mm/hr)	Measured Attenuation (dB)	Predicted Attenuation (dB)
Kjeller, Norway (60.0N, 0)	11.6	H	22	0.001	75.0	12.6	16.0
				0.01	40.0	6.9	7.5
				0.1	15.0	2.1	2.9
				1.0	3.0	0.8	1.0
Yonaguni, Japan (24.3N, 0.20)	12.1	V	58	0.001	117.1	16.7	23.3
				0.01	82.8	14.3	10.9
				0.1	37.8	8.1	4.2
				1.0	5.6	2.7	1.4
Clarksburg, Md. (39.2N, 0.18)	19	H	47.5	0.001	—	—	64.0
				0.001	68.0	22.0	29.9
				0.1	17.0	11.0	11.6
				1.0	2.0	3.0	3.9
Blacksburg, Va. (37.2N, 0.30)	28.6	V	45	0.001	100.0	—	48.5
				0.01	36.0	24.0	22.7
				0.1	8.0	11.8	8.8
				1.0	1.0	5.0	2.9

Rain Depolarization

The determination of depolarization is important for communications, interference considerations, and remote-sensing applications, particularly for dual-polarization antennas. At gigahertz frequencies, depolarization can be caused by raindrops, snow, and ice particles, with raindrops being the major contributor [24]. Engineering modelings for estimating depolarizations due to snow and ice particles have yet to be established. Even for rain, CCIR models are rather divergent. The basic reason is that depolarization results from the nonspherical geometry of raindrops as they tilt in the air while falling under the influence of turbulent wind shear. This is a highly random meteorological process [25]. For first-order applications, a simple equation is selected here for estimating cross polarization [26], XPD in decibels, from a signal at a known level of attenuation, $L_R(R)$, given in the previous section:

$$XPD = 30\log f - 40\log(\cos\theta) - 20\log(\sin 2\tau) - 20\log[L_R(P)] \qquad (25)$$

where

f = frequency in gigahertz

θ = elevation angle in degrees

τ = polarization tilt angle in degrees with respect to the horizon (for circular polarization $\tau = 45°$, the term $20\log[\sin 2\tau]$ disappears)

Again, the XPD here refers to long-term averaged cross-polarization level at P percent of the time for an averaged year.

Transionospheric Propagation

Caused by solar radiation, the earth's ionosphere consists of several regions of ionizations located from about 50 km to 1000 km in altitude. For all practical communication purposes three regions of the ionosphere, D, E, and F (F1, F2), have been identified as shown in Fig. 3, where N_e is the electron density in electrons per cubic meter.

In each region the ionized medium is neither homogeneous in space nor stationary in time. Generally speaking, the background ionization has relatively regular diurnal, seasonal, and 11-year solar cycle variations, with many anomalies, and depends strongly on geographical location. An international reference ionosphere has been constructed [27] that can be used as a general reference. In addition, there are highly dynamic, small-scaled nonstationary structures known as *irregularities*. Both the background ionization and irregularities degrade radio waves. The charged particles in the ionosphere cause the refractive index to become frequency dependent, i.e., the medium becomes dispersive. For transionospheric propagation above the very high frequencies, when the effect of earth's magnetic field can be neglected, the refractive index n can be expressed as

$$n = \left(1 - \frac{80.6}{f^2}N_e\right)^{1/2} \qquad (26)$$

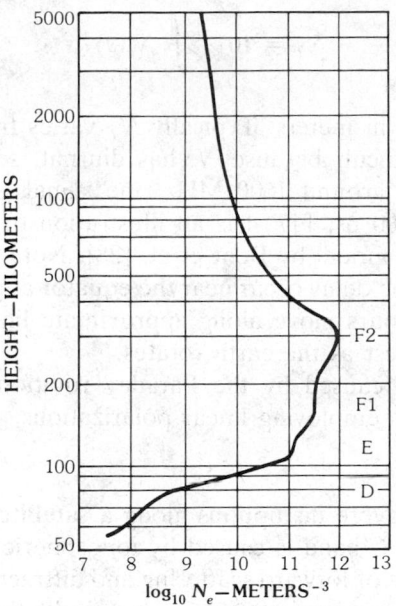

Fig. 3. Typical electron-density profile of the ionosphere.

where N_e is the electron density in electrons per cubic meter and f is the signal frequency in hertz. The density N_e has a regular part corresponding to the background density and a fluctuating part corresponding to the irregularities. The variation of the density will cause the refractive index to fluctuate, which in turn will affect the propagation of the radio waves. In general, for transionospheric propagation, the F2 layer is the most important region.

Group Delay

The presence of charged particles in the ionosphere slows down the propagation of radio signals along the satellite-earth path. The time delay in excess of the propagation time in free space is called the *group delay*. It is an important factor to be considered for digital communication systems. This quantity can be computed from the group velocity derived from (26):

$$\tau = 1.345 N_T / f^2 \qquad (27)$$

where

τ = delay time in nanoseconds in reference to propagation in a vacuum

f = frequency of propagations in gigahertz

N_T = integrations per 10^{16} electrons/m^2 of clectron density along the propagation path

The quantity N_T is called the *total electron content* (TEC) and can be evaluated by

$$N_T = 10^{-16} \int_s N_e(s)\, ds \tag{28}$$

where s is the ray path in meters. Typically N_T varies from 1 to 200. The exact evaluation of N_T is difficult because N_e has diurnal, seasonal, and solar cycle variations. For a band around 1600 MHz, the signal group delay varies from approximately 0.5 to 100 ns. Fig. 4 is an illustration of worldwide group delay distribution based on a model by Bent et al. [28]. Note from the figure that the regions of maximum time delay occur near the equator at plus and minus 15° to 20° of latitude. These contours move along approximate lines of constant magnetic latitude from east to west as the earth rotates.

The depolarization caused by the Faraday rotation may also cause signal degradation for systems employing linear polarizations.

Scintillation

One of the most severe disruptions along a satellite-earth propagation path for signals from vhf to C-band is caused by ionospheric scintillation. Principally through the mechanisms of forward scattering and diffraction, small-scale irregular structures in the ionization density produce the scintillation phenomenon in which the steady signal at the receiver is replaced by one which is fluctuating in amplitude, phase, and apparent direction of arrival. Depending on the modulation scheme of the system, various aspects of scintillation affect the system performance differently. The most commonly used parameter characterizing the intensity fluctuations is the scintillation index S_4, defined by

$$S_4 = \left(\frac{\langle I^2 \rangle - \langle I \rangle^2}{\langle I \rangle^2} \right)^{1/2} \tag{29}$$

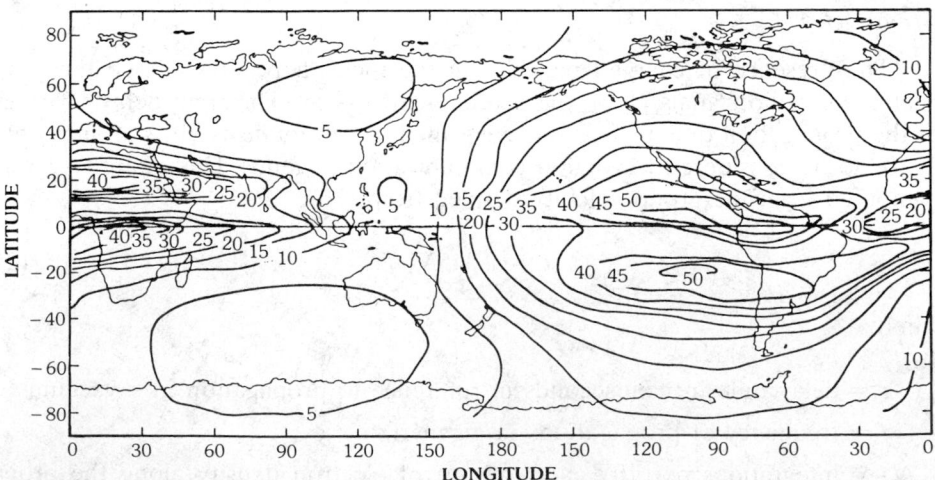

Fig. 4. Worldwide group delay in nanoseconds for a system operating frequency of 1.6 GHz at a universal time of 00 hours derived from the Bent model [28] for conditions appropriate to March 1968.

where I is the intensity of the signal and $\langle \ \rangle$ denotes averaging.

Under weak scintillation conditions which approximately apply when $S_4 \lesssim 0.5$, the scintillation index can be estimated from the following equation [8]:

$$S_4{}^2 = 8\pi L\lambda^2 r_e{}^2 \int\int_{-\infty}^{+\infty} S_N(\varkappa, 0) \sin^2(\varkappa^2/\varkappa_F{}^2)\, d^2\varkappa \tag{30}$$

where

$\qquad L = $ thickness in meters of the ionosphere layer

$\qquad r_e = $ classical radius of electron $= 2.818 \times 10^{-15}$ m

$\qquad \lambda = $ wavelength in meters

$\qquad S_N(\varkappa, \varkappa_z) = $ power spectrum of electron density fluctuations in the ionosphere, in meters^{-3}

$\qquad \varkappa_F = $ Fresnel wave number $= \sqrt{4\pi/\lambda z}$ in meters^{-1}

$\qquad z = $ height of the ionosphere layer in meters

The scintillation index S_4 is related to the peak-to-peak fluctuations of the intensity. The exact relation depends on the distribution of the intensity. Empirically, Chart 1 provides a convenient conversion between S_4 and the approximate peak-to-peak fluctuations P_{fluc} in decibels [29].

Chart 1. Empirical Conversion Table for Scintillation Indices

S_4	P_{fluc}
0.1	1.5
0.2	3.5
0.3	6
0.4	8.5
0.5	11
0.6	14
0.7	17
0.8	20
0.9	24
1.0	27.5

The intensity distribution is best described by the Nakagami distribution [8] for a wide range of S_4 values. As $S_4 \rightarrow 1.0$, it approaches the Rayleigh distribution. Occasionally, S_4 may exceed 1, reaching values as high as 1.5. This is due to focusing. For values less than 0.6, S_4 shows a consistent $f^{-1.5}$ frequency dependence for most multifrequency observations in the vhf and uhf bands [30]. Recent equatorial observations at gigahertz frequencies, however, suggested values higher than 1.5 for the spectral index [31]. As the scintillation becomes stronger such that S_4 exceeds 0.6, the spectral index decreases. This is due to the saturation of scintillation for Rayleigh fading under the strong influence of multiple scattering.

Another parameter that is important in system applications is the fade coherence time τ_c, which is defined as the time lag in seconds for the auto-correlation function of the intensity to drop to 50 percent of its maximum value. This coherence time varies from a few tenths of a second to about ten seconds, depending on the signal frequency and the ionospheric conditions [8, 31]. For a given frequency, strong scintillation corresponds to smaller τ_c. Time delays should be greater than τ_c to achieve any diversity gain.

The phase scintillations are in general dominated by slow fluctuations. They become important in systems that are required to maintain phase coherence over long time or spatial intervals.

Geographically, there are two intense zones of scintillations, one at high latitudes and the other centered within ±20° of the magnetic equator as shown in Fig. 5 [32]. Severe scintillations have been observed up to gigahertz frequencies in these two sectors, while in the middle latitudes scintillations mainly affect vhf signals. In all sectors there is a pronounced nighttime maximum of the activity as also indicated in Fig. 5. For equatorial gigahertz scintillation, peak activity around the vernal equinox and high activity at the autumnal equinox have been observed [31, 33].

To model the scintillation phenomenon one needs to understand the physical mechanisms that generate these irregularities. Current knowledge about these irregularities, however, does not provide sufficient basis for the construction of a comprehensive model. It has been suggested that spread-F irregularities are responsible for scintillations [34]. Spread-F irregularities are essentially evening and nighttime events in the F region. Depending on latitudinal locations the

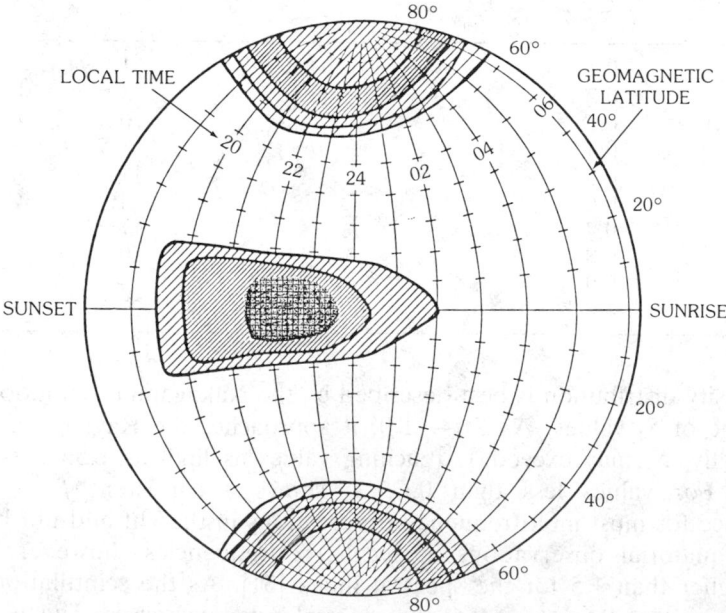

Fig. 5. Depth of scintillation fading (proportional to density of cross-hatching) during low to moderate solar activity. (*After Aarons [32], © 1982 IEEE*)

occurrence of spread F has distinctive patterns relating to seasonal variations and magnetic activities. Since spread F has been studied continuously from the early 1930s and enormous amounts of data are available, radio physicists have attempted to devise empirical models for predicting scintillations based on spread-F models. A step-by-step methodology consisting of inputs of morphological data of spread F and more than 50 numerical equations is available for statistical prediction of uhf scintillation [34]. A more recent model has been constructed based on numerous satellite propagation observations, particularly on observations performed in the DNA Wideband Satellite Experiment. Geographical, seasonal, diurnal, and solar activity as well as magnetic activity dependence are included in the model [35, 36, 37]. The models have been shown to be successful in predicting uhf scintillations.

As for scintillations at gigahertz frequencies, however, the available long-term observational data seem to indicate that additional efforts in modeling are needed [33, 35]. An example for an equatorial ionospheric path at 4 GHz is shown in Fig. 6. The annual occurrence statistics of peak-to-peak amplitude fluctuations, P_{fluc} in decibels, are given for two links. In the figure, P curves in solid lines are signal fluctuations as received from a satellite at the east direction of an approximately 20° elevation angle, while the I curves are from a satellite at the west direction at 30° elevation angle. Curves for different years when sunspot numbers (SSN) change from 10 to 165 are labeled. For link budget calculation, P_{fluc} is related to loss L_p given in (4) by

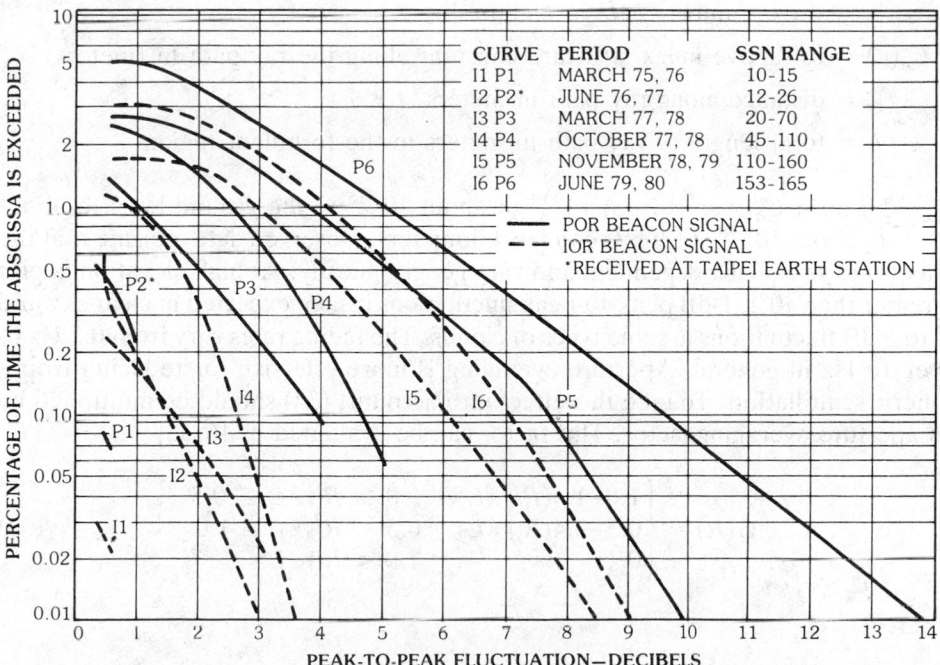

Fig. 6. Annual statistics of peak-to-peak fluctuations observed at Hong Kong earth station (curves I1, P1, I3–I6, P3–P6) and Taipei earth station (curves P2 and I2).

$$L_p = \frac{1}{\sqrt{2}} P_{\text{fluc}} \tag{31}$$

For scintillations at high latitude the CCIR document [35] should be used for further information.

For frequencies above C-band and propagation path at an elevation angle lower than 10° or so, troposphere-induced scintillations become a factor to be considered in computing the path loss. The tropospheric effects are produced by local atmospheric features such as high humidity gradients and temperature inversion layers leading to thin, horizontally stratified layers in which the refractive index is markedly different from the ambient value. Superimposed on these layers are fluctuations caused by internal waves or turbulence.

For a weak scintillation such that $S_4 < 0.6$, the rms fluctuations in log power can be estimated by the following equation assuming the scintillation is caused by Kolmogroff turbulence [6]:

$$\sigma_\chi^2 = 42.25 \, k^{7/6} \int_0^L C_n^2(\zeta) \, \zeta^{5/6} \, d\zeta \tag{32}$$

where

σ_χ = standard deviation of logarithm of received power related to S_4 by
$\quad S_4 = 4\sigma_\chi^2$

k = wave number, $2\pi/\lambda$ in meters^{-1}

$C_n(\zeta)$ = refractive index structure constant along the ray path in (meters)$^{-2/3}$

ζ = distance along ray path in meters

L = total length of ray path in meters in the turbulent region

The term C_n^2 varies from 10^{-14} to about 10^{-15} at the ground level to about 10^{-17} to about 10^{-18} at heights of a few kilometers. Observed data indicate that the effects are seasonally dependent and vary geographically. At high elevation angles (greater than 10°), 1-dB peak-to-peak fluctuations can be expected in clear sky and 2- to 6-dB fluctuations in some types of clouds. The fading rates vary from 0.5 Hz to over 10 Hz in general. Aperture averaging is more effective for reducing tropospheric scintillation. To take this effect into account, (32) should be multiplied by an aperture averaging factor. This factor can be evaluated by [6, 38]

$$G(R) = \begin{cases} 1.0 - 1.4(R/\sqrt{\lambda L}), & 0 \leq R/\sqrt{\lambda L} \leq 0.5 \\ 0.5 - 0.4(R/\sqrt{\lambda L}), & 0.5 < R/\sqrt{\lambda_L} \leq 1.0 \\ 0.1, & 1.0 < R/\sqrt{\lambda L} \end{cases} \tag{33}$$

where

R = effective radius in meters of circular antenna aperture, given by $\eta^{1/2}(D/2)$
$\quad$ where η is the antenna efficiency factor and D is the physical diameter of
$\quad$ the antenna in meters

L = slant distance in meters to height of a horizontal thin turbulent layer

$\quad = [h^2 + 2R_e h + (R_e \sin\theta)^2]^{1/2} - R_e \sin\theta$

h = height of layer in meters

θ = elevation angle

R_e = effective earth radius = 8.5×10^5 m

λ = operating wavelength in meters

3. Propagation beyond the Horizon via the Ionosphere

While for the satellite-earth propagation path the ionosphere causes additional path loss as discussed in Section 2, it provides the most reliable beyond-the-horizon transmission channels at frequency bands of hf and below and sometimes at the vhf band. When radio signals at hf or lower frequencies are transmitted from the earth surface (or aircraft) to the ionosphere, the refraction due to the ionosphere layers may be sufficient to bend the ray back to earth. In addition, irregular structures of the ionization density in the ionosphere may scatter radio-frequency energy and provide over-the-horizon transmission even at vhf band. In this section the various modes of propagation beyond the horizon via the ionosphere will be discussed.

Ionospheric Propagation at High Frequency

According to magnetoionic theory for radio waves propagating in the ionosphere, two characteristic modes exist. The refractive indices n for the two modes are given by the Appleton-Hartree formula [3, 4, 39]

$$n^2 = 1 - \frac{X}{U - Y_T^2/2(U - X) \pm [Y_T^2/4(U - X)^2 + Y_L^2]^{1/2}} \tag{34}$$

where

$X = 80.6\, N_e/f^2$

$Y_T = Y \sin\theta$

$Y_L = Y \cos\theta$

$Y = 2.8 \times 10^{10}\, B_0/f$

$U = 1 - j(\nu/2\pi f)$

N_e = electron density in electrons per cubic meter

$\quad f$ = frequency in hertz

$\quad \theta$ = angle between the propagation direction and the earth's magnetic field, Fig. 7

B_0 = earth magnetic flux density in webers per square meter

$\quad \nu$ = collisional frequency in hertz between electrons and molecules

The upper sign in (34) is referred to as ordinary mode, and the lower sign the extraordinary mode.

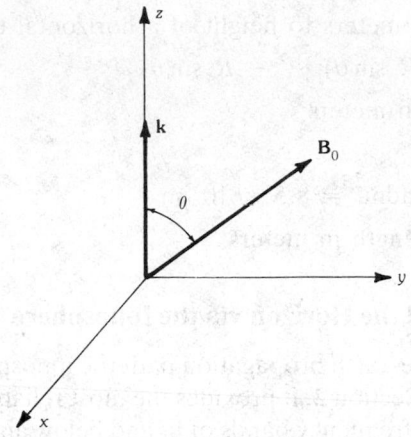

Fig. 7. Propagation geometry of radio waves in a magnetoplasma.

It is interesting to note that for $\theta = 90°$, if one neglects the collisions, the refractive index for the ordinary mode will decrease to zero as $X \to 1$, or as the signal frequency approaches the value

$$f_0 = \sqrt{80.6 N_{\max}} \tag{35}$$

where

f_0 = critical frequency in hertz

$N_{\max}$ = maximum electron density in electrons per cubic meter in the ionosphere

For a signal frequency f lower than the vertical frequency, a vertically propagated ordinary wave will be turned back by the ionosphere at a height where the electron density N_e is equal to $f^2/80.6$.

For an obliquely incident wave the maximum frequency below which the wave will be bent back to the earth is given by $f_0 \sec \phi_0$, where ϕ_0 is the angle the ray path makes with the vertical. This frequency is known as the theoretical maximum usable frequency, muf. It is this phenomenon that makes long-distance communication using high frequencies via the ionosphere possible. One-hop propagation usually can reach a maximum range of 4000 km. Several hops can be used for even longer distances. The critical frequency for reflection of the extraordinary wave is obtained from (34) by taking the lower sign. For the case $\theta = 90°$,

$$f_x \cong f_0 + f_H/2 \tag{36}$$

where

f_x = critical frequency for extraordinary wave in hertz

f_H = gyrofrequency for electrons = $2.84 \times 10^{10} B_0$ Hz

It is clear that the muf's for the two modes are different.

For a given propagation geometry the muf depends on many parameters, including diurnal and seasonal variation of ionization of the ionosphere, solar flux, magnetic-field disturbances, and propagation configuration with respect to the earth's magnetic field. Different muf's can be assigned to different ionization regions, such as the muf for the E region, muf for the E region irregularity known as "sporadic E," muf for the F region, etc. Fig. 8 shows a typical diurnal variation of critical frequencies. Both f_0 and f_x are included in the figure. In addition, f_z, corresponding to the critical frequency for the occasional z mode, and f_bE_s, corresponding to the critical frequency due to sporadic-E layers [39], are also shown in the figure. These critical frequencies are usually obtained from ionograms. An ionogram is a graph displaying the time delay for a sounding pulse to

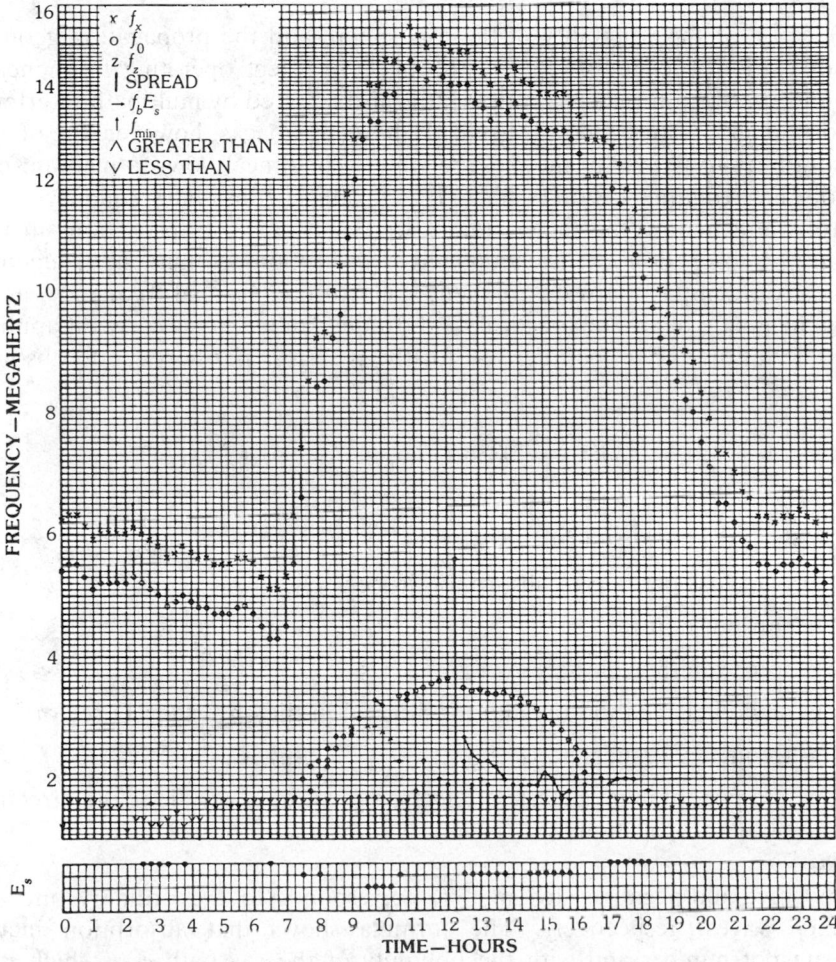

Fig. 8. Typical diurnal variation of critical frequencies at middle latitude: Washington, D.C., December 31, 1958. (*After Davies [39]*)

travel up to the ionosphere and back as a function of frequency. The sounding device is called the *ionosonde*, which essentially is a pulsed radar with a frequency-sweeping capability over a wide range of frequencies from 1 MHz up to 25 MHz in general. The frequency at which the pulse penetrates the ionosphere during vertical sounding is the critical frequency. Modern-day ionosondes can provide digitized output and additional information about the ionosphere.

The main layers in the ionosphere responsible for returning the hf radio signal are the E and F layers. F2 reflections are usually the most important. During the day, one-to-two–hop F-layer reflections are usually dominant. The F1 layer may be important for paths of 2000 km to 3500 km during the day. The E layer provides a useful propagation path during the day for distances up to 2000 km. Sporadic-E (E_s) layers may increase the muf to 150 MHz at times for ranges up to 2400 km. They can also provide a two-hop configuration together with F2-layer reflection with extended range since the D-region absorption can be avoided in the second hop.

Because of the variability of the ionosphere and the propagation geometry, hf propagation is characterized by the need to select operating frequencies as frequently as the ionosphere changes. It is also affected by multipath interference and ionospheric disturbances. Despite all these drawbacks, however, the hf circuit still continues to be one of the most reliable, economical, low- to medium-power long-distance communication systems. Due to a high collisional frequency between electrons and neutral molecules, radio waves suffer most absorptions in the D region. The absorption coefficient defined as absorption per unit length depends on the electron density, the wave frequency, electron collisional frequency, and the earth's magnetic field. For practical applications, an empirical formula applicable in middle latitudes has been derived for the path loss per hop due to absorption [39]:

$$L_a = 430(1 + 0.0035R)(\cos\chi)^{3/4}\sec\phi_D(f \pm f_L)^{-2} \tag{37}$$

where

L_a = path loss per hop to absorption in decibels

R = sunspot number

χ = solar zenith angle

ϕ_D = angle between the ray and the vertical in the absorbing D region

f = frequency in megahertz

f_L = component in megahertz of electron gyrofrequency along the direction of phase propagation

The positive-and-negative sign in (37) corresponds to the ordinary and extra-ordinary waves, respectively. The formula shows that absorption increases with sunspot number and with the obliquity of the ray path ϕ_D, which is proportional to the path length. After sunset $\chi > 90°$ and the absorption falls to low values. This is because of the disappearance of the D region after sunset. If all factors remain constant, absorption increases as the wave frequency decreases. The lowest usable frequency for a given link is called the luf.

The formula does not include the effect of the winter anomaly for which exceptionally high absorptions occur. Multiplicative factors must be used to the formula (37) for different months and geographical locations when estimating path loss during the winter anomaly period [39].

In the polar region extremely high D-region absorptions causing long-lasting blackouts of hf propagation are believed to be related closely to the energetic particle precipitation from the solar wind. These events are known as *polar cap absorptions* and are almost always preceded by major solar flare events. Another type of high-latitude absorption is the auroral blackout that occurs near the auroral zones during magnetic storms. Magnetic storms also affect the F region by reducing the critical frequency, thus lowering the muf. Other types of ionospheric disturbances include large-scale traveling ionospheric disturbances (TIDs) and small-scale irregularities. They also affect the propagation of hf waves and often cause deep fades in received signal strength.

Link budget computation for an hf circuit begins in the selection of the *optimum working frequency* (owf or FOT) which is usually chosen above the luf and just below the muf. To calculate the muf for a given path, prediction maps [40] are used. Great-circle paths can be drawn and overlaid on these maps to obtain the muf. These are monthly median values for a given hour. Day-to-day random variability of the muf can be taken into account by considering its random distribution as a function of season, solar activity, etc. With the operating frequency chosen, the path loss due to ionospheric propagation can be calculated as

$$L = L_a + L_g + L_e \tag{38}$$

where

L_a = path loss in decibels due to ionospheric absorption, computed using (37) or empirical curves

L_g = loss in decibels due to ground reflection (value ranges from 4 dB for land to 0.2 dB for ocean per reflection)

L_e = extra path loss in decibels due to defocusing of the rays upon refraction, polarization mismatch, and variability of the loss about monthly median values

Computerized prediction models are available as documented in CCIR publications [40, 41, 42]. These models can be used to estimate hf path loss at a given annual percentage of time for different locations and propagation geometries. The computer program based on [41] can be purchased from National Technical Information Services, US Department of Commerce.

Sky-Wave Propagation at Medium Frequency

Medium-frequency (mf) waves (300 kHz to 3 MHz) are primarily used for am broadcasting. From (37) we note that ionospheric D-region absorption during daytime renders medium frequencies useless for long-distance communication. After sunset, sky-wave propagation at medium frequencies becomes possible. The frequency band from 150 kHz to 1.5 MHz is used for maritime communications and navigation and medium-range broadcasting. For practical engineering evaluations at medium frequencies there are several methods recognized by the CCIR for distinctive regions of the world. The method recommended for northern America is documented here [43].

Let E be the annual median half-hourly field strength in decibels over 1 µV/m at the time of 6 hr after sunset at the receiver. Then

$$E = 10 \log P_t + G_t + (106.6 - 2 \sin \Phi) - 20 \log p - k_R p \times 10^{-3} \quad (39)$$

where

P_t = transmitted power in kilowatts

G_t = transmitting antenna gain in decibels

Φ = geomagnetic latitude in degrees of the path midpoint between transmitter and receiver

p = slant propagation path in kilometers

k_R = loss factor in decibels

It can be shown that

$$\Phi = \sin^{-1}[\sin \phi \sin 78.3° + \cos \phi \cos 78.3 \cos(\lambda - 291°)] \quad (40)$$

with ϕ and λ being the north latitude and east longitude of the midpoint. Furthermore,

$$p = \begin{cases} d, & d < 1000 \text{ km} \\ (d^2 - 40\,000)^{1/2}, & d > 1000 \text{ km} \end{cases} \quad (41)$$

where d is the ground distance, which can be found in the Solar Geographical Data Reports published by NOAA. Also,

$$k_R = bR \times 10^{-2} + K \quad (42)$$

$$b = \begin{cases} 0.4|\Phi| - 16, & \Phi > 45° \\ 0, & \Phi < 45° \end{cases} \quad (43)$$

where R is the smoothed Zurich sunspot number, usually between 20 and 150, which can be found in the Solar Geographical Data Reports published by NOAA. Also,

$$K = [0.0667 \times |\Phi| + 0.2] + 3 \tan^2(\Phi + 3°) \quad (44)$$

For local time other than 6 hr after local sunset the value of E can be adjusted by a reduction factor shown in Fig. 9. It has been found that the method provides a good agreement with the well-known FCC curves [44] as well as with measurement data from Europe and other parts of the world [45].

Ionospheric Waveguide-Mode Propagation at Very Low and Extremely Low Frequencies

For signals with frequencies below 30 kHz the wavelength becomes comparable to the scale height of the ionosphere, and the ionosphere behaves as a relatively well defined reflecting layer for these waves. This layer and the highly conducting surface of the earth (at these frequencies) constitute the boundaries for a spherical waveguide. Several special features stand out for the characteristic modes propagating in the ionosphere-earth waveguide. First, natural resonances can exist in this spherical shell structure. Indeed, these resonant oscillations have been observed near 8, 14, and 20 Hz and higher frequency bands. They are believed to be excited by naturally occurring lightning discharges and are known as Schumaun resonance [46]. Secondly, guided modes at vlf and elf bands have relatively low path attenuation, which is quite stable with time. Thirdly, good phase stability exists for the dominant mode. These last two points are in contrast with hf transmissions, which suffer from unreliability during ionospheric disturbances and rapid and deep fading. Therefore vlf and elf systems are attractive candidates for around-the-globe, highly reliable communication, navigation, and timing applications. Indeed, the modern Omega navigation system uses vlf transmissions at 10.2, 11.33, and 13.6 kHz [47]. The disadvantages of systems in these frequency bands are the large antenna installations and limited usable bandwidths. This is especially true due to the necessary high-Q antennas. Typically the bandwidths are limited to 20 to 150 Hz.

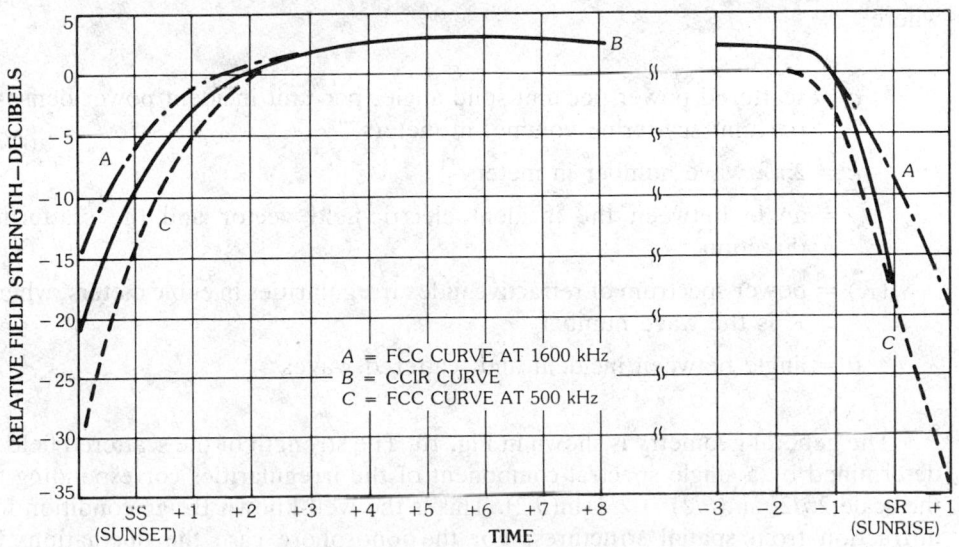

Fig. 9. Adjustment factor for F.

Very low and extremely low frequency propagation in the ionosphere-earth waveguide has been studied using different models [48]. The basic model is a spherical shell structure with a perfectly conducting boundary on the earth surface and an impedance boundary for the ionosphere. Refined models take into account the anisotropy of the ionosphere plasma, the terrain features on the surface, the penetration of the wave into the ionosphere, and the effect of propagation across a sunrise and sunset line. Predicted attenuation coefficients and phase propagation speeds from model computations compare rather favorably with measured data. These results are used in link budget computation for a given path. Since atmospheric noise is quite high in these frequency bands, it has to be included in the computation. A practical procedure for calculating power requirements for a vlf station operating over a specified path in the presence of a given level of background noise can be found in CCIR reports 265-24, 265-5 [46, 49].

Scatter Propagation at Very High and Ultrahigh Frequencies

At vhf and uhf frequencies the radio waves penetrate the ionospheric layers, and the over-the-horizon transmissions via the regular reflection of the ionosphere become impossible. However, irregular structures in the ionosphere are capable of scattering the wave energy back to the earth surface and yield much higher field strength. This is the basic mechanism for scatter propagation. It has been demonstrated that irregularities in the D region can be used to establish a scatter communication link for ranges up to a few thousand kilometers. The prediction of the field strength in such a link is based on the well-known Booker-Gordon formula for a scattering cross section given by [50]

$$\sigma = 2\pi k^4 \sin^2\chi \, S_n(2k \sin \theta/2) \tag{45}$$

where

σ = scattered power per unit solid angle, per unit incident power density, per unit scattering volume, in meters^{-1}

$k = 2\pi/\lambda$ wave number in meters^{-1}

χ = angle between the incident electric-field vector and the scattering direction

$S_n(K)$ = power spectrum of refractive index irregularities in cubic meters, where K is the wave number

θ = angle between incident and scattered waves

The general geometry is shown in Fig. 10. The strength of the scattered field is determined by a single spectral component of the irregularities corresponding to the scale $2\pi/2k \sin(\theta/2) = \lambda/2 \sin(\theta/2)$. This is the well-known Bragg condition for diffraction from spatial structures. For the ionosphere case the fluctuations in the refractive index are caused by electron density irregularities. The received scattered power is given by

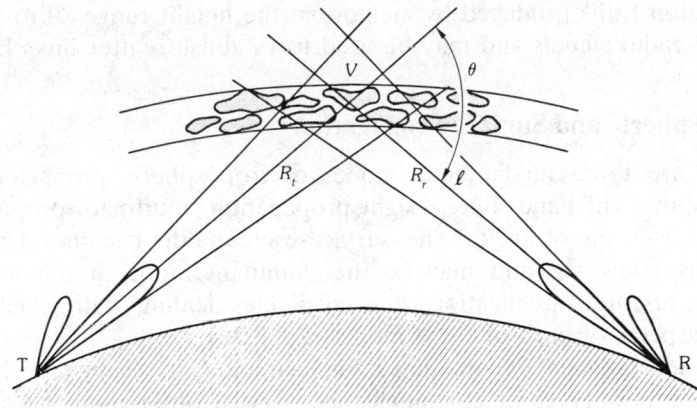

Fig. 10. Geometry of ionospheric propagation based on scattering mechanism.

$$P_r = \frac{\pi P_t G_t G_r \lambda^2 r_e^2}{2 R_t R_r^2} \sin^2\chi \cdot S_N(2k \sin\theta/2)\, V \qquad (46)$$

where

P_r = received power in watts

P_t = transmitted power in watts

G_t = antenna gain for the transmitting antenna

G_r = antenna gain for the receiving antenna

λ = wavelength in meters for the radio wave

R_t = distance in meters from the transmitter to scattering volume

R_r = distance in meters from the receiver to scattering volume

r_e = classical electron radius (2.82×10^{-15} m)

$S_N(k)$ = power spectrum of electron density fluctuations in meters^{-3}

V = common volume of the scattering region in cubic meters

Equation 46 can be compared to (2). From (46) it can be shown that the received power is proportional to $\lambda^m/\sin^p(\theta/2)$. The values of m and p depend on the spectral shape of the irregularity spectrum; observations show m varies between 7 and 9.5 while p is found to be in the range of 6 to 8.5. In deriving (45), the irregularities are assumed to be isotropic. This equation has to be modified if field aligned structures are responsible for the scattering.

At the equator, the F-region spread-F type of irregularities can support trans-equatorial vhf propagation at levels 30 to 40 dB above the level produced by usual scatter propagation. This usually occurs during the hours after local ionospheric sunset. It is closely related to the equatorial irregularities causing scintillations of the transionospheric signal as discussed in Section 2.

Ionization trails produced by meteors in the height range 80 to 120 km also scatter vhf radio signals and may be used to establish scatter links [39].

4. Tropospheric and Surface Propagation

There are three fundamental modes of tropospheric propagation for frequencies above vhf band: line-of-sight propagation, multipath propagation, and diffraction over an obstacle. The surface-wave mode becomes important for frequencies below vhf and may be the dominant mode at mf and lf bands. Numerous research publications are available dealing with theoretical and modeling aspects of each of these modes [2, 51, 52].

Rapid advancement of communication technology in recent years have opened the door for new types of communication service other than the conventional point-to-point communication. It becomes evident that propagation anomalies due to either tropospheric irregularities or surface irregularities affect each type of communication differently simply because of the unique requirement of each individual type. For instance, Doppler effects are much more serious for mobile unit communication than for fixed station communication. Among types of mobile unit communication, land- and sea-mobile units are sensitive to the surface-wave mode of propagation while air-mobile units are not. A broadcasting system involves point-to-area coverage, and so has completely different interference considerations than a point-to-point radio-relay system. A further complication is that even for one specific type of service, propagation degradations are highly region-dependent due to different meteorological conditions. For instance, the CCIR propagation document for a transhorizon radio-relay system [53] outlines the effect of nine climatological regions, including equatorial, continental subtropical, maritime subtropical, desert, Mediterranean, continental temperate, maritime temperate over land, and maritime temperate over sea and polar regions.

Obviously it is impractical to summarize here all engineering models for different communication services and applications at various climatological regions. The objective of this section is to provide a few sets of practical equations which enable one to gain an intuitive physical insight into propagation degradation due to the presence of the troposphere and the earth's surface. Such intuitive insight is always a prerequisite for an engineer to apply judiciously the specific modeling procedure among more than a dozen or so postulated by the CCIR, FCC, and others.

Refractive Index of the Atmosphere

The refractive index n in a medium is a function of ϵ_r and is given by (9). Typical values of ϵ_r and σ (in siemens per meter) relevant to troposphere and surface propagation are shown in Fig. 11. In the atmosphere, $\sigma \cong 0$; therefore

$$n = \sqrt{\epsilon_r} \tag{47}$$

In practice, the radio refractivity N is defined as

$$N = (n - 1) \times 10^6 \tag{48}$$

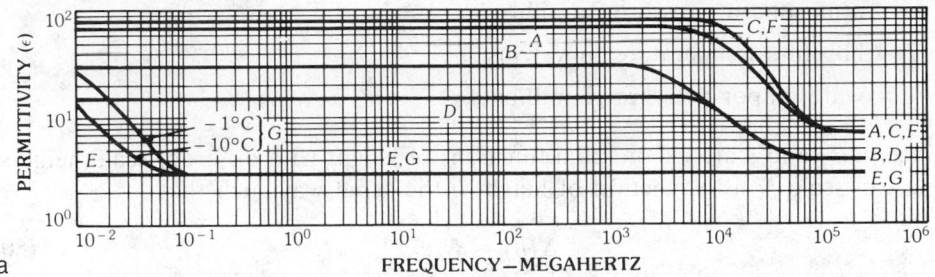

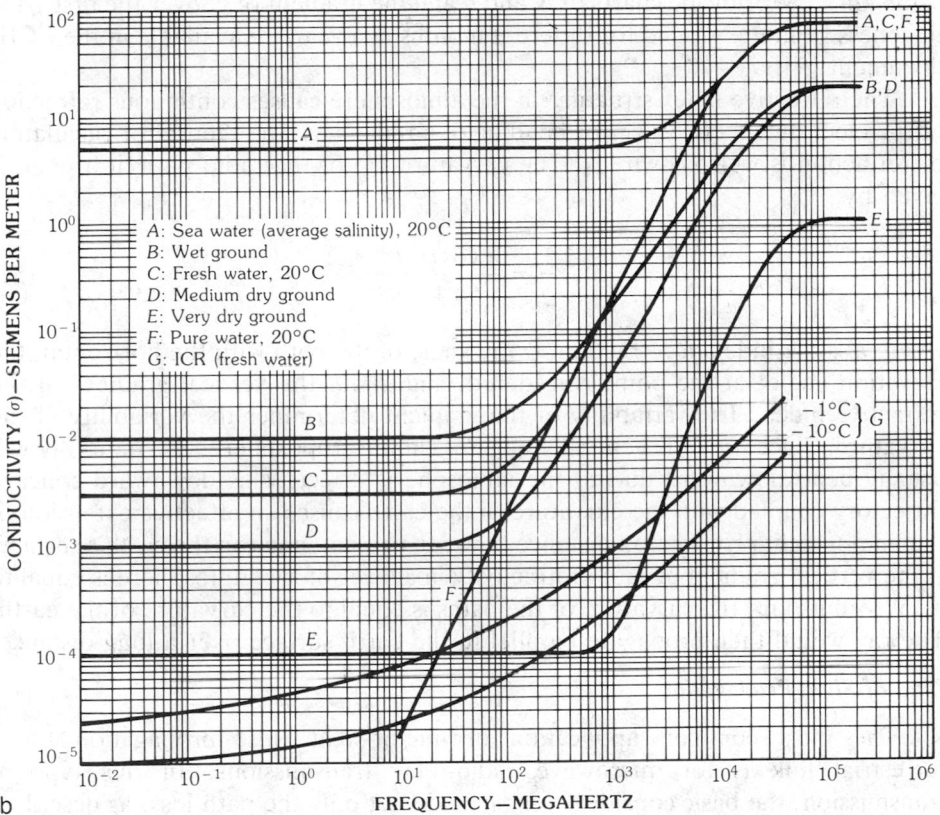

Fig. 11. Typical values of relative permittivity ϵ_r and conductivity σ for the troposphere propagation environment. (*a*) Relative permittivity. (*b*) Conductivity.

It is known that for all frequencies up to 100 GHz, the following approximations apply:

$$N = 77.6\frac{p}{T} + 3.73 \times 10^5 \frac{e}{T^2} \tag{49}$$

where

p = atmospheric pressure in millibars

T = absolute temperature in kelvins

e = water vapor pressure in millibars

Since p, T, and e are exponential functions of height with different scale heights, one can scale N as a function of height h the same way such that

$$N(h) = N_s e^{-bh} \tag{50}$$

The CCIR defines a "standard radio atmosphere" for which $N_s = 315$ units and $b = 0.136$ km^{-1}. Worldwide charts of N and b and the gradient of N over the first 1-km thickness from the ground for different months in a year are available in the CCIR document [54].

The refractive index structure in the atmosphere causes continuous refraction of the electromagnetic wave, or bending of a radio ray. The theory for calculation of refraction is well known [1]. The curvature $1/\varrho$ of the radio path at a point is given by

$$\frac{1}{\varrho} = \frac{-\cos\phi}{n} \frac{dn}{dh} \tag{51}$$

where n is the index of refraction, ϕ the angle of the path with the horizontal, the elevation angle, at the point considered, and dn/dh the vertical gradient of the refractive index. The variability of the gradient determines the variability of the curvature. Since dn/dh is generally negative in the troposphere, the radio ray will usually bend downward due to the atmospheric effect. This downward concave trajectory coupled with the curvature of the earth causes an effective extension of the line-of-sight propagation distance between two antennas on the earth's surface. If the vertical gradient of radio refractivity in a layer of the atmosphere is equal to $-157N$ units/km, the curvature of the paths is equal to the curvature of the earth. The ray would then propagate parallel to the earth surface over a long distance.

Line-of-Sight Propagation

The most common application of line-of-sight (los) propagation for a terrestrial link is for microwave radio-relay transmissions. In this type of transmission, the basic considerations include not only the path loss, as described in (6) and (7), but also the clearance of the path. An illustration is provided in Fig. 12a. An los straight path from transmitter T to receiver R may appear to have enough clearance over a hill with peak P. In reality, due to atmospheric refraction, the los ray simply hits point Q of the hill.

The amount of bending is a function of the vertical gradient dn/dh as given in (51). For practical computations the curvature of the rays can be taken into account by introducing an equivalent earth radius factor K. This factor, K, multiplied by the actual earth radius a_e, is the radius of an equivalent earth. When considering the path profiles, this equivalent earth curvature is such that the ray paths can be represented by straight lines. For the standard radio atmosphere, the effective earth radius factor $K = 4/3$. This concept was used in Fig. 1, where $R_e = \frac{4}{3} \times$

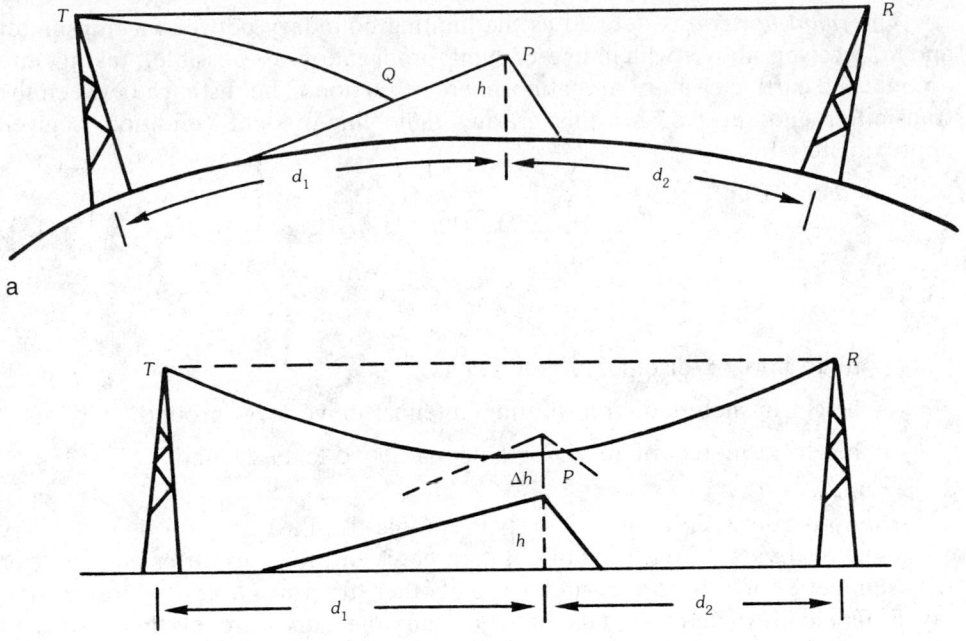

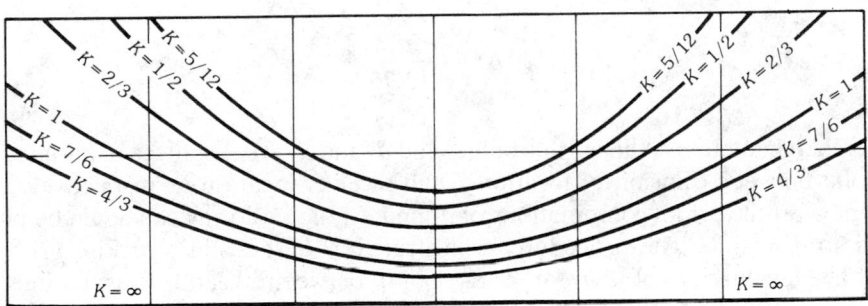

Fig. 12. Geometry of radio-wave propagation. (*a*) Propagation between a transmitter and a receiver of a radio wave in the troposphere. (*b*) Construction of the path profile. (*c*) Standard *K* curves.

earth's radius. Atmospheric refraction, however, may fluctuate, depending on a variety of factors including time of day, weather condition, the season, and geographical location. The relationship between K and ΔN (in N units), the mean refractivity change from the ground level to a 1-km altitude, is given by

$$K = \frac{1}{1 + 6.37\Delta N \times 10^{-3}} \qquad (52)$$

Data on the parameter ΔN have been compiled in a CCIR document [54]. Typically K varies from 4/3 to 5/12.

The *radio horizon* is defined as the limiting boundary between a transmitter and a receiver above which line-of-sight propagation is possible, taking into account the earth curvature and atmosphere refraction. The distance between the transmitter and receiver for this grazing radio line-of-sight condition is given approximately by

$$d_h = 3.57\sqrt{K}(\sqrt{h_1} + \sqrt{h_2}) \tag{53}$$

where

d_h = radio horizon distance in kilometers

h_1 = height in meters of transmitting antenna above level ground

h_2 = height in meters of receiving antenna above level ground

The procedure to construct a path profile is illustrated in Fig. 12b. A cross-sectional view of the topological map between the transmitter and receiver with silhouettes of hills, trees, buildings, and other structures that may obstruct the ray is first constructed. To take into account the curvature of the earth and atmospheric refraction, an earth's bulge should be added to the profile. The bulge is given by

$$\Delta h = \frac{d_1 d_2}{12.75\,K} \tag{54}$$

where Δh in meters is the earth's bulge, and d_1 and d_2 are distances in kilometers of a point between transmitter location T and receiver location R, respectively. With the new profile for all combinations of d_1 and d_2, the radio ray can again be plotted as a straight line. The procedure is illustrated as dashed lines in Fig. 12b.

The evaluation of Δh for every point between T and R is tedious. An alternative method is to bend the wave relative to the original topological cross section. This is shown in solid lines in Fig. 12b. As a matter of fact, the bending of waves for different K values can be prepared on a standard sheet as shown in Fig. 12c, which can then be used as an overlay on any cross-sectional view. This is indeed the method used by engineers [55].

The actual los propagation degradation is of course more than bending as mentioned above. The height of the transmitter and receiver towers may not be the same, thus the path may be a slant path. The true ray path can be determined only by a ray-tracing program [56] using a measured atmospheric refractive index profile as the time path length.

The path losses in the line-of-sight propagation situation are mainly due to hydrometeors and gaseous absorptions, as discussed previously in Section 2 for the satellite-earth propagation case. The basic principles in computing the path losses are the same. The main differences are the propagation geometry and the height region where the signal propagates. For gaseous absorption for a terrestrial path, the path loss is given by [57].

$$L_{\text{gas}} = \gamma_a d \tag{55}$$

where

L_{gas} = gaseous attenuation in decibels

γ_a = specific absorption loss in decibels per kilometer

$d = d_1 + d_2$, the path length in kilometers

Fig. 13, which is similar to a corresponding part of Fig. 2, shows γ_a by water vapor and oxygen for a water vapor density of 7.5 g/m^3 at an air temperature of 20°C and atmospheric pressure of 1000 mbar.

Similarly, for rain, the path loss is computed by

$$L_R = \gamma_R d\delta \tag{56}$$

where

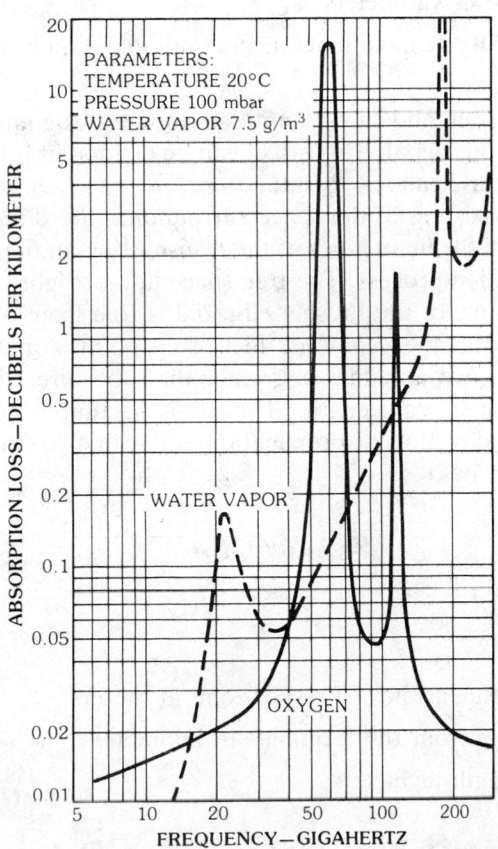

Fig. 13. Absorption by water vapor and oxygen in the standard radio atmosphere. (*After Brodhage and Hormuth [57], © Siemens Aktiengesellschaft, Berlin and Munich*)

L_R = attenuation in decibels due to rain

γ_R = specific attenuation in decibels per kilometer due to rain

d = path length in kilometers

δ = reduction factor for P percent of the time

The term γ_R can be computed by the same equation as in the case of the satellite-earth propagation path.

For the line-of-sight propagation geometry on the ground, the reduction factor is approximately given by

$$\delta = 61[1 - \exp(-d^{0.5}R^{0.7}/61)] d^{0.5}R^{0.7} \tag{57}$$

where

δ = reduction factor for P percent of the time

d = path length in kilometers

R = point rainfall rate in millimeters per hour at a required P percent of the time

In addition to rain and snow, scintillation due to fog and clear air turbulence may also degrade the signal. This effect can be estimated following the procedure discussed in Section 2, under "Scintillation."

All obstacles existing in the space surrounding the direct line-of-sight in the common volume of the main lobes of the transmitting and receiving antennas may affect the propagation process. The free-space line-of-sight propagation condition prevails if the volume of the Fresnel ellipsoid is free from obstacles. The Fresnel ellipsoid is defined as the geometric loci of all points for which the sum of the distances from the two antennas is greater than the direct distance between the antennas by one-half wavelength. The cross section of the Fresnel ellipsoid orthogonal to the direction of propagation is the first Fresnel zone whose linear dimension is given by

$$r_F = 31.6\sqrt{d_1 d_2 \lambda/(d_1 + d_2)} \tag{58}$$

where

r_F = linear dimension of Fresnel zone in meters

d_1, d_2 = distances from the terminals in kilometers

λ = wavelength in meters

Higher-order Fresnel zones are given by $r_F^{(n)} = \sqrt{n}r_F$.

Diffraction and reflection effects have to be taken into account if obstacles fall into the Fresnel zone along the line-of-sight path.

Multipath Propagation over an Earth Surface

Multipath propagation is important when a radio-wave transmission can no longer be considered a simple los transmission isolated from the reflecting surfaces in the terrain, buildings, and tropospheric layers. The reflected signals are able to reach the receiving antenna and interact with the los signal. This happens, for example, in the case of low elevation angle transmission and transmission involving wide antenna beams. The principal factor in multipath propagation is the relative phases of the waves that arrive at the receiving antenna via different paths. This is particularly significant at low elevation angles where the reflected wave from the surface may have the same amplitude as the direct wave along an los path. The phase differences depend on the relative positions of the reflectors, whether they are in the first Fresnel zone or higher-order Fresnel zones.

The primary type of multipath propagation is that of a two-path propagation, i.e., a direct wave and a specularly reflected wave from the surface. The geometry is provided in Fig. 1. A general equation for evaluating loss factor F [cf. (3) and (7)] is given by [58, 59]

$$F = 1 + \Gamma g \varrho_s D e^{-j\Delta} \tag{59}$$

Each of the five parameters Γ, g, ϱ_s, D, and Δ requires a few lines of explanation.

The first parameter Γ is the Fresnel reflection coefficient. It assumes a value either given by Γ_H or Γ_V, depending on whether the wave in consideration is horizontally or vertically polarized, where

$$\Gamma_H = \gamma_H e^{j\theta_H} = \frac{\sin\psi - \sqrt{\epsilon_r' - \cos^2\psi}}{\sin\psi + \sqrt{\epsilon_r' - \cos^2\psi}} \tag{60}$$

$$\Gamma_V = \gamma_V e^{j\theta_V} = \frac{\epsilon_r' \sin\psi - \sqrt{\epsilon_r' - \cos^2\psi}}{\epsilon_r' \sin\psi + \sqrt{\epsilon_r' - \cos^2\psi}} \tag{61}$$

where ϵ_r' is given in (9) and ψ is the grazing angle.

For example, according to Fig. 11, $\epsilon_r \cong 80$ and $\sigma \cong 4$ for seawater, the value of ϵ_r' at 1.54 GHz (L-band maritime communication) is $80 - j47$. Since $|\epsilon_r'| \gg 1$ for seawater, (60) yields $\gamma_H \cong 1$, $\theta_H \cong 180°$ for $0 < \psi < 10°$ for a wide frequency range from 100 MHz to at least 30 GHz; values of γ_V and θ_V are shown in Figs. 14a and 14b [52].

Propagation at low elevation angles often involves circular rather than linear polarization. In this case Γ should assume the value of either Γ_R or Γ_L, where

$$\Gamma_{R,L} = \frac{\Gamma_H \pm j\Gamma_V}{\sqrt{2}} \tag{62}$$

Magnitudes of $\Gamma_{R,L}$ at 1.54 GHz over the sea are shown in Fig. 15 [59].

The parameter g in (59) is the antenna pattern factor between the direct wave and reflected wave. Except for the phased array type of antenna, where g may be complex, the value of g can be considered as a scalar related to the antenna gain

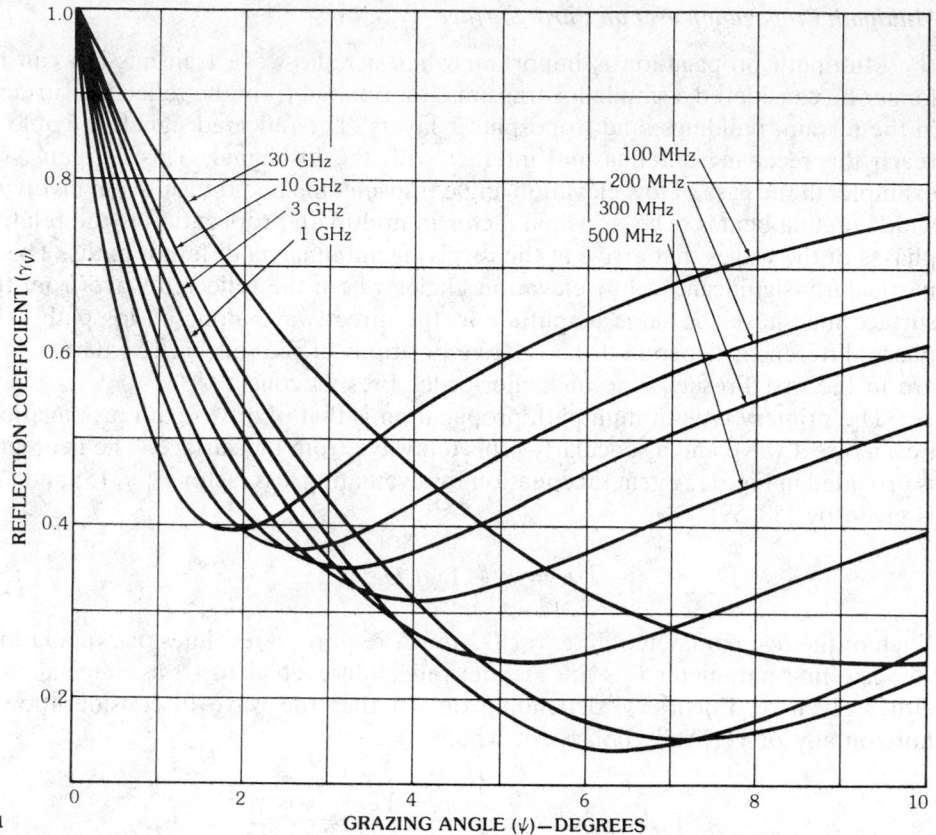

Fig. 14. Reflection coefficient γ_V and phase angle θ_V for vertically polarized wave over the ocean's surface. (*a*) Reflection coefficient. (*b*) Phase angle. (*After Blake [58]*)

function. For example, assume the antenna at point A (Fig. 1) is pointing along R_d toward point D, and the 3-dB beamwidth angle is ϕ_{3dB}. Then

$$g \cong \exp\left[-\left(\frac{\theta_d + \theta_r}{1.7\phi_{3dB}}\right)^2\right] \cong \exp\left[-\left(\frac{\psi}{\phi_{3dB}}\right)^2\right] \tag{63}$$

The last approximation in the above equation is valid for low grazing angles ($\psi \cong 10°$) only. The magnitudes of $\Gamma_{R,L}$ g represented as γ'_R and γ'_L according to (62) with $\phi_{3dB} = 7.22$ (14.4° beamwidth) are also shown in Fig. 15 [59].

Since the reflection coefficient Γ described above concerns only a smooth surface, a further correction factor to take into account surface roughness is required when the height deviations from a smooth surface over the area of the first Fresnel zone exceed the Rayleigh limit $\lambda/16 \sin \psi$. This is provided by ϱ_s in (59). Depending on the nature of surfaces (sea, land, ice, desert, forest, city, etc.) and roughness conditions (sinusoidal, layered, perfectly or partially conducting, hemispherical blobs, etc.), theoreticians have postulated numerous formulations for ϱ_s [5, 52, 58–64]. For the sea surface, a practical first-order equation for ϱ_s is

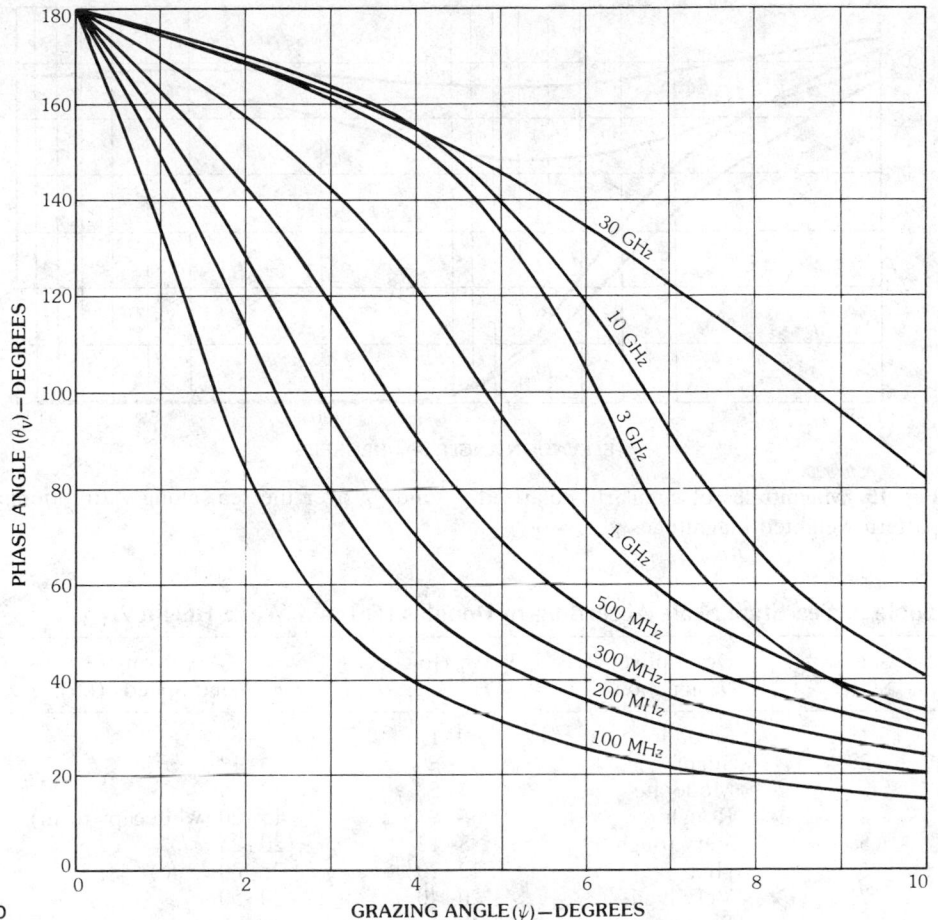

Fig. 14, *continued.*

$$\varrho_s = \exp\left[-2\left(\frac{2\pi\sigma_h \sin\psi}{\lambda}\right)^2\right] \tag{64}$$

where σ_h is the standard deviation of sea wave height, which can be related to the "significant height" $H_{1/3}$ generally used by oceanographers [65, 66]:

$$\sigma_h = \frac{1}{4}H_{1/3} \tag{65}$$

For example, for sea state 3, $H_{1/3}$ is about 4 ft or 1.23 m. At 1.54 GHz and 5° elevation angle, $\varrho_s \cong 0.23$. Table 3 provides general values for $H_{1/3}$.

The next factor in (59) is the divergence factor D, which considers the effect of a spherical earth. For most practical applications this factor can be approximated by [58]

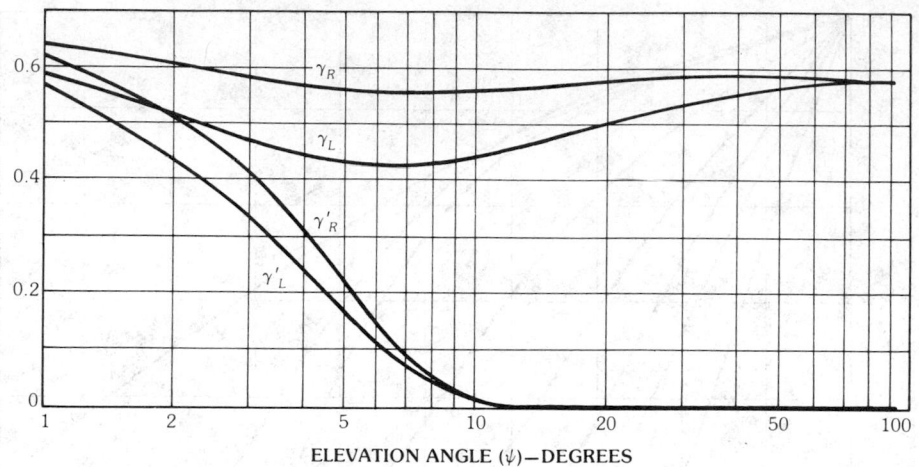

Fig. 15. Magnitudes of circularly polarized γ_R and γ_L over the sea, along with antenna-pattern weighted magnitudes $\gamma'_{R.L} = g\gamma_{R.L}$.

Table 3. Sea State Scale According to Douglas [66] and Wave Height $H_{1/3}$

Sea State	Descriptive Designation	Wave Height $H_{1/3}$ (ft)	Approximate Wind Speeds (kn)
1	Smooth	0–1	0–6
2	Slight	1–3	6–12
3	Moderate	3–5	12–15
4	Rough	5–8	15–20 (whitecaps form)
5	Very rough	8–12	20–25
6	High	12–20	25–30 (gale)
7	Very high	20–40	30–50
8	Precipitous	40	>50 (storm)

$$D \cong \left(1 + \frac{2G_1G_2}{a_eG\sin\psi}\right)^{-1/2} \tag{66}$$

where G_1, G_2, and G are shown in Fig. 1.

Factor Δ in (59) is the phase angle difference due to path length difference between the direct wave and the reflected wave, i.e.,

$$\Delta = \frac{2\pi}{\lambda}(R_1 + R_2 - R_d) \cong \frac{4\pi}{\lambda}h_1\psi \tag{67}$$

where the last approximation is valid at low grazing angles for a satellite-earth path with h_1 being the height of the transmitting antenna and ψ being the grazing angle in radians [59].

Equation 59 can now be employed for the evaluation of microwave fading. In addition to net attenuation, microwave signals under multipath conditions often fade with increasing peak-to-peak fluctuations as ψ decreases. The magnitude of fades can also be estimated by (59) by taking extreme values of F. Thus the peak-to-peak fluctuations of the signal power, ΔP in decibels, is

$$\Delta P = 20 \log \left(\frac{1 + |\Gamma g \varrho_s D|}{1 - |\Gamma g \varrho_s D|} \right) \tag{68}$$

The temporal variation of the signal is due to the fact that the random reflecting surface is moving or changing with respect to time, or the different paths are varying due to the varying condition of the atmosphere. Indeed, under these conditions, a Doppler spread will appear in the received signal when the input to the channel is a stable single-frequency signal.

In addition to the multipath propagation due to reflections from the earth's surface, tropospheric layers with refractive index structures substantially different from the ambient may provide reflections of the signal, giving rise to multipath propagation conditions. These layers usually are formed over flat ground, especially in humid areas, during calm weather, when the air temperature at higher levels is greater than that near the ground. Such layers can exist both above and below the direct path between the transmitting and receiving antennas. Total reflections, hence multipaths, can occur for small angles of incidence (less than about 0.5°). Sometimes when the conditions are favorable, waveguide modes can be set up between such a layer and the ground, or within the layer itself (elevated duct). Overreach propagation can be produced in such ducts. The maximum wavelength capable of propagating in a duct without loss is given approximately by [67]

$$\lambda_{\max} = 2.5 \, \Delta h \, (\Delta n)^{1/2} \tag{69}$$

where

$\lambda_{\max}$ = maximum wavelength in a duct in meters

Δh = depth of the duct in meters

Δn = difference of refractive index inside and outside the duct

Diffraction

Although, as discussed previously, under "Line-of-Sight Propagation," an los link should be designed for enough clearance by elevated antennas, this is not always possible. Obstacles such as hills, buildings, trees, etc., may be within the first Fresnel zone of the path or may even completely block the los path. The same can of course be said about transmissions under multipath conditions, as described in the preceding subsection, "Multipath Propagation over an Earth Surface." In these situations, wave energy can still be transmitted to the receiver through diffraction. But, in addition to the free-space path loss, the obstacle produces further losses due to the diffraction of waves over the obstacle. For transmissions at vhf and higher frequencies the losses can be evaluated by a simple CCIR model [68] as illustrated below.

For a single, general, blob-shaped obstacle as shown in Fig. 16, the free-space loss L_0 can still be calculated by (6), provided that the path length D is replaced by the actual path length $d_1 + d_2$. The additional medium loss L_p in decibels, due to the blob, can be considered as a sum of three terms:

$$L_p = F(v) + G(\varrho) + E(\chi) \tag{70}$$

where $F(v)$ is the Fresnel-Kirchhoff knife-edge loss given by Fig. 17 with the dimensionless parameter v given by

$$v = 2\sin(\theta/2)\left[\frac{2(d_1 + R\theta/2)(d_2 + R\theta/2)}{\lambda d}\right]^{1/2} \tag{71}$$

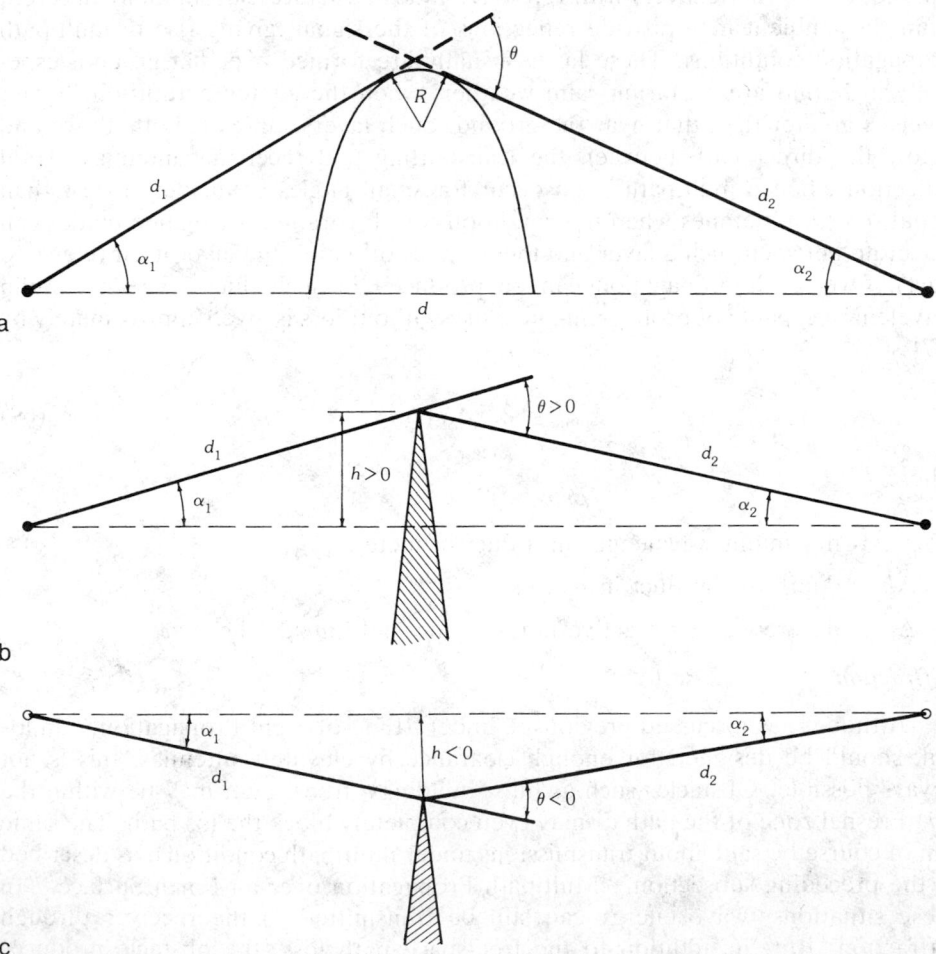

Fig. 16. Propagation geometry for diffraction over a blob. (*a*) Illustrating physical shape of blob. (*b*) Knife-edge diffraction for $h > 0$ and $\theta > 0$. (*c*) Knife-edge diffraction for $h < 0$ and $\theta < 0$.

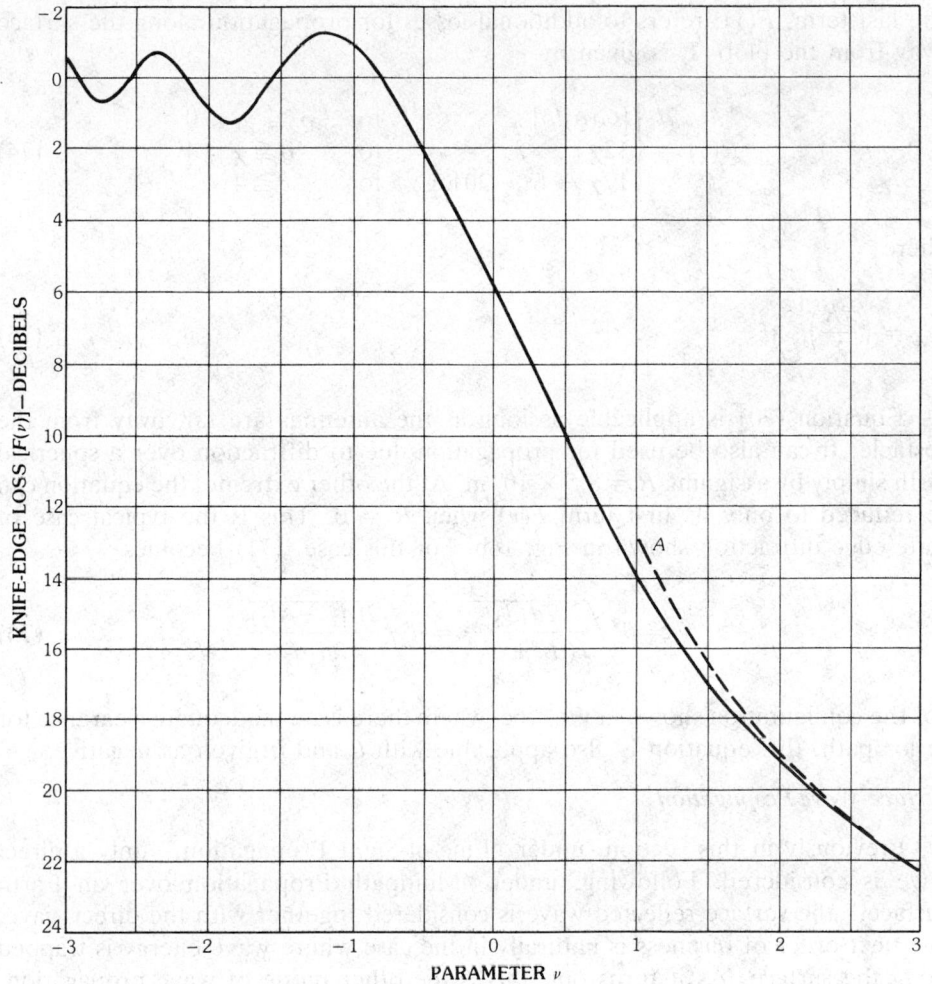

Fig. 17. Knife-edge–diffraction transmission loss relative to free space, where curve $A = 13 + 20 \log \nu$.

in which λ is the wavelength, d is the straight-line distance, and R is the radius of curvature. The term $G(\varrho)$ refers to a loss due to the curved surface. It can be calculated by

$$G(\varrho) = 7.2\varrho - 2.0\varrho^2 + 3.6\varrho^3 - 0.8\varrho^4 \tag{72}$$

with

$$\varrho^2 = \left(\frac{d_1 + d_2}{d_1 d_2} \right) \bigg/ \left[\left(\frac{\pi R}{\lambda} \right)^{1/3} \frac{1}{R} \right] \tag{73}$$

The last term, $E(\chi)$, refers to additional losses for propagation along the surface away from the blob. It is given by

$$E(\chi) = \begin{cases} [G(\varrho)/\varrho] & \text{for} \quad -\varrho \leqq \chi < 0 \\ 12\chi & \text{for} \quad 0 \leqq \chi < 4 \\ 17\chi - 6 - 20\log\chi & \text{for} \quad x \geqq 4 \end{cases} \qquad (74)$$

where

$$\chi = \left(\frac{\pi R}{\lambda}\right)^{1/3} \theta \qquad (75)$$

Equation (70) is applicable as long as the antennas are far away from the obstacle. It can also be used for propagation due to diffraction over a spherical earth simply by assigning $R = 8.5 \times 10^6$ m. At the other extreme, the equation can be reduced to only its first term $F(v)$ when $R = 0$. This is the typical case of knife-edge diffraction shown in Fig. 16b. For this case, (71) becomes

$$v = \theta\sqrt{\frac{2d_1 d_2}{\lambda(d_1 + d_2)}} = h\sqrt{\frac{2(d_1 + d_2)}{\lambda d_1 d_2}} \qquad (76)$$

For the configuration shown in Fig. 16c, where there is an inadequate clearance for an los path, this equation is also applicable with θ and h given as negative.

Surface-Wave Propagation

Previously in this section, under "Line-of-Sight Propagation," only a direct wave is considered. Following, under "Multipath Propagation over an Earth Surface," the surface-reflected wave is considered together with the direct wave. The next order of business is naturally in the case where wave energy is trapped along the surface. As it turns out, this is the other mode of wave propagation, known as *surface-wave* propagation. This mode arises because of the boundary between the two different media, the air and the earth. The wave is guided along the earth's surface while energy is abstracted from it to supply the losses in the ground.

The intensity of a surface wave decreases exponentially as height increases. The effect has been successfully modeled by an $\exp(-\alpha h/\lambda)$ factor where α is the attenuation factor, and h/λ is the height in wavelengths. Since there is always a practical height for the antenna of a communications system, it follows that the surface wave is important only for low frequencies where the wavelength is at least in the order of meters. This establishes the upper limit of frequencies (300 MHz) for surface waves.

The attenuation of the surface wave depends on the conductivity of the earth and the wave frequency. The attenuation is low at lower frequencies over a good conducting ground. Therefore, in practice, surface waves are most important in the mf and lf bands. Vertical polarization is always used in surface-wave propagation. This is due to the fact that at these frequencies the antenna heights are usually only a fraction of a wavelength. If horizontal polarization were used, the field induced in the earth would virtually cancel out with the field in the antenna. In practice, a surface wave can be conveniently excited by a vertical dipole antenna near the earth's surface. When both transmitting and receiving antennas are located right at the earth's surface, the surface wave is the major component of the wave energy, since the direct line-of-sight and ground-reflected components will nearly cancel each other.

The study of surface-wave propagation with a vertical dipole as a source in the presence of either a plane or a spherical earth is a classical problem. It was originally formulated by Sommerfeld in 1909 and revised by Norton in the late thirties for engineering applications [69], most of which are still being quoted in the current FCC documentations. Over the years, enormous new analytic approaches have been devised, most of which provide clearer ways of bookkeeping. Chapter 2 of Kerr's book [2], a textbook by Jordan and Balmain [70], and others [71, 72] provide excellent coverage on the subject. One aspect which often causes confusion is the definition and the context of the surface waves referred to in the literature. As stated above, under "Diffraction," the diffraction mechanism can also support wave propagation along the surface. This is true even in the case where the surface is perfectly conducting. Blob-shaped conductor and/or sinusoidal corrugations are known to enhance wave diffractions. In theoretical studies the trapped surface wave and the diffracted wave along the surface are difficult to separate mathematically. The term "surface wave" is thus sometimes used loosely for either type of these waves, or both.

Fig. 18 shows the ground-wave attenuation factor F for a vertical dipole over a plane earth in terms of a numerical distance given approximately by [69]

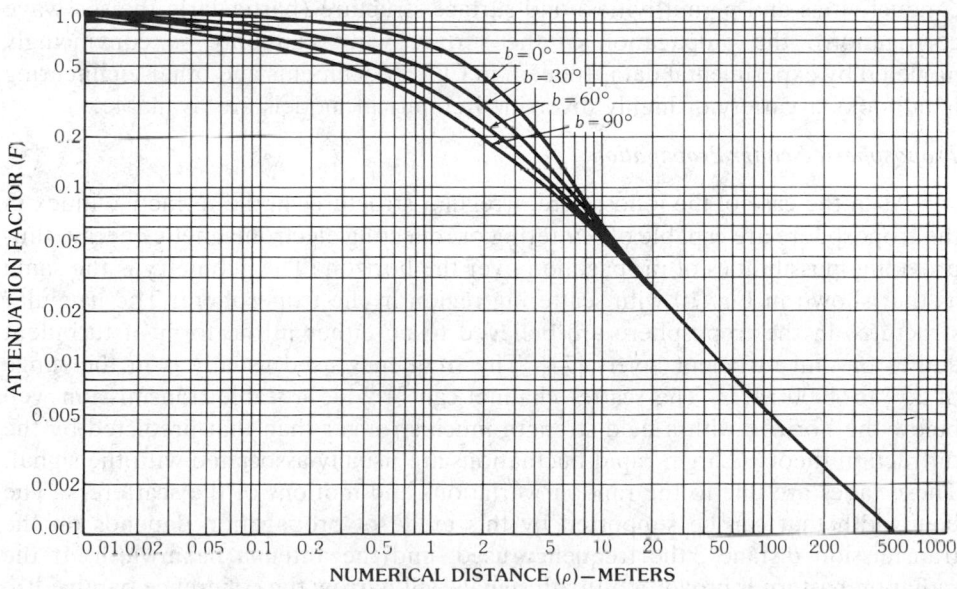

Fig. 18. Ground-wave attenuation factor F. (*After Jordan and Balmain [70]*, © *1968 Prentice-Hall, Inc., Englewood Cliffs, New Jersey; reprinted with permission*)

$$p = \frac{Dk}{2\chi}\cos b \qquad (77)$$

where

$$b = \tan^{-1}\left(\frac{\epsilon_r + 1}{\chi}\right)$$

$$\chi = \frac{\sigma}{\omega\epsilon_0} = \frac{18 \times 10^3 \sigma}{f_{MHz}}$$

σ = conductivity in siemens per meter of the earth

D = distance in meters between transmitting and receiving antennas

Both the transmitting and receiving antennas are on a plane earth surface.

For the case of spherical earth, the expression of F is very complicated. For hf applications the earth can often be approximated as perfectly conducting. Then again, for transmitter and receiver both on the surface separated by a distance D, the F factor is [73]

$$F \cong 2\left[\pi\left(\frac{k}{2R_e^2}\right)^{1/3} D\right]^{1/2} \exp[-\sqrt{3}/2(k/2R_e^2)^{1/3} D] \qquad (78)$$

where $(k/2R_e^2)^{1/3} D$ is a dimensionless quantity.

Finally, noting the fact that since a surface wave, either trapped or diffracted, depends on the surface condition (example: land, sea, forest, all having different conductivities and permittivities) and surface structure (particularly the sea wave corrugation), the propagation of the surface wave can only be convincingly modeled by experimental data. Indeed, in CCIR documents and other engineering handbooks a variety of highly diversified empirical models are available.

Tropospheric Scatter Propagation

As in the case of the ionosphere, irregular structures in the refractive index in the troposphere are capable of scattering or reflecting electromagnetic energy, thus providing possibilities of propagation over the horizon. The geometry is the same as that shown in Fig. 10 with scattering region in the troposphere. The irregular structures in the troposphere are believed to be either in the form of turbulent eddies or thin turbulent layers [74]. The frequency used usually is in the range of 150 to 4000 MHz. The scatter channel can provide a stable transmission well below the horizon with a field strength much stronger than that predicted by the diffraction theory. Large, rapid fluctuations are usually associated with the signal. These fades are due to the random variations and motions of the scatterers. The bandwidth that can be supported by this mode of propagation depends on the transmission distance, the frequency used, and the antenna beamwidth. If the radiation pattern is broad, multipath signals will narrow the coherence bandwidth. Therefore, to transmit broadband signals, it is necessary to use a narrow-beam antenna. For transmission over large distances (up to 1000 km), only narrow-band applications, such as small-capacity multiplex telephony signals, are practically possible. For shorter-range transmission (200 to 300 km), much broader band applications, including tv signals, can be supported.

The current understanding of the mechanisms which support the tropospheric scatter propagation is not yet complete. Experimental evidence indicates that both the volume scattering as described in Section 3 and partial reflections due to thin layers may be responsible for producing the received field. Because of the complexity of the mechanisms, results from theoretical analysis still do not allow for the quantitative predictions of the received fields and their fluctuations with the accuracy required for practical applications. Empirical procedures have been recommended by the CCIR [53] to calculate the path loss for troposcatter transmission. This procedure takes into account the meteorological condition of the troposphere as well as the altitude of the scattering volume. The attenuation of the medium due to scattering (in addition to the free-space path loss) can be estimated by the following procedure [53]:

Step 1

Construct the path profile by either of the two methods discussed previously in this section, under "Line-of-Sight Propagation," as shown in Fig. 19a.
Step 2

Calculate the scattering angle θ, using the formula in Fig. 19a.
Step 3

The long-term median troposcatter path loss (not exceeded for 50 percent of a year) may be estimated by the formula

$$L(50\%) = 30 \log f - 20 \log D + F(\theta D) - G_e + V \tag{79}$$

where

L = path loss in decibels

f = frequency in hertz

D = great-circle distance in kilometers

θ = scattering angle in radians

G_e = effective total combined antenna gain in decibels

V = factor for long- and short-term fading in different climate regions, may be up to ± 8 dB

$F(\theta D)$ = the attenuation function in decibels, given in Fig. 19b

NOTE: The different curves in Fig. 19b are for different surface refractivities N_s, which are different for different geographical regions. G_e represents the combined transmitting and receiving antenna gains minus their aperture-to-medium coupling losses.

5. Noise

To conclude this chapter, this section presents a short discussion on noise and its effect on a link. For a given link that is required to maintain a specified signal

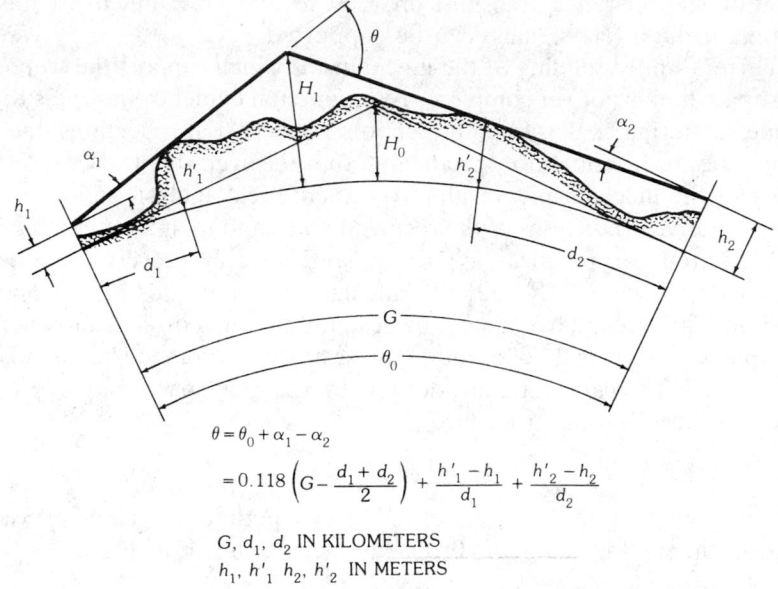

$$\theta = \theta_0 + \alpha_1 - \alpha_2$$

$$= 0.118\left(G - \frac{d_1 + d_2}{2}\right) + \frac{h'_1 - h_1}{d_1} + \frac{h'_2 - h_2}{d_2}$$

G, d_1, d_2 IN KILOMETERS

h_1, h'_1 h_2, h'_2 IN METERS

a

Fig. 19. Path profile and attenuation function for transhorizon propagation. (*a*) Path profile. (*After Picquenard [67], originally published by Macmillan Press, Ltd., London*) (*b*) Graph to obtain $F(\theta D)$. (*Courtesy NTIA, US Dept. of Commerce*)

carrier-to-noise ratio (cnr), the link budget is computed from the equation

$$\mathrm{cnr} = P_r - P_n \tag{80}$$

where

cnr = carrier-to-noise ratio in decibels

P_r = power available at receiving antenna in decibels referred to 1 W, given by (4)

P_n = total received noise power in decibels referred to 1 W

From (4) and (80), the required transmitter power for the link can be calculated. Noise power is generally referred to in terms of temperature, i.e., the noise temperature. Briefly, the noise temperature is related to the noise power P_n in decibels referred to 1 W by

$$P = 10\log(kTB) \tag{81}$$

where

k = Boltzmann's constant = 1.374×10^{-23} J/K

T = noise temperature in kelvins

B = bandwidth in hertz over which the energy is collected in the receiver, i.e., the passband of the receiver

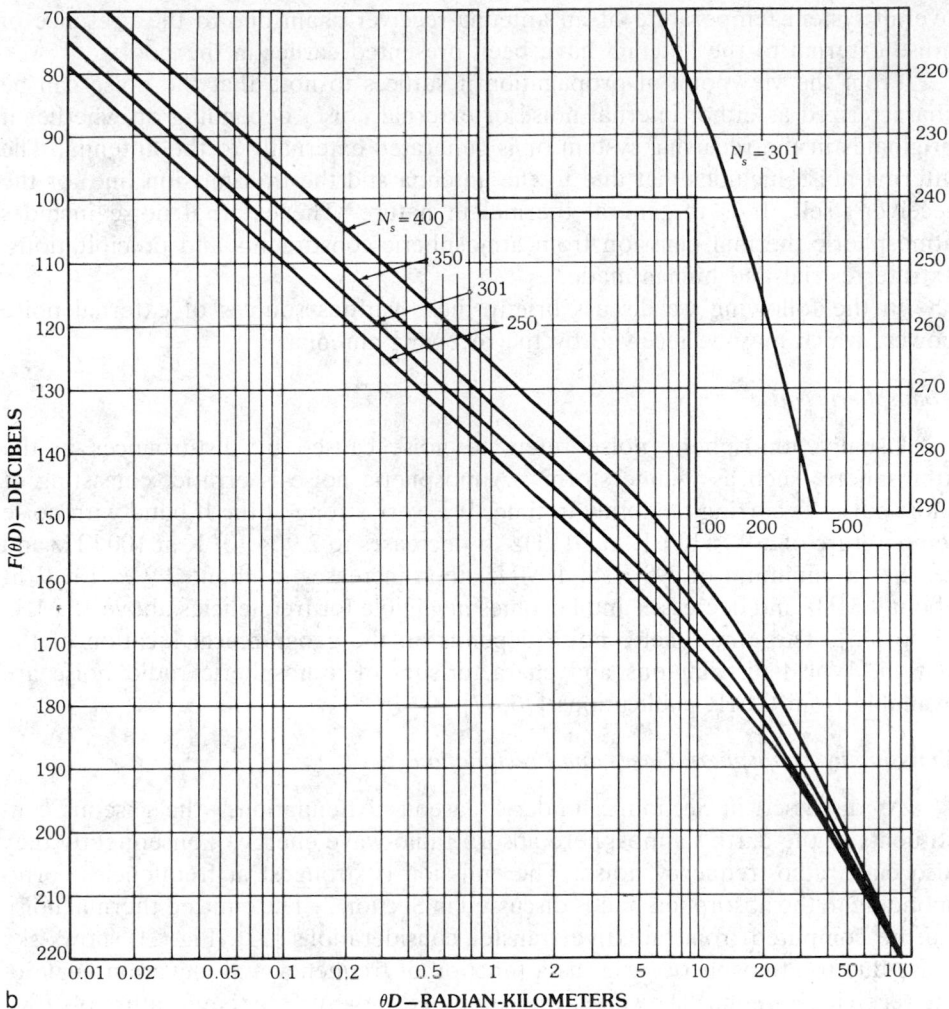

Fig. 19, *continued.*

Equation 81 is also often written as

$$P_n = -204 + 10\log(T/290) + 10\log B \qquad (82)$$

It is important to note that noise degradations of the system have to be addressed in a framework of the complete system. For instance, two shf systems can have the same gain over temperature (G/T) performance under clear sky conditions if one employs a large antenna while the other employs a small antenna but with a sensitive low-noise amplifier (lna). Under rainy conditions, however, the noise temperature degradation of the small antenna system is significantly worse than that of the large antenna system. Detailed formulas for the evaluation of overall system temperature of an antenna-receiver chain due to the presence of noise external to the antenna have been presented earlier in the book.

From the viewpoint of propagation it suffices to note that the noise can be characterized as either internal noise or external noise, depending on whether it originates in the receiving system or is generated externally to the antenna. The internal noise includes that due to the antenna and the transmission line, or the receiver itself. It is in general thermal in nature. The external noise includes atmospheric thermal emission from atmospheric constituents and precipitations, extraterrestrial and human-made.

In the following we discuss briefly these various sources of external noise power, which may be received by the receiving antenna.

Atmospheric Noise

Usually atmospheric noise refers to noise caused by disturbances in the atmosphere, such as thunderstorms. Atmospheric noise is erratic, consisting of short pulses occurring randomly in time. It is very strong at the lf band with noise temperature of 2.9×10^{16} K at 10 kHz. It decreases to 2.9×10^9 K at 100 kHz and reaches a minimum of 290 K at 1 MHz, then increases to about 2.9×10^5 K at about 9 MHz and decreases and becomes negligible for frequencies above 15 MHz or so [75]. The atmospheric noise depends on the geographical location of the station. World distributions and characteristics of atmospheric radio noise are available from CCIR publications [76, 77].

Emission by Atmospheric Gases and Precipitation

As discussed in Section 2, under "Gaseous Attenuation," the gaseous constituents of the earth's atmosphere absorb radio-wave energy. Consequently they also emit radio-frequency noise. The emission is strongest at frequencies corresponding to the absorption bands discussed in Section 2. The emitted thermal noise can be computed from radiative transfer considerations [22]. Fig. 20 shows sky noise due to atmospheric gases as a function of frequency and elevation angle of the receiving antenna. The curves are computed for a surface temperature of 20°C, surface pressure of 1 atm (101 325 Pa), and a surface water vapor concentration of 10 g/m³, corresponding to a relative humidity of 58 percent. Curves for other conditions can also be computed. If one avoids the absorption bands, the sky noise from atmospheric gases for elevation angles greater than 10° is usually less than 100 K in the frequency range of 1 to 100 GHz.

The increase in sky noise due to rain can also be computed in a similar manner. The sky noise temperature increases from about 60 K for a rain attenuation level of 1 dB to about 250 K for a 10-dB rain attenuation, then it saturates to about 270 K for 30-dB attenuation [22] at microwave frequencies.

Extraterrestrial Noise

The sources of extraterrestrial noise are the sun, the galaxy, the cosmic background, and other discrete sources. Fig. 21 gives the sky noise temperature due to the various sources [78]. Also included in the figure are curves for gaseous contributions of dry, clear skies which can be compared with curves shown in Fig. 20.

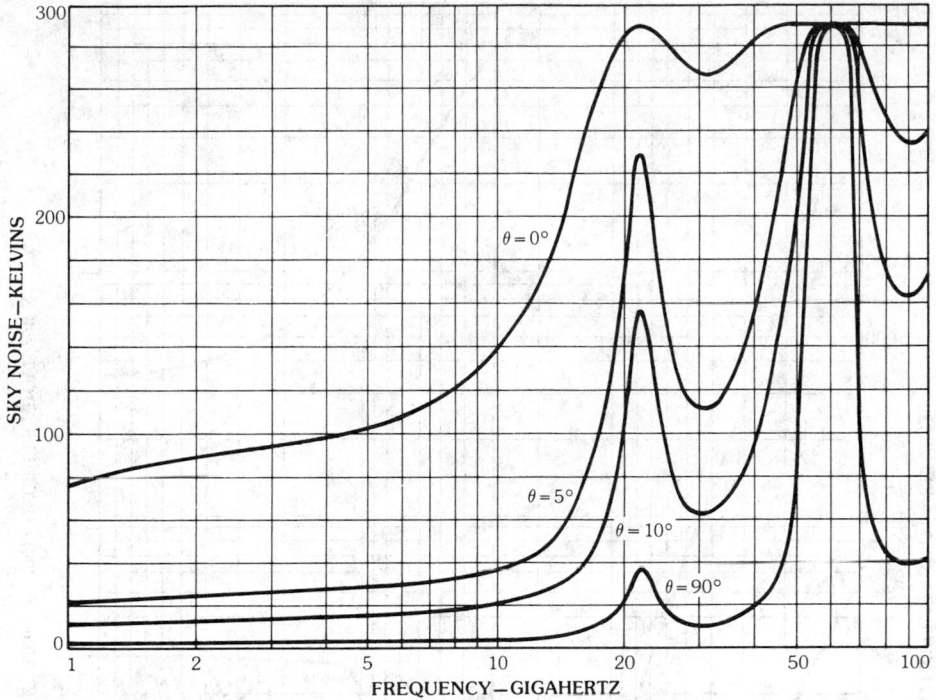

Fig. 20. Sky noise due to atmospheric gases as a function of frequency and elevation angle, for a surface temperature of 20°C, pressure of 1 atm (101 325 Pa), and water vapor concentration of 10 g/m^3. (*After Ippolito [22], © 1981 IEEE*)

Human-Made Noise

Human-made noise can arise from a number of sources, such as power lines, industrial machinery, ignition systems, etc. It varies with location and time and is difficult to estimate under general conditions. Limited observational data indicate that it decreases with frequency. Noise below 2 MHz is generally associated with power lines. At 20 MHz and above, ignition systems are the dominant contributors. An estimate for human-made noise in a business environment is included in Fig. 21. The curve can be extended to low frequencies below 1 MHz [75]. A similar trend is seen for other types of environment, such as residential, rural, and quiet rural areas. The noise levels for those areas are lower, with the quiet rural environment almost 25 dB down from the curve shown in Fig. 21.

From Fig. 21 it is apparent that for a satellite–earth station downlink, the sun constitutes the strongest source of noise and the antenna should be directed away from it. Over most of the uhf range, galactic noise is dominant, but human-made noise in business environments may dominate. Gaseous atmosphere and precipitation are the main contributors to the noise temperature at upper shf band.

For transhorizon communications below uhf frequencies, worldwide maps of radio noise levels were published by the CCIR [79] in 1963. Recently improved maps are available from the NTIA [80]. The improved maps provide corrections, ranging as high as 20 dB, to the 1-MHz noise level values given in the earlier CCIR maps.

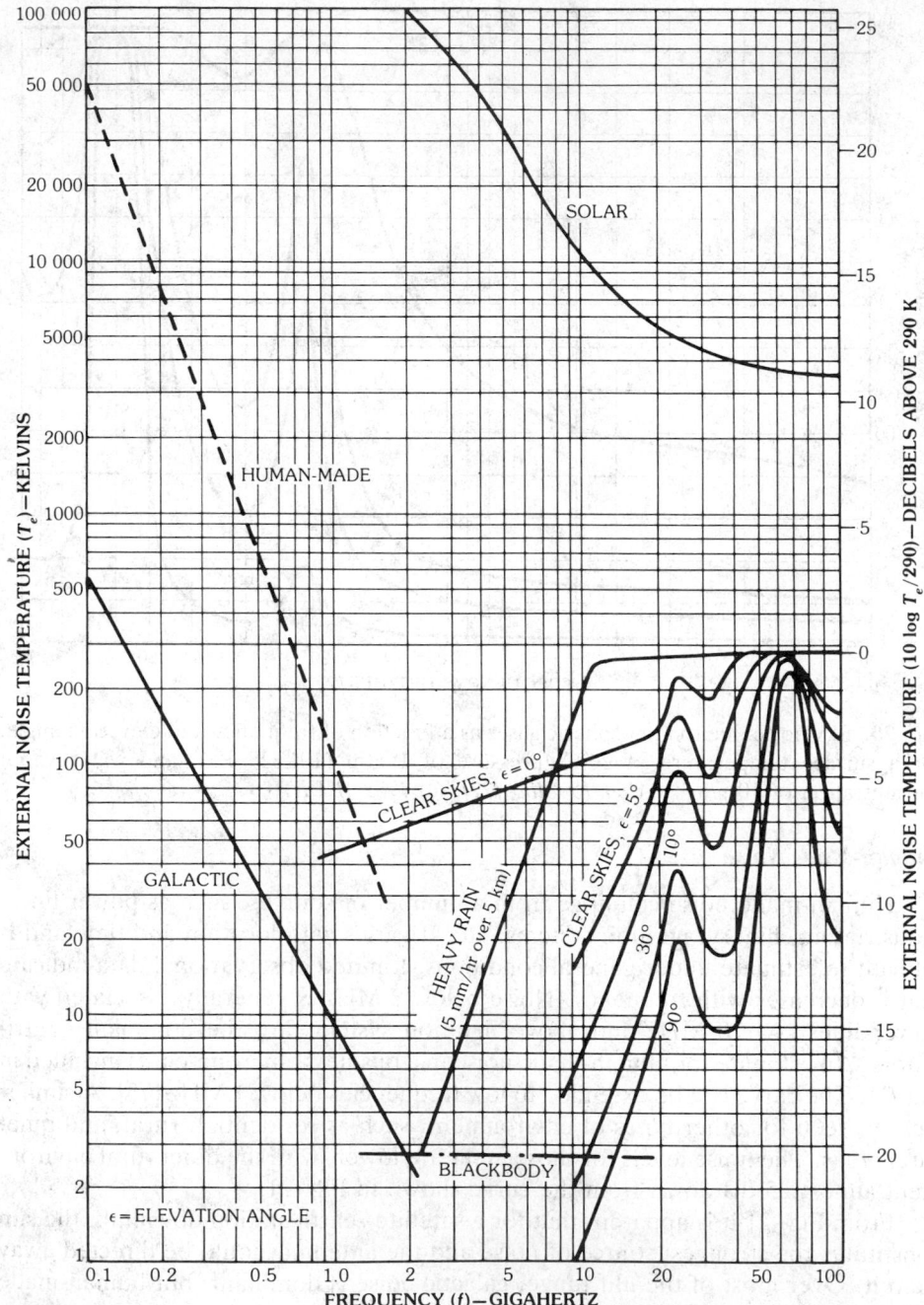

Fig. 21. Apparent temperatures of various external noise sources for downlink slant paths. (*After Dougherty [78]*)

6. References

[1] B. R. Bean and E. J. Dutton, *Radio Meteorology*, New York: Dover Publications, 1968.

[2] D. E. Kerr (ed.), *Propagation of Short Radio Waves*, MIT Radiation Lab Series, vol. 13, New York: McGraw-Hill Book Co., 1951.

[3] K. G. Budden, *Radio Waves in the Ionosphere*, Cambridge: Cambridge University Press, 1961.

[4] K. C. Yeh and C. H. Liu, *Theory of Ionospheric Waves*, New York: Academic Press, 1972.

[5] J. R. Wait, *Electromagnetic Waves in Stratified Media*, 2nd ed., New York: Pergamon Press, 1970.

[6] V. I. Tatarskii, "The effects of the turbulent atmosphere on wave propagation," Springfield, Va.: US Dept. of Commerce TT-68-50464, 1971.

[7] A. Ishimaru, *Wave Propagation and Scattering in Random Media*, vols. I and II, New York: Academic Press, 1978.

[8] K. C. Yeh and C. H. Liu, "Radio wave scintillations in the ionosphere," *Proc. IEEE*, vol. 70, pp. 324–360, 1982.

[9] D. C. Cox, H. W. Arnold, and H. H. Hoffman, "Observations of cloud-produced amplitude scintillation on 19- and 28-GHz earth-space paths," *Radio Sci.*, vol. 16, pp. 885–908, 1981.

[10] H. T. Dougherty and C. M. Rush, "Some propagational aspects of frequency allocation and frequency sharing," AGARD Conference No. 332, Propagation Aspects of Frequency Sharing, Interference and System Diversity, pp. 2.1–2.11, Paris, October 1982.

[11] CCIR Study Group 5 Doc. Report 719-1, Volume V, "Attenuation by atmospheric gases," XVth Plenary Assembly, Geneva, 1982.

[12] CCIR Study Group 5 Doc. Report 564-2, Volume V, "Propagation data required for space telecommunications systems," XVth Plenary Assembly, Geneva, 1982.

[13] D. V. Rogers, "Simple method for estimating 1- to 15-GHz atmospheric absorption," *COMSAT Tech. Rev.*, vol. 13, pp. 157–163, 1983.

[14] T. Oguchi, "Attenuation of electromagnetic waves due to rain with distorted raindrops," *J. Radio Res. Lab.* (Tokyo), vol. 11, pp. 19–37, 1964.

[15] R. K. Crane, "Propagation phenomena affecting satellite communications systems operating in the centimeter and millimeter wavelength bands," *Proc. IEEE*, vol. 59, pp. 173–188, 1971.

[16] D. C. Hogg and T. S. Chu, "The role of rain in satellite communication," *Proc. IEEE*, vol. 63, pp. 1308–1331, 1975.

[17] H. T. Dougherty and E. V. Dutton, "Estimating year-to-year variability of rainfall for microwave applications," *IEEE Trans. Comm.*, vol. COM-26, pp. 1321–1324, 1978.

[18] D. J. Fang and J. M. Harris, "Precipitation attenuation studies based on measurements of ATS-6 20/30 GHz beacon signals at Clarksburg, Md.," *IEEE Trans. Antennas Propag.*, vol. AP-27, pp. 1–11, 1979.

[19] CCIR Study Group 5 Doc. Report 721-1, Volume V, "Attenuation by hydrometeors, in particular, precipitations and other atmospheric particles," XVth Plenary Assembly, Geneva, 1982.

[20] Final Report of the Conference Preparatory Meeting (CPM) for RARC-82, chapter 3, "Radio propagation factors relating to frequencies near 12.5 and 17.5 GHz," Geneva, June–July 1982.

[21] Y. Otsu et al., "Propagation measurements and tv-reception tests with the Japanese broadcasting satellite for experimental purposes," *IEEE Trans. Broadcast Commun.*, vol. BC-25, p. 113, 1979.

[22] L. J. Ippolito, "Radio propagation for space communications systems," *IEEE Proc.*, vol. 69, pp. 697–727, 1981.

[23] P. N. Kumar, "Precipitation fade statistics for 19/29-GHz COMSAT beacon signals and 12-GHz radiometric measurements," *COMSAT Tech. Rev.*, vol. 12, pp. 1–27, 1982.

[24] D. C. Cox, "Depolarization of radio waves by atmospheric hydrometeors in earth-space paths: a review," *Radio Sci.*, vol. 16, pp. 781–812, 1981.

[25] D. J. Fang and C. H. Chen, "Propagation of centimeter/millimeter waves along a slant path through precipitation," *Radio Sci.*, vol. 17, pp. 989–1005, 1982.

[26] CCIR Study Group 5 Doc. Report No. 564, "Propagation data required for space communication system," XIVth Plenary Assembly, Kyoto, Japan, 1978.

[27] K. Rawer, "Intercomparison of different measuring techniques in the upper atmosphere: the international reference ionosphere," *Space Res.*, vol. XV, p. 212, 1975.

[28] R. B. Bent et al., "The development of a highly-successful worldwide empirical ionospheric model and its use in certain aspects of space communications and worldwide total content investigation," *Effects of the Ionosphere on Space Systems and Communications*, ed. by T. M. Goodman, Naval Research Lab, Washington, D.C. 20375, US Government Printing Office Stock No. 008-051-00064-0, 1981.

[29] H. E. Whitney, J. Aarons, R. S. Allen, and D. R. Seemann, "Estimation of the cumulative amplitude probability distribution function of ionospheric scintillations," *Radio Sci.*, vol. 7, pp. 1095–1104, 1972.

[30] R. Umeki, C. H. Liu, and K. C. Yeh, "Multifrequency studies of ionospheric scintillations," *Radio Sci.*, pp. 311–317, 1977.

[31] D. J. Fang and C. H. Liu, "A morphological study of gigahertz equatorial scintillations in the Asian region," *Radio Sci.*, vol. 18, pp. 241–252, 1983.

[32] J. Aarons, "Global morphology of ionospheric scintillations," *Proc. IEEE*, vol. 70, pp. 360–379, 1982.

[33] D. J. Fang and M. S. Pontes, "4/6 GHz ionospheric scintillation measurement at Hong Kong earth station during the peak of sunspot cycle 21," *COMSAT Tech. Rev.*, vol. 11, pp. 293–320, 1981.

[34] D. G. Singleton, "An improved ionospheric irregularity model," *Solar-Terrestrial Predictions Proc.*, US Department of Commerce, Document 003-017-00479-1, vol. 4, pp. D1–D15, 1980.

[35] CCIR Study Group 6 Doc. Report 263-4, Volume VI, "Ionospheric effects upon earth-space propagation," XVth Plenary Assembly, Geneva, 1982.

[36] E. J. Fremouw and C. L. Rino, "A signal-statistical and morphological model of ionospheric scintillation," *Proc. AGARD Conf. on Operational Modeling of Aerospace Propagation Environment*, pap. 18, Ottawa, Canada, 1978.

[37] J. Aarons, E. Mackenzie, and K. Bhavnani, "High latitude analytical formulas for scintillation levels," *Radio Sci.*, vol. 15, pp. 115–127, 1980.

[38] CCIR Study Group 5 Doc. Report 881, Volume V, "Effects of small-scale spatial or temporal variations of refraction on radio wave propagation," XVth Plenary Assembly, Geneva, 1982.

[39] K. Davies, "Ionospheric radio propagation," *National Bureau of Standards Monograph 80*, US Government Printing Office, Washington, D.C., November 1965.

[40] CCIR Study Group 6 Doc. Report 340-4, "CCIR atlas of ionospheric characteristics," XVth Plenary Assembly, Geneva, 1982. (This report is published separately from Volume VI.)

[41] CCIR Study Group 6 Doc. Supplement of Report 252-2, Volume VI, "Second CCIR computer-based interim method for estimating sky-wave field strength and transmission loss at frequencies between 2 and 30 MHz," and CCIR Report 571-1, "Comparisons between observed and predicted sky-wave signal intensities at frequencies between 2 and 30 MHz," XVth Plenary Assembly, Geneva, 1982.

[42] A. F. Barghausen et al., "Predicting long-term operational parameters of high frequency skywave telecommunication systems," *Tech. Rep. ERL 110 ITS 78* (US Environmental Science Services Administration).

[43] CCIR Study Group 6 Doc. Report 575-2, Volume VI, "Method for predicting skywave field-strengths at frequencies between 150 kHz and 1600 kHz," XVth Plenary Assembly, Geneva, 1982.

[44] *FCC Rules and Regulations*, pt. 73, Washington: US Government Printing Office.

[45] C. H. Wang, "Interference and sharing at medium frequency: skywave propagation

considerations," *AGARD Conf. No. 332, Propagation Aspects of Frequency Sharing, Interference and System Diversity*, Paris, October 1982.

[46] A. D. Watt, *VLF Radio Engineering*, New York: Pergamon Press, 1967.

[47] E. R. Swanson, "Omega," *Navigation*, vol. 18, 1972.

[48] J. Galejs, *Terrestrial Propagation of Long Electromagnetic Waves*, New York: Pergamon Press, 1972.

[49] CCIR Study Group 6 Doc. Report 265-5, Volume VI, "Sky-wave propagation and circuit performance at frequencies between about 30 kHz and 500 kHz," XVth Plenary Assembly, Geneva, 1982.

[50] H. G. Booker and W. E. Gordon, "A theory of radio scattering in the troposphere," *Proc. IRE*, vol. 38, p. 401, 1950.

[51] V. A. Fock, *Electromagnetic Diffraction and Propagation Problems*, New York: Pergamon Press, 1965.

[52] P. Beckman and A. Spizzichino, *The Scattering of Electromagnetic Waves from Rough Surfaces*, New York: Macmillan, 1963.

[53] CCIR Study Group 5 Doc. Report 239-4, Volume V, "Propagation data required for trans-horizon radio-relay system," XVth Plenary Assembly, Geneva, 1982.

[54] CCIR Study Group 5 Doc., Volume V, "Radiometeorological data," XVth Plenary Assembly, Geneva, 1982.

[55] GTE Lenkurt Publications, *Engineering Considerations for Microwave Communication Systems*, San Carlos, Calif.: GTE Lenkurt Incorporated, 1970.

[56] R. H. Ott, "Scattering, diffraction, propagation and refraction of acoustic and EM waves," *Remote Sensing of the Troposphere,*, Chapter 8, Boulder: Univ. of Colorado, and NOAA, June 13–30, 1972.

[57] H. Brodhage and W. Hormuth, *Planning and Engineering of Radio Relay Links*, New York: John Wiley & Sons, 1977, p. 87.

[58] L. V. Blake, *Radar Range-Performance Analysis*, Norwood, Massachusetts: Artech House, 1986.

[59] D. J. Fang and R. H. Ott, "A low elevation angle *L*-band maritime propagation measurement and modeling," *IEE Conf. Pub. 222*, pp. 45–50, Third International Conference on Satellite Systems for Mobile Communications and Navigations, London, 1983.

[60] S. O. Rice, "Reflection of electromagnetic waves from slightly rough surfaces," *Theory of Electromagnetic Waves*, ed. by M. Kline, New York: Interscience, pp. 351–378, 1951.

[61] H. Davis, "The reflection of electromagnetic waves from a rough surface," *Proc. IEEE*, vol. 101, pp. 209–214, 1954.

[62] D. E. Barrick, "Remote sensing of sea state by radar," *Remote Sensing of the Troposphere*, ed. by V. Derr, chapter 12, pp. 12-1 to 12-46, Washington: US Government Printing Office, 1972.

[63] D. J. Fang, "Scattering from a perfectly conducting sinusoidal surface," *IEEE Trans. Antennas Propag.*, vol. AP-20, pp. 388–390, 1972.

[64] R. H. Ott, "An alternative integral equation for propagation over irregular terrain, pt. II," *Radio Sci.*, vol. 6, no. 4, April 1971.

[65] B. Kinsman, *Wind Waves, Their Generation and Propagation on the Ocean Surface*, New York: Prentice-Hall, 1965.

[66] M. W. Long, *Reflectivity of Land and Sea*, Lexington, Massachusetts: D.C. Heath & Co., 1975.

[67] A. Picquenard, *Radio Wave Propagation*, New York: John Wiley & Sons, 1974.

[68] CCIR Study Group 5 Doc. Report 715-1, Volume V, "Propagation by diffraction," XVth Plenary Assembly, Geneva, 1982.

[69] K. A. Norton, "The propagation of radio waves over the surface of the earth and in the upper atmosphere," *Proc. IRE*, vol. 24, p. 1367, 1936; *Proc. IRE*, vol. 25, p. 1203, 1937; and *Proc. IRE*, vol. 25, p. 1192, 1937.

[70] E. C. Jordan and K. G. Balmain, *Electromagnetic Waves and Radiating Systems*, 2nd ed., New York: Prentice-Hall, 1958.

[71] J. R. Wait, "Electromagnetic Surface Waves," in *Advances in Radio Research*, ed. by J. A. Saxton, London and New York: Academic Press, 1964.

[72] H. M. Barlows and J. Brown, *Radio Surface Waves*, Oxford: Clarendon Press, 1962.

[73] R. H. Ott, "Fock currents for concave surface," *IEEE Trans. Antennas Propag.*, vol. AP-22, pp. 357–360, 1974; also see J. R. Wait, "EM surface waves," in *Advances in Radio Research*, ed. by J. Saxton, New York: Academic Press, 1964.

[74] F. DuCastel, *Tropospheric Radio Wave Propagation Beyond the Horizon*, New York: Pergamon Press, 1966.

[75] CCIR Study Group 1 Doc. Report 670, "Worldwide minimum external noise levels, 0.1 Hz to 100 GHz," XIVth Plenary Assembly, Kyoto, Japan, 1978.

[76] CCIR Study Group 6 Doc., Volume VI, "Characteristics and applications of atmospheric radio noise data," XVth Plenary Assembly, Geneva, 1982.

[77] CCIR Study Group 6 Doc., Volume VI, "Measurement of atmospheric radio noise from lightning," XVth Plenary Assembly, Geneva, 1982.

[78] H. T. Dougherty, "A consolidated model for uhf/shf telecommunication links between earth and synchronous satellites," *NTIA Rep. 80-45*, US Department of Commerce, August 1980.

[79] CCIR Study Group 6 Doc. Report 322-2, "World distribution and characteristics of atmospheric radio noise," Xth Plenary Assembly, Geneva, 1963.

[80] A. P. Spaulding and J. S. Washburn, "Atmospheric radio noise: worldwide levels and other characteristics," *NTIA Rep. 85-173*, Boulder, April 1985.

Chapter 30

Antenna Response to Electromagnetic Pulses

K. S. H. Lee

Dikewood, Division of Kaman Sciences Corporation

CONTENTS

Kelvin S. H. Lee received the BS, MS, and PhD degrees in engineering and applied science, in 1960, 1961, and 1963, respectively, from the California Institute of Technology.

Dr. Lee, Vice President of Kaman Sciences Corporation, Dikewood Division, is the Manager of Dikewood's California Research Branch in Santa Monica and is in charge of its R&D programs in electromagnetics.

He is the editor of the newly published book *EMP Interaction: Principles, Techniques, and Reference Data* (Hemisphere Publishing Corporation, New York, 1986).

He has been actively involved with system response to EMP, lighting, and high-power microwaves (HPM) in both the development and the experimental verification of analytical coupling models, in system-level testing, and in hardening design and hardness assessment of many military and commercial systems.

Dr. Lee has made numerous contributions to the electromagnetic theory involving the electrodynamics of moving media, the complex Doppler effect, electromagnetic shielding, reversible and irreversible power, dipole antennas, transient electromagnetic phenomena, and low-frequency approximations in electromagnetic theory.

1. High-Altitude Electromagnetic Pulses

Of the many different types of electromagnetic pulses (EMP) from a nuclear detonation the high-altitude EMP (HEMP) has received the most attention mainly because it covers a large area. HEMP is generally assumed to be a plane wave with the electric-field amplitude given by [1]

$$E_1(t) = E_0(e^{-\alpha t} - e^{-\beta t})u(t) \tag{1}$$

or

$$E_2(t) = E_0(e^{\alpha t} + e^{-\beta t})^{-1} \tag{2}$$

where u is the unit-step function, $1/\alpha \cong 250$ ns, $1/\beta \cong 2$ ns, and E_0 is tens of kilovolts per meter. The waveforms are plotted in Fig. 1 and their time- and frequency-

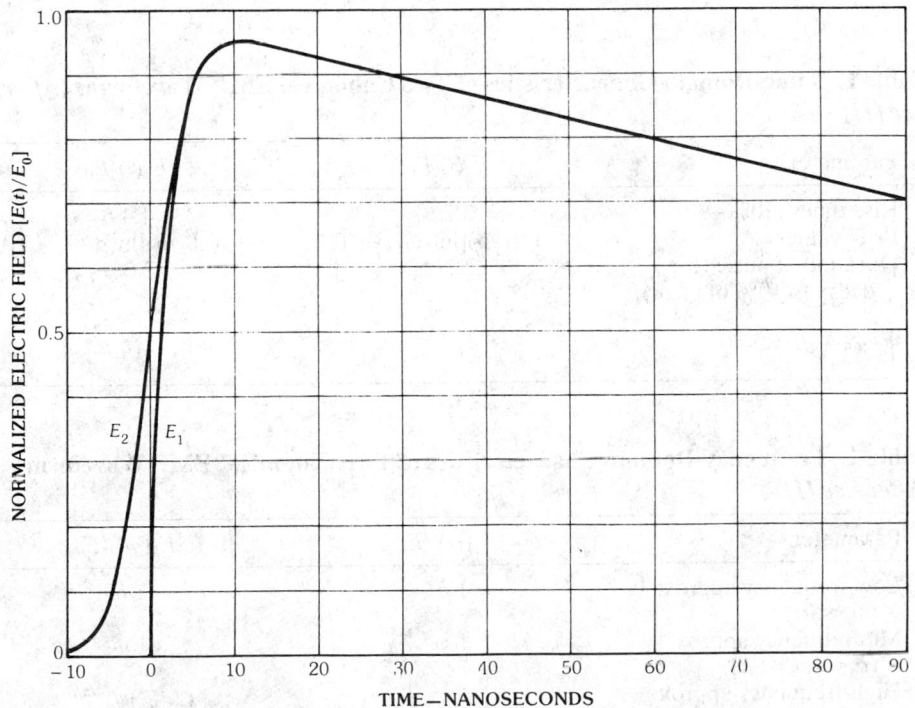

Fig. 1. Two canonical high-altitude EMP waveforms for which $\alpha = 4 \times 10^6$ s^{-1} and $\beta = 5 \times 10^8$ s^{-1}. (*After Lee [1]*)

domain characteristics are given in Tables 1 and 2. Note that E_1 has a discontinuous time derivative at $t = 0$, whereas the time derivative of E_2 at $t = 0$ is continuous. Both spectra, as shown in Fig. 2, have the first breakpoint at $\alpha/2\pi$ and roll off as $1/f$ until the second breakpoint at $\beta/2\pi$ is reached, from which point onward $E_1(\omega)$ decreases as $1/f^2$ whereas $E_2(\omega)$ drops off more rapidly. The 20-dB point of the half-width of both spectra is about 6 MHz.

2. The Effects of Antenna Mounting Structures

In most cases antennas are not directly exposed to the EMP plane-wave field described in the preceding section. The structures on which the antennas are mounted will modify not only the incident field but also the input impedances of the antennas. If the dimensions of the antenna are small compared with the local radii of curvature of the mounting structure, the effect of the mounting stucture can be approximated by that of an infinite ground plane so far as the antenna's input impedance is concerned. On the other hand, the *external* field to which the antenna is exposed contains the resonances of the mounting structure which can easily be excited by EMP. Figs. 3 and 4 are the scale-model measurements of the axial current density at the top and bottom midforward fuselages of the 747 and 707 aircraft [2]. The incident time-harmonic plane-wave field is striking from the top with **E** parallel to the fuselage. Fig. 5 shows the total current at three locations of a cylinder [3]. The resonances of the structures are obvious from these figures, which

Table 1. Time-Domain Characteristics of Two Canonical EMP Waveforms (*After Lee [1]*)

Parameter	$E_1(t)/E_0$	$E_2(t)/E_0$
Rise times (10%–90%)	$2.2/\beta$	$4.4/\beta$
Peak values	$1 + (\alpha/\beta)[\ln(\alpha/\beta) - 1]$	$1 + (\alpha/\beta)[\ln(\alpha/\beta) - 1]$
Duration of pulse (for decay to 10% of peak)	$2.3/\alpha$	$2.3/\alpha$
$\int_{-\infty}^{\infty} \cdots \, dt$	$1/\alpha$	$1/\alpha$

Table 2. Frequency-Domain Characteristics of Two Canonical EMP Waveforms (*After Lee [1]*)

Parameter	$|E_1(\omega)|/E_0$	$|E_2(\omega)|/E_0$
Low-frequency approx. ($\omega \ll \alpha$)	$1/\alpha$	$1/\alpha$
Midfrequency approx. ($\alpha \ll \omega \ll \beta$)	$1/\omega$	$1/\omega$
High-frequency approx. ($\omega \gg \beta$)	β/ω^2	$(2\pi/\beta)e^{-\omega\pi/\beta}$

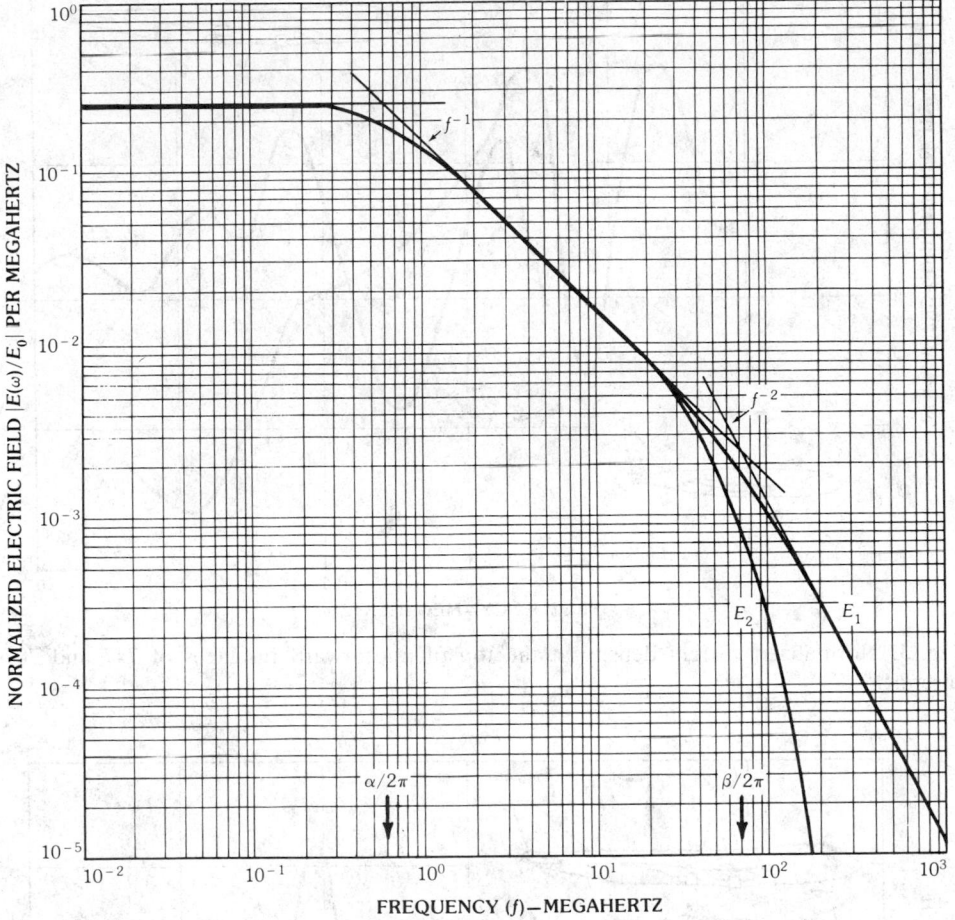

Fig. 2. Spectra of two canonical HEMP curves.

need be scaled by the EMP spectra given in Section 1 to obtain the excitation field for the mounted antenna. For example, take E_0 to be 50 kV/m in (1), and make use of Figs. 2 through 5 and the formula

$$I_{peak} = |\Delta\omega| \, |I(\omega_0)| \tag{3}$$

where $|I(\omega)|$ is the spectrum of $I(t)$ and I_{peak} is the peak amplitude of the damped sinusoid defined by

$$I(t) = I_{peak} \, e^{-|\Delta\omega| t/2} \sin\omega_0 t \tag{4}$$

with ω_0 being the resonance frequency and $|\Delta\omega|$ the 3-dB bandwidth. One obtains Table 3 for the first-resonance induced current or current density.

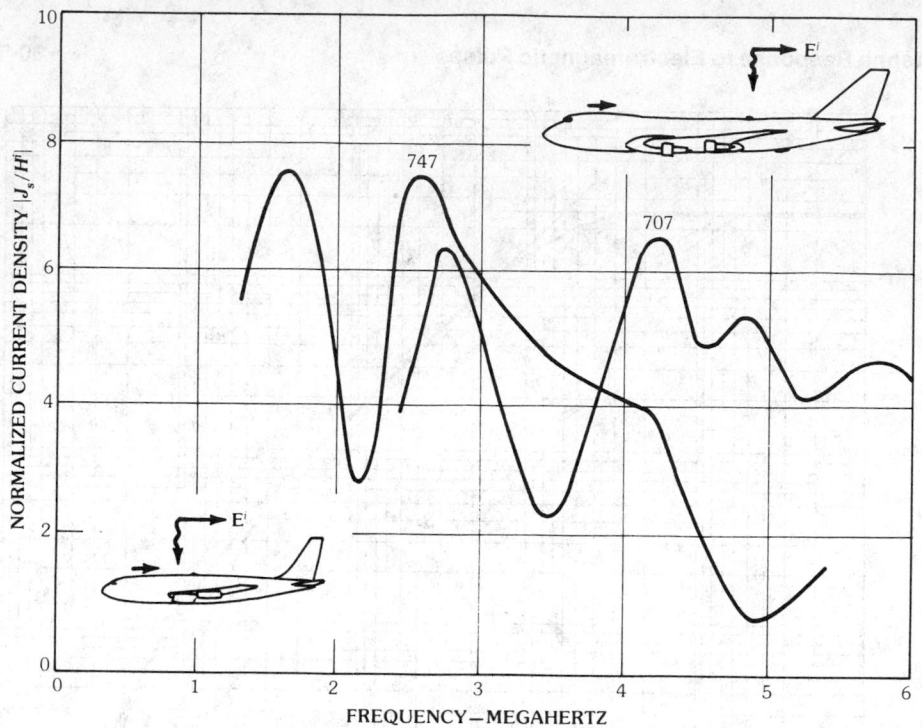

Fig. 3. Normalized current density at the top of midforward fuselages of 747 and 707 aircraft.

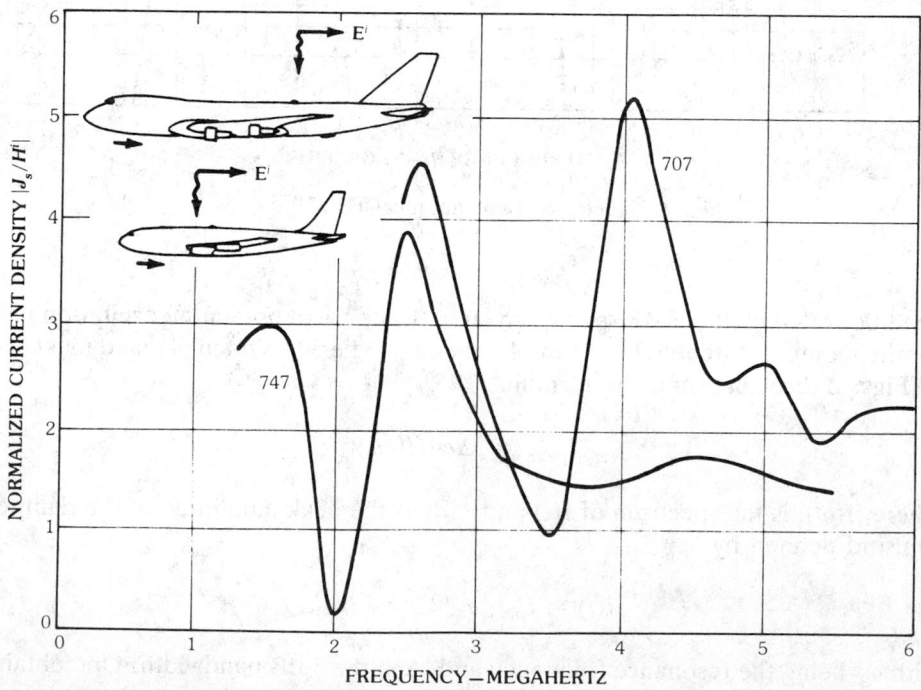

Fig. 4. Normalized current density at the bottom of midforward fuselages of 747 and 707 aircraft.

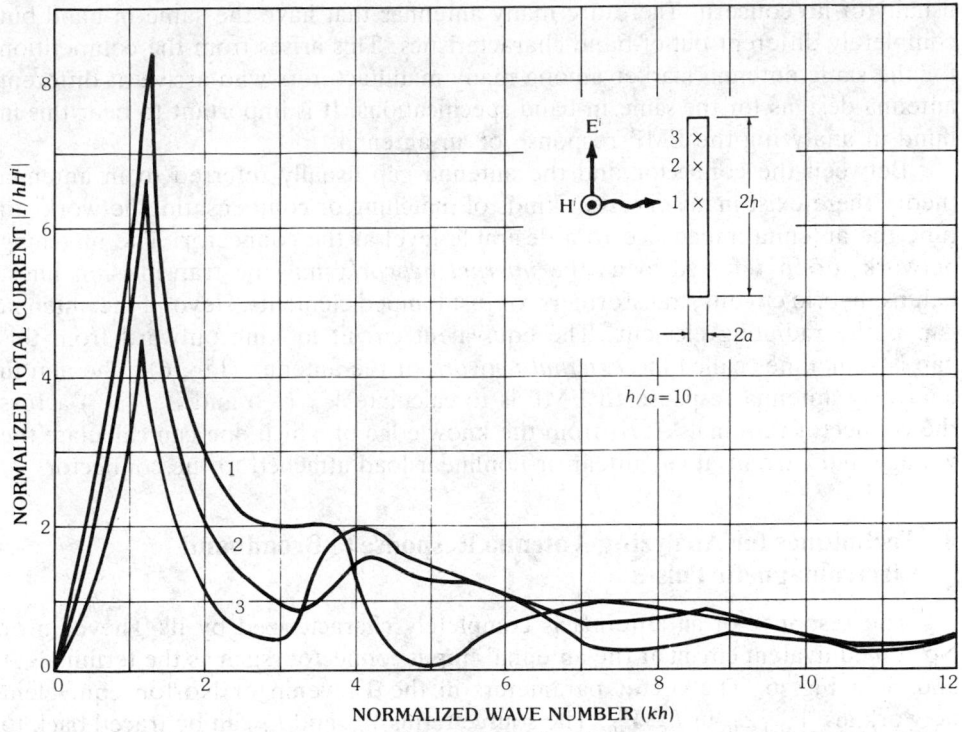

Fig. 5. Magnitude of total current induced on a cylinder.

Table 3. Characteristics of the First Damped Sinusoid Induced by a 50-kV/m High-Altitude EMP

Characteristic	747		707		Cylinder (2h = 10 m)		
	Top	Bottom	Top	Bottom	1	2	3
First resonance frequency (MHz)		1.60		2.65		11.20	
Bandwidth (MHz)		0.65		0.55		3.15	
Peak current density (A/m)	365	146	176	126			
Peak current (kA)					1.47	1.23	0.83

3. Principal Antenna Elements

The high-altitude EMP, as described in Section 1, has not only a high field strength but also a broad frequency spectrum. These two characteristics of EMP make it necessary to analyze the response of an antenna not only within its in-band frequencies but also in out-of-band frequencies in which the antenna's behavior is usually of no concern. There are many antennas that have the same in-band but completely different out-of-band characteristics. This arises from the competition for the same antenna market among many manufacturers who arrive at different antenna designs for the same in-band specifications. It is important to bear this in mind in analyzing the EMP response of an antenna.

Between the connector and the antenna gap usually referred to in antenna theory there exist many different kinds of matching or compensating networks to tune the antenna impedance to a desirable level at the connector. The matching network, often referred to as the *internal network*, may be transmission lines, baluns, hybrid circuits, transformers, or just lumped elements. Beyond the antenna gap is the radiating element. The equivalent circuit looking outward from this gap is sometimes called the *external network* of the antenna (Fig. 6). The aim in analyzing antenna response to EMP is to calculate V_{oc} (I_{sc}) and Z_{in} (Y_{in}) across the connector terminals A, B, from the knowledge of which one can calculate the voltage and current at any linear or nonlinear load attached to the connector.

4. Techniques for Analyzing Antenna Response to Broadband Electromagnetic Pulses

The response of an antenna is completely characterized by its Thevenin or Norton equivalent circuit at the antenna's input connector, such as the terminals A and B in Fig. 6. The circuit parameters in the Thevenin or Norton equivalent network are V_{oc}, Z_{in} or I_{sc}, Y_{in}. The source terms V_{oc} and I_{sc} can be traced back to the *induced* open-circuit voltage V_{ind} and short-circuit current I_{ind} across the antenna gap (Fig. 6). These latter source quantities can be factored into a product of effective height or area and the local electric or magnetic field on the mounting

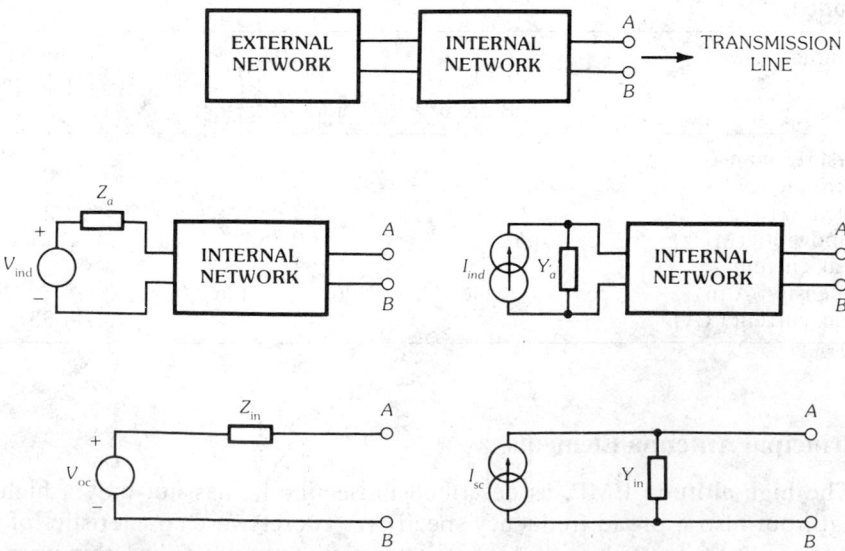

Fig. 6. Principal antenna components.

structure. Examples of the local field are given in Section 2. In this section techniques are given for calculating the circuit parameters of the external network shown in Fig. 6.

Low-Frequency Techniques

As far as their responses to EMP are concerned, most antennas can be considered electrically small. In Chart 1, formulas are given for the fundamental quantities of two basic dipole antennas. The definitions of these quantities are as follows:

$\mathbf{A}_e, \mathbf{A}_m$ = equivalent area of electric, magnetic dipole antenna [4]

$\mathbf{h}_e, \mathbf{h}_m$ = equivalent height of electric, magnetic dipole antenna [4]

V_e, V_m = equivalent volume of electric, magnetic dipole antenna [4]

$\mathbf{P}, \mathbf{M}$ = electric, magnetic polarizability tensor

$\mathbf{p}, \mathbf{m}$ = electric, magnetic dipole moment

$V_{\text{ind}}, I_{\text{ind}}$ = induced open-circuit voltage, short-circuit current

C_a, L_a = antenna input capacitance, inductance

Note that

$$\mathbf{A}_e \mathbf{h}_e = \mathbf{h}_e \mathbf{A}_e$$
$$\mathbf{A}_m \mathbf{h}_m = \mathbf{h}_m \mathbf{A}_m$$
$$V_{\text{ind}}^* I_{\text{ind}} = j\omega \mathbf{p} \cdot \mathbf{E}^* - j\omega\mu_0 \mathbf{m} \cdot \mathbf{H}^*$$
$$= j\omega(\epsilon_0 \mathbf{E} \cdot \mathbf{h}_e \mathbf{A}_e \cdot \mathbf{E}^* - \mu_0 \mathbf{H} \cdot \mathbf{h}_m \mathbf{A}_m \cdot \mathbf{H}^*)$$

$$(5)$$

Chart 1. Formulas for Electrically Small Electric and Magnetic Dipole Antennas

Rod	Loop

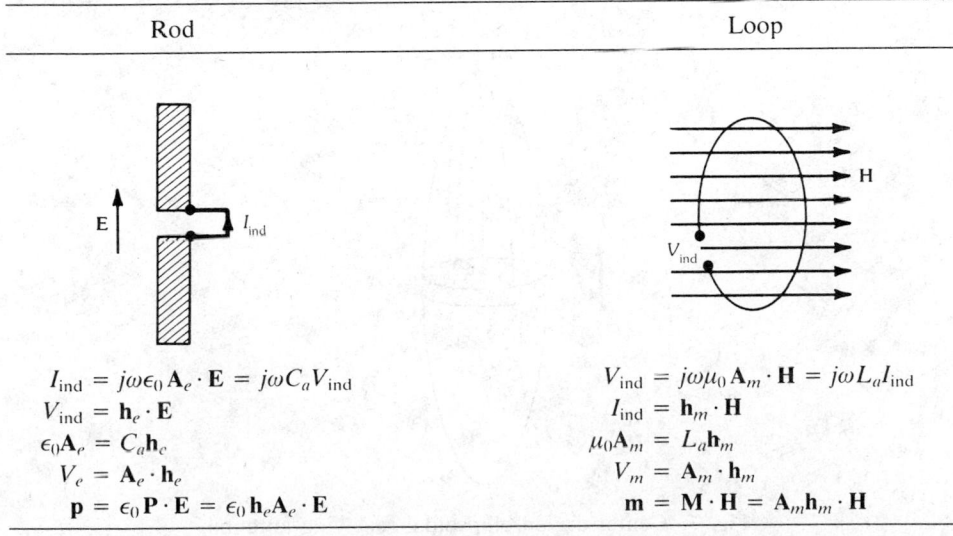

Rod	Loop
$I_{\text{ind}} = j\omega\epsilon_0 \mathbf{A}_e \cdot \mathbf{E} = j\omega C_a V_{\text{ind}}$	$V_{\text{ind}} = j\omega\mu_0 \mathbf{A}_m \cdot \mathbf{H} = j\omega L_a I_{\text{ind}}$
$V_{\text{ind}} = \mathbf{h}_e \cdot \mathbf{E}$	$I_{\text{ind}} = \mathbf{h}_m \cdot \mathbf{H}$
$\epsilon_0 \mathbf{A}_e = C_a \mathbf{h}_e$	$\mu_0 \mathbf{A}_m = L_a \mathbf{h}_m$
$V_e = \mathbf{A}_e \cdot \mathbf{h}_e$	$V_m = \mathbf{A}_m \cdot \mathbf{h}_m$
$\mathbf{p} = \epsilon_0 \mathbf{P} \cdot \mathbf{E} = \epsilon_0 \mathbf{h}_e \mathbf{A}_e \cdot \mathbf{E}$	$\mathbf{m} = \mathbf{M} \cdot \mathbf{H} = \mathbf{A}_m \mathbf{h}_m \cdot \mathbf{H}$

Since $V_e = \mathbf{h}_e \cdot \mathbf{A}_e$ and $V_m = \mathbf{h}_m \cdot \mathbf{A}_m$, it can be seen that V_e and V_m are, respectively, the measure of (maximum) electric and magnetic energy extracted from the external field that is stored in the antenna.

A useful quantity to measure the real power absorbed by an antenna is the effective receiving area, whose maximum value for an electrically small antenna is given by [5,6]

$$A_r = \frac{3\lambda^2}{8\pi} \quad \text{for a rod or slot antenna}$$

$$= \frac{3\lambda^2}{4\pi} \quad \text{for transmission through a slot antenna} \tag{6}$$

where λ is the wavelength of the incident wave. When A_r is multiplied by the time-average incident Poynting vector $\overline{S}_0$, one has the maximum time-average real power absorbed by the antenna. Combining (5) and (6), one has the following useful formula:

maximum time-average complex power absorption of an electrically

small antenna $= A_r \overline{S}_0 + 2j\omega(V_e \overline{W}_e - V_m \overline{W}_m)$ (7)

where $\overline{W}_e$ and $\overline{W}_m$ are the time-average electric and magnetic energy of the external wave.

For an electrically small ellipsoidal center-fed antenna as shown in Fig. 7, the normalized equivalent area is given by the formula [7]

$$\frac{A_e}{\pi bc} = \frac{(a^2 - b^2)\sqrt{a^2 - c^2}}{abc} \frac{1}{F(\phi/\alpha) - E(\phi/\alpha)} \tag{8}$$

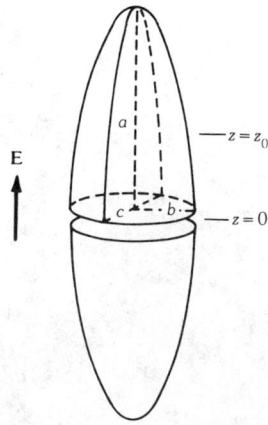

Fig. 7. Center-driven ellipsoidal receiving antenna.

where $a > b > c$, F and E are respectively the incomplete elliptic integrals of the first and second kind, and

$$\phi = \sin^{-1}\sqrt{1 - c^2/a^2}$$
$$\alpha = \sin^{-1}\sqrt{(a^2 - b^2)/(a^2 - c^2)}$$

Equation 8 is plotted in Fig. 8 for various b/a and c/a values.

When $c \ll b$, the ellipsoid degenerates into an elliptic disk or blade, which can be used to approximate most aircraft blade antennas. In this case, (8) becomes

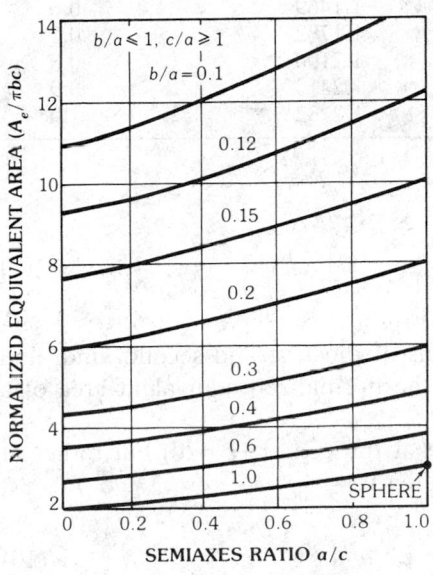

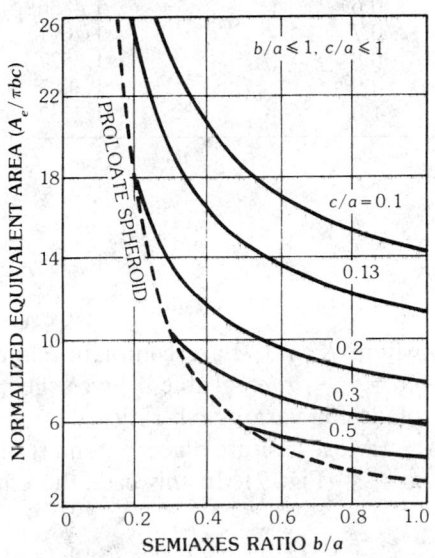

a   b

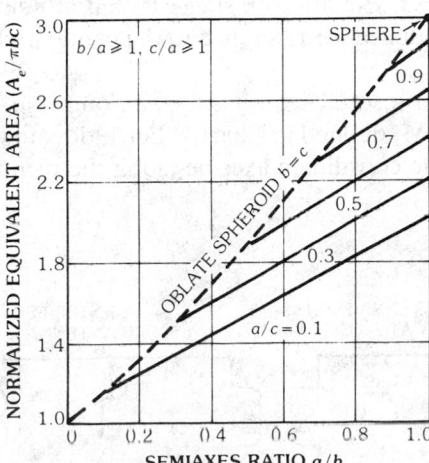

c

Fig. 8. Normalized equivalent area. (*a*) For various b/a. (*b*) For various c/a. (*c*) For various a/c.

Table 4. Normalized Equivalent Area of an Elliptic Disk (Blade Antenna)
(*After Lee [1]*)

b/a	$A_e/\pi ab$	$A_e/\pi ab$	a/b
10^{-3}	137.0980	1.0000	10^{-3}
10^{-2}	20.0328	1.0002	10^{-2}
0.1	3.6945	1.0112	0.1
0.2	2.4420	1.0324	0.2
0.3	1.9890	1.0581	0.3
0.4	1.7376	1.0864	0.4
0.5	1.5865	1.1162	0.5
0.6	1.4836	1.1469	0.6
0.7	1.4091	1.1782	0.7
0.8	1.3527	1.2100	0.8
0.9	1.3087	1.2413	0.9
1.0	1.2732	1.2732	1.0

$$A_e = \frac{\pi ab}{\sqrt{1-m}} \frac{m}{K(m) - E(m)} \tag{9}$$

where K and E are complete elliptic integrals of the first and second kind, and $m = 1 - b^2/a^2$. Table 4 gives the values of the normalized equivalent area of a blade for various b/a ratios.

Most aircraft blade antennas are fed not at the center ($z = 0$) but rather at $z = z_0$ (Fig. 7). In this case the equivalent area is

$$A_e(z_0) = A_e(1 - z_0^2/a^2) \tag{10}$$

where A_e is given by (9). Equation 10 suggests that an ideal transformer with $n = 1 - z_0^2/a^2$ can be introduced for reference to the center-fed quantities, as shown in Fig. 9.

There are certain useful relationships among the electric and magnetic polarizabilities, **P** and **M**, defined in Chart 1. Let V devote the geometric volume of the ellipsoid and let the coordinate axes be along the principal axes $(1, 2, 3)$ of the ellipsoid. Then [7]

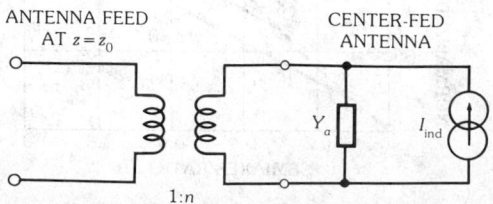

Fig. 9. Use of ideal transformer for referring to center-fed antenna parameters Y_a and I_{ind}.

$$P_{ij} = M_{ij} = 0, \qquad i \neq j$$

$$\frac{1}{P_{ii}} - \frac{1}{M_{ii}} = \frac{1}{V} \tag{11}$$

$$\sum_{i=1}^{3} \frac{1}{P_{ii}} = \frac{1}{V}$$

from which one can deduce

$$P_{11} = -2M_{33} \tag{12}$$

for a body rotationally symmetric about, say, the 3 axis, and

$$P_{11} = P_{22} = P_{33} = 3V$$

$$M_{11} = M_{22} = M_{33} = -\frac{3}{2}V \tag{13}$$

for a sphere.

Figs. 10 and 11 show, respectively, the normalized equivalent area and capacitance of an annular slot and center-fed spherical dipole antenna [8,9]. For the annular slot antenna one has the following limiting forms:

$$\frac{A_e}{\pi ab} \cong \frac{8}{\pi^2} + \frac{32}{\pi^4}\frac{a}{b}, \qquad a \ll b$$

$$\frac{C}{8\epsilon a} \cong \frac{1}{2}\ln\left(\frac{2\pi a}{b - a}\right) - 0.19, \qquad a \to b \tag{14}$$

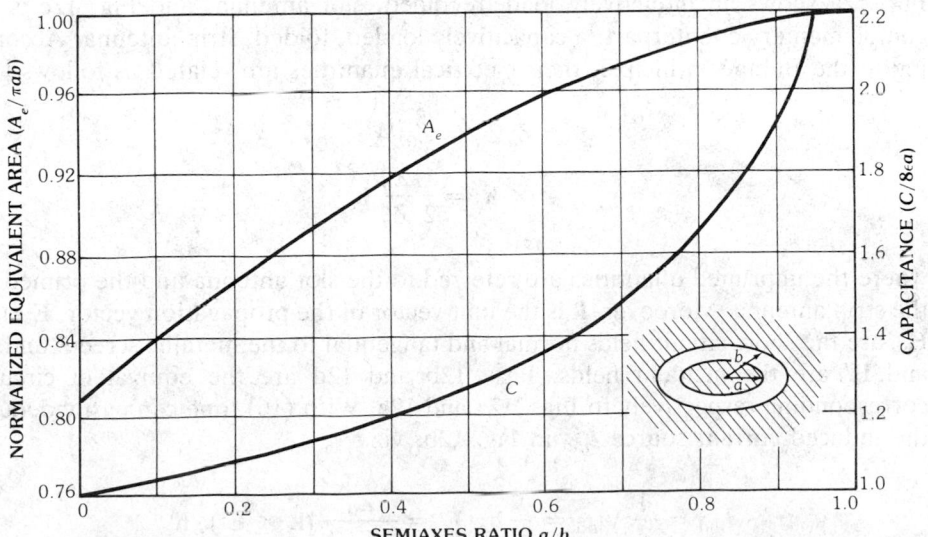

Fig. 10. Equivalent area and capacitance of annular slot antenna.

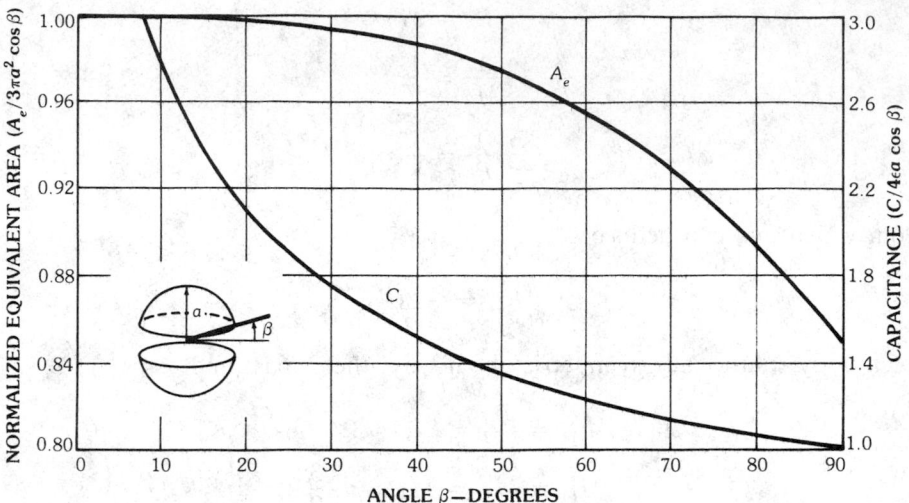

Fig. 11. Equivalent area and capacitance of spherical dipole antenna versus the gap half-angle.

For the spherical dipole antenna the limiting form is

$$\frac{C}{4\epsilon a} \cong 0.9872 - \ln\beta, \qquad \beta \leqq 10° \qquad (15)$$

The Babinet Principle

The Babinet principle is an indispensable tool for analyzing slot antennas. Fig. 12a shows an inductively loaded, folded, slot antenna, and Fig. 12c is its complementary counterpart, a capacitively loaded, folded, strip antenna. According to the Babinet principle, their electrical quantities are related as follows [1]:

$$Z_a Z_a' = Z_0^2/4$$

$$\hat{\mathbf{k}} \times \mathbf{h}_e = \frac{1}{2} \frac{Z_0}{Z_a'} \mathbf{h}_e' \qquad (16)$$

where the unprimed quantities are referred to the slot antenna and the primed to the strip antenna. Moreover, $\hat{\mathbf{k}}$ is the unit vector of the propagation vector, $\mathbf{E}_n$ and $\mathbf{H}_{sc}$ are the short-circuit fields normal and tangential to the metallic screen, and $\mathbf{E}^i$ and $\mathbf{H}^i$ are the incident fields. Figs. 12b and 12d are the equivalent circuits corresponding respectively to Figs. 12a and 12c. With (16), one can easily deduce the induced current source I_{ind} in Fig. 12b, viz.,

$$I_{\text{ind}} = \frac{1}{Z_a} V_{\text{ind}} = \frac{1}{Z_a} \mathbf{h}_e \cdot \mathbf{E}^i = \frac{Z_0}{2Z_a Z_a'} (\hat{\mathbf{k}} \times \mathbf{E}^i) \cdot \mathbf{h}_e'$$

$$= 2 \mathbf{H}^i \cdot \mathbf{h}_e' = \mathbf{h}_e' \cdot \mathbf{H}_{\text{sc}} \qquad (17)$$

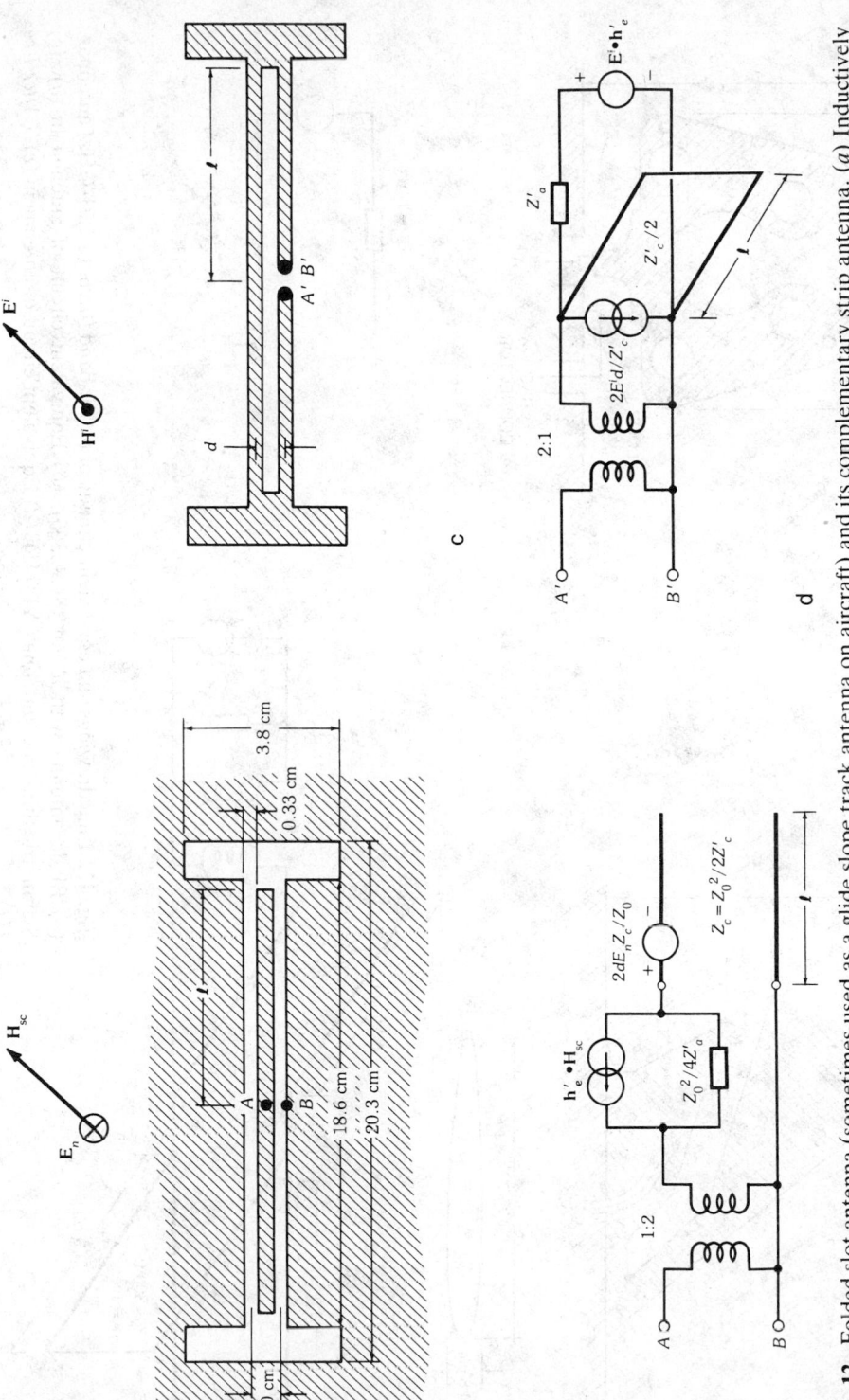

Fig. 12. Folded slot antenna (sometimes used as a glide slope track antenna on aircraft) and its complementary strip antenna. (*a*) Inductively loaded, folded, slot antenna. (*b*) Equivalent circuit of antenna in (*a*). (*c*) Capacitively loaded, folded, strip antenna. (*d*) Equivalent circuit of antenna in (*c*). (*After Lee [1]*)

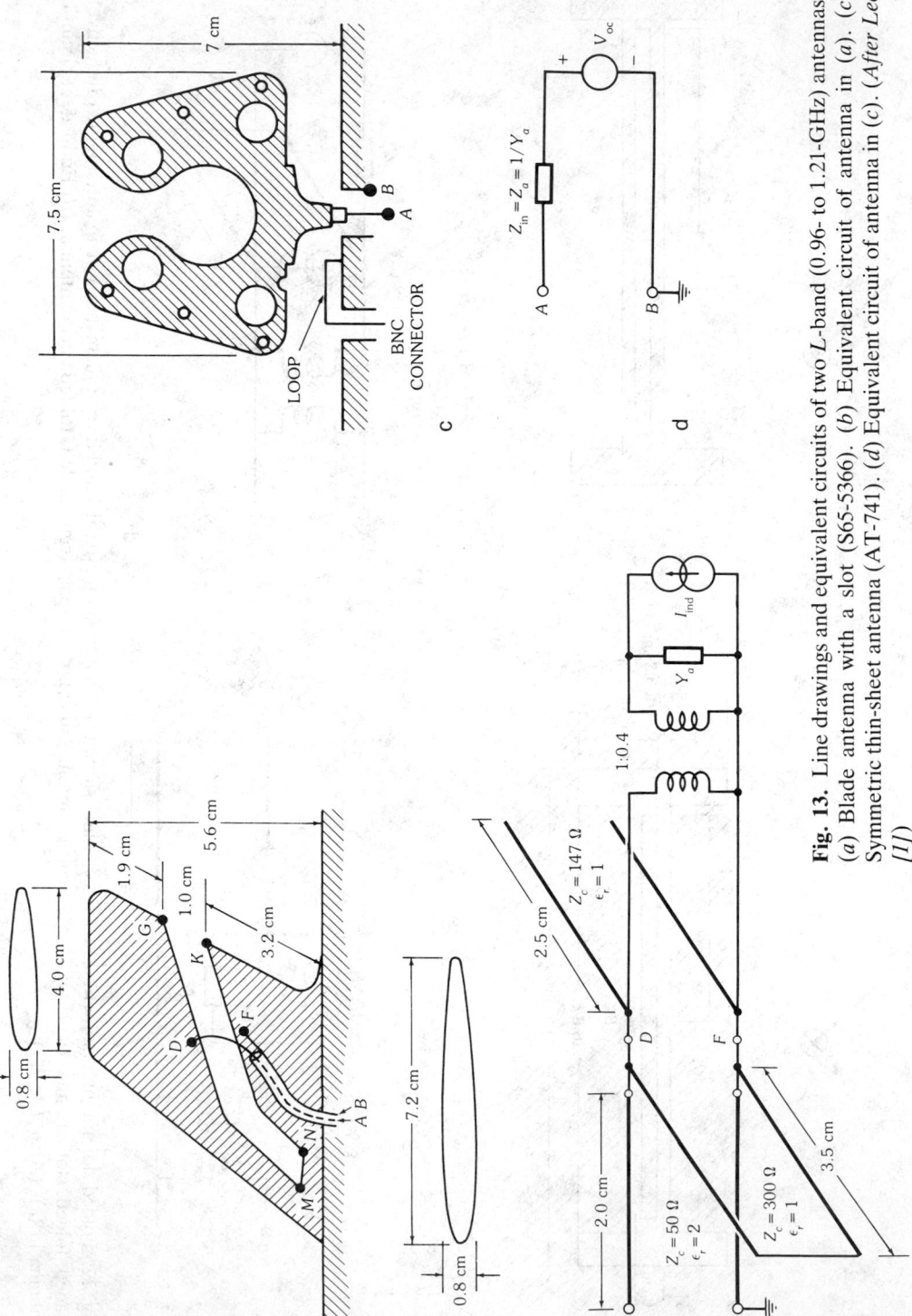

Fig. 13. Line drawings and equivalent circuits of two L-band (0.96- to 1.21-GHz) antennas. (*a*) Blade antenna with a slot (S65-5366). (*b*) Equivalent circuit of antenna in (*a*). (*c*) Symmetric thin-sheet antenna (AT-741). (*d*) Equivalent circuit of antenna in (*c*). (*After Lee [1]*)

5. EMP Responses of Aircraft Antennas

The EMP responses of many different types of aircraft antennas have been analyzed. Typical examples follow [1, 10–12].

L-Band Blade Antennas

Figs. 13a and 13c show two different antennas that both operate between 0.96 GHz and 1.21 GHz. Although their overall dimensions and in-band properties are similar, their out-of-band properties are quite different.

Fig. 13a is a blade antenna with a slot (the slot often being referred to as a notch). The equivalent circuit is given in Fig. 13b. The induced current I_{ind} and antenna admittance Y_a in the equivalent circuit of Fig. 13b are the quantities referred to the same blade antenna having a feeding gap across the entire blade at D, F. The two transmission lines with characteristic impedances 147 Ω and 300 Ω are due to the slot, whereas the transmission line between D, F and A, B represents the coaxial cable with a characteristic impedance of 50 Ω. The frequency variation of the antenna's input impedance is shown in Fig. 14a.

The interior structure of another L-band antenna is shown in Fig. 13c. The actual antenna element is a thin metal sheet cut out in a regular symmetric pattern. The holes cut out in the sheet have the effect of increasing the antenna inductance, since the current has to flow around the holes, thereby lengthening the current path. The input impedance is displayed in Fig. 14b, while the effective heights of both antennas are given in Fig. 15.

Both L-band antennas have roughly the same input impedance and effective height at in-band frequencies. The notch antenna has an inductive input impedance and a very small effective height for low frequencies, whereas the other blade antenna has a capacitance input impedance and considerable effective height for low frequencies.

UHF Communication Antennas

A very common uhf antenna is shown in Fig. 16a, and Fig. 16b is the corresponding equivalent circuit in which L is the inductance of the short stub. This stub not only provides a dc path for lightning protection between the upper part of the antenna and the aircraft skin but also mechanically fastens the two parts of the antenna. The different sections of the transmission lines and the end capacitance C_e are for tuning purposes, so that within the operating frequency band (225 to 400 MHz) the vswr is about 2:1 at the input connector. The capacitance C_p accounts for the two ends of the gap. Figs. 17a and 17b show the frequency variations of Z_{in} and h_e together with measured data for Z_{in}.

VHF Communication Antennas

A schematic diagram of a typical vhf communication antenna is shown in Figs. 18a and 18b. A careful examination of these figures reveals that Fig. 18c is the appropriate equivalent circuit for this antenna. The transmission lines with impedance $Z_c = 176$ Ω represent the empty portion of the slot, whereas the transmission line with impedance $Z_c/\sqrt{\epsilon_r} = 119$ Ω represents the dielectric-filled portion of the slot. The impedance Z_c can be estimated from the analysis of two

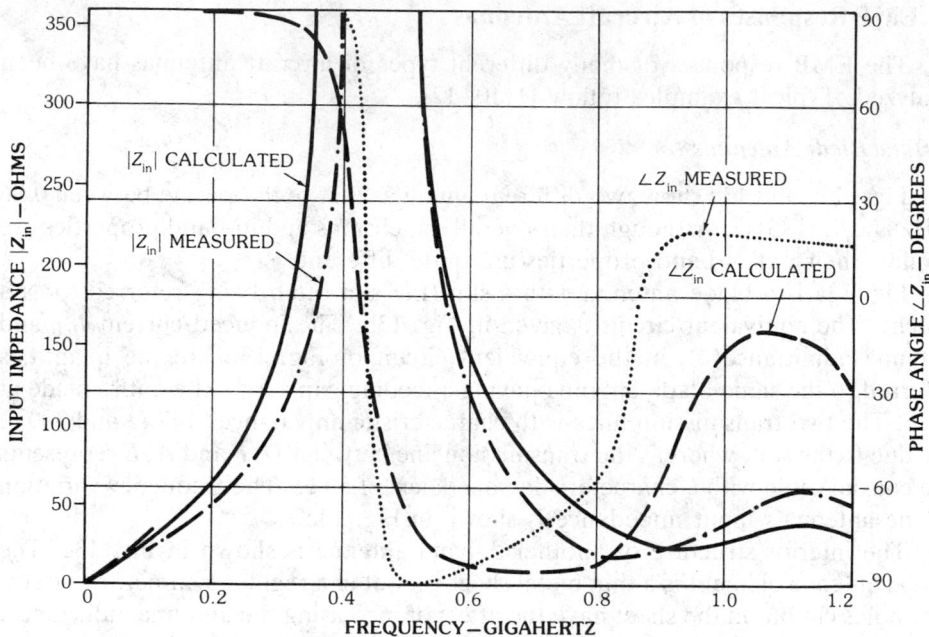

a

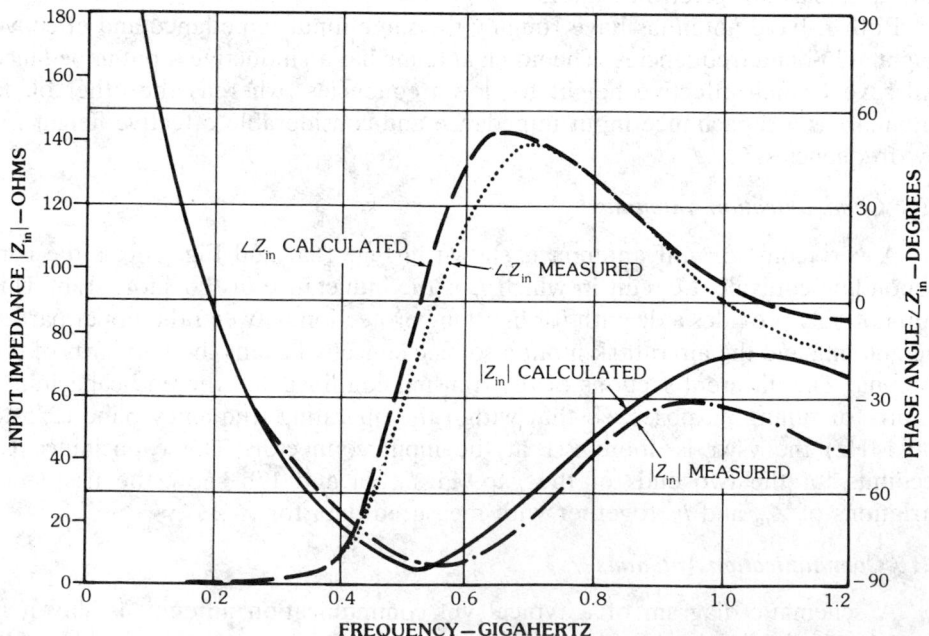

b

Fig. 14. Input impedance of two *L*-band (0.96- to 1.21-GHz) antennas. (*a*) Of blade antenna with a slot (S65-5366). (*b*) Of symmetric thin-sheet antenna (AT-741). (*After Lee [1]*)

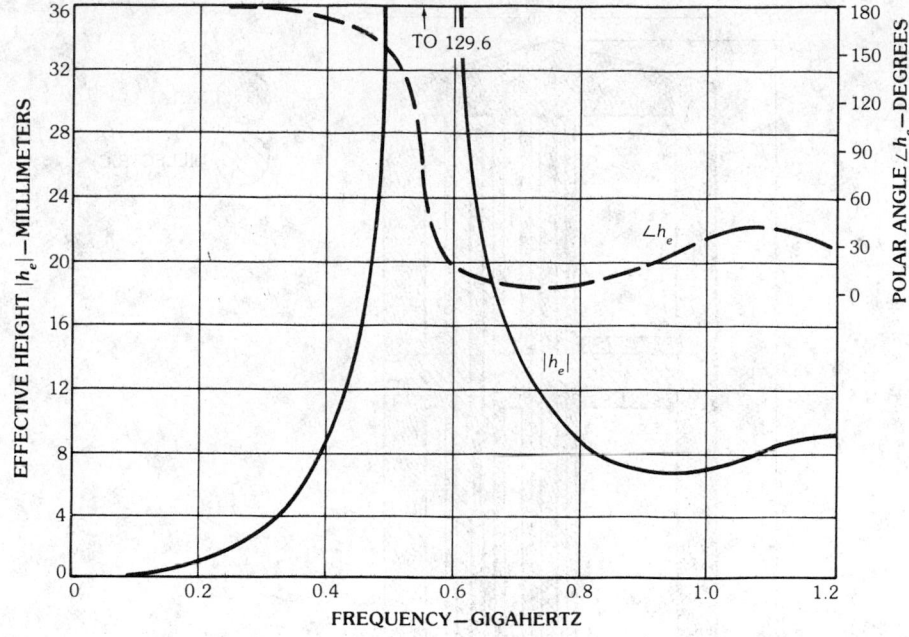

a

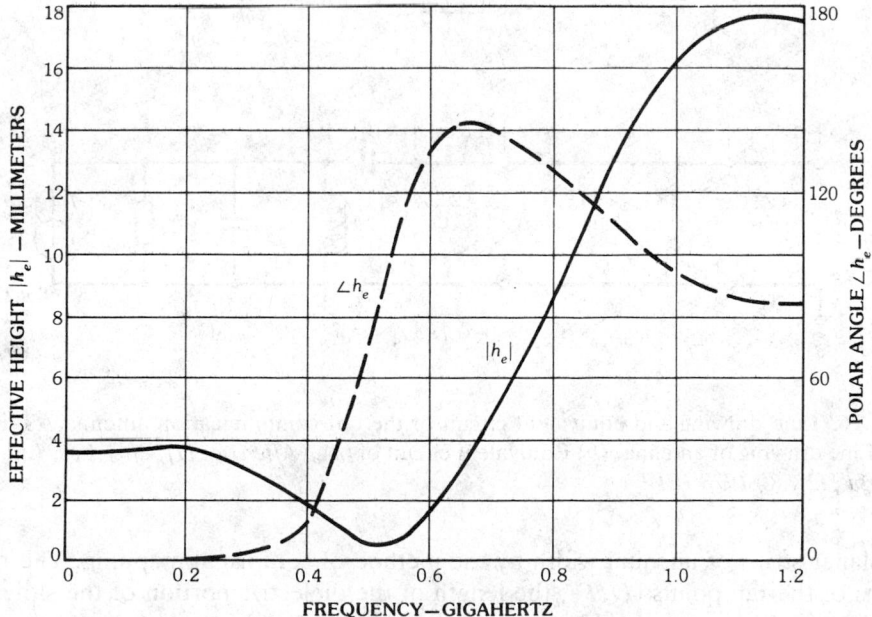

b

Fig. 15. Effective heights of two *L*-band (0.96- to 1.21-GHz) antennas. (*a*) Of blade antenna with a slot (S65-5366). (*b*) Of symmetric thin-sheet antenna (AT-741). (*After Lee [1]*)

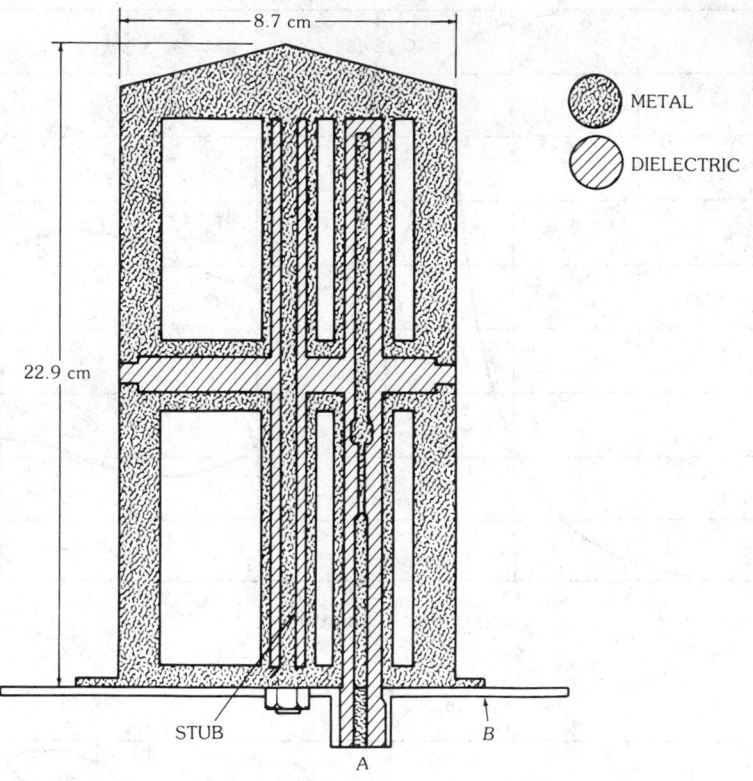

a

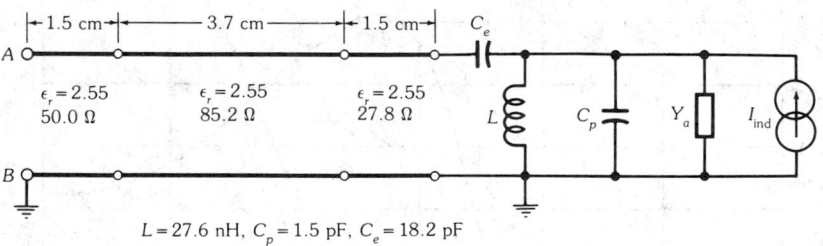

$L = 27.6$ nH, $C_p = 1.5$ pF, $C_e = 18.2$ pF

b

Fig. 16. Line drawing and equivalent circuit of the uhf communication antenna AT-1076. (*a*) Line drawing of antenna. (*b*) Equivalent circuit of (*a*). (*After Lee [1]; after Lee, Liu, and Marin [12], © 1978 IEEE*)

coplanar strips of unequal width by the method of conformal mapping. The positions of the tap points (D, F), the length of the dielectric portion of the slot, and the three added coaxial cables act as a matching network, so that the vswr is less than 1.75:1 at in-band frequencies (116 to 156 MHz). The ideal transformer in the equivalent circuit accounts for the fact that the short-circuit current between D and F is just a fraction of I_{ind}, which is the total induced current through a cross section of the antenna (located at D or F and being parallel to the antenna base)

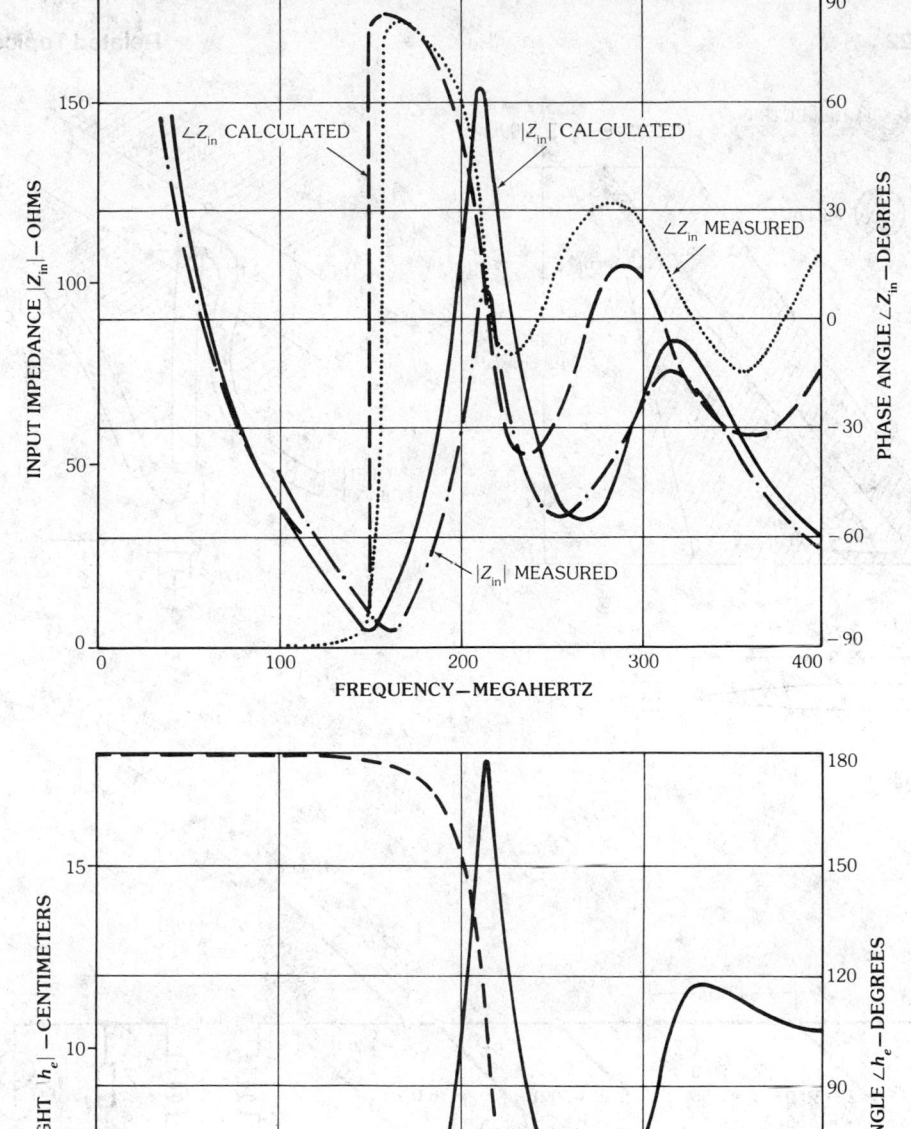

Fig. 17. Input impedance and effective height of the uhf communication antenna AT-1076 (225 to 400 MHz). (*a*) Input impedance. (*b*) Effective height. (*After Lee [1]; after Lee, Liu, and Marin [12]*, © *1978 IEEE*)

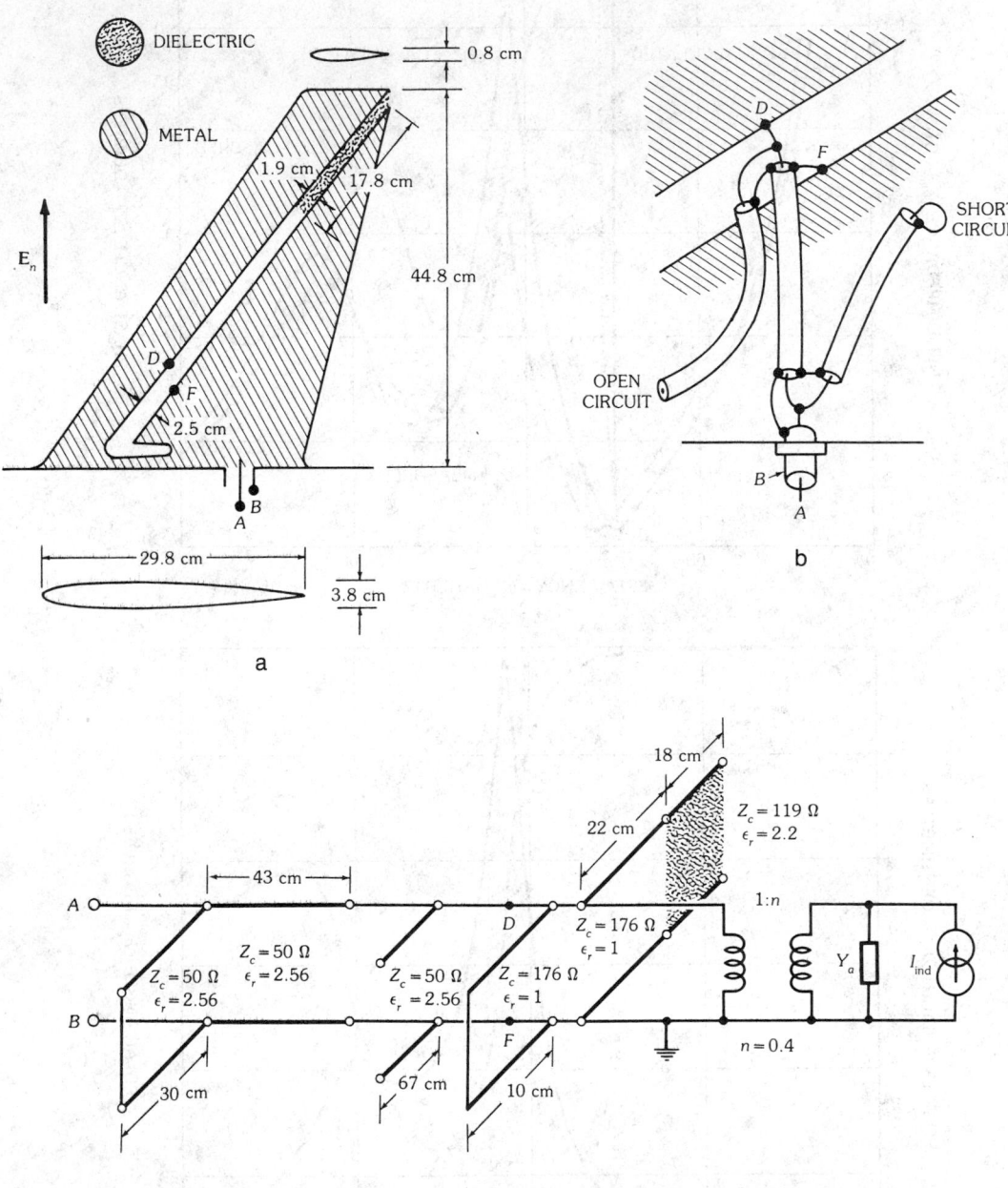

Fig. 18. Line drawings and equivalent circuit of the vhf communication antenna S65-8262-2. (*a*) Construction of antenna. (*b*) Antenna feed circuit. (*c*) Equivalent circuit of (*a*). (*After Lee [1]*)

when the slot is absent. The ideal transformer also accounts for the effects of the locations of D, F on the antenna's input admittance, since Y_a in Fig. 18c is the admittance across a gap bisecting the entire antenna at the location of the cross section discussed above. The transformer ratio n can be estimated by the method described in Section 4, under "Low-Frequency Techniques." Fig. 19 shows the calculated input impedance together with some measured data.

VHF Localizer Antennas

Figs. 20a and 20b show all the electrical elements of a localizer antenna and Fig. 20c is the equivalent circuit. The external elements Z_a and V_{ind} can be calculated from the theory of a circular loop with due account for factors of two resulting from the presence of the ground plane. The tuning capacitors at the ends of the loops (Fig. 20a) are represented by C in Fig. 20c. The coaxial cable wound around the center rods serves as a balun, converting a balanced signal into an unbalanced one for the terminals A, B and A', B within the in-band frequencies (108 to 112 MHz). The wires wrapped around the 100-Ω resistor can be identified as a hybrid circuit which splits an incoming signal into two equal parts for the two terminals A, B and A', B. Because of the close coupling between the wires, M can be taken equal to L. Fig. 21a the input impedance Z_{in} referred to A, B with A', B open circuited, or vice versa. Also shown in the figure are some measured data. Fig. 21b gives the frequency variation of V_{oc} across A, B or A', B for an incident plane wave with the electric-field vector perpendicular to the ground plane and the magnetic-field component B_0 perpendicular to the plane of the loop.

VHF Marker Beacon Antennas

The marker beacon antenna shown in Fig. 22a is flush-mounted at the bottom of the fuselage and operates around 75 MHz. The bowl can be approximated by a hemispherical indentation. The equivalent circuit is given by Fig. 22b in which L_1 and L_2 are respectively the inductances of the small and large loops formed by the metal rod and the feed wire. Figs. 23a and 23b give the frequency variations of Z_{in} and V_{oc}/B_0, where B_0 is the magnetic field tangential to the aircraft skin and perpendicular to the plane of the large loop, and V_{oc} and Z_{in} are the open-circuit voltage and input impedance across A, B.

HF Fixed-Wire Antennas

Two hf fixed-wire antennas on an aircraft are shown in Fig. 24a. A schematic drawing of the stick-model aircraft used in the calculations is shown in Fig. 24b. To find the input impedance between A, B when the port A', B is terminated by the impedance Z_L, let the antenna be driven with the voltage V between A, B. The induced currents on the antenna wires and the aircraft can be decomposed into (a) radiating currents on the aircraft, (b) a TEM mode such that the wire currents are of equal magnitude and in the same direction with the "return current" along the fuselage, and (c) a TEM mode such that the wire currents are of equal magnitudes but in opposite directions with no net current along the fuselage. The details of this decomposition can be found in reference 13. An equivalent circuit of the antenna is shown in Fig. 24c. The admittance Y_a is defined at an imaginary gap across the fuselage at the location of the antenna feed point, and can be calculated using a

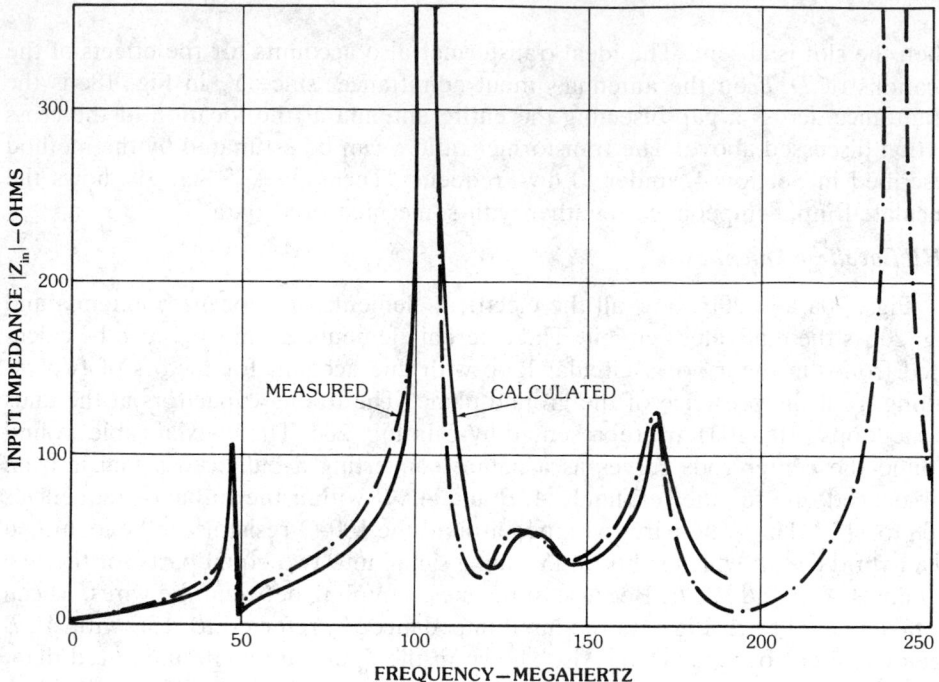

a

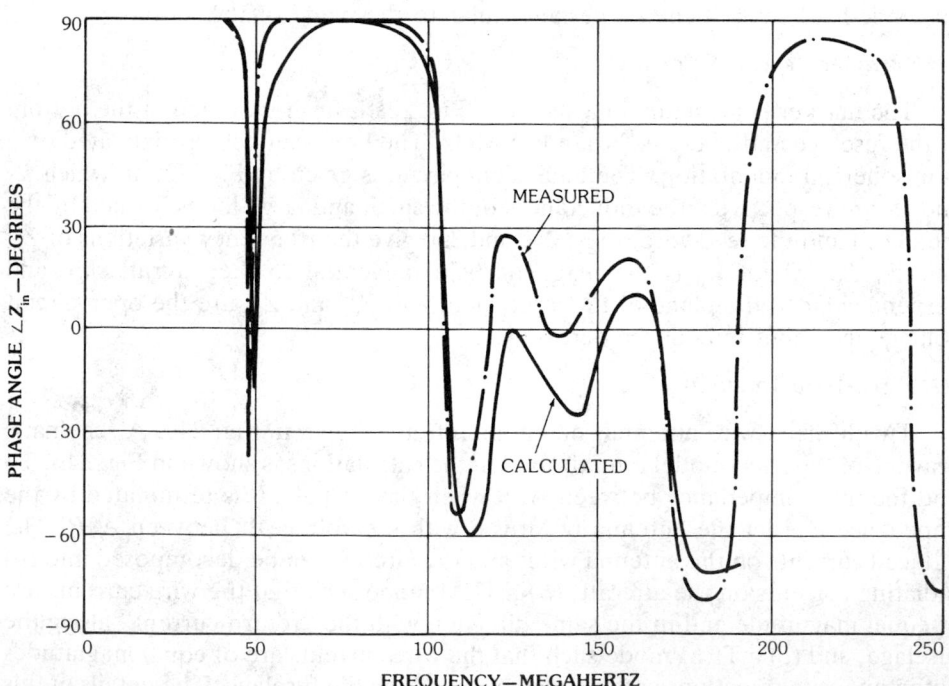

b

Fig. 19. Input impedance of the vhf communication antenna S65-8262-2 (116 to 156 MHz). (*a*) Magnitude of input impedance. (*b*) Phase angle of input impedance. (*After Lee [1]*)

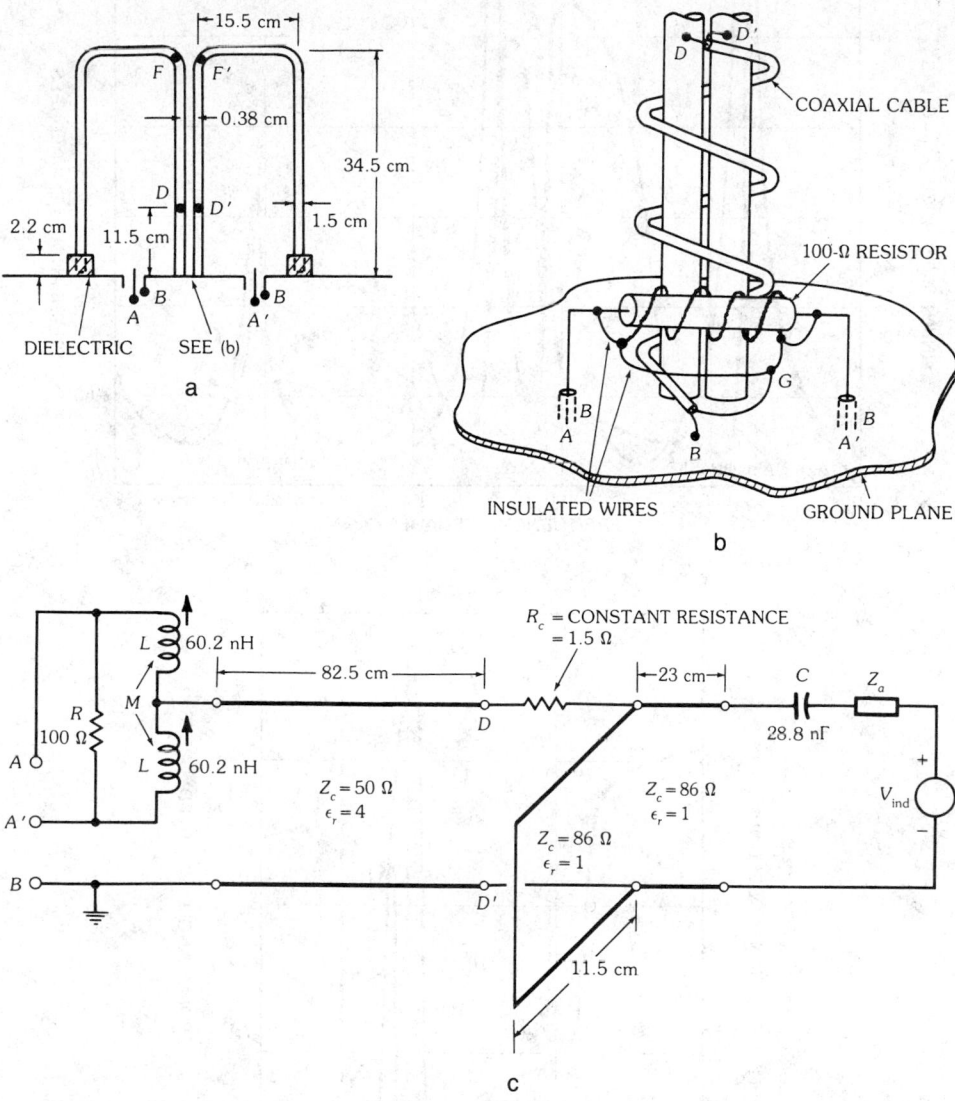

Fig. 20. Line drawings and equivalent circuit of the localizer antenna S65-147. (*a*) Loop structures. (*b*) Signal-splitting circuit. (*c*) Equivalent circuit of antenna in (*a*). (*After Lee [1]; after Lee, Liu, and Marin [12],* © *1978 IEEE*)

stick-model aircraft. The transmission line with the characteristic impedance Z_c' represents the first TEM mode described above, whereas the transmission line with impedance Z_c'' represents the second TEM mode. The ideal transformers account for the coupling between the TEM modes and the radiating currents. The current I_{ind} is induced on the fuselage by a plane wave at the location of the antenna feed gap and in the absence of the antenna wires (see Fig. 24b). The other current generators are found from the transmission-line theory [14].

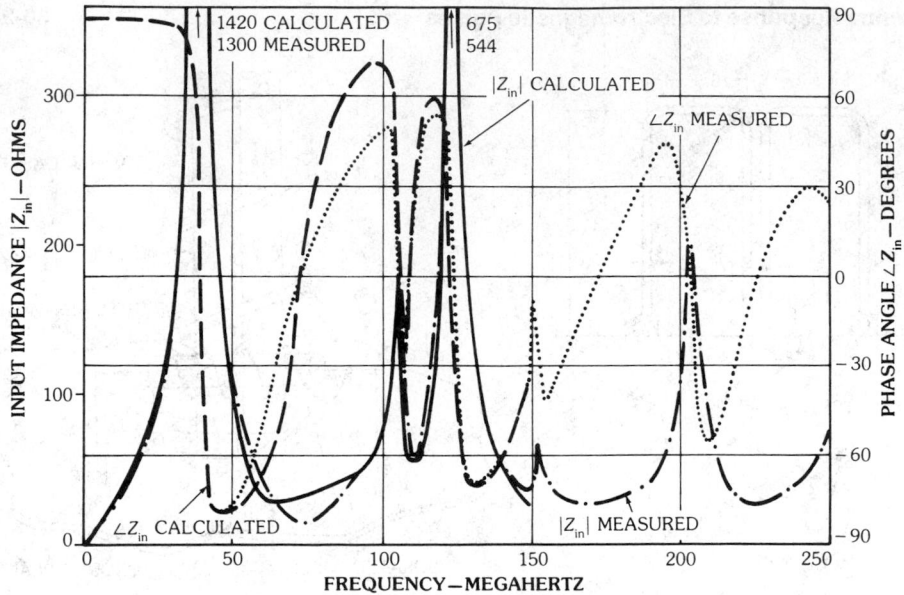

a

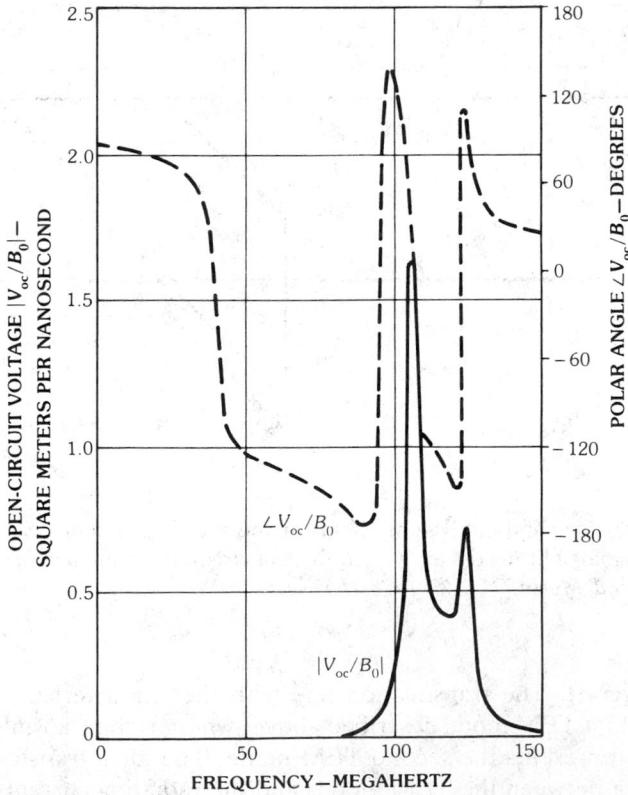

b

Fig. 21. Input impedance and open-circuit voltage of the localizer antenna S65-147 (108 to 112 MHz). (*a*) Input impedance. (*b*) Open-circuit voltage. (*After Lee [1]; after Lee, Liu, and Marin [12]*, © *1978 IEEE*)

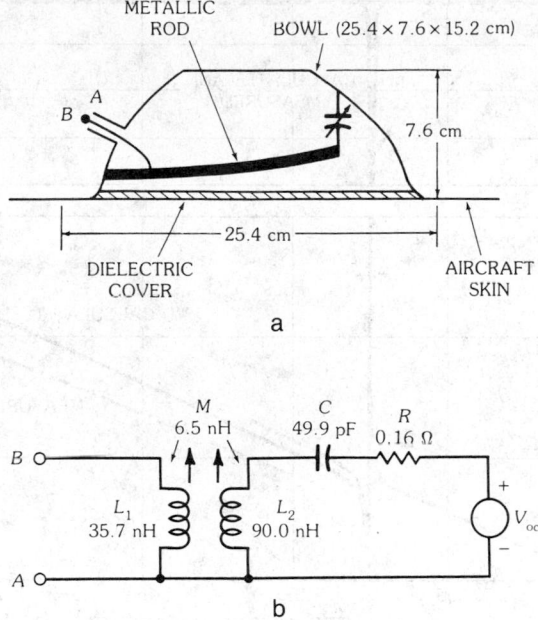

Fig. 22. Line drawing and equivalent circuit of the AT-536/ARN marker beacon antenna. (*a*) Antenna structure. (*b*) Equivalent circuit of antenna in (*a*). (*After Lee [1]; after Lee, Liu, and Marin [12], © 1978 IEEE*)

Fig. 25 gives the frequency variations of the input impedance and effective height. In this case the effective height relates the open-circuit voltage to the incident electric field, and is a function of the angle of incidence and polarization of the incident plane wave. In Fig. 25 the values $Z_L = 0$, $Z_L = \infty$, and $\theta = 120°$ have been used. Quantities with superscript "sc" correspond to $Z_L = 0$, and quantities with superscript "oc" correspond to $Z_L = \infty$. The variation of h_e with θ can be found in [13]. The transient response of this antenna to the high-altitude EMP waveform given by (1) can be found in reference 15.

VLF/LF Trailing-Wire Antennas

A vlf/lf (17- to 60-kHz) transmitting dual-wire antenna is shown in Fig. 26a. The response of this type of antenna to EMP can be obtained using two different steps. In the first step, which is valid in the frequency range where the aircraft is electrically small, one may model the aircraft by a capacitance to characterize its effect on the antenna response. The interaction between the two wires can be calculated by decomposing the wire currents into common-mode (antenna-proper) currents and differential-mode (transmission-line) currents. In the second step, which is valid in the frequency region around the first few aircraft resonances, the aircraft can be modeled by intersecting sticks. Fig. 26b is the equivalent circuit appropriate for the situation depicted in Fig. 26a. Explicit approximate formulas for I'_{ind}, I''_{ind}, Z'_a, h, Z''_a, Z_c, α, and β can be found in reference 16.

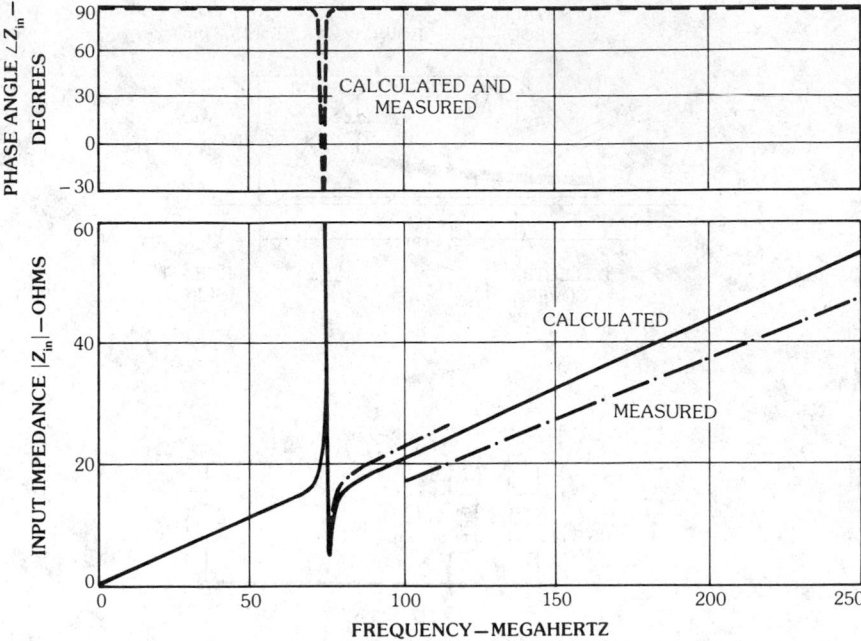

a

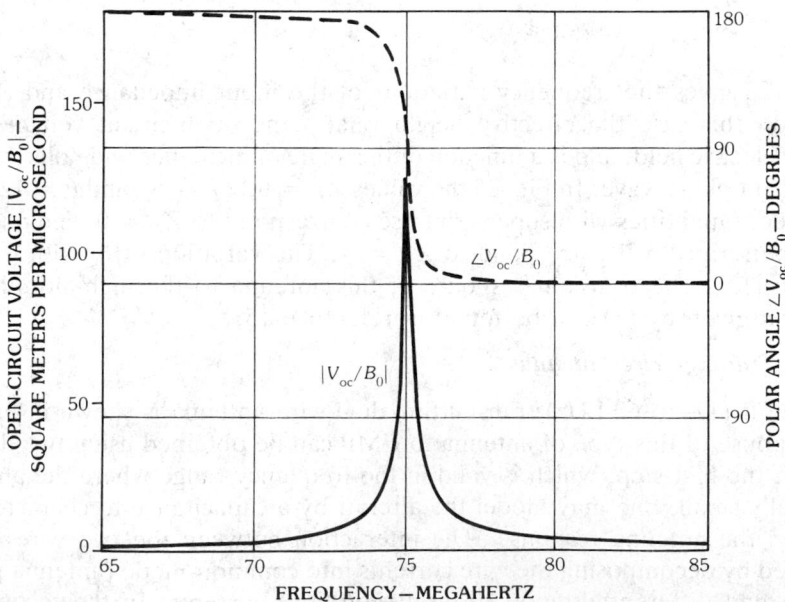

b

Fig. 23. Input impedance and relative open-circuit voltage of the AT-536/ARN marker beacon antenna (75 MHz). (*a*) Input impedance. (*b*) Open-circuit voltage. (*After Lee [1]; after Lee, Liu, and Marin [12],* © *1978 IEEE*)

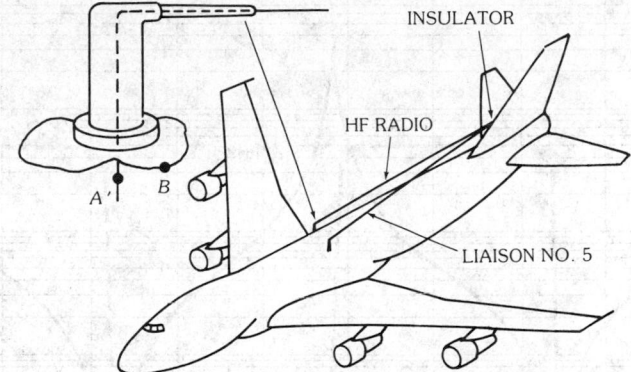

a

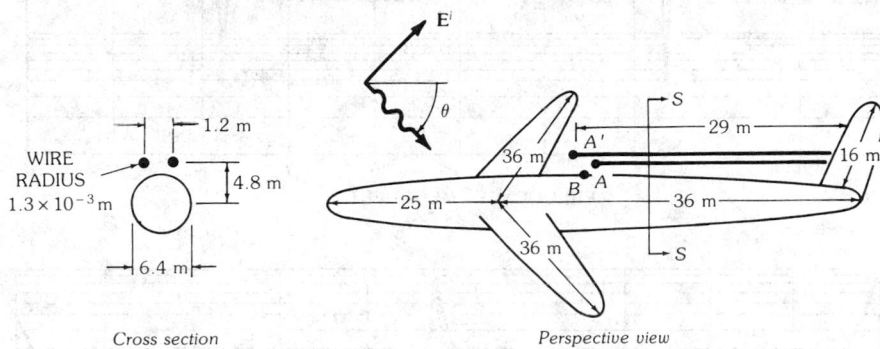

b

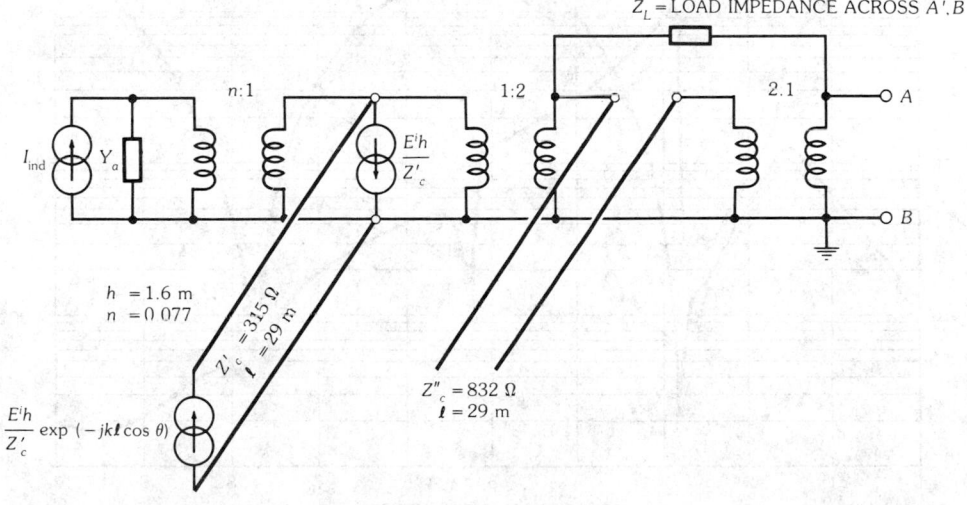

c

Fig. 24. Line drawings and equivalent circuit of the two hf fixed-wire antennas on the E-4 aircraft. (*a*) Aircraft with two fixed-wire hf antennas. (*b*) Wire position and incident electric field. (*c*) Equivalent circuit of antenna. (*After Lee [1]*)

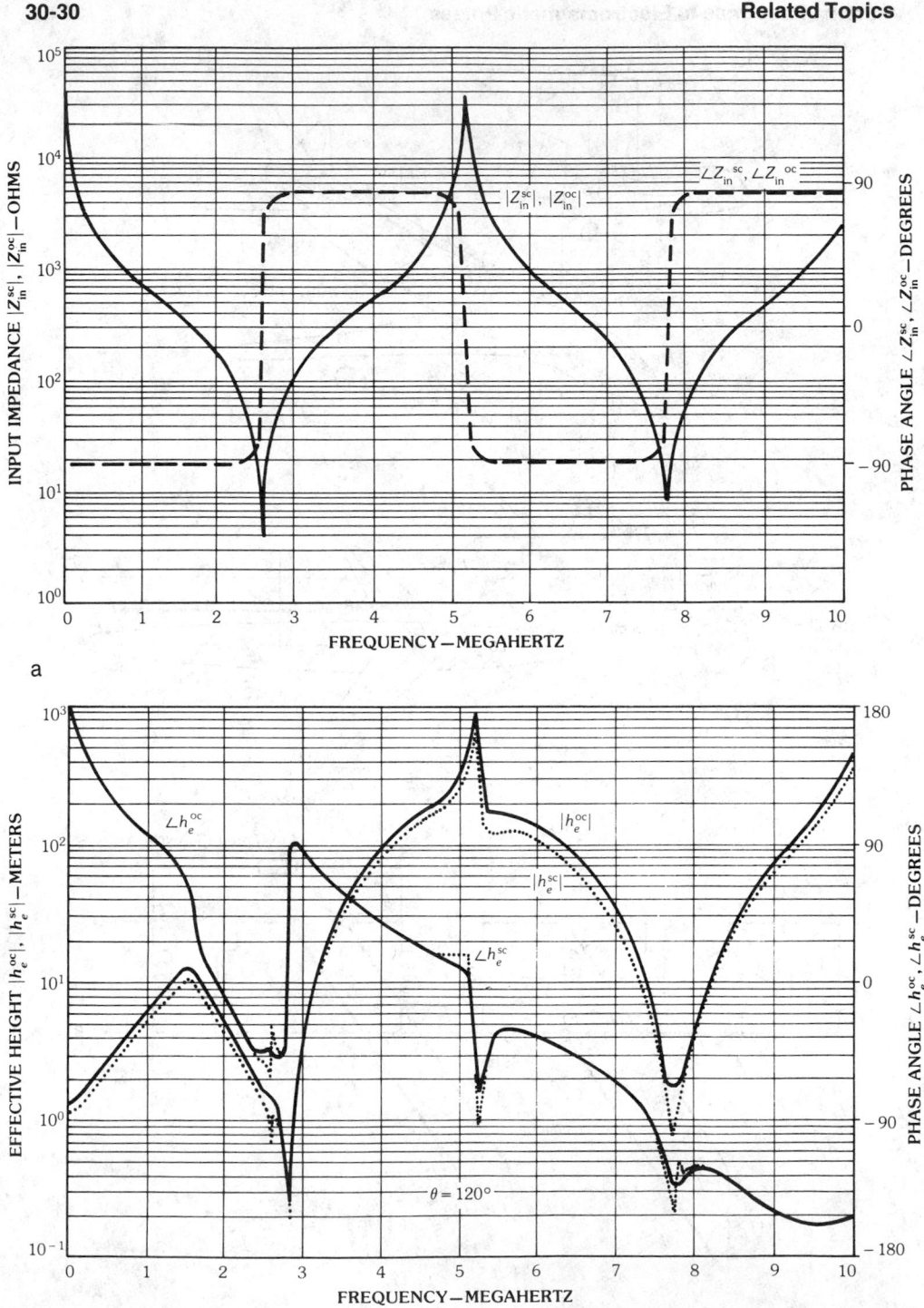

Fig. 25. Input impedance and effective height of the hf (2- to 30-MHz) fixed-wire antennas on the E-4 aircraft. (*a*) Input impedance. (*b*) Effective height. (*After Lee [1]*)

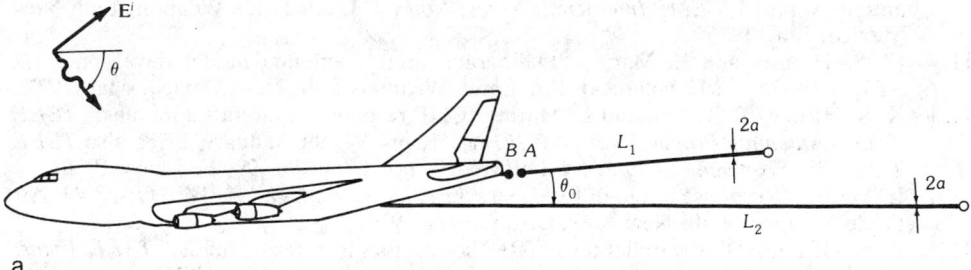

a

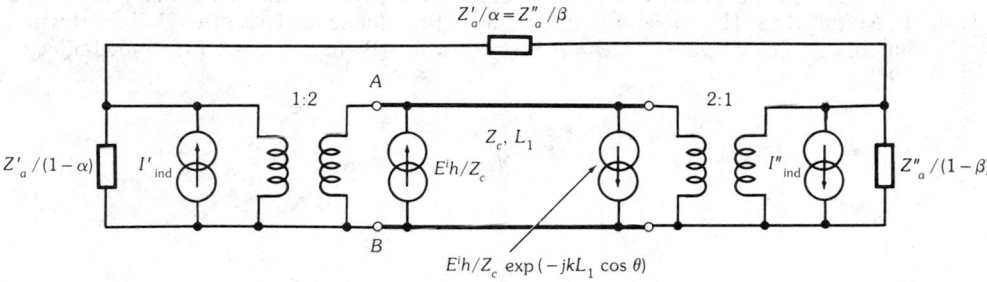

b

Fig. 26. Line drawing and equivalent circuit for the dual-wire antenna. (*a*) Aircraft with trailing-wire transmitting antenna. (*b*) Equivalent circuit of antenna in (*a*). (*After Lee [1]; after Marin, Lee, and Castillo [16],* © *1978 IEEE*)

6. References

[1] K. S. H. Lee, *EMP Interaction: Principles, Techniques, and Reference Data*, chapter 2.1, New York: Hemisphere Publishing Corp., 1986.

[2] V. V. Liepa, "Surface field measurements on scale model EC-135 and E-4 aircraft," *Interaction Application Memos 15 and 17*, Air Force Weapons Lab, New Mexico, March and August 1977.

[3] R. W. Sassman, "The current induced on a finite, perfectly conducting, solid cylinder in free space by an electromagnetic pulse," *EMP Interaction Notes, Note 11*, Air Force Weapons Lab, New Mexico, July 1967.

[4] C. E. Baum, "Parameters for some electrically-small electromagnetic sensors," *EMP Sensor and Simulation Notes, Note 38*, Air Force Weapons Lab, New Mexico, March 1967.

[5] J. D. Kraus, *Antennas*, chapter 3, New York: McGraw-Hill Book Co., 1950.

[6] K. S. H. Lee, "Maximum power penetration through an electrically small aperture," *IEEE Trans. Antennas Propag.*, vol. AP-31, no. 3, pp. 518–519, May 1983.

[7] K. S. H. Lee, "Electrically small ellipsoidal antennas," *EMP Sensor and Simulation Notes, Note 193*, Air Force Weapons Lab, New Mexico, February 1974.

[8] R. W. Latham and K. S. H. Lee, "Capacitance and equivalent area of a disk in a circular aperture," *EMP Sensor and Simulation Notes, Note 106*, Air Force Weapons Lab, New Mexico, May 1970.

[9] R. W. Latham and K. S. H. Lee, "Capacitance and equivalent area of a spherical dipole sensor," *EMP Sensor and Simulation Notes, Note 113*, Air Force Weapons Lab, New Mexico, July 1970.

[10] T. K. Liu, K. S. H. Lee, and L. Marin, "Broadband responses of deliberate aircraft

antennas, part I," *EMP Interaction Notes*, *Notes 228*, Air Force Weapons Lab, New Mexico, May 1975.

[11] K. S. H. Lee and L. Marin, "Deliberate aircraft antenna model development," *AFWL-TR-76-218* (2 volumes), Air Force Weapons Lab, New Mexico, May 1977.

[12] K. S. H. Lee, T. K. Liu, and L. Marin, "EMP response of aircraft antennas," *IEEE Trans. Antennas Propag.*, vol. AP-26, no. 1, pp. 94–99, January 1978; also *IEEE Trans. Electromagn. Compat.*, vol. EMC-20, no. 1, pp. 94–99, February 1978.

[13] L. Marin, "Response of the hf fixed-wire antenna on the E-4," *AFWL-TR-78-42*, Air Force Weapons Lab, New Mexico, February 1979.

[14] K. S. H. Lee, "Two parallel terminated conductors in external fields," *IEEE Trans. Electromagn. Compat.*, vol. EMC-20, no. 2, pp. 288–296, May 1978.

[15] G. Bedrosian and L. Marin, "Transient response of the hf fixed-wire antennas on the E-4," *AFWL-TR-78-43*, Air Force Weapons Lab, New Mexico, February 1979.

[16] L. Marin, K. S. H. Lee, and J. P. Castillo, "Broadband analysis of vlf/lf aircraft wire antennas," *IEEE Trans. Antennas Propag.*, vol. AP-26, no. 1, pp. 141–145, January 1978.

Chapter 31

Radome Electromagnetic Design

G. P. Tricoles
General Dynamics

CONTENTS

 **Gus Tricoles** received a BA in physics from the University of California, Los Angeles, in 1955, an MS in applied mathematics from San Diego State College in 1958, and an MS in physics in 1962 and a PhD in applied physics in 1970, both from the University of California, San Diego.

He was employed by General Dynamics from 1955 to 1959 and by Smyth Research Associates from 1960 to 1962; in 1962 he returned to General Dynamics, where he is now Engineering Manager of Electromagnetics in the Electronics Division in San Diego.

Dr. Tricoles' main work has been in propagation and microwave holography with applications to radomes, direction finding, and geological prospecting. He developed analytical methods for radome design and experimental, diagnostic methods for radomes. He has designed radomes for several aircraft and missiles, developed anisotropic radomes and guided-wave antennas, and modeled retinal receptors. He holds 14 patents and has published 20 papers and written many reports. He is a member of US Commissions A, B, and C of URSI, the IEEE Antenna Standards Committee, a Fellow of the Optical Society of America, and a Fellow of the Institute of Electrical and Electronic Engineers.

1. Introduction

Radomes are enclosures for antennas. Most radomes are hollow dielectric shells although some contain perforated metallic layers or metallic reinforcing structures. Radomes are used with large antennas on the earth's surface to reduce wind loading and to prevent accumulation of ice or snow; these radomes usually have spherical contours. Many aircraft and missiles have radomes; some are blunt, but a nose radome may be pointed to reduce aerodynamic drag. See Fig. 1. Although a radome may perform a useful function, it can degrade an electronic system connected to the enclosed antenna. Radomes reduce radar range because they reflect and attenuate incident waves. They also cause errors in determining the direction of a reflecting object or radiating source; in missiles, this directional error significantly affects the accuracy of guidance. Radomes also can increase radar clutter by reflections that are analogous to multipath. These effects on systems constitute the electromagnetic performance of a radome. A radome's electromagnetic performance is described by four parameters. They are as follows:

Radome Performance Parameters

Far-Field Radiation Patterns—Far-field patterns describe the angular distribution of the intensity radiated or received by an antenna. Radomes can change beamwidths, side lobe levels, directions of pattern maxima, values of maxima or minima, and axial ratio. Patterns of antennas in radomes are measured with conventional apparatus except for special fixtures to support the antenna and radome.

Power Transmittance—Power transmittance of a radome describes the intensity at the peak of the antenna pattern's main beam. It is the ratio of peak intensity received with the radome present to the intensity received when the radome is absent. Although transmittance can be measured by comparing intensities at the peaks of antenna patterns measured with and without the radome, accuracy is greater if the measurements are made with the enclosed antenna held fixed and receiving energy from a distant transmitter. The radome is pivoted about the enclosed antenna, and data are recorded continuously during the radome motion.

Boresight Error—Boresight error is an angular shift; it depends on the nature of the enclosed antenna. For monopulse antennas, it is the shift that the radome causes in the direction of the difference mode pattern's deepest minimum or the shift that is obtained by comparing the phases in a pair of antennas. For antennas with patterns that determine direction from the main beam peak, boresight error is the shift in direction of the main beam. Boresight error is usually measured by pivoting the radome about the fixed, enclosed receiving antenna. Several kinds of detection are used [1].

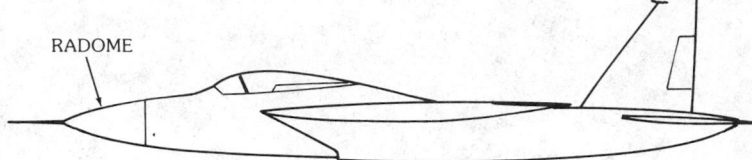

Fig. 1. Aircraft with nose radome.

Boresight Error Slope—Boresight error slope is the rate of change of boresight error with respect to the angle between the axes of the radome and the antenna. The antenna axis is some preferred direction, such as the monopulse null direction. The radome axis is an axis of revolution, if one exists, or some arbitrary direction. Boresight error slope is significant for guided missiles that are steered to maintain a constant target bearing.

Variables in the Radome Performance Parameters

The radome performance parameters depend on several variables, which may be classified as either constrained or design variables.

Constrained Variables—The parameters depend on some variables that are constrained by the system that uses the radome. The frequencies and polarization of the radiation have strong influence; so do the enclosed antenna's size, polarization, aperture distribution, position, and range of orientations. Radomes must obviously be big enough to provide antenna clearance, and many aircraft and missile radomes are pointed to reduce drag. In addition, the radome's environment is significant. Ground-based radomes are subjected to wind loads and weather. Aircraft and missile radomes must withstand aerodynamic loads and the high temperatures generated during flight. The temperatures and loads restrict the materials that can be used for fabrication. All these variables influence radome performance, but the constraints put them, to varying degrees, beyond a designer's control.

Design Variables—Although some variables are tightly constrained, values of several additional variables can be chosen in a design to achieve (or approach) desired performance. We call the latter class design variables. Some design variables describe a radome's composition; others, its configuration. The radome wall configuration is one of the most important design variables. Radome walls can consist of one homogeneous layer or of several; the dielectric constants and thicknesses can be chosen within restrictions imposed by material strength and temperatures. In addition, small local thickness variations or gradual tapers improve performance. Recently, materials with configurational anisotropy have been used to increase frequency bandwidths and to reduce effects produced by changing wave polarization [2,3]. Metallic layers also have been included in radomes to give protection from lightning or static electricity [4].

2. Physical Description of Radome Effects

This section describes what radomes do to electromagnetic waves. When a wave illuminates a radome, several mechanisms occur; these are refraction, reflection, attenuation, guided-wave excitation, and scattering by the tip or base. Because the mechanisms are diverse, occur simultaneously, and depend on frequency and polarization, radome analysis tends to be complicated.

Fields in Radome-Bounded Regions; Constituent Waves

Fig. 2 shows the kinds of waves that occur when a plane wave illuminates the outside of a pointed radome. The sketch is two-dimensional, so it does not show the effect of transverse radome curvature and wave polarization. Experiments, theory, and computation show that the field in a radome-bounded region consists of several constituent fields [5, 6]. These constituents interfere, so the antenna receives a complicated, nonplanar wave, even for specific polarization, frequency, and direction of the incident wave. One of the constituents is directly propagated through the illuminated side of the radome; it includes multiple internal reflections. A second constituent is a wave that propagates first through the illuminated side and is then reflected by the opposite side. Guided waves are excited; their propagation constants differ from those of the direct and reflected waves. A vertex may scatter a field. The base of the radome and the adjacent airframe also scatter, but absorbing material can suppress these fields, so they are omitted.

Knowledge about the constituent waves is useful. It helps in formulating and

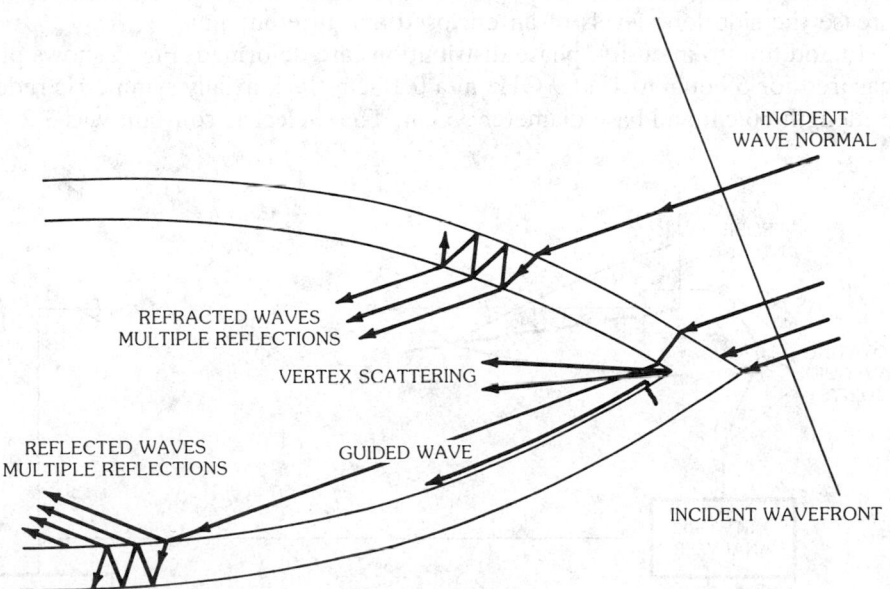

Fig. 2. A wave incident on a pointed radome produces several kinds of waves in the radome-bounded region.

evaluating numerical design methods, which usually are approximate because radomes often have shapes not defined by any separable coordinate surface. Knowing each wave's properties strengthens empirical design approaches, which may involve fabricating and testing several different radomes to connect configuration and measured performance. For example, identifying reflected waves helps locate reflecting regions, which can be modified to reduce antenna side lobe levels.

Field Distribution Measurement

Fig. 3 shows an arrangement of apparatus for measuring the field distribution in a radome-bounded region. A small probe scans the region and receives waves from a transmitting antenna which is connected to a signal generator. A coherent reference field also is supplied to the receiver, a network analyzer that measures phase and intensity.

Very nonuniform distributions can be produced. For example, Fig. 4 shows intensity measured with a small probe antenna for β, the angle between the radome axis and propagation direction of the incident wave, equal to 36°. (The angle β is called the *gimbal angle*.) The radome was pointed, with base diameter 33 cm and length 66 cm; the thickness was 0.49 cm. The dielectric constant was 4.0. The spacings of the fringes show they result from interference of direct (refracted) and reflected waves that are suggested in Fig. 2. Notice that the reflected-field magnitude is large, approximately that of the incident field. Large reflections can occur despite the high transmittance in the reflection-free region because incidence on the lower side of the radome in Fig. 2 is more nearly grazing. The dependence of reflection on incidence direction is described in Section 5. Radome reflections can increase the side lobe levels of an enclosed aperture antenna.

In addition to intensity, phase distributions are deformed. Fig. 5 shows phase measured for β equal to 4° at 9 GHz in a 0.15-cm-thick axially symmetric radome with length 66 cm and base diameter 33 cm. The dielectric constant was 3.2. The

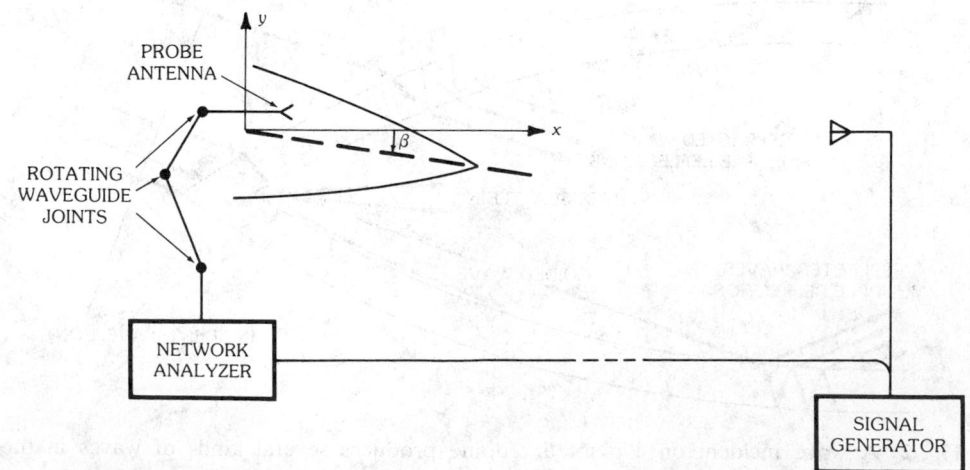

Fig. 3. Apparatus for measuring the field in a radome-bounded region.

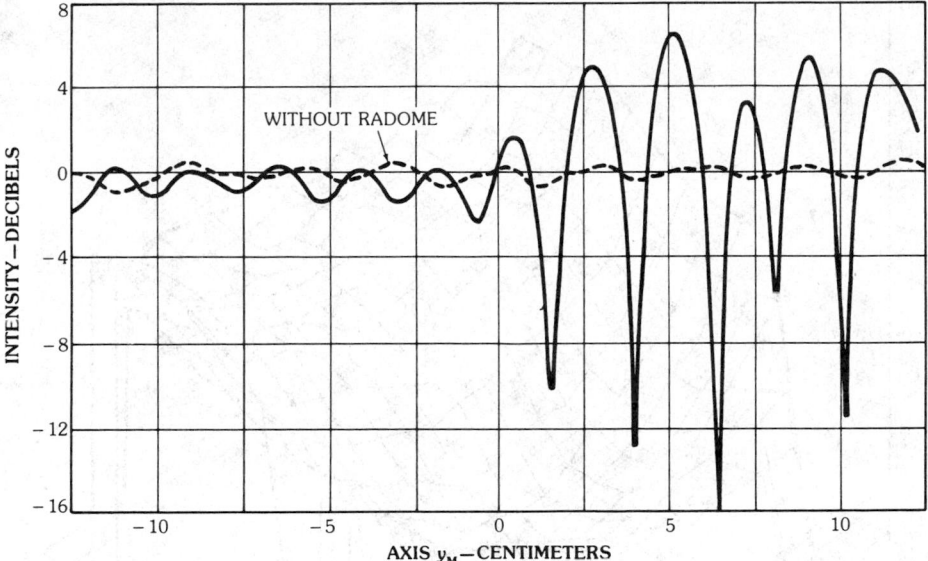

Fig. 4. Power transmittance in an axially symmetric shell measured at 16.5 GHz, with the probe on a plane perpendicular to the radome axis for $\beta = 36°$.

minimum occurs at the shadow of the tip. The distorted wavefronts cause boresight error.

Fig. 6 shows boresight error typical of pointed radomes. In this case the radome had a dielectric of constant of 5.5; length and base diameter respectively were 24.1 and 11.6 wavelengths at midband frequency. The *E*-plane is the plane that contains the electric field and the normal to the antenna. Note that boresight error depends on the orientation of polarization relative to the scan direction of the antenna.

Wavefront Aberrations and Boresight Error

Approximate diffraction analysis shows that radome boresight error is related to wavefront deformations that are asymmetric, like tilts. Although most practical radomes are three-dimensional, let us for the present consider Fig. 7, where the *z* axis points out of the page. For a fixed direction of an incident wave let the distorted wave phase Φ be expanded as a polynomial in the coordinates of Fig. 7 so that

$$\Phi = a + by + cy^2 + dy^3 + ez^2y \tag{1}$$

The phase Φ is measured over a plane; it is a variation from the incident wave's phase over that plane, and it has units of radians. The coefficient *a* has units of phase, *b* has dimension of reciprocal length, and so on. Some typical values of Φ are given in Section 5.

When (1) is used in a scalar diffraction integral for far-field patterns, the boresight error $\delta\beta$ is [7]

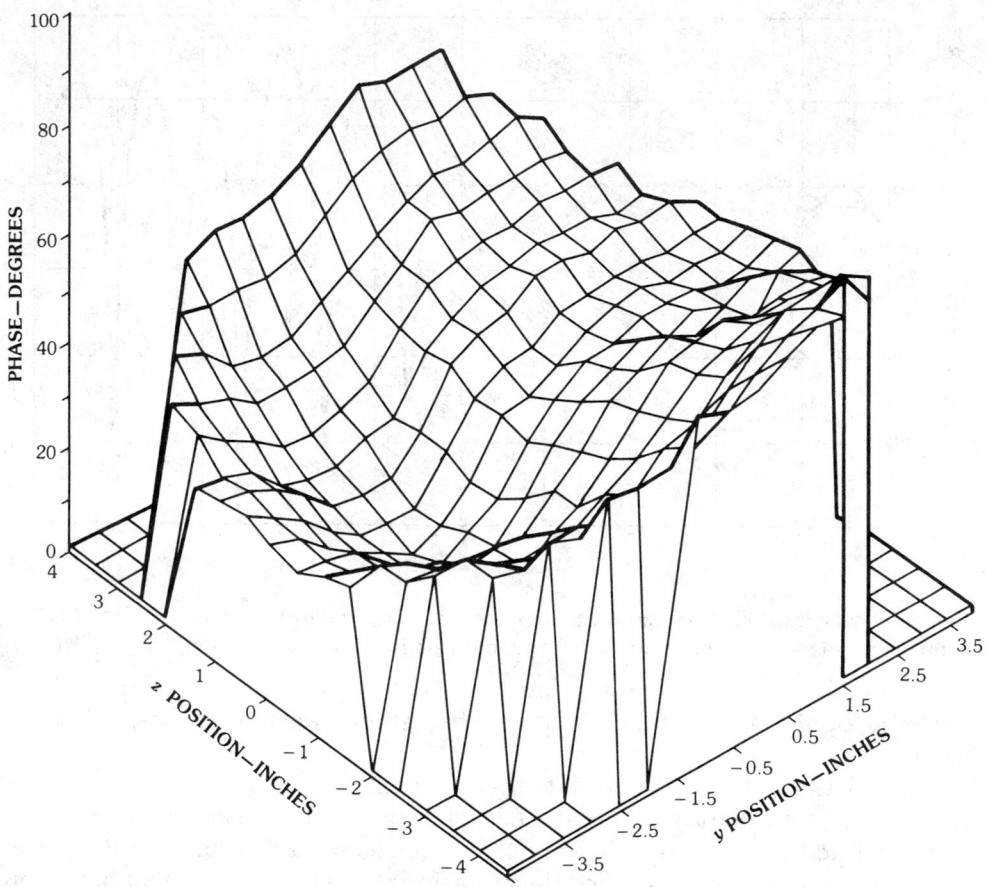

Fig. 5. Measured phase distribution in a thin radome.

$$\delta\beta = (\lambda/2\pi)[b + (e + 3d)R^2/6] \qquad (2)$$

where R is the radius of the aperture and λ is wavelength. Equation 2 was derived for conical scan antennas. It assumes no intensity variations accompany Φ and uniform antenna aperture distribution. It shows that error arises from linear tilts given by b and e, as well as cubic tilts given by d. The antenna radius also appears, suggesting that boresight error depends on antenna size for a specific radome; in fact, experience shows antenna size is significant.

Diffraction Methods for Radome Design

Although (2) gives a physical interpretation of boresight error, quantitative performance estimates of specific radome designs require computing the diffraction patterns of the enclosed antenna. Computers can do the calculations quickly and economically in most cases; however, accuracy depends on the theoretical basis of algorithms, especially for small antennas and radomes, so empirical design is common. Selection of a general design approach seems to depend on many fac-

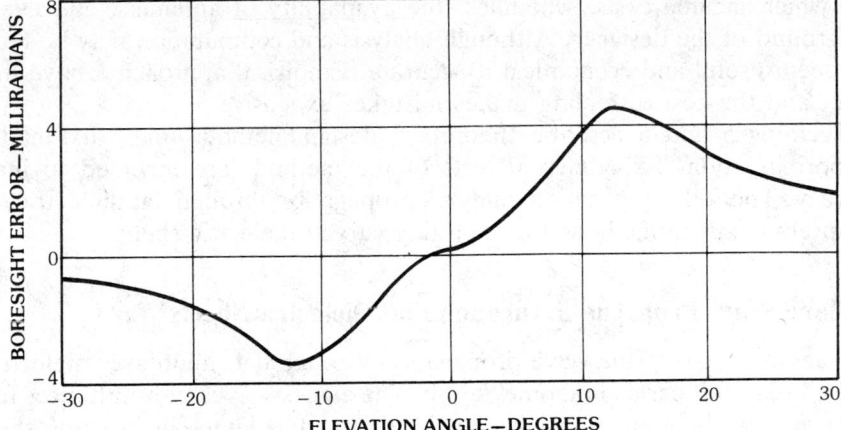

a

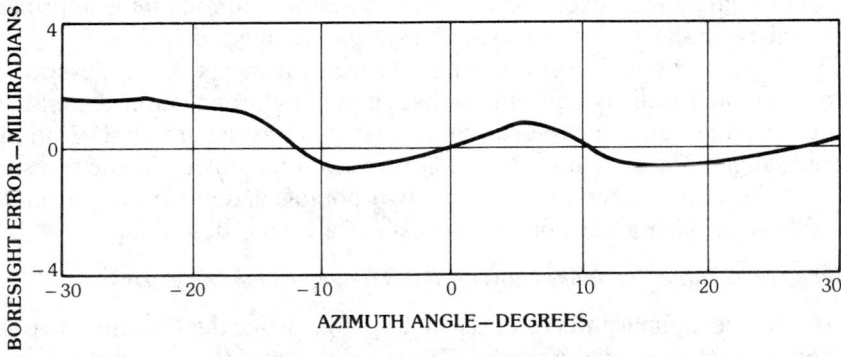

b

Fig. 6. Measured boresight error for a ceramic radome with thickness approximately 0.5λ, base diameter 11.6λ, and length 24.1λ, enclosing an antenna of diameter 10λ. (*a*) *E*-plane. (*b*) *H*-plane.

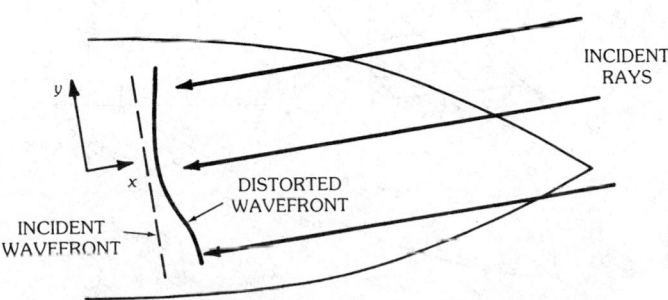

Fig. 7. Incident and distorted wavefronts in a two-dimensional view.

tors, which include costs, schedules, the availability of antennas, and even the background of the designer. Although analysis and computation may be tedious, they seem useful and economical if accurate. Empirical approaches have limited range, and the cost of tooling makes mistakes expensive.

Sections 5 and 6 describe theoretical design methods that vary in degree of approximation. Experimental tests of the methods are included to indicate accuracy. The following section analyzes propagation through flat dielectric sheets as a preliminary to methods for analyzing curved dielectric shells.

3. Plane Wave Propagation through Flat Dielectric Sheets

The analysis of plane wave propagation through flat, multilayer dielectrics of infinite breadth is basic to radome design. The analysis gives transmittances for use in approximate computations that analyze curved radomes by treating them as locally plane. The flat-sheet transmittances, combined with determinations of incidence angle, estimate power transmittance of curved radomes. The local flat-sheet approximation is suggested in Fig. 8; note that the radome is approximated by a number of sheets. The incidence angle is the angle α in Fig. 9.

This section summarizes propagation through flat sheets; it also describes some common radome wall configurations that show how propagation depends on dielectric constants and thicknesses of the layers, wavelength, polarization, and incidence angle. The section starts with formulas for the reflection and transmission at an infinite planar interface between two homogeneous regions; formulas for propagation through a flat homogenous sheet are then described.

Incidence on a Plane Boundary between Two Homogeneous Dielectrics

At a plane infinite interface, as in Fig. 10, an incident plane wave excites reflected and refracted plane waves. The amplitudes of these waves are given by well-known Fresnel formulas (see [8] and [9], pp. 38–41). The dielectrics are

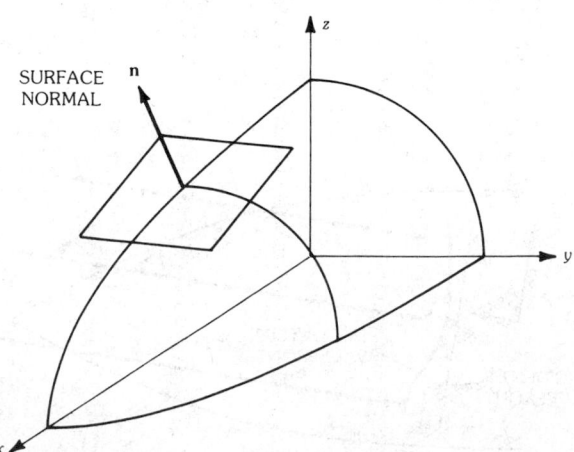

Fig. 8. Approximate local description of a radome by a flat sheet.

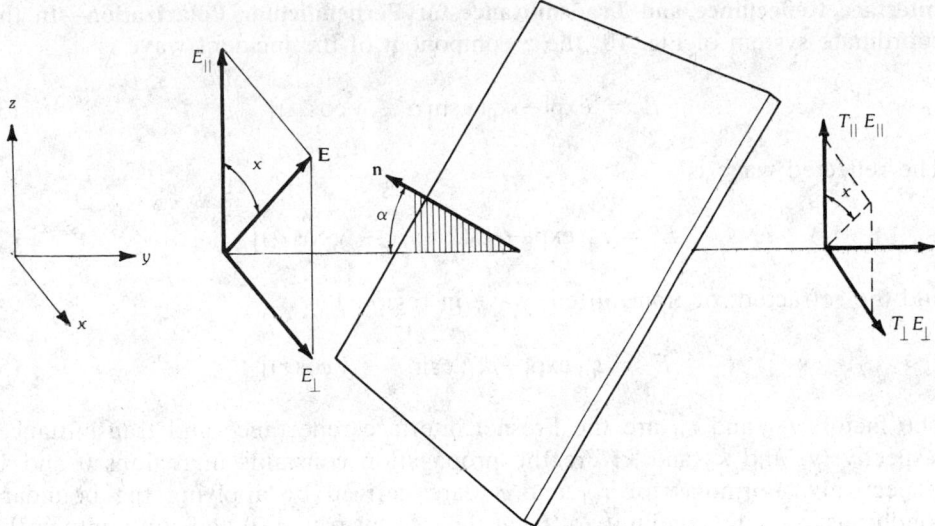

Fig. 9. Plane wave incident at angle α on a flat panel.

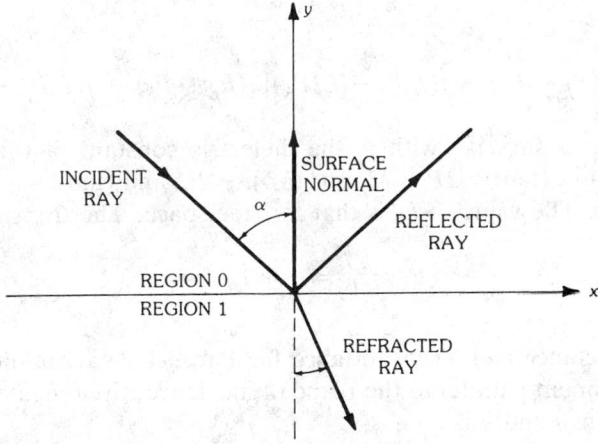

Fig. 10. Reflection and refraction at a plane interface between two homogeneous dielectric regions.

assumed homogeneous, isotropic, linear, and nonmagnetic. A rectangular electric-field component of an incident, linearly polarized wave is resolved into components parallel and perpendicular to the plane of incidence, which is defined with reference to Fig. 9 as the plane containing the surface normal **n** and the incident wave normal. Transmittances and reflectances for the two components differ because distinct boundary conditions apply. The name *parallel polarization* denotes the component parallel to the incidence plane; *perpendicular polarization* refers to the component orthogonal to the incidence plane.

Interface Reflectance and Transmittance for Perpendicular Polarization—In the coordinate system of Fig. 10, the z component of the incident wave is

$$E_z^i = \exp[-jk_0(x \sin \alpha - y \cos \alpha)] \tag{3}$$

The reflected wave is

$$E_z^r = r_{01} \exp[-jk_0(x \sin \alpha + y \cos \alpha)] \tag{4}$$

and the refracted, or transmitted, wave in region 1 is

$$E_z^t = t_{01} \exp[-jk_1(x \sin r - y \cos r)] \tag{5}$$

The factors r_{01} and t_{01} are the Fresnel interface reflectance and transmittance, respectively, and k_0 and k_1 are the propagation constants in regions 0 and 1, respectively. Formulas for r_{01} and t_{01} are derived by applying the boundary conditions, namely, continuity of E_z at the interface ($y = 0$) and continuity of the tangential component of the magnetic field **H**. This component is found by applying the Maxwell equation curl $\mathbf{E} = -j\omega\mu\mathbf{H}$ and taking the projection onto the interface. Because the derivation is standard, only the results are cited. For dielectrics that have loss tangents less than about 0.02, the reflectance between regions labeled j and k is

$$r_{jk} = [(P_j - P_k) - j(L_jP_j - L_kP_k)]/[(P_j + P_k) - j(L_jP_j + L_kP_k)] \tag{6}$$

where P_j is $(\varkappa_j - \sin^2\alpha)^{1/2}$ with $\varkappa_j$ the dielectric constant of the jth layer, i.e., $k_j = k_0\sqrt{\varkappa_j}$; L_j is $\varkappa_j(\tan\delta)_j/2P_j^2$; M_j is $(\tan\delta)_j - L_j$; and $(\tan\delta)_j$ is the loss tangent of the jth layer. The value of k_0 is that for free space. The transmittance is

$$t_{jk} = 1 + r_{jk} \tag{7}$$

Interface Reflectance and Transmittance for Parallel Polarization—For the electric-field component parallel to the plane of incidence, the reflectance for low-loss dielectric regions j and k is

$$\hat{r}_{jk} = [(\varkappa_jP_k - \varkappa_kP_j) - j(\varkappa_jM_kP_k - \varkappa_kM_kP_j)]/[(\varkappa_jP_k + \varkappa_kP_j) \\ - j(\varkappa_jM_jP_k + \varkappa_kM_kP_j)]^{-1} \tag{8}$$

The transmittance is

$$\hat{t}_{jk} = 1 + \hat{r}_{jk} \tag{9}$$

An important distinction exists between r_{01} and $\hat{r}_{01}$; for parallel polarization, $\hat{r}_{01}$ can vanish but r_{01}, for perpendicular polarization, cannot. The vanishing of $\hat{r}_{01}$ is well known; it is the Brewster phenomenon. From (8) and Snell's law, $\hat{r}_{01}$ is zero when the incidence angle is $\alpha_\beta = \tan^{-1}\sqrt{\varkappa}$, where $\varkappa$ is k_1/k_0; this angle is Brewster's angle. The Brewster phenomenon influences radome design because the parallel and perpendicular components are affected differently by radomes, as will be described in Section 5.

Incidence on Flat, Homogeneous, Dielectric Sheets

For plane wave incidence on a flat, homogeneous, isotropic, and nonmagnetic dielectric sheet, the transmitted field is the sum of multiple reflections as suggested by Fig. 11. For now, the polarization is assumed perpendicular or parallel. To be definite, consider first perpendicular polarization. The incident wave, with unit magnitude, produces a wave with magnitude $t_{12}t_{23}$; the Fresnel transmittances describe propagation through the interfaces. The incident wave also produces a second wave that is transmitted into the sheet, reflected twice, and then transmitted out of the slab; this wave has magnitude $t_{12}r_{23}r_{21}t_{23}$. This process continues, giving a series of terms. Initially assume the dielectric is lossless.

The phase of the various terms is evaluated at a plane at distance r from the origin. Let $E^i(0, d_2)$ be the incident field at $(0, d_2)$. With p_2 the path length in the sheet, the term with amplitude $t_{12}t_{23}$ changes phase by $k_2 p_2 + k_3(r_3 - p_2 \sin r_2 \sin r_3)$. Use the abbreviation Φ_2 for $k_2 p_2 - k_3 p_2 \sin r_2 \sin r_3$ so the phase of the first term is rewritten as $\Phi_2 + k_3 r_3$. For the second term the phase change is $3\Phi_2 + k_3 r_3$. Higher-order reflections have additional even multiples of Φ_2. The field propagated through the sheet is given by a geometric series which sums as

$$E_{13} = E^i(0, d_2) t_{12}t_{23} \exp(-j\Phi_2 - L_2\Phi_2) \exp(-jk_3 r)D_2^{-1} \tag{10}$$

where

$$D_2 = 1 - r_{21}r_{23} \exp(-j\Phi_2 - L_2\Phi_2) \tag{11}$$

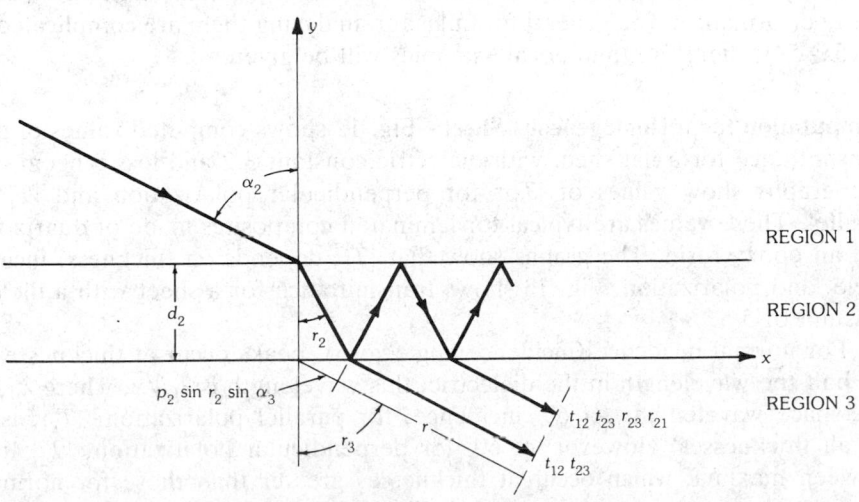

Fig. 11. Multiple reflection and transmission in a dielectric sheet.

It is convenient to express the formulas in terms of the angle of incidence. In particular, by Snell's law,

$$\Phi_2 = k_2 p_2 - k_3 p_2 \sin r_2 \sin r_3$$
$$= k_0 d_2 (\varkappa_2 - \varkappa_1 \sin^2 \alpha_2)^{1/2} \tag{12}$$

The transmittance is found from (10) normalized to (dividing by) the incident field at the reference plane. Thus the complex amplitude transmittance of the flat sheet is

$$T_{13} = t_{12} t_{23} D_2^{-1} \exp(-j\Phi_2 - L_2\Phi_2) \exp(jk_2 d_2 \cos \alpha_2) \tag{13}$$

The power transmittance (for $\tan \delta = 0$) is

$$|T_{13}|^2 = |t_{12} t_{23} D_2^{-1} \exp(-L_2\Phi_2)|^2 \tag{14}$$

The Fresnel reflectances and transmittances are given by (6) through (9).

The *insertion phase delay* is the phase change produced by the sheet; it is

$$\Delta\Phi = \arg(t_{12} t_{23} D_2^{-1}) - k_0 d_2 (\varkappa_2 - \varkappa_1 \sin^2 \alpha_2)^{1/2} + jk_2 d_2 \cos \alpha_2 \tag{15}$$

It can be measured with a microwave interferometer.

For parallel polarization, replace r_{21}, r_{32}, t_{12}, and t_{23} by $\hat{r}_{21}$, $\hat{r}_{32}$, $\hat{t}_{12}$, and $\hat{t}_{23}$. The reflectance is (for perpendicular polarization)

$$R_{13} = [r_{12} + r_{21} \exp(-j_2\Phi_2)] D_2^{-1} \tag{16}$$

Incidence on Multilayer Dielectric Sheets

Multilayer radome walls are sometimes used to obtain strength or broadband performance. The general formulas for analyzing them are complicated ([9], pp. 55–59); therefore, numerical examples will be given.

Computation for a Homogeneous Sheet—Fig. 12 shows computed values of power transmittance for a flat sheet with dielectric constant 3.2 and loss tangent 0.015. The graphs show values of $|T_\perp|^2$ for perpendicular polarization and $|T_\parallel|^2$ for parallel. These values are typical for laminated composites made of quartz fabric and an epoxy resin. The graphs show that $|T|^2$ depends on thickness, incidence angle, and polarization. Fig. 13 shows transmittance for a sheet with a dielectric constant of 5.5.

For normal incidence (incidence angle zero), peaks occur at thicknesses that are half the wavelength in the dielectric; this wavelength is $\lambda_0/\sqrt{\varkappa}$, where λ_0 is the free-space wavelength. At 60° incidence, for parallel polarization, $|T_\parallel|^2$ is high for all thicknesses. However, at 60° for perpendicular polarization, $|T_\perp|^2$ is low between maxima, which occur at thicknesses greater than those for normal incidence. The low transmittances restrict bandwidths of radomes.

Although graphs like those in Figs. 12 and 13 describe $|T|^2$ for flat sheets, the

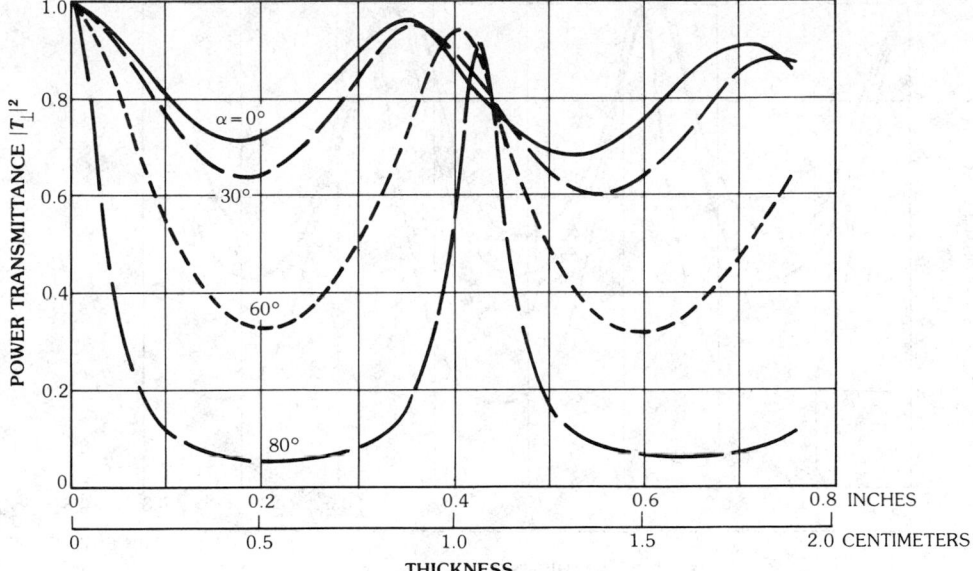

a

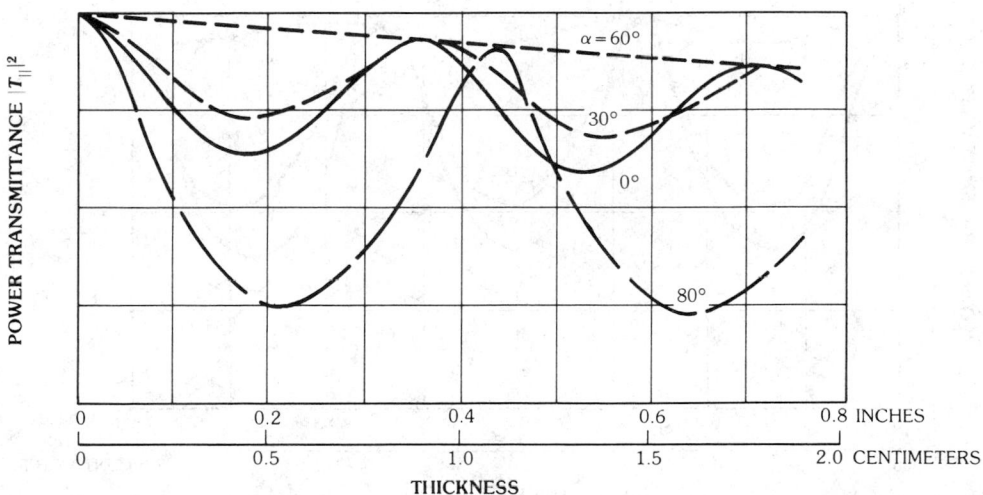

b

Fig. 12. Power transmittance of a flat sheet with dielectric constant 3.2 and loss tangent 0.015, at 9.375 GHz with incidence angle α. (*a*) For perpendicular polarization. (*b*) For parallel polarization.

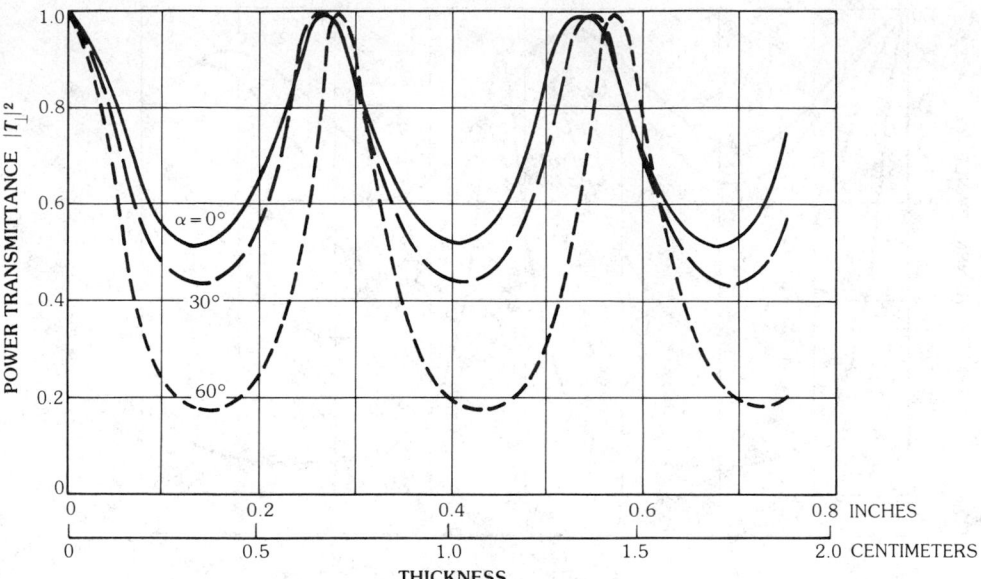

a

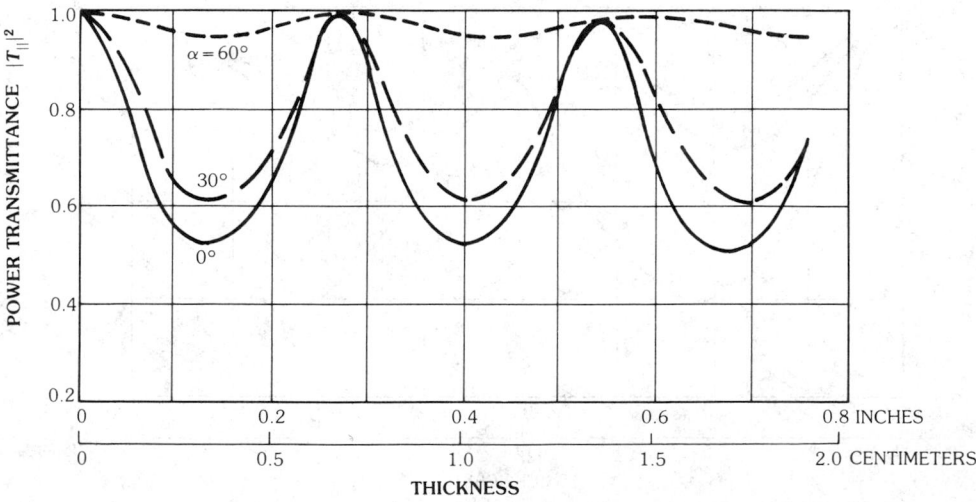

b

Fig. 13. Power transmittance of a flat sheet with dielectric constant 5.5 and loss tangent 0.0003, at 9.375 GHz with incidence angle α. (*a*) For perpendicular polarization. (*b*) For parallel polarization.

dependence on polarization and incidence angle will be shown, in Section 5, to be significant in the design of curved radomes.

An Approximation for Reflectances of a Single-Layer Sheet—An approximation to the reflectance in (16) provides some useful insight. For simplicity consider $\tan \delta = 0$. Suppose the flat sheet is in free space, so region 3 has the same properties as region 1. Thus $r_{23} = r_{21}$. Now electromagnetic theory shows that $r_{21} = -r_{12}$. If r_{12} has small absolute value relative to unit value, then the product $r_{21}r_{23}$ can be omitted in D_2, so from (16) the reflectance is approximately

$$R_{13}^a = r_{12}[1 - \exp(-j2\Phi_2)] \tag{17}$$

The power reflectance is approximately

$$|R_{13}^a|^2 = 2r_{12}{}^2(1 - \cos 2\Phi_2) \tag{18}$$

The power reflectance is least, and therefore transmittance is maximized, when 2Φ is $2\pi m$, where m is an integer.

Special Cases—From (18), $|R_{13}^a|^2$ vanishes for $\Phi_2 = 0$, and, from (12), when $d_2 = 0$. In fact $|R_{13}|^2$ is small when d_2 is small enough. A sheet or radome with small thickness is commonly called a *thin-wall design*. Fig. 12 shows the high transmittance for small d/λ_d.

From (18), $|R_{13}^a|^2$ vanishes for $\Phi_2 = \pi$, or, from (12), when

$$d_2' = \lambda/2\sqrt{\varkappa_2 - \varkappa_1 \sin^2 \alpha_2} \tag{19}$$

For thicknesses near d'_2 the reflectance is low. Sheets that satisfy this condition are called *half-wave radomes*. Fig. 12 shows the high transmittance for thickness approximately half the wavelength in the dielectric.

A-Sandwich—A common design (called an A-sandwich) consists of two thin layers, or skins, having high dielectric constant spaced by a third low dielectric constant layer called a *core*. The core thickness is approximately $\lambda_d/4$. The reflections from each sheet are identical, except for small propagation losses. Because of the quarter-wave spacing the field reflected from one thin sheet travels two quarter-wavelengths farther than the wave reflected from the other sheet; therefore, the two reflections are out of phase. In the single-layer, half-wave case, the negative sign of r_{21} (equal to $-r_{12}$) requires a path difference of two half-wavelengths for the reflections to cancel.

A-sandwich wall designs have some advantages over half-wave walls. They are light and have more broadband transmittance, for moderate incidence angles.

Fig. 14 shows computed transmittance of an A-sandwich as a function of frequency. In this figure the thicknesses are: skins, 0.035 in (0.89 μm); core, 0.25 in (6.35 mm). The dielectric constants are: skins, 3.2; core, 1.1. The loss tangents are: skins, 0.015; core, 0.005. Fig. 15 compares the transmittance of two A-sandwich flat sheets. In one case (Fig. 15a), the skins have the same thickness. In

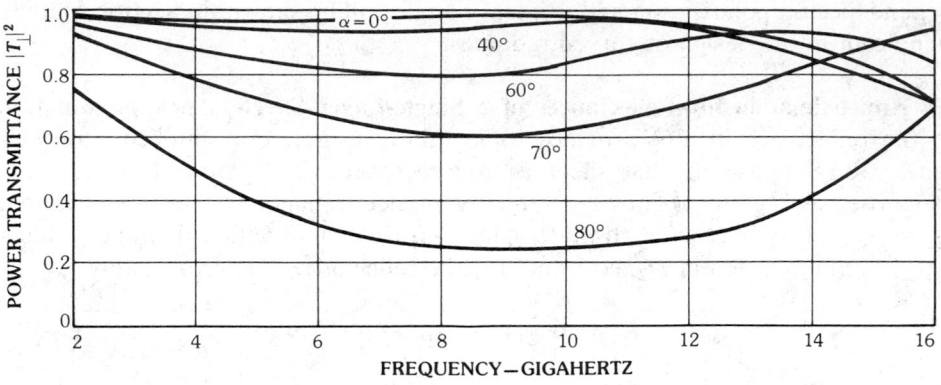

a

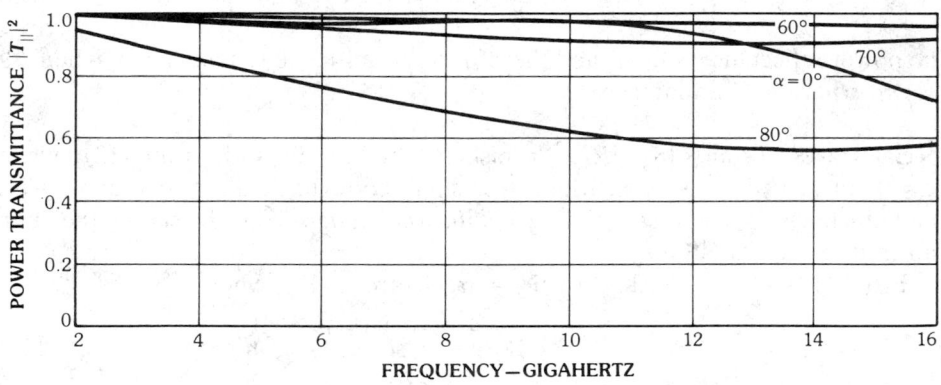

b

Fig. 14. Power transmittance of A-sandwich wall. The incidence angle is α. (*a*) For perpendicular polarization. (*b*) For parallel polarization.

the other case (Fig. 15b), the A-sandwich is asymmetric; the inner skin is thinner than the outer. The asymmetric design has more broadband performance.

4. General Aspects of Theoretical Design

Radome design is an analytic rather than synthetic process because antennas and radomes are complicated. They involve more variables than does synthesis of filters or one-dimensional antennas. Radome analysis starts with a candidate design, which is a statement of configuration and composition, and proceeds to evaluate performance (boresight error, transmittance, side lobe levels). The analysis may require iteration (sometimes called trial and error). The evaluation may be experimental, theoretical, or numerical; all three can be considered analytical.

Theoretical and numerical design methods evaluate performance parameters by computing the diffraction patterns of the enclosed antenna. The computations

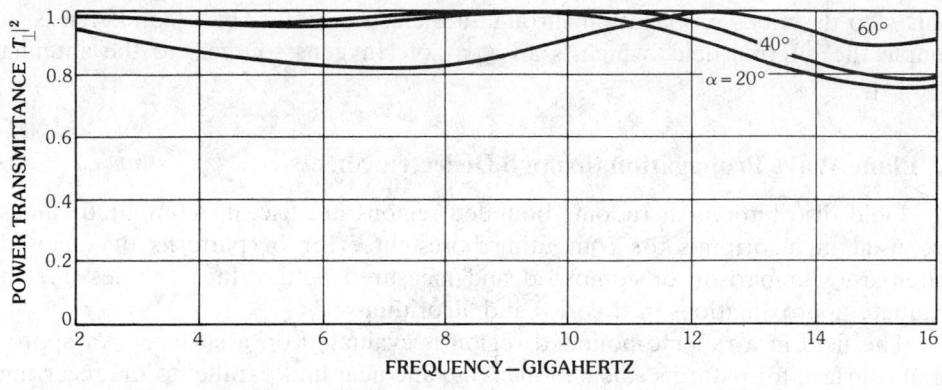

a

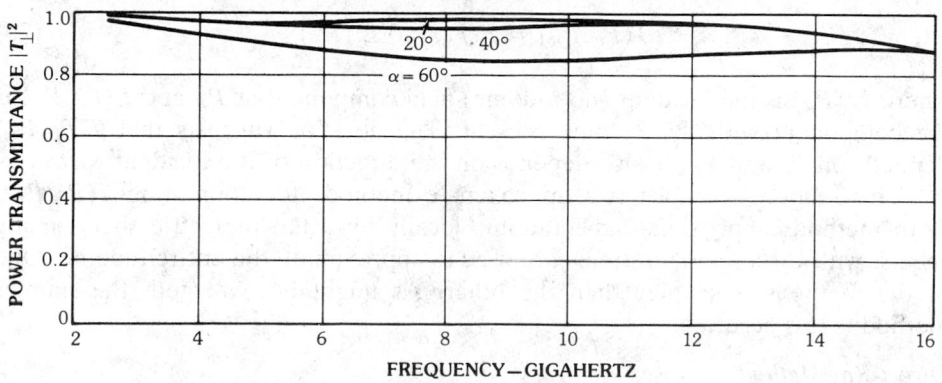

b

Fig. 15. Power transmittance of two A-sandwich panels for perpendicular polarization, with dielectric constants 2.85 (skins) and 1.15 (core), and loss tangents 0.003 (skins) and 0.004 (core). (*a*) Symmetric case: outer layers 0.04 in (1.016 mm) thick; central layer 0.30 in (7.62 mm) thick. (*b*) Asymmetric case: outer layers 0.04 in (1.016 mm) and 0.01 in (0.245 mm) thick; central layer 0.30 in (7.62 mm) thick.

are extensive because a pattern is necessary for each frequency, polarization direction, and orientation of the antenna relative to the radome. The antenna aperture distribution also is a variable. Monopulse antennas require three patterns for each frequency, polarization, and direction to describe the sum and two difference modes.

Receiving rather than transmitting operation is assumed in the following sections. Because of reciprocity, intensity patterns for reception and transmission are identical. Receiving operation justifies assuming a plane wave incident on the outside of a radome. This view seems simpler than analyzing the complicated field produced near a radome by a transmitting antenna.

Antenna pattern calculations involve two steps. The first determines the field produced in a radome-bounded region by a plane wave incident on the outside.

This step describes propagation through a dielectric shell. The second step is to couple the internal field, which is an array of Huygens sources, to the antenna [10, 11].

5. Plane Wave Propagation through Dielectric Sheets

Field distributions in radome-bounded regions are useful. Computed values are used in algorithms for computing boresight error or patterns of enclosed antennas. Comparison of computed and measured field values enables one to evaluate approximations in theories and algorithms.

The field in a radome-bounded region is evaluated on a surface. An appropriate surface for patterns calculations is a plane near and parallel to the receiving aperture; see Fig. 16. At a typical point P_R, the amplitude and phase of each rectangular field component are described by the complex-valued transmittance

$$T(P_0, P_R) = E(P_0, P_R)/E^i(P_R) \tag{20}$$

where $E^i(P_R)$ is the incident (no radome) field component at P_R and $E(P_0, P_R)$ is the component with the radome present. The notation suggests that $T(P_0, P_R)$ depends on P_0 and P_R; it also depends on the direction of the incident wave.

This section describes two approximate methods for computing $T(P_0, P_R)$. Both methods approximate the radome locally by a flat dielectric sheet; many sheets, with varying orientations, are used to approximate the entire radome. One of the methods is simpler than the other. As might be expected, the simpler method is less accurate.

Direct-Ray Method

This method [12] is the simpler. The simplification results from assuming that field propagation in the radome-bounded region is along rays parallel to the incident ray or wave normal. We call this model the *direct-ray method.** Fig. 16

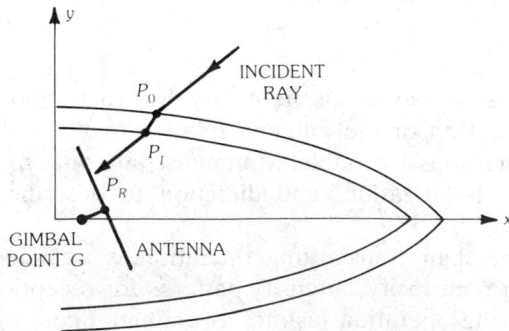

Fig. 16. A sketch of a two-dimensional direct-ray model.

---

*This name emphasizes the parallelism of the incident ray and propagated ray.

shows a direct ray. It intersects the radome at P_0, is refracted, leaves the radome at P_I, and from P_I to P_R is parallel to the incident ray. The parallelism depends on approximating a radome locally by a flat sheet as in Fig. 17. The flat sheet approximation implies multiple internal reflections, which also appear in Fig. 17. The incident ray that is refracted at P'_0, reflected at P'_I and P_0, and emergent at P_I is also a direct ray. The similarity between Figs. 17 and 11 is apparent; however, many radomes have both axial and circumferential curvature, so the approximating sheet depends on P_0 and P_R.

The thickness of the approximating sheet is defined to have the same ray path length as the distance $\overline{P_I P_0}$. This thickness is $\overline{P_I P_0} \cos r$ where r is the angle of refraction calculated by Snell's law with the radome's refractive index and the incidence angle at P_0. The difference between the actual thickness and $\overline{P_I P_0} \cos r$ is negligible in many cases where curvature is not too great. For example, boresight error calculations made with the thickness values showed differences of 5 percent at peak values of 3 mrad; in general, differences were less than 2 percent.

To determine the field at P_R, start with the incident field

$$\mathbf{E}^i(P_R) = E_0^i \exp[-j\mathbf{k} \cdot \mathbf{r}(P_R)] \mathbf{e} \tag{21}$$

where E_0^i is the scalar amplitude of the incident wave, which is assumed linearly polarized; j is $\sqrt{-1}$, k is the propagation constant; $\mathbf{k}$ is a vector with magnitude k and in the direction of the incident plane wave; $\mathbf{e}$ is a rectangular unit vector in the direction of the electric field; and $\mathbf{r}(P_R)$ is the position vector to P_R. At P_0 the incident field is resolved into components parallel and perpendicular to the plane of incidence, as suggested in Fig. 9. Let $\mathbf{1}_\perp$ and $\mathbf{1}_\parallel$ be unit vectors perpendicular and parallel, respectively, to the plane of incidence, which contains $\mathbf{k}$ and $\mathbf{n}$ and which is normal to the surface at P_0. Let $\mathbf{e} \cdot \mathbf{1}_\perp$ be α_p and $\mathbf{e} \cdot \mathbf{1}_\parallel$ be β_p. The incident field component is

$$E^i(P_0) = E_0^i[\alpha_p \mathbf{1}_\perp + \beta_p \mathbf{1}_\parallel] \tag{22}$$

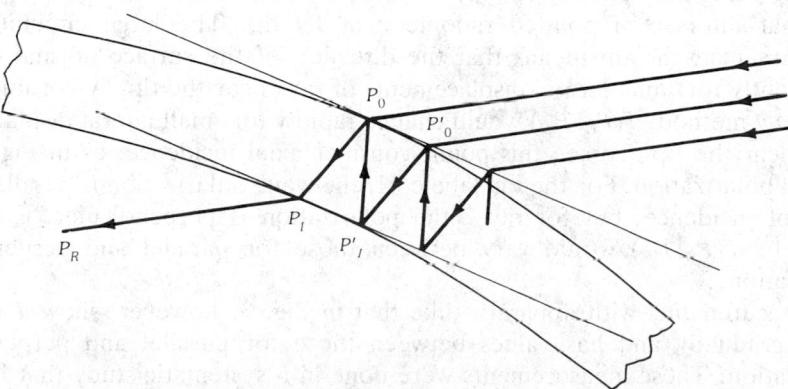

Fig. 17. An approximating flat sheet and multiple internal reflections.

(Note that α_p is not the incidence angle α.) The approximating thickness $\overline{P_I P_0} \cos r$ and the normal direction of the shell are used to compute transmittance and phase delays in (13) and (15) (for both parallel and perpendicular polarization). The field at P_I is then

$$\mathbf{E}'(P_I) = [\alpha_p E^i(P_0) \, |T_\perp| \, e^{-j\Delta_\perp}]\mathbf{1}_\perp + [\beta_p E^i(P_0) \, |T_\parallel| \, e^{-j\Delta_\parallel}]\mathbf{1}_\parallel \tag{23}$$

Assume that the receiving antenna responds to a linear field component that is parallel to the direction of a unit vector $\mathbf{1}_d$. The rectangular component of the electric field transmitted through the radome and in the direction of $\mathbf{1}_d$ is, from (23),

$$E_d^t(P_I) = |T_e| \, E^i(P_0) \, \exp[-j \, \arg(T_e)] \tag{24}$$

where

$$|T_e|^2 = (|T_\perp|\alpha_p\delta)^2 + (T_\parallel\beta_p\delta')^2 + 2|T_\perp| \, \alpha_p\beta_p\delta\delta'\cos\Delta \tag{25}$$

and δ is $\mathbf{1}_\perp \cdot \mathbf{1}_d$, δ^1 is $\mathbf{1}_\parallel \cdot \mathbf{1}_d$, and Δ is $\Delta\Phi_\perp - \Delta\Phi_\parallel$. In addition,

$$\arg(T_e) = \Delta_\parallel + \tan^{-1}\{|T_\perp|\alpha_p\delta|[|T_e|^2 - (|T_\perp| \, \delta_p \sin\Delta)]\}^{-1/2} \sin\Delta \tag{26}$$

At P_R the rectangular field component propagated through the radome and parallel to $\mathbf{1}_d$ is

$$E_d^t(P_R) = T_e \, E^i(P_R) \tag{27}$$

The direct-ray method is adequate for radomes with antennas that have diameters bigger than approximately ten wavelengths and lengths approximately twice the diameter. However, it is inaccurate for antennas with diameter only about five wavelengths in radomes and with length about ten wavelengths. The source of error appears to be in representing Huygens' principle by a set of parallel, direct rays and using only one ray $(\overline{P_0 P_R})$ to P_R. This approximation is questionable for rays that intersect a pointed radome near its tip. The large circumferential curvature near the tip means that the direction of the surface normal changes significantly for small lateral displacements of rays near the tip. According to the direct-ray method, $T(P_0, P_R)$ would change rapidly for small lateral displacements of P_0 near the tip. To see this point, consider axial incidence, as in Fig. 18 for vertical polarization. For the ray labeled 1, the wave polarization is parallel to the plane of incidence, but for ray 2 the polarization is perpendicular. Computed values of $T(P_0, P_R)$ would vary between those for parallel and perpendicular polarization.

Measurements with apparatus like that in Fig. 3, however, show $T_e(P_0, P_R)$ varies gradually and has values between those for parallel and perpendicular polarization. These measurements were done in a systematic study that involved three radomes of distinct sizes but with similar shapes and identical dielectric constants. Fig. 19 shows the radome profiles and gives dimensions. Fig. 20 shows

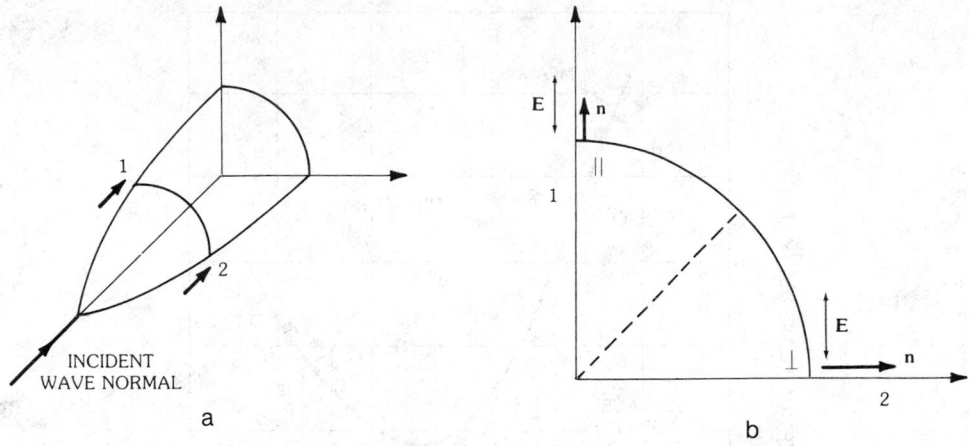

Fig. 18. Axially incident rays have distinct phase delays because the surface normal direction varies. (*a*) With axially incident rays 1 and 2. (*b*) Polarization of rays 1 and 2 with respect to surface normal.

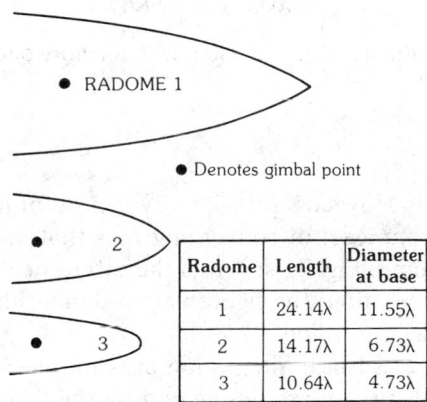

Radome	Length	Diameter at base
1	24.14λ	11.55λ
2	14.17λ	6.73λ
3	10.64λ	4.73λ

Fig. 19. Three radome configurations; all radomes had dielectric constant 5.55, and all were approximately 0.5λ thick.

insertion phase delay computed by the direct-ray method [$\arg(T_e)$ of (26)] for radome 2 of Fig. 19. The calculations differ appreciably from values measured with a small probe with aperture 1.6 by 1.0 wavelengths. The minimum in the values computed by the direct-ray method is the shadow of the tip.

For a given radome the deficiencies in the direct-ray method seem more significant for smaller antennas because the errors, in the shadow of the tip, extend over a larger portion of a small antenna.

The discrepancies between theory and measurement can be considered as caused by undersampling of the radome by the direct-ray method. The following subsection describes a method that more densely samples the tip region.

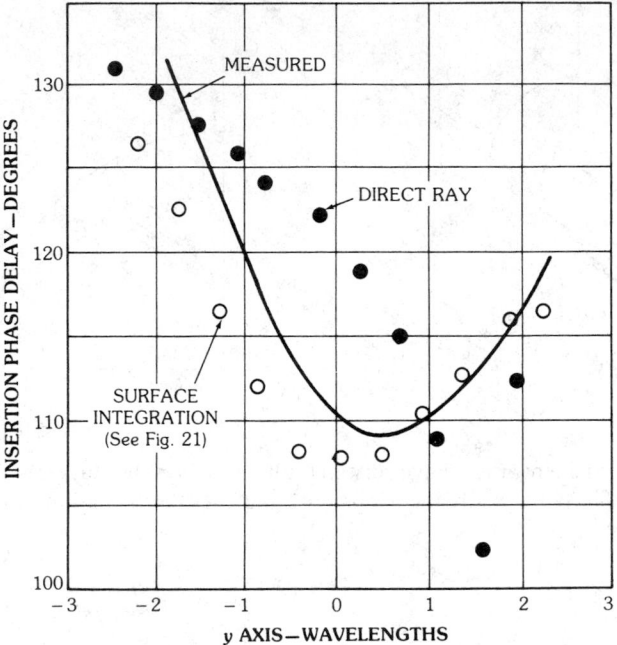

Fig. 20. Insertion phase delay of radome 2 for horizontal polarization.

Surface Integration

This method models Huygens' principle by representing the field at a typical point P_R as resulting from a set of converging rays that originate on a wavefront surface outside the radome. Fig. 21 suggests the difference between the direct-ray method and surface integration; the figure is two-dimensional but the converging rays arise from an area, not a line.

The transmittance for surface integration uses the flat sheet approximation at each of a set of points P_0; these are designated by the subscript m. The transmittance for surface integration is a sum corresponding to converging rays; it is

$$T_{\text{si}}(P_{0m}; P_R) = \sum_{M=1}^{M} T(P_{0m}, P_R) \tag{28}$$

The sum computationally represents an integral over the wavefront. The number of rays M and their distribution influence the transmittance. Determination of M and the distribution of rays is a truncation and sampling problem.

Comparison of measured and computed values showed that good accuracy was obtained by bounding the integration area to the first Fresnel zone centered about P_R. The field is then

$$E_d^t(P_R) = T_{\text{si}} E^i(P_R) \tag{29}$$

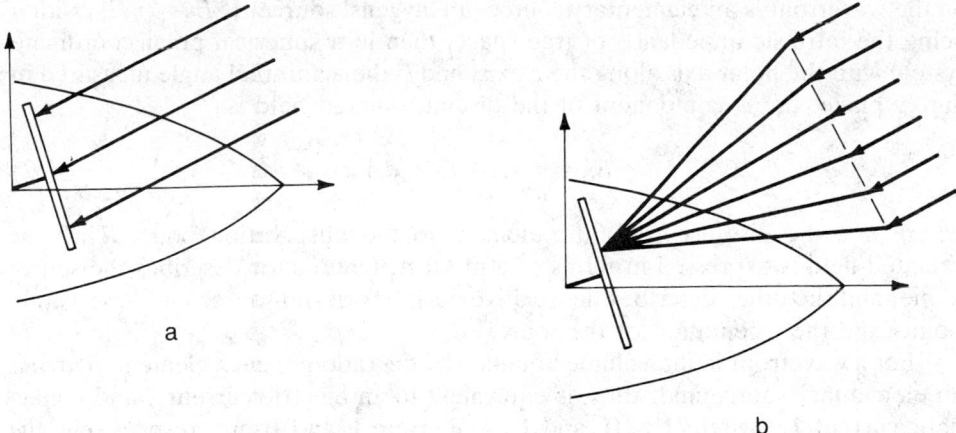

Fig. 21. Two methods for computing transmittance. (*a*) Direct-ray method. (*b*) Surface integration.

Fig. 20 shows phase delay values computed by surface integration. In this case and in many others, surface integration is more accurate than the direct-ray method. However, for sufficiently large antennas and radomes the direct-ray method gives useful predictions of boresight error. The direct-ray method has the advantage of lower computer cost than surface integration.

6. Antenna Patterns

Although Huygens' principle was generally described in Chapters 1 and 2, this section utilizes a special form suitable for computing patterns of antennas in a radome [13].

Pattern computations are based on considering the wavefront in the radome-bounded region as an array of secondary sources. Schelkunoff's induction theorem states that each secondary source is a combination of electric and magnetic current elements with moments that are proportional, respectively, to the magnetic and electric intensities tangential to the wavefront. For a wave in a homogeneous medium, the induction theorem becomes the equivalence theorem, and the densities of the magnetic current **M** and electric current **J** are

$$\mathbf{M} = \mathbf{E}^i \times \mathbf{1}_s \tag{30}$$

and

$$\mathbf{J} = \mathbf{1}_s \times \mathbf{H}^i \tag{31}$$

where $\mathbf{E}^i$ and $\mathbf{H}^i$ are, respectively, the electric and magnetic intensities of the incident plane wave, and $\mathbf{1}_s$ is a unit vector normal to the wavefront. For a linearly polarized wave with rectangular components E_x and H_y, each element of area dA

on the wavefront is an elementary source, a Huygens' source. If $E_x = \eta H_y$, with η being the intrinsic impedance of free space, then in a spherical polar coordinate system with the polar axis along the z axis, and ϕ the azimuthal angle measured in the xy plane, the ϕ component of the distant radiated field is

$$E_\phi = f(r, \phi) \, E \times dA \tag{32}$$

where r is the distance from the element to the observation point. Thus the radiated field is expressed in terms of a product. One factor describes the source alone, and the other describes the relative positions of the source and observation points and the orientation of the source.

For a wavefront in the volume bounded by the radome, each element of area is an elementary source and, thus, is equivalent to an electric current $\mathbf{J}$ and a magnetic current $\mathbf{J}_m$, where $\mathbf{J} \propto \mathbf{H}$, and $\mathbf{J}_m \propto \mathbf{E}$. Here $\mathbf{E}$ and $\mathbf{H}$ are, respectively, the electric and magnetic intensities tangential to the wavefront.

Now consider an aperture antenna receiving the wavefront. The antenna is connected to a detector by a waveguide. Let the field in the waveguide at P', due to an element of unit strength at a point P_R, be

$$dE_g = F(P_R, P')E_d^t(P_R) \, dx \, dy \tag{33}$$

where E_d^t is given by (27) or (29). The function F is the receiving aperture distribution, which can be measured, at least approximately [13].

The field due to the entire incident wave is

$$E_g = \int\int F(P_R, P')E^t(P_R) \, dx \, dy \tag{34}$$

The integral extends over that portion of the incident wavefront such that F has appreciable values. This surface is flat. The phase variation over this surface is given by that phase of T_{si} in (29) or T_e in (27).

Power patterns are obtained from $|E_g|^2$. Transmittance is obtained from peak values of the main beam, by comparing values with and without the radome.

Boresight error calculation depends on the antenna. For conical scan antennas, error is obtained from an intensity minimum determined from overlapping patterns for distinct beam direction. For amplitude comparison monopulse, boresight error is obtained from angular shifts of power pattern minima. For phase comparison monopulse, the shift is in the direction giving a quadrature relation of sum and difference patterns.

It is assumed that a reasonable approximation to F can be measured with a suitable probe for any particular receiving antenna. Call this approximation F_m.

In practice,

$$E_g = \sum_{j,k} F_m(P_R, P') \, E_d^t(P) \tag{35}$$

The sum in (35) is the evaluation of the integral for E_g. Dimensions are preserved on omitting the differentials by factoring the distance between equally spaced points P and P'. This distance is omitted for convenience. The preceding formulation is largely physical. It is suggested by and based on the induction theorem. Its practical use depends on knowing the function F_m. Some assumptions have been made on the nature of the secondary sources. It is assumed that for the wavefront propagated through the dielectric shell, $E = \eta H$ at any point; it is not assumed that E is a constant over some plane.

Another factor in actual cases is reflected energy. Fields reflected from the receiving antenna would change the field on the aperture plane, changing the strength of a secondary source. In addition, fields which enter the antenna after reflection from the radome must be considered; for example, if side lobe estimates are being made. Fig. 2 shows the path of a possible reflected ray which represents energy above that received from a collimated ray bundle entering the antenna after passing through the radome without reflection. Reflected rays may be very significant in some cases.

7. Examples of Boresight Error

For radome 1 of Fig. 19, Fig. 22 compares measured boresight error and values computed by the direct-ray method for the antenna axis rotated parallel and perpendicular to the plane of the electric field. Fig. 22 shows calculations made with an assumed uniform amplitude distribution that had a phase reversal on opposite sides of the aperture. Fig. 22 also shows calculations made with an aperture distribution that was measured with a small probe at a distance of four wavelengths and backward propagated to the aperture. The direct-ray method gives reasonable accuracy, and aperture distribution affects accuracy [14].

Fig. 23 shows boresight error for radome 3, the smallest in Fig. 19. The direct-ray method failed, but surface integration gave accurate predictions.

Surface integration also was applied to radome 1; it was more accurate than the direct-ray method.

Some alternative methods have been described [15–17].

8. The Moment Method

The analytical methods of Sections 5 and 6 are approximate. Propagation is described by direct rays or surface integration; both methods approximate the radome as locally plane. Guided waves are omitted. Coupling of the wavefront to the antenna is approximate.

A more accurate description is given by the moment method [18]. This method computes the field in a radome for a given incident field from the solution of an integral equation. The external fields, near or far, are computed from the internal field values.

In Richmond's formulation the total field $\mathbf{E}^T$ is defined as the sum of the incident (known) field and the scattered field $\mathbf{E}^s$; that is,

$$\mathbf{E}^T = \mathbf{E}^i + \mathbf{E}^s \tag{36}$$

The scattered field follows from vector and scalar potentials which are integrals over polarization currents. Thus

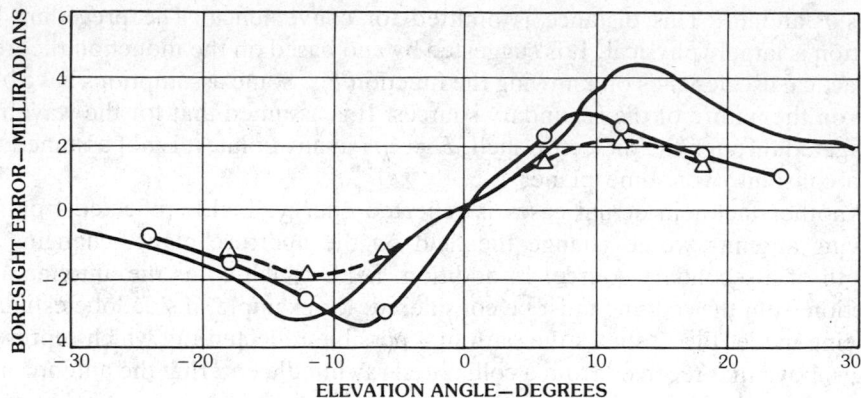

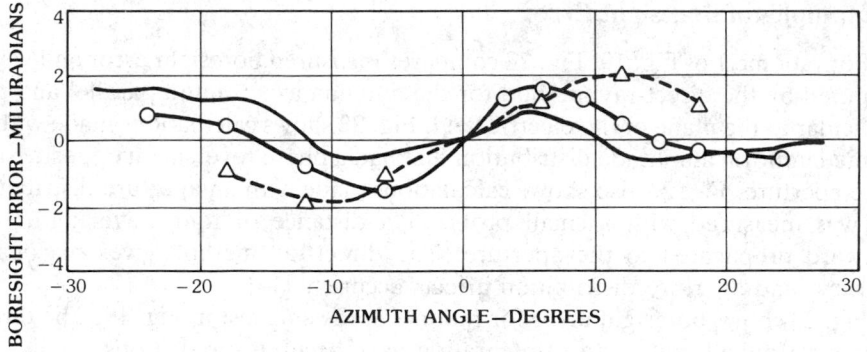

Fig. 22. In-plane boresight error of radome 1: measured (—); computed, uniform aperture amplitude ($\triangle$); computed, measured aperture distribution ($\bigcirc$).

$$\mathbf{E}^s = -j\omega\mathbf{A} - \operatorname{grad}\phi_s \tag{37}$$

$$\mathbf{A} = (\mu_0/4\pi)\int g\,\mathbf{J}\,dV \tag{38}$$

and

$$\phi_s = (4\pi\epsilon_0)^{-1}\int g\varrho\,dV \tag{39}$$

where ϱ is $-\epsilon_0\nabla\cdot(\varkappa - 1)\mathbf{E}$, $\mathbf{J}$ is $j\omega\epsilon_0(\varkappa - 1)\mathbf{E}$, and g is $r^{-1}\exp(-jkr)$. Notice that the unknown $\mathbf{E}^T$ appears in the potentials and thus in $\mathbf{E}^s$ by (37). When $\mathbf{E}^s$ from (37) is used in (36), the integral equation results.

To solve the integral equation the radome is subdivided into small volunes, or cells, and the field in each cell is assumed constant.

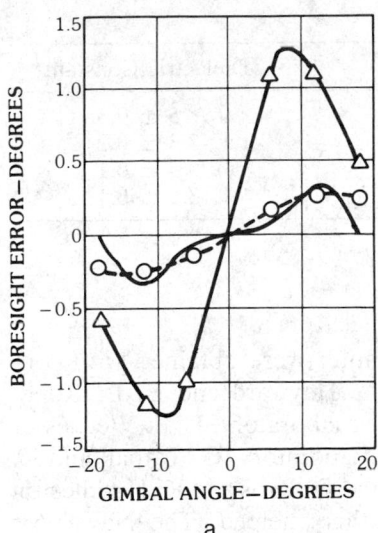
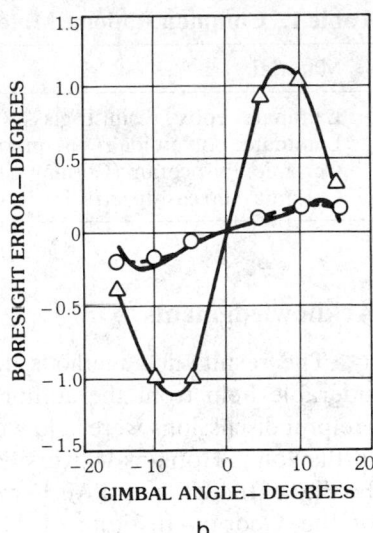

Fig. 23. Boresight error of radome 3: measured (—); computed direct-ray (△); computed surface integration (○). (*a*) *E*-plane. (*b*) *H*-plane.

Richmond gave explicit calculations for dielectric cylinders of arbitrary shape for both transverse electric and transverse magnetic fields. The arbitrarily shaped cylinders were represented by cells that were circular cylinders. The cells should be small relative to the wavelength to justify assuming that the field is constant in any cell. The integrals for $\mathbf{A}$ and ϕ_s reduce to sums of integrals; each integral is a sum that extends through a cell. The unknown $\mathbf{E}^T$ can be factored from under the integrals and appears as a factor. The integral equation is evaluated for each cell to reduce it to a set of simultaneous algebraic equations. Computations based on Richmond's method give good agreement with measured near fields; of course, cells must be small enough [19].

Recently the moment method has been applied to hollow cones [20, 21]. The cells are angular sectors of annuli, so ogival radomes can be analyzed. Measured near fields agreed well with calculations.

Although the moment method is accurate, current computers limit matrix inversion, so analysis is feasible for small cones; lengths are a few wavelengths. Iterative methods are being studied. Despite some limitations the moment method is powerful; it has estimated strengths of guided waves on dielectric slabs.

9. Comments on Materials

Although the selection of a material is significant for radome design and development, this chapter omits materials except to list some standard materials in Table 1. The dielectric constant values are approximate. The values depend on temperature. Descriptions of electromagnetic, thermal, and mechanical properties are available [22].

Table 1. Common Radome Materials

Material	Dielectric Constant
Laminate: epoxy resin, E-glass	5 ± 0.3
Laminate: polyamide resin, quartz fabric	3.3 ± 0.2
Ceramic: Pyroceram (Corning Glass Works)	5.5 ± 0.1
Ceramic: slip cast fused silica	3.2 ± 0.1

Acknowledgments

The results and methods described in this chapter were obtained with considerable help from the author's colleagues, R. A. Hayward and E. L. Rope. Helpful discussions were held with T. E. Fiscus of our laboratory, L. E. Weckesser of the Johns Hopkins University's Applied Physics Laboratory, G. Tatnall and K. Foulke of the US Naval Air Development Center, and D. Paris and K. Huddleston of the Georgia Institute of Technology; many others helped. The Naval Air Systems Command and James Willis of the Command supported much of the work.

10. References

[1] T. J. Lyon, "Operational radomes," chapter 7 in *Radome Engineering Handbook*, ed. by J. D. Walton, New York: Marcel Dekker, 1970.

[2] G. Tricoles and E. L. Rope, "Circular polarization radome technique," *Tech. Rep. AFAL-TR-69-164*, US Air Force Avionics Lab, Wright-Patterson AFB, Ohio 45433, August 1969.

[3] D. G. Bodnar and H. L. Bassett, "Analysis of an anisotropic dielectric radome," *IEEE Trans. Antennas Propag.*, vol. AP-23, pp. 841–846, 1975.

[4] E. A. Pelton and B. A. Munk, "A streamlined metallic radome," *IEEE Trans. Antennas Propag.*, vol. AP-22, pp. 799–803, 1974.

[5] G. Tricoles and E. L. Rope, "Guided waves in a dielectric slab, hollow wedge, and hollow cone," *J. Opt. Soc. Am.*, vol. 55, pp. 328–330, 1965.

[6] G. Tricoles and N. L. Farhat, "Microwave holography: application and techniques," *Proc. IEEE*, vol. 65, pp. 108–121, 1977.

[7] R. A. Hayward, E. L. Rope, and G. Tricoles, "Radome boresight error and its relation to wavefront distortion," *Proc. 13th Symp. on Electromagnetic Windows*, ed. by H. L. Bassett and J. M. Newton, pp. 87–89, Georgia Institute of Technology, Atlanta, 1976.

[8] J. A. Stratton, *Electromagnetic Theory*, New York: McGraw-Hill Book Co., 1941, pp. 492–497.

[9] M. Born and E. Wolf, *Principles of Optics*, New York: Pergamon Press, 1964.

[10] B. B. Baker and E. T. Copson, *The Mathematical Theory of Huygens' Principle*, 2nd ed., Oxford: Oxford University Press, 1953.

[11] S. A. Schelkunoff, "Kirchoff's formula, its vector analogue, and other field equivalence theorems," in *Theory of Electromagnetic Waves*, ed. by M. Kline, New York: Dover Publications, 1951, pp. 43–59.

[12] G. Tricoles, "Radiation patterns and boresight error of a microwave antenna enclosed in an axially symmetric dielectric shell," *J. Opt. Soc. Am.*, vol. 54, pp. 1094–1101, 1964.

[13] G. Tricoles, "Radiation patterns of a microwave antenna enclosed by a hollow dielectric wedge," *J. Opt. Soc. Am.*, vol. 53, pp. 545–557, 1963.

[14] R. A. Hayward, E. L. Rope, and G. Tricoles, "Accuracy of two methods for radome analysis." *Dig. 1979 IEEE Antennas Propag. Symp.*, pp. 598–601. IEEE cat. no. 79CH1456-3 AD.

[15] K. Siwiak et al., "Boresight error induced by missile radomes," *IEEE Trans. Antennas Propag.*, vol. AP-27, pp. 832–841, 1979.

[16] D. T. Paris, "Computer aided radome analysis," *IEEE Trans. Antennas Propag.*, vol. AP-18, pp. 7–15, 1970.

[17] D. G. Burks, E. R. Graf, and M. D. Fahey, "A high frequency analysis of radome-induced radar pointing error," *IEEE Trans. Antennas Propag.*, vol. AP-30, pp. 947–955, 1982.

[18] J. H. Richmond, "Scattering by a dielectric cylinder of arbitrary cross section shape," *IEEE Trans. Antennas Propag.*, vol. AP-13, pp. 334–341, 1965; "TE wave scattering by a dielectric cylinder of arbitrary cross section shape," *IEEE Trans. Antenna Propag.*, vol. AP-14, pp. 460–464, 1966.

[19] G. Tricoles, E. L. Rope, and R. A. Hayward, "Wave propagation through axially symmetric dielectric shells," *General Dynamics Electronics Division Report R-81-125*, final report for Contract N00019-79-C-0638, San Diego, June 1981.

[20] G. Tricoles, E. L. Rope, and R. A. Hayward, "Electromagnetic waves near dielectric structures," *General Dynamics Electronics Division Report R-83-047*, final report for Contract N00019-81-C-0389, San Diego, February 1983.

[21] G. Tricoles, E. L. Rope, and R. A. Hayward, "Electromagnetic analysis of radomes by the moment method," *Proc. Seventeenth Symp. Electromagnetic Windows*, ed. by H. Bussell, pp. 1–8, Georgia Institute of Technology, Atlanta, Georgia 30332.

[22] J. D. Walton, ed., *Radome Engineering Handbook*, New York: Marcel Dekker, 1970.

Chapter 32

Measurement of Antenna Radiation Characteristics on Far-Field Ranges

E. S. Gillespie

California State University, Northridge

CONTENTS

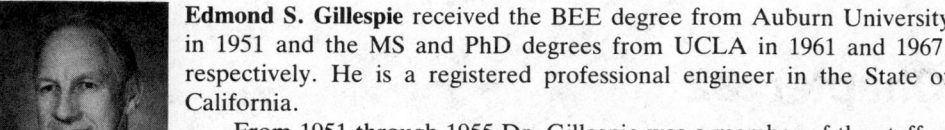

Edmond S. Gillespie received the BEE degree from Auburn University in 1951 and the MS and PhD degrees from UCLA in 1961 and 1967, respectively. He is a registered professional engineer in the State of California.

From 1951 through 1955 Dr. Gillespie was a member of the staff of the Sandia Corporation, Albuquerque, New Mexico, principally engaged in the development of airborne antennas. In 1955 he joined the Lockheed-California Company, where he was primarily concerned with the design of aircraft antennas and also directed the research activities of the antenna laboratory. During parts of 1961–1963, he was on leave from Lockheed attending UCLA, where he was appointed an associate in engineering. He taught courses in solid-state electronics, electromagnetics, and microwave engineering. In 1963 he left Lockheed, where he had held the position of Senior Research Specialist, and joined the UCLA Institute of Geophysics and Planetary Physics as a senior engineer. He was engaged in a study of the scattering of surface waves by obstacles. In 1965 he joined the faculty of the School of Engineering and Computer Science at the California State University, Northridge. Currently he is a professor of electrical engineering and Director of the Microwave and Antenna Engineering Program.

Dr. Gillespie is a Fellow of the IEEE, and he is a member of many technical societies and committees. He was responsible for the development of ANSI/IEEE Standard 149-1979, *Test Procedures for Antennas*, and IEEE Standard 145-1982, *Definitions of Terms for Antennas*.

1. Introduction

To assess the performance of a communication, radar, or telemetry system, it is necessary to know the characteristics of the antennas employed to within specified confidence limits. This usually requires a measurement program, the extent of which depends on the severity of the system requirements as well as economic factors. Chart 1 is a tabulation of *antenna** characteristics that are often measured.

For a simple line-of-sight communications system one might be required to measure only a limited number of amplitude patterns and the peak gain of the antennas. For a spacecraft antenna or an airborne radar antenna, however, an extensive measurement program encompassing many of the entries of Chart 1 may be required. Antennas which are designed to operate at frequencies above about 100 MHz usually can be tested at antenna test facilities away from their operational environment. There are exceptions such as large earth station and radioastronomy antennas which must be tested *in situ* [2]. On-site measurements of this type are beyond the scope of this chapter, although much that is presented is applicable. The characteristics of some types of antennas are greatly affected by their immediate environment. For example, a common situation where this occurs is when antennas are located on large supporting structures such as aircraft, spacecraft, or ships, the physical size of which may preclude their being tested at an antenna test facility. In those cases, scale-modeling techniques can be employed and the modeled structure can be treated as an integral part of the antenna being measured ([2], pp. 56–58). In this way the properties of the reduced-size model system can be measured at an antenna test facility.

The various measurements indicated by the entries in Chart 1 require specialized equipment, facilities, and personnel to perform them. This is particularly obvious for the major headings of the tabulation; however, it is also the case for the measurements listed under them. For example, consider those labeled as radiation characteristics. *Noise temperature*, *EIRP*, and *G/T* measurements are generally performed on-site and are, in fact, system measurements since they usually include amplifiers and transmitters along with the antennas being tested. Typically, there are separate, dedicated facilities, referred to as antenna ranges, for *boresight* and *radome* measurements, for *radar cross-section* measurements, and for amplitude and phase *radiation pattern* measurements. These ranges include a source antenna to illuminate the test antenna, positioners to support and orient them, the electronic instrumentation, and the physical space between and around the antennas.

---

*Throughout this chapter, the definitions of terms employed will conform to Institute of Electrical and Electronics Engineers' (IEEE) standards [1, 2], and those appearing in reference 1 will be italicized the first time they appear in the text.

Chart 1. Commonly Performed Antenna Measurements

1. Radiation Characteristics
   Amplitude and power radiation patterns
   Phase
   Directivity
   Gain
   Polarization
   Radiation efficiency
   Boresight
   Radome
   Equivalent isotropically radiated power (EIRP)
   Figure of merit (G/T)
   Noise temperature
   Passive intermodulation
   Scattering/radar cross section

2. Circuit Characteristics
   Self-impedance/reflection coefficient
   Mutual impedance

3. Power-Handling Capability
   Metallic and dielectric heating
   Voltage breakdown/corona
   Multipacting

4. Structural Integrity
   Vibration
   Surface deformation

5. Electromechanical
   Positioning accuracy
   Agility
   Et cetera

6. Environmental
   Wind
   Rain
   Salt spray
   Temperature
   Altitude
   Et cetera

This chapter is primarily concerned with *far-field* radiation pattern ranges and the measurements that are performed on them, namely, amplitude and *phase pattern*, *directivity*, *gain*, and *polarization* measurements. Far-field antenna ranges are those which are designed to produce a plane or quasi-plane wave over the *aperture* of the antenna being tested, thus allowing one to measure the far field of that antenna directly. This definition not only includes those ranges whose lengths or spacings between the source and test antenna are great enough so that far-field conditions are met, but also compact ranges. These latter ranges are ones in which the quasi-plane wave is achieved by collimating the field of a small source antenna by use of a large lens or reflector. Since the planar field exists in the aperture region of the *lens* or *reflector*, the required range length is reduced to relatively short distances [2, 3].

The success of any measurement is critically dependent on the design of the measurement system employed. For antenna measurements this includes not only the instrumentation but the space between and around the source and test antennas. Since the range length can be, in some cases, hundreds of meters in length, a substantial investment in land may be required in addition to that in instrumentation. Furthermore, the most serious sources of measurement error are introduced in the link between the source and test antennas. The causes of these errors include the finite range length, unwanted specular reflections from surfaces and objects, multipath between antennas, adverse weather conditions, and extraneous interfering signals. These effects cannot be completely eliminated; but, rather, the ranges should be designed in such a way as to reduce them to acceptable levels. Outdoor ranges pose particular difficulties because of weather conditions such as wind, rain, snow, and humidity, which cannot be controlled. However, detection systems can be installed to determine wind and moisture levels to alert the operator when measurements should be suspended. All other effects can be controlled by good antenna range design and management.

The design criteria for antenna ranges were fairly well established during World War II after the introduction of microwave radar systems. This work was summarized by Cutler, King, and Kock [4]. Hollis and his colleagues have refined and extended these criteria as well as antenna measurement techniques in general. Their work, including the work of other individuals, is presented in their landmark book, *Microwave Antenna Measurements* [5]. Johnson, Ecker, and Hollis [6], and Kummer and Gillespie [7], have published significant review papers on antenna measurements. IEEE Standard 149-1979, *Test Procedures for Antennas* [2], contains further extensions of these works and reference 8 is an excellent presentation on antenna measurements. The reader is referred to these sources for details that may be omitted here.

In this presentation the basic criteria for the design of antenna ranges are given. These form a guide for the range designer. No attempt has been made to present engineering details of range design, which are too varied and specialized to be included. The quality of an antenna range is dependent on the designer's knowledge and attention to operational details. When designing a test facility, one must literally visualize, step by step, the measurement procedures to be employed on a proposed range and provide a design that will allow the experimenter to perform measurements in an accurate and, where possible, convenient manner. Designing the range for anticipated future use is most important because of the large capital outlay that is required and the difficulty one has in making later modifications to the range.

The measurement procedures described are presented in the same manner, that is, only the basic principles are presented. Using them the experimenter can design the measurement procedure for the particular facility where the measurement is to be performed. The first step is usually the construction of an error budget based on a knowledge of the instrumentation employed, the physical layout of the antenna range, and the limits of uncertainty required for the measurement. This will guide the experimenter in the detail design of the experiment.

Because of the extreme measurement accuracy required by modern systems employing antennas and the resulting sophistication of the instrumentation required to achieve such results, the antenna measurements engineer must possess a broad knowledge of many fields. These include electromagnetics, electronics, digital and analog control, metrology, and data processing, to name a few. Like all engineering work the art is as important as the scientific aspects and it can only be obtained by experience.

2. Radiation Patterns and Pattern Cuts

The radiation pattern of an antenna is the spatial distribution of any of several quantities which characterize the field excited by the antenna. These quantities include *power flux density*, *field strength*, *radiation intensity*, *directivity*, gain, phase, and *polarization*. It is not practical to measure these quantities at every point in space, but, rather, one must sample them over a defined measurement surface. The most common surfaces are planar, cylindrical, and spherical. Any path over a measurement surface is called a *pattern cut* and the distribution of any of the above-mentioned quantities over a pattern cut is also called a radiation pattern. Practical considerations usually dictate that the cuts coincide with the natural coordinates of the measurement surface.

Planar and cylindrical measurement surfaces find their greatest use for measurements in the *near-field region* of an antenna (see Chapter 33), whereas the spherical surface can be employed for either near-field or far-field measurements. The spherical surface that is appropriate for far-field measurements is the *radiation sphere* of an antenna. The location of points of the radiation sphere are given in terms of the θ and ϕ coordinates of a standard spherical-coordinate system whose origin is located at the center of the sphere.

There are two standard types of pattern cuts over the radiation sphere. The path formed by the locus of points for which θ is a specified constant and ϕ is a variable is called a conical cut or ϕ-cut. The path formed by the locus of points for which ϕ is a specified constant and θ is the variable is called a great-circle cut or θ-cut (see Fig. 1).

In a practical measurement situation there must be a probe antenna to sample the field as depicted in Fig. 2. The probe's receiving coordinate system is the $x'y'z'$ Cartesian coordinate system as shown. Note that the z' axis is pointing in the radial direction which is determined by the values of θ and ϕ of the radiation sphere's coordinate system. The rotation of the probe about the z' axis is measured with respect to the unit vector $\hat{\theta}$ with the resulting angle designated as χ. The angles θ, ϕ, and χ are simply the Euler angles that relate the two coordinate systems. Continuous variations of θ or ϕ will yield radiation patterns, whereas a continuous variation of χ will result in a *polarization pattern*.

Among the decisions that an experimenter must make are the polarization of the probe and its path over the radiation sphere. In addition, the orientation of the antenna being tested with respect to the spherical-coordinate system of the radiation sphere must be specified. There are two common choices: the beam axis can be oriented toward one of the poles (usually the north pole) or the equator (usually either the x or the y direction) as shown in Fig. 3.

For directional antennas which are *linearly polarized*, the choice depends on the two orthogonally polarized field components to be measured over the radiation

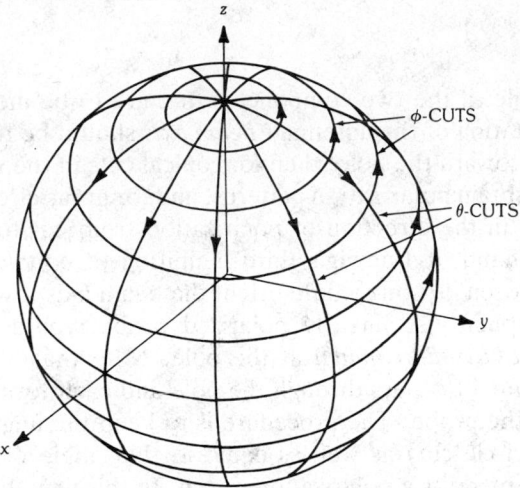

Fig. 1. An illustration of θ- and ϕ-cuts over the radiation sphere.

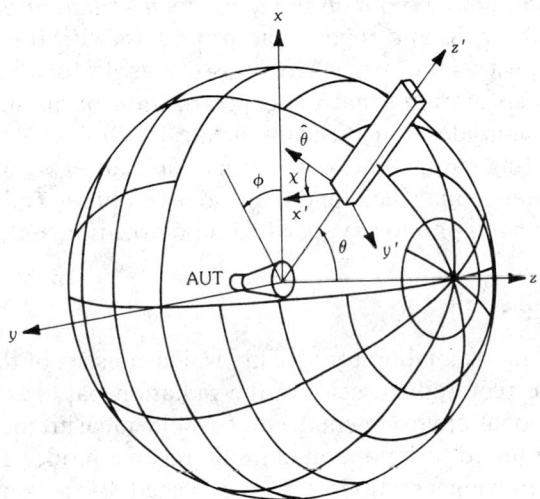

Fig. 2. An illustration of the coordinate systems of the antenna under test (AUT) and the probe used to sample the field over the radiation sphere.

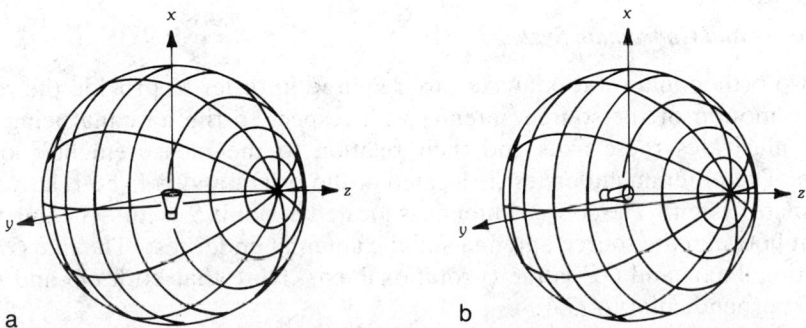

Fig. 3. Standard orientations of the antenna under test with respect to the spherical-coordinate system of the radiation sphere. (*a*) Equatorial orientation. (*b*) Polar orientation.

sphere. For example, if the two components that are to be measured are E_θ and E_ϕ, then the orientation of the antenna's *beam axis* should be toward the equator. If it were oriented toward the pole, then for conical cuts in the vicinity of the pole, one would simply obtain polarization patterns, and for great-circle cuts there would be drastic changes in the direction of polarization from cut to cut.

On the other hand, if Ludwig's third definition of *copolarization* and *cross polarization* are chosen [9], one would orient the beam axis toward the north pole of the radiation sphere. A linearly polarized probe would be employed and orientated for a *polarization match* at the pole. Note that all θ-cuts which are specified by the value of ϕ pass through the pole and each forms a different angle χ_0 with respect to the probe. The procedure is to keep the angle χ_0 constant over each θ-cut (great-circle cut) as well as equal to that angle obtained at the pole with the probe oriented for a polarization match. In this way, for each great-circle cut, there is a polarization match at $\theta = 0$. This procedure will result in a measurement of the *copolar radiation pattern* over the radiation sphere as specified by Ludwig's third definition. To obtain the *cross-polar radiation pattern*, one rotates the probe 90° at the pole and repeats the procedure with the probe oriented to $\chi_0 + 90°$ with respect to the axis where χ_0 is the angle for the copolar pattern.

It should be kept in mind that the copolarization of an antenna is a defined quantity. Systems considerations usually dictate its choice. With the availability of modern computing equipment, one can measure any two orthogonal far-field components of the antenna's field and their relative phases and then compute the components that correspond to any specified copolarization and cross polarization.

3. Antenna Ranges

An antenna range is a laboratory facility which consists of the instrumentation and physical space required to measure the radiation patterns of antennas away from their operational environments. For far-field measurements it is usual for the antenna being tested to be operated in the receive mode. The probe antenna discussed in the previous section is then replaced by a source (transmitting) antenna whose radiation pattern and polarization are defined by the particular measurement problem for which it is employed. Unless otherwise specified, the antenna under test will be considered to be operated in the receive mode for the remainder of this chapter.

Positioners and Coordinate Systems

Two orthogonal rotational axes are required in order to provide the required relative motion of the source antenna with respect to the antenna being tested. Fig. 4 illustrates these axes and their relation to the measurement coordinate system. The antenna under test is located at the origin and is fixed relative to the coordinate system. The source antenna is located at point S, with OS being the line of sight between the source antenna and the antenna under test. The line OA is the θ rotational axis and OZ is the ϕ rotational axis. Note that both OS and OZ are always perpendicular to OA.

There are two basic means of achieving the required motions about the θ and ϕ rotational axes. One approach is to move the source antenna physically over a

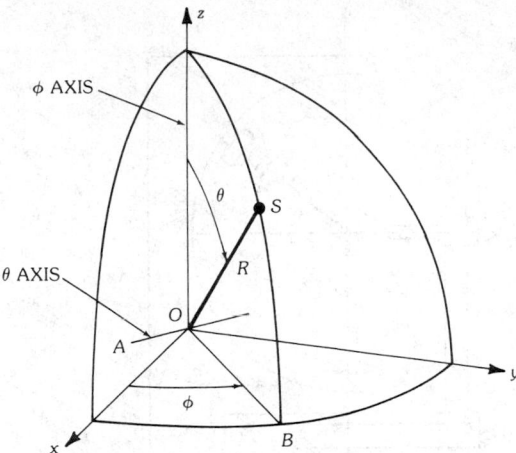

Fig. 4. The two orthogonal rotational axes required for a spherical-coordinate positioning system.

prescribed path. Ranges that employ this approach are called movable line-of-sight ranges. The second approach is one where the source antenna is fixed and the antenna under test is provided with a means of rotation about the two orthogonal axes, *OA* and *OZ*. Ranges employing this approach are called fixed line-of-sight ranges and they are the more common of the two.

The movable line-of-sight ranges can be realized by the use of a vertical arch that is either fixed or mounted on an azimuthal positioner. The antenna under test is usually mounted on an azimuthal positioner or a turntable so that its motion can be independent of that of the source antenna, as shown in Fig. 5a. There are two modes of operation: (1) The antenna under test is rotated incrementally in ϕ and the source antenna moves continuously over the arch for each θ-cut. (2) The source antenna is moved incrementally in θ and the antenna under test is rotated continuously for each ϕ-cut. If the arch is mounted on an azimuthal positioner then, with the antenna under test remaining fixed, the source antenna is moved incrementally in θ and the arch is rotated continuously for each ϕ-cut as depicted in Fig. 5b. An alternate approach is to use a gantry arrangement or simply a rotating arm to position the source antenna as shown in Fig. 5c. Note that the rotating arm could have been oriented so that the motion of the source antenna is in the horizontal plane. Also, not shown in Fig. 5 is the microwave-absorbing material that would be required to cover the reflecting surfaces of the positioners.

The movable line-of-sight ranges find their greatest use for testing vehicular antennas, for measuring the radiation patterns of scale-model antennas mounted on model ships, and for locating the *phase centers* of feed antennas for reflector antennas. The first two uses require very large arches of the type illustrated in Fig. 6 [10, 11, 12], which is known as an image-plane range (see discussion under "Outdoor Ranges"). A range employing a horizontal rotating arm is more appropriate for locating phase centers. To accomplish this, the antenna under test is moved longitudinally in increments until the measured phase is nearly constant over a cut through the peak of its *main beam*.

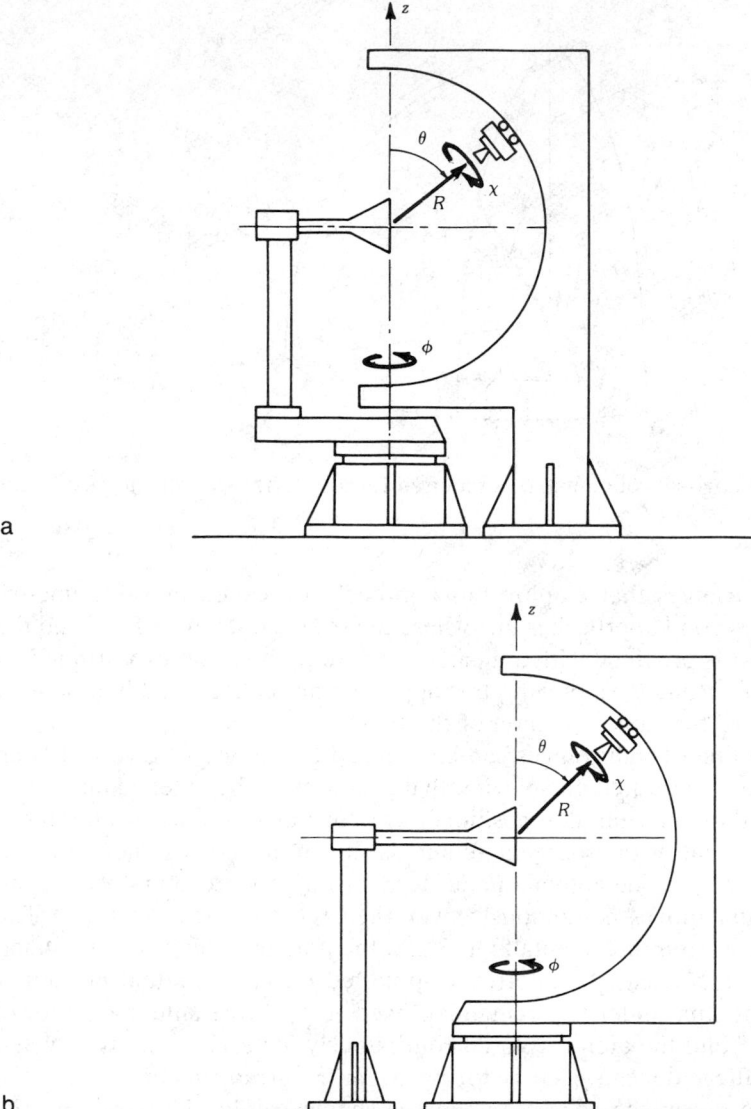

Fig. 5. Possible movable line-of-sight ranges. (*a*) AUT mounted on turntable. (*b*) With AUT fixed. (*c*) Source antenna on rotating arm.

The fixed line-of-sight ranges are by far the most commonly used ranges. They require a spherical-coordinate positioner for the antenna under test and a polarization positioner for the source antenna (unless its polarization can be controlled electronically). Each type of positioner has a coordinate system associated with it. The antenna under test should be mounted on the positioner in such a manner that its coordinate system properly coincides with that of the positioner; otherwise, errors will result.

There are three basic positioner configurations that provide the θ and ϕ

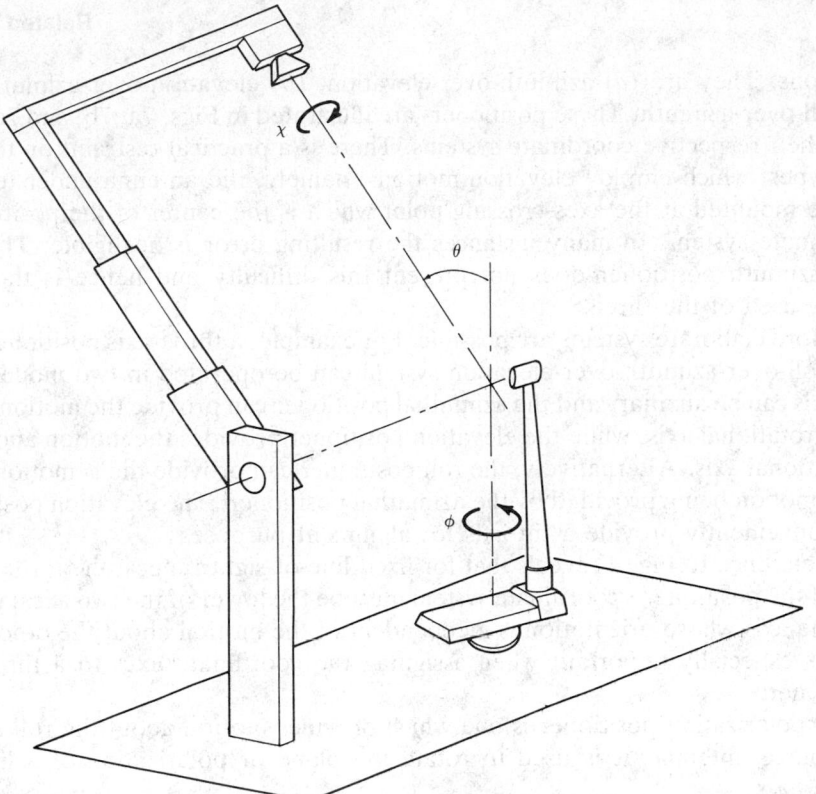

C

Fig. 5, *continued.*

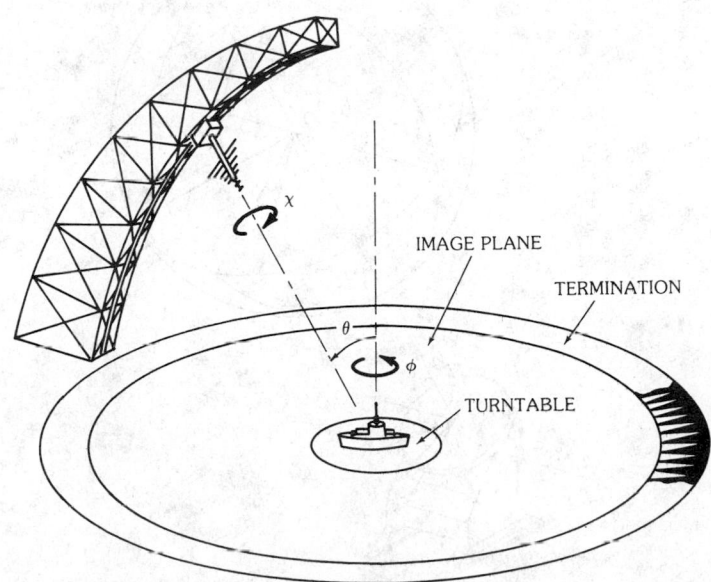

Fig. 6. An example of a movable line-of-sight range which employs a large arch. This is known as an image-plane range.

rotations. They are (*a*) azimuth-over-elevation, (*b*) elevation-over-azimuth, and (*c*) roll-over-azimuth. These positioners are illustrated in Figs. 7a, 7b, and 7c along with their respective coordinate systems. There is a practical restraint on the first two types, which employ elevation motions, namely, the antenna under test can not be mounted at the axes-crossing point which is the center of the positioner's coordinate system. In many instances the resulting error is negligible. The roll-over-azimuth positioner does not present this difficulty and hence is the most widely used of the three.

More elaborate systems are possible. For example, a three-axis positioner such as a roll-over-azimuth-over-elevation system can be operated in two modes. The roll axis can be auxiliary and the azimuthal positioner can provide the motion about the ϕ rotational axis, while the elevation positioner provides the motion about the θ rotational axis. Alternatively, the roll positioner can provide the ϕ motion, with the θ motion being provided by the azimuthal positioner. The elevation positioner can conveniently provide a tilt axis for alignment purposes.

Reference to Fig. 4 reveals that for fixed line-of-sight ranges, the θ rotational axis of the positioner's coordinate system must be the lower of the two axes; that is, it is the axis whose orientation is independent of the motion about the other axis. This is especially important when assigning the coordinate axes to a three-axis positioner.

A polarization positioner is one which provides motion about the roll axis of the source antenna. It is used to rotate the plane of polarization of a linearly

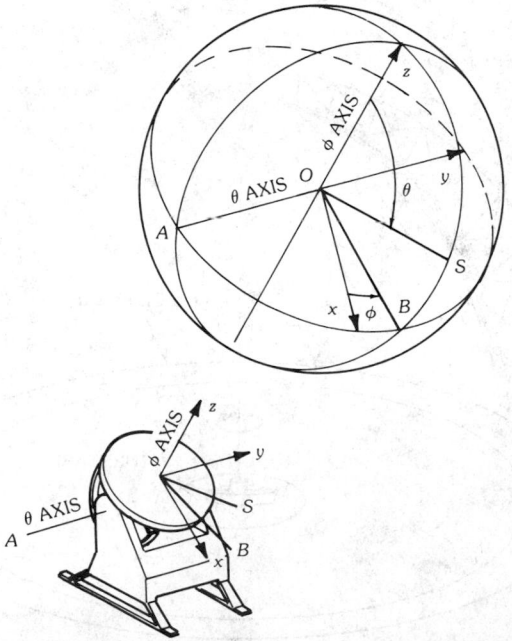

a

Fig. 7. The three basic spherical-coordinate positioners. (*a*) Azimuth-over-elevation positioner. (*b*) Elevation-over-azimuth positioner. (*c*) Roll-over-azimuth positioner.

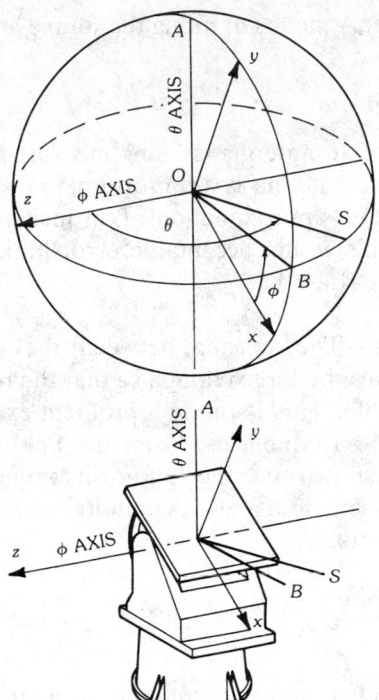

b

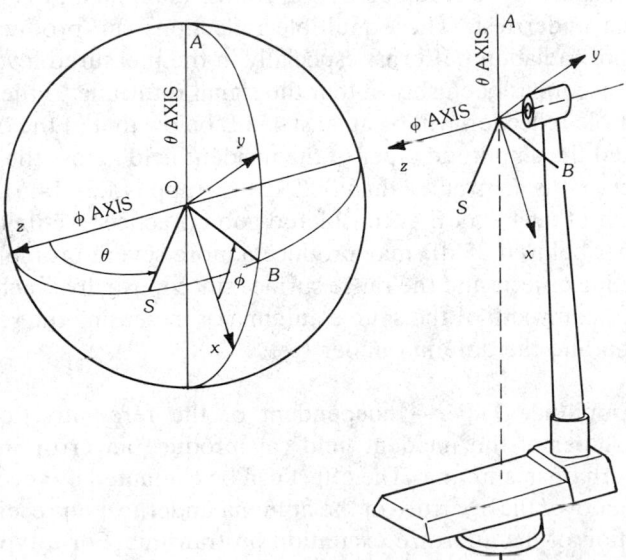

c

Fig. 7, *continued.*

polarized source antenna. A means of tilting the source antenna is usually provided for use in alignment.

Antenna Range Design Criteria

When the far field of an antenna is being measured in its receive mode, the ideal incident wave at the antenna is a uniform plane wave. Antenna ranges are designed to provide a close approximation of an incident plane wave at the test region. To accomplish this within acceptable error limits, the following must be considered ([5], pp. 14.1–14.41).

Reactive Field Coupling—The spacing between the source antenna and the antenna under test, R, must be large enough so that the reactive part of the field at the test region is negligible. This is rarely a problem except for electrically small antennas operating at low frequencies. From the field equations for a *Hertzian electric dipole*, it can be shown that the ratio of its inductive field to its radiation field is $\lambda/2\pi R$. This ratio is -36 dB at a separation of 10λ; this is the basis of the commonly accepted criterion

$$R \gtrsim 10\lambda \tag{1}$$

for antenna ranges to reduce reactive coupling to acceptable levels.

Reradiative Coupling—This type of coupling occurs because part of the energy received by the antenna under test is radiated back toward the source antenna. In turn, part of this energy is received by the source antenna and is reradiated back to the antenna under test. These multiple reflections can produce errors in the measurement of radiation patterns, especially in the measured level of *side lobes*. For a typical case it has been shown that the signal returning to the antenna under test after multiple reflection will be at least 45 dB below that of the original incident signal, provided the amplitude taper of the incident field across the aperture of the antenna under test is no greater than 0.25 dB ([5], pp. 14.3–14.5). An amplitude taper of no more than 0.25 dB is considered good practice. Actually, reducing the amplitude taper below 0.25 dB may produce a more severe problem of reflections from surrounding objects and the range surface since it usually is achieved by either increasing the beamwidth of the source antenna or increasing the separation of the source antenna and the antenna under test.

Transverse Amplitude Taper—Independent of the reradiative coupling effect, an amplitude taper of the incident field can produce an error in the measured radiation pattern of an antenna. The effect can be evaluated by recognizing that an amplitude taper over the aperture of the antenna under test on receive is analogous to a modification of the aperture excitation on transmit. For a typical microwave antenna with a circular aperture, Hollis, Lyon, and Clayton ([5], p. 14.13) have shown that a symmetrical amplitude taper of 0.5 dB produces an apparent reduction in gain of 0.15 dB and a taper of 0.25 dB produces a reduction in gain of 0.10 dB.

Longitudinal Amplitude Taper—A longitudinal amplitude taper has an effect similar to a transverse one; however, for an aperture-type antenna it is primarily the side lobes in the vicinity of 90° from the beam axis that are affected. The effect is most pronounced when high-gain end-fire antennas are being tested. In this case the main beam and the near-in side lobes are affected. If the range length is at least 10 times the length of the antenna (width in the case of an aperture antenna), the longitudinal taper will be less than 1 dB, which is considered to be satisfactory ([5], pp. 14.11–14.13, [13]).

Phase Taper—The measured radiation patterns of most antennas are considerably more sensitive to a deviation (from planar) of the phase of the incident wave over their apertures than they are to amplitude deviations. Thus the phase taper across the aperture of the antenna under test is the dominant factor in the selection of the separation distance R between the source and test antennas.

Hollis, Lyon, and Clayton ([5], pp. 14.5–14.7) have shown that for a typical source antenna whose aperture diameter is D meters, the phase front over its main beam, at least to the −0.5-dB level, is approximately spherical at range distances greater than about D^2/λ, where λ is the wavelength in meters. This means that the phase deviation over an aperture, in the absence of reflected signals, is determined solely by the range length R. With reference to Fig. 8, one finds that the phase deviation $\Delta\Psi$ across the test antenna's aperture is given by

$$\Delta\Psi = \frac{\pi D^2}{4\lambda R} \tag{2}$$

A criterion commonly employed is to limit this phase deviation to a maximum of $\pi/8$ rad. This leads to the restriction on range length,

$$R \geqq \frac{2D^2}{\lambda} \tag{3}$$

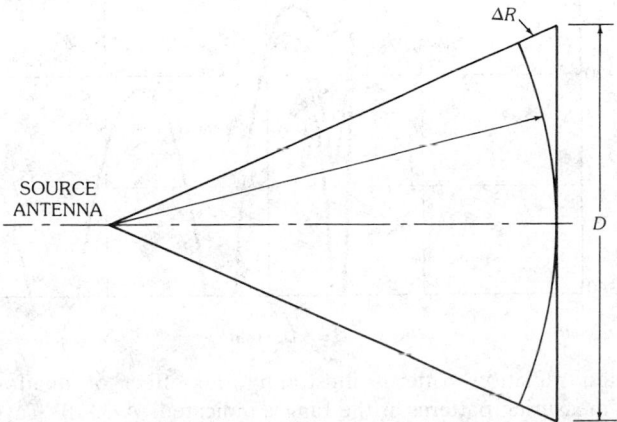

Fig. 8. A spherical phase front incident on a plane aperture.

While this criterion is adequate for many measurements, it is inadequate for others [14, 15, 16]. Equation 3 should therefore be interpreted simply as a reference criterion.

The effects of a phase taper on the measured radiation pattern of an antenna with a sum pattern include a change in levels of the first few side lobes and a filling of the nulls as shown in Fig. 9. As the phase taper is made more severe as a result of reducing the range length, the first side lobes on either side of the main beam begin to merge with the main beam, thereby forming a *shoulder lobe*. Further reduction in range length causes successive side lobes to merge with the main beam.

This effect is more pronounced for low and ultralow side lobe levels as demonstrated in Fig. 10. The distance at which the first side lobe begins to merge

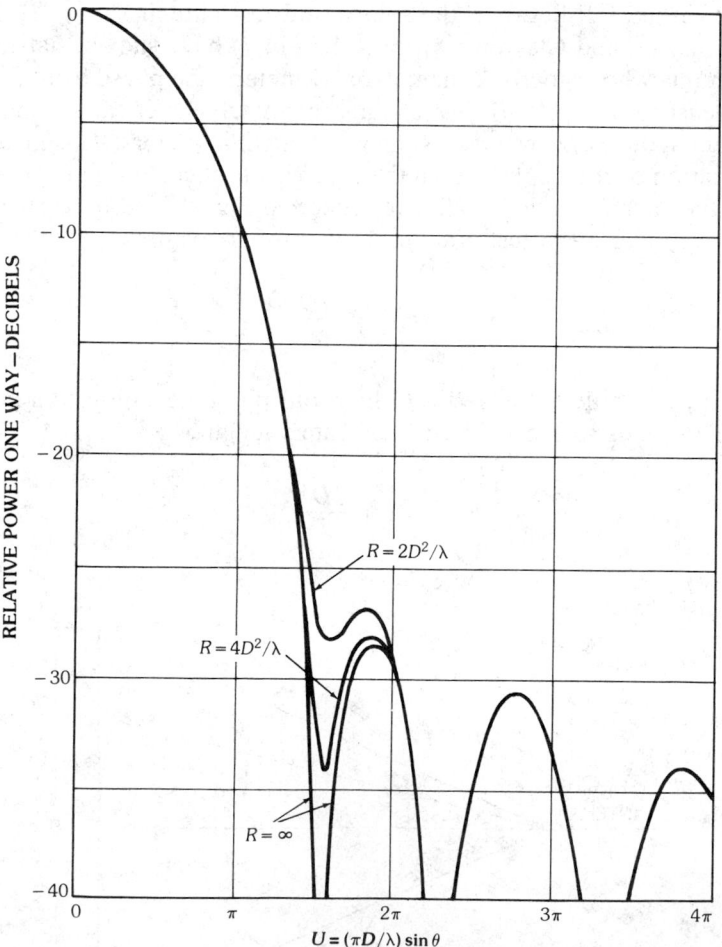

Fig. 9. Calculated radiation patterns illustrating the effect of quadratic phase errors encountered in measuring patterns at the ranges indicated. A 30-dB Taylor distribution is assumed. (*After ANSI/IEEE [2], © 1979 IEEE; after Hollis, Lyon, and Clayton [5], © 1970 Scientific-Atlanta*)

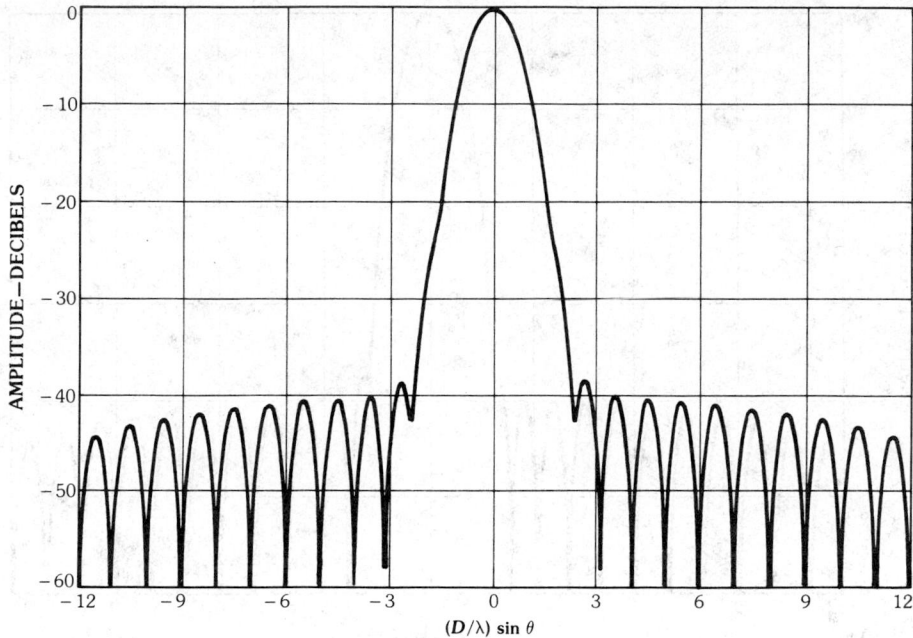

a

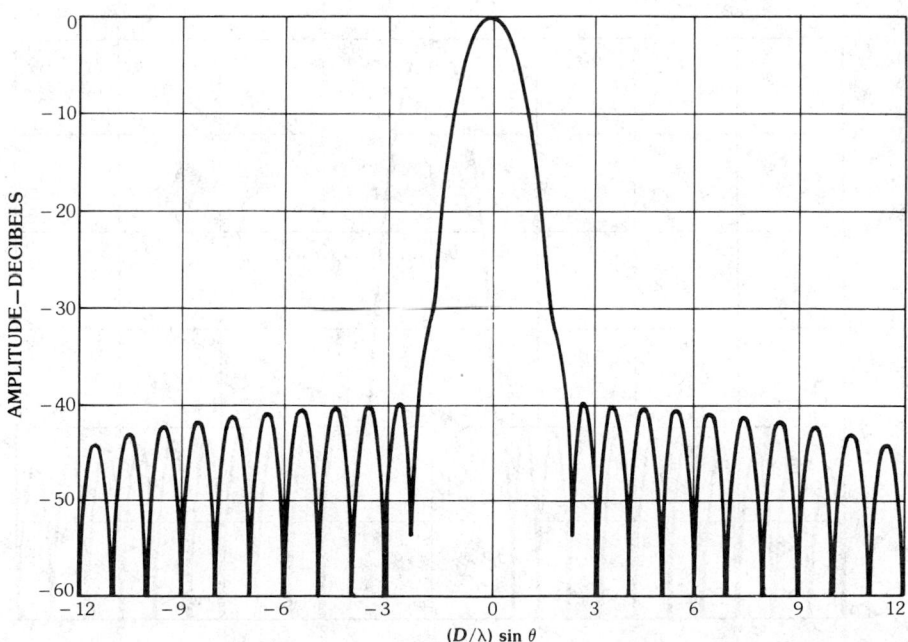

b

Fig. 10. Calculated radiation patterns illustrating the effect of quadratic phase errors encountered when measuring patterns at the ranges indicated. A 40-dB Taylor n distribution is assumed with $n = 11$. (a) $R = D^2/\lambda$. (b) $R = 2D^2/\lambda$. (c) $R = 4D^2/\lambda$. (d) $R = 6D^2/\lambda$. (*After Hansen [16], © 1984 IEEE*)

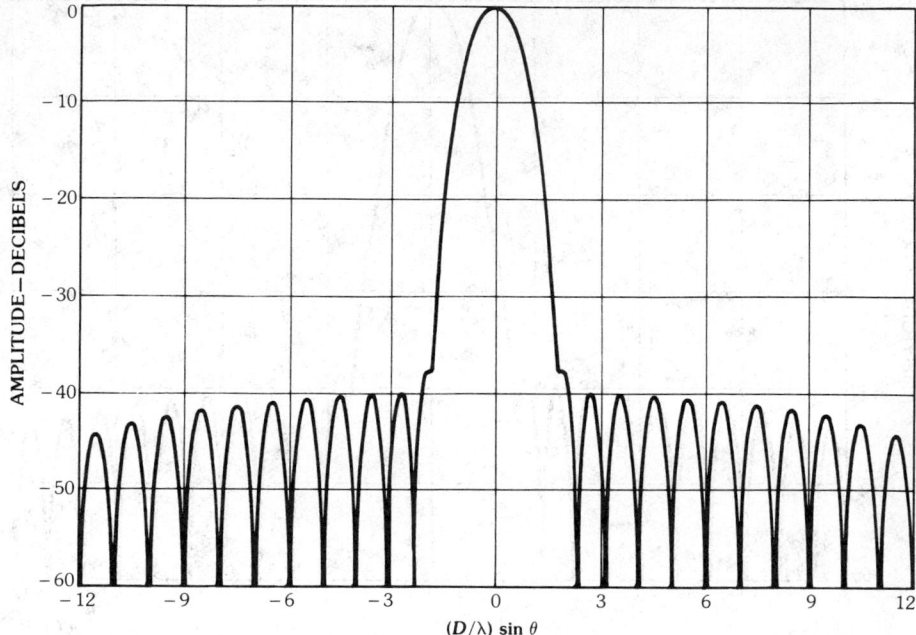

c

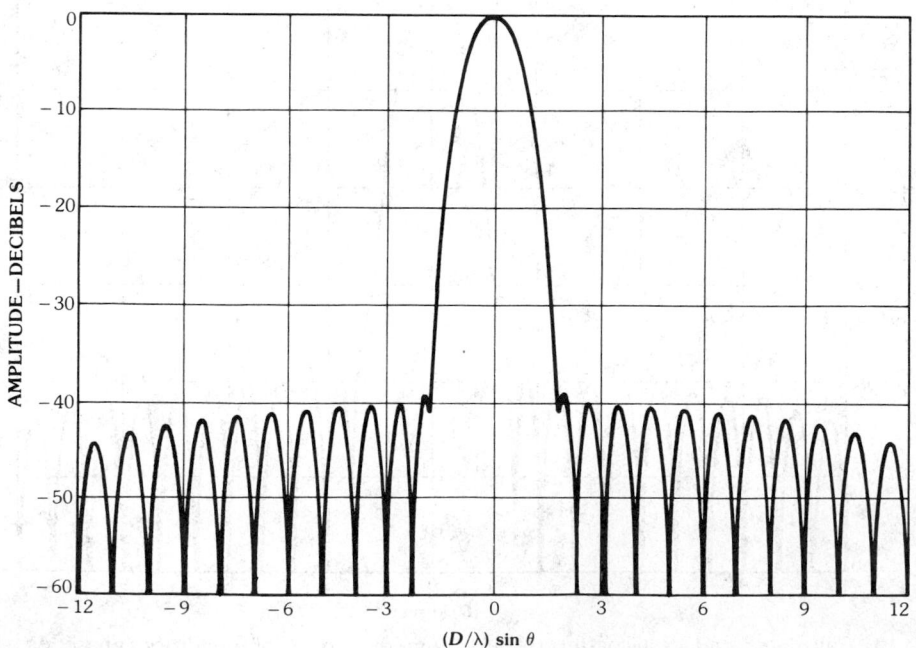

d

Fig. 10, *continued.*

with the main beam is dependent on the side lobe level of the antenna. In general, the lower the side lobe level of the antenna under test, the larger the range length must be made in order to prevent the merging of the first side lobes into the main beam [15, 16].

Shaped-beam antennas which consist of a reflector fed by an *array* also exhibit measured radiation pattern distortions as a result of finite range lengths [14]. These distortions decrease with increasing range lengths. As a rule of thumb the distance criterion for such antennas is usually taken to be about $8D^2/\lambda$ to $12D^2/\lambda$.

Whenever one designs a measurement program which involves the measurement of radiation patterns, the distance criterion should be based on the characteristics of the antenna to be tested.

Spatial Variation Caused by Reflections—Reflections of the radiated signal from the source antenna by range surfaces, nearby objects, or antenna positioners that reach the test region will cause a distortion of the incident field which illuminates the antenna under test. This can result in significant errors in radiation pattern measurements.

When an outdoor antenna range is designed, sufficient land should be cleared of reflecting objects so that it is principally the range surface (ground) and the positioner that must be considered. The latter problem can usually be reduced or eliminated with the use of a good-quality microwave-absorbing material covering the important reflecting surfaces.

The effect of reflections from the range surface can be reduced by proper range design, as will be discussed in the next section.

Interference from Sources Other Than the Range Source—Extraneous signals can produce errors in measurements if they are recorded along with the desired signal. Such signals can only be controlled by the proper selection of instrumentation. For example, a narrow-bandwidth receiver will discriminate against signals outside its beamwidth, whereas a simple detector and amplifier will not provide any discrimination.

Outdoor Ranges

There are three major types of outdoor ranges: elevated or free-space, ground-reflection, and slant ranges. In addition there are special-purpose ranges such as those used for measuring the radiation patterns of model ships and those used to measure the patterns of antennas on terrestrial vehicles [10, 11, 12]. These two particular examples of special-purpose ranges might be called image-plane or ground-plane ranges since the model ships and vehicles operate over an image plane which simulates seawater and earth, respectively.

Hollis, Lyon, and Clayton ([5], pp. 14.1–14.70) have provided detailed design and evaluation procedures for elevated and ground-reflection ranges. Much of their work appears in reference 2. The reader is referred to these references for design details not included here.

Elevated Antenna Ranges—An elevated range is one that simulates a free-space condition by judicious choices of the range length, height of the source and test antennas, and the radiation characteristics of the source antenna. In addition the reflections from the range surface are suppressed to be below an acceptable level.

The geometry of the elevated range is shown in Fig. 11. The range length R is usually determined by the maximum allowable phase taper of the incident field over the aperture of the test antenna as discussed under "Antenna Range Design Criteria." The value of R is usually chosen to be between $2D^2/\lambda$ and $12D^2/\lambda$, depending on the characteristics of the antenna being tested.

The height of the antenna under test should be chosen so that the main beam of the source antenna does not illuminate the range surface. If the 0.25-dB amplitude taper criterion is adhered to also, then the height of the antenna under test, h_r, should be at least four times its maximum dimension, D, i.e.,

$$h_r \gtrsim 4D \tag{4}$$

This height requirement is usually met by mounting the test and source antennas on towers; however, it can be achieved by mounting them on the roofs of neighboring buildings or other permanent structures, or by mounting them on adjacent hilltops.

If the range surface is irregular or if there are objects nearby that cannot be removed, one should construct *Fresnel zones* over a topographic map of the range surface in order to locate the specular regions from which significant reflections can occur. A plaster scaled model of the terrain is often employed for this purpose. Once the specular regions are located, remedial measures can be taken to reduce the effect of reflections on the measurements.

If the range surface is flat, diffraction fences can be employed to redirect the reflected signals away from the test region. There are two cautions that should be emphasized. First, the diffraction fences should not intercept the main beam of the source antenna. Secondly, the top edge of the fences should not be straight but rather serrated. The reason for this is to reduce the effect of knife-edge diffraction, which could actually cause the indirect signal to increase as a result of placing the fences on the range. In some measurement situations it is impractical to elevate the test antenna to a height of four times its largest dimension. In these cases a

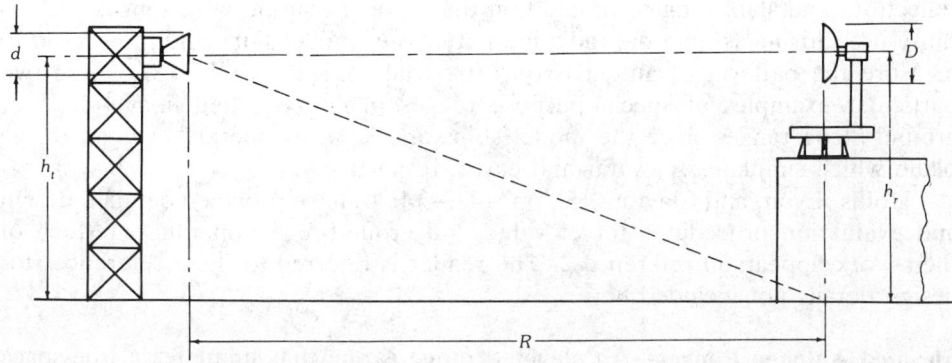

Fig. 11. An elevated range.

compromise must be made between (*a*) illuminating the range surface with energy contained in the main beam of the source antenna and (*b*) making the source antenna's beamwidth narrower. The former leads to increased reflections from the range surface, and the latter to an increased amplitude taper as well as increased radiation coupling. An error analysis is usually required in order to make a decision.

The height of the source antenna is usually chosen to be the same as that of the antenna being tested. If, however, one wishes to further reduce reflections from the range surface in order to measure near-in side lobes more accurately, then one could elevate the source antenna slightly higher than the test antenna. Some improvement in the measurement of a back lobe can be achieved by positioning the source antenna slightly lower than the antenna under test.

Finally, the source antenna should be chosen so that the amplitude taper over the antenna under test is no greater than 0.25 dB and the direction of its first null is toward the base of the test tower. This will ensure that the range surface intercepts only side lobe energy. Reducing the side lobe level of the source antenna will in turn reduce errors due to reflections.

Ground-Reflection Ranges—It should be apparent that for elevated ranges, if the range length is even moderately long, the source antenna must have a rather narrow beamwidth, which may not be easily achievable at lower frequencies (vhf-uhf). The restrictions on beamwidth can be relaxed for ground-reflection ranges which utilize the reflections from the range surface to approximate the desired plane wave at the antenna under test. Thus ground-reflection ranges find their greatest use in measurement of lower frequencies.

With reference to Fig. 12, the specularly reflected signal from the range surface combines with the direct signal from the source antenna to produce an interference pattern in the vertical direction at the test antenna. The heights of the source and test antennas are chosen so that the test antenna is centered on the first

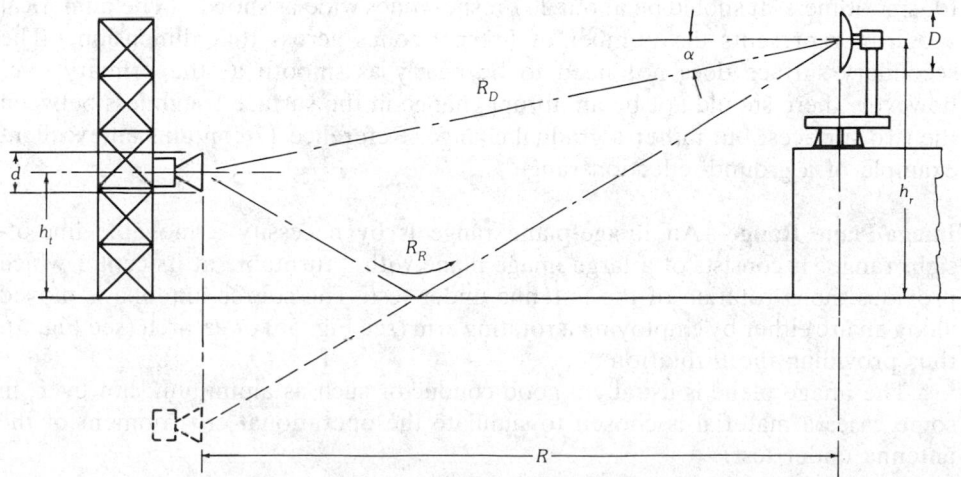

Fig. 12. A ground-reflection range.

interference lobe and the amplitude taper is no more than 0.25 dB. For a given frequency the approximate height of the source (transmitting) antenna, h_t, is given by

$$h_t \cong \frac{\lambda R}{4h_r} \tag{5}$$

If the 0.25-dB amplitude taper criterion is also adhered to, the height of the test antenna should be 3.3D, where D is the diameter of the antenna under test. Usually, it is recommended that the criterion be taken as

$$h_r \gtrsim 4D \tag{6}$$

since there are approximations in the derivation. Note that the height of the source antenna is frequency sensitive; that is, if the frequency changes, so must the height of the source antenna.

It is extremely important that the range surface be smooth so that energy is specularly reflected. This means that "smoothness" must be specified for the range surface. The Rayleigh criterion can be employed. A useful form of it is given by

$$\Delta h = \le \frac{\lambda}{M \sin \psi} \tag{7}$$

where Δh is the root-mean-squared deviation of the irregularities relative to a median surface, ψ is the angle of incidence measured from the horizontal, and M is a smoothness factor. Smoothness factors ranging from 8 to 32 or greater are commonly used. This corresponds to surfaces that range from tolerable to very smooth. A value of 20 is typical.

A possible ground-reflection range configuration is depicted in Fig. 13. It is the primary surface along the range axis that must have the most stringent requirement for smoothness. It should be about 20 Fresnel zones wide as shown. (The numerical subscript represents the number of Fresnel zones across that dimension.) The secondary surface does not need to be nearly as smooth as the primary one; however, there should not be an abrupt change in the surface roughness between the two surfaces, but rather a gradual change. Reference 17 contains an excellent example of a ground-reflection range.

Image-Plane Range—An image-plane range is by necessity a movable line-of-sight range. It consists of a large image plane with a turntable at its center which provides the ϕ rotation of the antenna under test. The source antenna is moved along an arc either by employing a rotating arm (see Fig. 5c) or an arch (see Fig. 6), thus providing the θ rotation.

The image plane is usually a good conductor such as aluminum; however, in some cases a material is chosen to simulate the operational environment of the antenna under test.

Two of the principal concerns in the design of an image-plane range are surface smoothness and edge diffraction around the image plane. The Rayleigh criterion,

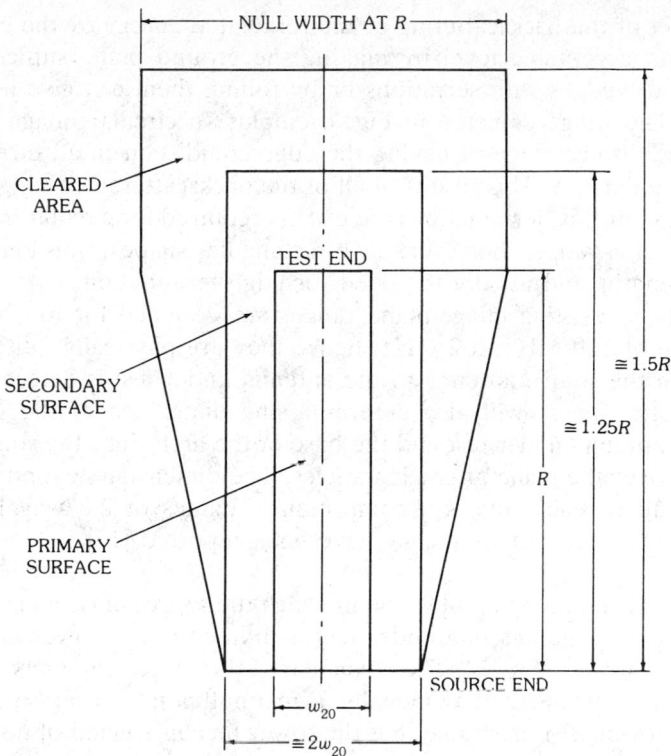

Fig. 13. A possible layout for a ground-reflection range. (*After ANSI/IEEE [2], © 1979 IEEE; after Hollis, Lyon, and Clayton [5], © 1970 Scientific-Atlanta*)

(7), is useful in establishing a limit on the root-mean-squared surface irregularities. Because the image plane is relatively small compared with a ground-reflection range, one can usually choose a larger value of smoothness factor for it. If the image plane is subjected to large extremes in temperature, there is the possibility that the surface will buckle, thus producing an unacceptable waviness in the surface. This should be taken into account when designing such a range.

Edge diffraction can also produce errors in image-plane measurements. It occurs in two ways. First, energy from the source antenna incident on the edges can be diffracted toward the antenna under test, and, secondly, energy reradiated by the antenna under test can be backscattered from the edges to the antenna under test.

The first effect can be minimized by designing the source antenna in such a way that, for a given-size image plane, the edge illumination is sufficiently low so that the diffracted signal at the antenna under test is at an acceptably low level. Since the beamwidth of the source antenna must also be broad enough to produce a wave with a sufficiently planar phase front over the antenna under test, a compromise must be made between edge illumination and the planarity of the incident phase front. It is good practice for the source antenna to extend over the image plane at grazing angles of incidence, thus eliminating the possibility of diffraction from the edge closest to the source antenna.

The effect of the backscattering of the reradiated energy by the edges can be reduced to an acceptable level by making the ground plane sufficiently large. Terminating the edges with serrations or by rolling them can give a further improvement. The range depicted in Fig. 6 employs a circular image plane. This shape has the disadvantage of having the edge equidistant in all directions from the antenna under test. This results in all of the backscattered energy returning in phase. Because of this, terminations are usually required for circular image planes. Most image-plane ranges, however, have rectangular shapes, thus eliminating the equal path lengths and usually the need for edge termination.

Many of the existing image-plane ranges are designed for use over the frequency range of 100 MHz to 2 GHz; hence they are physically quite large. The distance from the source antenna to the antenna under test is determined by the far-field criteria. These will also determine the dimension of the image plane between the antenna under test and the base of the arch since the source antenna must extend over the plane at grazing angles. The dimension beyond the antenna under test and to each side is a compromise. Values of 2.5 wavelengths to 8 wavelengths at the lowest frequency have been reported [12].*

Slant Range—A slant range [18] is one in which the source antenna is placed close to the ground and the antenna under test is mounted on a tower along with its positioner, as shown in Fig. 14. The beam axis of the source antenna points toward the center of the test antenna and ideally its first null points toward the base of the test antenna tower. It is desirable that the tower be constructed of nonconducting material.

If the antenna under test is highly directive and only the main beam and first few side lobes need to be measured, it is possible to reverse the positions of the source and test antennas. In this way the test antenna remains at ground level, which is a requirement for some types of antennas, such as satellite antennas, which must be protected by a radome housing from contamination.

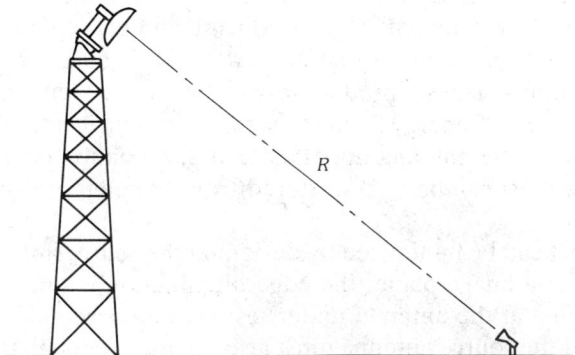

Fig. 14. A slant range.

*The US Naval Ocean Systems Center, San Diego, California, has a circular image-plane range that has a 160-ft (48.76-m) diameter.

Indoor Ranges

Environmental conditions such as adverse weather and electromagnetic interference can materially impede measurements made on outdoor ranges. This has led to the development of indoor ranges, all of which make use of microwave-absorbing material [18].

A room completely lined with absorbing material is called an anechoic chamber. Such chambers may also be electrically shielded to further prevent electromagnetic interference. There are two basic forms of chambers designed for antenna pattern measurements: the rectangular anechoic chamber and the tapered anechoic chamber.

These chambers must be long enough to satisfy the distance criteria presented under "Antenna Range Design Criteria." Therefore there is a restriction on their utility since it is impractical to construct a chamber of great length. One can, however, construct a partial anechoic chamber to house the antenna under test and its positioner by removing one wall of an otherwise enclosed chamber. In this way it can be combined with an outdoor range [19]. The distance criteria can be circumvented by the use of a compact range [3]. This type of range is one in which the antenna under test is illuminated by the collimated energy in the aperture of a larger point or line-focus reflector antenna.

Indoor ranges have other desirable features. For example, classified antennas can be tested without special shrouds covering them. Also, a relatively clean environment can be provided for satellite antennas.

Rectangular Chambers—A longitudinal-sectional view of a rectangular chamber is depicted in Fig. 15. The source antenna is located at the center of one of the end walls. The test antenna is located at the other end of the chamber at a point approximately equidistant from the side and back walls along the center line of the chamber.

Fig. 15. A longitudinal-sectional view of a rectangular anechoic chamber.

Although the chamber is completely lined with microwave-absorbing material, there will be reflections from the walls, floor, and ceiling. Of principal concern are the specular regions from which the reflected signals can reach the antenna under test. These regions are located midway between the source and test antennas on the side walls, floor, and ceiling. The center region of the back wall is also such a specular region.

Typically, absorbing materials produce the smallest reflection for normal incidence. As the angle of incidence θ_i increases from the normal, the reflection coefficient increases. It is good practice for the chamber width and height to be such that $\theta_i < 60°$. This requirement leads to the restriction that the length-to-width ratio of a chamber be about 2:1 ([8], pp. 669–671). In some instances it is necessary to increase this ratio to as much as 3:1 at the expense of higher levels of reflections from the walls, floor, and ceiling.

The region in which the test antenna is located is called the quiet zone. For a given chamber the size of the quiet zone depends on the specified or allowable deviation of the incident field from a uniform plane wave. The deviation is due to the finite distance from the source antenna and the specular reflections from the various surfaces of the chamber. There is no standardized figure of merit for anechoic chambers. What is often done is to establish the ratio of an "equivalent" reflected wave, which is the aggregate of all reflected waves incident on a probe antenna used to test the chamber, to the direct wave. The directivity of the probe antenna will affect the result [20].

Tapered Chambers—Rectangular anechoic chambers designed for frequencies below 1 GHz are very costly because of the increased size of the absorbing material. Furthermore, it is difficult to obtain accurate measurements in them. This is due in part to the fact that, at these frequencies, it is usually not possible to obtain a source antenna with a sufficiently narrow beam to avoid illumination of the walls, floor, and ceiling with the main beam.

The tapered anechoic chamber was introduced to overcome these difficulties. This type of chamber consists of a tapered section opening into a rectangular section. The taper has the shape of a pyramidal horn that tapers from a small source end to the large rectangular test region, as shown in Fig. 16. The tapered section is usually about twice the length of the rectangular section and the

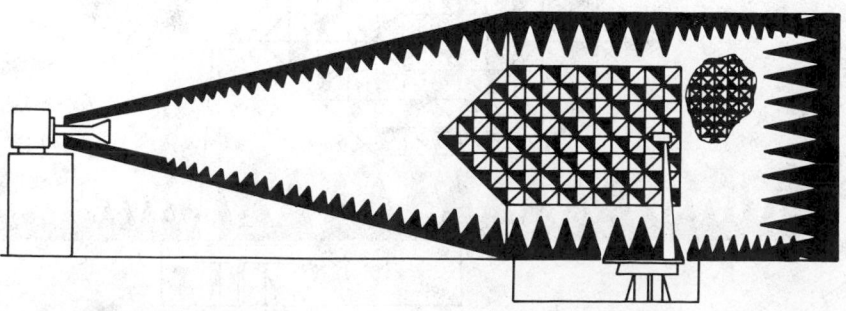

Fig. 16. A longitudinal-sectional view of a tapered anechoic chamber.

rectangular section is approximately cubical. The tapered chamber inherently requires less absorbing material; hence there is a substantial economic advantage compared with the rectangular chamber.

The low-frequency mode of operation requires the source antenna to be electrically very close to the apex of the tapered section. As a result, a slowly varying amplitude taper is produced in the test region. This can be understood by ray tracing. With reference to Fig. 17, note that the specular reflections that reach the test region occur close to the source antenna. The path lengths of the reflected signals are not very different, electrically, from that of the direct signal. The phasor sum of the reflected and direct signals thus produces the slowly varying amplitude variation shown. Another benefit is the fact that, as with the ground-reflection range, one is making use of the specularly reflected signal from the walls to create a constructive interference with the direct signal. This means that thinner absorbing material can be used over the walls in the tapered section.

The source antenna must be properly located to achieve the desired amplitude taper. If it is not close enough to the apex, deep nulls can appear in the test region. Furthermore, this location is frequency sensitive. Usually one must experimentally determine the proper location of the source antenna. This adds to the time required for a given measurement. When the source antenna is properly located the tapered chamber usually exhibits performance superior to that of the rectangular chamber at lower frequencies.

Because the source antenna must be moved closer to the apex as the frequency increases, the use of a log-periodic source antenna which has the reverse property is precluded. Also, as the frequency of operation increases, it becomes more difficult to place the source close enough to the apex to achieve proper operation. The chamber can then be operated in a free-space mode by moving the source antenna far enough forward so that it behaves as a rectangular chamber.

One can expect a certain amount of depolarization due to asymmetries in the chamber. This poses a problem when circularly polarized measurements are desired. One can adjust the polarization of the source antenna to compensate for the chamber at a single frequency of operation. For swept-frequency measurements one must sample the incident field and compensate for the axial ratio of the chamber [21].

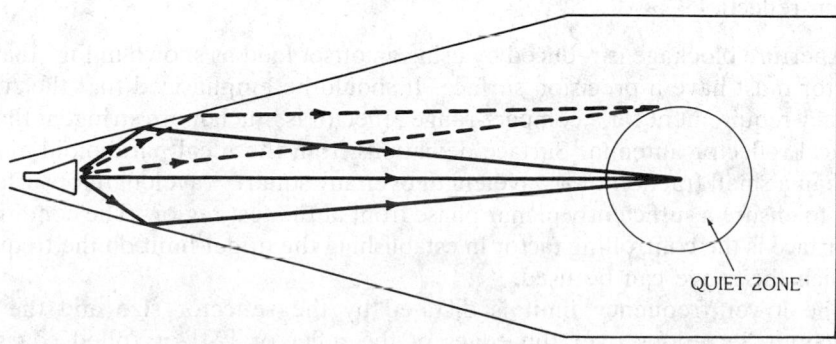

Fig. 17. Specular reflections that reach the quiet zone of a tapered anechoic chamber.

It should be noted that the field of the source antenna does not spread in the manner of a spherical wave; therefore one cannot make measurements based on the Friis transmission formula [22]. This means that absolute gain measurement cannot be made in a tapered chamber.

Compact Range—The distance criteria for the spacing between the source and test antennas are based on a maximum allowable deviation of the incident field's phase front over the test aperture from that of a plane wave. It is explicitly assumed that the incident phase front is spherical. Thus, for a given-size test aperture the spacing between the source and test antennas must be increased in order to decrease the phase deviation over the test aperture.

A large precision paraboloidal reflector [3, 6] or lens [23] provides a means of circumventing the distance criteria since either device can be used to convert a spherical phase front into a planar one as shown in Fig. 18. Because such a device is necessarily finite in size, its planar phase front exists only in a region near its aperture. This means that the test antenna will have to be located near the aperture of the reflector or lens if it is to be illuminated with a wave that exhibits a planar phase front. Thus, by using this technique a relatively short antenna test range can be constructed, hence the name compact range.

Practical considerations limit the usefulness of lenses for this purpose. One reason is that each lens exhibits an inherent mismatch at its surface, giving rise to unwanted reflections. The plano-convex dielectric lens shown in Fig. 18a is a case in point. To minimize this effect, a low-density dielectric must be used which results in a very thick, bulky structure that is difficult to manage. For these reasons, lenses are rarely employed for compact ranges, although they have found use, in a few cases, for anechoic chambers that are too short for a given measurement.

There are a number of factors that affect the performance of a compact range and which must be taken into account in the design and installation. They are:

> aperture blockage
> reflector surface accuracy
> reflector edge diffraction
> reflector focal length
> direct coupling between the feed and test antenna
> feed radiation pattern
> room reflections

Aperture blockage is reduced by using an offset feed as shown in Fig. 18a. The reflector must have a precision surface. It should be emphasized that the surface accuracy requirement for a compact-range reflector is much more stringent than for a typical reflector antenna. Surface deviations from the ideal paraboloid must be less than a small fraction of a wavelength over any square wavelength of surface in order to ensure a sufficiently planar phase front at the test region. The condition of the surface is the controlling factor in establishing the upper limit on the frequency at which the range can be used.

The lower frequency limit is dictated by the reflector size and the edge diffraction. Serrations over the edges of the reflector [24] or rolled edges [25] provide means of controlling edge diffraction effects. Serrations are designed in

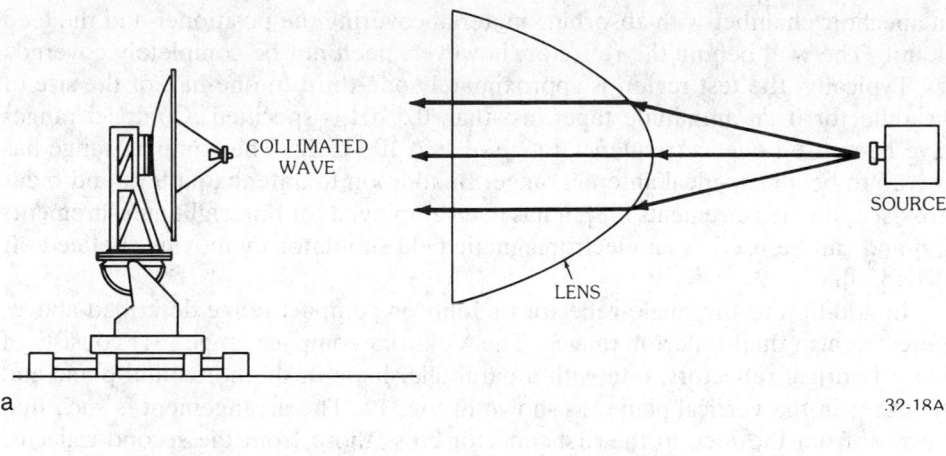

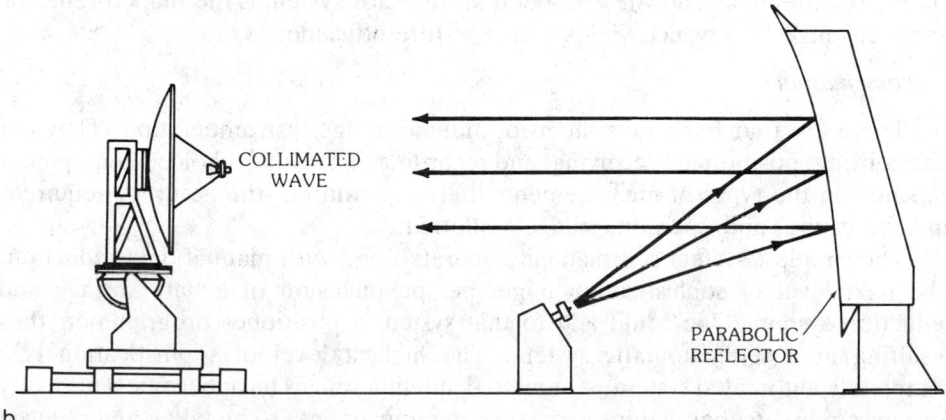

Fig. 18. Prime focus compact ranges. (*a*) Lens-type compact range. (*b*) Single-reflector or Johnson compact range.

such a manner that the cone of diffracted rays from the various edges of the serrations is in a direction away from the test antenna and feed. The appropriate shape of the rolled edge is elliptical; however, this shape should be blended with the parabolic shape of the reflector to minimize any discontinuities in the shape of the resulting surface [26]. In either case the effect of the two approaches is to direct energy from the feed incident at the edges away from the test antenna. This means that the absorbing material over the anechoic chamber in which the range is located must be designed to adequately absorb the redirected energy. Either approach when properly designed will yield excellent results for antenna pattern measurements. The rolled edge seems to yield a smoother amplitude variation over the test zone, which is required for low radar cross-section measurements [27].

The focal length of the reflector is chosen to be relatively long in order to reduce depolarization effects of the curved reflector surface. This also allows the feed to be located below the test antenna so that direct coupling between them can be controlled using absorbing material. The feed antenna must be designed to minimize side lobe radiation at wide angles. Finally, the range should be located in an anechoic chamber with absorbing material covering the positioner and the feed mount. The wall behind the reflector, however, need not be completely covered.

Typically, the test region is approximately one-third to one-half of the size of the reflector if an amplitude taper less than 0.5 dB is specified. Compact ranges have been used over a frequency range of 1 to 100 GHz. The compact range has proved to be a near-ideal antenna range. In addition to antenna pattern and radar cross-section measurements [28], it has been employed for boresight measurements [29] and can be used as an electromagnetic field simulator by moving the feed off focus [30].

In addition to the single-reflector or Johnson compact range described above, there are also dual-reflector ranges. The Vokurka compact range [31] consists of two cylindrical reflectors, one with a parabolic shape in the horizontal plane and the other in the vertical plane, as shown in Fig. 19. The arrangement is such that the rays from the feed to the first reflector cross those from the second reflector to the test antenna as shown. Another dual-reflector system is the shaped-reflector compact range [32], which yields high aperture efficiency.

Instrumentation

There are four basic subsystems of antenna range instrumentation. They are transmitting, positioning, receiving, and recording. The actual choice of equipment depends on the types of measurements that are required, the accuracy required, and the capital and operating budgets allotted.

The simplest system is a manually operated one with manual data reduction. The next level of sophistication might be the inclusion of a data storage and reduction system. One could add to that system a positioner programmer, thus resulting in a semiautomatic system. The highest level of sophistication is a completely automated system. Automated antenna ranges have become a necessity for measurements that require very large amounts of data to be taken and reduced efficiently and with a relatively high degree of accuracy. Simplified block diagrams of the manual and automated systems are given in Fig. 20.

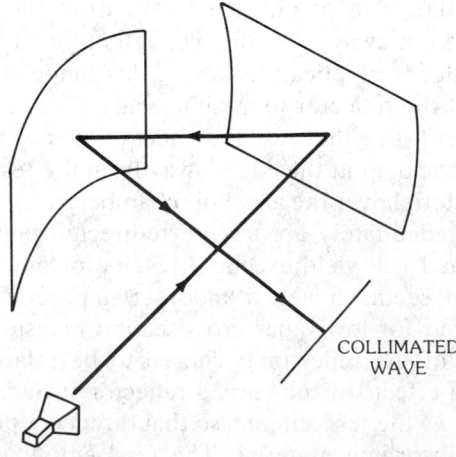

COLLIMATED
WAVE

Fig. 19. The Vokurka compact range.

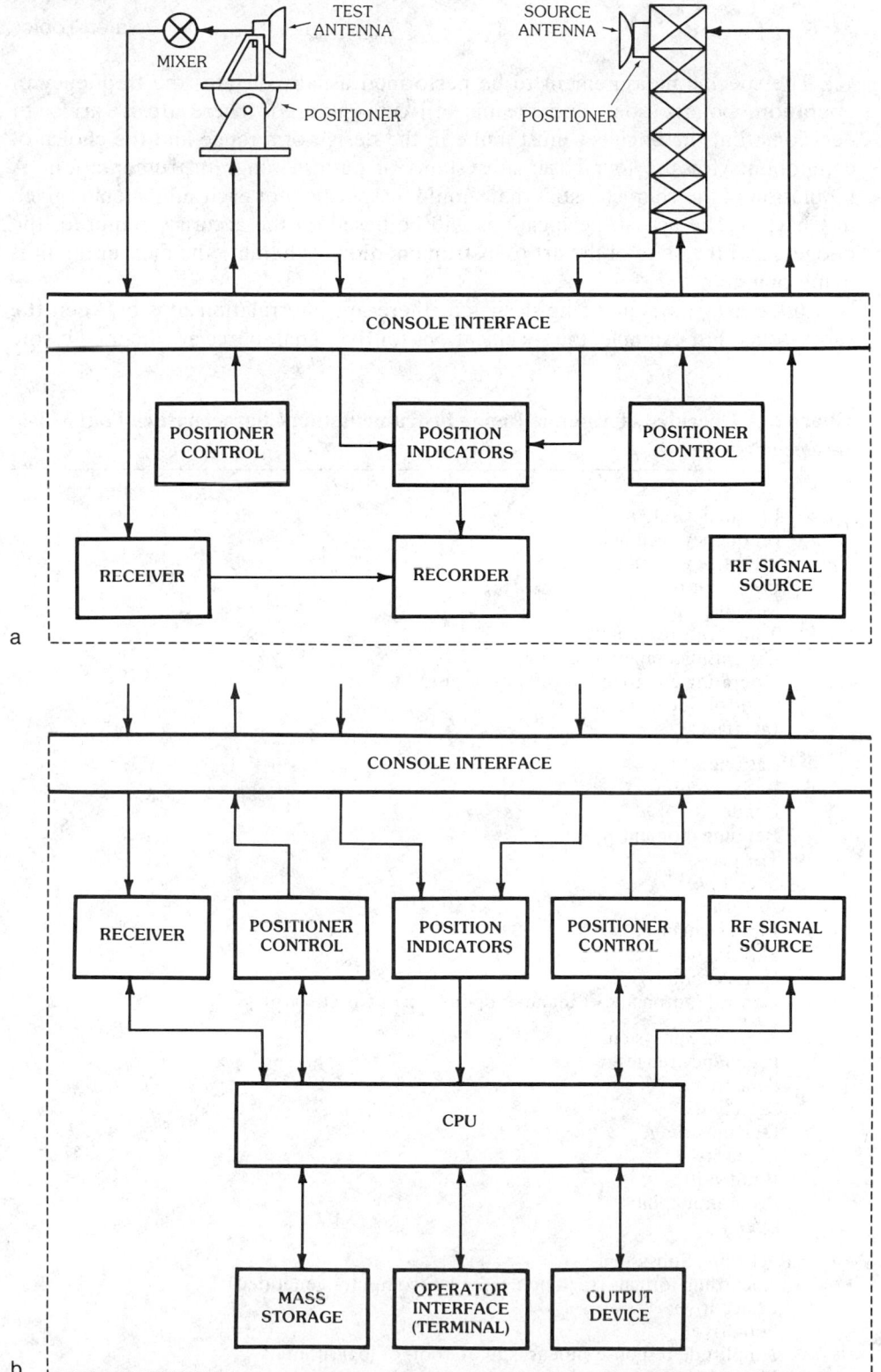

Fig. 20. Simplified block diagrams of manual and automated antenna instrumentation. (*a*) Manual system. (*b*) Automated system.

The specific measurement to be performed usually dictates the frequency of operation, polarizations, and the pattern cuts required. There are a variety of decisions that the engineer must make in the design of a range and the choice of components for the four basic subsystems of pattern range instrumentation. A tabulation of the characteristics that should be specified for each subsystem is given in Chart 2. The actual specifications will be based on the accuracy required, the budget, and the state of the art of instrumentation at the time the measurement is being planned.

Since it is a system being designed, there are interrelationships between the subsystems. For example, the specifications for the signal source are dictated by the

Chart 2. A Checklist of Antenna Range Instrumentation Characteristics That Must Be Specified

1. Signal Sources
    Frequency range
    Frequency accuracy
    Frequency stability
    Power output
    Modulation
    Temperature stability
    Operating temperature range
    Operating environment (indoor or outdoor)
    Control (local or remote)
    Interfaces

2. Positioners
    Axes required
    Load
    Bending moment
    Torque
    Speed
    Accuracy
    Limit switches
    Slip rings
    Rotary joints
    Control (automatic or manual; open loop or closed loop)

3. Receiving Subsystem
    Frequency range
    Sensitivity
    Selectivity
    Dynamic range
    Linearity
    Bandwidth
    Programmability
    Speed

4. Recording Subsystem
    Recording formats (rectangular, polar, computer generated)
    Chart drive
    Pen drive
    Amplitude response (linear, square root, or logarithmic)
    Phase response

choice of the receiver. If a simple detector-amplifier is employed for the receiver, then modulation would be required. Compared with the use of a superheterodyne receiver, higher output power would be required, but there could be a less stringent requirement for frequency stability.

Where high accuracy and precision are required, it is usually necessary to develop an error budget. It is at this point that trade-offs can be made in order to design a realistic experiment.

In addition to the basic elements of an antenna range instrumentation system, there is auxiliary equipment that must be available. This includes standard-gain antennas and range evaluation and alignment instrumentation.

Range Evaluation

An important part of antenna range measurements is the alignment and evaluation of the range. The first step in the procedure is that of alignment of the axes of the positioner. The vertical axis (*0* axis for the elevation-over-azimuth and the roll-over-azimuth positioners) can be aligned using a clinometer or a precision level. Also, a plumb bob can be used for the roll-over-azimuth positioner. The horizontal axis is usually aligned optically [33].

The source antenna must also be aligned. A convenient method of accomplishing this is by the use of a field probe ([5], pp. 14.53–14.54). A typical field probe is shown in Fig. 21. The probe is first mounted on the test antenna positioner in such a manner that it can be oriented for a horizontal pattern cut in a plane perpendicular to the range axis. The amplitude pattern obtained should have a clearly defined maximum. Equilevel points on either side of the maximum are located

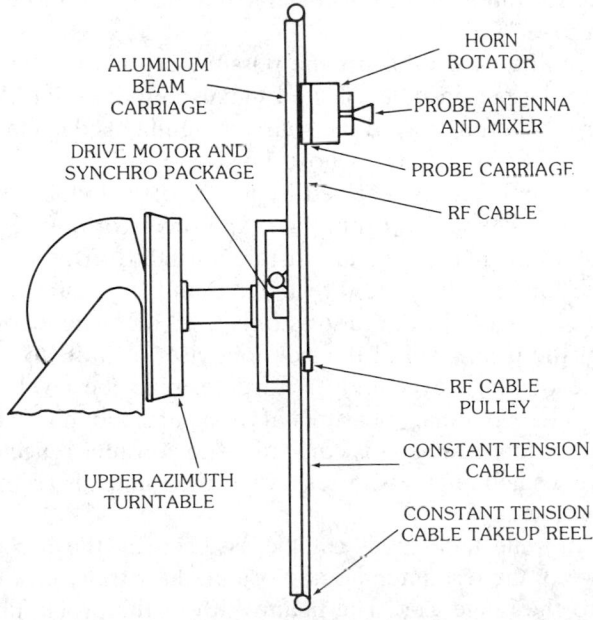

Fig. 21. A field-probe mechanism. (*After ANSI/IEEE [2], © 1979 IEEE; after Hollis, Lyon, and Clayton [5], © 1970 Scientific-Atlanta*)

for reference. The orientation of the source antenna is adjusted until the geometrical mean of these points is located on the range axis. The source antenna is then rotated 180° and another horizontal cut is made. If the pattern obtained is not symmetrical about the range axis, then the source antenna must be adjusted with respect to the polarization positioner until the pattern is symmetrical for both cases. In the case of a reflector antenna being used as a source antenna, the position of the feed might need adjusting.

For elevated ranges the procedure is repeated for the vertical direction. Ground-reflection ranges utilize the reflected signal from the range surface to obtain the vertical amplitude distribution. For this type of range the height of the source antenna is adjusted for symmetry in the vertical direction.

If there is a severe problem with range reflections, it may be necessary to take remedial measures prior to the completion of the source antenna alignment.

Once alignment has been accomplished, the state of the incident field over the test region must be determined. The principal reasons for the incident field to deviate from that predicted from an idealized range are reflections from the range surface, surrounding objects, the positioner, and the cables used for the test antenna. Sometimes signals from external sources also pose a problem.

To illustrate how a reflected signal can affect a radiation pattern, consider the situation depicted in Fig. 22. There is a direct signal from the source antenna and a reflected one from an angle measured from the range axis. The phasor sum of the two signals depends not only on their amplitudes but also their relative phase. The in-phase and out-of-phase conditions give the maximum and minimum levels. A plot of the errors due to the in-phase and out-of-phase conditions as a function of the decibel ratio E_R/E_D is given in Fig. 23. The plus and minus errors are essentially equal for ratios of -25 dB or less and are given in the logarithmic plot for ratios down to -75 dB.

The graph can be used to assess the possible errors in antenna pattern measurements due to the presence of reflected signals. Suppose that the main beam of the test antenna is pointing toward the source antenna, as depicted in Fig. 22a. At the same time a reflected signal, whose level is 35 dB below the direct signal received by the main beam, is received by a -25-dB side lobe. For this case the decibel ratio $E_R/E_D = -60$ dB. From the graph the error bounds in the measurement of the peak of the main beam are just under ± 0.01 dB. On the other hand, if the -25-dB side lobe receives the direct signal and the main beam the reflected signal, as shown in Fig. 22b, the ratio is only -10 dB. The error bounds on the measurement of the peak level of the side lobe are -3.3 dB to $+2.4$ dB.

The purpose of range evaluation is to determine the levels of the reflected signals and to take remedial measures if they are too high for the accuracy required. For an outdoor range it is convenient to consider reflected signals which arrive at relatively small angles with respect to the range axis separate from these which arrive at wide angles.

The near-axis reflected signals can be assessed by the use of a field probe mounted in place of the test antennas and oriented for transverse cuts in the plane perpendicular to the range axis. The beamwidth of the probe antenna should be broad enough to receive signals reflected from the specular region of the range surface.

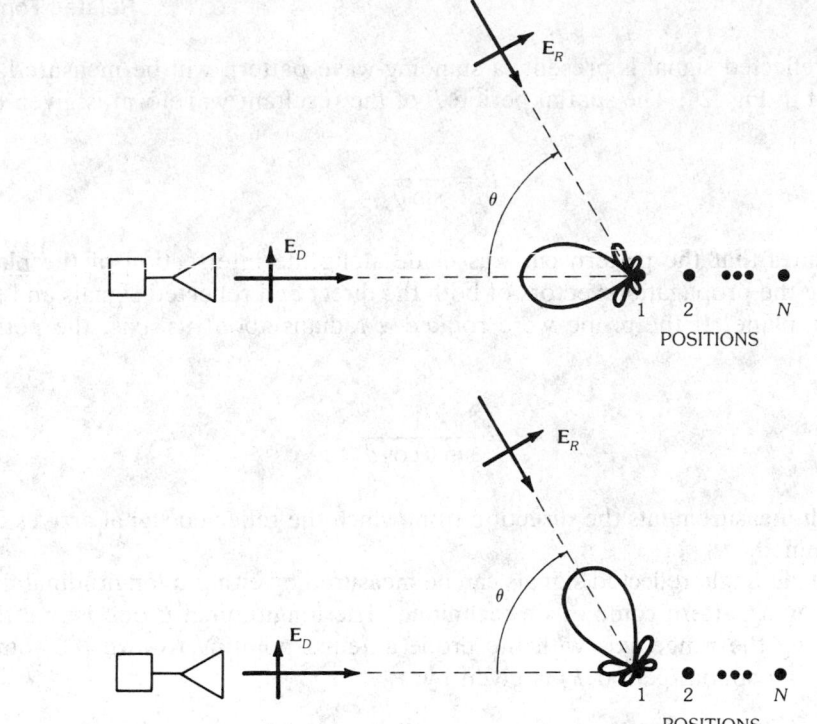

Fig. 22. Illustration of how the side lobe level of the test antenna is affected during antenna-pattern comparison measurement. (*a*) Test antenna pointing toward source. (*b*) Side lobe pointing toward source. (*After ANSI/IEEE [2], © 1979 IEEE*)

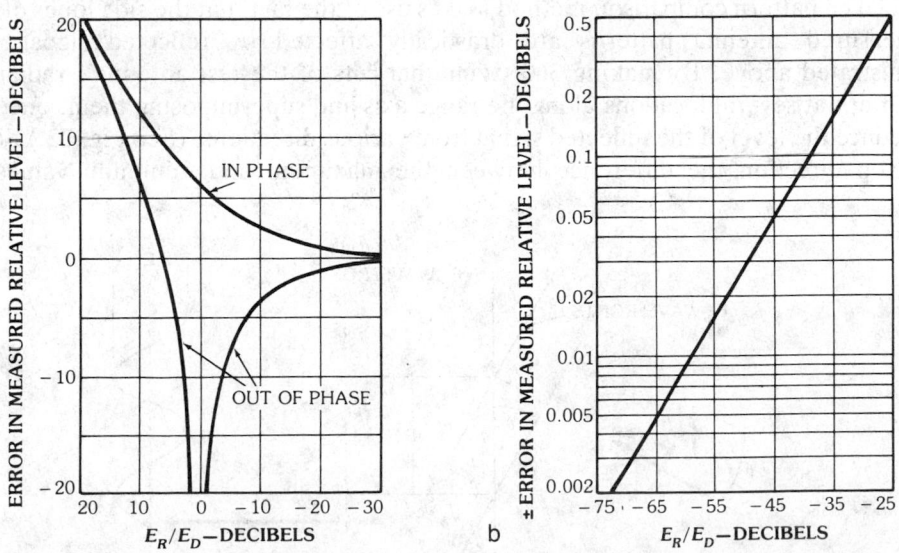

Fig. 23. Possible error in measured relative pattern level due to coherent extraneous signals. (*a*) Linear scales are employed for signal ratios of +20 to −30 dB. (*b*) Plus-or-minus errors are essentially equal for ratios of −25 dB or less, as indicated in this logarithmic plot for ratios down to −75 dB. (*After ANSI/IEEE [2], © 1979 IEEE; after Hollis, Lyon, and Clayton [5], © 1970 Scientific-Atlanta*)

If a reflected signal is present, a standing-wave pattern will be measured, as illustrated in Fig. 24. The spatial period P of the resultant waveform is given by

$$P = \frac{\lambda}{\sin \theta} \tag{8}$$

This assumes that the pattern cut was made along the intersection of the plane containing the propagation vectors of both the direct and reflected signals and the transverse plane. If the probe were rotated α radians about its axis, the period becomes

$$P = \frac{\lambda}{\sin \theta \cos \alpha} \tag{9}$$

From such measurements the direction from which the reflected signal arrives can be determined.

The wide-angle reflected signals can be measured by either a longitudinal field probe or by a pattern comparison technique. The longitudinal probe is one that moves along the range axis with the probe antenna pointing toward the source antenna. The spatial period P_l is given by

$$P_l = \frac{\lambda}{2 \sin^2(\theta/2)} \tag{10}$$

The direction from which the reflected signal arrives can be deduced from this equation.

The pattern comparison method makes use of the fact that the side lobes of the measured antenna patterns are drastically affected by reflected signals, as illustrated above. By making 360° azimuthal cuts of the test antenna's radiation pattern at several locations along the range axis and superimposing them, one can deduce the level of the reflected signal from various directions. (See Fig. 22.) For a given direction the difference between the maximum and minimum values in

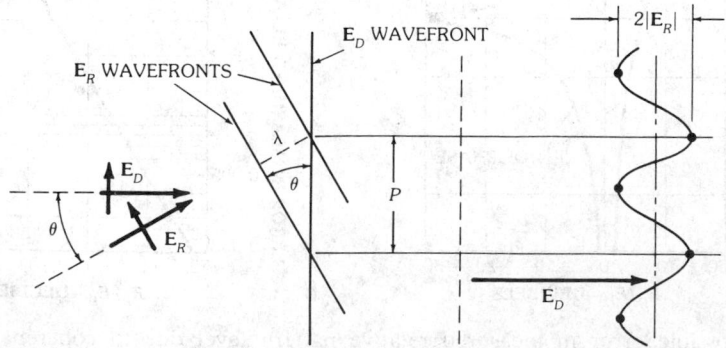

Fig. 24. Spatial interference pattern due to a reflected wave. (*After ANSI/IEEE [2]*, © *1979 IEEE; after Hollis, Lyon, and Clayton [5]*, © *1970 Scientific-Atlanta*)

decibels can be used to determine the level of the reflected signal from the graph of Fig. 25.* To illustrate this, consider the data shown in Fig. 26. Note that there is a variation of approximately 12 dB in the patterns at an azimuth angle of 120°. If the direct and major reflected waves were both received on the side lobe, this would mean that the reflected wave is 4.5 dB below the level of the direct wave. However, the reflected wave was received on the main beam and the direct wave on the side lobe. Since the variation occurred approximately 30 dB below the peak of the main beam, the reflected wave is at least −34.5 dB relative to the level of the direct wave.

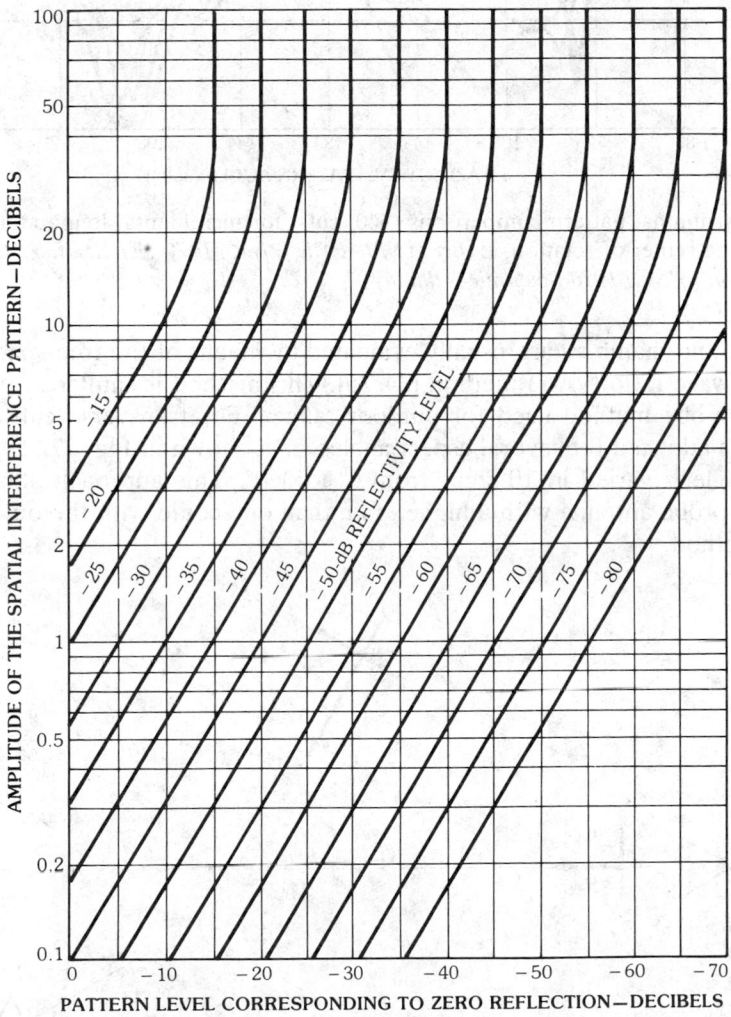

Fig. 25. Amplitude of spatial interference pattern for a given reflectivity level and antenna-pattern level. (*After ANSI/IEEE [2], © 1979 IEEE*)

*The curves plotted as straight lines below the 5.7-dB ordinate value are actually very slightly concave upward. All curves are correct as plotted above the 5.7-dB ordinate value.

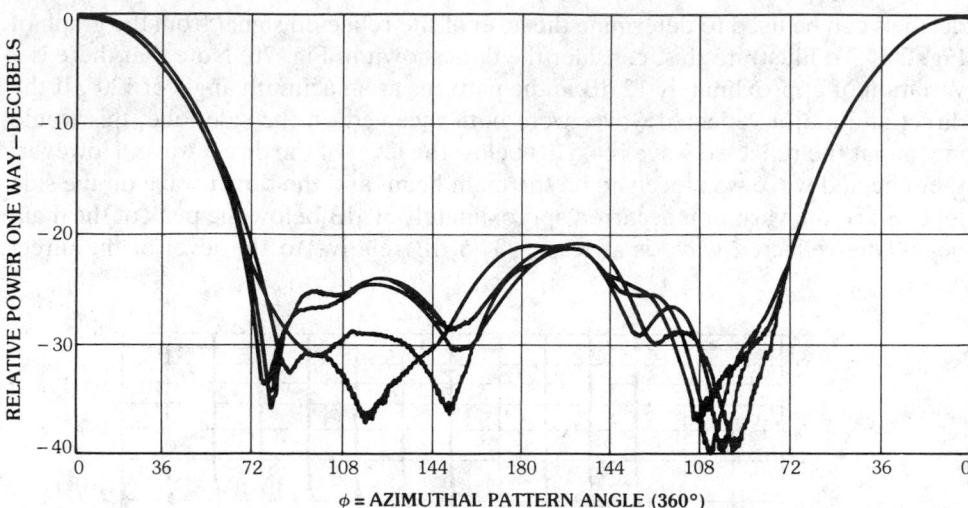

Fig. 26. Azimuthal pattern comparisons (360° cuts) for incremental longitudinal displacements of the center of rotation. (*After ANSI/IEEE [2], © 1979 IEEE; after Hollis, Lyon, and Clayton [5], © 1970 Scientific-Atlanta*)

Anechoic chambers are usually evaluated by means of the free-space voltage-standing-wave-ratio (vswr) method [34, 35]. This method is similar in principle to the field-probe method used for outdoor ranges. For transverse cuts the probe antenna is adjusted for several aspect angles ϕ, as shown in Fig. 27. Typically the aspect angle is varied in 10 steps from 0° to 360°. This approach allows one to choose a probe antenna with a higher gain than one could with the outdoor field-probe method.

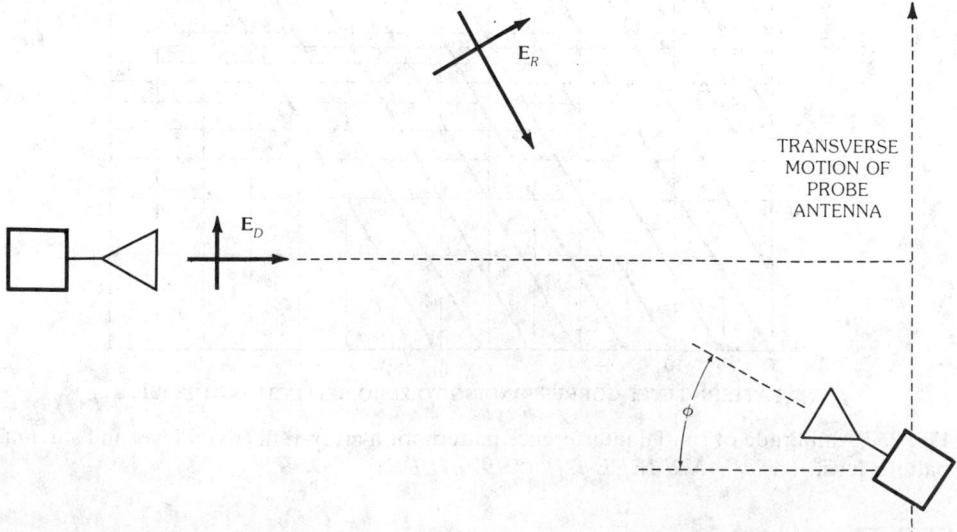

Fig. 27. Geometry for the free-space vswr method. (*After ANSI/IEEE [2], © 1979 IEEE*)

For a complete evaluation a horizontal vertical and longitudinal cut is required for each orientation of the probe antenna. Sometimes cuts in other directions are required.

Because of the six surfaces of the chamber it is desirable that the motion of the probe antenna be controlled at the base of the positioner. For example, the positioner can be mounted on rails at its base.

The antenna-pattern comparison method can also be applied; however, there is the possibility that the maximum reflectivity level will not be detected.

4. Measurement of Radiation Characteristics

The requirements of the system in which an antenna is to be incorporated usually dictate the extent of the measurements. In some instances only the principal-plane amplitude patterns and, perhaps, the peak gain of the antenna are all that are needed. Often, however, it is necessary to measure both the copolar and cross-polar amplitude patterns over the entire radiation sphere of the antenna. Sometimes a separate measurement of the antenna's polarization properties is also required. Both amplitude and phase patterns are needed for certain types of antennas, such as feeds for reflector and lens antennas and for antennas used as probes in near-field measurements where probe correction is required. Most systems require a knowledge of the gain of an antenna. Generally, for those cases where modeling techniques are employed, the gain can only be obtained from the directivity, which can be computed from the measured amplitude patterns and an estimation of the antenna's radiation efficiency. Conversely, there are cases where both the gain and directivity can be measured. The radiation efficiency of the antenna can then be determined from these results since it is the ratio of the gain to directivity.

Special measurements are usually required for antennas employed in radar systems. These include boresight and, for those cases where the antenna is enclosed within a radome, radome measurements. These measurements are beyond the scope of this chapter.

Amplitude Patterns and Directivity Measurements

The decisions that must be made before one begins a measurement program include which pattern cuts are required. As discussed in Section 2, this consideration influences the manner in which the antenna under test is mounted on the positioner. Other decisions include the frequencies of operation and the increments of the coordinate chosen as a parameter (ϕ increments for θ-cuts and θ increments for ϕ-cuts) and the mode of display.

Once the antenna is mounted on the positioner, it must be carefully aligned as described in Section 3, under "Range Evaluation." A few exploratory patterns should be measured before the production run begins. For example, if the general characteristics of the antenna's pattern are known, then a few cuts through the main beam will reveal whether or not there are any obvious problems. One might also perform a pattern comparison measurement using the antenna under test (see Section 3, under "Range Evaluation") to determine if the range reflections are at the same level as they were when the range was last evaluated.

The antenna's measured radiation patterns give the spatial distribution of relative power density contained in a specified polarization that would be obtained if the antenna operated in the transmit mode. The directivity of the antenna can be computed from the measured patterns provided that they are measured over the entire radiation sphere.

The directivity is usually determined for the direction of the peak value of the amplitude pattern (θ', ϕ'). From the definition of directivity, the maximum directivity, $D_m(\theta', \phi')$ can be written as

$$D_m(\theta', \phi') = \frac{\Phi_m(\theta', \phi')}{1/4\pi \int_0^\pi \left[\int_0^{2\pi} \Phi(\theta, \phi)d\phi \right] \sin\theta \, d\theta} \tag{11}$$

where $\Phi(\theta, \phi)$ is the radiation intensity as a function of direction, (θ, ϕ), and $\Phi_m(\theta', \phi')$ is its maximum value. Since only relative values of the radiation intensity are measured, it is necessary to normalize the radiation intensity by dividing by the value of the radiation intensity which corresponds to that used for the measured radiation patterns.

The radiation intensity can be decomposed into two components corresponding to the two orthogonal polarizations employed in the measurement of the radiation patterns. These are usually the copolarization and the cross polarization, in which case the normalization will be with respect to the peak radiation intensity of the copolarization.

This decomposition of the radiation intensity allows the directivity to be written as the sum of two partial directivities corresponding to the two orthogonal polarizations, that is,

$$\begin{aligned} D_m(\theta', \phi') &= D_1(\theta', \phi') + D_2(\theta', \phi') \\ &= \frac{\overline{\Phi}_1(\theta', \phi')}{1/4\pi(\overline{P}_{1t} + \overline{P}_{2t})} + \frac{\overline{\Phi}_2(\theta', \phi')}{1/4\pi(\overline{P}_{1t} + \overline{P}_{2t})} \end{aligned} \tag{12}$$

The subscripts 1 and 2 denote the two orthogonal polarizations; the bars indicate normalization, and $\overline{P}_{1t}$ and $\overline{P}_{2t}$ are given by

$$\overline{P}_{1t} = \int_0^\pi \left[\int_0^{2\pi} \overline{\Phi}_1(\theta, \phi)d\phi \right] \sin\theta \, d\theta \tag{13}$$

and

$$\overline{P}_{2t} = \int_0^\pi \left[\int_0^{2\pi} \overline{\Phi}_2(\theta, \phi)d\phi \right] \sin\theta \, d\theta \tag{14}$$

Note that both $\overline{P}_{1t}$ and $\overline{P}_{2t}$ are required in order to compute either $D_1(\theta', \phi')$ or $D_2(\theta', \phi')$.

Since the polarization of an antenna changes with direction, it is necessary to determine the partial directivities with respect to two specified orthogonal polarizations and then add them to obtain the (total) directivity of the antenna. It is desirable that the copolarization and cross polarization be the specified polarizations. If polarizations 1 and 2 are the copolarization and cross polarization, respectively, then in (12), $\overline{\Phi}(\theta', \phi') = 1$. Also, the normalized radiation intensities $\overline{\Phi}_1(\theta, \phi)$ and $\overline{\Phi}_2(\theta, \phi)$ are equal to the relative power patterns that are measured.

If the radiation patterns are measured using ϕ-cuts with the θ increments chosen to be π/M rad, the partial directivities are given by

$$D_1(\theta', \phi') = \frac{4M}{\sum\limits_{i=1}^{M} \left\{ \int_0^{2\pi} [\overline{\Phi}_1(\theta_i, \phi) + \overline{\Phi}_2(\theta_i, \phi)] d\phi \right\} \sin \theta_i} \tag{15}$$

$$D_2(\theta', \phi') = \frac{4M \overline{\Phi}_2(\theta', \phi')}{\sum\limits_{i=1}^{M} \left\{ \int_0^{2\pi} [\overline{\Phi}_1(\theta_i, \phi) + \overline{\Phi}_2(\theta_i, \phi)] d\phi \right\} \sin \theta_i} \tag{16}$$

where $\theta_i = i\pi/M$ rad.

If θ-cuts are employed with the ϕ increments chosen to be $2\pi/N$ rad, the partial directivities are given by

$$D_1(\theta', \phi') = \frac{2N}{\sum\limits_{j=1}^{N} \int_0^{\pi} [\overline{\Phi}_1(\theta, \phi_j) + \overline{\Phi}_2(\theta, \phi_j)] \sin \theta \, d\theta} \tag{17}$$

$$D_2(\theta', \phi') = \frac{2N \overline{\Phi}_2(\theta', \phi')}{\sum\limits_{j=1}^{N} \int_0^{\pi} [\overline{\Phi}_1(\theta, \phi_j) + \overline{\Phi}_2(\theta, \phi_j)] \sin \theta \, d\theta} \tag{18}$$

In this case each $\overline{\Phi}(\theta, \phi_j)$ must be multiplied by $\sin \theta$ during the process of integration.

These equations can be evaluated numerically using a computer [36]. On-line computers incorporated into the measurement system can be used to yield essentially real-time computations. Alternatively, the data can be stored and entered into a computer at a later time.

For pencil-beam antennas an estimate of the directivity can be obtained from the principal-plane patterns since it is inversely proportional to the *beam area* or *areal beamwidth*. The beam area is the product of the two half-power beamwidths, θ_{HP} and ϕ_{HP}. Therefore one can write

$$D_m(\theta', \phi') = \frac{K}{\theta_{HP} \phi_{HP}} \tag{19}$$

The value of K has been given as being between 25 000 and 41 253. For most cases a value of 32 400 yields an acceptable estimate of directivity [37, 38].

Antenna Gain Measurements

Antenna gain measurements are generally classified as absolute gain measurements or gain-transfer measurements ([2], pp. 94–112). Absolute gain measurements require no *a priori* knowledge of the gains of the antennas being tested, whereas the gain-transfer method requires the use of a gain standard whose gain is accurately known.

Absolute gain measurements have found their greatest use in the calibration of gain standards. The gain-transfer method is the more commonly employed method of measuring antenna gain.

Absolute Gain Measurements—Absolute gain measurements are based on the Friis transmission formula for coupling between two antennas. To illustrate this, consider two antennas, one of which is transmitting and the other receiving. They are separated by R meters as depicted in Fig. 28.

The power received by antenna B, P_R, is given by the Friis transmission formula

$$P_R = P_0 G_A G_B \left(\frac{\lambda}{4\pi R}\right)^2 q_B p \qquad (20)$$

where G_A and G_B are the gains of antennas A and B, respectively, P_0 is the power accepted by antenna A from its generator, q_B is the impedance mismatch factor evaluated at a reference plane between antenna B and its load, and p is the *polarization efficiency.**

This formulation implicitly contains the assumptions that free-space conditions prevail and that the separation R is such that the incident field produced by antenna A over the aperture of antenna B is planar.

The impedance mismatch factor is given by

$$q_B = \frac{(1 - |\Gamma_B|^2)(1 - |\Gamma_L|^2)}{|1 - \Gamma_B \Gamma_L|^2} \qquad (21)$$

where Γ_B and Γ_L are the reflection coefficients of antenna B and the load evaluated at a common reference plane. Note that both the amplitude and phase of Γ_B and Γ_L are required in order to evaluate q_B.

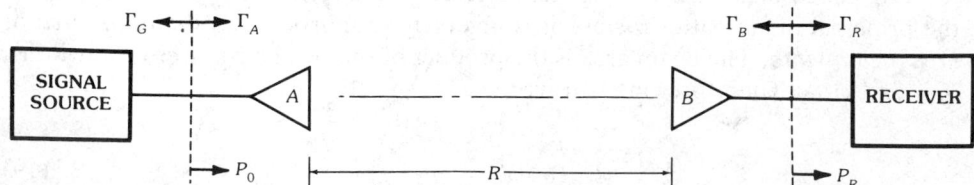

Fig. 28. Two-antenna system illustrating the Friis transmission formula.

---

*The polarization efficiency p is also called the *polarization mismatch factor*.

For measurements performed on a well-designed free-space range, one can usually assume that there are no depolarization effects. In that case the polarization efficiency p can be computed from a knowledge of the polarizations of the two antennas ([2], pp. 76–85). Various formulas for p are given in the following subsection on polarization measurements.

The accepted power P_0 cannot be measured directly. If one measures the available power, P_A, of the generator by conjugately matching a power meter to it, then

$$P_0 = P_A q_A \tag{22}$$

where

$$q_A = \frac{(1 - |\Gamma_G|^2)(1 - |\Gamma_A|^2)}{|1 - \Gamma_G\Gamma_A|^2} \tag{23}$$

Here Γ_G and Γ_A are the complex reflection coefficients of the generator and antenna A, respectively, measured at the same reference plane as shown in Fig. 28. Often a power meter which is matched to the transmission line is used to measure the power without a tuner to conjugately match it to the generator. This measurement will yield the line-matched power P_M and

$$P_0 = P_M q_A' \tag{24}$$

in which

$$q_A' = \frac{1 - |\Gamma_A|^2}{|1 - \Gamma_A\Gamma_G|^2} \tag{25}$$

In either case, if the generator were matched to the transmission line, the impedance mismatch factor becomes

$$q_A'' = 1 - |\Gamma_A|^2 \tag{26}$$

and

$$P_0 = P_A q_A'' \tag{27}$$

Here the power P_A is also known as the incident power at the antenna.

Since (22) contains all of the special cases, the Friis transmission formula will be written in terms of the available power P_A, that is,

$$P_R = P_A G_A G_B \left(\frac{\lambda}{4\pi R}\right)^2 q_A q_B p \tag{28}$$

Two-Antenna Method—For what follows, it is convenient to express (28) in logarithmic form which, when rearranged, becomes

$$(G_A)_{\mathrm{dB}} + (G_B)_{\mathrm{dB}} = 20\log\left(\frac{\lambda}{4\pi R}\right) - 10\log\left(\frac{P_A}{P_R}\right) - 10\log C \tag{29}$$

where

$$C = q_A q_B p \tag{30}$$

The expression C can be considered a correction factor which requires the measurement of Γ_G, Γ_A, Γ_B, Γ_R, and p. Of course, there will remain a residual measurement error in the measurement of these quantities.

If the gain of one of the antennas is known, then the gain of the other antenna can be determined from a measurement of λ, R, C, P_A, and P_R. If the gain of neither antenna is known but $G_A = G_B$, then the gains are given by

$$(G_A)_{\mathrm{dB}} = (G_B)_{\mathrm{dB}}$$
$$= 1/2\left[20\log\left(\frac{4\pi R}{\lambda}\right) - 10\log\left(\frac{P_A}{P_R}\right) - 10\log C\right] \tag{31}$$

This method is called the two-antenna method since only two antennas are required.

Three-Antenna Method—If the two antennas are not identical, or nearly so, and neither gain is known, the two-antenna method fails. In this case a third antenna is required in order to determine the gain of any one of the antennas. Since three antennas are required, the procedure is called the three-antenna method. Three sets of measurements are performed using the three possible combinations of antennas A, B, and C. The resulting set of equations is

$$(G_A)_{\mathrm{dB}} + (G_B)_{\mathrm{dB}} = 20\log\left(\frac{4\pi R}{\lambda}\right) - \left[10\log\left(\frac{P_A}{P_R}\right) + 10\log C\right]_{AB}$$
$$(G_A)_{\mathrm{dB}} + (G_C)_{\mathrm{dB}} = 20\log\left(\frac{4\pi R}{\lambda}\right) - \left[10\log\left(\frac{P_A}{P_R}\right) + 10\log C\right]_{AC} \tag{32}$$
$$(G_B)_{\mathrm{dB}} + (G_C)_{\mathrm{dB}} = 20\log\left(\frac{4\pi R}{\lambda}\right) - \left[10\log\left(\frac{P_A}{P_R}\right) + 10\log C\right]_{BC}$$

where the subscripts indicate the specific combinations of two antennas. The simultaneous solution of these equations gives the gains of all three antennas.

Insertion-Loss Method—There is an alternative method in which the insertion loss between the reference planes shown in Fig. 28 is measured [39]. This can be accomplished by first removing the antennas and connecting the load directly to the generator as shown in Fig. 29. The power transferred to the load, P_R^i, is measured. Secondly, the antennas are inserted and separated to a distance of R as before and the power transferred to the load for this case, P_R^f, is measured.

The sum of the two antennas' gains in decibels can be shown to be given by

$$(G_A)_{\mathrm{dB}} + (G_B)_{\mathrm{dB}} = 20\log\left(\frac{4\pi R}{\lambda}\right) + 10\log\left(\frac{P_R^f}{P_R^i}\right) + 10\log C'' \tag{33}$$

where

$$C'' = \frac{|1 - \Gamma_G\Gamma_A|^2|1 - \Gamma_B\Gamma_R|^2}{|1 - \Gamma_G\Gamma_R|^2(1 - |\Gamma_A|^2)(1 - |\Gamma_B|^2)} + p \tag{34}$$

Again, if one of the gains is known, the other can be computed using (33). If neither gain is known but G_A is known to equal G_B, then the gain of each antenna is equal to one-half the right side of (33). Three antennas are required as before if the gains are not equal and neither G_A or G_B is known.

One of the advantages of the insertion-loss method is that the measurement of the ratio of power (P_R^f/P_R^i) can be accomplished with the use of a precision attenuator. The attenuator can be placed in the line at either the generator or load end. It is adjusted so that the signal at the load is the same for both measurements. The difference in the settings for the direct measurement and measurement with antennas inserted is precisely the ratio of power in decibels.

The basic instrumentation required for the two- and three-antenna methods is shown in Fig. 30. The small squares at each device are included as a reminder of the need for connectors and adapters, which are potential sources of impedance mismatch. Furthermore, the devices themselves, as well as the antennas under test, are not perfectly matched at any given frequency within the band of operation; therefore tuners are often used. They must be used with care, however, since the resulting network becomes very narrow band. Thus a small amount of detuning will cause a great change in input impedance.

The transmitting and receiving systems should be bench tested and calibrated prior to assembly on the antenna range. Also, great care must be exercised in handling the interconnecting transmission lines. These must also be calibrated in order to account for losses and impedance mismatches. Wherever possible, transmission lines should be eliminated because they introduce calibration errors and they deteriorate or may be damaged through use. For example, the directional coupler might be connected directly to the transmitting antenna. Also, if power meters are used to measure the powers at the transmitting and receiving test ports, it may be possible to connect their power sensors directly to the directional coupler/attenuator and to the transmitting antenna, respectively.

The measurement system may be automated provided that the instruments employed are computer controllable [40]. Fig. 31 illustrates the basic instrumentation required. The system must be calibrated as in the manual case. With the automated system the calibration data are stored in the computer so that corrections can be made on the data. This type of system can easily be programmed to measure the gain over a band of frequencies.

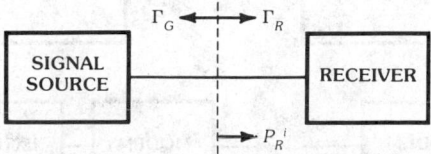

Fig. 29. An illustration of the direct measurement of power before the antennas are inserted in the system.

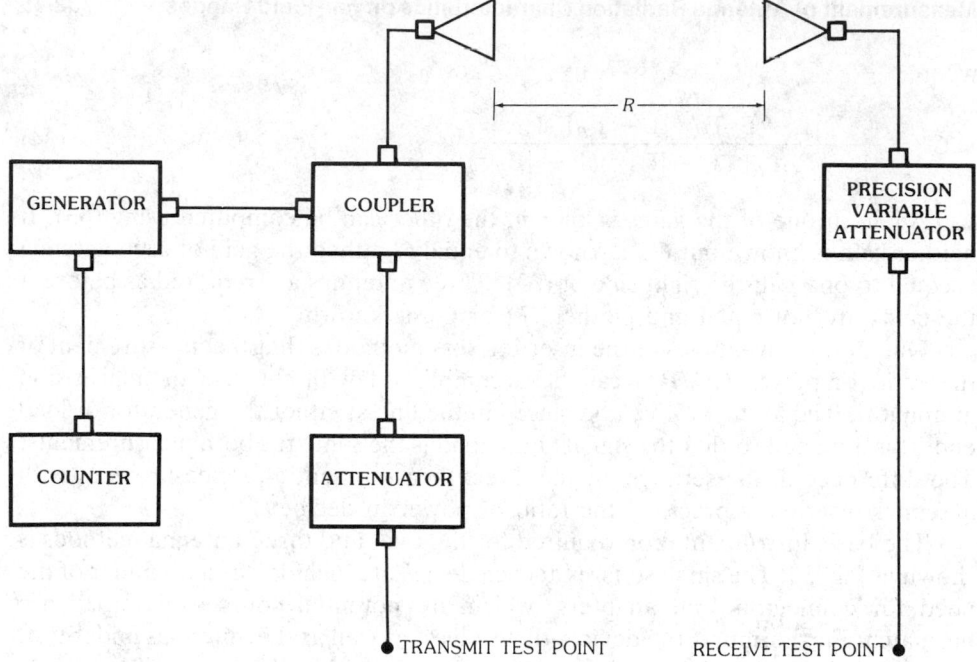

Fig. 30. Typical instrumentation for two-antenna and three-antenna methods of power-gain measurement.

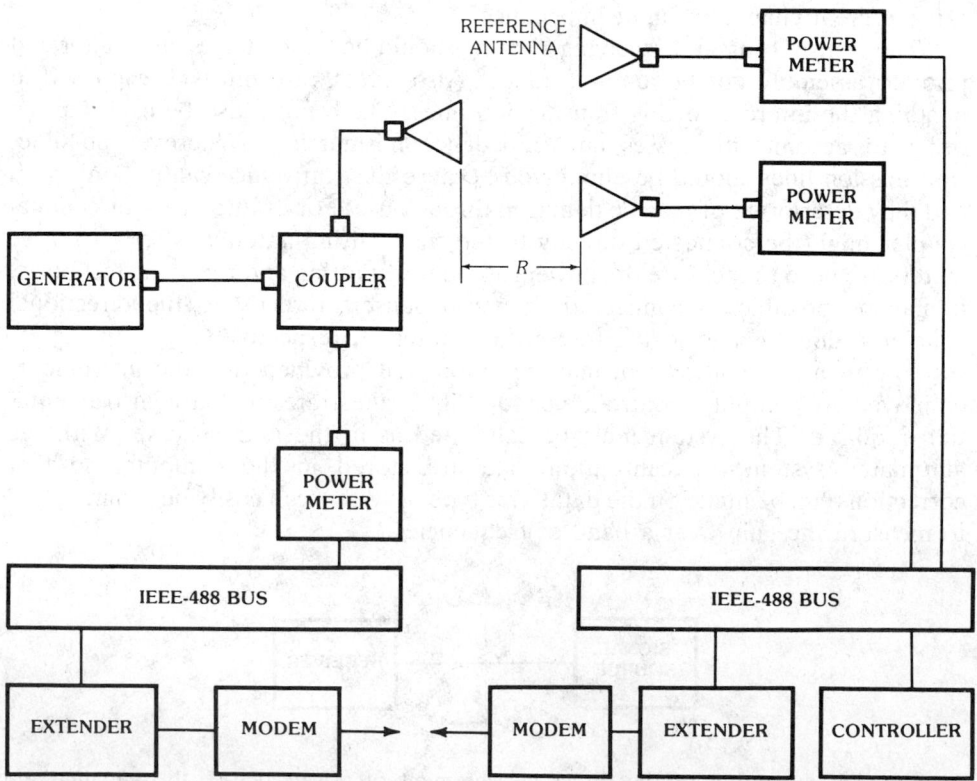

Fig. 31. Block diagram of an automated gain-measurement system.

Extrapolation Method—There are two significant sources of error in absolute gain measurements which can be avoided by the use of the extrapolation technique [41]. These are the errors caused by the finite spacing between the antennas and multipath interferences.

Because of finite spacing, the uncorrected measured value of an antenna's gain will be lower than its actual value because of the presence of near-field components which cause a broadening of the beam of the antenna.

Multipath interference causes an undulating received signal when the spacing between the transmitting and receiving antennas is varied. This results in the uncorrected measured gain exhibiting a similar characteristic.

The details of the extrapolation technique are given in references 8 and 41 and will not be given here. However, an examination of the principal result of the extrapolation theory yields insights into conventional far-field gain measurements. The extrapolation theory states that, for any two antennas used in the gain measurement, the received signal as a function of spacing can be represented by the double series

$$V(R) = C \sum_{p=0}^{\infty} \frac{\exp[-j(2p+1)kR]}{R^{2p+1}} \sum_{q=0}^{\infty} \frac{A_{pq}}{R^q} \qquad (35)$$

in which R is the spacing, k the free-space wave number, and C a constant which is determined by the incident signal at the transmitting antenna and reflection coefficients of the receiving antenna and its load. To interpret this equation it is convenient to rewrite it in the following manner:

$$V(R) = [\exp(-jkR)/R](A_{00} + A_{01}/R + A_{02}/R^2 + \cdots)$$
$$+ [\exp(-j3kR)/R^3](A_{10} + A_{11}/R + A_{12}/R^2 + \cdots) + \cdots \qquad (36)$$

The coefficients of $\exp(-jkR)/R$ are identified as the direct transmission terms with A_{00} being the one used to determine the far-field properties of the antenna, including its gain. The coefficients of $\exp(-j3kR)/R^3$ give the first-order reflection or multipath terms. All other terms represent higher-order reflections.

In a conventional far-field measurement of gain one must choose the spacing R to be sufficiently large to ensure that A_{00} is the only significant term. This results in very long ranges for some types of antennas. As one might expect, ground reflections will increase with increased spacing unless a corresponding increase in antenna height above the ground is provided.

The extrapolation procedure consists of the measurement of the received signal as a continuous function of the spacing between the transmitting and receiving antennas. The resulting cyclic variation in the received signal due to multipath is averaged out either mathematically or electronically. Next, a curve based on (36) is fit to the averaged measured data. From this result, A_{00} can be determined and hence the gain.

This method allows one to perform the actual measurement at spacings less than $2D^2/\lambda$, where D is the largest dimension of the antenna. The procedure is in effect an extrapolation of the fitted curve from the near field to the far field of the antenna.

Just as in the conventional far-field technique, the method can be employed as a two-antenna method or three-antenna method.

By measuring both the amplitude and phase of the received signal the method can be generalized to give both the gains and polarizations of the antennas under test. There is the restriction, however, that no two antennas be circularly polarized. If one antenna is circularly polarized, only that antenna's characteristics can be determined.

The extrapolation technique requires a special range consisting of two towers, at least one of which is movable. The range must be such that boresight between the transmitting and receiving antennas is maintained as their separation varies over the length of the range. Because of this requirement the extrapolation method seems to be only practical as a specialized technique for calibrating gain standards by a standards agency. Indeed, the technique is capable of yielding gain measurements with uncertainties as low as ±0.05 dB and more routine measurements to ±0.08 dB [7].

Absolute-Gain Measurement on a Ground-Reflection Range—At frequencies below about 1 GHz, ground-reflection ranges are commonly used. With some restrictions and modifications the two- or three-antenna method can be implemented on a ground-reflection range [42].

The method is limited to linearly polarized antennas which couple to the electric field only. It is recommended that the antennas be oriented for horizontal polarization in order to avoid the rapid variation of the earth's reflection coefficient as a function of the angle of incidence that is associated with vertical polarization.

The criteria for ground-reflection ranges, outlined in Section 3, under "Outdoor Ranges," should be satisfied. The geometry of the ground-reflection range for gain measurements is shown in Fig. 12. It is desirable that the range length R_0 be such that $R_0 \gg 2h_r$, where h_r is the height of the receiving antenna. When the height of the transmitting antenna is adjusted so that the field at the receiving antenna is at the first maximum closest to the ground, then the gain sum equation for the two-antenna or the three-antenna method can be modified to read

$$(G_A)_{\text{dB}} + (G_B)_{\text{dB}} = 20 \log\left(\frac{4\pi R_D}{\lambda}\right) - 10 \log\left(\frac{P_0}{P_r}\right)$$
$$- 20 \log\left[(D_A D_B)^{1/2} + \frac{rR_d}{R_R}\right] \qquad (37)$$

where D_A and D_B are the directivities along R_D relative to the peak directivities of antennas A and B, respectively. The factor r is to be determined. It is a function of the electrical and geometrical properties of the antenna range, the radiation patterns of the antennas, and the frequency of operation. Directivities D_A and D_B are obtained from amplitude patterns of the two antennas which should be measured prior to performing the gain measurement. The quantities R_D, R_R, and λ and the ratio P_0/P_r are measured directly. Once r is determined, the gain sum can be evaluated.

To obtain r, the preceding measurement is repeated, but this time with the height of the transmitting antenna adjusted to a position such that the field at the receiving antenna is at a minimum. To distinguish the quantities measured with this geometry from those of the previous one, let all the quantities associated with the latter geometry be represented by the same letters, except with primes. With this notation the equation for r can be written as

$$r = \left(\frac{R_R R_R'}{R_D R_D'}\right) \frac{[(P_r/P_r')(D_A' D_B')]^{1/2} R_D - (D_A D_B)^{1/2} R_D'}{(P_r/P_r')^{1/2} R_R + R_R'} \tag{38}$$

where R_R, R_D, and P_r are obtained with the transmitting antenna adjusted for a maximum signal at the receiving antenna, and R_R', R_D', and P_r' are obtained with the antenna adjusted for a minimum signal at the receiving antenna. The relative directivities D_A, D_A', D_B, and D_B' are obtained from the amplitude patterns of the two antennas.

The instrumentation for this measurement is essentially the same as that for the free-space range measurement (see Fig. 30). Accuracies of ± 0.3 dB are attainable with this method.

Gain-Transfer Method—The gain-transfer method, also called the gain-comparison method, requires the use of a gain standard with which the gain of the test antenna is compared. Once the comparison has been performed, the gain of the standard is said to have been transferred to the test antenna. The measurement can be performed on a free-space range or ground-reflection range in an anechoic chamber, or *in situ*; however, in any case the far-field distance criteria must be satisfied.

Typically the test antenna is operated in the receive mode. With it mounted and properly oriented, the received power P_T is measured. The test antenna is then replaced with a gain-standard antenna, leaving all other conditions the same. The received power P_S is then measured. The gain of the test antenna expressed in decibels is given by

$$(G_T)_{dB} = (G_S)_{dB} + 10\log\left(\frac{P_T}{P_S}\right) - 10\log\left(\frac{q_T}{q_S}\right) - 10\log\left(\frac{p_T}{p_S}\right) \tag{39}$$

in which G_S is the known gain of the gain-standard antenna, q_T and q_S are the impedance mismatch factors of the test antenna and the gain-standard antenna, respectively, and p_T and p_S are their polarization efficiencies.

The instrumentation for this method is essentially the same as that for the absolute gain measurement. A convenient method of achieving the exchange of the test antenna and the gain-standard antenna is to mount them back to back on the positioner so that rotating 180° in azimuth effects the positional exchange. Usually absorbing is required between them to reduce coupling.

Measurement of Circularly and Elliptically Polarized Antennas—In the preceding it was tacitly assumed that the source antenna, the test antenna, and the gain-standard antenna are nominally polarization matched; that is, the ratio p_T/p_S is very nearly equal to unity. Often, however, one must measure the gains of circularly or elliptically polarized antennas using a nominally linearly polarized gain-standard antenna.

This measurement can be accomplished by measuring the partial gains of the test antenna with respect to two orthogonal orientations of the gain-standard antenna. For example, the gain-standard and the source antennas are first oriented for vertical polarization and then for horizontal polarization. This results in the measurement of G_{TV} and G_{TH}, the partial gains with respect to vertical and horizontal polarization, respectively. The gain of the test antenna is then given by

$$(G_T)_{dB} = 10\log(G_{TV} + G_{TH}) \tag{40}$$

Note that the partial gains must be in ratio form when added; that is,

$$G_{TV} = G_{SV}\frac{P_{TV}\,q_{SV}\,p'_{SV}}{P_{SV}\,q_{TV}\,p'_{TV}}$$
$$G_{TH} = G_{SH}\frac{P_{TH}\,q_{SH}\,p'_{SH}}{P_{SH}\,q_{TH}\,p'_{TH}} \tag{41}$$

where the subscripts V and H indicate vertical and horizontal polarizations. The previously defined impedance mismatch factors apply here; however, the polarization efficiencies must be modified; hence, the primes.

Polarization mismatches enter in a more complicated fashion for this procedure. The factors p'_{TV}, p'_{TH}, p'_{SV}, and p'_{SH} are *not* the usual polarization efficiencies but rather factors that account for the deviations of the source and gain-standard antennas from linear polarization. This is best illustrated by an example.

Consider the case where the gain of a circularly polarized test antenna is being measured by this method. Suppose that the test and gain-standard antennas are purely circularly and linearly polarized, respectively, but the source antenna has a finite axial ratio. For this example p'_{SV} and p'_{SH} become the usual mismatch factors so that the primes can be dropped. The factors p'_{TV} and p'_{TH}, on the other hand, are the factors by which the usual mismatch factors deviate from 1/2. The reason for this is that a circularly polarized receiving antenna has a polarization efficiency of 1/2 when the incident wave is linearly polarized and the measurement requires a partial gain with respect to linear polarization. Table 1 gives the errors in measured

Table 1. Errors in the Measured Gain of a Purely Circularly Polarized Antenna Due to a Finite Axial Ratio of the Linearly Polarized Sampling Antenna

Sampling Antenna Axial Ratio (dB)	Measurement Error (dB)	
	Same Sense	Opposite Sense
20	+0.828	−0.915
25	+0.475	−0.503
30	+0.270	−0.279
35	+0.153	−0.156
40	+0.086	−0.109
45	+0.049	−0.049
50	+0.027	−0.028

gain for this example as a function of the source antenna axial ratio. Note that the error can be quite significant.

Polarization Measurements

The polarization of an antenna is defined as the polarization of the wave it radiates [1]. As discussed in Chapter 1, what is implied is that in a localized region in the far field of the antenna the radiated wave may be considered to be a uniform plane wave, and that the polarization of the antenna is the same as that of the plane wave. Furthermore, polarization is defined in such a way that it is dependent on the ratio and relative phase of any two orthogonal field components into which the plane wave is resolved. Therefore, polarization, like gain and directivity, is only a function of direction, (θ, ϕ). To completely specify an antenna's polarization, one would have to determine the distribution of its polarization over its radiation sphere.

Polarization can be specified in a number of different ways, the most common being by the *axial ratio r* and *tilt angle τ* of the polarization ellipse and the *sense of polarization*. The tilt angle requires the establishment of a reference direction in the *plane of polarization*, and because polarization is a function of direction, reference directions are required over the radiation sphere of the antenna. A common choice is in the direction of $\hat{\theta}$ as depicted in Fig. 32.

It is convenient to define a local Cartesian coordinate system $(1, 2, 3)$ at the point on the radiation sphere where the polarization is to be determined. One coordinate axis, 3, is the direction of propagation, thus placing the 1 and 2 axes in the plane of polarization. The orientation of the $(1,2)$ axes is arbitrary as long as their relationship to the (θ, ϕ) axes of the antenna's radiation sphere is known. Since the 1 axis will be the reference direction for measuring the tilt angle, it is desirable that it be in the $\hat{\theta}$ direction. For the antenna range measurements discussed here, the 1 axis will be chosen to be in the horizontal direction and the 2 axis in the vertical direction, which results in the horizontal direction being the reference direction for the tilt angle as shown in Fig. 33. In this figure $\mathbf{E}(t)$ is the instantaneous electric-field vector.

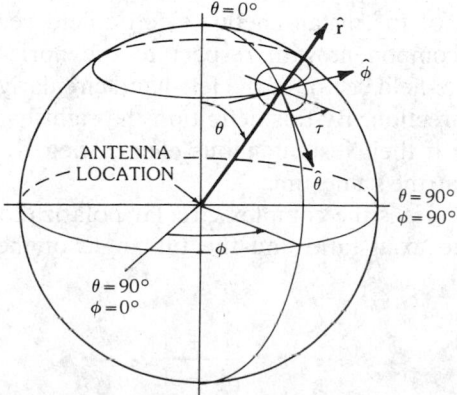

Fig. 32. The polarization ellipse shown in relation to the antenna's coordinate system.

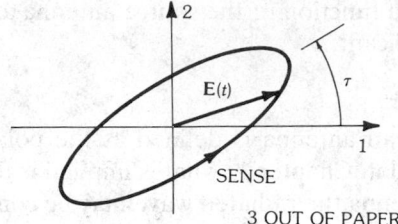

Fig. 33. A local coordinate system $(1, 2, 3)$ at a point (θ, ϕ) on the radiation sphere.

To express a plane wave in a form useful for antenna measurements, it is decomposed into the linear combination of two orthogonally polarized wave components. The orthogonal polarizations useful for antenna measurements are horizontal and vertical linear (E_H, E_V), diagonal linear $(E_{45°}, E_{135°})$, and right-hand and left-hand circular (E_R, E_L), where the symbol **E** denotes the electric-field vector and the subscript indicates the particular component. The polarization of a plane wave can be expressed in terms of *complex polarization ratios* as follows:

$$\hat{\varrho}_\ell = \varrho_\ell \exp(j\delta_\ell)$$
$$\hat{\varrho}_d = \varrho_d \exp(j\delta_d) \qquad (42)$$
$$\hat{\varrho}_c = \varrho_c \exp(j\delta_c)$$

where

$$\varrho_\ell = E_V/E_H$$
$$\varrho_d = E_{135°}/E_{45°} \qquad (43)$$
$$\varrho_c = E_R/E_L$$

and δ_ℓ, δ_d, and δ_c are the relative phases of the corresponding pairs of field components. Circular polarization components require a reference direction for the definition of phase. For our purposes the horizontal direction is chosen; hence δ_c is defined by the angle of the instantaneous electric-field vector of the right-hand circularly polarized component with respect to the horizontal direction at the instant that the electric-field vector of the left-hand circularly polarized component is in the horizontal direction. By this definition the right-hand and left-hand components are in phase if their instantaneous electric-field vectors are in the same horizontal direction at the same time.

Of particular interest is the complex circular polarization ratio $\hat{\varrho}_c$, because of its relationship to the axial ratio and the tilt angle of the wave's polarization, i.e.,

$$r = \frac{\varrho_c + 1}{\varrho_c - 1} \qquad (44)$$

and

$$\tau = \delta_c/2 \tag{45}$$

The sign of the denominator of (44) gives the sense of the polarization, with a positive value indicating right-hand sense and a negative value indicating a left-hand sense. Also, since the reference direction for defining phase is in the horizontal direction, the tilt angle is also referenced to the horizontal direction as required.

If linear polarization ratios are to be measured rather than circular ones, they can be converted to circular ones by use of

$$\hat{\varrho}_c = \varrho_c \exp(j\delta_c) = \frac{1 + j\hat{\varrho}_\ell}{1 - j\hat{\varrho}_\ell} \tag{46}$$

or, if diagonal components are employed,

$$\hat{\varrho}_c = \varrho_c \exp(j\delta_c) = -\frac{1 + j\hat{\varrho}_d}{j + \hat{\varrho}_d} \tag{47}$$

When making polarization measurements on an antenna range it is useful to establish not only the coordinate system for the antenna's polarization as defined above, but also the antenna's *receiving polarization*. Fig. 34 depicts these two coordinate systems. Note that the tilt angle of the antenna's polarization and that of the receiving polarization are supplements of one another; that is,

$$\tau_r = 180° - \tau_a$$

where τ_r and τ_a are the tilt angles of the receiving polarization and the (transmitting) polarization of the antenna, respectively. Note that the receiving polarization of the antenna is the polarization of an incident plane wave which is *polarization matched* to the antenna.

The receiving polarization of an antenna is introduced because, unlike the antenna's (transmitting) polarization, it is in the same coordinate system as an

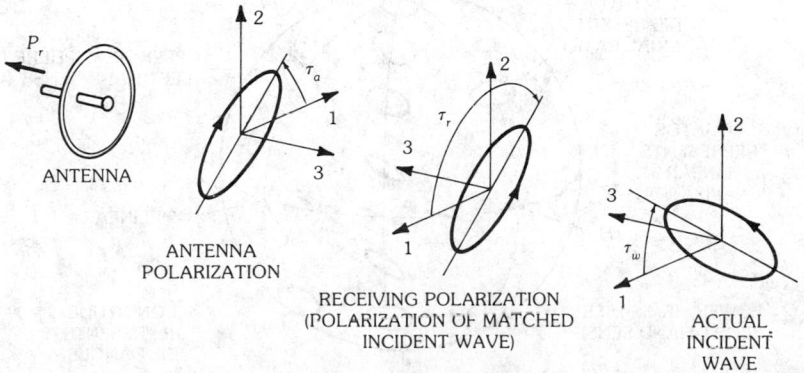

Fig. 34. The relationship between the antenna's (transmitting) polarization and receiving polarization.

incident plane wave of arbitrary polarization as shown in Fig. 34. It follows that the various mathematical representations of these polarizations will be in the same coordinate system and hence can be combined to give the coupling between the wave and the antenna. This is particularly important in polarization measurements since the antenna under test can be operated in either the transmit or receive mode.

The *Poincaré sphere* provides a means of visualizing polarization measurements. It is defined so that there is a one-to-one correspondence between all of the possible polarizations of a single-frequency plane wave and the points over its surface. This is illustrated in Fig. 35. Deschamps [43] introduced its use for antenna work. Hollis ([5], Chapters 3 and 10) greatly elaborated on its development and use, providing numerous relationships useful for polarization calculations and measurement. Much useful material appears in references 2 and 7. A thorough treatment is presented in reference 44. The reader is referred to these sources for detailed expositions on the Poincaré sphere and its use, as well as polarization theory and measurements.

Fig. 35 depicts the representation of all possible polarizations of a single-frequency plane wave in free space on the Poincaré sphere. The 1 axis is labeled H, which indicates horizontal polarization. The polarization of a specific plane wave, **W**, can be located on the sphere as shown in Fig. 36. Note the role the relative phases of the complex polarization ratios play in locating **W**. The polarization ratios are related to the angles α, β, and γ shown in Fig. 36 as follows:

$$\varrho_\ell = \tan\alpha$$
$$\varrho_d = \tan\beta \qquad\qquad (48)$$
$$\varrho_c = \tan\gamma$$

The polarization box ([5], pp. 3.23 to 3.24) is a useful tool in establishing further relationships between the various defined quantities. It is in fact an extension of

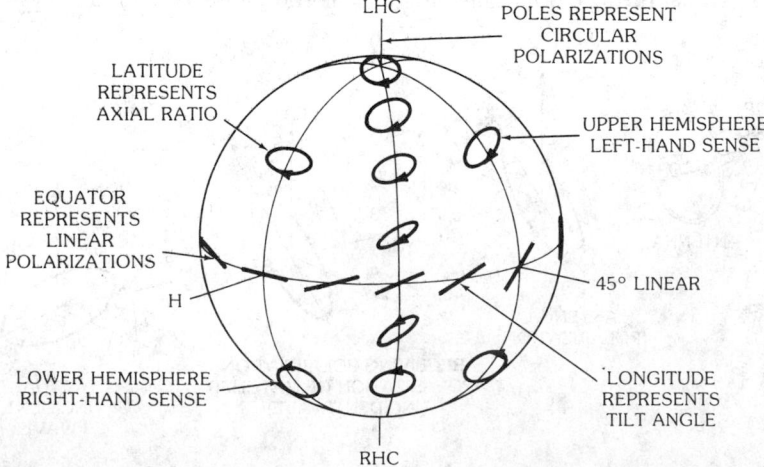

Fig. 35. Representation of polarization on Poincaré sphere.

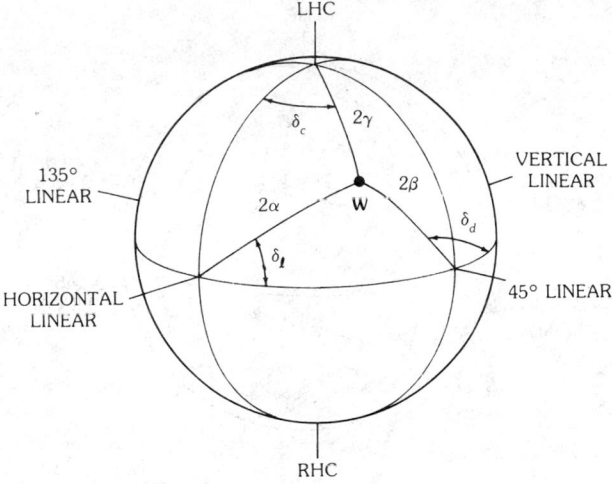

Fig. 36. The Poincaré sphere representation of the polarization of plane wave **W**.

the usual Stokes's parameters for a single-frequency wave of unit intensity. The polarization box is shown in Fig. 37 in relation to the Poincaré sphere.

The Poincaré sphere can also be used to determine the polarization efficiency when a plane wave is incident on an antenna. If the antenna's receiving polarization A_r is located on the sphere along with that of the incident wave, **W**, then the polarization efficiency is given by

$$p = \cos^2 \zeta \tag{49}$$

where ζ is one-half the angular separation of **W** and A_r on the Poincaré sphere as shown in Fig. 38. If the law of cosines for spherical triangles is applied to the one

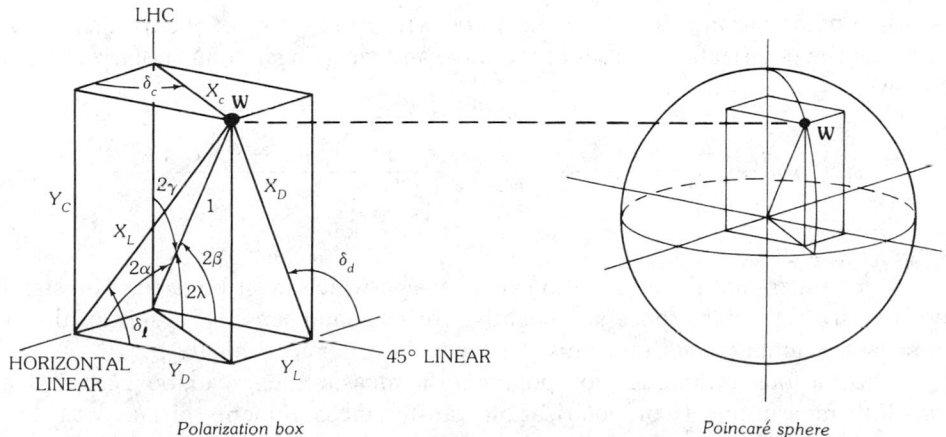

Polarization box   *Poincaré sphere*

Fig. 37. The polarization box and its relation to the Poincaré sphere. (*After ANSI/IEEE [2], © 1979 IEEE; after Hollis, Lyon, and Clayton [5], © 1970 Scientific-Atlanta*)

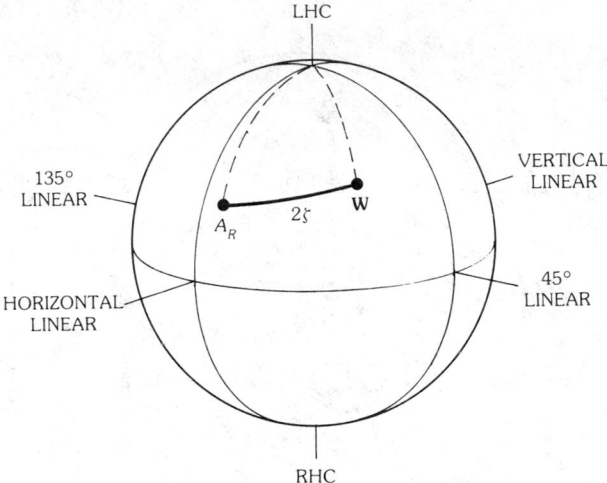

Fig. 38. The incident wave's polarization **W** and the antenna's receiving polarization A_r located on the Poincaré sphere.

formed by A_r, **W**, and the north pole of the sphere, a form for the polarization efficiency in terms of the polarization ratios results, i.e.,

$$p = \frac{1 + \varrho_w^2 \varrho_r^2 + 2\varrho_w \varrho_r \cos \Delta}{(1 + \varrho_w^2)(1 + \varrho_r^2)} \tag{50}$$

where, although the construction implies circular polarization ratios, ϱ_w and ϱ_r can be any of the three polarization ratios ϱ_ℓ, ϱ_d, or ϱ_c for the wave and antenna, respectively, and the angle Δ is the corresponding phase angle $[(\delta_\ell)_w - (\delta_\ell)_r]$, $[(\delta_d)_w - (\delta_d)_r]$, or $[(\delta_c)_w - (\delta_c)_r]$.

The polarization efficiency can also be written in terms of the axial ratios r_w and r_r of the polarization ellipses of the wave and receiving antenna polarizations as follows:

$$p = \frac{(1 + r_w^2)(1 + r_r^2) + 4r_w r_r + (1 - r_w^2)(1 - r_r^2) \cos \Delta_c}{2(1 + r_w^2)(1 + r_r^2)} \tag{51}$$

where $\Delta_c = 2(\tau_w - \tau_r)$.

Remember that the axial ratio carries a sign which must be used. The sign is positive for right-hand sense and negative for left-hand sense. If either axial ratio approaches infinity, the limit must be taken in order to evaluate p.

The various techniques for polarization measurement can be classified as absolute measurements or polarization-transfer measurements just as was done for gain measurements. In addition there are techniques which only yield partial polarization information.

Absolute measurements require no detailed *a priori* information about the polarizations of the antennas being tested. This method requires three antennas, no two of which can have circular or near-circular polarizations. This method can be combined with the three-antenna method of gain measurements and the extrapolation method to yield both gain and polarization of antennas [41].

The polarization-transfer method requires a polarization standard if highly accurate measurements are to be performed. The standard can be used as a sampling antenna or to calibrate the sampling antennas employed. Unfortunately, there is no universally accepted polarization standard. A carefully designed and fabricated linearly polarized antenna often can be used without proof of its polarization purity. When greater accuracy is required, the three-antenna absolute method can be used to calibrate the standard.

Generally, the sampling antennas used for polarization measurements are dual polarized with either orthogonal linear or orthogonal circular polarizations. With modern automated systems the antenna's polarization does not have to be purely linear or circular, provided the polarization is known with sufficient accuracy, since a computer can be programmed to make the necessary corrections. An alternate procedure is to adjust the polarization by use of a polarization-adjustment network [44]. The system can be designed to operate at radio, intermediate, or audio frequencies, and it can also be realized digitally.

The Phase-Amplitude Method of Polarization Measurement—Fig. 39 is a block diagram of the basic instrumentation required for the phase-amplitude method of polarization measurement. The dual-polarized sampling antenna and the two channels of the phase-amplitude receiver must be calibrated. As previously discussed this usually can be accomplished by employing a well-designed linearly polarized antenna as the standard.

First, assume that the sampling antenna has right- and left-hand circular polarizations. Then, with the antenna under test replaced by the linearly polarized standard, the amplitude responses at the outputs of the receiver should be equal and independent of the orientation of the standard in the transverse plane. The phase should be zero when the linearly polarized antenna is oriented for horizontal polarization. As the standard is rotated from the horizontal to vertical position in a positive angular direction with respect to the local coordinate system $(1, 2, 3)$, the output phase δ_c should increase from 0 to π rad. A helical antenna can be used to determine the senses of the polarization if required.

Once the system has been calibrated the linearly polarized standard antenna

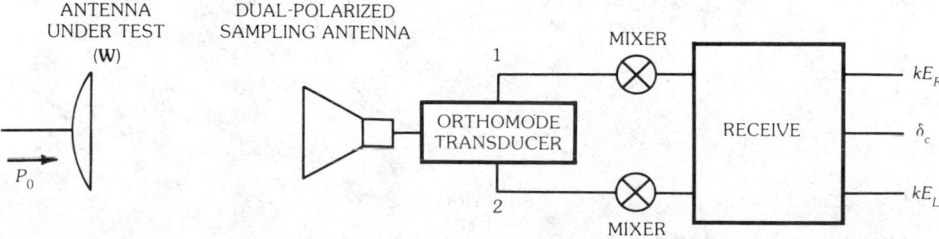

Fig. 39. Instrumentation for the phase-amplitude method of polarization measurement.

is replaced by the antenna under test. The ratio of the responses will give the circular polarization ratio, that is,

$$\varrho_c = \frac{|\hat{V}_R|}{|\hat{V}_L|} = \frac{E_R}{E_L} = \tan\gamma \qquad (52)$$

and the phase δ_c is given directly. The axial ratio r and the tilt angle τ are given by (44) and (45), respectively. The measurement is illustrated by means of the Poincaré sphere as shown in Fig. 40.

If extreme accuracy is required, the standard antenna should be measured using an absolute measurement technique so that its polarization is precisely known. However, usually polarization measurements are more limited by range errors than by the standard antenna, provided that care was exercised in the design of the standard.

Next, let the dual-polarized sampling antenna have horizontal and vertical linear polarizations (see Fig. 41). The accuracy of the calibration in this case is more sensitive to a deviation of the standard's polarization from pure linear than it was in the previous case, where circularly polarized sampling antennas were being used. Therefore, great care should be taken in the design and construction of the standard.

With the antenna under test replaced by the linearly polarized standard, the two polarizations of the sampling antenna are adjusted for optimum nulls when the standard is oriented for horizontal polarization and for vertical polarization, respectively.

Next, the standard is rotated to 45° with respect to the horizontal. The two responses $\hat{V}_V$ and $\hat{V}_H$ should be equal. Also the phase angle δ_ℓ should be zero for

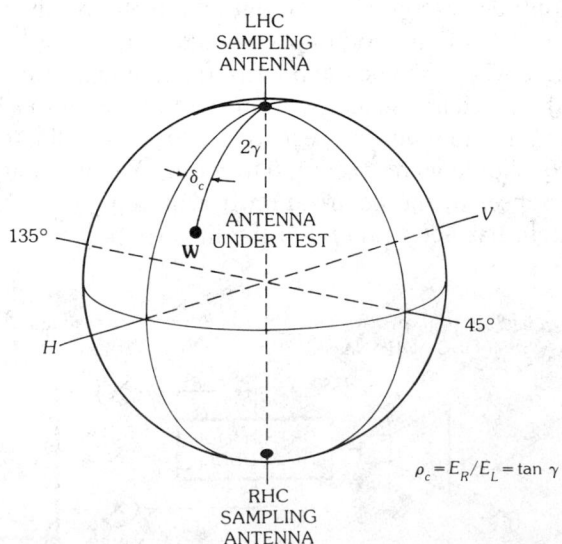

Fig. 40. Phase-amplitude measurements with circularly polarized sampling antennas.

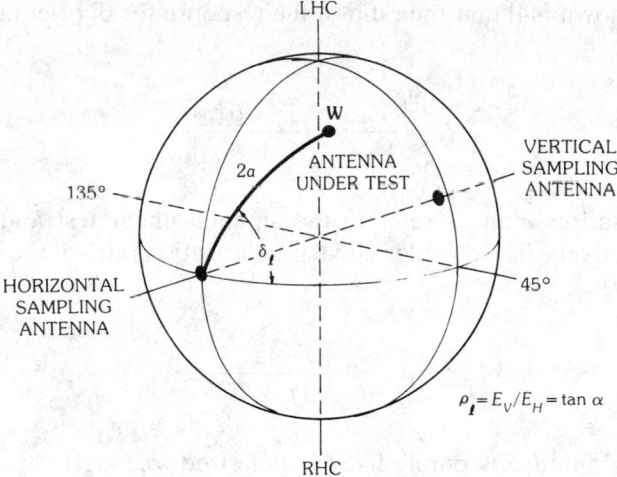

Fig. 41. Phase-amplitude measurements with linearly polarized sampling antennas.

all orientations of the standard. Finally, a helical antenna can be employed to check the sense of the phase angle.

The standard is replaced by the antenna under test. The ratio of the responses gives the linear polarization ratio ϱ_ℓ, i.e.,

$$\varrho_\ell = \frac{|\hat{V}_V|}{|\hat{V}_H|} = \frac{E_V}{E_H} = \tan \alpha \tag{53}$$

and the phase angle δ_ℓ is given directly. The complex circular polarization ratio and, hence, the axial ratio and tilt angle of the antenna under test can be computed with use of (46).

An alternate approach is to use a linearly polarized sampling antenna and a single-channel receiver. The two orthogonal polarizations are obtained by rotating the sampling antenna 90°, from horizontal to vertical, as illustrated in Fig. 42. In this figure $\hat{V}_{0°}$ is the relative response with the sampling antenna oriented to 0° and $\hat{V}_{90°}$ is the relative response with it oriented to 90°. For this measurement the complex circular polarization ratio of the sampling antenna must be accurately known.

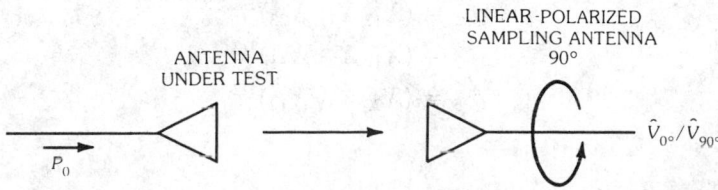

Fig. 42. The phase-amplitude method using a single linearly polarized sampling antenna; $\hat{V}_{0°}$ and $\hat{V}_{90°}$ are the relative responses when the sampling antenna is oriented at 0° and 90°, respectively.

It can be shown [44] that the ratio of the response for $0°$ orientation to that of $90°$ is given by

$$\hat{K} = \frac{\hat{V}_{0°}}{\hat{V}_{90°}} = j\frac{1 - \hat{\varrho}_{ca}\hat{\varrho}_{cs}}{1 + \hat{\varrho}_{ca}\hat{\varrho}_{cs}} \tag{54}$$

where the subscripts a and s refer to the antenna under test and the sampling antenna, respectively. The complex circular polarization ratio of the antenna under test is given by

$$\hat{\varrho}_{ca} = \frac{j - \hat{K}}{\hat{\varrho}_{cs}(j + \hat{K})} \tag{55}$$

If the sampling antenna is purely linearly polarized, $\varrho_{cs} = 1$.

Three-Antenna, Absolute Polarization Measurement—The three-antenna method does not require the use of a polarization standard; however, there is the restriction that no two of the antennas be circularly or near-circularly polarized; the third antenna's polarization can be arbitrary. A total of six measurements are required in order to determine the polarization of the three antennas. The method is illustrated in Fig. 43, where $\hat{V}_{12,0°}$ is the relative response at antenna 2 with its orientation at $0°$. The response is measured twice for each combination of antennas, once with the receiving antenna oriented for $0°$ and once with it rotated $90°$. The ratio of the two responses for each combination of antennas is determined. This gives the following three complex equations:

$$\hat{K}_{12} = j\frac{1 - \hat{\varrho}_{c1}\hat{\varrho}_{c2}}{1 + \hat{\varrho}_{c1}\hat{\varrho}_{c2}}$$

$$\hat{K}_{23} = j\frac{1 - \hat{\varrho}_{c2}\hat{\varrho}_{c3}}{1 + \hat{\varrho}_{c2}\hat{\varrho}_{c3}} \tag{56}$$

$$\hat{K}_{31} = j\frac{1 - \hat{\varrho}_{c3}\hat{\varrho}_{c1}}{1 + \hat{\varrho}_{c3}\hat{\varrho}_{c1}}$$

These equations can be combined to give

$$\hat{\varrho}_{c1} = (\hat{X}_{12}\hat{X}_{31}/\hat{X}_{23})^{1/2}$$
$$\hat{\varrho}_{c2} = (\hat{X}_{12}\hat{X}_{23}/\hat{X}_{31})^{1/2} \tag{57}$$
$$\hat{\varrho}_{c3} = (\hat{X}_{23}\hat{X}_{31}/\hat{X}_{12})^{1/2}$$

in which

$$\hat{X}_{mn} = \frac{1 + j\hat{K}_{mn}}{1 - j\hat{K}_{mn}}, \qquad mn = 12, 23, 31$$

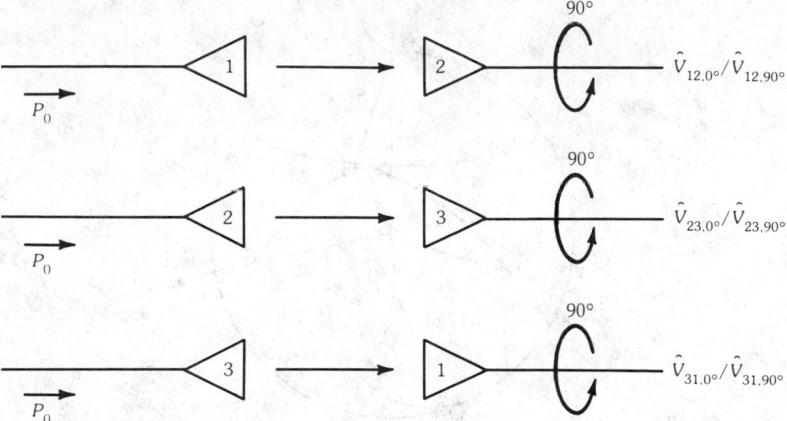

Fig. 43. The three-antenna absolute method of polarization measurement. (a) $\hat{V}_{12,0°}$ is the relative response at antenna 2 with its orientation at 0°. $\hat{V}_{12,90°}$ is the relative response with its orientation at 90°, etc.

There are two sets of solutions because of the square-root operation. One set is the negative of the other. This corresponds to an ambiguity in the resulting tilt angles, with those of one solution set rotated 90° from the other. This obliquity can be removed by a knowledge of at least one of the tilt angles. If none of the tilt angles is known, then an auxiliary measurement of one of the tilt angles may be required [44, 45].

Newell [46] has devised an improved procedure for this measurement. It entails rotating the receiving antenna from 0° to the angle that results in a minimum response, which is then recorded. The antenna is rotated 90° from that angle to where a maximum response is obtained. The ratio of the minimum to maximum responses is formed. This time the phase angle of the ratio is known to be 90°. This results in all of the $\hat{X}_{mn}$s being pure real; but then the phases of the circular polarization ratios are shifted by twice the total angular rotation from 0° to the angle where the maximum response is recorded. The ambiguity in the tilt angles remains, however.

The Multiple-Amplitude Component Method—Clayton and Hollis [47] have shown that the polarization of an antenna can be determined without measuring phase. It requires the measurement of the three polarization ratios, ϱ_ℓ, ϱ_d, and ϱ_c. From these data, 2α, 2β, and 2γ are determined [see (48)] since they define the loci of all possible polarizations on the Poincaré sphere which correspond to the polarization ratios ϱ_ℓ, ϱ_d, and ϱ_c, respectively. This is illustrated in Fig. 44.

The axial ratio r is determined by the use of (44). The phase angle of the complex circular polarization from which the tilt angle is determined is given by

$$\delta_c = \tan^{-1}(Y_D/Y_L) \qquad (58)$$

where, from the polarization box,

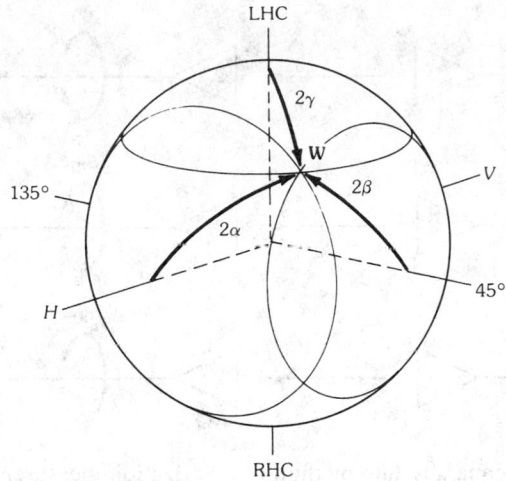

Fig. 44. Multiple-amplitude component method of polarization measurement.

$$Y_D = \frac{1 - \varrho_d^2}{1 + \varrho_d^2} \tag{59}$$

and

$$Y_L = \frac{1 - \varrho_\ell^2}{1 + \varrho_\ell^2} \tag{60}$$

A modified form of the multiple-amplitude component method attributed to Hollis [2] only requires the use of a linearly polarized sampling antenna which is used to measure ϱ_ℓ and ϱ_d. The method will yield the magnitude of the axial ratio and the tilt angle but not the sense of polarization. The tilt angle is determined from δ_c, which is computed using (58) as before, since only ϱ_ℓ and ϱ_d are required. The axial ratio is given by

$$r = -\cot \lambda \tag{61}$$

in which

$$\lambda = \tfrac{1}{2} \cos^{-1}(Y_L{}^2 + Y_D{}^2)^{1/2} \tag{62}$$

The terms Y_L and Y_D are obtained with use of (59) and (60). The square-root operation in the computation of λ results in the sense of polarization being indeterminate; however, the sense can be determined using a helical antenna.

Polarization Pattern Method—This method yields the tilt angle and the magnitude of the axial ratio, but it does not determine the sense of polarization. The antenna under test may be operated in either the receive or transmit mode. Assume that it is in the transmit mode. The method consists of plotting the magnitude of the voltage response $\hat{V}$ of a linearly polarized sampling antenna as its tilt angle is rotated in a plane perpendicular at the range axis. The graphical result is called a *polarization pattern*. A typical pattern is shown in Fig. 45. Note that the polarization ellipse of the antenna is inscribed in the pattern, it being tangent to the polarization pattern at the end of its major and minor axes.

Rotating-Source Method—In some cases a knowledge of the axial ratio over an entire pattern cut is required. Usually, it is a circularly polarized antenna for which this is a requirement. This can be accomplished by rotating the tilt angle of a linearly polarized sampling antenna operating as a source antenna while a pattern cut of the antenna under test is being recorded. The rate of rotation of the source antenna must be much greater than the angular rate of rotation of the positioner orientating the antenna under test. An *axial ratio pattern* results. A typical pattern is shown in Fig. 46. Note that an inner and an outer envelope can be discerned. At any given angle the axial ratio can be determined, it being the ratio of the outer to inner response levels. If the pattern is plotted in decibels, the axial ratios in decibels are simply the excursions of the responses recorded on the pattern. This method does not yield either tilt angle or sense of polarization.

5. Modeling and Model Measurements

Modeling techniques are used extensively in antenna measurements [2, 36, 48]. One reason for this is that it is often impractical or impossible to test an antenna in its operational environment. Antennas mounted on aircraft, ships, missiles, and satellites are examples of such situations. The design of the antennas for these vehicles usually begins in the preliminary design phase at which time only the external shape of the vehicle is available to the antenna designer. One of the first tasks is to determine the location of the various antennas subject to the structural and operational constraints imposed. While the experienced antenna engineer usually can predict the gross effect of the vehicular shape on the antenna's radia-

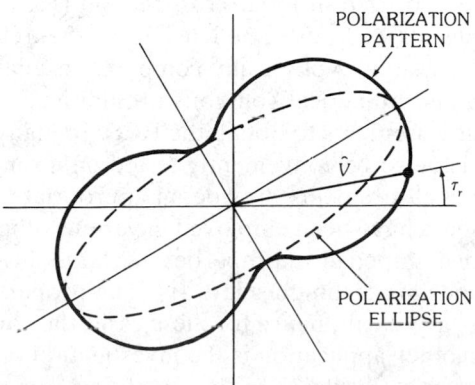

Fig. 45. The polarization pattern and its relation to the polarization ellipse of a wave. (*Courtesy Tecon Industries, Inc.*)

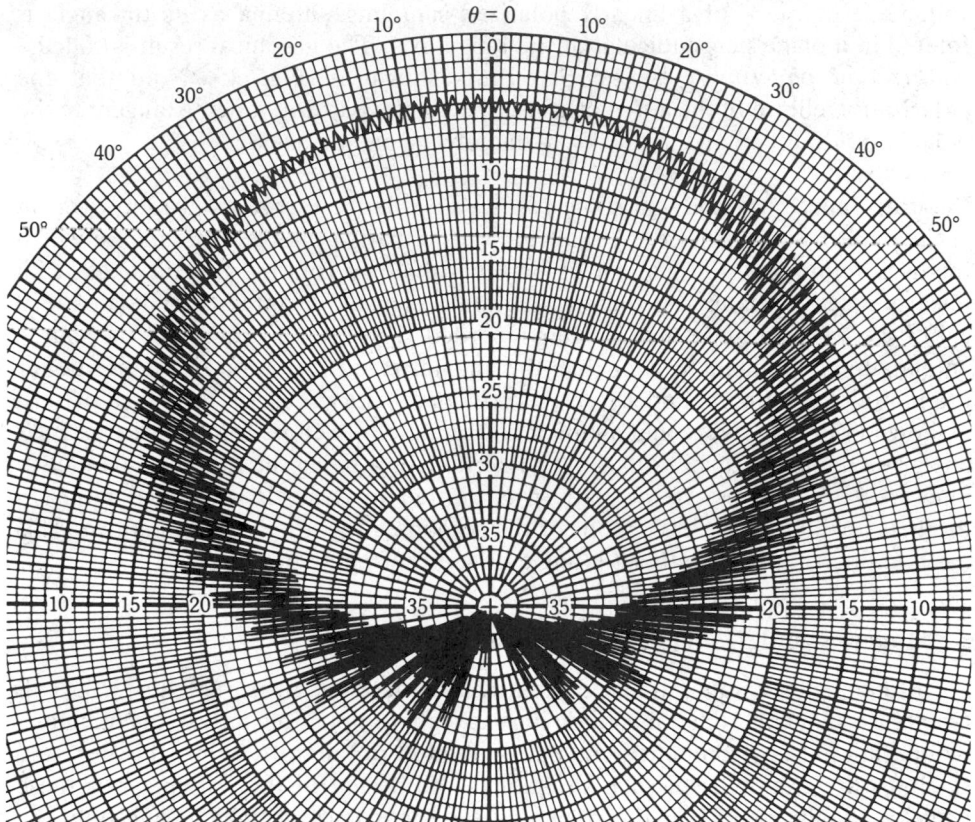

Fig. 46. Pattern of a circularly polarized test antenna taken with a rotating, linearly polarized, source antenna.

tion pattern, more precise information is ultimately required. The use of scale models provides a convenient means of obtaining such data.

Scale models are also used in radar cross-section (RCS) measurements [49]. Often such measurements are performed to verify theoretical computations of RCS; however, if the system is physically complex, an analytic solution may be impossible to obtain and numerical solutions prohibitively expensive. One must then rely solely on measurements to obtain the RCS. In many instances the targets to be tested are too large to be conveniently tested on a range or they are simply not available. Model measurements provide an appropriate solution.

Modeling techniques have been employed in various other situations. Both the radiation patterns and impedance properties of large hf antennas have been investigated using modeling techniques [50, 51]. The properties of these antennas are greatly affected by the earth; hence for these cases the earth is usually modeled or simulated also. Another application is the investigation of the effects that large metallic structures, such as bridges, towers, buildings, or other large antennas, have on communication or navigational systems [52, 53]. Scale-model systems are also used in the study of electromagnetic pulse effects [54], geological explorations

[55, 56], and radio-wave propagation [57, 58, 59]; these studies are beyond the scope of this chapter. References 60 and 61 contain thorough treatments of the use of modeling techniques for radio-wave propagation studies and experimental investigations of antennas in matter, respectively, and hence will not be considered here.

Modeling techniques find their greatest use in the study of physically large systems operating in free space by scaling the system to a smaller physical size. This allows controlled experiments to be performed on antenna ranges or in laboratories. Furthermore, large amounts of data can be conveniently obtained using scale models, whereas this is usually not possible if full-scale systems are tested. In almost all cases, however, it is desirable, if not required, that limited full-scale measurements be performed to substantiate the model measurements.

In addition to physical models there is widespread use of mathematical models. In the broadest sense of the word any mathematical equation that relates electromagnetic fields to sources may be considered a mathematical model. For our purpose a more restrictive usage is necessary. Here a mathematical model will refer to the case where a physical structure is described mathematically in order to compute electromagnetic fields in its presence. Perhaps "computer model" is a more apt term since the digital computer has made such modeling techniques possible. The method of moments [62, 63] and the geometrical theory of diffraction (GTD) [64] can be used to compute the fields and input impedances of antennas as well as RCS.

These methods find their greatest use in those cases where relatively simple shapes can be employed to represent the structures to be modeled. For example, an antenna mounted on the fuselage of an aircraft away from any of its complex structures, such as the cockpit, nose radar, or landing gear, lends itself to computer modeling [65–67]. The use of simple computer models to explore the sensitivity of the radiation pattern of the structure in various angular regions is very helpful in determining the requirements for a physical model or for a more sophisticated computer model.

Both physical and computer modeling are important and widely used tools in the study of electromagnetic systems.

Theory of Electromagnetic Models

Whenever the medium in which a field exists is linear, then Maxwell's equations for that medium will also be linear. It is the linearity of Maxwell's equations that suggests that scale models of electromagnetic systems are possible. Nonlinear modeling has been proposed [68]; however, its implementation has not been developed [54]. Only linear modeling will be considered here, which means that all systems having nonlinear or time-varying constituents are excluded.

The theory of electromagnetic models has long been known and used for antenna research and development [69–80]. Sinclair [81] laid the foundation for most modern work.* What follows was largely extracted from his work except that whereas he used time-domain quantities we shall use frequency-domain ones.

---

*References 70 and 81 contain rather complete bibliographies of early papers on modeling theory and techniques up to 1947.

To arrive at the condition for a scale model of an electromagnetic system: consider two electromagnetic systems: a full-scale, or actual, system and a scale model of that system. The complex electric- and magnetic-field vectors in the full-scale system shall be denoted as

$$\mathbf{E}(\omega, \mathbf{r}) \quad \text{and} \quad \mathbf{H}(\omega, \mathbf{r})$$

where ω is the angular frequency and $\mathbf{r}$ is the position vector. All variables in the model shall be denoted by primed quantities and they are related to the full-scale quantities in the following manner:

$$\mathbf{E}(\omega, \mathbf{r}) = \alpha \mathbf{E}'(\omega', \mathbf{r}')$$
$$\mathbf{H}(\omega, \mathbf{r}) = \beta \mathbf{H}'(\omega', \mathbf{r}') \tag{63}$$

The position vectors are related by

$$\mathbf{r} = p\mathbf{r}' \tag{64}$$

and the angular frequencies by

$$\omega = \omega'/\gamma \tag{65}$$

The factors α, β, p, and γ are the electric, magnetic, spatial, and temporal (inverse frequency) scale factors, respectively.* Four factors are required since there are four fundamental quantities required to express the electromagnetic fields, mass, length, time, and electric current.

Both sets of fields must satisfy Maxwell's equations and the constitutive relations. This leads to the relationship between the permittivity, permeability, and conductivity of the media that constitute the two systems.

In the full-scale system, Maxwell's equations can be written in the frequency domain as

$$\nabla \times \mathbf{H}(\omega, \mathbf{r}) = \sigma \mathbf{E}(\omega, \mathbf{r}) + j\omega\hat{\epsilon}\mathbf{E}(\omega, \mathbf{r})$$
$$\nabla \times \mathbf{E}(\omega, \mathbf{r}) = -j\omega\hat{\mu}\mathbf{H}(\omega, \mathbf{r}) \tag{66}$$

in which $\hat{\mu}$ and $\hat{\epsilon}$ are the complex permeability and permittivity, respectively. In the scale-model system they can be written as

$$\nabla' \times \mathbf{H}'(\omega', \mathbf{r}') = \sigma_m \mathbf{E}'(\omega', \mathbf{r}') + j\omega'\hat{\epsilon}_m\mathbf{E}'(\omega', \mathbf{r}')$$
$$\nabla' \times \mathbf{E}'(\omega', \mathbf{r}') = -j\omega'\hat{\mu}_m\mathbf{H}'(\omega', \mathbf{r}') \tag{67}$$

where $\hat{\mu}_m$ and $\hat{\epsilon}_m$ are the complex permeability and permittivity for the scaled system.

---

*King and Smith [61] and King [60] use scale factors that are reciprocal to the ones used here.

If the model is to be an exact simulation of the full-scale system, then the substitution for the scaled quantities given in (63)–(65) should transform (67) into (66). Since

$$\nabla' \times \mathbf{H}' = \frac{p}{\beta} \nabla \times \mathbf{H}$$

$$\nabla' \times \mathbf{E}' = \frac{p}{\alpha} \nabla \times \mathbf{E}$$

$$(68)$$

one obtains from the substitution

$$\nabla \times \mathbf{H} = \frac{\beta}{p\alpha} \sigma_m \mathbf{E} + j\omega \frac{\gamma\beta}{\alpha p} \hat{\epsilon}_m \mathbf{E}$$

$$\nabla \times \mathbf{E} = -j \frac{\gamma\alpha}{p\beta} \hat{\mu}_m \mathbf{H}$$

$$(69)$$

A comparison of (66) and (69) reveals that

$$\sigma_m = \frac{p\alpha}{\beta} \sigma$$

$$\hat{\epsilon}_m = \frac{p\alpha}{\beta\gamma} \hat{\epsilon}$$

$$\hat{\mu}_m = \frac{p\beta}{\alpha\gamma} \hat{\mu}$$

$$(70)$$

A tabulation of the equations relating the various quantities of a model system to the corresponding ones of the full-scale system is given in Table 2.

There are practical considerations that restrict one's choices for p, α, β, and

Table 2. The General Scale-Model System

Quantity	Transformation	Quantity	Transformation
Length	$\ell' = \ell/p$	Voltage	$V' = V/\alpha p$
Wavelength	$\lambda' = \lambda/p$	Current	$I' = I/\beta p$
Time	$t' = t/\gamma$	Power	$P' = P/\alpha\beta p^2$
Frequency	$f' = \gamma f$	Charge	$Q' = Q/\beta\gamma p$
Electric-field strength	$\mathbf{E}' = \mathbf{E}/\alpha$	Propagation constant	$k' = pk$
Magnetic field strength	$\mathbf{H}' = \mathbf{H}/\beta$	Resistance	$R' = \beta R/\alpha$
Conductivity	$\sigma' = p\alpha\sigma/\beta$	Reactance	$X' = \beta X/\alpha$
Permittivity	$\hat{\epsilon}' = p\alpha\hat{\epsilon}/\beta\gamma$	Impedance	$Z' = \beta Z/\alpha$
Permeability	$\hat{\mu}' = p\beta\hat{\mu}/\alpha\gamma$	Electric-flux density	$\mathbf{D}' = p\mathbf{D}/\beta\gamma$
Current density	$\mathbf{J}' = p\mathbf{J}/\beta$	Capacitance	$C' = \alpha C/\beta\gamma$
Power density	$\mathbf{S}' = \mathbf{S}/\alpha\beta$	Inductance	$L' = \beta L/\alpha\gamma$
Charge density	$\varrho' = p^2\varrho/\beta\gamma$	Magnetic-flux density	$\mathbf{B}' = p\mathbf{B}/\alpha\gamma$
Angular frequency	$\omega' = \gamma\omega$	Scattering cross section	$A' = A/p^2$
Phase velocity	$v' = \gamma v/p$	Antenna gain	$G' = G$

γ. Perhaps the most common restriction is imposed because it is, in general, impractical to scale the permeability of a good conductor or a dielectric. These materials generally have a permeability equal to that of free space. This implies that for such cases one must choose

$$p\beta/\gamma\alpha = 1 \tag{71}$$

There are also practical considerations that dictate the choices of p and γ. These include restrictions on the physical size of the model system and the availability of suitable equipment. It should be noted that p and γ can be chosen independently. Thus the ratio α/β usually cannot be independently chosen because of these restrictions.

Note that the ratio α/β is the scale factor for impedance. Of particular importance in modeling are surface impedances at the interfaces between two different media. With each vector component of the fields tangential to an interface denoted by a subscript t, one can write

$$Z_s = \frac{E_t}{H_t} = (\alpha/\beta)\frac{E'_t}{H'_t} = (\alpha/\beta)Z_{sm} \tag{72}$$

in which Z_s is the surface impedance in the full-scale system and Z_{sm} the corresponding one in the model system.

Wave impedances are important in modeling because there are instances where it is desirable to simply model the surface impedances over the surface bounding the region of interest rather than those media adjacent to it [60]. The uniqueness theorem [82] can be imposed to show that this procedure is valid.

A further restriction on the choices of p and γ occurs when the region of interest of the full-scale system is air and it is desired that the measurement on the model system also be in air. This results in

$$p\alpha/\gamma\beta = 1 \tag{73}$$

in addition to (71). A comparison of (71) and (73) reveals that

$$\alpha = \beta \tag{74}$$

$$p = \gamma \tag{75}$$

Therefore,

$$\hat{\epsilon}_m = \hat{\epsilon} \tag{76}$$

$$\hat{\mu}_m = \hat{\mu} \tag{77}$$

and hence

$$Z_s = Z_{sm} \tag{78}$$

However, the conductivity must be scaled, since from (70)

$$\sigma_m = p\sigma \tag{79}$$

Note that $\hat{\epsilon}$, $\hat{\mu}$, $\hat{\epsilon}_m$, and $\hat{\mu}_m$ are complex quantities.

Equation 74 states that the electric and magnetic scale factors are equal. This can be understood by recognizing that the equality results in the wave impedances being equal at corresponding points in the two systems. This is a consequence of making the media the same in both systems. The restriction imposed by (75), which states that the spatial and temporal scale factors are equal, results in the frequency of operation in the scaled system being inversely scaled from the full-scale system compared to the mechanical dimensions. That is, if the physical sizes of structures are reduced by a factor of p, then the frequency of operation is increased by a factor p.

Notice that both (74) and (75) merely require that the ratios α/β and p/γ be equal to unity. Thus the value of α can be left arbitrary and only p, hence γ, be specified. If this is done the model is said to be a *geometrical model*. The flux lines will be similar in the two systems but not the power levels. If both α and p, hence β and γ, are specified, the model is called an *absolute* one in which the power levels in the two systems are related by a scale factor given by the product of α and β. This means that for an absolute model, quantitative data can be obtained directly.

Table 3 gives the scaling relations for the simpler geometrical model system in which only one scale factor, p, needs to be specified. This system will yield all of the important parameters that are usually required for antenna and RCS investigations and is therefore the system most often employed.

Materials for Electromagnetic Models

Geometrical modeling is conceptionally straightforward. One first chooses an appropriate scale factor which can be such that the model is made smaller or larger than, or, in some cases, the same size as, the full-scale system. A common case is to reduce the physical dimensions by the scale factor. The frequency of operation will then be increased by the scale factor. All structures are constructed of materials which have the same complex permittivities and permeabilities as those of the full-scale system. The conductivities, however, are increased by the scale factor.

A difficulty arises because of the availability of materials. For example, if good conductors such as aluminum, copper, silver, or gold are to be modeled, it will not

Table 3. The Geometrical Scale-Model System

Quantity	Transformation	Quantity	Transformation
Length	$\ell' = \ell/p$	Permittivity	$\hat{\epsilon}' = \hat{\epsilon}$
Time	$t' = t/p$	Permeability	$\hat{\mu}' = \hat{\mu}$
Wavelength	$\lambda' = \lambda/p$	Phase velocity	$v' = v$
Capacitance	$C' = C/p$	Resistance	$R' = R$
Inductance	$L' = L/p$	Reactance	$X' = X$
Impedance	$Z' = Z$	Conductivity	$\sigma' = p\sigma$
Frequency	$f' = pf$	Propagation constant	$k' = pk$
Antenna gain	$G' = G$	Scattering cross section	$A' = A/p^2$

be possible to find conductors with higher conductivity. Also, it is difficult, and in some cases impossible, to find dielectric and magnetic materials that have complex permittivities and permeabilities which have both their real and imaginary parts at the scaled frequency equal to those in the full-scale system. This is especially true of magnetic materials.

Good Conductors and Dielectrics—In most cases the conductivities of good conductors do not have to be scaled in order to produce an acceptable model. To show this, consider a plane wave incident on the interface between a good nonmagnetic dielectric and a good nonmagnetic conductor as shown in Fig. 47. The angle of incidence is denoted by θ_i, the angle of reflection by θ_r, and that of transmission by θ_t.

Snell's law states that

$$\sin \theta_t = \frac{k_d}{k_m} \sin \theta_i \tag{80}$$

in which k_d and k_m are the propagation constants for the dielectric and conductor regions, respectively. They are given by

$$k_d = \omega \sqrt{\mu_0 \epsilon} \tag{81}$$

and

$$k_m = \sqrt{-j\omega\mu_0\sigma} \tag{82}$$

Substitution of (81) and (82) into Snell's law, (80), gives

$$\sin \theta_t \cong \sqrt{\frac{j\omega\epsilon}{\sigma}} \sin \theta_i \tag{83}$$

This is an extremely small quantity for good conductors even at microwave frequencies; hence, for all practical purposes the wave can be considered to prop-

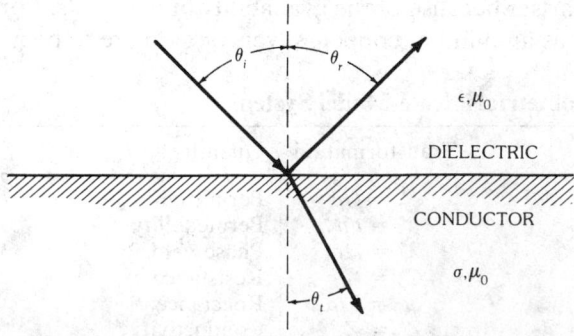

Fig. 47. The interface between a good nonmagnetic dielectric and a good nonmagnetic conductor.

agate normally ($\theta_t \cong 0$) into the conductor, regardless of the angle of incidence θ_i. This means that the surface impedance at the interface looking into the conductor is just the intrinsic impedance of the conductor, η_m; namely,

$$\eta_m \cong \left(\frac{\omega \mu_0}{2\sigma} \right)^{1/2} (1 + j) \tag{84}$$

which is independent of θ_i. In most cases η_m can be assumed to be equal to zero. One can conclude that scaling the conductivity σ is not necessary as long as a good nonmagnetic conductor is used for the model. For our purposes a good conductor is one which has a conductivity such that

$$\sigma \gg \omega \epsilon_0 \tag{85}$$

Copper, gold, silver, aluminum, and brass are examples of good conductors at microwave frequencies.

Similarly, it is often the case that only the real part of a good dielectric needs to be invariant to scaling in order to produce an acceptable model. A good dielectric here is one for which the real part of the complex permittivity is much larger than its imaginary part, i.e.,

$$\epsilon' \gg \epsilon'' \tag{86}$$

where ϵ' is the real part or dielectric constant and ϵ'' is the imaginary part.

The principal effect of not scaling the conductivities of good conductors and ignoring the loss factor of good dielectrics is that the system losses will not be properly scaled. This means that while the radiation patterns of the modeled antennas may be reproduced to an acceptable accuracy, the gain will not. One can, however, obtain the directivity by pattern integration. Therefore the data from such model measurements are qualitative; that is, at least one measurement on the full-scale system is required to obtain the gain. Alternatively, if the losses in the full-scale antenna can be determined, the gain can be obtained since it is the product of the radiation efficiency and the directivity.

Lossy Dielectrics—A great deal of ingenuity is required to model lossy dielectrics. It is indeed fortuitous for one to find a commercially available dielectric that satisfies a modeling need. Usually composite materials must be employed. These include mixtures of powered dielectric and conductive materials which behave as lossy dielectrics. To establish a nomenclature for describing such materials consider the complex permittivity of a homogeneous, lossy dielectric,

$$\hat{\epsilon} = \epsilon' - j\epsilon'' \tag{87}$$

The relative values of ϵ' and ϵ'' with respect to the free-space permittivity ϵ_0 are usually given in the literature. The quantity ϵ'/ϵ_0 is usually called the relative dielectric constant and ϵ''/ϵ_0 is usually called the loss factor. Often one uses the loss tangent in computing losses in dielectrics; it is defined as

$$\tan \delta = \frac{\epsilon''}{\epsilon'} \tag{88}$$

where δ is called the loss angle.

For composite materials there will be a conductivity as well as a complex permittivity. One can consider an equivalent conductivity that includes the dielectric's loss factor. To show this, consider the admittivity $\hat{y}(\omega)$ of the material,

$$
\begin{aligned}
\hat{y}(\omega) &= \sigma + j\omega\hat{\epsilon} \\
&= (\sigma + \omega\epsilon'') + j\omega\epsilon'
\end{aligned} \tag{89}
$$

The equivalent conductivity is defined as

$$\sigma_{eq} = \sigma + \omega\epsilon'' \tag{90}$$

Note that because of the factor ω multiplying ϵ'', ϵ_{eq} scales properly.

When a composite material is measured, it is more convenient to treat the equivalent conductivity as though it were a part of the loss factor; for example, one can write (89) as

$$
\begin{aligned}
\hat{y}(\omega) &= j\omega\left(\epsilon' - j\frac{\delta_{eq}}{\omega}\right) \\
&= j\omega(\epsilon' - j\epsilon''_{eq})
\end{aligned} \tag{91}
$$

where

$$\epsilon''_{eq} = \frac{\sigma + \omega\epsilon''}{\omega} \tag{92}$$

The equivalent loss factor in $\epsilon''_{eq}/\epsilon_0$ and the equivalent loss tangent is given by

$$\tan \delta_{eq} = \frac{\epsilon''_{eq}}{\epsilon'} \tag{93}$$

Generally these materials are measured using techniques that yield the complex permittivity, the imaginary part of which is the equivalent loss factor. Because of this, the adjective equivalent is usually dropped. A chart displaying the relative permittivity and loss tangent of the principal classes of solid materials at a frequency of 100 MHz and at room temperature is given in Fig. 48. Since the permittivities of most of these materials are reasonably independent of frequency up to the low microwave frequencies, these data can be considered typical for the classes of materials represented. For an exhaustive list of permittivities of specific materials, the reader is referred to references 83 and 84. Table 4 gives the conductivities of several conductors useful in modeling.

To obtain a specified complex permittivity it is usually necessary to use a mixture of dielectric and conductive materials. For example, a carbon-loaded

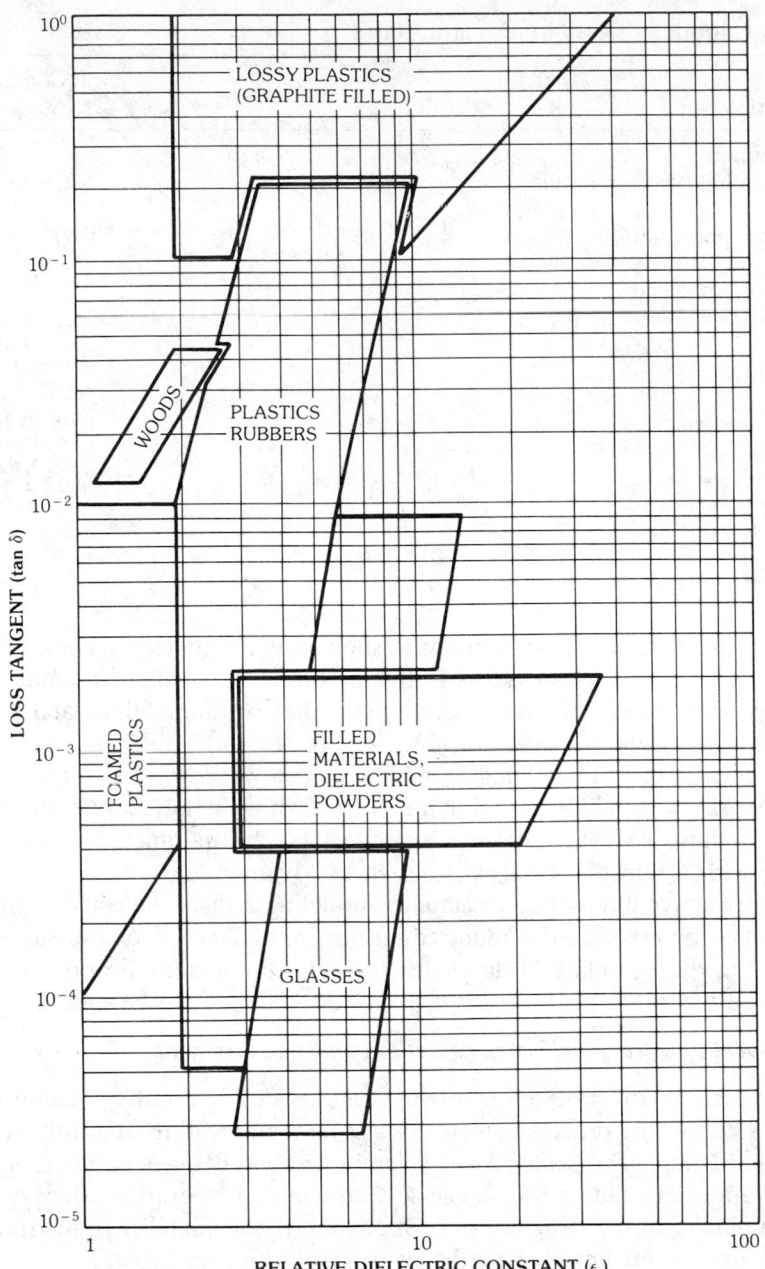

Fig. 48. The relative dielectric constant and loss tangent of solid materials at a frequency of 100 MHz and room temperature. (*After King and Smith [61], © 1981 MIT Press*)

polyurethane elastomer has been used to model land [85], as has a mixture of sand and carbon powder [58, 59].

There are a number of formulas for estimating the dielectric properties of mixtures [86, 87]; however, in all cases the complex permittivity must be measured.

Table 4. Conductivities of Metals and Alloys

Material	Conductivity (S/m at 20°C)
Aluminum	3.82×10^7
Aluminum, commercial alloys	$1.22–3.60 \times 10^7$
Brass	$1.2–1.6 \times 10^7$
Bronze, phosphor	$0.97–1.06 \times 10^7$
Carbon	$2.0–2.8 \times 10^4$
Copper, annealed	5.80×10^7
Copper, beryllium	1.72×10^7
Gold	4.10×10^7
Graphite	$0.7–1.2 \times 10^5$
Iron	1.03×10^7
Iron, gray cast	$0.05–0.20 \times 10^7$
Lead	0.457×10^7
Nickel	1.45×10^7
Silver	6.14×10^7
Steel	$0.5–1.0 \times 10^7$
Zinc	1.67×10^7

It is most useful to construct a graph to show how the dielectric constant and loss tangent vary with the constituents of the mixture. An excellent example has been given by Hall, Chambers, and McInnes [51]. They were modeling land using sand, salt, and water. Their results are given in Fig. 49 and Table 5.

Freshwater can be modeled by adding salt to water to obtain the scaled conductivity [61]. It should be noted that mixtures which include liquids are subject to change because of evaporation. This effect can be minimized by covering the mixture with a thin film of a plastic such as Mylar.

An attractive alternative to actually modeling a lossy dielectric is to simulate its presence by creating the required surface impedance over the surface corresponding to the boundary of the dielectric [88]. Of course, this approach can only be applied when the fields within the dielectric are not to be measured.

Scale-Model Construction, Instrumentation, and Measurement

A measurement program involving scale modeling requires careful planning. The first step in the process is to select a scale factor and to determine the extent of the modeling. The scale factor selection is based on a number of practical considerations, including the full-scale frequency of operation, the physical size and complexity of the structure to be modeled, the availability of instrumentation and any limitations imposed by the instrumentation, and cost.

The most convenient range of frequencies for model measurements is 100 MHz to 18 GHz. Instrumentation is usually readily available for these frequencies. Since radio frequencies for the system that require modeling are, in general, allocated internationally, it is not surprising that there is a striking uniformity throughout industry in the choice of scale factors for similar projects. For radiation pattern measurements, aircraft models range from 1/4 to 1/60 scale, ship models range from 1/48 to 1/200 scale with 1/48 or 1/50 being quite standard, and ground-based hf antennas are usually of the order of 1/100 scale.

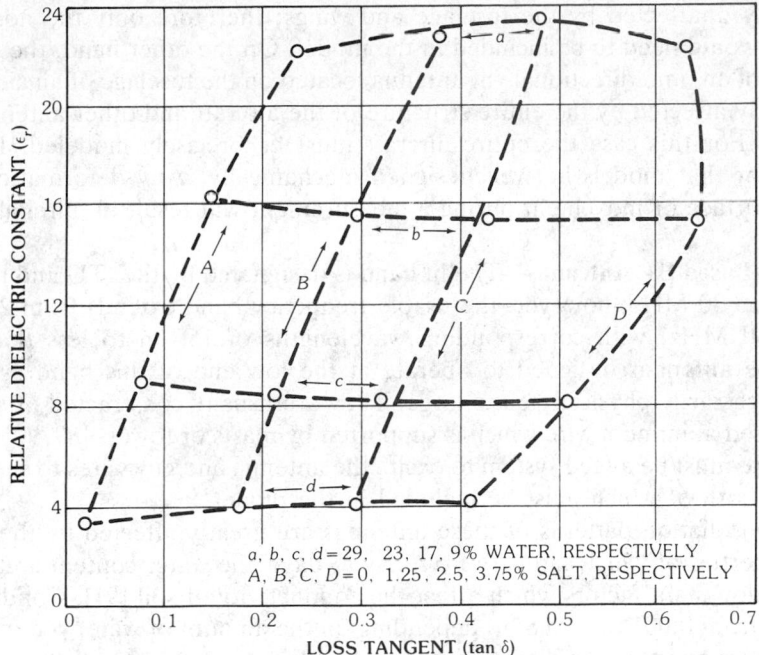

a, b, c, d = 29, 23, 17, 9% WATER, RESPECTIVELY
A, B, C, D = 0, 1.25, 2.5, 3.75% SALT, RESPECTIVELY

Fig. 49. Relative dielectric constant and loss tangent of various mixtures of water, salt, and sand. (*After Hall, Chambers, and McInnes [51], © 1978 IEE; published by Peter Peregrinus Ltd., London*)

Table 5. Comparison of the Dielectric Constant and Conductivity of a Sand, Salt, and Water, 1/80 Scale Simulation of Wet and Dry Ground With Those of the Full-Scale Ground (*After Hall, Chambers, and McInnes [51], © 1978 IEE; published by Peter Peregrinus Ltd., London*)

Ground Type	Wet	Dry
Full-scale (25 MHz)	$\epsilon_r = 23.0$ $\sigma = 5$ mS/m	$\epsilon_r = 5.0$ $\sigma = 1.0$ mS/m
Model (2 GHz)	$\epsilon_{rm} = 23.0$ $\sigma_m = 0.4$ S/m	$\epsilon_{rm} = 5.0$ $\sigma_m = 0.08$ S/m
Mixture (%) Sand Water Salt	 68.5 30.0 1.5	 87 12 1

The extent of the modeling refers to the amount of detail required for the model and how much of the antenna's environment must be included. This decision is based on the general radiation characteristics of the antenna to be tested and its location relative to other antennas, sharp discontinuities, and any protuberances on the surface of the structure on which it is mounted. For example, the radiation pattern of a directive antenna located on the nose bulkhead of an aircraft will be relatively unaffected by the fuselage and wings; therefore only the nose of the aircraft would need to be included in the model. On the other hand, the radiation pattern of an omnidirectional vhf antenna located on the fuselage of an aircraft will be greatly affected by the entire structure of the aircraft and other antennas in its vicinity. For this case the entire aircraft must be precisely modeled. It is very important that models be well designed mechanically. Any deformation of the model surface or movement during a measurement will result in unreliable data.

Ground-Based HF Antennas—The hf band is designated by the ITU and the IEEE to be 3 to 30 MHz; however, the usable frequency band extends from 2 MHz to above 30 MHz, with corresponding wavelengths of 150 m to less than 10 m. Directive antennas designed to operate at the low end of this band are, therefore, necessarily physically quite large. They are usually constructed of stranded, steel-cored aluminum wire which is supported by masts or towers [89, 90]. In addition there must be a feed system to excite the antenna and guy wires to support the towers, both of which must be included in the model.

The radiation patterns of these antennas are greatly affected by the ground, the properties of which can vary from day to day. The water content and salinity, are the dominant factors which affect the conductivity of soil [91]. Conductivities ranging from 0.001 to 0.05 S/m, depending on the amount of water present and its salinity, are normal. The dielectric constant is also affected by the water content; it varies from about 10 to 20. Because of the difficulties of modeling earth, scaled hf antennas are often tested on an image-plane range, the surface of which is constructed of a good conductor. The measured data are then adjusted to account for imperfect earth [89].

There are inherent difficulties in obtaining an exact scale model of an hf antenna. The principal reason is that when the wire diameters are exactly scaled, they become too fragile to be used. Also the ohmic loss becomes excessive. It is usual practice to use a copper wire of sufficient diameter to maintain mechanical integrity during the measurement process. A low−dielectric-constant foam can be used to support the modeled wires. The same situation occurs for guy wires and feed lines. Guy wires are usually designed to have little effect on the radiation pattern of the antenna; therefore one would not expect a problem because of the guy wire size.

The radiation pattern of some types of antennas is affected by the choice of wire diameter. The Yagi-Uda antenna as well as the log-periodic antenna are notable examples [51, 92]. If one is primarily interested in the effect of the antenna's environment on its radiation pattern, then it is appropriate to redesign the scaled model to account for the change in wire diameter so that its radiation pattern in free space duplicates that computed for the full-scale antenna alone. When the redesigned scaled antenna is placed in the modeled environment, the measured radiation patterns should replicate those of the full-scale system. Fig. 50 depicts an experimental apparatus for measuring the radiation pattern of an hf antenna where the earth is also simulated.

It is impractical to make input impedance measurements on a 1/100 scale-modeled hf antenna. Not only is the antenna's input impedance affected by the wire size and losses in the image plane, but also by the design of the feed of the

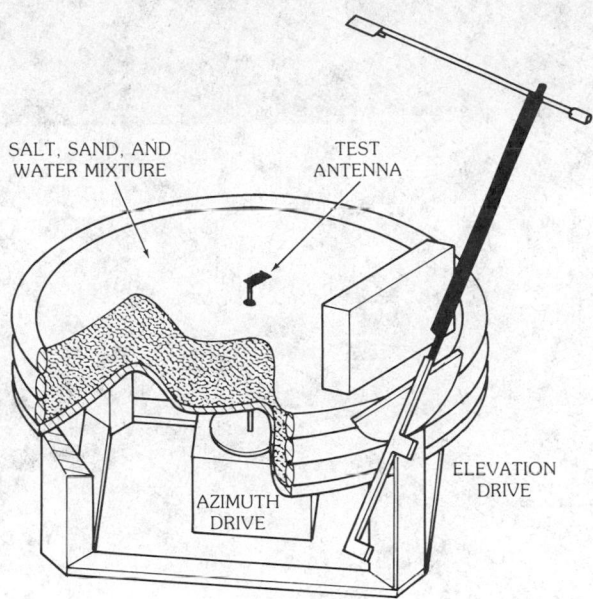

SALT, SAND, AND
WATER MIXTURE

TEST
ANTENNA

ELEVATION
DRIVE

AZIMUTH
DRIVE

Fig. 50. An apparatus for the simulation of ground and measurement of radiation patterns of antennas over ground. (*After Hall, Chambers, and McInnes [51],* © *1978 IEE; published by Peter Peregrinus Ltd., London*)

antenna. All of these are virtually impossible to precisely model with such a large scale factor; therefore impedance models are usually 1/2 to 1/7 of the full-scale antenna. It is often the case that designers of such structures make an engineering model with a scale factor within this range of values for mechanical and scheduling purposes. Such a model can also be used for impedance measurements.

Shipborne Antennas—The electronics systems aboard a large modern warship may require over 100 antennas, all of which must be located in the very restricted space of the ship's topside (79). This results in perhaps the most complex environment in which antenna engineers must place antennas. Modeling techniques are extensively used in the design of the hf antennas required and have been employed for omnidirectional vhf and uhf antennas as well [80]. The superstructure of a ship being physically complex, it must be modeled accurately since it provides many electrical paths commensurate with resonant lengths at hf. Model ships are usually scaled to 1/48 or 1/50 size. To obtain the precision required, the models must be constructed from the ship's actual plans and include all detail above the waterline of the full-scale ship. The construction is usually of wood covered with brass sheet. Turrets, masts, and other such objects are usually machined from brass. The model must include provisions for duplicating the motions of movable structures such as radar antennas, missile launchers, and landing-pad guardrails so that their effect on the characteristic of the antenna being tested can be measured. A strip of brass screen around the base of the model (the waterline of the full-scale ship) provides a means of making an electrical connection to the image-plane turntable. Figs. 51 and 52 picture models of ships illustrating the detail that is required.

Fig. 51. A 1/48 scale model ship for hf antenna radiation pattern measurements. (*Courtesy Naval Ocean Systems Center, San Diego, Calif.*)

Fig. 52. A scale model of a ship soldered to the turntable of an image-plane range. Note the gantry in the background used to rotate the source antenna. (*Courtesy Naval Sea Systems Command, Washington, D.C.*)

Like the model ground-based hf antenna, wire diameters cannot be modeled exactly; however, the radiation patterns for the types of wire antennas used aboard ships are not greatly affected by wire size.

Model shipborne antennas are tested on image-plane ranges such as that depicted in Fig. 6. In that figure a large arch is used to provide the elevation motion of the source antenna. An alternative approach is to use a large nonmetallic boom to provide the semicircular motion required for the source antenna [80, 94, 95]. A corner reflector can be used as a source antenna when propagation along the surface ($\theta = 90°$) is required [96].

The ground plane simulates the saltwater of the ocean. Because saltwater is, in fact, a good conductor, aluminum or wire mesh image planes are usually employed. While it is possible to model the ocean using composite materials, the cost can become prohibitive. This is particularly true in harsh climates where the surface is exposed to extremes in temperature which can cause damage to the surface.

By using a good conductor, such as aluminum, for the surface, the Brewster's angle phenomena will not be duplicated. This, however, is of little consequence at high frequencies since, for seawater, the Brewster's angle occurs at a grazing angle of less than 1°, as shown in Fig. 53.

The instrumentation for radiation patterns is the same as that for other ranges. Impedance measurements can also be performed on the modeled hf antennas. Usually a network analyzer is employed for this purpose. The model antenna is fed from below the image plane. Often all of the required instrumentation is also located under the image plane. Modeled shipborne antennas can also be used to perform electromagnetic compatibility studies.

Aircraft Antennas—Aircraft, like ships, carry many antennas. Table 6 lists the standard communication, navigational, and identification systems found on most aircraft. All of these systems require antennas. In addition, there are many more special-purpose antennas, especially on military aircraft.

For radiation pattern measurements, models 1/7 to 1/10 of the full-scale aircraft size are typical for those antennas which operate at frequencies above about 50 MHz. Models 1/50 or 1/60 full scale are typical for hf antennas. Model aircraft are usually fabricated from patternmaker's white pine or basswood [97]. The model contours are obtained from the aircraft designer's loft drawings photographically reduced to the proper scale and applied to metal sheets from which templates are made. Models are made hollow to accommodate the required instrumentation. The hollow interior also helps to reduce the weight of the model. Removable sections are required to allow access to the interior of the model. Where weight is an important factor the model can be fabricated from a glass cloth–reinforced plastic laminate with a honeycomb core. This type of construction is usually more expensive than wood since a plastic mold is required for the lay-up of the laminate. In some instances it may be advantageous to construct part of the model from wood and part from glass cloth–reinforced plastic laminate [48].

The model must be covered with a good conductor to simulate the metallic skin of the aircraft. Sheets of soft copper can be formed over the model, or the model can be flame-sprayed with copper or aluminum. A copper surface has the

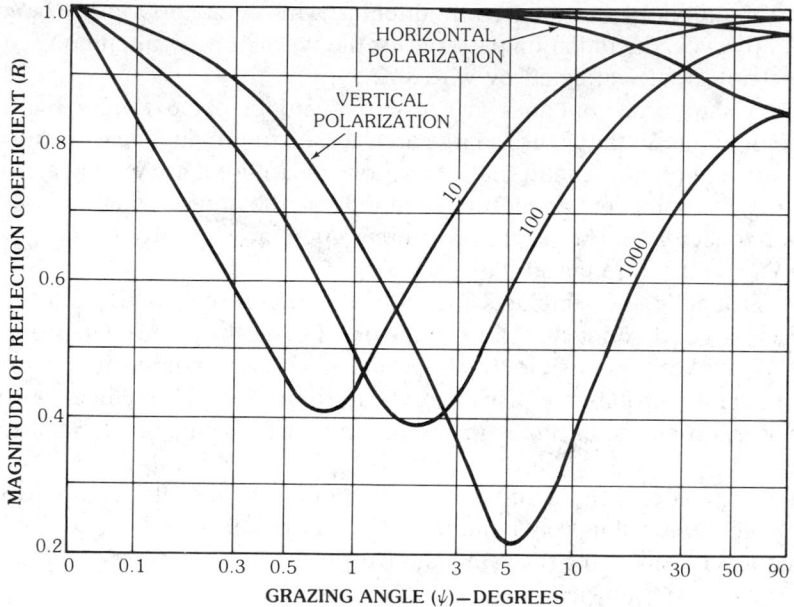

a

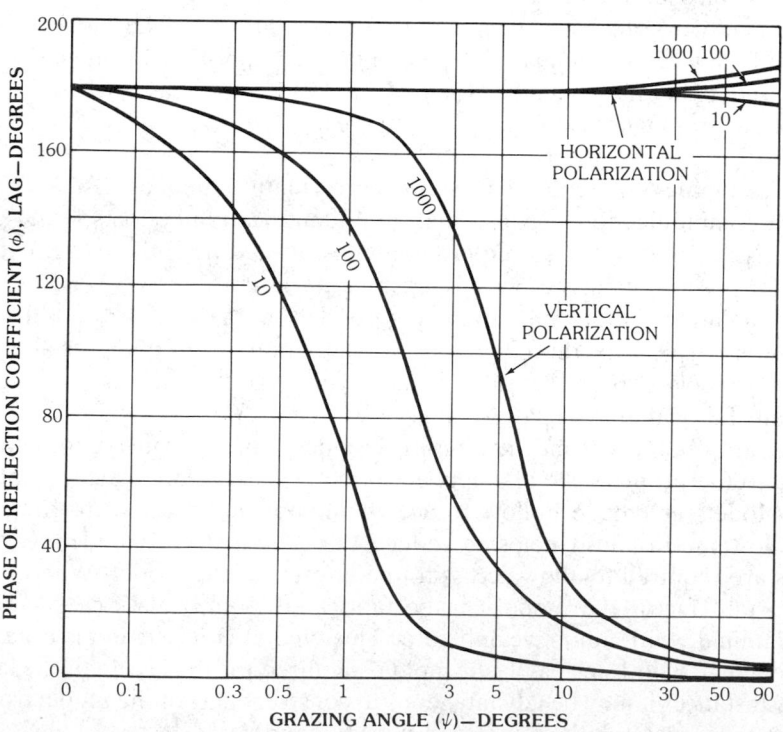

b

Fig. 53. The magnitude and phase of the reflection coefficient as a function of grazing angle for a plane wave incident on smooth seawater ($\epsilon_r = 80$ and $\sigma = 4.64$ S/m) at 10, 100, and 1000 MHz. (*a*) Magnitude of reflection coefficient. (*b*) Phase of reflection coefficient.

Table 6. Communication/Navigational/Identification Systems Requiring Antennas on Aircraft

System	Frequency Range (MHz)
Hf communication	2–30
Marker beacon	75
Localizer	108–112
VOR	108–122
Vhf communications	116–156
Uhf communications	225–400
Glide slope	329–335
Tacan	960–1020
DME	962–1024
Control radar beacon system	1030–1090
System	1227.60–1575.42
Radio altimeter	4300

advantage of being readily solderable, which is important when mounting antennas onto the model. A metal thickness of about 0.05 cm is usually satisfactory. A model aircraft for radiation pattern measurements is shown in Fig. 54.

Models that are physically large are more conveniently fabricated with wooden formers covered by thin plywood veneer and metal foil. Hot-dipped hard-ware cloth with 1/8-in (3.175-mm) mesh can also be used to simulate the metallic surface of the aircraft. Copper screen is not recommended because the screen wires are not bonded together, and therefore, after oxidation takes place, it will not adequately simulate a continuous conducting surface. If a section of the model has a constant cross section, then an aluminum sheet can be used.

As a rule of thumb, the dimensional tolerances required for the external surface of the model is a sixteenth of a wavelength at the highest model frequency used. This is usually about 0.064 cm. All protuberances on the full-scale vehicle which are a tenth of a wavelength or larger should be included in the model if they are in the vicinity of the antenna to be tested. These include other antennas, which should be appropriately terminated to simulate the actual full-scale environment of the test antenna. It is very important that all thin wires, sharp edges, and surface discontinuities be modeled if they are within two or three wavelengths of the antenna under test. Flaps, retracting landing gear, and other movable structures must be modeled.

Generally, if the dielectric in the full-scale system is electrically very thin, it can be neglected in the model; or, more properly stated, it can be modeled as free space (air). For example, an aircraft model for the study of scaled hf, vhf, or uhf antennas would not have to include the plastic canopy or nose radome. However, all metallic parts such as frames, pitot tubes, and lightning diverters, which are built into the canopy or radome, must be modeled along with the metallic structures enclosed by canopy and radome. One has to be especially careful if the enclosed parts are of resonant dimensions at the frequency of operation.

There are situations in which even electrically thin dielectrics affect the radiation pattern of antennas. These are cases where surface waves are excited or lens effects are prominent [98, 99]. One such example is a radome of circular cross

Fig. 54. A copper-clad model airplane mounted on a model tower in a large microwave anechoic chamber. (*Courtesy MBB, Munich, West Germany*)

section enclosing an end-fire surface-wave antenna [100]. If the average diameter of the radome is on the order of a wavelength or less, then it is, in fact, an integral part of the antenna even though it may be electrically thin and hence should be included in the model. The lenticularly shaped rotadome enclosing a uhf antenna is another such example [101].

Radomes are sometimes of a sandwich-type construction with a honeycomb core and glass fiber–reinforced epoxy or polyester laminates on either side. This type of radome can be satisfactorily modeled using a single sheet of dielectric which has the same electrical thickness [99], although, if the scale factor is not too great or the physical dimensions too small, one may model the sandwich construction using low-loss foam in place of the honeycomb for the model [101].

Aircraft antennas usually cannot be modeled exactly. This can be due to the complexity of their structures or a lack of detailed information from their manufacturers. To ensure that model antennas will reproduce the radiation characteristics of the full-scale antennas, a preliminary measurement, if possible, is advisable. For example, the radiation pattern of a stub-type antenna, which is to be modeled, can be measured with the antenna mounted on a flat, circular, ground plane. The measurement is repeated with the corresponding modeled antenna mounted on a scaled, circular, ground plane. A comparison of the patterns will reveal the adequacy of the scale-model antenna. Convenient sizes of the full-scale ground planes range from 0.85 to 3.5 m in diameter, depending on the frequency used. The modeled ground plane usually requires a quarter-wave choke on its periphery or a ring of absorbing material on its back side to reduce the currents that flow around its edge.

Once the model antenna's radiation patterns have been shown to replicate those of the full-scale antenna mounted on the flat ground plane, it is then ready for installation on the aircraft model. The use of a standard mounting system for model antennas is highly desirable. One approach is to use a standard brass receptacle which can be soldered in place on the model aircraft. The model antenna is mounted on a mating plug terminated with a miniature coaxial connector. The antenna-plug assembly can be inserted into the receptacle from the outside of the model. This method is most useful when one is making design changes in an antenna structure or when standard antennas are used on successive models. Fig. 55 depicts one type of receptacle and plug arrangement.

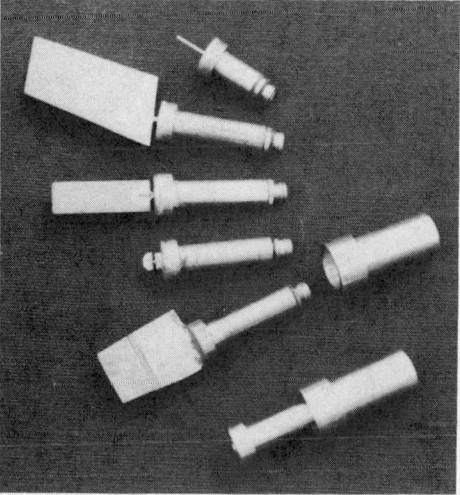

Fig. 55. A receptacle and plug assembly for mounting antennas onto model aircraft. (*Courtesy Lockheed-California Corp.*)

To perform radiation pattern measurements, the model aircraft is mounted on a special roll-over-azimuth positioner with a nonmetallic mast. These positioners are called model towers. The remote mixer or any other instrumentation that must be located at the test antenna is placed inside the model. Care has to be taken to mount the devices directly to the model structure so that there will be no unwanted movement. The coaxial cable between the remote mixer and the receiver exits the model at the point where the model is mounted to the model tower. The antenna being measured must be on the opposite side of the model from the mount. Care must be exercised to ensure that the cable does not affect the measurements. Two blade-type antennas mounted on a model aircraft are shown in Fig. 56.

A severe problem arises when the model aircraft is small compared with the wavelength at the test frequency. This situation arises when hf antennas are being modeled, in which case the model may be physically small also. Thus the head of the model tower and the cable from the model to the receiver can greatly affect the radiation patterns. To reduce the effect of the model tower head, the model can be mounted on a plastic rod and located above the model tower. The cable effect can be eliminated by use of a battery-operated transmitter located within the model. For this approach the receiver is connected to the source antenna. Another method is to use a fiber-optic waveguide between the model and the base of the model tower [97]. The optical transmitter must be mounted inside the model. Finally, one can use an amplitude-modulated rf signal of sufficient power to illuminate the model. The modulation frequency is usually 1 kHz. The detector is then located in the model aircraft. Instead of using coaxial cable to couple the audio signal to the amplifier, a carbon-loaded, high-impedance cable can be employed. This type of cable has little effect on the radiation pattern of the model. A system of this type requires higher power than does the conventional mixer-receiver system and does not have the selectivity of a receiver to discriminate against unwanted signals.

Fig. 56. Model blade antennas mounted on a model aircraft. (*Courtesy Northrop Corp.*)

It is standard practice to determine the directivity of the test antenna by pattern integration. The isotropic level can then be located on the graphical display of the radiation pattern. The radiation patterns measured on the model aircraft are ultimately compared to measurements made on the aircraft flying over a prescribed path.

With the exception of the hf antennas, impedance measurements are made

Fig. 57. A 1/24 scale-model helicopter mounted on a model tower. (*Courtesy National Research Council, Canada*)

at full scale on partial mock-ups of the aircraft. To measure the impedances of hf antennas, 1/4 to 1/7 scale models are usually employed. The model should be located at least a wavelength above the ground. All instrumentation must be located within the model and remotely controlled.

In addition to airplanes, helicopters, satellites, spacecraft, rockets, and projectiles of various types have systems that require antennas. Modeling techniques are often used in the design and study of these antennas [102]. Figs. 57 and 58 depict two such models.

Fig. 58. A 1/2.5 scale model of the German Communication Satellite DFS for testing the telemetry and telecommand antenna. The measurement was performed on the spherical near-field range of the Technical University of Denmark. (*Courtesy MBB, Munich, West Germany*)

6. References

[1] *IEEE Standard Definitions of Terms for Antennas*, IEEE Std. 145-1982, IEEE, New York, NY.

[2] *ANSI/IEEE Standard Test Procedures for Antennas*, ANSI/IEEE Std. 149-1979, IEEE, New York: John Wiley distributors.

[3] R. C. Johnson, H. A. Ecker, and R. A. Moore, "Compact range techniques and measurements," *IEEE Trans. Antennas Propag.*, vol. AP-17, pp. 563–576, September 1969.

[4] C. C. Cutler, A. P. King, and W. E. Kock, "Microwave antenna measurements," *Proc. IRE*, vol. 35, pp. 1462–1471, December 1947.

[5] J. S. Hollis, T. J. Lyon, and L. Clayton, Jr., eds., *Microwave Antenna Measurements*, Atlanta: Scientific-Atlanta, 1970.

[6] R. C. Johnson, H. A. Ecker, and J. S. Hollis, "Determination of far-field antenna patterns from near-field measurements," *Proc. IEEE*, vol. 61, pp. 1668–1694, December 1973.

[7] W. H. Kummer and E. S. Gillespie, "Antenna measurements—1978," *Proc. IEEE*, vol. 66, pp. 483–507, April 1978.

[8] J. Appel-Hansen, J. D. Dyson, E. S. Gillespie, and T. G. Hickman, "Antenna measurements," chapter 8 in *The Handbook of Antenna Design*, vol. 1, ed. by A. W. Rudge et al., Stevenage, Herts., UK: Peter Peregrinus, 1982, pp. 584–694.

[9] A. C. Ludwig, "The definition of cross-polarization," *IEEE Trans. Antennas Propag.*, vol. AP-21, pp. 116–119, January 1973.

[10] A. J. Lombardi and M. S. Polgar, Jr., "Wide band antenna facility," *Elec. Commun.*, vol. 49, pp. 94–98, January 1974.

[11] L. G. Sturgill and S. E. Thomas, "Scale model shipboard antenna measurements with a computer automated antenna analyzer system," *Proc. Antenna Meas. Techniques Assoc. Mtg.*, Atlanta, October 17–18, 1979.

[12] D. E. Fessenden and D. C. Portofee, "An antenna pattern arch for measurements over sea water," *Proc. Antenna Meas. Techniques Assoc. Mtg.*, Las Cruces, October 5–7, 1982.

[13] C. G. Bachman, H. E. King, and R. C. Hansen, "Techniques for measurements of reduced radar cross-sections, pt. 1," *Microwave J.*, vol. 6, pp. 61–67, January 1963.

[14] D. F. DiFonzo and W. J. English, "Far-field criteria for reflectors with phased array feeds," *1974 Intl. IEEE/AP-S Symp. Dig.*, IEEE, New York.

[15] P. S. Hacker and H. E. Schrank, "Range distance requirements for measuring low and ultra-low sidelobe antenna patterns," *IEEE Trans. Antennas Propag.*, vol. AP-30, pp. 956–966, September 1982.

[16] R. C. Hansen, "Measurement distance effects on low sidelobe patterns," *IEEE Trans. Antennas Propag.*, vol. AP-32, pp. 591–594, June 1984.

[17] D. E. Baker, "Development and evaluation of the 500m ground-reflection antenna test range of CSIR, Pretoria, South Africa," *Proc. Antenna Meas. Techniques Assoc. Mtg.*, San Diego, October 2–4, 1984.

[18] P. W. Arnold, "The slant antenna range," *IEEE Trans. Antennas Propag.*, vol. AP-14, pp. 658–659, September 1966.

[19] W. H. Emerson, "Electromagnetic wave absorbers and anechoic chambers through the years," *IEEE Trans. Antennas Propag.*, vol. 21, pp. 484–490, July 1973.

[20] J. A. Strom and W. G. Mavroides, "RADC electromagnetics test facility at Ipswich, Ma.," *Proc. 1982 Antenna Meas. Techniques Assoc. Mtg.*, Las Cruces, October 5–7, 1982.

[21] J. Appel-Hansen, "Reflectivity level of radio anechoic chamber," *IEEE Trans. Antennas Propag.*, vol. AP-21, pp. 490–498, July 1973.

[22] W. J. English and R. W. Gruner, "Precision polarization measurement techniques for circularly polarized antennas," *Dig. IEEE AP-S Intl. Symp. Antennas Propag.*, University of Illinois (IEEE catalog no. 75CH0965-9AP).

[23] J. R. Mentzer, *Scattering and Diffraction of Radar Waves*, London: Pergamon Press,

1955, pp. 124–126.

[24] D. W. Hess, F. G. Willwerth, and R. C. Johnson, "Compact range improvements and performance at 30 GHz," *Dig. Intl. Symp. Antennas Propag.*, Stanford Univ., June 20–22, 1977, pp. 264–267.

[25] W. D. Burnside, M. C. Gilreath, and B. Kent, "A rolled edge modification of compact range reflector," *Proc. Antenna Meas. Techniques Assoc. Mtg.*, San Diego, October 2–4, 1984.

[26] W. D. Burnside, A. K. Dominek, and R. Barger, "Blended surface concepts for a compact range reflector," *Proc. Antenna Meas. Techniques Assoc. Mtg.*, Melbourne, Fla., October 29–31, 1985.

[27] W. D. Burnside and L. Peters, Jr., "Target illumination requirements for low RCS target measurements," *Proc. Antenna Meas. Techniques Assoc. Mtg.*, Melbourne, Fla., October 29–31, 1985.

[28] D. W. Hess and R. C. Johnson, "Compact ranges provide accurate measurement of radar cross section," *Microwave Syst. News*, vol. 12, no. 9, September 1982.

[29] T. G. Hickmann and R. C. Johnson, "Boresight measurements utilizing a compact range," *Abstracts, Spring USNC/URSI Mtg.*, Washington, D.C., April 13–15, 1972.

[30] R. C. Rudduck, W. D. Burnside, M. C. Liang, and T. H. Lee, "The compact range as an electromagnetic field simulator," *Proc. Antenna Meas. Techniques Assoc. Mtg.*, Melbourne, Fla., October 29–31, 1985.

[31] V. J. Vokurka, "9.3 compact range," in *Satellite Communication Antenna Technology*, ed. by R. Mittra et al., Amsterdam: Elsevier Science Publishers B.V. (North-Holland), 1983.

[32] J. K. Conn, C. L. Armstrong, and L. S. Gans, "A dual shaped compact range for ehf antenna measurements," *Proc. Antenna Meas. Techniques Assoc. Mtg.*, San Diego, October 2–4, 1984.

[33] S. W. Zieg, "A precision optical range alignment technique," *Proc. Antenna Meas. Techniques Assoc. Mtg.*, Las Cruces, October 5–7, 1982.

[34] J. Appel-Hansen, "Reflectivity level of radio anechoic chambers," *IEEE Trans. Antennas Propag.*, vol. AP-21, pp. 490–498, July 1973.

[35] E. F. Buckley, "Outline of evaluation procedures for microwave anechoic chambers," *Microwave J.*, vol. 6, pp. 69–75, August 1963.

[36] C. A. Balanis, *Antenna Theory: Analysis and Design*, New York: Harper & Row, 1982, pp. 37–42.

[37] R. J. Stegen, "The gain-beamwidth product of an antenna," *IEEE Trans. Antennas Propag.*, vol. AP-12, pp. 505–506, July 1966.

[38] R. S. Elliott, *Antenna Theory and Design*, Englewood Cliffs: Prentice-Hall, 1981, pp. 205–207.

[39] R. R. Bowman, "Field strength above 1 GHz: measurement procedures for standard antennas," *Proc. IEEE*, vol. 55, pp. 981–990, June 1967.

[40] J. Bellamy, J. Hill, and S. Wilson, "Automatic gain measurement system," *Proc. Antenna Meas. Techniques Assoc. Mtg.*, September 27–29, 1983.

[41] A. C. Newell, R. C. Baird, and P. F. Wacker, "Accurate measurement of antenna gain and polarization at reduced distances by an extrapolation technique," *IEEE Trans. Antennas Propag.*, vol. AP-21, pp. 418–431, July 1973.

[42] L. H. Hemming and R. A. Heaton, "Antenna gain calibration on a ground reflection range," *IEEE Trans. Antennas Propag.*, vol. AP-21, pp. 532–538, July 1973.

[43] H. G. Booker, V. H. Rumsey, G. A. Deschamps, M. I. Kales, and J. I. Bohnert, "Techniques for handling elliptically polarized waves with special reference to antennas," *Proc. IRE*, vol. 39, pp. 533–552, May 1951.

[44] J. S. Hollis, et al. *Microwave Antenna Polarization Measurements*, to be published, Scientific-Atlanta, Inc.

[45] J. R. Jones and D. W. Hess, "Automated three-antenna polarization measurements using digital signal processing," *Proc. Antenna Meas. Techniques Assoc. Mtg.*, Melbourne, Fla., October 29–31, 1985.

[46] A. C. Newell, "Improved polarization measurements using a modified three-antenna

technique," *Dig. IEEE Intl. Antennas Propag. Symp.*, Urbana-Champaign, Ill., pp. 337–340, June 2–4, 1975.

[47] L. Clayton, Jr., and J. S. Hollis, "Antenna polarization analysis by amplitude measurement of multiple components," *Microwave J.*, vol. 8, pp. 35–41, January 1965.

[48] E. L. Kane, "Scale model aircraft antenna measurements," *Paper 7002*, Douglas Aircraft Company, Long Beach, 1980.

[49] C. K. Krichbaum, "Radar cross-section measurements," in *Radar Cross-section Handbook, Vol. 2*, ed. by George T. Ruck, New York: Plenum Press, 1970, pp. 893–935.

[50] E. L. Kilpatrick and H. V. Cottony, "Scaled model measurements of performance of an elevated sloping vee antenna," *ESSA Tech. Rep. ERL 72-ITS62*, US Department of Commerce, ESSA Research Laboratories, Boulder, 1968.

[51] P. S. Hall, B. Chambers, and P. A. McInnes, "Microwave modelling of h.f. Yagi antennas over imperfectly-conducting ground," *Proc. IEE*, vol. 125, no. 4, pp. 261–266, 1978.

[52] S. Toyada and H. Hashimoto, "Scattering characteristics of vhf television broadcasting waves by steel towers of overhead power transmission lines," *IEEE Trans. Electromagnetic Compat.*, vol. EMC-21, no. 1, pp. 62–65, 1979.

[53] V. V. Liepa and R. L. Frank, "Experimental scale model study of loran-C signals near bridges," *1983 Intl. Symp. Dig. Antennas Propag.*, vol. 2, pp. 545–547 (IEEE catalog no. 83CH1860-6).

[54] T. B. A. Senior and V. V. Liepa, "Scale modeling," in *EMP Interaction: Principles, Techniques and Reference Data*, ed. by K. S. H. Lee, Air Force Weapons Laboratory, EMP Interaction 2-1, AFWL-TR-80-402, Albuquerque, 1980.

[55] G. V. Keller and F. C. Frischknecht, *Electrical Methods in Geophysical Prospecting*, New York: Pergamon Press, 1966, pp. 295–299.

[56] F. C. Frischknecht, "Electromagnetic scale modelling," in *Electromagnetic Probing in Geophysics*, ed. by J. R. Wait, Boulder: Golem Press, 1971, pp. 265–320.

[57] E. Bahar, "Propagation of vlf radio waves in a model earth-ionosphere waveguide of arbitrary height and finite surface impedance boundary: theory and experiment," *Radio Sci.*, vol. 1, no. 8, pp. 925–938, 1966.

[58] D. E. Winder, I. C. Peden, and H. M. Swarm, "A 3 GHz scale model of submerged vlf antenna using lossy ceramic powder," *IEEE Trans. Antennas Propag.*, vol. AP-14, no. 4, pp. 507–509, 1966.

[59] G. E. Webber and I. C. Peden, "Vlf mode analysis using a ceramic dielectric model of the earth-ionosphere waveguide," *IEEE Trans. Antennas Propag.*, vol. AP-17, no. 5, pp. 613–620, 1969.

[60] R. J. King, "Physical modeling of EM wave propagation over the earth," *Radio Sci.*, vol. 19, no. 5, pp. 1103–1116, 1982.

[61] R. W. P. King and G. S. Smith, *Antennas in Matter*, Cambridge: MIT Press, 1981, pp. 727–818.

[62] R. F. Harrington, *Field Computations by Moment Methods*, New York: Macmillan, 1968.

[63] J. Perini and D. J. Buchanan, "Assessment of MOM techniques for shipboard applications," *IEEE Trans. Electromagnetic Compat.*, vol. EMC-L4, no. 1, pp. 32–39, 1982.

[64] J. B. Keller, "Geometrical theory of diffraction," *J. Opt. Soc. Am.*, vol. 52, no. 2, pp. 116–130, 1962.

[65] W. D. Burnside, M. C. Gilreath, R. J. Marhefka, and C. L. Yu, "A study of KC-135 aircraft antenna patterns," *IEEE Trans. Antennas Propag.*, vol. AP-23, no. 3, pp. 309–316, 1975.

[66] C. L. Yu, W. D. Burnside, and M. C. Gilreath, "Volumetric pattern analysis of airborne antennas," *IEEE Trans. Antennas Propag.*, vol. AP-26, no. 5, pp. 636–641, 1978.

[67] W. D. Burnside, N. Wang, and E. L. Pelton, "Near-field pattern analysis of airborne

antennas," *IEEE Trans. Antennas Propag.*, vol. AP-28, no. 3, pp. 318–327, 1980.

[68] R. K. Ritt, "The modeling of physical systems," *IEEE Trans. Antennas Propag.*, vol. AP-4, no. 3, pp. 216–218, 1956.

[69] J. Tykocinski-Tykociner, "Investigation of antennas by means of models," *Bulletin 22, no. 39*, University of Illinois, Urbana, 1925.

[70] G. Sinclair, E. C. Jordan, and E. W. Vaughn, "Measurement of aircraft-antenna patterns using models," *Proc. IRE*, vol. 35, no. 12, pp. 1451–1462, 1947.

[71] J. V. N. Granger and T. Morita, "Radio-frequency current distributions on aircraft structures," *Proc. IRE*, vol. 39, no. 8, pp. 932–938, 1951.

[72] E. F. Harris, "Investigating antennas for uhf mobiles," *Electronics*, vol. 25, no. 11, pp. 127–129, 1952.

[73] J. T. Bolljahn and R. F. Reese, "Electrically small antennas and the low-frequency aircraft problem," *IRE Trans. Antennas Propag.*, vol. AP-1, no. 2, pp. 46–54, 1953.

[74] W. Sichak and J. J. Nail, "Uhf omnidirectional antenna systems for large aircraft," *IRE Trans. Antennas Propag.*, vol. AP-2, no. 1, pp. 6–15, 1954.

[75] I. Carswell, "Current distribution on wing-cap and tail-cap antennas," *IRE Trans. Antennas Propag.*, vol. AP-3, no. 4, pp. 207–212, 1955.

[76] R. E. Webster, "20–70 mc monopole antennas on ground-based vehicles," *IRE Trans. Antennas Propag.*, vol. AP-5, no. 4, pp. 363–368, 1957.

[77] R. L. Tanner, "Shunt and notch-fed hf aircraft antennas," *IRE Trans. Antennas Propag.*, vol. AP-6, no. 1, pp. 35–43, 1958.

[78] R. A. Burberry, "Progress in aircraft aerials," *Proc. IEE*, vol. 109, pt. B, no. 48, pp. 431–444, 1962.

[79] P. E. Law, "Accommodating antenna systems in the ship design process," *Naval Engineer's J.*, vol. 91, no. 2, pp. 65–75, 1979.

[80] K. F. Woodman and B. E. Stemp, "Shipborne antenna modeling," *Microwave J.*, vol. 22, no. 4, pp. 73–81, 1979.

[81] G. Sinclair, "Theory of models of electromagnetic systems," *Proc. IRE*, vol. 36, no. 11, pp. 1364–1370, 1948.

[82] R. F. Harrington, *Time-Harmonic Electromagnetic Fields*, New York: McGraw-Hill Book Co., 1961, pp. 100–103.

[83] A. R. von Hippel, *Dielectric Materials and Applications*, Cambridge: MIT Press, 1954.

[84] E. C. Jordan, ed., *Reference Data for Engineers: Radio, Electronics, Computer and Communications*, 7th ed., Indianapolis: Howard W. Sams, Inc., 1985.

[85] J. F. Ramsey, B. V. Popovich, and J. F. Gobler, "Research on compact and efficient antennas," *Report ECOM-02111-F*, Airborne Instruments Laboratory, 1967.

[86] L. S. Taylor, "Dielectric properties of mixtures," *IEEE Trans. Antennas Propag.*, vol. AP-13, pp. 943–947, 1965.

[87] A. M. Shutko and E. M. Reutov, "Mixture formulas applied in estimation of dielectric and radioactive characteristics of soils and grounds at microwave frequencies," *IEEE Trans. Geosci. Remote Sensing*, vol. GE-20, pp. 29–32, 1982.

[88] R. J. King, D. V. Theil, and K. S. Park, "The synthesis of a surface reactance using artificial dielectric," *IEEE Trans. Antennas Propag.*, vol. AP-31, pp. 471–476, 1983.

[89] M. F. Radford, "High-frequency antennas," chapter 16 in *Handbook of Antenna Design*, vol. 2, ed. by A. W. Rudge et al., London: Peter Peregrinus, 1982, pp. 663–724.

[90] R. Wilensky and W. Wharton, "High-frequency Antennas," chapter 26 in *Antenna Engineering Handbook*, 2nd ed., ed. by R. C. Johnson and H. Jasik, New York: McGraw-Hill Book Co., 1984, pp. 26-1 to 26-41.

[91] D. Wobschall, "A theory of the complex dielectric permittivity of soil containing water: the semidisperse model," *IEEE Trans. Geosci. Electron.*, vol. GE-15, pp. 49–58, January 1977.

[92] E. D. Sharpe and R. L. Tanner, "Scale modeling of high-frequency antennas," *IEEE Trans. Antennas Propag.*, vol. AP-17, pp. 810–811, November 1969.

[93] P. E. Law, Jr., *Shipboard Antennas*, 2nd ed., Norwood, Massachusetts: Artech House, 1986.

[94] J. Y. Wong and J. C. Barnes, "Design and construction of a pattern range for testing high-frequency shipborne antennas," *Trans. Eng. Inst. Canada*, vol. 2, pp. 2–8, January 1958.

[95] J. Y. Wong and J. C. Barnes, "A new facility for measuring radiation patterns of model shipborne antennas," *The Engineering J.*, the Engineering Institute of Canada, pp. 3–4, June 1961.

[96] K. M. Keen, R. R. Grime, and B. E. Stemp, "Improvements to a surface-wave antenna measurement range with troublesome site effects," *Electron. Lett.*, vol. 18, pp. 439–440, May 27, 1982.

[97] C. D. Widell, "Modeling techniques," *Lecture Notes Microwave Antenna Measurements, Short Course*, California State University, Northridge, 1977.

[98] E. S. Gillespie, "The effect of radome geometry on the radiation pattern of a retarded wave antenna," *Proc. Ninth Ann. Symp. USAF Antenna Res. Dev.*, Univ. of Illinois, October 1959.

[99] E. S. Gillespie and R. S. Elliott, "The effects of a radome on a surface wave antenna," *LR 15,297*, Lockheed Aircraft Corp., Burbank, September 1961.

[100] D. D. Batson, "Retarded wave antenna for AN/APG-30 installed in model PLV-7," *LR 12,258*, Lockheed Aircraft Corp., Burbank, April 1957.

[101] I. Alne, R. Taron, E. S. Gillespie, and F. R. Zboril, "Evaluation of the WV-2E airframe interference on the rotadome uhf and IFF patterns, pt. 2," *LR 12,618*, Lockheed Aircraft Corp., Burbank.

[102] J. H. Lemanczyk, D. Fasold, and H.-J. Steiner, "Development and measurements of the TM-TC antenna for the DFS satellite," *Proc. Fourth IEEE Intl. Conf. Antennas Propag.*, Univ. of Warwich, Coventry, UK, April 1985.

Chapter 33

Near-Field Far-Field Antenna Measurements

Jørgen Appel-Hansen

Electromagnetics Institute, Technical University of Denmark

CONTENTS

 **Jørgen Appel-Hansen** was born in Farendløse, Denmark, on March 13, 1937. He received an MSc degree in electrical engineering in 1962 from the Technical University of Denmark, Lyngby. From 1962 to 1964 he was in the Royal Danish Army, where part of his time was spent as a teaching associate at the Physics Laboratory II, Technical University of Denmark. In 1964 he joined the Electromagnetics Institute, Technical University of Denmark, where he received the PhD degree in 1966. In 1967 he was a visiting research associate at the University of Michigan, Ann Arbor. From 1968 to 1978 he worked as head of the Radio Anechoic Chamber, Electromagnetics Institute.

Also, in 1972 he gave a postgraduate course in antenna measurements at the National Physical Laboratory, New Delhi, India. In May 1978, he was a visiting professor at Universitá di Napoli, Italy. In 1979 and 1981 he was a lecturer at the Microwave Antenna Measurements Short Course at the California State University, Northridge. He gave a series of lectures at the Beijing Municipal Institute of Labour Protection, the Shanghai Electrical Apparatus Research Institute, and Jiao Tong University, China, in 1984.

Dr. Appel-Larsen is currently an associate professor at the Electromagnetics Institute. He is also a consultant in design of anechoic chambers, test ranges and their instrumentation, and the implementation of measurement procedures.

1. Introduction

Over the past 15 years major efforts in antenna measurement studies have been invested in improving measurement accuracy and shortening the required test distance. It is interesting to note that some of the techniques seem to have managed to meet both objectives. As a result these techniques, in particular the scanning techniques and the compact range, have been implemented in various forms on several test ranges.

The purpose of this chapter is to give a concise introduction to established near-field far-field measurement techniques. These are here broadly defined as techniques which use test distances shorter than those obtained by using the conventional far-field criterion as derived in the next section. The presentation is based on a previous work [1] published in Rudge et al. [2]. The present work differs from the previous one in two respects. First, several mathematical derivations and discussions of literature earlier than 1980 in the previous work are shortened or omitted. This, it is hoped, will facilitate the reader's understanding of the basic ideas of the techniques treated. Second, the present chapter includes references to works carried out after 1980. This means that the following sections include the major progress in the near-field techniques over the past three years.

In Section 2 some general concepts are presented and the treated techniques are related to each other with respect to their test distance. Basically, far-field properties may be determined from knowledge of the current distribution on the antenna under test. Therefore, although current measurement techniques are not often used, these are touched upon briefly in Section 3. In Sections 4 to 8, the scanning techniques are dealt with in some detail. In comparison with the previous work, a more condensed mathematical notation is used and some error budgets are presented.

In Section 9 the plane-wave synthesis technique is described. The commonly used compact range, defocusing, and extrapolation techniques are outlined in Sections 10, 11, and 12, respectively.

2. General Concepts

The radiating properties of an antenna depend on the distance from the antenna structure to the considered field point. From an examination of the field from a Hertzian dipole, it turns out that the field expressions involve terms which vary as r^{-1} (radiated field), r^{-2} (inductive field), and r^{-3} (static field), where r is the distance between the dipole and the field point [3]. It is easily shown that for

$$r > \lambda/2\pi \tag{1}$$

the terms involving the inverse distance dependence r^{-1} are the largest ones. Since any antenna structure may be considered as composed of small dipoles, the region around an antenna structure may be divided into two major regions, viz., a reactive near-field region close to the antenna and a radiating-field region. The boundary between the two regions is determined by (1). An analysis of the instantaneous power flow density in space shows that this can be expressed as a sum of terms. Only the radiated field contributes a constant term, which gives the average power flow density away from the antenna. All types of fields contribute to the other terms which indicate sinusoidal varying power flow in space. Terms which are due to inductive and static fields can usually be neglected 10 wavelengths away from the antenna.

As the distance tends to infinity, the angular dependence of antenna properties becomes independent of distance. In the limit at infinite distance the properties are characterized as true far-field parameters. In practice there are several advantages to measuring the antenna properties at short distances. However, usually antennas are used for large distances. Therefore, measurements have to be carried out at such distances that the measured parameters agree with the far-field parameters within a certain specified accuracy. In each case the minimum distance depends on the type of antenna, its dimension, the wavelength λ, the measurement technique, the manner in which the data are dealt with, the measured property, and the specified accuracy. Various criteria for far-field measurements have obtained general acceptance [4]. As an example, taking the most popular criterion, the required distance for broadside major lobe pattern measurements on uniformly illuminated aperture antennas, which have a largest cross-sectional diameter D of several wavelengths, the minimum range R is given by

$$R = 2D^2/\lambda \tag{2}$$

Then, when the antenna is illuminated by a point source, the phase variation over the antenna aperture is less than $\pi/8$ rad. The criterion is easily derived from Fig. 1a and it is often referred to as the conventional far-field criterion. It subdivides the radiating-field region into two regions, the radiating near-field region and the radiating far-field region. The three major field regions are illustrated in Fig. 1b.

When the criterion (2) is satisfied, the measured data usually represent the far-field properties. However, at short wavelengths, demanded in particular for satellite communication, the criterion may require a test range which is long. This creates several disadvantages. Measurements may have to be carried out in open air with possible adverse weather conditions. They may also be disturbed by other signals, and obstacle reflections may give inaccurate results. Reduction of ground reflections requires high towers and special arrangements in the form of fences and absorbers around the specular reflection point. The long test distance requires high power or reduced dynamic range. The open-air range may also have to be placed inconveniently with respect to other activities. Thus, there is a need for indoor near-field measurements under fully controlled conditions. Mainly, two approaches have been followed to meet this objective. In one approach, test setups giving far-field data directly from measurements at short distances are invented. In

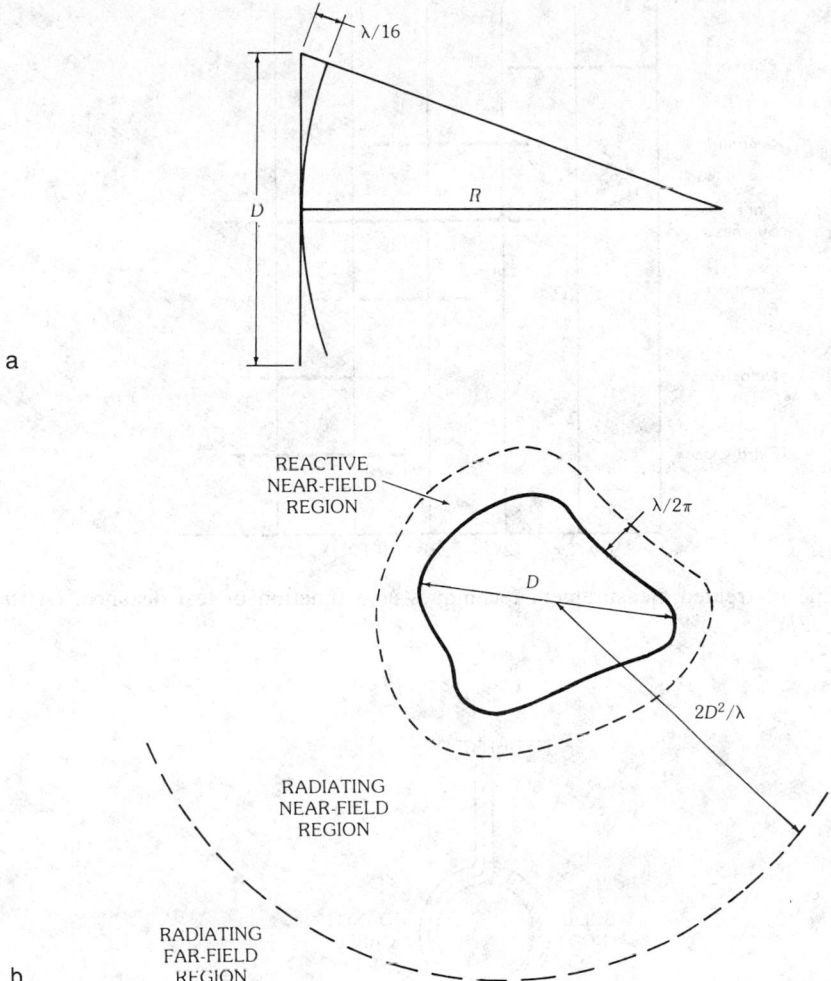

Fig. 1. Major field regions. (*a*) Deriving far-field criterion. (*b*) Three field regions.

the other approach the measured data are transformed mathematically to predict the true far-field data.

The measurement techniques to be described in this chapter are outlined in Fig. 2. The techniques are shown in a framework characterizing their typical test distances.

3. Current Distribution Measurements

Measurement of the surface current on an antenna structure is usually carried out with a probe placed in the reactive-field region as close to the surface as possible. For example, the magnetic field **H** may be measured in phase and amplitude with a circular loop as shown in Fig. 3. Then the surface current density **J** may be found from

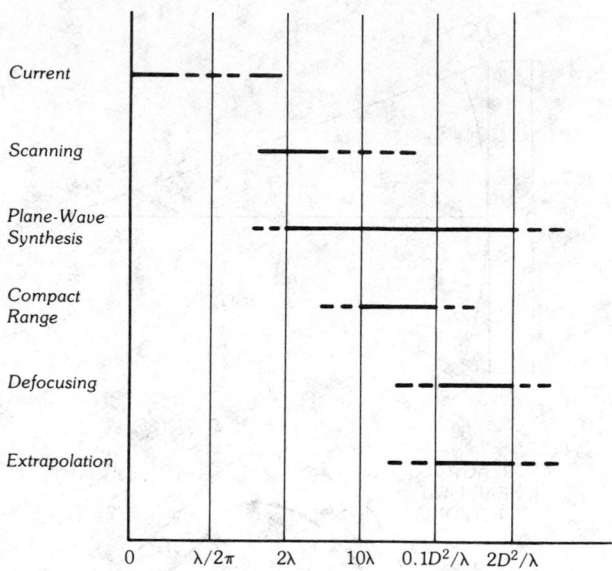

Fig. 2. Outline of treated measurement techniques as a function of test distance. (*After Rudge et al. [2]*)

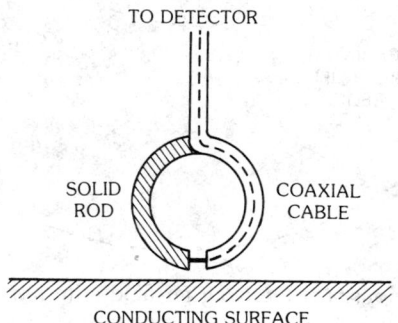

Fig. 3. Current measurement.

$$\mathbf{J} = \hat{\mathbf{n}} \times \mathbf{H} \tag{3}$$

where $\hat{\mathbf{n}}$ is a unit vector pointing perpendicularly out from the surface toward the loop. The loop should be sufficiently small so it does not disturb the current seriously. From the measured current distribution the radiating properties of the antenna structure can be calculated. A detailed discussion of various probes and supporting mechanisms may be found in Dyson's paper [5]. Current distribution measurements are not generally used for antennas. However, the current distribution measurement is mentioned here because it is the near-field technique which uses the shortest distance.

4. Planar Scanning

The simplest scanning technique to treat mathematically is the planar scanning. It was first described by Kerns and Dayhoff [6] using a scattering matrix formulation. A detailed discussion has been given recently [7]. By using the Lorentz reciprocity theorem Brown and Jull [8] treated the two-dimensional cylindrical case. Following the steps of this derivation Jensen [9] and Leach and Paris [10] derived the coupling equation for scanning on a sphere and cylinder, respectively. In Sections 4 to 8 the three types of scanning shall be introduced by following the scattering matrix formulation. This has also been adopted by Yaghjian [11] for the cylindrical case and by Wacker [12] and Larsen [13] in the spherical case. The equivalence between the Lorentz reciprocity theorem formulation and the scattering matrix formulation was demonstrated by Appel-Hansen [14]. The scattering matrix formulation is preferred here because all derivations are based on considerations of modes as for waveguide techniques. In the treatment of planar, cylindrical, and spherical techniques in Sections 4, 5, and 6, respectively, the same steps are followed in the exposition, only the scanning surface is changed in accordance with the technique. A condensed notation like the one introduced by Jensen [9] is made use of.

In planar scanning a probe antenna is moved in a plane situated in front of the antenna under test as shown in Fig. 4. The position of the probe is characterized by the position vector $\mathbf{r}_0$ with coordinates (x_0, y_0, z_0) in a rectangular xyz-coordinate system associated with the test antenna. During scanning, z_0 is kept constant and y_0 is varied for fixed values of x_0. This is illustrated in the figure. However, other types of scanning, such as plane-polar, are also possible [15, 16].

The field radiated by the test antenna and incident on the probe may be expressed as

$$\mathbf{E}(\mathbf{r}_0) = \int_{-\infty}^{\infty} \int_{-\infty}^{\infty} \sum_{s=1}^{2} b^s(k_x, k_y) \mathbf{f}_{sk_x k_y}(\mathbf{r}_0) dk_x dk_y \tag{4}$$

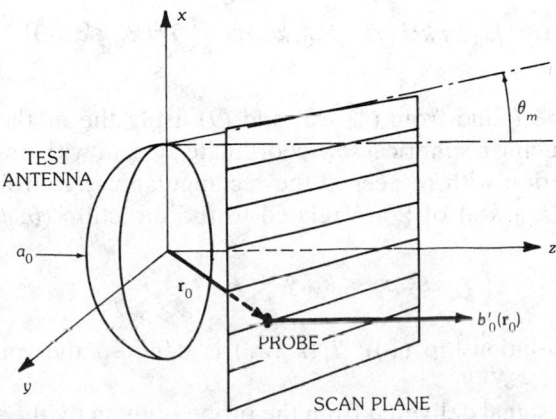

Fig. 4. Planar scanning.

where, for an arbitrary field point with position vector $\mathbf{r}$,

$$\mathbf{f}_{1k_xk_y}(\mathbf{r}) = j\frac{\hat{\mathbf{z}} \times \mathbf{k}}{|\hat{\mathbf{z}} \times \mathbf{k}|}e^{-j\mathbf{k}\cdot\mathbf{r}} \tag{5}$$

and

$$\mathbf{f}_{2k_xk_y}(\mathbf{r}) = -\frac{1}{k}\frac{\hat{\mathbf{z}} \times \mathbf{k}}{|\hat{\mathbf{z}} \times \mathbf{k}|} \times \mathbf{k}\,e^{-j\mathbf{k}\cdot\mathbf{r}} \tag{6}$$

are elementary planar vector wave functions. The time factor $e^{j\omega t}$ is omitted. The rectangular components (k_x, k_y, γ) of the propagation vector $\mathbf{k}$ are related through

$$\mathbf{k} = k_x\hat{\mathbf{x}} + k_y\hat{\mathbf{y}} + \gamma\hat{\mathbf{z}} \tag{7}$$

and

$$\gamma = \pm\sqrt{k^2 - k_x^2 - k_y^2} \tag{8}$$

where $k = \omega\sqrt{\mu\epsilon}$ as usual and $\hat{x}$, $\hat{y}$, and $\hat{z}$ are unit vectors of the xyz system. The sign is chosen so that γ is positive real in the visible region within the circle $k_x^2 + k_y^2 = k^2$ and negative imaginary outside this circle in the k_xk_y plane. Thus the incident field $\mathbf{E}(\mathbf{r}_0)$ consists of propagating and evanescent waves in the positive-z direction, i.e., from the test antenna toward the probe.

The coefficients $b^s(k_x, k_y)$ can be expressed as

$$b^s(k_x, k_y) = a_0\,T^s(k_x, k_y) \tag{9}$$

where a_0 is the amplitude of a single mode in the feed line to the test antenna and $T^s(k_x, k_y)$ are related to the far-field components of $\mathbf{E}(\mathbf{r})$ through

$$\mathbf{E}(\mathbf{r}) \cong ja_0\,2\pi\,k_{z0}[jT^1(k_{x0}, k_{y0})\hat{\boldsymbol{\phi}} - T^2(k_{x0}, k_{y0})\hat{\boldsymbol{\theta}}]\frac{e^{-jkr}}{r} \tag{10}$$

This result may be found from (4)–(6) and (9) using the method of steepest descents and introducing a spherical $r\theta\phi$-coordinate system with unit vectors $(\hat{\mathbf{r}}, \hat{\boldsymbol{\theta}}, \hat{\boldsymbol{\phi}})$ and usual orientation with respect to the rectangular xyz-coordinate system. The components (k_{x0}, k_{y0}, k_{z0}) of $\mathbf{k}$ are related to the direction (θ, ϕ) by

$$k_{x0}\hat{\mathbf{x}} + k_{y0}\hat{\mathbf{y}} + k_{z0}\hat{\mathbf{z}} = k\hat{\mathbf{r}} \tag{11}$$

Because of the relationship (10), $T^s(k_x, k_y)$ is denoted the transmitting characteristic.

The detected signal delivered from the probe is given by the so-called coupling equation

$$b_0'(\mathbf{r}_0) = a_0 \int_{-\infty}^{\infty} \int_{-\infty}^{\infty} \sum_{s=1}^{2} R'^s(k_x, k_y) T^s(k_x, k_y) e^{-j\mathbf{k}\cdot\mathbf{r}_0} dk_x dk_y \qquad (12)$$

From a consideration of (4), (9), and (12), it is seen that the $R'^s(k_x, k_y)$ express the sensitivity or the receiving characteristic of the probe to the elementary planar vector functions incident on the probe. Fourier inversion of (12) gives

$$\sum_{s=1}^{2} R'^s(k_x, k_y) T^s(k_x, k_y) = \frac{e^{j\gamma z_0}}{4\pi^2 a_0} \int_{-\infty}^{\infty} \int_{-\infty}^{\infty} b_0'(\mathbf{r}_0) e^{j(k_x x_0 + k_y y_0)} dx_0 dy_0 \qquad (13)$$

It can be shown that there exists the following reciprocity relation between $R'^s(k_x, k_y)$ and $T'^s(k_x, k_y)$:

$$\eta_0 R'^s(k_x, k_y) = -\eta\, T'^s(-k_x, -k_y) \qquad (14)$$

where η_0 is the admittance in the feed line and η is the free-space admittance. Thus, by measuring the far field of the probe and by making use of (10) and (14), $R'^s(k_x, k_y)$ may be calculated. This means that (13) is one equation in the two unknowns $T^1(k_x, k_y)$ and $T^2(k_x, k_y)$. An additional equation, which is needed in order to find $T^s(k_x, k_y)$, is obtained by repeating the measurements with another probe or the same probe rotated 90°. It is interesting to note that at a finite distance the received signal $b_0'(\mathbf{r}_0)$ in (12) is a Fourier transform of the product of true far-field characteristics of the antennas involved.

When $T^s(k_x, k_y)$ is determined, the gain in a direction characterized by (k_{x0}, k_{y0}, k_{z0}) is

$$G(k_{x0}, k_{y0}, k_{z0}) = 16\pi^3 k_{z0} [|T^1(k_{x0}, k_{y0})|^2 + |T^2(k_{x0}, k_{y0})|^2] \eta/\eta_0 \qquad (15)$$

In the above it has been assumed that impedance match exists in the feed lines of the probe and test antenna.

Measurement accuracy may be discussed from a consideration of (13). Evidently integration cannot be made over an infinite range in x_0 and y_0. The scan area has to be truncated. This means that the computed far-field pattern, e.g., in the xz plane as illustrated in Fig. 4, can only be trusted within an angle θ_m given by

$$\tan \theta_m = \frac{L_x - D_x}{2z_0} \qquad (16)$$

where L_x and D_x are the scan length and the antenna dimension in the x direction, respectively [17]. Therefore a single scan plane gives coverage only in a conical volume in front of the antenna. However, a combination of scan planes in a polyhedron type of scanning may be arranged to give spherical coverage for the predicted pattern.

Equation 12 shows that $b_0'(\mathbf{r}_0)$ is expressed as an integration in the $k_x k_y$ plane. A consideration of (5)–(8) demonstrates that outside the circle $k_x^2 + k_y^2 = k^2$, evanescent waves contribute to $b_0'(\mathbf{r}_0)$. This means that $b_0'(\mathbf{r}_0)$ is a Fourier transform of a bandlimited function of k_x and k_y. Therefore (13) can be written as a summation and the sampling spacing can be chosen in accordance with the sampling theory. The upper limit

$$\Delta x_0 = \Delta y_0 = \lambda/2 \tag{17}$$

is approached when $z_0 \to \infty$. In general, sampling spacings of $\lambda/3$ are sufficient for test distances z_0 equal to a few wavelengths [18].

5. Cylindrical Scanning

In comparison with a single planar scan giving only coverage in front of the antenna, cylindrical scanning giving azimuthal coverage in the far-field pattern may be preferred. In comparison with spherical scanning, cylindrical scanning may also be preferred when the antenna can only be rotated around a single axis for mechanical reasons. The cylindrical scanning may be implemented as exemplified in Fig. 5. Here, the test antenna is rotated around the z axis of an xyz-coordinate system while the probe is placed on a cylindrical surface at various levels relative to the xy plane.

The field radiated by the test antenna and incident on the probe at $\mathbf{r}_0$ may be expressed as

$$\mathbf{E}(\mathbf{r}_0) = \sum_{n=-\infty}^{\infty} \int_{-\infty}^{\infty} \sum_{s=1}^{2} b_n^s(\gamma) \mathbf{f}_{sn\gamma}^{(4)}(\mathbf{r}_0) d\gamma \tag{18}$$

where, at an arbitrary field point $\mathbf{r}$,

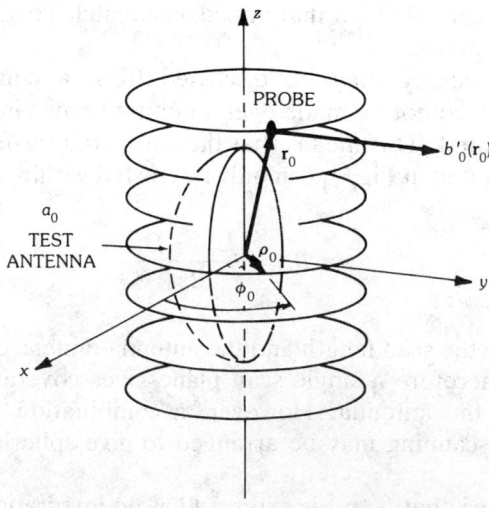

Fig. 5. Cylindrical scanning.

$$\mathbf{f}_{1n\gamma}^{(i)}(\mathbf{r}) = \left[\frac{jn}{\varrho} Z_n^{(i)}(\varkappa\varrho)\hat{\varrho} - \frac{\partial Z_n^{(i)}(\varkappa\varrho)}{\partial\varrho}\hat{\boldsymbol{\phi}} \right] e^{jn\phi} e^{-j\gamma z} \tag{19}$$

and

$$\mathbf{f}_{2n\gamma}^{(i)}(\mathbf{r}) = \left[-\frac{j\gamma}{k} \frac{\partial Z_n^{(i)}(\varkappa\varrho)}{\partial\varrho}\hat{\varrho} + \frac{n\gamma}{k\varrho} Z_n^{(i)}(\varkappa\varrho)\hat{\boldsymbol{\phi}} + \frac{\varkappa^2}{k} Z_n^{(i)}(\varkappa\varrho)\hat{\mathbf{z}} \right] e^{jn\phi} e^{-j\gamma z} \tag{20}$$

are the elementary cylindrical vector wave functions. The position vector $\mathbf{r}$ has cylindrical coordinates (ϱ,ϕ,z), n is any integer, $-\infty < \gamma < \infty$, $\varkappa = \sqrt{k^2 - \gamma^2}$ is chosen positive for $\gamma^2 < k^2$ and negative imaginary for $\gamma^2 > k^2$, $Z_n^{(i)}(\varkappa\varrho)$ is the cylinder functions $J_n(\varkappa\varrho)$, $N_n(\varkappa\varrho)$, $H_n^{(1)}(\varkappa\varrho)$, and $H_n^{(2)}(\varkappa\varrho)$ for $i = 1, 2, 3$, and 4, respectively.

The coefficients $b_n^s(\gamma)$ may be expressed by

$$b_n^s(\gamma) = T_n^s(\gamma)\, a_0 \tag{21}$$

where the transmitting characteristic $T_n^s(\gamma)$ of the test antenna is related to its far field at a position $\mathbf{r}$ with spherical coordinates (r,θ,ϕ) through

$$E_\phi(\mathbf{r}) \cong -\frac{a_0\, 2k\sin\theta\, e^{-jkr}}{r} \sum_{n=-\infty}^{\infty} j^n T_n^1(k\cos\theta)e^{jn\phi} \tag{22}$$

$$E_\theta(\mathbf{r}) \cong -\frac{a_0\, 2jk\sin\theta\, e^{-jkr}}{r} \sum_{n=-\infty}^{\infty} j^n T_n^2(k\cos\theta)e^{jn\phi} \tag{23}$$

where $E_\phi(\mathbf{r})$ and $E_\theta(\mathbf{r})$ are the ϕ and θ components of the far field $\mathbf{E}(\mathbf{r})$.

It can be shown that the signal received by the probe is

$$b_0'(\mathbf{r}_0) = a_0 \sum_{n=-\infty}^{\infty} \int_{-\infty}^{\infty} \sum_{s=1}^{2} R_n'^s(\gamma)\left[\sum_{m=-\infty}^{\infty} (-1)^n H_{m-n}^{(2)}(\varkappa\varrho_0)e^{jm\phi_0}\, e^{-j\gamma z_0}T_m^s(\gamma) \right] d\gamma \tag{24}$$

Fourier inversion of this equation readily gives

$$\sum_{s=1}^{2}\left[T_m^s(\gamma) \sum_{n=-\infty}^{\infty} (-1)^n H_{m-n}^{(2)}(\varkappa\varrho_0)R_n'^s(\gamma) \right] = \frac{1}{4\pi^2 a_0} \int_{-\infty}^{\infty}\int_{0}^{2\pi} b_0'(\mathbf{r}_0)e^{-jm\phi_0}\, e^{j\gamma z_0}\, d\phi_0 dz_0 \tag{25}$$

which is one equation in the two unknowns $T_m^s(\gamma)$, $s = 1, 2$. An additional equation is obtained as in the planar case.

Using (22) and (23) and the reciprocity relation

$$\eta_0 R_n'^s(\gamma) = (-1)^n \frac{4\pi\varkappa^2}{k}\, \eta\, T_{-n}'^s(-\gamma) \tag{26}$$

the receiving characteristic $R_n'^{s}(\gamma)$ of the probe can be found from far-field pattern measurements.

The upper limits for the sampling spacings $\Delta\phi_0$ and Δz_0 are determined from

$$\Delta\phi_0 = \lambda/2a \tag{27}$$

and

$$\Delta z_0 = \lambda/2 \tag{28}$$

where a is the radius of the smallest cylinder enclosing the test antenna. These limits are found from considerations analogous to the planar case discussed in the previous section and the spherical case discussed in the following section.

6. Spherical Scanning

While planar scanning requires at least four scans and cylindrical scanning at least two scans, the spherical scanning requires only a single scan over a sphere enclosing the test antenna to give spherical coverage. Mechanically, spherical scanning has the advantage that the test antenna may be mounted on a standard rotator, which can, for example, make a continuous scan in polar angle θ while ϕ is changed in steps as illustrated in Fig. 6. During measurements the probe is kept at a fixed position (see Fig. 9, in Section 8). Electrically, spherical scanning has the advantage that directivity can easily be determined from a single scan.

The field incident on the probe is expressed as

$$\mathbf{E}(\mathbf{r}_0) = \sum_{n=1}^{\infty} \sum_{m=-n}^{n} \sum_{s=1}^{2} b_{mn}^{s} \mathbf{f}_{smn}^{(4)}(\mathbf{r}_0) \tag{29}$$

where the spherical vector wave functions are

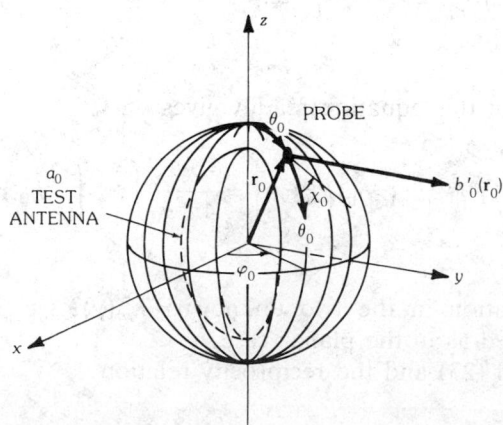

Fig. 6. Spherical scanning.

$$\mathbf{f}_{1mn}^{(i)}(\mathbf{r}) = z_n^{(i)}(kr)e^{jn\phi}\left[\frac{jm}{\sin\theta}P_n^m(\cos\theta)\hat{\theta} - \frac{\partial P_n^m(\cos\theta)}{\partial\theta}\hat{\phi}\right] \tag{30}$$

and

$$\mathbf{f}_{2mn}^{(i)}(\mathbf{r}) = \frac{n(n+1)}{kr}z_n^{(i)}(kr)\,P_n^m(\cos\theta)e^{jm\phi}\,\hat{\mathbf{r}} + \frac{1}{kr}\frac{\partial[rz_n^{(i)}(kr)]}{\partial r}\,e^{jm\phi}$$

$$\times\left[\frac{\partial P_n^m(\cos\theta)}{\partial\theta}\hat{\theta} + \frac{jm}{\sin\theta}P_n^m(\cos\theta)\hat{\phi}\right] \tag{31}$$

Here, $z_n^{(i)}(kr)$ are the spherical functions $j_n(kr)$, $n_n(kr)$, $h_n^{(1)}(kr)$, and $h_n^{(2)}(kr)$ for $i = 1, 2, 3$, and 4, respectively. The functions $P_n^m(\cos\theta)$ are associated Legendre functions, and n and m are integers, $0 \leqq n \leqq \infty$, $-n \leqq m \leqq n$. The coefficients b_{mn}^s are given by

$$b_{mn}^s = a_0\,T_{mn}^s \tag{32}$$

where T_{mn}^s characterizes the test antenna. The relations between T_{mn}^s and the far-field components are

$$E_\phi(\mathbf{r}) \cong j\frac{a_0 e^{-jkr}}{kr}\sum_{n=1}^{\infty}\sum_{m=-n}^{n}j^n\,e^{jm\phi}\left[-\frac{\partial P_n^m(\cos\theta)}{\partial\theta}\,T_{mn}^1\right.$$

$$\left. + \frac{m}{\sin\theta}P_n^m(\cos\theta)T_{mn}^2\right] \tag{33}$$

$$E_\phi(\mathbf{r}) \cong \frac{a_0 e^{-jkr}}{kr}\sum_{n=1}^{\infty}\sum_{m=-n}^{n}j^n\,e^{jm\phi}\left[-\frac{m}{\sin\theta}P_n^m(\cos\theta)T_{mn}^1\right.$$

$$\left. + \frac{\partial P_n^m(\cos\theta)}{\partial\theta}\,T_{mn}^2\right] \tag{34}$$

By transforming $\mathbf{f}_{smn}^{(4)}(\mathbf{r}_0)$ to a coordinate system associated with the probe, it is found that the coupling equation takes the form

$$b_0'(\mathbf{r}_0) = a_0\sum_{n=1}^{\infty}\sum_{m=-n}^{n}\sum_{\mu=-n}^{n}D_{\mu m}^{\mu n}(\phi_0, \theta_0, \chi_0)\,D(m, n, \mu) \tag{35}$$

where

$$D_{\mu m}^{\mu n}(\phi_0, \theta_0, \chi_0) = (-1)^{m+\mu}\sqrt{\frac{(n-\mu)!(n+m)!}{(n+\mu)!(n-m)!}}\,e^{jm\phi_0}\,d_{\mu m}^{(n)}(\theta_0)e^{j\mu\chi_0} \tag{36}$$

and $D(m, n, \mu)$ is the so-called coupling product given by

$$D(m, n, \mu) = \sum_{s=1}^{2}T_{mn}^s\sum_{\nu=(1,\mu)}^{\infty}[R_{\mu\nu}'^s(-1)^{n+\nu}A_{\mu\nu}^{\mu n} + R_{\mu\nu}'^{3-s}(-1)^{n+\nu+1}B_{\mu\nu}^{\mu n}] \tag{37}$$

The angle χ_0 is used to characterize the orientation of the probe with respect to the $\hat{\theta}_0$ vector. The coefficients $d_{\mu m}^{(n)}(\theta_0)$, $A_{\mu\nu}^{\mu n}$, and $B_{\mu\nu}^{\mu n}$ are expressed by summations involving the indices μ, ν, m, n, and $\sin(\theta_0/2)$, $\cos(\theta_0/2)$, $h_n^{(2)}(kr_0)$, and the Wigner $3-j$ symbols.

Using the orthogonality integral

$$\int_0^\pi d_{\mu m}^{(n)}(\theta_0) d_{\mu m}^{(n')}(\theta_0) \sin\theta_0 \, d\theta_0 = \frac{2}{2n+1}\delta_{nn'} \tag{38}$$

and (36), Fourier inversion of (35) can be made with respect to ϕ_0, θ_0, and χ_0. This gives

$$D(m,n,\mu) = \frac{(-1)^{\mu+m}(2n+1)}{a_0 \, 8\pi^2}\sqrt{\frac{(n+\mu)!(n-m)!}{(n-\mu)!(n+m)!}}\int_0^{2\pi}\int_0^\pi\int_0^{2\pi}$$
$$\times \, b_0'(r_0,\phi_0,\theta_0,\chi_0)e^{-jm\phi_0}d_{\mu m}^{(n)}(\theta_0)\sin\theta_0 \, e^{-j\mu\chi_0}d\phi_0 d\theta_0 d\chi_0 \tag{39}$$

When the far field of the probe is measured, its transmitting characteristic $T_{mn}'^s$ can be found from (33) and (34). Then, its receiving characteristic $R_{mn}'^s$ can be found from

$$\eta_0 R_{mn}'^s = (-1)^m \frac{\eta}{k^2}\frac{2\pi n(n+1)}{2n+1}T_{-mn}'^s \tag{40}$$

From using $R_{mn}'^s$ in (37), it is seen that (39) is one equation in the two unknowns T_{mn}^s, with $s = 1, 2$.

It turns out that measurements and calculations are facilitated by choosing a probe with $R_{\mu\nu}'^s \neq 0$ for μ equal to $+1$ and -1 only, e.g., an antenna with rotational symmetry as a linearly polarized circular conical horn fed with the TE_{11} mode. Then, measurements carried out for $\chi_0 = 0°$ and $90°$ are sufficient for making the integration over χ_0. Evaluation of $D(m,n,\mu)$ for μ equal to $+1$ and -1 gives the required two equations for determination of t_{mn}^s.

Due to the fact that the radiated field $\mathbf{E}(\mathbf{r}_0)$ can be truncated for $n = ka$, where a is the radius of the smallest sphere enclosing the antenna, $b_0'(\mathbf{r}_0)$ is a bandlimited function. Therefore the sampling spacings $\Delta\phi_0$ and $\Delta\theta_0$ should be chosen so that

$$\Delta\phi_0 < \lambda/2a \tag{41}$$

and

$$\Delta\theta_0 < \lambda/2a \tag{42}$$

The center of the smallest sphere should be chosen at the intersection between the axes of rotation (see Fig. 9, in Section 8). Therefore the maximum sampling spacing depends on the mounting of the test antenna as well as its size. It should be noted that in case sampling is chosen less than the theoretical maximum, possible modes with $n > ka$, which may be due to noise, may be removed by filtering.

A single computer program based on the coupling equation (35) may be developed to carry out the following major calculations: (*a*) inversion of the coupling equation to find the probe characteristics R'^s_{mn} from spherical scanning of the probe with another probe which can be approximated with a Hertzian dipole (nonprobe correction—this is an alternative to the use of (33) and (34) as described above; (*b*) inversion of the coupling equation to find the unknown characteristics T^s_{mn} from measurements on the unknown antenna with a probe with known R'^s_{mn}, i.e., probe correction; (*c*) the field intensity at any distance from an antenna with known T^s_{mn}; (*d*) the true far-field pattern by specifying the distance as infinite and using asymptotic forms for the mode functions. In the last two types of calculation the coupling equation is used with a probe specified as a Hertzian dipole.

7. Error Budgets for Scanning Ranges

The sources of errors in the determination of the test antenna characteristic T^s can be derived from a consideration of the inverse coupling equations (13), (25), and (39) and their background. Three types of sources exist:

1.   Discrepancies between the theoretical model and the actual experiment
2.   Inaccuracies in the measurement and digitizing of the quantities in the inverse equations
3.   Computations

The first source gives rise to errors due to the neglect of multiple reflections between the test antenna and the probe, multipath reflections from the lining of the test range and the applied instrumentation, and leakage signals. In the derived equations, reflections (mismatches) in the feed lines are also neglected. These, however, may easily be included by multiplying with proper factors. Furthermore, in the planar and cylindrical case there is a practical need for a finite scan area. Nonzero sampling spacing causes neglect of evanescent or higher-order modes. A special source of error is variation in the test antenna structure during measurements. Such variations may be due to temperature variations or deformations in case the antenna is rotated or otherwise moved during measurements.

The second source of error is related to inaccuracies in both the positioning of the probe and electrical characterization of the probe. It also includes inaccurate measurement of probe output signal, input signal to the test antenna, frequency, and reflection coefficients in the receiving system. In some cases determination of the permeability and permittivity of the transmission medium may also influence the determination of T^s.

Because of the development of computers the third source of error is in general insignificant. A particular error may arise in case the determinant of the two equations obtained by measuring two orthogonal components of the radiated field is close to zero.

An error budget for antenna measurements may be carried out in four steps:

1.   Specify the test antenna
2.   Specify the test range

3.   Study the test antenna in the test range
4.   Make the error budget

From the first two steps it can be decided whether the test range can handle the test antenna. In this respect, size of antenna and range, frequency, and instrumentation are important factors. When it is decided to start the test, the third step can be initiated. In particular, preliminary experiments, such as those indicating the error signal levels, may be carried out. A systematic analysis and classification of error signals in antenna measurements is given by Appel-Hansen [19]. It is characteristic that all types of error signals are present at the same time. However, each type has its own peculiarities which may be used in the evaluation of the influence of the individual signals. In near-field techniques it is convenient to classify the extraneous error signals into three types. The first type is due to reflections between the test antenna and the probe. These are referred to as *multiple reflections* and have a characteristic $\lambda/2$ period when the probe is moved longitudinally with respect to the test antenna. In particular, multiple reflections may be severe when the two antennas are pointing toward each other. They may be reduced by not measuring too close to the test antenna and by properly matching a probe having a low back-scattering cross section.

The second type of extraneous error signal is due to reflections from the lining of the test range and the instrumentation, which includes the test antenna support. These are referred to as *multipath reflections*. They often cause an irregular variation with change in probe position. They may be reduced by proper choice of probe directivity and of absorbers. Multipath reflections may be measured by moving the probe in various directions for several aspect angles as in the evaluation of the reflectivity level of anechoic chambers [20].

The third type is due to leakage from the transmitting system to the receiving system. Proper shielding and cabling, especially connectors, may reduce this source of error. This type of error signal may be discovered, for example, by probing a possible signal when the transmitting terminals are short-circuited.

To facilitate further insight, major details of error budgets for a planar scanning and a spherical scanning shall be outlined as examples. Additional information may be obtained from references 21 and 22 in the planar case and from reference 23 in the spherical case.

In the first budget the test antenna is a planar array for which the planar near-field technique is a candidate. Some of the details in the four steps of the error budget are the following:

1.   The test antenna specifications are:
         Maximum dimension $L_y = 538$ cm
         Minimum dimension $L_x = 202$ cm
         Wavelength $\lambda = 23.5$ cm
2.   The range specifications are:
         Maximum scan length $L_{max} = 640$ cm
         Transverse position accuracy $\Delta_x = \Delta_y = \pm 0.005$ cm
         Longitudinal position accuracy $\Delta_z = \pm 0.010$ cm
         Probe gain error $= \pm 0.12$ dB

The scanner is installed in a hall partly covered with absorbers.

3. The scan length may be obtained by a combination of two or several scan areas. An overlapping of the individual areas may secure proper transition between the patterns obtained from the various scans. The range distance z_0 is chosen to be 37.0 cm, i.e., between 1λ and 2λ. The truncation angle is about 50°. Preliminary transverse movements show a signal level of −35 dB at the edge of the scan plane relative to the maximum signal in the scan area. Longitudinal movement of the probe in front of the test antenna gives a peak-to-peak variation of about 1 dB. Since the distance between two consecutive peaks is $\lambda/2$, the oscillations must be associated with multiple reflections between the probe and test antennas.

4. Using upper-bound expressions [17], the five largest contributions in the error budget for gain measurements are:

Probe gain	±0.12 dB
Receiver nonlinearity	±0.12 dB
Multiple reflections	±0.10 dB
Normalization	±0.05 dB
Mismatch factors	±0.02 dB

It turns out that the position error is less than 0.01 dB. An analysis shows that the root-square sum of all errors is ±0.20 dB in X-band as well as in L-band.

The error budget for side lobe level measurements shows that area truncation error becomes important. It is the third largest component after that of multiple reflections, which is the predominant component, and probe gain. It is concluded that the error budgets give only an indication of expected errors. In particular, this is the case for the side lobe level budgets. However, the budgets probably give good information on the relative magnitudes of the different sources of error. More work seems appropriate in order to obtain agreement between error results derived from theory and from experiment.

1. The test antenna is a pyramidal standard-gain horn. The gain is determined in the frequency range 17.3 GHz to 18.1 GHz.

2. The test range is an anechoic chamber with a precision rotator as described in the next section. Some of the range specifications are as follows:

Positioner accuracy	±0.01°
Distance accuracy	±1 mm
Phase drift	±5°
Amplitude nonlinearity	±0.5 percent

    These values are based on measurements estimated or taken from manufacturers' indications.

3. When the test antenna is placed in the range, a multipath reflection level of −70 dB exists. The level of multiple reflections between the test antenna and the probe is −45 dB.

4. Introducing the individual errors into actually measured data, the five largest contributions for gain measurements are:

Multiple reflections	±0.045 dB
Amplitude nonlinearity	±0.025 dB

Asymmetry of probe	±0.02 dB
Internal loss calculation	{ +0.020 dB / −0.005 dB
Leakage	< ±0.010 dB

Multipath reflections, phase, positioning, distance, and amplitude drift errors contribute with less than ±0.003 dB each. The root-square-sum value of the upperbound errors is ±0.06 dB. The value of the theoretically determined internal loss is about 0.020 dB. This is subtracted from the measured directivity values oscillating between 24.647 dB and 24.938 dB. The so-determined gain as a function of frequency is shown in Fig. 7.

Finally, it should be mentioned that the error budgets may be improved by calibrating for systematic inaccuracies such as probe position, probe gain, and, especially, receiver nonlinearity. Part of the influence of the various extraneous reflections may be averaged out from repeated measurements at various range distances as in conventional pattern measurement [20]. The error budgets indicate that the scanning techniques are usually superior to conventional techniques with respect to measurement accuracy.

8. Scanning Ranges

The first commercially available scanner was of the planar type. It used an analog Fourier integral computer [24]. It has turned out, however, that test ranges

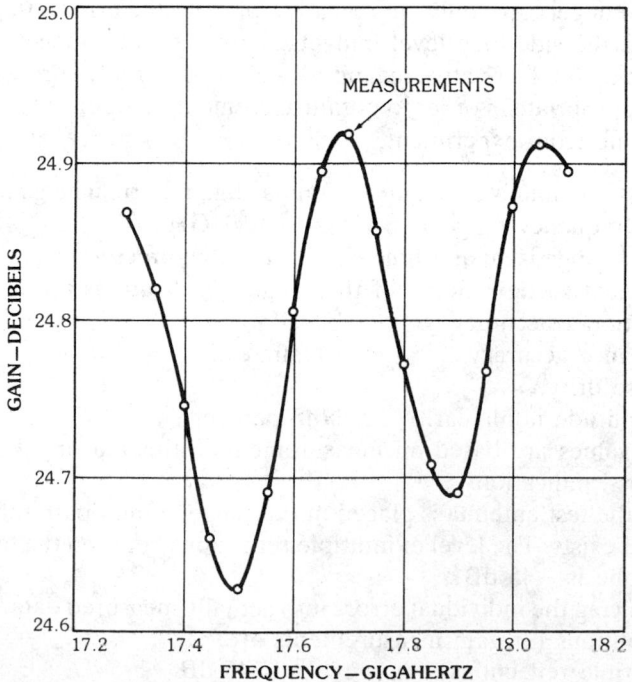

Fig. 7. Gain predicted from spherical near-field measurements. (*After Lemanczyk [23]*)

over the years have usually constructed their own planar scanners in accordance with their specific needs. A recent example is that of Borowick, Holley, Lange, Cummings, and Howard [25]. At present only spherical scanners are readily available. The success of the spherical technique is due to the fact that the mathematical problems have been solved during recent years in such a manner that minicomputers can handle the transformation of measured near-field data to predicted far-field behavior of the test antenna. Furthermore, the spherical-coordinate system is used in conventional far-field measurements. This means that the same instrumentation can be used for near-field work as well as for conventional far-field measurements.

Rectangular XY-scanners have been implemented in several places. They are based on experience obtained in particular at the National Bureau of Standards [26] and at the Georgia Institute of Technology [27]. Vertical as well as horizontal scanners have been built. The horizontal scanner may use existing turntables and rails for movement of the test antenna (see Fig. 8) [28]. Transverse movements from a few meters up to about ten meters have been arranged. Positioning and alignment are the critical works which often require the use of laser systems. When such systems are used, however, the remaining inaccurate positioning is usually an insignificant source of error, as shown in the previous section.

The most often used probe is an open-ended waveguide delivering its signal to a phase-amplitude receiver or a network analyzer. Due to the lengthy measure-

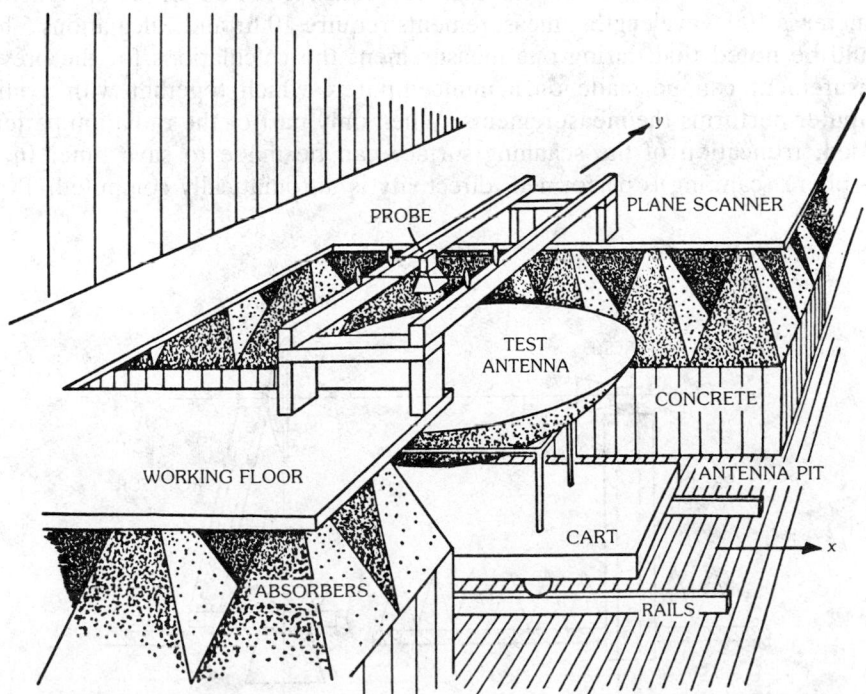

Fig. 8. Horizontal planar scanner setup in a radio anechoic chamber. (*After Anderson, Appel-Hansen, and Larsen [28]*)

ment procedure and in order to keep the mechanical source of error at an insignificant level, measurements in a temperature-stabilized environment may be demanded.

At the Technical University of Denmark a spherical near-field antenna test facility has been developed [29, 30]. In particular, precision measurements of directivity, gain, and copolar and cross-polar patterns can be carried out on microwave antennas of up to 200 wavelengths in electrical diameter and about 6 m in physical diameter. The test distance can be varied from 1.5 m to 5 m. The setup is illustrated in Fig. 9. The test antenna is mounted in a specially designed rotator. The resettability of the intersection between the vertical and horizontal rotation axes is ±0.1 mm. With no load, the intersection accuracy of the axes during rotation is within 0.05 mm. The probe is placed on a special probe tower and aligned accurately to point toward the intersection point. The setup is placed in a large anechoic chamber ($12 \times 14 \times 16$ m³) lined with 2-m-high pyramidal absorbers.

The probe antennas are conical horns with orthomode transducers for dual linear polarization. This means that during a spherical scan a single probe can be used to measure two orthogonal components simultaneously, i.e., two scans with two different orientations of the probe are not required. The measurements are facilitated by the use of a three-channel receiver, one channel for phase reference and two channels for the two polarizations. In the near-field to far-field transformation, correction is made for the receiving pattern of the probe as a routine.

For antennas with a 10-wavelength diameter the execution time is about 2 hr for measurements and a few minutes for transformation of data. When the diameter is 100 wavelengths, measurements require 10 hr and calculations 5 hr. It should be noted that during one measurement the calculations for the previous measurement can be made on a minicomputer, which together with a microcomputer performs the measurements. When only part of the radiation pattern is needed, truncation of the scanning surface can be made to save time. In case full-sphere scanning is performed, directivity is automatically computed. Typical

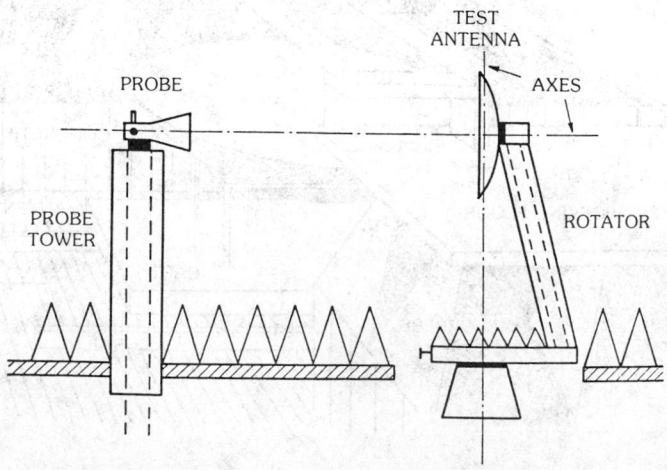

Fig. 9. Spherical near-field scanning.

directivity measurement accuracy is ± 0.10 dB (3σ). Determination of gain requires a substitution measurement with a standard-gain horn. The gain of this horn is determined by substracting analytically determined ohmic losses from measured directivity. Cross-polarization levels are measured with an accuracy of a few decibels down to about -55 dB. The use of accurate reference points makes determination of phase centers possible, for example.

9. Plane-Wave Synthesis Techniques

In general, attempts to make far-field measurements may be analyzed from a plane-wave illumination point of view [31]. This may be illustrated by a consideration of the three basic cases shown in Fig. 10. The first case is the simplest and used in the derivation of the conventional far-field criterion (2). Here, a single point source, as shown in Fig. 10a, illuminates the test volume in which the plane-wave quality is specified as an allowable phase error of 22.5°, corresponding to a path length difference of $\lambda/16$ (see Fig. 1).

In the second case, Fig. 10b, a simple array consisting of a few elements that are properly fed is used. In this manner a test volume of sufficient quality may be established at a shorter distance than required by the criterion (2). Lynggaard [32] studied an array of five pyramidal horn antennas theoretically and experimentally. The theory is based on a paper by Hald [33], who has studied the evaluation of the measurement error arising from imperfections in the plane wave. The compact range described in the next section may be considered as a continuous array of elements.

In the third case, Fig. 10c, a complete array of elements creating a perfect plane wave in a certain volume is thought of. It is probably impossible to build such an array. There is another approach, however. Instead of having a radiator at all positions, a single radiator steps from one position to the next. In this manner a

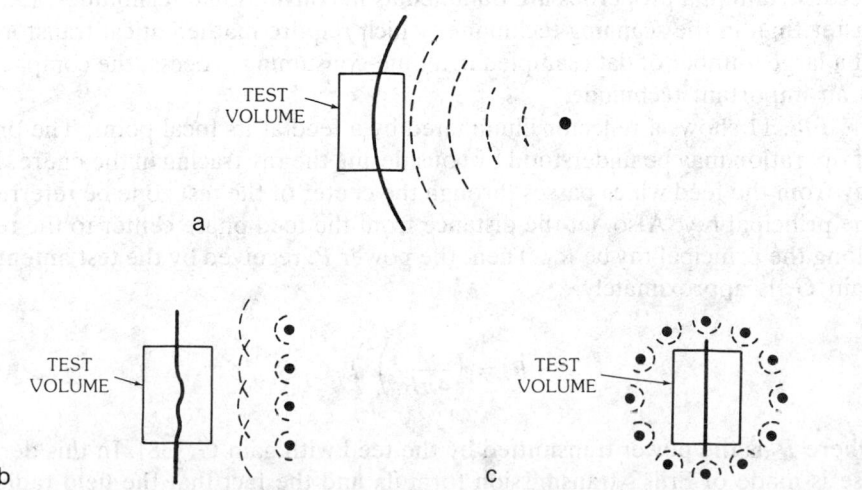

Fig. 10. Plane-wave synthesis. (*a*) Single point source. (*b*) Single array of sources. (*c*) Complete array of sources.

sequence of field contributions, each of which is separated in time, is generated. When a receiving test antenna is inserted in the test volume the response to each field contribution may be sampled. After the sampling, the responses may be multiplied with weighting functions. The result is that the responses added in this manner correspond to the response of the antenna to a synthesized plane-wave illuminator. By proper change of the weighting functions the direction of propagation of the plane wave can be changed and the complete antenna pattern can be obtained from a single scan. In fact, the procedure is analogous to the scanning techniques described in Sections 4 to 8. The approach has been studied by Martsafey [34] and Bennett and Schoessow [35] for the planar and spherical cases, respectively. Ludwig and Larsen [36] have used computer programs for spherical scanning to determine the size of the test volume.

An interesting aspect of the plane-wave synthesis technique is to create a plane wave field in the test volume and at the same time to minimize the illumination outside the test volume [37]. In this manner extraneous reflections may be reduced and the need for absorbers probably eliminated.

10. Compact Range

By *compact range* is usually meant a range which creates a plane wave field at distances which are considerably shorter than those needed under conventional far-field conditions. The required compact range is usually so short that the setup can be arranged indoors. Thus the compact range has all the advantages (controlled conditions, location, etc.) of indoor measurements. The compactness of the range may be obtained by using a set of sources which are properly fed as described in Section 9. Alternatively, the diverging rays from a single-source antenna may be collimated by using a lens or a reflector. Out of these two possibilities, the use of a parabolic reflector is the most widely implemented. Therefore this type of range is here designated "compact range" and will be treated in some detail. In particular, because antenna properties are obtained as in conventional techniques, i.e., much faster than in the scanning techniques which require mathematical transformation of a large number of data sampled in a time-consuming process, the compact range is an important technique.

Fig. 11 shows a reflector illuminated by a feed at its focal point. The principle of operation may be understood by considering the ray tracing in the figure. Let the ray from the feed which passes through the center of the test zone be referred to as the principal ray. Also, let the distance from the feed phase center to the reflector along the principal ray be R_0. Then, the power P_r received by the test antenna with gain G_r is approximately

$$P_r = \left(\frac{\lambda}{4\pi R_0}\right)^2 P_f G_f G_r \qquad (43)$$

where P_f is the power transmitted by the feed with gain G_f [38]. In this derivation use is made of Friis's transmission formula and the fact that the field radiated by the feed is only diverging over a distance R_0 and then collimated in a tube around the principal ray. Thus, in the test zone there is no inverse distance dependence

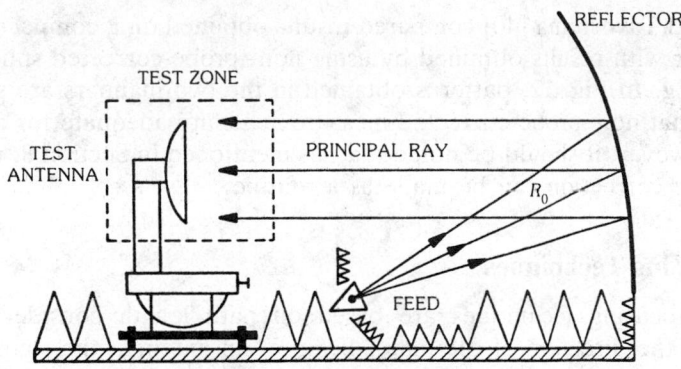

Fig. 11. Compact range.

as in conventional far-field measurements. Also, as opposed to conventional techniques, for fixed antenna dimensions the compact range does not require an increase in distance with an increase in frequency. This is due to the fact that the path lengths of the various rays are the same. As in conventional techniques, it is good practice to probe the field in the test zone to check amplitude and phase taper, and polarization conditions.

In the design and operation of the compact range, specific precautions have to be taken into account. The feed has to illuminate the reflector uniformly. Because the feed does not have a perfect phase center, the positioning of the feed ends up as an optimization problem. Aperture blockage is avoided by using asymmetrical illumination. Side lobes of the feed in the direction of the test antenna have to be minimized or screened with absorbers as illustrated in the figure. In this connection diffraction from the edge of the absorbers may cause adverse effects which also have to be minimized. The test antenna must not be placed too close to the reflector and R_0 must be chosen sufficiently large in order to avoid unwanted coupling between test antenna, probe, and reflector. The edge of the reflector has to be rolled and cut in an irregular manner in order not to give constructive interference of edge reflections. The reflector surface has to be smooth, for example, surface deviations less than $\lambda/100$ over any area of size λ^2. Also, in order to have small depolarization effects, a reflector with a long focal length has to be made.

Proper measurements require that the aperture of the reflector be about three times that of the test antenna. In the frequency range 2 GHz to 40 GHz, test zones with diameters of 1 m and length 1.5 m are typical. In order to increase the test zone diameter without increasing the overall dimensions, a compact range consisting of two crossed parabolic cylindrical reflectors has been suggested [39]. The focal lines of the reflectors are perpendicular to each other, and one reflector acts as a subreflector while the other acts as a main reflector. It is claimed that in the aperture of the main reflector a test zone superior to that of the single paraboloidal reflector with respect to amplitude and phase taper, cross polarization, and extraneous signal level is created. It would be interesting to obtain data for the test zone properties of the two types of compact ranges with the same probing mechanism.

Hess and Tavormina [40] compared results obtained on a compact range with one reflector with results obtained by using non–probe-corrected spherical near-field scanning. In Fig. 12, patterns obtained in the two manners are shown. It is concluded that non–probe-corrected measurements are adequate for many applications. However, it should be noted that, as mentioned in Section 8, ranges exist where probe correction can be made as a routine.

11. Defocusing Techniques

The defocusing techniques are based on path length considerations. For example, in the direction of the major beam of an antenna, the signals radiated from various parts of the antenna structure are usually in phase, i.e., there is no difference in the path lengths over which the signals propagate. It is said that the antenna is focused at infinity. In order to measure the antenna pattern at a finite

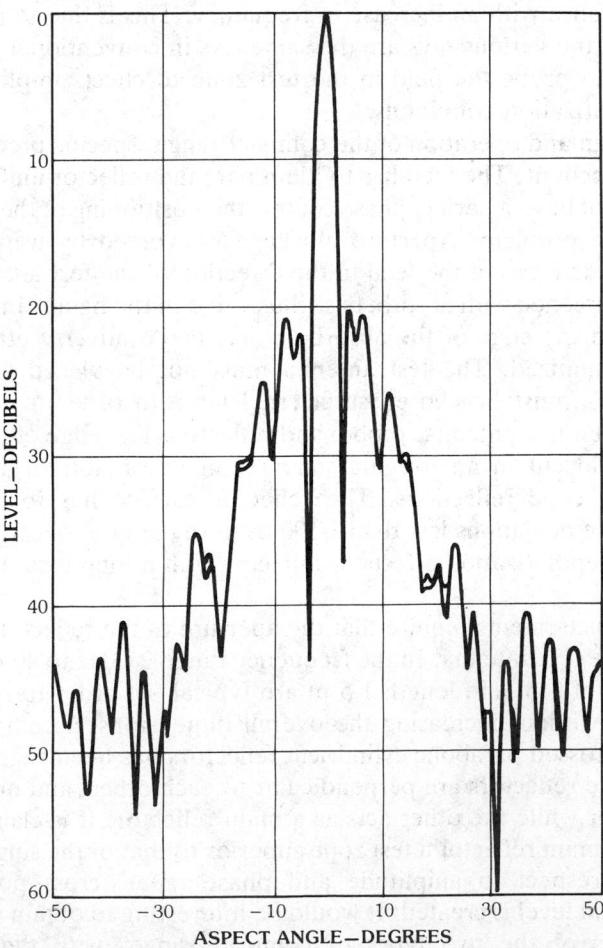

Fig. 12. Comparison of compact range and spherical near-field scanning. (*After Hess [41]*)

distance, the antenna structure is changed so that the path length differences occurring at infinity are established at the finite distance R, which is usually less than the conventional $2D^2/\lambda$ test distance. It is said that the antenna is defocused from infinity or focused at the finite distance. Besides changing the antenna structure physically, for example, bending a linear waveguide array, the defocusing may be done by modifying the feeding network or other components of the antenna.

The problems of the defocusing technique will be discussed in relation to the paraboloidal reflector shown in Fig. 13. The reflector has its focus at F and a focal length f. When a source is placed at F, there are no path length differences at infinity. This situation is approximated at a point F' by displacing the feed a distance ε to a point F'' along the axis and away from the vertex V of the reflector. However, the reflector is only approximately focused at F'. This is due to the fact that a focusing at F' would require an ellipsoidal reflector with foci at F' and F''. The simplest criterion used to find ε is to require that the two path lengths from F'' to F' via the vertex and the edge of the reflector are the same [42]. This is illustratted in the lower part of the figure. Another method is to make an approximate cancellation of the path length differences of all the rays arriving at F'. This gives

$$\varepsilon = \frac{1}{R}\left[f^2 + \left(\frac{D}{4}\right)^2\right] \qquad [44]$$

where $R = VF'$ and D is the diameter of the reflector. The approximation is valid for f/R small [43]. It should be noted that computations or experience with measurements of particular antenna parameters may be used in the choice of ε [44].

The defocusing is associated with a particular direction. In Fig. 13, defocusing is in the direction of the axis of the paraboloid. Therefore measurements can be trusted only in a certain region around the axis. Range distances should in general be chosen from $D^2/4\lambda$ up to $2D^2/\lambda$, depending on the required accuracy.

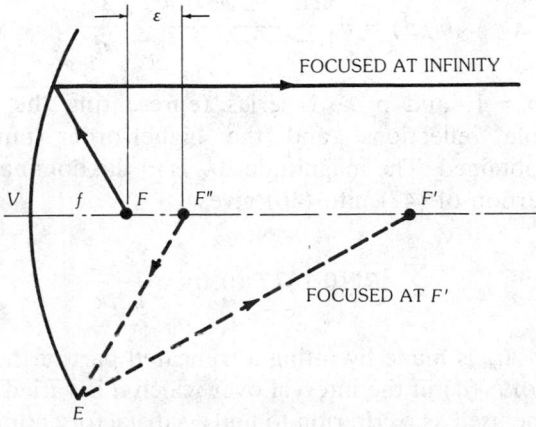

Fig. 13. Focusing at a finite distance.

12. Extrapolation Techniques

In the extrapolation technique two unknown antennas are arranged as in a conventional far-field measurement setup with the feature that the distance d between the antennas can be varied [45]. The theory developed in Section 4 on planar scanning may be used to describe the technique. The antennas are arranged as in Fig. 4 with $z_0 = d$ and $x_0 = y_0 = 0$. Let one of the two antennas be numbered 1 and be used as a transmitting antenna with transmitting characteristic $^1T^s(k_x, k_y)$. Let the other antenna be numbered 2 and be used as a probe with receiving characteristic $^2R'^s(k_x, k_y)$. Using the asymptotic form of the coupling equation (12), the signal received by antenna 2 is given by

$$b_0'(d) \cong j2\pi k a_0 \frac{e^{-jkd}}{d} \sum_{s=1}^{2} {}^2R'^s(0,0) \, {}^1T^s(0,0) \tag{45}$$

By theoretically extrapolating d to infinity we obtain one equation

$$\sum_{s=1}^{2} {}^2R'^s(0,0) \, {}^1T^s(0,0) = \lim_{d \to \infty} \left[\frac{b_0'(d) d \, e^{jkd}}{j2\pi k a_0} \right] \tag{46}$$

in the four unknowns $^2R'^s(0,0)$ and $^1T^s(0,0)$. By rotating antenna 2 about the direction to antenna 1 by 90°, an additional equation is obtained. Introducing an antenna number 3 and carrying out analogous experiments with the antenna pair 1 and 3 and the pair 2 and 3, it is understood [by making use of the reciprocity relation (14)] that six equations with the six unknown characteristics $^nT^s(0,0)$, for $s = 1, 2$ and $n = 1, 2, 3$, are obtained. The equations can in general be solved when no two antennas are circularly polarized. It should be noted that only those antenna characteristics along the axis between the antennas, usually in the main lobe directions, i.e., $(k_x, k_y) = (0,0)$, are measured.

To facilitate the solution, use is made of the series expression [46] for the received field:

$$b_0'(d) = a_0 \sum_{p=0}^{\infty} \frac{e^{-j(2p+1)kd}}{d^{2p+1}} \sum_{q=0}^{\infty} \frac{A_{pq}}{d^q} \tag{47}$$

For $p = 0$, $p = 1$, and $p > 1$ series representing the direct signal, the first-order multiple reflections, and the higher-order multiple reflections, respectively, are obtained. The magnitude A_{00} is in the dominant term for the far field. In fact, insertion of (47) into (46) gives

$$\sum_{s=1}^{2} {}^2R'^s(0,0) \, {}^1T^s(0,0) = \frac{A_{00}}{j2\pi k} \tag{48}$$

Determination of A_{00} is made by fitting a truncated part of the series (47) to the measured values of $b_0'(d)$ in the interval over which d is varied. Statistical tests of significance may be used as a criterion to find a satisfactory number of terms in the polynomial fitting. In actual measurements the ratio between the received and transmitted powers is measured as illustrated in Fig. 14. This means that $|A_{00}|^2$ is determined and not A_{00}. Therefore, in order to solve the system of six equations, phase differences are measured between the signals obtained for the two orientations of each pair of antennas.

The oscillations in Fig. 14 are due to multiple reflections, which may be averaged out by fitting processes. The same may be the case for ground reflections which have a longer period than the $\lambda/2$ characteristic for the multiple reflections. Repjar, Newell, and Baird [47] demonstrated a technique which overcomes unwanted oscillations due to ground reflections by including phase versus distance information. Hunter and Morgan [48] have combined the series expression (47) with theoretical correction factors in an extrapolation technique. The extrapolation technique is usually carried out for distances from $0.2D^2/\lambda$ and up to the conventional far-field criterion $2D^2/\lambda$. Newell, Baird, and Wacker [45] give further details on the extrapolation technique. In Fig. 15 their setup is shown.

In some recent investigations, Panicali and Nakamura [49] and Hollmann [50] have demonstrated that accurate results can be obtained by using the concept of an amplitude center, suggested by Appel-Hansen [51]. Scanning techniques and efficient methods of computation are discussed by Yaghjian [52]. Additional references to near-field far-field antenna measurements may be found in Hansen [53], which deals with spherical near-field measurements.

Acknowledgments

The author wishes to thank Johnny Vang Dahlberg and Steen Vedsegaard for reading and commenting on the manuscript. Thanks are also due to Flemming Holm Larsen and Jesper E. Hansen for many fruitful discussions on near-field techniques; to Mette Flagstad for creating most of the drawings; and to Ellinor Barfoed for typing the manuscript.

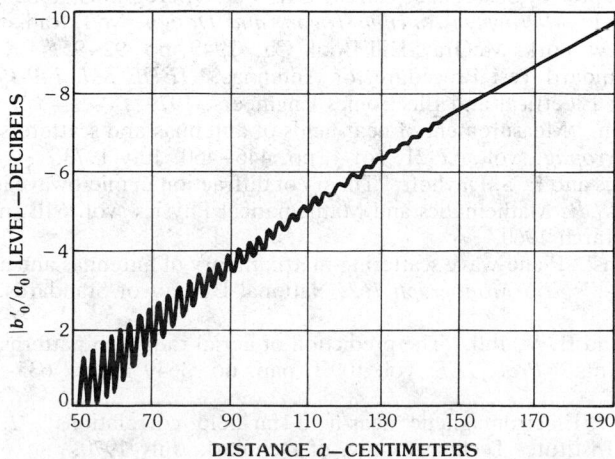

Fig. 14. Data for extrapolation to true far field. (*After Newell, Baird, and Wacker, © 1973 IEEE*)

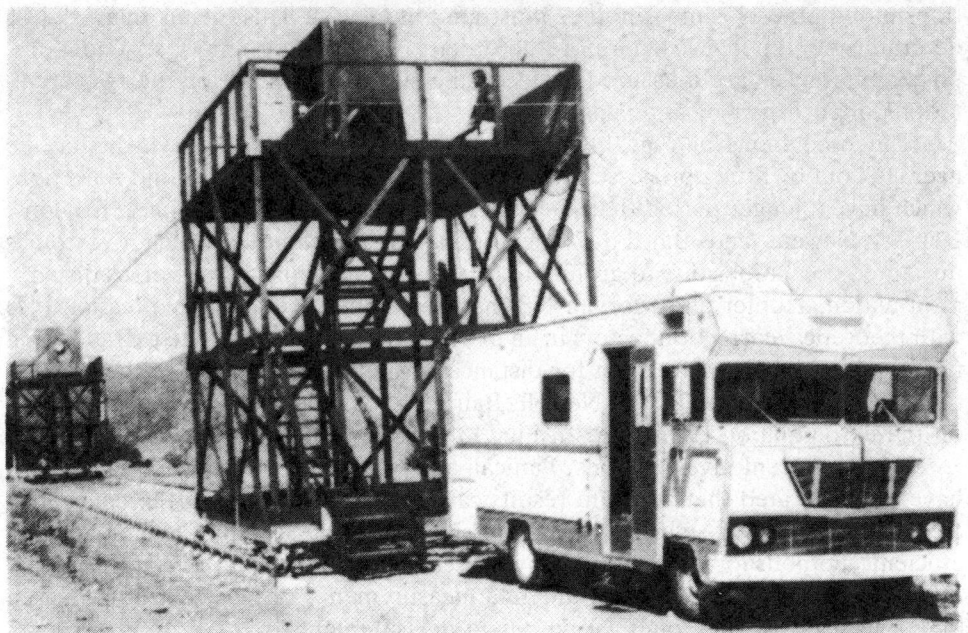

Fig. 15. Outdoor extrapolation range. (*After Newell, Baird, and Wacker [45],* © *1973 IEEE*)

13. References

[1]   J. Appel-Hansen, ed., *Antenna Measurements*, FN 36, vols. I and II, Electromagnetics Institute, Technical Univ. of Denmark, Lyngby, October 1980.

[2]   A. W. Rudge, K. Milne, A. D. Olver, and P. Knight, eds., *The Handbook of Antenna Design*, chapter 8, Stevenage, Herts., UK: Peter Peregrinus, 1982.

[3]   S. Silver, ed., *Microwave Antenna Theory and Design*, MIT Radiation Lab Series, vol. 12, New York: McGraw-Hill Book Co., 1949, pp. 92–95.

[4]   "IEEE Standard Test Procedure for Antennas," *IEEE Std. 149-1979*, pp. 18–20, Institute of Electrical and Electronics Engineers, 1979.

[5]   J. D. Dyson, "Measurement of near fields of antennas and scatterers," *IEEE Trans. Antennas Propag.*, vol. AP-21, no. 4, pp. 446–460, July 1973.

[6]   D. M. Kerns and E. S. Dayhoff, "Theory of diffraction in microwave interferometry," *J. Res. NBS B*, Mathematics and Mathematical Physics, vol. 64B, no. 1, pp. 1–13, January–March 1960.

[7]   D. M. Kerns, "Plane-wave scattering-matrix theory of antennas and antenna-antenna interactions," *NBS Monograph 162*, National Bureau of Standards, Boulder, June 1981.

[8]   J. Brown and E. V. Jull, "The prediction of aerial radiation patterns from near-field measurements," *Proc. IEE*, vol. 109B, pap. no. 3649 E, pp. 635–644, November 1961.

[9]   F. Jensen, "Electromagnetic near-field–far-field correlations," *LD 15*, Electromagnetics Institute, Technical Univ. of Denmark, July 1970.

[10] W. M. Leach, Jr., and D. T. Paris, "Probe compensated near-field measurements on a cylinder," *IEEE Trans. Antennas Propag.*, vol. AP-21, no. 4, pp. 435–445, July 1973.

[11] A. D. Yaghjian, "Near-field antenna measurements on a cylindrical surface: a source

scattering matrix formulation," *NBS Tech. Note 696*, National Bureau of Standards, Boulder, September 1977.

[12] P. F. Wacker, "Non-planar near-field measurements: spherical scanning," *NBSIR 75-809*, National Bureau of Standards, Boulder, June 1975.

[13] F. H. Larsen, "Probe correction of spherical near-field measurements," *Electron. Lett.*, vol. 13, no. 14, pp. 393–395, July 1977.

[14] J. Appel-Hansen, "On cylindrical near-field scanning techniques," *IEEE Trans. Antennas Propag.*, vol. AP-28, no. 2, pp. 231–234, March 1980.

[15] P. F. Wacker, "Plane-radial scanning techniques with probe correction; natural orthogonalities with respect to summation on planar measurement lattices," *1979 Intl. Symp. Dig. IEEE Antennas Propag. Soc.*, vol. 2, pp. 561–564, Seattle, June 1979.

[16] Y. Rahmat-Samii, V. Galindo-Israel, and R. Mittra, "A plane-polar approach for far-field construction from near-field measurements," *IEEE Trans. Antennas Propag.*, vol. AP-28, no. 2, pp. 216–230, March 1980.

[17] A. D. Yaghjian, "Upper-bound errors in far-field antenna parameters determined from planar near-field measurements," *NBS Tech. Note 667, Part 1: Analysis*, National Bureau of Standards, Boulder, October 1975.

[18] E. B. Joy and D. T. Paris, "Spatial sampling and filtering in near-field measurements," *IEEE Trans. Antennas Propag.*, vol. AP-20, no. 3, pp. 253–261, May 1972.

[19] J. Appel-Hansen, "Error signals in antenna and scattering measurements," *CPEM Dig. 1982*, pp. F4–6, Conference on Precision Electromagnetic Measurements, Univ. of Colorado, Boulder, June 28–July 1, 1982.

[20] J. Appel-Hansen, "Reflectivity level of radio anechoic chambers," *IEEE Trans. Antennas Propag.*, vol. AP-21, no. 4, pp. 490–498, July 1973.

[21] K. R. Carver and A. C. Newell, "A comparison of independent far-field and near-field measurements of large spaceborne planar arrays with controlled surface deformations," *1979 Intl. Symp. Dig. IEEE Antennas Propag. Soc.*, vol. 2, pp. 494–497, Seattle, June 1979.

[22] J. Appel-Hansen, F. Jensen, and A. C. Ludwig, "SAR antenna test techniques," *TICRA ApS Rep. S-117-02*, Copenhagen, January 1980.

[23] J. H. Lemanczyk, "Calibration report *P*-band standard gain horn Scientific-Atlanta Model 12-12 SN215," *R 265*, Electromagnetics Institute, Technical Univ. of Denmark, Lyngby, September 1982.

[24] L. Clayton, Jr., and J. S. Hollis, "Calculation of microwave antenna radiation patterns by the Fourier integral method," *The Essay*, no. 1, Scientific-Atlanta, Inc., March 1960.

[25] J. Borowick, A. E. Holley, W. Lange, R. Cummings, and R. W. Howard, "A near field antenna measurement system," *CPEM Dig. 1982*, p. F-10, Conference on Precision Electromagnetic Measurements, Univ. of Colorado, Boulder, June 28–July 1, 1982.

[26] P. F. Wacker and A. C. Newell, "Advantages and disadvantages of planar, circular cylindrical, and spherical scanning and description of the NBS antenna scanning facilities," *ESA Preprint SP127*, pp. 115–121, Workshop on Antenna Testing Techniques held at ESTEC, European Space Agency, Noordwijk, The Netherlands, June 6–8, 1977.

[27] E. B. Joy, W. M. Leach, Jr., G. P. Rodrigue, and D. T. Paris, "Application of probe-compensated near-field measurements," *IEEE Trans. Antennas Propag.*, vol. AP-26, no. 3, pp. 379–389, May 1978.

[28] J. Anderson, J. Appel-Hansen, and F. H. Larsen, "Horizontal planar scanner for near-field measurements," *Conf. Dig. IX Nat. Conv. Radio Sci. and the Nordic Antenna Seminar*, pp. 100–103, URSI Finnish National Committee, Otaniemi, Finland, October 27–29, 1976.

[29] J. Lemanczyk, F. H. Larsen, J. E. Hansen, and J. Aasted, "Calibration of standard gain antennas using a spherical near-field technique," *IEE Conf. Pub. No. 195*, pt. 1, pp. 241–245, Second International Conference on Antennas and Propagation, Univ. of York, Heslington, York, UK, April 13–16, 1981.

[30] J. E. Hansen et al., "Introduction to the TUD-ESA spherical near-field far-field antenna test facility," draft report, Electromagnetics Institute, Technical Univ. of Denmark, July 1, 1983.

[31] J. E. Hansen, "Spherical near-field testing of spacecraft antennas," *ESA Journal*, vol. 4, pp. 89–102, 1980.

[32] S. K. Lynggaard, "Antenna measurements using plane wave synthesis," MSc thesis (in Danish), Electromagnetics Institute, Technical Univ. of Denmark, Lyngby, August 1982.

[33] J. Hald, "Synthesis and evaluation of spherical quasi plane wave regions for antenna pattern measurements," *1981 Intl. Symp. Dig. IEEE Antennas Propag. Soc.*, vol. II, pp. 573–576, Los Angeles, June 16–19, 1981.

[34] V. V. Martsafey, "Measurement of electrodynamic antenna parameters by the method of synthesized aperture," *Radio Eng. Electron. Phys.*, vol. 13, no. 12, pp. 1869–1873, 1968.

[35] J. C. Bennett and E. P. Schoessow, "Antenna near-field/far-field transformation using a plane-wave synthesis technique," *Proc. IEE*, vol. 125, no. 3, pp. 179–184, March 1978.

[36] A. C. Ludwig and F. H. Larsen, "Spherical near-field measurements from a 'compact range' viewpoint," *IEE Conf. Pub. No. 195*, pt. 1, pp. 274–277, Second International Conference on Antennas and Propagation, Univ. of York, Heslington, York, UK, April 13–16, 1981.

[37] J. F. R. Pereira, A. P. Anderson, and J. C. Bennett, "A procedure for near field measurement of microwave antennas without anechoic environments," *IEE Conf. Pub. No. 219*, pt. 1, pp. 219–223, Third International Conference on Antennas and Propagation, Univ. of East Anglia, Norwich, UK, April 12–15, 1983.

[38] D. W. Hess and R. C. Johnson, "Compact ranges provide accurate measurement of radar cross-section," *Microwave Systems News*, vol. 12, no. 9, pp. 150–160, September 1982.

[39] V. J. Vokurka, "Compact-antenna range performance at 70 GHz," *1980 Intl. Symp. Dig. IEEE Antennas Propag. Symp.*, pp. 260–263, Université Laval, Quebec, vol. I, June 2–6, 1980.

[40] D. W. Hess and J. J. Tavormina, "Verification testing of a spherical near-field algorithm and comparison to compact range measurements," *1981 Intl. Symp. Dig. IEEE Antennas Propag. Soc.*, vol. I, pp. 242–271, Los Angeles, June 16–19, 1981.

[41] D. W. Hess, "Comparison of spherical near-field vs. compact range," private communication, Scientific-Atlanta, Inc., Atlanta, August 1981.

[42] S. T. Moseley, "On-axis defocus characteristics of the paraboloidal reflector," final report, USAF Contract AF30(602)-925, Department of Electrical Engineering, Syracuse Univ., August 1, 1954.

[43] D. K. Cheng and S. T. Moseley, "On-axis defocus characteristics of the paraboloidal reflector," *IRE Trans. Antennas Propag.*, vol. AP-3, pp. 214–216, October 1955.

[44] R. C. Johnson, H. A. Ecker, and J. S. Hollis, "Determination of far-field antenna patterns from near-field measurements," *Proc. IEEE*, vol. 61, no. 12, pp. 1668–1694, December 1973.

[45] A. C. Newell, R. C. Baird, and P. F. Wacker, "Accurate measurement of antenna gain and polarization at reduced distances by an extrapolation technique," *IEEE Trans. Antennas Propag.*, vol. AP-21, no. 4, pp. 418–431, July 1973.

[46] P. F. Wacker, "Theory and numerical techniques for accurate extrapolation of near-zone antenna and scattering measurements," unpublished report, National Bureau of Standards, Boulder, April 1972.

[47] A. Repjar, A. Newell, and R. Baird, "Antenna gain measurements by an extended version of the NBS extrapolation method," *CPEM Dig. 1982*, pp. F7–9, Conference on Precision Electromagnetic Measurements, Univ. of Colorado, Boulder, June 28–July 1, 1982.

[48] J. D. Hunter and I. G. Morgan, "Gain measurements in the National Measurement Laboratory anechoic chamber," *Dig. Seminar on Electromagnetic Antenna and*

Scattering Measurements, CSIRO Division of Applied Physics, pt. II, pp. 197–204, National Measurement Laboratory, Sydney, November 1982.

[49] A. R. Panicali and M. M. Nakamura, "On the amplitude center of radiating apertures," *IEEE Trans. Antennas Propag.*, vol. AP-33, no. 3, pp. 330–335, March 1985.

[50] H. Hollmann, "Gain measurements for pyramidal horns," a summary of "Gewinnmessung von Hornstrahlern," *Tech. Rep. 454 TB 42*, Forschungsinstitut beim FTZ, Deutsche Bundespost, Darmstadt, Germany, August 1987.

[51] J. Appel-Hansen, "Centers of structures in electromagnetism: a critical analysis," *IEEE Trans. Antennas Propag.*, vol. AP-30, no. 4, pp. 606–610, July 1982.

[52] A. D. Yaghjian, "An overview of near-field antenna measurements," *IEEE Trans. Antennas Propag.*, vol. AP-34, no. 1, pp. 30–45, January 1986.

[53] J. E. Hansen, ed., *Spherical Near-Field Antenna Measurements*, Stevenage, Herts., UK: Peter Peregrinus Ltd., to be published.

Appendixes

CONTENTS

Appendix A

Physical Constants, International Units, Conversion of Units, and Metric Prefixes

Yi-Lin Chen
*University of Illinois**

Physical Constants

Quantity	Symbol	Value
Speed of light in vacuum	c	$2.997\,925 \times 10^8\,\text{ms}^{-1}$
Electron charge	e	$1.602\,192 \times 10^{-19}\,\text{C}$
Electron rest mass	m_e	$9.109\,558 \times 10^{-31}\,\text{kg}$
Boltzmann constant	k	$1.380\,622 \times 10^{-23}\,\text{JK}^{-1}$
Dielectric constant in vacuum	ϵ_0	$8.854\,185 \times 10^{-12}\,\text{Fm}^{-1}$ $\cong (36\pi \times 10^9)^{-1}\,\text{Fm}^{-1}$
Permeability in vacuum	μ_0	$4\pi \times 10^{-7}\,\text{Hm}^{-1}$

International System of Units (SI Units): Basic Units

Quantity	Symbol	Units
Length	ℓ	meters (m)
Mass	m	kilograms (kg)
Time	t	seconds (s)
Electric current	I	amperes (A)
Temperature	T	kelvins (K)
Luminous intensity	I	candelas (cd)

---

*On leave from the Chinese Aeronautical Laboratory, Beijing, China, during 1983.

Derived Units in Electromagnetics

Quantity	Symbol	Units
Electric-field strength	**E**	volts per meter (V/m)
Magnetic-field strength	**H**	amperes per meter (A/m)
Electric-flux density	**D**	coulombs per meter squared (C/m^2)
Magnetic-flux density	**B**	teslas (T) = Wb/m^2
Electric-current density	**J**	amperes per meter squared (A/m^2)
Magnetic-current density	**K**	volts per meter squared (V/m^2)
Electric-charge density	ϱ	coulombs per meter cubed (C/m^3)
Magnetic-charge density	ϱ_m	webers per meter cubed (Wb/m^3)
Voltage	V	volts (V)
Electric current	I	amperes (A)
Dielectric constant (permittivity)	ϵ	farads/meter (F/m)
Permeability	μ	henrys/meter (H/m)
Conductivity	σ	siemens per meter (S/m) = $\mho/m$
Resistance	R	ohms (Ω)
Inductance	L	henrys (H)
Capacitance	C	farads (F)
Impedance	Z	ohms (Ω)
Admittance	Y	siemens (S) or mhos ($\mho$)
Power	P	watts (W)
Energy	W	joules (J)
Radiation intensity	I	watts per steradian (W/sr)
Frequency	f	hertz (Hz)
Angular frequency	ω	radians per second (rad/s)
Wavelength	λ	meters (m)
Wave number	k	1 per meter (m^{-1})
Phase shift constant	β	radians per meter (rad/m)
Attenuation factor	α	nepers per meter (Np/m)

Conversions of Units

Quantity	Symbol	SI Unit	CGS Electromagnetic Unit	Equivalent Number of CGS Electrostatic Unit
Electric charge	q	coulombs	10^{-1} abcoulomb	3×10^{9} statcoulombs
Current	I	amperes	10^{-1} abampere	3×10^{9} statamperes
Volume current density	$\mathbf{J}$	amperes/meter^2	10^{-5} abampere/centimeter^2	3×10^{5} statamperes/centimeter^2
Voltage	V	volts	10^{8} abvolts	$\frac{1}{3} \times 10^{-2}$ statvolt
Electric-field intensity	$\mathbf{E}$	volts/meter	10^{6} abvolts/cm	$\frac{1}{3} \times 10^{-4}$ statvolt/centimeter
Electric-flux density	$\mathbf{D}$	coulombs/meter^2	$4\pi \times 10^{-5}$ abcoulomb/centimeter^2	$12\pi \times 10^{5}$ statcoulombs/centimeter^2
Magnetic-field intensity	$\mathbf{H}$	amperes/meter	$4\pi \times 10^{-3}$ oersted	$12\pi \times 10^{7}$ oersteds
Magnetic-flux intensity	$\mathbf{B}$	webers/meter^2	10^{4} gausses	$\frac{1}{3} \times 10^{-6}$ gauss
Permittivity	ϵ	farads/meter	$4\pi \times 10^{-11}$ abfarad/centimeter	$36\pi \times 10^{9}$ statfarads/centimeter
Permeability	μ	henrys/meter	$\frac{1}{4\pi} \times 10^{7}$ gauss/oersted	$\frac{1}{36\pi} \times 10^{-13}$ gauss/oersted
Magnetic flux	Φ	webers	10^{8} gilberts	$\frac{1}{3} \times 10^{-2}$ gilbert
Resistance	R	ohms	10^{9} abohms	$\frac{1}{9} \times 10^{-11}$ statohm
Inductance	L	henrys	10^{9} abhenrys	$\frac{1}{9} \times 10^{-11}$ stathenry
Capacitance	C	farads	10^{-9} abfarad	9×10^{11} statfarads
Conductivity	σ	siemens/meter	10^{-11} absiemen/centimeter	9×10^{9} statsiemens/centimeter
Work	W	joules	10^{7} ergs	10^{7} ergs
Power	P	watts	10^{7} ergs/second	10^{7} ergs/second

Conversion of Length Units

Meters	Centimeters	Inches	Feet	Miles
1	100	39.37	3.281	6.214×10^{-4}
0.01	1	0.3937	3.281×10^{-2}	
0.0254	2.540	1	8.333×10^{-2}	
0.3048	30.48	12	1	1.894×10^{-4}
1609			5279	1

Metric Prefixes and Symbols*

Multiplication Factor	Prefix	Symbol
10^{18}	exa	E
10^{15}	peta	P
10^{12}	tera	T
10^{9}	giga	G
10^{6}	mega	M
10^{3}	kilo	k
10^{2}	hecto	h
10	deka	da
10^{-1}	deci	d
10^{-2}	centi	c
10^{-3}	milli	m
10^{-6}	micro	μ
10^{-9}	nano	n
10^{-12}	pico	p
10^{-15}	femto	f
10^{-18}	atto	a

*From *IEEE Standard Dictionary of Electrical and Electronics Terms*, p. 682, The Institute of Electrical and Electronics Engineers, Inc., 1984.

Appendix B

The Frequency Spectrum

Li-Yin Chen
*University of Illinois**

The wavelength of an electromagnetic wave in free space is $\lambda_0 = c/f$.

$$\lambda_0 = \frac{300\,000}{f(\text{kHz})}\,\text{m} = \frac{300}{f(\text{MHz})}\,\text{m} = \frac{30}{f(\text{GHz})}\,\text{cm}$$

$$= \frac{9.843 \times 10^5}{f(\text{kHz})}\,\text{ft} = \frac{9.843 \times 10^2}{f(\text{MHz})}\,\text{ft} = \frac{11.81}{f(\text{GHz})}\,\text{in}$$

The wave number of an electromagnetic wave in free space: $k_0 = \omega\sqrt{\mu_0\epsilon_0} = 2\pi f/c$.

$$k_0 = f(\text{Hz}) \times 2.0944 \times 10^{-8}\,\text{m}^{-1} = f(\text{kHz}) \times 2.0944 \times 10^{-5}\,\text{m}^{-1}$$

$$= f(\text{MHz}) \times 2.0944 \times 10^{-2}\,\text{m}^{-1} = f(\text{GHz}) \times 20.944\,\text{m}^{-1}$$

$$= f(\text{Hz}) \times 6.383 \times 10^{-9}\,\text{ft}^{-1} = f(\text{kHz}) \times 6.383 \times 10^{-6}\,\text{ft}^{-1}$$

$$= f(\text{MHz}) \times 6.383 \times 10^{-3}\,\text{ft}^{-1} = f(\text{GHz}) \times 6.383\,\text{ft}^{-1}$$

$$= f(\text{GHz}) \times 0.532\,\text{in}^{-1}$$

*On leave from the Chinese Aeronautical Laboratory, Beijing, China, during 1983.

Nomenclature of Frequency Bands

Adjectival Designation	Frequency Range	Metric Subdivision	Wavelength Range
elf: Extremely low frequency	30 to 300 Hz	Megametric waves	10 000 to 1000 km
vf: Voice frequency	300 to 3000 Hz		1000 to 100 km
vlf: Very low frequency	3 to 30 kHz	Myriametric waves	100 to 10 km
lf: Low frequency	30 to 300 kHz	Kilometric waves	10 to 1 km
mf: Medium frequency	300 to 3000 kHz	Hectrometric waves	1000 to 100 m
hf: High frequency	3 to 30 MHz	Decametric waves	100 to 10 m
vhf: Very high frequency	30 to 300 MHz	Metric waves	10 to 1 m
uhf: Ultrahigh frequency	300 to 3000 MHz	Decimetric waves	100 to 10 cm
shf: Superhigh frequency	3 to 30 GHz	Centimetric waves	10 to 1 cm
ehf: Extremely high frequency	30 to 300 GHz	Millimetric waves	10 to 1 mm
	300 to 3000 GHz	Decimillimetric waves	1 to 0.1 mm

Standard Radar-Frequency Letter Bands*

Band Designation	Nominal Frequency Range
hf	3–30 MHz
vhf	30–300 MHz
uhf	300–1000 MHz
L	1000–2000 MHz
S	2000–4000 MHz
C	4000–8000 MHz
X	8000–12 000 MHz
K_u	12.0–18 GHz
K	18–27 GHz
K_a	27–40 GHz
Millimeter	40–300 GHz

*Reprinted from ANSI/IEEE Std. 100-1984, *IEEE Standard Dictionary of Electrical and Electronics Terms*, © 1984 by The Institute of Electrical and Electronics Engineers, Inc., by permission of the IEEE Standards Department.

Television Channel Frequencies*

Channel Number[†]	Band (MHz)	Channel Number[†]	Band (MHz)	Channel Number[†]	Band (MHz)
2	54–60	29	560–566	57	728–734
3	60–66	30	566–572	58	734–740
4	66–72	31	572–578	59	740–746
5	76–82	32	578–584	60	746–752
6	82–88	33	584–590	61	752–758
7	174–180	34	590–596	62	758–764
8	180–186	35	596–602	63	764–770
9	186–192	36	602–608	64	770–776
10	192–198	37	608–614	65	776–782
11	198–204	38	614–620	66	782–788
12	204–210	39	620–626	67	788–794
13	210–216	40	626–632	68	794–800
14	470–476	41	632–638	69	800–806
15	476–482	42	638–644	70	806–812
16	482–488	43	644–650	71	812–818
17	488–494	44	650–656	72	818–824
18	494–500	45	656–662	73	824–830
19	500–506	46	662–668	74	830–836
20	506–512	47	668–674	75	836–842
21	512–518	48	674–680	76	842–848
22	518–524	49	680–686	77	848–854
23	524–530	50	686–692	78	854–860
24	530–536	51	692–698	79	860–866
25	536–542	52	698–704	80	866–872
26	542–548	53	704–710	81	872–878
27	548–554	54	710–716	82	878–884
28	554–560	55	716–722	83	884–890
		56	722–728		

*Note: The carrier frequency for the video portion is the lower frequency plus 1.25 MHz. The audio carrier frequency is the upper frequency minus 0.25 MHz. All channels have a 6-MHz bandwidth. For example, channel 2 video carrier is at 55.25 MHz and the audio carrier is at 59.75 MHz.

[†]Channels 2 through 13 are vhf; channels 14 through 83 are uhf. Channels 70 through 83 were withdrawn and reassigned to tv translator stations until licenses expire.

Appendix C

Electromagnetic Properties of Materials

Yi-Lin Chen
*University of Illinois**

Resistivities and Skin Depth of Metals and Alloys

Material	Resistivity* ($\mu\Omega$-cm)	Skin Depth[†] (μm at 1 GHz)
Aluminum	2.62	2.576
Brass (66% Cu, 34% Zn)	7.5	4.3586
Copper	1.7241	2.0898
Gold	2.44	2.4861
Iron	9.71	4.9594
Nickel	6.9	4.1807
Silver	1.62	2.0257
Steel (0.4–0.5% C, balance Fe)	13–22	5.7384–7.465
Steel, stainless (0.1% C, 18% Cr, 8% Ni, balance Fe)	90	15.0988
Tin	11.4	5.3737
Titanium	47.8	11.0036

*In solid form at 20°C; resistivity = (conductivity)$^{-1}$.
[†]Skin depth $\delta = (\pi\mu\sigma f)^{-1/2} = (20\pi)^{-1} [\sigma f(\text{GHz})]^{-1/2}$ m. The δs in the column are calculated at $f = 1$ GHz. For other frequencies, multiply them by $[f(\text{GHz})]^{-1/2}$.

*On leave from the Chinese Aeronautical Laboratory, Beijing, China, during 1983.

Characteristics of Insulating Materials*

Material Composition	T (°C)	Dielectric Constant† at (Frequency in Hertz)				Dissipation Factor† at (Frequency in Hertz)				Dielectric Strength in Volts/Mil at 25°C	DC Volume Resistivity in Ohm-cm at 25°C	Thermal Expansion (Linear) in Parts/°C	Softening Point in °C	Moisture Absorption in Percent
		10^4	10^6	3×10^9	2.5×10^{10}	10^4	10^6	3×10^9	2.5×10^{10}					
Ceramics:														
Aluminum oxide	25	8.80	8.80	8.79	—	0.00033	0.00030	0.0010	—	—	—	—	1400–1430	—
Barium titanate†	26	1143	167.7	600	100	0.0105	—	0.30	0.60	75	10^{12}–10^{13}	—	—	0.1
Calcium titanate	25	167.7	167.7	165	—	0.0002	0.0023	0.0023	—	100	10^{12}–10^{14}	—	1510	<0.1
Magnesium oxide	25	9.65	9.65	5.90	—	<0.0003	<0.0003	—	—	—	—	—	—	—
Magnesium silicate	25	5.97	5.96	5.90	5.90	0.0005	0.0004	0.0012	—	—	>10^{14}	9.2×10^{-6}	1350	0.1–1
Magnesium titanate	25	13.9	13.9	13.8	13.7	0.0004	0.0005	0.0017	0.0065	—	—	—	—	—
Oxides of aluminum, silicon, magnesium, calcium, barium	24	6.04	—	5.90	—	0.0011	—	0.0024	—	—	—	7.7×10^{-6}	1325	—
Porcelain (dry process)	25	5.08	5.04	—	—	0.0075	0.0078	—	—	—	—	—	—	—
Steatite 410	25	5.77	5.77	5.7	5.7	0.0007	0.0006	0.0006	—	—	—	—	—	—
Strontium titanate	25	232	232	—	—	0.0002	0.0001	0.00089	—	—	—	—	1510	0.1
Titanium dioxide (rutile)	26	100	100	100	100	0.0003	0.00025	—	—	100	10^{12}–10^{14}	—	—	—
Glasses:														
Iron-sealing glass	24	8.30	8.20	7.99	7.84	0.0005	0.0009	0.00199	0.0112	—	10^{10} at 250°	132×10^{-7}	484	poor
Soda-borosilicate	25	4.84	4.84	4.82	4.65	0.0036	0.0030	0.0054	0.0090	410 (0.25")	7×10^7 at 250°	50×10^{-7}	693	—
100% silicon dioxide (fused quartz)	25	3.78	3.78	3.78	3.78	0.0001	0.0002	0.00006	0.00025	—	>10^{18}	5.7×10^{-7}	1667	—
Plastics:														
Alkyd resin	25	4.76	4.55	4.50	—	0.0149	0.0138	0.0108	—	—	—	—	—	—
Cellulose acetate-butyrate, plasticized	26	3.30	3.08	2.91	—	0.018	0.017	0.028	—	—	—	11–17×10^{-5}	60–121	2.3
Cresylic acid–formaldehyde, 50% α-cellulose	25	4.51	3.85	3.43	3.21	0.036	0.055	0.051	0.038	250–400 (0.125")	—	—	>125	1.2
Cross-linked polystyrene	25	2.58	2.58	2.58	—	0.0016	0.0020	0.0019	—	1020 (0.033")	3×10^{12}	3×10^{-5}	—	—
Epoxy resin (Araldite CN-501)	25	3.62	3.35	3.09	—	0.019	0.034	0.027	—	405 (0.125")	>3.8×10^7	4.77×10^{-5}	109 (distortion)	0.14
Epoxy resin (Epon resin RN-48)	25	3.52	3.32	3.04	—	0.0142	0.0264	0.021	—	—	—	—	—	—
Foamed polystyrene, 0.25% filler	25	1.03	—	1.03	1.03	<0.0002	0.0001	0.0001	—	—	—	—	85	low
Melamine—formaldehyde, α-cellulose	24	7.00	6.0	4.93	—	0.041	0.085	0.103	—	300–400	—	—	99 (stable)	0.4–0.6
Melamine—formaldehyde, 55% filler	26	5.75	5.5	—	—	0.0115	0.020	—	—	—	10^{11}	1.7×10^{-5}	—	0.6
Phenol—formaldehyde (Bakelite BM 120)	25	4.36	3.95	3.70	3.55	0.0280	0.0380	0.0390	—	300 (0.125")	—	30–40×10^{-6}	<135 (distortion)	<0.6
Phenol—formaldehyde, 50% paper laminate	26	4.60	4.04	3.57	—	0.034	0.057	0.060	—	—	—	—	—	—
Phenol—formaldehyde, 65% mica, 4% lubricants	24	4.78	4.72	4.71	—	0.0082	0.0115	0.0126	—	—	—	—	—	—
Polycarbonate	—	2.96	—	—	—	0.010	0.0126	—	—	364 (0.125")	2×10^{16}	7×10^{-5}	135 (deflection)	—
Polychlorotrifluoroethylene	25	2.42	2.32	2.29	2.28	0.0082	0.0028	0.0053	—	—	10^{13}	—	—	—

Polyethylene	25	2.26	2.26	2.26	2.26	<0.0002	0.0002	0.00031	0.0006	1200 (0.033″)	10^{17}	19×10^{-5} (varies)	95–105 (distortion)	0.03
Polyethylene-terephthalate	—	2.98	—	—	—	0.016	—	—	—	4000 (0.002″)	—	—	—	—
Polyethylmethacrylate	22	2.55	2.52	2.51	2.5	0.0090	—	0.0075	0.0083	—	—	—	60 (distortion)	low
Polyhexamethylene-adipamide (nylon)	25	3.14	3.0	2.84	2.73	0.0218	0.0200	0.0117	0.0105	400 (0.125″)	8×10^{14}	10.3×10^{-5}	65 (distortion)	1.5
Polyimide	—	3.4	—	—	—	0.003	—	—	—	570	—	—	—	—
Polyisobutylene	25	2.23	2.23	2.23	—	0.0001	0.0003	0.00047	—	600 (0.010″)	—	—	25 (distortion)	low
Polymer of 95% vinyl-chloride, 5% vinyl-acetate	20	2.90	2.8	2.74	—	0.0150	0.0080	0.0059	—	—	—	—	—	—
Polymethyl methacrylate	27	2.76	—	2.60	—	0.0140	—	0.0057	—	990 (0.030″)	$>5 \times 10^{16}$	$8–9 \times 10^{-5}$	70–75 (distortion)	0.3–0.6
Polyphenylene oxide	—	2.55	—	2.55	—	0.0007	—	0.0011	—	500 (0.125″)	10^{17}	5.3×10^{-5}	195 (deflection)	—
Polypropylene	—	2.55	—	—	—	<0.0005	—	—	—	650 (0.125″)	6×10^{16}	$6–8.5 \times 10^{-5}$	99–116 (deflection)	—
Polystyrene	25	2.56	2.55	2.55	2.54	0.00007	<0.0001	0.00033	0.0012	500–700 (0.125″)	10^{18}	$6–8 \times 10^{-5}$	82 (distortion)	0.05
Polytetrafluoroethylene (Teflon)	22	2.1	2.1	2.1	2.08	<0.0002	<0.0002	0.00015	0.0006	1000–2000 (0.005″–0.012″)	10^{17}	9.0×10^{-5}	66 (distortion) (stable to 300)	0.00
Polyvinylcyclohexane	24	2.25	2.25	2.25	—	<0.0002	<0.0002	0.00018	—	—	—	—	—	—
Polyvinyl formal	26	2.92	2.80	2.76	2.7	0.019	0.013	0.0113	0.0115	860 (0.034″)	$>5 \times 10^{16}$	7.7×10^{-5}	190	1.3
Polyvinylidene fluoride	—	6.6	—	—	—	0.17	—	—	—	260 (0.125″)	2×10^{14}	12×10^{-5}	148 (deflection)	—
Urea-formaldehyde, cellulose	27	5.65	5.1	4.57	—	0.027	0.050	0.0555	—	375 (0.085″)	—	2.6×10^{-5}	152 (distortion)	2
Urethane elastomer	—	6.5–7.1	—	—	—	—	—	—	—	450–500 (0.125″)	2×10^{11}	$10–20 \times 10^{-5}$	—	—
Vinylidene–vinyl chloride copolymer	23	3.18	2.82	2.71	—	0.057	0.0180	0.0072	—	300 (0.125″)	$10^{14}–10^{16}$	15.8×10^{-5}	150	<0.1
100% aniline-formaldehyde (Dilectane-100)	25	3.58	3.50	3.44	—	0.0061	0.0033	0.0026	—	810 (0.068″)	10^{16}	5.4×10^{-5}	125	0.06–0.08
100% phenol-formaldehyde	24	5.4	4.4	3.64	—	0.060	0.077	0.052	—	277 (0.125″)	—	$8.3–13 \times 10^{-5}$	50 (distortion)	0.42
100% polyvinyl-chloride	20	2.88	2.85	2.84	—	0.0160	0.0081	0.0055	—	400 (0.125″)	10^{14}	6.9×10^{-5}	54 (distortion)	0.05–0.15
Organic Liquids:														
Aviation gasoline (100 octane)	25	1.94	1.94	1.92	—	—	0.0001	0.0014	—	—	—	—	—	—
Benzene (pure, dried)	25	2.28	2.28	2.28	2.28	<0.0001	<0.0001	<0.0001	<0.0001	—	—	—	—	—
Carbon tetrachloride	25	2.17	2.17	2.17	—	<0.00004	<0.0002	0.0004	—	—	—	—	—	—
Ethyl alcohol (absolute)	25	24.5	23.7	6.5	—	0.090	0.062	0.250	—	—	—	—	—	—
Ethylene glycol	25	41	41	12	—	0.030	0.045	1.00	—	—	—	—	—	—
Jet fuel (JP-3)	25	2.08	2.08	2.04	—	0.0001	—	0.0055	—	—	—	—	—	—
Methyl alcohol (absolute analytical grade)	25	31	31.0	23.9	—	0.20	0.038	0.64	—	—	—	—	—	—
Methyl or ethyl siloxane polymer (1000 cs)	22	2.78	—	2.74	—	<0.0003	—	0.0096	—	—	—	—	—	—
Monomeric styrene	22	2.40	2.40	2.40	—	<0.0003	—	0.0020	—	300 (0.100″)	3×10^{12}	—	—	0.06
Transil oil	26	2.22	2.20	2.18	—	<0.0005	0.0048	0.0028	—	300 (0.100″)	—	—	−40 (pour point)	—
Vaseline	25	2.16	2.16	2.16	—	<0.0001	<0.0004	0.00066	—	—	—	—	—	—

Characteristics of Insulating Materials* (cont'd.)

Material Composition	T(°C)	Dielectric Constant† at (Frequency in Hertz)				Dissipation Factor† at (Frequency in Hertz)				Dielectric Strength in Volts/Mil at 25°C	DC Volume Resistivity in Ohm-cm at 25°C	Thermal Expansion (Linear) in Parts/°C	Softening Point in °C	Moisture Absorption in Percent
		10^4	10^6	3×10^9	2.5×10^{10}	10^4	10^6	3×10^9	2.5×10^{10}					
Waxes:														
Beeswax, yellow	23	2.53	2.45	2.39	—	0.0092	0.0090	0.0075	—	—	—	—	45–64 (melts)	—
Dichloronaphthalenes	23	2.98	2.93	2.89	—	0.0003	0.0017	0.0037	—	—	—	—	35–63 (melts)	nil
Polybutene	25	2.34	2.30	2.27	—	0.00133	0.00133	0.0009	—	—	—	—	—	—
Vegetable and mineral waxes	25	2.3	2.3	2.25	—	0.0004	0.0004	0.00046	—	—	—	—	57	—
Rubbers:														
Butyl rubber	25	2.35	2.35	2.35	—	0.0010	0.0010	0.0009	—	—	—	—	—	—
GR-S rubber	25	2.90	2.82	2.75	—	0.0120	0.0080	0.0057	—	870 (0.040")	2×10^{15}	—	—	—
Gutta-percha	25	2.53	2.47	2.40	—	0.0042	0.0120	0.0060	—	—	10^{15}	—	—	—
Hevea rubber (pale crepe)	25	2.4	2.4	2.15	—	0.0018	0.0050	0.0030	—	—	—	—	—	—
Hevea rubber, vulcanized (100 pts pale crepe, 6 pts sulfur)	27	2.74	2.42	2.36	4.0	0.0446	0.0180	0.0047	0.025	300 (0.125")	8×10^{12}	—	—	—
Neoprene rubber	24	6.26	4.5	4.00	13.6	0.038	0.090	0.034	0.10	—	—	—	—	nil
Organic polysulfide, fillers	23	110	30	16	—	0.39	0.28	0.22	—	—	—	—	—	—
Silicone-rubber compound	25	3.20	3.16	3.13	—	0.0030	0.0032	0.0097	—	—	—	—	—	—
Woods:‡														
Balsa wood	26	1.37	1.30	1.22	—	0.0120	0.0135	0.100	—	—	—	—	—	—
Douglas fir	25	1.93	1.88	1.82	1.78	0.026	0.033	0.027	0.032	—	—	—	—	—
Douglas fir, plywood	25	1.90	—	—	1.6	0.0230	—	—	0.0220	—	—	—	—	—
Mahogany	25	2.25	2.07	1.88	1.6	0.025	0.032	0.025	0.020	—	—	—	—	—
Yellow birch	25	2.70	2.47	2.13	1.87	0.029	0.040	0.033	0.026	—	—	—	—	—
Yellow poplar	25	1.75	—	1.50	1.4	0.019	—	0.015	0.017	—	—	—	—	—
Miscellaneous:														
Amber (fossil resin)	25	2.65	—	2.6	—	0.0056	—	0.0090	—	2300 (0.125")	Very high	9.8×10^{-5}	200	—
DeKhotirsky cement	23	3.23	—	2.96	—	0.024	—	0.021	—	—	—	—	80–85	—
Gilsonite (99.9% natural bitumen)	26	2.58	2.56	—	—	0.0016	0.0011	0.0011	—	—	—	—	155	—
Shellac (natural XL)	28	3.47	3.10	2.86	—	0.031	0.030	0.0254	—	—	10^{16}	—	80	low after baking
Mica, glass-bonded	25	7.39	—	—	—	0.0013	—	—	—	—	—	—	—	—
Mica, glass, titanium dioxide	24	9.0	—	—	—	0.0026	—	0.0040	—	—	—	—	—	—
Ruby mica	26	5.4	5.4	5.4	—	0.0003	0.0002	0.0003	—	3800–5600 (0.040") 202 (0.125")	5×10^{13}	—	400	<0.5
Paper, royalgrey	25	2.99	2.77	2.70	—	0.038	0.066	0.056	—	—	—	—	—	—
Selenium (amorphous)	25	6.00	6.00	6.00	6.00	<0.0003	<0.0002	0.00018	0.0013	—	—	—	—	—
Asbestos fiber–chrysotile paper	25	3.1	—	—	—	0.025	—	0.0013	—	—	—	—	—	—
Sodium chloride (fresh crystals)	25	5.90	—	—	5.90	<0.0002	—	<0.0005	—	—	—	—	—	—

Soil, sandy dry	25	2.59	2.55	2.55	—	0.017	—	0.0062	—
Soil, loamy dry	25	2.53	2.48	2.44	—	0.018	—	0.0011	—
Ice (from pure distilled water)	−12	4.15	3.45	3.20	—	0.12	0.035	0.0009	—
Freshly fallen snow	−20	1.20	1.20	1.20	—	0.0215	—	0.00029	—
Hard-packed snow followed by light rain	−6	1.55	—	1.5	—	0.29	—	0.0009	—
Water (distilled)	25	78.2	78	76.7	34	0.040	0.005	0.157	0.2650	10^4

*Reproduced with permission of the publisher, Howard W. Sams & Company, Indianapolis, *Reference Data for Engineers: Radio, Electronics, Computer, and Communications*, 7th ed., by E. C. Jordan, ed., © 1985.

†The dissipation factor is defined as the ratio of the energy dissipated to the energy stored in the dielectric, or as the tangent of the loss angle. Dielectric constant and dissipation factor depend on electrical field strength.

‡Field perpendicular to grain.

Properties of Soft Magnetic Metals*

Name	Composition (%)	Permeability Initial	Permeability Maximum	Coercivity H_c (A/m)	Retentivity B_r (T)	B_{max} (T)	Resistivity ($\mu\Omega$-cm)
Ingot iron	99.8 Fe	150	5 000	80	0.77	2.14	10
Low carbon steel	99.5 Fe	200	4 000	100	—	2.14	12
Silicon iron, unoriented	3 Si, bal Fe	270	8 000	60	—	2.01	47
Silicon iron, grain oriented	3 Si, bal Fe	1 400	50 000	7	1.20	2.01	50
4750 alloy	48 Ni, bal Fe	11 000	80 000	2	—	1.55	48
4-79 Permalloy	4 Mo, 79 Ni, bal Fe	40 000	200 000	1	—	0.80	58
Supermalloy	5 Mo, 80 Ni, bal Fe	80 000	450 000	0.4	—	0.78	65
2V-Permendur	2V, 49 Co, bal Fe	800	8 000	160	—	2.30	40
Supermendur	2V, 49 Co, bal Fe	—	100 000	16	2.00	2.30	26
Metglas[†] 2605SC	$Fe_{81}B_{13.5}Si_{3.5}C_2$	—	210 000	14	1.46	1.60	125
Metglas[†] 2605S-3	$Fe_{79}B_{16}Si_5$	—	30 000	8	0.30	1.58	125

*Reproduced with permission of the publisher, Howard W. Sams & Company, Indianapolis, *Reference Data for Engineers: Radio, Electronics, Computer, and Communications*, 7th ed., by E. C. Jordan, ed., © 1985.
[†]Metglas is Allied Corporation's registered trademark for amorphous alloys.

Appendix D

Vector Analysis

Yi-Lin Chen
University of Illinois[*]

1. Change of Coordinate Systems

The transformations of the coordinate components of a vector **A** among the rectangular (x, y, z), cylindrical (θ, ϕ, z), and spherical (r, θ, ϕ) coordinates are given by the following relations (see Fig. 1):

$$A_x = A_\varrho \cos\phi - A_\phi \sin\phi = A_r \sin\theta\cos\phi + A_\theta \cos\theta\cos\phi - A_\phi \sin\phi$$
$$A_y = A_\varrho \sin\phi + A_\phi \cos\phi = A_r \sin\theta\sin\phi + A_\theta \cos\theta\sin\phi + A_\phi \cos\phi$$

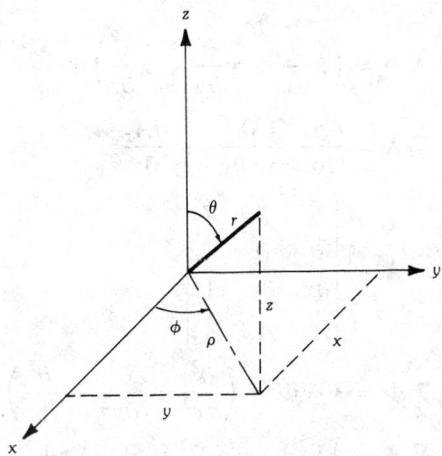

Fig. 1. Rectangular, cylindrical, and spherical coordinate systems.

---

*On leave from the Chinese Aeronautical Laboratory, Beijing, China, during 1983.

$$A_z = A_r \cos\theta - A_\theta \sin\theta$$
$$A_\varrho = A_x \cos\phi + A_y \sin\phi = A_r \sin\theta + A_\theta \cos\theta$$
$$A_\phi = -A_x \sin\phi + A_y \cos\phi$$
$$A_r = A_x \sin\theta \cos\phi + A_y \sin\theta \sin\phi + A_z \cos\theta = A_\varrho \sin\theta + A_z \cos\theta$$
$$A_\theta = A_x \cos\theta \cos\phi + A_y \cos\theta \sin\phi - A_z \sin\theta = A_\varrho \cos\theta - A_z \sin\theta$$

Differential element of volume:

$$dV = dx\,dy\,dz = \varrho\,d\varrho\,d\phi\,dz = r^2 \sin\theta\,dr\,d\theta\,d\phi$$

Differential element of vector area:

$$\mathbf{dS} = \hat{\mathbf{x}}\,dy\,dz + \hat{\mathbf{y}}\,dx\,dz + \hat{\mathbf{z}}\,dx\,dy$$
$$= \hat{\boldsymbol{\varrho}}\varrho\,d\phi\,dz + \hat{\boldsymbol{\phi}}\,d\varrho\,dz + \hat{\mathbf{z}}\varrho\,d\varrho\,d\phi$$
$$= \hat{\mathbf{r}}r^2 \sin\theta\,d\theta\,d\phi + \hat{\boldsymbol{\theta}}r \sin\theta\,dr\,d\phi + \hat{\boldsymbol{\phi}}r\,dr\,d\theta$$

Differential element of vector length:

$$d\boldsymbol{\ell} = \hat{\mathbf{x}}\,dx + \hat{\mathbf{y}}\,dy + \hat{\mathbf{z}}\,dz$$
$$= \hat{\boldsymbol{\varrho}}\,d\varrho + \hat{\boldsymbol{\phi}}\varrho\,d\phi + \hat{\mathbf{z}}\,dz$$
$$= \hat{\mathbf{r}}\,dr + \hat{\boldsymbol{\theta}}r\,d\theta + \hat{\boldsymbol{\phi}}r \sin\theta\,d\phi$$

2. ∇ Operator

In rectangular coordinates (x, y, z):

$$\nabla\Phi = \left(\hat{\mathbf{x}}\frac{\partial}{\partial x} + \hat{\mathbf{y}}\frac{\partial}{\partial y} + \hat{\mathbf{z}}\frac{\partial}{\partial z} \right)\Phi$$

$$\nabla\cdot\mathbf{A} = \frac{\partial A_x}{\partial x} + \frac{\partial A_y}{\partial y} + \frac{\partial A_z}{\partial z}$$

$$\nabla\times\mathbf{A} = \begin{vmatrix} \hat{\mathbf{x}} & \hat{\mathbf{y}} & \hat{\mathbf{z}} \\ \dfrac{\partial}{\partial x} & \dfrac{\partial}{\partial y} & \dfrac{\partial}{\partial z} \\ A_x & A_y & A_z \end{vmatrix}$$

$$\nabla^2\Phi = \nabla\cdot\nabla\Phi = \left(\frac{\partial^2}{\partial x^2} + \frac{\partial^2}{\partial y^2} + \frac{\partial^2}{\partial z^2} \right)\Phi$$

$$\nabla^2\mathbf{A} = \hat{\mathbf{x}}\nabla^2 A_x + \hat{\mathbf{y}}\nabla^2 A_y + \hat{\mathbf{z}}\nabla^2 A_z$$

In cylindrical coordinates (ϱ, ϕ, z):

$$\nabla\Phi = \left(\hat{\boldsymbol{\varrho}}\frac{\partial}{\partial\varrho} + \hat{\boldsymbol{\phi}}\frac{\partial}{\varrho\partial\phi} + \hat{\mathbf{z}}\frac{\partial}{\partial z} \right)\Phi$$

$$\nabla \cdot \mathbf{A} = \frac{1}{\varrho}\frac{\partial}{\partial \varrho}(\varrho A_\varrho) + \frac{1}{\varrho}\frac{\partial A_\phi}{\partial \phi} + \frac{\partial A_z}{\partial z}$$

$$\nabla \times \mathbf{A} = \frac{1}{\varrho}\begin{vmatrix} \hat{\varrho} & \varrho\hat{\phi} & \hat{z} \\ \dfrac{\partial}{\partial \varrho} & \dfrac{\partial}{\partial \phi} & \dfrac{\partial}{\partial z} \\ A_\varrho & \varrho A_\phi & A_z \end{vmatrix}$$

$$= \hat{\varrho}\left(\frac{1}{\varrho}\frac{\partial A_z}{\partial \phi} - \frac{\partial A_\phi}{\partial z}\right) + \hat{\phi}\left(\frac{\partial A_\varrho}{\partial z} - \frac{\partial A_z}{\partial \varrho}\right) + \hat{z}\left[\frac{1}{\varrho}\frac{\partial}{\partial \phi}(\varrho A_\phi) - \frac{1}{\varrho}\frac{\partial A_\varrho}{\partial \phi}\right]$$

$$\nabla^2\Phi = \frac{1}{\varrho}\frac{\partial}{\partial \varrho}\left(\varrho\frac{\partial \Phi}{\partial \varrho}\right) + \frac{1}{\varrho^2}\frac{\partial^2 \Phi}{\partial \phi^2} + \frac{\partial^2 \Phi}{\partial z^2}$$

$$\nabla^2\mathbf{A} = \nabla\nabla\cdot\mathbf{A} - \nabla\times\nabla\times\mathbf{A} \neq \hat{\varrho}\nabla^2 A_\varrho + \hat{\phi}\nabla^2 A_\phi + \hat{z}\nabla^2 A_z$$

In spherical coordinates (r, θ, ϕ):

$$\nabla\Phi = \left(\hat{r}\frac{\partial}{\partial r} + \hat{\theta}\frac{1}{r}\frac{\partial}{\partial \theta} + \hat{\phi}\frac{1}{r\sin\theta}\frac{\partial}{\partial \phi}\right)\Phi$$

$$\nabla\cdot\mathbf{A} = \frac{1}{r}\frac{\partial}{\partial r}(r^2 A_r) + \frac{1}{r\sin\theta}\frac{\partial}{\partial \theta}(A_\theta\sin\theta) + \frac{1}{r\sin\theta}\frac{\partial A_\phi}{\partial \phi}$$

$$\nabla\times\mathbf{A} = \frac{1}{r^2\sin\theta}\begin{vmatrix} \hat{r} & r\hat{\theta} & (r\sin\theta)\hat{\phi} \\ \dfrac{\partial}{\partial r} & \dfrac{\partial}{\partial \theta} & \dfrac{\partial}{\partial \phi} \\ A_r & rA_\theta & (r\sin\theta)A_\phi \end{vmatrix}$$

$$= \hat{r}\frac{1}{r\sin\theta}\left[\frac{\partial}{\partial \theta}(A_\phi\sin\theta) - \frac{\partial A_\theta}{\partial \phi}\right] + \hat{\theta}\frac{1}{r}\left[\frac{1}{\sin\theta}\frac{\partial A_r}{\partial \phi} - \frac{\partial}{\partial r}(rA_\phi)\right]$$

$$+ \hat{\phi}\frac{1}{r}\left[\frac{\partial}{\partial r}(rA_\phi) - \frac{\partial A_r}{\partial \theta}\right]$$

$$\nabla^2\Phi = \frac{1}{r^2}\frac{\partial}{\partial r}\left(r^2\frac{\partial \Phi}{\partial r}\right) + \frac{1}{r^2\sin\theta}\frac{\partial}{\partial \theta}\left(\sin\theta\frac{\partial \Phi}{\partial \theta}\right) + \frac{1}{r^2\sin^2\theta}\frac{\partial^2 \Phi}{\partial \phi^2}$$

$$\nabla^2\mathbf{A} = \nabla\nabla\cdot A - \nabla\times\nabla\times A \neq \hat{r}\nabla^2 A_r + \hat{\theta}\nabla^2 A_\theta + \hat{\phi}\nabla^2 A_\phi$$

3. Identities

$$\mathbf{a}\cdot\mathbf{b}\times\mathbf{c} = \mathbf{a}\times\mathbf{b}\cdot\mathbf{c} = \mathbf{b}\cdot\mathbf{c}\times\mathbf{a}$$

$$\mathbf{a}\times(\mathbf{b}\times\mathbf{c}) = (\mathbf{a}\cdot\mathbf{c})\mathbf{b} - (\mathbf{a}\cdot\mathbf{b})\mathbf{c}$$

$$(\mathbf{a}\times\mathbf{b})\cdot(\mathbf{c}\times\mathbf{d}) = \mathbf{a}\cdot\mathbf{b}\times(\mathbf{c}\times\mathbf{d}) = \mathbf{a}\cdot[(\mathbf{b}\cdot\mathbf{d})\mathbf{c} - (\mathbf{b}\cdot\mathbf{c})\mathbf{d}] = (\mathbf{a}\cdot\mathbf{c})(\mathbf{b}\cdot\mathbf{d}) - (\mathbf{a}\cdot\mathbf{d})(\mathbf{b}\cdot\mathbf{c})$$

$$(\mathbf{a}\times\mathbf{b})\times(\mathbf{c}\times\mathbf{d}) = (\mathbf{a}\times\mathbf{b}\cdot\mathbf{d})\mathbf{c} - (\mathbf{a}\times\mathbf{b}\cdot\mathbf{c})\mathbf{d}$$

$$\nabla(\Phi + \psi) = \nabla\Phi + \nabla\psi$$

$$\nabla(\Phi\psi) = \Phi\nabla\psi + \psi\nabla\Phi$$

$$\nabla\cdot(\mathbf{a} + \mathbf{b}) = \nabla\cdot\mathbf{a} + \nabla\cdot\mathbf{b}$$

$$\nabla \times (\mathbf{a} + \mathbf{b}) = \nabla \times \mathbf{a} + \nabla \times \mathbf{b}$$

$$\nabla \cdot (\Phi \mathbf{a}) = \mathbf{a} \cdot \nabla \Phi + \Phi \nabla \cdot \mathbf{a}$$

$$\nabla \times (\Phi \mathbf{a}) = \nabla \Phi \times \mathbf{a} + \Phi \nabla \times \mathbf{a}$$

$$\nabla (\mathbf{a} \cdot \mathbf{b}) = (\mathbf{a} \cdot \nabla) \mathbf{b} + (\mathbf{b} \cdot \nabla) \mathbf{a} + \mathbf{a} \times (\nabla \times \mathbf{b}) + \mathbf{b} \times (\nabla \times \mathbf{a})$$

$$\nabla \times (\mathbf{a} \times \mathbf{b}) = \mathbf{a} \nabla \cdot \mathbf{b} - \mathbf{b} \nabla \cdot \mathbf{a} + (\mathbf{b} \cdot \nabla) \mathbf{a} - (\mathbf{a} \cdot \nabla) \mathbf{b}$$

$$\nabla \cdot (\mathbf{a} \times \mathbf{b}) = \mathbf{b} \cdot \nabla \times \mathbf{a} - \mathbf{a} \cdot \nabla \times \mathbf{b}$$

$$\nabla \times \nabla \times \mathbf{a} = \nabla \nabla \cdot \mathbf{a} - \nabla^2 \mathbf{a}$$

$$\nabla \times \nabla \Phi \equiv 0$$

$$\nabla \cdot \nabla \times \mathbf{a} \equiv 0$$

$$\iiint_V \nabla \cdot \mathbf{a} \, dV = \oiint_S \mathbf{a} \cdot d\mathbf{S} \qquad \text{(Gauss's theorem)}$$

$$\iint_S \nabla \times \mathbf{a} \cdot d\mathbf{S} = \oint_c \mathbf{a} \cdot d\boldsymbol{\ell} \qquad \text{(Stokes's theorem)}$$

Green's first and second identities:

$$\iiint_V (\nabla \psi \cdot \nabla \Phi + \Phi \nabla^2 \psi) \, dV = \oiint_S \Phi \nabla \psi \cdot d\mathbf{S}$$

$$\iiint_V (\Phi \nabla^2 \psi - \psi \nabla^2 \Phi) \, dV = \oiint_S (\Phi \nabla \psi - \psi \nabla \Phi) \cdot d\mathbf{S}$$

$$\iiint_V (\nabla \times \mathbf{A} \cdot \nabla \times \mathbf{B} - \mathbf{A} \cdot \nabla \times \nabla \times \mathbf{B}) \, dV = \iiint_V (\nabla \cdot \mathbf{A} \times \nabla \times \mathbf{B}) \, dV$$

$$= \oiint_S \mathbf{A} \times \nabla \times \mathbf{B} \cdot d\mathbf{S}$$

$$\iiint_V (\mathbf{B} \cdot \nabla \times \nabla \times \mathbf{A} - \mathbf{A} \cdot \nabla \times \nabla \times \mathbf{B}) \, dV = \oiint_S (\mathbf{A} \times \nabla \times \mathbf{B} - \mathbf{B} \times \nabla \times \mathbf{A}) \cdot d\mathbf{S}$$

Appendix E

VSWR Versus Reflection Coefficient and Mismatch Loss

Yi-Lin Chen
University of Illinois[*]

The following relations are used in the construction of the vswr table below.

$$\text{vswr} = \frac{1 + |\Gamma|}{1 - |\Gamma|}, \qquad |\Gamma| = \frac{\text{vswr} - 1}{\text{vswr} + 1}$$

$$\text{mismatch loss (dB)} = -10 \log_{10}(1 - |\Gamma|^2)$$

VSWR Versus Reflection Coefficient (Γ) and Mismatch Loss

VSWR	$|\Gamma|$	Mismatch Loss (dB)	VSWR	$|\Gamma|$	Mismatch Loss (dB)
1.01	.0050	.0001	1.12	.0566	.0139
1.02	.0099	.0004	1.13	.0610	.0162
1.03	.0148	.0009	1.14	.0654	.0186
1.04	.0196	.0017	1.15	.0698	.0212
1.05	.0244	.0026	1.16	.0741	.0239
1.06	.0291	.0037	1.17	.0783	.0267
1.07	.0338	.0050	1.18	.0826	.0297
1.08	.0385	.0064	1.19	.0868	.0328
1.09	.0431	.0081	1.20	.0909	.0360
1.10	.0476	.0099	1.21	.0950	.0394
1.11	.0521	.0118	1.22	.0991	.0429

[*]On leave from the Chinese Aeronautical Laboratory, Beijing, China, during 1983.

VSWR Versus Reflection Coefficient (Γ) and Mismatch Loss (cont'd.)

VSWR	|Γ|	Mismatch Loss (dB)	VSWR	|Γ|	Mismatch Loss (dB)
1.23	.1031	.0464	1.73	.2674	.3222
1.24	.1071	.0501	1.74	.2701	.3289
1.25	.1111	.0540	1.75	.2727	.3357
1.26	.1150	.0579	1.76	.2754	.3425
1.27	.1189	.0619	1.77	.2780	.3493
1.28	.1228	.0660	1.78	.2806	.3561
1.29	.1266	.0702	1.79	.2832	.3630
1.30	.1304	.0745	1.80	.2857	.3698
1.31	.1342	.0789	1.81	.2883	.3767
1.32	.1379	.0834	1.82	.2908	.3837
1.33	.1416	.0880	1.83	.2933	.3906
1.34	.1453	.0927	1.84	.2958	.3976
1.35	.1489	.0974	1.85	.2982	.4046
1.36	.1525	.1023	1.86	.3007	.4116
1.37	.1561	.1072	1.87	.3031	.4186
1.38	.1597	.1121	1.88	.3056	.4257
1.39	.1632	.1172	1.89	.3080	.4327
1.40	.1667	.1223	1.90	.3103	.4398
1.41	.1701	.1275	1.91	.3127	.4469
1.42	.1736	.1328	1.92	.3151	.4540
1.43	.1770	.1382	1.93	.3174	.4612
1.44	.1803	.1436	1.94	.3197	.4683
1.45	.1837	.1490	1.95	.3220	.4755
1.46	.1870	.1546	1.96	.3243	.4827
1.47	.1903	.1602	1.97	.3266	.4899
1.48	.1935	.1658	1.98	.3289	.4971
1.49	.1968	.1715	1.99	.3311	.5043
1.50	.2000	.1773	2.00	.3333	.5115
1.51	.2032	.1831	2.05	.3443	.5479
1.52	.2063	.1890	2.10	.3548	.5844
1.53	.2095	.1949	2.15	.3651	.6212
1.54	.2126	.2009	2.20	.3750	.6582
1.55	.2157	.2069	2.25	.3846	.6952
1.56	.2188	.2130	2.30	.3939	.7324
1.57	.2218	.2191	2.35	.4030	.7696
1.58	.2248	.2252	2.40	.4118	.8069
1.59	.2278	.2314	2.45	.4203	.8441
1.60	.2308	.2377	2.50	.4286	.8814
1.61	.2337	.2440	2.55	.4366	.9186
1.62	.2366	.2503	2.60	.4444	.9557
1.63	.2395	.2566	2.65	.4521	.9928
1.64	.2424	.2630	2.70	.4595	1.0298
1.65	.2453	.2695	2.75	.4667	1.0667
1.66	.2481	.2760	2.80	.4737	1.1035
1.67	.2509	.2825	2.85	.4805	1.1402
1.68	.2537	.2890	2.90	.4872	1.1767
1.69	.2565	.2956	2.95	.4937	1.2131
1.70	.2593	.3022	3.00	.5000	1.2494
1.71	.2620	.3088	3.10	.5122	1.3215
1.72	.2647	.3155	3.20	.5238	1.3929

VSWR Versus Reflection Coefficient (Γ) and Mismatch Loss

VSWR	$|\Gamma|$	Mismatch Loss (dB)	VSWR	$|\Gamma|$	Mismatch Loss (dB)
3.30	.5349	1.4636	8.10	.7802	4.0754
3.40	.5455	1.5337	8.20	.7826	4.1170
3.50	.5556	1.6030	8.30	.7849	4.1583
3.60	.5652	1.6715	8.40	.7872	4.1992
3.70	.5745	1.7393	8.50	.7895	4.2397
3.80	.5833	1.8064	8.60	.7917	4.2798
3.90	.5918	1.8727	8.70	.7938	4.3196
4.00	.6000	1.9382	8.80	.7959	4.3591
4.10	.6078	2.0030	8.90	.7980	4.3982
4.20	.6154	2.0670	9.00	.8000	4.4370
4.30	.6226	2.1302	9.10	.8020	4.4754
4.40	.6296	2.1927	9.20	.8039	4.5135
4.50	.6364	2.2545	9.30	.8058	4.5513
4.60	.6429	2.3156	9.40	.8077	4.5888
4.70	.6491	2.3759	9.50	.8095	4.6260
4.80	.6552	2.4355	9.60	.8113	4.6628
4.90	.6610	2.4945	9.70	.8131	4.6994
5.00	.6667	2.5527	9.80	.8148	4.7356
5.10	.6721	2.6103	9.90	.8165	4.7716
5.20	.6774	2.6672	10.00	.8182	4.8073
5.30	.6825	2.7235	11.00	.8333	5.1491
5.40	.6875	2.7791	12.00	.8462	5.4665
5.50	.6923	2.8340	13.00	.8571	4.7625
5.60	.6970	2.8884	14.00	.8667	6.0399
5.70	.7015	2.9421	15.00	.8750	6.3009
5.80	.7059	2.9953	16.00	.8824	6.5472
5.90	.7101	3.0479	17.00	.8889	6.7804
6.00	.7143	3.0998	18.00	.8947	7.0017
6.10	.7183	3.1513	19.00	.9000	7.2125
6.20	.7222	3.2021	20.00	.9048	7.4135
6.30	.7260	3.2525	30.00	.9355	9.0354
6.40	.7297	3.3022	40.00	.9512	10.2145
6.50	.7333	3.3515	50.00	.9608	11.1411
6.60	.7368	3.4002	60.00	.9672	11.9045
6.70	.7403	3.4485	70.00	.9718	12.5536
6.80	.7436	3.4962	80.00	.9753	13.1182
6.90	.7468	3.5435	90.00	.9780	13.6178
7.00	.7500	3.5902	100.00	.9802	14.0658
7.10	.7531	3.6365	200.00	.9900	17.0330
7.20	.7561	3.6824	300.00	.9934	18.7795
7.30	.7590	3.7277	400.00	.9950	20.0217
7.40	.7619	3.7727	500.00	.9960	20.9865
7.50	.7647	3.8172	600.00	.9967	21.7754
7.60	.7674	3.8612	700.00	.9971	22.4428
7.70	.7701	3.9049	800.00	.9975	23.0212
7.80	.7727	3.9481	900.00	.9978	23.5315
7.90	.7753	3.9909	1000.00	.9980	23.9881
8.00	.7778	4.0334			

Appendix F

Decibels Versus Voltage and Power Ratios*

Yi-Lin Chen
University of Illinois[†]

The decibel chart below indicates decibels for any ratio of voltage or power up to 100 dB. For voltage ratios greater than 10 (or power ratios greater than 100) the ratio can be broken down into two products, the decibels found for each separately, the two results then added. For example, to convert a voltage ratio of 200:1 to dB, a 200:1 voltage ratio equals the product of 100:1 and 2:1. Now, 100:1 equals 40 dB; 21:1 equals 6 dB. Therefore a 200:1 voltage ratio equals 40 dB + 6 dB, or 46 dB.

$$dB = 20 \log_{10}(\text{voltage ratio}) = 10 \log_{10}(\text{power ratio})$$

*Reprinted with permission of *Microwave Journal*, from *The Microwave Engineer's Handbook and Buyer's Guide*, 1966 issue, © 1966 Horizon House–Microwave, Inc.
[†]On leave from the Chinese Aeronautical Laboratory, Beijing, China, during 1983.

Decibels Versus Voltage and Power Ratios

Voltage Ratio	Power Ratio	-dB +	Voltage Ratio	Power Ratio
1.0000	1.0000	0	1.000	1.000
.9886	.9772	.1	1.012	1.023
.9772	.9550	.2	1.023	1.047
.9661	.9333	.3	1.035	1.072
.9550	.9120	.4	1.047	1.096
.9441	.8913	.5	1.059	1.122
.9333	.8710	.6	1.072	1.148
.9226	.8511	.7	1.084	1.175
.9120	.8318	.8	1.096	1.202
.9016	.8128	.9	1.109	1.230
.8913	.7943	1.0	1.122	1.259
.8810	.7762	1.1	1.135	1.288
.8710	.7586	1.2	1.148	1.318
.8610	.7413	1.3	1.161	1.349
.8511	.7244	1.4	1.175	1.380
.8414	.7079	1.5	1.189	1.413
.8318	.6918	1.6	1.202	1.445
.8222	.6761	1.7	1.216	1.479
.8128	.6607	1.8	1.230	1.514
.8035	.6457	1.9	1.245	1.549
.7943	.6310	2.0	1.259	1.585
.7852	.6166	2.1	1.274	1.622
.7762	.6026	2.2	1.288	1.660
.7674	.5888	2.3	1.303	1.698
.7586	.5754	2.4	1.318	1.738

Voltage Ratio	Power Ratio	-dB +	Voltage Ratio	Power Ratio
.5309	.2818	5.5	1.884	3.548
.5248	.2754	5.6	1.905	3.631
.5188	.2692	5.7	1.928	3.715
.5129	.2630	5.8	1.950	3.802
.5070	.2570	5.9	1.972	3.890
.5012	.2512	6.0	1.995	3.981
.4955	.2455	6.1	2.018	4.074
.4898	.2399	6.2	2.042	4.169
.4842	.2344	6.3	2.065	4.266
.4786	.2291	6.4	2.089	4.365
.4732	.2239	6.5	2.113	4.467
.4677	.2188	6.6	2.138	4.571
.4624	.2138	6.7	2.163	4.677
.4571	.2089	6.8	2.188	4.786
.4519	.2042	6.9	2.213	4.898
.4467	.1995	7.0	2.239	5.012
.4416	.1950	7.1	2.265	5.129
.4365	.1905	7.2	2.291	5.248
.4315	.1862	7.3	2.317	5.370
.4266	.1820	7.4	2.344	5.495
.4217	.1778	7.5	2.371	5.623
.4169	.1738	7.6	2.399	5.754
.4121	.1698	7.7	2.427	5.888
.4074	.1660	7.8	2.455	6.026
.4027	.1622	7.9	2.483	6.166

Voltage Ratio	Power Ratio	-dB +	Voltage Ratio	Power Ratio
.2818	.07943	11.0	3.548	12.59
.2786	.07762	11.1	3.589	12.88
.2754	.07586	11.2	3.631	13.18
.2723	.07413	11.3	3.673	13.49
.2692	.07244	11.4	3.715	13.80
.2661	.07079	11.5	3.758	14.13
.2630	.06918	11.6	3.802	14.45
.2600	.06761	11.7	3.846	14.79
.2570	.06607	11.8	3.890	15.14
.2541	.06457	11.9	3.936	15.49
.2512	.06310	12.0	3.981	15.85
.2483	.06166	12.1	4.027	16.22
.2455	.06026	12.2	4.074	16.60
.2427	.05888	12.3	4.121	16.98
.2399	.05754	12.4	4.169	17.38
.2371	.05623	12.5	4.217	17.78
.2344	.05495	12.6	4.266	18.20
.2317	.05370	12.7	4.315	18.62
.2291	.05248	12.8	4.365	19.05
.2265	.05129	12.9	4.416	19.50
.2239	.05012	13.0	4.467	19.95
.2213	.04898	13.1	4.519	20.42
.2188	.04786	13.2	4.571	20.89
.2163	.04677	13.3	4.624	21.38
.2138	.04571	13.4	4.677	21.88

.7499	.5623	2.5	1.334	1.778	.3981	.1585	8.0	2.512	6.310	.2113	.04467	13.5	4.732	22.39
.7413	.5495	2.6	1.349	1.820	.3936	.1549	8.1	2.541	6.457	.2089	.04365	13.6	4.786	22.91
.7328	.5370	2.7	1.365	1.862	.3890	.1514	8.2	2.570	6.607	.2065	.04266	13.7	4.842	23.44
.7244	.5248	2.8	1.380	1.905	.3846	.1479	8.3	2.600	6.761	.2042	.04169	13.8	4.898	23.99
.7161	.5129	2.9	1.396	1.950	.3802	.1445	8.4	2.630	6.918	.2018	.04074	13.9	4.955	24.55
.7079	.5012	3.0	1.413	1.995	.3758	.1413	8.5	2.661	7.079	.1995	.03981	14.0	5.012	25.12
.6998	.4898	3.1	1.429	2.042	.3715	.1380	8.6	2.692	7.244	.1972	.03890	14.1	5.070	25.70
.6918	.4786	3.2	1.445	2.089	.3673	.1349	8.7	2.723	7.413	.1950	.03802	14.2	5.129	26.30
.6839	.4677	3.3	1.462	2.138	.3631	.1318	8.8	2.754	7.586	.1928	.03715	14.3	5.188	26.92
.6761	.4571	3.4	1.479	2.188	.3589	.1288	8.9	2.786	7.762	.1905	.03631	14.4	5.248	27.54
.6683	.4467	3.5	1.496	2.239	.3548	.1259	9.0	2.818	7.943	.1884	.03548	14.5	5.309	28.18
.6607	.4365	3.6	1.514	2.291	.3508	.1230	9.1	2.851	8.128	.1862	.03467	14.6	5.370	28.84
.6531	.4266	3.7	1.531	2.344	.3467	.1202	9.2	2.884	8.318	.1841	.03388	14.7	5.433	29.51
.6457	.4169	3.8	1.549	2.399	.3428	.1175	9.3	2.917	8.511	.1820	.03311	14.8	5.495	30.20
.6383	.4074	3.9	1.567	2.455	.3388	.1148	9.4	2.951	8.710	.1799	.03236	14.9	5.559	30.90
.6310	.3981	4.0	1.585	2.512	.3350	.1122	9.5	2.985	8.913	.1778	.03162	15.0	5.623	31.62
.6237	.3890	4.1	1.603	2.570	.3311	.1096	9.6	3.020	9.120	.1758	.03090	15.1	5.689	32.36
.6166	.3802	4.2	1.622	2.630	.3273	.1072	9.7	3.055	9.333	.1738	.03020	15.2	5.754	33.11
.6095	.3715	4.3	1.641	2.692	.3236	.1047	9.8	3.090	9.550	.1718	.02951	15.3	5.821	33.88
.6026	.3631	4.4	1.660	2.754	.3199	.1023	9.9	3.126	9.772	.1698	.02884	15.4	5.888	34.67
.5957	.3548	4.5	1.679	2.818	.3162	.1000	10.0	3.162	10.000	.1679	.02818	15.5	5.957	35.48
.5888	.3467	4.6	1.698	2.884	.3126	.09772	10.1	3.199	10.23	.1660	.02754	15.6	6.026	36.31
.5821	.3388	4.7	1.718	2.951	.3090	.09550	10.2	3.236	10.47	.1641	.02692	15.7	6.095	37.15
.5754	.3311	4.8	1.738	3.020	.3055	.09333	10.3	3.273	10.72	.1622	.02630	15.8	6.166	38.02
.5689	.3236	4.9	1.758	3.090	.3020	.09120	10.4	3.311	10.96	.1603	.02570	15.9	6.237	38.90
.5623	.3162	5.0	1.778	3.162	.2985	.08913	10.5	3.350	11.22	.1585	.02512	16.0	6.310	39.81
.5559	.3090	5.1	1.799	3.236	.2951	.08710	10.6	3.388	11.48	.1567	.02455	16.1	6.383	40.74
.5495	.3020	5.2	1.820	3.311	.2917	.08511	10.7	3.428	11.75	.1549	.02399	16.2	6.457	41.69
.5433	.2951	5.3	1.841	3.388	.2884	.08318	10.8	3.467	12.02	.1531	.02344	16.3	6.531	42.66
.5370	.2884	5.4	1.862	3.467	.2851	.08128	10.9	3.508	12.30	.1514	.02291	16.4	6.607	43.65

Decibels Versus Voltage and Power Ratios, (cont'd.)

Voltage Ratio	Power Ratio	-dB +	Voltage Ratio	Power Ratio
.1496	.02239	16.5	6.683	44.67
.1479	.02188	16.6	6.761	45.71
.1462	.02138	16.7	6.839	46.77
.1445	.02089	16.8	6.918	47.86
.1429	.02042	16.9	6.998	48.98
.1413	.01995	17.0	7.079	50.12
.1396	.01950	17.1	7.161	51.29
.1380	.01905	17.2	7.244	52.48
.1365	.01862	17.3	7.328	53.70
.1349	.01820	17.4	7.413	54.95
.1334	.01778	17.5	7.499	56.23
.1318	.01738	17.6	7.586	57.54
.1303	.01698	17.7	7.674	58.88
.1288	.01660	17.8	7.762	60.26
.1274	.01622	17.9	7.852	61.66
.1259	.01585	18.0	7.943	63.10
.1245	.01549	18.1	8.035	64.57
.1230	.01514	18.2	8.128	66.07
.1216	.01479	18.3	8.222	67.61
.1202	.01445	18.4	8.318	69.18
.1189	.01413	18.5	8.414	70.79
.1175	.01380	18.6	8.511	72.44
.1161	.01349	18.7	8.610	74.13
.1148	.01318	18.8	8.710	75.86
.1135	.01288	18.9	8.811	77.62
.1122	.01259	19.0	8.913	79.43
.1109	.01230	19.1	9.016	81.28
.1096	.01202	19.2	9.120	83.18
.1084	.01175	19.3	9.226	85.11
.1072	.01148	19.4	9.333	87.10
.1059	.01122	19.5	9.441	89.13
.1047	.01096	19.6	9.550	91.20
.1035	.01072	19.7	9.661	93.33
.1023	.01047	19.8	9.772	95.50
.1012	.01023	19.9	9.886	97.72
.1000	.01000	20.0	10.000	100.00
	10^{-3}	30		10^3
10^{-2}	10^{-4}	40	10^2	10^4
	10^{-5}	50		10^5
10^{-3}	10^{-6}	60	10^3	10^6
	10^{-7}	70		10^7
10^{-4}	10^{-8}	80	10^4	10^8
	10^{-9}	90		10^9
10^{-5}	10^{-10}	100	10^5	10^{10}

Index

Index